# VIRUSES OF PLANTS

## Descriptions and Lists from the VIDE Database

Edited by

Alan Brunt[1]
Karen Crabtree[2]
Michael Dallwitz[3]
Adrian Gibbs[2]
Leslie Watson[2]

[1] *Horticulture Research International, Wellesbourne, Warwick CV35 9EF, UK; and Imperial College of Science, Technology and Medicine, University of London, South Kensington, London SW7 2BB, UK*

[2] *Research School of Biological Sciences, Institute of Advanced Studies, Australian National University, GPO Box 475, Canberra, ACT 2601, Australia*

[3] *CSIRO Division of Entomology, GPO Box 1700, Canberra, ACT 2601, Australia*

*Present addresses:*
*Karen Crabtree: 65 Garden Road, Woolmer Green, Knebworth SG2 6JZ, UK*
*Leslie Watson: 78 Vancouver Street, Albany, WA 6330, Australia*

CAB INTERNATIONAL

18729827

CAB INTERNATIONAL
Wallingford
Oxon OX10 8DE
UK

Tel: +44 (0)1491 832111
Fax: +44 (0)1491 833508
E-mail: cabi@cabi.org
Telex: 847964 (COMAGG G)



A catalogue record for this book is available from the British Library.

ISBN 0 85198 794 X

Printed and bound in the UK at the University Press, Cambridge

# CONTENTS

# PREFACE

The VIDE (Virus Identification Data Exchange) project started in the early 1980s and, with support from the Rural Credit Development Fund of the Reserve Bank of Australia, produced and published an experimental database of viruses of legumes in 1983. Legumes were chosen, not just because of their economic importance, and that they are hosts to a large and representative collection of viruses, but also because the International Working Group on Legume Viruses comprised a group of experts willing to collate the information. The database was extended, with further support from the Fund, to produce the book 'Viruses of Plants in Australia' in 1988.

Since 1983 the project has also been generously supported by the Australian Centre for International Agricultural Research, and this work resulted in the publication of 'Viruses of Tropical Plants' in 1990 and, now, this book.

The VIDE project has relied on the help and encouragement of a very large number of colleagues around the world. Those who collated most of the data of each virus are specifically named in the description of that virus. Very large contributions were made by past members of the editorial team; K.F. Boswell, C. Büchen-Osmond and G.D. McLean. T.A. Paine of the CSIRO Division of Entomology provided invaluable help with the theory and practical use of the DELTA system. J.F. Antoniw of Rothamsted Experimental Station, U.K. very kindly provided the sequence database accession codes.

We are pleased to receive comments on all aspects of the VIDE project, to discuss additional applications of the database, to learn of errors in the data and to have offers of assistance in revising or adding new information. Contact the project co-ordinator: Adrian Gibbs, RSBS, ANU, PO Box 475, ACT, 2601, Australia; phone Intl. 61-6 249-4211; fax 61-6 249-4437; gibbs@rsbs-central.anu.edu.au.

The Editors
November, 1995

# 1. INTRODUCTION

This book contains information on all viruses of plants, not only those whose virions have been described, but also those like umbraviruses that have no virion protein genes of their own, and use the virion proteins of their symbiotic helper viruses. We have not included the many transmissible agents, that cause virus-like diseases, but whose infectious particles have not yet been characterized, however we have included the database accession numbers of the genomic sequences of satellite RNAs. The book also contains summaries of data for all the viruses that are definitive or tentative members of each accepted genus or 'group'.

## 1.1. DELTA and the VIDE database

This book is a product of the VIDE (Virus Identification Data Exchange) project (Boswell and Gibbs, 1983; Boswell and Gibbs, 1986; Boswell, Dallwitz, Gibbs and Watson, 1986; Brunt, Crabtree and Gibbs, 1990; Büchen-Osmond, Crabtree, Gibbs and McLean, 1988; Watson, Dallwitz, Gibbs and Pankhurst, 1988). The project has used the DELTA (DEscription Language for TAxonomy) system (Dallwitz, 1980; Dallwitz, Paine and Zurcher, 1993; also URL http://muse.bio.cornell.edu/delta/ or http://www.ento.csiro.au/delta/) for formatting, storing and manipulating data on all plant viruses. The data to be stored were defined in the character list (Section 4.2), which also served as a questionnaire for each virus. The data were then encoded, and manipulated using the DELTA programs, and finally translated and typeset in Postscript automatically using associated programs. They will be first available in the interactive key format, INTKEY, from early 1996, on InterNet from the BioWeb server (URL http://biology.anu.edu.au).

## 1.2. Notes on the data

### *1.2.1. Names*

Each virus species description (Section 3) is headed by a formal vernacular name which, in most instances, is the most commonly used name for that virus, or is that used in ICTV publications (Murphy *et al.*, 1995), or that listed in Steadman's ICTV Words (Calisher and Fauquet, 1992). The order of words in some names has been changed so that the host name appears first.

Many of the viruses described are definitive members of genera, whose names have been approved by the ICTV. The names of such viruses include their generic names (Appendix 2, character #541), in italics, in the style introduced by Fenner (1976). Thus 'American plum line pattern virus', for example, appears as 'Plum American line pattern *ilarvirus*'.

Some virus clusterings, although generally recognized by virologists, have not yet been given generic status with ICTV-approved names. For these we have used the most commonly used euphonious name, but without italics. However for those that have no generally agreed name we have coined one for use in this book. New names of this type are:

- bigeminivirus, monogeminivirus and hybrigeminivirus: the three genera of the *Geminiviridae*. The first includes the whitefly-transmitted geminiviruses that mostly have bipartite genomes, the second the leafhopper-transmitted geminiviruses with unipartite genomes and, mostly, monocotyledonous hosts, and the last includes viruses, such as beet curly top virus, which has intermediate properties and a genomic sequence that clearly shows it derived its replicase from a bigeminivirus and virion protein gene from a monogeminivirus (Howarth and Vandemark, 1989; Padidam, Beachy and Fauquet, 1995);
- ipomovirus: a genus of the *Potyviridae* with, at present, two species, both of which were isolated from sweet potato, *Ipomoea batatus*;
- macluravirus: a genus named after the natural host, *Maclura pomifera* (Moraceae), of its best known species;
- nanavirus: a group otherwise clumsily known as the 'subterranean clover stunt virus group'. The name, nanavirus, is derived from *nanos* Gr. meaning dwarf, as these viruses have among the smallest known virions and genome segment sizes, they stunt their hosts and, incidentally, the best known nanavirus infects bananas.
- ourmiavirus: named after the town in Iran where the melon infected with the type species was found;
- satellite RNA: sub-viral pathogens that produce no virions themselves, but are associated with, and depend on (but not *vice versa*), particular helper viruses;
- satellivirus: the satellite viruses, like that of tobacco necrosis virus, which depend on helper viruses, but which produce their own virions;
- varicosavirus: a group otherwise known as the 'lettuce big-vein virus group'. The name, varicosavirus, is derived from *varix* L. meaning 'abnormal dilatation or enlargement of a vein or artery' (Shorter Oxford Dictionary), and alludes to the disease caused by the type species.

Some viruses are only tentatively associated with particular genera or groups. The formal vernacular name at the start of the description of each of these viruses includes '(?)' placed before the generic component of the name. In addition there are several viruses with virions characteristic of the families Potyviridae or Rhabdoviridae, which have not yet been assigned to particular genera in these families. Their formal names contain '(?) potyvirus' or '(?) rhabdovirus' (N.B. no italics) respectively.

### *1.2.2. Descriptions*

The data for each virus description were mostly collated by the person named at the start of each description, however the editors have added some key references to some descriptions. The references given for each virus are indicative, certainly not exhaustive, and some are unfortunately incomplete; they are included to assist those wishing to find the major areas of published work on each species.

Where possible the data have been encoded numerically (i.e. as a series of characters and states). Viruses are, however, a particularly disparate group of pathogens, and this has made it impossible to define a set of genuinely homologous characters that apply to all plant viruses. As a result much information has been stored as 'text comments', associated with appropriate characters, and this is responsible for the many differences between the character list (Appendix 2) and the translated description of each species (Section 3).

Comments are removed during the conversion of the data for the INTKEY version, and, as this provides the 'generic summaries', all comments are missing from those summaries (Section 2). Also omitted from these summaries are data that are clearly specific to individual species, such as their geographic distribution and host range. The summaries give a list of the definitive and tentative species in the genus, and also the ranges and means for numerical characters, and the frequency of occurrence of each state of multistate characters.

## 1.3. References


Boswell, K.F. and Gibbs, A.J. (eds) (1983). *Viruses of Legumes 1983 — Descriptions and Keys from VIDE.* 139 pp. Australian National University, Canberra.

Boswell, K. and Gibbs, A. (1986). The VIDE data bank for plant viruses. In *Developments and Applications in Virus Testing* (ed R.A.C. Jones). pp. 283–287. Assoc. Applied Biologists, U.K.

Boswell, K.F., Dallwitz, M.J., Gibbs, A.J. and Watson, L. (1986). The VIDE (Virus Identification Data Exchange) project: a database for plant viruses. *Rev. Plant Path.* **65**: 221–231.

Büchen-Osmond, C., Crabtree, K., Gibbs, A. & McLean, G. (1988). *Viruses of Plants in Australia.* 590 pp. Australian National University, Canberra.

Brunt, A., Crabtree, K. & Gibbs, A. (1990). *Viruses of Tropical Plants: Descriptions and Lists from the VIDE Database.* 707 pp. C.A.B. International, U.K.

Calisher, C.H. and Fauquet, C.M. (1992) *Steadman's ICTV Virus Words.* 271 pp. Williams and Wilkins, Baltimore.

Dallwitz, M.J. (1980). A general system for coding taxonomic descriptions. *Taxon* **29**: 41–46.

Dallwitz, M.J., Paine, T.A. and Zurcher, E.J. (1993). *Delta User's Guide: A General System for Processing Taxonomic Descriptions.* 136 pp. CSIRO Division of Entomology, Canberra.

Fenner, F. (1976). Classification and Nomenclature of Viruses; 2nd Report of the International Committee on Taxonomy of Viruses. *Intervirology* **7**: 1–116.

Howarth, A.J. and Vandemark, G.J. (1989). Phylogeny of geminiviruses. *J. gen. Virol.* **70**: 2717–2727.

Murphy, F.A., Fauquet, C.M., Mayo, M.A., Jarvis, A.W., Ghabrial, S.A., Summers, M.D., Martelli, G.P. and Bishop, D.H.L. (1995). 6th Report of the International Committee on Taxonomy of Viruses. *Archiv. Virol.* Suppl. 10, Springer-Verlag, Wein, New York.

Padidam, M., Beachy, R.N. and Fauquet, C.M. (1995). Classification and identification of geminiviruses using sequence comparisons. *J. gen. Virol.* **76**: 249–263.

Watson, L., Dallwitz, M.J., Gibbs, A.J. and Pankhurst, R.J. (1988). Automated taxonomic descriptions. In *Prospects in Systematics* (ed D.L. Hawksworth). pp. 292–304. Clarendon Press, Oxford.

# 2. GENERIC SUMMARIES

## Alfamoviruses: *Bromoviridae*

*Definitive species* alfalfa mosaic alfamovirus.

**Natural host range and symptoms**

Symptoms persist, or disappear soon after infection.

**Transmission**

Transmitted by a vector; an insect; Aphididae. Transmitted in a non–persistent manner. Virus transmitted by mechanical inoculation; transmitted by grafting; not transmitted by contact between plants; transmitted by seed; transmitted by pollen to the seed.

**Experimental host range**

Many (>9) families susceptible.

**Properties of particles in sap**

TIP: 60–65 °C. LIV: 1–4 days. DEP: $\log_{10}$ minus 3–4. Infectivity of sap not changed by treatment with di–ethyl ether. Leaf sap contains many virions.

**Particle morphology**

Virions bacilliform; not enveloped; 30–56 nm in length; 18 nm wide.

**Physical properties**

Four sedimenting components in purified preparations; sedimentation coefficient 94 *S*; of the other(s) 66–82 *S*. Density 1.381–1.385 g $cm^{-3}$ in CsCl. Isoelectric point pH 4.6. $A_{260}/A_{280}$ ratio 1.7–1.8.

**Biochemical properties**

Virions contain 16 % nucleic acid; 84 % protein; 0 % lipid.

*Genome* consists of RNA; single–stranded; linear. Total genome size 8.27 kb. Genome of three parts; largest (or only) genome part 3.64 kb; the 2nd largest 2.59 kb; the 3rd largest 2.04 kb. Base composition 23 % G; 25 % A; 23 % C; 29 % U. 5' terminus of RNA has a methylated nucleotide cap. Infectivity retained when deproteinised with proteases, or decreased when deproteinised with proteases; retained when deproteinised with phenol or detergent. Poly A region absent. Additional factor required for infectivity. Genome has no *t*RNA–like activity.

*Non–genomic nucleic acid* found in the virions; is subgenomic *m*RNA. Sub–genomic *m*RNA found in infected cells.

*Virion protein(s)* one; $M_r$ 24250.

**Replication**

Replication does not depend on a helper virus.

**Cytopathology**

Virions found in cytoplasm, or in chloroplasts. Inclusions present in infected cells; are unusual in shape; they contain virions.

# Alphacryptoviruses: *Partitiviridae*

*Definitive species* alfalfa 1, beet 1, beet 2, beet 3, carnation 1, carrot temperate 1, carrot temperate 3, carrot temperate 4, cocksfoot, hop trefoil 1, hop trefoil 3, radish yellow edge, ryegrass, spinach temperate, vicia, white clover 1 and white clover 3 alphacryptoviruses. *Tentative species* amaryllis, avocado 3, carnation 2, fescue, garland chrysanthemum temperate, mibuna temperate, poinsettia, red pepper 1, red pepper 2, rhubarb temperate, santosai temperate and white mustard alphacryptoviruses.

**Natural host range and symptoms**

Symptoms disappear soon after infection (1/5), or none (4/5).

**Transmission**

Transmitted by means not involving a vector (7/7). Virus transmitted by mechanical inoculation (2/24), or not transmitted by mechanical inoculation (22/24); not transmitted by grafting (14/14); not transmitted by contact between plants (10/10); transmitted by seed (24/24); transmitted by pollen to the seed (6/6).

**Experimental host range**

Many (>9) families susceptible (1/9), or few (<3) families susceptible (8/9).

**Properties of particles in sap**

Leaf sap contains few virions (10/10).

**Particle morphology**

Virions isometric (26/26); not enveloped (22/22); 28–29.85–30 nm in diameter; 30 nm in length; rounded in profile (17/18), or angular in profile (3/18); with a conspicuous capsomere arrangement (2/12), or without a conspicuous capsomere arrangement (10/12).

**Physical properties**

One sedimenting component in purified preparations (8/8); sedimentation coefficient 118–119.5–120 *S*. Density 1.337–1.362–1.39 g $cm^{-3}$ in CsCl.

**Biochemical properties**

*Genome* consists of RNA (19/19); double-stranded (19/19); linear (6/6). Total genome size 2.99–3.986–7.1 kb. Genome of two parts (10/14), or of three parts (4/14); largest (or only) genome part 1.3–2.015–3.9 kb; the 2nd largest 1.25–1.729–3.2 kb; the 3rd largest 1.21–1.33–1.45 kb.

*Virion protein(s)* one (2/4), or two (2/4); $M_r$ 52500–59250–67000. $M_r$ of 2nd largest 53000–57000–61000.

**Replication**

Replication does not depend on a helper virus (2/2).

**Cytopathology**

Virions found in cytoplasm (3/3), or in nuclei (1/3). Inclusions present in infected cells (1/3), or absent from infected cells (2/3); are viroplasms (1/1); they contain virions (1/1).

# Badnaviruses

*Definitive species* banana streak, cacao swollen shoot, canna yellow mottle, commelina yellow mottle, dioscorea bacilliform, kalanchoe top-spotting, rice tungro bacilliform, schefflera ringspot and sugarcane bacilliform badnaviruses. *Tentative species* aucuba ringspot, dasheen (or Taro) bacilliform, mimosa striped chlorosis, rubus yellow net, sweet potato leaf curl, yam internal brown spot and yucca bacilliform badnaviruses.

**Natural host range and symptoms**

Symptoms persist (8/10), or vary seasonally (1/10), or disappear soon after infection (1/10).

**Transmission**

Transmitted by a vector (6/6); an insect (6/6); Aleyrodidae (1/6), or Aphididae (1/6), or Cicadellidae (1/6), or Pseudococcidae (3/6). Transmitted in a semi–persistent manner (3/4), or in a persistent manner (1/4). Virus retained when the vector moults (1/3), or lost by the vector when it moults (2/3); does not multiply in the vector (2/2); not transmitted congenitally to the progeny of the vector (3/3); requires, for vector transmission, a helper virus (1/1); transmitted by mechanical inoculation (2/10), or not transmitted by mechanical inoculation (8/10); transmitted by grafting (6/7), or not transmitted by grafting (1/7); not transmitted by contact between plants (6/6); transmitted by seed (2/6), or not transmitted by seed (4/6); transmitted by pollen to the seed (1/5), or not transmitted by pollen (4/5).

**Experimental host range**

Several (3–9) families susceptible (2/8), or few (<3) families susceptible (6/8).

**Properties of particles in sap**

TIP: 55–57.5–60 °C. LIV: 28–56.5–85 days. DEP: $\log_{10}$ minus 3–4 (1/1). Leaf sap contains few virions (8/8).

**Particle morphology**

Virions bacilliform (15/15); not enveloped (12/12); 80–139–400 nm in length; with a clear modal length (1/1); of 95 nm; 24–29.35–35 nm wide.

**Physical properties**

One sedimenting component in purified preparations (3/3); sedimentation coefficient 218–251.5–285 *S*. Density 1.312 g $cm^{-3}$ in CsCl.

**Biochemical properties**

*Genome* consists of DNA (5/5); double–stranded (4/4); circular (4/4). Total genome size 7.35–7.413–7.49 kb. Genome unipartite (4/4); largest (or only) genome part 7.35–7.683–8.3 kb.

*Virion protein(s)* one (1/1); $M_r$ 32000.

**Replication**

Replication does not depend on a helper virus (1/1).

**Cytopathology**

Virions found in cytoplasm (11/11), or in cell vacuoles (1/11). Inclusions present in infected cells (3/7), or absent from infected cells (4/7); are membranous bodies (1/2), or viroplasms (1/2); they contain virions (2/2).

# Betacryptoviruses: *Partitiviridae*

*Definitive species* carrot temperate 2, hop trefoil 2, red clover 2 and white clover 2 betacryptoviruses. *Tentative species* alfalfa 2 betacryptovirus.

**Transmission**

Transmitted by means not involving a vector (1/1). Virus not transmitted by mechanical inoculation; not transmitted by grafting; not transmitted by contact between plants (1/1); transmitted by seed.

**Experimental host range**

Few (<3) families susceptible (2/2).

**Properties of particles in sap**

Leaf sap contains few virions (1/1).

**Particle morphology**

Virions isometric; not enveloped (2/2); 30–36.4–38 nm in diameter; rounded in profile (2/3), or angular in profile (1/3); with a conspicuous capsomere arrangement (2/2).

**Physical properties**

One sedimenting component in purified preparations (1/1). Density 1.375 g $cm^{-3}$ in CsCl. $A_{260}/A_{280}$ ratio 1.31.

**Biochemical properties**

Virions contain 24 % nucleic acid; 76 % protein.

*Genome* consists of RNA (4/4); double–stranded (4/4); linear (2/2). Total genome size 4.31–4.803–5.15 kb. Genome of two parts (2/3), or of three parts (1/3); largest (or only) genome part 2.05–2.279–2.55 kb; the 2nd largest 1.65–2.04–2.4 kb; the 3rd largest 1.45 kb.

**Cytopathology**

Virions found in cytoplasm (1/1). Inclusions absent from infected cells (1/1).

## Bigeminiviruses: *Geminiviridae*

*Definitive species* abutilon mosaic, ageratum yellow vein, bean calico mosaic, bean golden mosaic, bhendi yellow vein mosaic, cassava African mosaic, cassava Indian mosaic, chino del tomate', cotton leaf crumple, cotton leaf curl, croton yellow vein mosaic, dolichos yellow mosaic, euphorbia mosaic, horsegram yellow mosaic, jatropha mosaic, lima bean golden mosaic, melon leaf curl, mung bean yellow mosaic, okra leaf-curl, pepper hausteco, pepper Texas, potato yellow mosaic, rhynchosia mosaic, serrano golden mosaic, squash leaf curl, tobacco leaf curl, tomato Australian leafcurl, tomato golden mosaic, tomato Indian leafcurl, tomato leaf crumple, tomato mottle, tomato yellow leaf curl, tomato yellow mosaic, watermelon chlorotic stunt and watermelon curly mottle bigeminiviruses. *Tentative species* bean distortion dwarf, cowpea golden mosaic, lupin leaf curl, pepper mild tigre', solanum apical leaf curling and soybean crinkle leaf bigeminiviruses.

**Natural host range and symptoms**

Symptoms persist (31/34), or vary cyclically over a few weeks (2/34), or vary seasonally (6/34).

**Transmission**

Transmitted by a vector (38/38); an insect (38/38); Aleyrodidae (38/38). Transmitted in a non–persistent manner (3/28), or in a semi–persistent manner (4/28), or in a persistent manner (22/28). Virus retained when the vector moults (9/14), or lost by the vector when it moults (5/14); does not multiply in the vector (12/12); not transmitted congenitally to the progeny of the vector (15/15); does not require a helper virus for vector transmission

(1/1); transmitted by mechanical inoculation (16/38), or not transmitted by mechanical inoculation (22/38); transmitted by grafting (29/29); not transmitted by contact between plants (21/21); not transmitted by seed (26/26); not transmitted by pollen (13/13).

**Experimental host range**

Many (>9) families susceptible (1/31), or several (3–9) families susceptible (8/31), or few (<3) families susceptible (22/31).

**Properties of particles in sap**

TIP: 40–50.16–60 °C. LIV: 0.01–14.27–72 days. DEP: $\log_{10}$ minus 1 (3/6), or 2–3 (2/6), or 4 (1/6). Infectivity of sap not changed by treatment with di–ethyl ether (2/2). Leaf sap contains few virions (19/20), or contains many virions (1/20).

**Particle morphology**

Virions geminate (33/33), or unusually shaped (1/33); not enveloped (19/19); 15–19.12–25 nm in diameter; 15–30.41–52 nm in length; rounded in profile (5/9), or angular in profile (5/9); with a conspicuous capsomere arrangement (1/12), or without a conspicuous capsomere arrangement (11/12).

**Physical properties**

One sedimenting component in purified preparations (6/9), or two sedimenting components in purified preparations (3/9), or three sedimenting components in purified preparations (1/9); sedimentation coefficient 69–78.6–90 *S*; of the other(s) 50–56.66–70 *S*. Density 1.717 g $cm^{-3}$ in CsCl. Isoelectric point pH 7. $A_{260}/A_{280}$ ratio 1.36–1.376–1.4.

**Biochemical properties**

Virions contain 18–20–22 % nucleic acid; 78–80–82 % protein; 0 % lipid.

*Genome* consists of DNA (23/23); single–stranded (22/22); circular (16/16). Total genome size 5.096–5.286–5.57 kb. Genome of two parts (15/15); largest (or only) genome part 2.588–2.678–2.87 kb; the 2nd largest 2.508–2.627–2.724 kb. Base composition 21–22.31–23.5 % G; 26–27.76–29 % A; 18.3–18.71–19.8 % C; 30.2–31.2–32.6 % T. Infectivity retained when deproteinised with proteases (2/4), or decreased when deproteinised with proteases (2/4); retained when deproteinised with phenol or detergent (5/5). Poly A region absent (3/3). Additional factor not required for infectivity (4/4). Genome has no *t*RNA–like activity (2/2).

*Non–genomic nucleic acid* found in the virions (2/6), or not found in the virions (4/6); is of an unusual type of structure (1/1). Sub–genomic *m*RNA found in infected cells (1/1).

*Virion protein(s)* one (6/7), or two (1/7); $M_r$ 27400–29950–34000. $M_r$ of 2nd largest 27000. Virion proteins not glycosylated (1/1).

*Virus–coded non–virion proteins* identified by genomic sequence analysis (2/2); five proteins found (2/2). $M_r$ 40100–40200–40300. $M_r$ of 2nd largest 32700–33200–33700. $M_r$ of 3rd 29300–29450–29600. $M_r$ of 4th 15600–15700–15800. $M_r$ of 5th and smaller 14500–14800–15100.

**Replication**

Genome replicates in nuclei (6/6). Replication does not depend on a helper virus (8/8).

**Cytopathology**

Virions found in cytoplasm (5/16), or in nuclei (15/16), or in nucleoli (2/16), or in chloroplasts (1/16), or in cell vacuoles (1/16). Inclusions present in infected cells (12/13), or absent from infected cells (1/13); are crystals in the nucleus (3/12), or amorphous X–bodies (2/12), or viroplasms (1/12), or unusual in shape (11/12); they contain virions (5/6), or they do not contain virions (2/6).

# Bromoviruses: *Bromoviridae*

*Definitive species* broad bean mottle, brome mosaic, cassia yellow blotch, cowpea chlorotic mottle, melandrium yellow fleck and spring beauty latent bromoviruses.

**Natural host range and symptoms**

Symptoms persist (5), or none (1).

**Transmission**

Transmitted by a vector (2/2); an insect (2/2); Coleoptera (2/2). Virus transmitted by mechanical inoculation; transmitted by grafting (2/2); transmitted by contact between plants (1/1); not transmitted by seed (5/5); not transmitted by pollen (1/1).

**Experimental host range**

Many (>9) families susceptible (1), or several (3–9) families susceptible (5).

**Properties of particles in sap**

TIP: 67–82.87–95 °C. LIV: 1–23.87–44 days. DEP: $\log_{10}$ minus 3–4 (2/4), or 5 (3/4). Infectivity of sap increased by treatment with di–ethyl ether (1/2), or not changed by treatment with di–ethyl ether (1/2). Leaf sap contains many virions (5/5).

**Particle morphology**

Virions isometric; not enveloped (5/5); 25–26.16–28 nm in diameter; rounded in profile (5/5); with a conspicuous capsomere arrangement (2/5), or without a conspicuous capsomere arrangement (3/5).

**Physical properties**

One sedimenting component in purified preparations (5/5), or three sedimenting components in purified preparations (1/5); sedimentation coefficient 85–87.06–90 *S*; of the other(s) 78.7 *S*. Density 1.35–1.356–1.36 g $cm^{-3}$ in CsCl. Isoelectric point pH 3.6–4.925–6.8. $A_{260}/A_{280}$ ratio 1.7–3.55–5.4.

**Biochemical properties**

Virions contain 20–21.92–23.7 % nucleic acid; 76.3–77.43–79 % protein; 0 % lipid.

*Genome* consists of RNA; single–stranded; linear (4/4). Total genome size 8.216–8.622–9.5 kb. Genome of three parts (5/5); largest (or only) genome part 3.158–3.372–3.9 kb; the 2nd largest 2.799–3.032–3.4 kb; the 3rd largest 2.117–2.296–2.6 kb. Base composition 24.4–25.95–28 % G; 24.9–26.05–27 % A; 19–20.62–22.2 % C; 24–27.42–29 % U. 5' terminus of RNA has a methylated nucleotide cap (3/3). Infectivity retained when deproteinised with proteases (1/1), or decreased when deproteinised with proteases (1/1); retained when deproteinised with phenol or detergent (3/3). Poly A region absent (1/1). Additional factor not required for infectivity (2/2). Genome has *t*RNA–like activity (3/3).

*Non–genomic nucleic acid* found in the virions (5/5); is subgenomic *m*RNA (4/4). Sub–genomic *m*RNA found in infected cells (3/3). 5 virus specified dsRNA species found in infected cells. Size of largest virus specified dsRNA 2.2 kbp; 2nd largest 2 kbp; 3rd largest 1.4 kbp; 4th largest 1.2 kbp; 5th largest 0.5 kbp.

*Virion protein(s)* one (5/5); $M_r$ 19600–20740–23000.

*Virus–coded non–virion proteins* isolated (1/1); one protein found (1/1). $M_r$ 35000.

**Replication**

Genome replicates in cytoplasm (1/1). Replication does not depend on a helper virus (5/5).

**Cytopathology**

Virions found in cytoplasm (5/5), or in nuclei (1/5), or in the perinuclear space (1/5), or in chloroplasts (1/5), or in cell vacuoles (1/5). Inclusions present in infected cells (3/4), or absent from infected cells (1/4); are crystals in the cytoplasm (1/3), or amorphous X–bodies (1/3), or unusual in shape (1/3); they contain virions (2/2).

# Bymoviruses: *Potyviridae*

*Definitive species* barley mild mosaic, barley yellow mosaic, oat mosaic, rice necrosis mosaic, wheat spindle streak mosaic and wheat yellow mosaic bymoviruses.

**Natural host range and symptoms**

Symptoms persist (2), or vary seasonally (4), or disappear soon after infection (1).

**Transmission**

Transmitted by a vector; a fungus; Plasmodiophorales. Virus transmitted by mechanical inoculation; not transmitted by grafting (1/1); not transmitted by contact between plants (5/5); not transmitted by seed; not transmitted by pollen (4/4).

**Experimental host range**

Several (3–9) families susceptible (1), or few (<3) families susceptible (5).

**Properties of particles in sap**

TIP: 44–51.25–65 °C. LIV: 1–4.333–14 days. DEP: $\log_{10}$ minus 1 (2), or 2–4 (4), or 5 (2). Infectivity of sap not changed by treatment with di–ethyl ether (1/1). Leaf sap contains few virions.

**Particle morphology**

Virions filamentous; not enveloped; usually flexuous; with no clear modal length (2), or with clear modal lengths (4); of 257–589.3–2000 nm; 12–13.83–16 nm wide. Axial canal obscure. Basic helix obscure (5/5).

**Physical properties**

One sedimenting component in purified preparations (1/2), or two sedimenting components in purified preparations (1/2). Density 1.28–1.299–1.32 g $cm^{-3}$ in CsCl. $A_{260}/A_{280}$ ratio 1.14.

**Biochemical properties**

Virions contain 5 % nucleic acid; 95 % protein.

*Genome* consists of RNA (3/3); single–stranded (3/3); linear (3/3). Total genome size 11.21–11.73–12 kb. Genome of two parts (3/3); largest (or only) genome part 7.632–7.877–8 kb; the 2nd largest 3.585–3.861–4 kb. Base composition 22.1 % G; 27.4 % A; 24 % C; 26.5 % U. Infectivity retained when deproteinised with phenol or detergent (1/1). Poly A region present (3/3). Additional factor not required for infectivity (2/2).

*Non–genomic nucleic acid* not found in the virions (2/2).

*Virion protein(s)* one (2/4), or two (1/4), or three (1/4); $M_r$ 30000–33180–36000. $M_r$ of 2nd largest 26500–28250–30000. $M_r$ of 3rd largest 27000.

*Virus–coded non–virion proteins* isolated (1/1); one protein found (1/1). $M_r$ 69000–70500–72000.

**Replication**

Replication does not depend on a helper virus (2/2).

### Cytopathology

Virions found in cytoplasm. Inclusions present in infected cells; are amorphous X–bodies (3), or membranous bodies (4), or pinwheels; they do not contain virions.

## Capilloviruses

*Definitive species* apple stem grooving, cherry A, citrus tatter leaf and lilac chlorotic leafspot capilloviruses. *Tentative species* nandina stem pitting capillovirus.

### Natural host range and symptoms

Symptoms persist (3/4), or vary seasonally (1/4).

### Transmission

Transmitted by means not involving a vector (1/1). Virus transmitted by mechanical inoculation (3/4), or not transmitted by mechanical inoculation (1/4); transmitted by grafting; transmitted by seed (2/3), or not transmitted by seed (1/3).

### Experimental host range

Several (3–9) families susceptible (3/4), or few (<3) families susceptible (2/4).

### Properties of particles in sap

TIP: 60–63.25–65 °C. LIV: 2–7–16 days. DEP: $\log_{10}$ minus 4 (2/2). Leaf sap contains few virions (3/3).

### Particle morphology

Virions filamentous; not enveloped (3/3); usually flexuous; with a clear modal length (4), or with no clear modal length (1); of 600–897.7–1536 nm; 12–13.87–19 nm wide. Axial canal obscure (3/3). Basic helix obvious (4/4), or obscure (1/4); pitch of basic helix 3.4–3.635–3.9 nm.

### Physical properties

One sedimenting component in purified preparations (2/2); sedimentation coefficient 96–104–112 *S*. Density 1.302 g $cm^{-3}$ in CsCl. Isoelectric point pH 4.3. $A_{260}/A_{280}$ ratio 1.22–1.46–1.69.

### Biochemical properties

*Genome* consists of RNA (3/3); single–stranded (2/2). Total genome size 6.5 kb. Genome unipartite (1/1). Base composition 20.7 % G; 30.8 % A; 20.7 % C; 29.8 % U. Poly A region present (2/2).

*Virion protein(s)* one (4/4); $M_r$ 24000–26250–27000.

*Virus–coded non–virion proteins* isolated (1/1); two proteins found (1/1). $M_r$ 239900. $M_r$ of 2nd largest 36000.

### Replication

Replication does not depend on a helper virus (1/1).

### Cytopathology

Virions found in cytoplasm (1/2), or in nuclei (1/2). Inclusions present in infected cells (2/2); are amorphous X–bodies (1/2), or viroplasms (1/2); they contain virions (2/2).

## Carlaviruses

*Definitive species* blueberry scorch, cactus 2, caper latent, carnation latent, chrysanthemum B, dandelion latent, elderberry, fig S, helenium S, honeysuckle latent, hop American latent, hop latent, hop mosaic, kalanchoe latent, lilac mottle, lily symptomless, mulberry latent, muskmelon vein necrosis, nerine latent, passiflora latent, pea streak, poplar mosaic, potato M, potato S, red clover vein mosaic, shallot latent and strawberry pseudo mild yellow edge carlaviruses. *Tentative species* anthriscus, arracacha latent, artichoke latent M, artichoke latent S, aster chlorotic stunt, butterbur mosaic, cardamine latent, cassava brown streak-associated, cassia mild mosaic, chicory yellow blotch, Chinese yam necrotic mosaic, cole latent, cowpea mild mottle, cynodon mosaic, dandelion, daphne S, eggplant mild mottle, fuchsia latent, garlic common latent, gentiana, helleborus mosaic, hydrangea latent, lilac ringspot, plantain 8, prunus S, Sint-Jan's onion latent, southern potato latent and white bryony mosaic carlaviruses.

**Natural host range and symptoms**

Symptoms persist (18/31), or vary cyclically over a few weeks (1/31), or vary seasonally (2/31), or disappear soon after infection (1/31), or none (10/31).

**Transmission**

Transmitted by a vector (29/29); an insect (29/29); Aleyrodidae (2/29), or Aphididae (27/29). Transmitted in a non–persistent manner (24/26), or in a semi–persistent manner (2/26). Virus lost by the vector when it moults (1/1); does not multiply in the vector (1/1); not transmitted congenitally to the progeny of the vector (1/1); transmitted by mechanical inoculation (41/45), or not transmitted by mechanical inoculation (4/45); transmitted by grafting (16/16); not transmitted by contact between plants (12/12); transmitted by seed (2/20), or not transmitted by seed (18/20); transmitted by pollen to the pollinated plant (1/7), or not transmitted by pollen (6/7).

**Experimental host range**

Many (>9) families susceptible (1/40), or several (3–9) families susceptible (18/40), or few (<3) families susceptible (21/40).

**Properties of particles in sap**

TIP: 50–68.5–85 °C. LIV: 0.5–4.852–21 days. DEP: $\log_{10}$ minus 2 (4/21), or 3 (12/21), or 4 (11/21), or 5 (5/21), or 6 (3/21), or 7 (1/21). Infectivity of sap not changed by treatment with di–ethyl ether (2/2). Leaf sap contains few virions (20/33), or contains many virions (13/33).

**Particle morphology**

Virions filamentous (49/49); not enveloped (40/40); usually straight (28/44), or usually flexuous (17/44); with a clear modal length (44/44); of 509–655.7–720 nm; 11–13.14–18 nm wide. Axial canal obvious (8/34), or obscure (26/34); 2–2.666–3.5 nm in diameter. Basic helix obvious (3/29), or obscure (27/29); pitch of basic helix 3.3–3.4–3.5 nm.

**Physical properties**

One sedimenting component in purified preparations (24/24); sedimentation coefficient 147.5–163.2–176 *S*. Density 1.28–1.306–1.32 g $cm^{-3}$ in CsCl. Density 1.29 g $cm^{-3}$ in sucrose. Isoelectric point pH 4.5. $A_{260}/A_{280}$ ratio 1.08–1.239–1.4.

**Biochemical properties**

Virions contain 2.5–5.703–8.5 % nucleic acid; 93.75–94.52–95 % protein; 0 % lipid.

*Genome* consists of RNA (20/20); single–stranded (20/20); linear (11/11). Total genome size 6.48–7.772–8.5 kb. Genome unipartite (19/19); largest (or only) genome part 6.48–7.759–8.535 kb. Base composition 31 % G; 24 % A; 23 % C; 22 % U. Infectivity retained when deproteinised with proteases (1/1); retained when deproteinised with phenol or detergent (4/4), or lost when deproteinised with phenol or detergent (1/4). Additional factor not required for infectivity (2/2).

*Non–genomic nucleic acid* not found in the virions (5/5). Sub–genomic *m*RNA not found in infected cells (1/1). 1 virus specified dsRNA species found in infected cells.

*Virion protein(s)* one (19/20), or two (1/20); $M_r$ 23200–32530–40000. $M_r$ of 2nd largest 24000.

**Replication**

Replication does not depend on a helper virus (13/13).

**Cytopathology**

Virions found in cytoplasm (23/23), or in chloroplasts (1/23), or in mitochondria (1/23). Inclusions present in infected cells (12/20), or absent from infected cells (8/20); are crystals in the cytoplasm (1/12), or amorphous X–bodies (2/12), or membranous bodies (3/12), or viroplasms (3/12), or unusual in shape (6/12); they contain virions (7/8), or they do not contain virions (1/8).

## Carmoviruses: *Tombusviridae*

*Definitive species* bean mild mosaic, cardamine chlorotic fleck, carnation mottle, cucumber leaf spot, cucumber soil-borne, galinsoga mosaic, hibiscus chlorotic ringspot, melon necrotic spot, pelargonium flower break and turnip crinkle carmoviruses. *Tentative species* Ahlum waterborne, blackgram mottle, cowpea mottle, elderberry latent, glycine mottle, johnsongrass chlorotic stripe, narcissus tip necrosis, pelargonium line pattern, plantain 6, saguaro cactus, tephrosia symptomless and Weddel waterborne carmoviruses.

**Natural host range and symptoms**

Symptoms persist (7/13), or vary seasonally (3/13), or disappear soon after infection (1/13), or none (4/13).

**Transmission**

Transmitted by a vector (6/10), or means not involving a vector (4/10); an insect (4/6), or a fungus (2/6); Coleoptera (3/3); Chytridiales (2/2). Transmitted in a non–persistent manner (1/1). Virus does not require a helper virus for vector transmission (1/1); transmitted by mechanical inoculation (19/20), or not transmitted by mechanical inoculation (1/20); transmitted by grafting (4/5), or not transmitted by grafting (1/5); transmitted by contact between plants (5/9), or not transmitted by contact between plants (4/9); transmitted by seed (4/11), or not transmitted by seed (7/11); not transmitted by pollen (1/1).

**Experimental host range**

Many (>9) families susceptible (2/19), or several (3–9) families susceptible (10/19), or few (<3) families susceptible (7/19).

**Properties of particles in sap**

TIP: 55–77.62–95 °C. LIV: 1–48.3–395 days. DEP: $\log_{10}$ minus 4 (3/15), or 5 (6/15), or 6 (9/15), or 7–8 (2/15). Infectivity of sap not changed by treatment with di–ethyl ether (3/3). Leaf sap contains few virions (3/14), or contains many virions (12/14).

**Particle morphology**

Virions isometric; not enveloped (19/19); 27–30.5–34 nm in diameter; rounded in profile (11/16), or angular in profile (5/16), or neither clearly angular or nor rounded in profile (1/16); with a conspicuous capsomere arrangement (11/16), or without a conspicuous capsomere arrangement (5/16).

**Physical properties**

One sedimenting component in purified preparations (17/19), or two sedimenting components in purified preparations (2/19); sedimentation coefficient 112–124.6–143 *S*; of the other(s) 48 *S*. Density 1.33–1.348–1.364 g $cm^{-3}$ in CsCl. Isoelectric point pH 3.5–4.833–5.5. $A_{260}/A_{280}$ ratio 1.48–1.548–1.64.

**Biochemical properties**

Virions contain 14–19.13–23 % nucleic acid; 77–80.93–86 % protein; 0 % lipid.

*Genome* consists of RNA; single–stranded; linear (7/7). Total genome size 3.94–4.45–5.3 kb. Genome unipartite (19/19); largest (or only) genome part 3.94–4.402–5.3 kb. Base composition 21.7–26.33–32 % G; 23–25.72–30 % A; 19–24.11–31.5 % C; 21–23.95–27 % U. 5' terminus of RNA has a methylated nucleotide cap (1/1). Infectivity retained when deproteinised with proteases (1/1); retained when deproteinised with phenol or detergent (2/2). Poly A region absent (3/3). Additional factor not required for infectivity (1/1). Genome has no *t*RNA–like activity (1/1).

*Non–genomic nucleic acid* found in the virions (3/5), or not found in the virions (2/5); is subgenomic *m*RNA (3/3). Sub–genomic *m*RNA found in infected cells (2/2). 3 virus specified dsRNA species found in infected cells. Size of largest virus specified dsRNA 4–4.2–4.4 kbp; 2nd largest 1.5–1.7–1.9 kbp; 3rd largest 1.35–1.525–1.7 kbp.

*Virion protein(s)* one (19/19); $M_r$ 36400–40610–47000. Virion proteins not glycosylated (1/1). Virion proteins not phosphorylated (1/1).

*Virus–coded non–virion proteins* isolated (2/3), or identified by genomic sequence analysis (1/3); two proteins found (2/3), or three proteins found (1/3), or four proteins found (1/3). $M_r$ 50000–71610–85830. $M_r$ of 2nd largest 26840–26940–27000. $M_r$ of 3rd 8000–9500–11000. $M_r$ of 4th 6900.

**Replication**

Genome replicates in cytoplasm (2/2). Replication does not depend on a helper virus (8/8). Acts as helper for a satellite RNA (1/1).

**Cytopathology**

Virions found in cytoplasm (11/11), or in nuclei (1/11), or in chloroplasts (1/11), or in mitochondria (2/11), or in cell vacuoles (4/11). Inclusions present in infected cells (5/11), or absent from infected cells (6/11); are crystals in the cytoplasm (2/5), or unusual in shape (3/5); they contain virions (3/5), or they do not contain virions (2/5).

# Caulimoviruses

*Definitive species* blueberry red ringspot, carnation etched ring, cauliflower mosaic, dahlia mosaic, figwort mosaic, horseradish latent, mirabilis mosaic, peanut chlorotic streak, soybean chlorotic mottle, sweet potato and thistle mottle caulimoviruses. *Tentative species* aquilegia necrotic mosaic, cassava vein mosaic, cestrum, petunia vein clearing, plantago 4 and strawberry vein banding caulimoviruses.

**Natural host range and symptoms**

Symptoms persist (11/12), or vary seasonally (2/12).

**Transmission**

Transmitted by a vector (8/9), or means not involving a vector (1/9); an insect (8/8); Aphididae (8/8). Transmitted in a non–persistent manner (1/5), or in a semi–persistent manner (4/5). Virus lost by the vector when it moults (3/3); does not multiply in the vector (4/4); not transmitted congenitally to the progeny of the vector (4/4); transmitted by mechanical inoculation (8/13), or not transmitted by mechanical inoculation (5/13); transmitted by grafting (8/8); transmitted by contact between plants (1/5), or not transmitted by contact between plants (4/5); not transmitted by seed (8/8); not transmitted by pollen (3/3).

**Experimental host range**

Several (3–9) families susceptible (2/10), or few (<3) families susceptible (8/10).

**Properties of particles in sap**

TIP: 75–81.5–90 °C. LIV: 1–8–32 days. DEP: $\log_{10}$ minus 3–4 (4/5). Infectivity of sap not changed by treatment with di–ethyl ether (3/3). Leaf sap contains few virions (10/11), or contains many virions (1/11).

**Particle morphology**

Virions isometric; not enveloped (14/14); 35–47.52–50 nm in diameter; rounded in profile (13/15), or angular in profile (2/15); with a conspicuous capsomere arrangement (3/10), or without a conspicuous capsomere arrangement (7/10).

**Physical properties**

One sedimenting component in purified preparations (11/12), or two sedimenting components in purified preparations (1/12); sedimentation coefficient 200–223.9–275 *S*; of the other(s) 212 *S*. Density 1.3–1.347–1.4 g $cm^{-3}$ in CsCl. Isoelectric point pH 4.

**Biochemical properties**

Virions contain 14.5–16.08–17 % nucleic acid; 83–84.41–85.9 % protein; 0 % lipid.

*Genome* consists of DNA (13/13); double–stranded (13/13); circular (10/10), or linear (2/10). Total genome size 6.8–7.771–8.175 kb. Genome unipartite (9/10), or of two parts (1/10); largest (or only) genome part 6.8–7.771–8.175 kb. Base composition of ds genome 28–35.5–43 % (G + C). Infectivity retained when deproteinised with proteases (2/3), or decreased when deproteinised with proteases (1/3); retained when deproteinised with phenol or detergent (4/4). Poly A region present (1/2), or absent (2/2). Genome has no *t*RNA–like activity (1/1).

*Non–genomic nucleic acid* not found in the virions (5/5). Sub–genomic *m*RNA found in infected cells (1/1).

*Virion protein(s)* one (6/7), or three (1/7); $M_r$ 32000–47500–75000. $M_r$ of 2nd largest 44000. $M_r$ of 3rd largest 33000.

**Replication**

Genome replicates in nuclei (3/3), or in cytoplasm (2/3). Replication does not depend on a helper virus (7/7).

**Cytopathology**

Virions found in cytoplasm (12/13), or in nuclei (3/13), or in the perinuclear space (1/13), or in unusual sites (1/13). Inclusions present in infected cells (15/15); are viroplasms (10/14), or unusual in shape (5/14); they contain virions (13/15), or they do not contain virions (2/15).

# Closteroviruses

*Definitive species* beet yellow stunt, beet yellows, broad bean severe chlorosis, burdock yellows, carnation necrotic fleck, citrus tristeza, clover yellows, grapevine stem pitting associated and wheat yellow leaf closteroviruses. *Tentative species* abutilon yellows, alligatorweed stunting, beet pseudo-yellows, carrot yellow leaf, cassia severe mosaic, cucumber chlorotic spot, dendrobium vein necrosis, diodia vein chlorosis, festuca necrosis, grapevine corky bark-associated, grapevine leafroll-associated, lettuce infectious yellows, peach yellow leaf, pineapple wilt-associated, sweet potato sunken vein and tomato infectious chlorosis closteroviruses.

**Natural host range and symptoms**

Symptoms persist (13/13), or vary seasonally (1/13), or disappear soon after infection (1/13).

**Transmission**

Transmitted by a vector (20/20); an insect (20/20); Aleyrodidae (6/19), or Aphididae (10/19), or Psyllidae (1/19), or Pseudococcidae (2/19). Transmitted in a non–persistent manner (2/14), or in a semi–persistent manner (12/14). Virus lost by the vector when it moults (4/4); does not multiply in the vector (4/4); not transmitted congenitally to the progeny of the vector (4/4); can help the vector transmission of another virus (1/1); transmitted by mechanical inoculation (9/19), or not transmitted by mechanical inoculation (10/19); transmitted by grafting (8/8); not transmitted by contact between plants (8/8); not transmitted by seed (7/7); not transmitted by pollen (2/2).

**Experimental host range**

Many (>9) families susceptible (2/18), or several (3–9) families susceptible (3/18), or few (<3) families susceptible (13/18).

**Properties of particles in sap**

TIP: 40–49.16–55 °C. LIV: 1–2–5 days. DEP: $\log_{10}$ minus 3 (1/5), or 4 (4/5), or 5 (2/5). Leaf sap contains few virions (12/13), or contains many virions (1/13).

**Particle morphology**

Virions filamentous (24/24); not enveloped (13/13); usually flexuous (23/23); with a clear modal length (13/17), or with no clear modal length (4/17); of 150–1504–2200 nm; 10–12.11–18 nm wide. Axial canal obvious (1/8), or obscure (7/8). Basic helix obvious (8/10), or obscure (2/10); pitch of basic helix 3–3.528–3.7 nm.

**Physical properties**

One sedimenting component in purified preparations (4/5), or two sedimenting components in purified preparations (1/5); sedimentation coefficient 110–125–140 *S*; of the other(s) 80–85–90 *S*. Density 1.325–1.332–1.34 g $cm^{-3}$ in CsCl. $A_{260}/A_{280}$ ratio 1.44–1.61–1.73.

**Biochemical properties**

Virions contain 5–5.075–5.15 % nucleic acid; 94.85 % protein; 0 % lipid.

*Genome* consists of RNA (7/7); single–stranded (7/7); linear (3/3). Total genome size 7.1–13.96–20 kb. Genome unipartite (5/5); largest (or only) genome part 7.1–13.22–20 kb. Base composition 23.4–25.9–28.4 % G; 24.4–25.35–26.3 % A; 21–22.95–24.9 % C; 22.3–25.8–29.3 % U. 5' terminus of RNA has a methylated nucleotide cap (1/1). Infectivity retained when deproteinised with phenol or detergent (1/1). Poly A region absent (2/2). Additional factor not required for infectivity (2/2). Genome has no *t*RNA–like activity (1/1).

*Non–genomic nucleic acid* not found in the virions (1/1). Sub–genomic *m*RNA found in infected cells (1/1). 1–3.5–6 virus specified dsRNA species found in infected cells. Size of largest virus specified dsRNA 4.6–10.9–15 kbp; 2nd largest 1.2–4.712–7 kbp; 3rd largest 0.4–3.037–5 kbp; 4th largest 2.7 kbp; 5th largest 1.6 kbp; 6th largest 1 kbp.

*Virion protein(s)* one (4/4); $M_r$ 22.3–20130–34000.

**Replication**

Replication does not depend on a helper virus (5/5). Acts as helper for another virus (1/1).

**Cytopathology**

Virions found in cytoplasm (9/9), or in nuclei (1/9). Inclusions present in infected cells (8/8); are crystals in the cytoplasm (2/8), or membranous bodies (3/8), or viroplasms (1/8), or unusual in shape (2/8); they contain virions (7/7).

## Comoviruses: *Comoviridae*

*Definitive species* bean pod mottle, bean rugose mosaic, broad bean stain, broad bean true mosaic, cowpea mosaic, cowpea severe mosaic, glycine mosaic, pea mild mosaic, potato Andean mottle, quail pea mosaic, radish mosaic, red clover mottle, squash mosaic and ullucus C comoviruses.

**Natural host range and symptoms**

Symptoms persist (7/10), or vary cyclically over a few weeks (3/10), or vary seasonally (2/10).

**Transmission**

Transmitted by a vector (10/10); an insect (10/10); Coleoptera (10/10). Transmitted in a non–persistent manner (1/2), or in a semi–persistent manner (1/2). Virus does not multiply in the vector (1/1); not transmitted congenitally to the progeny of the vector (1/1); transmitted by mechanical inoculation; transmitted by grafting (6/6); transmitted by contact between plants (1/2), or not transmitted by contact between plants (1/2); transmitted by seed (6/12), or not transmitted by seed (6/12); transmitted by pollen to the seed (1/3), or not transmitted by pollen (2/3).

**Experimental host range**

Many (>9) families susceptible (1), or several (3–9) families susceptible (7), or few (<3) families susceptible (6).

**Properties of particles in sap**

TIP: 55–66.1–80 °C. LIV: 2–23.71–112 days. DEP: $\log_{10}$ minus 3 (2), or 4 (9), or 5 (6), or 6 (3), or 7–8 (1). Infectivity of sap decreased by treatment with di–ethyl ether (1/6), or not changed by treatment with di–ethyl ether (5/6). Leaf sap contains few virions (1/12), or contains many virions (11/12).

**Particle morphology**

Virions isometric; not enveloped (13/13); 20–27.53–30 nm in diameter; rounded in profile (1/13), or angular in profile (13/13); with a conspicuous capsomere arrangement (3/11), or without a conspicuous capsomere arrangement (8/11).

**Physical properties**

Two sedimenting components in purified preparations (1), or three sedimenting components in purified preparations; sedimentation coefficient 112–118.5–130 *S*; of the other(s) 51–76.71–103 *S*. Density 1.288–1.401–1.463 g $cm^{-3}$ in CsCl. Isoelectric point pH 4.3–5.317–7. $A_{260}/A_{280}$ ratio 0.86–1.648–1.81.

**Biochemical properties**

Virions contain 0–17.87–38.6 % nucleic acid; 62–82.19–100 % protein; 0 % lipid.

*Genome* consists of RNA (13/13); single–stranded (13/13); linear (7/7). Total genome size 9.57–11.02–13.5 kb. Genome of two parts (12/12); largest (or only) genome part 6–6.757–8.5 kb; the 2nd largest 3.54–4.279–5 kb. Base composition 19.1–22.5–25 % G; 26.5–29.17–33.4 % A; 16.2–17.96–20.4 % C; 27–30.33–32.3 % U. 5' terminus of RNA has a VPg (4/4). Infectivity retained when deproteinised with proteases (2/2), or decreased when deproteinised with proteases (1/2); retained when deproteinised with phenol or detergent (5/5). Poly A region present (5/5). Genome has no *t*RNA–like activity (1/1).

*Non–genomic nucleic acid* not found in the virions (6/6). Sub–genomic *m*RNA not found in infected cells (1/1).

*Virion protein(s)* two (6/11), or three (5/11); $M_r$ 37500–40750–45100. $M_r$ of 2nd largest 20700–22320–24500. $M_r$ of 3rd largest 3538–11800–21000. Virion proteins glycosylated (2/2).

*Virus–coded non–virion proteins* isolated (1/1); six proteins found (1/1). $M_r$ 189900. $M_r$ of 2nd largest 51000. $M_r$ of 3rd 32000. $M_r$ of 4th 111900. $M_r$ of 5th and smaller 22000–63500–104900.

### Replication

Genome replicates in cytoplasm (1/1). Replication does not depend on a helper virus (10/10).

### Cytopathology

Virions found in cytoplasm (8/8), or in cell vacuoles (2/8). Inclusions present in infected cells (8/8); are crystals in the cytoplasm (4/8), or unusual in shape (8/8); they contain virions (5/5).

## Cucumoviruses: *Bromoviridae*

*Definitive species* cucumber mosaic, peanut stunt and tomato aspermy cucumoviruses.

### Natural host range and symptoms

Symptoms persist.

### Transmission

Transmitted by a vector; an insect; Aphididae. Transmitted in a non–persistent manner. Virus transmitted by mechanical inoculation; transmitted by grafting (1/1); not transmitted by contact between plants (1/1); transmitted by seed.

### Experimental host range

Many (>9) families susceptible.

### Properties of particles in sap

TIP: 50–56.66–70 °C. LIV: 1–3.5–10 days. DEP: $\log_{10}$ minus 3 (2), or 4, or 5–6 (2). Infectivity of sap not changed by treatment with di–ethyl ether (1/1). Leaf sap contains few virions (1/2), or contains many virions (2/2).

### Particle morphology

Virions isometric; not enveloped; 29–29.33–30 nm in diameter; rounded in profile, or angular in profile (1); with a conspicuous capsomere arrangement (1), or without a conspicuous capsomere arrangement (2).

### Physical properties

One sedimenting component in purified preparations, or four sedimenting components in purified preparations (1); sedimentation coefficient 98–116.6–152 *S*; of the other(s) 37–86–135 *S*. Density 1.357–1.365–1.367 g $cm^{-3}$ in CsCl. Isoelectric point pH 5.5–5.716–6.05. $A_{260}/A_{280}$ ratio 1.6–1.65–1.7.

**Biochemical properties**

Virions contain 16–18.4–21.2 % nucleic acid; 78.8–81.6–84 % protein; 0 % lipid.

*Genome* consists of RNA; single–stranded; linear (2/2). Total genome size 8.5–8.606–8.698 kb. Genome of three parts; largest (or only) genome part 3.355–3.384–3.41 kb; the 2nd largest 2.946–3.018–3.074 kb; the 3rd largest 2.186–2.199–2.214 kb. Base composition 23.1–23.83–24 % G; 23–24.8–26.2 % A; 21–21.83–23 % C; 29–29.51–30.1 % U. 5' terminus of RNA has a methylated nucleotide cap (1/2), or a VPg (1/2). Infectivity retained when deproteinised with proteases (1/2), or decreased when deproteinised with proteases (2/2); retained when deproteinised with phenol or detergent (2/2). Poly A region absent (2/2). Genome has *t*RNA–like activity (2/2).

*Non–genomic nucleic acid* found in the virions; is subgenomic *m*RNA, or satellite RNA. Sub–genomic *m*RNA found in infected cells (1/1).

*Virion protein(s)* one (1/1); $M_r$ 24240–24370–24500.

*Virus–coded non–virion proteins* isolated (1/2), or identified by genomic sequence analysis (1/2); one protein found (1/2), or three proteins found (1/2). $M_r$ 30000–33150–36300. $M_r$ of 2nd largest 31000. $M_r$ of 3rd 9900.

**Replication**

Replication does not depend on a helper virus (2/2).

**Cytopathology**

Virions found in cytoplasm, or in cell vacuoles (2). Inclusions present in infected cells; are crystals in the cytoplasm; they contain virions (2/2).

# Cytorhabdoviruses: *Rhabdoviridae*

*Definitive species* barley yellow striate mosaic, broad bean yellow vein, broccoli necrotic yellows, cereal northern mosaic, festuca leaf streak, lettuce necrotic yellows, sonchus and strawberry crinkle cytorhabdoviruses. *Tentative species* celery T, digitaria striate, ivy vein clearing, melon variegation, pelargonium vein clearing, pisum, raphanus, soursop yellow blotch and winter wheat Russian mosaic cytorhabdoviruses.

**Natural host range and symptoms**

Symptoms persist (6/7), or vary seasonally (1/7), or disappear soon after infection (1/7).

**Transmission**

Transmitted by a vector (7/7); an insect (7/7); Aphididae (3/7), or Delphacidae (4/7). Transmitted in a persistent manner (6/6). Virus retained when the vector moults (5/5); multiplies in the vector (5/5); transmitted congenitally to the progeny of the vector (3/5), or not transmitted congenitally to the progeny of the vector (2/5); transmitted by mechanical inoculation (9/15), or not transmitted by mechanical inoculation (6/15); transmitted by grafting (3/3); not transmitted by contact between plants (3/3); not transmitted by seed (8/8); not transmitted by pollen (2/2).

**Experimental host range**

Several (3–9) families susceptible (4/11), or few (<3) families susceptible (7/11).

### Properties of particles in sap

TIP: 50–52.78–60 °C. LIV: 0.1–1.328–5 days. DEP: $\log_{10}$ minus 2 (2/7), or 3–4 (3/7), or 5–6 (1/7). Infectivity of sap decreased by treatment with di–ethyl ether (3/3). Leaf sap contains few virions (8/9), or contains many virions (2/9).

### Particle morphology

Virions rhabdo– or bullet–shaped; enveloped (12/12); usually straight (3/3); with a clear modal length (12/13), or with no clear modal length (1/13); of 100–278.2–360 nm; 42–63.32–130 nm wide. Axial canal obvious (5/5); 14–20.5–35 nm in diameter. Basic helix obvious (6/7), or obscure (1/7); pitch of basic helix 4.2–4.633–5.5 nm.

### Physical properties

One sedimenting component in purified preparations (5/5); sedimentation coefficient 704–839.3–940 *S*. Density 1.19–1.194–1.2 g $cm^{-3}$ in sucrose.

### Biochemical properties

*Genome* consists of RNA (5/5); single–stranded (5/5); linear (1/1). Total genome size 10.5–12.34–14 kb. Genome unipartite (3/3); largest (or only) genome part 10.5–12.34–14 kb. Base composition 16.3 % G; 30.3 % A; 16.7 % C; 36.6 % U. Infectivity lost when deproteinised with phenol or detergent (2/2).

*Non–genomic nucleic acid* found in the virions (1/1); is *r*RNA (1/1). Sub–genomic *m*RNA found in infected cells (1/1).

*Virion protein(s)* three (1/3), or five (1/3), or six (1/3); $M_r$ 50000–108200–189900. $M_r$ of 2nd largest 49000–62160–76000. $M_r$ of 3rd largest 20000–40160–54000. $M_r$ of 4th largest 38000–39000–40000. $M_r$ of 5th largest 19000. Virion proteins glycosylated (4/4).

### Replication

Genome replicates in cytoplasm (1/1). Replication does not depend on a helper virus (3/3).

### Cytopathology

Virions found in cytoplasm, or in the perinuclear space (1). Inclusions present in infected cells (5/7), or absent from infected cells (2/7); are membranous bodies (2/5), or viroplasms (3/5), or unusual in shape (1/5); they contain virions (2/4), or they do not contain virions (2/4).

## Dianthoviruses

*Definitive species* carnation ringspot, red clover necrotic mosaic and sweet clover necrotic mosaic dianthoviruses. *Tentative species* furcraea necrotic streak dianthovirus.

### Natural host range and symptoms

Symptoms persist.

### Transmission

Transmitted by a vector (2/2); a nematode (1/1). Virus does not require a helper virus for vector transmission (1/1); transmitted by mechanical inoculation; transmitted by grafting (2/2); transmitted by contact between plants (1/1); not transmitted by seed (2/2).

### Experimental host range

Many (>9) families susceptible (1), or several (3–9) families susceptible (2), or few (<3) families susceptible (1).

**Properties of particles in sap**

TIP: 80–85.83–90 °C. LIV: 50–60.75–70 days. DEP: $\log_{10}$ minus 5 (2/3), or 6–7 (1/3). Infectivity of sap decreased by treatment with di–ethyl ether (1/2), or not changed by treatment with di–ethyl ether (1/2). Leaf sap contains few virions (1/3), or contains many virions (2/3).

**Particle morphology**

Virions isometric; not enveloped; 31–33.87–35 nm in diameter; rounded in profile (2), or angular in profile (2); with a conspicuous capsomere arrangement (1/3), or without a conspicuous capsomere arrangement (2/3).

**Physical properties**

One sedimenting component in purified preparations; sedimentation coefficient 126–132–135 *S*. Density 1.363–1.364–1.366 g $cm^{-3}$ in CsCl. Isoelectric point pH 4.5. $A_{260}/A_{280}$ ratio 1.64–1.72–1.8.

**Biochemical properties**

Virions contain 20–22.87–28 % nucleic acid; 72–77.12–80 % protein; 0 % lipid.

*Genome* consists of RNA; single–stranded; linear (2/2). Total genome size 5.25–5.382–5.6 kb. Genome of two parts; largest (or only) genome part 3.75–3.882–4 kb; the 2nd largest 1.448–1.499–1.6 kb. Base composition 26 % G; 27 % A; 23 % C; 24 % U. 5' terminus of RNA has a methylated nucleotide cap (1/1). Infectivity retained when deproteinised with phenol or detergent (2/2). Poly A region absent (1/1).

*Non–genomic nucleic acid* found in the virions (1/1). Sub–genomic *m*RNA found in infected cells (1/1).

*Virion protein(s)* one; $M_r$ 37000–38780–40800.

*Virus–coded non–virion proteins* identified by genomic sequence analysis (1/1); three proteins found (1/1). $M_r$ 99900. $M_r$ of 2nd largest 37000. $M_r$ of 3rd 34000.

**Replication**

Replication does not depend on a helper virus (2/2).

**Cytopathology**

Virions found in cytoplasm (3/3), or in nuclei (1/3). Inclusions present in infected cells (3), or absent from infected cells (1); are amorphous X–bodies (1/3), or unusual in shape (2/3); they contain virions (1/1).

# Enamoviruses

*Definitive species* pea enation mosaic enamovirus.

**Natural host range and symptoms**

Symptoms persist.

**Transmission**

Transmitted by a vector; an insect; Aphididae. Transmitted in a persistent manner. Virus retained when the vector moults; does not multiply in the vector; not transmitted congenitally to the progeny of the vector; can help the vector transmission of another virus; transmitted by mechanical inoculation; transmitted by seed.

**Experimental host range**

Several (3–9) families susceptible.

**Properties of particles in sap**

TIP: 55–65 °C. LIV: 4 days. DEP: $\log_{10}$ minus 4. Infectivity of sap not changed by treatment with di–ethyl ether. Leaf sap contains many virions.

**Particle morphology**

Virions isometric; not enveloped; 25–28 nm in diameter; rounded in profile, or angular in profile; without a conspicuous capsomere arrangement.

**Physical properties**

Two sedimenting components in purified preparations; sedimentation coefficient 112 *S*; of the other(s) 99 *S*. Density 1.436 g $cm^{-3}$ in CsCl. Isoelectric point pH 5–6. $A_{260}/A_{280}$ ratio 1.63.

**Biochemical properties**

Virions contain 28 % nucleic acid; 72 % protein; 0 % lipid.

*Genome* consists of RNA; single–stranded; linear. Total genome size 9.96 kb. Genome of two parts; largest (or only) genome part 5.71 kb; the 2nd largest 4.25 kb. Base composition 26.6 % G; 24.1 % A; 24.5 % C; 24.8 % U. 5' terminus of RNA has a VPg. Infectivity lost when deproteinised with proteases. Poly A region absent.

*Non–genomic nucleic acid* found in the virions; is satellite RNA.

*Virion protein(s)* three; $M_r$ 54000. $M_r$ of 2nd largest 21000. $M_r$ of 3rd largest 17500.

**Cytopathology**

Virions found in cytoplasm, or in nuclei, or in cell vacuoles. Inclusions present in infected cells; are unusual in shape; they contain virions.

# Fabaviruses: *Comoviridae*

*Definitive species* broad bean wilt and lamium mild mottle fabaviruses.

**Natural host range and symptoms**

Symptoms vary seasonally (1/1).

**Transmission**

Transmitted by a vector; an insect; Aphididae. Transmitted in a non–persistent manner. Virus transmitted by mechanical inoculation; not transmitted by seed (1/1).

**Experimental host range**

Many (>9) families susceptible (1), or several (3–9) families susceptible (1).

**Properties of particles in sap**

TIP: 55–57.5–60 °C. LIV: 3–3.5–4 days. DEP: $\log_{10}$ minus 4–5 (1/1). Leaf sap contains many virions (1/1).

**Particle morphology**

Virions isometric; not enveloped; 25–27.25–32 nm in diameter; angular in profile; without a conspicuous capsomere arrangement.

**Physical properties**

Three sedimenting components in purified preparations; sedimentation coefficient 126 *S*; of the other(s) 63–81.5–100 *S*. Density 1.39–1.415–1.44 g $cm^{-3}$ in CsCl. $A_{260}/A_{280}$ ratio 0.86–1.525–1.79.

**Biochemical properties**

Virions contain 0–16.5–33 % nucleic acid; 67–83.5–100 % protein; 0 % lipid.

*Genome* consists of RNA (1/1); single–stranded (1/1); linear (1/1). Total genome size 10.8 kb. Genome of two parts (1/1); largest (or only) genome part 6.3 kb; the 2nd largest 4.5 kb. Base composition 26 % G; 30 % A; 17.5 % C; 26.5 % U. Infectivity retained when deproteinised with phenol or detergent (1/1).

*Virion protein(s)* two; $M_r$ 42000–43500–45000. $M_r$ of 2nd largest 26000–27250–28500.

**Replication**

Replication does not depend on a helper virus.

**Cytopathology**

Virions found in cytoplasm. Inclusions present in infected cells (1/1); are crystals in the cytoplasm (1/1), or amorphous X–bodies (1/1); they contain virions (1/1).

# Fijiviruses: *Reoviridae*

*Definitive species* maize rough dwarf, oat sterile dwarf, pangola stunt, rice black-streaked dwarf and sugarcane Fiji disease fijiviruses. *Tentative species* garlic dwarf fijivirus.

**Natural host range and symptoms**

Symptoms persist.

**Transmission**

Transmitted by a vector (5/5); an insect (5/5); Delphacidae (5/5). Transmitted in a persistent manner (5/5). Virus retained when the vector moults (5/5); multiplies in the vector (5/5); transmitted congenitally to the progeny of the vector (2/4), or not transmitted congenitally to the progeny of the vector (2/4); does not require a helper virus for vector transmission (1/1); transmitted by mechanical inoculation (2/5), or not transmitted by mechanical inoculation (3/5); not transmitted by grafting (2/2); not transmitted by contact between plants (3/3); not transmitted by seed (4/4); not transmitted by pollen (3/3).

**Experimental host range**

Few (<3) families susceptible (5/5).

**Properties of particles in sap**

TIP: 60 °C. LIV: 6–6.5–7 days. DEP: $\log_{10}$ minus 5–6 (1/1). Leaf sap contains few virions (4/5), or contains many virions (4/5).

**Particle morphology**

Virions isometric; not enveloped (5/5); 54–67.5–80 nm in diameter; rounded in profile (5), or angular in profile (3); with a conspicuous capsomere arrangement (2/4), or without a conspicuous capsomere arrangement (2/4).

**Physical properties**

One sedimenting component in purified preparations (2/4), or two sedimenting components in purified preparations (3/4), or three sedimenting components in purified preparations (1/4), or four sedimenting components in purified preparations (1/4); sedimentation coefficient 400 *S*.

**Biochemical properties**

Virions 0 % lipid.

*Genome* consists of RNA; double–stranded; linear (4/4). Total genome size 23.89–26.25–28.91 kb. Genome of 10 parts; largest (or only) genome part 3.405–3.95–4.4 kb; the 2nd largest 3.2–3.459–3.9 kb; the 3rd largest 3.15–3.379–3.7 kb;

the 4th largest 2.6–3.231–3.7 kb; the 5th largest 2.55–2.975–3.3 kb; the 6th largest and other parts 2.293–2.488–2.9 kb. Additional factor required for infectivity (2/2).

*Non–genomic nucleic acid* not found in the virions (2/2).

*Virion protein(s)* three (1/2), or six (1/2); $M_r$ 138900–143200–149900. $M_r$ of 2nd largest 124900–125900–128900. $M_r$ of 3rd largest 36000–80250–122900. $M_r$ of 4th largest 110900. $M_r$ of 5th largest 97000. $M_r$ of 6th largest and smaller, 64000.

### Replication

Genome replicates in cytoplasm (1/3), or in cytoplasmic viroplasms (2/3). Replication does not depend on a helper virus (3/3).

### Cytopathology

Virions found in cytoplasm, or in cell vacuoles (1). Inclusions present in infected cells (5/5); are crystals in the cytoplasm (2), or crystals in the nucleus (1), or viroplasms (5), or unusual in shape (4); they contain virions.

## Furoviruses

*Definitive species* beet necrotic yellow vein, beet soil-borne, broad bean necrosis, oat golden stripe, peanut clump, potato mop-top, sorghum chlorotic spot and wheat soil-borne mosaic furoviruses. *Tentative species* chara australis, hypochoeris mosaic, nicotiana velutina mosaic and rice stripe necrosis furoviruses.

### Natural host range and symptoms

Symptoms persist (3/8), or vary seasonally (5/8).

### Transmission

Transmitted by a vector (7/9), or means not involving a vector (2/9); a fungus (7/7); Plasmodiophorales (6/6). Virus transmitted by mechanical inoculation (11/11); transmitted by grafting (3/3); transmitted by contact between plants (1/4), or not transmitted by contact between plants (3/4); transmitted by seed (2/5), or not transmitted by seed (3/5); not transmitted by pollen (1/1).

### Experimental host range

Several (3–9) families susceptible (7/11), or few (<3) families susceptible (4/11).

### Properties of particles in sap

TIP: 45–61.83–80 °C. LIV: 1–531–4000 days. DEP: $\log_{10}$ minus 1 (1/8), or 2–3 (4/8), or 4 (3/8), or 5 (1/8). Infectivity of sap not changed by treatment with di–ethyl ether (1/1). Leaf sap contains few virions (8/10), or contains many virions (2/10).

### Particle morphology

Virions rod–shaped; not enveloped (9/9); usually straight (11/11); with a clear modal length (8), or with no clear modal length (3), or with clear modal lengths (1); of 65–236.7–532 nm; 18–20.16–25 nm wide. Axial canal obvious (10/11), or obscure (1/11); 4.5 nm in diameter. Basic helix obvious (6/8), or obscure (2/8); pitch of basic helix 2.6–2.712–2.9 nm.

### Physical properties

One sedimenting component in purified preparations (1/9), or two sedimenting components in purified preparations (4/9), or three sedimenting components in purified preparations (3/9), or four sedimenting components in purified preparations (1/9), or five sedimenting components in purified preparations (1/9), or six sedimenting components in purified preparations (1/9), or eight sedimenting components in purified preparations

(1/9); sedimentation coefficient 155–206.8–236 *S*; of the other(s) 15–150.3–183 *S*. Density 1.32–1.32–1.321 g $cm^{-3}$ in CsCl. Isoelectric point pH 3.4–5–6.45. $A_{260}/A_{280}$ ratio 1.2.

**Biochemical properties**

Virions contain 4–4.75–5 % nucleic acid; 95–95.25–96 % protein; 0 % lipid.

*Genome* consists of RNA (10/10); single–stranded (10/10); linear (7/7). Total genome size 9.7–11.16–13.5 kb. Genome unipartite (1/7), or of two parts (5/7), or of four parts (1/7), or of five parts (1/7); largest (or only) genome part 6.2–7.342–11 kb; the 2nd largest 3–3.916–5.3 kb; the 3rd largest 1.8 kb; the 4th largest 1.5 kb; the 5th largest 1.45 kb. Base composition 24–24.25–24.5 % G; 26–27–28 % A; 15–17.5–20 % C; 27.5–30.75–34 % U. 5' terminus of RNA has a methylated nucleotide cap (1/1). Infectivity retained when deproteinised with phenol or detergent (3/3). Poly A region present (1/2), or absent (1/2). Additional factor not required for infectivity (1/1). Genome has no *t*RNA–like activity (1/1).

*Non–genomic nucleic acid* found in the virions (1/4), or not found in the virions (3/4). Sub–genomic *m*RNA found in infected cells (1/1). 3–3.5–4 virus specified dsRNA species found in infected cells. Size of largest virus specified dsRNA 4.1 kbp; 2nd largest 1.95 kbp; 3rd largest 1.68 kbp.

*Virion protein(s)* one (7/7); $M_r$ 18500–20890–24500.

*Virus–coded non–virion proteins* identified by genomic sequence analysis; seven proteins found (1/1). $M_r$ 236900. $M_r$ of 2nd largest 75000. $M_r$ of 3rd 42000. $M_r$ of 4th 31000. $M_r$ of 5th and smaller 25000.

**Replication**

Replication does not depend on a helper virus (4/4).

**Cytopathology**

Virions found in cytoplasm (7/7), or in cell vacuoles (1/7). Inclusions present in infected cells (5/8), or absent from infected cells (3/8); are crystals in the cytoplasm (1/4), or amorphous X–bodies (2/4), or membranous bodies (1/4), or viroplasms (1/4), or unusual in shape (2/4); they contain virions (4/4).

# Hordeiviruses

*Definitive species* anthoxanthum latent blanching, barley stripe mosaic, lychnis ringspot and poa semilatent hordeiviruses.

**Natural host range and symptoms**

Symptoms persist (3), or disappear soon after infection (1).

**Transmission**

Transmitted by means not involving a vector (1/1). Virus transmitted by mechanical inoculation; transmitted by grafting (1/1); transmitted by seed (2/3), or not transmitted by seed (1/3); transmitted by pollen to the pollinated plant (1/1).

**Experimental host range**

Many (>9) families susceptible (1/3), or several (3–9) families susceptible (1/3), or few (<3) families susceptible (1/3).

**Properties of particles in sap**

TIP: 60–65.5–68 °C. LIV: 15–69.25–120 days. DEP: $log_{10}$ minus 2 (1/2), or 3 (2/2), or 4 (1/2). Leaf sap contains few virions (1/1).

**Particle morphology**

Virions rod–shaped; not enveloped; usually straight; with a clear modal length (2), or with no clear modal length (1), or with clear modal lengths (1); of 100–144–175 nm; 18–23.25–25 nm wide. Axial canal obvious (2/2); 3–3.5–4 nm in diameter. Basic helix obvious (3/3); pitch of basic helix 2.5–2.55–2.6 nm.

**Physical properties**

Three sedimenting components in purified preparations (1/1); sedimentation coefficient 199 *S*; of the other(s) 166–180–194 *S*. Isoelectric point pH 4.5. $A_{260}/A_{280}$ ratio 0.99.

**Biochemical properties**

Virions contain 3.8–4.45–5 % nucleic acid; 95–95.5–96 % protein; 0 % lipid.

*Genome* consists of RNA (3/3); single–stranded (3/3); linear (2/2). Total genome size 9–11.62–15.6 kb. Genome of three parts (2/3), or of four parts (1/3); largest (or only) genome part 3.8–4.1–4.5 kb; the 2nd largest 3.289–3.529–4 kb; the 3rd largest 3–3.254–3.6 kb; the 4th largest 3.5 kb. Base composition 20.3–21.9–23.5 % G; 27–28.95–30.9 % A; 19.4–20.45–21.5 % C; 28–28.7–29.4 % U. 5' terminus of RNA has a methylated nucleotide cap (1/1). Poly A region present (1/1). Genome has *t*RNA–like activity (1/1).

*Non–genomic nucleic acid* found in the virions (1/1); is *t*RNA (1/1).

*Virion protein(s)* one (3/3); $M_r$ 21500–22500–23000. Virion proteins glycosylated (1/1).

*Virus–coded non–virion proteins* identified by genomic sequence analysis; three proteins found (1/1). $M_r$ 58080. $M_r$ of 2nd largest 17370. $M_r$ of 3rd 14110.

**Replication**

Replication does not depend on a helper virus (2/2).

**Cytopathology**

Virions found in cytoplasm (2/2), or in nuclei (1/2). Inclusions absent from infected cells (1/1).

# Hybrigeminiviruses: *Geminiviridae*

*Definitive species* beet curly top hybrigeminivirus. *Tentative species* tomato pseudo curly top hybrigeminivirus.

**Natural host range and symptoms**

Symptoms persist (1/1).

**Transmission**

Transmitted by a vector; an insect; Cicadellidae (1), or Membracidae (1). Transmitted in a semi–persistent manner (1), or in a persistent manner (1). Virus retained when the vector moults; does not multiply in the vector (1/1); not transmitted congenitally to the progeny of the vector (1/1); not transmitted by mechanical inoculation; transmitted by grafting (1/1); not transmitted by seed.

**Experimental host range**

Many (>9) families susceptible (1), or several (3–9) families susceptible (1).

**Properties of particles in sap**

TIP: 80 °C. LIV: 8 days. DEP: $\log_{10}$ minus 3 (1/1). Leaf sap contains few virions (1/1).

**Particle morphology**

Virions geminate; not enveloped; 18–20–22 nm in diameter; rounded in profile (1/1).

**Physical properties**

One sedimenting component in purified preparations (1/1), or two sedimenting components in purified preparations (1/1); sedimentation coefficient 82 *S*; of the other(s) 55–70.5–86 *S*.

**Biochemical properties**

*Genome* consists of DNA; single–stranded; circular. Total genome size 2.8 kb. Genome unipartite (1/1); largest (or only) genome part 2.8 kb.

**Cytopathology**

Virions found in nuclei. Inclusions present in infected cells (1/1); are viroplasms (1/1), or unusual in shape (1/1); they contain virions (1/1).

# Idaeoviruses

*Definitive species* raspberry bushy dwarf idaeovirus.

**Natural host range and symptoms**

Symptoms none.

**Transmission**

Transmitted by means not involving a vector. Virus transmitted by mechanical inoculation; transmitted by grafting; transmitted by seed; transmitted by pollen to the seed, or transmitted by pollen to the pollinated plant.

**Experimental host range**

Many (>9) families susceptible.

**Properties of particles in sap**

TIP: 65 °C. LIV: 4 days. DEP: $\log_{10}$ minus 4.

**Particle morphology**

Virions isometric; not enveloped; 33 nm in diameter; angular in profile; without a conspicuous capsomere arrangement.

**Physical properties**

One sedimenting component in purified preparations, or three sedimenting components in purified preparations; sedimentation coefficient 115 *S*. Density 1.37 g $cm^{-3}$ in CsCl.

**Biochemical properties**

Virions contain 24 % nucleic acid; 76 % protein.

*Genome* consists of RNA; single–stranded; linear. Total genome size 8.6 kb. Genome of three parts; largest (or only) genome part 5.4 kb; the 2nd largest 2.2 kb; the 3rd largest 1 kb. Poly A region absent.

*Non–genomic nucleic acid* found in the virions; is subgenomic *m*RNA.

*Virion protein(s)* one; $M_r$ 29000.

**Replication**

Replication does not depend on a helper virus.

# Ilarviruses: *Bromoviridae*

*Definitive species* apple mosaic, asparagus 2, blueberry necrotic shock, citrus leaf rugose, citrus variegation, elm mottle, humulus japonicus, hydrangea mosaic, lilac ring mottle, parietaria mottle, plum American line pattern, prune dwarf, prunus necrotic ringspot, spinach latent, tobacco streak and Tulare apple mosaic ilarviruses. *Tentative species* alstroemeria, apple necrosis, fragaria chiloensis, grapevine line pattern, lisianthus line pattern and sunflower ringspot ilarviruses.

**Natural host range and symptoms**

Symptoms persist (10/17), or vary cyclically over a few weeks (1/17), or vary seasonally (7/17), or disappear soon after infection (3/17).

**Transmission**

Transmitted by a vector (1/6), or means not involving a vector (5/6); an insect (1/1); Thysanoptera (1/1). Virus transmitted by mechanical inoculation (19/19); transmitted by grafting (15/15); transmitted by contact between plants (1/7), or not transmitted by contact between plants (6/7); transmitted by seed (13/15), or not transmitted by seed (2/15); transmitted by pollen to the seed (4/6), or transmitted by pollen to the pollinated plant (6/6).

**Experimental host range**

Many (>9) families susceptible (8/20), or several (3–9) families susceptible (8/20), or few (<3) families susceptible (4/20).

**Properties of particles in sap**

TIP: 45–57.35–66 °C. LIV: 0.0069–3.036–21 days. DEP: $\log_{10}$ minus 1 (1/13), or 2 (3/13), or 3 (9/13), or 4 (6/13), or 5–7 (1/13). Infectivity of sap increased by treatment with di–ethyl ether (1/4), or not changed by treatment with di–ethyl ether (3/4). Leaf sap contains few virions (7/8), or contains many virions (1/8).

**Particle morphology**

Virions isometric (18/20), or bacilliform (1/20), or unusually shaped (2/20); not enveloped (16/16); 19–27.46–36 nm in diameter; 20–39.16–54.5 nm in length; rounded in profile (14/16), or angular in profile (1/16), or neither clearly angular or nor rounded in profile (1/16); with a conspicuous capsomere arrangement (2/15), or without a conspicuous capsomere arrangement (13/15).

**Physical properties**

One sedimenting component in purified preparations (2/16), or two sedimenting components in purified preparations (2/16), or three sedimenting components in purified preparations (9/16), or four sedimenting components in purified preparations (3/16), or five sedimenting components in purified preparations (1/16); sedimentation coefficient 90–106.9–125 *S*; of the other(s) 72–92.33–114 *S*. Density 1.33–1.351–1.37 g $cm^{-3}$ in CsCl. Isoelectric point pH 4.6. $A_{260}/A_{280}$ ratio 1.31–1.49–1.67.

**Biochemical properties**

Virions contain 12–15.43–24 % nucleic acid; 76–84.56–88 % protein; 0 % lipid.

*Genome* consists of RNA (16/16); single–stranded (16/16); linear (7/7). Total genome size 7.717–9.86–12.22 kb. Genome of three parts (6/14), or of four parts (5/14), or of five parts (4/14); largest (or only) genome part 2.901–3.532–4.3 kb; the 2nd largest 2.366–3–3.7 kb; the 3rd largest 1.605–2.322–2.7 kb; the 4th largest 0.845–1.1–1.409 kb; the 5th largest 0.76–0.812–0.871 kb. Base composition 24–25.33–27 % G; 24–24.66–25 % A; 20–20.66–21 % C; 27–29.33–31 % U. 5' terminus of RNA has a methylated nucleotide cap (1/1). Infectivity retained when deproteinised with proteases (2/3), or lost when deproteinised with proteases (1/3); retained when deproteinised with phenol or detergent (5/6), or lost when deproteinised with phenol or detergent (1/6). Poly A region present (1/1). Additional factor required for infectivity (6/6). Genome has no *t*RNA–like activity (1/1).

*Non–genomic nucleic acid* found in the virions (6/6); is subgenomic *m*RNA (6/6). Sub–genomic *m*RNA found in infected cells (3/4), or not found in infected cells (1/4).

*Virion protein(s)* one (14/14); $M_r$ 19000–25220–30000. Virion proteins not glycosylated (1/1).

**Replication**

Replication does not depend on a helper virus (13/13).

**Cytopathology**

Virions found in cytoplasm (3/3), or in nuclei (1/3). Inclusions present in infected cells (1/5), or absent from infected cells (4/5); are unusual in shape (1/1); they do not contain virions (1/1).

## Ipomoviruses: *Potyviridae*

*Definitive species* sweet potato mild mottle ipomovirus. *Tentative species* sweet potato yellow dwarf ipomovirus.

**Natural host range and symptoms**

Symptoms persist.

**Transmission**

Transmitted by a vector; an insect; Aleyrodidae. Transmitted in a persistent manner. Virus transmitted by mechanical inoculation; transmitted by grafting; not transmitted by contact between plants (1/1); not transmitted by seed (1/1).

**Experimental host range**

Many (>9) families susceptible (1), or several (3–9) families susceptible (1).

**Properties of particles in sap**

TIP: 60–73.75–90 °C. LIV: 3–16.5–49 days. DEP: $\log_{10}$ minus 2–3 (1), or 6–7 (1).

**Particle morphology**

Virions filamentous; not enveloped (1/1); usually flexuous (1/1); with a clear modal length (1/1); of 750–812.5–950 nm.

**Physical properties**

One sedimenting component in purified preparations; sedimentation coefficient 155 *S*.

**Biochemical properties**

Virions contain 5 % nucleic acid; 95 % protein.

*Genome* consists of RNA (1/1); single–stranded (1/1).

*Virion protein(s)* one; $M_r$ 33000–35350–37700.

**Cytopathology**

Virions found in cytoplasm. Inclusions present in infected cells; are pinwheels; they do not contain virions (1/1).

# Luteoviruses

*Definitive species* barley yellow dwarf, bean leaf roll, beet mild yellowing, beet western yellows, carrot red leaf, groundnut rosette assistor, potato leafroll, solanum yellows, soybean dwarf, soybean Indonesian dwarf, strawberry mild yellow edge, subterranean clover red leafy and tobacco necrotic dwarf luteoviruses. *Tentative species* beet yellow net, celery yellow spot, citrus enation - woody gall, cotton anthocyanosis, cowpea stunt, filaree red leaf, grapevine ajinashika disease, millet red leaf, physalis mild chlorosis, physalis vein blotch, raspberry leaf curl, tobacco vein-distorting, tobacco yellow net and tobacco yellow vein assistor luteoviruses.

**Natural host range and symptoms**

Symptoms persist (19/20), or vary seasonally (4/20), or disappear soon after infection (2/20).

**Transmission**

Transmitted by a vector (25/26), or means not involving a vector (1/26); an insect (25/25); Aphididae (25/25). Transmitted in a persistent manner (25/25). Virus retained when the vector moults (14/14); does not multiply in the vector (11/11); not transmitted congenitally to the progeny of the vector (9/9); does not require a helper virus for vector transmission (2/8), or can help the vector transmission of another virus (7/8); not transmitted by mechanical inoculation (20/20); transmitted by grafting (10/12), or not transmitted by grafting (2/12); not transmitted by contact between plants (14/14); not transmitted by seed (16/16); not transmitted by pollen (12/12).

**Experimental host range**

Many (>9) families susceptible (1/23), or several (3–9) families susceptible (6/23), or few (<3) families susceptible (16/23).

**Properties of particles in sap**

TIP: 45–63.12–80 °C. LIV: 5–13.5–16 days. DEP: $\log_{10}$ minus 2 (2/3), or 3–4 (1/3). Infectivity of sap decreased by treatment with di–ethyl ether (1/2), or not changed by treatment with di–ethyl ether (1/2). Leaf sap contains few virions (13/13).

**Particle morphology**

Virions isometric (15/15); not enveloped (13/13); 23–26.36–30 nm in diameter; rounded in profile (7/14), or angular in profile (8/14); with a conspicuous capsomere arrangement (1/12), or without a conspicuous capsomere arrangement (11/12).

**Physical properties**

One sedimenting component in purified preparations (13/13); sedimentation coefficient 104–113.3–118 *S*. Density 1.38–1.4–1.42 g $cm^{-3}$ in CsCl. Isoelectric point pH 5.3. $A_{260}/A_{280}$ ratio 1.8.

**Biochemical properties**

Virions contain 28–30.31–37.8 % nucleic acid; 70–70.71–72 % protein; 0 % lipid.

*Genome* consists of RNA (11/11); single–stranded (11/11); linear (9/9). Total genome size 5.641–5.968–6.9 kb. Genome unipartite (9/9); largest (or only) genome part 5.6–6.044–6.9 kb. Base composition 24.6 % G; 29.6 % A; 23.8 % C; 22 % U. Infectivity retained when deproteinised with proteases (1/1); retained when deproteinised with phenol or detergent (1/1). Poly A region absent (4/4).

*Non–genomic nucleic acid* found in the virions (4/4); is subgenomic *m*RNA (1/4), or satellite RNA (2/4), or of an unusual type of structure (1/4). Sub–genomic *m*RNA found in infected cells (1/1).

*Virion protein(s)* one (6/10), or two (3/10), or four (1/10); $M_r$ 22600–31540–60000. $M_r$ of 2nd largest 7000–34000–55000. $M_r$ of 3rd largest 40000. $M_r$ of 4th largest 25000. Virion proteins not glycosylated (1/1). Virion proteins not phosphorylated (1/1).

*Virus–coded non–virion proteins* isolated (1/2), or identified by genomic sequence analysis (1/2); three proteins found (1/2), or six proteins found (1/2). $M_r$ 95000–109900–124900. $M_r$ of 2nd largest 69000–70000–71000. $M_r$ of 3rd 29000–31500–34000. $M_r$ of 4th 22000. $M_r$ of 5th and smaller 6000.

**Replication**

Genome replicates in cytoplasm (1/1). Replication does not depend on a helper virus (7/7).

**Cytopathology**

Virions found in cytoplasm (9/10), or in nuclei (5/10), or in cell vacuoles (6/10). Inclusions present in infected cells (3/9), or absent from infected cells (6/9); are crystals in the cytoplasm (2/3), or amorphous X–bodies (1/3), or membranous bodies (1/3); they contain virions (3/3).

# Machlomoviruses

*Definitive species* maize chlorotic mottle machlomovirus.

**Transmission**

Transmitted by a vector; an insect; Coleoptera, or Thysanoptera. Transmitted in a non–persistent manner. Virus transmitted by mechanical inoculation; transmitted by seed.

**Properties of particles in sap**

TIP: 80–85 °C. LIV: 12–33 days. DEP: $\log_{10}$ minus 6–8. Leaf sap contains few virions.

**Particle morphology**

Virions isometric; not enveloped; 30 nm in diameter; angular in profile; without a conspicuous capsomere arrangement.

**Physical properties**

One sedimenting component in purified preparations; sedimentation coefficient 109 *S*. Density 1.365 g $cm^{-3}$ in CsCl.

**Biochemical properties**

Virions contain 18 % nucleic acid; 82 % protein.

*Genome* consists of RNA; single–stranded; linear. Total genome size 4.4 kb. Genome unipartite; largest (or only) genome part 4.4 kb. Base composition 22–25 % G; 24–26 % A; 15–16 % C; 34–37 % U. 5' terminus of RNA has a methylated nucleotide cap. Poly A region absent.

*Non–genomic nucleic acid* found in the virions; is subgenomic *m*RNA. 2 virus specified dsRNA species found in infected cells. Size of largest virus specified dsRNA 4.4 kbp; 2nd largest 1.1 kbp.

*Virion protein(s)* one; $M_r$ 25100.

# Macluraviruses

*Definitive species* maclura mosaic and narcissus latent macluraviruses.

### Natural host range and symptoms

Symptoms persist, or vary seasonally (1).

### Transmission

Transmitted by a vector (1/1); an insect (1/1); Aphididae (1/1). Transmitted in a non–persistent manner (1/1). Virus transmitted by mechanical inoculation; transmitted by grafting (1/1); not transmitted by contact between plants (1/1); not transmitted by seed (1/1); not transmitted by pollen (1/1).

### Experimental host range

Several (3–9) families susceptible.

### Properties of particles in sap

TIP: 65–66.75–70 °C. LIV: 3–3.25–4 days. DEP: $\log_{10}$ minus 3–4. Infectivity of sap not changed by treatment with di–ethyl ether (1/1). Leaf sap contains few virions (1), or contains many virions (1).

### Particle morphology

Virions filamentous; not enveloped; usually straight (1), or usually flexuous (1); with a clear modal length; of 657–664.5–672 nm; 13–13.75–16 nm wide. Axial canal obscure. Basic helix obscure.

### Physical properties

One sedimenting component in purified preparations; sedimentation coefficient 155–156–158 *S*. Density 1.307–1.316–1.33 g $cm^{-3}$ in CsCl. $A_{260}/A_{280}$ ratio 1.04–1.065–1.09.

### Biochemical properties

Virions contain 3–4.25–5 % nucleic acid; 95–95.75–97 % protein; 0 % lipid.
*Genome* consists of RNA; single–stranded.
*Virion protein(s)* one; $M_r$ 32600–38800–45000.

### Replication

Replication does not depend on a helper virus (1/1).

### Cytopathology

Virions found in cytoplasm. Inclusions present in infected cells; are amorphous X–bodies (1), or pinwheels.

# Marafiviruses

*Definitive species* bermuda grass etched-line, maize rayado fino and oat blue dwarf marafivirus.

### Natural host range and symptoms

Symptoms persist (2/2), or vary seasonally (1/2).

### Transmission

Transmitted by a vector; an insect; Cicadellidae. Transmitted in a persistent manner. Virus retained when the vector moults (2/2); multiplies in the vector; not transmitted by mechanical inoculation; not transmitted by grafting (1/1); not transmitted by contact between plants (2/2); not transmitted by seed; not transmitted by pollen (1/1).

### Experimental host range

Several (3–9) families susceptible (1), or few (<3) families susceptible (2).

### Properties of particles in sap

TIP: 60–63.75–70 °C. LIV: 3–13.75–32 days. DEP: $\log_{10}$ minus 5 (1/1). Leaf sap contains few virions (1/2), or contains many virions (1/2).

### Particle morphology

Virions isometric; not enveloped; 28–30.16–33 nm in diameter; rounded in profile; with a conspicuous capsomere arrangement.

### Physical properties

Two sedimenting components in purified preparations (2), or three sedimenting components in purified preparations (1); sedimentation coefficient 119–119.3–120 *S*; of the other(s) 54–55.5–57 *S*. Density 1.28–1.367–1.46 g $cm^{-3}$ in CsCl. $A_{260}/A_{280}$ ratio 0.87–1.353–1.63.

### Biochemical properties

Virions contain 33 % nucleic acid; 67 % protein.

*Genome* consists of RNA; single–stranded; linear (2/2). Total genome size 6.7–6.98–7.5 kb. Genome unipartite; largest (or only) genome part 6.7–6.98–7.5 kb. Additional factor not required for infectivity (1/1).

*Virion protein(s)* one (2/2); $M_r$ 22000–24250–26500. Virion proteins not glycosylated (1/1).

### Replication

Replication does not depend on a helper virus (2/2).

### Cytopathology

Virions found in cytoplasm, or in cell vacuoles (2). Inclusions present in infected cells (1/2), or absent from infected cells (1/2); are crystals in the cytoplasm (1/1), or membranous bodies (1/1), or viroplasms (1/1); they contain virions (1/1).

## Monogeminiviruses: *Geminiviridae*

*Definitive species* chloris striate mosaic, datura striate mosaic, digitaria streak, maize streak, miscanthus streak, panicum streak, paspalum striate mosaic, sugarcane streak, tobacco yellow dwarf and wheat dwarf monogeminiviruses. *Tentative species* chickpea chlorotic dwarf monogeminivirus.

### Natural host range and symptoms

Symptoms persist (7/7).

### Transmission

Transmitted by a vector (10/10); an insect (10/10); Cicadellidae (10/10). Transmitted in a non–persistent manner (1/8), or in a semi–persistent manner (2/8), or in a persistent manner (6/8). Virus retained when the vector moults (5/5); does not multiply in the vector (5/5); not transmitted congenitally to the progeny of the vector (5/5); does not require a helper virus for vector transmission (1/1); transmitted by mechanical inoculation (1/10), or not transmitted by mechanical inoculation (9/10); transmitted by grafting (1/3), or not transmitted by grafting (2/3); not transmitted by contact between plants (5/5); transmitted by seed (1/6), or not transmitted by seed (5/6); not transmitted by pollen (4/4).

**Experimental host range**

Many (>9) families susceptible (1/9), or several (3–9) families susceptible (4/9), or few (<3) families susceptible (4/9).

**Properties of particles in sap**

TIP: 50–55–60 °C. LIV: 7–13.33–26 days. DEP: $\log_{10}$ minus 2–4 (1/1). Leaf sap contains few virions (4/7), or contains many virions (3/7).

**Particle morphology**

Virions geminate; not enveloped (6/6); 15–18.1–20 nm in diameter; 25–30–35 nm in length; angular in profile (5/5); with a conspicuous capsomere arrangement (1/5), or without a conspicuous capsomere arrangement (4/5).

**Physical properties**

One sedimenting component in purified preparations (5/6), or two sedimenting components in purified preparations (2/6); sedimentation coefficient 70–74.6–76 *S*; of the other(s) 50–52–54 *S*. Density 1.35 g $cm^{-3}$ in CsCl. $A_{260}/A_{280}$ ratio 1.1–1.283–1.4.

**Biochemical properties**

Virions contain 19–19.66–20 % nucleic acid; 80 % protein; 0 % lipid.

*Genome* consists of DNA; single–stranded; circular (9/9). Total genome size 2.672–2.752–2.9 kb. Genome unipartite (7/7); largest (or only) genome part 2.672–2.752–2.9 kb. Base composition 25.7–26.4–27.7 % G; 21.8–24.23–25.5 % A; 23.1–23.53–23.9 % C; 24.8–26.06–28.1 % T. Infectivity lost when deproteinised with proteases (1/1); retained when deproteinised with phenol or detergent (1/1). Poly A region absent (1/1). Additional factor required for infectivity (1/1). Genome has no *t*RNA–like activity (1/1).

*Non–genomic nucleic acid* not found in the virions (3/3). Sub–genomic *m*RNA found in infected cells (1/2), or not found in infected cells (1/2).

*Virion protein(s)* one (8/8); $M_r$ 26600–28120–32000.

*Virus–coded non–virion proteins* identified by genomic sequence analysis (1/1); one protein found (1/1). $M_r$ 10900.

**Replication**

Genome replicates in nuclei (4/4). Replication does not depend on a helper virus (4/4).

**Cytopathology**

Virions found in cytoplasm (3/8), or in nuclei (8/8), or in cell vacuoles (3/8). Inclusions present in infected cells (4/6), or absent from infected cells (2/6); are crystals in the nucleus (3/4), or viroplasms (1/4), or unusual in shape (2/4); they contain virions (3/3).

# Nanaviruses

*Definitive species* banana bunchy top, coconut foliar decay, faba bean necrotic yellows, milk vetch dwarf and subterranean clover stunt nanaviruses.

**Natural host range and symptoms**

Symptoms persist (4/4).

**Transmission**

Transmitted by a vector; an insect; Aphididae (4), or an unusual insect group (1). Transmitted in a semi–persistent manner (1), or in a persistent manner. Virus retained when the vector moults (4/4); does not multiply in the vector (1/1); not transmitted congenitally to the progeny of the vector (2/2); not transmitted by mechanical inoculation; transmitted by grafting (2/3), or not transmitted by grafting (1/3); not transmitted by contact between plants (2/2); not transmitted by seed (2/2); not transmitted by pollen (2/2).

**Experimental host range**

Few (<3) families susceptible (4/4).

**Properties of particles in sap**

Leaf sap contains few virions (3/4), or contains many virions (1/4).

**Particle morphology**

Virions isometric; not enveloped (4/4); 17–20.2–26 nm in diameter; rounded in profile (1), or angular in profile (4); without a conspicuous capsomere arrangement.

**Physical properties**

One sedimenting component in purified preparations (2/2). Density 1.34 g $cm^{-3}$ in CsCl. $A_{260}/A_{280}$ ratio 1.35–1.375–1.4.

**Biochemical properties**

Virions contain 16–16.5–17 % nucleic acid; 83–83.5–84 % protein.

*Genome* consists of DNA (4/4); single–stranded (4/4); circular (4/4). Total genome size 1.29–2.325–3.36 kb. Genome unipartite (1/4), or of three parts (1/4), or of 7 parts (2/4); largest (or only) genome part 0.89–1.147–1.3 kb; the 2nd largest 0.88–1.09–1.3 kb; the 3rd largest 0.88–1.025–1.171 kb; the 4th largest 0.87 kb. Poly A region present (1/1).

*Virion protein(s)* one (3/3); $M_r$ 19000–20330–22000.

**Cytopathology**

Virions found in cytoplasm (1/1), or in cell vacuoles (1/1). Inclusions absent from infected cells (2/2).

# Necroviruses

*Definitive species* tobacco necrosis necrovirus. *Tentative species* carnation yellow stripe and lisianthus necrosis necroviruses.

**Natural host range and symptoms**

Symptoms persist (2/2).

**Transmission**

Transmitted by a vector (2/2); a fungus (2/2); Chytridiales (2/2). Virus transmitted by mechanical inoculation; not transmitted by seed (1/1); not transmitted by pollen (1/1).

**Experimental host range**

Many (>9) families susceptible (2), or several (3–9) families susceptible (1).

**Properties of particles in sap**

TIP: 80–87.5–95 °C. LIV: 7–46.83–63 days. DEP: $\log_{10}$ minus 4 (1), or 7 (1), or 9–10 (1). Leaf sap contains few virions (1/2), or contains many virions (1/2).

**Particle morphology**

Virions isometric; not enveloped (2/2); 26–28.66–30 nm in diameter; rounded in profile (1/2), or angular in profile (1/2); without a conspicuous capsomere arrangement.

### Physical properties

One sedimenting component in purified preparations; sedimentation coefficient 99–118.6–136 *S*. Density 1.333–1.367–1.399 g $cm^{-3}$ in CsCl. Isoelectric point pH 4.5. $A_{260}/A_{280}$ ratio 1.61.

### Biochemical properties

Virions contain 18–18.5–19 % nucleic acid; 72–76.5–81 % protein; 0 % lipid.

*Genome* consists of RNA; single-stranded; linear (1/1). Total genome size 3.759–3.981–4.185 kb. Genome unipartite (2/2); largest (or only) genome part 3.759–3.879–4 kb. Base composition 24 % G; 27.5 % A; 22.5 % C; 25.5 % U. 5' terminus of RNA has an unusual type (1/1). Infectivity retained when deproteinised with phenol or detergent (1/1). Poly A region absent (1/1).

*Non–genomic nucleic acid* not found in the virions (1/1). Sub–genomic *m*RNA found in infected cells (1/1).

*Virion protein(s)* one; $M_r$ 27000–30830–35500.

### Replication

Replication does not depend on a helper virus (1/1).

### Cytopathology

Virions found in cytoplasm (2/2). Inclusions present in infected cells (2/2); are crystals in the cytoplasm (2/2), or membranous bodies (1/2), or unusual in shape (1/2); they contain virions (1/2), or they do not contain virions (1/2).

## Nepoviruses: *Comoviridae*

*Definitive species* arabis mosaic, arracacha A, artichoke Italian latent, artichoke yellow ringspot, blueberry leaf mottle, cacao necrosis, cassava green mottle, cherry leaf roll, cherry rasp leaf, chicory yellow mottle, crimson clover latent, cycas necrotic stunt, grapevine Bulgarian latent, grapevine chrome mosaic, grapevine fanleaf, hibiscus latent ringspot, lucerne Australian latent, mulberry ringspot, myrobalan latent ringspot, olive latent ringspot, peach rosette mosaic, potato black ringspot, potato U, raspberry ringspot, tobacco ringspot, tomato black ring and tomato ringspot nepoviruses. *Tentative species* arracacha B, artichoke vein banding, caraway latent, dogwood mosaic, eucharis mottle, lucerne Australian symptomless, peach enation, rubus Chinese seed-borne, satsuma dwarf, strawberry latent ringspot, sweet potato ringspot and tomato top necrosis nepoviruses.

### Natural host range and symptoms

Symptoms persist (15/29), or vary seasonally (7/29), or disappear soon after infection (8/29), or none (3/29).

### Transmission

Transmitted by a vector (18/20), or means not involving a vector (2/20); a nematode (17/17); Dorylamidae (11/11). Virus lost by the vector when it moults (1/1); does not multiply in the vector (1/1); not transmitted congenitally to the progeny of the vector (1/1); does not require a helper virus for vector transmission (2/2); transmitted by mechanical inoculation (38/38); transmitted by grafting (23/23); not transmitted by contact between plants (19/19); transmitted by seed (25/28), or not transmitted by seed (3/28); transmitted by pollen to the seed (7/9), or transmitted by pollen to the pollinated plant (4/9), or not transmitted by pollen (1/9).

**Experimental host range**

Many (>9) families susceptible (17/38), or several (3–9) families susceptible (19/38), or few (<3) families susceptible (2/38).

**Properties of particles in sap**

TIP: 50–60.63–75 °C. LIV: 1–15.9–100 days. DEP: $\log_{10}$ minus 2 (4/37), or 3 (21/37), or 4 (24/37), or 5 (16/37), or 6 (2/37). Infectivity of sap not changed by treatment with di–ethyl ether (7/7). Leaf sap contains few virions (19/26), or contains many virions (7/26).

**Particle morphology**

Virions isometric; not enveloped (36/36); 22–27.69–33 nm in diameter; rounded in profile (1/37), or angular in profile (37/37); with a conspicuous capsomere arrangement (4/26), or without a conspicuous capsomere arrangement (22/26).

**Physical properties**

One sedimenting component in purified preparations (3/35), or two sedimenting components in purified preparations (6/35), or three sedimenting components in purified preparations (26/35), or four sedimenting components in purified preparations (2/35); sedimentation coefficient 112–125.7–135 *S*; of the other(s) 40–78.71–128 *S*. Density 1.2–1.451–1.53 g $cm^{-3}$ in CsCl. Isoelectric point pH 4–5.437–7.5. $A_{260}/A_{280}$ ratio 0.65–1.493–1.97.

**Biochemical properties**

Virions contain 0–23.04–46 % nucleic acid; 54–77.64–100 % protein; 0 % lipid.

*Genome* consists of RNA (33/33); single–stranded (33/33); linear (18/18). Total genome size 10.1–12.82–16.9 kb. Genome of two parts (33/33); largest (or only) genome part 5.3–7.549–9.5 kb; the 2nd largest 3.774–5.214–7.5 kb. Base composition 21.6–24.42–26.6 % G; 21.4–23.98–26 % A; 19–22.07–25.6 % C; 27.9–29.54–31 % U. 5' terminus of RNA has a VPg (2/2). Infectivity decreased when deproteinised with proteases (1/6), or lost when deproteinised with proteases (5/6); retained when deproteinised with phenol or detergent (12/13), or lost when deproteinised with phenol or detergent (1/13). Poly A region present (3/3). Additional factor required for infectivity (1/6), or not required for infectivity (5/6). Genome has no *t*RNA–like activity (1/1).

*Non–genomic nucleic acid* found in the virions (5/13), or not found in the virions (8/13); is satellite RNA (5/5).

*Virion protein(s)* one (23/32), or two (8/32), or three (1/32), or four (1/32); $M_r$ 24000–51580–71000. $M_r$ of 2nd largest 4000–24170–65000. $M_r$ of 3rd largest 22000–39000–56000. $M_r$ of 4th largest 14300. Virion proteins not glycosylated (2/2). Virion proteins not phosphorylated (1/1).

**Replication**

Genome replicates in cytoplasm (2/2). Replication does not depend on a helper virus (18/18).

**Cytopathology**

Virions found in cytoplasm (18/21), or in nuclei (2/21), or in the perinuclear space (1/21), or in Golgi apparatus (1/21), or in cell vacuoles (5/21). Inclusions present in infected cells (17/20), or absent from infected cells (3/20); are crystals in the cytoplasm (1/17), or amorphous X–bodies (1/17), or membranous bodies (8/17), or viroplasms (1/17), or unusual in shape (12/17); they contain virions (12/13), or they do not contain virions (1/13).

# Nucleorhabdoviruses: *Rhabdoviridae*

*Definitive species* carrot latent, coriander feathery red vein, cow parsnip mosaic, cynodon chlorotic streak, datura yellow vein, eggplant mottled dwarf, maize mosaic, pittosporum vein yellowing, potato yellow dwarf, sonchus yellow net, sowthistle yellow vein, tomato vein clearing and wheat American striate mosaic nucleorhabdoviruses. *Tentative species* butterbur, caper vein yellowing, cereal chlorotic mottle, chondrilla juncea stunting, clover enation, coffee ringspot, croton vein yellowing, cynara, finger millet mosaic, gloriosa fleck, gomphrena, laburnum yellow vein, lotus streak, lucerne enation, maize Iranian mosaic, maize streak dwarf, passionfruit, pineapple chlorotic leaf streak, poplar vein yellowing, raspberry vein chlorosis, red clover mosaic, rice transitory yellowing, ryegrass bacilliform, sorghum stunt mosaic, sweet clover latent and thyme nucleorhabdoviruses.

### Natural host range and symptoms

Symptoms persist (18/20), or vary seasonally (2/20), or disappear soon after infection (1/20).

### Transmission

Transmitted by a vector (19/19); an insect (18/19), or a mite (1/19); Aphididae (7/17), or Cicadellidae (4/17), or Delphacidae (6/17). Transmitted in a persistent manner (13/13). Virus retained when the vector moults (7/8), or lost by the vector when it moults (1/8); multiplies in the vector (8/8); transmitted congenitally to the progeny of the vector (2/6), or not transmitted congenitally to the progeny of the vector (4/6); transmitted by mechanical inoculation (10/25), or not transmitted by mechanical inoculation (15/25); transmitted by grafting (11/11); transmitted by contact between plants (1/10), or not transmitted by contact between plants (9/10); not transmitted by seed (14/14); transmitted by pollen to the seed (1/7), or not transmitted by pollen (6/7).

### Experimental host range

Several (3–9) families susceptible (2/24), or few (<3) families susceptible (22/24).

### Properties of particles in sap

TIP: 40–54–75 °C. LIV: 0.12–7.564–44 days. DEP: $\log_{10}$ minus 2 (1/8), or 3 (5/8), or 4 (4/8), or 5 (3/8). Infectivity of sap decreased by treatment with di–ethyl ether (3/3). Leaf sap contains few virions (10/15), or contains many virions (5/15).

### Particle morphology

Virions bacilliform (1/37), or rhabdo– or bullet–shaped (37/37); enveloped (24/24); usually straight (9/9); with a clear modal length (28/28); of 95–243.2–500 nm; 43–77.67–100 nm wide. Axial canal obvious (13/18), or obscure (5/18); 4.5–17.25–30 nm in diameter. Basic helix obvious (10/13), or obscure (3/13); pitch of basic helix 4–4.437–5 nm.

### Physical properties

One sedimenting component in purified preparations (7/8), or two sedimenting components in purified preparations (2/8); sedimentation coefficient 774–922.1–1044 *S*. Density 1.165–1.181–1.2 g $cm^{-3}$ in sucrose.

### Biochemical properties

Virions contain 1–2.25–5 % nucleic acid; 68–76.5–80 % protein; 15–20.37–25 % lipid.

*Genome* consists of RNA (5/5); single–stranded (5/5); linear (4/4). Total genome size 7–11.84–14 kb. Genome unipartite (5/5); largest (or only) genome part 7–11.84–14 kb. Base composition 26.4 % G; 21.3 % A; 22.5 % C; 29.8 % U. Infectivity lost when deproteinised with proteases (4/4); lost when deproteinised with phenol or detergent (4/4). Poly A region absent (2/2). Additional factor required for infectivity (2/2). Genome has no *t*RNA–like activity (1/1).

*Non–genomic nucleic acid* not found in the virions (2/2). Sub–genomic *m*RNA found in infected cells (1/2), or not found in infected cells (1/2).

*Virion protein(s)* four (4/11), or five (6/11), or six (1/11); $M_r$ 200–118000–240900. $M_r$ of 2nd largest 78–62440–92000. $M_r$ of 3rd largest 56–45380–61000. $M_r$ of 4th largest 31–25670–45000. $M_r$ of 5th largest 37–23820–33000. Virion proteins glycosylated (7/7).

**Replication**

Genome replicates in nuclei (4/4), or in cytoplasm (1/4). Replication does not depend on a helper virus (8/8).

**Cytopathology**

Virions found in cytoplasm (13/36), or in nuclei (12/36), or in nucleoli (1/36), or in the perinuclear space (31/36), or in mitochondria (2/36), or in unusual sites (1/36). Inclusions present in infected cells (6/12), or absent from infected cells (6/12); are crystals in the nucleus (1/7), or membranous bodies (1/7), or viroplasms (4/7), or unusual in shape (3/7); they contain virions (5/6), or they do not contain virions (1/6).

## Oryzaviruses: *Reoviridae*

*Definitive species* echinochloa ragged stunt and rice ragged stunt oryzaviruses.

**Natural host range and symptoms**

Symptoms persist.

**Transmission**

Transmitted by a vector; an insect; Delphacidae. Transmitted in a persistent manner. Virus retained when the vector moults; multiplies in the vector; not transmitted congenitally to the progeny of the vector; does not require a helper virus for vector transmission (1/1); transmitted by mechanical inoculation (1), or not transmitted by mechanical inoculation (1); not transmitted by grafting (1/1); not transmitted by contact between plants; not transmitted by seed; not transmitted by pollen.

**Experimental host range**

Few (<3) families susceptible.

**Properties of particles in sap**

TIP: 70–75–80 °C. LIV: 3 days. DEP: $\log_{10}$ minus 6–7 (1/1). Leaf sap contains few virions, or contains many virions (1).

**Particle morphology**

Virions isometric; not enveloped; 65–72.5–80 nm in diameter; angular in profile; with a conspicuous capsomere arrangement (1), or without a conspicuous capsomere arrangement (1).

**Physical properties**

One sedimenting component in purified preparations (1), or two sedimenting components in purified preparations (1). $A_{260}/A_{280}$ ratio 1.8.

**Biochemical properties**

*Genome* consists of RNA; double–stranded; linear (1/1). Total genome size 26.64–26.65–26.66 kb. Genome of 10 parts; largest (or only) genome part

3.85–3.875–3.9 kb; the 2nd largest 3.82–3.86–3.9 kb; the 3rd largest 3.8 kb; the 4th largest 3.7–3.75–3.8 kb; the 5th largest 2.7–2.725–2.75 kb; the 6th largest and other parts 2.3–2.45–2.6 kb. Additional factor required for infectivity (1/1).

*Non–genomic nucleic acid* not found in the virions (1/1).

*Virion protein(s)* six (1), or 8 (1); $M_r$ 124900–125900–126900. $M_r$ of 2nd largest 97000–109900–122900. $M_r$ of 3rd largest 66000–84500–102900. $M_r$ of 4th largest 63000–63500–64000. $M_r$ of 5th largest 48000–49000–50000. $M_r$ of 6th largest and smaller, 34000–38500–43000.

**Replication**

Genome replicates in cytoplasmic viroplasms (1/1). Replication does not depend on a helper virus (1/1).

**Cytopathology**

Virions found in cytoplasm (1/1). Inclusions present in infected cells; are viroplasms; they contain virions.

# Ourmiaviruses

*Definitive species* cassava Ivorian bacilliform, epirus cherry, ourmia melon and pelargonium zonate spot ourmiaviruses. *Tentative species* olive latent 2 ourmiavirus.

**Natural host range and symptoms**

Symptoms persist (2/3), or disappear soon after infection (1/3).

**Transmission**

Virus transmitted by mechanical inoculation (4/4); transmitted by grafting (1/1); transmitted by seed (2/2); transmitted by pollen to the seed (1/1).

**Experimental host range**

Many (>9) families susceptible (1), or several (3–9) families susceptible (4).

**Properties of particles in sap**

TIP: 40–56–80 °C. LIV: 0.3–14.32–35 days. DEP: $\log_{10}$ minus 2–3 (1), or 5–6 (2). Leaf sap contains few virions (1/3), or contains many virions (2/3).

**Particle morphology**

Virions bacilliform; not enveloped (3/3); 26–41.6–76 nm in length; 18–21.1–26 nm wide.

**Physical properties**

One sedimenting component in purified preparations (1), or two sedimenting components in purified preparations (2), or three sedimenting components in purified preparations (3), or four sedimenting components in purified preparations (1); sedimentation coefficient 82–100–118 *S*; of the other(s) 77–81–90 *S*. Density 1.286–1.35–1.375 g $cm^{-3}$ in CsCl. $A_{260}/A_{280}$ ratio 0.62–1.105–1.58.

**Biochemical properties**

Virions contain 15–19.25–25 % nucleic acid; 75–80.33–85 % protein; 0 % lipid.

*Genome* consists of RNA; single–stranded; linear (2/2). Total genome size 1.58–6.826–10.65 kb. Genome of two parts (1), or of three parts (3), or of four parts (1); largest (or only) genome part 0.91–3–4.4 kb; the 2nd largest 0.35–2.106–3.3 kb; the 3rd largest 0.32–1.625–2.73 kb; the 4th largest 2.1 kb. Infectivity retained when deproteinised with phenol or detergent (1/1). Additional factor not required for infectivity (1/1).

*Non–genomic nucleic acid* found in the virions (1/1); is of unknown origin and function (1/1).

*Virion protein(s)* one; $M_r$ 21400–23340–26300.

**Replication**

Replication does not depend on a helper virus (1/1).

**Cytopathology**

Virions found in cytoplasm (2/2), or in nuclei (1/2), or in nucleoli (1/2). Inclusions absent from infected cells (1/1).

# Phytoreoviruses: *Reoviridae*

*Definitive species* clover wound tumor, rice dwarf and rice gall dwarf phytoreoviruses. *Tentative species* rice bunchy stunt and sweet potato phytoreoviruses.

**Natural host range and symptoms**

Symptoms persist (4/4).

**Transmission**

Transmitted by a vector (4/4); an insect (4/4); Cicadellidae (4/4). Transmitted in a persistent manner (4/4). Virus retained when the vector moults (3/3); multiplies in the vector (4/4); transmitted congenitally to the progeny of the vector (3/4), or not transmitted congenitally to the progeny of the vector (1/4); not transmitted by mechanical inoculation (4/4); not transmitted by contact between plants (3/3); not transmitted by seed (4/4); not transmitted by pollen (2/2).

**Experimental host range**

Many (>9) families susceptible (1/3), or few (<3) families susceptible (2/3).

**Properties of particles in sap**

TIP: 45–55.83–60 °C. LIV: 4–7–10 days. DEP: $\log_{10}$ minus 3–4 (1/2), or 6 (1/2). Infectivity of sap not changed by treatment with di–ethyl ether (1/1). Leaf sap contains few virions (1/2), or contains many virions (1/2).

**Particle morphology**

Virions isometric (4/4); not enveloped (2/2); 60–67–73 nm in diameter; angular in profile (3/3); with a conspicuous capsomere arrangement (2/3), or without a conspicuous capsomere arrangement (1/3).

**Physical properties**

One sedimenting component in purified preparations (1/3), or two sedimenting components in purified preparations (2/3); sedimentation coefficient 510 *S*. Density 1.26–1.385–1.46 g $cm^{-3}$ in CsCl. $A_{260}/A_{280}$ ratio 1.4–1.41–1.42.

**Biochemical properties**

Virions contain 20–21–22 % nucleic acid; 78–79–80 % protein.

*Genome* consists of RNA (3/3); double–stranded (3/3); linear (1/1). Total genome size 25.13–25.19–25.27 kb. Genome of 12 parts (3/3); largest (or only) genome part 4.131–4.302–4.423 kb; the 2nd largest 3.385–3.434–3.532 kb; the 3rd largest 3.195–3.201–3.205 kb; the 4th largest 2.468–2.511–2.565 kb; the 5th largest 2.47–2.551–2.613 kb; the 6th largest and other parts 1.699–1.715–1.726 kb. Base composition 21.8–22.3–22.8 % G; 27.8–28.1–28.4 % A; 21.4–21.5–21.6 % C; 28–28.1–28.2 % U. Base composition of ds genome 39–42.1–48.1 % (G + C). 5' terminus of RNA has an unusual type (1/1). Infectivity decreased when deproteinised with proteases (1/1); lost when deproteinised with phenol or detergent (2/2). Poly A region absent (1/1).

Sub–genomic *m*RNA not found in infected cells (1/1).

*Virion protein(s)* six (1/3), or 7 (2/3); $M_r$ 182900–196400–209900. $M_r$ of 2nd largest 164900–174900–184900. $M_r$ of 3rd largest 149900–157400–164900. $M_r$ of 4th largest 113900–128400–142900. $M_r$ of 5th largest 55000–87500–119900. $M_r$ of 6th largest and smaller, 45000–48250–56000.

**Replication**

Genome replicates in cytoplasm (1/1). Replication does not depend on a helper virus (2/2).

**Cytopathology**

Virions found in cytoplasm (3/3), or in cell vacuoles (1/3). Inclusions present in infected cells (3/3); are viroplasms (2/3), or unusual in shape (1/3); they contain virions (1/2), or they do not contain virions (1/2).

# Potexviruses

*Definitive species* asparagus 3, cactus X, cassava X, chicory X, clover yellow mosaic, commelina X, cymbidium mosaic, daphne X, foxtail mosaic, hydrangea ringspot, lily X, narcissus mosaic, nerine X, papaya mosaic, pepino mosaic, plantago asiatica mosaic, plantain X, potato aucuba mosaic, potato X, tulip X, viola mottle and white clover mosaic potexviruses. *Tentative species* artichoke curly dwarf, bamboo mosaic, burdock yellow mosaic, cassava Caribbean mosaic, cassava Colombian symptomless, cassava common mosaic, centrosema mosaic, clitoria mosaic, crotalaria spectabilis yellow mosaic, dioscorea latent, groundnut chlorotic spot, impatiens latent, lychnis, malva veinal necrosis, nandina mosaic, negro coffee mosaic, onion mite-borne latent, parsley 5, parsnip 3, rehmannia X, rhododendron necrotic ringspot, rhubarb 1, shallot mite-borne latent, Sieg River, silene X, smithiana, sonchus mosaic, strawberry mild yellow edge-associated, tamus latent, zygocactus Montana X and zygocactus potexviruses.

**Natural host range and symptoms**

Symptoms persist (23/29), or vary seasonally (3/29), or none (4/29).

**Transmission**

Transmitted by a vector (11/22), or means not involving a vector (11/22); an insect (8/10), or a mite (2/10); Aphididae (7/8), or an unusual insect group (1/8); Eriophyidae (2/2). Transmitted in a non–persistent manner (6/6). Virus requires, for vector transmission, a helper virus (2/3), or does not require a helper virus for vector transmission (1/3); transmitted by mechanical inoculation (43/46), or not transmitted by mechanical inoculation (3/46); transmitted by grafting (7/7); transmitted by contact between plants (5/11), or not transmitted by contact between plants (6/11); transmitted by seed (2/19), or not transmitted by seed (17/19); not transmitted by pollen (4/4).

**Experimental host range**

Many (>9) families susceptible (4/40), or several (3–9) families susceptible (23/40), or few (<3) families susceptible (13/40).

**Properties of particles in sap**

TIP: 55–69.3–100 °C. LIV: 0.25–80.59–365 days. DEP: $\log_{10}$ minus 3 (4/28), or 4 (7/28), or 5 (12/28), or 6 (14/28), or 7 (4/28), or 9 (2/28), or 10 (1/28). Infectivity of sap not changed by treatment with di–ethyl ether (4/4). Leaf sap contains few virions (10/31), or contains many virions (21/31).

**Particle morphology**

Virions filamentous (51/51); not enveloped (38/38); usually flexuous (45/45); with a clear modal length (50/50); of 445–523.1–775 nm; 11–12.8–15 nm wide. Axial canal obvious (2/35), or obscure (33/35); 3.4–6.3–12 nm in diameter. Basic helix obvious (8/27), or obscure (19/27); pitch of basic helix 2.8–3.331–3.5 nm.

**Physical properties**

One sedimenting component in purified preparations (17/22), or two sedimenting components in purified preparations (5/22); sedimentation coefficient 110–122–144 *S*; of the other(s) 102–111.6–122 *S*. Density 1.28–1.308–1.33 g $cm^{-3}$ in CsCl. Isoelectric point pH 4–4.75–5.3. $A_{260}/A_{280}$ ratio 1.18–1.229–1.4.

**Biochemical properties**

Virions contain 5–5.755–8 % nucleic acid; 92–94.14–95 % protein; 0 % lipid.

*Genome* consists of RNA (25/25); single–stranded (25/25); linear (16/16). Total genome size 5.845–6.725–8.1 kb. Genome unipartite (22/22); largest (or only) genome part 5.845–6.725–8.1 kb. Base composition 15.5–21.12–25 % G; 26.4–29.9–33.8 % A; 23.4–26.02–30.3 % C; 20.5–22.92–25.7 % U. 5' terminus of RNA has a methylated nucleotide cap (1/3), or an unusual type (2/3). Infectivity retained when deproteinised with proteases (1/1); retained when deproteinised with phenol or detergent (6/6). Poly A region present (5/5). Additional factor not required for infectivity (3/3).

*Non–genomic nucleic acid* found in the virions (3/9), or not found in the virions (6/9); is subgenomic *m*RNA (3/3). Sub–genomic *m*RNA found in infected cells (4/4). 4 virus specified dsRNA species found in infected cells. Size of largest virus specified dsRNA 6–11.55–17.1 kbp; 2nd largest 4.2–5.6–7 kbp; 3rd largest 1.7–3.7–5.7 kbp; 4th largest 0.5–2.55–4.6 kbp.

*Virion protein(s)* one (25/26), or two (1/26); $M_r$ 14000–24340–30000. $M_r$ of 2nd largest 30500. Virion proteins not glycosylated (2/2).

*Virus–coded non–virion proteins* isolated (2/4), or identified by genomic sequence analysis (2/4); one protein found (2/4), or three proteins found (1/4), or five proteins found (1/4). $M_r$ 115900–141800–159900. $M_r$ of 2nd largest 25340–45670–66000. $M_r$ of 3rd 11570–27780–44000. $M_r$ of 4th 11210. $M_r$ of 5th and smaller 8079.

**Replication**

Replication does not depend on a helper virus (19/19).

**Cytopathology**

Virions found in cytoplasm (26/26), or in nuclei (1/26), or in cell vacuoles (4/26). Inclusions present in infected cells (23/27), or absent from infected cells (4/27); are crystals in the cytoplasm (3/19), or crystals in the nucleus (1/19), or amorphous X–bodies (2/19), or viroplasms (1/19), or unusual in shape (15/19); they contain virions (19/20), or they do not contain virions (1/20).

# Potyviruses: *Potyviridae*

*Definitive species* alstroemeria mosaic, amaranthus leaf mottle, araujia mosaic, arracacha Y, artichoke latent, asparagus 1, banana bract mosaic, bean common mosaic necrosis, bean common mosaic, bean yellow mosaic, beet mosaic, bidens mosaic, bidens mottle, cardamom mosaic, carnation vein mottle, carrot thin leaf, cassava brown streak, cassia yellow spot, celery mosaic, chickpea bushy dwarf, chickpea distortion mosaic, clover yellow vein, commelina diffusa, commelina mosaic, cowpea green vein-banding, cowpea Moroccan aphid-borne mosaic, cowpea rugose mosaic, crinum mosaic, daphne Y, dasheen mosaic, datura Colombian, datura distortion mosaic, datura necrosis, datura shoestring, dendrobium mosaic, desmodium mosaic, dioscorea alata, dioscorea green banding mosaic, eggplant green mosaic, euphorbia ringspot, freesia mosaic, groundnut eyespot, guar symptomless, guinea grass mosaic, helenium Y, henbane mosaic, hippeastrum mosaic, hyacinth mosaic, iris fulva mosaic, iris mild mosaic, iris severe mosaic, johnsongrass mosaic, kennedya Y, leek yellow stripe, lettuce mosaic, lily mottle, maize dwarf mosaic, malva vein clearing, marigold mottle, narcissus yellow stripe, nerine, onion yellow dwarf, ornithogalum mosaic, papaya ringspot, parsnip mosaic, passiflora ringspot, passiflora South African, passionfruit woodiness, patchouli mosaic, pea mosaic, pea seed-borne mosaic, peanut green mosaic, peanut mottle, pepper Indian mottle, pepper mottle, pepper severe mosaic, pepper veinal mottle, plum pox, pokeweed mosaic, potato A, potato V, potato Y, primula mosaic, ranunculus mottle, sorghum mosaic, soybean mosaic, statice Y, sugarcane mosaic, sweet potato feathery mottle, sweet potato G, swordbean distortion mosaic, tamarillo mosaic, telfairia mosaic, tobacco etch, tobacco vein-banding mosaic, tobacco vein mottling, tobacco wilt, tomato Peru, tradescantia/Zebrina, tropaeolum 1, tropaeolum 2, tuberose, tulip band-breaking, tulip breaking, tulip chlorotic blotch, turnip mosaic, ullucus mosaic, vallota mosaic, vanilla mosaic, vanilla necrosis, voandzeia distortion mosaic, watermelon mosaic 1, watermelon mosaic 2, wild potato mosaic, wisteria vein mosaic, yam mosaic, zucchini yellow fleck and zucchini yellow mosaic potyviruses. *Tentative species* asystasia gangetica mottle, celery latent, datura mosaic, endive necrotic mosaic, kalanchoe mosaic, konjak mosaic, nasturtium mosaic, patchouli mottle, shallot yellow stripe, sweet potato vein mosaic and welsh onion yellow stripe potyviruses.

**Natural host range and symptoms**

Symptoms persist (86/99), or vary cyclically over a few weeks (3/99), or vary seasonally (9/99), or disappear soon after infection (7/99), or none (3/99).

**Transmission**

Transmitted by a vector (105/106), or means not involving a vector (1/106); an insect (105/105); Aphididae (105/105). Transmitted in a non–persistent manner (99/99). Virus does not require a helper virus for vector transmission (8/8), or can help the vector transmission of another virus (2/8); transmitted by mechanical inoculation (111/113), or not transmitted by mechanical inoculation (2/113); transmitted by grafting (30/32), or not transmitted by grafting (2/32); transmitted by contact between plants (2/38), or not transmitted by contact between plants (36/38); transmitted by seed (16/68), or not transmitted by seed (52/68); transmitted by pollen to the seed (3/20), or transmitted by pollen to the pollinated plant (1/20), or not transmitted by pollen (17/20).

**Experimental host range**

Many (>9) families susceptible (8/111), or several (3–9) families susceptible (53/111), or few (<3) families susceptible (50/111).

**Properties of particles in sap**

TIP: 44–59.1–90 °C. LIV: 0.125–7.283–64 days. DEP: $\log_{10}$ minus 1 (5/91), or 2 (23/91), or 3–4 (49/91), or 5 (25/91), or 6 (12/91), or 7 (2/91), or 8 (1/91). Infectivity of sap increased by treatment with di–ethyl ether (2/13), or decreased by treatment with di–ethyl ether (1/13), or not changed by treatment with di–ethyl ether (11/13). Leaf sap contains few virions (76/87), or contains many virions (12/87).

**Particle morphology**

Virions filamentous (120/120); not enveloped (99/99); usually flexuous (116/116); with a clear modal length (109/111), or with no clear modal length (2/111); of 650–758.4–900 nm; 11–12.86–20 nm wide. Axial canal obvious (5/84), or obscure (79/84); 2–2.5–3 nm in diameter. Basic helix obvious (5/71), or obscure (66/71); pitch of basic helix 3.3–3.38–3.4 nm.

**Physical properties**

One sedimenting component in purified preparations (61/61), or two sedimenting components in purified preparations (2/61); sedimentation coefficient 140–151.4–180 *S*; of the other(s) 150–155–160 *S*. Density 1.26–1.316–1.336 g $cm^{-3}$ in CsCl. Density 1.326–1.328–1.33 g $cm^{-3}$ in sucrose. Isoelectric point pH 4.3–4.65–5.3. $A_{260}/A_{280}$ ratio 0.77–1.417–2.9.

**Biochemical properties**

Virions contain 4–5.466–7 % nucleic acid; 93–94.63–96 % protein; 0 % lipid.

*Genome* consists of RNA (46/46); single–stranded (46/46); linear (25/25). Total genome size 9–10.08–12 kb. Genome unipartite (29/29); largest (or only) genome part 9–10.08–12 kb. Base composition 21–23.51–26 % G; 23–30.15–44 % A; 14.9–22.41–28 % C; 15.6–24.41–30.9 % U. 5' terminus of RNA has a VPg (4/4). Infectivity retained when deproteinised with proteases (1/3), or decreased when deproteinised with proteases (2/3), or lost when deproteinised with proteases (1/3); retained when deproteinised with phenol or detergent (7/7). Poly A region present (9/9). Additional factor not required for infectivity (5/5).

*Non–genomic nucleic acid* not found in the virions (7/7). Sub–genomic *m*RNA not found in infected cells (1/1).

*Virion protein(s)* one (45/49), or two (2/49), or three (1/49), or five (1/49); $M_r$ 28300–33940–54100. $M_r$ of 2nd largest 29000–34630–39900. $M_r$ of 3rd largest 27000–30500–34000. $M_r$ of 4th largest 33000. $M_r$ of 5th largest 32000.

*Virus–coded non–virion proteins* isolated (8/12), or identified by genomic sequence analysis (4/12); one protein found (2/12), or two proteins found (5/12), or three proteins found (1/12), or four proteins found (1/12), or five proteins found (1/12), or seven proteins found (2/12). $M_r$ 39000–63900–81000. $M_r$ of 2nd largest 49000–53750–70000. $M_r$ of 3rd 47000–50000–53000. $M_r$ of 4th 49000–59750–70000. $M_r$ of 5th and smaller 34000–45830–69000.

**Replication**

Genome replicates in cytoplasm (3/3). Replication does not depend on a helper virus (29/29). Acts as helper for another virus (1/1).

**Cytopathology**

Virions found in cytoplasm (78/79), or in nuclei (1/79), or in chloroplasts (1/79), or in Golgi apparatus (1/79), or in cell vacuoles (2/79), or in unusual sites (2/79). Inclusions present in infected cells (93/95), or absent from infected cells (2/95); are crystals in the cytoplasm (9/94), or crystals in the nucleus (8/94), or amorphous X–bodies (6/94), or membranous bodies (1/94), or viroplasms (2/94), or pinwheels (84/94), or unusual in shape (34/94); they contain virions (7/65), or they do not contain virions (59/65).

# Rymoviruses: *Potyviridae*

*Definitive species* agropyron mosaic, hordeum mosaic, oat necrotic mottle, ryegrass mosaic and wheat streak mosaic rymoviruses. *Tentative species* brome streak mosaic and spartina mottle rymoviruses.

**Natural host range and symptoms**

Symptoms persist (5/5), or vary seasonally (1/5).

**Transmission**

Transmitted by a vector (3/4), or means not involving a vector (1/4); a mite (3/3); Eriophyidae (3/3). Virus not transmitted congenitally to the progeny of the vector (1/1); transmitted by mechanical inoculation (6/6); not transmitted by contact between plants (2/2); transmitted by seed (1/4), or not transmitted by seed (3/4); not transmitted by pollen (3/3).

**Experimental host range**

Few (<3) families susceptible.

**Properties of particles in sap**

TIP: 50–52.8–60 °C. LIV: 1–8.75–56 days. DEP: $\log_{10}$ minus 3 (6/6), or 4 (1/6). Leaf sap contains few virions (2/4), or contains many virions (2/4).

**Particle morphology**

Virions filamentous; not enveloped (4/4); usually flexuous (6/6); with a clear modal length; of 693–709.8–760 nm; 11–13.7–15.5 nm wide. Axial canal obscure (4/4). Basic helix obscure (3/3).

**Physical properties**

One sedimenting component in purified preparations (5/5); sedimentation coefficient 150–161.5–166 *S*. Density 1.304–1.314–1.325 g $cm^{-3}$ in CsCl. $A_{260}/A_{280}$ ratio 1.33–1.35–1.37.

**Biochemical properties**

Virions contain 5.3 % nucleic acid; 0 % lipid.

*Genome* consists of RNA (5/5); single-stranded (5/5); linear (2/2). Total genome size 8–9–10 kb. Genome unipartite (4/4); largest (or only) genome part 8–9–10 kb. Base composition 31 % G; 23 % A; 24 % C; 22 % U. 5' terminus of RNA has a VPg (2/2). Infectivity retained when deproteinised with phenol or detergent (1/1). Poly A region present (1/1). Genome has no *t*RNA–like activity (1/1).

*Non–genomic nucleic acid* not found in the virions (1/1). Sub–genomic *m*RNA not found in infected cells (2/2).

*Virion protein(s)* one (5/5); $M_r$ 29200–38660–47000.

*Virus–coded non–virion proteins* isolated (2/2); one protein found (2/2). $M_r$ 68000–70000–72000.

**Replication**

Genome replicates in cytoplasm (2/2). Replication does not depend on a helper virus (3/3).

**Cytopathology**

Virions found in cytoplasm (6/6). Inclusions present in infected cells; are pinwheels, or unusual in shape (1); they do not contain virions (4/4).

# Satellite RNAs

*Definitive species* arabis mosaic, chicory yellow mottle, cucumber mosaic, grapevine fanleaf, strawberry latent ringspot, tobacco ringspot, tomato black ring and velvet tobacco mottle virus viroid-like satellite RNAs – no data, except sequence accession numbers stored.

# Satelliviruses

*Definitive species* maize white line mosaic, panicum mosaic, tobacco mosaic and tobacco necrosis satelliviruses.

**Transmission**

Transmitted by a vector (1/1); a fungus (1/1); Chytridiales (1/1). Virus transmitted by mechanical inoculation (2/3), or not transmitted by mechanical inoculation (1/3); not transmitted by seed (2/2).

**Experimental host range**

Several (3–9) families susceptible (1/1).

**Properties of particles in sap**

TIP: 95 °C.

**Particle morphology**

Virions isometric (3/3); not enveloped (3/3); 17 nm in diameter; rounded in profile (2/3), or neither clearly angular or nor rounded in profile (1/3); with a conspicuous capsomere arrangement (1/2), or without a conspicuous capsomere arrangement (1/2).

**Physical properties**

One sedimenting component in purified preparations; sedimentation coefficient 42–45.25–50 *S*. Density 1.355 g $cm^{-3}$ in CsCl. Isoelectric point pH 7. $A_{260}/A_{280}$ ratio 1.53–1.59–1.65.

**Biochemical properties**

Virions contain 20–22.5–25 % nucleic acid; 75–77.5–80 % protein.

*Genome* consists of DNA (1), or consists of RNA (3); single–stranded; linear (1/1). Total genome size 1.059–1.155–1.239 kb. Genome unipartite (3/3); largest (or only) genome part 1.059–1.155–1.239 kb. Base composition 22–23.7–24.4 % G; 26.1–27.3–29 % A; 20–21.15–22 % C; 25–27.6–29 % U. 5' terminus of RNA has an unusual type (1/1). Infectivity retained when deproteinised with proteases (1/1); retained when deproteinised with phenol or detergent (1/1). Poly A region absent (1/1). Additional factor not required for infectivity (1/1).

*Non–genomic nucleic acid* not found in the virions (1/1). Sub–genomic *m*RNA not found in infected cells (1/1). 1 virus specified dsRNA species found in infected cells. Size of largest virus specified dsRNA 1.1 kbp.

*Virion protein(s)* one (3/3); $M_r$ 17500–21730–24700.

*Virus–coded non–virion proteins* isolated (1/1); one protein found (1/1). $M_r$ 68000.

**Replication**

Genome replicates in cytoplasmic viroplasms (1/1). Replication depends on co–infection with a helper virus (1/1).

**Cytopathology**

Virions found in cytoplasm (3/3). Inclusions present in infected cells (1/3), or absent from infected cells (2/3); are crystals in the cytoplasm (1/1), or viroplasms (1/1), or unusual in shape (1/1); they contain virions (1/1).

# Sequiviruses: *Sequiviridae*

*Definitive species* dandelion yellow mosaic and parsnip yellow fleck sequiviruses.

**Natural host range and symptoms**

Symptoms persist.

**Transmission**

Transmitted by a vector; an insect; Aphididae. Transmitted in a semi–persistent manner (1/1). Virus lost by the vector when it moults (1/1); not transmitted congenitally to the progeny of the vector (1/1); requires, for vector transmission, a helper virus (1/1); transmitted by mechanical inoculation; not transmitted by seed (1/1); not transmitted by pollen (1/1).

**Experimental host range**

Several (3–9) families susceptible.

**Properties of particles in sap**

TIP: 57.5–61.87–65 °C. LIV: 1–3.25–7 days. DEP: $\log_{10}$ minus 3 (1), or 4, or 5 (1). Leaf sap contains few virions (1/1).

**Particle morphology**

Virions isometric; not enveloped (1/1); 31 nm in diameter; rounded in profile (1/1); without a conspicuous capsomere arrangement (1/1).

**Physical properties**

One sedimenting component in purified preparations (1), or two sedimenting components in purified preparations (1); sedimentation coefficient 153–156–159 *S*; of the other(s) 60 *S*. Density 1.297–1.393–1.49 g $cm^{-3}$ in CsCl. $A_{260}/A_{280}$ ratio 1.46–1.597–1.67.

**Biochemical properties**

Virions contain 40 % nucleic acid.

*Genome* consists of RNA; single–stranded. Total genome size 9.871 kb. Genome unipartite; largest (or only) genome part 9.871–9.935–10 kb.

*Virion protein(s)* three; $M_r$ 29000–30700–32000. $M_r$ of 2nd largest 25000–26370–29500. $M_r$ of 3rd largest 22700–24100–27000.

**Replication**

Replication does not depend on a helper virus (1/1).

**Cytopathology**

Virions found in cytoplasm (1/1). Inclusions present in infected cells (1/1); are membranous bodies (1/1); they contain virions (1/1).

# Sobemoviruses

*Definitive species* bean southern mosaic, blueberry shoestring, cocksfoot mottle, lucerne transient streak, rice yellow mottle, rottboelia yellow mottle, solanum nodiflorum mottle, sowbane mosaic, subterranean clover mottle, turnip rosette and velvet tobacco mottle sobemoviruses. *Tentative species* cocksfoot mild mosaic, cynosurus mottle, ginger chlorotic fleck, olive latent 1, panicum mosaic and ryegrass mottle sobemoviruses.

**Natural host range and symptoms**

Symptoms persist (14/16), or vary seasonally (2/16), or disappear soon after infection (1/16), or none (1/16).

**Transmission**

Transmitted by a vector (8/8); an insect (8/8); Aphididae (4/8), or Cicadellidae (1/8), or Coleoptera (6/8). Transmitted in a non–persistent manner (2/7), or in a semi–persistent manner (5/7). Virus lost by the vector when it moults (1/1); does not multiply in the vector (2/2); not transmitted congenitally to the progeny of the vector (2/2); does not require a helper virus for vector transmission (2/2); transmitted by mechanical inoculation (15/15); transmitted by grafting (3/3); not transmitted by contact between plants (2/2); transmitted by seed (4/11), or not transmitted by seed (8/11); transmitted by pollen to the seed (2/5), or transmitted by pollen to the pollinated plant (2/5), or not transmitted by pollen (3/5).

**Experimental host range**

Several (3–9) families susceptible (4), or few (<3) families susceptible (13).

**Properties of particles in sap**

TIP: 60–78.72–96 °C. LIV: 3–95.37–365 days. DEP: $\log_{10}$ minus 3–4 (1/10), or 5 (6/10), or 6 (3/10), or 7 (2/10), or 8 (3/10), or 9 (2/10). Infectivity of sap not changed by treatment with di–ethyl ether (3/3). Leaf sap contains few virions (2/11), or contains many virions (10/11).

**Particle morphology**

Virions isometric; not enveloped (12/12); 25–28.38–33 nm in diameter; rounded in profile (4/15), or angular in profile (12/15); with a conspicuous capsomere arrangement (2/10), or without a conspicuous capsomere arrangement (8/10).

**Physical properties**

One sedimenting component in purified preparations (16/16); sedimentation coefficient 104–112.5–120 *S*; of the other(s) 102 *S*. Density 1.314–1.367–1.399 g $cm^{-3}$ in CsCl. Density 1.1 g $cm^{-3}$ in sucrose. Isoelectric point pH 4.4–5.137–6. $A_{260}/A_{280}$ ratio 1.46–1.597–1.72.

**Biochemical properties**

Virions contain 17–21.5–28 % nucleic acid; 72–78.5–83 % protein; 0 % lipid.

*Genome* consists of RNA; single–stranded; linear (9/9). Total genome size 4.1–4.318–5.7 kb. Genome unipartite (15/15); largest (or only) genome part 4.1–4.317–5.7 kb. Base composition 24–25.8–29 % G; 21–23.98–26 % A; 22–25.3–28.4 % C; 21.6–24.85–27 % U. 5' terminus of RNA has a VPg (2/2). Infectivity retained when deproteinised with proteases (2/5), or lost when deproteinised with proteases (3/5); retained when deproteinised with phenol or detergent (5/5). Poly A region absent (3/3). Additional factor required for infectivity (2/3), or not required for infectivity (1/3). Genome has no *t*RNA–like activity (1/1).

*Non–genomic nucleic acid* found in the virions (5/9), or not found in the virions (4/9); is subgenomic *m*RNA (3/5), or satellite RNA (2/5), or of an unusual type of structure (1/5). Sub–genomic *m*RNA found in infected cells (2/3), or not found in infected cells (1/3). 3 virus specified dsRNA species found in infected cells. Size of largest virus specified dsRNA 4.6 kbp; 2nd largest 3 kbp; 3rd largest 1.3 kbp.

*Virion protein(s)* one (14/16), or two (2/16), or three (1/16); $M_r$ 24900–29850–37000. $M_r$ of 2nd largest 12000–25320–33000. $M_r$ of 3rd largest 31500.

*Virus–coded non–virion proteins* identified by genomic sequence analysis (1/1); three proteins found (1/1). $M_r$ 104900. $M_r$ of 2nd largest 78000. $M_r$ of 3rd 18000.

**Replication**

Replication does not depend on a helper virus (5/5). Acts as helper for a satellite virus (1/3), or a satellite RNA (2/3).

**Cytopathology**

Virions found in cytoplasm (12/13), or in nuclei (5/13), or in cell vacuoles (6/13). Inclusions present in infected cells (9/15), or absent from infected cells (6/15); are crystals in the cytoplasm (5/9), or membranous bodies (3/9), or unusual in shape (4/9); they contain virions (6/6).

# Tenuiviruses

*Definitive species* maize stripe, rice grassy stunt, rice hoja blanca and rice stripe tenuiviruses. *Tentative species* barley dubia, maize yellow stripe, oat pseudorosette, phleum green stripe, wheat European striate mosaic and wheat Iranian stripe tenuiviruses.

**Natural host range and symptoms**

Symptoms persist (5/5), or vary cyclically over a few weeks (1/5).

**Transmission**

Transmitted by a vector (9/9); an insect (9/9); Cicadellidae (1/9), or Delphacidae (8/9). Transmitted in a persistent manner (9/9). Virus retained when the vector moults (6/6); multiplies in the vector (5/5); transmitted congenitally to the progeny of the vector (4/5), or not transmitted congenitally to the progeny of the vector (1/5); transmitted by mechanical inoculation (1/8), or not transmitted by mechanical inoculation (7/8); not transmitted by contact between plants (6/6); not transmitted by seed (5/5); not transmitted by pollen (2/2).

**Experimental host range**

Few (<3) families susceptible (5/5).

**Properties of particles in sap**

TIP: 50–52.5–55 °C. LIV: 0.5 days. DEP: $\log_{10}$ minus 2–4 (1/2). Leaf sap contains few virions (1/3), or contains many virions (2/3).

**Particle morphology**

Virions filamentous (5/6), or rhabdo– or bullet–shaped (1/6); not enveloped (5/5); usually flexuous (3/3); with no clear modal length (2/2); of 950–1150–1350 nm; 3–5.2–8 nm wide. Axial canal obvious (1/2), or obscure (1/2). Basic helix obvious (1/1).

**Physical properties**

One sedimenting component in purified preparations (3/6), or two sedimenting components in purified preparations (1/6), or three sedimenting components in purified preparations (2/6), or four sedimenting components in purified preparations (2/6); sedimentation coefficient 87–137.5–187 *S*; of the other(s) 68–74–80 *S*. Density 1.28–1.283–1.288 g $cm^{-3}$ in CsCl. Isoelectric point pH 4.5.

**Biochemical properties**

Virions contain 5.2–9.733–12 % nucleic acid; 88–91.4–94.8 % protein; 0 % lipid.

*Genome* consists of RNA (4/4); single–stranded (4/4); linear (4/4). Total genome size 15.46–17.21–18.6 kb. Genome of four parts (3/4), or of five parts (1/4); largest (or only) genome part 4.39–8.077–9.8 kb; the 2nd largest 3.5–3.63–3.94 kb; the 3rd largest 2.3–2.7–3.64 kb; the 4th largest 1.99–2.462–3.49 kb; the 5th largest 1.32 kb. Base composition 18.1 % G; 26.3 % A; 17.8 % C; 37.8 % U. Additional factor not required for infectivity (1/1).

*Virion protein(s)* one (3/4), or two (1/4); $M_r$ 31000–32420–34000. $M_r$ of 2nd largest 229900.

*Virus–coded non–virion proteins* isolated (3/3); one protein found (3/3). $M_r$ 13800–15930–17500.

**Cytopathology**

Virions found in cytoplasm (5/5), or in nuclei (1/5). Inclusions present in infected cells (6/6); are crystals in the cytoplasm (3/6), or crystals in the nucleus (1/6), or amorphous X–bodies (3/6), or unusual in shape (2/6); they contain virions (2/5), or they do not contain virions (3/5).

# Tobamoviruses

*Definitive species* cucumber green mottle mosaic, frangipani mosaic, kyuri green mottle mosaic, odontoglossum ringspot, paprika mild mottle, pepper mild mottle, ribgrass mosaic, opuntia Sammons', sunn-hemp mosaic, tobacco mild green mosaic, tobacco mosaic, tomato mosaic and ullucus mild mottle tobamoviruses. *Tentative species* hibiscus yellow mosaic, maracuja mosaic, potato 14R and rose tobamoviruses.

**Natural host range and symptoms**

Symptoms persist (11/12), or vary seasonally (1/12), or disappear soon after infection (1/12).

**Transmission**

Transmitted by a vector (1/12), or means not involving a vector (12/12); an insect (1/1); Coleoptera (1/1). Virus transmitted by mechanical inoculation (16/16); transmitted by grafting (7/7); transmitted by contact between plants (7/7); transmitted by seed (5/7), or not transmitted by seed (2/7); not transmitted by pollen (3/3).

**Experimental host range**

Many (>9) families susceptible (4/16), or several (3–9) families susceptible (8/16), or few (<3) families susceptible (4/16).

**Properties of particles in sap**

TIP: 80–91.5–95 °C. LIV: 30–3898–35000 days. DEP: $\log_{10}$ minus 5 (2/11), or 6 (8/11), or 7 (7/11), or 8 (1/11), or 10 (1/11). Infectivity of sap not changed by treatment with di–ethyl ether (6/6). Leaf sap contains few virions (1/15), or contains many virions (14/15).

**Particle morphology**

Virions rod–shaped; not enveloped (15/15); usually straight (16/16); with a clear modal length (16), or with no clear modal length (1); of 297–303.3–320 nm; 15–17.26–18.2 nm wide. Axial canal obvious (14/16), or obscure (2/16); 2–3.257–4 nm in diameter. Basic helix obvious (11/12), or obscure (1/12); pitch of basic helix 2.3–2.314–2.4 nm.

### Physical properties

One sedimenting component in purified preparations (11/13), or two sedimenting components in purified preparations (1/13), or three sedimenting components in purified preparations (1/13); sedimentation coefficient 176–190.8–212 *S*; of the other(s) 35–87–119 *S*. Density 1.25–1.317–1.37 g $cm^{-3}$ in CsCl. Isoelectric point pH 3.16–3.853–4.64. $A_{260}/A_{280}$ ratio 0.63–1.113–1.33.

### Biochemical properties

Virions contain 5 % nucleic acid; 95 % protein; 0 % lipid.

*Genome* consists of RNA (12/12); single–stranded (12/12); linear (9/9). Total genome size 6.355–6.453–6.5 kb. Genome unipartite (11/11); largest (or only) genome part 6.355–6.453–6.5 kb. Base composition 23–24.32–25.8 % G; 24.6–27.92–29.8 % A; 18–19.02–20.6 % C; 26.3–28.72–31.6 % U. 5' terminus of RNA has a methylated nucleotide cap (3/3). Infectivity retained when deproteinised with proteases (2/5), or decreased when deproteinised with proteases (3/5); retained when deproteinised with phenol or detergent (7/7). Poly A region absent (3/3). Additional factor not required for infectivity (6/6). Genome has *t*RNA–like activity (3/3).

*Non–genomic nucleic acid* found in the virions (4/5), or not found in the virions (1/5); is subgenomic *m*RNA (4/4), or host genomic RNA (1/4). Sub–genomic *m*RNA found in infected cells (6/6).

*Virion protein(s)* one (12/14), or two (2/14); $M_r$ 17000–18820–26500. $M_r$ of 2nd largest 15900–16700–17500.

*Virus–coded non–virion proteins* identified by genomic sequence analysis (1/1); four proteins found (1/1). $M_r$ 182900. $M_r$ of 2nd largest 125900. $M_r$ of 3rd 54000. $M_r$ of 4th 30000.

### Replication

Genome replicates in cytoplasm (2/2). Replication does not depend on a helper virus (10/10).

### Cytopathology

Virions found in cytoplasm (12/12), or in nuclei (2/12), or in chloroplasts (2/12), or in cell vacuoles (3/12). Inclusions present in infected cells (12/12); are crystals in the cytoplasm (10/12), or amorphous X–bodies (5/12), or viroplasms (1/12), or unusual in shape (3/12); they contain virions (9/9), or they do not contain virions (1/9).

## Tobraviruses

*Definitive species* pea early browning, pepper ringspot and tobacco rattle tobraviruses. *Tentative species* harts tongue fern tobravirus.

### Natural host range and symptoms

Symptoms persist, or vary seasonally (1), or disappear soon after infection (1).

### Transmission

Transmitted by a vector (2/2); a nematode (2/2); Trichodoridae (2/2). Virus transmitted by mechanical inoculation; transmitted by grafting (2/2); not transmitted by contact between plants (2/2); transmitted by seed (2/2).

### Experimental host range

Many (>9) families susceptible (2), or several (3–9) families susceptible (1), or few (<3) families susceptible (1).

**Properties of particles in sap**

TIP: 74–80.33–85 °C. LIV: 40–90–183 days. DEP: $\log_{10}$ minus 5–6 (2/3). Infectivity of sap not changed by treatment with di–ethyl ether (2/2). Leaf sap contains few virions (3/3).

**Particle morphology**

Virions rod–shaped; not enveloped (3/3); usually straight; with a clear modal length; of 46–158.3–320 nm; 21–21.75–22 nm wide. Axial canal obvious (3), or obscure (1); 4 nm in diameter. Basic helix obvious (3), or obscure (1); pitch of basic helix 2.5 nm.

**Physical properties**

Two sedimenting components in purified preparations (3/3); sedimentation coefficient 286–297–305 *S*; of the other(s) 163–186.5–210 *S*. Density 1.306–1.31–1.324 g $cm^{-3}$ in CsCl. Density 1.27 g $cm^{-3}$ in sucrose. $A_{260}/A_{280}$ ratio 1.1–1.137–1.15.

**Biochemical properties**

Virions contain 5 % nucleic acid; 95 % protein; 0 % lipid.

*Genome* consists of RNA (3/3); single–stranded (3/3); linear (3/3). Total genome size 9.09–10.63–12 kb. Genome of two parts (3/3); largest (or only) genome part 7.272–7.557–7.9 kb; the 2nd largest 1.818–3.076–4.242 kb. Base composition 24–24.7–25.5 % G; 26–27.65–29.5 % A; 16.5–17.27–18 % C; 28.5–30.37–32 % U. 5' terminus of RNA has a methylated nucleotide cap (2/2). Infectivity retained when deproteinised with proteases (2/2); retained when deproteinised with phenol or detergent (2/2). Poly A region absent (2/2). Additional factor not required for infectivity (2/2). Genome has no *t*RNA–like activity (1/1).

*Non–genomic nucleic acid* found in the virions (2/2); is subgenomic *m*RNA (2/2). Sub–genomic *m*RNA found in infected cells (2/2).

*Virion protein(s)* one (3/3); $M_r$ 21000–22660–24000.

*Virus–coded non–virion proteins* isolated (1/2), or identified by genomic sequence analysis (1/2); two proteins found (1/2), or three proteins found (1/2). $M_r$ 169900–178400–186900. $M_r$ of 2nd largest 139900–140900–141900. $M_r$ of 3rd 30000.

**Replication**

Genome replicates in cytoplasm (1/1). Replication does not depend on a helper virus (2/2).

**Cytopathology**

Virions found in cytoplasm (2/2). Inclusions present in infected cells (2/2); are amorphous X–bodies (1/2), or unusual in shape (1/2); they contain virions (1/1).

## Tombusviruses: *Tombusviridae*

*Definitive species* artichoke mottled crinkle, carnation Italian ringspot, cucumber necrosis, cymbidium ringspot, eggplant mottled crinkle, grapevine Algerian latent, Lato River, Neckar River, pelargonium leaf curl, pepper Moroccan, petunia asteroid mosaic and tomato bushy stunt tombusviruses. *Tentative species* oat chlorotic stunt and Sitke waterborne tombusviruses.

**Natural host range and symptoms**

Symptoms persist (5/6), or vary cyclically over a few weeks (1/6), or vary seasonally (2/6).

**Transmission**

Transmitted by a vector (1/6), or means not involving a vector (6/6); a fungus (1/1); Chytridiales (1/1). Virus transmitted by mechanical inoculation (12/12); transmitted by grafting (6/6); transmitted by contact between plants (2/3), or not transmitted by contact between plants (1/3); transmitted by seed (1/4), or not transmitted by seed (3/4); transmitted by pollen to the seed (1/2), or not transmitted by pollen (1/2).

**Experimental host range**

Many (>9) families susceptible (3/12), or several (3–9) families susceptible (7/12), or few (<3) families susceptible (2/12).

**Properties of particles in sap**

TIP: 75–85–90 °C. LIV: 21–104.5–300 days. DEP: $\log_{10}$ minus 4–5 (2/6), or 6 (4/6), or 7 (1/6). Leaf sap contains many virions (7/7).

**Particle morphology**

Virions isometric; not enveloped (11/11); 28–30.57–35 nm in diameter; rounded in profile (4/7), or angular in profile (2/7), or neither clearly angular or nor rounded in profile (1/7); with a conspicuous capsomere arrangement (2/5), or without a conspicuous capsomere arrangement (3/5).

**Physical properties**

One sedimenting component in purified preparations (10/11), or two sedimenting components in purified preparations (1/11); sedimentation coefficient 130–134.2–140 *S*. Density 1.28–1.346–1.36 g $cm^{-3}$ in CsCl. Isoelectric point pH 3.9–4–4.1. $A_{260}/A_{280}$ ratio 1.64.

**Biochemical properties**

Virions contain 14–16.92–18 % nucleic acid; 82–82.66–84 % protein; 0 % lipid.

*Genome* consists of RNA (13/13); single–stranded (13/13); linear (5/5). Total genome size 4.114–4.673–4.8 kb. Genome unipartite (12/12); largest (or only) genome part 4.6–4.702–4.8 kb. Base composition 27.8–28.13–28.6 % G; 24.9–26.06–27 % A; 21.2–21.5–22 % C; 23–25.13–26.3 % U. 5' terminus of RNA has a methylated nucleotide cap (1/1). Infectivity retained when deproteinised with proteases (2/2); retained when deproteinised with phenol or detergent (5/5). Poly A region absent (5/5). Additional factor not required for infectivity (5/5). Genome has no *t*RNA–like activity (2/2).

*Non–genomic nucleic acid* found in the virions (4/5), or not found in the virions (1/5); is subgenomic *m*RNA (4/4), or satellite RNA (2/4), or of an unusual type of structure (2/4). Sub–genomic *m*RNA found in infected cells (3/3). 3–3.5–4 virus specified dsRNA species found in infected cells. Size of largest virus specified dsRNA 4.7–4.716–4.733 kbp; 2nd largest 2.1–2.144–2.188 kbp; 3rd largest 0.9–0.918–0.936 kbp; 4th largest 0.62 kbp; 5th largest 0.499 kbp.

*Virion protein(s)* one; $M_r$ 35800–40960–48200. Virion proteins not glycosylated (2/2). Virion proteins not phosphorylated (2/2).

*Virus–coded non–virion proteins* isolated (1/1); three proteins found (1/1). $M_r$ 92000. $M_r$ of 2nd largest 33000. $M_r$ of 3rd 22000.

**Replication**

Genome replicates in cytoplasm (2/2). Replication does not depend on a helper virus (4/4). Acts as helper for a satellite RNA (2/2).

**Cytopathology**

Virions found in cytoplasm (11/11), or in nuclei (4/11), or in nucleoli (2/11), or in the perinuclear space (1/11), or in chloroplasts (2/11), or in mitochondria (2/11), or in cell vacuoles (5/11). Inclusions present in infected cells (6/7), or absent from infected cells (1/7); are crystals in the cytoplasm (3/6), or crystals in the nucleus (1/6), or membranous bodies (2/6), or unusual in shape (3/6); they contain virions (1/5), or they do not contain virions (4/5).

# Tospoviruses: *Bunyaviridae*

*Definitive species* impatiens necrotic spot, peanut yellow spot and tomato spotted wilt tospoviruses. *Tentative species* groundnut ringspot and tomato chlorotic spot tospoviruses.

**Natural host range and symptoms**

Symptoms persist (2/2).

**Transmission**

Transmitted by a vector (4/4); an insect (4/4); Thysanoptera (4/4). Transmitted in a semi–persistent manner (1/2), or in a persistent manner (1/2). Virus retained when the vector moults (2/2); multiplies in the vector (1/1); not transmitted congenitally to the progeny of the vector (2/2); transmitted by mechanical inoculation (2/2); transmitted by grafting (2/2); not transmitted by contact between plants (2/2); not transmitted by seed (2/2); not transmitted by pollen (1/1).

**Experimental host range**

Many (>9) families susceptible (1/2), or several (3–9) families susceptible (1/2).

**Properties of particles in sap**

TIP: 40–43.75–45 °C. LIV: 0.2–0.6–1 days. DEP: $\log_{10}$ minus 2–3 (1/2). Infectivity of sap decreased by treatment with di–ethyl ether (2/2). Leaf sap contains few virions (2/2).

**Particle morphology**

Virions isometric (1/2), or unusually shaped (1/2); enveloped (2/2); 70–82.5–90 nm in diameter; rounded in profile (1/1); without a conspicuous capsomere arrangement (2/2).

**Physical properties**

One sedimenting component in purified preparations (2/2); sedimentation coefficient 550 *S*. Density 1.21 g $cm^{-3}$ in sucrose.

**Biochemical properties**

Virions contain 5 % nucleic acid; 70 % protein; 20 % lipid.

*Genome* consists of RNA (2/2); single–stranded (2/2); linear (2/2). Total genome size 17.2–17.4–17.6 kb. Genome of three parts (1/1); largest (or only) genome part 8.8–8.848–8.897 kb; the 2nd largest 5.4 kb; the 3rd largest 2.916–3.208–3.5 kb. Base composition 16.2 % G; 31.6 % A; 19.3 % C; 32.9 % U. Infectivity lost when deproteinised with proteases (2/2); lost when deproteinised with phenol or detergent (2/2). Poly A region absent (1/1). Additional factor required for infectivity (1/2), or not required for infectivity (1/2).

*Non–genomic nucleic acid* not found in the virions (1/1). Sub–genomic *m*RNA found in infected cells (1/1).

*Virion protein(s)* four (2/2); $M_r$ 72000–96000–119900. $M_r$ of 2nd largest 64000–71000–78000. $M_r$ of 3rd largest 58000–60000–62000. $M_r$ of 4th largest 27000–29000–31000. Virion proteins glycosylated (2/2). Virion proteins phosphorylated (1/1).

**Replication**

Genome replicates in cytoplasm (1/1). Replication does not depend on a helper virus (2/2).

**Cytopathology**

Virions found in cytoplasm (2/2). Inclusions present in infected cells (1/2), or absent from infected cells (1/2); are unusual in shape (1/1).

# Trichoviruses

*Definitive species* apple chlorotic leaf spot, heracleum latent and potato T trichoviruses. *Tentative species* cherry mottle leaf, grapevine A and grapevine B trichoviruses.

**Natural host range and symptoms**

Symptoms persist (1/3), or vary seasonally (1/3), or disappear soon after infection (1/3).

**Transmission**

Transmitted by a vector (3/3); an insect (2/3), or a mite (1/3); Aphididae (1/2), or Pseudococcidae (1/2); Eriophyidae (1/1). Transmitted in a semi–persistent manner (1/1). Virus lost by the vector when it moults (1/1); does not multiply in the vector (1/1); requires, for vector transmission, a helper virus (1/1); transmitted by mechanical inoculation (5), or not transmitted by mechanical inoculation (1); transmitted by grafting (5/5); not transmitted by contact between plants (2/2); transmitted by seed (1/3), or not transmitted by seed (2/3); transmitted by pollen to the seed (1/1).

**Experimental host range**

Many (>9) families susceptible (1/4), or several (3–9) families susceptible (2/4), or few (<3) families susceptible (1/4).

**Properties of particles in sap**

TIP: 40–54–65 °C. LIV: 1–2.75–10 days. DEP: $\log_{10}$ minus 4–5 (2/3). Infectivity of sap decreased by treatment with di–ethyl ether (1/1). Leaf sap contains few virions (2/4), or contains many virions (2/4).

**Particle morphology**

Virions filamentous; not enveloped (4/4); usually flexuous; with a clear modal length (4/4); of 637–909.5–1800 nm; 10.2–11.74–12.5 nm wide. Axial canal obscure (3/3). Basic helix obvious (5/5); pitch of basic helix 3.1–3.59–3.9 nm.

**Physical properties**

One sedimenting component in purified preparations (4/4); sedimentation coefficient 96–97–99 *S*. $A_{260}/A_{280}$ ratio 1.62.

**Biochemical properties**

Virions contain 95 % nucleic acid; 5 % protein; 0 % lipid.

*Genome* consists of RNA (4/4); single–stranded (4/4); linear (3/3). Total genome size 7.5–7.788–8.2 kb. Genome unipartite (3/3); largest (or only) genome part 7.5–7.651–7.9 kb. 5' terminus of RNA has a methylated nucleotide cap (1/1). Infectivity retained when deproteinised with proteases (1/1); retained when deproteinised with phenol or detergent (1/1). Poly A region present (2/2). Additional factor not required for infectivity (1/1).

*Non–genomic nucleic acid* not found in the virions (1/1). 5 virus specified dsRNA species found in infected cells. Size of largest virus specified dsRNA 7.56 kbp; 2nd largest 6.5 kbp; 3rd largest 5.5 kbp.

*Virion protein(s)* one (4/4); $M_r$ 20500–23620–27000. Virion proteins not glycosylated (1/1).

*Virus–coded non–virion proteins* isolated (1/1); two proteins found (1/1). $M_r$ 199900. $M_r$ of 2nd largest 45000.

**Replication**

Replication does not depend on a helper virus (3/3).

**Cytopathology**

Virions found in cytoplasm (2/2). Inclusions absent from infected cells (1/1).

# Tymoviruses

*Definitive species* abelia latent, belladonna mottle, cacao yellow mosaic, clitoria yellow vein, desmodium yellow mottle, dulcamara mottle, eggplant mosaic, erysimum latent, kennedya yellow mosaic, melon rugose mosaic, okra mosaic, ononis yellow mosaic, passionfruit yellow mosaic, physalis mosaic, plantago mottle, potato Andean latent, scrophularia mottle, turnip yellow mosaic, voandzeia necrotic mosaic and wild cucumber mosaic tymoviruses. *Tentative species* poinsettia mosaic tymovirus.

**Natural host range and symptoms**

Symptoms persist (18/20), or vary seasonally (1/20), or disappear soon after infection (1/20), or none (2/20).

**Transmission**

Transmitted by a vector (11/12), or means not involving a vector (2/12); an insect (10/10); Coleoptera (10/10). Transmitted in a non–persistent manner (3/5), or in a semi–persistent manner (2/5). Virus transmitted by mechanical inoculation (20/20); transmitted by grafting (8/9), or not transmitted by grafting (1/9); transmitted by contact between plants (1/7), or not transmitted by contact between plants (6/7); transmitted by seed (4/16), or not transmitted by seed (12/16); not transmitted by pollen (5/5).

**Experimental host range**

Many (>9) families susceptible (3), or several (3–9) families susceptible (9), or few (<3) families susceptible (9).

**Properties of particles in sap**

TIP: 55–70.62–90 °C. LIV: 2–22.1–100 days. DEP: $\log_{10}$ minus 2 (1), or 3 (2), or 4 (5), or 5 (4), or 6 (7), or 7 (5), or 8 (4), or 9 (1). Infectivity of sap not changed by treatment with di–ethyl ether (11/11). Leaf sap contains few virions (1/19), or contains many virions (18/19).

**Particle morphology**

Virions isometric; not enveloped (20/20); 25–28.03–32 nm in diameter; rounded in profile (18/20), or angular in profile (4/20); with a conspicuous capsomere arrangement (15/19), or without a conspicuous capsomere arrangement (4/19).

**Physical properties**

Two sedimenting components in purified preparations (20/20); sedimentation coefficient 106–114.2–125 *S*; of the other(s) 42–53.87–64 *S*. Density 1.26–1.358–1.45 g $cm^{-3}$ in CsCl. Isoelectric point pH 3.3–5.893–9.46. $A_{260}/A_{280}$ ratio 0.64–2.075–8.51.

**Biochemical properties**

Virions contain 23–34.73–39 % nucleic acid; 61–64.85–77 % protein; 0 % lipid.

*Genome* consists of RNA; single–stranded; linear (16/16). Total genome size 2–5.028–6.5 kb. Genome unipartite (16/16); largest (or only) genome part 2–5.028–6.5 kb. Base composition 13.6–16.35–20.5 % G; 17–21.91–24.1 % A; 31.3–35.55–41 % C; 22.15–26.24–29.4 % U. 5' terminus of RNA has a methylated nucleotide cap (5/6), or an unusual type (1/6). Infectivity retained when deproteinised with proteases (3/3); retained when deproteinised with phenol or detergent (10/10). Poly A region present (1/4), or absent (3/4). Additional factor not required for infectivity (5/5). Genome has *t*RNA–like activity (5/7), or has no *t*RNA–like activity (2/7).

*Non–genomic nucleic acid* found in the virions (4/4); is subgenomic *m*RNA (3/4), or of an unusual type of structure (1/4). Sub–genomic *m*RNA found in infected cells (4/4).

*Virion protein(s)* one (16/16); $M_r$ 19420–20600–22200. Virion proteins not glycosylated (3/3). Virion proteins not phosphorylated (3/3).

*Virus–coded non–virion proteins* identified by genomic sequence analysis (1/1); two proteins found (1/1). $M_r$ 48490. $M_r$ of 2nd largest 193800.

**Replication**

Genome replicates in association with chloroplasts (9/9). Replication does not depend on a helper virus (15/15).

**Cytopathology**

Virions found in cytoplasm (15/16), or in nuclei (5/16), or in nucleoli (1/16), or in chloroplasts (3/16), or in mitochondria (2/16), or in cell vacuoles (4/16). Inclusions present in infected cells (5/11), or absent from infected cells (6/11); are crystals in the cytoplasm (4/5), or crystals in the nucleus (2/5), or unusual in shape (2/5); they contain virions (5/5).

# Umbraviruses

*Definitive species* bean yellow vein banding, carrot mottle mimic, carrot mottle, carrot mottle mimic, groundnut rosette, lettuce speckles mottle and tobacco mottle umbraviruses. *Tentative species* sunflower crinkle, sunflower yellow blotch and tobacco yellow vein umbraviruses.

**Natural host range and symptoms**

Symptoms persist (3/4), or vary seasonally (1/4).

**Transmission**

Transmitted by a vector (7/7); an insect (7/7); Aphididae (7/7). Transmitted in a persistent manner (6/6). Virus retained when the vector moults (3/3); does not multiply in the vector (2/2); not transmitted congenitally to the progeny of the vector (2/2); requires, for vector transmission, a helper virus (7/7); transmitted by mechanical inoculation (7/7); transmitted by grafting (1/1); not transmitted by contact between plants (1/1); not transmitted by seed (2/2); not transmitted by pollen (1/1).

**Experimental host range**

Many (>9) families susceptible (1/7), or several (3–9) families susceptible (3/7), or few (<3) families susceptible (3/7).

**Properties of particles in sap**

TIP: 55–62.5–70 °C. LIV: 0.2–1.533–2 days. DEP: $\log_{10}$ minus 2 (1/2), or 3 (2/2). Infectivity of sap decreased by treatment with di–ethyl ether (1/2), or not changed by treatment with di–ethyl ether (1/2). Leaf sap contains few virions (1/1).

**Particle morphology**

Virions isometric (1/2), or unusually shaped (1/2); not enveloped (1/1); 50–70–90 nm in diameter; rounded in profile (1/1).

**Physical properties**

One sedimenting component in purified preparations (2/2); sedimentation coefficient 270 *S*. Density 1.15 g $cm^{-3}$ in CsCl.

**Biochemical properties**

*Genome* consists of RNA (4/4); single–stranded (4/4); linear (3/3). Total genome size 4.203–5.284–6.9 kb. Genome unipartite (3/3); largest (or only) genome part 2.05–3.667–4.75 kb. Base composition 27 % G; 25 % A; 25.5 % C; 22.4 % U. Infectivity retained when deproteinised with proteases (1/1); retained when deproteinised with phenol or detergent (1/1). Poly A region absent (1/1). Additional factor not required for infectivity (2/2). Genome has no *t*RNA–like activity (1/1).

*Non–genomic nucleic acid* found in the virions (2/2); is subgenomic *m*RNA (1/2), or satellite RNA (1/2). Sub–genomic *m*RNA found in infected cells (2/2). 2 virus specified dsRNA species found in infected cells. Size of largest virus specified dsRNA 4.2 kbp; 2nd largest 1.5 kbp.

*Virion protein(s)* none (1/1).

**Replication**

Genome replicates in cytoplasm (1/1). Replication depends on co–infection with a helper virus (1/2), or does not depend on a helper virus (1/2). Acts as helper for a satellite virus (1/1).

**Cytopathology**

Inclusions present in infected cells (1/1); are membranous bodies (1/1); they do not contain virions (1/1).

# Unassigned potyviruses: *Potyviridae*

*Tentative species* alstroemeria streak, amaranthus mosaic, amazon lily mosaic, anthoxanthum mosaic, aquilegia, bramble yellow mosaic, brinjal mild mosaic, broad bean V, bryonia mottle, canary reed mosaic, canavalia maritima mosaic, carrot mosaic, celery yellow mosaic, chickpea filiform, chilli veinal mottle, chrysanthemum spot, cocksfoot streak, cypripedium calceolus, datura innoxia Hungarian mosaic, dioscorea trifida, dock mottling mosaic, eggplant severe mottle, fern, fig, habenaria mosaic, holcus streak, isachne mosaic, launaea mosaic, melilotus mosaic, melon vein-banding mosaic, melothria mottle, mung bean mottle, narcissus degeneration, narcissus late season yellows, nerine Y, nothoscordum mosaic, palm mosaic, parsley green mottle, passionfruit Sri Lankan mottle, peanut top paralysis, pecteilis mosaic, pepper mild mosaic, perilla mottle, plantain 7, pleioblastus chino, poplar decline, primula mottle, shamrock chlorotic ringspot, soybean Z, sunflower mosaic, sweet potato latent, teasel mosaic, tomato mild mottle, trichosanthes mottle, watermelon Moroccan mosaic, white bryony, zinnia mild mottle and zoysia mosaic potyviruses.

# Unassigned rhabdoviruses: *Rhabdoviridae*

*Tentative species* asclepias, atropa belladonna, beet leaf curl, callistephus chinensis chlorosis, carnation, cassava symptomless, chrysanthemum vein chlorosis, citrus leprosis, colocasia bobone disease, cucumber toad-skin, euonymus fasciation, euonymus, gerbera symptomless, holcus lanatus yellowing, iris germanica leaf stripe, kenaf vein-clearing, lupin yellow vein, orchid fleck, parsley, passionfruit vein-clearing, patchouli mottle, peanut veinal chlorosis, pigeonpea proliferation, plantain mottle, ranunculus repens symptomless, saintpaulia leaf necrosis, sambucus vein clearing, sarracenia purpurea, soybean, strawberry latent C, vigna sinensis mosaic and wheat chlorotic spot rhabdoviruses.

## Varicosaviruses

*Definitive species* freesia leaf necrosis, lettuce big-vein and tobacco stunt varicosaviruses. *Tentative species* camellia yellow mottle varicosavirus.

### Natural host range and symptoms

Symptoms persist (2), or vary seasonally (3), or disappear soon after infection (1).

### Transmission

Transmitted by a vector (3/3); a fungus (3/3); Chytridiales (3/3). Virus does not require a helper virus for vector transmission (1/1); transmitted by mechanical inoculation (3), or not transmitted by mechanical inoculation (1); transmitted by grafting (3/3); not transmitted by contact between plants (3/3); not transmitted by seed (3/3); not transmitted by pollen (3/3).

### Experimental host range

Several (3–9) families susceptible (2), or few (<3) families susceptible (2).

### Properties of particles in sap

TIP: 40–47.5–50 °C. LIV: 1 days. DEP: $\log_{10}$ minus 1 (1/2), or 3–4 (1/2). Leaf sap contains few virions (3/3).

### Particle morphology

Virions rod–shaped (3/3); not enveloped (2/2); usually straight (2/2); with a clear modal length (2/2); of 120–268.3–360 nm; 18–22–30 nm wide. Axial canal obvious (3/3); 3–5.5–10 nm in diameter. Basic helix obvious (2/2); pitch of basic helix 5–6.666–10 nm.

### Biochemical properties

*Genome* consists of RNA (2/2); double–stranded (2/2); linear (2/2). Total genome size 11.98–12.74–13.5 kb. Genome of two parts (2/2); largest (or only) genome part 6.35–6.675–7 kb; the 2nd largest 5.63–6.065–6.5 kb.

*Virion protein(s)* one (2/2); $M_r$ 48000.

### Replication

Replication does not depend on a helper virus (2/2).

### Cytopathology

Virions found in cytoplasm (3/3). Inclusions present in infected cells (1/2), or absent from infected cells (1/2); are crystals in the cytoplasm (1/1); they contain virions (1/1).

## Waikaviruses: *Sequiviridae*

*Definitive species* anthriscus yellows, maize chlorotic dwarf and rice tungro spherical waikaviruses.

**Natural host range and symptoms**

Symptoms persist (2/2).

**Transmission**

Transmitted by a vector; an insect; Aphididae (1), or Cicadellidae (2). Transmitted in a semi–persistent manner. Virus lost by the vector when it moults (2/2); does not multiply in the vector (2/2); not transmitted congenitally to the progeny of the vector (2/2); can help the vector transmission of another virus (2/2); not transmitted by mechanical inoculation; not transmitted by contact between plants (2/2); not transmitted by seed (2/2); not transmitted by pollen (2/2).

**Experimental host range**

Few (<3) families susceptible.

**Properties of particles in sap**

LIV: 45 days. Leaf sap contains few virions (1/2), or contains many virions (1/2).

**Particle morphology**

Virions isometric; not enveloped; 29–29.66–30 nm in diameter; rounded in profile (2), or angular in profile (1); with a conspicuous capsomere arrangement (2), or without a conspicuous capsomere arrangement (1).

**Physical properties**

One sedimenting component in purified preparations (2), or two sedimenting components in purified preparations (1); sedimentation coefficient 175–179–183 *S*. Density 1.27–1.484–1.551 g $cm^{-3}$ in CsCl. $A_{260}/A_{280}$ ratio 0.73–1.28–1.83.

**Biochemical properties**

Virions contain 0–28.5–42 % nucleic acid; 58–71.5–100 % protein; 0 % lipid.

*Genome* consists of RNA; single–stranded; linear (1/1). Total genome size 10.6–10.63–10.67 kb. Genome unipartite; largest (or only) genome part 10.42–10.56–10.67 kb. Base composition 24 % G; 30 % A; 17 % C; 29 % U.

*Virion protein(s)* one (1), or two (1), or four (1); $M_r$ 23000–29330–35000. $M_r$ of 2nd largest 18000–23460–28400. $M_r$ of 3rd largest 24300. $M_r$ of 4th largest 22300.

**Replication**

Acts as helper for another virus (1/1).

**Cytopathology**

Virions found in cytoplasm. Inclusions present in infected cells (2), or absent from infected cells (1); are amorphous X–bodies (1/2), or viroplasms (2/2), or unusual in shape (2/2); they contain virions (2/2), or they do not contain virions (1/2).

# Ungrouped viruses

*Tentative species* alsike clover vein mosaic, apple stem pitting, barley mosaic, barley yellow streak mosaic, beet distortion mosaic, beet western yellows ST9-associated RNA, black raspberry necrosis, broad bean B, broad bean yellow ringspot, burdock mosaic, burdock mottle, cassia mosaic, cassia ringspot, celery yellow net, cereal flame chlorosis, citrus ringspot, clover mild mosaic, cucumber vein yellowing, dodonaea yellows-associated, grapevine fleck, grapevine stunt, guar top necrosis, gris Japanese necrotic ring, kalanchoe isometric, maize eyespot, maize line, maize mottle/chlorotic stunt, maize white line mosaic, malvastrum mottle, mimosa mosaic, oak ringspot, parsnip leafcurl, pea stem necrosis, pigeonpea sterility mosaic, purple granadilla mosaic, rice yellow stunt, soybean mild mosaic, soybean spherical, soybean yellow vein, strawberry mottle, strawberry pallidosis, thimbleberry ringspot, tulip halo necrosis, tulip mosaic, turnip vein-clearing, urd bean leaf crinkle, watercress yellow spot and wineberry latent viruses.

# All viruses in the VIDE database

### Natural host range and symptoms

Symptoms persist (469/585), or vary cyclically over a few weeks (15/585), or vary seasonally (83/585), or disappear soon after infection (42/585), or none (36/585).

### Transmission

Transmitted by a vector (473/531), or means not involving a vector (62/531); an insect (415/469), or a mite (9/469), or a nématode (20/469), or a fungus (25/469); Aleyrodidae (50/411), or Aphididae (256/411), or Cicadellidae (30/411), or Delphacidae (27/411), or Psyllidae (1/411), or Membracidae (1/411), or Pseudococcidae (6/411), or Coleoptera (36/411 ), or Thysanoptera (6/411), or an unusual insect group (2/411); Eriophyidae (6/8), or Tetranychidae (2/8); Dorylamidae (11/13), or Trichodoridae (2/13); Plasmodiophorales (15/24), or Chytridiales (9/24). Transmitted in a non–persistent manner (172/337), or in a semi–persistent manner (50/337), or in a persistent manner (118/337). Virus retained when the vector moults (72/98), or lost by the vector when it moults (26/98); multiplies in the vector (34/90), or does not multiply in the vector (56/90); transmitted congenitally to the progeny of the vector (14/87), or not transmitted congenitally to the progeny of the vector (73/87); requires, for vector transmission, a helper virus (13/45), or does not require a helper virus for vector transmission (22/45), or can help the vector transmission of another virus (13/45); transmitted by mechanical inoculation (561/743), or not transmitted by mechanical inoculation (182/743); transmitted by grafting (263/297), or not transmitted by grafting (34/297); transmitted by contact between plants (29/268), or not transmitted by contact between plants (239/268); transmitted by seed (138/441), or not transmitted by seed (304/441); transmitted by pollen to the seed (30/168), or transmitted by pollen to the pollinated plant (16/168), or not transmitted by pollen (133/168).

### Experimental host range

Many (>9) families susceptible (77/685), or several (3–9) families susceptible (262/685), or few (<3) families susceptible (347/685).

### Properties of particles in sap

TIP: 35–64.47–100 °C. LIV: 0.0069–130.6–35000 days. DEP: $\log_{10}$ minus 1–4–10. Infectivity of sap increased by treatment with di–ethyl ether (4/96), or decreased by treatment with di–ethyl ether (16/96), or not changed by treatment with di–ethyl ether (77/96). Leaf sap contains few virions (315/485), or contains many virions (181/485).

### Particle morphology

Virions isometric (285/812), or geminate (46/812), or bacilliform (23/812), or rod–shaped (42/812), or filamentous (334/812), or rhabdo– or bullet–shaped (80/812), or unusually shaped (6/812); enveloped (45/619), or not enveloped (574/619); 15–30.44–90 nm in diameter; 15–56.77–400 nm in length; rounded in profile (140/257), or angular in profile (128/257), or neither clearly angular or nor rounded in profile (4/257); with a conspicuous capsomere arrangement (67/216), or without a conspicuous capsomere arrangement (149/216); usually straight (81/356), or usually flexuous (276/356); with a clear modal length (361/388), or with no clear modal length (21/388), or with clear modal lengths (6/388); of 46–628.7–4000 nm; 3–27.07–130 nm wide. Axial canal obvious (69/259), or obscure (190/259); 2–8.354–35 nm in diameter. Basic helix obvious (80/222), or obscure (144/222); pitch of basic helix 2.3–3.618–10 nm.

**Physical properties**

One sedimenting component in purified preparations (298/433), or two sedimenting components in purified preparations (75/433), or three sedimenting components in purified preparations (67/433), or four sedimenting components in purified preparations (13/433), or five sedimenting components in purified preparations (2/433), or six sedimenting components in purified preparations (1/433), or eight sedimenting components in purified preparations (1/433); sedimentation coefficient 42–161–1044 *S*; of the other(s) 15–82.59–212 *S*. Density 1.15–1.358–1.717 g $cm^{-3}$ in CsCl. Density 1.1–1.217–1.33 g $cm^{-3}$ in sucrose. Isoelectric point pH 3.16–5.085–9.46. $A_{260}/A_{280}$ ratio 0.62–1.498–8.51.

**Biochemical properties**

Virions contain 0–15.95–95 % nucleic acid; 5–83.39–100 % protein; 0–0.7575–25 % lipid.

*Genome* consists of DNA (61/463), or consists of RNA (402/463); single–stranded (405/460), or double–stranded (55/460); circular (45/265), or linear (222/265). Total genome size 1.059–8.343–28.91 kb. Genome unipartite (225/377), or of two parts (94/377), or of three parts (29/377), or of four parts (11/377), or of five parts (6/377), or of 7 parts (3/377), or of 8 parts (1/377), or of 9 parts (1/377 ), or of 10 parts (9/377), or of 11 parts (1/377), or of 12 parts (4/377), or of 13 parts (1/377), or of 14 parts (1/377), or of 15 parts (1/377); largest (or only) genome part 0.89–5.979–20 kb; the 2nd largest 0.35–3.549–7.5 kb; the 3rd largest 0.32–2.422–3.8 kb; the 4th largest 0.845–2.26–3.8 kb; the 5th largest 0.76–2.174–3.3 kb; the 6th largest and other parts 0.25–2.102–2.9 kb. Base composition 13.6–23.14–32 % G; 17–26.06–44 % A; 14.9–23.8–41 % C; 15.6–26.95–37.8 % U; 24.8–28.63–32.6 % T. Base composition of ds genome 28–42.4–56.8 % (G + C). 5' terminus of RNA has a methylated nucleotide cap (25/47), or a VPg (16/47), or an unusual type (6/47). Infectivity retained when deproteinised with proteases (30/58), or decreased when deproteinised with proteases (15/58), or lost when deproteinised with proteases (18/58); retained when deproteinised with phenol or detergent (102/115), or lost when deproteinised with phenol or detergent (14/115). Poly A region present (36/83), or absent (48/83). Additional factor required for infectivity (17/72), or not required for infectivity (55/72). Genome has *t*RNA–like activity (14/34), or has no *t*RNA–like activity (20/34).

*Non–genomic nucleic acid* found in the virions (61/131), or not found in the virions (70/131); is subgenomic *m*RNA (40/57), or host genomic RNA (1/57), or *t*RNA (1/57), or *r*RNA (1/57), or satellite RNA (16/57), or of unknown origin and function (1/57), or of an unusual type of structure (6/57). Sub–genomic *m*RNA found in infected cells (45/56), or not found in infected cells (11/56). 1–3.104–14 virus specified dsRNA species found in infected cells. Size of largest virus specified dsRNA 1.1–7.078–17.1 kbp; 2nd largest 1.1–3.361–7 kbp; 3rd largest 0.4–2.451–5.7 kbp; 4th largest 0.5–1.924–4.6 kbp; 5th largest 0.499–0.8663–1.6 kbp; 6th largest 1 kbp.

*Virion protein(s)* one (320/393), or two (34/393), or three (15/393), or four (9/393), or five (8/393), or six (5/393), or 7 (2/393), or 8 (1/393 ), or none (1/393); $M_r$ 22.3–37520–240900. $M_r$ of 2nd largest 32–45580–229900. $M_r$ of 3rd largest 56–45380–164900. $M_r$ of 4th largest 31–44330–142900. $M_r$ of 5th largest 37–41630–119900. $M_r$ of 6th largest and smaller, 34000–47500–64000. Virion proteins glycosylated (16/31), or not glycosylated (15/31). Virion proteins phosphorylated (1/9), or not phosphorylated (8/9).

*Virus–coded non–virion proteins* isolated (27/46), or identified by genomic sequence analysis (19/46); one protein found (14/46), or two proteins found (11/46), or three proteins found (10/46), or four proteins found (3/46), or five proteins found (4/46), or six proteins found (2/46), or seven proteins found (3/46). $M_r$ 10900–88610–239900. $M_r$ of 2nd largest 17370–60450–193800. $M_r$ of 3rd 8000–31070–54000. $M_r$ of 4th 6900–40290–111900. $M_r$ of 5th and smaller 6000–29960–104900.

**Replication**

Genome replicates in nuclei (17/54), or in cytoplasm (27/54), or in association with chloroplasts (9/54), or in cytoplasmic viroplasms (4/54). Replication depends on co–infection with a helper virus (2/246), or does not depend on a helper virus (244/246). Acts as helper for a satellite virus (2/10), or a satellite RNA (5/10), or another virus (3/10).

**Cytopathology**

Virions found in cytoplasm (409/462), or in nuclei (81/462), or in nucleoli (7/462), or in the perinuclear space (39/462), or in chloroplasts (14/462), or in mitochondria (9/462), or in Golgi apparatus (2/462), or in cell vacuoles ( 61/462), or in unusual sites (5/462). Inclusions present in infected cells (347/436), or absent from infected cells (89/436); are crystals in the cytoplasm (65/343), or crystals in the nucleus (23/343), or amorphous X–bodies (35/343), or membranous bodies (34/343), or viroplasms (51/343), or pinwheels (125/343), or unusual in shape (142/343); they contain virions (155/262), or they do not contain virions (111/262).

**Taxonomy and Relationships**

*Alfamovirus: Bromoviridae* (1/864), or *Alphacryptovirus: Partitiviridae* (29/864), or *Badnavirus* (16/864), or *Betacryptovirus: Partitiviridae* (5/864), or bigeminivirus: *Geminiviridae* (41/864), or *Bromovirus: Bromoviridae* (6/864), or *Bymovirus: Potyviridae* (6/864), or *Capillovirus* (5 /864), or *Carlavirus* (55/864), or *Carmovirus: Tombusviridae* (22/864), or *Caulimovirus* (17/864), or *Closterovirus* (25/864), or *Comovirus: Comoviridae* (14/864), or *Cucumovirus: Bromoviridae* (3/864), or *Cytorhabdovirus: Rhabdoviridae* ( 17/864), or *Dianthovirus* (4 /864), or *Enamovirus* (1/864), or *Fabavirus: Comoviridae* (2/864), or *Fijivirus: Reoviridae* (6/864), or *Furovirus* (12/864), or *Hordeivirus* (4/864), or hybrigeminivirus: *Geminiviridae* (2/864), or *Idaeovirus* (1 /864), or *Ilarvirus: Bromoviridae* (22/864), or ipomovirus: *Potyviridae* (2/864), or *Luteovirus* (27/864), or *Machlomovirus* (1/864), or macluravirus (2/864), or *Marafivirus* (3/864), or monogeminivirus: *Geminiviridae* (11 /864), or nanavirus (5/864 ), or *Necrovirus* (3/864), or *Nepovirus: Comoviridae* (39/864), or *Nucleorhabdovirus: Rhabdoviridae* (39/864), or *Oryzavirus: Reoviridae* (2/864), or ourmiavirus (5/864), or *Phytoreovirus: Reoviridae* (5/864), or *Potexvirus* (53/864), or *Potyvirus: Potyviridae* (1 29/864), or *Rymovirus: Potyviridae* (7/864), or satellite RNA (8/864), or satellivirus (4/864), or *Sequivirus: Sequiviridae* (2/864), or *Sobemovirus* (17/864), or *Tenuivirus* (10/864), or *Tobamovirus* (17/864 ), or *Tobravirus* (4/864), or *Tombusvirus: Tombusviridae* (14/864), or *Tospovirus: Bunyaviridae* (5/864), or *Trichovirus* (6/864), or *Tymovirus* (21/864), or *Umbravirus* (10/864), or unassigned potyvirus (58/864), or unassigned rhabdovirus ( 32/864), or varicosavirus (4/864), or *Waikavirus: Sequiviridae* (3/864).

# 3. VIRUS SPECIES DESCRIPTIONS

## Abelia latent *tymovirus*

**Data collated by A.J. Gibbs, 1992.**

First reported in *Abelia grandiflora*; from Maryland, U.S.A.; by Waterworth *et al.* (1975).

**Natural host range and symptoms**

Symptoms none.
*Abelia grandiflora* — symptomless.

**Transmission**

Virus transmitted by mechanical inoculation.

**Geographical distribution**

Spreads in the North American region; the USA (eastern).

**Experimental host range**

Many (>9) families susceptible. Experimentally infected plants mostly show local lesions, mosaic or no symptoms.

Diagnostically susceptible

*Chenopodium quinoa* — chlorotic local lesions, bright systemic chlorosis.
*Datura stramonium* and *Glycine max* — yellow mosaic.
*Atropa belladonna, Brassica campestris* spp. *pekinensis* and *Vicia faba* — systemic, but symptomless.
*Nicotiana clevelandii* and *N. glutinosa* — chlorotic local lesions, systemic mosaic.
*Lycopersicon esculentum* — systemic mosaic.

Diagnostically insusceptible

*Capsicum frutescens, Cucurbita pepo, Dianthus barbatus, Spinacia oleracea* and *Zea mays*.

Dicotyledonous Hosts (***Susceptible*** *or Insusceptible)*

***Amaranthus caudatus***. *Antirrhinum majus*. *Beta vulgaris*. ***Brassica campestris*** ssp. ***pekinensis***. *Brassica campestris* ssp. *rapa*. *Calendula officinalis*. *Capsicum frutescens*. ***Catharanthus roseus***. ***Celosia argentea***. ***Chenopodium amaranticolor***. ***Chenopodium murale***. ***Chenopodium quinoa***. *Cucumis sativus*. *Cucurbita maxima*. *Cucurbita pepo*. ***Datura metel***. ***Datura stramonium***. *Dianthus barbatus*. ***Glycine max***. ***Gomphrena globosa***. ***Gossypium hirsutum***. ***Hibiscus esculentus***. *Ipomoea nil*. ***Lobelia erinus***. ***Lycopersicon esculentum***. ***Momordica balsamina***. ***Nicotiana clevelandii***. ***Nicotiana glutinosa***. ***Nicotiana rustica***. ***Petunia hybrida***. *Phaseolus vulgaris*. ***Physalis peruviana***. ***Sesamum indicum***. *Sesbania exaltata*. ***Solanum melongena***. ***Solanum tuberosum***. *Spinacia oleracea*. ***Tetragonia expansa***. *Theobroma cacao*. ***Tithonia speciosa***. ***Torenia fournieri***. ***Vicia faba***. *Vigna unguiculata*. *Zinnia elegans*.

Monocotyledonous Hosts (***Susceptible*** *or Insusceptible)*

***Hordeum vulgare***.

Additional hosts

*Acer palmatum.*

Sources of host-range data

Waterworth *et al.* (1975).

**Properties of particles in sap**

TIP: 74–78 °C. LIV: more than 50 days. DEP: $\log_{10}$ minus 8–9. Infectivity of sap not changed by treatment with di–ethyl ether. Leaf sap contains many virions.

**Purification**

Method: Waterworth *et al.* (1975).

**Particle morphology**

Virions isometric; not enveloped; 25 nm in diameter; rounded in profile.

**Physical properties**

Two sedimenting components in purified preparations; sedimentation coefficient of the fastest 114 *S* (B); of the other(s) 54 *S* (T). $A_{260}/A_{280}$ ratio 1.71; (B).

**Biochemical properties**

Virions contain 39 % nucleic acid; 61 % protein; 0 % lipid (B).

*Genome* consists of RNA; single–stranded; linear. Base composition 13.6 % G; 23.8 % A; 33.5 % C; 29.1 % U. Infectivity retained when deproteinised with phenol or detergent.

**Taxonomy and Relationships**

*Tymovirus.*

Virus(es) with serologically related virions: eggplant mosaic (SDI 2–5), Andean potato latent (3–7), belladonna mottle (4–6) and physalis mosaic (4–5).

**References**

Waterworth, H.E., Kaper, J.M. and Koenig, R. (1975). *Phytopathology* **65**: 891.

# Abutilon mosaic bigeminivirus

**Data collated by H. Jeske, 1987 and F.J. Morales, 1988.**

*Synonyms:* bean dwarf mosaic geminivirus, infectious chlorosis of Malvaceae, malvaceous chlorosis geminivirus (Christie *et al.*, 1986), possibly bean chlorotic mottle virus. *Standard acronym:* AbMV. *Strains:* abutilon mosaic A West Indies virus, abutilon mosaic B Brazil virus.

First reported in *Malva parviflora*; from Germany; by Baur (1906); see also Abouzid and Jeske (1986).

**Natural host range and symptoms**

Symptoms vary seasonally (depending on light intensity).

*Abutilon* spp., *Sida* spp., *Hibiscus* spp., *Gossypium hirsutum* — yellow mosaic.

*Malva* sp., *Phaseolus vulgaris* — mosaic.

**Transmission**

Transmitted by a vector; an insect; *Bemisia tabaci*; Aleyrodidae. Transmitted in a persistent manner. Virus retained when the vector moults; does not multiply in the vector; not transmitted congenitally to the progeny of the vector; transmitted by mechanical inoculation; transmitted by grafting; not transmitted by contact between plants; not transmitted by seed; not transmitted by pollen.

**Geographical distribution**

Probably distributed worldwide (in ornamental plants). Spreads in the South and Central American region; Argentina.

**Experimental host range**

Several (3–9) families susceptible. Experimentally infected plants mostly show yellow mosaic, vein clearing.

Diagnostically susceptible

*Malva parviflora, Phaseolus vulgaris* cv. Top Crop — systemic mosaic, vein clearing.

*Phaseolus vulgaris* cv. Alubia — systemic, severe stunting.

*Sida* spp. — systemic golden mosaic.

Maintenance and propagation hosts

*Malva parviflora, M. sylvestris, Phaseolus vulgaris* cvs Top Crop, Imuna, Stringless Green Refugee and Amanda, *Sida micrantha, S. rhombifolia.*

Assay hosts (**L**ocal lesions or **W**hole plants)

*Malva* spp. (W), *Phaseolus vulgaris* cvs Top Crop, Alubia and Imuna (W).

Dicotyledonous Hosts *(**Susceptible** or Insusceptible)*

***Arachis hypogaea*. *Cucumis melo*. *Cucumis sativus*.** *Cucurbita pepo*. ***Cyamopsis tetragonoloba*. *Datura stramonium*. *Datura tatula*. *Glycine max*. *Gossypium hirsutum*. *Hibiscus esculentus*.** *Lupinus albus*. ***Lycopersicon esculentum*. *Momordica balsamina*. *Nicandra physaloides*. *Nicotiana glutinosa*. *Nicotiana tabacum*. *Phaseolus lunatus*. *Phaseolus vulgaris*. *Solanum tuberosum*.**

Sources of host-range data

Costa (1955); Costa and Carvalho (1960*a*, 1960*b*).

**Properties of particles in sap**

TIP: 45–55 °C. LIV: 0.25–2 days. DEP: $\log_{10}$ minus 1–2. Infectivity of sap not changed by treatment with di–ethyl ether. Leaf sap contains few virions. Electron microscopy: glutaraldehyde fixation needed before negative staining.

**Purification**

Method: Abouzid and Jeske (1986); Morales *et al.* (1988).

**Particle morphology**

Virions geminate; not enveloped; 20 nm in diameter; the dimer is 29–33 nm in length; rounded in profile; without a conspicuous capsomere arrangement.

**Physical properties**

One sedimenting component in purified preparations, or two sedimenting components in purified preparations; sedimentation coefficient about 90 *S*; of the other(s) about 70 *S*. Density in $Cs_2SO_4$ 1.30 g $cm^{-3}$.

**Biochemical properties**

Virions contain 18 % nucleic acid; 82 % protein.

*Genome* consists of DNA; single–stranded; circular; of two parts; largest (or only) genome part the larger 2.663 kb; the 2nd largest 2.636 kb.

*Sequence database accession code(s):*

S67640 Em(40)_vi:S67640 Gb(84)_vi:S67640S1 Ac1...Av1 (DNA A) Abutilon mosaic virus, mRNA, 4 genes, 399 nt, segment 1 of 2. 11/93 399bp.
S67693 Em(40)_vi:S67693 Gb(84)_vi:S67640S2 Ac1...Av1 (DNA A) Abutilon mosaic virus, mRNA, 4 genes, 304 nt, segment 2 of 2. 11/93 304bp
S67833 Em(40)_vi:S67833 Gb(84)_vi:S67833S1 Bc1, Bv1 (DNA B) Abutilon mosaic virus, mRNA, 206 nt, segment 1 of 3. 11/93 206bp.
S67843 Em(40)_vi:S67843 Gb(84)_vi:S67833S2 Bc1, Bv1 (DNA B) Abutilon mosaic virus, mRNA, 224 nt, segment 2 of 3. 11/93 224bp.
S67851 Em(40)_vi:S67851 Gb(84)_vi:S67833S3 Bc1, Bv1 (DNA B) Abutilon mosaic virus, mRNA, 249 nt, segment 3 of 3. 11/93 249bp
X15983 Em(40)_vi:GEABMVA Gb(84)_vi:GEABMVA Abutilon mosaic virus subgenome DNA A. 9/93 2,629bp.
X15984 Em(40)_vi:GEABMVB Gb(84)_vi:GEABMVB Abutilon mosaic virus subgenome DNA B. 11/90 2,581bp.
M88179 Em(40)_vi:GEBDMARAL Gb(84)_vi:BDMARAL13A Bean dwarf mosaic virus coat protein (AR1) gene, putative replicative protein (AL1) gene comp
M88180 Em(40)_vi:GEBDMBRL1 Gb(84)_vi:BDMBRL1A Bean dwarf mosaic virus (BR1 and BL1) genes, complete cds. 9/92 2,576bp. 9 sequences.

*Virion protein(s)* two; $M_r$ of the larger 28000. $M_r$ of 2nd largest 27000. Method of preparation: Abouzid and Jeske (1986); Morales *et al.* (1988).

**Replication**

Genome replicates in nuclei (multiplication depending on daylength). Replication does not depend on a helper virus.

**Cytopathology**

Virions found in leaves, roots, vascular parenchyma, mesophyll, phloem and companion cells; in cytoplasm, in nuclei, in chloroplasts, and in cell vacuoles. Inclusions present in infected cells; are crystals in the nucleus, viroplasms, and unusual in shape; fibrillar bodies; they contain virions.

**Taxonomy and Relationships**

Bigeminivirus: *Geminiviridae*.

Virus(es) with serologically related virions: bean golden mosaic, African cassava mosaic and mung bean yellow mosaic viruses.

Genome homologous with those of bean golden mosaic, tomato golden mosaic and African cassava mosaic bigeminiviruses. Best tests for diagnosis: often in mixed infections with the closely related bean golden mosaic virus, making identification difficult.

**Comments**

The bean dwarf mosaic strain destroyed over 40,000 hectares of beans in 1978–81 in north west Argentina, where it causes 'achaparramiento' disease.

**References**

Abouzid, A. and Jeske, H. (1986). *J. Phytopath.* **115**: 344.
Baur, E. (1906). *Sitzber. Klg. preuss. Akad. Wiss.* **1**: 11.
Christie, R.G., Ko, N.J., Falk, B.W., Hiebert, E., Lastra, R., Bird, J. and Kim, K.S. (1986). *Phytopathology* **76**: 124.
Cooper, J.I. (1979). In: *Virus Diseases of Trees and Shrubs*. Inst. Terr. Ecol., Oxford, U.K.
Costa, A.S. (1975). In: *Tropical Diseases of Legumes*, p. 27; eds J. Bird and K. Maramorosch. Academic Press, New York.

Costa, A.S. (1955). *Phytopath. Z.* **24**: 97.
Costa, A.S. and Carvalho, A.M.B. (1960*a*). *Phytopath. Z.* **37**: 259.
Costa, A.S. and Carvalho, A.M.B. (1960*b*). *Phytopath. Z.* **38**: 129.
Gröning, B.R., Abouzid, A. and Jeske, H. (1987). *Proc. Natl. Acad. Sci. U.S.A.* **84**: 8996.
Jeske, H. (1980). *Eur. J. Cell Biol.* **22**: 549.
Jeske, H. (1986). *J. Phytopath.* **115**: 243.
Jeske, H. and Abouzid, A. (1986). *Mitt. biol. BundesAnst. Land. Forstw.* **232**: 403.
Jeske, H. and Schuchalter-Eicke, G. (1984). *Phytopath. Z.* **109**: 353.
Jeske, H. and Werz, G. (1978). *Phytopath. Z.* **91**: 1.
Jeske, H. and Werz, G. (1980*a*). *Phytopath. Z.* **97**: 43.
Jeske, H. and Werz, G. (1980*b*). *Virology* **106**: 155.
Jeske, H., Menzel, D. and Werz, G. (1977). *Phytopath. Z.* **89**: 289.
Morales, F., Niessen, A., Ramirez, C. and Castano, M. (1988). *Phytopathology* **78**: 858.
Orlando, A. and Silberschmidt, K. (1946). *Arch. Inst. Biol., Sao Paulo.* **17**: 1.
Schuchalter-Eicke, G. and Jeske, H. (1983). *Phytopath. Z.* **108**: 172.

# Abutilon yellows (?) *closterovirus*

**Data collated by A.A. Brunt, 1992.**

First reported in *Abutilon* spp.; from the U.S.A.; by Duffus *et al.* (1986).

### Natural host range and symptoms

*Abutilon* spp. — conspicuous leaf chlorosis.

### Transmission

Transmitted by a vector; an insect; *Trialeurodes abutilonea*; Aleyrodidae. Transmitted in a semi–persistent manner.

### Geographical distribution

Spreads in the North American region; the USA (California).

### Particle morphology

Virions filamentous; usually flexuous.

### Taxonomy and Relationships

Possible *Closterovirus*.

The virions and type of vector of this virus are like those of lettuce infectious yellows, beet pseudo-yellows and diodia yellow vein viruses, which may form a sub-group of closteroviruses (Duffus *et al.*, 1986).

### References

Duffus, J.E., Larsen, R.C. and Liu, H.Y. (1986). *Phytopathology* **76**: 97.

# Ageratum yellow vein bigeminivirus

**Data collated by S.M. Wong, 1993.**

*Standard acronym:* AYVV.

First reported in *Ageratum conyzoides*; from Singapore; by Wong *et al.* (1993).

**Natural host range and symptoms**

*Ageratum conyzoides* — vein yellowing.

**Transmission**

Transmitted by a vector; an insect; *Bemisia tabaci*; Aleyrodidae. Not transmitted by aphids. Virus not transmitted by mechanical inoculation; not transmitted by contact between plants; not transmitted by seed; not transmitted by pollen.

**Geographical distribution**

Spreads in India, Singapore, and Thailand.

**Experimental host range**

Several (3–9) families susceptible.

Diagnostically susceptible

*Asystasia nemorum* — leaf downcurling, chlorosis and mosaic.
*Emilia sonchifolia* — leaf downcurling, vein yellowing.
*Oxalis barrelieri* — vein yellowing, leaf downcurling, leaf distortion.
*Phyllanthus debilis* — vein yellowing.
*Synedrella nodiflora* — vein yellowing, mosaic, stunting.

Diagnostically insusceptible

*Carica papaya, Citrullus lanatus, Cleome rutidosperma, Cucumis melo, Eclipta prostrata.*

Maintenance and propagation hosts

*Ageratum conyzoides.*

Assay hosts (**L**ocal lesions or **W**hole plants)

*Ageratum conyzoides* (W).

Dicotyledonous Hosts *(**Susceptible** or Insusceptible)*

*Carica papaya. Citrullus lanatus. Glycine max. Phaseolus vulgaris.*

Comments on host-range

AYVV seems not to be transmitted by whitefly to ornamental *Ageratum* hybrids.

Sources of host-range data

Tan and Wong (1993).

**Properties of particles in sap**

Leaf sap contains few virions. Electron microscopy: best by ISEM.

**Particle morphology**

Virions geminate; not enveloped; 20 nm in diameter; 30 nm in length; with a conspicuous capsomere arrangement.

**Biochemical properties**

*Genome* consists of DNA; single–stranded; circular; of two parts; largest (or only) genome part 2.741 kb.

*Sequence database accession code(s):* EMBL X74516.

*Non–genomic nucleic acid* not found in the virions.

**Replication**

Replication does not depend on a helper virus.

**Cytopathology**

Virions found in cytoplasm and in nuclei. Inclusions present in infected cells; are unusual in shape; fibrillar rings.

**Taxonomy and Relationships**

Bigeminivirus: *Geminiviridae.*

Virus(es) with serologically related virions: cassava African mosaic virus.

**References**

Tan, H.N.P. and Wong, S.M. (1993). *J. Phytopathology* **139**: 165.

Wong, S.M., Swanson, M.M. and Harrison, B.D. *Plant Path.* **42**: 137.

# Agropyron mosaic *rymovirus*

**Data collated by W.G. Langenberg and N.L. Robertson, 1991.**

*Synonyms:* agropyron green mosaic virus, agropyron streak mosaic virus, agropyron yellow mosaic virus, couch grass streak mosaic virus. *Standard acronym:* AgMV. *Strains:* PV75 severe isolate, PV101, mild isolate.

First reported in *Agropyron repens*; from the U.S.A.; by McKinney, 1937.

**Natural host range and symptoms**

Symptoms persist and vary seasonally.

*Agropyron repens, Triticum aestivum, T. durum* — light green/yellow mosaic; leaves may become chlorotic and plants stunted.

**Transmission**

Transmitted by a vector; a mite; *Abacarus hystrix*; Eriophyidae. Not transmitted by *Aculus mckenziei, Eriophyes tulipae.* Virus transmitted by mechanical inoculation; not transmitted by seed; not transmitted by pollen.

**Ecology and control**

Studies reported by Slykhuis (1962; 1969).

**Geographical distribution**

Spreads in the Eurasian region and the North American region; Canada, Finland, the USA.

**Experimental host range**

Few (<3) families susceptible. Experimentally infected plants mostly show chlorotic mottle and streaks.

Diagnostically susceptible

*Agropyron repens, Elytrigia intermedium, Lophopyrum elongatum, Secale cereale, Triticum aestivum, T. durum* — light green/yellow mosaic in leaves; plants may die.

Diagnostically insusceptible

*Avena sativa, Bromus inermis, Echinochloa crus-galli, Hordeum jubatum, Lolium perenne.*

Maintenance and propagation hosts

*Agropyron repens, Triticum aestivum.*

Assay hosts (**L**ocal lesions or **W**hole plants)

*Triticum aestivum* (W).

Monocotyledonous Hosts *(**Susceptible** or Insusceptible)*

***Agropyron repens***. *Avena sativa*. *Bromus inermis*. ***Bromus secalinus***. *Dactylis glomerata*. *Echinochloa crus–galli*. ***Elytrigia intermedium***. ***Hordeum vulgare***. ***Lolium multiflorum***. *Lolium perenne*. ***Lophopyrum elongatum***. ***Poa pratensis***. ***Secale cereale***. ***Setaria viridis***. ***Triticum aestivum***. ***Triticum durum***.

Sources of host-range data

Slykhuis (1961; 1973); Slykhuis and Bell (1966).

**Properties of particles in sap**

TIP: 50 °C. LIV: 2–4 days (at 20°C). DEP: $\log_{10}$ minus 3. Leaf sap contains many virions.

**Purification**

Method: Staples and Brakke (1963); Slykhuis and Bell (1966).

**Particle morphology**

Virions filamentous; not enveloped; usually flexuous; with a clear modal length; of 700–760 nm; 15 nm wide. Axial canal obscure.

**Physical properties**

One sedimenting component in purified preparations; sedimentation coefficient 165 *S*.

**Biochemical properties**

*Genome* consists of RNA; single–stranded; linear; unipartite; largest (or only) genome part 8–10 kb. Genomic nucleic acid isolated by Langenberg and Robertson: polyadenylated full-length RNA genomes were obtained from partially purified virions with proteinase K and SDS, phenol/chloroform extraction, ethanol precipitation, and oligo-d(T) cellulose chromatography. 5' terminus of RNA has a VPg. Poly A region the 3' end has a poly (A) tail.

*Non–genomic nucleic acid* not found in the virions. Sub–genomic *m*RNA not found in infected cells.

*Virion protein(s)* one; coat protein.

*Virus–coded non–virion proteins* isolated (Langenberg (1991)); one protein found. $M_r$ 68000–72000; pinwheels.

**Replication**

Genome replicates presumably in cytoplasm. Coat protein *m*RNA translated in the cytoplasm. Replication does not depend on a helper virus.

**Cytopathology**

Virions found in leaves, roots, mesophyll, epidermis and the meristems; in cytoplasm. Inclusions present in infected cells; are pinwheels; they do not contain virions.

**Taxonomy and Relationships**

*Rymovirus: Potyviridae.*

Virus(es) with serologically related virions: hordeum mosaic and wheat streak mosaic viruses. Virus(es) with serologically unrelated virions: oat necrotic mottle and ryegrass mosaic viruses.

The 3' ends of the genomes of agropyron mosaic and hordeum mosaic viruses are 73% homologous. Best tests for diagnosis: use ISEM and immunogold labelling.

**References**


Langenberg, W.G. (1991). *J. gen. Virol.* **72**: 493.
McKinney, H.H. (1937). *U.S. Dept Agric. Circular* No. 442, 23 pp.
Robertson, N.L., French, R. and Langenberg, W.G. (1990). *Phytopathology* **80**: 1018.
Slykhuis, J.T. (1961). *Can. J. Pl. Sci.* **41**: 211.
Slykhuis, J.T. (1962). *Can. J. Bot.* **40**: 1439.
Slykhuis, J.T. (1969). *Phytopathology* **59**: 29.
Slykhuis, J.T. (1973). *CMI/AAB Descr. Pl. Viruses* No. 118, 4 pp.
Slykhuis, J.T. and Bell, W. (1966). *Can. J. Bot.* **44**: 1191.
Staples, R. and Brakke, M. (1963). *Phytopathology* **53**: 969.


# Ahlum waterborne (?) *carmovirus*

**Data collated by A.A. Brunt, 1994.**

*Standard acronym:* AWBV.

First reported in water from River Ahlum; from Germany; by Yi *et al.* (1992).

**Geographical distribution**

Found, but with no evidence of spread, in Germany.

**Experimental host range**

Few (<3) families susceptible.

Diagnostically susceptible

*Chenopodium amaranticolor, C. quinoa, Nicotiana benthamiana* and *N. clevelandii.*

Maintenance and propagation hosts

*Nicotiana benthamiana.*

Dicotyledonous Hosts *(**Susceptible** or Insusceptible)*

***Chenopodium amaranticolor*. *Chenopodium quinoa*. *Nicotiana benthamiana*. *Nicotiana clevelandii*.**

**Purification**

Method: Gallitelli *et al.* (1985).

**Particle morphology**

Virions isometric; not enveloped; 33 nm in diameter.

**Biochemical properties**

*Genome* consists of RNA; single–stranded; unipartite; largest (or only) genome part 4.7 kb.

*Virion protein(s)* one; $M_r$ 37800; coat protein.

**Cytopathology**

Virions found in cytoplasm.

**Taxonomy and Relationships**

Possible *Carmovirus: Tombusviridae.*

Virus(es) with serologically unrelated virions: Bean mild mosaic, blackgram mottle, carnation mottle, cucumber leaf spot, cucumber soil-borne, elderberry latent, galinsoga mosaic, glycine mottle, hibiscus chlorotic ringspot, melon necrotic spot, narcissus tip necrosis, pelargonium flower break, saguaro cactus, tephrosia symptomless and turnip crinkle carmoviruses and 12 tombusviruses.

In Northern blot hybridization tests, the genomic RNA showed some homology with that of Weddel waterborne carmovirus.

**References**

Gallitelli, D., Hull, R. and Koenig, R. (1985). *J. gen. Virol.* **66**: 1523.

Yi, L., Lesemann, D.-E., Koenig, R., Rüdel, M. and Pfeilstetter, E. (1992). *J. Phytopath.* **134**: 121.

# Alfalfa 1 *alphacryptovirus*

**Data collated by G.P. Accotto, 1991.**

*Synonyms:* alfalfa temperate virus (in part). *Standard acronym:* ACV-1.

First reported in *Medicago sativa*; from Japan; by Natsuaki *et al.* (1986); Luisoni *et al.* (1987).

**Natural host range and symptoms**

*Medicago sativa* — symptomless.

**Transmission**

Transmitted by means not involving a vector. Virus not transmitted by mechanical inoculation; not transmitted by grafting; not transmitted by contact between plants; transmitted by seed; transmitted by pollen to the seed.

**Geographical distribution**

Spreads in the Eurasian region and the Middle East. Found, but with no evidence of spread, in the Eurasian region, the Middle East and Japan.

Maintenance and propagation hosts

*Medicago sativa.*

Dicotyledonous Hosts *(**Susceptible** or Insusceptible)*

***Medicago sativa*.**

**Properties of particles in sap**

Leaf sap contains few virions.

**Purification**

Method: Accotto *et al.* (1990).

**Particle morphology**

Virions isometric; not enveloped; 30 nm in diameter; rounded in profile; without a conspicuous capsomere arrangement.

**Physical properties**

One sedimenting component in purified preparations. Density 1.34–1.35 g $cm^{-3}$ in CsCl.

**Biochemical properties**

*Genome* consists of RNA; double–stranded; linear; of two parts; largest (or only) genome part the larger 1.9 kb (pairs); the 2nd largest 1.77 kb (pairs). Genomic nucleic acid isolated by phenol/chloroform extraction in the presence of 0.5% SDS.

*Virion protein(s)* two; $M_r$ of the larger 67000; P1 protein — possible RNA-dependent RNA polymerase, obtained from *in vitro* translation of the large RNA segment. $M_r$ of 2nd largest 53000; P2 coat protein. Method of preparation: Accotto *et al.* (1990).

**Replication**

Replication does not depend on a helper virus.

**Cytopathology**

Virions found in all parts of the host plant; in cytoplasm. Inclusions absent from infected cells.

**Taxonomy and Relationships**

*Alphacryptovirus: Partitiviridae.*

Virus(es) with serologically related virions: hop trefoil 1 alphacryptovirus. Virus(es) with serologically unrelated virions: beet 1, carnation 1, radish yellow edge, red clover, vicia and white clover 1 alphacryptoviruses.

**References**


Accotto, G.P., Marzachi, C., Luisoni, E. and Milne, R.G. (1990). *J. gen. Virol.* **71**: 433.

Boccardo, G., Lisa, V., Luisoni, E. and Milne, R.G. (1987). *Adv. Virus Res.* **32**: 171.

Luisoni, E., Milne, R.G., Accotto, G.P. and Boccardo, G. (1987). *Intervirology* **28**: 144.

Natsuaki, K.T., Natsuaki, T., Okuda, S., Teranaka, M., Yamashita, S. and Doi, Y. (1984). *J. Agric. Sci.* **29**: 49.

Natsuaki, T., Natsuaki, K.T., Okuda, S., Teranaka, M., Milne, R.G., Boccardo, G. and Luisoni, E. (1986). *Intervirology* **25**: 69.


# Alfalfa 2 (?) *betacryptovirus*

**Data collated by A.A. Brunt, 1991.**

*Standard acronym:* ACV-2.

First reported in *Medicago sativa*; from Japan; by Boccardo *et al.* (1987).

**Natural host range and symptoms**

*Medicago sativa* — symptomless.

**Transmission**

Virus not transmitted by mechanical inoculation; not transmitted by grafting; transmitted by seed.

**Geographical distribution**

Found, but with no evidence of spread, in Japan and Italy.

Dicotyledonous Hosts *(**Susceptible** or Insusceptible)*

***Medicago sativa*.**

**Particle morphology**

Virions isometric; 38 nm in diameter; angular in profile.

**Taxonomy and Relationships**

Possible *Betacryptovirus: Partitiviridae.*

**Comments**

The virus was first found occurring in complex with alfalfa 1 alphacryptovirus.

**References**

Accotto, G.P., Marzachi, C. Luisoni, E. and Milne, R.G. (1990). *J. gen. Virol.* **71**: 433.

Boccardo, G., Lisa, V., Luisoni, E. and Milne, R.G. (1987). *Adv. Virus Res.* **32**: 171.

# Alfalfa mosaic *alfamovirus*

**Data collated by L. van Vloten-Doting, 1981. Revised 1987 by A.J. Gibbs.**

*Synonyms:* lucerne mosaic virus, potato calico virus. *Standard acronym:* AMV.

First reported in *Medicago sativa*; from the U.S.A.; by Weimer (1931).

**Natural host range and symptoms**

Symptoms persist, or disappear soon after infection.
The virus naturally infects many species including:
*Caryopteris incana* — white fleck.
*Cicer arietinum* — wilting.
*Lycopersicon esculentum* — severe necrosis.
*Apium graveolens, A. graveolens* var. *rapaceum, Lactuca sativa, Malva parviflora, Trifolium incarnatum, T. repens, Viburnum opulus* — mosaic.
*Medicago sativa* — mosaic, mottle, malformation, but often symptomless during summer.
*Nicotiana tabacum, Lupinus* spp. — mosaic and mottle.
*Phaseolus vulgaris, Vigna unguiculata, V. radiata, Astragalus glyciphyllos, Dolichos lablab, Glycine max, Lens esculenta, Capsicum annuum* — necrotic or chlorotic local lesions, sometimes mosaic.
*Philadelphus* sp. — white mottle.
*Pisum sativum* — necrosis.
*Solanum tuberosum* — calico and tuber necrosis.

**Transmission**

Transmitted by a vector; an insect; *Myzus persicae* and at least 13 other species; Aphididae. Transmitted in a non–persistent manner. Virus transmitted by mechanical inoculation; transmitted by grafting; not transmitted by contact between plants; transmitted by seed (50% in alfalfa seeds from individual infected plants and up to 10% in commercial seed); transmitted by pollen to the seed.

**Ecology and control**

Studies reported by Gibbs (1962); Garran and Gibbs (1982); Edwardson and Christie (1986).

**Geographical distribution**

Probably distributed worldwide.

**Experimental host range**

Many (>9) families susceptible. Experimentally infected plants mostly show necrotic local lesions, mottles or ringspots.

Diagnostically susceptible

*Chenopodium amaranticolor, C. quinoa* — chlorotic local lesions; systemic chlorotic and necrotic flecks which distinguish this virus from cucumber mosaic virus.

*Ocimum basilicum* — systemic yellow mosaic.

*Phaseolus vulgaris* — most strains induce necrotic local lesions, some induce chlorotic local lesions or a systemic mottle, vein necrosis and leaf malformation.

*Pisum sativum* — local lesions and/or wilting of inoculated leaves; systemic stem necrosis and plant death.

*Vicia faba* — black necrotic local lesions, stem necrosis and plant death, a few isolates induce systemic mottle.

*Vigna unguiculata* — necrotic local lesions, not systemic, some isolates induce various systemic symptoms.

*Nicotiana tabacum* — necrotic and chlorotic local lesions (some strains induce no local symptoms); systemic mottle, vein banding, ringspots, rarely malformation.

Maintenance and propagation hosts

*Nicotiana glutinosa, N. tabacum.*

Assay hosts (**L**ocal lesions or **W**hole plants)

*Chenopodium amaranticolor* (L,W), *C. quinoa* (L,W), *Phaseolus vulgaris* (L,W), *Vigna unguiculata* (L).

Dicotyledonous Hosts *(**Susceptible** or Insusceptible)*

***Ageratum conyzoides. Amaranthus caudatus. Amaranthus retroflexus. Antirrhinum majus. Apium graveolens. Arachis hypogaea.*** *Atriplex hortensis.* ***Beta vulgaris. Brassica campestris*** ssp. ***rapa.*** *Brassica oleracea* ssp. *capitata.* ***Calendula officinalis. Capsicum annuum. Capsicum frutescens.*** *Cassia tora.* ***Catharanthus roseus. Celosia argentea. Cheiranthus cheiri. Chenopodium album. Chenopodium amaranticolor.*** *Chenopodium capitatum. Chenopodium foetidum.* ***Chenopodium murale. Chenopodium quinoa. Cicer arietinum. Cichorium endiva. Coriandrum sativum. Crotalaria spectabilis. Cucumis melo. Cucumis sativus. Cucurbita pepo. Cyamopsis tetragonoloba.*** *Datura stramonium.* ***Daucus carota*** (var. *sativa*). ***Dianthus barbatus. Dianthus caryophyllus. Emilia sagittata. Fagopyrum esculentum. Glycine max. Gomphrena globosa. Helianthus annuus. Hibiscus esculentus.*** *Hyoscyamus niger.* ***Lactuca sativa. Lathyrus odoratus. Lens esculenta. Linum usitatissimum.*** *Lotus corniculatus.* ***Lupinus albus. Lycopersicon esculentum. Macroptilium lathyroides. Matthiola incana. Medicago hispida. Medicago sativa. Melilotus alba.*** *Nicandra physaloides.* ***Nicotiana bigelovii. Nicotiana clevelandii. Nicotiana debneyi. Nicotiana glutinosa. Nicotiana megalosiphon. Nicotiana rustica. Nicotiana sylvestris. Nicotiana tabacum. Ocimum basilicum. Petunia hybrida. Phaseolus lunatus. Phaseolus vulgaris. Physalis floridana. Physalis peruviana. Phytolacca americana. Pisum sativum.*** *Sesbania exaltata.* ***Solanum demissum. Solanum melongena. Solanum nigrum. Solanum nodiflorum. Solanum rostratum. Solanum tuberosum. Sonchus oleraceus. Spinacia oleracea. Stellaria media. Tetragonia expansa. Trifolium dubium. Trifolium hybridum. Trifolium incarnatum. Trifolium pratense. Trifolium repens. Trifolium subterraneum. Tropaeolum majus.*** *Verbesina encelioides.* ***Vicia faba.*** *Vicia sativa. Vicia villosa.* ***Vigna radiata. Vigna unguiculata. Vigna unguiculata*** ssp. ***sesquipedalis. Zinnia elegans.***

Sources of host-range data

Thornberry (1966); Hull (1969); Schmelzer *et al.* (1973).

**Properties of particles in sap**

TIP: 60–65 °C. LIV: 1–4 days. DEP: $\log_{10}$ minus 3–4. Infectivity of sap not changed by treatment with di–ethyl ether. Leaf sap contains many virions. Electron microscopy: most negative stains disrupt virions unless they are first fixed with formaldehyde. The virions of some strains are stable in AM (Price and McLean, 1982).

**Purification**

Method: Van Vloten-Doting and Jaspars (1972).

**Particle morphology**

Virions bacilliform; not enveloped; mostly 30, 35, 43, and 56 nm in length; 18 nm wide.

**Physical properties**

Four sedimenting components in purified preparations (at least); sedimentation coefficient of the fastest 94 *S* (B); of the other(s) 82 *S* (M), or 78 *S* (Ta), or 66 *S* (Tb; minor components of 53 *S* and 60 *S* found in some isolates). Density 1.381 g $cm^{-3}$ in CsCl ($T_a$), or 1.385 g $cm^{-3}$ in CsCl (B). Density in $Cs_2SO_4$ 1.278 g $cm^{-3}$. Isoelectric point pH 4.6. $A_{260}/A_{280}$ ratio 1.7–1.8.

**Biochemical properties**

Virions contain 16 % nucleic acid; 84 % protein; 0 % lipid.

*Genome* consists of RNA; single–stranded; linear; of three parts; largest (or only) genome part the largest, RNA-1 3.64 kb; the 2nd largest 2.59 kb (RNA-2); the 3rd largest 2.04 kb (RNA-3). Base composition 23 % G; 25 % A; 23 % C; 29 % U. 5' terminus of RNA has a methylated nucleotide cap. Infectivity retained when deproteinised with proteases, or decreased when deproteinised with proteases; retained when deproteinised with phenol or detergent (provided RNA-4 is present). Poly A region absent. Additional factor required for infectivity; coat protein *m*RNA or RNA-4 of 0.88 kb. RNA-1 and RNA-2 alone can infect protoplasts. Genome has no *t*RNA–like activity. Nucleotide sequence references: Cornelissen *et al.* (1983); Collot *et al.* (1977); Barker *et al.* (1983); Ravelonandro *et al.* (1984); Gunn and Symons (1980); Koper-Zwarthoff *et al.* (1980); Houwing and Jaspars (1982).

*Sequence database accession code(s):*

J02001 Gb(84)_vi:MAARNA23 alfalfa mosaic virus (Q strain) rna2 3' end. 2/85 228bp.

J02003 Em(40)_vi:ALRNA3 Gb(84)_vi:MAARNA3 alfalfa mosaic virus rna 3 35kd protein leader sequence. 4/90 318bp.

J02005 Gb(84)_vi:MAARNA35 alfalfa mosaic virus (strain 425) rna3 5' end. 2/85 101bp.

K02702 Gb(84)_vi:MAACG2Z Alfalfa mosaic virus (strain 425 Leiden) RNA 2 of complete genome. 9/88 2,593bp.

K02703 Em(40)_vi:ALMRNA3 Gb(84)_vi:MAACG3Z Alfalfa mosaic virus (strain 425 Madison) RNA 3 of complete genome. 4/90 2,037bp

K03542 Em(40)_vi:MAARNA3L Gb(84)_vi:MAARNA3L Alfalfa mosaic virus RNA 3 encoding viral coat protein, complete.B. 4/90 2,142bp.

L00161 Gb(84)_vi:MAARNA33 Alfalfa mosaic virus (strain Q) RNA 3, 3' end. 8/86 230bp

L00162 Em(40)_vi:ALMAARNA4 Gb(84)_vi:MAARNA4 Alfalfa mosaic virus (strain 425 Leiden) RNA 4 encoding viral coat protein. 5/94 964bp.

L00163 Em(40)_vi:ALMAACG1Z Gb(84)_vi:MAACG1Z Alfalfa mosaic virus (strain 425 Leiden) RNA 1 of complete genome. 5/94 3,644bp.

L00164 Gb(84)_vi:MAARNA13 Alfalfa mosaic virus (strain Q) RNA 1. 8/86 226bp.
M10826 Em(40)_vi:MAARNA01 Gb(84)_vi:MAARNA4AX Alfalfa mosaic virus (A1MV) RNA 4, 3' terminal fragment 29C. 7/91 91bp.
M10851 Em(40)_vi:MAARNA4A Gb(84)_vi:MAARNA4A Alfalfa mosaic virus RNA 4, 5' terminal region. 7/89 74bp.
M25004 Em(40)_vi:ALMAARNAA Gb(84)_vi:MAARNAA Alfalfa mosaic virus RNA 3 or 4, 3' end. 4/92 113bp.
M25005 Em(40)_vi:ALMAARNAB Gb(84)_vi:MAARNAB Alfalfa mosaic virus RNA 2, 3' end. 4/92 103bp.
M25006 Em(40)_vi:ALMAARNAC Gb(84)_vi:MAARNAC Alfalfa mosaic virus RNA 1, 3' end. 4/92 110bp.
M25452 Em(40)_vi:ALMAARNA1 Gb(84)_vi:MAARNA4D Alfalfa mosaic virus RNA 4 RNA fragment. 4/92 62bp.
M35975 Em(40)_vi:ALMAARNA Gb(84)_vi:MAARNA1A Alfalfa mosaic virus (strain AlMV-S) 5' end of RNA-1. 12/90 163bp.
M35976 Em(40)_vi:ALMAAR01 Gb(84)_vi:MAARNA1B Alfalfa mosaic virus (strain AlMV-B) 5' end of RNA-1. 12/90 115bp.
M36389 Em(40)_vi:ALMAAR02 Gb(84)_vi:MAARNA2A Alfalfa mosaic virus (strain AlMV-S) 5' end of RNA-2. 12/90 108bp.
M36390 Em(40)_vi:ALMAAR03 Gb(84)_vi:MAARNA2B Alfalfa mosaic virus (strain AlMV-B) 5' end of RNA-2. 12/90 109bp.
M36391 Em(40)_vi:ALMAAR04 Gb(84)_vi:MAARNA3B Alfalfa mosaic virus (strain AlMV-S) 5' end of RNA-3. 12/90 305bp.
M36392 Em(40)_vi:ALMAAR05 Gb(84)_vi:MAARNA3C Alfalfa mosaic virus (strain AlMV-B) 5' end of RNA-3. 12/90 290bp.
M59241 Em(40)_vi:ALMAA32K Gb(84)_vi:MAA32KDMP Alfalfa mosaic virus 32 kDa movement protein and coat protein RNA, complete cds. 8/92 2,188bp.
S55890 Em(40)_vi:S55890 Gb(84)_vi:S55890 RNA-3 coat protein homolog, alfalfa mosaic virus RNA-3 32K protein homolog (RNA-2) raspberry
U12509 Gb(84)n:AMU12509 Alfalfa mosaic virus NZ1 RNA4 coat protein mRNA, complete cds. 8/94 876bp.
U12510 Gb(84)n:AMU12510 Alfalfa mosaic virus NZ2 RNA4 coat protein mRNA, complete cds. 8/94 876bp.
V00044 Em(40)_vi:ALALM1 Gb(84)_vi:ALALM1 5' end of alfalfa mosaic virus RNA 1. 5/94 61bp.
V00045 Em(40)_vi:ALALM2 Gb(84)_vi:ALALM2 5' end of alfalfa mosaic virus RNA 2. 5/94 13bp.
V00046 Em(40)_vi:ALALM3 Gb(84)_vi:ALALM3 5' end of alfalfa mosaic virus RNA 3. 5/94 101bp.
V00047 Em(40)_vi:ALALM4 Gb(84)_vi:ALALM4 Intercistronic junction in alfalfa mosaic virus RNA 3. 5/94 122bp.
V00048 Em(40)_vi:ALALM5 Gb(84)_vi:ALALM5 alfalfa mosaic virus RNA 4 coding for the coat protein. 5/94 881bp.
V00049 Em(40)_vi:ALALM6 Gb(84)_vi:ALALM6 3' end of alfalfa mosaic virus RNA 1. 7/91 226bp.
V00050 Em(40)_vi:ALALM7 Gb(84)_vi:ALALM7 3' end of alfalfa mosaic virus RNA 2. 7/91 228bp.
V00051 Em(40)_vi:ALALM8 Gb(84)_vi:ALALM8 3' end of alfalfa mosaic virus RNA 3. 7/91 230bp.
V00052 Em(40)_vi:ALAM01 Gb(84)_vi:ALAM01 Alfalfa mosaic virus RNA 1 fragment. (Obtained after digestion with ribonuclease T1.). 5/94 8

V00053 Em(40)_vi:ALAM02 Gb(84)_vi:ALAM02 Alfalfa mosaic virus RNA 1 fragment. (Obtained after digestion with ribonuclease T1.). 5/94 6

V00054 Em(40)_vi:ALAM03 Gb(84)_vi:ALAM03 Alfalfa mosaic virus RNA 1 fragment. (Obtained after digestion with ribonuclease T1.). 5/94 4

V00055 Em(40)_vi:ALAM04 Gb(84)_vi:ALAM04 Alfalfa mosaic virus RNA 1 fragment. (Obtained after digestion with ribonuclease T1.). 5/94 4

V00056 Em(40)_vi:ALAM05 Gb(84)_vi:ALAM05 Alfalfa mosaic virus RNA 1 fragment. (Obtained after digestion with ribonuclease T1.). 5/94 4

V00057 Em(40)_vi:ALAM06 Gb(84)_vi:ALAM06 Alfalfa mosaic virus RNA 1 fragment. (Obtained after digestion with ribonuclease T1.). 5/94 3

V00058 Em(40)_vi:ALAM07 Gb(84)_vi:ALAM07 Alfalfa mosaic virus RNA 1 fragment. (Obtained after digestion with ribonuclease T1.). 5/94 3

V00059 Em(40)_vi:ALAM08 Gb(84)_vi:ALAM08 Alfalfa mosaic virus RNA 1 fragment. (Obtained after digestion with ribonuclease T1.). 5/94 3

V00060 Em(40)_vi:ALAM09 Gb(84)_vi:ALAM09 Alfalfa mosaic virus RNA 1 fragment. (Obtained after digestion with ribonuclease T1.). 5/94 2

V00061 Em(40)_vi:ALAM10 Gb(84)_vi:ALAM10 Alfalfa mosaic virus RNA 1 fragment. (Obtained after digestion with ribonuclease T1.). 5/94 2

V00062 Em(40)_vi:ALAM11 Gb(84)_vi:ALAM11 Alfalfa mosaic virus RNA 1 fragment. (Obtained after digestion with ribonuclease A.). 5/94 25

V00063 Em(40)_vi:ALAM12 Gb(84)_vi:ALAM12 Alfalfa mosaic virus RNA 1 fragment. (Obtained after digestion with ribonuclease A.). 5/94 19

V00064 Em(40)_vi:ALAM13 Gb(84)_vi:ALAM13 Alfalfa mosaic virus RNA 1 fragment. (Obtained after digestion with ribonuclease A.). 5/94 19

V00065 Em(40)_vi:ALAM14 Gb(84)_vi:ALAM14 Alfalfa mosaic virus RNA 1 fragment. (Obtained after digestion with ribonuclease A.). 5/94 18

V00066 Em(40)_vi:ALAM15 Gb(84)_vi:ALAM15 Alfalfa mosaic virus RNA 1 fragment. (Obtained after digestion with ribonuclease A.). 5/94 18

V00067 Em(40)_vi:ALAM16 Gb(84)_vi:ALAM16 Alfalfa mosaic virus RNA 1 fragment. (Obtained after digestion with ribonuclease A.). 5/94 15

V00068 Em(40)_vi:ALAM17 Gb(84)_vi:ALAM17 Alfalfa mosaic virus RNA 1 fragment. (Obtained after digestion with ribonuclease T1.). 5/94 9

X00819 Em(40)_vi:ALAM19 Gb(84)_vi:ALAM19 Alfalfa mosaic virus (strain S) complete RNA 3 sequence. 9/93 2,055bp.

X01572 Em(40)_vi:A1MVRNA2 Gb(84)_vi:A1MVRNA2 Alfalfa mosaic virus (A1M4) RNA 2. 7/91 2,593bp.

M28374 Em(43)_vi:Maatbts7a Gb(89)_vi:Maatbts7a Alfalfa mosaic virus (clone 143) temperature-sensitive mutant Tbts7 RNA3 (coat protein-encodi

M28375 Em(43)_vi:Maatbts7b Gb(89)_vi:Maatbts7b Alfalfa mosaic virus (clone 112) temperature-sensitive mutant Tbts7 RNA3 (coat protein-encodi

U12509 Em(43)_vi:Am12509 Gb(89)_vi:Amu12509 Alfalfa mosaic virus NZ1 RNA4 coat protein mRNA, complete cds. 8/94 876bp.

U12510 Em(43)_vi:Am12510 Gb(89)_vi:Amu12510 Alfalfa mosaic virus NZ2 RNA4 coat protein mRNA, complete cds. 8/94 876bp.

*Non–genomic nucleic acid* found in the virions; is subgenomic *m*RNA (of coat protein gene). Sub–genomic *m*RNA found in infected cells; coat protein *m*RNA also present in virion preparations.

*Virion protein(s)* one; $M_r$ 24250. Method of preparation: Kruseman *et al.* (1971). Amino acid sequence: Van Beynum (1977) for AMV-425 and Castel (1979) for AMV-VRU.

**Replication**

Coat protein *m*RNA translated in the cytoplasm. Replication does not depend on a helper virus.

**Cytopathology**

Virions found in all parts of the host plant; in cytoplasm and in chloroplasts; cytoplasmic invaginations in the chloroplasts. Inclusions cytoplasmic present in infected cells; are unusual in shape; vacuolated bodies derived from spheroidal and granular bodies, vacuolated or non-vacuolated bodies containing nucleoprotein and hexagonal crystals; they contain virions (aggregated). Also tubular bodies are found in the nucleus (Hull *et al.*, 1969).

**Taxonomy and Relationships**

*Alfamovirus: Bromoviridae* (the type species).

No other viruses are definitive alfamoviruses. However, sunflower ringspot virus (J. Dale, pers. comm.) and a virus isolated from *Mercurialis perennis* in the U.K. (A.A. Brunt, pers. comm.) have virions of similar shape, but are not serologically related. Cassava Ivorian bacilliform virus (Fargette *et al.*, 1991), and allied viruses, is unrelated. Some alfalfa mosaic virus genes are homologous to those of the ilar-, bromo- and cucumoviruses.

**References**


Barker, R.F., Jarvis, N.P., Thompson, D.U., Loesch-Fries, L.S. and Hall, T.C. (1983). *Nucl. Acids Res.* **11**: 2881.

Bos, L. and Jaspars, E.M.J. (1971). *CMI/AAB Descr. Pl. Viruses* No. 46, 4 pp.

Castel, A., Kraal, B., de Graaf, J.M. and Bosch, L. (1979). *Eur. J. Biochem.* **102**: 125.

Collot, D., Dupin, A. and Duranton, H. (1977). *Biochim. biophys. Acta* **492**: 260.

Cornelissen, B.J.C., Brederode, F.Th., Veenemann, G.H., van Boom, J.H. and Bol, J.F. (1983). *Nucl. Acids Res.* **11**: 3019.

Edwardson, J.R. and Christie, R.G. (1986). *Fla Agric. Exp. Stn Monogr.* No. 14, p. 30.

Garran, J. and Gibbs, A.J. (1982). *Aust. J. agric. Res.* **33**: 657.

Gibbs, A.J. (1962). *Pl. Path.* **11**: 167.

Gunn, M.R. and Symons, R.H. (1980). *FEBS Lett.* **109**: 145.

Houwing, C.J. and Jaspars, E.M.J. (1982). *Biochemistry* **21**: 3408.

Hull, R. (1969). *Adv. Virus Res.* 15, p. 365.

Hull, R., Hills, G.J. and Plaskitt, (1969). *J. Ultrastruct. Res.* **24**: 465.

Jaspars E.M.J. and Bos, L. (1980). *CMI/AAB Descr. Pl. Viruses* No. 229, 7 pp.

Koper-Zwarthoff, E.C., Brederode, F.Th., Veenemann, G. and van Boom, J.H. (1980). *Nucl. Acids Res.* **8**: 5635.

Kruseman, J., Jaspars, E.M.J., Bol, J.F., Brederode, F.Th. and Veldstra, H. (1971). *Biochemistry* **10**: 447.

Price, L.K. and McLean, G.D. (1982). *Australas. Pl. Path.* **11**: 48.

Ravelonandro, M., Pinck, M. and Pinck, L. (1984). *Biochimie* **66**: 395.

Schmelzer, K., Schmidt, H.B. and Beczner, L. (1973). *Biol. Zbl.* **92**: 211.

Thornberry, H.H. (1966). In: *Index of Plant Virus Diseases*. U.S. Dept Agric. Hdbk No. 307, p.264.

Van Beynum, G.M.A., de Graaf, J.M., Castel, A., Kraal, B. and Bosch, L. (1977). *Eur. J. Biochem.* **72**: 63.

Van Vloten-Doting, L. and Jaspars, E.M.J. (1972). *Virology* **48**: 699.

Weimer, J.L. (1931). *Phytopathology* **21**: 122.

# Alligatorweed stunting (?) *closterovirus*

**Data collated by A.A. Brunt, 1991.**

*Standard acronym:* AWSV.

First reported in *Alternanthera philoxeroides*; from Florida, U.S.A.; by Hill and Zettler (1973).

**Natural host range and symptoms**

*Alternanthera philoxeroides* — leaf reddening and malformation; severe stunting.

**Transmission**

Virus transmitted by mechanical inoculation.

**Geographical distribution**

Spreads in the North American region; the USA (Florida).

**Particle morphology**

Virions filamentous; usually flexuous; with a clear modal length; of 1717 nm.

**Cytopathology**

Virions found in leaves and phloem; in cytoplasm.

**Taxonomy and Relationships**

Possible *Closterovirus*.

**References**

Hill, H.R. and Zettler, F.W. (1973). *Phytopathology* **63**: 443.

Milne, R.G. (1988). In: *The Plant Viruses*, Vol. 4, The Filamentous Plant Viruses, p.3; ed. R.G. Milne. Plenum Press, New York.

# Alsike clover vein mosaic virus

**Data collated by K. Lindsten, 1981. Checked 1986.**

*Synonyms:* clover Alsike vein mosaic virus.

First reported in *Trifolium hybridum*; from Sweden; by Gerhardson and Lindsten (1971).

**Natural host range and symptoms**

Symptoms vary seasonally.

*Trifolium hybridum* — vein chlorosis, malformation.

**Transmission**

Virus transmitted by mechanical inoculation; not transmitted by seed; not transmitted by pollen.

**Geographical distribution**

Spreads in Denmark, Finland, Norway, and Sweden.

**Experimental host range**

Few (<3) families susceptible. Experimentally infected plants mostly show necrotic local lesions, not systemic.

Diagnostically susceptible

*Chenopodium amaranticolor, C. quinoa, Phaseolus vulgaris, Vigna unguiculata* — chlorotic local lesions; not systemic.

*Lathyrus odoratus* — chlorotic local lesions, wilting, death.

*Pisum sativum* — necrotic local lesions; leaf malformation, death.
*Trifolium hybridum, T. incarnatum* — systemic mosaic, chlorosis.
*Vicia faba* — necrotic and chlorotic local lesions; usually not systemic.

Maintenance and propagation hosts

*Trifolium hybridum.*

Assay hosts (**L**ocal lesions or **W**hole plants)

*Chenopodium quinoa* (L).

Dicotyledonous Hosts (***Susceptible*** *or Insusceptible)*

*Beta vulgaris.* ***Chenopodium amaranticolor.*** *Chenopodium quinoa. Cucumis sativus. Datura stramonium. Glycine max. Gomphrena globosa.* ***Lathyrus odoratus.*** *Lycopersicon esculentum. Medicago sativa. Nicotiana clevelandii. Nicotiana glutinosa. Nicotiana tabacum. Petunia hybrida.* ***Phaseolus vulgaris.*** *Physalis floridana.* ***Pisum sativum.*** *Spinacia oleracea. Tetragonia expansa.* ***Trifolium hybridum. Trifolium incarnatum.*** *Trifolium pratense. Trifolium repens.* ***Vicia faba. Vicia sativa. Vicia villosa. Vigna unguiculata.*** *Zinnia elegans.*

Sources of host-range data

Gerhardson *et al.* (1971).

**Properties of particles in sap**

TIP: 76 °C. LIV: 64 days. DEP: $\log_{10}$ minus 7. Leaf sap contains many virions.

**Particle morphology**

Virions isometric; not enveloped; 30 nm in diameter.

**Physical properties**

One sedimenting component in purified preparations; sedimentation coefficient 123 *S*.

**Taxonomy and Relationships**

Virus(es) with serologically unrelated virions: red clover necrotic mosaic, red clover mottle, alfalfa mosaic and arabis mosaic viruses.

**References**

Gerhardson, B. and Lindsten, K. (1971). *Phytopath. Z.* **72**: 76.

# Alstroemeria (?) *ilarvirus*

**Data collated by S. Phillips, 1987.**

First reported in *Alstroemeria caryophylla* 'Ligtu' hybrids; from England; by Brunt and Phillips (1981).

**Natural host range and symptoms**

*Alstroemeria caryophylla* 'Ligtu' hybrids — faint chlorosis.

**Transmission**

Virus transmitted by mechanical inoculation; not transmitted by contact between plants.

**Geographical distribution**

Found, but with no evidence of spread, in England.

**Experimental host range**

Several (3–9) families susceptible (but only 6 families have been tested). Experimentally infected plants mostly show local lesions, systemic chlorosis, mosaic, mosaic line patterns.

Diagnostically susceptible

*Chenopodium quinoa, C. amaranticolor, C. murale, Gomphrena globosa* — local lesions; severe systemic chlorosis and malformation.

*Nicotiana megalosiphon* — local etched rings; systemic chlorosis or white necrotic line patterns.

*Petunia hybrida* — systemic mosaic.

*Nicotiana clevelandii* — systemic mild chlorotic vein-banding.

Diagnostically insusceptible

*Lycopersicon esculentum, Nicotiana rustica, Brassica juncea.*

Maintenance and propagation hosts

*Nicotiana clevelandii, N. megalosiphon, N. benthamiana, Chenopodium quinoa, Petunia hybrida.*

Assay hosts (**L**ocal lesions or **W**hole plants)

*Chenopodium quinoa* (L and W), *Nicotiana megalosiphon* (L and W), *Phaseolus vulgaris* (L).

Dicotyledonous Hosts *(**Susceptible** or Insusceptible)*

*Brassica juncea*. ***Chenopodium amaranticolor***. ***Chenopodium murale***. ***Chenopodium quinoa***. ***Datura stramonium***. ***Gomphrena globosa***. *Lycopersicon esculentum*. ***Nicotiana benthamiana***. ***Nicotiana clevelandii***. ***Nicotiana debneyi***. ***Nicotiana glutinosa***. ***Nicotiana megalosiphon***. *Nicotiana rustica*. ***Nicotiana tabacum***. ***Petunia hybrida***. ***Phaseolus vulgaris***. ***Pisum sativum***. ***Tetragonia expansa***. ***Trifolium incarnatum***. ***Vigna unguiculata***.

Sources of host-range data

Phillips and Brunt (1986).

**Properties of particles in sap**

TIP: 60 °C. DEP: $\log_{10}$ minus 4 (purified preparation). Leaf sap contains few virions. Electron microscopy: usually only seen in purified preparations.

**Purification**

Method: Phillips and Brunt (1986).

**Particle morphology**

Virions isometric; not enveloped; 25–32 nm in diameter; rounded in profile; without a conspicuous capsomere arrangement.

**Replication**

Replication does not depend on a helper virus.

**Cytopathology**

Virions found in all parts of the host plant.

**Taxonomy and Relationships**

Possible *Ilarvirus: Bromoviridae.*

Virus(es) with serologically related virions: prunus necrotic ringspot, spinach latent, hydrangea mosaic and tobacco streak (dewberry latent isolate) viruses. Virus(es) with serologically unrelated virions: asparagus 2, crimson clover latent, elm mottle and tobacco streak (asparagus and dahlia isolates) viruses.

**References**

Brunt, A.A. and Phillips, S. (1981). *Rep. Glasshouse Crops Res. Inst.* 1979, p. 151.
Phillips, S. and Brunt, A.A. (1986). *Acta Hort.* **177**: 227.

# Alstroemeria mosaic *potyvirus*

**Data collated by S. Phillips, 1986.**

*Standard acronym:* AlMV.

First reported in *Alstroemeria psittacina*; from England; by Brunt (1975).

**Natural host range and symptoms**

Symptoms persist.

*Alstroemeria caryophylla* and *Alstroemeria psittacina* — some flower break and variable leaf chlorosis.

**Transmission**

Transmitted by a vector; an insect; *Myzus persicae*; Aphididae. Transmitted in a non–persistent manner. Virus transmitted by mechanical inoculation; not transmitted by contact between plants; not transmitted by seed; not transmitted by pollen.

**Geographical distribution**

Spreads in the Eurasian region and the North American region; the Netherlands and the USA.

**Experimental host range**

Several (3–9) families susceptible. Experimentally infected plants mostly show faint chlorotic local lesions, mild systemic chlorosis.

Diagnostically susceptible

*Nicotiana clevelandii, N. megalosiphon, N. benthamiana* — faint chlorotic lesions and mild systemic veinal chlorosis.

*Chenopodium quinoa, C. murale, C. capitatum, Tetragonia expansa* — chlorotic local lesions, not systemic.

Diagnostically insusceptible

*Narcissus pseudonarcissus, Cucumis sativus, Belamcanda chinensis, Trifolium incarnatum, Solanum tuberosum.*

Maintenance and propagation hosts

*Nicotiana clevelandii, N. benthamiana, N. megalosiphon.*

Assay hosts (Local lesions or Whole plants)

*Nicotiana clevelandii* (W), *Chenopodium quinoa* (L), *C. murale* (L).

Dicotyledonous Hosts *(**Susceptible** or Insusceptible)*

*Amaranthus caudatus. Beta vulgaris. Brassica campestris* ssp. *pekinensis. Capsicum frutescens. Chenopodium amaranticolor.* ***Chenopodium capitatum. Chenopodium murale. Chenopodium quinoa.*** *Cucumis sativus. Datura stramonium. Gomphrena globosa. Hyoscyamus niger. Lycopersicon esculentum.* ***Nicandra physaloides. Nicotiana benthamiana. Nicotiana clevelandii.*** *Nicotiana debneyi. Nicotiana glutinosa.* ***Nicotiana megalosiphon*** (*Nicotiana plumbaginafolia* **S**). *Nicotiana rustica.* ***Nicotiana sylvestris.***

***Nicotiana tabacum***. ***Petunia hybrida***. *Phaseolus vulgaris. Physalis floridana. Solanum demissum. Solanum tuberosum.* ***Tetragonia expansa***. *Trifolium incarnatum. Vicia faba. Vigna unguiculata.*

Monocotyledonous Hosts *(**Susceptible** or Insusceptible)*

*Belamcanda chinensis. Lilium formosanum. Narcissus pseudonarcissus.*

Sources of host-range data

Phillips and Brunt (1986).

**Properties of particles in sap**

TIP: 65 °C. DEP: $\log_{10}$ minus 4. Leaf sap contains few virions.

**Purification**

Method: Phillips and Brunt (1986).

**Particle morphology**

Virions filamentous; not enveloped; usually flexuous; with a clear modal length; of 745–750 nm. Axial canal obscure. Basic helix obscure.

**Physical properties**

One sedimenting component in purified preparations; sedimentation coefficient 145 *S*. Density 1.296 g $cm^{-3}$ in CsCl.

**Biochemical properties**

Virions contain 5 % nucleic acid; 95 % protein; 0 % lipid.

*Genome* consists of RNA; single–stranded; unipartite. Genomic nucleic acid isolated by Brunt *et al.* (1980).

*Non–genomic nucleic acid* not found in the virions.

*Virion protein(s)* one; $M_r$ 32500; coat protein. Method of preparation: Brunt *et al.* (1980).

**Replication**

Replication does not depend on a helper virus.

**Cytopathology**

Virions found in mesophyll, epidermis and all parts of the host plant; in cytoplasm. Inclusions present in infected cells; are pinwheels and unusual in shape; conical or cylindrical.

**Taxonomy and Relationships**

*Potyvirus: Potyviridae.*

Virus(es) with serologically related virions: iris severe mosaic (SDI 2), narcissus late season yellows (SDI 2), potato Y (SDI 3–7), clover yellow vein (SDI 7), watermelon mosaic 2 (SDI 7) and pea mosaic (SDI 9) viruses. Virus(es) with serologically unrelated virions: asparagus I, bean yellow mosaic (crocus and gladiolus isolates), carnation vein mottle, datura shoestring, henbane mosaic, hippeastrum mosaic, iris mild mosaic, lettuce mosaic, narcissus degeneration, narcissus yellow stripe, nerine yellow stripe, passiflora woodiness, pepper veinal mottle, potato Y (veinal necrosis strain), tobacco etch and turnip mosaic viruses.

Best tests for diagnosis: ISEM with 'decoration'.

**References**

Brunt, A.A. (1975). *Rep. Glasshouse Crops Res. Inst. for 1974*: 117.

Brunt, A.A. Barton, R.J., Phillips, S. and Lana, A.O. (1980). *Ann. appl. Biol.* **96**: 37.

Maat, D.Z. (1977). *Rep. Res. Inst. Pl. Prot.* for 1976: p.63.

Phillips, S. and Brunt, A.A. (1986). *Acta Hort.* **177**: 227.

# Alstroemeria streak (?) potyvirus

**Data collated by A.A. Brunt, 1994; S.M. Wong, 1995.**

*Standard acronym:* AlSV.

First reported in *Alstroemeria* sp.; from New York, U.S.A.; by Wong *et al.* (1992).

**Natural host range and symptoms**

Symptoms persist. Symptoms streaking and stunting.
*Alstroemeria* — streaking, stunting and mosaic of leaves.

**Transmission**

Transmitted by a vector; an insect; *Macrosiphum euphorbiae*; Aphididae. Virus transmitted by mechanical inoculation; not transmitted by contact between plants.

**Geographical distribution**

Found, but with no evidence of spread, in Netherlands, UK and USA.

**Experimental host range**

Few (<3) families susceptible. Experimentally infected plants mostly show chlorotic streaking.

Diagnostically susceptible

*Chenopodium amaranticolor, Hyoscyamus niger, Tetragonia expansa* — local chlorotic lesions, not systemic.

*Nicotiana benthamiana, N. clevelandii, N. hesperis* 67A, *N. occidentalis* P1 — local chlorotic lesions, systemic chlorosis.

*Nicotiana occidentalis* — systemic chlorosis. *Nicotiana occidentalis* Japan — systemic chlorotic streaks parallel to midrib.

Diagnostically insusceptible

*Nicotiana glutinosa, Chenopodium quinoa.*

Maintenance and propagation hosts

*Nicotiana occidentalis* Japan.

Assay hosts (**L**ocal lesions or **W**hole plants)

*Nicotiana occidentalis* Japan (W).

Dicotyledonous Hosts *(**Susceptible** or Insusceptible)*

***Chenopodium amaranticolor***. *Chenopodium quinoa*. ***Hyoscyamus niger***. *Nicotiana glutinosa*.

**Properties of particles in sap**

Leaf sap contains few virions.

**Purification**

Method: Wong *et al.* (1992).

**Particle morphology**

Virions filamentous; not enveloped.

**Physical properties**

One sedimenting component in purified preparations. $A_{260}/A_{280}$ ratio 1.34.

**Biochemical properties**

Genome unipartite.

*Virion protein(s)* one; $M_r$ 30000; coat protein.

**Cytopathology**

Virions found in cytoplasm. Inclusions present in infected cells; are pinwheels (and laminated aggregates).

**Taxonomy and Relationships**

Possible unassigned potyvirus.

Virus(es) with serologically related virions: alstroemeria mosaic virus.

**References**

De Blank, C.M., van Zaayen, A. and Bouwen, I. (1994). *Acta Hort.* **377**: 199.
Wong, S.M., Chang, C.G., Reiser, R.A. and Horst, R.K. (1994). *Acta. Hort.* **377**: 63.
Wong, S.M., Reiser, R.A. and Horst, R.K. (1992). *Phytopathology* **82**: 722.
Van Zaayen, A., de Blank, C. and Bouwen, I. (1994). *Europ. J. Pl. Path.* **100**: 85.

# Amaranthus leaf mottle *potyvirus*

**Data collated by S. Phillips and O. Lovisolo, 1992.**

*Standard acronym:* AmLMV. *Strains:* type and *Cirsium* (Casetto *et al.*, 1986).

First reported in *Amaranthus deflexus*; from Turin, Italy; by Lovisolo and Lisa (1976).

**Natural host range and symptoms**

Symptoms leaf mottling.

*Amaranthus deflexus* — leaf mottling, blistered mosaic and growth reduction.

*Cirsium arvense* — leaf mottling.

**Transmission**

Transmitted by a vector; an insect; *Macrosiphum euphorbiae, Myzus persicae*; Aphididae. Transmitted in a non–persistent manner. Virus transmitted by mechanical inoculation.

**Geographical distribution**

Spreads in the Eurasian region; Italy, Morocco, and Spain.

**Experimental host range**

Several (3–9) families susceptible. Experimentally infected plants mostly show local chlorotic lesions; systemic chlorosis.

Diagnostically susceptible

*Amaranthus caudatus, A. hybridus, A. tricolor, Gomphrena globosa, Nicotiana clevelandii* — systemic leaf chlorosis.

*Chenopodium quinoa, C. amaranticolor* — chlorotic local lesions, systemic mottling. *Phaseolus vulgaris* cv. Saxa — chlorotic local lesions, not systemic.

Diagnostically insusceptible

*Cucurbita pepo, Cucumis sativus, Nicotiana glutinosa, Pisum sativum, Vigna unguiculata.*

Maintenance and propagation hosts

*Chenopodium quinoa, Nicotiana clevelandii.*

Assay hosts (**L**ocal lesions or **W**hole plants)

*Chenopodium quinoa* (L), *C. amaranticolor* (L).

Dicotyledonous Hosts *(**Susceptible** or Insusceptible)*

***Amaranthus caudatus***. ***Amaranthus retroflexus***. ***Atriplex hortensis***. *Beta vulgaris*. *Capsicum annuum*. *Celosia cristata*. ***Chenopodium album***. ***Chenopodium amaranticolor***. ***Chenopodium murale***. ***Chenopodium quinoa***. *Cucumis sativus*. *Cucurbita pepo*. *Datura stramonium*. ***Gomphrena globosa***. ***Lupinus albus***. *Lycopersicon esculentum*. ***Nicotiana clevelandii***. ***Nicotiana debneyi***. *Nicotiana glutinosa*. ***Nicotiana rustica***. ***Nicotiana tabacum***. ***Ocimum basilicum***. ***Petunia hybrida***. ***Phaseolus vulgaris***. *Pisum sativum*. ***Spinacia oleracea***. ***Tetragonia expansa***. *Trifolium pratense*. ***Vicia faba***. *Vigna unguiculata*.

Sources of host-range data

Lovisolo and Lisa (1976 and 1979); Casetto *et al.* (1986).

**Purification**

Method: Lovisolo and Lisa (1979).

**Particle morphology**

Virions filamentous; not enveloped; usually flexuous; with a clear modal length; of 780 nm; 12 nm wide. Axial canal obscure. Basic helix obscure.

**Physical properties**

One sedimenting component in purified preparations. $A_{260}/A_{280}$ ratio 1.16.

**Replication**

Replication does not depend on a helper virus.

**Cytopathology**

Virions found in leaves.

**Taxonomy and Relationships**

*Potyvirus: Potyviridae.*

Virus(es) with serologically related virions: plum pox and bean yellow mosaic viruses, also Lisianthus potyvirus (Lisa *et al.*, 1990), peanut mottle (Rajeshwari *et al.*, 1983) and zucchini yellow mosaic (Lisa and Lecoq, 1984) viruses distantly. Virus(es) with serologically unrelated virions: clover yellow vein (English isolate and pea necrosis strain), lettuce mosaic, watermelon mosaic 2, wisteria vein mosaic, celery mosaic, henbane mosaic, potato Y, turnip mosaic and papaya ringspot viruses.

The *Cirsium* strain infects *Cirsium arvense, Lactuca sativa, Physalis floridana* and *Trigonella foenum-graecum.* Best tests for diagnosis: ISEM and decoration.

**References**

Casetta, A., D'Agostino, G. and Conti, M. (1986). *Inform. Fitopatol.* **36**:43

Lisa, V., Dellavalle, G., Vaira, A.M., Milne, R.G., Masenga, V., Boccardo, G. and D'Aquilio, M. (1990).

Lisa, V. and Lecoq, H. (1984). *CMI/AAB Descr. Pl. Viruses* No. **282**, 4pp.

Lovisolo, O., Caciagli, P. and Conti, M. (1980). *Proc. 5th Congr. Un. Phytopath. Medit.* Patras, Greece, pp.22–24.

Lovisolo, O. and Lisa, V. (1976). *Agric. Scient.* **39**: 553.

Lovisolo, O. and Lisa, V. (1979). *Phytopath. Medit.* **18**: 89.

Rajeshwari, R., Iizuka, N., Nolt, B.L. and Reddy, D.V.R. (1983). *Plant Path.* **32**: 197.

# Amaranthus mosaic (?) potyvirus

**Data collated by V.R. Mali, 1987.**

*Synonyms:* pigweed mosaic virus (Singh *et al.*, 1972). *Strains:* Coimbatore strain (Govindaswamy *et al.*, 1967).

First reported in *Amaranthus blitum* and *A. viridis*; from Delhi, India; by Phatak (1965).

## Natural host range and symptoms

Symptoms persist.

*Amaranthus blitum, A. viridis* — systemic blotchy mosaic and mottle, sometimes with severe vein yellowing.

*A. caudatus, A. gangeticus* — systemic mosaic, leaves puckered.

## Transmission

Virus transmitted by mechanical inoculation; transmitted by grafting.

## Ecology and control

Studies reported by Govindaswamy *et al.* (1967); Mali (1985); Phatak (1965); Singh *et al.* (1972).

## Geographical distribution

Spreads in the Central Asian region; India (Coimbatore (Tamil Nadu) and Delhi (Andhra Pradesh)).

## Experimental host range

Few (<3) families susceptible. Experimentally infected plants mostly show mosaics and mottling.

Diagnostically susceptible

*Amaranthus cruentus, A. dilatum, A. gangeticus, A. leucarpus, A. mangoostanus* — systemic mosaic and mottling.

Diagnostically insusceptible

*Chenopodium album, C. amaranticolor, Capsicum annuum, Nicotiana glutinosa, N. tabacum, Nicandra physalodes, Solanum nigrum.*

Maintenance and propagation hosts

*Amaranthus viridis, A. caudatus* and *A. gangeticus.*

Assay hosts (**L**ocal lesions or **W**hole plants)

*Amaranthus caudatus* (W).

Dicotyledonous Hosts *(**Susceptible** or Insusceptible)*

***Amaranthus caudatus***. *Beta vulgaris. Capsicum annuum. Chenopodium album. Chenopodium amaranticolor. Chenopodium murale. Cucumis melo. Cucurbita moschata. Cyamopsis tetragonoloba. Datura stramonium.* ***Gomphrena globosa***. *Helianthus annuus. Nicandra physaloides. Nicotiana glutinosa. Nicotiana tabacum.* ***Petunia hybrida***. *Solanum melongena. Solanum nigrum.*

Sources of host-range data

Govindaswamy *et al.* (1967). Phatak (1965). Singh *et al.* (1972).

## Properties of particles in sap

TIP: 55–60 °C. LIV: 1 days. DEP: $\log_{10}$ minus 1–2. Leaf sap contains few virions.

**Particle morphology**

Virions filamentous; not enveloped; usually flexuous; with a clear modal length; of 700–728 nm; 14 nm wide.

**Taxonomy and Relationships**

Unassigned potyvirus.

Phatak (1965) reported a mosaic disease of *Amaranthus blitum* and *A. viridis* from Delhi, India in 1965 and called it amaranthus mosaic virus. This has been regarded as the type culture of AMV. Subsequently a variant has been reported by Govindaswamy *et al.* (1967) from Coimbatore, Tamil Nadu, India in naturally infected *A. caudatus* and *A. gangeticus*. The Coimbatore isolate differs from the type culture (strain) in infecting *Gomphrena globosa* and *Petunia hybrida*. Singh *et al.* (1972) reported its virion morphology and it has since been considered a member of the potyvirus group (Mali, 1985), but its possible relationship with amaranthus leaf mottle potyvirus (Hollings and Brunt, 1981; Lovisolo and Lisa, 1976) needs to be tested.

**References**

Govindasamy, C.V., Mariappan, V., Kandaswamy, T.K. and Padmanabham, C. (1967). *Madras Agric. J.* **54**: 35.

Hollings, M. and Brunt, A.A. (1981). *CMI/AAB Descr. Pl. Viruses* No. 245, 7 pp.

Lovisolo, O. and Lisa, V. (1976). *4th Congr. Medit. Phytopath. Union, Poljopr Znanst Smotra* **39**: 553.

Mali, V.R. (1985). *Int. J. Trop. Pl. Dis.* **3**: 93.

Phatak, H.C. (1965). *Curr. Sci.* **34**: 645.

Singh, S., Varma, A., Chenulu, V.V. and Verma, V.S. (1972). *Phytopath. Z.* **75**: 82.

# Amaryllis (?) *alphacryptovirus*

**Data collated by A.A. Brunt, 1994.**

First reported in *Amaryllis* sp.; from India; by Raizada *et al.* (1983).

**Natural host range and symptoms**

Symptoms *Amaryllis* sp. — symptomless.

**Geographical distribution**

Found, but with no evidence of spread, in India.

**Taxonomy and Relationships**

Possible *Alphacryptovirus: Partitiviridae*.

Best tests for diagnosis: ELISA.

**References**

Raizada, R.K., Srivasta, K.M. and Singh, B.P. (1983). *Ind. J. Exp. Biol.* **21**: 299.

# Amazon lily mosaic (?) potyvirus

**Data collated by A.A. Brunt, 1994.**

*Standard acronym:* ALiMV.

First reported in *Eucharis grandiflora*; from Japan; by Terami *et al.* (1993).

**Natural host range and symptoms**

*Eucharis grandiflora* — leaf mosaic.

# Amazon lily mosaic (?) potyvirus

**Data collated by A.A. Brunt, 1994.**

*Standard acronym:* ALiMV.

First reported in *Eucharis grandiflora*; from Japan; by Terami *et al.* (1993).

**Natural host range and symptoms**

*Eucharis grandiflora* — leaf mosaic.

**Transmission**

Transmitted by a vector; an insect; *Myzus persicae*; Aphididae. Transmitted in a non–persistent manner. Virus lost by the vector when it moults; transmitted by mechanical inoculation.

**Geographical distribution**

Geographical distribution probably restricted. Spreads in the Eastern Asian region; Japan.

**Experimental host range**

Few (<3) families susceptible. Experimentally infected plants mostly show mosaic.

Diagnostically susceptible

*Chenopodium amaranticolor, C. quinoa* — necrotic local lesions, not systemic.
*Nicotiana benthamiana* — systemic leaf mosaic.

Maintenance and propagation hosts

*Nicotiana benthamiana.*

Assay hosts (**L**ocal lesions or **W**hole plants)

*Chenopodium amaranticolor* (L), *C. quinoa* (L).

Dicotyledonous Hosts *(**Susceptible** or Insusceptible)*

***Chenopodium amaranticolor***. ***Chenopodium quinoa***. ***Nicotiana benthamiana***.

**Purification**

Method: Terami *et al.* (1993).

**Particle morphology**

Virions filamentous; not enveloped; usually flexuous; with a clear modal length; of 760 nm.

**Physical properties**

$A_{260}/A_{280}$ ratio 1.32.

**Biochemical properties**

*Genome* consists of RNA; single–stranded; linear; unipartite.
*Virion protein(s)* one; $M_r$ 35000; coat.

**Cytopathology**

Virions found in cytoplasm. Inclusions present in infected cells; are pinwheels.

**Taxonomy and Relationships**

Possible unassigned potyvirus.

Virus(es) with serologically related virions: alstroemeria mosaic, hippeastrum mosaic, hyacinth mosaic, onion yellow dwarf, ornithogalum mosaic, potato Y, watermelon mosaic-2 and zucchini yellow mosaic potyviruses.

**References**

Terami, F., Honda, Y. and Fukumoto, F. (1993). *Ann. Phytopath. Soc. Japan* **59**: 334.

## Anthoxanthum latent blanching *hordeivirus*

**Data collated by A.A. Brunt, 1991.**

*Standard acronym:* ALBV.

First reported in *Anthoxanthum odoratum*; from Aberystwyth, Wales, U.K; by Catherall and Chamberlain (1981*a* and *b*).

**Natural host range and symptoms**

Symptoms persist.

*Anthoxanthum odoratum* — inconspicuous chlorotic stripes in leaves, or symptomless infection.

**Transmission**

Virus transmitted by mechanical inoculation.

**Geographical distribution**

Spreads in the Eurasian region; the UK (Wales and England).

**Experimental host range**

Few (<3) families susceptible.

Diagnostically susceptible

*Anthoxanthum odoratum* — inconspicuous chlorotic leaf stripes in some plants.
*Avena sativa, Hordeum vulgare* — symptomless systemic infection.
*Chenopodium amaranticolor, C. quinoa* — local lesions; no systemic infection.
*Lagurus ovatus* — inconspicuous leaf chlorosis, soon followed by wilting and death of leaves.
*Zea mays* — brown streaks in systemically infected leaves.

Diagnostically insusceptible

*Beta vulgaris, Dactylis glomerata, Festuca elatior, Lolium perenne, Triticum aestivum.*

Maintenance and propagation hosts

*Anthoxanthum odoratum.*

Assay hosts (**L**ocal lesions or **W**hole plants)

*Anthoxanthum odoratum* (W).

Dicotyledonous Hosts *(**Susceptible** or Insusceptible)*

***Chenopodium amaranticolor*. *Chenopodium quinoa*.**

Monocotyledonous Hosts *(**Susceptible** or Insusceptible)*

***Anthoxanthum odoratum*. *Avena sativa*. *Hordeum vulgare*. *Lagurus ovatus*.** *Lolium perenne*. *Triticum aestivum*. ***Zea mays*.**

**Particle morphology**

Virions rod–shaped; not enveloped; usually straight; with clear modal lengths; of 124, 142, and 167 nm; 22 nm wide.

**Cytopathology**

Virions found in leaves; in cytoplasm (in aggregates).

**Taxonomy and Relationships**

*Hordeivirus*.

Virus(es) with serologically related virions: barley stripe mosaic virus.

**References**

Catherall, P.L. and Chamberlain, J.A. (1981*a*). *Rep. Welsh Pl. Breeding Stn.* 1980, p.153.

Catherall, P.L. and Chamberlain, J.A. (1981*b*). *Proc. 3rd Conf. Virus Dis. Gramineae in Europe*, p. 83.

Edwards, M.L., Kelley, S.E., Arnold, M.K. and Cooper, J.I. (1989). *Pl. Path.* **38**: 209.

# Anthoxanthum mosaic (?) potyvirus

**Data collated by A.A. Brunt, 1991.**

*Standard acronym:* AntMV.

First reported in *Anthoxanthum odoratum*; from Wales, U.K.; by Catherall (1967).

**Natural host range and symptoms**

Symptoms persist.

*Anthoxanthum odoratum* — mild mosaic.

**Transmission**

Not transmitted by *Aulacorthum circumflexum, Macrosiphum (Sitobion) avenae, M. (S.) fragariae, Acyrthosiphon (Metopolophium) dirhodum, Myzus persicae, Rhopalosiphum padi*. Virus transmitted by mechanical inoculation; not transmitted by seed.

**Geographical distribution**

Spreads in the Eurasian region; the UK (Wales).

**Experimental host range**

Few (<3) families susceptible.

Diagnostically susceptible

*Anthoxanthum odoratum, Avena sativa, A. fatua, A. strigosa, Hordeum vulgare, Lagurus ovatus, Triticum aestivum* — mild mosaic.

Diagnostically insusceptible

*Dactylis glomerata, Festuca pratensis, Lolium perenne, Phleum pratense, Zea mays*.

Maintenance and propagation hosts

*Anthoxanthum odoratum, Avena sativa, A. fatua, Hordeum vulgare, Triticum aestivum*.

Assay hosts (**L**ocal lesions or **W**hole plants)

*Anthoxanthum odoratum* (W).

Monocotyledonous Hosts *(**Susceptible** or Insusceptible)*

***Anthoxanthum odoratum*. *Avena sativa*. *Avena strigosa*. *Hordeum vulgare*. *Lagurus ovatus*. *Triticum aestivum*.**

**Particle morphology**

Virions filamentous; with a clear modal length; of 750 nm; 12 nm wide.

**Taxonomy and Relationships**

Unassigned potyvirus.

**References**

Catherall, P.L. (1967). *Rep. Welsh Pl. Breed. Stn* 1966, p. 116.
Catherall, P.L. (1970). *Pl. Path.* **19**: 125.

# Anthriscus (?) *carlavirus*

**Data collated by P. van Dijk and L. Bos, 1991.**

*Standard acronym:* AntV.

First reported in *Anthriscus sylvestris*; from the Netherlands; by Van Dijk and Bos (1989).

**Natural host range and symptoms**

Symptoms none.
*Anthriscus sylvestris, Angelica sylvestris, Carum carvi* — symptomless.

**Transmission**

Transmitted by a vector; an insect; *Cavariella aegopodii*; Aphididae. Transmitted in a non–persistent manner. Virus transmitted by mechanical inoculation.

**Geographical distribution**

Found, but with no evidence of spread, in the Netherlands.

**Experimental host range**

Few (<3) families susceptible. Experimentally infected plants mostly show mottles, mosaics or symptomless.

Diagnostically susceptible

*Ammi majus* — interveinal chlorosis and systemic yellow mosaic.
*Anthriscus cerefolium* — interveinal yellowing or mottling, necrotic striping and leaf rolling.
*Apium graveolens* — chlorotic local lesions by some isolates.
*Chenopodium amaranticolor, C. quinoa* — pin-point local lesions.

Diagnostically insusceptible

*Nicotiana benthamiana, Phaseolus vulgaris.*

Maintenance and propagation hosts

*Ammi majus, Nicotiana pavicola.*

Assay hosts (**L**ocal lesions or **W**hole plants)

*Ammi majus* (L), *Chenopodium amaranticolor* (L), *C. quinoa* (L).

Dicotyledonous Hosts *(**Susceptible** or Insusceptible)*

***Anthriscus cerefolium*. *Apium graveolens*.** *Beta vulgaris. Celosia argentea.* ***Chenopodium amaranticolor*. *Chenopodium quinoa*.** *Coriandrum sativum. Cucumis sativus. Gomphrena globosa. Lycopersicon esculentum. Nicotiana benthamiana. Nicotiana clevelandii. Nicotiana glutinosa. Nicotiana megalosiphon. Nicotiana occidentalis. Nicotiana tabacum. Phaseolus vulgaris.*

Additional hosts

*Pastinaca sativa.*

Sources of host-range data

Van Dijk and Bos (1989).

**Properties of particles in sap**

Leaf sap contains many virions.

**Particle morphology**

Virions filamentous; not enveloped; usually flexuous; with a clear modal length; of 690 nm.

**Replication**

Replication does not depend on a helper virus.

**Cytopathology**

Virions found in leaves.

**Taxonomy and Relationships**

Possible *Carlavirus*.

Best tests for diagnosis: electron microscopy.

**References**

Van Dijk, P. and Bos, L. (1989). *Neth. J. Pl. Path.* **95** Suppl. 2: 34pp.

## Anthriscus yellows *waikavirus*

**Data collated by A.A. Brunt, 1991.**

*Standard acronym:* AYV.

First reported in *Anthriscus sylvestris*; from the U.K.; by Murant and Goold (1968).

**Natural host range and symptoms**

*Anthriscus sylvestris* — no conspicuous symptoms.

**Transmission**

Transmitted by a vector; an insect; *Cavariella aegopodii*; Aphididae. Transmitted in a semi–persistent manner. Virus can help the vector transmission of another virus (parsnip yellow fleck sequivirus); not transmitted by mechanical inoculation.

**Geographical distribution**

Spreads in the Eurasian region; the UK.

**Experimental host range**

Few (<3) families susceptible.

Diagnostically susceptible

*Anthriscus cerefolium* — systemically infected leaves faintly chlorotic, then reddish purple and slightly distorted; plants stunted.

*Coriandrum sativum* — slight yellowing of systemically infected leaves; plants stunted.

Diagnostically insusceptible

*Bupleurum rotundifolium, Daucus carota, Eryngium planum, Pastinaca sativa, Pimpinella anusum, Trinia glauca.*

Maintenance and propagation hosts

*Anthriscus cerefolium.*

Assay hosts (**L**ocal lesions or **W**hole plants)
*Anthriscus cerefolium* (W).

Dicotyledonous Hosts *(**Susceptible** or Insusceptible)*
***Anthriscus cerefolium*. *Coriandrum sativum*.** *Daucus carota. Pastinaca sativa.*

Sources of host-range data
Murant and Goold (1968).

**Properties of particles in sap**

LIV: purified virions in 1 mM $CaCl_2$ or less remain intact for 45 days (at 4°C). Leaf sap contains few virions.

**Purification**

Method: Hemida *et al.* (1989).

**Particle morphology**

Virions isometric; not enveloped; 29 nm in diameter; rounded in profile; without a conspicuous capsomere arrangement.

**Physical properties**

Two sedimenting components in purified preparations. Density 1.52 g $cm^{-3}$ in CsCl (B), or 1.27 g $cm^{-3}$ in CsCl (T). $A_{260}/A_{280}$ ratio 1.83 (B), or 0.73 (T).

**Biochemical properties**

Virions contain 42 % nucleic acid (B), or 0 % nucleic acid (T); 58 % protein (B), or 100 % protein (T).

*Genome* consists of RNA; single–stranded; unipartite; largest (or only) genome part 10.6 kb.

*Virion protein(s)* four; $M_r$ of the largest 35000; coat protein. $M_r$ of 2nd largest 28400; coat. $M_r$ of 3rd largest 24300; coat. $M_r$ of 4th largest 22300; coat.

**Replication**

Acts as helper for another virus; a helper for parsnip yellow fleck sequivirus.

**Cytopathology**

Virions found in phloem; in cytoplasm. Inclusions absent from infected cells.

**Taxonomy and Relationships**

*Waikavirus: Sequiviridae.*

Virus(es) with serologically unrelated virions: maize chlorotic dwarf and rice tungro spherical viruses.

**References**


Elnager, S. and Murant, A.F. (1976*a*). *Ann. appl. Biol.* **84**: 153.
Elnager, S. and Murant, A.F. (1976*b*). *Ann. appl. Biol.* **84**: 169.
Hemida, S.K., Murant, A.F. and Duncan, G.H. (1989). *Ann. appl. Biol.* **114**: 71.
Murant, A.F. and Goold, R.A. (1968). *Ann. appl. Biol.* **62**: 123.
Murant, A.F. and Roberts, I.M. (1977). *Ann. appl. Biol.* **85**: 403.
Murant, A.F., Roberts, I.M. and Elnager, S. (1976). *J. gen. Virol.* **53**: 321.

# Apple chlorotic leaf spot *trichovirus*

**Data collated by R.M. Lister, 1987.**

*Synonyms:* pear ring pattern mosaic virus (Cropley, 1969), apple latent virus type 1, plum pseudopox virus, quince stunt virus. *Standard acronym:* ACSLV. *Strains:* peach strain, apple strain, (Chairez and Lister; 1973).

First reported in *Malus* spp.; from the U.S.A.; by Mink and Shay (1959; 1962); Lister *et al.* (1965).

**Natural host range and symptoms**

Symptoms vary seasonally.
*Malus platycarpa* — line patterns and chlorotic rings.
*Malus sylvestris* cv. R12740–7A — chlorotic leaf spots and stem pitting.
*Malus sylvestris* cv. Spy — chlorotic spots and stem pitting, chlorosis and stunting.
*Malus sylvestris* cv. Virginia Crab — stem pitting.
*Cydonia oblonga* (Quince) C7/1 — chlorotic rings and spots.
*Pyronia veitchii* — chlorosis and stunting.
*Pyrus communis* — ring patterns, mosaic.
*Crataegus* spp. — chlorotic ring spots.
*Prunus armeniaca* — ring patterns, mosaic.
*Prunus spinosa* — chlorotic ringspot patterns.
*Prunus persica* — dark green mottle.
*Prunus domestica* — bark split and pseudopox.
*Prunus armeniaca* — stock scion incompatibility and pseudopox (Bar-Joseph *et al.*, 1979; Cropley, 1968; Cropley, 1969; Dunez *et al.*, 1972; Lister and Bar-Joseph, 1981; Luckwill and Campbell, 1963; Marenaud, 1968; Pena-Iglesias and Gonsalves, 1975; Posnette and Ellenberger, 1957; Yanase, 1974; Yanase *et al.*, 1975).

**Transmission**

Transmitted by eudorylamoid nematodes (Fritzsche and Kegler, 1968); an unconfirmed report. Virus transmitted by mechanical inoculation; transmitted by grafting (possibly by root-grafts; Gilmer *et al.*, 1971); not transmitted by seed.

**Ecology and control**

Studies reported by Bar-Joseph *et al.* (1979); Lister and Bar-Joseph (1981).

**Geographical distribution**

Probably distributed worldwide (and probably wherever apples are cultivated). Spreads in the Eastern Asian region, the Eurasian region, the North American region, and the Pacific region; Australia, China, and New Zealand.

**Experimental host range**

Several (3–9) families susceptible.

Diagnostically susceptible

*Malus sylvestris* cv. R12740–7A — asymmetric blotches and malformation of young leaves.

*M. platycarpa* (Long Ashton clone) — irregular chlorotic rings and line patterns, small leaves shed prematurely, malformation.

*Chenopodium quinoa* — large etched local lesions, becoming necrotic, systemic chlorotic spotting, mottling, ring and line patterns, chlorosis and necrosis.

*C. amaranticolor* — chlorotic local lesions, surrounded by necrotic dots.

*Phaseolus vulgaris* cvs Pinto, Bountiful, Kinghorn — purple-brown necrotic spots or rings (1–3 mm), not systemic.

Diagnostically insusceptible

*Nicotiana glutinosa*. Apple stem grooving capillovirus, but not apple chlorotic leafspot trichovirus causes Virginia Crab stem grooving symptoms in *Malus*.

Maintenance and propagation hosts

*Chenopodium quinoa*.

Assay hosts (**L**ocal lesions or **W**hole plants)

*Chenopodium quinoa* (L), *Phaseolus vulgaris* (L).

Dicotyledonous Hosts ***(Susceptible** or Insusceptible)*

***Chenopodium amaranticolor*. *Chenopodium quinoa*. *Malus platycarpa*. *Malus sylvestris*. *Phaseolus vulgaris*. *Prunus armeniaca*. *Prunus persica*.**

Comments on host-range

For symptoms in other species and horticultural varieties of *Malus* and woody Rosaceae, see Luckwill and Campbell (1963); Cropley (1969); Sweet (1980).

Sources of host-range data

Barnett and Murant (1971); Cropley (1968); Lister *et al.* (1965); Nemec (1967); Saksena and Mink (1969); Sweet (1980).

**Properties of particles in sap**

TIP: 52–55 °C (Lister *et al.*, 1965). LIV: 1 days (at 20°C), or 10 days (at 4°C). DEP: $\log_{10}$ minus 4. Leaf sap contains few virions. Electron microscopy: best stained with UA or UF; breakage occurs in PTA (Lister *et al.*, 1965).

**Purification**

Method: de Sequeira and Lister (1969); Lister and Hadidi (1971).

**Particle morphology**

Virions filamentous; not enveloped; usually flexuous; with a clear modal length; of 720–740 nm; 12 nm wide. Axial canal obscure (laterally). Basic helix obvious; pitch of basic helix 3.8–3.9 nm.

**Physical properties**

One sedimenting component in purified preparations; sedimentation coefficient 96 *S*. Density 1.27 g $cm^{-3}$ in $Cs_2SO_4$ (Bar-Joseph *et al.*, 1974).

**Biochemical properties**

*Genome* consists of RNA; single–stranded; unipartite; largest (or only) genome part 7.555 kb (Lister and Bar-Joseph, 1981; Bar-Joseph *et al.*, 1979). Genomic nucleic acid isolated by Lister and Bar-Joseph, (1981); Bar-Joseph *et al.* (1974; 1979). 5' terminus of RNA has a methylated nucleotide cap (probably). Poly A region present; 3' terminus.

*Sequence database accession code(s):*

D14996 Em(40)_vi:APC2165K Gb(84)_vi:APCCOMS Apple chlorotic leaf spot virus, complete sequence. 2/94 7,552bp.

M58152 Em(40)_vi:APCCG Gb(84)_vi:APCCG Apple chlorotic leaf spot virus, complete genome. 8/91 7,555bp. 2 sequences.

5 virus specified dsRNA species found in infected cells. Size of largest virus specified dsRNA 7.56 kbp (pairs); 2nd largest 6.5 kbp (pairs); 3rd largest 5.5 kbp (pairs).

*Virion protein(s)* one (Lister and Bar-Joseph, 1981; Bar-Joseph *et al.*, 1979); $M_r$ 23500. Method of preparation: Chairez and Lister (1973); Bar-Joseph *et al.* (1979).

*Virus–coded non–virion proteins* isolated (Chairez and Lister (1973); Lister and Bar-Joseph (1981)); two proteins found. $M_r$ *c.* 200000; polymerase module. $M_r$ of 2nd largest 45000; movement protein, possibly.

**Replication**

Replication does not depend on a helper virus.

**Cytopathology**

Virions found in all parts of the host plant probably.

**Taxonomy and Relationships**

*Trichovirus* (the type species).

Best tests for diagnosis: Virus identified by serological tests, electron microscopy, symptoms in *Chenopodium quinoa, C. amaranticolor* and *Phaseolus vulgaris* cvs Kinghorn, Pinto and Bountiful, and symptoms in graft infected Russian apple R-127407A.

**Comments**

The virus was originally considered a definitive closterovirus, then became a tentative closterovirus, but is now the type species of the *Trichovirus* genus (Martelli *et al.*, 1994).

**References**


Bar-Joseph, M., Garnsey, S.M. and Gonsalves, D. (1979). *Adv. Virus Res.* **25**: 93.

Cadman, C.H. (1963). *Pl. Dis. Reptr* **47**: 459.

Cropley, R. (1968). *Ann. appl. Biol.* **61**: 361.

Cropley, R. (1969). *Comm. Bur. Hort. Pl. Crops. Tech. Commun.* No. **30**, *Suppl.* 2/3/4: 10.

de Sequeira, O.A. and Lister, R.M. (1969). *Phytopathology* **59**: 1740.

Fritzsche, R. and Kegler, H. (1968). *TagBer. dt. Akad. Landw. Berlin* **97**: 289.

German, S., Candresse, T. Lanneau, M., Pernollet, J.C. and Dunez, J. (1990). *Virology* **179**: 104.

German, S., Candresse, T., Le Gall, O., Lanneau, M. and Dunez, J. (1992). *J. gen. Virol.* **73**: 767.

Lister, R.M. (1970). *CMI/AAB Descr. Pl. Viruses* No. 30, 4 pp.

Lister, R.M., Bancroft, J.B. and Nadakavukaren, M.J. (1965). *Phytopathology* **55**: 859.

Lister, R.M. and Bar-Joseph, M. (1981). Closteroviruses. In: *Handbook of Plant Virus Infections and Comparative Diagnosis*, p. 809; ed. E. Kurstak. Elsevier/North-Holland Biomedical Press.

Luckwill, L.C. and Campbell, A.I. (1963). *Commonw. Bur. Hort. Pl. Crops. Tech. Commun.* No. **30**: 789.

Martelli, G.P., Candresse, T. and Namba, S. (1994). *Arch. Virol.* **134**: 451.

Nemec, S. (1967). *Pl. Dis. Reptr* **51**: 283.

Okhi, S.T., Yoshikawa, N., Inouye, N., and Inouye, T. (1989). *Ann. Phytopath. Soc. Japan* **55**: 245.

Saksena, K.N. and Mink, G.I. (1969). *Phytopathology* **59**: 84.

Sato, K., Yoshikawa, N. and Takahashi, T. (1993). *J. gen. Virol.* **74**: 1927.

Sweet, J.B. (1980). *Ann. appl. Biol.* **94**: 83.

# Apple mosaic *ilarvirus*

**Data collated by R.W. Fulton, 1985.**

*Synonyms:* birch line pattern virus, birch ringspot virus, Dutch plum line pattern virus, European plum line pattern virus, hop A virus, horse chestnut yellow mosaic virus, rose mosaic virus (Thomas, 1980; 1981; 1982; 1984a and b). *Standard acronym:* ApMV. *Strains:* mild apple mosaic virus, severe apple mosaic virus.

First reported in Rose and *Malus domestica*; from the U.S.A.; by White (1928) and Bradford and Joley (1933).

**Natural host range and symptoms**

Symptoms persist.
*Malus domestica* and *Malus* spp. — mosaic.
*Rosaceae* (many species) — necrotic ringspots.
*Humulus* spp. — mottling.

**Transmission**

Virus transmitted by mechanical inoculation; transmitted by grafting (of roots); possibly not transmitted by seed; but probably transmitted by pollen to the pollinated plant.

**Geographical distribution**

Probably distributed worldwide.

**Experimental host range**

Many (>9) families susceptible. Experimentally infected plants mostly show necrotic and chlorotic local lesions, systemic mottles, mosaics and ringspots.

Diagnostically susceptible

*Amaranthus gangeticus, Teramnus uncinatus* — small brown local lesions.
*Catharanthus roseus, Vigna unguiculata* — systemic chlorotic lines and rings.
*Chenopodium quinoa* — mottling.
*Crotalaria juncea* — veinal necrosis.
*Cucumis sativus* — chlorotic local lesions, systemic chlorosis, stunting.
*Cyamopsis tetragonoloba* — necrotic local lesions.
*Torenia fournieri* — bright yellow mosaic.

Diagnostically insusceptible

*Beta vulgaris, Datura stramonium, Helianthus annuus, Lactuca sativa, Nicotiana tabacum.*

Maintenance and propagation hosts

*Catharanthus roseus, Cucumis sativus, Phaseolus vulgaris* (only certain isolates), *Vigna unguiculata* (only certain isolates).

Assay hosts (**L**ocal lesions or **W**hole plants)

*Cyamopsis tetragonoloba* (L), *Cucumis sativus*, (W).

Dicotyledonous Hosts ***(Susceptible** or Insusceptible)*

***Amaranthus retroflexus***. *Beta vulgaris*. ***Cassia occidentalis***. ***Catharanthus roseus***. ***Chenopodium murale***. ***Chenopodium quinoa***. ***Crotalaria spectabilis***. ***Cucumis melo***. ***Cucumis sativus***. ***Cucurbita maxima***. *Cucurbita pepo*. *Cyamopsis tetragonoloba*. *Datura stramonium*. ***Fragaria vesca***. ***Gomphrena globosa***. *Helianthus annuus*. ***Hyoscyamus niger***. ***Ipomoea nil***. *Lactuca sativa*. ***Melilotus alba***. ***Momordica balsamina***. ***Nicandra***

physaloides. ***Nicotiana benthamiana***. *Nicotiana clevelandii*. ***Nicotiana debneyi***. ***Nicotiana glutinosa***. ***Nicotiana megalosiphon***. ***Nicotiana rustica***. *Nicotiana tabacum*. ***Phaseolus vulgaris*** (some isolates). ***Phlox drummondii***. ***Ricinus communis***. *Sesbania exaltata*. ***Torenia fournieri***. ***Tropaeolum majus***. ***Vigna unguiculata***.

Sources of host-range data

Fulton (1952); Kristensen and Thomson (1963); Posnette and Ellenberger (1963).

**Properties of particles in sap**

TIP: 54 °C (in extracts stabilized with mercaptoethanol). LIV: 0.1–0.2 days (2–4 hours in buffered extracts, 2–5 minutes in crude sap). DEP: $\log_{10}$ minus 3 (in stabilized 2-mercapto-ethanol sap diluted in buffer). Leaf sap contains few virions. Electron microscopy: treat with 1% glutaraldehyde before negative staining.

**Particle morphology**

Virions isometric; not enveloped; 25 and 29 nm in diameter (and probably a third size); rounded in profile (and pleomorphic); without a conspicuous capsomere arrangement.

**Physical properties**

Three sedimenting components in purified preparations; sedimentation coefficient of the fastest 117 *S*; of the other(s) 88 and 95 *S*.

**Biochemical properties**

Virions contain 16 % nucleic acid (all components); 84 % protein; 0 % lipid.

*Genome* consists of RNA; single–stranded; of three parts. Genomic nucleic acid isolated by Gonsalves and Fulton (1977). Additional factor required for infectivity (coat protein).

*Sequence database accession code(s):* apple mosaic ilarvirus

L03726 Em(40)_vi:AMQCOATPA Gb(84)_vi:AMQCOATPA Apple mosaic virus coat protein mRNA, complete cds. 11/92 944bp.

U03857 Em(40)_vi:AM03857 Gb(84)_vi:AMU03857 Apple mosaic virus PV 32 putative coat protein mRNA, complete cds. 6/94 891bp.

U15608 Em(43)_vi:Am15608 Gb(89)_vi:Amu15608 Apple mosaic virus RNA-3 putative cell-to-cell movement protein gene and coat protein gene.

*Non–genomic nucleic acid* found in the virions; is subgenomic *m*RNA; for the coat protein. Sub–genomic *m*RNA found in infected cells.

*Virion protein(s)* one; $M_r$ 25000. Method of preparation: Gonsalves and Fulton (1977).

**Replication**

Replication does not depend on a helper virus.

**Taxonomy and Relationships**

*Ilarvirus: Bromoviridae* (subgroup 3).

Virus(es) with serologically related virions: prunus necrotic ringspot virus, but distantly. Virus(es) with serologically unrelated virions: tobacco streak, Tulare apple mosaic, prune dwarf, citrus variegation, citrus leaf rugose, asparagus 2 and elm mottle viruses.

**References**

Baur, F. (1907). *Ber. dt. Bot. Ges.* **25**: 410.

Bradford, F.C. and Joley, L. (1933). *J agric. Res.* **46**: 901.

Callaghan, K.L. (1962). *Rep. Forest Ent. Path. Branch, Canada Dep. Forestry* 1962, p. 42.
de Sequeira, O.A. (1967). *Virology* **31**: 314.
Fulton, R.W. (1952). *Phytopathology* **42**: 413.
Fulton, R.W. (1967). *Phytopathology* **57**: 1197.
Fulton, R.W. (1968). *Phytopathology* **58**: 635.
Fulton, R.W. (1972). *CMI/AAB Descr. Pl. Viruses* No. 83, 4 pp.
Gilmer, R.M. (1956). *Phytopathology* **46**: 127.
Gonsalves, D. and Fulton, R.W. (1977). *Virology* **81**: 398.
Gotlieb, A.R. and Berbee, J.G. (1973). *Phytopathology* **63**: 1470.
Kristensen, H.R. and Thomsen, A. (1963). *Phytopathol. Medit.* **2**: 97.
Posnette, A.F. and Ellenberger, C.E. (1963). *Ann. appl. Biol.* **45**: 74.
Schmelzer, K. (1968). In: *Pflanzliche Virologie*. Vol. 2, part 2, p. 281; ed M. Klinkowski, Acad. Verlag, Berlin.
Thomas, B.J.(1980). *Ann. appl. Biol.* **94**: 91.
Thomas, B.J. (1981). *Ann. appl. Biol.* **98**: 419.
Thomas, B.J. (1982). *Ann. appl. Biol.* **100**: 129.
Thomas, B.J. (1984). *Ann. appl. Biol.* **105**: 213.
Thomas, B.J. (1984). *Pl. Path.* **33**: 155.
White, R.P. (1928). *Pl. Dis. Reptr.* **12**: 33.

# Apple necrosis (?) *ilarvirus*

**Data collated by A.A. Brunt, 1992.**

First reported in *Malus* spp.; from Japan; by Y. Doi (in Hull *et al.*, 1989).

### Natural host range and symptoms

*Malus* spp. — leaf necrosis.

### Geographical distribution

Spreads in the Eastern Asian region; Japan.

### Taxonomy and Relationships

Possible *Ilarvirus: Bromoviridae.*

### References

Hull, R., Brown, F. and Payne, C. (1989). In: *Virology: Directory and Dictionary of Animal, Bacterial and Plant Viruses*, p.11. Macmillan Press, London.

# Apple stem grooving *capillovirus*

**Data collated by R.M. Lister, 1986.**

*Synonyms:* Virginia Crab stem grooving virus, chenopodium dark green epinasty virus, brown line disease virus. *Standard acronym:* ASGV. *Strains:* apple latent virus Type 2 isolate C–431 (Lister *et al.*, 1965); E-36.

First reported in *Malus sylvestris* cv. Virginia Crab; from the U.S.A.; by Lister *et al.* (1965); de Sequeira (1965 and 1967); Waterworth and Gilmer (1969).

### Natural host range and symptoms

Symptoms in Virginia Crab apple persist.
*Malus sylvestris* cv. Virginia Crab — stem grooves, abnormal graft union.

**Transmission**

Transmitted by means not involving a vector. Virus transmitted by mechanical inoculation (from apple, especially in spring, by inoculating extracts from buds, young leaves or petals ground in 0.05 M phosphate buffer, pH 7–8. Additives such as 2% (w/v) nicotine base or 2% (w/v) PVP or hide powder help but are not essential); transmitted by grafting; transmitted by seed (in *Chenopodium quinoa*).

**Ecology and control**

Studies reported by Mink and Shay (1962). The disease has been eliminated from some apple scions by heat therapy (30 days at 36°C; de Sequeira and Posnette, 1969).

**Geographical distribution**

Probably distributed worldwide. Spreads in Australia, China, India, Italy, Japan, the Netherlands, New Zealand, Portugal, and the UK (probably occurs wherever apple trees are cultivated).

**Experimental host range**

Several (3–9) families susceptible. Experimentally infected plants mostly show chlorotic or necrotic local lesions, systemic mosaics or necrosis.

Diagnostically susceptible

*Malus sylvestris* cv. Virginia Crab — grooves in the wood. Strain 'E-36' causes graft union to be brown, fragile, swollen and sunken areas on the stem.

*Chenopodium quinoa* — etched necrotic local lesions, systemic chlorotic rings and mottle, tip malformed; some isolates induce epinasty.

*Nicotiana glutinosa* — systemic yellow mosaic, line patterns.

*Phaseolus vulgaris* — purple brown spots, chlorotic local lesions, systemic necrosis.

Maintenance and propagation hosts

*Chenopodium quinoa* — for propagation. *Nicotiana glutinosa, Phaseolus vulgaris* — for maintaining cultures.

Assay hosts (**L**ocal lesions or **W**hole plants)

*Chenopodium quinoa* (L), *Phaseolus vulgaris* cv. Pinto (L).

Dicotyledonous Hosts *(**Susceptible** or Insusceptible)*

***Chenopodium quinoa***. ***Malus sylvestris***. ***Nicotiana glutinosa***. ***Phaseolus vulgaris***.

Additional hosts

Reported to infect about 20 species, usually symptomlessly, in 9 dicotyledonous families: Aizoaceae, Amaranthaceae, Chenopodiaceae, Cucurbitaceae, Lamiaceae, Fabaceae, Rosaceae, Scrophulariaceae, Solanaceae.

Sources of host-range data

Lister *et al.* (1965); de Sequeira (1967); Waterworth and Gilmer (1969); Lister *et al.* (1967); de Sequeira and Cropley (1968); de Sequeira and Posnette (1969); de Sequeira and Lister (1969*b*); Barnett and Murant (1971).

**Properties of particles in sap**

TIP: 60–63 °C. LIV: 2 days. DEP: $\log_{10}$ minus 4. Leaf sap contains few virions. Electron microscopy: best stained with UA or UF; breakage occurs in PTA (Lister *et al.*, 1965; de Sequeira and Lister 1969*b*).

**Purification**

Method: Lister *et al.* (1965); de Sequeira and Lister (1969); Lister (1970).

**Particle morphology**

Virions filamentous; not enveloped; usually flexuous; with a clear modal length; of 600–700 nm; 12 nm wide. Axial canal obscure. Basic helix obvious; pitch of basic helix 3.4–3.7 nm (Lister and Bar-Joseph, 1981).

**Physical properties**

One sedimenting component in purified preparations; sedimentation coefficient 112 *S*. Isoelectric point pH 4.3 (ionic strength of 0.1).

**Biochemical properties**

*Genome* consists of RNA; single–stranded; unipartite. Poly A region present; 3́ terminus.

*Sequence database accession code(s):*

D14995 Em(40)_vi:ASG241KP Gb(84)_vi:ASG241KP Apple stem grooving virus genome, complete sequence. 1/94 6,496bp.

S47260 Em(40)_un:S47260 orf1, orf2 apple stem grooving virus ASGV, Genomic RNA Complete, 6496 nt. 12/93 6,496bp. 2 sequences.

*Virion protein(s)* one; $M_r$ 27000.

*Virus–coded non–virion proteins* isolated; two proteins found. $M_r$ 240000; probably polymerase. $M_r$ of 2nd largest 36000; possibly movement protein.

**Cytopathology**

Virions found in cambium, probably.

**Taxonomy and Relationships**

*Capillovirus.*

Virus(es) with serologically related virions: potato virus T (Salazar and Harrison, 1977; 1978). Virus(es) with serologically unrelated virions: apple chlorotic leaf spot (Lister *et al.*, 1965; de Sequeira, 1967). In cross-protection experiments in Virginia Crab apple the virus seemed to be unrelated to apple stem grooving virus (de Sequeira and Cropley, 1968).

Best tests for diagnosis: this virus is not transmitted to *Solanum tuberosum*, like potato virus T (Salazar and Harrison, 1978), nor to *Nicotiana glutinosa* like apple chlorotic leaf spot virus. Symptoms in *Chenopodium quinoa, Phaseolus vulgaris*, Russian apple R-12740–7A and Virginia Crab distinguish apple chlorotic leaf spot and apple stem grooving viruses (de Sequeira and Cropley, 1968; de Sequeira and Posnette, 1967). *Chenopodium quinoa* is best for isolating this virus because other apple viruses such as tobacco mosaic and apple chlorotic leaf spot virus replicates poorly in that species and their local lesions develop 2–3 days later. When both viruses are present, symptoms are much more severe.

**References**


Bar-Joseph, M., Garnsey, S.M. and Gonsalves, D. (1979). *Adv. Virus Res.* **25**: 93.

de Sequeira, O.A. (1965). *Zast. Bilja* **16**: 247.

de Sequeira, O.A. (1967). *Ann. appl. Biol.* **60**: 59.

de Sequeira, O.A. and Cropley, R. (1968). *TagBer. dt. Akad. Landw. Berlin* **97**: 35.

de Sequeira, O.A. and Lister, R.M. (1969*a*). *Phytopathology* **59**: 572.

de Sequeira, O.A. and Lister, R.M. (1969*b*). *Phytopathology* **59**: 1740.

de Sequeira, O.A. and Posnette, A.F. (1969). *Commonw. Bur. Hort. Pl. Crops. Tech. Commun.* No. **30**, Suppl. 2/3/4, 76*a*.

Hull, R., Brown, F. and Payne, C. (1989). *Directory and Dictionary of Animal, Bacterial and Plant Viruses*, p. 76. MacMillan Reference Books, London.

Lister, R.M. (1970). *CMI/AAB Descr. Pl. Viruses* No. 31, 4 pp.

Lister, R.M., Bancroft, J.B. and Nadakavukaren, M.J. (1965). *Phytopathology* **55**: 859.
Lister, R.M., Bancroft, J.B. and Shay, J.R. (1967). *Phytopathology* **57**: 819.
Lister, R.M. and Bar-Joseph, M. (1981). Closteroviruses. In: *Handbook of Plant Virus Infections and Comparative Diagnosis*, p. 809; ed. E. Kurstak. Elsevier/North-Holland, Biomedical Press, Amsterdam.
Ohki, S.T., Yoshikawa, N., Inouye, N. and Inouye, T. (1989). *Ann. Phytopath. Soc. Japan* **55**: 245.
Waterworth, H.E. and Gilmer, R.M. (1969). *Phytopathology* **59**: 334
Yoshikawa, N. and Takahashi, T. (1988). *J. gen. Virol.* **69**: 241.
Yoshikawa, N. and Takahashi, T. (1992). *J. gen. Virol.* **73**: 1313.
Yoshikawa, N. Sasaki, E., Kato, M. and Takahashi, T. (1992). *Virology* **191**: 98.

# Apple stem pitting virus

**Data collated by H. Koganezawa, 1992.**

*Synonyms:* apple spy 227 epinasty and decline virus (Gilmer. 1962; Mink *et al.*, 1971; Desvignes and Savio, 1975), hawthorn ring pattern mosaic virus, pear necrotic spot virus (Yanase *et al.*, 1989; Koganezawa and Yanase, 1990), pear stony pit virus (Koganezawa and Yanase, 1990; Németh, 1986; Van der Meer, 1986), pear vein yellows virus (Yanase *et al.*, 1989; Van der Meer, 1986). *Strains:* mild and severe strains recognised.

First reported in *Malus sylvestris*; from the U.S.A.; by Smith (1954).

### Natural host range and symptoms

Symptoms persist.

*Malus sylvestris* — commercial cultivars show decline when grafted to sensitive rootstocks such as Virginia Crab and *M. sieboldii.*

*Malus sieboldii* — dieback, inner bark necrosis.

*Malus sieboldii* var. *arborescens* — leaf epinasty, inner bark necrosis.

*Pyrus communis* — vein yellowing.

*Crataegus* spp., *Sorbus mitchelli* — latent infection.

### Transmission

Transmitted by means not involving a vector. Virus transmitted by mechanical inoculation; transmitted by grafting; not transmitted by contact between plants; not transmitted by seed; not transmitted by pollen.

### Ecology and control

Studies reported by Welsh and Nyland (1965); Mink *et al.* (1973); Cutting and Montgomery (1973); Fridlund (1980*a*); Németh (1986).

### Geographical distribution

Spreads in Australia, Austria, Bulgaria, Canada, China, the former Czechoslovakia, Denmark, France, Germany, Greece, Hungary, Israel, Italy, Japan, Korea Republic, the Netherlands, New Zealand, Norway, Poland, Portugal, Romania, South Africa, Spain, Sweden, Switzerland, the UK, the USA, the former USSR, Uruguay, the former Yugoslavia.

### Experimental host range

Several (3–9) families susceptible.

Diagnostically susceptible

*Nicotiana occidentalis* 37B — vein yellowing and leaf necrosis.
*N. occidentalis* subsp. *obliqua* — necrotic local lesions.
*Malus sylvestris* cv. Spy 227 — leaf epinasty and decline.
*M. sylvestris* cv. Virginia Crab — stem pitting.
*M. sylvestris* cvs Radiant or Sparkler — leaf epinasty.
*M. sieboldii* MO65 — dieback, inner bark necrosis and necrotic leaf spot.
*Pyrus communis* cv. Nouveau Poiteau — necrotic leaf spot.
*Pyronia veitchii* — chlorotic spot on leaves and epinasty.

Diagnostically insusceptible

*Nicotiana tabacum, Phaseolus vulgaris.*

Maintenance and propagation hosts

Any susceptible *Malus* sp., *Nicotiana occidentalis* 37B.

Assay hosts (**L**ocal lesions or **W**hole plants)

*Nicotiana occidentalis* subsp. *obliqua* (L), *Malus sylvestris* cvs Virginia Crab, Spy 227, Radiant, Sparkler (W), *M. sieboldii* MO65 (W).

Dicotyledonous Hosts *(**Susceptible** or Insusceptible)*

***Celosia cristata. Chenopodium murale. Chenopodium quinoa***, or *Chenopodium quinoa*. ***Cucumis sativus. Gomphrena globosa. Malus platycarpa.*** *Nicotiana bigelovii. Nicotiana glutinosa. Nicotiana megalosiphon. Nicotiana rustica. Nicotiana tabacum. Phaseolus vulgaris.* ***Physalis floridana***, or *Physalis floridana.* ***Sesamum indicum. Tetragonia expansa.***

Sources of host-range data

Mink *et al.* (1971); Yanase (1974); Sweet (1980); Fridlund (1980*b*); Németh (1986); Van Dijk *et al.* (1987); Stouffer (1989); Yanase *et al.* (1989); Koganezawa and Yanase (1990).

**Properties of particles in sap**

TIP: 55–60 °C. LIV: 0.3–1 days. DEP: $\log_{10}$ minus 2–3. Leaf sap contains few virions.

**Purification**

Method: Koganezawa and Yanase (1990).

**Particle morphology**

Virions filamentous; not enveloped; usually flexuous; with a clear modal length; of 800 nm; 12–15 nm wide. Axial canal obscure. Basic helix obscure.

**Physical properties**

One sedimenting component in purified preparations (isopycnic centrifugation), or three sedimenting components in purified preparations (sucrose density gradient).

**Biochemical properties**

*Genome* consists of RNA; single–stranded; linear; unipartite. Genomic nucleic acid isolated by Koganezawa and Yanase (1990).

*Sequence database accession code(s):*

D21829 Em(40)_vi:AITASPV Gb(84)_vi:AITASPV Apple stem pitting virus (ASPV) RNAs for putative viral RNA-dependent RNA polymerase, triple

D21828 Em(40)_vi:PAVYV Gb(84)_vi:PAVYV Pear vein yellows-associated virus (PVYV) RNA for coat protein, complete cds. 6/94 1,598bp. 2 sequence.

*Virion protein(s)* one; $M_r$ 48000; coat protein.

**Replication**

Replication does not depend on a helper virus.

**Cytopathology**

Virions found in all parts of the host plant; in cytoplasm. Inclusions absent from infected cells. Other cellular changes: disorganisation of chloroplasts (Koganezawa and Yanase, 1990).

**Taxonomy and Relationships**

See Welsh and Uyemoto (1980). Best tests for diagnosis: The virus can be distinguished from others with filamentous virions with similar length by finding end-to-end aggregates of virions in leaf dip preparations by electron microscopy. Also by grafting to woody indicator plants.

**Comments**

Although filamentous virions are associated with apple stem pitting disease, this virus has not yet been transmitted back to apple. Pear vein yellows virus, pear stony pit virus, pear necrotic spot virus and quince sooty ringspot virus are probably identical with apple stem pitting virus (Desvignes and Savio, 1975; Kegler *et al.*, 1979; Yanase *et al.*, 1989).

**References**


Cutting, C.V. and Montgomery, H.B.S. (1973). *More and Better Fruit with East Malling and Long Ashton Research Stations*, U.K.

Desvignes, J.C. and Savio, A. (1975). *Acta Hort.* **44**: 139.

Fridlund, P.R. (1980*a*). *Plant Dis.* **64**: 826.

Fridlund, P.R. (1980*b*). *Acta Phytopath. Acad. Sci. Hung.* **15**: 153.

Gilmer, R.M. (1962). *Phytopathology* **52**: 1027.

Kegler, H., Verderevskaja, T. and Fuchs, E. (1979). *Arch. Gartenb.* **27**: 325.

Koganezawa, H. and Yanase, H. (1990). *Plant Dis.* **74**: 610.

Machita, I. (1991). *Ann. Phytopath. Soc. Japan* **57**: 90.

Mink, G.I., Shay, J.R., Gilmer, R.M. and Stouffer, R.F. (1971). *Search Agric.* **1**: 9.

Mink, G.I., Yanase, H., Yamaguchi, Y. and Sawamura, K. (1973). *Hort. Sci.* **8**: 296.

Németh, M. (1986). In: *Virus, Mycoplasma and Rickettsia Diseases of Fruit Trees*, 841 pp. Akademia Kiado, Budapest.

Smith, W.W. (1954). *Proc. Am. Soc. Hort. Sci.* **63**: 101.

Stouffer, R.F. (1989). In: *Virus and Virus-Like Diseases of Pome Fruits and Similar Noninfectious Disorders*, p. 138; ed. P.R. Fridlund. Washington State University Coop. Ext. Spec. Publ.

Sweet, J.B. (1980). *J. hort. Sci.* **55**: 103.

Van Dijk, P., van der Meer, F.A. and Piron, P.G.M. (1987). *Neth. J. Pl. Path.* **93**: 73.

Welsh, M.F. and Nyland, G. (1965). *Can. J. Pl. Sci.* **45**: 443.

Welsh, M.F. and Uyemoto, J.K. (1980). *Phytopathology* **70**: 349.

Yanase, H. (1974). *Bull. Fruit Tree Res. Stn*, Ser. C **1**: 47.

Yanase, H., Koganezawa, H. and Fridlund, P.R. (1989). *Acta Hort.* **235**: 157.

# Aquilegia necrotic mosaic (?) *caulimovirus*

**Data collated by A.A. Brunt, 1992.**

*Standard acronym:* ANMV.

First reported in *Aquilegia* sp.; from Japan; by Y. Doi (in Hull *et al.* 1989).

**Natural host range and symptoms**

*Aquilegia* sp. — leaf chlorosis and necrosis.

**Geographical distribution**

Spreads in the Eastern Asian region; Japan.

**Particle morphology**

Virions isometric; not enveloped; 50 nm in diameter; rounded in profile.

**Cytopathology**

Inclusions present in infected cells; are viroplasms; they contain virions.

**Taxonomy and Relationships**

Possible *Caulimovirus.*

**References**

Hull, R., Brown, F. and Payne, C.C. (1989). In: *Virology: Directory and Dictionary of Animal, Bacterial and Plant Viruses*, p.11. MacMillan Press, London.

# Aquilegia (?) potyvirus

**Data collated by A.A. Brunt, 1991.**

*Standard acronym:* AqV.

First reported in *Aquilegia* sp.; from Italy; by Marani and Pisi (1976).

**Natural host range and symptoms**

Symptoms persist.
*Aquilegia* sp. — necrotic ringspots in leaves, and flower deformation.

**Transmission**

Virus transmitted by mechanical inoculation.

**Geographical distribution**

Spreads in the Eurasian region; Italy.

**Experimental host range**

Few (<3) families susceptible.

Diagnostically susceptible

*Chenopodium amaranticolor, C. quinoa* — necrotic local lesions; systemic chlorosis and/or necrotic spotting.

*Petunia hybrida, Phaseolus vulgaris* — symptomless systemic infection.

Diagnostically insusceptible

*Cucumis sativus, Gomphrena globosa, Nicotiana tabacum, Ocimum basilicum, Vigna unguiculata.*

Maintenance and propagation hosts

*Chenopodium quinoa.*

Assay hosts (Local lesions or Whole plants)

*Chenopodium amaranticolor* (L), *C. quinoa* (L).

Dicotyledonous Hosts *(**Susceptible** or Insusceptible)*

***Chenopodium amaranticolor*. *Chenopodium quinoa*. *Petunia hybrida*. *Phaseolus vulgaris*.**

**Properties of particles in sap**

TIP: 60 °C. LIV: 8 days. DEP: $\log_{10}$ minus 6.

**Particle morphology**

Virions filamentous; not enveloped; usually flexuous; with no clear modal length; of 700–800 nm. Axial canal obscure. Basic helix obscure.

**Cytopathology**

Inclusions present in infected cells; are pinwheels; they do not contain virions.

**Taxonomy and Relationships**

Unassigned potyvirus.

**References**

Marani, F. and Pisi, A. (1976). *Phytopathol. Medit.* **15**: 137.

# Arabis mosaic *nepovirus*

**Data collated by A.F. Murant, 1984. Revised 1985.**

*Synonyms:* ash ring and line pattern virus (Lana and Agrios, 1973), raspberry yellow dwarf virus (Cadman, 1960; Harrison, 1958), Rhabarber-Mosaik-Virus, rhubarb mosaic virus, forsythia yellow net virus, probably jasmine yellow blotch virus (Cooper and Sweet, 1976). *Standard acronym:* ArMV. *Strains:* hop nettlehead virus, isolates from hop are serologically identical to type strain, but differ in host range and symptoms.

First reported in *Arabis hirsuta*; from England; by Smith and Markham (1944).

**Natural host range and symptoms**

Symptoms disappear soon after infection (but plants may remain stunted).

*Anagallis arvensis, Apium graveolens* var. *dulce*, *Arabis hirsuta, Arabis* sp., *Armoracia rusticana, Asparagus officinalis, Astilbe* sp., *Bellis perennis, Beta vulgaris, Buxus sempervirens, Capsella bursa-pastoris, Carum segetum, Chamaecyparis lawsoniana, Cucumis sativus, Cucurbita pepo, Cyphomandra betacea, Daphne mezereum, Daucus carota, Delphinium* sp., *Dianthus caryophyllus, Euonymus europaea, Forsythia intermedia, Fragaria vesca, Fraxinus excelsior, Humulus lupulus, Jasminum officinale, Lactuca sativa, Lamium amplexicaule, Ligustrum vulgare, Melilotus officinalis, Mentha arvensis, Narcissus* sp., *Phaseolus multiflorus, Plantago lanceolata, Polygonum aviculare, P. persicaria, Prunus avium, P. domestica, P. persica, Ranunculus repens, Rheum rhaponticum, Ribes* sp., *Rosa* spp., *Rubus idaeus, R. procerus, Sambucus nigra, Senecio vulgaris, Solanum nigrum, Stellaria media, Syringa vulgaris, Taraxacum officinale, Trifolium repens, Tulipa* spp. *Urtica dioica, U. urens, Vitis vinifera* — mosaics, mottling and chlorotic ringspots and sometimes necrosis.

**Transmission**

Transmitted by a vector; a nematode; *Xiphinema bakeri, X. coxi, X. diversicaudatum*; Dorylamidae. Principal natural vector(s): *Xiphinema diversicaudatum*. Virus transmitted by mechanical inoculation; transmitted by grafting; not transmitted by contact between plants; transmitted by seed (in 20 species of 14 families; up to 100% infection; Murant, 1983).

**Geographical distribution**

Spreads in the African region, the Eastern Asian region, the Eurasian region, the North American region, and the Pacific region; Australia, Belgium, Bulgaria, Canada, the former Czechoslovakia, Denmark, Finland, France, Germany, Hungary, Ireland, Italy, Japan, Luxembourg, the Netherlands, New Zealand, Norway, Poland, Romania, South Africa, Sweden, Switzerland, Turkey, the UK, the former USSR, the former Yugoslavia. Found, but with no evidence of spread, in North America and Australia.

**Experimental host range**

Many (>9) families susceptible. Experimentally infected plants mostly show necrotic or chlorotic local lesions and mosaics, ringspots or mottles. Plants may recover.

Diagnostically susceptible

*Chenopodium amaranticolor, C. quinoa* — local lesions; systemic chlorotic mottling. *Cucumis sativus* — local lesions, systemic chlorosis. *Nicotiana tabacum* cv. White Burley — local lesions; systemic chlorotic spots, rings and lines. *Petunia hybrida* — local lesions or necrotic rings; systemic chlorotic rings or lines.

Maintenance and propagation hosts

*Nicotiana clevelandii, Petunia hybrida.*

Assay hosts (**L**ocal lesions or **W**hole plants)

*Chenopodium amaranticolor* (L), *Cucumis sativus* (W) for use in nematode transmission tests as a bait plant.

Dicotyledonous Hosts *(**Susceptible** or Insusceptible)*

*Amaranthus caudatus. Amaranthus retroflexus.* ***Antirrhinum majus. Apium graveolens. Bellis perennis. Beta vulgaris. Brassica campestris. Brassica campestris*** ssp. ***pekinensis. Brassica campestris*** ssp. ***rapa. Brassica juncea. Brassica oleracea*** ssp. ***botrytis. Calendula officinalis. Capsella bursa-pastoris.*** *Celosia argentea.* ***Chenopodium album. Chenopodium amaranticolor. Chenopodium foetidum. Chenopodium murale. Chenopodium quinoa. Cucumis melo. Cucumis sativus. Cucurbita pepo. Cyamopsis tetragonoloba. Datura stramonium. Daucus carota. Delphinium hybridum. Dianthus barbatus. Dianthus caryophyllus. Emilia sagittata. Fagopyrum esculentum. Glycine max. Gomphrena globosa. Lactuca sativa. Lathyrus odoratus. Lobelia erinus. Lycopersicon esculentum.*** *Matthiola incana.* ***Myosotis sylvatica. Nicandra physaloides. Nicotiana clevelandii. Nicotiana glutinosa. Nicotiana rustica. Nicotiana sylvestris. Nicotiana tabacum. Ocimum basilicum. Petunia hybrida. Phaseolus vulgaris. Phlox drummondii. Physalis floridana. Pisum sativum. Senecio vulgaris.*** *Solanum nodiflorum.* ***Solanum tuberosum. Spinacia oleracea. Stellaria media. Tetragonia expansa. Torenia fournieri. Trifolium repens. Tropaeolum majus. Vicia faba. Vigna unguiculata. Zinnia elegans.***

Monocotyledonous Hosts *(**Susceptible** or Insusceptible)*

***Narcissus pseudonarcissus.***

**Properties of particles in sap**

TIP: 55–61 °C. LIV: 7–14 days. DEP: $\log_{10}$ minus 3–5. Infectivity of sap not changed by treatment with di–ethyl ether. Leaf sap contains few virions.

**Purification**

Method: Harrison and Nixon (1960).

**Particle morphology**

Virions isometric; not enveloped; 25–27 nm in diameter (in negative stain); angular in profile; without a conspicuous capsomere arrangement.

**Physical properties**

Three sedimenting components in purified preparations; sedimentation coefficient of the fastest 126 *S* (B); of the other(s) 93 *S* (M), or 53 *S* (T). Density 1.51 g $cm^{-3}$ in CsCl ($B_2$), or 1.5 g $cm^{-3}$ in CsCl ($B_1$), or 1.43 g $cm^{-3}$ in CsCl (M). $A_{260}/A_{280}$ ratio 1.68 (B), or 1.5 (M), or 0.73 (T).

**Biochemical properties**

Virions contain 46 % nucleic acid (B), or 22 % nucleic acid (M), or 0 % nucleic acid (T); 54 % protein (B), or 78 % protein (M), or 100 % protein (T); 0 % lipid.

*Genome* consists of RNA; single–stranded; linear; of two parts; largest (or only) genome part the larger 9 kb; the 2nd largest 4.1 kb. Base composition 21.6 % G; 24.9 % A; 25.6 % C; 27.9 % U. Infectivity decreased when deproteinised with proteases; retained when deproteinised with phenol or detergent.

*Sequence database accession code(s):*

D10086 Em(40)_vi:AMVCP Gb(84)_vi:AMVCP Arabis mosaic virus RNA-2, 3' terminal region. 5/92 2,406bp.

X55460 Em(40)_vi:NEAMCOAT Gb(84)_vi:NEAMCOAT Arabis Mosaic virus gene (RNA) coding for coat protein. 1/91 1,515bp.

X81814 Em(43)_vi:Amvrna2u Gb(89)_vi:Amvrna2u Arabis mosaic virus RNA for polyprotein P2-U. 3/95 3,852bp.

X81815 Em(43)_vi:Amvrna2l Gb(89)_vi:Amvrna2l Arabis mosaic virus RNA for polyprotein P2-L. 3/95 3,711bp.

*Non–genomic nucleic acid* found in the virions; is satellite RNA (in some isolates only).

*Virion protein(s)* one; $M_r$ 54000.

**Replication**

Genome replicates in cytoplasmic inclusion bodies. Replication does not depend on a helper virus.

**Cytopathology**

Virions found in epidermis, phloem and all parts of the host plant; in cytoplasm. Inclusions cytoplasmic present in infected cells; are amorphous X–bodies and unusual in shape; granular bodies and hollow spheroids or concentric spheres; they contain virions (aggregated). Other cellular changes: tubules containing virions aligned in rows and found in plasmodesmata (as in other nepovirus infections), but the spherical inclusions in the phloem are found only in arabis mosaic virus infections.

**Taxonomy and Relationships**

*Nepovirus: Comoviridae.*

Virus(es) with serologically related virions: grapevine fanleaf and dogwood mosaic viruses. Other nepoviruses are unrelated.

**Comments**

AMV is one of the viruses that causes lilac yellow ring symptoms (Novak, 1969).

**References**

Abu-Salih, H.S., Murant, A.F. and Daft, M.J. (1968). *J. gen. Virol.* **3**: 299.

Brown, D.J.F. and Trudgill, D.L. (1983). *Rev. de Némat.* **6**: 229.

Cadman, C.H. (1960). *Virology* **11**: 653.

Clark, M.F. and Adams, A.N. (1977). *J. gen. Virol.* **34**: 475.
Cooper, J.I. and Sweet, J.B. (1976). *Forestry* **49**: 73.
Dodd, S.M. and Robinson, D.J. (1984). *J. gen. Virol.* **65**: 1731.
Harrison, B.D. (1958). *Ann. appl. Biol.* **46**: 221.
Harrison, B.D. and Nixon, H.L. (1960). *Virology* **12**: 104.
Jha, A. (1961). *J. hort. Sci* **36**: 219.
Lana, A.O. and Agrios, G.N. (1973). *Phytopathology* **63**: 203.
Murant, A.F. (1970). *CMI/AAB Descr. Pl. Viruses* No. 16, 4 pp.
Murant, A.F. (1985). In: *Virus Diseases of Small Fruits*, ed. R.H. Converse. U.S. Dep. Agric. Hdbk No. 631, p. 204.
Murant, A.F. (1983). *Seed Sci. Technol.* **11**: 973.
Novak (1969). *Proc. 6th. Conf. Czech. Pl. Virologists, Olomone 1967*, p.289.
Roberts, I.M. and Brown, D.J.F. (1980). *Ann. appl. Biol.* **96**: 187.
Smith, K.M. and Markham, R. (1944). *Phytopathology* **34**: 324.
Thresh, J.M., Adams, A.N., Barbara, D.J. and Clark, M.F. (1977). *Ann. appl. Biol.* **87**: 57.

## Arabis mosaic satellite RNA

*Sequence database accession code(s):*
D00664 Em(40)_vi:AMVSRNA Gb(84)_vi:AMVSRNA Arabis mosaic virus satellite RNA (RNA-3). 3/91 1,104bp.
M21212 Em(40)_vi:SAM300CG Gb(84)_vi:SAM300CG Satellite arabis mosaic virus, complete genome. 7/89 300bp.

**Taxonomy and Relationships**

Satellite RNA.

## Araujia mosaic *potyvirus*

**Data collated by A.A. Brunt, and E. Hiebert, 1987.**

*Standard acronym:* ArjMV.

First reported in *Araujia angustifolia* and *Morrenia odorata*; from Argentina; by Charudattan *et al.* (1976).

**Natural host range and symptoms**

Symptoms persist.
*Araujia angustifolia, Morrenia odorata* — mosaics.

**Transmission**

Transmitted by a vector; an insect; *Aphis nerii, Aphis spiraecola, Myzus persicae*; Aphididae. Transmitted in a non–persistent manner. Virus transmitted by mechanical inoculation; not transmitted by contact between plants.

**Ecology and control**

Studies reported by Charudattan *et al.* (1980).

**Geographical distribution**

Spreads in the South and Central American region; Argentina.

**Experimental host range**

Few (<3) families susceptible. Experimentally infected plants mostly show mosaics, malformation.

Diagnostically susceptible

*Araujia angustifolia, A. hortorum, A. sericofera, Matelea floridana, Morrenia brachystephana, Sarcostemma clausum* — systemic mosaic and malformation.

*Hoya carnosa, H. coronaria* — symptomless.

Diagnostically insusceptible

*Chenopodium amaranticolor, Cucumis sativus, Phaseolus vulgaris, Nicotiana benthamiana, Nicotiana tabacum.*

Maintenance and propagation hosts

*Morrenia odorata, Araujia angustifolia, Araujia sericofera.*

Assay hosts (**L**ocal lesions or **W**hole plants)

*Morrenia odorata* (W).

Dicotyledonous Hosts *(**Susceptible** or Insusceptible)*

*Apium graveolens. Arachis hypogaea. Beta vulgaris. Brassica campestris* ssp. *rapa. Brassica juncea. Brassica oleracea* ssp. *botrytis. Brassica oleracea* ssp. *capitata. Capsicum annuum. Capsicum frutescens. Carica papaya. Chenopodium album. Chenopodium amaranticolor. Cichorium endiva. Citrullus lanatus. Citrus aurantifolia. Citrus aurantium. Citrus limon. Citrus medica. Citrus paradisi. Citrus sinensis. Coriandrum sativum. Cucumis melo. Cucumis sativus. Cucurbita maxima. Datura stramonium. Daucus carota. Glycine max. Gomphrena globosa. Gossypium hirsutum. Helianthus annuus. Lactuca sativa. Lespedeza stipulacea. Lycopersicon esculentum. Melilotus alba. Nicotiana benthamiana. Nicotiana tabacum. Phaseolus lunatus. Phaseolus vulgaris. Pisum sativum. Poncirus trifoliata. Raphanus sativus. Ricinus communis. Solanum melongena. Solanum tuberosum. Spinacia oleracea. Tetragonia expansa. Trifolium hybridum. Vigna radiata. Vigna unguiculata. Zinnia elegans.*

Monocotyledonous Hosts *(**Susceptible** or Insusceptible)*

*Agropyron repens. Allium cepa. Avena sativa. Bromus inermis. Bromus secalinus. Caladium hortulanum. Echinochloa crus–galli. Hippeastrum hybridum. Lolium multiflorum. Lolium perenne. Narcissus pseudonarcissus. Panicum miliaceum. Philodendron selloum. Secale cereale. Setaria italica. Triticum aestivum. Zea mays.*

Sources of host-range data

Charudattan *et al.* (1980).

**Properties of particles in sap**

TIP: 56–58 °C. LIV: 2–2.5 days. DEP: $\log_{10}$ minus 2–3. Leaf sap contains few virions.

**Purification**

Method: Hiebert and Charudattan (1984).

**Particle morphology**

Virions filamentous; not enveloped; usually flexuous; with a clear modal length; of 741 nm. Axial canal obscure. Basic helix obscure.

**Biochemical properties**

*Genome* consists of RNA; single–stranded.

*Virion protein(s)* one; $M_r$ 32000; coat protein.

*Virus–coded non–virion proteins* identified by genomic sequence analysis (Hiebert and Charudattan (1984)); five proteins found. $M_r$ of the largest 81000; helper component related. $M_r$ of 2nd largest 70000; cylindrical inclusion. $M_r$ of 3rd 53000; nuclear inclusion. $M_r$ of 4th 49000; nuclear inclusion. $M_r$ of 5th and smaller 40000; function unknown.

### Cytopathology

Virions found in mesophyll and epidermis. Inclusions present in infected cells; are pinwheels; they do not contain virions.

### Taxonomy and Relationships

*Potyvirus: Potyviridae.*

Virus(es) with serologically unrelated virions: bean common mosaic, bidens mottle, blackeye cowpea mosaic, dasheen mosaic, lettuce mosaic, papaya ringspot, pepper mottle, potato Y, soybean mosaic, tobacco etch, tobacco veinal mottle, tobacco etch and turnip mosaic viruses.

The 49kDa and 53kDa proteins produced by *in vitro* translation react with antisera to tobacco etch (TEV) nuclear proteins and the 32kDaprotein with antiserum to TEV capsid protein; the cylindrical inclusion protein of araujia mosaic is serologically related to that of dasheen mosaic virus but not TEV; and the 81kDaprotein of araujia mosaic virus reacted with antiserum to the helper component of tobacco vein mottling virus.

### Comments

Charudattan *et al.* (1980) suggest that the virus might be used to control *Morrenia* weeds in citrus orchards in Florida, U.S.A.

### References

Charudattan, R., Zettler, F.W., Cordo, H.A. and Christie, R.G. (1976). *Proc. Am. Phytopath. Soc.* **3**: 272.

Charudattan, R., Zettler, F.W., Cordo, H.A. and Christie, R.G. (1980). *Phytopathology* **70**: 909.

Hiebert, E. and Charudattan, R. (1984). *Phytopathology* **74**: 642.

## Arracacha A *nepovirus*

**Data collated by R.A.C. Jones, 1984. Revised 1989.**

*Standard acronym:* AVA.

First reported in *Arracacia xanthorrhiza* (arracacha); from the Peruvian Andes; by Jones and Kenten (1978).

### Natural host range and symptoms

Symptoms persist.

*Arracacia xanthorrhiza* — yellow mosaic.

### Transmission

Transmitted by a vector (probably); transmitted by vegetative propagation; a nematode (probably). Virus transmitted by mechanical inoculation; not transmitted by contact between plants; transmitted by seed (readily in *Nicotiana clevelandii*).

### Geographical distribution

Spreads in the South and Central American region; Peru.

**Experimental host range**

Many (>9) families susceptible. Experimentally infected plants mostly show local lesions, mosaics, mottling and may recover.

Diagnostically susceptible

*Chenopodium murale, C. quinoa, Nicotiana clevelandii* — local lesions or rings; systemic mosaics and mottling.

*Tetragonia expansa* — systemic mottling and necrosis.

Diagnostically insusceptible

*Brassica rapa, Phaseolus vulgaris, Vigna unguiculata, Capsicum annuum, Solanum melongena.*

Maintenance and propagation hosts

*Chenopodium quinoa, Nicotiana clevelandii* (best).

Assay hosts (**L**ocal lesions or **W**hole plants)

*Chenopodium murale* (L, W), *C. quinoa* (L, W).

Dicotyledonous Hosts ***(Susceptible** or Insusceptible)*

***Amaranthus caudatus. Anthriscus cerefolium. Beta vulgaris. Brassica campestris*** ssp. ***pekinensis.*** *Brassica campestris* ssp. *rapa.* ***Brassica oleracea*** ssp. ***capitata*** (cv. Greyhound). *Capsicum annuum.* ***Chenopodium amaranticolor. Chenopodium foliosum. Chenopodium murale. Chenopodium quinoa. Clitoria ternatea. Coriandrum sativum. Cucumis sativus. Cucurbita pepo. Cyamopsis tetragonoloba. Datura stramonium.*** *Daucus carota.* ***Gomphrena globosa. Hyoscyamus niger. Lycopersicon esculentum. Montia perfoliata. Nicandra physaloides. Nicotiana bigelovii. Nicotiana clevelandii. Nicotiana debneyi. Nicotiana glutinosa. Nicotiana occidentalis. Nicotiana tabacum.*** *Ocimum basilicum. Petroselinum crispum.* ***Petunia hybrida.*** *Phaseolus acutifolius. Phaseolus vulgaris.* ***Physalis floridana. Physalis peruviana. Sesamum indicum.*** *Solanum chacoense.* ***Solanum demissum.*** *Solanum melongena.* ***Solanum tuberosum. Tetragonia expansa.*** *Vigna unguiculata. Vigna unguiculata* ssp. *cylindrica. Zinnia elegans.*

Sources of host-range data

Jones and Kenten (1978).

**Properties of particles in sap**

TIP: 65–70 °C. LIV: 15 days. DEP: $\log_{10}$ minus 5.

**Purification**

Method: Jones and Kenten (1978).

**Particle morphology**

Virions isometric; 26 nm in diameter; angular in profile.

**Physical properties**

Three sedimenting components in purified preparations; sedimentation coefficient of the fastest 125 *S* (B); of the other(s) 92 *S* (M), or 50 *S* (T). Density 1.32 g $cm^{-3}$ in CsCl (T), or 1.45 g $cm^{-3}$ in CsCl (M), or 1.52 g $cm^{-3}$ in CsCl (B). $A_{260}/A_{280}$ ratio 0.65 (T), or 1.5 (M), or 1.85 (B).

**Biochemical properties**

Virions contain 43.5 % nucleic acid (B), or 32.5 % nucleic acid (M), or 0 % nucleic acid (T); 56.5 % protein (B), or 67.5 % protein (M), or 100 % protein (T); 0 % lipid.

*Genome* consists of RNA; single–stranded; of two parts; largest (or only) genome part the larger 8 kb; the 2nd largest 4.5 kb. Genomic nucleic acid isolated by Jones and Kenten (1980).

*Virion protein(s)* one; $M_r$ 53000. Method of preparation: Jones and Kenten (1978).

**Replication**

Replication does not depend on a helper virus.

**Taxonomy and Relationships**

*Nepovirus: Comoviridae.*

Virus(es) with serologically unrelated virions: potato black ringspot, arabis mosaic, artichoke Italian latent, cherry leaf roll, cacao necrosis, grapevine Bulgarian latent, grapevine chrome mosaic, grapevine fan leaf, mulberry ringspot, myrobalan latent ringspot, potato U, peach rosette mosaic, raspberry ringspot, strawberry latent ringspot, tobacco ringspot, tomato black ring (Scottish and English strains) and tomato ringspot viruses (Jones and Kenten, 1978, 1980; Jones *et al.*, 1982).

Best tests for diagnosis: arracacha A and B viruses are readily distinguishable by serology and by inoculation to *Chenopodium quinoa, Tetragonia expansa, Nicotiana clevelandii* and *Vigna unguiculata.*

**References**

Jones, R.A.C. and Kenten, R.H. (1978). *Ann. appl. Biol.* **90**: 85.
Jones, R.A.C. and Kenten, R.H. (1980). *CMI/AAB Descr. Pl. Viruses* No. 216, 4 pp.
Jones, R.A.C., Fribourg, C.E. and Koenig, R. (1983). *Phytopathology* **73**: 195.

# Arracacha B (?) *nepovirus*

**Data collated by R.A.C. Jones, 1984. Revised 1989.**

*Standard acronym:* AVB. *Strains:* Type strain (T), Oca strain (O).

First reported in *Arracacia xanthorrhiza* (arracacha); from the Peruvian Andes; by Jones and Kenten (1978).

**Natural host range and symptoms**

Symptoms none.

*Arracacia xanthorrhiza, Oxalis tuberosa, Solanum tuberosum* — symptomless.

**Transmission**

Virus transmitted by mechanical inoculation; transmitted by grafting; not transmitted by contact between plants; transmitted by seed (5% of strain T in *Chenopodium quinoa*; 12% strain O in *Solanum tuberosum*); transmitted by pollen to the seed (in *Solanum tuberosum*; Jones, 1982).

**Ecology and control**

Studies reported by Jones (1981*a, b*; 1982) indicate that the virus is disseminated in vegetative propagules (in root pieces of arracacha).

**Geographical distribution**

Spreads in the South and Central American region (where arracacha, oca and potato are grown; Jones and Kenten, 1983; Atkey and Brunt, 1982); Bolivia and Peru.

**Experimental host range**

Several (3–9) families susceptible. Experimentally infected plants mostly show vein clearing then systemic mosaics, mottles, ringspots or necrosis and may recover.

Diagnostically susceptible

*Chenopodium amaranticolor* — local lesions; systemic mosaic or mottle and may recover.

*C. murale* — systemic necrosis.

*Cucumis sativus* — systemic mosaic then recovery.

*Tetragonia expansa* — systemic mosaic or mottle.

Diagnostically insusceptible

*Brassica rapa, Phaseolus vulgaris, Capsicum annuum, Datura metel, Hyoscyamus niger.*

Maintenance and propagation hosts

*Chenopodium quinoa, Tetragonia expansa.*

Assay hosts (**L**ocal lesions or **W**hole plants)

*Chenopodium amaranticolor* (L,W), *C. murale* (W).

Dicotyledonous Hosts (***Susceptible*** *or Insusceptible)*

***Amaranthus caudatus. Anthriscus cerefolium.*** *Brassica campestris* ssp. *pekinensis. Brassica campestris* ssp. *rapa.* Cv. Golden Acre *Brassica oleracea* ssp. *capitata. Capsicum annuum.* ***Chenopodium amaranticolor. Chenopodium foetidum. Chenopodium murale. Chenopodium quinoa. Coriandrum sativum. Cucumis sativus. Cyamopsis tetragonoloba.*** *Datura metel.* ***Datura stramonium. Gomphrena globosa.*** *Hyoscyamus niger. Ipomoea setosa.* ***Lycopersicon esculentum.*** *Lycopersicon pimpinellifolium.* ***Montia perfoliata. Nicandra physaloides. Nicotiana bigelovii. Nicotiana clevelandii. Nicotiana debneyi. Nicotiana glutinosa. Nicotiana occidentalis. Nicotiana rustica. Nicotiana tabacum. Petroselinum crispum. Petunia hybrida.*** *Phaseolus vulgaris.* ***Physalis floridana. Physalis peruviana. Solanum chacoense.*** *Solanum melongena.* ***Solanum tuberosum***, or *Solanum tuberosum.* ***Tetragonia expansa. Vigna unguiculata. Vigna unguiculata*** ssp. ***cylindrica. Vigna unguiculata*** ssp. ***sesquipedalis.***

Sources of host-range data

Kenten and Jones (1979); Jones and Kenten (1981).

**Properties of particles in sap**

TIP: 65–70 °C. LIV: 12–19 days. DEP: $\log_{10}$ minus 3–5.

**Purification**

Method: Kenten and Jones (1979).

**Particle morphology**

Virions isometric; not enveloped; 26 nm in diameter; angular in profile.

**Physical properties**

One sedimenting component in purified preparations; sedimentation coefficient 126 *S*. Density 1.481 and 1.492 g $cm^{-3}$ in CsCl (for strain T; and 1.484 and 1.495 g $cm^{-3}$ for strain O).

**Biochemical properties**

Virions contain 40–41 % nucleic acid; 59–60 % protein; 0 % lipid.

*Virion protein(s)* two (in ratio 3:1, larger:smaller); $M_r$ of the larger 26000. $M_r$ of 2nd largest 20000 (strain T), or 22000 (strain O). Method of preparation: Kenten and Jones (1979).

### Replication

Replication does not depend on a helper virus.

### Taxonomy and Relationships

Possible *Nepovirus: Comoviridae.*

Virus(es) with serologically related virions: strains T and O are serologically distantly related (SDI 6–7), and induce distinct symptoms in most indicator hosts (Jones and Kenten, 1981). Virus(es) with serologically unrelated virions: Andean potato latent, Andean potato mottle, arracacha A, carnation Italian ringspot, carnation mottle, carnation ringspot, cherry rasp leaf, cymbidium ringspot, narcissus tip necrosis, pelargonium flower break, pelargonium leaf curl, pelargonium ringspot, pelargonium ring pattern, red clover necrotic mosaic, saguaro cactus, southern bean mosaic, sowbane mosaic, tomato bushy stunt, turnip crinkle, turnip rosette and turnip yellow mosaic viruses.

Best tests for diagnosis: See Jones and Kenten (1983). *Chenopodium murale* is the best diagnostic host.

### References

Atkey, P.T. and Brunt, A.A. (1982). *Phytopath. Z.* **103**: 294.
Jones, R.A.C. (1981*a*). *Plant Dis.* **65**: 753.
Jones, R.A.C. (1981*b*). In: *Pests, Pathogens and Vegetation*, p. 89; ed. J.M. Thresh. Pitman, London.
Jones, R.A.C. (1982). *Ann. appl. Biol.* **100**: 315.
Jones, R.A.C. and Kenten, R.H. (1978). *Ann. appl. Biol.* **90**: 85.
Jones, R.A.C. and Kenten, R.H. (1981). *Phytopath. Z.* **100**: 88.
Jones, R.A.C. and Kenten, R.H. (1983). *CMI/AAB Descr. Pl. Viruses* No. 270, 4 pp.
Kenten, R.H. and Jones, R.A.C. (1979). *Ann. appl. Biol.* **93**: 31.

## Arracacha latent (?) *carlavirus*

**Data collated by S. Phillips, 1986.**

*Standard acronym:* ALV.

First reported in *Arracacia xanthorrhiza*; from Littlehampton, U.K. in material from South America.

### Natural host range and symptoms

*Arracacia xanthorrhiza* — latent infection.

### Transmission

Virus transmitted by mechanical inoculation.

### Geographical distribution

Spreads in the South and Central American region; Bolivia and Peru.

### Experimental host range

Several (3–9) families susceptible. Experimentally infected plants mostly show local lesions, systemic chlorosis.

Diagnostically susceptible

*Nicotiana clevelandii* — systemic chlorotic vein-banding.
*Gomphrena globosa* — red local lesions; not systemic.
*Chenopodium quinoa* — chlorotic local lesions; not systemic.
*Nicotiana megalosiphon* — symptomless local and systemic infection.

Diagnostically insusceptible

*Trifolium incarnatum, Vigna unguiculata, Pisum sativum.*

Maintenance and propagation hosts

*Nicotiana clevelandii.*

Assay hosts (**L**ocal lesions or **W**hole plants)

*Chenopodium quinoa* (L), *Nicotiana clevelandii* (W).

Dicotyledonous Hosts ***(Susceptible** or Insusceptible)*

*Brassica campestris* ssp. *pekinensis*. ***Chenopodium amaranticolor***. ***Chenopodium quinoa***. *Cucumis sativus*. *Datura stramonium*. ***Gomphrena globosa***. *Lycopersicon esculentum*. ***Nicotiana clevelandii***. ***Nicotiana megalosiphon***. *Phaseolus vulgaris*. *Pisum sativum*. ***Tetragonia expansa***. *Trifolium incarnatum*. *Vicia faba*. *Vigna unguiculata*.

**Properties of particles in sap**

Leaf sap contains few virions.

**Purification**

Method: extract *Nicotiana* leaves with 0.5 M $KPO_4$ + 0.5% (w/v) $Na_2SO_3$ (pH 8.0), add 5% Triton X-100 and then differentially centrifuge twice.

**Particle morphology**

Virions filamentous; not enveloped; usually straight; with a clear modal length; of 650 nm. Basic helix obscure.

**Replication**

Replication does not depend on a helper virus.

**Cytopathology**

Virions found in all parts of the host plant.

**Taxonomy and Relationships**

Probably *Carlavirus*.

Virus(es) with serologically related virions: lily symptomless (SDI 3), carnation latent (SDI 4) and chrysanthemum B (SDI 7) viruses. Virus(es) with serologically unrelated virions: narcissus latent, potato S, elder carlavirus, red clover vein mosaic and honeysuckle latent viruses.

## Arracacha Y *potyvirus*

**Data collated by A.A. Brunt and S. Phillips, 1986.**

*Standard acronym:* AVY.

First reported in *Arracacia xanthorrhiza*; from Littlehampton, U.K. in material from South America.

**Natural host range and symptoms**

Symptoms persist.

*Arracacia xanthorrhiza* — leaf chlorosis.

**Transmission**

Transmitted by a vector; an insect; *Myzus persicae, Aphis gossypii*; Aphididae. Transmitted in a non–persistent manner. Virus transmitted by mechanical inoculation;

not transmitted by contact between plants; not transmitted by seed; not transmitted by pollen.

**Geographical distribution**

Spreads in the South and Central American region; Peru.

**Experimental host range**

Several (3–9) families susceptible. Experimentally infected plants mostly show local lesions, chlorosis.

Diagnostically susceptible

*Chenopodium amaranticolor, C. quinoa* — chlorotic or necrotic local lesions; not systemic.

*Nicotiana clevelandii* — systemic veinal chlorosis.

*N. benthamiana* — systemic chlorosis and malformation.

*N. tabacum, N. sylvestris* — chlorotic local lesions; not systemic.

Diagnostically insusceptible

*Vigna unguiculata, Gomphrena globosa, Ullucus tuberosus.*

Maintenance and propagation hosts

*Nicotiana clevelandii Nicotiana benthamiana.*

Assay hosts (**L**ocal lesions or **W**hole plants)

*Chenopodium quinoa* (L), *Chenopodium amaranticolor* (L), *Nicotiana clevelandii* (W), *Nicotiana benthamiana* (W).

Dicotyledonous Hosts *(**Susceptible** or Insusceptible)*

*Brassica campestris* ssp. *pekinensis*. ***Chenopodium amaranticolor***. ***Chenopodium murale***. ***Chenopodium quinoa***. *Cucumis sativus. Datura metel. Datura stramonium. Gomphrena globosa. Lycopersicon esculentum.* ***Nicotiana benthamiana***. ***Nicotiana clevelandii***. ***Nicotiana tabacum***. *Phaseolus vulgaris. Pisum sativum. Solanum demissum.* ***Tetragonia expansa***. *Trifolium incarnatum. Vicia faba. Vigna unguiculata.*

**Properties of particles in sap**

TIP: 65–70 °C (in *Nicotiana clevelandii* sap). LIV: 32 days. DEP: $\log_{10}$ minus 3. Leaf sap contains few virions.

**Purification**

Method: A.A. Brunt and R.S. Phillips: blend the leaves in 0.067 M $PO_4$ + 0.1% mercapto acetic acid, clarify with 8.0% butanol and then differentially centrifuge twice.

**Particle morphology**

Virions filamentous; not enveloped; usually flexuous; with a clear modal length; of 750 nm; 12 nm wide. Axial canal obscure. Basic helix obscure.

**Replication**

Replication does not depend on a helper virus.

**Cytopathology**

Virions found in all parts of the host plant.

**Taxonomy and Relationships**

*Potyvirus: Potyviridae.*

Virus(es) with serologically related virions: datura shoestring (SDI 6), sweet potato feathery mottle (SDI 2–3), ullucus mosaic (SDI 7) and watermelon mosaic 2 (SDI 7) viruses. Virus(es) with serologically unrelated virions: bean common mosaic, bean yellow mosaic, beet mosaic, bidens mottle, carnation vein mottle, clover yellow vein, commelina mosaic, freesia mosaic, henbane mosaic, hippeastrum mosaic, iris mild mosaic, lettuce mosaic, passion fruit woodiness, peanut mottle, pepper veinal mottle, potato Y, tobacco etch, tulip breaking and turnip mosaic viruses.

## Artichoke curly dwarf (?) *potexvirus*

**Data collated by A.A. Brunt, 1995.**

*Standard acronym:* ACDV.

First reported in *Cynara scolymus*; from California, USA; by Leach and Oswald (1950).

**Natural host range and symptoms**

Symptoms persist.

*Cynara scolymus* — veinal necrosis, leaf distortion and stunting.

**Transmission**

Virus transmitted by mechanical inoculation.

**Geographical distribution**

Spreads in the North American region; the USA.

**Experimental host range**

Few (<3) families susceptible. Experimentally infected plants mostly show Leaf distortion and stunting.

Diagnostically susceptible

*Callistephus chinensis, Helichrysum bracteatum* — chlorotic local lesions, systemic leaf distortion.

*Centaurea cyanus, Cynara cardunculus, C. scolymus* — systemic veinal necrosis, leaf distortion and stunting.

*Helianthus annuus* — chlorotic local lesions, symptomless systemic infection.

*Tagetes* — symptomless local and systemic infection.

*Zinnia elegans* — necrotic local lesions, systemic veinal necrosis, leaf distortion and stunting.

Diagnostically insusceptible

*Calendula officinalis, Cichorium endiva, Cosmos bipinnatus, Lactuca sativa, Taraxacum officinale.*

Assay hosts (**L**ocal lesions or **W**hole plants)

*Zinnia elegans* (L).

Dicotyledonous Hosts ***(Susceptible** or Insusceptible)*

*Calendula officinalis. Chrysanthemum morifolium. Cichorium endiva.* ***Helianthus annuus****. Lactuca sativa.* ***Zinnia elegans****.*

Comments on host-range

The virus seems only to infect species of the Asteraceae; it failed to infect unnamed species in 18 other families.

Sources of host-range data

Leach and Oswald (1950), Morton (1961).

**Properties of particles in sap**

TIP: 55–60 °C. LIV: 2–3 days (at 19°C). DEP: $\log_{10}$ minus 3–4.

**Particle morphology**

Virions filamentous; not enveloped; with a clear modal length; of 582 nm.

**Taxonomy and Relationships**

Possible *Potexvirus*.

**Comments**

Although first reported in 1950, the virus was probably present in California two or more decades earlier (Morton, 1961).

**References**

Leach, L.D. and Oswald, J.W. (1950). *Phytopathology* **40**: 967.
Morton, D.J. (1961). *Phytopathology* **51**: 731.

# Artichoke Italian latent *nepovirus*

**Data collated by C. Büchen-Osmond and D. Gallitelli, 1987.**

*Standard acronym:* AILV.

First reported in *Cynara scolymus*; from Italy; by Marjorana and Rana (1970); Vovlas *et al.* (1971).

**Natural host range and symptoms**

*Cynara scolymus* — usually latent, plants stunted and yellow.
*Cichorium intybus* — chlorotic mottle and bright yellow spots.
*Pelargonium zonale* — severely malformed leaves, plant stunted.
*Vitis vinifera* — fanleaf-like symptoms, stunted.
*Gladiolus* sp., *Sonchus arvensis* — yellow rings and line pattern.
*Crepis neglecta, Helminthia echioides, Hypochoeris aetensis, Lactuca virosa, Urospermum dalechampii, Lamium amplexicaule, Sonchus* spp. — symptomless.

**Transmission**

Transmitted by a vector; a nematode; *Longidorus apulus* (Lamberti and Bleve Zacheo, 1977); Dorylamidae. Virus transmitted by mechanical inoculation.

**Ecology and control**

Studies reported by Vovlas and Roca (1975); Roca *et al.* (1975).

**Geographical distribution**

Spreads in the Eurasian region; Bulgaria and Italy.

**Experimental host range**

Many (>9) families susceptible. Experimentally infected plants mostly show ring local lesions, then systemic mosaic, ringspots, recovery.

Diagnostically susceptible

*Phaseolus vulgaris* cv. La Victoire — chlorotic and necrotic local lesions; tip necrosis or mottle; then recovery.
*Gomphrena globosa* — pale necrotic, ring local lesions; sometimes tip necrosis.
*Cucumis sativus* — chlorotic/necrotic local lesions; severe systemic mosaic, and necrosis.

*Nicotiana tabacum* cv. White Burley — necrotic ring local lesions; systemic ringspots and line patterns.

Maintenance and propagation hosts

*Cucurbita pepo, Chenopodium quinoa, Phaseolus vulgaris.*

Assay hosts (Local lesions or Whole plants)

*Phaseolus vulgaris* cv. La Victoire (L); *Gomphrena globosa* (L).

Dicotyledonous Hosts *(**Susceptible** or Insusceptible)*

***Chenopodium quinoa. Cucurbita pepo. Gomphrena globosa. Phaseolus vulgaris. Vitis vinifera.***

Sources of host-range data

Martelli and Rana (1975); Vovlas *et al.* (1971); Vovlas and Roca (1975); Vovlas (1974); Jankulova *et al.* (1976); Savino *et al.* (1977); Quacquarelli *et al.* (1976*a*); Majorana and Rana (1970).

**Properties of particles in sap**

TIP: 55–60 °C. LIV: 3–14 days (at 4 °C). DEP: $\log_{10}$ minus 2–5.

**Purification**

Method: Dunn and Hitchborn (1965); Jankulova *et al.* (1976); Savino *et al.* (1977).

**Particle morphology**

Virions isometric; not enveloped; 30 nm in diameter.

**Physical properties**

Three sedimenting components in purified preparations; sedimentation coefficient of the fastest 121 *S* (B); of the other(s) 96 *S* (M), or 55 *S* (T).

**Biochemical properties**

Virions contain 41 % nucleic acid (B), or 34 % nucleic acid (M), or 0 % nucleic acid (T); 59 % protein (B), or 66 % protein (M), or 100 % protein (T).

*Genome* consists of RNA; single–stranded; linear; of two parts; largest (or only) genome part the larger 7.75 kb (RNA-1; present in B virions); the 2nd largest 4.75 kb (RNA-2; present in M virions). Genomic nucleic acid isolated by Jankulova *et al.* (1976); Quacquarelli *et al.* (1976*b*); Savino *et al.* (1977); Diener and Schneider (1968). Base composition 24 % G; 26 % A; 19 % C; 31 % U.

*Virion protein(s)* one; $M_r$ 54000; coat protein. Method of preparation: Jankulova *et al.* (1976); Savino *et al.* (1977).

**Cytopathology**

Virions found in leaves, roots and parenchyma; in cell vacuoles. Inclusions absent from infected cells. Other cellular changes: rows of membraneous tubules in parenchyma cells (Vovlas *et al.*, 1971).

**Taxonomy and Relationships**

*Nepovirus: Comoviridae.*

Virus(es) with serologically unrelated virions: grapevine fanleaf, arabis mosaic, tomato black ring, tobacco ringspot, tomato ringspot, peach rosette mosaic, raspberry ringspot, strawberry latent ringspot, grapevine chrome mosaic, cacao necrosis, cherry leaf roll, grapevine Bulgarian latent and myrobalan latent ringspot viruses (Vovlas *et al.*, 1971; Savino *et al.*, 1977).

The physico-chemical characters and hydrodynamic behaviour of the virions have been used to assign it to a subgroup containing tomato black ring and related viruses (Martelli, 1975; Quacquarelli *et al.* 1967*b*).

**References**

Diener, T.O. and Schneider, I.R. (1968). *Arch. Biochim. Biophys.* **124**: 401.
Dunn, D.B. and Hitchborn, J.H. (1965). *Virology* **25**: 171.
Jankulova, M., Savino, V., Gallitelli, D., Quacquarelli, A. and Martelli, G.P. (1976). *Abstr. Proc. 6th Conf. Int. Counc. Grapevine Viruses, Cordova 1976*, p.24.
Lamberti, F. and Bleve Zacheo (1977). *Nematol. Medit.* **5**: 73.
Majorana, G. and Rana, G.L. (1970). *Phytopath. Medit.* **9**: 193.
Martelli, G.P. (1975). In: *Nematode Vectors of Plant Viruses*, p. 223; eds F. Lamberti, C.E. Taylor and J.W. Seinhorst. Plenum Press, New York.
Martelli, G.P., Rana, G.L. and Savino, V. (1977). *CMI/AAB Descr. Pl. Viruses* No. 176, 4 pp.
Quacquarelli, A., Rana, G.L. and Martelli, G.P. (1976). *Poljopr. znanst. Smotra* **39**: 561.
Quacquarelli, A., Gallitelli, D., Savino, V., Piazzolla, P. and Martelli, G.P. (1976*b*). *Abstr. Proc. 6th Conf. Int. Counc. Grapevine Viruses, Cordova 1976*, p.10.
Rana, G.L. and Roca, F. (1975). *Atti. 2nd Congr. Int. Carciofo, Bari 1973*, p.855.
Roca, F., Martelli, G.P., Lamberti, F. and Rana, G.L. (1975). *Nematol. Medit.* **3**: 91.
Savino, V., Gallitelli, D., Jankulova, M. and Rana, G.L. (1977). *Phytopathol. Medit.* **16**: 41.
Vovlas, C. (1974). *Phytopathol. Medit.* **13**: 139.
Vovlas, C. and Roca, F. (1975). *Nematol. Medit.* **4**: 23.
Vovlas, C., Martelli, G.P. and Quacquarelli, A. (1971). *Phytopathol. Medit.* **10**: 244.

# Artichoke latent M (?) *carlavirus*

**Data collated by D. Gallitelli, 1991.**

*Standard acronym:* ArLVM.

First reported in *Cynara scolymus*; from Apulia, southern Italy; by Di Franco *et al.* (1989).

**Natural host range and symptoms**

Symptoms usually none.
*Cynara scolymus* — symptomless.

**Transmission**

Virus not transmitted by mechanical inoculation.

**Geographical distribution**

Spreads in the Eurasian region; Italy (Apulia).

Sources of host-range data

Di Franco *et al.* (1989).

**Properties of particles in sap**

Leaf sap contains many virions.

**Purification**

Method: Di Franco *et al.* (1989).

**Particle morphology**

Virions filamentous; usually flexuous; with a clear modal length; of 697 nm. Axial canal obvious. Basic helix obscure.

**Physical properties**

Density 1.31 g $cm^{-3}$ in CsCl.

**Biochemical properties**

*Genome* consists of RNA; single–stranded; linear; unipartite; largest (or only) genome part 7.5 kb.

*Non–genomic nucleic acid* not found in the virions.

*Virion protein(s)* one; $M_r$ 31000; coat protein. Method of preparation: Laemmli (1970).

**Cytopathology**

Virions found in cytoplasm. Inclusions present in infected cells; are membranous bodies; they do not contain virions.

**Taxonomy and Relationships**

Possible *Carlavirus*.

Virus(es) with serologically related virions: carnation latent and poplar mosaic viruses. Virus(es) with serologically unrelated virions: artichoke latent S, caper latent, potato M, potato S, shallot latent, chrysanthemum B, helenium S and pea streak viruses.

Best tests for diagnosis: serology.

**References**

Di Franco, A., Gallitelli, D., Vovlas, C. and Martelli, G.P. (1989). *J. Phytopath.* **127**: 265.

Laemmli, U.K. (1970). *Nature, Lond.* **227**: 680.

## Artichoke latent *potyvirus*

**Data collated by D. Gallitelli, 1991.**

*Synonyms:* artichoke Californian latent virus, probably artichoke potyvirus (Schwarz, 1977). *Standard acronym:* ArLV.

First reported in *Cynara scolymus*; from California, U.S.A.; by Costa *et al.* (1959).

**Natural host range and symptoms**

Symptoms persist, vary seasonally, and none. Symptoms stunting and yellow flecking of plants.

*Cynara scolymus, C. cardunculus* — latent. May cause symptoms in mixed infections (Rana *et al.*, 1982).

*Petunia hybrida* — systemic mosaic, stunting.

**Transmission**

Transmitted by a vector; an insect; *Myzus persicae, Aphis fabae, Brachycaudus cardui*; Aphididae. Transmitted in a non–persistent manner. Virus transmitted by mechanical inoculation; transmitted by grafting; not transmitted by seed.

**Geographical distribution**

Spreads in the Eurasian region, the Mediterranean region, the Middle East, the North American region, and the Pacific region; Albania, Algeria, Australia (Victoria), Egypt,

France, Greece, Israel, Italy, Lebanon, Malta, Morocco, Spain, Tunisia, Turkey, the USA (California).

**Experimental host range**

Several (3–9) families susceptible. Experimentally infected plants mostly show systemic mottling, mosaic, necrosis.

Diagnostically susceptible

*Gomphrena globosa* — red-ring local lesions.
*Datura stramonium* — systemic mottle.
*Chenopodium quinoa, C. amaranticolor* — necrotic local lesions.
*Nicotiana benthamiana, N. clevelandii* — severe systemic mosaic.

Diagnostically insusceptible

*Brassica chinensis, Beta vulgaris, Capsicum annuum, Lycopersicon esculentum, Pisum sativum.*

Maintenance and propagation hosts

*Nicotiana benthamiana, N. clevelandii.*

Assay hosts (**L**ocal lesions or **W**hole plants)

*Chenopodium quinoa* (L), *C. amaranticolor* (L), *Gomphrena globosa* (L), *Nicotiana benthamiana* (W), *N. clevelandii* (W).

Dicotyledonous Hosts *(**Susceptible** or Insusceptible)*

*Beta vulgaris*. *Brassica campestris* ssp. *chinensis*. *Brassica campestris* ssp. *napus*. *Brassica campestris* ssp. *rapa*. ***Chenopodium album***. ***Chenopodium amaranticolor***. ***Chenopodium ambrosioides***. ***Chenopodium murale***. ***Chenopodium quinoa***. *Cucumis sativus*. ***Datura stramonium***. *Daucus carota*. ***Gomphrena globosa***. ***Helianthus annuus***. *Lactuca sativa*. ***Lens esculenta***. *Lupinus albus*. *Lycopersicon esculentum*. ***Nicotiana bigelovii***. ***Nicotiana clevelandii***. ***Petunia hybrida***. ***Phaseolus vulgaris***. *Solanum melongena*. *Tropaeolum majus*. *Vigna unguiculata*. *Zinnia elegans*.

**Properties of particles in sap**

TIP: 60–70 °C. LIV: 3 days. DEP: $\log_{10}$ minus 4–5. Leaf sap contains few virions.

**Purification**

Method: Thompson *et al.* (1988).

**Particle morphology**

Virions filamentous; not enveloped; usually flexuous; with a clear modal length; of 725–746 nm; 12 nm wide. Axial canal obscure. Basic helix obscure.

**Physical properties**

One sedimenting component in purified preparations; sedimentation coefficient 145 *S*. Density 1.31 g $cm^{-3}$ in CsCl. $A_{260}/A_{280}$ ratio 1.2.

**Biochemical properties**

Virions contain 6 % nucleic acid.

*Genome* consists of RNA; single–stranded; linear; unipartite; largest (or only) genome part 9 kb. Genomic nucleic acid isolated by Gallitelli: phenol-SDS. 5' terminus of RNA has possible a VPg. Infectivity lost when deproteinised with proteases; retained when deproteinised with phenol or detergent. Poly A region present. Additional factor not required for infectivity.

*Non–genomic nucleic acid* not found in the virions.

*Virion protein(s)* one; $M_r$ 33000; coat protein. Method of preparation: Laemmli (1970).

**Replication**

Replication does not depend on a helper virus.

**Cytopathology**

Virions found in leaves, mesophyll, epidermis, meristems, vascular parenchyma and companion cells; in cytoplasm. Inclusions present in infected cells; are pinwheels.

**Taxonomy and Relationships**

*Potyvirus: Potyviridae.*

Virus(es) with serologically unrelated virions: bean common mosaic, bean yellow mosaic, beet mosaic, clover yellow vein, dasheen mosaic, lettuce mosaic, papaya ringspot, potato Y, soybean mosaic, turnip mosaic, watermelon mosaic 2 and zucchini yellow fleck viruses.

Best tests for diagnosis: Serology (ISEM), molecular probes and cytopathology.

**Comments**

This virus is frequently found in mixed infections with other viruses with filamentous virions.

**References**

Costa, A.S., Duffus, J.E., Morton, D.J., Yarwood, C.E. and Bardin, R. (1959). *Phytopathology* **49**: 49.

Laemmli, U.K. (1970). *Nature, Lond.* **227**: 680.

Rana, G.L., Russo, M., Gallitelli, D. and Martelli, G.P. (1982). *Ann. appl. Biol.* **101**: 279.

Schwarz (1977). *B.Sc. Thesis*, La Trobe University, Melbourne, Australia.

Thompson, S., Frazier, S.S. and Barnden, K.L. (1988). *J. virol. Meth.* **20**: 57.

# Artichoke latent S (?) *carlavirus*

**Data collated by D. Gallitelli, 1991.**

*Standard acronym:* ArLVS.

First reported in *Cynara scolymus*; from northern Italy from *in vitro* meristem culture; by Rana *et al.* (1989).

**Natural host range and symptoms**

Symptoms none.

*Cynara scolymus* — symptomless.

**Transmission**

Not transmitted by *Myzus persicae*. Virus transmitted by mechanical inoculation.

**Geographical distribution**

Found, but with no evidence of spread, in northern Italy.

**Experimental host range**

Few (<3) families susceptible. Experimentally infected plants mostly show chlorotic to necrotic local lesions and systemic leaf chlorosis.

Diagnostically susceptible

*Chenopodium amaranticolor, C. quinoa* — chlorotic or necrotic local lesions.

*Nicotiana benthamiana* — systemic leaf chlorosis, leaf rolling.
*Zinnia elegans* — local and systemic mottling.

Diagnostically insusceptible

*Cucumis sativus, Gomphrena globosa, Nicotiana clevelandii, N. megalosiphon, Cynara scolymus* cv. Molese.

Maintenance and propagation hosts

*Nicotiana benthamiana.*

Assay hosts (**L**ocal lesions or **W**hole plants)

*Chenopodium amaranticolor* (L), *C. quinoa* (L), *Nicotiana benthamiana* (W).

Dicotyledonous Hosts *(**Susceptible** or Insusceptible)*

*Celosia cristata.* ***Chenopodium amaranticolor***. ***Chenopodium quinoa***. *Cucumis sativus. Gomphrena globosa.* ***Nicotiana benthamiana***. *Nicotiana clevelandii. Nicotiana megalosiphon. Phaseolus vulgaris.*

Additional hosts

*Cynara scolymus* cv. Molese.

Sources of host-range data

Rana *et al.* (1989).

**Purification**

Method: Rana *et al.* (1989); Gallitelli and Di Franco (1987).

**Particle morphology**

Virions filamentous; usually flexuous; with a clear modal length; of 664 nm; 12 nm wide. Axial canal obscure. Basic helix obvious.

**Physical properties**

One sedimenting component in purified preparations; sedimentation coefficient 150 *S*.

**Biochemical properties**

*Genome* consists of RNA; single–stranded; linear; unipartite; largest (or only) genome part 6.6 kb.

*Non–genomic nucleic acid* not found in the virions.

*Virion protein(s)* one; $M_r$ 29000; coat protein. Method of preparation: Laemmli (1970).

**Cytopathology**

Virions found in leaves and mesophyll; in cytoplasm, in chloroplasts, and in mitochondria. Inclusions present in infected cells; are crystals in the cytoplasm and membranous bodies; they contain virions.

**Taxonomy and Relationships**

Possible *Carlavirus.*

Virus(es) with serologically related virions: poplar mosaic virus. Virus(es) with serologically unrelated virions: potato S, potato M, carnation latent, elderberry, helleborus, lily symptomless, pea streak, shallot latent, caper latent, nerine latent and chrysanthemum B viruses.

Best tests for diagnosis: serology.

**References**

Gallitelli, D. and Di Franco, A. (1987). *J. Phytopath.* **119**: 97.

Laemmli, U.K. (1970). *Nature, Lond.* **227**: 680.
Rana, G.L., Piazzolla, P., Lafortezza, R. and Greco, N. (1989). *J. Phytopath.* **125**: 289.

# Artichoke mottled crinkle *tombusvirus*

**Data collated by D. Gallitelli, 1987. Revised 1991.**

*Standard acronym:* AMCV.

First reported in *Cynara scolymus*; from Italy; by Russo *et al.* (1967, 1968).

**Natural host range and symptoms**

Symptoms persist and vary seasonally.

*Cynara scolymus* — chlorotic blotching and puckering, flower malformation and colour breaking; plants stunted.

**Transmission**

Virus transmitted by mechanical inoculation.

**Geographical distribution**

Spreads in Italy (in the south).

**Experimental host range**

Several (3–9) families susceptible. Experimentally infected plants mostly show necrotic local lesions, systemic mosaic, necrosis and death.

Diagnostically susceptible

*Chenopodium quinoa, Ocimum basilicum* — necrotic local lesions.
*Nicotiana clevelandii, N. benthamiana* — systemic mosaic, necrosis, death.
*Gomphrena globosa* — red-ringed local lesions.

Diagnostically insusceptible

*Nicotiana tabacum* cvs Samsun, Xanthi, *N. rustica*.

Maintenance and propagation hosts

*Nicotiana clevelandii, N. benthamiana.*

Assay hosts (**L**ocal lesions or **W**hole plants)

*Chenopodium quinoa* (L), *Gomphrena globosa* (L), *Nicotiana clevelandii* (W), *Nicotiana benthamiana* (W).

Dicotyledonous Hosts (***Susceptible*** *or Insusceptible)*

***Chenopodium amaranticolor*. *Chenopodium quinoa*. *Cucumis sativus*. *Cucurbita pepo*. *Gomphrena globosa*. *Nicotiana benthamiana*. *Nicotiana clevelandii*. *Nicotiana glutinosa*.** *Nicotiana tabacum*. *Phaseolus vulgaris*. ***Vigna unguiculata*.**

**Properties of particles in sap**

TIP: 85–90 °C. LIV: 84 days. DEP: $\log_{10}$ minus 6. Leaf sap contains many virions.

**Purification**

Method: Gallitelli *et al.* (1985).

**Particle morphology**

Virions isometric; not enveloped; 30 nm in diameter; rounded in profile; with a conspicuous capsomere arrangement.

**Physical properties**

One sedimenting component in purified preparations; sedimentation coefficient 132 *S*. Density 1.35 g $cm^{-3}$ in CsCl.

**Biochemical properties**

Virions contain 18 % nucleic acid; 82 % protein; 0 % lipid.

*Genome* consists of RNA; single–stranded; linear; unipartite; largest (or only) genome part 4.7 kb. Genomic nucleic acid isolated by Gallitelli *et al.* (1985); SDS. Infectivity retained when deproteinised with proteases; retained when deproteinised with phenol or detergent. Poly A region absent. Additional factor not required for infectivity.

*Sequence database accession code(s):*

X16060 Em(40)_vi:TOAMCVCP Gb(84)_vi:TOAMCVCP Artichoke mottled crinkle virus (AMCV) genomic RNA for coat protein. 9/93 1,718bp.

X51456 Em(40)_vi:CRTRNA Gb(84)_vi:CRTRNA Artichoke mottled crinkle virus genomic RNA 3' terminal region. 9/93 2,200bp.

X62493 Em(40)_vi:AMCVCG Gb(84)_vi:AMCVCG Artichoke Mottled Crinkle Virus complete genome. 5/92 4,789bp. 3 sequences.

*Features of the genome:* in the 3'-terminal region of the genome (*c.* 2.2 kb) there are 3 ORF's and a non coding region of 349 nucleotides. $ORF_1$ 41kDaprotein, nucleotides 79–1242. $ORF_2$ 22kDaprotein, nucleotides 1283–1849. $ORF_3$ 19kDaprotein, nucleotides 1315–1830.

*Non–genomic nucleic acid* found in the virions; is subgenomic *m*RNA and satellite RNA. Sub–genomic *m*RNA found in infected cells; about 2.1 kb and 1 kb.

*Virion protein(s)* one; $M_r$ 41049; coat protein. Method of preparation: Laemmli (1970).

**Replication**

Genome replicates in cytoplasm. Replication does not depend on a helper virus. Acts as helper for a satellite RNA (0.7 kb).

**Cytopathology**

Virions found in leaves, roots, mesophyll, epidermis, meristems, vascular parenchyma, xylem, phloem and companion cells; in cytoplasm, in nuclei, and in cell vacuoles. Inclusions present in infected cells; are crystals in the cytoplasm and crystals in the nucleus; they contain virions.

**Taxonomy and Relationships**

*Tombusvirus: Tombusviridae.*

Virus(es) with serologically related virions: several tombusviruses; see Koenig and Gibbs, and Martelli *et al.* (1988 and 1989).

**References**


Gallitelli, D., Hull, R. and Koenig, R. (1985). *J. gen. Virol.* **66**: 1523.

Gallitelli, D. and Hull, R. (1985). *J. gen. Virol.* **66**: 1533.

Grieco, F. and Gallitelli, D. (1990). *Nucl. Acids Res.* **18**: 1300.

Koenig, R. and Gibbs, A.J. (1986). *J. gen. Virol.* **67**: 75.

Laemmli, U.K. (1970). *Nature, Lond.* **227**: 680.

Martelli, G.P., Gallitelli, D. and Russo, M. (1988). In: *The Plant Viruses*, Vol. 3, Polyhedral Virions with Monopartite RNA Genomes, p.13.; ed. R. Koenig. Plenum Press, New York.

Martelli, G.P., Russo, M. and Gallitelli, D. (1989). *AAB Descr. Pl. Viruses* No. 352, 8pp.

Quacquarelli, A., Martelli, G.P. and Russo, M. (1966). *Phytopathol. Medit.* **5**: 164.

Russo, M., Martelli, G.P. and Quacquarelli, A. (1967). *Virology* **33**: 555.
Russo, M., Martelli, G.P. and Quacquarelli, A. (1968). *Virology* **34**: 679.

# Artichoke vein banding (?) *nepovirus*

**Data collated by D. Gallitelli, 1987.**

*Standard acronym:* AVBV.

First reported in *Cynara scolymus*; from Apulia, Italy; by Gallitelli *et al.* (1978).

**Natural host range and symptoms**

Symptoms vary seasonally.

*Cynara scolymus* cvs Bayrombasa, Sakiz, Mazzaferrata — vein chlorosis.

**Transmission**

Virus transmitted by mechanical inoculation.

**Geographical distribution**

Found, but with no evidence of spread, in Europe (Apulia, Italy.

**Experimental host range**

Several (3–9) families susceptible. Experimentally infected plants mostly show local lesions, severe systemic mosaic.

Diagnostically susceptible

*Chenopodium amaranticolor, C. quinoa* — small chlorotic or necrotic local lesions, severe mosaic and tip necrosis.

*Phaseolus vulgaris* — pinpoint necrotic local lesions, mosaic, malformation of trifoliate leaves.

Maintenance and propagation hosts

*Chenopodium quinoa, Phaseolus vulgaris.*

Assay hosts (**L**ocal lesions or **W**hole plants)

*Ocimum basilicum* (L), *Chenopodium quinoa* (W), *Phaseolus vulgaris* (W).

Dicotyledonous Hosts *(**Susceptible** or Insusceptible)*

*Amaranthus caudatus. Antirrhinum majus. Apium graveolens. Beta vulgaris. Brassica campestris* ssp. *pekinensis. Brassica oleracea* ssp. *botrytis. Calendula officinalis. Carthamus tinctorius. Celosia argentea. Celosia cristata.* ***Chenopodium amaranticolor. Chenopodium ambrosioides. Chenopodium murale. Chenopodium quinoa. Cucumis sativus. Cucurbita pepo.*** *Datura metel. Datura stramonium. Dianthus barbatus. Dolichos biflorus.* ***Gomphrena globosa.*** *Helianthus annuus. Lactuca sativa. Lathyrus odoratus. Lupinus albus. Lycopersicon esculentum.* ***Nicotiana benthamiana. Nicotiana clevelandii. Nicotiana glutinosa. Nicotiana rustica.*** *Nicotiana tabacum.* ***Ocimum basilicum. Petunia hybrida. Phaseolus vulgaris.*** *Physalis floridana. Pisum sativum. Saponaria vaccaria. Sinapis alba. Solanum melongena. Spinacia oleracea. Tetragonia expansa. Tropaeolum majus.* ***Vicia faba. Vigna radiata. Vigna unguiculata. Zinnia elegans.***

Sources of host-range data

Gallitelli *et al.* (1978).

**Properties of particles in sap**

TIP: 55 °C. LIV: 3 days. DEP: $\log_{10}$ minus 5. Leaf sap contains many virions.

**Purification**

Method: homogenize infected leaves in 100 mM neutral phosphate buffer. Clarify with 10% Mg-activated bentonite. Differential centrifugation.

**Particle morphology**

Virions isometric; not enveloped; 30 nm in diameter; angular in profile; without a conspicuous capsomere arrangement.

**Physical properties**

Three sedimenting components in purified preparations; sedimentation coefficient of the fastest 124 *S* (B); of the other(s) 92 *S* (M), or 56 *S* (T). Density virions are unstable even after fixation with formaldehyde.

**Biochemical properties**

Virions contain 37 % nucleic acid (B), or 24 % nucleic acid (M), or 0 % nucleic acid (T); 63 % protein (B), or 76 % protein (M), or 100 % protein (T).

*Genome* consists of RNA; single–stranded; linear; of two parts; largest (or only) genome part the larger 7.8 kb; the 2nd largest 4.5 kb. Genomic nucleic acid isolated by : using phenol-SDS. Infectivity retained when deproteinised with phenol or detergent.

*Non–genomic nucleic acid* not found in the virions.

*Virion protein(s)* four; $M_r$ of the largest 27000. $M_r$ of 2nd largest 24000. $M_r$ of 3rd largest 22000. $M_r$ of 4th largest 14300. Method of preparation: Laemmli (1970); Chu and Francki (1979).

**Cytopathology**

Virions found in mesophyll; in cytoplasm. Inclusions present in infected cells; are membranous bodies and unusual in shape; vesicles containing finely stranded material resembling nucleic acid; they do not contain virions.

**Taxonomy and Relationships**

Possible *Nepovirus: Comoviridae.*

Virus(es) with serologically unrelated virions: arabis mosaic, artichoke Italian latent, artichoke yellow ringspot, cherry leaf roll, cherry rasp leaf, chicory yellow mottle, cacao necrosis, grapevine Bulgarian latent, grapevine chrome mosaic, grapevine fanleaf, myrobalan latent ringspot, olive latent ringspot, peach rosette, raspberry ringspot, strawberry latent ringspot, tobacco ringspot and tomato ringspot.

Best tests for diagnosis: use ISEM. In nature artichoke vein banding virus may be mixed with artichoke latent potyvirus which has filamentous virions or with artichoke mottled crinkle or artichoke Italian latent viruses which are serologically distinct.

**References**


Chu, P.W.G. and Francki, R.I.B. (1979). *Virology* **93**: 398.

Gallitelli, D., Martelli, G.P. and Rana, G.L. (1984). *CMI/AAB Descr. Pl. Viruses* No. 285, 4 pp.

Gallitelli, D., Rana, G.L. and Di Franco, A. (1978). *Phytopathol. Medit.* **17**: 1.

Laemmli, U.K. (1970). *Nature, Lond.* **277**: 680.

# Artichoke yellow ringspot *nepovirus*

**Data collated by D. Gallitelli, 1987.**

*Standard acronym:* AYRSV.

First reported in *Cynara cardunculus* (cardoon) and *Cynara scolymus* (artichoke); from Greece (north-east Poleponnesus) and Italy (Sicily); by Kyriakopoulou and Bem (1973); Rana *et al.* (1978).

**Natural host range and symptoms**

Symptoms persist.

*Cynara scolymus, C. cardunculus, Nicotiana tabacum* — chlorotic blotches, rings and lines.

*Anethum graveolens, Foeniculum* sp., *Nicotiana glauca, Phaseolus vulgaris, Reseda alba, Stellaria media, Vicia faba* — chlorotic mottled and malformed leaves, plants stunted.

**Transmission**

Virus transmitted by mechanical inoculation; transmitted by seed; transmitted by pollen to the seed (in *Datura stramonium, Nicotiana clevelandii, N. glutinosa, N. tabacum, Petunia hybrida*), or transmitted by pollen to the pollinated plant (in *Nicotiana clevelandii*).

**Geographical distribution**

Spreads in the Eurasian region; Greece and Italy.

**Experimental host range**

Many (>9) families susceptible.

Diagnostically susceptible

*Gomphrena globosa* — reddish local lesions; systemic yellow ring and line patterns.

*Chenopodium quinoa* — chlorotic/necrotic local lesions; systemic mottling and tip necrosis.

*Phaseolus vulgaris* — reddish necrotic local lesions; systemic mosaic and leaf deformation.

*Cucumis sativus* — chlorotic local lesions in cotyledons then mosaic, malformation and enations.

*Nicotiana tabacum* — local and systemic chlorotic rings, line and oakleaf patterns.

Maintenance and propagation hosts

*Gomphrena globosa, Phaseolus vulgaris, Chenopodium quinoa, Cucumis sativus.*

Assay hosts (**L**ocal lesions or **W**hole plants)

*Chenopodium quinoa* (L); *Cucumis sativus* (L); *Canavalia ensiformis* (L).

Dicotyledonous Hosts *(**Susceptible** or Insusceptible)*

***Chenopodium quinoa. Cucumis sativus. Gomphrena globosa. Nicotiana clevelandii. Nicotiana glutinosa. Nicotiana rustica. Nicotiana tabacum. Petunia hybrida. Phaseolus vulgaris. Stellaria media. Vicia faba.***

Sources of host-range data

Kyriakopoulou and Bem (1973); Rana *et al.* (1978; 1980).

**Properties of particles in sap**

TIP: 60 °C. LIV: 2 days. DEP: $\log_{10}$ minus 5. Leaf sap contains few virions.

**Purification**

Method: Rana *et al.* (1980).

**Particle morphology**

Virions isometric; not enveloped; 30 nm in diameter; angular in profile.

**Physical properties**

Two sedimenting components in purified preparations (T and B). Density 1.2–1.3 g $cm^{-3}$ in CsCl (T), or 1.4–1.53 g $cm^{-3}$ in CsCl (B).

**Biochemical properties**

*Genome* consists of RNA; single–stranded; of two parts; largest (or only) genome part the larger 6.3 kb (RNA-1); the 2nd largest 5.3 kb (RNA-2). Genomic nucleic acid isolated by Reijnders *et al.* (1974).

*Non–genomic nucleic acid* not found in the virions.

*Virion protein(s)* one; $M_r$ 53000.

**Cytopathology**

Virions found in leaves, roots, mesophyll and epidermis; in cytoplasm (stacked layers or between membranes or within tubular structures), or in nuclei (aggregates of empty virions). Inclusions present in infected cells; are unusual in shape; tangled membranes and vesicles containing finely stranded material resembling nucleic acid; they contain virions. Other cellular changes: cell wall abnormalities (callose deposition and wall outgrowths) and paramural bodies; dictyosomes are also affected and may take part in some steps of virus multiplication.

**Taxonomy and Relationships**

*Nepovirus: Comoviridae.*

Virus(es) with serologically unrelated virions: 13 other known members of the genus.

$M_r$ of its RNA-2 suggests artichoke yellow ringspot virus is in the cherry leaf roll subgroup of nepoviruses (Martelli *et al.*, 1978). Best tests for diagnosis: the symptoms shown by infected artichoke and cardoon plants are distinctive and unlike those caused by other viruses (Martelli *et al.*, 1981).

**References**


Kyriakopoulou, P.E. and Bem, F. (1973). *Symp. Agric. Res., Athens, 1973.* p.64.

Martelli, G.P., Quacquarelli, A., Gallitelli, D., Savino, V. and Piazzolla, P. (1978). *Phytopathol. Medit.* **17**: 145.

Martelli, G.P., Russo, M. and Rana, G.L. (1981). *Atti 3° Congr. Int. Carciofo, Bari, 1979*, 895.

Rana, G.L., Kyriakopoulou, P.E. and Martelli, G.P. (1985). *CMI/AAB Descr. Pl. Viruses* No. 271, 4 pp.

Rana, G.L., Rosciglione, B. and Cannizzaro, G. (1978). *Phytopathol. Medit.* **17**: 63.

Rana, G.L., Gallitelli, D., Kyriakopoulou, P., Russo, M. and Martelli, G.P. (1980). *Ann. appl. Biol.* **96**: 177.

Reijnders, L., Aalbers, A.M.J., van Kammen, A. and Thuring, R.W.J. (1974). *Virology* **60**: 515.

# Asclepias (?) rhabdovirus

**Data collated by A.A. Brunt, 1989.**

First reported in *Asclepias curassavica*; from Ecuador; by Dollet and Gargani, 1989.

**Natural host range and symptoms**

*Asclepias curassavica* — diffuse leaf chlorosis and mosaic.

**Geographical distribution**

Spreads in the South and Central American region; Ecuador.

**Particle morphology**

Virions rhabdo– or bullet–shaped; enveloped; of 300 nm; 50 nm wide.

**Taxonomy and Relationships**

Unassigned rhabdovirus.

**Comments**

This rhabdovirus sometimes occurs together with a flagellate protozoan (*Phytomonas* sp.).

**References**

Dollet, M. and Gargani, D. (1989). *J. Phytopath.* **125**: 269.

# Asparagus 1 *potyvirus*

**Data collated by A. Bertaccini and W.E. Howell, 1991.**

*Synonyms:* asparagus virus B. *Standard acronym:* AV-1.

First reported in *Asparagus officinalis*; from Germany; by Hein (1960).

**Natural host range and symptoms**

Symptoms none.
*Asparagus officinalis*—symptomless.

**Transmission**

Transmitted by a vector; an insect; *Aphis craccivora, Myzus persicae*; Aphididae. Principal natural vector(s): *Myzus persicae*. Not transmitted by *Aphis gossypii, Macrosiphum euphorbiae*. Transmitted in a non–persistent manner. Virus transmitted by mechanical inoculation; not transmitted by contact between plants; not transmitted by seed; not transmitted by pollen.

**Ecology and control**

Studies reported by Yang and Clore (1976); Howell and Mink (1985); Bertaccini *et al.* (1990).

**Geographical distribution**

Probably distributed worldwide (wherever asparagus is grown commercially). Spreads in the Eastern Asian region, the Eurasian region, and the North American region; Germany, Italy, Japan, the UK, and the USA.

**Experimental host range**

Several (3–9) families susceptible. Experimentally infected plants mostly show necrotic local lesions.

Diagnostically susceptible

*Chenopodium album, C. amaranticolor, C. quinoa, Tetragonia expansa* — chlorotic or necrotic local lesions.

Diagnostically insusceptible

*Beta vulgaris, Cucumis sativus, Glycine max, Nicotiana tabacum, Phaseolus vulgaris, Vigna unguiculata.*

Maintenance and propagation hosts

*Asparagus officinalis, Chenopodium amaranticolor, C. quinoa, Tetragonia expansa.*

Assay hosts (**L**ocal lesions or **W**hole plants)

*Chenopodium amaranticolor* (L), *Chenopodium quinoa* (L), *Tetragonia expansa* (L).

Dicotyledonous Hosts *(**Susceptible** or Insusceptible)*

*Amaranthus retroflexus. Apium graveolens. Beta vulgaris. Brassica campestris* ssp. *rapa. Capsicum frutescens. Celosia cristata.* ***Chenopodium amaranticolor. Chenopodium capitatum****. Chenopodium murale.* ***Chenopodium quinoa****. Cucumis sativus. Cucurbita pepo. Datura stramonium. Daucus carota. Dianthus caryophyllus. Glycine max.* ***Gomphrena globosa***, or *Gomphrena globosa. Lactuca sativa. Lycopersicon esculentum. Nicotiana benthamiana. Nicotiana clevelandii. Nicotiana glutinosa. Nicotiana sylvestris. Nicotiana tabacum. Ocimum basilicum. Petunia hybrida. Phaseolus vulgaris. Pisum sativum. Sesamum indicum. Solanum tuberosum.* ***Tetragonia expansa****. Vicia faba. Vigna unguiculata. Vigna unguiculata* ssp. *sesquipedalis. Zinnia elegans.*

Monocotyledonous Hosts *(**Susceptible** or Insusceptible)*

*Allium cepa. Allium fistulosum. Zea mays.*

Additional hosts

*Asparagus officinalis* (**S**); *Allium tuberosum* (**S**); *Lilium elegans* (**I**).

Sources of host-range data

Hein (1960); Mink and Uyeda (1977); Fujisawa *et al.* (1983).

**Properties of particles in sap**

TIP: 50–55 °C. LIV: 2–11 days. DEP: $\log_{10}$ minus 3–4. Leaf sap contains few virions.

**Purification**

Method: Fujisawa *et al.* (1983); Howell and Mink (1985).

**Particle morphology**

Virions filamentous; not enveloped; usually flexuous; with a clear modal length; of 740 nm; 13 nm wide. Axial canal obscure. Basic helix obscure.

**Physical properties**

One sedimenting component in purified preparations; sedimentation coefficient 146 *S*. $A_{260}/A_{280}$ ratio 1.24.

**Biochemical properties**

Virions contain 6 % nucleic acid; 94 % protein.
*Genome* consists of RNA; single–stranded.

**Replication**

Replication does not depend on a helper virus.

### Cytopathology

Virions found in leaves, stems, roots and mesophyll; in cytoplasm. Inclusions present in infected cells; are pinwheels; they do not contain virions.

### Taxonomy and Relationships

*Potyvirus: Potyviridae.*

Virus(es) with serologically related virions: bean yellow mosaic, lettuce mosaic and turnip mosaic viruses. Virus(es) with serologically unrelated virions: beet mosaic, iris mild mosaic and potato Y viruses.

Best tests for diagnosis: asparagus 1 virus occurs naturally in asparagus plants that often are infected with tobacco streak, asparagus 2 or cucumber mosaic viruses. Few, if any, symptoms are caused. Asparagus 1 virus is readily separated from the others because its host range is limited and it causes only necrotic local lesions in *Chenopodium quinoa* and no symptoms in *Cucumis sativus, Phaseolus vulgaris* or *Nicotiana tabacum.*

### Comments

Asparagus infected with asparagus 1 and 2 viruses are significantly more susceptible to damage caused by *Fusarium oxysporum* f. sp. *asparagi.*

### References

Bertaccini, A., Marani, F. and Passarelli, V. (1984). *Atti Giornate Fitopatol., 1984, Sorrento*, Vol. 3, p. 437.
Bertaccini, A., Giuncheoli, L. and Poggi Pollini, C. (1990). *Acta Hort.* **271**: 279.
Evans, T.A. and Stephens, C.T. (1989). *Phytopathology* **79**: 253.
Fujisawa, I., Goto, T., Tsuchizaki, T. and Chizuka, N. (1983). *Ann. Phytopath. Soc. Japan* **49**: 299.
Gröschel, H. and Jan-Ladwig, R. (1977). *Phytopath. Z.* **88**: 183.
Hein, A. (1960). *Phytopath. Z.* **67**: 217.
Hein, A. (1969). *Z. PflKrankh. PflPath. PflSchutz.* **76**: 395.
Howell, W.E. and Mink, G.I. (1985). *Plant Dis.* **69**: 1044.
Mink, G.I. and Uyeda, I. (1977). *Pl. Dis. Reptr* **61**: 398.
Montasser, M.S. and Davis, R.F. (1987). *Plant Dis.* **71**: 497.
Yang, H.J. (1979). *Hort. Sci.* **14**: 734.
Yang, H.J. and Clore, W.J. (1976). *Hort. Sci.* **12**: 140.

## Asparagus 2 *ilarvirus*

**Data collated by A. Bertaccini, 1991.**

*Synonyms:* asparagus virus C, asparagus latent virus. *Standard acronym:* AV-2.

First reported in *Asparagus officinalis*; from Denmark and Germany; by Paludan (1964) and Groschel (1976) respectively.

### Natural host range and symptoms

Symptoms persist.

*Asparagus officinalis* — stunting, decline.

### Transmission

Not transmitted by *Myzus persicae*. Virus transmitted by mechanical inoculation; transmitted by grafting; transmitted by seed; transmitted by pollen to the seed and transmitted by pollen to the pollinated plant.

**Ecology and control**

Studies reported by Bertaccini *et al.* (1990); Yang and Clore (1976).

**Geographical distribution**

Spreads in the Eurasian region and the North American region. Found, but with no evidence of spread, in Europe and North America; Denmark, Germany, Italy, the UK and the USA.

**Experimental host range**

Several (3–9) families susceptible. Experimentally infected plants mostly show necrotic local lesions and rings, vein chlorosis and mottling.

Diagnostically susceptible

*Chenopodium amaranticolor, C. quinoa* — local necrotic spots, systemic necrosis.
*Gomphrena globosa* — white spots with red borders.
*Nicotiana tabacum* cv. Samsun — local necrotic rings and systemic vein yellowing and necrotic lines.
*Nicotiana tabacum* cv. White Burley — local necrotic rings.

Diagnostically insusceptible

*Cucumis melo, Datura stramonium, Lactuca sativa.*

Maintenance and propagation hosts

*Chenopodium amaranticolor* (L), *C. quinoa* (L), *Nicotiana tabacum* cv. Samsun (W).

Dicotyledonous Hosts *(**Susceptible** or Insusceptible)*

***Beta vulgaris*. *Chenopodium amaranticolor*. *Chenopodium murale*. *Chenopodium quinoa*.** *Cucumis melo*. ***Cucumis sativus*.** *Datura stramonium*. ***Glycine max*. *Gomphrena globosa*.** *Lactuca sativa*. ***Nicotiana glutinosa*. *Nicotiana glutinosa* × *clevelandii*. *Nicotiana tabacum*. *Ocimum basilicum*. *Petunia hybrida*. *Phaseolus vulgaris*. *Vigna unguiculata*.**

Sources of host-range data

Giunchedi *et al.* (1987); Uyeda and Mink (1981).

**Properties of particles in sap**

TIP: 45 °C. LIV: 2 days. DEP: $\log_{10}$ minus 3.

**Purification**

Method: Giunchedi *et al.* (1987); Phillips and Brunt (1985); Uyeda and Mink (1981).

**Particle morphology**

Virions isometric; not enveloped; neither clearly angular or nor rounded in profile; without a conspicuous capsomere arrangement.

**Physical properties**

Three sedimenting components in purified preparations; sedimentation coefficient of the fastest 90 *S* (±2); of the other(s) 104 *S* (±2), or 95 *S*. $A_{260}/A_{280}$ ratio 1.31–1.36.

*Sequence database accession code(s):*

X86352 Em(43)_vi:Aviirna3 Gb(89)_vi:Aviirna3 Asparagus virus II RNA3 ORF1 & ORF2. 5/95 2,307bp.

**Cytopathology**

Inclusions absent from infected cells.

**Taxonomy and Relationships**

*Ilarvirus: Bromoviridae* (subgroup 2).

Virus(es) with serologically related virions: citrus leaf rugose, citrus variegation, elm mottle and Tulare apple mosaic viruses. Virus(es) with serologically unrelated virions: tobacco streak viruses.

**References**

Bertaccini, A., Marani, F. and Passarelli, V. (1984). *Atti giornate Fitopatol. 1984, Sorrento*, Vol. 30, p. 437.
Bertaccini, A., Macri, S., Poggi Pollini, C. and Martini, L. (1988). *Inform. Fitopatol.* **38**: 39.
Bertaccini, A., Giunchedi, L. and Poggi Pollini, C. (1990). *Acta Hort.* **271**: 279.
Giunchedi, L., Poggi Pollini, C., Bertaccini, A. and Marani, F. (1987). *Phytopath. Medit.* **26**: 117.
Groschel, H. (1976). *Ph.D. Thesis*, Tech. Univ. Hanover, pp.103.
Hartung, A.C., Evans, T.A. and Stephens, C.T. (1985). *Plant Dis.* **6**: 501.
Mink, G.I. and Uyeda, I. (1977). *Pl. Dis. Reptr* **61**: 398.
Paludan, N. (1964). *Maanedsovers. PlSygd.* **407**: 11.
Phillips, S. and Brunt, A.A. (1985). *Pl. Path.* **34**: 440.
Uyeda, I. and Mink, G.I. (1981). *Phytopathology* **71**: 1264.
Uyeda, I. and Mink, G.I. (1983). *Phytopathology* **73**: 47.
Uyeda, I. and Mink, G.I. (1984). *CMI/AAB Descr. Pl. Viruses* No.288, 4 pp.
Yang, H.J. and Clore, W.J. (1976). *Hort. Science* **12**: 140.

# Asparagus 3 *potexvirus*

**Data collated by I. Fujisawa, 1990.**

*Standard acronym:* AV-3.

First reported in *Asparagus* sp.; from Japan; by Fujisawa (1986).

**Natural host range and symptoms**

Symptoms vary seasonally.

*Asparagus* spp. — faint leaf chlorosis.

**Transmission**

Transmitted by means not involving a vector. Not transmitted by aphids. Virus transmitted by mechanical inoculation; not transmitted by seed.

**Geographical distribution**

Spreads in the Eastern Asian region; Japan.

**Experimental host range**

Several (3–9) families susceptible. Experimentally infected plants mostly show mosaic, mild mosaic and necrotic local lesions.

Diagnostically susceptible

*Chenopodium quinoa, Nicotiana clevelandii* — mosaic.

*Spinacia oleracea* — symptomless systemic infection.

*Nicotiana tabacum* cv. Bright Yellow, *Vicia faba* — necrotic local lesions; not systemic.

*Nicotiana debneyi, Glycine max* — mild mosaic.

Diagnostically insusceptible

*Allium cepa, A. fistulosum, Lycopersicon esculentum, Brassica rapa, Cucurbita pepo.*

Maintenance and propagation hosts

*Nicotiana debneyi, N. megalosiphon, N. clevelandii.*

Assay hosts (**L**ocal lesions or **W**hole plants)

*Chenopodium amaranticolor* (L), *Vicia faba* (L), *Sesamum indicum* (L), *Nicotiana benthamiana* (W).

Dicotyledonous Hosts *(**Susceptible** or Insusceptible)*

***Amaranthus retroflexus*. *Beta vulgaris*.** *Brassica campestris* ssp. *pekinensis*. *Brassica campestris* ssp. *rapa*. *Brassica oleracea* ssp. *capitata*. *Capsicum annuum*. ***Chenopodium amaranticolor*. *Chenopodium murale*. *Chenopodium quinoa*.** *Cucumis melo*. *Cucumis sativus*. *Cucurbita maxima*. *Cucurbita pepo*. ***Datura metel*. *Datura stramonium*.** *Dianthus caryophyllus*. ***Glycine max*. *Gomphrena globosa*.** *Lycopersicon esculentum*. ***Nicotiana glutinosa*. *Nicotiana megalosiphon*. *Nicotiana rustica*. *Nicotiana tabacum*.** *Petunia hybrida*. *Phaseolus vulgaris*. ***Pisum sativum*. *Sesamum indicum*. *Solanum tuberosum*. *Spinacia oleracea*. *Tetragonia expansa*. *Vicia faba*. *Vigna angularis*. *Vigna unguiculata*** ssp. ***sesquipedalis*. *Zinnia elegans*.**

Monocotyledonous Hosts *(**Susceptible** or Insusceptible)*

*Hordeum vulgare*. *Triticum aestivum*. *Zea mays*.

**Properties of particles in sap**

TIP: 60 °C. LIV: 23 days. DEP: $\log_{10}$ minus 4. Leaf sap contains many virions. Electron microscopy: use glutaraldehyde or osmic acid to fix.

**Purification**

Method: Fujisawa (1986).

**Particle morphology**

Virions filamentous; not enveloped; usually flexuous; with a clear modal length; of 580 nm; 13 nm wide. Axial canal obscure. Basic helix obscure.

**Physical properties**

One sedimenting component in purified preparations. $A_{260}/A_{280}$ ratio 1.18–1.19.

**Replication**

Replication does not depend on a helper virus.

**Cytopathology**

Virions found in leaves and mesophyll; in cytoplasm. Inclusions absent from infected cells.

**Taxonomy and Relationships**

*Potexvirus*.

Virus(es) with serologically related virions: cactus X and narcissus mosaic viruses. Virus(es) with serologically unrelated virions: white clover mosaic, clover yellow mosaic, potato X and tulip X viruses.

Best tests for diagnosis: bioassay on test plants: *Sesamum indicum, Vicia faba*.

**References**

Fujisawa, I. (1986). *Ann. Phytopath. Soc. Japan* **52**: 2.

Hull, R., Brown, F. and Payne, C. (1989). In: *Virology: Directory and Dictionary of Animal, Bacterial and Plant Viruses*. Macmillan Press, U.K.

# Aster chlorotic stunt (?) *carlavirus*

**Data collated by L.W. Stobbs, 1991.**

First reported in *Aster novae-angliae*; from Jordan Station, Ontario, Canada; by Stobbs and van Schagen (1991).

**Natural host range and symptoms**

Symptoms persist. Symptoms systemic chlorosis, stunting, chlorotic local lesions.

*Aster* spp. — systemic chlorosis and stunting, curling of apical leaves, often elongated with necrotic flecking. Rosetting caused by shortened internodes; irregular chlorotic local lesions, often with necrotic centres, on basal leaves.

**Transmission**

Transmitted by a vector; an insect; *Myzus persicae, Aphis gossypii*; Aphididae. Transmitted in a non–persistent manner. Virus transmitted by mechanical inoculation; not transmitted by seed.

**Geographical distribution**

Spreads in the North American region; Canada (southern Ontario).

**Experimental host range**

Few (<3) families susceptible. Experimentally infected plants mostly show systemic chlorosis, stunting, chlorotic local lesions.

Diagnostically susceptible

*Aster novae-angliae, A. novi-belgii* — systemic chlorosis, necrotic flecking; stunting.

*Chenopodium amaranticolor, C. quinoa* — chlorotic local lesions, vein-netting and chlorotic spotting.

Diagnostically insusceptible

*Helenium autumnale, Impatiens wallerana, Dianthus barbatus, D. caryophyllus.*

Maintenance and propagation hosts

*Aster* spp., *Chenopodium amaranticolor, C. quinoa.*

Assay hosts (**L**ocal lesions or **W**hole plants)

*Chenopodium amaranticolor* (W).

Dicotyledonous Hosts *(**Susceptible** or Insusceptible)*

***Chenopodium amaranticolor***. ***Chenopodium quinoa***. *Cucumis sativus. Dahlia pinnata. Datura stramonium. Dianthus barbatus. Dianthus caryophyllus. Gomphrena globosa. Lycopersicon esculentum. Nicotiana clevelandii. Nicotiana glutinosa. Nicotiana rustica. Nicotiana tabacum. Papaver nudicaule. Petunia hybrida. Phaseolus vulgaris. Primula malacoides. Salvia splendens. Vigna unguiculata. Zinnia elegans.*

Additional hosts

***Achillea filipendulina, A. ptarmica, Ajuga reptans, Althaea rosea, Aurinia saxatilis compactum, Begonia x tubohybrida, Callistephus chinensis, Chrysanthemum maximum, Coreopsis tinctoria, Gazania rigens, Gerbera jamesonii, Helenium autumnale, Helianthus maximilianii, Impatiens wallerana, Iris pumila, Liatris spicata, Myosotis scorpioides, Pelargonium x domesticum, Portulaca grandiflora, Saxifraga bergenia cordifolia*** and ***Tagetes erecta***.

Sources of host-range data

Stobbs and van Schagen (1991).

### Properties of particles in sap

TIP: 50–60 °C. LIV: 5 days (at 25°C). DEP: $\log_{10}$ minus 6–7. Leaf sap contains many virions.

### Purification

Method: Stobbs and van Schagen (1991).

### Particle morphology

Virions filamentous; not enveloped; usually flexuous; with a clear modal length; of 610–620 nm; 13 nm wide. Axial canal obvious. Basic helix obscure.

### Physical properties

One sedimenting component in purified preparations; sedimentation coefficient 160 *S*. Density 1.29 g $cm^{-3}$ in sucrose. $A_{260}/A_{280}$ ratio 1.38 (±0.05).

### Biochemical properties

*Genome* consists of RNA; single–stranded; linear; unipartite; largest (or only) genome part 8 kb. Infectivity retained when deproteinised with proteases; retained when deproteinised with phenol or detergent. Additional factor not required for infectivity.

*Virion protein(s)* one; $M_r$ 35400. Method of preparation: Stobbs and van Schagen (1991).

### Cytopathology

Virions found in all parts of the host plant; in cytoplasm.

### Taxonomy and Relationships

Possible *Carlavirus*.

### References

Stobbs, L.W. and van Schagen, J.G. (1991). *Plant Dis.* **75**: 421.

## Asystasia gangetica mottle (?) potyvirus

**Data collated by J.-C.Thouvenel, 1986. Revised 1989.**

*Standard acronym:* AGMoV.

First reported in *Asystasia gangetica*; from the Ivory coast; by Thouvenel *et al.* (1982).

### Natural host range and symptoms

Symptoms persist.

*Asystasia gangetica* — vein mottle, lines and ringspots.

### Transmission

Transmitted by a vector; an insect; *Aphis spiraecola*; Aphididae. Transmitted in a non–persistent manner. Virus transmitted by mechanical inoculation; transmitted by grafting; not transmitted by seed.

### Geographical distribution

Spreads in the African region; the Congo, Cote d'Ivoire, and Guinea.

### Experimental host range

Several (3–9) families susceptible. Experimentally infected plants mostly show local lesions, mosaics, ringspots.

Diagnostically susceptible

*Chenopodium amaranticolor, Tetragonia expansa* — necrotic local lesions; not systemic.

*Nicotiana benthamiana* — systemic mosaic, leaf curling.

*Nicotiana megalosiphon* — systemic mottling.

*Nicotiana tabacum* cvs Samsun NN and Xanthi-*nc* — necrotic ringspots and line pattern.

*Anthirrhinum majus, Phaseolus vulgaris* — systemic chlorosis.

*Sesamum indicum* — systemic chlorotic spotting.

Diagnostically insusceptible

*Cucurbita* spp.,*Cucumis* spp., *Nicotiana glutinosa, Vigna unguiculata.*

Maintenance and propagation hosts

*Nicotiana benthamiana.*

Assay hosts (**L**ocal lesions or **W**hole plants)

*Nicotiana benthamiana* (W), *Chenopodium amaranticolor* (L).

Dicotyledonous Hosts *(**Susceptible** or Insusceptible)*

***Antirrhinum majus***. *Arachis hypogaea*. ***Chenopodium amaranticolor***. *Cucumis melo*. *Cucumis sativus*. *Cucurbita maxima*. *Cucurbita moschata*. *Cucurbita pepo*. *Hibiscus esculentus*. *Momordica balsamina*. ***Nicotiana benthamiana***. ***Nicotiana clevelandii***. ***Nicotiana megalosiphon***. ***Nicotiana tabacum***. ***Phaseolus vulgaris***. ***Sesamum indicum***. ***Solanum nigrum***. ***Tetragonia expansa***. *Vigna unguiculata*. *Vigna unguiculata* ssp. *cylindrica*.

Monocotyledonous Hosts *(**Susceptible** or Insusceptible)*

*Zea mays.*

Sources of host-range data

Thouvenel *et al.* (1988).

**Properties of particles in sap**

TIP: 70–75 °C. LIV: 3 days. DEP: $\log_{10}$ minus 3–4. Infectivity of sap not changed by treatment with di–ethyl ether. Leaf sap contains few virions.

**Purification**

Method: Thouvenel *et al.* (1982).

**Particle morphology**

Virions filamentous; not enveloped; usually flexuous; with a clear modal length; of 750 nm; 13 nm wide. Axial canal obscure. Basic helix obscure.

**Physical properties**

One sedimenting component in purified preparations.

**Biochemical properties**

Virions contain 5 % nucleic acid; 95 % protein.

*Virion protein(s)* one; $M_r$ 33000; coat protein. Amino acid composition: Fauquet and Thouvenel (1987).

**Cytopathology**

Virions found in all parts of the host plant. Inclusions present in infected cells; are pinwheels.

**Taxonomy and Relationships**

Possible *Potyvirus: Potyviridae.*

Virus(es) with serologically unrelated virions: 30 other group members tested.

**References**

Fauquet, C. and Thouvenel, J.-C. (1987). In: *Plant Viral Diseases in the Ivory Coast*, pp. 228; Initiations, Documentations, Techniques. No.46. ORSTOM. Paris.

Thouvenel, J.-C. Fauquet, C. and Monsarrat, A. (1982). *C.r. Acad. Sci. Paris* **295**: 213.

Thouvenel, J.-C., Fauquet, C. and Fargette, D. (1988). *Ann. appl. Biol.* **112**: 127.

# Atropa belladonna (?) rhabdovirus

**Data collated by A.A. Brunt, 1991.**

*Standard acronym:* AtBV.

First reported in *Atropa belladonna*; from Germany; by Lesemann (1972).

**Natural host range and symptoms**

*Atropa belladonna* — symptoms unknown (found only in complex with other viruses).

**Geographical distribution**

Found, but with no evidence of spread, in Germany.

**Particle morphology**

Virions rhabdo– or bullet–shaped; with a clear modal length; of 110 nm; 55 nm wide.

**Taxonomy and Relationships**

Unassigned rhabdovirus.

Best tests for diagnosis: often found in plants also containing belladonna mottle tymovirus and henbane mosaic potyvirus but can be distinguished from these most quickly by electron microscopy.

**References**

Lesemann, D.-E. (1972). *Phytopath. Z.* **73**: 83.

# Aucuba ringspot (?) *badnavirus*

**Data collated by S. Yamashita, 1991.**

*Standard acronym:* AuBV.

First reported in *Aucuba japonica*; from Japan; by Kusunoki (1980).

**Natural host range and symptoms**

Symptoms persist.

*Aucuba japonica* — yellow ringspots and mosaic.

**Transmission**

Virus not transmitted by mechanical inoculation; transmitted by grafting; not transmitted by contact between plants; not transmitted by seed; not transmitted by pollen.

**Geographical distribution**

Found, but with no evidence of spread, in Japan and the U.K.

**Experimental host range**

Few (<3) families susceptible.

Diagnostically susceptible

*Aucuba japonica* — yellow spots and mosaic.

Dicotyledonous Hosts *(**Susceptible** or Insusceptible)*

*Beta vulgaris. Brassica campestris* ssp. *pekinensis. Brassica oleracea* ssp. *botrytis. Catharanthus roseus. Chenopodium amaranticolor. Chenopodium murale. Chenopodium quinoa. Cucumis melo. Cucumis sativus. Datura stramonium. Daucus carota. Glycine max. Gomphrena globosa. Lactuca sativa. Lycopersicon esculentum. Nicotiana glutinosa. Nicotiana rustica. Nicotiana tabacum. Petunia hybrida. Phaseolus vulgaris. Pisum sativum. Raphanus sativus. Solanum melongena. Spinacia oleracea. Tetragonia expansa. Trifolium repens. Vicia faba. Vigna unguiculata.*

Monocotyledonous Hosts *(**Susceptible** or Insusceptible)*

*Hordeum vulgare. Oryza sativa. Triticum aestivum. Zea mays.*

**Properties of particles in sap**

Leaf sap contains few virions.

**Particle morphology**

Virions bacilliform; not enveloped; *c.* 130 nm in length; 30 nm wide.

**Cytopathology**

Virions found in the leaves and all parts of the host plant; in cytoplasm. Inclusions absent from infected cells.

**Taxonomy and Relationships**

Possibly *Badnavirus.*

**References**

Kusunoki, M. (1980). *Ann. Phytopath. Soc. Japan* **46**: 414.

Yamashita, S. (1983). In: *Handbook of Plant Viruses in Japan*, p. 204; eds K. Yora, Y. Saito, Y. Doi, T Inouye and K. Tomaru. Asakura Shoten, Tokyo, Japan.

Yamashita, S., Doi, Y., Maeda, H. and Yora, K. (1982). *Ann. Phytopath. Soc. Japan* **48**: 395.

Yamashita, S., Natsuaki, T., Doi, Y. and Yora, K. (1985). *Ann. Phytopath. Soc. Japan* **51**: 642.

# Avocado 3 (?) *alphacryptovirus*

**Data collated by A.A. Brunt 1994.**

First reported in *Persea americana*; from the U.S.A.; by Jordan *et al.* (1983).

**Natural host range and symptoms**

Symptoms *Persea americana* — symptomless.

**Geographical distribution**

Found, but with no evidence of spread, in U.S.A.

**Taxonomy and Relationships**

Possible *Alphacryptovirus: Partitiviridae.*

**References**

Jordan, R.L. Dodds, J.A. and Ohr, H. (1983). *Phytopathology* **73**: 1130.

# Bamboo mosaic (?) *potexvirus*

**Data collated by M.T. Lin, 1984.**

*Standard acronym:* BaMV.

First reported in *Bambusa multiplex* and *B. vulgaris*; from Brasilia, Brazil; by Lin *et al.* (1977).

**Natural host range and symptoms**

Symptoms persist.

Ten species of bamboo, *Bambusa beecheyana, B. beecheyana* cv. *pubescens, B. edulis, B. multiplex, B. oldhamii, B. vulgaris, B. vulgaris* cv. *striata, Dendrocalamus latiflorus, D. latiflorus* cv. Mei-nung, *Phyllostachys nigra* — mosaic, necrotic streaks on culms, vascular discoloration.

**Transmission**

Transmitted by means not involving a vector. Virus transmitted by mechanical inoculation.

**Geographical distribution**

Spreads in the Pacific region; the Philippines and Taiwan. Found, but with no evidence of spread, in Brazil.

**Experimental host range**

Several (3–9) families susceptible. Experimentally infected plants mostly show mosaics, stunting.

Diagnostically susceptible

*Gomphrena globosa, Chenopodium amaranticolor* — necrotic local lesions; not systemic.

*Bambusa vulgaris* cv. Vittatu, *Dendrocalamus latiflorus* cv. Mei-nung — systemic mosaic.

Diagnostically insusceptible

*Phaseolus vulgaris, Capsicum annuum, Lycopersicon esculentum, Nicotiana tabacum* cv. Turkish, *Oryza sativa.*

Maintenance and propagation hosts

*Bambusa multiplex, B. vulgaris, Dendrocalamus latiflorus.*

Assay hosts (**L**ocal lesions or **W**hole plants)

*Gomphrena globosa* (L), *Chenopodium amaranticolor* (L), *Bambusa vulgaris* cv. Vittatu (W), *Dendrocalamus latiflorus* cv. Mei-nung (W).

Dicotyledonous Hosts *(**Susceptible** or Insusceptible)*

*Brassica oleracea* ssp. *botrytis. Capsicum annuum. Capsicum frutescens.* ***Chenopodium amaranticolor***. *Chenopodium quinoa. Datura metel. Datura stramonium.* ***Gomphrena globosa***. *Lactuca sativa. Lycopersicon esculentum. Nicotiana glutinosa. Nicotiana tabacum. Phaseolus vulgaris. Raphanus sativus. Vigna unguiculata.*

Monocotyledonous Hosts *(**Susceptible** or Insusceptible)*

*Oryza sativa. Zea mays.*

Sources of host-range data

Lin *et al.* (1977; 1979).

**Properties of particles in sap**

TIP: 65–70 °C. DEP: $\log_{10}$ minus 5–6. Leaf sap contains many virions.

**Purification**

Method: Lin *et al.* (1977).

**Particle morphology**

Virions filamentous; not enveloped; with a clear modal length; of 490 nm; 15 nm wide. Axial canal obscure.

*Sequence database accession code(s):*

X57581 Gb(84)_vi:BOMVCPA Bamboo mosaic virus coat protein gene. 6/94 729bp

D26017 Em(43)_vi:Am1bamvgr Gb(89)_vi:Am1bamvgr bamboo mosaic virus BaMV gRNA gene. 10/94 6,366bp.

L22762 Em(43)_vi:Bbssatpro Gb(89)_vi:Bbssatpro Bamboo mosaic virus satellite RNA, complete cds. 10/94 836bp.

X57581 Em(43)_vi:Bomvcpa Gb(89)_vi:Bomvcpa Bamboo mosaic virus coat protein gene. 6/94 729bp.

**Cytopathology**

Virions found in all parts of the host plant; in cytoplasm and in cell vacuoles; vacuoles and attached to chloroplasts, as well as occurring abundantly in cytoplasm. Inclusions present in infected cells; are crystals in the cytoplasm and crystals in the nucleus; they do not contain virions.

**Taxonomy and Relationships**

Possible *Potexvirus*.

**References**


Kitajima, E.W., Lin, M.T., Cupertino, F.P. and Costa, C.L. (1977). *Phytopath. Z.* **90**: 180.

Lin, M.T. and Chen, C.C. (1991). *Phytopathology* **81**: 1551.

Lin, M.T., Chen, C.C., Kiang, T. and Lin, W.C. (1979). *Taiwan Forest Res. Inst. Bull.* No. 317.

Lin, M.T., Kitajima, E.W., Cupertino, F.P. and Costa, C.L. (1977). *Phytopathology* **67**: 1439.


# Banana bract mosaic *potyvirus*

**Data collated by A.A. Brunt, 1991.**

*Synonyms:* banana potyvirus.

First reported in *Musa sapientum*; from the Philippines; by Bateson and Dale, (1995).

**Natural host range and symptoms**

*Musa sapientum* — chlorosis, especially obvious in bracts.

**Geographical distribution**

Spreads in the South and Central American region; the Philippines.

**Particle morphology**

Virions filamentous; usually flexuous; with a clear modal length; of 750 nm.

### Taxonomy and Relationships

*Potyvirus: Potyviridae.*

Virus(es) with serologically related virions: the virus reacts with antiserum prepared to the conserved core regions of a mixture of potyviruses. Virus(es) with serologically unrelated virions: sugarcane mosaic potyvirus.

The coat protein gene of the virus is 71% identical to that of maize dwarf mosaic virus, and its 3'UTR is most similar to that of ornithogalum mosaic virus (40%).

### References

Bateson, M.F. and Dale, J.L. (1995). *Arch. Virol.* **140**: 515.

## Banana bunchy top nanavirus

**Data collated by J.L. Dale, 1986. Revised 1991.**

*Synonyms:* abaca bunchy top virus. *Standard acronym:* BBTV.

First reported in *Musa* spp.; from Fiji; by Magee (1953).

### Natural host range and symptoms

Symptoms persist. Symptoms chlorosis, stunting.
*Musa* spp. — chlorosis, stunting, death.

### Transmission

Transmitted by a vector; an insect; *Pentalonia nigronervosa*; Aphididae. Transmitted in a persistent manner. Virus probably retained when the vector moults; not transmitted congenitally to the progeny of the vector; not transmitted by mechanical inoculation.

### Ecology and control

Studies reported by Magee (1940 and 1953).

### Geographical distribution

Spreads in the African region and the Pacific region; Australia, Burundi, Egypt, Gabon, the Philippines, and Taiwan.

### Experimental host range

Few (<3) families susceptible. Experimentally infected plants mostly show chlorosis, stunting.

Diagnostically susceptible

*Musa sapientum* cv. Cavendish — chlorosis, stunting.

Maintenance and propagation hosts

*Musa* spp.

### Properties of particles in sap

Leaf sap contains few virions.

### Particle morphology

Virions isometric; not enveloped; 18–20 nm in diameter; rounded in profile; without a conspicuous capsomere arrangement.

### Biochemical properties

*Genome* consists of DNA; single–stranded; circular; of three parts (J. Dale, personal communication); largest (or only) genome part the largest 1.3 kb; the 2nd largest 1.3 kb; the 3rd largest 1.171 kb. Genomic nucleic acid isolated by Dale *et al.* (1986).

*Sequence database accession code(s):*

L32166 Gb(84)n:BYTV1 Banana bunchy top virus (BBTV DNA I) V1 and C1-C3 genes, complete cds's. 8/94 1,106bp.

L32167 Gb(84)n:BYTV2 Banana bunchy top virus (BBTV DNA II) V1, V2, C1 and C2 genes, complete cds's. 8/94 1,096bp.

S56276 Em(40)_vi:S56276 Gb(84)_vi:S56276 orf V1 (component 1) banana bunchy top virus BBTV, Genomic, 1111 nt. 1/94 1,111bp.

U02312 Em(40)_vi:BBU02312 Gb(84)_vi:U02312 Banana bunchy top virus component 2. 6/94 1,095bp

L41574 Gb(89)n:Bytorfaa Banana bunchy top virus gene, complete cds. 6/95 1,075bp.

L41575 Gb(89)n:Bytorfab Banana bunchy top virus gene, complete cds. 6/95 1,043bp.

L41576 Gb(89)n:Byttataaa Banana bunchy top virus gene, TATA and polyA signals. 6/95 1,060bp.

L41577 Gb(89)n:Bytorfac Banana bunchy top virus gene, complete cds. 6/95 1,089bp.

L41578 Gb(89)n:Bytorfad Banana bunchy top virus gene, complete cds. 6/95 1,018bp.

U12586 Em(43)_vi:Bb12586 Gb(89)_vi:Bbu12586 Banana bunchy top virus DNA I II ORF V1 and ORF C1 genes, complete cds. 2/95 1,111bp.

U12587 Em(43)_vi:Bb12587 Gb(89)_vi:Bbu12587 Banana bunchy top virus DNA IV ORF V1 and ORF C1 genes, complete cds. 2/95 1,127bp.

U18077 Em(43)_vi:Bb18077 Gb(89)_vi:Bbu18077 Banana bunchy top virus putative replicase and hypothetical 13.8 kDa protein, component 1 DNA

U18078 Em(43)_vi:Bb18078 Gb(89)_vi:Bbu18078 Banana bunchy top virus unknown protein, component 3 DNA, complete cds. 12/94 1,057bp.

U18079 Em(43)_vi:Bb18079 Gb(89)_vi:Bbu18079 Banana bunchy top virus unknown protein, component 4 DNA, complete cds. 12/94 1,017bp.

*Virion protein(s)* one; $M_r$ 20000; virion protein.

## Cytopathology

Virions found in phloem.

## Taxonomy and Relationships

Nanavirus.

In nucleic acid hybridization tests, its genome reacts distantly with that of abaca bunchy top virus. It was originally thought to be a luteovirus: infected plants sometimes contain dsRNA, but these may come from a contaminant virus.

## References

Dale, J.L. (1987). *Adv. Virus Res.* **33**: 301.

Magee, C.J.P. (1940). *J. Aust. Inst. Agric. Sci.* **6**: 109.

Magee, C.J.P. (1953). *J. Proc. R. Soc. N.S.W.* **87**: 3.

# Banana streak *badnavirus*

**Data collated by A.A. Brunt, S. Molyneux, and B.E. Lockhart, 1987.**

*Standard acronym:* BSV.

First reported in *Musa* spp.; from Morocco; by Lockhart (1986).

**Natural host range and symptoms**

*Musa* spp. — chlorotic leaf streaks, necrosis.

**Transmission**

Virus not transmitted by mechanical inoculation.

**Geographical distribution**

Spreads in Morocco.

**Experimental host range**

Several (3–9) families susceptible. Experimentally infected plants mostly show chlorotic and necrotic streaks.

Diagnostically susceptible

*Musa* spp. — chlorotic and necrotic leaf streaks.

Diagnostically insusceptible

*Hordeum vulgare, Nicotiana benthamiana, N. glutinosa, N. tabacum, Zea mays.*

Dicotyledonous Hosts *(**Susceptible** or Insusceptible)*

*Nicotiana benthamiana. Nicotiana × edwardsonii. Nicotiana glutinosa. Nicotiana tabacum.*

Monocotyledonous Hosts *(**Susceptible** or Insusceptible)*

*Avena sativa. Hordeum vulgare. Zea mays.*

Sources of host-range data

Lockhart (1986).

**Purification**

Method: Lockhart (1986).

**Particle morphology**

Virions bacilliform; 119 nm in length; *c.* 30 nm wide.

**Taxonomy and Relationships**

*Badnavirus.*

Virus(es) with serologically unrelated virions: cacao swollen shoot and rice tungro bacilliform viruses.

**References**

Lockhart, B.E. (1986). *Phytopathology* **76**: 995.

Lockhart, B.E.L. and Autrey, J.C. (1988). *Plant Dis.* **72**: 230.

# Barley dubia (?) *tenuivirus*

**Data collated by A.A. Brunt, 1991.**

First reported in *Hordeum vulgare*; from Germany; by Kempiak (1972).

### Natural host range and symptoms

*Hordeum vulgare*.

### Transmission

Transmitted by a vector; an insect; *Javesella pellucida, J. dubia*; Delphacidae. Transmitted in a persistent manner.

### Taxonomy and Relationships

Possibly *Tenuivirus*.
See Gingery (1988).

### References

Gingery, R.E. (1988). In: *The Plant Viruses*, Vol. 4, The Filamentous Plant Viruses, p. 303; ed. R.G. Milne. Plenum Press, New York.
Kempiak, G. (1972). *TagBer. dt. Akad. Landw. Berlin* **121**: 99.

# Barley mild mosaic *bymovirus*

**Data collated by M.J. Adams, 1990.**

*Synonyms:* barley yellow mosaic virus mechanically transmissible strain, BaYMV — Streatley strain. *Standard acronym:* BaMMV. *Strains:* two serotypes have been reported from Japan.

First reported in *Hordeum vulgare*; from Germany; by Huth *et al.* (1984).

### Natural host range and symptoms

Symptoms vary seasonally (in winter or early spring only). Symptoms yellow streaks, brown necrosis.

*Hordeum vulgare* — yellow streaks, especially in younger leaves and sometimes in leaf sheath, also brown necrotic patches. Leaves may roll making the plant look 'spiky'.

### Transmission

Transmitted by a vector; a fungus; *Polymyxa graminis* (Mastigomycotina); Plasmodiophorales. Principal natural vector(s): *Polymyxa graminis*. Virus transmitted by mechanical inoculation; not transmitted by contact between plants; not transmitted by seed; not transmitted by pollen.

### Ecology and control

Studies reported by Huth (1989).

### Geographical distribution

Spreads in Belgium, France, Germany, Italy, Japan, the Netherlands, and the UK.

### Experimental host range

Few (<3) families susceptible. Experimentally infected plants mostly show yellow streaks.

Diagnostically susceptible

*Hordeum vulgare, Lagarus ovatus, Eremopyrum hirsutum, Aegilops kotschyi, A. searsii, A. speltoides* — systemic leaf streaking.

Diagnostically insusceptible

*Avena sativa, Oryza sativa, Secale cereale, Triticum aestivum, Zea mays.*

Maintenance and propagation hosts

*Hordeum vulgare* (but some cultivars are immune).

Assay hosts (**L**ocal lesions or **W**hole plants)

*Hordeum vulgare* (W).

Monocotyledonous Hosts *(**Susceptible** or Insusceptible)*

*Agropyron repens. Agrostis tenuis. Anthoxanthum odoratum. Avena sativa. Bromus inermis. Bromus mollis. Dactylis glomerata. Echinochloa crus–galli. Elymus junceus. Festuca pratensis.* ***Hordeum vulgare. Lagurus ovatus.*** *Lolium multiflorum. Lolium perenne. Lolium temulentum. Oryza sativa. Panicum maximum. Phalaris arundinacea. Poa pratensis. Secale cereale. Setaria viridis. Triticum aestivum. Zea mays.*

Sources of host-range data

Proeseler (1988).

**Properties of particles in sap**

DEP: $\log_{10}$ minus 4. Leaf sap contains few virions.

**Purification**

Method: Huth *et al.* (1984); Adams *et al.* (1987); Pröls *et al.* (1990).

**Particle morphology**

Virions filamentous; not enveloped; slightly usually flexuous; with clear modal lengths (two); of 270–290 nm, or 570–600 nm; 12 nm wide (PTA), or 15 nm wide (UA). Axial canal obscure. Basic helix obscure.

**Physical properties**

Density 1.314 g $cm^{-3}$ in CsCl.

**Biochemical properties**

*Genome* consists of RNA; single–stranded; linear; of two parts; largest (or only) genome part the larger 8 kb (approximately); the 2nd largest *c.* 4 kb. Genomic nucleic acid isolated by Batista *et al.* (1989). Infectivity retained when deproteinised with phenol or detergent. Poly A region present; found at the 3' end of both genome segments. Additional factor not required for infectivity.

*Sequence database accession code(s):*

D10947 Em(40)_vi:BMMNA1 Gb(84)_vi:BMMNA1 Barley mild mosaic virus RNA 1, polyprotein precursor, partial sequence. 1/93 3,464bp.

D10949 Em(40)_vi:BMMKA1 Gb(84)_vi:BMMKA1 Barley mild mosaic virus RNA 1, polyprotein precursor, partial sequence. 1/93 3,946bp.

X69204 Em(40)_vi:BMPOLYPRO Gb(84)_vi:BMPOLYPRO Barley Mild Mosaic Virus sequence for polyprotein, 3' terminus. 11/92 2,095bp

X74095 Em(40)_vi:BAMMVRNA1 Gb(84)_vi:BAMMVRNA1 Barley mild mosaic virus (BaMMV) RNA-1 (3' region) ORF for coat protein. 8/93 1,476bp

X75933 Em(43)_vi:Bammvrna2 Gb(89)_vi:Bammvrna2 Barley mild mosaic virus complete sequence of RNA2. 11/94 3,524bp.

X82625 Em(43)_vi:Bmmvp1p2 Gb(89)_vi:Bmmvp1p2 Barley mild mosaic virus mRNA for RNA2 polyprotein. 4/95 2,428bp.

*Non–genomic nucleic acid* not found in the virions.

*Virion protein(s)* one; $M_r$ 35000. Method of preparation: Huth *et al.* (1984); Ehlers and Paul (1986).

*Virus–coded non–virion proteins* isolated (Ehlers (1987)); one protein found. $M_r$ 72000; possibly pinwheel inclusion protein.

**Replication**

Replication does not depend on a helper virus.

**Cytopathology**

Virions found in leaves, roots, mesophyll, epidermis, vascular parenchyma and the companion cells; in cytoplasm. Inclusions present in infected cells; are membranous bodies and pinwheels; they do not contain virions (but virions may be attached to pinwheels).

**Taxonomy and Relationships**

*Bymovirus: Potyviridae.*

Virus(es) with serologically unrelated virions: barley yellow mosaic, wheat spindle streak mosaic (syn. wheat yellow mosaic) and oat mosaic viruses.

The virus was reported to have virions serologically related to those of bean yellow mosaic potyvirus and those of one isolate of turnip mosaic potyvirus (Stanarius *et al.*, 1989) but was not confirmed by Chen and Adams (1991). Best tests for diagnosis: longitudinal streaks on unfurling leaves in winter or early spring distinguish this from barley yellow dwarf luteovirus. Serological tests or sequencing are necessary to distinguish it from barley yellow mosaic bymovirus.

**References**

Adams, M.J. (1991). *Pl. Path.* **40**: 53.
Adams, M.J., Jones, P. and Swaby, A.G. (1987). *Ann. appl. Biol.* **110**: 321.
Batista, M.F., Antoniw, J.F., Swaby, A.G., Jones, P. and Adams, M.J. (1989). *Pl. Path.* **38**: 226.
Chen, J. and Adams, M.J. (1991). *Pl. Path.* **40**: 226.
Ehlers, V. (1987). *Ph.D. Thesis*, Technischen Universitat Carlo-Wilhelmina, Braunschweig, Germany.
Huth, W. (1988). In: *Viruses with Fungal Vectors*, p. 61; eds J.I. Cooper and M.J.C. Asher. Association of Applied Biologists, Wellesbourne, U.K.
Huth, W. (1989). *EPPO Bulletin* **19**: 547.
Huth, W. and Adams, M.J. (1990). *Intervirology* **31**: 38.
Huth, W., Lesemann, D.-E. and Paul, H.L. (1984). *Phytopath. Z.* **111**: 37.
Proeseler, G. (1988). *Arch. Phytopath. Pflanzen.* **24**: 267.
Pröls, M., Davidson, A., Schell, J. and Steinbiss, H.-H. (1990). *J. Phytopath.* **130**: 249.
Stanarius, A., Proeseler, G. and Richter, J. (1989). *Arch. Phytopath. Pflanzen.* **25**: 303.

# Barley mosaic virus

**Data collated by S.L. Bhattiprolu, 1991.**

First reported in *Hordeum vulgare*; from Delhi, India; by Dhanraj and Raychaudhuri (1969).

**Natural host range and symptoms**

Symptoms persist. Symptoms mosaic and stunting.

*Hordeum vulgare* — mosaic and stunting.

**Transmission**

Transmitted by a vector; an insect; *Rhopalosiphum maidis*; Aphididae. Virus transmitted by mechanical inoculation; transmitted by seed.

**Geographical distribution**

Found, but with no evidence of spread, in India.

**Experimental host range**

Few (<3) families susceptible. Experimentally infected plants mostly show mosaic.

Diagnostically susceptible

*Avena sativa, Hordeum vulgare, Triticum aestivum* — mosaic in systemically infected leaves.

Maintenance and propagation hosts

*Avena sativa, Hordeum vulgare, Triticum aestivum.*

Assay hosts (**L**ocal lesions or **W**hole plants)

*Avena sativa* (W), *Hordeum vulgare* (W), *Triticum aestivum* (W).

Monocotyledonous Hosts *(**Susceptible** or Insusceptible)*

***Avena sativa. Hordeum vulgare. Triticum aestivum.***

Sources of host-range data

Dhanraj and Raychaudhuri (1969).

**Properties of particles in sap**

TIP: 53–55 °C. LIV: 0.33 days. DEP: $\log_{10}$ minus 2–3. Leaf sap contains few virions.

**Particle morphology**

Virions isometric; 40 nm in diameter; rounded in profile.

**Cytopathology**

Virions found in leaves.

**References**

Dhanraj, K.S. and Raychaudhuri, S.P. (1969). *Pl. Dis. Reptr* **53**: 766.

# Barley stripe mosaic *hordeivirus*

**Data collated by C. Büchen-Osmond, 1987.**

*Synonyms:* barley false stripe virus, possibly barley yellow stripe virus, barley mild stripe virus, oat stripe mosaic virus. *Standard acronym:* BSMV. *Strains:* type strain, Russian, Norwich (ND 18), Canadian severe, Argentine mild (AM), Rothamsted (R).

First reported in *Hordeum vulgare* and *Triticum aestivum*; from the U.S.A.; by McKinney (1951).

**Natural host range and symptoms**

Symptoms persist.
*Hordeum vulgare, Triticum aestivum* — mild stripe mosaic to lethal necrosis.

**Transmission**

Transmitted by means not involving a vector. Virus transmitted by mechanical inoculation; transmitted by seed (up to 90–100%); transmitted by pollen to the pollinated plant.

**Ecology and control**

Studies reported by McKinney (1953); Gold *et al.* (1954); Gardner (1967); Sprague *et al.* (1963); McNeal *et al.* (1976); Sandfaer (1971).

**Geographical distribution**

Probably distributed worldwide. Spreads in the Eurasian region, the North American region, and the Pacific region; Australia, China, the UK, the USA, and the former USSR.

**Experimental host range**

Several (3–9) families susceptible. Experimentally infected plants mostly show stripe mosaic in monocotyledonous species, and chlorotic local lesions in dicotyledonous species.

Diagnostically susceptible

*Hordeum vulgare, Triticum aestivum, Avena sativa* — systemic stripe mosaic.
*Chenopodium amaranticolor, C. quinoa, C. album* — large chlorotic local lesions; not systemic.
*Beta vulgaris* — chlorotic local lesions; not systemic.
*Zea mays* — systemic stripe mosaic.
*Spinacia oleracea* — systemic mosaic.
*Nicotiana tabacum* cv. Samsun — local chlorotic lesions; not systemic.

Maintenance and propagation hosts

*Hordeum vulgare, Triticum aestivum.*

Assay hosts (**L**ocal lesions or **W**hole plants)

*Chenopodium amaranticolor* (L); *Chenopodium quinoa* (L).

Dicotyledonous Hosts ***(Susceptible** or Insusceptible)*

***Beta vulgaris**. **Chenopodium album**. **Chenopodium amaranticolor**. **Chenopodium quinoa**. **Nicotiana tabacum**. **Spinacia oleracea**.*

Monocotyledonous Hosts ***(Susceptible** or Insusceptible)*

***Agropyron elongatum**. **Agropyron intermedium**. **Anthoxanthum cristatum**. **Anthoxanthum odoratum**. **Avena sativa**. **Bromus secalinus**. **Bromus tectorum**. **Dactylis glomerata**. **Echinochloa crus–galli**. **Eragrostis cilianensis**. **Festuca pratensis**. **Hordeum vulgare**. **Lagurus ovatus**. **Lolium multiflorum**. **Lolium perenne**. **Lolium temulentum**. **Lolium persicum**. **Oryza sativa**. **Oryzopsis miliacea**. **Panicum capillare**. **Panicum miliaceum**. **Phalaris arundinacea**. **Phalaris paradoxa**. **Phleum arenarium**. **Phleum pratense**. **Poa annua**. **Poa pratensis**. **Secale cereale**. **Setaria italica**. **Setaria macrostachia**. **Setaria viridis**. **Sorghum bicolor**. **Triticum aestivum**. **Triticum durum**. **Zea mays**.*

Sources of host-range data

Nitzany and Gerechter (1962); Ohmann-Kreutzberg (1962); McKinney and Greeley (1965); Kirstensen and Engsbro (1970); Novikov and Atabekov (1970); Polak and Slykhuis (1972).

**Properties of particles in sap**

TIP: 60–68 °C (McKinney, 1951; Kassanis and Slykhuis, 1959). LIV: 15–22 days. DEP: $\log_{10}$ minus 2–4.

**Purification**

Method: Novikov (1970).

**Particle morphology**

Virions rod–shaped; not enveloped; usually straight; with a clear modal length; of 112–150 nm (Gold *et al.*, 1954; Harrison *et al.*, 1965); 18–24 nm wide. Axial canal obvious; 3–4 nm in diameter (Gibbs *et al.*, 1963; Harrison *et al.*, 1965; Finch, 1966). Basic helix obvious; pitch of basic helix 2.5–2.6 nm.

**Physical properties**

Three sedimenting components in purified preparations; sedimentation coefficient of the fastest 199 *S*; of the other(s) 166–194 *S* (Brakke and Palomar, 1976; Harrison *et al.*, 1965). Isoelectric point pH 4.5. $A_{260}/A_{280}$ ratio 0.99 (corrected).

**Biochemical properties**

Virions contain 3.8–4 % nucleic acid; 96 % protein; 0 % lipid.

*Genome* consists of RNA; single–stranded; linear; of three parts (RNA-a, RNA-b and RNA-g are separately encapsidated); largest (or only) genome part the largest 4 kb; the 2nd largest 3.289 kb; the 3rd largest 3.164 kb (or smaller depending on the strain. The Agentine strain has another RNA component of 2.5 kb). Genomic nucleic acid isolated by Lane (1974); Gustafson and Armour (1986). Base composition 20.3–23.5 % G; 27–30.9 % A; 19.4–21.5 % C; 28–29.4 % U (Atabekov and Novikov, 1971). 5' terminus of RNA has a methylated nucleotide cap. Poly A region present; of variable lengths and between the coding region and the 3' *t*RNA-like structure in each genomic RNA. Genome has *t*RNA–like activity. Genome accepts and accepts tyrosine. Nucleotide sequence references: Gustafson and Armour (1986).

*Sequence database accession code(s):*

J04342 Em(40)_vi:HOMBSARN Gb(84)_vi:MBSARNA Barley stripe mosaic virus alpha RNA, complete cds. 11/89 3,768bp.

M11509 Em(40)_vi:HOBSMG1 Gb(84)_vi:MBSRNAGSA Barley stripe mosaic virus (ND18) gamma-subgenomic RNA, 3' end. 4/90 156bp.

M11510 Em(40)_vi:HOBSMG2 Gb(84)_vi:MBSRNAGSB Barley stripe mosaic virus (Type) gamma-subgenomic RNA, 3' end. 4/90 159bp.

M11511 Em(40)_vi:HOBSMG3 Gb(84)_vi:MBSRNAG Barley stripe mosaic virus gamma-RNA, 3' end. 4/90 92bp.

M16576 Em(40)_vi:HOMBS Gb(84)_vi:MBSRNAGT Barley stripe mosaic virus (Type) RNA-gamma segment. 7/89 3,164bp.

M16577 Em(40)_vi:HOMBSRNA Gb(84)_vi:MBSRNAGND Barley stripe mosaic virus (ND18) RNA-gamma segment. 7/89 2,791bp.

M28702 Em(40)_sy:CVRVPS Gb(84)_sy:SYNBSMRVPS Barley stripe mosaic virus, 3' end/E.coli T7 promoter recombinant DNA. 2/90 54bp

M38631 Em(40)_vi:MBSRNAL Gb(84)_vi:MBSRNAL Barley stripe mosaic virus 5' leader sequence of RNA. 5/94 91bp.

M38633 Em(40)_vi:MBSRNAN Gb(84)_vi:MBSRNAN Barley stripe mosaic virus 5' leader sequence of RNA, CV17. 5/94 88bp.

M59831 Em(40)_vi:MBSRNAM Gb(84)_vi:MBSRNAM Barley stripe mosaic virus 5' leader sequence of RNA. 5/94 91bp.

U05346 Em(40)_vi:BS05346 Gb(84)_vi:BSU05346 Barley stripe mosaic virus ND18 alpha protein gene, partial cds. 3/94 823bp

U05347 Em(40)_vi:BS05347 Gb(84)_vi:BSU05347 Barley stripe mosaic virus CV42 RNA-alpha segment alpha protein gene, complete cds. 3/94 3,78

U13916 Gb(84)n:BSU13916 Barley stripe mosaic virus CV17 RNA gamma segment, complete sequence. 10/94 3,164bp.

U13917 Gb(84)n:BSU13917 Barley stripe mosaic virus ND18 RNA gamma segment, complete sequence. 10/94 2,790bp.

U13918 Gb(84)n:BSU13918 Barley stripe mosaic virus Type RNA gamma segment, complete sequence. 10/94 3,168bp.

X01025 Em(40)_vi:HOBSMVRN Gb(84)_vi:HOBSMVRN Barley stripe mosaic virus RNA2 3'-terminal non coding region. 7/89 238bp.

X03854 Em(40)_vi:HOBMSVRB Gb(84)_vi:HOBMSVRB Barley stripe mosaic virus (BSMV, Type strain) genomic RNA beta. 9/93 3,289bp.

X05214 Em(40)_pl:ZMBS1TRA Gb(84)_pl:ZMBS1TRA Maize transposon Bs1 terminal repeat in Adh1 gene following barley stripe mosaic virus infect

X52774 Em(40)_vi:BSMVRNA3 Gb(84)_vi:BSMVRNA3 Barley stripe mosaic virus (BSMV) third genomic component RNA 3. 5/93 3,164bp.

X52775 Em(40)_vi:BSMVRNA1 Gb(84)_vi:BSMVRNA1 Barley stripe mosaic virus 3' region of the first genomic component RNA1. 5/93 872bp. 20 sequences.

*Features of the genome:* RNA-a has one ORF that encodes the 130 kDa (? replicase); RNA-b encodes the 22 kDa (coat protein) and the 60, 17 and 14 kDa (movement proteins); RNA-g has 2 ORFs and encodes the 87 kDa (? polymerase) and 17 kDa (regulator of the translation of RNA-b genes.

*Non–genomic nucleic acid* found in the virions; is *t*RNA; generated from the 3'-end of genome and of 1.830 and 0.788 kb.

*Virion protein(s)* one; $M_r$ 21500; it is coat protein. Method of preparation: Atabekov *et al.* (1968); Gumpf and Hamilton (1968). Amino acid sequence: Gustafson and Armour (1986). Amino acid composition: Atabekov *et al.* (1968); Gumpf and Hamilton (1968). Virion proteins glycosylated; are are the coat proteins (Gumpf *et al.*, 1977; Partridge *et al.*, 1976).

*Virus–coded non–virion proteins* identified by genomic sequence analysis (Gustafson and Armour (1986)); at least three proteins found. $M_r$ 58087. $M_r$ of 2nd largest 17378. $M_r$ of 3rd 14119.

**Replication**

Replication does not depend on a helper virus.

**Cytopathology**

Virions found in all parts of the host plant; in cytoplasm and in nuclei. Inclusions absent from infected cells. Other cellular changes: peripheral vesicles in chloroplasts (Carroll, 1970).

**Taxonomy and Relationships**

*Hordeivirus* (the type species).

Virus(es) with serologically related virions: lychnis ringspot virus, but distantly (Gibbs *et al.*, 1963).

The genomic RNA do not hybridize with those of lychnis ringspot and poa semilatent hordeiviruses.

**References**

Atabekov, J.G. and Novikov, V.K. (1971). *CMI/AAB Descr. Pl. Viruses* No. 68, 4 pp.

Atabekov, J.G. and Novikov, V.K. (1989). *CMI/AAB Descr. Pl. Viruses* No. 344, 6 pp.

Atabekov, J.G., Novikov, V.K., Kiselev, N.A., Kaftanova, A.S. and Egporov, A.M. (1968). *Virology* **36**: 620.

Brakke, M.K. and Palomar, M.K. (1976). *Virology* **71**: 255.

Carroll, T.W. (1970). *Virology* **42**: 1015.

Carroll, T.W. (1986). In: *The Plant Viruses*. Vol. 2, The Rod-Shaped Plant Viruses, p. 273; eds M.H.V. van Regenmortel and H. Fraenkel-Conrat. Plenum Press, New York.

Edwards, M.C., Petty, I.T.D. and Jackson, A.O. (1992). *Virology* **189**: 389.

Finch, J.T. (1966). *Nature, Lond.* **212**: 349.

Gibbs, A.J., Kassanis, B., Nixon, H.L. and Woods, R.D. (1963). *Virology* **20**: 194.

Gold, A.H., Suneson, C.A. Houston, B.R. and Oswald, J.W. (1954). *Phytopathology* **44**: 115.

Gumpf, D.J., Cunningham, D.S., Heick, J.A. and Shannon, L.M. (1977). *Virology* **78**: 328.

Gumpf, D.J. and Hamilton, R.I. (1968). *Virology* **35**: 87.

Gustafson, G. and Armour, S.L. (1986). *Nucl. Acids Res.* **14**: 3895.

Harrison, B.D., Nixon, H.L. and Woods, R.D. (1965). *Virology* **26**: 284.

Jackson, A.O. and Lane, L.C. (1981). In: *Handbook of Plant Virus Infections and Comparative Diagnosis*, Hordeiviruses, p. 565; ed. E. Kurstak. Elsevier/North-Holland Biomedical Press, Amsterdam.

Jackson, A.O., Hunter, B.G. and Gustafson, G.D. (1989). *Ann. Rev. Phytopath.* **27**: 95.

Jackson, A.O., Petty, I.T.D., Jones, R.W., Edwards, M.C. and French, R. (1991). *Semin. Virol.* **2**: 107.

Jackson, A.O., Petty, I.T.D., Jones, R.W., Edwards, M.C. and French, R. (1991). *Can. J. Pl. Path.* **13**: 163.

Kassanis, B. and Slykhuis, J.T. (1959). *Ann. appl. Biol.* **47**: 254.

Kirsten, H.R. and Engsbro, B. (1970). *Tidsskr. PlAvl.* **74**: 326.

Kiselev, N.A., Atabekov, J.G., Kaftanova, A.S. and Novikov, V.K. (1966). *Biokhimiya* **31**: 670.

McKinney, H.H. (1951). *Phytopathology* **41**: 563.

McKinney, H.H. (1953). *Pl. Dis. Reptr* **37**: 292.

McKinney, H.H and Greeley, L.W. (1953). *Tech. Bull. U.S. Dep. Agric.* **1324**: 84 pp.

McNeal, F.H., Berg, M.A. and Carroll, T.W. (1976). *Pl. Dis. Reptr* **60**: 730.

Na-Sheng, L. and Langenberg, W.G. (1985). *Virology* **142**: 291.

Nitzany, F.E. and Gerechter, E.K. (1962). *Phytopathol. Medit.* **2**: 11.

Novikov, V.K. (1970). *Ph.D. Thesis*, Moscow State University, USSR.

Novikov, V.K. and Atabekov, J.G. (1970). *Virology* **41**: 101.

Ohmann-Kreutzberg, G. (1962). *Phytopath. Z.* **47**: 1.

Partridge, J.E., Shannon, L.M. and Gumpf, D.J. (1976). *Biochim. biophys. Acta* **451**: 470.

Partridge, J.E., Shannon, L.M., Gumpf, D.J. and Colbaugh, P. (1974). *Nature, Lond* **247**: 491.

Petty, I.T.D. and Jackson, A.O. (1990). *Virology* **179**: 712.

Petty, I.T.D., Edwards, M.C. and Jackson, A.O. (1990). *Proc. Natl. Acad. Sci. USA* **87**: 8894.

Petty, I.T.D., French, R., Jones, R.W. and Jackson, A.O. (1990). *EMBO J.* **9**: 33453.

Polak, Z. and Slykhuis, J.T. (1972). *Can. J. Bot.* **50**: 263.

Sandfaer, J. (1971). *Danish A.E.C. Risol. Rep. 230.*
Sprague, G.F., McKinney, H.H. and Greeley, L.W. (1963). *Science, N.Y.* **141**: 1052.
Veerisetty, V. (1978). *Virology* **84**: 523.

# Barley yellow dwarf *luteovirus*

**Data collated by W.F. Rochow, 1986; R.J. Sward and P.M. Waterhouse, 1987.**

*Synonyms:* cereal yellow dwarf virus, hordeum nanescens virus, oat red leaf virus, rice giallume virus, ryegrass chlorotic streak virus (Clark and Christensen, 1972; Catherall, 1974), possibly maize leaf fleck virus (Stoner, 1952). *Standard acronym:* BYDV-MAV, BYDV-PAV, BYDV-RGV, BYDV-RMV, BYDV-RPV, BYDV-SGV. *Strains:* Subgroup 1: (1) BYDV-MAV — transmitted by ***Macrosiphum (Sitobion) avenae.*** (2) BYDV-PAV — transmitted by *Rhopalosiphum **padi*** and by *Macrosiphum **avenae**.* (3) BYDV-SGV — transmitted by ***Schizaphis** graminum.* Subgroup 2: (4) BYDV-RMV — transmitted by ***Rhopalosiphum maidis.*** (5) BYDV-RPV — Transmitted by ***Rhopalosiphum padi.***

First reported in *Avena sativa, Hordeum vulgare* and *Triticum aestivum*; from California, U.S.A.; by Oswald and Houston (1951; 1953*a*; 1953*b*).

**Natural host range and symptoms**

Symptoms persist, or vary seasonally, or disappear soon after infection (cool temperatures and supplementary light are needed for conspicuous symptoms to develop in plants grown in greenhouses).

*Hordeum vulgare* — chlorotic leaf symptoms.

*Avena sativa, A. byzantina* — floret blast, chlorotic and red leaves.

*Triticum aestivum* — leaf chlorosis.

*Oryza sativa* — yellow to orange leaves (rice giallume disease).

*Lolium perenne, L. multiflorum, Cereale secale, Zea mays* — often symptomless, but sometimes purple or crimson leaf streaking.

**Transmission**

Transmitted by a vector; an insect; *Aulocorthum circumflexum* (synonyms *Amphorophora circumflexa, Macrosiphum circumflexa, Myzus circumflexus, Aulacorthum circumflexum, Siphonophora circumflexa*); *Ceruraphis eriophori*; *Acyrthosiphon (Metopolophium) dirhodum* (synonyms *Acyrthosiphon dirhodum, Macrosiphum dirhodum*); *Myzus persicae*; *Rhopalosiphum insertum* (synonyms *Rhopalosiphum annuae, Rhopalosiphum fitchii*); *Rhopalosiphum maidis* (synonym *Aphis maydis*); *Rhopalosiphum padi* (synonyms *Anuraphis padi, Rhopalosiphum prunifoliae*); *Rhopalosiphum rufiabdominalis*; *Schizaphis graminum* (synonym *Toxoptera graminum*); *Sipha agropyrella*; *Macrosiphum (Sitobion) avenae* (synonym *Macrosiphum granarium*); *Macrosiphum (Sitobion) fragariae*); *Macrosiphum (Sitobion) avenae miscanthi, Sitobion avenae miscanthi* (synonyms *Sitobion avenae* ssp. *miscanthi, Macrosiphum avenae miscanthi cf.* Jedlinski, 1981).

Subgroup 1: (1) BYDV-MAV is transmitted usually by *Macrosiphum (Sitobion) avenae*, but rarely by *Rhopalosiphum padi, R. maidis* and *Schizaphis graminum.* (2) BYDV-PAV is transmitted usually by *R. padi* and *M.(S.) avenae*, erratically by *Sch. graminum*, but rarely by *R. maidis.* (3) BYDV-SGV is transmitted regularly by *Sch. graminum*, but rarely if at all by *M.(S.) avenae, R. padi* or *R. maidis.*

Subgroup 2: (4) BYDV-RMV is regularly transmitted by *R. maidis*, but infrequently by *R. padi, M.(S.) avenae* and *Sch. graminum.* (5) BYDV-RPV is transmitted regularly by *R. padi*, erratically by *Sch. graminum*, but rarely by *R. maidis* and *M.(S.) avenae*;

Aphididae. Transmitted in a persistent manner. Virus retained when the vector moults; does not multiply in the vector; not transmitted congenitally to the progeny of the vector; not transmitted by mechanical inoculation; not transmitted by contact between plants; not transmitted by seed; not transmitted by pollen.

**Ecology and control**

Studies reported by Bruehl (1961); Jedlinski and Brown (1965); Rochow (1963; 1969; 1970; 1979); Rochow and Muller (1974); Paliwal (1982); Plumb (1976); Smith and Plumb (1981); Watson and Mulligan (1960); Waterhouse (1986); Zhang *et al.* (1983). See also Büchen-Osmond *et al.* (1988).

Cultivars tolerant of the virus are important for controlling the virus, which is probably the most widely distributed virus affecting cereals. Losses great from early infections.

**Geographical distribution**

Probably distributed worldwide.

**Experimental host range**

Few (<3) families susceptible. Experimentally infected plants mostly show chlorosis, reddening, stunting.

Diagnostically susceptible

*Hordeum vulgare* cv. Black Hulls, *Avena sativa* cvs Clintland, Clintland 64 and Blenda (spring oat), *Avena byzantina* cvs Coast Black and California Red (winter oat) — leaf chlorosis.

*Avena byzantina* cv. Black: strains BYDV-RMV and BYDV-RPV — little effect; BYDV-MAV — moderate effect; BYDV-PAV — severe effect.

Maintenance and propagation hosts

*Avena byzantina* cv. Coast Black.

Assay hosts (**L**ocal lesions or **W**hole plants)

*Avena sativa* (W); *Avena byzantina* (W); *Hordeum vulgare* (W).

Monocotyledonous Hosts (***Susceptible*** *or Insusceptible)*

***Agropyron intermedium***, or *Agropyron intermedium*. ***Agrostis alba***, or *Agrostis alba*. ***Anthoxanthum odoratum***. ***Avena byzantina***. ***Avena sativa***. ***Avena strigosa***. ***Bromus inermis***, or *Bromus inermis*. ***Bromus mollis***. ***Bromus racemosus***. ***Bromus secalinus***. ***Bromus tectorum***. ***Chloris gayana***. ***Dactylis glomerata***, or *Dactylis glomerata*. ***Festuca pratensis***, or *Festuca pratensis*. ***Hordeum vulgare***. ***Lolium perenne***, or *Lolium perenne*. ***Lolium temulentum***. ***Oryza sativa***. ***Panicum dichotomiflorum***. ***Paspalum dilatatum***. ***Phalaris arundinacea***. ***Phleum pratense***, or *Phleum pratense*. ***Poa annua***. ***Poa pratensis***. ***Poa trivialis***, or *Poa trivialis*. ***Secale cereale***. ***Triticum aestivum***. ***Triticum durum***. ***Zea mays***.

Additional hosts

At least 100 monocotyledonous species are susceptible.

Sources of host-range data

Bruehl (1961); Rochow (1961); Slykhuis (1967).

**Properties of particles in sap**

TIP: 60–75 °C (Heagy and Rochow, 1965). Leaf sap contains few virions.

**Purification**

Method: Rochow and Brakke (1964); Hebert (1963); Hewish and Shukla (1983); D'Arcy *et al.* (1983); Hammond *et al.* (1983); Waterhouse *et al.* (1987).

**Particle morphology**

Virions isometric; not enveloped; 25 nm in diameter (negatively stained or in sections; Jensen, 1969), or 30 nm in diameter (approximately in shadowed preparations; Rochow and Brakke, 1964); rounded in profile; without a conspicuous capsomere arrangement.

**Physical properties**

Usually one sedimenting component in purified preparations; sedimentation coefficient 115–118 *S* (for BYDV-MAV). Density 1.4–1.405 g $cm^{-3}$ in CsCl (for BYDV-PAV). Density 1.335 g $cm^{-3}$ in $Cs_2SO_4$ for BYDV-PAV.

**Biochemical properties**

Virions contain 28 % nucleic acid (in BYDV-MAV-like isolate; Paliwal, 1978); 72 % protein; 0 % lipid.

*Genome* consists of RNA; single–stranded; linear; unipartite; largest (or only) genome part 5.6 kb (BYDV-RPV; 5.677 for BYDV-PAV). Genomic nucleic acid isolated by Brakke and Rochow (1974). Base composition 24.6 % G; 29.6 % A; 23.8 % C; 22 % U. Poly A region absent. Nucleotide sequence references: Miller *et al.* (1987).

*Sequence database accession code(s):*

D10206 Em(40)_vi:BYDNYRPV Gb(84)_vi:BYDNYRPV Barley yellow dwarf virus genomic RNA for coat protein, complete cds. 11/93 5,600bp.

D11028 Em(40)_vi:BYDMAV Gb(84)_vi:BYDMAV Barley yellow dwarf virus (serotype MAV) genome, coat protein. 1/93 5,273bp.

D11032 Em(40)_vi:BYDPAV Gb(84)_vi:BYDPAV Barley yellow dwarf virus (serotype PAV) genome, coat protein. 1/93 5,179bp.

L10356 Em(40)_vi:BYDCOATP Gb(84)_vi:BYDCOATP Barley yellow dwarf virus coat protein gene, complete cds. 2/93 606bp.

L12757 Em(40)_vi:BYDCPA Gb(84)_vi:BYDCPA Barley yellow dwarf virus capsid protein mRNA, 3'end cds. 3/93 513bp.

L12758 Em(40)_vi:BYDCPB Gb(84)_vi:BYDCPB Barley yellow dwarf virus capsid protein mRNA, 3'end cds. 3/93 519bp.

L12759 Em(40)_vi:BYDSEQA Gb(84)_vi:BYDSEQA Barley yellow dwarf virus ORF 2 mRNA, 3'end cds. 3/93 363bp.

L19471 Em(40)_vi:BYDCOAT Gb(84)_vi:BYDCOAT Barley yellow dwarf virus coat protein mRNA, complete cds, and genome-linked proten VPg mRNA,

L19504 Em(40)_vi:BYDCOATPR Gb(84)_vi:BYDCOATPRO Barley yellow dwarf luteovirus coat protein gene; 17 kDa genome-linked viral protein gene. 6/

L25299 Em(40)_vi:BYDCG Gb(84)_vi:BYDCG Barley yellow dwarf virus complete genome. 11/93 5,723bp.

M21347 Em(40)_vi:BYDPCT Gb(84)_vi:BYDPCT Barley yellow dwarf virus coat protein gene, complete cds. 7/89 603bp.

M63666 Em(40)_vi:BYDSATRN Gb(84)_vi:BYDSATRNA Barley yellow dwarf virus satellite RNA. 4/91 322bp.

U06865 Gb(84)n:BYU06865 Barley yellow dwarf virus NY-SGV putative coat protein and putative readthrough protein RNA,

U06866 Gb(84)n:BYU06866 Barley yellow dwarf virus TX-SGV putative coat protein and putative readthrough protein RNA,

U12928 Gb(84)n:BYU12928 Barley yellow dwarf virus PAV serotype coat protein and readthrough protein genes, partial cd

X07653 Em(40)_vi:BYDVPAV Gb(84)_vi:BYDVPAV Barley yellow dwarf virus (BYDV-PAV) genomic RNA. 9/93 5,677bp.

X17259 Em(40)_vi:BYDRPVCP Gb(84)_vi:BYDRPVCP Barley yellow dwarf virus (isolate NY-RPV) coat protein gene. 1/93 615bp.

X17260 Em(40)_vi:BYDMAVCP Gb(84)_vi:BYDMAVCP Barley yellow dwarf virus (isolate MAV-PS1) coat protein gene. 1/93 600bp.

X17261 Em(40)_vi:BYDPAVCP Gb(84)_vi:BYDPAVCP Barley yellow dwarf virus (isolate P-PAV) coat protein gene. 1/93 603bp.

X53174 Em(40)_vi:LUBYDCP Gb(84)_vi:LUBYDCP Barley yellow dwarf virus genomic RNA for capsid protein. 9/93 655bp.

X56050 Em(40)_vi:BYDCPG Gb(84)_vi:BYDCPG Barley yellow dwarf virus gene for capsid protein. 1/92 984bp

X80050 Gb(84)n:BYDV13T Barley yellow dwarf Virus RNA (13t) for 3' terminal. 8/94 855bp.

Z14123 Em(40)_vi:BYDVCP Gb(84)_vi:BYDVCP Barley yellow dwarf virus gene for capsid protein. 7/92 647bp.

*Non–genomic nucleic acid* found in the virions (Miller *et al.*, 1987); is satellite RNA. Sub–genomic *m*RNA found in infected cells (Gerlach *et al.*, 1987).

*Virion protein(s)* one; $M_r$ 23500 (MAV), or 24400 (PAV), or 24450 (RPV) (Scalla and Rochow, 1977; Hammond *et al.*, 1983). Method of preparation: Scalla and Rochow (1977).

*Virus–coded non–virion proteins* identified by genomic sequence analysis; six proteins found. $M_r$ 95000; probably a polymerase, read through protein. $M_r$ of 2nd largest 69000; possibly with a transport function. $M_r$ of 3rd 34000; probably a polymerase. $M_r$ of 4th 22000; coat protein. $M_r$ of 5th and smaller 6000; possibly VPg; 17000 internal gene coding for coat protein.

**Replication**

Replication does not depend on a helper virus.

**Cytopathology**

Virions found in phloem (Jensen, 1969); in cytoplasm and in nuclei; and around plasmodesmata. Viruses of different BYDV subgroups alter the nucleus in different ways (Gill and Chong, 1975, 1976, 1979*a*, 1979*b*). Other cellular changes: the degeneration of phloem tissue to which virions are confined.

Subgroup 1 viruses initially cause malformation and lobing of the nucleus, in which, later, masses of densely stained material accumulate, before it disintegrates (Gill and Chong, 1975).

Subgroup 2 viruses induce changes that are quite different. Heterochromatin disappears during the later phases of infection, and the nucleus devlops a double membrane with defunct nucleoplasm.

**Taxonomy and Relationships**

*Luteovirus* (BYDV-MAV, the type species).

Virus(es) with serologically related virions: BYDV-MAV, BYDV-PAV and BYDV-SGV belong to one group and are related to soybean dwarf virus, whereas BYDV-RPV (including ryegrass chlorotic streak isolates), BYDV-RGV and BYDV-RMV belong to a second which is closely related to beet western yellows virus (BWYV; Duffus and Rochow, 1978; Rochow and Duffus, 1978; Martin and D'Arcy, 1990), potato leafroll and carrot red leaf viruses, and more distantly to pea enation mosaic and southern bean mosaic sobemovirus. BYDV-RPV and BWYV are serologically close, induce similar cytological effects and have similar genome sequences, which are related to those of carmoviruses, dianthoviruses and tombusviruses.

Best tests for diagnosis: field symptoms easily confused with nutrient or water deficiencies. The disease resembles aster yellows MLO in small grains (Banttari, 1965; Gill *et al.*, 1969).

**References**


Aapola, A.I.E. (1968). *Ph.D. Thesis*, Cornell University, U.S.A.

Aapola, A.I.E. and Rochow, W.F. (1968). *Phytopathology* **58**: 398.

Banttari, E.E. (1965). *Phytopathology* **55**: 838.

Barbara, D.J., Kawata, E.E., Ueng, P.P., Lister, R.M. and Larkins, B.A. (1987). *J. gen. Virol.* **68**: 2419.

Brakke, M.K. and Rochow, W.F. (1974). *Virology* **61**: 240.

Bruehl, G.W. (1961). *Am. Phytopathol. Soc. Monograph* No. 1.

Büchen-Osmond, C., Crabtree, K.J., Gibbs, A.J. and McLean, G.D. (1988). In: *Viruses of Plants in Australia*, p. 402, Aust. Nat. Univ., Canberra.

Catherall, P.L. (1974). *Pl. Path.* **23**: 116.

Clark, M.F. and Christensen, M.J. (1972). *N.Z. J. agric. Res.* **15**: 179.

D'Acry, C.J., Hewings, A.D., Burnett, P.A. and Jedlinski, H. (1983). *Phytopathology* **73**: 755.

Duffus, J.E. and Rochow, W.F. (1978). *Phytopathology* **68**: 45.

Gerlach, W., Miller, A. and Waterhouse, P.M. (1987). *BYDV Newsl.* **1**: 17.

Gill, C.C. and Chong, J. (1975). *Virology* **66**: 440.

Gill, C.C. and Chong, J. (1975). *Virology* **75**: 33.

Gill, C.C. and Chong, J. (1975). *Phytopathology* **69**: 363.

Gill, C.C. and Chong, J. (1975). *Virology* **95**: 59.

Gill, C.C., Westdal, P.H. and Richardson, H.P. (1975). *Phytopathology* **59**: 527.

Hammond, J., Lister, R.M. and Foster, J.E. (1983). *J. gen. Virol.* **64**: 667.

Heagy, J. and Rochow, W.F. (1965). *Phytopathology* **55**: 809.

Hebert, T.T. (1963). *Phytopathology* **53**: 362.

Hewish, D.R. and Shukla, D.D. (1983). *J. virol. Meth.* **7**: 223.

Jedlinski, H. (1972). *Pl. Dis. Reptr* **56**: 230.

Jedlinski, H. and Brown, C.M. (1969). *Virology* **38**: 83.

Lister, R.M. and Rochow, W.F. (1979). *Phytopathology* **62**: 921.

Lister, R.M. and Sward, R.J. (1988). *Phytopathology* **78**: 766.

Martin, R.R. and D'Arcy, C.J. (1990). *Intervirology* **31**: 23.

Miller, W.A., Waterhouse, P.M., Gerlach, W.L. and Helms, K. (1987*a*). *Abstr. 7th. Conf. Virol., Edmonton, Canada*, p. 329.

Miller, A., Waterhouse, P.M., Gerlach, W.L. and Helms, K. (1987*b*). *Phytopathology* **77**: 1704.

Orlob, G.B., Arny, D.C. and Medler, J.T. (1961). *Phytopathology* **51**: 515.

Oswald, J.W. and Houston, B.R. (1951). *Pl. Dis. Reptr* **35**: 471.

Oswald, J.W. and Houston, B.R. (1953*a*). *Phytopathology* **43**: 128.

Oswald, J.W. and Houston, B.R. (1953*b*). *Phytopathology* **43**: 309.

Paliwal, Y.C. (1982). *Can. J. Pl. Pathol.* **4**: 367.

Plumb, R.T. (1976). *Ann. appl. Biol.* **83**: 53.

Rochow, W.F. (1961). *Adv. Agron.* **13**: 217.

Rochow, W.F. (1963). *Phytopathology* **53**: 355.

Rochow, W.F. (1969). In: *Viruses, Vectors and Vegetation*, p. 175; ed. K. Maramorosch. Interscience, New York.

Rochow, W.F. (1970). *Science, N.Y.* **167**: 875.
Rochow, W.F. (1970). *CMI/AAB Descr. Pl. Viruses* No. 32, 4 pp.
Rochow, W.F. (1982). *Phytopathology* **72**: 302.
Rochow, W.F. (1986). In: *Methods of Enzymatic Analysis*, Vol. XI. p. 420; eds H.U. Bergmeyer and M. Grassl. Chemié Verlag, Weinheim, Germany.
Rochow, W.F. and Ball, E.W. (1967). *Virology* **33**: 359.
Rochow, W.F. and Brakke, M.K. (1964). *Virology* **24**: 310.
Rochow, W.F. and Carmichael, L.E. (1979). *Virology* **95**: 415.
Rochow, W.F. and Duffus, J.E. (1978). *Phytopathology* **68**: 51.
Rochow, W.F. and Muller, I. (1974). *Pl. Dis. Reptr* **58**: 472.
Scalla, R. and Rochow, W.F. (1977). *Virology* **78**: 576.
Slykhuis, J.T. (1967). *Rev appl. Mycol.* **46**: 401.
Smith, P.R. and Plumb, R.T. (1981). *Aust. J. agric. Res.* **32**: 249.
Smith, P.R. and Sward, R. (1982). *Aust. J. agric. Res.* **33**: 179.
Stoner, W.N. (1952). *Phytopathology* **42**: 683.
Sward, R.J. and Kollmorgen, J.F. (1986). *Aust. J. agric. Res.* **37**: 11.
Sward, R.J. and Lister, R.M. (1987). *Aust. J. agric. Res.* **38**: 821.
Sward, R.J. and Lister, R.M. (1988). *Aust. J. agric. Res.* **39**: 375.
Timian, R.G. (1964). *Phytopathology* **54**: 910.
Ueng, P.P., Vincent, J.R., Kawata, E.E., Lei, C.-H., Lister, R.M. and Larkins, B.A. (1992). *J. gen. Virol.* **73**: 487.
Vincent, J.R., Lister, R.M. and Larkins, B.A. (1991). *J. gen. Virol.* **72**: 2347.
Waterhouse, P.M. (1985). *Australas. Pl. Path.* **14**: 32.
Waterhouse, P.M., Gerlach, W.R. and Miller, W.A. (1986). *J. gen. Virol.* **67**: 1273.
Watson, M.A. and Mulligan, T.E. (1960). *Ann. appl. Biol.* **48**: 711.
Zhang, Q.F., Guan, W.N., Ren, Z.Y. and Zhu, X.S. (1983). *Plant Dis.* **67**: 895.

# Barley yellow mosaic *bymovirus*

**Data collated by M.J. Adams, 1991.**

*Standard acronym:* BaYMV. *Strains:* six reported from Japan (Kashiwazaki *et al.*. 1989*a*); the M and Streatley strains in Europe are now recognised as strains of barley mild mosaic bymovirus. Two strains reported from Germany (Huth, 1989*a*) and elsewhere in Europe.

First reported in *Hordeum vulgare*; from Japan; by Ikata and Kawai (1940); Miyamoto (1958); Inouye (1964).

### Natural host range and symptoms

Symptoms vary seasonally. Symptoms yellow streaks especially on youngest leaves in winter or spring.

*Hordeum vulgare* — yellow streaking, especially of younger leaves and sometimes of leaf sheath, also brown necrotic patches. Leaves may roll so the plants appear 'spiky'. Vigour reduced.

### Transmission

Transmitted by a vector; a fungus; *Polymyxa graminis*; Plasmodiophorales. Virus transmitted by mechanical inoculation; not transmitted by contact between plants; not transmitted by seed; not transmitted by pollen.

### Ecology and control

Studies reported by Huth (1989*b*); Usugi (1988).

**Geographical distribution**

Spreads in the Eastern Asian region and the Eurasian region; Belgium, China, France, Germany, Greece, Japan, and the UK.

**Experimental host range**

Few (<3) families susceptible. Experimentally infected plants mostly show yellow leaf streaking.

Diagnostically susceptible

*Hordeum vulgare* — systemic leaf streaking.

Diagnostically insusceptible

*Avena sativa, Oryza sativa, Secale cereale, Triticum aestivum, Zea mays.*

Maintenance and propagation hosts

*Hordeum vulgare.*

Assay hosts (**L**ocal lesions or **W**hole plants)

*Hordeum vulgare* (W).

Dicotyledonous Hosts (***Susceptible*** *or Insusceptible)*

*Chenopodium amaranticolor. Chenopodium quinoa. Nicotiana tabacum.*

Monocotyledonous Hosts (***Susceptible*** *or Insusceptible)*

*Avena sativa.* ***Hordeum vulgare****. Lolium multiflorum. Oryza sativa. Triticum aestivum. Zea mays.*

Sources of host-range data

Miyamoto (1958); Ruan and Jin (1983).

**Properties of particles in sap**

DEP: $\log_{10}$ minus 2. Leaf sap contains few virions.

**Purification**

Method: Huth *et al.* (1984); Pröls *et al.* (1990).

**Particle morphology**

Virions filamentous; not enveloped; usually flexuous; with clear modal lengths (two); of 270–290 nm, or 570–600 nm; 12–15 nm wide. Axial canal obscure. Basic helix obscure.

**Physical properties**

Density 1.32 g $cm^{-3}$ in CsCl. $A_{260}/A_{280}$ ratio 1.14.

**Biochemical properties**

Virions contain 5 % nucleic acid; 95 % protein.

*Genome* consists of RNA; single–stranded; linear; of two parts; largest (or only) genome part the larger 7.632 kb; the 2nd largest 3.585 kb. Genomic nucleic acid isolated by Kashiwazaki *et al.* (1989*b*). Base composition 22.1 % G; 27.4 % A; 24 % C; 26.5 % U. Poly A region present; at the 3' terminus of both RNAs. Additional factor not required for infectivity. Nucleotide sequence references: Kashiwazaki *et al.* (1990); Kashiwazaki *et al.* (1991); Davidson *et al.* (1991).

*Sequence database accession code(s):*

D00544 Em(40)_vi:BYMCP Gb(84)_vi:BYMCP Barley yellow mosaic virus capsid protein gene, 3' end. 8/92 1,370bp.

D01091 Em(40)_vi:BYMR1P Gb(84)_vi:BYMR1P Barley yellow mosaic virus RNA 1 encoding polyprotein. 8/92 7,632bp.

D01092 Em(40)_vi:BYMR2P Gb(84)_vi:BYMR2P Barley yellow mosaic virus RNA 2 encoding polyprotein. 8/92 3,585bp.

D01099 Em(40)_vi:BYMGR2P Gb(84)_vi:BYMGR2P Barley yellow mosaic virus RNA 2 encoding polyprotein. 1/94 3,585bp.

X69757 Em(40)_vi:BAYMVRNA1 Gb(84)_vi:BAYMVRNA1 Barley yellow mosaic virus RNA for polyprotein RNA1. 1/93 7,648bp.

X69836 Em(40)_vi:BYMV270KD Gb(84)_vi:BYMV270KD Barley Yellow Mosaic Virus-2 DNA for 70 kDa protein. 3/93 621bp.

Z24677 Em(40)_vi:BYCOAPROA Gb(84)_vi:BYCOAPROA Barley yellow mosaic virus coat protein mRNA, complete CDS. 7/93 1,313bp. 7 sequences.

*Features of the genome:*

RNA 1: single ORF encoding a 270 kDa protein which is processed to produce the capsid protein (c-terminal) and at least 3 non-structural proteins.

RNA 2: single ORF encoding a 98 kDa protein; function not yet determined.

*Non–genomic nucleic acid* not found in the virions.

*Virion protein(s)* one; $M_r$ 32234. Method of preparation: Huth *et al.* (1984); Usugi *et al.* (1989). Amino acid sequence: Kashiwazaki *et al.* (1989*b*).

**Replication**

Replication does not depend on a helper virus.

**Cytopathology**

Virions found in leaves, roots, mesophyll, epidermis, vascular parenchyma and companion cells; in cytoplasm. Inclusions present in infected cells; are amorphous X–bodies, membranous bodies, and pinwheels; they do not contain virions.

**Taxonomy and Relationships**

*Bymovirus: Potyviridae.*

Virus(es) with serologically related virions: rice necrosis mosaic, oat mosaic, wheat yellow mosaic and wheat spindle streak mosaic viruses. Virus(es) with serologically unrelated virions: barley mild mosaic virus.

Best tests for diagnosis: leaf streaking and unfurling distinguish this from barley yellow dwarf luteovirus. Serological tests required to distinguish it from barley mild mosaic bymovirus with which it is sometimes associated.

**References**


Davidson, A.D., Pröls, M., Schell, J. and Steinbiss, H.-H. (1991). *J. gen. Virol.* **72**: 989.

Huth, W. (1988). In: *Viruses with Fungal Vectors*, p. 61; eds J.I. Cooper and M.J.C. Asher, Association of Applied Biologists.

Huth, W. (1989*a*). *Nach. dtsch. PflSchutz.* **41**: 6.

Huth, W. (1989*b*). *EPPO Bulletin* **19**: 547.

Huth, W. and Adams, M.J. (1990). *Intervirology* **31**: 38.

Huth, W., Lesemann, D.-E. and Paul, H.L. (1984). *Phytopath. Z.* **111**: 37.

Ikata, S. and Kawai, I. (1940). *Noji Kairyo Shiryo* **154**: 1.

Inouye, T. (1964). *Nogaku Kenkyu* **50**: 117.

Kashiwazaki, S., Ogawa, K., Usugi, T., Omura, T. and Tsuchizaki, T. (1989*a*). *Ann. Phytopath. Soc. Japan* **55**: 16.

Kashiwazaki, S., Hayano, Y., Minobe, Y., Omura, T., Hibino, H. and Tsuchizaki, T. (1989*b*). *J. gen. Virol.* **70**: 3015.

Kashiwazaki, S., Minobe, Y., Omura, T. and Hibino, H. (1990). *J. gen. Virol.* **71**: 2781.

Kashiwazaki, S., Minobe, Y. and Hibino, H. (1991). *J. gen. Virol.* **72**: 995.

Miyamoto, Y. (1958). *Ann. Phytopath. Soc. Japan* **23**: 199.

Pröls, M., Davidson, A., Schell, J. and Steinbiss, H.-H. (1990). *J. Phytopath.* **130**: 249.

Ruan, Y. and Jin, D. (1983). *Acta phytopath. Sin.* **13**: 49.

Usugi, T. (1988). In: *Viruses with Fungal Vectors*, p. 213; eds J.I. Cooper and M.J.C. Asher, Association of Applied Biologists.

Usugi, T., Kashiwazaki, S., Omura, T. and Tsuchizaki, T. (1989). *Ann. Phytopath. Soc. Japan* **55**: 26.

# Barley yellow streak mosaic virus

**Data collated by N.L. Robertson, 1991.**

First reported in *Hordeum vulgare*; from Conrad, Montana, U.S.A.; by Robertson and Carroll (1988).

**Natural host range and symptoms**

Symptoms vary cyclically over a few weeks and vary seasonally. Symptoms light green to yellow dashes, streaks and stripes parallel to the mid-rib, producing mosaic pattern.

*Echinochloa crus-galli, Hordeum vulgare, Lolium persicum, Setaria viridis, Triticum aestivum* — mosaic and banding.

**Transmission**

Transmitted by a vector; a mite; *Petrobia latens*; Tetranychidae. Not transmitted by *Endria inimica, Macrosteles fascifrons, Balchutha neglecta, Psammotettix alienus, P. lividellus, Hebecephalus occidentalis, Dicraneura probicarneola, Amblysellus grex, Auridius auratus, Sorhoanus uhleri, Athysanella attenuata, A. acuticauda, A. occidentalis, Rhopalosiphum padi, R. maidis, Macrosiphum (Sitobion) avenae, Limothrips denticornis, Anapothrips obscurus brachypterus, Thrips tabaci, Frankliniella triici, F. fusca.* Virus transmitted by mechanical inoculation; not transmitted by contact between plants; not transmitted by seed; not transmitted by pollen.

**Ecology and control**

Studies reported by Robertson (1987); Robertson and Carroll (1988).

**Geographical distribution**

Spreads in the North American region; Canada and the USA.

**Experimental host range**

Several (3–9) families susceptible. Experimentally infected plants mostly show mosaics.

Diagnostically susceptible

*Chenopodium quinoa* — few necrotic local lesions.

*Nicotiana benthamiana* — chlorotic and necrotic local lesions and rings; new leaves show systemic symptoms of vein-chlorosis and yellow mosaics. Severely diseased plants were reduced in height and leaf size. Remission occurs on plants with faint chlorotic spots or mild mosaics.

*Echinochloa crus-galli, Hordeum vulgare, Lolium persicum, Setaria viridis, Triticum aestivum* — mosaic of streaks and dashes, some banding; systemic.

Diagnostically insusceptible

*Chenopodium amaranticolor, Gomphrena globosa, Nicotiana clevelandii, N. debneyi, N. glutinsoa.*

Maintenance and propagation hosts

*Hordeum vulgare, Nicotiana benthamiana.*

Assay hosts (**L**ocal lesions or **W**hole plants)

*Chenopodium quinoa* (L), *Hordeum vulgare* (W), *Nicotiana benthamiana* (W).

Dicotyledonous Hosts *(**Susceptible** or Insusceptible)*

*Brassica campestris* ssp. *chinensis*. *Capsicum annuum*. *Chenopodium amaranticolor*. ***Chenopodium quinoa***. *Cucumis melo*. *Cucumis sativus*. *Datura stramonium*. *Daucus carota*. *Gomphrena globosa*. ***Nicotiana benthamiana***. *Nicotiana clevelandii*. *Nicotiana* × *edwardsonii*. *Nicotiana debneyi*. *Nicotiana glutinosa*. *Nicotiana megalosiphon*. *Nicotiana rustica*. *Nicotiana sylvestris*. *Petunia hybrida*. *Physalis floridana*.

Monocotyledonous Hosts *(**Susceptible** or Insusceptible)*

*Agropyron repens*. *Avena sativa*. ***Echinochloa crus–galli***. ***Hordeum vulgare***. ***Lolium persicum***. ***Setaria viridis***. ***Triticum aestivum***.

Comments on host-range

Susceptible monocotyledons are only infected by the mite vector, not by sap inoculation, whereas the reverse is true for *Nicotiana benthamiana*.

Sources of host-range data

Robertson and Carroll (1988; 1991.

**Properties of particles in sap**

Leaf sap contains few virions, or contains many virions. Electron microscopy: aldehyde fixation helps to keep virions intact.

**Purification**

Method: Robertson and Carroll (1991).

**Particle morphology**

Virions unusually shaped; enveloped; usually flexuous; with no clear modal length; of 127–4000 nm; 52–81 nm wide. Basic helix obscure.

**Biochemical properties**

*Genome* consists of RNA; single–stranded; linear; unipartite; largest (or only) genome part 10 kb. Genomic nucleic acid isolated by Robertson and Carroll (1991). Infectivity lost when deproteinised with phenol or detergent.

*Virion protein(s)* two (at least); $M_r$ of the larger 100. $M_r$ of 2nd largest 32. Method of preparation: Robertson and Carroll (1991).

**Replication**

Genome replicates in cytoplasm.

**Cytopathology**

Virions found in leaves, roots, mesophyll, epidermis, vascular parenchyma, xylem, phloem, companion cells and all parts of the host plant; in cytoplasm. Inclusions present in infected cells; are crystals in the cytoplasm.

**Taxonomy and Relationships**

Best tests for diagnosis: electron microscopy of leaf dips; infected leaf sap mechanically inoculated to *Nicotiana benthamiana* seedlings.

**References**

Robertson, N.L. (1987). *Ph.D. Thesis*, Montana State University, U.S.A.

Robertson, N.L. and Carroll, T.W. (1988). *Science* **240**: 1188.

Robertson, N.L. and Carroll, T.W. (1989). *J. Ultrastruct. and Mol. Struct. Res.* **102**: 139.

Robertson, N.L. and Carroll, T.W. (1991). *Plant Dis.* **75**:: 839.

# Barley yellow striate mosaic *cytorhabdovirus*

**Data collated by R.G. Milne and M. Conti, 1985.**

*Synonyms:* cereal striate mosaic virus (Matthews, 1982). *Standard acronym:* BYSMV. *Strains:* maize sterile stunt virus (Greber, 1982; 1983*a* and *b*; 1984*a* and *b*; Milne *et al.*, 1986), Moroccan wheat rhabdovirus (Milne *et al.*, 1986), wheat chlorotic streak virus (Signoret *et al.* 1976), wheat chlorotic streak mosaic virus (Milne *et al.*, 1986; Signoret *et al.*, 1972; 1977).

First reported in *Triticum durum, T. compactum, T. aestivum*; from northern Italy; by Conti (1969*a*); Signoret *et al.* (1972).

**Natural host range and symptoms**

*Agropyron repens, Triticum durum, T. compactum, T. aestivum, Zea mays* — chlorotic stripes or mosaic.

*Avena sativa, Secale cereale* — yellow mosaic, red basal leaves.

*Setaria viridis, Setaria* spp. — chlorotic streaks and stripes in leaves; stunting.

**Transmission**

Transmitted by a vector; an insect; *Laodelphax striatellus* (Conti 1969*a*, 1969*b*, 1974, 1980; Conti and Plumb, 1977; Leclant and Signoret, 1976; Signoret *et al.*, 1976; Signoret *et al.*, 1977). *Javesella pellucida* transmits the virus experimentally (Conti, 1980); Delphacidae. Transmitted in a persistent manner. Virus retained when the vector moults; multiplies in the vector; transmitted congenitally to the progeny of the vector; not transmitted by mechanical inoculation; not transmitted by seed (Conti, 1980).

**Ecology and control**

Studies reported by Conti (1969*a*); Signoret *et al.* (1976); Conti and Milne, (1977); Amici *et al.* (1978); Conti (1980) Lockhart and Elyamani (1983).

**Geographical distribution**

Spreads in the African region, the Eurasian region, the Middle East, and the Pacific region; Australia, France, Iran, Italy, and Morocco.

**Experimental host range**

Few (<3) families susceptible. Experimentally infected plants mostly show chlorosis, reddening.

Diagnostically susceptible

*Avena sativa* cv. Alba — yellow mosaic, reddening of basal leaves.

*Bromus aleutensis, B. arvensis, B. rigidus* (=*B. maximus*) — stunting.

*Hordeum vulgare* cv. Delisa — bright yellow mosaic.

*Setaria italica* — chlorotic stripes.

Diagnostically insusceptible

*Digitaria sanguinalis, Echinochloa crus-galli, Festuca pratensis, Holcus lanatus, Phleum pratense.*

Maintenance and propagation hosts

*Hordeum vulgare.*

Assay hosts (**L**ocal lesions or **W**hole plants)

*Hordeum vulgare* (W).

Monocotyledonous Hosts *(**Susceptible** or Insusceptible)*

***Agropyron repens. Avena sativa. Bromus mollis. Dactylis glomerata.*** *Echinochloa crus–galli. Festuca pratensis.* ***Hordeum vulgare.*** *Phleum pratense.* ***Secale cereale. Setaria italica. Triticum aestivum. Zea mays.***

Additional hosts

Transmitted by the plant hopper vector to 26 of 44 species in Gramineae only (Conti, 1980), in the genera *Lagurus* (Agrostideae), *Avena* and *Trisetum* (Aveneae); *Briza, Bromus, Dactylis* and *Poa* (Festuceae); *Hordeum, Lolium, Triticum* and *Secale* (Andropogoneae); *Zea* (Maydeae) and *Panicum* and *Secale* (Paniceae).

Sources of host-range data

Conti (1969*a*); Signoret *et al.*, (1976); Conti and Milne (1977); Amici *et al.*, (1978); Conti (1980); Lockhart and Elyamani (1983).

**Properties of particles in sap**

TIP: 50–60 °C. LIV: 1–2 days (at 22°C and 2–4 days at 5°C (Conti, 1980)). DEP: $\log_{10}$ minus 2–3. Leaf sap contains few virions. Electron microscopy: fix in glutaraldehyde; unfixed virions disrupt in neutral PTA and UA.

**Purification**

Method: Luisoni (1969); Milne and Conti (1986).

**Particle morphology**

Virions rhabdo– or bullet–shaped; enveloped; with a clear modal length; of 270 nm (Conti, 1969*a*), but longer in thin sections or UA (Conti and Appiano, 1974; Appiano and Conti, 1973; Conti and Plumb, 1977)); 55 nm wide. Basic helix of nucleocapsid obvious (after envelope has been removed with chloroform, 0.2% Nonidet P 40 or 0.2% Triton X-100); pitch of basic helix 4.4 nm.

**Physical properties**

One sedimenting component in purified preparations.

**Biochemical properties**

*Genome* consists of RNA; single–stranded (Bassi *et al.*, 1980).

**Replication**

Replication does not depend on a helper virus.

**Cytopathology**

Virions found in phloem; in cytoplasm; within membrane-bound cisternae, from which the virus buds, but not found in nucleus or perinuclear space. Inclusions present in infected cells; are viroplasms.

**Taxonomy and Relationships**

*Cytorhabdovirus: Rhabdoviridae.*

Virus(es) with serologically related virions: Moroccan wheat rhabdovirus, wheat chlorotic streak, maize sterile stunt (SDI:2), northern cereal mosaic (SDI:6), digitaria striate and wheat rosette stunt (SDI:6) viruses. Maize sterile stunt (MSSV) is a serotype of BYSMV (2 SDI units) (Greber, 1984*b*). Virus(es) with serologically unrelated virions: festuca leaf streak virus.

**References**

Amici, A., Faoro, F. and Tornaghi, R. (1978). *Riv. Patol. veg. S IV* **14**: 85.

Appiana, A. and Conti, M. (1974). *J. Submicrosc. Cytol.* **6**: 105.

Bassi, M., Barbieri, N., Appiano, A., Conti, M., D'Agostino, G. and Caciagli, P. (1980). *J. Submicrosc. Cytol.* **12**: 201.

Conti, M. (1969*a*). *Phytopath. Z.* **66**: 275.

Conti, M. (1969*b*). *Ric. Sci.* **39**: 701.

Conti, M. (1972). *Phytopath. Z.* **73**: 39.

Conti, M. (1974). *Microbiologica* **11**: 49.

Conti, M. (1980). *Ann. appl. Biol.* **95**: 83.

Conti, M. and Appiano, A. (1973). *J. gen. Virol.* **21**: 315.

Conti, M. and Milne, R.G. (1977). *Annls. Phytopath.* **9**: 255.

Conti, M. and Plumb, R.T. (1977). *J. gen. Virol.* **34**: 107.

Greber, R.S. (1979). *Aust. J. agric. Res.* **30**: 43.

Greber, R.S. (1982). *Aust. J. agric. Res.* **33**: 13.

Greber, R.S. (1983*a*). *Aust. Microbiol.* **4**: 100.

Greber, R.S. (1983*b*). In: *Proc. Int. Maize Virus Disease Colloq. and Workshop, Wooster, Ohio State Univ., 1982*, pp. 206–209; eds D. Gordon, J. Knoke, L. Nault and R. Ritter.

Greber, R.S. (1984*a*). In: *Maize Virus Disease Newsl.*, Vol. 1 p. 46; ed. L. Autrey. Ohio State Univ., Wooster.

Greber, R.S. (1984*b*). In: *Current Topics in Vector Research*, Vol. 2, p 141; ed. K. Harris. Praeger Scientific, New York,

Izadpanah, K., Ebrahim-Nesbak, F. and Afsharifar, A.R. (1991). *J. Phytopath.* **131**: 290.

Leclant, F. and Signoret, P.A. (1976). *C.R. IV Journées Phytiatr. Phytopharm. Circum-Méditerr. Montpellier 1974*, pp. 215–217. Rulliere-Libeccio, Avignon.

Lockhart, B.E.L. and Elyamani, M. (1983). In: *Proc. Int. Maize Virus Dis. Colloq. and Wksh., Wooster, Ohio State Univ., 1982*, pp. 127–129; eds D. Gordon, J. Knoke, L. Nault and R. Ritter.

Luisoni, E. (1969). *Ric. Sci.* **39**: 708.

Matthews, R.E.F. (1982). *Intervirology* **17**: 1.

Milne, R.G. and Conti, M. (1986). *CMI/AAB Descr. Pl. Viruses* No. 312, 5 pp.

Milne, R.G., Masenga, V. and Conti, M. (1986). *Intervirology* **25**: 83.

Signoret, P.A. Conti, M., Leclant, F., Alliot, B. and Giannotti, J. (1977). *Annls. Phytopath.* **9**: 381.

Signoret, P.A., Giannotti, J. and Alliot, B. (1972). *Annls. Phytopath.* **4**: 45.

Signoret, P.A., Leclant, F. and Alliot, B. (1976). *Poljopr. znanst. Smotra 39* **49**: 183.

# Bean calico mosaic bigeminivirus

**Data collated by J.K. Brown, 1990.**

*Standard acronym:* BCaMV.

First reported in *Phaseolus vulgaris*; from Mexico; by Brown *et al.* (1988).

## Natural host range and symptoms

Symptoms persist and vary seasonally.
*Phaseolus vulgaris* cv. Red Kidney, *Phaseolus lunatus* — leaf curl, yellow mosaic.
*Malva parviflora* — yellow mosaic, leaves stunted and puckered.

## Transmission

Transmitted by a vector; an insect; *Bemisia tabaci*; Aleyrodidae. Transmitted in a persistent manner. Virus does not multiply in the vector; not transmitted congenitally to the progeny of the vector; transmitted by mechanical inoculation; transmitted by grafting; not transmitted by contact between plants; not transmitted by seed.

## Ecology and control

Studies reported by Brown *et al.* (1989).

## Geographical distribution

Spreads in the North American region; Mexico.

## Experimental host range

Several (3–9) families susceptible. Experimentally infected plants mostly show mosaics, mottles, ringspots or necrosis first on youngest leaves as vein clearing.

Diagnostically susceptible

*Phaseolus vulgaris* cvs Red Kidney, Top Crop, *Phaseolus lunatus* — yellow mosaic, stunted.

*Phaseolus aureus, P. acutifolius, Glycine max* cv. Frost Beater — systemic vein clearing and necrosis, mosaic curled leaves.

*Vigna unguiculata* cv. Californian Blackeye — malformed leaves and recovery.

Diagnostically insusceptible

*Gossypium hirsutum* cv. DP90, *Datura stramonium, Lactuca sativa* cv. Empire, *Cucurbita maxima* cv. Big Max, *Capsicum annuum* cv. Anaheim.

Maintenance and propagation hosts

*Phaseolus vulgaris* cvs Red Kidney, Top Crop. *Phaseolus lunatus* cv. Henderson Bush.

Assay hosts (**L**ocal lesions or **W**hole plants)

*Phaseolus vulgaris* cv. Red Kidney (W), *P. vulgaris* cv. Top Crop (W), *P. lunatus* cv. Henderson Bush (W), *Nicotiana benthamiana* (W).

Dicotyledonous Hosts *(**Susceptible** or Insusceptible)*

*Amaranthus retroflexus. Arachis hypogaea. Beta vulgaris. Brassica campestris* ssp. *chinensis. Brassica campestris* ssp. *rapa. Capsella bursa–pastoris. Capsicum annuum. Cassia obtusifolia. Chenopodium album. Chenopodium capitatum. Chenopodium murale. Cicer arietinum. Citrullus lanatus. Cucumis melo. Cucumis sativus. Cucurbita maxima. Cyamopsis tetragonoloba. Datura stramonium.* ***Glycine max****. Gossypium hirsutum. Helianthus annuus. Hibiscus cannabinus. Lactuca sativa. Lycopersicon esculentum.* ***Nicotiana benthamiana*****.** ***Phaseolus acutifolius*****.** ***Phaseolus lunatus*****.** ***Phaseolus vulgaris*****.**

*Physalis floridana. Pisum sativum. Solanum melongena. Sonchus oleraceus. Spinacia oleracea.* ***Vigna radiata. Vigna unguiculata.*** *Zinnia elegans.*

Monocotyledonous Hosts *(**Susceptible** or Insusceptible)*

*Zea mays.*

Sources of host-range data

Brown *et al.* (1988); Brown *et al.* (1989).

**Properties of particles in sap**

Leaf sap contains few virions. Electron microscopy: fix with 0.25% glutaraldehyde.

**Purification**

Method: Brown and Nelson (1989).

**Particle morphology**

Virions geminate; not enveloped; 20 nm in diameter; 30 nm in length; rounded in profile, or angular in profile; without a conspicuous capsomere arrangement.

**Physical properties**

One sedimenting component in purified preparations.

**Biochemical properties**

Virions contain 20 % nucleic acid; 80 % protein.

*Genome* consists of DNA; single–stranded; circular; of two parts; largest (or only) genome part the larger 2.6–2.7 kb; the 2nd largest 2.6–2.7 kb. Genomic nucleic acid isolated by Brown: resuspend purified virions in Tris-EDTA (10mM each) plus 0.1M NaCl, pH 8.0, 1% SDS; add 50ug/ml proteinase K; incubate 42°C 1 hour; phenol:$CHCl_3$: (50:48:2) extract; precipitate with ethanol at — 20°C overnight.

*Sequence database accession code(s):*

L22757 Em(40)_vi:BCOORGRPL Gb(84)_vi:BCOORGRPL Bean calico mosaic virus origin of replication. 1/94 169bp.

L22758 Em(40)_vi:BCOREPR Gb(84)_vi:BCOREPR Bean calico mosaic virus replicative protein gene, 5' end. 1/94 659bp.

L27264 Em(40)_vi:GEBCOAC1A Gb(84)_vi:BCOAC1A Bean calico mosaic geminivirus replication-associated protein (AC1) gene, partial cds. 1/94 2

L27266 Em(40)_vi:GETLGBC1A Gb(84)_vi:TLGBC1A Bean calico mosaic virus movement protein (BC1) gene, partial cds. 1/94 204bp.

U00120 Em(40)_vi:BCMOVCR Gb(84)_vi:BCMOVCR Phaseolus vulgaris calico mosaic virus origin of replication sequence. 6/94 169bp.

U00121 Em(40)_vi:BCMAL1 Gb(84)_vi:BCMAL1 Phaseolus vulgaris calico mosaic virus replicative protein (AL1) gene, partial cds. 6/94 659b 6 sequences.

Method of preparation: Brown and Nelson (1989).

**Replication**

Replication does not depend on a helper virus.

**Cytopathology**

Inclusions present in infected cells; are unusual in shape.

**Taxonomy and Relationships**

Bigeminivirus: *Geminiviridae.*

Virus(es) with serologically related virions: bean golden mosaic and watermelon curly mottle viruses. Virus(es) with serologically unrelated virions: five other legume geminiviruses which induce similar symptoms in common hosts.

Best tests for diagnosis: use whitefly (*Bemisia tabaci*) transmission to *Malva parviflora* to separate from other viruses commonly found in mixed infections in Mexico.

**References**

Brown, J.K. and Nelson, M.R. (1989). *Ann. appl. Biol.* **115**: 243.
Brown, J.K., Jimenez-Garcia, E. and Nelson, M.R. (1988). *Phytopathology* **78**: 1579.
Brown, J.K., Chapman, M.A. and Nelson, M.R. (1990). *Plant Dis.* **73**: **74**: 81.

# Bean common mosaic necrosis *potyvirus*

*Synonyms:* bean common mosaic virus — serotype A. *Standard acronym:* BCMNV.

**Particle morphology**

Virions of 810–818 nm.

**Taxonomy and Relationships**

*Potyvirus: Potyviridae.*

# Bean common mosaic *potyvirus*

**Data collated by L. Bos, 1981. Revised 1987. Condensed by A.J. Gibbs, 1995.**

*Synonyms:* bean common mosaic virus — serotype B, bean mosaic virus, bean western mosaic virus (Bos, 1964), mungbean mosaic virus (Abu Kassim, 1981; Kaiser and Mossahebi, 1974; Rao *et al.*, 1986). *Standard acronym:* BCMV. *Strains:* Many strains of the virus have been distinguished (Drijfhout *et al.*, 1978). Those once grouped as serotype A are now considered isolates of a separate potyvirus species, bean common necrosis virus, and several viruses, once considered to be distinct, have now been shown to be strains of this virus (McKern *et al.*, 1992). The latter include azuki bean mosaic, blackeye cowpea mosaic, cowpea (aphid-borne) mosaic, cowpea (blackeye) mosaic, cowpea vein-banding mosaic, peanut blotch and peanut stripe viruses and some isolates from soybean.

First reported in *Phaseolus vulgaris*; from New York, U.S.A.; by Stewart and Reddick (1917); Pierce (1934).

**Natural host range and symptoms**

Symptoms persist.

*Phaseolus vulgaris, Phaseolus coccineus* — tolerant cultivars develop mosaic and malformation, sensitive ones show rugosity of lower leaves, mosaic, malformation of leaves and pods, those with the dominant necrosis gene develop vein necrosis, 'black root' and death (especially above 26°C).

**Transmission**

Transmitted by a vector; an insect; *Acyrthosiphon pisum, Aphis craccivora, A. fabae, Myzus persicae* and other spp.; Aphididae. Transmitted in a non–persistent manner. Virus transmitted by mechanical inoculation; transmitted by seed (up to 83% in *Phaseolus vulgaris* and from 7–22% in tepary bean); transmitted by pollen to the seed.

**Ecology and control**

Studies reported by Drijfhout (1978) on breeding for resistance.

**Geographical distribution**

Probably distributed worldwide (in *Phaseolus* beans wherever they are grown). Spreads in China and the USA.

**Experimental host range**

Several (3–9) families susceptible (strain specific response). Experimentally infected plants mostly show mosaics, local lesions, necrosis — very variable.

Diagnostically susceptible

*Chenopodium quinoa* — faint chlorotic local lesions developing into green rings; not systemic.

*Macroptilium lathyroides* — necrotic local lesions; systemic necrosis.

*Phaseolus vulgaris* cvs Dubbele Witte, Stingless Green Refugee — green vein-banding, malformed leaves (Morales and Bos, 1987).

*Pisum sativum, Vicia faba* — symptomless.

Diagnostically insusceptible

*Cucumis sativus, Medicago sativa, Nicotiana tabacum, N. glutinosa, Pisum sativum.*

Maintenance and propagation hosts

*Phaseolus vulgaris*; cultivars Dubbele Witte and Stringless Green Refugee susceptible to all known strains.

Assay hosts (**L**ocal lesions or **W**hole plants)

*Chenopodium amaranticolor, C. quinoa, Phaseolus vulgaris; Pisum sativum, Vicia faba* latent or no infection.

Dicotyledonous Hosts *(**Susceptible** or Insusceptible)*

*Arachis hypogaea. Bauhinia purpurea.* ***Cajanus cajan****. Capsicum annuum.* ***Cassia tora****.* ***Chenopodium amaranticolor****.* ***Chenopodium quinoa****.* ***Cicer arietinum****.* ***Crotalaria spectabilis****. Cucumis sativus.* ***Cyamopsis tetragonoloba****. Datura stramonium.* ***Glycine max****.* ***Gomphrena globosa****. Lablab purpureus. Lactuca sativa. Lathyrus odoratus.* ***Lens esculenta****.* ***Lupinus albus****.* ***Lupinus angustifolius****. Lycopersicon esculentum.* ***Macroptilium lathyroides****. Medicago sativa.* ***Melilotus alba****.* ***Nicotiana benthamiana****.* ***Nicotiana clevelandii****. Nicotiana glutinosa. Nicotiana tabacum.* ***Phaseolus lathyroides****.* ***Phaseolus lunatus****.* ***Phaseolus vulgaris****. Pisum sativum.* ***Sesbania exaltata****.* ***Tetragonia expansa****. Trifolium hybridum.* ***Trifolium incarnatum****. Trifolium pratense. Trifolium repens.* ***Trifolium subterraneum****.* ***Vicia faba****.* ***Vicia sativa****.* ***Vicia villosa****.* ***Vigna angularis****.* ***Vigna radiata****.* ***Vigna unguiculata****. Vigna unguiculata* ssp. *sesquipedalis. Zinnia elegans.*

Sources of host-range data

Zaumeyer and Goth (1964); Thornberry (1966); Edwardson (1974); Drijfhout *et al.* (1977).

**Properties of particles in sap**

TIP: 60 °C. LIV: 1–4 days. DEP: $\log_{10}$ minus 3–4. Leaf sap contains few virions.

**Purification**

Method: Morales (1979).

**Particle morphology**

Virions filamentous; not enveloped; usually flexuous; of 847–886 nm; 12–15 nm wide. Axial canal obscure. Basic helix obscure.

**Physical properties**

One sedimenting component in purified preparations; sedimentation coefficient 154–158 *S*. Density 1.31–1.32 g $cm^{-3}$ in CsCl.

**Biochemical properties**

Virions contain 5 % nucleic acid; 95 % protein.

*Genome* consists of RNA; single–stranded; linear; unipartite; largest (or only) genome part 10 kb. Genomic nucleic acid isolated by Bravo (1984).

*Sequence database accession code(s):*

L11890 Em(40)_vi:BCMCOATP Gb(84)_vi:BCMCOATP Bean common mosaic virus coat protein gene, 3' end. 3/93 1,185bp.

L12740 Em(40)_vi:BCMCPA Gb(84)_vi:BCMCPA Bean common mosaic virus coat protein mRNA, complete cds. 3/93 1,550bp.

L15332 Em(40)_vi:BCMCP Gb(84)_vi:BCMCP Bean common mosaic virus coat protein gene, 3' end. 5/93 1,144bp.

L19472 Em(40)_vi:BCMCOATA Gb(84)_vi:BCMCOATA Bean common mosaic virus coat protein mRNA, 3' end. 6/93 1,139bp.

L19473 Em(40)_vi:BCMCOATB Gb(84)_vi:BCMCOATB Bean common mosaic virus coat protein mRNA, 3' end. 6/93 1,127bp.

L19474 Em(40)_vi:BCMCOATC Gb(84)_vi:BCMCOATC Bean common mosaic virus coat protein mRNA, 3' end. 6/93 1,174bp.

L19539 Em(40)_vi:BCMCTP Gb(84)_vi:BCMCTP Bean common mosaic virus coat protein mRNA sequence, 3' end. 6/93 1,123bp.

L21766 Em(40)_vi:BCMCOAT Gb(84)_vi:BCMCOAT Bean common mosaic virus coat protein mRNA, 3' end cds. 7/93 1,190bp.

L21767 Em(40)_vi:BCMCOATPR Gb(84)_vi:BCMCOATPRO Bean common mosaic virus coat protein mRNA, 3' end. 7/93 1,183bp.

S66251 Em(40)_vi:S66251 Gb(84)_un:S66251S1 polymerase, coat protein bean common mosaic virus BCMV, NL1, Genomic, 945 nt, segment 1 of 2

S66252 Em(40)_vi:S66252 Gb(84)_un:S66252S1 polymerase, coat protein bean common mosaic virus BCMV, NY15, Genomic, 945 nt, segment 1 of

S66274 Em(40)_vi:S66274 Gb(84)_un:S66274S1 polymerase, coat protein bean common mosaic virus BCMV, NL3, Genomic, 867 nt, segment 1 of 2

S66275 Em(40)_vi:S66275 Gb(84)_un:S66251S2 polymerase, coat protein bean common mosaic virus BCMV, NL1, Genomic, 253 nt, segment 2 of 2

S66276 Em(40)_vi:S66276 Gb(84)_un:S66274S2 polymerase, coat protein bean common mosaic virus BCMV, NL3, Genomic, 240 nt, segment 2 of 2

S66279 Em(40)_vi:S66279 Gb(84)_un:S66252S2 polymerase, coat protein bean common mosaic virus BCMV, NY15, Genomic, 256 nt, segment 2 of

Z15057 Em(40)_vi:BCMVCP Gb(84)_vi:BCMVCP Bean common mosaic virus gene for coat protein (partial). 2/94 1,119bp.

Z17203 Em(40)_vi:BCMVNL3CP Gb(84)_vi:BCMVNL3CP Bean common mosaic virus NL-3 coat protein gene. 2/94 1,030bp.

U19287 Em(43)_vi:Bc19287 Gb(89)_vi:Bcu19287 Bean common mosaic virus polyprotein gene, complete cds. 1/95 9,612bp.

U20818 Em(43)_vi:Bc20818 Gb(89)_vi:Bcu20818 Bean common mosaic virus polyprotein gene, partial cds. 3/95 1,291bp.

*Virion protein(s)* one; $M_r$ 32000–35000 (sometimes two of 32000 and 34000, are found). Method of preparation: Morales (1979); Bravo (1984).

**Cytopathology**

Virions found in leaves and stems including the apical meristems; in cytoplasm. Inclusions present in infected cells; are unusual in shape; cylindrical *i.e.* pinwheels with associated scrolls (Edwardson's Group I).

**Taxonomy and Relationships**

*Potyvirus: Potyviridae.*

Virus(es) with serologically related virions: 17 potyviruses, including potato Y, watermelon mosaic 2, bean yellow mosaic, blackeye cowpea mosaic and soybean mosaic viruses.

**Isolates originally called azuki bean mosaic virus** (Matsumoto, 1922) were first reported in Japan inducing mosaic and green vein banding in *Vigna angularis.* They are found throughout the East Asian region. They probably differ from type isolates in infecting *Nicotiana tabacum*, but neither *Chenopodium amaranticolor* nor *Vicia faba.*

References.

Hampton, R., Beczner, L., Hagedorn, D., Bos, L., Inouye, T., Barnett, O., Musil, M. and Meiners, J. (1978). *Phytopathology* **68**: 989.

Matsumoto, T. (1922). *Byochugai Zasshi* **9**: 517.

**Isolates originally called blackeye cowpea mosaic virus** (Anderson, 1995), cowpea (aphid-borne) mosaic virus, cowpea (blackeye) mosaic virus and cowpea vein-banding mosaic virus were first reported in the U.S.A. inducing mosaics, mottles, streaks and leaf malformation in *Crotalaria spectabilis* and *Vigna unguiculata.* They are found worldwide, and also do not infect *Vicia faba*. Some gene sequences have been determined.

*Sequence database accession codes*:

S66253 Em(40)_vi:S66253 Gb(84)_un:S66253S1 polymerase, coat protein blackeye cowpea mosaic virus BlCMV, W, Genomic, 945 nt, segment 1.

S66280 Em(40)_vi:S66280 Gb(84)_un:S66253S2 polymerase, coat protein blackeye cowpea mosaic virus BlCMV, W, Genomic, 254 nt, segment 2, 2 sequences.

References

Anderson, C.W. (1955). *Pl. Dis. Reptr* **39**: 349.

Dijkstra, J., Bos, L., Bouwmeester, H.J., Hadiastono, T. and Lohuis, H. (1987). *Neth. J. Pl. Path.* **93**: 115.

Lovisolo, O. and Conti, M. (1966). *Neth. J. Pl. Path.* **72**: 265.

Tsuchizaki, T., Senboku, T., Iwaki, M., Pholauporn, S., Srithongchi, W., Deema, N. and Ching, A.O. (1984). *Ann. Phytopath. Soc. Japan* **50**: 461.

**Isolates originally called peanut stripe virus** (Demski *et al.*, 1984), peanut chlorotic ring virus (Demski *et al.*, 1988; Fukumoto *et al.*, 1986; Wongkaew and Dollet, 1990), peanut mild mottle virus (Zeyong *et al.*, 1983; Demski *et al.*, 1988), peanut mosaic virus and sesame yellow mosaic virus were first found in Georgia, U.S.A. in *Arachis hypogaea*, which is not infected by type isolates. They cause severe mosaic, striping and stunting in several other legumes including *Lupinus albus*, *Glycine max* and *Sesamum* spp., and are transmitted not only by *Aphis craccivora*, but also *A. glycines* and *Rhopalosiphum maidis*. Some gene sequences have been determined.

*Sequence database accession codes*:

U05771 Em(40)_vi:PS05771 Gb(84)_vi:PSU05771 Peanut stripe virus, complete genome. 2/94 10,062bp.

X63559 Em(40)_vi:PSTVCP Gb(84)_vi:PSTVCP Peanut stripe virus gene for capsid protein. 2/94 1,367bp

Z21700 Em(40)_vi:PSV3TERM Gb(84)_vi:PSV3TERM Peanut stripe virus 370 capsid protein. 8/93 1,388bp. 3 sequences.

References

Demski, J.W., Reddy, D.V.R., Sowel, G. (1984). *FAO Pl. Prot. Bull.* **32**: 114.

Demski, J.W., Reddy, D.V.R., Wongkaew, S., Iwaki, M., Seleh, N. and Xu, Z. (1988). *Phytopathology* **78**: 631.

Fukumoto, F., Thongmeearkom, P., Iwaki, M., Choopanya, D., Sarindu, N., Deema, N. and Tsuchizaki, T. (1986). *Tech. Bull. Trop. Agric. Res. Center, Japan* **21**: 150.

Wongkaew, S. and Dollet, M. (1990). *Oleagineux* **45**: 268.

Zeyong, X., Ziling, Y. and Jialing, L. (1983). *Plant Dis.* **67**: 1029. Peptide analysis and immunoblot tests show that this virus is very closely related to blackeye cowpea mosaic potyvirus (D. Shukla, personal communication).

**References**

Abu Kassim, A.B. (1981). *Malaysian Agric. J.* **53**: 29. See *Rev. Pl. Path.* **62**: 119.
Bercks, R. (1959). *Phytopath. Z.* **35**: 105.
Bos, L. (1964). *Neth. J. Pl. Path.* **70**: 161.
Bravo (1984). *M.Sc. Thesis*, Univ. Valle, 126 pp.
Drijfhout, E. (1978). *Agric. Res. Rep. Wageningen*, 872.
Drijfhout, E. and Bos, L. (1977). *Neth. J. Pl. Path.* **83**: 13.
Drijfhout, E., Silbernagel, M.J. and Burke, D.W. (1978). *Neth. J. Pl. Path.* **84**: 13.
Edwardson, J.R. (1974). *Fla Agric. Exp. Stn Monog.* No. 5, p. 19.
Kaiser, W.J. and Mossahebi, G.H. (1974). *Phytopathology* **64**: 1209.
Koenig, R., Lesemann, D.-E. and Vetten, H.J. (1991). *19th Ann. Newsl. for 1990 of the ISHS Veg. Virus Working Grp*, p., 15.
Lovisolo, O. (1971). *CMI/AAB Descr. Pl. Viruses* No. 73, 4 pp.
Morales, F.J. (1979). *Turrialba* **29**: 320.
Morales, F.J. and Bos, L. (1987). *AAB Descr. Pl. Viruses* No. 337.
McKern, N.M., Mink, G.I., Barnett, O.W., Mishra, A., Whittaker, L.A., Silbernagel, M.J., Ward, C.W. and Shukla, D.D. (1992). *Phytopathology* **82**: 923.
Pierce, W.H. (1934). *Phytopathology* **24**: 87.
Quantz, L. (1961). *Phytopath. Z.* **43**: 79.
Rao, G.P., Pandey, A.K. and Shukla, K. (1986). *Indian Perfumer* **30**: 483.
Reddick, D. and Stewart, V.B. (1918). *Phytopathology* **8**: 530.
Stewart, V.B. and Reddick, D. (1917). *Phytopathology* **7**: 61.
Thornberry, H.H. (1966). In: *Index of Plant Virus Diseases*. U.S. Dep. Agric. Hdbk No. 307.
Zaumeyer, W.J. and Goth, R.W. (1964). *Phytopathology* **54**: 1378.

# Bean distortion dwarf (?) bigeminivirus

**Data collated by A.A. Brunt, 1992.**

*Standard acronym:* BDDV.

First reported in *Phaseolus vulgaris*; from China; by Xi *et al.* (1982).

**Natural host range and symptoms**

Symptoms persist.

*Phaseolus vulgaris* — small distorted leaves, systemic necrosis and stunting.

**Transmission**

Transmitted by a vector; an insect; *Bemisia tabaci*; Aleyrodidae. Virus transmitted by mechanical inoculation.

**Geographical distribution**

Spreads in China.

**Experimental host range**

Few (<3) families susceptible.

Diagnostically susceptible

*Phaseolus vulgaris, Vigna unguiculata* — mild systemic chlorosis, production of small distorted leaves, systemic necrosis and plant stunting.

Maintenance and propagation hosts

*Phaseolus vulgaris, Vigna unguiculata.*

Assay hosts (**L**ocal lesions or **W**hole plants)

*Phaseolus vulgaris* (W).

Dicotyledonous Hosts *(**Susceptible** or Insusceptible)*

***Capsicum annuum. Datura stramonium. Nicotiana glutinosa. Nicotiana tabacum. Phaseolus vulgaris. Vigna unguiculata.***

Sources of host-range data

Xi *et al.* (1982).

**Particle morphology**

Virions geminate; not enveloped; 20 nm in diameter.

**Physical properties**

$A_{260}/A_{280}$ ratio 1.36.

**Biochemical properties**

*Genome* consists of DNA; single–stranded.

**Cytopathology**

Virions found in leaves and mesophyll; in cytoplasm.

**Taxonomy and Relationships**

Possible bigeminivirus: *Geminiviridae.*

**Comments**

The reported size of the virions is unusually small for a geminivirus.

**References**

Xi, Z., Xu, S. and Mang, K. (1982). *Acta Microbiol. Sin.* **22**: 293.

# Bean golden mosaic bigeminivirus

**Data collated by S. Haber and A.J. Howarth, 1983. Revised A.J. Howarth, 1990.**

*Synonyms:* virus del mosaico dorado. *Standard acronym:* BGMV. *Strains:* isolates from Venezuela and Puerto Rico differ in DNA hybridization analyses.

First reported in *Phaseolus vulgaris*; from Costa Rica; by Galvez and Castano (1976).

**Natural host range and symptoms**

Symptoms persist.

*Macroptilium lathyroides, Phaseolus vulgaris, P. lunatus, Malvastrum coromandelianum* — vein yellow net, interveinal chlorosis.

**Transmission**

Transmitted by a vector; an insect; *Bemisia tabaci* (*Sida* race); Aleyrodidae. Transmitted in a persistent manner. Virus retained when the vector moults; does not multiply in the vector; not transmitted congenitally to the progeny of the vector; some isolates transmitted by mechanical inoculation; transmitted by grafting; not transmitted by contact between plants; not transmitted by seed; not transmitted by pollen.

**Geographical distribution**

Spreads in Brazil, the Dominican Republic, Guatemala, Jamaica, Puerto Rico, and Venezuela.

**Experimental host range**

Few (<3) families susceptible. Experimentally infected plants mostly show chlorosis.

Diagnostically susceptible

*Phaseolus vulgaris* cv. Top Crop, *Phaseolus lunatus* — systemic vein chlorosis, golden mosaic.

Diagnostically insusceptible

*Glycine max, Rhynchosia minima* (distinguishes the virus from *Rhynchosia* mosaic).

Maintenance and propagation hosts

*Phaseolus vulgaris* cv. Top Crop.

Assay hosts (**L**ocal lesions or **W**hole plants)

*Phaseolus vulgaris* cvs Top Crop and Red Kidney (W) and Red Mexican 35 (L).

Dicotyledonous Hosts *(**Susceptible** or Insusceptible)*

***Cajanus cajan*. *Glycine max*. *Macroptilium lathyroides*. *Phaseolus acutifolius*. *Phaseolus lunatus*. *Phaseolus vulgaris*.**

Additional hosts

***Malvastrum coromandelianum*, *Pachyrhizus erosus*, *Phaseolus angularis*, *Phaseolus coccineus*, *Phaseolus longipedunculatus*** and ***Phaseolus mungo***.

Sources of host-range data

Galvez and Castano (1976); J. Bird (unpublished).

**Properties of particles in sap**

TIP: 50–55 °C. LIV: 72 days. DEP: $\log_{10}$ minus 2. Leaf sap contains few virions. Electron microscopy: aldehyde fixation essential unless PTA used at pH 4.

**Purification**

Method: Goodman *et al.* (1977); Goodman and Bird (1978).

**Particle morphology**

Virions geminate; not enveloped; 18–20 nm in diameter; dimers 30 nm in length; angular in profile; without a conspicuous capsomere arrangement.

**Physical properties**

One sedimenting component in purified preparations; sedimentation coefficient 69 *S*. Density 1.717 g $cm^{-3}$ in CsCl (without aldehyde fixation). Density 1.31 in $Cs_2SO_4$. Isoelectric point pH 7.

**Biochemical properties**

Virions contain 20 % nucleic acid; 80 % protein; 0 % lipid.

*Genome* consists of DNA; single–stranded; circular; of two parts; largest (or only) genome part the larger 2.647 kb; the 2nd largest 2.587 kb. Base composition 21 % G; 28.1 % A; 18.3 % C; 32.6 % T. Infectivity decreased when deproteinised with proteases (proteinase K has no effect on protoplast infectivity); retained when deproteinised with phenol or detergent. Poly A region absent. Additional factor not required for infectivity. Genome has no *t*RNA–like activity. Nucleotide sequence references: Howarth *et al.* (1985).

*Sequence database accession code(s):*

D00200 Em(40)_vi:GEMBGACG Gb(84)_vi:MBGACG Bean golden mosaic virus (BGMV), DNA A, complete sequence. 4/90 2,585bp

D00201 Em(40)_vi:GEMBGBCG Gb(84)_vi:MBGBCG Bean golden mosaic virus (BGMV), DNA B, complete genome. 1/91 2,647bp.

L01635 Em(40)_vi:GEMBGCOAR Gb(84)_vi:MBGCOAREPA Bean golden mosaic geminivirus type II coat protein (AR1) gene complete cds, (AL2 and AL3)

L01636 Em(40)_vi:GEMBGCOA1 Gb(84)_vi:MBGCOAREPB Bean golden mosaic geminivirus type II (BR1 and BL1) genes, complete cds. 11/92 2,608bp.

L34266 Gb(84)_vi:MBGAC1A Bean golden mosaic geminivirus (clone pCR1) AC1 gene, partial cds. 7/94 222bp.

L34267 Gb(84)_vi:MBGAC1B Bean golden mosaic geminivirus (clone pCR1) AC1 gene, partial cds. 7/94 222bp.

L34268 Gb(84)_vi:MBGAC1C Bean golden mosaic geminivirus (clone pGA3) AC1 gene, partial cds. 7/94 222bp.

L34269 Gb(84)_vi:MBGAC1D Bean golden mosaic geminivirus (clone pJA1) AC1 gene, partial cds. 7/94 222bp

M10070 Em(40)_vi:GEBGMV1 Gb(84)_vi:MBGCG1Z Bean golden mosaic virus, DNA 1 of complete genome. 1/91 2,646bp.

M10080 Em(40)_vi:GEBGMV2 Gb(84)_vi:MBGCG2Z Bean golden mosaic virus, DNA 2 of complete genome. 4/90 2,587bp.

M88686 Em(40)_vi:GEMBGARAL Gb(84)_vi:MBGARAL Bean golden mosaic geminivirus viral coat protein (AR1) gene, complete cds; putative replicat

M88687 Em(40)_vi:GEMBGBRBL Gb(84)_vi:MBGBRBL Bean golden mosaic geminivirus BR1 gene, complete cds and BL1 gene, complete cds. 9/92 2,580b

M91604 Em(40)_vi:GEMBGCGA Gb(84)_vi:MBGCGA Bean golden mosaic virus coat protein gene, complete cds; putative replication protein gene

M91605 Em(40)_vi:GEMBGCGB Gb(84)_vi:MBGCGB Bean golden mosaic virus BR1 and BL1 genes, complete cds's. 9/92 2,596bp.

S66724 Em(40)_vi:S66724 Gb(84)_vi:S66724 (clone DR-1 hypervariable region) bean golden mosaic geminivirus, Genomic, 203 nt. 1/94 203

S66726 Em(40)_vi:S66726 Gb(84)_vi:S66726 (clone DR-2 hypervariable region) bean golden mosaic geminivirus, Genomic, 203 nt. 1/94 203

S66729 Em(40)_vi:S66729 Gb(84)_vi:S66729 (clone DR-3 hypervariable region) bean golden mosaic geminivirus, Genomic, 203 nt. 1/94 203

S66730 Em(40)_vi:S66730 Gb(84)_vi:S66730 (clone DR-4 hypervariable region) bean golden mosaic geminivirus, Genomic, 203 nt. 1/94 203

S66732 Em(40)_vi:S66732 Gb(84)_vi:S66732 (clone DR-5 hypervariable region) bean golden mosaic geminivirus, Genomic, 203 nt. 1/94 203

S66733 Em(40)_vi:S66733 Gb(84)_vi:S66733 (DNA-B, clone pDRB-PCR hypervariable region) bean golden mosaic geminivirus, Genomic, 203 nt.

*Non–genomic nucleic acid* not found in the virions.

*Virion protein(s)* one; $M_r$ 27400. Method of preparation: Goodman *et al.* (1980). Amino acid sequence: inferred from nucleotide sequence.

*Virus–coded non–virion proteins* identified by genomic sequence analysis; five proteins found. $M_r$ of the largest 40100. $M_r$ of 2nd largest 32700. $M_r$ of 3rd 29600. $M_r$ of 4th 15600. $M_r$ of 5th and smaller 14500.

**Replication**

Genome replicates probably in nuclei. Coat protein *m*RNA translated in the cytoplasm. Replication does not depend on a helper virus.

**Cytopathology**

Virions found in leaves, mesophyll, phloem and stems; in nuclei and in nucleoli. Inclusions present in infected cells; are unusual in shape; hexagonally packed crystalline arrays or loose aggregates of virion-like particles in the nucleus. Other cellular changes: the occurrence of characteristic fibrillar rings in nuclei, after nucleoli undergo hypertrophy.

**Taxonomy and Relationships**

Bigeminivirus: *Geminiviridae.*

Virus(es) with serologically related virions: abutilon mosaic, African cassava mosaic, squash leaf curl and tomato golden mosaic viruses.

**References**


Bird, J., Perez, J.E., Alconero, R., Vakili, N.G. and Melandez, P.L. (1972). *J. agric. Univ. Puerto Rico* **56**: 64.

Galvez, G.E. and Castano, M. (1976). *Turrialba* **26**: 205.

Goodman, R.M. (1977). *Virology* **83**: 171.

Goodman, R.M. and Bird, J. (1978). *CMI/AAB Descr. Pl. Viruses* No. 192, 4 pp.

Goodman, R.M., Bird, J. and Thongmeearkom, P. (1977). *Phytopathology* **67**: 37.

Goodman, R.M., Shock, T.L., Haber, S., Browning, K.S. and Bowers, G.R. (1980). *Virology* **106**: 168.

Haber, S., Howarth, A.J. and Goodman, R.M. (1983). *Virology* **129**: 517.

Howarth, A.J. and Vandemark, G.J. (1989). *J. gen. Virol.* **70**: 2717.

Howarth, A.J., Caton, J., Bossert, M. and Goodman, R.M. (1985). *Proc. Natl. Acad. Sci. U.S.A.* **82**: 3572.

Kim, K.S., Shock, T.L. and Goodman, R.M. (1978). *Virology* **89**: 22.


# Bean leaf roll *luteovirus*

**Data collated by J.W. Ashby, 1981. Revised 1989 by R.R. Martin and for chickpea stunt strain by Y.L. Nene, 1981 and revised 1983.**

*Synonyms:* alfalfa Michigan luteovirus (Thottappilly *et al.* 1977), legume yellows virus, pea leaf roll virus, pea tip yellowing virus, pea top yellows virus. *Standard acronym:* BLRV. *Strains:* chickpea stunt virus.

First reported in *Pisum sativum* and *Vicia faba*; from Germany; by Quantz and Volk (1954).

**Natural host range and symptoms**

Symptoms persist, or disappear soon after infection.

*Medicago sativa* — usually symptomless.

*Pisum sativum* — chlorotic, leaves rolled downward.

*Trifolium repens* — yellowed, rolled leaves.

*Vicia faba* — chlorotic leaves and flowers abscissed.

*Phaseolus vulgaris, Cicer arietinum, Vigna unguiculata, Lens esculenta* — leaves yellowed.

**Transmission**

Transmitted by a vector; an insect; *Acyrthosiphon pisum, Myzus persicae, Macrosiphum euphorbiae, Megoura viciae, Aphis craccivora* (chickpea stunt strain); Aphididae. Transmitted in a persistent manner. Virus retained when the vector moults; does not multiply in the vector; can help the vector transmission of another virus (bean yellow vein banding umbravirus); not transmitted by mechanical inoculation; transmitted by grafting; not transmitted by seed; not transmitted by pollen.

**Geographical distribution**

Spreads in Ethiopia, Germany, Iran, Lebanon, Pakistan, Sudan, Syria, Tunisia, Turkey, the UK, and the USA (California, the Midwest and Northwest).

**Experimental host range**

Few (<3) families susceptible. Experimentally infected plants mostly show chlorosis, leaf rolling.

Diagnostically susceptible

*Medicago hispida* — systemic chlorosis.

*Pisum sativum* cv. Onyx — tip leaves chlorotic, leaves rolled (cv. Kelvedon Wonder is insusceptible).

*Trifolium subterraneum* — red leaf margins.

*Vicia faba* — interveinal chlorosis, leaves rolled upward.

Diagnostically insusceptible

*Beta vulgaris, Capsella bursa-pastoris, Lactuca sativa, Pisum sativum* cv. Kelvedon Wonder.

Maintenance and propagation hosts

*Pisum sativum* cvs Galaxy, Onyx, Eurofin, *Vicia faba, Medicago sativa.*

Assay hosts (**L**ocal lesions or **W**hole plants)

*Pisum sativum* cv. Onyx (W), *Trifolium subterraneum* cv. Bacchus Marsh (W).

Dicotyledonous Hosts *(**Susceptible** or Insusceptible)*

***Arachis hypogaea***. ***Astragalus sinicus***. *Beta vulgaris*. *Brassica campestris* ssp. *napus*. *Capsella bursa–pastoris*. ***Cicer arietinum***. ***Glycine max***, or *Glycine max*. *Gomphrena globosa*. *Lablab purpureus*. *Lactuca sativa*. ***Lathyrus odoratus***. ***Lens esculenta***, or *Lens esculenta*. *Linum usitatissimum*. *Lotus corniculatus*. *Lupinus albus*. ***Lupinus cosentinii***. ***Medicago hispida***. ***Medicago sativa***. *Melilotus alba*. ***Phaseolus vulgaris***, or *Phaseolus vulgaris*. ***Pisum sativum***. *Senecio vulgaris*. *Stellaria media*. *Tetragonia expansa*. ***Trifolium dubium***. ***Trifolium hybridum***. ***Trifolium incarnatum***. ***Trifolium pratense***, or *Trifolium pratense*. ***Trifolium repens***. ***Trifolium subterraneum***. ***Vicia faba***, or *Vicia faba*. ***Vicia sativa***. ***Vigna unguiculata***. *Zinnia elegans*.

Sources of host-range data

Cockbain and Gibbs (1973), Johnstone *et al.*, (1984).

**Properties of particles in sap**

Leaf sap contains few virions.

**Purification**

Method: Takanami and Kubo (1979); Ashby and Huttinga (1979).

**Particle morphology**

Virions isometric; not enveloped; 27 nm in diameter; rounded in profile; without a conspicuous capsomere arrangement.

**Physical properties**

One sedimenting component in purified preparations. Density 1.32 in $Cs_2SO_4$.

**Biochemical properties**

Virions contain 37.8 % nucleic acid.

*Genome* consists of RNA; single–stranded; linear; unipartite; largest (or only) genome part 6 kb. Genomic nucleic acid isolated by R.R. Martin; use hot phenol. Poly A region absent.

*Sequence database accession code(s):*

X53865 Em(40)_vi:LUBLRVCP Gb(84)_vi:LUBLRVCP Bean leaf roll virus gene for coat protein. 9/93 591bp.

U15978 Em(43)_vi:Bl15978 Gb(89)_vi:Blu15978 Bean leaf roll virus coat protein mRNA, complete cds. 11/94 591bp.

*Virion protein(s)* one; $M_r$ 23000 (type strain), or 30000–35000 (chickpea stunt strain).

**Cytopathology**

Virions found in phloem parenchyma mainly in necrotic sieve-tube cells; in cytoplasm. Inclusions absent from infected cells. Other cellular changes: necrosis and collapse of phloem.

**Taxonomy and Relationships**

*Luteovirus.*

Virus(es) with serologically related virions: soybean dwarf, beet western yellows, potato leafroll, carrot red leaf and barley yellow dwarf (RPV) viruses.

The chickpea stunt strain of the virus (Kaiser and Danesh, 1966) has been recorded in the same geographical areas as the type strain. Hosts recorded for the chickpea stunt strain but not for the type strain are: *Chenopodium amaranticolor* (**I**); *Cucumis sativus* (**I**); *Cyamopsis tetragonoloba* (**I**); *Datura stramonium* (**I**); *Nicotiana glutinosa* (**I**); *Nicotiana tabacum* (**I**); *Phaseolus lunatus* (**S**); *Vicia villosa* (**S**); *Vigna radiata* (**I**) (Anon., 1978/79). Best tests for diagnosis: ELISA tests with MAbs very effective.

**References**

Anon. (1978–79). *Rep. Int. Crops Res. Inst. for Semi-Arid Tropics* 1978–79: 30.

Ashby, J.W. (1984). *CMI/AAB Descr. Pl. Viruses* No. 286, 5 pp.

Ashby, J.W. and Huttinga, H. (1979). *Neth. J. Pl. Path.* **85**: 113.

Cockbain, A.J. and Gibbs, A.J. (1973). *Ann. appl. Biol.* **73**: 177.

D'Arcy, C.J., Martin, R.R. and Torrance, L. (1988). *Phytopathology* **78**: 1537.

Fukano and Yokoyama (1951). *Kyushu Nogyo Kenkyu* **10**: 133.

Inouye, T. (1971). *Ber. Ohara Inst. Landw. Biol.* **15**: 69.

Inouye, T. and Asatari (1968). *Ann. Phytopath. Soc. Japan* **34**: 317.

Johnstone, G.R., Ashby, J.W., Gibbs, A.J., Duffus, J.E., Thottapilly, G. and Fletcher, J.D. (1984). *Neth J. Pl. Path.* **90**: 107.

Kaiser, W.J. (1972). *FAO Pl. Prot. Bull.* **20**: 127.

Kaiser, W.J. and Danesh, D. (1966). *Phytopathology* **61**: 453.

Kaiser, W.J and Danesh, D.(1971). *Phytopathology* **61**: 372.

Quantz, L. and Volk, J. (1954). *NachrBl. dtsch. PflSchutz., Braunschweig* **6**: 177.

Takanami, Y. and Kubo, S. (1979). *J. gen. Virol.* **44**: 153.

Thottappilly, G., Kao, Y.J., Hooper, G.R. and Bath, J.E. (1977). *Phytopathology* **67**: 1451.

# Bean mild mosaic *carmovirus*

**Data collated by H. Waterworth, 1981. Revised by A.V. Karasev, 1991.**

*Standard acronym:* BMMV.

First reported in *Phaseolus vulgaris*; from the U.S.A.; by Waterworth *et al.* (1977).

**Natural host range and symptoms**

*Phaseolus vulgaris* — mild chlorotic vein banding and mosaic.

**Transmission**

Transmitted by a vector; an insect; *Epilachna varivestis, Diabrotica undecimpunctata howardi, D. balteata, Ceratoma ruficornis, Gynandrobrotica variabilis*; Coleoptera. Virus transmitted by mechanical inoculation.

**Geographical distribution**

Spreads in the South and Central American region; Colombia and El Salvador.

**Experimental host range**

Several (3–9) families susceptible. Experimentally infected plants mostly show chlorosis, mosaic.

Diagnostically susceptible

*Cyamopsis tetragonoloba, Lablab purpureus, Phaseolus vulgaris* cvs 27R, Top Crop — mild systemic chlorotic mosaic.

Diagnostically insusceptible

*Arachis hypogaea, Cicer arietinum, Pisum sativum, Phaseolus coccineus, P. lunatus, Vicia faba, Vigna radiata, V. unguiculata.*

Maintenance and propagation hosts

*Phaseolus vulgaris* cvs Top Crop, Pinto.

Assay hosts (**L**ocal lesions or **W**hole plants)

*Phaseolus vulgaris* cvs Top Crop (W), Pinto (W).

Dicotyledonous Hosts *(**Susceptible** or Insusceptible)*

*Amaranthus caudatus. Antirrhinum majus. Arachis hypogaea. Beta vulgaris. Brassica campestris* ssp. *rapa. Calendula officinalis. Capsicum annuum. Capsicum frutescens. Chenopodium quinoa. Cicer arietinum. Cucumis sativus. Cucurbita maxima. Datura stramonium. Dianthus barbatus.* ***Glycine max. Gomphrena globosa.*** *Lens esculenta. Lycopersicon esculentum.* ***Macrotyloma uniflorum.*** *Momordica balsamina. Nicotiana glutinosa. Nicotiana tabacum. Petunia hybrida. Phaseolus lunatus.* ***Phaseolus vulgaris.*** *Physalis floridana. Pisum sativum. Raphanus sativus. Sesamum indicum.* ***Sesbania exaltata.*** *Tetragonia expansa. Torenia fournieri. Vicia faba. Vigna radiata. Vigna unguiculata. Zinnia elegans.*

Monocotyledonous Hosts *(**Susceptible** or Insusceptible)*

*Hordeum vulgare. Zea mays.*

Sources of host-range data

Waterworth *et al.* (1977); Waterworth (1981).

**Properties of particles in sap**

TIP: 84 °C. LIV: 42 days. DEP: $\log_{10}$ minus 8.

**Purification**

Method: Waterworth *et al.* (1977); Karasev *et al.* (1989).

**Particle morphology**

Virions isometric; not enveloped; 27–30 nm in diameter; angular in profile; with a conspicuous capsomére arrangement.

**Physical properties**

One sedimenting component in purified preparations; sedimentation coefficient 127 *S*. Density 1.336 g $cm^{-3}$ in CsCl (Hari *et al.*, 1989). $A_{260}/A_{280}$ ratio 1.52 (Karasev *et al.*, 1989).

**Biochemical properties**

Virions contain 20 % nucleic acid; 80 % protein.

*Genome* consists of RNA; single–stranded; linear; unipartite; largest (or only) genome part 4.2 kb. Base composition 21.7 % G; 25.8 % A; 31.5 % C; 21 % U. 5' terminus of RNA has a methylated nucleotide cap (only a portion of the RNA). Poly A region absent.

*Non–genomic nucleic acid* found in the virions; is subgenomic *m*RNA. Sub–genomic *m*RNA found in infected cells.

*Virion protein(s)* one; $M_r$ 40000; coat protein.

*Virus–coded non–virion proteins* isolated; two proteins found, or four proteins found. $M_r$ of the largest 79000. $M_r$ of 2nd largest 27000. $M_r$ of 3rd 8000. $M_r$ of 4th 6900.

**Taxonomy and Relationships**

*Carmovirus: Tombusviridae.*

Virus(es) with serologically unrelated virions: bean pod mottle, southern bean mosaic, cowpea chlorotic mottle, bean rugose mosaic, red clover mottle, broad bean true mosaic, cowpea mosaic, radish mosaic and squash mosaic viruses.

**References**

Hampton, R.O. and Hancock, C.L. (1981). *Phytopathology* **71**: 223.
Hari, V., Gadh, I.P.S., Baunoch, D. and Das, P. (1989). *Virus Genes* **2**: 211.
Karasev, A.V., Chirkov, S.N., Kaftanova, A.S., Miroshnichenko, N.A., Surgucheva, N.A. and Fedotina, V.L. (1989). *Intervirology* **30**: 285.
Meiners, J.P., Waterworth, H.E., Lawson, R.H. and Smith, F.F. (1977). *Phytopathology* **67**: 163.
Scott, H.A. and Phatak, H.C. (1979). *Phytopathology* **69**: 346.
Waterworth, H.E. (1981). *CMI/AAB Descr. Pl. Viruses* No. 231, 3 pp.
Waterworth, H.E., Meiners, J.P., Lawson, R.H. and Smith, F.F. (1977). *Phytopathology* **67**: 169.

# Bean pod mottle *comovirus*

**Data collated by J.P. Fulton, 1981. Revised 1985.**

*Synonyms:* pod mottle virus. *Standard acronym:* BPMV. *Strains:* desmodium virus (Lee and Walters, 1970).

First reported in *Phaseolus vulgaris* cv. Tendergreen; from Charleston, U.S.A; by Zaumeyer and Thomas (1948).

**Natural host range and symptoms**

Symptoms persist.

*Phaseolus vulgaris* — severe mottling, malformed leaves and pods.

*Glycine max* — mottling and streaking of seed coats.

*Desmodium paniculatum* — mottling.

**Transmission**

Transmitted by a vector; an insect; *Ceratoma trifurcata, Diabrotica balteata, D. undecimpunctata howardii, Colaspis flavida, C. lata, Epicauta vittata, Epilachna varivestis*; Coleoptera. Virus transmitted by mechanical inoculation; transmitted by grafting; not transmitted by seed; not transmitted by pollen.

**Geographical distribution**

Spreads in the North American region.

**Experimental host range**

Few (<3) families susceptible. Experimentally infected plants mostly show mottling, malformation.

Diagnostically susceptible

*Glycine max* — severe mottling and malformation of leaves, pods and seed coats.

*Phaseolus vulgaris* cvs Black Valentine, Bountiful, Tendergreen, Cherokee Wax — severe leaf mottling, malformation.

*Stizolobium deeringianum, Trifolium incarnatum, Lespedeza striata* — mottling.

Maintenance and propagation hosts

*Glycine max, Phaseolus vulgaris* cvs Black Valentine, Cherokee Wax, Bountiful.

Assay hosts (Local lesions or Whole plants)

*Phaseolus vulgaris* cvs Pinto, Kentucky Wonder, Tendergreen (L).

Dicotyledonous Hosts *(**Susceptible** or Insusceptible)*

*Arachis hypogaea. Brassica oleracea* ssp. *botrytis. Capsicum frutescens.* ***Cassia occidentalis. Catharanthus roseus. Chenopodium quinoa.*** *Crotalaria spectabilis. Cucumis sativus. Datura stramonium.* ***Glycine max.*** *Lactuca sativa.* ***Lens esculenta. Lespedeza stipulacea.*** *Lycopersicon esculentum. Medicago sativa. Melilotus alba. Nicotiana glutinosa. Nicotiana tabacum. Petunia hybrida.* ***Phaseolus lunatus***, or *Phaseolus lunatus.* ***Phaseolus vulgaris. Pisum sativum***, or *Pisum sativum.* ***Sesbania exaltata. Trifolium incarnatum.*** *Trifolium pratense. Trifolium repens. Vicia faba. Vigna angularis.* ***Vigna radiata. Vigna unguiculata***, or *Vigna unguiculata. Vigna unguiculata* ssp. *sesquipedalis.*

Sources of host-range data

Skotland (1958); Thornberry (1966); Hampton *et al.* (1978).

**Properties of particles in sap**

TIP: 70–75 °C. LIV: 62–93 days. DEP: $\log_{10}$ minus 4–5. Leaf sap contains many virions.

**Purification**

Method: Bancroft (1962).

**Particle morphology**

Virions isometric; not enveloped; 28 nm in diameter; angular in profile; without a conspicuous capsomere arrangement.

**Physical properties**

Three sedimenting components in purified preparations; sedimentation coefficient of the fastest 112 *S* (B); of the other(s) 91 *S* (M), or 54 *S* (T). Isoelectric point pH 4.8 (slow), or 5.3 (fast). $A_{260}/A_{280}$ ratio 1.77 (B), or 1.7 (M).

**Biochemical properties**

Virions contain 37 % nucleic acid (B), or 30 % nucleic acid (M), or 0 % nucleic acid (T); 63 % protein (B), or 70 % protein (M), or 100 % protein (T); 0 % lipid.

*Genome* consists of RNA; single–stranded; linear; of two parts; largest (or only) genome part the larger 8 kb; the 2nd largest 4.8 kb. Base composition 19.1 % G; 33.4 % A; 16.7 % C; 30.8 % U.

*Sequence database accession code(s):*

M62738 Em(40)_vi:COMBPMCO Gb(84)_vi:BPMCOAT Bean pod mottle virus coat protein gene, complete cds, complete middle component (M) RNA. 4/9 1 sequence.

**Cytopathology**

Virions found in all parts of the host plant. Inclusions present in infected cells; are crystals in the cytoplasm and unusual in shape; masses of clustered membranous vesicles and bundles of filamentous structures; they contain virions. Other cellular changes: prominent cell wall protrusions containing virions embedded in tubules.

**Taxonomy and Relationships**

*Comovirus: Comoviridae.*

Virus(es) with serologically related virions: cowpea mosaic, radish mosaic, squash mosaic and red clover mottle viruses (Bruening, 1978).

**References**


Bancroft, J.B. (1962). *Virology* **16**: 419.

Bruening, G. (1978). *CMI/AAB Descr. Pl. Viruses* No. 199, 4 pp.

Hampton, R., Beczner, L., Hagedorn, D., Bos, L., Inouye, T., Barnett, O., Musil, M. and Meiners, J. (1978). *Phytopathology* **68**: 989.

Lee, F.N. and Walters, H.J. (1970). *Phytopathology* **60**: 585.

Semancik, J.S. (1972). *CMI/AAB Descr. Pl. Viruses* No. 108, 4 pp.

Skotland, C.B. (1958). *Pl. Dis. Reptr* **42**: 1155.

Thornberry, H.H. (1966). In: *Index of Plant Virus Diseases.* U.S. Dep. Agric. Hdbk No. 307.

Zaumeyer, W.J. and Thomas, H.R. (1948). *J. Agric. Res.* **77**: 81.


# Bean rugose mosaic *comovirus*

**Data collated by P.W.G. Chu, 1985. Revised 1989.**

*Synonyms:* virus del mosaico rugoso de frijol, virus ampollado del frijol, bean mosaico-em-desenho. *Standard acronym:* BRMV. *Strains:* three strains that differ biologically but are serologically close. 1. Type strain (from Costa Rica and Guatemala). 2. Ampollado strain (VAF) severe strain. 3. BMDV strain.

First reported in *Phaseolus vulgaris*; from Costa Rica; by Gamez (1971; 1972).

**Natural host range and symptoms**

Symptoms persist.

*Phaseolus vulgaris* — severe mosaic, mottling, malformation of leaves and pods (type strain); or severe stunting and blistering of leaves (Ampollado strain); or symmetric mosaic patterns on leaves (BMDV strain).

**Transmission**

Transmitted by a vector; an insect; *Ceratoma ruficornis, Diabrotica balteata, D. adelpha*; Coleoptera. Transmitted in a semi–persistent manner. Virus does not multiply in the vector; not transmitted congenitally to the progeny of the vector; transmitted by mechanical inoculation; not transmitted by contact between plants; not transmitted by seed (in *Phaseolus vulgaris*).

**Ecology and control**

Studies reported by Gamez (1972, 1982); Fulton and Scott (1977); Fulton *et al.* (1981); Cupertino *et al.* (1982); Gamez and Moreno (1983).

**Geographical distribution**

Spreads in Brazil, Colombia, Costa Rica, El Salvador, Guatemala, and Panama.

**Experimental host range**

Few (<3) families susceptible. Experimentally infected plants mostly show severe systemic mosaics and malformation.

Diagnostically susceptible

*Chenopodium amaranticolor* — necrotic local lesions; not systemic.

*Phaseolus vulgaris* cv. Plentiful — severe mosaic, rugosity, mottling and malformation of leaves and pods. Some cultivars develop necrotic local lesions, not systemic.

*Vicia faba* — mild mottle.

Diagnostically insusceptible

*Cucumis sativus, Phaseolus vulgaris* cvs Kentucky Wonder, Potomac, Tendergreen and *Vigna unguiculata* cv. Early Ramshorn *Nicotiana tabacum* cv. White Burley.

Maintenance and propagation hosts

*Phaseolus vulgaris* cvs Plentiful, Col. 109-R, Mex. 27-R.

Assay hosts (**L**ocal lesions or **W**hole plants)

*Phaseolus vulgaris* cvs TCA–Pijao, Top Crop, Pinto III, Jamapa (L).

Dicotyledonous Hosts *(**Susceptible** or Insusceptible)*

*Chenopodium album*. ***Chenopodium amaranticolor***. *Cucumis sativus*. *Nicotiana tabacum*. *Petunia hybrida*. ***Phaseolus vulgaris***. *Physalis floridana*. *Raphanus sativus*. ***Vicia faba***. *Zinnia elegans*.

Comments on host-range

Host differences alone do not distinguish this virus from closely related comoviruses such as bean pod mottle and cowpea mosaic viruses.

Sources of host-range data

Gamez (1972, 1982).

**Properties of particles in sap**

TIP: 65–75 °C. LIV: 2–4 days. DEP: $\log_{10}$ minus 4–5. Infectivity of sap not changed by treatment with di–ethyl ether. Leaf sap contains many virions.

**Purification**

Method: Bancroft (1962); Gamez (1982); Van Kammen and de Jager (1978).

**Particle morphology**

Virions isometric; not enveloped; 25–30 nm in diameter; angular in profile.

**Physical properties**

Two sedimenting components in purified preparations, or three sedimenting components in purified preparations; sedimentation coefficient of the fastest 113 *S* (B); of the other(s) 97 *S* (M), or 59 *S* (T). Density 1.45 g $cm^{-3}$ in CsCl (B), or 1.414 g $cm^{-3}$ in CsCl (M2), or 1.401 g $cm^{-3}$ in CsCl (M1), or 1.293 g $cm^{-3}$ in CsCl (T).

**Biochemical properties**

Virions contain 36 % nucleic acid (B), or 29 % nucleic acid (M), or 0 % nucleic acid (T); 64 % protein (B), or 71 % protein (M), or 100 % protein (T); 0 % lipid. Also coat proteins contain polysaccharides.

*Genome* consists of RNA; single–stranded; linear; of two parts; largest (or only) genome part the larger 6.75 kb; the 2nd largest 4.5 kb. Genomic nucleic acid isolated by Ramirez *et al.* (1987). 5' terminus of RNA has a VPg. Poly A region present; presumably on the 3' end.

*Non–genomic nucleic acid* not found in the virions.

*Virion protein(s)* three; $M_r$ of the largest 41000. $M_r$ of 2nd largest 23000. $M_r$ of 3rd largest 21000. Method of preparation: Acosta *et al.* (1986); Ramirez *et al.* (1987). Virion proteins glycosylated.

**Replication**

Replication does not depend on a helper virus.

**Cytopathology**

Virions found in mesophyll, epidermis and trichomal cells; in cytoplasm and in cell vacuoles; and in the plasmodesmata. Virions tend to aggregate in areas rich in membranous vesicles and tubules and are sometimes surrounded by membranes. Inclusions present in infected cells; are crystals in the cytoplasm and unusual in shape; and sometimes in dense amorphous material; they contain virions. Other cellular changes: collapsed cells with densely staining vacuoles, cytoplasm very vesiculated.

**Taxonomy and Relationships**

*Comovirus: Comoviridae.*

Virus(es) with serologically related virions: bean pod mottle, cowpea severe mosaic, and cowpea mosaic viruses. Virus(es) with serologically unrelated virions: quail pea mosaic virus.

Best tests for diagnosis: host range does not distinguish the virus from bean pod mottle and cowpea mosaic viruses, serological tests are also required; *Phaseolus vulgaris* cv. Tendergreen and *Vigna unguiculata* cv. Early Ramshorn insusceptible to the virus but not to bean pod mottle and cowpea mosaic comovirus.

**References**

Acosta, O., Alegria, A. and Lastra, R. (1986). *Phytopathology* **76**: 1182.

Bancroft, J.B. (1962). *Virology* **16**: 419.

Cupertino, F.P., Costa, A.S., Lin, M.T. and Kitajima, E.W. (1982). *Fitopatol. Bras.* **7**: 269.

Fulton, J.P. and Scott, H.A. (1977). *Fitopatol. Bras.* **2**: 9.

Fulton, J.P. and Scott, H.A. (1979). *Phytopathology* **69**: 305.

Fulton, J.P., Scott, H.A. and Gamez, R. (1981). In: *Vectors of Plant Pathogens*. pp. 115–132; eds K. Maramarosch and K.F. Harrison. Academic Press, New York.
Galvez, G.E., Cardenas, M., Kitajima, E., Diaz, A.J. and Nieto, M.P. (1977). *Turrialba* **27**: 343.
Gamez, R. (1971). In: *Inst. Interamericano Sci. Agric. 17th Reunion Ann., Panama*, 2 pp.
Gamez, R. (1972). *Turrialba* **22**: 249.
Gamez, R. (1982). *CMI/AAB Descr. Pl. Viruses* No. 246, 4 pp.
Gamez, R. and Moreno, R.A. (1983). In: *Plant Virus Epidemiology*, pp. 103–113; eds R.T. Plumb and J.M. Thresh. Blackwell Scientific Publications, Oxford.
Lin, M.T., Gamez, R. and Kitajima, E.W. (1981). *Fitopatol. Bras.* **6**: 293.
Ramirez, P., Espinoza, A.M., Fuentes, A.L. and Leon, P. (1987). *Phytopathology* **77**: 1317.
Stace-Smith, R. (1981). In: *Handbook of Plant Virus Infections and Comparative Diagnosis*, pp. 171–195; ed. E. Kurstak. Elsevier/North Holland Biomedical Press, Amsterdam.
Van Kammen, A. and de Jager, C.P. (1978). *CMI/AAB Descr. Pl. Viruses* No. 197, 6 pp.

# Bean southern mosaic *sobemovirus*

**Data collated by O.P. Sehgal, 1980. Revised 1984.**

*Synonyms:* southern bean mosaic virus, southern bean mosaic virus 1. *Standard acronym:* SBMV. *Strains:* cowpea strain (strain C), Ghana strain (strain G), severe bean mosaic strain or Mexican strain (strain M) (Tremaine and Hamilton 1983), type strain (strain B).

First reported in *Phaseolus vulgaris*; from samples from Louisiana and California, U.S.A.; by Zaumeyer and Harter (1943).

## Natural host range and symptoms

Symptoms persist.

*Phaseolus vulgaris, Vigna unguiculata, V. mungo* and *Glycine max* — mosaic and/or mottle, and stunting (especially in *Vigna unguiculata*).

## Transmission

Transmitted by a vector; an insect; Chrysomelidae: *Ceratoma trifurcata, Epilachna variestis*; Coleoptera. Transmitted in a semi–persistent manner. Virus transmitted by mechanical inoculation; transmitted by grafting; transmitted by seed (3–7% in *V. unguiculata* cv. Early Wilt Resistant Ramshorn); transmitted by pollen to the seed and transmitted by pollen to the pollinated plant.

## Geographical distribution

Spreads in the African region, the North American region, and the South and Central American region; France.

## Experimental host range

Few (<3) families susceptible.

Diagnostically susceptible

*Glycine max* — systemic vein clearing and mild mottling.
*Phaseolus lunatus* — necrotic local lesions; no systemic infection.
*Phaseolus vulgaris* cv. Pinto — local lesions systemic mottling, leaf curling, stunting.

*Vigna unguiculata* — chlorotic spots, systemic vein clearing and banding, leaf deformation, stunting; cv. Clay is a local lesion host of the cowpea strain.

*Vigna unguiculata* ssp. *sesquipedalis* — local lesions followed by systemic chlorosis and vein clearing.

Diagnostically insusceptible

*Cassia tora, Cicer arietinum, Lupinus alba, Melilotus alba.*

Maintenance and propagation hosts

*Phaseolus vulgaris* cv. Bountiful (bean strain) *Vigna unguiculata* cv. Blackeye (cowpea strain).

Assay hosts (**L**ocal lesions or **W**hole plants)

*Phaseolus vulgaris* cv. Pinto (bean strain) (L), *Vigna unguiculata* cv. Clay (cowpea strain) (L).

Dicotyledonous Hosts *(**Susceptible** or Insusceptible)*

*Amaranthus retroflexus. Arachis hypogaea. Brassica campestris. Cajanus cajan. Capsicum frutescens. Chenopodium album. Chenopodium amaranticolor. Chenopodium quinoa. Cicer arietinum. Crotalaria spectabilis. Cucumis sativus. Cucurbita pepo.* ***Cyamopsis tetragonoloba****. Datura stramonium.* ***Glycine max****. Helianthus annuus. Lactuca sativa. Lens esculenta. Lotus corniculatus. Lycopersicon esculentum. Macroptilium lathyroides. Medicago sativa. Nicotiana glutinosa. Nicotiana tabacum. Petunia hybrida.* ***Phaseolus lunatus****.* ***Phaseolus vulgaris****.* ***Pisum sativum***, or *Pisum sativum. Sesbania exaltata. Tetragonia expansa. Trifolium incarnatum. Trifolium pratense. Trifolium repens. Vicia faba.* ***Vigna radiata****.* ***Vigna unguiculata***, or *Vigna unguiculata.* ***Vigna unguiculata*** ssp. ***sesquipedalis****. Zinnia elegans.*

Sources of host-range data

Zaumeyer and Harter (1943); Shepherd and Fulton (1962); Lamptey *et al.* (1974).

**Properties of particles in sap**

TIP: 90–95 °C. LIV: 20–165 days. DEP: $\log_{10}$ minus 5–6. Infectivity of sap not changed by treatment with di–ethyl ether. Leaf sap contains many virions.

**Purification**

Method: Hull (1977); Sehgal and Hsu (1977).

**Particle morphology**

Virions isometric; not enveloped; 28 nm in diameter; rounded in profile; without a conspicuous capsomere arrangement.

**Physical properties**

One sedimenting component in purified preparations; sedimentation coefficient 115 *S*. Density 1.36 g $cm^{-3}$ in CsCl. Density 1.1 g $cm^{-3}$ in sucrose. Isoelectric point pH 5.5.

**Biochemical properties**

Virions contain 21 % nucleic acid; 79 % protein; 0 % lipid.

*Genome* consists of RNA; single–stranded; linear; unipartite; largest (or only) genome part 4.195 kb. Genomic nucleic acid isolated by Weber and Sehgal (1982). Base composition 24 % G; 26 % A; 23 % C; 27 % U. 5' terminus of RNA has a VPg. Infectivity lost when deproteinised with proteases; retained when deproteinised with phenol or detergent. Poly A region absent. Additional factor required for infectivity; a genome-linked protein. Nucleotide sequence references: Wu *et al.* (1987) for the cowpea strain.

*Sequence database accession code(s):*

J02355 Em(40)_vi:SOB3 Gb(84)_vi:MSBB3 southern bean mosaic virus(bean strain)rna, 3' terminus. 4/90 393bp.

J02356 Em(40)_vi:SOC3 Gb(84)_vi:MSBC3 southern bean mosaic virus(cowpea strain)rna, 3' terminus. 4/90 400bp.

L34672 Gb(84)_vi:MSBP5P Southern bean mosaic virus p11.5 protein gene, complete cds, p96 protein gene, complete cds,

M23021 Gb(84)_vi:MSBMVCCG Southern bean mosaic virus complete genome. 7/94 4,194bp.

*Non–genomic nucleic acid* found in the virions; is subgenomic *m*RNA (including autonomised coat cistrons with $M_r$ of 300,000–400,000 d).

*Virion protein(s)* two; $M_r$ of the larger 30000. $M_r$ of 2nd largest 12000. Method of preparation: Hsu *et al.* (1977); Denloye *et al.* (1978). Amino acid sequence: Hermondson *et al.* (1982).

## Replication

Replication does not depend on a helper virus.

## Cytopathology

Virions found in mesophyll; in cytoplasm and in nuclei. Inclusions present in infected cells; are unusual in shape; crystalline aggregates in cowpea strain; they contain virions.

## Taxonomy and Relationships

*Sobemovirus* (the type species).

Virus(es) with serologically related virions: none.

Bean (type) strain infects most *Phaseolus vulgaris* cultivars systemically (but local lesions only in cv. Pinto), but it does not infect cowpea. Only the bean strain infects *Phaseolus lunatus*. The cowpea strain strain infects most cowpea cultivars systemically but not *Phaseolus vulgaris* except cv. Pinto, in which it is symptomless. The Ghana strain infects many cowpea cultivars and induces local or systemic symptomless infection in some cultivars of *Phaseolus vulgaris*, but small necrotic spots in *Vigna radiata*. The severe bean mosaic strain or Mexican strain induces more severe symptoms in *Phaseolus vulgaris* than the bean strain, and also infects cowpea (Tremaine and Hamilton 1983).

## References


Abad-Zapatero, C., Abad-Meguid, S.S., Johnon, J.E., Leslie, A.G.W., Rayment, I., Rossmann, M.G., Suck, D. and Tsukihara, T. (1980). *Nature, Lond.* **286**: 33.

Denloye, A.O., Homer, R.B. and Hull, R.(1978). *J. gen. Virol.* **41**: 77.

Ghabrial, S.A., Shepherd, R.J. and Grogan, R.G. (1967). *Virology* **33**: 17.

Ghosh, A., Dasgupta, R., Salerno-Rife, T., Rutgers, T. and Kaesberg, P. (1979). *Nucl. Acids Res.* **7**: 2137.

Ghosh, A., Rutgers, T., Ke-Qiang, M. and Kaesberg, P.(1981). *J. Virol.* **39**: 87.

Gorbalenya, A.E., Koonin, E.V., Blinov, V.M. and Donchenko, A.P. (1988). *FEBS Lett.* **236**: 2987.

Hermondson, M.A., Abad-Zapatero, C., Abdel-Meguid, S.S., Pundak, S., Rossmann, M.G. and Tremaine, J.H. (1982). *Virology* **39**: 133.

Hsu, C.H., White, J.A. and Sehgal, O.P. (1977). *Virology* **81**: 471.

Hull, R. (1977). *Virology* **79**: 50.

Lamptey, P.N.L. and Hamilton, R.I. (1974). *Phytopathology* **64**: 1100.

Rossmann, M.G., Abad-Zapatero, C., Hermondson, M.A. and Erikson, J.W. (1983). *J. mol. Biol.* **166**: 37.

Rutgers, T., Salerno-Rife, T. and Kaesberg, P. (1980). *Virology* **104**: 506.

Salerno-Rife, T., Rutgers, T. and Kaesberg, P. (1980). *J. Virol.* **34**: 51.

Sehgal, O.P. and Hsu, C.H. (1977). *Virology* **77**: 1.

Sehgal, O.P. and Hsu, C.H. (1976). *Stad. Symp.* **8**: 57.
Shepherd, R.J. (1971). *CMI/AAB Descr. Pl. Viruses* No. 57, 4 pp.
Shepherd, R.J. and Fulton, R.W. (1962). *Phytopathology* **52**: 489.
Tremaine, J.H. and Hamilton, R.I. (1983). *CMI/AAB Descr. Pl. Viruses* No. 274, 6 pp.
Veerisetty, V. and Sehgal, O.P. (1980). *Phytopathology* **70**: 58.
Wu, S., Rinehart, C.A. and Kaesberg, P. (1987). *Virology* **161**: 73.
Zaumeyer, W.J. and Harter, L.L. (1943). *J. Agric. Res.* **67**: 305.

# Bean yellow mosaic *potyvirus*

**Data collated by O.W. Barnett and Xu Zeyong, 1982. Revised 1987 by O.W. Barnett; data on Croatian clover strain collated by N. Taraku, 1983.**

*Synonyms:* bean virus 2, canna mosaic virus (Brierley and Smith, 1948), gladiolus mosaic virus, gloriosa stripe mosaic virus. *Standard acronym:* BYMV. *Strains:* Croatian clover potyvirus, Croatian clover mosaic virus (Taraku *et al.* 1977), dipladenia mosaic virus (Paludan *et al.*, 1988), pea mosaic 1 virus, pea 2 virus, pisum 2 virus, pea common mosaic virus. There are many other strains, Scott's isolate, B-25 and isolates from *Gladiolus* spp. are considered typical isolates, pea mosaic virus (PMV) 204 and other isolates from clovers are more distant.

First reported in *Phaseolus vulgaris*; from the U.S.A. and the Netherlands; by Doolittle and Jones (1925).

### Natural host range and symptoms

Symptoms vary seasonally.

*Phaseolus vulgaris* and *Pisum sativum* — tip necrosis, mosaic (strains B25 and Scott).

*Gladiolus* sp., *Freesia* sp., *Lupinus luteus, Robinia pseudo-acacia, Trigonella foenum-graecum, Vicia sativa, Papaver somniferum* — streaks, mosaics, mottling and leaf malformation.

*Arachis hypogaea* — leaf mottling.

*Crotalaria spectabilis, Glycine max, Canna spp., Trifolium subterraneum, Lisianthus russelianus, Vicia faba, Medicago sativa, Trifolium hybridum, T. vesiculatum, T. incarnatum, T. pratense, T. repens* — mosaics.

### Transmission

Transmitted by a vector; an insect; more than 20 spp. including *Acyrthosiphon pisum, Macrosiphum euphorbiae, Myzus persicae, Aphis fabae*; Aphididae. Transmitted in a non–persistent manner. Virus transmitted by mechanical inoculation; transmitted by seed (to 3%).

### Geographical distribution

Probably distributed worldwide.

### Experimental host range

Several (3–9) families susceptible. Experimentally infected plants mostly show mosaic.

Diagnostically susceptible

B-25 or Scott strains.

*Gomphrena globosa* — necrotic local lesions; not systemic.

*Chenopodium amaranticolor, C. quinoa* — necrotic local lesions with red border; some systemic chlorotic vein banding and spots.

*Phaseolus vulgaris* — necrotic or chlorotic local lesions; systemic mosaic.
*Pisum sativum* — Perfection types not infected; systemic mosaic in others.
*Vicia faba* — mild systemic mosaic.

Diagnostically insusceptible

*Cucumis sativus* (**I**), *Trifolium repens* (**I**) Scott strain.

Maintenance and propagation hosts

*Pisum sativum, Vicia faba, Phaseolus vulgaris, Nicotiana clevelandii.*

Assay hosts (**L**ocal lesions or **W**hole plants)

*Chenopodium amaranticolor* (L), *C. quinoa* (L).

Dicotyledonous Hosts (***Susceptible*** *or Insusceptible)*

***Amaranthus caudatus***. *Amaranthus retroflexus*. *Beta macrocarpa*. ***Cajanus cajan***. *Cassia tora*. *Cheiranthus cheiri*. ***Chenopodium album***. ***Chenopodium amaranticolor***. *Chenopodium quinoa*. ***Cicer arietinum***. ***Crotalaria retusa***. ***Crotalaria spectabilis***. *Cucumis sativus*. ***Cucurbita maxima***. *Cucurbita pepo*. ***Cyamopsis tetragonoloba***. *Datura stramonium*. *Datura tatula*. ***Glycine max***. ***Gomphrena globosa***. ***Lathyrus odoratus***. ***Lens esculenta***. ***Lupinus albus***. ***Lupinus angustifolius***. *Lycopersicon esculentum*. ***Macrotyloma uniflorum***. ***Medicago sativa***. ***Melilotus alba***. ***Nicotiana clevelandii***. ***Nicotiana debneyi***. ***Nicotiana glutinosa***. ***Nicotiana rustica***. ***Nicotiana sylvestris***. ***Nicotiana tabacum***. ***Petunia hybrida***. ***Phaseolus lunatus***. ***Phaseolus vulgaris***. *Phlox drummondii*. ***Physalis floridana***. ***Phytolacca americana***. ***Pisum sativum***. ***Spinacia oleracea***. ***Stellaria media***. ***Tetragonia expansa***. ***Trifolium hybridum***. ***Trifolium incarnatum***. ***Trifolium pratense***. ***Trifolium repens***. ***Trifolium subterraneum***. ***Vicia faba***. ***Vicia sativa***. ***Vicia villosa***. ***Vigna angularis***. ***Vigna radiata***. ***Vigna unguiculata***. *Vigna unguiculata* ssp. *sesquipedalis*. ***Zinnia elegans***.

Sources of host-range data

Bos *et al.* (1974); Edwardson (1974); Ghabriel *et al.* (1977); Hampton *et al.* (1978); Horváth (1979*b*); Johnstone and McLean (1987); Komuro *et al.* (1968); Lindsten *et al.* (1976); Lisa *et al.* (1987); McLaughlin (1988); Moghal *et al.* (1976); Randles *et al.* (1980); Taylor *et al.* (1986); Thornberry (1966).

**Properties of particles in sap**

TIP: 65 °C. LIV: 2–7 days. DEP: $\log_{10}$ minus 3–5. Leaf sap contains few virions.

**Purification**

Method: coat protein degrades if not purified rapidly (Reddick and Barnett, 1983).

**Particle morphology**

Virions filamentous; not enveloped; usually flexuous; with a clear modal length; of 750 nm ( longer in presence of divalent cations); 12–15 nm wide. Axial canal obscure. Basic helix obvious; pitch of basic helix 3.4 nm.

**Physical properties**

One sedimenting component in purified preparations; sedimentation coefficient 151 *S*.

**Biochemical properties**

Virions contain 5 % nucleic acid; 95 % protein; 0 % lipid.

*Genome* consists of RNA; single–stranded; linear; unipartite; largest (or only) genome part 10 kb.

*Sequence database accession code(s):*

D00490 Em(40)_vi:BYMVCP Gb(84)_vi:MBGVCP Bean yellow mosaic virus, coat protein gene, complete cds. 8/90 1,015bp.

D00604 Em(40)_vi:MBGCAP Gb(84)_vi:MBGCAP Bean yellow mosaic virus capsid protein gene, complete cds. 8/91 1,392bp.

D28819 Em(40)_vi:MBGMB625 Gb(84)_vi:MBGMB625 Bean yellow mosaic virus gene. 4/94 4,306bp.

S77515 Em(40)_vi:S77515 Gb(84)_vi:S77515 viral coat protein (3' region) bean yellow mosaic virus BYMV, strain S, Genomic RNA, 1019 nt

X53684 Em(40)_vi:POBYMV Gb(84)_vi:POBYMV Bean yellow mosaic virus (BYMV) mRNA for coat protein. 9/93 993bp.

X63358 Em(40)_vi:BYMV3G Gb(84)_vi:BYMV3G Bean yellow mosaic virus 3'part of genome. 6/93 4,320bp.

X81124 Gb(84)n:BYMVCPNIB Bean yellow mosaic virus genomic RNA for coat protein. 8/94 1,239bp.

X81124 Em(43)_vi:Bymvcpnib Gb(89)_vi:Bymvcpnib Bean yellow mosaic virus genomic RNA for coat protein. 8/94 1,239bp.

Amino acid composition: Randles *et al.*, (1980) for bean yellow mosaic virus-S and -Q.

*Virus–coded non–virion proteins* isolated (Nagel *et al.* (1983) for nuclear inclusion protein); two proteins found. $M_r$ of the largest 56000. $M_r$ of 2nd largest 51000.

**Cytopathology**

Virions found in all parts of the host plant; in cytoplasm. Inclusions present in infected cells; are crystals in the nucleus, crystals in the cytoplasm, and pinwheels (Edwardson's Group II); they do not contain virions (although sometimes banded bodies are found in the cytoplasm that consist of virions).

**Taxonomy and Relationships**

*Potyvirus: Potyviridae.*

Virus(es) with serologically related virions: clover yellow vein and bean common mosaic viruses.

The Croatian clover strain infects *Nicotiana megalosiphon* and *Chenopodium murale*, and is serologically distinct from type strains. With random primed *c*DNA and S1 nuclease, several bean yellow mosaic potyvirus strains show close homology, but no hybridization was detected between bean yellow mosaic strains and clover yellow vein, bean yellow mosaic (PMV) 204–1 or sweet pea mosaic viruses (Barnett *et al.*, 1987).

**References**


Barnett, O.W., Randles, J.W. and Burrows, P.M. (1987). *Phytopathology* **77**: 791.

Bays, D.C. and Demski, J.W. (1986). *Plant Dis.* **70**: 667.

Bos, L. (1970). *Neth. J. Pl. Path.* **76**: 8.

Bos, L. (1970). *CMI/AAB Descr. Pl. Viruses* No. 40, 4 pp.

Bos, L., Kowalska, C. and Maat, D.Z. (1974). *Neth. J. Pl. Path.* **80**: 173.

Bos, L., Lindsten, K. and Maat, D.Z. (1977). *Neth. J. Pl. Path.* **83**: 97.

Brierley, P. and Smith, F.F. (1948). *Phytopathology* **38**: 230.

Doolittle, S.P. and Jones, F.R. (1925). *Phytopathology* **15**: 763.

Edwardson, J.R. (1974). *Fla Agric. Exp. Stn Monog.* No. 5, p. 22.

Ghabrial, S.A., Pickard, C.M. and Stuckey, R.E. (1977). *Pl. Dis. Reptr* **61**: 690.

Hampton, R., Beczner, L., Hagedorn, P., Bos, L., Inouye, T., Barnett, O.W., Musil, M. and Meiners, J. (1978). *Phytopathology* **68**: 989.

Horváth, J. (1979*a*). *Acta Phytopath. Acad. Sci. Hung.* **14**: 157.

Horváth, J. (1979*b*). *Acta Phytopath. Acad. Sci. Hung.* **14**: 147.

Jones, R.T. and Diachun, S. (1977). *Phytopathology* **67**: 831.
Johnstone, G.R. and McLean, G.D. (1987). *Ann. appl. Biol.* **110**: 421.
Komuro, Y. and Iwaki, M. (1968). *Ann. Phytopath. Soc. Japan* **34**: 7.
Lindsten, K., Brishammas, S. and Tominius, K. (1976). *Meddn St. VäxtskAnst.* **16**: 289.
Lisa, V. and Dellavalle, G. (1987). *Pl. Path.* **36**: 214.
McLaughlin, M.R. (1988). *Plant Dis.* **72**: 539.
Moghal, S.M. and Francki, R.I.B. (1976). *Virology* **73**: 350.
Moghal, S.M. and Francki, R.I.B. (1981). *Virology* **112**: 210.
Nagel, J., Zettler, F.W. and Hiebert, E. (1983). *Phytopathology* **73**: 449.
Paludan, N., Heide, M., Begtrup, J. and Borkhardt, B. (1988). *Acta Hort.* **234**: 53.
Pierce, W.H. (1934). *Phytopathology* **24**: 87.
Randles, J.W., Davies, J.W., Gibbs, A.J. and Hatta, T. (1980). *Aust. J. biol. Sci.* **33**: 245.
Reddick, B.B. and Barnett, O.W. (1983). *Phytopathology* **73**: 1506.
Schmelzer, K., Schmidt, H.E. and Beczner, L. (1973). *Biol. Zbl.* **92**: 211.
Taraku, N., Juretič, N. and Miličič, D. (1977). *Acta bot. Croat.* **36**: 47
Taylor, N.L., Ghabrial, S.A., Diachun, S. and Cornelius, P.L. (1986). *Crop Science* **26**: 68.
Thornberry, H.H. (1966). In: *Index of Plant Virus Diseases* U.S. Dep. Agric. Hdbk No. 307.
Van Regenmortel, M.H.V. (1964). *Virology* **23**: 495.
Wetter, C. (1960). *Arch. Microbiol.* **37**: 278.

## Bean yellow vein banding *umbravirus*

**Data collated by K. Boswell, 1985. From Cockbain 1983.**

*Synonyms:* possibly pea silky green virus (Cockbain *et al.*, 1986). *Standard acronym:* BYVBV.

First reported in *Vicia faba*; from southern England; by Cockbain (1978).

### Natural host range and symptoms

*Vicia faba* — yellow vein-banding, mottling; symptoms enhanced when co-infecting with pea enation mosaic enamovirus.

### Transmission

Transmitted by a vector; an insect; *Acyrthosiphon pisum, Myzus persicae*; Aphididae. Not transmitted by *Aphis fabae*. Transmitted in a persistent manner. Virus requires, for vector transmission, a helper virus (pea enation mosaic enamovirus in the field; or bean leaf roll luteovirus in glasshouse tests); transmitted by mechanical inoculation.

### Geographical distribution

Spreads in the UK.

### Experimental host range

Many (>9) families susceptible.

Diagnostically susceptible

*Chenopodium amaranticolor, C. quinoa* — local chlorotic or necrotic lesions; no symptoms.

*Nicotiana clevelandii, N. glutinosa, Pisum sativum, Trifolium incarnatum, Vicia faba, V. sativa* — no local symptoms; systemic mottling.

Dicotyledonous Hosts (***Susceptible*** *or Insusceptible)*

***Vicia faba*.**

**Particle morphology**

Virions isometric; not enveloped; 50–90 nm in diameter; rounded in profile.

**Taxonomy and Relationships**

*Umbravirus*.

**References**

Cockbain, A.J. (1978). *Rep. Rothamsted Exp. Stn.* for 1977, p. 221.

Cockbain, A.J. (1983). *The Faba Bean (Vicia faba): A Basis for Improvement*; ed. P.D. Hebblethwaite. Butterworths, Sevenoaks, U.K.

Cockbain, A.J. Jones, P. Woods, R.D. (1986). *Ann. appl. Biol.* **108**: 59.

# Beet 1 *alphacryptovirus*

**Data collated by R.F. White, 1986.**

*Synonyms:* beet temperate virus. *Standard acronym:* BCV-1.

First reported in *Beta vulgaris*; from the U.K.

**Natural host range and symptoms**

Symptoms none.

*Beta vulgaris* — no symptoms.

**Transmission**

Transmitted by means not involving a vector. Virus not transmitted by mechanical inoculation; not transmitted by grafting; not transmitted by contact between plants; transmitted by seed (up to 100% in sugar beet); transmitted by pollen to the seed.

**Geographical distribution**

Spreads in the Eurasian region. Found, but with no evidence of spread, in Australian, the Eurasian region, Japan and the U.S.A.

**Experimental host range**

Few (<3) families susceptible.

Maintenance and propagation hosts

*Beta vulgaris*.

Assay hosts (**L**ocal lesions or **W**hole plants)

None.

Dicotyledonous Hosts (***Susceptible*** *or Insusceptible)*

***Beta vulgaris***. *Chenopodium amaranticolor. Chenopodium capitatum. Chenopodium foetidum. Chenopodium hybridum. Chenopodium murale. Chenopodium quinoa. Cucumis sativus. Gomphrena globosa. Nicotiana clevelandii. Nicotiana glutinosa. Nicotiana tabacum. Petunia hybrida. Vigna unguiculata* ssp. *sesquipedalis*.

**Properties of particles in sap**

Leaf sap contains few virions.

**Purification**

Method: Kassanis *et al.* (1977); Accotto and Boccardo (1986).

**Particle morphology**

Virions isometric; not enveloped; 30 nm in diameter; rounded in profile, or angular in profile; without a conspicuous capsomere arrangement.

**Physical properties**

Usually one sedimenting component in purified preparations; sedimentation coefficient 120 *S*. Density 1.36–1.38 g $cm^{-3}$ in CsCl.

**Biochemical properties**

*Genome* consists of RNA; double–stranded; linear; of two parts; largest (or only) genome part the larger 2.06 kb (pairs); the 2nd largest 1.42 kb (pairs).

*Virion protein(s)* one; $M_r$ 52500–54500. Method of preparation: Accotto and Boccardo (1986).

**Cytopathology**

Virions found in all parts of the host plant.

**Taxonomy and Relationships**

*Alphacryptovirus: Partitiviridae.*

Virus(es) with serologically unrelated virions: alfalfa 1, beet 2, carnation 1, carrot 1, fescue, garland chrysanthemum temperate, hop trefoil 1, radish yellow edge, red clover 1, rhubarb temperate, ryegrass, spinach temperate, vicia and white clover 1 and 3 alphacryptoviruses.

The genomic dsRNAs have no significant sequence homology with those of beet 2 alphacryptovirus. Best tests for diagnosis: cDNA probes can be used to detect the virus in beet plants.

**References**

Accotto, G.P. and Boccardo, G. (1986). *J. gen. Virol.* **67**: 367.

Accotto, G.P., Brisco, M.J. and Hull, R. (1987). *J. gen. Virol.* **68**: 1417.

Antoniw, J.F., Linthorst, H.J.M., White, R.F. and Bol, J.F. (1986). *J. gen. Virol.* **67**: 2047.

Kassanis, B., White, R.F. and Woods, R.D. (1977). *Phytopath. Z.* **90**: 350.

Kassanis, B., Russell, G.E. and White, R.F. (1978). *Phytopath. Z.* **91**: 6.

Natsuaki, T., Natsuaki, K.T., Okuda, S., Teranaka, M., Milne, R.G., Boccardo, G. and Luisoni, E. (1986). *Intervirology* **31**: 69.

Xie, W., Antoniw, J.F. and White, J.F. (1989). *Pl. Path.* **38**: 527.

# Beet 2 *alphacryptovirus*

**Data collated by R.F. White, 1986.**

*Synonyms:* beet cryptic virus. *Standard acronym:* BCV-2.

First reported in *Beta vulgaris*; from the U.K.; by Kassanis *et al.* (1977).

**Natural host range and symptoms**

Symptoms none.

*Beta vulgaris* — no symptoms.

**Transmission**

Transmitted by means not involving a vector. Virus not transmitted by mechanical inoculation; not transmitted by grafting; not transmitted by contact between plants; transmitted by seed (up to 100% in sugar beet); transmitted by pollen to the seed.

**Geographical distribution**

Spreads in the Eurasian region. Found, but with no evidence of spread, in Australia, the Eurasian region, Japan and the U.S.A.

**Experimental host range**

Few (<3) families susceptible.

Maintenance and propagation hosts

*Beta vulgaris*.

Assay hosts (**L**ocal lesions or **W**hole plants)

None.

Dicotyledonous Hosts *(**Susceptible** or Insusceptible)*

***Beta vulgaris***. *Chenopodium amaranticolor*. *Chenopodium capitatum*. *Chenopodium foetidum*. *Chenopodium hybridum*. *Chenopodium murale*. *Chenopodium quinoa*. *Cucumis sativus*. *Gomphrena globosa*. *Nicotiana clevelandii*. *Nicotiana glutinosa*. *Nicotiana tabacum*. *Petunia hybrida*. *Vigna unguiculata* ssp. *sesquipedalis*.

**Properties of particles in sap**

Leaf sap contains few virions.

**Purification**

Method: Kassanis *et al.* (1977); Accotto and Boccardo (1986).

**Particle morphology**

Virions isometric; not enveloped; 30 nm in diameter; rounded in profile, or angular in profile; without a conspicuous capsomere arrangement.

**Physical properties**

Usually one sedimenting component in purified preparations; sedimentation coefficient 120 *S*. Density 1.36–1.38 g $cm^{-3}$ in CsCl.

**Biochemical properties**

*Genome* consists of RNA; double–stranded; of two parts; largest (or only) genome part the larger 1.74 kb (pairs); the 2nd largest 1.32 kb (pairs).

*Virion protein(s)* one; $M_r$ 52500–54500. Method of preparation: Accotto and Boccardo (1986).

**Cytopathology**

Virions found in all parts of the host plant.

**Taxonomy and Relationships**

*Alphacryptovirus: Partitiviridae.*

Virus(es) with serologically unrelated virions: alfalfa 1, carnation 1, carrot 1, fescue cryptic, garland chrysanthemum temperate, hop trefoil 1, radish yellow edge, red clover 1, rhubarb temperate, ryegrass, spinach temperate, vicia and white clover cryptic 1 and 3 alphacryptoviruses.

The genomic dsRNAs have no detectable homology with those of beet 1 alphacryptovirus. Best tests for diagnosis: cDNA probes can be used to detect this virus in infected beet plants.

**References**

Accotto, G.P. and Boccardo, G. (1986). *J. gen. Virol.* **67**: 367.

Antoniw, J.F., Linthorst, H.J.M., White, R.F. and Bol, J.F. (1989). *J. gen. Virol.* **67**: 2047.

Kassanis, B., White, R.F. and Woods, R.D. (1977). *Phytopath. Z.* **90**: 350.
Kassanis, B., Russell, G.E. and White, R.F. (1978). *Phytopath. Z.* **91**: 6.
Natsuaki, T., Natsuaki, K.T., Okuda, S., Teranaka, M., Milne, R.G., Boccardo, G. and Luisoni, E. (1986). *Intervirology* **31**: 69.
Xie, W., Antoniw, J.F. and White, R.F. (1989). *Pl. Path.* **38**: 527.

# Beet 3 *alphacryptovirus*

**Data collated by A.A. Brunt, 1994.**

*Standard acronym:* BCV-3.

First reported in *Beta vulgaris* var *cicla*; from Italy, in plants grown from Japanese seed; by Accotto *et al.* (1987), Xie *et al.* (1993).

**Natural host range and symptoms**

Symptoms none.
*Beta vulgaris* var *cicla* cv. Fudanso — symptomless.

**Transmission**

Virus not transmitted by mechanical inoculation; not transmitted by grafting; transmitted by seed.

**Geographical distribution**

Found, but with no evidence of spread, in Japan, Italy and the U.K.

**Experimental host range**

Many (>9) families susceptible.

Maintenance and propagation hosts

*Beta vulgaris* var. *cicla*.

Dicotyledonous Hosts *(**Susceptible** or Insusceptible)*

***Beta vulgaris***.

**Purification**

Method: Kassanis *et al.* (1977), Xie *et al.* (1993).

**Particle morphology**

Virions isometric; not enveloped; 30 nm in diameter; rounded in profile.

**Biochemical properties**

*Genome* consists of RNA; double–stranded; of two parts; largest (or only) genome part 1.74 kb (pairs); the 2nd largest 1.6 kb (pairs).

*Sequence database accession code(s):*

S63913 Em(40)_vi:S63913 Gb(84)_vi:S63913 putative RNA-dependent RNA polymerase (double-stranded RNA2) beet cryptic virus 3 BCV3, Genome 1 sequence.

**Taxonomy and Relationships**

*Alphacryptovirus: Partitiviridae.*
Virus(es) with serologically unrelated virions: beet 1 and 2 *alphacryptoviruses*.
Viral dsRNAs have no homology with those beet 1 and 2 *alphacryptoviruses*.

**References**

Accotto, G.P., Brisco, M.J. and Hull, R. (1987). *J. gen. Virol.* **68**: 1417.
Kassanis, B., White, R.F. and Woods, R.D. (1977). *Phytopath. Z.* **90**: 350.

Xie, W.S., Antoniw, J.F., White, R.F. and Woods, R.D. (1993). *Pl. Path.* **42**: 465.

# Beet curly top hybrigeminivirus

**Data collated by K. Boswell, 1985.**

*Synonyms:* sugarbeet curly top virus, sugarbeet curly-leaf virus, tomato yellows virus, western yellow blight virus. *Standard acronym:* BCTV. *Strains:* tomato yellows virus. Many stable strains described (Thomas and Mink, 1979).

First reported in *Beta vulgaris*; from Western U.S.A; by Ball (1909).

**Natural host range and symptoms**

Symptoms persist.

*Spinacia oleracea*, Cucurbitaceae, *Phaseolus vulgaris*, *Linum* spp., *Capsicum* spp., *Lycopersicon esculentum, Solanum tuberosum*, and many other species — vein-clearing, swelling and malformation in young leaves; rigid, dwarfed, yellowed, twisted and malformed leaves; axillary buds stimulated, phloem necrosis, exudation of fluid from phloem; death of plant. Some hosts symptomless.

*Beta vulgaris* — leaf rolling, vein clearing, leaves become dark, dull green in colour, thick crisp and brittle.

**Transmission**

Transmitted by a vector; an insect; *Circulifer tenellus* in N. America, *C. tenellus, C. opacipennis* in Mediterranean Basin; Cicadellidae. Transmitted in a persistent manner. Virus retained when the vector moults; does not multiply in the vector; not transmitted congenitally to the progeny of the vector; not transmitted by mechanical inoculation (unless special procedures used); transmitted by grafting (by dodder from plants that are not hosts of the vector); not transmitted by seed.

**Geographical distribution**

Spreads in the African region, the Eurasian region, the Mediterranean region, the North American region, and the South and Central American region (in the arid and semi-arid regions of the west); Argentina, Bolivia, Brazil, Canada, Costa Rica, Cyprus, Egypt, India, Iran, Italy, Mexico, Puerto Rico, Spain, Turkey, the USA, and Uruguay.

**Experimental host range**

Many (>9) families susceptible.

Diagnostically susceptible

*Beta vulgaris* — vein-clearing and upward and inward rolling of leaves. Vein swelling and galling to give rough abaxial leaf surface. Phloem exudate on petioles and large leaf veins, phloem necrosis in roots.

*Cucumis sativus* — seedlings killed. Older plants stunted and malformed, leaves rolled upward.

*Phaseolus vulgaris* cv. Bountiful — first trifoliate leaf epinastic. Buds of survivors proliferate, leaflets brittle, curled and crinkled.

Maintenance and propagation hosts

*Beta vulgaris* for maintenance cultures. *Phaseolus vulgaris* cvs Romano, Bountiful, and *Nicotiana tabacum* cv. Turkish for propagation.

Assay hosts (**L**ocal lesions or **W**hole plants)

No local-lesion host. *Beta vulgaris* (W).

Dicotyledonous Hosts *(**Susceptible** or Insusceptible)*

***Amaranthus caudatus. Amaranthus retroflexus. Apium graveolens. Atriplex hortensis. Bellis perennis. Beta macrocarpa. Beta patellaris. Beta vulgaris. Brassica campestris*** *ssp.* ***rapa. Calendula officinalis. Capsella bursa–pastoris. Capsicum frutescens. Catharanthus roseus.*** *Celosia argentea.* ***Chenopodium album. Chenopodium amaranticolor. Chenopodium capitatum. Chenopodium murale. Cicer arietinum. Citrullus lanatus. Coriandrum sativum. Cucumis sativus. Cucurbita maxima. Cucurbita moschata. Cucurbita pepo.*** *Datura stramonium.* ***Dianthus barbatus. Dianthus caryophyllus. Euphorbia marginata. Fagopyrum esculentum. Gomphrena globosa. Gossypium hirsutum. Hibiscus esculentus. Ipomoea setosa. Lactuca sativa. Lathyrus odoratus. Linum usitatissimum.*** *Lycopersicon esculentum. Lycopersicon pimpinellifolium.* ***Matthiola incana. Medicago hispida. Medicago sativa. Melilotus alba. Nicotiana benthamiana. Nicotiana bigelovii. Nicotiana clevelandii. Nicotiana debneyi. Nicotiana glutinosa. Nicotiana rustica. Nicotiana sylvestris. Nicotiana tabacum. Papaver nudicaule. Petunia hybrida.*** *Phaseolus lunatus.* ***Phaseolus vulgaris. Phlox drummondii. Physalis floridana. Physalis peruviana. Phytolacca americana. Pisum sativum. Raphanus sativus. Ricinus communis. Salvia splendens. Senecio vulgaris. Solanum demissum. Solanum melongena. Solanum nigrum.*** *Solanum tuberosum. Sonchus oleraceus.* ***Spinacia oleracea. Stellaria media. Trifolium hybridum. Trifolium incarnatum. Trifolium pratense. Trifolium repens. Tropaeolum majus. Vicia faba. Vicia sativa. Vicia villosa. Vigna angularis. Vigna unguiculata. Viola cornuta. Zinnia elegans.***

Comments on host-range

No monocotyledonous species found to be susceptible.

Sources of host-range data

Bennet (1971).

**Properties of particles in sap**

TIP: 80 °C. LIV: 8 days. DEP: $\log_{10}$ minus 3. Leaf sap contains few virions.

**Purification**

Method: Mink and Thomas (1974); Mumford (1974).

**Particle morphology**

Virions geminate; not enveloped; 18–22 nm in diameter; rounded in profile.

**Physical properties**

One sedimenting component in purified preparations, or two sedimenting components in purified preparations; sedimentation coefficient 82 *S* (for single virions); of the other(s) 86 *S*, or 55 *S* (Egbert *et al.* 1976).

**Biochemical properties**

*Genome* consists of DNA; single–stranded; circular; unipartite; largest (or only) genome part 2.8 kb.

*Sequence database accession code(s):*

M24597 Gb(84)_vi:BCTCG028A Beet curly top virus (clone pBCT028) DNA, complete genome. 9/89 2,994bp.

M74562 Em(40)_vi:BCTSTLPA Gb(84)_vi:BCTSTLPA Beet curly top virus stem loop. 1/92 58bp.

M74563 Em(40)_vi:BCTSTLPB Gb(84)_vi:BCTSTLPB Beet curly top virus stem loop. 1/92 60bp.

U02311 Em(40)_vi:BCU02311 Gb(84)_vi:BCU02311 Beet curly top virus CFH complete genome. 1/94 2,927bp

X04144 Em(40)_vi:GEBCTV Gb(84)_vi:GEBCTV Beet curly top virus (BCTV) virion DNA. 7/89 2,993bp.

S72015 Gb(89)_un:S72015 orf L1 (intergenic region) (beet curly top virus BCTV, Worland, Genomic, 106 nt). 12/94 106bp.

**Cytopathology**

Virions found in phloem; in nuclei.

**Taxonomy and Relationships**

Hybrigeminivirus: *Geminiviridae* (type species).

**References**


Ball, E.D. (1909). *Bull. Bur. Ent. U.S. Dep. Agric.* **66**: 33.
Bennett, C.W. (1971). *Monogr. Am. Phytopath. Soc. No.* **7**: 81.
Mink, G.I. and Thomas, P.E. (1974). *Phytopathology* **64**: 140.
Mumford, D.L. (1974). *Phytopathology* **64**: 136.
Mumford, D.L. (1977). *Phytopathology* **67**: 949.
Stanley, J., Markham, P.G., Callis, R.J. and Pinner, M.S. (1986). *EMBO J.* **5**: 1761.
Thomas, P.E. and Mink, G.I. (1979). *CMI/AAB Descr. Pl. Viruses* No. 210, 6 pp.


# Beet distortion mosaic virus

**Data collated by A.A. Brunt, 1991.**

First reported in *Beta vulgaris*; from the U.S.A.; by Liu *et al.* (1987).

**Natural host range and symptoms**

*Beta vulgaris* — malformed and chlorotic leaves.

**Transmission**

Transmitted by a vector; probably a fungus; probably *Polymyxa betae*; Plasmodiophorales. Virus transmitted by mechanical inoculation.

**Geographical distribution**

Spreads in the USA (California).

**Particle morphology**

Virions filamentous; usually flexuous; with a clear modal length; of 200–2400 nm; 12 nm wide.

**Taxonomy and Relationships**

Virus(es) with serologically unrelated virions: wheat spindle streak mosaic virus.

**References**


Liu, H.Y., Duffus, J.E. and Genk, J.S. (1987). *Phytopathology* **77**: 1732.


# Beet leaf curl (?) rhabdovirus

**Data collated by Sc.H. Kleinhempel, 1987.**

*Synonyms:* beet leaf crinkle virus, beta virus 3, Rübenkräusel-Virus, sugarbeet leafcurl rhabdovirus. *Standard acronym:* BLCV.

First reported in *Beta vulgaris*; from Germany; by Wille (1928).

**Natural host range and symptoms**

Symptoms persist.

*Beta vulgaris* — vein deformation, leaves curved towards the apex.

**Transmission**

Transmitted by a vector; an insect; Heteroptera, Lygaeidae: *Piesma quadratum*. Transmitted in a persistent manner. Virus retained when the vector moults; multiplies in the vector; not transmitted congenitally to the progeny of the vector; not transmitted by mechanical inoculation; transmitted by grafting; not transmitted by contact between plants.

**Ecology and control**

Studies reported by Proeseler (1980).

**Geographical distribution**

Spreads in the Eurasian region and the Mediterranean region; the former Czechoslovakia, Germany, Poland, Turkey, and the former USSR.

**Experimental host range**

Few (<3) families susceptible. Experimentally infected plants mostly show vein clearing, leaf curling, stunting.

Diagnostically susceptible

*Beta vulgaris, Chenopodium quinoa, Spinacia oleracea, Tetragonia expansa* — vein clearing, leaf curling, stunting.

Maintenance and propagation hosts

*Beta vulgaris.*

Assay hosts (**L**ocal lesions or **W**hole plants)

*Beta macrocarpa* (W), *Atriplex patula* (W), *Chenopodium murale* (W), *Spinacia tetranda* (W).

Dicotyledonous Hosts *(**Susceptible** or Insusceptible)*

***Beta macrocarpa***. ***Beta vulgaris***. ***Chenopodium album***. ***Chenopodium amaranticolor***. ***Chenopodium capitatum***. ***Chenopodium foliosum***. ***Chenopodium murale***. ***Chenopodium quinoa***. ***Spinacia oleracea***. ***Tetragonia expansa***.

Sources of host-range data

Proeseler (1966); Schmutterer and Ehrhardt (1966).

**Properties of particles in sap**

TIP: 54–58 °C. LIV: less than 1 days. DEP: $\log_{10}$ minus 4–5.

**Purification**

Method: Eisbein and Proll (1978).

**Particle morphology**

Virions rhabdo– or bullet–shaped; enveloped; of 225–350 nm (longer in roots); 80 nm wide.

**Physical properties**

One sedimenting component in purified preparations.

### Cytopathology

Virions found in leaves, roots mesophyll and vascular parenchyma; in cytoplasm, in the perinuclear space, and in cell vacuoles. Inclusions present in infected cells; are crystals in the cytoplasm; they contain virions.

### Taxonomy and Relationships

Unassigned rhabdovirus.

### References

Boening, K. (1928). *Anz. Schädlingsk.* **4**: 8.
Eisbein, K. and Proll, E. (1978). *Arch. Phytopathol. Pflanzenschutz.* **14**: 81.
Proeseler, G. (1966). *Phytopath. Z.* **56**: 191.
Proeseler, G. (1980). In: *Vectors of Plant Pathogens*, p. 97; eds K.F. Harris and K. Maramorosch. Academic Press, New York.
Proeseler, G. (1983). *CMI/AAB Descr. Pl. Viruses* No. 268, 3 pp.
Schmutterer, H. and Ehrhardt, P. (1966). *Z. PflKrankh. PflPath. PflSchutz.* **73**: 271.
Wille, J. (1928). *Arb. biol. BundesAnst. Land. Forstw.* **16**: 115.

## Beet mild yellowing *luteovirus*

**Data collated by H.G. Smith, 1991.**

*Synonyms:* beet western yellows. *Standard acronym:* BMYV.

First reported in *Beta vulgaris*; from England; by Russel (1958).

### Natural host range and symptoms

Symptoms persist. Symptoms yellowing of older leaves, which become curled, thickened and brittle. Some reddening of older leaves of weed species.

*Beta* species — golden yellowing of older leaves, which become thick and brittle.

*Capsella bursa-pastoris* — yellowing of older leaves, curling and reddening of leaves.

*Stellaria media* — yellowing of older leaves.

*Lamium amplexicaule* — yellowing of older leaves.

*Senecio vulgaris, Veronia* sp. — yellowing of older leaves, reddening of some older leaves.

### Transmission

Transmitted by a vector; an insect; *Myzus persicae, M. (Sciamyzus) ascolonicus, M. (Nectarosiphon) certus, Macrosiphum euphorbiae, Acyrthosiphon (Aulocorthum) solani, Myzus (Phorodon) humuli, Brachycaudus helichrysi*; Aphididae. Transmitted in a persistent manner. Virus retained when the vector moults; does not multiply in the vector; not transmitted congenitally to the progeny of the vector; not transmitted by mechanical inoculation; not transmitted by grafting; not transmitted by contact between plants; not transmitted by seed; not transmitted by pollen.

### Ecology and control

Studies reported by Dunning *et al.* (1988).

### Geographical distribution

Spreads in the Eurasian region.

**Experimental host range**

Several (3–9) families susceptible. Experimentally infected plants mostly show yellowing or reddening of older leaves, which also may become thickened and brittle. Curling of leaves. Stunting.

Diagnostically susceptible

*Capsella bursa-pastoris* and
*Spinacia oleracea* — yellowing of older leaves, curling and thickening.
*Montia perfoliata* — reddening of older leaves.
*Senecio vulgaris*.

Diagnostically insusceptible

*Brassica* spp., *Lactuca sativa, Nicotiana clevelandii, Pisum sativum, Trifolium incarnatum*.

Maintenance and propagation hosts

*Beta vulgaris, Capsella bursa-pastoris, Montia perfoliata*.

Assay hosts (**L**ocal lesions or **W**hole plants)

*Beta vulgaris* (W), *Capsella bursa-pastoris* (W), *Montia perfoliata* (W).

Dicotyledonous Hosts *(**Susceptible** or Insusceptible)*

*Amaranthus caudatus*. *Amaranthus retroflexus*. *Atriplex hortensis*. ***Beta macrocarpa***. ***Beta patellaris***. ***Beta vulgaris***. *Brassica campestris*. *Brassica campestris* ssp. *chinensis*. *Brassica campestris* ssp. *napus*. *Brassica campestris* ssp. *pekinensis*. *Brassica campestris* ssp. *rapa*. *Brassica juncea*. *Brassica oleracea* ssp. *botrytis*. *Brassica oleracea* ssp. *capitata*. ***Capsella bursa–pastoris***. *Celosia argentea*. *Celosia cristata*. *Chenopodium album*. *Chenopodium amaranticolor*. *Chenopodium ambrosioides*. *Chenopodium capitatum*. *Chenopodium foetidum*. *Chenopodium foliosum*. *Chenopodium hybridum*. *Chenopodium quinoa*. *Chrysanthemum morifolium*. ***Crambe abyssinica***. *Cucumis melo*. ***Cucumis sativus***. *Datura metel*. *Datura stramonium*. *Emilia sagittata*. ***Gomphrena globosa***. *Hyoscyamus niger*. *Lactuca sativa*. *Melandrium album*. *Melilotus alba*. ***Montia perfoliata***. *Nicotiana clevelandii*. *Nicotiana sylvestris*. *Nicotiana tabacum*. *Petunia hybrida*. ***Physalis floridana***. *Pisum sativum*. *Prunus persica*. *Raphanus sativus*. ***Senecio vulgaris***. ***Sinapis alba***. *Solanum nigrum*. *Sonchus oleraceus*. ***Spinacia oleracea***. ***Stellaria media***. ***Tetragonia expansa***. *Trifolium incarnatum*. *Trifolium pratense*. *Trifolium repens*. *Trifolium subterraneum*. *Vicia faba*. ***Viola cornuta***. ***Zinnia elegans***.

Sources of host-range data

Peters (1988).

**Properties of particles in sap**

TIP: 65 °C. LIV: 16 days. Leaf sap contains few virions.

**Purification**

Method: Govier (1985).

**Particle morphology**

Virions isometric; not enveloped; 26 nm in diameter; angular in profile; without a conspicuous capsomere arrangement.

**Physical properties**

One sedimenting component in purified preparations; sedimentation coefficient 116 *S*. Density 1.42 g $cm^{-3}$ in CsCl. $A_{260}/A_{280}$ ratio 1.8.

**Biochemical properties**

*Genome* consists of RNA; single–stranded. Genomic nucleic acid isolated by Veidt *et al.* (1988).

**Cytopathology**

Virions found in leaves, roots and phloem. Inclusions absent from infected cells.

**Taxonomy and Relationships**

*Luteovirus.*

Virus(es) with serologically related virions: beet western yellows virus, closely.

The virus is considered by some virologists to be a synonym of beet western yellows luteovirus. However, those working closely with the virus consider it to be a distinct pathogen. Best tests for diagnosis: ELISA using polyclonal antiserum to distinguish it from beet yellows virus, with which it is frequently associated in field-grown sugar beet.

**References**

Govier, D.A. (1985). *Ann. appl. Biol.* **107**: 439.

Peters, D. (1988) In: *Virus Yellows Monograph* pp. 87–117. IIRB, Rue Montoyer, 47, B–1040, Brussels, Belgium.

Russel, G.E. (1958). *Ann. appl. Biol.* **46**: 393.

Veidt, I., Lot, H., Leiser, M., Scheidecker, D., Guilley, H., Richards, K. and Jonard, G. (1988). *Nucl. Acids Res.* **16**: 9917.

# Beet mosaic *potyvirus*

**Data collated by G.D. Heathcote and R.D. Woods, 1982. Revised by F. Nienhaus, 1987 and I. Fujisawa, 1988. Revised by H.G. Smith and A.V. Karasev, 1991.**

*Synonyms:* spinach mosaic virus, sugarbeet mosaic virus. *Standard acronym:* BtMV. *Strains:* dioscorea alata ring mottle virus, dioscorea alata ringspot virus.

First reported in *Beta vulgaris*; from Braunschweig, Germany; by Schneider and Mundry (1956); Zimmer and Brandes (1956).

**Natural host range and symptoms**

Symptoms persist. Symptoms flecking then mosaic, malformation.

*Beta maritima, Melilotus indicus, Trifolium incarnatum*—vein yellowing or netting and leaf malformation.

*Beta vulgaris* — small local lesions on tip leaves, then mottling and malformation.

*Spinacia oleracea* — flecking, yellowing.

*Chenopodium album* — local lesions and curling of leaves.

*Amaranthus retroflexus, Sonchus arvensis, Melilotus indica* — vein chlorosis, mottling.

*Trifolium incarnatum* — vein chlorosis, mottling, necrosis.

**Transmission**

Transmitted by a vector; an insect; more than 28 spp. including *Myzus persicae, Aphis fabae, Rhopalosiphum padi, Acyrthosiphon (Metopolophium) dirhodum, Macrosiphum (Sitobion) avenae*; Aphididae. Principal natural vector(s): *Myzus persicae, Aphis fabae.* Transmitted in a non–persistent manner. Virus transmitted by mechanical inoculation; transmitted by grafting; not transmitted by contact between plants; not transmitted by seed; not transmitted by pollen.

**Ecology and control**

Studies reported by Smith (1957).

**Geographical distribution**

Probably distributed worldwide (in all major beet growing areas (Russell 1971)).

**Experimental host range**

Several (3–9) families susceptible. Experimentally infected plants mostly show mosaic, necrotic local lesions.

Diagnostically susceptible

*Amaranthus caudatus, A. retroflexus, Beta patellaris* — necrotic local lesions.

*Gomphrena globosa, Chenopodium quinoa, Spinacia oleracea*—chlorotic local lesions.

*B. vulgaris* — systemic mosaic.

*Nicotiana quadrivalvis, N. clevelandii, Phacelia campanularia, Pisum sativum, Samolus parviflorus.*

*Stellaria media* — necrotic spots; systemic mosaic.

*Glycine max* — systemic mosaic.

Maintenance and propagation hosts

*Beta vulgaris, Spinacia oleracea, Stellaria media.*

Assay hosts (**L**ocal lesions or **W**hole plants)

*Amaranthus caudatus* (L), Kazakh isolate (W), *A. retroflexus* (L), *Gomphrena globosa* (L), *Beta patellaris* (L), *B. vulgaris* (W), *Chenopodium quinoa* (L), *Spinacia oleracea* (W).

Dicotyledonous Hosts *(**Susceptible** or Insusceptible)*

***Amaranthus caudatus***. ***Amaranthus retroflexus***. *Apium graveolens*. ***Atriplex hortensis***. ***Beta macrocarpa***. ***Beta patellaris***. ***Beta vulgaris***. *Brassica campestris*. *Brassica campestris* ssp. *pekinensis*. *Brassica juncea*. *Brassica oleracea* ssp. *botrytis*. *Brassica oleracea* ssp. *capitata*. *Calendula officinalis*. ***Capsella bursa-pastoris***. *Capsicum frutescens*. ***Celosia argentea***. ***Celosia cristata***. *Cheiranthus cheiri*. ***Chenopodium album***. ***Chenopodium amaranticolor***. ***Chenopodium capitatum***. ***Chenopodium murale***. ***Chenopodium quinoa***. *Chrysanthemum morifolium*. *Cucumis sativus*. ***Cucurbita pepo***. *Dahlia pinnata*. *Datura metel*. *Datura stramonium*. *Daucus carota*. ***Dianthus barbatus***. ***Glycine max***. ***Gomphrena globosa***. ***Hibiscus esculentus***. ***Hyoscyamus niger***. *Lactuca sativa*. *Lathyrus odoratus*. *Lycopersicon esculentum*. *Matthiola incana*. *Medicago sativa*. *Melilotus alba*. *Nicandra physaloides*. ***Nicotiana clevelandii***, or *Nicotiana clevelandii*. *Nicotiana debneyi*. ***Nicotiana glutinosa***, or *Nicotiana glutinosa*. *Nicotiana rustica*. *Nicotiana sylvestris*. *Nicotiana tabacum*. *Petunia hybrida*. *Phaseolus vulgaris*. *Phlox drummondii*. *Physalis floridana*. ***Phytolacca americana***. ***Pisum sativum***. *Raphanus sativus*. ***Senecio vulgaris***. *Solanum melongena*. *Solanum nigrum*. *Solanum tuberosum*. *Sonchus oleraceus*. ***Spinacia oleracea***. ***Stellaria media***. ***Tetragonia expansa***. ***Trifolium incarnatum***. *Trifolium pratense*. *Trifolium repens*. ***Vicia faba***. *Vigna unguiculata*. ***Zinnia elegans***.

Sources of host-range data

Bennett (1949); Smith (1957); Thornberry (1966); Russell (1971); Edwardson (1974); Rogov *et al.* (1991).

**Properties of particles in sap**

TIP: 55–60 °C. LIV: 1–2 days. DEP: $\log_{10}$ minus 3. Leaf sap contains few virions.

**Purification**

Method: Chod and Polak (1969); Fujisawa *et al.* (1983); Rogov *et al.* (1991); Grüntzig and Fuchs (1979*a*).

**Particle morphology**

Virions filamentous; not enveloped; usually flexuous; of 695–770 nm; 13 nm wide. Axial canal obscure. Basic helix obscure; pitch of basic helix 3.4 nm.

**Physical properties**

Of the other(s) 150–160 *S*. Density 1.31 g $cm^{-3}$ in CsCl. $A_{260}/A_{280}$ ratio 1.22 to 1.25 (Fujisawa *et al.*, 1983); 1.23 to 1.40 (Rogov *et al.*, 1991).

**Biochemical properties**

*Genome* consists of RNA; single–stranded; linear; unipartite; largest (or only) genome part 10 kb. Genomic nucleic acid isolated by Karasev: two phenol/chloroform extractions (pH 8–8.5).

*Non–genomic nucleic acid* not found in the virions.

*Virion protein(s)* one; $M_r$ 34700; coat protein.

**Cytopathology**

Virions found' in all parts of the host plant; in cytoplasm, in nuclei, and in chloroplasts. Inclusions present in infected cells; are crystals in the cytoplasm, crystals in the nucleus, amorphous X–bodies, pinwheels, and unusual in shape; bundle type (Hoefert, 1969); they do not contain virions. Other cellular changes: enlarged and malformed nucleoli in *Beta vulgaris* and *Gomphrena globosa*.

**Taxonomy and Relationships**

*Potyvirus: Potyviridae.*

Virus(es) with serologically related virions: bean yellow mosaic, soybean mosaic and potato Y viruses, but distantly. Virus(es) with serologically unrelated virions: plum pox potyvirus.

Dioscorea alata ring mottle virus strain differences include: purified virions sediment as 1 component, sedimentation coefficient 162 **S**, density in CsCl 1.321 g $cm^{-3}$, nucleic acid 5.1%, genome is ssRNA, in 1 part, total size 10kb. Method of protein preparation reported by Laemmli and Favre (1973), 2 species found, Mr 36000 and 35000.

**References**


Bennett, C.W. (1949). *Phytopathology* **39**: 669.

Chod, J. and Polak, Z. (1969). *Biologia Pl.* **11**: 324.

Edwardson, J.R. (1974). *Fla Agric. Exp. Stn Monog.* No. 5.

Fujisawa, I., Tsuchizaki, T. and Iizuka, N. (1983). *Ann. Phytopath. Soc. Japan* **49**: 22.

Grüntzig, M. and Fuchs, E. (1979*a*). *Arch. Phytopathol. Pflanzenschutz* **15**: 89.

Grüntzig, M. and Fuchs, E. (1979*b*). *Arch. Phytopathol. Pflanzenschutz* **15**: 153.

Hoefert, L.L. (1969). *Virology* **37**: 498.

Katis, N. and Gibson, R.W. (1984). *Pl. Path.* **33**: 425.

Laemmli, U.K. and Favre, M. (1973). *J. mol. Biol.* **80**: 575.

Rogov, V.V., Bobkova, A.F., Karasev, A.V., Agranovsky, A.A. and Gorbunova, N.I. (1989). *Doklady VASChNIL*, No. 8, p. 7.

Rogov, V.V., Karasev, A.V., Agranovsky, A.A. and Gorbunova, N.I. (1991). *Pl. Path.* **40** in press.

Russell, G.E. (1971). *CMI/AAB Descr. Pl. Viruses* No. 53, 3 pp.

Schneider, F. and Mundry, K.W. (1956). *Z. Naturf.* **11**: 393.

Thornberry, H.H. (1966). In: *Index of Plant Virus Diseases.* U.S. Dep. Agric. Hdbk No. 307.
Tulegenov, T.A. (1977). *Vest. Sel'.khoz. Nauki, Kazakhstana* **20**: 122.
Zimmer, K. and Brandes, J. (1956). *Phytopath. Z.* **26**: 439.

# Beet necrotic yellow vein *furovirus*

**Data collated by K.E. Richards, 1991.**

*Synonyms:* possibly beet yellow vein virus. *Standard acronym:* BNYVV. *Strains:* YS, CS, CR and NS.

First reported in *Beta vulgaris*; from Italy; by Canova (1966); Matsuda *et al.* (1965); Tamada (1975).

### Natural host range and symptoms

Symptoms persist. Symptoms rootlet proliferation, necrosis, stunting or death.

*Beta vulgaris* — rootlet proliferation from main tap root; necrosis of vascular bundle; rootlet necrosis; stunting or death of infected plant. Occasional bright yellow invasive chlorotic local lesions form on veins in the head.

*Beta macrocarpa*
*Spinacia oleracea.*

### Transmission

Transmitted by a vector; a fungus; *Polymyxa betae*; Plasmodiophorales. Not transmitted by *Polymyxa graminis, Myzus persicae, Acyrthosiphon (Aulacorthum) solani.* Virus transmitted by mechanical inoculation; not transmitted by contact between plants; not transmitted by seed; not transmitted by pollen.

### Geographical distribution

Probably distributed worldwide (where there are major sugarbeet industries). Spreads in the Eastern Asian region, the Eurasian region, the Mediterranean region, and the North American region; Austria, Belgium, Bulgaria, China, the former Czechoslovakia, France, Germany, Greece, Hungary, Italy, Japan, Mongolia, the Netherlands, Poland, Romania, St. Vincent and Grenadines, Switzerland, Turkey, the UK, the USA, the former USSR, the former Yugoslavia. Found, but with no evidence of spread, in the U.K.

### Experimental host range

Few (<3) families susceptible. Experimentally infected plants mostly show chlorotic local lesions.

Diagnostically susceptible

*Beta vulgaris* — rootlet proliferation from main tap root; necrosis of vascular bundle; rootlet necrosis; stunting or death of infected plant. Occasional bright yellow invasive chlorotic local lesions form in veins.

*Chenopodium amaranticolor, C. quinoa, Tetragonia expansa* — chlorotic local lesions.

*Beta macrocarpa, Spinacia oleracea* — chlorotic local lesions, chlorotic mottle, yellow mosaic and flecks; systemic infection. Plants become stunted and die prematurely.

Maintenance and propagation hosts

*Chenopodium quinoa, Tetragonia expansa.*

Assay hosts (**L**ocal lesions or **W**hole plants)

*Chenopodium quinoa* (L), *Tetragonia expansa* (L).

Dicotyledonous Hosts *(**Susceptible** or Insusceptible)*

*Amaranthus retroflexus. Antirrhinum majus. Arachis hypogaea. Astragalus sinicus.* ***Beta macrocarpa. Beta patellaris. Beta vulgaris.*** *Brassica campestris* ssp. *chinensis. Brassica campestris* ssp. *napus. Brassica campestris* ssp. *pekinensis. Brassica oleracea* ssp. *botrytis. Brassica oleracea* ssp. *capitata. Capsicum annuum. Celosia cristata.* ***Chenopodium album. Chenopodium amaranticolor. Chenopodium murale. Chenopodium quinoa.*** *Citrullus lanatus. Cucumis melo. Cucumis sativus. Cucurbita moschata. Datura metel. Datura stramonium. Glycine max.* ***Gomphrena globosa.*** *Lactuca sativa. Lycopersicon esculentum. Medicago sativa. Melilotus alba. Nicandra physaloides. Nicotiana clevelandii. Nicotiana debneyi. Nicotiana glutinosa. Nicotiana sylvestris. Nicotiana tabacum. Petunia hybrida. Phaseolus vulgaris. Physalis floridana. Pisum sativum. Raphanus sativus. Senecio vulgaris. Sesamum indicum. Solanum nigrum. Solanum tuberosum.* ***Spinacia oleracea. Tetragonia expansa.*** *Trifolium hybridum. Trifolium incarnatum. Trifolium pratense. Trifolium repens. Vicia faba. Vicia sativa. Vicia villosa. Vigna unguiculata. Vigna unguiculata* ssp. *sesquipedalis. Zinnia elegans.*

Monocotyledonous Hosts *(**Susceptible** or Insusceptible)*

*Hordeum vulgare. Oryza sativa. Secale cereale. Zea mays.*

Sources of host-range data

Tamada and Baba (1973).

**Properties of particles in sap**

TIP: 65–70 °C. LIV: 5–8 days. DEP: $\log_{10}$ minus 4–5. Leaf sap contains few virions.

**Purification**

Method: Putz and Kuszala (1978).

**Particle morphology**

Virions rod–shaped; not enveloped; usually straight; with a clear modal length; of 70, 100, 265, and 390 nm; 20 nm wide. Axial canal obscure. Basic helix obvious; pitch of basic helix 2.6 nm.

**Physical properties**

Four sedimenting components in purified preparations, or five sedimenting components in purified preparations (in some Japanese isolates).

**Biochemical properties**

Virions contain 5 % nucleic acid; 95 % protein.

*Genome* consists of RNA; single–stranded; linear; of four parts, or of five parts (in some Japanese isolates); largest (or only) genome part the largest 6.7 kb; the 2nd largest 4.7 kb; the 3rd largest 1.8 kb; the 4th largest 1.5 kb; the 5th largest 1.45 kb (in some Japanese isolates). Genomic nucleic acid isolated by Richards: phenol extraction. Base composition 24 % G; 26 % A; 15 % C; 34 % U. 5' terminus of RNA has a methylated nucleotide cap. Infectivity retained when deproteinised with phenol or detergent. Poly A region present; 3' poly (A) tail on all 4 genome components. Additional factor not required for infectivity. Genome has no *t*RNA–like activity. Nucleotide sequence references: Richards: Genbank accession Numbers RNA 1 D00115; RNA 2 X04197; RNA 3 M36894; RNA 4 see Bouzoubaa *et al.* (1985, 1986).

*Sequence database accession code(s):*

D00115 Gb(84)_vi:BNYVVR1 Beet necrotic yellow vein mosaic virus RNA-1. 12/87 6,746bp.

M36894 Em(40)_vi:BNYVVR3A Gb(84)_vi:BNYVVR3A Beet necrotic yellow vein mosaic virus F2 RNA-3. 8/90 1,775bp.

M36895 Em(40)_vi:BNYVVR3B Gb(84)_vi:BNYVVR3B Beet necrotic yellow vein mosaic virus G1 RNA-3. 8/90 1,420bp

M36896 Em(40)_vi:BNYVVR4A Gb(84)_vi:BNYVVR4A Beet necrotic yellow vein mosaic virus F2 RNA-4. 8/90 1,108bp.

M36897 Em(40)_vi:BNYVVR4B Gb(84)_vi:BNYVVR4B Beet necrotic yellow vein mosaic virus G1 RNA-4. 8/90 1,370bp

X04197 Em(40)_vi:TOBNYVV2 Gb(84)_vi:TOBNYVV2 Beet necrotic yellow vein virus RNA-2. 9/93 4,612bp.

X05147 Em(40)_vi:NCBNYVV1 Gb(84)_vi:NCBNYVV1 Beet Necrotic Yellow Vein Virus complete RNA-1 with 237K ORF. 9/93 6,746bp.

X75574 Em(40)_vi:BNYVMTP2 Gb(84)_vi:BNYVMTP2 Beet necrotic yellow vein mosaic virus (Yu2) genomic RNA for transport proteins. 6/94 1,922bp.

X75575 Em(40)_vi:BNYVMTP1 Gb(84)_vi:BNYVMTP1 Beet necrotic yellow vein mosaic virus (RG1)genomic RNA for transport proteins. 6/94 1,922bp.

S71490 Gb(89)_un:S71490 CP=coat protein (beet necrotic yellow vein virus BNYVV, NM, Genomic RNA, 633 nt). 1/95 633bp.

*Non–genomic nucleic acid* not found in the virions. Sub–genomic *m*RNA found in infected cells. 4 virus specified dsRNA species found in infected cells.

*Virion protein(s)* one; $M_r$ 21000; and it is coat protein. Method of preparation: Putz (1977).

*Virus–coded non–virion proteins* identified by genomic sequence analysis; seven proteins found. $M_r$ of the largest 237000; replicase. $M_r$ of 2nd largest 75000; vector interactions. $M_r$ of 3rd 42000; cell to cell movement. $M_r$ of 4th 31000; vector interactions. $M_r$ of 5th and smaller 25000 (, 15000, 13000); cell to cell movement.

## Replication

Replication does not depend on a helper virus.

## Cytopathology

Virions found in roots of *Beta vulgaris*; in cytoplasm. Inclusions absent from infected cells.

## Taxonomy and Relationships

*Furovirus.*

Virus(es) with serologically unrelated virions: peanut clump, potato mop-top and soil-borne wheat mosaic viruses.

RNA 3 and 4 components from different strains produce different symptom phenotypes: YS (yellow spot) RNA 1+2+3+4. CS (chlorotic spot) and CR (chlorotic ring): RNA 1+2+4. NS (necrotic spot) RNA 1+2+deleted RNAs. P42, P13 and P15 are homologous to genes in carlaviruses, potexviruses and barley stripe mosaic hordeivirus.

## References


Bouzoubaa, S., Quillet, L., Guilley, H., Jonard, G. and Richards, K. (1987). *J. gen. Virol.* **68**: 615.

Brunt, A.A. and Richards, K.E. (1989). *Adv. Virus Res.* **36**: 1.

Brunt, A.A. and Shikata, E. (1986). In: *The Plant Viruses*, Vol. 2, The Rod-Shaped Plant Viruses, p. 385; eds M.H.V. van Regenmortel and H. Fraenkel-Conrat. Plenum Press, New York.

Canova, A. (1966). *Inf. Fitopatol.* **16**: 235.

Hill, S.A. and Torrance, L. (1989). *Pl. Path.* **38**: 114.

Masuda, T., Kagawa, K. and Kanzawa, K. (1969). *Bull. Sugar Beet Res. Suppl.* **11**: 77.

Morozov, S.Y., Dolja, V.V. and Atabekov, J.G. (1989). *J. Mol. Evol.* **29**: 52.

Putz, C. (1977). *J. gen. Virol.* **35**: 397.
Putz, C. and Kuszala, M. (1978). *Annls. Phytopath.* **10**: 247.
Quillet, L., Guilley, H., Jonard, G. and Richards, K. (1989). *Virology* **172**: 293.
Steven, A.C., Trus, B.L., Putz, C. and Wurtz, M. (1981). *Virology* **113**: 428.
Tamada, T. (1975). *CMI/AAB Descr. Pl. Viruses* No. 144, 4 pp.
Tamada, T. and Abe, H. (1989). *J. gen. Virol.* **70**: 3391.
Tamada, T. and Baba, T. (1973). *Ann. Phytopath. Soc. Japan* **39**: 325.
Tamada, T. and Kusumi, T. (1991). *J. gen. Virol.* **72**: 1497.
Tamada, T., Shirako, Y., Abe, H., Saito, M., Kiguchi, T. and Harada, T. (1989). *J. gen. Virol.* **70**: 3399.

# Beet pseudo-yellows (?) *closterovirus*

**Data collated by G.R. Johnstone, 1987.**

*Synonyms:* cucumber yellows virus (Yamashita *et al.*, 1979), muskmelon yellows virus (Liu and Duffus, 1990; Lot *et al.*, 1982). *Standard acronym:* BPYV.

First reported in *Beta vulgaris*; from Salinas, California, U.S.A.; by Duffus (1965).

**Natural host range and symptoms**

Symptoms persist.

*Beta vulgaris, Lactuca sativa, Cichorium endiva* — yellowing, thickening, brittleness of leaves.

*Capsella bursa-pastoris, Cucumis sativus* — yellowing, thickening, brittleness and down-curling of leaf margins.

*Taraxacum officinalis* — interveinal reddening and chlorosis.

*Conium maculatum* — symptoms not described.

**Transmission**

Transmitted by a vector; an insect; *Trialeurodes vaporariorum*; Aleyrodidae. Transmitted in a semi–persistent manner. Virus does not multiply in the vector; not transmitted congenitally to the progeny of the vector; not transmitted by mechanical inoculation; not transmitted by contact between plants; not transmitted by seed.

**Geographical distribution**

Spreads in Australia (Tasmania), France, Japan, the Netherlands, the USA (California).

**Experimental host range**

Many (>9) families susceptible. Experimentally infected plants mostly show yellowing, reddening, curling.

Diagnostically susceptible

*Beta vulgaris, Capsella bursa-pastoris, Lactuca sativa* — yellowing, thickening, brittleness of leaves.

*Taraxacum officinale, Chenopodium capitatum* — interveinal reddening and chlorosis.

*Linum usitatissimum, Nicotiana glutinosa, Cucumis melo* — interveinal yellowing.

Diagnostically insusceptible

*Datura stramonium, Lycopersicon esculentum, Physalis floridana, Tetragonia expansa*.

Assay hosts (**L**ocal lesions or **W**hole plants)

*Capsella bursa-pastoris* (W), *Lactuca sativa* (W), *Taraxacum officinale* (W).

Dicotyledonous Hosts (***Susceptible*** *or Insusceptible)*

***Beta macrocarpa***. ***Beta vulgaris***. ***Capsella bursa–pastoris***. ***Chenopodium album***. ***Chenopodium amaranticolor***. ***Chenopodium capitatum***. ***Chenopodium murale***. ***Cichorium endiva***. *Citrullus lanatus*. ***Conium maculatum***. ***Cucumis melo***. ***Cucumis sativus***. ***Cucurbita moschata***. *Cucurbita pepo*. *Datura stramonium*. ***Daucus carota***. *Gossypium hirsutum*. *Helianthus annuus*. *Hibiscus esculentus*. *Ipomoea nil*. ***Lactuca sativa***. ***Linum usitatissimum***. *Lycopersicon esculentum*. *Medicago sativa*. ***Montia perfoliata***. *Nicandra physaloides*. ***Nicotiana clevelandii***. ***Nicotiana glutinosa***. ***Nicotiana tabacum***. *Phaseolus vulgaris*. *Physalis floridana*. *Raphanus sativus*. ***Senecio vulgaris***. ***Sonchus oleraceus***. ***Spinacia oleracea***. *Tetragonia expansa*. ***Zinnia elegans***.

Additional hosts

Additional Hosts: *Althaea roseus* (**I**), *Aquilegia* sp. (**S**), *Callistephus chinensis* (**S**), *Erodium cicutarium* (**S**), *Geranium dissectum* (**S**), *Ipomoea purpurea* (**I**), *Ipomoea tricolor* (**I**), *Lavatera assurgentiflora* (**I**), *Linum grandiflorum* (**S**), *Malva parviflora* (**S**), *Malva sylvestris* (**I**), *Physalis ixocarpa* (**S**), *Physalis wrightii* (**S**), *Sisymbrium irio* (**I**), *Solanum dulcamara* (**S**), *Thlaspi arvense* (**I**), *Urtica californica* (**S**).

Sources of host-range data

Duffus (1965).

**Replication**

Replication does not depend on a helper virus.

**Taxonomy and Relationships**

Possible *Closterovirus*.

Beet pseudo-yellows virus seems similar to cucumber yellows virus, from Japan and also transmitted by *Trialeurodes vaporariorum*, but has a host range restricted to cucurbits and has flexuous virions of *c*. 1000 nm long (Yamashita *et al*., 1979).

**References**

Duffus, J.E. (1965). *Phytopathology* **55**: 450.
Duffus, J.E. (1973). *Adv. Virus Res.* **18**: 347.
Duffus, J.E. and Johnstone, G.R. (1981). *Aust. Pl. Path.* **10**: 68.
Lot, H., Delecolle, B. and Lecoq, H. (1982). *Acta Hort.* **127**: 175.
Lot, H., Onillon, J.C. and Lecoq, H. (1980). *Revue hort.* **209**: 31.
Liu, H.Y. and Duffus, J.E. (1990). *Phytopathology* **80**: 866.
Van Dorst, H.J.M., Huijberts, N. and Bos, L. (1980). *Neth. J. Pl. Path.* **86**: 311.
Yamashita, S., Doi, Y., Yora, K. and Yoshino, M. (1979). *Ann. Phytopath. Soc. Japan* **45**: 484.

# Beet soil-borne *furovirus*

**Data collated by C. Henry, 1991.**

*Synonyms:* Norfolk virus, Barney Patch virus. *Standard acronym:* BSBV. *Strains:* similar viruses have been found in Germany and the U.S.A., but it is uncertain if they are strains.

First reported in *Beta vulgaris*; from Barney, Norfolk, England; by Ivanovic and Macfarlane (1982).

**Natural host range and symptoms**

*Beta vulgaris.*

**Transmission**

Transmitted by probably a vector; probably a fungus; *Polymyxa betae*; Plasmodiophorales. Virus transmitted by mechanical inoculation; may be transmitted by contact between plants (but not easily).

**Geographical distribution**

Found, but with no evidence of spread, in Sweden, Germany, Belgium and England. Possibly found in the U.S.A. but not yet confirmed.

**Experimental host range**

Few (<3) families susceptible. Experimentally infected plants mostly show local lesions and necrotic ringspots.

Diagnostically susceptible

*Beta vulgaris* cv. Monoire — chlorotic rings.
*Chenopodium quinoa, C. amaranticolor* — chlorotic or necrotic ringspots.
*Spinacia oleracea* — chlorotic spots sometimes.

Diagnostically insusceptible

*Cucumis sativus, Avena sativa, Phaseolus vulgaris, Nicotiana tabacum.*

Maintenance and propagation hosts

*Chenopodium quinoa.*

Assay hosts (**L**ocal lesions or **W**hole plants)

*Chenopodium quinoa* (L), *C. amaranticolor* (L).

Dicotyledonous Hosts *(**Susceptible** or Insusceptible)*

*Amaranthus retroflexus.* ***Beta macrocarpa. Beta vulgaris.*** *Brassica campestris* ssp. *napus. Capsella bursa–pastoris.* ***Chenopodium album. Chenopodium amaranticolor. Chenopodium capitatum. Chenopodium foetidum. Chenopodium foliosum. Chenopodium murale. Chenopodium quinoa.*** *Datura stramonium. Dianthus barbatus. Dianthus caryophyllus. Gomphrena globosa. Lycopersicon esculentum. Montia perfoliata. Nicotiana benthamiana. Nicotiana clevelandii. Nicotiana debneyi. Nicotiana tabacum. Petunia hybrida. Phaseolus vulgaris. Physalis floridana. Pisum sativum. Raphanus sativus.* ***Spinacia oleracea.*** *Tetragonia expansa. Vicia faba. Zinnia elegans.*

Monocotyledonous Hosts *(**Susceptible** or Insusceptible)*

*Avena sativa. Zea mays.*

Sources of host-range data

Henry *et al.* (1986).

**Properties of particles in sap**

TIP: 65 °C. LIV: 2 days. DEP: $\log_{10}$ minus 2–2. Leaf sap contains few virions.

**Particle morphology**

Virions rod–shaped; not enveloped; usually straight; with a clear modal length; of 65, 150, and 300 nm; 19 nm wide. Axial canal obvious; *c.* 4.5 nm in diameter. Basic helix obscure.

**Biochemical properties**

*Genome* consists of RNA; single–stranded.

3 virus specified dsRNA species found in infected cells. Size of largest virus specified dsRNA 4.1 kbp; 2nd largest 1.95 kbp; 3rd largest 1.68 kbp.

**Cytopathology**

Virions found in roots.

**Taxonomy and Relationships**

*Furovirus.*

Virus(es) with serologically unrelated virions: beet necrotic yellow vein, potato mop-top, wheat soil-borne mosaic and hypochoeris mosaic viruses.

Does not appear to have sequence homology with beet necrotic yellow vein or potato mop-top viruses. Best tests for diagnosis: rod-shaped; few virions in sap; does not infect tobaccos or legumes; host range is restricted to Chenopodiaceae; does not cross-react with antisera to beet necrotic yellow vein virus.

**References**

Henry, C. and Hutchinson (1989). *Aspects Appl. Biol.* **22**: 109.

Henry, C., Jones, R.A.C. and Coutts, R.H.A. (1986). *Pl. Path.* **38**: 585.

Ivanovic, M. and Macfarlane, I. (1982). *Ann. Rep. Rothamsted Exp. Stn 1982* p. 190.

Ivanovic, M., Macfarlane, I. and Wood (1983). *Ann. rep. for Rothamsted Exp. Stn.* 1983 p. 189.

Lesemann, D.-E., Koenig, R., Lindsten, K. and Henry, C. (1989). *EPPO Bulletin* **19**: 539.

Verhoyen, M., van den Bossche, M. and van Steyvoort, L. (1987). *Revue Agric., Brux.* **40**: 1463.

# Beet western yellows *luteovirus*

**Data collated by J.E. Duffus and G.R. Johnstone, 1983.**

*Synonyms:* beet mild yellowing virus (Russell, 1958), malva yellows virus (Costa *et al.*, 1959), pea leaf roll virus — New Zealand, radish yellows virus (Duffus, 1960; 1961), turnip mild yellows virus. *Standard acronym:* BWYV.

First reported in *Beta vulgaris, Lactuca sativa, Spinacia oleracea, Raphanus sativus*; from California, U.S.A; by Duffus (1960; 1961).

**Natural host range and symptoms**

Symptoms persist, or vary seasonally.

More than 150 species in 23 dicotyledonous families susceptible including such economically important species as:

*Beta vulgaris* (sugar and table beet), *Spinacia oleracea, Helianthus annuus, Lactuca sativa, Brassica napo-brassica, B. napus, B. nigra, B. oleracea* vars *botrytis, capitata, B. rapa, Raphanus sativus, Crambe abyssinica, Citrullus lanatus, Cucumis sativus, Cucurbita pepo, Cicer arietinum, Glycine max, Pisum sativum, Trifolium subterraneum, Vicia faba, Phlox drummondii, Capsicum annuum, Lycopersicon esculentum* — mild chlorotic spotting, yellowing, thickening and brittleness of older leaves.

**Transmission**

Transmitted by a vector; an insect; *Aphis craccivora, A. gossypii, Acyrthosiphon (Aulacorthum) solani, Brachycaudus helichrysi, Brevicoryne brassicae, Macrosiphum euphorbiae, Myzus (Sciamyzus) ascalonicus, M. ornatus, M. persicae, Myzus (Phorodon) humuli*; Aphididae. Principal natural vector(s): *Myzus persicae*. Transmitted in a persistent manner. Virus retained when the vector moults; does not multiply in the vector;

not transmitted congenitally to the progeny of the vector; does not require a helper virus for vector transmission and can help the vector transmission of another virus (lettuce speckles mottle umbravirus); not transmitted by mechanical inoculation; not transmitted by grafting; not transmitted by contact between plants; not transmitted by seed; not transmitted by pollen.

**Geographical distribution**

Probably distributed worldwide.

**Experimental host range**

Many (>9) families susceptible. Experimentally infected plants mostly show yellowing, reddening, stunting.

Diagnostically susceptible

*Gomphrena globosa, Crambe abyssinica, Trifolium subterraneum, Montia perfoliata* — systemic leaf reddening.

*Lactuca sativa, Capsella bursa-pastoris, Pisum sativum* — systemic leaf yellowing.

*Arachis hypogaea, Glycine max* cv. Shirotsuronoko, *Lens esculenta* — stunting.

*Senecio vulgaris, Spinacia oleracea.*

Diagnostically insusceptible

*Chenopodium capitatum, Lactuca sativa, Medicago sativa, Nicotiana clevelandii, Phaseolus vulgaris, Pisum sativum, Trifolium hybridum, T. incarnatum.*

Maintenance and propagation hosts

*Beta vulgaris, Capsella bursa-pastoris, Crambe abyssinica, Montia perfoliata, Raphanus sativus, Physalis floridana.*

Assay hosts (**L**ocal lesions or **W**hole plants)

*Beta vulgaris* (W), *Senecio vulgaris* (W), *Capsella bursa-pastoris* (W), *Crambe abyssinica* (W), *Montia perfoliata* (W).

Dicotyledonous Hosts *(**Susceptible** or Insusceptible)*

*Amaranthus caudatus*. ***Amaranthus retroflexus***. ***Arachis hypogaea***. *Atriplex hortensis*. ***Beta macrocarpa***. ***Beta patellaris***. ***Beta vulgaris***. *Brassica campestris*. *Brassica campestris* ssp. *chinensis*. ***Brassica campestris*** ssp. ***napus***, or *Brassica campestris* ssp. *napus*. ***Brassica campestris*** ssp. ***pekinensis***, or *Brassica campestris* ssp. *pekinensis*. ***Brassica campestris*** ssp. ***rapa***, or *Brassica campestris* ssp. *rapa*. ***Brassica juncea***, or *Brassica juncea*. ***Brassica oleracea*** ssp. ***botrytis***, or *Brassica oleracea* ssp. *botrytis*. ***Brassica oleracea*** ssp. ***capitata***, or *Brassica oleracea* ssp. *capitata*. ***Capsella bursa–pastoris***. ***Capsicum annuum***. ***Celosia argentea***, or *Celosia argentea*. *Celosia cristata*. ***Cheiranthus cheiri***. *Chenopodium album*. *Chenopodium amaranticolor*. *Chenopodium ambrosioides*. ***Chenopodium capitatum***, or *Chenopodium capitatum*. *Chenopodium foetidum*. *Chenopodium foliosum*. *Chenopodium hybridum*. *Chenopodium quinoa*. *Chrysanthemum morifolium*. ***Cicer arietinum***. ***Citrullus lanatus***. *Citrullus vulgaris*. ***Crambe abyssinica***. *Cucumis melo*. ***Cucumis sativus***, or *Cucumis sativus*. ***Cucurbita moschata***. ***Cucurbita pepo***. *Datura metel*. *Datura stramonium*. *Emilia sagittata*. ***Glycine max***. ***Gomphrena globosa***. ***Helianthus annuus***. ***Hibiscus esculentus***. *Hyoscyamus niger*. *Lablab purpureus*. ***Lactuca sativa***, or *Lactuca sativa*. ***Lathyrus odoratus***. ***Lens esculenta***. *Linum usitatissimum*. ***Lupinus albus***. ***Lupinus cosentinii***. ***Lycopersicon esculentum***. ***Matthiola incana***. ***Medicago hispida***. *Medicago sativa*. *Melandrium album*. *Melilotus alba*. ***Montia perfoliata***. ***Nicandra physaloides***. ***Nicotiana bigelovii***. ***Nicotiana clevelandii***, or *Nicotiana clevelandii*. *Nicotiana sylvestris*. *Nicotiana tabacum*. ***Petroselinum crispum***. ***Petunia hybrida***, or *Petunia hybrida*. *Phaseolus* vulgaris. ***Phlox drummondii***. ***Physalis floridana***. ***Pisum sativum***, or *Pisum sativum*. *Prunus persica*. ***Raphanus sativus***. ***Senecio vulgaris***. ***Sinapis alba***. *Solanum nigrum*. ***Solanum tuberosum***. ***Sonchus oleraceus***, or *Sonchus oleraceus*. ***Spinacia oleracea***. ***Stellaria media***. ***Tetragonia expansa***. *Trifolium hybridum*. ***Trifolium incarnatum***, or *Trifolium incarnatum*. ***Trifolium pratense***, or *Trifolium pratense*. ***Trifolium repens***, or *Trifolium repens*. ***Trifolium subterraneum***, or *Trifolium subterraneum*. ***Tropaeolum majus***. *Verbesina encelioides*. ***Vicia faba***, or *Vicia faba*. ***Vicia sativa***. ***Vigna unguiculata***. ***Viola cornuta***. ***Zinnia elegans***.

Monocotyledonous Hosts *(**Susceptible** or Insusceptible)*

***Arundo donax***.

Additional hosts

***Arctotheca calendula, Calandrinia caulescens, Clarkia pulchella, Coronopus didymus, Erodium moschatum, Euphorbia peplus, Gazania rigens, Lactuca serriola, Lepidium campestre, Malva parviflora, Papaver rhoeas, Sinapis arvensis*** and ***Sonchus asper***.

Sources of host-range data

Duffus (1960, 1964, 1973); Johnstone *et al.* (1983); Duffus and Rochow (1978); Duffus and Johnstone (1984).

**Properties of particles in sap**

TIP: 65 °C. LIV: 16 days. Leaf sap contains few virions.

**Purification**

Method: Duffus (1981); D'Arcy *et al.* (1983); Govier (1985).

**Particle morphology**

Virions isometric; not enveloped; 26 nm in diameter; angular in profile; without a conspicuous capsomere arrangement.

**Physical properties**

One sedimenting component in purified preparations; sedimentation coefficient 116 *S*. Density 1.42 g $cm^{-3}$ in CsCl. Density 1.36 in $Cs_2SO_4$. $A_{260}/A_{280}$ ratio 1.8.

**Biochemical properties**

Virions contain *c.* 30 % nucleic acid; *c.* 70 % protein; 0 % lipid.

*Genome* consists of RNA; single–stranded; linear; unipartite; largest (or only) genome part 5.641 kb. Genomic nucleic acid isolated by Veidt *et al.* (1988).

*Sequence database accession code(s):*

D00337 Em(40)_vi:BWY Beet western yellows virus, 3'end of virus genome. 11/92 109bp.

D13750 Em(40)_vi:BWY3E Gb(84)_vi:BWY3E Beet western yellows virus, 3'end of virus genome. 12/92 109bp.

L04281 Em(40)_vi:BWYRNA Gb(84)_vi:BWYRNA Beet western yellows virus genomic RNA sequence, open reading frame 1,2,3, and 4. 5/93 2,843b

X13062 Em(40)_vi:BWYV3 Gb(84)_vi:BWYV3 Beet Western Yellows Virus (BWYV) genomic RNA 3'-end. 9/93 2,208bp.

X13063 Em(40)_vi:BWYVFL1 Gb(84)_vi:BWYVFL1 Beet western yellows virus (BWYV-FL1) genomic RNA. 9/93 5,641bp.

L39952 Gb(89)_vi:Bwycoat Beet western yellows virus RNA. 2/95 1,041bp.

L39982 Em(43)_vi:Lebwycp31 Gb(89)_vi:Bwycp31a Beet western yellows luteovirus (strain bwyv-1, isolate 31) coat protein RNA fragment. 5/95 2

L40009 Gb(89)_vi:Bwycoata Beet western yellows virus coat protein. 2/95 205bp.
L40010 Gb(89)_vi:Bwycoatb Beet western yellows virus coat protein. 2/95 703bp.
L40011 Gb(89)_vi:Bwycoatc Beet western yellows virus coat protein. 2/95 878bp.
L40012 Gb(89)_vi:Bwycoatd Beet western yellows virus coat protein. 2/95 877bp.
L40013 Gb(89)_vi:Bwycoate Beet western yellows virus coat protein. 2/95 878bp.
L40014 Gb(89)_vi:Bwycoatf Beet western yellows virus coat protein. 2/95 878bp.
L40015 Gb(89)_vi:Bwycoatg Beet western yellows virus coat protein. 2/95 878bp.
L40016 Gb(89)_vi:Bwycoath Beet western yellows virus coat protein. 2/95 878bp.
L40017 Gb(89)_vi:Bwycoati Beet western yellows virus coat protein. 2/95 878bp.
L40018 Gb(89)_vi:Bwycoatj Beet western yellows virus coat protein. 2/95 972bp.
L40019 Em(43)_vi:Lebwycoat Gb(89)_vi:Bwycoatk Beet western yellows virus coat protein. 5/95 1,154bp.
L40020 Gb(89)_vi:Bwycoatl Beet western yellows virus coat protein. 2/95 1,041bp.

*Virion protein(s)* two; $M_r$ of the larger 56000. $M_r$ of 2nd largest 24000. Method of preparation: Falk and Duffus (1984).

**Replication**

Replication does not depend on a helper virus.

**Cytopathology**

Virions found in leaves, roots, phloem and stems; in nuclei. Inclusions absent from infected cells.

**Taxonomy and Relationships**

*Luteovirus.*

Virus(es) with serologically related virions: barley yellow dwarf-RPV and beet mild yellowing viruses are closely related; soybean dwarf, subterranean clover red leaf, potato leafroll are more distantly related and bean leaf roll virus is very distantly related. Beet mild yellowing virus is so close that some consider it to be the same species.

**References**


Casper, R. (1988). In: *The Plant Viruses*, Vol. 3; Polyhedral Virions with Monopartite RNA Genomes, p. 235, ed. R. Koenig, Plenum Press, New York.

Chin, L.-S., Forster, J.L. and Falk, B.W. (1993). *Virology* **192**: 473.

Costa, A.S., Duffus, J.E. and Bardin, R. (1959). *J. Am. Soc. Sug. Beet Technol.* **10**: 371.

D'Arcy, C.J., Hewings, A.D., Burnett, P.A. and Jedlinski, H. (1983). *Phytopathology* **73**: 755.

Duffus, J.E. (1960). *Phytopathology* **50**: 389.

Duffus, J.E. (1961). *Phytopathology* **51**: 605.

Duffus, J.E. (1964). *Phytopathology* **54**: 736.

Duffus, J.E. (1972). *CMI/AAB Descr. Pl. Viruses* No. 89, 4 pp.

Duffus, J.E. (1973). *Adv. Virus Res.* **18**: 347.

Duffus, J.E. (1977). In: *Aphids as Virus Vectors*, p. 361; ed. K.F. Harris and K. Maramarosch. Academic Press, New York.

Duffus, J.E. (1981). *Phytopathology* **71**: 193.

Duffus, J.E. and Rochow, W.F. (1978). *Phytopathology* **68**:45.

Duffus, J.E. and Russel, G.E. (1975). *Phytopathology* **65**: 811.

Falk, B.W. and Duffus, J.E. (1984). *Phytopathology* **74**: 1224.

Gold, A.H. and Duffus, J.E. (1967). *Virology* **31**: 308.

Govier, G.A. (1985). *Ann. appl. Biol.* **107**: 439.

Johnstone, G.R. and Duffus, J.E. (1984). *Aust. J. agric. Res.* **35**: 821.

Johnstone, G.R., Ashby, J.W., Gibbs, A.J., Duffus, J.E., Thottappilly, G. and Fletcher, J.D. (1984). *Neth. J. Pl. Path.* **90**: 107.
Rochow, W.F. and Duffus, J.E. (1981). In: *Handbook of Plant Virus Infections and Comparative Diagnosis*, p. 147; ed. E. Kurstak. Elsevier/North Holland Biomedical Press, Amsterdam.
Russell, G.E. (1958). *Ann. appl. Biol.* **46**: 393.
Veidt, I., Lot, H., Leiser, M., Scheidecker, D., Guilley, H., Richards, K. and Jonard, G. (1988). *Nucl. Acids Res.* **16**: 9917.

# Beet western yellows ST9-associated RNA virus

**Data collated by M.J. Gibbs (1995).**

First reported in *Brassica oleracea* var. *botrytis* (broccoli); from California, U.S.A.; by Falk and Duffus (1984).

**Transmission**

Transmitted by a vector; an insect; *Myzus persicae*; Aphididae. Transmitted in a persistent manner (probably). Virus requires, for vector transmission, a helper virus (beet western yellows luteovirus). The virus genome is probably transcapsidated in the virions of the helper virus (Falk and Duffus, 1984). Virus transmitted by mechanical inoculation (when transcripts from a DNA encoding the genome of the virus were inoculated to young plants of shepherd's purse (*Capsella bursa-pastoris*).

**Geographical distribution**

Found, but with no evidence of spread, in California, U.S.A.

**Experimental host range**

Few (<3) families susceptible. Experimentally infected plants mostly show stunting, yellowing and twisting.

Diagnostically susceptible

*Capsella bursa-pastoris* — lethal yellowing and necrosis, but causes milder yellowing, twisting and stunting when with beet western yellows luteovirus.

Diagnostically insusceptible

*Cichorium endiva, C. intybus, Lactuca sativa, Nicotiana clevelandii* and *Solanum nigrum.*

Maintenance and propagation hosts

*Capsella bursa-pastoris.*

Assay hosts (**L**ocal lesions or **W**hole plants)

*Capsella bursa-pastoris* (W).

Dicotyledonous Hosts *(**Susceptible** or Insusceptible)*

***Brassica oleracea*** ssp. ***botrytis***. *Cichorium endiva. Lactuca sativa. Nicotiana clevelandii.* ***Raphanus sativus***. *Solanum nigrum.*

**Biochemical properties**

*Genome* consists of RNA; single–stranded; linear; unipartite; largest (or only) genome part 2.843 kb. Infectivity retained when deproteinised with phenol or detergent (Passmore *et al.*, 1993). Poly A region absent. Additional factor not required for infectivity. Nucleotide sequence references: Chin *et al.*, 1993.

*Sequence database accession code(s):* L04281.

*Features of the genome:* Four ORFs. Only ORF3 encodes a protein with recognizable function, the polymerase. ORF2 may be translated after *t*RNA suppression of the amber stop codon of ORF1, but sequence analysis suggests that the reported sequence may be in error and that ORFs 2 and 3 may be a single ORF (Gibbs, 1995). ORF4 is probably translated from a subgenomic RNA.

Sub–genomic *m*RNA found in infected cells; *c.* 400 bp long and co-terminal with the 3' end. 2 virus specified dsRNA species found in infected cells (one from the helper virus and of *c.* 2.9 kb pairs).

**Replication**

Replication does not depend on a helper virus; ST9 a-virus does not produce its own virion proteins or virions, however plants co-infected with it and its helper produce 10 times more virions than the helper does alone (Sanger *et al.*, 1994).

**Taxonomy and Relationships**

This virus is not sufficiently similar to viruses in any of the currently recognized groups to be comfortably placed with them. Its polymerase protein shows affinities with those of the carmoviruses, dianthoviruses, type 1 and 2 luteoviruses, machlomoviruses, necroviruses, tombusviruses, umbraviruses and pea enation mosaic virus RNA-2.

**Comments**

This organism has features of a satellite RNA, but can replicate independently, and therefore is included as a virus.

**References**

Chin, L.-S., Forster, J.L., Falk, B.W. and Bruening, G. (1993). *Proc. Natl. Acad. Sci. USA* **90**: 10168.

Falk, B.W. and Duffus, J.E. (1984). *Phytopathology* **74**: 1224.

Gibbs, M.J. (1995). *D. Phil. thesis*, Oxford University.

Sanger, M. Passmore, B., Falk, B.W., Bruening, G., Ding, B. and Lucas, W.J. (1994). *Virology* **200**: 48.

# Beet yellow net (?) *luteovirus*

**Data collated by A.A. Brunt, 1991.**

*Standard acronym:* BYNV.

First reported in *Beta vulgaris*; from Belgium; U.S.A.; by Roland (1948); and by Sylvester (1948).

**Natural host range and symptoms**

*Beta vulgaris* — conspicuous veinal chlorosis.

**Transmission**

Transmitted by a vector; an insect; *Myzus persicae*; Aphididae. Transmitted in a persistent manner. Virus not transmitted by mechanical inoculation; not transmitted by seed.

**Geographical distribution**

Spreads in Belgium and the USA (California).

**Experimental host range**

Experimentally infected plants mostly show vein clearing.

Diagnostically susceptible

*Beta vulgaris, Beta vulgaris* var. *cicla* — veinal chlorosis.

Maintenance and propagation hosts

*Beta vulgaris.*

Assay hosts (**L**ocal lesions or **W**hole plants)

*Beta vulgaris* (W).

Dicotyledonous Hosts *(**Susceptible** or Insusceptible)*

***Beta vulgaris.***

**Taxonomy and Relationships**

Possible *Luteovirus.*

**Comments**

Pathogenicity tests indicate that this virus is distinct from tobacco yellow net (?) *luteovirus.*

**References**

Roland, G. (1948). *Parasitica* **4**: 152.
Sylvester, E.S. (1948). *Phytopathology* **38**: 429.
Sylvester, E.S. (1949). *Phytopathology* **39**: 117.
Sylvester, E.S. (1956). *J. econ. Ent.* **49**: 789.
Sylvester, E.S. (1958). *J. econ. Ent.* **51**: 812.
Waterhouse, P.M., Gildow, F.E. and Johnstone, G.R. (1988). *CMI/AAB Descr. Pl. Viruses* No. 339, 9 pp.

## Beet yellow stunt *closterovirus*

**Data collated by D.M. Wright, 1991.**

*Standard acronym:* BYSV.

From the U.S.A. in California; by Duffus (1971).

**Natural host range and symptoms**

Symptoms persist.
*Beta vulgaris* — yellowing, stunting.
*Lactuca sativa* — severe stunting and chlorosis, severe necrosis of phloem tissue.
*Sonchus oleraceus* — red or yellow interveinal colouration.

**Transmission**

Transmitted by a vector; an insect; *Hyperomyzus lactucae, Myzus persicae, Macrosiphum euphorbiae*; Aphididae. Principal natural vector(s): *Hyperomyzus lactucae.* Transmitted in a semi–persistent manner. Virus lost by the vector when it moults; not transmitted congenitally to the progeny of the vector; not transmitted by mechanical inoculation; not transmitted by contact between plants; not transmitted by seed.

**Geographical distribution**

Found, but with no evidence of spread, in the U.K. in Scotland.

Dicotyledonous Hosts *(**Susceptible** or Insusceptible)*

***Beta vulgaris. Sonchus oleraceus.***

**Particle morphology**

Virions filamentous.

*Sequence database accession code(s):*

L20761 Gb(84)_vi:BEYRDRP Beet yellow stunt virus RNA-dependent RNA polymerase gene, 3' end and heat shock protein 70 g.

**Cytopathology**

Virions found in roots.

**Taxonomy and Relationships**

*Closterovirus.*

**References**

Duffus, J E. (1979). *CMI/AAB Descr. Pl. Viruses* No. 207, 4 pp.

# Beet yellows *closterovirus*

**Data collated by H. Smith and A. Karasev, 1991.**

*Synonyms:* sugarbeet yellows closterovirus. *Standard acronym:* BYV. *Strains:* strains of differing pathogenicity are known to exist.

First reported in *Beta vulgaris*; from Belgium; by Roland (1936).

**Natural host range and symptoms**

Symptoms persist. Symptoms yellowing and premature senescence of older leaves.

*Beta vulgaris* — vein clearing, then leaves become thick, brittle, yellow with necrotic spots.

**Transmission**

Transmitted by a vector; an insect; *Myzus persicae, Aphis fabae*; Aphididae. Not transmitted by *Rhopalosiphum padi, Macrosiphum (Sitobion) avenae*. Transmitted in a semi–persistent manner. Virus lost by the vector when it moults; does not multiply in the vector; not transmitted congenitally to the progeny of the vector; transmitted by mechanical inoculation (with difficulty); transmitted by grafting; not transmitted by contact between plants; not transmitted by seed; not transmitted by pollen.

**Ecology and control**

Studies reported by Hull (1968); Harrington *et al.* (1989).

**Geographical distribution**

Probably distributed worldwide (wherever *Beta vulgaris* is grown). Spreads in China.

**Experimental host range**

Several (3–9) families susceptible. Experimentally infected plants mostly show yellowing or reddening, necrotic spots.

Diagnostically susceptible

*Tetragonia expansa* — stunting, vein clearing, yellowing.

*Montia perfoliata* — necrotic spots, reddening and yellowing of older leaves.

*Beta vulgaris* — vein clearing then yellowing, thickening and brittleness of leaves; necrotic spots.

*Spinacia oleracea, Stellaria media* — yellowing.

Diagnostically insusceptible

*Brassica* spp. *Lactuca sativa, Pisum sativum, Nicotiana tabacum, Plantago lanceolata.*

Maintenance and propagation hosts

*Montia perfoliata, Tetragonia expansa, Beta vulgaris.*

Assay hosts (**L**ocal lesions or **W**hole plants)

*Beta vulgaris* (L and W), *Montia perfoliata* (L and W), *Chenopodium foliosum* (L and W), *C. quinoa* (L).

Dicotyledonous Hosts *(**Susceptible** or Insusceptible)*

***Atriplex hortensis***. *Brassica campestris*. *Brassica campestris* ssp. *chinensis*. *Brassica campestris* ssp. *napus*. *Brassica campestris* ssp. *pekinensis*. *Brassica campestris* ssp. *rapa*. *Brassica juncea*. *Brassica oleracea* ssp. *botrytis*. *Brassica oleracea* ssp. *capitata*. *Capsella bursa–pastoris*. ***Celosia cristata***. ***Chenopodium album***. ***Chenopodium amaranticolor***. ***Chenopodium ambrosioides***. ***Chenopodium capitatum***. ***Chenopodium foliosum***. ***Chenopodium hybridum***. ***Chenopodium quinoa***. *Chrysanthemum morifolium*. *Hyoscyamus niger*. ***Lactuca sativa***. ***Montia perfoliata***. ***Nicotiana clevelandii***. *Nicotiana tabacum*. *Petunia hybrida*. *Pisum sativum*. *Raphanus sativus*. *Senecio vulgaris*. *Sinapis alba*. *Solanum nigrum*. *Sonchus oleraceus*. ***Spinacia oleracea***. ***Stellaria media***. ***Tetragonia expansa***. ***Trifolium hybridum***. *Trifolium incarnatum*. *Trifolium pratense*. *Trifolium repens*. *Trifolium subterraneum*. *Vicia faba*. *Zinnia elegans*.

Sources of host-range data

Russell (1965).

**Properties of particles in sap**

TIP: 55 °C. LIV: 1 days. DEP: $\log_{10}$ minus 4. Leaf sap contains few virions.

**Purification**

Method: Kassanis *et al.* (1977); Karasev *et al.* (1989).

**Particle morphology**

Virions filamentous; not enveloped; usually flexuous; with a clear modal length; of 1250 nm; 10 nm wide. Axial canal obscure. Basic helix obvious; pitch of basic helix 3–3.4 nm.

**Physical properties**

Two sedimenting components in purified preparations; sedimentation coefficient of the fastest 110 *S*; of the other(s) 80–90 *S*. Density 1.34 g $cm^{-3}$ in CsCl (fixed). $A_{260}/A_{280}$ ratio 1.44–1.54.

**Biochemical properties**

Virions contain 5 % nucleic acid.

*Genome* consists of RNA; single–stranded; linear; unipartite; largest (or only) genome part 14.5 kb. Genomic nucleic acid isolated by Karasev: two phenol/SDS extractions. Base composition 23.4 % G; 26.3 % A; 21 % C; 29.3 % U. 5' terminus of RNA has a methylated nucleotide cap. Infectivity retained when deproteinised with phenol or detergent. Poly A region absent. Additional factor not required for infectivity. Genome has no *t*RNA–like activity. Nucleotide sequence references: Karasev: 3'-terminal half (6746 nt). EMBL AC. No. X53462.

*Sequence database accession code(s):*

M59452 Em(40)_vi:BYVVCPG Gb(84)_vi:BYVVCPG Beet yellows virus coat protein gene, complete cds, and two ORFs. 8/91 2,724bp.

X53462 Em(40)_vi:CLBYV3PH Gb(84)_vi:CLBYV3PH Beet yellows virus genome 3'-proximal half, for capsid protein HSP70 related protein, RNA-dep

X73475 Em(40)_vi:BYVMBPA Gb(84)_vi:BYVMBPA Beet yellows virus (isolate German BYV-G) genes for putative membrane-binding protein, heat s

X73476 Em(40)_vi:BYVUAA Gb(84)_vi:BYVUAA Beet yellows virus (isolate Ukrainian BYV-U) complete genome. 2/94 15,480bp. 4 sequences.

*Features of the genome:* incomplete ORF 1 (nt 3–656) encodes putative polymerase (219 AA). ORF 2 (nt 710–871) encodes 6.4kDa protein (AA 1–54). ORF 3 (nt 874–2667) encodes 65kDa homologue of HSP70 heat shock proteins (AA 1–598). ORF 4 (nt 2576–4234) encodes 64kDa protein (AA1–553). ORF 5 (nt 4185–4832) encodes 24kDa protein, a diverged copy of capsid protein (AA 1–216). ORF 6 (nt 4906–5517) encodes 22.3kDa capsid protein (AA 1–204). ORF 7 (nt 5517–6056) encodes 20kDa protein (AA 1–180). ORF 8 (nt 6034–6564) encodes 21kDa protein (AA 1–177).

*Non–genomic nucleic acid* not found in the virions. Sub–genomic *m*RNA found in infected cells. 6 virus specified dsRNA species found in infected cells (at least, including genome-size dsRNA). Size of largest virus specified dsRNA 14.5 kbp; 2nd largest 6.3 kbp; 3rd largest 4.8 kbp; 4th largest 2.7 kbp; 5th largest 1.6 kbp; 6th largest 1 kbp.

*Virion protein(s)* one; $M_r$ 22.3 (KDa); capsid protein. Method of preparation: Carpenter *et al.* (1977). Amino acid composition: Carpenter *et al.* (1977).

## Replication

Replication does not depend on a helper virus.

## Cytopathology

Virions found in leaves, roots, mesophyll, vascular parenchyma, phloem, companion cells and all parts of the host plant. Inclusions present in infected cells; are crystals in the cytoplasm; they contain virions.

## Taxonomy and Relationships

*Closterovirus.*

Its 6.4kDaprotein is apparently related to 12kDaproteins of potex and carlaviruses. The putative polymerase is related to polymerases of tricornaviruses (bromo and cucumoviruses). Its capsid protein is a member of monophyletic family of capsid proteins encoded by filamentous plant viruses, and its 65kDaprotein is a direct homologue of cell HSP70 heat shock proteins. Best tests for diagnosis: ELISA using polyclonal antiserum to beet yellows closterovirus is routinely used to identify the virus and distinguish it from beet mild yellowing virus which often occurs with it in sugar beet crops.

## References


Agranovsky, A.A., Boyko, V.P., Karasev, A.V., Lunina, N.A., Koonin, E.V. and Dolja, V.V. (1991). *J. gen. Virol.* **72**: 15.

Agranovsky, A.A., Boyko, V.P., Karasev, A.V., Koonin, E.V. and Dolja, V.V. (1991). *J. mol. Biol.* **217**: 603.

Bar-Joseph, M. and Hull, R. (1974). *Virology* **62**: 552.

Carpenter, J.M., Kassasnis, B. and White, R.F. (1977). *Virology* **77**: 101.

Dolja, V.V., Karasev, A.V. and Agranovsky, A.A. (1990). In: *New Aspects of Positive-Strand RNA Viruses*, Amer. Soc. Micro., Washington D.C., U.S.A., p. 31.

Esau, K. (1960). *Virology* **11**: 317.

Harrington, R., Dewar, A.M. and George, B. (1989). *Ann. appl. Biol.* **114**: 459.

Heathcote, G.D. (1988). In: *Virus Yellows Monograph* IIRB Pests and Diseases Study Group, Rue Montoyer 47, B–1040, Brussels, Belgium.
Hull, R. (1968). *Pl. Path.* **17**: 1.
Karasev, A.V., Agranovsky, A.A., Rogov, V.V., Miroschnichenko, N.A., Dolja, V.V. and Atabekov, J.G. (1989). *J. gen. Virol.* **70**: 241.
Kassanis, B., Carpenter, J.M., White, R.F. and Woods, R.D. (1977). *Virology* **77**: 95.
Russell, G.E. (1965). *Ann. appl. Biol.* **55**: 245.

# Belladonna mottle *tymovirus*

**Data collated by A.J. Gibbs, 1991.**

*Synonyms:* Tollkirschenscheckungs virus. *Standard acronym:* BeMV.

First reported in *Atropa belladonna*; from Germany; by Bode and Marcus (1959).

**Natural host range and symptoms**

Symptoms persist. Symptoms mottling.
*Atropa belladonna* — mottling, distortion.

**Transmission**

Transmitted by a vector; an insect; *Epithrix atropae* (Jankulova *et al.*, 1968); Coleoptera. Not transmitted by aphids. Virus transmitted by mechanical inoculation; not transmitted by contact between plants; not transmitted by seed.

**Geographical distribution**

Spreads in Bulgaria and Germany.

**Experimental host range**

Few (<3) families susceptible. Experimentally infected plants mostly show necrosis, mosaics.

Diagnostically susceptible

*Chenopodium quinoa* — local lesions, not systemic.
*Datura stramonium* — necrotic local lesions, mosaic.
*Nicotiana tabacum* — ring local lesions, vein clearing, mosaic.

Maintenance and propagation hosts

*Nicotiana tabacum.*

Assay hosts (**L**ocal lesions or **W**hole plants)

*Nicotiana tabacum* (L).

Dicotyledonous Hosts *(**Susceptible** or Insusceptible)*

*Antirrhinum majus. Arachis hypogaea. Bellis perennis. Beta vulgaris. Brassica campestris* ssp. *pekinensis. Catalpa bignonioides.* ***Catharanthus roseus. Chenopodium quinoa.*** *Cucumis sativus.* ***Datura stramonium.*** *Daucus carota. Delphinium hybridum. Eucalyptus cloesiana. Euphorbia marginata. Fagopyrum esculentum. Fragaria vesca. Gossypium hirsutum.* ***Gypsophila elegans.*** *Helianthus annuus. Ipomoea nil. Lactuca sativa. Lobelia erinus.* ***Nicotiana tabacum.*** *Papaver nudicaule. Pastinaca sativa. Petroselinum crispum. Pisum sativum. Podolepis robusta. Salvia splendens.* ***Sinningia speciosa.*** *Tropaeolum majus. Valeriana officinalis. Vigna radiata. Viola cornuta.*

Additional hosts

Additional hosts: *Leptosiphon* spp. (**S**).

Sources of host-range data

Paul *et al.* (1968); Guy *et al.* (1984).

**Properties of particles in sap**

TIP: 80 °C. LIV: 5–20 days. DEP: $\log_{10}$ minus 6–7. Infectivity of sap not changed by treatment with di–ethyl ether. Leaf sap contains many virions.

**Purification**

Method: Steere's method.

**Particle morphology**

Virions isometric; not enveloped; 27 nm in diameter; rounded in profile; with a conspicuous capsomere arrangement.

**Physical properties**

Two sedimenting components in purified preparations; sedimentation coefficient of the fastest 113 *S*; of the other(s) 53 *S*. Isoelectric point pH 9.12.

**Biochemical properties**

Virions contain 37 % nucleic acid; 63 % protein.

*Genome* consists of RNA; single–stranded; linear; unipartite; largest (or only) genome part 6.3 kb. Base composition 17.5 % G; 22.8 % A; 32.8 % C; 26.9 % U. 5' terminus of RNA has a methylated nucleotide cap. Infectivity retained when deproteinised with proteases; retained when deproteinised with phenol or detergent. Poly A region absent. Additional factor not required for infectivity. Genome has *t*RNA–like activity. Nucleotide sequence references: Ding *et al.* (1990): EMBL X54529.

*Sequence database accession code(s):*

X54529 Em(40)_vi:BMVVP Gb(84)_vi:BMVVP Belladonna mottle virus VP gene for virion protein and ORF. 9/93 615bp.

*Features of the genome:* Untranslated 5' terminal region, overlapping gene of about 1800 nucleotides followed 7 nucleotides (to 3') by replicase gene of about 5400 nucleotides, then after the conserved tymobox promoter region by the virion protein gene of 567 nucleotides.

Sub–genomic *m*RNA found in infected cells; mRNA of virion protein.

*Virion protein(s)* one; $M_r$ 19784; coat protein. Amino acid sequence: Ding *et al.* (1990). Virion proteins not glycosylated. Virion proteins not phosphorylated.

**Replication**

Genome replicates in association with chloroplasts (in vesicles in the chloroplast external membrane). Coat protein *m*RNA translated in the cytoplasm. Replication does not depend on a helper virus.

**Cytopathology**

Virions found in all parts of the host plant; in cytoplasm. Inclusions absent from infected cells.

**Taxonomy and Relationships**

*Tymovirus.*

Virus(es) with serologically related virions: Andean potato latent, dulcamara mottle, eggplant mosaic, ononis yellow mosaic and scrophularia mottle viruses.

**References**

Bode, O. and Marcus, O. (1959). *Proc. 4th Int. Cong. Crop Protect. Hamburg* **1**: 375.

Ding, S.W., Howe, J., MacKenzie, A., Skotnicki, M. and Gibbs, A.J. (1990). *Nucl. Acids Res.* **18**: 6138.

Guy, P.L., Dale, J.L., Adena, M.A. and Gibbs, A.J. (1984). *Pl. Path.* **33**: 337.
Koenig, R. (1976). *Virology* **72**: ;1.
Paul, H.L. (1971). *CMI/AAB Descr. Pl. Viruses* No. 52, 3 pp.
Paul, H.L., Bode, O., Jankulova, M. and Brandes, J. (1968). *Phytopath. Z.* **61**: 342.
Peter, R., Peter, C., Dupin, A. and Witz, J. (1989). *C.r. Acad. Sci., Paris* **309**: 599.

# Bermuda grass etched-line *marafivirus*

**Data collated by A.A Brunt and B.E.L. Lockhart, 1987.**

*Standard acronym:* BELV.

First reported in *Cynodon dactylon* and *Sorghum halepense*; from Morocco; by Lockhart *et al.* (1985).

**Natural host range and symptoms**

Symptoms persist and vary seasonally.
*Cynodon dactylon* — white etched lines and spots in leaves; plants stunted.
*Sorghum halepense* — chlorotic streaks in leaves.

**Transmission**

Transmitted by a vector; an insect; *Aconurella prolixa*; Cicadellidae. Transmitted in a persistent manner. Virus multiplies in the vector; not transmitted by mechanical inoculation; not transmitted by seed.

**Ecology and control**

Studies reported by Lockhart *et al.* (1985).

**Geographical distribution**

Spreads in Morocco.

**Experimental host range**

Few (<3) families susceptible. Experimentally infected plants mostly show streaks.

Diagnostically susceptible

*Avena sativa, Cynodon dactylon, Triticum durum, Zea mays, Sorghum bicolor, S. halepense* — chlorotic spots and lines.

Diagnostically insusceptible

*Agrostis alba, Agropyron repens, Hordeum vulgare, Lolium perenne, Saccharum officinarum, Setaria viridis.*

Maintenance and propagation hosts

*Cynodon dactylon, Zea mays.*

Assay hosts (**L**ocal lesions or **W**hole plants)

*Cynodon dactylon* (W), *Zea mays* (W).

Monocotyledonous Hosts ***(Susceptible** or Insusceptible)*

*Agropyron repens. Agrostis alba.* ***Avena sativa. Bromus inermis. Cynodon dactylon.*** *Echinochloa crus–galli. Hordeum vulgare. Lolium perenne. Panicum dichotomiflorum. Phalaris arundinacea. Saccharum officinarum. Secale cereale. Setaria viridis.* ***Sorghum bicolor. Triticum durum. Zea mays.***

Additional hosts

*Festuca arundinacea.*

Sources of host-range data
    Lockhart *et al.* (1985).

**Properties of particles in sap**

Leaf sap contains many virions.

**Purification**

Method: Lockhart *et al.* (1985).

**Particle morphology**

Virions isometric; not enveloped; *c.* 30 nm in diameter; rounded in profile; with a conspicuous capsomere arrangement.

**Physical properties**

Two sedimenting components in purified preparations; sedimentation coefficient of the fastest 119 *S* ((B)); of the other(s) 57 *S* ((T)). Density 1.44 g $cm^{-3}$ in CsCl ((B)), or 1.29 g $cm^{-3}$ in CsCl ((T)). Density in $Cs_2SO_4$ 1.40 g/ml (B) and 1.26 g/ml (T). $A_{260}/A_{280}$ ratio 1.53 (B), or 0.88 (T).

**Biochemical properties**

*Genome* consists of RNA; single–stranded; linear; unipartite; largest (or only) genome part 7.5 kb. Genomic nucleic acid isolated by Lockhart: use standard phenol-SDS procedure.

*Virion protein(s)* one; $M_r$ 26500; coat protein. Virion proteins not glycosylated.

**Replication**

Replication does not depend on a helper virus.

**Cytopathology**

Virions found in leaves, mesophyll, epidermis, vascular parenchyma and all parts of the host plant; in cytoplasm and in cell vacuoles. Inclusions absent from infected cells.

**Taxonomy and Relationships**

*Marafivirus* (the type species).

Virus(es) with serologically related virions: maize rayado fino and oat blue dwarf viruses.

**References**

Khaless, N. (1988). *M.Sc. Thesis*, University of Minnesota, 53 pp.

Lockhart, B.E.L., Khaless, N., Lennon, A.M. and El Maataoui, M. (1985). *Phytopathology* **75**: 1258.

# Bhendi yellow vein mosaic bigeminivirus

**Data collated by V. Muniyappa, 1991.**

First reported in *Abelmoschus esculentus*; from Bombay, India; by Kulkarni (1924).

**Natural host range and symptoms**

Symptoms persist. Symptoms vein clearing followed by yellowing and thickening, leaves and fruit reduced in size.

*Abelmoschus esculentus* — vein clearing, yellowing and thickening.

**Transmission**

Transmitted by a vector; an insect; *Bemisia tabaci*; Aleyrodidae. Transmitted in a persistent manner. Virus retained when the vector moults; does not multiply in the vector;

not transmitted congenitally to the progeny of the vector; not transmitted by mechanical inoculation; transmitted by grafting; not transmitted by contact between plants; not transmitted by seed; not transmitted by pollen.

**Ecology and control**

Studies reported by Sastry and Singh (1973).

**Geographical distribution**

Spreads in India.

**Experimental host range**

Few (<3) families susceptible. Experimentally infected plants mostly show leaf curling and crinkling.

Diagnostically susceptible

*Althaea rosea* — leaf curling and crinkling.

Diagnostically insusceptible

*Ageratum conyzoides, Cajanus cajan, Croton bonplandianum, Glycine max, Lablab purpureus.*

Maintenance and propagation hosts

*Abelmoschus esculentus.*

Assay hosts (**L**ocal lesions or **W**hole plants)

*Abelmoschus esculentus* (W).

Sources of host-range data

Rao (1986).

**Particle morphology**

Virions geminate; 16–18 nm in diameter; dimers 30 nm in length.

**Taxonomy and Relationships**

Bigeminivirus: *Geminiviridae.*

Virus(es) with serologically related virions: African cassava mosaic virus.

**References**

Harrison, B.D., Muniyappa, V., Swanson, M.M., Roberts, I.M. and Robinson, D.J. (1991). *Ann. appl. Biol.* **118**: 595.

Kulkarni, G.S. (1924). *Poona Agric. Coll. Mag.* **6**: 6.

Rao, M. (1986). *M.Sc. Thesis*, Univ. Agric. Sci., Bangalore, India, 74 pp.

Sastry, K.S.M. and Singh, S.J. (1973). *Indian Phytopath.* **26**: 129.

Singh, A.B., Joshi, B.S., Khanna, P. and Gupta, P.S. (1962). *Ind. J. Genet. Pl. Breeding* **22**: 137.

Uppal, B.N., Varma, P.M. and Capoor, S.P. (1940). *Curr. Sci.* **9**: 227.

Varma, P.M. (1955). *Indian J. agric. Sci.* **25**: 293.

# Bidens mosaic *potyvirus*

**Data collated by E.W. Kitajima, 1991.**

*Standard acronym:* BiMV.

First reported in *Bidens pilosa*; from Brazil; by Kitajima *et al.* (1961).

**Natural host range and symptoms**

Symptoms persist. Symptoms mosaics.
*Bidens pilosa, Helianthus annuum* — mosaic, necrosis.
*Lactuca sativa, Coreopsis lanceolata* — mosaic.

**Transmission**

Transmitted by a vector; an insect; *Aphis coreopsidis, Myzus persicae, Dactynotus* spp.; Aphididae. Principal natural vector(s): *Aphis coreopsidis*. Transmitted in a non–persistent manner. Virus transmitted by mechanical inoculation; transmitted by grafting; not transmitted by seed.

**Geographical distribution**

Spreads in Brazil.

**Experimental host range**

Several (3–9) families susceptible. Experimentally infected plants mostly show mosaics, local necrotic lesions, local and systemic chlorotic spots.

Diagnostically susceptible

*Nicotiana tabacum* cv. White Burley — chlorotic local and systemic spotting.
*Zinnia elegans* — systemic mosaic.
*Lactuca sativa, Pisum sativum* — mosaic.
*Helianthus annuum* — systemic chlorotic and necrotic spotting, mosaics.
*Callistephus chinensis* — systemic vein clearing and mosaic.
*Acanthospermum hispidum* — systemic chlorotic spots.
*Bidens pilosa* — mosaic and vein banding.
*Chenopodium amaranticolor* — necrotic local lesions.
*C. quinoa* — chlorotic local lesions.

Diagnostically insusceptible

*Phaseolus vulgaris, Cucurbita maxima, Euphorbia prunifolia, Triticum aestivum, Gomphrena globosa.*

Maintenance and propagation hosts

*Lactuca sativa, Bidens pilosa, Helianthus annuum.*

Assay hosts (**L**ocal lesions or **W**hole plants)

*Chenopodium quinoa* (L), *C. amaranticolor* (L), *Lactuca sativa* (W), *Helianthus annuus* (W).

Dicotyledonous Hosts *(**Susceptible** or Insusceptible)*

*Ageratum conyzoides. Antirrhinum majus. Arachis hypogaea. Brassica oleracea* ssp. *capitata. Capsicum annuum.* ***Cassia occidentalis***. ***Chenopodium amaranticolor***. *Chenopodium capitatum. Chenopodium murale.* ***Chenopodium quinoa***. *Citrullus lanatus. Coriandrum sativum. Cucumis melo. Cucurbita maxima. Datura stramonium. Galinsoga parviflora. Glycine max. Gomphrena globosa.* ***Helianthus annuus***. ***Lactuca sativa***. ***Lupinus albus***. ***Lupinus angustifolius***. *Lycopersicon esculentum. Medicago sativa.* ***Nicotiana debneyi***. ***Nicotiana glutinosa***. ***Nicotiana rustica***. ***Nicotiana tabacum***.

*Phaseolus vulgaris*. ***Physalis floridana***. ***Pisum sativum***. *Raphanus sativus*. ***Solanum nigrum***. *Tetragonia expansa*. *Trifolium pratense*. *Trifolium repens*. ***Zinnia elegans***.

Monocotyledonous Hosts (***Susceptible*** *or Insusceptible)*

*Hordeum vulgare*. *Triticum aestivum*.

Sources of host-range data

Kuhn *et al.* (1980).

**Properties of particles in sap**

TIP: 55–60 °C. LIV: 5 days. DEP: $\log_{10}$ minus 3. Leaf sap contains few virions.

**Purification**

Method: Marinho and Kitajima (1989).

**Particle morphology**

Virions filamentous; not enveloped; usually flexuous; with a clear modal length; of 730 nm; 15 nm wide. Axial canal obscure. Basic helix obscure.

**Cytopathology**

Virions found in leaves, mesophyll and the epidermis; in cytoplasm. Inclusions present in infected cells; are pinwheels. Other cellular changes: Kuhn *et al.* (1982).

**Taxonomy and Relationships**

*Potyvirus: Potyviridae*.

**References**

Kitajima, E.W., Costa, C.L. and Carvalho (1961). *Bragantia* **20**: 503.
Kuhn, G.B., Lin, M.T. and Costa, C.L. (1980). *Fitopatol. Bras.* **5**: 39.
Kuhn, G.L., Lin, M.T. and Kitajima, E.W. (1982). *Fitopatol. Bras.* **7**: 185.
Marinho, V.L.A. and Kitajima, E.W. (1989). *Fitopatol. Bras.* **14**: 91.
Rodrigues, M.G.R., Kitajima, E.W. and Marinho, V.L.A. (1985). *Fitopatol. Bras.* **10**: 305.

# Bidens mottle *potyvirus*

**Data collated by S.R. Christie, 1982. Revised by J.R. Edwardson, 1989.**

*Standard acronym:* BiMoV.

First reported in *Bidens pilosa, Lepidium virginicum*; from Florida, U.S.A; by Christie (1968).

**Natural host range and symptoms**

Symptoms persist.

*Erigeron* sp., *Bidens* sp., *Senecio* sp. *Cichorium endiva, Lactuca sativa* — mild chlorotic mottling.

*Lepidium virginicum, Lupinus angustifolius* — chlorotic mottling.

*Fittonia vershaffeltii* var. *argyroneura* — leaf malformation.

*Rudbeckia hirta hybridum, Zinnia elegans, Ageratum conyzoides* — leaf chlorosis.

**Transmission**

Transmitted by a vector; an insect; *Myzus persicae*; Aphididae. Transmitted in a non–persistent manner. Virus transmitted by mechanical inoculation; not transmitted by seed.

**Geographical distribution**

Spreads in the USA.

**Experimental host range**

Several (3–9) families susceptible. Experimentally infected plants mostly show systemic mosaics, mottles, ringspots or necrosis.

Diagnostically susceptible

*Chenopodium amaranticolor* — chlorotic local lesions; not systemic.

*C. quinoa* — chlorotic local lesions; systemic chlorotic spots and streaks.

*Helianthus annuus, Zinnia elegans, Nicotiana clevelandii, N. x edwardsonii, Lactuca sativa, Cyamopsis tetragonoloba* — systemic leaf mottling, malformation and stunting.

Diagnostically insusceptible

*Lactuca sativa* cv. Valmaine, Brassica campestris (perviridis), *Pisum sativum.*

Maintenance and propagation hosts

*Nicotiana clevelandii, N. x edwardsonii.*

Assay hosts (**L**ocal lesions or **W**hole plants)

*Chenopodium .amaranticolor* (L), *C. quinoa* (L), *Zinnia elegans, Pisum sativum, Nicotiana x edwardsonii.*

Dicotyledonous Hosts *(**Susceptible** or Insusceptible)*

***Ageratum conyzoides. Calendula officinalis. Chenopodium amaranticolor. Chenopodium quinoa. Cichorium endiva. Cyamopsis tetragonoloba.*** *Gomphrena globosa.* ***Helianthus annuus. Lactuca sativa. Lupinus angustifolius.*** *Lycopersicon esculentum.* ***Nicotiana clevelandii.*** *Nicotiana glutinosa. Nicotiana tabacum.* ***Petunia hybrida.*** *Phaseolus vulgaris. Pisum sativum. Raphanus sativus. Vigna unguiculata.* ***Zinnia elegans.***

Sources of host-range data

Christie (1968); Purcifull *et al.* (1971).

**Properties of particles in sap**

TIP: 50–55 °C. LIV: 16 days. DEP: $\log_{10}$ minus 3. Leaf sap contains few virions.

**Purification**

Method: Hiebert and McDonald (1973).

**Particle morphology**

Virions filamentous; not enveloped; usually flexuous; with a clear modal length; of 720 nm. Axial canal obscure. Basic helix obvious (in uranyl formate).

**Physical properties**

One sedimenting component in purified preparations.

*Virion protein(s)* one; $M_r$ 33000. Method of preparation: Hiebert and McDonald (1973).

**Cytopathology**

Virions found in mesophyll and parenchyma cells throughout the host; in cytoplasm. Inclusions present in infected cells; are pinwheels; they do not contain virions.

**Taxonomy and Relationships**

*Potyvirus: Potyviridae.*

Virus(es) with serologically related virions: potato virus Y.

**References**

Christie, S.R. (1968). *Pl. Dis. Reptr* **52**: 763.
Christie, S.R. and Edwardson, J.R. (1977). *Fla Agric. Exp. Stn Monog.* No. 9, p. 55.
Edwardson, J.R. (1974). *Fla Agric. Exp. Stn Monog.* No. 4, p. 398.
Edwardson, J.R., Purcifull, D.E., Christie, R.G. and Christie, S.R. (1976). *Pl. Dis. Reptr* **60**: 776.
Hiebert, E. and McDonald, J.G. (1973). *Virology* **56**: 349.
Kitajima, E.W., Carvalho, A.M.B. and Costa, A.S. (1961). *Bragantia* **20**: 503.
Kuhn, G.B., Lin, M.T. and Kitajima, E.W. (1978). *Fitopatol. Bras.* **3**: 93.
Kuhn, G.B., Lin, M.T. and Costa, C.L. (1980). *Fitopatol. Bras.* **5**: 39.
Logan, A.E., Zettler, F.W. and Christie, S.R. (1984). *Plant Dis.* **68**: 260.
Purcifull, D.E. (1973). *Virology* **55**: 275.
Purcifull, D.E., Christie, S.R., Zitter, T.A. and Bassett, M.J. (1971). *Pl. Dis. Reptr* **55**: 1061.
Purcifull, D.E., Christie, S.R. and Zitter, T.A. (1976). *CMI/AAB Descr. Pl. Viruses* No. 161, 4 pp.
Purcifull, D.E. and Zitter, T.A. (1973). *Proc. Fla Stn Hort. Soc.* **86**: 143 and 165.
Shepherd, J.E., Secor, G.A. and Purcifull, D.E. (1974). *Virology* **58**: 464.
Zettler, F.W., Lima, J.A.A. and Zurawski, D. (1977). *Proc. Am. Phytop. Soc.* **4**: 121.

# Blackgram mottle (?) *carmovirus*

**Data collated by H.C. Phatak, 1983.**

*Standard acronym:* BMoV.

First reported in *Vigna mungo*; from New Delhi, India; by Phatak (1974).

**Natural host range and symptoms**

Symptoms persist.
*Vigna mungo* — mottling, puckering, vein clearing and malformation of leaves.

**Transmission**

Transmitted by a vector; an insect; *Ceratoma trifurcata, Epilachna varivestis*; Coleoptera. Virus does not require a helper virus for vector transmission; transmitted by mechanical inoculation; transmitted by seed (8% in *Vigna mungo*).

**Geographical distribution**

Spreads in the Pacific region; Australia, India, the Philippines, and Thailand.

**Experimental host range**

Few (<3) families susceptible. Experimentally infected plants mostly show necrotic local lesions, mosaics.

Diagnostically susceptible

*Cyamopsis tetragonoloba, Phaseolus lunatus, Phaseolus vulgaris* cv. Pinto, *Vicia faba* — necrotic local lesions; not systemic.
*P. vulgaris* cv. Black Valentine — systemic mosaic, leaf malformation.
*Vigna radiata* — necrotic local lesions; systemic mosaic.

Diagnostically insusceptible

*Chenopodium* spp., *Glycine max, Sesamum indicum, Nicotiana clevelandii.*

Maintenance and propagation hosts

*Cyamopsis tetragonoloba, Phaseolus vulgaris* cv. Black Valentine.

Assay hosts (**L**ocal lesions or **W**hole plants)

*Cyamopsis tetragonoloba* (L), *Macrotyloma uniflorum* (L), *Phaseolus lunatus* (L), *P. vulgaris* cvs Pinto, Puregold (L).

Dicotyledonous Hosts *(**Susceptible** or Insusceptible)*

*Chenopodium amaranticolor. Chenopodium quinoa. Cyamopsis tetragonoloba. Datura stramonium.* ***Glycine max.*** *Gomphrena globosa.* ***Lablab purpureus. Macroptilium lathyroides. Macrotyloma uniflorum.*** *Nicotiana clevelandii. Nicotiana glutinosa. Nicotiana rustica. Ocimum basilicum.* ***Phaseolus lunatus. Phaseolus vulgaris. Pisum sativum. Sesamum indicum. Vigna radiata. Vigna unguiculata.***

Sources of host-range data

Phatak (1974).

**Properties of particles in sap**

TIP: 90–92 °C. LIV: 45 days. DEP: $\log_{10}$ minus 5. Infectivity of sap not changed by treatment with di–ethyl ether. Leaf sap contains many virions.

**Purification**

Method: Scott and Phatak (1979).

**Particle morphology**

Virions isometric; not enveloped; 28 nm in diameter; angular in profile; with a conspicuous capsomere arrangement.

**Physical properties**

One sedimenting component in purified preparations; sedimentation coefficient 122 *S*. Density 1.364 g $cm^{-3}$ in CsCl.

**Biochemical properties**

*Genome* consists of RNA; single–stranded. Base composition 24.9 % G; 25.9 % A; 25.2 % C; 24 % U.

*Non–genomic nucleic acid* not found in the virions.

*Virion protein(s)* one; $M_r$ 38200. Method of preparation: Scott and Phatak (1979).

**Cytopathology**

Virions found in mesophyll, epidermis and in the lumen of xylem; in cytoplasm and in mitochondria. Inclusions absent from infected cells. Other cellular changes: cell membrane proliferation in association with dictyosomes and endoplasmic reticulum, accumulation of Golgi vesicles.

**Taxonomy and Relationships**

Possible *Carmovirus: Tombusviridae.*

Virus(es) with serologically unrelated virions: southern bean mosaic (type and cowpea strain), broad bean mottle, cowpea severe mosaic, bean pod mottle, cowpea chlorotic mottle, quail pea mosaic, cowpea mottle, bean mild mottle and more than 35 other isometric viruses.

**References**


Phatak, H.C. (1974). *Seed Sci. Technol.* **2**: 3.

Scott, H.A. and Hoy, J.W. (1981). *CMI/AAB Descr. Pl. Viruses* No. 237, 4 pp.

Scott, H.A. and Phatak, H.C. (1979). *Phytopathology* **69**: 346.

# Black raspberry necrosis virus

**Data collated by R. Stace-Smith, 1991.**

*Synonyms:* black raspberry mild mosaic virus, raspberry black necrosis virus. *Standard acronym:* BRNV.

First reported in *Rubus idaeus*; from Canada; by Stace-Smith (1955).

**Natural host range and symptoms**

Symptoms disappear soon after infection and none. Symptoms mild leaf mottle.

*Rubus idaeus, fruticosus, procerus, ursinus, lasiocarpus, phoenicolasius, loganobaccus* — symptomless.

*R. occidentalis* — stem necrosis, leaf mottle.

*R. henryi* — stem necrosis.

**Transmission**

Transmitted by a vector; an insect; *Amphorophora agathonica, A. idaei, A. rubicumberlandii, Illinoia rubicola*; Aphididae. Transmitted in a semi–persistent manner. Virus lost by the vector when it moults; does not multiply in the vector; not transmitted congenitally to the progeny of the vector; transmitted by mechanical inoculation; transmitted by grafting; not transmitted by contact between plants; not transmitted by seed; not transmitted by pollen.

**Ecology and control**

Studies reported by Jones and Jennings (1986).

**Geographical distribution**

Spreads in Australia, Canada, France, Germany, Italy, the Netherlands, New Zealand, Norway, Poland, and the USA. Found, but with no evidence of spread, in Australia and New Zealand.

**Experimental host range**

Few (<3) families susceptible.

Diagnostically susceptible

*Chenopodium amaranticolor, C. quinoa, Gomphrena globosa, Nicotiana debneyi, Petunia hybrida, Spinacia oleracea* — local chlorotic lesions.

Maintenance and propagation hosts

*Rubus idaeus, R. occidentalis.*

Assay hosts (**L**ocal lesions or **W**hole plants)

*Rubus occidentalis* (W), *R. henryi* (W).

Dicotyledonous Hosts *(**Susceptible** or Insusceptible)*

***Chenopodium amaranticolor*. *Chenopodium murale*. *Chenopodium quinoa*. *Gomphrena globosa*. *Nicotiana debneyi*. *Petunia hybrida*.**

Sources of host-range data

Jones and Murant (1972); Murant *et al.* (1976).

**Properties of particles in sap**

DEP: $\log_{10}$ minus 1–2. Leaf sap contains few virions.

**Purification**

Method: Murant *et al.* (1976).

**Particle morphology**

Virions isometric; not enveloped; 25–30 nm in diameter; rounded in profile; without a conspicuous capsomere arrangement.

**Physical properties**

One sedimenting component in purified preparations.

**Cytopathology**

Virions found in the leaves; in unusual sites.

**Taxonomy and Relationships**

Best tests for diagnosis: aphid transmission to *Rubus* indicator hosts.

**References**

Jones, A.T. and Jennings, D.L. (1986). *Ann. appl. Biol.* **96**: 59.
Jones, A.T. and Murant, A.F. (1972). *Pl. Path.* **21**: 166.
Murant, A.D., Jones, A.T. and Roberts, I.M. (1976). *Acta Hort.* **66**: 39.
Stace-Smith, R. (1955). *Can. J. Bot.* **33**: 314.
Stace-Smith, R. and Jones, A.T. (1987). In: *Virus Diseases of Small Fruits*; ed. R.H. Converse, U.S. Dep. Agric. Hdbk No. 631, p. 178.

# Blueberry leaf mottle *nepovirus*

**Data collated by D.C. Ramsdell, 1987.**

*Standard acronym:* BLMV. *Strains:* blueberry leaf mottle virus — M1 and blueberry leaf mottle virus — NY.

First reported in *Vaccinium corymbosum*; from Michigan, U.S.A.; by Ramsdell and Stace-Smith (1979).

**Natural host range and symptoms**

*Vaccinium corymbosum* cv. Rubel — stem dieback and poor regrowth, hence stunting. Leaves malformed and mottled.

*V. corymbosum* cv. Jersey — stem dieback. Regrowth stunted. Leaves rosetted and mottled.

**Transmission**

Virus transmitted by mechanical inoculation; transmitted by grafting; not transmitted by contact between plants; not transmitted by seed; transmitted by pollen to the pollinated plant (by honeybees).

**Ecology and control**

Studies reported by Childress and Ramsdell (1986*a*); Childress and Ramsdell (1986*b*); Childress and Ramsdell (1987).

**Geographical distribution**

Spreads in the USA (Michigan). Found, but with no evidence of spread, in New York, USA (grapevine) and in Pleven, Bulgaria.

**Experimental host range**

Several (3–9) families susceptible. Experimentally infected plants mostly show local lesions, mottle, tip necrosis.

Diagnostically susceptible

*Chenopodium quinoa, C. album* — chlorotic local lesions, systemic mottle and tip necrosis.

*C. amaranticolor* — systemic mottle.

*Nicotiana clevelandii* — systemic necrosis.

Maintenance and propagation hosts

*Chenopodium amaranticolor, C. quinoa* respectively.

Assay hosts (**L**ocal lesions or **W**hole plants)

*Chenopodium quinoa* (L), *C. amaranticolor* (W), *Nicotiana clevelandii* (W).

Dicotyledonous Hosts *(**Susceptible** or Insusceptible)*

*Cassia occidentalis. Catharanthus roseus.* ***Chenopodium album. Chenopodium amaranticolor. Chenopodium capitatum. Chenopodium foetidum. Chenopodium murale. Chenopodium quinoa. Cucumis sativus.*** *Cucurbita maxima. Cucurbita pepo.* ***Cyamopsis tetragonoloba.*** *Datura stramonium. Dolichos biflorus.* ***Gomphrena globosa. Lycopersicon esculentum.*** *Momordica balsamina.* ***Nicotiana benthamiana. Nicotiana clevelandii. Nicotiana debneyi. Nicotiana glutinosa. Nicotiana rustica. Nicotiana sylvestris. Nicotiana tabacum.*** *Petunia hybrida.* ***Phaseolus vulgaris.*** *Sesbania exaltata. Tithonia speciosa. Vigna unguiculata. Vigna unguiculata* ssp. *cylindrica. Zinnia elegans.*

Additional hosts

Additional non-hosts: *Plantago major, P. lanceolata.*

Sources of host-range data

Ramsdell and Stace-Smith (1979).

**Properties of particles in sap**

TIP: 65–70 °C. LIV: 14–21 days (at 4°C), or 5–7 days (at 20°C). DEP: $\log_{10}$ minus 4–5. Leaf sap contains few virions.

**Purification**

Method: Ramsdell and Stace-Smith (1983).

**Particle morphology**

Virions isometric; not enveloped; 28 nm in diameter; angular in profile; without a conspicuous capsomere arrangement.

**Physical properties**

Three sedimenting components in purified preparations; sedimentation coefficient of the fastest 128 *S* (B); of the other(s) 120 *S* (M), or 53 *S* (T). Density 1.497 g $cm^{-3}$ in CsCl (B), or 1.471 g $cm^{-3}$ in CsCl (M). Isoelectric point pH 7.5.

**Biochemical properties**

*Genome* consists of RNA; single–stranded; linear; of two parts; largest (or only) genome part the larger 7.6 kb; the 2nd largest 7 kb. Genomic nucleic acid isolated by Ramsdell and Stace-Smith (1981).

*Sequence database accession code(s):*

U20621 Em(43)_vi:Bl20621 Gb(89)_vi:Blu20621 Blueberry leaf mottle virus segment RNA2 polyprotein gene, partial cds. 5/95 3,082bp.

U20622 Em(43)_vi:Bl20622 Gb(89)_vi:Blu20622 Blueberry leaf mottle virus segment RNA1 polyprotein gene, partial cds. 5/95 1,908bp.

*Non–genomic nucleic acid* not found in the virions.

$M_r$ 54000; coat protein. Method of preparation: Ramsdell and Stace-Smith (1981).

**Replication**

Replication does not depend on a helper virus.

**Cytopathology**

Virions found in leaves, roots and blossoms; in cytoplasm; the cytoplasm of parenchyma cells close to endoplasmic reticulum but not with other organelles. Inclusions absent from infected cells.

**Taxonomy and Relationships**

*Nepovirus: Comoviridae.*

Virus(es) with serologically related virions: grapevine Bulgarian latent virus is distantly related.

**References**

Childress, A.M. and Ramsdell, D.C. (1986*a*). *Acta Hort.* **186**: 87.
Childress, A.M. and Ramsdell, D.C. (1986*b*). *Phytopathology* **76**: 1333.
Childress, A.M. and Ramsdell, D.C. (1987). *Phytopathology* **77**: 167.
Ramsdell, D.C. (1987). In: *Virus Diseases of Small Fruits*; ed. R.H. Converse. U.S. Dep. Agric. Hdbk No. 631, p. 112.
Ramsdell, D.C. and Stace-Smith, R. (1979). *Acta Hort.* **95**: 37.
Ramsdell, D.C. and Stace-Smith, R. (1980). *Proc. 7th Meeting Int. Counc. for the Study of Viruses and Virus-Like Dis. of Grapevine, Niagara Falls, Canada.*
Ramsdell, D.C. and Stace-Smith, R. (1981). *Phytopathology* **71**: 468.
Ramsdell, D.C. and Stace-Smith, R. (1983). *CMI/AAB Descr. Pl. Viruses* No. 267, 4 pp.
Uyemoto, J.K., Taschenberg, E.F. and Hummer, D.K. (1977). *Pl. Dis. Reptr* **61**: 949.

# Blueberry necrotic shock *ilarvirus*

**Data collated by R.R. Martin, 1991.**

*Standard acronym:* BlShV.

First reported in *Vaccinium corymbosum*; from Washington State, U.S.A.; by MacDonald *et al.* (1991).

**Natural host range and symptoms**

Symptoms vary seasonally. Symptoms leaf and flower necrosis.

*Vaccinium corymbosum* — leaf and flower necrosis during bloom, plants show symptoms for 1–4 years, then recover and remain symptomless and produce good quality fruit.

**Transmission**

Virus spreads quickly in a crop. Removal of infected plants does not slow the spread of the virus significantly. Not transmitted by *Fimbriaphis fimbriata, Myzus persicae.* Virus transmitted by mechanical inoculation; transmitted by grafting; not transmitted by contact between plants.

**Geographical distribution**

Spreads in the USA (Oregon and Washington State).

**Experimental host range**

Few (<3) families susceptible.

Diagnostically susceptible

*Nicotiana benthamiana, N. clevelandii, N. sylvestris, N. tabacum* cvs Samsun and Havana-425 — necrotic local lesions; systemic necrosis.

*Vaccinium corymbosum* — necrosis of leaves and flowers.

Diagnostically insusceptible

*Chenopodium quinoa, Cucumis sativus, Gomphrena globosa, Petunia hybrida, Phaseolus vulgaris.*

Maintenance and propagation hosts

*Nicotiana benthamiana, N. clevelandii, Vaccinium corymbosum.*

Assay hosts (**L**ocal lesions or **W**hole plants)

*Nicotiana clevelandii* (W).

Dicotyledonous Hosts *(**Susceptible** or Insusceptible)*

*Brassica juncea. Chenopodium amaranticolor. Chenopodium quinoa. Cucumis sativus. Gomphrena globosa. Lycopersicon esculentum.* ***Nicotiana benthamiana. Nicotiana clevelandii. Nicotiana sylvestris. Nicotiana tabacum.*** *Petunia hybrida. Phaseolus vulgaris.*

Sources of host-range data

MacDonald *et al.* (1991).

**Properties of particles in sap**

Leaf sap contains few virions.

**Purification**

Method: MacDonald *et al.* (1991).

**Particle morphology**

Virions isometric; not enveloped; 27 nm in diameter; angular in profile; without a conspicuous capsomere arrangement.

**Physical properties**

Two sedimenting components in purified preparations. $A_{260}/A_{280}$ ratio 1.46–1.6 (uncorrected for light scattering).

**Biochemical properties**

*Genome* consists of RNA; single–stranded; linear; of four parts; largest (or only) genome part 2.901 kb; the 2nd largest 2.366 kb; the 3rd largest 1.605 kb; the 4th largest 0.845 kb. Genomic nucleic acid isolated by proteinase K digest of coat protein.

*Virion protein(s)* one; $M_r$ 27300; coat protein. Method of preparation: MacDonald *et al.* (1991).

**Cytopathology**

Virions found in leaves, roots and mesophyll; in cytoplasm.

**Taxonomy and Relationships**

*Ilarvirus: Bromoviridae* (subgroup 2).

Virus(es) with serologically related virions: apple mosaic and prunus necrotic ringspot viruses. Virus(es) with serologically unrelated virions: asparagus 2, citrus leaf rugose, citrus variegation, elm mosaic, elm mottle, Humulus japonicus, prune dwarf, spinach latent, tobacco streak and Tulare apple mosaic viruses.

**References**

MacDonald, S.G., Martin, R.R. and Bristow, P.R. (1989). *Acta Hort.* **241**: 295.
MacDonald, S.G., Martin, R.R. and Bristow, P.R. (1991). *Phytopathology* **81**: 210.

# Blueberry red ringspot *caulimovirus*

**Data collated by A.A. Brunt, 1991.**

*Standard acronym:* BRRV.

First reported in *Vaccinium corymbosum*; from the U.S.A.; by Hutchinson and Varney (1954).

**Natural host range and symptoms**

Symptoms persist.

*Vaccinium australe, V. corybosum* and possibly *V. macrocarpa* — red ringspots in leaves and older stems; earlier, but decreased, fruit yield.

**Transmission**

Virus not transmitted by mechanical inoculation; transmitted by grafting.

**Geographical distribution**

Spreads in the USA.

**Experimental host range**

Few (<3) families susceptible.

Diagnostically susceptible

*Vaccinium corymbosum* — red ringspots in mature leaves of graft-inoculated plants.

Maintenance and propagation hosts

*Vaccinium corymbosum.*

**Purification**

Method: Gillet and Ramsdell (1984).

**Particle morphology**

Virions isometric; not enveloped; 42–46 nm in diameter; rounded in profile; without a conspicuous capsomere arrangement.

**Physical properties**

Two sedimenting components in purified preparations; sedimentation coefficient 275 *S*; of the other(s) 212 *S*. Density 1.3 and 1.4 g $cm^{-3}$ in CsCl. Isoelectric point pH 4.

**Biochemical properties**

*Genome* consists of DNA; double–stranded.

*Virion protein(s)* one; $M_r$ 44000; coat protein.

**Cytopathology**

Virions found in leaves and the mesophyll; in nuclei and in unusual sites; but most are within cytoplasmic inclusions. Inclusions present in infected cells; are viroplasms (enclosed within endoplasmic reticulum); they contain virions.

**Taxonomy and Relationships**

*Caulimovirus.*

Virus(es) with serologically unrelated virions: cauliflower mosaic, carnation etched ring, dahlia mosaic and figwort mosaic viruses.

**References**

Gillet, J.M. and Ramsdell, D.C. (1984). *Phytopathology* **74**: 862.

Gillet, J.M. and Ramsdell, D.C. (1988). *AAB Descr. Pl. Viruses* No. 327, 4 pp.

Hepp, R.F. and Converse, R.H. (1987). *Plant Dis.* **71**: 536.

Hutchinson, M.T. and Varney, E.H. (1954). *Pl. Dis. Reptr* **38**: 260.

Kim, K.S., Ramsdell, D.C., Gillet, J.M. and Fulton, J.P. (1981). *Phytopathology* **71**: 673.

# Blueberry scorch *carlavirus*

**Data collated by A.A. Brunt and R.R. Martin, 1991.**

*Standard acronym:* BlSV. *Strains:* Sheep Pen Hill virus.

First reported in *Vaccinium corymbosum*; from Washington State, U.S.A.; by Martin and Bristow (1988).

**Natural host range and symptoms**

Symptoms persist. Symptoms flower necrosis and leaf chlorosis and necrosis.

*Vaccinium corymbosum* — marginal necrosis or chlorosis of leaves, flowers aborted; some cultivars remain symptomless.

**Transmission**

Virus not transmitted by mechanical inoculation; transmitted by grafting; not transmitted by contact between plants.

**Geographical distribution**

Spreads in the USA (Oregon and Washington).

**Experimental host range**

Few (<3) families susceptible.

Diagnostically susceptible

*Vaccinium corymbosum* — marginal leaf chlorosis.

Maintenance and propagation hosts

*Vaccinium corymbosum.*

Assay hosts (**L**ocal lesions or **W**hole plants)

*Vaccinium corymbosum* cv. Pemberton (W).

Dicotyledonous Hosts *(**Susceptible** or Insusceptible)*

*Beta vulgaris. Brassica campestris* ssp. *rapa. Brassica juncea. Catharanthus roseus. Chenopodium amaranticolor. Chenopodium murale.* ***Chenopodium quinoa****. Cucumis sativus. Datura stramonium. Gomphrena globosa. Lactuca sativa. Nicotiana benthamiana. Nicotiana clevelandii. Nicotiana glutinosa. Nicotiana rustica. Nicotiana sylvestris. Nicotiana tabacum. Petunia hybrida. Phaseolus vulgaris. Pisum sativum. Trifolium incarnatum. Trifolium repens. Verbesina encelioides. Vigna unguiculata. Zinnia elegans.*

Comments on host-range

Only blueberry infected by most strains, one infects *Chenopodium quinoa*.

Sources of host-range data

Martin and Bristow (1988).

**Properties of particles in sap**

Leaf sap contains few virions.

**Purification**

Method: Martin and Bristow (1988).

**Particle morphology**

Virions filamentous; not enveloped; usually straight; with a clear modal length; of 690 nm; 14 nm wide. Axial canal obscure. Basic helix obscure.

**Physical properties**

One sedimenting component in purified preparations. Density 1.28 g $cm^{-3}$ in CsCl. $A_{260}/A_{280}$ ratio 1.2.

**Biochemical properties**

*Genome* consists of RNA; single–stranded; linear; unipartite; largest (or only) genome part 8.4 kb. Genomic nucleic acid isolated by Martin and Bristow (1988).

*Sequence database accession code(s):*

L25658 Em(40)_vi:BLUORF Gb(84)_vi:BLUORF Blueberry scorch carlavirus mRNA. 4/94 8,512bp. 1 sequence.

*Virion protein(s)* one; $M_r$ 35200; coat protein. Method of preparation: Martin and Bristow (1988).

**Cytopathology**

Virions found in leaves, roots, mesophyll and epidermis; in cytoplasm. Inclusions present in infected cells; are viroplasms; they contain virions.

**Taxonomy and Relationships**

*Carlavirus.*

Virus(es) with serologically related virions: carnation latent, chrysanthemum B, dandelion, elderberry, helenium S, potato M, potato S, pea streak and red clover vein mosaic carlaviruses.

Best tests for diagnosis: serological tests, ELISA is best. Due to the low pH of blueberry leaf sap, it is necessary to ensure the pH of homogenized leaves is above pH 6.0 by using a strong buffer (10 × PBS or 0.5% nicotine).

**References**


MacDonald, S.G., Martin, R.R. and Bristow, P.R. (1989). *Acta Hort.* **241**: 295.

Martin, R.R. and Bristow, P.R. (1988). *Phytopathology* **78**: 1636.

Martin, R.R. and MacDonald, S.G. (1988). *Can. Pl. Dis. Surv.* **68**: 98.


# Blueberry shoestring *sobemovirus*

**Data collated by D.C. Ramsdell, 1987.**

*Standard acronym:* BSSV.

First reported in *Vaccinium* spp.; from New Jersey, U.S.A.; by Varney (1957).

**Natural host range and symptoms**

Symptoms persist and vary seasonally.

*Vaccinium corymbosum, V. angustifolium* — "shoestring" disease; reddish streaks on stems, narrow strap-like leaves, occasional flower-breaking, severe yield reduction.

**Transmission**

Transmitted by a vector; an insect; *Masonaphis pepperi*; Aphididae. Transmitted in a semi–persistent manner (probably). Virus not transmitted congenitally to the progeny of the vector; transmitted by mechanical inoculation; transmitted by grafting; not transmitted by contact between plants; not transmitted by seed; not transmitted by pollen.

**Ecology and control**

Studies reported by Morimoto and Ramsdell (1985); Morimoto *et al.* (1985).

**Geographical distribution**

Spreads in the USA (in Michigan and New Jersey).

**Experimental host range**

Few (<3) families susceptible. Experimentally infected plants mostly show malformation, streaking.

Diagnostically susceptible

*Vaccinium* spp. — oak leaf pattern on leaves malformed shoestring leaves, red streaks on stems; blossoms pink.

Maintenance and propagation hosts

*Vaccinium corymbosum.*

Sources of host-range data

Ramsdell (1979*a*).

**Properties of particles in sap**

Leaf sap contains few virions.

**Purification**

Method: Ramsdell (1979*a*); Ramsdell (1979*b*).

**Particle morphology**

Virions isometric; not enveloped; 27 nm in diameter; rounded in profile; without a conspicuous capsomere arrangement.

**Physical properties**

One sedimenting component in purified preparations; sedimentation coefficient 120 *S*. Density 1.392 g $cm^{-3}$ in CsCl. Density 1.273 in $Cs_2SO_4$ and 1.355 g/$cm^3$. Isoelectric point pH 4.6–4.7.

**Biochemical properties**

Virions contain 20 % nucleic acid (in the single component); 80 % protein (in the single component).

*Genome* consists of RNA; single–stranded; linear; unipartite; largest (or only) genome part 4.2 kb. Genomic nucleic acid isolated by Ramsdell (1979*b*). Poly A region absent. Additional factor not required for infectivity. Genome has no *t*RNA–like activity.

*Non–genomic nucleic acid* not found in the virions. Sub–genomic *m*RNA not found in infected cells.

*Virion protein(s)* one; $M_r$ 30000; coat protein. Method of preparation: Ramsdell (1979*b*).

**Replication**

Replication does not depend on a helper virus.

### Cytopathology

Virions found in mesophyll, epidermis, meristem, xylem, phloem and palisade parenchyma; in cytoplasm, in nuclei, and in cell vacuoles (in small cytoplasmic vacuoles). Inclusions absent from infected cells. Other cellular changes: rod-like materials in palisade cells (Hartman *et al.*, 1973).

### Taxonomy and Relationships

*Sobemovirus.*

Virus(es) with serologically unrelated virions: 33 viruses with isometric virions (Ramsdell, 1979a).

Best tests for diagnosis: look for reddish streaks on stems.

### References


Converse, R.H., Martin, R.R. and Spiegel, S. (1987). In: *Virus Diseases of Small Fruits*; ed. R.H. Converse. U.S. Dep. Agric. Hdbk No. 631.

Fraser, N.W. (1970). In: *Virus and Virus-Like Diseases of Small Fruit and Grapevines*; ed. N.W. Fraser. University of California, Berkeley, California, U.S.A..

Gillet, J.M., Morimoto, K.M., Ramsdell, D.C., Baker, K.K. Chaney, W.G. and Esselman, W.J. (1982). *Acta Hort.* **129**: 25.

Hancock, J.F., Morimoto, K.M., Pritts, M.P. and Ramsdell, D.C. (1986). *Hort. Science* **21**: 1059.

Hancock, J.F., Morimoto, K.M., Roeche, J.M., Schulte, N.L. and Ramsdell, D.C. (1986). *Fruit Var. J.* **40**: 56.

Hartman, J.X., Bath, J.E. and Hooper, G.R. (1973). *Phytopathology* **63**: 432.

Klomparens, K.L., Petersen, J.F., Ramsdell, D.C. and Chaney, W.G. (1986). *J. Electron Micro. Tech.* **4**: 47.

Lesney, M.S. and Ramsdell, D.C. (1976). *Acta Hort.* **66**: 105.

Lesney, M.S., Ramsdell, D.C. and Sun, M. (1978). *Phytopathology* **68**: 295.

Morimoto, K.M. and Ramsdell, D.C. (1985). *Phytopathology* **75**: 1217.

Morimoto, K.M., Ramsdell, D.C., Gillet, J.M. and Chaney, W.G. (1985). *Phytopathology* **75**: 709.

Ramsdell, D.C. (1979*a*). *CMI/AAB Descr. Pl. Viruses* No. 204, 4 pp.

Ramsdell, D.C. (1979*b*). *Phytopathology* **69**: 1087.

Schulte, N.L. (1985). *J. Am. Soc. Hort. Sci.* **110**: 343.

Varney, E.H. (1957). *Phytopathology* **47**: 307.


## Bramble yellow mosaic (?) potyvirus

**Data collated by A.A. Brunt, 1991.**

*Standard acronym:* BrmYMV.

First reported in *Rubus rigidus* (wild trailing blackberry); from South Africa; by Engelbrecht (1963).

### Natural host range and symptoms

Symptoms persist.

*Rubus rigidus* — yellow mosaic and line patterns in early spring; later leaves symptomless.

**Transmission**

Virus transmitted by mechanical inoculation; transmitted by grafting; transmitted by seed.

**Geographical distribution**

Found, but with no evidence of spread, in South Africa.

**Experimental host range**

Several (3–9) families susceptible.

Diagnostically susceptible

*Chenopodium murale, C. quinoa* — local lesions; systemic necrotic ringspots or chlorotic mottling.

*Gomphrena globosa, Nicotiana tabacum* — chlorotic or necrotic local lesions; not systemic.

Diagnostically insusceptible

*Cucumis sativus, Petunia hybrida, Phaseolus vulgaris.*

Maintenance and propagation hosts

*Chenopodium murale.*

Assay hosts (**L**ocal lesions or **W**hole plants)

*Chenopodium murale* (L/W).

**Properties of particles in sap**

TIP: 55 °C. LIV: 8 days (at 20°C). DEP: $\log_{10}$ minus 4.

**Particle morphology**

Virions filamentous; not enveloped; usually flexuous; with a clear modal length; of 730 nm. Axial canal obscure. Basic helix obscure.

**Taxonomy and Relationships**

Unassigned potyvirus.

**References**

Engelbrecht, D.J. (1963). *S. Afr. J. Lab. clin. Med.* **9**: 140.
Engelbrecht, D.J. (1976). *Acta Hort.* **66**: 79.
Engelbrecht, D.J. and van der Walt, W.J.K. (1974). *Phytophylactica* **6**: 311.

# Brinjal mild mosaic (?) potyvirus

**Data collated by A.A. Brunt, 1991.**

First reported in *Solanum melongena*; from India (Aligarh); by Naqvi and Mahmood (1976).

**Natural host range and symptoms**

Symptoms persist.
*Solanum melongena* — mild leaf chlorosis.

**Transmission**

Virus transmitted by mechanical inoculation.

**Geographical distribution**

Spreads in India.

**Experimental host range**

Few (<3) families susceptible.

Diagnostically susceptible

*Datura metel, D. stramonium, Lycopersicon esculentum, Nicotiana glutinosa, Petunia hybrida, Physalis peruviana, Solanum melongena* — systemic chlorotic vein-banding, mottling and some distortion.

Diagnostically insusceptible

*Chenopodium amaranticolor, Gomphrena globosa, Nicotiana rustica, Solanum tuberosum, Vigna unguiculata.*

Dicotyledonous Hosts *(**Susceptible** or Insusceptible)*

*Chenopodium amaranticolor. **Datura metel. Datura stramonium.** Gomphrena globosa. **Lycopersicon esculentum.** Nicotiana rustica. **Petunia hybrida. Physalis floridana. Solanum melongena.** Solanum tuberosum. Vigna unguiculata.*

**Properties of particles in sap**

TIP: 65 °C. LIV: 4 days. DEP: $\log_{10}$ minus 4.

**Purification**

Method: Naqvi and Mahmood (1978).

**Particle morphology**

Virions filamentous; usually flexuous; with a clear modal length; of 675–760 nm; 13 nm wide.

**Physical properties**

One sedimenting component in purified preparations; sedimentation coefficient 168 *S*.

**Biochemical properties**

*Genome* consists of RNA; single–stranded.

**Taxonomy and Relationships**

Unassigned potyvirus.

Virus(es) with serologically unrelated virions: potato Y and tobacco etch viruses.

**References**

Naqvi, Q.A. and Mahmood, K. (1976). *Indian Phytopath.* **29**: 138.

Naqvi, Q.A. and Mahmood, K. (1978). *Phytopath. Z.* **93**: 86.

# Broad bean B virus

**Data collated by K. Boswell, 1985.**

First reported in *Vicia faba*; from Colmar, France; by Putz and Kuszala (1973).

**Natural host range and symptoms**

Symptoms persist.

*Vicia faba* — stunting.

**Geographical distribution**

Spreads in France.

**Experimental host range**

Several (3–9) families susceptible. Experimentally infected plants mostly show stunting.

Assay hosts (**L**ocal lesions or **W**hole plants)

*Vicia faba.*

Dicotyledonous Hosts *(**Susceptible** or Insusceptible)*

***Chenopodium amaranticolor. Chenopodium quinoa. Nicotiana clevelandii. Pisum sativum. Vicia faba. Vigna unguiculata.***

Sources of host-range data

Putz and Kuszala (1973).

**Particle morphology**

Virions isometric; not enveloped; 25–30 nm in diameter.

**References**

Putz, C. and Kuszala, M. (1973). *Annls. Phytopath.* **5**: 447.

# Broad bean mottle *bromovirus*

**Data collated by A.J. Gibbs, 1981, Revised 1985.**

*Standard acronym:* BBMV.

First reported in *Vicia faba*; from the U.K; by Bawden *et al.* (1951).

**Natural host range and symptoms**

Symptoms persist.
*Vicia faba* — leaf mottling and malformation.

**Transmission**

Transmitted by a vector; an insect; *Acalymma trivittata, Colaspis flavida, Diabrotica undecimpunctata, Sitona lineata*; Coleoptera. Virus transmitted by mechanical inoculation; transmitted by grafting; not transmitted by seed.

**Geographical distribution**

Spreads in Portugal, Sudan, and the UK.

**Experimental host range**

Several (3–9) families susceptible. Experimentally infected plants mostly show local lesions, mosaics, mottles.

Diagnostically susceptible

*Chenopodium amaranticolor* — chlorotic local lesions; not systemic.
*Cyamopsis psoraloides* — necrotic local lesions or streaks.
*Phaseolus vulgaris* cv. Canadian Wonder — chlorotic local lesions; systemic mottle and vein yellowing or netting.
*Pisum sativum* — lethal systemic wilt.
*Vicia faba* — systemic blotchy mottle.
*Nicotiana clevelandii* — mild mottle.

Diagnostically insusceptible

*Nicotiana tabacum.*

Maintenance and propagation hosts

*Vicia faba, Nicotiana clevelandii.*

Assay hosts (**L**ocal lesions or **W**hole plants)

*Cyamopsis psoraloides* (L).

Dicotyledonous Hosts *(**Susceptible** or Insusceptible)*

***Chenopodium album**. **Chenopodium amaranticolor**. **Chenopodium hybridum**. **Chenopodium quinoa**. Cucumis sativus. Datura stramonium. **Glycine max**. **Gomphrena globosa**. Lablab purpureus. **Lathyrus odoratus**. **Lupinus albus**. Lycopersicon pimpinellifolium. **Medicago sativa**. **Melilotus alba**. **Nicotiana clevelandii**. Nicotiana glutinosa. Nicotiana tabacum. **Phaseolus vulgaris**. **Pisum sativum**. Spinacia oleracea. **Trifolium hybridum**. **Trifolium incarnatum**. **Trifolium pratense**. **Trifolium repens**. **Trifolium subterraneum**. **Vicia faba**. Vicia sativa. **Vicia villosa**. Vigna radiata. Vigna unguiculata.*

Sources of host-range data

Thornberry (1966); Lane (1974; 1981).

**Properties of particles in sap**

TIP: 95 °C. LIV: 20 days. DEP: $\log_{10}$ minus 3. Infectivity of sap not changed by treatment with di–ethyl ether. Leaf sap contains many virions.

**Purification**

Method: Lane (1974).

**Particle morphology**

Virions isometric; not enveloped; 26 nm in diameter; rounded in profile; without a conspicuous capsomere arrangement.

**Physical properties**

One sedimenting component in purified preparations; sedimentation coefficient 85 *S* (pH 7). Density 1.36 g $cm^{-3}$ in CsCl (pH 6 or lower). Isoelectric point pH 5.6 (and 0.1 ionic strength). $A_{260}/A_{280}$ ratio 5.4.

**Biochemical properties**

Virions contain 21–23 % nucleic acid; 77–79 % protein; 0 % lipid. Also spermidine.

*Genome* consists of RNA; single–stranded; linear; of three parts; largest (or only) genome part the largest 3.158 kb; the 2nd largest 2.799 kb; the 3rd largest 2.293 kb. Base composition 25 % G; 27 % A; 19 % C; 29 % U. 5' terminus of RNA has a methylated nucleotide cap. Infectivity retained when deproteinised with phenol or detergent. Genome has *t*RNA–like activity. Genome accepts tyrosine.

*Sequence database accession code(s):*

K01776 Em(40)_vi:BRRNA1 Gb(84)_vi:BBMRNA1 Broad bean mottle virus RNA 1, 3' end. 6/94 200bp.

K01777 Em(40)_vi:BRRNA2 Gb(84)_vi:BBMRNA2 Broad bean mottle virus rna 2, 3' end. 6/94 214bp.

K01778 Em(40)_vi:BRRNA3 Gb(84)_vi:BBMRNA3 Broad bean mottle virus rna 3/rna 4, 3' end. 6/94 266bp.

M60291 Em(40)_vi:BBM3ACT Gb(84)_vi:BBM3ACT Broad bean mottle virus 3a and coat proteins, complete cds. 4/94 2,293bp.

M64713 Em(40)_vi:BBMRNA2Q Gb(84)_vi:BBMRNA2Q Broad bean mottle virus RNA 2, complete cds. 5/91 2,799bp.

M65138 Em(40)_vi:BBMIAP Gb(84)_vi:BBMIAP Broad bean mottle virus (BBMV) 1a protein gene, complete cds. 8/92 3,158bp.

U24495 Em(43)_vi:Bb24495 Gb(89)_vi:Bbu24495 Broad bean mottle virus Type strain 2a protein gene, complete cds. 5/95 2,834bp.

U24496 Em(43)_vi:Bb24496 Gb(89)_vi:Bbu24496 Broad bean mottle virus Mo strain 2a protein gene, complete cds. 5/95 2,917bp.

*Non-genomic nucleic acid* found in the virions. Sub-genomic *m*RNA found in infected cells.

*Virion protein(s)* one; $M_r$ 20000. Method of preparation: Gibbs (1972): 1M NaCl and reducing agent.

**Replication**

Coat protein *m*RNA translated in the cytoplasm (cycloheximide sensitive). Replication does not depend on a helper virus.

**Cytopathology**

Virions found in all parts of the host plant; in cytoplasm. Inclusions present in infected cells; are amorphous X-bodies; they contain virions.

**Taxonomy and Relationships**

*Bromovirus: Bromoviridae.*

Virus(es) with serologically related virions: brome mosaic and cowpea chlorotic mottle viruses.

**References**

Bawden, F.C., Chaudhuri, R.P. and Kassanis, B. (1951). *Ann. appl. Biol.* **38**: 774.

Dzianott, A.M. and Bujarski, J.J. (1991). *Virology* **185**: 553.

Gibbs, A.J. (1972). *CMI/AAB Descr. Pl. Viruses* No. 101, 4 pp.

Lane, L.C. (1974). *Adv. Virus Res.* **19**: 151.

Lane, L.C. (1979). *CMI/AAB Descr. Pl. Viruses* No. 215, 4 pp.

Lane, L.C. (1981). In: *Handbook of Pl. Virus Infections and Comparative Diagnosis*, p.333; ed. E. Kurstak. Elsevier/North Holland Biomedical Press, Amsterdam.

Thornberry, H.H. (1966). In: *Index of Plant Virus Diseases*. U.S. Dep. Agric. Hdbk No. 307, p.235.

# Broad bean necrosis *furovirus*

**Data collated by T. Inouye, 1981. Revised 1984.**

*Standard acronym:* BBNV.

First reported in *Vicia faba*; from Japan; by Fukano and Yokoyama (1951).

**Natural host range and symptoms**

Symptoms vary seasonally.

*Vicia faba* — necrotic and chlorotic spots, rings, streaks.

**Transmission**

Virus transmitted by mechanical inoculation.

**Geographical distribution**

Spreads in Japan.

**Experimental host range**

Several (3–9) families susceptible. Experimentally infected plants mostly show necrotic or chlorotic local lesions; systemic mosaic.

Diagnostically susceptible

*Chenopodium amaranticolor, C. quinoa* — chlorotic local lesions or ringspots.

*Lathyrus odoratus, Pisum sativum, vicia faba* — necrotic local lesions; systemic mosaic.

*Phaseolus vulgaris* cv. Yamashiro Kurosando — necrotic local lesions.

*Nicotiana clevelandii* — chlorotic local lesions.

Diagnostically insusceptible

*Cucumis sativus, Glycine max, Nicotiana tabacum.*

Maintenance and propagation hosts

*Lathyrus odoratus, Pisum sativum, Vicia faba.*

Assay hosts (**L**ocal lesions or **W**hole plants)

*Chenopodium amaranticolor* (L), *C. quinoa* (L).

Dicotyledonous Hosts *(**Susceptible** or Insusceptible)*

*Astragalus sinicus. Cassia tora.* ***Chenopodium amaranticolor. Chenopodium quinoa.*** *Crotalaria spectabilis. Cucumis sativus. Cucurbita maxima. Glycine max. Gomphrena globosa. Lactuca sativa.* ***Lathyrus odoratus.*** *Medicago sativa.* ***Melilotus alba. Nicotiana clevelandii.*** *Nicotiana glutinosa.* ***Nicotiana rustica.*** *Nicotiana tabacum. Petunia hybrida.* ***Phaseolus vulgaris. Pisum sativum.*** *Raphanus sativus. Spinacia oleracea.* ***Tetragonia expansa.*** *Trifolium hybridum. Trifolium incarnatum. Trifolium pratense. Trifolium repens. Trifolium subterraneum.* ***Vicia faba.*** *Vicia sativa. Vicia villosa. Vigna angularis. Vigna radiata. Vigna unguiculata. Vigna unguiculata* ssp. *sesquipedalis. Zinnia elegans.*

Monocotyledonous Hosts *(**Susceptible** or Insusceptible)*

*Hordeum vulgare. Triticum aestivum.*

Sources of host-range data

Inouye *et al.* (1968).

**Properties of particles in sap**

TIP: 55–60 °C. LIV: 8 days. DEP: $\log_{10}$ minus 3–4. Leaf sap contains few virions.

**Purification**

Method: Nakasone and Inouye (1977).

**Particle morphology**

Virions rod–shaped; not enveloped; usually straight; with a clear modal length; of 250 nm (& 150); 25 nm wide. Axial canal obvious. Basic helix obvious.

**Cytopathology**

Virions found in all parts of the host plant; in cytoplasm. Inclusions present in infected cells; are amorphous X–bodies; they contain virions.

**Taxonomy and Relationships**

*Furovirus.*

**References**

Fukano, H. and Yokoyama, S. (1951). *Kyushu Nogyo Kenkyu* **10**: 133.

Inouye, T. (1971). *Ber. Ohara Inst. Landw. Biol.* **15**: 69.
Inouye, T. and Asatari, M. (1968). *Ann. Phytopath. Soc. Japan* **34**: 317.
Inouye, T. and Nakasone, W. (1980). *CMI/AAB Descr. Pl. Viruses* No. 223, 4 pp.
Nakasone, W. and Inouye, T. (1977). *Ann. Phytopath. Soc. Japan* **43**: 100.

# Broad bean severe chlorosis *closterovirus*

**Data collated by G. Thottappilly, 1983. Revised 1989.**

First reported in *Vicia faba*; from Michigan, U.S.A; by Thottappilly *et al.* (1975).

**Natural host range and symptoms**

*Vicia faba* — chlorosis, necrosis, leaf malformation.

**Transmission**

Transmitted by a vector; an insect; *Acyrthosiphon pisum*; Aphididae. Transmitted in a non–persistent manner. Virus transmitted by mechanical inoculation.

**Geographical distribution**

Spreads in the USA (Michigan).

**Experimental host range**

Few (<3) families susceptible.

Diagnostically susceptible

*Pisum sativum* cv. Midfreezer — systemic chlorosis, mosaic; then necrosis and plant dies.

Maintenance and propagation hosts

*Pisum sativum* cv. Midfreezer.

Assay hosts (**L**ocal lesions or **W**hole plants)

*Pisum sativum* cv. Midfreezer (W).

Dicotyledonous Hosts *(**Susceptible** or Insusceptible)*

***Chenopodium amaranticolor*. *Chenopodium quinoa*.** *Cucumis sativus. Gomphrena globosa. Lycopersicon esculentum. Nicotiana glutinosa. Nicotiana tabacum.* ***Phaseolus vulgaris*. *Pisum sativum***, or *Pisum sativum.* ***Trifolium incarnatum*. *Vicia faba*.** *Vigna unguiculata.*

Sources of host-range data

Thottappilly *et al.* (1975).

**Properties of particles in sap**

TIP: 45–50 °C. LIV: 1 days (12–20 hours). DEP: $\log_{10}$ minus 3–4.

**Particle morphology**

Virions filamentous; not enveloped; usually flexuous; with a clear modal length; of 650–1640 nm (2 major peaks 650–820 and 1360–1640 nm : 4 minor peaks 2180–2450, 2970–3180, 3650–3850 and 4500–4700 nm); 14 nm wide.

**Taxonomy and Relationships**

*Closterovirus.*

**References**

Thottappilly, G., Harris, K.F. and Bath, J.E. (1975). *Z. PflKrankh. PflPath. PflSchutz.* **79**: 686.

# Broad bean stain *comovirus*

**Data collated by A.J. Gibbs, 1981. Revised 1987.**

*Synonyms:* virus de la mosaique de la fève, MF virus, broad bean F1 virus (Gibbs *et al.*, 1968), broad bean Evesham stain virus. *Standard acronym:* BBSV. *Strains:* pea green mottle virus.

First reported in *Vicia faba*; from the U.K.; by Lloyd *et al.* (1965).

**Natural host range and symptoms**

Symptoms vary cyclically over a few weeks, or vary seasonally.
*Vicia faba* — systemic mottling, necrosis and malformation.

**Transmission**

Transmitted by a vector; an insect; *Apion vorax, Sitona* spp.; Coleoptera. Virus transmitted by mechanical inoculation; transmitted by grafting; transmitted by seed (up to 10% in some *Vicia faba* cultivars).

**Geographical distribution**

Spreads in the African region and the Eurasian region. Found, but with no evidence of spread, in Australia from imported seed.

**Experimental host range**

Several (3–9) families susceptible. Experimentally infected plants mostly show mottle, mosaic, necrosis.

Diagnostically susceptible

*Phaseolus vulgaris* cv. Tendergreen — systemic mottle.
*Pisum sativum, Vicia faba* — systemic mottle; necrosis.

Diagnostically insusceptible

*Chenopodium amaranticolor, Phaseolus vulgaris* cv. Pinto, *Nicotiana clevelandii.*

Maintenance and propagation hosts

*Phaseolus vulgaris* cv. Tendergreen, *Pisum sativum, Vicia faba.*

Assay hosts (**L**ocal lesions or **W**hole plants)

*Phaseolus vulgaris* cv. Tendergreen (L), *Vicia faba* (W).

Dicotyledonous Hosts *(**Susceptible** or Insusceptible)*

*Cassia occidentalis.* ***Chenopodium album.*** *Chenopodium amaranticolor.* ***Chenopodium capitatum. Chenopodium murale. Chenopodium quinoa. Cicer arietinum.*** *Cyamopsis tetragonoloba. Datura stramonium.* ***Glycine max.*** *Gomphrena globosa.* ***Lathyrus odoratus.*** *Lens esculenta. Lotus corniculatus. Medicago sativa.* ***Melilotus alba.*** *Nicotiana clevelandii.* ***Nicotiana debneyi.*** *Nicotiana glutinosa. Nicotiana megalosiphon. Nicotiana rustica. Nicotiana sylvestris. Nicotiana tabacum.* ***Phaseolus vulgaris. Pisum sativum. Trifolium hybridum. Trifolium incarnatum.*** *Trifolium pratense. Trifolium repens.* ***Vicia faba.*** *Vigna radiata.* ***Vigna unguiculata.***

Sources of host-range data

Thornberry (1966); Moghal and Francki (1974); Musil *et al.* (1978); Valenta *et al.* (1979).

**Properties of particles in sap**

TIP: 60–65 °C. LIV: 31 days. DEP: $\log_{10}$ minus 3. Infectivity of sap not changed by treatment with di–ethyl ether. Leaf sap contains many virions.

**Purification**

Method: Gibbs *et al.* (1968).

**Particle morphology**

Virions isometric; not enveloped; 28 nm in diameter; angular in profile; without a conspicuous capsomere arrangement.

**Physical properties**

Three sedimenting components in purified preparations; sedimentation coefficient of the fastest 127 *S* (B); of the other(s) 100 *S* (M), or 60 *S* (T).

**Biochemical properties**

Virions contain 34 % nucleic acid (B), or 25 % nucleic acid (M), or 0 % nucleic acid (T); 66 % protein (T), or 75 % protein (M), or 100 % protein (T); 0 % lipid.

*Genome* consists of RNA; single–stranded; of two parts; largest (or only) genome part the larger 6.75 kb; the 2nd largest 4.5 kb. Base composition 23.3 % G; 26.5 % A; 18.4 % C; 31.9 % U. Infectivity retained when deproteinised with phenol or detergent.

**Replication**

Replication does not depend on a helper virus.

**Cytopathology**

Virions found in all parts of the host plant; in cytoplasm. Inclusions present in infected cells; are unusual in shape; rows of virions in tubules, finger-like outgrowths at plasmodesmata, vacuolate-vesiculate bodies and virus aggregates (Russo *et al.*, 1982).

**Taxonomy and Relationships**

*Comovirus: Comoviridae.*

Virus(es) with serologically related virions: cowpea mosaic, cowpea severe mosaic, glycine mosaic, red clover mottle and squash mosaic viruses, but distantly (Beczner *et al.*, 1983).

**References**


Beczner, L., Musil, M. and Kowalska, C. (1983). *Zesz. probl. Poste. Nauk rolm.* **291**: 21.

Cockbain, A.J., Cook, S.M. and Bowen, R.(1975). *Ann. appl. Biol.* **81**: 331.

Gibbs, A.J. and Smith, H.G. (1970). *CMI/AAB Descr. Pl. Viruses* No. 29, 3 pp.

Gibbs, A.J., Giussani-Belli, G. and Smith, H.G. (1968). *Ann. appl. Biol.* **61**: 99.

Jones, A.T. and Barker, H. (1976). *Ann. appl. Biol.* **83**: 231.

Lloyd, A.T.E., Smith, H.G. and Jones, L.H. (1965). *Hort. Res.* **5**: 13.

Moghal, S.M. and Francki, R.I.B. (1974). *Aust. J. biol. Sci.* **27**: 341.

Musil, M., Leskova, O. and Kleja, S. (1978). *Ochr. Rost.* **14**, 161.

Russo, M., Castellano, M.A. and Martelli, G.P. (1982). *J. Submicrosc. Cytol.* **14**: 149.

Valenta, V., Gressnerova, M., Marcinka, K. and Nermut, M.V. (1979). *Acta Virol., Prague* **13**: 422.

# Broad bean true mosaic *comovirus*

**Data collated by A.J. Gibbs, 1981. Revised 1987.**

*Synonyms: Vicia* virus 1, Echtes Ackerbohnenmosaikvirus. *Standard acronym:* BBTMV.

First reported in *Vicia faba*; from Germany; by Quantz (1953).

**Natural host range and symptoms**

Symptoms vary cyclically over a few weeks.
*Vicia faba* — mild mosaic and malformation, necrosis.

**Transmission**

Transmitted by a vector; an insect; *Apion vorax*, *Sitonia* spp. and other weevils; Coleoptera. Virus transmitted by mechanical inoculation; transmitted by seed.

**Geographical distribution**

Spreads in the African region and the Eurasian region; China, Morocco, and the UK. Found, but with no evidence of spread, in Australia in crops grown from imported seed.

**Experimental host range**

Few (<3) families susceptible.

Diagnostically susceptible

*Pisum sativum* — systemic chlorosis, shoot necrosis.
*Vicia faba* — mottle, mosaic and shoot necrosis.

Diagnostically insusceptible

*Chenopodium amaranticolor, Nicotiana clevelandii.*

Maintenance and propagation hosts

*Pisum sativum, Vicia faba.*

Assay hosts (**L**ocal lesions or **W**hole plants)

*Pisum sativum* (W), *Vicia faba* (W).

Dicotyledonous Hosts *(**Susceptible** or Insusceptible)*

*Antirrhinum majus.* ***Cicer arietinum***. *Cucumis sativus. Datura stramonium. Glycine max. Lathyrus odoratus.* ***Lens esculenta***. *Lotus corniculatus.* ***Lupinus albus***. ***Melilotus alba***. *Nicotiana clevelandii. Nicotiana glutinosa. Nicotiana tabacum. Phaseolus lunatus.* ***Phaseolus vulgaris***. ***Pisum sativum***. *Spinacia oleracea. Stellaria media. Trifolium hybridum.* ***Trifolium incarnatum***. *Trifolium pratense.* ***Trifolium repens***. ***Vicia faba***. ***Vicia sativa***. ***Vicia villosa***.

Sources of host-range data

Quantz (1953).

**Properties of particles in sap**

TIP: 65–75 °C. LIV: 6–7 days. DEP: $\log_{10}$ minus 4. Infectivity of sap not changed by treatment with di–ethyl ether. Leaf sap contains many virions.

**Purification**

Method: Blevings and Stace-Smith (1976).

**Particle morphology**

Virions isometric; not enveloped; 28 nm in diameter; angular in profile; without a conspicuous capsomere arrangement.

**Physical properties**

Three sedimenting components in purified preparations (in reducing buffer); sedimentation coefficient of the fastest 116 *S* (B); of the other(s) 95 *S* (M), or 59 *S* (T). $A_{260}/A_{280}$ ratio 1.81 (B), or 1.76 (M).

**Biochemical properties**

Virions contain 35 % nucleic acid (B), or 26 % nucleic acid (M), or 0 % nucleic acid (T); 65 % protein (B), or 74 % protein (M), or 100 % protein (T).

*Genome* consists of RNA; single–stranded; linear. Base composition 22.8 % G; 26.5 % A; 18.4 % C; 32.3 % U. Infectivity retained when deproteinised with phenol or detergent.

*Non–genomic nucleic acid* not found in the virions.

*Virion protein(s)* two; $M_r$ of the larger 37500. $M_r$ of 2nd largest 24500. Method of preparation: Blevings and Stace-Smith (1976).

**Replication**

Replication does not depend on a helper virus.

**Cytopathology**

Virions found in all parts of the host plant; in cytoplasm. Inclusions present in infected cells; are unusual in shape; are tubules from plasmodesmata containing rows of virus virions, also vesiculated bodies and virus aggregates (Russo *et al.*, 1982).

**Taxonomy and Relationships**

*Comovirus: Comoviridae.*

Virus(es) with serologically related virions: broad bean stain, glycine mosaic, radish mosaic and squash mosaic viruses.

**References**

Blevings, S. and Stace-Smith, R. (1976). *J. gen. Virol.* **31**: 199.
Cockbain, A.J., Cook, S.M. and Bowen, R. (1975). *Ann. appl. Biol.* **81**: 331.
Cockbain, A.J., Bowen, R. and Vorra-Urai, S. (1976). *Ann. appl. Biol.* **84**: 321.
Gibbs, A.J. and Paul, H.L. (1970). *CMI/AAB Descr. Pl. Viruses* No. 20, 4 pp.
Gibbs, A.J., Giussani-Belli, G. and Smith, H.G. (1968). *Ann. appl. Biol.* **61**: 99.
Jones, A.T. (1978). *Ann. appl. Biol.* **88**: 137.
Quantz, L. (1953). *Phytopath. Z.* **20**: 421.
Randles, J.W. and Dube, A.J. (1977). *Australas. Pl. Path. Soc. Newsl.* **6**: 37.
Russo, M., Castellano, M.A. and Martelli, G.P. (1982). *J. Submicrosc. Cytol.* **14**: 149.

# Broad bean V (?) potyvirus

**Data collated by Boswell from Cockbain 1983.**

First reported in *Vicia faba*; from Spain; by Rubio-Huertos *et al.* (1967).

**Natural host range and symptoms**

*Vicia faba* — mosaic.

**Transmission**

Transmitted by a vector; an insect; *Aphis fabae*; Aphididae. Virus transmitted by mechanical inoculation.

Dicotyledonous Hosts *(**Susceptible** or Insusceptible)*

***Nicotiana glutinosa****. Nicotiana tabacum.* ***Petunia hybrida. Phaseolus vulgaris. Vicia faba****.*

**Particle morphology**

Virions filamentous; not enveloped; usually flexuous; of 657–750 nm.

**Taxonomy and Relationships**

Unassigned potyvirus.

Best tests for diagnosis: distinguishable from bean yellow mosaic by the ability to infect *Nicotiana glutinosa, N. tabacum* and *Petunia hybrida.*

**References**

Cockbain, A.J. (1983). *The Faba Bean (Vicia faba) A Basis for Improvement*; ed. P.D. Hebblethwaite. Butterworths.

Rubio-Huertos, M., Santaolalla and Moreno, S. (1967). *An. Edafol. Agrobiol.* **26**: 763.

# Broad bean wilt *fabavirus*

**Data collated by J.W. Bowyer, 1980.**

*Synonyms:* catalpa chlorotic leaf spot virus (Schmelzer, 1970), tropaeolum ringspot virus (Cook and Gibbs, 1971), nasturtium ringspot virus, Ringmosaikvirus der Kapuzinerkresse (Frowd and Tomlinson, 1972), petunia ringspot virus (Hull and Plaskitt, 1973; 1974), P.O. pea streak virus, parsley virus 3. *Standard acronym:* BBWV.

First reported in *Vicia faba*; from Melbourne, Australia; by Stubbs (1947).

**Natural host range and symptoms**

Symptoms vary seasonally.

*Vicia faba, Pisum sativum* — vein clearing, mottling and necrosis of shoot apex, plant wilts, mottled, malformed and stunted.

*Spinacia oleracea* — wilting, shoot die-back, leaves mottled.

*Petroselinum crispum* — symptomless.

*Tropaeolum majus* — mosaic.

*Petunia hybrida* — line-ring spots and oak leaf patterns.

*Plantago lanceolata* — mottling, ringspots.

**Transmission**

Transmitted by a vector; an insect; *Acyrthosiphon pisum, Aphis craccivora, A. faba, A. nasturtii, Macrosiphum euphorbiae, M. solanifolii, M. persicae*; Aphididae. Transmitted in a non–persistent manner. Virus transmitted by mechanical inoculation; not transmitted by seed.

**Geographical distribution**

Spreads in the African region, the Eurasian region, the Middle East, the North American region, and the Pacific region; Australia and China. Found, but with no evidence of spread, in Argentina.

**Experimental host range**

Many (>9) families susceptible. Experimentally infected plants mostly show necrotic local lesions, ringspots, mosaic and systemic necrosis.

Diagnostically susceptible

*Chenopodium amaranticolor, C. quinoa* — chlorotic local lesions; tip epinasty, leaves malformed and mottled.

*Spinacia oleracea* — systemic wilt, necrosis and die-back.

*Vicia faba* — apical necrosis, wilting and mosaic.

*Vigna unguiculata* cv. Blackeye — red or chlorotic local lesions; systemic mosaic and malformation.

*Datura stramonium* — coalescent, necrotic local lesions; systemic ringspots and line patterns.

*Nicotiana x edwardsonii* — necrotic local lesions; systemic necrosis.

*N. glutinosa* — systemic mosaic with some ringspots.

*N. tabacum* cv. White Burley — large ringspots; systemic leaf chlorosis.

*Petunia hybrida* —systemic line and oak leaf patterns and ringspots.

Maintenance and propagation hosts

*Chenopodium quinoa, Pisum sativum, Vicia faba, Nicotiana x edwardsonii.*

Assay hosts (**L**ocal lesions or **W**hole plants)

*Chenopodium amaranticolor* (L), *C. quinoa* (L), *Phaseolus vulgaris* (Moroccan isolate only), *Vigna unguiculata* cv. Blackeye (L), *Datura stramonium* (L).

Dicotyledonous Hosts *(**Susceptible** or Insusceptible)*

***Antirrhinum majus. Beta vulgaris. Brassica campestris. Brassica campestris*** **ssp.** ***rapa.*** *Brassica oleracea* ssp. *botrytis. Cajanus cajan.* ***Calendula officinalis. Capsella bursa-pastoris.*** *Capsicum annuum.* ***Capsicum frutescens.*** *Chenopodium amaranticolor.* ***Chenopodium hybridum. Chenopodium quinoa.*** *Cucumis melo.* ***Cucumis sativus. Cyamopsis tetragonoloba. Datura metel. Datura stramonium. Glycine max. Gomphrena globosa.*** *Helianthus annuus.* ***Lactuca sativa. Lathyrus odoratus. Lupinus angustifolius.*** *Lycopersicon esculentum. Lycopersicon pimpinellifolium.* ***Matthiola incana. Nicandra physaloides. Nicotiana clevelandii. Nicotiana × edwardsonii. Nicotiana debneyi. Nicotiana glutinosa. Nicotiana rustica. Nicotiana sylvestris. Nicotiana tabacum. Ocimum basilicum. Petroselinum crispum. Petunia hybrida.*** *Phaseolus lunatus.* ***Phaseolus vulgaris. Physalis floridana. Physalis peruviana. Pisum sativum.*** *Raphanus sativus.* ***Solanum nigrum.*** *Solanum tuberosum. Sonchus oleraceus.* ***Spinacia oleracea. Tropaeolum majus. Vicia faba. Vigna unguiculata.*** *Zinnia elegans.*

Sources of host-range data

Gracia and Feldman (1976); Schmelzer and Stahl (1977).

**Properties of particles in sap**

TIP: 55–60 °C. LIV: 3–4 days. DEP: $\log_{10}$ minus 4–5. Leaf sap contains many virions.

**Purification**

Method: Doel (1975).

**Particle morphology**

Virions isometric; not enveloped; 25 nm in diameter, or 28–32 nm in diameter (for parsley virus 3); angular in profile; without a conspicuous capsomere arrangement.

**Physical properties**

Three sedimenting components in purified preparations; sedimentation coefficient of the fastest 126 *S* (B); of the other(s) 100 *S* (M), or 63 *S* (T). Density in the presence of

2.34 w/v Igepon $T_{73}$ 1.39 g $cm^{-3}$ in CsCl (M), or 1.44 g $cm^{-3}$ in CsCl (B). $A_{260}/A_{280}$ ratio 1.69 (M), or 1.79 (B).

**Biochemical properties**

Virions contain 33 % nucleic acid (B), or 22 % nucleic acid (M), or 0 % nucleic acid (T); 67 % protein (B), or 78 % protein (M), or 100 % protein (T); 0 % lipid.

*Genome* consists of RNA; single–stranded; linear; of two parts; largest (or only) genome part the larger 6.3 kb; the 2nd largest 4.5 kb. Base composition 26 % G; 30 % A; 17.5 % C; 26.5 % U. Infectivity retained when deproteinised with phenol or detergent.

*Virion protein(s)* two; $M_r$ of the larger 42000. $M_r$ of 2nd largest 26000. Method of preparation: Doel (1975); Brown and Hull (1973).

**Replication**

Replication does not depend on a helper virus.

**Cytopathology**

Virions found in mesophyll and epidermis; in cytoplasm. Inclusions present in infected cells; are crystals in the cytoplasm (Russo *et al.*, 1975), or amorphous X–bodies (Russo and Martelli, 1975); they contain virions (the inclusions are unlike those induced by members of the comovirus group, yet are similar to those induced by some tymoviruses; Edwardson and Christie, 1986).

**Taxonomy and Relationships**

*Fabavirus: Comoviridae* (the Type species of the genus).

Virus(es) with serologically related virions: all isolates are serologically closely related and indistinguishable. Virus(es) with serologically unrelated virions: four comoviruses.

**References**


Brown, F. and Hull, R. (1973). *J. gen. Virol.* **20**: 43.

Cook, S.M. and Gibbs, A.J. (1971). *Rep. Rothamsted Exp. Stn for 1970*, p. 124.

Doel, T.R. (1975). *J. gen. Virol.* **26**: 95.

Edwardson, J.R. and Christie, R.G. (1986). *Fla Agric. Res. Stn Monog.* No. 14, p. 95.

Frowd, J.A. and Tomlinson, J.A. (1972). *Ann. appl. Biol.* **72**: 189.

Govier, D.A. (1988). *Ann. appl. Biol.* **113**: 287.


Govier, D.A. (1988). *Ann. appl. Biol.* **113**: 287.


Gracia, O. and Feldman, J.M. (1976). *Phytopath. Z.* **85**: 227.

Harrison, B.D. (1976). *Rep. Scottish Hort. Res. Inst.* 1976.

Hull, R. and Plaskitt, A. (1973/74). *Intervirology* **2**: 352.

Lockhart, B.E.L. and Fischer, H.U. (1977). *Phytopath. Z.* **88**: 209.

Russo, M. and Martelli, G.P. (1975). *J. Submicrosc. Cytol.* **7**: 335.

Sahambi, H.S., Milne, R.G., Cook, S.M., Gibbs, A.J. and Woods, R.D. (1973). *Phytopath. Z.* **76**: 158.

Schmelzer, K. (1970). *Phytopath. Z.* **67**: 285.

Schmelzer, K. and Stahl, P. (1977). *Zbl. Bakt. ParasitKde.* Abt. 2, **132**: 123.

Schroeder, W.T. and Provvidenti, R. (1970). *Phytopathology* **60**: 1405.

Shukla, D.D., Teakle, D.S. and Gough, K.H. (1980). *Plant Dis.* **64**: 802.

Stubbs, L.L. (1947). *J. Dep. Agric. Vic.* **46**: 323.

Stubbs, L.L. (1960). *Aust. J. agric. Res.* **11**: 734.

Taylor, R.H. and Stubbs, L.L. (1968). *Aust. J. biol. Sci.* **21**: 929.

Taylor, R.H. and Stubbs, L.L. (1972). *CMI/AAB Descr. Pl. Viruses* No. 81, 4 pp.

Tian, W.H., Wang, X.F., Pei, M.Y. and Xie, D.Z. (1982). *Acta Phytopath. Sin.* **9**: 153.
Uyemoto, J.K. and Provvidenti, R. (1974). *Phytopathology* **64**: 1547.
Xi, Z.X., Xu, S.H. and Mang, K.Q. (1982). *Acta Phytopath. Sin.* **12**: 38.
Xu, Z.G., Cockbain, A.J., Woods, R.D. and Govier, D.A. (1988). *Ann. appl. Biol.* **113**: 287.
Xu, Z.G., Pu, Z.C., Cao, Q., Fang, Z.D. and Cockbain, A.J. (1985). *J. Nanjing Agric. Univ. 1985*, pp. 42–48.

# Broad bean yellow ringspot virus

**Data collated by K.T. Natsuaki, 1981. Revised 1984.**

First reported in *Vicia faba*; from Japan.

**Natural host range and symptoms**

Symptoms persist.
*Vicia faba* —chlorotic local lesions and ringspots.

**Transmission**

Virus not transmitted by mechanical inoculation.

**Geographical distribution**

Spreads in Japan.

Dicotyledonous Hosts *(**Susceptible** or Insusceptible)*

***Vicia faba***.

**Properties of particles in sap**

Leaf sap contains few virions.

**Particle morphology**

Virions isometric; not enveloped; 28 nm in diameter; rounded in profile; without a conspicuous capsomere arrangement.

**Cytopathology**

Virions found in mesophyll, epidermis and phloem; in cytoplasm, in nuclei, and in cell vacuoles. Inclusions present in infected cells; are unusual in shape; oval or round inclusions, 1–2 μ; they contain virions. Other cellular changes: swelling of the nucleus.

**Taxonomy and Relationships**

Virus(es) with serologically unrelated virions: cucumber mosaic virus (Japanese Y strain).

**References**

Natsuaki, K.T. (1979). *Ann. Phytopath. Soc. Japan* **45**: 84.

# Broad bean yellow vein *cytorhabdovirus*

**Data collated by K.T. Natsuaki, 1981. Revised 1984.**

First reported in *Vicia faba*; from Japan; by Natsuaki (1981).

**Natural host range and symptoms**

Symptoms persist.
*Vicia faba* —chlorotic veins.

**Transmission**

Virus not transmitted by mechanical inoculation.

**Geographical distribution**

Spreads in Japan.

Dicotyledonous Hosts *(**Susceptible** or Insusceptible)*

***Vicia faba*.**

**Properties of particles in sap**

Leaf sap contains few virions.

**Particle morphology**

Virions rhabdo– or bullet–shaped; enveloped; of 230–250 nm; 110–130 nm wide.

**Cytopathology**

Virions found in mesophyll, epidermis and phloem; in cytoplasm. Inclusions present in infected cells; are unusual in shape; oval or rounded inclusions, 0.05–2.0 µm; they do not contain virions. Other cellular changes: vesiculation of cytoplasm. Virions bud from cytoplasm.

**Taxonomy and Relationships**

*Cytorhabdovirus: Rhabdoviridae.*

**References**

Natsuaki, K.T. (1981). *Ann. Phytopath. Soc. Japan* **47**: 410.

# Broccoli necrotic yellows *cytorhabdovirus*

**Data collated by C. Büchen-Osmond, 1986.**

*Standard acronym:* BNYV.

First reported in *Brassica oleracea* var. *botrytis* (cauliflower-headed broccoli); from England; by Hills and Campbell (1968).

**Natural host range and symptoms**

*Brassica oleracea* ssp. *botrytis* (broccoli), *Brassica oleracea* var. *gemmifera* (brussel sprouts) —symptomless.

**Transmission**

Transmitted by a vector; an insect; *Brevicoryne brassicae* (Tomlinson, Webb and Faithfull, 1972); Aphididae. Virus transmitted by mechanical inoculation; not transmitted by seed.

**Geographical distribution**

Spreads in Australia and the UK.

**Experimental host range**

Several (3–9) families susceptible. Experimentally infected plants mostly show chlorosis, necrosis, sometimes symptomless.

Diagnostically susceptible

*Brassica oleracea* — vein-clearing and leaf-rolling, later symptomless.

*Datura stramonium* — chlorotic and necrotic local lesions, then systemic mosaic and necrosis of veins.

*Nicotiana glutinosa* —sometimes local lesions, then systemic mosaic.

Maintenance and propagation hosts

*Datura stramonium.*

Assay hosts (**L**ocal lesions or **W**hole plants)

*Datura stramonium* (L/W).

Dicotyledonous Hosts *(**Susceptible** or Insusceptible)*

***Brassica oleracea*** ssp. ***botrytis. Datura stramonium. Nicotiana glutinosa.***

**Properties of particles in sap**

TIP: 50 °C. LIV: 0.5–0.8 days (12–18 hours) at 23°C), or 2 days (at 4°C). DEP: $\log_{10}$ minus 3–4. Infectivity of sap decreased by treatment with di–ethyl ether (Lin and Campbell, 1972). Electron microscopy: preparations stained with PTA-K have bullet-shaped virions.

**Purification**

Method: Lin and Campbell (1972).

**Particle morphology**

Virions rhabdo– or bullet–shaped; enveloped; with a clear modal length; of 266 nm (in PTA-K), or 275 nm (in UA); 66 nm wide (in PTA-K), or 75 nm wide (in UA). Basic helix obvious; pitch of basic helix 4.5 nm.

**Physical properties**

One sedimenting component in purified preparations; sedimentation coefficient 874 *S* (±41). Density 1.183–1.195 g/cm$^3$ in potassium tartrate (Lin and Campbell, 1972).

**Biochemical properties**

*Genome* consists of RNA; single–stranded; linear.

**Cytopathology**

Virions found in all parts of the host plant; in cytoplasm (in membrane-bounded sacs; Hill and Campbell (1968)). Inclusions absent from infected cells. Other cellular changes: mitochondria swollen with few cristae (Hill and Campbell, 1968).

**Taxonomy and Relationships**

*Cytorhabdovirus: Rhabdoviridae.*

Virus(es) with serologically unrelated virions: lettuce necrotic yellows and barley yellow striate mosaic viruses.

Virions of broccoli necrotic yellows and lettuce necrotic yellows viruses are similar in size and intracellular location but these viruses differ in host range, host reaction, geographical distribution and are serologically distinct.

**References**

Campbell, R.N. and Lin, M.T. (1972). *CMI/AAB Descr. Pl. Viruses* No. 85, 4 pp.

Garret, R.G. and O'Loughlin, G.T. (1976). *Virology* **76**: 653.

Hills, G.J. and Campbell, R.N. (1968). *J. Ultrastruct. Res.* **24**: 134.

Lin, M.T. and Campbell, R.N. (1972). *Virology* **48**: 30.

Tomlinson, J.A., Webb, M.J.W. and Faithfull, E.M. (1972). *Ann. appl. Biol.* **71**: 127.

# Brome mosaic *bromovirus*

**Data collated by C. Büchen-Osmond, 1987.**

*Synonyms:* weidelgrasmosaikvirus, ryegrass streak virus, trespengrasmosaikvirus. *Standard acronym:* BMV.

First reported in *Bromus inermis*; from the U.S.A.; by McKinney *et al.* (1942).

**Natural host range and symptoms**

Symptoms persist.
*Bromus inermis* — mosaic.

**Transmission**

In laboratory tests, nematodes of the genus *Xiphinema* transmitted the virus (Schmidt *et al.*, 1963; Fritzsche, 1975). Attempts to transmit the virus with aphids and mites have been unsuccessful (Lane, 1974). Virus transmitted by mechanical inoculation; not transmitted by seed (Lane, 1974).

**Geographical distribution**

Spreads in the Eurasian region (eastern); Australia, South Africa, and the USA (central).

**Experimental host range**

Several (3–9) families susceptible. Experimentally infected plants mostly show local lesions, systemic necrosis.

Diagnostically susceptible

*Zea mays* —lesions then systemic necrosis and death.
*Chenopodium amaranticolor, C. hybridum, C. quinoa* — local lesions; not systemic.

Maintenance and propagation hosts

*Hordeum vulgare*.

Assay hosts (**L**ocal lesions or **W**hole plants)

*Chenopodium hybridum* (L).

Dicotyledonous Hosts *(**Susceptible** or Insusceptible)*

***Chenopodium amaranticolor*. *Chenopodium hybridum*. *Chenopodium quinoa*.**

Monocotyledonous Hosts *(**Susceptible** or Insusceptible)*

***Bromus inermis*. *Hordeum vulgare*. *Zea mays*.**

Sources of host-range data

Lane (1974); Rochow (1959).

**Properties of particles in sap**

TIP: 80 °C. LIV: 28 days (or longer). DEP: $\log_{10}$ minus 4–5. Leaf sap contains many virions.

**Purification**

Method: Lane and Kaesberg (1971); Lane (1977).

**Particle morphology**

Virions isometric; not enveloped; 26 nm in diameter; rounded in profile; with a conspicuous capsomere arrangement (Bancroft *et al.*, 1967).

**Physical properties**

One sedimenting component in purified preparations; sedimentation coefficient 87.3 *S* (at pH 3–6); of the other(s) 78.7 *S* (at pH 7; Incardona and Kaesberg, 1964). Density 1.35 g $cm^{-3}$ in CsCl (Bancroft, 1971). Isoelectric point pH 6.8 (Rice and Horst, 1972).

**Biochemical properties**

Virions contain 22 % nucleic acid; 78 % protein; 0 % lipid.

*Genome* consists of RNA; single–stranded; linear; of three parts; largest (or only) genome part the largest 3.234 kb (RNA-1); the 2nd largest 2.865 kb (RNA-2); the 3rd largest 2.117 kb (RNA-3). Genomic nucleic acid isolated by Bockstahler and Kaesberg (1962; 1965). Base composition 28 % G; 27 % A; 21 % C; 24 % U. 5' terminus of RNA has a methylated nucleotide cap (of 7-methyl guanosine). Genome has *t*RNA–like activity. Genome accepts tyrosine. Nucleotide sequence references: Bastin *et al.* (1976); P. Kaesberg in Lane (1977); DasGupta *et al.* (1975).

*Sequence database accession code(s):*

J02042 Em(40)_vi:MBRCG3Z Gb(84)_vi:MBRCG3Z Brome mosaic virus (Russian strain) RNA 3 (and RNA 4), of complete genome. 6/94 2,111bp.

K01774 Em(40)_vi:BRMBRR05 Gb(84)_vi:MBRRNA1 Brome mosaic virus (Russian strain) RNA 1, 3' end. 6/94 260bp.

K01775 Em(40)_vi:BRMBRR06 Gb(84)_vi:MBRRNA2 Brome mosaic virus (Russian strain) RNA 2, 3' end. 6/94 265bp

K02706 Gb(84)_vi:MBRCG1Z Brome mosaic virus RNA 1 of complete genome. 11/85 3,234bp.

K02707 Gb(84)_vi:MBRCG2Z Brome mosaic virus RNA 2 of complete genome. 11/85 2,865bp.

M10570 Em(40)_vi:BRMBRR04 Gb(84)_vi:MBRRNA4A Brome mosaic virus (BMV) RNA 4, 3' end. 7/89 227bp.

M12024 Em(40)_vi:BRMBRTRL Gb(84)_vi:MBRTRLA Brome mosaic virus 3' terminal tRNA-like structure. 4/90 131bp

M12675 Em(40)_vi:BRMBRR03 Gb(84)_vi:MBRRNA3B Brome mosaic virus RNA encoding coat protein, partial cd. 4/90 53bp.

M18690 Em(40)_vi:BRMBRR02 Gb(84)_vi:MBRRNA3A Brome mosaic virus (BMV) RNA 3, 3' end. 7/89 227bp.

M19306 Em(40)_vi:BRMBRRNA Gb(84)_vi:MBRRNA1A Brome mosaic virus (BMV) RNA 1, 3' end. 7/89 170bp.

M19307 Em(40)_vi:BRMBRR01 Gb(84)_vi:MBRRNA2A Brome mosaic virus (BMV) RNA 2, 3' end. 7/89 196bp.

M19550 Em(40)_sy:AGRNA3C Gb(84)_sy:SYNRNA3C Synthetic Brome mosaic virus (BMV) RNA 3' end/CAT gene from pB3CA42, partial cds. 4/90 50bp.

M19551 Em(40)_sy:AGRNA3D Synthetic brome mosaic virus (BMV) RNA 3' with CAT gene from plasmid pB3CA61, partial cd. 7/8

M19552 Em(40)_sy:AGRNA3E Gb(84)_sy:SYNRNA3E Synthetic brome mosaic virus (BMV) RNA 3' with CAT gene from plasmid pB3CA71, partial cd. 4/9

M19553 Em(40)_sy:AGRNA301 Gb(84)_sy:SYNRNA3D Synthetic brome mosaic virus (BMV) RNA 3' with CAT gene from plasmid pB3CA61, partial cds. 4/

M25172 Em(40)_vi:BRMBR5RN Gb(84)_vi:MBR5RNA3 Brome mosaic virus RNA-3, 5' end. 1/90 164bp.

V00099 Em(40)_vi:BRBMV3 Gb(84)_vi:BRBMV3 Brome mosaic virus complete sequence of RNA 3 encoding cistron 3a and the coat protein. 6/94

X01678 Em(40)_vi:BRBMV2 Gb(84)_vi:BRBMV2 Brome mosaic virus (BMV) RNA2. 9/93 2,865bp

X02380 Em(40)_vi:BRBMV1 Gb(84)_vi:BRBMV1 Brome mosaic virus (BMV) RNA1 sequence. 9/93 3,234bp.
X58456 Em(43)_vi:Bmv1aprot Gb(89)_vi:Bmv1aprot Brome mosaic virus mRNA for 1a protein. 10/94 3,234bp.
X58457 Em(43)_vi:Bmv2aprot Gb(89)_vi:Bmv2aprot Brome mosaic virus mRNA for 2a protein. 10/94 2,867bp.
X58458 Em(43)_vi:Bmv3aprot Gb(89)_vi:Bmv3aprot Brome mosaic virus mRNA for 3a protein and coat protein. 10/94 2,114bp.

*Non–genomic nucleic acid* found in the virions; is subgenomic *m*RNA; for the coat protein; it is called RNA-4 and is found in the virions containing RNA-3. Sub–genomic *m*RNA found in infected cells.

*Virion protein(s)* one; $M_r$ 20300; coat protein. Method of preparation: Stubbs and Kaesberg (1964). Amino acid composition: Stubbs and Kaesberg (1964); Ahlquist *et al.* (1981).

*Virus–coded non–virion proteins* isolated (Hariharasubramanian *et al.* (1973)); one protein found. $M_r$ 35000; membrane-bound RNA polymerase.

**Replication**

Replication does not depend on a helper virus.

**Cytopathology**

Virions found in leaves and mesophyll; in cytoplasm, in the perinuclear space, and in chloroplasts (within invaginations; Paliwal, 1970). Inclusions present in infected cells; are crystals in the cytoplasm. Other cellular changes: virions which can associate in a helical array around the outside of membraneous tubules and are about 30 nm in diameter (Burges *et al.*, 1974).

**Taxonomy and Relationships**

*Bromovirus: Bromoviridae* (the type species).

Virus(es) with serologically related virions: cowpea chlorotic mottle virus, but distantly (Scott and Slack, 1971).

RNA-1 and RNA-2 from brome mosaic virus and RNA-3 from cowpea chlorotic mottle virus form a viable genetic hybrid, although this replicates more slowly and has a narrower host range (Bancroft, 1972). This indicates their close relationship. Brome mosaic and broad bean mottle viruses also are related as shown by their physical properties of their virions and similar coat protein composition. Best tests for diagnosis: Brome mosaic virus is distinguishable from most other viruses of Gramineae by its symptoms in maize and by its ability to infect several non-graminaceous hosts.

**References**

Ahlquist, P., Luckow and Kaesberg, P. (1981). *J. mol. Biol.* **153**: 23.
Bancroft, J.B. (1970). *CMI/AAB Descr. Pl. Viruses* No. 3, 3 pp.
Bancroft, J.B. (1971). *Virology* **45**: 830.
Bancroft, J.B. (1972). *J. gen. Virol.* **14**: 223.
Bancroft, J.B., Hills, G.J. and Markham, R. (1967). *Virology* **31**: 354.
Bastin, M., DasGupta, R., Hall, T.C. and Kaesberg, P. (1976). *J. mol. Biol.* **103**: 737.
Bockstahler, L.E. and Kaesberg, P. (1962). *Biophys. J.* **2**: 1.
Bockstahler, L.E. and Kaesberg, P. (1965). *J. mol. Biol.* **13**: 127.
Burgess, J., Motoyoshi, F. and Fleming, E.N. (1974). *Planta* **117**: 133.
DasGupta, R., Shih, D.S., Saris, C. and Kaesberg, P. (1972). *Biochem. J.* **129**: 969.
Fritzsche, R. (1975). *Archiv. Phytopath. PflSchutz.* **11**: 197.
Hariharasubramanian, V., Hadidi, A., Singer, B. and Fraenkel-Conrat, H. (1973). *Virology* **54**: 190.

Incardona, N.L. and Kaesberg, P. (1964). *Biophys. J.* **4**: 11.
Lane, L.C. (1974). *Adv. Virus Res.* **19**: 151.
Lane, L.C. (1977). *CMI/AAB Descr. Pl. Viruses* No. 183, 4 pp.
Lane, L.C. and Kaesberg, P. (1971). *Nature, New Biol.* **232**: 40.
McKinney, H.H., Fellows, H. and Johnston, C.O. (1942). *Phytopathology* **32**: 331.
Paliwal, Y.C. (1970). *J. Ultrastruct. Res.* **30**: 491.
Rice, R.H. and Horst, J. (1972). *Virology* **49**: 602.
Rochow, W.F. (1959). *Phytopathology* **49**: 126.
Schmidt, H.B., Fritsche, R. and Lehmann, W. (1963). *Naturw.* **50**: 386.
Stubbs, J.D. and Kaesberg, P. (1964). *J. mol. Biol.* **8**: 314.
Von Wechmar, M.B. and van Regenmortel, M.H.V. (1968). *Virology* **34**: 36.

# Brome streak mosaic (?) *rymovirus*

**Data collated by A.A. Brunt, 1995.**

*Standard acronym:* BStV.

First reported in *Bromus mollis, Hordeum murinum*; from Former Yugoslavia; by Miličič *et al.* (1980, 1982).

**Natural host range and symptoms**

*Bromus mollis, Hordeum murinum* — chlorotic leaf streaking.

**Geographical distribution**

Spreads in the former Yugoslavia.

**Experimental host range**

Few (<3) families susceptible.

Diagnostically susceptible

*Avena, Bromus, Hordeum, Triticum* and *Vulpia* spp. — chlorotic leaf streaking.

Maintenance and propagation hosts

*Bromus mollis* (W).

Monocotyledonous Hosts ***(Susceptible** or Insusceptible)*

***Bromus mollis*. *Chrysalidocarpus lutescens*. *Hordeum vulgare*. *Triticum aestivum*.**

**Properties of particles in sap**

LIV: 5 days (or more). DEP: $\log_{10}$ minus 3.

**Purification**

Method: Miličič *et al.* (1982).

**Particle morphology**

Virions filamentous; usually flexuous; with a clear modal length; of 693 nm.

*Sequence database accession code(s):*

X78485 Gb(89)n:Bsmpnicp Brome Streak Mosaic Potyvirus mRNA for viral polymerase, coat protein. 6/95 3,044bp.

**Cytopathology**

Inclusions present in infected cells; are pinwheels.

**Taxonomy and Relationships**

Possible *Rymovirus: Potyviridae*.

**References**

Miličič, D., Kujundzic, M., Wrischer, M. and Plavsic, B. (1980). *Acta bot. Croat.* **39**: 27.

Miličič, D., Mamula, D. and Plazibat, M. (1982). *Acta bot. Croat.* **41**: 7.

# Bryonia mottle (?) potyvirus

**Data collated by A.A. Brunt, 1994.**

*Standard acronym:* BryMV.

First reported in *Bryonia dioica*; from Morocco; by Lockhart and Fischer (1979).

**Taxonomy and Relationships**

Possible unassigned potyvirus.

**References**

Lockhart, B.E.L. and Fischer, H.U. (1979). *Phytopath. Z.* **96**: 244.

# Burdock mosaic virus

**Data collated by A.A. Brunt, 1992.**

First reported in *Arctium lappa*; from Japan; by Oshima and Goto (1955).

**Natural host range and symptoms**

*Arctium lappa* — yellow leaf mottling and mosaic.

**Transmission**

Transmitted by a vector; an insect; Aphididae. Transmitted in a semi–persistent manner. Virus transmitted by mechanical inoculation.

**Geographical distribution**

Spreads in Japan.

**Experimental host range**

Few (<3) families susceptible.

Diagnostically susceptible

*Arctium lappa* — systemic leaf chlorosis.

Maintenance and propagation hosts

*Arctium lappa*.

Assay hosts (Local lesions or Whole plants)

*Arctium lappa* (W).

**Properties of particles in sap**

TIP: 60 °C. LIV: 2 days.

**Particle morphology**

Virions isometric; 30 nm in diameter; rounded in profile.

**References**

Inouye, T. and Mitsuhata, K. (1971). *Agric. Res.* **54**: 1.

Oshima, N. and Goto, T. (1955). *Rep. Hokkaido Agric. Res. Stn No. 68*, p. 55.

# Burdock mottle virus

**Data collated by A.A. Brunt, 1992.**

First reported in *Arctium lappa*; from Japan; by Inouye (1973).

**Natural host range and symptoms**

*Arctium lappa* — faint leaf chlorosis.

**Transmission**

Virus transmitted by mechanical inoculation.

**Geographical distribution**

Spreads in Japan.

**Experimental host range**

Few (<3) families susceptible.

Diagnostically susceptible

*Chenopodium quinoa, C. murale, Nicotiana clevelandii, N. rustica* — systemic leaf chlorosis.

Maintenance and propagation hosts

*Chenopodium quinoa, Nicotiana clevelandii.*

Assay hosts (**L**ocal lesions or **W**hole plants)

*Chenopodium amaranticolor* (L).

Dicotyledonous Hosts *(**Susceptible** or Insusceptible)*

***Chenopodium amaranticolor***. ***Chenopodium murale***. ***Chenopodium quinoa***. ***Nicotiana clevelandii***. ***Nicotiana rustica***.

**Properties of particles in sap**

TIP: 65 °C. LIV: 5 days.

**Particle morphology**

Virions rod–shaped; usually straight; with a clear modal length; of 250 nm; 17 nm wide. Basic helix obvious; pitch of basic helix 2.3 nm.

**Cytopathology**

Virions found in leaves and mesophyll; in cytoplasm. Inclusions present in infected cells; are viroplasms.

**References**

Inouye, T. (1973). *Ber. Ohara Inst., Okayama Univ.* **15**: 207.

# Burdock yellow mosaic (?) *potexvirus*

**Data collated by L.W. Stobbs, 1991.**

*Strains:* 1 isolate.

First reported in *Arctium minus*; from the Niagara region, Ontario, Canada; by Stobbs and van Schagen (1988).

**Natural host range and symptoms**

Symptoms persist. Symptoms bright yellow mosaic and stunting.

*Arctium minus* — stunting and bright yellow mosaic. Symptoms more pronounced in plants growing in sunny positions than those in full shade.

**Transmission**

Transmitted by means not involving a vector (possible). Virus transmitted by mechanical inoculation; transmitted by contact between plants; not transmitted by seed.

**Geographical distribution**

Spreads in Canada.

**Experimental host range**

Several (3–9) families susceptible. Experimentally infected plants mostly show local lesions, latent and systemic infection.

Diagnostically susceptible

*Chenopodium amaranticolor, Gomphrena globosa, Nicotiana tabacum* — local lesions.

*Chenopodium murale* — latent infection.

*C. quinoa* — systemic infection.

Maintenance and propagation hosts

*Arctium minus, Chenopodium quinoa* for propagation.

Assay hosts (**L**ocal lesions or **W**hole plants)

*Chenopodium amaranticolor* (L), *Gomphrena globosa* (L), *Nicotiana tabacum* cv. Harrow Velvet (L).

Dicotyledonous Hosts *(**Susceptible** or Insusceptible)*

*Beta vulgaris. Brassica campestris* ssp. *rapa. Capsicum annuum.* ***Chenopodium amaranticolor. Chenopodium murale. Chenopodium quinoa.*** *Cucumis sativus. Datura stramonium.* ***Gomphrena globosa.*** *Lycopersicon esculentum. Nicotiana clevelandii. Nicotiana glutinosa. Nicotiana rustica.* ***Nicotiana tabacum.*** *Petunia hybrida. Phaseolus vulgaris. Spinacia oleracea. Vigna unguiculata. Zinnia elegans.*

**Properties of particles in sap**

TIP: 58 °C. LIV: 91 days. DEP: $\log_{10}$ minus 6. Leaf sap contains many virions.

**Purification**

Method: Stobbs and van Schagen (1988).

**Particle morphology**

Virions filamentous; not enveloped; usually flexuous; with a clear modal length; of 525–530 nm; 13 nm wide. Axial canal obvious.

**Physical properties**

One sedimenting component in purified preparations. Density 1.304 g $cm^{-3}$ in CsCl. Isoelectric point pH 4–4.5. $A_{260}/A_{280}$ ratio 1.22.

**Biochemical properties**

*Genome* consists of RNA; single–stranded.

*Virion protein(s)* two; $M_r$ of the larger 28300. $M_r$ of 2nd largest 30500. Method of preparation: Allen and Dias (1977).

**Cytopathology**

Virions found in all parts of the host plant; in cytoplasm. Inclusions absent from infected cells.

**Taxonomy and Relationships**

Possible *Potexvirus*.

Virus(es) with serologically related virions: clover yellow mosaic and potato aucuba mosaic viruses.

Best tests for diagnosis: symptoms in *Arctium minus* are diagnostic for this virus.

**References**

Allen, W.R. and Dias, H.F. (1977). *Can. J. Bot.* **55**: 1028.

Stobbs, L.W. and van Schagen, J.G. (1988). *Plant Dis.* **72**: 314.

# Burdock yellows *closterovirus*

**Data collated by A.A. Brunt, 1991.**

*Standard acronym:* BuYV.

First reported in *Arctium lappa*; from Japan; by Inouye and Mitsuhata (1971).

**Natural host range and symptoms**

*Arctium lappa* — faint chlorosis, often marginal, or symptomless. Infected plants often also contain burdock mosaic virus.

**Transmission**

Transmitted by a vector; an insect; *Dactynotus gobonis*; Aphididae. Transmitted in a semi–persistent manner. Virus transmitted by mechanical inoculation (with difficulty).

**Geographical distribution**

Spreads in Japan.

**Experimental host range**

Few (<3) families susceptible. Experimentally infected plants mostly show yellows.

Diagnostically susceptible

*Nicotiana clevelandii* — systemic leaf chlorosis.

Maintenance and propagation hosts

*Nicotiana clevelandii*.

Assay hosts (**L**ocal lesions or **W**hole plants)

*Nicotiana clevelandii* (W).

Dicotyledonous Hosts *(**Susceptible** or Insusceptible)*

***Nicotiana clevelandii***.

**Properties of particles in sap**

TIP: 45–50 °C. LIV: 1–2 days. Leaf sap contains few virions.

**Particle morphology**

Virions filamentous; not enveloped; usually flexuous; with a clear modal length; of 1725 nm; 12 nm wide. Axial canal obscure. Basic helix obvious; pitch of basic helix 3.6 nm.

**Physical properties**

$A_{260}/A_{280}$ ratio 1.73.

**Cytopathology**

Virions found in leaves, mesophyll and vascular parenchyma; in cytoplasm. Inclusions present in infected cells; are viroplasms; they contain virions. Other cellular changes: necrosis of phloem cells.

**Taxonomy and Relationships**

*Closterovirus.*

Virus(es) with serologically unrelated virions: carnation necrotic fleck, citrus tristeza and wheat yellow leaf viruses.

**References**

Inouye, T. and Mitsuhata, K. (1971). *Agric. Res.* **54**: 1.
Nakano, M. and Inouye, T. (1980). *Ann. Phytopath. Soc. Japan* **46**: 7.

# Butterbur mosaic (?) *carlavirus*

**Data collated by A.A. Brunt, 1992.**

*Standard acronym:* ButMV.

First reported in *Petasites officinalis*; from Japan; by Tochihara and Tamura (1976).

**Natural host range and symptoms**

Symptoms persist.

*Petasites officinalis* — conspicuous leaf chlorosis and plant stunting.

**Transmission**

Transmitted by a vector; an insect; *Myzus persicae*; Aphididae. Transmitted in a non–persistent manner. Virus transmitted by mechanical inoculation.

**Geographical distribution**

Spreads in Japan.

**Experimental host range**

Several (3–9) families susceptible.

Diagnostically susceptible

*Beta vulgaris, Chenopodium amaranticolor, C. quinoa* — local lesions.

*Calendula officinalis, Petasites officinalis, Petunia hybrida, Zinnia elegans* — systemic leaf chlorosis.

Maintenance and propagation hosts

*Calendula officinalis, Petunia hybrida.*

Assay hosts (**L**ocal lesions or **W**hole plants)

*Chenopodium amaranticolor* (L), *C. quinoa* (L).

Dicotyledonous Hosts *(**Susceptible** or Insusceptible)*

***Calendula officinalis***. ***Chenopodium amaranticolor***. ***Chenopodium quinoa***. ***Petunia hybrida***. ***Zinnia elegans***.

**Particle morphology**

Virions filamentous; not enveloped; usually straight; with a clear modal length; of 670 nm; 13 nm wide. Axial canal obscure. Basic helix obscure.

**Taxonomy and Relationships**

Possible *Carlavirus.*

Virus(es) with serologically related virions: carnation latent virus.

**References**

Tochihara, H. and Tamura, M. (1976). *Ann. Phytopath. Soc. Japan* **42**: 533.

# Butterbur (?) *nucleorhabdovirus*

**Data collated by A.A. Brunt, 1992.**

First reported in *Petasites officinalis*; from Japan; by Yamashita *et al.* (1982).

**Natural host range and symptoms**

*Petasites officinalis* — symptomless infection.

**Transmission**

Virus transmitted by mechanical inoculation.

**Geographical distribution**

Spreads in Japan.

**Experimental host range**

Few (<3) families susceptible.

Diagnostically susceptible

*Nicotiana glutinosa.*

Maintenance and propagation hosts

*Nicotiana glutinosa.*

Dicotyledonous Hosts *(**Susceptible** or Insusceptible)*

***Nicotiana glutinosa*.**

**Particle morphology**

Virions rhabdo– or bullet–shaped; with a clear modal length; of 230–250 nm; 80 nm wide. Axial canal obvious; 4.5 nm in diameter. Basic helix obvious; pitch of basic helix 4.5 nm.

**Cytopathology**

Virions found in leaves and mesophyll; in nuclei and in the perinuclear space.

**Taxonomy and Relationships**

Possible *Nucleorhabdovirus: Rhabdoviridae.*

**References**

Yamashita, S., Maeda, H., Doi, Y. and Yora, K. (1982). *Ann. Phytopath. Soc. Japan* **48**: 395.

# Cacao necrosis *nepovirus*

**Data collated by A.A. Brunt, 1984.**

*Synonyms:* cacao swollen shoot virus — strain S. *Standard acronym:* CNV.

First reported in *Theobroma cacao*; from Ghana; by Kenten (1972).

**Natural host range and symptoms**

*Theobroma cacao* — systemic, necrotic or chlorotic spots, streaks and mosaic.

**Transmission**

Transmitted by a vector; a nematode (possibly). Virus transmitted by mechanical inoculation; transmitted by grafting; not transmitted by contact between plants; transmitted by seed (up to 24% in *Glycine max, Phaseolus lunatus* and *P. vulgaris*).

**Geographical distribution**

Spreads in Ghana and Nigeria.

**Experimental host range**

Several (3–9) families susceptible. Experimentally infected plants mostly show necrotic local lesions, systemic mottling.

Diagnostically susceptible

*Beta vulgaris* — red local lesions.

*Chenopodium amaranticolor, C. quinoa* — systemic leaf chlorosis.

*Phaseolus vulgaris* — chlorotic local lesions or ringspots; systemic vein-banding and mottling.

Diagnostically insusceptible

*Spinacia oleracea, Trifolium repens, Vicia faba, Datura stramonium.*

Maintenance and propagation hosts

*Phaseolus vulgaris.*

Assay hosts (**L**ocal lesions or **W**hole plants)

*Beta vulgaris* (L), but less sensitive than *Phaseolus vulgaris* (L).

Dicotyledonous Hosts *(**Susceptible** or Insusceptible)*

***Beta vulgaris***. *Brassica campestris* ssp. *pekinensis*. *Brassica campestris* ssp. *rapa*. *Capsicum annuum*. ***Chenopodium amaranticolor***. ***Chenopodium quinoa***. ***Cucumis sativus***. *Datura stramonium*. *Daucus carota*. ***Glycine max***. ***Gomphrena globosa***. *Hyoscyamus niger*. *Lactuca sativa*. *Lycopersicon esculentum*. ***Nicotiana clevelandii***. ***Nicotiana glutinosa***. ***Nicotiana tabacum***. *Pastinaca sativa*. ***Petunia hybrida***. ***Phaseolus vulgaris***. *Physalis floridana*. *Pisum sativum*. *Raphanus sativus*. ***Tetragonia expansa***. ***Theobroma cacao***. *Trifolium repens*. *Vicia faba*. ***Vigna unguiculata***.

Monocotyledonous Hosts *(**Susceptible** or Insusceptible)*

*Triticum aestivum.*

Sources of host-range data

Kenten (1972).

**Properties of particles in sap**

TIP: 65 °C. LIV: 4–7 days. DEP: $\log_{10}$ minus 3–4. Leaf sap contains few virions.

**Purification**

Method: Kenten (1972).

**Particle morphology**

Virions isometric; not enveloped; 24–26 nm in diameter (a minor component of 20–30 *S* has isometric virions 12 nm diameter); angular in profile; without a conspicuous capsomere arrangement.

**Physical properties**

Four sedimenting components in purified preparations (one of which may be polymerized coat protein); sedimentation coefficient of the fastest 129 *S* (B); of the other(s) 101 *S* (M), or 54 *S* (T). $A_{260}/A_{280}$ ratio 1.78 (B), or 1.63 (M; both corrected for scattering).

**Biochemical properties**

Virions contain 38 % nucleic acid (B), or 26 % nucleic acid (M), or 0 % nucleic acid (T); 62 % protein (B), or 74 % protein (M), or 100 % protein (T); 0 % lipid.

*Genome* consists of RNA; single–stranded; linear; of two parts.

*Virion protein(s)* one; $M_r$ 60000.

**Taxonomy and Relationships**

*Nepovirus: Comoviridae.*

Virus(es) with serologically related virions: tomato black ring virus, but distantly, and grapevine chrome mosaic virus, very distantly. Virus(es) with serologically unrelated virions: arabis mosaic, cherry leaf roll, strawberry latent ringspot, tobacco ringspot and tomato ringspot viruses.

**References**

Kenten, R.H. (1972). *Ann. appl. Biol.* **71**: 119.

Kenten, R.H. (1977). *CMI/AAB Descr. Pl. Viruses* No. 173, 4 pp.

Owusu, G.K. (1971). *Trop. Agric., Trin.* **48**: 133.

Thresh, J.M. and Tinsley, T.W. (1959). *Tech. Bull. W. Afr. Cocoa Res. Inst.* No. 7, p. 27.

# Cacao swollen shoot *badnavirus*

**Data collated by A.A. Brunt, 1986.**

*Synonyms:* theobroma virus 1, cacao swollen shoot virus, cacao mottle leaf virus. *Standard acronym:* CSSV. *Strains:* Many variants have been recorded; initially named alphabetically, they are now named by where they were found. Well-known strains include Bisa, Bosumtwe, Ikiri, Kpeve, Mampong, New Juaben, Nsaba and Offa Igbo.

First reported in *Theobroma cacao* (cocoa); from Ghana; by Posnette (1940); Brunt and Kenten (1960).

**Natural host range and symptoms**

Symptoms persist (in some and disappears in others).

*Theobroma cacao* — interveinal chlorosis, swollen stems and roots caused by many strains, a few cause root swelling only, and some cause no symptoms.

*Ceiba pentandra, Cola chlamydantha, Cola gigantea* var. *glabrescens, Sterculia tragacantha* — transient leaf chlorosis.

*Adansonia digitata* — leaf chlorosis.

**Transmission**

Transmitted by a vector; an insect; *Planococcoides njalensis, Planococcus citri, Planococcus kenyae, Planococcus* sp. *celtis, Dysmicoccus brevipes, Ferrisia virgata, Pseudococcus longispinus* (only Mampong strain); Pseudococcidae. Transmitted in a semi–persistent manner. Virus retained when the vector moults; does not multiply in the vector; not transmitted congenitally to the progeny of the vector; transmitted by mechanical inoculation (with difficulty); transmitted by grafting; not transmitted by contact between plants; not transmitted by seed; not transmitted by pollen.

**Ecology and control**

Studies reported by Ollenu *et al.* (1989).

**Geographical distribution**

Spreads in Cote d'Ivoire, Ghana, Nigeria, Sierra Leone, Sri Lanka, and Togo (and, possibly, Trinidad).

**Experimental host range**

Several (3–9) families susceptible. Experimentally infected plants mostly show chlorosis, necrosis, swelling.

Diagnostically susceptible

*Theobroma cacao* (Amelonado) — systemic red vein-banding, then leaf chlorosis.
*Corchorus* spp. — lethal systemic necrosis.
*Adansonia digitata* — systemic mottling and stunting.
*Ceiba pentandra, Ceiba chlamydantha, Cola gigantea* — transient leaf chlorosis.
*Bombax brevicuspe* — chlorosis.

Diagnostically insusceptible

*Nicotiana tabacum, Phaseolus vulgaris, Chenopodium amaranticolor, Chenopodium quinoa, Nicotiana clevelandii.*

Maintenance and propagation hosts

*Theobroma cacao, Adansonia digitata, Bombax brevicuspe.*

Assay hosts (**L**ocal lesions or **W**hole plants)

*Theobroma cacao* (W). Cocoa cotyledons are readily infected by viruliferous mealybugs or by mechanical inoculation.

Dicotyledonous Hosts *(**Susceptible** or Insusceptible)*

***Adansonia digitata***. *Arachis hypogaea. Capsicum annuum. Cassia occidentalis. Cassia tora.* ***Ceiba pentandra***. *Chenopodium amaranticolor. Chenopodium foetidum. Chenopodium quinoa. Crotalaria spectabilis. Cucumis sativus. Datura stramonium. Glycine max. Gomphrena globosa. Hibiscus esculentus. Nicotiana clevelandii. Nicotiana glutinosa. Nicotiana tabacum. Petunia hybrida. Phaseolus lunatus. Phaseolus vulgaris. Ricinus communis. Solanum melongena.* ***Theobroma cacao***. *Vigna unguiculata.*

Sources of host-range data

Legg and Lovi (1968); Legg and Lovi (1969); Tinsley and Wharton (1958).

**Properties of particles in sap**

TIP: 55–60 °C. LIV: 28–85 days (at 2°C). DEP: $\log_{10}$ minus 3–4. Leaf sap contains few virions.

**Purification**

Method: Adomako *et al.* (1983); Lot *et al.* (1991), Hagen *et al.* (1993).

**Particle morphology**

Virions bacilliform; not enveloped; 130 nm in length; 28 nm wide.

**Physical properties**

One sedimenting component in purified preparations; sedimentation coefficient 218 *S*.

**Biochemical properties**

*Genome* consists of DNA; double–stranded; circular; unipartite; largest (or only) genome part 7.4 kb. Nucleotide sequence references: Lot *et al.* (1991).

*Sequence database accession code(s):*

L14546 Em(40)_vi:CSHCG Gb(84)_vi:CSHCG Cacao swollen shoot virus polyprotein gene, complete circular genome. 1/94 7,161bp. 1 sequence.

**Cytopathology**

Virions found in all parts of the host plant; in cytoplasm. Inclusions absent from infected cells.

**Taxonomy and Relationships**

*Badnavirus.*

Virus(es) with serologically unrelated virions: banana streak, rice tungro bacilliform, taro bacilliform, yam internal brown spotting and yucca bacilliform viruses.

**References**


Adomako, D., Lesemann, D.E. and Paul, H.L. (1983). *Ann. appl. Biol.* **103**: 109.

Brunt, A.A. (1970). *CMI/AAB Descr. Pl. Viruses* No. 10, 4 pp.

Brunt, A.A. and Kenten, R.H. (1960). *Virology* **12**: 328.

Brunt, A.A. and Kenten, R.H. (1962). *Ann. appl. Biol.* **50**: 749.

Brunt, A.A. and Kenten, R.H. (1963). *Virology* **19**: 388.

Brunt, A.A., Kenten, R.H. and Nixon, H.L. (1964). *J. gen. Microbiol.* **36**: 303.

Hagen, L.S., Jacquemond, M., Lepingle, A., Lot, H. and Tepfer, M. (1993). *J. gen. Virol.*

Kenten, R.H. and Legg, J.T. (1971). *Ann. appl. Biol.* **67**: 195.

Legg, J.T. and Lovi, N.K. (1968). *Rep. W. Afr. Cocoa Res. Inst.* for **1966–1967**: 22.

Legg, J.T. and Lovi, N.K. (1969). *Rep. W. Afr. Cocoa Res. Inst.* for **1967–1968**: 29.

Lot, H., Djekpor, E. and Jacquemond, M. (1991). *J. gen. Virol.* **72**: 1735.

Ollenu, L.A.A., Owusu, G.K. and Thresh, J.M. (1989). *Cocoa Growers Bull.* **42**: 25.

Partiot, M., Amefia, Y.K., Djiekpor, E.K. and Baker, K.A. (1978). *Cafe, Cacao, The* **23**: 217.

Posnette, A.F. (1940). *Trop. Agric., Trin.* **17**: 98.

Posnette, A.F. (1947). *Ann. appl. Biol.* **34**: 388.

Posnette, A.F. (1950). *Ann. appl. Biol.* **37**: 378.

Posnette, A.F. and Robertson, N.F. (1950). *Ann. appl. Biol.* **37**: 363.

Posnette, A.F., Robertson, N.F. and Todd, J.M. (1950). *Ann. appl. Biol.* **37**: 229.

Posnette, A.F. and Todd, J.M. (1955). *Ann. appl. Biol.* **43**: 443.

Sagemann, W., Lesemann, D.-E., Paul, H.L., Adomako, D. and Owusu, G.K. (1985). *J. Phytopath.* **114**: 79.

Sagemann, W., Paul, H.L., Adomako, D. and Owusu, G.K. (1983). *Phytopath. Z.* **106**: 281.

Tinsley, T.W. and Wharton, A.L. (1958). *Ann. appl. Biol.* **46**: 1.


# Cacao yellow mosaic *tymovirus*

**Data collated by A.A. Brunt, 1985.**

*Standard acronym:* CYMV.

First reported in *Theobroma cacao*; from Giehuna, Sierra Leone; by Blencowe *et al.* (1963).

**Natural host range and symptoms**

Symptoms persist.

*Theobroma cacao* — leaves with yellow blotches or mosaic.

**Transmission**

Transmitted by a vector. Virus transmitted by mechanical inoculation; transmitted by grafting; not transmitted by seed.

**Geographical distribution**

Spreads in Sierra Leone.

**Experimental host range**

Several (3–9) families susceptible. Experimentally infected plants mostly show local lesions, mosaic.

Diagnostically susceptible

*Chenopodium amaranticolor, C. quinoa* — local lesions; systemic chlorosis.
*Tetragonia expansa* — chlorotic local lesions; not systemic.
*Nicandra physalodes, Nicotiana clevelandii, Catharanthus roseus* — systemic chlorosis.

Diagnostically insusceptible

*Gomphrena globosa, Phaseolus vulgaris, Datura stramonium, Nicotiana glutinosa, Physalis floridana.*

Maintenance and propagation hosts

*Catharanthus roseus, Chenopodium quinoa, Nicandra physalodes, Nicotiana clevelandii, Theobroma cacao.*

Assay hosts (**L**ocal lesions or **W**hole plants)

*Chenopodium amaranticolor* (L), *C. quinoa* (L), *Tetragonia expansa* (L), *Nicotiana clevelandii* (W).

Dicotyledonous Hosts *(**Susceptible** or Insusceptible)*

***Adansonia digitata***. *Arachis hypogaea*. ***Beta vulgaris***. *Brassica campestris* ssp. *pekinensis*. *Calendula officinalis*. *Capsicum frutescens*. ***Catharanthus roseus***. ***Ceiba pentandra***. ***Chenopodium amaranticolor***. ***Chenopodium quinoa***. ***Cucumis melo***. *Cucumis sativus*. ***Cucurbita pepo***. *Datura stramonium*. *Daucus carota*. *Glycine max*. *Gomphrena globosa*. *Lycopersicon esculentum*. ***Nicotiana clevelandii***. *Nicotiana glutinosa*. ***Nicotiana tabacum***. *Petunia hybrida*. *Phaseolus lunatus*. *Phaseolus vulgaris*. *Phlox drummondii*. *Physalis floridana*. *Physalis peruviana*. *Raphanus sativus*. *Solanum melongena*. ***Tetragonia expansa***. ***Theobroma cacao***. *Trifolium incarnatum*. ***Vigna unguiculata***. *Zinnia elegans*.

Monocotyledonous Hosts *(**Susceptible** or Insusceptible)*

*Avena sativa*. *Dactylis glomerata*. *Hordeum vulgare*. *Lolium perenne*. *Zea mays*.

Sources of host-range data

Brunt *et al.* (1965).

**Properties of particles in sap**

TIP: 60–65 °C. LIV: 16–32 days. DEP: $\log_{10}$ minus 4. Leaf sap contains many virions.

**Purification**

Method: Brunt *et al.* (1965).

**Particle morphology**

Virions isometric; not enveloped; 28 nm in diameter; angular in profile; with a conspicuous capsomere arrangement.

**Physical properties**

Two sedimenting components in purified preparations; sedimentation coefficient 108 *S* ((B)); of the other(s) 49 *S* ((T)). Isoelectric point pH 3.6.

**Biochemical properties**

Virions contain 38 % nucleic acid (B); 62 % protein (B).

*Genome* consists of RNA; single–stranded; linear; unipartite; largest (or only) genome part 6 kb. Base composition 16 % G; 22 % A; 33 % C; 29 % U.

*Sequence database accession code(s):*

X54354 Em(40)_vi:CYMVP Gb(84)_vi:CYMVP Cacao yellow mosaic virion protein gene. 9/93 681bp. 1 sequence.

*Virion protein(s)* one; $M_r$ 19788; coat.

**Cytopathology**

Virions found in cytoplasm (as paracrystals), or in nuclei (but only empty virions). Other cellular changes: small regular vesicles formed by invagination of the outer chloroplast membrane. Mitochondria swollen.

**Taxonomy and Relationships**

*Tymovirus*.

Virus(es) with serologically related virions: clitoria yellow vein, desmodium yellow mottle, and okra mosaic viruses are related closely; scrophularia mottle, Andean potato latent and eggplant mosaic viruses are distantly related. Virus(es) with serologically unrelated virions: dulcamara mottle, ononis yellow mosaic, belladonna mottle, turnip yellow mosaic and wild cucumber mosaic viruses.

**References**

Blencowe, J.W., Brunt, A.A., Kenten, R.H. and Lovi, N.K. (1963). *Trop. Agric., Trin.* **40**: 233.

Brunt, A.A. (1970). *CMI/AAB Descr. Pl. Viruses* No. 11, 4 pp.

Brunt, A.A., Kenten, R.H., Gibbs, A.J. and Nixon, H.L. (1965). *J. gen. Microbiol.* **38**: 81.

Koenig, R. (1976). *Virology* **72**: 1.

Lesemann, D.-E. (1977). *Phytopath. Z.* **90**: 315.

# Cactus 2 *carlavirus*

**Data collated by A.A. Brunt, 1989.**

*Standard acronym:* CV-2.

From the Federal Republic of Germany; by Brandes and Wetter (1959).

**Natural host range and symptoms**

Cultivated cacti — symptomless.

**Transmission**

Virus transmitted by mechanical inoculation.

**Geographical distribution**

Found, but with no evidence of spread, in Germany.

**Experimental host range**

Few (<3) families susceptible. Experimentally infected plants mostly show no symptoms.

Diagnostically susceptible

*Chenopodium quinoa* — symptomless systemic infection.

Maintenance and propagation hosts

*Chenopodium quinoa.*

Assay hosts (**L**ocal lesions or **W**hole plants)

*Chenopodium quinoa* (W).

Dicotyledonous Hosts *(**Susceptible** or Insusceptible)*

***Chenopodium quinoa.***

**Particle morphology**

Virions filamentous; not enveloped; usually straight; with a clear modal length; of 650 nm.

**Taxonomy and Relationships**

*Carlavirus.*

Virus(es) with serologically related virions: passiflora latent, potato S, potato M, chrysanthemum B, red clover vein mosaic and carnation latent viruses.

**References**

Brandes, J. and Wetter, C. (1959). *Virology* **8**: 99.

Brandes, J. and Wetter, C. (1963/64). *Phytopath. Z.* **49**: 61.

# Cactus X *potexvirus*

**Data collated by R. Koenig, 1987.**

*Synonyms:* barrel cactus virus (Attathom *et al*, 1978; Milne, 1988). *Standard acronym:* CVX. *Strains:* Chessin, Miličič, Braunschweig.

First reported in Cacti; from Germany; by Amelunxen (1958).

**Natural host range and symptoms**

Symptoms none.

*Opuntia monacantha Austrocylindropuntia cylindrica, Pereskia saccarosa, Schlumbergera bridgesii, Epiphyllum* spp., *Cereus* spp., *Echinopsis* spp., *Zygocactus* spp., *Ferocactus acanthodes, Echinocereus procumbens* — usually infected symptomlessly.

**Transmission**

Virus transmitted by mechanical inoculation; transmitted by grafting; transmitted by contact between plants; not transmitted by seed.

**Geographical distribution**

Probably distributed worldwide (in cultivated cacti). Spreads in the North American region; the USA.

**Experimental host range**

Several (3–9) families susceptible. Experimentally infected plants mostly show necrotic local lesions, mottle.

Diagnostically susceptible

*Amaranthus caudatus* — necrotic local lesions; rarely systemic.

*Gomphrena globosa, Chenopodium amaranticolor* — necrotic local lesions; not systemic.

*C. quinoa* — systemic mottle.

Diagnostically insusceptible

*Nicotiana tabacum, Phaseolus* spp., *Cucumis sativus.*

Maintenance and propagation hosts

*Chenopodium quinoa.*

Assay hosts (**L**ocal lesions or **W**hole plants)

*Gomphrena globosa* (L), *Chenopodium quinoa* (W).

Dicotyledonous Hosts *(**Susceptible** or Insusceptible)*

***Amaranthus caudatus***. ***Amaranthus retroflexus***. ***Beta vulgaris***. ***Celosia argentea***. *Chenopodium album*. *Chenopodium amaranticolor*. ***Chenopodium foetidum***. ***Chenopodium murale***. ***Chenopodium quinoa***. *Cucumis sativus*. ***Gomphrena globosa***. *Nicotiana tabacum*. ***Ocimum basilicum***. ***Opuntia vulgaris***. *Phaseolus vulgaris*.

Sources of host-range data

Plese and Miličič (1966).

**Properties of particles in sap**

TIP: 82 °C. LIV: 28 days (or longer). DEP: $\log_{10}$ minus 5. Infectivity of sap not changed by treatment with di–ethyl ether. Leaf sap contains many virions.

**Purification**

Method: Koenig and Lesemann (1983).

**Particle morphology**

Virions filamentous; not enveloped; usually flexuous; with a clear modal length; of 520 nm; 13 nm wide. Axial canal obscure.

**Physical properties**

One sedimenting component in purified preparations; sedimentation coefficient 120 *S*.

**Biochemical properties**

Virions contain 5 % nucleic acid; 95 % protein; 0 % lipid.

*Genome* consists of RNA; single–stranded; linear; unipartite; largest (or only) genome part 6.5 kb. Genomic nucleic acid isolated by Attathom *et al.* (1978); Koenig (1971). Base composition 20.8 % G; 26.4 % A; 29.2 % C; 23.6 % U. Infectivity retained when deproteinised with phenol or detergent.

*Virion protein(s)* one; $M_r$ 20000. Method of preparation: Attathom *et al.* (1978). Amino acid composition: Attathom *et al.* (1978).

**Replication**

Replication does not depend on a helper virus.

**Cytopathology**

Virions found in all parts of the host plant; usually in cytoplasm, or in cell vacuoles (occasionally). Inclusions cytoplasmic present in infected cells; are unusual in shape; banded bodies or paracrystalline structures; they contain virions.

**Taxonomy and Relationships**

*Potexvirus*.

Virus(es) with serologically unrelated virions: zygocactus X potexvirus.

**References**


Amelunxen, F. (1958). *Protoplasma* **49**: 140.
Attathom, S., Weathers, L.G. and Gumpf, D.J. (1978). *Pl. Dis. Reptr.* **62**: 228.
Attathom, S., Weathers, L.G. and Gumpf, D.J. (1978). *Phytopathology* **68**: 1401.
Bercks, R. (1971). *CMI/AAB Descr. Pl. Viruses* No. 58, 3 pp.
Koenig, R. (1978). *J. gen. Virol.* **40**: 309.
Koenig, R. and Lesemann, D.-E. (1983). *CMI/AAB Descr. Pl. Viruses* No. 265, 3 pp.
Milne, R.G. (1988). In: *The Plant Viruses*, Vol. 4; *The Filamentous Viruses*, p.6; ed Milne, R.G., Plenum, New York.
Plese, N. and Miličič, D. (1966). *Phytopath. Z.* **55**: 197
Ready, K.F.M. and Bancroft, J.B. (1985). *Virology* **141**: 302.


# Callistephus chinensis chlorosis (?) rhabdovirus

**Data collated by A.A. Brunt, 1991.**

*Synonyms:* Callistephus chinensis chlorosis virus (Francki *et al.*, 1985). *Standard acronym:* CCCV.

**Natural host range and symptoms**

*Callistephus chinensis* — leaf chlorosis.

**Geographical distribution**

Spreads in the South and Central American region; Brazil.

**Particle morphology**

Virions rhabdo– or bullet–shaped; with a clear modal length; of 200–300 nm; 60–80 nm wide.

**Taxonomy and Relationships**

Unassigned rhabdovirus.

**References**


Francki, R.I.B., Milne, R.G. and Hatta, T. (1985). In: *Atlas of Plant Viruses*, Vol. 1, p. 73. CRC Press, Boca Raton, Florida.
Kitajima, E.W. and Costa, A.S. (1979). *Fitopatol. Bras.* **4**: 55.


# Camellia yellow mottle (?) varicosavirus

**Data collated by C. Hiruki, 1991.**

*Synonyms:* camellia leaf yellow mottle virus, camellia infectious variegation virus (Milbrath and McWhorter, 1946; Gailhofer *et al.*, 1988), camellia colour-breaking virus (Gailhofer *et al.*, 1988), camellia variegation virus.

First reported in *Camellia* spp.; from the U.S.A; by Milbrath and McWhorter (1946); Plakidas (1954).

**Natural host range and symptoms**

Symptoms persist, or vary seasonally, or disappear soon after infection. Symptoms chlorotic leaf flecks and flower breaking.

*Camellia japonica, C. sasanqua* — chlorotic leaf flecks and flower breaking.

**Transmission**

Virus not transmitted by mechanical inoculation; transmitted by grafting; not transmitted by contact between plants; not transmitted by pollen.

**Geographical distribution**

Probably distributed worldwide (wherever camellias are grown).

**Experimental host range**

Few (<3) families susceptible. Experimentally infected plants mostly show chlorotic flecks in leaves and flower breaking.

Diagnostically susceptible

*Camellia japonica, C. sasanqua* — chlorotic flecks and mottle, flower breaking.

Maintenance and propagation hosts

*Camellia japonica, C. sasanqua.*

Assay hosts (**L**ocal lesions or **W**hole plants)

*Camellia japonica* (W), *C. sasanqua* (W).

**Properties of particles in sap**

Leaf sap contains few virions.

**Particle morphology**

Virions rod–shaped; of 120–170 nm; 30 nm wide. Axial canal obvious; 10 nm in diameter. Pitch of basic helix 10 nm.

**Cytopathology**

Virions found in leaves, mesophyll, petals and stamens; in cytoplasm.

**Taxonomy and Relationships**

Possible varicosavirus.

**References**

Gailhofer, M., Thaler, I. and Miličič, D. (1988). *Acta Hort.* **234**: 385.
Hiruki, C. (1985). *Acta Hort.* **164**: 55.
Milbrath, J.A. and McWhorter, F.P. (1946). *Am. Camellia Soc. Yrbk.*, p. 51.
Miličič, D., Thaler, I. and Gailhofer, M. (1986). *Acta bot. Croat.* **45**: 1.
Miličič, D., Thaler, I. and Gailhofer, M. (1988). *Phytopathol. Medit.* **27**: 119.
Plakidas, A.G. (1954). *Phytopathology* **44**: 14.
Plakidas, A.G. (1962). *Phytopathology* **52**: 77.

# Canary reed mosaic (?) potyvirus

**Data collated by A.A. Brunt, 1991.**

*Standard acronym:* CRMV.

First reported in *Phalaris arundinacea*; from Japan; by Toriyama and Yora (1972).

**Natural host range and symptoms**

Symptoms persist.
*Phalaris arundinacea* — mosaic symptoms in leaves.

**Transmission**

Transmitted by a vector; an insect; *Myzus persicae*; Aphididae. Transmitted in a non–persistent manner. Virus transmitted by mechanical inoculation.

**Geographical distribution**

Spreads in Japan.

**Experimental host range**

Few (<3) families susceptible.

Diagnostically susceptible

*Lolium multiflorum, Phalaris arundinacea* — mosaic leaf symptoms.

Maintenance and propagation hosts

*Lolium multiflorum, Phalaris arundinacea.*

Assay hosts (Local lesions or Whole plants)

*Lolium multiflorum* (W), *Phalaris arundinacea* (W).

**Properties of particles in sap**

TIP: 50 °C. LIV: 1 days. DEP: $\log_{10}$ minus 5.

**Particle morphology**

Virions filamentous; not enveloped; usually straight; with no clear modal length; of 700–800 nm.

**Taxonomy and Relationships**

Unassigned potyvirus.

**References**

Toriyama, S. and Yora, K. (1972). In: *Virus Diseases of Wild Grasses and Cereal Crops in Japan*. University of Tokyo Press, Tokyo, Japan.

# Canavalia maritima mosaic (?) potyvirus

**Data collated by K. Boswell, 1985.**

*Standard acronym:* CnMMV.

First reported in *Canavalia maritima*; from Puerto Rico; by Rodriguez *et al.* (1975).

**Natural host range and symptoms**

*Canavalia maritima* — mosaic, mottling, vein banding and leaf malformation.

**Transmission**

Transmitted by a vector; an insect; *Aphis craccivora, A. gossypii, Myzus persicae*; Aphididae. Not transmitted by *Dactynotus ambrosiae*. Virus transmitted by mechanical inoculation.

**Geographical distribution**

Spreads in Puerto Rico.

**Experimental host range**

Few (<3) families susceptible. Experimentally infected plants mostly show mosaic.

Diagnostically susceptible

*Canavalia maritima* — systemic mosaic, mottling and leaf malformation.
*Phaseolus vulgaris* cvs Bountiful, Top Crop, *Vigna unguiculata* cv. Early Ramshorn.

Diagnostically insusceptible

*Phaseolus vulgaris* cv. Green Refugee, *Vigna unguiculata* cv. Blackeye.

Dicotyledonous Hosts *(**Susceptible** or Insusceptible)*

*Capsicum annuum*. *Cassia occidentalis*. ***Chenopodium album***. ***Chenopodium amaranticolor***. ***Chenopodium quinoa***. *Cucumis sativus*. *Datura stramonium*. ***Glycine max***. *Gomphrena globosa*. *Medicago sativa*. *Nicotiana glutinosa*. *Nicotiana tabacum*. ***Phaseolus lunatus***. ***Phaseolus vulgaris***, or *Phaseolus vulgaris*. *Pisum sativum*. ***Spinacia oleracea***. *Trifolium pratense*. *Trifolium repens*. ***Vicia faba***. ***Vigna angularis***. ***Vigna radiata***. ***Vigna unguiculata***, or *Vigna unguiculata*.

Sources of host-range data

Rodriguez *et al.* (1975).

**Properties of particles in sap**

TIP: 55–60 °C. DEP: $\log_{10}$ minus 3. Leaf sap contains many virions.

**Particle morphology**

Virions filamentous; not enveloped.

**Cytopathology**

Virions found in all parts of the host plant; in cytoplasm. Inclusions present in infected cells; are pinwheels.

**Taxonomy and Relationships**

Unassigned potyvirus.

Virus(es) with serologically unrelated virions: bean yellow mosaic and bean common mosaic viruses.

**References**

Rodriguez, R.L., Bird, J., Monllor, A.C., Waterworth, H.E., Kimura, M. and Maramorosch, K. (1975). In: *Tropical Diseases of Legumes*, eds J. Bird and K. Maramorosch. Academic Press, New York.

## Canna yellow mottle *badnavirus*

**Data collated by A.A. Brunt and S. Yamashita, 1987.**

*Synonyms:* canna mottle virus. *Standard acronym:* CaYMV.

First reported in *Canna* spp.; from Ibaragi, Japan; by Yamashita *et al.* (1979).

**Natural host range and symptoms**

Symptoms persist.

*Canna* spp. — veinal yellowing or necrosis and mottling.

**Geographical distribution**

Spreads in Japan and the USA.

**Experimental host range**

Experimentally infected plants mostly show chlorosis, mottle.

Diagnostically susceptible

*Canna* spp. — systemic veinal mottling and chlorosis.

Maintenance and propagation hosts

*Canna* spp.

**Properties of particles in sap**

Leaf sap contains few virions.

### Particle morphology

Virions bacilliform; not enveloped; 120–130 nm in length; 28 nm wide.

### Cytopathology

Virions found in vascular parenchyma; in cytoplasm. Inclusions present in infected cells; are membranous bodies; they contain virions.

### Taxonomy and Relationships

Probably *Badnavirus*.

### References

Lockhart, B.E.L. (1988). *Acta Hort.* **234**: 72.

Yamashita, S., Natsuaki, T., Doi, Y. and Yora, K. (1979). *Ann. Phytopath. Soc. Japan* **45**: 85.

Yamashita, S., Natsuaki, T., Doi, Y. and Yora, K. (1985). *Ann. Phytopath. Soc. Japan* **51**: 642.

## Caper latent *carlavirus*

**Data collated by D. Gallitelli, 1991.**

*Synonyms:* caper vein banding virus. *Standard acronym:* CapLV.

First reported in *Capparis spinosa*; from Apulia, southern Italy; by Gallitelli and Di Franco (1987).

### Natural host range and symptoms

Symptoms persist. Symptoms vein yellowing.

*Capparis spinosa* — vein yellowing in plants also infected with caper vein yellowing rhabdovirus (Di Franco and Gallitelli, 1985).

### Transmission

Virus transmitted by mechanical inoculation; transmitted by grafting.

### Geographical distribution

Spreads in Italy (Apulia).

### Experimental host range

Several (3–9) families susceptible. Experimentally infected plants mostly show chlorotic or necrotic local lesions, systemic mosaic and leaf malformation.

Diagnostically susceptible

*Chenopodium amaranticolor* — pinpoint local lesions.
*C. quinoa, Nicotiana glutinosa* — chlorotic local lesions.
*Capparis spinosa* var. *rupestris* — latent.
*Phaseolus aureus* — red local lesions.
*P. vulgaris* — necrotic local lesions.
*Petunia hybrida* — chlorotic ringspots.
*Nicotiana clevelandii, N. megalosiphon, N. benthamiana* — systemic mosaic.

Diagnostically insusceptible

*Nicotiana tabacum* cv. Samsun, White Burley, *Gomphrena globosa, Celosia cristata.*

Maintenance and propagation hosts

*Nicotiana benthamiana, N. clevelandii.*

Assay hosts (**L**ocal lesions or **W**hole plants)

*Chenopodium amaranticolor* (L), *C. quinoa* (L), *Phaseolus vulgaris* (L), *Nicotiana megalosiphon* (W), *N. benthamiana* (W).

Dicotyledonous Hosts *(**Susceptible** or Insusceptible)*

*Celosia cristata*. ***Chenopodium amaranticolor***. ***Chenopodium quinoa***. *Gomphrena globosa*. ***Nicotiana benthamiana***. ***Nicotiana clevelandii***. ***Nicotiana glutinosa***. ***Nicotiana megalosiphon***. *Nicotiana rustica*. *Nicotiana tabacum*. ***Petunia hybrida***. ***Phaseolus vulgaris***. ***Vigna radiata***.

**Properties of particles in sap**

Leaf sap contains many virions.

**Purification**

Method: sap-clarification with Triton X-100, differential centrifugation, centrifugation through a 30% sucrose or 49% CsCl.

**Particle morphology**

Virions filamentous; not enveloped; usually flexuous; of 662 nm. Axial canal obscure.

**Physical properties**

One sedimenting component in purified preparations; sedimentation coefficient 168 *S*. Density 1.31 g $cm^{-3}$ in CsCl.

**Biochemical properties**

*Genome* consists of RNA; single–stranded; linear; unipartite; largest (or only) genome part 8 kb. Genomic nucleic acid isolated by Gallitelli: phenol SDS.

*Non–genomic nucleic acid* not found in the virions.

*Virion protein(s)* one; $M_r$ 35000; coat protein. Method of preparation: Laemmli (1970).

**Replication**

Replication does not depend on a helper virus.

**Cytopathology**

Virions found in leaves and mesophyll; in cytoplasm. Inclusions present in infected cells; are membranous bodies; they contain virions.

**Taxonomy and Relationships**

*Carlavirus*.

Virus(es) with serologically related virions: helenium S virus. Virus(es) with serologically unrelated virions: carnation latent, poplar mosaic, potato M, potato S, shallot latent, chrysanthemum B and pea streak viruses.

Best tests for diagnosis: inoculation onto *Capparis spinosa*.

**References**

Di Franco, A. and Gallitelli, D. (1985). *Phytopathol. Medit.* **24**: 234.

Gallitelli, D. and Di Franco, A. (1987). *J. Phytopath.* **119**: 97.

Laemmli, U.K. (1970). *Nature, Lond.* **227**: 680.

# Caper vein yellowing (?) *nucleorhabdovirus*

**Data collated by A.A. Brunt, 1991.**

First reported in *Capparis spinosa* cv. rupestris; from Italy; by Di Franco and Gallitelli (1985).

**Natural host range and symptoms**

*Capparis spinosa* — vein yellowing. Infected plants often also contain caper latent (?) *carlavirus*.

**Geographical distribution**

Spreads in Italy.

**Particle morphology**

Virions rhabdo- or bullet-shaped; of *in vivo* 200–300 nm; 80–90 nm wide.

**Cytopathology**

Virions found in nuclear envelope of mesophyll, vascular parenchyma and sieve tubes; in cytoplasm and in the perinuclear space.

**Taxonomy and Relationships**

Possible *Nucleorhabdovirus: Rhabdoviridae.*

**References**

Di Franco, A.P. and Gallitelli, D. (1985). *Phytopathol. Medit.* **24**: 234.

# Caraway latent (?) *nepovirus*

**Data collated by P. van Dijk and L. Bos, 1991.**

*Standard acronym:* CawLV.

First reported in *Carum carvi*; from the Netherlands; by Van Dijk and Bos (1989).

**Natural host range and symptoms**

Symptoms usually none.
*Carum carvi* — symptomless.

**Transmission**

Not transmitted by *Cavariella aegopodii.* Virus transmitted by mechanical inoculation.

**Geographical distribution**

Found, but with no evidence of spread, in the Netherlands.

**Experimental host range**

Many (>9) families susceptible. Experimentally infected plants mostly show chlorotic or necrotic local lesions followed by chlorotic and often necrotic systemic symptoms.

Diagnostically susceptible

*Celosia argentea, Chenopodium amaranticolor, C. quinoa* — local and systemic necrosis.

*Nicotiana clevelandii* — necrotic local lesions and systemic etching and chlorosis.

*Cyamopsis tetragonoloba, Phaseolus vulgaris, Pisum sativum* — necrotic local lesions without systemic symptoms.

*Gomphrena globosa* — chlorotic /necrotic local lesions and chlorotic systemic symptoms.

*Nicotiana glutinosa* — chlorotic local lesions without systemic symptoms.

*Apium graveolens* var. *rapaceum* — systemic vein mosaic followed by latent infection.

Diagnostically insusceptible

*Catharanthus roseus, Vicia faba, Vigna unguiculata, Pimpinella anisum.*

Maintenance and propagation hosts

*Chenopodium quinoa, Nicotiana benthamiana, N. clevelandii.*

Assay hosts (**L**ocal lesions or **W**hole plants)

*Chenopodium quinoa* (W), *Cyamopsis tetragonoloba* (L), *Nicotiana clevelandii* (W), *Phaseolus vulgaris* (L).

Dicotyledonous Hosts *(**Susceptible** or Insusceptible)*

***Apium graveolens**. **Beta vulgaris**. Catharanthus roseus. **Celosia argentea**. **Chenopodium amaranticolor**. **Chenopodium quinoa**. **Cucumis sativus**. **Cyamopsis tetragonoloba**. **Datura stramonium**. **Gomphrena globosa**. **Nicotiana benthamiana**. **Nicotiana clevelandii**. **Nicotiana glutinosa**. **Nicotiana rustica**. **Phaseolus vulgaris**. **Pisum sativum**. Vicia faba. Vigna unguiculata. **Zinnia elegans**.*

Additional hosts

***Ammi majus, Carum carvi**, Catharanthus roseus, Pimpinella anisum.*

Sources of host-range data

Van Dijk and Bos (1989).

**Properties of particles in sap**

LIV: 9–13 days. DEP: $\log_{10}$ minus 3–4. Leaf sap contains many virions.

**Particle morphology**

Virions isometric; not enveloped; 25 nm in diameter.

**Replication**

Replication does not depend on a helper virus.

**Cytopathology**

Virions found in leaves.

**Taxonomy and Relationships**

Possible *Nepovirus: Comoviridae.*

Best tests for diagnosis: electron microscopy, serology, inoculation to *Chenopodium quinoa* and *C. amaranticolor* and estimation of longevity *in vitro* to distinguish caraway latent virus from heracleum latent, cow parsnip mosaic, tobacco streak, parsnip leafcurl, arabis mosaic and HV3 viruses.

**References**

Van Dijk, P. and Bos, L. (1989). *Neth. J. Pl. Path.* **95**: 34.

# Cardamine chlorotic fleck *carmovirus*

**Data collated by A.A. Brunt, 1992 and M.L. Skotnicki, 1993.**

First reported in *Cardamine lilacina*; from Australia; by Skotnicki *et al.* (1992, 1993).

**Natural host range and symptoms**

*Cardamine lilacina* — chlorotic leaf flecking.

**Transmission**

Virus transmitted by mechanical inoculation.

**Geographical distribution**

Spreads in Australia.

Diagnostically susceptible

*Arabidopsis thaliana, Brassica rapa, Cardamine lilacina, Iberis coronaria* — chlorotic leaf flecking.

Diagnostically insusceptible

*Brassica campestris* ssp. *pekinensis*, *Phaseolus vulgaris* and *Pisum sativum*.

**Purification**

Method: Skotnicki *et al.* (1992).

**Particle morphology**

Virions isometric; not enveloped; 30 nm in diameter.

**Biochemical properties**

*Genome* consists of RNA; single–stranded; unipartite; largest (or only) genome part 4.1 kb.

*Sequence database accession code(s):*

L16015 Em(40)_vi:CDKCG Gb(84)_vi:CDKCG Cardamine chlorotic fleck carmovirus genome, including replicase, coat protein, p8 protein, 1 sequence.

**Taxonomy and Relationships**

*Carmovirus: Tombusviridae.*

Virus(es) with serologically unrelated virions: turnip crinkle virus.

The viral RNA shows significant sequence homology with those of turnip crinkle and carnation mottle viruses but not with those of cymbidium ringspot, tomato bushy stunt, turnip yellow mosaic and southern bean mosaic viruses.

**References**

Skotnicki, M.L., Mackenzie, A.M., Torronen, M., Brunt, A.A. and Gibbs, A.J. (1992). *Australas. Pl. Path.* **21**: 120;

Skotnicki, M.L., Mackenzie, A.M., Torronen, M. and Gibbs, A.J. (1993). *J. gen. Virol.* **74**: 1933.

# Cardamine latent (?) *carlavirus*

**Data collated by P.L. Guy, 1987.**

*Standard acronym:* CaLV.

First reported in *Cardamine* spp.; from the Snowy Mountains, NSW, Australia; by Guy and Gibbs (1985).

**Natural host range and symptoms**

*Cardamine lilacina* ssp. B and C — latent.

**Transmission**

Virus transmitted by mechanical inoculation.

**Geographical distribution**

Spreads in Australia (southern New South Wales, but probably not in Tasmania).

**Experimental host range**

Few (<3) families susceptible. Experimentally infected plants mostly show local lesions vein-clearing.

Diagnostically susceptible

*Chenopodium quinoa* — large local lesions; not systemic.
*Sinapis alba* cv. White Mustard — systemic vein clearing with leaves malformed.

Diagnostically insusceptible

*Brassica campestris* ssp. *pekinensis*.

Maintenance and propagation hosts

*Cardamine lilacina* spp. B and C.

Assay hosts (**L**ocal lesions or **W**hole plants)

*Chenopodium quinoa* (L); *Sinapis alba* (W).

Dicotyledonous Hosts (***Susceptible*** *or Insusceptible)*

*Antirrhinum majus. Arachis hypogaea. Bellis perennis. Brassica campestris* ssp. *pekinensis. Catalpa bignonioides.* ***Chenopodium quinoa****. Clarkia amoena. Cucumis sativus. Delphinium hybridum. Eucalyptus cloesiana. Euphorbia marginata. Fagopyrum esculentum. Gomphrena globosa. Hibiscus esculentus. Nicotiana clevelandii. Phaseolus vulgaris.* ***Sinapis alba****.*

**Properties of particles in sap**

Leaf sap contains few virions.

**Particle morphology**

Virions filamentous; not enveloped; usually straight; with a clear modal length; of 650 nm. Axial canal obscure. Basic helix obscure.

**Taxonomy and Relationships**

Possible *Carlavirus*.
Virus(es) with serologically related virions: lily symptomless virus.

**References**

Guy, P.L. and Gibbs, A.J. (1985). *Australas. Pl. Path.* **34**: 532.

# Cardamom mosaic *potyvirus*

**Data collated by D. Gonsalves, 1986.**

*Standard acronym:* CdMV.

First reported in *Elettaria cardamomum*; from U.S.A.; by Gonsalves *et al.* (1986).

**Natural host range and symptoms**

Symptoms persist.
*Elettaria cardamomum* — mosaic, possibly the same as the 'Katte' or marble disease described in India.

**Transmission**

Transmitted by a vector; an insect; *Pentalonia nigronervosa*; Aphididae. Transmitted in a non–persistent manner. Virus not transmitted by mechanical inoculation; transmitted by grafting; not transmitted by contact between plants; not transmitted by seed.

**Geographical distribution**

Spreads in Guatemala and India (possibly).

**Experimental host range**

Few (<3) families susceptible. Experimentally infected plants mostly show mosaic.

Diagnostically susceptible

*Elettaria cardamomum* — systemic mosaic.

Maintenance and propagation hosts

*Elettaria cardamomum.*

**Purification**

Method: Gonsalves *et al.* (1986).

**Particle morphology**

Virions filamentous; not enveloped; usually flexuous; of 700–720 nm. Basic helix obscure.

**Biochemical properties**

*Genome* consists of RNA; single–stranded; linear; unipartite.

$M_r$ 37500 (also fragments of 32200 and 29000); capsid. Method of preparation: Gonsalves *et al.* (1986).

**Cytopathology**

Virions found in all parts of the host plant; in cytoplasm. Inclusions present in infected cells; are pinwheels.

**Taxonomy and Relationships**

*Potyvirus: Potyviridae.*

Virus(es) with serologically related virions: zucchini yellow mosaic, papaya ringspot and several legume potyviruses.

**References**

Gonsalves, D., Trujillo, E. and Hoch, H.C. (1986). *Plant Dis.* **70**: 65.

# Carnation 1 *alphacryptovirus*

**Data collated by R.G. Milne, V. Lisa and C. Marzachí, 1991.**

*Standard acronym:* CCV-1.

First reported in *Dianthus caryophyllus*; from Italy; by Lisa *et al.* (1980); Lisa *et al.* (1981*a*, *b*).

**Natural host range and symptoms**

Symptoms none.

*Dianthus caryophyllus* — symptomless.

**Transmission**

Transmitted by means not involving a vector. Virus not transmitted by mechanical inoculation; not transmitted by grafting; not transmitted by contact between plants; transmitted by seed (probably, and also in cuttings).

**Geographical distribution**

Probably distributed worldwide (wherever carnations are grown).

**Experimental host range**

Few (<3) families susceptible.

Maintenance and propagation hosts

*Dianthus caryophyllus.*

Dicotyledonous Hosts (***Susceptible*** *or Insusceptible)*

***Dianthus caryophyllus.***

**Properties of particles in sap**

Leaf sap contains few virions. Electron microscopy: ISEM effective.

**Purification**

Method: Lisa *et al.* (1981). Grind leaves in 0.5M phosphate buffer, pH 7, containing 0.02 M $Na_2SO_3$, 0.005 M EDTA and 0.01 M DIECA. Shake with equal volume of $CHCl_3$, separate the aqueous phase by low speed centrifugation, and pellet the virus at 78000 x g for 150 mins. Resuspend, band in a CsCl density gradient.

**Particle morphology**

Virions isometric; not enveloped; 29 nm in diameter; rounded in profile; without a conspicuous capsomere arrangement.

**Physical properties**

One sedimenting component in purified preparations. Density 1.337 g $cm^{-3}$ in CsCl.

**Biochemical properties**

*Genome* consists of RNA; double–stranded; linear; of three parts. Genomic nucleic acid isolated by Lisa *et al.* (1981). Additional factor Marzachí *et al.* (1988).

**Replication**

Replication does not depend on a helper virus.

**Cytopathology**

Virions found in leaves, roots, mesophyll and probably all parts of the host plant. Inclusions absent from infected cells.

**Taxonomy and Relationships**

*Alphacryptovirus: Partitiviridae.*

Virus(es) with serologically unrelated virions: white clover 1 and 2, beet 1 and 2, ryegrass, all three hop trefoil viruses and poinsettia cryptoviruses.

No named or clearly identified strains, but one isolate from Israel had serologically related, but distinct, virions. Best tests for diagnosis: the dsRNA pattern is diagnostic. The small isometric virions can be confused with those of carnation mottle virus, especially its attenuated forms; however, serology, especially ISEM plus decoration, can distinguish the two and identify mixtures. *N.B.* all antisera to other carnation viruses may contain antibodies to carnation cryptic virus, as most carnations are infected.

**References**

Boccardo, G., Lisa, V., Luisoni, E. and Milne, R.G. (1987). *Adv. Virus Res.* **32**: 171.

Francki, R.I.B., Milne, R.G. and Hatta, T. (1985). In: *An Atlas of Plant Viruses* Vol. 2. CRC Press, Boca Raton, Florida, U.S.A.

Lisa, V., Luisoni, E. and Milne, R.G. (1980). *Acta Hort.* **110**: 175.

Lisa, V., Boccardo, G. and Milne, R.G. (1981). *Virology* **115**: 410.

Lisa, V., Luisoni, E. and Milne, R.G. (1981). *Ann. appl. Biol.* **98**: 431.

Lisa, V., Luisoni, E. and Milne, R.G. (1986). *CMI/AAB Descr. Pl. Viruses* No. 315, 4 pp.

Marzachí, C., Milne, R.G. and Boccardo, G. (1988). *Virology* **165**: 115.

# Carnation 2 (?) *alphacryptovirus*

*Standard acronym:* CCV-2.

First reported in *Dianthus caryophyllus*; from Italy; by Boccardo *et al.* (1987).

**Particle morphology**

Virions isometric.

**Taxonomy and Relationships**

Possible *Alphacryptovirus: Partitiviridae.*

**References**

Boccardo, G., Lisa, V., Luisoni, E. and Milne, R.G. (1987). *Adv. Virus Res.* **32**: 171.

# Carnation etched ring *caulimovirus*

**Data collated by J. Hammond, R.H. Lawson and J.R. Moran, 1987.**

*Standard acronym:* CERV.

First reported in *Dianthus caryophyllus*; from England; by Hollings and Stone (1960).

**Natural host range and symptoms**

Symptoms vary seasonally.

*Dianthus caryophyllus* — often symptomless but sometimes necrotic flecks and lines.

**Transmission**

Transmitted by a vector; an insect; *Myzus persicae*; Aphididae. Transmitted in a semi–persistent manner. Virus lost by the vector when it moults; does not multiply in the vector; not transmitted congenitally to the progeny of the vector; transmitted by mechanical inoculation; transmitted by grafting; not transmitted by contact between plants; not transmitted by seed.

**Ecology and control**

Studies reported by Hakkaart and Jordanova (1968); it is more difficult to eliminate by heat therapy and meristem culture than other carnation viruses.

**Geographical distribution**

Probably distributed worldwide.

**Experimental host range**

Few (<3) families susceptible. Experimentally infected plants mostly show necrotic flecks, lines and chlorotic blotches.

Diagnostically susceptible

*Saponaria vaccaria* cv. Pink Beauty — red ringspots or necrotic flecks, spots, lines and rings.

*Dianthus caryophyllus* cv. Joker — typical "etched ring" symptoms.

*Silene armeria* — systemic necrotic lines and blotches.

Diagnostically insusceptible

*Chenopodium quinoa, Nicotiana tabacum.*

Maintenance and propagation hosts

*Dianthus caryophyllus, Saponaria vaccaria* (Hsu and Lawson, 1985).

Assay hosts (**L**ocal lesions or **W**hole plants)

No reliable host is known) *Saponaria vaccaria* cv. Pink beauty, can be used if plants are grown under controlled light and temperature, and flowering is prevented (Hearon and Lawson, 1981).

Dicotyledonous Hosts *(**Susceptible** or Insusceptible)*

*Chenopodium amaranticolor. Chenopodium quinoa.* ***Dianthus caryophyllus.*** *Nicotiana clevelandii. Nicotiana rustica. Nicotiana tabacum.* ***Saponaria vaccaria. Silene armeria.***

**Properties of particles in sap**

TIP: 80–85 °C. DEP: $\log_{10}$ minus 3–4. Leaf sap contains few virions. Electron microscopy: ISEM best using UA.

**Purification**

Method: Hull *et al.* (1976).

**Particle morphology**

Virions isometric; not enveloped; 45 nm in diameter; angular in profile; with a conspicuous capsomere arrangement.

**Physical properties**

One sedimenting component in purified preparations; sedimentation coefficient 206 *S*.

**Biochemical properties**

*Genome* consists of DNA; double–stranded; circular and linear (also super coiled); unipartite; largest (or only) genome part 7.932 kb (pairs). Genomic nucleic acid isolated by Lawson and Civerolo (1976, 1978); Hull and Donson (1982). Infectivity retained when deproteinised with proteases; retained when deproteinised with phenol or detergent. Genome has no *t*RNA–like activity.

*Sequence database accession code(s):*

X04658 Em(40)_vi:CERVDNA Gb(84)_vi:CERVDNA Carnation etched ring virus (CERV) DNA. 9/93 7,932bp. 1 sequence.

*Non–genomic nucleic acid* not found in the virions. Sub–genomic *m*RNA found in infected cells.

*Virus–coded non–virion proteins* Lawson and Civerolo (1978).

**Replication**

Replication does not depend on a helper virus.

**Cytopathology**

Virions found in all parts of the host plant; in cytoplasm and in nuclei. Inclusions present in infected cells; are viroplasms; they contain virions.

**Taxonomy and Relationships**

*Caulimovirus.*

Virus(es) with serologically related virions: cauliflower mosaic, sweet potato caulimo-like and dahlia mosaic viruses (Hollings and Stone, 1969).

Distant homologies in nucleic acid hybridization tests with other caulimoviruses (Hull and Donson, 1982; Donson and Hull, 1983). Best tests for diagnosis: difficult to detect in carnations.

**References**

Donson, J. and Hull, R. (1983). *J. gen. Virol.* **64**: 2281.
Hakkaart, F.A. and Jordanova, J. (1968). *Neth. J. Pl. Path.* **74**: 146.
Hearon, S.S. and Lawson, R.H. (1981). *Phytopathology* **71**: 645.
Hollings, M. and Stone, O.M. (1960). *Rep. Glasshouse Crops Res. Inst. 1960*, p. 94.
Hollings, M. and Stone, O.M. (1969). *Rep. Glasshouse Crops Res. Inst. 1968*, p. 102.
Hsu, H.T. and Lawson, R.H. (1985). *Phytopathology* **75**: 778.
Hull, R. and Donson, J. (1982). *J. gen. Virol.* **60**: 125.
Hull, R., Shepherd, R.J. and Harvey, J.D. (1976). *J. gen. Virol.* **31**: 9246.
Lawson, R.H. and Civerolo, E.L. (1976). *Acta Hort.* **59**: 49.
Lawson, R.H. and Civerolo, E.L. (1977). *CMI/AAB Descr. Pl. Viruses* No. 182, 4 pp.
Lawson, R.H. and Civerolo, E.L. (1978). *Phytopathology* **68**: 181.

# Carnation Italian ringspot *tombusvirus*

**Data collated by A.A. Brunt, 1995.**

*Synonyms:* tomato bushy stunt virus — carnation strain (Martelli *et al.* 1971). *Standard acronym:* CIRV.

First reported in *Dianthus caryophyllus*; from the U.K. in plants from Italy and the U.S.A.; by Hollings *et al.* (1970).

**Natural host range and symptoms**

*Dianthus caryophyllus* — transient chlorotic spots and rings in leaves and slight stunting.

**Transmission**

Virus transmitted by mechanical inoculation; transmitted by grafting.

**Geographical distribution**

Spreads in Germany, Italy, the UK, and the USA.

**Experimental host range**

Many (>9) families susceptible.

**Purification**

Method: Hollings *et al.* (1970).

**Particle morphology**

Virions isometric; 30 nm in diameter.

**Physical properties**

One sedimenting component in purified preparations; sedimentation coefficient 135 *S*. Density 1.36 g $cm^{-3}$ in CsCl (unfixed).

**Biochemical properties**

*Genome* consists of RNA; single–stranded; unipartite; largest (or only) genome part 4.7 kb.

*Sequence database accession code(s):*

X85215 Em(43)_vi:Cirv12345 Gb(89)_vi:Cirv12345 Carnation Italian ringspot virus genomic sequence for ORF1,2,3,4,5. 3/95 4,760bp.

*Virion protein(s)* one; $M_r$ 37000; coat protein.

### Cytopathology

Virions found in cytoplasm, in nuclei, and in mitochondria.

### Taxonomy and Relationships

*Tombusvirus: Tombusviridae.*

Virus(es) with serologically related virions: Tomato bushy stunt virus.

The viral genome has sequence homology with that of tomato bushy stunt virus.

### Comments

The virus occurs in some rivers in Germany.

### References

Buttner, C., Jacobi, V. and Koenig, R. (1987). *J. Phytopathology* **118**: 131.

Gallitelli, D., Hull, R. and Koenig, R. (1985). *J. gen. Virol.* **66**: 1523.

Hayes, R.J., Buck, K.W. and Brunt, A.A. (1984). *J. gen. Virol.* **65**: 1239.

Hollings, M., Stone, O.W. and Boutell, G. (1970). *Ann. appl. Biol.* **65**: 299.

Koenig, R. and Lesemann, D.-E. (1985). *Phytopath. Z.* **112**: 105.

Koenig, R. and Gibbs, A.J. (1986). *J. gen. Virol.* **67**: 75.

Martelli, G.P., Gallitelli, D. and Russo, M. (1988). In: *The Plant Viruses*; 3, *Polyhedral Virions with Monopartite RNA Genomes*, pp. 13–72. Plenum Press, New York.

Martelli, G.P., Quacquarelli, A. and Russo, M. (1971). *CMI/AAB Descr. Pl. Viruses* No. 69, 4pp.

# Carnation latent *carlavirus*

**Data collated by J.R. Moran, 1987.**

*Synonyms:* dulcamara A virus, dulcamara B virus (Phillips and Brunt, 1981). *Standard acronym:* CLV.

First reported in *Dianthus caryophyllus*; from England; by Kassanis (1955).

### Natural host range and symptoms

*Dianthus caryophyllus* — few or no symptoms.

### Transmission

Transmitted by a vector; an insect; *Myzus persicae*; Aphididae. Transmitted in a non–persistent manner. Virus transmitted by mechanical inoculation (but sometimes is difficult to transmit from carnation because its sap contains inhibitors of infection); not transmitted by contact between plants; not transmitted by seed.

### Geographical distribution

Probably distributed worldwide (wherever carnations are grown).

### Experimental host range

Several (3–9) families susceptible. Experimentally infected plants mostly show local lesions, mottle.

Diagnostically susceptible

*Chenopodium amaranticolor*, *C. quinoa* — small chlorotic local lesions; systemic mottle.

Maintenance and propagation hosts

*Dianthus barbatus*.

Assay hosts (**L**ocal lesions or **W**hole plants)

*Chenopodium amaranticolor* (L); *C. quinoa* (L).

Dicotyledonous Hosts *(**Susceptible** or Insusceptible)*

***Beta vulgaris*. *Chenopodium amaranticolor*. *Chenopodium quinoa*. *Dianthus barbatus*. *Dianthus caryophyllus*. *Gynura aurantiaca*. *Nicotiana clevelandii*.**

**Properties of particles in sap**

TIP: 60–65 °C. LIV: 2–3 days. DEP: $\log_{10}$ minus 3–4. Leaf sap contains few virions.

**Purification**

Method: Wetter and Paul (1961).

**Particle morphology**

Virions filamentous; not enveloped; usually straight (or slightly curved); with a clear modal length; of 650 nm; 12 nm wide. Axial canal obscure. Basic helix obscure; pitch of basic helix 3.3 nm.

**Physical properties**

One sedimenting component in purified preparations; sedimentation coefficient 167 *S*.

**Biochemical properties**

Virions contain 6 % nucleic acid.

*Genome* consists of RNA; single–stranded; unipartite. Genomic nucleic acid isolated by Wetter and Milne (1981). Nucleotide sequence references: Haylor *et al.* (1990). The nucleotide sequence of the 3' proximal region of the genomic RNA has been reported.

*Sequence database accession code(s):*

X55897 Em(40)_vi:CLVTUCRF Gb(84)_vi:CLVTUCRF Carnation latent virus (CLV) 3' terminal ORF. 9/93 393bp. 1 sequence.

*Virion protein(s)* one; $M_r$ 32000. Method of preparation: Paul (1973).

**Taxonomy and Relationships**

*Carlavirus* (type member).

Virus(es) with serologically related virions: cactus 2, chrysanthemum B, helenium S, hop latent, lilac mottle, lily symptomless, mulberry latent, muskmelon vein necrosis, nerine latent, passiflora latent, pea streak, potato M, potato S and red clover vein mosaic viruses. Virus(es) with serologically unrelated virions: narcissus latent, poplar mosaic, shallot latent and honeysuckle latent viruses.

Best tests for diagnosis: use ISEM decoration to distinguish carnation latent from carnation vein mottle virus.

**References**


Haylor, M.T., Brunt, A.A. and Coutts, R.H.A. (1990). *Nucl. Acids Res.* **18**: 6127.

Kassanis, B. (1954). *Nature, Lond.* **173**: 109.

Kassanis, B. (1955). *Ann. appl. Biol.* **43**: 103.

Kassanis, B. (1958). *Ann. appl. Biol.* **43**: 103.

Paul, H.L. (1974). *Phytopath. Z.* **80**: 330.

Phillips, S. and Brunt, A.A. (1981). *Rep. Glasshouse Crops Res. Inst. 1980*, p. 152.

Wetter, C. (1971). *CMI/AAB Descr. Pl. Viruses* No. 61, 3 pp.

Wetter, C. and Milne, R.G. (1981). In: *Handbook of Plant Virus Infections and Comparative Diagnosis*, p 695; ed. E. Kurstak. Elsevier, Amsterdam.

Wetter, C. and Paul, H.L. (1961). *Phytopath. Z.* **43**: 207.

# Carnation mottle *carmovirus*

**Data collated by J.H. Tremaine and J.R. Moran, 1985.**

*Standard acronym:* CarMV. *Strains:* PSR, PR4 and also attenuated forms.

First reported in *Dianthus* spp.; from the U.K.; by Kassanis (1955).

**Natural host range and symptoms**

*Dianthus caryophyllus, D. barbatus, Saponaria vaccaria* — mild mottle.

*Begonia elatior, Begonia* × *cheimantha, Daphne odorata, Dianthus chinensis, D. superbus, Saponaria officinalis* — mottle.

**Transmission**

Transmitted by means not involving a vector. Virus transmitted by mechanical inoculation; transmitted by grafting; transmitted by contact between plants; not transmitted by seed.

**Geographical distribution**

Probably distributed worldwide.

**Experimental host range**

Many (>9) families susceptible. Experimentally infected plants mostly show chlorotic and necrotic local lesions.

Diagnostically susceptible

*Chenopodium amaranticolor* — small chlorotic local lesions.

*Chenopodium quinoa, Atriplex hortensis* — chlorotic spots, systemic chlorotic flecks, spots and malformation.

*Gomphrena globosa* — fawn necrotic dots.

*Tetragonia expansa* — small chlorotic and/or necrotic local lesions.

Maintenance and propagation hosts

*Dianthus caryophyllus, Chenopodium quinoa.*

Assay hosts (**L**ocal lesions or **W**hole plants)

*Chenopodium amaranticolor* (L); *C. quinoa* (L).

Dicotyledonous Hosts (***Susceptible*** *or Insusceptible)*

*Ageratum conyzoides*. ***Amaranthus caudatus***. ***Antirrhinum majus***. *Apium graveolens* (var. *dulce*). ***Atriplex hortensis***. *Bellis perennis*. ***Beta vulgaris***. *Brassica campestris* ssp. *rapa*. *Brassica oleracea* ssp. *capitata*. *Calendula officinalis*. *Capsicum frutescens*. ***Catharanthus roseus***. ***Celosia argentea***. ***Chenopodium amaranticolor***. ***Chenopodium quinoa***. *Chrysanthemum morifolium*. *Crotalaria spectabilis*. ***Cucumis sativus***. *Cucurbita pepo*. *Dahlia pinnata*. *Datura stramonium*. ***Dianthus barbatus***. ***Dianthus caryophyllus***. *Emilia sagittata*. *Fagopyrum esculentum*. *Glycine max*. ***Gomphrena globosa***. *Helianthus annuus*. *Lactuca sativa*. *Lathyrus odoratus*. ***Lunaria annua***. ***Lycopersicon esculentum***. ***Medicago sativa***. *Myosotis sylvatica*. ***Nicandra physaloides***. ***Nicotiana clevelandii***. *Nicotiana glutinosa*. *Nicotiana tabacum*. ***Petunia hybrida***. *Phaseolus vulgaris*. *Physalis floridana*. ***Phytolacca americana***. *Pisum sativum*. ***Primula malacoides***. ***Saponaria vaccaria***. *Senecio vulgaris*. ***Silene armeria***. *Solanum melongena*. *Solanum nigrum*. *Solanum tuberosum* (*Solanum quercifolium* (**I**)). ***Spinacia oleracea***. ***Stellaria media***. ***Tetragonia expansa***. ***Torenia fournieri***. ***Trifolium repens***. *Tropaeolum majus*. *Vicia faba*. *Vigna unguiculata*. *Zinnia elegans*.

Monocotyledonous Hosts *(**Susceptible** or Insusceptible)*

*Avena sativa.*

Additional hosts

***Ajuga reptans, Amsonia tabernae-montana, Anagallis arvensis, Aquilegia coerulea, Araujia sericofera, Begonia semperflorens, Berberis darwinii, Borago officinalis, Bryonia dioica, Calandrinia grandiflora, Callistephus chinensis, Campanula persicifolia, Caryopteris x clandonensis, Clematis vitalba, Chrysanthemum parthenium, Digitalis purpurea, Dipsacus fullonum*** var. ***sylvestris, Freesia refracta, Fumaria officinalis, Genista hispanica, Glechoma hederacea, Impatiens balsamina, Limonium latifolium, Lonicera periclymenum, Lupinus polyphyllus, Lychnis chalcedonica, L. dioica, Lythrum salicaria, Mimulus luteus, Moricanda charantia, Mycelis muralis, Oxalis corniculata, Papaver rhoeas, Passiflora coerulea, Plantago major, P. lanceolata, Portulaca oleracea, Pulicaria dysenterica, Ranunculus repens, Reseda odorata, Rumex obtusifolia, Sonchus asper, Streptocarpus kewensis, Tanacetum vulgare, Trichosanthes anguina, Verbena hybrida, Veronica chamaedrys*** and ***Zebrina pendula***.

Sources of host-range data

Hollings and Stone (1964).

**Properties of particles in sap**

TIP: 90 °C. LIV: 70 days, or 395 days (for attenuated forms). DEP: $\log_{10}$ minus 5. Infectivity of sap not changed by treatment with di–ethyl ether. Leaf sap contains many virions.

**Purification**

Method: Moran *et al.* (1985).

**Particle morphology**

Virions isometric; not enveloped; 34 nm in diameter; rounded in profile; without a conspicuous capsomere arrangement.

**Physical properties**

One sedimenting component in purified preparations; sedimentation coefficient 122 *S*. Isoelectric point pH 5.2.

**Biochemical properties**

Virions contain 20 % nucleic acid; 80 % protein.

*Genome* consists of RNA; single–stranded; linear; unipartite; largest (or only) genome part 4.003 kb. Genomic nucleic acid isolated by Carrington and Morris (1985). Base composition 27 % G; 30 % A; 19 % C; 24 % U. Infectivity retained when deproteinised with proteases. Poly A region absent (Guilley *et al.*, 1985). Additional factor not required for infectivity. Genome has no *t*RNA–like activity. Nucleotide sequence references: Guilley *et al.* (1985).

*Sequence database accession code(s):*

X02986 Em(40)_vi:XXCARMV Gb(84)_vi:XXCARMV Carnation mottle virus (CarMV) complete genomic RNA sequence. 9/93 4,003bp. 1 sequence.

*Non–genomic nucleic acid* found in the virions; is subgenomic *m*RNA; two major species, the smaller for the coat protein *m*RNA is 1.6 kb long and codes for 347 amino acid residues. The function of the larger one, 1.75 kb, is unknown (Guilley *et al.*, 1985).

*Virion protein(s)* one (plus a fragment); $M_r$ 37787 (plus 26841); coat protein. Method of preparation: Tremaine (1970). Amino acid sequence: Guilley *et al.* (1985). Amino acid composition: Tremaine (1970).

*Virus–coded non–virion proteins* identified by genomic sequence analysis; two proteins found. $M_r$ of the largest 85831. $M_r$ of 2nd largest 26841.

**Replication**

Replication does not depend on a helper virus.

**Cytopathology**

Virions found in all parts of the host plant. Inclusions absent from infected cells.

**Taxonomy and Relationships**

Type species of *Carmovirus: Tombusviridae.*

Virus(es) with serologically unrelated virions: sowbane mosaic, broad bean mottle, tobacco ringspot, tomato bushy stunt, carnation ringspot, turnip crinkle, carnation Italian ringspot, tobacco necrosis, cymbidium ringspot, pelargonium leaf curl and raspberry ringspot viruses.

The coat protein has amino acid sequence homology with those of turnip crinkle and tomato bushy stunt viruses (Carrington and Morris, 1986), but they are not serologically related.

**References**

Carrington, J.C, and Morris, T.J. (1984). *Virology* **139**: 22.
Carrington, J.C. and Morris, T.J. (1985). *Virology* **144**: 1.
Carrington, J.C. and Morris, T.J. (1986). *Virology* **150**: 196.
Guilley, H., Carrington, J.C., Balàzs, E., Jonard, G., Richards, K. and Morris, T.J. (1985). *Nucl. Acids Res.* **13**: 6663.
Hollings, M. and Stone, O.M. (1964). *Ann. appl. Biol.* **53**: 103.
Hollings, M. and Stone, O.M. (1970). *CMI/AAB Descr. Pl. Viruses* No. 7, 4 pp.
Kassanis, B. (1955). *Ann. appl. Biol.* **43**: 103.
Kemp, W.G. and Fazekas, L.J. (1966). *Can. J. Bot.* **44**: 1261.
Koenig, R. and Lesemann, D.E. (1985). *Phytopath. Z.* **112**: 105.
Moran, J.R., Wilson, J.M., Garrett, R.G. and Smith, P.R. (1985). *Pl. Path.* **34**: 467.
Tremaine, J.H. (1970). *Virology* **42**: 611.

# Carnation necrotic fleck *closterovirus*

**Data collated by J.R. Moran and A.A. Brunt, 1987.**

*Synonyms:* carnation yellow fleck virus (Bar-Joseph *et al.*, 1976), carnation streak virus. *Standard acronym:* CNFV.

First reported in *Dianthus caryophyllus*; from Japan; by Inouye and Mitsuhata (1973).

**Natural host range and symptoms**

*Dianthus caryophyllus* — mottling, streaking becoming necrotic; grey streaks or red necrotic flecks.

**Transmission**

Transmitted by a vector; an insect; Aphididae. Transmitted in a semi–persistent manner. Virus lost by the vector when it moults; does not multiply in the vector; not transmitted congenitally to the progeny of the vector; transmitted by mechanical inoculation (with difficulty); not transmitted by contact between plants.

**Ecology and control**

Studies reported by Smookler and Loebenstein (1974); Poupet *et al.* (1975).

**Geographical distribution**

Probably distributed worldwide (wherever carnations are grown).

**Experimental host range**

Few (<3) families susceptible. Experimentally infected plants mostly show vein clearing and flecking.

Diagnostically susceptible

*Dianthus chinensis, D. barbatus* — vein clearing and flecking after 3–5 weeks.

Diagnostically insusceptible

*Chenopodium quinoa.*

Maintenance and propagation hosts

*Dianthus barbatus.*

Assay hosts (**L**ocal lesions or **W**hole plants)

*Dianthus barbatus* (L).

Dicotyledonous Hosts *(**Susceptible** or Insusceptible)*

*Beta vulgaris. Chenopodium amaranticolor. Chenopodium capitatum. Chenopodium hybridum.* ***Dianthus barbatus****. Gomphrena globosa.* ***Montia perfoliata****. Nicotiana clevelandii.* ***Silene armeria****. Tetragonia expansa.*

Additional hosts

Additional Hosts: *Dianthus chinensis* (**S**); *Lychnis chalcedonia* (**S**).

Sources of host-range data

Bar-Joseph *et al.* (1979).

**Properties of particles in sap**

TIP: 40–45 °C. LIV: 2–4 days. DEP: $\log_{10}$ minus 4. Leaf sap contains many virions. Electron microscopy: virions fragile and do not stain well in carnation sap.

**Purification**

Method: Bar-Joseph and Smookler (1976); Inouye (1976).

**Particle morphology**

Virions filamentous; not enveloped; usually flexuous; with no clear modal length; of 1400–1500 nm; 12 nm wide. Basic helix obvious; pitch of basic helix 3.4 nm.

**Physical properties**

One sedimenting component in purified preparations; sedimentation coefficient 122–128 *S* (80 *S* for broken virions). Density 1.325 g $cm^{-3}$ in CsCl.

**Biochemical properties**

Virions contain 5.15 % nucleic acid; 94.85 % protein; 0 % lipid.

*Genome* consists of RNA; single–stranded; unipartite; largest (or only) genome part 12.8 kb. Genomic nucleic acid isolated by Bar-Joseph and Smookler (1976); Bar-Joseph *et al.* (1976).

*Virion protein(s)* one; $M_r$ 23500. Method of preparation: Bar-Joseph *et al.* (1979); Short *et al.* (1977). Amino acid composition: Bar-Joseph *et al.* (1979); Short *et al.* (1977).

**Cytopathology**

Virions found in epidermis and phloem mainly in the parts of the plant that show symptoms. Inclusions present in infected cells; are membranous bodies; they contain virions.

**Taxonomy and Relationships**

*Closterovirus.*

Virus(es) with serologically related virions: beet yellows and wheat yellow leaf viruses. Virus(es) with serologically unrelated virions: citrus tristeza and heracleum latent viruses and all capilloviruses.

Best tests for diagnosis: virions of CNFV are usually found in large numbers in carnation sap and are easily detected by electron microscopy.

**References**

Bar-Joseph, M., Inouye, T. and Sutton, J. (1976). *Pl. Dis. Reptr* **60**: 851.
Bar-Joseph, M., Garnsey, S.M. and Gonsalves, D. (1979). *Adv. Virus Res.* **25**: 93.
Bar-Joseph, M., Loebenstein, G. and Cohen, J. (1976). *Proc. 7th. Conf. Int. Org. Citrus Virol.*, p. 39.
Bar-Joseph, M. and Smookler, M. (1976). *Phytopathology* **66**: 835.
Inouye, T. (1974). *CMI/AAB Descr. Pl. Viruses* No. 136, 3 pp.
Inouye, T. and Mitsuhata, K. (1973). *Ber. Ohara Inst. Landw. Biol.* **15**: 195.
Poupet, A., Cardin, L., Marais, A. and Cadilhac, B. (1975). *Ann. Phytopath.* **7**: 277.
Short, M.N., Hull, R., Bar-Joseph, M. and Rees, M.W. (1977). *Virology* **77**: 408.
Smookler, M. and Loebenstein, G. (1974). *Phytopathology* **7**: 277.

# Carnation (?) rhabdovirus

**Data collated by A.A. Brunt, 1990.**

*Synonyms:* carnation bacilliform virus.

First reported in *Dianthus caryophyllus*; from New Zealand; by Bennett and Milne (1976).

**Natural host range and symptoms**

*Dianthus* sp.

**Geographical distribution**

Spreads in New Zealand.

Dicotyledonous Hosts ***(Susceptible** or Insusceptible)*

***Dianthus barbatus*.**

**Particle morphology**

Virions rhabdo– or bullet–shaped; of 260 nm; 55 nm wide (in leaf sections).

**Taxonomy and Relationships**

Unassigned rhabdovirus.

**References**

Bennett, P.R. and Milne, K.S. (1976). *Acta Hort.* **59**: 61.

# Carnation ringspot *dianthovirus*

**Data collated by J.H. Tremaine and J.R. Moran, 1985.**

*Synonyms:* Anjermozaick virus. *Standard acronym:* CRSV. *Strains:* N, A and R.

First reported in *Dianthus* spp.; from the U.K.; by Kassanis (1955).

**Natural host range and symptoms**

Symptoms persist.

*Dianthus caryophyllus, D. barbatus* — leaf mottling and ringspotting, plant stunting and distortion, flower distortion.

**Transmission**

Transmitted by a vector; a nematode (Fritsche *et al.* (1979), Trudgill *et al.* (1983)); *Longidorus elongatus, L. macrosoma and Xiphinema diversicaudatum*; reported, but questioned. Virus transmitted by mechanical inoculation; transmitted by grafting; transmitted by contact between plants; not transmitted by seed.

**Geographical distribution**

Probably distributed worldwide (wherever carnations are grown). Spreads in the Central Asian region, the Eurasian region, and the North American region; India. Found, but with no evidence of spread, in Australia.

**Experimental host range**

Several (3–9) families susceptible. Experimentally infected plants mostly show chlorotic and necrotic local lesions rings and flecks, sometimes systemic.

Diagnostically susceptible

*Dianthus barbatus, Gomphrena globosa, Phaseolus vulgaris* — local and systemic chlorotic and necrotic lesions rings and flecks.

*Chenopodium amaranticolor, C. quinoa, Tetragonia expansa* — necrotic lesions, usually not systemic.

*Vigna unguiculata* — necrotic local lesions, systemic mottling, necrotic flecking and distortion.

Maintenance and propagation hosts

*Dianthus barbatus, Nicotiana clevelandii, Phaseolus vulgaris, Vigna unguiculata.*

Assay hosts (**L**ocal lesions or **W**hole plants)

*Chenopodium amaranticolor* (L); *C. quinoa* (L); *Vigna unguiculata*(L).

Dicotyledonous Hosts *(**Susceptible** or Insusceptible)*

*Ageratum conyzoides*. ***Amaranthus caudatus***. ***Antirrhinum majus***. *Apium graveolens*. *Atriplex hortensis*. *Brassica campestris* ssp. *pekinensis*. *Brassica oleracea* ssp. *capitata*. ***Capsicum frutescens***. *Catharanthus roseus*. *Cheiranthus cheiri*. ***Chenopodium amaranticolor***. ***Chenopodium quinoa***. *Chrysanthemum morifolium*. ***Cichorium endiva***. *Citrus limon*. ***Crotalaria spectabilis***. ***Cucumis sativus***. ***Datura stramonium***. ***Dianthus barbatus***. ***Dianthus caryophyllus***. ***Emilia sagittata***. ***Fagopyrum esculentum***. *Glycine max*. ***Gomphrena globosa***. ***Hyoscyamus niger***. ***Lactuca sativa***. ***Lathyrus odoratus***. ***Lobelia erinus***. *Lycopersicon esculentum*. *Matthiola incana*. *Melilotus alba*. ***Myosotis sylvatica***. ***Nicandra physaloides***. ***Nicotiana debneyi***. ***Nicotiana glutinosa***. ***Nicotiana rustica***. ***Nicotiana tabacum***. *Petunia hybrida*. ***Phaseolus vulgaris***. ***Phlox drummondii***. ***Physalis floridana***. ***Phytolacca americana***. *Pisum sativum*. *Primula malacoides*. *Senecio vulgaris*. *Solanum tuberosum*. ***Spinacia oleracea***. ***Stellaria media***. ***Tetragonia expansa***. ***Torenia fournieri***. *Trifolium repens*. *Tropaeolum majus*. ***Vicia faba***. ***Vigna unguiculata***. ***Zinnia elegans***.

Monocotyledonous Hosts *(**Susceptible** or Insusceptible)*

*Zea mays*.

Additional hosts

*Alonsoa warscewiczii, Callistephus chinensis, Convolvulus tricolor, Fumaria officinalis, Gypsophila paniculata, Lamium maculatum, Lythrum salicaria, Portulaca oleracea, Saintpaulia ionantha, Salpiglossis sinuata, Verbena hybrida.*

Insusceptible hosts: *Anthriscus sylvestris, Begonia semperflorens, Ceratostigma plumbaginoides, Geranium pratense, Geum chiloense, Hypericum perforatum, Lavatera trimestris, Mimulus luteus, Mycelis muralis, Passiflora coerulea, Plantago lanceolata, Pelargonium hortum, Ranunculus repens, Reseda odorata, Rheum rhapontium, Taraxacum officinale, Zebrina pendula.*

Sources of host-range data

Tremaine and Dodds (1985).

**Properties of particles in sap**

TIP: 80 °C. LIV: 50–60 days. DEP: $\log_{10}$ minus 5. Infectivity of sap not changed by treatment with di–ethyl ether. Leaf sap contains few virions. Electron microscopy: virions best stained with UA, pH 5.0.

**Purification**

Method: Tremaine *et al.* (1976); Kalmakoff and Tremaine (1967).

**Particle morphology**

Virions isometric; not enveloped; 34 nm in diameter; rounded in profile; without a conspicuous capsomere arrangement.

**Physical properties**

One sedimenting component in purified preparations (but there are probably two classes of virions, one containing RNA-1, the other probably three copies of RNA-2); sedimentation coefficient 135 *S*. Density 1.366 g $cm^{-3}$ in CsCl (when unfixed at pH 5; virions swell at pH 7, but stabilized by divalent cations). Isoelectric point pH 4.5.

**Biochemical properties**

Virions contain 20.5 % nucleic acid; 79.5 % protein; 0 % lipid.

*Genome* consists of RNA; single–stranded; linear; of two parts; largest (or only) genome part the larger 3.89 kb (RNA-1); the 2nd largest 1.45 kb (RNA-2). Genomic nucleic acid isolated by Dodds *et al.* (1977). Base composition 26 % G; 27 % A; 23 % C; 24 % U. 5' terminus of RNA has a methylated nucleotide cap. Infectivity retained when deproteinised with phenol or detergent. Poly A region absent.

*Sequence database accession code(s):*

L18870 Em(40)_vi:CIRRNA1A Gb(84)_vi:CIRRNA1A Carnation ringspot virus fusion protein polymerase (p88K), p27K, p57K, capsid protein (p37K)

M88589 Em(40)_vi:CIRCRSV Gb(84)_vi:CIRCRSV Carnation ringspot virus CRSV movement protein mRNA, complete cds. 1/93 1,394bp 2 sequences.

*Non–genomic nucleic acid* found in the virions. Sub–genomic *m*RNA found in infected cells (Lommel, 1983).

*Virion protein(s)* one; $M_r$ 37000 (339 amino acid residues); coat protein. Method of preparation: Tremaine and Dodds (1985).

*Virus–coded non–virion proteins* identified by genomic sequence analysis (Tremaine *et al.* (1984)); three proteins found. $M_r$ of the largest 100000. $M_r$ of 2nd largest 37000 (coat protein). $M_r$ of 3rd 34000; transport protein.

**Replication**

Replication does not depend on a helper virus.

**Cytopathology**

Virions found in all parts of the host plant; in cytoplasm and in nuclei. Inclusions absent from infected cells.

**Taxonomy and Relationships**

*Dianthovirus* (the type species).

Virus(es) with serologically related virions: red clover necrotic mosaic and sweet clover necrotic mosaic viruses, but remotely.

Tremaine *et al.* (1976, 1983, 1984) and Tremaine and Ronald (1976) have described strains N, A and R that differ in the physico-chemical properties and composition of their virions.

**References**

Chen, M.H., Hiruki, C. and Okuro, T. (1984). *Can. J. Pl. Pathol.* **6**: 191.
Dodds, J.A., Tremaine, J.H. and Ronald, W.P. (1977). *Virology* **83**: 322.
Fritsche, R., Kegler, H., Thiele ? and Gruber ?. (1979). *Arch. Phytopath. PflSchutz.* **15**: 177.
Hiruki, C. (1987). *Adv. Virus Res.* **33**: 257.
Hiruki, C., Rao, A.L.N., Furuya, Y. and Figueiredo, G. (1984). *J. gen. Virol.* **65**: 2273.
Hiruki, C., Rao, D.V., Chen, M.H., Okuno, T. and Figueiredo, G. (1984). *Phytopathology* **74**: 482.
Hollings, M. and Stone, O.M. (1969). *Ann. appl. Biol.* **56**: 73.
Hollings, M. and Stone, O.M. (1977). *CMI/AAB Descr. Pl. Viruses* No. 21, 3 pp.
Kalmakoff, J. and Tremaine, J.H. (1967). *Virology* **33**: 10.
Kassanis, B. (1955). *Ann. appl. Biol.* **43**: 103.
Lommel, S.A. (1983). *Ph.D. Thesis*, University California, Berkeley, California, U.S.A.
Lommel, S. and Morris, J. (1982). *Phytopathology* **72**: 955.
Lommel, S., McCain, and Morris, J. (1982). *Phytopathology* **72**: 1018.
Lommel, S., McCain, Mayhew, D.E. and Morris, J. (1983). *Plant Dis.* **67**: 53.
Tremaine, J.H. and Dodds, J.A. (1985). *CMI/AAB Descr. Pl. Viruses* No. 308, 5 pp.
Tremaine, J.H. and Ronald, W.P. (1976). *J. gen. Virol.* **30**: 299.
Tremaine, J.H., Ronald, W.P. and Valcic, A. (1976). *Phytopathology* **66**: 34.
Tremaine, J.H., Ronald, W.P. and McGauley, E.M. (1983). *Phytopathology* **73**: 1241.
Tremaine, J.H., Ronald, W.P. and McGauley, E.M. (1984). *Phytopathology* **74**: 161.
Trudgill, D.L., Brown, D.J.F. and McNamara, D. (1983). *Revue Nématol.* **6**: 133.

# Carnation vein mottle *potyvirus*

**Data collated by J.R. Moran, M. Verhoyen and G. van Caneghem, 1987.**

*Standard acronym:* CVMV.

First reported in *Dianthus caryophyllus*; from the U.S.A.; by Kassanis (1954).

**Natural host range and symptoms**

*Dianthus caryophyllus* — chlorotic and darker green spots, flecks and mottling. Flower breaking and malformation.

*Dianthus barbatus* — mottle.

**Transmission**

Transmitted by a vector; an insect; Aphididae. Transmitted in a non–persistent manner. Virus transmitted by mechanical inoculation; not transmitted by contact between plants; not transmitted by seed.

**Ecology and control**

Studies reported by Hollings *et al.* (1977); Poupet *et al.* (1981). Stocks freed from virus by tip propagation and heat treatment.

**Geographical distribution**

Probably distributed worldwide (wherever carnations are grown).

**Experimental host range**

Few (<3) families susceptible. Experimentally infected plants mostly show chlorotic and necrotic local lesions rarely systemic.

Diagnostically susceptible

*Chenopodium amaranticolor* — chlorotic and necrotic local lesions; not systemic.

*Chenopodium quinoa* — chlorotic spots, local lesions; systemic spotting.

*Silene pendula* — necrotic local lesions, systemic chlorotic or necrotic spots (Tochihara *et al.*, 1975).

Maintenance and propagation hosts

*Chenopodium quinoa, Dianthus barbatus.*

Assay hosts (**L**ocal lesions or **W**hole plants)

*Chenopodium quinoa* (L), *Chenopodium amaranticolor* (L).

Dicotyledonous Hosts *(**Susceptible** or Insusceptible)*

***Amaranthus caudatus***. *Antirrhinum majus. Apium graveolens. Arachis hypogaea.* ***Atriplex hortensis***. ***Beta vulgaris***. *Brassica juncea. Brassica oleracea* ssp. *capitata. Calendula officinalis. Capsicum frutescens. Catharanthus roseus.* ***Celosia argentea***. *Cheiranthus cheiri.* ***Chenopodium amaranticolor***. *Chenopodium capitatum.* ***Chenopodium murale***. ***Chenopodium quinoa***. *Chrysanthemum morifolium. Cucumis sativus. Cucurbita maxima. Datura stramonium. Emilia sagittata. Fagopyrum esculentum. Glycine max.* ***Gomphrena globosa***, or *Gomphrena globosa. Helianthus annuus. Hibiscus esculentus. Lactuca sativa. Lathyrus odoratus. Nicotiana debneyi. Nicotiana glutinosa. Nicotiana glutinosa* × *clevelandii. Nicotiana rustica. Nicotiana tabacum. Ocimum basilicum. Petunia hybrida. Phaseolus vulgaris. Phlox drummondii. Physalis floridana. Phytolacca americana. Pisum sativum. Primula malacoides.* ***Saponaria vaccaria***. *Senecio vulgaris.* ***Silene armeria***. *Solanum melongena. Solanum tuberosum.* ***Spinacia oleracea***. ***Stellaria media***. ***Tetragonia expansa***. *Torenia fournieri. Trifolium incarnatum. Tropaeolum majus. Verbesina encelioides. Vicia faba. Vigna unguiculata. Zinnia elegans.*

Monocotyledonous Hosts *(**Susceptible** or Insusceptible)*

*Avena sativa. Zea mays.*

Additional hosts

*Ageratum houstonianum, Alonsoa warscewiczii, Anchusa azurea, Aquilegia coerulea, Asclepias curassavica, Begonia semperflorens, Callistephus chinensis, Calystegia sepium, Convolvulus arvensis, Convolvulus tricolor, Digitalis purpurea, Dipsacus sylvestris, Geranium robertianum, Gomphocarpus fruticosus, Lychnis alba,* ***Lychnis chalcedonica***, *Lychnis dioica, Lythrum salicaria, Mimulus luteus, Mirabilis jalapa, Nicotiana megalosiphon, Oenothera biennis, Oxalis corniculata, Papaver rhoeas, Pelargonium domesticum,* ***Plantago lanceolatum, Plantago major***, *Plantago maritima, Polygonum aviculare, Polygonum persicaria,* ***Portulaca oleracea***, *Reseda odorata, Rheum rhaponticum,* ***Rumex obtusifolius, Senecio cruentus, Silene cucubalis, Silene pendula*** and *Verbena hybrida.*

Sources of host-range data

Hollings and Stone (1977); Tochihara *et al.* (1975).

**Properties of particles in sap**

TIP: 55–65 °C. LIV: 2–10 days. DEP: $\log_{10}$ minus 2–5. Leaf sap contains many virions.

**Purification**

Method: Hollings and Stone (1971); Borges *et al.* (1981).

**Particle morphology**

Virions filamentous; not enveloped; usually flexuous; of 790 nm; 12 nm wide. Axial canal obvious.

**Physical properties**

One sedimenting component in purified preparations; sedimentation coefficient 144 *S*.

**Biochemical properties**

*Genome* consists of RNA; single–stranded. Base composition 21 % G; 25 % A; 28 % C; 29 % U.

*Virion protein(s)* one; $M_r$ 32000–34800.

**Cytopathology**

Virions found in all parts of the host plant; in unusual sites; in a single, thin layer between two membranes just inside the tonoplast. Inclusions present in infected cells; are pinwheels and unusual in shape; dense bands, loops and circles (Weintraub, 1970) or tubular structures (Bergtrup, 1976); they do not contain virions. Other cellular changes: elongated nuclei with abnormally electron dense areas of chromatin.

**Taxonomy and Relationships**

*Potyvirus: Potyviridae.*

Virus(es) with serologically related virions: bean yellow mosaic, pepper veinal mottle, pea mosaic, freesia streak, turnip mosaic (cabbage black ring strain) and watermelon mosaic 2 viruses (Borges, 1981). Virus(es) with serologically unrelated virions: turnip mosaic (tigridia mosaic strain), narcissus degeneration, clover yellow vein, narcissus yellow stripe, hippeastrum mosaic, potato veinal necrosis, iris mosaic, potato Y, sugar beet mosaic, tobacco severe etch and lettuce mosaic viruses.

Best tests for diagnosis: for electron microscopy use leaf sap, virus aggregates are found in plant membranes. CVMV can be distinguished from carnation latent virus either by ISEM decoration or by inoculation to *Chenopodium amaranticolor*, which carnation latent infects systemically, but CVMV does not.

**References**

Begtrup, J. (1976). *Phytopath. Z.* **86**: 127.

Borges, M. de L.V., Sequeira, J.C. and Louro, D. (1981). *Bolm Soc. broteriana* **53**: 933.

Hollings, M. and Stone, O.M. (1977). *CMI/AAB Descr. Pl. Viruses* No. 78, 4 pp.

Hollings, M., Stone, O.M., Atkey, P.T. and Barton, R.J. (1977). *Ann. appl. Biol.* **85**: 59.

Hooper, G.R. and Weise, M.V. (1972). *Virology* **47**: 664.

Kassanis, B. (1954). *Nature, Lond.* **173**: 1097.

Poupet, A., Jacquemont, R., Beck, D., Onesto, J.P. and Delomon, D. (1981). *C. r. Séanc. Acad. Agric. Fr.* **65**: 427.

Tochihara, H., Idei, T., Yabuki, S. and Fukumoto, F. (1975). *Ann. Phytopath. Soc. Japan* **41**: 390.
Weintraub, M. and Ragetli, H.W.S. (1970). *Virology* **40**: 868.

# Carnation yellow stripe (?) *necrovirus*

**Data collated by D. Gallitelli, 1991.**

*Standard acronym:* CYSV.

First reported in *Dianthus caryophyllus*; from Apulia, southern Italy; by Rana *et al.* (1977).

**Natural host range and symptoms**

*Dianthus caryophyllus* cv. Pink Shocking—yellow leaf stripes.

**Transmission**

Virus transmitted by mechanical inoculation.

**Geographical distribution**

Spreads in Italy.

**Experimental host range**

Several (3–9) families susceptible. Experimentally infected plants mostly show chlorotic or necrotic local lesions.

Diagnostically susceptible

*Chenopodium quinoa, C. amaranticolor, C. album* — small chlorotic local lesions, becoming necrotic.
*Chenopodium ambrosioides* — bright red local lesions.
*Gomphrena globosa* — red-rimmed necrotic local lesions.
*Vigna unguiculata* — systemic mosaic, chlorotic vein-banding.
*Saponaria vaccaria* — chlorotic local lesions.

Diagnostically insusceptible

*Hordeum vulgare, Cucumis sativus, Phaseolus aureus, Nicotiana benthamiana, N. rustica.*

Maintenance and propagation hosts

*Vigna unguiculata.*

Assay hosts (**L**ocal lesions or **W**hole plants)

*Vigna unguiculata* (W), *Chenopodium quinoa* (L), *C. amaranticolor* (L), *Gomphrena globosa* (L), *Dianthus barbatus* (L).

Dicotyledonous Hosts *(**Susceptible** or Insusceptible)*

***Chenopodium album*. *Chenopodium amaranticolor*. *Chenopodium ambrosioides*. *Chenopodium quinoa***. *Cucumis sativus*. *Cucurbita pepo*. *Dianthus barbatus*. ***Gomphrena globosa*. *Lupinus albus***. *Nicotiana benthamiana*. ***Nicotiana clevelandii***. *Nicotiana rustica*. ***Vicia faba*. *Vigna unguiculata***.

Monocotyledonous Hosts *(**Susceptible** or Insusceptible)*

*Avena sativa*. *Hordeum vulgare*. *Triticum durum*. *Zea mays*.

Sources of host-range data

Gallitelli *et al.* (1979).

**Properties of particles in sap**

TIP: 80 °C. LIV: 60 days. DEP: $\log_{10}$ minus 7. Leaf sap contains many virions.

**Purification**

Method: Gallitelli *et al.* (1979).

**Particle morphology**

Virions isometric; 30 nm in diameter; angular in profile; without a conspicuous capsomere arrangement.

**Physical properties**

One sedimenting component in purified preparations; sedimentation coefficient 99–105 *S*. Density 1.369 g $cm^{-3}$ in CsCl. $A_{260}/A_{280}$ ratio 1.61.

**Biochemical properties**

Virions contain 18 % nucleic acid; 72 % protein; 0 % lipid.

*Genome* consists of RNA; single–stranded; unipartite; largest (or only) genome part 4 kb. Genomic nucleic acid isolated by Gallitelli: phenol-SDS.

*Virion protein(s)* one; $M_r$ 27000; coat protein. Method of preparation: Laemmli (1970).

**Cytopathology**

Virions found in leaves and mesophyll; in cytoplasm. Inclusions present in infected cells; are crystals in the cytoplasm and membranous bodies; they do not contain virions.

**Taxonomy and Relationships**

Possible *Necrovirus*.

Virus(es) with serologically related virions: tobacco necrosis virus (grape strain).

Best tests for diagnosis: inoculation onto *Chenopodium quinoa* and serology.

**References**


Gallitelli, D., Castellano, M.A., Di Franco, A. and Rana, G.L. (1979). *Phytopathol. Medit.* **18**: 31.

Laemmli, U.K. (1970). *Nature, Lond.* **227**: 680.

Rana, G.L., Di Franco, A. and Castellano, M.A. (1977). *Ann. Fac. Agr. Univ. Bari,* **29**: 397.


# Carrot latent *nucleorhabdovirus*

**Data collated by A.A. Brunt, 1991.**

*Standard acronym:* CtLtV.

First reported in *Daucus carota*; from Japan; by Ohki *et al.* (1978).

**Natural host range and symptoms**

*Daucus carota* — symptomless infection.

**Transmission**

Transmitted by a vector; an insect; *Brachycaudus (Semiaphis) heraclei*; Aphididae. Virus not transmitted by mechanical inoculation.

**Geographical distribution**

Spreads in Japan.

Diagnostically susceptible

*Daucus carota* — symptomless infection.

Maintenance and propagation hosts

*Daucus carota*.

Dicotyledonous Hosts *(**Susceptible** or Insusceptible)*

***Daucus carota***.

**Particle morphology**

Virions rhabdo– or bullet–shaped; enveloped; with a clear modal length; of 220 nm (in sections); 70 nm wide (*in vivo*).

**Cytopathology**

Virions found in all parts of the host plant; in nuclei.

**Taxonomy and Relationships**

Possible *Nucleorhabdovirus: Rhabdoviridae*.

**References**

Ohki, S.T., Doi, Y. and Yora, K. (1978). *Ann. Phytopath. Soc. Japan* **44**: 22.

# Carrot mosaic (?) potyvirus

**Data collated by A.A. Brunt, 1991.**

*Standard acronym:* CtMV.

First reported in *Daucus carota*; from Czechoslovakia; by Chod (1965*a*).

**Natural host range and symptoms**

Symptoms persist.

*Daucus carota* — chlorotic spots and mosaic of leaves, and yellow spots on petioles.

**Transmission**

Transmitted by a vector; an insect; *Acyrthosiphon pisum, Cavariella aegopodii, Myzus persicae*; Aphididae. Virus transmitted by mechanical inoculation.

**Geographical distribution**

Spreads in the former Czechoslovakia.

Diagnostically susceptible

*Vigna unguiculata* — chlorotic flecks and chlorosis of leaves.

*Chenopodium quinoa* — local chlorotic lesions.

Dicotyledonous Hosts *(**Susceptible** or Insusceptible)*

***Apium graveolens*. *Chenopodium amaranticolor*. *Chenopodium quinoa*. *Datura stramonium*. *Daucus carota*. *Lycopersicon esculentum*. *Medicago sativa*. *Phaseolus vulgaris***.

**Properties of particles in sap**

TIP: 65 °C. LIV: 4 days (at 20–22°C). DEP: $\log_{10}$ minus 3.

**Particle morphology**

Virions filamentous; not enveloped; usually flexuous; with a clear modal length; of 740 nm; 15 nm wide.

**Taxonomy and Relationships**

Unassigned potyvirus.

**References**

Chod, J. (1965*a*). *Biologia Pl.* **7**: 463.
Chod, J. (1965*b*). *Ochr. Rost.* **1**: 49.
Chod, J. (1966). *Biologia Pl.* **8**: 53.

# Carrot mottle mimic *umbravirus*

**Data collated by M.J. Gibbs, 1995.**

First reported in *Daucus carota*; from New South Wales, Australia; by Gibbs (1991; 1995).

**Natural host range and symptoms**

Symptoms persist.

*Daucus carota* — systemic yellowing and mottling, stunting.

**Transmission**

Transmitted by a vector; an insect; probably *Cavariella aegopodii*; Aphididae. Transmitted in a persistent manner. Virus requires, for vector transmission, a helper virus (carrot red leaf luteovirus). The virus genome is probably transcapsidated in the virions of the helper virus. Virus transmitted by mechanical inoculation.

**Geographical distribution**

Found, but with no evidence of spread, in Australia, New Zealand and the U.S.A.; but may have been confused with carrot mottle umbravirus.

**Experimental host range**

Few (<3) families susceptible. Experimentally infected plants mostly show vein necrosis.

Diagnostically susceptible

*Nicotiana megalosiphon* — necrotic ringspots, systemic necrotic and chlorotic lesions, mostly along major veins.

Maintenance and propagation hosts

*Nicotiana megalosiphon, N. clevelandii.*

Assay hosts (**L**ocal lesions or **W**hole plants)

*Nicotiana megalosiphon* (W), *N. clevelandii* (W).

Dicotyledonous Hosts *(**Susceptible** or Insusceptible)*

***Anthriscus cerefolium*. *Coriandrum sativum*. *Daucus carota*. *Nicotiana benthamiana*. *Nicotiana clevelandii*. *Nicotiana glutinosa*.**

**Properties of particles in sap**

Leaf sap contains few virions (and these have the coat protein of carrot red leaf luteovirus).

**Biochemical properties**

*Genome* consists of RNA; single–stranded; linear; unipartite; largest (or only) genome part 4.203 kb. Genomic nucleic acid isolated by as dsRNA by Franklin column chromatography using Whatman CF-11 fibrous cellulose powder; as described for carrot mottle umbravirus by Murant *et al.* (1985). Base composition 27 % G; 25 % A; 25.5 %

C; 22.4 % U. Infectivity retained when deproteinised with proteases. Poly A region absent. Additional factor not required for infectivity. Genome has no *t*RNA–like activity. Nucleotide sequence references: Gibbs (1995).

*Features of the genome:* Four ORFs. ORF1 encodes a protein that may be part of the viral replicase, ORF2 encodes the RNA dependent RNA polymerase, ORF3 encodes a protein of unknown function and ORF4 encodes the movement protein gene. ORF2 overlaps ORF1 and is translated after a -1 ribosomal frameshift, ORFs 3&4 also overlap and are translated from a sub-genomic RNA.

*Non–genomic nucleic acid* found in the virions; is subgenomic *m*RNA. Sub–genomic *m*RNA found in infected cells; *c.* 1.5kb co-terminal with 3' end of the genome. 2 virus specified dsRNA species found in infected cells. Size of largest virus specified dsRNA 4.2 kbp; 2nd largest *c.* 1.5 kbp.

*Virion protein(s)* none.

**Cytopathology**

Inclusions present in infected cells; are membranous bodies; they do not contain virions.

**Taxonomy and Relationships**

*Umbravirus.*

Genome sequence comparisons show the virus to be most closely related to RNA-2 of pea enation mosaic enamovirus, in addition the polymerase sequences show affinities to carmoviruses, dianthoviruses, type 1 and 2 luteoviruses, necroviruses, machlomoviruses, tombusviruses, beet western yellows ST9 associated-RNA and oat chlorotic stunt virus. Comparisons of movement protein sequences show affinities with bromo- and cucumoviruses. Best tests for diagnosis: often confused with carrot mottle umbravirus, distinguished by infection of *Nicotiana megalosiphon* and *N. benthamiana*; carrot mottle mimic virus induces a mild chlorotic mottle in *N. benthamiana* and severe symptoms in *N megalosiphon*, carrot mottle virus induces a more severe chlorotic mottle in *N. benthamiana*, but no symptoms in *N. megalosiphon.*

**References**


Gibbs, M.J. (1991). *Grad. Dip. Thesis*, Aust. Nat. Univ.

Gibbs, M.J. (1995). *D. Phil. Thesis*, Univ. Oxford.

Murant, A.F., Waterhouse, P.M., Raschke, J.H. and Robinson, D.J. (1985). *J. gen. Virol.* **66**: 1575.


# Carrot mottle *umbravirus*

**Data collated by P.M. Waterhouse, 1986.**

*Synonyms:* carrot motley dwarf virus (Stubbs, 1948; 1952). *Standard acronym:* CMoV.

First reported in *Daucus carota*; from England; by Watson *et al.* (1964).

**Natural host range and symptoms**

*Daucus carota, Anethum graveolens, Anthriscus sylvestris* — symptomless but, with CRLV, induces yellowing and reddening.

**Transmission**

Transmitted by a vector; an insect; *Cavariella aegopodii*; Aphididae. Transmitted in a persistent manner. Virus retained when the vector moults; does not multiply in the vector; not transmitted congenitally to the progeny of the vector; requires, for vector transmission, a helper virus (carrot red leaf luteovirus); transmitted by mechanical inoculation; not transmitted by seed; not transmitted by pollen.

**Geographical distribution**

Spreads in Australia, Canada, Germany, Japan, New Zealand, the UK, and the USA.

**Experimental host range**

Several (3–9) families susceptible. Experimentally infected plants mostly show local lesions, mottling.

Diagnostically susceptible

*Chenopodium quinoa* — local lesions.
*Nicotiana clevelandii* — systemic vein etching.
*Phaseolus vulgaris* cv. 'The Prince' — brown local lesions.
*Coriandrum sativum* — systemic mottle.

Diagnostically insusceptible

*Lactuca sativa*.

Maintenance and propagation hosts

*Nicotiana clevelandii*.

Assay hosts (**L**ocal lesions or **W**hole plants)

*Chenopodium quinoa* (L); *Nicotiana clevelandii* (W); *Phaseolus vulgaris* (L).

Dicotyledonous Hosts *(**Susceptible** or Insusceptible)*

***Apium graveolens***. ***Chenopodium amaranticolor***. ***Chenopodium capitatum***. ***Chenopodium quinoa***. ***Coriandrum sativum***. ***Daucus carota***. ***Gomphrena globosa***. ***Nicotiana benthamiana***. ***Nicotiana clevelandii***. ***Nicotiana glutinosa***. ***Nicotiana rustica***. ***Petroselinum crispum***. ***Phaseolus vulgaris***.

Sources of host-range data

Waterhouse (1981); Watson *et al.* (1964); Murant *et al.* (1969, 1973); Murant (1975); Frowd and Tomlinson (1972).

**Properties of particles in sap**

TIP: 70 °C. LIV: 0.2–1 days. DEP: $\log_{10}$ minus 3. Infectivity of sap not changed by treatment with di–ethyl ether.

**Purification**

Method: Murant *et al.* (1969).

**Particle morphology**

Virions unusually shaped; virions not identified; vesicles reported, but unconfirmed.

**Physical properties**

One sedimenting component in purified preparations; sedimentation coefficient 270 *S*. Density 1.15 g $cm^{-3}$ in CsCl.

**Biochemical properties**

*Genome* consists of RNA; single–stranded; linear; unipartite; largest (or only) genome part 4.75 kb.

**Replication**

Replication does not depend on a helper virus.

### Cytopathology

Other cellular changes: outgrowths from plasmodesmatal, seen by light microscopy.

### Taxonomy and Relationships

*Umbravirus*.

CMotV resembles viruses such as lettuce speckles mottle and groundnut rosette and other sap-transmitted components of luteovirus-containing virus symbioses; the name umbravirus has been proposed for these viruses. Preliminary sequence information indicates a relationship with the carmo/tombus/luteovirus cluster (M.J. Gibbs, unpublished text). Best tests for diagnosis: Lettuce speckles mottle virus infects manually inoculated lettuce, CMotV does not.

### Comments

Carrot motley dwarf virus (Stubbs, 1948; 1952) is a complex of carrot red leaf and carrot mottle viruses.

### References

Frowd, J.A. and Tomlinson, J.A. (1972). *Ann. appl. Biol.* **72**: 177.
Halk, E.L., Robinson, D.J. and Murant, A.F. (1979). *J. gen. Virol.* **45**: 383.
Murant, A.F. (1974). *CMI/AAB Descr. Pl. Viruses* No. 137, 4 pp.
Murant, A.F. (1975). *Can. Pl. Dis. Surv.* **55**: 103.
Murant, A.F., Goold, R.A., Roberts, I.M. and Cathro, J. (1969). *J. gen. Virol.* **4**: 329.
Murant, A.F., Roberts, I.M. and Goold, R.A. (1973). *J. gen. Virol.* **21**: 269.
Murant, A.F., Waterhouse, P.M., Raschke, J.H. and Robinson, D.J. (1985). *J. gen. Virol.* **66**: 1575.
Stubbs, L.L. (1948). *Aust. J. sci. Res.* **1**: 303.
Stubbs, L.L. (1952). *Aust. J. sci. Res.* **5**: 399.
Waterhouse, P.M. (1981). *Ph.D. Thesis*, University of Dundee, U.K., 244 pp.
Watson, M.A., Serjeant, E.P. and Lennon, E.A. (1964). *Ann. appl. Biol.* **54**: 153.

## Carrot red leaf *luteovirus*

**Data collated by P.M. Waterhouse, 1986.**

*Synonyms:* carrot motley dwarf virus (Stubbs, 1948; 1952). *Standard acronym:* CtRLV.

First reported in *Daucus carota*; from England; by Watson *et al.* (1964).

### Natural host range and symptoms

Symptoms persist and vary seasonally.

*Daucus carota, Anethum graveolens* — chlorosis or reddening of older leaves, but mostly symptomless.

*Anthriscus sylvestris* — reddening of leaf margins.

### Transmission

Transmitted by a vector; an insect; *Cavariella aegopodii*; Aphididae. Transmitted in a persistent manner. Virus retained when the vector moults; does not multiply in the vector; not transmitted congenitally to the progeny of the vector; can help the vector transmission of another virus (carrot mottle umbravirus); not transmitted by mechanical inoculation; not transmitted by contact between plants; not transmitted by seed; not transmitted by pollen.

### Geographical distribution

Spreads in Australia, Canada, Germany, Japan, New Zealand, the UK, and the USA.

**Experimental host range**

Few (<3) families susceptible. Experimentally infected plants mostly show yellowing, reddening.

Diagnostically susceptible

*Anthriscus cerefolium, Apium leptophyllum, Coriandrum sativum* — yellowing and reddening of older leaves.

Diagnostically insusceptible

*Apium graveolens, Petroselinum crispum.*

Maintenance and propagation hosts

*Anthriscus cerefolium.*

Assay hosts (**L**ocal lesions or **W**hole plants)

*Anthriscus cerefolium* (W); *Apium leptophyllum* (W); *Coriandrum sativum* (W).

Dicotyledonous Hosts *(**Susceptible** or Insusceptible)*

*Apium graveolens. Capsella bursa–pastoris. Chenopodium capitatum. Chenopodium murale.* ***Coriandrum sativum****. Datura stramonium.* ***Daucus carota****. Lactuca sativa. Nicotiana clevelandii. Petroselinum crispum. Raphanus sativus. Senecio vulgaris. Trifolium pratense.*

**Properties of particles in sap**

Leaf sap contains few virions.

**Purification**

Method: Waterhouse and Murant (1981).

**Particle morphology**

Virions isometric; not enveloped; 25 nm in diameter; rounded in profile, or angular in profile.

**Physical properties**

One sedimenting component in purified preparations; sedimentation coefficient 104 *S*. Density 1.403 g $cm^{-3}$ in CsCl.

**Biochemical properties**

Virions contain 28 % nucleic acid; 72 % protein; 0 % lipid.

*Genome* consists of RNA; single–stranded; linear; unipartite; largest (or only) genome part 5.75 kb. Genomic nucleic acid isolated by Murant *et al.* (1985).

*Virion protein(s)* four; $M_r$ of the largest 60000; component 1. $M_r$ of 2nd largest 50000; minor component 2. $M_r$ of 3rd largest 40000; minor component 3. $M_r$ of 4th largest 25000; coat protein (major component). Method of preparation: Murant *et al.* (1985).

**Replication**

Replication does not depend on a helper virus.

**Cytopathology**

Virions found in phloem and companion cells; in cytoplasm and in cell vacuoles. Inclusions present in infected cells; are amorphous X–bodies and membranous bodies; they contain virions.

**Taxonomy and Relationships**

*Luteovirus.*

Virus(es) with serologically related virions: barley yellow dwarf-RPV and beet western yellows viruses moderately closely; tobacco necrotic dwarf, potato leafroll and bean leaf roll viruses distantly; and soybean dwarf, barley yellow dwarf-MAV viruses very distantly.

Best tests for diagnosis: This virus is the only luteovirus known to infect Apiaceae and has a very narrow host range. It is usually found in a complex with CMotV, which unlike CRLV is sap transmitted.

**Comments**

Carrot motley dwarf virus (Stubbs, 1948; 1952) is a complex of carrot red leaf and carrot mottle viruses.

**References**

Murant, A.F. (1974). *CMI/AAB Descr. Pl. Viruses* No. 137, 4 pp.

Murant, A.F. and Roberts, I.M. (1979). *Ann. appl. Biol.* **92**: 343.

Murant, A.F., Waterhouse, P.M., Raschke, J.H. and Robinson, D.J. (1985). *J. gen. Virol.* **66**: 1575.

Stubbs, L.L. (1948). *Aust. J. sci. Res.* **1**: 303.

Stubbs, L.L. (1952). *Aust. J. Sci. Res.* **5**: 399.

Stubbs, L.L., Smith, P.R. and O'Loughlin, G.T. (1983). *Abstr. 4th Int. Congr. Pl. Path. Melbourne, Australia, 1983*, p. 123.

Waterhouse, P.M. (1981). *Ph.D. Thesis*, University of Dundee, U.K., pp. 244.

Waterhouse, P.M. and Murant, A.F. (1982). *Ann. appl. Biol.* **97**: 191.

Waterhouse, P.M. (1985). *Australas. Pl. Path.* **14**: 32.

Waterhouse, P.M. and Murant, A.F. (1982). *CMI/AAB Descr. Pl. Viruses* No. 249, 4 pp.

Watson, M.A., Serjeant, E.P. and Lennon, E.A. (1964). *Ann. appl. Biol.* **54**: 153.

# Carrot temperate 1 *alphacryptovirus*

**Data collated by A.A. Brunt, 1991.**

*Standard acronym:* CTeV-1.

First reported in *Daucus carota*; from Japan; by Natsuaki *et al.* (1983).

**Natural host range and symptoms**

*Daucus carota* — symptomless.

**Transmission**

Virus not transmitted by mechanical inoculation; transmitted by seed.

**Geographical distribution**

Found, but with no evidence of spread, in Japan.

Dicotyledonous Hosts *(**Susceptible** or Insusceptible)*

***Daucus carota***.

**Particle morphology**

Virions isometric; not enveloped; 30 nm in diameter.

**Biochemical properties**

*Genome* consists of RNA; double–stranded; of three parts.

**Taxonomy and Relationships**

*Alphacryptovirus: Partitiviridae*.

The dsRNA genome of the virus shows no significant homology with those of carrot temperate 2, 3, 4 and 5 cryptoviruses. Best tests for diagnosis: the virus often occurs in mixed infections with other carrot temperate cryptoviruses.

**References**

Boccardo, G., Lisa, V., Luisoni, E. and Milne, R.G. (1987). *Adv. Virus Res.* **32**: 171.
Natsuaki, T., Yamashita, S., Doi, Y., Okuda, S. and Teranaka, M. (1983). *Ann. Phytopath. Soc. Japan* **49**: 593.
Natsuaki, T., Natsuaki, K.T., Okuda, S., Teranaka, M., Milne, R.G., Boccardo, G. and Luisoni, E. (1986). *Intervirology* **25**: 69.

# Carrot temperate 2 *betacryptovirus*

**Data collated by A.A. Brunt, 1991.**

*Standard acronym:* CTeV-2.

First reported in *Daucus carota*; from Japan; by Natsuaki *et al.* (1990).

**Natural host range and symptoms**

*Daucus carota* — symptomless infection.

**Transmission**

Virus not transmitted by mechanical inoculation; not transmitted by grafting; transmitted by seed.

**Geographical distribution**

Found, but with no evidence of spread, in Japan.

Dicotyledonous Hosts *(**Susceptible** or Insusceptible)*

***Daucus carota.***

**Particle morphology**

Virions isometric; 30 nm in diameter.

**Biochemical properties**

*Genome* consists of RNA; double–stranded; of two parts; largest (or only) genome part the larger 2.55 kb (pairs); the 2nd largest 2.4 kb (pairs).

**Taxonomy and Relationships**

*Betacryptovirus: Partitiviridae.*

The viral dsRNA genome has no detectable sequence homology with those of carrot temperate 1, 3 and 4 cryptoviruses. Best tests for diagnosis: the virus often occurs in mixed infections with other carrot temperate cryptoviruses.

**References**

Natsuaki, T., Muroi, Y., Okuda, S. and Teranaka, M. (1990). *Ann. Phytopath. Soc. Japan* **56**: 354.

# Carrot temperate 3 *alphacryptovirus*

**Data collated by A.A. Brunt, 1991.**

*Standard acronym:* CTeV-3. *Strains:* carrot temperate 5 cryptovirus.

First reported in *Daucus carota*; from Japan; by Natsuaki *et al.* (1990).

**Natural host range and symptoms**

*Daucus carota* — symptomless infection.

**Transmission**

Virus not transmitted by mechanical inoculation; not transmitted by grafting; transmitted by seed.

**Geographical distribution**

Found, but with no evidence of spread, in Japan.

Dicotyledonous Hosts *(**Susceptible** or Insusceptible)*

***Daucus carota*.**

**Particle morphology**

Virions isometric; 30 nm in length.

**Biochemical properties**

*Genome* consists of RNA; double–stranded; of two parts; largest (or only) genome part the larger 1.58 kb (pairs); the 2nd largest 1.41 kb (pairs).

**Taxonomy and Relationships**

*Alphacryptovirus: Partitiviridae.*

The viral dsRNAs hybridized strongly with that of carrot temperate 5 cryptovirus, but they showed no significant sequence homology with those of carrot temperate 1, 2 and 4 cryptoviruses. Best tests for diagnosis: the virus often occurs in mixed infections with other carrot temperate cryptoviruses.

**References**

Natsuaki, T., Muroi, Y., Okuda, S. and Teranaka, M. (1990). *Ann. Phytopath. Soc. Japan* **56**: 354.

# Carrot temperate 4 *alphacryptovirus*

**Data collated by A.A. Brunt, 1991.**

*Standard acronym:* CTeV-4.

First reported in *Daucus carota*; from Japan; by Natsuaki *et al.* (1990).

**Natural host range and symptoms**

*Daucus carota* — symptomless infection.

**Transmission**

Virus not transmitted by mechanical inoculation; not transmitted by grafting; transmitted by seed.

**Geographical distribution**

Found, but with no evidence of spread, in Japan.

Dicotyledonous Hosts *(**Susceptible** or Insusceptible)*

***Daucus carota*.**

**Particle morphology**

Virions isometric; 30 nm in diameter.

**Biochemical properties**

*Genome* consists of RNA; double–stranded; of two parts; largest (or only) genome part the larger 1.95 kb (pairs); the 2nd largest 1.8 kb (pairs).

**Taxonomy and Relationships**

*Alphacryptovirus: Partitiviridae.*

The viral dsRNA showed no significant homology with those of carrot temperate 1, 2 and 3 cryptoviruses. Best tests for diagnosis: the virus often occurs in mixed infections with other carrot temperate cryptoviruses.

**References**

Natsuaki, T., Muroi, Y., Okuda, S. and Teranaka, M. (1990). *Ann. Phytopath. Soc. Japan* **56**: 354.

# Carrot thin leaf *potyvirus*

**Data collated by W.E. Howell and A.A. Brunt, 1991.**

*Standard acronym:* CTLV.

First reported in *Daucus carota*; from Washington State, U.S.A.; by Howell and Mink (1976).

**Natural host range and symptoms**

Symptoms persist. Symptoms malformation, leaf narrowing and some vein clearing.

*Daucus carota, Conium maculatum* — symptomless.

*Daucus carota* var. *sativum* — leaflet lobes malformed and narrow, vein clearing and some chlorotic spots.

**Transmission**

Transmitted by a vector; an insect; *Cavariella aegopodii, Myzus persicae*; Aphididae. Transmitted in a non–persistent manner. Virus transmitted by mechanical inoculation; not transmitted by contact between plants; not transmitted by seed.

**Ecology and control**

Studies reported by Howell and Mink (1977*a* and *b*).

**Geographical distribution**

Spreads in the USA.

**Experimental host range**

Several (3–9) families susceptible. Experimentally infected plants mostly show systemic thread like leaflets, twisted leaves, vein clearing in Apiaceae; chlorotic and necrotic local lesions in *Chenopodium*.

Diagnostically susceptible

*Anthriscus cerefolium* — thread like leaflets.

*Apium australe* — chlorosis and death of older leaves.

*Chenopodium quinoa* — chlorotic and necrotic local lesions.

*C. amaranticolor* — chlorotic and red ring local lesions.

*Coriandrum sativum, Daucus carota* var. *sativum* — thread like leaflets, vein clearing.

*Pastinaca sativa, Petroselinum hortense, Zinnia elegans* — symptomless.

Diagnostically insusceptible

*Apium graveolens* var. dulce, *Cucumis sativus, Datura stramonium, Nicotiana tabacum, Phaseolus vulgaris.*

Maintenance and propagation hosts

*Chenopodium quinoa, Coriandrum sativum, Daucus carota.*

Assay hosts (**L**ocal lesions or **W**hole plants)

*Chenopodium amaranticolor* (L), *C. quinoa* (L), *Coriandrum sativum* (W), *Daucus carota* (W).

Dicotyledonous Hosts *(**Susceptible** or Insusceptible)*

***Anthriscus cerefolium***. *Apium graveolens. Arachis hypogaea.* ***Atriplex hortensis***. *Beta vulgaris. Brassica oleracea* ssp. *botrytis. Capsella bursa–pastoris. Capsicum frutescens. Catharanthus roseus.* ***Chenopodium amaranticolor***. ***Chenopodium ambrosioides***. ***Chenopodium foliosum***. ***Chenopodium murale***. ***Chenopodium quinoa***. ***Cichorium endiva***. ***Conium maculatum***. ***Coriandrum sativum***. *Cucumis melo. Cucumis sativus. Cucurbita maxima. Datura stramonium.* ***Daucus carota***. *Dianthus caryophyllus. Glycine max. Gomphrena globosa. Hyoscyamus niger. Lactuca sativa.* ***Lathyrus odoratus***. *Lycopersicon esculentum. Nicandra physaloides.* ***Nicotiana clevelandii***. *Nicotiana glutinosa. Nicotiana tabacum.* ***Pastinaca sativa***. *Petunia hybrida. Phaseolus vulgaris. Physalis floridana.* ***Pisum sativum***, or *Pisum sativum. Raphanus sativus. Sesbania exaltata. Solanum melongena. Solanum tuberosum. Spinacia oleracea. Tetragonia expansa. Trifolium pratense. Vicia sativa.* ***Zinnia elegans***.

Sources of host-range data

Howell and Mink (1976; 1977*a*).

**Properties of particles in sap**

TIP: 50–55 °C. LIV: 2 days. DEP: $\log_{10}$ minus 5–6. Leaf sap contains few virions.

**Purification**

Method: Howell and Mink (1976).

**Particle morphology**

Virions filamentous; not enveloped; usually flexuous; with a clear modal length; of 736 nm; 11 nm wide. Axial canal obscure.

**Physical properties**

One sedimenting component in purified preparations, or two sedimenting components in purified preparations. $A_{260}/A_{280}$ ratio 1.18.

**Biochemical properties**

Virions contain 4 % nucleic acid; 96 % protein.

*Genome* consists of RNA; single–stranded.

**Replication**

Replication does not depend on a helper virus.

**Cytopathology**

Virions found in leaves, stems and roots. Inclusions present in infected cells; are pinwheels.

**Taxonomy and Relationships**

*Potyvirus: Potyviridae.*

Virus(es) with serologically unrelated virions: bean common mosaic, bean yellow mosaic, celery mosaic, clover yellow vein, iris mild mosaic, parsnip mosaic, pea seed-borne mosaic, peanut mottle, potato Y, soybean mosaic, sugarcane mosaic, tobacco etch and turnip mosaic viruses.

**References**

Howell, W.E. (1974). *Proc. Am. Phytopath. Soc.* **1**: 38.
Howell, W.E. and Mink, G.I. (1976). *Phytopathology* **66**: 949.
Howell, W.E. and Mink, G.I. (1976). *Pl. Dis. Reptr* **60**: 1047.
Howell, W.E. and Mink, G.I. (1977*a*). *Pl. Dis. Reptr* **61**: 217.
Howell, W.E. and Mink, G.I. (1977*b*). *Pl. Dis. Reptr* **61**; 841.
Howell, W.E. and Mink, G.I. (1979). *Pl. Dis. Reptr* **63**: 989.
Howell, W.E. and Mink, G.I. (1980). *CMI/AAB Descr. Pl. Viruses* No. 218, 3 pp.

# Carrot yellow leaf (?) *closterovirus*

**Data collated by P. van Dijk and L. Bos, 1991.**

*Synonyms:* hogweed 6 virus, heracleum 6 capillovirus. *Standard acronym:* CYLV.

First reported in *Daucus carota* ssp. *sativus*; from Japan; by Yamashita *et al.* (1976).

**Natural host range and symptoms**

Symptoms yellowing in cultivated *Daucus carota* and no visible symptoms in wild umbelliferae.

*Daucus carota* — yellowing and reddening.

*Heracleum sphondylium, Anthriscus sylvestris* — symptomless.

**Transmission**

Transmitted by a vector; an insect; *Cavariella aegopodii, C. pastinacae, C. theobaldi*; Aphididae. Transmitted in a semi–persistent manner. Virus can help the vector transmission of another virus (heracleum latent (?) *closterovirus*); transmitted by mechanical inoculation.

**Geographical distribution**

Found, but with no evidence of spread, in Japan, the Netherlands and the U.K.

**Experimental host range**

Few (<3) families susceptible. Experimentally infected plants mostly show yellowing and necrosis.

Diagnostically susceptible

*Anthriscus cerefolium, Coriandrum sativum, Pimpinella anisum* — vein necrosis, severe leaf rolling and yellowing, leading to plant death.

*Nicotiana benthamiana* — necrotic local lesions and systemic lethal necrosis and yellowing.

*N. clevelandii* — necrotic local lesions without systemic symptoms.

Diagnostically insusceptible

*Chenopodium amaranticolor, C. quinoa, Nicotiana glutinosa, N. occidentalis, Phaseolus vulgaris.*

Maintenance and propagation hosts

*Anthriscus cerefolium* (W), *Coriandrum sativum* (W), *Nicotiana benthamiana* (W).

Dicotyledonous Hosts *(**Susceptible** or Insusceptible)*

***Anthriscus cerefolium***. *Chenopodium amaranticolor. Chenopodium capitatum. Chenopodium foetidum. Chenopodium murale. Chenopodium quinoa.* ***Coriandrum sativum***. *Cucumis sativus.* ***Daucus carota***. ***Heracleum sphondylium***. ***Nicotiana benthamiana***. ***Nicotiana clevelandii***. *Nicotiana glutinosa. Nicotiana megalosiphon. Nicotiana occidentalis. Phaseolus vulgaris. Pisum sativum. Vicia faba.*

Additional hosts

***Anthriscus sylvestris, Heracleum sphondylium, Pimpinella anisum***.

Sources of host-range data

Van Dijk and Bos (1989).

**Properties of particles in sap**

Leaf sap contains few virions. Electron microscopy: virions visible without fixation and staining by PTA.

**Purification**

Method: Murant *et al.* (1984).

**Particle morphology**

Virions filamentous; not enveloped; usually flexuous; with a clear modal length; of 1650 nm (±45 nm); 12 nm wide. Basic helix obvious; pitch of basic helix 3.7 nm.

**Replication**

Replication does not depend on a helper virus. Acts as helper for another virus; a helper for heracleum latent virus. Acts as a helper by possibly phenotypic mixing.

**Cytopathology**

Virions found in leaves and phloem; in cytoplasm and in nuclei. Other cellular changes: characteristic vesicles and morphological changes of mitochondria in phloem cells (Yamashita *et al.*, 1976).

**Taxonomy and Relationships**

Possible *Closterovirus*.

Best tests for diagnosis: transfer by aphid to *Anthriscus cerefolium* or *Coriandrum sativum*, and test these on *Chenopodium quinoa*, which is susceptible to heracleum latent virus but not CYLV, and *Nicotiana benthamiana*, which is susceptible to CYLV not heracleum latent virus. CYLV has flexuous virions, anthriscus yellows virus has isometric ones.

**References**

Bem, F. and Murant, A.F. (1979). *Ann. appl. Biol.* **92**: 237.

Murant, A.F. (1981). *Ann. Rep. Scottish Hort. Res. Inst.* 1980, p. 102.

Murant, A.F. (1983). *Ann. Rep. Scottish. Crop Res. Inst.* 1982, p. 191.

Murant, A.F., Duncan, G.H. and Raschké, J.H. (1984). *Ann. Rep. Scottish Crop Res. Inst.* 1983, p. 188.

Van Dijk, P. and Bos, L. (1985). *Neth. J. Pl. Path.* **91**: 169.

Van Dijk, P. and Bos, L. (1989). *Neth. J. Pl. Path.* **95(2)**: 34.

Yamashita, S., Ohki, S.T., Doi, Y. and Yora, K. (1976). *Ann. Phytopath. Soc. Japan* **42**: 382.

# Cassava African mosaic bigeminivirus

**Data collated by C. Büchen-Osmond, 1987. Revised by B.D. Harrison, 1989.**

*Synonyms:* cassava latent virus, cassava mosaic virus, cassava African mosaic virus. *Standard acronym:* ACMV. *Strains:* type strain isolate 844 (Bock *et al.*, 1978); Kenya coast (C) strain (Bock *et al.*, 1981), Angola defective isolates (Sequeira and Harrison, 1982).

First reported in *Manihot esculenta*; by Warburg (1894); see also Storey and Nichols (1938); Harrison *et al.* (1977); Bock *et al.* (1978).

### Natural host range and symptoms

Symptoms persist and vary cyclically over a few weeks.

*Manihot esculenta, Jatropha multifida* — severe mosaic.

*Hewittia sublobata* — mosaic (probably the natural host of the Kenya coast strain; Bock *et al.*, 1981).

*Laportea* (=*Fluerya*) *aestuans* — bright chlorosis (possibly the natural host in Nigeria; Anon., 1979).

*Manihot glaziovii* — mosaic.

### Transmission

Transmitted by a vector; an insect; *Bemisia tabaci* (Chant, 1958; Dubern, 1979); Aleyrodidae. Transmitted in a persistent manner. Virus retained when the vector moults; not transmitted congenitally to the progeny of the vector (Dubern, 1979); transmitted by mechanical inoculation (with difficulty to cassava (Bock and Woods, 1983) but more readily to *Nicotiana benthamiana*); transmitted by grafting; not transmitted by seed.

### Ecology and control

Studies reported by Bock and Guthrie (1982); Bock (1984); Robertson (1984). Incidence often exceeds 80%, but only a quarter of the cuttings of infected susceptible cassava cultivars yield healthy plants (Bock, 1983). Virus-free cassava plants can be produced by heat treatment and meristem-tip culture. Some cassava genotypes resist infection. Two methods of control are advocated; in Kenya (Bock and Guthrie, 1982; Bock, 1983) and the Ivory Coast (Fargette *et al.*, 1985), virus-free planting material of cultivars with moderate resistance to infection is used, whereas in Nigeria cultivars with superior resistance to, and tolerance of, infection are favoured (Hahn *et al.*, 1980).

### Geographical distribution

Spreads in the African region; Angola, Kenya, Nigeria, and the Seychelles.

### Experimental host range

Few (<3) families susceptible. Experimentally infected plants mostly show chlorotic local lesions, systemic mosaic, leaf curling.

Diagnostically susceptible

*Manihot esculenta* — bright mosaic 3–5 weeks after inoculation by whitefly or by grafting. Leaves with symptoms may alternate with leaves showing few or no symptoms.

*Nicotiana benthamiana* — diffuse chlorotic lesions then systemic leaf curling, malformation and blotching.

*Nicotiana clevelandii* — the type strain may induce chlorotic local lesions then severe malformation, and irregular yellow veinbanding and blotching. The Kenya coast strain is milder and less readily transmitted.

*Datura stramonium* — type strain induces chlorotic and necrotic local lesions then systemic veinbanding, leaf curling and malformation.

Diagnostically insusceptible

*Cucurbita pepo, Chenopodium amaranticolor, Gomphrena globosa, Phaseolus vulgaris.*

Maintenance and propagation hosts

*Manihot esculenta, Nicotiana benthamiana.*

Assay hosts (**L**ocal lesions or **W**hole plants)

*Datura stramonium* (L) for type strain; *Nicotiana benthamiana* (W) for Kenya coast strain.

Dicotyledonous Hosts (***Susceptible*** *or Insusceptible)*

*Atriplex hortensis. Beta vulgaris. Capsicum frutescens. Chenopodium amaranticolor. Chenopodium quinoa. Citrullus lanatus. Cucumis melo. Cucumis sativus. Cucurbita maxima. Cucurbita pepo.* ***Datura ferox****. Datura metel.* ***Datura stramonium****. Gomphrena globosa. Gossypium hirsutum. Ipomoea setosa. Lycopersicon esculentum.* ***Manihot esculenta. Nicandra physaloides. Nicotiana benthamiana. Nicotiana clevelandii. Nicotiana debneyi. Nicotiana glutinosa. Nicotiana rustica. Nicotiana tabacum****. Petunia hybrida. Phaseolus lathyroides. Phaseolus vulgaris. Physalis peruviana. Ricinus communis. Solanum melongena.* ***Solanum nigrum****. Solanum tuberosum. Spinacia oleracea.*

Monocotyledonous Hosts (***Susceptible*** *or Insusceptible)*

*Zea mays.*

Sources of host-range data

Storey and Nichols (1938); Bock and Woods (1983); Bock (1983); Bock *et al.* (1978); Bock *et al.* (1981); Anon. (1979); Dubern (1979).

**Properties of particles in sap**

TIP: 55 °C. LIV: 2–4 days. DEP: $\log_{10}$ minus 3 (type strain in sap of *Nicotiana clevelandii*). Infectivity of sap not changed by treatment with di–ethyl ether. Leaf sap contains few virions. Electron microscopy: virions best detected by immunosorbent electron microscopy.

**Purification**

Method: Sequeira and Harrison (1982).

**Particle morphology**

Virions geminate; not enveloped; 20 nm in diameter; the dimer is 30 nm in length; slightly angular in profile; without a conspicuous capsomere arrangement.

**Physical properties**

Two sedimenting components in purified preparations; sedimentation coefficient of the fastest (main) component 76 *S*; of the other(s) 50 *S*.

**Biochemical properties**

Virions contain 22 % nucleic acid; 78 % protein.

*Genome* consists of DNA; single–stranded; circular; of two parts; largest (or only) genome part the larger 2.779 kb (DNA-1); the 2nd largest 2.724 kb (DNA-2). Genomic nucleic acid isolated by Harrison *et al.* (1977); Sequeira *et al.* (1983). Base composition 23.5 % G (DNA-1 24.6%, DNA-2 22.3%); 27.7 % A (DNA-1 26.9%, DNA-2 28.8%); 18.6 % C (DNA-1 19.9%, DNA-2 17.3%); 30.2 % T (DNA-1 28.5%, DNA-2 31.9%). Infectivity decreased when deproteinised with proteases; retained when deproteinised with phenol or detergent (isolated DNA is 1–50% as infective as the same amount of

DNA in virions). Poly A region absent. Additional factor not required for infectivity (cloned DNA is infective). Genome has no *t*RNA–like activity. Nucleotide sequence references: Stanley and Gay (1983); Stanley (1983).

*Sequence database accession code(s):*

J02057 Em(40)_vi:GE1G Gb(84)_vi:CLV1G Cassava latent virus (West Kenyan 844), DNA 1 of complete genome. 4/90 2,779bp.

J02058 Em(40)_vi:GE2G Gb(84)_vi:CLV2G Cassava latent virus (west kenyan 844), dna 2 of complete genome. 4/90 2,724bp.

X17095 Em(40)_vi:GEN1G Gb(84)_vi:GEN1G African cassava mosaic virus (Nigerian strain) DNA 1. 9/93 2,781bp.

X17096 Em(40)_vi:GEN2G Gb(84)_vi:GEN2G African cassava mosaic virus (Nigerian strain) DNA-2. 9/93 2,725bp.

X68318 Em(40)_vi:ACPPCR6 Gb(84)_vi:ACPPCR6 African Cassava Mosaic Virus gene for coat protein (clone pPCR6). 7/93 852bp.

X68319 Em(40)_vi:ACPPCR5 Gb(84)_vi:ACPPCR5 African Cassava Mosaic Virus gene for coat protein (clone pPCR5). 7/93 852bp.

X68320 Em(40)_vi:ACPPCR4 Gb(84)_vi:ACPPCR4 African Cassava Mosaic Virus gene for coat protein (clone pPCR4). 7/93 852bp. 5 sequences.

*Features of the genome:* the two circular genome DNA species share a sequence of 193 nucleotides which contains a stem and loop structure of 33 nucleotides. Transcription in both directions. DNA-1 encodes one protein of more than 15 kDain the plus (virion DNA) sense and three proteins more than 15 kDain the minus sense. DNA-2 encodes one protein more than 15 kDain each sense. Possible promoter and terminator and polyadenylation signals are found in all six ORFs. RNA transcripts of the expected sizes are found.

*Non–genomic nucleic acid* found in the virions; Non-genomic nucleic acid is derived from DNA-2. Sub–genomic *m*RNA found in infected cells.

*Virion protein(s)* one; $M_r$ 31800 (±700 (Sequeira, 1982) or 30200 when translated from the nucleotide sequence (Stanley and Gay, 1983)). Method of preparation: Bock *et al.* (1977); Sequeira (1982). Amino acid sequence: Stanley *et al.* (1985). Virion proteins not glycosylated (Sequeira, 1982).

*Virus–coded non–virion proteins* identified by genomic sequence analysis; five proteins found. $M_r$ 40300. $M_r$ of 2nd largest 33700; 2L1, which assists virus movement. $M_r$ of 3rd 29300; 2R1, assisting virus movement. $M_r$ of 4th 15800. $M_r$ of 5th and smaller 15100.

## Replication

Replication does not depend on a helper virus.

## Cytopathology

Virions found in mesophyll, epidermis, phloem and companion cells; in nuclei. Inclusions present in infected cells; are unusual in shape; a granular material; they contain virions. Other cellular changes: the production of fibrillar rings in the nuclei of phloem parenchyma cells (Horvat and Verhoyen, 1981; Adejare and Coutts, 1982).

## Taxonomy and Relationships

Bigeminivirus: *Geminiviridae.*

Virus(es) with serologically related virions: bean golden mosaic (SDI:2) (Sequeira and Harrison, 1982), squash leaf curl (Cohen *et al.*, 1983), tomato golden mosaic (SDI:2) (Stein *et al.*, 1983), euphorbia mosaic (Roberts *et al.*, 1984) and many other whitefly transmitted geminiviruses (M. Aiton and B.D. Harrison, unpublished information).

There are sequence homologies between the larger genome part (DNA-1) of African cassava mosaic virus and the DNA of bean golden mosaic, tomato golden mosaic, tobacco leaf curl and tomato leaf curl viruses (Roberts *et al.*, 1984). DNA-1 of African cassava mosaic virus shows clear sequence similarities with bean golden mosaic virus in coding regions, but little in non-coding regions (Hamilton *et al.*, 1984; Harrison, 1985). No serological relationship or genome homology has been found between African cassava mosaic virus and any monogeminiviruses.

The close serological relationship between biologically distinct geminiviruses can lead to difficulties in virus identification, especially when plant species not previously recorded as hosts are infected. However, African cassava mosaic virus does not infect *Phaseolus vulgaris* or *Cucurbita pepo* whereas bean golden mosaic and squash leaf curl viruses do. In hybridization tests, African cassava mosaic virus DNA-2 probes do not react with extracts of plants infected with bean golden mosaic, tobacco golden mosaic or tobacco leaf curl and allied geminiviruses (Roberts *et al.*, 1984). Best tests for diagnosis: African cassava mosaic virus is the only geminivirus reported from cassava in Africa. However, in South America, similar symptoms are caused by cassava common mosaic potexvirus which also causes local lesions in *Chenopodium amaranticolor* and *Gomphrena globosa* and is not transmitted by *Bemisia tabaci* (Costa and Kitajima, 1972). Viruses associated with two other mosaic diseases of cassava in Colombia are not sap-transmissible to *Nicotiana benthamiana*. Indian cassava mosaic virus is now considered to be a separate geminivirus.

**References**


Adejare, G.O. and Coutts, R.H.A. (1982). *Phytopath. Z.* **103**: 87.

Anon. (1979). *Rep. Int. Inst. Trop. Agric. 1978*, p.107.

Bock, K.R. (1983). In: *Plant Virus Epidemiology*, p. 337; eds R.T. Plumb and J.M. Thresh. Oxford, Blackwell Scientific Publications.

Bock, K.R. (1984). *ODA Crop Virol. Res. Proj. Kenya Agric. Res. Inst., Final Report.* Overseas Development Administration, London.

Bock, K.R. and Guthrie, E.J. (1982). *Trop. Pest Manag.* **28**: 219.

Bock, K.R. and Harrison, B.D. (1985). *CMI/AAB Descr. Pl. Viruses* No. 297, 6 pp.

Bock, K.R. and Woods, R.D. (1983). *Plant Dis.* **67**: 994.

Bock, K.R., Guthrie, E.J. and Figueiredo, G. (1981). *Ann. appl. Biol.* **99**: 151.

Bock, K.R., Guthrie, E.J. and Meredith, G. (1978). *Ann. appl. Biol.* **90**: 361.

Bock, K.R., Guthrie, E.J., Meredith, G. and Barker, H. (1977). *Ann. appl. Biol.* **85**: 305.

Chant, S.R. (1958). *Ann. appl. Biol.* **46**: 210.

Cohen, S., Duffus, J.E., Larsen, R.C., Liu, H.Y. and Flock, R.A. (1983). *Phytopathology* **73**: 1669.

Costa, A.S. and Kitajima, E.W. (1972). *CMI/AAB Descr. Pl. Viruses* No. 90, 4 pp.

Dubern, J. (1979). *Phytopath. Z.* **96**: 25.

Fargette, D., Fauquet, C. and Thouvenel, J.-C. (1985). *Ann. appl. Biol.* **106**: 285.

Hahn, S.K., Terry, E.R. and Leuschner, K. (1980). *Euphytica* **29**: 673.

Hamilton, W.D.O., Stein, V.E., Coutts, R.H.A. and Buck, K.W. (1984). *EMBO J.* **3**: 2197.

Harrison, B.D. (1985). *Ann. Rev. Phytopath.* **23**: 55.

Harrison, B.D., Barker, H., Bock, K.R., Guthrie, E.J., Meredith, G. and Atkinson, M. (1977). *Nature, Lond.* **270**: 760.

Horvat, F. and Verhoyen, M. (1981). *Parasitica* **37**: 27119.

Kaiser, W.J. and Louie, R. (1982). *Plant Dis.* **66**: 475.

Roberts, I.M., Robinson, D.J. and Harrison, B.D. (1984). *J. gen. Virol.* **65**: 1723.

Robertson, I.A.D. (1984). *Cassava Whitefly Proj. Crop Virol. Res. Proj. Kenya Agric. Res. Inst., Final Report.* Overseas Development Administration, London.
Sequeira, J.C. (1982). *Ph.D. Thesis*, University of Dundee, U.K.
Sequeira, J.C. and Harrison, B.D. (1982). *Ann. appl. Biol.* **101**: 33.
Sequeira, J.C., Harrison, B.D. and Duncan, G.H. (1983). *Rep. Scottish Crop Res. Inst.* 1982, p. 193.
Stanley, J. and Gay, M.R. (1983). *Nature, Lond* **301**: 260.
Stein, V.E., Coutts, R.H.A. and Buck, K.W. (1983). *J. gen. Virol.* **64**: 2493.
Storey, H.H. and Nichols, R.F.W. (1938). *Ann. appl. Biol.* **25**: 790.
Townsend, R., Stanley, J., Curson, S.J. and Short, M.N. (1985). *EMBO J.* **4**: 33.
Warburg, O. (1894). *Mitt. Deutsch. Schutzgeb.* **7**: 131.

# Cassava brown streak-associated (?) *carlavirus*

**Data collated by A.A. Brunt, 1990.**

*Standard acronym:* CBSaV.

First reported in *Manihot esculenta*; from East Africa; by Storey (1936).

**Natural host range and symptoms**

*Manihot esculenta* — mixed infections (with cassava brown streak potyvirus) cause mottling and brown streaked stems.

**Transmission**

Transmitted by a vector; an insect; *Bemisia tabaci*; Aleyrodidae. Virus transmitted by mechanical inoculation.

**Ecology and control**

Studies reported by Jennings (1960).

**Geographical distribution**

Spreads in the African region (in the east).

**Experimental host range**

Several (3–9) families susceptible. Experimentally infected plants mostly show systemic mosaics, mottles, ringspots or necrosis.

Diagnostically susceptible

*Nicotiana benthamiana, N. debneyi* — systemic mosaic and mottling.
*Chenopodium quinoa, C. amaranticolor* — local infection.

Maintenance and propagation hosts

*Nicotiana benthamiana, N. debneyi.*

Dicotyledonous Hosts *(**Susceptible** or Insusceptible)*

***Adansonia digitata**. **Chenopodium amaranticolor**. **Chenopodium quinoa**. **Manihot esculenta**. **Nicotiana benthamiana**. **Nicotiana debneyi**.*

**Properties of particles in sap**

Leaf sap contains many virions.

**Particle morphology**

Virions filamentous; not enveloped; usually flexuous; with a clear modal length; of 650–690 nm.

**Taxonomy and Relationships**

Possible *Carlavirus*.

Virus(es) with serologically related virions: cowpea mild mottle (?) *carlavirus*.

**References**

Jennings, D.C. (1960). *Eur. J. exp. Agric.* **28**: 261.
Kitajima, E.W. and Costa, A.S. (1964). *E. Afr. agric. For. J.* **30**: 28.
Lennon, A.M., Aiton, M.M. and Harrison, B.D. (1986). *Rep. Scottish Crop Res. Inst.* 1985, p. 168.
Storey. H.H. (1936). *E. Afr. agric. J.* **2**: 34.

# Cassava brown streak *potyvirus*

**Data collated by A.A. Brunt, 1990.**

First reported in *Manihot esculenta*; from East Africa; by Storey (1936).

**Natural host range and symptoms**

Symptoms persist.

*Manihot esculenta* — mixed infections with cassava brown streak (?) *carlavirus* give mottling and brown streaked stems.

**Geographical distribution**

Spreads in the African region (in the east).

**Experimental host range**

Few (<3) families susceptible.

Diagnostically susceptible

*Nicotiana benthamiana, N. debneyi.*

Maintenance and propagation hosts

*Nicotiana benthamiana.*

Dicotyledonous Hosts *(**Susceptible** or Insusceptible)*

***Manihot esculenta*. *Nicotiana benthamiana*. *Nicotiana debneyi*.**

**Particle morphology**

Virions filamentous; not enveloped; usually flexuous; with a clear modal length; of 750 nm.

**Taxonomy and Relationships**

*Potyvirus: Potyviridae.*

Virus(es) with serologically related virions: wild potato mosaic potyvirus.

**References**

Kitajima, E.W. and Costa, A.S. (1964). *E. Afr. agric. For. J.* **30**: 28.
Lennon, A.M., Aiton, M.M. and Harrison, B.D. (1986). *Rep. Scottish Crop Res. Inst.* 1985, p. 168.
Storey, H.H. (1936). *Rep. E. Afr. agric. Res. Inst.*.

# Cassava Caribbean mosaic (?) *potexvirus*

**Data collated by A.A. Brunt, 1990.**

First reported in *Manihot esculenta*; from Colombia; by Lennon *et al.* (1986).

**Natural host range and symptoms**

*Manihot esculenta* — mosaic.

**Transmission**

Virus transmitted by mechanical inoculation.

**Geographical distribution**

Spreads in Colombia.

**Experimental host range**

Experimentally infected plants mostly show local lesions or symptomless.

Diagnostically susceptible

*Chenopodium quinoa* — local lesions.
*Nicotiana benthamiana* — symptomless; not systemic.

Dicotyledonous Hosts *(**Susceptible** or Insusceptible)*

***Chenopodium quinoa*. *Manihot esculenta*. *Nicotiana benthamiana*.**

**Particle morphology**

Virions filamentous; with a clear modal length; of 460 nm.

**Taxonomy and Relationships**

Possible *Potexvirus*.

Virus(es) with serologically related virions: cassava X potexvirus is distantly related. Virus(es) with serologically unrelated virions: cassava Colombian symptomless potexvirus (Lennon *et al.*, 1987).

**References**

Lennon, A.M., Aiton, M.M. and Harrison, B.D. (1986). *Rep. Scottish Crop Res. Inst.* 1985, p. 167.
Lennon, A.M., Aiton, M.M. and Harrison, B.D. (1987). *Rep. Scottish Crop Res. Inst.* 1986, p. 181.

# Cassava Colombian symptomless (?) *potexvirus*

**Data collated by A.A. Brunt, 1990.**

First reported in *Manihot esculenta*; from Colombia; by Lennon *et al.* (1987).

**Natural host range and symptoms**

*Manihot esculenta* — systemic, but symptomless.

**Transmission**

Virus transmitted by mechanical inoculation.

**Geographical distribution**

Spreads in Colombia.

Diagnostically susceptible

*Chenopodium* spp., *Nicotiana benthamiana* — symptomless, not systemic.

Dicotyledonous Hosts (***Susceptible*** *or Insusceptible)*

***Manihot esculenta*. *Nicotiana benthamiana*.**

**Properties of particles in sap**

Leaf sap contains few virions.

**Particle morphology**

Virions filamentous; with a clear modal length; of 460 nm; 13 nm wide.

*Virion protein(s)* one; $M_r$ 26000; coat protein.

**Taxonomy and Relationships**

Possible *Potexvirus*.

Virus(es) with serologically related virions: narcissus mosaic virus, but only distantly related to cassava common mosaic and cassava X potexviruses. Virus(es) with serologically unrelated virions: nine other potexviruses (Aiton and Harrison, 1988).

Best tests for diagnosis: this virus sometimes occurs with cassava X potexvirus or in plants with frogskin or Caribbean mosaic symptoms but apparently is not the cause of these diseases.

**References**

Aiton, M.M. and Harrison, B.D. (1988). *Rep. Scottish Crop Res. Inst.* 1987, p. 192.

Lennon, A.M., Aiton, M.M. and Harrison, B.D. (1987). *Rep. Scottish Crop Res. Inst.* 1986, p. 181.

# Cassava common mosaic (?) *potexvirus*

**Data collated by E.W. Kitajima, 1985. Revised by F.W. Zettler, 1989.**

*Synonyms:* Brazilian cassava common mosaic virus. *Standard acronym:* CsCMV. *Strains:* chaya strain.

First reported in *Manihot esculenta*; from Sao Paulo, Brazil; by Costa and Normanha (1940).

**Natural host range and symptoms**

Symptoms persist.

*Cnidoscolus acontifolius, Manihot esculenta* — mosaic.

**Transmission**

Transmitted by means not involving a vector; disseminated in infected cuttings. Virus transmitted by mechanical inoculation; transmitted by grafting; not transmitted by seed; not transmitted by pollen.

**Geographical distribution**

Spreads in Brazil, Colombia, Mexico, Peru, Taiwan, and the USA (Florida), Cote d'Ivoire (possibly).

**Experimental host range**

Several (3–9) families susceptible. Experimentally infected plants mostly show chlorotic or necrotic local lesions, not systemic.

Diagnostically susceptible

*Gomphrena globosa, Chenopodium amaranticolor, C. murale, C. quinoa, Gossypium hirsutum, Manihot esculenta, Ricinus communis* — local lesions; not systemic.

*Cassia occidentalis, Datura stramonium* — chlorotic or necrotic local lesions; not systemic (Chaya strain).

*Nicotiana benthamiana* — chlorotic local lesions; mild systemic chlorosis.

Diagnostically insusceptible

*Nicotiana glutinosa, N. tabacum, Zea mays.*

Maintenance and propagation hosts

*Euphorbia prunifolia, Manihot esculenta.*

Assay hosts (**L**ocal lesions or **W**hole plants)

*Gomphrena globosa* (L), *Chenopodium amaranticolor* (L), *C. quinoa* (L).

Dicotyledonous Hosts *(**Susceptible** or Insusceptible)*

***Chenopodium album***. ***Chenopodium amaranticolor***. *Chenopodium murale*. ***Chenopodium quinoa***, or *Chenopodium quinoa*. *Chrysanthemum morifolium*. *Cucurbita pepo*. *Datura metel*. *Datura stramonium*. *Euphorbia marginata*. *Glycine max*. ***Gomphrena globosa***. ***Gossypium hirsutum***. *Lycopersicon esculentum*. *Nicandra physaloides*. ***Nicotiana benthamiana***. *Nicotiana clevelandii*. *Nicotiana × edwardsonii*. *Nicotiana debneyi*. *Nicotiana glutinosa*. *Nicotiana rustica*. *Nicotiana tabacum*. *Phaseolus vulgaris*. *Physalis floridana*. *Pisum sativum*. ***Ricinus communis***. *Vigna unguiculata*.

Monocotyledonous Hosts *(**Susceptible** or Insusceptible)*

*Zea mays*.

**Properties of particles in sap**

TIP: 65–70 °C. LIV: 128 days. DEP: $\log_{10}$ minus 5–6. Infectivity of sap not changed by treatment with di–ethyl ether. Leaf sap contains few virions.

**Purification**

Method: Nolt *et al.* (1991).

**Particle morphology**

Virions filamentous; not enveloped; usually flexuous; with a clear modal length; of 495–525 nm; 15 nm wide. Axial canal obscure. Basic helix obscure.

**Physical properties**

One sedimenting component in purified preparations; sedimentation coefficient 126 *S*.

**Biochemical properties**

Virions contain 5 % nucleic acid.

*Genome* consists of RNA; single–stranded; linear (probably); unipartite; largest (or only) genome part 6 kb.

*Sequence database accession code(s):*

U23414 Em(43)_vi:Cc23414 Gb(89)_vi:Ccu23414 Cassava common mosaic virus Brazilian strain, complete genome. 5/95 6,376bp.

U23415 Em(43)_vi:Cc23415 Gb(89)_vi:Ccu23415 Cassava common mosaic virus Brazilian strain defective RNA, complete sequence. 5/95 1,094bp.

*Virion protein(s)* one; $M_r$ 21000.

**Replication**

Replication does not depend on a helper virus.

**Cytopathology**

Virions found in all parts of the host plant; in cytoplasm. Inclusions present in infected cells; are unusual in shape; fibrous masses; they contain virions.

**Taxonomy and Relationships**

Possible *Potexvirus*.

Virus(es) with serologically related virions: potato X and tulip X potexviruses. Virus(es) with serologically unrelated virions: cactus X, cassava X, hydrangea ringspot, narcissus mosaic and nerine X potexviruses.

Chaya strain, unlike type strain, infects *Chenopodium quinoa* and induces nuclear inclusions.

**References**

Costa, A.S. and Kitajima, E.W. (1972). *CMI/AAB Descr. Pl. Viruses* No. 90, 4 pp.
Costa, A.S. and Normanha, E.S. (1940). *J. Agron. (Piracicuba)* **3**: 239.
Elliott, M.S. and Zettler, F.W. (1987). *Plant Dis.* **71**: 353.
Kitajima, E.W., Wetter, C., Oliveira, A.R., Silva, D.M. and Costa, A.S. (1965). *Bragantia* **24**: 260.
Nolt, B.L., Velasco, A.C. and Pineda, B. (1991). *Ann. appl. Biol.* **118**: 105.
Silva, D.M., Kitajima, E.W. and Oliveira, A.R. (1963). *Cienc. Cult.* **15**: 304.
Nolt, B.L., Velasco, A.C. and Pineda, B. (1991). *Ann. appl. Biol.* **105**: .

# Cassava green mottle *nepovirus*

**Data collated by A.M. Lennon, M.M. Aiton and B.D. Harrison, 1986.**

*Standard acronym:* CGMV.

First reported in *Manihot esculenta*; from in a sample from the Solomon Islands examined in Scotland; by Lennon *et al.* (1985).

**Natural host range and symptoms**

Symptoms disappear soon after infection.
*Manihot esculenta* — leaf mottling.

**Transmission**

Transmitted by means not involving a vector. Virus transmitted by mechanical inoculation.

**Geographical distribution**

Spreads in the Pacific region; the Solomon Islands.

**Experimental host range**

Many (>9) families susceptible. Experimentally infected plants mostly show mottles, both in inoculated and other leaves, mosaic.

Diagnostically susceptible

*Manihot esculenta* — local and systemic mottle.
*Chenopodium quinoa* — necrotic local lesions; systemic necrosis.
*Nicotiana clevelandii* — necrotic and systemic spots.
*Phaseolus vulgaris, Cucumis sativus* — systemic mottling.
*Ricinis communis* — necrotic local lesions, mosaic.
*Ipomoea batatas* — symptomless systemic infection.
*Solanum tuberosum* — local and systemic black necrotic spots.

Diagnostically insusceptible

*Zea mays, Arachis hypogaea, Hordeum vulgare, Triticum aestivum, Brassica oleracea.*

Maintenance and propagation hosts

*Nicotiana clevelandii.*

Assay hosts (**L**ocal lesions or **W**hole plants)

*Chenopodium quinoa* (L), *Chenopodium amaranticolor* (L).

Dicotyledonous Hosts *(**Susceptible** or Insusceptible)*

***Apium graveolens***. *Arachis hypogaea*. *Brassica campestris* ssp. *rapa*. *Brassica oleracea* ssp. *botrytis*. ***Catharanthus roseus***. ***Chenopodium amaranticolor***. ***Chenopodium murale***. ***Chenopodium quinoa***. ***Cucumis sativus***. ***Datura stramonium***. ***Daucus carota***. ***Gomphrena globosa***. ***Lactuca sativa***. ***Lycopersicon esculentum***. ***Nicotiana benthamiana***. ***Nicotiana clevelandii***. ***Nicotiana debneyi***. ***Nicotiana glutinosa***. ***Nicotiana rustica***. ***Nicotiana tabacum***. ***Petunia hybrida***. ***Phaseolus vulgaris***. *Pisum sativum*. ***Ricinus communis***. ***Solanum tuberosum***. ***Tetragonia expansa***. ***Vicia faba***.

Monocotyledonous Hosts *(**Susceptible** or Insusceptible)*

*Hordeum vulgare*. *Triticum aestivum*. *Zea mays*.

**Properties of particles in sap**

TIP: 60–65 °C. LIV: 12 days (or more). DEP: $\log_{10}$ minus 5–6. Infectivity of sap not changed by treatment with di–ethyl ether. Leaf sap contains few virions.

**Particle morphology**

Virions isometric; not enveloped; 27 nm in diameter; rounded in profile, or angular in profile; without a conspicuous capsomere arrangement.

**Physical properties**

Two sedimenting components in purified preparations, or three sedimenting components in purified preparations. $A_{260}/A_{280}$ ratio 1.8–1.9 (uncorrected).

**Biochemical properties**

*Genome* consists of RNA; single–stranded; linear; of two parts; largest (or only) genome part the larger 9.5 kb; the 2nd largest 7.4 kb. Infectivity retained when deproteinised with phenol or detergent.

*Virion protein(s)* one (or more); $M_r$ 53000.

**Replication**

Replication does not depend on a helper virus.

**Cytopathology**

Virions found in all parts of the host plant; in cytoplasm. Inclusions present in infected cells; are unusual in shape. Other cellular changes: the virus occasionally being found in tubules.

**Taxonomy and Relationships**

*Nepovirus: Comoviridae.*

Virus(es) with serologically unrelated virions: arabis mosaic, arracacha A, artichoke Italian latent, artichoke vein-banding, cherry leaf roll, cacao necrosis, grapevine Bulgarian latent, grapevine fanleaf, grapevine Hungarian chrome mosaic, hibiscus latent ringspot, lucerne Australian latent, mulberry ringspot, olive latent ringspot, peach rosette mosaic, potato black ringspot, potato U, raspberry ringspot, strawberry latent ringspot, tobacco ringspot, tomato black ring, tomato necrosis and tomato ringspot viruses.

**References**

Lennon, A.M., Aiton, M.M. and Harrison, B.D. (1987). *Ann. appl. Biol.* **10**: 545.

Lennon, A.M., Aiton, M.M. and Harrison, B.D. (1985). *Rep. Scottish Crop Res. Inst.* 1985, p. 167.

# Cassava Indian mosaic bigeminivirus

**Data collated by V. Muniyappa, 1992.**

*Synonyms:* Indian cassava mosaic virus. *Standard acronym:* ICMV.

First reported in *Manihot esculenta* — cassava; from India; by Abraham (1956).

**Natural host range and symptoms**

Symptoms vary seasonally. Symptoms mosaic, chlorosis and deformation of leaves.

*Manihot esculenta* — mosaic and irregular chlorotic patches on leaves with distortion and stunting. Enations sometimes, on abaxial surfaces of leaves.

*M. glaziovii* — leaf mosaic and deformation.

**Transmission**

Transmitted by a vector; an insect; *Bemisia tabaci*; Aleyrodidae. Transmitted in a persistent manner. Virus retained when the vector moults; does not multiply in the vector; not transmitted congenitally to the progeny of the vector; transmitted by mechanical inoculation; transmitted by grafting; not transmitted by contact between plants; not transmitted by seed.

**Ecology and control**

Studies reported by Alagianagalingam *et al.* (1966), Malathi *et al.* (1985).

**Geographical distribution**

Spreads in India and Sri Lanka.

**Experimental host range**

Few (<3) families susceptible. Experimentally infected plants mostly show chlorotic lesions, systemic mosaic, vein thickening, leaf curling deformation and stunting.

Diagnostically susceptible

*Datura stramonium, Nicandra physaloides, Nicotiana benthamiana, N. debneyi, N. glutinosa, N. occidentalis, N. rustica, N. sylvestris, N. tabacum* (cv. Jaysree) and *Petunia hybrida* — vein-clearing, chlorotic lesions, mosaic, curling and stunting.

Diagnostically insusceptible

*Abelmoschus esculentus, Chenopodium quinoa, Croton bonplaudianum, Euphorbia geniculata, Gomphrena globosa.*

Maintenance and propagation hosts

*Manihot esculenta, M. glaziovii, Nicandra physaloides, Nicotiana tabacum* cv. Jayasree and *N. hybrida*.

Assay hosts (**L**ocal lesions or **W**hole plants)

*Nicotiana benthamiana* (W), *N. glutinosa* (W), *N. hybrida* (W), *N. tabacum* (cvs. Jaysree and Samsun) (W).

Dicotyledonous Hosts *(**Susceptible** or Insusceptible)*

*Ageratum conyzoides. Arachis hypogaea. Cajanus cajan. Capsicum annuum. Chenopodium album. Chenopodium amaranticolor. Chenopodium murale. Chenopodium quinoa. Citrullus lanatus. Cucumis sativus. Cucurbita pepo. Cyamopsis tetragonoloba. Datura metel.* ***Datura stramonium****. Dolichos biflorus. Glycine max. Gomphrena globosa.*

*Gossypium hirsutum. Lycopersicon esculentum. Macrotyloma uniflorum.* ***Manihot esculenta. Nicandra physaloides. Nicotiana benthamiana. Nicotiana bigelovii. Nicotiana clevelandii. Nicotiana debneyi. Nicotiana glutinosa. Nicotiana megalosiphon. Nicotiana occidentalis. Nicotiana rustica. Nicotiana sylvestris. Nicotiana tabacum. Petunia hybrida.*** *Phaseolus lunatus. Phaseolus vulgaris. Sesamum indicum. Solanum melongena. Solanum tuberosum. Vigna radiata. Vigna unguiculata. Vigna unguiculata* ssp. *cylindrica. Vigna unguiculata* ssp. *sesquipedalis. Zinnia elegans.*

Sources of host-range data

Mathew (1988), Mathew and Muniyappa (1991), Mathew and Muniyappa (1993).

**Properties of particles in sap**

TIP: 55–60 °C. LIV: 3 days (at room temperature). DEP: $\log_{10}$ minus 3–4. Leaf sap contains few virions.

**Purification**

Method: Mathew and Muniyappa, (1992).

**Particle morphology**

Virions geminate; 16–18 nm in diameter; 30 nm in length.

**Physical properties**

$A_{260}/A_{280}$ ratio 1.4.

**Biochemical properties**

*Genome* consists of DNA; single–stranded; circular.

*Sequence database accession code(s):*

Z24758 Em(40)_vi:ICMVCGA Gb(84)_vi:ICMVCGA Indian cassava mosaic virus encoding AR0 complete CDS. 3/94 2,815bp.

Z24759 Em(40)_vi:ICMVBR1A Gb(84)_vi:ICMVBR1A Indian cassava mosaic virus BR1 and BL1 genes, complete CDS's. 2/94 2,645bp. 2 sequences.

*Virion protein(s)* one; $M_r$ 34000; coat protein.

**Cytopathology**

Virions found in leaves, phloem; in cytoplasm and in nuclei.

**Taxonomy and Relationships**

Bigeminivirus: *Geminiviridae.*

**Comments**

Virus isolates from cassava (*Manihot esculenta*) in India and Sri Lanka, although earlier tentatively regarded as Group C isolates of African cassava mosaic geminivirus, are now considered to be those of a distinct virus and named Indian cassava mosaic geminivirus (Harrison *et al.*, 1986; Harrison and Robinson, 1988). The virus is transmitted by *Bemisia tabaci*, induces conspicuous leaf symptoms in infected plants and causes ultrastructural effects typical of geminiviruses (Roberts, 1989). The virus is serologically more closely related to some other geminiviruses than to African cassava mosaic geminivirus.

**References**

Abraham, A. (1956). *Farm Bulletin No. 17. Ind. Counc. Agric. Res. New Delhi*, 20 pp.

Alagianagalingam, M.N. and Ramakrishnan, K. (1966). *South Ind. Hort* **14**: 71.

Malathi, V.G., Narr, N.G. and Santha, P. (1985). *Tech. Bull. Ser. 5. Central Tuber Crops Res. Inst.* Trivandrum, India. 18pp.

Harrison, B.D. and Robinson, D.J. (1988). *Phil. Trans. Roy. Soc.* London **3321**: 447.
Harrison, B.D., Lennon, A.M., Massalski, P.R., Robinson, D.J. and Thomas, J.E. (1986). *Rep. Scottish Crop Res. Inst.* 1985, p. 179.
Mathew, A.V. (1988) Ph.D. Thesis, 237 pp. Univ. Agric. Sci., Bangalore, India.
Mathew, A.V. and Muniyappa, V. (1991). *Fitopatol. Bras.* **16**: 46.
Mathew, A.V. and Muniyappa, V. (1992). *J. Phytopathology* **135**: 299.
Mathew, A.V. and Muniyappa, V. (1993). *Indian Phytopathology* **46**: 16
Roberts, I.M. (1989). *J. gen. Virol.* **70**: 2729.

# Cassava Ivorian bacilliform ourmiavirus

**Data collated by A.A. Brunt, 1990. Revised 1992.**

First reported in *Manihot esculenta*; from Ivory Coast; by Aiton *et al.* (1988).

**Natural host range and symptoms**

*Manihot esculenta* — symptomless.

**Transmission**

Not transmitted by *Myzus persicae*. Virus transmitted by mechanical inoculation.

**Geographical distribution**

Spreads in Cote d'Ivoire.

**Experimental host range**

Several (3–9) families susceptible. Experimentally infected plants mostly show systemic chlorosis.

Diagnostically susceptible

*Chenopodium amaranticolor* — slight distortion of young systemically infected leaves.

*Chenopodium murale, C. quinoa* — chlorotic or necrotic local lesions; systemic necrosis.

*Gomphrena globosa, Tetragonia expansa* — chlorotic or necrotic local lesions; symptomless systemic infection.

*Nicotiana benthamiana* — symptomless or faint systemic mottle.

*Nicotiana clevelandii, N. rustica, N. tabacum* cv. Samsun,
*Phaseolus vulgaris, Senecio cruentus* — symptomless systemic infection.

Diagnostically insusceptible

*Cucumis sativus, Datura stramonium, Nicotiana debneyi, Physalis floridana, Vigna unguiculata.*

Maintenance and propagation hosts

*Chenopodium quinoa.*

Assay hosts (**L**ocal lesions or **W**hole plants)

*Chenopodium murale* (L), *C. amaranticolor* (W), *C. quinoa* (W).

Dicotyledonous Hosts *(**Susceptible** or Insusceptible)*

*Apium graveolens. Brassica campestris* ssp. *rapa. Catharanthus roseus.* ***Chenopodium amaranticolor. Chenopodium murale. Chenopodium quinoa.*** *Coriandrum sativum. Cucumis sativus. Datura stramonium. Daucus carota.* ***Gomphrena globosa.*** *Lactuca sativa. Lycopersicon chilense. Lycopersicon esculentum.* ***Manihot esculenta. Nicotiana benthamiana. Nicotiana clevelandii.*** *Nicotiana debneyi.* ***Nicotiana*** rustica. ***Nicotiana tabacum***. *Pastinaca sativa*. ***Petunia hybrida***. ***Phaseolus vulgaris***. *Physalis floridana*. *Pisum sativum*. ***Tetragonia expansa***. *Vicia faba*. *Vigna unguiculata*.

Comments on host-range

Additional Hosts: species of the Aizoaceae, Amaranthaceae, Chenopodiaceae, Compositae, Leguminosae, Solanaceae.

Sources of host-range data

Aiton *et al.* (1988).

**Properties of particles in sap**

TIP: 55 °C. DEP: $\log_{10}$ minus 3.

**Purification**

Method: Fargette *et al.* (1991).

**Particle morphology**

Virions bacilliform; not enveloped; 42 nm in length, or 49 nm in length, or 76 nm in length; 18 nm wide.

**Physical properties**

Three sedimenting components in purified preparations to four sedimenting components in purified preparations.

**Biochemical properties**

Virions contain 25 % nucleic acid; 75 % protein; 0 % lipid.

*Genome* consists of RNA; single–stranded; of three parts; largest (or only) genome part 3.94 kb; the 2nd largest 3.03 kb; the 3rd largest 2.73 kb.

*Virion protein(s)* one; $M_r$ 22000; coat protein.

**Taxonomy and Relationships**

Ourmiavirus.

Virus(es) with serologically unrelated virions: alfalfa mosaic alfamovirus.

**References**


Aiton, M.M., Roberts, I.M. and Harrison, B.D. (1988). *Rep. Scottish Crop Res. Inst.* 1987, p. 191.

Fargette, D., Roberts, I.M. and Harrison, B.D. (1989). *Rep. Scottish Crop Res. Inst. for 1988*, p. 179.

Fargette, D., Roberts, I.M. and Harrison, B.D. (1991). *Ann. appl. Biol.* **119**: 303.


# Cassava symptomless (?) rhabdovirus

**Data collated by A.A. Brunt, 1989.**

*Standard acronym:* CsSLV.

First reported in *Manihot* spp.; from Brazil; by Kitajima and Costa (1979).

**Natural host range and symptoms**

*Manihot* spp. — symptomless.

**Geographical distribution**

Spreads in Brazil.

**Taxonomy and Relationships**

Unassigned rhabdovirus.

**References**

Kitajima, E.W. and Costa, A.S. (1979). *Fitopatol. Bras.* **4**: 55.

# Cassava vein mosaic (?) *caulimovirus*

**Data collated by E.W. Kitajima, 1985.**

*Standard acronym:* CVMV.

First reported in *Manihot esculenta*; from Sao Paulo, Brazil; by Costa and Normanha (1940).

**Natural host range and symptoms**

Symptoms persist.
*Manihot esculenta* — vein mosaic.

**Transmission**

Transmitted by means not involving a vector. Virus transmitted by mechanical inoculation; transmitted by grafting; not transmitted by seed; not transmitted by pollen.

**Geographical distribution**

Spreads in Brazil.

**Experimental host range**

Few (<3) families susceptible. Experimentally infected plants mostly show vein mosaic.

Diagnostically susceptible
*Manihot esculenta* — vein mosaic.

Maintenance and propagation hosts
*Manihot esculenta*.

Assay hosts (**L**ocal lesions or **W**hole plants)
*Manihot esculenta* (W).

**Properties of particles in sap**

Infectivity of sap not changed by treatment with di–ethyl ether. Leaf sap contains few virions.

**Purification**

Method: Lin and Kitajima (1980).

**Particle morphology**

Virions isometric; not enveloped; 45–50 nm in diameter; rounded in profile; without a conspicuous capsomere arrangement.

**Physical properties**

One sedimenting component in purified preparations; sedimentation coefficient 246 *S*.

**Biochemical properties**

*Genome* consists of DNA; double–stranded; circular. Nucleotide sequence references: R.J. Shepherd (personal communication).
*Sequence database accession code(s):*
U20341 Em(43)_vi:Cv20341 Gb(89)_vi:Cvu20341 Cassava vein mosaic virus, complete genome. 4/95 8,158bp.

### Replication

Replication does not depend on a helper virus.

### Cytopathology

Virions found in all parts of the host plant; in cytoplasm. Inclusions present in infected cells; are viroplasms (but not as dense as those of other caulimoviruses); they contain virions.

### Taxonomy and Relationships

Possible *Caulimovirus*.

Virus(es) with serologically related virions: none.

### References

Costa, A.S. and Normanha, E.S. (1940). *J. Agron.* **3**: 239.
Kitajima, E.W. and Costa, A.S. (1966). *Bragantia* **25**: 211.
Lin, M.T. and Kitajima, E.W. (1980). *Fitopatol. Bras.* **5**: 419.

# Cassava X *potexvirus*

**Data collated by A.A. Brunt, 1990.**

*Standard acronym:* CsVX.

First reported in *Manihot esculenta*; from Colombia; by Lennon *et al.* (1986).

### Natural host range and symptoms

*Manihot esculenta* — systemic and symptomless.

### Transmission

Virus transmitted by mechanical inoculation.

### Geographical distribution

Spreads in Colombia.

### Experimental host range

Experimentally infected plants mostly show local lesions, chlorosis.

Diagnostically susceptible

*Nicotiana benthamiana* — local lesions and severe systemic chlorosis.

Dicotyledonous Hosts *(**Susceptible** or Insusceptible)*

***Manihot esculenta. Nicotiana benthamiana.***

### Particle morphology

Virions filamentous; of *c.* 480 nm.

### Biochemical properties

*Genome* consists of RNA; single–stranded; unipartite; largest (or only) genome part 6.5 kb.

*Virion protein(s)* one; $M_r$ 26000.

### Taxonomy and Relationships

*Potexvirus.*

Virus(es) with serologically unrelated virions: cassava common mosaic potexvirus.

**References**

Lennon, A.M., Aiton, M.M. and Harrison, B.D. (1986). *Rep. Scottish Crop Res. Inst.* 1985, p. 167.

# Cassia mild mosaic (?) *carlavirus*

**Data collated by M.T. Lin, 1983, revised E.W. Kitajima, 1991.**

*Synonyms:* cassia mottle carlavirus. *Standard acronym:* CasMMV.

First reported in *Cassia sylvestris*; from Brasilia, Brazil; by Lin *et al.* (1977).

**Natural host range and symptoms**

Symptoms vary cyclically over a few weeks.
*Cassia sylvestris* — mild chlorosis.
*C. macranthera* — chlorotic local lesions, vein clearing and interveinal chlorosis.

**Transmission**

Virus transmitted by mechanical inoculation; not transmitted by contact between plants; not transmitted by seed; not transmitted by pollen.

**Geographical distribution**

Spreads in Brazil.

**Experimental host range**

Few (<3) families susceptible. Experimentally infected plants mostly show necrotic or chlorotic local lesions, mosaics or chlorosis.

Diagnostically susceptible

*Gomphrena globosa* — necrotic local lesions; not systemic

*Cassia coluteoides, C. sylvestris, C. macranthera* — chlorotic local lesions; systemic chlorosis.

*C. obtusifolia, C. occidentalis* — necrotic and chlorotic local lesions; systemic wilting, mosaic, interveinal chlorosis, pods and seeds abnormal.

Diagnostically insusceptible

*Chenopodium amaranticolor, C. quinoa, Medicago sativa, Phaseolus vulgaris, Pisum sativum.*

Maintenance and propagation hosts

*Cassia obtusifolia, C. occidentalis.*

Assay hosts (**L**ocal lesions or **W**hole plants)

*Gomphrena globosa* (L), *Cassia obtusifolia* (L), *C. occidentalis* (L).

Dicotyledonous Hosts *(**Susceptible** or Insusceptible)*

***Apium graveolens*. *Arachis hypogaea*. *Brassica campestris*** ssp. ***rapa*. *Brassica oleracea*** ssp. ***capitata*. *Capsicum annuum*.** *Cassia occidentalis*. ***Chenopodium amaranticolor*. *Chenopodium murale*. *Chenopodium quinoa*. *Cichorium endiva*. *Citrullus lanatus*. *Cucumis melo*. *Cucumis sativus*. *Cucurbita pepo*. *Datura stramonium*. *Glycine max*.** *Gomphrena globosa*. ***Helianthus annuus*. *Lactuca sativa*. *Lupinus albus*. *Lupinus angustifolius*. *Lycopersicon esculentum*. *Medicago sativa*. *Nicandra physaloides*. *Nicotiana debneyi*. *Nicotiana glutinosa*. *Nicotiana sylvestris*. *Nicotiana tabacum*. *Phaseolus vulgaris*. *Physalis floridana*. *Pisum sativum*. *Raphanus sativus*. *Solanum melongena*. *Solanum tuberosum*. *Sonchus oleraceus*. *Tetragonia expansa*. *Vigna unguiculata*. *Zinnia elegans*.**

Monocotyledonous Hosts *(**Susceptible** or Insusceptible)*

***Sorghum bicolor. Zea mays.***

Sources of host-range data

Lin *et al.* (1977).

**Properties of particles in sap**

TIP: 55–60 °C. DEP: $\log_{10}$ minus 5–6. Leaf sap contains many virions.

**Purification**

Method: Veerisetty and Brakke (1978); Tremaine and Hamilton (1983).

**Particle morphology**

Virions filamentous; not enveloped; usually flexuous; with a clear modal length; of 640 nm; 15 nm wide. Axial canal obscure. Basic helix obscure.

**Cytopathology**

Virions found in leaves; in cytoplasm. Inclusions present in infected cells; are unusual in shape; are loose aggregates of elongated virions up to 1μ in size.

**Taxonomy and Relationships**

Possible *Carlavirus.*

**References**


Lin, M.T., Kitajima, E.W., Cupertino, F.P. and Costa, C.L. (1977). *Fitopatol. Bras.* **2**: 87.

Lin, M.T., Kitajima, E.W., Cupertino, F.P. and Costa, C.L. (1977). *Pl. Dis. Reptr* **63**: 501.

Lin, M.T., Kitajima, E.W. and Costa, C.L. (1980). *Plant Dis.* **64**: 587.

Munro, D. (1981). *Aust. Pl. Path. Soc. Newsl.* **10**: 61.

Tremaine, J.H. and Hamilton, R.I. (1983). *CMI/AAB Descr. Pl. Viruses* No. 274, 3 pp.

Veerisetty, V. and Brakke, M.K. (1978). *Phytopathology* **68**: 59.


# Cassia mosaic virus

**Data collated by R.J. van Velsen, 1984.**

First reported in *Cassia tora, C. occidentalis, Crotalaria anagyroides*; by Van Velsen (1961).

**Natural host range and symptoms**

Symptoms persist.

*Cassia tora, C. occidentalis, Crotalaria anagyroides* — mosaic.

**Transmission**

Virus transmitted by mechanical inoculation; transmitted by grafting; not transmitted by seed.

**Geographical distribution**

Spreads in Papua New Guinea.

**Experimental host range**

Few (<3) families susceptible. Experimentally infected plants mostly show mosaic.

Diagnostically susceptible

*Cassia tora, C. occidentalis, Crotalaria anagyroides* — systemic mosaic.

Maintenance and propagation hosts

*Cassia tora, C. occidentalis.*

Dicotyledonous Hosts (***Susceptible*** *or Insusceptible)*

*Arachis hypogaea. Capsicum annuum.* ***Cassia occidentalis. Cassia tora.*** *Chenopodium amaranticolor. Crotalaria spectabilis. Cucumis sativus. Datura stramonium. Lablab purpureus. Lycopersicon esculentum. Medicago sativa. Nicotiana glutinosa. Nicotiana rustica. Nicotiana sylvestris. Nicotiana tabacum. Petunia hybrida. Phaseolus lunatus. Phaseolus vulgaris. Physalis floridana. Pisum sativum. Solanum nigrum. Trifolium pratense. Trifolium repens. Trifolium subterraneum. Vicia faba. Vigna unguiculata. Vigna unguiculata* ssp. *sesquipedalis.*

Sources of host-range data

Van Velsen (1961).

**Properties of particles in sap**

TIP: 62 °C. LIV: 6 days. DEP: $\log_{10}$ minus 3.

**References**

Van Velsen, R.J. (1961). *Papua New Guin. agric. J.* **14**: 124.

# Cassia ringspot virus

**Data collated by B.R. Singh, 1983. Revised 1984.**

*Synonyms:* coffee senna ringspot mosaic virus.

First reported in *Cassia occidentalis*; from Kanpur, India; by Mathur and Singh (1969).

**Natural host range and symptoms**

Symptoms persist.
*Cassia occidentalis* — variable; mosaics, ringspots and necrosis.

**Transmission**

Not transmitted by *Myzus persicae, Aphis gossypii, A. craccivora, Bemisia tabaci.* Virus transmitted by mechanical inoculation; transmitted by grafting; not transmitted by contact between plants.

**Geographical distribution**

Spreads in India.

**Experimental host range**

Experimentally infected plants mostly show variable; local lesions, mosaics, mottles.

Diagnostically susceptible

*Arachis hypogaea* — systemic mosaic.
*Cassia occidentalis* — mild necrotic and chlorotic mosaic and mottle.
*C. tora* — mosaic, mottle, vein banding.
*Pisum sativum* — mosaic, mottle and wilting.
*Trifolium alexandrinum* — mosaic, mottle.
*Vigna mungo, V. radiata* — mosaic, mottle.
*V. unguiculata* cv. T2, *V. unguiculata* cv. 5269 — necrotic local lesions.

Maintenance and propagation hosts

*Arachis hypogaea, Cassia occidentalis, C. tora, Trifolium alexandrinum, Vigna unguiculata* cv. T2.

Assay hosts (**L**ocal lesions or **W**hole plants)

*Arachis hypogaea, Cassia occidentalis, C. tora, Trifolium alexandrinum, Vigna unguiculata* cv. T2.

Dicotyledonous Hosts *(**Susceptible** or Insusceptible)*

***Arachis hypogaea**. Cajanus cajan. Capsicum annuum. **Cassia occidentalis**. **Cassia tora**. Chenopodium amaranticolor. Crotalaria retusa. Cucumis sativus. Cucurbita pepo. Cyamopsis tetragonoloba. Datura stramonium. Glycine max. Gomphrena globosa. Helianthus annuus. Hibiscus esculentus. Lablab purpureus. Lycopersicon esculentum. Nicandra physaloides. Nicotiana glutinosa. Nicotiana tabacum. **Pisum sativum**. Raphanus sativus. Ricinus communis. Solanum melongena. Solanum nigrum. Spinacia oleracea. **Vigna radiata**. **Vigna unguiculata**. **Vigna unguiculata*** ssp. ***sesquipedalis**. Zinnia elegans.*

Monocotyledonous Hosts *(**Susceptible** or Insusceptible)*

*Zea mays.*

Sources of host-range data

Mathur and Singh (1969).

**Properties of particles in sap**

TIP: 80 °C. LIV: 0.75 days. DEP: $\log_{10}$ minus 5–6.

**Comments**

Some virologists are as yet unconvinced that cassia ringspot virus is a distinct virus.

**References**

Mathur, A.K. and Singh B.R. (1969). *Indian Phytopath.* **25**: 314.

# Cassia severe mosaic (?) *closterovirus*

**Data collated by R.G.T. Hicks, 1991.**

First reported in *Cassia corymbosa*; from Bath, U.K.; by Ngamyeesoon and Hicks (1989).

**Natural host range and symptoms**

Symptoms persist. Symptoms leaf mosaic, mottling and malformation.

*Cassia corymbosa* — mosaic, chlorotic spotting, flecking, vein yellowing, leaf malformation, yellow blotching and mottling.

**Transmission**

By vegetative propagation. Not transmitted by *Myzus persicae*. Virus transmitted by mechanical inoculation; transmitted by grafting; not transmitted by seed.

**Geographical distribution**

Spreads in the UK.

**Experimental host range**

Several (3–9) families susceptible. Experimentally infected plants mostly show chlorotic local lesions, systemic mottle or mosaic.

Diagnostically susceptible

*Chenopodium amaranticolor, C. quinoa* — chlorotic local lesions, systemic flecks and lines.

*Cassia occidentalis* — local systemic mild mottle/mosaic; plants die.

*Nicotiana clevelandii* — chlorotic local lesions, systemic mottle.

*N. tabacum* cv. White Burley — faint chlorotic local lesions; not systemic.

*Phaseolus vulgaris* cv. Prince — chlorotic local lesions/vein yellowing; systemic chlorotic flecks and vein yellowing.

*Pisum sativum* cv. Monteith — systemic chlorotic mosaic and leaf malformation; plants die.

*Trifolium repens* cv. White Clover — systemic vein necrosis.

*Vigna sesquipedalis* — necrotic local lesions; systemic necrotic spots and vein necrosis.

Diagnostically insusceptible

*Brassica pekinensis, Cassia obtusifolia, Cucumis sativus* cv. Parisienne Pickling, *Nicotiana glutinosa, N. rustica.*

Maintenance and propagation hosts

*Nicotiana clevelandii, Pisum sativum* cv. Monteith (for purification).

Assay hosts (**L**ocal lesions or **W**hole plants)

*Chenopodium quinoa* (L).

Dicotyledonous Hosts ***(Susceptible** or Insusceptible)*

*Cassia obtusifolia*. ***Cassia occidentalis***. ***Cassia tora***. ***Chenopodium album***. ***Chenopodium amaranticolor***. ***Chenopodium foetidum***. ***Chenopodium murale***. ***Chenopodium quinoa***. *Cucumis sativus*. ***Glycine max***, or *Glycine max*. ***Nicotiana clevelandii***. *Nicotiana debneyi*. *Nicotiana glutinosa*. ***Nicotiana megalosiphon***. *Nicotiana rustica*. *Nicotiana sylvestris*. ***Nicotiana tabacum***. ***Phaseolus vulgaris***. ***Pisum sativum***, or *Pisum sativum*. ***Trifolium repens***. ***Vicia faba***. ***Vigna unguiculata***, or *Vigna unguiculata*. ***Vigna unguiculata*** ssp. ***sesquipedalis***.

Sources of host-range data

Ngamyeesoon (1989); Ngamyeesoon and Hicks (1988).

**Properties of particles in sap**

TIP: 50–55 °C. LIV: less than 1 days. DEP: $\log_{10}$ minus 4–5. Leaf sap contains few virions.

**Purification**

Method: purification largely unsuccessful due to fragility of virions.

**Particle morphology**

Virions filamentous; not enveloped; usually flexuous; with no clear modal length; of 150–2168 nm; 10–12 nm wide. Axial canal obscure. Basic helix obscure.

1–3 virus specified dsRNA species found in infected cells (major, also 8 minor). Size of largest virus specified dsRNA 13–15 kbp; 2nd largest 1.2–1.3 kbp; 3rd largest 0.4–0.5 kbp.

**Cytopathology**

Virions found in leaves, epidermis and veins.

**Taxonomy and Relationships**

Possible *Closterovirus*.

Best tests for diagnosis: virion properties distinguish this virus from a carlavirus isolated from *Cassia*, and which does not infect *Chenopodium amaranticolor, C. quinoa, Phaseolus vulgaris* and *Pisum sativum* (Lin *et al.*, 1979).

**References**

Ngamyeesoon, N. (1989). *Ph.D. Thesis*, University of Bath, U.K. 343 pp.
Ngamyeesoon, N. and Hicks, R.G.T. (1988). *Acta Hort.* **234**: 45.

# Cassia yellow blotch *bromovirus*

**Data collated by J.L. Dale, 1984. Revised 1987.**

*Standard acronym:* CYBV.

First reported in *Cassia pleurocarpa*; from Mitchell, Queensland; by Dale *et al.* (1984).

**Natural host range and symptoms**

Symptoms persist.
*Cassia pleurocarpa* — yellow leaf blotches.

**Transmission**

Virus transmitted by mechanical inoculation.

**Geographical distribution**

Spreads in Australia.

**Experimental host range**

Several (3–9) families susceptible. Experimentally infected plants mostly show chlorotic local lesions, mottle, tip necrosis.

Diagnostically susceptible

*Chenopodium amaranticolor, C. quinoa* — chlorotic local lesions; not systemic.
*Cassia hirsuta, C. occidentalis, Trifolium incarnatum* — systemic chlorotic mottle.
*Cicer arietinum* — tip necrosis.
*Nicotiana clevelandii* — systemic green mottle.

Diagnostically insusceptible

*Arachis hypogaea, Pisum sativum, Vigna radiata.*

Maintenance and propagation hosts

*Nicotiana clevelandii.*

Assay hosts (**L**ocal lesions or **W**hole plants)

*Chenopodium amaranticolor* (L).

Dicotyledonous Hosts *(**Susceptible** or Insusceptible)*

*Arachis hypogaea. Cajanus cajan.* ***Cassia occidentalis. Cassia tora. Chenopodium amaranticolor. Chenopodium quinoa.*** *Cicer arietinum. Clitoria ternatea.* ***Cucumis sativus.*** *Datura stramonium.* ***Desmodium tortuosum. Glycine max. Gomphrena globosa.*** *Helianthus annuus.* ***Lathyrus odoratus. Lupinus angustifolius. Macroptilium lathyroides.*** *Medicago sativa.* ***Nicotiana clevelandii. Phaseolus vulgaris.*** *Pisum sativum.* ***Trifolium incarnatum.*** *Trifolium pratense. Vigna radiata. Vigna unguiculata* ssp. *sesquipedalis.*

Monocotyledonous Hosts *(**Susceptible** or Insusceptible)*

*Avena sativa. Hordeum vulgare. Sorghum bicolor. Triticum aestivum.* ***Zea mays.***

Sources of host-range data

Dale *et al.* (1984).

**Properties of particles in sap**

Leaf sap contains many virions. Electron microscopy: best with UA or AM.

**Purification**

Method: Bancroft *et al.* (1967); Dale *et al.* (1984).

**Particle morphology**

Virions isometric; not enveloped; 26 nm in diameter; rounded in profile; without a conspicuous capsomere arrangement.

**Physical properties**

One sedimenting component in purified preparations (at pH 7); sedimentation coefficient 85 *S*. Isoelectric point pH 3.6.

**Biochemical properties**

*Genome* consists of RNA; single–stranded; of three parts; largest (or only) genome part the largest 3.4 kb; the 2nd largest 3 kb; the 3rd largest 2.3 kb. Genomic nucleic acid isolated by Dale *et al.* (1984). Base composition 24.4 % G; 24.9 % A; 22.2 % C; 28.5 % U. Infectivity retained when deproteinised with phenol or detergent. Additional factor not required for infectivity.

*Non–genomic nucleic acid* found in the virions; is subgenomic *m*RNA.

*Virion protein(s)* one; $M_r$ 20800. Amino acid composition: Dale *et al.* (1984).

**Replication**

Replication does not depend on a helper virus.

**Cytopathology**

Virions found in all parts of the host plant; in cytoplasm and in nuclei. Inclusions absent from infected cells.

**Taxonomy and Relationships**

*Bromovirus: Bromoviridae.*

Virus(es) with serologically related virions: broad bean mottle, brome mosaic and cowpea chlorotic mottle viruses.

**References**

Bancroft, J.B., Hills, G.J. and Markham, R. (1967). *Virology* **31**: 354.

Dale, J.L., Gibbs, A.J. and Behncken, G.M. (1984). *J. gen. Virol.* **65**: 281.

## Cassia yellow spot *potyvirus*

**Data collated by A.A. Brunt and O. Lovisolo, 1992.**

*Synonyms:* cassia yellow blotch virus. *Standard acronym:* CasYSV.

First reported in *Cassia hoffmanseggi*; from Brazil (Pernambuco State); by Paguio and Kitajima (1981).

**Natural host range and symptoms**

Symptoms persist.

*Cassia hoffmanseggi* — chlorotic leaf blotching.

**Transmission**

Transmitted by a vector; an insect; *Myzus persicae, Uroleucon ambrosiae*; Aphididae. Transmitted in a non–persistent manner. Virus transmitted by mechanical inoculation; transmitted by seed (in *Cassia occidentalis* (Souto *et al.*, 1990)).

**Geographical distribution**

Found, but with no evidence of spread, in Brazil (Pernambuco State).

**Experimental host range**

Several (3–9) families susceptible. Experimentally infected plants mostly show systemic mosaic, crinkle and growth reduction, necrotic local lesions.

Diagnostically susceptible

*Cassia occidentalis, Phaseolus vulgaris, Canavalia ensiformis* — systemic mosaic, crinkle and growth reduction.

*Vigna unguiculata* (some cvs), *Sesamum indicum* — mosaic.

*Gycine max* — latent.

*Chenopodium amaranticolor* — necrotic local lesions.

Diagnostically insusceptible

*Gomphrena globosa, Cucumis melo, Pisum sativum, Nicotiana glutinosa, N. tabacum.*

Maintenance and propagation hosts

*Phaseolus vulgaris.*

Assay hosts (**L**ocal lesions or **W**hole plants)

*Phaseolus vulgaris* (L and W), *C. amaranticolor* (L), *Canavalia ensiformis* (W), *Crotalaria juncea* (W).

Dicotyledonous Hosts *(**Susceptible** or Insusceptible)*

*Arachis hypogaea. Capsicum annuum.* ***Cassia occidentalis****. Cassia tora.* ***Chenopodium amaranticolor****.* ***Crotalaria spectabilis****. Cucumis melo. Cucumis sativus.* ***Cucurbita pepo****.* ***Glycine max****. Gomphrena globosa. Lactuca sativa. Nicotiana glutinosa. Nicotiana tabacum.* ***Phaseolus lunatus****.* ***Phaseolus vulgaris***, or *Phaseolus vulgaris. Pisum sativum. Raphanus sativus.* ***Sesamum indicum****.* ***Vigna radiata****.* ***Vigna unguiculata****. Zinnia elegans.*

Additional hosts

***Canavalia ensiformis***; ***Crotalaria juncea***; *Lupinus luteus.*

Sources of host-range data

Souto (1990).

**Particle morphology**

Virions filamentous; not enveloped; usually flexuous; with a clear modal length; of 700 nm. Axial canal obscure.

**Cytopathology**

Virions found in leaves and mesophyll; in cytoplasm. Inclusions present in infected cells; are pinwheels.

**Taxonomy and Relationships**

*Potyvirus: Potyviridae.*

Virus(es) with serologically related virions: bean common mosaic, canavalia acronecrosis, carnation vein mottle and passionfruit woodiness viruses. Virus(es) with serologically unrelated virions: blackeye cowpea mosaic, papaya ringspot and watermelon mosaic viruses (Lovisolo and Kitajima, 1992; Lovisolo *et al.*, 1991).

**Comments**

This virus was first designated cassia yellow blotch virus (Paguio and Kitajima, 1980 and 1981) but, to avoid confusion with cassia yellow blotch bromovirus, its name was later changed (Souto *et al.*, 1989).

**References**

Costa, C.L., Kitajima, E.W. and Marinho, V.L.A. (1989). *Fitopatol. Bras.* **14**: 115.
Lovisolo, O. and Kitajima, E.W. (1992). *Arch. Virol., Suppl.* **5**: in press.
Lovisolo, O., De Sa, P.B., Lima, J.A.A., Costa, C.L. and Kitajima, E.W. (1991). *Fitopatol. Bras.* **16**: 24.
Lovisolo, O., Marinho, V.L.A., Souto, E., Costa, C.L. and Kitajima, E.W. (1989). *Fitopatolo. Bras.* **14**: 119.
Paguio, D.R. and Kitajima, E.W. (1980). *Fitopatol. Bras.* **5**: 436.
Paguio, D.R. and Kitajima, E.W. (1981). *Fitopatol. Bras.* **6**: 187.
Souto, E.R. (1990). *Inst. Cienc. Biol. Univ. de Brasilia*, 93 pp.
Souto, E.R. and Kitajima, E.W. (1991). *Fitopatol. Bras.* **16**: in press.
Souto, E.R. and Kitajima, E.W. (1991). *Fitopatol. Bras.* **16**: 256.
Souto, E.R., Costa, C.L. and Kitajima, E.W. (1990). *Fitopatol. Bras.* **15**: 136.
Souto, E.R., Kitajima, E.W., Marinho, V.L.A. and Costa, A.F. (1988). *Fitopatol. Bras.* **13**: 145.
Souto, E.R., Marinho, V.L.A., Oliveira, C.R.B. and Kitajima, E.W. (1989). *Fitopatol. Bras.* **14**: 114.

# Cauliflower mosaic *caulimovirus*

**Data collated by R.G. Garrett, 1982.**

*Synonyms:* brassica virus 3, broccoli mosaic virus, cabbage mosaic virus, cabbage virus B. *Standard acronym:* CaMV.

First reported in *Brassica campestris* and *B. oleracea*; from the U.S.A.; by Tompkins (1937).

**Natural host range and symptoms**

Symptoms persist.

*Arabidopsis thaliana*, *Brassica* spp., *Raphanus* spp. and other species of the Brassicaceae, Resedaceae — vein-clearing or banding mosaic.

**Transmission**

Transmitted by a vector; an insect; *Brevicoryne brassicae*, *Myzus persicae* and at least 25 other species; Aphididae. Transmitted in a semi–persistent manner. Virus lost by the vector when it moults; does not multiply in the vector; not transmitted congenitally to the progeny of the vector; transmitted by mechanical inoculation.

**Geographical distribution**

Probably distributed worldwide.

**Experimental host range**

Few (<3) families susceptible. Experimentally infected plants mostly show vein clearing, mosaic.

Diagnostically susceptible

*Brassica oleracea* ssp. *botrytis, B. campestris* — vein clearing, mosaic.
*Brassica campestris* ssp. *napus* — mosaic.

Maintenance and propagation hosts

*Brassica campestris* cv. Just Right.

Assay hosts (**L**ocal lesions or **W**hole plants)

*Brassica campestris* cv. Just Right.

Dicotyledonous Hosts ***(Susceptible** or Insusceptible)*

***Brassica campestris. Brassica campestris*** ssp. ***chinensis. Brassica campestris*** ssp. ***napus. Brassica campestris*** ssp. ***pekinensis. Brassica campestris*** ssp. ***rapa. Datura stramonium.*** *Nicotiana bigelovii.* ***Nicotiana clevelandii.*** *Nicotiana debneyi. Nicotiana glutinosa.*

Sources of host-range data

Tompkins (1937); Broadbent and Tinsley (1953).

**Properties of particles in sap**

TIP: 75–80 °C. LIV: 7–15 days. DEP: $\log_{10}$ minus 3. Infectivity of sap not changed by treatment with di–ethyl ether. Leaf sap contains few virions. Electron microscopy: no special treatment required but virions flatten.

**Purification**

Method: Gardner and Shepherd (1980).

**Particle morphology**

Virions isometric; not enveloped; 50 nm in diameter; rounded in profile; without a conspicuous capsomere arrangement.

**Physical properties**

One sedimenting component in purified preparations; sedimentation coefficient 208 *S* (Hull *et al.*, 1976). Density 1.37 g $cm^{-3}$ in CsCl.

**Biochemical properties**

Virions contain 16–17 % nucleic acid; 83–84 % protein; 0 % lipid.

*Genome* consists of DNA; double–stranded; circular (linear molecules found but not infective); unipartite; largest (or only) genome part 8.032 kb (pairs). Genomic nucleic acid isolated by Shepherd *et al.* (1970). Base composition of ds genome 43 % (G + C). Poly A region absent. Nucleotide sequence references: Franck *et al.* (1980); Gardner *et al.* (1981).

*Sequence database accession code(s):*

A00001 Em(40)_vi:A00001 Gb(84)_pat:A00001 Cauliflower mosaic virus satellite cDNA. 1/93 335bp.

D00335 Em(40)_vi:MCAB1G6 Gb(84)_vi:MCAB1G6 Cauliflower mosaic virus (strain Bari 1), gene VI and flanks. 3/94 1,908bp

J02046 Em(40)_vi:MCA1841 Gb(84)_vi:MCA1841 Cauliflower mosaic virus (strain CM1841), complete genome. 2/92 8,031bp.

J02048 Gb(84)_vi:MCASTRAS Cauliflower mosaic virus (Strasbourg strain), complete genome. 2/85 8,024bp.

L09053 Em(40)_vi:MCAMIBP. Gb(84)_vi:MCAMIBP. Cauliflower mosaic virus major inclusion body protein (VI) gene, complete cds. 11/93 3,036bp.

M10166 Em(40)_vi:MCAIBP Gb(84)_vi:MCA4184P Cauliflower mosaic virus IBP. gene promoter region. 7/89 92bp
M10326 Em(40)_vi:MCA41841 Gb(84)_vi:MCA41841 Cauliflower mosaic virus (strain CM4–184) 5' region, segment 1. 7/89 69bp.
M10327 Em(40)_vi:MCA41842 Gb(84)_vi:MCA41842 Cauliflower mosaic virus (strain CM4–184) 5' region, segment 2. 7/89 135bp.
M10376 Em(40)_vi:MCACGDH Gb(84)_vi:MCACGDH Cauliflower mosaic virus (altered virulence isolate D/H), complete genome. 5/94 8,016bp.
M10385 Em(40)_vi:MCA4184D Gb(84)_vi:MCA4184D Cauliflower mosaic virus (strain CM4–184) deletion mutant DNA. 7/89 90bp.
M13776 Em(40)_vi:MCAIS Gb(84)_vi:MCAIS Cauliflower mosaic virus (CaMV) in vitro initiation sites. 7/89 320bp.
M14412 Em(40)_sy:AG5MCA Gb(84)_sy:SYNTRN5MCA Synthetic gene of Cauliflower mosaic virus promoter and Transposon Tn5 kanamycin resistant ge
M14790 Gb(84)_vi:MCAREPREC Cauliflower mosaic virus (strain CM4–184) U5 region DNA. 5/87 2,539bp.
M16239 Em(40)_vi:MCA4184P Gb(84)_sy:SYN4184P1 Cauliflower mosaic virus IBP. gene promoter region with oligomer BamHI linker. 7/89 67bp
M17415 Em(40)_vi:MCAATFA Gb(84)_vi:MCAATFA Cauliflower mosaic virus (CAMV) BstEII/XhoI aphid transmissibility fragment DNA. 7/89 1,556bp.
M19736 Em(40)_sy:AG35S Gb(84)_sy:SYNMCA35S Cauliflower mosaic virus (CaMV) 35S promoter gene under Tn10-encoded tet repressor-operator c
M19740 Em(40)_vi:MCACPPG2 Gb(84)_vi:MCACPPG2 Cauliflower mosaic virus viable recombinant DNA, 3' flank to ORF (VIII,IV,V,V*) region. 7/89
M19741 Em(40)_vi:MCACPPG1 Gb(84)_vi:MCACPPG1 Cauliflower mosaic virus viable recombinant ORF VIII mRNA, 3' end; ORF V* mRNA, complete cds;
M23620 Em(40)_vi:MCAMV6 Gb(84)_vi:MCAMV6 Cauliflower mosaic virus (CaMV) gene six protein gene, complete cds. 12/93 1,904bp.
M32808 Em(40)_vi:MCAORFIX Gb(84)_vi:MCAORFIX1 Cauliflower mosaic virus unidentified protein gene, partial cds. 6/92 74bp.
M32809 Em(40)_vi:MCAAAF1 Gb(84)_vi:MCAAAF1 Cauliflower mosaic virus aphid acquisition factor 5' end. 6/92 229bp.
M32810 Em(40)_vi:MCAORFX2 Gb(84)_vi:MCAORFX2 Cauliflower mosaic virus unidentified protein gene, 5' end. 6/92 178bp.
M32811 Em(40)_vi:MCAMPX Gb(84)_vi:MCAMPX Cauliflower mosaic virus matrix protein (put.), partial cds. 6/92 972bp.
M32812 Em(40)_vi:MCACTCTP Gb(84)_vi:MCACTCTP Cauliflower mosaic virus cell-to-cell transport protein, 5' end. 6/92 221bp.
M32813 Em(40)_vi:MCAAAFCP Gb(84)_vi:MCAAAFCP Cauliflower mosaic virus aphid acquisition factor (3' end), and coat protein (5' end) genes.
M32814 Em(40)_vi:MCACPX Gb(84)_vi:MCACPX Cauliflower mosaic virus coat protein, partial cds. 6/92 106bp
M32815 Em(40)_vi:MCARTRA Gb(84)_vi:MCARTRA Cauliflower mosaic virus reverse transcriptase (V) gene, partial cds. 6/92 467bp.
M37581 Em(40)_vi:MCAORF01 Gb(84)_vi:MCAORFIIPV Cauliflower mosaic virus ORF II protein gene, complete cds. 9/90 649bp.
M37582 Em(40)_vi:MCAORFII Gb(84)_vi:MCAORFIICM Cauliflower mosaic virus ORF II protein gene, complete cds. 9/90 648bp.
M74305 Em(40)_sy:AGCAMVCM5 Gb(84)_sy:SYNCAMVCM5 Synthetic expression vector pUC18cpPRV, constructed from cauliflower mosaic virus 35S gene-re

M74306 Em(40)_sy:AGCAMVCM3 Gb(84)_sy:SYNCAMVCM3 Synthetic expression vector pUC18cpPRV constructed from cauliflower mosaic virus 35S gene-reg

M90541 Em(40)_vi:MCAMATCAP Gb(84)_vi:MCAMATCAP Cauliflower mosaic virus movement protein, aphid transmission protein, DNA binding protein, c

M90542 Em(40)_vi:MCAMADCRI Gb(84)_vi:MCAMADCRI Cauliflower mosaic virus movement protein gene, aphid transmission protein gene, DNA binding

M90543 Em(40)_vi:MCACOMGEN Gb(84)_vi:MCACOMGEN Cauliflower mosaic virus movement protein gene, aphid transmission protein gene, DNA binding

M94887 Em(40)_vi:MCAORFS Gb(84)_vi:MCAORFS Cauliflower mosaic virus DNA sequence, ORFs I, II, and VII. 10/92 1,692bp.

S51055 Em(40)_vi:S51055 Gb(84)_vi:S51055 19S (promoter) cauliflower mosaic virus, Genomic, 240 nt. 11/93 240bp.

S51061 Em(40)_vi:S51061 Gb(84)_vi:S51061 35S (promoter) cauliflower mosaic virus, Genomic, 210 nt. 11/93 210bp.

V00140 Em(40)_vi:CAMVG1 Gb(84)_vi:CAMVG1 Cauliflower mosaic virus genome. 9/93 8,031bp.

V00141 Em(40)_vi:CAMVG2 Gb(84)_vi:CAMVG2 Cauliflower mosaic virus genome. 9/93 8,024bp.

X02606 Em(40)_vi:CACM4184 Gb(84)_vi:CACM4184 Cauliflower mosaic virus (CM4184) R, U5 and flanking regions. 7/86 2,539bp.

X04879 Em(40)_sy:ARCAMVPR Gb(84)_sy:ARCAMVPR CaMV promoter in pMON273 vector (transgenic petunia) CaMV = Cauliflower Mosaic Virus 35S prom

X14911 Em(40)_vi:CAMVI Gb(84)_vi:CAMVI Cauliflower mosaic virus gene VI for the inclusion body matrix protein. 9/93 3,131bp.

X53860 Em(40)_vi:CMVVIG Gb(84)_vi:CMVVIG Cauliflower mosaic virus (CaMV) gene VI for viroplasmin. 9/93 2,283bp.

X79465 Em(43)_vi:Cmvdna Gb(89)n:Cmvdna Cauliflower Mosaic Virus strain B29 complete DNA sequence. 5/95 8,031bp.

*Non–genomic nucleic acid* not found in the virions.

*Virion protein(s)* one; $M_r$ 44000, or 58000. Method of preparation: Al Ani *et al.* (1979); Hahn and Shepherd (1980). Amino acid composition: Brunt *et al.* (1975); protein is basic because of relatively high lysine and arginine content.

**Replication**

Genome replicates in nuclei. Coat protein *m*RNA translated in the cytoplasm. Replication does not depend on a helper virus.

**Cytopathology**

Virions found in all parts of the host plant; in cytoplasm; plasmodesmata. Inclusions present in infected cells; are unusual in shape; electron-dense matrices; they contain virions. Other cellular changes: deeply lobed nuclei, cell wall protrusions associated with vesicles and convoluted tubules.

**Taxonomy and Relationships**

*Caulimovirus* (the type species).

Virus(es) with serologically related virions: dahlia mosaic, strawberry vein-banding and carnation etched ring viruses.

**References**

Al Ani, R., Pfeiffer, P. and Lebeurier, G. (1979). *Virology* **93**: 188.

Broadbent, L. and Tinsley, T.W. (1953). *Pl. Path.* **2**: 88.

Brunt, A.A., Barton, R.J., Tremaine, J.H. and Stace-Smith, R. (1975). *J. gen. Virol.* **27**: 101.
Cheng, R.H., Olron, N.H. and Barker, T.S. (1992). *Virology* **186**: 655.
Covey, S.N. and Hull, R. (1985). *Oxford Surveys of Pl. Mol. Cell Biol.* **2**: 339.
Donson, J. and Hull, R. (1983). *J. gen. Virol.* **64**: 2281.
Franck, A., Guilley, H., Jonard, G., Richards, K. and Hirth, L. (1980). *Cell* **21**: 285.
Gardner, R.C. and Shepherd, R.J. (1980). *Virology* **106**: 159.
Gardner, R.C., Howarth, A.J., Hahn, P., Brown-Luedi, M., Shepherd, R.J. and Messing, J. (1981). *Nucl. Acids Res.* **9**: 2871.
Hahn, P. and Shepherd, R.J. (1980). *Virology* 107: 295.
Hull, R. and Covey, S.N. (1983). *Trends Biochem. Sci.* **8**: 119.
Kruse, J., Timmians, P. and Witz, J. (1987). *Virology* **159**: 166.
Meagher, R.B., Shepherd, R.J. and Boyer, H.W. (1977). *Virology* **80**: 362.
Pfeiffer, P. and Hohn, T. (1983). *Cell* **33**: 781.
Shepherd, R.J. (1970). *CMI/AAB Descr. Pl. Viruses* No. 24, 4 pp.
Shepherd, R.J. (1981). *CMI/AAB Descr. Pl. Viruses* No. 243, 6 pp.
Tompkins, C.M. (1937). *J. Agric. Res.* **55**: 33.

# Celery latent (?) potyvirus

**Data collated by E. Luisoni, 1991.**

First reported in *Apium graveolens* var. *dulce*; from Italy, near Turin; by Brandes and Luisoni (1966); Luisoni (1966).

### Natural host range and symptoms

Symptoms none.

*Apium graveolens* var. *dulce, A. graveolens* var. *rapaceum* — symptomless.

### Transmission

Transmitted by means not involving a vector. Not transmitted by *Acyrthosiphon pisum, Aphis fabae, Cavariella aegopodii, Macrosiphum euphorbiae, Myzus persicae, Aulacorthum circumflexum*. Virus transmitted by mechanical inoculation; transmitted by grafting; transmitted by seed.

### Ecology and control

Studies reported by Verhoyen *et al.* (1976); Bos *et al.* (1978). The virus was detected in seedlings of commercial cultivars of celery and celeriac. The virus seemed to reduce the yield and the use of virus-free mother plants for seed production was advised. The availability of antiserum facilitates this control.

### Geographical distribution

Spreads in Belgium, Italy, and the Netherlands.

### Experimental host range

Several (3–9) families susceptible. Experimentally infected plants mostly show no symptoms. Some *Chenopodium* and *Amaranthus* species and *Atriplex hortensis* give local lesions and generally systemic rings, flecks, deformations and dwarfing.

Diagnostically susceptible

*Apium graveolens dulce, A. graveolens rapaceum* — no visible symptoms.

*Anthriscus cerefolium, Coriandrum sativum* — symptomless.

*Chenopodium quinoa* — local lesions, chlorotic flecks and rings with slight leaf deformation.

*Chenopodium amaranticolor* — local lesions; less distinct systemic symptoms.

*Amaranthus hybridus hypochondriacus* — necrotic local lesions; sometimes severe systemic symptoms.

Diagnostically insusceptible

*Anethum graveolens, Daucus carota, Pastinaca sativa, Petroselinum crispum, Nicotiana rustica.*

Maintenance and propagation hosts

*Apium graveolens dulce* (for maintaining the virus long term), *Chenopodium quinoa* (for maintaining and propagating the virus short term).

Assay hosts (**L**ocal lesions or **W**hole plants)

*Chenopodium quinoa* (W), *Chenopodium amaranticolor* (L), *Amaranthus hybridus hypochondriacus* (L).

Dicotyledonous Hosts *(**Susceptible** or Insusceptible)*

***Amaranthus caudatus***. ***Amaranthus retroflexus***. ***Anthriscus cerefolium***. ***Apium graveolens***. ***Atriplex hortensis***. ***Beta vulgaris***. *Brassica campestris* ssp. *chinensis*. *Calendula officinalis*. ***Capsicum annuum***. ***Chenopodium album***. ***Chenopodium amaranticolor***. ***Chenopodium foetidum***. ***Chenopodium quinoa***. *Conium maculatum*. ***Coriandrum sativum***. *Cucumis sativus*. *Cucurbita pepo*. *Datura metel*. *Datura stramonium*. *Daucus carota*. *Glycine max*. ***Gomphrena globosa***. *Heracleum sphondylium*. ***Hyoscyamus niger***. *Lactuca sativa*. *Lycopersicon esculentum*. ***Nicotiana clevelandii***. ***Nicotiana debneyi***. ***Nicotiana glutinosa***. *Nicotiana glutinosa* × *clevelandii*. ***Nicotiana megalosiphon***. *Nicotiana rustica*. ***Nicotiana tabacum***. *Ocimum basilicum*. *Pastinaca sativa*. *Petroselinum crispum*. ***Petunia hybrida***. ***Phaseolus vulgaris***. *Physalis floridana*. ***Pisum sativum***. *Solanum nigrum*. ***Spinacia oleracea***. ***Tetragonia expansa***. ***Trifolium incarnatum***. *Trifolium pratense*. *Trifolium repens*. ***Vicia faba***. *Vigna unguiculata*. ***Zinnia elegans***.

Additional hosts

***Amaranthus hybridus hypochondriacus, Trigonella foenum-graecum***.

Sources of host-range data

Bos *et al.* (1978); Brandes and Luisoni (1966); Luisoni (1966); Verhoyen *et al.* (1976).

**Properties of particles in sap**

TIP: 54 °C. LIV: 10 days. DEP: $\log_{10}$ minus 4. Leaf sap contains few virions.

**Purification**

Method: Bos *et al.* (1978).

**Particle morphology**

Virions filamentous; not enveloped; usually flexuous; with a clear modal length; of 860 nm. Axial canal obvious.

**Physical properties**

One sedimenting component in purified preparations; sedimentation coefficient 161 *S*.

**Cytopathology**

Virions found in leaves; in cytoplasm. Inclusions absent from infected cells. Other cellular changes: bundles of virions in the cytoplasm, often near or parallel to the tonoplast (Verhoyen *et al.*, 1976; Bos *et al.*, 1978).

**Taxonomy and Relationships**

*Potyvirus: Potyviridae* (possibly).

Best tests for diagnosis: ISEM and decoration is diagnostic. Susceptibility of *C. quinoa* is not as some strains of celery mosaic virus.

**References**

Bos, L., Diaz-Ruiz, J.R. and Maat, D.Z. (1978). *Neth. J. Pl. Path.* **84**: 61.

Brandes, J. and Luisoni, E. (1966). *Phytopath. Z.* **57**: 277.

Luisoni, E. (1966). *Atti Accad. Sci., Torino* **100**: 541.

Luisoni, E. and Lisa, V. (1969). *Annls. Phytopath.* **1**: 375.

Verhoyen, M., Esparza-Duque, J., Matthieu, J.L. and Horvat, F. (1976). *Parasitica* **32**: 158.

# Celery mosaic *potyvirus*

**Data collated by R.N. Campbell, 1986.**

*Synonyms:* apium virus 1, celery western mosaic virus, celery ringspot virus (Hollings, 1964; Pemberton and Frost, 1986). *Standard acronym:* CeMV. *Strains:* celery crinkle-leaf strain (Freitag and Severin, 1945a), poison hemlock ringspot virus (Freitag and Severin, 1945a and b; Sutabutra and Campbell, 1971).

First reported in *Apium graveolens* var. *dulce*; from California, U.S.A.; by Severin and Freitag (1938).

**Natural host range and symptoms**

Symptoms persist.

*Apium graveolens* var. *dulce* — mottling, malformation of leaflets.
*Conium maculatum* — systemic mottle and malformation.
*Pastinaca sativa* — crinkly leaves and mottle.
*Petroselinum crispum* — chlorosis and necrotic spots.
*Coriandrum sativum* — chlorosis, stunted.
*Daucus carota* — mosaic.
*Anethum graveolens* — stunted and malformed.

**Transmission**

Transmitted by a vector; an insect; *Myzus persicae, Aphis middletonii, A. ferruginea-striata; A. apigraveolens, A. apii, A. gossypii, A. rumicis, Cavariella capreae, Myzus circumflexus, M. convolvuli, Rhopalosiphum melliferum; Cavariella aegopodii, C. pastinacae* (Severin and Freitag, 1968; Walkey *et al.*, 1970); Aphididae. Transmitted in a non–persistent manner. Virus transmitted by mechanical inoculation; not transmitted by contact between plants; not transmitted by seed.

**Ecology and control**

Studies reported by Milbrath in California, U.S.A.. Controlled by a celery-free period (Milbrath, 1948). In Britain, the virus infects wild hemlock and other weeds (Walkey, 1970), thus a celery-free period may not control the virus there.

**Geographical distribution**

Probably distributed worldwide. Spreads in Argentina (Gracia and Feldman, 1977), Australia (western Australia), Canada (Kemp and Froud, 1975), the former Czechoslovakia (Chod, 1984)), France (Marchoux *et al.*, 1969), Germany (Brandes and Luisoni, 1966), Italy (Avgelis and Quacquarelli, 1972), Japan (Iwaki and Komuro, 1970),

New Zealand (Fry and Proctor, 1968), the UK (Walkey *et al.*, 1970), the USA (in the western states and Florida (Purcifull and Shepard, 1967)).

**Experimental host range**

Few (<3) families susceptible. Experimentally infected plants mostly show mottle leaf malformation.

Diagnostically susceptible

*Apium graveolens* var. *dulce* — mottled and malformed leaflets.
*Conium maculatum* — mottled and leaves malformed (Sutabutra, 1968).
*Pastinaca sativa* — leaves malformed.
*Daucus carota* var. *sativa* chlorotic spots.
*Coriandrum sativum* — systemic chlorosis.

Maintenance and propagation hosts

*Apium graveolens* cv. Utah 10B.

Assay hosts (**L**ocal lesions or **W**hole plants)

*Apium graveolens* (W), *Chenopodium amaranticolor* (L), *C. quinoa* (L).

Dicotyledonous Hosts *(**Susceptible** or Insusceptible)*

***Apium graveolens*. *Chenopodium amaranticolor*. *Chenopodium quinoa*. *Conium maculatum*. *Coriandrum sativum*. *Daucus carota*. *Heracleum sphondylium*. *Pastinaca sativa*. *Petroselinum crispum*.**

Sources of host-range data

Severin and Freitag (1938); Sutabutra and Campbell (1971); Frowd and Tomlinson (1970); Walkey *et al.* (1970); Wolf (1969); Wolf and Schmelzer (1972).

**Properties of particles in sap**

TIP: 55–60 °C. LIV: 6 days. DEP: $\log_{10}$ minus 2–3. Leaf sap contains few virions.

**Purification**

Method: Shepard and Grogan (1967*b*).

**Particle morphology**

Virions filamentous; not enveloped; usually flexuous; with a clear modal length; of 780 nm; 15 nm wide (Walkey *et al.*, 1970). Axial canal obscure.

**Biochemical properties**

Virions contain 5.5 % nucleic acid; 94.5 % protein; 0 % lipid.

*Genome* consists of RNA; single–stranded; linear; unipartite; largest (or only) genome part 9.5–11 kb. Infectivity decreased when deproteinised with proteases; retained when deproteinised with phenol or detergent.

*Virion protein(s)* one.

**Replication**

Replication does not depend on a helper virus.

**Cytopathology**

Virions found in all parts of the host plant; in cytoplasm. Inclusions present in infected cells; are pinwheels and unusual in shape; tubules in cytoplasm (Purcifull and Shepard, 1967).

**Taxonomy and Relationships**

*Potyvirus: Potyviridae.*

Virus(es) with serologically related virions: tobacco etch and potato Y (Shepard *et al.*, 1974) viruses.

*Pastinaca sativa* is not infected by the common strain, but is susceptible to the crinkle leaf strain (Freitag and Severin, 1945a). *Daucus carota* var. *sativa* cultivars, Western Red and Early Nantes, are not infected by mechanical inoculation with British isolates (Walkey *et al.*, 1970).

**References**


Avgelis, A. and Quacquarelli, A. (1972). *Phytopathol. Medit.* **11**: 124.
Brandes, J. and Luisoni, E. (1966). *Phytopath. Z.* **57**: 277.
Chod, J. (1984). *Ochr. Rost.* **20**: 91.
Freitag, J.H. and Severin, H.H.P. (1945a). *Hilgardia* **16**: 361.
Freitag, J.H. and Severin, H.H.P. (1945b). *Hilgardia* **16**: 389.
Fry, P.R. and Proctor, C.H. (1968). *N.Z. comml. Grow.* **24**: 23.
Gracia, O. and Feldman, J.M. (1977). *Pl. Dis. Reptr* **61**: 905.
Hari, V. (1981). *Virology* **112**: 391.
Hollings, M. (1964). *J. hort. Sci.* **39**: 130.
Iwaki, M. and Komuro, Y. (1970). *Ann. Phytopathol. Proc. Japan* **36**: 36.
Kemp, W.G. and Frowd, J.A. (1975). *Pl. Dis. Reptr* **59**: 50.
Marchoux, G., Navatel, J.C., Rougier, J. and Duteil, M. (1969). *Annls. Phytopath.* **1**: 227.
Milbrath, D.G. (1948). *Bull. Calif. Dep. Agric.* **37**: 3.
Pemberton, A.W. and Frost, R.R. (1974). *Pl. Path.* **23**: 20.
Pemberton, A.W. and Frost, R.R. (1986). *Ann. appl. Biol.* **108**: 319.
Purcifull, D.E. and Shepard, J.F. (1967). *Pl. Dis. Reptr* **51**: 502.
Severin, H.H.P. and Freitag, J.H. (1938). *Hilgardia* **11**: 493.
Shepard, J.F. and Grogan, R.G. (1967*a*). *Phytopathology* **57**: 1104.
Shepard, J.F. and Grogan, R.G. (1967*b*). *Phytopathology* **57**: 1136.
Shepard, J.F. and Grogan, R.G. (1971). *CMI/AAB Descr. Pl. Viruses* No. 50, 3 pp.
Shepard, J.F., Secor, G.A. and Purcifull, D.E. (1974). *Virology* **58**: 464.
Sutabutra, T. and Campbell, R.N. (1971). *Pl. Dis. Reptr* **55**: 328.
Walkey, D.G.A., Tomlinson, J.A. and Frowd, J.A. (1970). *Pl. Dis. Reptr* **54**: 370.
Wolf, P. (1969). *TagBer. dt. Akad. Wiss., Berlin* **11**: 293.
Wolf, P. and Schmelzer, K. (1972). *Zbl. Bakt. ParasitKde.* Abt. 2, **127**: 665.


# Celery T (?) *cytorhabdovirus*

**Data collated by A.A. Brunt, 1994.**

First reported in *Apium graveolens* var. *rapaceum* (celeriac); from Turin, Italy; by D'Agostino *et al.* (1992).

**Natural host range and symptoms**

*Apium graveolens* var. *rapaceum*, but only found in plants also infected with celery mosaic potyvirus, so the symptoms caused by CTV alone are unknown.

**Transmission**

Virus not transmitted by mechanical inoculation; transmitted by grafting.

**Geographical distribution**

Spreads in the Mediterranean region; Italy.

**Experimental host range**

Few (<3) families susceptible.

Dicotyledonous Hosts *(**Susceptible** or Insusceptible)*

***Apium graveolens*.**

**Particle morphology**

Virions rhabdo– or bullet–shaped; with a clear modal length; of 360 nm; 65–70 nm wide.

**Cytopathology**

Virions found in Virions found within cisternae of the endoplasmic reticulum; in cytoplasm.

**Taxonomy and Relationships**

Probably *Cytorhabdovirus: Rhabdoviridae*.

**References**

D'Agostino, G., Luisoni, E. and Lovisolo, O. (1992). *Proc. Eur. EM* (Granada, Spain) **3**: 427.

# Celery yellow mosaic (?) potyvirus

**Data collated by A.A. Brunt, 1991.**

*Standard acronym:* CeYMV.

From Brazil; by Kitajima and Costa (1968).

**Natural host range and symptoms**

Symptoms persist.

*Apium graveolens* — conspicuous leaf chlorosis (except in cv. Florida 683 strain K which is tolerant).

**Transmission**

Transmitted by a vector; an insect; *Myzus persicae*; Aphididae. Transmitted in a non–persistent manner. Virus transmitted by mechanical inoculation.

**Geographical distribution**

Spreads in Brazil.

**Experimental host range**

Few (<3) families susceptible.

Diagnostically susceptible

*Apium graveolens, Coriandrum sativum, Petroselinum sativum* — systemic leaf chlorosis.

*Chenopodium amaranticolor, C. quinoa* — local lesions; no systemic infection.

Diagnostically insusceptible

*Conium maculatum, Datura stramonium, Glycine max, Gomphrena globosa, Nicotiana tabacum.*

Maintenance and propagation hosts

*Apium graveolens, Coriandrum sativum, Petroselinum sativum.*

Assay hosts (**L**ocal lesions or **W**hole plants)

*Chenopodium amaranticolor* (L), *C. quinoa* (L).

Dicotyledonous Hosts *(**Susceptible** or Insusceptible)*

***Apium graveolens***. *Beta vulgaris*. *Brassica campestris* ssp. *chinensis*. *Brassica juncea*. *Brassica oleracea* ssp. *botrytis*. *Brassica oleracea* ssp. *capitata*. *Calendula officinalis*. *Capsicum annuum*. *Cassia occidentalis*. ***Chenopodium amaranticolor***. *Chenopodium murale*. ***Chenopodium quinoa***. *Conium maculatum*. ***Coriandrum sativum***. *Cucumis melo*. *Cucumis sativus*. *Cucurbita pepo*. *Dahlia pinnata*. *Datura stramonium*. *Daucus carota*. *Dianthus caryophyllus*. *Gomphrena globosa*. *Helianthus annuus*. *Hibiscus esculentus*. *Lactuca sativa*. *Lycopersicon esculentum*. *Nicandra physaloides*. *Nicotiana debneyi*. *Nicotiana glutinosa*. *Nicotiana tabacum*. ***Petroselinum crispum***. *Physalis floridana*. *Pisum sativum*. *Raphanus sativus*. *Solanum nigrum*. *Sonchus oleraceus*. *Spinacia oleracea*. *Vigna unguiculata*. *Zinnia elegans*.

Monocotyledonous Hosts *(**Susceptible** or Insusceptible)*

*Zea mays*.

Sources of host-range data

De Olivera and Kitajima (1981).

**Properties of particles in sap**

TIP: 55 °C. LIV: 6 days. DEP: $\log_{10}$ minus 4.

**Particle morphology**

Virions filamentous; not enveloped; usually flexuous; with a clear modal length; of 750 nm.

**Physical properties**

One sedimenting component in purified preparations; sedimentation coefficient 133 *S*.

**Biochemical properties**

Virions contain 5.5 % nucleic acid; 94.5 % protein.

**Cytopathology**

Virions found in leaves and mesophyll; in cytoplasm and in nuclei. Inclusions present in infected cells; are viroplasms, pinwheels, and unusual in shape; are fibrous proteinaceous nuclear inclusions.

**Taxonomy and Relationships**

Unassigned potyvirus.

Virus(es) with serologically related virions: celery mosaic virus.

**References**

Da Cruz, N.D., Medina, D.M., Kitajima, E.W. and Costa, A.S. (1972). *Bragantia* **31**: 217.

de Olivera, M.L. and Kitajima, E.W. (1981). *Fitopatol. Bras.* **6**: 47.

Kitajima, E.W. and Costa, A.S. (1968). *Bragantia* **27**: 7.

Kitajima, E.W. and Costa, A.S. (1968). *Bragantia* **27**: 9.

Kitajima, E.W. and Costa, A.S. (1978). *Fitopatol. Bras.* **3**: 287.

# Celery yellow net (?) virus

**Data collated by A.A. Brunt, 1991.**

First reported in *Apium graveolens*; from the U.K.; by Hollings (1964).

**Natural host range and symptoms**

Symptoms persist.

*Apium graveolens* — yellow flecking and vein banding, leaves malformed.

**Transmission**

Not transmitted by *Myzus persicae*. Virus transmitted by mechanical inoculation; transmitted by grafting; not transmitted by seed.

**Geographical distribution**

Spreads in the UK.

**Experimental host range**

Few (<3) families susceptible.

Diagnostically susceptible

*Apium graveolens* — yellow vein-netting vein-banding, necrotic spotting and malformation.

*Petroselinum crispum* — green systemic spotting then chlorotic or necrotic spots and rings.

*Daucus carota, Apium repens* — transient spots or systemic yellow-netting.

Diagnostically insusceptible

*Chenopodium amaranticolor, Cucumis sativus, Nicotiana tabacum, Petunia hybrida, Vigna unguiculata.*

Maintenance and propagation hosts

*Apium graveolens, Petroselinum crispum.*

Assay hosts (Local lesions or Whole plants)

*Apium graveolens* (W).

**Properties of particles in sap**

TIP: 45 °C. LIV: 2 days (at 18°C). DEP: $\log_{10}$ minus 2.

**References**

Hollings, M. (1964). *J. hort. Sci.* **39**: 130.

Pemberton, A.W. and Frost, R.R. (1986). *Ann. appl. Biol.* **108**: 319.

# Celery yellow spot (?) *luteovirus*

**Data collated by A.A. Brunt, 1991.**

*Standard acronym:* CeYSV.

First reported in *Apium graveolens*; from California, U.S.A.; by Severin and Freitag (1945).

**Natural host range and symptoms**

Symptoms persist.

*Apium graveolens* — pale green spots and stripes, with yellow spots and stripes along veins.

*Conium maculatum* — symptomless infection.

*Pastinaca sativa* — mild leaf chlorosis.

**Transmission**

Transmitted by a vector; an insect; *Hyadaphis foeniculi*; Aphididae. Not transmitted by *Myzus persicae*. Transmitted in a persistent manner. Virus not transmitted by mechanical inoculation (usually).

**Geographical distribution**

Spreads in the UK and the USA (in California).

**Experimental host range**

Few (<3) families susceptible. Experimentally infected plants mostly show chlorotic spots and stripes, or mild leaf chlorosis.

Diagnostically susceptible

*Apium graveolens* — systemic yellow spots and stripes along the veins.
*Conium maculatum* — symptomless systemic infection.
*Pastinaca sativum* — faint systemic leaf chlorosis.

Diagnostically insusceptible

*Anthriscus cerefolium, Coriandrum sativum, Cucumis sativus, Daucus carota, Foeniculum vulgare, Lycopersicon esculentum, Nicotiana glutinosa, N. tabacum, Petroselinum crispum, Petunia hybrida, Phaseolus vulgaris, Tetragonia expansa.*

Maintenance and propagation hosts

*Apium graveolens, Pastinaca sativa.*

Assay hosts (**L**ocal lesions or **W**hole plants)

*Apium graveolens* (W), *Pastinaca sativa* (W).

**Taxonomy and Relationships**

Possible *Luteovirus*.

**References**

Freitag, J.H. and Severin, H.H.P. (1945). *Hilgardia* **16**: 375.
Hollings, M. (1964). *J. hort. Sci.* **39**: 130.
Pemberton, A.W. and Frost, R.R. (1986). *Ann. appl. Biol.* **108**: 319.
Severin, H.H.P. and Freitag, J.H. (1938). *Hilgardia* **11**: 493.
Waterhouse, P.M., Gildow, F.E. and Johnstone, G.R. (1988). *AAB Descr. Pl. Viruses* No. 339, 9 pp.

# Centrosema mosaic (?) *potexvirus*

**Data collated by A.J. Gibbs, 1985. Revised 1987.**

*Standard acronym:* CenMV.

First reported in *Centrosema* and *Crotalaria* spp. and other tropical legumes; from Papua New Guinea; by Van Velsen and Crowley (1962).

**Natural host range and symptoms**

Symptoms persist.

*Centrosema pubescens, Crotalaria anagyroides, C. retusa, C. goreensis, C. mucronata, Calopogonium mucunoides* (which sets no seeds), *Desmodium distortum* — vein-clearing, mosaic, leaf malformation.

**Transmission**

Transmitted by a vector; an insect; Heteroptera: *Lygaeidae* and *Nysius* spp.; an unusual insect group. Transmitted in a non–persistent manner. Virus transmitted by mechanical inoculation; transmitted by grafting; not transmitted by contact between plants; not transmitted by seed.

**Geographical distribution**

Spreads in Papua New Guinea.

**Experimental host range**

Few (<3) families susceptible. Experimentally infected plants mostly show mosaic.

Diagnostically susceptible

*Crotalaria spectabilis, C. juncea, C. retusa, C. anagyroides, Centrosema pubescens* and *Trifolium subterraneum* — mosaic.

Diagnostically insusceptible

*Nicotiana tabacum, Phaseolus vulgaris, Pisum sativum, Vicia faba, Vigna unguiculata.*

Maintenance and propagation hosts

*Crotalaria anagyroides.*

Assay hosts (**L**ocal lesions or **W**hole plants)

*Crotalaria anagyroides* (W).

Dicotyledonous Hosts *(**Susceptible** or Insusceptible)*

*Arachis hypogaea. Chenopodium amaranticolor. Cicer arietinum.* ***Crotalaria retusa. Crotalaria spectabilis.*** *Cucumis sativus. Datura stramonium. Glycine max. Lupinus albus. Lupinus angustifolius. Lycopersicon esculentum. Medicago sativa. Nicotiana glutinosa. Nicotiana rustica. Nicotiana tabacum. Petunia hybrida. Phaseolus lunatus. Phaseolus vulgaris. Pisum sativum. Sesbania exaltata. Trifolium hybridum. Trifolium incarnatum. Trifolium pratense. Trifolium repens.* ***Trifolium subterraneum.*** *Vicia faba. Vigna unguiculata. Vigna unguiculata* ssp. *cylindrica. Vigna unguiculata* ssp. *sesquipedalis.*

**Properties of particles in sap**

TIP: 55–58 °C. LIV: 0.25 days. DEP: $\log_{10}$ minus 3–4.

**Purification**

Method: Crowley and Francki, 1963.

**Particle morphology**

Virions filamentous; not enveloped; usually flexuous; with a clear modal length; of 580 nm. Axial canal obscure.

**Taxonomy and Relationships**

Possible *Potexvirus.*

**References**

Crowley, N.C. and Francki, R.I.B. (1963). *Aust. J. biol. Sci.* **16**: 468.

Van Velsen, R.J. and Crowley, N.C. (1962). *Aust. J. agric. Sci.* **13**: 220.

# Cereal chlorotic mottle (?) *nucleorhabdovirus*

**Data collated by R.S. Greber, 1980.**

*Standard acronym:* CerCMV. *Strains:* CCMV-Morocco.

First reported in *Zea mays*; from Brisbane, Australia; by Greber (1977; 1979).

## Natural host range and symptoms

Symptoms persist.

*Avena sativa, Digitaria ciliaris, Dinebra retroflexa, Echinochloa colona, Hordeum vulgare, Leptochloa filiformis, Setaria verticillata, Triticum aestivum, Urochloa panicoides, Zea mays* — chlorotic streaking, variable stunting and sterility.

## Transmission

Transmitted by a vector; an insect; *Nesoclutha pallida, Cicadula bimaculata, Cicadulina bipunctata* subsp. *bipunctella*; Cicadellidae. Transmitted in a persistent manner. Virus retained when the vector moults; multiplies in the vector; not transmitted congenitally to the progeny of the vector; not transmitted by mechanical inoculation; not transmitted by seed.

## Geographical distribution

Spreads in Australia and Morocco.

## Experimental host range

Few (<3) families susceptible (only Gramineae). Experimentally infected plants mostly show chlorotic striations.

Diagnostically susceptible

*Dinebra retroflexa* — chlorotic streaking.
*Leptochloa filiformis* — chlorotic and necrotic streaking.
*Triticum* spp. and x *Triticosecale* — chlorotic streaking.
*Zea mays* — chlorotic streaks, but cvs differ in susceptibility.

Diagnostically insusceptible

*Oryza sativa, Sorghum bicolor.*

Maintenance and propagation hosts

*Dinebra retroflexa.*

Monocotyledonous Hosts (***Susceptible*** *or Insusceptible)*

***Avena sativa***. *Chloris gayana*. *Digitaria decumbens*. ***Dinebra retroflexa***. ***Echinochloa colonum***. ***Eleusine coracana***. ***Eragrostis cilianensis***. ***Hordeum vulgare***. ***Leptochloa filiformis***. *Oryza sativa*. *Paspalum dilatatum*. *Sorghum bicolor*. ***Triticum aestivum***. ***Zea mays***.

Sources of host-range data

Greber (1979).

## Properties of particles in sap

Leaf sap contains many virions. Electron microscopy: use PTA-H.

## Purification

Method: Jackson and Christie (1977); Greber and Gowanlock (1979).

## Particle morphology

Virions rhabdo– or bullet–shaped; enveloped; with a clear modal length; of 214 nm; 75 nm wide. Axial canal obscure.

### Physical properties

One sedimenting component in purified preparations. Density 1.165 g $cm^{-3}$ in sucrose.

*Virion protein(s)* five; $M_r$ of the largest 82000. $M_r$ of 2nd largest 52000. $M_r$ of 3rd largest 45500. $M_r$ of 4th largest 31800. $M_r$ of 5th largest 28900. Method of preparation: Laemmli (1970).

### Replication

Replication does not depend on a helper virus.

### Cytopathology

Virions found in mesophyll, phloem and companion cells; in the perinuclear space; cytoplasmic vesicles.

### Taxonomy and Relationships

Possible *Nucleorhabdovirus: Rhabdoviridae.*

Virus(es) with serologically unrelated virions: 8 rhabdoviruses of *Poaceae*, except for Moroccan cereal chlorotic mottle virus.

### References

Greber, R.S. (1977). *Aust. Pl. Path. Soc. Newsl.* **6**: 17.
Greber, R.S. (1979). *Aust. J. agric. Res.* **30**: 433.
Greber, R.S. (1982). *CMI/AAB Descr. Pl. Viruses* No. 251, 4 pp.
Greber, R.S. and Gowanlock, D.H. (1979). *Aust. J. biol. Sci.* **32**: 399.
Jackson, A.O. and Christie, S.R. (1977). *Virology* **77**: 344.
Laemmli, U.K. (1970). *Nature, Lond.* **227**: 680.
Lockhart, B.E.L. (1986). *Plant Dis.* **70**: 912.

# Cereal flame chlorosis (?) virus

**Data collated by S. Haber, 1991.**

*Strains:* flame chlorosis virus (FCV) is associated with distinct dsRNA patterns in barley, bread and durum wheat and oat. There may be barley, wheat and oat strains of FCV.

First reported in *Hordeum vulgare* cv. Argyle; from Manitoba, Canada near Newdale; by Haber *et al.* (1990).

### Natural host range and symptoms

Symptoms persist. Symptoms flame-like chlorosis, stunting, prostrate growth.

*Avena sativa, Hordeum vulgare, Triticum aestivum, T. durum* — defined chlorosis radiating in a flame-like pattern from the leaf base. Plants stunted and prostrate.

### Transmission

Transmitted in soil; probably by fungi. Not transmitted by *Polymyxa graminis*. Virus not transmitted by mechanical inoculation; not transmitted by contact between plants; not transmitted by seed; not transmitted by pollen.

### Ecology and control

Studies reported by the disease is found often in fields frequently sown with spring cereals, particularly barley.

### Geographical distribution

Spreads in Canada (Manitoba).

**Experimental host range**

Few (<3) families susceptible. Experimentally infected plants mostly show variegated yellow-whitish chlorosis.

Diagnostically susceptible

*Avena sativa, Hordeum vulgare, Triticum aestivum* — variegated yellow-whitish to pale-green chlorosis.

Maintenance and propagation hosts

*Hordeum vulgare* is the best, but may also use *Triticum aestivum* and *Avena sativa.*

Monocotyledonous Hosts *(**Susceptible** or Insusceptible)*

***Avena sativa. Hordeum vulgare. Triticum aestivum. Triticum durum.***

Sources of host-range data

Haber and Barr (1990).

**Biochemical properties**

*Genome* consists of RNA; double–stranded; linear; of 7 parts to of 15 parts (depending on isolate and variant); largest (or only) genome part the largest 3.5 kb (pairs); the 6th largest and other parts 0.25 kb (pairs; but varies in different isolates). Genomic nucleic acid isolated by Haber and Barr (1990); Haber *et al.* (1990). Base composition of ds genome 56.8 % (G + C).

*Features of the genome:* from sequencing of cDNA clones at least 1 ORF has been found; it encodes a putative 108 amino-acid histone-like polypeptide. Known sequence is in EMBL Data Library (Ref. DS7101) X59248 HORDEUM, FLAME CHLOROSIS DNA; putative 108 amino-acid basic polypeptide has possible (distant) similarity with tymovirus nucleotide-binding fold.

**Cytopathology**

Inclusions present in infected cells; are unusual in shape; they do not contain virions. Other cellular changes: malformation and vesiculation of peripheries of chloroplasts, mitochondria and leukoplasts (Haber *et al.*, 1990).

**Taxonomy and Relationships**

Effects on chloroplasts resemble those caused by tombusviruses and the vesicular inclusions resemble those of tymoviruses. The pattern of numerous dsRNAs resembles those of some mycoviruses.

**Comments**

CFCV may have arisen or spread recently as it was not seen before 1985. This corresponds with increasing frequency of barley cultivation. Since 1985 it has been found at greater distances from the site where it was orginally found; it is expected that it will soon be found in border areas of North Dakota and Minnesota.

**References**


Haber, S. (1990). *Can. Pl. Dis. Surv.* **70**: 50.

Haber, S. (1991). *Can. Pl. Dis. Surv.* **7**: in press.

Haber, S. and Barr, D.J. (1990). In: German Phytomedical Series Vol. 1. *Proc. First Symp. of Int. Working Group on Plant Viruses with Fungal Vectors*, ed. R. Koenig, p. 169.

Haber, S. and Platford, R.G. (1991). *Can. Pl. Dis. Surv.* **71**: 79.

Haber, S., Kim, W., Gillespie, R. and Tekauz, A. (1990). *J. Phytopath.* **129**: 245.

Haber, S., Barr, D.J.S. and Platford, R.G. (1991). *Can. J. Pl. Pathol.* **13**: in press.

# Cereal northern mosaic *cytorhabdovirus*

**Data collated by S. Toriyama, 1987.**

*Synonyms:* northern cereal mosaic virus, wheat rosette stunt virus (Lundsgaard *et al.*, 1984; Gong *et al.*, 1985; Milne *et al.*, 1986). *Standard acronym:* NCMV. *Strains:* barley yellow striate mosaic cytorhabdovirus.

First reported in *Hordeum vulgare*; from Hokkaido, Japan; by Ito and Fukushi (1944*a*).

**Natural host range and symptoms**

*Hordeum vulgare.*

**Transmission**

Transmitted by a vector; an insect; Delphacidae. Transmitted in a persistent manner. Virus retained when the vector moults; multiplies in the vector; not transmitted congenitally to the progeny of the vector; not transmitted by mechanical inoculation; not transmitted by contact between plants; not transmitted by seed.

**Geographical distribution**

Spreads in China and Japan.

**Experimental host range**

Few (<3) families susceptible.

Monocotyledonous Hosts *(**Susceptible** or Insusceptible)*

*Agrostis palustris*. ***Anthoxanthum odoratum***. ***Avena byzantina***. ***Avena sativa***. ***Avena strigosa***. *Bromus inermis*. *Cynodon dactylon*. *Dactylis glomerata*. ***Echinochloa crus–galli***. *Eleusine coracana*. ***Eragrostis cilianensis***. ***Hordeum vulgare***. ***Lolium multiflorum***. ***Lolium perenne***. *Oryza sativa*. ***Panicum maximum***. *Phleum pratense*. ***Poa annua***. *Poa compressa*. *Poa pratensis*. *Poa trivialis*. ***Secale cereale***. ***Setaria italica***. ***Setaria viridis***. ***Triticum aestivum***. ***Zea mays***.

Additional hosts

***Alopecurus aequalis, A. geniculatus, A. japonica, A. pratensis, Avena fatua, Beckmannia syzgachne, Calamagrostis epigejoo, Chloris viagara, Diplachne fusca***, *Eleusine indica*, ***Eragrostis multicaulis, E. poaeoides, E. pilosa, Festuca elation***, *F. ovina, F. rubra*, ***Holcus lanatus***, *H. mollis, Leersia japonica*, ***Panicum bisulcatum, Pennisetum alopecuroides, Phleum paniculata, Polypogon fugux***, *Setaria faberii*, ***S. faberii***, *Sorghum vulgare* and *Sporobolus elongatus.*

Sources of host-range data

Ito and Fukushi (1944*a*); Ito and Fukushi (1944*b*); Shinkai (1955); Liu and Chen (1982); Lin *et al.* (1983).

**Properties of particles in sap**

TIP: 50–55 °C. LIV: 4–5 days (at 5°C). DEP: $\log_{10}$ minus 3 (sap), or 4 (insect extracts). Infectivity of sap decreased by treatment with di–ethyl ether. Leaf sap contains many virions.

**Purification**

Method: Toriyama: grind tissues with glycine at pH 7, Celite filtration, sucrose density gradient centrifugation, zone electrophoresis.

**Particle morphology**

Virions rhabdo– or bullet–shaped; enveloped; with a clear modal length; of 350 nm; 68 nm wide. Axial canal obvious. Basic helix obvious (if stained with UA); pitch of basic helix 5.5 nm.

**Physical properties**

One sedimenting component in purified preparations.

**Biochemical properties**

Also phospho-lipids and sterols. Composition of sterols 10.7%, cholesterol 12.9%, campesterol 25.6%, stigmasterol 50.8% β-sitosterol.

*Genome* consists of RNA; single–stranded; unipartite; largest (or only) genome part 10.5 kb. Base composition 16.3 % G; 30.3 % A; 16.7 % C; 36.6 % U (for wheat rosette stunt strain). Infectivity lost when deproteinised with phenol or detergent.

*Virion protein(s)* six; $M_r$ (wheat rosette stunt strain only) of the largest(L) 140000. $M_r$ of 2nd largest 60000–63000. $M_r$ of 3rd largest 46000–47000. $M_r$ of 4th largest 40000. $M_r$ of 5th largest 19000. Virion proteins glycosylated.

**Replication**

Genome replicates in cytoplasm.

**Cytopathology**

Virions found in leaves, roots, mesophyll, epidermis, meristem, vascular parenchyma, phloem, companion cells and in developing xylem cells; in cytoplasm. Inclusions present in infected cells; are viroplasms; they do not contain virions. Other cellular changes: suppression of the elongation of cells in wheat and barley: the proportion of glucose decreases and that of xylose arabinose increases.

**Taxonomy and Relationships**

*Cytorhabdovirus: Rhabdoviridae.*

Virus(es) with serologically related virions: wheat rosette stunt virus is closely related and barley yellow striate mosaic virus is distantly related. Virus(es) with serologically unrelated virions: festuca leaf streak virus.

Best tests for diagnosis: After one day acquisition feeding and about 10 days latent period, transmit the virus by *Laodelphax striatellus* to wheat (or barley) and japonica type rice varieties. Ceral northern mosaic virus infects wheat (or barley), but not rice. Infected wheat (or barley) plants, first show chlorotic striping, then develop many tillers, one by one, to give stunted rosettes.

**References**


Anon. (1979). *Scientia Sin.* **22**: 573.

Gong, Z. and Zhen, Q. (1985). *Acta Biochim. Biophys. Sin.* **17**: 650.

Gong, Z., Zhen, Q., Peng, T. and Shikata, E. (1985). *Chinese J. Virol.* **1**: 257.

Ishii, M. (1966). *Res. Bull. Hokkaido natn agric. Exp. Stn* **89**: 49.

Ishii, M. (1967). *Jap. J. appl. Ent. Zool.* **11**: 191.

Ishii, M. and Matsumoto (1964). *Ann. Phytopath. Soc. Japan* **29**: 280.

Ito and Fukushi, T. (1944*a*). *J. Sapporo Soc. Agric. For.* **36**: 62.

Ito and Fukushi, T. (1944*b*). *J. Sapporo Soc. Agric. For.* **36**: 65.

Ito and Fukushi, T. (1956). *Ann. Phytopath. Soc. Japan* **21**: 3.

Kung and Zheng (1985). *Acta Biochim. Biophys. Sin.* **17**: 248.

Lee and Shikata, E. (1977). *Korean J. Pl. Prot.* **16**: 87.

Lin, Ruan and Jin (1983). *Scientia agric. Sin.* **64**: 1983.

Liu and Chen (1982). *Scientia agric. Sin.* **8**: 1982.

Lu, Shikata, E. and Murayama (1968). *Mem. Fac. Agric. Hokkaido Univ.* **6**: 335.
Lundsgaard, T. (1984). *Intervirology* **22**: 50.
Lundsgaard, T., Tien, P. and Toriyama, S. (1984). *Phytopath. Z.* **111**: 232.
Mikoshiba, Nemoto and Gomi (1984). *Ann. Phytopath. Soc. Japan* **50**: 108.
Milne, R.G., Masenga, V. and Conti, M. (1986). *Intervirology* **25**: 83.
Murayama and Lu (1967). *J. Fac. Agric. Hokkaido Univ.* **55**: 182.
Ogawa and Moichi (1984). *Ann. Phytopath. Soc. Japan* **50**: 108.
Shinkai, A. (1955).. *Ann. Phytopath. Soc. Japan* **19**:1 40.
Shinkai, A. (1960). In: *Plant Virus Diseases* p. 275; eds J. Hidaka, T. Hirai, D. Murayama, and K. Yora. Asakura Pub. Co. Ltd., Tokyo.
Shinkai, A. (1966). *Ann. Phytopath. Soc. Japan* **32**: 317.
Shinkai, A. (1968). *Shokubutsu Boeki* **22**: 445.
Shinkai, A. (1969). *Shokubutsu Boeki* **23**: 380.
Shinkai, A., Kureha and Suzuki (1968). *Ann. Phytopath. Soc. Japan* **34**: 376.
Shikata, E. and Lu (1967). *Proc. Japan Acad.* **43**: 918.
Shirako, Y. and Ehara, Y. (1985). *Phytopathology* **75**: 453.
Toriyama, S. (1972). *Virus* **22**: 114.
Toriyama, S. (1976*a*). *Ann. Phytopath. Soc. Japan* **42**: 494.
Toriyama, S. (1976*b*). *Ann. Phytopath. Soc. Japan* **42**: 563.
Toriyama, S. (1977). *Ann. Phytopath. Soc. Japan* **43**: 126.
Toriyama, S. (1986). *CMI/AAB Descr. Pl. Viruses* No. 322, 5 pp.
Toriyama, S. and Kuwata, S. (1973). *Ann. Phytopath. Soc. Japan* **39**: 136.
Toriyama, S. and Peters, D. (1981). *J. gen. Virol.* **56**: 59.
Toriyama, S. and Yora, K. (1969). *Ann. Phytopath. Soc. Japan* **35**: 360.
Yamada and Shikata, E. (1969). *J. Fac. Agric. Hokkaido Univ.* **56**: 91.
Yamada, Shikata, E. and Murayama (1967). *Ann. Phytopath. Soc. Japan* **33**: 97.
Zheng, Xu, Zhu, Shen and Kung (1983). *Acta Biochim Biophys. Sin* **15**: 561.

# Cestrum (?) *caulimovirus*

**Data collated by A.A. Brunt, 1991.**

*Standard acronym:* CV.

First reported in *Cestrum parqui*; from Italy; by Ragozzino (1974).

### Natural host range and symptoms

*Cestrum parqui* — chlorotic oak-leaf patterns and malformation.

### Transmission

Virus not transmitted by mechanical inoculation; transmitted by grafting.

### Geographical distribution

Found, but with no evidence of spread, in Italy.

### Particle morphology

Virions isometric; 35–40 nm in diameter.

### Taxonomy and Relationships

Possible *Caulimovirus*.

### References

Hull, R. (1984). *CMI/AAB Descr. Pl. Viruses* No. 295, 4 pp.
Ragozzino, A. (1974). *Ann. Fac. Sci. agr. Univ. Napoli IV* **8**: 249.

# Chara australis (?) *furovirus*

**Data collated by A.J. Gibbs and M. Torronen, 1987 and 1991.**

*Synonyms:* Chara corallina virus. *Standard acronym:* ChAV.

First reported in *Chara australis*; from the A.C.T., Australia; by Gibbs *et al.* (1975).

**Natural host range and symptoms**

*Chara australis* — plants symptomless but with inclusions.

**Transmission**

Virus not transmitted by contact between plants.

**Ecology and control**

Studies reported by Skotnicki *et al.* (1976).

**Geographical distribution**

Spreads in Australia (south-eastern rivers, but not found in the United Kingdom (4spp.) nor in a worldwide collection of charophyte species held in Lubbock, Texas (V. Proctor and A.J. Gibbs, unpublished data)).

**Experimental host range**

Few (<3) families susceptible.

Diagnostically susceptible

*Chara australis* — infected by injection become chlorotic and die.

Diagnostically insusceptible

*Nicotiana clevelandii, N. glutinosa, Brassica campestris* spp. pekinensis, *Phaseolus vulgaris.*

Maintenance and propagation hosts

*Chara australis.*

Assay hosts (**L**ocal lesions or **W**hole plants)

*Chara australis.*

Dicotyledonous Hosts *(**Susceptible** or Insusceptible)*

*Brassica campestris* ssp. *pekinensis. Brassica campestris* ssp. *rapa. Chenopodium amaranticolor. Chenopodium quinoa. Nicotiana clevelandii. Nicotiana glutinosa. Petunia hybrida. Phaseolus vulgaris. Pisum sativum. Spinacia oleracea. Vicia faba.*

Monocotyledonous Hosts *(**Susceptible** or Insusceptible)*

*Avena sativa. Triticum aestivum.*

Sources of host-range data

Gibbs *et al.* (1975).

**Properties of particles in sap**

Leaf sap contains many virions.

**Purification**

Method: blend *Chara* plants in 1–0.5 vol. of 50 mM phosphate buffer pH 7.5 containing 100 mM urea and 5 mM disodium EDTA. Filter. Stir in 2% w/v sodium chloride 4% w/v polyethylene glycol 6000 at 4°C overnight. Centrifuge at 3000 g for 30 minutes. Collect pellet. Resuspend in original buffer containing 8.0% butanol at room temperature. Emulsify with 10% chloroform. Centrifuge and collect supernatant. Purify further by centrifuging in buffered sucrose gradients.

**Particle morphology**

Virions rod–shaped; usually straight; with a clear modal length; of 532 nm; 18 nm wide. Axial canal obvious. Pitch of basic helix 2.75 nm.

**Physical properties**

One sedimenting component in purified preparations, or two sedimenting components in purified preparations (in some preparations); sedimentation coefficient 230 *S*. Isoelectric point pH 3.4–3.7.

**Biochemical properties**

Virions contain 5 % nucleic acid; 95 % protein.

*Genome* consists of RNA; single–stranded; linear; unipartite; largest (or only) genome part 11 kb. Base composition 24.5 % G; 28 % A; 20 % C; 27.5 % U.

*Non–genomic nucleic acid* not found in the virions.

*Virion protein(s)* one; $M_r$ 18700; coat protein with approximately 170 residues. Amino acid composition: Skotnicki *et al.* (1976).

**Replication**

Replication does not depend on a helper virus.

**Cytopathology**

Virions found in all parts of the host plant; in cytoplasm and in cell vacuoles. Inclusions present in infected cells (in the cytoplasm, are banana-shaped paracrystals); they contain virions.

**Taxonomy and Relationships**

Possible *Furovirus*.

Virus(es) with serologically related virions: tobacco mosaic and odontoglossum ringspot tobamoviruses, but very distantly.

The nucleotide sequence of the replicase motif has closest homology to that of beet necrotic yellow vein furovirus.

**Comments**

The host of this virus was incorrectly identified first as *Chara corallina*.

**References**

Gibbs, A.J., Skotnicki, A.H., Gardiner, J.E., Walker, E.S. and Hollings M. (1975). *Virology* **64**: 571.

Skotnicki, A.H., Gibbs, A.J. and Wrigley, N.G. (1976). *Virology* **75**: 457.

# Cherry A *capillovirus*

**Data collated by A.A. Brunt.**

First reported in *Prunus cerasus*; from Germany; by Jelkmann (1995).

**Natural host range and symptoms**

*Prunus cerasus* — unknown.

**Transmission**

Virus transmitted by grafting.

**Geographical distribution**

Spreads in the Eurasian region. Found, but with no evidence of spread, in Germany.

**Particle morphology**

Virions filamentous; usually flexuous; with no clear modal length; 12 nm wide.

**Biochemical properties**

*Genome* consists of RNA. Base composition 20.7 % G; 30.8 % A; 20.7 % C; 29.8 % U. Poly A region present; 3'-terminus.

*Virion protein(s)* one; $M_r$ 24000; coat protein. Amino acid sequence: Jelkmann (1995).

**Taxonomy and Relationships**

*Capillovirus.*

Virus(es) with serologically unrelated virions: apple stem grooving, apple chlorotic leafspot, apple stem pitting and cherry mottle viruses.

**References**

Jelkmann, W. (1995). *J. gen. Virol.* **76**: 2015.

# Cherry leaf roll *nepovirus*

**Data collated by A.T. Jones 1986.**

*Synonyms:* ash mosaic virus (Nienhaus and Hamacher, 1990), sambucus ringspot and yellow net virus. *Standard acronym:* CLRV. *Strains:* birch ring and line pattern virus (Cooper and Atkinson, 1975); blackberry and red raspberry strains (Cropley and Tomlinson, 1971; Jones and Wood, 1978); berteroa ringspot virus (Lockhart, 1977); elm mosaic strain (Varney and Moore, 1952; Jones and Murant, 1971); dogwood ringspot strain (Waterworth and Lawson, 1973); golden elderberry strain (Hansen (1967); Hansen and Stace-Smith, 1971; Jones and Murant, 1971); red elder ringspot strain (Schmelzer, 1972*b*); rhubarb strain (Tomlinson and Walkey, 1967); type (cherry) strain (Cropley, 1961); walnut black line virus (Cooper and Edwards, 1980; De Zoeten, 1982; Mircetich *et al.*, 1980; Rowhani *et al.*, 1985), walnut line pattern and mosaic virus (Christoff, 1958), walnut yellow mosaic virus, walnut ringspot and walnut yellow vein strains (Savino *et al.*, 1977).

First reported in *Ulmus americana* (American elm), *Prunus avium* (cherry) and *Juglans regia* (walnut). Symptoms of walnut blackline disease, now known to be caused by cherry leaf roll first described in 1933 by Schuster and Miller; by Swingle *et al.* (1941; 1943); Posnette and Cropley (1955); Schuster and Miller (1933).

**Natural host range and symptoms**

Symptoms persist and vary seasonally.

*Juglans regia* — leaf pattern and black line.

*Ulmus americana* — chlorotic mosaic, ring pattern and dieback.

*Prunus avium* — leaf rolling and death.

*Betula* spp., *Sambucus* spp., *Rubus* spp., *Cornus florida* — chlorotic ringspot, leaf patterns and/or yellow vein netting.

*Ptelea trifoliata* — yellow spotting.

*Olea europea, Rheum rhaponticum, Berteroa incana, Delphinium elatum, Rumex obtusifolius*—symptomless.

**Transmission**

Transmitted by a vector; *Xiphinema coxi, X. diversicaudatum* and *X. vuittenezi* (cherry isolates — Fritzsche and Kegler (1964); Flegg (1969), but not the golden elderberry and rhubarb isolates). Not transmitted by *X. americanum, X. bakeri,*

*Longidorus elongatus, L. leptocephalus, L. macrosoma* or *Paralongidorus maximus* (Van Hoof, 1971; Jones *et al.*, 1981). The elm mosaic strain is not transmitted by *X. americanum* (Fulton and Fulton, 1970) or by the aphid *Myzus persicae* (Ford *et al.*, 1972). Virus transmitted by mechanical inoculation; transmitted by grafting; not transmitted by contact between plants.

**Geographical distribution**

Spreads in the Eurasian region and the North American region; Australia, China, New Zealand, Turkey, and the former USSR.

**Experimental host range**

Many (>9) families susceptible. Experimentally infected plants mostly show chlorotic or necrotic local lesions, systemic necrosis or mosaic.

Diagnostically susceptible

*Chenopodium amaranticolor, C. quinoa* — chlorotic or necrotic local lesions; systemic mottle, necrosis and malformation.

*Cucumis sativus* — chlorotic local lesions, occasional systemic mosaic.

*Nicotiana rustica, N. tabacum* cvs White Burley, Xanthi-*nc* — necrotic local lesions and rings; systemic necrotic or chlorotic rings.

Maintenance and propagation hosts

*Nicotiana rustica, N. tabacum* cvs White Burley, Xanthi-*nc* for maintaining virus cultures; *Nicotiana clevelandii* and *Chenopodium quinoa* for propagation and purification of virions.

Assay hosts (Local lesions or Whole plants)

*Nicotiana rustica, N. tabacum* cvs White Burley, Xanthi-*nc* (L), *Chenopodium amaranticolor* (L).

Dicotyledonous Hosts *(**Susceptible** or Insusceptible)*

*Amaranthus caudatus.* ***Antirrhinum majus. Apium graveolens.*** *Arachis hypogaea.* ***Atriplex hortensis. Beta macrocarpa. Beta vulgaris. Capsicum frutescens.*** *Cassia tora.* ***Catharanthus roseus***, or *Catharanthus roseus.* ***Celosia argentea. Celosia cristata. Chenopodium amaranticolor. Chenopodium foetidum. Chenopodium murale. Chenopodium quinoa.*** *Citrullus lanatus.* ***Cucumis melo. Cucurbita maxima. Cucurbita pepo***, or *Cucurbita pepo. Datura metel. Datura stramonium. Dianthus barbatus.* ***Glycine max. Gomphrena globosa. Helianthus annuus. Lathyrus odoratus. Lavatera trimestris. Lobelia erinus.*** *Lycopersicon esculentum.* ***Momordica balsamina. Nicotiana benthamiana. Nicotiana clevelandii. Nicotiana debneyi. Nicotiana megalosiphon. Nicotiana occidentalis. Nicotiana rustica. Nicotiana tabacum. Ocimum basilicum. Petunia hybrida. Phaseolus vulgaris. Phlox drummondii. Physalis floridana. Physalis peruviana. Pisum sativum. Prunus cerasifera. Prunus persica. Raphanus sativus. Rubus idaeus. Sambucus racemosus.*** *Saponaria vaccaria.* ***Senecio vulgaris. Sesamum indicum.*** *Solanum rostratum.* ***Spinacia oleracea. Stellaria media. Torenia fournieri. Tropaeolum majus.*** *Vigna unguiculata.* ***Zinnia elegans.***

Additional hosts

All isolates have a wide experimental host range in more than 36 plant families, including the monocotyledons *Tinantia erecta* and *Commelina* spp.

Sources of host-range data

Schmelzer (1966); Hansen and Stace-Smith (1971); Horváth (1979); Tomlinson and Walkey (1967); Lockhart (1977); Ahmed and Bailiss (1975); Walkey and Cooper (1973).

**Properties of particles in sap**

TIP: 50–60 °C. LIV: 4–16 days. DEP: $\log_{10}$ minus 3–5. Infectivity of sap not changed by treatment with di–ethyl ether. Leaf sap contains many virions.

**Purification**

Method: Jones (1973); Cooper and Atkinson (1975), Walkey *et al.* (1973).

**Particle morphology**

Virions isometric; not enveloped; 28 nm in diameter; angular in profile; without a conspicuous capsomere arrangement.

**Physical properties**

Three sedimenting components in purified preparations (Jones, 1973; Haber and Hamilton, 1980); sedimentation coefficient of the fastest 128 *S* (B); of the other(s) 115 *S* (M), or 52 *S* (T). Density 1.5 g $cm^{-3}$ in CsCl (B), or 1.47 g $cm^{-3}$ in CsCl (M).

**Biochemical properties**

Virions contain 46 % nucleic acid (B), or 41 % nucleic acid (M), or 0 % nucleic acid (T); 54 % protein (B), or 59 % protein (M), or 100 % protein (T).

*Genome* consists of RNA; single–stranded; linear; of two parts (both required for infectivity; Jones and Mayo, 1972; Jones and Duncan, 1980; Haber and Hamilton, 1980); largest (or only) genome part the larger 9 kb (RNA-1); the 2nd largest 7.5 kb (RNA-2). Genomic nucleic acid isolated by Jones and Mayo (1972); Walkey *et al.* (1973); Murant *et al.* (1981). Base composition RNA-1 and RNA-2, 26.5 and 26.6 % G; 21.4 and 22 % A; 22.3 and 22.8 % C; 29.1 and 29.2 % U. Infectivity retained when deproteinised with phenol or detergent. Poly A region present (in both RNA species; Massalski, 1984). Additional factor required for infectivity; a genome-linked protein of estimated M.Wt 3500 (C.U.T. Hellen and J.I. Cooper, personal communication), although not yet known if required for infectivity.

*Sequence database accession code(s):*

S63537 Em(40)_vi:S63537 Gb(84)_vi:S63537 polyprotein (RNA-2) cherry leaf roll nepovirus CLRV, birch isolate I2, Genomic RNA, 1920 nt

S84124 Em(40)_vi:S84124 Gb(84)_vi:S84124 (3' terminal region, RNA-1) cherry leaf roll nepovirus, Genomic RNA, 1743 nt. 1/94 1,743bp.

S84125 Em(40)_vi:S84125 Gb(84)_vi:S84125 (3' terminal region, RNA-2) cherry leaf roll nepovirus, Genomic RNA, 1805 nt. 1/94 1,805bp.

S84126 Em(40)_vi:S84126 Gb(84)_vi:S84126 (3' terminal region, R25) cherry leaf roll nepovirus, Genomic RNA, 1182 nt. 1/94 1,182bp.

U24694 Em(43)_vi:Cl24694 Gb(89)_vi:Clu24694 Cherry leaf roll virus RNA2, 3' end. 5/95 1,565bp.

Z34265 Em(43)_vi:Clrvrn Cherry leaf roll virus (walnut) genomic RNA (1588bp). 11/94 1,588bp.

*Virion protein(s)* one; $M_r$ 54000. Method of preparation: Jones and Mayo (1972); Walkey *et al.* (1973); Murant *et al.* (1981). Amino acid composition: Walkey *et al.* (1973) (five strains).

**Replication**

Replication does not depend on a helper virus.

**Cytopathology**

Virions found in leaves, roots, meristem and within tubules in pollen, ovules and mature seeds (Walkey and Webb, 1968; 1970; Jones *et al.*, 1973); in cytoplasm. Inclusions present in infected cells; are membranous bodies (connected with endoplasmic reticulum and ribosomes; Jones *et al.*, 1973); they contain virions. Other cellular changes: projections from cell walls (Jones *et al.*, 1973; Cooper and Atkinson, 1975). Birch leaves showing symptoms often contain chloroplasts with large densely stained plastoglobuli and poorly defined lamellae (Cooper and Atkinson, 1975).

**Taxonomy and Relationships**

*Nepovirus: Comoviridae.*

Virus(es) with serologically unrelated virions: grapevine Bulgarian latent, lucerne Australian latent, peach rosette mosaic (Jones and Duncan, 1980), arabis mosaic, grapevine fanleaf, raspberry ringspot, strawberry latent ringspot, tobacco ringspot, tomato black ring, tomato ringspot and tomato top necrosis viruses (Cropley, 1961; Hansen and Stace-Smith, 1971; Jones and Murant, 1971).

*Gomphrena globosa* is only susceptible to some strains. Virions of elm and dogwood strains are less stable than most others (Jones, 1976). Most strains only produce M and B components, and the ratios of the components differ between strains. Varney and Moore (1952) found that plants infected with tomato ringspot virus resisted infection with the elm mosaic strain. Jones (1976) confirmed this using eight strains of cherry leaf roll virus in *Nicotiana tabacum* cv. Xanthi-*nc* but found none of these protected plants against infection by tomato ringspot virus.

**References**


Ahmed, A.H. and Bailiss, K.W. (1975). *J. hort. Sci.* **50**: 47.

Callahan, K.L. (1957*a*). *Diss. Abstr.* **17**: 1861.

Callahan, K.L. (1957*b*). *Phytopathology* **47**: 5.

Christoff, A. (1958). *Phytopath. Z.* **31**: 381.

Cooper, J.I. (1976). *Mitt. biol. BundesAnst. Land. Forstw.* **170**: 17.

Cooper, J.I. and Atkinson, J.A. (1975). *Forestry* **48**: 193.

Cooper, J.I. and Edwards, M.L. (1975). *Forestry* **53**: 41.

Cooper, J.I. and Edwards, M.L. (1980). *Forestry* **53**: 41.

Cooper, J.I., Massalski, P.R. and Edwards, M.L. (1984). *Ann. appl. Biol.* **105**: 55.

Cropley, R. (1961). *Ann. appl. Biol.* **49**: 524.

Cropley, R. and Tomlinson, J.A. (1971). *CMI/AAB Descr. Pl. Viruses* No. 80, 4 pp.

de Zoeten, G.A. (1982). *Phytopathology* **72**: 1261.

Flegg, J.J.M. (1969). *Rep. E. Malling Res. Stn.* 1968, p. 155.

Ford, R.E., Moline, H.E., McDaniel, G.L., Mayhew, D.E. and Epstein, A.H. (1972). *Phytopathology* **62**: 987.

Fritzsche, R. and Kegler, H. (1964). *Naturw.* **51**: 299.

Fulton, J.P. and Fulton, R.W. (1970). *Phytopathology* **60**: 114.

Haber, S. and Hamilton, R.I. (1980). *J. gen. Virol.* **50**: 377.

Hansen, A.J. (1967). *Phytopathology* **57**: 813.

Hansen, A.J. and Stace-Smith, R. (1971). *Phytopathology* **61**: 1222.

Horváth, J. (1979). *Acta phytopath. Acad. Sci. Hung.* **14**: 319.

Jones, A.T. (1973). *Ann. appl. Biol.* **74**: 211.

Jones, A.T. (1976). *Poljopr. Znanst. Smotra* **39**: 527.

Jones, A.T. (1985). *CMI/AAB Descr. Pl. Viruses* No. 306, 6 pp.

Jones, A.T. and Duncan, G.H. (1980). *J. gen. Virol.* **50**: 269.

Jones, A.T. and Mayo, M.A. (1972). *J. gen. Virol.* **16**: 349.

Jones, A.T. and Murant, A.F. (1971). *Ann. appl. Biol.* **69**: 11.

Jones, A.T. and Wood, G.A. (1978). *Pl. Dis. Reptr* **62**: 835.

Jones, A.T., Kinninmonth, A.M., and Roberts, I.M. (1973). *J. gen. Virol.* **18**: 61.

Jones, A.T., McElroy, F.D. and Brown, D.J.F. (1981). *Ann. appl. Biol.* **99**: 143.

Lister, R.M. and Murant, A.F. (1967). *Ann. appl. Biol.* **59**: 49.

Lockhart, B.E.L. (1977). *Phytopathology* **4**: 90.
Massalski, P.R. (1984). *Ph.D. Thesis*, Univ. Oxford, U.K. 246 pp.
Massalski, P.R. and Cooper, J.I. (1984). *Pl. Path.* **33**: 255.
Mircetich, S.M., Sanborn, R.R. and Ramos, D.E. (1980). *Phytopathology* **70**: 962.
Murant, A.F. (1983). *Seed Sci. Technol.* **11**: 973.
Murant, A.F., Taylor, M., Duncan, G.H. and Raschké, J.H. (1981). *J. gen. Virol.* **53**: 321.
Nienhaus, F. and Hamacher, J. (1990). *Allg. Forstzeitg.* **16**: 385.
Novak, J. and Lanzova, J. (1975). *Biologia Pl.* **17**: 226.
Posnette, A.F. and Cropley, R. (1955). *Rep. E. Malling Res. Stn.* 1954, p. 126.
Quacquarelli, A. and Savino, V. (1977). *Phytopathol. Medit.* **16**: 154.
Rowhani, A., Mircetich, S.M., Shepherd, R.J. and Cucuzza, J.D. (1985). *Phytopathology* **75**: 48.
Savino, V. and Gallitelli, D. (1981). *Phytopathol. Medit.* **20**: 202.
Savino, V., Quacquarelli, A., Gallitelli, D., Piazzolla P. and Martelli, G.P. (1977). *Phytopathol. Medit.* **16**: 96.
Schimanski, H.H. and Schmelzer, K. (1972). *Zbl. Bakt. ParasitKde.* Abt. 2, **127**: 673.
Schimanski, H.H., Schmelzer, K. and Albrecht, H.J. (1975). *Arch. Phytopath. PflSchutz.* **11**: 329.
Schimanski, H.H., Schmelzer, K. and Albrecht, H.J. (1976). *Zbl. Bakt. ParasitKde.* Abt. 2, **131**: 117.
Schimanski, H.H., Albrecht, H.J. and Kegler, H. (1980). *Arch. PflSchutz.* **3**: 231.
Schmelzer, K. (1965). *Zast. Bilja* **85–88**: 485.
Schmelzer, K. (1966). *Phytopath. Z.* **55**: 317.
Schmelzer, K. (1972). *Zbl. Bakt. ParasitKde.* Abt. 2, **127**: 140.
Schuster, C.E. and Miller, P.W. (1933). *Phytopathology* **23**: 408.
Swingle, R.V., Tilford, P.E. and Irish, C.F. (1941). *Phytopathology* **31**: 22.
Swingle, R.V., Tilford, P.E. and Irish, C.F. (1943). *Phytopathology* **33**: 1196.
Tomlinson, J.A. and Walkey, D.G.A. (1967). *Ann. appl. Biol.* **59**: 415.
Van Hoof, H.A. (1971). *Neth. J. Pl. Path.* **77**: 30.
Varney, E.H. and Moore, J.D. (1952). *Phytopathology* **42**: 476.
Walkey, D.G.A. and Cooper, J.I. (1973). *Rep. nat. Veg. Res. Stn.* 1972, p. 100.
Walkey, D.G.A. and Webb, M.J.W. (1968). *J. gen. Virol.* **3**: 311.
Walkey, D.G.A., Stace-Smith, R. and Tremaine, J.H. (1973). *Phytopathology* **63**: 566.
Waterworth, H.E. and Lawson, R.H. (1973). *Phytopathology* **63**: 141.

# Cherry mottle leaf (?) *trichovirus*

**Data collated by Delano James, 1993.**

*Synonyms: Cerasuvirus maculans* virus, *Marmor cerasae* virus, *Marmor cerasi* virus, *Prunivir cerasi* virus, Prunus virus 1.

First reported in *Prunus avium* (cherry); from Oregon, USA; by Zeller (1934).

**Natural host range and symptoms**

Symptoms persist. Symptoms irregular chlorotic mottle and distortion of leaves, especially terminal leaves.

*Prunus avium* — irregular chlorotic mottling and distortion of foliage, especially terminal leaves.

**Transmission**

Transmitted by a vector; a mite; *Eriophyes inaequalis*; Eriophyidae. Virus transmitted by mechanical inoculation; transmitted by grafting.

**Geographical distribution**

Spreads in Belgium, Canada, the former Czechoslovakia, Italy, Poland, Romania, South Africa, and the USA.

**Experimental host range**

Few (<3) families susceptible.

Diagnostically susceptible

*Chenopodium amaranticolor* — local lesions, systemic leaf distortion, epinasty, chlorosis and stunting.

*C. quinoa* — local chlorotic spots and stunting.

Diagnostically insusceptible

*Chenopodium foetida, Cucumis sativus, Gomphrena globosa, Nicotiana clevelandii* and *N. benthamiana*.

Maintenance and propagation hosts

*Chenopodium quinoa*.

Assay hosts (**L**ocal lesions or **W**hole plants)

*Chenopodium amaranticolor* (L), *C. quinoa* (W).

Dicotyledonous Hosts *(**Susceptible** or Insusceptible)*

***Chenopodium amaranticolor***. *Chenopodium foetidum*. ***Chenopodium quinoa***. *Cucumis sativus*. *Datura stramonium*. *Gomphrena globosa*. *Lycopersicon esculentum*. *Nicotiana benthamiana*. *Nicotiana clevelandii*. *Nicotiana glutinosa*. *Nicotiana occidentalis*. ***Prunus armeniaca***. ***Prunus cerasifera***. ***Prunus mahaleb***. ***Prunus persica***. ***Prunus serrulata***.

Sources of host-range data

Cheney and Parish (1976), James (1992), Nemeth (1986).

**Properties of particles in sap**

TIP: 50–55 °C. LIV: 1 days. Leaf sap contains many virions.

**Purification**

Method: James and Mukerji (1993).

**Particle morphology**

Virions filamentous; not enveloped; usually flexuous; with a clear modal length; of 760 nm; 10.2 nm wide. Axial canal obscure. Basic helix obvious; pitch of basic helix 3.1 nm.

**Physical properties**

One sedimenting component in purified preparations. $A_{260}/A_{280}$ ratio 1.62.

**Biochemical properties**

*Genome* consists of RNA; single–stranded; linear. Genomic nucleic acid isolated by James and Mukerjee, 1993.

*Virion protein(s)* one; $M_r$ 20500; coat protein.

**Replication**

Replication does not depend on a helper virus.

### Cytopathology

Virions found in leaves.

### Taxonomy and Relationships

*Trichovirus.*

Virus(es) with serologically unrelated virions: apple chlorotic leafspot virus, several grapevine viruses (GLRV-1, GLRV-3, NY-1 and GVA) and heracleum latent virus.

Best tests for diagnosis: ELISA, western blot using mAbs.

### References

Cheney, P.W. and Parish, C.L. (1976). *U.S. Dep. Agric. Hdbk* No. **437**, pp. 216.

James, D. (1992). *Acta Hort.* **309**: 39.

James, D. and Mukerji, S. (1993). *Plant Disease* **77**: 271. McCarty, H.R. (1935). *Northwest Assoc. Hort. Ent. Pl. Pathol.* **1**: 5.

Nemeth, M. (1986). In: *Virus, Mycoplasma and Rickettsia Diseases of Fruit Trees.* pp. 296–8. Akademia Kiado, Budapest.

Reeves, E.L. (1941). *J. Agric. Res.* **62**: 555.

Zeller, S.M. (1934). *Rep. Oregon State Hort. Soc.* **26**: 92.

## Cherry rasp leaf *nepovirus*

### Data collated by A.J. Hansen, 1986.

*Synonyms:* flat apple virus. *Standard acronym:* CRLV.

First reported in *Prunus avium*; from the USA; by Bodine and Newton (1942).

### Natural host range and symptoms

Symptoms persist.

*Prunus avium, P. maheleb* — affected cherry branches are very frost sensitive (Bodine *et al.*, 1951; Hansen *et al.*, 1974; Wagnon *et al.*, 1968) enations, stunting.

*Prunus persica* — enations, stunting, decline.

*Malus sylvestris* — enations, flat fruit.

*Balsamorhiza sagittata, Taraxacum officinalis, Plantago major* — symptomless.

### Transmission

Transmitted by a vector; a nematode; *Xiphinema americana* (Nyland *et al.*, 1969; Hansen *et al.*, 1974); Dorylamidae. Virus transmitted by mechanical inoculation; transmitted by grafting; transmitted by seed (10–20%).

### Ecology and control

Studies reported by Luepschen *et al.* (1974); Wagnon *et al.* (1968); Parish and Cheney (1974).

### Geographical distribution

Spreads in the North American region and the Pacific region; Canada, New Zealand, South Africa, and the USA. Found, but with no evidence of spread, in the U.K. (Jones *et al.*, 1985), China, Australia and New Zealand and possibly South Africa. *Prunus avium* raspleaf symptoms have been found in many countries associated with one or another of seven viruses (Hansen *et al.*, 1974).

### Experimental host range

Several (3–9) families susceptible. Experimentally infected plants mostly show necrotic or chlorotic local lesions, systemic mottle.

Diagnostically susceptible

*Cucumis sativus* — chlorotic local lesions; systemic mottle.

*Cyamopsis tetragonoloba* — necrotic local lesions; not systemic.

*Chenopodium quinoa* — mottle and vein clearing.

*Chenopodium amaranticolor* — systemic mottle.

*Vigna unguiculata, Physalis floridana, Sesbania exalta* — local lesions but unreliable.

Maintenance and propagation hosts

*Chenopodium quinoa, Cucumis sativus.*

Assay hosts (**L**ocal lesions or **W**hole plants)

*Cucumis sativus* (L), *Cyamopsis tetragonoloba* (L).

Dicotyledonous Hosts *(**Susceptible** or Insusceptible)*

***Chenopodium amaranticolor. Chenopodium quinoa. Cucumis sativus. Cyamopsis tetragonoloba. Malus sylvestris. Physalis floridana. Prunus mahaleb. Prunus persica. Sesbania exaltata. Vigna unguiculata.***

Sources of host-range data

Hansen *et al.* (1974).

**Properties of particles in sap**

TIP: 58 °C. LIV: 7 days (at 4°C). DEP: $\log_{10}$ minus 4 (Stace-Smith and Hansen, 1976).

**Purification**

Method: Stace-Smith and Hansen (1976).

**Particle morphology**

Virions isometric; not enveloped; 30 nm in diameter; angular in profile; without a conspicuous capsomere arrangement.

**Physical properties**

Three sedimenting components in purified preparations; sedimentation coefficient of the fastest 128 *S* (B); of the other(s) 96 *S* (M), or 56 *S* (T).

**Biochemical properties**

*Genome* consists of RNA; single–stranded; of two parts; largest (or only) genome part the larger 6.5 kb; the 2nd largest 4.8 kb.

*Virion protein(s)* two; $M_r$ of the larger 24000. $M_r$ of 2nd largest 22500. Method of preparation: Stace-Smith and Hansen (1976).

**Replication**

Replication does not depend on a helper virus.

**Cytopathology**

Virions found in parenchyma and plasmodesmata (Stace-Smith and Hansen, 1976); in cytoplasm. Inclusions present in infected cells; are unusual in shape (tubular structures); they contain virions.

**Taxonomy and Relationships**

Possible *Nepovirus: Comoviridae.*

Virus(es) with serologically unrelated virions: tobacco ringspot, tomato ringspot, cherry leaf roll, peach rosette mosaic, raspberry ringspot, tomato black ring and arabis mosaic viruses (Hansen *et al.*, 1974).

Cherry rasp leaf differs from Eola rasp leaf and other cherry diseases that cause leaf enations.

**References**


Bodine, E.W. and Newton, N. (1942). *Phytopathology* **31**: 333.

Bodine, E.W., Blodgett, E.C. and Lott, T.B. (1951). *Hdbk. U.S. Dep. Agric.* **10**: 71.

Hansen, A.J., Nyland, G., McElroy, F.D. and Stace-Smith, R. (1974). *Phytopathology* **64**: 721.

Jones, A.T., Mayo, M.A. and Henderson, S.J. (1985). *Ann. appl. Biol.* **106**: 101.

Luepschen, N.S., Harder, H.H., Rohrback, K.G. and Sisson, M.A. (1974). *Pl. Dis. Reptr* **58**: 26.

Nyland, G., Lownsbery, B.F., Lowe, S.K. and Mitchell, J.F. (1969). *Phytopathology* **59**: 1111.

Parish, C.L. and Cheney, P.W. (1974). *Proc. Am. phytopath. Soc.* **1**: 52.

Stace-Smith, R. and Hansen, A.J. (1976). *Acta Hort.* **67**: 193.

Stace-Smith, R. and Hansen, A.J. (1976). *CMI/AAB Descr. Pl. Viruses* No. 159, 4 pp.

Wagnon, H.F., Traylor, J., Willams, H.E. and Weiner, A.C. (1968). *Pl. Dis. Reptr* **52**: 618.


# Chickpea bushy dwarf *potyvirus*

**Data collated by A.J. Gibbs, 1990.**

*Standard acronym:* CpBDV.

First reported in *Cicer arietinum*; from India (Andra Pradesh); by Anjaiah *et al.* (1989).

**Natural host range and symptoms**

Symptoms persist.

*Cicer arietinum* — small chlorotic brittle leaves, axillary buds activated but stunted giving bushy dwarf symptoms.

**Experimental host range**

Several (3–9) families susceptible. Experimentally infected plants mostly show mosaics, mottles, ringspots or necrosis.

Diagnostically susceptible

*Cicer arietinum* — chlorotic bushy dwarf.

*Chenopodium amaranticolor, C. murale, Cyamopsis tetragonoloba* — chlorotic local lesions; not systemic.

*Canavalia ensiformis, Chenopodium quinoa, Gossypium herbaceum, Nicotiana benthamiana, Phaseolus vulgaris* cvs Bountiful, Pinto — systemic necrosis.

*Macrotyloma uniflorum, Cassia obtusifolia* — necrotic local lesions; not systemic.

Diagnostically insusceptible

*Arachis hypogaea, Cucumis sativus, Nicotiana glutinosa, Pisum sativum* cv. Bonneville, *Phaseolus vulgaris* cv. Top Crop.

Maintenance and propagation hosts

*Cicer arietinum*.

Assay hosts (**L**ocal lesions or **W**hole plants)

*Chenopodium amaranticolor* (L).

Dicotyledonous Hosts *(**Susceptible** or Insusceptible)*

*Arachis hypogaea*. ***Cassia obtusifolia***. ***Chenopodium amaranticolor***. ***Chenopodium murale***. ***Chenopodium quinoa***. *Cucumis sativus*. ***Cyamopsis tetragonoloba***. *Glycine max*. ***Macrotyloma uniflorum***. *Nicotiana clevelandii*. *Nicotiana glutinosa*. *Nicotiana rustica*. *Petunia hybrida*. ***Phaseolus lunatus***, or *Phaseolus lunatus*. *Pisum sativum*. *Vigna radiata*. *Vigna unguiculata*.

Sources of host-range data

Anjaiah *et al.* (1989).

**Properties of particles in sap**

Leaf sap contains few virions.

**Purification**

Method: Anjaiah *et al.* (1989).

**Particle morphology**

Virions filamentous; not enveloped; usually flexuous; with a clear modal length; of 700–720 nm; 13–15 nm wide. Axial canal obscure.

**Physical properties**

One sedimenting component in purified preparations.

**Biochemical properties**

*Genome* consists of RNA; single–stranded; linear; unipartite; largest (or only) genome part 10 kb. Genomic nucleic acid isolated by Anjaiah *et al.* (1989).

*Virion protein(s)* one; $M_r$ 32500 (, sometimes degraded to 29000); virion coat protein. Method of preparation: Anjaiah *et al.* (1989).

**Taxonomy and Relationships**

*Potyvirus: Potyviridae*.

Virus(es) with serologically related virions: azuki bean mosaic, peanut mottle and soybean mosaic virus; peanut stripe and blackeye cowpea mosaic are more distantly related. Virus(es) with serologically unrelated virions: bean yellow mosaic virus.

Best tests for diagnosis: In India, chickpea may be also infected with alfalfa mosaic, bean yellow mosaic, cucumber mosaic and chickpea stunt viruses. In the U.S.A., chickpea filiform potyvirus (Kaiser *et al.*, 1988) also infects chickpea, and bean yellow mosaic and beet western yellow luteovirus are found in chickpea in other parts of the world. Serological and host range tests distinguishes readily between these viruses.

**References**

Anjaiah, V., Reddy, D.V.R., Mahohar, S.K., Naidu, R.A., Nene, Y.L. and Ratna, A.S. (1989). *Pl. Path.* **38**: 520.

Kaiser, W.J., Wyatt, S.D., Hannan, R.M. and Cody, Y. (1988). *Plant Dis.* **72**: 70.

# Chickpea chlorotic dwarf (?) monogeminivirus

**Data collated by A.A. Brunt, 1995.**

First reported in *Cicer arietinum*; from India; by Horn *et al.* (1993).

**Natural host range and symptoms**

*Cicer arietinum* — reddening of leaflets, phloem browning and severe stunting.

**Transmission**

Transmitted by a vector; an insect; *Orosius orientalis*; Cicadellidae. Virus not transmitted by mechanical inoculation.

**Geographical distribution**

Spreads in India.

**Experimental host range**

Several (3–9) families susceptible.

Diagnostically susceptible

*Beta vulgaris, Lens esculentus, Pisum sativum, Datura stramonium, Nicotiana benthamiana, N. tabacum* — systemic leaf chlorosis, sometimes leaf mottling and stunting.

*Cicer arietinum* — leaf yellowing and reddening, leaf rolling, phloem browning and stunting.

Diagnostically insusceptible

*Arachis hypogaea, Cajanus cajan, Cucumis sativus, Glycine max, Nicotiana occidentalis, Vigna unguiculata.*

Maintenance and propagation hosts

*Cicer arietinum, Pisum sativum.*

Assay hosts (**L**ocal lesions or **W**hole plants)

*Cicer arietinum* (W), *Pisum sativum* (W).

Dicotyledonous Hosts *(**Susceptible** or Insusceptible)*

*Arachis hypogaea.* ***Beta vulgaris****. Cajanus cajan.* ***Cicer arietinum****. Cucumis sativus.* ***Datura stramonium****. Glycine max.* ***Lens esculenta. Lycopersicon esculentum****. Medicago sativa.* ***Nicotiana benthamiana. Nicotiana glutinosa****. Nicotiana occidentalis. Nicotiana rustica.* ***Nicotiana tabacum. Phaseolus vulgaris. Pisum sativum****. Solanum melongena.* ***Vicia faba****. Vigna radiata. Vigna unguiculata.*

**Purification**

Method: Horn *et al.* (1993).

**Particle morphology**

Virions geminate; 15 nm in diameter; dimers 25 nm in length.

**Physical properties**

$A_{260}/A_{280}$ ratio 1.4.

**Biochemical properties**

*Genome* consists of DNA; single–stranded; circular; largest (or only) genome part 2.9 kb.

*Virion protein(s)* one; $M_r$ 32000; coat protein.

**Taxonomy and Relationships**

Possible monogeminivirus: *Geminiviridae.*

Virus(es) with serologically related virions: Tobacco yellow dwarf. Virus(es) with serologically unrelated virions: Cassava African mosaic, squash leaf curl, chloris striate, maize streak and wheat dwarf viruses.

**References**

Horn, N.M., Reddy, S.V., Roberts, I.M. and Reddy, D.V.R. (1993). *Ann. appl. Biol.* **122**: 467.

## Chickpea distortion mosaic *potyvirus*

**Data collated by V.R. Mali, 1987.**

First reported in *Cicer arietinum* L.; from Parbhani, Maharashtra State, India.

**Natural host range and symptoms**

*Cicer arietinum* — mosaic, leaf malformation, stunting.

**Transmission**

Transmitted by a vector; an insect; *Aphis craccivora* Koch., *A. gossypii* Glover, *Myzus persicae* Sulz.; Aphididae. Transmitted in a non–persistent manner. Virus transmitted by mechanical inoculation; transmitted by grafting; not transmitted by contact between plants; not transmitted by seed.

**Ecology and control**

Studies reported by Mali (1980, 1985, 1988*a*); Mali and Vyanjane (1980*a*, 1980*c*, 1981); Polajiwar (1985).

**Geographical distribution**

Spreads in India (Maharashtra State).

**Experimental host range**

Several (3–9) families susceptible.

Diagnostically susceptible

*Canavalia ensiformis* — systemic mosaic and leaf malformation.

*Dolichos biflorus* — vein necrosis.

*Glycine max, Phaseolus lunatus, Trigonella foenum graecum, Vigna unguiculata, Nicotiana benthamiana* — systemic mosaic, mottling.

*P. vulgaris* — chlorotic or necrotic local lesions, systemic mosaic and leaf malformation.

Diagnostically insusceptible

*Chenopodium album, C. amaranticolor, C. quinoa, Arachis hypogaea, Vicia faba.*

Maintenance and propagation hosts

*Cicer arietinum, Canavalia ensiformis, Phaseolus lunatus, P. vulgaris.*

Assay hosts (**L**ocal lesions or **W**hole plants)

*Phaseolus vulgaris* (W), *P. lunatus* (W), *Cicer arietinum* (W).

Dicotyledonous Hosts *(**Susceptible** or Insusceptible)*

*Arachis hypogaea. Beta vulgaris. Brassica oleracea* ssp. *botrytis. Cajanus cajan. Capsicum annuum.* ***Cassia occidentalis****. Chenopodium album. Chenopodium amaranticolor. Chenopodium murale. Chenopodium quinoa.* ***Cicer arietinum****. Cucumis sativus. Cucurbita pepo.* ***Cyamopsis tetragonoloba****. Datura stramonium.* ***Dolichos biflorus****.* ***Glycine max****. Helianthus annuus. Lablab purpureus. Lactuca sativa.* ***Lens esculenta****. Linum usitatissimum. Luffa acutangula. Lycopersicon esculentum.* ***Medicago sativa****.* ***Nicotiana benthamiana****. Nicotiana glutinosa. Nicotiana rustica. Nicotiana sylvestris. Nicotiana tabacum. Ocimum basilicum. Petunia hybrida.* ***Phaseolus lunatus****.*

***Phaseolus vulgaris*. *Pisum sativum*.** *Raphanus sativus*. *Ricinus communis*. *Sesamum indicum*. *Solanum melongena*. *Solanum nigrum*. *Spinacia oleracea*. ***Tetragonia expansa*.** *Vicia faba*. ***Vigna radiata*. *Vigna unguiculata*. *Vigna unguiculata*** ssp. ***cylindrica*. *Vigna unguiculata*** ssp. ***sesquipedalis*.**

Sources of host-range data

Mali (1985); Mali and Vyanjane (1980*a*); Mali and Vyanjane (1980*b*); Polajiwar (1985); Mali *et al.* (1988*a*).

**Properties of particles in sap**

TIP: 65–70 °C. LIV: 2.5 days. DEP: $\log_{10}$ minus 5–6. Leaf sap contains few virions.

**Purification**

Method: Mali *et al.* (1985, 1988*a*).

**Particle morphology**

Virions filamentous; not enveloped; usually flexuous; with a clear modal length; of 750 nm; 11 nm wide. Axial canal obscure. Basic helix obscure.

**Cytopathology**

Inclusions present in infected cells; are pinwheels (associated with scrolls, but not laminated aggregates in chickpea infected cells. Inclusions like those of subdivision I of the potyvirus group (Edwardson, 1974; Mali *et al.*, 1988*b*)).

**Taxonomy and Relationships**

*Potyvirus: Potyviridae.*

Virus(es) with serologically related virions: bean common mosaic, bean yellow mosaic, peanut mottle, soybean mosaic, wisteria vein mosaic and swordbean distortion mosaic viruses. Virus(es) with serologically unrelated virions: blackeye cowpea mosaic, celery mosaic, cowpea aphid-borne mosaic, datura distortion mosaic, papaya ringspot, potato Y, watermelon mosaic 1 and watermelon mosaic 2 viruses.

Best tests for diagnosis: chickpea is known to be naturally infected with potyviruses like bean yellow mosaic (BYMV) (Erwin and Snyder, 1958; Kaiser and Danesh, 1971), and its strains chickpea narrow leaf virus (CNLV) (Chalam *et al.*, 1985) and chickpea filiform virus (CFV) (Kaiser *et al.*, 1988). CpDMV is distinct from BYMV in host range and inclusion (CCI) morphology which for the latter are type II inclusions (Edwardson *et al.*, 1987).

**References**


Chalam, T.V., Reddy, D.V.R., Subbayya, J., Nene, Y.L. and Beniwal, S.P.S. (1985). *Indian Phytopath.* **38**(4): 708.

Edwardson, J.R. (1974). *Fla Agric. Exp. Stn Monog.* No. 4, p. 1.

Edwardson, J.R., Christie, R.G. and Ko, N.J. (1984). *Phytopathology* **74**: 1111.

Erwin, D.C. and Snyder, W.C. (1958). *Calif. Agric.* **13**: 6.

Hollings, M. and Brunt, A.A. (1981). *CMI/AAB Descr. Pl. Viruses* No. 245, 7 pp.

Kaiser, W.J. and Danesh, D. (1971). *Phytopathology* **61**: 453.

Kaiser, W.J., Wyatt, S.D., Hannam, R.M. and Cody, T. (1988). *Plant Dis.* **72**: 70.

Mali, V.R. (1979). *Indian Phytopath.* **33**: 155.

Mali, V.R. (1980). *Indian J. Mycol. & Pl. Pathol.* **10**: 198.

Mali, V.R. (1985). *Int. J. Tropical Pl. Diseases* **3**: 93.

Mali, V.R. and Vyanjane, N.T. (1980*a*). *Curr. Sci.* **49**: 599.

Mali, V.R. and Vyanjane, N.T. (1980*b*). *Indian Phytopath.* **33**: 953.

Mali, V.R. and Vyanjane, N.T. (1980*c*). *Indian J. Mycol & Pl. Pathol.* **10**: 198.

Mali, V.R. and Vyanjane, N.T. (1981). *Indian Phytopath.* **34**: 109.

Mali, V.R., Markham, P.G., Wells, B., Pinner, M.S. and Patil, F.S. (1990). *Indian Phytopath.* **43**: 340.

Mali, V.R., Nirmal, D.D., Mundhe, G.E., Vyanjane, N.T. and Raut, K.G. (1985). *Indian Phytopath.* **38**: 2.

Mali, V.R., Polajiwar, B.B., Nirmal, D.D. and Patil, F.S. (1988*a*). *Indian Phytopath.* **41**: 336.

Polajiwar, B.B. (1985). *M.Sc. Thesis*, M.A.U., Parbhani, M.S., India 79 pp.

# Chickpea filiform (?) potyvirus

**Data collated by P. Jeyanandarajah, 1992.**

*Standard acronym:* CpFV.

First reported in *Cicer arietinum*; from the U.S.A.; by Kaiser *et al.* (1988).

**Natural host range and symptoms**

Symptoms persist.

*Cicer arietinum* — narrow, deformed leaflets, axillary buds proliferate and become stunted; fewer pods with small misshapen seeds.

**Transmission**

Transmitted by a vector; an insect; *Acyrthosiphon pisum, Myzus persicae*; Aphididae. Transmitted in a non–persistent manner. Virus transmitted by mechanical inoculation; not transmitted by seed; not transmitted by pollen.

**Geographical distribution**

Found, but with no evidence of spread, in the U.S.A. (Washington State), in chickpea originating from India.

**Experimental host range**

Several (3–9) families susceptible.

Diagnostically susceptible

*Cicer arietinum* — filiform leaves, wilting and necrosis.

*Lens culinaris, Trigonella foenum-graecum* — systemic leaf chlorosis, leaf curling and stunting.

*Nicotiana clevelandii* — symptomless systemic infection.

*Chenopodium amaranticolor, C. quinoa, Gomphrena globosa* — local lesions; no systemic infection.

Diagnostically insusceptible

*Cassia occidentalis, Cucumis sativus, Nicotiana glutinosa, N. tabacum, Vigna unguiculata.*

Maintenance and propagation hosts

*Cicer arietinum.*

Assay hosts (**L**ocal lesions or **W**hole plants)

*Chenopodium quinoa* (L).

Dicotyledonous Hosts *(**Susceptible** or Insusceptible)*

*Cassia occidentalis.* ***Chenopodium amaranticolor.*** ***Chenopodium quinoa.*** ***Cicer arietinum.*** *Cucumis sativus. Cyamopsis tetragonoloba. Glycine max.* ***Gomphrena globosa.*** ***Lens esculenta.*** *Medicago sativa.* ***Nicotiana clevelandii.*** *Nicotiana glutinosa.*

*Nicotiana tabacum. Phaseolus lunatus. Phaseolus vulgaris. Tetragonia expansa. Trifolium repens. Vicia faba. Vicia villosa. Vigna radiata. Vigna unguiculata.*

Sources of host-range data

Kaiser *et al.* (1988).

**Purification**

Method: Kaiser *et al.* (1988).

**Particle morphology**

Virions filamentous; of 770 nm.

**Cytopathology**

Virions found in leaves.

**Taxonomy and Relationships**

Unassigned potyvirus.

Virus(es) with serologically related virions: bean yellow mosaic and blackeye cowpea mosaic viruses. Virus(es) with serologically unrelated virions: clover yellow vein, pea mosaic and pea seedborne mosaic viruses.

**References**

Kaiser, W.J., Wyatt, S.D., Hannan, R.M. and Cody, Y. (1988). *Plant Dis.* **72**: 70.

# Chicory X *potexvirus*

**Data collated by D. Gallitelli, 1991.**

*Standard acronym:* ChVX.

First reported in *Cichorium intybus*; from Apulia, southern Italy; by Gallitelli and Di Franco (1982).

**Natural host range and symptoms**

Symptoms persist.

*Cichorium intybus* — leaf yellowing.

**Transmission**

Virus transmitted by mechanical inoculation.

**Geographical distribution**

Spreads in Italy (Apulia).

**Experimental host range**

Few (<3) families susceptible. Experimentally infected plants mostly show necrotic local lesions.

Diagnostically susceptible

*Chenopodium amaranticolor, C. quinoa* — necrotic local lesions.

*Gomphrena globosa* — red-rimmed lesions.

Diagnostically insusceptible

*Phaseolus vulgaris, Nicotiana tabacum, N. benthamiana, N. glutinosa, Cucumis sativus.*

Maintenance and propagation hosts

*Chenopodium amaranticolor.*

Assay hosts (**L**ocal lesions or **W**hole plants)

*Chenopodium amaranticolor* (L), *C. quinoa* (L), *Gomphrena globosa* (L).

Dicotyledonous Hosts *(**Susceptible** or Insusceptible)*

*Beta vulgaris. Celosia argentea.* ***Chenopodium amaranticolor. Chenopodium quinoa.*** *Cucumis sativus. Cucurbita pepo. Datura stramonium.* ***Gomphrena globosa.*** *Lycopersicon esculentum. Nicotiana benthamiana. Nicotiana clevelandii. Nicotiana glutinosa. Nicotiana megalosiphon. Nicotiana rustica. Nicotiana tabacum. Ocimum basilicum. Petunia hybrida. Phaseolus vulgaris. Vicia faba. Zinnia elegans.*

Sources of host-range data

Gallitelli and Di Franco (1982).

**Properties of particles in sap**

TIP: 55–60 °C. LIV: 15 days. DEP: $\log_{10}$ minus 6–7. Leaf sap contains few virions.

**Purification**

Method: Hammond and Hull (1981).

**Particle morphology**

Virions filamentous; usually flexuous; with a clear modal length; of 553 nm. Axial canal obscure. Basic helix obvious.

**Physical properties**

One sedimenting component in purified preparations; sedimentation coefficient 115 *S*. Density 1.304 g $cm^{-3}$ in CsCl. $A_{260}/A_{280}$ ratio 1.19.

**Biochemical properties**

Virions contain *c.* 5 % nucleic acid.

*Genome* consists of RNA; single–stranded; linear; unipartite; largest (or only) genome part 7 kb. Genomic nucleic acid isolated by phenol-SDS.

*Non–genomic nucleic acid* not found in the virions.

*Virion protein(s)* one; $M_r$ 25000; coat protein. Method of preparation: Laemmli (1970).

**Replication**

Replication does not depend on a helper virus.

**Cytopathology**

Virions found in leaves and mesophyll; in cytoplasm. Inclusions present in infected cells; are unusual in shape; fibrous aggregates of virions; they contain virions.

**Taxonomy and Relationships**

*Potexvirus.*

Virus(es) with serologically unrelated virions: potato X, cactus X, viola mottle, pepino mosaic, clitoria mosaic, clover yellow mosaic and white clover mosaic viruses.

Best tests for diagnosis: inoculation onto *Chenopodium amaranticolor* and serology.

**References**

Gallitelli, D. and Di Franco, A. (1982). *Phytopath. Z.* **105**: 120.

Hammond, J. and Hull, R. (1981). *J. gen. Virol.* **54**: 75.

Laemmli, U.K. (1970). *Nature, Lond.* **227**: 680.

# Chicory yellow blotch (?) *carlavirus*

**Data collated by A.A. Brunt, 1991.**

*Synonyms:* chicory yellows virus, chicory blotch virus. *Standard acronym:* ChYBV.

First reported in *Cichorium intybus*; from the former Czechoslovakia; by Kvičala (1956).

**Natural host range and symptoms**

Symptoms persist.
*Cichorium intybus* — vein-clearing, yellow flecks, necrosis and leaf distortion.

**Transmission**

Virus transmitted by mechanical inoculation.

**Geographical distribution**

Found, but with no evidence of spread, in Czechoslovakia.

**Experimental host range**

Few (<3) families susceptible.

Diagnostically susceptible

*Beta vulgaris, Chenopodium giganteum, C. quinoa, Gomphrena globosa* — local lesions; no systemic infection.
*Cichorium intybus* — conspicuous leaf chlorosis.

Diagnostically insusceptible

*Amaranthus caudatus, Cucumis sativus.*

Maintenance and propagation hosts

*Cichorium intybus.*

Assay hosts (**L**ocal lesions or **W**hole plants)

*Chenopodium giganteum* (L), *C. quinoa* (L).

Dicotyledonous Hosts *(**Susceptible** or Insusceptible)*

*Amaranthus caudatus*. ***Beta vulgaris***. ***Chenopodium quinoa***. *Cucumis sativus*. ***Gomphrena globosa***.

**Properties of particles in sap**

TIP: 70 °C. LIV: 1 days (at 18–22°C).

**Particle morphology**

Virions filamentous; not enveloped; usually straight; with a clear modal length; of 657 nm; 11 nm wide. Axial canal obscure.

**Taxonomy and Relationships**

Possible *Carlavirus*.

**References**

Brčak, J. (1960). *Biol. Pl.. Acad. Sci. bohemsl.* **2**: 244.
Brčak, J. (1961). *Biol. Pl.. Acad. Sci. bohemsl.* **3**: 215.
Brčak, J. and Cech, M. (1962). *Phytopath. Z.* **45**: 335.
Kvičala, A. (1956). *Sbornik CSAZV (Rost. vyr.)* **29**: 617.

# Chicory yellow mottle *nepovirus*

**Data collated by A.A. Brunt, 1992.**

*Synonyms:* parsley carrot leaf virus (Avgelis and Quacquarelli, 1974). *Standard acronym:* ChYMV.

First reported in *Cichorium intybus*; from Italy; by Vovla *et al.* (1971).

**Natural host range and symptoms**

Symptoms persist.
*Cichorium intybus* — ringspots, line pattern and bright yellow mottling of leaves.
*Petroselinum crispum* — reduction of leaf laminae.

**Transmission**

Possibly by nematodes. Virus transmitted by mechanical inoculation; transmitted by seed.

**Geographical distribution**

Spreads in Italy.

**Experimental host range**

Many (>9) families susceptible.

Diagnostically susceptible

*Cucurbita pepo* — chlorotic local lesions; systemic chlorosis, ringspots and necrosis.
*Phaseolus vulgaris* — chlorotic local lesions; severe systemic mottling.
*Chenopodium amaranticolor, C. quinoa* — chlorotic or necrotic local lesions; apical necrosis.

Diagnostically insusceptible

*Capsicum annuum, Datura metel, Lycopersicon esculentum, Tropaeolum majus, Ocimum basilicum.*

Maintenance and propagation hosts

*Cucurbita pepo.*

Assay hosts (**L**ocal lesions or **W**hole plants)

*Chenopodium quinoa* (W).

Dicotyledonous Hosts (***Susceptible*** *or Insusceptible)*

***Amaranthus caudatus***. ***Apium graveolens***. ***Brassica campestris*** ssp. ***rapa***. ***Calendula officinalis***. *Capsicum annuum*. ***Celosia cristata***. ***Cheiranthus cheiri***. ***Chenopodium amaranticolor***. ***Chenopodium quinoa***. ***Cucumis sativus***. ***Cucurbita pepo***. *Datura metel*. ***Datura stramonium***. *Daucus carota*. *Dianthus barbatus*. ***Gomphrena globosa***. ***Lactuca sativa***. ***Lathyrus odoratus***. ***Lupinus albus***. *Lycopersicon esculentum*. ***Nicotiana benthamiana***. ***Nicotiana clevelandii***. ***Nicotiana glutinosa***. ***Nicotiana rustica***. ***Nicotiana tabacum***. *Ocimum basilicum*. *Pastinaca sativa*. ***Petroselinum crispum***. ***Petunia hybrida***. ***Phaseolus vulgaris***. *Pisum sativum*. *Raphanus sativus*. ***Sinapis alba***. *Solanum melongena*. *Tropaeolum majus*. *Vicia faba*.

Sources of host-range data

Vovlas *et al.* (1971); Avgelis and Quacquarelli (1974*a*).

**Properties of particles in sap**

TIP: 56 °C. LIV: 4 days. DEP: $\log_{10}$ minus 5.

**Purification**

Method: Vovlas *et al.* (1971).

**Particle morphology**

Virions isometric; not enveloped; 30 nm in diameter; angular in profile; with a conspicuous capsomere arrangement.

**Physical properties**

More than one sedimenting component in purified preparations; sedimentation coefficient 128 *S* (B); of the other(s) 116 *S* (M), or 51 *S* (T).

**Biochemical properties**

Virions contain 42 % nucleic acid (B), or 39 % nucleic acid (M), or 0 % nucleic acid (T); 58 % protein (B), or 61 % protein (M), or 100 % protein (T).

*Genome* consists of RNA; single–stranded; of two parts; largest (or only) genome part 7.27 kb; the 2nd largest 6.06 kb.

*Virion protein(s)* one; $M_r$ 54000; coat protein.

**Taxonomy and Relationships**

*Nepovirus: Comoviridae.*

Virus(es) with serologically unrelated virions: tomato ringspot virus.

**References**


Avgelis, A. and Quacquarelli, A. (1974*a*). *Phytopath. Medit.* **13**: 1.

Avgelis, A. and Quacquarelli, A. (1974*b*). *Phytopath. Medit.* **13**: 97.

Piazzolla, P., Gallitelli, D., Vovlas, C. and Quacquarelli, A. (1979). *Phytopath. Medit.* **17**: 149.

Quacquarelli, A., Martelli, G.P. and Vovlas, C. (1974). *CMI/AAB Descr. Pl. Viruses* No. 132 4 pp.

Quacquarelli, A., Piazzolla, P. and Vovlas, C. (1972*a*). *J. gen. Virol.* **17**: 147.

Quacquarelli, A., Piazzolla, P. and Vovlas, C. (1972*b*). *Phytopath. Medit.* **11**: 207.

Quacquarelli, A., Vovlas, C., Piazzolla, P., Russo, M. and Martelli, G.P. (1972*c*). *Phytopath. Medit.* **11**: 180.

Quacquarelli, A., Piazzolla, P., Vovlas, C. and Martelli, G.P. (1973). *Mikrobiologiya* **10**: 15.

Vovlas, C. (1973). *Phytopath. Medit.* **12**: 102.

Vovlas, C. (1978). *Phytopath. Medit.* **17**: 193.

Vovlas, C., Martelli, G.P. and Quacquarelli, A. (1974). *Phytopath. Medit.* **13**: 179.

Vovlas, C., Martelli, G.P. and Quacquarelli, A. (1971). *Phytopath. Medit.* **10**: 244.


# Chicory yellow mottle satellite RNA

*Sequence database accession code(s):*

D00685 Em(40)_vi:UTSSR Gb(84)_vi:CHYTSSR Chicory yellow mottle virus isolate T small satellite RNA. 6/91 457bp.

D00686 Em(40)_vi:UCLSR Gb(84)_vi:CHYCLSR Chicory yellow mottle virus isolate C large satellite RNA. 6/91 1,165bp.

D00721 Em(40)_vi:CHYSATS1 Gb(84)_vi:CHYSATS1 Chicory yellow mottle virus satellite RNA S1. 9/92 457bp.

D00722 Em(40)_vi:CHYSATL1 Gb(84)_vi:CHYSATL1 Chicory yellow mottle virus satellite RNA L1. 12/91 1,145bp. 4 sequences.

**Taxonomy and Relationships**

Satellite RNA.

## Chilli veinal mottle (?) potyvirus

**Data collated by A.A. Brunt, 1989. Revised 1992.**

*Standard acronym:* ChiVMV.

First reported in *Capsicum annuum*; from Malayasia; by Burnett (1947).

**Natural host range and symptoms**

Symptoms veinal mottle, malformation.

*Capsicum annuum, C. frutescens* — dark green mottling adjacent to main leaf veins, reduction in leaf area and distortion; plants stunted (severity dependent upon cultivar and duration of infection).

*Capsicum* spp. — veinal mottle, malformation.

**Transmission**

Transmitted by a vector; an insect; *Aphis craccivora, A. gossypii, A. spiraecola, Myzus persicae, Toxoptera citricidus, Hysteroneura setarieae, Rhopalosiphum maidis*; Aphididae. Transmitted in a non–persistent manner. Virus transmitted by mechanical inoculation; transmitted by grafting; not transmitted by seed.

**Geographical distribution**

Spreads in Korea D.P.R. (North), Korea Republic, Malaysia, the Philippines, Taiwan, and Thailand.

**Experimental host range**

Few (<3) families susceptible.

Diagnostically susceptible

*Capsicum annuum* — systemic vein clearing, followed by leaf mottling.

Diagnostically insusceptible

*Chenopodium quinoa, Datura stramonium, Gomphrena globosa, Cucumis sativus.*

Maintenance and propagation hosts

*Capsicum annuum, Nicandra physalodes, Nicotiana megalosiphon, Physalis floridana, Nicotiana tabacum.*

Assay hosts (**L**ocal lesions or **W**hole plants)

*Capsicum annuum* (W).

Dicotyledonous Hosts *(**Susceptible** or Insusceptible)*

*Ageratum conyzoides. Arachis hypogaea. Brassica juncea. Brassica oleracea* ssp. *capitata.* ***Capsicum annuum. Capsicum frutescens.*** *Cassia occidentalis. Catharanthus roseus. Chenopodium amaranticolor. Chenopodium capitatum. Chenopodium quinoa. Cucumis sativus. Datura stramonium.* ***Hyoscyamus niger.*** *Luffa acutangula. Lycopersicon esculentum.* ***Nicandra physaloides.*** *Nicotiana glutinosa.* ***Nicotiana megalosiphon.*** *Nicotiana rustica.* ***Nicotiana tabacum.*** *Passiflora edulis. Phaseolus vulgaris.* ***Physalis floridana.*** *Solanum melongena. Solanum nigrum. Tetragonia expansa. Vigna unguiculata. Vigna unguiculata* ssp. *sesquipedalis. Zinnia elegans.*

Monocotyledonous Hosts *(**Susceptible** or Insusceptible)*

*Zea mays.*

**Properties of particles in sap**

TIP: 60 °C. LIV: 7 days. DEP: $\log_{10}$ minus 4.

**Particle morphology**

Virions filamentous; not enveloped; usually flexuous; with a clear modal length; of 750 nm; 12 nm wide.

**Cytopathology**

Inclusions present in infected cells; are pinwheels.

**Taxonomy and Relationships**

Unassigned potyvirus.

Virus(es) with serologically unrelated virions: potato Y, tobacco etch, pepper veinal mottle and pepper mottle (S.K. Green, personal communication).

**References**

Abu Kassim, A.B. (1986). In: *Food and Fertiliser Technology Centre Book Series* No. 33, p. 3.

Burnett, F. (1947). *Rep. Agric. Malaya 1946*, 85 pp.

Ong, C.A., Varghese, G. and Poh, T.W. (1979). *MARDI Res. Bull.* **7**: 78.

# Chinese yam necrotic mosaic (?) *carlavirus*

**Data collated by A.A. Brunt and S. Molyneux, 1986 and F. Nienhaus, 1987.**

*Standard acronym:* ChYNMV.

First reported in *Dioscorea batatus f.typica*; from Japan; by Fukumoto and Tochihara (1978).

**Natural host range and symptoms**

Symptoms persist.

*Dioscorea batatus* — chlorotic and necrotic spots or netting.

**Transmission**

Transmitted by a vector; an insect; *Myzus persicae* and *Aphis gossypii*; Aphididae. Transmitted in a non–persistent manner. Virus transmitted by mechanical inoculation; not transmitted by contact between plants.

**Geographical distribution**

Spreads in Japan.

**Experimental host range**

Experimentally infected plants mostly show chlorosis.

Diagnostically susceptible

*Dioscorea batatus* — chlorosis and necrosis.

Maintenance and propagation hosts

*Dioscorea batatus*.

Assay hosts (**L**ocal lesions or **W**hole plants)

*Dioscorea batatus* (W).

**Properties of particles in sap**

Leaf sap contains few virions.

**Purification**

Method: Fukumoto and Tochihara (1978).

**Particle morphology**

Virions filamentous; of 660 nm; 12–13 nm wide.

**Taxonomy and Relationships**

Possible *Carlavirus*.

Best tests for diagnosis: The virus can be detected and identified rapidly by electroblot immuno-assay.

**References**

Fukumoto, F. and Tochihara, H. (1978). *Ann. Phytopath. Soc. Japan* **44**: 1.
Shirako, Y. and Ehara, Y. (1986). *Ann. Phytopath. Soc. Japan* **52**: 453.

# Chino del tomaté bigeminivirus

**Data collated by J.K. Brown, 1990.**

*Standard acronym:* CdTV.

First reported in *Lycopersicon esculentum*; from Mexico; by Brown and Hine (1984).

**Natural host range and symptoms**

Symptoms persist and vary seasonally.

*Lycopersicon esculentum* — leaf mottling, yellowing, curling.

*Capsicum annuum* — vein-puckering and striping on stems and fruit. This virus together with pepper mild tigré virus and another uncharacterised virus induces symptoms in peppers a disease described as tigré, in Mexico and south Texas.

*Malva parviflora* — leaf mottling, yellowing and malformation.

*Datura stramonium* — severe mosaic and leaf malformation.

**Transmission**

Transmitted by a vector; an insect; *Bemisia tabaci*; Aleyrodidae. Transmitted in a persistent manner. Virus does not multiply in the vector; not transmitted congenitally to the progeny of the vector; not transmitted by mechanical inoculation; transmitted by grafting; not transmitted by contact between plants; not transmitted by seed.

**Ecology and control**

Studies reported by Brown and Nelson (1988).

**Geographical distribution**

Spreads in Mexico and the USA.

**Experimental host range**

Several (3–9) families susceptible. Experimentally infected plants mostly show systemic mosaics, mottles, ringspots or necrosis and leaf curling.

Diagnostically susceptible

*Lycopersicon esculentum* cv. Pole Boy, *Malva parviflora, Nicotiana benthamiana, Datura stramonium, Lens culinaris* cv. Chilean Lentil 78 — systemic mottling and leaf curling.

Diagnostically insusceptible

*Lactuca sativa* cv. Empire, *Citrullus lanatus* cv. Charleston Gray, *Pisum sativum, Gossypium hirsutum* cv. DP90, *Solanum melongena* cv. Black Beauty.

Maintenance and propagation hosts

*Lycopersicon esculentum* cv. Pole Boy, *Datura stramonium, Nicotiana benthamiana.*

Assay hosts (**L**ocal lesions or **W**hole plants)

*Lycopersicon esculentum* (W), *Datura stramonium* (W), *Capsicum annuum* (W), *Lens culinaris* (W).

Dicotyledonous Hosts *(**Susceptible** or Insusceptible)*

*Beta vulgaris. Capsella bursa–pastoris.* ***Capsicum annuum. Capsicum frutescens.*** *Chenopodium album. Cicer arietinum. Citrullus lanatus. Cucumis melo. Cucumis sativus. Cucurbita maxima.* ***Datura stramonium. Datura tatula.*** *Daucus carota. Gomphrena globosa. Gossypium hirsutum. Helianthus annuus. Lactuca sativa.* ***Lens esculenta. Lycopersicon esculentum. Lycopersicon pimpinellifolium.*** *Medicago sativa.* ***Nicotiana benthamiana. Nicotiana clevelandii. Nicotiana glutinosa. Nicotiana rustica. Nicotiana tabacum. Phaseolus vulgaris.*** *Physalis peruviana. Pisum sativum. Raphanus sativus. Solanum melongena. Solanum tuberosum. Sonchus oleraceus. Spinacia oleracea. Vicia faba. Vigna unguiculata. Zinnia elegans.*

Monocotyledonous Hosts *(**Susceptible** or Insusceptible)*

*Zea mays.*

Sources of host-range data

Brown and Nelson (1988).

**Properties of particles in sap**

Leaf sap contains few virions. Electron microscopy: labile even after fixation with buffered (pH 7) 0.25% glutaraldehyde.

**Particle morphology**

Virions geminate; 20 nm in diameter; 30 nm in length.

**Biochemical properties**

Virions contain 20 % nucleic acid; 80 % protein.

*Genome* consists of DNA; single–stranded; circular; of two parts; largest (or only) genome part the larger 2.6–2.7 kb; the 2nd largest 2.6–2.7 kb.

**Taxonomy and Relationships**

Bigeminivirus: *Geminiviridae*.

**References**

Brown, J.K. and Hine, R.B. (1984). *Phytopathology* **74**: 844.
Brown, J.K. and Nelson, M.R. (1988). *Plant Dis.* **72**: 866.
Brown, J.K., Goldstein, D.E. and Nelson, M.R. (1986). *Phytopathology* **76**: 862.

# Chloris striate mosaic monogeminivirus

**Data collated by R. Greber, 1986.**

*Synonyms:* Australian wheat striate mosaic virus. *Standard acronym:* CSMV. *Strains:* Chloris (C) and Microlaena.

First reported in *Chloris gayana*; from Australia; by Grylls (1963).

**Natural host range and symptoms**

Gramineae including *Chloris gayana, Ixophorus unisetus* and *Dactylis glomerata* and cereals such as *Triticum* spp., *Avena sativa, Hordeum vulgare, Zea mays* — streaking or mosaic.

**Transmission**

Transmitted by a vector; an insect; *Nesoclutha (obscura) pallida*; Cicadellidae. Transmitted in a persistent manner. Virus possibly transmitted by seed.

**Geographical distribution**

Spreads in Australia.

**Experimental host range**

Few (<3) families susceptible. Experimentally infected plants mostly show chlorotic streaks.

Diagnostically susceptible

*Chloris gayana, Hordeum vulgare* — yellow or grey streaks.

Maintenance and propagation hosts

*Chloris gayana* — for maintaining cultures by vegetative propagation, the leaves are a good source of virus.

Assay hosts (**L**ocal lesions or **W**hole plants)

*Chloris gayana* (W) and the vector *Nesoclutha pallida* assayed by feeding on *Chloris gayana* (Grylls and Waterford, 1976; Francki *et al.*, 1979).

Monocotyledonous Hosts *(**Susceptible** or Insusceptible)*

***Avena sativa. Chloris gayana. Dactylis glomerata. Eleusine coracana. Hordeum vulgare. Triticum aestivum. Zea mays.***

Sources of host-range data

Grylls (1963).

**Properties of particles in sap**

Leaf sap contains many virions. Electron microscopy: use UA not PTA-H.

**Purification**

Method: Francki *et al.* (1979).

**Particle morphology**

Virions geminate; 18 nm in diameter; 30 nm in length (Hatta and Francki, 1979).

**Physical properties**

One sedimenting component in purified preparations. Density 1.35 g $cm^{-3}$ in CsCl (after fixation in glutaraldehyde).

**Biochemical properties**

Virions contain 19 % nucleic acid.

*Genome* consists of DNA; single–stranded; circular; unipartite; largest (or only) genome part 2.75 kb.

*Sequence database accession code(s):*

M20021 Em(40)_vi:GECHMGEN Gb(84)_vi:CHMGEN Chloris striate mosaic virus DNA, complete genome. 7/89 2,750bp. 1 sequence.

*Virion protein(s)* one; $M_r$ 28000; coat protein.

**Cytopathology**

Virions found in leaves, mesophyll, meristem and phloem (Francki *et al.*, 1979); in nuclei; sometimes in the cytoplasm and cell vacuole (Francki *et al.*, 1979).

**Taxonomy and Relationships**

Monogeminivirus: *Geminiviridae*.

Virus(es) with serologically unrelated virions: bean golden mosaic virus, cassava African mosaic virus (Francki *et al.*, 1979, 1980).

**References**

Derrick, K.S. and Brlansky, R.H. (1976). *Phytopathology* **66**: 815.
Francki, R.I.B. (1980). *CMI/AAB Descr. Pl. Viruses* No. 221, 4 pp.
Francki, R.I.B., Hatta, T., Grylls, N.E. and Grivell, C.J. (1979). *Ann. appl. Biol.* **91**: 51.
Francki, R.I.B., Hatta, T., Boccardo, G. and Randles, J.W. (1980). *Virology* **101**: 233.
Grylls, N.E. (1963). *Aust. J. agric. Res.* **14**: 143.
Grylls, N.E. and Waterford, C.J. (1976). *Aust. Pl. Path. Soc. Newsl.* **5**(Suppl.): 89.
Hatta, T. and Francki, R.I.B. (1979). *Virology* **92**: 428.

# Chondrilla juncea stunting (?) *nucleorhabdovirus*

**Data collated by A.A. Brunt, 1991.**

First reported in *Chondrilla juncea*; from Australia; by Hasan *et al.* (1973).

**Natural host range and symptoms**

*Chondrilla juncea* — leaf chlorosis and distortion, reduced seed production and plant stunting.

**Transmission**

Virus not transmitted by mechanical inoculation.

**Particle morphology**

Virions rhabdo– or bullet–shaped; with a clear modal length; of 142 nm (*in vivo*); 58 nm wide (*in vivo*). Axial canal obvious.

**Cytopathology**

Virions found in all parts of the host plant, especially the companion cells; in the perinuclear space.

**Taxonomy and Relationships**

Possible *Nucleorhabdovirus: Rhabdoviridae*.

**References**

Hasan, S., Gionnotti, J. and Vago, C. (1973). *Phytopathology* **63**: 791.

# Chrysanthemum B *carlavirus*

**Data collated by J.R. Moran, 1987.**

*Synonyms:* chrysanthemum virus Q, chrysanthemum dwarf mottle virus, chrysanthemum mild mosaic virus (Hollings, 1957), chrysanthemum necrotic mottle virus, gynura latent virus. *Standard acronym:* CVB. *Strains:* chrysanthemum vein mottle virus (Hakkaart and Maat, 1974).

First reported in *Chrysanthemum morifolium*; from the Netherlands; by Noordam (1952).

**Natural host range and symptoms**

Symptoms persist.

*Chrysanthemum morifolium* — sometimes mottling, vein clearing or necrotic streaks, but often symptomless.

*Gynura aurantiaca* — symptomless infection.

**Transmission**

Transmitted by a vector; an insect; *Myzus persicae, Macrosiphum euphorbiae, Acyrthosiphon (Aulacorthum) solani, Coloradoa rufomaculata* and *Macrosiphoniella (Pyrethromyzus) sanborni* (Hollings, 1957); Aphididae. Transmitted in a non–persistent manner. Virus transmitted by mechanical inoculation; transmitted by grafting; not transmitted by contact between plants; not transmitted by seed.

**Ecology and control**

Studies reported by Hollings (1957).

**Geographical distribution**

Probably distributed worldwide (wherever chrysanthemums are grown).

**Experimental host range**

Few (<3) families susceptible. Experimentally infected plants mostly show chlorotic local lesions, systemic mottle.

Diagnostically susceptible

*Nicotiana glutinosa, Petunia hybrida, Tetragonia expansa* — chlorotic spots.
*Nicotiana clevelandii* — systemic mottling, vein clearing.

Diagnostically insusceptible

*Chenopodium quinoa.*

Maintenance and propagation hosts

*Nicotiana clevelandii.*

Assay hosts (**L**ocal lesions or **W**hole plants)

*Petunia hybrida* (L) but not reliable.

Dicotyledonous Hosts *(**Susceptible** or Insusceptible)*

*Ageratum conyzoides. Amaranthus caudatus. Antirrhinum majus. Beta vulgaris. Brassica campestris* ssp. *pekinensis. Brassica campestris* ssp. *rapa. Brassica oleracea* ssp. *botrytis. Brassica oleracea* ssp. *capitata. Capsicum annuum. Celosia argentea. Cheiranthus cheiri. Chenopodium album. Chenopodium amaranticolor.* ***Chrysanthemum morifolium****. Cucumis sativus. Cucurbita pepo. Dahlia pinnata. Datura stramonium. Dianthus barbatus. Fagopyrum esculentum. Gomphrena globosa.* ***Gynura aurantiaca****. Helianthus annuus. Lactuca sativa. Lavatera trimestris. Lycopersicon esculentum. Matthiola incana. Medicago sativa. Melilotus alba. Myosotis sylvatica. Nicandra physaloides.* ***Nicotiana glutinosa****.* ***Nicotiana rustica****. Nicotiana tabacum. Phaseolus vulgaris. Phlox drummondii. Physalis floridana. Phytolacca americana. Pisum sativum. Raphanus sativus. Salvia splendens. Solanum tuberosum. Spinacia oleracea. Stellaria media.* ***Tetragonia expansa****. Trifolium repens. Tropaeolum majus.* ***Vicia faba****. Vigna unguiculata. Zinnia elegans.*

Additional hosts

*Ageratum conyzoides, Alonsoa warscewiczii, Anchusa azurea, Apium graveolens, Aquilegia coerulea,* ***Aster amellus****, Brassica oleracea* var. *gemmifera, Brassica oleracea* var. *italica*, *Calendula chrysantha*, *Callistephus chinensis*, *Calystegia sepium*, ***Chrysanthemum carinatum***, Convolvulus tricolor, Delphinium grandiflorum, Epilobium montanum, Geum chiloensis, Helichrysum bracteatum, Lactuca muralis, Lobularia maritima, Lupinus polyphyllus, Lythrum salicaria, Mimulus luteus, ***Nicotiana affinis***, *Papaver rhoeas*, *Papaver somniferum*, ***Passiflora coerulea***, *Pelargonium domesticum*, *Plantago lanceolata*, *Portulaca oleracea*, *Primula sinensis*, *Primula malacoides*, *Ranunculus ficaria*, *Ranunculus repens*, *Saintpaulia ionantha*, *Salpiglossis sinuata*, *Schizanthus pinnatus*, ***Senecio cruentus***, ***Tagetes erecta***, ***Taraxacum officinale*** and *Verbena hybrida*.

Sources of host-range data

Hollings (1957).

**Properties of particles in sap**

TIP: 70–75 °C. LIV: 2–6 days. DEP: $\log_{10}$ minus 2–4. Leaf sap contains few virions.

**Purification**

Method: Hollings *et al.* (1970).

**Particle morphology**

Virions filamentous; not enveloped; usually straight; with a clear modal length; of 685 nm.

**Physical properties**

One sedimenting component in purified preparations; sedimentation coefficient 168 *S*.

**Biochemical properties**

*Genome* consists of RNA; single–stranded; unipartite; largest (or only) genome part 7.5 kb.

*Sequence database accession code(s):*

S60150 Em(40)_vi:S60150 Gb(84)_vi:S60150 ORF1...ORF6 (3' terminal reigon) chrysanthemum virus B CVB, Genomic RNA, 6 genes, 3426 nt. 1 sequence.

**Cytopathology**

Virions found in all parts of the host plant; in cytoplasm. Inclusions absent from infected cells.

**Taxonomy and Relationships**

*Carlavirus*.

Virus(es) with serologically related virions: carnation latent, potato S and potato M and helenium S viruses.

**References**

Hakkaart, F.A. and Maat, D.Z. (1974). *Neth. J. Pl. Path.* **80**: 97.

Hollings, M. (1957). *Ann. appl. Biol.* **45**: 589.

Hollings, M. (1972). *CMI/AAB Descr. Pl. Viruses* No. 110, 4 pp.

Hollings, M., Stone, O.M., Thorne, J.M. and Atkey, P.T. (1970). *Rep. Glasshouse Crops Res. Inst.* 1969, p. 128.

Noordam, D. (1952). *Tijdsch. PlZiekt.* **58**: 121.

# Chrysanthemum spot (?) potyvirus

**Data collated by A.A. Brunt, 1994.**

First reported in *Chrysanthemum frutescens*; from Italy; by Bertaccini *et al.* (1994).

**Geographical distribution**

Spreads in Italy.

**Taxonomy and Relationships**

Possible unassigned potyvirus.

**References**

Bertaccini, A., Bellardi, M.G., Marani, F. and Rabiti, A. (1994). *Acta Hort.* **377**: 107.

# Chrysanthemum vein chlorosis (?) rhabdovirus

**Data collated by A.A. Brunt, 1991.**

*Standard acronym:* CVCV.

First reported in *Chrysanthemum* spp.; from Brazil; by Kitajima and Costa (1979).

**Natural host range and symptoms**

*Chrysanthemum* spp. — chlorotic banding of leaf veins.

**Geographical distribution**

Spreads in Brazil.

**Particle morphology**

Virions rhabdo– or bullet–shaped; with a clear modal length; of 200–300 nm; 60–80 nm wide.

**Taxonomy and Relationships**

Possible unassigned rhabdovirus.

**References**

Kitajima, E.W. and Costa, A.S. (1979). *Fitopatol. Bras.* **4**: 55.

# Citrus enation – woody gall (?) *luteovirus*

**Data collated by C. Büchen-Osmond, 1987. Revised by R.F. Lee, 1989.**

First reported in Citrus species as *Citrus limon, C. aurantifolia* and *C. paradisi*; from California, U.S.A.; by Wallace and Drake (1953).

**Natural host range and symptoms**

Symptoms persist.

*Citrus limon* — abaxial enations on main leaf veins of seedlings and woody galls in older trees.

*C. aurantifolia, C. paradisi, C. sinensis* — enations on leaves of seedling.

*C. reticulata* — few or no symptoms.

Also found in *Ficus* sp., *Ulmus* sp., *Morus* sp., *Rosa* sp., *Prunus domestica, Viburnum* sp., *Eucalyptus* sp. in coastal regions of N.S.W., Australia — woody galls.

**Transmission**

Transmitted by a vector; an insect; *Toxoptera citricidus*, *Myzus persicae* and *Aphis gossypii*; Aphididae. Transmitted in a persistent manner. Virus retained when the vector moults; transmitted by grafting (of buds).

**Geographical distribution**

Spreads in Australia (the central coast region of N.S.W., citrus trees are found far from citrus plantations. Such feral trees are often infected with woody gall, and similar galls are occasionally seen on other species), South Africa and the USA (in California and most citrus growing countries).

**Experimental host range**

Several (3–9) families susceptible.

Diagnostically susceptible

*Citrus limon* — enation in seedlings and galls after 18 months.

Maintenance and propagation hosts

*Citrus limon*.

Assay hosts (**L**ocal lesions or **W**hole plants)

*Citrus limon* (W).

Dicotyledonous Hosts *(**Susceptible** or Insusceptible)*

***Citrus aurantifolia***. ***Citrus aurantium***. ***Citrus limon***. ***Citrus paradisi***. ***Citrus sinensis***. ***Morus alba***. ***Prunus domestica***.

Sources of host-range data

Fraser and Broadbent (1979).

**Particle morphology**

Virions isometric; 24–26 nm in diameter (Hooper *et al.*, 1974; Maharaj and da Graca, 1988).

**Cytopathology**

Virions found in phloem, companion cells and the bark in cambium.

**Taxonomy and Relationships**

Possible *Luteovirus*.

**References**


Fraser, L.R. (1959). *Proc. Linn. Soc. N.S.W.* **LXXIV**: 332.

Fraser, L.R. and Broadbent, P. (1979). "Virus and Related Diseases of Citrus in New South Wales", p. 50. *Dep. Agric. N.S.W.* Surrey Beatty and Sons Pty Ltd, Australia.

Hooper, G.R., Spink, G.C. and Myers, R. (1974). *Virology* cited by P.R. Desjardins *et al.* (1974) *Proc. 6th Conf. Int. Org. Citrus Virol.*, Univ. California, Riverside, U.S.A.

Maharaj, S.B. and da Graca, J.V. (1988). *Phytophylactica* **20**: 357.

McLean, A.P.D. (1954). *S. Afr. J. Sci.* **50**: 147.

Wallace, J.M. and Drake, R.J. (1953). *Calif. Citrogr.* **38**: 180.

Wallace, J.M. and Drake, R.J. (1960). *Pl. Dis. Reptr* **44**: 580.

Wallace, J.M. and Drake, R.J. (1961). *Pl. Dis. Reptr* **45**: 682.

# Citrus leaf rugose *ilarvirus*

**Data collated by P. Barkley, 1987. Checked by R.F. Lee, 1989.**

*Synonyms:* citrus psorosis virus complex, citrus crinkly leaf ilarvirus. *Standard acronym:* CiLRV.

First reported in *Citrus limon*; from the U.S.A.; by Fawcett (1936).

## Natural host range and symptoms

Symptoms persist, or vary seasonally.

*Citrus aurantifolia, C. limon, C. paradisi* — flecking of young leaves and malformation of old leaves (particularly of Eureka lemons), fruit small and lumpy.

*Phaseolus vulgaris* cv. Red Kidney — necrotic local lesions, no systemic infection.

## Transmission

Transmitted by means not involving a vector. Virus transmitted by mechanical inoculation; transmitted by grafting; possibly transmitted by seed.

## Geographical distribution

Spreads in the USA (California). Found, but with no evidence of spread, in Australia.

## Experimental host range

Few (<3) families susceptible.

Diagnostically susceptible

*Citrus limon* Eureka lemon — asteroid local lesions.
*Citrus sinensis, C. limon, C. reticulata* — crinkling and flecking.
*Vigna unguiculata* — vein clearing, malformation.
*Crotalaria spectabilis* — systemic chlorotic veinbanding.

Maintenance and propagation hosts

Any *Citrus* species.

Assay hosts (**L**ocal lesions or **W**hole plants)

*Vigna unguiculata* cv. 'Blackeye' No. 5. (L).

Dicotyledonous Hosts *(**Susceptible** or Insusceptible)*

***Catharanthus roseus**. **Citrus aurantifolia**. **Citrus limon**. **Citrus sinensis**. **Crotalaria spectabilis**. **Vigna unguiculata**.*

## Purification

Method: Yot-Dauthy and Bové (1968); Garnsey (1975).

## Particle morphology

Virions unusually shaped; irregularly isometric virions; 25–32 nm in diameter (Yot-Dauthy and Bové, 1968).

## Physical properties

Four sedimenting components in purified preparations; sedimentation coefficient 105 *S*; of the other(s) 98 *S* (also 89 and 79). $A_{260}/A_{280}$ ratio 1.4–1.46 (mixed components).

## Biochemical properties

*Genome* consists of RNA; single–stranded; linear; of three parts; largest (or only) genome part 3.099 kb; the 2nd largest 2.817 kb; the 3rd largest 1.972 kb.

*Sequence database accession code(s):*

U17390 Em(43)_vi:Cl17390 Gb(89)_vi:Clu17390 Citrus leaf rugose virus RNA 3 movement protein and coat protein genes, complete cds. 4/95 2,

U17726 Em(43)_vi:Cl17726 Gb(89)_vi:Clu17726 Citrus leaf rugose virus RNA-2, polymerase gene, complete cds. 2/95 2,990bp.

*Non–genomic nucleic acid* found in the virions; is subgenomic *m*RNA; of the coat protein.

*Virion protein(s)* one; $M_r$ 26000; coat protein.

**Taxonomy and Relationships**

*Ilarvirus: Bromoviridae* (subgroup 2).

**References**

Attathom, S., Weathers, L.G. and Greer, F.C. (1975). *Proc. Am. Phytopath. Soc.* **2**: 111.

Fawcett, H.S. (1936). In: *Citrus Diseases and their Control*. McGraw Hill, New York, London, 656 pp.

Garnsey, S.M. (1975). *Phytopathology* **65**: 50.

Gonsalves, D. and Garnsey, S.M. (1974). *Virology* **61**: 343.

Gonsalves, D. and Garnsey, S.M. (1975a). *Virology* **64**: 23.

Gonsalves, D. and Garnsey, S.M. (1975b). *Virology* **67**: 311.

Gonsalves, D. and Garnsey, S.M. (1975c). *Virology* **67**: 319.

Gonsalves, D. and Garnsey, S.M. (1976). *CMI/AAB Descr. Pl. Viruses* No. 164, 4pp.

Yot-Dauthy, D. and Bové, J.M. (1968). *Proc. 4th Conf. Int. Org. Citrus Virol.*, p. 255.

# Citrus leprosis (?) rhabdovirus

**Data collated by E.W. Kitajima, 1985. Revised by R.F. Lee, 1989.**

*Standard acronym:* CiLV.

First reported in *Citrus sinensis*; from Sao Paolo, Brazil; by Bitancourt (1940).

**Natural host range and symptoms**

Symptoms disappear soon after infection.

*Citrus sinensis* — large chlorotic blotches on fruits, leaves and twigs.

**Transmission**

Transmitted by a vector; a mite; *Brevipalpus phoenicis*; Tetranychidae. Virus not transmitted by mechanical inoculation; transmitted by grafting; not transmitted by seed; not transmitted by pollen.

**Geographical distribution**

Spreads in Argentina, Brazil, and the USA.

**Experimental host range**

Few (<3) families susceptible.

Diagnostically susceptible

*Citrus* sp.

Assay hosts (**L**ocal lesions or **W**hole plants)

*Citrus* sp. (W).

Dicotyledonous Hosts *(**Susceptible** or Insusceptible)*

***Citrus aurantifolia*. *Citrus sinensis*.**

**Particle morphology**

Virions rhabdo– or bullet–shaped; not enveloped; of 100–110 nm; 30 nm wide.

**Cytopathology**

Virions found in leaves; in nuclei (commonly associated with its envelope). Inclusions present in infected cells; are viroplasms; they do not contain virions.

**Taxonomy and Relationships**

Possible unassigned rhabdovirus.

**References**


Bitancourt, A.A. (1940). *Biologico* **6**: 39.

Chagas, C.M. and Rossetti, V. (1980). *Fitopatol. Bras.* **5**: 211.

Fawcett, H.S. (1936). In: *Citrus Diseases and their Control*, 2nd ed. McGraw Hill, New York.

Kitajima, E.W., Muller, G.W., Costa, A.S. and Yuki, W. (1972). *Virology* **50**: 254.

Knorr, C.L. (1950). *Phytopathology* **40**: 15.

Knorr, L.C. (1968). *Proc. 4th Conf. Int. Org. Citrus Virol.*, Univ. Florida, Gainesville, p. 332.


# Citrus ringspot virus

**Data collated by K.S. Derrick, 1989.**

*Synonyms:* citrus psorosis virus complex — A and B (Whiteside *et al.*, 1988; da Graca *et al.*, 1991).

First reported in *Citrus* sp.; from Florida and California, U.S.A.; by Swingle and Webber (1896); Fawcett (1933); Wallace and Drake (1968).

**Natural host range and symptoms**

Symptoms persist.

*Citrus* sp. — epinasty, chlorotic flecks, leaf mottling, ringspots or large, irregular chlorotic patterns on mature leaves.

**Transmission**

Possibly by a vector as, in Argentina spread is not related to grafting. Virus transmitted by mechanical inoculation; transmitted by grafting.

**Ecology and control**

Studies reported by in Compendium of Citrus Diseases (1988).

**Geographical distribution**

Spreads in Argentina (controlled by using virus-free budwood).

**Experimental host range**

Several (3–9) families susceptible. Experimentally infected plants mostly show systemic mosaics, mottles, ringspots or necrosis.

Diagnostically susceptible

*Chenopodium quinoa, Phaseolus vulgaris* cv. Red Kidney — local lesions.

*Zinnia elegans* cv. Thembelina — symptomless.

Maintenance and propagation hosts

*Citrus* sp.

Assay hosts (**L**ocal lesions or **W**hole plants)

*Chenopodium quinoa* (L), *Phaseolus vulgaris* (L).

Dicotyledonous Hosts *(**Susceptible** or Insusceptible)*

***Catharanthus roseus. Chenopodium amaranticolor. Chenopodium quinoa. Citrus aurantifolia. Citrus aurantium. Citrus limon. Citrus medica. Citrus paradisi. Citrus sinensis. Crotalaria spectabilis. Cucumis sativus. Gomphrena globosa. Helianthus annuus. Nicotiana benthamiana. Nicotiana clevelandii. Nicotiana megalosiphon. Nicotiana rustica. Nicotiana tabacum. Petunia hybrida. Phaseolus vulgaris. Pisum sativum. Poncirus trifoliata. Sesamum indicum. Zinnia elegans.***

Sources of host-range data

Timmer *et al.* (1978).

**Properties of particles in sap**

LIV: 2 days (at 4°C). DEP: $\log_{10}$ minus 2. Infectivity of sap decreased by treatment with di–ethyl ether. Leaf sap contains few virions.

**Purification**

Method: Derrick *et al.* (1988).

**Particle morphology**

Virions filamentous; they vary in length; not enveloped; usually flexuous (extremely); of 300–500 nm, or 1500–2500 nm; *c.* 8–10 nm wide.

**Physical properties**

Two sedimenting components in purified preparations.

**Biochemical properties**

*Genome* probably consists of RNA; single–stranded; of two parts.

*Virion protein(s)* one; $M_r$ 48000; probably coat protein. Method of preparation: Derrick *et al.* (1988).

**Cytopathology**

Virions found in leaves (isolated from leaf tissue).

**Taxonomy and Relationships**

Best tests for diagnosis: see Derrick *et al.* (1988).

**References**


Compendium of Citrus Diseases (1988). *Amer. Phytopath. Soc.*. Eds J.O. Whiteside, S.M. Garnsey and L.W. Timmer. APS Press.

Da Graca, J.V., Lee, R.F., Moreno, P., Civerolo, E.L. and Derrick, K.S. (1991). *Plant Dis.* **75**: 613.

Derrick, K.S., Brlansky, R.H., Da Graca, J.V., Lee, R.F., Timmer, L.W. and Nguyen, T.K. (1988). *Phytopathology* **78**: 1298.

Derrick, K.S., Lee, R.F., Hewitt, B.G., Barthe, G.A. and Da Graca, J.V. (1991). *Proc. 11th Conf. Int. Org. Citrus Virol.*

Fawcett, H.S. (1933). *Phytopathology* **23**: 930.

Swingle, W.T. and Webber, H.J. (1896). *U.S. Dep. Agr. Div. Veg. Phys. Path. Bull.* **8**: 42 pp.

Timmer, L.W., Garnsey, S.M. and McRitchie, J.J. (1978). *Pl. Dis. Reptr* **62**: 1054.

Wallace, J.M. and Drake, R.J. (1968). *Proc. 4th Conf. Int. Org. Citrus Virol.* Ed. J.F.L. Childs. Florida Press, Gainesville, Florida, U.S.A..

Whiteside, J.O., Garnsey, S.M. and Timmer, L.W. (1988). In: *Compendium Citrus Diseases*, pp. 88. Amer. Phytopath. Soc., St. Paul, Minnesota, U.S.A.

# Citrus tatter leaf *capillovirus*

**Data collated by C. Büchen-Osmond, 1987. Revised by R.F. Lee, 1989.**

*Synonyms:* citrange stunt virus (Roistacher, 1988). *Standard acronym:* CTLV.

First reported in *Citrus limon* cv. Meyer lemon; from California, U.S.A.; by Wallace and Drake (1962).

**Natural host range and symptoms**

Symptoms vary seasonally.
*Citrus limon* cv. Meyer — tatter leaf; blotchy and malformed leaves.
*C. excelsa* — tatter leaf or symptomless.
*C. sinensis* — symptomless.
*Poncirus trifoliata* — interveinal chlorosis.

**Transmission**

Probably is moved and dispersed only in infected budwood. Virus transmitted by mechanical inoculation; transmitted by grafting; transmitted by seed (in lily).

**Ecology and control**

Studies reported by Takahara *et al.* (1988); Koizumi (1984); Navarro *et al.* (1989); Miyakawa (1980); Roistacher *et al.* (1980).

**Geographical distribution**

Spreads in the African region, the Eastern Asian region, the North American region, and the Pacific region; Australia, China, Japan, Korea D.P.R. (North), Korea Republic, South Africa, Taiwan, and the USA. Found, but with no evidence of spread, in N.S.W., Australia.

**Experimental host range**

Several (3–9) families susceptible to few (<3) families susceptible. Experimentally infected plants mostly show tatter leaf, systemic mottle.

Diagnostically susceptible

*Citrus excelsa* — tatter leaf.
*Citrus trifoliata* — interveinal chlorosis.
*Nicotiana clevelandii, N. glutinosa* — chlorotic veinal blotching
*Chenopodium quinoa* — chlorotic local lesions, systemic mottle and curling.
*Citrus sinensis* x *Poncirus trifoliata* cv. Rusk *Poncirus trifoliata* x *Citrus paradisi* cv. Swingle — leaf malformation and twisted stems.

Maintenance and propagation hosts

*Citrus excelsa, Nicotiana clevelandii, N. glutinosa, Vigna unguiculata, Rusk citrange.*

Assay hosts (**L**ocal lesions or **W**hole plants)

*Citrus excelsa* (W), *Chenopodium quinoa* (L).

Dicotyledonous Hosts *(**Susceptible** or Insusceptible)*

*Chenopodium amaranticolor*. ***Chenopodium quinoa***. ***Citrus aurantifolia***. ***Citrus limon***. ***Citrus medica***. ***Citrus sinensis***. *Crotalaria spectabilis*. ***Nicotiana clevelandii***. ***Nicotiana glutinosa***. *Nicotiana sylvestris*. *Nicotiana tabacum*. ***Poncirus trifoliata***. *Vigna unguiculata*.

Sources of host-range data

Garnsey (1974); Fraser and Broadbent (1979); Marais and Lee (1986).

**Purification**

Method: Semancik and Weathers (1965).

**Particle morphology**

Virions filamentous; usually flexuous; with a clear modal length; of 650 nm; 19 nm wide. Basic helix obvious; pitch of basic helix 3.4 nm.

**Biochemical properties**

*Genome* consists of RNA; single–stranded.

*Sequence database accession code(s):*

D14455 Gb(84)n:CTLVCC Citrus tatter leaf virus RNA for viral polymerase protein, cell to cell movement protein and

D16368 Em(40)_vi:CTLA Gb(84)_vi:CTLA Citrus tatter leaf virus RNA. 3/94 2,956bp.

D14455 Em(43)_vi:Ctlvcc Gb(89)_vi:Ctlvcc Citrus tatter leaf virus RNA for viral polymerase protein, cell to cell movement protein.

*Virion protein(s)* one; $M_r$ 27000.

**Taxonomy and Relationships**

*Capillovirus*.

Virus(es) with serologically related virions: apple stem grooving virus.

**Comments**

Recent tests indicate that this virus is probably a closely related strain of apple stem grooving *capillovirus* rather than a distinct virus (Ohira *et al.*, 1994).

**References**

Catara, A. and Wallace, J.M. (1970). *Phytopathology* **70**: 737.

Fulton, R.W. (1966). *Phytopathology* **56**: 575.

Garnsey, L.M. (1974). *Proc. 6th Conf. Int. Org. Citrus Virol.*, Univ. Calif., Riverside, U.S.A., p. 137.

Inouye, N., Maeda, T. and Mitsuhata, K. (1989). *Ann. Phytopath. Soc. Japan* **45**: 712.

Koizumi, M. (1984). *Proc. 9th Conf. Int. Org. Citrus Virol.*, Univ. Calif., Riverside, U.S.A., p. 229.

Marais, L.J. and Lee, R.F. (1986). *Plant Dis.* **70**: 892.

Miyakawa, T. (1980). *Proc. 8th Conf. Int. Org. Citrus Virol.*, Univ. Calif., Riverside, U.S.A., p. 220.

Navarro, L., Civerolo, E.L., Juarez, J. and Garnsey, S.M. (1989). *Abs 11th Conf. Int. Org. Citrus Virol.*, Orlando, Florida, U.S.A.

Nishio, T., Kawai, A., Takahashi, T., Namba, S. and Yamashita, S. (1989). *Ann. Phytopath. Soc. Japan* **55**: 254.

Ohki, S.T., Yoshikawa, N., Inouye, N. and Inouye, T. (1989). *Ann. Phytopath. Soc. Japan* **55**: 245.

Ohira, K., Ito, T., Kawai, A., Namba, S., Kusumi, T. and Tsuchizaki, T. (1994). *Virus Genes* **8**: 169.

Roistacher, C.N. (1988). *Proc. 10th Conf. Int. Org. Citrus Virol.*, Univ. Calif., Riverside, U.S.A., p. 353.

Semancik, J.S. and Weathers, L.G. (1965). *Phytopathology* **55**: 1354.

Wallace, J.M. and Drake, R.J. (1962). *Pl. Dis. Reptr* **46**: 21407.

Wallace, J.M. and Drake, R.J. (1968). *Proc. 4th Conf. Int. Org. Citrus Virol.*, Univ. Florida, Gainsville, p. 177.
Weathers, C.E. and Semancik, J.S. (1963). *Phytopathology* **55**: 1081.

# Citrus tristeza *closterovirus*

**Data collated by C. Büchen-Osmond, 1986; D.J. Gumpf, 1987, U.S.A. Revised by R.F. Lee, 1989.**

*Synonyms:* citrus quick decline virus (Fawcett and Wallace, 1946), grapevine A virus, hassaku dwarf virus, lime dieback closterovirus. *Standard acronym:* CTV. *Strains:* seedling yellows virus (Fraser, 1952), grapefruit stem pitting virus (Oberholzer *et al.*, 1946), grapefruit stunt bush virus (Anon, 1950), lime die-back virus (Hughes and Lister, 1949), Ellendale mandarin decline virus (Stubbs, 1952).

First reported in *Citrus sinensis* and *C. aurantifolia*; from Italy and California, U.S.A.; by Meneghini (1946); Fawcett and Wallace (1946); Kitajima *et al.* (1964).

### Natural host range and symptoms

Symptoms persist.

*Citrus* ssp. grafted onto *Citrus aurantifolia* (sour orange) root stock — quick decline, pitted stems.

*C. paradisi* (grapefruit) — stunt.
*C. aurantifolia* (lime) — die-back.
*C. aurantifolia* (Seville orange) — seedling yellows.
*C. reticulata* (mandarin) — decline.
*Aeglopsis chevalieri, Afraegle paniculata, Pamburus missiones, Passiflora gracilis.*

### Transmission

Transmitted by a vector; an insect; *Toxoptera citricidus* (Meneghini, 1946; Costa and Grant, 1951), *Aphis gossypii, A. spiraecola* and *Toxoptera aurantii* (Dickson *et al.*, 1951; Norman and Grant, 1956); Aphididae. Transmitted in a non–persistent manner. Virus transmitted by mechanical inoculation; transmitted by grafting; not transmitted by seed.

### Ecology and control

Studies reported by Costa and Muller (1980).

### Geographical distribution

Probably distributed worldwide (whereever citrus is grown). Spreads in the African region, the Eastern Asian region, the Eurasian region, the Mediterranean region, the North American region, and the Pacific region; Algeria, American Samoa, Antigua and Barbuda, Argentina, Australia, Belize, Bermuda, Bolivia, Brazil, Brunei Darussalam, Cameroon, the Central African Republic, Chad, China, Colombia, Costa Rica, Cyprus, the Dominican Republic, Ecuador, Egypt, El Salvador, Ethiopia, Fiji, French Polynesia, Gabon, Ghana, Guyana, India, Indonesia, Iran, Israel, Italy, Jamaica, Japan, Kenya, Korea Republic, Malaysia, Mauritius, Morocco, Mozambique, Nepal, Netherlands Antilles, New Caledonia, New Zealand, Nicaragua, Nigeria, Pakistan, Panama, Paraguay, Peru, the Philippines, Portugal, Puerto Rico, Saudi Arabia, Spain, Sri Lanka, Suriname, Taiwan, Tanzania, Thailand, Trinidad and Tobago, Turkey, the USA, Uganda, Uruguay, Venezuela, Vietnam, Zaire, Zambia, Western Samoa, the former Yugoslavia, Zimbabwe.

### Experimental host range

Few (<3) families susceptible.

Diagnostically susceptible

*Citrus aurantifolia* cvs West Indian lime, Key lime, Mexican lime, Galego lime, Kagzi lime — chlorotic vein flecks, leaves cupped or canoe-shaped; stem pitting (Wallace and Drake, 1951).

*C. sinensis* on rootstock *C. aurantifolia* — sudden wilt, decline and death, overgrowth on bud union and 'honeycombing' just below union.

*C. sinensis, C. paradisi* — stem pitting.

Maintenance and propagation hosts

*Citrus aurantifolia, C. exelsa, C. sinensis.*

Assay hosts (**L**ocal lesions or **W**hole plants)

*Citrus aurantifolia* (W).

Dicotyledonous Hosts *(**Susceptible** or Insusceptible)*

***Citrus aurantifolia*. *Citrus aurantium*. *Citrus limon*. *Citrus medica*. *Citrus paradisi*. *Citrus sinensis*.**

Sources of host-range data

Wallace and Drake (1951); Knorr (1956); Müller *et al.* (1974).

**Properties of particles in sap**

Leaf sap contains few virions.

**Purification**

Method: Bar-Joseph *et al.* (1970).

**Particle morphology**

Virions filamentous; not enveloped; usually flexuous; with a clear modal length; of 2000 nm; 12 nm wide.

**Physical properties**

One sedimenting component in purified preparations; sedimentation coefficient 140 *S* (±10). Density 1.257 g/ml.

**Biochemical properties**

*Genome* consists of RNA; single–stranded; unipartite; largest (or only) genome part 17–20 kb. Genomic nucleic acid isolated by Bar-Joseph *et al.* (1985).

*Sequence database accession code(s):*

L12175 Em(40)_vi:CTVCOATB Gb(84)_vi:CTVCOATB Citrus tristeza virus coat protein mRNA, complete cds. 12/93 672bp.

M76485 Em(40)_vi:CTVCOAT Gb(84)_vi:CTVCOAT Citrus tristeza virus coat protein gene, complete cds. 7/92 1,080bp.

S67800 Em(40)_vi:S67800 Gb(84)_vi:S67800 coat protein citrus tristeza virus CTV, ssp VT, Ach-T isolate, Genomic RNA, 672 nt. 4/94 67

U02547 Em(40)_vi:CT02547 Gb(84)_vi:CTU02547 Citrus tristeza virus Hsp70 homolog mRNA, partial sequence. 5/94 7,292bp.

X75433 Em(40)_vi:GVMPCP Gb(84)_vi:GVMPCP Grapevine virus A genomic RNA, MP and CP genes. 5/94 1,883bp.

U16304 Em(43)_vi:Ct16304 Gb(89)_vi:Ctu16304 Citrus tristeza virus complete genome. 5/95 19,296bp.

U17265 Em(43)_vi:Ct17265 Gb(89)_vi:Ctu17265 Citrus tristeza virus p18, p13, p20, and p23.5 genes, complete cds, and coat protein gene, pa

U20531 Em(43)_vi:Ct20531 Gb(89)_vi:Ctu20531 Citrus tristeza virus defective RNA D–CTV-RNA. 4/95 2,424bp.

Method of preparation: Lee *et al.* (1988). Amino acid composition: Lee *et al.* (1988).

**Replication**

Replication does not depend on a helper virus.

**Cytopathology**

Virions found in phloem; in cytoplasm. Inclusions present in infected cells; are crystals in the cytoplasm; they contain virions.

**Taxonomy and Relationships**

*Closterovirus.*

**References**


Anon (1950). *Agric. Gaz. N.S.W.* **61**: 365.

Bar-Joseph, M., Loebenstein, G. and Cohen, J. (1970). *Phytopathology* **60**: 75.

Bar-Joseph, M., Gumpf, D.J., Dodds, J.A., Rosner, A. and Ginsberg, I. (1985). *Phytopathology* **75**: 195.

Brlansky, R.H., Garnsey, S.M., Lee, R.F. and Purcifull, D.E. (1984). *Proc. 9th Conf. Int. Org. Citrus Virol.*, p. 337. Univ. California, Riverside.

Brlansky, R.H., Lee, R.F. and Garnsey, S.M. (1988). *Plant Dis.* **72**: 1039.

Costa, A.S. and Grant, T.J. (1951). *Phytopathology* **41**: 105.

Costa, A.S. and Müller, G.W. (1980). *Plant Dis.* **64**: 538.

Dickson, C.R., Flock, R.A. and Johnson, M.M. (1951). *Calif. Citrogr.* **36**: 135.

Fawcett, H.S. and Wallace, J.M. (1946). *Calif. Citrogr.* **32**: 50.

Fraser, L.R. (1952). *Agric. Gaz. N.S.W.* **63**: 125.

Grant, T.J. and Higgins, R.P. (1957). *Phytopathology* **47**: 272.

Hughes, W.A. and Lister, C.A. (1949). *Nature, Lond.* **164**: 880.

Knorr, L.C. (1956). *Phytopathology* **46**: 557.

Lee, R.F., Garnsey, S.M., Brlansky, R.H. and Goheen, A.C. (1987). *Phytopathology* **77**: 543.

Lee, R.F., Calvert, L.A., Nagel, J. and Hubbard, J.D. (1988). *Phytopathology* **78**: 1221.

Meneghini, M. (1946). *Biologico* **12**: 285.

Müller, G.W., Costa, A.S., Kitajima, E.W. and Camargo, I.J.B. (1974). *Proc. 6th Conf. Int. Org. Citrus Virol.*, p. 75. Univ. California, Riverside.

Norman, P.A. and Grant, T.J. (1956). *Proc. Fla Stn hort Soc.* **69**: 38.

Oberholzer, P.C.F., Mathews, I. and Stiemie, S.F. (1949). *Sci. Bull. Dep. Agric. S. Afr.*, 287.

Price, W.C. (1970). *CMI/AAB Descr. Pl. Viruses* No. 33, 3 pp.

Stubbs, L.L. (1952). *J. Dep. Agric. Vic.* **50**: 124.

Wallace, J.M. and Drake, R.J. (1951). *Calif. Citrogr.* **36**: 136.


# Citrus variegation *ilarvirus*

**Data collated by P. Barkley, 1987. Checked by R.F. Lee, 1989.**

*Synonyms:* citrus psorosis virus complex — infectious variegation component. *Standard acronym:* CVV. *Strains:* CVV-I, CVV-II.

First reported in *Citrus limon*; from California, U.S.A.; by Fawcett and Klotz (1939).

**Natural host range and symptoms**

Symptoms persist and vary seasonally (with growth flushes).

*Citrus aurantifolia, C. limon* and other species — malformation and young-leaf flecking (psorosis): more severe in sour orange and lemon.

**Transmission**

Virus transmitted by mechanical inoculation; transmitted by grafting; possibly transmitted by seed.

**Ecology and control**

Studies reported by Wallace (1978).

**Geographical distribution**

Probably distributed worldwide (occurs in most citrus growing countries, but is not common). Spreads in Australia and the USA.

**Experimental host range**

Several (3–9) families susceptible.

Diagnostically susceptible

*Citrus aurantifolia, C. limon* cv. Eureka — leaf flecking, mosaic and malformation.

Maintenance and propagation hosts

Any *Citrus* species.

Assay hosts (**L**ocal lesions or **W**hole plants)

*Vigna unguiculata* (W), *Capsicum frutescens* var. grossum (W).

Dicotyledonous Hosts *(**Susceptible** or Insusceptible)*

***Capsicum frutescens*. *Citrus aurantium*. *Citrus limon*. *Vigna unguiculata*.**

**Particle morphology**

Virions unusually shaped; irregularly isometric; 26–35 nm in diameter.

*Sequence database accession code(s):*

U17389 Em(43)_vi:Cv17389 Gb(89)_vi:Cvu17389 Citrus variegation virus RNA 3 movement protein and coat protein genes, complete cds. 5/95 2.

**Taxonomy and Relationships**

*Ilarvirus: Bromoviridae* (subgroup 2).

Virus(es) with serologically related virions: citrus leaf rugose virus (but distinct).

**References**

Calvert, L.A., Lee, R.F. and Hiebert, E. (1987). *Proc. 10th Conf. Int. Org. Citrus Virol.*, p. 114.

Corbett, M.K. and Grant, T.J. (1967). *Phytopathology* **57**: 137.

Davino, M. and Garnsey, S.M. (1980). *Proc. 9th Conf. Int. Org. Citrus Virol.*, p. 196.

Desjardins, P.R. and French, J.V. (1970). *Virology* **40**: 746.

Fawcett, H.S. and Klotz, L.J. (1939). *Phytopathology* **29**: 911.

Garnsey, S.M. (1974). *Proc. 6th Conf. Int. Org. Citrus Virol.*, p. 169.

Gonsalves, D. and Garnsey, S.M. (1976). *Proc. 7th Conf. Int. Org. Citrus Virol.*, p. 109.

Rana, G.L., Quacquarelli, A. and Martelli, G.P. (1974). *Proc. 6th Crop Protect. Conf. Int. Org. Citrus Virol.*, p.165.

Wallace, J.M. (1978). In: *Citrus Industry* Vol. 4, p. 72.

# Clitoria mosaic (?) *potexvirus*

**Data collated by K.M. Srivastava and R.K. Raizada, 1984.**

First reported in *Clitoria ternatea*; from Lucknow, India; by Srivastava and Singh (1978).

**Natural host range and symptoms**

Symptoms persist.
*Clitoria ternatea, Crotalaria spectabilis* — systemic mosaic.

**Transmission**

Virus transmitted by mechanical inoculation; transmitted by grafting; not transmitted by seed.

**Geographical distribution**

Spreads in India.

**Experimental host range**

Few (<3) families susceptible. Experimentally infected plants mostly show necrotic lesions, systemic mosaic or veinal necrosis.

Diagnostically susceptible

*Chenopodium amaranticolor* — necrotic local lesions.
*Crotalaria spectabilis* — systemic mosaic.
*Vigna unguiculata* — necrotic local lesions, systemic veinal necrosis.

Diagnostically insusceptible

*Cucumis sativus, Arachis hypogaea, Datura stramonium, Nicotiana glutinosa.*

Maintenance and propagation hosts

*Vigna unguiculata.*

Assay hosts (**L**ocal lesions or **W**hole plants)

*Chenopodium amaranticolor* (L).

Dicotyledonous Hosts *(**Susceptible** or Insusceptible)*

*Ageratum conyzoides. Amaranthus caudatus. Antirrhinum majus. Arachis hypogaea. Beta vulgaris. Brassica campestris. Brassica campestris* ssp. *rapa. Brassica juncea. Cajanus cajan. Calendula officinalis. Capsicum annuum. Cassia bicapsularis. Cassia occidentalis. Catharanthus roseus.* ***Chenopodium amaranticolor****. Chrysanthemum morifolium. Citrullus lanatus.* ***Clitoria ternatea****.* ***Crotalaria spectabilis****. Cucumis melo. Cucumis sativus. Cucurbita maxima. Cucurbita moschata. Cucurbita pepo. Dahlia pinnata.* ***Datura metel****. Daucus carota. Dianthus barbatus. Glycine max. Gomphrena globosa. Helianthus annuus. Lablab purpureus. Lycopersicon esculentum. Momordica balsamina. Nicotiana glutinosa. Nicotiana rustica. Nicotiana tabacum. Ocimum basilicum. Petunia hybrida. Phaseolus lunatus. Phlox drummondii. Raphanus sativus. Salvia splendens.* ***Solanum nigrum****. Vicia faba. Vigna radiata. Viola cornuta. Zinnia elegans.*

Monocotyledonous Hosts *(**Susceptible** or Insusceptible)*

*Oryza sativa. Saccharum officinarum. Zea mays.*

Sources of host-range data

Srivastava *et al.* (1978).

**Properties of particles in sap**

TIP: 70 °C. LIV: 0.75 days. DEP: $\log_{10}$ minus 3–4.

**Purification**

Method: Srivastava and Raizada: partially purified by differential centrifugation.

**Particle morphology**

Virions filamentous; not enveloped; usually flexuous; with a clear modal length; of 510 nm; 12 nm wide. Axial canal obscure. Basic helix obscure.

**Biochemical properties**

*Genome* consists of RNA; single–stranded.

**Cytopathology**

Virions found in all parts of the host plant.

**Taxonomy and Relationships**

Possible *Potexvirus*.
Virus(es) with serologically related virions: potato virus X.

**References**

Srivastava, K.M. and Singh, B.P. (1978). *Indian Phytopath.* **31**: 248.

# Clitoria yellow vein *tymovirus*

**Data collated by K. Bock, 1980. Revised 1983.**

*Standard acronym:* CYVV.

First reported in *Clitoria ternatea*; from Kenya; by Bock *et al.* (1977*a*).

**Natural host range and symptoms**

Symptoms persist.
*Abrus precatorius* — mottle.
*Clitoria ternatea* — systemic vein chlorosis.

**Transmission**

Virus transmitted by mechanical inoculation; not transmitted by seed.

**Geographical distribution**

Spreads in Kenya.

**Experimental host range**

Several (3–9) families susceptible. Experimentally infected plants mostly show necrotic or chlorotic local lesions, systemic vein chlorosis or mosaic.

Diagnostically susceptible

*Bauhinia purpurea*, *Cajanus cajan* — vein netting.
*Clitoria ternatea* — systemic mosaic and vein netting.
*Glycine max* — systemic vein netting, chlorosis.
*Phaseolus vulgaris* cv. Long Tom — necrotic or chlorotic local lesions, systemic vein netting.
*Vigna radiata* — necrotic local lesions.
*V. unguiculata* — systemic mottle.
*Voandzeia subterranea* — systemic mottle.
*Datura metel* — systemic vein netting.
*Nicotiana clevelandii* — systemic vein netting.

Maintenance and propagation hosts

*Nicotiana clevelandii.*

Assay hosts (**L**ocal lesions or **W**hole plants)

*Glycine max, Phaseolus vulgaris, Datura metel, Nicotiana clevelandii.*

Dicotyledonous Hosts *(**Susceptible** or Insusceptible)*

*Adansonia digitata. **Arachis hypogaea. Bauhinia purpurea**. Beta vulgaris. Cajanus cajan. Capsicum annuum. Capsicum frutescens. **Cassia occidentalis**. Ceiba pentandra. Chenopodium amaranticolor. Chenopodium quinoa. **Cicer arietinum**. Citrullus lanatus. **Clitoria ternatea**. Crotalaria spectabilis. Cucumis melo. Cucumis sativus. Cucurbita maxima. Cucurbita pepo. Datura ferox. **Datura metel**. Datura stramonium. **Desmodium tortuosum**. Glycine max. Gomphrena globosa. Gossypium hirsutum. **Hibiscus esculentus**. Ipomoea setosa. Lycopersicon esculentum. Macroptilium lathyroides. Medicago sativa. Melilotus alba. Nicandra physaloides. **Nicotiana clevelandii**. Nicotiana glutinosa. Nicotiana tabacum. Petunia hybrida. Phaseolus lunatus. **Phaseolus vulgaris**. Physalis floridana. **Pisum sativum**. Solanum melongena. **Solanum nigrum**. Spinacia oleracea. Theobroma cacao. Trifolium repens. Trifolium subterraneum. Vicia faba. **Vigna angularis**. **Vigna radiata**. **Vigna unguiculata**.*

Sources of host-range data

Bock *et al.* (1977*a*; 1977*b*).

**Properties of particles in sap**

TIP: 72 °C. LIV: 21 days. DEP: $\log_{10}$ minus 8. Leaf sap contains many virions.

**Purification**

Method: Bock *et al.* (1977*a*).

**Particle morphology**

Virions isometric; not enveloped; 28 nm in diameter; rounded in profile; with a conspicuous capsomere arrangement.

**Physical properties**

Two sedimenting components in purified preparations; sedimentation coefficient of the fastest 109 *S*; of the other(s) 50 *S*. $A_{260}/A_{280}$ ratio 1.69 (B; uncorrected).

**Biochemical properties**

Virions contain 37 % nucleic acid.

*Genome* consists of RNA; single–stranded.

*Sequence database accession code(s):*

M15963 Em(40)_vi:TYMTY3CY Gb(84)_vi:CYE3CYVV Clitoria yellow vein virus 3' end which can form tRNA-like structure. 11/90 120bp.

M58311 Em(40)_vi:CYERRLT Gb(84)_vi:CYERRLT Clitoria yellow vein virus (CYVV) 3' terminus tRNA-like structure. 8/91 83bp. 2 sequences.

*Virion protein(s)* one; $M_r$ 19429. Method of preparation: Bock *et al.* (1977*a*). Amino acid sequence: from the VP gene; (A.M. Mackenzie, pers. comm.).

**Taxonomy and Relationships**

*Tymovirus.*

**References**

Bock, K.R., Guthrie, E.J. and Meredith, G. (1977*a*). *Ann. appl. Biol.* **85**: 97.

Bock, K.R. and Guthrie, E.J. (1977*b*). *CMI/AAB Descr. Pl. Viruses* No. 171, 3 pp.

Koenig, R. (1976). *Virology* **72**: 1.

# Clover enation (?) *nucleorhabdovirus*

**Data collated by P.W.G. Chu, 1985.**

*Synonyms:* clover enation mosaic virus. *Standard acronym:* ClEV.

First reported in *Trifolium repens*; from Northern Italy; by Rubio-Huertos and Bos (1969).

## Natural host range and symptoms

Symptoms persist.
*Trifolium repens* — enations on main veins of leaves.

## Transmission

Virus not transmitted by mechanical inoculation; transmitted by grafting.

## Geographical distribution

Spreads in Italy (the north).

## Experimental host range

Experimentally infected plants mostly show tumours.

Diagnostically susceptible

*Vicia faba* — vein clearing, then vein swelling to give chains of spindle-shaped tumours.

Dicotyledonous Hosts *(**Susceptible** or Insusceptible)*

***Trifolium incarnatum*. *Trifolium pratense*. *Trifolium repens*. *Vicia faba*.**

## Particle morphology

Virions rhabdo– or bullet–shaped; enveloped; with a clear modal length; of 200 nm; 80 nm wide. Axial canal obvious.

## Cytopathology

Virions found in in leaf tissue with enations; in cytoplasm, in nuclei, and in mitochondria. Inclusions present in infected cells; are crystals in the nucleus (arrays of tubular virions), or membranous bodies (formed by the endoplasmic reticulum); they contain virions. Other cellular changes: enations on the abaxial surface of the main leaf veins, containing tumorous virus-containing cells with enlarged nuclei and nucleoli surrounded by normal virus-free cells.

## Taxonomy and Relationships

Possible *Nucleorhabdovirus: Rhabdoviridae*.

## Comments

A MLO is thought to cause a similar disease.

## References

Bogatyrenko, T.N. (1977). *Biol. Nauki* **20**: 120.
Francki, R.I.B. (1973). *Adv. Virus Res.* **18**: 257.
Matthews, R.E.F. (1982). *2nd Report of ICTV.*
Rubio-Huertos, M. and Bos, L. (1969). *Neth J. Pl. Path.* **75**: 329.

# Clover mild mosaic virus

**Data collated by B. Gerhardson. 1982. Revised 1983.**

First reported in *Trifolium pratense*; from Sweden; by Gerhardson (1977).

**Natural host range and symptoms**

Symptoms persist.
*Trifolium pratense, T. hybridum* — mosaic and malformation.

**Transmission**

Transmitted by a vector; an insect; Aphididae. Transmitted in a semi–persistent manner. Virus transmitted by mechanical inoculation.

**Geographical distribution**

Spreads in Sweden.

**Experimental host range**

Few (<3) families susceptible.

Diagnostically susceptible

*Phaseolus vulgaris* — necrotic local lesions.
*Pisum sativum* — systemic mosaic, ringspots, leaf shape malformation.
*Trifolium hybridum, Vicia faba* — systemic mosaic, leaf malformation.

Maintenance and propagation hosts

*Pisum sativum, Trifolium hybridum, Vicia faba.*

Assay hosts (**L**ocal lesions or **W**hole plants)

*Phaseolus vulgaris, Pisum sativum, Vicia faba.*

Dicotyledonous Hosts *(**Susceptible** or Insusceptible)*

*Antirrhinum majus. Beta vulgaris. Brassica campestris* ssp. *pekinensis. Chenopodium amaranticolor. Chenopodium quinoa. Cucumis sativus. Gomphrena globosa. Lactuca sativa.* ***Lathyrus odoratus****. Lycopersicon esculentum. Medicago sativa. Melilotus alba. Nicotiana clevelandii. Nicotiana glutinosa. Nicotiana tabacum. Petunia hybrida.* ***Phaseolus vulgaris***, or *Phaseolus vulgaris. Physalis floridana.* ***Pisum sativum****. Spinacia oleracea. Tetragonia expansa.* ***Trifolium hybridum. Trifolium incarnatum. Trifolium pratense****. Trifolium repens.* ***Tropaeolum majus. Vicia faba. Vicia sativa. Vicia villosa****. Zinnia elegans.*

Sources of host-range data

Gerhardson (1977).

**Properties of particles in sap**

TIP: 56–60 °C. LIV: 7–14 days. DEP: $\log_{10}$ minus 3. Infectivity of sap not changed by treatment with di–ethyl ether. Leaf sap contains few virions.

**Purification**

Method: Gerhardson (1977).

**Particle morphology**

Virions isometric; not enveloped; 27–28 nm in diameter; rounded in profile; without a conspicuous capsomere arrangement.

**Physical properties**

Four sedimenting components in purified preparations; sedimentation coefficient of the fastest 121 *S*; of the other(s) 101 *S*, or 82 *S*, or 54 *S*.

**Taxonomy and Relationships**

Virus(es) with serologically unrelated virions: red clover mottle, broad bean stain, broad bean true mosaic, broad bean wilt, alsike clover vein mosaic and arabis mosaic viruses.

**References**

Gerhardson, B. (1977). *Phytopath. Z.* **89**: 116.

# Clover wound tumor *phytoreovirus*

**Data collated by L.M. Black, 1981. Revised 1984 and 1986.**

*Synonyms:* wound tumor virus, clover big vein virus, *Aureogenus magnivena* virus, *Trifoliumvirus nervicrassans* virus. *Standard acronym:* WTV. *Strains:* NJ and WT.

First reported in *Melilotus officinalis* clone C10, and the vector *Agallia constricta*; from the U.S.A; by Black (1944).

**Natural host range and symptoms**

Symptoms persist.

*Trifolium incarnatum, Melilotus officinalis* — irregular vein enlargement on the abaxial surface of leaves, depression of veins on the adaxial surface, root tumours at bases of lateral roots.

*Catharanthus roseus* — no symptoms described (Hillman *et al.*, 1991).

**Transmission**

Transmitted by a vector; an insect; *Agalliopsis novella, Agallia constricta, A. quadripunctata*; Cicadellidae. Not transmitted by *Aceratagallia sanguinolenta*. Transmitted in a persistent manner. Virus retained when the vector moults; multiplies in the vector; transmitted congenitally to the progeny of the vector (2% of vector eggs); not transmitted by mechanical inoculation; not transmitted by seed.

**Geographical distribution**

Spreads in the North American region.

**Experimental host range**

Many (>9) families susceptible. Experimentally infected plants mostly show tumours.

Diagnostically susceptible

*Rumex acetosa* — tumours at bases of lateral roots.

*Melilotus officinalis.*

*Trifolium incarnatum* — irregular vein enlargement on abaxial surface of leaf, depression of veins on adaxial surface.

Maintenance and propagation hosts

*Melilotus officinalis, Trifolium incarnatum.*

Assay hosts (**L**ocal lesions or **W**hole plants)

*Trifolium incarnatum* (W), and quantitative assays using vector cell monolayers.

Dicotyledonous Hosts *(**Susceptible** or Insusceptible)*

***Capsella bursa-pastoris*. *Catharanthus roseus*. *Dianthus barbatus*. *Lycopersicon esculentum*.** *Melilotus alba*. ***Nicotiana glutinosa*. *Nicotiana tabacum*. *Rumex acetosa*. *Tetragonia expansa*. *Torenia fournieri*. *Trifolium incarnatum*. *Trifolium pratense*.**

Sources of host-range data

Black (1945), and L.M. Black (pers. comm.).

**Properties of particles in sap**

TIP: 60 °C. LIV: 1 year at 0°C. DEP: $\log_{10}$ minus 6.

**Purification**

Method: Reddy and Black (1973).

**Particle morphology**

Virions isometric; 73 nm in diameter (but when the outer amorphous layer is removed a shell of distinct capsomeres, 59–63 nm in diameter, is revealed, and within that a core without spikes *c*.50 nm in diameter); angular in profile; without a conspicuous capsomere arrangement (unless the outer shell is removed).

**Physical properties**

One sedimenting component in purified preparations; sedimentation coefficient 510 *S* (Kalmakoff *et al.*, 1969). Density 1.39 g $cm^{-3}$ in CsCl.

**Biochemical properties**

Virions contain 22 % nucleic acid; 78 % protein.

*Genome* consists of RNA; double–stranded; of 12 parts (named S1–10); largest (or only) genome part the largest 4.131 kb (pairs); the 2nd largest 3.532 kb (pairs); the 3rd largest 3.205 kb (pairs); the 4th largest 2.565 kb (pairs); the 5th largest 2.613 kb (pairs); the 6th largest and other parts 1.726 kb (pairs; also 1.700, 1.472, 1.182, 1.192, 1.128 and 0.851 kb pairs). Base composition of ds genome 39 % (G + C). Infectivity decreased when deproteinised with proteases (after removal of 2 virion proteins with trypsin or chymotrypsin); lost when deproteinised with phenol or detergent.

*Sequence database accession code(s):*

J03020 Em(40)_vi:WTVDI Gb(84)_vi:WTVDI Wound tumor virus genome segment S5 encoding polypeptide P5, complete cds. 7/89 2,613bp.

J04344 Em(40)_vi:WTVP8SP Gb(84)_vi:WTVP8SP Wound tumor virus genomic segment S8 encoding P8 structural polypeptide, complete cds. 4/90 1

M11133 Em(40)_vi:WTVS12 Gb(84)_vi:WTVS12 Plant wound tumor virus (reovius), segment S12. 7/89 851bp.

M24114 Em(40)_vi:WTVPNS11 Gb(84)_vi:WTVPNS11A Wound Tumor Virus genomic segment S10 encoding Pns11 nonstructural polypeptide, complete cds.

M24115 Em(40)_vi:WTVPNS10 Gb(84)_vi:WTVPNS10A Wound Tumor Virus genomic segment S9 encoding Pns10 nonstructural polypeptide, complete cds.

M24116 Em(40)_vi:WTVPNS7A Gb(84)_vi:WTVPNS7A Wound Tumor Virus genomic segment S6 encoding Pns7 nonstructural polypeptide, complete cds. 7

M24117 Em(40)_vi:WTVPNS4A Gb(84)_vi:WTVPNS4A Wound Tumor Virus genomic segment S4 encoding Pns4 nonstructural polypeptide, complete cds. 7

M77019 Em(40)_vi:WTVNJORF1 Gb(84)_vi:WTVNJORF1 Wound tumor virus ORF1 mRNA, complete cds. 10/91 1,726bp.

M77020 Em(40)_vi:WTVNJORF2 Gb(84)_vi:WTVNJORF2 Wound tumor virus ORF2 mRNA, complete cds. 12/91 1,128bp.

M77021 Em(40)_vi:WTVNJORF3 Gb(84)_vi:WTVNJORF3 Wound tumor virus ORF3 and ORF4 mRNA, complete cds. 12/91 851bp.

X14218 Em(40)_vi:WTVS7P6 Gb(84)_vi:WTVS7P6 Wound tumor virus genomic RNA segment S7 encoding structural protein P6. 9/93 1,726bp

X14219 Em(40)_vi:WTVS11 Gb(84)_vi:WTVS11 Wound tumor virus RNA segment S11 for structural protein P9. 9/93 1,128bp. 12 sequences.

*Virion protein(s)* 7.

**Replication**

Coat protein *m*RNA translated in the cytoplasm only. Replication does not depend on a helper virus.

**Cytopathology**

Virions found in all parts of the host plant; in cytoplasm. Inclusions present in infected cells; are unusual in shape; globular bodies.

**Taxonomy and Relationships**

*Phytoreovirus: Reoviridae* (the type species).

**References**

Bils, R.F. and Hall, C.E. (1962). *Virology* **17**: 123.
Black, L.M. (1944). *Proc. Am. phil. Soc.* **88**: 132.
Black, L.M. (1970). *CMI/AAB Descr. Pl. Viruses* No. 34, 4 pp.
Black, L.M. (1979). *Adv. Virus Res.* **25**: 191.
Black, L.M. (1982). In: *Molecular Biology of Plant Tumors*, p. 69; eds. G. Kahl and J. Schell. Academic Press, New York.
Black, D.R. and Knight, C.A. (1970). *J. Virol.* **6**: 194.
Brakke, M.K., Vatter, A.E. and Black, L.M. (1954). *N.Y. Symp. in Biol.* **6**: 137.
Gomatos, P.J. and Tamm, I. (1963). *Proc. Natl. Acad. Sci. U.S.A.* **50**: 878.
Hillman, B.I., Anzola, J.V., Halpern, B.T., Cavileer, T.D. and Nuss, D.L. (1991). *Virology* **185**: 896.
Kalmakoff, J., Lewandowski, L.J. and Black, D.R. (1969). *J. Virol.* **4**: 851.
Kleinschmidt, A.K., Dunnebacke, T.H., Spendlove, R.S., Schaffer, F.L. and Whitcomb, R.F. (1964). *J. mol. Biol.* **10**: 282.
Maramorosch, K., Shikata, E., Hirumi, H. and Granados, R.R. (1968). *Nat. Can. Inst. Monog.* **31**: 493.
Nuss, D.L. (1984). *Adv. Virus Res.* **29**: 57.
Nuss, D.L. and Peterson, A.J. (1980). *J. Virol.* **34**: 532.
Reddy, D.V.R. and Black, L.M. (1973). *Virology* **54**: 150.
Tomita, K.I. and Rich, A. (1964). *Nature, Lond.* **201**: 1160.

# Clover yellow mosaic *potexvirus*

**Data collated by O.W. Barnett, R.H. Baum, S. Scott, B. Reddick and J.C. Tu, 1981. Revised 1983.**

*Synonyms:* pea mottle virus. *Standard acronym:* ClYMV. *Strains:* broad bean mild mosaic virus (Yu, 1979).

First reported in *Trifolium repens*; from the U.S.A; by Johnson (1942); Pratt (1961).

**Natural host range and symptoms**

Symptoms persist.

*Trifolium* spp., *Pisum sativum, Chenopodium album, Malus pumila, Medicago alba, M. sativa, Vicia sativa, Stellaria media* — mosaic with necrotic streaks.

**Transmission**

Not transmitted by *Acyrthosiphon pisum, Anuraphis bakeri* (Pratt 1961; Purcifull 1968). Virus transmitted by mechanical inoculation.

**Geographical distribution**

Spreads in Canada (in the south-west), the USA (in the north-west).

**Experimental host range**

Several (3–9) families susceptible. Experimentally infected plants mostly show chlorotic or necrotic local lesions, systemic mosaic.

Diagnostically susceptible

*Chenopodium quinoa, Gomphrena globosa* — systemic mosaic.
*Medicago sativa* — local lesions (some strains only).
*Pisum sativum* — necrotic local lesions, systemic mosaic.
*Vigna unguiculata* — local infection.

Maintenance and propagation hosts

*Phaseolus vulgaris, Pisum sativum.*

Assay hosts (**L**ocal lesions or **W**hole plants)

*Gomphrena globosa, Chenopodium quinoa.*

Dicotyledonous Hosts *(**Susceptible** or Insusceptible)*

***Amaranthus caudatus***. *Antirrhinum majus*. ***Arachis hypogaea***. ***Beta vulgaris***. *Capsicum annuum*. *Cassia occidentalis*. ***Catharanthus roseus***. ***Chenopodium album***. ***Chenopodium amaranticolor***. ***Chenopodium quinoa***. ***Cucumis sativus***, or *Cucumis sativus*. ***Cucurbita maxima***. *Cucurbita pepo*. *Datura stramonium*. ***Glycine max***. ***Gomphrena globosa***. ***Helianthus annuus***. ***Lablab purpureus***. ***Lactuca sativa***. ***Lathyrus odoratus***. *Lespedeza stipulacea*. *Lycopersicon esculentum*. ***Medicago sativa***. ***Melilotus alba***. *Nicotiana clevelandii*. *Nicotiana* × *edwardsonii*. ***Nicotiana debneyi***. *Nicotiana glutinosa*. ***Nicotiana megalosiphon***. *Nicotiana rustica*. *Nicotiana tabacum*. *Petunia hybrida*. ***Phaseolus lunatus***. ***Phaseolus vulgaris***. ***Pisum sativum***. *Raphanus sativus*. ***Solanum melongena***. ***Spinacia oleracea***. ***Stellaria media***. ***Tithonia speciosa***. ***Torenia fournieri***. ***Trifolium hybridum***. ***Trifolium incarnatum***. ***Trifolium pratense***. ***Trifolium repens***. ***Vicia faba***. ***Vicia sativa***. ***Vicia villosa***. ***Vigna radiata***. ***Vigna unguiculata***. ***Vigna unguiculata*** ssp. ***sesquipedalis***. *Zinnia elegans*.

Monocotyledonous Hosts *(**Susceptible** or Insusceptible)*

*Lilium formosanum*. *Zea mays*.

Sources of host-range data

Pratt (1961); Hampton *et al.* (1978*a*); Rao *et al.* (1980).

**Properties of particles in sap**

TIP: 60–65 °C. LIV: 188 days (or more). DEP: $\log_{10}$ minus 5–6. Leaf sap contains many virions.

**Purification**

Method: Barnett and Gibson (1975); Rao *et al.* (1980); Welsh *et al.* (1973).

**Particle morphology**

Virions filamentous; not enveloped; usually flexuous; with a clear modal length; of 540 nm; 13 nm wide.

**Physical properties**

One sedimenting component in purified preparations; sedimentation coefficient 132 *S*. Isoelectric point pH 5.3.

**Biochemical properties**

*Genome* consists of RNA; single–stranded; linear; largest (or only) genome part 7.015 kb.

*Sequence database accession code(s):*

D00485 Em(40)_vi:CYMVCP Gb(84)_vi:CYMVCP Clover yellow mosaic potexvirus genomic RNA, 3'-terminal region. 4/90 1,051bp.

D29630 Em(40)_vi:CYMRNA Gb(84)_vi:CYMRNA Clover yellow mosaic potexvirus genomic RNA, complete sequence. 3/94 7,015bp.

M63511 Em(40)_vi:CYMCPFP Gb(84)_vi:CYMCPFP Clover yellow mosaic potexvirus non-structural/coat protein fusion product RNA, complete cds.

M63512 Em(40)_vi:CYMCPFPD Gb(84)_vi:CYMCPFPD Clover yellow mosaic potexvirus non-structural/coat protein fusion product RNA, complete cds.

M63513 Em(40)_vi:CYMCPFPG Gb(84)_vi:CYMCPFPG Clover yellow mosaic potexvirus non-structural/coat protein fusion product RNA, complete cds.

M63514 Em(40)_vi:CYMCPFPS Gb(84)_vi:CYMCPFPS Clover yellow mosaic potexvirus non-structural/coat protein fusion product RNA, complete cds. 6 sequences.

**Cytopathology**

Virions found in mesophyll and parenchyma cells; in cytoplasm. Inclusions present in infected cells; are unusual in shape; banded and of complex structures; they contain virions.

**Taxonomy and Relationships**

*Potexvirus*.

Virus(es) with serologically related virions: white clover mosaic, cactus X and hydrangea ringspot viruses, but distantly.

Best tests for diagnosis: often found with white clover mosaic potexvirus (WCMV) but causes systemic mosaic in *Chenopodium quinoa*, which WCMV infects only locally. By contrast WCMV infects *Vigna unguiculata* systemically, but CYMV only locally.

**References**

Agrawal, H., Bos, L. and Chessin, M. (1962). *Phytopathology* **62**: 517.

Barnett, O.W. and Gibson, P.B. (1975). *Crop Sci.* **15**: 32.

Bos, L. (1973). *CMI/AAB. Descr. Pl. Viruses* No. 111, 4 pp.

Ford, R.E. (1973). *Phytopathology* **63**: 926.

Hampton, R., Beczner, L., Hagedorn, D., Bos, L., Inouye, T., Barnett, O., Musil, M. and Meiners, J. (1978). *Phytopathology* **68**: 989.

Johnson, F. (1942). *Phytopathology* **32**: 103.

Pratt, M.J. (1961). *Can. J. Bot.* **39**: 655.

Pratt, M.J. (1967). *Can. J. Pl. Sci.* **47**: 289.

Pratt, M.J. and Reichmann, M.E. (1961). *Proc. Can. Phytop. Soc.* **28**: 13.

Purcifull, D.E. (1968). *Pl. Dis. Reptr.* **52**: 759.

Purcifull, D.E. and Shepherd, R.J. (1964). *Phytopathology* **54**: 1102.

Rao, D.V., Hiruki, C. and Matsumoto, T. (1980). *Phytopath. Z.* **98**: 260.

Sit, T.L., White, K.A., Holy, S., Padmanabhan, U., Eweida, M., Heibert, M., Mackie, G.A. and Abouhaidar, M. *J. gen. Virol.* **71**: 1990.

Welsh, M.F., Stace-Smith, R. and Brennan, E. (1973). *Phytopathology* **63**: 50.

Yu, Y. (1979). In: *Vicia Faba Diseases*, Scientific Press, Peking (in Chinese).

# Clover yellow vein *potyvirus*

**Data collated by O.W. Barnett, R.H. Baum, S. Scott, B. Reddick, 1981. Revised 1987; and on the broad bean mild mosaic strain by Tien Po, 1983.**

*Synonyms:* dendrobium mosaic virus, pea mottle virus (Pratt, 1961; Johnson, 1942), pea necrosis virus (Bos *et al.*, 1977), probably alsike clover mosaic virus, pea western ringspot virus. *Standard acronym:* ClYVV. *Strains:* broad bean mild mosaic (Yu, 1939; 1979), CYVV Hollings and CYVV-Pratt are 'typical' isolates. CYVV-PQ and other Australian isolates are distantly related to the type isolates.

First reported in *Trifolium repens*; from the U.K; by Hollings and Nariani (1965).

**Natural host range and symptoms**

Symptoms vary seasonally.

*Coriandrum sativum, Daucus carota* (wild and domestic), *Pisum sativum, Trifolium hybridum, T. incarnatum, T. pratense, T. repens, T. subterraneum, T. vesiculosum* — mosaics, mottles or streaks, vein yellowing or netting.

*Limonium sinnatum* — (Lawson *et al.*, 1985).

*Lupinus luteus, Lupinus* sp. — (Abu-Samah *et al.*, 1983).

*Glycine max* — (Jones *et al.*, 1977).

**Transmission**

Transmitted by a vector; an insect; *Acyrthosiphon (Aulacorthum) solani, Macrosiphum euphorbiae, Myzus persicae* (Singh *et al.*, 1971); Aphididae. Transmitted in a non–persistent manner. Virus transmitted by mechanical inoculation.

**Geographical distribution**

Probably distributed worldwide (wherever white clover occurs).

**Experimental host range**

Several (3–9) families susceptible. Experimentally infected plants mostly show chlorotic or necrotic local lesions systemic mosaic or necrosis.

Diagnostically susceptible

CYVV-Pratt strain):

*Chenopodium quinoa* — chlorotic or necrotic local lesions; systemic mosaic and necrosis.

*Phaseolus vulgaris* — chlorotic local lesions; systemic mosaic.

*Nicotiana clevelandii* — chlorotic or necrotic local lesions; systemic mosaic and necrosis.

*N. tabacum* — necrotic ringspots; not systemic.

*Trifolium repens* — mosaic and chlorotic banding or symptomless.

Diagnostically insusceptible

*Cucumis sativus*, but susceptible to pea necrosis strain.

Maintenance and propagation hosts

*Pisum sativum, Nicotiana clevelandii.*

Assay hosts (**L**ocal lesions or **W**hole plants)

*Chenopodium amaranticolor, C. quinoa, Nicotiana clevelandii.*

Dicotyledonous Hosts *(**Susceptible** or Insusceptible)*

***Amaranthus caudatus*. *Antirrhinum majus*.** *Apium graveolens*. ***Arachis hypogaea*. *Atriplex hortensis*. *Beta vulgaris*.** *Capsicum annuum*. ***Cassia occidentalis*.** *Catharanthus* roseus. ***Celosia argentea***. ***Chenopodium album***. ***Chenopodium amaranticolor***. ***Chenopodium quinoa***. ***Coriandrum sativum***. ***Crotalaria spectabilis***. ***Cucumis sativus***, or *Cucumis sativus*. *Cucurbita maxima*. ***Cucurbita pepo***, or *Cucurbita pepo*. *Datura metel*. *Datura stramonium*. ***Daucus carota***. ***Glycine max***. ***Gomphrena globosa***. ***Helianthus annuus***. *Lactuca sativa*. ***Lathyrus odoratus***. *Lespedeza stipulacea*. *Lycopersicon esculentum*. ***Medicago sativa***. ***Melilotus alba***. ***Nicandra physaloides***. ***Nicotiana clevelandii***. ***Nicotiana × edwardsonii***. ***Nicotiana debneyi***. *Nicotiana glutinosa*. ***Nicotiana megalosiphon***. ***Nicotiana occidentalis***. *Nicotiana rustica*. ***Nicotiana tabacum***. *Ocimum basilicum*. *Pastinaca sativa*. *Petroselinum crispum*. *Petunia hybrida*. *Phaseolus lunatus*. ***Phaseolus vulgaris***. *Phlox drummondii*. ***Pisum sativum***. *Solanum melongena*. *Solanum tuberosum*. ***Spinacia oleracea***. ***Tetragonia expansa***. *Tithonia speciosa*. ***Torenia fournieri***. ***Trifolium hybridum***. ***Trifolium incarnatum***. ***Trifolium pratense***. ***Trifolium repens***. ***Trifolium subterraneum***. ***Vicia faba***. ***Vicia villosa***. ***Vigna unguiculata***. *Vigna unguiculata* ssp. *sesquipedalis*. ***Zinnia elegans***.

Monocotyledonous Hosts (***Susceptible*** *or Insusceptible*)

*Zea mays*.

Sources of host-range data

Singh *et al.* (1971); Edwardson and Christie (1986); Bos *et al.* (1974); Lindsten *et al.* (1976); Hampton *et al.* (1978); Howell *et al.* (1981); Munro (1981).

**Properties of particles in sap**

TIP: 65 °C. LIV: 2–8 days. DEP: $\log_{10}$ minus 3–5. Leaf sap contains few virions. Electron microscopy: different lengths are obtained in different media.

**Purification**

Method: Reddick and Barnett (1983), for an example which uses, PEG, Triton X-100 and $Cs_2SO_4$ gradients; complete rapidly to avoid coat protein degradation.

**Particle morphology**

Virions filamentous; not enveloped; usually flexuous; with a clear modal length; of 760 nm (divalent cations may increase lengths); 12–15 nm wide. Axial canal obscure. Basic helix obvious (in uranyl formate).

**Physical properties**

One sedimenting component in purified preparations; sedimentation coefficient 159.5 *S*.

**Biochemical properties**

Virions contain 5 % nucleic acid; 95 % protein; 0 % lipid.

*Genome* consists of RNA; single–stranded; linear; unipartite; largest (or only) genome part 9.5 kb. Poly A region present.

*Sequence database accession code(s):*

D00605 Em(40)_vi:CYVCAP Gb(84)_vi:CYVCAP Clover yellow vein virus capsid protein gene, complete cds. 3/91 1,357bp.

S77521 Em(40)_vi:S77521 Gb(84)_vi:S77521 viral coat protein (3' region) clover yellow vein virus CYVV, strain B, Genomic RNA, 1018 nt 2 sequences.

*Virus–coded non–virion proteins* isolated (Nagel *et al.* (1983)); two proteins found. $M_r$ of the largest 59000. $M_r$ of 2nd largest 49000.

**Cytopathology**

Virions found in all parts of the host plant; in cytoplasm. Inclusions present in infected cells; are unusual in shape; small crystals in close association with the nucleus, or large and irregular shaped in the nucleus, and cylindrical in the cytoplasm consisting of pinwheels, bundles and laminated aggregates (Edwardson's Group II); they do not contain virions.

**Taxonomy and Relationships**

*Potyvirus: Potyviridae.*

Virus(es) with serologically related virions: bean yellow mosaic, pea mosaic, blackeye cowpea mosaic and soybean mosaic viruses.

The broad bean mild mosaic strain in China gives mottled elongated leaves in *Vicia faba* (Yu, 1979), and is transmitted in the non-persistent manner by *Aphis rumicis* and *Macrosiphum pisum*, unlike the type strain it infects *Trifolium pratense* and *T. repens*. With random primed *c*DNA and S1 nuclease, several CYVV strains show close homology but no hybridization was detected between CYVV strains and bean yellow mosaic, BYMV (PMV) 204–1, or sweet pea mosaic virus. Best tests for diagnosis: often confused with bean yellow mosaic potyvirus especially in white clover.

**Comments**

Pea western ringspot virus is probably a mixed infection with cucumber mosaic virus; R.O. Hampton, personal communication).

**References**


Abu-Samah, N. and Randles, J.W. (1983). *Ann. appl. Biol.* **103**: 97.

Barnett, O.W., Randles, J.W. and Burrows, P.M. (1987). *Phytopathology* **77**: 791.

Bos, L., Lindsten, K. and Maat, D.Z. (1977). *Neth. J. Pl. Path.* **83**: 97.

Edwardson, J.R. (1974). *Fla Agric. Exp. Stn Monog.* No. 5.

Edwardson, J.R. and Christie, R.G. (1986). *Fla Agric. Exp. Stn Monog.* No. 14, p. 410.

Gibbs, A.J., Hecht Poinar, E., Woods, R.D. and McKee, R.K. (1966). *J. gen. Microbiol.* **44**: 177.

Hampton, R., Beczner, L., Hagedorn, D., Bos, L., Inouye, T., Barnett, O., Musil, M. and Meiners, J. (1978). *Phytopathology* **68**: 989.

Hollings, M. and Nariani, T.K. (1965). *Ann. appl. Biol.* **56**: 99.

Hollings, M. (1965). *Ann. appl. Biol.* **55**: 447.

Hollings, M. and Stone, O.M. (1974). *Rep. Glasshouse Crops Res. Inst.*

Hollings, M. and Stone, O.M. (1974). *CMI/AAB Descr. Pl. Viruses* No. 131, 4 pp.

Horváth, J. (1979). *Acta Phytopath. Acad. Sci. Hung.* **14**: 157.

Howell, W.E. and Mink, G.I. (1981). *Pl. Dis. Reptr* **65**: 277.

Johnson, F. (1942). *Phytopathology* **32**: 103.

Jones, R.T. and Diachun, S. (1977). *Phytopathology* **67**: 831.

Jones, R.T. and Gordon, D.T. (1974). *Proc. Phytop. Soc.* 1–116–107.

Jones, R.T. and Diachun, S. (1976). *Pl. Dis. Reptr* **60**: 690.

Lawson, R.H., Brannigan, M.D. and Foster, J. (1985). *Phytopathology* **75**: 899.

Lindsten, K.S., Brishammer, S. and Tomenius, K. (1976). *Meddn. St. VäxtskAnst.* **16**: 289.

Munro, D. (1981). *Aust. Pl. Path. Soc. Newsl.* **10**: 61.

Nagel, J., Zettler, F.W., and Hiebert, E. (1983). *Phytopathology* **73**: 449.

Pratt, M. (1961). *Can. J. Bot.* **39**: 655.

Singh, R.P. and Lopez-Abella, D. (1971). *Phytopathology* **61**: 333.

Varma, A., Gibbs, A.J., Woods, R.D. and Finch, J.T. (1968). *J. gen. Virol.* **2**: 107.

Yu, T.F. (1939). *Phytopathology* **29**: 448.

Yu, T.F. (1979). *Vicia faba Diseases* Scientific Press, Peking. (in Chinese).

# Clover yellows *closterovirus*

**Data collated by S.T. Ohki, 1981. Revised 1984.**

*Standard acronym:* CYV.

First reported in *Trifolium repens*; from Tokyo, Japan; by Ohki *et al.*, 1976.

## Natural host range and symptoms

Symptoms persist.
*Trifolium repens, T. incarnatum* — yellowing or reddening.

## Transmission

Transmitted by a vector; an insect; *Aphis craccivora*; Aphididae. Transmitted in a semi–persistent manner. Virus not transmitted by mechanical inoculation; not transmitted by contact between plants.

## Geographical distribution

Spreads in Japan.

## Experimental host range

Few (<3) families susceptible. Experimentally infected plants mostly show yellowing, reddening.

Diagnostically susceptible

*Pisum sativum, Vicia faba* — systemic yellowing.

Maintenance and propagation hosts

*Pisum sativum, Vicia faba.*

Assay hosts (**L**ocal lesions or **W**hole plants)

*Pisum sativum, Vicia faba.*

Dicotyledonous Hosts *(**Susceptible** or Insusceptible)*

*Beta vulgaris. Chenopodium amaranticolor. Gomphrena globosa. Nicotiana glutinosa. Nicotiana tabacum.* ***Pisum sativum. Trifolium incarnatum. Trifolium repens.*** *Vicia faba. Vigna unguiculata.*

Sources of host-range data

Ohki *et al.* (1976).

## Properties of particles in sap

Leaf sap contains few virions.

## Particle morphology

Virions filamentous; not enveloped; usually flexuous; with a clear modal length; of 1700 nm; 12 nm wide. Axial canal obscure. Basic helix obvious; pitch of basic helix 3.7 nm.

## Cytopathology

Virions found in phloem; in cytoplasm. Inclusions present in infected cells; are unusual in shape; banded bodies and containing vesicles fibril; they contain virions. Other cellular changes: necrosis and vesicles in phloem cells.

## Taxonomy and Relationships

*Closterovirus.*

## References

Ohki, S.T., Doi, Y. and Yora, K. (1976). *Ann. Phytopath. Soc. Japan* **42**: 313.

# Cocksfoot *alphacryptovirus*

**Data collated by A.A. Brunt, 1994.**

First reported in *Dactylis glomerata*; from Scotland, U.K.; by Torrance *et al.* (1994).

**Natural host range and symptoms**

Symptoms none.

*Dactylis glomerata* — symptomless.

**Transmission**

Transmitted by means not involving a vector. Virus not transmitted by mechanical inoculation; not transmitted by grafting; not transmitted by contact between plants; transmitted by seed.

**Geographical distribution**

Found, but with no evidence of spread, in U.K.

**Experimental host range**

Few (<3) families susceptible.

Maintenance and propagation hosts

*Dactylis glomerata.*

Monocotyledonous Hosts *(**Susceptible** or Insusceptible)*

***Dactylis glomerata.***

**Properties of particles in sap**

Leaf sap contains few virions.

**Purification**

Method: Torrance *et al.* (1994).

**Particle morphology**

Virions isometric; 30 nm in diameter; rounded in profile.

**Biochemical properties**

*Genome* consists of RNA; double–stranded; of two parts; largest (or only) genome part 2 kb (pairs); the 2nd largest 1.85 kb (pairs).

**Taxonomy and Relationships**

Possible *Alphacryptovirus: Partitiviridae.*

Virus(es) with serologically related virions: ryegrass, beet 1 and 2 alphacryptoviruses. Virus(es) with serologically unrelated virions: carnation and vicia alphacryptoviruses.

**References**

Torrance, L., Jones, A.T. and Duncan, G.H. (1994). *Ann. appl. Biol.* **124**: 579.

# Cocksfoot mild mosaic (?) *sobemovirus*

**Data collated by W. Huth, 1991.**

*Standard acronym:* CMMV. *Strains:* brome stem leaf mottle virus (Catherall and Chamberlain, 1975 and 1977; Paul *et al.*, 1980), cocksfoot necrosis and mosaic virus (Hariri and Lapierre, 1977), festuca mottle virus (Catherall and Chamberlain, 1977; Gibson and Boyes, 1974; Paul *et al.*, 1980), holcus transitory mottle virus (Catherall and

Chamberlain, 1975; Paul *et al.*, 1980), phleum mottle virus (Catherall, 1966; Catherall and Chamberlain, 1975; Paul *et al.*, 1973/74; Paul *et al.*, 1980).

First reported in *Phleum pratense, Dactylis glomerata*; from the U.K. and Germany; by Carr (1966); Huth (1968).

**Natural host range and symptoms**

Symptoms persist and vary seasonally. Symptoms light green streaks.

*Agrostis stolonifera, Bromus mollis, Dactylis glomerata, Festuca pratensis, Holcus lanatus, Phleum pratense, Poa trivialis* — faint leaf chlorosis.

**Transmission**

Transmitted by Aphididae and Coleoptera (inefficiently, and probably non-specifically). Virus transmitted by mechanical inoculation; not transmitted by contact between plants; not transmitted by seed; not transmitted by pollen.

**Geographical distribution**

Spreads in Canada, the former Czechoslovakia, Denmark, France, Germany, Norway, and the UK.

**Experimental host range**

Few (<3) families susceptible. Experimentally infected plants mostly show light green streaks.

Diagnostically susceptible

*Hordeum vulgare* — light green streaks.

*Avena sativa, Dactylis glomerata, Festuca pratense, Secale cereale, Setaria italica* — diffuse systemic mild mosaics.

Diagnostically insusceptible

*Agropyron repens, Anthoxanthum odoratum, Lolium perenne.*

Maintenance and propagation hosts

*Setaria italica.*

Assay hosts (**L**ocal lesions or **W**hole plants)

*Dactylis glomerata, Festuca pratense, Hordeum vulgare, Secale cereale, Setaria italica.*

Monocotyledonous Hosts *(**Susceptible** or Insusceptible)*

*Agropyron intermedium. Agropyron repens.* ***Avena sativa. Avena strigosa. Bromus inermis. Bromus mollis. Bromus racemosus. Bromus secalinus. Bromus tectorum. Dactylis glomerata. Festuca pratensis. Hordeum vulgare. Lamarkia aurea. Lolium multiflorum.*** *Lolium perenne.* ***Lolium temulentum. Lolium persicum. Panicum miliaceum. Paspalum membranaceum. Phalaris arundinacea. Phleum arenarium.*** *Poa annua.* ***Poa compressa. Poa trivialis. Secale cereale. Setaria italica. Setaria viridis. Triticum aestivum. Zea mays.***

**Properties of particles in sap**

TIP: 80–85 °C. LIV: 3 days. DEP: $log_{10}$ minus 5. Leaf sap contains many virions.

**Purification**

Method: Huth *et al.* (1970).

**Particle morphology**

Virions isometric; 28 nm in diameter; angular in profile.

**Physical properties**

One sedimenting component in purified preparations; sedimentation coefficient 105 *S* (±1). $A_{260}/A_{280}$ ratio 1.65 (±0.02).

**Biochemical properties**

Virions contain 23.5 % nucleic acid; 76.5 % protein.

*Genome* consists of RNA; single–stranded; linear; unipartite (sometimes also a minor component of 1.5 kb); largest (or only) genome part 5.7 kb. Base composition 26.6 % G; 23.4 % A; 28.4 % C; 21.6 % U.

*Virion protein(s)* one; $M_r$ 24900; coat protein.

**Cytopathology**

Inclusions absent from infected cells.

**Taxonomy and Relationships**

Possible *Sobemovirus*.

The virus is distantly serologically related to molinia streak virus (Querfurth and Bercks, 1976), which is now considered to be a strain of panicum mosaic sobemovirus (Paul *et al.*, 1980). There have also been unconfirmed and disputed reports of serological cross-reactions with virions of tobacco mosaic virus and some viruses of the turnip yellow mosaic virus group. Best tests for diagnosis: cocksfoot mottle and cocksfoot streak viruses are often found in mixed infections in *Dactylis glomerata* and the quickest diagnostic method is by serological testing.

**References**

A'Brook, J. and Benigno, D.A. (1972). *Ann. appl. Biol.* **72**: 169.
Benigno, D.A. (1972). *Ann. appl. Biol.* **72**: 43.
Benigno, D.A. and A.'Brook, J. (1972). *Pl. Path.* **21**: 142.
Bercks, R. and Querfurth, G. (1971). *Phytopath. Z.* **72**: 354.
Bercks, R. and Querfurth, G. (1972). *Phytopath. Z.* **75**: 215.
Carr, A.J.H. (1966). *Rep. Welsh Pl. Breed. Stn* 1965, p. 90.
Carr, A.J.H. (1967). *Rep. Welsh Pl. Breed. Stn* 1966, p. 110.
Catherall, P.L. (1966). *Rep. Welsh Pl. Breeding Stn.* 1965, p. 94.
Catherall, P.L. (1982). *Rep. Welsh Pl. Breed. Stn.* 1981, p. 161.
Catherall, P.L. and Chamberlain, J.A. (1975). *Pl. Path.* **24**: 217.
Catherall, P.L. and Chamberlain, J.A. (1977). *Ann. appl. Biol.* **87**: 147.
Chamberlain, J.A. and Catherall, P.L. (1976). *J. gen. Virol.* **30**: 41.
Chamberlain, J.A. and Catherall, P.L. (1976). *Ann. appl. Biol.* **83**: 475.
Gibson, R.W. and Boyes, D. (1974). *Rep. Rothamsted Exp. Stn.* 1973, p. 137.
Hariri, D. and Lapierre, H. (1977). *Annls. Phytopath.* **9**: 281.
Hariri, D. and Lapierre, H. (1979). *Annls. Phytopath.* **11**: 246.
Hull, R. (1977). *J. gen. Virol.* **36**: 289.
Huth, W. (1968). *Phytopath. Z.* **62**: 300.
Huth, W. and Paul, H.L. (1972). *CMI/AAB Descr. Pl. Viruses* No. 107, 4 pp.
Huth, W., Brandes, J. and Paul, H.L. (1970). *Phytopath. Z.* **68**: 367.
Paul, H.L. (1974). *Acta Biol. Jugosl. Ser. B. Microb.* **11**: 193.
Paul, H.L. and Huth, W. (1970). *Phytopath. Z.* **69**: 1.
Paul, H.L., Huth, W. and Querfurth, G. (1973/74). *Intervirology* **2**: 253.
Paul, H.L., Querfurth, G. and Huth, W. (1980). *J. gen. Virol.* **47**: 67.
Querfurth, G. and Bercks, R. (1976). *Phytopath. Z.* **85**: 193.

# Cocksfoot mottle *sobemovirus*

**Data collated by P. Jeyanandarajah, 1991.**

*Synonyms:* cocksfoot necrotic mosaic virus. *Standard acronym:* CoMV.

First reported in *Dactylis glomerata*; from the U.K; by Serjeant (1963).

**Natural host range and symptoms**

Symptoms persist. Symptoms streaking and mottling of leaves.

*Dactylis glomerata, Triticum aestivum* — conspicuous yellow streaking and mottling of leaves.

**Transmission**

Transmitted by a vector; an insect; Chrysomelid beetles, *Lema melanopa, L. lichenis*; Coleoptera. Transmitted in a semi–persistent manner. Virus transmitted by mechanical inoculation; not transmitted by seed.

**Geographical distribution**

Spreads in France, Germany, Japan, New Zealand, and the UK.

**Experimental host range**

Few (<3) families susceptible. Experimentally infected plants mostly show yellow streaking and mottling with necrosis.

Diagnostically susceptible

*Avena sativa, Dactylis glomerata, Hordeum vulgare, Triticum aestivum* — systemic leaf streaking, mottling and necrosis.

Diagnostically insusceptible

*Musa balbisiana, Oryza sativa, Poa pratensis, Sorghum halepense, Zea mays.*

Maintenance and propagation hosts

*Dactylis glomerata.*

Dicotyledonous Hosts *(**Susceptible** or Insusceptible)*

*Chenopodium amaranticolor. Chenopodium hybridum. Chenopodium quinoa. Cucumis sativus. Nicotiana clevelandii. Nicotiana glutinosa. Nicotiana tabacum. Petunia hybrida. Phaseolus vulgaris. Vicia faba.*

Monocotyledonous Hosts *(**Susceptible** or Insusceptible)*

*Agropyron repens. Anthoxanthum odoratum.* ***Avena sativa****. Bromus mollis.* ***Dactylis glomerata****.* ***Hordeum vulgare****. Lilium formosanum. Lolium perenne. Oryza sativa. Poa pratensis. Poa trivialis.* ***Triticum aestivum****. Zea mays.*

**Properties of particles in sap**

TIP: 65 °C. LIV: 4–6 days (at 20°C). DEP: $\log_{10}$ minus 3.

**Particle morphology**

Virions isometric; 30 nm in diameter; angular in profile.

**Physical properties**

Sedimentation coefficient 118 *S*. Density 1.39 g $cm^{-3}$ in CsCl. $A_{260}/A_{280}$ ratio 1.6.

**Biochemical properties**

Virions contain 25 % nucleic acid; 75 % protein.

*Genome* consists of RNA; single–stranded; unipartite; largest (or only) genome part 4.35 kb.

*Sequence database accession code(s):*
Z36903 Gb(89)n:Cmvrep Cocksfoot mottle virus gene for putative replicase. 6/95 2,838bp.
*Virion protein(s)* one; $M_r$ 31000; coat protein.

### Cytopathology

Virions found in all parts of the host plant.

### Taxonomy and Relationships

*Sobemovirus.*

Virus(es) with serologically related virions: cynosurus mottle virus (New Zealand isolate). Virus(es) with serologically unrelated virions: cynosurus mottle virus (European isolate).

The genome of a New Zealand isolate is reported to have 5–8% nucleotide sequence homology with that of cynosurus mottle virus.

### Comments

Although the virus alone will not infect *Cynosurus cristatus*, it induces systemic infection when present together with cynosurus mottle (?) *sobemovirus.*

### References


Catherall, P.L. (1970). *CMI/AAB Descr. Pl. Viruses* No. 23, 3 pp.
Hariri, D. and Lapierre, H. (1978). *C.N.R.A., I.N.R.A.*, Versailles, France, p. 281.
Hull, R. (1977*a*).*J. gen. Virol.* **36**: 289.
Hull, R. (1977*b*). *Virology* **79**: 50.
Hull, R. (1977*c*). *Virology* **79**: 58.
Huth, W. (1968). *Phytopath. Z.* **62**: 300.
Mohamed, N.A. (1980). *J. agric. Res.* **23**: 273.
Mohamed, N.A. and Mossop, D.W. (1981). *J. gen. Virol.* **55**: 63.
Rabenstein, F. and Schmidt, H.B. (1979). *Arch. Phytopath. Pflanzenschutz* **15**: 351.
Serjeant, E.P. (1963). *Rep. Rothamsted Exp. Stn.* 1962, p. 112.
Serjeant, E.P. (1964). *Pl. Path.* **13**: 23.
Serjeant, E.P. (1967). *Ann. appl. Biol.* **59**: 31.
Taliansky, M.E., Malyshenko, S.I., Pshannikova, E.S. and Atabekov, J.G. (1982). *Virology* **122**: 327.
Toriyama, S. (1982). *Ann. Phytopath. Soc. Japan* **48**: 514.


## Cocksfoot streak (?) potyvirus

**Data collated by A.A. Brunt, 1991.**

*Synonyms:* orchard grass mosaic virus (McKinney, 1956), cocksfoot mosaic virus (Slykhuis, 1957). *Standard acronym:* CSV.

First reported in *Dactylis glomerata*; from England; by Smith (1952); Storey (1952).

### Natural host range and symptoms

Symptoms persist.
*Dactylis glomerata* — chlorotic streaks on leaves; decreased tillering.

### Transmission

Transmitted by a vector; an insect; *Hyalopteroides humulus, Macrosiphum avenae, M. euphorbiae, Acyrthosiphon (Metopolophium) dirhodum, Myzus persicae*; Aphididae.

Transmitted in a non–persistent manner. Virus transmitted by mechanical inoculation; not transmitted by seed.

**Geographical distribution**

Spreads in the Eurasian region; Canada, the UK, and the USA.

**Experimental host range**

Few (<3) families susceptible.

Diagnostically susceptible

*Dactylis glomerata* — chlorotic streaks in leaves; tillering and seed set reduced.
*Lolium multiflorum, L. perenne* — pale green leaf streaking.
*Paspalum membranaceum, Setaria macrostachia* — conspicuous yellow leaf streaking.
*Phalaris paradoxa, Lamarkia aurea* — yellow leaf spotting.

Maintenance and propagation hosts

*Dactylis glomerata.*

Assay hosts (**L**ocal lesions or **W**hole plants)

*Dactylis glomerata* (W), *Paspalum membranaceum* (W).

Monocotyledonous Hosts *(**Susceptible** or Insusceptible)*

***Anthoxanthum cristatum***. *Avena sativa*. ***Avena strigosa***. ***Bromus mollis***. ***Dactylis glomerata***. ***Lagurus ovatus***. ***Lamarkia aurea***. ***Lolium multiflorum***. ***Lolium perenne***. ***Paspalum membranaceum***. ***Phalaris paradoxa***. ***Setaria macrostachia***. ***Setaria viridis***.

Sources of host-range data

Ohmann-Kreutzberg (1963).

**Properties of particles in sap**

TIP: 55 °C. LIV: 16 days. DEP: $\log_{10}$ minus 3. Leaf sap contains few virions.

**Particle morphology**

Virions filamentous; not enveloped; usually flexuous; with a clear modal length; of 750 nm; 13 nm wide. Axial canal obscure. Basic helix obscure.

**Taxonomy and Relationships**

Unassigned potyvirus.
Virus(es) with serologically unrelated virions: anthoxanthum mosaic potyvirus.

**References**

Brandes, J. (1959). *Phytopath. Z.* **35**: 205.
Catherall, P.L. (1971). *CMI/AAB Descr. Pl. Viruses* No. 59, 3 pp.
Catherall, P.L. and Griffiths, E. (1966). *Ann. appl. Biol.* **57**: 141.
McKinney, H.H. (1956). *Pl. Dis. Reptr* **40**: 524.
Ohmann-Kreutzberg, G. (1963). *Phytopath. Z.* **47**: 113.
Slykhuis, J. (1957). *Proc. Canad. phytopath. Soc.* **25**: 17.
Slykhuis, J. (1958). *FAO Pl. Prot. Bull.* **6**: 129.
Smith, K.M. (1952). *Pl. Path.* **1**: 118.
Storey, I.F. (1952). *Pl. Path.* **1**: 101.
Watson, M.A. and Mulligan, T.E. (1960). *Rep. Rothamsted Exp. Stn for 1959*, p. 101.

# Coconut foliar decay virus nanavirus

**Data collated by J.W. Randles, 1989.**

*Standard acronym:* CFDV.

First reported in *Cocos nucifera*; from Saraoutou, Vanuatu; by Calvez *et al.* (1980).

**Natural host range and symptoms**

*Cocos nucifera* — necrosis.

**Transmission**

Transmitted by a vector; an insect; Cixiidae: *Myndus crudus*; an unusual insect group. Transmitted in a semi–persistent manner, or in a persistent manner. Virus not transmitted by mechanical inoculation; not transmitted by grafting (coconuts will not graft).

**Geographical distribution**

Spreads in Vanuatu.

**Experimental host range**

Experimentally infected plants mostly show progressive yellowing and necrosis.

Diagnostically susceptible

*Cocos nucifera.*

Maintenance and propagation hosts

*Cocos nucifera.*

Assay hosts (**L**ocal lesions or **W**hole plants)

*Cocos nucifera* cv. Malayan Red Dwarf (W).

Monocotyledonous Hosts *(**Susceptible** or Insusceptible)*

***Cocos nucifera***.

**Properties of particles in sap**

Electron microscopy: virions not seen in sap extracts.

**Purification**

Method: Randles and Hanold (1989).

**Particle morphology**

Virions isometric; 20 nm in diameter; angular in profile; without a conspicuous capsomere arrangement.

**Physical properties**

Density 1.21–1.25 g/ml in Nycodenz; 1.30 g/ml in $CS_2SO_4$.

**Biochemical properties**

*Genome* consists of DNA; single–stranded; circular; possibly unipartite; largest (or only) genome part 1.29 kb.

*Sequence database accession code(s):*

M29963 Em(40)_vi:CFDCG Gb(84)_vi:CFDCG Coconut foliar decay virus, complete genome. 7/90 1,291bp. 1 sequence.

**Cytopathology**

Virions found in leaves, roots and meristem.

**Taxonomy and Relationships**

Nanavirus.

Resembles subterranean clover stunt, banana bunchy top, milk vetch dwarf and faba bean necrotic yellows viruses, and also porcine circovirus, in the size of its circular single stranded DNA genome; but the latter does not hybridize with that of subterranean clover stunt virus in dot blot hybridization tests. Best tests for diagnosis: isolate and characterize DNA.

**References**

Calvez, C., Renard, J.L. and Marty, G. (1980). *Oleagineux* **35**: 443.
Randles, J.W. and Hanold, D. (1989). *Intervirology* **30**: 177.
Rohde, W., Randles, J.W. and Langridge, P. (1990). *Virology* **176**: 648.

# Coffee ringspot (?) *nucleorhabdovirus*

**Data collated by C.M. Chagas, 1985.**

*Standard acronym:* CoRSV. *Strains:* Coffee ringspot from the Philippines is reported to be seed-borne, that from Brazil is not.

First reported in *Coffea arabica*; from Brazil; by Kitajima and Costa (1972).

**Natural host range and symptoms**

Symptoms persist (while viruliferous mites present).
*Coffea arabica* — chlorotic lesions.

**Transmission**

Transmitted by a vector; a mite; Tenuipalpidae: *Brevipalpus phoenicis*. Virus transmitted by mechanical inoculation; not transmitted by seed.

**Geographical distribution**

Spreads in Brazil and the Philippines.

**Experimental host range**

Few (<3) families susceptible. Experimentally infected plants mostly show chlorotic or necrotic local lesions.

Diagnostically susceptible

*Chenopodium amaranticolor, C. quinoa* — necrotic local lesions.
*Coffea arabica* — chlorotic spots on leaves about 40 days after infestation with viruliferous mites.

Diagnostically insusceptible

*Gomphrena globosa, Cucumis sativus, Phaseolus vulgaris, Nicotiana glutinosa, N. tabacum.*

Dicotyledonous Hosts *(**Susceptible** or Insusceptible)*

*Capsicum annuum. Cassia occidentalis. Catharanthus roseus.* ***Chenopodium amaranticolor***. ***Chenopodium quinoa***. *Cucumis sativus. Cucurbita pepo. Datura stramonium. Gomphrena globosa. Lycopersicon esculentum. Nicandra physaloides. Nicotiana glutinosa. Nicotiana tabacum. Phaseolus vulgaris. Pisum sativum. Solanum tuberosum. Tetragonia expansa. Tropaeolum majus. Zinnia elegans.*

Monocotyledonous Hosts *(**Susceptible** or Insusceptible)*

*Zea mays.*

Additional hosts

*Diodia teres, Genipa americana, Hamelia patens*, and *Menetia cordifolia*.

**Properties of particles in sap**

TIP: 70–75 °C. LIV: 19–22 days at 4 °C. DEP: $\log_{10}$ minus 4–5.

**Particle morphology**

Virions rhabdo– or bullet–shaped; enveloped; with a clear modal length; of 178–224 nm (in thin sections); 59–76 nm wide (in thin sections). Axial canal obvious; 8 nm in diameter.

**Cytopathology**

Virions found in lesions of leaves and in fruit pulp; in nuclei and in the perinuclear space.

**Taxonomy and Relationships**

Possible *Nucleorhabdovirus: Rhabdoviridae*.

**References**

Chagas, C.M. (1978). *Ph.D. Thesis,* Univ. Sao Paolo, Brazil.
Chagas, C.M. (1980). *Phytopath. Z.* **99**: 301.
Chagas, C.M., July, J.R. and Alba, A.P.C. (1981). *Phytopath. Z.* **102**:
Kitajima, E.W: and Costa, A.S. (1972). *Cienc. Cult.* **24**: 542.

# Cole latent (?) *carlavirus*

**Data collated by E.W. Kitajima, 1991.**

*Standard acronym:* CoLV.

First reported in *Brassica oleracea* var. *acephala*; from Brazil; by Kitajima *et al.* (1970).

**Natural host range and symptoms**

Symptoms none.

*Brassica oleracea* — usually symptomless.

**Transmission**

Transmitted by a vector; an insect; *Myzus persicae*, but transmits infrequently; Aphididae. Transmitted in a non–persistent manner. Virus transmitted by mechanical inoculation; transmitted by grafting; not transmitted by seed.

**Geographical distribution**

Spreads in Brazil (in the states of San Paulo and the Federal District).

**Experimental host range**

Several (3–9) families susceptible. Experimentally infected plants mostly show latent infection, some show local lesions.

Diagnostically susceptible

*Gomphrena globosa, Chenopodium amaranticolor, C. quinoa, Malva parviflora* — local lesions.

*Brassica oleracea* var. *acephala, B. oleracea* var. *botrytis, Raphanus sativus, Citrullus lanatus, Glycine max, Solanum melongena* — latent.

Diagnostically insusceptible

*Dianthus caryophyllus, Lactuca sativa, Zea mays, Phaseolus vulgaris, Capsicum annuum.*

Maintenance and propagation hosts

*Brassica oleracea* var. *acephala, Chenopodium quinoa.*

Assay hosts (**L**ocal lesions or **W**hole plants)

*Chenopodium amaranticolor* (L), *C. quinoa* (L), *Gomphrena globosa* (L).

Dicotyledonous Hosts *(**Susceptible** or Insusceptible)*

*Beta vulgaris.* ***Brassica campestris*** ssp. ***rapa***. ***Brassica juncea***. ***Brassica oleracea*** ssp. ***botrytis***. ***Brassica oleracea*** ssp. ***capitata***. *Capsicum annuum.* ***Chenopodium amaranticolor***. *Chenopodium murale.* ***Chenopodium quinoa***. *Cichorium endiva.* ***Citrullus lanatus***. *Coriandrum sativum. Cucumis melo. Cucumis sativus. Cucurbita maxima. Datura stramonium. Daucus carota. Dianthus caryophyllus. Glycine max.* ***Gomphrena globosa***. *Hibiscus esculentus. Lactuca sativa. Lycopersicon esculentum. Nicandra physaloides. Nicotiana tabacum. Passiflora edulis. Petunia hybrida. Phaseolus vulgaris. Pisum sativum.* ***Rubus idaeus***. ***Solanum melongena***. *Solanum tuberosum.*

Monocotyledonous Hosts *(**Susceptible** or Insusceptible)*

*Triticum aestivum. Zea mays.*

Sources of host-range data

Mello *et al.* (1987).

**Properties of particles in sap**

TIP: 75–80 °C. LIV: 2–4 days. DEP: $\log_{10}$ minus 3–4. Leaf sap contains few virions.

**Purification**

Method: extraction, clarification, precipitation with PEG 6000, differential centrifugation (2 cycles).

**Particle morphology**

Virions filamentous; not enveloped; usually straight; with a clear modal length; of 650 nm; 15 nm wide. Axial canal obscure. Basic helix obscure.

**Biochemical properties**

Virions contain 8.5 % nucleic acid.

**Cytopathology**

Virions found in leaves, mesophyll and epidermis; in cytoplasm. Inclusions present in infected cells; are unusual in shape; are masses of virions.

**Taxonomy and Relationships**

Possible *Carlavirus*.

Virus(es) with serologically related virions: potato S virus.

Best tests for diagnosis: *Chenopodium amaranticolor* and *Gomphrena globosa* are susceptible.

**References**

Costa, A.S., Kitajima, E.W., Costa, C.L. and Nagai, H. (1972). *Rev. Olericult.* **12**: 82.

Kitajima, E.W., Camargo, I.J. and Costa, A.S. (1970). *Bragantia* **29**: 181.

Mello, S.M., Cupertino, F.P., Kitajima, E.W. and Lin, M.T. (1987). *Fitopatol. Bras.* **12**: 353.

Mello, S.M. (1981). *M.Sc. Dissertation*, 56 pp., Univ. Brasilia.

# Colocasia bobone disease (?) rhabdovirus

**Data collated by R. Plumb, 1984. Revised by A.A. Brunt, 1990.**

*Standard acronym:* CBDV.

First reported in *Colocasia esculenta*; from in samples from the Solomon Islands, South Pacific; by James *et al.* (1973).

**Natural host range and symptoms**

Symptoms disappear soon after infection.

*Colocasia esculenta* — thickened malformed and brittle leaves, severe stunting.

**Transmission**

Transmitted by a vector; an insect; *Tarophagus proserpina*; Delphacidae. Transmitted in a persistent manner. Virus not transmitted by mechanical inoculation; not transmitted by contact between plants; not transmitted by seed; not transmitted by pollen.

**Geographical distribution**

Spreads in Papua New Guinea and the Solomon Islands.

**Experimental host range**

Few (<3) families susceptible.

Diagnostically susceptible

*Colocasia esculenta* — malformation, puckering and thickening of leaves followed by recovery.

*Philodendron selloum* — leaf chlorosis.

Diagnostically insusceptible

No natural infection recorded in *Xanthosoma* sp., *Caladium* sp., *Cyrtosperma* sp., *Alocasia* sp, and not transmitted to *Tetragonia expansa.*

Maintenance and propagation hosts

*Colocasia esculenta.*

Assay hosts (**L**ocal lesions or **W**hole plants)

*Colocasia esculenta* (W).

Dicotyledonous Hosts *(**Susceptible** or Insusceptible)*

*Tetragonia expansa.*

Monocotyledonous Hosts *(**Susceptible** or Insusceptible)*

***Colocasia esculenta***.

Sources of host-range data

Gollifer *et al.* (1977); Shaw *et al.* (1979).

**Properties of particles in sap**

Leaf sap contains few virions (generally, but often aggregated).

**Purification**

Method: James *et al.* (1973).

**Particle morphology**

Virions rhabdo– or bullet–shaped; not enveloped; with a clear modal length; of 300–335 nm; 50–55 nm wide. Axial canal obvious (can be seen in broken virions); 7 nm in diameter. Basic helix obvious; pitch of basic helix 4.5 nm.

**Physical properties**

Sedimentation coefficient 285 *S* (±15).

**Cytopathology**

Virions found in mesophyll and phloem sieve tubes; in the perinuclear space. Inclusions present in infected cells; are viroplasms; they contain virions. Other cellular changes: formation of crystalline arrays of virions.

**Taxonomy and Relationships**

Possible unassigned rhabdovirus.

Isolates that differ in virulence are found in different Pacific Island countries.

**Comments**

"Male" taro plants in Papua New Guinea and Solomon Islands also containing taro badnavirus develop a very severe disease described as "alomae".

**References**

Gollifer, D.E. and Brown, J.E. (1972). *Pl. Dis. Reptr* **56**: 597.

Gollifer, D.E., Jackson, G.V.H., Dabek, A.J., Plumb, R.T. and May, Y.Y. (1977). *PANS* **23**: 171.

Jackson, G.H.V. (1980). *Diseases and Pests of Taro*. South Pacific Commission, Noumea.

James, M., Kenten, R.H. and Woods, R.D. (1973). *J. gen. Virol.* **21**: 145.

Kenten, R.H. and Woods, R.D. (1973). *PANS* **19**: 38.

Shaw, D.E., Plumb, R.T. and Jackson, G.V.H. (1979). *Papua New Guinea agric. J.* **30**: 71.

# Commelina diffusa *potyvirus*

**Data collated by F.W. Zettler, 1989.**

First reported in *Commelina diffusa* Burm.; from Florida, U.S.A.; by Morales and Zettler (1977).

**Natural host range and symptoms**

Symptoms persist.

*Commelina diffusa* — mosaic.

*Rhoeo spathacea* (discolor) — mild mosaic.

**Transmission**

Transmitted by a vector; an insect; *Aphis gossypii, Myzus persicae*; Aphididae. Transmitted in a non–persistent manner. Virus transmitted by mechanical inoculation; not transmitted by grafting; not transmitted by contact between plants; not transmitted by seed.

**Ecology and control**

Studies reported by Morales and Zettler (1977); Baker and Zettler (1988).

**Geographical distribution**

Spreads in Puerto Rico and the USA (Florida).

**Experimental host range**

Few (<3) families susceptible. Experimentally infected plants mostly show systemic mosaics, mottles, ringspots or necrosis.

Diagnostically susceptible

*Commelina diffusa* — mosaic.
*Rhoeo spathacea* — mild mosaic.

Diagnostically insusceptible

*Commelina benghalensis, Zebrina pendula, Commelina erecta, Murdannia nudiflora, Aneilema aequinoctiale.*

Maintenance and propagation hosts

*Commelina diffusa, Rhoeo spathacea.*

Assay hosts (**L**ocal lesions or **W**hole plants)

*Commelina diffusa* (W).

Dicotyledonous Hosts *(**Susceptible** or Insusceptible)*

*Apium graveolens. Capsicum annuum. Chenopodium amaranticolor. Cucumis sativus. Cucurbita maxima. Datura stramonium. Daucus carota. Gomphrena globosa. Nicotiana × edwardsonii. Nicotiana tabacum. Phaseolus vulgaris. Pisum sativum. Vigna unguiculata.*

Monocotyledonous Hosts *(**Susceptible** or Insusceptible)*

*Belamcanda chinensis. Philodendron selloum. Zea mays.*

Additional hosts

***Commelina diffusa*** and ***Rhoeo spathacea***.

Sources of host-range data

Morales and Zettler (1977); Baker and Zettler (1988).

**Properties of particles in sap**

TIP: 50–60 °C. LIV: 12–20 days. DEP: $\log_{10}$ minus 1–2. Leaf sap contains few virions.

**Purification**

Method: F.W. Zettler: several methods have been tried and been unsuccessful, due to the viscosity of *Commelina diffusa* extracts.

**Particle morphology**

Virions filamentous; not enveloped; usually flexuous; with a clear modal length; of 734 nm; *c.* 13 nm wide. Axial canal obscure. Basic helix obscure.

**Cytopathology**

Virions found in leaves, epidermis, companion cells, probably roots, probably mesophyll, vascular parenchyma and probably phloem; in cytoplasm (probably). Inclusions present in infected cells; are pinwheels; they contain virions.

**Taxonomy and Relationships**

*Potyvirus: Potyviridae.*

Virus(es) with serologically unrelated virions: bidens mottle, blackeye cowpea mosaic, dasheen mosaic, lettuce mosaic, pepper mottle, potato Y, tobacco etch and turnip mosaic viruses.

Best tests for diagnosis: can be distinguished from cucumber mosaic virus in naturally infected plants by the symptoms (Morales and Zettler, 1977).

**References**

Baker, C.A. and Zettler, F.W. (1988). *Plant Dis.* **72**: 513.

Morales, F.J. and Zettler, F.W. (1977). *Phytopathology* **67**: 839.

# Commelina mosaic *potyvirus*

**Data collated by F.J. Morales, 1984.**

*Standard acronym:* ComMV.

First reported in *Commelina* spp.; from Florida, U.S.A; by Morales and Zettler (1977).

### Natural host range and symptoms

Symptoms persist.
*Commelina diffusa* — systemic mottling and mosaic.
*Rhoeo spathacea.*

### Transmission

Transmitted by a vector; an insect; *Aphis gossypii, Myzus persicae*; Aphididae. Transmitted in a non–persistent manner. Virus transmitted by mechanical inoculation.

### Geographical distribution

Spreads in the USA.

Assay hosts (**L**ocal lesions or **W**hole plants)

*Commelina diffusa* (W).

Dicotyledonous Hosts *(**Susceptible** or Insusceptible)*

*Apium graveolens. Capsicum annuum. Chenopodium amaranticolor. Cucumis sativus. Cucurbita maxima. Datura stramonium. Daucus carota. Gomphrena globosa. Nicotiana benthamiana. Nicotiana tabacum. Phaseolus vulgaris. Pisum sativum. Vigna unguiculata.*

Monocotyledonous Hosts *(**Susceptible** or Insusceptible)*

*Belamcanda chinensis. Philodendron selloum. Zea mays.*

Sources of host-range data

Morales and Zettler (1977).

### Properties of particles in sap

TIP: 50–60 °C. LIV: less than 1 days. DEP: $\log_{10}$ minus 1–2.

### Particle morphology

Virions filamentous; not enveloped; usually flexuous; with a clear modal length; of 707–808 nm.

### Cytopathology

Virions found in all parts of the host plant; in cytoplasm. Inclusions present in infected cells; are pinwheels; they do not contain virions.

### Taxonomy and Relationships

*Potyvirus: Potyviridae.*

Virus(es) with serologically unrelated virions: bidens mottle, blackeye cowpea mosaic, dasheen mosaic, lettuce mosaic, pepper mottle, potato Y, tobacco etch and turnip mosaic viruses.

### References

Morales, F.J. and Zettler, F.W. (1977). *Phytopathology* **67**: 839.

# Commelina X *potexvirus*

**Data collated by A.A. Brunt, 1987.**

*Standard acronym:* ComVX.

First reported in *Commelina diffusa*; from the U.K., but probably originated from Mexico or South America; by Stone (1980).

**Natural host range and symptoms**

*Commelina diffusa*—necrotic streaks, malformation.

**Transmission**

Virus transmitted by mechanical inoculation; not transmitted by seed.

**Experimental host range**

Several (3–9) families susceptible. Experimentally infected plants mostly show symptomless or chlorotic or necrotic local lesions, symptomless infection.

Diagnostically susceptible

*Datura stramonium, Nicotiana clevelandii, N. debneyi, N. glutinosa, Trifolium incarnatum*—symptomless local and systemic infection.

*Chenopodium quinoa, C. murale, Tetragonia expansa* — chlorotic or necrotic local lesions.

*Nicotiana tabacum* — symptomless local infection.

Diagnostically insusceptible

*Chenopodium amaranticolor, Petunia hybrida, Phaseolus vulgaris, Vicia faba, Vigna unguiculata.*

Maintenance and propagation hosts

*Nicotiana clevelandii.*

Assay hosts (**L**ocal lesions or **W**hole plants)

*Chenopodium murale* (L).

Dicotyledonous Hosts *(**Susceptible** or Insusceptible)*

*Chenopodium amaranticolor*. ***Chenopodium capitatum***. ***Chenopodium murale***. ***Chenopodium quinoa***. ***Datura stramonium***. ***Glycine max***. ***Gomphrena globosa***. ***Lycopersicon esculentum***. ***Nicotiana clevelandii***. ***Nicotiana debneyi***. ***Nicotiana glutinosa***. ***Nicotiana megalosiphon***. ***Nicotiana tabacum***. *Ocimum basilicum*. *Petunia hybrida*. *Phaseolus vulgaris*. ***Tetragonia expansa***. ***Trifolium incarnatum***. *Vicia faba*. *Vigna unguiculata*.

Sources of host-range data

Stone (1980).

**Purification**

Method: Stone (1980).

**Particle morphology**

Virions filamentous; usually flexuous; with a clear modal length; of 550 nm. Axial canal obscure. Basic helix obscure.

**Taxonomy and Relationships**

*Potexvirus.*

Virus(es) with serologically related virions: cactus X, hydrangea ringspot, lily X, narcissus mosaic, nerine X and papaya mosaic viruses.

**References**

Stone, O.M. (1980). *Acta Hort.* **110**: 59.

# Commelina yellow mottle *badnavirus*

**Data collated by A.A.Brunt, 1988.**

*Standard acronym:* ComYMV.

First reported in *Commelina diffusa*; from Guadeloupe; by Migliori and Lastra (1978).

**Natural host range and symptoms**

*Commelina diffusa* — found only as mixed infections.

**Transmission**

Virus not transmitted by mechanical inoculation; transmitted by grafting.

**Geographical distribution**

Spreads in Guadeloupe and Sudan (Migliori and Lastra, 1978; B.E.L. Lockhart, unpublished data).

**Properties of particles in sap**

Leaf sap contains few virions.

**Particle morphology**

Virions bacilliform; not enveloped; 133 nm in length; 24 nm wide.

**Biochemical properties**

*Genome* consists of DNA; double–stranded; circular; unipartite.
*Sequence database accession code(s):*
X52938 Em(40)_vi:CYMVCG Gb(84)_vi:CYMVCG Commelina yellow mottle virus complete circular genome. 9/93 7,489bp. 1 sequence.

**Cytopathology**

Virions found in mesophyll; in cytoplasm.

**Taxonomy and Relationships**

*Badnavirus* ( the type species).

**References**

Medberry, S.L., Lockhart, B.E.L. and Olszewski, N.E. (1990). *Nucl. Acids Res.* **18**: 5505.
Migliori, A. and Lastra, R. (1978). *Annls. Phytopath.* **10**: 467.

# Coriander feathery red vein *nucleorhabdovirus*

**Data collated by A.A. Brunt, 1991.**

*Standard acronym:* CFRVV.

First reported in *Coriandrum sativum, Pastinaca sativa*; from the U.S.A.; by Misari and Sylvester (1983).

**Natural host range and symptoms**

*Coriandrum sativum.*
*Pastinaca sativa* — symptomless infection.

## Transmission

Transmitted by a vector; an insect; *Hydaphis foeniculi, Myzus persicae*; Aphididae. Not transmitted by *Aphis fabae, A. helianthi heraciella, Acyrthosiphon (Aulacorthum) solani, Cavariella aegopodii, Chaetosiphon thomasi jacobi, Dysaphis apiifolia, Hydaphis erysimi, Hyperomyzus lactucae, Myzus ornatus, Aulocorthum circumflexum*. Transmitted in a persistent manner. Virus multiplies in the vector; transmitted congenitally to the progeny of the vector; transmitted by mechanical inoculation; not transmitted by seed.

## Geographical distribution

Spreads in the USA (California).

## Experimental host range

Few (<3) families susceptible.

Diagnostically susceptible

*Anthriscus sylvestris, Daucus carota, Foeniculum vulgare, Petroselinum crispum* — symptomless infection.

*Apium graveolens* — systemic mottling and chlorotic spotting.

*Coriandrum sativum* — systemic chlorotic vein-banding and, later, red vein-banding.

*Nicotiana clevelandii, N. clevelandii x glutinosa* — systemic veinal chlorosis and, later, interveinal chlorosis.

Diagnostically insusceptible

*Chenopodium quinoa, Cucumis sativus, Phaseolus vulgaris, Physalis floridana, Nicotiana tabacum.*

Maintenance and propagation hosts

*Apium graveolens, Coriandrum sativum, Nicotiana glutinosa.*

Assay hosts (**L**ocal lesions or **W**hole plants)

*Coriandrum sativum* (W), *Apium graveolens* (W).

Dicotyledonous Hosts *(**Susceptible** or Insusceptible)*

*Capsicum frutescens. Chenopodium amaranticolor. Chenopodium quinoa. Cucumis sativus.* ***Daucus carota****. Fragaria vesca. Gomphrena globosa. Lactuca sativa. Lycopersicon esculentum.* ***Nicotiana clevelandii****.* ***Nicotiana glutinosa****.* ***Nicotiana glutinosa* × *clevelandii****. Nicotiana tabacum.* ***Pastinaca sativa****.* ***Petroselinum crispum****. Petunia hybrida. Phaseolus vulgaris. Physalis floridana. Sonchus oleraceus. Vigna unguiculata.*

Sources of host-range data

Misari and Sylvester (1983).

## Particle morphology

Virions rhabdo– or bullet–shaped; enveloped; with a clear modal length; of 216 nm (in sections); 75 nm wide (in sections). Axial canal obvious. Basic helix obvious.

## Cytopathology

Virions found in leaves and mesophyll; in nuclei and in the perinuclear space.

## Taxonomy and Relationships

Probably *Nucleorhabdovirus: Rhabdoviridae.*

Virus(es) with serologically unrelated virions: panicum mosaic satellite and tobacco necrosis satellite viruses.

**References**

Misari, S.M. and Sylvester, E.S. (1983). *Hilgardia* **51**: 1.

# Cotton anthocyanosis (?) *luteovirus*

**Data collated by A.A. Brunt, 1989.**

*Standard acronym:* CAV.

First reported in *Gossypium barbadense*; from Brazil; by Costa (1957).

**Natural host range and symptoms**

Symptoms persist.

*Gossypium barbadense, Hibiscus cannabinus, H. esculentus, Sida rhombifolia, S. micrantha* — reddening of mature leaves.

**Transmission**

Transmitted by a vector; an insect; *Aphis gossypii*; Aphididae. Transmitted in a persistent manner. Virus not transmitted by mechanical inoculation; transmitted by grafting; not transmitted by contact between plants; not transmitted by seed.

**Ecology and control**

Studies reported by Costa (1957).

**Geographical distribution**

Spreads in Brazil.

**Experimental host range**

Few (<3) families susceptible. Experimentally infected plants mostly show reddening.

Diagnostically susceptible

*Gossypium barbadense, Hibiscus cannabinus, H. esculentum, Sida rhombifolia, S. micrantha* — reddening of mature leaves.

Maintenance and propagation hosts

*Gossypium hirsutum, Hibiscus cannabinus, Hibiscus esculentus.*

Assay hosts (**L**ocal lesions or **W**hole plants)

*Gossypium barbadense* (W), *Hibiscus cannabinus* (W), *H. esculentus* (W).

Dicotyledonous Hosts *(**Susceptible** or Insusceptible)*

***Hibiscus cannabinus*. *Hibiscus esculentus*.**

**Taxonomy and Relationships**

Possible *Luteovirus.*

See Waterhouse *et al.* (1988).

**References**

Costa, A.S. (1957). *Phytopath. Z.* **28**: 167.

Waterhouse, P.M., Gildow, F.E. and Johnstone, G.R. (1988). *AAB Descr. Pl. Viruses* No. 339, 9 pp.

# Cotton leaf crumple bigeminivirus

**Data collated by J.K. Brown, 1987 and A.A. Brunt, 1986.**

*Standard acronym:* CLCrV. *Strains:* Mild and severe strains recognised (Erwin and Meyer, 1961).

First reported in *Gossypium hirsutum*; from California, U.S.A.; by Dickson *et al.* (1954).

**Natural host range and symptoms**

Symptoms persist.

Species of *Gossypium, Abutilon, Althaea, Hibiscus, Malva* of *Castanospermum, Glycine, Phaseolus, Vicia*—mosaic and malformation.

**Transmission**

Transmitted by a vector; an insect; *Bemisia tabaci*; Aleyrodidae. Transmitted in a semi–persistent manner. Virus lost by the vector when it moults; does not multiply in the vector; not transmitted congenitally to the progeny of the vector; not transmitted by mechanical inoculation; transmitted by grafting; not transmitted by contact between plants; not transmitted by seed; not transmitted by pollen.

**Ecology and control**

Studies reported by Brown *et al.* (1983); Brown *et al.* (1987); Duffus and Flock (1982); Erwin and Meyer (1961); Brown *et al.* (1986).

**Geographical distribution**

Spreads in the Middle East; India, Mexico, and the USA.

**Experimental host range**

Few (<3) families susceptible. Experimentally infected plants mostly show chlorosis and malformation.

Diagnostically susceptible

*Malva parviflora, Gossypium hirsutum* 'Delta Pine 70', *Phaseolus angularis* (Azuki bean) and *Castanospermum australe* (Delgado bean) — systemic chlorosis and malformation.

Diagnostically insusceptible

*Cucumis melo, Hibiscus esculentum, Spinacia oleracea, Nicotiana glutinosa, Zinnia elegans.*

Maintenance and propagation hosts

*Gossypium hirsutum, Malva parviflora* and a useful purification host is *Phaseolus vulgaris.*

Assay hosts (**L**ocal lesions or **W**hole plants)

*Gossypium hirsutum* (W), *Malva parviflora* (W).

Dicotyledonous Hosts *(**Susceptible** or Insusceptible)*

*Cucumis melo*. ***Glycine max***. ***Gossypium hirsutum***. ***Hibiscus cannabinus***. *Hibiscus esculentus*. *Nicotiana glutinosa*. ***Phaseolus acutifolius***. ***Phaseolus vulgaris***. *Spinacia oleracea*. *Zinnia elegans*.

Sources of host-range data

Brown and Nelson (1984); Brown *et al.* (1986).

**Properties of particles in sap**

Leaf sap contains few virions. Electron microscopy: fixation with glutaraldehyde is essential. UA is better than PTA.

**Purification**

Method: Brown and Nelson (1984).

**Particle morphology**

Virions geminate; not enveloped; 17–20 nm in diameter; dimers 30–32 nm in length.

**Cytopathology**

Virions found in phloem. Inclusions present in infected cells (in phloem and parenchyma only); are unusual in shape; spherical or amorphous. They are more numerous in younger leaves. Other cellular changes: hypertrophy of cells containing inclusions.

**Taxonomy and Relationships**

Bigeminivirus: *Geminiviridae*.

**References**

Brown, J.K. and Nelson, M.R. (1984). *Phytopathology* **74**: 987.
Brown, J.K. and Nelson, M.R. (1987). *Plant Dis.* **71**: 522.
Brown, J.K., Mihail, J.D. and Nelson, M.R. (1987). *Plant Dis.* **71**: 699.
Brown, J.K., Butler, G.D. and Nelson, M.R. (1983). *Phytopathology* **73**: 787.
Brown, J.K., Nelson, M.R. and Lambe, R.C. (1986). *Plant Dis.* **70**: 981.
Dickson, R.C., Johnson, M.M. and Laird, E. (1954). *Phytopathology* **44**: 479.
Duffus, J.E. and Flock, R.A. (1982). *Calif. Agric.* **36**: 4.
Erwin, D.C. and Meyer, R. (1961). *Phytopathology* **51**: 472.
Tsao, P.M. (1963). *Phytopathology* **53**: 243.

# Cotton leaf curl bigeminivirus

**Data collated by A.A. Brunt, 1989.**

*Standard acronym:* CLCuV.

First reported in *Gossypium barbadense*; from Nigeria; by Farquharson (1912).

**Natural host range and symptoms**

Symptoms persist.

*Gossypium barbadense, Hibiscus esculentus, Corchorus fasicularis* — chlorosis, thickened and malformed veins, petioles and leaves.

**Transmission**

Transmitted by a vector; an insect; *Bemisia tabaci*; Aleyrodidae. Transmitted in a persistent manner. Virus not transmitted congenitally to the progeny of the vector; not transmitted by mechanical inoculation; transmitted by grafting; not transmitted by contact between plants; not transmitted by seed.

**Ecology and control**

Studies reported by Abu Salih (1965); El Nur (1967); El Nur and Abu Salih (1966); Golding (1930); Kirkpatrick (1931); Nour and Nour (1964); Tarr (1951, 1957).

**Geographical distribution**

Spreads in Nigeria and Sudan.

**Experimental host range**

Few (<3) families susceptible. Experimentally infected plants mostly show leaf curling and malformation.

Diagnostically susceptible

*Gossypium barbadense, Hibiscus cannabinus, H. esculentus, H. subdarifa, Althaea rosea, Sida alba* — leaf curling.

Maintenance and propagation hosts

*Gossypium barbadense, Hibiscus esculentus, H. cannabinus.*

Assay hosts (**L**ocal lesions or **W**hole plants)

*Gossypium barbadense* (W), *Hibiscus esculentus* (W).

Dicotyledonous Hosts *(**Susceptible** or Insusceptible)*

***Corchorus spp.. Hibiscus cannabinus. Hibiscus esculentus.***

Sources of host-range data

Nour and Nour (1964); Tarr (1957).

**Taxonomy and Relationships**

Bigeminivirus: *Geminiviridae.*

**References**

Abu Salih, H.S. (1965). *Rep. Res. Div. Minist. Agric. Sudan* 1964–65, p. 74.
El Nur, E. (1967). *Rep. Res. Div. Minist. Agric. Sudan* 1966–67.
El Nur, E. and Abu Salih, H.S. (1966). *Rep. Res. Div. Minist. Agric. Sudan* 1965–66.
Farquharson, C.O. (1912). *Rep. Agric. Dep. Nigeria.*
Golding, F.D. (1930). *Emp. Cott. Grow. Rev.* **7**: 120.
Kirkpatrick, T.W. (1931). *Bull. ent. Res.* **12**: 323.
Massay, R.E. and Andrews, F.W. (1932). *Emp. Cott. Grow. Rev.* **9**: 32.
Nour, M.A. and Nour, J.J. (1964). *Emp. Cott. Grow. Rev.* **41**: 27.
Tarr, S.A.J. (1957). *Leaf Curl Disease of Cotton. Comm. Myc. Inst.*, Kew, U.K.
Tarr, S.A.J. (1957). *Emp. Cott. Grow. Rev.* **34**: 258.
Tarr, S.A.J. (1964). *Misc. Publs. Commonw. Mycol. Inst.* **18**: 23.

# Cow parsnip mosaic *nucleorhabdovirus*

**Data collated by A.A. Brunt, 1994.**

*Standard acronym:* CowPMV.

**Taxonomy and Relationships**

Possible *Nucleorhabdovirus: Rhabdoviridae.*

**References**

Dijk, P. and Bos, L. (1989). *Plant Path.* **95(2)**: 34.
Francki, R.I.B., Milne, R.G. and Hatta, T. (1985). In: 434,03 *Atlas of Plant Viruses*, Vol. 1, p. 73. CRC Press, Boca Raton, Florida.

# Cowpea chlorotic mottle *bromovirus*

**Data collated by L.C. Lane, 1981.**

*Synonyms:* bean yellow stipple virus, virus del moteado amarillo. *Standard acronym:* CCMV.

First reported in *Phaseolus vulgaris*; from Illinois, U.S.A; by Bancroft (1971); Ahlquist *et al.* (1981).

**Natural host range and symptoms**

Symptoms persist.
*Vigna unguiculata, Glycine max* — bright leaf mottle.

**Transmission**

Transmitted by a vector; an insect; *Ceratoma ruficornis, C. trifurcata, Diabrotica balteata, D. undecimpunctata*; Coleoptera. Virus transmitted by mechanical inoculation; not transmitted by seed; not transmitted by pollen.

**Geographical distribution**

Spreads in Costa Rica and the USA.

**Experimental host range**

Several (3–9) families susceptible. Experimentally infected plants mostly show necrotic local lesions or streaks, mosaics.

Diagnostically susceptible

*Chenopodium hybridum, C. quinoa, Glycine max* — necrotic local lesions or streaks.
*Vigna unguiculata* — systemic mosaic, mottle, streaks.

Maintenance and propagation hosts

*Chenopodium quinoa, Phaseolus vulgaris, Vigna unguiculata, Nicotiana benthamiana.*

Assay hosts (**L**ocal lesions or **W**hole plants)

*Chenopodium hybridum* (L), *Glycine max* (L).

Dicotyledonous Hosts *(**Susceptible** or Insusceptible)*

*Arachis hypogaea*. ***Beta vulgaris***. ***Cajanus cajan***. *Capsicum frutescens*. ***Cassia occidentalis***. ***Cassia tora***. ***Chenopodium album***. ***Chenopodium amaranticolor***. ***Chenopodium hybridum***. ***Chenopodium quinoa***. *Crotalaria spectabilis*. ***Cucumis sativus***. ***Cucurbita pepo***. ***Cyamopsis tetragonoloba***. *Desmodium tortuosum*. ***Glycine max***. ***Gomphrena globosa***. ***Lablab purpureus***. ***Lupinus albus***. *Lycopersicon esculentum*. ***Macroptilium lathyroides***. ***Nicotiana benthamiana***. *Nicotiana glutinosa*. ***Nicotiana tabacum***. ***Petunia hybrida***. ***Phaseolus lunatus***. ***Phaseolus vulgaris***. *Physalis peruviana*. ***Pisum sativum***. ***Vicia faba***. ***Vigna unguiculata***. ***Vigna unguiculata*** ssp. ***cylindrica***. ***Vigna unguiculata*** ssp. ***sesquipedalis***. *Zinnia elegans*.

Monocotyledonous Hosts *(**Susceptible** or Insusceptible)*

***Zea mays***.

Sources of host-range data

Kuhn (1964); Lane (1974, 1981).

**Properties of particles in sap**

TIP: 67–76 °C. LIV: 1–44 days. DEP: $\log_{10}$ minus 3–5. Infectivity of sap increased by treatment with di–ethyl ether. Leaf sap contains many virions.

**Purification**

Method: Lane (1974).

**Particle morphology**

Virions isometric; not enveloped; 26 nm in diameter; rounded in profile; with a conspicuous capsomere arrangement.

**Physical properties**

One sedimenting component in purified preparations, or three sedimenting components in purified preparations (1 component at pH 3–6, 3 components at pH 7); sedimentation coefficient 88 *S* (at pH 3–6). Density 1.36 g $cm^{-3}$ in CsCl. Isoelectric point pH 3.7. $A_{260}/A_{280}$ ratio 1.7.

**Biochemical properties**

Virions contain 23.7 % nucleic acid; 76.3 % protein; 0 % lipid.

*Genome* consists of RNA; single–stranded; linear; of three parts (but encapsidated with *m*RNA of virion proteins); largest (or only) genome part the largest 3.171 kb; the 2nd largest 3.1 kb; the 3rd largest 2.173 kb. Base composition 26.4 % G; 25.3 % A; 20.3 % C; 28.2 % U. 5' terminus of RNA has a methylated nucleotide cap. Infectivity retained when deproteinised with proteases, or decreased when deproteinised with proteases; retained when deproteinised with phenol or detergent. Poly A region absent. Genome has *t*RNA–like activity. Genome accepts tyrosine. Nucleotide sequence references: Gunn *et al.* (1980); Ahlquist *et al.* (1981).

*Sequence database accession code(s):*

J02052 Em(40)_vi:CORNA3 Gb(84)_vi:MCCRNA3 Cowpea chlorotic mottle virus rna 3, complete coat protein cds and complete rna 4 span. 6/94

K01779 Em(40)_vi:CORNA1 Gb(84)_vi:MCCRNA1 Cowpea chlorotic mottle virus rna 1, 3' end. 6/94 208bp.

K01780 Em(40)_vi:CORNA2 Gb(84)_vi:MCCRNA2 Cowpea chlorotic mottle virus rna 2, 3' end. 6/94 234bp.

M18658 Em(40)_vi:BRMCCRNA Gb(84)_vi:MCCRNA4 Cowpea chlorotic mottle virus (CCMV) RNA 4, 3' end. 7/89 166bp.

M28817 Gb(84)_vi:MCCRNAA2 Cowpea chlorotic mottle virus RNA2, encoding 2a protein, complete cds. 3/90 2,774bp.

M28818 Em(40)_vi:MCCRNAA3 Gb(84)_vi:MCCRNAA3 Cowpea chlorotic mottle virus RNA3, encoding 3a protein and coat protein, complete cds. 4/90

M65139 Em(40)_vi:MCCP1A Gb(84)_vi:MCCP1A Cowpea chlorotic mottle virus (CCMV) 1a protein gene, complete cds. 8/92 3,171bp. 7 sequences.

*Non–genomic nucleic acid* found in the virions; is subgenomic *m*RNA (for the coat protein, and named RNA-4). Sub–genomic *m*RNA found in infected cells; the coat protein *m*RNA.

*Virion protein(s)* one; $M_r$ 19600. Method of preparation: Lane: salt dissociation with NaCl or $CaCl_2$. Amino acid composition: Casey (1979); Rees *et al.* (1979).

**Replication**

Genome replicates probably in cytoplasm (but there is no direct evidence). Coat protein *m*RNA translated in the cytoplasm. Replication does not depend on a helper virus.

**Cytopathology**

Virions found in all parts of the host plant; in cytoplasm and in cell vacuoles. Inclusions present in infected cells; are unusual in shape; of irregular or crystalloid structure; they contain virions. Other cellular changes: proliferation of endoplasmic reticulum.

**Taxonomy and Relationships**

*Bromovirus: Bromoviridae.*

Virus(es) with serologically related virions: brome mosaic virus, but distantly; broad bean mottle virus is very distantly related.

**References**

Ahlquist, P., Dasgupta, R. and Kaesberg, P. (1981). *Cell* **23**: 183.
Bancroft, J.B. (1971). *CMI/AAB Descr. Pl. Viruses* No. 49, 4 pp.
Bancroft, J.B., Hiebert, E., Rees, M.W. and Markham, R. (1968). *Virology* **34**: 224.
Casey, R. (1979). *J. gen. Virol.* **43**: 143.
Christie, R.G. and Edwardson, J.R. (1977). *Fla Agric. Exp. Stn Monog.* No. 9, p. 115.
Dzianott, A.M. and Bujarski, J.J. (1991). *Virology* **185**: 553.
Gunn, M.R. and Symons, R.H. (1980). *FEBS. Lett.* **115**: 77.
Kuhn, C.W. (1964). *Phytopathology* **54**: 739.
Lane, L.C. (1974). *Adv. Virus Res.* **19**: 151.
Lane, L.C. (1981). In: *Handbook of Plant Virus Infections and Comparative Diagnosis*, p. 333; ed. E. Kurstak. Elsevier/North Holland Biomedical Press, Amsterdam.
Langenberg, W.G. (1979). *J. Ultrastruct. Res.* **66**: 120.
Malorni, M.C. (1979). *Rep. John Innes Inst.* 1979, p.101.
Rees, M.W. and Short, M.N. (1980). *Rep. John Innes Inst.* 1979, p.102.

# Cowpea golden mosaic (?) bigeminivirus

**Data collated by H.W. Rossel and A.A. Brunt, 1989.**

*Standard acronym:* CPGMV.

First reported in *Vigna unguiculata*; from northern and southern Nigeria; by Rossel (1977*a* and *b*).

**Natural host range and symptoms**

Symptoms persist.

*Vigna unguiculata* — bright yellow mosaic.

**Transmission**

Transmitted by a vector; an insect; *Bemisia tabaci*; Aleyrodidae. Transmitted in a persistent manner. Virus not transmitted by mechanical inoculation; transmitted by grafting; not transmitted by contact between plants; not transmitted by seed; not transmitted by pollen.

**Ecology and control**

Studies reported by Rossel and Thottappilly (1985) who found that wild cowpeas are likely perennial sources of infection for cowpea crops in Nigeria. See also Anno-Nyako *et al.* (1983); Vetten and Allen (1983).

**Geographical distribution**

Spreads in India, Kenya, Mozambique, Niger, Nigeria, Pakistan, and Tanzania.

**Experimental host range**

Few (<3) families susceptible. Experimentally infected plants mostly show systemic mosaics, mottles, ringspots or necrosis.

Diagnostically susceptible

*Vigna unguiculata* — yellow mosaic.

Diagnostically insusceptible

*Phaseolus lunatus.*

Maintenance and propagation hosts

*Vigna unguiculata.*

Assay hosts (**L**ocal lesions or **W**hole plants)

*Vigna unguiculata* (W).

Dicotyledonous Hosts ***(Susceptible** or Insusceptible)*

*Phaseolus lunatus.* ***Vigna unguiculata. Vigna unguiculata*** ssp. ***sesquipedalis.***

Additional hosts

***Vigna unguiculata*** subsp. ***dekindtiana*** var. ***dekindtiana*** (wild perennial cowpea).

Sources of host-range data

Rossel and Thottappilly (1985).

**Taxonomy and Relationships**

Bigeminivirus: *Geminiviridae.*

The failure of reciprocal transmission tests using grafting and *Bemisia tabaci* indicate that the virus is distinct from the whitefly-transmitted virus inducing similar symptoms in *Phaseolus lunatus* (Lima bean) in Nigeria (Williams, 1976).

**Comments**

Rossel and Thottappilly (1985) note that similar cowpea diseases occur in Pakistan, East Africa and Niger (Ahmad, 1978; Sharma and Varma, 1976; Singh and Allen, 1979).

**References**

Ahmad, M. (1978). *Pl. Dis. Reptr* **62**: 224.

Anno-Nyako, F.O., Vetten, H.J., Allen, D.J. and Thottappilly, G. (1983). *Ann. appl. Biol.* **102**: 319.

Rossel, H.W. (1977*a*). *Tropical Grain Legume Bulletin* **8**: 41.

Rossel, H.W. (1977*b*). *Rep. Int. Inst. Trop. Agric.* 1977, pp. 28, 89.

Rossel, H.W. and Thottappilly, G. (1985). *Virus Diseases of Important Food Crops in Tropical Africa*: Ibadan, Nigeria, Int. Inst. Trop. Agric., p. 35.

Sharma, S.R. and Varma, A. (1976). *Indian Phytopath.* **29**: 421.

Singh, S.R. and Allen, D.J. (1979). In: *Cowpea Pests and Diseases*. Ibadan, Nigeria. Int. Inst. Trop. Agric.

Vetten, H.J. and Allen, D.J. (1983). *Ann. appl. Biol.* **102**: 219.

Williams, R.J. (1976). *Pl. Dis. Reptr* **60**: 853.

# Cowpea green vein-banding *potyvirus*

**Data collated by M.T. Lin, 1984.**

*Standard acronym:* CGVBV.

First reported in *Vigna unguiculata*; from Brazil; by Lin *et al.* (1979*a*).

**Natural host range and symptoms**

Symptoms persist.

*Vigna unguiculata* — green vein banding.

**Transmission**

Transmitted by a vector; an insect; *Myzus persicae*; Aphididae. Transmitted in a non–persistent manner. Virus transmitted by mechanical inoculation; transmitted by grafting; not transmitted by contact between plants; transmitted by seed.

**Geographical distribution**

Spreads in Brazil.

**Experimental host range**

Few (<3) families susceptible. Experimentally infected plants mostly show chlorotic local lesions, mosaic, leaf curling.

Diagnostically susceptible

*Chenopodium amaranticolor, C. quinoa* — chlorotic local lesions; not systemic.
*Phaseolus acutifolius* — systemic vein clearing, mosaic, leaf curling.
*P. arborigenus* — local vein necrosis.
*P. vulgaris* cv. Black Turtle Soup — local and systemic infection.
*P. vulgaris* cv. Bountiful — chlorotic local lesions.
*Vigna unguiculata* cv. TVu 4536 — systemic green vein banding.

Diagnostically insusceptible

*Glycine max* cv. Bragg, *Medicago sativa* cv. du Puits, *Phaseolus vulgaris* cv. Pinto III, *Pisum sativum* cv. Dark Skin Perfection, *Datura stramonium.*

Maintenance and propagation hosts

*Phaseolus vulgaris* cv. Rico 23, *Vigna unguiculata* cv. Sempre Verde, TVu 4536.

Assay hosts (**L**ocal lesions or **W**hole plants)

*Chenopodium amaranticolor* (L), *C. quinoa* (L), *Phaseolus vulgaris* cv. Rico 23 (W).

Dicotyledonous Hosts *(**Susceptible** or Insusceptible)*

*Antirrhinum majus. Brassica oleracea* ssp. *capitata. Capsicum annuum.* ***Chenopodium amaranticolor****. Chenopodium murale.* ***Chenopodium quinoa****. Cichorium endiva. Cucumis sativus. Cucurbita maxima. Datura stramonium. Glycine max.* ***Gomphrena globosa****. Helianthus annuus. Lycopersicon esculentum. Medicago sativa. Nicotiana × edwardsonii. Nicotiana glutinosa. Nicotiana tabacum. Petunia hybrida.* ***Phaseolus lunatus****.* ***Phaseolus vulgaris****. Physalis floridana. Pisum sativum. Raphanus sativus. Solanum melongena. Solanum nigrum.* ***Vigna unguiculata****.*

Monocotyledonous Hosts *(**Susceptible** or Insusceptible)*

*Hordeum vulgare. Zea mays.*

**Properties of particles in sap**

TIP: 50–55 °C. DEP: $\log_{10}$ minus 3–4. Leaf sap contains few virions.

**Particle morphology**

Virions filamentous; not enveloped; usually flexuous; with a clear modal length; of 780 nm; 12 nm wide. Axial canal obscure.

**Cytopathology**

Virions found in all parts of the host plant; in cytoplasm. Inclusions present in infected cells; are pinwheels; they do not contain virions.

**Taxonomy and Relationships**

*Potyvirus: Potyviridae.*

Virus(es) with serologically related virions: cowpea rugose mosaic virus is related closely. Virus(es) with serologically unrelated virions: bean common mosaic, bean yellow mosaic, blackeye cowpea mosaic and soybean mosaic viruses.

**References**

Lin, M.T., Anjos, J.R.N., Kitajima, E.W. and Rios, G.P. (1979*a*). *Fitopatol. Bras.* **4**: 120.

Lin, M.T., Anjos, J.R.N., Kitajima, E.W. and Rios, G.P. (1979*b*). *Fitopatol. Bras.* **4**: 203.

# Cowpea mild mottle (?) *carlavirus*

**Data collated by A.A. Brunt, 1980. Revised 1989 by N.M. Horn.**

*Synonyms:* bean angular mosaic virus (Costa *et al.*, 1983; Gaspar *et al.*, 1985), voandezia mosaic virus, groundnut crinkle virus, probably Psophocarpus necrotic mosaic virus (Fauquet and Thouvenel, 1987; Fortuner *et al.*, 1979). *Standard acronym:* CPMMV. *Strains:* tomato pale chlorosis virus (Antignus and Cohen, 1987; Cohen and Antignus, 1982), voandzeia mosaic virus (Fauquet and Thouvenel, 1987).

First reported in *Vigna unguiculata*; from Tafo, Ghana; by Brunt *et al.* (1973).

**Natural host range and symptoms**

Symptoms vary seasonally.

*Vigna unguiculata* cv. Blackeye — chlorotic blotches, systemic mottling and malformation.

*Arachis hypogaea* — necrotic local lesions, chlorotic rings, systemic chlorosis, rolling, veinal necrosis.

*Canavalia ensiformis, Glycine max* — vein mosaic and chlorosis, apical necrosis and malformation.

*Voandzeia subterranea, Lycopersicon esculentum* — mottling.

**Transmission**

Transmitted by a vector; an insect; *Bemisia tabaci*; Aleyrodidae. Transmitted in a semi–persistent manner. Virus transmitted by mechanical inoculation; some isolates transmitted by seed (and others not; Horn *et al.*, 1991; Jeyanandarajah and Brunt, 1993).

**Geographical distribution**

Spreads in Brazil, Cote d'Ivoire, Egypt, Fiji, Ghana, India, Indonesia, Israel, Kenya, Malaysia, Nigeria, the Solomon Islands, Tanzania, Thailand, and Yemen.

**Experimental host range**

Several (3–9) families susceptible. Experimentally infected plants mostly show chlorotic local lesions, systemic mottle or not systemic.

Diagnostically susceptible

*Beta vulgaris, Chenopodium murale, C. quinoa* — chlorotic local lesions.

*Arachis hypogaea, Cajanus cajan, Canavalia ensiformis, Glycine max, Vigna unguiculata, Nicotiana clevelandii* — systemic mottle.

*Tetragonia expansa* — chlorotic local lesions; not systemic.

Maintenance and propagation hosts

*Cajanus cajan, Canavalia ensiformis, Glycine max, Phaseolus vulgaris, Nicotiana clevelandii.*

Assay hosts (Local lesions or Whole plants)

*Beta vulgaris, Chenopodium quinoa.*

Dicotyledonous Hosts *(**Susceptible** or Insusceptible)*

*Ageratum conyzoides.* ***Arachis hypogaea.*** *Atriplex hortensis.* ***Beta vulgaris.*** *Brassica campestris* ssp. *pekinensis. Brassica juncea. Brassica oleracea* ssp. *capitata.* ***Cajanus cajan.*** *Capsicum annuum. Catharanthus roseus. Celosia argentea.* ***Chenopodium amaranticolor. Chenopodium foetidum. Chenopodium murale. Chenopodium quinoa.*** *Cicer arietinum. Cucumis sativus. Cucurbita moschata. Dahlia pinnata. Datura stramonium.* ***Glycine max. Gomphrena globosa.*** *Helianthus annuus. Lens esculenta.* ***Lycopersicon esculentum.*** *Nicandra physaloides.* ***Nicotiana clevelandii.*** *Nicotiana glutinosa. Nicotiana megalosiphon. Nicotiana rustica. Nicotiana tabacum. Ocimum basilicum. Petunia hybrida.* ***Phaseolus lunatus. Phaseolus vulgaris.*** *Physalis floridana. Phytolacca americana. Pisum sativum. Senecio vulgaris. Solanum melongena. Solanum tuberosum.* ***Spinacia oleracea. Tetragonia expansa. Theobroma cacao.*** *Torenia fournieri.* ***Trifolium incarnatum.*** *Tropaeolum majus. Verbesina encelioides. Vicia faba. Vigna radiata.* ***Vigna unguiculata.*** *Zinnia elegans.*

Monocotyledonous Hosts *(**Susceptible** or Insusceptible)*

*Zea mays.*

Sources of host-range data

Brunt *et al.* (1973).

**Properties of particles in sap**

TIP: 65–70 °C. LIV: 4 days. DEP: $\log_{10}$ minus 3–4. Infectivity of sap not changed by treatment with di–ethyl ether. Leaf sap contains few virions.

**Purification**

Method: Brunt *et al.* (1973).

**Particle morphology**

Virions filamentous; not enveloped; usually straight; with a clear modal length; of 650 nm; 13 nm wide. Axial canal obscure; 2–3 nm in diameter. Basic helix obscure.

**Physical properties**

One sedimenting component in purified preparations; sedimentation coefficient 165 *S*.

**Biochemical properties**

Virions contain 5 % nucleic acid; 95 % protein; 0 % lipid.

*Genome* consists of RNA; single–stranded. Additional factor not required for infectivity.

*Virion protein(s)* one; $M_r$ 35700.

**Cytopathology**

Virions found in mesophyll, epidermis and palisade parenchyma; in cytoplasm. Inclusions present in infected cells; are unusual in shape; of brush-like structure; apparently they contain virions.

**Taxonomy and Relationships**

Possible *Carlavirus*.

Virus(es) with serologically unrelated virions: 12 recognised members of the carlavirus group (Iwaki *et al.*, 1982; Adams and Barbara, 1982; Brunt *et al.*, 1983).

**References**

Adams, A.N. and Barbara, D.J. (1982). *Ann. appl. Biol.* **101**: 495.
Anon. (1987). *Rep. Int. Inst. Trop. Agric.* 1986, Ibadan.
Antignus, Y. and Cohen, S. (1987). *Ann. appl. Biol.* **110**: 563.
Brunt, A.A. and Kenten, R.H. (1973). *Ann. appl. Biol.* **74**: 67.
Brunt, A.A. and Kenten, R.H. (1974). *CMI/AAB Descr. Pl. Viruses* No. 140, 4 pp.
Brunt, A.A. and Phillips, S. (1981). *Trop. Agric. Trinidad* **58**: 177.
Brunt, A.A., Atkey, P.T. and Woods, R.D. (1983). *Intervirology* **20**: 137.
Cohen, S. and Antignus, Y. (1982). *Phytoparasitica* **10**: 101.
Costa, A.S., Gaspar, J.O. and Vega, J. (1983). *Fitopatol. Bras.* **8**: 325.
Dubera, J. and Dollet, M. (1981). *Phytopath. Z.* **101**: 337.
Fauquet, C. and Thouvenel, J.-C. (1987). In: *Plant Viral Diseases in the Ivory Coast*, No. 46, p. 132. ORSTOM, Paris.
Fortuner, R., Fauquet, C. and Lourd, M. (1979). *Pl. Dis. Reptr* **63**: 194.
Gaspar, J.O., Beriam, L.O.S., Alves, M.N., Oliveira, A.R. and Costa, A.S. (1985). *Fitopatol. Bras.* **10**: 195.
Horn, N.M., Nasin, S. and Yuliantoro, B. (1989). *Neth. J. Pl. Path.* **97** 125.
Iizuka, N., Rajeshwari, R., Reddy, D.V.R., Goto, T., Muniyappa, V., Bharathan, N. and Ganekar, A.M. (1984). *Phytopath. Z.* **109**: 245.
Iwaki, M., Thongmeearkom, P., Prommin, M., Honda, Y. and Hibi, T. (1982). *Plant Dis.* **66**: 365.
Jeyanandarajah, P. and Brunt, A.A. (1993). *J. Phytopathology* **137**: 148.
Mink, G.I. (1972). *CMI/AAB Descr. Pl. Viruses* No. 92, 4 pp.
Mali, V.R., Nirmal, D.D., Nolt, B.L., Rajeshwar, R., Bharathan, N. and Reddy, D.V.R. (1987). *Indian Phytopath.* **40** 22.
Mink, G.I. and Keswani, C.L. (1987). *Pl. Dis. Reptr.* **71**: 557.
Sun, M.K.C. and Hebert, T.T. (1972). *Phytopathology* **62**: 832.
Thouvenel, J.C., Fauquet, C. and Monsarret, A. (1982). *Plant Dis.* **66**: 336.
Walkey, D.G.A., Alhubaishi, A.A. and Webb, M.J.W. (1990). *Trop. Pest Manag.* **36**: 195.

# Cowpea Moroccan aphid-borne mosaic *potyvirus*

**Data collated by A.A. Brunt, 1994.**

*Synonyms:* possibly passiflora South African potyvirus (Brand *et al.*, 1993; McKern *et al.*, 1994).

First reported in *Vigna unguiculata*; from Morocco; by Fischer and Lockhart (1976).

### Natural host range and symptoms

*Vigna unguiculata* — severe leaf chlorosis (symptoms even more severe when co-infected with cucumber mosaic virus.

### Transmission

Transmitted by a vector; an insect; *Myzus persicae*; Aphididae. Transmitted in a non–persistent manner. Virus does not require a helper virus for vector transmission; transmitted by mechanical inoculation; transmitted by seed.

### Geographical distribution

Spreads in the African region; Morocco and South Africa.

**Particle morphology**

Virions filamentous; usually flexuous; with a clear modal length; of 765 nm.

**Taxonomy and Relationships**

*Potyvirus: Potyviridae.*

**Comments**

The virus was originally called cowpea aphid-borne mosaic virus, like that described by Lovisolo and Conti (1966). The latter was probably an isolate of bean common mosaic virus, but is no longer available for comparative tests, and hence a *nomen nudum*. This virus is best differentiated from the latter by adding "Moroccan" to its name.

**References**

Brand, R.J., Burger, J.T. and Rybicki, E.P. (1993). *Arch. Virol.* **128**: 29.
Fischer, H.U. and Lockhart, B.E.L. (1976). *Phytopath. Z.* **84**: 43.
Lovisolo, O. and Conti, M. (1966) *Neth. J. Pl. Path.* **72**: 265.
McKern, N.M., Strike, P.M., Barnett, O.W., Dijkstra, J., Shukla, D.D. and Ward, C.M. (1994). *Arch. Virol.* **136**: 207.

# Cowpea mosaic *comovirus*

**Data collated by C.P. de Jager, 1980. Revised 1983. Revised by G. Bruening.**

*Synonyms:* cowpea yellow mosaic virus. *Standard acronym:* CPMV. *Strains:* cowpea mosaic comovirus SB strain and cowpea mosaic comovirus Vu strain (Agrawal, 1964).

First reported in *Vigna unguiculata*; from Nigeria; by Chant (1959).

**Natural host range and symptoms**

Symptoms persist.

*Vigna unguiculata, Cajanus cajan, Crotalaria juncea* — mosaic, vein yellowing, leaf malformation.

**Transmission**

Transmitted by a vector; an insect; *Ootheca mutabilis, Paraluperodes quaternus, Nematocerus acerbus, Ceratoma variegata, C. ruficornis, C. trifurcata, Diabrotica balteata, D. undecimpunctata howardi, D. virgifera, Acalymma vittatum*; Coleoptera. Virus transmitted by mechanical inoculation; transmitted by grafting; transmitted by seed (1–5%).

**Geographical distribution**

Spreads in Cuba, Kenya, Nigeria, Suriname, and Tanzania. Found, but with no evidence of spread, in the U.S.A.

**Experimental host range**

Several (3–9) families susceptible. Experimentally infected plants mostly show necrotic or chlorotic local lesions and systemic mosaic, necrotic or chlorotic streaks.

Diagnostically susceptible

*Chenopodium amaranticolor* — necrotic local lesions, systemic chlorotic spots and streaks, and apical deformation.

*Canavalia ensiformis.*

*Phaseolus vulgaris* cv. Pinto — necrotic or chlorotic local lesions.

*Vigna unguiculata* cv. Blackeye — chlorotic local lesions, systemic mosaic.

*Vigna unguiculata* cv. Chinese Red x Iron — pinpoint necrotic local lesions.

Diagnostically insusceptible

*Nicotiana glutinosa.*

Maintenance and propagation hosts

*Vigna unguiculata* cvs Blackeye Early Ramshorn, Blackeye 5.

Assay hosts (**L**ocal lesions or **W**hole plants)

*Chenopodium amaranticolor (L), Canavalia ensiformis (L), Phaseolus vulgaris* cvs Pinto (L), Scotia (L), *Vigna unguiculata* cv. Early Red (L).

Dicotyledonous Hosts (***Susceptible*** *or Insusceptible)*

***Antirrhinum majus***. *Apium graveolens. Arachis hypogaea. Beta vulgaris.* ***Cajanus cajan***. *Capsicum frutescens.* ***Cassia tora***. ***Chenopodium amaranticolor***. ***Chenopodium quinoa***. *Crotalaria retusa.* ***Crotalaria spectabilis***. *Cucumis sativus. Cucurbita pepo.* ***Cyamopsis tetragonoloba***. ***Datura stramonium***. ***Glycine max***. ***Gomphrena globosa***. ***Lablab purpureus***. *Lathyrus odoratus. Lens esculenta. Lupinus albus. Lycopersicon esculentum. Lycopersicon pimpinellifolium. Macrotyloma uniflorum. Medicago sativa. Melilotus alba.* ***Nicotiana clevelandii***. *Nicotiana glutinosa. Nicotiana rustica.* ***Nicotiana tabacum***. ***Petunia hybrida***. ***Phaseolus lunatus***. ***Phaseolus vulgaris***. *Physalis peruviana.* ***Pisum sativum***. *Solanum melongena. Solanum tuberosum.* ***Spinacia oleracea***. *Trifolium incarnatum. Trifolium pratense. Trifolium repens. Tropaeolum majus.* ***Vicia faba***. *Vicia sativa. Vicia villosa.* ***Vigna angularis***. ***Vigna radiata***. ***Vigna unguiculata***. ***Vigna unguiculata*** ssp. ***cylindrica***. ***Vigna unguiculata*** ssp. ***sesquipedalis***. ***Zinnia elegans***.

Additional hosts

***Stizolobium deeringianum***.

Sources of host-range data

Thornberry (1966); Swaans and van Kammen (1973); Hampton *et al.* (1978).

**Properties of particles in sap**

TIP: 55–65 °C. LIV: 4–10 days. DEP: $\log_{10}$ minus 5. Infectivity of sap not changed by treatment with di–ethyl ether. Leaf sap contains many virions.

**Purification**

Method: Klootwijk *et al.* (1977).

**Particle morphology**

Virions isometric; not enveloped; 20–24 nm in diameter; angular in profile; without a conspicuous capsomere arrangement.

**Physical properties**

Three sedimenting components in purified preparations; sedimentation coefficient of the fastest 119 *S* (B); of the other(s) 98 *S* (M), or 60 *S* (T). Density 1.447 g $cm^{-3}$ in CsCl (B), or 1.406 g $cm^{-3}$ in CsCl ($M_2$), or 1.395 g $cm^{-3}$ in CsCl ($M_1$), or 1.288 g $cm^{-3}$ in CsCl (T). Isoelectric point pH 5–7. $A_{260}/A_{280}$ ratio 1.74 (B), or 1.61 (M).

**Biochemical properties**

Virions contain 34.5 % nucleic acid (B), or 24 % nucleic acid (M), or 0 % nucleic acid (T); 65.5 % protein (B), or 76 % protein (M), or 100 % protein (T). Also 1.9% carbohydrates and polyamines: spermidine 5.05 µg/mg, spermine 0.17 µg/mg.

*Genome* consists of RNA; single–stranded; linear; of two parts; largest (or only) genome part the larger 6.6 kb; the 2nd largest 3.8 kb. Base composition 23.5 % G; 27.9 % A; 18.1 % C; 30.5 % U. Infectivity retained when deproteinised with proteases, or decreased when deproteinised with proteases; retained when deproteinised with phenol or detergent. Poly A region present (El Manna and Bruening, 1973). Genome has no *t*RNA–like activity. Nucleotide sequence references: Van Wezenbeek *et al.* (1983); Lomonossoff and Shanks (1983).

*Sequence database accession code(s):*

J02064 Em(40)_vi:COBRNA Gb(84)_vi:MCPBRNA cowpea mosaic virus bottom comp. (b-rna); rgn of 1st aug. 4/90 149bp.

J02065 Em(40)_vi:COMRNA Gb(84)_vi:MCPMRNA cowpea mosaic virus middle comp. (b-rna); rgn of 1st aug. 4/90 325bp.

M10534 Em(40)_vi:COMMCPM5 Gb(84)_vi:MCPM5E Cowpea mosaic virus M RNA, 5' end. 7/89 87bp.

M10535 Em(40)_vi:COMMCPB5 Gb(84)_vi:MCPB5E Cowpea mosaic virus B genmic RNA, 5' end. 7/89 79bp.

M25438 Em(40)_vi:COMMCPRN Gb(84)_vi:MCPRNA3A Cowpea mosaic virus M RNA 3' terminal sequence. 7/90 80bp.

M25439 Gb(84)_vi:MCPRNA3B Cowpea mosaic virus B RNA 3' terminal sequence. 9/90 80bp.

X00206 Em(40)_vi:COCPMVB Gb(84)_vi:COCPMVB Cowpea mosaic virus bottom component RNA (B RNA). 9/93 5,889bp

X00729 Em(40)_vi:COCPMVM Gb(84)_vi:COCPMVM Cowpea mosaic virus (CPMV) middle-component RNA (M RNA). 9/93 3,481bp. 8 sequences.

*Non–genomic nucleic acid* not found in the virions. Sub–genomic *m*RNA not found in infected cells.

*Virion protein(s)* three (one of these is derived from another by proteolytic trimming of the carboxyl terminus — Kridl and Bruening, 1983); $M_r$ 40700. $M_r$ of 2nd largest 23708. $M_r$ of 3rd largest 3538. Method of preparation: Kridl and Bruening (1983). Amino acid sequence: Van Wezenbeek *et al.* (1983). Virion proteins glycosylated; are covalently linked carbohydrates (Partridge *et al.*, 1974).

**Replication**

Genome replicates in cytoplasm (in membranous structures). Coat protein *m*RNA translated in the cytoplasm. Replication does not depend on a helper virus.

**Cytopathology**

Virions found in mesophyll; in cytoplasm. Inclusions present in infected cells; are crystals in the cytoplasm and unusual in shape; electron dense material within vacuolated vesicles or membranous tubules; they contain virions. Other cellular changes: membranous tubules containing rows of virion-like particles that cross large vacuoles or are embedded in enlarged plasmodesmata.

**Taxonomy and Relationships**

*Comovirus: Comoviridae* (type species of the genus — SB isolate).

Virus(es) with serologically related virions: cowpea severe mosaic, glycine mosaic (GW strain), broad bean true mosaic, bean rugose mosaic and red clover mottle viruses.

**References**

Agrawal, H.O. (1964). *Meded. Landb. Hoogesch. Wagen.* **64**: 1.

Agrawal, H.O. and Maat, D.Z. (1964). *Nature, Lond.* **202**: 674.

Beier, H., Bruening, G., Russell, M.L. and Tucjer, C.L. (1979). *Virology* **95**: 165.

Beier, H., Siler, D.J., Russell, M.L. and Bruening, G. (1977). *Phytopathology* **67**: 917.

Bruening, G. (1969). *Virology* **56**: 577.

Bruening, G. and Agrawal, H.O. (1967). *Virology* **32**: 306.

Chant, S.R. (1959). *Ann. appl. Biol.* **77**: 565.

Davies, J.E., Stanley, J. and van Kammen, A. (1979). *Nucl. Acids Res.* **7**: 493.

El Manna, M.M. and Bruening, G. (1973). *Virology* **56**: 198.
Geelen, J.L.M.C., van Kammen, A. and Verduin, B.J.M. (1972). *Virology* **49**: 205.
Hampton, R., Beczner, L., Hagedorn, D., Bos, L., Inouye, T., Barnett, O., Musil, M. and Meiners, J. (1978). *Phytopathology* **68**: 989.
Klootwijk, J., Klein, I., Zabel, P., and van Kammen, A. (1977). *Cell* **11**: 73.
Kridl, J.C. and Bruening, G. (1983). *Virology* **129**: 369.
Lomonossoff, G.P. and Shanks, M. (1983). *EMBO J.* **2**: 2253
Partridge, J.E., Shannon, L.M., Gumpf, D.J. and Colbaugh, P. (1974). *Nature, Lond.* **247**: 391.
Siler, D.J., Babcock, J. and Bruening, G. (1976). *Virology* **71**: 560.
Stanley, J. and van Kammen, A. (1979). *Eur. J. Biochem.* **101**: 45.
Swaans, H. and van Kammen, A. (1973). *Neth. J. Pl. Path.* **79**: 257.
Thornberry, H.H. (1966). In: *Index of Plant Virus Diseases.* U.S. Dep. Agric. Hdbk No. 307, p.247.
Van Kammen, A. (1971). *CMI/AAB Descr. Pl. Viruses* No. 47, 4 pp.
Van Kammen, A. and de Jager, C.P. (1978). *CMI/AAB Descr. Pl. Viruses* No. 197, 6 pp.
Van Lent, J.W.M., Wellink, J. and Goldbach, R. (1990). *J. gen. Virol.* **71**: 219.
Van Wezenbeck, P., Verrer, J., Harmsen, J., Vos, P. and van Kammen, A. (1983). *EMBO J.* **2**: 941.
Wellink, J., van Lent, J.W.M. and Goldbach, R. (1988). *J. gen. Virol.* **69**: 751.
Wu, G.J. and Bruening, G. (1971). *Virology* **46**: 596.
Zimmern, D. (1984). *EMBO J.* **3**: 855.

# Cowpea mottle (?) *carmovirus*

**Data collated by R. Bozarth, 1980. Revised 1990.**

*Standard acronym:* CPMoV.

First reported in *Vigna unguiculata*; from Ibadan, Nigeria; by Shoyinka *et al.* (1978).

### Natural host range and symptoms

Symptoms persist.
*Vigna unguiculata* — mottling or mosaic.
*Voandzeia subterranea* — green mottle, shortening of petioles.

### Transmission

Transmitted by a vector; an insect; *Ootheca mutabilis, Paraluperodes lineata*; Coleoptera. Virus transmitted by mechanical inoculation; transmitted by seed (10% in *V. unguiculata*).

### Geographical distribution

Spreads in Nigeria.

### Experimental host range

Several (3–9) families susceptible. Experimentally infected plants mostly show chlorotic local lesions, vein necrosis, mottling, malformation.

Diagnostically susceptible

*Chenopodium amaranticolor, C. quinoa* — diffuse chlorotic local lesions; not systemic.

*Cajanus cajan, Vigna unguiculata* — severe systemic chlorotic or green mottle, leaf malformation. Necrosis in youngest primary leaves.

*Glycine max, Macroptilium lathyroides, Phaseolus acutifolius, P. lunatus, V. radiata, V. mungo, P. vulgaris* — systemic mosaic.

Maintenance and propagation hosts

*Vigna unguiculata.*

Assay hosts (**L**ocal lesions or **W**hole plants)

*Chenopodium amaranticolor (L), C. quinoa (L), Vigna unguiculata (L).*

Dicotyledonous Hosts *(**Susceptible** or Insusceptible)*

*Atriplex hortensis.* ***Cajanus cajan***. ***Chenopodium amaranticolor***. *Chenopodium murale.* ***Chenopodium quinoa***. *Cucumis sativus. Datura stramonium.* ***Glycine max***. *Lycopersicon esculentum.* ***Macrotyloma uniflorum***. *Nicotiana clevelandii. Nicotiana glutinosa.* ***Nicotiana tabacum***, or *Nicotiana tabacum. Petunia hybrida.* ***Phaseolus lunatus***. ***Phaseolus vulgaris***. *Tetragonia expansa. Vicia faba.* ***Vigna radiata***. ***Vigna unguiculata***. *Zinnia elegans.*

Sources of host-range data

Shoyinka *et al.* (1978).

**Properties of particles in sap**

TIP: 60 °C. LIV: 1 days. DEP: $\log_{10}$ minus 6 (x 2). Leaf sap contains many virions.

**Purification**

Method: Shoyinka *et al.* (1978).

**Particle morphology**

Virions isometric; not enveloped; 30 nm in diameter; rounded in profile; with a conspicuous capsomere arrangement.

**Physical properties**

One sedimenting component in purified preparations; sedimentation coefficient 122 *S*. Density 1.3492 g $cm^{-3}$ in CsCl.

**Biochemical properties**

Virions contain 20 % nucleic acid; 80 % protein; 0 % lipid.

*Genome* consists of RNA; single–stranded; unipartite; largest (or only) genome part 3.94 kb.

*Sequence database accession code(s):*

M65155 Em(40)_vi:MCCCP Gb(84)_vi:MCCCP Cowpea mottle virus capsid protein gene. 8/91 1,104bp. 1 sequence.

3 virus specified dsRNA species found in infected cells. Size of largest virus specified dsRNA 4.4 kbp; 2nd largest 1.9 kbp; 3rd largest 1.35 kbp.

*Virion protein(s)* one; $M_r$ 44500 (±1500); coat protein. Method of preparation: Bozarth and Shoyinka (1979).

*Virus–coded non–virion proteins* isolated; three proteins found. $M_r$ of the largest 50000. $M_r$ of 2nd largest 27000. $M_r$ of 3rd 11000.

**Replication**

Replication does not depend on a helper virus.

**Taxonomy and Relationships**

Possible *Carmovirus: Tombusviridae.*

Virus(es) with serologically unrelated virions: almond mosaic, apple mosaic, arabis mosaic, artichoke crinkle, bean mild mosaic, bean pod mottle, bean rugose mosaic, broad bean mottle virus, brome mosaic, carnation ringspot, cowpea severe mosaic, cucumber necrosis, cynosurus mottle, eggplant mosaic, elm mosaic, hibiscus chlorotic ringspot, okra mosaic, pelargonium ringspot, southern bean mosaic, sowbane mosaic, squash mosaic virus, tobacco ringspot, tobacco streak and tomato ringspot viruses.

**References**

Bozarth, R.F. and Shoyinka, S.A. (1979). *CMI/AAB Descr. Pl. Viruses* No. 212, 3 pp.

Kim, J.W. (1989). *Ph.D. Thesis*. Indiana State University, Terra Haute, Indiana, U.S.A.

Shoyinka, S.A., Bozarth, R.F., Reese, J. and Rossel, H.W. (1978). *Phytopathology* **68**: 693.

# Cowpea rugose mosaic *potyvirus*

**Data collated by E.W. Kitajima, 1991.**

*Standard acronym:* CPRMV.

First reported in *Vigna unguiculata* syn. *sinensis*; from Piaui State, Brazil; by Santos (1981); Santos *et al.* (1981).

**Natural host range and symptoms**

Symptoms persist. Symptoms rugose mosaic.
*Vigna unguiculata* syn. *sinensis.*

**Transmission**

Transmitted by a vector (probably); an insect; *Myzus persicae, Aphis neri*; Aphididae. Not transmitted by *Aphis coreopsidis, A. citricola, Dactynotus* sp. Transmitted in a non–persistent manner. Virus transmitted by mechanical inoculation; not transmitted by seed.

**Geographical distribution**

Spreads in Brazil (Praui).

**Experimental host range**

Several (3–9) families susceptible.

Diagnostically susceptible

*Vigna unguiculata, Phaseolus vulgaris, Pisum sativum, Chenopodium amaranticolor, C. quinoa, C. murale, Gomphrena globosa, Centrosema* sp.

Diagnostically insusceptible

*Glycine max, Crotalaria paulina, Nicotiana tabacum, Sonchus oleraceus, Zea mays.*

Maintenance and propagation hosts

*Vigna unguiculata* syn. *sinensis, Phaseolus vulgaris.*

Assay hosts (**L**ocal lesions or **W**hole plants)

*Chenopodium quinoa* (L), *Vigna unguiculata* syn. *sinensis* (W).

Dicotyledonous Hosts *(**Susceptible** or Insusceptible)*

*Cassia obtusifolia.* ***Chenopodium amaranticolor*****.** ***Chenopodium murale*****.** ***Chenopodium quinoa*****.** *Cichorium endiva. Datura stramonium.* ***Gomphrena globosa*****.** *Lactuca sativa. Lycopersicon esculentum. Nicandra physaloides. Nicotiana rustica. Nicotiana tabacum. Solanum melongena. Sonchus oleraceus.*

Monocotyledonous Hosts ***(Susceptible** or Insusceptible)*

*Zea mays.*

**Properties of particles in sap**

TIP: 60–65 °C. LIV: 7–9 days. DEP: $\log_{10}$ minus 3–4. Leaf sap contains few virions.

**Purification**

Method: Kitajima: extraction in phosphate buffer + EDTA + $Na_2SO_3$. Clarification in 8% n-butanol. Differential centrifugation, then into sucrose cushion, and finally density gradient. Yields *c.* 6 mg per kg.

**Particle morphology**

Virions filamentous; not enveloped; usually flexuous; with a clear modal length; of 730 nm; 15 nm wide. Axial canal obscure. Basic helix obscure.

**Cytopathology**

Virions found in leaves, mesophyll and epidermis; in cytoplasm. Inclusions present in infected cells; are pinwheels.

**Taxonomy and Relationships**

*Potyvirus: Potyviridae.*

Virus(es) with serologically unrelated virions: soybean mosaic, blackeye cowpea mosaic, turnip mosaic, bean common mosaic and watermelon mosaic 1 viruses.

**References**

Santos, A.A. (1981). *M.Sc. Dissertation*, Univ. Brasilia.

Santos, A.A., Lin, M.T. and Kitajima, E.W. (1981). *Phytopathology* **71**: 890.

## Cowpea severe mosaic *comovirus*

**Data collated by C.P. de Jager, 1980. Revised 1983. Revised 1989 by G. Bruening.**

*Synonyms:* Puerto Rico cowpea mosaic virus. *Standard acronym:* CPSMV. *Strains:* see Beier *et al.* (1977); Agrawal (1964).

First reported in *Vigna unguiculata*; from Louisiana, Arkansas and Indiana, U.S.A; by Perez and Cortez-Monllor (1970).

**Natural host range and symptoms**

Symptoms persist.

*Calopogonium mucunoides, Canavalia ensiformis, Centrosema pubescens, Crotalaria juncea, Desmodium canescens, Glycine max, Macroptilium lathyroides, Phaseolus vulgaris, Psophocarpus tetragonolobus, Vigna radiata, V. unguiculata* — chlorotic lesions, concentric ringspots often becoming necrotic; vein clearing, then mosaic, and malformation and blistering of younger leaves.

**Transmission**

Transmitted by a vector; an insect; *Ceratoma arcuata, C. ruficornis, C. trifurcata, C. variegata, Chalcodermus bimaculatus, Diabrotica balteata, D. speciosa, D. virgifera, D. undecimpunctata, Diphaulaca* sp., *Epilachna varivestis, Acalymma vittatum*; Coleoptera. Virus transmitted by mechanical inoculation; transmitted by seed (10% in *V. unguiculata* ssp. *sesquipedalis*, 8% in *V. unguiculata*, varies with strain and cultivar); transmitted by pollen to the seed.

**Geographical distribution**

Spreads in the North American region and the South and Central American region. Found, but with no evidence of spread, in Peru.

**Experimental host range**

Several (3–9) families susceptible.

Diagnostically susceptible

*Chenopodium amaranticolor, Phaseolus vulgaris* cv. Pinto — necrotic local lesions; not systemic.

*Vigna unguiculata* — systemic mosaic, vein yellowing or yellow netting, leaf malformation.

Maintenance and propagation hosts

*Phaseolus vulgaris* cv. Prince, *Vigna unguiculata* cvs Blackeye, Early Ramshorn, Blackeye 5, Pituiba.

Assay hosts (**L**ocal lesions or **W**hole plants)

*Chenopodium amaranticolor* (L), *C. quinoa* (L), *Canavalia ensiformis* (W), *Phaseolus vulgaris* cvs Pinto, Scotia, (L) *Vigna unguiculata* cv. Arlington (W).

Dicotyledonous Hosts *(**Susceptible** or Insusceptible)*

***Cajanus cajan*** (cvs Khaki, Saragateado, Amarillo, 6 Selections). ***Cassia occidentalis***. ***Catharanthus roseus***. ***Chenopodium album***. ***Chenopodium amaranticolor***. ***Chenopodium capitatum***. ***Chenopodium quinoa***. *Cyamopsis tetragonoloba*. ***Datura stramonium***. ***Glycine max*** (8 cultivars). ***Gomphrena globosa***. ***Macroptilium lathyroides***. ***Nicotiana clevelandii***. ***Nicotiana glutinosa***. ***Nicotiana tabacum*** (cvs Samsun NN, Ambalena, Virginia 115, Wisonsin Havana 425). ***Petunia hybrida***. ***Phaseolus vulgaris*** (18/25 cultivars). ***Pisum sativum*** (cv. Dwarf Telephone). ***Sesbania exaltata***. ***Trifolium incarnatum***. ***Vigna radiata***. ***Vigna unguiculata*** (cvs Going, Early Red, PI 221731, Blackeye Early Ramshorn).

Comments on host-range

Cowpea severe mosaic comovirus has a wider host range among cowpea cvs than cowpea mosaic comovirus (Beier *et al.*, 1977).

Sources of host-range data

Perez and Cortez-Monllor (1970); Swaans and van Kammen (1973).

**Properties of particles in sap**

TIP: 65–70 °C. LIV: 5 days. DEP: $\log_{10}$ minus 4–5. Leaf sap contains many virions.

**Purification**

Method: Klootwijk *et al.* (1977).

**Particle morphology**

Virions isometric; not enveloped; 25 nm in diameter; angular in profile; without a conspicuous capsomere arrangement.

**Physical properties**

Three sedimenting components in purified preparations; sedimentation coefficient of the fastest 115 *S* (B); of the other(s) 95 *S* (M), or 58 *S* (T).

**Biochemical properties**

Virions contain 36 % nucleic acid (B), or 25 % nucleic acid (M), or 0 % nucleic acid (T); 64 % protein (B), or 75 % protein (M), or 100 % protein (T).

*Genome* consists of RNA; single–stranded; of two parts; largest (or only) genome part 6 kb (RNA-1); the 2nd largest 3.73 kb (RNA-2). 5' terminus of RNA has a VPg (Daubert and Bruening, 1979). Infectivity retained when deproteinised with phenol or detergent. Poly A region present.

*Sequence database accession code(s):*

M83309 Em(40)_vi:COMCPSRNA Gb(84)_vi:CPSRNA2A Cowpea severe mosaic virus RNA-2 region. 7/92 3,732bp

M83830 Em(40)_vi:COMCPSRN1 Gb(84)_vi:CPSRNA1A Cowpea severe mosaic virus complete RNA1 sequence (isolated from Vigna unguiculata). 4/93 5,9 2 sequences.

*Non–genomic nucleic acid* not found in the virions.

*Virion protein(s)* three; $M_r$ of the largest 39000. $M_r$ of 2nd largest 22000. $M_r$ of 3rd largest *c.* 4000.

**Replication**

Replication does not depend on a helper virus.

**Cytopathology**

Virions found in mesophyll; in cytoplasm. Inclusions present in infected cells; are crystals in the cytoplasm and unusual in shape; vacuolated structures adjacent to the nucleus.

**Taxonomy and Relationships**

*Comovirus: Comoviridae.*

Virus(es) with serologically related virions: bean pod mottle virus is related closely; Andean potato mottle, red clover mottle, pea green mottle and cowpea mosaic viruses are related distantly; radish mosaic, quail pea mosaic, bean rugose mosaic, pea mild mosaic and broad bean stain viruses are very distantly related.

**References**


Agrawal, H.O. (1964). *Meded. Landb. Hoogesch. Wagen.* **64**: 1.

Agrawal, H.O. and Maat, D.Z. (1964). *Nature, Lond.* **202**: 674.

Beier, H., Siler, D.J., Russell, M.L. and Bruening, G. (1977). *Phytopathology* **67**: 917.

Daubert, S. and Bruening, G. (1979). *Virology* **98**: 246.

de Jager, C.P. (1979). *CMI/AAB Descr. Pl. Viruses* No. 209, 5 pp.

Klootwijk, J., Klein, I., Zabel, P. and van Kammen, A. (1977). *Cell* **11**: 73.

Lin, M.T., Anjos, J.R.N. and Rios, G.P. (1981). *Phytopathology* **71**: 435.

Perez, J.E. and Cortez-Monllor, A. (1970). *Pl. Dis. Reptr* **54**: 212.

Smith, C.E. (1924). *Science* **60**: 268.

Swaans, H. and van Kammen, A. (1973). *Neth. J. Pl. Path.* **79**: 257.


# Cowpea stunt (?) *luteovirus*

**Data collated by A.A. Brunt, and M. Kameya-Iwaki, 1987.**

First reported in *Vigna sesquipedalis* and *V. unguiculata*; from Indonesia; by Anon. (1976).

**Natural host range and symptoms**

Symptoms persist.

*Vigna sesquipedalis, V. unguiculata* — stunting.

**Transmission**

Transmitted by a vector; an insect; *Aphis craccivora*, but not *Myzus persicae* or *Acyrthosiphon (Aulacorthum) solani*; Aphididae. Transmitted in a persistent manner. Virus not transmitted by mechanical inoculation.

**Ecology and control**

Studies reported by Tsuchizaki *et al.* (1986).

**Geographical distribution**

Spreads in Indonesia and Thailand.

**Experimental host range**

Several (3–9) families susceptible.

Diagnostically susceptible

*Vigna sesquipedalis, V. unguiculata, Glycine max, Phaseolus vulgaris, Pisum sativum, Psophocarpus tetragonolobus, Vigna radiata* — small, malformed leaves and severe stunting.

Maintenance and propagation hosts

*Vigna sesquipedalis, V. unguiculata.*

Assay hosts (**L**ocal lesions or **W**hole plants)

*Vigna sesquipedalis* (W), *Vigna unguiculata* (W).

Dicotyledonous Hosts *(**Susceptible** or Insusceptible)*

*Chenopodium amaranticolor. Chenopodium quinoa.* ***Glycine max.*** *Gomphrena globosa. Nicotiana glutinosa. Petunia hybrida.* ***Phaseolus vulgaris.*** *Physalis floridana.* ***Pisum sativum. Vigna radiata. Vigna unguiculata. Vigna unguiculata*** ssp. ***sesquipedalis.***

Sources of host-range data

Tsuchizaki *et al.* (1986).

**Taxonomy and Relationships**

Possible *Luteovirus.*

**References**


Anon. (1976). In: *Japan-Indonesia Food Crop Research Program Report 1970–1975*, p. 171. Japan Int. Coop. Agency, Tokyo.

Tsuchizaki, T. (1985). *Nettai Noken Shuho* **53**: 143.

Tsuchizaki, T., Senboku, T., Iwaki, M., Kiratiya-Angul, S., Srithongchai, W. and Deema, N. (1986). *Tech. Bull. Trop. Agr. Res. Center, Japan* **21**: 219.


# Crimson clover latent *nepovirus*

**Data collated by A.J. Cockbain, 1982. Revised 1983.**

*Synonyms:* clover Crimson latent virus. *Standard acronym:* CCLV.

First reported in *Trifolium incarnatum*; from Hertfordshire, U.K; by Kenten *et al.* (1980).

**Natural host range and symptoms**

*Trifolium incarnatum* — symptomless.

**Transmission**

Not transmitted by *Myzus persicae*. Virus transmitted by mechanical inoculation; not transmitted by contact between plants; transmitted by seed (97% in *T. incarnatum*, 20% in *C. quinoa*).

**Geographical distribution**

Found, but with no evidence of spread, in seed from France, Italy, The Netherlands, U.K. and the U.S.A.

**Experimental host range**

Few (<3) families susceptible.

Diagnostically susceptible

*Chenopodium amaranticolor* — systemic mosaic, leaf malformation.
*C. quinoa* — chlorotic local lesions, systemic mosaic, ringspots, leaf malformation.

Diagnostically insusceptible

*Cucumis sativus, Phaseolus vulgaris, Pisum sativum, Vicia faba, Nicotiana clevelandii, N. glutinosa, N. tabacum, Petunia hybrida.*

Maintenance and propagation hosts

*Chenopodium album, C. amaranticolor, C. quinoa, Trifolium incarnatum.*

Assay hosts (**L**ocal lesions or **W**hole plants)

*Chenopodium amaranticolor, C. quinoa.*

Dicotyledonous Hosts *(**Susceptible** or Insusceptible)*

*Beta vulgaris. Brassica campestris* ssp. *pekinensis.* ***Chenopodium album. Chenopodium amaranticolor.*** *Chenopodium foetidum. Chenopodium murale.* ***Chenopodium quinoa.*** *Cucumis sativus. Glycine max. Gomphrena globosa. Lycopersicon esculentum. Medicago sativa. Nicotiana clevelandii. Nicotiana glutinosa. Nicotiana tabacum. Petunia hybrida. Phaseolus vulgaris. Physalis floridana. Pisum sativum. Tetragonia expansa.* ***Trifolium incarnatum.*** *Trifolium pratense. Trifolium repens. Trifolium subterraneum. Vicia faba. Vigna unguiculata. Vigna unguiculata* ssp. *sesquipedalis.*

Sources of host-range data

Kenten *et al.* (1980).

**Properties of particles in sap**

TIP: 60–65 °C. LIV: 20–30 days. DEP: $\log_{10}$ minus 2–3. Leaf sap contains few virions.

**Purification**

Method: Kenten *et al.* (1980).

**Particle morphology**

Virions isometric; not enveloped; 26 nm in diameter; angular in profile; without a conspicuous capsomere arrangement.

**Physical properties**

Three sedimenting components in purified preparations; sedimentation coefficient of the fastest 122 *S* (B); of the other(s) 101 *S* (M), or 52 *S* (T). Density 1.5 g $cm^{-3}$ in CsCl (B), or 1.44 g $cm^{-3}$ in CsCl (M).

**Biochemical properties**

Virions contain 41 % nucleic acid (B), or 34 % nucleic acid (M), or 0 % nucleic acid (T); 59 % protein (B), or 66 % protein (M), or 100 % protein (T).

*Genome* consists of RNA; single–stranded; of two parts; largest (or only) genome part the larger 7.1 kb; the 2nd largest 5.2 kb. Additional factor not required for infectivity.

*Virion protein(s)* one; $M_r$ 52000.

**Taxonomy and Relationships**

*Nepovirus: Comoviridae.*

Virus(es) with serologically unrelated virions: arabis mosaic, arracacha A, artichoke Italian latent, lucerne Australian latent, bean leaf roll, bean pod mottle, beet cryptic, broad bean stain, broad bean mottle, broad bean true mosaic, broad bean wilt, cherry leaf roll, cherry rasp leaf, cacao necrosis, grapevine Bulgarian latent, grapevine chrome mosaic, grapevine fanleaf, hibiscus latent ringspot, mulberry ringspot, myrobalan latent ringspot, pea enation mosaic, peach rosette mosaic, potato black ringspot, raspberry ringspot, red clover mottle, red clover necrotic mosaic, southern bean mosaic, strawberry latent ringspot, tobacco ringspot, tomato black ring and tomato ringspot viruses.

**References**

Kenten, R.H., Cockbain, A.J. and Woods, R.D. (1980). *Ann. appl. Biol.* **96**: 79.

# Crinum mosaic *potyvirus*

**Data collated by R.D. Pares, 1985. Revised 1990.**

*Standard acronym:* CriMV.

First reported in *Crinum* spp.; from New South Wales, Australia; by Pares and Bertus (1978).

**Natural host range and symptoms**

Symptoms persist. Symptoms longitudinal streaking, yellow mosaic and general chlorosis.

*Crinum* spp. — pale streaking, then mosaic and general chlorosis.

**Geographical distribution**

Spreads in Australia and Fiji (Brunt, 1980).

**Experimental host range**

Few (<3) families susceptible.

Diagnostically insusceptible

*Chenopodium quinoa, Capsicum annuum, Nicotiana clevelandii, Vicia faba, Lycopersicon esculentum.*

Dicotyledonous Hosts (***Susceptible** or Insusceptible*)

*Capsicum annuum. Chenopodium amaranticolor. Chenopodium quinoa. Cucumis sativus. Datura metel. Datura stramonium. Gomphrena globosa. Lycopersicon esculentum. Nicotiana clevelandii. Nicotiana glutinosa. Nicotiana rustica. Nicotiana sylvestris. Nicotiana tabacum. Physalis peruviana. Trifolium subterraneum. Vicia faba. Vigna unguiculata.*

**Properties of particles in sap**

DEP: $\log_{10}$ minus 2–4. Leaf sap contains many virions.

**Particle morphology**

Virions filamentous; not enveloped; usually flexuous; with a clear modal length; of 780 nm; 11 nm wide. Axial canal obscure. Basic helix obscure.

**Cytopathology**

Virions found in leaves, mesophyll and epidermis; in cytoplasm. Inclusions present in infected cells; are pinwheels and unusual in shape; and tubular structures; they contain virions. Other cellular changes: pinwheel and tubular inclusions, large masses of virions, nuclei with invaginations containing mitochondria.

**Taxonomy and Relationships**

*Potyvirus: Potyviridae.*

**References**

Brunt, A.A. (1980). *UNDP/FAO Survey of Plant Virus Diseases in Fiji.*
Pares, R.D. and Bertus, A.L. (1978). *Phytopath. Z.* **91**: 170.

# Crotalaria spectabilis yellow mosaic (?) *potexvirus*

**Data collated by A.A. Brunt, 1987.**

First reported in *Crotalaria spectabilis*; from Anambra State, Nigeria; by Igwegbe (1982).

**Natural host range and symptoms**

Symptoms persist.
*Crotalaria spectabilis* — vein clearing, yellow mosaic or general chlorosis.

**Transmission**

Transmitted by a vector; an insect; *Myzus persicae*; Aphididae. Transmitted in a non–persistent manner. Virus transmitted by mechanical inoculation; not transmitted by seed.

**Geographical distribution**

Spreads in Nigeria.

**Experimental host range**

Few (<3) families susceptible.

Diagnostically susceptible

*Crotalaria spectabilis* — vein yellowing, yellow mosaic and mottling.
*Calopogonium mucunoides* — systemic chlorotic mottling and leaf curling.
*Canavalia ensiformis* — interveinal systemic leaf chlorosis and malformation.
*Phaseolus lunatus* — systemic mosaic.
*Vigna unguiculata* — vein chlorosis and mottling.
*Glycine max* — systemic vein chlorosis and mosaic.

Diagnostically insusceptible

*Cajanus cajan, Crotalaria juncea, Cyamopsis tetragonoloba, Nicotiana glutinosa, Chenopodium quinoa.*

Maintenance and propagation hosts

*Calopogonium mucunoides, Crotalaria spectabilis.*

Assay hosts (**L**ocal lesions or **W**hole plants)

*Calopogonium mucunoides* (W).

Dicotyledonous Hosts *(**Susceptible** or Insusceptible)*

*Arachis hypogaea. Brassica oleracea* ssp. *botrytis. Brassica oleracea* ssp. *capitata. Cajanus cajan. Capsicum annuum. Capsicum frutescens. Catharanthus roseus. Celosia argentea. Chenopodium amaranticolor. Chenopodium quinoa.* ***Crotalaria spectabilis****. Cucumis sativus. Cyamopsis tetragonoloba.* ***Glycine max****. Gomphrena globosa. Lycopersicon esculentum. Nicotiana clevelandii. Nicotiana glutinosa. Nicotiana tabacum. Phaseolus lathyroides.* ***Phaseolus lunatus****.* ***Phaseolus vulgaris***, or *Phaseolus vulgaris. Physalis floridana. Pisum sativum. Sesamum indicum. Sesbania exaltata. Solanum melongena. Solanum nigrum.* ***Vigna unguiculata****. Vigna unguiculata* ssp. *cylindrica.*

Additional hosts

***Vigna unguiculata*** cv. California Blackeye; *Vigna unguiculata* cvs Dinner and New Era.

Sources of host-range data

Igewegbe (1982).

**Properties of particles in sap**

TIP: 55–60 °C. LIV: 1–2 days (at 20–22°C). DEP: $\log_{10}$ minus 3–4.

**Particle morphology**

Virions filamentous; not enveloped; usually flexuous; with a clear modal length; of 500 nm. Axial canal obscure; 12 nm in diameter. Basic helix obscure.

**Taxonomy and Relationships**

Possible *Potexvirus.*

**Comments**

This virus resembles centrosema mosaic virus, but does not infect *Centrosema pubescens*, and also, unlike most potexviruses, is reported to be transmitted by aphids.

**References**

Igwegbe, E.C.K. (1982). *Plant Dis.* **66**: 74.
Van Velsen, R.J. and Crowley, N.C. (1981). *Nature, Lond.* **189**: 858.
Van Velsen, R.J. and Crowley, N.C. (1962). *Aust. J. agric. Res.* **13**: 220.

# Croton vein yellowing (?) *nucleorhabdovirus*

**Data collated by A.A. Brunt, 1992.**

*Synonyms:* croton (?) rhabdovirus.

First reported in *Codiaeum variegatum*; from Italy; by Bellardi *et al.* (1991).

**Natural host range and symptoms**

Symptoms persist.
*Codiaeum variegatum* — yellow and/or pink vein banding and stunting.

**Transmission**

Virus transmitted by mechanical inoculation; transmitted by grafting.

**Geographical distribution**

Spreads in Italy.

**Experimental host range**

Few (<3) families susceptible.

Diagnostically susceptible

*Chenopodium amaranticolor* — systemic leaf chlorosis and distortion.

*Nicotiana glutinosa* — chlorotic local lesions; systemic vein clearing and distortion, and plant stunting.

Diagnostically insusceptible

*Chenopodium quinoa, Gomphrena globosa, Phaseolus vulgaris, Nicotiana clevelandii, N. tabacum.*

Maintenance and propagation hosts

*Nicotiana glutinosa.*

Assay hosts (**L**ocal lesions or **W**hole plants)

*Nicotiana glutinosa* (W).

Dicotyledonous Hosts *(**Susceptible** or Insusceptible)*

*Apium graveolens. Capsicum annuum.* ***Chenopodium amaranticolor****. Chenopodium murale. Chenopodium quinoa. Gomphrena globosa. Lycopersicon esculentum. Nicotiana clevelandii.* ***Nicotiana glutinosa****. Nicotiana tabacum. Ocimum basilicum. Phaseolus vulgaris. Tetragonia expansa.*

Sources of host-range data

Bertaccini and Bellardi (1992).

**Particle morphology**

Virions rhabdo– or bullet–shaped; with a clear modal length; of 180–240 nm; 70–80 nm wide.

**Replication**

Genome replicates in nuclei.

**Cytopathology**

Virions found in leaves and mesophyll; in nuclei and in the perinuclear space. Other cellular changes: : nuclei contain chromatin-rich clumps and less electron dense granular material.

**Taxonomy and Relationships**

Possible *Nucleorhabdovirus: Rhabdoviridae.*

**References**

Bellardi, M.G., Bertaccini, A., Villani, A. and Vicchi, V. (1991). *Plant Dis.* **75**: 862.

Bertaccini, A. and Bellardi, M.G. (1992). *Plant Path.* **41**: 79.

# Croton yellow vein mosaic bigeminivirus

**Data collated by V. Muniyappa, 1991.**

*Standard acronym:* CYVMV.

First reported in *Croton bonplandianum*; from India; by Varma (1963).

**Natural host range and symptoms**

Symptoms persist and vary seasonally. Symptoms vein clearing, vein yellowing, thickening of veins and curling.

*Croton bonplandianum* — vein clearing and yellowing, thickening of veins and leaf curling.

**Transmission**

Transmitted by a vector; an insect; *Bemisia tabaci*; Aleyrodidae. Transmitted in a persistent manner. Virus retained when the vector moults; does not multiply in the vector; not transmitted congenitally to the progeny of the vector; not transmitted by mechanical inoculation; transmitted by grafting; not transmitted by contact between plants; not transmitted by seed; not transmitted by pollen.

**Geographical distribution**

Spreads in India.

**Experimental host range**

Many (>9) families susceptible. Experimentally infected plants mostly show vein clearing, malformation, yellowing and thickening, leaf curling, enations and dark green veins.

Diagnostically susceptible

*Croton bonplandianum* — yellow vein.
*Beta vulgaris, Spinacia oleracea* — vein thickening.
*Carica papaya, Lycopersicon esculentum, Phaseolus vulgaris* — leaf curling.
*Raphanus sativus* — curling and rolling of leaves.

Diagnostically insusceptible

*Abelmoschus esculentus, Arachis hypogaea, Cucumis sativus, Euphorbia geniculata, Vigna unguiculata.*

Maintenance and propagation hosts

*Croton bonplandianum, Nicandra physalodes, Nicotiana hybrida, N. tabacum* cvs Samsun and Jayasree.

Assay hosts (**L**ocal lesions or **W**hole plants)

*Croton bonplandianum* (W), *Datura stramonium* (W), *Nicotiana tabacum* cv. Jayasree (W).

Dicotyledonous Hosts *(**Susceptible** or Insusceptible)*

***Ageratum conyzoides*. *Beta vulgaris*. *Carica papaya*. *Chenopodium amaranticolor*. *Datura stramonium*. *Lycopersicon esculentum*. *Nicotiana benthamiana*. *Nicotiana clevelandii*. *Nicotiana glutinosa*. *Nicotiana rustica*. *Nicotiana sylvestris*. *Nicotiana tabacum*. *Petunia hybrida*. *Phaseolus vulgaris*. *Spinacia oleracea*. *Zinnia elegans*.**

Sources of host-range data

Mandal and Muniyappa (1991).

**Properties of particles in sap**

Leaf sap contains many virions.

**Purification**

Method: Mandal and Muniyappa (1991).

**Particle morphology**

Virions geminate; 18 nm in diameter; dimers 30 nm in length.

**Physical properties**

One sedimenting component in purified preparations.

**Taxonomy and Relationships**

Bigeminivirus: *Geminiviridae*.

Virus(es) with serologically related virions: African cassava mosaic and Indian cassava mosaic viruses.

Croton yellow mosaic virus has sequence similarity to African cassava mosaic geminivirus.

**References**

Harrison, B.D., Muniyappa, V., Swanson, M.M., Roberts, I.M. and Robinson, D.J. (1991). *Ann. appl. Biol.* **118**: 595.

Mandal, B. and Muniyappa, V. (1991). *Ann. appl. Biol.* submitted.

Mandal, B. and Muniyappa, V. (1991). *Fitopatol. Bras.* (in press).

Muniyappa, V. and Veeresh, G.K. (1984). *Proc. Indian Acad. Sci. (Anim. Sci.* **93**: 397.

Varma, P.M. (1963). *Bull. Nat. Inst. Sci. India* **24**: 11.

# Cucumber chlorotic spot (?) *closterovirus*

**Data collated by A.A. Brunt, 1995.**

First reported in *Cucumis sativus*; from Southern France; by Woudt *et al.* (1993).

**Natural host range and symptoms**

*Cucumis sativus* — chlorotic spotting of younger leaves; interveinal chlorosis to complete chlorosis of older leaves.

**Transmission**

Transmitted by a vector; an insect; *Trialeurodes vaporariorum*; Aleyrodidae.

**Geographical distribution**

Spreads in France.

Diagnostically susceptible

*Cucumis sativus* — systemic chlorotic spotting and chlorosis.

Dicotyledonous Hosts *(**Susceptible** or Insusceptible)*

***Cucumis sativus*.**

**Particle morphology**

Virions filamentous; usually flexuous.

**Biochemical properties**

*Genome* consists of RNA; single–stranded; unipartite. Poly A region absent.

**Cytopathology**

Virions found in cytoplasm.

**Taxonomy and Relationships**

Possible *Closterovirus*.

The genome has some sequence homology with that of beet yellows virus.

**References**

Woudt, L.P., de Rover, A.P., de Haan, P.T. and van Grisven, M.Q.J.M. (1993). *Abstr. 9th Int. Congr. Virol., Glasgow*, P60–26.

# Cucumber green mottle mosaic *tobamovirus*

**Data collated by R.I.B. Francki, 1988.**

*Synonyms:* cucumber green mottle mosaic virus, tobacco mosaic virus watermelon strain — W, cucumber virus 3 (Ainsworth, 1935), cucumber virus 4 (Ainsworth, 1935; Hollings *et al.*, 1975), bottlegourd Indian mosaic virus, cucumis virus 2. *Standard acronym:* CGMMV.

First reported in *Cucumis sativus*; from Great Britain; by Ainsworth (1935).

**Natural host range and symptoms**

Symptoms persist.

*Cucumis sativus, Citrullus vulgaris, Lagenaria siceraria* (bottlegourd) — mosaic.

**Transmission**

Transmitted by means not involving a vector, or a vector; an insect; *Raphidopalpa fevicolis* (Rao and Varma, 1984); Coleoptera. Virus transmitted by mechanical inoculation.

**Geographical distribution**

Spreads in the Eurasian region; India, Japan, and the UK.

**Experimental host range**

Few (<3) families susceptible.

Diagnostically susceptible

*Cucumis sativus, Citrullus vulgaris* — systemic mosaic.

Diagnostically insusceptible

*Datura stramonium, Petunia hybrida.*

Maintenance and propagation hosts

*Cucumis sativus.*

Assay hosts (**L**ocal lesions or **W**hole plants)

*Chenopodium amaranticolor* (L), *Cucumis sativus* (W).

Dicotyledonous Hosts (***Susceptible*** *or Insusceptible)*

*Capsicum annuum*. ***Chenopodium murale***. ***Citrullus lanatus***. *Nicotiana glutinosa*. ***Nicotiana tabacum***.

**Properties of particles in sap**

TIP: 80–90 °C. LIV: more than 240 days (at 20°C). DEP: $\log_{10}$ minus 6. Leaf sap contains many virions.

**Purification**

Method: Tung and Knight (1972); Francki and McLean (1968).

**Particle morphology**

Virions rod–shaped; not enveloped; usually straight; with a clear modal length; of *c* 300 nm; *c.* 15 nm wide. Axial canal obvious. Basic helix obvious.

**Biochemical properties**

Virions contain *c.* 5 % nucleic acid; *c.* 95 % protein.

*Genome* consists of RNA; single–stranded; linear; unipartite; largest (or only) genome part 6.5 kb. Genomic nucleic acid isolated by Peden and Symons (1973). Base composition 23.2 % G (CGMMV-W, 25.8% (CV4), 25.5% (CV3)); 24.6 % A (CGMMV-

W, 25.8% (CV4), 25.8% (CV3)); 20.6 % C (CGMMV-W, 19.3% (CV4), 18.3% (CV3)); 31.6 % U (CGMMV-W, 29.5% (CV4), 30.8% (CV3)). Infectivity decreased when deproteinised with proteases; retained when deproteinised with phenol or detergent. Poly A region absent. Additional factor not required for infectivity. Genome has *t*RNA–like activity. Genome accepts histidine. Nucleotide sequence references: Meshi *et al.* (1983).

*Sequence database accession code(s):*

D12505 Em(40)_vi:MCGSHCG Gb(84)_vi:MCGSHCG Cucumber green mottle mosaic virus (CGMMV) complete genome. 10/92 6,421bp.

J02054 Gb(84)_vi:MCGCPW cucumber green mottle mosaic virus coat protein gene. 9/83 1,071bp.

J04322 Em(40)_vi:TOBMCG30 Gb(84)_vi:MCG30KP Cucumber green mottle mosaic virus 30K protein gene, complete cds, and 180K and coat protein 3 sequences.

*Features of the genome:* the genomic RNA has been sequenced (Ugaki *et al.*, 1991).

*Non–genomic nucleic acid* found in the virions; is subgenomic *m*RNA. Sub–genomic *m*RNA found in infected cells.

*Virion protein(s)* one; $M_r$ 17261; coat protein. Method of preparation: Fraenkel-Conrat (1957); Tung and Knight (1972). Amino acid sequence: Meshi *et al.* (1983). Amino acid composition: Nozu *et al.* (1971); Tung and Knight (1972).

## Replication

Replication does not depend on a helper virus.

## Cytopathology

Virions found in leaves, mesophyll, epidermis, vascular parenchyma, xylem, phloem, companion cells and all parts of the host plant; in cytoplasm and in cell vacuoles. Inclusions present in infected cells; are crystals in the cytoplasm; they contain virions. Other cellular changes: vesiculation of mitochondria.

## Taxonomy and Relationships

*Tobamovirus.*

Virus(es) with serologically related virions: frangipani mosaic, kyuri green mottle mosaic, ribgrass mosaic, tobacco mosaic, tomato mosaic and tobacco mild green mosaic viruses. Virus(es) with serologically unrelated virions: odontoglossum ringspot, sunn-hemp mosaic viruses.

The host ranges of cucumber green mottle mosaic virus and kyuri green mottle mosaic virus are similar, and their coat proteins have greater amino acid sequence similarity than either has to those of other tobamoviruses, which are serologically more distantly related.

## References


Ainsworth, G.C. (1935). *Ann. appl. Biol.* **22**: 55.

Fraenkel-Conrat, H. (1957). *Virology* **4**: 1.

Francki, R.I.B. and McLean, G.D. (1968). *Aust. J. biol. Sci.* **21**: 1311.

Francki, R.I.B., Hu, J. and Palukaitis, P. (1986). *Intervirology* **26**: 156.

Hollings, M., Komuro, Y. and Tochihara, H. (1975). *CMI/AAB Descr. Pl. Viruses* No. 154, 4 pp.

Meshi, T., Kiyama, R. Ohmi, T. and Okada, Y. (1983). *Virology* **127**: 54.

Nozu, Y., Tochihara, H., Komuro, Y. and Okada, Y. (1971). *Virology* **45**: 577.

Peden, K.W.C. and Symons, R.H. (1973). *Virology* **53**: 487.

Rao, D.C. and Varma, A. (1984). *Phytopath. Z.* **109**: 325.

Tung, J.S. and Knight, C.A. (1972). *Virology* **48**: 574.

Ugaki, M., Toriyama, M., Kakutani, T., Hidaka, S., Kiguchi, T., Nagata, R., Sato, T., Motoyoshi, F. and Nishiguchi, M. (1991). *J. gen. Virol.* **72**: 1487.

# Cucumber leaf spot *carmovirus*

**Data collated by Sc. H. Kleinhempel and I. Weber, 1987.**

*Synonyms:* cucumber fruit streak virus (Gallitelli *et al.*, 1983; Weber *et al.*, 1986). *Standard acronym:* CLSV. *Strains:* cucumber leaf spot (LS), cucumber fruit streak (FS).

First reported in *Cucumis sativus*; from Germany; by Weber *et al.* (1982).

**Natural host range and symptoms**

Symptoms disappear soon after infection.
*Cucumis sativus* — chlorotic spots with necrotic centres.

**Transmission**

Transmitted by a vector; a fungus; *Olpidium radicale*; Chytridiales. Virus transmitted by mechanical inoculation; not transmitted by contact between plants; transmitted by seed.

**Ecology and control**

Studies reported by Weber *et al.* (1982); Gallitelli *et al.* (1983).

**Geographical distribution**

Spreads in Germany, Greece, Jordan, and the UK.

**Experimental host range**

Several (3–9) families susceptible.

Diagnostically susceptible

*Celosia argentea*, *Gomphrena globosa*, *Chenopodium murale*, *C. quinoa*, *Nicotiana clevelandii*, *Petunia hybrida* — necrotic local lesions.

Diagnostically insusceptible

*Phaseolus vulgaris* var. 'Pinto', *Avena sativa*, *Datura stramonium*, *Lycopersicon esculentum*, *Nicotiana glutinosa*.

Maintenance and propagation hosts

*Cucumis sativus, Nicotiana megalosiphon.*

Assay hosts (**L**ocal lesions or **W**hole plants)

*Chenopodium quinoa* (L).

Dicotyledonous Hosts (***Susceptible*** *or Insusceptible)*

*Apium graveolens*. *Brassica oleracea* ssp. *botrytis*. *Capsicum annuum*. ***Celosia argentea***. ***Celosia cristata***. ***Chenopodium amaranticolor***. ***Chenopodium capitatum***. ***Chenopodium murale***. ***Chenopodium quinoa***. ***Cucumis melo***. ***Cucumis sativus***. ***Cucurbita maxima***. ***Cucurbita pepo***. *Datura stramonium*. ***Gomphrena globosa***. *Lactuca sativa*. ***Luffa acutangula***. *Lycopersicon esculentum*. ***Nicotiana clevelandii***. *Nicotiana glutinosa*. ***Nicotiana megalosiphon***. *Nicotiana tabacum*. ***Ocimum basilicum***. ***Petunia hybrida***. *Phaseolus vulgaris*. *Physalis floridana*. *Pisum sativum*. *Raphanus sativus*. *Solanum melongena*. ***Spinacia oleracea***. *Vicia faba*. ***Vigna unguiculata***. ***Zinnia elegans***.

Monocotyledonous Hosts (***Susceptible*** *or Insusceptible)*

*Avena sativa*. *Triticum aestivum*.

Sources of host-range data

Weber *et al.* (1982); Gallitelli *et al.* (1983).

**Properties of particles in sap**

TIP: 80–85 °C. LIV: 20 days. DEP: $\log_{10}$ minus 6–7. Leaf sap contains few virions (naturally), or contains many virions (artificially). Electron microscopy: ISEM is recommended.

**Purification**

Method: Weber *et al.* (1982).

**Particle morphology**

Virions isometric; not enveloped; *c.* 28 nm in diameter; rounded in profile; with a conspicuous capsomere arrangement.

**Physical properties**

One sedimenting component in purified preparations; sedimentation coefficient 132 *S* (FS). Density 1.342 g $cm^{-3}$ in CsCl (LS), or 1.35 g $cm^{-3}$ in CsCl (FS).

**Biochemical properties**

Virions contain 20 % nucleic acid; 80 % protein.

*Genome* consists of RNA; single–stranded; linear; unipartite; largest (or only) genome part 5.3 kb (LS), or 4.6 kb (FS). Genomic nucleic acid isolated by Leiser *et al.* (1981).

*Virion protein(s)* one; $M_r$ 44000 (LS), or 47000 (FS).

**Replication**

Replication does not depend on a helper virus.

**Cytopathology**

Virions found in leaves, mesophyll and phloem; in cytoplasm. Inclusions present in infected cells; are crystals in the cytoplasm; they contain virions. Other cellular changes: mitochondria maybe deformed and the Golgi increased to form many small vesicles containing fine fibrils.

**Taxonomy and Relationships**

*Carmovirus: Tombusviridae.*

Virus(es) with serologically unrelated virions: carnation mottle, cocksfoot mild mosaic, cucumber necrosis, melon necrotic spot, narcissus tip necrosis, tomato bushy stunt, tobacco necrosis (Weber *et al.*, 1982), artichoke mottled crinkle, cymbidium ringspot, carnation mottle, cucumber necrosis, pelargonium leaf curl, southern bean mosaic, sowbane mosaic, tobacco necrosis and tomato bushy stunt viruses (Gallitelli *et al.*, 1983).

Best tests for diagnosis: ELISA.

**References**


Campbell, R.N., Lecoq, H., Wipf-Scheibel, C. and Sim, S.T. (1991). *J. gen. Virol.* **72**: 3115.

Gallitelli, D., Vovlas, C. and Avgelis, A. (1983). *Phytopath. Z.* **106**: 149.

Leiser, R.M., Stanarius, A., and Stanarius, Th. (1981). *Arch. Phytopath. PflSchutz.* **17**: 299.

Weber, I. (1986). *CMI/AAB Descr. Pl. Viruses* No. 319, 4 pp.

Weber, I., Proll, E., Ostermann, W.D., Leiser, R.M., Stanarius, A. and Kegler, H. (1982). *Arch. Phytopath. PflSchutz.* **18**: 137.

Weber, I., Stenarius, A. and Kalinina, I. (1986). *Arch. Phytopath. PflSchutz.* **22**: 169.

# Cucumber mosaic *cucumovirus*

**Data collated by R.I.B. Francki, 1980; N. Habili, 1987.**

*Synonyms:* banana infectious chlorosis virus, coleus mosaic virus (Creager, 1945; Holcomb and Valverde, 1991), cowpea banding mosaic virus, cowpea ringspot virus, cucumber virus 1, lily ringspot virus (Brierley and Travis, 1958), pea top necrosis virus, peanut yellow mosaic virus, southern celery mosaic virus (Doolittle, 1916; Price, 1935; Wellman, 1934), soybean stunt virus (Hanada and Tochihara, 1982), spinach blight virus, tomato fern leaf virus, pea western ringspot virus. *Standard acronym:* CMV. *Strains:* numerous, the better known include: A–CMV, E–CMV, L–CMV, N–CMV, P–CMV, Z–CMV and WAI/WAII; there seem to be two antigenic groups, ToRS and DTL (Devergne, 1975).

First reported in *Cucumis sativus*; from the U.S.A; by Price (1934).

## Natural host range and symptoms

Symptoms persist.

Very many hosts (Kaper and Waterworth, 1981) including:

*Cucumis sativus* and many other cucurbits — mosaics and stunting, reduced fruit yield.

*Lycopersicon esculentum* — mosaic, reduction of leaf laminae ("fernleaf") and stunting.

*Spinacia oleracea* — severe chlorosis and stunting.

## Transmission

Transmitted by a vector; an insect; more than 60 spp. including *Acyrthosiphon pisum, Aphis craccivora* and *Myzus persicae*; Aphididae. Transmitted in a non–persistent manner. Virus transmitted by mechanical inoculation; transmitted by seed (in 19 species, but in variable extents).

## Geographical distribution

Probably distributed worldwide.

## Experimental host range

Many (>9) families susceptible.

Diagnostically susceptible

*Chenopodium amaranticolor, C. quinoa* — chlorotic local lesions.

*Cucumis sativus* — systemic mosaic.

*Vigna unguiculata* — necrotic local lesions.

*Lycopersicon esculentum, N. x edwardsonii, N. glutinosa, N. tabacum* — symptoms depend on virus strain.

Maintenance and propagation hosts

*Cucumis sativus, Nicotiana clevelandii, N. glutinosa, N. tabacum.*

Assay hosts (**L**ocal lesions or **W**hole plants)

*Chenopodium amaranticolor* (L), *C. quinoa* (L), *Vigna unguiculata* (L).

Dicotyledonous Hosts *(**Susceptible** or Insusceptible)*

***Amaranthus caudatus*. *Amaranthus retroflexus*. *Antirrhinum majus*. *Apium graveolens*. *Atriplex hortensis*. *Beta vulgaris*. *Brassica campestris*** ssp. ***napus*. *Brassica campestris*** ssp. ***pekinensis*. *Brassica campestris*** ssp. ***rapa*. *Brassica juncea*. *Calendula officinalis*. *Capsella bursa–pastoris*. *Capsicum annuum*. *Capsicum frutescens*. *Catharanthus roseus*.** *Celosia argentea*. ***Cheiranthus cheiri*. *Chenopodium album*.**

***Chenopodium amaranticolor. Chenopodium foetidum. Chenopodium hybridum. Chenopodium murale. Chenopodium quinoa. Chrysanthemum morifolium. Cicer arietinum. Cichorium endiva. Citrullus lanatus.*** *Citrullus vulgaris.* ***Coriandrum sativum. Crotalaria spectabilis. Cucumis melo. Cucumis sativus. Cucurbita maxima. Cucurbita moschata. Cucurbita pepo.*** *Dahlia pinnata.* ***Datura ferox. Datura metel. Datura stramonium. Datura tatula. Daucus carota. Emilia sagittata. Fagopyrum esculentum. Glycine max. Gomphrena globosa. Gossypium hirsutum. Helianthus annuus. Hibiscus esculentus. Hyoscyamus niger. Ipomoea nil.*** *Lactuca sativa. Lathyrus odoratus.* ***Lens esculenta. Lotus corniculatus. Lupinus albus. Lupinus angustifolius. Lycopersicon esculentum. Lycopersicon pimpinellifolium. Matthiola incana. Medicago sativa. Melilotus alba. Momordica balsamina.*** *Nicandra physaloides.* ***Nicotiana bigelovii. Nicotiana clevelandii. Nicotiana debneyi. Nicotiana glutinosa. Nicotiana megalosiphon. Nicotiana rustica. Nicotiana sylvestris. Nicotiana tabacum.*** *Petunia hybrida.* ***Phaseolus lunatus. Phaseolus vulgaris. Physalis floridana. Physalis peruviana. Phytolacca americana. Pisum sativum. Raphanus sativus. Rumex acetosa. Senecio vulgaris. Solanum demissum. Solanum melongena. Solanum nigrum. Solanum nodiflorum. Solanum rostratum. Solanum tuberosum. Sonchus oleraceus.*** *Spinacia oleracea. Stellaria media.* ***Tetragonia expansa. Trifolium hybridum. Trifolium incarnatum. Trifolium pratense. Trifolium repens. Tropaeolum majus. Verbesina encelioides. Vicia faba. Vicia sativa. Vigna radiata. Vigna unguiculata. Vigna unguiculata*** ssp. ***sesquipedalis.*** *Zinnia elegans.*

Sources of host-range data

Douine *et al.* (1979).

**Properties of particles in sap**

TIP: 55–70 °C. LIV: 1–10 days. DEP: $\log_{10}$ minus 3–6. Infectivity of sap not changed by treatment with di–ethyl ether. Leaf sap contains few virions, or contains many virions. Electron microscopy: freeze-dry preparations for shadowing, or fix with aldehyde for negative staining with PTA.

**Purification**

Method: Lot *et al.* (1972) and modified by Peden and Symons in 1973 (Clark *et al.*, 1974).

**Particle morphology**

Virions isometric; not enveloped; 29 nm in diameter; rounded in profile; without a conspicuous capsomere arrangement.

**Physical properties**

One sedimenting component in purified preparations; sedimentation coefficient 99 *S*. Density 1.367 g $cm^{-3}$ in CsCl (after fixation with formaldehyde). Isoelectric point pH 5.5 (Q strain). $A_{260}/A_{280}$ ratio 1.7 (corrected).

**Biochemical properties**

Virions contain 18 % nucleic acid; 82 % protein; 0 % lipid.

*Genome* consists of RNA; single–stranded; linear; of three parts; largest (or only) genome part the largest 3.389 kb; the 2nd largest 3.035 kb; the 3rd largest 2.197 kb. Genomic nucleic acid isolated by Gould and Symons (1977). Base composition 24 % G; 23 % A; 23 % C; 30 % U. 5' terminus of RNA has a methylated nucleotide cap. Infectivity retained when deproteinised with proteases, or decreased when deproteinised with proteases; retained when deproteinised with phenol or detergent. Poly A region absent. Genome has *t*RNA–like activity. Genome accepts tyrosine.

*Sequence database accession code(s):*

D00355 Em(40)_vi:CUCMVRN2 Gb(84)_vi:MCVRN2 Cucumber mosaic Virus Fny strain RNA2. 8/90 3,050bp.

D00356 Em(40)_vi:CUMCVFRN Gb(84)_vi:MCVFRNA1 (Fny strain) RNA1. 3/91 3,357bp.

D00385 Em(40)_vi:CUMCVO3 Gb(84)_vi:MCVO3, strain O, RNA 3, virus genome. 4/90 2,217bp.

D00462 Em(40)_vi:MCVRNA3C Gb(84)_vi:MCVRNA3C RNA 3, coat protein gene and flanks. 3/91 1,255bp.

D00463 Em(40)_vi:MCVRNA3W Gb(84)_vi:MCVRNA3WL RNA 3, coat protein gene and flanks. 3/91 1,267bp.

D10209 Em(40)_vi:MCVORNA2 Gb(84)_vi:MCVORNA2 (CMV) RNA 2. 10/92 3,049bp.

D10538 Em(40)_vi:MCVCP3A1 Gb(84)_vi:MCVCP3A1 (Fny-CMV) RNA 3, complete genome. 6/92 2,216bp.

D10539 Em(40)_vi:MCVCP3A2 Gb(84)_vi:MCVCP3A2 (M-CMV) RNA 3, complete genome. 6/92 2,215bp.

D10544 Em(40)_vi:MCVCPFC Gb(84)_vi:MCVCPFC (CMV-FC) coat protein, complete cds. 6/92 657bp.

D10545 Em(40)_vi:MCVCPP6 Gb(84)_vi:MCVCPP6 (CMV-P6) coat protein, complete cds. 6/92 657bp.

D12499 Em(40)_vi:MCVYRNA3 Gb(84)_vi:MCVYRNA3 RNA3, 3a protein and coat protein genes, complete cds. 9/92 2,217bp.

D12537 Em(40)_vi:MCVRNA1 Gb(84)_vi:D12537 RNA1, complete sequence. 10/92 3,361bp.

D12538 Em(40)_vi:MCVRNA2 Gb(84)_vi:D12538 RNA2, complete sequence. 10/92 3,051bp.

D16403 Em(40)_vi:MCVSVR1 Gb(84)_vi:MCVL1 RNA for viral replicase subunit, complete cds. 11/93 3,359bp

D16405 Em(40)_vi:MCVSVR2 Gb(84)_vi:MCVL3 RNAs for movement protein and viral coat protein, complete cds. 11/93 2

D16406 Em(40)_vi:MCVPMPVCP Gb(84)_vi:MCVL2 RNA for viral replicase subunit, complete cds. 11/93 3,047bp

D28559 Em(40)_vi:MCVKN Gb(84)_vi:MCVKN RNA. 5/94 405bp.

D28778 Em(40)_vi:MCV1A1 Gb(84)_vi:MCV1A1 RNA for 1a. 3/94 3,358bp.

D28779 Em(40)_vi:MCV2A2 Gb(84)_vi:MCV2A2 RNA for 2a and 2b. 3/94 3,042bp.

D28780 Em(40)_vi:MCV3A3 Gb(84)_vi:MCV3A3 RNA for 3a and coat protein. 3/94 2,214bp

J02059 Em(40)_vi:MCVRNA3 Gb(84)_vi:MCVRNA3 (Q strain) RNA 3, complete segment including RNA 4 cds. 5/92 2,193bp.

K01781 Gb(84)_vi:MCVRNA1 (Q strain) RNA 1, 3' end. 1/85 271bp.

K01782 Gb(84)_vi:MCVRNA2 (Q strain) RNA 2, 3' end. 1/85 271bp.

L15336 Em(40)_vi:MCV3APCOA Gb(84)_vi:MCV3APCOAT 3A protein and coat protein mRNA (RNA3) complete cds. 5/93 2,209bp.

L36251 Gb(84)n:MCVRNA3KOR RNA3 encoding movement protein (MP) and capsid protein (CP) genes, comp

L36525 Gb(84)n:MCVCPB coat protein (CP) gene, complete cds. 9/94 894bp.

M21464 Em(40)_vi:CUMCVRNA Gb(84)_vi:MCVRNA3A RNA 3 containing 3A protein and coat protein, complete cds. 7/89 2,197b

M22710 Em(40)_vi:CUMCVCPA Gb(84)_vi:MCVCPA coat protein gene, complete cds. 4/90 1,362bp.

M29129 Em(40)_vi:CUMCVCOP Gb(84)_vi:MCVCOPR coat protein (segment 4) mRNA, partial cds. 2/90 106bp.

M57602 Em(40)_vi:MCVYCNS Gb(84)_vi:MCVYCNS Cucumber mosaic virus Y strain, 3A protein, coat protein mRNA, complete cds. 9/92 2,217bp.

M98499 Em(40)_vi:MCVCPAA Gb(84)_vi:MCVCPAA coat protein (CP) RNA, complete cds. 8/92 657bp.

M98500 Em(40)_vi:MCVCPAB Gb(84)_vi:MCVCPAB coat protein (CP) RNA, complete cds. 8/92 657bp.

M98501 Em(40)_vi:MCVCPAC Gb(84)_vi:MCVCPAC coat protein (CP) RNA, complete cds. 8/92 657bp.

S70105 Em(40)_un:S70105 Gb(84)_vi:S70105 cp=coat protein CMV, subgroup II, WL, Genomic, 1360 nt. 2/93 1,360bp.

U10922 Gb(84)_vi:CMU10922 DKRD coat protein gene, complete cds. 6/94 834bp.

U10923 Gb(84)_vi:CMU10923 SP103 coat protein gene, complete cds. 6/94 841bp.

U10924 Gb(84)_vi:CMU10924 SP104 coat protein gene, complete cds. 6/94 835bp.

X00817 Em(40)_vi:CURNA01 Gb(84)_vi:CURNA01 strain Y (CMV-Y) RNA 5 sequence. 5/92 368bp.

X00985 Em(40)_vi:CVRNA02 Gb(84)_vi:CVRNA02 RNA 2 (Q-CMV RNA 2). 9/93 3,035bp.

X01333 Em(40)_vi:CURNA1 Gb(84)_vi:CURNA1 (CMV) RNA1 3' terminal region. 11/85 270bp.

X01334 Em(40)_vi:CURNA2 Gb(84)_vi:CURNA2 (CMV) RNA2 3' terminal region. 5/92 270bp

X01335 Em(40)_vi:CURNA3 Gb(84)_vi:CURNA3 (CMV) RNA3 3' terminal region. 11/85 267bp.

X01336 Em(40)_vi:CURNA4 Gb(84)_vi:CURNA4 (CMV) RNA4 3' terminal region. 11/85 267bp.

X02733 Em(40)_vi:CURNA1Q Gb(84)_vi:CURNA1Q strain Q (Q-CMV) RNA 1 sequence. 9/93 3,389bp.

X16386 Em(40)_vi:CUMCVRN4 Gb(84)_vi:CUMCVRN4 (CMV) RNA4 gene for coat protein. 9/93 1,007bp.

X65017 Em(40)_vi:CMVCOPR Gb(84)_vi:CMVCOPR genomic RNA for coat protein. 6/92 963bp

X77855 Em(40)_vi:CMVCOATP Gb(84)_vi:CMVCOATP (As) RNA4 mRNA for coat protein. 2/94 1,043bp.

X81161 Gb(84)n:CMVT8920 T/89/20 genomic RNA. 8/94 450bp.

X81162 Gb(84)n:CMVT8943 T/89/43 genomic RNA. 8/94 450bp.

X81163 Gb(84)n:CMVT9020 T/90/20 genomic RNA. 8/94 450bp.

X81164 Gb(84)n:CMVT9023 T/90/23 genomic RNA. 8/94 450bp.

X81165 Gb(84)n:CMVT9104 T/91/04 genomic RNA. 8/94 450bp.

X81166 Gb(84)n:CMVT9102 T/91/02 genomic RNA. 8/94 450bp

X81167 Gb(84)n:CMVT9107 T/91/07 genomic RNA. 8/94 450bp.

X81168 Gb(84)n:CMVT9109 T/91/09 genomic RNA. 8/94 450bp.

X81169 Gb(84)n:CMVT9224 T/92/24 genomic RNA. 8/94 450bp.

X81170 Gb(84)n:CMV920 T/92/09 genomic RNA. 8/94 450bp.

Z12818 Em(40)_vi:CMV3ACP Gb(84)_vi:CMV3ACP genes for 3a protein and coat protein. 2/94 2,199bp.

Z21863 Em(40)_vi:CM2BPRO Gb(84)_vi:CM2BPRO RNA 4A. 3/94 682bp.

D42079 Em(43)_vi:Mcvr3mpcp Gb(89)_vi:Mcvr3mpcp1 Cucumber mosaic virus gene for movement protein and viral coat protein. 11/94 2,238bp.

D42080 Em(43)_vi:Mcvr3mpc1 Gb(89)_vi:Mcvr3mpcp2 Cucumber mosaic virus gene for movement protein and viral coat protein. 11/94 2,214bp.

D42081 Em(43)_vi:Mcvst3acp Gb(89)_vi:Mcvst3acp Cucumber mosaic virus gene for movement protein and coat protein. 11/94 334bp.

D42082 Em(43)_vi:Mcvsrmpcp Gb(89)_vi:Mcvsrmpcpa Cucumber mosaic virus gene for movement potein and coat protein. 11/94 334bp.

D43798 Em(43)_vi:Mcv3apa Gb(89)_vi:Mcv3apa Cucumber mosaic virus mRNA for 3a protein. 12/94 840bp.

D43799 Em(43)_vi:Mcvr1pb Gb(89)_vi:Mcvr1pb Cucumber mosaic virus mRNA for RNA1 protein. 12/94 1,137bp.

D43800 Em(43)_vi:Mcvcp Gb(89)_vi:Mcvcp Cucumber mosaic virus mRNA for coat protein. 12/94 733bp.

D49496 Em(43)_vi:Mcv3acp Gb(89)_vi:Mcv3acp Cucumber mosaic virus gene for 3a and cp. 3/95 2,205bp.

L36251 Em(43)_vi:Mcvrna3ko Gb(89)_vi:Mcvrna3kor Cucumber mosaic virus RNA3 encoding movement protein (MP) and capsid protein (CP) genes.

L36525 Em(43)_vi:Mcvcpb Gb(89)_vi:Mcvcpb Cucumber mosaic virus coat protein (CP) gene, complete cds. 10/94 894bp.

L40953 Em(43)_vi:Mcvcopra Gb(89)_vi:Mcvcopra Cucumber mosaic virus coat protein, complete cds. 4/95 757bp.

S72187 Gb(89)_un:S72187 orf2a, orf2b (cucumber mosaic cucumovirus CMV, K, Genomic RNA, nt). 10/94 3,048bp.

U20668 Em(43)_vi:Cm20668 Gb(89)_vi:Cmu20668 Cucumber mosaic virus truncated protein 3a and coat protein mRNA, complete cds. 5/95 1,906bp.

U22821 Em(43)_vi:Cm22821 Gb(89)_vi:Cmu22821 Cucumber mosaic virus Ny coat protein gene, complete cds. 4/95 1,033bp.

U22822 Em(43)_vi:Cm22822 Gb(89)_vi:Cmu22822 Cucumber mosaic virus Sn coat protein gene, complete cds. 4/95 1,041bp.

X81161 Em(43)_vi:Cmvt8920 Gb(89)_vi:Cmvt8920 Cucumber mosaic Virus T/89/20 genomic RNA. 8/94 450bp.

X81162 Em(43)_vi:Cmvt8943 Gb(89)_vi:Cmvt8943 Cucumber mosaic Virus T/89/43 genomic RNA. 8/94 450bp.

X81163 Em(43)_vi:Cmvt9020 Gb(89)_vi:Cmvt9020 Cucumber mosaic Virus T/90/20 genomic RNA. 8/94 450bp.

X81164 Em(43)_vi:Cmvt9023 Gb(89)_vi:Cmvt9023 Cucumber mosaic Virus T/90/23 genomic RNA. 8/94 450bp. X81165 Em(43)_vi:Cmvt9104 Gb(89)_vi:Cmvt9104 Cucumber mosaic Virus T/91/04 genomic RNA. 8/94 450bp.

X81166 Em(43)_vi:Cmvt9102 Gb(89)_vi:Cmvt9102 Cucumber mosaic Virus T/91/02 genomic RNA. 8/94 450bp.

X81167 Em(43)_vi:Cmvt9107 Gb(89)_vi:Cmvt9107 Cucumber mosaic Virus T/91/07 genomic RNA. 8/94 450bp.

X81168 Em(43)_vi:Cmvt9109 Gb(89)_vi:Cmvt9109 Cucumber mosaic Virus T/91/09 genomic RNA. 8/94 450bp.

X81169 Em(43)_vi:Cmvt9224 Gb(89)_vi:Cmvt9224 Cucumber mosaic Virus T/92/24 genomic RNA. 8/94 450bp.

X81170 Em(43)_vi:Cmv920 Gb(89)_vi:Cmv920 Cucumber mosaic Virus T/92/09 genomic RNA. 8/94 450bp.

*Non–genomic nucleic acid* found in the virions; is subgenomic *m*RNA (of coat protein), or satellite RNA (in some isolates). Sub–genomic *m*RNA found in infected cells; coat protein *m*RNA, which is also the encapsidated RNA-4 of 1.027 kb.

*Virion protein(s)* one; $M_r$ 24247; coat protein. Method of preparation: Habili and Francki (1974). Amino acid sequence: Davies and Symons (1987). Amino acid composition: Mossop (1978).

*Virus–coded non–virion proteins* identified by genomic sequence analysis (Clark *et al.* (1974); Kumarasamy and Symons (1979); Gill *et al.* (1981)); three proteins found. $M_r$ of the largest 36300 (791 residues); RNA-1 translation product from Q-CMV, putative viral RNA replicase component (Rezaian *et al.*, 1985). $M_r$ of 2nd largest 31000 (333 residues; RNA-2 translation product from Q-CMV, putative viral RNA polymerase (Rezaian *et al.*, 1985)). $M_r$ of 3rd 9900 (353 residues); 3*a* protein, possible movement protein (Davies and Symons, 1987).

## Replication

Replication does not depend on a helper virus.

## Cytopathology

Virions found in all parts of the host plant; in cytoplasm. Inclusions present in infected cells; are crystals in the cytoplasm (that are often rhomboidal, hexagonal or roughly spherical and may appear as solid hollow structures); they contain virions.

## Taxonomy and Relationships

*Cucumovirus: Bromoviridae* (the type species).

Virus(es) with serologically related virions: peanut stunt and tomato aspermy viruses, but distantly.

Sequence homologies indicate that the virus is distantly related to brome mosaic virus, and also, in sequences in RNA-1 and RNA-2, to alfalfa mosaic and tobacco mosaic viruses (Rezaian *et al.*, 1985; Davies and Symons, 1987).

## Comments

Pea western ringspot virus is probably a mixture of cucumber mosaic and clover yellow vein viruses.

## References


Brierley, P. and Travis, R.V. (1958). *Pl. Dis. Reptr* **42**: 1034.

Clark, G.L., Peden, K.W.C. and Symons, R.H. (1974). *Virology* **62**: 434.

Creager, D.B. (1945). *Phytopathology* **35**: 223.

Devergne, J.-C. (1975). *Meded. Facult. Landb. Rijks-Univ. Gent* **40**: 19.

Doolittle, S.P. (1916). *Phytopathology* **6**: 145.

Douine, L., Quiot, J.B., Marchoux, G. and Archange, P. (1979). *Annls. Phytopath.* **11**: 439.

Finch, J.T., Klug, A. and van Regenmortel, M.H.V. (1967). *J. mol. Biol.* **24**: 303.

Francki, R.I.B., Mossop, D.W. and Hatta, T. (1979). *CMI/AAB Descr. Pl. Viruses* No. 213, 6 pp.

Gibbs, A.J. and Harrison, B.D. (1970). *CMI/AAB Descr. Pl. Viruses* No. 1, 4 pp.

Gill, D.S., Kumarasamy, R. and Symons, R.H. (1981). *Virology* **113**: 1.

Habili, N. and Francki, R.I.B. (1974). *Virology* **57**: 392.

Hanada, K. and Tochihara, H. (1982). *Phytopathology* **72**: 761.

Holcomb, G.E. and Valverde, R.A. (1991). *Plant Dis.* **75**: 1183.

Jagger, I.C. (1916). *Phytopathology* **6**: 148.

Kaper, J.M. and Waterworth, H.C. (1981). In: *Plant Virus Infections: Comparative Diagnosis*, pp.257–332; ed. E. Kurstak. Elsevier/North Holland, Amsterdam.

Komuro, Y. (1973). *Viruses in Vegetables*. Seibundo Shinkosha, Tokyo.
Kumarasamy, R. and Symons, R.H. (1979). *Virology* **96**: 622.
Lot, H., Jonard, G. and Richards, K. (1977). *FEBS Lett.* **80**: 395.
Mossop, D.W. (1978). *Ph.D. Thesis*, University of Adelaide, Australia.
Price, W.C. (1934). *Phytopathology* **24**: 743.
Price, W.C. (1935). *Phytopathology* **25**: 947.
Rezaian, M.A., Williams, R.H.V. and Symons, R.H. (1985). *Eur. J. Biochem.* **150**: 331.
Wellman, F.L. (1934). *Phytopathology* **24**: 695.

# Cucumber mosaic virus satellite RNA

*Sequence database accession code(s):*

A10086 Em(40)_vi:A10086 Gb(84)_pat:A10086 satellite I 17N. 2/94 335bp
D00541 Em(40)_vi:MCVSATRA Gb(84)_vi:MCVSATRA satellite RNA, complete cds. 3/91 338bp.
D00542 Em(40)_vi:MCVSATY Gb(84)_vi:MCVSATY (CMV) satellite RNA, complete sequence. 5/92 369bp.
D00699 Em(40)_vi:CUMCVS10 Gb(84)_vi:MCVSATR satellite RNA. 3/91 338bp.
D10036 Em(40)_vi:MCVS19SAT Gb(84)_vi:MCVS19SAT (CMV) satellite RNA. 5/92 370bp.
D10037 Em(40)_vi:MCVT73SAT Gb(84)_vi:MCVT73SAT (CMV) satellite RNA. 5/92 335bp.
D10038 Em(40)_vi:MCVYSAT Gb(84)_vi:MCVYSAT2 (CMV) satellite RNA, complete sequence. 5/92 369bp.
D10039 Em(40)_vi:MCVT43SAT Gb(84)_vi:MCVT43SAT (CMV) satellite RNA. 9/93 384bp
D11048 Em(40)_vi:MCVPKO2S8 Gb(84)_vi:MCVPKO2S1 satellite RNA. 3/93 368bp
D11049 Em(40)_vi:MCVPKO2S7 Gb(84)_vi:MCVPKO2S2 satellite RNA. 3/93 368bp
D11050 Em(40)_vi:MCVPKO2S2 Gb(84)_vi:MCVPKO2S3 satellite RNA. 3/93 368bp
D11051 Em(40)_vi:MCVPKO2S3 Gb(84)_vi:MCVPKO2S4 satellite RNA. 3/93 368bp
D11052 Em(40)_vi:MCVPKO2S4 Gb(84)_vi:MCVPKO2S5 satellite RNA. 3/93 368bp
D11053 Em(40)_vi:MCVPKO2S5 Gb(84)_vi:MCVPKO2S6 satellite RNA. 3/93 368bp
J02060 Em(40)_vi:CUSATRNA Gb(84)_vi:MCVSATRNA satellite rna(sat–rna). 4/90 336bp.
J02061 Em(40)_vi:CUSATRN5 Gb(84)_vi:MCVSATRN5 -associated satellite RNA 5 ((1)caRNA5). 4/90 334bp.
M10686 Em(40)_vi:CUMCVSAT Gb(84)_vi:MCVSATRNB -associated RNA 5 (CARNA 5), complete sequence. 7/89 336bp.
M14934 Em(40)_vi:CUMCVS01 Gb(84)_vi:MCVSATRNC (strain S)-associated satellite (S)CARNA 5 RNA, complete cds. 7/89 339b
M16585 Em(40)_vi:CUMCVSRG Gb(84)_vi:MCVSRG G-sat RNA. 7/89 332bp.
M16586 Em(40)_vi:CUMCVSRB Gb(84)_vi:MCVSRB1 B1-sat RNA. 7/89 337bp.
M16587 Em(40)_vi:CUMCVS02 Gb(84)_vi:MCVSRB2 B2-sat RNA. 7/89 340bp.

M16588 Em(40)_vi:CUMCVS03 Gb(84)_vi:MCVSRB3 B3-sat RNA. 7/89 337bp

M16589 Em(40)_vi:CUMCVSRW Gb(84)_vi:MCVSRWL1 WL1-sat RNA. 7/89 336bp

M16590 Em(40)_vi:CUMCVS08 Gb(84)_vi:MCVSRWL2 WL2-sat RNA. 7/89 342bp

M18867 Em(40)_vi:CUMCVSRI Gb(84)_vi:MCVSRI17A (CMV) satellite I17N(1) RNA. 12/91 335bp.

M18868 Em(40)_vi:CUMCVS05 Gb(84)_vi:MCVSRI17B (CMV) satellite I17N(2) RNA. 12/91 335bp.

M18869 Em(40)_vi:CUMCVSRR Gb(84)_vi:MCVSRRA (CMV) satellite R(1) RNA. 12/91 307bp.

M18870 Em(40)_vi:CUMCVS06 Gb(84)_vi:MCVSRRB (CMV) satellite R(2) RNA. 12/91 331bp.

M18871 Em(40)_vi:CUMCVS07 Gb(84)_vi:MCVSRRC (CMV) satellite R(3) RNA. 12/91 333bp.

M18872 Em(40)_vi:CUMCVSRD Gb(84)_vi:MCVSRD (CMV) satellite D RNA. 12/91 335bp

M20350 Em(40)_vi:CUMCVC5T Gb(84)_vi:MCVC5TNA CARNA 5 gene with ORFs I, IIA, and IIB, 5' end, clone D. 7/89 335bp.

M20352 Em(40)_vi:CUMCVC01 Gb(84)_vi:MCVC5TNB CARNA 5 gene with ORFs I, IIA, and IIB, 5' end, clone Sq10. 7/89 335bp.

M20353 Em(40)_vi:CUMCVC02 Gb(84)_vi:MCVC5TNC CARNA 5 gene with ORFs I, IIA, and IIB, 5' end, clone Ch20. 7/89 335bp.

M20354 Em(40)_vi:CUMCVC03 Gb(84)_vi:MCVC5TND CARNA 5 gene with ORFs I, IIA, and IIB, 5' end, clone X2nT3. 7/89 334bp.

M20355 Em(40)_vi:CUMCVC04 Gb(84)_vi:MCVC5TNE CARNA 5 gene with ORFs I, IIA, and IIB, 5' end, clone X7. 7/89 334bp.

M20356 Em(40)_vi:CUMCVC05 Gb(84)_vi:MCVC5TNF CARNA 5 gene with ORFs I, IIA, and IIB, 5' end, clone X12. 7/89 336bp.

M20357 Em(40)_vi:CUMCVC06 Gb(84)_vi:MCVC5TNG CARNA 5 gene with ORFs I, IIA, and IIB, 5' end, clone X15. 7/89 336bp.

M20360 Em(40)_vi:CUMCVC07 Gb(84)_vi:MCVC5TNH CARNA 5 gene with ORFs I, IIA, and IIB, 5' end, clone X2c. 7/89 338bp.

M20844 Em(40)_vi:CUMCVSRN Gb(84)_vi:MCVSRNAE E-satellite (E-satRNA) RNA. 7/89 339bp.

M20845 Em(40)_vi:CUMCVS09 Gb(84)_vi:MCVSRNAY OY-satellite (OY2-satRNA) RNA. 7/89 386bp.

M21742 Em(40)_vi:CUMCVS04 Gb(84)_vi:MCVSRDC5 (D) CARNA 5-sat RNA. 7/89 58bp.

M21743 Em(40)_vi:CUMCVSR1 Gb(84)_vi:MCVSR1C5 (1)CARNA 5-sat RNA. 7/89 59bp.

M21744 Em(40)_vi:CUMCVSRS Gb(84)_vi:MCVSRSC5 (S) CARNA 5-sat RNA. 7/89 61bp.

M21745 Em(40)_vi:CUMCVSRY Gb(84)_vi:MCVSRY Y-sat RNA. 7/89 59bp.

M30584 Em(40)_vi:MCVDSAT4 Gb(84)_vi:MCVDSAT4A (CMV) D-satellite RNA, clone pDsat4. 4/90 335bp.

M30585 Em(40)_vi:MCVDSAT5 Gb(84)_vi:MCVDSAT5A (CMV) D-satellite RNA, clone pDsat5. 4/90 334bp.

M30586 Em(40)_vi:MCVDSAT6 Gb(84)_vi:MCVDSAT6A (CMV) D-satellite RNA, clone pDsat6. 4/90 335bp.

M30587 Em(40)_vi:MCVDSAT9 Gb(84)_vi:MCVDSAT9A (CMV) D-satellite RNA, clone pDsat9. 4/90 335bp.

M30588 Em(40)_vi:MCVWLSA4 Gb(84)_vi:MCVWLSA47A (CMV) WL-satellite RNA, clone pWLsat47. 4/90 335bp.

M30589 Em(40)_vi:MCVWLS21 Gb(84)_vi:MCVWLS210A (CMV) WL- satellite RNA, clone pWLsat210. 4/90 333bp.

M30590 Em(40)_vi:MCVWLS32 Gb(84)_vi:MCVWLS324A (CMV) WL-satellite RNA, clone pWLsat324. 4/90 315bp

M30591 Em(40)_vi:MCVB1SAT Gb(84)_vi:MCVB1SAT1A (CMV) B-1-satellite RNA, clones pBsat 1,3,5. 4/90 338bp.

M30592 Em(40)_vi:MCVBSA2A Gb(84)_vi:MCVBSA2A (CMV) B-1-satellite RNA, clone pBsat2. 4/90 338bp.

M30593 Em(40)_vi:MCVBSA4A Gb(84)_vi:MCVBSA4A (CMV) B-1-satellite RNA, clone pBsat4. 4/90 338bp.

M64284 Em(40)_vi:SCBSATRNA Gb(84)_vi:SCBSATRNA Satellite (strain Ix) RNA. 12/91 334bp

S61869 Em(40)_vi:S61869 Gb(84)_vi:S61869 (satellite RNA) CMV, satellite RNA, Genomic Mutagenesis, 58 nt. 1/94

S68622 Gb(84)n:S68622 (satellite RNA D27- CARNA-5) CMV, -associated RNA-

S68624 Gb(84)n:S68624 (satellite RNA J876- CARNA-5) CMV-associated RNA

X51465 Em(40)_vi:CUCMWSAT Gb(84)_vi:CUCMWSAT WL3-satellite RNA. 9/90 340bp.

X53534 Em(40)_vi:CUMVSAT Gb(84)_vi:CUMVSAT satellite RNA. 9/90 339bp.

X54065 Em(40)_vi:CMVFNY1 Gb(84)_vi:CMVFNY1 Fny1-satellite RNA. 5/92 337bp.

X54066 Em(40)_vi:CMVFNY2 Gb(84)_vi:CMVFNY2 Fny2-satellite RNA. 5/92 337bp.

X57582 Gb(84)_vi:CMVSATA satellite RNA. 6/94 334bp.

X65455 Em(40)_vi:CMVSAT Gb(84)_vi:CMVSAT satellite RNA. 5/94 390bp

X69136 Em(40)_vi:CMVSATRNA Gb(84)_vi:CMVSATRNA (CMV) satellite RNA. 1/93 339bp

S68622 Gb(89)_vi:S68622 (satellite RNA D27-CARNA-5) (cucumber mosaic virus CMV, cucumber mosaic virus-associated RNA-

S68624 Gb(89)_vi:S68624 (satellite RNA J876-CARNA-5) (cucumber mosaic virus CMV, cucumber mosaic virus-associated RNA

X57582 Em(43)_vi:Cmvsata Gb(89)_vi:Cmvsata Cucumber mosaic virus satellite RNA. 6/94 334bp.

X86408 Em(43)_vi:Cmvt8asat Gb(89)_vi:Cmvt8asat Cucumber mosaic virus (CMV) satellite RNA (strain T8A). 4/95 340bp.

X86409 Em(43)_vi:Cmvt1sat Gb(89)_vi:Cmvt1sat Cucumber mosaic virus (CMV) satellite RNA (strain T1). 4/95 338bp.

X86410 Em(43)_vi:Cmvt2sat Gb(89)_vi:Cmvt2sat Cucumber mosaic virus (CMV) satellite RNA (strainT2). 4/95 334bp.

X86411 Em(43)_vi:Cmvt7sat Gb(89)_vi:Cmvt7sat Cucumber mosaic virus (CMV) satellite RNA (strain ). 4/95 335bp.

X86412 Em(43)_vi:Cmvt22sat Gb(89)_vi:Cmvt22sat Cucumber mosaic virus (CMV) satellite RNA (strain T22). 4/95 339bp.

X86413 Em(43)_vi:Cmvt16sat Gb(89)_vi:Cmvt16sat Cucumber mosaic virus (CMV) satellite RNA (strain T16). 4/95 335bp.

X86414 Em(43)_vi:Cmvt6sat Gb(89)_vi:Cmvt6sat Cucumber mosaic virus (CMV) satellite RNA (strain T6). 4/95 335bp.

X86415 Em(43)_vi:Cmvt4sat Gb(89)_vi:Cmvt4sat Cucumber mosaic virus (CMV) satellite RNA (strain T4). 4/95 340bp.
X86416 Em(43)_vi:Cmvt12sat Gb(89)_vi:Cmvt12sat Cucumber mosaic virus (CMV) satellite RNA (strain T12). 4/95 334bp.
X86417 Em(43)_vi:Cmvto79sa Gb(89)_vi:Cmvto79sa Cucumber mosaic virus (CMV) satellite RNA (strain To79). 4/95 336bp.
X86418 Em(43)_vi:Cmvt57sat Gb(89)_vi:Cmvt57sat Cucumber mosaic virus (CMV) satellite RNA (strain T57). 4/95 334bp.
X86419 Em(43)_vi:Cmvt27sat Gb(89)_vi:Cmvt27sat Cucumber mosaic virus (CMV) satellite RNA (strain ). 4/95 334bp.
X86420 Em(43)_vi:Cmvt8sat Gb(89)_vi:Cmvt8sat Cucumber mosaic virus (CMV) satellite RNA (strain ). 4/95 335bp.
X86421 Em(43)_vi:Cmvto80sa Gb(89)_vi:Cmvto80sa Cucumber mosaic virus (CMV) satellite RNA (strain TO80). 4/95 334bp.
X86422 Em(43)_vi:Cmvto77sa Gb(89)_vi:Cmvto77sa Cucumber mosaic virus (CMV) satellite RNA (strain TO77). 4/95 334bp.
X86423 Em(43)_vi:Cmvt59sat Gb(89)_vi:Cmvt59sat Cucumber mosaic virus (CMV) satellite RNA (strain T59). 4/95 334bp.
X86424 Em(43)_vi:Cmvng71sa Gb(89)_vi:Cmvng71sa Cucumber mosaic virus (CMV) satellite RNA (strain NG71). 4/95 338bp.
X86425 Em(43)_vi:Cmvcesat Gb(89)_vi:Cmvcesat Cucumber mosaic virus (CMV) satellite RNA (strain CE). 4/95 336bp.
X86426 Em(43)_vi:Cmvpgsat Gb(89)_vi:Cmvpgsat Cucumber mosaic virus (CMV) satellite RNA (strain PG). 4/95 334bp.
X86708 Em(43)_vi:Cmvt18sat Gb(89)_vi:Cmvt18sat Cucumber mosaic virus (CMV) satellite RNA (strain T18). 4/95 333bp.

**Taxonomy and Relationships**

Satellite RNA.

# Cucumber necrosis *tombusvirus*

**Data collated by D.M. Rochon, 1990.**

*Standard acronym:* CuNV.

First reported in *Cucumis sativus*; from Harrow, Ontario, Canada; by McKeen (1959).

## Natural host range and symptoms

Symptoms vary seasonally. Symptoms malformed chlorotic or purplish leaves, which dessicate and die. Fruits mottled.

*Cucumis sativus* — malformed chlorotic leaves.

## Transmission

Transmitted by a vector and means not involving a vector (in water and soil); a fungus; *Olpidium radicale*; Chytridiales. Principal natural vector(s): *Olpidium radicale*. Not transmitted by *Aphis gossypii*. Virus transmitted by mechanical inoculation; transmitted by contact between plants; not transmitted by seed; not transmitted by pollen.

## Ecology and control

Studies reported by Bos *et al.* (1984).

## Geographical distribution

Spreads in Canada (Ontario).

## Experimental host range

Several (3–9) families susceptible. Experimentally infected plants mostly show necrotic local lesions and systemic necrosis.

Diagnostically susceptible

*Nicotiana clevelandii* — necrotic local lesions, systemic necrosis.
*Nicotiana benthamiana* — systemic necrosis.
*Chenopodium amaranticolor* — small necrotic local lesions, no systemic infection.
*Vigna unguiculata* — brown necrotic local lesions, no systemic infection.
*Gomphrena globosa* — grey necrotic local lesions, no systemic infection.
*Nicotiana tabacum* — grey necrotic spots, no systemic infection.

Maintenance and propagation hosts

*Nicotiana clevelandii, Cucumis sativus.*

Assay hosts (**L**ocal lesions or **W**hole plants)

*Chenopodium amaranticolor* (L), *Vigna unguiculata* (L), *Gomphrena globosa* (L).

Dicotyledonous Hosts *(**Susceptible** or Insusceptible)*

***Chenopodium amaranticolor*. *Chenopodium quinoa*. *Cucumis sativus*. *Gomphrena globosa*. *Nicotiana benthamiana*. *Nicotiana clevelandii*. *Nicotiana tabacum*. *Phaseolus vulgaris*. *Vigna unguiculata*.**

Sources of host-range data

Dias and McKeen (1972).

## Properties of particles in sap

TIP: 75–80 °C (10 min). LIV: 30 days. DEP: $\log_{10}$ minus 4–5. Leaf sap contains many virions.

## Purification

Method: Tremaine *et al.* (1983).

## Particle morphology

Virions isometric; not enveloped; 30 nm in diameter; angular in profile.

## Physical properties

One sedimenting component in purified preparations; sedimentation coefficient 133 *S*. Density 1.33 g $cm^{-3}$ in CsCl. Isoelectric point pH 3.9 (pH).

## Biochemical properties

Virions contain 16 % nucleic acid; 84 % protein; 0 % lipid.

*Genome* consists of RNA; single–stranded; linear; unipartite; largest (or only) genome part 4.7 kb. Genomic nucleic acid isolated by Rochon: phenol or chloroform extraction in buffer containing 10 mm EDTA, Tris–HCl pH 8.9, 1% SDS. Infectivity retained when deproteinised with phenol or detergent. Poly A region absent. Additional factor not required for infectivity. Genome has no *t*RNA–like activity.

*Sequence database accession code(s):*

M25270 Gb(84)_vi:CNVGENO Cucumber necrosis virus, complete genome. 4/91 4,701bp.

S64508 Em(40)_vi:S64508 Gb(84)_vi:S64508 (defective interfering RNA DI15) cucumber necrosis virus CNV, Genomic RNA, 622 nt. 12/93 62.

U16698 Em(43)_vi:Cn16698 Gb(89)_vi:Cnu16698 Cucumber necrosis virus defective interfering RNA isolate C1. 11/94 432bp.

U16699 Em(43)_vi:Cn16699 Gb(89)_vi:Cnu16699 Cucumber necrosis virus defective interfering RNA isolate C12. 11/94 432bp.
U16700 Em(43)_vi:Cn16700 Gb(89)_vi:Cnu16700 Cucumber necrosis virus defective interfering RNA isolate C2. 11/94 432bp.
U16701 Em(43)_vi:Cn16701 Gb(89)_vi:Cnu16701 Cucumber necrosis virus defective interfering RNA isolate C39. 11/94 594bp.
U16702 Em(43)_vi:Cn16702 Gb(89)_vi:Cnu16702 Cucumber necrosis virus defective interfering RNA isolate C40. 11/94 411bp.
U16703 Em(43)_vi:Cn16703 Gb(89)_vi:Cnu16703 Cucumber necrosis virus defective interfering RNA isolate C46. 11/94 416bp.
U16704 Em(43)_vi:Cn16704 Gb(89)_vi:Cnu16704 Cucumber necrosis virus defective interfering RNA isolate C61. 11/94 594bp.
U16705 Em(43)_vi:Cn16705 Gb(89)_vi:Cnu16705 Cucumber necrosis virus defective interfering RNA isolate C62. 11/94 597bp.
U16706 Em(43)_vi:Cn16706 Gb(89)_vi:Cnu16706 Cucumber necrosis virus defective interfering RNA isolate C66. 11/94 414bp.

*Features of the genome:* there are 5 ORFs; p33, p92, p41, p21 and p20. p33 and p92 are translated from the full length genomic RNA. p92 would be translated if suppression of the amber codon of p33 were to occur. p41, p21 and p20 are translated from subgenomic RNAs. p41 is coat protein, p92 is possibly a replicase protein, p33, p21 and p20 functions are unknown.

*Non–genomic nucleic acid* found in the virions; is subgenomic *m*RNA and of an unusual type of structure; defective interfering RNA. Sub–genomic *m*RNA found in infected cells; two subgenomic RNAs found in infected cells, 2.1 and 0.9 kb, and there are possibly others. 3 virus specified dsRNA species found in infected cells (also several minor ones). Size of largest virus specified dsRNA 4.7 kbp; 2nd largest 2.1 kbp; 3rd largest 0.9 kbp.

*Virion protein(s)* one; $M_r$ 41000; coat protein. Amino acid sequence: Rochon and Tremaine (1989). Amino acid composition: Rochon and Tremaine (1989). Virion proteins not glycosylated. Virion proteins not phosphorylated.

## Replication

Replication does not depend on a helper virus.

## Cytopathology

Virions found in mesophyll; in cytoplasm. Inclusions present in infected cells; are membranous bodies and unusual in shape; multivesicular bodies; they do not contain virions.

## Taxonomy and Relationships

*Tombusvirus: Tombusviridae.*

Virus(es) with serologically unrelated virions: no serological relationship to several other tombusviruses.

Its possible replicase has similar amino acid sequence to the replicases of carnation mottle carmovirus, barley yellow dwarf luteovirus, tobacco necrosis *necrovirus*, red clover necrotic mosaic dianthovirus, maize chlorotic mottle machlomovirus and hepatitis C virus.

## References


Dias, H.F. and McKeen, C.D. (1972). *CMI/AAB Descr. Pl. Viruses* No. 82, 4 pp.
Johnston, J.C. and Rochon, D.M. (1990). *J. gen. Virol.* **71**: 2233.
McKeen, C.D. (1959). *Can. J. Bot.* **37**: 913.
Riviere, C.J. and Rochon, D.M. (1990). *J. gen. Virol.* **71**: 1887.

Rochon, D.M. and Tremaine, J.H. (1988). *Virology* **69**: 395.
Rochon, D.M. and Tremaine, J.H. (1989). *Virology* **169**: 251.
Rochon, D.M., Johnston, J.C. and Riviere, C.J. (1991). *Can. J. Pl. Pathol.* **13**: 162.

# Cucumber soil-borne *carmovirus*

**Data collated by K.M. Makkouk, 1987.**

*Standard acronym:* CuSBV.

First reported in *Cucumis sativus*; from Jieh, south of Beirut, Lebanon; by Koenig *et al.* (1983).

**Natural host range and symptoms**

*Cucumis sativus* — present in roots only, no leaf symptoms.

**Transmission**

Virus transmitted by mechanical inoculation.

**Geographical distribution**

Found, but with no evidence of spread, in Lebanon.

**Experimental host range**

Few (<3) families susceptible.

Diagnostically susceptible

*Chenopodium quinoa, C. amaranticolor, Datura stramonium, Gomphrena globosa, Nicotiana glutinosa, Beta vulgaris, Vigna unguiculata* Black eye No. 5 — local lesions not systemic.

Diagnostically insusceptible

*Physalis floridana, Vicia faba, Spinacia oleracea, Capsicum annuum* cv. Yolo Wonder, *Solanum melongena* cv. Black Beauty, *Vigna unguiculata* cv. California Black Eye No. 5.

Assay hosts (**L**ocal lesions or **W**hole plants)

*Chenopodium quinoa (L), C. amaranticolor (L), Datura stramonium (L), Gomphrena globosa (L), Nicotiana glutinosa (L), Beta vulgaris (L), Vigna unguiculata* cv. Black Eye No. 5 (L).

Dicotyledonous Hosts *(**Susceptible** or Insusceptible)*

***Cucumis sativus*** (root only).

Sources of host-range data

Koenig *et al.* (1983).

**Properties of particles in sap**

Leaf sap contains many virions.

**Purification**

Method: Koenig *et al.* (1983).

**Particle morphology**

Virions isometric; not enveloped; 31 nm in diameter; rounded in profile and angular in profile (occasionally); with a conspicuous capsomere arrangement.

**Physical properties**

One sedimenting component in purified preparations; sedimentation coefficient 120 *S*. Density 1.343 g $cm^{-3}$ in CsCl.

**Biochemical properties**

*Genome* consists of RNA; single–stranded; unipartite; largest (or only) genome part 4.75 kb (also a satellite RNA of 0.5 kb). Genomic nucleic acid isolated by Koenig *et al.* (1983). Base composition 32 % G; 23 % A; 22 % C; 23 % U. Infectivity retained when deproteinised with phenol or detergent.

*Virion protein(s)* one; $M_r$ 41400; coat protein. Method of preparation: Koenig *et al.* (1983).

**Replication**

Acts as helper for a satellite RNA.

**Cytopathology**

Virions found in epidermis; in cytoplasm. Inclusions present in infected cells; are unusual in shape; thin fibrils about 9 nm thick; they do not contain virions. Other cellular changes: small vesicles with a single membrane and containing fibrils resembling nucleic acid.

**Taxonomy and Relationships**

*Carmovirus: Tombusviridae.*

**References**

Koenig, R., Lesemann, D.-E., Huth, W. and Makkouk, K.M. (1983). *Phytopathology* **73**: 515.

# Cucumber toad-skin (?) rhabdovirus

**Data collated by A.A. Brunt, 1995.**

First reported in *Cucumis sativus*; from France; by Lecoq (1982).

**Natural host range and symptoms**

*Cucumis sativus* — vein-clearing, severe leaf crinkling and stunting (toad skin disease).

**Geographical distribution**

Spreads in France and Greece (possibly).

Dicotyledonous Hosts *(**Susceptible** or Insusceptible)*

***Cucumis sativus*.**

**Taxonomy and Relationships**

Unassigned rhabdovirus.

**References**

Katis, N., Avgelis, A. and Lecoq, H. (1995). *Proc. 8th Conf. on Virus Dis. Vegs*, p.140 Prague, July, 1995.

Lecoq, H. (1982). *Rev. Hort.* **223**: 15.

# Cucumber vein yellowing virus

**Data collated by A.A. Brunt and E. Tanne, 1988.**

*Standard acronym:* CVYV.

First reported in *Cucumis* spp.; from Israel; by Cohen and Nitzany (1960).

**Natural host range and symptoms**

Symptoms persist.

*Cucumis sativus, Citrullus vulgaris* — vein clearing, leaf chlorosis and necrosis.

**Transmission**

Transmitted by a vector; an insect; *Bemisia tabaci*; Aleyrodidae. Transmitted in a semi–persistent manner. Virus transmitted by mechanical inoculation; transmitted by grafting; not transmitted by contact between plants.

**Geographical distribution**

Spreads in Israel.

**Experimental host range**

Few (<3) families susceptible.

Diagnostically susceptible

*Cucumis sativus, Citrullus vulgaris* — systemic leaf chlorosis.

Diagnostically insusceptible

*Cucurbita moschata, Ecballium elaterium, Nicotiana glutinosa, Pisum sativum, Chenopodium amaranticolor.*

Maintenance and propagation hosts

*Cucumis sativus, Citrullus vulgaris.*

Assay hosts (**L**ocal lesions or **W**hole plants)

*Cucumis sativus* (W).

Dicotyledonous Hosts *(**Susceptible** or Insusceptible)*

*Chenopodium amaranticolor.* ***Citrullus lanatus***. ***Cucumis sativus***. *Cucurbita moschata. Nicotiana glutinosa. Pisum sativum.*

**Properties of particles in sap**

TIP: 52 °C. LIV: less than 1 days. DEP: $\log_{10}$ minus 3.

**Purification**

Method: Sela *et al.* (1980).

**Particle morphology**

Virions filamentous; not enveloped; usually flexuous; of 740–800 nm.

**Physical properties**

One sedimenting component in purified preparations; sedimentation coefficient 220 *S*.

**Biochemical properties**

*Genome* consists of DNA; double–stranded.

*Virion protein(s)* one; $M_r$ 39000; coat protein. Method of preparation: Sela *et al.* (1980).

**References**

Cohen, S. and Nitzany, F.E. (1960). *Phytopathol. Medit.* **1**: 44.
Cohen, S. and Nitzany, F.E. (1963). *Phytopathology* **53**: 193.
Harpaz, I. and Cohen, S. (1965). *Phytopath. Z.* **54**: 240.
Sela, I., Assouline, I., Tanne, E., Cohen, S. and Marco, S. (1980). *Phytopathology* **70**: 226.

# Cycas necrotic stunt *nepovirus*

**Data collated by M. Kusunoki, K. Hanada and M. Kameya, 1987.**

*Standard acronym:* CNSV.

First reported in *Cycas revoluta*; from Chiba Prefecture, Japan; by Kusunoki *et al.* (1975).

**Natural host range and symptoms**

Symptoms persist and vary seasonally.

*Cycas revoluta* — twisted young leaves and chlorotic or necrotic spots of mature leaves, stunting.

*Daphne odora* — mild mosaic.

*Aucuba japonica* — vein clearing.

*Gladiolus spp.* — symptomless.

**Transmission**

Transmitted by possibly a vector; a nematode. Virus transmitted by mechanical inoculation; transmitted by seed (about 30% in *Chenopodium amaranticolor* and over 80% in *C. serotinum*).

**Geographical distribution**

Spreads in Japan.

**Experimental host range**

Several (3–9) families susceptible.

Diagnostically susceptible

*Tetragonia expansa, Chenopodium amaranticolor, C. quinoa, C. serotinum* — chlorotic local lesions or ringspots, then systemic mottling and necrosis.

*Celosia cristata* — systemic mottling.

*Gomphrena globosa, Beta vulgaris* — latent infection.

*Spinacia oleracea* — latent infection.

*Cycas revoluta* — chlorotic spotting, then systemic mottling and necrosis.

Diagnostically insusceptible

*Cucumis sativus, Cucurbita moschata, Glycine max, Nicotiana clevelandii, Lycopersicon esculentum, Nicotiana tabacum, N. glutinosa, Phaseolus vulgaris, Pisum sativum, Catharanthus roseus, Datura stramonium.*

Maintenance and propagation hosts

*Chenopodium serotinum, C. quinoa, C. amaranticolor, Celosia cristata.*

Assay hosts (**L**ocal lesions or **W**hole plants)

*Chenopodium quinoa* (L and W), *C. amaranticolor* (L) and then (W), *C. murale* (L) and then (W), *C. serotinum* (W), *Tetragonia expansa* (L).

Dicotyledonous Hosts *(**Susceptible** or Insusceptible)*

***Beta vulgaris***. *Brassica campestris* ssp. *chinensis*. *Brassica oleracea* ssp. *botrytis*. *Catharanthus roseus*. ***Celosia cristata***. ***Chenopodium amaranticolor***. ***Chenopodium murale***. ***Chenopodium quinoa***. *Cucumis sativus*. *Cucurbita moschata*. *Datura stramonium*. *Dianthus barbatus*. *Dianthus caryophyllus*. *Fagopyrum esculentum*. *Glycine max*. ***Gomphrena globosa***. *Helianthus annuus*. *Lactuca sativa*. *Lathyrus odoratus* (*Lycium chinense* **S**). *Lupinus albus*. *Lycopersicon esculentum*. *Nicotiana clevelandii*. *Nicotiana glutinosa*. *Nicotiana rustica*. *Nicotiana tabacum*. *Petunia hybrida*. *Phaseolus vulgaris*. *Physalis floridana*. *Pisum sativum*. *Raphanus sativus*. ***Spinacia oleracea***. ***Tetragonia expansa***. *Trifolium pratense*. *Trifolium repens*. *Vicia faba*. *Vigna unguiculata*. *Vigna unguiculata* ssp. *sesquipedalis*. *Zinnia elegans*.

Additional hosts

***Aucuba japonica, Cycas revoluta, Daphne odora***.

Sources of host-range data

Kusunoki *et al.* (1986).

**Properties of particles in sap**

TIP: 60–65 °C. LIV: 18–24 days. DEP: $\log_{10}$ minus 3. Leaf sap contains many virions.

**Purification**

Method: Kusunoki *et al.* (1986).

**Particle morphology**

Virions isometric; not enveloped; 28 nm in diameter; angular in profile; with a conspicuous capsomere arrangement.

**Physical properties**

Three sedimenting components in purified preparations; sedimentation coefficient of the fastest 112 *S* (B); of the other(s) 85 *S* (M). Density 1.472 g $cm^{-3}$ in CsCl (B), or 1.404 g $cm^{-3}$ in CsCl (M). $A_{260}/A_{280}$ ratio 1.67 (B), or 1.41 (M), or 1 (T).

**Biochemical properties**

Virions contain 24 % nucleic acid (B), or 12 % nucleic acid (M), or 0 % nucleic acid (T); 76 % protein (B), or 88 % protein (M), or 100 % protein (T).

*Genome* consists of RNA; single–stranded; linear; of two parts; largest (or only) genome part the larger 8.1 kb; the 2nd largest 4.75 kb. Genomic nucleic acid isolated by Hanada *et al.* (1986). Infectivity retained when deproteinised with phenol or detergent. Additional factor not required for infectivity.

*Non–genomic nucleic acid* not found in the virions.

*Virion protein(s)* three; $M_r$ of the largest 71000. $M_r$ of 2nd largest 65000. $M_r$ of 3rd largest 56000.

**Replication**

Genome replicates in cytoplasm. Replication does not depend on a helper virus.

**Cytopathology**

Virions found in leaves; in cytoplasm, in Golgi apparatus, and in cell vacuoles; in plasmodesmata. Inclusions present in infected cells; are viroplasms (crystalline aggregates of virions and capsid protein); they contain virions. Other cellular changes: vesicular structures also occur in cytoplasm.

**Taxonomy and Relationships**

*Nepovirus: Comoviridae.*

Virus(es) with serologically unrelated virions: arabis mosaic, cacao necrosis, cherry leaf roll, crimson clover latent, grapevine chrome mosaic, mulberry ringspot, myrobalan latent, tobacco ringspot, tomato ringspot and tomato black ring viruses.

Best tests for diagnosis: use gel diffusion serology; species of *Fabaceae* and *Solanaceae* not infected despite 'nepovirus-like' symptoms in *Chenopodium* plants.

**References**


Hanada, K., Kusunoki, M. and Iwaki, M. (1986). *Ann. Phytopath. Soc. Japan* **52**: 422.

Kusunoki, M., Nagai, Y., Yamashita, S., Doi, Y. and Yora, K. (1975). *Ann. Phytopath. Soc. Japan* **41**: 285.

Kusunoki, M., Hanada, K., Iwaki, M., Chang, M.U., Doi, Y. and Yora, K. (1986). *Ann. Phytopath. Soc. Japan* **52**: 302.


# Cymbidium mosaic *potexvirus*

**Data collated by R.I.B. Francki, 1984.**

*Synonyms:* orchid mosaic virus. *Standard acronym:* CymMV.

First reported in *Cymbidium* spp.; from California, U.S.A.; by Jensen (1950).

**Natural host range and symptoms**

Symptoms vary seasonally.
*Cymbidium* spp. — mosaic, necrosis.
*Cattleya* spp. — mosaic, flower necrosis.
*Phalaenopsis* spp.—mosaic, water soaked local lesions.
*Vanda* spp. — chlorotic flecks.
*Epidendrum* spp., *Laelia* spp., *Laeliocattleya* spp., *Oncidium* spp., *Zygopetalum* spp., *Vanilla fragrans* — also naturally infected.

**Transmission**

Transmitted by means not involving a vector. Virus transmitted by mechanical inoculation; transmitted by contact between plants.

**Geographical distribution**

Probably distributed worldwide.

**Experimental host range**

Several (3–9) families susceptible.

Diagnostically susceptible

*Chenopodium amaranticolor, Datura stramonium* — blotchy local lesions developing slowly; not systemic.
*Cassia occidentalis* — small local lesions; not systemic.

Diagnostically insusceptible

*Beta vulgaris, Nicotiana glutinosa, N. tabacum, Petunia hybrida, Zea mays.*

Maintenance and propagation hosts

*Datura stramonium, Cymbidium* sp.

Assay hosts (**L**ocal lesions or **W**hole plants)

*Chenopodium amaranticolor* (L), *Cassia occidentalis* (L), *Datura stramonium*, (L).

Dicotyledonous Hosts (***Susceptible*** *or Insusceptible)*

*Apium graveolens. Beta vulgaris. Brassica campestris* ssp. *napus. Brassica campestris* ssp. *pekinensis. Brassica oleracea* ssp. *botrytis. Capsicum frutescens. Catharanthus roseus.* ***Chenopodium amaranticolor. Cucumis sativus****. Cucurbita maxima.* ***Datura stramonium****. Dianthus caryophyllus. Glycine max.* ***Gomphrena globosa****. Lactuca sativa. Lathyrus odoratus. Lycopersicon esculentum. Nicotiana glutinosa. Nicotiana tabacum. Ocimum basilicum. Petunia hybrida. Phaseolus vulgaris. Raphanus sativus. Trifolium incarnatum.* ***Tropaeolum majus****. Vigna unguiculata.* ***Zinnia elegans****.*

Monocotyledonous Hosts (***Susceptible*** *or Insusceptible)*

***Oryza sativa****. Zea mays.*

Comments on host-range

Some early reports of host range were probably of odontoglossum ringspot tobamovirus.

Sources of host-range data

Jensen (1955); Jensen and Gold (1955); Murakishi (1958); Kado and Jensen (1964); Faccioli and Marani (1979).

**Properties of particles in sap**

TIP: 60–70 °C. LIV: 25 days. DEP: $\log_{10}$ minus 6–7. Leaf sap contains many virions.

**Purification**

Method: Francki and McLean (1968).

**Particle morphology**

Virions filamentous; not enveloped; usually flexuous; with a clear modal length; of 480 nm; 13 nm wide. Axial canal obscure. Basic helix obvious; pitch of basic helix 2.8 nm.

**Physical properties**

One sedimenting component in purified preparations; sedimentation coefficient 121 *S*.

**Biochemical properties**

Virions contain 5.6 % nucleic acid; 94 % protein; 0 % lipid.

*Genome* consists of RNA; single–stranded; linear; unipartite; largest (or only) genome part 8.1 kb. Genomic nucleic acid isolated by Frowd and Tremaine (1977). Base composition 21.1 % G; 28.9 % A; 24.4 % C; 25.6 % U.

*Sequence database accession code(s):*

X62133 Em(40)_vi:CMVIRCP Gb(84)_vi:CMVIRCP Cymbidium mosaic virus gene for coat protein. 6/92 1,290bp.

X62663 Em(40)_vi:CMV11KD Gb(84)_vi:CMV11KD Cymbidium mosaic virus RNA for 11Kd protein. 6/92 276bp.

X62664 Em(40)_vi:CMV14KD Gb(84)_vi:CMV14KD Cymbidium mosaic virus RNA for 14Kd protein. 6/92 339bp

X62665 Em(40)_vi:CMVCP Gb(84)_vi:CMVCP Cymbidium mosaic virus RNA for coat protein. 10/93 749bp. 4 sequences.

*Virion protein(s)* one; $M_r$ 27600. Method of preparation: Francki and McLean (1968). Amino acid composition: Frowd and Tremaine (1977).

**Replication**

Replication does not depend on a helper virus.

**Cytopathology**

Virions found in leaves; in cytoplasm. Inclusions cytoplasmic present in infected cells; are unusual in shape; banded bodies; they contain virions. Other cellular changes: osmiophilic globules in chloroplasts.

**Taxonomy and Relationships**

*Potexvirus.*

Virus(es) with serologically unrelated virions: white clover mosaic and potato X viruses.

**References**


Faccioli, G. and Marani, F. (1979). *Phytopathol. Medit.* **18**: 21.

Francki, R.I.B. (1970). *CMI/AAB Descr. Pl. Viruses* No. 27, 3 pp.

Francki, R.I.B. and McLean, G.D. (1968). *Aust. J. biol. Sci.* **21**: 1311.

Francki, R.I.B., Randles, J.W., Chambers, T.C. and Wilson, S.B. (1966). *Virology* **28**: 729.

Frowd, J.A. and Tremaine, J.H. (1977). *Phytopathology* **67**: 43.

Hanchey, P., Livingstone, C.H. and Reeves, F.B. (1975). *Physiol. Pl. Path.* **6**: 227.

Jensen, D.D. (1950). *Phytopathology* **40**: 966.

Jensen, D.D. (1955). *Phytopathology* **41**: 401.

Jensen, D.D. and Gold, A.H. (1955). *Phytopathology* **45**: 327.

Kado, C.I. and Jensen, D.D. (1964). *Phytopathology* **54**: 974.

Murakishi, H.H. (1958). *Phytopathology* **48**: 132.

Pearson, M.N., Brunt, A.A. and Pone, S.P. (1990). *J. Phytopath.* **128**: 46.


# Cymbidium ringspot *tombusvirus*

**Data collated by K. Boswell. 1985.**

*Standard acronym:* CymRSV. *Strains:* cymbidium and clover.

First reported in *Cymbidium* spp.; from Sussex, U.K; by Hollings and Stone (1963).

**Natural host range and symptoms**

Symptoms persist.

*Cymbidium* spp. — chlorotic ring-mottling, more intense when mixed with cymbidium mosaic virus.

*Trifolium repens* — chlorotic flecking, mottling.

**Transmission**

Transmitted by means not involving a vector. Virus transmitted by mechanical inoculation; transmitted by contact between plants; not transmitted by seed (in *Trifolium incarnatum, Phaseolus vulgaris* cv. The Prince, *Nicotiana clevelandii*).

**Geographical distribution**

Spreads in the UK.

**Experimental host range**

Many (>9) families susceptible.

Diagnostically susceptible

*Chenopodium amaranticolor, C. murale, C. quinoa* — chlorotic or necrotic local lesions; not systemic.

*Emilia sagittata* — chlorotic local lesions; systemic yellow spots and rings, leaf malformation and necrosis.

*Phaseolus vulgaris* cv. The Prince — whitish necrotic local lesions with *Cymbidium* strain, semi-necrotic local lesions with clover strain; systemic chlorosis and mottling (both strains).

*Vigna unguiculata* cv. Blackeye — necrotic whitish and brown local lesions; not systemic.

*Nicotiana clevelandii* — diffuse chlorotic, and some brown local lesions, necrosis spreads killing leaf; systemic chlorotic and necrotic, speckling, severe puckering, stunting.

*Nicotiana glutinosa, N. tabacum* cvs White Burley, Judy's Pride, Havana 425, Xanthi-*nc* — brown necrotic local lesions; not systemic.

Maintenance and propagation hosts

*Nicotiana clevelandii* virus is best stored as freeze-dried leaves.

Assay hosts (**L**ocal lesions or **W**hole plants)

*Chenopodium quinoa* (L), *Phaseolus vulgaris* (L).

Dicotyledonous Hosts *(**Susceptible** or Insusceptible)*

*Ageratum conyzoides*. ***Amaranthus caudatus***. ***Antirrhinum majus***. *Apium graveolens*. ***Arachis hypogaea***. ***Atriplex hortensis***. *Bellis perennis*. ***Beta vulgaris***. *Brassica campestris* ssp. *pekinensis*. *Brassica campestris* ssp. *rapa*. *Brassica oleracea* ssp. *capitata*. ***Capsicum frutescens***. *Catharanthus roseus*. ***Celosia argentea***. *Cheiranthus cheiri*. ***Chenopodium amaranticolor***. ***Chenopodium murale***. ***Chenopodium quinoa***. *Chrysanthemum morifolium*. *Citrus limon*. *Cucumis melo*. ***Cucumis sativus***. ***Datura stramonium***. *Dianthus barbatus*. ***Emilia sagittata***. ***Fagopyrum esculentum***. ***Glycine max***. ***Gomphrena globosa***. ***Helianthus annuus***. *Hibiscus esculentus*. *Hyoscyamus niger*. *Ipomoea nil* (cv. Scarlett o'Hara). ***Lactuca sativa***. ***Lathyrus odoratus***. ***Lupinus angustifolius***. ***Lycopersicon esculentum***. ***Myosotis sylvatica***. ***Nicandra physaloides***. ***Nicotiana benthamiana***. *Nicotiana clevelandii*. ***Nicotiana debneyi***. ***Nicotiana glutinosa***. ***Nicotiana megalosiphon***. ***Nicotiana rustica***. ***Nicotiana tabacum***. ***Ocimum basilicum***. ***Petunia hybrida***. ***Phaseolus vulgaris***. ***Phlox drummondii***. ***Physalis floridana***. ***Phytolacca americana***. ***Pisum sativum***. ***Saponaria vaccaria***. ***Senecio vulgaris***. *Solanum melongena*. ***Solanum nigrum***. *Solanum tuberosum*. ***Spinacia oleracea***. ***Stellaria media***. ***Tetragonia expansa***. ***Torenia fournieri***. ***Trifolium incarnatum***. ***Trifolium repens***. ***Tropaeolum majus***. ***Vicia faba***. ***Vigna unguiculata***. ***Zinnia elegans***.

Monocotyledonous Hosts *(**Susceptible** or Insusceptible)*

*Avena sativa*. *Dactylis glomerata*. *Hordeum vulgare*. *Phleum pratense*. *Triticum aestivum*. *Zea mays*.

Sources of host-range data

Hollings *et al.* (1977).

**Properties of particles in sap**

TIP: 85–90 °C. LIV: 300 days. DEP: $\log_{10}$ minus 5–6. Leaf sap contains many virions.

**Purification**

Method: Hollings *et al.* (1977).

**Particle morphology**

Virions isometric; not enveloped; 30 nm in diameter.

**Physical properties**

One sedimenting component in purified preparations; sedimentation coefficient 136–139 *S*. Density 1.36 g $cm^{-3}$ in CsCl.

**Biochemical properties**

Virions contain 14–15 % nucleic acid.

*Genome* consists of RNA; single–stranded; unipartite; largest (or only) genome part 4.733 kb. Base composition 27.8 % G; 24.9 % A; 21.3 % C; 26.1 % U.

*Sequence database accession code(s):*

D00719 Em(40)_vi:CRVDIR Gb(84)_vi:CRVDIR Cymbidium ringspot virus (CyRSV) defective interfering (DI) RNA. 3/92 499bp.

D00720 Em(40)_vi:CRVSATR Gb(84)_vi:CRVSATR Cymbidium ringspot virus (CyRSV) satellite RNA. 3/92 621bp.

S43932 Em(40)_un:S43932 Gb(84)_vi:S43932 (defective interfering RNA) cymbidium ringspot tombusvirus CyRSV, DI-371, Other Genetic Mate

X15511 Em(40)_vi:CRTV1 Gb(84)_vi:CRTV1 Cymbidium ringspot tombusvirus RNA. 9/93 4,733bp. 4 sequences.

*Virion protein(s)* one; $M_r$ 43600.

**Cytopathology**

Virions found in leaves, roots, stems and flowers.

**Taxonomy and Relationships**

*Tombusvirus: Tombusviridae.*

Virus(es) with serologically related virions: some tombusviruses. Virus(es) with serologically unrelated virions: some tombusviruses (Koenig and Gibbs, 1986).

**References**


Dalmay, T., Rubino, L., Burgyan, J., Kollar, A. and Russo, M. (1993). *Virology* **194**: 697.

Grieco, F., Burgyan, J. and Russo, M. (1989). *Nucl. Acids Res.* **17**: 6383.

Hollings, M. and Stone, O.M. (1963). *Rep. Glasshouse Crops Res. Inst.* 1962, p. 90.

Hollings, M. and Stone, O.M. (1977). *CMI/AAB Descr. Pl. Viruses* No. 178, 4 pp.

Hollings, M., Stone, O.M. and Barton, R.J. (1977). *Ann. appl. Biol.* **85**: 233.

Koenig, R. and Gibbs, A.J. (1986). *J. gen. Virol.* **67**: 75.

Rubino, L., Burgyan, J., Grieco, F. and Russo, M. (1989). *J. gen. Virol.* **71**: 1655.


# Cynara (?) *nucleorhabdovirus*

**Data collated by A.A. Brunt, 1992.**

*Standard acronym:* CraV.

First reported in *Cynara scolymus*; from Italy; by Russo *et al.* (1975).

**Natural host range and symptoms**

*Cynara scolymus, C. syriaca* — symptomatology unknown (virus found only in complex with other viruses).

**Transmission**

Virus transmitted by mechanical inoculation.

**Geographical distribution**

Spreads in Italy and Spain (probably).

**Experimental host range**

Few (<3) families susceptible.

Diagnostically susceptible

*Nicotiana benthamiana* — systemic leaf chlorosis.

Maintenance and propagation hosts

*Nicotiana benthamiana.*

Assay hosts (**L**ocal lesions or **W**hole plants)

*Nicotiana benthamiana* (W).

Dicotyledonous Hosts *(**Susceptible** or Insusceptible)*

***Nicotiana benthamiana*.**

**Particle morphology**

Virions rhabdo– or bullet–shaped; enveloped; with a clear modal length; of 260 nm (*in vivo*); 73 nm wide (*in vivo*). Axial canal obvious; 20 nm in diameter.

**Cytopathology**

Virions found in leaves, mesophyll and the xylem; in cytoplasm.

**Taxonomy and Relationships**

Probably *Nucleorhabdovirus: Rhabdoviridae.*

**References**

Pena-Iglesias, A., Rubio-Huertos, M., Moreno, S. and Martin, R. (1972). *Annl. I.N.I.A.* **2**: 123.

Russo, M., Martelli, G.P. and Rana, G.L. (1975). *Phytopath. Z.* **83**: 223.

# Cynodon chlorotic streak *nucleorhabdovirus*

**Data collated by A.A. Brunt and B.E. Lockhart, 1989.**

First reported in *Zea mays*; from Morocco; by Lockhart and Elyamani (1982).

**Natural host range and symptoms**

Symptoms persist.
*Cynodon dactylon, Zea mays* — chlorotic streaking.

**Transmission**

Transmitted by a vector; an insect; *Toya propinqua*; Delphacidae. Virus not transmitted by mechanical inoculation; not transmitted by contact between plants.

**Ecology and control**

Studies reported by Lockhart *et al.* (1985).

**Geographical distribution**

Spreads in France, Iran, Jordan, Morocco, Spain, and Tunisia.

**Experimental host range**

Few (<3) families susceptible.

Diagnostically susceptible

*Cynodon dactylon, Zea mays* — chlorotic streaking.

Maintenance and propagation hosts

*Cynodon dactylon, Zea mays.*

Assay hosts (**L**ocal lesions or **W**hole plants)

*Cynodon dactylon* (W), *Zea mays* (W).

Monocotyledonous Hosts *(**Susceptible** or Insusceptible)*

***Cynodon dactylon*. *Zea mays*.**

Sources of host-range data

Lockhart *et al.* (1985).

**Purification**

Method: Lockhart *et al.* (1985).

**Particle morphology**

Virions rhabdo– or bullet–shaped; enveloped; usually straight; with a clear modal length; of 240 nm (in sap), or 280 nm (in thin sections); 72 nm wide (in sap), or 80 nm wide (in thin sections). Axial canal obvious.

**Cytopathology**

Virions found in leaves and roots; in the perinuclear space.

**Taxonomy and Relationships**

*Nucleorhabdovirus: Rhabdoviridae.*

Virus(es) with serologically unrelated virions: maize mosaic, cereal chlorotic mottle, maize sterile stunt, barley yellow striate mosaic and festuca leaf streak viruses.

**References**


Izadpanah, K. (1989). *Proc. 9th Pl. Prot. Congr. Iran*, p. 161. Ferdowsi Univ., Mashad, Iran.

Lockhart, B.E. and Elyamani, M. (1982). In: *Proc. Int. Maize Virus Dis. Colloq. and Wksh.*; eds D.T. Gordon, J.K. Knocke, L.R. Nault and R.M. Ritter. Ohio Agric. Res. and Dev. Centre, Wooster, 261 pp.

Lockhart, B.E., Kharless, N., El Maataoui, M. and Lastra, R. (1985). *Phytopathology* **75**: 1094.


# Cynodon mosaic (?) *carlavirus*

**Data collated by A.A. Brunt, 1990.**

*Standard acronym:* CynMV.

First reported in *Cynodon dactylon*; from India; by Bhargava *et al.* (1971).

**Natural host range and symptoms**

*Cynodon dactylon* — chlorotic streaking and mosaic.

**Transmission**

Transmitted by a vector; an insect; Aphididae. Virus transmitted by mechanical inoculation.

**Geographical distribution**

Found, but with no evidence of spread, in India.

Diagnostically susceptible

*Zea mays, Sorghum bicolor, Pennisetum pedicillatum*—systemic chlorotic mottling.

Monocotyledonous Hosts *(**Susceptible** or Insusceptible)*

***Sorghum bicolor*. *Zea mays*.**

**Particle morphology**

Virions filamentous; with a clear modal length; of 509–632 nm.

**Taxonomy and Relationships**

Possible *Carlavirus.*

**References**

Bhargava, K.S., Joshi, R.D. and Rishi, N. (1971). *Indian Phytopath.* **24**: 119.
Koenig, R. (1982). *CMI/AAB Descr. Pl. Viruses* No. 259, 4 pp.

## Cynosurus mottle (?) *sobemovirus*

**Data collated by A.A. Brunt, 1992.**

*Synonyms:* lolium mottle virus (A'Brook, 1972). *Standard acronym:* CnMoV.

First reported in *Cynosurus cristatus, Lolium perenne*; from the U.K.; by A'Brook (1972).

**Natural host range and symptoms**

Symptoms persist.
*Agrostis tenuis, Cynosurus cristatus* — severe leaf mottling and necrosis.
*Agrostis stolonifera* — dark green leaf mottling.
*Lolium perenne, Lolium perenne x. L. multiflorum* — chlorotic leaf mottling.

**Transmission**

Transmitted by a vector; an insect; *Rhopalosiphum padi* (New Zealand isolate); Aphididae. Transmitted in a non–persistent manner. Virus transmitted by mechanical inoculation.

**Geographical distribution**

Spreads in Germany, New Zealand, and the UK.

**Experimental host range**

Few (<3) families susceptible.

Diagnostically susceptible

*Agrostis puchella, Bromus secalinus, Cynosurus cristatus, Lagurus ovatus, Phleum arenarium, Triticum aestivum* — severe systemic leaf mottling and, often, necrosis.

*Avena fatua, A. sativa, Hordeum vulgare, Secale cereale* — mild systemic leaf mottling.

Diagnostically insusceptible

*Agropyron repens, Bromus mollis, Dactylis glomerata, Lolium multiflorum, Setaria italica.*

Maintenance and propagation hosts

*Avena sativa, Triticum aestivum.*

Assay hosts (**L**ocal lesions or **W**hole plants)

*Avena sativa* (W), *Cynosurus cristatus* (W), *Triticum aestivum* (W).

Monocotyledonous Hosts *(**Susceptible** or Insusceptible)*

*Agropyron repens*. ***Agrostis tenuis***. ***Avena sativa***. *Bromus mollis*. ***Bromus secalinus***. *Echinochloa colonum*. *Echinochloa crus–galli*. *Festuca pratensis*. ***Hordeum vulgare***. ***Lagurus ovatus***. *Lamarkia aurea*. *Lolium multiflorum*. *Lolium perenne*. *Oryza sativa*.

*Panicum miliaceum*. ***Phleum arenarium***. ***Poa annua***, or *Poa annua*. ***Secale cereale***. *Setaria italica*. *Setaria viridis*. ***Triticum aestivum***. *Zea mays*.

Sources of host-range data

Catherall *et al.* (1977); Mohamed (1978*a*).

**Properties of particles in sap**

TIP: 75 °C. DEP: $\log_{10}$ minus 8.

**Purification**

Method: Mohamed (1978*b*).

**Particle morphology**

Virions isometric; not enveloped; 28 nm in diameter; angular in profile.

**Physical properties**

One sedimenting component in purified preparations; sedimentation coefficient 113 *S*. Density 1.399 g $cm^{-3}$ in CsCl. Density 1.332 g $cm^{-3}$ in $Cs_2SO_4$. $A_{260}/A_{280}$ ratio 1.72.

**Biochemical properties**

Virions contain 22.5 % nucleic acid; 77.5 % protein.

*Genome* consists of RNA; single–stranded; largest (or only) genome part 4.3 kb. Genomic nucleic acid isolated by Mohamed (1978*b*). Base composition 24.2 % G; 24.3 % A; 26.2 % C; 25.3 % U.

*Virion protein(s)* one; $M_r$ 30500; and it is coat protein.

**Cytopathology**

Virions found in leaves and mesophyll; in cytoplasm and in cell vacuoles. Inclusions absent from infected cells.

**Taxonomy and Relationships**

Possible *Sobemovirus*.

Virus(es) with serologically related virions: cocksfoot mottle virus.

The viral RNA has 5–8% nucleotide sequence homology with that of cocksfoot mottle virus.

**References**

A'Brook, J. (1972). *Pl. Path.* **32**: 118.

Catherall, P.L., Andrews, P.A. and Chamberlain, J.A. (1977). *Ann. appl. Biol.* **87**: 233.

Hull, R. (1977). *J. gen. Virol.* **36**: 289.

Hull, R. (1988). In: *The Plant Viruses*, Vol. 3, Polyhedral Virions with Monopartite RNA Genomes, p. 113; ed. R. Koenig. Plenum Press, New York.

Huth, W. and Paul, H.L. (1977). *Annls. Phytopath.* **9**: 293.

Mohamed, N.A. (1978*a*). *N.Z. Jl agric. Res.* **21**: 709.

Mohamed, N.A. (1978*b*). *J. gen. Virol.* **40**: 379.

Mohamed, N.A. and Mossop, D.W. (1981). *J. gen. Virol.* **55**: 63.

Paul, H.L., Querfurth, G. and Huth, W. (1980). *J. gen. Virol.* **47**: 67.

# Cypripedium calceolus (?) potyvirus

**Data collated by A.A. Brunt, 1991.**

*Standard acronym:* CypCV.

First reported in *Cypripedium calceolus*; from Germany; by Lesemann and Vetten (1985).

**Natural host range and symptoms**

Symptoms persist.

*Cypripedium calceolus* — chlorotic and necrotic streaking and leaf deformation.

**Geographical distribution**

Spreads in Germany.

**Particle morphology**

Virions filamentous; not enveloped; usually flexuous; with a clear modal length; of 774 nm. Axial canal obscure. Basic helix obscure.

**Cytopathology**

Inclusions present in infected cells; are pinwheels; they do not contain virions.

**Taxonomy and Relationships**

Unassigned potyvirus.

Virus(es) with serologically related virions: bean yellow mosaic, tobacco etch, tulip breaking and turnip mosaic viruses.

**References**

Lesemann, D.-E. and Vetten, H.J. (1985). *Acta Hort.* **164**: 45.

# Dahlia mosaic *caulimovirus*

**Data collated by J. Hammond, 1986; A.A. Brunt, 1987.**

*Synonyms:* dahlia virus 1. *Standard acronym:* DMV.

First reported in *Dahlia pinnata*; from Germany; by Brandenburg (1928).

**Natural host range and symptoms**

Symptoms persist.

*Dahlia pinnata* — chlorotic vein-banding, mosaic and leaf malformation.

**Transmission**

Transmitted by a vector; an insect; 16 species, notably *Aphis fabae, Myzus persicae* and *Macrosiphum euphorbiae* (Heinze, 1951, 1952; Brierley and Smith, 1950); Aphididae. Transmitted in a non–persistent manner (in an atypical manner, because the virus can be retained by both feeding and by starved *M. persicae* (Brierley and Smith, 1950)). Virus transmitted by mechanical inoculation; not transmitted by seed.

**Ecology and control**

Studies reported by Brierley (1933; 1951); Brierley and Smith (1950); Heinze (1951; 1952). Dahlia mosaic is prevalent because it is transmitted effectively by several aphid species common on dahlias, and cultivars with few or no symptoms are important reservoirs of infection.

**Geographical distribution**

Probably distributed worldwide.

**Experimental host range**

Several (3–9) families susceptible.

Diagnostically susceptible

*Verbesina encelioides* — chlorotic local lesions then vein-banding, chlorosis and slight malformation.

*Dahlia pinnata* — systemic chlorotic vein-banding, then chlorosis and stunting.

*Ageratum conyzoides* cv. Blue Ball — chlorotic local lesions, then systemic chlorotic vein-banding and chlorosis.

*Zinnia elegans* — symptomless inoculated leaves, then slight systemic malformation, with transient chlorosis.

*Amaranthus caudatus* — few chlorotic local lesions, then mild systemic mottle.

*Chenopodium capitatum* — no symptoms in inoculated leaves, then conspicuous systemic chlorotic vein-banding and chlorosis.

Maintenance and propagation hosts

*Verbesina encelioides* and *Zinnia elegans*.

Assay hosts (**L**ocal lesions or **W**hole plants)

No reliable local lesion host, but *Verbesina encelioides* is best for whole plant assay.

Dicotyledonous Hosts *(**Susceptible** or Insusceptible)*

***Ageratum conyzoides*. *Amaranthus caudatus*. *Chenopodium capitatum*. *Dahlia pinnata*. *Verbesina encelioides*. *Zinnia elegans*.**

Sources of host-range data

Brunt (1971*a*).

**Properties of particles in sap**

TIP: 75–80 °C. LIV: 2–4 days (at 18°C), or 16–32 days (at 2°C; Brierley and Smith, 1950; Brunt, 1971). DEP: $\log_{10}$ minus 4. Infectivity of sap not changed by treatment with di–ethyl ether. Leaf sap contains few virions. Electron microscopy: use neutral 2% PTA, in other stains such as UA or UF the virions are swollen or disrupted, even after fixing.

**Purification**

Method: Brunt (1971*b*).

**Particle morphology**

Virions isometric; not enveloped; 48–50 nm in diameter; angular in profile.

**Physical properties**

One sedimenting component in purified preparations; sedimentation coefficient 254 *S*.

**Biochemical properties**

Virions contain 14.5–15 % nucleic acid (Goméc, 1973); 85–85.5 % protein; 0 % lipid.

*Genome* consists of DNA; double–stranded; circular; unipartite; largest (or only) genome part *c.* 6.8 kb (pairs).

*Non–genomic nucleic acid* not found in the virions.

*Virion protein(s)* one.

**Replication**

Replication does not depend on a helper virus.

**Cytopathology**

Virions found in mesophyll, epidermis, xylem, phloem sieve tubes and enlarged plasmodesmata (Kitajima and Lauritis, 1969); in cytoplasm and in the perinuclear space; associated with dictyosomes or Golgi cisternae (Kitajima *et al.*, 1969). Inclusions present in infected cells (Brunt, 1969; Kitajima and Lauritis, 1969; Kitajima *et al.*, 1969); are unusual in shape; electron-dense amorphous often vacuolated matrix; they contain virions (in great numbers and apparently hollow, 40–45 nm in diameter).

**Taxonomy and Relationships**

*Caulimovirus.*

Virus(es) with serologically related virions: cauliflower mosaic (Brunt, 1966; 1969) and carnation etched ring viruses (Hollings and Stone, 1969).

Antisera to dahlia mosaic and cauliflower mosaic virus differ little in homologous and heterologous titres in agar gel (0.8%) double diffusion tests. However neither virus infects the natural host of the other. Best tests for diagnosis: tobacco streak, tobacco rattle, tomato spotted wilt and cucumber mosaic viruses also infect dahlia. Unlike dahlia mosaic, however, each of these viruses is readily isolated and identified serologically and/or by the reactions of *Chenopodium quinoa, C. amaranticolor, Nicotiana glutinosa, N. tabacum* and *Vigna unguiculata.*

**References**

Brandenburg, E. (1928). *Forsch. Geb. Pflkrankh., Berlin* **5**: 39.
Brierley, P. (1933). *Contr. Boyce Thompson Inst. Pl. Res.* **5**: 235.
Brierley, P. (1951). *Pl. Dis. Reptr* **35**: 405.
Brierley, P. and Smith, F.F. (1950). *Pl. Dis. Reptr* **34**: 363.
Brunt, A.A. (1966). *Virology* **28**: 778.
Brunt, A.A. (1969). *Rep. Glasshouse Crops Res. Inst.* 1968, 104.
Brunt, A.A. (1971*a*). *Ann. appl. Biol.* **67**: 357.
Brunt, A.A. (1971*b*). *CMI/AAB Descr. Pl. Viruses* No. 51, 4 pp.
Goméc, B. (1973). *Ph.D. Thesis*, Univ. Calif. Davis, California, U.S.A.
Heinze, K. (1951). *Mitt. biol. BundesAnst. Land. Forstw.* 71, 126 pp.
Heinze, K. (1952). *Z. PflKrankh. PflPath. PflSchutz.* **59**: 3.
Hollings, M. and Stone, O.M. (1969). *Rep. Glasshouse Crops Res. Inst.* for 1968, p. 102.
Kitajima, E.W. and Lauritis, J.A. (1969). *Virology* **37**: 681.
Kitajima, E.W., Lauritis, J.A. and Swift, H. (1969). *Virology* **39**: 240.
Richins, R.D. and Shepherd, R.J. (1983). *Virology* **124**: 208.

# Dandelion (?) *carlavirus*

**Data collated by A.A. Brunt, 1994.**

First reported in *Taraxacum officinale*; from the Netherlands; by (Dijkstra *et al.*, 1985).

**Particle morphology**

Virions filamentous.

**Taxonomy and Relationships**

Possible *Carlavirus.*

**Comments**

Possibly a strain of dandelion latent *carlavirus* (Milne, 1988).

**References**

Dijkstra, J., Clement, Y. and Lohuis, H. (1985). *Neth. J. Pl. Path.* **91**: 77

John, L.J. (1982). *Phytopathology* **72**: 1239.

Milne, R.G. (1988). in *The Plant Viruses*, Vol.3; *The Filamentous Viruses*, p.11; ed. R.G. Milne. Plenum, New York.

# Dandelion latent *carlavirus*

**Data collated by R. Stace-Smith, 1991.**

*Standard acronym:* DaLV. *Strains:* dandelion carlavirus, possibly (Dijkstra *et al.*, 1985).

First reported in *Taraxacum officinale*; from British Columbia; by Johns (1982).

**Natural host range and symptoms**

Symptoms none.

*Taraxacum officinale.*

**Transmission**

Transmitted by probably a vector; probably an insect; *Myzus persicae* transmits in experiments; the natural vector is unknown; Aphididae. Transmitted in a non–persistent manner. Virus transmitted by mechanical inoculation; transmitted by grafting; not transmitted by contact between plants; not transmitted by seed; not transmitted by pollen.

**Geographical distribution**

Spreads in Canada.

**Experimental host range**

Few (<3) families susceptible. Experimentally infected plants mostly show chlorotic local lesions and systemic mottle.

Diagnostically susceptible

*Chenopodium quinoa, C. amaranticolor* — chlorotic local lesions, systemic mottle.
*Gomphrena globosa* — local lesions; no systemic symptoms.

Diagnostically insusceptible

*Cucumis sativus, Nicotiana tabacum, Lycopersicon esculentum, Vigna unguiculata* syn. *sinensis.*

Maintenance and propagation hosts

*Chenopodium quinoa.*

Assay hosts (**L**ocal lesions or **W**hole plants)

*Chenopodium quinoa* (L).

Dicotyledonous Hosts *(**Susceptible** or Insusceptible)*

*Brassica juncea. Capsicum frutescens.* ***Chenopodium quinoa****. Cucumis sativus. Datura stramonium. Helianthus annuus. Lycopersicon esculentum. Nicotiana debneyi. Nicotiana glutinosa. Nicotiana tabacum. Phaseolus vulgaris. Pisum sativum. Solanum tuberosum. Verbesina encelioides. Vicia faba. Vigna unguiculata.*

Sources of host-range data

John (1982).

**Properties of particles in sap**

TIP: 75–80 °C. LIV: 5 days (at 23°C). DEP: $\log_{10}$ minus 5–5. Leaf sap contains many virions.

**Purification**

Method: John (1982).

**Particle morphology**

Virions filamentous; not enveloped; usually flexuous; with a clear modal length; of 640 nm; 12–13 nm wide. Axial canal obvious. Basic helix obscure.

**Physical properties**

One sedimenting component in purified preparations. $A_{260}/A_{280}$ ratio 1.4.

**Biochemical properties**

Virions contain 2.5 % nucleic acid.

*Genome* consists of RNA; single–stranded.

**Cytopathology**

Virions found in leaves.

**Taxonomy and Relationships**

*Carlavirus.*

Virus(es) with serologically related virions: potato S, helenium S, chrysanthemum B and narcissus latent viruses. Virus(es) with serologically unrelated virions: carnation latent, pea streak, poplar mosaic, potato M and red clover vein mosaic viruses.

**Comments**

A latent virus of dandelion in the Netherlands is possibly a strain of this virus (Milne, 1988).

**References**


Dijkstra, J., Clement, Y. and Lohuis, H. (1985). *Neth. J. Pl. Path.* **91**: 77

John, L.J. (1982). *Phytopathology* **72**: 1239.

Milne, R.G. (1988). in *The Plant Viruses*, Vol.3; *The Filamentous Viruses*, p.11; ed. R.G. Milne. Plenum, New York.


## Dandelion yellow mosaic *sequivirus*

**Data collated by A.A. Brunt and L. Bos, 1992.**

*Synonyms:* lettuce necrosis virus (Hein, 1963). *Standard acronym:* DaYMV.

First reported in *Taraxacum officinale*; from the U.K.; by Kassanis (1944); Bos (1983).

**Natural host range and symptoms**

Symptoms persist. Symptoms chlorotic rings, interveinal necrosis, distortion and stunting.

*Lactuca sativa* — veinal chlorosis and necrosis, and veinal and interveinal necrotic etching; leaf narrowing and curling and severe plant stunting.

*Taraxacum officinale* — conspicuous yellow rings and mottling.

**Transmission**

Transmitted by a vector; an insect; *Acyrthosiphon (Aulacorthum) solani, Myzus ornatus, Myzus ascalonicus, Myzus persicae*; Aphididae. Not transmitted by *Aphis fabae,*

*Macrosiphum euphorbiae, Nasonovia ribicola, Aulocorthum circumflexum, Pemphigus bursarius*. Virus transmitted by mechanical inoculation.

**Ecology and control**

Studies reported by Kassanis (1947).

**Geographical distribution**

Spreads in the former Czechoslovakia, Germany, the Netherlands, Norway, and the UK.

**Experimental host range**

Several (3–9) families susceptible. Experimentally infected plants mostly show chlorotic or necrotic local lesions, systemic chlorosis, necrosis, leaf curling and stunting or symptomless infection.

Diagnostically susceptible

*Chenopodium amaranticolor, C. quinoa* — numerous small chlorotic local lesions; systemic stippling.
*Chenopodium murale* — small local necrotic lesions.
*Lactuca sativa* — chlorotic mosaics, leaf curling and plant stunting.
*Nicotiana benthamiana* — systemic chlorosis (some isolates only).
*N. clevelandii* — systemic chlorotic rings (some isolates only).

Diagnostically insusceptible

*Pisum sativum, Sonchus oleraceus, Vigna unguiculata.*

Maintenance and propagation hosts

*Chenopodium quinoa, Lactuca sativa, Nicotiana clevelandii.*

Assay hosts (**L**ocal lesions or **W**hole plants)

*Chenopodium murale* (L), *C. amaranticolor* (L), *C. quinoa* (L), *Nicotiana clevelandii* (W), *Gomphrena globosa* (L).

Dicotyledonous Hosts *(**Susceptible** or Insusceptible)*

*Apium graveolens. Calendula officinalis. Cheiranthus cheiri.* ***Chenopodium amaranticolor. Chenopodium murale. Chenopodium quinoa. Cichorium endiva. Cucumis sativus. Gomphrena globosa. Lactuca sativa. Nicotiana benthamiana. Nicotiana clevelandii. Nicotiana glutinosa. Nicotiana rustica. Nicotiana sylvestris. Pastinaca sativa.*** *Phaseolus vulgaris. Pisum sativum. Senecio vulgaris. Sonchus oleraceus.* ***Spinacia oleracea.*** *Vigna unguiculata.*

Additional hosts

*Lactuca serriola, Lactuca virosa.*

Sources of host-range data

Kassanis (1947); Bos *et al.* (1983).

**Properties of particles in sap**

TIP: 60–65 °C. LIV: 1 days. DEP: $\log_{10}$ minus 4–5. Leaf sap contains few virions. Electron microscopy: ISEM.

**Purification**

Method: Bos *et al.* (1983).

**Particle morphology**

Virions isometric; not enveloped; rounded in profile; without a conspicuous capsomere arrangement.

**Physical properties**

One sedimenting component in purified preparations; sedimentation coefficient 159 *S*. Density 1.42 in $Cs_2SO_4$. $A_{260}/A_{280}$ ratio 1.67.

**Biochemical properties**

*Genome* consists of RNA; single–stranded; unipartite; largest (or only) genome part 10 kb.

*Virion protein(s)* three; $M_r$ 29000, or 32000. $M_r$ of 2nd largest 25000, or 29500. $M_r$ of 3rd largest 24000, or 27000.

**Replication**

Replication does not depend on a helper virus.

**Cytopathology**

Virions found in leaves; in cytoplasm.

**Taxonomy and Relationships**

*Sequivirus: Sequiviridae.*

Virus(es) with serologically unrelated virions: parsnip yellow fleck virus.

**References**

Blystad, D.-R. (1989). *Norw. J. agric. Sci.* **3**: 373.

Bos, L., Huijberts, N., Huttinga, H. and Maat, D. (1983). *Neth. J. Pl. Path.* **89**: 207.

Hein, A. (1963). *NachrBl. dtsch. Pflanzenschutzd. Braunschweig* **15**: 17.

Hemida, S.K. and Murant, A.F. (1989). *Ann. appl. Biol.* **114**: 87.

Kassanis, B. (1944). *Nature, Lond.* **154**: 16.

Kassanis, B. (1947). *Ann. appl. Biol.* **34**: 412.

Kristensen, H.R., Tapio, E. and Lihnell, D. (1965). *List of Plant Viruses Occurring in the Nordic Countries*, p. 25.

Murant, A.F. (1988). In: *The Plant Viruses*; Vol. 3, Polyhedral Virions with Monopartite RNA Genomes, p. 273; ed. R. Koenig. Plenum Press, New York.

Vetten, H.J., Breyel, E., Lesemann, D.-E., Maiss, E. and Weidemann, H.L. (1985). *Phytoparasitica* **13**: 271.

# Daphne S (?) *carlavirus*

**Data collated by H.F. Nielson and K.S. Milne, 1991.**

*Synonyms:* daphne leaf distortion S carlavirus. *Standard acronym:* DVS.

First reported in *Daphne* spp.; from New Zealand; by Forster and Milne.

**Natural host range and symptoms**

Symptoms none. Symptoms sometimes downcurling of leaf edges.

*Daphne odora* cv. Leucanthe — malformation and downcurling of mature leaves.

*Daphne odora* cvs Leucanthe Variegata, Rubra, Rubra variegata, *D. burkwoodii* cvs Somerset, Variegata, *D. cneorum* cvs Major, Variegata, *D. x. dauphinii, D. genkwa, D. giraldii, D. laureola, D. x. neopolitana, D. retusa* — symptomless.

**Transmission**

Transmitted by a vector; an insect; *Myzus persicae*; Aphididae. Transmitted in a non–persistent manner. Virus not transmitted by mechanical inoculation; transmitted by grafting; not transmitted by contact between plants.

**Geographical distribution**

Spreads in New Zealand.

**Experimental host range**

Few (<3) families susceptible.

Diagnostically susceptible

*Daphne* spp. — symptomless.

Maintenance and propagation hosts

*Daphne odora* cvs Leucanthe or Rubra.

Dicotyledonous Hosts *(**Susceptible** or Insusceptible)*

*Antirrhinum majus. Beta vulgaris. Capsicum frutescens. Chenopodium amaranticolor. Chenopodium quinoa. Cucumis sativus. Cucurbita maxima. Dianthus barbatus. Gomphrena globosa. Nicotiana clevelandii. Nicotiana glutinosa. Nicotiana rustica. Nicotiana tabacum. Petunia hybrida. Phaseolus vulgaris. Pisum sativum. Spinacia oleracea. Vicia faba. Vigna unguiculata.*

Sources of host-range data

Forster (1974); Forster and Milne (1978).

**Properties of particles in sap**

Leaf sap contains few virions. Electron microscopy: flowers or young leaves are good sources of virions.

**Purification**

Method: Forster and Milne (1978).

**Particle morphology**

Virions filamentous; not enveloped; usually straight, or usually flexuous; with a clear modal length; of 720 nm; 13 nm wide. Axial canal obvious. Basic helix obscure.

**Physical properties**

One sedimenting component in purified preparations; sedimentation coefficient 174 *S* (±1). $A_{260}/A_{280}$ ratio 1.35.

1 virus specified dsRNA species found in infected cells.

**Replication**

Replication does not depend on a helper virus.

**Cytopathology**

Virions found in cytoplasm.

**Taxonomy and Relationships**

Possible *Carlavirus*.

Virus(es) with serologically related virions: carnation latent and potato S viruses.

Best tests for diagnosis: flowers or young leaves are good sources of virions for examination by electron microscopy. This virus, which has virions 720 nm long, commonly occurs in mixed infections with daphne Y virus, which has slightly longer (730 nm), less rigid and narrower virions. If, however, all virions appear the same, have a length of *c.*700–750 nm and can be mechanically transmitted to *C. quinoa* inducing chlorotic local lesions in 4–6 days, and scattered systemic blotches in 10–28 days then the virus is probably the potyvirus; the carlavirus is not sap transmissible, and other viruses found in daphne but not readily seen in the electron microscope give quite different reactions when sap transmitted to *C. quinoa.*

**References**

Forster, R.L.S. (1974). *M.Hort.Sci Thesis*, Massey University, New Zealand 137 pp.
Forster, R.L.S. and Milne, K.S. (1975). *N.Z. Jl agric. Res.* **18**: 391.
Forster, R.L.S. and Milne, K.S. (1978). *N.Z. Jl agric. Res.* **21**: 131.

# Daphne X *potexvirus*

**Data collated by R.L.S. Forster, 1986.**

*Synonyms:* daphne latent X potexvirus. *Standard acronym:* DVX.

First reported in *Daphne cneorum*; from New Zealand; by Forster and Milne (1978*b*).

**Natural host range and symptoms**

Symptoms none.
*Daphne cneorum, D. odora* — symptomless.

**Transmission**

Transmitted by means not involving a vector; Not transmitted between peas by the aphid *Myzus persicae.* Virus transmitted by mechanical inoculation; not transmitted by seed.

**Ecology and control**

Studies reported by Forster and Milne (1976; 1978). In New Zealand, no *Daphne cneorum* plants have been found free of daphne X virus out of more than 100 tested. *D. cneorum* plants imported from West Germany free of daphne X virus have been propagated and used to replace infected stock plants in nurseries.

**Geographical distribution**

Spreads in Australia and New Zealand.

**Experimental host range**

Several (3–9) families susceptible.

Diagnostically susceptible

*Cucumis sativus* — sunken chlorotic local lesions; then systemic faint chlorotic spotting, but later leaves are symptomless.
*Dolichos biflorus* — small necrotic local lesions; not systemic.
*Gomphrena globosa* — necrotic local lesions, often with red haloes; not systemic.
*Nicotiana clevelandii* — chlorotic and necrotic local lesions; systemic chlorotic rings or mottle.
*Pisum sativum* — necrotic local lesions; then systemic chlorotic mottle and sometimes stem blackening.

Diagnostically insusceptible

*Vicia faba, Spinacia oleracea.*

Maintenance and propagation hosts

*Nicotiana clevelandii* — inoculated leaves are also good source for purification.

Assay hosts (**L**ocal lesions or **W**hole plants)

*Cucumis sativus* (L), *Dolichos biflorus* (L), *Gomphrena globosa* (L).

Dicotyledonous Hosts *(**Susceptible** or Insusceptible)*

***Chenopodium quinoa*. *Cucumis melo*. *Cucumis sativus*. *Cucurbita maxima*. *Datura stramonium*. *Dolichos biflorus*. *Gomphrena globosa*. *Nicotiana clevelandii*. *Pisum sativum*.** *Spinacia oleracea*. *Vicia faba*.

**Properties of particles in sap**

TIP: 80 °C. LIV: 35 days. DEP: $\log_{10}$ minus 5. Leaf sap contains many virions.

**Purification**

Method: Forster and Milne (1978*c*).

**Particle morphology**

Virions filamentous; not enveloped; usually flexuous; with a clear modal length; of 500 nm; 12 nm wide. Axial canal obscure. Basic helix obvious.

**Physical properties**

One sedimenting component in purified preparations; sedimentation coefficient 110 *S*. Density 1.28 g $cm^{-3}$ in CsCl.

**Biochemical properties**

Virions contain 6 % nucleic acid; 94 % protein; 0 % lipid.

*Genome* consists of RNA; single–stranded; linear; unipartite; largest (or only) genome part 6.5 kb. Genomic nucleic acid isolated by Guilford and Forster (1986). Infectivity retained when deproteinised with phenol or detergent. Poly A region present; at the 3'–terminal and 200 bases long. Additional factor not required for infectivity.

*Non–genomic nucleic acid* found in the virions; is subgenomic *m*RNA. Sub–genomic *m*RNA found in infected cells; Five subgenomic RNAs were detected in infected tissues (4.9; 4.0; 2.1; 1.4 and 0.8 Kb). The 0.8 Kb subgenomic RNA is the *m*RNA for the coat protein (Guilford and Forster, 1986).

*Virion protein(s)* one; $M_r$ 23000.

**Replication**

Replication does not depend on a helper virus.

**Cytopathology**

Virions found in leaves; in cytoplasm. Inclusions present in infected cells; are crystals in the cytoplasm; they contain virions.

**Taxonomy and Relationships**

*Potexvirus*.

Virus(es) with serologically unrelated virions: clover yellow mosaic, narcissus mosaic, potato X and white clover mosaic viruses, although preliminary tests (Milne and Forster, 1976) had suggested a distant relationship to white clover mosaic virus.

Best tests for diagnosis: Only two other viruses infecting daphne have flexuous filamentous virions; daphne S virus has virions *c.* 720 nm long (Forster and Milne, 1978*a*), and daphne Y virus has virions *c.* 730 nm long (Forster and Milne, 1976). Unlike daphne X virus, these viruses do not infect *Cucumis sativus* or *Gomphrena globosa*.

**References**

Forster, R.L.S. and Milne, K.S. (1976). *N.Z. Jl agric. Res.* **19**: 359.
Forster, R.L.S. and Milne, K.S. (1978*a*). *N.Z. Jl agric. Res.* **21**: 131.
Forster, R.L.S. and Milne, K.S. (1978*b*). *N.Z. Jl agric. Res.* **21**: 137.

Forster, R.L.S. and Milne, K.S. (1978*c*). *CMI/ABB Descr. Pl. Viruses* No. 195, 3 pp.
Guilford, P.J. and Forster, R.L.S. (1986). *J. gen. Virol.* **67**: 83.
Milne, K.S. and Forster, R.L.S. (1976). *Acta Hort.* **59**: 95.

# Daphne Y *potyvirus*

**Data collated by A.A. Brunt, 1991.**

*Synonyms:* daphne chlorotic mosaic Y potyvirus. *Standard acronym:* DVY.

First reported in *Daphne odora, D. cneorum, D. retusa*; from New Zealand; by Forster and Milne (1975).

**Natural host range and symptoms**

*Daphne odora, D. cneorum, D. retusa, D. mezereum* — leaf chlorosis and distortion, decreased flower quality.

**Transmission**

Transmitted by a vector; an insect; *Myzus persicae*; Aphididae. Transmitted in a non–persistent manner. Virus transmitted by mechanical inoculation.

**Geographical distribution**

Spreads in Australia, the Netherlands, New Zealand, and the UK.

**Experimental host range**

Few (<3) families susceptible.

Diagnostically susceptible

*Chenopodium album, C. amaranticolor, Spinacia oleracea* — chlorotic and/or necrotic local lesions; no systemic infection.

*Chenopodium quinoa* — chlorotic local lesions; systemic chlorotic flecks or blotches.

*Nicotiana clevelandii* — symptomless local infection or etched necrotic rings; symptomless systemic infection.

Diagnostically insusceptible

*Cucumis sativus, Gomphrena globosa, Nicotiana tabacum, Phaseolus vulgaris, Vigna unguiculata.*

Maintenance and propagation hosts

*Chenopodium quinoa.*

Assay hosts (**L**ocal lesions or **W**hole plants)

*Chenopodium amaranticolor* (L), *C. quinoa* (L).

Dicotyledonous Hosts *(**Susceptible** or Insusceptible)*

*Antirrhinum majus. Beta vulgaris. Brassica campestris* ssp. *pekinensis. Cassia occidentalis.* ***Chenopodium album. Chenopodium amaranticolor. Chenopodium quinoa.*** *Cucumis melo. Cucumis sativus. Cucurbita maxima. Dianthus barbatus. Matthiola incana.* ***Nicotiana clevelandii.*** *Nicotiana glutinosa. Nicotiana tabacum. Petunia hybrida. Phaseolus vulgaris. Pisum sativum.* ***Spinacia oleracea.*** *Vicia faba. Vigna unguiculata. Zinnia elegans.*

Sources of host-range data

Forster and Milne (1976).

**Properties of particles in sap**

TIP: 65 °C. LIV: 8–10 days (at 20°C). DEP: $\log_{10}$ minus 5.

**Purification**

Method: Forster and Milne (1976).

**Particle morphology**

Virions filamentous; not enveloped; usually flexuous; with a clear modal length; of 733 nm; 13 nm wide. Axial canal obscure. Basic helix obscure.

**Physical properties**

One sedimenting component in purified preparations; sedimentation coefficient 149 *S*.

**Cytopathology**

Virions found in leaves and mesophyll; in cytoplasm. Inclusions present in infected cells; are pinwheels; they do not contain virions.

**Taxonomy and Relationships**

*Potyvirus: Potyviridae.*

Virus(es) with serologically unrelated virions: bean yellow mosaic, lettuce mosaic, maize dwarf mosaic, passionfruit woodiness, potato Y, tobacco etch, tamarillo mosaic, turnip mosaic and watermelon mosaic 2 viruses.

**References**

Bos, L. (1971). *Jversl. Inst. PlZt.* 1971, p. 84.
Forster, R.L.S. and Milne, K.S. (1975). *N.Z. Jl agric. Res.* **18**; 391.
Forster, R.L.S. and Milne, K.S. (1976). *N.Z. Jl agric. Res.* **19**: 359.
Sutton, J. and Taylor, R.M. (1973). *Rep. Vic. Res. Inst.* 1972–1973, p. 12.

# Dasheen mosaic *potyvirus*

**Data collated by F.W. Zettler and E. Hiebert, 1986; B. Rodoni, 1986.**

*Standard acronym:* DsMV.

First reported in *Zantedeschia elliottina* corms imported into the U.S.A. from Holland; from Alachua County, Florida, U.S.A.; by Zettler *et al.* (1970).

**Natural host range and symptoms**

Symptoms vary seasonally.

*Aglaonema, Alocasia, Amorphophallus, Arisaema, Caladium, Cyrtosperma* spp. — mosaic.

*Cryptocoryne, Dieffenbachia, Philodendron, Richardia, Zantedeschia* spp.—mosaic and leaf malformation.

*Colocasia, Xanthosoma* spp. — mosaic, chlorotic feathering.

**Transmission**

Transmitted by a vector; an insect; *Myzus persicae, Aphis craccivora* (Morales and Zettler, 1977) and *Aphis gossypii* (Gollifer *et al.*, 1977), but apparently not by *Pentalonia nigronervosa* (Morales and Zettler, 1977) or *Rhopalosiphum padi* (Gollifer *et al.*, 1977); Aphididae. Transmitted in a non–persistent manner. Virus transmitted by mechanical inoculation; not transmitted by contact between plants; not transmitted by seed (Hartman *et al.*, 1972; Zettler and Abo El-Nil, 1979); not transmitted by pollen.

**Ecology and control**

Studies reported by Abo El-Nil and Zettler (1974); Hartman and Zettler (1972); Debrot and Ordosgoitti (1974); Alconero and Zettler (1971); Zettler *et al.* (1970), reported that dasheen mosaic virus is widespread in *Aglaonema, Caladium, Colocasia, Dieffenbachia, Xanthosoma* and *Zantedeschia* spp. Tissue culture techniques give virus-free plants of *Caladium, Colocasia, Dieffenbachia* and *Xanthosoma*, and these can be used for propagation (Hartman, 1974*b*, Zettler *et al.*, 1980). Attempts to eliminate dasheen mosaic virus from *Xanthosoma* corms by heat treatment were unsuccessful (Alconero, 1972).

**Geographical distribution**

Probably distributed worldwide (especially in tropical and subtropical regions). Spreads in China.

**Experimental host range**

Many (>9) families susceptible.

Diagnostically susceptible

*Philodendron selloum* — systemic mosaic, vein clearing, malformation.
*Caladium hortulanum, Colocasia* spp., *Xanthosoma* spp. — systemic mosaic.
*Dieffenbachia* spp. — systemic mosaic, epinasty, systemic and local necrosis.
*Philodendron verrucosum* — local necrosis; not systemic.
*Zantedeschia* spp. — systemic mosaic, malformation.

Diagnostically insusceptible

*Nicotiana x edwardsonii, N. benthamiana, Vigna unguiculata.*

Maintenance and propagation hosts

*Philodendron selloum; Caladium hortulanum* (certain cultivars); *Xanthosoma* spp. (certain cultivars); *Colocasia* spp. (certain cultivars); *Zantedeschia* spp., *Richardia* spp.

Assay hosts (**L**ocal lesions or **W**hole plants)

*Philodendron verrucosum* (L), *Philodendron selloum* (W) — test plant for aphid transmission. The following species gave local lesions in Italy, but not outside Europe (Rana *et al.*, 1983): *Chenopodium amaranticolor, C. quinoa, C. ambrosioides, Nicotiana benthamiana, Saponaria vaccaria* cv. Pink Beauty.

Dicotyledonous Hosts *(**Susceptible** or Insusceptible)*

*Amaranthus caudatus. Apium graveolens. Beta vulgaris. Brassica oleracea* ssp. *botrytis. Capsicum annuum. Carica papaya. Cassia occidentalis.* ***Chenopodium amaranticolor***, or *Chenopodium amaranticolor.* ***Chenopodium ambrosioides***. ***Chenopodium quinoa***, or *Chenopodium quinoa. Crotalaria spectabilis. Cucumis sativus. Cucurbita moschata. Cucurbita pepo. Dahlia pinnata. Datura stramonium. Gomphrena globosa. Lupinus angustifolius. Lycopersicon esculentum.* ***Nicotiana benthamiana***, or *Nicotiana benthamiana. Nicotiana clevelandii. Nicotiana debneyi. Nicotiana glutinosa. Nicotiana glutinosa × clevelandii. Nicotiana megalosiphon. Nicotiana rustica. Nicotiana tabacum* (*N. langsdorffii* (I) (Van der Meer, 1985)). *Petunia hybrida. Phaseolus vulgaris. Physalis floridana. Pisum sativum.* ***Saponaria vaccaria***. *Spinacia oleracea.* ***Tetragonia expansa***, or *Tetragonia expansa. Tropaeolum majus. Vicia faba. Vigna unguiculata. Zinnia elegans.*

Monocotyledonous Hosts *(**Susceptible** or Insusceptible)*

***Caladium hortulanum**. **Colocasia esculenta**. **Dieffenbachia picta**. **Philodendron selloum**. **Philodendron verrucosum**. Sorghum bicolor. **Xanthosoma caracu**. **Zantedeschia sp.**. Zea mays.*

Additional hosts

***Aglaonema*** spp., ***Alocasia*** spp., ***Amorphophallus*** spp., ***Arisaema*** spp., *Brassica acephala*, ***Cryptocoryne*** spp., ***Cyrtosperma*** spp., ***Dieffenbachia*** spp., ***Philodendron*** spp., ***Ricardia*** spp. and ***Spathiphyllum*** spp.

Comments on host-range

There are conflicting reports on the susceptibility of non-aroids; in European studies some are susceptible, but, in tests Florida and Venezuela, non-aroids were not infected (Zettler and Hartman, 1986).

Sources of host-range data

Chase and Zettler (1982); Debrot and Ordosgoitti (1974); Hartman (1974*a*); Hartman and Zettler (1972); Rana *et al.* (1983); Rodrigues *et al.* (1984); Shanmuganathan (1980); Tooyama (1975*a*; 1975*b*); Van der Meer (1985); Wisler *et al.* (1978); Zettler and Hartman (1986); Zettler *et al.* (1970).

**Properties of particles in sap**

TIP: 60–65 °C. LIV: 3–4 days (at 26°C (Alconero, 1972)). DEP: $\log_{10}$ minus 2. Leaf sap contains few virions. Electron microscopy: see Chase and Zettler (1982), p. 892.

**Purification**

Method: Moghal and Francki (1977); Abo El-Nil *et al.* (1977).

**Particle morphology**

Virions filamentous; not enveloped; usually flexuous; with a clear modal length; of 750 nm (approximately). Axial canal obscure. Basic helix obscure.

**Physical properties**

One sedimenting component in purified preparations. Density 1.26 g $cm^{-3}$ in CsCl. $A_{260}/A_{280}$ ratio 2.38 (specific absorbance at 260 nm).

**Biochemical properties**

Virions contain 5 % nucleic acid; 95 % protein; 0 % lipid.

*Genome* consists of RNA; single–stranded; linear; unipartite; largest (or only) genome part 10 kb. Genomic nucleic acid isolated by Dougherty and Hiebert (1980).

*Sequence database accession code(s):*

S55485 Em(40)_vi:S55485 coat protein (3' untranslated region) dasheen mosaic potyvirus DMV, LA isolate, mRNA Partial

S66478 Em(40)_vi:S66478 Gb(84)_un:S66478 (3' region) dasheen mosaic virus DMV, isolate LA, Genomic, 252 nt. 1/94 252bp.

U00122 Em(40)_vi:DPU00122 Gb(84)_vi:DMU00122 Dasheen mosaic potyvirus capsid protein gene, complete cds, RNA polymerase gene, partial cds.

U08124 Em(40)_vi:DM08124 Gb(84)_vi:DMU08124 Dasheen mosaic virus RNA polymerase (NIb) gene, partial cds, and capsid protein (CP) gene, co 4 sequences.

*Virion protein(s)* one (; many virion species detected in SDS-PAGE probably due to proteolysis); $M_r$ 29000; capsid protein. Method of preparation: Morales, Zettler and Hiebert (unpublished); Hiebert *et al.* (1984).

*Virus–coded non–virion proteins* identified by genomic sequence analysis (Hiebert *et al.* (1984)); four proteins found. $M_r$ of the largest 69000; cylindrical inclusion protein.

$M_r$ of 2nd largest 56000; nuclear inclusion-like protein, polymerase. $M_r$ of 3rd 47000; nuclear inclusion-like protein, protease. $M_r$ of 4th 51000; helper protein for aphid transmisson.

**Replication**

Replication does not depend on a helper virus.

**Cytopathology**

Virions found in mesophyll, epidermis and probably phloem; in cytoplasm. Inclusions present in infected cells; are pinwheels; they do not contain virions (but virions may be associated with pinwheel inclusions).

**Taxonomy and Relationships**

*Potyvirus: Potyviridae.*

Virus(es) with serologically related virions: blackeye cowpea mosaic, tobacco etch, and araujia mosaic viruses (Hiebert and Charudattan, 1984; Abo El-Nil *et al.*, 1977). Virus(es) with serologically unrelated virions: potato Y and turnip mosaic viruses.

The cylindrical inclusion protein (pinwheel) of dasheen mosaic virus is serologically related to that of araujia mosaic virus (Hiebert and Charudattan, 1984). The helper component protein of tobacco vein mottling potyvirus is related to *in vitro* synthesized gene products of dasheen mosaic virus (Hiebert *et al.*, 1984). Best tests for diagnosis: Filamentous viruses of *Anthurium andraeanum* in Venezuela (Herold, 1967) and *Zantedeschia aethiopica* in the Soviet Union (Kolbasina and Protsenko, 1973) have been reported, but their relationship to dasheen mosaic virus has not been determined. Cucumber mosaic virus (Lovisolo and Conti, 1969), tomato spotted wilt virus (Tomkins and Severin, 1950), and two bacilliform viruses infecting taro (James *et al.*, 1973) can readily be distinguished from dasheen mosaic virus by virion morphology, host range, serology and/or mode of transmission. Banana bunchy top virus in *Colocasia* (Ram and Summanwar, 1984) and tobacco necrosis virus in *Dieffenbachia* (Paludan and Begtrup, 1982) can be similarly distinguished.

**References**


Abo El-Nil, M.M. and Zettler, F.W. (1976). *Pl. Dis. Reptr* **60**: 281.

Abo El-Nil, M.M., Zettler, F.W. and Hiebert, E. (1977). *Phytopathology* **67**: 1445.

Alconero, R. (1972). *Pl. Dis. Reptr* **56**: 320.

Alconero, R. and Zettler, F.W. (1971). *Pl. Dis. Reptr* **55**: 506.

Chase, A.R. and Zettler, F.W. (1982). *Plant Dis.* **66**: 891.

Christie, R.G. and Edwardson, J.R. (1977). *Fla Agric. Exp. Stn Monog.* No. 9, p. 150.

Debrot, E.A. and Ordosgoitti, A. (1974). *Pl. Dis. Reptr* **58**: 1032.

Dougherty, W.G. and Hiebert, E. (1980). *Virology* **101**: 466.

Gollifer, D.E., Jackson, G.V.H., Dabek, A.J., Plumb, R.T. and May, Y.Y. (1977). *Pesticide Articles and News Summaries* **23**: 171.

Hakkaart, F.A. and Waterreus, H.A.J.I. (1976). *Acta Hort.* **59**: 175.

Hartman, R.D. (1974*a*). *Ph.D. Thesis*, Univ. Florida, U.S.A..

Hartman, R.D. (1974*b*). *Phytopathology* **64**: 237.

Hartman, R.D. (1972). *Phytopathology* **62**: 804.

Hartman, R.D., Zettler, F.W., Knauss J.F. and Hawkins, E.M. (1972). *Proc. Fla State Hort. Soc.* **85**: 404.

Herold, F. (1967). *Phytopathology* **57**: 8.

Hiebert, E. and Charudattan, R. (1984). *Phytopathology* **74**: 642.

Hiebert, E., Thornbury, D.W. and Pirone, T.P. (1984). *Virology* **135**: 1.

Hill, S.A. and Wright, D.M. (1980). *Pl. Path.* **29**: 143.

James, M., Kenten, R.H. and Woods, R.D. (1973). *J. gen. Virol.* **21**: 145.

Kolbasina, N.I. and Protsenko, A.E. (1973). *Byull. glavn. bot. Sada Leningrad* **88**: 98.
Lovisolo, O. and Conti, M. (1969). *Annls. Phytopath.* **1**: 367.
Moghal, S.M. and Francki, R.I.B. (1977). *Virology* **73**: 352.
Morales, F.J. and Zettler, F.W. (1977). *Fitopatol. Columbiana* **6**: 134.
Paludan, N. and Begtrup, J. (1982). *Danish Res. Serv. for Plant and Soil Sci. Report* No. 1612, p. 399.
Ram, R.D. and Summanwar, A.S. (1984). *Curr. Sci.* **53**: 145.
Rana, G.L., Vovlas, C. and Zettler, F.W. (1983). *Plant Dis.* **67**: 1121.
Rodoni, B. (1986). *B.Sc. (Hons) Report*, Burnley, P.R.I., Vic. Dep. Agric.
Rodrigues, M.G.R., Kitajima, E.W. and Lin, M.T. (1984). *Fitopatol. Bras.* **9**: 291.
Shanmuganathan, N. (1980). *F.A.O. Pl. Prot. Bull.* **18**: 29.
Tompkins, C.M. and Severin, H.H.P. (1950). *Hilgardia* **20**: 207.
Tooyama, A. (1975*a*). *Ann. Phytopath. Soc. Japan* **41**: 504.
Tooyama, A. (1975*b*). *Ann. Phytopath. Soc. Japan* **41**: 506.
Van der Meer, F.W. (1985). *Phytophylactica* **17**: 95.
Wisler, G.C., Zettler, F.W., Hartman, R.D. and McRitchie, J.J. (1978). *Proc. Fla State Hort. Soc.* **91**: 237.
Zettler, F.W. and Hartman, R.D. (1986). *Food and Fertilizer Technology Center* Ext. Bull. 233.
Zettler, F.W., Abo El-Nil, M.M. and Hartman, R.D. (1978). *CMI/AAB Descr. Pl. Viruses* No. 191, 4 pp.
Zettler, F.W., Foxe, M.J., Hartman, R.D., Edwardson, R.D. and Christie, R.G. (1970). *Phytopathology* **60**: 983.
Zettler, F.W., Knauss, J.F., Taylor-Knauss, M.E., Hartman, R.D. and Chase, A.R. (1980). *Acta Hort.* **110**: 259.

# Dasheen bacilliform (?) *badnavirus*

**Data collated by R.D. Woods, 1984.**

*Synonyms:* colocasia badnavirus, taro bacilliform virus. *Standard acronym:* CBV.

First reported in *Colocasia esculenta*; from the Solomon Islands; by Kenten and Woods (1973).

### Natural host range and symptoms

Symptoms disappear soon after infection.
*Colocasia esculenta*—vein clearing and slight leaf malformation.

### Transmission

Transmitted by a vector; an insect; *Planococcus citri*; Pseudococcidae. Virus not transmitted by mechanical inoculation; not transmitted by grafting; not transmitted by contact between plants.

### Geographical distribution

Spreads in the Cook Islands, Fiji, the Solomon Islands, Vanuatu, and Western Samoa.

Diagnostically susceptible

*Colocasia esculenta, Alocasia macrorrhiza, Xanthosoma* sp. — transient vein-clearing and slight leaf malformation.

### Properties of particles in sap

Leaf sap contains few virions.

**Purification**

Method: James *et al.* (1973).

**Particle morphology**

Virions bacilliform; not enveloped; 125 nm in length; 28 nm wide.

**Physical properties**

One sedimenting component in purified preparations; sedimentation coefficient 285 *S*.

**Cytopathology**

Virions found in all parts of the host plant and often in phloem sieve tubes; in cytoplasm and in cell vacuoles.

**Taxonomy and Relationships**

Possible *Badnavirus*.

Virus(es) with serologically unrelated virions: cacao swollen shoot badnavirus.

Best tests for diagnosis: "Male" taro plants in Papua New Guinea and the Solomon Islands also containing taro bobone rhabdovirus develop very severe "alomae" symptoms.

**References**

Gollifer, D.E., Jackson, G.H.V., Dabek, A.J., Plumb. R.T. and May, Y.Y. (1977). *PANS* **23**: 171.

Jackson, G.H.V. (1980). *Diseases and Pests of Taro* South Pacific Commission, Noumea.

James, M., Kenten, R.H. and Woods, R.D. (1973). *J. gen. Virol.* **21**: 145.

Kenten, R.H. and Woods, R.D. (1973). *PANS* **19**: 38.

# Datura Colombian *potyvirus*

**Data collated by R.P. Kahn, 1987.**

*Synonyms:* Colombian datura virus. *Standard acronym:* CDV.

First reported in *Datura sanguinea, Datura candida*; from *Datura* cuttings imported from the Sibundoy Valley, Putumato, Colombia; by Kahn and Bartels (1968).

**Natural host range and symptoms**

Symptoms persist, or disappear soon after infection.

*Datura candida, D. sanguinea* — veinbanding, chlorotic flecking followed by mottling.

**Transmission**

Transmitted by a vector; an insect; *Myzus persicae*; Aphididae. Transmitted in a non–persistent manner. Virus transmitted by mechanical inoculation; transmitted by grafting.

**Geographical distribution**

Spreads in Colombia.

**Experimental host range**

Few (<3) families susceptible.

Diagnostically susceptible

*Datura metel* — wilting and death.

*Nicotiana tabacum* cv. Virginia — chlorotic and necrotic spots.

*Nicotiana tabacum* cv. Samsun — chlorotic or necrotic spots, veinal necrosis.
*Solanum tuberosum* x *S. demissum* 'A-6' — necrotic spots.
*Datura stramonium* — chlorotic spots, occasionally.
*Gomphrena globosa* var. *rubra* — red bordered local lesions.

Diagnostically insusceptible

*Capsicum annuum.*

Maintenance and propagation hosts

*Nicotiana tabacum* cv. Samsun.

Assay hosts (**L**ocal lesions or **W**hole plants)

*Datura metel* (W); *Nicotiana tabacum* cv. Samsun (W).

Dicotyledonous Hosts *(**Susceptible** or Insusceptible)*

*Capsicum annuum.* ***Datura metel. Datura stramonium. Gomphrena globosa. Nicotiana tabacum. Solanum demissum* × *tuberosum.***

Sources of host-range data

Kahn and Bartels (1968).

**Properties of particles in sap**

TIP: 55–60 °C. LIV: 6 days (or more at 20–25°C in crude sap). DEP: $\log_{10}$ minus 5. Leaf sap contains few virions.

**Purification**

Method: Kahn and Bartels (1968).

**Particle morphology**

Virions filamentous; not enveloped; usually flexuous; with a clear modal length; of ±28 721 nm.

**Taxonomy and Relationships**

*Potyvirus: Potyviridae.*

Virus(es) with serologically related virions: henbane mosaic and tobacco etch viruses. Virus(es) with serologically unrelated virions: potato A and Y viruses.

**References**

Kahn, R.P. and Bartels, R. (1968). *Phytopathology* **58**: 587.

# Datura distortion mosaic *potyvirus*

**Data collated by V.R. Mali, 1987. Revised 1989.**

*Standard acronym:* DDMV.

First reported in *Datura fastuosa* L.; from Maharashtra State, India; by Mali *et al.* (1985).

**Natural host range and symptoms**

Symptoms persist.

*Datura fastuosa, D. alba* — mosaic and blistering of leaves, small fruits, no spines, stunting.

**Transmission**

Transmitted by a vector; an insect; *Aphis craccivora, A. gossypii, A. nerii, Dactynotus sonchi, Myzus persicae*, but not *Melanaphis sacchari*; Aphididae. Transmitted in a non–persistent manner. Virus transmitted by mechanical inoculation; transmitted by grafting; not transmitted by contact between plants; not transmitted by seed.

**Ecology and control**

Studies reported by Mali *et al.* (1985); Game (1988).

**Geographical distribution**

Spreads in India (Maharashtra State).

**Experimental host range**

Few (<3) families susceptible.

Diagnostically susceptible

*Datura ferox* — systemic vein clearing, mosaic, leaf malformation and shoe-stringing.

*Datura metel* — veinal necrosis, then systemic mosaic and malformation.

*D. stramonium* — systemic veinal chlorosis.

*Nicotiana clevelandii* — chlorotic local lesions, then systemic vein clearing, epinasty and leaf malformation.

*N. glutinosa, N. sylvestris* — chlorotic local lesions, then systemic mosaic and leaf malformation.

*N. megalosiphon* — chlorotic local lesions, systemic mottle.

*N. tabacum* cv. White Burley — systemic mottle.

*N. tabacum* cv. Xanthi,

*Solanum nigrum* — systemic mosaic, blistering.

*Petunia hybrida, Physalis floridana* — systemic vein clearing and mosaic.

Diagnostically insusceptible

*Lycopersicon esculentum, Nicotiana rustica, N. tabacum* cv. Samsun, *Solanum melongena, S. tuberosum, Capsicum pendulum, Chenopodium album, C. amaranticolor, C. quinoa, C. murale.*

Maintenance and propagation hosts

*Datura fastuosa, D. metel, Nicotiana megalosiphon, N. tabacum* L. cv. Xanthi.

Assay hosts (**L**ocal lesions or **W**hole plants)

*Datura fastuosa* (W), *D. metel* (W), *Nicotiana glutinosa* (W).

Dicotyledonous Hosts *(**Susceptible** or Insusceptible)*

*Arachis hypogaea.* ***Capsicum annuum***. *Carica papaya. Cassia occidentalis. Cassia tora. Chenopodium album. Chenopodium amaranticolor. Chenopodium murale. Chenopodium quinoa. Cicer arietinum. Cucumis sativus. Cucurbita moschata.* ***Datura ferox***. ***Datura metel***. ***Datura stramonium***. *Luffa acutangula.* ***Lycopersicon esculentum***. ***Nicotiana clevelandii***. ***Nicotiana glutinosa***. ***Nicotiana megalosiphon***. *Nicotiana rustica.* ***Nicotiana sylvestris***. ***Nicotiana tabacum***, or *Nicotiana tabacum.* ***Petunia hybrida***. *Phaseolus vulgaris.* ***Physalis floridana***. *Pisum sativum. Solanum melongena.* ***Solanum nigrum***. *Solanum tuberosum. Tetragonia expansa. Vigna unguiculata. Vigna unguiculata* ssp. *sesquipedalis. Zinnia elegans.*

Additional hosts

*Amaranthus polygamus, Bidens pilosa, Canavalia ensiformis, Capsicum pendulum, Commelina benghalensis, Crotalaria juncea,* ***Datura alba, Datura fastuosa, Petunia axillaris, Petunia violacea, Physalis minima***, *Solanum xanthocarpus* and *Vigna mungo.*

Sources of host-range data

Mali *et al.* (1985); Game (1988).

**Properties of particles in sap**

TIP: 55–60 °C. LIV: 4 days. DEP: $\log_{10}$ minus 2–3. Leaf sap contains few virions.

**Purification**

Method: Mali *et al.* (1985).

**Particle morphology**

Virions filamentous; not enveloped; usually flexuous; with a clear modal length; of ±11 770 nm; 11 nm wide. Axial canal obscure. Basic helix obscure.

**Cytopathology**

Inclusions cytoplasmic cylindrical present in infected cells; are pinwheels and unusual in shape; scrolls and long straight laminated aggregates.

**Taxonomy and Relationships**

*Potyvirus: Potyviridae.*

Virus(es) with serologically related virions: henbane mosaic virus. Virus(es) with serologically unrelated virions: potato A, potato Y, bean common mosaic, bean yellow mosaic, peanut mottle and soybean mosaic viruses.

Best tests for diagnosis: The virus is a potyvirus and hence distinct from tobacco mosaic (TMV) (Hotin and Procenko, 1973), potato virus M (PVM) (Habib, 1982), cucumber mosaic (CMV) (Suteri and Bala, 1982) and tobacco streak (TSV) (Blakeslee, 1921; Edwardson and Purcifull, 1974). *Datura* species are known to be infected at least by ten potyviruses such as CDV (Kahn and Bartels, 1968), DNV (Badami and Kassanis, 1959), DMV (Quereshi and Mahmood, 1968), datura 437 virus (Damsteegt, 1974), DSV (Giri and Agrawal, 1971), HMV (Bradley, 1952; Lovisolo and Bartels, 1970), HDMV (Peralta *et al.*, 1981), PMeV (Nelson and Wheeler, 1974), PVY (Blaszczak, 1976; Rao and Yaraguntaiah, 1976) and TEV (Chester, 1937; Klesser, 1966; Zitter, 1971). DDMV differs from DMV, DNV and DSV in aphid transmissibility. Although DDMV is serologically related to HMV, it differs in forming type-III CCI inclusions (HMV induces type-II CCI inclusions), in virion size and host range. HMV has longer virions (800–900 nm) and causes more severe symptoms in *Datura stramonium* than DDMV. In CCI morphology, DDMV is also distinct from CDV, HDMV, datura 437 virus, PMeV, PVY and TEV (Edwardson *et al.*, 1984). DDMV differs from all ten potyviruses in host range and symptoms. It differs from DSV, PMeV and PVY, in infecting *Datura stramonium* and from HMV, TEV, DMV and HDMV in its mild symptoms in *Datura stramonium*. It differs from HMV, TEV and PVY, DDMV in not infecting *Chenopodium amaranticolor* and *Chenopodium quinoa*, and is distinct from PMeV in infecting *Datura metel* and from HDMV in not infecting *Nicotiana rustica* and *N. tabacum* cv. Samsun. Therefore, DDMV is considered to be a distinct member of the potyvirus group (Mali, 1985) and in subdivision III of Edwardson (1974). Few comparisons have been made between DDMV and other similar viruses reported from India (Capoor and Varma, 1948, 1951; Garga, 1958; Verma and Verma, 1963).

**References**

Badami, R.S. and Kassanis, B. (1959). *Ann. appl. Biol.* **47**: 90.

Blakeslee, A.F. (1921). *J. Genet.* **11**: 17.

Blaszczak, W. (1976). *Rezn. nauk. roln. seria. E.T. 6, Z* **2**: 69.

Bradley, R.H.E. (1952). *Ann. appl. Biol.* **39**: 78.

Capoor, S.P. and Varma, P.M. (1948). *Curr. Sci.* **17**: 151.

Capoor, S.P. and Varma, P.M. (1951). *Indian J. agric. Sci.* **22**: 303.
Chester, K.S. (1937). *Phytopathology* **27**: 903.
Damsteegt, V.D. (1974). *Proc. Am. Phytopath. Soc.* **1**: 50.
Edwardson, J.R. (1974). *Fla Agric. Exp. Stn Monog.* No. 4, p. 398.
Edwardson, J.R. and Purcifull, D.E. (1974). *Phytopathology* **64**: 1322.
Edwardson, J.R., Christie, R.G. and Ko, N.J. (1984). *Phytopathology* **74**: 1111.
Game (1988). *M.Sc. Thesis*, Maharashtra Agric. Univ., Parbhani, India, 50 pp.
Garga, R.P. (1958). *Indian Phytopath.* **11**: 133.
Giri, L. and Agrawal, H.O. (1971). *Phytopath. Z.* **70**: 81.
Habibi, N.M. (1982). *Egypt. J. Bot.* **23**: 163.
Hotin, Y.A. and Procenko, A.I. (1973). *Byull. glavn. bot. Sada., Leningrad* **87**: 113.
Kahn, R.P. and Bartels, R. (1962). *Phytopathology* **48**: 587.
Kahn, R.P. and Bartels, R. (1968). *Pl. Dis. Reptr* **48**: 603.
Klesser, P.J. (1966). *Bothalia* **9**: 173.
Lovisolo, O. and Bartels, R. (1970). *Phytopath. Z.* **69**: 73.
Mali, V.R. (1985). *Int. J. Trop. Pl. Dis.* **3**: 93.
Mali, V.R., Nirmal, D.D., Patel, K.V. and Vyanjane, N.T. (1985). *Indian Phytopath.* **38**: 413.
Matthews, R.E.F. (1982). *Intervirology* **17**: 11.
Nelson, M.R. and Wheeler, R.E. (1978). *Phytopathology* **68**: 979.
Peralta, E.L., Beczner, L. and Dezsery, M. (1981). *Acta Phytopath. Acad. Sci. Hung.* **16**: 85.
Quereshi, S. and Mahmood, K. (1978). *Phytopath. Z.* **93**: 113.
Rao, R.D.V.J.P. and Yaraguntaiah, R.C. (1976). *Curr. Sci.* **49**: 467.
Suteri, B.D. and Bala, S. (1982). *Sci. Cult.* **48**: 384.
Varma, G.S. and Verma, H.N. (1963). *Indian Phytopath.* **16**: 366.
Yaraguntaiah, R.C. and Govindu, H.C. (1972). *Mysore J. Agric. Sci.* **6**: 352.
Zitter, T.A. (1973). *Pl. Dis. Reptr* **57**: 991.

# Datura innoxia Hungarian mosaic (?) potyvirus

**Data collated by A.A. Brunt, 1991.**

*Standard acronym:* HDIV.

First reported in *Datura innoxia*; from Hungary; by Peralta *et al.* (1981).

**Natural host range and symptoms**

Symptoms persist.
*Datura innoxia* — mosaic leaf symptoms.

**Transmission**

Virus transmitted by mechanical inoculation.

**Geographical distribution**

Spreads in Hungary (Budapest).

**Experimental host range**

Few (<3) families susceptible.

Diagnostically susceptible

*Datura innoxia, D. stramonium, Petunia hybrida, Physalis floridana, Nicotiana tabacum* — chlorosis of systemically infected leaves which is especially conspicuous in *D. innoxia*.

Diagnostically insusceptible

*Capsicum annuum, Chenopodium quinoa, Cucumis sativus, Gomphrena globosa, Solanum melongena.*

Maintenance and propagation hosts

*Datura innoxia.*

Assay hosts (**L**ocal lesions or **W**hole plants)

*Datura innoxia* (W).

Dicotyledonous Hosts *(**Susceptible** or Insusceptible)*

*Capsicum annuum. Capsicum frutescens. Chenopodium quinoa. Cucumis sativus.* ***Datura stramonium****. Gomphrena globosa.* ***Lycopersicon esculentum****.* ***Nicotiana debneyi****.* ***Nicotiana glutinosa****.* ***Nicotiana megalosiphon****.* ***Nicotiana rustica****.* ***Nicotiana sylvestris****.* ***Nicotiana tabacum****.* ***Petunia hybrida****.* ***Physalis floridana****. Solanum demissum × tuberosum. Solanum melongena. Solanum tuberosum.*

**Properties of particles in sap**

TIP: 60 °C. LIV: 3 days. DEP: $\log_{10}$ minus 5.

**Particle morphology**

Virions filamentous; not enveloped; usually flexuous; with a clear modal length; of 750 nm.

**Physical properties**

$A_{260}/A_{280}$ ratio 1.22.

**Cytopathology**

Inclusions present in infected cells; are pinwheels; they do not contain virions.

**Taxonomy and Relationships**

Unassigned potyvirus.

Virus(es) with serologically unrelated virions: potato Y and tobacco etch viruses.

**References**

Peralta, E.L., Beczner, L. and Dezsery, M. (1981). *Acta Phytopath. Acad. Sci. Hung.* **16**: 85.

# Datura mosaic (?) potyvirus

**Data collated by V.R. Mali, 1987.**

*Standard acronym:* DTMV.

First reported in *Datura metel* L.; from Aligarh, India; by Quereshi and Mahmood (1978).

**Natural host range and symptoms**

Symptoms persist.

*Datura metel* — mosaic, blistering and shoe-stringing.

**Transmission**

Some species of aphid have been tested and found not to be vectors. Virus transmitted by mechanical inoculation; transmitted by grafting; not transmitted by contact between plants; not transmitted by seed.

**Ecology and control**

Studies reported by Quereshi and Mahmood (1978).

**Geographical distribution**

Spreads in India.

**Experimental host range**

Few (<3) families susceptible.

Diagnostically susceptible

*Datura innoxia, D. metel* — necrotic local lesions, mosaic, blistering, reduced fruits, no spines.

*D. stramonium* — necrotic local lesions, flowering checked, systemic mosaic.

*Nicotiana glutinosa* — mosaic, leaf malformation and stunting.

*N. tabacum* — mosaic, leaf malformation, flowering checked.

*Petunia hybrida* — mosaic, leaf reduction and malformation, stunting.

Diagnostically insusceptible

*Capsicum annuum, Lycopersicon esculentum, Solanum luteum, S. melongena, S. tuberosum, Chenopodium album, C. amaranticolor, C. ambrosioides, C. murale.*

Maintenance and propagation hosts

*Datura metel.*

Assay hosts (**L**ocal lesions or **W**hole plants)

*Datura metel* (L).

Dicotyledonous Hosts *(**Susceptible** or Insusceptible)*

*Antirrhinum majus. Beta vulgaris. Brassica campestris. Brassica campestris* ssp. *pekinensis. Brassica oleracea* ssp. *botrytis. Brassica oleracea* ssp. *capitata. Calendula officinalis. Capsicum annuum. Catharanthus roseus. Chenopodium album. Chenopodium amaranticolor. Chenopodium ambrosioides. Chenopodium murale. Chenopodium quinoa. Cucumis melo. Cucumis sativus. Cucurbita pepo. Cyamopsis tetragonoloba. Dahlia pinnata.* ***Datura metel. Datura stramonium.*** *Daucus carota. Dianthus barbatus. Glycine max. Gomphrena globosa. Helianthus annuus. Hibiscus esculentus. Lablab purpureus.* ***Lycopersicon esculentum. Nicotiana glutinosa. Nicotiana tabacum. Petunia hybrida.*** *Phaseolus vulgaris. Phlox drummondii.* ***Physalis peruviana.*** *Pisum sativum. Raphanus sativus. Solanum melongena.* ***Solanum nigrum.*** *Solanum tuberosum. Sonchus oleraceus. Spinacia oleracea. Stellaria media. Tetragonia expansa. Tropaeolum majus. Vicia faba. Zinnia elegans.*

Sources of host-range data

Quereshi and Mahmood (1978).

**Properties of particles in sap**

TIP: 45–50 °C. LIV: 1 days. DEP: $\log_{10}$ minus 6–7.

**Purification**

Method: Quereshi and Mahmood (1978).

**Particle morphology**

Virions filamentous; not enveloped; usually flexuous; with a clear modal length; of 712 nm; 11–13 nm wide. Axial canal obscure. Basic helix obscure.

**Physical properties**

One sedimenting component in purified preparations.

**Taxonomy and Relationships**

Probably *Potyvirus: Potyviridae.*

Virus(es) with serologically unrelated virions: tobacco etch virus.

Datura mosaic virus is a potyvirus, but its relationships with others are uncertain (Hollings and Brunt, 1981). Best tests for diagnosis: Datura shoestring virus (DSV) differs from DMV in its failure to infect *Datura stramonium, D. innoxia* and *Solanum nigrum*. Datura distortion mosaic virus (DDMV) differs from DMV in its failure to induce local lesions in *Datura* species and in being transmissible by aphids. *Datura* species are known to be naturally infected by other potyviruses such as datura necrosis virus (DNV) (Badami and Kassanis, 1959), datura Colombian virus (DCV) (Kahn and Bartels, 1968), datura shoestring virus (DSV) (Giri and Agrawal, 1971) and datura distortion mosaic virus (DDMV) (Mali *et al.*, 1985). However, they differ in various properties; DNV infects *Lycopersicon esculentum*, which is insusceptible to DMV; DCV incites few or no symptoms in *Datura stramonium, D. metel* and *Nicotiana tabacum* var. Samsun, is transmitted by *Myzus persicae*, and has a greater LIV; DMV incites severe symptoms in *Datura* species, is not transmitted by aphids and has a short LIV; DMV, unlike DSV, induces local lesions in *Datura* species; DSV, unlike DMV, induces lesions in *Nicotiana rustica* and does not infect *Datura stramonium, D. innoxia* and *Solanum nigrum*. DDMV differs from DMV in being aphid transmitted and able to infect *Capsicum annuum* but not cause local lesions in *Datura* species (Mali, 1985). The virion morphology of other Indian viruses infecting *Datura* (Capoor and Verma, 1948; 1951; Garga, 1958; Verma and Verma, 1963; Yaraguntaiah and Govindu, 1973) is not known.

**References**


Badami, R.S. and Kassanis, B. (1959). *Ann. appl. Biol.* **47**: 90.
Capoor, S.P. and Varma, P.M. (1948). *Curr. Sci.* **17**: 151.
Capoor, S.P. and Varma, P.M. (1951). *Indian J. agric. Sci.* **22**: 303.
Garga, R.P. (1958). *Indian Phytopath.* **11**: 133.
Giri, L. and Agrawal, H.O. (1971). *Phytopath. Z.* **70**: 81.
Hollings, M. and Brunt, A.A. (1981). *CMI/AAB Descr. Pl. Viruses* No. 245, 7 pp.
Kahn, R.P. and Bartels, R. (1962). *Phytopathology* **48**: 587.
Mali, V.R. (1985). *Int. J. Trop. Pl. Dis.* **3**: 93.
Mali, V.R., Nirmal, D.D., Patel, K.V. and Vyanjane, N.T. (1985). *Indian Phytopath.* **38**: 413.
Quereshi, S. and Mahmood, K. (1978). *Phytopath. Z.* **93**: 113.
Verma, G.S. and Verma, H.N. (1963). *Indian Phytopath.* **16**: 366.
Yaraguntaiah, R.C. and Govindu, H.C. (1972). *Mysore J. Agric. Sci.* **6**: 352.


# Datura necrosis *potyvirus*

**Data collated by A.A. Brunt and A. Varma, 1987.**

*Standard acronym:* DNV.

First reported in *Solanum jasminoides*; from England from plants imported from India; by Badami and Kassanis (1959).

**Natural host range and symptoms**

*Solanum jasminoides* — symptoms unknown all the naturally infected plants examined also contained potato Y and tobacco wilt viruses.

**Transmission**

Virus transmitted by mechanical inoculation.

**Geographical distribution**

Spreads in India.

**Experimental host range**

Several (3–9) families susceptible.

Diagnostically susceptible

*Nicotiana tabacum*—vein banding and mosaic.
*Nicotiana glutinosa* — mottling.
*Datura tatula* — local lesions and necrosis.
*Solanum jasminoides* — mottling.

Diagnostically insusceptible

*Solanum tuberosum, Cucumis sativus, Phaseolus vulgaris, Gomphrena globosa.*

Maintenance and propagation hosts

*Nicotiana tabacum, N. glutinosa.*

Assay hosts (**L**ocal lesions or **W**hole plants)

*Nicotiana tabacum* (W), *N. glutinosa* (W).

Dicotyledonous Hosts *(**Susceptible** or Insusceptible)*

*Cucumis sativus.* ***Datura tatula.*** *Gomphrena globosa.* ***Hyoscyamus niger. Lycopersicon esculentum. Nicotiana glutinosa. Nicotiana tabacum.*** *Phaseolus vulgaris.* ***Physalis floridana.*** *Solanum tuberosum.*

**Properties of particles in sap**

TIP: 60–65 °C. LIV: 8–12 days. DEP: $\log_{10}$ minus 4. Leaf sap contains few virions.

**Particle morphology**

Virions filamentous; not enveloped; usually flexuous; with a clear modal length; of 750 nm. Axial canal obscure. Basic helix obscure.

**Taxonomy and Relationships**

*Potyvirus: Potyviridae.*

Virus(es) with serologically unrelated virions: potato Y and tobacco wilt viruses.

**References**

Badami, R.S. and Kassanis, B. (1959). *Ann. appl. Biol.* **47**: 90.

# Datura shoestring *potyvirus*

**Data collated by M. Weintraub, 1984.**

*Standard acronym:* DSTV.

First reported in *Solanum jasminoides*; from Simla, India; by Giri and Agrawal (1971).

**Natural host range and symptoms**

*Solanum jasminoides* — mottling and leaf malformation.

**Transmission**

Virus transmitted by mechanical inoculation.

**Geographical distribution**

Spreads in India.

**Experimental host range**

Few (<3) families susceptible.

Diagnostically susceptible

*Datura metel* and *N. glutinosa* — systemic mosaic, leaf blistering.

*N. rustica* — necrotic local lesions, then vein and stem necrosis, mottling, necrosis and plant death.

*N. tabacum* cv. White Burley — necrotic-chlorotic local lesions, then vein necrosis and plant death.

Diagnostically insusceptible

*Chenopodium amaranticolor, Datura stramonium, D. tatula, Solanum demissum, Physalis floridana.*

Maintenance and propagation hosts

*Datura metel, Nicotiana glutinosa.*

Assay hosts (**L**ocal lesions or **W**hole plants)

*Datura metel* (W), *Nicotiana glutinosa* (W), *N. rustica* (W), *N. tabacum* cv. White Burley (L, W).

Dicotyledonous Hosts *(**Susceptible** or Insusceptible)*

*Amaranthus caudatus. Atriplex hortensis. Capsicum annuum. Chenopodium amaranticolor. Chenopodium quinoa. Cucumis sativus.* ***Datura metel****. Datura stramonium. Datura tatula. Gomphrena globosa. Lycopersicon esculentum. Nicotiana debneyi.* ***Nicotiana glutinosa****.* ***Nicotiana rustica****.* ***Nicotiana tabacum****. Phaseolus vulgaris. Physalis floridana. Pisum sativum. Solanum demissum. Solanum nigrum. Solanum nodiflorum. Solanum tuberosum. Vicia faba. Vigna unguiculata.*

Sources of host-range data

Giri and Agrawal (1971).

**Properties of particles in sap**

TIP: 60–65 °C. LIV: 1–2 days. DEP: $\log_{10}$ minus 4. Leaf sap contains many virions. Electron microscopy: 2% PTA-Na, pH 7.

**Purification**

Method: Brunt and Stace-Smith (1973).

**Particle morphology**

Virions filamentous; not enveloped; usually flexuous; with a clear modal length; of 720–750 nm; 16 nm wide. Axial canal obscure. Basic helix usually obscure (but sometimes visible in thin sections).

**Physical properties**

One sedimenting component in purified preparations; sedimentation coefficient 143 *S*. Density 1.33 g $cm^{-3}$ in CsCl.

*Virion protein(s)* one; $M_r$ 36600 (degrades to 27400). Method of preparation: Agrawal and Tremaine (1972).

**Cytopathology**

Virions found in leaves and mesophyll; in cytoplasm. Inclusions present in infected cells; are crystals in the cytoplasm, amorphous X–bodies, pinwheels, and unusual in shape; tubules in the nucleus; they do not contain virions (although long helical structures are found near virus aggregates). Other cellular changes: drastic reduction of leaf lamina.

**Taxonomy and Relationships**

*Potyvirus: Potyviridae.*

Virus(es) with serologically related virions: pepper veinal mottle virus. Virus(es) with serologically unrelated virions: potato virus Y.

**References**


Agrawal, H.O. and Tremaine, J.H. (1972). unpublished data.

Brunt, A.A. and Stace-Smith, R. (1973). unpublished data.

Giri, L. and Agrawal, H.O. (1971). *Phytopath. Z.* **70**: 81.

Pushkarnath, M.J. (1952). *Curr. Sci.* **21**: 249.

Weintraub, M., Agrawal, H.O. and Ragetli, H.W.J. (1973). *Can. J. Bot.* **51**: 855.


# Datura striate mosaic monogeminivirus

**Data collated by A.A. Brunt, 1995.**

*Synonyms:* Digitaria didactyla striate mosaic virus (Pinner *et al.*, 1992).

First reported in *Digitaria didactyla*; from Queensland, Australia; by Greber (1989).

**Natural host range and symptoms**

*Digitaria didactyla* — striate mosaic.

**Transmission**

Transmitted by a vector; an insect; *Nesoclutha pallida*; Cicadellidae. Not transmitted by *Cicadulina bimaculata*. Transmitted in a persistent manner. Virus not transmitted by mechanical inoculation.

**Geographical distribution**

Geographical distribution probably restricted. Spreads in Australia.

**Experimental host range**

Few (<3) families susceptible. Experimentally infected plants mostly show Striate mosaic.

Diagnostically susceptible

*Dactyloctenium aegyptium, Leptochloa filiformis, Triticum aestivum* — leaf streaking.

Diagnostically insusceptible

*Bromus catharticus, Digitaria ciliaris, Lolium multiflorum, Paspalum dilatatum.*

Maintenance and propagation hosts

*Triticum aestivum* cv. Gatcher.

Assay hosts (**L**ocal lesions or **W**hole plants)

*Triticum aestivum* cv. Gatcher (W).

Monocotyledonous Hosts *(**Susceptible** or Insusceptible)*

***Avena sativa. Chloris gayana. Hordeum vulgare. Leptochloa filiformis. Lolium multiflorum.*** *Panicum miliaceum. Paspalum dilatatum.* ***Setaria italica. Triticum aestivum. Zea mays.***

Sources of host-range data

Greber (1989).

**Purification**

Method: Pinner *et al.* (1992).

**Particle morphology**

Virions geminate; not enveloped; 18 nm in diameter; dimers 30 nm in length.

**Biochemical properties**

*Genome* consists of DNA; single–stranded.

**Cytopathology**

Virions found in cytoplasm and in nuclei.

**Taxonomy and Relationships**

Monogeminivirus: *Geminiviridae.*

Virus(es) with serologically related virions: Chloris striate mosaic, paspalum striate mosaic, distantly and digitaria streak, wheat dwarf, maize streak, panicum streak and sugarcane streak very distantly.

**References**

Greber, R.S. (1989). *Ann. appl. Biol.* **114**: 471.

Pinner, M.S., Markham, P.G., Rybicki, E.P. and Greber, R.S. (1992). *Pl. Path.* **41**: 618.

# Datura yellow vein *nucleorhabdovirus*

**Data collated by J.E. Thomas, 1990.**

*Standard acronym:* DYVV. *Strains:* Thunbergia rhabdovirus.

First reported in *Datura stramonium*; from south-east Queensland, Australia; by Thomas and Behncken (1983).

**Natural host range and symptoms**

Symptoms persist. Symptoms vein yellowing and leaf malformation.

*Datura stramonium* — vein yellowing, leaf and flower bud malformation.

*Thunbergia alata* — vein clearing and yellowing, leaf malformation.

*Lycopersicon esculentum* — vein clearing and chlorosis, mottling and leaf malformation.

**Transmission**

Not transmitted by *Myzus persicae, Macrosiphum euphorbiae, Aphis gossypii, Acyrthosiphon (Aulacorthum) solani, Orosius argentatus, Austroasca alfalfae, Henosepilachna* sp. Virus not transmitted by mechanical inoculation; transmitted by grafting; not transmitted by contact between plants; not transmitted by seed.

**Geographical distribution**

Spreads in Australia.

**Experimental host range**

Few (<3) families susceptible. Experimentally infected plants mostly show vein yellowing.

Diagnostically susceptible

*Datura stramonium* — vein yellowing, leaf and flower bud malformation.

*Thunbergia alata* — vein yellowing and leaf malformation.

*Lycopersicon esculentum* — vein clearing and leaf malformation.

*Solanum melongena* — vein etching.
*Nicotiana glutinosa* — vein yellowing.
*Nicotiana tabacum* — symptomless.

Diagnostically insusceptible

*Capsicum annuum, Petunia hybrida, Nicotiana rustica, N. physalodes, Solanum tuberosum* cv. Sebago.

Maintenance and propagation hosts

*Datura stramonium, Lycopersicon esculentum.*

Assay hosts (**L**ocal lesions or **W**hole plants)

*Datura stramonium* (W).

Dicotyledonous Hosts *(**Susceptible** or Insusceptible)*

*Capsicum annuum.* ***Datura stramonium. Lycopersicon esculentum.*** *Nicandra physaloides.* ***Nicotiana glutinosa.*** *Nicotiana rustica.* ***Nicotiana tabacum. Solanum melongena.*** *Solanum tuberosum.*

**Properties of particles in sap**

Leaf sap contains few virions.

**Purification**

Method: Thomas and Dietzgen (1991).

**Particle morphology**

Virions rhabdo– or bullet–shaped; enveloped; usually straight; with a clear modal length; of 166 nm; 77 nm wide. Basic helix obscure.

*Virion protein(s)* four; $M_r$ of the largest 78000; G protein. $M_r$ of 2nd largest 47000; N protein. $M_r$ of 3rd largest 41000; $M_1$ protein. $M_r$ of 4th largest 36000; $M_2$ protein. Method of preparation: Thomas and Dietzgen (1991). Virion proteins glycosylated; are the G protein is glycosylated and binds to concanavalin A (lectin).

**Cytopathology**

Virions found in leaves; in the perinuclear space. Inclusions absent from infected cells.

**Taxonomy and Relationships**

*Nucleorhabdovirus: Rhabdoviridae.*

Virus(es) with serologically related virions: sonchus yellow net virus. Virus(es) with serologically unrelated virions: potato yellow dwarf, potato chlorotic stunt, tomato vein yellowing, maize sterile stunt, cereal chlorotic mottle, digitaria striate, festuca leaf streak, lettuce necrotic yellows, northern cereal mosaic, maize mosaic and barley yellow striate mosaic viruses.

Best tests for diagnosis: virion shape in negatively stained sap, characteristic symptoms in *Datura stramonium*, detection of viral protein from crude sap in Western immunoblots and gel diffusion serology.

**References**

Dietzgen, R.G. and Thomas, J.E. (1990). *Abstr. 8th Int. Congr. Virol., Berlin.* p. 444.

Thomas, J.E. and Behncken, G.M. (1983). *Abstr. 4th Int. Congr. Pl. Path., Melbourne.* p. 123.

Thomas, J.E. and Dietzgen, R.G. (1991). *Ann. appl. Biol.* **118**: 339.

# Dendrobium mosaic *potyvirus*

*Standard acronym:* DeMV.

**Natural host range and symptoms**

*Dendrobium* spp.
*Sequence database accession code(s):*
U23564 Em(43)_vi:Dm23564 Gb(89)_vi:Dmu23564 Dendrobium mosaic potyvirus coat protein gene, partial cds. 5/95 1,143bp.

**Taxonomy and Relationships**

*Potyvirus: Potyviridae.*

**Comments**

Synonym of clover yellow vein virus.

**References**

Francki, R.I.B., Milne, R.G. and Hatta, T. (1985). In: *Atlas of Plant Viruses* Vol. 2. p. 183. CRC Press, Boca Raton, Florida.

# Dendrobium vein necrosis (?) *closterovirus*

**Data collated by D.E. Lesemann, 1989.**

*Standard acronym:* DVNV.

First reported in *Dendrobium phalaenopsis*; from Germany; by Lesemann (1977).

**Natural host range and symptoms**

Symptoms persist.
*Dendrobium phalaenopsis* — vein necrosis in leaves, petals and sepals.

**Geographical distribution**

Found, but with no evidence of spread, in Germany, but only in two glasshouses.

**Properties of particles in sap**

Leaf sap contains few virions. Electron microscopy: mount sap in 2% PTA, pH 7.

**Particle morphology**

Virions filamentous; not enveloped; usually flexuous; very with a clear modal length; of 1865 nm; 10–12 nm wide. Axial canal obvious (but indistinct). Basic helix obvious.

**Cytopathology**

Virions found in leaves, mesophyll, epidermis and parenchyma surrounding the vascular bundle and in the petals and sepals of the flowers; in cytoplasm. Inclusions present in infected cells; are unusual in shape; are bundles of virions; they contain virions. Other cellular changes: peripheral vesiculation of mitochondria.

**Taxonomy and Relationships**

Possible *Closterovirus*.

**References**

Lesemann, D.-E. (1977). *Phytopath. Z.* **89**: 330.

# Desmodium mosaic *potyvirus*

**Data collated by J.R. Edwardson, 1981.**

*Standard acronym:* DesMV.

First reported in *Desmodium canum*; from Florida, U.S.A; by Edwardson *et al.* (1970).

**Natural host range and symptoms**

Symptoms persist.
*Desmodium canum* — mosaic, leaf malformation.

**Transmission**

Transmitted by a vector; an insect; *Aphis craccivora, Myzus persicae*; Aphididae. Virus transmitted by mechanical inoculation; transmitted by seed (8%).

**Geographical distribution**

Spreads in the USA.

**Experimental host range**

Few (<3) families susceptible.

Diagnostically susceptible

*Cyamopsis tetragonoloba* — faint systemic mosaic.
*Desmodium canum* — systemic mosaic, leaf shape malformation.
*Nicotiana benthamiana* — systemic vein clearing.

Diagnostically insusceptible

*Gomphrena globosa, Chenopodium amaranticolor, Cucumis sativus, Cassia tora, Phaseolus vulgaris, Datura stramonium, Nicotiana clevelandii.*

Maintenance and propagation hosts

*Desmodium canum.*

Assay hosts (**L**ocal lesions or **W**hole plants)

*Cyamopsis tetragonoloba* (W), *Desmodium canum* (W), *Nicotiana benthamiana* (W).

Dicotyledonous Hosts *(**Susceptible** or Insusceptible)*

*Apium graveolens. Capsicum annuum. Cassia tora. Chenopodium amaranticolor. Cucumis sativus. Cucurbita maxima. Cucurbita pepo.* ***Cyamopsis tetragonoloba****. Datura stramonium. Gomphrena globosa.* ***Nicotiana benthamiana****. Nicotiana clevelandii. Nicotiana tabacum. Phaseolus vulgaris. Pisum sativum. Vicia faba. Vigna unguiculata.*

Sources of host-range data

Edwardson *et al.* (1970); Edwardson (1974*b*).

**Properties of particles in sap**

Leaf sap contains few virions.

**Particle morphology**

Virions filamentous; not enveloped; usually flexuous; with a clear modal length; of 775 nm. Axial canal obscure. Basic helix obscure.

**Biochemical properties**

*Genome* consists of RNA; single–stranded.

**Cytopathology**

Virions found in mesophyll and epidermis. Inclusions present in infected cells; are pinwheels (with scrolls attached); they do not contain virions.

**Taxonomy and Relationships**

*Potyvirus: Potyviridae.*

**References**

Edwardson, J.R. (1974*a*). *Fla Agric. Exp. Stn Monog.* No. 4, p. 398.
Edwardson, J.R. (1974*b*). *Fla Agric. Exp. Stn Monog.* No. 5, p. 48.
Edwardson, J.R., Purcifull, D.E., Zettler, F.W., Christie, R.G. and Christie, S.R. (1970). *Pl. Dis. Reptr* **54**: 161.

# Desmodium yellow mottle *tymovirus*

**Data collated by H.A. Scott, 1980. Revised 1984.**

*Standard acronym:* DYMV.

First reported in *Desmodium laevigatum* and *D. paniculatum*; from Arkansas, U.S.A; by Walters and Scott (1968).

**Natural host range and symptoms**

Symptoms persist.

*Desmodium laevigatum, D. paniculatum* — yellow mottle and leaf malformation.

**Transmission**

Virus transmitted by mechanical inoculation.

**Geographical distribution**

Spreads in the USA.

**Experimental host range**

Few (<3) families susceptible (only Leguminosae infected).

Diagnostically susceptible

*Desmodium tortuosum* — necrotic local lesions.
*Phaseolus vulgaris* cvs Black Valentine, Bountiful, Pinto — local lesions.
*P. vulgaris* cv. Great Northern,
*Vigna unguiculata* cv. Monarch — systemic mottle.

Diagnostically insusceptible

*P. vulgaris* cv. Red Kidney, Small White.

Maintenance and propagation hosts

*Phaseolus vulgaris* cv. Great Northern.

Assay hosts (**L**ocal lesions or **W**hole plants)

*Desmodium tortuosum* (L).

Dicotyledonous Hosts ***(Susceptible** or Insusceptible)*

*Antirrhinum majus. Brassica campestris* ssp. *pekinensis. Brassica campestris* ssp. *rapa. Brassica juncea. Brassica oleracea* ssp. *capitata. Calendula officinalis. Chenopodium album. Chenopodium hybridum. Chenopodium quinoa. Cyamopsis tetragonoloba. Datura stramonium.* ***Desmodium tortuosum****. Dianthus barbatus. Glycine max. Gomphrena globosa.* ***Lathyrus odoratus****. Lens esculenta.* ***Lespedeza stipulacea****. Lotus corniculatus. Lupinus albus. Lupinus angustifolius.* ***Macroptilium lathyroides. Macrotyloma uniflorum****. Matthiola incana. Medicago sativa. Nicotiana glutinosa. Nicotiana rustica. Nicotiana tabacum. Petunia hybrida.* ***Phaseolus vulgaris. Pisum sativum****. Ricinus communis. Sesamum indicum.* ***Sesbania exaltata. Trifolium*** incarnatum. *Trifolium pratense*. ***Trifolium repens***. ***Vigna radiata***. ***Vigna unguiculata***. ***Vigna unguiculata*** ssp. ***sesquipedalis***. *Zinnia elegans*.

Sources of host-range data

Walters and Scott (1972).

**Properties of particles in sap**

TIP: 70 °C. LIV: 38 days. DEP: $\log_{10}$ minus 7. Leaf sap contains many virions.

**Purification**

Method: Walters and Scott (1972).

**Particle morphology**

Virions isometric; not enveloped; 30 nm in diameter; rounded in profile; with a conspicuous capsomere arrangement.

**Physical properties**

Two sedimenting components in purified preparations; sedimentation coefficient of the fastest 114 *S* (B); of the other(s) 54 *S* (T).

**Biochemical properties**

Virions contain 35 % nucleic acid (B); 65 % protein (B); 0 % lipid.

*Genome* consists of RNA; single–stranded; linear; unipartite; largest (or only) genome part 6.3 kb. Base composition 16.5 % G; 22.5 % A; 37.2 % C; 23.8 % U. Infectivity retained when deproteinised with phenol or detergent.

**Replication**

Replication does not depend on a helper virus.

**Cytopathology**

Virions found in leaves; in cytoplasm, in nuclei, and in cell vacuoles. Inclusions present in infected cells; are crystals in the cytoplasm, crystals in the nucleus, and unusual in shape; crystals in the vacuole; they contain virions. Other cellular changes: vesicles in the peripheries of chloroplasts opening to cytoplasm.

**Taxonomy and Relationships**

*Tymovirus*.

Virus(es) with serologically related virions: kennedya yellow mosaic, okra mosaic, cocoa yellow mosaic and clitoria yellow vein viruses are related closely; turnip yellow mosaic, Andean potato latent, dulcamara mottle, ononis yellow mosaic, scrophularia mottle and wild cucumber mosaic viruses are distantly related.

**References**


Koenig, R. (1976). *Virology* **72**: 1.

Koenig, R. and Givord, L. (1974). *Virology* **58**: 119.

Scott, H.A. (1976). *CMI/AAB Descr. Pl. Viruses* No. 168, 3 pp.

Scott, H.A. and Moore, B.J. (1972). *Virology* **50**: 613.

Walters, H.J. and Scott, H.A. (1972). *Phytopathology* **62**: 125.

Walters, H.J. and Scott, H.A. (1968). *Phytopathology* **58**: 1071.

# Digitaria streak monogeminivirus

**Data collated by P.G. Markham, P.M. Mullineaux, G.P. Accotto, 1991.**

*Standard acronym:* DSV.

First reported in *Digitaria setigera* (syn. *D. sanguinalis*); from Vanuatu; by Dollet *et al.* (1986).

**Natural host range and symptoms**

Symptoms persist. Symptoms white or chlorotic streaks or local lesions.

*Digitaria setigera* — narrow white streaks or local lesions, usually 1–3 mm long. Longer wider chlorotic often yellow local lesions with diffuse margins.

**Transmission**

Transmitted by a vector; an insect; *Nesoclutha declivata*; Cicadellidae. Not transmitted by *Cicadulina mbila, C. triangula, C. chinai, C. bipunctella, Nesoclutha pallida*. Transmitted in a persistent manner. Virus does not multiply in the vector; not transmitted congenitally to the progeny of the vector; not transmitted by mechanical inoculation; not transmitted by grafting; not transmitted by contact between plants; not transmitted by seed; not transmitted by pollen.

**Geographical distribution**

Spreads in Vanuatu.

**Experimental host range**

Several (3–9) families susceptible. Experimentally infected plants mostly show narrow, short, white or chlorotic streaks or local lesions with some showing diffuse edges to local lesions, but most local lesions clearly defined.

Diagnostically susceptible

*Avena sativa, Digitaria setigera, Hordeum vulgare, Zea mays* — chlorotic streaks and local lesions, severe deformation of plant.

Diagnostically insusceptible

*Nicotiana tabacum*.

Maintenance and propagation hosts

*Digitaria setigera*.

Assay hosts (**L**ocal lesions or **W**hole plants)

*Digitaria setigera* (W).

Monocotyledonous Hosts *(**Susceptible** or Insusceptible)*

***Avena sativa*. *Hordeum vulgare*. *Zea mays*.**

Additional hosts

*Digitaria setigera*.

Sources of host-range data

Dollet *et al.* (1986); Donson *et al.* (1988); Pinner and Markham (1990); Creissen *et al.* (1990).

**Properties of particles in sap**

Leaf sap contains few virions. Electron microscopy: ISEM for negative staining.

**Purification**

Method: resuspend virus in 0.01M sodium acetate buffer pH 4.5, purify through sucrose cushion (30%).

**Particle morphology**

Virions geminate; not enveloped; 18 nm in diameter; 30 nm in length; angular in profile; with a conspicuous capsomere arrangement.

**Physical properties**

One sedimenting component in purified preparations. $A_{260}/A_{280}$ ratio 1.1.

**Biochemical properties**

Virions contain 20 % nucleic acid; 80 % protein.

*Genome* consists of DNA; single–stranded; circular; unipartite; largest (or only) genome part 2.701 kb. Genomic nucleic acid isolated by Dollet *et al.* (1986). Base composition 25.8 % G; 25.5 % A; 23.9 % C; 24.8 % T. Genome has no *t*RNA–like activity. Nucleotide sequence references: Donson *et al.* (1987).

*Sequence database accession code(s):*

M23022 Gb(84)_vi:DIGGENOME Digitaria streak virus, complete genome. 3/92 2,701bp. 1 sequence.

*Features of the genome:* 4 ORFs; V1, V2 (coat protein), C1 and C2. Core promoter sequences located at coordinates 2671 (virion sense) and 2496 (complementary sense). Main stem loop structure in intergenic region (coordinates 2542–2581). Polyadenylation sequences at 1064 (virions strand) and 1084 (complementary strand).

Transcripts; 2 abundant virion-sense transcripts (1+, 2+). 1+ 5' end at 2700–3' end at *c.* 1060. 2+ 5' end at 170–3' end at *c.* 1060. 1+ may direct V1 translation, 2+ may direct V2 translation. Rare transcripts on complementary sense. 3 transcripts (1-, 2- and 4-). 4- is spliced between coordinates 1759 and 1666. Intron sequences typical for plant leads to fusion of C1:C2 ORF's which synthesise 41kDa protein. 2- unspliced equivalent would synthesise, C1 product only (30.5kDa).

5' ends 1– 2650, 2– 2467 and 2476 (2 ends), 4– 2467 and 2476 (2 ends).

3' ends for all transcripts; major end at 1063, minor end at 1067.

*Non–genomic nucleic acid* not found in the virions.

*Virion protein(s)* one; $M_r$ 26600; it is ORF V2 and functions as coat protein. Method of preparation: Dollet *et al.* (1986). Amino acid sequence: Donson *et al.* (1987).

**Replication**

Genome replicates in nuclei.

**Cytopathology**

Virions found in leaves, mesophyll, meristems and probably the roots; in nuclei and in cell vacuoles. Inclusions present in infected cells; are crystals in the nucleus and unusual in shape; typically in the formation of large irregular sheets of virions regularly arranged in rows; they contain virions.

**Taxonomy and Relationships**

Monogeminivirus: *Geminiviridae.*

Virus(es) with serologically related virions: maize streak, panicum streak and sugarcane streak viruses (Pinner and Markham, 1988). Virus(es) with serologically unrelated virions: chloris striate mosaic and paspalum striate mosaic viruses.

There is 74% homology between the coat proteins of maize streak and digitaria streak viruses. Best tests for diagnosis: vector specific transmission; indirect ELISA test on sap from infected plants, followed by dot-blot nucleic acid hybridization.

**References**

Accotto, G.P., Donson, J. and Mullineaux, P.M. (1989). *EMBO J.* **8**: 1033.

Dollet, M., Accotto, G.P., Lisa, V., Menissier, J. and Boccardo, G. (1986). *J. gen. Virol.* **67**: 933.

Donson, J., Accotto, G.P., Boulton, M.I., Mullineaux, P.M. and Davies, J.W. (1987). *Virology* **161**: 160.

Donson, J., Gunn, H.V., Woolston, C.J., Pinner, M.S., Boulton, M.I., Mullineaux, P.M. and Davies, J.W. (1988). *Virology* **162**; 248.

Julia, J.F. and Dollet, M. (1989). *J. Phytopath.* **127**: 42.

Mullineaux, P.M., Guerineau, F. and Accotto, G.P. (1990). *Nucl. Acids Res.* **18**: 7259.

Pinner, M.S. and Markham, P.G. (1990). *J. gen. Virol.* **71**: 1635.

# Digitaria striate (?) *cytorhabdovirus*

**Data collated by R.S. Greber, 1987.**

*Standard acronym:* DiSV.

First reported in *Digitaria decumbens* and *Digitaria ciliaris*; from the coastal areas of northern and southern Queensland; by Greber (1972).

**Natural host range and symptoms**

Symptoms persist.

*Digitaria ciliaris, D. decumbens, D. sanguinalis, Brachiaria milliiformis* — chlorotic spots and stripes.

**Transmission**

Transmitted by a vector; an insect; *Sogatella kolophon*; Delphacidae. Transmitted in a persistent manner. Virus retained when the vector moults; multiplies in the vector; not transmitted by contact between plants; not transmitted by seed; not transmitted by pollen.

**Geographical distribution**

Spreads in Australia.

**Experimental host range**

Few (<3) families susceptible.

Diagnostically susceptible

*Avena sativa, Digitaria ciliaris, Digitaria decumbens, Dinebra retroflexa, Eleusine indica, Hordeum vulgare, Lolium vulgare.*

Diagnostically insusceptible

*Echinochloa colonum, Zea mays* and *Oryza sativa.*

Maintenance and propagation hosts

*Digitaria ciliaris, D. decumbens.*

Assay hosts (**L**ocal lesions or **W**hole plants)

*Digitaria ciliaris* (W).

Monocotyledonous Hosts *(**Susceptible** or Insusceptible)*

***Avena sativa***. *Chloris gayana*. ***Digitaria decumbens***. ***Dinebra retroflexa***. *Echinochloa colonum*. ***Hordeum vulgare***. *Leptochloa filiformis*. ***Lolium multiflorum***. *Oryza sativa*. *Sorghum bicolor*. *Zea mays*.

Sources of host-range data

Greber (1979).

**Properties of particles in sap**

Leaf sap contains few virions.

**Purification**

Method: Greber: for nucleocapsids extract *Digitaria ciliaris* in 0.1 M phosphate and thioglycollate, pH 7.6. Clarify in 25% $CHCl_3$ + 5% butanol. Purify and concentrate by high speed and sucrose gradient centrifugation.

**Particle morphology**

Virions rhabdo– or bullet–shaped; enveloped; with a clear modal length; of 280 nm; 55 nm wide. Axial canal obvious; 14 nm in diameter. Basic helix obvious; pitch of basic helix 4.2 nm.

$M_r$ 50000; N protein. Virion proteins glycosylated; are Only the G-protein.

**Replication**

Replication does not depend on a helper virus.

**Cytopathology**

Virions found in leaves, mesophyll, phloem and companion cells; in cytoplasm. Inclusions absent from infected cells.

**Taxonomy and Relationships**

Possible *Cytorhabdovirus: Rhabdoviridae.*

Virus(es) with serologically related virions: barley yellow striate mosaic and maize sterile stunt viruses. Virus(es) with serologically unrelated virions: American wheat striate mosaic, cereal chlorotic mottle, cynodon rhabdovirus, festuca rhabdovirus, lettuce necrotic yellows and maize mosaic viruses.

**References**

Greber, R.S. (1972). *Aust. Pl. Path. Soc. Newsl.* **1**: 27.
Greber, R.S. (1976). *Aust. Pl. Path. Soc. Newsl.* **5**: 228.
Greber, R.S. (1979). *Aust. J. agric. Res.* **30**: 43.
Greber, R.S. (1984). In: *Current Topics in Vector Research*, Vol. 2, p. 141; ed. K. Harris. Praeger Scientific, New York.

# Diodia vein chlorosis (?) *closterovirus*

**Data collated by A.A. Brunt, 1992.**

*Standard acronym:* DVCV.

First reported in *Diodia virginiana*; from the U.S.A. (Arkansas); by Larsen *et al.* (1991).

**Natural host range and symptoms**

Symptoms persist.

*Diodia virginiana* — chlorotic vein-banding and general chlorosis.

**Transmission**

Transmitted by a vector; an insect; *Trialeurodes abutilonea*; Aleyrodidae. Virus not transmitted by mechanical inoculation; transmitted by grafting.

**Geographical distribution**

Spreads in the USA (Arkansas).

**Experimental host range**

Few (<3) families susceptible.

Diagnostically susceptible

*Diodia virginiana* — chlorotic vein-banding.

Maintenance and propagation hosts

*Diodia virginiana*.

Assay hosts (**L**ocal lesions or **W**hole plants)

*Diodia virginiana* (W).

Dicotyledonous Hosts *(**Susceptible** or Insusceptible)*

*Beta vulgaris. Chenopodium murale. Chenopodium quinoa. Cucumis sativus. Cucurbita maxima. Datura stramonium. Lactuca sativa. Nicotiana tabacum. Phaseolus vulgaris. Spinacia oleracea.*

Sources of host-range data

Larsen *et al.* (1991).

**Particle morphology**

Virions filamentous; not enveloped; usually flexuous; with no clear modal length; 12 nm wide.

3 virus specified dsRNA species found in infected cells. Size of largest virus specified dsRNA 4.6 kbp; 2nd largest 4.3 kbp; 3rd largest 1.9 kbp.

**Cytopathology**

Virions found in leaves and vascular parenchyma; in cytoplasm. Other cellular changes: the presence of virions in phloem parenchyma and sieve elements; virions are associated with membranous vesicles containing fibrils. Cytoplasm also contains greatly proliferated tubular membranes and vacuoles have double membrane-bound bodies.

**Taxonomy and Relationships**

Possible *Closterovirus*.

Virus(es) with serologically unrelated virions: beet yellows and lettuce infectious yellows viruses.

**Comments**

A virus with similar virions and transmissible by *Trialeurodes abutilonea* occurs in *Diodia maritima* in California, U.S.A. (Duffus *et al.*, 1986). Its possible relationship to diodia vein chlorosis virus has yet to be determined.

**References**

Duffus, J.E., Larsen, R.C. and Lui, H.Y. (1986). *Phytopathology* **76**: 97.

Larsen, R.C., Kim, K.S. and Scott, H.A. (1991). *Phytopathology* **81**: 227.

## Dioscorea alata *potyvirus*

**Data collated by A. Porth and F. Nienhaus, 1986.**

First reported in *Dioscorea alata*; from Togo; by Reckhaus (1980).

**Natural host range and symptoms**

*Dioscorea alata* — flecking.

**Transmission**

Virus not transmitted by mechanical inoculation.

**Geographical distribution**

Spreads in Togo.

Diagnostically insusceptible

*Gomphrena globosa, Chenopodium amaranticolor, Cucumis sativus, Phaseolus vulgaris, Nicotiana benthamiana.*

Maintenance and propagation hosts

*Dioscorea alata.*

Dicotyledonous Hosts *(**Susceptible** or Insusceptible)*

*Chenopodium amaranticolor. Chenopodium quinoa. Cucumis sativus. Gomphrena globosa. Nicotiana benthamiana. Nicotiana clevelandii. Nicotiana tabacum. Phaseolus vulgaris. Physalis floridana.*

**Purification**

Method: Lisa *et al.* (1981); Koenig *et al.* (1973).

**Particle morphology**

Virions filamentous; not enveloped; usually flexuous; with a clear modal length; of 805 nm (when stained with UA).

*Virion protein(s)* five; $M_r$ of the largest 36000. $M_r$ of 2nd largest 35000. $M_r$ of 3rd largest 34000. $M_r$ of 4th largest 33000. $M_r$ of 5th largest 32000. Method of preparation: Laemmli and Favre (1973).

**Cytopathology**

Virions found in leaves; in cytoplasm. Inclusions present in infected cells; are pinwheels and unusual in shape; and scrolls; they do not contain virions.

**Taxonomy and Relationships**

*Potyvirus: Potyviridae.*

Virus(es) with serologically related virions: dioscorea greenbanding mosaic virus; yam virus-Nigeria isolate and yam mosaic viruses are distantly related. Virus(es) with serologically unrelated virions: beet mosaic (yam isolate) and beet mosaic viruses.

**References**

Porth, A. (1986). *Ph.D. Thesis*, Univ. Bonn, Federal Republic of Germany, 167 pp.

Reckhaus, P. (1980). *Ph.D. Thesis*, Univ. Bonn, Federal Republic of Germany.

# Dioscorea bacilliform *badnavirus*

*Standard acronym:* DBV.

**Taxonomy and Relationships**

*Badnavirus.*

**References**

D'A. Hughes, J. (1986). *Ph.D. Thesis*, p.160. Univ. Reading, UK.

# Dioscorea green banding mosaic *potyvirus*

**Data collated by A. Porth and F. Nienhaus, 1986.**

*Strains:* Nigerian yam virus.

First reported in *Dioscorea rotundata*; from Togo; by Reckhaus (1979, 1980); Reckhaus and Nienhaus (1981).

**Natural host range and symptoms**

Symptoms persist.
*Dioscorea rotundata* — chlorotic flecks, greenbanding, and leaf malformation.

**Transmission**

Transmitted by a vector; an insect; *Aphis faba* and *Myzus persicae*; Aphididae. Transmitted in a non–persistent manner. Virus transmitted by mechanical inoculation.

**Geographical distribution**

Spreads in the African region (in the west); Nigeria and Togo.

**Experimental host range**

Several (3–9) families susceptible.

Diagnostically susceptible
*Nicotiana benthamiana* — chlorotic ringspots.

Maintenance and propagation hosts
*Nicotiana benthamiana.*

Dicotyledonous Hosts *(**Susceptible** or Insusceptible)*

*Celosia argentea. Chenopodium album.* ***Chenopodium amaranticolor****. Chenopodium ambrosioides. Chenopodium foliosum.* ***Chenopodium quinoa****. Gossypium hirsutum.* ***Nicotiana benthamiana****.* ***Nicotiana clevelandii****. Nicotiana megalosiphon.* ***Nicotiana occidentalis****. Nicotiana tabacum.*

Monocotyledonous Hosts *(**Susceptible** or Insusceptible)*

*Sorghum bicolor.*

**Purification**

Method: Lisa *et al.* (1981); Koenig *et al.* (1973).

**Particle morphology**

Virions filamentous; not enveloped; usually flexuous; with a clear modal length; of 732 nm (when stained with PTA–Na), or 769 nm (when stained with UA); 13–14 nm wide.

**Physical properties**

One sedimenting component in purified preparations; sedimentation coefficient 156 *S* (±1). Density 1.336 g $cm^{-3}$ in CsCl.

**Biochemical properties**

Virions contain 5.4 % nucleic acid.
*Genome* consists of RNA; single–stranded; unipartite; largest (or only) genome part 10.4 kb. Genomic nucleic acid isolated by Hull and Lane (1973).
*Virion protein(s)* one; $M_r$ 36000. Method of preparation: Laemmli and Favre (1973).

**Cytopathology**

Virions found in epidermis; in cytoplasm. Inclusions present in infected cells; are pinwheels and unusual in shape; and laminated aggregates; they do not contain virions. Other cellular changes: aggregated mitochondria, deformed chloroplasts and tubular bundles containing virions.

**Taxonomy and Relationships**

*Potyvirus: Potyviridae.*

Virus(es) with serologically related virions: yam mosaic virus. Virus(es) with serologically unrelated virions: beet mosaic and dioscorea alata ring mottle potyviruses.

**References**

Hull, R. and Lane, L.C. (1973). *Virology* **55**: 1.

Koenig, R., Lesemann, D.E., Brunt, A.A. and Kühne, H. (1973). *Intervirology* **1**: 348.

Laemmli, U.K. and Favre, M. (1973). *J. mol. Biol.* **80**: 575.

Lisa, V., Boccardo, G., D'Agostino, G., Dellavalle, G. and D'Aquilio, M. (1981). *Phytopathology* **71**: 667.

Porth, A. (1986). *Ph.D. Thesis*, Univ. Bonn, Federal Republic of Germany.

Reckhaus, P. (1979). *Z. PflKrankh. PflPath. PflSchutz.* **86**: 763.

Reckhaus, P. (1980). *Ph.D. Thesis*, Univ. Bonn, Federal Republic of Germany, pp 103.

Reckhaus, P. and Nienhaus, F. (1981). *Z. PflKrankh. PflPath. PflSchutz.* **88**: 492.

# Dioscorea latent (?) *potexvirus*

**Data collated by S. Phillips, A.A. Brunt, 1985 and F. Nienhaus, 1987.**

*Standard acronym:* DLV.

First reported in *Dioscorea composita, D. floribunda*; from Puerto Rico; by Waterworth *et al.* (1974).

**Natural host range and symptoms**

*Dioscorea composita, D. floribunda* — symptomless.

**Transmission**

Virus transmitted by mechanical inoculation; not transmitted by contact between plants; not transmitted by seed; not transmitted by pollen.

**Geographical distribution**

Spreads in Puerto Rico.

**Experimental host range**

Several (3–9) families susceptible.

Diagnostically susceptible

*Nicotiana megalosiphon, N. benthamiana* — symptomless local and systemic infection.

*Philodendron speciosum, Phaseolus vulgaris* cv. The Prince, *Vigna unguiculata* cv. Blackeye, *Datura metel, D. stramonium, Nicotiana clevelandii, N. sylvestris, N. tabacum* cv. Dutch A — local symptomless infection only.

Diagnostically insusceptible

*Solanum tuberosum, Commelina diffusa, Chenopodium quinoa, Trifolium incarnatum.*

Maintenance and propagation hosts

*Nicotiana benthamiana, N. megalosiphon.*

Assay hosts (**L**ocal lesions or **W**hole plants)

*Nicotiana benthamiana* (W), *N. megalosiphon* (W).

Dicotyledonous Hosts *(**Susceptible** or Insusceptible)*

*Brassica campestris* ssp. *pekinensis. Capsicum annuum. Chenopodium capitatum. Chenopodium murale. Chenopodium quinoa. Cucumis sativus.* ***Datura metel. Datura stramonium.*** *Gomphrena globosa.* ***Nicotiana benthamiana. Nicotiana clevelandii.*** *Nicotiana glutinosa.* ***Nicotiana megalosiphon.*** *Nicotiana sylvestris. Nicotiana tabacum.* ***Phaseolus vulgaris.*** *Solanum tuberosum.* ***Tetragonia expansa.*** *Trifolium incarnatum. Vicia faba.* ***Vigna unguiculata.*** *Zinnia elegans.*

Monocotyledonous Hosts *(**Susceptible** or Insusceptible)*

*Colocasia esculenta. Oryza sativa. Philodendron selloum. Zea mays.*

Additional hosts

Additional Hosts: *Philodendron speciosum* (**S**). Virus symptomless in *Nicotiana plumbaginifolia* and *Dioscorea bulbifera*, but does not infect *D. alata*.

Sources of host-range data

Phillips *et al.* (1986).

## Properties of particles in sap

TIP: 75–80 °C (in *Nicotiana megalosiphon* sap). LIV: 365 days (or more). DEP: $\log_{10}$ minus 6 (in *Nicotiana megalosiphon* sap). Leaf sap contains many virions.

## Purification

Method: Phillips *et al.* (1986).

## Particle morphology

Virions filamentous; not enveloped; usually flexuous; with a clear modal length; of 445 nm; 11 nm wide. Axial canal obscure. Basic helix obscure.

## Physical properties

Two sedimenting components in purified preparations (monomers and dimers); sedimentation coefficient of the fastest 131 *S*; of the other(s) 111 *S*. Density 1.33 g $cm^{-3}$ in CsCl (unfixed).

## Biochemical properties

Virions contain 5 % nucleic acid; 95 % protein; 0 % lipid.

*Genome* consists of RNA; single–stranded; unipartite; largest (or only) genome part 7.5 kb. Genomic nucleic acid isolated by Phillips *et al.* (1986).

*Non–genomic nucleic acid* not found in the virions.

*Virion protein(s)* one; $M_r$ 24900 (but reduced to 23100 by proteolysis). Method of preparation: Phillips *et al.* (1986).

## Replication

Replication does not depend on a helper virus.

## Cytopathology

Virions found in all parts of the host plant; in cytoplasm. Inclusions present in infected cells; are amorphous X–bodies and viroplasms; they contain virions.

**Taxonomy and Relationships**

Possible *Potexvirus*.

Virus(es) with serologically related virions: commelina X and lily X potexviruses. Virus(es) with serologically unrelated virions: cactus X, cymbidium mosaic, hydrangea ringspot, narcissus mosaic, nerine X, papaya mosaic, pepino mosaic, potato X, potato aucuba mosaic, viola mottle and white clover mosaic viruses.

**References**

Hearon, S.S., Corbett, M.K., Lawson, R.H., Gillespie, A.G. and Waterworth, H.E. (1978). *Phytopathology* **68**: 1137.

Phillips, S., Piggott, J.D'A. and Brunt A.A. (1986). *Ann. appl. Biol.* **109**: 137.

Waterworth, H.E., Lawson, R.H. and Kahn, R.P. (1974). *J. Agric. Univ. P. Rico* **58**: 351.

# Dioscorea trifida (?) potyvirus

**Data collated by A. Porth and F. Nienhaus, 1986.**

*Standard acronym:* DTV.

First reported in *Dioscorea trifida*; from Guadeloupe; by Migliori and Cadilhac (1976).

**Natural host range and symptoms**

*Dioscorea trifida* — mosaic, chlorotic flecking, vein-banding and leaf deformation.

**Transmission**

Transmitted by a vector; an insect; *Aphis gossypii*; Aphididae. Transmitted in a non–persistent manner. Virus transmitted by mechanical inoculation.

**Geographical distribution**

Spreads in Guadeloupe.

Diagnostically susceptible

*Dioscorea alata, D. cayenensis* — mosaic, chlorotic flecking, vein-banding and leaf deformation.

**Purification**

Method: Migliori and Cadilhac (1976).

**Particle morphology**

Virions filamentous; not enveloped; usually flexuous; with a clear modal length; of 760 nm.

**Cytopathology**

Virions found in all parts of the host plant; in cytoplasm. Inclusions present in infected cells; are pinwheels and unusual in shape; and laminated aggregates; they do not contain virions.

**Taxonomy and Relationships**

Unassigned potyvirus.

Best tests for diagnosis: cucumber mosaic virus often occurs with the virus (Migliori and Cadilhac, 1976).

**References**

Marchoux, G., Edwige, S. and Migliori, A. (1979). *Ann. Phytopath.* **11**: 535.

Migliori, A. and Cadilhac, B. (1976). *Ann. Phytopath.* **8**: 73.

# Dock mottling mosaic (?) potyvirus

**Data collated by A.A. Brunt, 1991.**

*Synonyms:* probably dock mosaic virus (Grainger and Cockerham, 1930; dock mosaic viruses reported in the U.S.A. (Fernow, 1948) and New Zealand (Chamberlain and Matthews, 1948) could also be synonyms, but tests of possible affinity have not been made). *Standard acronym:* DMMV.

First reported in *Rumex* spp.; from Hungary; by Juretič *et al.* (1976).

**Natural host range and symptoms**

Symptoms persist.
*Rumex* spp. — chlorotic mottling of leaves.

**Transmission**

Transmitted by a vector; an insect; *Myzus persicae*; Aphididae. Transmitted in a non–persistent manner. Virus transmitted by mechanical inoculation.

**Geographical distribution**

Spreads in Hungary, New Zealand, and the UK (possibly).

**Experimental host range**

Few (<3) families susceptible.

Diagnostically susceptible

*Rumex conglomeratus, R. obtusifolius, R. patientia, R. pulcher* — leaf mosaic.

Diagnostically insusceptible

*Chenopodium amaranticolor, C. quinoa, Nicotiana megalosiphon, N. tabacum.*

Maintenance and propagation hosts

*Rumex obtusifolius.*

Dicotyledonous Hosts *(**Susceptible** or Insusceptible)*

*Chenopodium amaranticolor. Chenopodium quinoa. Nicotiana megalosiphon. Nicotiana tabacum.*

**Properties of particles in sap**

TIP: 60 °C. LIV: 1 days.

**Particle morphology**

Virions filamentous; not enveloped; usually flexuous; with a clear modal length; of 700 nm.

**Cytopathology**

Inclusions present in infected cells; are pinwheels.

**Taxonomy and Relationships**

Unassigned potyvirus.

**References**


Chamberlain, E.E. and Matthews, R.E.F. (1948). *N.Z. J. Sci. Tech.* **29**: 286.

Fernow, K.H. (1925). *Cornell Univ. Agric. Exp. Stn Memoir 26.*

Grainger, J. and Cockherham, G. (1930). *Proc. Leeds Phil. Lit. Soc.* (Sci. Sect.) **2**: 120.

Juretič, N., Wrischer, M. and Miličič, D. (1976). *Agric. consp. Scientif.* **39**: 565.

# Dodonaea yellows-associated virus

**Data collated by A.A. Brunt, 1991.**

First reported in *Dodonaea viscosa*; from Hawaii; by Gardner and Kageler (1984).

**Natural host range and symptoms**

Symptoms persist.

*Dodonaea viscosa* — pendulous 'witches brooms' formed of abnormally long proliferated branches, the leaves on which are small and chlorotic.

**Transmission**

Virus not transmitted by mechanical inoculation.

**Geographical distribution**

Spreads in the USA (Hawaii).

**Purification**

Method: Borth *et al.* (1990).

**Particle morphology**

Virions filamentous; not enveloped; usually flexuous; with a clear modal length; of 700 nm; 16 nm wide.

1 virus specified dsRNA species found in infected cells. Size of largest virus specified dsRNA 3 kbp.

**Comments**

The symptoms in *Dodonaea viscosa* resemble those induced by mycoplasma-like organisms (MLO) in other hosts. Although no MLO were detected in affected plants, further work is necessary to elucidate the etiology of the disease.

**References**

Borth, W.B., Gardener, D.E. and German, T.L. (1990). *Plant Dis.* **74**: 434.

Gardener, D.E. and Kageler, V.A.D. (1984). *Hawaii Bot. Soc. Newsl.* **23**: 7.

# Dogwood mosaic (?) *nepovirus*

**Data collated by O.W. Barnett, 1991.**

*Synonyms:* arabis mosaic virus (first considered to be a strain of this virus, but now thought to be a separate viral species; Barnett, 1981; Barnett and Baxter, 1976; Handley and Barnett, 1951; Reddick *et al.*, 1980).

First reported in *Cornus florida*; from Clemson, South Carolina, U.S.A.; by Barnett and Baxter (1976).

**Natural host range and symptoms**

Symptoms vary seasonally. Symptoms mosaic.

*Cornus florida* — symptomless or mottling or mosaic.

**Transmission**

Not transmitted by *Xiphinema diversicaudatum*. Virus transmitted by mechanical inoculation; transmitted by grafting.

**Ecology and control**

Studies reported by Barnett *et al.* (1989).

**Geographical distribution**

Spreads in the USA (South Carolina).

**Experimental host range**

Many (>9) families susceptible. Experimentally infected plants mostly show mosaic and/or ringspots.

Diagnostically susceptible

*Chenopodium amaranticolor* — chlorotic local lesions, then systemic chlorotic spots, line patterns, mosaic.

*C. quinoa* — chlorotic local lesions becoming necrotic, then systemic mosaic and tip necrosis.

*Vigna unguiculata* subsp. *unguiculata, Lupinus albus* — chlorotic ringspots and systemic mosaic.

*Pisum sativum* — chlorotic local lesions, then systemic mosaic.

*Torenia fournieri* — systemic chlorosis and veinal necrosis.

*Nicotiana occidentalis* — local lesions; systemic mosaic.

*N. tabacum* — chlorotic local lesions, then systemic in some cultivars.

*Petunia hybrida* — necrotic ringspots; systemic necrotic spots.

*Lactuca sativa* — systemic mosaic.

Diagnostically insusceptible

*Arabis alpina, Fragaria vesca, Phaseolus lunatus, Zinnia elegans, Citrullus lanatus* var. *lanatus*.

Maintenance and propagation hosts

*Nicotiana clevelandii, N. benthamiana, Chenopodium quinoa, Gomphrena globosa.*

Assay hosts (**L**ocal lesions or **W**hole plants)

*Chenopodium quinoa* (L or W), *Nicotiana clevelandii* (W).

Dicotyledonous Hosts *(**Susceptible** or Insusceptible)*

*Anthriscus cerefolium*. ***Antirrhinum majus***. *Apium graveolens*. ***Beta patellaris***. ***Beta vulgaris***. *Brassica campestris* ssp. *pekinensis*. *Capsicum annuum*. ***Catharanthus roseus***. ***Celosia argentea***. ***Celosia cristata***. *Chenopodium album*. ***Chenopodium amaranticolor***. ***Chenopodium foetidum***. ***Chenopodium murale***. ***Chenopodium quinoa***. *Citrullus lanatus*. ***Cucumis sativus***. *Cucurbita pepo*. ***Datura stramonium***. *Desmodium tortuosum*. *Dolichos biflorus*. *Fragaria vesca*. ***Glycine max***. ***Gomphrena globosa***. ***Lactuca sativa***. ***Lupinus albus***. ***Lycopersicon esculentum***. *Medicago sativa*. *Momordica balsamina*. ***Nicandra physaloides***. ***Nicotiana clevelandii***. ***Nicotiana megalosiphon***. ***Nicotiana occidentalis***. ***Nicotiana sylvestris***. ***Nicotiana tabacum***. ***Petunia hybrida***. *Phaseolus lunatus*. ***Phaseolus vulgaris***, or *Phaseolus vulgaris*. ***Phlox drummondii***. ***Pisum sativum***. ***Spinacia oleracea***. ***Tetragonia expansa***. ***Torenia fournieri***. ***Trifolium pratense***, or *Trifolium pratense*. ***Verbesina encelioides***. ***Vicia faba***. ***Vigna unguiculata***. ***Vigna unguiculata*** ssp. ***cylindrica***. ***Zinnia elegans***, or *Zinnia elegans*.

Monocotyledonous Hosts *(**Susceptible** or Insusceptible)*

*Zea mays*.

Additional hosts

*Arabis alpina, Brassica nigra*, ***Chenopodium botrys, Dolichos lablab***, *Lagenaria siceraria*, ***Lupinus hirsutus, Lupinus luteus***, *Medicago lupulina, Mirabilis jalapa*, ***Nicotiana longiflora, Nicotiana palmeri, Physalis alkekengi***, *Pueraria lobata* and *Ipomoea sloteri*.

**Properties of particles in sap**

TIP: 50 °C. LIV: 8 days. DEP: $\log_{10}$ minus 4.

**Purification**

Method: Barnett *et al.* (1989).

**Particle morphology**

Virions isometric; not enveloped; 27 nm in diameter; angular in profile; without a conspicuous capsomere arrangement.

**Physical properties**

Three sedimenting components in purified preparations; sedimentation coefficient of the fastest 125 *S* (B); of the other(s) 54 *S* (T). $A_{260}/A_{280}$ ratio 1.63 (B), or 1.5 (M), or 0.88 (T).

**Biochemical properties**

*Genome* consists of RNA; single–stranded; linear; of two parts; largest (or only) genome part the larger 8.2 kb; the 2nd largest 3.9 kb. Genomic nucleic acid isolated by Barnett *et al.* (1989). Additional factor not required for infectivity.

*Non–genomic nucleic acid* not found in the virions.

*Virion protein(s)* one; $M_r$ 54100.

**Replication**

Replication does not depend on a helper virus.

**Cytopathology**

Virions found in leaves and roots.

**Taxonomy and Relationships**

*Nepovirus: Comoviridae.*

Virus(es) with serologically related virions: arabis mosaic and grapevine fanleaf viruses. Virus(es) with serologically unrelated virions: artichoke Italian latent, cherry leaf roll, grapevine chrome mosaic, raspberry ringspot, strawberry latent ringspot, tobacco ringspot and tomato ringspot viruses.

Dogwood mosaic virus is serologically more distant from arabis mosaic virus than is grapevine fanleaf virus. Best tests for diagnosis: cucumber mosaic and broad bean wilt viruses are found in *Cornus florida* in the USA with the virus. It, unlike cucumber mosaic virus, infects cowpea systemically and, unlike broad bean wilt, induces local lesions in cucumber. Serological tests are best for diagnosis.

**References**

Barnett, O.W. (1981). *Abstr. 5th Int. Congr. Virol., Strasbourg*, France.

Barnett, O.W. and Baxter, L.W. (1976). *Proc. Am. Phytopathol. Soc.* **3**: 249.

Barnett, O.W., Reddick, B.B., Burrows, P.M. and Baxter, L.W. (1989). *Phytopathology* **79**: 951.

Handley, M.K. and Barnett, O.W. (1981). *Phytopathology* **71**: 223.

Reddick, B.B., Barnett, O.W. and Baxter, L.W. (1979). *Pl. Dis. Reptr* **63**: 529.

Reddick, B.B., Barnett, O.W. and Baxter, L.W. (1980). *Ornamentals South* **2**: 4.

# Dolichos yellow mosaic bigeminivirus

**Data collated by V. Muniyappa, 1991.**

*Standard acronym:* DoYMV.

First reported in *Lablab purpureus*; from Poona, India; by Capoor and Varma (1950*b*).

## Natural host range and symptoms

Symptoms persist. Symptoms yellow or green mosaic with pitting.
*Lablab purpureus* — yellow mosaic.

## Transmission

Transmitted by a vector; an insect; *Bemisia tabaci*; Aleyrodidae. Transmitted in a semi–persistent manner. Virus retained when the vector moults; does not multiply in the vector; not transmitted congenitally to the progeny of the vector; not transmitted by mechanical inoculation; transmitted by grafting.

## Ecology and control

Studies reported by Subramanian *et al.* (1971).

## Geographical distribution

Spreads in India.

## Experimental host range

Few (<3) families susceptible. Experimentally infected plants mostly show yellow mosaic.

Maintenance and propagation hosts

*Lablab purpureus*.

Assay hosts (**L**ocal lesions or **W**hole plants)

*Lablab purpureus* (W).

Dicotyledonous Hosts (***Susceptible** or Insusceptible)*

***Lablab purpureus***.

## Particle morphology

Virions geminate; not enveloped; 20 nm in diameter; dimers 30 nm in length.

## Taxonomy and Relationships

Bigeminivirus: *Geminiviridae*.
Virus(es) with serologically related virions: African cassava mosaic virus.

## References


Capoor, S.P. and Varma, P.M. (1950*b*). *Curr. Sci.* **19**: 248.

Chinnadurai, G. and Nair, M.C. (1971). *Curr. Sci.* **40**: 18.

Harrison, B.D., Muniyappa, V., Swanson, M.M., Roberts, I.M. and Robinson, D.J. (1991). *Ann. appl. Biol.* **118**: 595.

Subramanian, K.S., Samuel, G.S., Janarthana, R. and Kandaswamy, T.C. (1971). *South Indian Hort.* **19**: 96.

Yaraguntaiah, R.C. and Govindu, H.C. (1964). *Curr. Sci.* **33**: 271.

# Dulcamara mottle *tymovirus*

**Data collated by A.J. Gibbs, 1991.**

*Standard acronym:* DuMV.

First reported in *Solanum dulcamara*; from the U.K. in Hertfordshire; by Gibbs *et al.* (1966).

### Natural host range and symptoms

Symptoms persist. Symptoms mottle.
*Solanum dulcamara* — mild mottle.

### Transmission

Transmitted by a vector; an insect; *Psylloides affinis*; Coleoptera. Transmitted in a semi–persistent manner. Virus transmitted by mechanical inoculation; not transmitted by contact between plants; transmitted by seed.

### Ecology and control

Studies reported by Gibbs *et al.* (1966).

### Geographical distribution

Spreads in the UK.

### Experimental host range

Several (3–9) families susceptible. Experimentally infected plants mostly show local lesions, mosaics, mottles, distortion.

Diagnostically susceptible

*Datura stramonium* — local lesions, mottle.
*Lycopersicon esculentum* — mottle.
*Nicotiana clevelandii, N. glutinosa* — local lesions, mosaic, distortion.
*Solanum nigrum* — mosaic.

Diagnostically insusceptible

*Solanum tuberosum* several cultivars.

Maintenance and propagation hosts

*Nicotiana clevelandii, N. glutinosa.*

Assay hosts (**L**ocal lesions or **W**hole plants)

*Nicotiana glutinosa* (L/W).

Dicotyledonous Hosts *(**Susceptible** or Insusceptible)*

*Antirrhinum majus. Arachis hypogaea. Bellis perennis. Beta vulgaris. Brassica campestris* ssp. *pekinensis. Catalpa bignonioides.* ***Catharanthus roseus****. Chenopodium amaranticolor. Chenopodium foetidum. Chenopodium quinoa. Cucumis sativus. Cucurbita pepo.* ***Datura stramonium****. Daucus carota. Delphinium hybridum. Eucalyptus cloesiana.* ***Euphorbia marginata****. Fagopyrum esculentum. Fragaria vesca. Gossypium hirsutum. Gypsophila elegans. Helianthus annuus.* ***Ipomoea nil****. Lactuca sativa. Lobelia erinus.* ***Lycopersicon esculentum****. Myosotis sylvatica.* ***Nicotiana clevelandii****. Nicotiana debneyi.* ***Nicotiana glutinosa****. Nicotiana tabacum. Papaver nudicaule. Pastinaca sativa. Petroselinum crispum. Petunia hybrida. Pisum sativum. Podolepis robusta. Salvia splendens.* ***Sinningia speciosa****.* ***Solanum nigrum****. Trifolium incarnatum. Tropaeolum majus. Vicia faba.* ***Vigna radiata****. Viola cornuta.*

Additional hosts

***Leptosiphon*** spp.

Sources of host-range data

Gibbs *et al.* (1966); Guy *et al.* (1984).

**Properties of particles in sap**

TIP: 65–70 °C. LIV: 2–7 days. DEP: $\log_{10}$ minus 3–4. Infectivity of sap not changed by treatment with di–ethyl ether. Leaf sap contains many virions.

**Particle morphology**

Virions isometric; not enveloped; 28 nm in diameter; rounded in profile; with a conspicuous capsomere arrangement.

**Physical properties**

Two sedimenting components in purified preparations; sedimentation coefficient of the fastest 121 *S*; of the other(s) 55 *S*. Isoelectric point pH 8.15.

**Biochemical properties**

Virions contain 37.5 % nucleic acid; 62.5 % protein.

*Genome* consists of RNA; single–stranded; linear; unipartite; largest (or only) genome part 6 kb. Base composition 16.7 % G; 22.6 % A; 32.2 % C; 28.5 % U. 5' terminus of RNA has a methylated nucleotide cap. Infectivity retained when deproteinised with proteases; retained when deproteinised with phenol or detergent. Poly A region present; 3' terminal poly A tract. Additional factor not required for infectivity. Genome has no *t*RNA–like activity.

Sub–genomic *m*RNA found in infected cells; mRNA for virion protein.

Virion proteins not glycosylated. Virion proteins not phosphorylated.

**Replication**

Genome replicates in association with chloroplasts (in vesicles in the peripheral membrane). Coat protein *m*RNA translated in the cytoplasm. Replication does not depend on a helper virus.

**Cytopathology**

Virions found in all parts of the host plant; in cytoplasm. Inclusions absent from infected cells.

**Taxonomy and Relationships**

*Tymovirus.*

Virus(es) with serologically related virions: Andean potato latent, belladonna mottle, eggplant mosaic and ononis yellow mosaic viruses.

**References**

Gibbs, A.J., Hecht-Poinar, E., Woods, R.D. and McKee, R.K. (1966). *J. gen. Microbiol.* **44**: 177.

Guy, P.L., Dale, J.L., Adena, M.A. and Gibbs, A.J. (1984). *Pl. Path.* **33**: 337.

# Echinochloa ragged stunt *oryzavirus*

**Data collated by C.C. Chen, 1991.**

*Standard acronym:* ERSV.

First reported in *Echinochloa crus-galli* var. *oryzicola*; from Taiwan; by Chen *et al.* (1980).

**Natural host range and symptoms**

Symptoms persist. Symptoms severe stunting, darker green leaves, serrated margins and veinal swellings.

*Echinochloa crus-galli* var. *oryzicola, Setaria italica, Triticum aestivum* — tillering, severe dwarfing, malformation of leaves with serrated margins and veinal swellings. Plants appear greener and fail to head, or show a delay in heading time leading to empty grains.

**Transmission**

Transmitted by a vector; an insect; *Sogatella vibix, S. longifurcifera*; Delphacidae. Not transmitted by *Nephotettix cincticeps, N. nigropictus, Nilaparvata lugens, Laodelphax striatellus, Sogatella furcifera, Recilia dorsalis*. Transmitted in a persistent manner. Virus retained when the vector moults; multiplies in the vector; not transmitted congenitally to the progeny of the vector; transmitted by mechanical inoculation; not transmitted by grafting; not transmitted by contact between plants; not transmitted by seed; not transmitted by pollen.

**Geographical distribution**

Spreads in Taiwan.

**Experimental host range**

Few (<3) families susceptible. Experimentally infected plants mostly show severe dwarfing, leaf malformation, ragged leaf, vein-swelling on abaxial surface of leaf sheaths, and a dark green plant colour.

Diagnostically susceptible

*Alopecurus aequalis, Echinochloa crus-galli* var. *formosensis, Eleusine coracana, Lolium multiflorum, Oryza sativa, Setaria italica, Triticum aestivum* — excessive tillering, dwarfing and production of malformed leaves with serrated margins and veinal swellings. Leaves darker green.

*Imperata cylindrica* var. *major* — leaf yellowing and small vein swellings.

*Zea mays* — milder symptoms and recovery after 2–3 weeks.

Diagnostically insusceptible

*Cynodon dactylon, Eleusine indica, Hordeum vulgare, Leersia hexandra, Sorghum bicolor.*

Maintenance and propagation hosts

*Echinochloa crus-galli, Lolium multiflorum, Setaria italica.*

Assay hosts (Local lesions or Whole plants)

*Alopecurus aequalis* (W), *Echinochloa crus-galli* (W), *Eleusine coracana* (W), *Lolium multiflorum* (W), *Setaria italica* (W).

Monocotyledonous Hosts *(**Susceptible** or Insusceptible)*

*Cynodon dactylon. Echinochloa colonum.* ***Echinochloa crus–galli. Eleusine coracana. Imperata cylindrica. Lolium multiflorum. Setaria italica.*** *Sorghum bicolor.* ***Triticum aestivum. Zea mays.***

Sources of host-range data

Chen *et al.* (1986).

**Properties of particles in sap**

TIP: 70–80 °C. LIV: 3 days. DEP: $\log_{10}$ minus 6–7. Leaf sap contains few virions. Electron microscopy: fix with glutaraldehyde.

**Purification**

Method: Chen *et al.* (1988).

**Particle morphology**

Virions isometric; not enveloped; 80 nm in diameter; angular in profile; with a conspicuous capsomere arrangement.

**Physical properties**

Two sedimenting components in purified preparations. $A_{260}/A_{280}$ ratio 1.8.

**Biochemical properties**

*Genome* consists of RNA; double–stranded; linear; of 10 parts (named S1–10); largest (or only) genome part the largest 3.85 kb (pairs); the 2nd largest 3.82 kb (pairs); the 3rd largest 3.8 kb (pairs); the 4th largest 3.7 kb (pairs); the 5th largest 2.7 kb (pairs); the 6th largest and other parts 2.6 kb (pairs; also 1.95, 1.85, 1.20 and 1.17 kb pairs). Genomic nucleic acid isolated by Chen *et al.* (1989).

*Virion protein(s)* six (4 major, 2 minor); $M_r$ of the largest 127000 (major). $M_r$ of 2nd largest 123000 (major). $M_r$ of 3rd largest 103000 (minor). $M_r$ of 4th largest 63000 (major). $M_r$ of 5th largest 50000 (minor). $M_r$ of 6th largest and smaller, 34000 (major). Method of preparation: Chen *et al.* (1989).

**Cytopathology**

Virions found in leaves and phloem. Inclusions present in infected cells; are viroplasms; they contain virions.

**Taxonomy and Relationships**

*Oryzavirus: Reoviridae.*

Best tests for diagnosis: ELISA.

**References**


Chen, C.C. (1985). *Ph.D. Thesis*, 150 pp. Grad. Inst. Pl. Path., Nat. Chung-Hsing Univ.

Chen, C.C., Chen, M.J. and Chiu, R.J. (1986). *Pl. Prot. Bull.* (Taiwan) **28**: 371.

Chen, C.C., Chen, M.J., Chiu, R.J. and Hsu, H.T. (1989). *Phytopathology* **79** ;235.

Chen, C.C., Chiu, R.J., Chen, M.J., Wang, Y.S. and Ko, W.H. (1980). *Pl. Prot. Bull. (Taiwan)* **22**: 425.

Chen, C.C., Hsu, Y.H., Chen, M.J. and Chiu, R.J. (1989). *Intervirology* **30**: 278.

Chen, C.C., Hsu, Y.H., Liu, M.C., Chen, M.J. and Chiu, R.J. (1988). *Pl. Prot. Bull. (Taiwan)* **30**: 420.


# Eggplant green mosaic *potyvirus*

**Data collated by A.A. Brunt, D.E. Lesemann and R. Koenig, 1988.**

*Standard acronym:* EGMV.

First reported in *Solanum melongena*; from southern Nigeria; by Ladipo (1976).

**Natural host range and symptoms**

Symptoms persist.

*Solanum melongena* — green mosaic.

**Transmission**

Virus transmitted by mechanical inoculation.

**Geographical distribution**

Spreads in Nigeria.

**Experimental host range**

Few (<3) families susceptible.

Diagnostically susceptible

*Lycopersicon esculentum* — mild systemic chlorosis.

*Nicandra physalodes* — necrotic local lesions, then systemic chlorosis and necrotic spotting.

*Nicotiana megalosiphon, N. occidentalis, N. rustica, N. sylvestris* — necrotic local lesions, then systemic necrotic spotting and veinal necrosis.

*N. tabacum* cvs White Burley, Xanthi-*nc*. — systemic chlorotic spotting and vein necrosis.

*N. glutinosa* — systemic vein clearing and mottling.

*Petunia hybrida* — chlorotic local lesions, then systemic veinal chlorosis and necrosis.

*Physalis peruviana* — chlorotic rings and lesions, systemic mottling.

Diagnostically insusceptible

*Chenopodium amaranticolor, Cucumis sativus, Datura stramonium, Gomphrena globosa, Vigna unguiculata.*

Maintenance and propagation hosts

*Nicotiana glutinosa, N. clevelandii.*

Assay hosts (**L**ocal lesions or **W**hole plants)

*Nicotiana megalosiphon* (L), *N. sylvestris* (L), *N. occidentalis* (L), *N. rustica* (L).

Dicotyledonous Hosts *(**Susceptible** or Insusceptible)*

*Capsicum annuum. Capsicum frutescens. Catharanthus roseus. Chenopodium album. Chenopodium amaranticolor. Cucumis sativus. Datura stramonium. Gomphrena globosa.* ***Lycopersicon esculentum. Nicandra physaloides. Nicotiana benthamiana. Nicotiana clevelandii. Nicotiana debneyi. Nicotiana glutinosa. Nicotiana megalosiphon. Nicotiana occidentalis. Nicotiana rustica. Nicotiana sylvestris. Nicotiana tabacum. Petunia hybrida. Physalis floridana. Physalis peruviana. Solanum melongena. Solanum nigrum.*** *Vigna unguiculata. Vigna unguiculata* ssp. *cylindrica.*

Sources of host-range data

Ladipo (1976); Ladipo *et al.* (1988).

**Properties of particles in sap**

TIP: 60 °C. LIV: 4 days (at 25–28°C). DEP: $\log_{10}$ minus 4. Leaf sap contains few virions.

**Purification**

Method: Ladipo *et al.* (1976).

**Particle morphology**

Virions filamentous; not enveloped; usually flexuous; with a clear modal length; of 820 nm. Axial canal obscure. Basic helix obscure.

**Cytopathology**

Virions found in leaves; in cytoplasm. Inclusions present in infected cells; are pinwheels; they do not contain virions. Other cellular changes: masses of endoplasmic reticulum, microbodies, lipid droplets, mitochondria and abnormal vesicles in cytoplasm.

**Taxonomy and Relationships**

*Potyvirus: Potyviridae.*

Virus(es) with serologically related virions: dioscorea green-banding, groundnut eyespot, zucchini yellow mosaic and an unnamed potyvirus from tomato in Taiwan. Virus(es) with serologically unrelated virions: datura Colombian, henbane mosaic, pepper mottle, pepper veinal mottle, tomato (Peru), potato A, potato Y, tobacco etch and wild potato mosaic viruses.

**References**

Ladipo, J.L. (1976). *Pl. Dis. Reptr* **60**: 1068.

Ladipo, J.L., Lesemann, D.E. and Koenig, R. (1988). *J. Phytopath.* **121**: 159.

# Eggplant mild mottle (?) *carlavirus*

**Data collated by A.A. Brunt, 1994.**

*Standard acronym:* EMMV.

First reported in *Solanum melongena*; from Al-Sharja; by Khalil and Nelson (1977).

**Natural host range and symptoms**

Symptoms persist. Symptoms mild leaf mottling.

*Solanum melongena* — mild leaf mottling.

**Transmission**

Not transmitted by *Myzus persicae*. Virus transmitted by mechanical inoculation; not transmitted by seed.

**Geographical distribution**

Spreads in the Middle East.

**Experimental host range**

Several (3–9) families susceptible. Experimentally infected plants mostly show etching and necrosis.

Diagnostically susceptible

*Solanum melongena* — systemic mild mottling.

*Nicotiana tabacum* cvs Havana 425 and Xanthi-*nc* — necrotic etching of leaves, stunting, plant dies sometimes.

Diagnostically insusceptible

*Capsicum annuum, Lycopersicon esculentum, Solanum tuberosum.*

Maintenance and propagation hosts

*Solanum melongena, Nicotiana tabacum* cv. Havana 425, *N. xanthi*-nc.

Assay hosts (**L**ocal lesions or **W**hole plants)

*Chenopodium amaranticolor* (L).

Dicotyledonous Hosts *(**Susceptible** or Insusceptible)*

*Capsicum annuum*. ***Chenopodium amaranticolor***. *Lycopersicon esculentum*. ***Nicotiana tabacum***. ***Solanum melongena***. *Solanum tuberosum*.

**Properties of particles in sap**

TIP: 70 °C. LIV: 6 days. DEP: $\log_{10}$ minus 4.

**Purification**

Method: Khalil *et al.* (1982).

**Particle morphology**

Virions filamentous; not enveloped; usually straight; with a clear modal length; of 693 nm. Axial canal obscure.

**Cytopathology**

Virions found in cytoplasm. Inclusions absent from infected cells.

**Taxonomy and Relationships**

Possibly *Carlavirus*.

Virus(es) with serologically unrelated virions: pea streak, potato S and red clover vein mosaic carlaviruses.

**References**

Khalil, J.A. (1978). Ph.D. Dissertation, 69 pp. University of Arizona.

Khalil, J.A. and Nelson, M.R. (1977). *Proc. Am. Phytopath. Soc.* **4**: 160.

Khalil, J.A., Nelson, M.R. and Wheeler, R.E. (1982). *Phytopathology* **72**: 1064.

# Eggplant mosaic *tymovirus*

**Data collated by C. Büchen-Osmond, 1987.**

*Synonyms:* abelia latent virus (Waterworth *et al.*, 1975). *Standard acronym:* EMV.

First reported in *Solanum melongena*; from Trinidad, West Indies; by Ferguson (1951); Dale (1954).

**Natural host range and symptoms**

Symptoms persist.

*Solanum melongena* — mosaic.

*Solanum seaforthianum* — mottle, veinbanding.

*Lycopersicon esculentum* — mosaic.

**Transmission**

Transmitted by a vector; an insect; *Epitrix* sp.; Coleoptera. Virus transmitted by mechanical inoculation; transmitted by seed.

**Ecology and control**

Studies reported by Gibbs *et al.* (1966); Debrot *et al.* (1977).

**Geographical distribution**

Spreads in Trinidad and Tobago.

**Experimental host range**

Few (<3) families susceptible.

Diagnostically susceptible

*Chenopodium amaranticolor, C. quinoa* — chlorotic or necrotic local lesions, then systemic mottle or mosaic.

*Cucumis sativus* — symptomless, local infection only.

*Nicotiana clevelandii* — necrotic local lesions, then systemic mottling and, sometimes, necrosis, stunting.

*Nicotiana glutinosa* — chlorotic or necrotic local lesions, then systemic chlorotic or necrotic lesions or flecks and malformation.

Diagnostically insusceptible

*Phaseolus vulgare* cvs The Prince, Bountiful, Stringless Green Pod and Top Crop.

Maintenance and propagation hosts

*Nicotiana clevelandii, N. glutinosa.*

Assay hosts (**L**ocal lesions or **W**hole plants)

*Chenopodium amaranticolor* (L), *C. hybridum* (L), *Vigna unguiculata* cv. Blackeye (L).

Dicotyledonous Hosts *(**Susceptible** or Insusceptible)*

*Apium graveolens*. ***Capsicum annuum***. ***Chenopodium album***. ***Chenopodium amaranticolor***. ***Chenopodium hybridum***. ***Chenopodium quinoa***. ***Cucumis sativus***, or *Cucumis sativus*. *Cucurbita moschata*. ***Datura stramonium***. ***Gomphrena globosa***. ***Lycopersicon esculentum***. ***Lycopersicon pimpinellifolium***. ***Nicandra physaloides***. ***Nicotiana clevelandii***. ***Nicotiana debneyi***. ***Nicotiana glutinosa***. *Nicotiana rustica*. *Nicotiana sylvestris*. *Nicotiana tabacum*. ***Petunia hybrida***. *Phaseolus vulgaris*. ***Physalis floridana***. ***Solanum melongena***. ***Solanum tuberosum***. ***Vigna unguiculata***. *Zinnia elegans*.

Sources of host-range data

Dale (1954); Debrot *et al.* (1977); Gibbs *et al.* (1966); Gibbs and Harrison (1969).

**Properties of particles in sap**

TIP: 80–90 °C. LIV: 7 days. DEP: $\log_{10}$ minus 7.

**Purification**

Method: Gibbs and Harrison (1973).

**Particle morphology**

Virions isometric; not enveloped; 30 nm in diameter; rounded in profile; with a conspicuous capsomere arrangement (capsid has 32 subunits).

**Physical properties**

Two sedimenting components in purified preparations; sedimentation coefficient of the fastest 111 *S* (B); of the other(s) 55 *S* (T). Isoelectric point pH 9.46.

**Biochemical properties**

Virions contain 36 % nucleic acid (B); 64 % protein (B); 0 % lipid.

*Genome* consists of RNA; single–stranded; linear; unipartite; largest (or only) genome part 6.33 kb. Genomic nucleic acid isolated by Bercks *et al.* (1971). Genome sequence reported by Osorio-Keese *et al.* (1989).

*Sequence database accession code(s):*

J04374 Em(40)_vi:MTYRPVP Gb(84)_vi:EMVRPVP Eggplant mosaic virus genome. 5/90 6,331bp

M10868 Em(40)_vi:TYMTY301 Gb(84)_vi:EMV3EMVA Eggplant mosaic virus RNA from the 3' terminus. 7/89 59bp.

M15285 Em(40)_vi:TYMTY3EM Gb(84)_vi:EMV3EMV Eggplant mosaic virus 3' end which can form tRNA-like structure. 11/90 79bp

M58314 Em(40)_vi:EMVRRLS Gb(84)_vi:EMVRRLS Eggplant mosaic virus (EMV) 3' terminus tRNA-like structure. 8/91 80bp. 4 sequences.

*Virion protein(s)* one; $M_r$ 19815; coat protein (Dupin *et al.*, 1984). Method of preparation: Bercks *et al.* (1971). Amino acid composition: Gibbs and Harrison (1973) and Dupin *et al.* (1984).

**Replication**

Genome replicates in association with chloroplasts (in vesicles of the membrane). Replication does not depend on a helper virus.

**Cytopathology**

Virions found in all parts of the host plant; in cytoplasm, in nuclei, in chloroplasts, in mitochondria, and in cell vacuoles. Inclusions present in infected cells; are unusual in shape; crystals in the vacuole; they contain virions.

**Taxonomy and Relationships**

*Tymovirus.*

Virus(es) with serologically related virions: Andean potato latent virus is closely related; belladonna mottle, dulcamara mottle, ononis yellow mosaic and scrophularia mottle viruses are more distantly related (Bercks *et al.*, 1971).

**References**

Dale, W.T. (1954). *Ann. appl. Biol.* **41**: 240.

Debrot, E.A., Lastra, R. and de Uzcátegui, R.C. (1977). *Pl. Dis. Reptr* **61**: 628.

Dupin, A., Peter, R., Collot, D., Das, B.C., Peter, C., Bouillon, P. and Duranton, H. (1984). *C.R. hebd. Séanc. Acad. Sci., Paris, Serie III* No. **8**: 24.

Ferguson, I.A.C. (1951). *Pl. Dis. Reptr* **35**: 102.

Gibbs, A.J. and Harrison, B.D. (1969). *Ann. appl. Biol.* **64**: 225.

Gibbs, A.J. and Harrison, B.D. (1973). *CMI/AAB Descr. Pl. Viruses* No. 124, 4 pp.

Gibbs, A.J., Hecht-Poinar, E., Woods, R.D. and McKee, R.K. (1966). *J. gen. Microbiol.* **44**: 177.

Osorio-Keese, M., Keese, P. and Gibbs, A.J. (1989). *Virology* **172**: 547.

Waterworth, H.E., Kaper, J.M. and Koenig, R. (1975). *Phytopathology* **65**: 891.

# Eggplant mottled crinkle *tombusvirus*

**Data collated by K.M. Makkouk, 1987.**

*Standard acronym:* EMCV.

First reported in *Solanum melongena*; from Jieh, south of Beirut, Lebanon; by Makkouk *et al.* (1981).

**Natural host range and symptoms**

Symptoms persist.

*Solanum melongena* — mottling and crinkling of leaves.

**Transmission**

Virus transmitted by mechanical inoculation; transmitted by grafting.

**Geographical distribution**

Spreads in Lebanon.

**Experimental host range**

Several (3–9) families susceptible.

Diagnostically susceptible

*Chenopodium quinoa, C. amaranticolor, Datura stramonium, Gomphrena globosa, Nicotiana glutinosa, Phaseolus vulgaris* cv. Bountiful — local lesions.

Maintenance and propagation hosts

*Nicotiana clevelandii.*

Assay hosts (**L**ocal lesions or **W**hole plants)

*Chenopodium amaranticolor* (L), *C. quinoa* (L), *Datura stramonium* (L), *Gomphrena globosa* (L), *Nicotiana glutinosa* (L), *Phaseolus vulgaris* cv. Bountiful (L).

Dicotyledonous Hosts *(**Susceptible** or Insusceptible)*

***Chenopodium amaranticolor**. **Chenopodium quinoa**. **Cucumis sativus**. **Cucurbita pepo**. **Datura stramonium**. **Nicotiana clevelandii**. **Nicotiana glutinosa**. **Nicotiana tabacum**. **Phaseolus vulgaris**. **Physalis floridana**. **Solanum melongena**. **Vigna unguiculata**.*

**Purification**

Method: Makkouk *et al.* (1981).

**Particle morphology**

Virions isometric; not enveloped; 28–30 nm in diameter; without a conspicuous capsomere arrangement.

**Physical properties**

One sedimenting component in purified preparations; sedimentation coefficient 132 *S*.

*Virion protein(s)* one; $M_r$ 41000.

**Cytopathology**

Virions found in leaves; in cytoplasm and in chloroplasts (but possibly after chloroplast membranes disintegrate). Inclusions present in infected cells; are unusual in shape; multivesicular cytoplasmic inclusions. Other cellular changes: an increase in the volume of cytoplasm.

**Taxonomy and Relationships**

*Tombusvirus: Tombusviridae.*

Virus(es) with serologically related virions: pelargonium leaf curl, tomato bushy stunt (strain BS-3) and petunia asteroid mosaic viruses.

Best tests for diagnosis: host range, as a large number of indicator plants produce local lesions, and serology.

**References**

Makkouk, K.M., Koenig, R. and Lesemann, D.-E. (1981). *Phytopathology* **71**: 572.

# Eggplant mottled dwarf *nucleorhabdovirus*

**Data collated by A.A. Brunt, 1989.**

*Synonyms:* tomato vein chlorosis virus, tomato yellow vein virus, tomato vein yellowing virus (Adam *et al.*, 1987; Castellano and Martelli, 1987; El Maataoui *et al.*, 1985), tomato vein clearing virus. *Standard acronym:* EMDV.

First reported in *Solanum melongena*; from southern Italy; by Martelli (1969).

**Natural host range and symptoms**

Symptoms persist.
*Solanum melongena* — mottling, vein clearing and leaf malformation.
*Hibiscus rosa-sinensis* — vein yellowing and leaf malformation.
*Lycopersicon esculentum, Solanum nigrum* — yellow vein banding.
*Solanum sodomaeum* — severe leaf chlorosis.
*Lonicera* sp — yellow vein banding.

**Transmission**

Virus transmitted by mechanical inoculation; transmitted by grafting; not transmitted by seed.

**Ecology and control**

Studies reported by the virus is widespread in some Mediterranean countries, but its vector is unknown. In Morocco, *Solanum sodomaeum* is probably an important perennial source of infection.

**Geographical distribution**

Spreads in Algeria, Greece, Italy, Morocco, Portugal, Tunisia, and Turkey (and the Canary Islands).

**Experimental host range**

Several (3–9) families susceptible.

Diagnostically susceptible

*Solanum melongena, Nicotiana benthamiana, N. clevelandii, N. langsdorffi, N. glutinosa, N. megalosiphon, N. rustica, N. tabacum* — chlorotic local lesions, then systemic chlorosis.

*Chenopodium amaranticolor, C. quinoa, Gomphrena globosa* — local lesions, not systemic.

Diagnostically insusceptible

*Lactuca sativa, Datura stramonium, Physalis floridana, Phaseolus vulgaris, Cucumis sativus.*

Maintenance and propagation hosts

*Solanum melongena, Nicotiana glutinosa, N. tabacum, N. benthamiana, N. clevelandii.*

Assay hosts (**L**ocal lesions or **W**hole plants)

*Nicotiana tabacum* (L), *N. glutinosa* (L).

Dicotyledonous Hosts *(**Susceptible** or Insusceptible)*

***Capsicum annuum*. *Chenopodium amaranticolor*. *Chenopodium quinoa*.** *Cucumis sativus. Cucurbita pepo.* ***Datura metel*.** *Datura stramonium.* ***Gomphrena globosa*.** *Lactuca sativa.* ***Lycopersicon esculentum*. *Lycopersicon pimpinellifolium*. *Nicandra physaloides*. *Nicotiana benthamiana*. *Nicotiana clevelandii*. *Nicotiana* × *edwardsonii*.**

***Nicotiana debneyi*. *Nicotiana glutinosa*. *Nicotiana megalosiphon*. *Nicotiana rustica*. *Nicotiana sylvestris*. *Nicotiana tabacum*.** *Ocimum basilicum*. ***Petunia hybrida*.** *Phaseolus vulgaris*. ***Physalis floridana*. *Solanum melongena***, or *Solanum melongena*. ***Solanum nigrum*.** *Vigna unguiculata*.

Comments on host-range

The Italian tomato isolate has a narrower host range than the Italian eggplant and Moroccan tomato isolates and fails also to infect eggplant (Castellano and Martelli, 1987). El Maataoui *et al.* (1985) also report that 32 unnamed dicotyledonous species are insusceptible.

Sources of host-range data

Martelli and Rana (1970); El Maataoui *et al.* (1985); Castellano and Martelli (1987).

**Properties of particles in sap**

TIP: 54 °C. LIV: 30–44 days (at 4°C). DEP: $\log_{10}$ minus 3–4. Infectivity of sap decreased by treatment with di–ethyl ether. Leaf sap contains few virions.

**Purification**

Method: El Maataoui *et al.* (1985).

**Particle morphology**

Virions rhabdo– or bullet–shaped; enveloped; of 220–230 nm; 80–90 nm wide. Axial canal obvious; 18 nm in diameter. Basic helix obvious; pitch of basic helix 4.5 nm.

*Virion protein(s)* five; $M_r$ of the largest 100000 (or more); L protein. $M_r$ of 2nd largest 81000–86000; G protein. $M_r$ of 3rd largest 56000–61000; N protein. $M_r$ of 4th largest 27000–32000; $M_1$ protein. $M_r$ of 5th largest 21000–25000; $M_2$ protein.

**Cytopathology**

Virions found in leaves, vascular parenchyma, flowers (except stamens) and pericarp; in cytoplasm (within membranous cisternae attached to the nuclear membrane), or in the perinuclear space. Inclusions absent from infected cells. Other cellular changes: nuclei which develop uniformly granular nucleoplasm and loose chromatin. The chloroplasts and mitochondria are also altered.

**Taxonomy and Relationships**

*Nucleorhabdovirus: Rhabdoviridae.*

Virus(es) with serologically unrelated virions: potato yellow dwarf (SYDV and CYDV serotypes), broccoli necrotic yellows and sonchus yellow net viruses.

The tomato vein clearing rhabdovirus occurring in Japan (Kano *et al.*, 1985) shows some similarities to eggplant mottled dwarf virus, but tests for possible relatedness have yet to be made. The yellow vein disease of tomatoes in Nigeria (Ladipo, 1977) is induced by an uncharacterised sap-transmissible virus which also shows some similarities to eggplant mottled dwarf rhabdovirus.

**References**


Adam, G., Chagas, C.M. and Lesemann, D.E. (1987). *Phytopathology* **120**: 31.

Castellano, M.A. and Martelli, G.P. (1987). *Phytopathol. Medit.* **26**: 46.

Cherif, C. and Martelli, G.P. (1985). *FAO Pl. Prot. Bull.* **33**: 166.

Dale, J.L. and Peters, D. (1981). *Intervirology* **16**: 86.

El Maataoui, M., Lockhart, B.E.L. and Lesemann, D.E. (1985). *Phytopathology* **75**: 109.

Kano, T.S., Namba, S., Yamashita, Y., Doi, Y. and Yora, K. (1985). *Ann. Phytopath. Soc. Japan* **51**: 606.

Ladipo, J.L. (1977). *Pl. Dis. Reptr* **61**: 958.
Lockhart, B.E.L. (1987). *Plant Dis.* **71**: 731.
Peters, D. (1981). *CMI/AAB Descr. Pl. Viruses* No. 244, 4 pp.
Plavsic, B., Eric, Z. and Miličič, D. (1984). *Phytopathol. Medit.* **23**: 49.
Martelli, G.P. (1969). *J. gen. Virol.* **5**: 319.
Martelli, G.P. and Castellano, M.A. (1970). *Phytopathol. Medit.* **9**: 39.
Martelli, G.P. and Cherif, C. (1987). *J. Phytopath.* **119**: 32.
Martelli, G.P. and Cirulli, M. (1969). *Annls. Phytopath.* **1**: 393.
Martelli, G.P. and Hmadi, A. (1986). *Pl. Path.* **35**: 595.
Martelli, G.P. and Rana, G.L. (1970). *Phytopathol. Medit.* **9**: 187.
Martelli, G.P., Yilmaz, M.A. and Baloglu, S. (1984). *Phytopathol. Medit.* **23**: 9.
Russo, M. and Martelli, G.P. (1972). *Phytopathol. Medit.* **11**: 136.

# Eggplant severe mottle (?) potyvirus

**Data collated by A.A. Brunt, 1991.**

*Synonyms:* eggplant severe mosaic virus. *Standard acronym:* ESMV. *Strains:* eggplant and tomato strains.

First reported in *Solanum melongena*; from Nigeria; by Ladipo *et al.* (1988).

**Natural host range and symptoms**

Symptoms persist.

*Lycopersicon esculentum* — conspicuous leaf mosaic.

*Solanum melongena* — severe mottling, blistering and distortion of leaves, and abnormal serration of leaf margins (but cv. New York is resistant to infection).

**Transmission**

Transmitted by a vector; an insect; *Aphis craccivora, Myzus persicae*; Aphididae. Transmitted in a non–persistent manner. Virus transmitted by mechanical inoculation.

**Geographical distribution**

Spreads in Nigeria.

**Experimental host range**

Few (<3) families susceptible.

Diagnostically susceptible

*Capsicum annuum, Datura metel, Nicotiana megalosiphon, Physalis floridana* — systemic vein-clearing and/or vein-banding.

*Datura stramonium, Lycopersicon esculentum, Nicotiana clevelandii, Nicotiana benthamiana* — systemic leaf mottling.

*Nicotiana sylvestris, N. tabacum* — diffuse local lesions; no systemic infection.

*Nicotiana occidentalis* — local chlorotic spots and rings; systemic leaf mottling.

Diagnostically insusceptible

*Arachis hypogaea, Cajanus cajan, Dolichos biflorus, Vigna unguiculata.*

Maintenance and propagation hosts

*Solanum melongena* (eggplant strain), *Nicotiana megalosiphon, N. occidentalis* (tomato strain).

Assay hosts (**L**ocal lesions or **W**hole plants)

*Nicotiana occidentalis* (W), *Solanum melongena* (W).

Dicotyledonous Hosts (***Susceptible*** or *Insusceptible*)

*Arachis hypogaea*. *Cajanus cajan*. ***Capsicum annuum***. ***Capsicum frutescens***. ***Datura metel***. ***Datura stramonium***. *Dolichos biflorus*. ***Lycopersicon esculentum***. ***Nicotiana benthamiana***. ***Nicotiana clevelandii***. ***Nicotiana debneyi***. ***Nicotiana glutinosa***. ***Nicotiana megalosiphon***. ***Nicotiana occidentalis***. ***Nicotiana rustica***. ***Nicotiana sylvestris***. ***Nicotiana tabacum***. ***Physalis floridana***. ***Physalis peruviana***. ***Solanum melongena***. ***Vigna unguiculata***.

Sources of host-range data

Ladipo *et al.* (1988).

**Properties of particles in sap**

TIP: 55–65 °C. LIV: 7–9 days (eggplant strain). DEP: $\log_{10}$ minus 5.

**Cytopathology**

Virions found in leaves; in cytoplasm. Inclusions present in infected cells; are pinwheels; they do not contain virions. Other cellular changes: enlarged, unusually vesiculated and disrupted mitochondria.

**Taxonomy and Relationships**

Unassigned potyvirus.

Virus(es) with serologically related virions: dioscorea green banding mosaic, eggplant green mosaic, groundnut eyespot, peanut stripe (mungbean isolate), pepper veinal mottle and telfairia mosaic viruses. Virus(es) with serologically unrelated virions: amaranthus leaf mottle, bean common mosaic, bean yellow mosaic, beet mosaic, celery mosaic, Colombian datura, henbane mosaic, pea seed-borne mosaic, pepper mild mottle, pepper mottle, Peru tomato, potato A, potato Y, soybean mosaic, sugarcane mosaic, sweet potato feathery mottle, tobacco etch, turnip mosaic, watermelon mosaic 2, wild potato mosaic and zucchini yellow mosaic viruses.

**References**

Ladipo, J.L., Lesemann, D.-E. and Koenig, R. (1988). *J. Phytopath.* **122**: 359.

# Elderberry *carlavirus*

**Data collated by A.A. Brunt, 1992.**

*Synonyms:* elderberry A virus (Van Lent *et al.*, 1980). *Standard acronym:* EV.

First reported in *Sambucus canadensis*; from Scotland in plants imported from Canada; by Jones (1970).

**Natural host range and symptoms**

*Sambucus canadensis, S. nigra, S. racemosa* — probably symptomless but, when present in complex, it probably intensifies the symptoms induced by cherry leaf roll nepovirus.

**Transmission**

Transmitted by a vector; an insect; *Myzus persicae*; Aphididae. Not transmitted by *Aphis fabae*. Transmitted in a non–persistent manner. Virus transmitted by mechanical inoculation; transmitted by grafting.

**Geographical distribution**

Found, but with no evidence of spread, in Canada, the Netherlands and the U.K. and possibly also the former Czechoslovakia, U.S.A. and Japan.

**Experimental host range**

Several (3–9) families susceptible.

Diagnostically susceptible

*Amaranthus caudatus, Vicia faba, Vigna unguiculata* cv. Blackeye — local and systemic infection.

*Arachis hypogaea, Beta vulgaris, Celosia argentea, Phaseolus vulgaris* cv. Processor — symptomless local infection.

*Chenopodium amaranticolor, C. quinoa* — local lesions in late summer or early autumn; symptomless systemic infection during other seasons.

*Crotalaria juncea* — chlorotic mottling, veinal necrosis and epinasty of systemically infected leaves.

*Gomphrena globosa* — local lesions, no systemic infection.

Diagnostically insusceptible

*Chenopodium album, Cucumis sativus, Datura stramonium, Lycopersicon esculentum, Nicotiana clevelandii.*

Maintenance and propagation hosts

*Gomphrena globosa, Sambucus nigra, S. racemosa.*

Assay hosts (**L**ocal lesions or **W**hole plants)

*Gomphrena globosa* (L).

Dicotyledonous Hosts ***(Susceptible** or Insusceptible)*

***Amaranthus caudatus. Arachis hypogaea. Beta vulgaris.*** *Capsicum annuum. Catharanthus roseus.* ***Celosia argentea.*** *Chenopodium album.* ***Chenopodium amaranticolor.*** *Chenopodium foetidum. Chenopodium hybridum.* ***Chenopodium quinoa.*** *Cucumis sativus.* ***Cyamopsis tetragonoloba.*** *Datura metel. Datura stramonium. Dianthus barbatus.* ***Gomphrena globosa.*** *Lycopersicon esculentum. Melilotus alba. Nicotiana benthamiana. Nicotiana clevelandii. Nicotiana glutinosa. Nicotiana megalosiphon. Nicotiana rustica. Nicotiana tabacum. Petunia hybrida.* ***Phaseolus vulgaris.*** *Phlox drummondii. Physalis floridana. Pisum sativum. Solanum melongena. Solanum nigrum.* ***Solanum rostratum.*** *Solanum tuberosum. Trifolium hybridum. Trifolium incarnatum. Trifolium pratense. Trifolium repens.* ***Vicia faba. Vigna unguiculata.*** *Zinnia elegans.*

Sources of host-range data

Van Lent *et al.* (1980).

**Properties of particles in sap**

TIP: 75 °C. LIV: 4 days (at 20°C). DEP: $\log_{10}$ minus 4. Leaf sap contains few virions.

**Purification**

Method: Veerisetty and Brakke (1978).

**Particle morphology**

Virions filamentous; not enveloped; usually straight; with a clear modal length; of 678 nm; 12 nm wide. Axial canal obscure. Basic helix obscure.

**Physical properties**

One sedimenting component in purified preparations. Density 1.315 g $cm^{-3}$ in CsCl. $A_{260}/A_{280}$ ratio 1.17.

*Virion protein(s)* one; $M_r$ 31000; and it is coat protein.

**Cytopathology**

Virions found in leaves; in cytoplasm. Inclusions absent from infected cells.

**Taxonomy and Relationships**

*Carlavirus.*

Virus(es) with serologically related virions: carnation latent virus.

**Comments**

Carlaviruses reported to occur in *Sambucus* spp. in the former Czechoslovakia (Brčak, 1964; Brčak and Polák, 1966), the U.S.A. (Uyemoto and Gilmer, 1971; Uyemoto *et al.*, 1971) and probably also Japan (Y. Doi, in Hull *et al.*, 1989) might also be related to elderberry carlavirus, but their affinities have not yet been reported.

**References**

Brčak, J. (1964). *Proc. 5th Conf. Czechosl. Pl. Virologists, Prague 1962*, p. 280.
Brčak, J. and Polák, Z. (1966). *Meded. Rijksfak. Landb. Gent* **31**: 397.
Brunt, A.A. and Stace-Smith, R. (1972). *Rep. Glasshouse Crops Res. Inst.* 1971, p. 121.
Dijkstra, J. and van Lent, J.W.M. (1983). *CMI/AAB Descr. Pl. Viruses* No. 263, 4 pp.
Hull, R., Brown, F. and Payne, C.C. (1989). In: *Virology: Directory and Dictionary of Animal, Bacterial and Plant Viruses*, p. 325. MacMillan Reference Books, London, U.K.
Jones, A.T. (1970). *Rep. Scot. Crops Res. Inst.* for 1969, p. 58.
Jones, A.T. (1972). *Ann. appl. Biol.* **70**: 49.
Uyemoto, J.K. and Gilmer, R.M. (1971). *Phytopathology* **61**: 132.
Uyemoto, J.K., Gilmer, R.M. and Williams, E. (1971). *Pl. Dis. Reptr* **55**: 913.
Van Lent, J.W.M., Wit, A.J. and Dijkstra, J. (1980). *Neth. J. Pl. Path.* **86**: 117.
Veerisetty, V. and Brakke, M.K. (1978). *Phytopathology* **68**: 59.

# Elderberry latent (?) *carmovirus*

**Data collated by R. Stace-Smith, 1991.**

*Synonyms:* pelargonium ringspot virus (Hollings, 1957; A.T. Jones and A.A. Brunt, unpublished data). *Standard acronym:* ELV.

First reported in *Sambucus canadensis*; from Scotland, U.K.; by Jones (1972).

**Natural host range and symptoms**

Symptoms usually none. Symptoms sometimes line patterns.

*Sambucus canadensis* — symptomless.

*Pelargonium hortorum* — occasional line patterns.

**Transmission**

Vectorless soil transmission. Not transmitted by *Macrosiphum euphorbiae, Myzus persicae*. Virus transmitted by mechanical inoculation; transmitted by grafting; not transmitted by contact between plants; not transmitted by seed; not transmitted by pollen.

**Geographical distribution**

Found, but with no evidence of spread, in the U.K. and in Europe, Canada and the U.S.A.

**Experimental host range**

Several (3–9) families susceptible. Experimentally infected plants mostly show symptomless or chlorotic local lesions.

Diagnostically susceptible

*Chenopodium amaranticolor, C. quinoa* — chlorotic local lesions, sometimes systemic spots.

*Cucumis sativus, Nicotiana clevelandii, N. tabacum* cv. Xanthi, *Petunia hybrida* — symptomless.

Diagnostically insusceptible

*Brassica pekinensis, Phaseolus vulgaris, Pisum sativum.*

Maintenance and propagation hosts

*Chenopodium quinoa* (L).

Assay hosts (**L**ocal lesions or **W**hole plants)

*Chenopodium quinoa.*

Dicotyledonous Hosts *(**Susceptible** or Insusceptible)*

*Anthriscus cerefolium.* ***Beta vulgaris.*** *Brassica campestris* ssp. *pekinensis.* ***Chenopodium amaranticolor. Chenopodium quinoa. Cucumis sativus. Datura stramonium.*** *Daucus carota.* ***Gomphrena globosa. Lycopersicon esculentum. Nicotiana benthamiana. Nicotiana clevelandii. Nicotiana debneyi. Nicotiana glutinosa. Nicotiana rustica. Nicotiana tabacum.*** *Pastinaca sativa.* ***Petunia hybrida.*** *Phaseolus vulgaris. Pisum sativum. Raphanus sativus.* ***Spinacia oleracea.***

Sources of host-range data

Jones (1972).

**Properties of particles in sap**

TIP: 65 °C. LIV: 7 days. DEP: $log_{10}$ minus 5–6.

**Purification**

Method: Jones (1972).

**Particle morphology**

Virions isometric; not enveloped; 30 nm in diameter; rounded in profile; without a conspicuous capsomere arrangement.

**Physical properties**

Two sedimenting components in purified preparations; sedimentation coefficient 112 *S* (major, B); of the other(s) 48 *S* (minor, T). Density 1.36 g $cm^{-3}$ in CsCl. Isoelectric point pH 4.8. $A_{260}/A_{280}$ ratio 1.62.

**Biochemical properties**

Virions contain 23 % nucleic acid; 77 % protein.

*Genome* consists of RNA; single–stranded; unipartite; largest (or only) genome part 4.7 kb. Genomic nucleic acid isolated by Mayo and Jones (1973).

*Virion protein(s)* one; $M_r$ 40000; coat protein. Method of preparation: Mayo and Jones (1973).

**Cytopathology**

Virions found in leaves.

**Taxonomy and Relationships**

Possible *Carmovirus: Tombusviridae*.

Best tests for diagnosis: ELISA, serology.

**References**

Ellis, P.J., Converse, R.H. and Stace-Smith, R. (1992). *Acta Hort* **308**: 69.
Hollings, M. (1957). *Pl. Path.* **6**: 17.
Jones, A.T. (1972). *J. gen. Virol.* **70**: 49.
Jones, A.T. (1974). *CMI/AAB Descr. Pl. Viruses* No. 127, 4 pp.
Mayo, M.A. and Jones, A.T. (1973). *J. gen. Virol.* **19**: 245.
Morris, T.J. and Carrington, J.C. (1988). In: *The Plant Viruses*, Vol. 3, Polyhedral Virions with Monopartite RNA Genomes, p. 73: ed. R. Koenig. Plenum Press, New York.

# Elm mottle *ilarvirus*

**Data collated by A.T. Jones, H. Kleinhempel and K. Kontzog, 1987.**

*Synonyms:* possibly hydrangea mosaic virus (Jones, 1985), lilac streak mosaic virus (Smolak and Novak, 1950), lilac white mosaic virus. *Standard acronym:* EMoV.

First reported in *Ulmus carpinifolia*; from Germany; by Schmelzer *et al.* (1966a).

**Natural host range and symptoms**

Symptoms persist (but restricted to a few branches).

*Hydrangea macrophylla* — chlorotic mosaic.

*Syringa vulgaris* — white mosaic.

*Ulmus carpinifolia* — light green blotches, diffuse mottling.

*Ulmus glabra* — mottling, banded or 'oak-leaf' pattern.

**Transmission**

Virus transmitted by mechanical inoculation; transmitted by grafting; transmitted by seed (Jones and Mayo 1973).

**Ecology and control**

Studies reported by Schmelzer *et al.* (1966); Schmelzer (1969).

**Geographical distribution**

Probably distributed worldwide. Found, but with no evidence of spread, in Germany and Scotland; similar leaf symptoms in elm seen in other countries (Schmelzer *et al.*, 1966).

**Experimental host range**

Many (>9) families susceptible.

Diagnostically susceptible

*Chenopodium quinoa* — chlorotic or necrotic local lesions, then systemic mottling and malformation.

*Cucumis sativus* — chlorotic blotches, severe systemic mosaic.

*Nicotiana megalosiphon* — chlorotic or necrotic blotches systemic crinkling.

Diagnostically insusceptible

*Arachis hypogaea, Brassica oleracea, Digitalis purpurea, Gypsophila perfoliata, Reseda alba.*

Maintenance and propagation hosts

*Chenopodium quinoa, Nicotiana tabacum* for propagation. *Petunia hybrida* for maintenance.

Assay hosts (**L**ocal lesions or **W**hole plants)

*Chenopodium quinoa* (W), *Phaseolus vulgaris* (L), *Vigna sinensis* (L).

Dicotyledonous Hosts (***Susceptible*** *or Insusceptible)*

***Amaranthus retroflexus. Antirrhinum majus.*** *Arachis hypogaea.* ***Atriplex hortensis. Beta vulgaris.*** *Brassica oleracea* ssp. *capitata.* ***Catharanthus roseus. Celosia argentea. Chenopodium album. Chenopodium amaranticolor. Chenopodium foetidum. Chenopodium murale. Chenopodium quinoa.*** *Citrullus lanatus.* ***Cucumis sativus. Cucurbita maxima. Cucurbita pepo.*** *Dianthus barbatus.* ***Helianthus annuus. Lathyrus odoratus. Lavatera trimestris. Lobelia erinus. Lycopersicon esculentum. Matthiola incana. Nicotiana clevelandii. Nicotiana glutinosa. Nicotiana megalosiphon. Nicotiana tabacum. Ocimum basilicum. Petunia hybrida. Phaseolus vulgaris. Phlox drummondii. Physalis peruviana. Senecio vulgaris. Spinacia oleracea. Stellaria media.*** *Tropaeolum majus.* ***Vicia faba. Vigna unguiculata. Zinnia elegans.***

Additional hosts

For other susceptible and insusceptible hosts see Schmelzer (1969).

Sources of host-range data

Schmelzer (1969).

**Properties of particles in sap**

TIP: 58–62 °C. LIV: 6 days. DEP: $\log_{10}$ minus 3–4. Infectivity of sap increased by treatment with di–ethyl ether. Leaf sap contains few virions. Electron microscopy: stain with UF.

**Purification**

Method: Schmelzer (1969); Jones and Mayo (1973).

**Particle morphology**

Virions isometric; not enveloped; 25–30 nm in diameter; rounded in profile; without a conspicuous capsomere arrangement.

**Physical properties**

Three sedimenting components in purified preparations; sedimentation coefficient of the fastest 101 *S* (B); of the other(s) 88 *S* (M), or 83 *S* (T). Density 1.35 g $cm^{-3}$ in CsCl (Jones and Mayo, 1973).

**Biochemical properties**

*Genome* consists of RNA; single–stranded; linear; of five parts; largest (or only) genome part the largest 4 kb; the 2nd largest 3.6 kb; the 3rd largest 2.5 kb; the 4th largest 1 kb; the 5th largest 0.8 kb. Genomic nucleic acid isolated by Jones and Mayo (1973). Infectivity retained when deproteinised with proteases.

*Virion protein(s)* one; $M_r$ 25000. Method of preparation: Jones and Mayo (1973).

**Replication**

Replication does not depend on a helper virus.

**Cytopathology**

Virions found in pollen and seed.

**Taxonomy and Relationships**

*Ilarvirus: Bromoviridae* (subgroup 2).

Virus(es) with serologically related virions: asparagus II, citrus leaf rugose, citrus variegation and hydrangea mosaic viruses. Virus(es) with serologically unrelated virions: tobacco streak, prunus necrotic ringspot, apple mosaic and prune dwarf viruses.

**References**

Jones, A.T. (1974). *CMI/AAB Descr. Pl. Viruses* No. 139, 4pp.
Jones, A.T. (1985). *Rep. Scottish Crop Res. Inst.* 1984, p. 190.
Jones, A.T. and Mayo, M.A. (1973). *Ann. appl. Biol.* **75**: 347.
Schmelzer, K. (1969). *Phytopath. Z.* **64**: 39.
Schmelzer, K. (1977). In: *Pflanzliche Virologie* Vol. 3. p. 276; ed. M. Klinkowski. Berlin, Akademie-Verlag.
Schmelzer, K., Schmidt, H.E. and Schmidt, H.B. (1966a). *Arch. Gartenb.* **14**: 303.
Schmelzer, K., Schmidt, H.E. and Schmidt, H.B. (1966b). *Arch. Forstw.* **15**: 107.
Smolak, J. and Novak, H.B. (1950). *Ochr. Rost.* **23**: 285.
Thomas, B.J., Barton, R.J. and Tuszynski, A. (1983). *Ann. appl. Biol.* **103**: 261.

# Endive necrotic mosaic (?) *potyvirus*

**Data collated by A.A. Brunt, 1995.**

First reported in *Cichorium endiva*; from Germany; by Vetten *et al.* (1995).

**Natural host range and symptoms**

*Cichorium endiva, Lactuca sativa* (some crisphead cvs only) — severe mosaic and necrosis.

**Transmission**

Transmitted by a vector; an insect; *Myzus persicae*; Aphididae. Virus transmitted by mechanical inoculation.

**Geographical distribution**

Spreads in Germany.

Diagnostically susceptible

*Chenopodium quinoa* — local lesions.
*Cichorium endiva, Lactuca sativa* — systemic chlorosis.

Maintenance and propagation hosts

*Cichorium endiva, Lactuca sativa.*

Assay hosts (**L**ocal lesions or **W**hole plants)

*Chenopodium quinoa* (L), *Cichorium endiva* (W), *Lactuca sativa* (W).

Dicotyledonous Hosts ***(Susceptible** or Insusceptible)*

***Chenopodium quinoa*. *Cichorium endiva*. *Lactuca sativa*.**

**Particle morphology**

Virions filamentous; usually flexuous; with a clear modal length; of 820 nm.
*Virion protein(s)* one; $M_r$ 33000; coat protein.

**Cytopathology**

Inclusions are crystals in the nucleus and pinwheels (scrolls).

**Taxonomy and Relationships**

Possible *Potyvirus: Potyviridae.*

Virus(es) with serologically unrelated virions: Bidens mottle, lettuce mosaic, turnip mosaic and more than 50 other potyviruses.

**References**

Vetten, H.J., Lesemann, D.-E. and Schönfelder, M. (1995). *Proc. 8th Conf. Virus Dis. Veg., Prague* (July 9–15, 1995), p. 113.

# Epirus cherry ourmiavirus

**Data collated by A. Avgelis, 1991.**

*Standard acronym:* EpCV.

First reported in *Prunus avium*; from Giannina, Epirus, Greece; by Avgelis *et al.* (1988).

**Natural host range and symptoms**

Symptoms persist. Symptoms rasp-leaf symptoms.

*Prunus avium*—enations on abaxial leaf surfaces, stunting.

**Transmission**

Not transmitted by *Myzus persicae*. Virus transmitted by mechanical inoculation; transmitted by seed.

**Geographical distribution**

Found, but with no evidence of spread, in Epirus, Greece.

**Experimental host range**

Several (3–9) families susceptible. Experimentally infected plants mostly show chlorotic or necrotic local lesions or ringspots, then systemic mottle or chlorotic or necrotic ringspots.

Diagnostically susceptible

*Cucumis sativus* — chlorotic local lesions.

*Datura stramonium* — systemic mottle and ringspotting.

*Gomphrena globosa* — necrotic local lesions, then systemic necrotic ring spots and malformation.

*Hibiscus esculentus* — chlorotic local lesions, then systemic yellow ringspots.

*Nicotiana tabacum* cv. Xanthi-*nc*. — local and systemic necrotic ringspots.

*Petunia hybrida* — local and systemic chlorotic ring spots.

*Phaseolus vulgaris* — reddish local lesions.

*Vicia faba* — reddish local lesions, then systemic chlorotic mottle and necrotic spots.

*Vigna unguiculata* — reddish local lesions, then systemic necrotic spots and tip necrosis.

Diagnostically insusceptible

*Capsicum annuum, Cucurbita pepo, Nicotiana glutinosa, Ocimum basilicum, Zinnia elegans*.

Maintenance and propagation hosts

*Gomphrena globosa, Nicotiana benthamiana, Vicia faba.*

Assay hosts (**L**ocal lesions or **W**hole plants)

*Chenopodium amaranticolor* (L), *Gomphrena globosa* (L and W), *Hibiscus esculentus* (W), *Phaseolus vulgaris* (L), *Vicia faba* (W).

Dicotyledonous Hosts *(**Susceptible** or Insusceptible)*

***Beta vulgaris**. Brassica oleracea* ssp. *botrytis. Brassica oleracea* ssp. *capitata. Capsicum annuum. Capsicum frutescens. **Celosia cristata. Chenopodium amaranticolor. Chenopodium quinoa. Cicer arietinum.** Citrullus lanatus. Cucumis melo. Cucumis metuliferus. **Cucumis sativus.** Cucurbita maxima. Cucurbita pepo. **Datura metel. Datura stramonium. Gomphrena globosa.** Helianthus annuus. **Hibiscus esculentus. Lactuca sativa.** Lens esculenta. **Luffa acutangula. Lupinus albus. Lycopersicon esculentum.** Medicago sativa. **Momordica balsamina. Nicotiana benthamiana. Nicotiana clevelandii.** Nicotiana glutinosa. **Nicotiana megalosiphon. Nicotiana rustica. Nicotiana tabacum.** Ocimum basilicum. Petroselinum crispum. **Petunia hybrida. Phaseolus vulgaris. Physalis floridana. Pisum sativum.** Raphanus sativus. Solanum melongena. Solanum tuberosum. **Spinacia oleracea. Vicia faba. Vigna radiata. Vigna unguiculata.** Zinnia elegans.*

Monocotyledonous Hosts *(**Susceptible** or Insusceptible)*

*Allium cepa. Zea mays.*

Sources of host-range data

Avgelis *et al.* (1989).

**Properties of particles in sap**

TIP: 50 °C. LIV: 15 days. DEP: $\log_{10}$ minus 5. Leaf sap contains few virions. Electron microscopy: fix with glutaraldehyde.

**Purification**

Method: Avgelis *et al.* (1989).

**Particle morphology**

Virions bacilliform; 26 nm in length, or 34 nm in length, or 43 nm in length; 18 nm wide.

**Physical properties**

Three sedimenting components in purified preparations. $A_{260}/A_{280}$ ratio 0.62–0.64 (unfractionated preparations).

**Biochemical properties**

Virions contain 15 % nucleic acid; 85 % protein.

*Genome* consists of RNA; single–stranded; of three parts; largest (or only) genome part the largest 2.45 kb; the 2nd largest 1.05 kb; the 3rd largest 1 kb. Genomic nucleic acid isolated by Boccardo *et al.* (1985); Barba *et al.* (1991).

*Virion protein(s)* one; $M_r$ 21400; coat protein. Method of preparation: Boccardo and Milne (1975).

**Cytopathology**

Virions found in the leaves.

**Taxonomy and Relationships**

Ourmiavirus.

The size and shape of the virions of epirus cherry and melon Ourmia viruses (Lisa *et al.*, 1988) are like those of alfalfa mosaic virus. However epirus cherry virus is not an alfamovirus but seems to be related to cassava Ivorian bacilliform virus (Aiton *et al.*, 1988), olive latent 2, ourmia melon and pelargonium zonate spot viruses, but no serological relationship between their virions has been shown. Best tests for diagnosis: electron microscopy.

**References**

Avgelis, A., Barba, M. and Rumbos, J. (1988). *Acta Hort.* **235**: 245.

Avgelis, A., Barba, M. and Rumbos, J. (1989). *P. Phytopath.* **126**: 51.

Aiton, M.M., Lennon, A.M., Roberts, I.M. and Harrison, B.D. (1988). *Abstr. 5th Int. Congr. Pl. Path. Kyoto, Japan*, p. 43.

Barba, M., Riccioni, L. and Avgelis, A. (1991). *J. Phytopath.* in press.

Boccardo, G. and Milne, R.G. (1975). *Virology* **68**: 29.

Boccardo, G., Milne, R.G., Luisoni, E., Lisa, V. and Accotto, G.P. (1985). *Virology* **147**: 29.

# Erysimum latent *tymovirus*

**Data collated by A.J. Gibbs, 1991.**

*Standard acronym:* ErLV.

First reported in *Erysimum helveticum, E. perovskianum, E. pulchellum, E. sylvestre, E. crepidifolium, Fibigia clypata, Arabis ludoviciana, Barbarea vulgaris*; from Germany.

**Natural host range and symptoms**

Symptoms persist, or none. Symptoms mosaics.

*Erysimum* spp. — latent, mosaics.

**Transmission**

Transmitted by a vector; an insect; *Phyllotreta* spp.; Coleoptera. Transmitted in a semi–persistent manner. Virus transmitted by mechanical inoculation; not transmitted by contact between plants; not transmitted by seed; not transmitted by pollen.

**Geographical distribution**

Spreads in Germany.

**Experimental host range**

Many (>9) families susceptible. Experimentally infected plants mostly show local lesions, vein clearing, mosaics.

Diagnostically susceptible

*Brassica campestris* ssp. *napus, B. campestris* ssp. *pekinensis* — local lesions and mosaic.

*Cucumis sativus, Nicotiana glutinosa* — mottle. *Capsella bursa-pastoris* — mosaic.

*Ocimum basilicum, Sinapis alba* — local lesions, mosaic.

*Stellaria media, Lunaria annua* — vein clearing and mottling.

*Erysimum* spp. — latent.

Diagnostically insusceptible

*Chenopodium quinoa, Nicotiana clevelandii, Nicotiana tabacum, Phaseolus vulgaris, Pisum sativum.*

Maintenance and propagation hosts

*Brassica campestris* ssp. *pekinensis, B. juncea, B. campestris* ssp. *napus, Lunaria annua.*

Assay hosts (**L**ocal lesions or **W**hole plants)

*Brassica campestris* ssp. *napus* (L).

Dicotyledonous Hosts (***Susceptible*** or *Insusceptible)*

*Anthriscus cerefolium. Antirrhinum majus. Apium graveolens. Arachis hypogaea. Bellis perennis. Beta vulgaris.* ***Brassica campestris*** ssp. ***chinensis. Brassica campestris*** ssp. ***napus. Brassica campestris*** ssp. ***pekinensis. Brassica campestris*** ssp. ***rapa. Brassica juncea. Brassica oleracea*** ssp. ***botrytis.*** *Brassica oleracea* ssp. *capitata.* ***Capsella bursa-pastoris.*** *Catalpa bignonioides.* ***Catharanthus roseus. Celosia argentea.*** *Cheiranthus cheiri. Chenopodium amaranticolor. Chenopodium foetidum. Chenopodium murale. Chenopodium quinoa.* ***Cucumis sativus.*** *Delphinium hybridum. Eucalyptus cloesiana. Euphorbia marginata. Fagopyrum esculentum. Fragaria vesca. Gossypium hirsutum. Gypsophila elegans. Helianthus annuus. Ipomoea nil. Lactuca sativa. Lobelia erinus. Matthiola incana. Myosotis sylvatica. Nicotiana clevelandii.* ***Nicotiana glutinosa. Nicotiana megalosiphon.*** *Nicotiana tabacum.* ***Ocimum basilicum.*** *Papaver nudicaule. Pastinaca sativa. Petroselinum crispum. Petunia hybrida. Phaseolus vulgaris. Phlox drummondii. Pisum sativum. Podolepis robusta. Raphanus sativus. Salvia splendens.* ***Sinapis alba.*** *Sinningia speciosa. Spinacia oleracea.* ***Stellaria media.*** *Tropaeolum majus. Vicia faba. Vigna unguiculata.* ***Viola cornuta.***

Additional hosts

***Arabidopsis thaliana, Camelina sativa, Reseda odorata, Solanum integrifolium, Trigonella caerulae*** and *Plantago major.*

**Properties of particles in sap**

TIP: 76–78 °C. LIV: 21 days. DEP: $\log_{10}$ minus 6. Infectivity of sap not changed by treatment with di–ethyl ether. Leaf sap contains many virions.

**Purification**

Method: Shukla and Gough (1980).

**Particle morphology**

Virions isometric; 27 nm in diameter; rounded in profile; with a conspicuous capsomere arrangement.

**Physical properties**

Two sedimenting components in purified preparations; sedimentation coefficient of the fastest 113 *S*; of the other(s) 59 *S*. Density 1.26 g $cm^{-3}$ in CsCl, or 1.42 g $cm^{-3}$ in CsCl. Isoelectric point pH 5.4. $A_{260}/A_{280}$ ratio 1.76 (B).

**Biochemical properties**

Virions contain 32 % nucleic acid; 68 % protein.

*Genome* consists of RNA; single–stranded; linear; unipartite; largest (or only) genome part 6.034 kb. Base composition 15.6 % G; 23.9 % A; 34.5 % C; 26 % U. 5' terminus of RNA has a methylated nucleotide cap. Infectivity retained when deproteinised with proteases; retained when deproteinised with phenol or detergent. Poly A region absent. Additional factor not required for infectivity. Genome has no *t*RNA–like activity.

*Non–genomic nucleic acid* found in the virions; is subgenomic *m*RNA. Sub–genomic *m*RNA found in infected cells; virion proteins.

*Virion protein(s)* one; $M_r$ 21483; coat protein. Method of preparation: Srifah *et al.* (1990). Amino acid sequence: Srifah *et al.* (1990). Virion proteins not glycosylated. Virion proteins not phosphorylated.

*Virus–coded non–virion proteins* identified by genomic sequence analysis; two proteins found. $M_r$ of the largest 48497; overlapping protein. $M_r$ of 2nd largest 193811; replicase protein.

### Replication

Genome replicates in association with chloroplasts (in vesicles in the outer membrane). Coat protein *m*RNA translated in the cytoplasm. Replication does not depend on a helper virus.

### Cytopathology

Virions found in all parts of the host plant; in cytoplasm. Inclusions absent from infected cells.

### Taxonomy and Relationships

*Tymovirus*.

Virus(es) with serologically related virions: Andean potato latent, ononis yellow mosaic and okra mosaic viruses.

Genome sequence homologous with all other tymoviruses but is the most distinct of the group.

### References

Gough, K., Lilley, G.G., Shukla, D.D. and Woods, E.F. (1982). *Aust. J. biol. Sci.* **35**: 5.

Shukla, D.D. and Gough, K. (1980). *CMI/AAB Descr. Pl. Viruses* No. 222, 4 pp.

Shukla, D.D., Koenig, R., Gough, K.H., Huth, H. and Lesemann, D.-E. (1980). *Phytopathology* **70**: 382.

Shukla, D.D., Proll, E., Schmelzer, K. and Schmidt, H.B. (1973). *Acta phytopath. Acad. Sci. Hung.* **8**: 91.

Srifah, P., Keese, P., Shukla, D.D. and Gibbs, A.J. (1990). *J. gen. Virol.* **71**: 3085.

## Eucharis mottle (?) *nepovirus*

**Data collated by A.A. Brunt, 1987.**

First reported in *Eucharis candida*; from the U.S.A. in plants imported from Peru; by Kahn *et al.* (1962).

### Natural host range and symptoms

Symptoms persist.

*Eucharis candida* — leaf mottling.

### Transmission

Virus transmitted by mechanical inoculation; transmitted by seed.

### Geographical distribution

Spreads in Peru.

### Experimental host range

Many (>9) families susceptible.

Diagnostically susceptible

*Celosia cristata* — chlorotic local lesions, then systemic vein-banding and malformation.

*Chenopodium amaranticolor* — local and systemic ringspots.

*Cucumis sativus* — chlorotic and necrotic local lesions, then systemic mottling.

*Datura stramonium* — local green rings, becoming systemic.

*Gomphrena globosa* — systemic vein-banding and rings.

*Nicotiana tabacum* — mostly symptomless.

*Phaseolus vulgaris* — local lesions then systemic necrosis.
*Tetragonia expansa* — white necrotic rings.
*Vigna unguiculata* — necrotic local lesions, then systemic necrosis.

Maintenance and propagation hosts

*Nicotiana tabacum, Gomphrena globosa.*

Assay hosts (**L**ocal lesions or **W**hole plants)

*Chenopodium amaranticolor* (L), *Vigna unguiculata* (L).

Dicotyledonous Hosts ***(Susceptible** or Insusceptible)*

***Celosia cristata. Chenopodium amaranticolor. Cucumis sativus. Cucurbita maxima. Datura stramonium. Glycine max. Gomphrena globosa. Nicotiana bigelovii. Nicotiana tabacum. Phaseolus vulgaris. Tetragonia expansa. Vigna unguiculata.***

Sources of host-range data

Kahn *et al.* (1962).

**Properties of particles in sap**

TIP: in sap of *Gomphrena globosa* or tobacco 65–75 °C. LIV: 6–10 days. DEP: $\log_{10}$ minus 2–3.

**Purification**

Method: Kahn *et al.* (1962).

**Particle morphology**

Virions isometric; not enveloped; 26 nm in diameter; angular in profile.

**Taxonomy and Relationships**

Possible *Nepovirus: Comoviridae.*

Virus(es) with serologically related virions: tobacco ringspot virus (several isolates). Virus(es) with serologically unrelated virions: arabis mosaic, the beet ringspot strain of tomato black ring and raspberry ringspot viruses.

Kahn *et al.* (1962) considered eucharis mottle virus to be a strain of tobacco ringspot virus. However, as its virions are serologically distinct and the two viruses fail to cross-protect and are unable to form pseudo-recombinants, it is considered to be a separate virus species (Gooding, 1970; Salazar and Harrison, 1978).

**References**

Gooding, G.V. (1970). *Phytopathology* **60**: 708.
Kahn, R.F., Scott, H.A. and Monroe, R.L. (1962). *Phytopathology* **52**: 1211.
Salazar, L.F. and Harrison, B.D. (1978). *Ann. appl. Biol.* **90**: 387.

# Euonymus fasciation (?) rhabdovirus

**Data collated by A.A. Brunt, 1991.**

*Standard acronym:* EFV.

First reported in *Euonymus japonica*; from France; by Codaccioni and Cossard (1975).

**Natural host range and symptoms**

*Euonymus japonica, E. japonica* var. *microphylla* — fasciated stems.

**Geographical distribution**

Found, but with no evidence of spread, in France.

**Particle morphology**

Virions rhabdo– or bullet–shaped.

**Taxonomy and Relationships**

Possible unassigned rhabdovirus.

**Comments**

It is not known if this virus is related to a rhabdovirus that occurs in *Euonymus japonica* with vein-yellowing and chlorotic leaf spotting, but no fasciation, in the formar Yugoslavia (Pleše and Erič, 1980).

**References**

Codaccioni, M. and Cossard, C. (1975). *C.r. hebd. Seanc. Acad. Sci.* **280**: 1497.
Codaccioni, M. and Cossard, C. (1977). *C.r. hebd. Seanc. Acad. Sci.* **284**: 701.
Pleše, N. and Erič, Z. (1980). *Acta Phytopath. Acad. Sci. Hung.* **15**: 291.

# Euonymus (?) rhabdovirus

**Data collated by A.A. Brunt, 1991.**

First reported in *Euonymus japonica*; from the formar Yugoslavia; by Pleše and Erič (1980).

**Natural host range and symptoms**

Symptoms persist.

*Euonymus japonica, E. japonica* var. *microphylla* — vein-yellowing and chlorotic spotting.

**Geographical distribution**

Found, but with no evidence of spread, in Yugoslavia.

**Particle morphology**

Virions rhabdo– or bullet–shaped; with a clear modal length; of 300 nm (in sections); 70 nm wide (in sections).

**Cytopathology**

Virions found in cytoplasm and in the perinuclear space.

**Taxonomy and Relationships**

Possible unassigned rhabdovirus.

**Comments**

It is not known whether a rhabdovirus from fasciated *Euonymus japonica* in France (Codaccioni and Cossard, 1975) is related to this virus.

**References**

Codaccioni, M. and Cossard, C. (1975). *C.r. hebd. Acad. Séanc., Paris* **280**: 1497.
Pleše, N. and Erič, Z. (1980). *Acta Phytopath. Acad. Sci. Hung.* **15**: 291.

# Euphorbia mosaic bigeminivirus

**Data collated by A.A. Brunt, 1990.**

*Standard acronym:* EuMV.

First reported in *Euphorbia prunifolia*; from Sao Paulo, Brazil; by Costa and Bennett (1950).

**Natural host range and symptoms**

Symptoms persist.
*Euphorbia prunifolia* — golden mosaic and leaf malformation.
*Euphorbia heterophylla* — mosaic.

**Transmission**

Transmitted by a vector; an insect; *Bemisia tabaci* (nymphs and adults); Aleyrodidae. Transmitted in a persistent manner. Virus transmitted by mechanical inoculation; transmitted by grafting; not transmitted by contact between plants; not transmitted by seed.

**Geographical distribution**

Spreads in Brazil, Puerto Rico, and the USA (Florida), Venezuela.

**Experimental host range**

Few (<3) families susceptible.

Diagnostically susceptible

*Euphorbia prunifolia* — chlorosis.
*Datura stramonium* — chlorotic local lesions and systemic chlorosis.
*Datura metel* — local and systemic chlorotic spotting.
*Nicotiana benthamiana, N. glutinosa* — systemic crinkling.

Diagnostically insusceptible

*Chenopodium quinoa, Glycine max, Gomphrena globosa, Nicandra physaloides, Nicotiana tabacum.*

Maintenance and propagation hosts

*Datura stramonium. Nicotiana benthamiana, N. glutinosa.*

Assay hosts (**L**ocal lesions or **W**hole plants)

*Datura stramonium* (W), *Nicotiana benthamiana* (W), *N. glutinosa* (W).

Dicotyledonous Hosts *(**Susceptible** or Insusceptible)*

*Chenopodium amaranticolor. Chenopodium quinoa.* ***Datura metel. Datura stramonium. Fagopyrum esculentum.*** *Glycine max. Gomphrena globosa. Lycopersicon esculentum. Lycopersicon pimpinellifolium.* ***Nicandra physaloides. Nicotiana benthamiana.*** *Nicotiana debneyi.* ***Nicotiana glutinosa.*** *Nicotiana megalosiphon. Nicotiana tabacum. Phaseolus lunatus. Phaseolus vulgaris.*

Sources of host-range data

Costa and Bennett (1950); Jaramillo and Lastra (1986).

**Properties of particles in sap**

TIP: in *Nicotiana benthamiana* sap 44–46 °C. LIV: 6–7 days. DEP: $\log_{10}$ minus 1. Leaf sap contains few virions.

**Purification**

Method: Jaramillo and Lastra (1986).

**Particle morphology**

Virions geminate; not enveloped; 18 nm in diameter; dimers 32 nm in length.

**Biochemical properties**

*Genome* consists of DNA; single–stranded; of two parts; largest (or only) genome part the larger 2.87 kb; the 2nd largest 2.7 kb. Genomic nucleic acid isolated by Jaramillo and Lastra (1986).

*Virion protein(s)* one; $M_r$ 32000; coat protein. Method of preparation: Jaramillo and Lastra (1986).

**Cytopathology**

Virions found in leaves, mesophyll and epidermis; in nuclei. Inclusions present in infected cells; are unusual in shape; fibrillar bodies in nuclei; they do not contain virions. Other cellular changes: unusual for a geminivirus, cytoplasmic inclusions.

**Taxonomy and Relationships**

Bigeminivirus: *Geminiviridae.*

Virus(es) with serologically unrelated virions: bean golden mosaic, African cassava mosaic, tomato golden mosaic and lima bean golden mosaic geminiviruses.

The Brazilian isolate is more stable than the Venezuelan isolate.

**References**


Bird, J. and Maramorosch, K. (1978). *Adv. Virus Res.* **22**: 55.

Bird, J., Sanchez, J., Rodriguez, R.L. and Julia, F.J. (1975). In: *Tropical Diseases of Legumes*, p. 325; eds J. Bird and K. Maramorosch. Academic Press, New York.

Costa, A.S. (1965). *FAO Pl. Prot. Bull.* **13**: 1.

Costa, A.S. (1976). *Ann. Rev. Phytopath.* **16**: 429.

Costa, A.S. and Bennett, C.W. (1950). *Phytopathology* **40**: 266.

Costa, A.S. and Carvalho, A.M.B. (1960). *Phytopath. Z.* **38**: 129.

Debrot, E.A. and Centeno, F. (1983). *Agron. trop.* **33**:.

Jaramillo, S. and Lastra, R. (1986). *J. Phytopath.* **115**: 193.

Kim, K.S. and Flores, E.M. (1979). *Phytopathology* **69**: 980.

Kim, K.S. and Fulton, R.W. (1984). *Phytopathology* **74**: 236.

Matys, J.C., Silva, D.M., Oliveira, A.R. and Costa, A.S. (1976). *Fitopatol. (Lima)* **11**: 22.


# Euphorbia ringspot *potyvirus*

**Data collated by D.-E. Lesemann, 1991.**

*Standard acronym:* EuRV.

First reported in *Euphorbia milii x lophogona*; from western Germany; by Bode and Lesemann (1976).

**Natural host range and symptoms**

Symptoms persist. Symptoms chlorotic spots, leaf and flower malformation.

*Euphorbia milii x lophogona* — ringshaped chlorotic spots, leaf and flower deformations.

*Euphorbia loricata* — ringspots, malformation.

**Transmission**

Transmitted by a vector; an insect; *Myzus persicae*; Aphididae. Transmitted in a non–persistent manner. Virus transmitted by mechanical inoculation; transmitted by grafting.

**Geographical distribution**

Found, but with no evidence of spread, in Germany and Denmark.

**Experimental host range**

Few (<3) families susceptible.

Diagnostically susceptible

*Euphorbia milii x lophogona* — chlorotic spots, leaf and flower malformation.
*Chenopodium quinoa* Danish isolate — local white rings, not systemic.
*Euphorbia milii* — chlorotic ringspots.

Diagnostically insusceptible

*Datura stramonium, Gomphrena globosa, Nicotiana clevelandii, N. glutinosa, Tetragonia expansa.*

Maintenance and propagation hosts

*Euphorbia milii, E. milii x lophogona.*

**Properties of particles in sap**

Leaf sap contains few virions.

**Particle morphology**

Virions filamentous; not enveloped; usually flexuous; with a clear modal length; of 750 nm. Axial canal obscure. Basic helix obscure.

**Cytopathology**

Virions found in leaves and epidermis; in cytoplasm. Inclusions present in infected cells; are pinwheels.

**Taxonomy and Relationships**

*Potyvirus: Potyviridae.*

**References**

Bode, O. and Lesemann, D.-E. (1976). *Acta Hort.* **59**: 161.
Paludan, N. (1981). *Rep. National Inst. Pl. Path. Denmark* **528**: 74.

# Faba bean necrotic yellows nanavirus

**Data collated by K.M. Makkouk, 1994.**

First reported in *Vicia faba*; from Lattakia, Syria; by Makkouk *et al.* (1991).

**Natural host range and symptoms**

Symptoms persist. Symptoms yellowing and stunting.
*Cicer arietinum, Lens culinaris* — yellowing and stunting.
*Phaseolus vulgaris* — stunting, yellowing and rosetting.
*Vicia faba* — stunting, yellowing and leaf necrosis.

**Transmission**

Transmitted by a vector; an insect; *Acyrthosiphon pisum, Aphis craccivora, Aphis fabae*; Aphididae. Not transmitted by *Myzus persicae*. Transmitted in a persistent manner.

Virus retained when the vector moults; not transmitted by mechanical inoculation; not transmitted by contact between plants; not transmitted by seed; not transmitted by pollen.

**Geographical distribution**

Spreads in Egypt, Ethiopia, Jordan, Lebanon, Morocco, Syria, and Turkey. Found, but with no evidence of spread, in Morocco.

**Experimental host range**

Few (<3) families susceptible. Experimentally infected plants mostly show stunting and yellowing, few pods, rolling of youngest leaves.

Diagnostically susceptible

*Cicer arietinum, Lens culinaris, Phaseolus vulgaris, Pisum sativum, Vicia faba, V. sativa, V. palaestina* — leaf yellowing and plant stunting.

*Medicago hispida, Trifolium incarnatum* — leaf yellowing and reddening, plant stunting.

Diagnostically insusceptible

*Beta vulgaris, Brassica napus, Capsella bursa-pastoris, Cucumis sativus, Glycine max, Spinacia oleracea.*

Maintenance and propagation hosts

*Lens culinaris, Pisum sativum, Vicia faba.*

Assay hosts (**L**ocal lesions or **W**hole plants)

*Pisum sativum* (W), *Vicia sativum* (W).

Dicotyledonous Hosts *(**Susceptible** or Insusceptible)*

*Antirrhinum majus. Beta vulgaris. Brassica campestris* ssp. *napus. Capsella bursa–pastoris.* ***Cicer arietinum***. *Crambe abyssinica. Cucumis sativus. Cucurbita pepo. Glycine max.* ***Lathyrus odoratus***. ***Lens esculenta***. *Lycopersicon esculentum.* ***Medicago hispida***. *Medicago sativa. Melilotus alba.* ***Melilotus officinalis***. *Ocimum basilicum.* ***Phaseolus vulgaris***. ***Pisum sativum***. *Spinacia oleracea. Trifolium hybridum.* ***Trifolium incarnatum***. *Trifolium pratense.* ***Trifolium subterraneum***. ***Vicia faba***. ***Vicia sativa***. *Vicia villosa. Vigna unguiculata.*

Sources of host-range data

Katul *et al.*, 1993.

**Properties of particles in sap**

Leaf sap contains few virions. Electron microscopy: ISEM.

**Purification**

Method: Katul *et al.*, 1993.

**Particle morphology**

Virions isometric; not enveloped; 18 nm in diameter; angular in profile; without a conspicuous capsomere arrangement.

**Physical properties**

One sedimenting component in purified preparations. $A_{260}/A_{280}$ ratio 1.4.

**Biochemical properties**

*Genome* consists of DNA; single–stranded; circular; of 7 parts; largest (or only) genome part each *c.* 0.97–1.25 kb. Genomic nucleic acid isolated by Katul *et al.*, 1995.

Poly A region present; in the gene that probably encodes the replicase. Nucleotide sequence references: Katul *et al.*, 1993.

*Virion protein(s)* one; $M_r$ 22000; coat. Method of preparation: Katul *et al.*, 1993.

**Cytopathology**

Virions found in leaves, roots and phloem.

**Taxonomy and Relationships**

Nanavirus.

Best tests for diagnosis: Distinguished most quickly from bean leafroll luteovirus by the TBIA test.

**References**


Franz, A., Makkouk, K.M. and Vetten, H.J. (1995). *J. Phytopath.* in press.

Katul, L., Vetten, H.J., Maiss, E., Makkouk, K.M., Lesemann, D.E. and Caspar, R. (1993). *Ann. appl. Biol.* **123**: 629.

Katul, L., Maiss, E. and Vetten, H.J. (1995). *J. gen. Virol.* **76**: 475.

Makkouk, K.M., Kumari, S.G., Katul, L. and Caspar, R. (1991). *Abstr. 4th Arab Cong. Pl. Prot., Cairo, Egypt.* 1991, p.204.


# Fern (?) potyvirus

**Data collated by A.A. Brunt, 1991.**

First reported in *Dryopteris felix-mas* and *Polypodium vulgare*; from Germany; by Nienhaus *et al.* (1974).

**Natural host range and symptoms**

*Dryopteris felix-mas, Polypodium vulgare.*

**Transmission**

Transmitted by a vector; an insect; *Myzus persicae*; Aphididae. Transmitted in a non–persistent manner.

**Geographical distribution**

Found, but with no evidence of spread, in Germany.

**Particle morphology**

Virions filamentous; of 785 nm; 12–14 nm wide.

**Taxonomy and Relationships**

Unassigned potyvirus.

Virus(es) with serologically unrelated virions: potato Y potyvirus.

**References**


Nienhaus, F., Mack, C. and Schinzer, U. (1974). *Z. PflKrankh. PflPath. PflSchutz.* **81**: 533.

# Fescue (?) *alphacryptovirus*

**Data collated by A.A. Brunt, 1991.**

*Standard acronym:* FCV.

First reported in *Festuca pratensis*; from Italy; by Boccardo *et al.* (1983).

**Natural host range and symptoms**

*Festuca pratensis* — symptomless.

**Transmission**

Virus not transmitted by mechanical inoculation; not transmitted by grafting; transmitted by seed.

**Geographical distribution**

Found, but with no evidence of spread, in Italy.

**Particle morphology**

Virions isometric; not enveloped; 30 nm in diameter; rounded in profile; with a conspicuous capsomere arrangement.

**Biochemical properties**

*Genome* consists of RNA; double–stranded.

**Taxonomy and Relationships**

Possible *Alphacryptovirus: Partitiviridae.*

**Comments**

The isolate orginally studied is possibly a complex of two or more viruses.

**References**

Boccardo, G., Lisa, V. and Milne, R.G. (1983). In: *Double-Stranded RNA Viruses*, p. 425; eds R.W. Compans and D.H.L. Bishop. Elsevier, New York.

# Festuca leaf streak *cytorhabdovirus*

**Data collated by T. Lundsgaard, 1991.**

*Standard acronym:* FLSV.

First reported in *Festuca gigantea*; from Denmark; by Lundsgaard and Albrechtsen (1976).

**Natural host range and symptoms**

Symptoms persist.

*Festuca gigantea* — chlorotic streaks of some veins of mature leaves.

**Transmission**

Not transmitted by *Javesella pellucida, Laodelphax striatellus, Psammotettix alienus.* Virus not transmitted by mechanical inoculation; not transmitted by seed.

**Geographical distribution**

Found, but with no evidence of spread, in Denmark and Germany.

Diagnostically susceptible

*Vigna unguiculata* syn. *sinensis* (*N.B.* protoplast infection).

**Properties of particles in sap**

Leaf sap contains few virions. Electron microscopy: virions are disrupted when extracted in 2% PTA. Virions are preserved by 2% AM extraction or aldehyde fixation prior to PTA staining.

**Purification**

Method: Lundsgaard (1987) .

**Particle morphology**

Virions rhabdo– or bullet–shaped; enveloped; usually straight; with a clear modal length; of 286 nm; 61 nm wide. Axial canal obvious; 15 nm in diameter. Basic helix obvious; pitch of basic helix 4.7 nm.

**Physical properties**

One sedimenting component in purified preparations; sedimentation coefficient 704 *S*. Density 1.194 g $cm^{-3}$ in sucrose.

**Biochemical properties**

*Genome* probably consists of RNA; single–stranded; unipartite; largest (or only) genome part 14 kb. Genomic nucleic acid isolated by Lundsgaard (1987).

*Non–genomic nucleic acid* found in the virions; is probably *r*RNA.

*Virion protein(s)* three; $M_r$ of the largest 58000; G protein. $M_r$ of 2nd largest 49000; N protein. $M_r$ of 3rd largest 20000; M protein. Method of preparation: Lundsgaard (1987). Virion proteins glycosylated; are it is G protein.

**Cytopathology**

Virions found in leaves roots, mesophyll and epidermis; in cytoplasm. Inclusions present in infected cells; are membranous bodies and viroplasms; they contain virions.

**Taxonomy and Relationships**

*Cytorhabdovirus: Rhabdoviridae.*

Virus(es) with serologically unrelated virions: Northern cereal mosaic, barley yellow striate mosaic, cereal chlorotic mottle, maize sterile stunt, digitaria striate and Moroccan wheat viruses (Lundsgaard, 1984; Greber, 1990).

Best tests for diagnosis: serology.

**References**

Greber, R.S. (1990). *Ann. appl. Biol.* **116**: 259.

Lundsgaard, T. (1984). *Intervirology* **22**: 50.

Lundsgaard, T. (1987). *J. gen. Virol.* **68**: 931.

Lundsgaard, T. (1991). *J. Phytopath.* (in press).

Lundsgaard, T. and Albrechtsen, S.E. (1976). *Phytopath. Z.* **87**: 12.

Lundsgaard, T. and Albrechtsen, S.E. (1979). *Phytopath. Z.* **94**: 112.

Van Beek, N.A.M., Lohuis, D., Dijkstra, J. and Peters, D. (1985). *J. gen. Virol.* **66**: 2485.

# Festuca necrosis (?) *closterovirus*

**Data collated by A.A. Brunt, 1991.**

*Standard acronym:* FNV.

First reported in *Festuca pratensis, Lolium multiflorum*; from Germany; by Schmidt *et al.* (1963).

**Natural host range and symptoms**

Symptoms persist.
*Festuca pratensis, Lolium multiflorum* — death of stems and plants.

**Transmission**

Transmitted by a vector; an insect; *Rhopalosiphum padi*; Aphididae.

**Geographical distribution**

Spreads in Germany.

**Experimental host range**

Few (<3) families susceptible.

Diagnostically susceptible

*Festuca pratensis, Lolium multiflorum* — leaf and stem necrosis.

Maintenance and propagation hosts

*Festuca pratensis, Lolium multiflorum.*

Assay hosts (**L**ocal lesions or **W**hole plants)

*Festuca pratensis* (W), *Lolium multiflorum* (W).

Monocotyledonous Hosts *(**Susceptible** or Insusceptible)*

***Festuca pratensis*. *Lolium multiflorum*.**

**Properties of particles in sap**

Leaf sap contains few virions.

**Particle morphology**

Virions filamentous; not enveloped; usually flexuous; with a clear modal length; of 1725 nm; 18 nm wide. Axial canal obscure. Basic helix obscure.

**Taxonomy and Relationships**

Possible *Closterovirus*.

**References**

Schmidt, H.B., Richter, J., Hertzsch, W. and Klinkowski, M. (1963). *Phytopath. Z.* **47**: 66.

# Fig (?) potyvirus

**Data collated by A.A. Brunt, 1991.**

*Standard acronym:* FicCV.

First reported in *Ficus carica*; from Yugoslavia; by Grbelja (1983).

**Natural host range and symptoms**

Symptoms persist.
*Ficus carica* — leaf chlorosis and malformation.

**Transmission**

Transmitted by a vector; an insect; *Myzus persicae*; Aphididae. Virus transmitted by mechanical inoculation.

**Geographical distribution**

Spreads in the former Yugoslavia.

**Experimental host range**

Few (<3) families susceptible.

Diagnostically susceptible

*Ficus carica, Nicotiana megalosiphon, N. sylvestris* — chlorosis and systemic leaf malformation.

Maintenance and propagation hosts

*Nicotiana megalosiphon, N. sylvestris.*

Assay hosts (**L**ocal lesions or **W**hole plants)

*Nicotiana megalosiphon* (W), *N. sylvestris* (W).

Dicotyledonous Hosts *(**Susceptible** or Insusceptible)*

***Nicotiana megalosiphon**. **Nicotiana sylvestris**.*

**Particle morphology**

Virions filamentous; not enveloped; usually flexuous; with a clear modal length; of 750–800 nm.

**Cytopathology**

Inclusions present in infected cells; are pinwheels; they do not contain virions.

**Taxonomy and Relationships**

Unassigned potyvirus.

**References**

Grbelja, E. (1983). *Acta bot. Croat.* **42**: 11.

# Fig S *carlavirus*

**Data collated by A.A. Brunt, 1991.**

First reported in *Ficus carica*; from Japan; by Y. Doi in Hull *et al.*, 1989.

**Taxonomy and Relationships**

*Carlavirus.*

**References**

Hull, R., Brown, F. and Payne, C. (1989). *Directory and Dictionary of Animal, Bacterial and Plant Viruses*, p. 76. MacMillan Reference Books, London.

# Figwort mosaic *caulimovirus*

**Data collated by A.A. Brunt, 1991.**

*Synonyms:* figwort mottle virus. *Standard acronym:* FMV.

First reported in *Scrophularia californica*; from the U.S.A.; by Handley *et al.* (1982).

**Natural host range and symptoms**

Symptoms persist.
*Scrophularia californica*—leaf chlorosis.

**Transmission**

Transmitted by a vector; an insect; Aphididae.

**Geographical distribution**

Spreads in the USA.

**Particle morphology**

Virions isometric; not enveloped; 50 nm in diameter; rounded in profile.

**Biochemical properties**

*Genome* consists of DNA; double–stranded; circular; unipartite; largest (or only) genome part 7.743 kb (pairs).

*Sequence database accession code(s):*

M59930 Em(40)_vi:FMVGVI Gb(84)_vi:FMVGVI Figwort mosaic virus gene VI, complete cds. 5/92 1,617bp.

X06166 Em(40)_vi:CAFMVXX Gb(84)_vi:CAFMVXX Figwort mosaic virus (FMV) genome. 9/93 7,743bp.

X16673 Em(40)_vi:CAFMV34P Gb(84)_vi:CAFMV34P Figwort mosaic virus DNA for 34S promoter region. 6/90 554bp.

X52754 Em(40)_vi:FMVVI Gb(84)_vi:FMVVI Figwort mosaic virus gene VI for viral inclusion body protein. 4/92 26bp. 4 sequences.

**Cytopathology**

Virions found in leaves and mesophyll; in cytoplasm. Inclusions present in infected cells; are viroplasms; they contain virions.

**Taxonomy and Relationships**

*Caulimovirus.*

The viral DNA has little or no sequence homology with that of cauliflower mosaic virus.

**References**

Handley, M.K., Duffus, J.E. and Shepherd, R.J. (1982). *Phytopathology* **72**: 952.

Hull, R. and Donson, J. (1982). *J. gen. Virol.* **60**: 125.

# Filaree red leaf (?) *luteovirus*

**Data collated by A.A. Brunt, 1991.**

*Standard acronym:* FLRV.

First reported in *Erodium moschatum*; from California, U.S.A.; by Frazier (1951).

**Natural host range and symptoms**

Symptoms persist.

*Erodium moschatum, E. circutarium, E. botrys* — vein clearing, immature leaves cupped and chlorotic; older leaves red and brittle. Plants stunted, no flowers.

**Transmission**

Transmitted by a vector; an insect; *Macrosiphum gei, Acyrthosiphon pelargonii zerozalphum, Acyrthosiphon (Aulocorthum) solani*; Aphididae. Not transmitted by *Myzus persicae, M. ornatus, Chaetosiphon fragaraefolii*. Transmitted in a persistent manner. Virus not transmitted by mechanical inoculation.

**Geographical distribution**

Spreads in the USA (California).

**Experimental host range**

Few (<3) families susceptible.

Diagnostically susceptible

*Erodium moschatum, E. circutarium, E. botrys* — younger leaves chlorotic and older leaves reddened.

Maintenance and propagation hosts

*Erodium moschatum.*

Assay hosts (**L**ocal lesions or **W**hole plants)

*Erodium moschatum* (W).

Sources of host-range data

Frazier (1951).

**Particle morphology**

Virions isometric; 28 nm in diameter; rounded in profile.

**Taxonomy and Relationships**

Possible *Luteovirus.*

**References**

Frazier, N.W. (1951). *Phytopathology* **41**: 221.

Waterhouse, P.M., Gildow, F.E. and Johnstone, G.R. (1988). *AAB Descr. Pl. Viruses* No. 339, 9 pp.

# Finger millet mosaic (?) *nucleorhabdovirus*

**Data collated by A.A. Brunt, 1987.**

*Synonyms:* eleusine mosaic virus, millet finger mosaic virus. *Standard acronym:* FMMV.

First reported in *Eleusine coracana*; from India; by Rao *et al.* (1965).

**Natural host range and symptoms**

*Eleusine coracana* — streaking, mosaic, mottling and general chlorosis.

**Transmission**

Transmitted by a vector; an insect; *Sogatella longifurcifera*; Delphacidae. Virus not transmitted by mechanical inoculation.

**Ecology and control**

Studies reported by Govindu *et al.* (1976).

**Geographical distribution**

Spreads in India (the south).

**Experimental host range**

Few (<3) families susceptible.

Diagnostically susceptible

*Eleusine caracana* — chlorotic streaks then mottling and general leaf chlorosis.

Maintenance and propagation hosts

*Eleusine caracana.*

Assay hosts (**L**ocal lesions or **W**hole plants)

*Eleusine caracana* (W), for insect transmission tests.

**Particle morphology**

Virions rhabdo– or bullet–shaped; with a clear modal length; of 285 nm (in sectioned cells); 80 nm wide.

**Cytopathology**

Virions found in all parts of the host plant; in the perinuclear space.

**Taxonomy and Relationships**

Possible *Nucleorhabdovirus: Rhabdoviridae.*

**References**

Govindu, H.C., Yaraguntaiah, R.C. and Keshavamurthy, K. (1976). *Pesticides* **10**: 21.

Maramorosch, K., Govindu, H.C. and Kondo, F. (1977). *Pl. Dis. Reptr* **61**: 1029.

Rao, D.G., Varma, P.M. and Capoor, S.P. (1965). *Indian Phytopath.* **18**: 139.

# Foxtail mosaic *potexvirus*

**Data collated by A.A. Brunt, 1991.**

*Standard acronym:* FoMV.

First reported in *Setaria italica* and *S. viridis*; from the U.S.A.; by Paulsen and Sill (1969).

**Natural host range and symptoms**

Symptoms persist.
*Setaria italica, S. viridis* — leaf mosaic.

**Transmission**

Virus not transmitted by mechanical inoculation; transmitted by seed.

**Geographical distribution**

Spreads in the USA.

**Experimental host range**

Many (>9) families susceptible.

Diagnostically susceptible

*Gomphrena globosa, Pisum sativum*—necrotic local lesions; not systemic.
*Chenopodium amaranticolor, Nicotiana clevelandii, Tetragonia expansa* — necrotic local lesions; systemic mosaic.
*Avena sativa, Hordeum vulgare, Triticum aestivum* — systemic mosaic.

Diagnostically insusceptible

*Arrhenatherum elatius, Lolium perenne, Sorghum halepense.*

Maintenance and propagation hosts

*Avena sativa, Hordeum vulgare, Triticum aestivum.*

Assay hosts (**L**ocal lesions or **W**hole plants)

*Chenopodium amaranticolor* (L).

Dicotyledonous Hosts (***Susceptible*** *or Insusceptible)*

***Beta vulgaris. Brassica juncea. Catharanthus roseus. Chenopodium album. Chenopodium amaranticolor. Chenopodium ambrosioides. Chenopodium capitatum. Chenopodium foetidum. Chenopodium hybridum. Chenopodium murale. Chenopodium quinoa. Crotalaria spectabilis. Cucumis sativus. Cyamopsis***

tetragonoloba. ***Glycine max. Gomphrena globosa. Lycopersicon esculentum. Nicandra physaloides. Nicotiana clevelandii. Nicotiana megalosiphon. Nicotiana rustica. Nicotiana tabacum. Physalis floridana. Spinacia oleracea. Tetragonia expansa. Torenia fournieri. Vigna unguiculata. Zinnia elegans.***

Monocotyledonous Hosts *(**Susceptible** or Insusceptible)*

***Agropyron elongatum. Avena sativa. Bromus inermis. Cynodon dactylon. Hordeum vulgare. Lagurus ovatus. Oryza sativa. Panicum capillare. Panicum maximum. Panicum miliaceum. Phalaris arundinacea. Phleum pratense. Poa pratensis. Secale cereale. Setaria italica. Setaria viridis. Sorghum bicolor. Triticum aestivum. Triticum durum. Zea mays.***

Sources of host-range data

Paulsen and Niblett (1977).

**Properties of particles in sap**

TIP: 70 °C. LIV: 46 days (in sap), or 105 days (in buffered leaf extracts). DEP: $\log_{10}$ minus 7.

**Purification**

Method: Bancroft *et al.* (1979).

**Particle morphology**

Virions filamentous; not enveloped; usually flexuous; with a clear modal length; of 500 nm. Axial canal obscure. Pitch of basic helix 3.5 nm.

**Physical properties**

Two sedimenting components in purified preparations; sedimentation coefficient of the fastest 144 *S* (probably dimers); of the other(s) 122 *S* (monomers). $A_{260}/A_{280}$ ratio 1.2.

**Biochemical properties**

Virions contain 7 % nucleic acid; 93 % protein.

*Genome* consists of RNA; single–stranded; unipartite; largest (or only) genome part 6.8 kb.

*Sequence database accession code(s):*

M62730 Em(40)_vi:FXMCGA Gb(84)_vi:FXMCGA Foxtail mosaic virus, complete genome. 12/91 6,151bp. 1 sequence.

Amino acid composition: Short (1981). Virion proteins not glycosylated.

**Cytopathology**

Virions found in leaves and mesophyll; in cytoplasm.

**Taxonomy and Relationships**

*Potexvirus.*

Virus(es) with serologically related virions: narcissus mosaic and viola mottle viruses. Virus(es) with serologically unrelated virions: cactus X, clover yellow mosaic, papaya mosaic, potato X and white clover mosaic viruses.

**References**

Bancroft, J.E., Abouhaidar, M. and Erickson, J.W. (1979). *Virology* **98**: 121.
Paulsen, A.Q. and Niblett, C.L. (1977). *Phytopathology* **67**: 1346.
Paulsen, A.Q. and Sill, W.H. (1969). *Phytopathology* **59**: 1043.
Richardson, J.F., Tollin, P. and Bancroft, J.B. (1981). *Virology* **112**: 34.
Short, M.N. (1981). *Rep. John Innes Inst. 1980*, p. 126.

Short, M.N. (1983). *CMI/AAB Descr. Pl. Viruses* No. 264, 3 pp.

# Fragaria chiloensis (?) *ilarvirus*

**Data collated by J.D. Postman, 1994.**

First reported in *Fragaria chiloensis*; from the USDA National Germplasm Repository, Corvallis, Oregon in strawberry plants collected in Chile; by Spiegel *et al.* (1993).

**Natural host range and symptoms**

*Fragaria chiloensis* and *Fragaria* sp. — symptomless.

**Transmission**

Transmitted by means not involving a vector. Virus transmitted by mechanical inoculation; transmitted by grafting; not transmitted by contact between plants; transmitted by seed (30–50% in *Fragaria chiloensis*).

**Geographical distribution**

Spreads in Chile.

**Experimental host range**

Few (<3) families susceptible.

Diagnostically susceptible

*Chenopodium amaranticolor, C. quinoa, Cucumis sativus* — systemic mottle, chlorotic patches and stunting.

Diagnostically insusceptible

*Datura stramonium, Gomphrena stramonium, Nicotiana* (several species) and *Phaseolus vulgaris.*

Maintenance and propagation hosts

*Fragaria* spp.

Assay hosts (**L**ocal lesions or **W**hole plants)

*Chenopodium quinoa* (W).

Dicotyledonous Hosts *(**Susceptible** or Insusceptible)*

***Chenopodium amaranticolor***. ***Chenopodium quinoa***. ***Cucumis sativus***. *Datura stramonium. Gomphrena globosa. Nicotiana benthamiana. Nicotiana clevelandii. Nicotiana glutinosa. Nicotiana rustica. Nicotiana sylvestris. Nicotiana tabacum. Petunia hybrida. Phaseolus vulgaris.*

Sources of host-range data

Spiegel *et al.*, 1993.

**Particle morphology**

Virions isometric; 21.4 nm in diameter (18.0 - 24.3 nm); 54.5 nm in length (maximum).

**Biochemical properties**

*Genome* consists of RNA; single–stranded; of four parts; largest (or only) genome part 3.7 kb; the 2nd largest 2.65 kb; the 3rd largest 2.6 kb; the 4th largest 1.2 kb.

*Virion protein(s)* one; $M_r$ 28000; coat.

**Taxonomy and Relationships**

Possible *Ilarvirus: Bromoviridae.*

Virus(es) with serologically related virions: asparagus virus 3, lilac ring mottle. Virus(es) with serologically unrelated virions: apple mosaic, blueberry shock, citrus leaf rugose, elm mosaic, elm mottle, prune dwarf, prunus necrotic ringspot, spinach latent, tobacco stunt, Tulare apple mosaic.

Best tests for diagnosis: ELISA.

**References**

Spiegel, S., Martin, R.R., Leggett, F., ter Borg, M. and Postman, J. (1993). *Phytopathology* **83**: 991.

# Frangipani mosaic *tobamovirus*

**Data collated by A. Varma, 1984.**

*Synonyms:* temple tree mosaic virus, champa mosaic virus. *Standard acronym:* FrMV. *Strains:* Adel (Adelaide), Ald (Allahabad), Del (Delhi).

First reported in *Plumeria acutifolia, P. alba*; from India and Australia; by Francki *et al.* (1971); Varma and Gibbs (1978).

**Natural host range and symptoms**

Symptoms persist.

*Plumeria acutifolia* — mosaic, ringspots, vein banding, bronzing.

*P. alba* — ringspots, leaf malformation, necrosis.

**Transmission**

Transmitted by means not involving a vector. Virus transmitted by mechanical inoculation.

**Geographical distribution**

Spreads in Australia, India, and Kenya.

**Experimental host range**

Several (3–9) families susceptible.

Diagnostically susceptible

*Datura stramonium* — chlorotic or black necrotic local lesions; most strains not systemic.

*Nicotiana glutinosa* — chlorotic local lesions, not systemic.

*Nicotiana tabacum* cvs Samsun, Virginia Gold, White Burley — chlorotic or necrotic ringspots, then systemically when grown at 35°C.

*Nicotiana x edwardsonii* — faint chlorotic lesions which become necrotic when plants are grown at 35°C. Not systemic.

Diagnostically insusceptible

*Chenopodium amaranticolor, Cucumis sativus, Phaseolus vulgaris.*

Maintenance and propagation hosts

*Nicotiana glutinosa.*

Assay hosts (**L**ocal lesions or **W**hole plants)

*Datura stramonium* (L).

Dicotyledonous Hosts *(**Susceptible** or Insusceptible)*

*Chenopodium amaranticolor*. ***Chenopodium quinoa***. *Cucumis sativus*. *Cucurbita pepo*. ***Datura stramonium***. ***Nicotiana clevelandii***. ***Nicotiana glutinosa***. ***Nicotiana tabacum***. *Phaseolus vulgaris*.

Comments on host-range

FMV requires high temperatures for replication. More species become infected at temperatures above 25°C than below. Symptoms appear in 3–6 days at 35°C, but after 2 weeks or more at 22°C.

Sources of host-range data

Varma and Gibbs (1978).

**Properties of particles in sap**

TIP: 95 °C. LIV: 70 days (or more). DEP: $\log_{10}$ minus 5. Leaf sap contains many virions.

**Purification**

Method: Varma *et al.* (1970).

**Particle morphology**

Virions rod–shaped; not enveloped; usually straight; with a clear modal length; of 297 nm; 17 nm wide. Axial canal obvious.

**Physical properties**

One sedimenting component in purified preparations; sedimentation coefficient 188 *S*.

**Biochemical properties**

Virions contain 5 % nucleic acid; 95 % protein; 0 % lipid.

*Genome* consists of RNA; single–stranded; linear; unipartite.

*Virion protein(s)* one. Method of preparation: Francki *et al.* (1971). Amino acid composition: Francki *et al.* (1971).

**Replication**

Replication does not depend on a helper virus.

**Cytopathology**

Virions found in mesophyll; in cytoplasm. Inclusions present in infected cells; are crystals in the cytoplasm; they contain virions.

**Taxonomy and Relationships**

*Tobamovirus*.

Virus(es) with serologically related virions: cucumber green mottle mosaic and sunn-hemp mosaic viruses are closely related; tobacco mosaic virus-type strain, tobacco mild green mosaic (U2), tomato mosaic and ribgrass mosaic viruses are more distantly related (Francki *et al.*, 1971). Nucleic hybridization tests have confirmed the relatedness of tomato mosaic and tobacco mild green mosaic viruses (Palukaitis and Symons, 1980; Palukaitis *et al.*, 1981). Virus(es) with serologically unrelated virions: Sammons' opuntia virus.

**References**


Francki, R.I.B., Zaitlin, M. and Grivell, C.J. (1971). *Aust. J. biol. Sci.* **24**: 815.

Palukaitis, P. and Symons, R.H. (1980). *Virology* **107**: 354.

Varma, A. and Gibbs, A.J. (1978). *CMI/AAB Descr. Pl. Viruses* No. 196, 4 pp.

Palukaitis, P., Randles, J.W., Tian, Y.C., Kang, L.Y. and Tien, P. (1981). *Intervirology* **16**: 136.

Varma, A., Gibbs, A.J. and Woods, R.D. (1970). *J. gen. Virol.* **8**: 21.

# Freesia leaf necrosis varicosavirus

**Data collated by D. Peters, 1991; I. Bouwen, 1994; A.A. Brunt, 1995.**

First reported in *Freesia refracta*; from the Netherlands; by Van Dorst (1973).

**Natural host range and symptoms**

Symptoms persist. Symptoms Chlorotic flecks, spots and necrosis. More severe when plants are stressed by bright light or waterlogging.

*Freesia* spp. — chlorotic flecks and spots in leaves, which become necrotic. Flowers and corms unaffected.

**Transmission**

Transmitted by a vector; a fungus; *Olpidium brassicae*; Chytridiales. Virus transmitted by mechanical inoculation (with difficulty); not transmitted by seed.

**Ecology and control**

Studies reported by The virus occurs in soil and is spread by infected corms, hence it is controlled by soil treatment and corm choice (Hakkaart, 1971, 1975; Van Dorst, 1973, 1975, 1988).

**Geographical distribution**

Spreads in the Eurasian region; Germany, Ireland, Italy, the Netherlands, and the UK.

**Experimental host range**

Few (<3) families susceptible.

Diagnostically susceptible

*Chenopodium quinoa, C. amaranticolor* — pinpoint necrotic local lesions with chlorotic halo. Not systemic.

*Nicotiana hesperis* (Accession 67A) — chlorotic local, and sometimes systemic, spots, centres becoming necrotic and white.

*Nicotiana occidentalis* (Accession P1) — chlorotic spots and rings, sometimes systemic vein chlorosis.

Diagnostically insusceptible

*Echinops exaltatus, Nicotiana glutinosa, N. tabacum* cv. White Burley, *Phaseolus vulgaris* cv. Bataaf, *Vicia faba* cv. Compacta.

Maintenance and propagation hosts

*Freesia refracta.*

Assay hosts (**L**ocal lesions or **W**hole plants)

*Freesia refracta* (W), *Chenopodium quinoa* (L), *Nicotiana hesperis* (L), *N. occidentalis* (L).

Dicotyledonous Hosts *(**Susceptible** or Insusceptible)*

*Ageratum conyzoides. Amaranthus caudatus. Antirrhinum majus. Atriplex hortensis. Bellis perennis. Beta vulgaris. Capsicum annuum. Catharanthus roseus.* ***Chenopodium amaranticolor***. *Chenopodium murale.* ***Chenopodium quinoa***. *Cucumis sativus. Datura stramonium. Gomphrena globosa. Lactuca sativa. Lathyrus odoratus. Lycopersicon* esculentum. *Nicotiana clevelandii. Nicotiana debneyi. Nicotiana glutinosa.* ***Nicotiana occidentalis***. *Nicotiana rustica. Nicotiana tabacum. Papaver nudicaule. Petunia hybrida. Physalis floridana. Pisum sativum. Raphanus sativus. Salvia splendens. Sonchus oleraceus. Spinacia oleracea. Torenia fournieri. Verbesina encelioides. Vicia faba.*

**Taxonomy and Relationships**

Possible varicosavirus.

**Comments**

This virus, together with freesia mosaic *potyvirus*, causes freesia streak disease (Brunt, 1974; Van Dorst, 1973).

**References**


Bellardi, M.G. and Bertaccini, A. (1989). *Adv. Hort. Sci.* **3**: 29.
Bouwen, I. (1994). *Acta. Hort.* in press.
Brunt, A.A. (1970). *Rep. Glasshouse Crops Res. Inst.* 1969, p. 131.
Brunt, A.A. (1974). *Rep. Glasshouse Crops Res. Inst.* 1973, p. 117.
Caspar, R. and Brunt, A.A. (1971). *Nachbl. PflSchutz.*, Braunschweig **23**: 89.
Foxe, M.J. and Wilson, U.E. (1985). *Acta Hort.* **164**: 291.
Hakkaart, F.A. (1971). *Proc. 7th Conf. Czech. Pl. Virol.* 1971: 231.
Hakkaart, F.A. (1975). *Jaarvers. Inst. Pl. Onderz.* 1974, p. 70.
Van Dorst, H.J.M. (1973). *Neth. J. Pl. Path.* **:79** 130.
Van Dorst, H.J.M. (1975). *Neth. J. Pl. Path.* **81** 45.
Van Dorst, H.J.M. (1988). In: *Viruses with Fungal Vectors*, p. 315 (eds J.I. Cooper and M.J.C. Asher). Association of Applied Biologists, Wellesbourne, U.K.


# Freesia mosaic *potyvirus*

**Data collated by S. Phillips, 1985.**

*Synonyms:* freesia streak virus. *Standard acronym:* FreMV.

First reported in *Freesia refracta* Klatt; from Lisse, The Netherlands; by Van Koot *et al.* (1954).

**Natural host range and symptoms**

Symptoms persist.
*Freesia refracta* Klatt — mild chlorosis symptomless.

**Transmission**

Transmitted by a vector; an insect; *Macrosiphum euphorbiae, Myzus persicae*; Aphididae. Transmitted in a non–persistent manner. Virus transmitted by mechanical inoculation; not transmitted by contact between plants; not transmitted by seed; not transmitted by pollen.

**Geographical distribution**

Spreads in Australia, Ireland, the Netherlands, and the UK. Found, but with no evidence of spread, in New Zealand.

**Experimental host range**

Few (<3) families susceptible.

Diagnostically susceptible

None reported.

Diagnostically insusceptible

*Phaseolus vulgaris, Chenopodium quinoa, Trifolium incarnatum, Vicia faba.*

Maintenance and propagation hosts

*Freesia refracta.*

Assay hosts (**L**ocal lesions or **W**hole plants)

None reported.

Dicotyledonous Hosts *(**Susceptible** or Insusceptible)*

*Brassica campestris* ssp. *pekinensis. Brassica juncea. Chenopodium amaranticolor. Chenopodium murale. Chenopodium quinoa. Cucumis sativus. Datura stramonium. Gomphrena globosa. Nicotiana benthamiana. Nicotiana clevelandii. Nicotiana glutinosa. Nicotiana rustica. Nicotiana tabacum. Petunia hybrida. Phaseolus vulgaris. Pisum sativum. Silene armeria. Trifolium incarnatum. Verbesina encelioides. Vicia faba. Vigna unguiculata.*

Monocotyledonous Hosts *(**Susceptible** or Insusceptible)*

*Lilium formosanum.*

**Properties of particles in sap**

Leaf sap contains many virions.

**Purification**

Method: Van Koot *et al.* (1954).

**Particle morphology**

Virions filamentous; not enveloped; usually flexuous; with a clear modal length; of 825 nm; 12 nm wide. Axial canal obscure.

**Replication**

Replication does not depend on a helper virus.

**Cytopathology**

Virions found in all parts of the host plant; in cytoplasm. Inclusions absent from infected cells.

**Taxonomy and Relationships**

*Potyvirus: Potyviridae.*

**Comments**

This virus, together with freesia leaf necrosis virus is the cause of a disease originally named freesia streak (Brunt, 1974; van Dorsk, 1973).

**References**

Brunt, A.A. (1974). *Rep. Glasshouse Crops Res. Inst.* 1973, p. 117.

Foxe, M.J. and Wilson, U.E. (1985). *Acta Hort.* **164**: 291.

Van Dorst, H.J.M. (1973). *Neth. J. Pl. Path.* **79**: 130.

Van Koot, Y., van Slogteren, D.H.M., Cremer, M.C. and Camfferman, J. (1954). *Tijdschr. Plziekt.* **60**: 157.

# Fuchsia latent (?) *carlavirus*

**Data collated by R. Stace-Smith, 1991.**

*Synonyms:* fuchsia S virus. *Standard acronym:* FLV.

First reported in *Fuchsia speciosa* × *hybrida*; from British Columbia, Canada; by John *et al.* (1980).

**Natural host range and symptoms**

Symptoms none.
*Fuchsia* spp. — symptomless.

**Geographical distribution**

Found, but with no evidence of spread, in most *Fuchsia* cultivars.

**Experimental host range**

Few (<3) families susceptible.

Diagnostically susceptible

*Fuchsia* spp. — symptomless.

Diagnostically insusceptible

*Chenopodium quinoa, Gomphrena globosa, Cucumis sativus.*

Maintenance and propagation hosts

*Fuchsia* spp.

Dicotyledonous Hosts *(**Susceptible** or Insusceptible)*

*Chenopodium amaranticolor. Chenopodium quinoa. Cucumis sativus. Datura stramonium. Gomphrena globosa. Lycopersicon esculentum. Nicotiana clevelandii. Nicotiana debneyi. Nicotiana glutinosa. Vigna unguiculata.*

**Particle morphology**

Virions filamentous; not enveloped; usually flexuous; of 650 nm; 12 nm wide. Axial canal obvious. Basic helix obscure.

**Cytopathology**

Virions found in leaves and phloem parenchyma.

**Taxonomy and Relationships**

Possible *Carlavirus.*

**References**

John, L.J., Stace-Smith, R. and Kadota, D.Y. (1980). *Acta Hort.* **110**: 195.

# Furcraea necrotic streak (?) *dianthovirus*

**Data collated by A.A. Brunt, 1987. Revised by F.J. Morales and M. Castâno, 1992.**

*Synonyms:* cabuya necrotic streak mosaic virus. *Standard acronym:* FNSV.

First reported in *Furcraea* spp.; from Colombia; by Gálvez *et al.* (1976).

**Natural host range and symptoms**

Symptoms persist. Symptoms chlorotic and necrotic streaks.
*Furcraea* spp., *Yucca* spp. — chlorotic and then necrotic streaks.

**Transmission**

The virus may have a soil-inhabiting vector. Not transmitted by *Planococcus citri, Saissetia caffea.* Virus transmitted by mechanical inoculation; transmitted by grafting.

**Geographical distribution**

Spreads in Colombia.

**Experimental host range**

Few (<3) families susceptible.

Diagnostically susceptible

*Agave* spp. — reddish streaks, necrosis.
*Furcraea* spp. — chlorotic and necrotic streaking.

Diagnostically insusceptible

*Phaseolus vulgaris, Physalis floridana, Nicotiana tabacum* cv. White Burley, *N. glutinosa, Chenopodium amaranticolor.*

Maintenance and propagation hosts

*Agave* spp., *Furcraea macrophylla.*

Assay hosts (**L**ocal lesions or **W**hole plants)

*Furcraea* spp. (W).

Dicotyledonous Hosts *(**Susceptible** or Insusceptible)*

*Arachis hypogaea. Brassica campestris* ssp. *rapa. Chenopodium amaranticolor. Chenopodium murale. Cucumis sativus. Hyoscyamus niger. Lycopersicon esculentum. Nicotiana clevelandii. Nicotiana glutinosa. Nicotiana tabacum. Phaseolus vulgaris. Physalis floridana. Pisum sativum. Vicia faba. Vigna unguiculata.*

Monocotyledonous Hosts *(**Susceptible** or Insusceptible)*

*Allium cepa. Allium porrum. Avena sativa. Colocasia esculenta. Hordeum vulgare. Philodendron selloum. Triticum aestivum.*

Sources of host-range data

Dabek and Castano (1978).

**Properties of particles in sap**

Leaf sap contains many virions.

**Purification**

Method: Gálvez *et al.* (1977).

**Particle morphology**

Virions isometric; not enveloped; 34 nm in diameter; angular in profile; with a conspicuous capsomere arrangement.

**Physical properties**

One sedimenting component in purified preparations. $A_{260}/A_{280}$ ratio 1.64.

**Biochemical properties**

Virions contain 28 % nucleic acid; 72 % protein.

*Genome* consists of RNA; single–stranded; linear; of two parts; largest (or only) genome part 4 kb; the 2nd largest 1.6 kb. Genomic nucleic acid isolated by Morales *et al.* (1992).

*Virion protein(s)* one; $M_r$ 40000; coat protein.

**Cytopathology**

Virions found in leaves, roots and the mesophyll; in cytoplasm. Inclusions present in infected cells; are unusual in shape; electron dense tubular inclusions; they contain virions.

**Taxonomy and Relationships**

*Dianthovirus.*

Virus(es) with serologically related virions: red clover necrotic mosaic virus. Virus(es) with serologically unrelated virions: bean pod mottle, bean rugose mosaic, bean yellow stipple, carnation ringspot and southern bean mosaic viruses.

The genomic RNAs of this virus cross-hybridizes with cDNAs encoding the two genomic RNAs of red clover necrotic mosaic virus.

**References**

Dabek, A.J. and Castâno, J.J. (1978). *Phytopath. Z.* **92**: 57.

Castâno, J.J. (1990). *Fitopatol. Colombiana* **14**: 24.

Gálvez, G.E., Castâno, M., Vesga, B. and Kitajima, E.W. (1976). *Proc. Am. Phytopath. Soc.* **3**: 296.

Gálvez, G.E., Castâno, M., Vesga, B. and Kitajima, E.W. (1977). *Pl. Dis. Reptr* **61**: 1017.

Morales, F.J., Castâno, M., Calvert, L.A. and Arroyave, S. (1992). *J. Phytopath.* **134**: 247.

# Galinsoga mosaic *carmovirus*

**Data collated by A.J. Gibbs, 1986.**

*Standard acronym:* GaMV.

First reported in *Galinsoga parviflora*; from Queensland, Australia; by Behncken (1970).

**Natural host range and symptoms**

Symptoms persist.

*Galinsoga parviflora* — mosaic.

**Transmission**

Without a vector through soil. Virus transmitted by mechanical inoculation; not transmitted by contact between plants; not transmitted by seed.

**Geographical distribution**

Spreads in Australia.

**Experimental host range**

Several (3–9) families susceptible.

Diagnostically susceptible

*Chenopodium amaranticolor, C. quinoa* — small necrotic local lesions, not systemic.

*Galinsoga parviflora* — chlorotic or necrotic local lesions, systemic mosaic; leaf malformation.

*Phaseolus vulgaris* cvs Bountiful, Pinto — reddish-brown necrotic lesions; not systemic.

*Spinacia oleracea* — systemic chlorotic spots and vein necrosis, leaves produced later symptomless.

Diagnostically insusceptible

*Cucumis sativus, Datura stramonium, Lycopersicon esculentum, Trifolium repens, Vicia faba.*

Maintenance and propagation hosts

*Chenopodium amaranticolor, Phaseolus vulgaris* (several cultivars).

Assay hosts (**L**ocal lesions or **W**hole plants)

*Chenopodium amaranticolor* (L); *Chenopodium quinoa* (L); *Phaseolus vulgaris* (L).

Dicotyledonous Hosts (***Susceptible*** *or Insusceptible)*

***Calendula officinalis***. *Capsella bursa–pastoris*. ***Chenopodium album***. ***Chenopodium amaranticolor***. ***Chenopodium quinoa***. ***Cichorium endiva***. *Cucumis sativus*. *Datura stramonium*. ***Daucus carota***. ***Emilia sagittata***. ***Glycine max***. ***Gomphrena globosa***. ***Helianthus annuus***. ***Lablab purpureus***. ***Lactuca sativa***. *Lycopersicon esculentum*. ***Melilotus alba***. ***Nicotiana clevelandii***. ***Nicotiana glutinosa***. ***Nicotiana tabacum***. ***Petunia hybrida***. ***Phaseolus vulgaris***. ***Sonchus oleraceus***. ***Spinacia oleracea***. *Stellaria media*. ***Trifolium incarnatum***. *Trifolium repens*. *Vicia faba*. ***Vigna radiata***. ***Vigna unguiculata***. *Vigna unguiculata* ssp. *cylindrica*. ***Zinnia elegans***.

Additional hosts

***Emilia sonchifolia***.

Sources of host-range data

Behncken (1970).

**Properties of particles in sap**

TIP: 75–80 °C. LIV: 42 days (and longer). DEP: $\log_{10}$ minus 6. Infectivity of sap not changed by treatment with di–ethyl ether. Leaf sap contains many virions.

**Purification**

Method: Behncken (1970).

**Particle morphology**

Virions isometric; not enveloped; 34 nm in diameter; rounded in profile; without a conspicuous capsomere arrangement.

**Physical properties**

Two sedimenting components in purified preparations; sedimentation coefficient 118 *S*; of the other(s) 48 *S* (minor component of some preparations).

**Biochemical properties**

Virions contain 22 % nucleic acid; 78 % protein; 0 % lipid.

*Genome* consists of RNA; single–stranded; probably linear; unipartite; largest (or only) genome part 4.4 kb. Base composition 24.8 % G; 28.1 % A; 21.8 % C; 25.3 % U. Infectivity retained when deproteinised with phenol or detergent.

*Non–genomic nucleic acid* not found in the virions.

*Virion protein(s)* one; $M_r$ 36400; coat protein. Method of preparation: Skotnicki and Gibbs (1981). Amino acid composition: Skotnicki and Gibbs (1981).

**Replication**

Replication does not depend on a helper virus.

**Cytopathology**

Virions found in all parts of the host plant; in cytoplasm (but not in the nucleus or perinuclear space). Inclusions present in infected cells; are unusual in shape (multivesicular bodies derived from mitochondria); they do not contain virions. Other cellular changes: multivesicular bodies similar to those found in cells infected with tomato bushy stunt virus.

**Taxonomy and Relationships**

*Carmovirus: Tombusviridae.*

Virus(es) with serologically unrelated virions: tomato bushy stunt and glycine mottle viruses.

**References**

Behncken, G.M. (1970). *Aust. J. biol. Sci.* **23**: 497.
Behncken, G.M., Francki, R.I.B. and Gibbs, A.J. (1982). *CMI/AAB Descr. Pl. Viruses* No. 252, 4 pp.
Hatta, T. and Francki, R.I.B. (1981). *J. Ultrastruct. Res.* **74**: 116.
Shukla, D.D., Shanks, G.J., Teakle, D.S. and Behncken, G.M. (1979). *Aust. J. biol. Sci.* **32**: 267.
Skotnicki, A. and Gibbs, A.J. (1981). *Australas. Pl. Path.* **10**: 27.

# Garland chrysanthemum temperate (?) *alphacryptovirus*

**Data collated by A.A. Brunt, 1991.**

*Synonyms:* garland chrysanthemum temperate (?) cryptovirus. *Standard acronym:* GCTV.

First reported in *Chrysanthemum coronarium*; from Japan; by Natsuaki *et al.* (1983a).

**Natural host range and symptoms**

*Chrysanthemum coronarium* — symptomless.

**Transmission**

Virus not transmitted by mechanical inoculation; transmitted by seed.

**Geographical distribution**

Found, but with no evidence of spread, in Japan.

**Particle morphology**

Virions isometric; not enveloped; 30 nm in diameter.

**Taxonomy and Relationships**

Possible *Alphacryptovirus: Partitiviridae.*

**References**

Boccardo, G., Lisa, V., Luisoni, E. and Milne, R.G. (1987). *Adv. Virus Res.* **32**: 171.
Natsuaki, T., Yamashita, S., Doi, Y., Okuda, S. and Teranaka, M. (1983*a*). *Ann. Phytopath. Soc. Japan* **49**: 593.
Natsuaki, T., Yamashita, S., Doi, Y., Okuda, S. and Teranaka, M. (1983*b*). *Ann. Phytopath. Soc. Japan* **49**: 709.

# Garlic common latent (?) *carlavirus*

**Data collated by A.A. Brunt, 1995.**

*Synonyms:* garlic latent virus — France (Delacolle and Lot, 1981), garlic latent virus — Germany (Graichen and Leistner, 1987).

First reported in *Allium sativum* (garlic), *A. ampeloprasum* var. *holmense* (great headed garlic); *A. ampeloprasum* var. *porrum* (leek); from the Netherlands; by Van Dijk (1993).

**Natural host range and symptoms**

*Allium sativum, A. ampeloprasum* var. *holmense, A. ampeloprasum* var. *porrum* — no obvious symptoms.

**Transmission**

Virus transmitted by mechanical inoculation.

**Geographical distribution**

Probably distributed worldwide. Spreads in Argentina, the former Czechoslovakia, France, Germany, Israel, Japan, and the UK.

Diagnostically susceptible

*Celosia argentea* — chlorotic and/or necrotic local lesions; no systemic infection.
*Chenopodium amaranticolor* — faint green local rings; no systemic infection.
*Chenopodium murale, C. quinoa* — chlorotic etched local lesions; no systemic infection.
*Nicotiana occidentalis* — no local symptoms, systemic vein necrosis.

Diagnostically insusceptible

*Vicia faba.*

Maintenance and propagation hosts

*Allium ampeloprasum* var. *porrum, Nicotiana occidentalis.*

Assay hosts (**L**ocal lesions or **W**hole plants)

*Chenopodium murale* (L), *C. quinoa* (L), *Allium ampeloprasum* var. *porrum* (W).

Dicotyledonous Hosts *(**Susceptible** or Insusceptible)*

***Celosia argentea. Chenopodium amaranticolor. Chenopodium murale. Chenopodium quinoa. Nicotiana occidentalis.*** *Vicia faba.*

Monocotyledonous Hosts *(**Susceptible** or Insusceptible)*

***Allium porrum.***

**Particle morphology**

Virions filamentous; of 650 nm.
*Sequence database accession code(s):*
D11161 Em(40)_vi:GCVRRCP1 Gb(84)_vi:GCVRRCP Garlic mosaic virus genes for RNA replicase and coat protein. 10/92 3,459bp.
X67134 Em(40)_vi:GMVRNA Gb(84)_vi:GMVRNA Garlic mosaic virus genomic RNA. 7/92 1,972bp. 2 sequences.

**Taxonomy and Relationships**

Possible *Carlavirus.*
Virus(es) with serologically unrelated virions: shallot latent virus.

**References**

Cadilhac, B., Quiot, J.B., Marrou, J. and Leroux, J.P. (1976). *Ann. Phytopath.* **8**: 65.
Conci, V.C., Nome, S.F. and Milne, R.G. (1992). *Plant Dis.* **76**: 594.
Delacolle, K. and Lot, H. (1981). *Agronomie* **1**: 763.
Francki, R.I.B., Milne, R.G. and Hatta, T. (1985). In: *An Atlas of Plant Viruses* Vol. 2, p. 173. CRC Press, Boca Raton, Florida.
Graichen, K. and Leistner, H.-U. (1987). *Arch. Phytopath. PflSchutz.* **23**: 165.

Van Dijk, P. (1993). *Neth. J. Pl. Path.* **99**: 233.
Walkey, D.G.A., Webb, M.J.W., Bolland, C.J. and Miller, A. (1987). *J. hort. Sci.* **62**: 211.

# Garlic dwarf (?) *fijivirus*

**Data collated by A.A. Brunt, 1994.**

First reported in *Allium sativum*; from Southern France; by Lot *et al.* (1993).

**Natural host range and symptoms**

Symptoms persist. Symptoms dwarfing and leaf thickening.

*Allium sativum* — severe dwarfing, leaves dark green with swollen veins and enations.

**Geographical distribution**

Geographical distribution probably restricted. Spreads in the Eurasian region; France.

**Purification**

Method: Lot *et al.* (1993).

**Particle morphology**

Virions isometric; not enveloped; 65–70 nm in diameter; rounded in profile.

**Biochemical properties**

*Genome* consists of RNA; double–stranded; probably of 10 parts; largest (or only) genome part 4 kb (pairs); the 2nd largest 3.5 kb (pairs); the 3rd largest 3.35 kb (pairs); the 4th largest 3.25 kb (pairs); the 5th largest 2.7 kb (pairs); the 6th largest and other parts 2.3 kb (pairs; also 1.60, 1.40 and 1.30 kb pairs).

**Cytopathology**

Virions found in phloem; in cytoplasm. Inclusions are viroplasms; they contain virions.

**Taxonomy and Relationships**

Possible *Fijivirus: Reoviridae.*

Virus(es) with serologically unrelated virions: fiji disease, maize rough dwarf and oat sterile dwarf viruses.

**References**

Lot, H., Dellacolle, B., Boccardo, G., Marachi, C. and Milne, R.G. (1993). *Plant Path.* **43**.

# Gentiana (?) *carlavirus*

**Data collated by A.A. Brunt, 1991.**

*Standard acronym:* GenV.

First reported in *Gentiana sino-ornata*; from Germany; by Koenig (1985).

**Natural host range and symptoms**

*Gentiana sino-ornata.*

**Taxonomy and Relationships**

Possible *Carlavirus.*

**References**

Koenig, R. (1985). *Acta Hort.* **164**: 21.

# Gerbera symptomless (?) rhabdovirus

*Synonyms:* gerbera latent virus. *Standard acronym:* GerSLV.

First reported in *Gerbera* spp.; from Japan; by Chang *et al.* (1976).

**Natural host range and symptoms**

*Gerbera* spp. — symptomless.

**Taxonomy and Relationships**

Possible unassigned rhabdovirus.

**References**

Chang, M.U., Doi, Y. and Yora, K. (1976). *Ann. Phytopath. Soc. Japan* **42**: 383.
Francki, R.I.B., Milne, R.G. and Hatta, T. (1985). In: *An Atlas of Plant Viruses* Vol. 1, p. 73. CRC Press, Boca Raton, Florida.

# Ginger chlorotic fleck (?) *sobemovirus*

**Data collated by J.E. Thomas, 1984. Revised 1987.**

*Standard acronym:* GCFV.

First reported in *Zingiber officinale*; from New Zealand and Queensland, Australia; by Thomas (1986).

**Natural host range and symptoms**

Symptoms persist.
*Zingiber officinale*—chlorotic leaf flecking.

**Transmission**

Virus transmitted by mechanical inoculation.

**Geographical distribution**

Found, but with no evidence of spread, in Australia, where it was isolated from ginger imported from India, Malaysia, Mauritius and Thailand, but all infected material was destroyed and there was no evidence that it became established in Australia. Its presence has not yet been confirmed in the countries of origin of the infected ginger.

**Experimental host range**

Few (<3) families susceptible.

Diagnostically susceptible

*Zingiber officinale* — systemic chlorotic flecking.

Diagnostically insusceptible

*Gomphrena globosa, Chenopodium quinoa, Cucumis sativus* cv. Supermarket, *Phaseolus vulgaris* cv. Bountiful, *Zea mays* cv. Iochief.

Maintenance and propagation hosts

*Zingiber officinale.*

Assay hosts (**L**ocal lesions or **W**hole plants)

*Zingiber officinale* (W). No local lesion host.

Dicotyledonous Hosts *(**Susceptible** or Insusceptible)*

*Chenopodium amaranticolor. Chenopodium quinoa. Cucumis sativus. Gomphrena globosa. Nicotiana tabacum. Phaseolus vulgaris. Vigna unguiculata.*

Monocotyledonous Hosts *(**Susceptible** or Insusceptible)*

*Oryza sativa. Sorghum bicolor. Zea mays.*

Additional hosts

*Elettaria cardamomum, Hedychium garderanum* and *Musa acuminata* cv. Cavendish.

Sources of host-range data

Thomas (1986).

**Properties of particles in sap**

Leaf sap contains many virions. Electron microscopy: virions are unstable in PTA, but stable in UA and AM.

**Purification**

Method: Thomas (1986).

**Particle morphology**

Virions isometric; 28–33 nm in diameter (from side to side, 4 nm less than from vertex to vertex); angular in profile; without a conspicuous capsomere arrangement.

**Physical properties**

One sedimenting component in purified preparations; sedimentation coefficient 111 *S* (in 10 mM phosphate pH 7, 20°C). Density 1.355 g $cm^{-3}$ in CsCl (fixed). Density in $Cs_2SO_4$ 1.297 g $cm^{-3}$ at pH 4.9 and 1.293 g $cm^{-3}$ at pH 6.9.

**Biochemical properties**

Virions contain 20 % nucleic acid; 80 % protein.

*Genome* consists of RNA; single–stranded; unipartite (but with two minor components); largest (or only) genome part the largest 4.2 kb. Genomic nucleic acid isolated by Thomas (1986).

*Virion protein(s)* two (the smaller probably a degradation product of the larger); $M_r$ of the larger 29000. $M_r$ of 2nd largest 27300. Method of preparation: Thomas (1986): SDS digestion.

**Cytopathology**

Virions found in leaves and mesophyll; in cytoplasm. Inclusions present in infected cells; are membranous bodies.

**Taxonomy and Relationships**

Possible *Sobemovirus*.

Virus(es) with serologically unrelated virions: carnation mottle, cocksfoot mild mottle (syn. phleum mottle), cocksfoot mottle, cucumber leaf spot (cucumber fruit streak strain), cucumber mosaic, cymbidium ringspot, elm latent, galinsoga mosaic, glycine yellow mosaic, hibiscus chlorotic ringspot, lucerne transient streak, narcissus tip necrosis, pelargonium flower break, pelargonium line pattern, pelargonium vein netting, red clover necrotic mosaic, saguaro cactus, solanum nodiflorum mottle, southern bean mosaic, sowbane mosaic, tobacco necrosis, turnip crinkle, turnip rosette and velvet tobacco mottle viruses.

Best tests for diagnosis: easily diagnosed by serological methods.

**References**

Thomas, J.E. (1986). *Ann. appl. Biol.* **108**: 43.

Thomas, J.E. (1988). *AAB Descr. Pl. Viruses* No. 328, 4 pp.

## Gloriosa fleck (?) *nucleorhabdovirus*

**Data collated by A.A. Brunt, 1991.**

First reported in *Gloriosa rothschildiana*; from Japan; by Araki *et al.* (1985).

**Natural host range and symptoms**

*Gloriosa rothschildiana* — leaf flecking.

**Geographical distribution**

Spreads in Japan.

**Particle morphology**

Virions rhabdo– or bullet–shaped; of 316 nm; 69 nm wide.

**Cytopathology**

Inclusions are viroplasms (in the nucleus).

**Taxonomy and Relationships**

Possible *Nucleorhabdovirus: Rhabdoviridae.*

**References**

Araki, M., Yamashita, S., Doi, Y. and Yora, K. (1985). *Ann. Phytopath. Soc. Japan* **51**: 632.

## Glycine mosaic *comovirus*

**Data collated by G.M. Behncken, 1980. Revised 1983.**

*Standard acronym:* GMV.

First reported in *Glycine clandestina*; from the Blue Mountains, N.S.W., Australia; by Bowyer *et al.* (1980).

**Natural host range and symptoms**

Symptoms vary cyclically over a few weeks and vary seasonally.

*Glycine clandestina, G. tabacina* — green mosaic, leaves malformation.

**Transmission**

Virus transmitted by mechanical inoculation.

**Geographical distribution**

Spreads in Australia.

**Experimental host range**

Few (<3) families susceptible.

Diagnostically susceptible

*Chenopodium amaranticolor, C. quinoa* — chlorotic lesions; then systemic mosaic, vein yellowing or netting.

*Glycine max, G. soja* — systemic mosaic.

*Macrotyloma uniflorum* — necrotic lesions; then systemic lethal necrosis.

*Phaseolus vulgaris* cvs Bountiful, Black Turtle Soup — necrotic lesions; then systemic mottling in cv. Bountiful.

*Pisum sativum* — systemic mosaic.

Maintenance and propagation hosts

*Chenopodium quinoa, Phaseolus vulgaris* cv. Bountiful, *Pisum sativum* cv. Greenfeast.

Assay hosts (**L**ocal lesions or **W**hole plants)

*Chenopodium amaranticolor* (L), *C. quinoa* (L), *Phaseolus vulgaris* cv. Bountiful (L).

Dicotyledonous Hosts ***(Susceptible** or Insusceptible)*

***Cajanus cajan**. Chenopodium amaranticolor. **Chenopodium quinoa**. **Cicer arietinum**. Cucumis sativus. Datura metel. Datura stramonium. **Glycine max**. Gomphrena globosa. Lupinus albus. Lupinus angustifolius. **Macroptilium lathyroides**. **Macrotyloma uniflorum**. Medicago sativa. Nicotiana clevelandii. Nicotiana glutinosa. Nicotiana tabacum. **Phaseolus vulgaris**. **Pisum sativum**. Trifolium incarnatum. Trifolium pratense. Trifolium repens. Trifolium subterraneum. Vicia faba. Vigna radiata. **Vigna unguiculata**.*

Sources of host-range data

Bowyer *et al.* (1980).

**Properties of particles in sap**

TIP: 55–60 °C. LIV: 7–8 days. DEP: $\log_{10}$ minus 4–5. Leaf sap contains many virions.

**Purification**

Method: Bowyer *et al.* (1980).

**Particle morphology**

Virions isometric; not enveloped; 28 nm in diameter; angular in profile; without a conspicuous capsomere arrangement.

**Physical properties**

Three sedimenting components in purified preparations; sedimentation coefficient of the fastest 130 *S* (B); of the other(s) 60 *S* (T), or 103 *S* (M). $A_{260}/A_{280}$ ratio 1.63 (M), or 1.75 (B).

**Biochemical properties**

Virions contain 38.6 % nucleic acid (B), or 26.4 % nucleic acid (M), or 0 % nucleic acid (T); 63.2 % protein (B), or 73.6 % protein (M), or 100 % protein (T).

*Virion protein(s)* two (major and 1 minor); $M_r$ of the largest 42000. $M_r$ of 2nd largest 21500. $M_r$ of 3rd largest 19000. Method of preparation: Bowyer *et al.* (1980).

**Replication**

Replication does not depend on a helper virus.

**Cytopathology**

Virions found in mesophyll; in cytoplasm. Inclusions present in infected cells; are unusual in shape; large aggregates, often in crystalline array, associated with amorphous, membrane-bound bodies; they contain virions. Other cellular changes: cytoplasmic vesiculation.

**Taxonomy and Relationships**

*Comovirus: Comoviridae.*

Virus(es) with serologically related virions: broad bean stain, broad bean true mosaic and squash mosaic viruses to both strains; cowpea mosaic (Sb and Ark), bean pod mottle and red clover mottle viruses are related to the GW strain.

**References**

Bowyer, J.W., Dale, J.L. and Behncken, G.M. (1980). *Ann. appl. Biol.* **95**: 385.

# Glycine mottle (?) *carmovirus*

**Data collated by G.M. Behncken, 1981. Revised 1983.**

*Standard acronym:* GMoV.

First reported in *Glycine tomentella*; from from Herberton, Queensland, Australia; by Behncken and Dale (1984).

**Natural host range and symptoms**

*Glycine tomentella* — systemic mottle.

**Transmission**

Virus transmitted by mechanical inoculation.

**Geographical distribution**

Found, but with no evidence of spread, in Australia.

**Experimental host range**

Few (<3) families susceptible.

Diagnostically susceptible

*Cyamopsis tetragonoloba*—necrotic local lesions.
*Glycine max, Phaseolus vulgaris* — chlorotic local lesions.
*Glycine soja* — systemic mottle.
*Macrotyloma uniflorum* — necrotic or chlorotic local lesions.

Diagnostically insusceptible

*Chenopodium quinoa, Pisum sativum.*

Maintenance and propagation hosts

*Glycine soja.*

Assay hosts (Local lesions or Whole plants)

*Cyamopsis tetragonoloba* (L), *Glycine max* (L), *Macrotyloma uniflorum* (L), *Phaseolus vulgaris* (L).

Dicotyledonous Hosts *(**Susceptible** or Insusceptible)*

***Cassia occidentalis***. *Chenopodium amaranticolor*. *Chenopodium quinoa*. *Cicer arietinum*. *Clitoria ternatea*. *Cucumis sativus*. ***Cyamopsis tetragonoloba***. ***Glycine max***. ***Lathyrus odoratus***. ***Lupinus albus***. *Macroptilium lathyroides*. ***Macrotyloma uniflorum***. *Medicago sativa*. *Nicotiana clevelandii*. *Nicotiana tabacum*. ***Phaseolus vulgaris***. *Pisum sativum*. ***Trifolium incarnatum***. *Trifolium pratense*. *Trifolium repens*. ***Vigna radiata***.

Sources of host-range data

Behncken and Dale (1984).

**Properties of particles in sap**

TIP: 80–85 °C. LIV: 60 days (or more). DEP: $\log_{10}$ minus 5–6. Electron microscopy: use 2% UA for negative staining.

**Purification**

Method: Behncken and Dale (1984).

**Particle morphology**

Virions isometric; not enveloped; 30 nm in diameter; rounded in profile; with a conspicuous capsomere arrangement.

**Physical properties**

One sedimenting component in purified preparations; sedimentation coefficient 138 *S*.

**Biochemical properties**

Virions contain 20 % nucleic acid.

*Genome* consists of RNA; single–stranded; unipartite; largest (or only) genome part 4.2 kb.

*Virion protein(s)* one (plus 2 fragments); $M_r$ 39000 (plus two of 37500 and 31000); coat protein. Method of preparation: Dale and Gibbs (1976).

**Cytopathology**

Virions found in all parts of the host plant; in cytoplasm and in nuclei (especially associated with the nucleolus and cytoplasmic invaginations of chloroplasts). Other cellular changes: vesicles in peroxisomes closely associated with chloroplasts. Number of vesicles are variable and are located around microbodies.

**Taxonomy and Relationships**

Possible *Carmovirus: Tombusviridae.*

Virus(es) with serologically unrelated virions: artichoke mottled crinkle, carnation Italian ringspot, cymbidium ringspot, galinsoga mosaic, pelargonium leaf curl, petunia asteroid mosaic, saguaro cactus, tephrosia symptomless, tomato bushy stunt and turnip crinkle viruses.

**References**

Behncken, G.M. and Dale, J.L. (1984). *Intervirology* **21**: 159.

Dale, J.L. and Gibbs, A.J. (1976). *Aust. J. biol. Sci.* **29**: 397.

# Gomphrena (?) *nucleorhabdovirus*

*Standard acronym:* GoV.

First reported in *Gomphrena globosa*; from Brazil; by Kitajima and Costa (1966).

**Natural host range and symptoms**

*Gomphrena* spp.

**Taxonomy and Relationships**

Possible *Nucleorhabdovirus: Rhabdoviridae.*

This virus is similar to lettuce necrotic yellows rhabdovirus (Kitajima and Costa, 1966).

**References**

Kitajima, E.W. and Costa, A.S. (1966). *Virology* **29**: 523.

# Grapevine A (?) *trichovirus*

**Data collated by A.A. Brunt, 1995.**

First reported in *Vitis vinifera*; from Italy; by Conti *et al.* (1980).

**Natural host range and symptoms**

*Vitis vinifera* — leaf rolling, reddening of leaf margins and petioles, and localised swelling and cracking of stems (probably Kober stem grooving disease).

**Transmission**

Transmitted by a vector; an insect; *Pseudococcus longispinus, Planococcus citri, P. ficus*; Pseudococcidae. Virus transmitted by mechanical inoculation.

**Geographical distribution**

Found, but with no evidence of spread, in France, Italy, Israel, Switzerland, South Africa, Tunisia and Yemen.

Diagnostically susceptible

*Nicotiana benthamiana, N. clevelandii* — narrow leaflets, vein-banding of systemically infected leaves, followed by mottling and puckering.

*Vitis vinifera* — leaf rolling and reddening (in autumn).

Diagnostically insusceptible

*Chenopodium amaranticolor, C. quinoa, Cucurbita pepo, Cucumis sativus, Nicotiana glutinosa, N. rustica, N. tabacum, N. rustica, Phaseolus vulgaris, Pisum sativum.*

Maintenance and propagation hosts

*Nicotiana benthamiana, N. clevelandii.*

Assay hosts (**L**ocal lesions or **W**hole plants)

*Nicotiana benthamiana* (W), *N. clevelandii* (W).

Dicotyledonous Hosts *(**Susceptible** or Insusceptible)*

*Chenopodium amaranticolor. Chenopodium quinoa. Cucumis sativus. Cucurbita pepo.* ***Nicotiana benthamiana. Nicotiana clevelandii.*** *Nicotiana glutinosa. Nicotiana rustica. Nicotiana tabacum. Phaseolus vulgaris. Pisum sativum.* ***Vitis vinifera.***

**Particle morphology**

Virions filamentous; usually flexuous; of 800 nm; 12 nm wide. Basic helix obvious; pitch of basic helix 3.8 nm.

**Cytopathology**

Virions found in phloem cells; in cytoplasm. Other cellular changes: Rosciglione *et al.* (1983).

**Taxonomy and Relationships**

Possible *Trichovirus*.

**References**


Boccardo, G. and d'Aquilo, M. (1981). *J. gen. Virol.* **53**: 179.

Boscia, D., Aslouj, E., Elicio, V., Savino, V., Castellano, M.A. and Martelli, G.P. (1992). *Arch. Virol.* **127**: 185.

Chevalier, S., Greif, C., Clauzel, J.-M., Walter, B. and Fritsch, C. (1995). *J. Phytopathology* **143**: 369.

Conti, M., Milne, R.G., Luisoni, E. and Boccardo, G. (1980). *Phytopathology* **70**: 394.

Englebrecht, D.J. and Kasdorf, G.G.F. (1985). *Phytopath. medit.* **24**: 101.

Milne, R.G., Conti, M., Lesemann, D.-E., Stellmach, G., Tanne, E. and Cohen, J. (1984). *Phytopath. Z.* **110**: 360.
Minafra, A., Saldarelli, F., Grieco, F. and Martelli, G.P. (1994). *Arch. Virol.* **137**: 249.
Rosciglione, B., Castellano, M.A., Martelli, G.P., Savino, V. and Cannizzaro, G. (1983). *Vitis* **22**: 331.

# Grapevine ajinashika disease (?) *luteovirus*

**Data collated by S. Holland, 1986. Revised by A.A. Brunt, 1991.**

*Standard acronym:* GAV.

First reported in *Vitis vinifera*; from Japan; by Namba *et al.* (1979).

**Natural host range and symptoms**

Symptoms persist.

*Vitis vinifera* — no leaf symptoms, fruit are unpalatable because of small sugar content.

**Transmission**

Transmitted by means not involving a vector. Virus not transmitted by mechanical inoculation; transmitted by grafting; not transmitted by contact between plants.

**Geographical distribution**

Found, but with no evidence of spread, in Japan.

**Experimental host range**

Few (<3) families susceptible.

Diagnostically susceptible

*Vitis vinifera* — no leaf symptoms, fruit unpalatable.

Maintenance and propagation hosts

*Vitis vinifera.*

Dicotyledonous Hosts *(**Susceptible** or Insusceptible)*

*Chenopodium amaranticolor. Chenopodium quinoa. Cucumis sativus. Cucurbita moschata. Datura stramonium. Lactuca sativa. Lycopersicon esculentum. Nicotiana glutinosa. Nicotiana rustica. Nicotiana tabacum. Petunia hybrida. Vigna unguiculata* ssp. *sesquipedalis.* ***Vitis vinifera*****.**

**Properties of particles in sap**

Leaf sap contains few virions.

**Purification**

Method: Namba *et al.* (1991).

**Particle morphology**

Virions isometric; not enveloped; 28 nm in diameter; angular in profile; without a conspicuous capsomere arrangement.

**Physical properties**

One sedimenting component in purified preparations; sedimentation coefficient 110 *S*. Density 1.38 g $cm^{-3}$ in CsCl.

**Biochemical properties**

Virions contain 30 % nucleic acid.

*Genome* consists of RNA; single–stranded; largest (or only) genome part 6.8 kb.

*Virion protein(s)* one; $M_r$ 23000; coat protein.

**Cytopathology**

Virions found in phloem and companion cells in sieve tubes; in cytoplasm, in nuclei, and in cell vacuoles. Inclusions present in infected cells; are crystals in the cytoplasm; they contain virions. Other cellular changes: necrosis of phloem cells.

**Taxonomy and Relationships**

Possible *Luteovirus*.

Virus(es) with serologically unrelated virions: barley yellow dwarf and potato leafroll viruses.

**References**

Namba, S., Yamashita, S., Doi, Y. and Yora, K. (1979). *Ann. Phytopath. Soc. Japan* **45**: 70.

Namba, S., Boschia, D., Yamashita, S., Tsuchizaki, T. and Gonsalves, D. (1991). *Plant Dis.* **75**: 1249.

# Grapevine Algerian latent *tombusvirus*

**Data collated by A.A. Brunt, 1992.**

*Standard acronym:* GALV.

First reported in *Vitis vinifera*; from Algeria; by Gallitelli *et al.* (1989).

**Natural host range and symptoms**

*Vitis vinifera* — no conspicuous symptoms.

**Transmission**

Transmitted by means not involving a vector.

**Geographical distribution**

Found, but with no evidence of spread, in Algeria, Germany and Italy.

**Experimental host range**

Few (<3) families susceptible.

Diagnostically susceptible

*Chenopodium amaranticolor, C. quinoa* — local lesions.

*Nicotiana benthamiana* — systemic chlorosis and necrosis.

Diagnostically insusceptible

*Nicotiana clevelandii.*

Maintenance and propagation hosts

*Nicotiana benthamiana.*

Assay hosts (**L**ocal lesions or **W**hole plants)

*Chenopodium quinoa* (L), *C. amaranticolor* (L).

Dicotyledonous Hosts *(**Susceptible** or Insusceptible)*

***Chenopodium amaranticolor**. **Chenopodium quinoa**. **Nicotiana benthamiana**. Nicotiana clevelandii.*

**Particle morphology**

Virions isometric; not enveloped; 33 nm in diameter.

**Biochemical properties**

*Genome* consists of RNA; single–stranded; unipartite; largest (or only) genome part 4.6 kb.

*Virion protein(s)* one; $M_r$ 35900; coat protein.

**Cytopathology**

Virions found in leaves and mesophyll; in cytoplasm. Inclusions absent from infected cells. Other cellular changes: the virus induces the formation of membrane associated vesicles at the periphery of peroxisomes which possibly contain dsRNA. It also sometimes induces the formation of rod-like structures in cytoplasm and in the stroma of chloroplasts and mitochondria.

**Taxonomy and Relationships**

*Tombusvirus: Tombusviridae.*

Virus(es) with serologically related virions: eggplant mottled crinkle tombusvirus.

The viral RNA has some sequence homology with those of cymbidium ringspot, petunia asteroid mosaic and Sikte waterborne tombusviruses.

**Comments**

The virus has only been found in river water in Germany and in Sicily.

**References**

Cannizzaro, G., Rosciglione, B. and Castellano, M.A. (1990). *Inform. Fitopatol.* **40**: 55.

Gallitelli, D., Martelli, G.P. and di Franco, A. (1989). *Phytoparasitica* **17**: 61.

Yi, L., Lesemann, D.-E., Koenig, R., Rüdel, M. and Pfeilstetter, E. (1992). *J. Phytopath.* **134**: 121.

# Grapevine B (?) *trichovirus*

**Data collated by A.A. Brunt, 1995.**

First reported in *Vitis vinifera*; from Switzerland, Italy; by Gugerli *et al.* (1984); Milne *et al.* (1984).

**Natural host range and symptoms**

*Vitis vinifera.*

**Transmission**

Virus not transmitted by mechanical inoculation; transmitted by grafting.

**Geographical distribution**

Found, but with no evidence of spread, in Italy, Switzerland and, possibly, Israel.

Dicotyledonous Hosts *(**Susceptible** or Insusceptible)*

***Vitis vinifera.***

**Particle morphology**

Virions filamentous; usually flexuous; of 1800 nm.

**Taxonomy and Relationships**

Possible *Trichovirus.*

**References**

Conti, M. and Milne, R.G. (1985). *Phytopath. medit.* **24**: 110.

Gugerli, P., Brugger, J.J. and Bovey, R. (1984). *Rev. Suisse Vitic. Arb. Hort.* **16**: 299.

Martelli, G.P. (1986). *FAO Pl. Prot. Bull.* **34**: 25.

Martelli, G.P. and Prota, U. (1985). *Ital. Agric.* **122**: 201.

Milne, R.G., Conti, M., Lesemann, D.-E., Stellmach, G., Tanne, E. and Cohen, J. (1984). *Phytopath. Z.* **110**: 360.

# Grapevine Bulgarian latent *nepovirus*

**Data collated by S. Holland, 1986.**

*Standard acronym:* GBLV. *Strains:* several distinct strains are known (Martelli *et al.*, 1978*a*; Gallitelli *et al.*, 1983).

First reported in *Vitis vinifera*; from Bulgaria; by Martelli *et al.* (1977).

**Natural host range and symptoms**

Symptoms persist.

*Vitis vinifera* — symptomless, when infected with European strains (Martelli *et al.*, 1977).

*Vitis lumbrusca* cv. Concord — delayed budbreak, irregular elongation of the shoots, pale green foliage and straggly fruit clusters, when infected with New York strain (Uyemoto *et al.*, 1977).

**Transmission**

Transmitted by means not involving a vector. Virus transmitted by mechanical inoculation; transmitted by grafting; not transmitted by contact between plants; transmitted by seed (New York strain *c.* 5% in *Vitis lambrusca* and *c.* 12% in *Chenopodium quinoa*; Uyemoto *et al.*, 1977).

**Ecology and control**

Studies reported by Martelli *et al.* (1978).

**Geographical distribution**

Found, but with no evidence of spread, in Bulgaria (Martelli *et al.*, 1977, 1978*a*), Yugoslavia (Dimitrijevic, 1985), Portugal (Gallitelli *et al.*, 1983) and the U.S.A. (Uyemoto *et al.*, 1977).

**Experimental host range**

Several (3–9) families susceptible.

Diagnostically susceptible

*Chenopodium quinoa* — chlorotic or necrotic local lesions, then severe systemic mottling and necrotic flecking.

*Chenopodium amaranticolor* — systemic mottling and leaf malformation.

*Gomphrena globosa* — reddish lesions, then malformation of upper leaves.

*Nicotiana clevelandii* — necrotic lesions; then systemic chlorosis and stunting.

*Phaseolus vulgaris* — systemic mottling (Portuguese strain).

*Nicotiana megalosiphon* — local and systemic chlorotic rings (Bulgarian strain).

Diagnostically insusceptible

*Datura stramonium, Cucurbita pepo, Pisum sativum, Vigna unguiculata, Zinnia elegans.*

Maintenance and propagation hosts

*Chenopodium quinoa, Gomphrena globosa, Nicotiana clevelandii.*

Assay hosts (**L**ocal lesions or **W**hole plants)

*Chenopodium quinoa* (L); *Chenopodium amaranticolor* (W).

Dicotyledonous Hosts *(**Susceptible** or Insusceptible)*

***Amaranthus retroflexus*. *Atriplex hortensis*. *Beta vulgaris*. *Chenopodium album*. *Chenopodium amaranticolor*. *Chenopodium ambrosioides*. *Chenopodium capitatum*. *Chenopodium hybridum*. *Chenopodium murale*. *Chenopodium quinoa*. *Cucumis sativus*.** *Cucurbita pepo*. *Datura stramonium*. ***Gomphrena globosa*. *Nicotiana benthamiana*. *Nicotiana clevelandii*. *Nicotiana rustica*. *Nicotiana tabacum*. *Phaseolus vulgaris*.** *Pisum sativum*. *Vigna unguiculata* ssp. *sesquipedalis*. ***Vitis labrusca*. *Vitis rupestris*. *Vitis vinifera*.** *Zinnia elegans*.

Sources of host-range data

Martelli *et al.*, 1977; 1978*a*; Gallitelli *et al.*, 1983.

**Properties of particles in sap**

TIP: 65–70 °C. LIV: 15–20 days (at 22°C). DEP: $\log_{10}$ minus 5–6 (Martelli *et al.*, 1977; 1987*a*). Leaf sap contains few virions.

**Purification**

Method: Martelli *et al.* (1977).

**Particle morphology**

Virions isometric; not enveloped; 30 nm in diameter; angular in profile; without a conspicuous capsomere arrangement.

**Physical properties**

Three sedimenting components in purified preparations; sedimentation coefficient of the fastest 127 *S* ($B_2$); of the other(s) 120 *S* ($B_1$), or 52 *S* (T). Density 1.489 g $cm^{-3}$ in CsCl ($B_2$), or 1.479 g $cm^{-3}$ in CsCl ($B_1$). $A_{260}/A_{280}$ ratio 1.7 (B), or 0.8 (T); (Martelli *et al.*, 1977; Gallitelli *et al.*, 1983).

**Biochemical properties**

Virions contain 41 % nucleic acid ($B_2$), or 39 % nucleic acid ($B_1$), or 0 % nucleic acid (T; Reichmann, 1965); 59 % protein ($B_2$), or 61 % protein ($B_1$), or 100 % protein (T).

*Genome* consists of RNA; single–stranded; linear; of two parts; largest (or only) genome part 6.2 kb ($B_2$); the 2nd largest 5.9 kb ($B_1$). Genomic nucleic acid isolated by Quacquarelli *et al.* (1972) — freezing and thawing; Martelli *et al.* (1977) — heating at 68°C for 90 seconds. Additional factor not required for infectivity.

*Non–genomic nucleic acid* found in the virions (of some isolates); is satellite RNA (of *c.*1.4 kb; Gallitelli *et al.* 1983).

*Virion protein(s)* one; $M_r$ 54000 (Martelli *et al.*, 1977; Gallitelli *et al.*, 1983). Method of preparation: Agrawal and Tremaine (1972).

**Cytopathology**

Virions found in mesophyll in herbaceous plants and grapevines; in cytoplasm. Inclusions present in infected cells; are unusual in shape; membranous bodies and tubules; they contain virions. Other cellular changes: cell wall outgrowths centered on plasmodesmata and engulfing virus-containing tubules (Martelli *et al.*, 1982).

### Taxonomy and Relationships

*Nepovirus: Comoviridae.*

Virus(es) with serologically related virions: blueberry leaf mottle virus, but distantly (Ramsdell and Stace-Smith, 1982; Gallitelli *et al.*, 1983). Virus(es) with serologically unrelated virions: many other nepoviruses.

Different strains are serologically distinguishable from one another in gel diffusion tests (Martelli *et al.*, 1978*a*; Uyemoto *et al.*, 1977; Gallitelli *et al.*, 1983). Best tests for diagnosis: Grapevine Bulgarian latent virus, like most other nepoviruses, infects *Nicotiana clevelandii* but grapevine fleck virus does not. Grapevine Bulgarian latent is serologically distinct and also has a much narrower host range than most nepoviruses.

### References

Agrawal, H.D. and Tremaine, J.H. (1972). *Virology* **47**: 8.

de Mendonça, A., de Sequeira, O.A., Mota, M., Pereira, A.N. and Simoes, V. (1982). *Proc. ICVG, Niagara Falls, 1980*, p. 245.

Gallitelli, D., Savino, V. and de Sequeira, O.A. (1983). *Phytopathol. Medit.* **22**: 27.

Martelli, G.P., Gallitelli, D. Abracheva, A., Savino, V. and Quacquarelli, A. (1978*a*). *Proc. 6th Meeting ICVG, Cordova 1976*, Monogr. INIA **18**: 136.

Martelli, G.P., Quacquarelli, A. and Gallitelli, D. (1978*b*). *CMI/AAB Descr. Pl. Viruses* No. 186, 4 pp.

Martelli, G.P., Di Franco, A., Russo, M. and Savino, V. (1982). *Proc. ICVG, Niagara Falls 1980*, p. 251.

Quacquarelli, A., Piazzolla, P. and Vovlas, C. (1972). *J. gen. Virol.* **17**: 147.

Reichmann, M.E. (1965). *Virology* **25**: 166.

Russo, M., Martelli, G.P. and Savino, V. (1982). *Proc. 7th Meeting ICVG. Niagara Falls 1980*, p. 251.

Uyemoto, J.K., Taschenberg, C.F. and Humer, D.K. (1977). *Pl. Dis. Reptr* **61**: 949.

## Grapevine chrome mosaic *nepovirus*

**Data collated by G.P. Martelli, 1987; S. Holland, 1986.**

*Synonyms:* Hungarian chrome mosaic virus. *Standard acronym:* GCMV.

First reported in *Vitis vinifera*; from Hungary; by Martelli *et al.* (1965).

### Natural host range and symptoms

Symptoms persist.

*Vitis vinifera* and interspecific hybrids — green mottling or yellow blotching of leaves, leaf and cane malformation, stunting, no fruit.

*Apium graveolens* — yellow veins.

### Transmission

Virus-infected vines distributed in patches in the field. Reports of transmission by *Xiphinema index* (Mali, 1976) are doubtful. Virus transmitted by mechanical inoculation; transmitted by grafting (Martelli *et al.*, 1965; Lehoczky and Sarospataki, 1967); not transmitted by contact between plants.

### Geographical distribution

Spreads in Hungary. Found, but with no evidence of spread, in Yugoslavia (Saric and Hranuelli, 1979) and England (Hollings *et al.*, 1969).

**Experimental host range**

Several (3–9) families susceptible. Experimentally infected plants mostly show local lesions, systemic mosaic, stunting; solanaceous plants usually remain symptomless.

Diagnostically susceptible

*Chenopodium quinoa* — small necrotic lesions; then severe systemic chlorosis, necrotic speckling, apical necrosis.

*Gomphrena globosa* — chlorotic to reddish lesions, then transient systemic mosaic.

*Datura stramonium* — inoculated leaves symptomless; then systemic transient yellow spots.

*Phaseolus vulgaris* — symptomless inoculated leaves; systemic mosaic, yellow-green rings and specks with some necrotic blotches.

*Vitis vinifera* cv. Jubileum 75, Pinot noir — chlorosis and apical necrosis, severe stunting (Lehoczky, 1985).

Maintenance and propagation hosts

*Chenopodium quinoa*, *Phaseolus vulgaris*.

Assay hosts (**L**ocal lesions or **W**hole plants)

*Chenopodium quinoa* (L).

Dicotyledonous Hosts *(**Susceptible** or Insusceptible)*

***Apium graveolens***. ***Chenopodium amaranticolor***. ***Chenopodium foetidum***. ***Chenopodium quinoa***. ***Cucumis sativus***. ***Cucurbita pepo***. ***Datura stramonium***. ***Gomphrena globosa***. ***Nicotiana benthamiana***. ***Nicotiana clevelandii***. ***Nicotiana glutinosa***. *Nicotiana megalosiphon*. ***Nicotiana tabacum***. ***Petunia hybrida***. ***Phaseolus vulgaris***. ***Vigna unguiculata***. ***Vitis vinifera***.

Sources of host-range data

Martelli (1965); Martelli and Quacquarelli (1972*a*).

**Properties of particles in sap**

TIP: 60–62 °C. LIV: 7 days (at 22°C). DEP: $\log_{10}$ minus 3–4 (Martelli *et al.*, 1965). Leaf sap contains few virions. Electron microscopy: see Martelli (1965); Martelli and Quacquarelli (1972*a*).

**Purification**

Method: Martelli and Quacquarelli (1972*a*): Chloroform-butanol followed by differential and density gradient centrifugation.

**Particle morphology**

Virions isometric; not enveloped; 30 nm in diameter; angular in profile.

**Physical properties**

Two sedimenting components in purified preparations; sedimentation coefficient of the fastest 117 *S* (B); of the other(s) 92 *S* (M). Density 1.486 g $cm^{-3}$ in CsCl (B), or 1.416 g $cm^{-3}$ in CsCl (M).

**Biochemical properties**

Virions contain 40 % nucleic acid (B), or 31 % nucleic acid (M); 60 % protein (B), or 69 % protein (M); 0 % lipid.

*Genome* consists of RNA; single–stranded; linear; of two parts; largest (or only) genome part 7.2 kb; the 2nd largest 4.4 kb.

*Sequence database accession code(s):*

X15163 Em(40)_vi:GCMVRNA2 Gb(84)_vi:GCMVRNA2 Hungarian grapevine chrome mosaic virus RNA 2. 9/93 4,441bp.

X15346 Em(40)_vi:GCMVRNA1 Gb(84)_vi:GCMVRNA1 Hungarian grapevine chrome mosaic nepovirus RNA1 sequence. 9/93 7,212bp. 2 sequences.

**Taxonomy and Relationships**

*Nepovirus: Comoviridae.*

Virus(es) with serologically related virions: cacao necrosis (Kenten, 1972) and tomato black ring viruses (Kerlan *et al.*, 1982), but distantly.

Distorting and chromogenic strains are antigenically close (Martelli *et al.*, 1965). An isolate of chrome mosaic virus has been associated with celery yellow vein disease.

**References**


Brault, V., Hibrand, L., Candresse, T., Le Gall, O. and Dunez, J. (1989). *Nucl. Acid Res.* **17**: 7809.

Le Gall, O., Candresse, T., Brault, V. and Dunez, J. (1989). *Nucl. Acid Res.* **17**: 7795.

Hollings, M., Stone, O.M. and Martelli, G.P. (1969). *Rep. Glasshouse Crops Res. Inst.* 1968, p. 102.

Kenten, R.H. (1972). *Ann. appl. Biol.* **71**: 119.

Kerlan, C., Mille, B., Dietenne and Dunez, J. (1982). *Annls. Virol. Inst. Pasteur* **133**E: 3.

Kolber, M., Beczner, L., Pacsa, S. and Lehoczky, J. (1985). *Phytopathol. Medit.* **24**: 135.

Lehoczky, J. (1985). *Phytopathol. Medit.* **24**: 129.

Lehoczky, J. and Sarospataki (1967). *Acta Phytopath. Acad. Sci. Hung.* **2**: 309.

Mali, V.R. (1976). *Indian Phytopath.* **26**: 363.

Martelli, G.P. (1965). *Proc. Int. Conf. Virus Vector Peren. Hosts, Univ. Cal. Agr. Sci., Davis*, 402.

Martelli, G.P. and Quacquarelli, A. (1972*a*). *Annls. Phytopath.* hors series 1972, 123.

Martelli, G.P. and Quacquarelli, A. (1972*b*). *CMI/AAB Descr. Pl. Viruses* No. 103, 4 pp.

Martelli, G.P., Quacquarelli, A. and Lehoczky, J. (1965). *Proc. Int. Conf. Virus Vector Peren. Hosts, Univ. Cal. Agr. Sci., Davis*, 389.

Russo, M., Martelli, G.P. and Savino, V. (1982). *Proc. 7th Meeting ICVG, Niagara Falls 1980*, 251.

Saric, A. and Hranuelli (1979). *Proc. Conf. Excoriosis Virus Dis. Grapevine, Mostar 1977*, 149.


# Grapevine corky bark-associated (?) *closterovirus*

**Data collated by A.A. Brunt (1995).**

First reported in *Vitis vinifera*; from the U.S.A.; by Beukman and Goheen (1996).

**Natural host range and symptoms**

*Vitis vinifera* — in intolerant cultivars, delayed bud opening, reduced vigour and yield, swelling of trunk and, sometimes, thick and rough spongy bark; woody cylinder of rootstock and/or scion pitted and/or grooved. Some cultivars and rootstocks are tolerant and produce no symptoms.

**Transmission**

Transmitted by a vector; an insect; Psyllidae. Not transmitted by *Planococcus ficus*. Virus not transmitted by mechanical inoculation; transmitted by grafting.

**Geographical distribution**

Spreads in the African region.

**Experimental host range**

Few (<3) families susceptible.

Diagnostically susceptible

*Vitis berlandieri* × Conderc 1613 (LN33) — swelling of basal internodes due to proliferation of secondary phloem, and stem grooving.

*Vitis berlandieri* × *V. riparia* (Kober SBB) — symptomless.

*Vitis rupestris* — stem grooving.

Maintenance and propagation hosts

*Vitis vinifera* (LN33) and *V. rupestris.*

Dicotyledonous Hosts ***(Susceptible** or Insusceptible)*

***Vitis rupestris*. *Vitis vinifera*.**

**Particle morphology**

Virions filamentous; usually flexuous; of 1400–2000 nm.

**Taxonomy and Relationships**

Possible *Closterovirus.*

**References**


Beukman, E.F. and Goheen, A.C. (1966). *Proc. Int. Conf. Virus Vectors Perenn. Hosts and Vitis*, Davis 1965, p.164.

Candresse, T. and Martelli, G.P. (1995). *Arch. Virol.*, Suppl. **10**: 461.

Frison, A.E. and Ikin, R. (1991). *FAO/IBPGR Tech. Guidelines for Safe Mut of Grapevine Germplasm*, p.28. FAO/IBPGR, Rome.

Graniti, A. and Martelli, G.P. (1966). *Proc. Int. Conf. Virus Vectors Perenn Hosts and Vitis*, Davis 1965, p.168.

Namba, S., Boscia, D., Azzam, O., Maixner, M., Hu, J.S., Golino, D. and Gonsalves, D. (1991). *Proc. 10th Meeting I.C.V.G.*, Volos 1990, p.10.

Savino, D., Boscia, D. and Martelli, G.P. (1989). *Proc. 9th Meeting, I.C.V.G.*, Kiryat Anavim 1987, p.91.

Tanne, E., Ben-Dov, Y. and Raccah, B. (1989). *Phytoparasitica* **17**: 55.


# Grapevine fanleaf *nepovirus*

**Data collated by G.P. Martelli, 1987; S.A. Holland, 1986.**

*Synonyms:* grapevine arricciamento virus, grapevine court noué virus, grapevine infectious degeneration virus, grapevine Reisigkrankheit virus, grapevine roncet virus, grapevine urticado virus. *Standard acronym:* GFLV. *Strains:* Fanleaf strain (Hewitt, 1950*b*); yellow mosaic strain (Hewitt, 1950*a*) and Dias (1950); veinbanding strain (Goheen and Hewitt, 1962) and Martelli and Hewitt (1963*a*).

First reported in *Vitis vinifera*; from Austria; by Rathay (1883).

**Natural host range and symptoms**

Symptoms persist and vary seasonally.

*Vitis vinifera, V. rupestris* and many other *Vitis* ssp. and interspecific hybrids — systemic green or yellow mosaic, ring and line patterns and flecks, leaf and nodal malformation.

**Transmission**

Transmitted by a vector; a nematode; *Xiphinema index* (Hewitt *et al.*, 1958) and *X. italiae* (Cohn *et al.*, 1970). Virus does not require a helper virus for vector transmission; transmitted by mechanical inoculation; transmitted by grafting; not transmitted by contact between plants; transmitted by seed (rarely if ever transmitted to embryos of *Vitis vinifera*, virus abundant in endosperm of infected viruses but not in embryo. Transmitted by seed in *Chenopodium amaranticolor* (Dias, 1963), *C. quinoa* (Brückbauer and Rüdel, 1961*a*), *Glycine max* (Cory and Hewitt, 1968); not transmitted by pollen (although found in pollen of *Vitis*).

**Ecology and control**

Studies reported by Martelli (1978); Raski *et al.* (1983); Esmenjavd (1983).

**Geographical distribution**

Probably distributed worldwide (in all areas where *Vitis vinifera* and American hybrid rootstocks are grown).

**Experimental host range**

Several (3–9) families susceptible.

Diagnostically susceptible

*Vitis* ssp. — green or yellow systemic mosaic, rings, line patterns and flecks.

*Chenopodium amaranticolor* and *C. quinoa* — systemic mottling and leaf malformation.

*Cucumis sativus, Phaseolus vulgaris* cv. Bountiful — systemic chlorotic or necrotic mosaic, mottling, flecking or ringspots. Immune to some strains.

*Gomphrena globosa* — systemic chlorotic spots, leaves twisted.

*Nicotiana benthamiana* — systemic mottling, leaf malformation, stunting.

*Nicotiana clevelandii* — systemic mottling and stunting.

Maintenance and propagation hosts

*Chenopodium amaranticolor; C. quinoa; Gomphrena globosa, Phaseolus vulgaris.* Good sources for virus purification are *C. quinoa, N. clevelandii, N. benthamiana. G. globosa* is good for maintaining cultures.

Assay hosts (**L**ocal lesions or **W**hole plants)

*Chenopodium amaranticolor* (W); *Gomphrena globosa* (W); *Phaseolus vulgaris* cv. Bountiful (W). No satisfactory local lesion hosts are known.

Dicotyledonous Hosts *(**Susceptible** or Insusceptible)*

***Amaranthus caudatus*. *Beta vulgaris*. *Chenopodium album*. *Chenopodium amaranticolor*. *Chenopodium ambrosioides*. *Chenopodium capitatum*. *Chenopodium foetidum*. *Chenopodium foliosum*. *Chenopodium hybridum*. *Chenopodium murale*. *Chenopodium quinoa*. *Cucumis sativus*. *Cucurbita maxima*. *Glycine max*. *Gomphrena globosa*. *Nicotiana benthamiana*. *Nicotiana bigelovii*. *Nicotiana clevelandii***, or *Nicotiana clevelandii*. ***Nicotiana debneyi*. *Nicotiana glutinosa***, or *Nicotiana glutinosa*. ***Nicotiana megalosiphon*. *Nicotiana rustica*. *Nicotiana tabacum*. *Petunia hybrida*. *Phaseolus vulgaris***. *Poncirus trifoliata*. *Prunus americana* × *salicina*. *Prunus armeniaca*. *Prunus cerasifera*. *Prunus domestica*. *Prunus mahaleb*. *Prunus persica*. *Prunus serrulata*. *Rubus idaeus*. *Rubus occidentalis*. ***Vicia faba*. *Vigna unguiculata*. *Vitis labrusca*. *Vitis rupestris*. *Vitis vinifera***.

Sources of host-range data

Brückbauer and Rüdel (1961*b*); Cadman *et al.* (1960); Dias (1963); Martelli and Hewitt (1963*a*); Taylor and Hewitt (1964); Vuittenez (1963).

**Properties of particles in sap**

TIP: 60–65 °C. LIV: 15–30 days (at 20°C). DEP: $log_{10}$ minus 3–4 (Brückbauer and Rüdel, 1961*b*; Cadman *et al.*, 1960; Dias, 1963; Taylor and Hewitt, 1964). Leaf sap contains few virions.

**Purification**

Method: Dias and Harrison (1963); Martelli and Hewitt (1963*b*); Taylor and Hewitt (1964).

**Particle morphology**

Virions isometric; not enveloped; 30 nm in diameter; angular in profile; without a conspicuous capsomere arrangement.

**Physical properties**

Three sedimenting components in purified preparations; sedimentation coefficient of the fastest 120 *S* (B); of the other(s) 86 *S* (M), or 50 *S* (T). Density 1.49 g $cm^{-3}$ in CsCl (B), or 1.41 g $cm^{-3}$ in CsCl (M), or 1.31 g $cm^{-3}$ in CsCl (T). Isoelectric point pH 4. $A_{260}/A_{280}$ ratio 1.67 (B), or 1.58 (M), or 0.73 (T; Quacquarelli *et al.*, 1976).

**Biochemical properties**

Virions contain 42 % nucleic acid (B), or 30 % nucleic acid (M), or 0 % nucleic acid (T); 58 % protein (B), or 70 % protein (M), or 100 % protein (T).

*Genome* consists of RNA; single–stranded; linear; of two parts; largest (or only) genome part the larger 7.342 kb; the 2nd largest 3.774 kb (Quacquarelli *et al.*, 1976). Genomic nucleic acid isolated by Diener and Schneider (1968); Quacquarelli *et al.* (1972; 1976): single-phase phenol/SDS; freezing and thawing or heating for 90 secs at 66°C. Infectivity retained when deproteinised with phenol or detergent.

*Sequence database accession code(s):*

D00915 Em(40)_vi:GFLRNA1 Gb(84)_vi:GFLRNA1 Grapevine fanleaf virus (GFLV) RNA1 for polyprotein. 4/94 7,342bp.

S38553 Em(40)_vi:S38553 Gb(84)_vi:S38553 RNA1=VPg (253K polyprotein) grapevine fanleaf nepovirus, Genomic RNA, 84 nt. 1/94 84bp.

X16907 Em(40)_vi:NEGFLVR2 Gb(84)_vi:NEGFLVR2 Grapevine Fanleaf virus (GFLV-F13) RNA2 encoding capsid protein. 9/93 3,774bp.

X60775 Em(40)_vi:GVFVCP Gb(84)_vi:GVFVCP Grapevine fanleaf virus genomic RNA for capsid protein. 10/91 2,305bp.

U11768 Em(43)_vi:Gf11768 Gb(89)_vi:U11768 Grapevine fanleaf virus coat protein gene, partial cds. 10/94 1,515bp.

U11770 Em(43)_vi:Gf770 Gb(89)_vi:Gfu11770 Grapevine fanleaf virus RNA2, deletion mutant. 10/94 501bp.

*Virion protein(s)* one; $M_r$ 54000 (Quacquarelli *et al.*, 1976). Method of preparation: Agrawal and Tremaine (1972): boil in presence of 1% SDS, 1% 2-mercaptoethanol and 4 M urea.

**Replication**

Replication does not depend on a helper virus (Morris-Krsinich *et al.*, 1983).

**Cytopathology**

Virions found in roots, mesophyll and vascular parenchyma in herbaceous hosts or grapevines (Gerola *et al.*, 1969; Saric and Wrischer, 1975; Kalasjan *et al.*, 1979); in cytoplasm, or in nuclei (where empty capsids may accumulate (Savina *et al.*, 1985)). Inclusions present in infected cells; are membranous bodies and unusual in shape (crystalline aggregates in the cytoplasm and tubules); they contain virions. Other cellular changes: leaf deformations that are open marginal, petiolar sinuses and prominent marginal teeth. Canes have unevenly spaced or double nodes and are flat. Cell walls have outgrowths (Saric and Wrischer, 1975; Kalasjan *et al.*, 1979) and trabeculae in phloem and xylem tissues (Gifford *et al.*, 1956).

**Taxonomy and Relationships**

*Nepovirus: Comoviridae.*

Virus(es) with serologically related virions: different strains of grapevine fanleaf virus are antigenically close (Cadman *et al.*, 1960; Barabino, 1963; Dias and Harrison, 1963; Martelli and Hewitt, 1963*b*; Taylor and Hewitt, 1964), and all are distantly related to arabis mosaic virus. Virus(es) with serologically unrelated virions: tomato ringspot and tobacco ringspot viruses.

Best tests for diagnosis: serology must be used to identify fanleaf virus. Seedlings or rooted cuttings of *V. vinifera* and *V. rupestris* are good bait plants in studies with nematode vectors.

**References**


Agrawal, H.D. and Tremaine, J.H. (1972). *Virology* **47**: 8.

Barabino, G. (1963). *Hort. Res.* **3**: 27.

Bercks, R. (1968). *Weinb. Kell.* **14**: 151.

Bovey, R., Brugger, J.J. and Gugerli, P. (1982). *Proc. 7th Meeting ICVG, Niagara Falls 1980*, p. 251.

Brückbauer, H. and Rüdel, M. (1961*a*). *Wein-Wiss.* **16**: 177.

Brückbauer, H. and Rüdel, M. (1961*b*). *Wein-Wiss.* **16**: 197.

Cadman, C.H., Dias, H.F. and Harrison, B.D. (1960). *Nature, Lond.* **187**: 577.

Cohn, E., Tanne, E. and Nitzany, F.E. (1970). *Phytopathology* **60**: 181.

Cory, L. and Hewitt, W.B. (1968). *Phytopathology* **58**: 1316.

Dias, H.F. (1963). *Ann. appl. Biol.* **51**: 85.

Dias, H.F. and Harrison, B.D. (1963). *Ann. appl. Biol.* **51**: 97.

Diener, T.O. and Schneider, I.R. (1968). *Arch. Biochim. Biophys.* **124**: 401.

Esmenjavd, D. (1983). *Prog. Agric. Viticole* **100**: 514.

Gerola, F.M., Bassi, M. and Belli, G. (1971). *G. Bot. Ital.* **103**: 271.

Gifford, E.M., Hewitt, W.B., Graham, A.D. and Lamoureux, C.H. (1956). *Bull. Calif. Dep. Agric.* **45**: 268.

Goheen, A.C. and Hewitt, W.B. (1962). *Am. J. Enol. Vitic.* **13**: 73.

Hewitt, W.B. (1950*a*). *Bull. Calif. Dep. Agric.* **39**: 6.

Hewitt, W.B. (1950*b*). *Bull. Calif. Dep. Agric.* **39**: 62.

Hewitt, W.B., Martelli, G., Dias, H.F. and Taylor, R.H. (1970). *CMI/AAB Descr. Pl. Viruses* No. 28, 4 pp.

Kalasjan, J.A., Litvak, L.A. and Marinesku, V. (1979). *Arch. Phytopath. Pflschutz* **6**: 373.

Martelli, G.P. (1978). *Nematol. Medit.* **6**: 1.

Martelli, G.P. and Hewitt, W.B. (1963*a*). *Phytopathol. Medit.* **2**: 275.

Martelli, G.P. and Hewitt, W.B. (1963*b*). *Phytopathol. Medit.* **2**: 285.

Morris-Krsinich, B.A.M., Foster, R.L.S. and Mossop, D.W. (1983). *Virology* **130**: 523.

Quacquarelli, A., Piazzolla, P. and Vovlas, C. (1972). *J. gen. Virol.* **17**: 147.

Quacquarelli, A., Gallitelli, D., Savino, V. and Martelli, G.P. (1976). *J. gen. Virol.* **32**: 349.

Raski, D.J., Goheen, A.C., Lider, L.A. and Meredith, C.P. (1983). *Plant Dis.* **67**: 335.
Rathay, E. (1883). *Önol. Lehranstalt Klosterneuburg*, 3.
Ritzenhaler, C., Viry, M., Pinck, M., Margis, R., Fuchs, M. and Pinck, L. (1991). *J. gen. Virol.* **172**: 2357.
Russo, M. (1985). *Phytopathol. Medit.* **24**: 144.
Russo, M., Martelli, G.P. and Savino, V. (1982). *Proc. 7th Meeting ICVG, Niagara Falls 1980*, p. 251.
Saric, A. and Wrischer, M. (1975). *Phytopath. Z.* **84**: 97.
Savino, V., Cherif, C. and Martelli, G.P. (1985). *Phytopathol. Medit.* **24**: 29.
Taylor, R.H. and Hewitt, W.B. (1964). *Aust. J. agric. Res.* **15**: 571.
Vuittenez, A. (1963). *C. r. hebd. Seanc. Acad. Agric. Fr.* **49**: 795.

# Grapevine fanleaf satellite RNA

*Sequence database accession code(s):*
D00442 Em(40)_vi:GFLRNA3 Gb(84)_vi:GFLRNA3 Grapevine fanleaf virus, satellite RNA (RNA3), complete sequence. 3/91 1,114bp.

**Taxonomy and Relationships**

Satellite RNA.

# Grapevine fleck virus

**Data collated by A.A. Brunt, 1992.**

*Synonyms:* grapevine marbrure virus (Vuittenez *et al.*, 1966), grapevine phloem-limited isometric virus (Boulila *et al.*, 1990). *Standard acronym:* GFkV.

First reported in *Vitis vinifera*; from France; by Vuittenez *et al.* (1966).

**Natural host range and symptoms**

*Vitis vinifera* — symptomless infection in most European cultivars and American *Vitis* species and their hybrids.

**Transmission**

Virus not transmitted by mechanical inoculation; transmitted by grafting.

**Geographical distribution**

Spreads in South Africa. Found, but with no evidence of spread, in European countries and the U.S.A.

**Experimental host range**

Few (<3) families susceptible.

Diagnostically susceptible

*Vitis rupestris* — scattered clearing of veinlets and leaf deformation with severe strains.

Maintenance and propagation hosts

*Vitis rupestris*.

**Particle morphology**

Virions isometric; not enveloped; 30 nm in diameter.

**Biochemical properties**

*Genome* consists of RNA; single–stranded.

**Taxonomy and Relationships**

Best tests for diagnosis: by grafting to *Vitis vinifera* and serological methods (ELISA and ISEM).

**References**

Boscia, D., Martelli, G.P., Savino, V. and Castellano, M.A. (1991). *Vitis* (in press).
Boulila, M., Boscia, D., di Terlizzi, B., Castellano, M.A., Minafra, A., Savino, V. and Martelli, G.P. (1990). *J. Phytopath.* **129**: 151.
Engelbrecht, D.J. and Kasdorf, G.G.F. (1990). *Phytophylactica* **22**: 347.
Hewitt, W.B., Goheen, A.C., Cory, L. and Luhn, C. (1972). *Annls. Phytopath.* **23**: 43.
Vuittenez, A., Legin, R. and Kuszala, J. (1966). *Annls. Épiphyt.* **17**: 67.

# Grapevine leafroll-associated (?) *closteroviruses*

**Data collated by A.A. Brunt 1995.**

**Natural host range and symptoms**

Five viruses, designated grapevine leafroll-associated 1,2,3,4 and 5 closteroviruses have been associated with grapevine leafroll disease (Candresse and Martelli, 1995), but none has been fully characterized.

*Vitis vinifera* — rolling and discoloration of leaves (reddish-purple in red-fruited, and yellowish in white-fruited cultivars), especially conspicuous in mate summer and early autumn in intolerant cultivars. Symptomless infection in some American *Vitis* spp. and hybrids.

**Transmission**

Transmitted by *Pseudococcus longispinus, Planococcus ficus* (GLRaV–3). Virus transmitted by grafting.

**Geographical distribution**

Probably distributed worldwide.

Diagnostically susceptible

*Vitis vinifera* cvs Cabernet, Pinot Noir, Mission, Gamay — leafrolling and discoloration.

Dicotyledonous Hosts *(**Susceptible** or Insusceptible)*

***Vitis vinifera.***

**Particle morphology**

Virions filamentous; usually flexuous; of 1800–2200 nm.

*Sequence database accession code(s):*

X75448 Em(40)_vi:GVBMPCP Gb(84)_vi:GVBMPCP Grapevine virus B genomic RNA for MP and CP genes. 5/94 2,161bp.

U22158 Gb(89)_vi:Glu22158 Grapevine leafroll associated virus type III p20 protein, partial cds. 3/95 1,012bp.

U22170 Em(43)_vi:Gl22170 Gb(89)_vi:Glu22170 Grapevine leafroll associated virus RNA6 gene, partial cds. 5/95 965bp.

**Taxonomy and Relationships**

Possible *Closterovirus*.

**References**

Candresse, T. and Martelli, G.P. (1995). *Arch. Virol.*, Suppl. **10**: 461.
Faoro, F., Tornaghi, R., Cinquanta, S. and Belli, G. (1992). *Riv. Pat. Veg. SV* **2**: 67.
Goheen, A.C., Harmon, F.N. and Weinberger, J.H. (1958). *Phytopathology* **48**: 51.
Gugerli, P., Brugger, J.J. and Bovey, R. (1984). *Rev. Suisse Vitic. Arb. Hort.* **16**; 299.
Hu, J.S., Gonsalves, D. and Teliz, D. (1990). *J. Phytopathology* **128**: 1.
Mossop, D.W., Elliot and Richards (1985). *N.Z. Jl agric. Res.* **28**: 419
Rezaian, M.A., Krake, L.R., Canying, Q. and Hazzalin, C.A. (1991). *J. Virol. Methods* **31**: 325.
Rosciglione, B. and Gugerli, P. (1989). *Proc. 9th Meeting I.C.V.G.*, Kiryat Anavim 1987, p.67.
Tanne, E., Ben-Dov, Y. and Raccah, B. (1989). *Proc. 9th Meeting I.C.V.G.*, Kiryat Anavim 1987, p.71.
Teliz, D., Tanne, E., Gonsalves, D. and Zee, F. (1987). *Plant Dis.* **71**: 704.
Zimmermann, D., Bass, P., Legin, R. and Walter, B. (1990). *Phytopathology* **130**: 205.

# Grapevine line pattern (?) *ilarvirus*

**Data collated by A.A. Brunt, 1992.**

First reported in *Vitis vinifera*; from Hungary; by Lehoczky *et al.* (1989).

**Natural host range and symptoms**

Symptoms persist.
*Vitis vinifera* — chlorotic spots, blotches, rings or line patterns.

**Transmission**

Virus transmitted by mechanical inoculation; transmitted by grafting.

**Geographical distribution**

Spreads in Hungary.

**Experimental host range**

Few (<3) families susceptible.

Diagnostically susceptible

*Cucumis sativus, Nicotiana glutinosa* — chlorotic local lesions, systemic mottle.

Maintenance and propagation hosts

*Cucumis sativus, N. glutinosa.*

Assay hosts (**L**ocal lesions or **W**hole plants)

*Cucumis sativus* (W), *Nicotiana glutinosa* (W).

Dicotyledonous Hosts *(**Susceptible** or Insusceptible)*

***Cucumis sativus*. *Nicotiana glutinosa*. *Vitis vinifera*.**

**Particle morphology**

Virions isometric; not enveloped; 24 nm in diameter.

**Taxonomy and Relationships**

*Ilarvirus: Bromoviridae* (possible).

**References**

Lehoczhy, J., Bosca, D., Burgyan, J., Castellano, M.A., Beczner, L. and Farkas, G. (1989). *Proc. 9th Meeting ICVG, Kiryat Anavim, 1987*, pp. 23–30.

# Grapevine stem pitting associated *closterovirus*

**Data collated by S. Holland, 1986.**

*Synonyms:* grapevine virus A, grapevine corky bark virus, grapevine stem-pitting virus.

First reported in *Vitis vinifera*; from Taranto, Italy; by Ciccarone (1961).

**Natural host range and symptoms**

Symptoms persist.

*Vitis vinifera* — pits and grooves in trunk and *legno riccio*, rugose wood, possibly caused.

**Transmission**

Transmitted by a vector (very rarely); an insect; Pseudococcidae. Virus transmitted by mechanical inoculation (only to *Nicotiana glutinosa*); transmitted by grafting; not transmitted by contact between plants; not transmitted by seed; not transmitted by pollen.

**Ecology and control**

Studies reported by Rosciglione *et al.* (1983).

**Geographical distribution**

Spreads in the Middle East; Italy, South Africa, and the USA (California).

**Experimental host range**

Few (<3) families susceptible.

Diagnostically susceptible

*Vitis rupestris* var. *rupestris* cv. St. George — pits and grooves in trunk.

Maintenance and propagation hosts

*Nicotiana glutinosa.*

Dicotyledonous Hosts ***(Susceptible** or Insusceptible)*

*Chenopodium amaranticolor. Chenopodium quinoa.* ***Nicotiana clevelandii.*** *Nicotiana glutinosa.* ***Vitis labrusca. Vitis rupestris. Vitis vinifera.***

**Properties of particles in sap**

TIP: 50 °C. LIV: 4–5 days. DEP: $\log_{10}$ minus 5. Leaf sap contains few virions (when isolated from *Nicotiana clevelandii*). Electron microscopy: when stained in UA the pitch of the basic helix can be measured.

**Purification**

Method: Milne *et al.* (1980).

**Particle morphology**

Virions filamentous; not enveloped; usually flexuous; with a clear modal length; of 800 nm; 11–12 nm wide. Basic helix obvious; pitch of basic helix 3.7 nm.

**Physical properties**

One sedimenting component in purified preparations (in CsCl).

**Biochemical properties**

*Genome* consists of RNA; single–stranded; linear. Genomic nucleic acid isolated by Boccardo and D'Aquilo (1981). Base composition 28.4 % G; 24.4 % A; 24.9 % C; 22.3 % U. Additional factor not required for infectivity.

Method of preparation: Boccardo and D'Aquilo (1981).

**Replication**

Replication does not depend on a helper virus.

**Cytopathology**

Virions found in phloem, but are very difficult to extract from host tissue.

**Taxonomy and Relationships**

*Closterovirus*.

Virus(es) with serologically related virions: citrus tristeza virus, but distantly (Conti *et al.*, 1980). Virus(es) with serologically unrelated virions: apple chlorotic leafspot, beet yellows, carnation yellow fleck, heracleum latent, lilac chlorotic leafspot, apple stem grooving and potato T viruses.

**References**

Boccardo and D'Aquilo (1981). *J. gen. Virol.* **53**: 179.
Ciccarone (1961). *Hort. Agric.* **98**: 901.
Conti, M. (1980).
Rosciglione, B. (1983). *Vitis* **22**: 331.

# Grapevine stunt virus

**Data collated by S. Yamashita, 1991.**

First reported in *Vitis vinifera*; from Japan, in Okayama Prefecture; by Namba *et al.* (1981).

**Natural host range and symptoms**

Symptoms vary seasonally. Symptoms leaf rolling and stunting in spring.

*Vitis vinifera, V. thumbergii* — leaf rolling and stunting.

**Transmission**

Transmitted by a vector and means not involving a vector; an insect; *Arboridia apicalis*; Cicadellidae. Not transmitted by *Myzus persicae*. Transmitted in a persistent manner. Virus not transmitted by mechanical inoculation; transmitted by grafting; not transmitted by contact between plants; not transmitted by seed.

**Ecology and control**

Studies reported by Yamashita: control by vector insect control and use of virus-free stock for grafting.

**Geographical distribution**

Spreads in Japan.

**Experimental host range**

Few (<3) families susceptible. Experimentally infected plants mostly show leaf rolling and stunting.

Diagnostically susceptible

*Vitis vinifera* cv. Campbell Early, *V. thumbergii* — leaf rolling and stunting.

Maintenance and propagation hosts

*Vitis vinifera* cv. Campbell Early.

Dicotyledonous Hosts (***Susceptible*** *or Insusceptible)*

*Brassica oleracea* ssp. *botrytis. Brassica oleracea* ssp. *capitata. Chenopodium amaranticolor. Chenopodium quinoa. Cucumis sativus. Datura stramonium. Glycine max. Gomphrena globosa. Lycopersicon esculentum. Nicotiana glutinosa. Nicotiana rustica. Nicotiana tabacum. Petunia hybrida. Phaseolus vulgaris. Physalis floridana. Raphanus sativus. Solanum melongena. Tetragonia expansa. Vigna unguiculata.* ***Vitis vinifera***.

Additional hosts

***Vitis thunbergii*** and ***Vitis vinifera*** cv. Campbell Early.

Sources of host-range data

Hatamoto *et al.* (1982).

**Properties of particles in sap**

Leaf sap contains few virions.

**Purification**

Method: Namba *et al.* (1981 and 1986).

**Particle morphology**

Virions isometric; not enveloped; 25 nm in diameter; angular in profile; without a conspicuous capsomere arrangement.

**Replication**

Replication does not depend on a helper virus.

**Cytopathology**

Virions found in leaves and phloem; in cytoplasm. Inclusions present in infected cells; are unusual in shape; are small vesicles containing fibrous material in the cytoplasm and vacuole. Other cellular changes: phloem necrosis, development of small vesicles and the accumulation of starch grains in chloroplasts in the mesophyll cells.

**Taxonomy and Relationships**

Best tests for diagnosis: induction of characteristic symptoms in *Vitis vinifera* cv. Campbell Early and *Vitis thumbergii* and leafhopper transmission.

**References**


Hatamoto, M., Fujii, S., Namba, S., Yamashita, S. and Doi, Y. (1982). *Ann. Phytopath. Soc. Japan* **48**: 396.

Namba, S. (1983). In: *"Handbook of Plant viruses in Japan"*, p. 344. Eds K. Yora, Y. Saito, Y. Doi, T. Inouye and K. Tomaru. Asakura Shoten, Japan.

Namba, S., Iwanami, T., Yamashita, S. and Doi, Y. (1986). *A.A.T.C. Tech. Bull.* **33**: 109.

Namba, S., Yamashita, S., Doi, Y. and Yora, K. (1981). *Ann. Phytopath. Soc. Japan* **47**: 137.

# Groundnut chlorotic spot (?) *potexvirus*

**Data collated by A.A. Brunt, 1990.**

First reported in *Arachis hypogaea*; from Ivory Coast; by Fauquet *et al.* (1985).

**Natural host range and symptoms**

Symptoms persist.

*Arachis hypogaea* — chlorotic spotting and mottling, ring spotting and line patterns in young leaves and green vein-banding in older leaves.

**Transmission**

Transmitted by a vector; an insect; *Aphis craccivora* and *A. spiraecola*; Aphididae. Transmitted in a non–persistent manner. Virus transmitted by mechanical inoculation.

**Geographical distribution**

Spreads in Cote d'Ivoire.

**Experimental host range**

Few (<3) families susceptible. Experimentally infected plants mostly show systemic mosaics, mottles, ringspots or necrosis.

Diagnostically susceptible

*Arachis hypogaea.*
*Chenopodium amaranticolor* — red necrotic local lesions; not systemic.
*Physalis floridana* — green systemic mosaic.
*Nicotiana benthamiana* — leaf malformation.
*Nicotiana megalosiphon* — systemic chlorotic spotting.

Maintenance and propagation hosts

*Arachis hypogaea, Nicotiana megalosiphon, Physalis floridana.*

Assay hosts (**L**ocal lesions or **W**hole plants)

*Chenopodium amaranticolor* (L).

Dicotyledonous Hosts *(**Susceptible** or Insusceptible)*

***Arachis hypogaea. Chenopodium amaranticolor. Nicotiana benthamiana. Nicotiana megalosiphon.***

Sources of host-range data

Fauquet *et al.* (1985).

**Properties of particles in sap**

TIP: 55–60 °C. LIV: 5 days (at 24°C), or 14 days (at 4°C). DEP: $\log_{10}$ minus 5–6. Leaf sap contains many virions.

**Purification**

Method: Fauquet *et al.* (1985).

**Particle morphology**

Virions filamentous; not enveloped; usually flexuous; with a clear modal length; of 456 nm (±22); 13 nm wide (±1). Axial canal obscure. Basic helix obscure.

**Taxonomy and Relationships**

Possible *Potexvirus*.

Virus(es) with serologically unrelated virions: cactus X, clover yellow mosaic, cymbidium mosaic, dioscorea latent, narcissus mosaic, papaya mosaic, potato X and white clover mosaic viruses.

**Comments**

The virus differs from better characterised potexviruses in being transmitted in the non-persistent manner by aphids.

**References**

Fauquet, C., Thouvenel, J.-C. and Fargette, D. (1985). *C.r. hebd. Seanc. Acad. Sci., Paris, Serie III* No. 17: 773.

# Groundnut eyespot *potyvirus*

**Data collated by K. Boswell, 1985.**

*Standard acronym:* GEV.

First reported in *Arachis hypogaea* (unnamed cultivar); from Korhogo, North of the Ivory Coast; by Dubern and Dollet (1978).

**Natural host range and symptoms**

*Arachis hypogaea* — dark green spots with chlorotic halos and dark green line patterns on leaflets.

*Physalis floridana* — mild mosaic leaf symptoms.

**Transmission**

Transmitted by a vector; an insect; *Aphis craccivora*; Aphididae. Transmitted in a non–persistent manner. Virus transmitted by mechanical inoculation.

**Geographical distribution**

Spreads in Cote d'Ivoire.

**Experimental host range**

Several (3–9) families susceptible.

Maintenance and propagation hosts

*Arachis hypogaea, Physalis floridana.*

Assay hosts (**L**ocal lesions or **W**hole plants)

No suitable local lesion host. *Arachis hypogaea* cv. Te3 (W) used for *in vitro* properties and host range studies. *A. hypogaea* or *Physalis floridana* used as source for purification.

Dicotyledonous Hosts (***Susceptible*** *or Insusceptible)*

***Antirrhinum majus***. ***Arachis hypogaea***. ***Beta vulgaris***. *Cajanus cajan. Capsicum annuum. Capsicum frutescens. Cassia occidentalis. Chenopodium amaranticolor. Chenopodium murale. Chenopodium quinoa. Clitoria ternatea. Datura metel. Datura stramonium. Gomphrena globosa. Gossypium hirsutum. Helianthus annuus. Hibiscus esculentus.* ***Lycopersicon esculentum***. *Macroptilium lathyroides. Medicago sativa. Melilotus alba. Nicotiana clevelandii. Nicotiana × edwardsonii. Nicotiana glutinosa. Nicotiana megalosiphon. Nicotiana rustica. Nicotiana tabacum.* ***Petunia hybrida***. *Phaseolus vulgaris.* ***Physalis floridana***. *Ricinus communis. Solanum nigrum.* ***Tetragonia expansa***. ***Torenia fournieri***. *Trifolium repens. Vicia faba. Vigna unguiculata. Zinnia elegans.*

Monocotyledonous Hosts (***Susceptible*** *or Insusceptible)*

*Oryza sativa. Zea mays.*

Comments on host-range

Dioscoraceae and Passifloraceae not susceptible.

Sources of host-range data

Dubern and Dollet (1980).

**Properties of particles in sap**

TIP: 44 °C. LIV: 0.125 days (3 hours). DEP: $\log_{10}$ minus 3–4. Leaf sap contains few virions.

**Purification**

Method: Dubern and Dollet (1980).

**Particle morphology**

Virions filamentous; not enveloped; usually flexuous; with a clear modal length; of 755 nm; 12.5 nm wide.

**Physical properties**

One sedimenting component in purified preparations.

**Biochemical properties**

Virions contain 5.5–6.5 % nucleic acid (from $A_{260}/A_{280}$ ratio (Layne 1977)).

**Cytopathology**

Virions found in all parts of the host plant; in cytoplasm. Inclusions present in infected cells; are pinwheels; they do not contain virions.

**Taxonomy and Relationships**

*Potyvirus: Potyviridae.*

Virus(es) with serologically related virions: guinea grass mosaic, iris fulva mosaic, passionfruit ringspot, pepper veinal mottle (Ivory Coast and Ghana strains) and wisteria vein mosaic viruses. Virus(es) with serologically unrelated virions: pepper mottle, bean common mosaic, soybean mosaic, bean yellow mosaic, bidens mottle, clover yellow vein, Colombian datura virus, blackeye cowpea mosaic, henbane mosaic, hippeastrum mosaic, lettuce mosaic, parsnip mosaic, pea seed-borne mosaic, peanut mottle, potato A, potato Y, sugarcane mosaic, ryegrass mosaic, tobacco etch, turnip mosaic and yam mosaic viruses.

**References**

Dubern, J. (1981). *Phytopath. Z.* **100**: 237.
Dubern, J. and Dollet, M. (1978). *Oleagineux* **33**: 175.
Dubern, J. and Dollet, M. (1980). *Ann. appl. Biol.* **96**: 193.
Layne, E. (1977). In: *Methods in Enzymology* Vol. 3, p. 447; ed. S.P. Colowick and N.O. Kaplan. Academic Press, New York.

# Groundnut ringspot (?) *tospovirus*

**Data collated by A.A. Brunt, 1995.**

First reported in *Arachis hypogaea*; from South Africa; by Klesser (1966).

**Natural host range and symptoms**

*Arachis hypogaea* — ringspots in leaves.

**Transmission**

Transmitted by a vector; an insect; *Frankliniella occidentalis*; Thysanoptera.

**Geographical distribution**

Spreads in Argentina, Brazil, and South Africa.

**Taxonomy and Relationships**

Possible *Tospovirus: Bunyaviridae*.

**References**

De Avila, A.C., Huguenot, R., Resende, R. de O., Kitajima, E.W., Goldbach, R.W. and Peters, D. (1990). *J. gen. Virol.* **71**: 2801.

De Avila, A.C., de Haan, P., Kitajima, E.W., Kormelink, R., Resende, R. de O., Goldbach, R.W. and Peters, D. (1992). *J. Phytopathology* **134**: 133.

De Avila, A.C., de Haan, P., Kormelink, R., Resende, R. de O., Goldbach, R.W. and Peters, D. (1993). *J. gen. Virol.* **74**: 153.

Dewey, R.A., Semorile, L., Grau, O., de Avila, A.C. and Kitajima, E.W. (1995). *J. Phytopathology* **143**: in press.

Klesser, P.J. (1966). *S. Afr. J. agric. Sci.* **9**: 711.

# Groundnut rosette assistor *luteovirus*

**Data collated by R. Hull, 1981. Revised by D.V.R. Reddy in 1989.**

*Standard acronym:* GRAV.

First reported in *Arachis hypogaea* (groundnut rosette disease); from Tanzania and later in Malawi, Zimbabwe and Nigeria; by Zimmerman (1907).

**Natural host range and symptoms**

Symptoms persist.

*Arachis hypogaea* — symptomless infection.

**Transmission**

Transmitted by a vector; an insect; *Aphis craccivora, A. gossypii*; Aphididae. Transmitted in a persistent manner. Virus retained when the vector moults; does not multiply in the vector; not transmitted congenitally to the progeny of the vector; can help the vector transmission of another virus (groundnut rosette (?) *umbravirus*); not transmitted by mechanical inoculation; transmitted by grafting; not transmitted by contact between plants; not transmitted by seed; not transmitted by pollen.

**Ecology and control**

Studies reported by Storey and Bottomley (1928); Storey and Ryland (1955); Davies (1975); Adams (1967); Gibbons (1977); Reddy (1984).

**Geographical distribution**

Spreads in the African region (in those countries south of the Sahara. It is the most important virus disease of groundnuts in Africa). Found, but with no evidence of spread, in India and the Phillipines.

**Experimental host range**

Few (<3) families susceptible.

Diagnostically susceptible

*Arachis hypogaea* — latent.

Maintenance and propagation hosts

*Arachis hypogaea*.

Assay hosts (**L**ocal lesions or **W**hole plants)

*Arachis hypogaea* (W).

Dicotyledonous Hosts *(**Susceptible** or Insusceptible)*

***Arachis hypogaea.***

Sources of host-range data

Zimmerman (1907); Hull (1972).

**Properties of particles in sap**

Leaf sap contains few virions.

**Purification**

Method: Rajeshwari and Murant (1988).

**Particle morphology**

Virions isometric; not enveloped; 28 nm in diameter; rounded in profile; with a conspicuous capsomere arrangement.

**Physical properties**

One sedimenting component in purified preparations. Density in caesium sulphate 1.34.

**Biochemical properties**

*Genome* consists of RNA; single–stranded; linear; unipartite; largest (or only) genome part 6.9 kb. Genomic nucleic acid isolated by Rajeshwari and Murant (1988).

*Sequence database accession code(s):*

S54327 Em(40)_un:S54327 Gb(84)_vi:S54327 nucleoprotein (S RNA) groundnut ringspot virus, isolate SA–05, Genomic RNA, 928 nt. 3/93 92 1 sequence.

*Non–genomic nucleic acid* found in the virions; is of an unusual type of structure; found in virions and are groundnut rosette virus nucleic acid and its satellite RNA.

*Virion protein(s)* one; $M_r$ 24000. Virion proteins not glycosylated. Virion proteins not phosphorylated.

**Replication**

Genome replicates in cytoplasm. Replication does not depend on a helper virus.

**Cytopathology**

Virions found in roots and phloem; in cytoplasm.

**Taxonomy and Relationships**

*Luteovirus.*

Virus(es) with serologically related virions: bean leaf roll, beet western yellows and potato leafroll viruses.

A MAb for the virions of potato leafroll virus reacted with those of groundnut rosette assistor virus. Best tests for diagnosis: serology using luteovirus antisera.

**References**

Hull, R. (1967). *Ann. appl. Biol.* **60**: 199.
Hull, R. (1968). *Ann. appl. Biol.* **62**: 139.
Hull, R. (1972). *Ann. appl. Biol.* **71**: 135.
Rajeshwari, R. and Murant, A.F. (1988). *Ann. appl. Biol.* **112**: 403.
Zimmerman, A. (1907). *Pflanzer* **3**: 129.

# Groundnut rosette *umbravirus*

**Data collated by D.V.R. Reddy, 1989.**

*Standard acronym:* GRV. *Strains:* Different diseases called groundnut chlorotic rosette, groundnut green rosette and groundnut mosaic rosette caused by complexes of different strains of the virus, its assistor luteovirus and satellite RNAs.

First reported in *Arachis hypogaea*; from Tanzania; by Zimmerman (1907).

**Natural host range and symptoms**

Symptoms persist.

*Arachis hypogaea*, different isolates give:

chlorotic rosette — faint mottling of youngest leaflets, then yellowing, curling and malformation of leaflets, some veinbanding and blotching.

green rosette — mild mottling and flecking, but mostly dark green, severe stunting.

mosaic rosette — green blotching and severe chlorosis, but less severe rosetting than with green rosette. There are three components associated with all three rosette diseases;

1) Groundnut rosette assistor virus (GRAV): a luteovirus which gives no overt symptoms in groundnut.

2) Groundnut rosette virus: a virus with a single-stranded RNA genome, which becomes packaged in the GRAV virions, and thus depends on GRAV for aphid transmission, but produces no overt symptoms in groundnut.

3) Groundnut rosette satellites: satellite RNAs that control the symptoms and cause the different types of rosette (chlorotic, green and mosaic).

**Transmission**

Transmitted by a vector; an insect; Aphididae. Transmitted in a persistent manner. Virus retained when the vector moults; does not multiply in the vector; not transmitted congenitally to the progeny of the vector; requires, for vector transmission, a helper virus (groundnut rosette assistor luteovirus); transmitted by mechanical inoculation; transmitted by grafting; not transmitted by contact between plants; not transmitted by seed.

**Ecology and control**

Studies reported by Storey and Bottomley (1928); Storey and Ryland (1955); Davies (1975); Adams (1967); Gibbons (1977); Reddy (1984).

**Geographical distribution**

Spreads in the African region (in all countries south of the Sahara).

**Experimental host range**

Several (3–9) families susceptible.

Diagnostically susceptible

*Chenopodium amaranticolor* — chlorotic local lesions, not systemic.

icotiana clevelandii — necrotic rings, then systemic curling and malformation.

*Arachis hypogaea* — see 'Natural host and symptoms'.

Diagnostically insusceptible

*Cucurbita pepo* cv. Eyes, *Pisum sativum, Phaseolus vulgaris* cvs Black turtle soup, Bountiful, Great Northern and Pinto, *Vigna unguiculata.*

Maintenance and propagation hosts

*Arachis hypogaea, Nicotiana clevelandii.*

Assay hosts (**L**ocal lesions or **W**hole plants)

*Chenopodium amaranticolor* (L), *C. murale* (L), *C. quinoa* (L).

Dicotyledonous Hosts *(**Susceptible** or Insusceptible)*

*Anthriscus cerefolium*. ***Arachis hypogaea***. *Beta vulgaris*. *Brassica campestris*. *Calendula officinalis*. *Catharanthus roseus*. ***Chenopodium amaranticolor***. ***Chenopodium murale***. ***Chenopodium quinoa***. *Coriandrum sativum*. *Cucurbita pepo*. *Cyamopsis tetragonoloba*. *Datura stramonium*. *Daucus carota*. *Dolichos biflorus*. ***Glycine max***. *Gomphrena globosa*. *Hibiscus esculentus*. *Hyoscyamus niger*. *Lathyrus odoratus*. *Medicago sativa*. ***Nicotiana benthamiana***. ***Nicotiana clevelandii***. ***Nicotiana* × *edwardsonii***. *Nicotiana glutinosa*. ***Nicotiana rustica***. *Nicotiana tabacum*. *Pastinaca sativa*. *Petunia hybrida*. *Phaseolus vulgaris*. *Physalis floridana*. *Pisum sativum*. ***Trifolium incarnatum***. ***Trifolium repens***. *Voandzeia subterranea*. *Zinnia elegans*.

Sources of host-range data

Okusanya and Watson (1966); Dubern (1980); Reddy *et al.* (1985).

**Properties of particles in sap**

Infectivity of sap decreased by treatment with di–ethyl ether.

**Purification**

Method: Reddy *et al.* (1985).

**Physical properties**

One sedimenting component in purified preparations.

**Biochemical properties**

*Genome* consists of RNA; single–stranded; linear; unipartite; largest (or only) genome part 2.05 kb. Genomic nucleic acid isolated by Reddy *et al.* (1985); Murant *et al.* (1988). Infectivity retained when deproteinised with phenol or detergent. Additional factor not required for infectivity.

*Sequence database accession code(s):*

Z29702 Em(40)_vi:GRVSATR1 Gb(84)_vi:GRVSATR1 Groundnut rosette virus Malawi chlorotic (MC3A) satellite RNA. 2/94 901bp.

Z29703 Em(40)_vi:GRVSATR2 Gb(84)_vi:GRVSATR2 Groundnut rosette virus Malawi chlorotic (MC3B) satellite RNA. 2/94 901bp.

Z29704 Em(40)_vi:GRVSATR3 Gb(84)_vi:GRVSATR3 Groundnut rosette virus Nigerian green (NG3A) satellite RNA. 2/94 898bp.

Z29705 Em(40)_vi:GRVSATR4 Gb(84)_vi:GRVSATR4 Groundnut rosette virus Nigerian green (NG3B) satellite RNA. 2/94 899bp.

Z29706 Em(40)_vi:GRVSATR5 Gb(84)_vi:GRVSATR5 Groundnut rosette virus Nigerian mild (NM3A) satellite RNA. 2/94 896bp.

Z29707 Em(40)_vi:GRVSATR6 Gb(84)_vi:GRVSATR6 Groundnut rosette virus Nigerian mild (NM3B) satellite RNA. 2/94 896bp.

Z29708 Em(40)_vi:GRVSATR7 Gb(84)_vi:GRVSATR7 Groundnut rosette virus Nigerian mild (NM3C) satellite RNA. 2/94 895bp.

Z29709 Em(40)_vi:GRVSATR8 Gb(84)_vi:GRVSATR8 Groundnut rosette virus Nigerian mild (NM3D) satellite RNA. 2/94 895bp.

Z29710 Em(40)_vi:GRVSATR9 Gb(84)_vi:GRVSATR9 Groundnut rosette virus Yellow blotch (YB3A) satellite RNA. 2/94 901bp.

Z29711 Em(40)_vi:GRVSATR0 Gb(84)_vi:GRVSATR0 Groundnut rosette virus Yellow blotch (YB3B) satellite RNA. 2/94 902bp. 10 sequences.

*Non–genomic nucleic acid* found in the virions; is satellite RNA. Sub–genomic *m*RNA found in infected cells.

**Replication**

Genome replicates in cytoplasm. Replication depends on co–infection with a helper virus; it requires groundnut rosette assistor virus for aphid transmission, and the satellite RNA determines the symptoms. Acts as helper for a satellite virus; a helper for groundnut satellite RNA, which has regions of homology with the host DNA, and is 2.7kb.

**Taxonomy and Relationships**

*Umbravirus*.

Best tests for diagnosis: diagnostic hosts; gel electrophoresis of double-stranded RNA.

**References**

Adams, A.N. (1967). *J. agric. Res.* **5**: 1.
Davies, J.C. (1975). *Trop. Agric.* **52**: 359.
Dubern, J. (1980). *Phytopath. Z.* **99**: 318.
Gibbons, R.W. (1977). In: *Diseases Pests and Weeds in Tropical Crops*, Groundnut rosette virus, p. 19; ed. P. Dorey. Verlag.
Murant, A.F. *et al.* (1988). *J. gen. Virol.* **69**: 1479.
Okusanya, B.A.M. and Watson, M.A. (1966). *Ann. appl. Biol.* **58**: 377.
Reddy, D.V.R. (1984). In: *Compendium of Peanut Diseases. Amer. Phytopath. Soc.* Groundnut rosette virus, p. 51.
Reddy, D.V.R., Murant, A.F. and Duncan, G.H. (1985). *Ann. appl. Biol.* **107**: 57.
Reddy, D.V.R., Raschke, J.H. and Mayo, M.A. (1985). *Ann. appl. Biol.* **107**: 65.
Storey, H.H. and Bottomley, A.M. (1928). *Ann. appl. Biol.* **15**: 26.
Storey, H.H. and Ryland, A.K. (1955). *Ann. appl. Biol.* **43**: 423.
Zimmerman, A. (1907). *Pflanzer* **3**: 129.

## Guar symptomless *potyvirus*

**Data collated by G.M. Behncken, 1983.**

*Standard acronym:* GSLV.

First reported in *Cyamopsis tetragonoloba*; from seed stocks from India, Pakistan, U.S.A.; by Hansen and Lesemann (1978).

**Natural host range and symptoms**

Symptoms disappear soon after infection.

*Cyamopsis tetragonoloba* — mild green mottle or symptomless.

**Transmission**

Transmitted by a vector; an insect; *Myzus persicae*; Aphididae. Transmitted in a non–persistent manner. Virus transmitted by mechanical inoculation; transmitted by seed (up to 70% in commercial seeds lots).

**Geographical distribution**

Probably distributed worldwide. Found, but with no evidence of spread, in Australia, India, Pakistan and the U.S.A.

**Experimental host range**

Few (<3) families susceptible.

Diagnostically susceptible

*Chenopodium amaranticolor* — necrotic local lesions.
*C. quinoa, Phaseolus vulgaris* cv. Bountiful — chlorotic local lesions.
*Glycine soja, Macroptilium lathyroides* — systemic mosaic.
*Macrotyloma uniflorum* — necrotic local lesions and systemic lethal necrosis.

Assay hosts (**L**ocal lesions or **W**hole plants)

*Chenopodium amaranticolor* (L), *Phaseolus vulgaris* cv. Bountiful (L).

Dicotyledonous Hosts *(**Susceptible** or Insusceptible)*

***Chenopodium album***. ***Chenopodium amaranticolor***. *Chenopodium foetidum*. ***Chenopodium murale***. ***Chenopodium quinoa***. *Cucumis sativus*. *Cucurbita maxima*. ***Cyamopsis tetragonoloba***. *Lycopersicon esculentum*. ***Macroptilium lathyroides***. *Macrotyloma uniflorum*. *Nicotiana clevelandii*. *Nicotiana tabacum*. ***Phaseolus vulgaris***. *Pisum sativum*. *Vicia faba*. ***Vigna unguiculata***.

Sources of host-range data

Hansen and Lesemann (1978).

**Properties of particles in sap**

Leaf sap contains few virions.

**Particle morphology**

Virions filamentous; not enveloped; usually flexuous; of 760 nm; 12 nm wide.

**Cytopathology**

Virions found in all parts of the host plant; in cytoplasm. Inclusions present in infected cells; are pinwheels (of the scroll-type); they do not contain virions. Other cellular changes: cell wall proliferations and enlarged mitochondria.

**Taxonomy and Relationships**

*Potyvirus: Potyviridae*.

**References**

Hansen, A.J. and Lesemann, D.E. (1978). *Phytopathology* **68**: 841.

# Guar top necrosis virus

**Data collated by J.P. Varma, 1981.**

*Strains:* type and attenuated.

First reported in *Cyamopsis tetragonoloba*; from Lucknow, India; by Verma *et al.* (1962).

**Natural host range and symptoms**

Symptoms persist.

*Cyamopsis tetragonoloba* — chlorotic leaf mottling and yellowing, chlorotic depressions then systemic necrosis and bronzing killing plant.

**Transmission**

Virus transmitted by mechanical inoculation.

**Geographical distribution**

Spreads in India.

**Experimental host range**

Few (<3) families susceptible.

Diagnostically susceptible

*Glycine max* — systemic necrosis, starting at tips and margins of leaves and spreading to whole plant.

*Phaseolus vulgaris* — systemic chlorotic mottling and stippling.

*Vigna radiata* — systemic leaf curling and tip necrosis.

*V. unguiculata* — chlorotic local lesions.

*Datura stramonium* — chlorotic mottling, leaves leathery and rough.

*Nicotiana tabacum* cv. White Burley — systemic mosaic.

*Petunia hybrida* — slight mottle.

*Solanum nigrum* — systemic mosaic, mottle.

Maintenance and propagation hosts

*Datura stramonium, Nicotiana tabacum* cv. White Burley, *Petunia hybrida, Solanum nigrum.*

Assay hosts (**L**ocal lesions or **W**hole plants)

*Glycine max* (W), *Phaseolus vulgaris* (W), *Vigna radiata* (W), *V. unguiculata* (L).

Dicotyledonous Hosts *(**Susceptible** or Insusceptible)*

*Antirrhinum majus. Beta vulgaris. Brassica oleracea* ssp. *capitata. Capsicum frutescens. Catharanthus roseus. Cucumis sativus.* ***Cyamopsis tetragonoloba. Datura stramonium. Glycine max.*** *Lycopersicon esculentum. Macrotyloma uniflorum. Nicotiana glutinosa.* ***Nicotiana tabacum***, or *Nicotiana tabacum.* ***Petunia hybrida. Pisum sativum.*** *Raphanus sativus.* ***Solanum nigrum.*** *Vigna radiata. Vigna unguiculata. Zinnia elegans.*

Sources of host-range data

Verma *et al.* (1962).

**Properties of particles in sap**

TIP: 70 °C. LIV: 7 days. DEP: $\log_{10}$ minus 3.

**Taxonomy and Relationships**

The attenuated strain causes mottling, leaf malformation and stunting.

**References**

Verma, G.S., Varma, J.P. and Saxena, P.N. (1962). *Proc. Natl. Acad. Sci. India* **32B**: 287.

# Guinea grass mosaic *potyvirus*

**Data collated by J.C. Thouvenel, 1984.**

*Standard acronym:* GGMV. *Strains:* guinea grass mosaic virus — strain B (from maize), guinea grass mosaic virus — strain D (from pearl millet), brachiaria strain (Morales *et al.*, 1994).

First reported in *Panicum maximum*; from the Ivory Coast; by Thouvenel *et al.* (1976).

**Natural host range and symptoms**

Symptoms persist.

*Panicum maximum, Pennisetum americanum, Zea mays* — green mosaic.

**Transmission**

Transmitted by a vector (both strains B and D); an insect; *Hysteroneura setariae, Rhopalosiphum maidis*; Aphididae. Transmitted in a non–persistent manner. Virus does not require a helper virus for vector transmission; transmitted by mechanical inoculation; not transmitted by seed.

**Geographical distribution**

Spreads in Brazil, Colombia, and Cote d'Ivoire.

**Experimental host range**

Few (<3) families susceptible.

Diagnostically susceptible

*Panicum maximum, Bromus macrostachys* — systemic green mosaic (strains A and D).

*Avena fatua, Eleusine coracana* — systemic green mosaic (strain B).

*Setaria italica, Bromus sterilis* — systemic green mosaic (strains A and B).

*Zea mays* — systemic green mosaic (strains A, B and D).

Diagnostically insusceptible

*Dactylis glomerata, Digitaria sanguinalis, Hordeum vulgare, Oryza sativa, Triticum durum.*

Maintenance and propagation hosts

*Panicum maximum* (strain A), *Pennisetum americanum* (strain D), *Zea mays* (strain B).

Assay hosts (**L**ocal lesions or **W**hole plants)

*Setaria italica* (W) for strains A and B.

Monocotyledonous Hosts *(**Susceptible** or Insusceptible)*

*Avena sativa. Dactylis glomerata.* ***Echinochloa crus-galli.*** *Eleusine coracana. Hordeum vulgare. Oryza sativa. Panicum capillare. Saccharum officinarum.* ***Setaria italica. Stenotaphrum secundatum.*** *Triticum aestivum.* ***Zea mays.***

Sources of host-range data

Thouvenel *et al.* (1976); Lamy *et al.* (1979); Kukla *et al.* (1984).

**Properties of particles in sap**

TIP: 50–55 °C. LIV: 1 days. DEP: $\log_{10}$ minus 2–3. Infectivity of sap not changed by treatment with di–ethyl ether. Leaf sap contains few virions.

**Purification**

Method: Thouvenel *et al.* (1976).

**Particle morphology**

Virions filamentous; not enveloped; usually flexuous (virions straighter in 50mM $MgCl_2$); with a clear modal length; of 815 nm; 15 nm wide. Axial canal obscure. Basic helix obscure.

**Physical properties**

Isoelectric point pH 4.7 (strains A and D), or 5.3 (strain B).

**Biochemical properties**

Virions contain 6 % nucleic acid; 94 % protein; 0 % lipid.

*Genome* consists of RNA; single–stranded. Infectivity retained when deproteinised with phenol or detergent.

*Virion protein(s)* one; $M_r$ 32500 (strains A and D), or 34500 (strain B).

**Cytopathology**

Virions found in all parts of the host plant; in cytoplasm. Inclusions present in infected cells; are pinwheels; they do not contain virions.

**Taxonomy and Relationships**

*Potyvirus: Potyviridae.*

Virus(es) with serologically related virions: pepper veinal mottle and yam mosaic viruses. Virus(es) with serologically unrelated virions: sugarcane mosaic and maize dwarf mosaic strains A and B.

Specific diagnostic hosts for identification of strain B (maize strain) are *Avena fatua, A. paniculata, A. strigosa, Bromus macrostachys, Eleusine coracana, E. tocusa, Lolium multiflorum, Panicum capillare, Sorghum vulgare.*

**References**

Kukla, B., Thouvenel, J.C. and Fauquet, C. (1984). *Phytopath. Z.* **109**: 65.
Lamy, D., Thouvenel, J.C. and Fauquet, C. (1979). *Ann. appl. Biol.* **93**: 37.
Morales, F.J., Castano, M., Velasco, A.C. and Arroyave, J. (1994). *Plant Dis.* **78**: 425.
Thouvenel, J.C., Givord, L. and Pfeiffer, P. (1976). *Phytopathology* **66**: 954.
Thouvenel, J.C., Fauquet, C. and Lamy, D. (1978). *CMI/AAB Descr. Pl. Viruses* No. 190, 3 pp.

# Habenaria mosaic (?) potyvirus

**Data collated by A.A. Brunt, 1991.**

*Standard acronym:* HaMV.

First reported in *Habenaria radiata*, from Japan; by Inouye (1983).

**Natural host range and symptoms**

Symptoms persist.

*Habenaria radiata* — conspicuous leaf chlorosis.

**Transmission**

Transmitted by a vector; an insect; *Myzus persicae*; Aphididae. Not transmitted by *Aphis gossypii*. Transmitted in a non–persistent manner. Virus transmitted by mechanical inoculation.

**Geographical distribution**

Spreads in Japan.

**Experimental host range**

Several (3–9) families susceptible.

Diagnostically susceptible

*Habenaria radiata* — conspicuous systemic leaf chlorosis.

*Chenopodium amaranticolor, C. quinoa, Spinacia oleracea, Tetragonia expansa, Gomphrena globosa* — local chlorotic lesions.

Maintenance and propagation hosts

*Habenaria radiata.*

Assay hosts (**L**ocal lesions or **W**hole plants)

*Habenaria radiata* (W), *Chenopodium amaranticolor* (L), *C. quinoa* (L).

Dicotyledonous Hosts *(**Susceptible** or Insusceptible)*

***Chenopodium amaranticolor***. ***Chenopodium quinoa***. ***Gomphrena globosa***. ***Spinacia oleracea***.

Additional hosts

***Habenaria radiata***.

**Properties of particles in sap**

TIP: 65 °C. LIV: 8–16 days (at 20°C). DEP: $\log_{10}$ minus 6.

**Particle morphology**

Virions filamentous; usually flexuous; with a clear modal length; of 750 nm; 13 nm wide.

**Cytopathology**

Virions found in leaves and mesophyll; in cytoplasm. Inclusions present in infected cells; are pinwheels; they do not contain virions.

**Taxonomy and Relationships**

Unassigned potyvirus.

**References**

Inouye, N. (1983). *Abstr. 4th Int. Congr. Pl. Path.* 467, p. 117.

# Harts tongue fern (?) *tobravirus*

**Data collated by A.A. Brunt, 1991.**

*Synonyms:* fern tobravirus.

First reported in *Phyllitis scolopendrium*; from the U.K.; by Hull (1968).

**Natural host range and symptoms**

Symptoms persist. Symptoms mottling and deformation of fronds.

*Pellaea falcata, Phyllitis scolopendrium, Polystichum falcatum, Pteris cretica* var. *albo-lineatum, P. 'Childsii'* — mottling and deformation of fronds.

**Transmission**

Virus transmitted by mechanical inoculation.

**Geographical distribution**

Found, but with no evidence of spread, in the U.K.

**Experimental host range**

Few (<3) families susceptible.

Diagnostically susceptible

*Phyllitis scolopendrium* — mottling of fronds.

Diagnostically insusceptible

*Cucumis sativus, Gomphrena globosa, Nicotiana clevelandii, N. tabacum, Vigna unguiculata*.

**Particle morphology**

Virions rod–shaped; usually straight; with a clear modal length; of 135 and 320 nm; 22 nm wide. Axial canal obscure. Basic helix obscure.

**Taxonomy and Relationships**

Possible *Tobravirus*.

**References**

Hull, R. (1968). *Virology* **35**: 333.

## Helenium S *carlavirus*

**Data collated by A.A. Brunt, 1991.**

*Standard acronym:* HVS. *Strains: Helenium* and *Impatiens* strains.

First reported in *Helenium amarum*; from Germany; by Kuschki *et al.* (1978).

**Natural host range and symptoms**

Symptoms none.

*Helenium amarum* hybrids, *Impatiens holstii* — no conspicuous symptoms, but the virus is often found in complex with helenium Y potyvirus in *H. amarum* plants with leaf chlorosis and flower breaking.

**Transmission**

Transmitted by a vector; an insect; *Myzus persicae* (*Impatiens* isolate); Aphididae. Transmitted in a non–persistent manner. Virus transmitted by mechanical inoculation; not transmitted by seed.

**Geographical distribution**

Spreads in Germany and the USA.

**Experimental host range**

Few (<3) families susceptible.

Diagnostically susceptible

*Chenopodium album, C. amaranticolor, C. quinoa* — chlorotic local lesions; systemic veinal chlorosis.

*Gomphrena globosa, Tetragonia expansa* (*Impatiens* isolates) — local lesions; no systemic infection.

*Beta vulgaris* var. *cicla*, *B. patellaris* — local symptomless infection; no systemic infection.

Diagnostically insusceptible

*Cucumis sativus, Glycine max, Chrysanthemum segetum, Nicotiana clevelandii, Zinnia elegans.*

Maintenance and propagation hosts

*Chenopodium album, C. amaranticolor, C. quinoa.*

Assay hosts (**L**ocal lesions or **W**hole plants)

*Chenopodium album* (L,W), *C. amaranticolor* (L,W), *C. quinoa* (L,W).

Dicotyledonous Hosts *(**Susceptible** or Insusceptible)*

*Ageratum conyzoides. Amaranthus caudatus. Antirrhinum majus. Apium graveolens. Arachis hypogaea. Atriplex hortensis.* ***Beta vulgaris****. Brassica campestris* ssp. *rapa. Brassica juncea. Brassica oleracea* ssp. *capitata. Capsicum annuum. Cheiranthus cheiri.* ***Chenopodium album. Chenopodium amaranticolor. Chenopodium quinoa.*** *Chrysanthemum morifolium. Citrullus lanatus. Cucumis sativus. Cucurbita maxima. Cucurbita pepo. Dahlia pinnata. Datura metel. Daucus carota. Dianthus barbatus.*

*Glycine max*. ***Gomphrena globosa***. *Gynura aurantiaca*. *Helianthus annuus*. *Hibiscus esculentus*. *Lactuca sativa*. *Lupinus albus*. *Lycopersicon pimpinellifolium*. *Matthiola incana*. *Medicago sativa*. *Melilotus alba*. *Nicotiana clevelandii*. *Nicotiana debneyi*. *Nicotiana glutinosa*. *Nicotiana megalosiphon*. *Nicotiana tabacum*. *Ocimum basilicum*. *Pastinaca sativa*. *Petunia hybrida*. *Phaseolus vulgaris*. *Pisum sativum*. *Solanum melongena*. *Spinacia oleracea*. ***Tetragonia expansa***. *Trifolium hybridum*. *Trifolium incarnatum*. *Tropaeolum majus*. *Verbesina encelioides*. *Vicia faba*. *Vigna unguiculata*. *Zinnia elegans*.

Monocotyledonous Hosts *(**Susceptible** or Insusceptible)*

*Dactylis glomerata*. *Hordeum vulgare*. *Zea mays*.

Sources of host-range data

Kuschki *et al.* (1978); Koenig *et al.* (1983).

**Properties of particles in sap**

TIP: 70–72 °C. LIV: 21 days. DEP: $\log_{10}$ minus 5. Leaf sap contains few virions.

**Purification**

Method: Koenig *et al.* (1983).

**Particle morphology**

Virions filamentous; not enveloped; usually straight; with a clear modal length; of 640 nm (in *Impatiens* isolate), or 676 nm (in PTA), or 693 nm (in UA). Axial canal obscure. Basic helix obscure.

**Physical properties**

One sedimenting component in purified preparations; sedimentation coefficient 160 *S*. $A_{260}/A_{280}$ ratio 1.1.

**Biochemical properties**

*Genome* consists of RNA; single–stranded; linear; unipartite; largest (or only) genome part 7.5 kb.

*Sequence database accession code(s):*

D10454 Em(40)_vi:HENSG Gb(84)_vi:HENSG Helenium virus S (HelVS) RNA, 3'terminal region encoding coat protein. 7/92 1,389bp.

S71594 Gb(89)_un:S71594 12K, 7K=coat protein Helenium virus S HelVS, Genomic RNA, 381 nt). 10/94 381bp.

*Virion protein(s)* one; $M_r$ 31000; coat protein.

**Cytopathology**

Virions found in leaves and mesophyll; in cytoplasm. Inclusions present in infected cells; are viroplasms and unusual in shape; aggregates of virions in parallel array associated with endoplasmic reticulum and ribosomes.

**Taxonomy and Relationships**

*Carlavirus*.

Virus(es) with serologically related virions: carnation latent, chrysanthemum B, dandelion latent, hop latent, hop mosaic, lily symptomless and potato M viruses. Virus(es) with serologically unrelated virions: poplar mosaic and potato S viruses.

*Helenium* and *Impatiens* strains differ in host range and transmissibility by aphids.

**References**

Koenig, R. and Lesemann, D.-E. (1983). *CMI/AAB Descr. Pl. Viruses* No. 265, 3 pp.

Koenig, R., Lesemann, D.-E., Lockhart, B.E., Betzold, J.A. and Weidemann, H.L. (1983). *Phytopath. Z.* **106**: 133.

Kuschki, G.H., Koenig, R., Düvel, D. and Kühne, H. (1978). *Phytopathology* **68**: 1407.

# Helenium Y *potyvirus*

**Data collated by A.A. Brunt, 1991.**

*Standard acronym:* HVY.

First reported in *Helenium amarum*; from Germany; by Kuschki *et al.* (1978).

**Natural host range and symptoms**

Symptoms persist.

*Helenium amarum* hybrids — leaf chlorosis, flower breaking and stunting.

**Transmission**

Virus transmitted by mechanical inoculation.

**Geographical distribution**

Spreads in Germany.

**Experimental host range**

Few (<3) families susceptible.

Diagnostically susceptible

*Chenopodium album, C. amaranticolor, C. quinoa* — chlorotic local lesions; systemic chlorotic spotting and leaf distortion in *C. quinoa*.

*Gomphrena globosa* — chlorotic and, later, necrotic local lesions; no systemic infection.

Diagnostically insusceptible

*Chrysanthemum segetum, Cucumis sativus, Helianthus annuus, Nicotiana clevelandii, Phaseolus vulgaris.*

Maintenance and propagation hosts

*Chenopodium album, C. amaranticolor, C. quinoa.*

Assay hosts (**L**ocal lesions or **W**hole plants)

*Chenopodium quinoa* (L or W).

Dicotyledonous Hosts (***Susceptible*** *or Insusceptible)*

*Ageratum conyzoides. Amaranthus caudatus. Antirrhinum majus. Apium graveolens. Arachis hypogaea. Atriplex hortensis. Beta vulgaris. Brassica campestris* ssp. *rapa. Brassica oleracea* ssp. *capitata. Capsicum annuum. Cheiranthus cheiri.* ***Chenopodium album. Chenopodium amaranticolor. Chenopodium quinoa****. Citrullus lanatus. Cucumis sativus. Cucurbita maxima. Cucurbita pepo. Dahlia pinnata. Datura metel. Daucus carota. Dianthus barbatus. Glycine max.* ***Gomphrena globosa****. Helianthus annuus. Lactuca sativa. Lupinus albus. Lycopersicon pimpinellifolium. Matthiola incana. Nicotiana clevelandii. Nicotiana debneyi. Nicotiana glutinosa. Nicotiana megalosiphon. Nicotiana tabacum. Ocimum basilicum. Pastinaca sativa. Petunia hybrida. Phaseolus vulgaris. Pisum sativum. Solanum melongena. Spinacia oleracea. Tetragonia expansa. Trifolium hybridum. Tropaeolum majus. Verbesina encelioides. Vicia faba. Vigna unguiculata. Zinnia elegans.*

Monocotyledonous Hosts *(**Susceptible** or Insusceptible)*

*Dactylis glomerata. Hordeum vulgare. Zea mays.*

**Properties of particles in sap**

TIP: 60 °C. DEP: $\log_{10}$ minus 4.

**Particle morphology**

Virions filamentous; not enveloped; usually flexuous; with a clear modal length.

**Cytopathology**

Inclusions present in infected cells; are pinwheels; they do not contain virions.

**Taxonomy and Relationships**

*Potyvirus: Potyviridae.*

Virus(es) with serologically related virions: bean yellow mosaic, lettuce mosaic and potato Y viruses. Virus(es) with serologically unrelated virions: bidens mottle, celery mosaic, plum pox, potato A and tulip breaking viruses.

**References**

Kuschki, G.H., Koenig, R., Düvel, D. and Kühne, K. (1978). *Phytopathology* **68**: 1407.

## Helleborus mosaic (?) *carlavirus*

**Data collated by A.A. Brunt, 1991.**

*Standard acronym:* HeMV.

First reported in *Helleborus niger*; from Germany; by Koenig (1985).

**Natural host range and symptoms**

*Helleborus niger* — symptomless.

**Geographical distribution**

Found, but with no evidence of spread, in Germany.

**Taxonomy and Relationships**

Possible *Carlavirus.*

Virus(es) with serologically related virions: chrysanthemum B and helenium S viruses.

**References**

Koenig, R. (1985). *Acta Hort.* **164**: 21.

## Henbane mosaic *potyvirus*

**Data collated by R. Plumb, 1991.**

*Synonyms:* hyoscyamus virus I, hyoscyamus virus III, atropa mild mosaic virus (Govier and Woods, 1971; Harrison and Roberts, 1971). *Standard acronym:* HMV. *Strains: alkekengi* strain (HMV–A).

From Bedfordshire, U.K.; by Hamilton (1932).

**Natural host range and symptoms**

Symptoms persist and vary seasonally. Symptoms mosaics and malformation.

*Hyoscyamus niger* — vein clearing, green/yellow mosaic.

*Datura stramonium, D. inermis* — mosaic, malformation.
*Physalis alkekengi, Atropa belladonna* — mild mosaic.

**Transmission**

Transmitted by a vector; an insect; *Myzus persicae, M. ascalonicus, Aulacorthum circumflexum, Macrosiphum euphorbiae*; Aphididae. Principal natural vector(s): *Myzus persicae*. Not transmitted by *Psylloides hyoscyami*. Transmitted in a non–persistent manner. Virus probably does not require a helper virus for vector transmission; transmitted by mechanical inoculation.

**Ecology and control**

Studies reported by Hamilton (1937).

**Geographical distribution**

Spreads in Germany, Hungary, Italy, and the UK.

**Experimental host range**

Few (<3) families susceptible. Experimentally infected plants mostly show mild to severe mosaics, some leaf malformation.

Diagnostically susceptible

*Datura stramonium* — severe mosaic.
*Hyoscyamus niger* — mosaic.
*Nicotiana glutinosa, N. tabacum* cvs White Burley and Samsun, *Physalis floridana* — mild mosaic.
*Nicotiana clevelandii* — necrotic local lesions and then mosaic.

Diagnostically insusceptible

*Amaranthus* spp. *Cucumis sativus, Spinacia oleracea, Chenopodium quinoa, Tetragonia expansa.*

Maintenance and propagation hosts

*Nicotiana tabacum* cv. Xanthi-*nc*.

Assay hosts (**L**ocal lesions or **W**hole plants)

*Nicotiana sylvestris* (L and W), *Datura stramonium* (W), *Chenopodium amaranticolor* (L), *Solanum demissum x. S. tuberosum* ‘A–6’ (L).

Dicotyledonous Hosts *(**Susceptible** or Insusceptible)*

*Amaranthus caudatus. Atriplex hortensis. Brassica campestris* ssp. *rapa.* ***Capsicum annuum***, or *Capsicum annuum. Capsicum frutescens.* ***Celosia argentea. Celosia cristata.*** *Chenopodium album.* ***Chenopodium amaranticolor***, or *Chenopodium amaranticolor. Chenopodium ambrosioides. Chenopodium foliosum. Chenopodium murale.* ***Chenopodium quinoa***, or *Chenopodium quinoa. Crambe abyssinica. Cucumis sativus.* ***Datura metel. Datura stramonium. Datura tatula.*** *Emilia sagittata. Gomphrena globosa.* ***Hyoscyamus niger. Lycopersicon esculentum. Nicandra physaloides. Nicotiana bigelovii. Nicotiana clevelandii. Nicotiana debneyi. Nicotiana glutinosa. Nicotiana megalosiphon. Nicotiana rustica. Nicotiana sylvestris. Nicotiana tabacum.*** *Ocimum basilicum.* ***Petunia hybrida.*** *Phaseolus vulgaris.* ***Physalis floridana. Solanum demissum × tuberosum. Solanum nigrum. Solanum tuberosum.*** *Spinacia oleracea. Tetragonia expansa. Tropaeolum majus. Vicia faba. Vigna unguiculata.*

Sources of host-range data

Lovisolo and Bartels (1970); Salamon (1989).

**Properties of particles in sap**

TIP: 60 °C. LIV: 2–3 days. DEP: $\log_{10}$ minus 6. Leaf sap contains few virions.

**Purification**

Method: Govier and Plumb (1972).

**Particle morphology**

Virions filamentous; not enveloped; with no clear modal length; of 800–900 nm; 12 nm wide. Axial canal obscure.

**Physical properties**

One sedimenting component in purified preparations; sedimentation coefficient 160 *S*. $A_{260}/A_{280}$ ratio 1.1.

**Biochemical properties**

*Genome* consists of RNA; single–stranded. Base composition 26 % G; 29 % A; 27 % C; 18 % U.

**Replication**

Genome replicates in cytoplasm. Acts as helper for another virus. Acts as a helper by aiding transmission of potato aucuba mosaic virus; presumably through a helper component.

**Cytopathology**

Virions found in leaves, mesophyll and the epidermis; in cytoplasm. Inclusions present in infected cells; are crystals in the cytoplasm and pinwheels; they do not contain virions. Other cellular changes: mitochondrial aggregates found in some plants.

**Taxonomy and Relationships**

*Potyvirus: Potyviridae.*

Virus(es) with serologically related virions: potato Y, potato A, tobacco etch and tulip breaking viruses; Columbian datura and pokeweed mosaic viruses are distantly related.

**References**


Bartels, R. (1964). *Phytopath. Z.* **49**: 257.

Bartels, R. (1971). *Phytopath. Z.* **76**: 87.

Bode, O., Brandes, J. and Paul, H.L. (1969). *Biol. Bund. für Land. Fortwirt. in Berlin and Braunschweig, 1968*, p. 61.

Bradley, R.H.E. (1952). *Ann. appl. biol.* **39**: 78.

Doncaster, J.P. and Kassanis, B. (1946). *Ann. appl. Biol.* **33**: 66.

Govier, D.A. and Plumb, R.T. (1972). *CMI/AAB Descr. Pl. Viruses* No. 95.

Govier, D.A. and Woods, R.D. (1971). *J. gen. virol.* **13**: 127.

Hamilton, M.A. (1932). *Ann. appl. Biol.* **19**: 550.

Hamilton, M.A. (1937). *Ann. appl. Biol.* **24**: 557.

Harrison, B.D. and Roberts, I.M. (1971). *J. gen. Virol.* **10**: 71.

Horvath, J. (1988). *Acta Phytopath. Ent. Hung.* **23**: 143.

Horvath, J., Salamon, P., Wolf, I. and Kölber, M. (1988). *Potato Research* **31**: 311.

Kahn, R.P. and Bartels, R. (1968). *Phytopath. Z.* **58**: 587.

Kitajima, E.W. and Lovisolo, O. (1972). *J. gen. Virol.* **16**: 265.

Lovisolo, O. and Bartels, R. (1970). *Phytopath. Z.* **69**: 189.

Mamula, D., Stefanac, Z., Thaler, I. and Gailhofer, M. (1988). *Acta bot. Croat.* **47**: 15.

Plumb, R.T. and Vince, D.A. (1971). *J. gen. Virol.* **13**: 357.

Salamon, P. (1989). *Novenyvedelem* **25**: 337.

Watson, M.A. (1968). *Rep. Rothamsted Exp. Stn.* 1967, p.123.
Watson, M.A. and Roberts, F.M. (1939). *Proc. Royal Soc. London, B* **127**: 543.

# Heracleum latent *trichovirus*

**Data collated by F. Bem, 1988.**

*Synonyms:* HV2. *Standard acronym:* HLV.

First reported in *Heracleum sphondylium*; from the U.K.; by Murant and Goold (1972).

**Natural host range and symptoms**

*Heracleum sphondylium, Apiaceae* — occurs in hogweed causing no symptoms.

**Transmission**

Transmitted by a vector; an insect; *Cavariella aegopodii, C. pastinacae, C. theobaldi*; Aphididae. Transmitted in a semi–persistent manner. Virus lost by the vector when it moults; does not multiply in the vector; requires, for vector transmission, a helper virus (carrot yellow leaf (?) *closterovirus*); transmitted by mechanical inoculation; transmitted by grafting; not transmitted by contact between plants; not transmitted by seed.

**Ecology and control**

Studies reported by Bem and Murant (1979); Murant and Duncan (1984).

**Geographical distribution**

Spreads in the Netherlands and the UK.

**Experimental host range**

Many (>9) families susceptible.

Diagnostically susceptible

*Chenopodium quinoa* — necrotic local lesions, systemic epinasty and stunting.
*C. amaranticolor* — systemic leaf curling and vein-clearing.
*Coriandrum sativum* — systemic interveinal yellowing or symptomless.

Diagnostically insusceptible

*Nicotiana clevelandii, N. tabacum* cvs Samsun and White Burley, *Cyamopsis tetragonoloba.*

Maintenance and propagation hosts

*Chenopodium quinoa.*

Assay hosts (**L**ocal lesions or **W**hole plants)

*Chenopodium murale* (L), *C. quinoa* (W), *Coriandrum sativum* (W).

Dicotyledonous Hosts *(**Susceptible** or Insusceptible)*

***Amaranthus caudatus***. ***Anthriscus cerefolium***. ***Apium graveolens***. ***Atriplex hortensis***. ***Beta vulgaris***. *Brassica campestris* ssp. *pekinensis*. *Brassica campestris* ssp. *rapa*. *Capsicum annuum*. *Catharanthus roseus*. ***Chenopodium album***. ***Chenopodium amaranticolor***. *Chenopodium ambrosioides*. ***Chenopodium capitatum***. ***Chenopodium foetidum***. ***Chenopodium foliosum***. ***Chenopodium murale***. ***Chenopodium quinoa***. ***Coriandrum sativum***. ***Cucumis sativus***. *Cucurbita maxima*. *Cucurbita pepo*. *Cyamopsis tetragonoloba*. *Datura metel*. *Datura stramonium*. ***Daucus carota***. *Dianthus barbatus*. *Fragaria vesca*. ***Gomphrena globosa***. ***Heracleum sphondylium***. ***Lycopersicon chilense***. *Lycopersicon esculentum*. *Malus sylvestris*. *Nicandra physaloides*. *Nicotiana benthamiana*. *Nicotiana clevelandii*. *Nicotiana debneyi*. *Nicotiana glutinosa*. *Nicotiana* *glutinosa* × *clevelandii*. *Nicotiana rustica*. *Nicotiana tabacum*. *Ocimum basilicum*. ***Pastinaca sativa***. ***Petroselinum crispum***. *Petunia hybrida*. *Phaseolus vulgaris*. *Physalis floridana*. *Physalis peruviana*. *Pisum sativum*. *Prunus persica*. *Sesbania exaltata*. *Solanum chacoense*. ***Solanum demissum***. ***Solanum melongena***. ***Spinacia oleracea***. ***Stellaria media***. ***Tetragonia expansa***. *Trifolium pratense*. *Vicia faba*. ***Vigna unguiculata***. *Vigna unguiculata* ssp. *cylindrica*. *Vigna unguiculata* ssp. *sesquipedalis*.

Monocotyledonous Hosts *(**Susceptible** or Insusceptible)*

*Allium porrum*. *Avena sativa*. *Hordeum vulgare*. *Setaria italica*. ***Zea mays***.

Additional hosts

*Amaranthus hybridus* cv. *hypochondriacus*, ***Anethum graveolens***, *Anthriscus sylvestris*, *Apium nodiflorum*, *Brassica perviridis*, ***Browallia speciosa***, ***Celosia plumosa***, ***Montia perfoliata***, *Eryngium planum*, ***Fedia cornucopiae***, *Foeniculum vulgare*, *Levisticum officinale*, *Nicotiana forgetiana*, *Nicotiana knightiana*, *Nicotiana paniculata*, *Nicotiana repanda*, *Oenanthe pimpinelloides*, *Phaseolus mungo*, *Pimpinella anisum*, *Pimpinella saxifraga*, ***Scandix pecten-veneris***, *Senecio cineraria*, ***Solanum acaule***, ***Solanum commersonii***, ***Solanum spegazzinii***, *Triticum vulgare*, *Urtica pilulifera* and ***Valerianella carinata***.

Sources of host-range data

Bem and Murant (1979).

**Properties of particles in sap**

TIP: 40–50 °C (in *Chenopodium quinoa* sap, clarified with bentonite). LIV: 1–2 days (at 22°C, in *C. quinoa* sap). DEP: $\log_{10}$ minus 4–5 (in *C. quinoa* sap). Infectivity of sap decreased by treatment with di–ethyl ether (or abolished). Leaf sap contains many virions. Electron microscopy: intact virions are seen when 2% PTA, pH7 is used, but disrupted ones seen in AM.

**Purification**

Method: Bem and Murant (1979).

**Particle morphology**

Virions filamentous; not enveloped; usually flexuous; with a clear modal length; of 730 nm; 12.5 nm wide. Basic helix obvious; pitch of basic helix 3.8 nm.

**Physical properties**

One sedimenting component in purified preparations; sedimentation coefficient 96 *S*. Density 1.24 in $Cs_2SO_4$.

**Biochemical properties**

Virions 0 % lipid.

*Genome* consists of RNA; single–stranded; linear; unipartite; largest (or only) genome part 7.9 kb. Genomic nucleic acid isolated by Bem and Murant (1979); Murant *et al.* (1979). Infectivity retained when deproteinised with proteases; retained when deproteinised with phenol or detergent. Additional factor not required for infectivity.

*Non–genomic nucleic acid* not found in the virions.

*Virion protein(s)* one; $M_r$ 23500; coat protein. Method of preparation: Bem and Murant (1979). Virion proteins not glycosylated.

**Replication**

Replication does not depend on a helper virus.

**Cytopathology**

Virions found in leaves, roots, phloem and companion cells; in cytoplasm. Inclusions absent from infected cells.

**Taxonomy and Relationships**

*Trichovirus.*

Virus(es) with serologically unrelated virions: apple chlorotic leafspot, apple stem grooving, beet yellow stunt, carnation necrotic fleck, citrus tristeza, beet yellows, lilac chlorotic leaf spot, potato T and wheat yellow leaf viruses.

Best tests for diagnosis: gives epinasty and stunting in *Chenopodium quinoa.*

**References**


Bem, F. and Murant, A.F. (1979). *Ann. appl. biol.* **92**: 243.

Bem, F. and Murant, A.F. (1979). *J. gen. Virol.* **44**: 817.

Murant, A.F. (1983). *Abstr. 4th Int. Congr. Pl. Path.*, p. 121.

Murant, A.F. and Duncan, G.H. (1984). *Abstr. 6th Int. Congr. Virol. Sendai, Japan*, p. 328.

Murant, A.F. and Goold, R.A. (1972). *Rep. Scottish Hort. Res. Inst.* 1971, p. 63.

Murant, A.F., Taylor, M. and Duncan, G.H. (1979). *Rep. Scottish Hort. Res. Inst.* 1978, p. 98.


## Hibiscus chlorotic ringspot *carmovirus*

**Data collated by C. Büchen-Osmond, 1987. Revised by R. Bonfiglioli and K. Gibb, 1991.**

*Standard acronym:* HCRSV.

First reported in *Hibiscus rosa-sinensis*; from the U.S.A.; by Waterworth *et al.* (1976).

**Natural host range and symptoms**

Symptoms persist.

*Hibiscus rosa-sinensis* — mottling or chlorotic spots, rings or vein-banding. Very variable.

*Hibiscus manihot* — mosaic.

**Transmission**

Transmitted by means not involving a vector; is not transmitted by *Myzus persicae*, nor by *Epilachna varivestis*. Virus transmitted by mechanical inoculation.

**Ecology and control**

Studies reported by Waterworth (1980).

**Geographical distribution**

Spreads in Australia, El Salvador, Fiji, the Solomon Islands, Thailand, and the USA.

**Experimental host range**

Several (3–9) families susceptible.

Diagnostically susceptible

*Chenopodium quinoa* — chlorotic local lesions, not systemic.

*Cyamopsis tetragonoloba, Dolichos biflorus, Gossypium hirsutum* — necrotic local lesions, not systemic.

*Hibiscus cannabinus* (kenaf) — ringspots, local lesions, then systemic mottle.

*Phaseolus vulgaris* cv. Pinto — small local lesions, not systemic.

Maintenance and propagation hosts

*Hibiscus cannabinus* — for purification. *Althaea* spp. and *Hibiscus rosa-sinensis* — for maintaining cultures.

Assay hosts (Local lesions or Whole plants)

*Cyamopsis tetragonoloba* (L); *Dolichos biflorus* (L). *Hibiscus cannabinus* (L) is the most sensitive host.

Dicotyledonous Hosts *(**Susceptible** or Insusceptible)*

***Chenopodium quinoa*. *Cyamopsis tetragonoloba*. *Dolichos biflorus*. *Gossypium hirsutum*. *Hibiscus cannabinus*. *Phaseolus vulgaris*.**

Sources of host-range data

Waterworth *et al.* (1976).

**Properties of particles in sap**

TIP: 72 °C. LIV: 30 days (at 20°C). DEP: $\log_{10}$ minus 8. Leaf sap contains many virions.

**Purification**

Method: Waterworth *et al.* (1976); Jones and Behncken (1980).

**Particle morphology**

Virions isometric; not enveloped; 28 nm in diameter; angular in profile; with a conspicuous capsomere arrangement (negative stain apparently penetrates about one virion in five — Waterworth *et al.*, 1976; Jones and Behncken, 1980).

**Physical properties**

One sedimenting component in purified preparations; sedimentation coefficient 118 *S* (Jones and Behncken, 1980; Waterworth *et al.*, 1976). Density 1.35 g $cm^{-3}$ in CsCl.

**Biochemical properties**

Virions contain 14 % nucleic acid; 86 % protein; 0 % lipid.

*Genome* consists of RNA; single–stranded; unipartite; largest (or only) genome part 4.4 kb. Base composition 23.8 % G; 25.5 % A; 26.4 % C; 24.3 % U.

*Virion protein(s)* one; $M_r$ 44100; coat protein.

**Replication**

Replication does not depend on a helper virus.

**Cytopathology**

Virions found in mesophyll; in cytoplasm and in cell vacuoles (of broken cells). Inclusions absent from infected cells.

**Taxonomy and Relationships**

*Carmovirus: Tombusviridae.*

Virus(es) with serologically unrelated virions: 45 viruses with isometric virions (Waterworth *et al.*, 1976), including okra mosaic virus (Givord and Hirth, 1973) and hibiscus latent ringspot virus (Brunt *et al.*, 1980). Isolates from the U.S.A. and El Salvador (Waterworth *et al.*, 1976), Australia (Jones and Behncken, 1980), Fiji (Brunt, 1980) and Thailand (H.C. Phatak, unpublished data) are serologically indistinguishable.

Hibiscus chlorotic ringspot virus may be related to, or the same, as an agent that induces a mosaic disease of *Hibiscus rosa-sinensis* in Hawaii (Hendrix and Murakishi, 1951). Best tests for diagnosis: hibiscus chlorotic ringspot virus is just one of several viruses that infect Malvaceae in nature. Others include

(1) okra mosaic tymovirus which causes a mosaic in *Hibiscus esculentus* (Givord and Hirth, 1973).

(2) hibiscus latent ringspot nepovirus which sediments as 3 components and causes leaf chlorosis in *Hibiscus rosa-sinensis* (Brunt *et al.*, 1980).

(3) malva yellows luteovirus is an aphid-transmitted virus found in many species that hibiscus chlorotic ringspot does not infect (Costa *et al.*, 1959).

(4) chlorosis in Malvaceae is caused by abutilon mosaic bigeminivirus and is transmitted by whitefly (Flores and Silberschmidt, 1967; Costa, 1954).

(5) yellow vein mosaic of *Hibiscus esculentus* (bhendi) (Capoor and Varma, 1950), of *H. esculentus* (okra) and of *Gossypium hirstutum* (cotton) (Santry and Singh, 1973).

(6) hibiscus line pattern which is not transmissible by sap inoculation (Wolfswinkel, 1966).

(7) leaf curl virus which induces different symptoms (Makherjie and Raychaudhuri, 1964).

**References**

Brunt, A.A. (1980). *UNDP/FAO Survey of Plant Virus Diseases in Fiji.*

Brunt, A.A., Barton, R.J., Phillips, S. and Lana, A.O. (1980). *Ann. appl. Biol.* **96**: 37.

Capoor, S.P. and Varma, P.M. (1950). *Ind. J. agric. Sci.* **20**: 217.

Costa, A.S. (1954). *Bragantia* **13**: 23.

Costa, A.S., Duffus, J.E. and Bardin, R. (1959). *J. Am. Soc. Sug. Beet Technol.* **10**: 371.

Flores, E. and Silberschmidt, K. (1967). *Phytopath. Z.* **60**: 181.

Givord, L. and Hirth, L. (1973). *Ann. appl. Biol.* **74**: 359.

Hendrix, J.W. and Murakishi, H. (1951). *Rep. Hawaii Univ. agric. Exp. Stn 1948–1950*, p. 123.

Jones, D.R. and Behncken, G.M. (1980). *Aust. Plant Path.* **9**: 4.

Lana, A.O. (1974). *Pl. Dis. Reptr* **58**: 1040.

Mukherjee, A.K. and Raychaudhuri, P. (1964). *Indian J. Hort.* **21**: 176.

Sastry, K.S.M. and Singh, S.J. (1973). *Indian Phytopath.* **26**: 136.

Van Velsen, R.J. (1967). *Papua New Guin. agric. J.* **19**: 10.

Waterworth, H.E. (1980). *CMI/AAB Descr. Pl. Viruses* No. 227, 4 pp.

Waterworth, H.E., Lawson, R.H. and Monroe, R.L. (1976). *Phytopathology* **66**: 570.

Wolfswinkel, L.D. (1966). *S. Afr. J. agric. Sci.* **9**: 483.

# Hibiscus latent ringspot *nepovirus*

**Data collated by A.A. Brunt, 1984.**

*Standard acronym:* HLRSV.

First reported in *Hibiscus rosa-sinensis*; from Ibadan, Nigeria; by Lana (1974).

**Natural host range and symptoms**

Symptoms disappear soon after infection.

*Hibiscus rosa-sinensis* and *H. cannabinus* — leaf chlorosis but usually symptomless.

**Transmission**

Transmitted by a vector; a nematode (probably). Virus transmitted by mechanical inoculation; transmitted by grafting; not transmitted by contact between plants; not transmitted by seed (in limited tests with seed from *Nicotiana clevelandii* and *Chenopodium quinoa*).

**Geographical distribution**

Spreads in Italy and Nigeria.

**Experimental host range**

Several (3–9) families susceptible.

Diagnostically susceptible

*Gomphrena globosa* — grey necrotic local lesions; not systemic.
*Chenopodium murale* — necrotic lesions, not systemic.
*C. quinoa* — chlorotic local lesions; systemic chlorosis.
*Trifolium incarnatum* — severe systemic chlorosis.
*Hibiscus cannabinus* — systemic leaf chlorosis.
*Nicotiana clevelandii* — systemic leaf chlorosis.

Diagnostically insusceptible

*Phaseolus vulgaris, Hibiscus esculentus* (okra), *Capsicum annuum, Lycopersicon esculentum, Physalis floridana.*

Maintenance and propagation hosts

*Hibiscus cannabinus, Nicotiana clevelandii.*

Assay hosts (**L**ocal lesions or **W**hole plants)

*Chenopodium murale* (L), *C. quinoa* (L,W), *Hibiscus cannabinus* (W).

Dicotyledonous Hosts *(**Susceptible** or Insusceptible)*

*Adansonia digitata. Beta vulgaris. Brassica campestris* ssp. *pekinensis. Brassica oleracea* ssp. *botrytis. Capsicum annuum. Capsicum frutescens. Catharanthus roseus.* ***Chenopodium amaranticolor. Chenopodium murale. Chenopodium quinoa.*** *Citrullus lanatus. Cucumis melo. Cucurbita pepo. Dahlia pinnata. Glycine max.* ***Gomphrena globosa. Gossypium hirsutum.*** *Helianthus annuus. Hibiscus esculentus. Lactuca sativa. Lycopersicon esculentum.* ***Nicandra physaloides. Nicotiana benthamiana. Nicotiana clevelandii. Nicotiana × edwardsonii. Nicotiana glutinosa.*** *Nicotiana megalosiphon. Nicotiana sylvestris.* ***Nicotiana tabacum. Petunia hybrida.*** *Phaseolus vulgaris. Physalis floridana. Pisum sativum. Sesamum indicum. Solanum demissum. Solanum melongena. Solanum nigrum.* ***Tetragonia expansa.*** *Theobroma cacao.* ***Trifolium incarnatum.*** *Trifolium repens. Vicia faba. Vigna unguiculata. Zinnia elegans.*

Monocotyledonous Hosts *(**Susceptible** or Insusceptible)*

*Zea mays.*

Sources of host-range data

Brunt *et al.* (1980).

**Properties of particles in sap**

TIP: 65 °C (in *Nicotiana clevelandii* sap). LIV: 21–28 days (in *N. clevelandii* sap). DEP: $\log_{10}$ minus 4–5 (in *N. clevelandii* sap). Leaf sap contains few virions.

**Particle morphology**

Virions isometric; not enveloped; 28 nm in diameter; angular in profile; without a conspicuous capsomere arrangement.

**Physical properties**

Three sedimenting components in purified preparations; sedimentation coefficient of the fastest 132 *S* (B); of the other(s) 114 *S* (M), or 51 *S* (T). Density 1.52 g $cm^{-3}$ in CsCl

(B), or 1.49 g $cm^{-3}$ in CsCl, or 1.32 g $cm^{-3}$ in CsCl (M; unfixed). $A_{260}/A_{280}$ ratio 1.85 (B), or 1.78 (M; corrected for light scattering).

**Biochemical properties**

Virions contain 42 % nucleic acid (B), or 38 % nucleic acid (M), or 0 % nucleic acid (T); 58 % protein (B), or 62 % protein (M), or 100 % protein (T); 0 % lipid.

*Genome* consists of RNA; single–stranded; of two parts; largest (or only) genome part the larger 7 kb; the 2nd largest 6.2 kb. Genomic nucleic acid isolated by Brunt *et al.* (1980).

*Virion protein(s)* one; $M_r$ 53600.

**Taxonomy and Relationships**

*Nepovirus: Comoviridae.*

Virus(es) with serologically unrelated virions: arabis mosaic, arracacha A, artichoke Italian latent, cherry leaf roll, cherry rasp leaf, cacao necrosis, raspberry ringspot, strawberry latent ringspot, tobacco ringspot, tomato black ring and tomato ringspot viruses.

**References**

Brunt, A.A., Barton, R.J., Phillips, S. and Lana, A.O. (1980). *Ann. appl. Biol.* **96**: 37.

Brunt, A.A., Barton, R.J. and Phillips, S. (1981). *CMI/AAB Descr. Pl. Viruses* No. 233, 3 pp.

Lana, A.O. (1974). *Pl. Dis. Reptr* **58**: 1040.

# Hibiscus yellow mosaic (?) *tobamovirus*

**Data collated by A.A. Brunt, 1992.**

First reported in *Hibiscus* sp.; from Japan; by Kashiwazaki *et al.* (1982).

**Natural host range and symptoms**

*Hibiscus* sp. — yellow leaf mottling.

**Transmission**

Virus transmitted by mechanical inoculation.

**Geographical distribution**

Spreads in Japan.

**Experimental host range**

Few (<3) families susceptible.

Diagnostically susceptible

*Chenopodium amaranticolor, Petunia hybrida* — local lesions.

Assay hosts (**L**ocal lesions or **W**hole plants)

*Chenopodium amaranticolor* (L).

Dicotyledonous Hosts *(**Susceptible** or Insusceptible)*

***Chenopodium amaranticolor*. *Petunia hybrida*.**

**Particle morphology**

Virions rod–shaped; usually straight; with a clear modal length; of 300 nm; 18 nm wide. Axial canal obscure; 2.3 nm in diameter. Basic helix obvious; pitch of basic helix 2.3 nm.

**Physical properties**

One sedimenting component in purified preparations; sedimentation coefficient 196 *S*.

**Cytopathology**

Virions found in leaves and mesophyll. Inclusions present in infected cells; are viroplasms.

**Taxonomy and Relationships**

Possible *Tobamovirus*.

**References**

Kashiwazaki, S., Yamashita, S. and Doi, Y. (1982). *Ann. Phytopath. Soc. Japan* **48**: 395.

# Hippeastrum mosaic *potyvirus*

**Data collated by S. Phillips, 1985.**

*Synonyms:* amaryllis mosaic virus. *Standard acronym:* HiMV. *Strains:* none reported.

First reported in *Hippeastrum hybridum*; from Baarn, the Netherlands; by Kunkel (1922); Brants and van den Heuvel (1965).

**Natural host range and symptoms**

Symptoms persist.

*Hippeastrum hybridum, H. equestre, Eucharis grandiflora*—chlorotic streaking of leaves and flower stalks.

**Transmission**

Transmitted by a vector; an insect; *Aphis fabae, A. gossypii, Myzus persicae*; Aphididae. Transmitted in a non–persistent manner. Virus transmitted by mechanical inoculation; not transmitted by contact between plants; not transmitted by seed; not transmitted by pollen.

**Geographical distribution**

Spreads in the former Czechoslovakia, Fiji, Japan, the Netherlands, South Africa, the UK, and the USA.

**Experimental host range**

Several (3–9) families susceptible.

Diagnostically susceptible

*Hippeastrum hybridum* — irregular green mosaic on leaves and flower stalks.
*Chenopodium quinoa* — necrotic local lesions; not systemic.
*Hyoscyamus niger* — numerous chlorotic lesions; not systemic.
*Nicotiana clevelandii* — white etched rings or chlorotic lesions; not systemic.
*Nicotiana tabacum* cv. Xanthi — white etched rings or chlorotic lesions; not systemic.

Diagnostically insusceptible

*Gomphrena globosa, Tetragonia expansa, Phaseolus vulgaris.*

Maintenance and propagation hosts

*Nicotiana clevelandii, Hippeastrum hybridum, Nicotiana tabacum* cv. Xanthi, *Chenopodium quinoa.*

Assay hosts (**L**ocal lesions or **W**hole plants)

*Hyoscyamus niger* (L), *Chenopodium quinoa* (L), *Hippeastrum hybridum* (W).

Dicotyledonous Hosts (***Susceptible*** *or Insusceptible*)

*Amaranthus caudatus. Beta vulgaris. Brassica campestris* ssp. *pekinensis. Calendula officinalis. Capsicum annuum. Catharanthus roseus. Celosia argentea. Chenopodium amaranticolor. Chenopodium capitatum.* ***Chenopodium quinoa****. Cichorium endiva. Citrullus lanatus. Cucumis sativus. Cucurbita pepo. Dahlia pinnata.* ***Datura stramonium****. Emilia sagittata. Fagopyrum esculentum. Gomphrena globosa. Helianthus annuus.* ***Hyoscyamus niger****. Lactuca sativa. Lycopersicon esculentum. Momordica balsamina. Nicandra physaloides. Nicotiana benthamiana.* ***Nicotiana clevelandii****. Nicotiana glutinosa. Nicotiana megalosiphon.* ***Nicotiana rustica****.* ***Nicotiana tabacum****.* ***Petunia hybrida****. Phaseolus vulgaris. Phlox drummondii. Physalis floridana. Pisum sativum. Solanum melongena. Spinacia oleracea. Tetragonia expansa. Trifolium incarnatum. Trifolium repens. Tropaeolum majus. Vicia faba. Vigna unguiculata. Zinnia elegans.*

Monocotyledonous Hosts (***Susceptible*** *or Insusceptible*)

*Belamcanda chinensis.* ***Hippeastrum hybridum****. Hordeum vulgare. Zea mays.*

Sources of host-range data

Brants and van den Heuvel (1965); Jayasinghe and Dijkstra (1979).

**Properties of particles in sap**

TIP: 60 °C (in *Hippeastrum hybridum* sap). LIV: 2 days (in *H. hybridum* sap). DEP: $\log_{10}$ minus 4. Infectivity of sap not changed by treatment with di–ethyl ether. Leaf sap contains few virions.

**Purification**

Method: Brunt (1973); Jayasinghe and Dijkstra (1979).

**Particle morphology**

Virions filamentous; not enveloped; usually flexuous; with a clear modal length; of 782 nm; 12 nm wide. Axial canal obscure. Basic helix obscure.

**Physical properties**

One sedimenting component in purified preparations; sedimentation coefficient 153 *S*. Density 1.31 g $cm^{-3}$ in CsCl (unfixed). $A_{260}/A_{280}$ ratio 1.21 (corrected for scattering).

**Biochemical properties**

Virions contain 4.5 % nucleic acid; 95.5 % protein; 0 % lipid.

*Virion protein(s)* one; $M_r$ 32000. Method of preparation: Brunt *et al.* (1975).

**Cytopathology**

Virions found in all parts of the host plant; in cytoplasm (as membrane-associated aggregates). Inclusions present in infected cells; are pinwheels and unusual in shape; and bundles near the nucleus. Other cellular changes: chloroplasts being smaller in infected cells.

**Taxonomy and Relationships**

*Potyvirus: Potyviridae.*

Virus(es) with serologically unrelated virions: bean yellow mosaic, carnation vein mottle, clover yellow vein, iris mild mosaic, lettuce mosaic, tobacco etch, potato Y and turnip mosaic viruses. Reported relationships with other potyviruses have not been confirmed (Brunt and Phillips, unpublished results).

**Comments**

*Gomphrena globosa* and *Tetragonia expansa* have been reported as local lesion hosts of hippeastrum mosaic virus, but are not, however they are susceptible to hippeastrum latent virus, a carlavirus which is closely related to nerine latent and carnation latent viruses, and which frequently occurs, together with hippeastrum mosaic virus, in hippeastrums.

**References**

Brants, D.H. and van den Heuvel, J. (1965). *Neth. J. Pl. Path.* **71**: 145.
Brunt, A.A. (1973). *Rep. Glasshouse Crops Res. Inst.* 1972, p. 103.
Brunt, A.A. (1973). *CMI/AAB Descr. Pl. Viruses* No. 117, 4 pp.
Brunt, A.A., Barton, R.J., Tremaine, J.H. and Stace-Smith, R. (1975). *J. gen. Virol.* **27**: 101.
Iwaki, M. (1967). *Ann. Phytopath. Soc. Japan* **33**: 237.
Jayasinghe, U. and Dijkstra, J. (1979). *Neth. J. Pl. Path.* **85**: 47.
Kunkel, L.O. (1922). *Science* **55**: 73.

# Holcus lanatus yellowing (?) rhabdovirus

**Data collated by A.A. Brunt, 1991.**

*Standard acronym:* HLYV.

First reported in *Holcus lanatus*; from Italy; by Amici *et al.* (1978).

**Natural host range and symptoms**

*Holcus lanatus* — clear leaf chlorosis.

**Particle morphology**

Virions rhabdo– or bullet–shaped.

**Taxonomy and Relationships**

Possible unassigned rhabdovirus.

**References**

Amici, A., Faoro, F. and Tornaghi, R. (1978). *Riv. Patol. Veg.* **14**: 85.

# Holcus streak (?) potyvirus

**Data collated by A.A. Brunt, 1991.**

*Standard acronym:* HSV.

First reported in *Holcus mollis* and *Holcus lanatus*; from Wales, U.K; by Catherall and Chamberlain (1975).

**Natural host range and symptoms**

Symptoms persist.

*Holcus lanatus, H. mollis*—yellow streaking of younger leaves and rusty brown streaks and mottle of flag leaves.

**Transmission**

Virus transmitted by mechanical inoculation.

**Geographical distribution**

Spreads in the UK (Wales).

**Experimental host range**

Few (<3) families susceptible.

Diagnostically susceptible

*Holcus lanatus, H. mollis, Avena sativa, A. strigosa, Bromus mollis, B. macrostachia, Phalaris paradoxa* — grey or chlorotic leaf streaking.

Maintenance and propagation hosts

*Holcus lanatus, H. mollis.*

Assay hosts (**L**ocal lesions or **W**hole plants)

*Holcus lanatus* (W), *H. mollis* (W), *Avena sativa* (W).

Monocotyledonous Hosts ***(Susceptible*** *or Insusceptible)*

***Avena sativa. Avena strigosa. Bromus mollis. Phalaris paradoxa.***

**Particle morphology**

Virions filamentous; usually flexuous; with a clear modal length; of 750 nm; 12 nm wide.

**Taxonomy and Relationships**

Unassigned potyvirus.

**References**

Catherall, P.L. and Chamberlain, J.A. (1975). *Pl. Path.* **24**: 247.

# Honeysuckle latent *carlavirus*

**Data collated by F.A. van der Meer, 1987. Revised by A.A. Brunt, 1991.**

*Synonyms:* lonicera latent virus (Brunt *et al.*, 1980; Brunt and van der Meer, 1984; Van der Meer *et al.*, 1980). *Standard acronym:* HnLV.

First reported in *Lonicera japonica L. periclymenum*; from the U.K. and the Netherlands; by Brunt and Thomas (1976); Van der Meer *et al.* (1980).

**Natural host range and symptoms**

*Lonicera americana, L. brownii* cvs Dropmore Scarlet, *Fuchsioides* and *Punica, L. caprifolium, L. heckottii* cv. Goldflame, *L. henryi, L. japonica* cvs Halliana and Reticulata, *L. periclymenum* cvs Belgica, Belgica Elect and Seotina, *L. telmanniana* — symptomless.

**Transmission**

Transmitted by a vector; an insect; *Hydaphis foeniculi*; Aphididae. Not transmitted by *Myzus persicae.* Transmitted in a non–persistent manner. Virus transmitted by mechanical inoculation (but not between plants of *Lonicera* species); transmitted by grafting; not transmitted by contact between plants; not transmitted by seed.

**Ecology and control**

Studies reported by Brunt *et al.* (1980) and Van der Meer *et al.* (1985) showed that honeysuckle plants free of honeysuckle latent virus can be obtained by rooting small shoots (*c.* 1 cm long) from infected plants which have been grown for 6–8 weeks at alternating temperatures of 16 hours at 37°C and 8 hours at 26°C (Brunt *et al.*, 1980). Ecological studies by Van der Meer *et al.* (1980); (1984).

**Geographical distribution**

Spreads in the Netherlands and the UK.

**Experimental host range**

Several (3–9) families susceptible. Experimentally infected plants mostly show local chlorotic lesions.

Diagnostically susceptible

*Nicotiana megalosiphon, N. occidentalis* — local lesions and systemic mosaic.
*Nicotiana clevelandii* — faint local lesions and systemic vein chlorosis.
*Nicotiana glutinosa* — chlorotic local lesions.
*Gomphrena globosa* — red local lesions.
*Chenopodium quinoa* — local lesions then systemic mosaic.

Diagnostically insusceptible

*N. tabacum* cv. White Burley, *N. rustica, Vigna unguiculata, Phaseolus vulgaris.*

Maintenance and propagation hosts

*Nicotiana clevelandii, N. megalosiphon, N. occidentalis.*

Assay hosts (**L**ocal lesions or **W**hole plants)

*Chenopodium murale* (L), *Nicotiana megalosiphon* (L and W), *N. occidentalis* (L and W).

Dicotyledonous Hosts ***(Susceptible** or Insusceptible)*

*Beta macrocarpa. Brassica campestris* ssp. *pekinensis. Calendula officinalis.* ***Chenopodium amaranticolor. Chenopodium murale. Chenopodium quinoa. Cucumis sativus. Gomphrena globosa. Nicotiana clevelandii. Nicotiana glutinosa. Nicotiana megalosiphon. Nicotiana occidentalis.*** *Nicotiana rustica. Nicotiana tabacum.* ***Petunia hybrida.*** *Phaseolus vulgaris. Solanum demissum. Solanum demissum × tuberosum. Tetragonia expansa. Trifolium incarnatum. Vicia faba. Vigna unguiculata.*

Comments on host-range

Sap from infected *Lonicera* does not infect *Chenopodium quinoa*, but that from infected *Nicotiana megalosiphon* does. In later serial transfers from *C. quinoa* to *C. quinoa* the virus becomes adapted to *C. quinoa* and causes local lesions and systemic mosaic (Van der Meer *et al.*, 1980).

Sources of host-range data

Brunt *et al.* (1980); Van der Meer *et al.* (1980).

**Properties of particles in sap**

TIP: 75–85 °C. LIV: 2–8 days (at 20°C). DEP: $\log_{10}$ minus 3–4. Leaf sap contains few virions.

**Purification**

Method: Brunt *et al.* (1980).

**Particle morphology**

Virions filamentous; not enveloped; usually flexuous; with a clear modal length; of 650 nm; 13 nm wide.

**Physical properties**

One sedimenting component in purified preparations; sedimentation coefficient 157 *S*. Density 1.31 g $cm^{-3}$ in CsCl. $A_{260}/A_{280}$ ratio 1.12.

**Biochemical properties**

Virions contain 5 % nucleic acid; 95 % protein.

*Genome* consists of RNA; single–stranded; linear; unipartite; largest (or only) genome part 8 kb.

*Virion protein(s)* one; $M_r$ 31000; coat protein.

**Replication**

Replication does not depend on a helper virus.

**Taxonomy and Relationships**

*Carlavirus.*

Virus(es) with serologically related virions: shallot latent, poplar mosaic and possibly red clover vein mosaic viruses. Virus(es) with serologically unrelated virions: carnation latent, chrysanthemum B, cowpea mild mottle, elderberry A, hippeastrum latent, hop latent, lily symptomless, nerine latent, narcissus latent, passiflora latent, pea streak, potato M, potato S and white bryony mosaic viruses.

Honeysuckle latent virus does not infect *Vigna unguiculata* and *Phaseolus vulgaris*, whereas poplar mosaic virus induces local lesions in inoculated leaves of these species (Van der Meer *et al.*, 1980). Honeysuckle latent virus infects *Nicotiana clevelandii* systemically (Van der Meer *et al.*, 1980), whereas shallot latent virus does not (Bos *et al.*, 1978). Best tests for diagnosis: honeysuckle latent virus may occur in mixed infections with tobacco leaf curl geminivirus (Osaki *et al.*, 1977), cucumber mosaic virus (Lihnell, 1951; Brunt and Thomas, 1976; Van der Meer, 1984) or with an unidentified virus with isometric virions (Huttinga, 1977); honeysuckle latent virus can be distinguished from these by the size and shape of its virions.

**References**

Bos, L., Huttinga, H. and Maat, D.Z. (1978). *Neth. J. Pl. Path.* **84**: 227.

Brunt, A.A. and Thomas, B.J. (1976). *Rep. Glasshouse Crops Res. Inst.* 1975, p. 124.

Brunt, A.A., Phillips, S. and Thomas, B.J. (1980). *Acta Hort.* **110**: 205.

Brunt, A.A. and van der Meer, F.A. (1984). *CMI/AAB Descr. Pl. Viruses* No. 289, 4 pp.

Huttinga, H. (1977). *Rep. Res. Inst. Pl. Prot., Wageningen* 1976, p. 31.

Lihnell, D. (1951). *Vaxteskyddsnotiser* **15**: 52.

Osaki, T., Kobatake, H. and Inouye, T. (1979). *Ann. Phytopath. Soc. Japan* **45**: 62.

Van der Meer, F.A. (1985). *Rep. Res. Inst. Pl. Prot. Wageningen* 1984, p. 35.

Van der Meer, F.A., Maat, D.Z. and Vink, J. (1980). *Neth. J. Pl. Path.* **86**: 69.

# Hop American latent *carlavirus*

**Data collated by A.N. Adams, 1986.**

*Synonyms:* American hop latent virus, New Zealand hop virus. *Standard acronym:* AHLV.

First reported in *Humulus lupulus*; from Washington State, U.S.A.; by Probasco and Skotland (1976).

**Natural host range and symptoms**

*Humulus lupulus* — symptomless.

**Transmission**

Transmitted by a vector; an insect; *Myzus (Phorodon) humuli* experimental vector, no information on natural spread; Aphididae. Transmitted in a non–persistent manner. Virus transmitted by mechanical inoculation.

**Geographical distribution**

Spreads in the North American region; the USA. Found, but with no evidence of spread, in the U.K., Germany and Australia where it has occurred only in recent imports from the U.S.A. and has probably been eliminated. Also found in New Zealand.

**Experimental host range**

Several (3–9) families susceptible. Experimentally infected plants mostly show necrotic and chlorotic local lesions, systemic veinbanding and mottle.

Diagnostically susceptible

*Datura stramonium* — necrotic spots and systemic ringspots.
*Chenopodium quinoa* — chlorotic spots, systemic vein-banding and mottle.
*Humulus lupulus* — usually symptomless, may show faint chlorotic ringspots.

Diagnostically insusceptible

*Nicotiana clevelandii.*

Maintenance and propagation hosts

*Chenopodium quinoa.*

Assay hosts (**L**ocal lesions or **W**hole plants)

*Datura stramonium* (L).

Dicotyledonous Hosts *(**Susceptible** or Insusceptible)*

***Chenopodium album***. ***Chenopodium amaranticolor***. ***Chenopodium foetidum***. ***Chenopodium murale***. ***Chenopodium quinoa***. *Cucumis sativus*. ***Datura stramonium***. *Gomphrena globosa. Helianthus annuus. Lactuca sativa. Lycopersicon esculentum. Medicago sativa. Nicotiana bigelovii. Nicotiana debneyi. Nicotiana glutinosa. Nicotiana rustica. Nicotiana tabacum*, or ***Nicotiana tabacum***. *Petunia hybrida*. ***Phaseolus vulgaris***. *Pisum sativum. Raphanus sativus. Trifolium hybridum. Trifolium pratense. Trifolium repens*. ***Vicia faba***.

Monocotyledonous Hosts *(**Susceptible** or Insusceptible)*

*Allium cepa. Avena sativa.*

Sources of host-range data

Adams and Barbara (1982).

**Properties of particles in sap**

Leaf sap contains few virions.

**Purification**

Method: Adams and Barbara (1982).

**Particle morphology**

Virions filamentous; not enveloped; usually straight (or slightly flexuous); with a clear modal length; of 678 nm; 15 nm wide. Axial canal obscure.

**Physical properties**

One sedimenting component in purified preparations.

**Biochemical properties**

Virions contain 6 % nucleic acid; 94 % protein; 0 % lipid.

*Genome* consists of RNA; single–stranded; linear; unipartite; largest (or only) genome part 8.45 kb. Genomic nucleic acid isolated by Adams and Barbara (1982).

*Virion protein(s)* one; $M_r$ 33000; coat. Method of preparation: Adams and Barbara (1980).

**Replication**

Replication does not depend on a helper virus.

**Taxonomy and Relationships**

*Carlavirus.*

Virus(es) with serologically related virions: nerine latent virus (distantly); hop latent, hop mosaic and chrysanthemum B viruses (very distantly). Virus(es) with serologically unrelated virions: carnation latent, lily symptomless, potato virus S, poplar mosaic, helenium S and narcissus latent viruses.

**References**

Adams, A.N. and Barbara, D.J. (1980). *Ann. appl. Biol.* **96**: 201.
Adams, A.N. and Barbara, D.J. (1982). *Ann. appl. Biol.* **101**: 483.
Barbara, D.J. and Adams, A.N. (1983). *CMI/AAB Descr. Pl. Viruses* No. 262, 3 pp.
Probasco, E.G. and Skotland, C.B. (1976). *Proc. Am. Phytopath. Soc.* **3**: 319.

# Hop latent *carlavirus*

**Data collated by A.N. Adams, 1986.**

*Standard acronym:* HpLV.

First reported in *Humulus japonicus, Humulus lupulus*; from Aschersleben, Germany; by Schmidt *et al.* (1966).

**Natural host range and symptoms**

Symptoms persist.

*Humulus lupulus* — systemic chlorotic flecking, but usually symptomless.

**Transmission**

Transmitted by a vector; an insect; *Myzus (Phorodon) humuli*; Aphididae. Transmitted in a non–persistent manner. Virus transmitted by mechanical inoculation; transmitted by grafting; not transmitted by seed.

**Geographical distribution**

Spreads in the Eurasian region; Australia, China, and the USA. Found, but with no evidence of spread, in New Zealand. Probably occurs worldwide as it is latent and planting material of hops has been distributed to all parts of the world.

**Experimental host range**

Few (<3) families susceptible.

Diagnostically susceptible

No reliable diagnostic host is known, as most susceptible plants show no symptoms.

Diagnostically insusceptible

*Nicotiana clevelandii*, which is however susceptible to hop mosaic carlavirus but shows no symptoms.

Maintenance and propagation hosts

*Humulus lupulus.*

Assay hosts (**L**ocal lesions or **W**hole plants)

*Phaseolus vulgaris* cv. Kinghorn wax (L) — local lesions, but unreliable.

Dicotyledonous Hosts *(**Susceptible** or Insusceptible)*

*Apium graveolens.* ***Atriplex hortensis.*** *Beta vulgaris. Brassica campestris* ssp. *pekinensis. Brassica oleracea* ssp. *capitata. Capsicum annuum. Catharanthus roseus.* ***Chenopodium album.*** *Chenopodium amaranticolor. Chenopodium capitatum. Chenopodium foetidum.* ***Chenopodium hybridum. Chenopodium murale. Chenopodium quinoa.*** *Cucumis sativus. Cucurbita pepo. Datura stramonium. Gomphrena globosa. Helianthus annuus. Lycopersicon esculentum. Medicago sativa. Nicotiana bigelovii. Nicotiana clevelandii. Nicotiana debneyi. Nicotiana glutinosa. Nicotiana rustica. Nicotiana tabacum. Petunia hybrida.* ***Phaseolus vulgaris.*** *Physalis floridana. Pisum sativum. Raphanus sativus. Solanum melongena. Trifolium hybridum. Trifolium pratense. Vicia faba.* ***Vigna unguiculata.***

Monocotyledonous Hosts *(**Susceptible** or Insusceptible)*

*Allium cepa. Avena sativa.*

Sources of host-range data

Probasco and Skotland (1978); Adams and Barbara (1982).

**Properties of particles in sap**

DEP: $\log_{10}$ minus 2–3. Infectivity of sap not changed by treatment with di–ethyl ether. Leaf sap contains few virions (usually).

**Purification**

Method: Probasco and Skotland (1978); Adams and Barbara (1982).

**Particle morphology**

Virions filamentous; not enveloped; usually straight (or slightly flexuous); with a clear modal length; of 674 nm; 14.4 nm wide. Axial canal obscure.

**Physical properties**

One sedimenting component in purified preparations; sedimentation coefficient 176 *S*.

**Biochemical properties**

Virions contain 6 % nucleic acid; 94 % protein; 0 % lipid.

*Genome* consists of RNA; single–stranded; linear; unipartite; largest (or only) genome part 8.2 kb. Genomic nucleic acid isolated by Adams and Barbara (1982).

*Virion protein(s)* one; $M_r$ 33000. Method of preparation: Adams and Barbara (1980).

**Replication**

Replication does not depend on a helper virus.

**Taxonomy and Relationships**

*Carlavirus.*

Virus(es) with serologically related virions: carnation latent, hop mosaic, lily symptomless, nerine latent and helenium S viruses are moderately related; American hop latent and potato M viruses are more distantly related. Virus(es) with serologically unrelated virions: potato S and poplar mosaic viruses.

**References**

Adams, A.N. and Barbara, D.J. (1980). *Ann. appl. Biol.* **96**: 201.
Adams, A.N. and Barbara, D.J. (1982). *Ann. appl. Biol.* **101**: 483.
Barbara, D.J. and Adams, A.N. (1983). *CMI/AAB Descr. Pl. Viruses* No. 261, 4 pp.
Probasco, E.G. and Skotland, C.B. (1978). *Phytopathology* **68**: 277.

# Hop mosaic *carlavirus*

**Data collated by A.N. Adams, 1986.**

*Standard acronym:* HpMV.

First reported in *Humulus lupulus*; from England; by Salmon (1923); see Bock (1967).

**Natural host range and symptoms**

Symptoms persist.

*Humulus lupulus* — mosaic, chlorotic vein-banding; infected plants climb poorly. Many cultivars are symptomless.

*Urtica urens* — no symptoms.

**Transmission**

Transmitted by a vector; an insect; *Myzus (Phorodon) humuli, Macrosiphum euphorbiae, Myzus persicae*; Aphididae. Transmitted in a non–persistent manner. Virus transmitted by mechanical inoculation; transmitted by grafting; not transmitted by seed.

**Geographical distribution**

Spreads in the Eurasian region; Australia, China, and the USA. Found, but with no evidence of spread, in New Zealand. Hop mosaic virus probably occurs worldwide as it is latent, and planting material of hops has been distributed worldwide.

**Experimental host range**

Several (3–9) families susceptible.

Diagnostically susceptible

*Humulus lupulus*, Wye Golding varieties only — chlorotic vein-banding and mosaic, tendency to fall from the support.

*Nicotiana clevelandii* — symptomless systemic infection.

Maintenance and propagation hosts

*Nicotiana clevelandii.*

Assay hosts (**L**ocal lesions or **W**hole plants)

*Nicotiana clevelandii* (W) — no symptoms. *Humulus lupulus* Golding varieties (W).

Dicotyledonous Hosts *(**Susceptible** or Insusceptible)*

*Ageratum conyzoides. Amaranthus caudatus. Arachis hypogaea. Atriplex hortensis. Beta vulgaris. Brassica campestris* ssp. *pekinensis. Brassica oleracea* ssp. *capitata. Cajanus cajan.* ***Chenopodium album. Chenopodium amaranticolor. Chenopodium foetidum. Chenopodium murale. Chenopodium quinoa.*** *Cucumis sativus. Cucurbita maxima. Cucurbita pepo. Datura stramonium. Daucus carota. Glycine max. Gomphrena globosa. Lactuca sativa.* ***Lathyrus odoratus.*** *Lupinus albus. Lycopersicon esculentum. Melilotus alba. Nicandra physaloides.* ***Nicotiana clevelandii. Nicotiana debneyi.*** *Nicotiana glutinosa. Nicotiana sylvestris. Nicotiana tabacum. Petunia hybrida.* ***Phaseolus vulgaris.*** *Phlox drummondii. Physalis floridana.* ***Pisum sativum.*** *Solanum* melongena. *Solanum nigrum. Tetragonia expansa. Trifolium repens. Vicia faba. Vigna unguiculata.*

Monocotyledonous Hosts *(**Susceptible** or Insusceptible)*

*Allium cepa.*

Sources of host-range data

Bock (1967); Adams and Barbara (1980).

**Properties of particles in sap**

Leaf sap contains few virions (when from hop, more when from *Nicotiana clevelandii*).

**Purification**

Method: Adams and Barbara (1980).

**Particle morphology**

Virions filamentous; not enveloped; usually straight (or slightly flexuous); with a clear modal length; of 651 nm; 13.8 nm wide. Axial canal obscure.

**Physical properties**

One sedimenting component in purified preparations.

**Biochemical properties**

Virions contain 6 % nucleic acid; 94 % protein; 0 % lipid.

*Genome* consists of RNA; single–stranded; linear; unipartite; largest (or only) genome part 8.5 kb. Genomic nucleic acid isolated by Adams and Barbara (1980).

*Virion protein(s)* one; $M_r$ 34000. Method of preparation: Adams and Barbara (1980). Amino acid composition: Yu and Liu (1987).

**Replication**

Replication does not depend on a helper virus.

**Cytopathology**

Virions found in all parts of the host plant; in cytoplasm (the plants examined by Macovei and Titu (1976) were probably also infected with hop latent virus).

**Taxonomy and Relationships**

*Carlavirus.*

Virus(es) with serologically related virions: helenium S, carnation latent and hop latent viruses, but distantly; American hop latent, nerine latent, lily symptomless, potato S, potato M and poplar mosaic viruses are very distantly related. Virus(es) with serologically unrelated virions: narcissus latent virus.

**References**


Adams, A.N. and Barbara, D.J. (1980). *Ann. appl. Biol.* **96**: 201.

Barbara, D.J. and Adams, A.N. (1981). *CMI/AAB Descr. Pl. Viruses* No. 241, 3 pp.

Bock, K.R. (1967). *Rep. East Malling Res. Stn.* 1966, p. 163.

Macovei, A. and Titu, H. (1976). *Rev. Roum. Biol. — Biol. Veg. (Bucarest)* **21**: 71.

Salmon, E.S. (1923). *J. Minist. Agric.* **29**: 1.

Salmon, E.S. and Ware, W.M. (1928). *Ann. appl. Biol.* **15**: 342.

Thresh, J.M., Adams, A.N., Barbara, D.J. and Clark, M.F. (1977). *Ann. appl. Biol.* **87**: 57.

Yu, J. and Liu, Y. (1987). *Pl. Path.* **36**: 38.

# Hop trefoil 1 *alphacryptovirus*

**Data collated by E. Luisoni, 1991.**

*Standard acronym:* HTCV-1.

First reported in *Medicago lupulina*; plants grown from seeds from Poland and Denmark; from Italy; by Boccardo *et al.* (1987); Luisoni *et al.* (1987).

**Natural host range and symptoms**

*Medicago lupulina* — symptomless.

**Transmission**

Transmitted by means not involving a vector. Virus not transmitted by mechanical inoculation; not transmitted by grafting; not transmitted by contact between plants; transmitted by seed.

**Geographical distribution**

Spreads in Italy (plants from seeds from Denmark and Poland and presumed widespread elsewhere).

**Experimental host range**

Few (<3) families susceptible.

Diagnostically susceptible

*Medicago lupulina* — symptomless.

Maintenance and propagation hosts

*Medicago lupulina.*

Sources of host-range data

Boccardo *et al.* (1987); Luisoni *et al.* (1987).

**Properties of particles in sap**

Leaf sap contains few virions (very few). Electron microscopy: use ISEM; to aid adhesion of negative stain use carbon-coated Formvar films that have been glow-discharged in air at 0.1 torr.

**Purification**

Method: Luisoni *et al.* (1987).

**Particle morphology**

Virions isometric; not enveloped; 30 nm in diameter; rounded in profile; without a conspicuous capsomere arrangement.

**Physical properties**

One sedimenting component in purified preparations. Density 1.35–1.37 g $cm^{-3}$ in CsCl.

**Biochemical properties**

*Genome* consists of RNA; double–stranded; linear; of two parts; largest (or only) genome part the larger 1.95 kb (pairs); the 2nd largest 1.75 kb (pairs). Genomic nucleic acid isolated by Boccardo *et al.* (1985); Luisoni *et al.* (1987).

**Taxonomy and Relationships**

*Alphacryptovirus: Partitiviridae.*

Virus(es) with serologically related virions: alfalfa 1 alphacryptovirus. Virus(es) with serologically unrelated virions: beet 1 and 2, carnation 1, poinsettia, radish yellow edge, ryegrass, vicia and white clover 3 cryptoviruses.

No serological cross-reaction between the virions of species of the two cryptovirus groups. Best tests for diagnosis: ISEM and decoration may distinguish between hop trefoil 2 betacryptovirus and hop trefoil 1 and 3 alphacryptoviruses. The detection, by PAGE, of two dsRNAs similar to those of the virus can confirm serological results.

**References**

Boccardo, G., Milne, R.G., Luisoni, E., Lisa, V. and Accotto, G.P. (1985). *Virology* **147**: 29.

Boccardo, G., Lisa, V., Luisoni, E. and Milne, R.G. (1987). *Adv. Virus Res.* **32**: 171.

Luisoni, E., Milne, R.G., Accotto, G.P. and Boccardo, G. (1987). *Intervirology* **28**: 144.

# Hop trefoil 2 *betacryptovirus*

**Data collated by E. Luisoni, 1991.**

*Standard acronym:* HTCV-2.

First reported in *Medicago lupulina*; from Italy, in plants grown from seeds imported from Poland and Denmark; by Boccardo *et al.* (1987); Luisoni *et al.* (1987).

**Natural host range and symptoms**

*Medicago lupulina* — symptomless.

**Transmission**

Transmitted by means not involving a vector. Virus not transmitted by mechanical inoculation; not transmitted by grafting; not transmitted by contact between plants; transmitted by seed.

**Geographical distribution**

Found, but with no evidence of spread, in in Italy, in plants grown from seeds from Denmark and Poland, probably widespread elsewhere.

**Experimental host range**

Few (<3) families susceptible.

Maintenance and propagation hosts

*Medicago lupulina.*

Sources of host-range data

Boccardo *et al.* (1987); Luisoni *et al.* (1987).

**Properties of particles in sap**

Leaf sap contains few virions (very few). Electron microscopy: use ISEM. To aid negative stain adhesion of purified preparations, it is best to use carbon-coated Formvar-film that has been treated with glow-discharge in air at 0.1 torr.

**Purification**

Method: Luisoni *et al.* (1987).

**Particle morphology**

Virions isometric; not enveloped; 38 nm in diameter; rounded in profile; with a conspicuous capsomere arrangement.

**Biochemical properties**

*Genome* consists of RNA; double–stranded; linear; of three parts (in the mixture hop trefoil cryptic viruses 2 and 3); largest (or only) genome part the largest 2.05 kb (pairs); the 2nd largest 1.65 kb (pairs); the 3rd largest 1.45 kb (pairs). Genomic nucleic acid isolated by Boccardo *et al.* (1985); Luisoni *et al.* (1987).

**Taxonomy and Relationships**

*Betacryptovirus: Partitiviridae.*

Virus(es) with serologically related virions: red clover 2 and white clover 2 betacryptoviruses.

No serological relation to alphacryptoviruses. Best tests for diagnosis: ISEM and decoration may distinguish between hop trefoil 2 betacryptovirus and hop trefoil 1 and 3 alphacryptoviruses. ELISA may be useful but it does not distinguish between 1, 2 and 3. The detection by PAGE of dsRNAs similar to those found for the mixture hop trefoil cryptic viruses 2 and 3 can confirm serology.

**References**


Boccardo, G., Milne, R.G., Luisoni, E., Lisa, V. and Accotto, G.P. (1985). *Virology* **147**: 29.

Boccardo, G., Lisa, V., Luisoni, E. and Milne, R.G. (1987). *Adv. Virus Res.* **32**: 171.

Luisoni, E., Milne, R.G., Accotto, G.P. and Boccardo, G. (1987). *Intervirology* **28**: 144.


# Hop trefoil 3 *alphacryptovirus*

**Data collated by E. Luisoni, 1991.**

*Standard acronym:* HTCV-3.

First reported in *Medicago lupulina*; in plants grown from seeds from Poland and Denmark; from Italy; by Boccardo *et al.* (1987); Luisoni *et al.* (1987).

**Natural host range and symptoms**

*Medicago lupulina* — symptomless.

**Transmission**

Transmitted by means not involving a vector. Virus not transmitted by mechanical inoculation; not transmitted by grafting; not transmitted by contact between plants; transmitted by seed.

**Geographical distribution**

Found, but with no evidence of spread, in Italy, but probably widespread elsewhere.

Maintenance and propagation hosts

*Medicago lupulina.*

Sources of host-range data

Boccardo *et al.* (1987); Luisoni *et al.* (1987).

**Properties of particles in sap**

Leaf sap contains few virions (very few). Electron microscopy: use ISEM; to aid negative stain adhesion use carbon-coated Formvar films treated with glow-discharge in air at 0.1 torr.

**Purification**

Method: Luisoni *et al.* (1987).

**Particle morphology**

Virions isometric; not enveloped; 30 nm in diameter; rounded in profile; without a conspicuous capsomere arrangement.

**Physical properties**

One sedimenting component in purified preparations. Density 1.35–1.37 g $cm^{-3}$ in CsCl.

**Biochemical properties**

*Genome* consists of RNA; double–stranded; linear; of three parts (in the mixture hop trefoil cryptic viruses 2 and 3); largest (or only) genome part the largest 2.05 kb (pairs); the 2nd largest 1.65 kb (pairs); the 3rd largest 1.45 kb (pairs). Genomic nucleic acid isolated by Boccardo *et al.* (1985); Luisoni *et al.* (1987).

**Taxonomy and Relationships**

*Alphacryptovirus: Partitiviridae.*

Virus(es) with serologically related virions: vicia cryptic virus. Virus(es) with serologically unrelated virions: alfalfa 1, beet 1 and 2, carnation, poinsettia, radish yellow edge, ryegrass and white clover 3 cryptoviruses.

Best tests for diagnosis: ISEM and decoration may distinguish between hop trefoil 2 betacryptovirus and hop trefoil 1 and 3 alphacryptoviruses. The detection, by PAGE, of dsRNAs similar to those found for the mixture hop trefoil cryptic viruses 2 and 3 can confirm serological results.

**References**

Boccardo, G., Milne, R.G., Luisoni, E., Lisa, V. and Accotto, G.P. (1985). *Virology* **147**: 29.

Boccardo, G., Lisa, V., Luisoni, E. and Milne, R.G. (1987). *Adv. Virus Res.* **32**: 171.

Luisoni, E., Milne, R.G., Accotto, G.P. and Boccardo, G. (1987). *Intervirology* **28**: 144.

# Hordeum mosaic *rymovirus*

**Data collated by W.G. Langenberg and N.L. Robertson, 1991.**

*Standard acronym:* HoMV.

First reported in *Hordeum jubatum*; from Alberta, Canada; by Slykhuis and Bell (1966).

**Natural host range and symptoms**

Symptoms persist. Symptoms mottle.

*Hordeum jubatum, H. vulgare* — mottle.

**Transmission**

Assumed to be transmitted by eriophyid mites. Not transmitted by *Eriophyes tulipae, Abacarus hystrix, Aculus mckenziei, Rhopalosiphum padi, R. maidis, Macrosiphum (Sitobion) avenae, Myzus persicae.* Virus transmitted by mechanical inoculation; not transmitted by seed; not transmitted by pollen.

**Geographical distribution**

Spreads in Canada.

**Experimental host range**

Few (<3) families susceptible. Experimentally infected plants mostly show chlorotic mottles; streaking; stunting; leaf rolling.

Diagnostically susceptible

*Agropyron trachycaulum, Avena sativa, Echinochloa crus-galli, Elymus canadensis, E. virginicus, Hordeum jubatum, H. vulgare, Lolium multiflorum, Poa palustris, Secale cereale, Triticum aestivum, T. durum* — diffuse chlorotic mottle; streaks at base of young leaves; moderate stunting and inrolling of leaves only with severe isolates.

Diagnostically insusceptible

*Agropyron repens, Bromus inermis, Lolium perenne, Lophopyrum elongatum.*

Maintenance and propagation hosts

*Hordeum jubatum, Lolium multiflorum, Triticum aestivum.*

Assay hosts (**L**ocal lesions or **W**hole plants)

*Hordeum jubatum* (W), *Lolium multiflorum* (W), *Triticum aestivum* (W).

Monocotyledonous Hosts *(**Susceptible** or Insusceptible)*

*Agropyron repens*. ***Avena sativa***. *Dactylis glomerata*. ***Echinochloa crus–galli***. *Elytrigia intermedium*. ***Hordeum vulgare***. ***Lolium multiflorum***. *Lolium perenne*. *Lophopyrum elongatum*. ***Secale cereale***. ***Triticum aestivum***. ***Triticum durum***.

Sources of host-range data

Slykhuis and Bell (1966).

**Properties of particles in sap**

TIP: 50 °C. LIV: 8 days. DEP: $\log_{10}$ minus 3. Leaf sap contains many virions.

**Purification**

Method: Slykhuis and Bell (1966).

**Particle morphology**

Virions filamentous; not enveloped; usually flexuous; with a clear modal length; of 701 nm; 15.5 nm wide. Axial canal obscure. Basic helix obscure.

**Physical properties**

One sedimenting component in purified preparations.

**Biochemical properties**

*Genome* consists of RNA; single–stranded; linear; unipartite; largest (or only) genome part 8–10 kb. Genomic nucleic acid isolated by (Robertson *et al.*, 1990); polyadenylated full-length RNA genomes were obtained from partially purified virions with proteinase K and SDS, phenol/chloroform extraction, ethanol precipitation, and oligo–d(T) cellulose chromatography. 5' terminus of RNA has a VPg. Poly A region present; the genome has a poly A tail of at least 33 residues attached to the 3' non-coding region. Genome has no *t*RNA–like activity.

Sub–genomic *m*RNA not found in infected cells.

*Virion protein(s)* one; coat protein. Method of preparation: Slykhuis and Bell (1966).

*Virus–coded non–virion proteins* isolated (Langenberg (1991)); one protein found. $M_r$ 68000–72000; pinwheel inclusion protein.

**Replication**

Genome replicates in cytoplasm. Coat protein *m*RNA translated in the cytoplasm. Replication does not depend on a helper virus.

**Cytopathology**

Virions found in leaves, roots, mesophyll, epidermis and the meristem; in cytoplasm. Inclusions present in infected cells; are pinwheels; they do not contain virions.

**Taxonomy and Relationships**

*Rymovirus: Potyviridae.*

Virus(es) with serologically related virions: agropyron mosaic virus. Virus(es) with serologically unrelated virions: oat necrotic mottle, ryegrass mosaic and wheat streak mosaic viruses.

The genomes of hordeum mosaic and agropyron mosaic viruses are 73% homologous at their 3' termini.

**References**

Langenberg, W.G. (1991). *J. gen. Virol.* **72**: 493.
Langenberg, W.G. and Schroeder, H.F. (1974). *J. gen. Virol.* **23**: 51.
Robertson, N.L., French, R. and Langenberg, W.G. (1990). *Phytopathology* **80**: 1018.
Slykhuis, J.T. and Bell, W. (1966). *Can. J. Bot.* **44**: 1191.

# Horsegram yellow mosaic bigeminivirus

**Data collated by V. Muniyappa. 1983. Revised 1984. Revised 1989 by D.V.R. Reddy.**

*Standard acronym:* HgYMV.

First reported in *Macrotyloma uniflorum*; from India; by Muniyappa *et al.* (1976).

**Natural host range and symptoms**

Symptoms persist.

*Cajanus cajan, Glycine max, Indigofera hirsuta, Macrotyloma uniflorum, Phaseolus lunatus, P. vulgaris, Vigna mungo, V. radiata* — mosaic, mottling and yellowing.

**Transmission**

Transmitted by a vector; an insect; *Bemisia tabaci*; Aleyrodidae. Transmitted in a non–persistent manner. Virus not transmitted by mechanical inoculation; not transmitted by contact between plants; not transmitted by seed; not transmitted by pollen.

**Geographical distribution**

Spreads in India.

**Experimental host range**

Few (<3) families susceptible.

Diagnostically susceptible

*Glycine max* — systemic mottle, mosaic, pod and seed malformation.

*Macrotyloma uniflorum, Phaseolus lunatus, P. vulgaris* — systemic mottle, mosaic, yellowing, pod and seed malformation.

Diagnostically insusceptible

*Macrotyloma uniflorum* cv. BGM.

Assay hosts (**L**ocal lesions or **W**hole plants)

*Glycine max* (W), *Macrotyloma uniflorum* (W), *Phaseolus lunatus* (W), *Phaseolus vulgaris* (W).

Dicotyledonous Hosts ***(Susceptible** or Insusceptible)*

***Cajanus cajan***. *Cicer arietinum*. ***Clitoria ternatea***. ***Glycine max***. *Lablab purpureus*. *Lens esculenta*. *Lycopersicon esculentum*. ***Macroptilium lathyroides***. ***Macrotyloma uniflorum***. ***Phaseolus lunatus***. ***Phaseolus vulgaris***. *Pisum sativum*. *Solanum tuberosum*. ***Vigna angularis***. ***Vigna radiata***. *Vigna unguiculata*.

Sources of host-range data

Muniyappa *et al.* (1976; 1977).

**Particle morphology**

Virions geminate; 15–18 nm in diameter; dimers 30 nm in length.

**Taxonomy and Relationships**

Bigeminivirus: *Geminiviridae.*

**References**

Muniyappa, V., Reddy, D.V.R. and Shivashankar, G. (1976). *Mysore J. agric. Sci.* **10**: 611.

Muniyappa, V., Setty, B.A.S. and Shivashankar, G. (1977). *Indian J. Genet.* **37**: 488.

Muniyappa, V., Rajeshwari, R., Bharathan, N., Reddy, D.V.R. and Nolt, B.L. (1987). *J. Phytopath.* **119**: 81.

# Horseradish latent *caulimovirus*

**Data collated by A.A. Brunt, 1991.**

*Standard acronym:* HRLV.

First reported in *Armoracia rusticana*; from the U.S.A., but in plants from Denmark; by Richins and Shepherd (1986).

**Natural host range and symptoms**

*Armoracia rusticana* — no conspicuous symptoms in American cultivars.

**Transmission**

Transmitted by a vector; an insect; *Myzus persicae*; Aphididae. Virus transmitted by mechanical inoculation.

**Geographical distribution**

Spreads in Denmark.

**Experimental host range**

Few (<3) families susceptible.

Diagnostically susceptible

*Brassica campestris* ssp. *rapa* — local chlorotic lesions; mild systemic chlorotic mottling and faint chlorotic banding of main veins.

*Matthiola incarna* — mild chlorotic mottling of systemically infected leaves.

Diagnostically insusceptible

*Beta vulgaris, Chenopodium quinoa, Cucumis sativus, Nicotiana tabacum, Phaseolus vulgaris.*

Maintenance and propagation hosts

*Brassica campestris* cv. Just Right.

Assay hosts (**L**ocal lesions or **W**hole plants)

*Brassica campestris* cv. Just Right (W).

Dicotyledonous Hosts *(**Susceptible** or Insusceptible)*

*Beta vulgaris*. ***Brassica campestris***. ***Brassica campestris*** ssp. ***napus***. ***Brassica campestris*** ssp. ***pekinensis***. ***Brassica campestris*** ssp. ***rapa***. *Brassica juncea*. *Chenopodium amaranticolor*. *Chenopodium quinoa*. *Cucumis melo*. *Cucumis sativus*.

*Datura stramonium. Lactuca sativa. Lycopersicon esculentum.* ***Malus sylvestris****. Nicotiana bigelovii. Nicotiana glutinosa. Phaseolus vulgaris. Pisum sativum. Vigna unguiculata. Zinnia elegans.*

Sources of host-range data

Richins and Shepherd (1986).

**Purification**

Method: Richins and Shepherd (1986).

**Particle morphology**

Virions isometric; not enveloped; 50 nm in diameter; rounded in profile.

**Physical properties**

One sedimenting component in purified preparations; sedimentation coefficient 200 *S*.

**Biochemical properties**

*Genome* consists of DNA; double–stranded; circular and linear; unipartite; largest (or only) genome part 8.031 kb (pairs).

*Virion protein(s)* one; $M_r$ 40000.

**Taxonomy and Relationships**

*Caulimovirus*.

Virus(es) with serologically related virions: cauliflower mosaic virus.

**References**

Richins, R.D. and Shepherd, R.J. (1986). *Phytopathology* **76**: 749.

Shalla, T.A., Shepherd, R.J. and Peterson, L.J. (1980). *Virology* **102**: 381.

# Humulus japonicus *ilarvirus*

**Data collated by A.N. Adams, 1991.**

*Standard acronym:* HjV.

First reported in *Humulus japonicus*; from the U.K. in seedlings grown from seed imported from the Peoples Republic of China; by Adams *et al.* (1989).

**Natural host range and symptoms**

*Humulus japonicus, H. lupulus* — symptomless.

**Transmission**

Virus transmitted by mechanical inoculation; transmitted by seed.

**Ecology and control**

Studies reported by Adams *et al.* (1989).

**Geographical distribution**

Spreads in China (probably; it was detected by serology in a few samples from China). Found, but with no evidence of spread, in samples imported to the U.K., but is now thought to have been eradicated.

**Experimental host range**

Many (>9) families susceptible. Experimentally infected plants mostly show necrotic local lesions, chlorotic mottle or mosaic in systemic leaves.

Diagnostically susceptible

*Chenopodium quinoa* — necrotic local lesions, systemic chlorosis and necrosis.

*Chenopodium amaranticolor* — pin-point necrotic local lesions, systemic chlorosis.

*Nicotiana clevelandii* — etched necrotic local lesions, systemic chlorosis and necrosis.

Diagnostically insusceptible

*Lactuca sativa, Brassica oleracea, Vigna unguiculata, Nicotiana glutinosa.*

Maintenance and propagation hosts

*Chenopodium quinoa.*

Assay hosts (**L**ocal lesions or **W**hole plants)

*Chenopodium quinoa* (L) and *C. amaranticolor* (L).

Dicotyledonous Hosts *(**Susceptible** or Insusceptible)*

***Amaranthus caudatus. Beta vulgaris.*** *Brassica oleracea* ssp. *capitata. Capsella bursa-pastoris. Capsicum annuum.* ***Catharanthus roseus. Celosia cristata. Chenopodium album. Chenopodium amaranticolor. Chenopodium foetidum. Chenopodium foliosum. Chenopodium murale. Chenopodium quinoa. Cucumis melo. Cucumis sativus. Cucurbita maxima. Helianthus annuus. Humulus lupulus.*** *Lactuca sativa. Lycopersicon esculentum.* ***Nicandra physaloides. Nicotiana clevelandii.*** *Nicotiana glutinosa.* ***Nicotiana tabacum. Petunia hybrida.*** *Raphanus sativus. Vicia faba. Vigna unguiculata.* ***Zinnia elegans.***

Monocotyledonous Hosts *(**Susceptible** or Insusceptible)*

*Allium cepa. Allium porrum.*

Sources of host-range data

Adams *et al.* (1989).

**Properties of particles in sap**

LIV: less than 0.25 days. Infectivity of sap not changed by treatment with di–ethyl ether. Leaf sap contains many virions.

**Purification**

Method: Adams *et al.* (1989).

**Particle morphology**

Virions isometric; not enveloped; 24–33 nm in diameter; rounded in profile; with a conspicuous capsomere arrangement.

**Physical properties**

Four sedimenting components in purified preparations. Density 1.364 g $cm^{-3}$ in CsCl (fixed with 2% formaldehyde).

**Biochemical properties**

*Genome* consists of RNA; single–stranded; of four parts; largest (or only) genome part the largest 3.393 kb; the 2nd largest 2.958 kb; the 3rd largest 2.113 kb; the 4th largest 1.099 kb. Genomic nucleic acid isolated by Adams *et al.* (1989). Infectivity retained when deproteinised with phenol or detergent.

*Virion protein(s)* one; $M_r$ 26350. Method of preparation: Adams *et al.* (1989).

**Replication**

Replication does not depend on a helper virus.

**Cytopathology**

Virions found in leaves.

**Taxonomy and Relationships**

*Ilarvirus: Bromoviridae* (subgroup 9).

Virus(es) with serologically related virions: prunus necrotic ringspot virus, but distantly. Virus(es) with serologically unrelated virions: apple mosaic, prune dwarf, American plum line pattern, tobacco streak, asparagus 2, citrus leaf rugose, citrus variegation, elm mottle and lilac ring mottle viruses.

Best tests for diagnosis: its only known natural hosts are species of *Humulus*. Test plant reactions could be confused with those of prunus necrotic ringspot virus which is widespread in hops and hence diagnosis is best confirmed by serology. Clear results obtained with ISEM and decoration.

**References**

Adams, A.N., Clark, M.F. and Barbara, D.J. (1989). *Ann. appl. Biol.* **114**: 497.

# Hyacinth mosaic *potyvirus*

**Data collated by S. Phillips, 1985.**

*Standard acronym:* HyaMV.

First reported in *Hyacinthus orientalis* L.; from Bulgaria and Oregon, U.S.A.; by Atanasoff (1928); Smith and Brierley (1944).

**Natural host range and symptoms**

Symptoms persist.

*Hyacinthus orientalis* L. — leaf chlorosis, stem spotting, occasional flower break.

**Transmission**

Transmitted by a vector; an insect; *Myzus persicae, Macrosiphum solanifolii*; Aphididae. Transmitted in a non–persistent manner. Virus transmitted by mechanical inoculation; not transmitted by contact between plants; not transmitted by seed; not transmitted by pollen.

**Geographical distribution**

Spreads in Australia, Bulgaria, the Netherlands, the UK, and the USA.

**Experimental host range**

Several (3–9) families susceptible.

Diagnostically insusceptible

*Chenopodium quinoa, Lilium formosanum, Nicotiana clevelandii, Phaseolus vulgaris, Tetragonia expansa.*

Maintenance and propagation hosts

*Hyacinthus orientalis.*

Dicotyledonous Hosts *(**Susceptible** or Insusceptible)*

*Chenopodium amaranticolor. Chenopodium murale. Chenopodium quinoa. Gomphrena globosa. Nicotiana clevelandii. Nicotiana glutinosa. Nicotiana tabacum. Petunia hybrida. Phaseolus vulgaris. Trifolium incarnatum. Vicia faba.*

Monocotyledonous Hosts *(**Susceptible** or Insusceptible)*

*Lilium formosanum.*

**Properties of particles in sap**

Leaf sap contains few virions.

**Purification**

Method: Phillips: extract from infected hyacinth leaves and flower stems by blending in 0.4 M Na citrate + 0.2 M urea + 10 mM DIECA + 0.2% (w/v) hemicellulase + 0.1% (v/v) mercapto-acetic acid (pH 6.9) (1 g : 4 ml). Incubate at 37°C for 1 hour. Add Triton X-100 (2% v/v of extractant) and stir for 30 minutes. Leave at 2°C overnight. Filter through cheesecloth and centrifuge (20 minutes at 12,000 g; 75 minutes at 65,000 g). Resuspend pellets in 0.2 M Na–citrate + 0.1 M urea + 0.1% (v/v) mercapto-acetic acid (pH 6.6, 1 ml/1 g of plant material). Leave at 2°C overnight. Give a second cycle of differential centrifugation and resuspend pellets in a 33 mM phosphate buffer (pH 7.3).

**Particle morphology**

Virions filamentous; not enveloped; usually flexuous; with a clear modal length; of 740 nm. Axial canal obscure. Basic helix obscure.

**Taxonomy and Relationships**

*Potyvirus: Potyviridae.*

Virus(es) with serologically related virions: ornithogalum mosaic, scilla mosaic, freesia mosaic and iris severe mosaic viruses. Virus(es) with serologically unrelated virions: bean yellow mosaic, hippeastrum mosaic, iris mild mosaic, narcissus yellow stripe, onion yellow dwarf, potato Y, tulip breaking and turnip mosaic viruses.

**References**

Atanasoff, D. (1928). *Bull. Soc. Bot. Bulgarie* **2**: 51.
Derks, A.F.L.M. and Vink-van den Abeele, J.L. (1980). *Acta Hort.* **109**: 495.
Smith, F.F. and Brierley, P. (1944). *Phytopathology* **34**: 497.

# Hydrangea latent (?) *carlavirus*

**Data collated by A.A. Brunt, 1991.**

*Standard acronym:* HdLV.

First reported in *Hydrangea macrophylla*; from the U.S.A.; by Allen *et al.* (1985).

**Natural host range and symptoms**

*Hydrangea macrophylla.*

**Particle morphology**

Virions filamentous; slightly usually flexuous; with a clear modal length; of 600–700 nm.

**Taxonomy and Relationships**

Possible *Carlavirus.*

**References**

Allen, T.C., McMorran, J.P. and Lawson, R.H. (1985). *Acta Hort.* **164**: 85.

# Hydrangea mosaic *ilarvirus*

**Data collated by B.J. Thomas, 1991.**

*Standard acronym:* HdMV.

First reported in *Hydrangea macrophylla*; from West Sussex, U.K.; by Thomas *et al.* (1983).

## Natural host range and symptoms

Symptoms vary seasonally. Symptoms mottles and mosaic.

*Hydrangea macrophylla*—light and dark green mottle and mosaics on leaves.

## Transmission

Not transmitted by *Myzus persicae*. Virus transmitted by mechanical inoculation; transmitted by grafting; transmitted by seed.

## Geographical distribution

Found, but with no evidence of spread, in U.K. (West Sussex, Guernsey, the Channel Islands).

## Experimental host range

Several (3–9) families susceptible. Experimentally infected plants mostly show transitory systemic chlorosis.

Diagnostically susceptible

*Chenopodium quinoa* — diffuse chlorotic local lesions; transitory systemic chlorotic local lesions.

*Gomphrena globosa, Nicotiana benthamiana, N. clevelandii* — systemic but symptomless.

*Chenopodium amaranticolor* — small chlorotic local lesions; not systemic.

Diagnostically insusceptible

*Beta vulgaris, Cucumis sativus, Nicotiana glutinosa, Pisum sativum, Vigna unguiculata.*

Maintenance and propagation hosts

*Chenopodium quinoa.*

Assay hosts (**L**ocal lesions or **W**hole plants)

*Chenopodium quinoa* (W).

Dicotyledonous Hosts *(**Susceptible** or Insusceptible)*

*Beta vulgaris. Brassica campestris* ssp. *pekinensis. Brassica oleracea* ssp. *botrytis.* ***Chenopodium amaranticolor****. Chenopodium murale.* ***Chenopodium quinoa****. Cucumis sativus. Datura metel.* ***Gomphrena globosa. Nicotiana benthamiana. Nicotiana clevelandii****. Nicotiana glutinosa. Nicotiana tabacum. Pisum sativum. Trifolium pratense. Vicia faba. Vigna unguiculata.*

Sources of host-range data

Thomas *et al.* (1983).

## Properties of particles in sap

TIP: 50 °C. LIV: 21 days (at 4°C). DEP: $\log_{10}$ minus 3. Infectivity of sap not changed by treatment with di–ethyl ether. Leaf sap contains few virions. Electron microscopy: virions are unstable in phosphate buffer and must be fixed in aldehyde. UA can be used without fixing purified preparations.

**Purification**

Method: Thomas *et al.* (1983).

**Particle morphology**

Virions isometric; not enveloped; 28–30 nm in diameter; 30–38 nm in length; rounded in profile; with a conspicuous capsomere arrangement.

**Physical properties**

Three sedimenting components in purified preparations; sedimentation coefficient of the fastest 105 *S* ((B)); of the other(s) 86 and 97 *S* (;(T) and (M) respectively). Density 1.37 g $cm^{-3}$ in CsCl (necessary to fix with aldehydes). $A_{260}/A_{280}$ ratio 1.32 (±0.44).

**Biochemical properties**

Virions contain 24 % nucleic acid; 76 % protein.

*Genome* consists of RNA; single–stranded; of five parts; largest (or only) genome part the largest 4.032 kb (RNA-1); the 2nd largest 3.484 kb (RNA-2); the 3rd largest 2.677 kb (RNA-3); the 4th largest 1.161 kb (RNA-4); the 5th largest 0.871 kb (RNA-5). Genomic nucleic acid isolated by methods described by Rogerson (1975); Wilcockson and Hull (1974). Infectivity retained when deproteinised with phenol or detergent. Additional factor required for infectivity; infectivity depends on RNAs 1–3 (total genome) plus RNA-5 or coat protein.

*Virion protein(s)* one; $M_r$ 26400 (±0.4); coat protein. Method of preparation: Laemmli (1970). Virion proteins not glycosylated.

**Replication**

Replication does not depend on a helper virus.

**Cytopathology**

Virions found in leaves, stamens, gynaecium and petals.

**Taxonomy and Relationships**

*Ilarvirus: Bromoviridae* (subgroup 8).

Virus(es) with serologically unrelated virions: apple mosaic, elm mottle, prunus necrotic ringspot, asparagus 2, spinach latent and tobacco streak viruses. Hydrangea mosaic may best be regarded as a strain of elm mottle ilarvirus (Jones, 1985).

Best tests for diagnosis: in *Hydrangea*, leaf symptoms differ from the malformations, ringspots or chlorosis caused by hydrangea ringspot virus, cucumber mosaic, tobacco ringspot, tomato ringspot, tomato black ring, arabis mosaic or tobacco rattle viruses. Virion instability, shape and small numbers in leaves is characteristic.

**References**

Jones, A.T. (1985). *Rep. Scottish Crop Res. Inst.* 1984, p.190.

Laemmli, U.K. (1970). *Nature, Lond.* **227**: 680.

Rogerson, A.A. (1975). *Ann. Biochem.* **65**: 34.

Thomas, B.J., Barton, R.J. and Tuszynski, A. (1983). *Ann. appl. Biol.* **103**: 261.

Wilcockson, J. and Hull, R. (1974). *J. gen. Virol.* **23**: 107.

# Hydrangea ringspot *potexvirus*

**Data collated by A.A. Brunt, 1994.**

*Standard acronym:* HRSV.

First reported in *Hydrangea macrophylla*; from the U.S.A.; by Brierley (1954).

**Natural host range and symptoms**

*Hydrangea macrophylla* — often symptomless, but leaves sometimes crinkled and distorted, fewer florets, plants stunted.

**Transmission**

Transmitted by means not involving a vector. Virus transmitted by mechanical inoculation; transmitted by grafting; not transmitted by seed.

**Geographical distribution**

Probably distributed worldwide. Spreads in the Eurasian region, the North American region, and the Pacific region; Belgium, Canada, the former Czechoslovakia, Denmark, France, Germany, Ireland, Italy, New Zealand, the UK, and the USA.

**Experimental host range**

Several (3–9) families susceptible. Experimentally infected plants mostly show chlorotic and necrotic local lesions.

Diagnostically susceptible

*Chenopodium amaranticolor, C. quinoa, Gomphrena globosa* — local chlorotic or necrotic lesions, not systemic.

*Primula malacoides* — systemic, but symptomless.

Maintenance and propagation hosts

*Primula malacoides*.

Assay hosts (**L**ocal lesions or **W**hole plants)

*Chenopodium amaranticolor* (L), *C. quinoa* (L) and *Gomphrena globosa* (L).

Dicotyledonous Hosts *(**S**usceptible or **I**nsusceptible)*

***Chenopodium amaranticolor*. *Chenopodium quinoa*. *Gomphrena globosa*. *Primula malacoides*.**

**Properties of particles in sap**

TIP: 70–75 °C. LIV: 14–21 days. DEP: $\log_{10}$ minus 5. Leaf sap contains few virions.

**Purification**

Method: Bercks and Brandes, (1961); Koenig, (1973).

**Particle morphology**

Virions filamentous; not enveloped; usually flexuous; with a clear modal length; of 490 nm. Axial canal obscure.

**Biochemical properties**

Virions contain 5 % nucleic acid; 95 % protein; 0 % lipid.

*Genome* consists of RNA; single–stranded; unipartite; largest (or only) genome part 7 kb.

*Virion protein(s)* one; $M_r$ 22000; coat.

**Cytopathology**

Virions found in cytoplasm. Inclusions present in infected cells.

**Taxonomy and Relationships**

*Potexvirus.*

Virus(es) with serologically related virions: cactus X, clover yellow mosaic, potato X and white clover mosaic potexviruses, distantly.

**References**

Bercks, R. and Brandes, J. (1961). *Phytopath. Z.* **42**: 45.
Brandes, J. and Bercks, R. (1965). *Adv. Virus Res.* **11**: 1.
Brierley, P. (1954). *Phytopathology* **44**: 696.
Brierley, P. and Lorentz, P. (1957). *Phytopathology* **47**: 39.
Hollings, M. (1958). *J. Hort. Sci.* **33**: 257.
Koenig, R. (1970). *NachrBl. dtsch. PflSchutz. Stuttg.* **22**: 34.
Koenig, R. (1973). *CMI/AAB Descr. Pl. Viruses* No. 114, 3pp.
Sisler, H.D., Chang, S.C., Reagan, R.L. and Brierley, P. (1957). *Phytopathology* **47**: 491.

# Hypochoeris mosaic (?) *furovirus*

**Data collated by R. Stace-Smith, 1987.**

*Synonyms:* cat's ear yellow spot virus (Schmidt, 1977; Brunt and Stace-Smith, 1983).
*Standard acronym:* HyMV.

First reported in *Hypochoeris radiata* (cat's ear); from Western Canada; by Brunt and Stace-Smith (1978*a*; 1978*b*).

**Natural host range and symptoms**

Symptoms vary seasonally (and they are absent or less conspicuous at higher temperatures (20–25°C and above)).

*Hypochoeris radiata* — faint leaf chlorosis.

*Leontodon autumnalis* — chlorotic spots in young leaves.

**Transmission**

Transmitted by means not involving a vector. Virus transmitted by mechanical inoculation.

**Geographical distribution**

Spreads in Australia, Canada, and Germany (possibly).

**Experimental host range**

Several (3–9) families susceptible.

Diagnostically susceptible

*Chenopodium amaranticolor* and *C. quinoa* — chlorotic local lesions and veinal necrosis, then systemic chlorosis.

*Chenopodium murale* — necrotic local lesions.

*Montia perfoliata*, isolate from western Canada — systemic severe chlorosis.

*Nicotiana clevelandii* — local grey necrotic lesions or oakleaf pattern, followed by veinal necrosis and death, systemic faint chlorosis in abnormally small leaves and severe stunting.

*Tetragonia expansa* — few local chlorotic lesions.

Maintenance and propagation hosts

*Nicotiana clevelandii.*

Assay hosts (**L**ocal lesions or **W**hole plants)

*Chenopodium amaranticolor* (L), *Chenopodium quinoa* (L).

Dicotyledonous Hosts *(**Susceptible** or Insusceptible)*

***Chenopodium amaranticolor*. *Chenopodium murale*. *Chenopodium quinoa*. *Cucumis sativus*. *Gomphrena globosa*. *Montia perfoliata*. *Nicotiana clevelandii*. *Tetragonia expansa*.**

Sources of host-range data

Brunt and Stace-Smith (1978*a*); Singh and McDonald (1980); Greber and Finlay (1981).

**Properties of particles in sap**

TIP: 45–50 °C. LIV: 1–2 days. DEP: $\log_{10}$ minus 2–3.

**Purification**

Method: Brunt and Stace-Smith (1978*a*).

**Particle morphology**

Virions rod–shaped; not enveloped; usually straight; with a clear modal length; of 220–240 nm (and 120–140 nm); 22 nm wide. Axial canal obvious. Basic helix obvious.

**Physical properties**

Six sedimenting components in purified preparations; sedimentation coefficient of the fastest 173 *S*; of the other(s) 15, 45, 71, 114, and 143 *S*.

**Biochemical properties**

*Genome* consists of RNA; single–stranded; linear.

$M_r$ 24500; coat protein.

**Replication**

Replication does not depend on a helper virus.

**Cytopathology**

Virions found in mesophyll; in cytoplasm. Inclusions cytoplasmic present in infected cells; are membranous bodies and unusual in shape; crystals; occasionally they contain virions (within the crystals, but more frequently adjacent to the enclosing membrane (Brunt and Stace-Smith, 1978*b*)).

**Taxonomy and Relationships**

Possible *Furovirus*.

An isolate from Australia, unlike the type isolate, infects *Cucumis sativus* in which it induces chlorotic local lesions followed by systemic chlorotic mottling and leaf malformation. An isolate from eastern Canada differs from the type isolate in infecting *Gomphrena globosa* to cause chlorotic local lesions and veinal necrosis, then systemic chlorotic mottling and leaf malformation. Hypochoeris mosaic virus has virions of similar morphology, size and properties to those of broad bean necrosis (Inouye and Asatani, 1968), beet necrotic yellow vein (Tamada and Baba, 1973), Nicotiana velutina mosaic (Randles *et al.*, 1976), peanut clump (Thouvenel *et al.*, 1976), potato mop-top (Harrison and Jones, 1970) and wheat soil-borne mosaic viruses (Brakke *et al.*, 1965), and to those of defective strains of tobacco mosaic virus (Kassanis and Woods, 1969). It also resembles some of those viruses in the intracellular location and appearance of its virions (Brunt and Stace-Smith, 1978*b*). Although hypochoeris mosaic virus is apparently serologically distinct, its similarities to these labile viruses suggests that it too may be a furovirus. It may also, like these viruses, be transmitted by a root-infecting fungus, but such transmission tests have not yet been reported.

**References**


Brakke, M.K., Estes, A.P. and Schuster, M.L. (1965). *Phytopathology* **55**: 79.
Brunt, A.A. and Stace-Smith, R. (1978*a*). *Ann. appl. Biol.* **90**: 205.
Brunt, A.A. and Stace-Smith, R. (1978*b*). *J. gen. Virol.* **39**: 63.
Brunt, A.A. and Stace-Smith, R. (1983). *CMI/AAB Descr. Pl. Viruses* No. 273, 4 pp.
Greber, R.S. and Finlay, J.R. (1981). *Australas. Pl. Path.* **10**: 30.
Harrison, B.D. and Jones, R.A.C. (1970). *Ann. appl. Biol.* **65**: 393.
Inouye, T. and Asatani, M. (1968). *Ann. Phytopath. Soc. Japan* **34**: 317.
Kassanis, B. and Woods, R.D. (1969). *Ann. appl. Biol.* **64**: 213.
Randles, J.W., Harrison, B.D. and Roberts, I.M. (1976). *Ann. appl. Biol.* **84**: 193.
Schmidt, H.E. (1977). In: *Pflanzliche Virologie*, Vol. 4, p. 517; eds K. Schmelzer and D. Spaar. Berlin, Akademie-Verlag.
Singh, R.P. and McDonald, J.G. (1980). *Canad. Pl. Dis. Surv.* **60**: 4.
Tamada, T. and Baba, T. (1973). *Ann. Phytopath. Soc. Japan* **39**: 325.
Thouvenel, J.-C., Dollet, M. and Fauquet, C. (1976). *Ann. appl. Biol.* **84**: 311.


## Impatiens latent (?) *potexvirus*

**Data collated by A.A. Brunt, 1991.**

*Standard acronym:* ILV.

First reported in *Impatiens holstii*; from Minnesota, U.S.A.; by Lockhart and Betzold (1980).

**Natural host range and symptoms**

*Impatiens holstii* — symptomless infection.

**Transmission**

Not transmitted by *Myzus persicae*. Virus transmitted by mechanical inoculation.

**Geographical distribution**

Spreads in the USA.

**Experimental host range**

Several (3–9) families susceptible.

Diagnostically susceptible

*Gomphrena globosa, Tetragonia expansa* — local lesions; not systemic.

*Chenopodium album, C. amaranticolor, C. quinoa* — local lesions; then systemic mottling.

*Beta vulgaris* — symptomless systemic infection.

Diagnostically insusceptible

*Cucumis sativus, Phaseolus vulgaris, Vigna unguiculata, Nicotiana clevelandii.*

Maintenance and propagation hosts

*Chenopodium quinoa.*

Assay hosts (**L**ocal lesions or **W**hole plants)

*Chenopodium quinoa* (L).

Dicotyledonous Hosts *(**Susceptible** or Insusceptible)*

*Antirrhinum majus.* ***Beta vulgaris.*** ***Chenopodium album.*** ***Chenopodium amaranticolor.*** ***Chenopodium quinoa.*** *Cucumis sativus.* ***Gomphrena globosa.*** *Melilotus alba. Nicotiana clevelandii. Phaseolus vulgaris.* ***Tetragonia expansa.*** *Trifolium incarnatum. Vigna unguiculata.*

**Properties of particles in sap**

TIP: 68–72 °C. LIV: 21–28 days (at 22–24°C).

**Particle morphology**

Virions filamentous; usually flexuous; with a clear modal length; of 520 nm. Axial canal obscure. Basic helix obscure.

**Cytopathology**

Virions found in leaves and mesophyll.

**Taxonomy and Relationships**

Possible *Potexvirus.*

Virus(es) with serologically unrelated virions: potato X, cymbidium mosaic, clover yellow mosaic, papaya mosaic and nandina mosaic viruses.

**References**

Lockhart, B.E.L. and Betzold, J.A. (1980). *Acta Hort.* **110**: 81.

# Impatiens necrotic spot *tospovirus*

**Data collated by A.A. Brunt, 1995.**

*Standard acronym:* INSV.

First reported in *Impatiens* sp.; from the U.S.A.; by Law and Moyer (1990).

**Natural host range and symptoms**

*Impatiens* sp. — most leaves symptomless, some necrotic spots with necrotic rings in young leaves.

**Transmission**

Transmitted by a vector; an insect; *Frankliniella occidentalis*; Thysanoptera.

**Geographical distribution**

Spreads in Belgium, Germany, the Netherlands, Poland, the UK, and the USA.

*Sequence database accession code(s):*

D00914 Em(40)_vi:IMNNCG Gb(84)_vi:IMNNUCL Impatiens necrotic spot virus (INSV) S RNA, 3'-noncoding region and N gene. 11/92 1,144bp.

M74904 Em(40)_vi:IMNGNSM Gb(84)_vi:IMNGNSM Impatiens necrotic spot virus G2-G1 polyprotein and nonstructural protein of the middle RNA,

X66972 Em(40)_vi:INSVNSP Gb(84)_vi:INSVNSP Impatiens necrotic spot virus gene for nucleocapsid and non-structural proteins. 11/93 2,992b 3 sequences.

**Taxonomy and Relationships**

*Tospovirus: Bunyaviridae.*

**References**

De Avila, A.C., de Haan, P., Kitajima, E.W., Kormelink, R., Resende, R. de O., Goldbach, R.W. and Peters, D. (1992). *J. Phytopathology* **134**: 133.

De Avila, A.C., de Haan, P., Kormelink, R., Resende, R. de O., Goldbach, R.W. and Peters, D. (1993). *J. gen. Virol.* **74**: 153.

De Haan, P., de Avila, A.C., Kormelink, R., Westerbroek, A., Gielen, J.J.L., Peters, D. and Goldbach, R. (1992). *FEBS Letts* **306**: 27.

Law, M.D. and Moyer, J.W. (1990). *J. gen. Virol.* **71**: 933.

Law, M.D., Speck, J. and Moyer, J.W. (1991). *J. gen. Virol.* **72**: 2597.

Law, M.D., Speck, J. and Moyer, J.W. (1992). *Virology* **188**: 732.

## Iris fulva mosaic *potyvirus*

**Data collated by O.W. Barnett, 1991.**

*Standard acronym:* IFMV.

First reported in an Apogon iris hybrid (*Iris fulva x I. brevicaulis*); from Massachusetts, U.S.A; by Travis (1957).

**Natural host range and symptoms**

Symptoms persist. Symptoms faint mosaic.

*Iris fulva* and hybrids with *Iris fulva* — faint mosaic.

**Transmission**

Transmitted by a vector; an insect; *Myzus persicae, Macrosiphum euphorbiae, Aphis fabae*; Aphididae. Not transmitted by *Aphis craccivora*. Transmitted in a non–persistent manner. Virus transmitted by mechanical inoculation; not transmitted by seed.

**Geographical distribution**

Spreads in the USA.

**Experimental host range**

Few (<3) families susceptible. Experimentally infected plants mostly show mosaic, necrosis, malformation of leaves and seed pods.

Diagnostically susceptible

*Iris fulva* and hybrids, *Amaranthus caudatus, Belamcanda chinensis, Chenopodium quinoa, Iris sibirica.*

Diagnostically insusceptible

*Gladiolus* sp. hybrids, *Iris spuria, Neomarica gracilis, Nicotiana clevelandii, Tetragonia expansa.*

Maintenance and propagation hosts

*Belamcanda chinensis* (W).

Dicotyledonous Hosts *(**Susceptible** or Insusceptible)*

***Amaranthus caudatus*****.** ***Chenopodium quinoa*****.** *Gomphrena globosa. Nicotiana clevelandii. Nicotiana tabacum. Phaseolus vulgaris. Pisum sativum. Tetragonia expansa. Vigna unguiculata.*

Monocotyledonous Hosts *(**Susceptible** or Insusceptible)*

***Belamcanda chinensis*****.**

Additional hosts

***Iris fulva*** and ***Iris sibirica***.

Sources of host-range data

Barnett (1986); Barnett and Alper (1977).

**Properties of particles in sap**

TIP: 50 °C. LIV: 2 days. DEP: $\log_{10}$ minus 4. Leaf sap contains few virions.

**Purification**

Method: Barnett (1986).

**Particle morphology**

Virions filamentous; not enveloped; usually flexuous; with a clear modal length; of 767 nm. Axial canal obscure. Basic helix obscure.

**Physical properties**

One sedimenting component in purified preparations; sedimentation coefficient 141 *S*. $A_{260}/A_{280}$ ratio 1.19.

**Biochemical properties**

Virions contain 6 % nucleic acid; 94 % protein.

*Genome* consists of RNA; single–stranded; linear; unipartite; largest (or only) genome part 9.4 kb. Genomic nucleic acid isolated by Reddick and Barnett (1983). Additional factor not required for infectivity.

*Non–genomic nucleic acid* not found in the virions.

*Virion protein(s)* one; $M_r$ 32000. Method of preparation: Barnett: heat in phosphate, mercaptoethanol, SDS and electrophoresis on SDS gels.

**Replication**

Replication does not depend on a helper virus.

**Cytopathology**

Virions found in leaves; in cytoplasm and in unusual sites; aggregate mitochondria. Inclusions present in infected cells; are pinwheels; they do not contain virions. Other cellular changes: mitochondrial aggregates.

**Taxonomy and Relationships**

*Potyvirus: Potyviridae.*

Virus(es) with serologically related virions: iris virus 1 (Lisa, 1980). Virus(es) with serologically unrelated virions: iris severe mosaic and iris mild mosaic viruses.

Best tests for diagnosis: serology in EM (decoration). Pinwheel inclusion morphology easily separates it from iris severe mosaic virus.

**References**

Barnett, O.W. (1986). *AAB Descr. Pl. Viruses* No. 310, 4 pp.

Barnett, O.W. and Alper, M. (1977). *Phytopathology* **67**: 448.

Lisa, V. (1980). *Acta Hort.* **110**: 39.

Reddick, B.B. and Barnett, O.W. (1983). *Phytopathology* **73**: 1506.

Travis, R.V. (1957). *Phytopathology* **47**: 454.

## Iris germanica leaf stripe (?) rhabdovirus

**Data collated by A.A. Brunt, 1991.**

*Standard acronym:* IGLSV.

First reported in *Iris germanica*; from Spain; by Rubio-Huertos (1978).

**Natural host range and symptoms**

*Iris germanica* — conspicuous yellow leaf stripes.

**Geographical distribution**

Spreads in Spain.

**Particle morphology**

Virions rhabdo– or bullet–shaped.

**Taxonomy and Relationships**

Possible unassigned rhabdovirus.

**References**

Rubio-Huertos, M. (1978). *Phytopath. Z.* **92**: 294.

## Iris Japanese necrotic ring virus

**Data collated by A.A. Brunt, 1992.**

First reported in *Iris kaempferi*; from Japan; by Yasukawa *et al.* (1991).

**Natural host range and symptoms**

*Iris kaempferi* — necrotic spindle-shaped necrotic streaks or rings in leaves.

**Transmission**

Not transmitted by *Myzus persicae, Rhopalosiphum padi*. Virus transmitted by mechanical inoculation.

**Geographical distribution**

Spreads in Japan.

**Experimental host range**

Few (<3) families susceptible.

Diagnostically susceptible

*Iris kaempferi* — necrotic streaks and rings in systemically infected leaves.
*Iris laevigata, I. sanguinea* — symptomless systemic infection.

Maintenance and propagation hosts

*Iris kaempferi.*

Assay hosts (**L**ocal lesions or **W**hole plants)

*Iris kaempferi* (W).

Comments on host-range

The virus failed to infect any of 40 other species in 13 families.

**Particle morphology**

Virions isometric; not enveloped; 35 nm in diameter; rounded in profile.

**Cytopathology**

Virions found in leaves and mesophyll; in cytoplasm.

**References**

Yasukawa, K., Osaki, T. and Inouye, T. (1991). *Bull. Univ. Osaka, Ser. B* **43**: 21.

# Iris mild mosaic *potyvirus*

**Data collated by C.J. Asjes, 1986.**

*Synonyms:* iris mosaic virus, iris latent mosaic virus. *Standard acronym:* IMMV.

First reported in *Iris* spp.; from Lisse, the Netherlands; by Van Slogteren (1958); Loebenstein and Alper (1963).

**Natural host range and symptoms**

Symptoms vary cyclically over a few weeks.

*Iris* spp.; bulbous irises, *Iris xiphoides* (English iris), *I. xiphium* (Spanish iris), *I. tingitana x I. xiphium* (Dutch iris), *I. reticulata* and *I. danfordiae* — mild mosaic.

**Transmission**

Transmitted by a vector; an insect; *Aphis gossypii*, *Macrosiphum euphorbiae* and *Myzus persicae*, but not by *Macrosiphoniella sanborni*; Aphididae. Transmitted in a non–persistent manner. Virus not transmitted by seed.

**Geographical distribution**

Probably distributed worldwide (most important cultivars are totally infected (Van Slogteren, 1963; Loebenstein and Alper, 1963; Lawson, 1967; Asjes, 1979)).

**Experimental host range**

Several (3–9) families susceptible.

Diagnostically susceptible

*Iris* ssp. — green mosaic of leaves, more intense on spathes (Van Slogteren, 1963; Loebenstein and Alper, 1963; Lawson, 1967; Asjes, 1979).

*Chenopodium quinoa* — chlorotic local lesions, not systemic.

*Nicotiana clevelandii* — chlorotic lesions.

*Tetragonia expansa* — circular chlorotic local lesions with dark grey necrotic margins, not systemic.

*Belamcanda chinensis* — (Hammond, unpublished data).

Maintenance and propagation hosts

*Iris* spp.

Assay hosts (**L**ocal lesions or **W**hole plants)

*Tetragonia expansa* (L); *Chenopodium quinoa* (L).

Dicotyledonous Hosts *(**Susceptible** or Insusceptible)*

***Chenopodium amaranticolor*. *Chenopodium quinoa*. *Nicotiana clevelandii*. *Nicotiana megalosiphon*. *Tetragonia expansa*.**

Sources of host-range data

Asjes (1979).

**Properties of particles in sap**

TIP: probably 65–70 °C. LIV: 3–4 days (at 20°C and 16–32 days at 2°C. Sap lyophilized with dextrose and peptone remains infective for at least 2 years (Brunt, 1968)). DEP: $\log_{10}$ minus 3–4. Leaf sap contains many virions.

**Purification**

Method: Asjes (1979); Brunt (1973).

**Particle morphology**

Virions filamentous; usually flexuous; with a clear modal length; of 750 nm; 12 nm wide. Axial canal obvious (when stained with UF).

**Physical properties**

One sedimenting component in purified preparations; sedimentation coefficient 142–147 *S* (but the virions may form faster sedimenting aggregates).

**Biochemical properties**

Virions contain 5–6 % nucleic acid; 95 % protein; 0 % lipid.

**Cytopathology**

Virions found in all parts of the host plant; in cytoplasm. Inclusions present in infected cells; are unusual in shape; primarily pinwheels with laminated aggregates and the lamellae close to virions.

**Taxonomy and Relationships**

*Potyvirus: Potyviridae.*

Virus(es) with serologically related virions: bean yellow mosaic virus. Virus(es) with serologically unrelated virions: iris severe mosaic (Van Slogteren, 1963; Lawson, 1967), bearded iris mosaic (Barnett *et al.*, 1971) and potato Y viruses.

**References**


Asjes, C.J. (1979). *Neth. J. Pl. Path.* **85**: 269.

Barnett, O.W., de Zoeten, G.A. and Gaard, G. (1971). *Phytopathology* **61**: 926.

Brunt, A.A. (1968). *Ann. appl. Biol.* **61**: 187.

Brunt, A.A. (1973). *CMI/AAB Descr. Pl. Viruses* No. 116, 4 pp.

Brunt, A.A. (1986). *CMI/AAB Descr. Pl. Viruses* No. 324, 5 pp.

Derks, A.F.L.M., Hollinger, T.C. and Vink-van den Abeele, J.L. (1985). *Acta. Hort.* **164**: 309.

Hammond, J., Brunt, A.A., Derks, A.F.L.M., Inouye, N., Barnett, O.W., Allen, T.C. and Lawson, R.H. (1985). *Acta Hort.* **164**: 395.

Lawson, R.H. (1967). *Pl. Dis. Reptr* **51**: 952.

Loebenstein, G. and Alper, M. (1963). *Phytopathology* **53**: 349.

Van Slogteren, D.H.M. (1958). *Versl. Werkzaamh. Bloemboll.* Lisse, 1957.

Van Slogteren, D.H.M. (1963). *Versl. Werkzaamh. Bloemboll.* Lisse, 1962.


# Iris severe mosaic *potyvirus*

**Data collated by C.J. Asjes, 1986; J. Hammond, 1987.**

*Synonyms:* bearded iris mosaic virus (Brunt *et al.*, 1988), iris grijs virus, iris stripe virus, iris yellow mosaic virus. *Standard acronym:* ISMV. *Strains:* type strain, avirulent strain, virulent strain (Barnett *et al.*, 1986).

First reported in *Iris* spp., *Crocus vernus* cv. King of the Blues; from the Netherlands; by Atanasoff (1928); Anon. (1961).

**Natural host range and symptoms**

Symptoms persist and vary cyclically over a few weeks.

*Crocus vernus* — leaf mosaic; striping of flowers, if not white or yellow.

*Belamcanda chinensis* — severe chlorosis.

*Iris* spp. — leaf striping, stunting, flower breaking. Plants die prematurely.

**Transmission**

Transmitted by a vector; an insect; *Macrosiphum euphorbiae* and *Myzus persicae*, but not by *Myzus pelargonii, M. circumflexum* or *Anuraphis tulipae* and *Rhopalosiphum tulipaella* (Brierley and McWhorter, 1934; Brierley and Smith, 1948; Travis, 1957; Barnett *et al.*, 1971; Asjes, 1979); Aphididae. Transmitted in a non–persistent manner. Virus transmitted by mechanical inoculation; not transmitted by grafting; not transmitted by contact between plants.

**Ecology and control**

Studies reported by Deutsch and Loebenstein (1967); the spread of the virus in bulbous iris can be partially controlled by spraying crops with mineral oils or pyrethroid and perimicarb insecticides (Asjes, 1985).

**Geographical distribution**

Spreads in Australia, the former Czechoslovakia, Denmark, Japan, the Netherlands, Sweden, the UK, and the USA (and probably wherever bulbous and rhizomatous irises and crocus are grown).

**Experimental host range**

Few (<3) families susceptible.

Diagnostically susceptible

*Belamcanda chinensis, Iris × hollandica.* No herbaceous hosts.

Diagnostically insusceptible

*Chenopodium quinoa, C. foetidum, Freesia refracta* cv. Imperial, *Nicotiana clevelandii.*

Maintenance and propagation hosts

*Crocus vernus, Iris* sp.

Assay hosts (**L**ocal lesions or **W**hole plants)

None.

Dicotyledonous Hosts *(**Susceptible** or Insusceptible)*

*Chenopodium foetidum. Chenopodium quinoa. Nicotiana clevelandii. Nicotiana debneyi. Nicotiana megalosiphon. Nicotiana tabacum.*

Monocotyledonous Hosts *(**Susceptible** or Insusceptible)*

***Belamcanda chinensis*.**

Sources of host-range data

Derks *et al.* (1985), Hammond *et al.* (1985).

**Properties of particles in sap**

TIP: 50 °C. LIV: 1–2 days (Barnett *et al.*, 1971). DEP: $\log_{10}$ minus 3. Leaf sap contains many virions.

**Purification**

Method: Asjes (1979). Homogenize in phosphate buffer, 6.7mM; pH 7.2; (w/v = 1/1) containing 0.1% thioglycollic acid, or in 0.2 M boric acid plus 0.05 M sodium borate pH 7.2; (w/v = 1/1). Extract can be frozen (-20°C) or not before the mixing with n-butanol/chloroform (w/v = 1/1; 30 minutes). The aqueous phase is differentially centrifuged and the pellet resuspended in phosphate or borate buffer.

**Particle morphology**

Virions filamentous; usually flexuous; with a clear modal length; of 750 nm; 12 nm wide. Axial canal obvious.

**Physical properties**

One sedimenting component in purified preparations; sedimentation coefficient 150 *S* (Barnett *et al.*, 1986).

**Biochemical properties**

Virions contain 5 % nucleic acid; 95 % protein.

*Sequence database accession code(s):*

X75939 Em(40)_vi:ISMVCP Gb(84)_vi:ISMVCP Iris severe mosaic potyvirus gene for coat protein. 12/93 2,510bp. 1 sequence.

*Virion protein(s)* one; $M_r$ 33000. Method of preparation: Alper *et al.* (1985); Hammond and Lawson (1984).

*Virus–coded non–virion proteins* isolated (Alper *et al.* (1984); Hammond and Lawson (1985)); one protein found. $M_r$ 71000.

**Replication**

Replication does not depend on a helper virus.

**Cytopathology**

Virions found in leaves and petals; scattered throughout in cytoplasm (or sometimes in membrane-enclosed masses). Inclusions present in infected cells; are pinwheels and unusual in shape; bundles and laminated aggregates (Barnett *et al.*, 1971; Lisa, 1980); they do not contain virions.

**Taxonomy and Relationships**

*Potyvirus: Potyviridae.*

Virus(es) with serologically unrelated virions: iris mild mosaic, iris fulva mosaic, bean yellow mosaic, turnip mosaic, freesia mosaic and four other viruses (Barnett *et al.*, 1971; Brunt, 1975).

Best tests for diagnosis: Rhizomatous irises infected with iris severe mosaic virus in Britain also contain cucumber mosaic virus (Brunt, 1968; Barnett *et al.*, 1971) or, rarely, broad bean wilt virus (Bailiss *et al.*, 1975). Tobacco ringspot virus also occurs occasionally in rhizomatous iris in the U.S.A. and Britain (Travis, 1957; Travis and Brierley, 1957; Brunt, 1972) and in bulbous iris in the Netherlands (Asjes, 1979). These five viruses may be distinguished from iris severe mosaic virus (ISMV) by their isometric or rod-shaped virions, their host ranges or by serological tests. ISMV and three other potyviruses are known to infect bulbous and rhizomatous irises (Brunt and Phillips, 1980; Inouye and Mitsuhata, 1978); these three (bean yellow mosaic, iris mild mosaic and turnip mosaic viruses) are all readily sap-transmitted to dicotyledonous test plants; moreover, and are serologically distinct from ISMV. Two other viruses with filamentous virions, narcissus latent carlavirus (Brunt, 1977; Asjes, 1979) and lilac chlorotic leafspot closterovirus (Lisa, 1980), also infect irises but are readily distinguished from ISMV.

**References**


Ainsworth, G.C. (1936). *Rep. exp. Res. Stn Cheshunt* 1936, p. 62.
Alper, M., Salomon, R. and Loebenstein, G. (1984). *Phytopathology* **74**: 960.
Anon. (1961). *Tuinbouwgids*, p. 468.
Asjes, C.J. (1979). *Neth. J. Pl. Path.* **85**: 269.
Atanasoff, D. (1928). *Izv. bul. bot. Druzh.* **2**: 51.
Bailiss, K.W., Brunt, A.A. and Dale, J. (1975). *Pl. Path.* **24**: 60.
Barnett, O.W., Brunt, A.A. and Dercks, A.F.L.M. (1987). *CMI/AAB Descr. Pl. Viruses* (in press).
Barnett, O.W., de Zoeten, G.A. and Gaard, G. (1971). *Phytopathology* **61**: 926.
Brierley, P. and McWhorter, F.P. (1934). *Phytopathology* **2**: 24.
Brierley, P. and McWhorter, F.P. (1936). *J. agric. Res.* **53**: 621.
Brierley, P. and Smith, F.E. (1948). *Phytopathology* **38**: 574.
Brunt, A.A. (1975). *Acta Hort.* **47**: 45.
Brunt, A.A. (1977). *Ann appl. Biol.* **87**: 355.
Brunt, A.A. and Phillips, S. (1980). *Acta Hort.* **109**: 503.
Brunt, A.A., Derks, A.F.L.M. and Barnett, O.W. (1988). *AAB Descr. Pl. Viruses* No. 338, 4 pp.
Carne (1927). *J. Proc. R. Soc. West Aust.* **14**: 23.
Derks, A.F.L.N. and Hollinger, T.C. (1986). *Acta Hort.* **177**:
Derks, A.F.L.N., Hollinger, T.C. and Vink-van den Abeele, J.L. (1985). *Acta Hort.* **164**: 309.
Deutsch, M. and Loebenstein, G. (1967). *Pl. Dis. Reptr* **51**: 318.
Fukushi, T. (1932). *Trans. Sapporo Nat. Hist. Soc.* **12**: 130.
Griffiths (1928). *US Dep. Agric. Circ.* **25**: 16.
Hammond, J. and Lawson, R.H. (1985). *Acta Hort.* **164**: 225.
Hammond, J., Brunt, A.A., Derks, A.F.L.N., Inouye, T., Barnett, O.W., Allen, T.C. and Lawson, R.H. (1985). *Acta Hort.* **164**: 395.
Hammond, J., Lawson, R.H. and Hsu, T. (1985). *Phytopathology* **75**: 1353.
Hsu, H.T., Franssen, J.M., Hammond, J., Dercks, A.F.L.N. and Lawson, R.H. *Phytopathology* (1986). **76**: 1132.
Ingelström (1938). *Växts kyddnotiser Vaxtkyddanst. Stockh.* **2**: 22.
Inouye, T. and Mitsuhata, K. (1978). *Nogaku Kenkyu* **57**: 1.
Lisa, V. (1980). *Acta Hort.* **110**: 39.
Loebenstein, G. and Alper, M. (1963). *Phytopathology* **53**: 349.
Neergaard (1939). *Rep. Phytopath. Lab. 1938–39*, p. 16.
Travis, R.V. (1957). *Phytopathology* **47**: 454.
Travis, R.V. and Brierley, P. (1957). *Pl. Dis. Reptr* **41**: 524.
Van Slogteren, D.H.M. (1958). *Jversl. Lab. Bloemboll.* Lisse, 1957, p. 67.
Van Slogteren, D.H.M. (1963). *Jversl. Lab. Bloemboll.* Lisse, 1963, p. 80.


# Isachne mosaic (?) potyvirus

**Data collated by A.A. Brunt, 1991.**

*Standard acronym:* IsaMV.

First reported in *Isachne globosa*; from Japan; by Toriyama and Yora (1972).

**Natural host range and symptoms**

Symptoms persist.
*Isachne globosa* — leaf chlorosis.

**Geographical distribution**

Spreads in Japan.

**Particle morphology**

Virions filamentous; not enveloped; usually flexuous; with a clear modal length; of 750 nm.

**Cytopathology**

Inclusions present in infected cells; are pinwheels; they do not contain virions.

**Taxonomy and Relationships**

Unassigned potyvirus.

**References**

Toriyama, S. and Yora, K. (1972). In: *Virus Diseases of Wild Grasses and Cereal Crops in Japan*. University of Tokyo Press, Tokyo, Japan.

# Ivy vein clearing (?) *cytorhabdovirus*

**Data collated by A.A. Brunt, 1991.**

*Standard acronym:* IVCV.

First reported in *Hedera helix*; from southern Italy; by Russo *et al.* (1979).

**Natural host range and symptoms**

Symptoms vary seasonally.

*Hedera helix* — transitory chlorosis of leaf veins, some chlorotic blotching in spring.

**Transmission**

Virus transmitted by mechanical inoculation.

**Geographical distribution**

Spreads in Italy.

**Experimental host range**

Several (3–9) families susceptible.

Diagnostically susceptible

*Gomphrena globosa, Nicotiana benthamiana* — chlorotic or necrotic local lesions; transitory systemic vein clearing.

*Nicotiana clevelandii, N. tabacum* cv. Xanthi — systemic ringspots.

*Nicotiana glutinosa, N. megalosiphon* — systemic vein clearing.

*Datura stramonium* — ring-like local lesions; not systemic.

Diagnostically insusceptible

*Chenopodium quinoa, Cucumis sativus, Nicotiana rustica, N. tabacum, Solanum melongena.*

Maintenance and propagation hosts

*Nicotiana benthamiana, N. clevelandii.*

Assay hosts (**L**ocal lesions or **W**hole plants)

*Gomphrena globosa* (L/W), *Nicotiana benthamiana* (W), *N. clevelandii* (W).

**Properties of particles in sap**

TIP: 50 °C. LIV: 0.25 days (6–9 hours). DEP: $\log_{10}$ minus 5. Leaf sap contains few virions.

**Particle morphology**

Virions rhabdo– or bullet–shaped; usually straight; of 100–200 nm (in sap), or 180–320 nm (in vivo); 60–70 nm wide (in sap), or 45–50 nm wide (in vivo).

**Cytopathology**

Virions found in mesophyll, vascular parenchyma and the phloem; in cytoplasm.

**Taxonomy and Relationships**

Possible *Cytorhabdovirus: Rhabdoviridae.*

**References**

Castellano, M.A. and Rana, G.L. (1981). *Phytopathol. Medit.* **20**: 199.
Russo, M., Castellano, M.A. and Martelli, G.P. (1979). *Phytopath. Z.* **96**: 122.

# Jatropha mosaic bigeminivirus

**Data collated by K.S. Kim, 1986.**

*Standard acronym:* JMV.

First reported in *Jatropha gossypifolia*; from Puerto Rico; by Bird (1957).

**Natural host range and symptoms**

. Symptoms persist.

*Jatropha gossypifolia, J. podagrica, J. multifida, Croton lobatus, Jacquemontia tamnifolia, Phaseolus vulgaris* — yellow mosaics.

**Transmission**

Transmitted by a vector; an insect; *Bemisia tabaci*; Aleyrodidae. Transmitted in a semi–persistent manner. Virus lost by the vector when it moults; not transmitted by mechanical inoculation; transmitted by grafting; not transmitted by seed.

**Geographical distribution**

Spreads in Kenya and Puerto Rico.

**Particle morphology**

Virions geminate; not enveloped; 15–18 nm in diameter; 15 and 30 nm in length; rounded in profile; without a conspicuous capsomere arrangement.

**Biochemical properties**

*Genome* consists of DNA; single–stranded.

**Replication**

Genome replicates in nuclei.

**Cytopathology**

Virions found in leaves, phloem and companion cells; in nuclei; in sieve tubes. Inclusions present in infected cells; are crystals in the nucleus, or amorphous X–bodies (in cytoplasm), or unusual in shape; fibrillar bodies; they contain virions (in the nuclear inclusions), or they do not contain virions (in the cytoplasmic inclusions). Other cellular changes: the nucleolar segregation of granular and fibrillar components (Kim *et al.*, 1986).

**Taxonomy and Relationships**

Bigeminivirus: *Geminiviridae*.

**References**

Bird, J. (1957). *P. R. Agric. Exp. Stn Tech. Pap.*, 22.

Kim, K.S., Bird, J., Rodriguez, R.L., Martin, E.M. and Escudero, J. (1986). *Phytopathology* **76**: 80.

# Johnsongrass chlorotic stripe (?) *carmovirus*

**Data collated by W. Huth, 1991; K. Izadpanah, 1992.**

First reported in *Sorghum halepense* (Johnson grass); from Fars Province, Iran; by Izadpanah (1982).

**Natural host range and symptoms**

Symptoms persist. Symptoms stripe mosaic.

*Sorghum halepense* — chlorotic leaf stripes, dwarfing, fewer seeds.

**Transmission**

Transmitted by means not involving a vector. Not transmitted by *Empoasca* spp., *Laodelphax striatellus, Neoaliturus haematoceps, Unkanodes tanasijevici, Eriophyes tulipae* or *Rhopalosiphum* spp. Virus not transmitted by mechanical inoculation; not transmitted by contact between plants; not transmitted by seed (although present in the cotyledon).

**Geographical distribution**

Spreads in Iran.

**Properties of particles in sap**

Leaf sap contains many virions.

**Purification**

Method: Izadpanah (1988).

**Particle morphology**

Virions isometric; not enveloped; 30 nm in diameter; rounded in profile (with a knobby surface); with a conspicuous capsomere arrangement.

**Physical properties**

One sedimenting component in purified preparations (with a minor second component). Density 1.359 g $cm^{-3}$ in CsCl. $A_{260}/A_{280}$ ratio 1.48.

**Biochemical properties**

*Genome* consists of RNA; single-stranded; linear; unipartite; largest (or only) genome part 4.65 kb. Base composition 27.5 % G; 23.5 % A; 26 % C; 23 % U.

*Virion protein(s)* one; $M_r$ 41000; coat protein. Method of preparation: Izadpanah *et al.* (1992).

**Cytopathology**

Virions found in leaves and mesophyll; in cytoplasm, in chloroplasts, in mitochondria, and in cell vacuoles. Inclusions absent from infected cells. Other cellular changes: swollen endoplasmic reticulum or nuclear envelope containing fibrils and vesicles.

**Taxonomy and Relationships**

Possible *Carmovirus: Tombusviridae.*

No serological relationship to any of 30 isometric viruses tested. Best tests for diagnosis: JGCSMV and sugarcane mosaic potyvirus infect Johnson grass in the Fras Province of Iran. JGCSMV induces more obvious symptoms.

**References**

Izadpanah, K. (1982). *Iran J. Pl. Path.* **18**: 3–7.
Izadpanah, K. (1988). *J. Phytopathology* **121**: 209.
Izadpanah, K., Huth, W., Lesemann, D.-E. and Vetten H.J. (1992). *J. Phytopathology* **137**: 105.

# Johnsongrass mosaic *potyvirus*

**Data collated by A.J. Gibbs, 1990.**

*Synonyms:* maize dwarf mosaic virus — strain O (McDaniel and Gordon, 1985; Shukla *et al.*, 1989), sugarcane mosaic virus — Australian Johnson grass virus (Shukla *et al.*, 1987), maize dwarf mosaic virus — Kansas I strain (McKern *et al.*, 1990). *Standard acronym:* JGMV.

First reported in *Sorghum halepense* and *Zea mays*; from Australia; by Taylor and Pares (1968).

**Natural host range and symptoms**

Symptoms persist.

*Sorghum halepense* (Johnson grass) — systemic mosaic.

*Zea mays* — systemic mosaic.

*Brachiaria miliiformis, Cenchrus ciliaris, Panicum miliaceum, Paspalum orbiculare, Pennisetum typhoideum, Sorghum almum, S. laxiflorum, S. macrospermum, S. miliaceum, S. stipoideum, S. sudanense, S. verticilliflorum, S. vulgare* — systemic chlorotic mosaic mottling, necrosis, stunting.

**Transmission**

Transmitted by a vector; an insect; *Aphis craccivora, A. gossypii, Rhopalosiphum maidis* but not *Macrosiphum euphorbiae, M. miscanthi, Rhopalosiphum padi*; Aphididae. Transmitted in a non–persistent manner. Virus transmitted by mechanical inoculation; not transmitted by contact between plants; not transmitted by seed.

**Geographical distribution**

Spreads in Australia and the USA.

**Experimental host range**

Several (3–9) families susceptible. Experimentally infected plants mostly show systemic mosaics, mottles, ringspots or necrosis.

Diagnostically susceptible

*Zea mays* cv. Iochief — mosaic.

*Sorghum vulgare* cv. Yates NK22OY — necrosis or mosaic.

*S. vulgare* cv. Atlas — mosaic.

*Avena sativa* — mosaic.

Diagnostically insusceptible

*Phaseolus vulgaris* cv. Bountiful, *Chenopodium quinoa, Saccharum officinarum, Nicotiana clevelandii, Cucumis sativus.*

Maintenance and propagation hosts

*Sorghum vulgare, Zea mays* cv. Iochief (Sweet corn).

Assay hosts (**L**ocal lesions or **W**hole plants)

*Zea mays* cv. Iochief (W), *Sorghum verticilliflorum* (W).

Dicotyledonous Hosts *(**Susceptible** or Insusceptible)*

*Brassica campestris* ssp. *rapa. Chenopodium album. Chenopodium amaranticolor. Chenopodium quinoa. Cicer arietinum. Cucumis sativus. Datura stramonium. Dolichos biflorus. Glycine max. Gomphrena globosa. Lycopersicon esculentum. Medicago sativa. Nicotiana clevelandii. Nicotiana glutinosa. Nicotiana tabacum. Phaseolus lathyroides. Pisum sativum. Vigna unguiculata.*

Monocotyledonous Hosts *(**Susceptible** or Insusceptible)*

*Agropyron repens. Allium cepa.* ***Avena sativa****. Bromus inermis. Bromus secalinus. Cynodon dactylon. Dactylis glomerata.* ***Echinochloa crus-galli****.* ***Eleusine coracana****.* ***Eragrostis cilianensis****. Hordeum vulgare. Lolium perenne. Oryza sativa. Oryzopsis miliacea.* ***Panicum capillare****.* ***Panicum miliaceum****. Phalaris arundinacea. Phleum pratense. Poa pratensis. Secale cereale.* ***Setaria italica****.* ***Setaria viridis****.* ***Sorghum bicolor****. Triticum aestivum.* ***Zea mays****.*

Sources of host-range data

Teakle and Grylls (1973); McDaniel and Gordon (1985).

**Properties of particles in sap**

TIP: 60 °C. LIV: 2 days. DEP: $\log_{10}$ minus 4. Leaf sap contains few virions.

**Purification**

Method: Teakle and Grylls (1973); McDaniel and Gordon (1985).

**Particle morphology**

Virions filamentous; usually flexuous; with a clear modal length; of 773–778 nm. Axial canal obscure. Basic helix obscure.

**Physical properties**

One sedimenting component in purified preparations.

*Sequence database accession code(s):*

U07217 Em(40)_vi:JM07217 Gb(84)_vi:JMU07217 Johnsongrass mosaic virus strain Maize dwarf O polyprotein (nuclear inclusion II protein, coa

U07218 Em(40)_vi:JM07218 Gb(84)_vi:JMU07218 Johnsongrass mosaic virus strain Maize dwarf KS1 polyprotein (nuclear inclusion II protein, c

Z26920 Em(40)_vi:JGMVVPS Gb(84)_vi:JGMVVPS Johnson grass mosaic virus gene for protease 1 and 3, helper component 6K protein, coat prote 3 sequences.

*Virion protein(s)* one; $M_r$ 33505; coat protein.

**Cytopathology**

Virions found in all parts of the host plant; in cytoplasm. Inclusions present in infected cells; are pinwheels; they do not contain virions.

**Taxonomy and Relationships**

*Potyvirus: Potyviridae.*

Virion protein distantly homologous to that of maize dwarf mosaic potyvirus B and more distantly related to other potyviruses. Using electro-blot immunoassay with cross-absorbed polyclonal antibodies against the N-terminus of the virion proteins show that four distinct potyviruses infect sugarcane, maize and sorghum. Johnson grass mosaic

virus (synonyms SCMV–JG and maize dwarf mosaic virus strain O) is one, the others are:
1) Maize dwarf mosaic potyvirus (syn; MDMV–A, MDMV–D, MDMV–E and MDMV–F);
2) Sugarcane mosaic potyvirus (syn; MDMV–B, SCMV–A, SCMV–B, SCMV–D, SCMV–E, SCMV–Sc, SCMV–Bc and SCMV–Sabi);
3) Sorghum mosaic potyvirus (syn; SCMV–H, SCMV–I and SCMV–M). (Shukla *et al.*, 1989).

**References**

McDaniel, L.L and Gordon, D.T. (1985). *Plant Dis.* **69**: 602. McKern, N.M, Whittaker, L.A., Strike, P.M., Ford, R.E., Jensen, S.G. and Shukla, D.D. (1990). *Phytopathology* **80**: 907.
Shukla, D.D. and Gough, K.H. (1984). *Plant Dis.* **68**: 204.
Shukla, D.D., Gough, K.H. and Ward, C.M. (1987). *Arch. virol.* **96**: 59.
Shukla, D.D., Tosic, M., Jilka, J., Ford, R.E., Toler, R.W. and Langham, M.A.C. (1989). *Phytopathology* **79**: 223.
Taylor, R.H. and Pares, R.D. (1968). *Aust. J. agric. Res.* **19**: 767.
Teakle, D.S. and Grylls, N.E. (1973). *Aust. J. agric. Res.* **24**: 465.

# Kalanchoe isometric virus

**Data collated by M.L. Izaguirre-Mayoral, 1991.**

*Synonyms:* IVLP. *Strains:* Venezuelan.

First reported in *Kalanchoe pinnata, K. tubiflora, K. daigremontiana*; from Edo Miranda, Venezuela; by Izaguirre-Mayoral *et al.* (1990).

**Natural host range and symptoms**

Symptoms vary seasonally. Symptoms sunken tissues and necrosis.

*Kalanchoe pinnata, K. tubiflora, K. daigremontiana* — sunken tissues between leaf veins, sometimes necrosis.

**Transmission**

Virus transmitted by mechanical inoculation.

**Geographical distribution**

Spreads in Venezuela.

**Experimental host range**

Several (3–9) families susceptible. Experimentally infected plants mostly show chlorotic and necrotic local lesions, chlorotic rings.

Diagnostically susceptible

*Chenopodium amaranticolor* — chlorotic local lesions.
*Chenopodium quinoa, Cucumis sativus* var. Palomar — necrotic local lesions.
*Datura stramonium* — chlorotic local lesions, chlorotic rings.
*Nicotiana benthamiana* — systemic mosaic.
*N. tabacum* cvs Samsun, White Burley — chlorotic rings and systemic mosaic.
*Phaseolus vulgaris* cv. Top crop — systemic mosaic.
*P. vulgaris* cv. Red Kidney — chlorotic local lesions and systemic mosaic.
*Tetragonia expansa* — chlorotic rings.

Diagnostically insusceptible

*Cucumis melo, C. sativus* var. Ashley, *Cucurbita maxima, Lycopersicon esculentum, Vigna unguiculata.*

Maintenance and propagation hosts

*Nicotiana benthamiana.*

Assay hosts (**L**ocal lesions or **W**hole plants)

*Chenopodium amaranticolor* (L), *C. quinoa* (L), *Cucumis sativus* var. Palomar (L).

Dicotyledonous Hosts *(**Susceptible** or Insusceptible)*

***Chenopodium amaranticolor***. ***Chenopodium quinoa***. *Cucumis melo*. *Cucurbita maxima*. ***Datura stramonium***. *Lycopersicon esculentum*. ***Nicotiana benthamiana***. *Nicotiana glutinosa*. ***Phaseolus vulgaris***. ***Tetragonia expansa***. *Vigna unguiculata*.

Monocotyledonous Hosts *(**Susceptible** or Insusceptible)*

*Zea mays.*

Sources of host-range data

Izaguirre-Mayoral *et al.* (1990).

**Properties of particles in sap**

TIP: 50 °C. LIV: 0.2 days (20 hours). Leaf sap contains few virions.

**Purification**

Method: Izaguirre-Mayoral *et al.* (1990).

**Particle morphology**

Virions isometric; not enveloped; 35 nm in diameter; angular in profile; without a conspicuous capsomere arrangement.

**Physical properties**

One sedimenting component in purified preparations. Density 1.32 g $cm^{-3}$ in CsCl. $A_{260}/A_{280}$ ratio 1.56.

**Biochemical properties**

Virions contain 9–10 % nucleic acid.

*Genome* consists of DNA; single–stranded; unipartite; largest (or only) genome part 3.5 kb.

*Non–genomic nucleic acid* not found in the virions.

*Virion protein(s)* one; $M_r$ 46000.

**Cytopathology**

Virions found in leaves. Inclusions absent from infected cells. Other cellular changes: severely damaged chloroplasts of different forms, randomly distributed in the cytoplasm and bigger than those of non-infected leaves. Starch grains occupied 78% of the chloroplast internal volume during the night as well as in the light period, and the number of thylakoids per grana reduced to 5±2.

**References**

Izaguirre-Mayoral, M.L., Carballo, O. and Gil, F. (1990). *J. Phytopath.* **130**: 303.

# Kalanchoe latent *carlavirus*

**Data collated by S.L. Bhattiprolu, 1992.**

*Synonyms:* kalanchoe I virus. *Standard acronym:* KLV. *Strains:* KV–1 and KV–2.

First reported in *Kalanchoe blossfeldiana*; from the U.S.A.; by Hearon (1981).

**Natural host range and symptoms**

*Kalanchoe blossfeldiana* — symptomless infection.

**Transmission**

Virus transmitted by mechanical inoculation; transmitted by grafting.

**Ecology and control**

Studies reported by Paludan (1985). Kalanchoe latent virus 1 was eliminated from the mother plants by meristem tip culture combined with previous heat treatment.

**Experimental host range**

Few (<3) families susceptible.

Diagnostically susceptible

*Chenopodium amaranticolor, C. album, C. quinoa* — local lesions, systemic yellow vein-banding and leaf distortion with strain KV2).

*Chenopodium album, C. murale, Tetragonia expansa* — local lesions; no systemic infection.

Diagnostically insusceptible

*Datura stramonium, Gomphrena globosa, Petunia hybrida, Nicotiana clevelandii, Phaseolus vulgaris.*

Maintenance and propagation hosts

*Chenopodium quinoa.*

Assay hosts (**L**ocal lesions or **W**hole plants)

*Chenopodium quinoa* (L and W), *Tetragonia expansa* (L).

Dicotyledonous Hosts *(**Susceptible** or Insusceptible)*

*Beta vulgaris. Calendula officinalis. Capsicum annuum. Capsicum frutescens. Cassia occidentalis.* ***Chenopodium album*****.** ***Chenopodium amaranticolor*****.** ***Chenopodium capitatum*****.** ***Chenopodium murale*****.** ***Chenopodium quinoa*****.** *Cucumis sativus. Cucurbita pepo. Dahlia pinnata. Datura metel. Datura stramonium. Fagopyrum esculentum. Glycine max. Gomphrena globosa. Hibiscus cannabinus. Hibiscus esculentus. Lycopersicon esculentum. Nicotiana benthamiana. Nicotiana clevelandii. Nicotiana debneyi. Nicotiana glutinosa. Nicotiana glutinosa × clevelandii. Nicotiana megalosiphon. Nicotiana rustica. Nicotiana tabacum. Petunia hybrida. Phaseolus vulgaris. Pisum sativum. Plantago rugelii. Saponaria vaccaria. Solanum melongena. Solanum rostratum. Spinacia oleracea.* ***Tetragonia expansa*****.** *Tropaeolum majus. Vicia faba. Vigna unguiculata. Zinnia elegans.*

Comments on host-range

Only the strain KV–2 induces systemic infection in *Chenopodium amaranticolor* and *C. quinoa.*

Sources of host-range data

Hearon (1982; 1984).

**Properties of particles in sap**

TIP: 75–80 °C. LIV: 3–5 days. Leaf sap contains many virions.

**Purification**

Method: Hearon (1984).

**Particle morphology**

Virions filamentous; not enveloped; usually flexuous; with a clear modal length; of 620–650 nm. Axial canal obscure. Basic helix obscure.

**Physical properties**

One sedimenting component in purified preparations; sedimentation coefficient 170 *S*. Density 1.3 g $cm^{-3}$ in CsCl. $A_{260}/A_{280}$ ratio 1.32–1.38.

**Biochemical properties**

Virions contain 5 % nucleic acid; 95 % protein.

*Genome* consists of RNA; single–stranded. Infectivity retained when deproteinised with phenol or detergent (KV–1), or lost when deproteinised with phenol or detergent (KV–2).

*Virion protein(s)* one; $M_r$ 33000–34000; it is coat protein. Method of preparation: Hearon (1982).

**Replication**

Replication does not depend on a helper virus.

**Cytopathology**

Virions found in leaves, mesophyll and the vascular parenchyma; in cytoplasm. Inclusions present in infected cells; are viroplasms; they contain virions. Other cellular changes: fusiform inclusions consisting of aggregates of parallel virions and pockets of cytoplasmic materials, vesicles, or microbodies; virions also aggregate along organelle membranes and the tonoplast. Bands of virions and 'fingerprint like' inclusions also in cells within the local lesions (Hearon, 1982, 1984).

**Taxonomy and Relationships**

*Carlavirus.*

Virus(es) with serologically related virions: carnation latent, potato S, lily symptomless and chrysanthemum B carlaviruses. Virus(es) with serologically unrelated virions: elderberry, potato M and red clover vein mosaic carlaviruses.

**References**

Hearon, S.S. (1981). *Phytopathology* **71**: 767.
Hearon, S.S. (1982). *Phytopathology* **72**: 838.
Hearon, S.S. (1984). *Phytopathology* **74**: 670.
Paludan, N. (1985). *Tidsskr. PlAvl.* **89**: 191.

# Kalanchoe mosaic (?) potyvirus

**Data collated by A.A. Brunt, 1995.**

First reported in *Kalanchoe blossfeldiana*; from Denmark; by Paludan (1989); Husted *et al.* (1994).

**Natural host range and symptoms**

*Kalanchoe blossfeldiana* — mosaic of light and dark green areas ("green island mosaic disease").

**Transmission**

Transmitted by a vector; an insect; *Myzus persicae*; Aphididae. Transmitted in a non–persistent manner. Virus transmitted by mechanical inoculation; transmitted by grafting.

**Geographical distribution**

Spreads in the Eurasian region; Denmark. Found, but with no evidence of spread, in USA, probably.

**Experimental host range**

Few (<3) families susceptible. Experimentally infected plants mostly show local lesions.

Diagnostically susceptible

*Chenopodium amaranticolor, C. quinoa* — chlorotic and necrotic local lesions; not systemic.

Diagnostically insusceptible

*Beta vulgaris, Brassica campestris* ssp. *pekinensis, Cucurbita pepo, Nicotiana benthamiana, N. occidentalis, N. tabacum, Phaseolus vulgaris* and *Pisum sativum.*

Maintenance and propagation hosts

*Chenopodium quinoa, Kalanchoe blossfeldiana.*

Assay hosts (**L**ocal lesions or **W**hole plants)

*Chenopodium quinoa* (L), *Kalanchoe blossfeldiana* (W).

Dicotyledonous Hosts *(**Susceptible** or Insusceptible)*

*Beta vulgaris. Brassica campestris* ssp. *napus. Brassica campestris* ssp. *pekinensis.* ***Chenopodium amaranticolor*. *Chenopodium quinoa*.** *Cucurbita pepo. Nicotiana benthamiana. Nicotiana clevelandii. Nicotiana occidentalis. Nicotiana tabacum. Phaseolus vulgaris. Pisum sativum. Saponaria vaccaria.*

**Purification**

Method: Husted *et al.* (1994).

**Particle morphology**

Virions filamentous; usually flexuous; with a clear modal length; of 724 nm.

**Biochemical properties**

*Genome* consists of RNA; single–stranded. Poly A region present. Nucleotide sequence references: Husted *et al.* (1994).

*Virion protein(s)* one; $M_r$ 33000; coat protein.

**Taxonomy and Relationships**

Possible *Potyvirus: Potyviridae.*

Virus(es) with serologically unrelated virions: bean common mosaic, bean yellow mosaic, carnation vein mottle, celery mosaic, dasheen mosaic, leek yellow stripe, onion yellow dwarf, pea seed-borne mosaic, potato Y and papaya ringspot viruses.

**References**

Hearn, S.S. (1981). *Phytopathology* **71**: 767.

Husted, K., Bech, K., Albrechtsen, M. and Borkhardt, B. (1994). *Phytopathology* **84**: 161.

Paludan, N. (1989). *Rep. Res. Centre Pl. Prot. Denmark* **106**: 20.

# Kalanchoe top-spotting *badnavirus*

**Data collated by A.A. Brunt, 1992.**

*Synonyms:* kalanchoe 'commelina yellow mottle badnavirus'. *Standard acronym:* KTSV.

First reported in *Kalanchoe blossfeldiana*; from the U.S.A.; by Hearon and Locke (1984).

**Natural host range and symptoms**

Symptoms persist.

*Kalanchoe blossfeldiana* — sunken circular yellow spots and occasional leaf deformation in intolerant cultivars.

**Transmission**

Transmitted by a vector; an insect; *Planococcus citri*; Pseudococcidae. Virus transmitted by mechanical inoculation; transmitted by seed; transmitted by pollen to the seed.

**Geographical distribution**

Spreads in the UK and the USA.

**Experimental host range**

Few (<3) families susceptible.

Diagnostically susceptible

*Kalanchoe blossfeldiana* — circular yellow spots in systemically infected leaves.

Diagnostically insusceptible

*Chenopodium quinoa, Cucumis sativus, Nicotiana benthamiana, N. glutinosa, Phaseolus vulgaris.*

Maintenance and propagation hosts

*Kalanchoe blossfeldiana.*

Assay hosts (**L**ocal lesions or **W**hole plants)

*Kalanchoe blossfeldiana* (W).

Dicotyledonous Hosts *(**Susceptible** or Insusceptible)*

*Chenopodium quinoa. Cucumis sativus. Nicotiana benthamiana. Nicotiana glutinosa. Phaseolus vulgaris. Pisum sativum.*

**Purification**

Method: Lockhart and Ferji (1988).

**Particle morphology**

Virions bacilliform; not enveloped; 130 nm in length; 30 nm wide.

**Biochemical properties**

*Genome* consists of DNA; double–stranded; circular; unipartite; largest (or only) genome part 7.35 kb.

**Cytopathology**

Virions found in leaves and mesophyll; in cytoplasm.

**Taxonomy and Relationships**

*Badnavirus.*

Virus(es) with serologically unrelated virions: banana streak, cacao swollen shoot, commelina yellow mottle, dioscorea bacilliform, rice tungro bacilliform and sugarcane viruses.

The viral genome shows weak homology with those of banana streak and badnaviruses.

**References**

Hearon, S.S. and Locke, J.C. (1984). *Phytopathology* **74**: 347.
Lockhart, B.E. and Ferji, Z. (1988). *Acta Hort.* **234**: 73.

# Kenaf vein-clearing (?) rhabdovirus

**Data collated by A.A. Brunt.**

First reported in *Hibiscus cannabinus*; from Italy; by Riccioni *et al.* (1991).

**Natural host range and symptoms**

Symptoms *Hibiscus cannabinus* — chlorotic vein banding.

**Geographical distribution**

Found, but with no evidence of spread, in Italy.

**Particle morphology**

Virions rhabdo– or bullet–shaped.

**Taxonomy and Relationships**

Unassigned rhabdovirus.

**References**

Riccioni, L., Saponara, C. and Barba, M. (1991). *Phytopath. medit.* **30**: 128.

# Kennedya Y *potyvirus*

**Data collated by A.J. Gibbs, 1987.**

*Standard acronym:* KVY.

First reported in *Kennedya rubicunda*; from the south coast of N.S.W., Australia; by Dale *et al.* (1975).

**Natural host range and symptoms**

Symptoms persist.
*Kennedya rubicunda* — mild marble mosaic.

**Transmission**

Virus transmitted by mechanical inoculation; not transmitted by seed.

**Geographical distribution**

Spreads in Australia (the south eastern region).

Diagnostically susceptible

*Phaseolus vulgaris* cv. Pinto — local lesions, tip blight.
*Glycine max* — systemic mosaic.

Diagnostically insusceptible

*Pisum sativum, Nicotiana glutinosa.*

Maintenance and propagation hosts

*Soja max.*

Assay hosts (**L**ocal lesions or **W**hole plants)

*Phaseolus vulgaris* cv. Pinto (L).

Dicotyledonous Hosts (***Susceptible*** *or Insusceptible)*

***Kennedya rubicunda***. ***Nicotiana glutinosa***. ***Phaseolus vulgaris***. *Pisum sativum*.

Sources of host-range data

Dale *et al.* (1975).

**Properties of particles in sap**

Leaf sap contains few virions.

**Particle morphology**

Virions filamentous; usually flexuous; with a clear modal length; of 710 nm; *c.* 12 nm wide. Axial canal obscure. Basic helix obscure.

**Cytopathology**

Virions found in leaves; in cytoplasm. Inclusions present in infected cells; are pinwheels; they do not contain virions.

**Taxonomy and Relationships**

*Potyvirus: Potyviridae.*

**References**

Dale, J.L., Gardiner, J.E. and Gibbs, A.J. (1975). *Aust. Pl. Path. Soc. Newsl.* **4**: 13.

# Kennedya yellow mosaic *tymovirus*

**Data collated by A.J. Gibbs, 1981. Revised 1989.**

*Standard acronym:* KYMV. *Strains:* Port Douglas (Mount Jukes), Jervis Bay, Wapengo; these being the type members of three geographically separate strains found in north Queensland, central coast N.S.W. and south coast N.S.W. respectively.

First reported in *Kennedya rubicunda*; from samples from Murrays Beach, Jervis Bay, Australia; by Dale *et al.* (1975, 1976); Gibbs (1978).

**Natural host range and symptoms**

Symptoms persist (and are often bright).

*Kennedya rubicunda, Desmodium triflorum, D. scorpiurus, Indigofera australis* — mild chlorotic local lesions, systemic blotchy yellow mosaic.

**Transmission**

Virus transmitted by mechanical inoculation; not transmitted by contact between plants; not transmitted by seed; not transmitted by pollen.

**Geographical distribution**

Spreads in Australia (the eastern coastal region).

**Experimental host range**

Few (<3) families susceptible.

Diagnostically susceptible

*Glycine max* — blotchy systemic yellow mosaic.

*Kennedya rubicunda* — systemic yellow mosaic.

*Phaseolus vulgaris* cv. Redland Pioneer, *Pisum sativum* — chlorotic local lesions; then systemic yellow mosaic.

*Vigna radiata* — necrotic local lesions; systemic leaf malformation.
*Datura stramonium, Nicotiana glutinosa* — chlorotic local lesions; not systemic.

Diagnostically insusceptible

*Lycopersicon esculentum, Nicotiana clevelandii.*

Maintenance and propagation hosts

*Pisum sativum* cv. Early Massey, *Glycine max.*

Assay hosts (**L**ocal lesions or **W**hole plants)

*Phaseolus vulgaris* cv. Redland Pioneer (L), *Pisum sativum* cv. Massey (W), *Vigna radiata* (L).

Dicotyledonous Hosts (***Susceptible*** *or Insusceptible)*

***Cajanus cajan***. *Chenopodium amaranticolor. Chenopodium quinoa.* ***Clitoria ternatea***. *Cucumis sativus.* ***Datura stramonium***. ***Glycine max***. *Gomphrena globosa. Lotus corniculatus.* ***Lupinus angustifolius***. *Lycopersicon esculentum.* ***Macrotyloma uniflorum***. *Medicago sativa. Nicotiana clevelandii.* ***Nicotiana glutinosa***. ***Phaseolus vulgaris***. ***Pisum sativum***. *Trifolium incarnatum. Trifolium pratense. Trifolium repens. Vicia faba. Vicia sativa.* ***Vigna angularis***. ***Vigna radiata***. ***Vigna unguiculata***. ***Vigna unguiculata*** ssp. ***sesquipedalis***.

Sources of host-range data

Dale and Gibbs (1976); Gibbs (1978) and unpublished data.

**Properties of particles in sap**

TIP: 65–70 °C. LIV: 10–100 days. DEP: $\log_{10}$ minus 5–6. Infectivity of sap not changed by treatment with di–ethyl ether. Leaf sap contains many virions.

**Purification**

Method: Gibbs (1978).

**Particle morphology**

Virions isometric; not enveloped; 28 nm in diameter; rounded in profile; with a conspicuous capsomere arrangement.

**Physical properties**

Two sedimenting components in purified preparations; sedimentation coefficient of the fastest; B 110 *S*; of the other(s) 54 *S*. Isoelectric point pH 3.3–3.57. $A_{260}/A_{280}$ ratio 0.81 (T), or 1.69 (B).

**Biochemical properties**

Virions contain 34.1 % nucleic acid; 65.9 % protein; 0 % lipid.

*Genome* consists of RNA; single–stranded; linear; unipartite; largest (or only) genome part 6.362 kb. Base composition 15.1 % G; 22.5 % A; 39.4 % C; 23 % U. 5' terminus of RNA has a methylated nucleotide cap. Infectivity retained when deproteinised with phenol or detergent. Poly A region absent. Additional factor not required for infectivity. Nucleotide sequence references: Ding *et al.* (1990).

*Sequence database accession code(s):*

D00637 Em(40)_vi:MTYCG01 Gb(84)_vi:KYVCGA Kennedya yellow mosaic virus genome, complete sequence. 11/90 6,362bp.

M15286 Em(40)_vi:TYMTY3KY Gb(84)_vi:KYV3KYMV Kennedya yellow mosaic virus 3' end which can form a tRNA-like structure. 11/90 83bp.

M58310 Em(40)_vi:KYVRRLT Gb(84)_vi:KYVRRLT Kennedya yellow mosaic virus 3' terminal tRNA-like structure. 8/91 84bp. 3 sequences.

*Non–genomic nucleic acid* found in the virions; is of an unusual type of structure (*m*RNA like).

*Virion protein(s)* one; $M_r$ 19602.

**Replication**

Genome replicates in association with chloroplasts (in the vesicles). Coat protein *m*RNA translated in the cytoplasm. Replication does not depend on a helper virus.

**Cytopathology**

Virions found in all parts of the host plant; in cytoplasm. Inclusions absent from infected cells. Other cellular changes: vesiculation of the chloroplasts outer membrane.

**Taxonomy and Relationships**

*Tymovirus*.

Virus(es) with serologically related virions: turnip yellow mosaic, clitoria yellow vein and okra mosaic, closely; desmodium yellow mottle and cacao yellow mosaic viruses more distantly.

*C*DNA-RNA hybridization tests correlate with the serological relationships, but also show an unexpected close relationship with eggplant mosaic tymovirus, which is serologically very distantly related to kennedya yellow mosaic virus.

**References**

Blok, J., Gibbs, A.J. and Mackenzie, A. (1987). *Arch. Virol.* **96**: 225.
Dale, J.L., Gardiner, J.E. and Gibbs, A.J. (1975). *Aust. Pl. Path. Soc. Newsl.* **4**: 13.
Dale, J.L. and Gibbs, A.J. (1976). *Aust. J. biol. Sci.* **29**: 397.
Ding, S., Keese, P. and Gibbs, A.J. (1990). *J. gen. Virol.* **71**: 295.
Gibbs, A.J. (1978). *CMI/AAB Descr. Pl. Viruses* No. 193, 4 pp.

# Konjak mosaic (?) *potyvirus*

**Data collated by A.A. Brunt, 1994.**

*Standard acronym:* KMV.

First reported in *Amorphophallus konjac*; from Gunma Prefecture, Japan; by Shimoyama *et al*. (1992a).

**Natural host range and symptoms**

*Amorphophallus konjac* — leaf mosaic.

**Transmission**

Transmitted by a vector; an insect; *Aphis gossypii*; Aphididae. Transmitted in a non–persistent manner. Virus transmitted by mechanical inoculation.

**Geographical distribution**

Spreads in the Eastern Asian region; Japan.

**Experimental host range**

Few (<3) families susceptible.

Diagnostically susceptible

*Amorphophallus konjac, Philodendron oxycardium, P. selloum* — leaf mosaic.

Diagnostically insusceptible

*Chenopodium amaranticolor, C. quinoa, Phaseolus vulgaris, Nicotiana benthamiana, N. clevelandii, N. glutinosa, N. tabacum.*

Maintenance and propagation hosts

*Amorphophallus konjac.*

Assay hosts (**L**ocal lesions or **W**hole plants)

*Amorphophallus konjac* (W), *Philodendron selloum* (W).

Dicotyledonous Hosts *(**Susceptible** or Insusceptible)*

*Chenopodium amaranticolor. Chenopodium quinoa. Nicotiana benthamiana. Nicotiana clevelandii. Nicotiana glutinosa. Nicotiana tabacum. Phaseolus vulgaris.*

**Properties of particles in sap**

TIP: 55–60 °C. LIV: 2–4 days (at 18C). DEP: $\log_{10}$ minus 2–3.

**Purification**

Method: Shimoyama *et al.* (1992b).

**Particle morphology**

Virions filamentous; not enveloped; usually flexuous; with a clear modal length; of 800 nm.

**Biochemical properties**

*Genome* consists of RNA; single–stranded; linear.
*Virion protein(s)* one; $M_r$ 35500; coat.

**Taxonomy and Relationships**

Possible *Potyvirus: Potyviridae.*

Virus(es) with serologically related virions: soybean mosaic and watermelon mosaic-2 potyviruses. Virus(es) with serologically unrelated virions: bean yellow mosaic, bean common mosaic (blackeye cowpea mosaic strain), dasheen mosaic, potato Y, turnip mosaic and zucchini yellow mosaic potyviruses.

**References**

Shimoyama, J., Kameya-Iwaki, M., Hanada, K. and Gunyi, T. (1992a). *Ann. Phytopath. Soc. Japan* **58**: 706.

Shimoyama, J., Hanada, K., Tsuda, S., Kameya-Iwaki, M. and Gunyi, T. (1992b). *Ann. Phytopath. Soc. Japan* **58**: 713.

# Kyuri green mottle mosaic *tobamovirus*

**Data collated by R.I.B. Francki, 1988.**

*Synonyms:* Japanese CV3, cucumber green mottle mosaic virus - cucumber strain or strain C (Francki *et al.*, 1986). *Standard acronym:* KGMMV. *Strains:* C and Y isolates or strains.

First reported in *Cucumis sativus*; from Japan; by Inouye *et al.* (1967*a*).

**Natural host range and symptoms**

Symptoms persist.

*Cucumis sativus* — mosaic, severe fruit mottling and malformation.

**Transmission**

Transmitted by means not involving a vector. Virus transmitted by mechanical inoculation.

**Geographical distribution**

Spreads in Japan.

**Experimental host range**

Several (3–9) families susceptible.

Diagnostically susceptible

*Cucumis sativus* — mottling, and fruit malformation.

*Datura stramonium, Petunia hybrida* — diffuse chlorotic local lesions; not systemic.

Maintenance and propagation hosts

*Cucumis sativus.*

Assay hosts (**L**ocal lesions or **W**hole plants)

*Datura stramonium* (L), *Cucumis sativus* (W).

Dicotyledonous Hosts *(**Susceptible** or Insusceptible)*

*Beta vulgaris. Catharanthus roseus. Chenopodium amaranticolor.* ***Cucumis melo. Cucumis sativus. Cucurbita maxima. Cucurbita moschata. Cucurbita pepo. Datura stramonium.*** *Glycine max. Gomphrena globosa. Lactuca sativa. Lycopersicon esculentum. Nicotiana glutinosa. Nicotiana rustica.* ***Nicotiana tabacum. Petunia hybrida.*** *Phaseolus vulgaris. Raphanus sativus. Spinacia oleracea. Tetragonia expansa. Vicia faba. Zinnia elegans.*

Sources of host-range data

Inouye *et al.* (1967*b*).

**Properties of particles in sap**

Leaf sap contains many virions. Electron microscopy: no special treatments needed.

**Purification**

Method: Tung and Knight (1972); Francki and McLean (1968).

**Particle morphology**

Virions rod–shaped; not enveloped; usually straight; with a clear modal length; of *c.* 300 nm; *c.* 15 nm wide. Axial canal obvious. Basic helix obvious.

**Biochemical properties**

Virions contain *c.* 5 % nucleic acid; *c.* 95 % protein.

*Genome* consists of RNA; single–stranded; linear; unipartite; largest (or only) genome part 6.5 kb. Genomic nucleic acid isolated by Peden and Symons (1973). Infectivity retained when deproteinised with phenol or detergent. Additional factor not required for infectivity.

*Non–genomic nucleic acid* found in the virions; is subgenomic *m*RNA. Sub–genomic *m*RNA found in infected cells.

*Virion protein(s)* one; $M_r$ 17000; coat protein. Method of preparation: Fraenkel-Conrat (1957); Tung and Knight (1972). Amino acid sequence: Nozu *et al.* (1971); Tung and Knight (1972); Kurachi *et al.* (1972).

**Replication**

Replication does not depend on a helper virus.

**Cytopathology**

Virions found in leaves, mesophyll, epidermis, vascular parenchyma, xylem, phloem, companion cells and all parts of the host plant; in cytoplasm and in cell vacuoles.

Inclusions present in infected cells; are crystals in the cytoplasm; they contain virions. Other cellular changes: vesiculation of mitochondria.

**Taxonomy and Relationships**

*Tobamovirus.*

Virus(es) with serologically related virions: cucumber green mottle mosaic (SDI:6), ribgrass mosaic (SDI:5), frangipani mosaic (SDI:5), odontoglossum ringspot (SDI:4), tobacco mosaic (SDI:7) and tobacco mild green mosaic (SDI:6) viruses. Virus(es) with serologically unrelated virions: sunn-hemp mosaic.

Biologically kyuri green mottle mosaic and cucumber green mottle mosaic viruses are similar but have little genomic sequence homology and are serologically distantly related.

**Comments**

Although long considered to be a strain of cucumber green mottle mosaic tobamovirus, this virus has been shown to be a distinct tobamovirus (Francki *et al.*, 1986).

**References**

Fraenkel-Conrat, H. (1957). *Virology* **4**: 1.

Francki, R.I.B. and McLean, G.D. (1968). *Aust. J. biol. Sci.* **21**: 1311.

Francki, R.I.B., Hu, J. and Palukaitis, P. (1986). *Intervirology* **26**: 156.

Inouye, T., Inouye, N., Asatani, M. and Mitsuhata, K. (1967*a*). *Nogaku Kenkyu* **51**: 175.

Inouye, T., Inouye, N., Asatani, M. and Mitsuhata, K. (1967*b*). *Ber. Ohara Inst. landw. Biol.* **14**: 49.

Kurachi, K., Funatsu, G., Funatsu, M. and Hidaka, S. (1972). *Agric. Biol. Chem.* **36**: 1109.

Nozu, Y., Tochihara, H., Komuro, Y. and Okada, Y. (1971). *Virology* **45**: 577.

Okada, Y. (1986). In: *The Plant Viruses.* Vol. 2, The Rod-Shaped Plant Viruses, p. 267; eds M.H.V. van Regenmortel and H. Fraenkel-Conrat. Plenum Press, New York.

Peden, K.W.C. and Symons, R.H. (1973). *Virology* **53**: 487.

Tung, J.S. and Knight, C.A. (1972). *Virology* **48**: 574.

# Laburnum yellow vein (?) *nucleorhabdovirus*

**Data collated by A.A. Brunt, 1991.**

*Synonyms:* laburnum anagyroides virus.

First reported in *Laburnum anagyroides*; from the U.K; by Schultz and Harrap (1975).

**Natural host range and symptoms**

Symptoms persist.

*Laburnum anagyroides* — chlorotic banding of veins.

**Geographical distribution**

Spreads in the Eurasian region; the UK (probably).

Diagnostically insusceptible

*Chenopodium amaranticolor, C. quinoa, Cucumis sativus, Nicotiana clevelandii, N. tabacum.*

**Particle morphology**

Virions rhabdo– or bullet–shaped; enveloped; usually straight; with a clear modal length; of 250 nm; 95 nm wide.

**Cytopathology**

Virions found in leaves and mesophyll; in the perinuclear space.

**Taxonomy and Relationships**

Possible *Nucleorhabdovirus: Rhabdoviridae.*

**References**

Schultz, M.G. and Harrap, K.A. (1975). *Ann. appl. Biol.* **79**: 247.

## Lamium mild mottle *fabavirus*

**Data collated by S. Phillips and O. Lovisolo, 1992.**

*Standard acronym:* LMMV.

First reported in *Lamium alba and Lamium purpureum*; from Cambridge, U.K.; by Lovisolo (1958).

**Natural host range and symptoms**

Symptoms mild mosaic or symptomless.

*Lamium album, L. purpureum, Marrubium peregrinum, M. vulgare* — mild mosaic or symptomless.

**Transmission**

Transmitted by a vector; an insect; *Cryptomyzus alboapicalis, Myzus persicae*; Aphididae. Transmitted in a non–persistent manner. Virus transmitted by mechanical inoculation.

**Geographical distribution**

Spreads in the UK (Cambridge area).

**Experimental host range**

Several (3–9) families susceptible. Experimentally infected plants mostly show chlorotic or necrotic lesions followed by systemic infection, chlorosis or line patterns.

Diagnostically susceptible

*Chenopodium quinoa, Datura stramonium, Phaseolus vulgaris, Nicotiana clevelandii, N. tabacum* cv. Samsun — chlorotic or necrotic local lesions; systemic chlorosis.

*Nicotiana rustica* — systemic line patterns.

*Vicia faba* — local lesions; no systemic infection.

Diagnostically insusceptible

*Capsicum annuum, Lycopersicon esculentum, Vigna unguiculata.*

Maintenance and propagation hosts

*Chenopodium quinoa, Nicotiana clevelandii, N. rustica.*

Assay hosts (**L**ocal lesions or **W**hole plants)

*Chenopodium quinoa* (L).

Dicotyledonous Hosts ***(Susceptible** or Insusceptible)*

***Amaranthus caudatus***. *Beta vulgaris*. *Capsicum annuum*. ***Chenopodium album***. ***Chenopodium amaranticolor***. ***Chenopodium foliosum***. ***Chenopodium quinoa***. ***Datura stramonium***. ***Gomphrena globosa***. *Lycopersicon esculentum*. ***Nicotiana clevelandii***. ***Nicotiana glutinosa***. ***Nicotiana megalosiphon***. ***Nicotiana rustica***. ***Nicotiana tabacum***. ***Ocimum basilicum***. ***Petunia hybrida***. ***Phaseolus vulgaris***. ***Physalis floridana***. *Solanum melongena*. ***Tetragonia expansa***. ***Tropaeolum majus***. ***Vicia faba***. *Vigna unguiculata*.

Additional hosts

*Ballota nigra* (**S**); *Brassica chinensis* (**I**); *Coleus blumei* (**I**); *Lamium album* (**S**); *Lamium purpureum* (**S**); *Marrubium pereprinum* (**S**); *Marrubium vulgare* (**S**); *Pelargonium zonale* (**I**); *Primula obconica* (**S**); *Solanum capricastrum* (**I**).

Sources of host-range data

Lovisolo (1958; 1960*a*; 1960*b*); Lisa *et al.* (1982).

**Purification**

Method: Lisa *et al.* (1982).

**Particle morphology**

Virions isometric; not enveloped; 25–27 nm in diameter; angular in profile; without a conspicuous capsomere arrangement.

**Physical properties**

Three sedimenting components in purified preparations. $A_{260}/A_{280}$ ratio 0.86 (T), or 1.67 (M), or 1.76 (B).

**Biochemical properties**

Genomic nucleic acid isolated by Lisa *et al.* (1982).

*Virion protein(s)* two; $M_r$ 45000. $M_r$ of 2nd largest 28500. Method of preparation: Lisa *et al.* (1982).

**Replication**

Replication does not depend on a helper virus.

**Cytopathology**

Virions found in leaves; in cytoplasm.

**Taxonomy and Relationships**

*Fabavirus: Comoviridae.*

Virus(es) with serologically related virions: broad bean wilt virus serotype I.

Not related to six comoviruses.

**References**


Lisa, V., Luisoni, E., Boccardo, G., Milne, R.G. and Lovisolo, O. (1982). *Ann. appl. Biol.* **100**: 467.

Lovisolo, O. (1958). *Boll. Staz. Pat. Veg., Roma* **15**: 89.

Lovisolo, O. (1960*a*). *Boll. Staz. Pat. Veg., Roma* **17**: 193.

Lovisolo, O. (1960*b*). *Not. Mal. Piante* **53–54**: 233.

# Lato River *tombusvirus*

**Data collated by C. Vovlas, 1991.**

*Standard acronym:* LRV.

From water from Lato river, near Apulia, Italy; by Vovlas (1986).

### Transmission

Most probably spread in irrigation water. Virus transmitted by mechanical inoculation.

### Geographical distribution

Spreads in Italy.

### Experimental host range

Several (3–9) families susceptible. Experimentally infected plants mostly show local lesions, systemic mosaic, and death.

Diagnostically susceptible

*Chenopodium quinoa, Gomphrena globosa* — local lesions.
*Nicotiana benthamiana, N. clevelandii* — systemic mosaic.

Diagnostically insusceptible

*Nicotiana glutinosa, Chenopodium amaranticolor.*

Maintenance and propagation hosts

*Nicotiana benthamiana, N. clevelandii.*

Assay hosts (**L**ocal lesions or **W**hole plants)

*Chenopodium quinoa* (L), *Gomphrena globosa* (L), *Nicotiana benthamiana* (W), *N. clevelandii* (W).

Dicotyledonous Hosts *(**Susceptible** or Insusceptible)*

*Chenopodium amaranticolor*. ***Gomphrena globosa***. ***Nicotiana benthamiana***. ***Nicotiana clevelandii***.

### Purification

Method: Blend infected leaves of *Nicotiana benthamiana* in phosphate buffer (pH 7.2). Clarify with magnesium bentonite.

### Particle morphology

Virions isometric; 30 nm in diameter (also some 17 nm in diameter); rounded in profile; with a conspicuous capsomere arrangement.

### Physical properties

Two sedimenting components in purified preparations. Density 1.28 g $cm^{-3}$ in CsCl, or 1.33 g $cm^{-3}$ in CsCl.

### Biochemical properties

Virions contain 18 % nucleic acid; 82 % protein.

*Genome* consists of RNA; single–stranded; linear; unipartite; largest (or only) genome part 4.7 kb. Genomic nucleic acid isolated by Vovlas: phenol–SDS. Infectivity retained when deproteinised with phenol or detergent. Poly A region absent. Additional factor not required for infectivity.

*Non–genomic nucleic acid* found in the virions; is subgenomic *m*RNA.

*Virion protein(s)* one (plus a fragment); $M_r$ 38000 (plus 29000); coat protein. Method of preparation: Laemmli (1970).

**Replication**

Replication does not depend on a helper virus.

**Cytopathology**

Virions found in leaves and all parts of the host plant; in cytoplasm and in cell vacuoles.

**Taxonomy and Relationships**

*Tombusvirus: Tombusviridae.*

Virus(es) with serologically related virions: Moroccan pepper, artichoke mottled crinkle and Neckar River tombusviruses. Virus(es) with serologically unrelated virions: cymbidium ringspot and grapevine Algerian latent viruses.

**Comments**

Purified preparations, after cool storage, contain 30 nm diameter virions and also some about 17 nm in diameter that contain no RNA. The small virions are formed of the coat protein of the larger; such virions are not produced by other tombusviruses.

**References**


Gallitelli, D., Di Franco, A., Vovlas, C. and Martelli, G.P. (1989). *J. Phytopath.* **126**: 213.

Laemmli, U.K. (1970). *Nature, Lond.* **227**: 680.

Vovlas, C. (1986). *Proc. 8th Wksh. Epidem. Pl. Virus Dis.*, Orlando, Florida, U.S.A. p. 13.

Vovlas, C. and Di Franco, A. (1987). *Inform. Fitopatol.* **3**: 55.


# Launaea mosaic (?) potyvirus

**Data collated by A.A. Brunt, 1991.**

First reported in *Launaea naudicaulis*; from India; by Padma *et al.* (1973).

**Natural host range and symptoms**

Symptoms persist.

*Launaea asplenfolia, L. naudicaulis* — leaf chlorosis.

**Transmission**

Transmitted by a vector; an insect; *Myzus persicae*; Aphididae. Virus transmitted by grafting.

**Geographical distribution**

Spreads in India (New Delhi and Uttar Pradesh).

**Particle morphology**

Virions filamentous; not enveloped; usually flexuous; with a clear modal length; of 812 nm.

**Taxonomy and Relationships**

Unassigned potyvirus.

**References**


Naqvi, S.Q. and Mahmood, K. (1976). *Curr. Sci.* **45**: 152.

Padma, R., Verma, V.S. and Singh, S. (1973). *Z. PflKrankh. PflPath. Pflschutz.* **80**: 68.

Verma, V.S. and Singh, S. (1974). *Biol. Plant.* **16**: 373.

Verma, V.S. and Singh, S. (1975). *Zbl. Bakt. Parasit. Inf. and Hyg. II* **130**: 121.

# Leek yellow stripe *potyvirus*

**Data collated by L. Bos, 1987.**

*Standard acronym:* LYSV.

First reported in *Allium porrum*; from Germany and several European countries; by Bos *et al.* (1978).

**Natural host range and symptoms**

Symptoms persist.
*Allium porrum* — yellow leaf striping.
*Allium cepa* var. *ascalonicum, A. cepa* — rarely infected.

**Transmission**

Transmitted by a vector; an insect; Aphididae. Transmitted in a non–persistent manner. Virus transmitted by mechanical inoculation; not transmitted by contact between plants; not transmitted by seed.

**Ecology and control**

Studies reported by Bos *et al.* (1975).

**Geographical distribution**

Spreads in the Eurasian region; New Zealand.

**Experimental host range**

Several (3–9) families susceptible.

Diagnostically susceptible

*Allium porrum* — yellow leaf stripes.
*Chenopodium amaranticolor, C. quinoa* — chlorotic local lesions, becoming green rings when leaves senesce.
*Celosia argentea* — brown necrotic local lesions.

Diagnostically insusceptible

*Allium fistulosum, Cucumis sativus, Nicotiana clevelandii, N. megalosiphon, Petunia hybrida.*

Maintenance and propagation hosts

*Allium porrum.*

Assay hosts (**L**ocal lesions or **W**hole plants)

*Chenopodium quinoa* (L), *C. amaranticolor* (L).

Dicotyledonous Hosts *(**Susceptible** or Insusceptible)*

*Apium graveolens. Brassica oleracea* ssp. *botrytis.* ***Celosia argentea. Chenopodium album. Chenopodium amaranticolor.*** *Chenopodium ambrosioides. Chenopodium capitatum.* ***Chenopodium foetidum.*** *Chenopodium foliosum.* ***Chenopodium murale. Chenopodium quinoa.*** *Cucumis sativus. Datura stramonium. Gomphrena globosa. Lactuca sativa. Lycopersicon esculentum. Nicotiana clevelandii.* ***Nicotiana debneyi.*** *Nicotiana glutinosa. Nicotiana megalosiphon. Nicotiana tabacum. Petunia hybrida. Phaseolus vulgaris. Pisum sativum. Silene armeria. Sinapis alba. Spinacia oleracea. Tetragonia expansa. Trifolium pratense. Trifolium repens. Vicia faba. Zinnia elegans.*

Monocotyledonous Hosts *(**Susceptible** or Insusceptible)*

*Allium ascalonicum. Allium cepa. Allium fistulosum.* ***Allium porrum.***

Sources of host-range data

Verhoyen and Horvat (1973); Bos *et al.* (1978); Graichen (1978).

**Properties of particles in sap**

TIP: 50–60 °C. LIV: 3–4 days. DEP: $\log_{10}$ minus 2–3. Leaf sap contains many virions.

**Purification**

Method: Huttinga (1975); Bos *et al.* (1978).

**Particle morphology**

Virions filamentous; not enveloped; usually flexuous; with a clear modal length; of 820 nm. Axial canal obscure. Basic helix obscure.

**Physical properties**

One sedimenting component in purified preparations. Density 1.326 g $cm^{-3}$ in sucrose.

*Virion protein(s)* one; $M_r$ 34000; coat protein. Method of preparation: Huttinga (1975).

**Cytopathology**

Virions found in leaves. Inclusions present in infected cells; are pinwheels and unusual in shape; large fibrous or granular cytoplasmic inclusion bodies.

**Taxonomy and Relationships**

*Potyvirus: Potyviridae.*

Virus(es) with serologically related virions: onion yellow dwarf virus.

Best tests for diagnosis: onions infected by leek yellow stripe virus (LYSV) with difficulty, and leek is very resistant to onion yellow dwarf virus (OYDV). *Allium fistulosum* is immune to LYSV but infected symptomlessly by OYDV. LYSV gives local lesions in *Chenopodium amaranticolor* and *C. quinoa*. LYSV differs from shallot latent carlavirus in virion size and morphology and in forming larger and later local lesions on *Chenopodium amaranticolor* and *C. quinoa*.

**References**

Bos, L. (1981). *CMI/AAB Descr. Pl. Viruses* No. 240, 4 pp.
Bos, L., Huijberts, N., Huttinga, H. and Maat, D.Z. (1978). *Neth. J. Pl. Path.* **84**: 185.
Bremer, H. (1937). *Phytopath. Z.* **10**: 79.
Graichen, K. (1978). *Arch. Phytopath. PflSchutz.* **14**: 1.
Kupke, F.W. (1957). *Rhein. Monatschr. Gem. Obst. Gartenb.* **45**: 173.
Mohamed, N.A. and Young, B.R. (1980). *N.Z. Jl agric. Res.* **23**: 129.
Verhoyen, M. and Horvat, F. (1973). *Parasitica* **29**: 16.

# Lettuce big-vein varicosavirus

**Data collated by S. Kuwata, 1991.**

*Standard acronym:* LBVV.

First reported in *Lactuca sativa*; from Imperial Valley, California, U.S.A.; by Jagger and Chandler (1934).

**Natural host range and symptoms**

Symptoms vary seasonally. Symptoms vein yellowing then chlorotic or white vein banding.

*Lactuca sativa* — chlorotic or white vein banding.
*Sonchus oleraceus* — symptomless or vein banding.
*Sonchus asper* — yellow vein-banding.

**Transmission**

Transmitted by a vector; a fungus; *Olpidium brassicae*; Chytridiales. Not transmitted by *Myzus persicae, Hyperomyzus lactuca, Xiphinema* spp., *Longidorus* spp. Virus transmitted by mechanical inoculation; transmitted by grafting; not transmitted by contact between plants; not transmitted by seed; not transmitted by pollen.

**Ecology and control**

Studies reported by Allen (1948); Tomlinson and Faithfull (1979).

**Geographical distribution**

Spreads in Australia, the former Czechoslovakia, Italy, Japan, the Netherlands, New Zealand, the UK, and the USA.

**Experimental host range**

Several (3–9) families susceptible. Experimentally infected plants mostly show chlorotic or white vein-banding.

Diagnostically susceptible

*Chenopodium amaranticolor*, *C. quinoa* — chlorotic local lesions.
*Lactuca sativa* — chlorotic or white vein-banding.
*Nicotiana benthamiana, N. clevelandii* — symptomless.
*Nicotiana occidentalis* — chlorotic local lesions and systemic chlorosis.
*Sonchus asper* — yellow vein-banding.
*Sonchus oleracea* — slight vein-banding.
*Tetragonia expansa* — green rings after leaf senescence.

Diagnostically insusceptible

*Chenopodium murale, Nicotiana tabacum.*

Maintenance and propagation hosts

*Lactuca sativa, Nicotiana occidentalis.*

Assay hosts (**L**ocal lesions or **W**hole plants)

*Chenopodium quinoa* (L).

Dicotyledonous Hosts *(**Susceptible** or Insusceptible)*

***Chenopodium amaranticolor***. *Chenopodium murale*. ***Chenopodium quinoa***. ***Lactuca sativa***. ***Nicotiana benthamiana***. ***Nicotiana clevelandii***. ***Nicotiana occidentalis***. *Nicotiana tabacum*. ***Sonchus oleraceus***. ***Tetragonia expansa***.

Sources of host-range data

Campbell (1965); Huijberts *et al.* (1990).

**Properties of particles in sap**

TIP: 50 °C. LIV: 1 days. DEP: $\log_{10}$ minus 1. Leaf sap contains few virions. Electron microscopy: aldehyde fixation essential.

**Purification**

Method: Kuwata *et al.* (1983).

**Particle morphology**

Virions rod–shaped; not enveloped; usually straight; with a clear modal length; of 320–360 nm; 18 nm wide. Axial canal obvious; 3 nm in diameter. Basic helix obvious; pitch of basic helix 5 nm.

**Physical properties**

Density 1.27 g/cm$^3$ in $Cs_2SO_4$.

**Biochemical properties**

*Genome* consists of RNA; double–stranded; linear; of two parts; largest (or only) genome part the larger 7 kb (pairs); the 2nd largest 6.5 kb (pairs). Genomic nucleic acid isolated by Kuwata: proteinase K hydrolysis then phenol.

*Virion protein(s)* one; $M_r$ 48000; coat protein.

**Replication**

Replication does not depend on a helper virus.

**Cytopathology**

Virions found in leaves, roots, mesophyll and vascular parenchyma; in cytoplasm. Inclusions present in infected cells; are crystals in the cytoplasm; they contain virions.

**Taxonomy and Relationships**

Varicosavirus.

The virions of this virus are closely serological related to those of tobacco stunt virus, but as they differ in host range, they are considered to be separate species, rather than related strains (Kuwata and Kubo, 1986). Best tests for diagnosis: soil transmissibility. Serological tests such as ISEM and ELISA.

**References**

Allen, M.W. (1948). *Phytopathology* **38**: 616.
Campbell, R.N. (1962). *Nature, Lond.* **195**: 675.
Campbell, R.N. (1985). *Can. J. Bot.* **63**: 2288.
Campbell, R.N. and Grogan, R.G. (1964). *Phytopathology* **54**: 681.
Campbell, R.N., Grogan, R.G. and Purcifull, D.E. (1961). *Virology* **15**: 82.
Huijberts, N., Blystad, D.-R. and Bos, L. (1990). *Ann. appl. Biol.* **116**: ;463.
Kuwata, S. and Kubo, S. (1986). *AAB Descr. Pl. Viruses* No. 313.
Kuwata, S., Kubo, S., Yamashita, S. and Doi, Y. (1983). *Ann. Phytopath. Soc. Japan* **49**: 246.
Tomlinson, J.A. and Faithfull, E.M. (1979). *Ann. appl. Biol.* **93**: 13.
Tomlinson, J.A. and Garrett, R.G. (1962). *Nature, Lond.* **194**: 249.
Tomlinson, J.A. and Garrett, R.G. (1964). *Ann. appl. Biol.* **54**: 45.
Vetten, H.J., Lesemann, D.-E. and Dalchow, J. (1987). *J. Phytopath.* **120**: 53.

# Lettuce infectious yellows (?) *closterovirus*

**Data collated by J.E. Duffus and S. Molyneux, 1986.**

*Standard acronym:* LIYV.

First reported in lettuce, sugarbeet, cantaloupe, watermelon, melons, squash and carrots; from California and Arizona, U.S.A.; by Duffus *et al.* (1982).

**Natural host range and symptoms**

Symptoms persist.

*Lactuca sativa, Citrullus lanatus, Cucumis melo, Cucurbita pepo, C. maxima, C. moschata, Beta vulgaris, Daucus carota* — severe chlorosis and/or reddening, rolling vein-clearing and brittleness of leaves, stunting.

**Transmission**

Transmitted by a vector; an insect; *Bemisia tabaci*; Aleyrodidae. Transmitted in a semi–persistent manner. Virus does not multiply in the vector; not transmitted by mechanical inoculation.

**Ecology and control**

Studies reported by Duffus *et al.* (1986).

**Geographical distribution**

Spreads in the USA (California and Arizona).

**Experimental host range**

Many (>9) families susceptible (45 spp. in 15 families).

Diagnostically susceptible

*Beta vulgaris, Chenopodium quinoa, Lactuca sativa, Brassica pekinensis Citrullus lanatus, Cucumis melo, Cucumis sativus, Trifolium subterraneum Malva parviflora, Nicotiana clevelandii* — severe chlorosis, and/or reddening, rolling, vein-clearing and brittleness of leaves; stunting.

Diagnostically insusceptible

*Capsicum annuum, Nicotiana tabacum, Phaseolus vulgaris, Vicia faba, Datura stramonium.*

Maintenance and propagation hosts

*Lactuca sativa, Malva parviflora, Nicotiana clevelandii.*

Assay hosts (**L**ocal lesions or **W**hole plants)

*Lactuca sativa* (W), *Nicotiana clevelandii* (W), *Malva parviflora* (W).

Dicotyledonous Hosts *(**Susceptible** or Insusceptible)*

*Bellis perennis*. ***Beta macrocarpa***. ***Beta vulgaris***. *Brassica campestris* ssp. *napus*. ***Brassica campestris*** ssp. ***pekinensis***. ***Brassica campestris*** ssp. ***rapa***. *Brassica oleracea* ssp. *botrytis*. ***Capsella bursa–pastoris***. *Capsicum annuum*. ***Carthamus tinctorius***. ***Chenopodium capitatum***. ***Chenopodium murale***. ***Chenopodium quinoa***. ***Cichorium endiva***. ***Citrullus lanatus***. *Coriandrum sativum*. ***Cucumis melo***. ***Cucumis sativus***. ***Cucurbita maxima***. ***Cucurbita moschata***. ***Cucurbita pepo***. *Cyamopsis tetragonoloba*. *Datura metel*. *Datura stramonium*. ***Daucus carota***. ***Dianthus caryophyllus***. *Gossypium hirsutum*. ***Helianthus annuus***. *Hibiscus esculentus*. ***Lactuca sativa***. *Lycopersicon esculentum*. *Medicago sativa*. *Nicandra physaloides*. ***Nicotiana clevelandii***. *Nicotiana tabacum*. *Ocimum basilicum*. ***Petunia hybrida***. *Phaseolus vulgaris*. *Physalis floridana*. ***Pisum sativum***. ***Raphanus sativus***. *Sesamum indicum*. ***Sonchus oleraceus***. ***Spinacia oleracea***. ***Trifolium subterraneum***. ***Tropaeolum majus***. *Vicia faba*. ***Zinnia elegans***.

Monocotyledonous Hosts *(**Susceptible** or Insusceptible)*

*Allium cepa*.

Sources of host-range data

Duffus *et al.* (1982); Duffus *et al.* (1986).

**Properties of particles in sap**

Leaf sap contains few virions.

**Purification**

Method: Duffus *et al.* (1986).

**Particle morphology**

Virions filamentous; usually flexuous; of 1800–2000 nm; 13–14 nm wide.

**Biochemical properties**

*Genome* consists of RNA; single–stranded.

*Sequence database accession code(s):*

U05242 Em(40)_vi:LI05242 Gb(84)_pl:LIU05242 Lettuce infectious yellows virus coat protein gene, complete cds. 2/94 791bp.

U15440 Em(43)_vi:Li15440 Gb(89)_vi:Liu15440 Lettuce infectious yellows virus RNA 1 methyltransferase and RNA-dependent RNA polymerase.

U15441 Em(43)_vi:Li15441 Gb(89)_vi:Liu15441 Lettuce infectious yellows virus RNA 2 heat-shock protein, coat protein, and duplicated coat.

**Cytopathology**

Virions found in vascular parenchyma. Inclusions present in infected cells; are membranous bodies; they contain virions.

**Taxonomy and Relationships**

Possible *Closterovirus*.

Virus(es) with serologically unrelated virions: beet pseudoyellows virus.

**References**

Duffus, J.E., Larsen, R.C. and Liu, H.Y. (1986). *Phytopathology* **76**: 97.

Duffus, J.E., Mayhew, D.E. and Flock, R.A. (1982). *Phytopathology* **72**: 963.

Houk, M.S. and Hoefert, L.L. (1983). *Phytopathology* **73**: 790.

## Lettuce mosaic *potyvirus*

**Data collated by J.A. Tomlinson, 1982. Revised 1984.**

*Synonyms:* lactuca virus 1. *Standard acronym:* LMV. *Strains:* three pathotypes of the virus are recognised.

First reported in *Lactuca sativa*; from Florida, U.S.A.

**Natural host range and symptoms**

Symptoms persist.

*Lactuca* spp., *Senecio vulgaris, Sonchus* spp. — necrotic or chlorotic local lesions and streaking, then mosaic, vein yellowing, leaf malformation.

*Cicer arietinum* — leaf chlorosis.

*Pisum sativum* — chlorotic mottling.

**Transmission**

Transmitted by a vector; an insect; *Aphis gossypii, Macrosiphum euphorbiae, M. gei, Myzus persicae*; Aphididae. Transmitted in a non–persistent manner. Virus transmitted by mechanical inoculation; transmitted by seed (3–10%); transmitted by pollen to the seed.

**Geographical distribution**

Probably distributed worldwide (wherever *Lactuca sativa* is grown).

**Experimental host range**

Several (3–9) families susceptible.

Diagnostically susceptible

*Gomphrena globosa* — necrotic local lesions; not systemic.

*Chenopodium amaranticolor, C. quinoa* — necrotic local lesions; then systemic mosaic, rugosity.

*Carthamnus tinctorius, Nicotiana benthamiana* — systemic mosaic.

Maintenance and propagation hosts

*Chenopodium quinoa, Lactuca sativa.*

Assay hosts (**L**ocal lesions or **W**hole plants)

*Gomphrena globosa* (L), *Chenopodium amaranticolor* (L), *C. quinoa* (L).

Dicotyledonous Hosts *(**Susceptible** or Insusceptible)*

***Amaranthus caudatus***. *Beta vulgaris*, or ***Beta vulgaris***. ***Capsella bursa–pastoris***. ***Chenopodium album***. ***Chenopodium amaranticolor***. ***Chenopodium capitatum***. ***Chenopodium murale***. ***Chenopodium quinoa***. ***Cichorium endiva***. *Cucumis sativus*. *Datura stramonium*. *Gomphrena globosa*. ***Lactuca sativa***. ***Lathyrus odoratus***. *Lycopersicon esculentum*. ***Nicotiana benthamiana***. ***Nicotiana clevelandii***. *Nicotiana glutinosa*. *Nicotiana tabacum*. *Petunia hybrida*. *Phaseolus vulgaris*. ***Pisum sativum***. ***Senecio vulgaris***. *Sonchus oleraceus*. ***Spinacia oleracea***. ***Stellaria media***. ***Tetragonia expansa***. *Trifolium pratense*. *Trifolium repens*. ***Zinnia elegans***.

Sources of host-range data

Costa and Duffus (1958); Thornberry (1966); Edwardson (1974).

**Properties of particles in sap**

TIP: 55–60 °C. LIV: 1–2 days. DEP: $\log_{10}$ minus 2–3. Leaf sap contains few virions.

**Purification**

Method: Moghal and Francki (1976).

**Particle morphology**

Virions filamentous; not enveloped; usually flexuous; with a clear modal length; of 750 nm; 15 nm wide. Axial canal obvious. Basic helix obscure.

**Physical properties**

One sedimenting component in purified preparations; sedimentation coefficient 143 *S*. Density 1.33 g $cm^{-3}$ in sucrose.

**Biochemical properties**

*Genome* consists of RNA; single–stranded.

*Sequence database accession code(s):*

X65652 Em(40)_vi:LMVPCP Gb(84)_vi:LMVPCP Lettuce Mosaic Virus 3'end of polyprotein gene for polymerase and coat protein. 5/92 1,679bp.

U24661 Em(43)_vi:Lm24661 Gb(89)_vi:Lmu24661 Lettuce mosaic virus isolate Asgrow coat protein (CP) mRNA, partial cds. 5/95 571bp. U24662 Em(43)_vi:Lm24662 Gb(89)_vi:Lmu24662 Lettuce mosaic virus isolate Firestone coat protein (CP) mRNA, partial cds. 5/95 571bp.

U24663 Em(43)_vi:Lm24663 Gb(89)_vi:Lmu24663 Lettuce mosaic virus isolate Gould coat protein (CP) mRNA, partial cds. 5/95 571bp.

U24664 Em(43)_vi:Lm24664 Gb(89)_vi:Lmu24664 Lettuce mosaic virus isolate TA Gazania coat protein (CP) mRNA, partial cds. 5/95 571bp.

U24665 Em(43)_vi:Lm24665 Gb(89)_vi:Lmu24665 Lettuce mosaic virus isolate Hartnell coat protein (CP) mRNA, partial cds. 5/95 571bp.

U24666 Em(43)_vi:Lm24666 Gb(89)_vi:Lmu24666 Lettuce mosaic virus isolate Hitchcock coat protein (CP) mRNA, partial cds. 5/95 571bp.
U24667 Em(43)_vi:Lm24667 Gb(89)_vi:Lmu24667 Lettuce mosaic virus isolate Hunter coat protein (CP) mRNA, partial cds. 5/95 571bp.
U24668 Em(43)_vi:Lm24668 Gb(89)_vi:Lmu24668 Lettuce mosaic virus isolate Peto coat protein (CP) mRNA, partial cds. 5/95 571bp.
U24669 Em(43)_vi:Lm24669 Gb(89)_vi:Lmu24669 Lettuce mosaic virus isolate LMV-P coat protein (CP) mRNA, partial cds. 5/95 571bp.
U24670 Em(43)_vi:Lm24670 Gb(89)_vi:Lmu24670 Lettuce mosaic virus isolate Romaine coat protein (CP) mRNA, partial cds. 5/95 1,123bp.
U24671 Em(43)_vi:Lm24671 Gb(89)_vi:Lmu24671 Lettuce mosaic virus isolate Spreckles coat protein (CP) mRNA, partial cds. 5/95 571bp.
$M_r$ 33000. Method of preparation: Moghal and Francki (1976).

**Cytopathology**

Virions found in all parts of the host plant; in cytoplasm. Inclusions present in infected cells; are pinwheels and unusual in shape; and bundles; they do not contain virions.

**Taxonomy and Relationships**

*Potyvirus: Potyviridae.*

Virus(es) with serologically related virions: bean yellow mosaic, clover yellow vein, potato Y and tobacco etch potyviruses.

**References**


Bosque-Perez, N.A. and Buddenhagen, I.W. (1989). *Plant Dis.* **73**: 608.
Costa, A.S. and Duffus, J.E. (1958). *Pl. Dis. Reptr* **42**: 583.
Edwardson, J.R. (1974). *Fla Agric. Exp. Stn Monog.* No. 5, p. 56.
Huttinga, H. and Mosch, W.H.M. (1974). *Neth. J. Pl. Path.* **80**: 19.
Moghal, S.M. and Francki, R.I.B. (1976). *Virology* **73**: 350.
Moghal, S.M. and Francki, R.I.B. (1981). *Virology* **112**: 210.
Pink, D.A.C., Kostova, D. and Walkey, D.G.A. (1992). *Plant Path.* **41**: 5.
Provvidenti, R. (1973). *Pl. Dis. Reptr* **57**: 688.
Saric, A. and Wrischer, M. (1977). *Phytopath. Z.* **90**: 27.
Thornberry, H.H. (1966). In: *Index of Plant Virus Diseases* U.S. Dep. Agric. Hdbk No. 307.
Tomlinson, J.A. (1970). *CMI/AAB Descr. Pl. Viruses* No. 9, 4 pp.


# Lettuce necrotic yellows *cytorhabdovirus*

**Data collated by J.W. Randles, 1980. Revised 1987.**

*Standard acronym:* LNYV. *Strains:* severe (SE 3) and mild strains reported.

First reported in *Lactuca sativa*; from Victoria, Australia; by Stubbs *et al.* (1963).

**Natural host range and symptoms**

Symptoms persist, or disappear soon after infection (symptomless).

*Lactuca sativa* — leaves first dull green, then chlorotic, outer leaves flaccid, bronzed and sometimes necrotic; plants stunted, and late infected plants have heads with internal necrosis.

*Embergeria megalocarpa, Pterolobium tingitana, Sonchus oleraceus, S. hydrophilus* — symptomless.

**Transmission**

Transmitted by a vector; an insect; *Hyperomyzus lactucae*; Aphididae. Transmitted in a persistent manner. Virus retained when the vector moults; multiplies in the vector; transmitted congenitally to the progeny of the vector; transmitted by mechanical inoculation; not transmitted by seed.

**Ecology and control**

Studies reported by Randles (1983).

**Geographical distribution**

Spreads in Australia. Found, but with no evidence of spread, in New Zealand.

**Experimental host range**

Several (3–9) families susceptible.

Diagnostically susceptible

*Lactuca sativa* — systemic yellowing, leaf malformation.

*Datura stramonium* — faint systemic interveinal chlorosis.

*Nicotiana glutinosa* — chlorotic local lesions; systemic yellowing, rugosity, leaf malformation.

Maintenance and propagation hosts

*Lactuca sativa, Nicotiana glutinosa.*

Assay hosts (**L**ocal lesions or **W**hole plants)

*Nicotiana glutinosa.*

Dicotyledonous Hosts *(**Susceptible** or Insusceptible)*

*Brassica oleracea* ssp. *botrytis*. ***Datura stramonium***. ***Gomphrena globosa***. ***Lactuca sativa***. ***Lycopersicon esculentum***. ***Nicotiana clevelandii***. ***Nicotiana glutinosa***. ***Nicotiana glutinosa* × *clevelandii***. *Nicotiana tabacum*. ***Petunia hybrida***. ***Sonchus oleraceus***. ***Spinacia oleracea***. *Trifolium subterraneum*. *Vicia faba*. *Vigna unguiculata*. *Zinnia elegans*.

Sources of host-range data

Stubbs *et al.* (1963).

**Properties of particles in sap**

TIP: 52 °C. LIV: 0.3–1 days. DEP: $\log_{10}$ minus 2. Infectivity of sap decreased by treatment with di–ethyl ether. Leaf sap contains few virions, or contains many virions. Electron microscopy: aldehyde fixation necessary.

**Purification**

Method: McLean and Francki (1967).

**Particle morphology**

Virions rhabdo– or bullet–shaped; enveloped; with a clear modal length; of 227 nm (in negative stain), or 360 nm (in tissue sections); 52 nm wide (in tissue sections), or 66 nm wide (in negative stain). Axial canal obvious. Basic helix obscure; pitch of basic helix 4.5 nm.

**Physical properties**

One sedimenting component in purified preparations; sedimentation coefficient 940 *S*. Density 1.19–1.2 g $cm^{-3}$ in sucrose. Density in Nycodenz, *c.* 1.135 g $cm^{-3}$.

**Biochemical properties**

*Genome* consists of RNA; single–stranded; unipartite; largest (or only) genome part 12.54 kb. Infectivity lost when deproteinised with phenol or detergent.

*Sequence database accession code(s):*

L24364 Em(40)_vi:LYLRNAA Gb(84)_vi:LYLRNAA Lettuce necrotic yellows virus RNA, 3' UTR. 5/94 84bp.

L24365 Em(40)_vi:LYLRNAB Gb(84)_vi:LYLRNAB Lettuce necrotic yellows virus RNA, 5' UTR. 6/94 187bp.

L30103 Gb(84)n:RHWNUPR Lettuce necrotic yellows virus nucleocapsid gene, complete cds. 8/94 1,533bp.

L30103 Em(43)_vi:Rhwnupr Gb(89)_vi:Rhwnupr Lettuce necrotic yellows virus nucleocapsid gene, complete cds. 1/95 1,533bp.

Sub–genomic *m*RNA found in infected cells (Randles *et al.*, 1972)).

*Virion protein(s)* five (Laemmli, 1970); $M_r$ of the largest 180000–190000; L protein. $M_r$ of 2nd largest 76000; G protein. $M_r$ of 3rd largest 54000; N protein. $M_r$ of 4th largest 38000; NS protein. $M_r$ of 5th largest 19000; M protein. Method of preparation: Francki *et al.* (1975). Virion proteins glycosylated; are G protein.

**Replication**

Replication does not depend on a helper virus.

**Cytopathology**

Virions found in mesophyll, epidermis, xylem, phloem and hair cells; in cytoplasm. Inclusions present in infected cells; are membranous bodies; they contain virions.

**Taxonomy and Relationships**

*Cytorhabdovirus: Rhabdoviridae.*

Virus(es) with serologically unrelated virions: sowthistle yellow vein virus.

**References**

Chu, P.W.G. and Francki, R.I.B. (1982). *Ann appl. Biol.* **100**: 149.

Crowley, N.C., Harrison, B.D. and Francki, R.I.B. (1965). *Virology* **26**: 290.

Dale, J.L. and Peters, D. (1981). *Intervirology* **16**: 86.

Francki, R.I.B. and Randles, J.W. (1970). *CMI/AAB Descr. Pl. Viruses* No. 26, 4 pp.

Francki, R.I.B. and Randles, J.W. (1975). In: *Negative stranded viruses*, p. 223. Academic Press.

McLean, G.D. and Francki, R.I.B. (1967). *Virology* **31**: 585.

Randles, J.W. (1983). In: *Current Topics in Vector Research*, Vol. 1, p. 169; ed. K.F. Harris. Praeger Scientific, New York.

Randles, J.W. and Coleman, D.F. (1972). *Physiol. Pl. Path.* **2**: 247.

Stubbs, L.L. and Grogan, R.G. (1963). *Aust. J. agric. Res.* **14**: 439.

# Lettuce speckles mottle *umbravirus*

**Data collated by A.A. Brunt, 1992.**

*Standard acronym:* LSMV.

First reported in *Beta vulgaris, Lactuca sativa, Spinacia oleracea*; from the U.S.A.; by Falk *et al.* (1978; 1979).

**Natural host range and symptoms**

Symptoms vary seasonally.

*Beta vulgaris, Lactuca sativa, Spinacia oleracea* — in mixed infections, the virus enhances the symptoms induced by beet western yellows luteovirus; the outer leaves of infected plants, especially those of lettuce, develop numerous angular chlorotic spots to produce the so-called speckles disease. Symptoms of the complex become less conspicuous as ambient temperatures rise.

**Transmission**

Transmitted by a vector; an insect; *Acyrthosiphon (Aulacorthum) solani, Brevicoryne brassicae, Myzus persicae*; Aphididae. Principal natural vector(s): *Myzus persicae*. Transmitted in a persistent manner. Virus requires, for vector transmission, a helper virus (beet western yellows luteovirus); transmitted by mechanical inoculation.

**Geographical distribution**

Spreads in the USA (California).

**Experimental host range**

Several (3–9) families susceptible.

Diagnostically susceptible

*Beta vulgaris, Spinacia oleracea, Nicotiana glutinosa, Physalis floridana* — no conspicuous systemic infection.

*Chenopodium quinoa* — local lesions.

*Nicotiana clevelandii* — faint interveinal chlorotic spotting of systemically infected leaves (especially under cool shady conditions).

Diagnostically insusceptible

*Beta macrocarpa, Chenopodium amaranticolor, Datura stramonium, Nicotiana tabacum, Petunia hybrida.*

Maintenance and propagation hosts

*Lactuca sativa, Nicotiana clevelandii.*

Assay hosts (**L**ocal lesions or **W**hole plants)

*Chenopodium quinoa* (L), *Nicotiana clevelandii* (W).

Dicotyledonous Hosts *(**Susceptible** or Insusceptible)*

*Apium graveolens. Beta macrocarpa.* ***Beta vulgaris. Brassica campestris*** ssp. ***napus***. *Brassica campestris* ssp. *pekinensis. Brassica juncea.* ***Capsella bursa–pastoris***. *Chenopodium amaranticolor.* ***Chenopodium capitatum. Chenopodium murale. Chenopodium quinoa***. *Cichorium endiva. Conium maculatum. Coriandrum sativum. Datura stramonium. Daucus carota. Gomphrena globosa.* ***Hyoscyamus niger. Lactuca sativa***. *Lycopersicon esculentum. Nicandra physaloides.* ***Nicotiana clevelandii. Nicotiana glutinosa. Nicotiana glutinosa*** × ***clevelandii***. *Nicotiana tabacum. Petunia hybrida.* ***Physalis floridana***. *Raphanus sativus. Sonchus oleraceus.* ***Spinacia oleracea***.

Comments on host-range

The host range of the virus is wider when viruliferous aphids are used to infect plants.

Sources of host-range data

Falk *et al.* (1979).

**Properties of particles in sap**

LIV: 2 days (but twice as long in sap from plants also infected with beet yellows luteovirus).

**Biochemical properties**

*Genome* consists of RNA; single–stranded.

**Taxonomy and Relationships**

*Umbravirus.*

**Comments**

The virus probably lacks a functional coat protein and, in mixed infections, is probably encapsidated in that of beet western yellow luteovirus (its helper virus). It thus resembles carrot mottle, groundnut rosette, tobacco mottle and tobacco yellow vein viruses.

**References**

Falk, B.W., Morris, T.J. and Duffus, J.E. (1978). *Proc. Am. Phytopath. Soc.* **5**: 97.
Falk, B.W., Duffus, J.E. and Morris, T.J. (1977). *Phytopathology* **69**: 612.

# Lilac chlorotic leafspot *capillovirus*

**Data collated by A.A. Brunt, 1991.**

*Standard acronym:* LCLV.

First reported in *Syringa vulgaris*; from south east England; by Brunt (1978).

**Natural host range and symptoms**

Symptoms persist.

*Syringa vulgaris* — chlorotic local lesions, which may coalesce, enlarge and produce indistinct line patterns.

**Transmission**

Not transmitted by *Aphis gossypii, Macrosiphoniella sanborni, Myzus ligustri, M. persicae.* Virus transmitted by mechanical inoculation; transmitted by grafting; not transmitted by seed.

**Geographical distribution**

Found, but with no evidence of spread, in south east England.

**Experimental host range**

Several (3–9) families susceptible.

Diagnostically susceptible

*Beta vulgaris, Chenopodium quinoa* — irregular necrotic local lesions, then systemic leaf chlorosis.

*Chenopodium amaranticolor, C. capitatum, C. murale* — few necrotic local lesions; not systemic.

*Nicotiana clevelandii, N. glutinosa, N. clevelandii x N. glutinosa, N. rustica* — faint systemic leaf chlorosis.

*Phaseolus vulgaris* — systemic chlorotic local lesions.

Diagnostically insusceptible

*Cucumis sativus, Datura stramonium, Glycine max, Nicotiana megalosiphon, Vigna unguiculata.*

Maintenance and propagation hosts

*Nicotiana clevelandii, N. glutinosa, N. rustica, Chenopodium quinoa.*

Assay hosts (**L**ocal lesions or **W**hole plants)

*Chenopodium quinoa* (L).

Dicotyledonous Hosts (***Susceptible*** *or Insusceptible)*

***Amaranthus retroflexus. Beta macrocarpa. Beta vulgaris.*** *Brassica campestris* ssp. *pekinensis. Cassia occidentalis. Catharanthus roseus.* ***Chenopodium amaranticolor. Chenopodium capitatum. Chenopodium murale. Chenopodium quinoa.*** *Citrus aurantifolia. Cucumis melo. Cucurbita pepo. Datura stramonium. Dianthus barbatus. Dianthus caryophyllus. Emilia sagittata. Fagopyrum esculentum. Glycine max. Gomphrena globosa. Helianthus annuus. Hyoscyamus niger. Lactuca sativa. Malus sylvestris. Montia perfoliata.* ***Nicotiana benthamiana. Nicotiana clevelandii. Nicotiana glutinosa. Nicotiana glutinosa*** × ***clevelandii.*** *Nicotiana megalosiphon.* ***Nicotiana rustica. Nicotiana tabacum. Petunia hybrida. Phaseolus vulgaris.*** *Pisum sativum.* ***Silene armeria. Tetragonia expansa. Trifolium incarnatum.*** *Trifolium repens. Tropaeolum majus.* ***Vicia faba.*** *Vigna unguiculata.*

Monocotyledonous Hosts (***Susceptible*** *or Insusceptible)*

*Avena sativa. Hordeum vulgare. Triticum aestivum. Zea mays.*

Sources of host-range data

Brunt (1978).

**Properties of particles in sap**

TIP: 65 °C. LIV: 8–16 days (at 20°C). DEP: $\log_{10}$ minus 4. Leaf sap contains few virions.

**Purification**

Method: Brunt (1978).

**Particle morphology**

Virions filamentous; not enveloped; usually flexuous; with a clear modal length; of 1536 nm; 12.5 nm wide. Axial canal obscure. Basic helix occasionally obvious, or obscure (usually); pitch of basic helix 3.69 nm.

**Physical properties**

One sedimenting component in purified preparations; sedimentation coefficient 96 *S*. Density 1.302 g $cm^{-3}$ in CsCl. $A_{260}/A_{280}$ ratio 1.69.

*Virion protein(s)* one; $M_r$ 27000; coat protein.

**Cytopathology**

Virions found in leaves, mesophyll and the vascular parenchyma; in cytoplasm. Inclusions present in infected cells; are viroplasms; they contain virions. Other cellular changes: enlarged and sinuous thylakoids in chloroplasts near inclusions.

**Taxonomy and Relationships**

*Capillovirus.*

Virus(es) with serologically unrelated virions: beet yellow stunt, beet yellows, carnation necrotic fleck and apple chlorotic leafspot viruses.

**References**

Brunt, A.A. (1978). *Ann. appl. Biol.* **88**: 383.

Brunt, A.A. and Stace-Smith, R. (1978). *J. gen. Virol.* **39**: 63.

# Lilac mottle *carlavirus*

**Data collated by A.A. Brunt, 1991.**

*Standard acronym:* LiMV.

First reported in *Syringa oblata affinis*; from the U.S.A. (in plants from China); by Waterworth (1972).

**Natural host range and symptoms**

Symptoms persist.
*Syringa oblata affinis* — leaf chlorosis and mottling.

**Transmission**

Transmitted by a vector; an insect; *Myzus persicae*; Aphididae. Transmitted in a non–persistent manner. Virus transmitted by mechanical inoculation.

**Geographical distribution**

Found, but with no evidence of spread, in the U.S.A.

**Experimental host range**

Several (3–9) families susceptible.

Diagnostically susceptible

*Chenopodium amaranticolor* — local lesions; not systemic.
*Chenopodium quinoa* — local lesions; systemic chlorosis.
*Nicotiana megalosiphon, Petunia hybrida, Physalis franchetii* — systemic chlorosis.

Diagnostically insusceptible

*Cucumis sativus, Datura stramonium, Capsicum annuum, Nicotiana tabacum, Glycine max.*

Maintenance and propagation hosts

*Petunia hybrida, Chenopodium quinoa.*

Assay hosts (**L**ocal lesions or **W**hole plants)

*Chenopodium amaranticolor* (L), *C. quinoa* (L and W) *Petunia hybrida* (W).

**Properties of particles in sap**

TIP: 67–70 °C. LIV: 0.5–1 days. DEP: $\log_{10}$ minus 5. Leaf sap contains few virions.

**Purification**

Method: Waterworth (1972).

**Particle morphology**

Virions filamentous; not enveloped; usually straight; with a clear modal length; of 575–610 nm. Axial canal obscure. Basic helix obscure.

**Taxonomy and Relationships**

*Carlavirus.*

Virus(es) with serologically related virions: carnation latent, chrysanthemum B and red clover vein mosaic viruses. Virus(es) with serologically unrelated virions: potato S virus.

**Comments**

A carlavirus tentatively named lilac ringspot virus occurs in lilacs in Japan (Y. Doi, in Hull *et al.*, 1989). Its relationship to lilac mottle and other carlaviruses has not yet been determined.

**References**

Hull, R., Brown, F. and Payne, C. (1989). *Directory and Dictionary of Animal, Bacterial and Plant Viruses*, p.119. MacMillan Reference Books, London.

Waterworth, H.E. (1972). *Pl. Dis. Reptr* **56**: 923.

# Lilac ring mottle *ilarvirus*

**Data collated by F.A. van der Meer, 1987.**

*Standard acronym:* LRMV.

First reported in *Syringa vulgaris*; from the Netherlands; by Van der Meer (1976).

**Natural host range and symptoms**

Symptoms vary cyclically over a few weeks and vary seasonally (infected plants show symptoms in spring; cvs Mme Florent Stepman and Mme Lemoine show clear symptoms).

*Syringa vulgaris* — faint rings and line patterns.

**Transmission**

Virus transmitted by mechanical inoculation; transmitted by grafting; transmitted by contact between plants; transmitted by seed (in *Chenopodium quinoa, C. amaranticolor*, but not in lilac (Van der Meer, 1976)).

**Ecology and control**

Studies reported by Van der Meer (1976); Van der Meer *et al.* (1976).

**Geographical distribution**

Found, but with no evidence of spread, in Canada, the Netherlands and the U.K.

**Experimental host range**

Several (3–9) families susceptible.

Diagnostically susceptible

*Chenopodium quinoa* — systemic epinasty and tip necrosis.
*Celosia plumosa* — chlorotic flecks, then systemic chlorosis.
*Nicotiana megalosiphon* — necrotic local lesions, then systemic mottling.
*Nicotiana tabacum* — transient wilting of the first leaves to be systemically infected.

Diagnostically insusceptible

*Cucumis sativus, Amaranthus caudatus, Capsicum annuum, Lycopersicon esculentum.*

Maintenance and propagation hosts

*Chenopodium quinoa.*

Assay hosts (**L**ocal lesions or **W**hole plants)

*Chenopodium quinoa* (W).

Dicotyledonous Hosts *(**Susceptible** or Insusceptible)*

*Amaranthus caudatus. Capsicum annuum.* ***Catharanthus roseus. Celosia argentea. Chenopodium album. Chenopodium amaranticolor. Chenopodium foetidum. Chenopodium quinoa.*** *Cucumis sativus.* ***Gomphrena globosa.*** *Lycopersicon esculentum.* ***Nicotiana clevelandii. Nicotiana glutinosa. Nicotiana megalosiphon. Nicotiana rustica. Nicotiana tabacum. Phaseolus vulgaris. Vigna unguiculata.***

Sources of host-range data

Van der Meer (1976).

**Properties of particles in sap**

TIP: 63–65 °C. LIV: 0.2 days (at 20°C). DEP: $\log_{10}$ minus 4 (in sap of *Chenopodium quinoa*). Infectivity of sap not changed by treatment with di–ethyl ether. Leaf sap contains few virions. Electron microscopy: fix virions with 2% glutaraldehyde.

**Purification**

Method: Van der Meer *et al.* (1976).

**Particle morphology**

Virions isometric; not enveloped; 27 nm in diameter; rounded in profile; without a conspicuous capsomere arrangement.

**Physical properties**

Two sedimenting components in purified preparations; sedimentation coefficient of the fastest (B) component 98 *S*; of the other(s) 83 *S* (M). Density 1.345 g $cm^{-3}$ in CsCl. Density 1.265 in $Cs_2SO_4$. $A_{260}/A_{280}$ ratio 1.43–1.46.

**Biochemical properties**

Virions contain 13–14 % nucleic acid; 86–87 % protein.

*Genome* consists of RNA; single–stranded; linear; of three parts; largest (or only) genome part the largest 3.324 kb; the 2nd largest 3.183 kb; the 3rd largest 2.535 kb. Genomic nucleic acid isolated by Huttinga and Mosch (1976). Infectivity retained when deproteinised with phenol or detergent. Additional factor required for infectivity; is coat protein plus RNA 1, 2 and 3.

*Sequence database accession code(s):*

U17391 Em(43)_vi:Lr17391 Gb(89)_vi:Lru17391 Lilac ring mottle virus RNA 3 movement protein and coat protein genes, complete cds. 5/95.

*Non–genomic nucleic acid* found in the virions; is subgenomic *m*RNA (RNA-4 1.042 kb). Sub–genomic *m*RNA found in infected cells.

*Virion protein(s)* one; $M_r$ 27500; coat protein. Method of preparation: Huttinga and Mosch (1976).

**Replication**

Replication does not depend on a helper virus.

**Taxonomy and Relationships**

*Ilarvirus: Bromoviridae* (subgroup 7).

Virus(es) with serologically unrelated virions: apple mosaic, elm mottle, prunus necrotic ringspot, spinach latent and tobacco streak viruses.

Serological tests quickly distinguish lilac ring mottle virus from related viruses. Best tests for diagnosis: ten other sap transmissible viruses are known to infect lilac (Van der Meer, 1980, 1981; Waterworth, 1972), thus identification of lilac ring mottle virus by testing in standard indicator plants, like *Chenopodium* and *Nicotiana* species, is unreliable and serological tests are required.

**References**

Bos, L., Huttinga, H. and Maat, D.Z. (1980). *Neth. J. Pl. Path.* **86**: 79.

Fulton, R.W. (1983). *CMI/AAB Descr. Pl. Viruses* No. 275, 3 pp.

Huttinga, H. and Mosch, W.H.M. (1976). *Acta Hort.* **59**: 113.

Huttinga, H., Maat, D.Z. and van der Meer, F.A. (1976). *Acta Bot. neerl.* **25**: 251.

Van der Meer, F.A. (1976). *Acta Hort.* **59**: 105.

Van der Meer, F.A. (1980). *Acta Hort.* **110**: 211.
Van der Meer, F.A. and Huttinga, H. (1979). *CMI/AAB Descr. Pl. Viruses* No. 201, 4 pp.
Van der Meer, F.A., Huttinga, H. and Maat, D.Z. (1976). *Neth. J. Pl. Path.* **82**: 67.
Waterworth, H.E. (1972). *Pl. Dis. Reptr* **56**: 923.

# Lilac ringspot (?) *carlavirus*

**Data collated by A.A. Brunt.**

*Standard acronym:* LacRSV.

First reported in *Syringa vulgaris*; from Japan; by Y. Doi, in Hull *et al.*, 1989.

**Taxonomy and Relationships**

Possible *Carlavirus.*

**References**

Hull, R., Brown, F. and Payne, C. (1989). *Directory and Dictionary of Animal, Bacterial and Plant Viruses*, p.119. MacMillan Reference Books, London.

# Lily mottle *potyvirus*

**Data collated by A.A. Brunt, 1994.**

*Synonyms:* tulip breaking virus — lily strain (Alper *et al.*, 1982; Maeda *et al.*, 1984).

First reported in *Lilium* sp.; from the U.S.A.; by Brierley and Smith (1944); Dekker *et al.* (1993).

**Geographical distribution**

Spreads in Germany, Israel, the Netherlands, and the USA.

**Taxonomy and Relationships**

*Potyvirus: Potyviridae.*

**References**

Alper, M., Koenig, R., Lesemann, D.-E. and Loebenstein, G. (1982). *Phytoparasitica* **10**: 193.
Brierley, P. and Smith, F.F. (1944). *Phytopathology* **34**: 718.
Dekker, E.L., Derks, A.F.L.M., Asjes, C.J., Lemmers, M.E.F., Bol, J.F. and Langeveld, S.A. (1993). *J. gen. Virol.* **74**: 881.
Derks, A.F.L.M., Lemmers, M.E.C. and van Gemmen, B.A. (1994). *Acta Hort.* **377**: 281.
Maeda, T., Inouye, N. and Mitsuhata, K. (1984). *Nogaku Kenkyu* **60**: 135.

# Lily symptomless *carlavirus*

**Data collated by C.J. Asjes, 1986.**

*Synonyms:* lily curl stripe virus, lily virus, lily streak virus. *Standard acronym:* LSV. *Strains:* alstroemeria carlavirus (Brunt and Phillips, 1981; Derks *et al.*, 1991; Phillips and Brunt, 1982).

First reported in *Lilium* spp.; from Oregon, U.S.A.; by Brierley and Smith (1944).

**Natural host range and symptoms**

Symptoms vary seasonally.
The only known natural hosts are from the Liliaceae, and all are infected systemically.
*Lilium longiflorum* — curling and striping in cool conditions.

**Transmission**

Transmitted by a vector; an insect; *Myzus persicae*; Aphididae. Transmitted in a non–persistent manner. Virus transmitted by mechanical inoculation.

**Geographical distribution**

Probably distributed worldwide.

**Experimental host range**

Few (<3) families susceptible.

Diagnostically susceptible

*Lilium longiflorum* (Asiatic hybrids) — usually symptomless, but chlorotic striping and leaf curling in cool conditions.

Maintenance and propagation hosts

Asiatic hybrids of lily.

Assay hosts (**L**ocal lesions or **W**hole plants)

*Lilium longiflorum* (W), *Tulipa gesneriana* cultivars (W).

Monocotyledonous Hosts *(**Susceptible** or Insusceptible)*

***Lilium longiflorum***.

Additional hosts

***Alstroemeria*** spp. and ***Lilium*** spp.

Sources of host-range data

Brierley and Smith (1944); McWhorter and Allen (1964, 1967); Derks and Asjes (1975).

**Purification**

Method: Civerolo *et al.* (1968); Derks and Asjes (1975); Derks and Vink-van den Abeele (1980).

**Particle morphology**

Virions filamentous; usually flexuous; with a clear modal length; of 640 nm; 17–18 nm wide. Axial canal obvious.

**Physical properties**

One sedimenting component in purified preparations; sedimentation coefficient 172 *S*.

**Biochemical properties**

Virions contain 8.3 % nucleic acid (Civerolo *et al.*, 1968).
Genome unipartite; largest (or only) genome part 8 kb.
*Sequence database accession code(s):*
X15343 Em(40)_vi:POLS3TER Gb(84)_vi:POLS3TER Lily symptomless virus RNA 3-terminal region encoding 25kD, 12kD, 7kD, 16kDa proteins and 32kD.
D43801 Em(43)_vi:Llvcp Gb(89)_vi:Llvcp Lily symptomless virus mRNA for coat protein. 12/94 876bp.

**Cytopathology**

Virions found in all parts of the host plant; in cytoplasm (except possibly the meristematic region). Inclusions absent from infected cells.

**Taxonomy and Relationships**

*Carlavirus.*

Virus(es) with serologically related virions: Alstroemeria carlavirus closely, but distinct, also chrysanthemum B, potato M, potato S, carnation latent and passiflora latent viruses.

**Comments**

Lily streak disease is now known to be caused by tulip breaking potyvirus alone or mixed with LSV.

**References**


Allen, T.C. (1971). *Lily Yrbk., N. Am. Lily Soc.* **24**: 29.

Allen, T.C. (1972). *CMI/AAB Descr. Pl. Viruses* No. 96, 4 pp.

Asjes, C.J., de Vos, N. and van Slogteren, D.H.M. (1973). *Neth. J. Pl. Path.* **79**: 23.

Brierley, P. and Smith, F.F. (1944). *Phytopathology* **34**: 529.

Brunt, A.A. and Phillips, S. (1981). *Rep. Glasshouse Crops Res. Inst.* 1979, p. 151.

Civerolo, E.L., Semancik, J.S. and Weathers, L.G. (1968). *Phytopathology* **58**: 1481.

Derks, A.F.L.M. and Asjes, C.J. (1975). *Neth. J. Pl. Path.* **81**: 14.

Derks, A.F.L.M. and Vink-van den Abeele, J.J. (1980). *Neth. J. Pl. Path.* **86**: 239.

Derks, A.F.L.M., Lemmers, M.E.C. and Hollinger, Th.C. (1982). *Rep. Glasshouse Crops Res. Inst.* 1981, 142.

McWhorter, F.P. and Allen, T.C. (1964). *Nature, Lond.* **204**: 604.

McWhorter, F.P. and Allen, T.C. (1967). In: *Easter Lilies*, p.111; eds. D.C. Kiplinger and R.W Langhans. The New York and Ohio Lily Schools.

Phillips, S. and Brunt, A.A. (1986). *Acta Hort.* **177**: 227.


# Lily X *potexvirus*

**Data collated by S. Phillips, 1986.**

*Synonyms:* lily potexvirus. *Standard acronym:* LVX.

First reported in *Lilium formosanum*; from England; by Stone (1976; 1980).

**Natural host range and symptoms**

*Lilium formosanum* — symptomless.

**Transmission**

Virus transmitted by mechanical inoculation; not transmitted by contact between plants.

**Geographical distribution**

Spreads in the UK.

**Experimental host range**

Several (3–9) families susceptible.

Diagnostically susceptible

*Tetragonia expansa* — chlorotic local lesions, becoming green rings.
*Chenopodium quinoa* — minute chlorotic local lesions.
*C. murale* — chlorotic local lesions.

Diagnostically insusceptible

*Trifolium incarnatum, Nicotiana megalosiphon, Nicotiana tabacum.*

Maintenance and propagation hosts

*Chenopodium quinoa, Tetragonia expansa.*

Assay hosts (**L**ocal lesions or **W**hole plants)

*Tetragonia expansa* (L), *Chenopodium quinoa* (L), *Chenopodium murale* (L).

Dicotyledonous Hosts *(**Susceptible** or Insusceptible)*

*Brassica campestris* ssp. *pekinensis*. *Chenopodium amaranticolor*. ***Chenopodium capitatum***. ***Chenopodium murale***. ***Chenopodium quinoa***. *Datura stramonium*. *Glycine max*. ***Gomphrena globosa***. *Lycopersicon esculentum*. ***Nicotiana benthamiana***. ***Nicotiana clevelandii***. *Nicotiana debneyi*. *Nicotiana glutinosa*. *Nicotiana megalosiphon*. *Nicotiana tabacum*. *Ocimum basilicum*. *Petunia hybrida*. *Phaseolus vulgaris*. ***Tetragonia expansa***. *Trifolium incarnatum*. *Vicia faba*. *Vigna unguiculata*.

Additional hosts

Some hosts, recorded as insusceptible by Stone (1980), have subsequently been infected using concentrated purified virion preparations.

Sources of host-range data

Stone (1980).

**Properties of particles in sap**

Electron microscopy: usually necessary to use ISEM to detect virions in sap.

**Purification**

Method: Stone (1980).

**Particle morphology**

Virions filamentous; not enveloped; usually flexuous; with a clear modal length; of 470 nm; 13 nm wide. Axial canal obscure. Basic helix obscure.

**Biochemical properties**

Virions contain 5 % nucleic acid; 95 % protein; 0 % lipid.

*Genome* consists of RNA; single--stranded; linear; unipartite; largest (or only) genome part 7 kb.

*Sequence database accession code(s):*

X15342 Em(40)_vi:POLX3TERM Gb(84)_vi:POLX3TERM Lily virus X RNA 3' terminal region encoding 24kDa and 12KD proteins, 22kDa coat protein. 5/91 1 sequence.

**Replication**

Replication does not depend on a helper virus.

**Cytopathology**

Virions found in all parts of the host plant.

**Taxonomy and Relationships**

*Potexvirus*.

Virus(es) with serologically related virions: dioscorea latent (SDI 9) and commelina X (SDI 10) viruses. Virus(es) with serologically unrelated virions: cactus X, cymbidium mosaic, hydrangea ringspot, narcissus mosaic, nerine X (Agapanthus strain), papaya mosaic (papaya and ullucus strains), potato aucuba mosaic, potato X and white clover mosaic viruses.

ISEM with 'decoration' rapidly distinguishes this virus from other potexviruses. Best tests for diagnosis: ISEM.

**Comments**

Stone (1980) estimated the length of the virions to be about 550 nm, but more recent estimates are of about 470 nm, which is shorter than that recorded for most potexviruses (usually 480–580 nm), but is longer than that recorded for dioscorea latent virus (445 nm), with which this virus shows some serological affinity.

**References**

Stone, O.M. (1976). *Rep. Glasshouse Crops Res. Inst.* 1975, p. 122.
Stone, O.M. (1980). *Acta Hort.* **110**: 59.

# Lima bean golden mosaic bigeminivirus

**Data collated by A.A. Brunt, 1989.**

*Synonyms:* bean lima golden mosaic virus. *Standard acronym:* LGMV.

First reported in
*Phaseolus lunatus*; by Williams, 1976.

**Natural host range and symptoms**

Symptoms persist. Symptoms yellow mosaic.
*Phaseolus lunatus* — yellow mosaic.

**Transmission**

Transmitted by a vector; an insect;
*Bemisia tabaci*; Aleyrodidae. Transmitted in a persistent manner.

**Geographical distribution**

Spreads in Nigeria.

**Taxonomy and Relationships**

Bigeminivirus: *Geminiviridae*.

The possible relationship of this virus to those causing similar diseases in lima beans in India (Capoor and Varma, 1948), Puerto Rico (Bird *et al.*, 1972), Brazil (Costa, 1965), Colombia (Galvez and Cardenas, 1980) and Costa Rica (Gamez, 1971) has not been determined.

**References**

Bird, J., Perez, J.E., Alconero, R., Vakili, N.G. and Melendez, P.L. (1972). *J. Agric. Univ. P.R.* **56**: 64.
Capoor, S.P. and Varma, P.M. (1948). *Curr. Sci.* **17**: 152.
Costa, A.S. (1965). *F.A.O. Pl. Prot. Bull.* **13**: 1.
Galvez, G.E. and Cardenas, M.R. (1980). In: *Bean Production Problems*, pp. 261–289; eds H.F. Schwartz and G.E. Galvez. Colombia: C.I.A.T.
Gamez, R. (1971). *Turrialba* **21**: 22.
Williams, R.J. (1976). *Pl. Dis. Reptr* **60**: 853.

# Lisianthus line pattern (?) *ilarvirus*

**Data collated by A.A. Brunt, 1994.**

First reported in *Eustoma grandiflorum* (synonym *Lisianthus russellianum*; from Italy; by Lisa *et al.* (1990).

**Natural host range and symptoms**

Symptoms *Eustoma grandiflorum* — necrotic line patterns.

**Geographical distribution**

Spreads in Italy.

**Taxonomy and Relationships**

Possible *Ilarvirus: Bromoviridae.*

**References**

Lisa, V., Dellavalle, G., Vaira, A.M., Milne, R.G., Masenga, V., Boccardo, G. and d'Aquilio, M. (1990). *Abstr. 23rd Hort. Congr. Florence*, 1990, Vol.1, No.2196, p.537.

Lisa, V., Vaira, A.M., d'Aquilio, M., Dellavalle, G., Masenga, V., Milne, R.G. and Boccardo, G. (1994). *Acta Hort.* **377**: 81.

# Lisianthus necrosis (?) *necrovirus*

**Data collated by P. Jeyanandarajah, 1991.**

*Standard acronym:* LNV.

First reported in *Eustoma russellianum*; from Japan; by Iwaki *et al.* (1985).

**Natural host range and symptoms**

Symptoms persist.

*Eustoma russellianum* — necrotic lesions and rings, tip necrosis and, in some cultivars, colour breaking.

**Transmission**

Transmitted by a vector; a fungus; *Olpidium* sp.; Chytridiales. Virus transmitted by mechanical inoculation.

**Geographical distribution**

Spreads in Japan.

**Experimental host range**

Many (>9) families susceptible. Experimentally infected plants mostly show necrotic local lesions.

Diagnostically susceptible

*Gomphrena globosa, Nicotiana clevelandii* — local lesions, systemic necrotic spots and/or general necrosis.

*Chenopodium amaranticolor, C. quinoa, Cucumis sativus, Nicotiana glutinosa, N. tabacum, Phaseolus vulgaris, Sesamum indicum, Vigna unguiculata* — necrotic local lesions; not systemic.

Diagnostically insusceptible

*Brassica campestris* ssp. *rapa, Capsicum annuum, Dianthus caryophyllus, Lycopersicon esculentum, Nicotiana debneyi.*

Maintenance and propagation hosts

*Nicotiana clevelandii, Eustoma russellianum.*

Assay hosts (**L**ocal lesions or **W**hole plants)

*Chenopodium amaranticolor* (L), *Phaseolus vulgaris* (L), *Nicotiana tabacum* cv. Xanthi (L), *N. clevelandii* (W).

Dicotyledonous Hosts *(**Susceptible** or Insusceptible)*

*Brassica campestris* ssp. *rapa*. *Capsicum annuum*. ***Chenopodium amaranticolor***. ***Chenopodium quinoa***. ***Cucumis sativus***. ***Cucurbita maxima***. ***Cucurbita moschata***. ***Cucurbita pepo***. *Dianthus caryophyllus*. ***Gomphrena globosa***. *Lycopersicon esculentum*. ***Nicotiana clevelandii***. *Nicotiana debneyi*. ***Nicotiana glutinosa***. ***Nicotiana tabacum***. ***Petunia hybrida***. ***Phaseolus vulgaris***. ***Sesamum indicum***. ***Tetragonia expansa***. ***Vigna unguiculata*** ssp. ***sesquipedalis***. ***Zinnia elegans***.

Sources of host-range data

Iwaki *et al.* (1987).

**Properties of particles in sap**

TIP: 90–95 °C. LIV: more than 63 days. DEP: $\log_{10}$ minus 9–10.

**Purification**

Method: Iwaki *et al.* (1987).

**Particle morphology**

Virions isometric; not enveloped; 30 nm in diameter; without a conspicuous capsomere arrangement.

**Physical properties**

One sedimenting component in purified preparations; sedimentation coefficient 136 *S*. Density 1.333 g $cm^{-3}$ in CsCl. $A_{260}/A_{280}$ ratio 1.61.

**Biochemical properties**

*Genome* consists of RNA; single–stranded.

*Virion protein(s)* one; $M_r$ 35500; coat protein.

**Taxonomy and Relationships**

Possible *Necrovirus*.

Virus(es) with serologically unrelated virions: tobacco necrosis, cucumber necrosis and melon necrotic spot viruses.

**References**

Iwaki, M., Maria, E.R.A., Hanada, K., Onogi, S. and Zenbayashi, R. (1985). *Ann. Phytopath. Soc. Japan* **52**: 355.

Iwaki, M., Hanada, K., Maria, E.R.A. and Onogi, S. (1987). *Phytopathology* **77**: 867.

# Lotus streak (?) *nucleorhabdovirus*

**Data collated by A.A. Brunt and S. Yamashita, 1987.**

First reported in *Nelumbo nucifera* (East Indian Lotus); from Okayama, Japan; by Yamashita *et al.* (1978).

**Natural host range and symptoms**

Symptoms persist.

*Nelumbo nucifera* — chlorotic spots or ring spots, streaking of roots.

**Transmission**

Transmitted by a vector; an insect; *Rhopalosiphum nymphaeae*; Aphididae. Transmitted in a persistent manner.

**Geographical distribution**

Spreads in Japan.

Maintenance and propagation hosts

*Nelumbo nucifera.*

**Properties of particles in sap**

Leaf sap contains few virions. Electron microscopy: necessary to prefix with glutaraldehyde or osmium tetroxide.

**Purification**

Method: Yamashita *et al.* (1985).

**Particle morphology**

Virions rhabdo– or bullet–shaped; with a clear modal length; of 300–340 nm; 90 nm wide. Axial canal obscure. Basic helix obvious; pitch of basic helix 4.5 nm.

**Cytopathology**

Virions found in leaves and shoots; in cytoplasm and in nuclei. Inclusions present in infected cells; are viroplasms (in the nucleus); they contain virions (of nucleoprotein without envelope).

**Taxonomy and Relationships**

Possible *Nucleorhabdovirus: Rhabdoviridae.*

**References**

Yamashita, S., Okamoto, Y., Fujii, S., Doi, Y. and Yora, K. (1978). *Ann. Phytopath. Soc. Japan* **44**: 61.

Yamashita, S., Doi, Y., Yora, K., Okamoto, Y. and Fujii, S. (1985). *Ann. Phytopath. Soc. Japan* **51**: 627.

# Lucerne Australian latent *nepovirus*

**Data collated by A.T. Jones, 1980. Revised 1987.**

*Synonyms:* lucerne latent virus. *Standard acronym:* LALV. *Strains:* lucerne strain, white clover strain.

First reported in *Medicago sativa*; from eastern Victoria, Australia; by Taylor and Smith (1971).

**Natural host range and symptoms**

Symptoms vary seasonally.
*Medicago sativa* — symptomless.
*Trifolium repens* — chlorotic line patterns or symptomless.

**Transmission**

Virus transmitted by mechanical inoculation; transmitted by seed (*i.e.* 1% in *C. quinoa* and lucerne); transmitted by pollen to the seed.

**Ecology and control**

Studies reported by Jones *et al.* (1979).

**Geographical distribution**

Spreads in Australia and New Zealand.

**Experimental host range**

Several (3–9) families susceptible.

Diagnostically susceptible

*Gomphrena globosa* — systemic mottle, vein yellowing.
*Chenopodium amaranticolor* — systemic mottle.
*C. quinoa* — necrotic local lesions; systemic mottling or streaking.
*Pisum sativum* — necrotic local lesions; symptomless systemic infection.

Maintenance and propagation hosts

*Chenopodium quinoa, Pisum sativum.*

Assay hosts (**L**ocal lesions or **W**hole plants)

*Chenopodium amaranticolor (W), C. quinoa (L,W).*

Dicotyledonous Hosts *(**Susceptible** or Insusceptible)*

*Arachis hypogaea*. ***Cajanus cajan***. *Capsicum annuum*. ***Chenopodium album***. ***Chenopodium amaranticolor***. ***Chenopodium foetidum***. *Chenopodium murale*. *Chenopodium quinoa*. ***Cicer arietinum***. *Crotalaria spectabilis*. ***Cucumis sativus***, or *Cucumis sativus*. *Cucurbita maxima*. ***Datura stramonium***. *Glycine max*. ***Gomphrena globosa***. *Ipomoea setosa*. *Lablab purpureus*. *Lotus corniculatus*. ***Lupinus albus***. ***Lupinus angustifolius***. *Lycopersicon esculentum*. ***Medicago sativa***. ***Nicotiana clevelandii***. ***Nicotiana glutinosa***. ***Nicotiana rustica***. ***Nicotiana tabacum***. *Phaseolus lunatus*. ***Phaseolus vulgaris***, or *Phaseolus vulgaris*. *Physalis floridana*. ***Pisum sativum***. *Solanum melongena*. ***Spinacia oleracea***. ***Tetragonia expansa***. ***Trifolium incarnatum***. ***Trifolium pratense***. ***Trifolium repens***. ***Trifolium subterraneum***, or *Trifolium subterraneum*. ***Vicia faba***. ***Vicia sativa***. ***Vigna unguiculata***. *Zinnia elegans*.

Sources of host-range data

Blackstock (1978); Jones *et al.* (1979); Forster *et al.* (1985).

**Properties of particles in sap**

TIP: 50–55 °C. LIV: 8 days. DEP: $\log_{10}$ minus 3–4. Leaf sap contains few virions.

**Purification**

Method: Jones *et al.* (1979).

**Particle morphology**

Virions isometric; not enveloped; 24–27 nm in diameter; angular in profile; without a conspicuous capsomere arrangement.

**Physical properties**

Three sedimenting components in purified preparations; sedimentation coefficient of the fastest 133 *S* (B); of the other(s) 128 *S* (M), or 56 *S* (T). $A_{260}/A_{280}$ ratio 1.71 (unfractionated M and B).

**Biochemical properties**

Virions contain 41 % nucleic acid (B), or 39 % nucleic acid (M), or 0 % nucleic acid (T); 59 % protein (B), or 61 % protein (M), or 100 % protein (T).

*Genome* consists of RNA; single–stranded; of two parts; largest (or only) genome part the larger 7.9 kb; the 2nd largest 6.8 kb.

*Virion protein(s)* one; $M_r$ 55000. Method of preparation: Jones *et al.* (1979).

**Taxonomy and Relationships**

*Nepovirus: Comoviridae.*

**References**

Blackstock, J.M. (1978). *Aust. J. agric. Res.* **29**: 291.
Forster, R.L.S. and Morris-Krsinich, B. (1985). *Ann. appl. Biol.* **107**: 449.
Jones, A.T. and Forster, R.L.S. (1980). *CMI/AAB Descr. Pl. Viruses* No. 225, 4 pp.
Jones, A.T., Forster, R.L.S. and Mohamed, N.A. (1979). *Ann. appl. Biol.* **92**: 49.
Taylor, R.H. and Smith, P.R. (1971). *Rep. Vic. Pl. Res. Inst.* **5**: 20.

# Lucerne Australian symptomless (?) *nepovirus*

**Data collated by A.T. Jones, 1987.**

*Synonyms:* lucerne Australian latent virus SM strain. *Standard acronym:* LASV.

First reported in *Medicago sativa*; from Melbourne, Australia; by Blackstock (1978).

**Natural host range and symptoms**

Symptoms none (but infection persists).
*Medicago sativa* — symptomless.

**Transmission**

Virus transmitted by mechanical inoculation; transmitted by seed (up to 6% in *Chenopodium quinoa*).

**Geographical distribution**

Spreads in Australia.

**Experimental host range**

Several (3–9) families susceptible.

Diagnostically susceptible

*Chenopodium amaranticolor* — faint systemic mottle.
*C. murale* — chlorotic or necrotic local lesions; then systemic necrotic spots and epinasty of apical leaves.
*C. quinoa* — faint systemic mottle.
*Gomphrena globosa* — red rings spots; not systemic.
*Nicotiana benthamiana, N. clevelandii, N. tabacum* cvs White Burley, Samsun NN, *Pisum sativum, Spinacia oleracea* — symptomless.
*Nicotiana debneyi* and *Petunia hybrida* — local symptomless infection.

Diagnostically insusceptible

*Brassica rapa, B. nigra, Catharanthus roseus, Cucumis sativus, Datura stramonium, Lycopersicon esculentum, Nicotiana glutinosa, N. rustica, Phaseolus vulgaris, Tetragonia expansa, Vicia faba.*

Maintenance and propagation hosts

*Chenopodium quinoa.*

Assay hosts (**L**ocal lesions or **W**hole plants)

*Chenopodium quinoa* (L); *C. murale* (L).

Dicotyledonous Hosts *(**Susceptible** or Insusceptible)*

*Brassica campestris* ssp. *rapa*. ***Chenopodium amaranticolor***. ***Chenopodium murale***. ***Chenopodium quinoa***. *Cucumis sativus*. *Datura stramonium*. ***Gomphrena globosa***.

*Lycopersicon esculentum*. ***Nicotiana benthamiana***. ***Nicotiana clevelandii***. ***Nicotiana debneyi***. *Nicotiana glutinosa*. *Nicotiana rustica*. ***Nicotiana tabacum***. ***Petunia hybrida***. *Phaseolus vulgaris*. ***Pisum sativum***. ***Spinacia oleracea***. *Tetragonia expansa*. *Vicia faba*.

**Properties of particles in sap**

TIP: 50–55 °C. LIV: greater than 24 days. DEP: $\log_{10}$ minus 2–3. Leaf sap contains few virions. Electron microscopy: most virions are either penetrated by stain or disrupted, when stained with 2% AM, MT or UA.

**Purification**

Method: Remah *et al.* (1986).

**Particle morphology**

Virions isometric; not enveloped; 26–28 nm in diameter; angular in profile.

**Physical properties**

One sedimenting component in purified preparations; sedimentation coefficient 130 *S*.

**Biochemical properties**

*Genome* consists of RNA; single–stranded; of two parts (plus 2 minor uncharacterized components of lengths 0.85 kb and 0.56 kb); largest (or only) genome part the larger 7 kb; the 2nd largest 3.9 kb. Genomic nucleic acid isolated by Remah *et al.* (1986). Infectivity lost when deproteinised with proteases.

*Virion protein(s)* two; $M_r$ of the larger 40000. $M_r$ of 2nd largest 26000. Method of preparation: Remah *et al.* (1986).

**Replication**

Replication does not depend on a helper virus.

**Taxonomy and Relationships**

Possible *Nepovirus: Comoviridae*.

Virus(es) with serologically unrelated virions: arracacha B (strains O and T), bean pod mottle, broad bean mottle, broad bean true mosaic, broad bean stain, broad bean wilt (serotype 1 and 2), cowpea mosaic, crimson clover latent, cucumber mosaic, lucerne transient streak, rubus Chinese seed-borne, strawberry latent ringspot and tobacco ringspot viruses.

The virus is similar to strawberry latent ringspot (Murant, 1974) and its relative rubus Chinese seed-borne virus (Barbara *et al.*, 1985), which are tentative members of the nepovirus group (Harrison and Murant, 1977; Barbara *et al.*, 1985).

**References**

Barbara, D.J., Ashby, S.C. and McNamara, D.G. (1985). *Ann. appl. Biol.* **107**: 45.
Blackstock, J.M. (1978). *Aust. J. agric. Res.* **29**: 291.
Harrison, B.D. and Murant, A.F. (1977). *CMI/AAB Descr. Pl. Viruses* No. 185, 4 pp.
Murant, A.F. (1974). *CMI/AAB Descr. Pl. Viruses* No. 126, 4 pp.
Remah, A., Jones, A.T. and Mitchell, M.J. (1986). *Ann. appl. Biol.* **109**: 307.

# Lucerne enation (?) *nucleorhabdovirus*

**Data collated by P.A. Signoret, B.C. Alliot, 1981.**

*Standard acronym:* LEV.

First reported in *Medicago sativa*; from France; by Alliot *et al.* (1972); Leclant *et al.* (1973).

**Natural host range and symptoms**

Symptoms persist.
*Medicago sativa, Vicia faba* — enations on main leaf veins.

**Transmission**

Transmitted by a vector; an insect; *Aphis craccivora*; Aphididae. Transmitted in a persistent manner. Virus transmitted by grafting; not transmitted by seed; not transmitted by pollen.

**Geographical distribution**

Spreads in France (the south and the Mediterranean). Found, but with no evidence of spread, in Morocco, Spain, Italy, Yugoslavia, Romania.

**Experimental host range**

Few (<3) families susceptible.

Diagnostically susceptible

*Vicia faba* — stem enations.

Dicotyledonous Hosts (***Susceptible*** *or Insusceptible)*

***Medicago sativa***. ***Vicia faba***.

**Particle morphology**

Virions rhabdo– or bullet–shaped; enveloped; of 250 nm; 82–89 nm wide.

**Cytopathology**

Virions found in phloem; in the perinuclear space. Inclusions absent from infected cells.

**Taxonomy and Relationships**

*Nucleorhabdovirus: Rhabdoviridae.*

**References**

Alliot, B. and Signoret, P.A. (1972). *Phytopath. Z.* **74**: 69.
Alliot, B., Giannotti, J. and Signoret, P.A. (1972). *C. r. Acad. Sci. Paris* **274**: 1974.
Leclant, F., Alliot, B. and Signoret, P.A. (1973). *Annls. Phytop.* **5(4)**: 441.

# Lucerne transient streak *sobemovirus*

**Data collated by R.L.S. Forster, 1982. Revised 1983.**

*Standard acronym:* LTSV.

First reported in *Medicago sativa*; from Victoria, Australia; by Blackstock (1974; 1978).

**Natural host range and symptoms**

Symptoms disappear soon after infection.
*Medicago sativa* — vein yellowing.

**Transmission**

Virus transmitted by mechanical inoculation; not transmitted by seed; not transmitted by pollen.

**Geographical distribution**

Spreads in Australia, Canada, and New Zealand.

**Experimental host range**

Several (3–9) families susceptible.

Diagnostically susceptible

*Chenopodium amaranticolor* — necrotic local lesions; systemic chlorotic blotching (Australian, but not Ne Zealand isolates).

*Chenopodium quinoa* — necrotic and chlorotic local lesions; systemic mosaic.

*Medicago scutellata* — necrotic local lesions.

*Pisum sativum* — necrotic local lesions (New Zealand isolates), or symptomless (Australian isolates); no systemic infection.

*Nicotiana clevelandii* — chlorotic local lesions; systemic mosaic.

Diagnostically insusceptible

*Cucumis sativus, Gomphrena globosa, Phaseolus vulgaris, Nicotiana glutinosa, Vigna unguiculata.*

Maintenance and propagation hosts

*Chenopodium quinoa, Nicotiana clevelandii.*

Assay hosts (**L**ocal lesions or **W**hole plants)

*Chenopodium amaranticolor* (L), *C. quinoa* (L), *Pisum sativum* (L).

Dicotyledonous Hosts *(**Susceptible** or Insusceptible)*

*Arachis hypogaea. Brassica campestris* ssp. *pekinensis. Calendula officinalis. Capsicum annuum. Capsicum frutescens.* ***Chenopodium album. Chenopodium amaranticolor. Chenopodium quinoa. Cicer arietinum.*** *Crotalaria spectabilis. Cucumis sativus. Cucurbita maxima. Datura stramonium. Dianthus barbatus. Glycine max. Gomphrena globosa. Lablab purpureus. Lactuca sativa. Lotus corniculatus.* ***Lupinus albus. Lupinus angustifolius.*** *Lycopersicon esculentum. Macroptilium lathyroides. Macrotyloma uniflorum.* ***Medicago sativa. Nicotiana clevelandii.*** *Nicotiana glutinosa.* ***Nicotiana rustica.*** *Nicotiana sylvestris.* ***Nicotiana tabacum. Petunia hybrida.*** *Phaseolus lunatus. Phaseolus vulgaris. Physalis floridana.* ***Pisum sativum.*** *Solanum melongena. Solanum nigrum.* ***Spinacia oleracea. Trifolium incarnatum.*** *Trifolium pratense. Trifolium repens. Trifolium subterraneum.* ***Vicia faba***, or *Vicia faba.* ***Vicia sativa.*** *Vigna angularis. Vigna unguiculata. Vigna unguiculata* ssp. *cylindrica.* ***Zinnia elegans.***

Sources of host-range data

Blackstock (1978); Forster and Jones (1979).

**Properties of particles in sap**

TIP: 70 °C. LIV: 28 days. DEP: $\log_{10}$ minus 5. Leaf sap contains few virions, or contains many virions.

**Purification**

Method: Forster and Jones (1979).

**Particle morphology**

Virions isometric; not enveloped; 27–28 nm in diameter; angular in profile; without a conspicuous capsomere arrangement (ring-like subunits on some virions after freeze-dehydration and staining).

**Physical properties**

One sedimenting component in purified preparations; sedimentation coefficient 112 *S*. Density 1.37 g $cm^{-3}$ in CsCl.

**Biochemical properties**

Virions contain 18 % nucleic acid; 82 % protein; 0 % lipid.

*Genome* consists of RNA; single–stranded; linear; unipartite; largest (or only) genome part 4.23 kb. 5' terminus of RNA has a VPg. Infectivity lost when deproteinised with proteases; retained when deproteinised with phenol or detergent. Poly A region absent.

*Sequence database accession code(s):*

D00341 Em(40)_vi:MTSCA Gb(84)_vi:MTSCA Lucerne transient streak virus(LTSV-C) RNA2, complete. 4/90 322bp.

X01984 Em(40)_vi:LTSVARN2 Gb(84)_vi:LTSVARN2 Lucerne transient streak virus (LTSV-A) RNA 2 sequence. 7/93 324bp.

X01985 Em(40)_vi:LTSVNRN2 Gb(84)_vi:LTSVNRN2 Lucerne transient streak virus (LTSV-N) RNA2 sequence. 7/93 324bp. 3 sequences.

*Non–genomic nucleic acid* found in the virions; is subgenomic *m*RNA and satellite RNA. Sub–genomic *m*RNA found in infected cells; codes for the coat protein and has $M_r$ 35000.

*Virion protein(s)* one (with a minor component that is probably a hydrolysis product of it); $M_r$ 32000; coat protein. $M_r$ of 2nd largest 29000. Method of preparation: Forster and Jones (1979).

*Virus–coded non–virion proteins* identified by genomic sequence analysis; three proteins found. $M_r$ of the largest 105000. $M_r$ of 2nd largest 78000. $M_r$ of 3rd 18000.

**Replication**

Replication does not depend on a helper virus. Acts as helper for a satellite RNA.

**Cytopathology**

Virions found in all parts of the host plant; in cytoplasm and in cell vacuoles. Inclusions absent from infected cells.

**Taxonomy and Relationships**

*Sobemovirus.*

Virus(es) with serologically related virions: subterranean clover mottle virus, but distantly.

**References**

Blackstock, J.M. (1974). *Aust. Pl. Path. Soc. Newsl.* **3**: 6.

Blackstock, J.M. (1978). *Aust. J. agric. Res.* **29**: 291.

Forster, R.L.S. and Jones, A.T. (1979). *Ann. appl. Biol.* **93**: 181.

Forster, R.L.S. and Jones, A.T. (1980). *CMI/AAB Descr. Pl. Viruses* No. 224, 4 pp.

Jones, A.T., Mayo, M.A. and Duncan G.H. (1983). *J. gen. Virol.* **64**: 1167.

Morris-Krsinich, B.A.M., Forster, R.L.S. and Mossop, D.W. (1983). *Virology* **124**: 349.

Tien, P., Davies, C., Hatta, T. and Francki, R.I.B. (1981). *FEBS Lett.* **132**: 353.

# Lupin leaf curl (?) bigeminivirus

**Data collated by A.A. Brunt, 1991.**

*Standard acronym:* LLCV.

First reported in *Lupinus hartwegii*; from India; by Verma (1974).

**Natural host range and symptoms**

Symptoms persist.

*Lupinus hartwegii* — thickening of veins, downcurling of leaflets and petioles, greatly decreased flowering and severe plant stunting.

**Transmission**

Transmitted by a vector; an insect; *Bemisia tabaci*; Aleyrodidae. Virus not transmitted by mechanical inoculation.

**Geographical distribution**

Spreads in India.

**Experimental host range**

Few (<3) families susceptible.

Diagnostically susceptible

*Lupinus hartwegii* — downcurling of leaves and petioles, thickening of veins and stunting.

Assay hosts (**L**ocal lesions or **W**hole plants)

*Lupinus hartwegii* (W).

**Taxonomy and Relationships**

Possible bigeminivirus: *Geminiviridae*.

**Comments**

This uncharacterized virus has been included by the ICTV as a possible bigeminivirus (Francki *et al.*, 1991).

**References**

Francki, R.I.B., Fauquet, C.M., Knudson, D.L. and Brown, F. (1991). *Fifth Rep. ICTV. Arch Virol.* Suppl. 2, p. 176.

Jones, R.A.C. and McLean, G.D. (1989). *Ann. appl. Biol.* **114**: 609.

Verma, V.S. (1974). *Gartenbauwiss.* **39**: 53.

# Lupin yellow vein (?) rhabdovirus

**Data collated by A.A. Brunt, 1991.**

*Synonyms:* lupin rhabdovirus. *Standard acronym:* LYVV.

First reported in *Lupinus polyphyllus x L. arboreus*; from the U.K.; by Tomlinson *et al.* (1972).

**Natural host range and symptoms**

*Lupinus polyphyllus x L. arboreus* — chlorotic banding of leaf veins.

**Geographical distribution**

Spreads in the UK.

**Particle morphology**

Virions rhabdo– or bullet–shaped.

**Taxonomy and Relationships**

Possible unassigned rhabdovirus.

**References**

Tomlinson, J.A., Webb, M.J.W. and Faithfull, E.M. (1972). *Ann. appl. Biol.* **71**: 127.

# Lychnis (?) *potexvirus*

**Data collated by A.A. Brunt, 1994.**

First reported in *Lychnis alba*; from Canada; by R.I. Hamilton in Koenig (1985).

**Taxonomy and Relationships**

Possible *Potexvirus*.
Virus(es) with serologically related virions: narcissus mosaic virus.

**References**

Koenig, R. (1985). *Acta Hort.* **164**: 21.

# Lychnis ringspot *hordeivirus*

**Data collated by A.O. Jackson, 1991.**

*Standard acronym:* LRSV. *Strains:* Salinas isolate, European isolate (Hungary), Mentha.

First reported in *Lychnis divaricata*; from Salinas, California, U.S.A. in seed imported from Europe.

**Natural host range and symptoms**

Symptoms persist. Symptoms mosaic, ringspot, necrosis.

*Lychnis divaricata, L. chalcedonica, L. coronaria, L. haageana* — systemic ringspots, mosaic and necrosis.

**Transmission**

Virus transmitted by mechanical inoculation; probably transmitted by grafting; transmitted by seed.

**Geographical distribution**

Spreads in Hungary.

**Experimental host range**

Many (>9) families susceptible. Experimentally infected plants mostly show ringspots and mosaic.

Diagnostically susceptible

*Capsicum frutescens, Chenopodium album, C. amaranticolor, Nicotiana tabacum* — chlorotic or necrotic local lesions.

*Callistephus chinensis, Nicotiana clevelandii, N. glutinosa, Petunia hybrida, Tetragonia expansa* — systemic.

Diagnostically insusceptible

*Centaurea cyanus, Lactuca sativa, Phaseolus vulgaris, Vigna sinensis, Lycopersicon esculentum.*

Maintenance and propagation hosts

*Spinacia oleracea.*

Assay hosts (**L**ocal lesions or **W**hole plants)

*Chenopodium amaranticolor* (L).

Dicotyledonous Hosts *(**Susceptible** or Insusceptible)*

***Beta patellaris. Capsella bursa–pastoris. Chenopodium album. Chenopodium amaranticolor. Chenopodium capitatum. Chenopodium murale. Gypsophila elegans.*** *Lactuca sativa.* ***Nicotiana clevelandii. Nicotiana glutinosa. Nicotiana rustica. Nicotiana tabacum. Petunia hybrida.*** *Phaseolus vulgaris. Phytolacca americana. Solanum nigrum.* ***Spinacia oleracea. Stellaria media. Tetragonia expansa.*** *Theobroma cacao.*

**Purification**

Method: Hunter *et al.* (1989).

**Particle morphology**

Virions rod–shaped; not enveloped; usually straight; with no clear modal length; of 100–175 nm; 25 nm wide. Basic helix obvious.

**Biochemical properties**

Virions contain 5 % nucleic acid; 95 % protein; 0 % lipid.

*Genome* consists of RNA; single–stranded; linear; of three parts; largest (or only) genome part the largest 3.8 kb; the 2nd largest 3.3 kb; the 3rd largest 3 kb. Genomic nucleic acid isolated by Hunter *et al.* (1989).

*Virion protein(s)* one; $M_r$ 23000. Method of preparation: SDS gel electrophoresis.

**Replication**

Replication does not depend on a helper virus.

**Taxonomy and Relationships**

*Hordeivirus.*

Virus(es) with serologically unrelated virions: barley stripe mosaic and poa semilatent viruses.

No detectable cross-hybridization between lychnis ringspot virus, barley stripe mosaic virus and poa semilatent virus genomes and cDNAs transcribed from them. Best tests for diagnosis: ELISA.

**References**

Beczner, L., Hamilton, R.I. and Rochon, D.M. (1992). *Intervirology* **33**: 49.

Bennett, C.W. (1959). *Phytopathology* **49**: 708.

Hunter, B.G., Smith, J., Fattouh, F. and Jackson, A.O. (1989). *Intervirology* **30**: 18.

Jackson, A.O. and Lane, L.C. (1981). In: *Handbook of Virus Infections and Comparative Diagnosis*, p. 565; ed. E. Kurstak, Elsevier/North Holland Biomedical Press, Amsterdam.

# Maclura mosaic macluravirus

**Data collated by N. Plese, 1991.**

*Standard acronym:* MacMV.

First reported in *Maclura pomifera*; from Zagreb, Croatia, Yugoslavia; by Plese and Miličič (1973).

**Natural host range and symptoms**

Symptoms persist and vary seasonally. Symptoms mosaic.

*Maclura pomifera* — interveinal mosaic, occasional vein-banding and leaf distortion especially in spring and autumn (Plese and Miličič, 1973; Plese and Stefanac, 1976).

**Transmission**

Experimentally transmitted by the aphid *Myzus persicae* and is non-persistent. Virus transmitted by mechanical inoculation; transmitted by grafting.

**Geographical distribution**

Spreads in the former Yugoslavia.

**Experimental host range**

Several (3–9) families susceptible. Experimentally infected plants mostly show local mild symptoms or no symptoms.

Diagnostically susceptible

*Amaranthus caudatus, Chenopodium album* — chlorotic local lesions.
*C. amaranticolor* — circular necrotic local lesions.
*C. quinoa* — latent local infection.
*Nicotiana clevelandii* — mild systemic mosaic.
*N. megalosiphon* — latent systemic infection.
*Tetragonia expansa* — diffuse chlorotic local lesions.

Diagnostically insusceptible

*Brassica campestris* ssp. *rapa*, *Datura stramonium, Petunia hybrida, Phaseolus vulgaris.*

Maintenance and propagation hosts

*Nicotiana clevelandii, Tetragonia expansa.*

Assay hosts (**L**ocal lesions or **W**hole plants)

*Chenopodium amaranticolor* (L).

Dicotyledonous Hosts *(**Susceptible** or Insusceptible)*

***Amaranthus caudatus***. ***Atriplex hortensis***. *Brassica campestris* ssp. *rapa*. ***Celosia cristata***. ***Chenopodium album***. ***Chenopodium amaranticolor***. ***Chenopodium ambrosioides***. ***Chenopodium foetidum***. ***Chenopodium foliosum***. ***Chenopodium murale***. ***Chenopodium quinoa***. *Cucumis sativus*. *Datura stramonium*. ***Maclura pomifera***. ***Nicotiana clevelandii***. ***Nicotiana megalosiphon***. ***Nicotiana tabacum***. *Ocimum basilicum*. *Petunia hybrida*. *Phaseolus vulgaris*. ***Tetragonia expansa***. ***Vigna unguiculata***.

Sources of host-range data

Plese and Miličič (1973); Plese and Stefanac (1976); Plese *et al.* (1979).

**Properties of particles in sap**

TIP: 65–67 °C. LIV: 3 days (at 20°C). DEP: $\log_{10}$ minus 3–4. Leaf sap contains few virions.

**Purification**

Method: Plese *et al.* (1979).

**Particle morphology**

Virions filamentous; not enveloped; usually flexuous; with a clear modal length; of 672 nm; 13–16 nm wide. Axial canal obscure. Basic helix obscure.

**Physical properties**

One sedimenting component in purified preparations; sedimentation coefficient 155 *S*. Density 1.307 g $cm^{-3}$ in CsCl. $A_{260}/A_{280}$ ratio 1.04.

**Biochemical properties**

Virions contain 3–4 % nucleic acid; 96–97 % protein.

*Genome* consists of RNA; single–stranded.

*Virion protein(s)* one; $M_r$ 45000. Method of preparation: Koenig *et al.* (1970); Koenig (1972).

**Cytopathology**

Virions found in leaves; in cytoplasm. Inclusions present in infected cells; are amorphous X–bodies and pinwheels.

**Taxonomy and Relationships**

Macluravirus.

Virus(es) with serologically related virions: bean yellow mosaic virus (Plese *et al.*, 1979). Weak reactions were observed with antisera to poplar mosaic and narcissus latent carlaviruses (Plese *et al.*, 1979; Mowat *et al.*, 1991). Virus(es) with serologically unrelated virions: bean common mosaic, bidens mottle, clover yellow vein, gloriosa stripe mosaic, lettuce mosaic, narcissus yellow stripe, pea necrosis, pea seed borne mosaic, plum pox, potato Y, tobacco etch and turnip mosaic viruses.

**References**


Koenig, R. (1972). *Virology* **50**: 263.

Koenig, R., Stegemann, H., Francksen, H. and Paul, H.L. (1970). *Biochim. biophys. Acta* **207**: 184.

Koenig, R., Lesemann, D.-E. and Plese, N. (1981). *CMI/AAB Descr. Pl. Viruses* No. 239, 3 pp.

Mowat, W.P., Dawson, S., Duncan, G.H. and Robinson, D.J. (1991). *Ann. appl. Biol.* **119**: 31.

Plese, N. and Miličič, D. (1973). *Phytopath. Z.* **77**: 178.

Plese, N. and Stefanac, Z. (1976). *Mitt. Biol. BundesAnst. Land. Forstw.* **170**: 47.

Plese, N. and Wrischer, M. (1978). *Acta bot. Croat.* **37**: 47.

Plese, N., Koenig, R., Lesemann, D.-E. and Bozarth, R.F. (1979). *Phytopathology* **69**: 471.


# Maize chlorotic dwarf *waikavirus*

**Data collated by R.E. Gingery, 1984.**

*Synonyms:* Ohio corn stunt agent. *Standard acronym:* MCDV.

First reported in *Zea mays*; from Ohio, U.S.A; by Bradfute *et al.* (1972).

**Natural host range and symptoms**

Symptoms persist.

*Sorghum halepense* — mild vein clearing.

*Zea mays* — stunting, reddening and yellowing, tertiary vein banding.

**Transmission**

Transmitted by a vector; an insect; *Graminella nigrifrons* most important, but also *Graminella sonora and Exitianus exitiosus*; Cicadellidae. Transmitted in a semi–persistent manner. Virus lost by the vector when it moults; does not multiply in the vector; not transmitted congenitally to the progeny of the vector; not transmitted by mechanical inoculation; not transmitted by contact between plants; not transmitted by seed; not transmitted by pollen.

**Geographical distribution**

Spreads in the USA (the south).

**Experimental host range**

Few (<3) families susceptible.

Diagnostically susceptible

*Digitaria sanguinalis Panicum miliaceum, Pennisetum glaucum, Sorghum vulgare* var. *sudanense* — symptoms mild or absent.

*Setaria faberii, S. lutescens.*

*Sorghum bicolor, S. halepense* — mild vein clearing.

*Triticum aestivum* — symptomless.

*Zea mays* — stunting, yellowing and reddening, tertiary vein banding.

Diagnostically insusceptible

*Avena sativa, Hordeum vulgare, Oryza sativa, Saccharum officinarum, Secale cereale.*

Maintenance and propagation hosts

*Zea mays.*

Assay hosts (**L**ocal lesions or **W**hole plants)

*Zea mays* (W).

Monocotyledonous Hosts *(**Susceptible** or Insusceptible)*

*Agropyron repens. Avena sativa. Cynodon dactylon. Dactylis glomerata. Eragrostis cilianensis. Hordeum vulgare. Lolium perenne. Oryza sativa. Paspalum dilatatum. Phleum pratense. Saccharum officinarum. Secale cereale.* ***Sorghum bicolor****. Stenotaphrum secundatum. Triticum aestivum.* ***Zea mays****.*

Sources of host-range data

Nault *et al.* (1976).

**Purification**

Method: Louie *et al.* (1974); Gingery (1976).

**Particle morphology**

Virions isometric; not enveloped; 30 nm in diameter; angular in profile; with a conspicuous capsomere arrangement.

**Physical properties**

One sedimenting component in purified preparations; sedimentation coefficient 183 $S$ ($\pm$ 6 $S$). Density 1.507 g cm$^{-3}$ in CsCl (unfixed).

**Biochemical properties**

Virions contain 36 % nucleic acid; 64 % protein; 0 % lipid.

*Genome* consists of RNA; single–stranded; linear; unipartite; largest (or only) genome part 10.67 kb. Genomic nucleic acid isolated by Gingery (1976). Base composition 24 % G; 30 % A; 17 % C; 29 % U.

*Virion protein(s)* two; $M_r$ of the larger 30000; coat protein. $M_r$ of 2nd largest 18000.

**Cytopathology**

Virions found in leaves and phloem but less frequently in bundle sheath cells; in cytoplasm. Inclusions present in infected cells; are viroplasms and unusual in shape; striated sheets and dense granular bodies; they contain virions (only in the dense granular inclusions). Other cellular changes: numerous cytoplasmic vesicles.

**Taxonomy and Relationships**

*Waikavirus: Sequiviridae.*

Virus(es) with serologically related virions: rice tungro *waikavirus*.

**References**


Bradfute, O.E., Gingery, R.E., Gordon, D.T. and Nault, L.R. (1972). *J. Cell Biol.* **55**: 25*a*.

Derrick, K.S. and Brlansky, R.H. (1976). *Phytopathology* **66**: 815.

Ge, X., Gordon, D.T. and Gingery, R.E. (1989). *Phytopathology* **79**: 1157.

Ge, X., Gordon, D.T. and Gingery, R.E. (1989). *Phytopathology* **79**: 1195.

Gingery, R.E. (1976). *Virology* **73**: 311.

Gingery, R.E. (1988). In: *The Plant Viruses*, Vol.3, pp.259–272; Ed. R. Koenig. Plenum Press, New York.

Gingery, R.E. (1989). *Phytopathology* **79**: 1157.

Gingery, R.E., Bradfute, D.E., Gordon, D.T. and Nault, L.R. (1978). *CMI/AAB Descr. Pl. Viruses* No. 194, 4 pp.

Gingery, R.E., Gordon, D.T., Nault, L.R. and Bradfute, O.E. (1981). In: *Handbook of Plant Virus Infections and Comparative Diagnosis*, p. 19; ed. E. Kurstak. Elsevier/North Holland Biomedical Press.

Gordon, D.T. and Nault, L.R. (1977). *Phytopathology* **67**: 27.

Hunt, R.E., Nault, L.R. and Gingery, R.E. (1988). *Phytopathology* **78**: 498.

Louie, R., Knoke, J.K. and Gordon, D.T. (1974). *Phytopathology* **64**: 1455.

Maroon, C.M., Gordon, D.T. and Gingery, R.E. (1989). *Phytopathology* **79**: 1157.

Nault, L.R., Gordon, D.T., Robertson, D.C. and Bradfute, D.E. (1976). *Pl. Dis. Reptr* **60**: 374.

Nault, L.R., Styer, W.E., Knoke, J.K. and Pitre, H.N. (1973). *J. Econ. Ent.* **66**: 1271.

Pirone, T.P., Bradfute, O.E., Freytag, P.H., Lung, M.C.Y. and Poneleit, G.G. (1972). *Pl. Dis. Reptr* **56**: 652.

Reeves, J.T., Jackson, A.O., Paschke, J.D. and Lister, R.M. (1978). *Pl. Dis. Reptr* **62**: 667.


# Maize chlorotic mottle *machlomovirus*

**Data collated by A.A. Brunt and D.T. Gordon, 1987. Revised 1991 by A.A. Brunt.**

*Standard acronym:* MCMV. *Strains:* Kansas serotypes 1 and 2, Peru serotype.

First reported in *Zea mays*; from Peru; by Hebert and Castillo (1973).

**Natural host range and symptoms**

*Zea mays* — systemic chlorotic leaf mottling; in mixtures with any of several potyviruses also causes necrosis, stunting and death.

**Transmission**

Transmitted by a vector; an insect; *Diabrotica* spp. and thrips; Coleoptera, or Thysanoptera. Transmitted in a non–persistent manner. Virus transmitted by mechanical inoculation; transmitted by seed (detected in 17 of 42000 seedlings from 25 seed lots).

**Ecology and control**

Studies reported by Nault *et al.* (1978).

**Geographical distribution**

Spreads in Argentina, Mexico, Peru, and the USA (Kansas, Nebraska and Hawaii).

Diagnostically susceptible

*Zea mays* — systemic chlorotic leaf mottling.
*Triticum aestivum* — systemic chlorotic leaf spotting and mottling.

Diagnostically insusceptible

*Sorghum bicolor.*

Maintenance and propagation hosts

Maize hybrids H84, N28 HT, N.28 and Oh 28, maize hybrid PM 205, *Zea mays* cv. Aristogold Bantam Evergreen; see Uyemoto *et al.* (1980).

Assay hosts (**L**ocal lesions or **W**hole plants)

Maize hybrids H84, N28 HT, N.28 and Oh 28 (W), maize hybrid PM 205 (W), *Zea mays* cv. Aristogold Bantam Evergreen (W).

Monocotyledonous Hosts *(**Susceptible** or Insusceptible)*

***Bromus inermis*. *Bromus mollis*. *Bromus racemosus*. *Bromus secalinus*. *Bromus tectorum*. *Hordeum vulgare*. *Panicum capillare*. *Panicum dichotomiflorum*. *Panicum maximum*. *Panicum miliaceum*.** *Sorghum bicolor.* ***Triticum aestivum*. *Zea mays*.**

**Properties of particles in sap**

TIP: 80–85 °C (in maize sap). LIV: 12–33 days. DEP: $\log_{10}$ minus 6–8. Leaf sap contains few virions.

**Purification**

Method: R.E. Gingery in Gordon *et al.* (1984); virions are stabilized by divalent cations.

**Particle morphology**

Virions isometric; not enveloped; *c.* 30 nm in diameter (in PTA–K, but 33 nm in UA); angular in profile; without a conspicuous capsomere arrangement.

**Physical properties**

One sedimenting component in purified preparations; sedimentation coefficient 109 *S*. Density 1.365 g $cm^{-3}$ in CsCl.

**Biochemical properties**

Virions contain 18 % nucleic acid; 82 % protein.

*Genome* consists of RNA; single–stranded; linear; unipartite; largest (or only) genome part 4.4 kb. Genomic nucleic acid isolated by Lommel *et al.* (1991). Base composition 22–25 % G; 24–26 % A; 15–16 % C; 34–37 % U. 5' terminus of RNA has a methylated nucleotide cap. Poly A region absent.

*Sequence database accession code(s):* Em_vi:MCMVXX — genome, 4,437bp.

*Features of the genome:* the genome has four ORFs encoding proteins of 32 kDa, 50 kDa (possibly the polymerase), 9kDa and 25.1 kDa (coat protein).

*Non–genomic nucleic acid* found in the virions; is subgenomic *m*RNA (encodes coat protein). 2 virus specified dsRNA species found in infected cells. Size of largest virus specified dsRNA 4.4 kbp (pairs); 2nd largest 1.1 kbp (pairs).

*Virion protein(s)* one; $M_r$ 25100 (238 amino acid residues); coat protein.

**Taxonomy and Relationships**

*Machlomovirus* (a monotypic genus).

Virus(es) with serologically unrelated virions: brome mosaic, cocksfoot mild mottle, cocksfoot mottle, cynosurus mottle, holcus transitory mottle, maize chlorotic dwarf, maize rayado fino, maize streak, maize stripe, maize white line, molinia streak, panicum mosaic, phleum mottle, rice yellow mottle and sowbane mosaic viruses.

The virus resembles carnation mottle and turnip crinkle carmoviruses in the amino acid sequence of some of its proteins, in its genome organisation and gene expression strategy, but has an additional ORF and lacks a second sub-genomic RNA.

**References**


Goldberg, K.B. and Brakke, M.K. (1987). *Phytopath.* **77**: 162.

Gordon, D.T., Bradfute, O.E., Gingery, R.E., Nault, L.R. and Uyemoto, J.K. (1984). *AAB Descr. Pl. Viruses* No. 284, 5 pp.

Hebert, T.T. and Castillo (1973). *Abstr. 2nd Int. Congr. Pl. Path., Minneapolis,* No. 72.

Jaing, X.Q., Wilkinson, D.R. and Berry, J.A. (1990). *Phytopath.* **80**: 1060.

Jensen, S.G., Wysong, D.S., Ball, E.M. and Higley, P.M. (1991). *Plant Dis.* **75**: 497.

Lommel, S.A., Kendall, T.L., Siu, N.F. and Nutter, R.C. (1991). *Phytopathology* **81**: 819.

Lommel, S.A., Kendall, T.L., Xiong, Z. and Nutter, R.C. (1991). *Virology* **181**: 382.

Nault, L.R., Styer, W.E., Coffey, M.E., Gordon, D.T., Negi, L.S. and Niblett, C.L. (1978). *Phytopathology* **68**: 1071.

Nutter, R.C., Sheeks, K., Panganiban, L.C. and Lommel, S.A. (1989). *Nucleic Acids Res.* **17**: 3163.

Uyemoto, J.K., Bockelman, D.L. and Claflin, L.E. (1980). *Plant Dis.* **64**: 99.


# Maize dwarf mosaic *potyvirus*

**Data collated by A.J. Gibbs, 1990.**

*Synonyms:* MDMV–A, MDMV–D, MDMV–E, MDMV–F. *Standard acronym:* MDMV.

First reported in *Zea mays*; from Ohio, U.S.A.; by Williams and Alexander, 1965; Dale, 1965.

**Natural host range and symptoms**

Symptoms persist.

*Sorghum halepense* — mosaic.

*Zea mays* — mosaic, stunting.

**Transmission**

Transmitted by a vector; an insect; several species; Aphididae. Transmitted in a non–persistent manner. Virus transmitted by mechanical inoculation; not transmitted by contact between plants; transmitted by seed (maize); not transmitted by pollen.

**Geographical distribution**

Spreads in China, South Africa, and the USA.

**Experimental host range**

Several (3–9) families susceptible.

Diagnostically susceptible

*Panicum miliaceum, Zea mays* — mosaic, reddening and dwarfing.

*Sorghum halepense, S. bicolor* — mosaic.

Diagnostically insusceptible

*Setaria adhaerans, Hordeum vulgare, Saccharum officinarum* (some cultivars).

Maintenance and propagation hosts

*Zea mays* (sweet corn).

Assay hosts (**L**ocal lesions or **W**hole plants)

*Zea mays* (sweet corn) (W).

Monocotyledonous Hosts *(**Susceptible** or Insusceptible)*

*Anthoxanthum odoratum*. ***Arundo donax***. *Avena sativa*. ***Bromus mollis***. ***Bromus secalinus***. ***Bromus tectorum***. ***Chloris gayana***. ***Cynodon dactylon***. *Dactylis glomerata*. ***Echinochloa crus–galli***. ***Eleusine coracana***. *Hordeum vulgare*. ***Lagurus ovatus***. *Lolium perenne*. *Lolium temulentum*. ***Oryza sativa***. ***Panicum capillare***. ***Panicum maximum***. ***Panicum miliaceum***. ***Paspalum dilatatum***. ***Phalaris paradoxa***. *Poa pratensis*. ***Rottboellia exaltata***. ***Saccharum officinarum***. ***Sacciolepis indica***. *Secale cereale*. ***Setaria italica***. ***Setaria viridis***. ***Sorghum bicolor***. *Triticum aestivum*. ***Zea mays***.

Comments on host-range

Differential hosts for different isolates recorded by Rosenkranz (1987).

Sources of host-range data

Rosenkranz (1987).

**Properties of particles in sap**

TIP: 56 °C. LIV: 1–2 days. DEP: $\log_{10}$ minus 2–4. Infectivity of sap not changed by treatment with di–ethyl ether. Leaf sap contains few virions.

**Purification**

Method: Shukla *et al.* (1989*a*).

**Particle morphology**

Virions filamentous; not enveloped; usually flexuous; with a clear modal length; of 770 nm.

*Sequence database accession code(s):*

D00949 Em(40)_vi:MDMCP Gb(84)_vi:MDMCP Maize dwarf mosaic virus gene for coat protein, 3'–terminal region. 2/93 1,377bp.

U07216 Em(40)_vi:MD07216 Gb(84)_vi:MDU07216 Maize dwarf mosaic virus MDMV-A polyprotein (nuclear inclusion II protein, coat protein) gene 2 sequences.

**Cytopathology**

Virions found in leaves, mesophyll and all parts of the host plant; in cytoplasm. Inclusions present in infected cells; are pinwheels and unusual in shape; bundles, scrolls and tubes; type I inclusions (Edwardson, 1974); they do not contain virions.

**Taxonomy and Relationships**

*Potyvirus: Potyviridae.*

This virus is one of a complex of potyviruses infecting tropical grasses. These have been shown by immunoblot and virion protein sequencing to be isolates of four species; maize dwarf mosaic, sugarcane mosaic, sorghum mosaic and Johnson grass mosaic. For details check the descriptions of the other three species (Shukla *et al.*, 1989).

**References**

Dale, J.L. (1965). *Pl. Dis. Reptr* **49**: 202.

Edwardson, J.R. (1974). *Fla Agric. Exp. Stn Monog. No. 4.*

Rosenkranz, E. (1980). *Phytopathology* **70**: 1056.

Rosenkranz, E. (1987). *Phytopathology* **77**: 598.

Shukla, D.D., Tosic, M., Jilka, J., Ford, R.E., Toler, R.W. and Langham, M.A.C. (1989). *Phytopathology* **79**: 223.

Williams, L.E. and Alexander, L.J. (1965). *Phytopathology* **55**: 802.

# Maize eyespot virus

**Data collated by A.A. Brunt, S. Molyneux and C. Fauquet, 1987.**

First reported in *Zea mays*; from Ivory Coast; by Fauquet and Thouvenel (1982).

**Natural host range and symptoms**

Symptoms persist.

*Zea mays* — chlorotic 'eyespotting', and chlorosis.

**Transmission**

Virus transmitted by mechanical inoculation.

**Geographical distribution**

Spreads in Cote d'Ivoire.

**Experimental host range**

Few (<3) families susceptible.

Diagnostically susceptible

*Zea mays* — chlorotic 'eyespotting', and chlorosis.
*Avena sativa, Bromus sterilis* — symptomless systemic infection.
*Hordeum vulgare* — systemic chlorotic rings and mosaic.
*Triticum aestivum, T. durum, Nicotiana megalosiphon* — symptomless.

Diagnostically insusceptible

*Chenopodium amaranticolor, C. quinoa, C. murale, Oryza sativa.*

Maintenance and propagation hosts

*Zea mays.*

Assay hosts (**L**ocal lesions or **W**hole plants)

*Zea mays* (W).

Dicotyledonous Hosts *(**Susceptible** or Insusceptible)*

*Chenopodium amaranticolor. Chenopodium murale. Chenopodium quinoa.* ***Nicotiana glutinosa* × *clevelandii*. *Nicotiana megalosiphon*. *Torenia fournieri*.**

Monocotyledonous Hosts *(**Susceptible** or Insusceptible)*

***Avena sativa*. *Hordeum vulgare*.** *Oryza sativa.* ***Triticum aestivum*. *Triticum durum*.**

Sources of host-range data

Fauquet and Thouvenel (1982).

**Properties of particles in sap**

TIP: 80–85 °C. LIV: more than 20 days. DEP: $\log_{10}$ minus 7–8. Leaf sap contains few virions.

**Purification**

Method: Fauquet and Thouvenel (1982).

**Particle morphology**

Virions isometric; 26 nm in diameter.

**Biochemical properties**

Virions contain 26 % nucleic acid; 74 % protein.

**Taxonomy and Relationships**

Virus(es) with serologically unrelated virions: rice yellow mottle, tobacco necrosis, maize stripe, tephrosia latent, southern bean mosaic, broad bean stain, pea enation mosaic, squash mosaic, cowpea mosaic, tobacco ringspot, tomato ringspot, cucumber mosaic, peanut stunt, tomato aspermy and red clover mottle viruses.

**References**

Fauquet, C. and Thouvenel, J.-C. (1982). *C. R. Acad. Sci. Paris* **295**: 293.

# Maize Iranian mosaic (?) *nucleorhabdovirus*

**Data collated by K. Izadpanah, 1991.**

*Synonyms:* Iranian Shiraz maize mosaic virus.

First reported in *Zea mays*; from Shiraz, Iran; by Izadpanah and Parvin (1979).

**Natural host range and symptoms**

Symptoms persist. Symptoms fine chlorotic striping on veins.

*Echinochloa* spp. — chlorotic streaking.

*Sorghum sudanense* — reddish striping of leaves.

*Zea mays* — fine chlorotic or necrotic striping on leaf and sheath veins, stunting and ear abortion wehen infected early.

**Transmission**

Transmitted by a vector; an insect; *Unkanodes tanasijevici, Laodelphax striatellus, Peregrinus maidis*; Delphacidae. Principal natural vector(s): *Unkanodes tanasijevici, Laodelphax striatellus*. Not transmitted by a number of aphid and cicadellid species. Transmitted in a persistent manner. Virus retained when the vector moults; multiplies in the vector; not transmitted by mechanical inoculation; not transmitted by contact between plants; not transmitted by seed.

**Ecology and control**

Studies reported by crops planted late avoid infection (Izadpanah, 1991).

**Geographical distribution**

Spreads in Iran.

**Experimental host range**

Few (<3) families susceptible. Experimentally infected plants mostly show chlorotic leaf streaking, mosaic.

Diagnostically susceptible

*Echinochloa* sp., *Hordeum vulgare, Sorghum sudanense, Zea mays* — chlorotic striping.

Maintenance and propagation hosts

*Zea mays.*

Assay hosts (**L**ocal lesions or **W**hole plants)

*Zea mays* (W).

Monocotyledonous Hosts *(**Susceptible** or Insusceptible)*

***Hordeum vulgare. Zea mays.***

**Properties of particles in sap**

Leaf sap contains few virions. Electron microscopy: PTA and UA.

**Purification**

Method: Izadpanah (1989).

**Particle morphology**

Virions rhabdo– or bullet–shaped; enveloped; usually straight; of 180 nm; 80 nm wide. Basic helix obscure.

**Physical properties**

Two sedimenting components in purified preparations.

*Virion protein(s)* six; proteins G, L, N, NS, M1 and M2 have been reported (Gomez-Luengo and Gordon, 1987). Virion proteins glycosylated.

**Cytopathology**

Virions found in leaves, mesophyll, epidermis, vascular parenchyma and phloem (Ammar *et al.*, 1987); in cytoplasm and in the perinuclear space. Other cellular changes: enlargement of nucleoli, vesiculation of the cytoplasm, tubular structures and degeneration of chloroplasts (Ammar *et al.*, 1987).

**Taxonomy and Relationships**

Possible *Nucleorhabdovirus: Rhabdoviridae.*

Virus(es) with serologically unrelated virions: barley yellow striate mosaic, cereal chlorotic mottle, cynodon chlorotic streak, maize mosaic, maize sterile stunt, festuca leaf streak, Moroccan wheat, northern cereal mosaic, wheat chlorotic streak and wheat rosette stunt viruses.

Best tests for diagnosis: Iranian maize mosaic (IMMV) and maize rough dwarf (MRDV) viruses are often found together in the same field and sometimes the same plant. MRDV induces small galls on the leaf veins, whereas IMMV induces chlorotic streaks. Also, sugarcane mosaic virus may occur in the same field but it produces mosaics.

**References**


Ammar, E.D., Gomez-Luengo, R.G. and Gordon, D.T. (1987). *Phytopathology* **77**:; 1732.

Gomez-Luengo, R.G. and Gordon, D.T. (1987). *Phytopathology* **77**: 1705.

Izadpanah, K. (1989). *J. Phytopath.* **126**: 43.

Izadpanah, K. (1991). *J. Phytopath.* **131**: 290.

Izadpanah, K. and Parvin, S. (1979). *Iran. J. Pl. Path.* **15**: 53.

Izadpanah, K., Ahmadi, A.A., Parvin, S. and Jafari, S.A. (1983). *J. Phytopath.* **107**: 283.

Milne, R.G., Masenga, V. and Conti, M. (1986). *Intervirology* **25**: 83.

Nault, L.R. and Ammar, E.D. (1989). *Ann. Rev. Ent.* **34**: 503.

# Maize line virus

**Data collated by A.A. Brunt and K. Bock, 1987.**

First reported in *Zea mays* (in complex with maize stripe tenuivirus); from Tanzania; by Storey (1936).

**Natural host range and symptoms**

Symptoms persist.

*Zea mays* — yellow striping along leaf veins but no apical bending or stunting, like that caused by maize stripe virus.

**Transmission**

Transmitted by a vector; an insect; *Peregrinus maidis*; Delphacidae. Transmitted in a persistent manner. Virus not transmitted by mechanical inoculation; not transmitted by contact between plants.

**Ecology and control**

Studies reported by Kulkarni (1973); Smith (1936).

**Geographical distribution**

Spreads in Kenya and Tanzania (symptoms are swamped by maize streak virus, which is widespread, so this virus may occur more commonly (Kulkarni, 1973)).

**Experimental host range**

Few (<3) families susceptible.

Diagnostically susceptible

*Zea mays* (hybrid 511, hybrid 613B, Wisconsin 641AA) — small chlorotic spots at leaf bases, followed by vein chlorosis and mottling.

Diagnostically insusceptible

*Chenopodium amaranticolor, Chenopodium quinoa, Glycine max, Petunia hybrida, Phaseolus vulgaris.*

Maintenance and propagation hosts

*Zea mays.*

Assay hosts (**L**ocal lesions or **W**hole plants)

*Zea mays* (W).

Dicotyledonous Hosts *(**Susceptible** or Insusceptible)*

*Chenopodium amaranticolor. Chenopodium quinoa. Glycine max. Petunia hybrida. Phaseolus vulgaris.*

Monocotyledonous Hosts *(**Susceptible** or Insusceptible)*

***Zea mays.***

Sources of host-range data

Kulkarni (1973); Storey (1936).

**Properties of particles in sap**

Leaf sap contains few virions.

**Purification**

Method: Kulkarni (1973).

**Particle morphology**

Virions isometric; not enveloped; 34 nm in diameter; rounded in profile; without a conspicuous capsomere arrangement.

**Taxonomy and Relationships**

Virus(es) with serologically unrelated virions: cucumber mosaic, maize mottle chlorotic stunt and maize streak viruses.

Best tests for diagnosis: Maize line virus sometimes occurs together with maize streak and/or sugarcane mosaic viruses.

**References**

Kulkarni, H.Y. (1973). *Ann. appl. Biol.* **75**: 205.
Storey, H.H. (1936). *E. Afr. Agric. J.* **1**: 333.

# Maize mosaic *nucleorhabdovirus*

**Data collated by R. Lastra, 1985.**

*Synonyms:* virus del enanismo rayado maiz, maize stripe Indian virus (Sharma and Payak, 1983), sorghum chlorosis virus, zea mays virus. *Standard acronym:* MMV. *Strains:* MMV — raya fina, MMV — raya gruesa.

First reported in *Zea mays*; from Colombia; by Lastra and Acosta (1979).

**Natural host range and symptoms**

Symptoms persist.

*Rottboellia exaltata, Setaria vulpiseta, Zea mays* — green-yellow spotting, striping and vein-banding, some apical bending.

*Sorghum bicolor* — interveinal chlorotic streaking, stunting and severe yield loss.

**Transmission**

Transmitted by a vector; an insect; *Peregrinus maidis*; Delphacidae. Transmitted in a persistent manner. Virus retained when the vector moults; multiplies in the vector; not transmitted congenitally to the progeny of the vector; not transmitted by mechanical inoculation; not transmitted by contact between plants; not transmitted by seed; not transmitted by pollen.

**Geographical distribution**

Spreads in possibly Australia, Colombia, Costa Rica, Fiji, India, Mauritius, Mexico, Peru, Spain, Tanzania, and the USA (and the Caribbean Islands).

**Experimental host range**

Few (<3) families susceptible.

Maintenance and propagation hosts

*Zea mays*.

Assay hosts (**L**ocal lesions or **W**hole plants)

*Zea mays* (W).

Monocotyledonous Hosts *(**Susceptible** or Insusceptible)*

***Sorghum bicolor*. *Zea mays*.**

**Properties of particles in sap**

Leaf sap contains many virions.

**Purification**

Method: Lastra and Acosta (1979).

**Particle morphology**

Virions rhabdo– or bullet–shaped; enveloped; with a clear modal length; of 225 nm (242 nm in thin sections); 90 nm wide (48m in thin sections).

**Physical properties**

One sedimenting component in purified preparations; sedimentation coefficient 774 *S*. Density 1.181 g $cm^{-3}$ in sucrose.

**Biochemical properties**

*Genome* consists of RNA; single–stranded; unipartite; largest (or only) genome part 12.54 kb. Genomic nucleic acid isolated by Lastra and Acosta (1979). Infectivity lost when deproteinised with proteases; lost when deproteinised with phenol or detergent. Additional factor required for infectivity; probably a polymerase.

Non–genomic nucleic acid N protein.

*Virion protein(s)* five. $M_r$ of 2nd largest 75000; 0 protei. $M_r$ of 3rd largest 50000. $M_r$ of 4th largest 45000; NS protein. $M_r$ of 5th largest 33000; M protein. Method of preparation: Autrey (1983); Falk and Tsai (1983). Virion proteins probably glycosylated.

**Replication**

Genome replicates in nuclei. Coat protein *m*RNA translated in probably in cytoplasm. Replication does not depend on a helper virus.

**Cytopathology**

Virions found in all parts of the host plant; in the perinuclear space; and the cores bud through the inner nuclear membrane. Inclusions present in infected cells; are unusual in shape; aggregates; they contain virions.

**Taxonomy and Relationships**

*Nucleorhabdovirus: Rhabdoviridae.*

**References**


Autrey, L.C. (1983). *Proc. Int. Maize Virus Disease Wksh., 1982*, p. 167. Wooster, Ohio, Agric. Res. and Dev. Center.

Christie, R.G. and Edwardson, J.R. (1977). *Fla Agric. Exp. Stn Monog.* No. 9.

Falk, B.W. and Tsai, J.H. (1983). *Phytopathology* **73**: 1536.

Herold, F. (1972). *CMI/AAB Descr. Pl. Viruses* No. 94, 4 pp.

Herold, F. and Munz, K. (1965). *Virology* **25**: 412.

Herold, F. and Munz, K. (1967). *J. gen. Virol.* **1**: 227.

Herold, F., Bergold, G.H. and Weibel, J. (1960). *Virology* **12**: 335.

Kunkel, L.O. (1921). *Bull. Hawaiian Sug. Pltrs. Assoc. Exp. Stn Bot.* Ser. **3**: 44.

Lastra, R.G. and Acosta, J.M. (1979). *Intervirology* **11**: 215.

Lastra, R.G. and Corballo, O. (1983). *Proc. Int. Maize Disease Wksh, 1982*, p. 83. Wooster, Ohio, Agric. Res. and Dev. Center.

Naidan, R.A., Harikrishnan, R., Manohar, S.K., Reddy, D.V.R., Rakna, A.S., King, S.B. and Bandyopadhyay, R. (1989). *Ann. appl. Biol.* **114**: 301.

Sharma, R.C. and Payak, M.M. (1983). *Proc. Int. Maize Virus Dis. Colloq. Wksh.*, Wooster, Ohio 1982, p. 186.

Tsai, J.H. and Falk, B.W. (1988). In: *Advances in Vector Research* Vol 5. Springer-Verlag. New York.

# Maize mottle/chlorotic stunt virus

**Data collated by A.A. Brunt, 1989.**

*Synonyms:* maize mottle virus.

First reported in *Zea mays*; from Tanzania; by Storey (1937).

**Natural host range and symptoms**

Symptoms persist.

*Zea mays* — transient leaf mottling in young leaves and, in susceptible cultivars, conspicuous chlorosis of older leaves, stunting of plants and sometimes tassel abortion, loss of seed production and bending of shoots.

**Transmission**

Transmitted by a vector; an insect; *Cicadulina mbila, C. zeae, C. storeyi* and *C. triangula*; Cicadellidae. Transmitted in a persistent manner. Virus not transmitted by mechanical inoculation; not transmitted by contact between plants.

**Ecology and control**

Studies reported by Storey (1937); Rossel and Thottappilly (1985).

**Geographical distribution**

Spreads in Nigeria, Rwanda, Sao Tome and Principe, Tanzania, Togo, Zambia, and Zimbabwe.

**Experimental host range**

Few (<3) families susceptible.

Diagnostically susceptible

*Zea mays* — transient chlorosis in tip leaves, then bright chlorosis of mature leaves.

Diagnostically insusceptible

*Pennisetum typhoides.*

Maintenance and propagation hosts

*Zea mays.*

Assay hosts (**L**ocal lesions or **W**hole plants)

*Zea mays* (W).

Monocotyledonous Hosts *(**Susceptible** or Insusceptible)*

***Zea mays.***

Additional hosts

*Pennisetum typhoides* (millet) is insusceptible and is thus a useful host for rearing non-viruliferous colonies of the virus vectors.

Sources of host-range data

Storey (1937); Rossel and Thottappilly (1985).

**Particle morphology**

Virions isometric; not enveloped; 40 nm in diameter; rounded in profile.

**Comments**

The Nigerian virus may be close to that called maize mottle virus in Tanzania by Storey (1937). However only modern varieties are available and so this cannot be proven. Therefore, despite historical precedence, the name maize mottle/chlorotic stunt virus should be used and "maize mottle virus" recognised as a synonym (Rossel and Thottappilly, 1985).

**References**

Rossel, H.W. (1984). *Maize Virus Disease Newsl.*.

Rossel, H.W. and Thottappilly, G. (1983). In: *Proc. Int. Maize Virus Dis. Coll. and Wksh.*, pp. 158–160. Wooster, Ohio Agric. Res. and Dev. Center.

Rossel, H.W. and Thottappilly, G. (1985). *Virus Diseases of Important Food Crops in Tropical Africa*, pp. 49–51. I.I.T.A., Ibadan.

Rossel, H.W., Buddenhagen, I.W. and Thottappilly, G. (1980). *I.I.T.A. Res. Briefs* **1**: 2.

Storey, H.H. (1937). *Ann. appl. Biol.* **24**: 87.

# Maize rayado fino *marafivirus*

**Data collated by A.A. Brunt and R. Gamez, 1988.**

*Standard acronym:* MRFV. *Strains:* Maize rayado Colombian strain (Martinez-Lopez, 1977), Brazilian corn streak strain (Gámez, 1980).

First reported in *Zea mays*; from El Salvador; by Ancalmo and Davies (1961); but initially described as a strain of corn stunt, until studied fully by Gamez (1969).

**Natural host range and symptoms**

*Zea mays* ssp. *mays* (maize), *Zea mays* ssp. *mexicana* (teosinte) — chlorotic vein stippling and striping especially in new cultivars.

**Transmission**

Transmitted by a vector; an insect; *Dalbulus maidis*; Cicadellidae. Transmitted in a persistent manner. Virus retained when the vector moults; multiplies in the vector; not transmitted by mechanical inoculation; not transmitted by grafting; not transmitted by contact between plants; not transmitted by seed.

**Ecology and control**

Studies reported by Gamez (1973, 1983); Gamez and Leon (1985, 1988); Gamez and Saavedra (1986).

**Geographical distribution**

Spreads in Argentina, Brazil, Colombia, Costa Rica, Mexico, Peru, Venezuela, and the USA (in the south).

**Experimental host range**

Few (<3) families susceptible.

Diagnostically susceptible

*Zea mays* — chlorotic spotting of base and veins of young leaves, then chlorotic striping. Root development poor, yield decreased.

Maintenance and propagation hosts

*Zea mays* (Tico H1 and Tico H4).

Assay hosts (**L**ocal lesions or **W**hole plants)

*Zea mays* (W). For insect transmission studies.

Monocotyledonous Hosts *(**Susceptible** or Insusceptible)*

***Zea mays***.

**Properties of particles in sap**

TIP: , determined by feeding vectors in sap through membranes, 60–65 °C. LIV: 3–4 days (at 20°C; 4 to 5 days at 3°C). Leaf sap contains few virions.

**Purification**

Method: Gamez *et al.* (1977); Leon and Gamez (1981).

**Particle morphology**

Virions isometric; not enveloped; 30–33 nm in diameter; rounded in profile; with a conspicuous capsomere arrangement.

**Physical properties**

Two sedimenting components in purified preparations; sedimentation coefficient of the fastest 120 *S* (B); of the other(s) 54 *S* (T). Density 1.28 g $cm^{-3}$ in CsCl (T), or 1.46 g $cm^{-3}$ in CsCl (B). Density 1.24 (T), 1.37 (B) in $Cs_2$ $SO_4$. $A_{260}/A_{280}$ ratio 0.87 (T), or 1.58 (B).

**Biochemical properties**

Virions contain 33 % nucleic acid; 67 % protein.

*Genome* consists of RNA; single–stranded; unipartite; largest (or only) genome part 6.7 kb.

*Virion protein(s)* one; $M_r$ 22000; coat protein; in virions of some isolates there is a small amount of a 28kDa protein. Method of preparation: Gamez and Leon (1988).

**Cytopathology**

Virions found in leaves, mesophyll, epidermis, vascular parenchyma and phloem; in cytoplasm and in cell vacuoles (in loose aggregates); occasionally in sieve tubes in loose aggregates or large crystalline arrays within the membrane bound vesicles.

**Taxonomy and Relationships**

*Marafivirus* (the type species).

Virus(es) with serologically unrelated virions: Bermuda grass etched line and oat blue dwarf viruses.

Best tests for diagnosis: This virus is frequently in mixed infections with MLOs and spiroplasms; these infections induce the corn stunt disease.

**References**

Ancalmo, O. and Davies, W.C. (1961). *Pl. Dis. Reptr* **45**: 281.

Espinoza, A.M. and Gamez, R. (1980). *Turrialba* **30**: 413.

Falk, B.W. and Tsai, J.H. (1986). *Intervirology* **25**: 111.

Gamez, R. (1969). *Pl. Dis. Reptr* **53**: 929.

Gamez, R. (1973). *Ann. appl. Biol.* **73**: 285.

Gamez, R. (1980). *CMI/AAB Descr. Pl. Viruses* No. 220, 4 pp.

Gamez, R. (1983). In: *Plant Virus Epidemiology*, p. 267; eds R.T. Plumb and J.M. Thresh. Blackwell Scientific, Oxford, England.

Gamez, R. and Leon, P. (1985). In: *Leafhoppers and Planthoppers*, p. 331; eds L.R. Nault and Rodriguez. J. Wiley and Sons Inc., New York,.

Gamez, R. and Saavedra, R. (1986). In: *Plant Virus Epidemics*, p. 315; eds G. Maclean, R. Garrett and W. Ruesnik. Academic Press, Melbourne.

Gamez, R., Kosuka, Y. and Fukuoka, T. (1977). *Rev. Biol. Trop.* **25**: 151.

Gamez, R., Kitajima, E.W. and Lin, M.T. (1979). *Pl. Dis. Reptr* **63**: 830.

Gamez, R. and Leon, P. (1986). In: *The Plant Viruses*, Vol. 3, Polyhedral Virions with Monopartite RNA Genomes, p. 213; ed. R. Koenig. Plenum Press, New York.

Gingery, R.E., Gordon, D.T. and Nault, L.R. (1982). *Phytopathology* **72**: 1313.
Kitajima, E.W. and Gamez, R. (1977). *Turrialba* **27**: 71.
Kitajima, E.W., Yano, T. and Costa, A.S. (1976). *Cienc. Cult.*, S. Paulo **28**: 427.
Lastra, R. and de Uscategui, R.C. (1980). *Turrialba* **30**: 405.
Leon, P. and Gamez, R. (1986). *Rev. Biol. Trop.* **34**: 111.
Martinez-Lopez, G. (1977). *Proc. int. Maize Viruses Diseases Coloq. Wksh.*, p. 20. Ohio Agric. Res. Dev. Center.
Ramirez, P. and Leon, P. (1988). *J. gen. Virol.* **69**: 757.
Rivera, C. and Gamez, R. (1986). *Intervirology* **25**: 76.
Tsai, J.H. (1986). *Intervirology* **25**: 111.

# Maize rough dwarf *fijivirus*

**Data collated by R.G. Milne, 1984.**

*Synonyms:* virus del nanismo ruvido del mais. *Standard acronym:* MRDV. *Strains:* type strain (TS), cereal tillering disease strain (CTD) (Boccardo and Milne, 1984; Lindsten and Gerhardson, 1971), mal de Rio Cuarto (MR4) (Boccardo and Milne, 1984; Nome *et al.*, 1981).

First reported in *Zea mays*; from Israel; by Harpaz (1959).

### Natural host range and symptoms

Symptoms persist.

*Avena sativa* (CTD strain), *Cynodon dactylon, Digitaria sanguinalis, Echinochloa crus-galli, Hordeum vulgare* (CTD strain), *Lolium perenne, Setaria verticillata, Triticum aestivum* (CTD strain), *Zea mays* (dent, not flint cultivars) — stunting, dark green colour, enations on veins of leaves and sheaths, longitudinal splitting of roots, flowers suppressed.

### Transmission

Transmitted by a vector; an insect; *Delphacodes propinqua, Dicranotropis hamata, Laodelphax striatellus, Javasella pellucida, Sogatella vibix*; Delphacidae. Transmitted in a persistent manner. Virus retained when the vector moults; multiplies in the vector; transmitted congenitally to the progeny of the vector (one unconfirmed report); transmitted by mechanical inoculation (injection; but very inefficient); not transmitted by grafting; not transmitted by contact between plants; not transmitted by seed; not transmitted by pollen.

### Geographical distribution

Spreads in Argentina, the former Czechoslovakia, France, Israel, Italy, Norway, Spain, Sweden, and the former Yugoslavia.

### Experimental host range

Few (<3) families susceptible.

Diagnostically susceptible

*Hordeum vulgare* cv. Delesa — stunting.

*Zea mays* American yellow dent hybrids; Wisconsin 641 AA — stunting, dark green colour, pale yellow enations on abaxial surface of leaves and leaf sheaths.

Diagnostically insusceptible

*Saccharum* spp., *Sorghum halepense* (susceptible to the Chinese virus), *Stipa bromoides*, probably all non-Gramineae.

Maintenance and propagation hosts

*Hordeum vulgare, Triticum aestivum, Zea mays.*

Assay hosts (**L**ocal lesions or **W**hole plants)

*Hordeum vulgare* (W), *Triticum aestivum* (W), *Zea mays* (W).

Monocotyledonous Hosts (***Susceptible*** *or Insusceptible)*

***Avena sativa***. *Chloris gayana*. ***Cynodon dactylon***. ***Echinochloa crus–galli***. ***Hordeum vulgare***. ***Lolium perenne***. ***Oryza sativa***. *Saccharum officinarum*. ***Secale cereale***. ***Setaria italica***. ***Triticum aestivum***. ***Zea mays***.

Sources of host-range data

Milne and Lovisolo (1977).

**Properties of particles in sap**

Leaf sap contains few virions, or contains many virions (few virions in whole leaf sap, many in sap from enations). Electron microscopy: for whole virions avoid neutral PTA, use 2% aqueous UA. No special treatment for subvirions.

**Purification**

Method: Milne *et al.* (1973); Boccardo and Milne (1981).

**Particle morphology**

Virions isometric; not enveloped; 70 nm in diameter; rounded in profile, or angular in profile; with a conspicuous capsomere arrangement.

**Physical properties**

Two sedimenting components in purified preparations (whole and subvirions); sedimentation coefficient 400 *S* (for subviral non-infectious virions).

**Biochemical properties**

Virions 0 % lipid.

*Genome* consists of RNA; double–stranded; linear; of 10 parts (named S1–10); largest (or only) genome part the largest 3.9 kb (pairs); the 2nd largest 3.2 kb (pairs); the 3rd largest 3.15 kb (pairs); the 4th largest 3.15 kb (pairs); the 5th largest 3.1 kb (pairs); the 6th largest and other parts 2.293 kb (pairs; also 2.198, 1.735, 1.728 and 1.715 kb pairs). Genomic nucleic acid isolated by Boccardo and Milne (1975). Additional factor required for infectivity; proteins from the virions; isolated nucleic acid is not infectious.

*Sequence database accession code(s):*

X55701 Em(40)_un:MRDVS6 Gb(84)_un:MRDVS6 Maize rough dwarf virus (MRDV) mRNA for genomic segment S6. 3/91 2,193bp. 1 sequence.

*Non–genomic nucleic acid* not found in the virions.

*Virion protein(s)* six; $M_r$ of the largest 139000. $M_r$ of 2nd largest 125000. $M_r$ of 3rd largest 123000. $M_r$ of 4th largest 111000. $M_r$ of 5th largest 97000. $M_r$ of 6th largest and smaller, 64000. Method of preparation: Boccardo and Milne (1975).

**Replication**

Genome replicates in cytoplasmic viroplasms. Coat protein *m*RNA translated in the cytoplasm. Replication does not depend on a helper virus.

**Cytopathology**

Virions found in phloem enations; in cytoplasm. Inclusions present in infected cells; are crystals in the cytoplasm, or viroplasms (containing immature virions), or unusual in shape; tubular structures some 100 nm diameter; they contain virions. Other cellular changes: cell wall proliferation and cell enlargement.

**Taxonomy and Relationships**

*Fijivirus: Reoviridae.*

Virus(es) with serologically related virions: The virus is so closely related to rice black streaked dwarf virus, that it may be the same species, it is more distantly related to pangola stunt virus. Virus(es) with serologically unrelated virions: oat sterile dwarf, rice ragged stunt and sugarcane Fiji disease viruses.

**Comments**

The virus has been reported from China, but it may be that the virus was incorrectly identified, and was rice black-streaked dwarf virus, which is common in the region.

**References**


Boccardo, G. and Milne, R.G. (1975). *Virology* **68**: 79.

Boccardo, G. and Milne, R.G. (1981). *J. virol. Meth.* **3**: 109.

Boccardo, G. and Milne, R.G. (1984). *CMI/AAB Descr. Pl. Viruses* No. 294, 7 pp.

Conti, M. (1984). In: *Current Topics in Vector Research*; ed. K.F. Harris. Praeger Scientific, New York.

Francki, R.I.B. and Boccardo, G. (1983). In: *The Reoviridae*. p. 155; ed. K.W. Joklik. Plenum Press, New York.

Francki, R.I.B., Milne, R.G. and Hatta, T. (1984). In: *An Atlas of Plant Viruses.* Vol. 1. CRC Press, Boca Raton, Florida.

Harpaz, I. (1959). *Nature, Lond.* **184**: 77.

Lindsten, K. and Conti, M. (1977). *Annls. Phytopath.* **9**: 30.

Lindsten, K. and Gerhardson, B. (1971). *Nat. Swed. Inst. Pl. Prot. Count. S.* **14**: 285.

Lovisolo, O. (1971). *CMI/AAB Descr. Pl. Viruses* No. 72, 4 pp.

Marzachi, C., Boccardo, G. and Nuss, D.L. (1991). *Virology* **180**: 518.

Milne, R.G. and Lovisolo, O. (1977). *Adv. Virus Res.* **21**: 267.

Milne, R.G. and Luisoni, E. (1977). *Virology* **80**: 12.

Milne, R.G., Conti, M. and Lisa, V. (1973). *Virology* **53**: 130.

Nome, S.F., Lenardon, S.L., Raju, B.C., Laguna, I.G., Lowe, S.K. and Docampo, D. (1981). *Phytopath. Z.* **101**: 7.


# Maize streak dwarf (?) *nucleorhabdovirus*

**Data collated by A.J. Gibbs, 1990.**

First reported in *Zea mays*; from China; by Peng *et al.* (1979).

**Natural host range and symptoms**

Symptoms persist.

*Zea mays* — chlorotic streaking of leaf, husk and sheath.

**Transmission**

Transmitted by a vector; an insect; *Laodelphax striatellus* (both adults and nymphs); Delphacidae. Transmitted in a persistent manner. Virus retained when the vector moults.

**Geographical distribution**

Spreads in China (Gansu Province).

**Experimental host range**

Few (<3) families susceptible.

Diagnostically susceptible

*Zea mays* — chlorotic streaking of leaves and sheaths.

Maintenance and propagation hosts

*Zea mays.*

Assay hosts (**L**ocal lesions or **W**hole plants)

*Zea mays* (W).

Monocotyledonous Hosts *(**Susceptible** or Insusceptible)*

***Zea mays.***

**Properties of particles in sap**

Leaf sap contains few virions.

**Particle morphology**

Virions rhabdo– or bullet–shaped; enveloped; with a clear modal length; of 150–220 nm (and 200–250 nm); 43–64 nm wide (and 70–80 nm).

**Cytopathology**

Virions found in leaves; in the perinuclear space (attached to the inner nuclear membrane); the endoplasmic reticulum.

**Taxonomy and Relationships**

Possible *Nucleorhabdovirus: Rhabdoviridae.*

**References**

Peng, C.M., Chen, T.Y., Shen, C.Y., Zhu, F.C. and Chin, M.C. (1979). *Acta Biochim. Biophys. Sin.* **11**: 187.

Tsai, J.H., Zhu, F.C. and Zhu, X.S. (1984). *Maize Virus Dis. Newsl.* **1**: 21.

# Maize streak monogeminivirus

**Data collated by M.B. von Wechmar, 1985. Revised by P.G. Markham and M.I. Boulton, 1989.**

*Synonyms:* bajra streak virus (Seth *et al.*, 1972*a* & 1972*b*), cereal African streak virus, maize streak A virus. *Standard acronym:* MSV. *Strains:* maize strain, *Panicum maximum* strain, sugarcane strain, *Digitaria setigera* strain.

First reported in *Zea mays*; from South Africa; by Fuller (1901); Bock *et al.* (1974); Storey (1925); Rose (1978); Gorter (1953).

**Natural host range and symptoms**

Symptoms persist.

*Avena sativa, Brachiaria reptens, Cenchrus echinatus, Chloris gayana, Coix lacryma-jobi, Digitaria abyssinica, Digitaria velutina, Eleusine africana, Eleusine indica, Hordeum vulgare, Panicum maximum, Pennisetum purpureum, Rottboellia exaltata, Saccharum* spp., *Secale cereale, Setaria homonyma, Setaria pallidifusca, Setaria verticillata, Sorghum bicolor, Sorghum sudanense, Triticum aestivum, Urochloa panicoides, Urochloa trichopus, Zea mays* — chlorotic or white streaking or lesions.

**Transmission**

Transmitted by a vector; an insect; *Cicadulina mbila, C. arachidis, C. bipunctella, C. triangula, C. bimaculata, C. similis, C. latens, C. ghaurii, C. parazeae*; Cicadellidae. Transmitted in a persistent manner. Virus retained when the vector moults; does not multiply in the vector; not transmitted congenitally to the progeny of the vector; not transmitted by mechanical inoculation; not transmitted by contact between plants; not transmitted by seed; not transmitted by pollen.

**Ecology and control**

Studies reported by Rose (1973*a*; *b*; 1974; 1978).

**Geographical distribution**

Spreads in the African region; India, Madagascar, Reunion, and Yemen.

**Experimental host range**

Many (>9) families susceptible. Experimentally infected plants mostly show streaks and spots.

Diagnostically susceptible

*Zea mays* — systemic chlorotic streaking.

Diagnostically insusceptible

*Zea perennis, Zea diploperennis, Pennisetum americanum.*

Maintenance and propagation hosts

*Hordeum vulgare, Triticum aestivum, Zea mays.*

Assay hosts (**L**ocal lesions or **W**hole plants)

*Hordeum vulgare* (W), *Triticum aestivum* (W), *Zea mays* cv. Golden Cross Bantam (W).

Monocotyledonous Hosts *(**Susceptible** or Insusceptible)*

***Agropyron repens. Avena sativa.*** *Bromus inermis.* ***Chloris gayana***, or *Chloris gayana.* ***Cynodon dactylon***, or *Cynodon dactylon. Dactylis glomerata. Digitaria decumbens.* ***Digitaria longiflora.*** *Digitaria valida. Echinochloa colonum. Echinochloa crus–galli.* ***Eleusine coracana. Hordeum vulgare. Lolium multiflorum. Lolium perenne.*** *Panicum maximum.* ***Panicum miliaceum.*** *Paspalum dilatatum. Phalaris arundinacea. Phleum pratense. Poa compressa. Poa pratensis. Poa trivialis.* ***Rottboellia exaltata. Setaria italica***, or *Setaria italica.* ***Setaria viridis. Sorghum bicolor***, or *Sorghum bicolor. Stenotaphrum secundatum.* ***Triticum aestivum. Zea mays.***

Additional hosts

***Aegilops cylindrica, Aegilops triaristata, Aegilops triucialis, Aegilops umbellilata***, *Agropyron cristatum* or ***Agropyron sibiricum (=Agropyron cristatum)***, *Agrostis canina*, ***Agrostis gigantea***, *Agrostis palustris, Agrostis stolonifera, Agrostis tenuis, Alopecurus aequalis*, ***Alopecurus pratensis, Andropogon gerardi***, *Andropogon hallii, Arrhenatherum elatius*, ***Avena fatua, Bothriochloa alta***, *Bothriochloa barbinodis, Bouteloua gracilis, B. hirsuta*, ***Brachiaria erusaeformis, Brachiaria reptens***, *Bromus carinatus, B. erectus*, ***B. uniloides, Calamagrostis canadensis***, *Cenchrus ciliaris*, ***C. echinatus, C. argentina***, *C. canterai*, ***C. cucullata***, *C. distichophylla*, ***C. gayana, C. radiata, C. submutica, C. virgata, Coix lacryma-jobi, Cymbopogon distans, C. schoenanthus***, *Dactyloctenium gigantea*, ***Digitaria enantha, D. horizontalis, D. marginata, D. milanjiana, D. sanguinalis, D. ternata, Echinochloa polystachya, Eleusine africana, E. indica***, *Elymus canadensis*, ***Eragrostis abyssinica, E. aspera, E. ciliaris, E. curvula, E. valida***, *Eremochloa ophiuroides*, ***Euchlaena mexicana***, *E. perennis*, ***Festuca elatior*** var.***Arundinacea***, *F. ovina*, , *F. rubra*, ***Glyceria fluitans, Heteropogon contortus, Holcus lanatus***, *Hordeum jubatum*, ***Hyparrhenia rufa, Imperata arundinacea***, *Leersia hexandra*, ***Leptochloa virgata, Lolium multiflorum, L. subulatum, Panicum bergii, P. coloratum***, *P. coloratum* var. *Makarikari*, ***P. hallii, P. texanum, P. virgatum, Paspalum almum, P. notatum, P. scrobiculatum, P. urvillei***, *Pennisetum americanum*, ***P. clandestinum*** (or *P. clandestinum)*, *P. macrorum*, ***P. purpureum***, *P. typhoides, Phalaris tuberosa, Phleum pratense*, ***Schedonnardus paniculatus, Schizachrium scoparium, Setaria chevalieri,***

***Setaria faberi, S. homonyma, S. pallidifusca***, *S. nigrirostris, S. sphacelata*, ***Sorghastrum nutans***, *Sorghum halepense, S. miliaceum, S. propinguum*, ***S. sudanense***, *S. vulgare*, ***Trichachne californica***, *Tripsacum dactyloides*, ***Trichloris crinita, T. pluriflora, Urochloa helopus, U. panicoides, U. trichopus, Vaseyochloa multinervosa***.

Comments on host-range

All isolates tested infect *Zea mays* cv. Golden Cross Bantam.

Sources of host-range data

Bock *et al.* (1974); Rose (1978); Damsteegt (1983); Storey (1925); Storey and McClean (1930).

**Properties of particles in sap**

LIV: more than 7 days. Leaf sap contains many virions. Electron microscopy: UA is best. ISEM is most reliable.

**Purification**

Method: von Wechmar and Milne (1982).

**Particle morphology**

Virions geminate; not enveloped; 18 nm in diameter; 30 nm in length; angular in profile; without a conspicuous capsomere arrangement.

**Physical properties**

Two sedimenting components in purified preparations (in sucrose (Howell, 1984). However, one could be of damaged virion as most geminiviruses sediment as a single component); sedimentation coefficient of the fastest 76 *S*; of the other(s) 54 *S*.

**Biochemical properties**

*Genome* consists of DNA; single–stranded; circular; unipartite; largest (or only) genome part 2.689 kb (Nigerian isolate, Mullineaux, 1984; 2.690, South African isolate, Lazarowitz, 1988; 2.681, Kenyan isolate, Howell, 1984). Base composition 25.7 % G; 25.4 % A; 23.6 % C; 25.3 % T. Infectivity retained when deproteinised with phenol or detergent (via agroinfection). Poly A region absent (but does have polyadenylated transcripts). Additional factor required for infectivity (coat protein is essential for transmission by leafhoppers, but cloned DNA can infect by agroinfection). Nucleotide sequence references: Mullineaux *et al.* (1984); Howell (1984); Lazarowitz (1988).

*Sequence database accession code(s):*

K02026 Gb(84)_vi:MZS Maize streak virus, complete genome. 6/88 2,687bp.

M26262 Em(40)_vi:GEMZSMSV Gb(84)_vi:MZSMSV Maize streak virus putative primer of second-strand synthesis. 2/90 81bp.

S54420 Em(40)_vi:S54420 Gb(84)_vi:S54420 (upstream activating sequence, promoter) maize streak virus, Genomic, 134 nt. 11/93 134bp.

X01089 Em(40)_vi:GEMSVKV Gb(84)_vi:GEMSVKV Maize streak virus (Kenyan isolate) virion strand (MSV-K). 7/89 2,681bp.

X01633 Em(40)_vi:GEMSV01 Gb(84)_vi:GEMSV01 Maize streak virus DNA sequence (MSV). 9/93 2,687bp. Y00514 Em(40)_vi:GEMSVSXX Gb(84)_vi:GEMSVSXX Maize streak virus (MSV-S) infectious genome. 9/93 2,690bp.

U20768 Em(43)_vi:Ms20768 Gb(89)_vi:Msu20768 Maize streak virus replication associated protein gene, partial cds. 5/95 227bp.

U20769 Gb(89)n:Msu20769 Maize streak virus RepB (C2) gene, partial cds. 6/95 227bp.

U20870 Em(43)_vi:Ms20870 Gb(89)_vi:Msu20870 Maize streak virus RepB (C2) gene, partial cds. 4/95 251bp.

U20871 Em(43)_vi:Ms20871 Gb(89)_vi:Msu20871 Maize streak virus strain Wheat/Eleusine replication-associated protein RepB (C2) gene, parti

U20893 Em(43)_vi:Ms20893 Gb(89)_vi:Msu20893 Maize streak virus movement protein (V1) gene, partial cds. 3/95 303bp.

U20904 Em(43)_vi:Ms20904 Gb(89)_vi:Msu20904 Maize streak virus replication-associated protein RepB (C2) gene, partial cds. 5/95 227bp.

U20905 Em(43)_vi:Ms20905 Gb(89)_vi:Msu20905 Maize streak virus strain Tas replication-associated protein RepB (C2) gene, partial cds. 5/9

X71953 Em(43)_vi:Gemsvcpa Gb(89)_vi:Gemsvcpa Maize streak virus V2 gene for coat protein (clone Y852). 11/94 992bp.

X71954 Em(43)_vi:Gemsvcpb Gb(89)_vi:Gemsvcpb Maize streak virus V2 gene for coat protein (clone N64). 11/94 990bp.

X71955 Em(43)_vi:Gemsvcpc Gb(89)_vi:Gemsvcpc Maize streak virus V2 gene for coat protein (clone Y511). 11/94 990bp.

X71956 Em(43)_vi:Gemsvcpd Gb(89)_vi:Gemsvcpd Maize streak virus V2 gene for coat protein (clone N58). 11/94 992bp.

X71957 Em(43)_vi:Gemsvcpe Gb(89)_vi:Gemsvcpe Maize streak virus V2 gene for coat protein (clone Y861). 11/94 990bp.

X71958 Em(43)_vi:Gemsvcpf Gb(89)_vi:Gemsvcpf Maize streak virus V2 gene for coat protein (clone Y831). 11/94 991bp.

X71959 Em(43)_vi:Gemsvcpg Gb(89)_vi:Gemsvcpg Maize streak virus V2 gene for coat protein (clone N74). 11/94 991bp.

X71960 Em(43)_vi:Gemsvcph Gb(89)_vi:Gemsvcph Maize streak virus V2 gene for coat protein (clone Y845). 11/94 917bp.

X71961 Em(43)_vi:Gemsvcpi Gb(89)_vi:Gemsvcpi Maize streak virus V2 gene for coat protein (clone Y946). 11/94 992bp.

X71962 Em(43)_vi:Gemsvcpj Gb(89)_vi:Gemsvcpj Maize streak virus V2 gene for coat protein (clone Y9614). 11/94 992bp.

X71963 Em(43)_vi:Gemsvcpk Gb(89)_vi:Gemsvcpk Maize streak virus V2 gene for coat protein (clone Y803). 11/94 990bp.

X71964 Em(43)_vi:Gemsvcpl Gb(89)_vi:Gemsvcpl Maize streak virus V2 gene for coat protein (clone O154). 11/94 992bp.

*Features of the genome:* promoters, stem loops and ORF information see Mullineaux *et al.* (1984).

Sub–genomic *m*RNA found in infected cells (Morris-Krsinich *et al.*, 1985).

*Virion protein(s)* one; $M_r$ 26900; coat protein.

*Virus–coded non–virion proteins* identified by genomic sequence analysis (Mullineaux *et al.* (1984)); one protein found. $M_r$ 10900.

## Replication

Genome replicates in nuclei. Replication does not depend on a helper virus.

## Cytopathology

Virions found in leaves, mesophyll, epidermis, meristem, vascular parenchyma, companion cells and all parts of the host plant; in cytoplasm, in nuclei, and in cell vacuoles. Inclusions present in infected cells; are crystals in the nucleus and viroplasms; they contain virions. Other cellular changes: changes in chloroplasts and other organelles destroyed.

**Taxonomy and Relationships**

Monogeminivirus: *Geminiviridae* (type species).

Virus(es) with serologically unrelated virions: African cassava mosaic and chloris striate mosaic viruses.

Best tests for diagnosis: use *Cicadulina mbila* to transmit to *Zea mays* cv. Golden Bantam.

**References**


Bock, K. (1974). *CMI/AAB Descr. Pl. Viruses* No. 133, 4 pp.

Bock, K.R., Guthrie, E.J. and Woods, R.D. (1974). *Ann. appl. Biol.* **77**: 289.

Boulton, M.I., King, D.I., Donson, J. and Davies, J.W. (1991). *Virology* **183**: 114.

Damsteegt, V.D. (1983). *Plant Dis.* **67**: 734.

Fuller, C. (1901). In: *1st Rep. Gov. Entomologist, Natal*, pp. 17–18.

Gorter, G.J.M.A. (1953). *Union S. Afr. Sci. Bull.*

Howarth, A.J. and Vandemark, G.J. (1989). *J. gen. Virol.* **70**: 2717.

Howell, S.H. (1984). *Nucl. Acids Res.* **12**:

Lazarowitz, S.G. (1988). *Nucl. Acids Res.* **16**: 229.

Mullineaux, P.M., Donson, J., Morris-Krsinich, B.A.M., Boulton, M. and Davies, J.W. (1984). *EMBO J.* **3**: 3063.

Pinner, M.S., Markham, P.G., Markham, R.H. and Dekker, E.L. (1988). *Pl. Path.* **37**: 74.

Rose, D.J.W. (1973*a*). *Bull. ent. Res.* **62**: 477.

Rose, D.J.W. (1973*b*). *Bull. ent. Res.* **62**: 497.

Rose, D.J.W. (1974). *Ann. appl. Biol.* **76**: 199.

Rose, D.J.W. (1978). *Ann. Rev. Ent.* **23**: 259.

Seth, M.L., Raychaudhuri, S.P. and Singh, D.V. (1972*a*). *Curr. Sci.* **41**: 684.

Seth, M.L., Raychaudhuri, S.P. and Singh, D.V. (1972*b*). *Pl. Dis. Reptr* **56**: 424.

Storey, H.H. (1924). *Nature, Lond.* **114**; 245.

Storey, H.H. (1925). *Science Bull.* **39**: 30 pp.

Storey, H.H. and McClean, A.P.D. (1930). *Ann. appl. Biol.* **17**: 691.

Von Wechmar, M.B. and Milne, R.G. (1983). *Phytopathology* **72**: 164.


# Maize stripe *tenuivirus*

**Data collated by R.E. Gingery, 1984.**

*Synonyms:* maize chlorotic stripe virus (Autrey, 1983; Gingery and Autrey, 1984), maize hoja blanca virus (Gingery *et al.*, 1979; Trujillo *et al.*, 1974; Lastra and Carballo, 1985), sorghum chlorosis virus (Mendoza Robles, 1986; Peterschmitt *et al.*, 1991). *Standard acronym:* MSpV.

First reported in *Zea mays*; from Florida, U.S.A; by Gingery *et al.* (1981).

**Natural host range and symptoms**

Symptoms persist (chlorosis becomes progressively more severe).

*Rottboellia exaltata, Zea mays* — stunting, chlorotic leaf striping and apical bending.

*Sorghum bicolor* — chlorotic stripes and bands in leaves, little or no panicle forms, stunting.

**Transmission**

Transmitted by a vector; an insect; *Peregrinus maidis*; Delphacidae. Transmitted in a persistent manner. Virus retained when the vector moults; multiplies in the vector; transmitted congenitally to the progeny of the vector (via eggs); not transmitted by mechanical inoculation; not transmitted by contact between plants; not transmitted by seed; not transmitted by pollen.

**Geographical distribution**

Spreads in Australia, Botswana, Guadeloupe, India, Kenya, Mauritius, Nigeria, Peru, the Philippines, Reunion, Sao Tome and Principe, the USA, and Venezuela.

**Experimental host range**

Few (<3) families susceptible.

Diagnostically susceptible

*Rottboellia exaltata, Zea mays* — systemic chlorotic striping, apical bending.
*Sorghum bicolor* — variable.

Diagnostically insusceptible

*Avena sativa, Oryza sativa, Triticum aestivum.*

Maintenance and propagation hosts

*Zea mays.*

Assay hosts (**L**ocal lesions or **W**hole plants)

*Zea mays* (W).

Monocotyledonous Hosts *(**Susceptible** or Insusceptible)*

*Avena sativa. Chloris gayana. Eleusine coracana.* ***Hordeum vulgare****. Oryza sativa. Saccharum officinarum.* ***Secale cereale****.* ***Sorghum bicolor****. Triticum aestivum.* ***Zea mays****.*

Sources of host-range data

Gingery *et al.* (1981); Greber (1981).

**Purification**

Method: Gingery *et al.* (1981).

**Particle morphology**

Virions rhabdo– or bullet–shaped; Virions are flexuous filaments 3–4 nm wide, or spiral-shaped filaments formed from coils of the 3 nm filaments; not enveloped.

**Physical properties**

Usually four sedimenting components in purified preparations (sometimes more); sedimentation coefficient 187 *S*. Density 1.28 g $cm^{-3}$ in CsCl.

**Biochemical properties**

Virions contain 5.2 % nucleic acid (±0.3); 94.8 % protein; 0 % lipid.

*Genome* consists of RNA; single–stranded; linear; of five parts; largest (or only) genome part the largest 9.12 kb; the 2nd largest 3.58 kb; the 3rd largest 2.38 kb; the 4th largest 2.23 kb; the 5th largest 1.32 kb. Base composition 18.1 % G; 26.3 % A; 17.8 % C; 37.8 % U. Additional factor not required for infectivity.

*Sequence database accession code(s):*

L13438 Em(40)_vi:MZSNCPNS4 Gb(84)_vi:MZSNCPNS4P Maize stripe virus noncapsid protein (NCP); nonstructural protein (NS4) genes. 5/93 2,227bp.

L13446 Em(40)_vi:MZSNSRNA5 Gb(84)_vi:MZSNSRNA5P Maize stripe virus nonstructural protein RNA5, complete cds. 5/93 1,317bp.

M57426 Em(40)_vi:MZSRNA3 Gb(84)_vi:MZSRNA3 Maize stripe virus RNA 3 nonstructural protein (NS-3) and nucleocapsid protein (N) genes, com

M60181 Em(40)_vi:MZSMNCP Gb(84)_vi:MZSMNCP Maize stripe virus major noncapsid protein mRNA, complete cds. 2/91 720bp.

S40180 Em(40)_un:S40180 Gb(84)_vi:S40180 RNA4: NCP=non-capsid protein, NS4=non-structural protein NS4 maize stripe virus MStV, Genomi

S58504 Em(40)_vi:S58504 Gb(84)_vi:S58504 NS5=44.2 kda protein (RNA5) maize stripe virus MStV, Genomic RNA, 1317 nt. 1/94 1,317bp. 6 sequences.

*Virion protein(s)* one; $M_r$ 32700 (±500).

*Virus–coded non–virion proteins* isolated; one protein found. $M_r$ 16500 (±300).

**Cytopathology**

Virions found in mesophyll and phloem; in cytoplasm. Inclusions present in infected cells; are unusual in shape; amorphous, semi-opaque, and filamentous electron dense laminated aggregates; they do not contain virions (apparently the filamentous electron dense inclusions contain non-capsid protein. No nucleoprotein has been shown in either type of inclusion). Other cellular changes: induction of smaller vacuoles and increasing amounts of electron dense cytoplasm after 4–5 weeks apparently due to the large volume of amorphous semi-opaque inclusions.

**Taxonomy and Relationships**

*Tenuivirus.*

Virus(es) with serologically related virions: rice stripe virus is closely related; rice grassy stunt virus is distantly related. Virus(es) with serologically unrelated virions: rice hoja blanca and European wheat striate mosaic viruses.

The maize stripe virus reported from India (Raychaudhuri *et al.*, 1977) is maize mosaic rhabdovirus (Sharma and Payak, 1983).

**References**


Ammar, E.D., Gingery, R.E. and Nault, L.R. (1985). *Phytopathology* **75**: 84.

Autrey, L.J.C. (1983). *Proc. Maize Virus Dis. Coll. and Wksh. 1982*, p. 167.

Capoor, S.P., Rao, D.G. and Varma, P.M. (1967). *Ind. J. Agric. Sci.* **38**: 198.

Falk, B.W. and Tsai, J.H. (1984). *Phytopathology* **74**: 909.

Gingery, R.E. (1983). *Proc. Int. Maize Virus Dis. Coll. and Wksh.*, 1982; eds D.T. Gordon, J.K. Knoke, L.R. Nault and R.M. Ritter. Wooster, Ohio Agric. Res. and Dev. Center.

Gingery, R.E. (1985). *CMI/AAB Descr. Pl. Viruses* No. 300, 4 pp.

Gingery, R.E. and Autrey, L.J.C. (1984). *Maize Virus Dis. Newsl.* **1**: 49.

Gingery, R.E., Nault, L.R., Tsai, J.H. and Lastra, R.J. (1979). *Pl. Dis. Reptr* **63**: 341.

Gingery, R.E., Nault, L.R. and Bradfute, O.E. (1981). *Virology* **112**: 99.

Gingery, R.E., Nault, L.R. and Yamashita, S. (1983). *J. gen. Virol.* **64**: 1765.

Greber, R.S. (1981). *Aust. J. agric. Res.* **32**: 27.

Huiet, L., Klassen, V., Tsai, J.H. and Falk, B.W. (1991). *Virology* **182**: 47.

Huiet, L., Tsai, J.H. and Falk, B.W. (1992). *J. gen. Virol.* **73**: 1603.

Huiet, L., Tsai, J.H. and Falk, B.H. (1993). *J. gen. Virol.* **74**: 459.

Kulkarni, H.Y. (1973). *Ann. appl. Biol.* **75**: 205.

Lastra, R. and Carballo, O. (1985). *Phytopath. Z.* **144**: 168.

Mendoza Robles, J.L. (1986). *Agric. Tecnica en Mexico* **12**: 313.

Peterschmitt, M., Ratna, A.S., Sacks, W.R., Reddy, D.V.R. and Mughogho, L.K. (1991). *Ann. appl. Biol.* **118**: 57.

Raychaudhuri, S.P., Seth, M.L., Renfro and Varma, A. (1977). *Proc. int. Maize Virus Dis. Colloq. Wksh.*, Wooster, Ohio, 1976: 186.

Sharma, R.C. and Payak, M.M. (1983). *Proc. int. Maize Virus Dis. Colloq. Wksh.*, Wooster, Ohio, 1982: 186.

Storey, H.H. (1936). *East Afr. Agric. J.* **1**: 333.

Trujillo, G.E., Acosta, J.M. and Pinero, A. (1974). *Pl. Dis. Reptr* **58**: 122.

Tsai, J.H. (1975). *Pl. Dis. Reptr* **59**: 830.
Tsai, J.H. and Falk, B.W. (1988). In: *Advances in Disease Vector Research*, Vol. 5. Springer-Verlag, New York.

## Maize white line mosaic satellivirus

**Data collated by A.A. Brunt, 1991.**

First reported in *Zea mays*; from the U.S.A.; by Gingery and Louie (1984).

**Natural host range and symptoms**

*Zea mays* — unknown (found only in complex with maize white line mosaic virus).

**Transmission**

Probably in soil.

**Geographical distribution**

Spreads in France and the USA (Ohio).

Monocotyledonous Hosts *(**Susceptible** or Insusceptible)*

***Zea mays***.

**Purification**

Method: Gingery and Louie (1984).

**Particle morphology**

Virions isometric; not enveloped; 17 nm in diameter; rounded in profile.

**Physical properties**

One sedimenting component in purified preparations; sedimentation coefficient 47 *S*. Density 1.355 g $cm^{-3}$ in CsCl.

**Biochemical properties**

*Genome* consists of RNA; single–stranded; unipartite; largest (or only) genome part 1.168 kb.

*Sequence database accession code(s):*

M55012 Em(40)_vi:SVMSECG Gb(84)_vi:SVMSECG Satellite virus of maize white line mosaic virus ORF, complete cds. 2/91 1,168bp. 1 sequence.

*Virion protein(s)* one; $M_r$ 24700; coat protein.

**Cytopathology**

Virions found in leaves and roots; in cytoplasm. Inclusions absent from infected cells.

**Taxonomy and Relationships**

Satellivirus.

Virus(es) with serologically unrelated virions: panicum mosaic satellite and tobacco necrosis satellite viruses.

**References**

Gingery, R.E. and Louie, R. (1984). *Phytopathology* **74**: 808.

# Maize white line mosaic virus

**Data collated by R.E. Gingery and R. Louie, 1985. Revised 1989 by R.E. Gingery.**

*Synonyms:* maize dwarf ringspot virus, possibly maize white line virus. *Standard acronym:* MWLMV.

First reported in *Zea mays*; from Vermont, U.S.A; by Gotlieb and Liese (1980).

**Natural host range and symptoms**

Symptoms persist.

*Digitaria sanguinalis, Panicum dichotomiflorum, Setaria faberi, S. viridis* — dark green mottle.

*Zea mays* — mottle, mosaic and characteristic chlorotic white lines, small ears with few or no kernels.

**Transmission**

Virus not transmitted by mechanical inoculation; not transmitted by contact between plants; not transmitted by seed; not transmitted by pollen.

**Geographical distribution**

Spreads in France, Italy, and the USA.

**Experimental host range**

Several (3–9) families susceptible.

Diagnostically susceptible

*Zea mays* — mottle, mosaic, discrete chlorotic white lines, stunting.

Maintenance and propagation hosts

*Zea mays*.

Assay hosts (**L**ocal lesions or **W**hole plants)

*Zea mays* (W).

Monocotyledonous Hosts ***(Susceptible** or Insusceptible)*

***Zea mays*.**

Sources of host-range data

Louie *et al.* (1982).

**Properties of particles in sap**

Leaf sap contains many virions.

**Purification**

Method: de Zoeten *et al.* (1980); Louie *et al.* (1982).

**Particle morphology**

Virions isometric; not enveloped; 30–35 nm in diameter; rounded in profile, or angular in profile; with a conspicuous capsomere arrangement.

**Physical properties**

One sedimenting component in purified preparations; sedimentation coefficient 117 *S*. Density 1.333 g $cm^{-3}$ in CsCl (unfixed), or 1.353 g $cm^{-3}$ in CsCl (when fixed in glutaraldehyde).

**Biochemical properties**

Virions contain 20.9 % nucleic acid; 79.1 % protein.

*Genome* consists of RNA; single–stranded; probably linear; unipartite; largest (or only) genome part 1.3–1.6 kb. Genomic nucleic acid isolated by de Zoeten *et al.* (1980). Base composition 30 % G; 19.7 % A; 26 % C; 24.3 % U.

*Virion protein(s)* one; $M_r$ 32000–35000. Method of preparation: de Zoeten *et al.* (1980).

**Cytopathology**

Virions found in all parts of the host plant; in cytoplasm. Inclusions absent from infected cells.

**Comments**

Satellite-like virions are associated with this virus.

**References**


Boothroyd, C.W. and Israel, H.W. (1980). *Plant Dis.* **64**: 218.

de Zoeten, G.A., Arny, D.C., Grau, C.R., Saad, S.M. and Gaard, G. (1980). *Phytopathology* **70**: 1019.

Gingery, R.E. and Louie, R. (1984). *Phytopathology* **74**: 808.

Gotlieb, A.R. and Liese, A.L. (1980). *Phytopathology* **70**: 462.

Lapierre, H., Moreau, J.C. and Molin, G. (1976). *Poljopr. Znan. Smotra. Agric. Conspect. Sci.* **39**: 187.

Louie, R., Gordon, D.T., Knoke, J.K., Gingery, R.E., Bradfute, O.E. and Lipps, P.E. (1982). *Plant Dis.* **66**: 167.

Louie, R., Gordon, D.T., Madden, L.V. and Knoke, J.K. (1983). *Plant Dis.* **67**: 371.


## Maize yellow stripe (?) *tenuivirus*

**Data collated by E.D. Ammar, 1989.**

*Synonyms:* maize yellow stripe virus, maize fine stripe virus, maize chlorotic stunt virus.

First reported in *Zea mays*; from Giza, Egypt; by Ammar *et al.* (1984).

**Natural host range and symptoms**

Symptoms persist, or vary cyclically over a few weeks.

*Zea mays, Sorghum vulgare, S. bicolor, Triticum aestivum, Hordeum vulgare* — yellow striping or chlorosis.

**Transmission**

Transmitted by a vector; an insect; *Cicadulina chinai*; Cicadellidae. Transmitted in a persistent manner. Virus retained when the vector moults; not transmitted by mechanical inoculation; not transmitted by contact between plants.

**Ecology and control**

Studies reported by Ammar *et al.* (1987); Ammar *et al.* (1989).

**Geographical distribution**

Spreads in Egypt.

**Experimental host range**

Few (<3) families susceptible. Experimentally infected plants mostly show systemic mosaics.

Diagnostically susceptible

*Zea mays, Sorghum vulgare, S. bicolor, Triticum aestivum, Hordeum vulgare* — yellow leaf striping.

Maintenance and propagation hosts

*Zea mays, Sorghum vulgare, S. bicolor, Triticum aestivum, Hordeum vulgare.*

Assay hosts (**L**ocal lesions or **W**hole plants)

*Zea mays* (W).

Monocotyledonous Hosts *(**Susceptible** or Insusceptible)*

***Hordeum vulgare*. *Sorghum bicolor*. *Triticum aestivum*. *Zea mays*.**

Sources of host-range data

Aboul-Ata and Ammar (1985); Ammar *et al.* (1987, 1989).

**Properties of particles in sap**

Leaf sap contains many virions. Electron microscopy: UA or PTA may be used without fixation.

**Purification**

Method: Ammar *et al.* (1990).

**Particle morphology**

Virions filamentous; usually flexuous; 5–7 nm wide. Axial canal obscure.

**Physical properties**

One sedimenting component in purified preparations; sedimentation coefficient 87 *S*.

*Virus–coded non–virion proteins* isolated (Ammar *et al.* (1990)); one protein found. $M_r$ 13800.

**Cytopathology**

Virions found in leaves, mesophyll, vascular parenchyma, phloem and companion cells; in cytoplasm. Inclusions present in infected cells; are amorphous X–bodies; they contain virions (helical filaments are found). Other cellular changes: tubular helical structures which are found abundantly in cytoplasmic inclusions, sometimes associated with mitochondria.

**Taxonomy and Relationships**

Possible *Tenuivirus*.

Virus(es) with serologically related virions: maize mottle — chlorotic stunt. Virus(es) with serologically unrelated virions: maize stripe, maize chlorotic dwarf, maize chlorotic mottle, maize dwarf mosaic (A, B or D), maize streak and maize mosaic viruses (Ammar *et al.*, 1990).

Like tenuiviruses it has thin filamentous virions and similar non-virion proteins, but differs in cytopathology and in being transmitted by a leafhopper and not a planthopper. Best tests for diagnosis: transmission with *Cicadulina chinai* (leafhoppers).

**References**


Aboul-Ata, A.E. and Ammar, E.D. (1985). *Abstr. 4th Int. Conf. Impact of Viral Dis. on the Devel. of African and Middle East Countries* 14–19 April 1985, Rabat, Morocco.

Ammar, E.D., Elnager, S., Tolba, A.E. and Aboul-Ata, A.E. (1984). *Abstr. 6th Cong. Medit. Phytopath. Union* 1–6 Oct., 1984, Cairo, Egypt.

Ammar, E.D., Aboul-Ata, A.E., El-Sheikh, M.A. and Sewify, G.H. (1987). *J. Phytopath.* **19**: 97.

Ammar, E.D., Elnager, S., Aboul-Ata, A.E. and Sewify, G.H. (1989). *J. Phytopath.* **126**: 246.

Ammar, E.D., Gingery, R.E., Gordon, D.T. and Aboul-Ata, A.E. (1990). *Phytopathology* **80**: 303.

# Malva vein clearing *potyvirus*

**Data collated by G.R. Johnstone, 1987.**

*Synonyms:* malva green mosaic virus, malva yellow vein mosaic virus, malva mosaic virus. *Standard acronym:* MVCV. *Strains:* Hein (1957) distinguishes two strains, that are the yellow vein mosaic and green mosaic strains of Costa and Duffus (1957).

First reported in *Malva sylvestris*; from Germany; by Hein (1957).

### Natural host range and symptoms

Symptoms persist.

*Malva borealis, M. neglecta, M. nicaensis, M. parviflora, M. rotundifolia, M. sylvestris, Lavatera assurgentiflora, L. cretia, L. trimestris* — vein clearing and yellow mosaic.

### Transmission

Transmitted by a vector; an insect; *Aphis umbrella* (syn. *Aphis malvae* Koch) and *Myzus persicae* are vectors, whereas *Acyrthosiphon solani, Coloradoa rufomaculata, Macrosiphum euphorbiae, Myzus ornatus, M. ascalonicus* and *Aulocorthum circumflexum* are non-vectors; Aphididae. Transmitted in a non–persistent manner. Virus transmitted by mechanical inoculation.

### Geographical distribution

Spreads in Australia (Tasmania), Brazil, the former Czechoslovakia, Germany, Israel, Italy, Portugal, the USA (California), the former USSR, the former Yugoslavia.

### Experimental host range

Few (<3) families susceptible (and only some genera of the Malvaceae are infected).

Diagnostically susceptible

*Malva parviflora, M. neglecta, M. sylvestris, M. rotundifolia, Lavatera assurgentiflora.*

Diagnostically insusceptible

*Chenopodium amaranticolor, Cucumis sativus, Datura stramonium, Gossypium hirsutum, Hibiscus esculentus.*

Assay hosts (**L**ocal lesions or **W**hole plants)

*Malva parviflora* (W); *M. neglecta* (W); *M. sylvestris* (W); *M. rotundifolia* (W); *Lavatera assurgentiflora* (W).

Dicotyledonous Hosts *(**Susceptible** or Insusceptible)*

*Amaranthus retroflexus. Beta vulgaris. Capsicum annuum. Chenopodium amaranticolor. Chenopodium quinoa. Coriandrum sativum. Cucumis sativus. Cucurbita pepo. Datura stramonium. Gomphrena globosa. Gossypium hirsutum. Hibiscus esculentus.* ***Lavatera trimestris****. Lycopersicon esculentum. Nicotiana glutinosa. Nicotiana sylvestris. Nicotiana tabacum. Phaseolus vulgaris. Stellaria media.*

Additional hosts

***Abutilon indicum, A. theophrastii, Althaea officinalis, A. sinensis****, A. rosea, Ameria vulgaris,* ***Anoda cristata, A. dillerriana, A. lavateroides****, Astrocarpa sesmoides, Beta maritima, Delphinium ajacis,* ***Gossypium barbadense, Hibiscus cannabinus, H.***

***sabdariffa, H. trionum, Kitaibelia vitifolia***, *Lamium purpureum*, ***Lavatera ambigua, Lavatera arborea, Malva alcea, M. crispa, M. meluca, M. moschata, M. pusilla, M. verticillata, Malvastrum capense***, *Melandrium nodiflorum*, ***Napea dioica***, *Salsola kali, Scabiosa columbaria, Schizanthus pinnatus*, ***Sida napaeae, S. rhombifolia, Sidalcea malvaefolia*** and ***Urocarpidum peruvianum***.

Sources of host-range data

Hein (1957); Horváth *et al.* (1979).

**Properties of particles in sap**

TIP: 60–65 °C. LIV: 1–8 days. DEP: $\log_{10}$ minus 2–3. Leaf sap contains few virions.

**Particle morphology**

Virions filamentous; not enveloped; usually flexuous; with a clear modal length; of 850–860 nm; 12 nm wide. Axial canal obscure. Basic helix obscure.

**Physical properties**

Presumably one sedimenting component in purified preparations.

**Biochemical properties**

Virions contain *c.* 5 % nucleic acid; *c.* 95 % protein.

*Genome* consists of RNA; single-stranded; linear; unipartite (presumably).

*Virus-coded non-virion proteins* isolated; probably three proteins found. $M_r$ coding for inclusion protein and helper component.

**Cytopathology**

Virions found in all parts of the host plant; in cytoplasm. Inclusions present in infected cells; are crystals in the cytoplasm, pinwheels, and unusual in shape; and scrolls; they do not contain virions.

**Taxonomy and Relationships**

*Potyvirus: Potyviridae.*

**References**

Costa, A.S. and Duffus, J.E. (1957). *Pl. Dis. Reptr* **41**: 1006

Hein, A. (1957). *Phytopath. Z.* **28**: 205.

Horváth, J., Mamula, D., Besada, W.H. and Juretič, N. (1979). *Phytopath. Z.* **95**: 51.

# Malva veinal necrosis (?) *potexvirus*

**Data collated by A.A. Brunt, 1990 and E.W. Kitajima, 1991.**

*Synonyms:* malva veinal chlorosis virus. *Standard acronym:* MVNV.

First reported in *Malva parviflora*; from Brazil; by Costa and Kitajima (1970).

**Natural host range and symptoms**

Symptoms persist.

*Malva parviflora*, other *Malva* spp. — veinal necrosis.

**Transmission**

Virus transmitted by mechanical inoculation.

**Geographical distribution**

Spreads in Brazil.

**Experimental host range**

Several (3–9) families susceptible. Experimentally infected plants mostly show systemic mosaics, mottles, ringspots or necrosis.

Diagnostically susceptible

*Malva parviflora, M. rotundifolia* — necrotic local lesions, then systemic veinal necrosis.

*Gomphrena globosa* — necrotic local lesions.

*Chenopodium amaranticolor, C. quinoa* — local lesions.

*Sesamum indicum, Tetragonia expansa.*

Maintenance and propagation hosts

*Malva parviflora.*

Assay hosts (**L**ocal lesions or **W**hole plants)

*Malva parviflora* (L,W), *Chenopodium amaranticolor* (L), *Gomphrena globosa* (L).

Dicotyledonous Hosts *(**Susceptible** or Insusceptible)*

***Chenopodium amaranticolor. Chenopodium quinoa.*** *Datura stramonium.* ***Gomphrena globosa.*** *Gossypium hirsutum. Hibiscus esculentus. Nicotiana glutinosa. Nicotiana tabacum. Physalis floridana.* ***Sesamum indicum.*** *Solanum tuberosum.* ***Tetragonia expansa.***

**Properties of particles in sap**

TIP: 70–75 °C. DEP: $\log_{10}$ minus 6. Leaf sap contains few virions.

**Particle morphology**

Virions filamentous; usually flexuous; with a clear modal length; of 525 nm; 13 nm wide. Basic helix obscure.

**Taxonomy and Relationships**

Possible *Potexvirus.*

**References**

Costa, A.S. and Kitajima, E.W. (1970). *Bragantia* **29**: 51.

# Malvastrum mottle virus

**Data collated by J.L. Dale, 1987.**

First reported in *Malvastrum coromandelianum*; from Queensland, Australia.

**Natural host range and symptoms**

Symptoms persist.

*Malvastrum coromandelianum* and *Euphorbia peplus* — leaf mottling.

**Transmission**

Virus transmitted by mechanical inoculation.

**Geographical distribution**

Spreads in Australia.

**Experimental host range**

Several (3–9) families susceptible.

Diagnostically susceptible

*Chenopodium amaranticolor* — necrotic local lesions, then systemic chlorotic spotting and malformation.

*C. quinoa* — chlorotic local lesions; then systemic chlorosis and malformation.

*Nicotiana clevelandii* — systemic chlorosis.

*Gomphrena globosa* — faint necrotic local lesions; systemic mottling.

Maintenance and propagation hosts

*Chenopodium amaranticolor.*

Assay hosts (**L**ocal lesions or **W**hole plants)

*Chenopodium amaranticolor* (L), *Gomphrena globosa* (L).

Dicotyledonous Hosts *(**Susceptible** or Insusceptible)*

***Chenopodium amaranticolor***. ***Chenopodium quinoa***. *Datura stramonium*. ***Gomphrena globosa***. *Helianthus annuus*. *Lycopersicon esculentum*. ***Nicotiana clevelandii***.

**Particle morphology**

Virions isometric; not enveloped; 28 nm in diameter; rounded in profile; without a conspicuous capsomere arrangement.

**Biochemical properties**

Genome of three parts (1 minor and 2 major); largest (or only) genome part the largest 1.78 kb; the 2nd largest 1.7 kb; the 3rd largest 0.47 kb.

# Maracuja mosaic (?) *tobamovirus*

**Data collated by D.-E. Lesemann, 1987.**

*Standard acronym:* MarMV.

First reported in *Passiflora edulis*; from Peru; by Fribourg *et al.* (1987).

**Natural host range and symptoms**

Symptoms persist.

*Passiflora edulis* — mosaic and leaf crinkling.

**Transmission**

Transmitted by means not involving a vector. Virus transmitted by mechanical inoculation; transmitted by contact between plants.

**Ecology and control**

Studies reported by Fribourg *et al.* (1987).

**Geographical distribution**

Spreads in Peru.

**Experimental host range**

Many (>9) families susceptible.

Diagnostically susceptible

*Passiflora edulis* — systemic leaf mosaic, crinkle.

*Nicotiana benthamiana* — necrotic local lesions; then systemic line patterns, mosaic.

*Chenopodium quinoa, C. amaranticolor, Luffa acutangula* — necrotic local lesions; not systemic.

*Nicotiana tabacum* — necrotic local lesions, necrotic rings and lines; not systemic.
*Cucumis sativus, Nicotiana glutinosa* — necrotic local lesions; not systemic.
*Antirrhinum majus* — symptomless local infection; not systemic.
*Datura stramonium* — chlorotic local lesions; not systemic.

Diagnostically insusceptible

*Brassica pekinensis, Capsicum annuum, Citrullus vulgaris, Phaseolus vulgaris* cv. Monroe, *Vigna unguiculata.*

Maintenance and propagation hosts

*Nicotiana benthamiana.*

Assay hosts (**L**ocal lesions or **W**hole plants)

*Chenopodium quinoa* (L).

Dicotyledonous Hosts (***Susceptible*** *or Insusceptible*)

***Antirrhinum majus***. *Brassica campestris* ssp. *pekinensis*. *Cajanus cajan*. *Capsicum annuum*. *Capsicum frutescens*. ***Chenopodium amaranticolor***. ***Chenopodium murale***. ***Chenopodium quinoa***. *Clitoria ternatea*. ***Cucumis sativus***. ***Cucurbita maxima***. ***Cucurbita moschata***. *Datura metel*. ***Datura stramonium***. *Dolichos biflorus*. ***Gomphrena globosa***. ***Helianthus annuus***. ***Luffa acutangula***. ***Lycopersicon chilense***. ***Lycopersicon esculentum***. *Lycopersicon pimpinellifolium*. *Matthiola incana*. *Montia perfoliata*. ***Nicandra physaloides***. ***Nicotiana benthamiana***. ***Nicotiana glutinosa***. ***Nicotiana tabacum***. ***Passiflora edulis***. *Phaseolus acutifolius*. *Phaseolus vulgaris*. ***Physalis floridana***. ***Salvia splendens***. *Sesamum indicum*. *Solanum tuberosum*. *Vigna unguiculata*. *Vigna unguiculata* ssp. *cylindrica*. ***Zinnia elegans***.

Sources of host-range data

Fribourg *et al.* (1987).

**Properties of particles in sap**

TIP: 90 °C. LIV: 90 days. DEP: $\log_{10}$ minus 10 (in *Chenopodium quinoa* sap). Leaf sap contains many virions.

**Purification**

Method: Fribourg *et al.* (1987); Fribourg and Nakashima (1984).

**Particle morphology**

Virions rod–shaped; not enveloped; usually straight; with a clear modal length; of 304 nm; 18 nm wide. Axial canal obvious.

**Physical properties**

One sedimenting component in purified preparations.

*Virion protein(s)* two; $M_r$ of the larger 17400. $M_r$ of 2nd largest 15900. Method of preparation: Fribourg *et al.* (1987).

**Cytopathology**

Virions found in mesophyll, epidermis, vascular parenchyma and xylem; in cytoplasm. Inclusions present in infected cells; are crystals in the cytoplasm, amorphous X–bodies, and unusual in shape; tonoplast-associated small vesicles; they contain virions (in the crystals), or they do not contain virions (in the others).

**Taxonomy and Relationships**

Possible *Tobamovirus*.

Virus(es) with serologically related virions: cucumber green mottle mosaic, frangipani mosaic, odontoglossum ringspot, tobacco mosaic and ullucus tobamoviruses, but distantly. Virus(es) with serologically unrelated virions: ribgrass mosaic and Sammons' opuntia tobamoviruses.

**References**

Fribourg, C.E. and Nakashima, J. (1984). *Phytopathology* **74**: 1363.

Fribourg, C.E., Koenig, R. and Lesemann, D.-E. (1987). *Phytopathology* **77**: 486.

# Marigold mottle *potyvirus*

**Data collated by A.A. Brunt, 1991.**

*Standard acronym:* MaMoV.

First reported in *Tagetes erecta*; from India; by Naqvi *et al.* (1981).

**Natural host range and symptoms**

Symptoms persist.

*Tagetes erecta* — severe leaf mottling, small malformed flowers, stunting and death.

**Transmission**

Transmitted by a vector; an insect; *Myzus persicae*; Aphididae. Not transmitted by *Aphis gossypii, Brevicoryne brassicae*. Transmitted in a non–persistent manner. Virus transmitted by mechanical inoculation.

**Geographical distribution**

Spreads in India.

**Experimental host range**

Several (3–9) families susceptible.

Diagnostically susceptible

*Amaranthus caudatus, Calendula officinalis, Antirrhinum majus, Capsicum annuum, Datura metel, Nicotiana glutinosa, N. tabacum, Petunia hybrida* — chlorosis of systemically infected leaves.

*Chenopodium amaranticolor, Tetragonia expansa*—chlorotic local lesions; no systemic infection.

Diagnostically insusceptible

*Chenopodium album, C. murale, Cucumis sativus, Cucurbita pepo, Physalis peruviana.*

Maintenance and propagation hosts

*Datura metel, Nicotiana glutinosa.*

Assay hosts (**L**ocal lesions or **W**hole plants)

*Chenopodium amaranticolor* (L), *Datura metel* (W), *Nicotiana glutinosa* (W), *Tetragonia expansa* (L).

Dicotyledonous Hosts ***(Susceptible** or Insusceptible)*

*Ageratum conyzoides. Amaranthus caudatus.* ***Antirrhinum majus. Beta vulgaris.*** *Brassica oleracea* ssp. *botrytis. Cajanus cajan.* ***Calendula officinalis. Capsicum annuum. Cassia tora.*** *Catharanthus roseus. Celosia cristata. Chenopodium album.* ***Chenopodium amaranticolor. Chenopodium ambrosioides.*** *Chenopodium murale. Cucumis sativus. Cucurbita moschata. Cucurbita pepo. Cyamopsis tetragonoloba. Dahlia pinnata.* ***Datura metel.*** *Daucus carota. Dianthus barbatus. Gomphrena globosa.* ***Lycopersicon esculentum. Nicotiana glutinosa. Petunia hybrida.*** *Phlox drummondii.*

*Physalis peruviana. Pisum sativum. Raphanus sativus. Solanum melongena.* ***Solanum nigrum****. Sonchus oleraceus.* ***Stellaria media****.* ***Tetragonia expansa****. Tropaeolum majus. Vigna unguiculata. Zinnia elegans.*

**Properties of particles in sap**

TIP: 50 °C. LIV: 6 days (at 20–25°C). DEP: $\log_{10}$ minus 2.

**Particle morphology**

Virions filamentous; not enveloped; usually flexuous; with a clear modal length; of 675 nm. Axial canal obscure.

**Physical properties**

One sedimenting component in purified preparations; sedimentation coefficient 152 *S*. $A_{260}/A_{280}$ ratio 1.28.

**Cytopathology**

Inclusions present in infected cells; are pinwheels; they do not contain virions.

**Taxonomy and Relationships**

*Potyvirus: Potyviridae.*

Virus(es) with serologically related virions: datura mosaic and tobacco etch viruses. Virus(es) with serologically unrelated virions: brinjal mild mosaic and melilotus mosaic viruses.

**References**

Naqvi, Q.A., Hadi, S. and Mahmood, K. (1981). *Plant Dis.* **65**: 271.

# Melandrium yellow fleck *bromovirus*

**Data collated by J. Horváth, 1987.**

*Standard acronym:* MYFV.

First reported in *Melandrium album*; from Hungary; by Hollings and Stone (1978).

**Natural host range and symptoms**

Symptoms persist.
*Melandrium album* — systemic vein clearing and yellow flecks.

**Transmission**

Virus transmitted by mechanical inoculation; not transmitted by seed.

**Geographical distribution**

Spreads in Hungary.

**Experimental host range**

Many (>9) families susceptible.

Diagnostically susceptible

*Chenopodium quinoa, Phaseolus vulgaris* — necrotic local lesions.
*Nicotiana clevelandii* — chlorotic/necrotic local lesions, then systemic mottle.
*Gomphrena globosa* — necrotic local lesions, then systemic mottle.
*Datura stramonium* — chlorotic spots, not systemic.
*Nicotiana glutinosa* — chlorotic local lesions.
*Tetragonia expansa* — necrotic local lesions, then systemic vein yellowing.
*Lycium barbarum* — chlorotic/necrotic local lesions.
*Physalis aequata* — chlorotic/necrotic systemic vein clearing, mosaic.

Diagnostically insusceptible

*Dianthus caryophyllus* cv. Joker, *Brassica pekinensis, B. oleracea, Crambe abyssinica, Trifolium incarnatum.*

Maintenance and propagation hosts

*Nicotiana clevelandii, Gomphrena globosa, Melandrium album, Tropaeolum majus, Physalis franchetti.*

Assay hosts (**L**ocal lesions or **W**hole plants)

*Chenopodium quinoa* (L), *C. amaranticolor* (L), *Phaseolus vulgaris* (L), *Emilia flammea* (L), *Phaseolus tuberosus* (L), *Gomphrena globosa* (W), *Pisum sativum* (W), *Nolana prostrata* (W).

Dicotyledonous Hosts *(**Susceptible** or Insusceptible)*

***Calendula officinalis***. ***Chenopodium amaranticolor***. ***Chenopodium hybridum***. ***Chenopodium murale***. ***Chenopodium quinoa***. *Crambe abyssinica*. ***Cucumis sativus***. ***Cucurbita maxima***. ***Cucurbita moschata***. ***Datura ferox***. ***Datura metel***. ***Datura stramonium***. ***Dianthus caryophyllus***. ***Gomphrena globosa***. ***Helianthus annuus***. ***Melandrium album***. ***Nicandra physaloides***. ***Nicotiana clevelandii***. ***Nicotiana glutinosa***. ***Nicotiana glutinosa × clevelandii***. ***Nicotiana tabacum***. ***Petunia hybrida***. ***Phaseolus acutifolius***. ***Phaseolus lunatus***. ***Phaseolus vulgaris***. ***Physalis floridana***. ***Tetragonia expansa***. *Trifolium incarnatum*. ***Tropaeolum majus***. ***Vicia faba***.

Sources of host-range data

Horváth *et al.* (1988).

**Properties of particles in sap**

TIP: 85 °C. LIV: 25 days. DEP: $\log_{10}$ minus 5.

**Purification**

Method: Hollings *et al.* (1981).

**Particle morphology**

Virions isometric; 25 nm in diameter.

**Biochemical properties**

Virions contain 20 % nucleic acid.
*Genome* consists of RNA; single–stranded.

**Cytopathology**

Virions found in leaves, roots and stems.

**Taxonomy and Relationships**

*Bromovirus: Bromoviridae.*

**References**

Hollings, M. and Horváth, J. (1978). *Rep. Glasshouse Crops Res. Inst.* 1977, p.129.
Hollings, M. and Horváth, J. (1981). *CMI/AAB Descr. Pl. Viruses* No. 236, 4 pp.
Hollings, M., Horváth, J. and Besada, W.H. (1981). *3rd Int. Congr. Pl. Path., München*, 1978 p.26.
Horváth, J., Hollings, M. and Barton, R.J. (1988). *Acta Phytopath. Ent.* **23**: 449.

# Melilotus mosaic (?) potyvirus

**Data collated by A.A. Brunt, 1991.**

*Standard acronym:* MeMV.

First reported in *Melilotus indica*; from India (Aligarh); by Naqvi and Mahmood (1978).

**Natural host range and symptoms**

Symptoms persist.
*Melilotus indica* — leaf mosaic, mottling and stunting.

**Transmission**

Not transmitted by *Aphis craccivora, A. gossypii*. Virus transmitted by mechanical inoculation; not transmitted by seed.

**Geographical distribution**

Spreads in India.

**Experimental host range**

Several (3–9) families susceptible.

Diagnostically susceptible

*Chenopodium amaranticolor, Tetragonia expansa, Verbena hybrida* — local lesions.
*Cajanus cajan, Crotalaria medicaginea, Datura stramonium, Nicotiana glutinosa, N. tabacum, Petunia hybrida* — systemic leaf chlorosis, necrosis and distortion.

Maintenance and propagation hosts

*Nicotiana glutinosa.*

Assay hosts (**L**ocal lesions or **W**hole plants)

*Chenopodium amaranticolor* (L).

Dicotyledonous Hosts *(**Susceptible** or Insusceptible)*

***Cajanus cajan*. *Chenopodium amaranticolor*. *Datura stramonium*. *Lycopersicon esculentum*. *Nicotiana glutinosa*. *Nicotiana rustica*. *Nicotiana tabacum*. *Petunia hybrida*. *Solanum tuberosum*. *Tetragonia expansa*.**

Sources of host-range data

Naqvi and Mahmood (1978).

**Properties of particles in sap**

TIP: 60 °C. LIV: 4 days (at 25°C). DEP: $\log_{10}$ minus 4.

**Purification**

Method: Naqvi and Mahmood (1978).

**Particle morphology**

Virions filamentous; usually flexuous; of 600–620 nm; 12 nm wide.

**Physical properties**

Sedimentation coefficient 118 *S*.

**Taxonomy and Relationships**

Unassigned potyvirus.
Virus(es) with serologically related virions: tobacco etch virus.

**Comments**

Although reported to be serologically related to tobacco etch virus, the virus has unusually short virions for a potyvirus.

**References**

Naqvi, Q.A. and Mahmood, K. (1978). *Phytopath. Z.* **93**: 249.

# Melon leaf curl bigeminivirus

**Data collated by A.A. Brunt, S. Molyneux and J.E. Duffus, 1986.**

*Standard acronym:* MLCV.

First reported in *Cucumis melo*; from Imperial Valley, California; by Duffus *et al.* (1985).

**Natural host range and symptoms**

Symptoms persist.
*Cucumis melo* — leaf curling and chlorosis.

**Transmission**

Transmitted by a vector; an insect; *Bemisia tabaci*; Aleyrodidae. Transmitted in a non–persistent manner. Virus transmitted by mechanical inoculation.

**Ecology and control**

Studies reported by Duffus *et al.* (1985).

**Geographical distribution**

Spreads in the USA (California).

Diagnostically susceptible

*Cucumis melo, Citrullus lanatus, Cucumis sativus, Cucurbita maxima, Cucurbita moschata, Phaseolus vulgaris* — chlorosis and leaf malformation.

Maintenance and propagation hosts

*Cucumis melo, Citrullus lanatus, Cucumis sativus* (non hosts of similar squash leaf curl virus).

Assay hosts (**L**ocal lesions or **W**hole plants)

*Cucumis melo* (W), *Citrullus lanatus* (W), *Cucumis sativus* (W).

Dicotyledonous Hosts *(**Susceptible** or Insusceptible)*

***Citrullus lanatus*. *Cucumis melo*. *Cucumis sativus*. *Cucurbita maxima*. *Cucurbita moschata*. *Phaseolus vulgaris*.**

**Particle morphology**

Virions geminate; 22 nm in length.

**Taxonomy and Relationships**

Bigeminivirus: *Geminiviridae*.
Virus(es) with serologically related virions: squash leaf curl virus (by ELISA but not agar gel diffusion tests).

**References**

Duffus, J.E., Liu, H.Y. and Johns, M.R. (1985). *Phytopathology* **75**: 1312.

# Melon necrotic spot *carmovirus*

**Data collated by T. Hibi, 1986.**

*Standard acronym:* MNSV. *Strains:* Japanese isolate, Californian isolate, European isolate.

First reported in *Cucumis melo*; from Japan; by Kishi (1966).

**Natural host range and symptoms**

Symptoms persist.
*Cucumis melo* — necrotic spotting.
*Cucumis sativus* — chlorotic spotting.

**Transmission**

Transmitted by a vector; a fungus; *Olpidium radicale*; Chytridiales. Virus transmitted by mechanical inoculation; transmitted by contact between plants; transmitted by seed (10–40% in melon).

**Geographical distribution**

Spreads in Japan, the Netherlands, the UK, and the USA.

**Experimental host range**

Few (<3) families susceptible.

Diagnostically susceptible

*Cucumis melo* — necrotic local lesions then, occasionally, systemic necrosis.
*Cucumis sativus* — chlorotic local lesions then, occasionally, systemic chlorosis or spotting.
*Citrullus lanatus* — local lesions, not systemic.
*Vigna unguiculata ssp. sesquipedalis* — necrotic spots; not systemic.
*Cucurbita moschata, Lagenaria siceraria* — local lesions; not systemic.

Diagnostically insusceptible

*Chenopodium amaranticolor, Chenopodium quinoa, Nicotiana glutinosa.*

Maintenance and propagation hosts

*Cucumis melo.*

Assay hosts (**L**ocal lesions or **W**hole plants)

*Cucumis melo* (L) and (W) *Cucumis sativus* (L).

Dicotyledonous Hosts *(**Susceptible** or Insusceptible)*

*Apium graveolens. Beta vulgaris. Brassica campestris* ssp. *rapa. Brassica oleracea* ssp. *capitata. Capsicum annuum. Chenopodium amaranticolor. Chenopodium quinoa.* ***Citrullus lanatus. Cucumis melo. Cucumis sativus.*** *Cucurbita maxima.* ***Cucurbita moschata.*** *Cucurbita pepo. Glycine max. Gomphrena globosa. Lactuca sativa. Nicotiana glutinosa. Phaseolus vulgaris. Pisum sativum. Spinacia oleracea. Trifolium repens. Vicia faba.* ***Vigna unguiculata.***

Monocotyledonous Hosts *(**Susceptible** or Insusceptible)*

*Allium cepa.*

Sources of host-range data

Hibi and Furuki (1985); Furuki (1981).

**Properties of particles in sap**

TIP: 60 °C. LIV: 9–32 days. DEP: $\log_{10}$ minus 4–5. Leaf sap contains few virions.

**Purification**

Method: Hibi and Furuki (1985).

**Particle morphology**

Virions isometric; not enveloped; 30 nm in diameter; angular in profile; with a conspicuous capsomere arrangement.

**Physical properties**

One sedimenting component in purified preparations; sedimentation coefficient 134 *S*. Density 1.33–1.34 g $cm^{-3}$ in CsCl (unfixed).

**Biochemical properties**

Virions contain 17.8 % nucleic acid; 82.2 % protein.

*Genome* consists of RNA; single–stranded; linear; unipartite; largest (or only) genome part 4.3 kb. Genomic nucleic acid isolated by Hibi *et al.* (1980).

*Sequence database accession code(s):*

D00562 Em(40)_vi:MNSCP Gb(84)_vi:MNSCP Melon necrotic spot virus coat protein gene, complete cds. 3/92 1,173bp.

D12536 Em(40)_vi:MNSCG1 Gb(84)_vi:D12536 Melon necrotic spot virus genomic RNA, almost complete sequence. 10/92 4,262bp.

M29671 Em(40)_vi:MNSCG Gb(84)_vi:MNSCG Melon necrotic spot virus, complete genome. 10/92 4,266bp.

D29662 Em(43)_vi:Mnscp1 Gb(89)_vi:Mnscp1 Melon necrotic spot virus (strain NH) gene for coat protein. 4/95 1,173bp.

D29663 Em(43)_vi:Mnscp2 Gb(89)_vi:Mnscp2 Melon necrotic spot virus (strain S) gene for coat protein. 4/95 1,173bp.

$M_r$ 46000; coat protein. Method of preparation: Bos *et al.* (1984).

**Replication**

Genome replicates in cytoplasm. Replication does not depend on a helper virus.

**Cytopathology**

Virions found in leaves, roots, stems, petioles, fruits and seeds; in cytoplasm and in cell vacuoles. Inclusions present in infected cells; are crystals in the cytoplasm; they contain virions.

**Taxonomy and Relationships**

*Carmovirus: Tombusviridae.*

Virus(es) with serologically unrelated virions: cucumber necrosis, wild cucumber mosaic, carnation mottle, tobacco necrosis, turnip yellow mosaic, carnation ringspot, cucumber mosaic, pelargonium leaf curl, squash mosaic, cucumber soil-borne, cucumber leaf spot and tomato bushy stunt viruses.

**References**

Bos, L., van Dorst, H.J.M., Huttinga, H. and Maat, D.Z. (1984). *Neth. J. Pl. Path.* **90**: 55.

Furuki, I. (1981). *Tech. Bull. Shizuoka Agric. Exp. Stn* **14**: 94.

Hibi, T., Furuki, I., Honda, Y., Saito, Y. and Komuro, Y. (1980). *Ann. Phytopath. Soc. Japan* **46**: 419.

Hibi, T. and Furuki, I. (1985). *CMI/AAB Descr. Pl. Viruses* No. 302, 4 pp.

Kishi, K. (1966). *Ann. Phytopath. Soc. Japan* **32**: 138.

Riviere, C.J. and Rochon, D.M. (1990). *J. gen. Virol.***69**: 395.

Tomlinson, J.A. and Thomas, B.J. (1986). *Ann. appl. Biol.* **108**: 71.

# Melon Ourmia ourmiavirus

**Data collated by A.A. Brunt, 1990.**

*Synonyms:* Ourmia melon virus. *Standard acronym:* OuMV.

First reported in *Cucumis melo*; from Ourmia, northern Iran; by Lisa *et al.* (1988).

**Natural host range and symptoms**

Symptoms persist.
*Cucumis melo* — chlorotic spots and irregular ringspots, puckering.

**Geographical distribution**

Spreads in Iran.

**Experimental host range**

Many (>9) families susceptible. Experimentally infected plants mostly show systemic mosaics, mottles, ringspots or necrosis.

Diagnostically susceptible

*Chenopodium quinoa, Gomphrena globosa* — necrotic local lesions.
*Nicotiana clevelandii, N. benthamiana, N. megalosiphon* — chlorotic local lesions; systemic necrotic mosaic.
*Petunia hybrida, Callistephus chinensis, Cucumis sativus, Cucumis melo* — severe systemic chlorosis.

Diagnostically insusceptible

*Brassica oleracea, Pisum sativum, Solanum tuberosum, Trifolium repens.*

Maintenance and propagation hosts

*Cucumis melo, Nicotiana benthamiana, N. clevelandii, Petunia hybrida.*

Assay hosts (**L**ocal lesions or **W**hole plants)

*Chenopodium quinoa* (L), *Gomphrena globosa* (L).

Dicotyledonous Hosts *(**Susceptible** or Insusceptible)*

***Apium graveolens*. *Calendula officinalis*. *Capsicum annuum*. *Chenopodium amaranticolor*. *Chenopodium quinoa*. *Cucumis melo*. *Cucumis metuliferus*. *Cucumis sativus*. *Datura stramonium*. *Dianthus barbatus*. *Gomphrena globosa*. *Luffa acutangula*. *Lycopersicon esculentum*. *Nicotiana benthamiana*. *Nicotiana clevelandii*. *Nicotiana glutinosa*. *Nicotiana megalosiphon*. *Nicotiana tabacum*. *Ocimum basilicum*. *Petunia hybrida*. *Phaseolus vulgaris*. *Phlox drummondii*. *Raphanus sativus*. *Spinacia oleracea*. *Tetragonia expansa*. *Vicia faba*. *Vigna unguiculata*. *Zinnia elegans*.**

Sources of host-range data

Lisa *et al.* (1988).

**Properties of particles in sap**

TIP: 70–80 °C. LIV: 35 days (more than at 20–25°C). DEP: $\log_{10}$ minus 5–6.

**Purification**

Method: Lisa *et al.* (1988).

**Particle morphology**

Virions bacilliform; not enveloped; 30 nm in length, or 37 nm in length; 18.5 nm wide.

**Physical properties**

Two sedimenting components in purified preparations (in sucrose), or one sedimenting component in purified preparations (in caesium chloride). Density 1.375 g $cm^{-3}$ in CsCl.

**Biochemical properties**

*Genome* consists of RNA; single–stranded; linear; of three parts; largest (or only) genome part the largest 0.91 kb; the 2nd largest 0.35 kb; the 3rd largest 0.32 kb. Genomic nucleic acid isolated by Lisa *et al.* (1988).

*Virion protein(s)* one (and a fragment); $M_r$ 26300 (and 23300); coat protein. Method of preparation: Lisa *et al.* (1988).

**Taxonomy and Relationships**

Ourmiavirus (the type species).

Virus(es) with serologically unrelated virions: alfalfa mosaic, olive latent 2, cassava Ivorian bacilliform and pelargonium zonate spot viruses.

**Comments**

Ourmiavirus has been proposed as the name of a viral genus containing this and the similar, but serologically unrelated, viruses listed above (Accotto *et al.*, 1990).

**References**

Accotto, G.P., Riccioni, L., Barba, M., Lisa, V. and Boccardo, G. (1990). *8th Int. Congr. Virology*, Berlin, 1990; No. P79–025, p. 448.

Lisa, V., Milne, R.G., Acotto, G.P., Boccardo, G. and Caciagli, P. (1988). *Ann. appl. Biol.* **112**: 291.

## Melon rugose mosaic *tymovirus*

**Data collated by P. Jones, 1984. Revised 1989.**

First reported in *Citrullus lanatus*; from the Peoples Democratic Republic of Yemen; by Jones (1981).

**Natural host range and symptoms**

Symptoms persist.

*Citrullus lanatus* — yellowing, rugose mosaic.

*Cucumis melo* — severe mosaic and rugosity.

**Transmission**

Virus transmitted by mechanical inoculation; not transmitted by seed.

**Geographical distribution**

Spreads in Yemen.

**Experimental host range**

Few (<3) families susceptible.

Diagnostically susceptible

*Citrullus lanatus, Cucumis melo, C. sativus, Cucurbita pepo* — systemic mosaic.

Diagnostically insusceptible

*Chenopodium amaranticolor, Brassica chinensis, Datura stramonium, Nicotiana clevelandii, N. tabacum* cv. Xanthi-*nc*.

Maintenance and propagation hosts

*Citrullus lanatus, Cucumis melo.*

Assay hosts (**L**ocal lesions or **W**hole plants)

*Citrullus lanatus* (W), *Cucumis melo* (W), *C. sativus* (W), *Cucurbita pepo* (W).

Dicotyledonous Hosts *(**Susceptible** or Insusceptible)*

*Chenopodium amaranticolor*. ***Citrullus lanatus***. ***Cucumis melo***. ***Cucumis sativus***. ***Cucurbita pepo***. *Datura stramonium*. *Nicotiana clevelandii*. *Nicotiana tabacum*.

Sources of host-range data

Jones *et al.* (1986).

**Properties of particles in sap**

TIP: 55–60 °C. LIV: 21 days (or more). DEP: $\log_{10}$ minus 6. Leaf sap contains many virions.

**Purification**

Method: Jones *et al.* (1986).

**Particle morphology**

Virions isometric; not enveloped; 32 nm in diameter; rounded in profile, or angular in profile.

**Physical properties**

Two sedimenting components in purified preparations; sedimentation coefficient of the fastest 118 *S*; of the other(s) 64 *S*. Density 1.39 g $cm^{-3}$ in CsCl.

**Biochemical properties**

Virions contain 23 % nucleic acid (B); 77 % protein (B); 0 % lipid.

*Genome* consists of RNA; single–stranded; linear; unipartite; largest (or only) genome part 2.1 kb.

*Virion protein(s)* one; $M_r$ 22000; coat protein.

**Replication**

Replication does not depend on a helper virus.

**Taxonomy and Relationships**

*Tymovirus*.

Virus(es) with serologically related virions: belladonna mottle, dulcamara mottle, Scrophularia mottle and Andean potato latent viruses, but distantly.

Best tests for diagnosis: mechanical inoculation to watermelon, sweetmelon and squash.

**References**

Jones, P. (1981). *Ann. Rep. Rothamsted Exp. Stn.*

Jones, P., Ba Angood S. and Carpenter J.M. (1986). *Ann. appl. Biol.* **108**: 303.

# Melon variegation (?) *cytorhabdovirus*

**Data collated by A.A. Brunt, 1991.**

*Standard acronym:* MVV.

First reported in *Cucumis melo* plants also containing watermelon mosaic potyvirus; from Spain; by Rubio-Huertos and Pena-Iglesias (1973).

**Natural host range and symptoms**

*Cucumis melo* — stunting, chlorosis of leaves and fruits.

**Geographical distribution**

Spreads in Spain.

**Particle morphology**

Virions rhabdo– or bullet–shaped; with a clear modal length; of 320 nm; 60 nm wide.

**Cytopathology**

Virions found in cytoplasm.

**Taxonomy and Relationships**

Possible *Cytorhabdovirus: Rhabdoviridae.*

**References**

Rubio-Huertos, M. and Pena-Iglesias, A. (1973). *Pl. Dis. Reptr* **57**: 649.

# Melon vein-banding mosaic (?) potyvirus

**Data collated by A.A.Brunt, 1994.**

*Standard acronym:* MVBMV.

First reported in *Cucumis melo*; from Taiwan; by Huang *et al.* (1993).

**Geographical distribution**

Spreads in Taiwan.

**Taxonomy and Relationships**

Possible unassigned potyvirus.

**References**

Huang, C.H., Chang, L. and Tsai, J.H. (1993). *Pl. Path.* **42**: 100.

# Melothria mottle (?) potyvirus

**Data collated by A.A. Brunt, 1994.**

First reported in *Melothria liukiuensis*; from Okinawa Prefecture, Japan; by Yonaha *et al.* (1993).

**Natural host range and symptoms**

*Melothria liukiuensis* — faint leaf mottling.

**Transmission**

Transmitted by a vector; an insect; Aphididae; *Myzus persicae*. Transmitted in a non–persistent manner.

**Geographical distribution**

Spreads in the Eastern Asian region; Japan.

**Experimental host range**

Several (3–9) families susceptible.

Diagnostically susceptible

*Melothria liukiuensis* — systemic leaf mottling.

*Cucurbita pepo, Nicotiana benthamiana* — local chlorotic spotting, systemic vein-clearing and mottling.

*Bryonopsis laciniosa, Citrullus lanatus, Cucumis sativus* and *Cucurbita maxima* — local necrotic and local and systemic chlorotic spotting.

*Sesamum indicum* — local necrotic spotting, systemic stem necrosis and wilting.

*Chenopodium amaranticolor, C. murale, C. quinoa, Gomphrena globosa* and *Spinacia oleracea* — local necrotic lesions, not systemic.

Diagnostically insusceptible

*Datura metel, D. stramonium, Nicotiana clevelandii, N. glutinosa, N. rustica.*

Maintenance and propagation hosts

*Nicotiana benthamiana.*

Assay hosts (**L**ocal lesions or **W**hole plants)

*Chenopodium amaranticolor* (L), *C. quinoa* (L), *Nicotiana benthamiana* (W).

Dicotyledonous Hosts *(**Susceptible** or Insusceptible)*

***Chenopodium amaranticolor***. ***Chenopodium murale***. ***Chenopodium quinoa***. ***Citrullus lanatus***. ***Cucumis melo***. ***Cucumis sativus***. ***Cucurbita moschata***. ***Cucurbita pepo***. *Datura stramonium*. ***Gomphrena globosa***. ***Nicotiana benthamiana***. *Nicotiana clevelandii*. *Nicotiana glutinosa*. *Nicotiana rustica*. ***Sesamum indicum***. ***Spinacia oleracea***.

**Properties of particles in sap**

TIP: 65 °C. LIV: 8–10 days. DEP: $\log_{10}$ minus 3–4.

**Purification**

Method: Yonaha *et al.* (1993).

**Particle morphology**

Virions filamentous; not enveloped; usually flexuous; with a clear modal length; of 725 nm; 12 nm wide.

**Taxonomy and Relationships**

Possible unassigned potyvirus.

Virus(es) with serologically unrelated virions: papaya ringspot, trichosanthes mottle, watermelon mosaic-2 and zucchini yellow mosaic viruses.

**References**

Yonaha, T., Nagata, M., Kawano, S. and Tamori, M. (1993). *Bull. Coll. Agric. Univ. Ryukus* **40**: 9–19.

# Mibuna temperate (?) *alphacryptovirus*

**Data collated by A.A. Brunt, 1991.**

*Standard acronym:* MTV.

First reported in *Brassica rapa var. laciniifolia*; from Japan; by Natsuaki *et al.* (1983).

**Natural host range and symptoms**

*Brassica rapa var. laciniifolia* — symptomless.

**Transmission**

Virus not transmitted by mechanical inoculation; transmitted by seed.

**Geographical distribution**

Found, but with no evidence of spread, in Japan.

**Particle morphology**

Virions isometric; not enveloped; 30 nm in diameter.

**Taxonomy and Relationships**

Possible *Alphacryptovirus: Partitiviridae*.

**References**

Boccardo, G., Lisa, V., Luisoni, E. and Milne, R.G. (1987). *Adv. Virus Res.* **32**: 171.

Natsuaki, T., Yamashita, S., Doi, Y., Okuda, S. and Teranaka, M. (1983). *Ann. Phytopath. Soc. Japan* **49**: 593.

Natsuaki, T., Natsuaki, K.T., Okuda, S., Teranaka, M., Milne, R.G., Boccardo, G. and Luisoni, E. (1986). *Intervirology* **25**: 69.

# Milk vetch dwarf nanavirus

**Data collated by T. Inouye, 1981. Revised 1984.**

*Standard acronym:* MVDV.

First reported in *Astragalus sinicus*; from Japan; by Ohki (1975).

**Natural host range and symptoms**

Symptoms persist.

*Astragalus sinicus, Glycine max, Phaseolus vulgaris, Pisum sativum, Vicia faba* — yellow dwarf and leaf rolling.

**Transmission**

Transmitted by a vector; an insect; *Aphis craccivora, A. gossypii, Acyrthosiphon pisum, Acyrthosiphon (Aulocorthum) solani*; Aphididae. Transmitted in a persistent manner. Virus retained when the vector moults; not transmitted by mechanical inoculation; transmitted by grafting.

**Geographical distribution**

Spreads in Japan.

**Experimental host range**

Few (<3) families susceptible.

Diagnostically susceptible

*Spinacia oleracea, Vigna unguiculata* ssp. *sesquipedalis* — leaf rolling, stunting.
*Astragalus sinicus, Vicia sativa* — yellow dwarf.
*Phaseolus vulgaris* — stunting.
*Pisum sativum, Vicia faba* — yellow dwarf, leaf roll.
*Datura stramonium* — interveinal yellowing, leaf rolling.
*Nicotiana rustica* — leaf rolling.

Maintenance and propagation hosts

*Pisum sativum, Vicia faba*.

Assay hosts (Local lesions or Whole plants)

*Astragalus sinicus* (W), *Pisum sativum* (W), *Vicia faba* (W).

Dicotyledonous Hosts *(**Susceptible** or Insusceptible)*

***Arachis hypogaea***. ***Astragalus sinicus***. ***Cassia tora***. *Chenopodium amaranticolor*. ***Crotalaria spectabilis***. *Cucumis sativus*. ***Datura stramonium***. ***Glycine max***. *Gomphrena globosa*. ***Lathyrus odoratus***. *Lycopersicon esculentum*. ***Medicago sativa***. ***Melilotus alba***. ***Nicotiana glutinosa***. ***Nicotiana rustica***. ***Nicotiana tabacum***. *Petunia hybrida*. ***Phaseolus vulgaris***. ***Pisum sativum***. ***Spinacia oleracea***. *Tetragonia expansa*. *Trifolium hybridum*. ***Trifolium incarnatum***. *Trifolium pratense*. *Trifolium repens*. ***Trifolium subterraneum***. ***Vicia faba***. ***Vicia sativa***. ***Vicia villosa***. ***Vigna angularis***. ***Vigna unguiculata***. ***Vigna unguiculata*** ssp. ***sesquipedalis***.

Sources of host-range data

Inouye *et al.* (1968).

**Properties of particles in sap**

Leaf sap contains few virions.

**Purification**

Method: Ohki *et al.* (1975).

**Particle morphology**

Virions isometric; not enveloped; 26 nm in diameter; angular in profile; without a conspicuous capsomere arrangement.

**Cytopathology**

Virions found in phloem; in cytoplasm and in cell vacuoles. Inclusions absent from infected cells. Other cellular changes: an increase of mitochondria.

**Taxonomy and Relationships**

Nanavirus.

Virus(es) with serologically related virions: Faba bean necrotic yellows virus, closely (Katul *et al.*, 1993. Virus(es) with serologically unrelated virions: This virus was previously considered to be a luteovirus, but is now known to share properties with subterranean clover stunt and related viruses. Its virions are very closely serologically related to those of faba bean necrotic yellows virus (Katul *et al.*, 1993) but, at present, they are listed as separate species, rather than related strains.

**References**


Asada, Ohki, S.T., Osaki, T. and Inouye, T. (1984). *Shokubutsu Boeki* **38**: 184.

Katul, L., van Vetten, H.J., Maiss, E., Makkouk, K.M., Lesemann, D.-E. and Casper, R. (1993) *Ann. appl. Biol.* **123**: 629.

Inouye *et al.* (1968). *Ann. Phytopath. Soc. Japan* **34**: 28.

Ohki, S.T. (1975). *Ann. Phytopath. Soc. Japan* **41**: 508.


# Millet red leaf (?) *luteovirus*

*Standard acronym:* MRLV.

First reported in *Setaria italica*; from the Peoples Republic of China; by Yu *et al.* (1957).

**Natural host range and symptoms**

*Setaria italica* — reddening of leaf blades and stems of purple stemmed cultivars and leaf yellowing in green stemmed cultivars.

**Transmission**

Transmitted by a vector; an insect; *Rhopalosiphum maidis, Macrosiphum (Sitobion) avenae* and *Schizaphis graminum*; Aphididae. Transmitted in a persistent manner. Virus not transmitted by seed.

**Geographical distribution**

Found, but with no evidence of spread, in the Peoples Republic of China (Yu *et al.*, 1957; 1958).

**Taxonomy and Relationships**

Possible *Luteovirus*.

Possibly a strain of barley yellow dwarf virus (Waterhouse *et al.*, 1988).

**References**

Waterhouse, P.M., Gildow, F.E. and Johnstone, G.R. (1988). *AAB Descr. Pl. Viruses* No. 339, 7 pp.

Yu, T.F., Pei, M.Y. and Hsu, H.K. (1957). *Acta Phytopath. Sin.* **3**: 1.

Yu, T.F., Hsu, H.K. and Pei, M.Y. (1958). *Acta Phytopath. Sin* **4**: 1.

# Mimosa mosaic virus

**Data collated by E.W. Kitajima, 1987.**

First reported in *Mimosa sensitiva*; from Belem, Brazil; by Marinho *et al.* (1985).

**Natural host range and symptoms**

Symptoms persist.

*Mimosa sensitiva* — mosaic.

**Transmission**

Transmitted by a vector; an insect; *Diabrotica speciosa*; Coleoptera. Transmitted in a semi–persistent manner. Virus does not multiply in the vector; transmitted by mechanical inoculation; transmitted by grafting; not transmitted by seed.

**Geographical distribution**

Spreads in Brazil.

**Experimental host range**

Few (<3) families susceptible.

Diagnostically susceptible

*Mimosa sensitiva* — systemic leaf chlorosis.

Maintenance and propagation hosts

*Mimosa sensitiva.*

Assay hosts (**L**ocal lesions or **W**hole plants)

*Mimosa sensitiva* (W).

Dicotyledonous Hosts *(**Susceptible** or Insusceptible)*

*Capsicum annuum. Capsicum frutescens. Cassia occidentalis. Cassia tora. Chenopodium amaranticolor. Chenopodium murale. Chenopodium quinoa. Cucumis melo. Cucumis sativus. Cucurbita maxima. Cucurbita pepo. Datura stramonium. Glycine max. Gomphrena globosa. Helianthus annuus. Lactuca sativa. Lycopersicon esculentum. Mimosa pudica. Nicandra physaloides. Nicotiana debneyi. Nicotiana glutinosa. Nicotiana rustica. Nicotiana tabacum. Petunia hybrida. Phaseolus lunatus. Phaseolus vulgaris.*

*Physalis floridana. Pisum sativum. Raphanus sativus. Solanum tuberosum. Vigna unguiculata* ssp. *sesquipedalis.*

Monocotyledonous Hosts *(**Susceptible** or Insusceptible)*

*Allium cepa. Triticum aestivum. Zea mays.*

**Properties of particles in sap**

Leaf sap contains many virions.

**Purification**

Method: Marinho *et al.* (1985).

**Particle morphology**

Virions isometric; not enveloped; 30 nm in diameter; angular in profile; without a conspicuous capsomere arrangement.

**Physical properties**

One sedimenting component in purified preparations. $A_{260}/A_{280}$ ratio 1.39.

**Biochemical properties**

Virions contain 20 % nucleic acid; 80 % protein.
*Genome* consists of RNA; single–stranded; linear.
*Virion protein(s)* one; $M_r$ 31000; coat protein.

**Cytopathology**

Virions found in all parts of the host plant; in cytoplasm and in cell vacuoles. Inclusions absent from infected cells.

**Taxonomy and Relationships**

Virus(es) with serologically unrelated virions: purple granadilla and maize rayado fino viruses and ilarviruses, sobemoviruses, cucumoviruses, comoviruses and tymoviruses.

**References**

Marinho, V.L.A., Crestani, D.A., Lin, M.T. and Kitajima, E.W. (1985). *Fitopatol. Bras.* **10**: 308.

# Mimosa striped chlorosis (?) *badnavirus*

**Data collated by P. Jeyanandarajah, 1991.**

*Synonyms:* mimosa bacilliform virus. *Standard acronym:* MBV.

First reported in *Albizzia julibrissin*; from the U.S.A.; by Martin and Kim (1987).

**Natural host range and symptoms**

Symptoms persist.
*Albizzia julibrissin* — chlorotic stripes parallel to or along leaf veins and interveinal chlorosis.

**Transmission**

Virus transmitted by grafting; transmitted by seed.

**Geographical distribution**

Found, but with no evidence of spread, in the U.S.A. (Arkansas).

**Particle morphology**

Virions bacilliform; with a clear modal length; of 95 nm (*in vivo*); 35 nm wide (*in vivo*).

**Biochemical properties**

*Genome* consists of DNA.

**Cytopathology**

Inclusions present in infected cells; are viroplasms; they contain virions.

**Taxonomy and Relationships**

Possible *Badnavirus*.

**References**

Martin, E.M. and Kim, K.S. (1987). *Phytopathology* **77**: 935.

# Mirabilis mosaic *caulimovirus*

**Data collated by R. Hull, 1987.**

*Standard acronym:* MiMV.

First reported in *Mirabilis nyctaginea*; from Illinois, U.S.A.; by Brunt and Kitajima (1973).

**Natural host range and symptoms**

Symptoms persist. Symptoms systemic mosaics, mottles, ringspots or necrosis.

*Mirabilis nyctaginea* — chlorotic vein banding of tip leaves and interveinal chlorosis of older leaves.

**Transmission**

Transmitted by a vector; an insect; *Myzus persicae*; Aphididae. Transmitted in a semi–persistent manner. Virus lost by the vector when it moults; does not multiply in the vector; not transmitted congenitally to the progeny of the vector; transmitted by mechanical inoculation; not transmitted by contact between plants; not transmitted by seed; not transmitted by pollen.

**Geographical distribution**

Spreads in the USA.

Diagnostically susceptible

*Mirabilis jalapa* — chlorotic vein banding of tip leaves and interveinal chlorosis of older leaves.

Maintenance and propagation hosts

*Mirabilis jalapa*.

Sources of host-range data

Brunt and Kitajima (1973).

**Properties of particles in sap**

Leaf sap contains many virions.

**Purification**

Method: Donson and Hull (1983); Richins and Shepherd (1983).

**Particle morphology**

Virions isometric; not enveloped; 50 nm in diameter; rounded in profile; without a conspicuous capsomere arrangement.

**Physical properties**

One sedimenting component in purified preparations; sedimentation coefficient 210–254 *S*. Density 1.32 g $cm^{-3}$ in CsCl.

**Biochemical properties**

Virions contain 17 % nucleic acid; 83 % protein; 0 % lipid.

*Genome* consists of DNA; double–stranded; circular; unipartite; largest (or only) genome part 7.8 kb (pairs). Genomic nucleic acid isolated by Donson and Hull (1983); Richins and Shepherd (1983). Infectivity retained when deproteinised with proteases; retained when deproteinised with phenol or detergent. Poly A region absent, or present; present in full length *c*RNA.

*Non–genomic nucleic acid* not found in the virions.

*Virion protein(s)* one; $M_r$ 32000; coat protein. Method of preparation: Brunt and Kitajima (1973).

**Replication**

Genome replicates presumably in nuclei and in cytoplasm. Replication does not depend on a helper virus.

**Cytopathology**

Virions found in mesophyll and epidermis; in cytoplasm. Inclusions present in infected cells; are viroplasms; they contain virions.

**Taxonomy and Relationships**

*Caulimovirus*.

Virus(es) with serologically unrelated virions: dahlia mosaic, cauliflower mosaic and carnation etched ring viruses.

**References**

Brunt, A.A. and Kitajima, E.W. (1973). *Phytopath. Z.* **76**: 265.

Donson, J. and Hull, R. (1983). *J. gen Virol.* **64**: 2281.

Richins, R.D. and Shepherd, R.J. (1983). *Virology* **124**: 208.

# Miscanthus streak monogeminivirus

**Data collated by A.A. Brunt and S. Yamashita, 1987.**

*Standard acronym:* MiSV.

First reported in *Miscanthus sacchariflorus*; from Chiba, Japan; by Yamashita *et al.* (1979).

**Natural host range and symptoms**

Symptoms persist.

*Miscanthus sacchariflorus*—chlorotic streaking, plant stunting.

**Transmission**

Virus not transmitted by mechanical inoculation.

**Geographical distribution**

Spreads in Japan.

Maintenance and propagation hosts

*Miscanthus saccariflorus* (naturally infected).

**Properties of particles in sap**

Leaf sap contains few virions.

**Purification**

Method: Yamashita *et al.* (1985).

**Particle morphology**

Virions geminate; 30 nm in length; angular in profile; without a conspicuous capsomere arrangement.

**Physical properties**

One sedimenting component in purified preparations; sedimentation coefficient 75 *S*. Density 1.35 g $cm^{-3}$ in CsCl.

**Biochemical properties**

*Genome* consists of DNA; single–stranded; circular; unipartite; largest (or only) genome part 2.672 kb.

*Sequence database accession code(s):*

D00800 Em(40)_vi:GEMSTGEN Gb(84)_vi:MSTGEN Miscanthus streak virus genome. 8/92 2,672bp

D01030 Em(40)_vi:GEMSTGVC Gb(84)_vi:MSTGVC Miscanthus streak virus ss-DNA, ORFs VO, V1, V2, C1, and C2, complete CDS. 12/91 2,672bp. 2 sequences.

*Virion protein(s)* one; $M_r$ 28000.

**Cytopathology**

Virions found in leaves and mesophyll; in cytoplasm, in nuclei, and in cell vacuoles (in geometrical crystalline arrays).

**Taxonomy and Relationships**

Monogeminivirus: *Geminiviridae*.

Virus(es) with serologically unrelated virions: maize streak, chloris striate mosaic, African cassava mosaic and tobacco leaf curl viruses.

The genome of the virus has clear sequence homology with the genomes of maize streak and wheat dwarf viruses.

**References**

Chatani, M., Matsumoto, Y., Mizuka, H., Ikegama, M., Boulton, M.I. and Davies, J.W. (1991). *J. gen. Virol.* **72**: 2325.

Ikegama, M., Yazaki, K., Chatani, M. and Mizuka, A. (1989). *Intervirology* **30**: 340.

Nonaka, N., Namba, S., Yamashita, S. and Doi, Y. (1985). *Ann. Phytopath. Soc. Japan* **51**: 58.

Yamashita, S., Doi, Y. and Yora, K. (1979). *Ann phytopath. Soc. Japan* **45**: 128.

Yamashita, S., Nonaka, N., Namba, N., Doi, Y. and Yora, K. (1985). *Ann. Phytopath. Soc. Japan* **51**: 582.

# Mulberry latent *carlavirus*

**Data collated by T. Tsuchizaki, 1987.**

*Standard acronym:* MLV.

First reported in *Morus alba* L.; from Japan; by Tsuchizaki (1976).

**Natural host range and symptoms**

*Morus alba* — latent.

**Transmission**

Virus transmitted by mechanical inoculation; transmitted by grafting; not transmitted by seed (in *Chenopodium quinoa*).

**Geographical distribution**

Spreads in Japan.

**Experimental host range**

Several (3–9) families susceptible.

Diagnostically susceptible

*Chenopodium quinoa* — local and systemic chlorotic spotting, mosaic and malformation.

*C. amaranticolor* — green ringspots; then symptomless systemic infection.

Diagnostically insusceptible

*Vigna unguiculata, Glycine max, Cucumis sativus, Lycopersicon esculentum, Phaseolus* spp.

Maintenance and propagation hosts

*Morus alba* (maintenance), *Chenopodium quinoa* (propagation).

Assay hosts (**L**ocal lesions or **W**hole plants)

*Chenopodium quinoa* (L and W).

Dicotyledonous Hosts *(**Susceptible** or Insusceptible)*

***Beta vulgaris***. *Brassica campestris* ssp. *rapa*. *Capsicum annuum*. ***Chenopodium amaranticolor***. ***Chenopodium quinoa***. *Cucumis sativus*. *Cucurbita maxima*. *Datura stramonium*. *Glycine max*. ***Gomphrena globosa***. *Humulus lupulus*. *Lycopersicon esculentum*. ***Morus alba***. ***Nicotiana clevelandii***. *Phaseolus vulgaris*. *Pisum sativum*. ***Sesamum indicum***. ***Spinacia oleracea***. *Trifolium subterraneum*. *Vicia faba*. *Vigna unguiculata*. *Vigna unguiculata* ssp. *sesquipedalis*. *Zinnia elegans*.

Monocotyledonous Hosts *(**Susceptible** or Insusceptible)*

*Zea mays*.

Sources of host-range data

Tsuchizaki (1976).

**Properties of particles in sap**

TIP: 55–65 °C. LIV: 3–7 days (at 20°C). DEP: $\log_{10}$ minus 3–4. Leaf sap contains few virions.

**Purification**

Method: Tsuchizaki (1976).

**Particle morphology**

Virions filamentous; not enveloped; usually straight; with a clear modal length; of 700 nm; 12 nm wide. Axial canal obscure. Basic helix obscure.

**Physical properties**

One sedimenting component in purified preparations.

**Cytopathology**

Virions found in mesophyll and epidermis; in cytoplasm. Inclusions absent from infected cells.

**Taxonomy and Relationships**

*Carlavirus.*

Virus(es) with serologically related virions: carnation latent virus.

Best tests for diagnosis: electron microscopic detection in sap possible.

**References**

Tsuchizaki, T. (1976). *Ann. Phytopath. Soc. Japan* **42**: 304.

# Mulberry ringspot *nepovirus*

**Data collated by A.A. Brunt and T. Tsuchizaki, 1987.**

*Standard acronym:* MRSV.

First reported in *Morus alba*; from Japan; by Tsuchizaki *et al.* (1971).

**Natural host range and symptoms**

Symptoms persist and vary seasonally.

*Morus alba* — mosaic and ringspotting, enations with some isolates.

**Transmission**

Transmitted by a vector; a nematode; *Longidorus martini*; Dorylamidae. Virus transmitted by mechanical inoculation; transmitted by grafting; not transmitted by contact between plants; transmitted by seed (*c.* 10% in *Glycine max*).

**Ecology and control**

Studies reported by Yagita and Komuro (1972).

**Geographical distribution**

Spreads in Japan.

**Experimental host range**

Several (3–9) families susceptible.

Diagnostically susceptible

*Morus alba* — mosaic, ringspotting, enations with some isolates.

*Glycine max, Pisum sativum* — systemic chlorotic spotting and mosaic.

*Vigna unguiculata* — necrotic spots or ringspots; systemic mosaic and/or necrosis of leaves and stems.

*Chenopodium quinoa* — faint chlorotic local lesions; then systemic mosaic.

Diagnostically insusceptible

*Nicotiana tabacum, Vicia faba, Lycopersicon esculentum, Chenopodium amaranticolor, Cucumis sativus.*

Maintenance and propagation hosts

*Vigna unguiculata, Glycine max.*

Assay hosts (**L**ocal lesions or **W**hole plants)

*Vigna unguiculata* (L).

Dicotyledonous Hosts *(**Susceptible** or Insusceptible)*

*Amaranthus retroflexus. Capsicum annuum. Catharanthus roseus. Chenopodium album. Chenopodium amaranticolor.* ***Chenopodium murale. Chenopodium quinoa.*** *Cucumis sativus. Daucus carota. Galinsoga parviflora.* ***Glycine max.*** *Gomphrena globosa. Hibiscus esculentus. Lycopersicon esculentum.* ***Morus alba. Nicotiana clevelandii.*** *Nicotiana tabacum. Petunia hybrida. Phaseolus vulgaris.* ***Pisum sativum. Sesamum indicum.*** *Solanum melongena. Solanum nigrum. Spinacia oleracea. Tetragonia expansa. Vicia faba.* ***Vigna unguiculata. Vigna unguiculata*** ssp. ***sesquipedalis.*** *Zinnia elegans.*

Monocotyledonous Hosts *(**Susceptible** or Insusceptible)*

*Zea mays.*

Sources of host-range data

Tsuchizaki *et al.* (1971); Tsuchizaki (1975).

**Properties of particles in sap**

TIP: 50–60 °C (in *Glycine max* sap). LIV: 3–5 days (at room temperature). DEP: $\log_{10}$ minus 3–4. Leaf sap contains few virions.

**Purification**

Method: Tsuchizaki *et al.* (1971).

**Particle morphology**

Virions isometric; not enveloped; 22–25 nm in diameter; angular in profile; with a conspicuous capsomere arrangement.

**Physical properties**

Three sedimenting components in purified preparations; sedimentation coefficient of the fastest 122 *S* ($B_1$ and $B_2$); of the other(s) 93 *S* (M), or 50 *S* (T). Density 1.504 g $cm^{-3}$ in CsCl ($B_2$), or 1.497 g $cm^{-3}$ in CsCl ($B_1$), or 1.413 g $cm^{-3}$ in CsCl (T).

**Biochemical properties**

Virions contain 41 % nucleic acid ($B_2$), or 40.1 % nucleic acid ($B_1$), or 29.5 % nucleic acid (M), or 0 % nucleic acid (T); 59 % protein ($B_2$), or 59.9 % protein ($B_1$), or 70.5 % protein (M), or 100 % protein (T).

*Genome* consists of RNA; single–stranded; linear; of two parts; largest (or only) genome part the larger 7.3 kb; the 2nd largest 3.9 kb. Genomic nucleic acid isolated by Hibi *et al.* (1984). Infectivity lost when deproteinised with phenol or detergent.

*Non–genomic nucleic acid* not found in the virions.

**Replication**

Replication does not depend on a helper virus.

**Cytopathology**

Virions found in mesophyll and epidermis; in cytoplasm. Inclusions present in infected cells; are membranous bodies (with vesicles). Other cellular changes: tubules containing a single row of virions in the cytoplasm.

**Taxonomy and Relationships**

*Nepovirus: Comoviridae.*

Virus(es) with serologically unrelated virions: arabis mosaic, cherry leaf roll, grapevine chrome mosaic, grapevine fanleaf, raspberry ringspot, strawberry latent ringspot, tobacco ringspot, tomato black ring and tomato ringspot viruses.

**References**

Hibi, T., Yagita, H. and Iwaki, M. (1984). *Ann. Phytopath. Soc. Japan* **50**: 445.
Tsuchizaki, T. (1975). *CMI/AAB Descr. Pl. Viruses* No. 142, 3 pp.
Tsuchizaki, T., Hibino, H. and Saito, Y. (1971). *Ann. Phytopath. Soc. Japan* **37**: 266.
Yagita, H. and Komuro, Y. (1972). *Ann. Phytopath. Soc. Japan* **38**: 275.

# Mung bean mottle (?) potyvirus

**Data collated by A.A. Brunt, 1991.**

*Standard acronym:* MMTV.

First reported in *Vigna radiata*; from Taiwan; by Sun *et al.* (1977).

**Natural host range and symptoms**

*Vigna radiata, V. unguiculata* ssp. *sesquipedalis, Chenopodium* spp. — leaf chlorosis and malformation.

**Transmission**

Virus transmitted by mechanical inoculation.

**Geographical distribution**

Spreads in Taiwan.

**Experimental host range**

Several (3–9) families susceptible.

Diagnostically susceptible

*Chenopodium amaranticolor, C. quinoa* — local lesions.

*Nicotiana glutinosa, Phaseolus acutifolius, P. vulgaris, Vigna radiata, V. unguiculata* ssp. *sesquipedalis* — systemic leaf chlorosis and deformation.

Maintenance and propagation hosts

*Vigna radiata, V. unguiculata* ssp. *sesquipedalis.*

Assay hosts (**L**ocal lesions or **W**hole plants)

*Vigna radiata* (W), *Vigna unguiculata* ssp. *sesquipedalis* (W).

Dicotyledonous Hosts ***(Susceptible** or Insusceptible)*

***Chenopodium amaranticolor*. *Chenopodium quinoa*. *Nicotiana glutinosa*. *Phaseolus vulgaris*. *Vigna radiata*. *Vigna unguiculata*** ssp. ***sesquipedalis***.

**Purification**

Method: Sun *et al.* (1977).

**Particle morphology**

Virions filamentous; not enveloped; usually flexuous; with a clear modal length; of 750 nm; 13 nm wide. Axial canal obscure.

**Cytopathology**

Inclusions present in infected cells; are pinwheels; they do not contain virions.

**Taxonomy and Relationships**

Unassigned potyvirus.

Virus(es) with serologically unrelated virions: bean common mosaic, bean yellow mosaic, henbane mosaic, peanut mottle, potato Y, soybean mosaic and tobacco etch viruses.

**References**

Sun, M., Lai, P.L. and Yan, L.P. (1977). *Proc. Am. Phytopath. Soc.* **4**: 91.

# Mung bean yellow mosaic bigeminivirus

**Data collated by J.P. Varma, 1981. Revised 1983.**

*Standard acronym:* MYMV.

First reported in *Vigna radiata*; from Delhi, India; by Nariani (1960).

**Natural host range and symptoms**

Symptoms persist.

*Vigna mungo, V. radiata* — faint yellow specks or spots and irregular blotching of tip leaves; slight leaf malformation.

**Transmission**

Transmitted by a vector; an insect; *Bemisia tabaci*; Aleyrodidae. Transmitted in a non–persistent manner. Virus not transmitted by mechanical inoculation; not transmitted by seed; not transmitted by pollen.

**Ecology and control**

Studies reported by two recessive genes for resistance identified in mungbean.

**Geographical distribution**

Spreads in Bangladesh, India, Pakistan, and the Philippines.

**Experimental host range**

Few (<3) families susceptible.

Diagnostically susceptible

*Eclipta alba, Xanthium strumarium, Cajanus cajan, Glycine max, Macrotyloma uniflorum, Macroptilium lathyroides, Phaseolus aconitifolia, Vigna mungo, V. radiata, Brachiaria remosa* — systemic mosaic and yellowing.

Maintenance and propagation hosts

*Cajanus cajan* (W), *Glycine max* (W), *Phaseolus aconitifolia* (W), *Vigna mungo* (W), *V. radiata* (W).

Assay hosts (**L**ocal lesions or **W**hole plants)

*Cajanus cajan, Glycine max, Phaseolus aconitifolia, Vigna radiata, V. mungo.*

Dicotyledonous Hosts *(**Susceptible** or Insusceptible)*

***Cajanus cajan***. ***Glycine max***. *Hibiscus esculentus*. ***Macroptilium lathyroides***. ***Macrotyloma uniflorum***. *Phaseolus lunatus*. ***Phaseolus vulgaris***. ***Vigna radiata***. *Vigna unguiculata*.

Sources of host-range data

Nariani (1960); Nene (1972).

*Sequence database accession code(s):*

D14703 Em(40)_vi:MYGDNA1 Gb(84)_vi:MYGDNA1 Mungbean yellow mosaic geminivirus DNA-1. 6/94 2,723bp

D14704 Em(40)_vi:MYGDNA2 Gb(84)_vi:MYGDNA2 Mungbean yellow mosaic geminivirus DNA-2. 6/94 2,675bp. 2 sequences.

**Taxonomy and Relationships**

Bigeminivirus: *Geminiviridae*.

**References**

Nariani, T.K. (1960). *Indian Phytopath.* **13**: 24.

Nene, Y.L. (1972). *G. B. Pant. Univ. Agric. and Tech. Bull.* **4**: 191.

Nene, Y.L. (1973). *Ind. Pl. Dis. Tep.* **57**: 463.

Premchand and Varma, J.P. (1980). *Indian Phytopath.* **33**: 48.

Varma, J.P., Kadian, O.P. and Singh, J.P. (1973). *Indian Phytopath.* **26**: 592.

Verma, R.P.S. and Singh, D.P. (1989). *Indian J. Genet. Plant Breed.* **49**: 321.

Williams, F.J., Grewal, J.S. and Amin, K.S. (1966). *Pl. Dis. Reptr* **52**: 300.

## Muskmelon vein necrosis *carlavirus*

**Data collated by A.A. Brunt and K.S. Milne, 1987.**

*Standard acronym:* MuVNV.

First reported in *Cucumis melo*; from California, U.S.A.; by Freitag (1952).

**Natural host range and symptoms**

Symptoms persist.

*Cucumis melo* — leaf chlorosis and vein necrosis.

*Pisum sativum* — leaf chlorosis.

**Transmission**

Transmitted by a vector; an insect; *Myzus persicae*; Aphididae. Transmitted in a non–persistent manner. Virus transmitted by mechanical inoculation; not transmitted by contact between plants.

**Ecology and control**

Studies reported by Freitag and Milne (1970).

**Geographical distribution**

Spreads in China and the USA (California).

**Experimental host range**

Several (3–9) families susceptible.

Diagnostically susceptible

*Cucumis sativus, Cucumis melo* — necrotic local lesions, then systemic leaf and vein necrosis.

*Vicia faba* — systemic mottling and stem necrosis.

*Medicago alba, Vicia sativa Trifolium pratense* — local lesions, then systemic chlorosis.

*Chenopodium amaranticolor, Gomphrena globosa, Chenopodium quinoa* — local lesions; not systemic.

Diagnostically insusceptible

*Phaseolus vulgaris, Gomphrena globosa, Nicotiana clevelandii, Glycine max, Nicotiana tabacum.*

Maintenance and propagation hosts

*Cucumis melo, Vicia faba.*

Assay hosts (**L**ocal lesions or **W**hole plants)

*Chenopodium amaranticolor* (L), *Gomphrena globosa* (L), *Medicago alba* (L), *Vicia sativa* (L), *Trifolium pratense* (L).

Dicotyledonous Hosts *(**Susceptible** or Insusceptible)*

*Beta vulgaris*. ***Chenopodium amaranticolor***. ***Chenopodium quinoa***. *Citrullus lanatus*. ***Cucumis melo***. ***Cucumis sativus***. *Cucurbita maxima*. *Cucurbita moschata*. *Cucurbita pepo*. ***Cyamopsis tetragonoloba***. *Datura stramonium*. *Dianthus barbatus*. *Dianthus caryophyllus*. *Glycine max*. ***Gomphrena globosa***. *Lactuca sativa*. ***Lathyrus odoratus***. *Lavatera trimestris*. *Luffa acutangula*. ***Medicago hispida***. *Medicago sativa*. ***Melilotus alba***. *Momordica balsamina*. *Nicotiana clevelandii*. *Nicotiana glutinosa*. *Nicotiana tabacum*. *Phaseolus vulgaris*. ***Pisum sativum***. ***Trifolium hybridum***. ***Trifolium pratense***. ***Trifolium repens***. *Vigna radiata*. *Vigna unguiculata*.

**Properties of particles in sap**

TIP: 50–55 °C. LIV: 2–7 days. DEP: $\log_{10}$ minus 3–4.

**Purification**

Method: Freitag and Milne (1970).

**Particle morphology**

Virions filamentous; not enveloped; usually straight; with a clear modal length; of 674 nm; 15 nm wide. Axial canal obscure. Basic helix obscure.

**Taxonomy and Relationships**

*Carlavirus*.

Virus(es) with serologically related virions: red clover vein mosaic carlavirus.

**References**


Freitag, J.H. (1952). *Phytopathology* **42**: 8.

Freitag, J.H. and Milne, K.S. (1970). *Phytopathology* **60**: 166.

Milne, K.S., Grogan, R.G. and Kimble, K.A. (1969). *Phytopathology* **59**: 819.


# Myrobalan latent ringspot *nepovirus*

**Data collated by R. Delbos, 1991.**

*Standard acronym:* MLRSV.

First reported in *Prunus cerasifera*; from France; by Dunez *et al.* (1971).

**Natural host range and symptoms**

*Prunus avium* — enations on undersides of leaves.

*Prunus persica* — severe rosetting.

**Transmission**

Transmitted by a vector; a nematode (probably). Virus transmitted by mechanical inoculation; transmitted by grafting; not transmitted by contact between plants.

**Geographical distribution**

Found, but with no evidence of spread, in France.

Diagnostically susceptible

*Chenopodium quinoa* — local lesions; systemic necrosis.
*Nicotiana tabacum* — necrotic spots and ringspots.

Maintenance and propagation hosts

*Chenopodium quinoa, Nicotiana clevelandii, Pisum sativum.*

Assay hosts (**L**ocal lesions or **W**hole plants)

*Chenopodium quinoa* (L), *Nicotiana tabacum* (L and W).

**Properties of particles in sap**

TIP: 72 °C. LIV: 100 days. DEP: $\log_{10}$ minus 4.

**Purification**

Method: clarify with butanol:chloroform, followed by differential centrifugation.

**Particle morphology**

Virions isometric; not enveloped; 28 nm in diameter; angular in profile.

**Physical properties**

Two sedimenting components in purified preparations; sedimentation coefficient of the fastest 115 *S* (B); of the other(s) 105 *S* (M). Density 1.5 g $cm^{-3}$ in CsCl (B), or 1.46 g $cm^{-3}$ in CsCl (M). $A_{260}/A_{280}$ ratio 1.73 (B), or 1.65 (M).

**Biochemical properties**

Virions contain 38 % nucleic acid (B), or 28 % nucleic acid (M), or 0 % nucleic acid (T); 62 % protein (B), or 72 % protein (M), or 100 % protein (T).

*Genome* consists of RNA; single–stranded; linear; of two parts; largest (or only) genome part the larger 7 kb; the 2nd largest 5.7 kb. Genomic nucleic acid isolated by Dunez *et al.* (1976): SDS followed by sucrose gradient centrifugation and ethanol precipitation. Infectivity lost when deproteinised with proteases; retained when deproteinised with phenol or detergent.

*Non–genomic nucleic acid* found in the virions; is of some isolates satellite RNA.

*Virion protein(s)* one; $M_r$ 53000; capsid protein. Virion proteins not glycosylated. Virion proteins not phosphorylated.

**Replication**

Replication does not depend on a helper virus.

**Cytopathology**

Virions found in leaves and roots.

**Taxonomy and Relationships**

*Nepovirus: Comoviridae.*

Virus(es) with serologically related virions: tomato black ring, but distantly. Virus(es) with serologically unrelated virions: arabis mosaic, artichoke Italian latent, cherry leaf roll, raspberry ringspot, strawberry latent ringspot, tobacco ringspot and tomato ringspot viruses (Delbos *et al.*, 1976).

**References**


Delbos, R., Dunez, J., Barrau, J. and Fisac, R. (1976). *Annls. Microbiol.* (Paris) **127A**: 101.

Dunez, J. and Delbos, R. (1976). *Mitt. Biol. Bund. fur Land. Fortw., Berlin* **170**: 9.

Dunez, J., Delbos, R. and Dupont (1975). *CMI/AAB Descr. Pl. Viruses* No. 160.

Dunez, J., Delbos, R., Desvignes, J.C., Marenaud, C., Kuszala, J. and Vuittenez, A. (1971). *Annls. Phytopath.*, p. 117.

# Nandina mosaic (?) *potexvirus*

**Data collated by A.A. Brunt, 1991.**

*Synonyms:* nandina (?) *potexvirus*. *Standard acronym:* NaMV.

First reported in *Nandina domestica*; from California, U.S.A.; by Moreno *et al.* (1976).

**Natural host range and symptoms**

Symptoms persist.
*Nandina domestica* — occasional mosaic leaf symptoms.

**Transmission**

Not transmitted by *Myzus persicae*. Virus transmitted by mechanical inoculation.

**Geographical distribution**

Spreads in the USA (in California).

**Experimental host range**

Several (3–9) families susceptible.

Diagnostically susceptible

*Gomphrena globosa, Chenopodium amaranticolor, Cassia occidentalis* — local lesions; not systemic.
*Nicotiana benthamiana, N. megalosiphon* — systemic mottling and malformation.
*Chenopodium quinoa, Spinacia oleracea* — local lesions; then systemic chlorosis.

Diagnostically insusceptible

*Cucumis sativus, Datura stramonium, Nicotiana glutinosa, N. tabacum, Phaseolus vulgaris.*

Maintenance and propagation hosts

*Nicotiana benthamiana.*

Assay hosts (**L**ocal lesions or **W**hole plants)

*Gomphrena globosa* (L).

Dicotyledonous Hosts *(**Susceptible** or Insusceptible)*

*Antirrhinum majus. Beta vulgaris. Brassica oleracea* ssp. *capitata. Capsicum annuum. Carica papaya.* ***Cassia occidentalis. Catharanthus roseus. Chenopodium amaranticolor. Chenopodium quinoa.*** *Crotalaria spectabilis. Cucumis sativus. Dahlia pinnata. Datura stramonium. Glycine max.* ***Gomphrena globosa.*** *Gynura aurantiaca. Lactuca sativa. Lathyrus odoratus. Lycopersicon esculentum. Melilotus alba.* ***Nicotiana benthamiana.*** *Nicotiana × edwardsonii. Nicotiana glutinosa.* ***Nicotiana megalosiphon.*** *Nicotiana rustica. Pastinaca sativa. Phaseolus vulgaris. Physalis floridana. Pisum sativum. Solanum melongena.* ***Spinacia oleracea.*** *Tetragonia expansa. Trifolium pratense. Trifolium repens. Vigna unguiculata. Zinnia elegans.*

Monocotyledonous Hosts *(**Susceptible** or Insusceptible)*

*Hordeum vulgare.*

**Purification**

Method: Zettler *et al.* (1980).

**Particle morphology**

Virions filamentous; usually flexuous; with a clear modal length; of 471 nm. Axial canal obscure. Basic helix obscure.

**Physical properties**

$A_{260}/A_{280}$ ratio 1.35.

*Virion protein(s)* one; $M_r$ 20500; coat protein.

**Cytopathology**

Virions found in leaves, mesophyll and the epidermis; in cytoplasm. Inclusions present in infected cells; are unusual in shape; aggregated virions; they contain virions.

**Taxonomy and Relationships**

Possible *Potexvirus*.

Virus(es) with serologically unrelated virions: clover yellow mosaic, cymbidium mosaic, hydrangea ringspot, papaya mosaic, potato X and white clover mosaic viruses.

**References**

Moreno, P., Attathom, S. and Weathers, L.G. (1976). *Proc. Am. Phytopath. Soc.* **3**: 319.

Santos-Pulgar, R. (1978). *Ph.D. Dissertation*, pp. 84.

Zettler, F.W., Hiebert, E., Christie, R.G. and Abo El-Nil, M.M. (1980). *Acta Hort.* **110**: 71.

# Nandina stem pitting (?) *capillovirus*

**Data collated by S.L. Bhattiprolu, 1991.**

*Standard acronym:* NSPV.

First reported in *Nandina domestica*; from Florida, U.S.A.; by Ahmed *et al.* (1982).

**Natural host range and symptoms**

Symptoms persist.

*Nandina domestica* — mosaic, leaf cupping and malformation; stem pitting and grooving.

**Transmission**

Not transmitted by *Myzus persicae*. Virus not transmitted by mechanical inoculation; transmitted by grafting.

**Geographical distribution**

Found, but with no evidence of spread, in the U.S.A.

**Experimental host range**

Few (<3) families susceptible.

Diagnostically susceptible

*Nandina domestica* — reddening and some inward curling of leaves.

Maintenance and propagation hosts

*Nandina domestica*.

Assay hosts (**L**ocal lesions or **W**hole plants)

*Nandina domestica* (W).

Dicotyledonous Hosts *(**Susceptible** or Insusceptible)*

*Chenopodium amaranticolor. Chenopodium quinoa. Gomphrena globosa. Nicotiana benthamiana. Vigna radiata.*

**Properties of particles in sap**

Leaf sap contains few virions.

**Purification**

Method: Ahmed *et al.* (1982); Lee *et al.* (1987).

**Particle morphology**

Virions filamentous; not enveloped; usually flexuous; with a clear modal length; of 755 nm (or 1600–1800nm; Lee *et al.*, 1987). Axial canal obscure. Basic helix obvious; pitch of basic helix 3.9 nm.

**Physical properties**

$A_{260}/A_{280}$ ratio 1.22–1.24.

**Replication**

Replication does not depend on a helper virus.

**Cytopathology**

Virions found in leaves, meristems and phloem; in nuclei. Inclusions present in infected cells; are amorphous X–bodies; they contain virions.

**Taxonomy and Relationships**

Possible *Capillovirus*.

**References**

Ahmed, N.A., Christie, S.R. and Zettler, F.W. (1982). *Phytopathology* **72**: 354.
Ahmed, N.A., Christie, S.R. and Zettler, F.W. (1982). *Phytopathology* **73**: 470.
Lee, R.F., Garnsey, S.M., Brlansky, R.H. and Goheen, A.C. (1987). *Phytopathology* **77**: 543.
Martelli, G.P., Candresse, T. and Namba, S. (1994). *Arch. Virol.* **134**: 451.

# Narcissus degeneration (?) potyvirus

**Data collated by A.A. Brunt.**

*Standard acronym:* NDV.

First reported in *Narcissus tazetta* cv. Grand Soleil d'Or; from England; by Brunt (1970).

**Taxonomy and Relationships**

Possible unassigned potyvirus.

**References**

Brunt, A.A. (1970). *Rep. Glasshouse Crops Res. Inst.* 1969, p.134.
Brunt, A.A. (1980). *Acta Hort.* **110**: 23.

# Narcissus late season yellows (?) potyvirus

**Data collated by A.A.Brunt.**

*Standard acronym:* NLSYV.

First reported in *Narcissus pseudonarcissus*; from the U.K.; by Brunt (1977; 1980).

**Taxonomy and Relationships**

Possible unassigned potyvirus.

**References**

Brunt, A.A. (1977). *Rep. Glasshouse Crops Res. Inst.* 1976, p.122.

Brunt, A.A. (1980). *Acta Hort.* **110**: 23.

Mowat, W.P., Duncan, G.H. and Dawson, S. (1988). *Ann. appl. Biol.* **113**: 531.

# Narcissus latent macluravirus

**Data collated by S. Phillips, 1985.**

*Synonyms:* gladiolus latent virus, gladiolus ringspot virus, iris mild mosaic virus (Derks *et al.*, 1985), iris mild yellow mosaic virus (Asjes, 1979; Derks *et al.*, 1985), irisbontvirus, narcissus mild mottle carlavirus. *Standard acronym:* NLV.

First reported in *Narcissus pseudonarcissus* cv. Golden Harvest; from Sussex, England; by Brunt and Atkey (1967).

**Natural host range and symptoms**

Symptoms persist.

*Gladiolus* cv. Scarlet Royal and *Iris xiphium* cv. *praecox x I. tingitana* (syn. *Iris hollandica*) bulbous iris — probably symptomless when present alone.

*Narcissus pseudonarcissus* — leaf tip chlorosis in some cultivars.

*Nerine sarniensis* cv. Rotherside — probably symptomless when present alone.

**Transmission**

Transmitted by a vector; an insect; *Acyrthosiphon pisum, Myzus persicae, Aphis gossypii*; Aphididae. Transmitted in a non–persistent manner. Virus transmitted by mechanical inoculation; not transmitted by contact between plants; not transmitted by seed; not transmitted by pollen.

**Geographical distribution**

Spreads in Australia, the Netherlands, and the UK.

**Experimental host range**

Several (3–9) families susceptible.

Diagnostically susceptible

*Nicotiana clevelandii* — systemic chlorotic spotting.

*Nicotiana benthamiana* and *N. megalosiphon* — faint chlorotic local lesions, then some faint systemic chlorotic vein banding.

*Phaseolus vulgaris* cv. The Prince — small chlorotic local lesions.

*Tetragonia expansa* — chlorotic local lesions, rarely systemic.

Diagnostically insusceptible

*Chenopodium quinoa, Dianthus barbatus, Vigna unguiculata, Trifolium incarnatum, Gomphrena globosa.*

Maintenance and propagation hosts

*Nicotiana clevelandii, N. benthamiana, N. megalosiphon.*

Assay hosts (**L**ocal lesions or **W**hole plants)

*Nicotiana clevelandii* (W), *N. megalosiphon* (W), *Tetragonia expansa* (L) (unreliable), *Phaseolus vulgaris* (L) (unreliable).

Dicotyledonous Hosts *(**Susceptible** or Insusceptible)*

***Amaranthus retroflexus. Antirrhinum majus. Arachis hypogaea. Atriplex hortensis. Beta macrocarpa. Beta vulgaris. Brassica campestris*** **ssp.** ***pekinensis. Brassica juncea.***

*Capsicum annuum. Chenopodium amaranticolor. Chenopodium capitatum. Chenopodium murale. Chenopodium quinoa. Cucumis sativus.* ***Datura metel. Datura stramonium.*** *Dianthus barbatus. Emilia sagittata. Fagopyrum esculentum. Gomphrena globosa. Hyoscyamus niger. Lycopersicon esculentum. Nicandra physaloides.* ***Nicotiana benthamiana. Nicotiana clevelandii. Nicotiana debneyi. Nicotiana glutinosa. Nicotiana megalosiphon. Nicotiana rustica. Nicotiana tabacum.*** *Petunia hybrida.* ***Phaseolus vulgaris.*** *Physalis floridana. Pisum sativum. Silene armeria. Solanum melongena. Spinacia oleracea.* ***Tetragonia expansa.*** *Trifolium incarnatum. Verbesina encelioides. Vicia faba. Vigna unguiculata.*

Monocotyledonous Hosts *(**Susceptible** or Insusceptible)*

*Belamcanda chinensis. Hippeastrum hybridum.* ***Narcissus pseudonarcissus.*** *Zea mays.*

Sources of host-range data

Brunt (1977).

**Properties of particles in sap**

TIP: 65–70 °C (in *Nicotiana clevelandii* sap). LIV: 3–4 days (in *Nicotiana clevelandii* sap). DEP: $\log_{10}$ minus 3–4 (in *Nicotiana clevelandii* sap). Infectivity of sap not changed by treatment with di–ethyl ether. Leaf sap contains many virions.

**Purification**

Method: Brunt (1977).

**Particle morphology**

Virions filamentous; not enveloped; usually straight; with a clear modal length; of 657 nm; 13 nm wide. Axial canal obscure. Basic helix obscure.

**Physical properties**

One sedimenting component in purified preparations; sedimentation coefficient 156–158 *S*. Density 1.32–1.33 g $cm^{-3}$ in CsCl (unfixed). $A_{260}/A_{280}$ ratio 1.09 (corrected).

**Biochemical properties**

Virions contain 5 % nucleic acid; 95 % protein; 0 % lipid.

*Genome* consists of RNA; single–stranded.

*Virion protein(s)* one; $M_r$ 32600. Method of preparation: Brunt *et al.* (1975).

**Replication**

Replication does not depend on a helper virus.

**Cytopathology**

Virions found in cytoplasm. Inclusions present in infected cells; are pinwheels.

**Taxonomy and Relationships**

Macluravirus.

Virus(es) with serologically related virions: lily symptomless virus, but distantly (SDI 6). Virus(es) with serologically unrelated virions: cactus 2, carnation latent, chrysanthemum B, cowpea mild mottle, elderberry, hippeastrum latent, hop latent, honeysuckle latent, passiflora latent, potato M, potato S, red clover vein mosaic and white bryony mosaic viruses.

**Comments**

Formerly assumed to be a carlavirus, however now known to induce cylindrical inclusions ('pinwheels') to form (Mowat *et al.*, 1991; Brunt *et al.*, 1994). Thus the virus shares properties with maclura mosaic virus which, it is suggested (Mowat *et al.*, 1991), should form a new genus, that has been tentatively named *macluravirus* (Shukla *et al.*, 1994).

**References**

Asjes, C.J. (1979). *Neth. J. Pl. Path.* **85**: 269.
Brunt, A.A. (1976). *Rep. Glasshouse Crops Res. Inst.* 1975, p. 122.
Brunt, A.A. (1976). *CMI/AAB Descr. Pl. Viruses* No. 170, 4 pp.
Brunt, A.A. (1977). *Ann. appl. Biol.* **87**: 355.
Brunt, A.A. and Atkey, P.T. (1967). *Rep. Glasshouse Crops Res. Inst.* 1966, p. 155.
Brunt, A.A., Barton, R.J., Tremaine, J.H. and Stace-Smith, R. (1975). *J. gen. Virol.* **27**: 101.
Brunt, A.A., Atkey, P.T., Frost, C. and Phillips, S. (1994). *Acta Hort.* (in press).
Derks, A.F.L.M., Hollinger, Th.C. and Vink-van den Abeele, J.L. (1985). *Acta Hort.* **164**: 309.
Mowat, W.P., Dawson, S., Duncan, G.H. and Robinson, D.J. (1991). *Ann. appl. Biol.* **119**: 31.
Shukla, D.S., Ward, C. and Brunt, A.A. (1994). *The Potyviridae*. CABI, Wallingford, U.K.

# Narcissus mosaic *potexvirus*

**Data collated by W.P. Mowat, 1986.**

*Standard acronym:* NMV. *Strains:* Lychnis potexvirus (R.I. Hamilton in Koenig, 1985).

First reported in *Narcissus pseudonarcissus*; from the Netherlands and the U.K.; by Van Slogteren and de Bruyn Ouboter (1946); Brunt (1966).

**Natural host range and symptoms**

*Narcissus pseudonarcissus* — symptomless infection.

**Transmission**

Virus transmitted by mechanical inoculation; not transmitted by contact between plants (between narcissus plants); not transmitted by seed.

**Geographical distribution**

Probably distributed worldwide (in all countries which have imported narcissus and nerine from Europe). Spreads in particularly the Netherlands and the UK.

**Experimental host range**

Several (3–9) families susceptible.

Diagnostically susceptible

*Chenopodium amaranticolor* — chlorotic local lesions.
*Nicotiana clevelandii* — symptomless systemic infection.
*Gomphrena globosa* — necrotic local lesions.

Diagnostically insusceptible

*Nicotiana tabacum* cv. White Burley, *N. glutinosa, Brassica pekinensis*.

Maintenance and propagation hosts

*Chenopodium quinoa, Nicotiana clevelandii* (not infected by some isolates).

Assay hosts (**L**ocal lesions or **W**hole plants)

*Chenopodium amaranticolor* (L).

Dicotyledonous Hosts *(**Susceptible** or Insusceptible)*

*Antirrhinum majus.* ***Apium graveolens. Atriplex hortensis.*** *Beta vulgaris. Brassica campestris* ssp. *pekinensis. Brassica juncea.* ***Catharanthus roseus. Chenopodium amaranticolor. Chenopodium quinoa.*** *Cucumis sativus.* ***Datura stramonium.*** *Fagopyrum esculentum.* ***Glycine max. Gomphrena globosa. Lactuca sativa.*** *Lycopersicon esculentum.* ***Medicago sativa. Nicandra physaloides.*** *Nicotiana glutinosa. Nicotiana rustica. Nicotiana tabacum.* ***Petunia hybrida.*** *Phaseolus vulgaris.* ***Pisum sativum.*** *Solanum tuberosum.* ***Tetragonia expansa. Trifolium pratense. Trifolium repens.*** *Vicia faba.*

Monocotyledonous Hosts *(**Susceptible** or Insusceptible)*

*Allium cepa. Lilium formosanum.* ***Narcissus pseudonarcissus. Nerine bowdenii.*** *Zea mays.*

Sources of host-range data

Brunt (1966).

**Properties of particles in sap**

TIP: 75 °C. LIV: 85 days (and more). DEP: $\log_{10}$ minus 6. Infectivity of sap not changed by treatment with di–ethyl ether. Leaf sap contains many virions.

**Purification**

Method: Mowat (1971).

**Particle morphology**

Virions filamentous; not enveloped; usually flexuous; with a clear modal length; of 550 nm; 13–14 nm wide. Axial canal obscure. Basic helix obscure; pitch of basic helix 3.4 nm.

**Physical properties**

Two sedimenting components in purified preparations; sedimentation coefficient of the fastest, the 550 nm virions 114 *S*.

**Biochemical properties**

Virions contain 6 % nucleic acid; 94 % protein; 0 % lipid.

*Genome* consists of RNA; single–stranded; linear; unipartite; largest (or only) genome part 6.955 kb. Genomic nucleic acid isolated by Short and Davies (1983).

*Sequence database accession code(s):*

D00405 Em(40)_vi:NMV Narcissus mosaic virus (NMV), complete genome. 8/90 6,955bp

D13747 Em(40)_vi:MNCCGAA Gb(84)_vi:MNCCGAA Narcissus mosaic virus (NMV), complete genome. 11/92 6,955bp. 2 sequences.

*Non–genomic nucleic acid* found in the virions; is subgenomic *m*RNA. Sub–genomic *m*RNA found in infected cells.

*Virion protein(s)* one; $M_r$ 24000. Method of preparation: Robinson *et al.* (1975). Amino acid composition: Short (1982).

*Virus–coded non–virion proteins* identified by genomic sequence analysis; three proteins found. $M_r$ of the largest 116000. $M_r$ of 2nd largest 66000. $M_r$ of 3rd 44000.

**Replication**

Replication does not depend on a helper virus.

**Cytopathology**

Virions found in epidermis; in cytoplasm, in nuclei, and in cell vacuoles. Inclusions present in infected cells; are unusual in shape; spindle-shaped; they contain virions.

**Taxonomy and Relationships**

*Potexvirus.*

Virus(es) with serologically unrelated virions: tulip X virus.

**References**


Bancroft, J.B., Hills, G.J. and Richardson, J.F. (1980). *J. gen. Virol.* **50**: 451.

Brunt, A.A. (1966). *Ann. appl. Biol.* **58**: 13.

Koenig, R. (1985). *Acta Hort.* **164**: 21.

Mowat, W.P. (1971). *CMI/AAB Descr. Pl. Viruses* No. 45, 3 pp.

Robinson, D.J., Hutcheson, A., Tollin, P. and Wilson, H.R. (1975). *J. gen. Virol.* **29**: 325.

Short, M.N. (1982). *Ph.D. Thesis*, University of East Anglia, U.K.

Short, M.N. and Davis, J.W. (1983). *Bioscience Rep.* **3**: 837.

Van Slogteren, D.H.M. and de Bruyn Ouboter, M.P. (1946). *Daffodil Tulip Yrbk.* **12**: 3.

Zuidema, D., Linthorst, H.J.M., Huisman, M.J., Asjes, C.J. and Bol, J.F. (1989). *J. gen. Virol.* **70**: 267.


# Narcissus tip necrosis (?) *carmovirus*

**Data collated by S. Phillips, 1992.**

*Standard acronym:* NTNV.

First reported in *Narcissus pseudonarcissus*; from the Netherlands; by Asjes (1972).

**Natural host range and symptoms**

Symptoms vary seasonally.

*Narcissus pseudonarcissus, N. tazetta, N. poeticus*—usually symptomless; sometimes general leaf chlorosis, elongate chlorotic areas near leaf tips, later becoming brown and necrotic, premature leaf senescence.

**Transmission**

Virus transmitted by mechanical inoculation; not transmitted by grafting.

**Geographical distribution**

Spreads in the Netherlands and the UK.

**Experimental host range**

Several (3–9) families susceptible. Experimentally infected plants mostly show small chlorotic local lesions 2mm; no systemic infection.

Diagnostically susceptible

*Chenopodium quinoa, C. murale* — chlorotic local lesions; no systemic infection.

*Tetragonia expansa* — chlorotic local lesions or green rings; no systemic infection.

*Chenopodium amaranticolor* — few pin prick chlorotic local lesions.

*Chenopodium capitatum, Nicotiana benthamiana* — symptomless local infection only.

Diagnostically insusceptible

*Nicotiana clevelandii, N. tabacum, Belamcanda chinensis.*

Maintenance and propagation hosts

*Chenopodium quinoa, C. capitatum, C. murale, Tetragonia expansa, Narcissus pseudonarcissus* cvs Barrett Browning, Goblet, Redmarley.

Assay hosts (**L**ocal lesions or **W**hole plants)

*Chenopodium quinoa* (L).

Dicotyledonous Hosts *(**Susceptible** or Insusceptible)*

***Chenopodium amaranticolor***. ***Chenopodium capitatum***. ***Chenopodium murale***. ***Chenopodium quinoa***. *Dahlia pinnata*. *Dianthus barbatus*. *Gomphrena globosa*. *Helianthus annuus*. ***Nicotiana benthamiana***. *Nicotiana glutinosa*. *Nicotiana megalosiphon*. *Nicotiana tabacum*. *Phaseolus vulgaris*. *Solanum demissum* × *tuberosum*. *Spinacia oleracea*. ***Tetragonia expansa***. *Trifolium incarnatum*. *Vicia faba*. *Vigna unguiculata* ssp. *sesquipedalis*.

Monocotyledonous Hosts *(**Susceptible** or Insusceptible)*

*Avena sativa*. *Triticum aestivum*.

Comments on host-range

Hosts become infected only by inoculation with purified virus preparations, never directly from infected *Narcissus* sap.

Sources of host-range data

S. Phillips and A. Brunt (unpublished data); Mowat *et al.* (1977).

**Properties of particles in sap**

TIP: 75 °C. Leaf sap contains few virions. Electron microscopy: ISEM.

**Purification**

Method: Mowat *et al.* (1977).

**Particle morphology**

Virions isometric; not enveloped; 30.7 nm in diameter; neither clearly angular or nor rounded in profile; without a conspicuous capsomere arrangement.

**Physical properties**

One sedimenting component in purified preparations; sedimentation coefficient 123 *S*. Density 1.356 g $cm^{-3}$ in CsCl. $A_{260}/A_{280}$ ratio 1.5–1.6.

**Biochemical properties**

Virions contain 16–20 % nucleic acid; 80–84 % protein; 0 % lipid.

*Genome* consists of RNA; single–stranded; unipartite; largest (or only) genome part 4.85 kb. Genomic nucleic acid isolated by Mowat *et al.* (1977); Murant *et al.* (1972).

*Virion protein(s)* one (plus a fragment); $M_r$ 42000 (plus 39000); coat protein. Method of preparation: Mayo *et al.* (1974); Brunt *et al.* (1975). Virion proteins not glycosylated. Virion proteins not phosphorylated.

**Replication**

Replication does not depend on a helper virus.

**Cytopathology**

Virions found in leaves, mesophyll, epidermis and meristems; in cytoplasm.

**Taxonomy and Relationships**

Possible *Carmovirus: Tombusviridae*.

Virus(es) with serologically unrelated virions: carnation mottle, elderberry latent, pelargonium flower break, saguaro cactus and turnip crinkle viruses.

**References**

Asjes, C.J. (1972). *Jversl. Lab. Bloemboll.* Lisse, 1970–1971 p. 45.
Asjes, C.J. (1974). *Jversl. Lab. Bloemboll.*, Lisse, 1973, p. 46.
Brunt, A.A. (1975). *Rep. Glasshouse Crops Res. Inst.* 1974, p. 116.
Brunt, A.A., Barton, R.J., Tremaine., J.H. and Stace-Smith, R. (1975). *J. gen. Virol.* **27**: 101.
Mayo, M.A., Robinson, D.J. and Perombelon, M.C.M. (1974). *J. gen. Microbiol.* **85**: 121.
Mowat, W.P. and Hutcheson, A.M. (1974). *Rep. Scottish Hort. Res. Inst.* 1973, p. 61.
Mowat, W.P., Asjes, C.J. and Brunt, A.A. (1976). *CMI/AAB Descr. Pl. Viruses* No. 166, 4 pp.
Mowat, W.P., Asjes, C.J. and Brunt, A.A. (1977). *Ann. appl. Biol.* **86**: 189.
Murant, A.F., Mayo, M.A., Harrison, B.D. and Goold, R.A. (1972). *J. gen. Virol.* **16**: 327.

# Narcissus yellow stripe *potyvirus*

*Synonyms:* narcissus degeneration potyvirus, narcissus yellow streak virus. *Standard acronym:* NYSV.

**Natural host range and symptoms**

*Narcissus* spp.

**Geographical distribution**

Spreads in the Eurasian region; possibly Australia.

**Taxonomy and Relationships**

*Potyvirus: Potyviridae.*

**Comments**

Often in mixed infections with narcissus latent virus (Mowat *et al.*, 1991).

**References**

Mowat, W.P., Dawson, S., Duncan, G.H. and Robinson, D.J. (1991). *Ann. appl. Biol.* **119**: in press.

# Nasturtium mosaic (?) *potyvirus*

**Data collated by A.A. Brunt, 1991.**

*Standard acronym:* NasMV.

First reported in *Tropaeolum majus*; from the U.S.A. and the U.K.; by Jensen (1950); Smith (1950).

**Natural host range and symptoms**

Symptoms persist.

*Tropaeolum majus* — vein clearing, mottling and slight leaf malformation; occasionally flower breaking.

**Transmission**

Transmitted by a vector; an insect; *Aphis fabae, Dysaphis apifolia, Myzus persicae, Aulocorthum circumflexum, Rhopalosiphum padi*; Aphididae. Transmitted in a non–persistent manner. Virus transmitted by mechanical inoculation.

**Geographical distribution**

Spreads in Brazil, South Africa, the UK, and the USA.

**Experimental host range**

Few (<3) families susceptible.

Diagnostically susceptible

*Tropaeolum majus* — severe leaf mosaic.
*Zinnia elegans* — mild leaf chlorosis.

Maintenance and propagation hosts

*Tropaeolum majus, Zinnia elegans.*

Assay hosts (**L**ocal lesions or **W**hole plants)

*Tropaeolum majus* (W).

**Properties of particles in sap**

TIP: 60 °C. LIV: 2–4 days.

**Particle morphology**

Virions filamentous; not enveloped; usually flexuous; with a clear modal length; of 750 nm.

**Taxonomy and Relationships**

Possible *Potyvirus: Potyviridae.*

Although there is no definite evidence, it is thought that the viruses in *Tropaeolum majus* in Brazil, South Africa, the U.K. and the U.S.A. are the same (DaGraca and Martin, 1977).

**References**

Da Graca, J.V. and Budnik (1993). *J. Phytopath.*.
Da Graca, J.V. and Martin, M.M. (1977). *Phytopath. Z.* **88**: 276.
Jensen, D.D. (1950). *Phytopathology* **40**: 967.
Silberschmidt, K. (1953). *Phytopathology* **43**: 304.
Smith, K.M. (1950). *J. Roy. hort. Soc.* **75**: 350.

# Neckar River *tombusvirus*

**Data collated by A.A. Brunt, 1991.**

*Standard acronym:* NRV.

From water from the River Neckar, Germany; by Koenig and Lesemann (1985).

**Transmission**

Virus transmitted by mechanical inoculation.

**Geographical distribution**

Found, but with no evidence of spread, in Germany.

**Experimental host range**

Several (3–9) families susceptible.

Diagnostically susceptible

*Nicotiana benthamiana, N. clevelandii* — conspicuous systemic leaf chlorosis.

*Chenopodium quinoa, Datura stramonium, Gomphrena globosa, Nicotiana tabacum, Petunia hybrida* — local lesions; no systemic infection.

Maintenance and propagation hosts

*Nicotiana benthamiana, N. clevelandii.*

Assay hosts (**L**ocal lesions or **W**hole plants)

*Chenopodium quinoa* (L).

Dicotyledonous Hosts *(**Susceptible** or Insusceptible)*

***Chenopodium quinoa*. *Datura stramonium*. *Gomphrena globosa*. *Nicotiana benthamiana*. *Nicotiana clevelandii*. *Nicotiana tabacum*. *Petunia hybrida*.**

Sources of host-range data

Koenig and Lesemann (1985).

**Properties of particles in sap**

Leaf sap contains many virions.

**Purification**

Method: Gallitelli *et al.* (1985).

**Particle morphology**

Virions isometric; not enveloped; 30 nm in diameter; rounded in profile.

**Physical properties**

One sedimenting component in purified preparations; sedimentation coefficient 133 *S*. Density 1.357 g $cm^{-3}$ in CsCl.

**Biochemical properties**

*Genome* consists of RNA; single–stranded; unipartite; largest (or only) genome part 4.7 kb.

*Virion protein(s)* one; $M_r$ 45000.

**Cytopathology**

Virions found in leaves and the mesophyll; in cytoplasm. Other cellular changes: vesicles at the periphery of microbodies.

**Taxonomy and Relationships**

*Tombusvirus: Tombusviridae.*

Virus(es) with serologically related virions: eggplant mottled crinkle, pelargonium leaf curl, Moroccan pepper, petunia asteroid mosaic, tomato bushy stunt and carnation Italian ringspot viruses. Virus(es) with serologically unrelated virions: cymbidium ringspot virus.

The viral genome has about 15% nucleotide sequence homology with that of tomato bushy stunt tombusvirus.

**References**

Gallitelli, D. and Russo, M. (1987). *J. Phytopath.* **119**: 106.

Koenig, R. and Gibbs, A.J. (1986). *J. gen. Virol.* **67**: 75.

Koenig, R. and Lesemann, D.-E. (1985). *Phytopath. Z.* **112**: 105.

# Negro coffee mosaic (?) *potexvirus*

**Data collated by A.A. Brunt, 1990.**

*Synonyms:* coffee Negro mosaic virus. *Standard acronym:* NeCMV.

First reported in *Cassia occidentalis*; from India; by Verma and Niazi (1974).

**Natural host range and symptoms**

*Cassia occidentalis* — mosaic on leaves.

**Transmission**

Transmitted by a vector; an insect; *Aphis solanella.*; Aphididae.

**Geographical distribution**

Spreads in India.

**Particle morphology**

Virions filamentous; with a clear modal length; of 550–580 nm.

**Taxonomy and Relationships**

Possible *Potexvirus*.

**References**

Verma, V.S. and Niazi, F.R. (1974). *Z. PflKrankh. PflPath. PflSchutz.* **81**: 608.

# Nerine latent *carlavirus*

**Data collated by S. Phillips, 1986.**

*Synonyms:* hippeastrum latent virus. *Standard acronym:* NeLV. *Strains:* nerine latent carlavirus, hippeastrum strain.

First reported in *Nerine bowdenii*; from England; by Brunt *et al.* (1970).

**Natural host range and symptoms**

*Nerine bowdenii, Hippeastrum hybridum* — symptomless.

**Transmission**

Transmitted by a vector; an insect; *Myzus persicae*; Aphididae. Transmitted in a non–persistent manner. Virus transmitted by mechanical inoculation; not transmitted by contact between plants; not transmitted by seed; not transmitted by pollen.

**Geographical distribution**

Spreads in the Netherlands and the UK.

**Experimental host range**

Several (3–9) families susceptible.

Diagnostically susceptible

*Chenopodium quinoa, C. murale* and *C. quinoa* — chlorotic local lesions, not systemic.

*Gomphrena globosa* — pinkish chlorotic local lesions only.

*Nicotiana clevelandii* — faint chlorotic local lesions, then systemic mild veinal chlorosis.

*Datura stramonium* — chlorotic local lesions, not systemic.

Diagnostically insusceptible

*Hyoscyamus niger, Trifolium incarnatum, Tetragonia expansa.*

Maintenance and propagation hosts

*Nicotiana clevelandii, Chenopodium quinoa, Nicotiana clevelandii x glutinosa hybrid.*

Assay hosts (**L**ocal lesions or **W**hole plants)

*Chenopodium quinoa, Chenopodium murale* and *Nicotiana clevelandii.*

Dicotyledonous Hosts *(**Susceptible** or Insusceptible)*

*Amaranthus caudatus. Brassica campestris* ssp. *pekinensis. Calendula officinalis. Capsicum annuum.* ***Chenopodium amaranticolor. Chenopodium capitatum. Chenopodium murale. Chenopodium quinoa.*** *Cucumis melo. Cucumis sativus. Cucurbita pepo. Dahlia pinnata.* ***Datura stramonium.*** *Dianthus caryophyllus. Emilia sagittata. Fagopyrum esculentum.* ***Gomphrena globosa.*** *Helianthus annuus. Hyoscyamus niger. Lactuca sativa. Lycopersicon esculentum. Momordica balsamina. Nicotiana benthamiana.* ***Nicotiana clevelandii.*** *Nicotiana glutinosa.* ***Nicotiana glutinosa × clevelandii.*** *Nicotiana megalosiphon. Nicotiana rustica. Nicotiana tabacum. Petunia hybrida. Phaseolus vulgaris. Pisum sativum. Tetragonia expansa. Trifolium incarnatum. Trifolium repens. Tropaeolum majus. Vigna unguiculata.*

Monocotyledonous Hosts *(**Susceptible** or Insusceptible)*

*Belamcanda chinensis.* ***Hippeastrum hybridum.*** *Hordeum vulgare.* ***Nerine bowdenii.*** *Zea mays.*

Sources of host-range data

Brolman-Hupkes (1975); Maat *et al.* (1978).

**Properties of particles in sap**

TIP: 70–80 °C. LIV: 1–2 days. DEP: $\log_{10}$ minus 3–4. Leaf sap contains many virions. Electron microscopy: ISEM may be necessary to detect virions in sap.

**Purification**

Method: Maat *et al.* (1978).

**Particle morphology**

Virions filamentous; not enveloped; usually straight; with a clear modal length; of 665 nm; 13 nm wide. Axial canal obscure. Basic helix obscure.

**Physical properties**

One sedimenting component in purified preparations; sedimentation coefficient 155 *S*. Density 1.298 g $cm^{-3}$ in CsCl (unfixed). $A_{260}/A_{280}$ ratio 1.08–1.19.

**Biochemical properties**

Virions contain 5 % nucleic acid; 95 % protein; 0 % lipid.

*Virion protein(s)* one; $M_r$ 32600; coat protein. Method of preparation: Brunt *et al.* (1975).

**Replication**

Replication does not depend on a helper virus.

**Cytopathology**

Virions found in all parts of the host plant; in cytoplasm.

**Taxonomy and Relationships**

*Carlavirus.*

Virus(es) with serologically related virions: carnation latent (SDI 3) and chrysanthemum B (SDI 10) viruses. Virus(es) with serologically unrelated virions: red clover vein mosaic, cowpea mild mottle, elderberry, honeysuckle latent, lily symptomless, narcissus latent, potato M and potato S viruses.

Best tests for diagnosis: ISEM with 'decoration' best for distinguishing from similar viruses. NLV together with hippeastrum mosaic potyvirus is common in *Hippeastrum hybridum* varieties, but the two viruses can be separated and identified by inoculation to *Gomphrena globosa* and *Hyoscyamus niger*.

**References**

Brolman-Hupkes (1975). *Neth. J. Pl. Path.* **81**: 226.

Brunt, A.A., Hollings, M. and Stone, O.M. (1970). *Rep. Glasshouse Crops Res. Inst.* 1969, p. 138.

Brunt, A.A., Barton, R.J., Tremaine, J.H. and Stace-Smith, R. (1975). *J. gen. Virol.* **27**: 101.

Maat, D.Z., Huttinga, H. and Hakkaart, F.A. (1978). *Neth. J. Pl. Path.* **84**: 47.

# Nerine *potyvirus*

**Data collated by S. Phillips, 1986.**

*Synonyms:* nerine yellow stripe virus. *Standard acronym:* NV.

First reported in *Nerine sarniensis*; from England; by Brunt *et al.* (1970).

**Natural host range and symptoms**

Symptoms persist.
*Nerine sarniensis* — chlorotic streaking.
*Nerine bowdenii* — chlorotic mosaic.

**Transmission**

Transmitted by a vector; an insect; Aphididae. Transmitted in a non–persistent manner. Virus transmitted by mechanical inoculation; not transmitted by contact between plants.

**Geographical distribution**

Spreads in the Netherlands and the UK.

**Experimental host range**

Several (3–9) families susceptible.

Diagnostically susceptible

*Nicotiana clevelandii* — systemic chlorotic vein-banding.
*Chenopodium quinoa* — chlorotic local lesions.
*Chenopodium amaranticolor* — small chlorotic local lesions.
*Chenopodium murale* — diffuse chlorotic local lesions.
*Hyoscyamus niger* — chlorotic local lesions.
*Nicotiana benthamiana* — chlorotic local lesions, then systemic chlorotic vein-banding.
*Tetragonia expansa* — green chlorotic rings.

Diagnostically insusceptible

*Alstroemeria caryophylla, Phaseolus vulgaris, Solanum tuberosum, Belamcanda chinensis, Trifolium incarnatum.*

Maintenance and propagation hosts

*Nicotiana clevelandii, Nicotiana benthamiana.*

Assay hosts (**L**ocal lesions or **W**hole plants)

*Nicotiana clevelandii* (W), *Nicotiana benthamiana* (W), *Chenopodium quinoa* (L), *Chenopodium amaranticolor* (L), *Hyoscyamus niger* (L).

Dicotyledonous Hosts *(**Susceptible** or Insusceptible)*

*Brassica campestris* ssp. *pekinensis. Calendula officinalis.* ***Chenopodium amaranticolor. Chenopodium capitatum. Chenopodium murale. Chenopodium quinoa.*** *Cucumis sativus. Dahlia pinnata. Datura metel. Gomphrena globosa. Helianthus annuus.* ***Hyoscyamus niger.*** *Lactuca sativa. Lycopersicon esculentum.* ***Nicotiana benthamiana. Nicotiana clevelandii. Nicotiana glutinosa. Nicotiana glutinosa × clevelandii. Nicotiana megalosiphon. Nicotiana rustica. Nicotiana sylvestris. Nicotiana tabacum.*** *Phaseolus vulgaris. Solanum demissum. Solanum tuberosum.* ***Tetragonia expansa.*** *Trifolium incarnatum. Vicia faba. Vigna unguiculata.*

Monocotyledonous Hosts *(**Susceptible** or Insusceptible)*

*Belamcanda chinensis. Lilium formosanum.*

**Properties of particles in sap**

TIP: 65 °C. Leaf sap contains few virions. Electron microscopy: ISEM best for finding virions in sap.

**Purification**

Method: Phillips: Homogenise infected leaves in 0.25M phosphate: 0.1M ascorbic acid: 0.14M mercaptoacetic acid (3:9:1 vols/wt; pH7.0), then ether and finally carbon tetrachloride (1:1 vols). Stir for 15 minutes then two cycles differential centrifugation.

**Particle morphology**

Virions filamentous; not enveloped; usually flexuous; with a clear modal length; of 780 nm; 12 nm wide. Axial canal obscure. Basic helix obscure.

**Replication**

Replication does not depend on a helper virus.

**Cytopathology**

Virions found in all parts of the host plant.

**Taxonomy and Relationships**

*Potyvirus: Potyviridae.*

Virus(es) with serologically related virions: potato Y (SDI 5), henbane mosaic (SDI 8), clover yellow vein, (SDI 9) and alstroemeria mosaic (SDI 10) viruses. Virus(es) with serologically unrelated virions: bean yellow mosaic (gladiolus and crocus strains), carnation vein mottle, freesia mosaic, hippeastrum mosaic, iris mild mosaic, narcissus degeneration, narcissus late season yellows, pea mosaic, pepper veinal mottle, tobacco severe etch, and watermelon mosaic viruses.

**References**

Brunt, A.A., Hollings, M. and Stone, O.M. (1970). *Rep. Glasshouse Crops Res. Inst.* 1969, p. 138.

Hakkaart, F.A., Maat, D.Z. and Quak (1975). *Acta Hort.* **47**: 51.

# Nerine X *potexvirus*

**Data collated by S. Phillips, 1986.**

*Synonyms:* agapanthus X virus. *Standard acronym:* NVX. *Strains:* Agapanthus, nerine.

First reported in *Nerine sarniensis*; from England; by Brunt *et al.* (1970).

**Natural host range and symptoms**

*Nerine sarniensis* (nerine strain) — symptomless.
*Agapanthus praecox ssp. orientalis* (agapanthus strain) — symptomless.

**Transmission**

Virus transmitted by mechanical inoculation; not transmitted by contact between plants; not transmitted by seed; not transmitted by pollen.

**Geographical distribution**

Spreads in the Netherlands and the UK.

**Experimental host range**

Several (3–9) families susceptible.

Diagnostically susceptible

*Chenopodium quinoa* — chlorotic or necrotic local lesions, then systemic chlorosis.

*Chenopodium amaranticolor, C. murale, C. capitatum, Gomphrena globosa* and *Amaranthus retroflexus* — chlorotic local lesions, not systemic.

*Tetragonia expansa* — etched chlorotic local lesions, not systemic. All tests done with the agapanthus strain.

Diagnostically insusceptible

*Lilium cv. Enchantment, Narcissus pseudonarcissus, Solanum tuberosum, Nicotiana clevelandii, Hippeastrum hybridum.*

Maintenance and propagation hosts

*Chenopodium quinoa.*

Assay hosts (Local lesions or Whole plants)

*Chenopodium quinoa* (L-W), *Chenopodium amaranticolor* (L) and *Chenopodium murale* (L).

Dicotyledonous Hosts ***(Susceptible** or Insusceptible)*

***Amaranthus retroflexus**. Arachis hypogaea. Brassica campestris* ssp. *pekinensis. Cajanus cajan. Calendula officinalis. Capsicum annuum.* ***Chenopodium amaranticolor. Chenopodium capitatum. Chenopodium murale. Chenopodium quinoa.*** *Cucumis sativus. Dahlia pinnata. Datura stramonium.* ***Gomphrena globosa.*** *Helianthus annuus. Lactuca sativa.* ***Lycopersicon esculentum.*** *Montia perfoliata.* ***Nicandra physaloides.*** *Nicotiana clevelandii.* ***Nicotiana debneyi.*** *Nicotiana glutinosa. Nicotiana tabacum. Petunia hybrida. Phaseolus vulgaris. Solanum demissum. Solanum tuberosum.* ***Tetragonia expansa. Trifolium incarnatum.*** *Verbesina encelioides. Vicia faba. Zinnia elegans.*

Monocotyledonous Hosts ***(Susceptible** or Insusceptible)*

*Allium cepa. Belamcanda chinensis. Hippeastrum hybridum. Lilium formosanum. Narcissus pseudonarcissus.*

Sources of host-range data

Phillips and Brunt (1980).

**Properties of particles in sap**

TIP: 95–100 °C. LIV: 365 days (at least). DEP: $\log_{10}$ minus 7 (all tests in *Chenopodium quinoa* sap; agapanthus strain). Leaf sap contains few virions.

**Purification**

Method: Maat (1976), for nerine strain; Phillips and Brunt (1980), agapanthus strain.

**Particle morphology**

Virions filamentous; not enveloped; usually flexuous; with a clear modal length; of 540 nm; 11 nm wide. Axial canal obscure. Basic helix obscure.

**Physical properties**

One sedimenting component in purified preparations; sedimentation coefficient 120 *S*. Density 1.31 g $cm^{-3}$ in CsCl (unfixed). $A_{260}/A_{280}$ ratio 1.18–1.4.

**Biochemical properties**

Virions contain 5 % nucleic acid; 95 % protein.

*Genome* consists of RNA; single–stranded; unipartite; largest (or only) genome part 7 kb. Genomic nucleic acid isolated by Brunt *et al.* (1980).

*Non–genomic nucleic acid* not found in the virions.

*Virion protein(s)* one; $M_r$ 25600; coat protein. Method of preparation: Brunt *et al.* (1975).

**Replication**

Replication does not depend on a helper virus.

**Cytopathology**

Virions found in all parts of the host plant.

**Taxonomy and Relationships**

*Potexvirus.*

Virus(es) with serologically related virions: commelina X (SDI 8), hydrangea ringspot (SDI 9), potato X (SDI 5) and viola mottle (SDI 9) viruses. Virus(es) with serologically unrelated virions: bamboo mosaic, cactus X, cymbidium mosaic, lily X, narcissus mosaic, papaya mosaic (type strain and ullucus strain), potato aucuba mosaic and white clover mosaic viruses.

Best tests for diagnosis: ISEM with 'decoration'.

**Comments**

The type strain from *Nerine sarniensis* is isolated and maintained in test plants with difficulty. Most data in this description are from work with the strain from *Agapanthus praecox* ssp. *orientalis*, which is serologically indistinguishable from the type strain, but is readily isolated, reaches a high concentration in test plants and is very stable *in vitro*.

**References**


Brunt, A.A., Hollings, M. and Stone, O.M. (1970). *Rep. Glasshouse Crops Res. Inst.* 1969, p. 138.

Brunt, A.A., Barton, R.J., Tremaine, J.H. and Stace-Smith, R. (1975). *J. gen. Virol.* **27**: 101.

Brunt, A.A., Barton, R.J., Phillips, S. and Lana, A.O. (1980). *Ann. appl. Biol.* **96**: 37.

Maat, D.Z. (1976). *Neth. J. Pl. Path.* **82**: 95.

Phillips, S. and Brunt, A.A. (1980). *Acta Hort.* **110**: 65.

# Nerine Y (?) potyvirus

**Data collated by A.A. Brunt, 1991.**

First reported in *Nerine sarniensis*; from the U.K.; by Brunt (1976).

### Natural host range and symptoms

*Nerine sarniensis* — severe leaf chlorosis.

### Geographical distribution

Spreads in New Zealand and the UK.

### Particle morphology

Virions filamentous; with a clear modal length; of 800 nm.

### Biochemical properties

*Genome* consists of RNA; single–stranded; unipartite; largest (or only) genome part 10 kb.

### Taxonomy and Relationships

Unassigned potyvirus.

A specific *c*DNA probe to the NVY genome did not hybridize with the genome of nerine yellow stripe, potato Y, hippeastrum mosaic and bean yellow mosaic viruses.

### References

Balasingam, G., Ellison, N., Milne, K.S. and Forster, R.L.S. (1988). *Acta Hort.* **234**: 267.

Brunt, A.A. (1976). *Rep. Glasshouse Crops Res. Inst.* 1976, p.123.

# Nicotiana velutina mosaic (?) *furovirus*

**Data collated by J.W. Randles, 1985.**

*Standard acronym:* NVMV.

First reported in *Nicotiana velutina*; from Lake Frome, Australia; by Randles *et al.* (1976).

### Natural host range and symptoms

*Nicotiana velutina* — bright yellow mosaic.

### Transmission

Transmitted by means not involving a vector. Virus transmitted by mechanical inoculation; transmitted by grafting; transmitted by seed (up to 72% in *Nicotiana glutinosa*).

### Geographical distribution

Spreads in Australia.

### Experimental host range

Several (3–9) families susceptible.

Diagnostically susceptible

*Nicotiana glutinosa* — systemic chlorosis.
*N. tabacum* — etched ringspots, then mild systemic mottling.
*N. debneyi* — necrosis; then systemic mosaic and malformation.
*Chenopodium amaranticolor, C. quinoa* — chlorotic local lesions; not systemic.
*Beta macrocarpa* — local lesions; systemic mosaic.
*Gomphrena globosa* — necrotic ringspots; systemic chlorosis.

Diagnostically insusceptible

*Tetragonia expansa.*

Maintenance and propagation hosts

*Nicotiana debneyi.*

Assay hosts (**L**ocal lesions or **W**hole plants)

*Chenopodium amaranticolor* (L), *C. quinoa* (L).

Dicotyledonous Hosts *(**Susceptible** or Insusceptible)*

***Beta macrocarpa*. *Chenopodium amaranticolor*. *Chenopodium quinoa*.** *Cucumis sativus. Dianthus barbatus. Lathyrus odoratus.* ***Nicotiana benthamiana. Nicotiana clevelandii. Nicotiana debneyi. Nicotiana glutinosa. Nicotiana rustica. Nicotiana sylvestris. Nicotiana tabacum.*** *Phaseolus vulgaris.* ***Spinacia oleracea.*** *Tetragonia expansa. Vicia faba.*

Monocotyledonous Hosts *(**Susceptible** or Insusceptible)*

*Triticum aestivum.*

**Properties of particles in sap**

TIP: 60–70 °C. LIV: 1–4 days. DEP: $\log_{10}$ minus 1–2. Leaf sap contains few virions.

**Purification**

Method: Randles (1978).

**Particle morphology**

Virions rod–shaped; not enveloped; usually straight; with no clear modal length; of 125–150 nm (range 100–175 nm); 18 nm wide. Axial canal obvious. Basic helix obvious; pitch of basic helix 2.9 nm.

**Physical properties**

Eight sedimenting components in purified preparations (a broad band (aggregates, not in separate components)); sedimentation coefficient of the fastest 155 *S* (and greater).

**Biochemical properties**

*Genome* consists of RNA; single–stranded; linear; of two parts; largest (or only) genome part the larger 8 kb (RNA 1); the 2nd largest 3 kb (RNA 2). Infectivity retained when deproteinised with phenol or detergent. Nucleotide sequence references: Randles and Rohde (1990).

*Sequence database accession code(s):*

D00906 Em(40)_vi:NVMV2 Gb(84)_vi:NVMV2 Nicotiana velutina mosaic virus RNA2. 2/93 2,325bp. 1 sequence.

*Features of the genome:* RNA 2 contains 4 ORF's; ORF 1 = protein 20K, ORF 2 = protein 39K, ORF 3 = protein 13K and one ORF is incomplete.

$M_r$ 21400. Method of preparation: Randles *et al.* (1976).

**Replication**

Replication does not depend on a helper virus.

**Cytopathology**

Virions found in locally or systemically infected tissue in thin sections. Inclusions absent from infected cells.

**Taxonomy and Relationships**

Possible *Furovirus*.

Virus(es) with serologically unrelated virions: potato mop top, soil-borne wheat mosaic, tobacco mosaic, beet necrotic yellow vein, barley stripe mosaic and tobacco rattle viruses viruses.

CDNA clones did not hybridize with tobacco mosaic, frangipani mosaic, odontoglossum ringspot, tomato mosaic, sunn-hemp mosaic, cucumber green mottle mosaic and Kyuri green mottle mosaic tobamoviruses or beet necrotic yellow vein and peanut clump furoviruses. Although the virus has a divided genome with two RNA's of a size comparable to those of furoviruses, its unique biological properties and lack of serological relationship with furoviruses suggest that it has no clear relationship to any recognised taxonomic group of plant viruses.

**References**

Randles, J.W. (1978). *CMI/AAB Descr. Pl. Viruses* No. 189, 4 pp.
Randles, J.W. and Rohde, W. (1990). *J. gen. Virol.* **71**: 1019.
Randles, J.W., Harrison, B.D. and Roberts, I.M. (1976). *Ann. appl. Biol.* **84**: 193.

# Nothoscordum mosaic (?) potyvirus

**Data collated by A.A. Brunt and P. Jeyanandarajah, 1991.**

*Standard acronym:* NoMV.

First reported in *Nothoscordum fragrans*; from Canary Islands; by McKinney (1929).

**Natural host range and symptoms**

Symptoms persist.
*Nothoscordum fragrans* — mosaic mottling.

**Geographical distribution**

Spreads in the USA (Louisiana, and the Canary Islands).

**Experimental host range**

Few (<3) families susceptible.

Diagnostically insusceptible

*Cucumis sativus, Datura stramonium, Nicotiana tabacum, Avena sativa, Zea mays.*

Maintenance and propagation hosts

*Nothoscordum fragrans* (W).

Dicotyledonous Hosts *(**Susceptible** or Insusceptible)*

*Beta vulgaris. Cucumis sativus. Datura stramonium. Lycopersicon esculentum. Nicotiana tabacum.*

Monocotyledonous Hosts *(**Susceptible** or Insusceptible)*

*Allium cepa. Avena sativa. Lilium formosanum.*

Sources of host-range data

McKinney (1929).

**Properties of particles in sap**

TIP: 58–60 °C. LIV: 17–20 days. DEP: $\log_{10}$ minus 5–6. Leaf sap contains few virions.

**Particle morphology**

Virions filamentous; not enveloped; usually flexuous; with a clear modal length; of 771 nm.

**Taxonomy and Relationships**

Unassigned potyvirus.

**References**

Gold, A.H., Scott, H.A. and McKinney, H.H. (1957). *Pl. Dis. Reptr* **41**: 250.
McKinney, H.H. (1929). *J. Agric. Res.* **39**: 557.
McKinney, H.H. (1950). *Phytopathology* **40**: 703.

# Oak ringspot virus

**Data collated by J.P. Fulton, K.S. Kim and O.W. Barnett, 1987.**

First reported in *Quercus marilandica* and *Q. velutina*; from Northwest Arkansas, U.S.A.; by Barnett (1971); Kim and Fulton (1973).

**Natural host range and symptoms**

Symptoms persist.

*Quercus marilandica, Q. velutina* — chlorotic ringspots and mosaic, vein-banding and oak leaf patterns, especially on older leaves.

**Transmission**

Virus not transmitted by mechanical inoculation; transmitted by grafting.

**Ecology and control**

Studies reported by Barnett (1971); Kim and Fulton (1973).

**Geographical distribution**

Spreads in the USA.

**Experimental host range**

Few (<3) families susceptible.

Diagnostically susceptible

*Quercus marilandica, Q. velutina.*

Maintenance and propagation hosts

*Quercus marilandica, Q. velutina.*

**Properties of particles in sap**

Leaf sap contains few virions (none found in extracts).

**Particle morphology**

Virions filamentous; not enveloped; usually flexuous; with no clear modal length; 13 nm wide. Axial canal obscure.

**Cytopathology**

Virions found in leaves and mesophyll; in cytoplasm. Inclusions present in infected cells; they contain virions. Other cellular changes: areas of vesiculation, prominent Golgi and paramural bodies, and cell wall protrusions.

**Taxonomy and Relationships**

Best tests for diagnosis: A virus-like disorder of overcup oak is not graft transmissible (Gregory and Seliskar, 1970), and no virions could be found (Neinhaus and Yarwood, 1972).

**Comments**

The relationship to viruses of oak in Europe (Blattny and Prochazkov, 1966; Schmelzer *et al.*, 1966) is not known.

**References**

Barnett, O.W. (1971). *Pl. Dis. Reptr* **55**: 411.
Blattny, C. and Prochazkova, Z. (1966). *Rev. Roumaine Biol. Ser. Bot.* **11(1/3)**: 39.
Gregory, G.F. and Seliskar, C.E. (1970). *Pl. Dis. Reptr* **54**: 844.
Kim, K.S. and Fulton, J.P. (1973). *Pl. Dis. Reptr* **57**: 1029.
Nienhaus, F. and Yarwood, C.E. (1972). *Phytopathology* **62**: 313.
Schmelzer, K., Schmidt, H.E. and Schmidt, H.B. (1966). *Arch. Forstw.* **15**: 107.

# Oat blue dwarf *marafivirus*

**Data collated by E.E. Banttari, R.J. Zeyen and R.G. Timian, 1991.**

*Synonyms:* flax crinkle virus. *Standard acronym:* OBDV.

First reported in *Avena sativa*; from Minnesota, U.S.A.; by Goto and Moore (1952).

**Natural host range and symptoms**

Symptoms persist. Symptoms stunting, enations, deep blue colour in *Avena* sp.

*Avena sativa* — stunting, short stiffened leaves, bluish leaves, enations along leaf veins, sterility of florets.

*Hordeum vulgare, Chrysanthemum morifolium, Hordeum jubatum, Spinacia oleracea, Stellaria media, Thlaspi arvense* — dwarfing and stiffening of leaves, enations.

*Linum grandiflorum, L. usitatissimum* — stunting, prominent enations along leaf veins.

**Transmission**

Transmitted by a vector; an insect; *Macrosteles fascifrons*; Cicadellidae. Not transmitted by *Macrosiphum avenae*. Transmitted in a persistent manner. Virus retained when the vector moults; multiplies in the vector; not transmitted by mechanical inoculation; not transmitted by contact between plants; not transmitted by seed; not transmitted by pollen.

**Ecology and control**

Studies reported by Banttari and Zeyen (1976); Long *et al.* (1975); Timian (1985); Westdal (1968).

**Geographical distribution**

Spreads in Canada, Finland, Sweden, and the USA.

**Experimental host range**

Several (3–9) families susceptible. Experimentally infected plants mostly show stunting, enations along leaf veins, bluish color, sterility of florets.

Diagnostically susceptible

*Avena sativa, Hordeum vulgare* — enations, blue colour, stunting, sterility, excess tillering.

*Chrysanthemum morifolium, Stellaria media* — enations.

*Linum grandiflorum, L. usitatissimum* — enations, less tillering, dimples on leaves, stunting.

*Spinacia oleracea* — wrinkling and cupping of leaves.

*Thlaspi arvense* — see Westdal (1968).

Diagnostically insusceptible

*Amaranthus retroflexus, Brassica campestris, Callistephus chinensis, Cucumis sativus, Zinnia elegans.*

Maintenance and propagation hosts

*Avena sativa, Hordeum vulgare, Linum usitatissimum.*

Assay hosts (**L**ocal lesions or **W**hole plants)

*Avena sativa* (W), *Hordeum vulgare* (W), *Linum usitatissimum* (W).

Dicotyledonous Hosts (***Susceptible*** *or Insusceptible)*

*Amaranthus retroflexus. Apium graveolens. Beta vulgaris. Brassica campestris* ssp. *napus. Brassica oleracea* ssp. *botrytis. Brassica oleracea* ssp. *capitata. Calendula officinalis. Capsicum frutescens. Carthamus tinctorius. Chenopodium album.* ***Chrysanthemum morifolium. Cichorium endiva.*** *Cucumis sativus.* ***Daucus carota.*** *Dianthus barbatus. Fagopyrum esculentum. Helianthus annuus. Lactuca sativa.* ***Linum usitatissimum.*** *Lycopersicon esculentum. Melilotus alba. Nicotiana rustica. Pastinaca sativa. Petroselinum crispum. Petunia hybrida. Phaseolus vulgaris. Raphanus sativus. Senecio vulgaris. Solanum melongena.* ***Spinacia oleracea. Stellaria media.*** *Trifolium pratense. Vicia sativa. Zinnia elegans.*

Monocotyledonous Hosts (***Susceptible*** *or Insusceptible)*

***Avena sativa. Hordeum vulgare. Secale cereale.*** *Setaria viridis.* ***Triticum aestivum. Triticum durum.*** *Zea mays.*

Additional hosts

***Avena fatua, Brassica kabir*** var. ***pinnatifida, Centaurea cyanus, Gaillardia aristata, Linum grandiflorum, L. usitatissimum, Spinacia oleracea*** and ***Thlaspi arvense.***

Sources of host-range data

Westdal (1968).

**Properties of particles in sap**

TIP: 60–70 °C. LIV: 16–32 days. DEP: $\log_{10}$ minus 5.

**Purification**

Method: Banttari and Zeyen (1969).

**Particle morphology**

Virions isometric; not enveloped; 28–30 nm in diameter; probably rounded in profile; with a conspicuous capsomere arrangement.

**Physical properties**

Three sedimenting components in purified preparations; sedimentation coefficient of the fastest 119 *S*. $A_{260}/A_{280}$ ratio 1.63.

**Biochemical properties**

*Genome* consists of RNA; single–stranded; linear; unipartite; largest (or only) genome part 6.74 kb. Genomic nucleic acid isolated by Pring *et al.* (1973). Additional factor not required for infectivity.

**Replication**

Genome replicates in phloem cells. Replication does not depend on a helper virus.

**Cytopathology**

Virions found in leaves, roots, meristems and phloem; in cytoplasm. Inclusions present in infected cells; are crystals in the cytoplasm, membranous bodies, and viroplasms; they contain virions. Other cellular changes: see Banttari and Zeyen (1971); Zeyen and Banttari (1972).

**Taxonomy and Relationships**

*Marafivirus*.

Virus(es) with serologically related virions: Bermuda grass etched line viruses. Virus(es) with serologically unrelated virions: Maize rayado fino virus.

Best tests for diagnosis: transmission to flax, barley and oats using the aster leafhopper. Crinkle in flax, leaf enations, serology, blue colour and dwarfing in leaves.

**References**


Atcham, T. and Banttari, E.E. (1986). *Phytopathology* **76**: 1062.
Azar, M. and Banttari, E.E. (1981). *Phytopathology* **71**: 858.
Banttari, E.E. (1967). *Phytopathology* **57**: 802.
Banttari, E.E. (1979). *Phytopathology* **69**: 1021.
Banttari, E.E. (1981). *Phytopathology* **71**: 1242.
Banttari, E.E. and Frederiksen, R.A. (1959). *Phytopathology* **49**: 539.
Banttari, E.E. and Moore, M.B. (1959). *Phytopathology* **49**: 533.
Banttari, E.E. and Moore, M.B. (1962). *Phytopathology* **52**: 897.
Banttari, E.E. and Zeyen, R.J. (1968). *Phytopathology* **58**: 1042.
Banttari, E.E. and Zeyen, R.J. (1969). *Phytopathology* **59**: 183.
Banttari, E.E. and Zeyen, R.J. (1970). *Phytopathology* **60**: 399.
Banttari, E.E. and Zeyen, R.J. (1971). *Phytopathology* **61**: 1249.
Banttari, E.E. and Zeyen, R.J. (1972). *Phytopathology* **62**: 745.
Banttari, E.E. and Zeyen, R.J. (1972). *Virology* **49**: 305.
Banttari, E.E. and Zeyen, R.J. (1973). *Ann. N.Y. Acad. Sci.* **225**: 503.
Banttari, E.E. and Zeyen, R.J. (1973). *CMI/AAB Descr. Pl. Viruses* No. 123.
Banttari, E.E. and Zeyen, R.J. (1976). *Phytopathology* **66**: 896.
Banttari, E.E. and Zeyen, R.J. (1979). In: *Leafhopper Vectors and Plant Disease Agents*, p. 327; eds K. Maramorosch and K.F. Harris. Academic Press, New York.
Conner, R. and Banttari, E.E. (1979). *Can. J. Pl. Pathol.* **1**: 111.
Fredericksen, R.A. and Goth, R.W. (1959). *Phytopathology* **49**: 538.
Goto, S. and Moore, M.B. (1952). *Pl. Dis. Reptr* **36**: 69.
Heagle, A.S. and Moore, M.B. (1969). *Pl. Dis. Reptr* **53**: 664.
Hoppe, W. and Vacke, J. (1972). *Polska Akad. Nauk* **133**: 123.
Hsu, T.P. and Banttari, E.E. (1979). *Phytopathology* **69**: 843.
Johnson, D.A., Tew, T.L. and Banttari, E.E. (1977). *Pl. Dis. Reptr* **61**: 280.
Leon, P. and Gamez, R. (1981). *J. gen. Virol.* **56**: 67.
Lockhart, B.E., Khaless, N., Lennon, A.M. and El Maataoui, M. (1985). *Phytopathology* **75**: 1258.
Long, D.L. and Timian, R.G. (1971). *Phytopathology* **61**: 1230.
Long, D.L. and Timian, R.G. (1972). *Phytopathology* **62**: 773.
Long, D.L. and Timian, R.G. (1975). *Phytopathology* **65**: 848.
Pring, D.R., Zeyen, R.J. and Banttari, E.E. (1972). *Phytopathology* **62**: 783.
Pring, D.R., Zeyen, R.J. and Banttari, E.E. (1973). *Phytopathology* **63**: 393.
Timian, R.G. (1983). *Barley Newsl.* **27**: 54.
Timian, R.G. (1985). *Plant Dis.* **69**: 706.

Timian, R.G. and Alm, K. (1973). *Phytopathology* **63**: 109.
Westdal, P.H. (1968). *Can. J. Bot.* **46**: 1431.
Zeyen, R.J. and Banttari, E.E. (1969). *Phytopathology* **59**: 1059.
Zeyen, R.J. and Banttari, E.E. (1972). *Can. J. Bot.* **50**: 2511.

# Oat chlorotic stunt (?) *tombusvirus*

**Data collated by A.A. Brunt, 1995.**

First reported in *Avena sativa*; from Wales; by Catherall (1986).

**Natural host range and symptoms**

*Avena sativa* — pale green to bright yellow leaf streaking, then necrosis and severe stunting, no viable seed produced.

**Transmission**

Virus transmitted by mechanical inoculation.

**Geographical distribution**

Spreads in the UK.

Monocotyledonous Hosts *(**Susceptible** or Insusceptible)*

***Avena sativa***.

**Properties of particles in sap**

Leaf sap contains many virions.

**Particle morphology**

Virions isometric; not enveloped; 35 nm in diameter; angular in profile.

**Biochemical properties**

*Genome* consists of RNA; single–stranded. Genomic nucleic acid isolated by Boonham *et al.* (1995).

*Virion protein(s)* one; $M_r$ 48200; coat protein.

**Taxonomy and Relationships**

Possible *Tombusvirus: Tombusviridae*.

The sequence of the coat protein places this virus closest to the tombusviruses, but the polymerase sequence places it outside the Tombusviridae.

**References**

Boonham, N., Henry, C.M. and Wood, K.R. (1995). *J. gen. Virol.* **76**: 2025.
Catherall, P.L. (1980). *Rep. Welsh Pl. Breeding Stn.* 1985, p. 132.

# Oat golden stripe *furovirus*

**Data collated by P. Jones, 1990.**

*Synonyms:* oat chlorotic stripe virus, oat tubular virus (Plumb and MacFarlane, 1978), soil-borne oat stripe virus. *Standard acronym:* OGSV.

First reported in *Avena sativa*; from Devon, England; by Plumb *et al.* (1977).

**Natural host range and symptoms**

Symptoms vary seasonally. Symptoms yellow striping on tip leaves when systemically infected.

*Avena sativa* — yellow striping on two youngest leaves.
Many plants may have infected roots, but show no systemic symptoms.

**Transmission**

Transmitted by a vector; a fungus (presumably); *Polymyxa graminis*; Plasmodiophorales. Principal natural vector(s): *Polymyxa graminis*. Virus transmitted by mechanical inoculation; not transmitted by contact between plants.

**Geographical distribution**

Spreads in France, the UK, and the USA (North Carolina).

**Experimental host range**

Several (3–9) families susceptible. Experimentally infected plants mostly show mottling; striping rare in glasshouse plants.

Diagnostically susceptible

*Avena sativa* — yellow mottling and stripes on leaves; systemic infection.
*Nicotiana clevelandii, Chenopodium amaranticolor* — necrotic local lesions, sometimes systemic in *N. clevelandii*.
*Nicotiana debneyi* — chlorotic local lesions.
*Nicotiana benthamiana*.

Diagnostically insusceptible

*Triticum* sp., *Hordeum vulgare, Phaseolus vulgaris*.

Maintenance and propagation hosts

*Avena* sp.

Dicotyledonous Hosts *(**Susceptible** or Insusceptible)*

***Chenopodium amaranticolor***. ***Nicotiana benthamiana***. ***Nicotiana clevelandii***. ***Nicotiana debneyi***. *Phaseolus vulgaris*.

Monocotyledonous Hosts *(**Susceptible** or Insusceptible)*

***Avena sativa***. *Hordeum vulgare*. *Triticum aestivum*.

Sources of host-range data

Plumb *et al.* (1977).

**Properties of particles in sap**

Leaf sap contains few virions.

**Purification**

Method: Adams *et al.* (1988).

**Particle morphology**

Virions rod–shaped; not enveloped; usually straight; with a clear modal length; of 150–300 nm; 20 nm wide. Axial canal obvious. Basic helix obscure.

**Physical properties**

Two sedimenting components in purified preparations; sedimentation coefficient of the fastest 218 *S*; of the other(s) 168 *S*. Density 1.321 g $cm^{-3}$ in CsCl.

**Biochemical properties**

*Genome* consists of RNA; single–stranded; linear; of two parts; largest (or only) genome part the larger 6.2 kb; the 2nd largest 3.5 kb. Genomic nucleic acid isolated by Jones: proteinase K incubation followed by phenol/chloroform and precipitation with ethanol.

*Virion protein(s)* one; $M_r$ 20000.

**Cytopathology**

Virions found in leaves and roots; in cytoplasm. Other cellular changes: disruption of organelles (Plumb *et al.*, 1977).

**Taxonomy and Relationships**

*Furovirus.*

Virus(es) with serologically related virions: soil borne wheat mosaic virus. Virus(es) with serologically unrelated virions: beet necrotic yellow vein, hypochoeris mosaic and potato mop top viruses.

Very distantly serologically related to potato M and red clover vein mosaic carlaviruses. Best tests for diagnosis: often found with oat mosaic virus. Distinguish by virion morphology in EM 'dip preparations'.

**References**

Adams, M.J., Jones, P. and Swaby, A.G. (1988). *Ann. appl. biol.* **112**: 285.

Chen, J. and Adams, M.J. (1991). *Pl. Path.* **40**: in press.

Hariri, D. and Lapierre, H. (1985). *Mitt. Biol. Bund. Land. Forst. Berlin-Dahlem* **228**: 72.

Macfarlane, I., Jenkins, J.E. and Melville, S.C. (1968). *Pl. Path.* **17**: 167.

Plumb, R.T. and Macfarlane, I. (1977). *Rep. Rothamsted Exp. Stn* p. 256.

Plumb, R.T. and MacFarlane, I. (1978). *Rep. Rothamsted Exp. Stn 1977* p. 212.

Plumb, R.T., Catherall, P.L., Chamberlain, J. and Macfarlane, I. (1977). *Annls. Phytopath.* **9**: 365.

# Oat mosaic *bymovirus*

**Data collated by T. Usugi, 1987.**

*Synonyms:* soil-borne oat mosaic virus (Macfarlane *et al.* 1968). *Standard acronym:* OMV. *Strains:* apical mosaic strain and eyespot strain.

First reported in *Avena sativa*; from the U.S.A.; by Atkinson (1945); McKinney (1946).

**Natural host range and symptoms**

Symptoms persist.

*Avena* spp. — mottling.

**Transmission**

Transmitted by a vector; a fungus; Plasmodiophorales; *Polymyxa graminis.* Virus transmitted by mechanical inoculation; not transmitted by seed.

**Ecology and control**

Studies reported by Toler and Hebert (1964); McKinney (1946).

**Geographical distribution**

Spreads in the UK and the USA.

**Experimental host range**

Several (3–9) families susceptible.

Diagnostically susceptible

*Avena sativa* — mottling, especially in the first leaves.

Diagnostically insusceptible

*Hordeum vulgare, Zea mays, Lolium* spp., *Oryza sativa.*

Maintenance and propagation hosts

*Avena sativa.*

Assay hosts (**L**ocal lesions or **W**hole plants)

*Avena sativa* (W).

Dicotyledonous Hosts *(**Susceptible** or Insusceptible)*

*Amaranthus caudatus. Antirrhinum majus. Beta vulgaris. Brassica oleracea* ssp. *botrytis. Capsicum annuum. Carthamus tinctorius. Chenopodium album. Chenopodium amaranticolor. Chenopodium hybridum. Chrysanthemum morifolium. Cicer arietinum. Cucumis melo. Cucumis sativus. Cucurbita pepo. Datura stramonium. Glycine max. Gomphrena globosa. Gossypium hirsutum. Hibiscus esculentus. Lactuca sativa. Lathyrus odoratus. Lycopersicon esculentum. Medicago sativa. Nicotiana glutinosa. Nicotiana sylvestris. Nicotiana tabacum. Petunia hybrida. Phaseolus acutifolius. Phaseolus lathyroides. Phaseolus lunatus. Phaseolus vulgaris. Phlox drummondii. Physalis peruviana. Pisum sativum. Raphanus sativus. Solanum melongena. Solanum tuberosum. Spinacia oleracea. Trifolium incarnatum. Trifolium repens. Vicia faba. Zinnia elegans.*

Monocotyledonous Hosts *(**Susceptible** or Insusceptible)*

*Allium cepa.* ***Avena byzantina. Avena sativa. Avena strigosa.*** *Hordeum vulgare. Oryza sativa. Secale cereale.* ***Triticum aestivum.*** *Tulipa gesneriana. Zea mays.*

Sources of host-range data

McKinney (1946); Toler and Hebert (1963).

**Properties of particles in sap**

TIP: 44–46 °C. LIV: 1–2 days. DEP: $\log_{10}$ minus 1–2 (using the leaf rubbing method), or 2–5 (using the air brush method). Infectivity of sap not changed by treatment with di–ethyl ether. Leaf sap contains few virions.

**Purification**

Method: Usugi (1981).

**Particle morphology**

Virions filamentous; not enveloped; usually flexuous; with no clear modal length; of 600–750 nm; 12–14 nm wide. Axial canal obscure. Basic helix obscure.

**Physical properties**

Density 1.3 g $cm^{-3}$ in CsCl.

**Biochemical properties**

*Genome* consists of RNA; single–stranded; linear; of two parts; largest (or only) genome part the larger 8 kb; the 2nd largest 4 kb. Genomic nucleic acid isolated by nucleic acid is extracted by adding 2% SDS to purified preparation. Poly A region present.

*Virion protein(s)* two; $M_r$ of the larger 30000. $M_r$ of 2nd largest 26500. Method of preparation: Usugi (1981).

**Cytopathology**

Virions found in all parts of the host plant; in cytoplasm. Inclusions present in infected cells; are pinwheels; they do not contain virions.

**Taxonomy and Relationships**

*Bymovirus: Potyviridae.*

Best tests for diagnosis: sap inoculation tests; only *Avena* species are susceptible, and develop mottling symptoms below 15°C.

**References**

Atkinson, M. (1945). *Pl. Dis. Reptr* **29**: 86.
Hebert, T.T. and Panizo, C.H. (1975). *CMI/AAB Descr. Pl. Viruses* No. 145, 4 pp.
Macfarlane, I., Jenkins, J.E.E. and Melville, S. (1968). *Pl. Path.* **17**: 167.
McKinney, H.H. (1946). *Phytopathology* **36**: 359.
Panizo, C.H. (1975). Ph.D. Thesis N. Carolina State University.
Toler, R.W. and Hebert, T.T. (1963). *Pl. Dis. Reptr* **47**: 58.
Toler, R.W. and Hebert, T.T. (1964). *Phytopathology* **54**: 428.
Usugi, T. (1981). *Ann. Phytopath. Soc. Japan* **47**: 581.

# Oat necrotic mottle *rymovirus*

**Data collated by A.A. Brunt, 1991.**

*Standard acronym:* ONMV.

First reported in *Avena sativa*; from Manitoba, Canada; by Gill and Westdal (1966).

**Natural host range and symptoms**

Symptoms persist.

*Avena sativa* — leaf chlorosis and, in spring-sown crops, abortion of florets, seed loss and plant stunting.

*Poa canadensis, P. pratensis* — mild leaf chlorosis or symptomless infection.

**Transmission**

Virus transmitted by mechanical inoculation; not transmitted by contact between plants.

**Ecology and control**

Studies reported by Gill and Westdal (1966); *Poa pratensis* and *P. canadensis* adjacent to crops are often infected and may be important sources of infection.

**Geographical distribution**

Spreads in Canada.

**Experimental host range**

Few (<3) families susceptible.

Diagnostically susceptible

*Avena sativa* — leaf chlorosis, sheath mottling or necrosis.

Diagnostically insusceptible

*Agropyron repens, Dactylis glomerata, Festuca elatior, F. rubra, Hordeum vulgare, Lolium perenne, Secale cereale, Triticum aestivum, Zea mays.*

Maintenance and propagation hosts

*Avena sativa* cv. Clintland–64.

Assay hosts (**L**ocal lesions or **W**hole plants)

*Avena sativa* cv. Clintland–64 (W).

Monocotyledonous Hosts (***Susceptible*** *or Insusceptible)*

*Agropyron repens. Agrostis alba. Agrostis palustris. Agrostis tenuis. Elytrigia intermedium. Lophopyrum elongatum.*

**Properties of particles in sap**

TIP: 50 °C. LIV: 3–7 days (at 23°C), or 42–56 days (at 4°C). DEP: $\log_{10}$ minus 3.

**Purification**

Method: Gill (1971).

**Particle morphology**

Virions filamentous; usually flexuous; with a clear modal length; of 720 nm; 11 nm wide.

**Physical properties**

$A_{260}/A_{280}$ ratio 1.33.

**Biochemical properties**

*Genome* consists of RNA; single–stranded.
*Virion protein(s)* one; $M_r$ 39800; coat protein.

**Cytopathology**

Virions found in cytoplasm. Inclusions present in infected cells; are pinwheels; they do not contain virions. Other cellular changes: the occurrence of deposits, sometimes extensive, on cell walls.

**Taxonomy and Relationships**

*Rymovirus: Potyviridae.*
Virus(es) with serologically related virions: wheat streak mosaic virus. Virus(es) with serologically unrelated virions: agropyron mosaic, hordeum mosaic and ryegrass mosaic viruses.

**References**

Gill, C.C. (1967). *Phytopathology* **57**: 302.
Gill, C.C. (1971). *J. gen. Virol.* **12**: 259.
Gill, C.C. (1974). *Can. J. Bot.* **52**: 621.
Gill, C.C. (1976*a*). *Phytopathology* **66**: 415.
Gill, C.C. (1976*b*). *CMI/AAB Descr. Pl. Viruses* No. 170, 4 pp.
Gill, C.C. (1980). *Can. J. Pl. Path.* **2**: 86.
Gill, C.C. and Westdal, P.H. (1966). *Can. Pl. Dis. Surv.* **46**: 18.

# Oat pseudorosette (?) *tenuivirus*

**Data collated by A.A. Brunt, 1991.**

*Synonyms:* oat pupation disease virus, oat Siberian mosaic virus.

First reported in *Avena sativa*; from USSR (Siberia); by Sukhov (1940).

**Natural host range and symptoms**

*Avena sativa* — light green stripes and spots in leaves and leaf sheaths, proliferic tillering, severe stunting and, often, sterility.

**Transmission**

Transmitted by a vector; an insect; *Laodelphax striatellus*; Delphacidae. Transmitted in a persistent manner. Virus not transmitted by mechanical inoculation; not transmitted by seed.

**Geographical distribution**

Spreads in the former USSR.

Diagnostically susceptible

*Avena sativa* — leaf chlorosis, excessive tillering and stunting.

**Taxonomy and Relationships**

Possible *Tenuivirus*.

**Comments**

Symptoms, persistent transmission by a delphacid planthopper vector and the possible presence of a non-capsid protein in infected plants suggest that the virus may be a tenuivirus.

**References**

Gingery, R.E. (1988). In: *The Plant Viruses*, Vol. 4, The Filamentous Viruses, p. 297; ed R.G. Milne, Plenum Press, New York.

Sukhov, K.S. and Vovk, A.M. (1938). *C.R. (Dokl.) Acad. Sci., URSS* **19**: 207.

Sukhov, K.S. (1940). *Mikrobiologiya* **9**: 188.

Sukhov, K.S., Vovk, A.M. and Alexeeva, T.S. (1943). *C.R. (Dokl.) Acad. Sci., URSS* **41**: 344.

# Oat sterile dwarf *fijivirus*

**Data collated by R.G. Milne, 1991.**

*Synonyms:* arrhenatherum blue dwarf virus (Milne and Lesemann, 1978; Muhle and Kempiak, 1971; Luisoni *et al.*, 1979), lolium enation virus (Lesemann and Huth, 1975; Milne and Lesemann, 1978). *Standard acronym:* OSDV.

First reported in *Avena sativa*; from the former Czechoslovakia and Sweden; by Prusa (1958); Lindsten (1959).

**Natural host range and symptoms**

Symptoms persist. Symptoms bushy, dwarf, dark green grass-like appearance; leaf malformation, yellowish or white enations (small galls) centred on veins or vein swellings on backs of leaves.

*Arrhenatherum elatius, Avena sativa, Hordeum vulgare, Lolium multiflorum, L. perenne* — dwarfing, excess tillering, flower suppression and retention of non-maturing juvenile phase; dark green colour of leaves, which may be notched, ragged or malformed; small enations on abaxial surfaces of leaves, nodes and spike.

**Transmission**

Transmitted by a vector; an insect; *Javesella pellucida, J. dubia, J. discolor, J. obscurella, Dicranotropis hamata*; Delphacidae. Principal natural vector(s): *Javesella pellucida*. Not transmitted by *Laodelphax striatellus*. Transmitted in a persistent manner. Virus retained when the vector moults; multiplies in the vector; does not require a helper virus for vector transmission; transmitted by mechanical inoculation; not transmitted by contact between plants; not transmitted by seed; not transmitted by pollen.

**Ecology and control**

Studies reported by Boccardo and Milne (1980); Lindsten (1979); Brčak (1979).

**Geographical distribution**

Spreads in the former Czechoslovakia, Finland, Germany, Norway, Poland, Sweden, and the UK.

**Experimental host range**

Few (<3) families susceptible. Experimentally infected plants mostly show bushy, dwarf, dark green grass-like appearance; leaf malformation, yellow or white enations on veins or vein swellings on backs of leaves.

Diagnostically susceptible

*Arrhenatherum elatius, Avena* spp., *Cynosurus cristatus, Hordeum* spp., *Lolium* spp., *Phalaris canariensis, Poa annua, Secale cereale, Triticum* spp., *Zea* spp. — bushy dwarf dark green grasslike appearance, leaf malformation, yellow or white enations on veins or vein swellings on backs of leaves.

Diagnostically insusceptible

All species except those from the Gramineae. Insusceptible Gramineae include *Agropyron repens, Agrostis gigantea, Bromus inermis, Digitaria sanguinalis, Echinochloa crus-galli.*

Maintenance and propagation hosts

*Lolium multiflorum.*

Assay hosts (**L**ocal lesions or **W**hole plants)

*Avena sativa* (W), *Lolium multiflorum* (W).

Monocotyledonous Hosts *(**Susceptible** or Insusceptible)*

*Agropyron repens.* ***Avena sativa***. *Bromus inermis. Echinochloa crus–galli.* ***Festuca pratensis***. ***Hordeum vulgare***. ***Lagurus ovatus***. ***Lolium multiflorum***. ***Lolium temulentum***. ***Poa annua***. ***Secale cereale***. ***Triticum aestivum***. ***Zea mays***.

Additional hosts

*Agrostis gigantea, Alopecurus myosuroides, Apera spica-venti*, ***Arrhenatherum elatius, Avena fatua, Cynosurus cristatus***, *Digitaria saguinalis*, ***Holcus lanatus*** and ***Phalaris canariensis***.

Sources of host-range data

Milne and Lovisolo (1977).

**Properties of particles in sap**

Leaf sap contains few virions. Electron microscopy: take material from enations only, or swollen leaf veins; use UA, not PTA, glutaraldehyde causes damage.

**Purification**

Method: Luisoni *et al.* (1979); Boccardo and Milne (1980). Only B-spiked subvirions (SVPs) have been purified. Use 0.4M phosphate buffer pH 7, shake with Freon, recover SVPs in aqueous phase; apply 2 cycles of differential centrifugation, and a final rate-zonal caesium sulphate density gradient.

**Particle morphology**

Virions isometric; subvirions are commonly seen, appearing as 50 nm diameter spheres with spikes; not enveloped; 65–70 nm in diameter (complete virions); rounded in profile.

**Physical properties**

One sedimenting component in purified preparations to four sedimenting components in purified preparations.

**Biochemical properties**

*Genome* consists of RNA; double–stranded; linear; of 10 parts (named S1–10); largest (or only) genome part the largest 4.4 kb (pairs); the 2nd largest 3.9 kb (pairs); the 3rd largest 3.7 kb (pairs); the 4th largest 3.7 kb (pairs); the 5th largest 3.3 kb (pairs); the 6th largest and other parts 2.9 kb (pairs; also 1.86, 1.83, 1.82 and 1.5 kb pairs). Genomic nucleic acid isolated by Luisoni *et al.* (1979).

**Replication**

Genome replicates in cytoplasm (of phloem and related cells of plants, and in the cytoplasm of many cell types in insects). Replication does not depend on a helper virus.

**Cytopathology**

Virions found in leaves, roots, phloem and companion cells; in cytoplasm; veins of infected plants are swollen or develop enations consisting of enlarged and proliferated phloem parenchyma. Most virions are found in these cells, and in the roots. Inclusions present in infected cells; are viroplasms and unusual in shape; crystalline arrays of complete virions and tubules or scroll-like structures composed of protein, usually containing virions; they contain virions. Other cellular changes: cell proliferation (Milne and Lovisolo, 1977; Milne, 1980).

**Taxonomy and Relationships**

*Fijivirus: Reoviridae.*

Virus(es) with serologically unrelated virions: sugarcane fiji disease, maize rough dwarf, rice black-streaked dwarf and pangola stunt viruses.

**References**


Boccardo, G. and Milne, R.G. (1980). *CMI/AAB Descr. Pl. Viruses* No. 217.

Boccardo, G. and Milne, R.G. (1984). *CMI/AAB Descr. Pl. Viruses* No. 294, 7pp.

Brčak, J. (1979). In: *Leafhopper Vectors and Plant Disease Agents* p. 97; eds K. Maramorosch and K. Harris. Academic Press, New York.

Lesemann, D.-E. and Huth, W. (1975). *Phytopath. Z.* **82**: 246.

Lindsten, K. (1959). *Phytopath. Z.* **35**: 420.

Lindsten, K. (1979). In: *Leafhopper Vectors and Plant Disease Agents.* p. 155; eds K. Maramorosch and K. Harris, Academic Press, New York.

Luisoni, E., Boccardo, G., Milne, R.G. and Conti, M. (1979). *J. gen. Virol.* **45**: 651.

Milne, R.G. (1980). *Microbiologica* **3**: 333.

Milne, R.G. and Lesemann, D.-E. (1978). *Virology* **90**: 299.

Milne, R.G. and Lovisolo, O. (1977). *Adv. Virus Res.* **21**: 267.

Milne, R.G. and Luisoni, E. (1977). *Virology* **80**: 12.

Milne, R.G., Lindsten, K. and Conti, M. (1975). *Ann. appl. Biol.* **79**: 371.

Muhle, E. and Kempiak, G. (1971). *Phytopath. Z.* **72**: 269.

Prusa, V. (1958). *Phytopath. Z.* **33**: 99.


# Odontoglossum ringspot *tobamovirus*

**Data collated by D.-E. Lesemann and P. Beetham, 1986.**

*Synonyms:* tobacco mosaic virus — orchid strain (Jensen and Gold, 1951; Kado *et al.*, 1968). *Standard acronym:* ORSV.

First reported in *Odontoglossum grande*; from the U.S.A.; by Jensen and Gold (1951).

**Natural host range and symptoms**

*Odontoglossum grande* — ringspots.

*Cymbidium* spp. — 'diamond' mottle.

More than 20 genera of orchids are reported to be infected by odontoglossum ringspot virus and show flower colour-breaking, chlorotic streaking, mosaic or necrosis. Symptomless in some cultivars.

**Transmission**

Transmitted by means not involving a vector. Virus transmitted by mechanical inoculation.

**Geographical distribution**

Probably distributed worldwide.

**Experimental host range**

Several (3–9) families susceptible.

Diagnostically susceptible

*Nicotiana tabacum* cv. Xanthi-*nc* — necrotic local lesions in older leaves, ring-like local lesions in younger leaves.

*Cassia occidentalis, Chenopodium amaranticolor* — pinpoint necrotic spots.

The symptoms induced in orchids are not diagnostic.

Maintenance and propagation hosts

*Nicotiana tabacum* cv. Samsun, *Nicotiana clevelandii*.

Assay hosts (**L**ocal lesions or **W**hole plants)

*Chenopodium quinoa* (L), *Nicotiana tabacum* cv. Xanthi-*nc* (L).

Dicotyledonous Hosts *(**Susceptible** or Insusceptible)*

***Beta vulgaris*. *Chenopodium amaranticolor*. *Chenopodium quinoa*. *Gomphrena globosa*. *Nicotiana clevelandii*. *Nicotiana glutinosa*. *Nicotiana tabacum*. *Tetragonia expansa*. *Zinnia elegans*.**

Monocotyledonous Hosts *(**Susceptible** or Insusceptible)*

***Cymbidium alexanderi*.**

Sources of host-range data

Paul *et al.* (1965); Lawson and Ali (1975).

**Properties of particles in sap**

TIP: 90 °C. DEP: $\log_{10}$ minus 6. Infectivity of sap not changed by treatment with di–ethyl ether. Leaf sap contains many virions.

**Particle morphology**

Virions rod–shaped; not enveloped; usually straight; with a clear modal length; of mostly 300 nm (but also shorter, broken virions); 18 nm wide. Axial canal obvious.

**Physical properties**

Two sedimenting components in purified preparations; sedimentation coefficient of the longer virions 212 *S*; of the other(s) 119 *S* (of the shorter virions). Density 1.25 g $cm^{-3}$ in CsCl (unfixed). $A_{260}/A_{280}$ ratio 0.99 (for predominantly longer virions), or 0.63 (for the shorter virions; both values corrected for scattering).

**Biochemical properties**

Virions contain 5 % nucleic acid.

*Sequence database accession code(s):*

X55295 Em(40)_vi:TOORCOAT Gb(84)_vi:TOORCOAT Odontoglossum ringspot virus coat protein gene. 12/91 891bp.

X55296 Em(40)_vi:ORVTRANS Gb(84)_vi:ORVTRANS Odontoglossum ringspot virus cell-to-cell transport gene. 12/90 912bp.

X80053 Gb(84)_vi:ORSPVCY Odontoglossum ringspot Virus-Cy RNA. 7/94 1,015bp.

X78966 Em(43)_vi:Orsvcp Gb(89)n:Orsvcp Odontoglossum Ringspot Virus (strain Cy) coat protein mRNA. 5/95 900bp.

X82130 Em(43)_vi:Orsvrna Gb(89)_vi:Orsvrna Odontoglossum ringspot virus RNA for replicase, movement protein and coat protein genes. 10/9.

*Virion protein(s)* one; $M_r$ 17598; coat protein. Method of preparation: Paul and Buchta (1971); Henning (1972). Amino acid composition: Henning (1972).

**Cytopathology**

Virions found in mesophyll, epidermis and vascular parenchyma; in chloroplasts and in cytoplasm (very occasionally). Inclusions present in infected cells; are crystals in the cytoplasm.

**Taxonomy and Relationships**

*Tobamovirus.*

Virus(es) with serologically related virions: cucumber green mottle mosaic, ribgrass mosaic, tomato mosaic, tobacco mosaic and tobacco mild green mosaic tobamoviruses (see Dubs and Regenmortel, 1990).

Local lesions caused by the type strain are larger than those of Corbett's (1967) isolate TMV-O. Best tests for diagnosis: this virus differs from all others known to infect orchids by its virion morphology, and by its ability to induce necrotic local lesions in *Nicotiana tabacum* cv. Xanthi.

**References**

Corbett, M.K. (1967). *Phytopathology* **57**: 164.

Dubs, M.C. and van Regenmortel, M.H.V. (1990). *Arch. Virol.* **115**: 239.

Edwardson, J.R. and Zettler, F.W. (1986). In: *The Plant Viruses*, Vol. 2, The Rod-Shaped Plant Viruses, pp. 233–247; eds M.H.V. van Regenmortel and H. Fraenkel-Conrat. Plenum Press, New York.

Hennig, B. (1972). In: *Atlas of Protein Sequence and Structure*, Vol. 5, D287; ed. M.O. Dayhoff. National Biomedical Research Foundation, Washington.

Jensen, D.D. and Gold, H.A. (1951). *Phytopathology* **41**: 648.

Kado, C.I., van Regenmortel, M.H.V. and Knight, C.A. (1968). *Virology* **34**: 17.

Lawson, R.H. and Ali, S. (1975). In: *Handbook on Orchid Pests and Diseases*, pp. 62–103. American Orchid Society, Cambridge, Mass., U.S.A.

Paul, H.L. (1975). *CMI/AAB Descr. Pl. Viruses* No. 155, 4 pp.

Paul, H.L. and Buchta, U. (1971). *J. gen. Virol.* **11**: 11.

Paul, H.L., Wetter, K., Wittmann, H.G. and Brandes, J. (1965). *Z. VererbLehre* **97**: 186.

# Okra leaf-curl bigeminivirus

**Data collated by L. Givord, 1991.**

*Synonyms:* bamia leaf-crinkle virus. *Standard acronym:* OLCV.

First reported in *Hibiscus esculentus* syn. *Abelmoschus esculentus*; from the Sudan; by Kirkpatrick (1930).

**Natural host range and symptoms**

Symptoms persist. Symptoms vein thickening, leaf crinkling and curling, vein enations, plants stunted and develop fewer and smaller pods.

*Hibiscus esculentus* — leaf crinkling, curling, plants stunted.

**Transmission**

Transmitted by a vector; an insect; *Bemisia tabaci*; Aleyrodidae. Transmitted in a persistent manner. Virus not transmitted by mechanical inoculation; transmitted by grafting; not transmitted by seed.

**Ecology and control**

Studies reported by Atiri and Fayoyin (1989).

**Geographical distribution**

Spreads in Chad, Cote d'Ivoire, Nigeria, and Sudan.

**Experimental host range**

Few (<3) families susceptible. Experimentally infected plants mostly show leaf curl.

Diagnostically susceptible

*Gossypium hirsutum, Lycopersicon esculentum* cv. Congo Jute, *Urena lobata* — leaf curl.

*Nicotiana tabacum* cvs Havana 24 — faint mottling.

Maintenance and propagation hosts

*Hibiscus esculentus.*

Assay hosts (**L**ocal lesions or **W**hole plants)

*Hibiscus esculentus* (W).

Dicotyledonous Hosts *(**Susceptible** or Insusceptible)*

***Gossypium hirsutum. Lycopersicon esculentum. Nicotiana tabacum.***

Sources of host-range data

Yassin and Nour (1965); Lana (1976).

**Properties of particles in sap**

Leaf sap contains few virions.

**Purification**

Method: grinding in phosphate buffer 0.2M pH 7.8, butanol/chloroform clarification, differential centrifugation, sucrose density gradient, yielding at 1–5 mg/kg.

**Particle morphology**

Virions geminate; 20 nm in diameter; rounded in profile.

**Taxonomy and Relationships**

Bigeminivirus: *Geminiviridae.*

Virus(es) with serologically related virions: African cassava mosaic virus.

**References**

Ahmed, A.H. (1978). *Acta Hort.* **84**: 65.

Atiri, G.I. and Fayoyin, G.A. (1989). *Ann. appl. Biol. (suppl.)* **10**: 152.

Bink, F.A. (1973). *Cot. Fib. Trop.* **28**: 365.

Fauquet, C. and Thouvenel, J.C. (1987). *Initiations Documentations Techniques* No. 46.

Kirkpatrick, T.W. (1930). *Bull. Ent. Res.* **21**; 127.

Kirkpatrick, T.W. (1931). *Bull. Ent. Res.* **22**: 323.
Lana, A.F. (1976). *PANS* **22**: 474.
Yassin, A.M. and Nour, M.A. (1965). *Ann. appl. Biol.* **56**: 207.

# Okra mosaic *tymovirus*

**Data collated by L. Givord, 1984. Revised 1991.**

*Synonyms:* virus de la mosaique du Gombo. *Standard acronym:* OkMV. *Strains:* VMG-CI (IC. OkMV) from *H. esculentus*, Côte d'Ivoire; VM-HR, from *H. rosa-sinensis*, Côte d'Ivoire; VMG-Nig (N.OkMV) from *H. esculentus* in Nigeria (Givord, 1977; Bozarth *et al.*, 1977).

First reported in *Hibiscus esculentus*; from Côte d'Ivoire; by Givord *et al.* (1972).

### Natural host range and symptoms

Symptoms persist (in all other families except Malvaceae). Symptoms mosaic, vein chlorosis and banding in Malvaceae. Other families show interveinal chlorosis, vein chlorosis and vein banding. In all families plants are often stunted.

*Abutilon hirtum* — yellow mosaic.
*Borreria intricans* — interveinal mosaic.
*Corchorus olitorius, Hibiscus sabdariffa* — vein chlorosis and vein banding.
*Hibiscus esculentus, H. rosa-sinensis* — mosaic and vein banding.
*Malvastrum coromandelianum, Sida acuta, S. rhomboidea* — yellow mosaic.
*Physalis angulata* — symptomless.
*S. linifolia* — mosaic and vein banding.
*Urena lobata* — some star-shaped light green areas.

### Transmission

Transmitted by a vector; an insect; *Podagrica decolorata* in Côte d'Ivoire (Givord and den Boer, 1980); *Podagrica uniforma, P. sjostedti* in Nigeria (Lana and Taylor, 1976; Atiri, 1984). Possibly *Bemisia tabaci* is a vector in Nigeria (Lana and Taylor, 1976), but needs confirmation; Coleoptera. Not transmitted by *Aphis gossypii, Chrysolagria cuprina, Lagria villosa, Medythia quaterna, Ootheca mutabilis, Nisotra dilecta.* Transmitted in a non–persistent manner. Virus transmitted by mechanical inoculation; transmitted by grafting; not transmitted by seed.

### Ecology and control

Studies reported by natural and synthetic chemicals are known for control (Atiri, 1991). Tolerant cultivars; Perkins Long Pod in Côte d'Ivoire (Givord, 1980), NHAE 621 and NHAE 117 in Nigeria (Atiri, 1983).

### Geographical distribution

Spreads in Cote d'Ivoire and Nigeria.

### Experimental host range

Many (>9) families susceptible. Experimentally infected plants mostly show regular vein chlorosis, large chlorotic bands along principal veins, spotting or dotting, chlorotic mosaic, chlorotic local lesions, stunting.

Diagnostically susceptible

*Chenopodium amaranticolor* — systemic chlorotic spotting and line patterns.
*Cucumis sativus* — large chlorotic local lesions, then systemic vein chlorosis.
*Arachis hypogaea* — dark and light green mosaic.

*Vigna unguiculata* — fine systemic vein chlorosis and mosaic.
*Nicotiana clevelandii* — systemic dark green and whitish mosaic, dwarfing.

Diagnostically insusceptible

*Zinnia elegans, Brassica campestris* ssp. *pekinensis, Cucurbita pepo, Phaseolus vulgaris, Datura stramonium.*

Maintenance and propagation hosts

*Cucumis sativus* in temperate regions. *Hibiscus esculentus* in tropical regions.

Assay hosts (**L**ocal lesions or **W**hole plants)

No reliable local lesion host known. *Hibiscus esculentus* (W).

Dicotyledonous Hosts ***(Susceptible*** *or Insusceptible)*

***Adansonia digitata***. *Amaranthus caudatus.* ***Amaranthus retroflexus***. *Antirrhinum majus.* ***Apium graveolens***. ***Arachis hypogaea***. *Beta vulgaris. Brassica campestris* ssp. *pekinensis. Brassica oleracea* ssp. *botrytis.* ***Cajanus cajan***. *Calendula officinalis.* ***Capsella bursa–pastoris***. *Capsicum annuum. Capsicum frutescens.* ***Catharanthus roseus***, or *Catharanthus roseus.* ***Ceiba pentandra***. *Celosia argentea. Cheiranthus cheiri.* ***Chenopodium album***. ***Chenopodium amaranticolor***. ***Chenopodium ambrosioides***. ***Chenopodium foetidum***. ***Chenopodium hybridum***. ***Chenopodium quinoa***. *Citrullus lanatus*, or ***Citrullus lanatus***. ***Corchorus spp.***. *Cucumis melo.* ***Cucumis sativus***. *Cucurbita maxima.* ***Cucurbita pepo***. *Datura metel. Datura stramonium.* ***Daucus carota***. ***Dianthus barbatus***. *Glycine max.* ***Gomphrena globosa***, or *Gomphrena globosa.* ***Gossypium hirsutum***, or *Gossypium hirsutum.* ***Hibiscus cannabinus***. ***Hibiscus esculentus***. *Hyoscyamus niger. Lactuca sativa.* ***Lathyrus odoratus***. ***Lavatera trimestris***. *Lycopersicon esculentum. Lycopersicon pimpinellifolium.* ***Matthiola incana***. *Medicago sativa. Melilotus alba.* ***Momordica balsamina***. *Nicandra physaloides.* ***Nicotiana benthamiana***. ***Nicotiana clevelandii***. ***Nicotiana* × *edwardsonii***. ***Nicotiana glutinosa***, or *Nicotiana glutinosa.* ***Nicotiana glutinosa* × *clevelandii***. ***Nicotiana megalosiphon***. ***Nicotiana rustica***, or *Nicotiana rustica. Nicotiana tabacum.* ***Passiflora edulis***. *Pastinaca sativa. Petunia hybrida. Phaseolus lathyroides.* ***Phaseolus lunatus***. ***Phaseolus vulgaris***, or *Phaseolus vulgaris.* ***Phlox drummondii***. *Physalis floridana. Physalis peruviana.* ***Pisum sativum***, or *Pisum sativum. Raphanus sativus. Ricinus communis. Senecio vulgaris.* ***Sesamum indicum***. ***Sinapis alba***. ***Solanum melongena***. ***Solanum nigrum***. *Solanum tuberosum. Sonchus oleraceus.* ***Spinacia oleracea***. ***Tetragonia expansa***. ***Theobroma cacao***. ***Trifolium incarnatum***. ***Trifolium pratense***. *Trifolium repens.* ***Tropaeolum majus***. ***Vicia faba***, or *Vicia faba.* ***Vigna radiata***. ***Vigna unguiculata***. ***Voandzeia subterranea***. *Zinnia elegans.*

Monocotyledonous Hosts ***(Susceptible*** *or Insusceptible)*

*Hordeum vulgare. Setaria italica. Triticum aestivum. Zea mays.*

Additional hosts

Additional Hosts: *Crotalaria juncea* (**S**).

Sources of host-range data

Givord (1977; 1979); Givord and Hirth (1973); Igwegbe (1983); Lana and Bozarth (1975).

**Properties of particles in sap**

TIP: 80 °C. LIV: 7–9 days. DEP: $\log_{10}$ minus 6. Leaf sap contains many virions.

**Purification**

Method: Givord and Hirth (1973).

**Particle morphology**

Virions isometric; not enveloped; 28 nm in diameter; rounded in profile; with a conspicuous capsomere arrangement.

**Physical properties**

Two sedimenting components in purified preparations; sedimentation coefficient of the fastest 106 *S* (B); of the other(s) 42 *S* (T). $A_{260}/A_{280}$ ratio 1.68 (B), or 0.89 (T).

**Biochemical properties**

Virions contain 32 % nucleic acid; 68 % protein; 0 % lipid.

*Genome* consists of RNA; single–stranded; unipartite (plus a sub-genomic RNA coding for coat protein found in Bottom and Top). Base composition 17.2 % G; 17.5 % A; 39.8 % C; 25.5 % U. Infectivity retained when deproteinised with phenol or detergent. Additional factor not required for infectivity. Genome has *t*RNA–like activity. Genome accepts and it is valine.

*Non–genomic nucleic acid* found in the virions; is subgenomic *m*RNA.

*Virion protein(s)* one. Amino acid composition: Paul *et al.* (1980).

**Replication**

Replication does not depend on a helper virus.

**Cytopathology**

Virions found in leaves and roots; in nuclei (containing masses of empty protein shells), or in mitochondria (containing vacuole-like vesicles with empty protein shell-like structures). Other cellular changes: the small double membrane vesicles, that form in the peripheries of chloroplasts, which become rounded, clumped, vacuolated and disorganised (Lesemann, 1977).

**Taxonomy and Relationships**

*Tymovirus.*

Virus(es) with serologically related virions: kennedya yellow mosaic, desmodium yellow mottle, clitoria yellow vein and cacao yellow mosaic viruses are closely related, and other tymoviruses are distantly serologically related (Koenig, 1976).

VMG–CI, VMG–Nig and VM.HR can be distinguished by five diagnostic species and by their serological interrelationships (Givord, 1977). IC–OkMV and N–OkMV can be distinguished on the basis of cross-serological reaction with some other tymoviruses and differential host susceptibility (Bozarth *et al.*, 1977).

**References**

Atiri, G.I. (1983). *Ann. appl. Biol.* **102** no Suppl.: 132.
Atiri, G.I. (1984). *Ann. appl. Biol.* **104**: 261.
Atiri, G.I., Ivbijaro, M.F. and Oladele, A.D. (1991). *Trop. Agric., Trin.* **68**: 178.
Bozarth, R.F., Lana, A.O., Koenig, R. and Reese, J. (1977). *Phytopathology* **67**: 735.
Genevaux, M., Pinck, M. and Duranton, H.M. (1976). *Ann. Microbiol.* **127**: 47.
Givord, L. (1977). *Annls. Phytopath.* **9**: 53.
Givord, L. (1978). *Pl. Dis. Reptr* **62**: 412.
Givord, L. (1979). *Agron. trop.* **34**: 88.
Givord, L. (1980). *Ph.D. Thesis,* University Louis Pasteur of Strasbourg, France.
Givord, L. and Den Boer, L. (1980). *Ann. appl. Biol.* **94**: 235.
Givord, L. and Hirth, L. (1973). *Ann. appl. Biol.* **74**: 359.
Givord, L. and Koenig, R. (1974). *CMI/AAB Descr. Pl. Viruses* No. 128, 4 pp.
Givord, L., Pfeiffer, P. and Hirth, L. (1972). *C. r. Acad. Sci. Paris* **275**: 1563.

Hirth, L. and Givord, L. (1988). In: *The Plant Viruses*. Vol. 3, Polyhedral virions with monopartite RNA genomes, p. 163; ed. R. Koenig, Plenum Press, New York.
Igwegbe, E.C.K. (1983). *Plant Dis.* **67**: 320.
Koenig, R. (1976). *Virology* **72**: 1.
Koenig, R. and Givord, L. (1974). *Virology* **58**: 119.
Lana, A.O. and Ajibola Taylor, T. (1976). *Ann. appl. Biol.* **82**: 361.
Lana, A.O. and Bozarth, R.F. (1975). *Phytopath. Z.* **83**: 77.
Lesemann, D.-E. (1977). *Phytopath. Z.* **90**: 315.
Marshall, B. and Matthews, R.E.F. (1981). *Virology* **110**: 253.
Paul, H.L., Gibbs, A. and Wittmann-Liebold, B. (1980). *Intervirology* **13**: 99.

# Olive latent ringspot *nepovirus*

**Data collated by D. Gallitelli, 1987.**

*Standard acronym:* OLRSV.

First reported in *Olea europaea*; from Lazio, Central Italy; by Savino *et al.* (1983).

**Natural host range and symptoms**

*Olea* spp. — symptomless.

**Transmission**

Virus transmitted by mechanical inoculation; transmitted by grafting.

**Geographical distribution**

Spreads in Italy (Lozio, central Italy).

**Experimental host range**

Several (3–9) families susceptible.

Diagnostically susceptible

*Chenopodium quinoa, C. amaranticolor* — chlorotic or necrotic local lesions, systemic mosaic, apical necrosis.

*Gomphrena globosa* — red-rimmed local lesions, tip leaves malformed.

Diagnostically insusceptible

*Nicotiana tabacum* cvs Samsun, White Barley, *N. glutinosa*, *Phaseolus aureus*.

Maintenance and propagation hosts

*Chenopodium quinoa, Nicotiana clevelandii.*

Assay hosts (**L**ocal lesions or **W**hole plants)

*Chenopodium quinoa* (W), *Chenopodium amaranticolor* (W), *Cucumis sativus* cv. Markerer (W).

Dicotyledonous Hosts *(**Susceptible** or Insusceptible)*

***Chenopodium amaranticolor*. *Chenopodium quinoa*. *Cucumis sativus*. *Gomphrena globosa*. *Nicotiana clevelandii*.** *Nicotiana glutinosa. Nicotiana tabacum.* ***Petunia hybrida*. *Phaseolus vulgaris*.**

Sources of host-range data

Savino *et al.* (1983).

**Properties of particles in sap**

TIP: 60 °C. LIV: 15 days. DEP: $\log_{10}$ minus 3.

**Purification**

Method: Gallitelli: homogenise symptom bearing *Chenopodium quinoa* leaves in 0.1M neutral phosphate buffer. Clarify the sap with Mg-activated bentonite. Purify by two cycles of centrifugation.

**Particle morphology**

Virions isometric; not enveloped; 28 nm in diameter; angular in profile; without a conspicuous capsomere arrangement.

**Physical properties**

Three sedimenting components in purified preparations; sedimentation coefficient of the fastest 132 *S* (B); of the other(s) 97 *S* (M), or 52 *S* (T). Density 1.52 g $cm^{-3}$ in CsCl ($B_2$), or 1.51 g $cm^{-3}$ in CsCl ($B_1$), or 1.43 g $cm^{-3}$ in CsCl (M), or 1.29 g $cm^{-3}$ in CsCl (T). $A_{260}/A_{280}$ ratio 1.8 (B), or 1.51 (M), or 1.03 (T).

**Biochemical properties**

Virions contain 45 % nucleic acid ($B_2$), or 43 % nucleic acid ($B_1$), or 32 % nucleic acid (M), or 0 % nucleic acid (T); 55 % protein ($B_2$), or 57 % protein ($B_1$), or 68 % protein (M), or 100 % protein (T).

*Genome* consists of RNA; single–stranded; of two parts; largest (or only) genome part the larger 7.5 kb; the 2nd largest 3.9 kb. Genomic nucleic acid isolated by Gallitelli; best means by phenol — SDS. Infectivity retained when deproteinised with phenol or detergent. Additional factor not required for infectivity.

*Virion protein(s)* one; $M_r$ 58000; a polymer of a monomeric form of 16kDa. Method of preparation: Gallitelli; boil for 3 minutes in 125mM Tris-Cl pH 6.8 containing 1% SDS, 2% beta-mercaptoethanol, 6M urea. Virion proteins not glycosylated.

**Cytopathology**

Virions found in leaves mesophyll and epidermis; in cytoplasm and in cell vacuoles. Inclusions present in infected cells; are crystals in the cytoplasm and membranous bodies; they contain virions. Other cellular changes: clumping of chloroplasts and finger-like outgrowths of the cell.

**Taxonomy and Relationships**

*Nepovirus: Comoviridae.*

Virus(es) with serologically unrelated virions: arabis mosaic, artichoke Italian latent, artichoke yellow ringspot, artichoke vein banding, cherry leaf roll, cherry rasp leaf, chicory yellow mottle, cacao necrosis, grapevine Bulgarian latent, grapevine fanleaf, grapevine chrome mosaic, myrobalan latent ringspot raspberry ringspot, strawberry latent ringspot, tobacco ringspot, tomato black ring and tomato ringspot viruses.

**References**

Gallitelli, D., Martelli, G.P. and Savino, V. (1985). *AAB Descr. Pl. Viruses* No. 301, 4 pp.

Savino, V., Gallitelli, D. and Barba, M. (1983). *Ann. appl. Biol.* **103**: 243.

# Olive latent 1 (?) *sobemovirus*

**Data collated by D. Gallitelli, 1991.**

*Standard acronym:* OLV-1.

First reported in *Olea europea*; from Apulia, southern Italy; by Gallitelli and Savino (1985).

**Natural host range and symptoms**

Symptoms none.

*Olea europea* — symptomless.

**Transmission**

Virus transmitted by mechanical inoculation.

**Geographical distribution**

Found, but with no evidence of spread, in Apulia, southern Italy.

**Experimental host range**

Several (3–9) families susceptible. Experimentally infected plants mostly show local lesions, but no systemic infection, except in *Nicotiana benthamiana*.

Diagnostically susceptible

*Nicotiana benthamiana* — systemic mosaic.

*Chenopodium amaranticolor, C. quinoa, Cucumis sativus, Nicotiana clevelandii, N. glutinosa, N. megalosiphon, N. tabacum, Gomphrena globosa* — necrotic local lesions; no systemic infection.

Diagnostically insusceptible

*Datura stramonium, Nicotiana rustica, Petunia hybrida.*

Maintenance and propagation hosts

*Nicotiana benthamiana.*

Assay hosts (**L**ocal lesions or **W**hole plants)

*Chenopodium quinoa* (L), *Nicotiana benthamiana* (W), *N. clevelandii* (L), *Cucumis sativus* (L), *Gomphrena globosa* (L).

Dicotyledonous Hosts *(**Susceptible** or Insusceptible)*

***Celosia cristata*. *Chenopodium amaranticolor*. *Chenopodium quinoa*. *Cucumis sativus*. *Cucurbita pepo*.** *Datura stramonium*. ***Gomphrena globosa*. *Momordica balsamina*. *Nicotiana clevelandii*. *Nicotiana glutinosa*. *Nicotiana megalosiphon*.** *Nicotiana rustica*. ***Nicotiana tabacum*.** *Petunia hybrida*. ***Phaseolus vulgaris*. *Vigna unguiculata*.**

Sources of host-range data

Gallitelli and Savino (1985).

**Properties of particles in sap**

Leaf sap contains many virions.

**Purification**

Method: Gallitelli and Savino (1985).

**Particle morphology**

Virions isometric; 30 nm in diameter; angular in profile; without a conspicuous capsomere arrangement.

**Physical properties**

One sedimenting component in purified preparations; sedimentation coefficient of the fastest 111 *S*. Density 1.314 g $cm^{-3}$ in CsCl. $A_{260}/A_{280}$ ratio 1.64.

**Biochemical properties**

Virions contain *c.* 17 % nucleic acid; 83 % protein; 0 % lipid.

*Genome* consists of RNA; single–stranded; linear; unipartite; largest (or only) genome part 4.1 kb. Genomic nucleic acid isolated by Gallitelli: phenol–SDS. Infectivity retained when deproteinised with proteases; retained when deproteinised with phenol or detergent.

*Non–genomic nucleic acid* not found in the virions.

*Virion protein(s)* one; $M_r$ 32000; coat protein. Method of preparation: Laemmli (1970).

**Cytopathology**

Virions found in leaves and mesophyll; in cytoplasm. Inclusions present in infected cells; are crystals in the cytoplasm and unusual in shape; vesiculated bodies and tubules; they contain virions. Other cellular changes: those described by Castellano *et al.* (1987).

**Taxonomy and Relationships**

Possible *Sobemovirus*.

**References**

Castellano, M.A., Di Franco, A. and Martelli, G.P. (1987). *J. Submicrosc. Cytol.* **19**: 495.

Gallitelli, D. and Savino, V. (1985). *Ann. appl. Biol.* **106**: 295.

Laemmli, U.K. (1970). *Nature, Lond.* **227**: 680.

# Olive latent 2 (?) ourmiavirus

**Data collated by M. Castellano, 1991.**

*Standard acronym:* OLV-2.

First reported in *Olea europaea*; from Apulia, southern Italy; by Savino *et al.* (1984).

**Natural host range and symptoms**

*Olea* spp. — symptomless.

**Transmission**

Virus transmitted by mechanical inoculation.

**Geographical distribution**

Found, but with no evidence of spread, in Apulia, southern Italy.

**Experimental host range**

Several (3–9) families susceptible.

Diagnostically susceptible

*Gomphrena globosa* — reddish necrotic local lesions then symptomless systemic infection.

*Chenopodium quinoa* — chlorotic and necrotic local lesions.

*Nicotiana megalosiphon, N. benthamiana* — mild mottling.

*Petunia hybrida, Nicotiana clevelandii, N. debneyi, N. tabacum, Phaseolus aureus* — symptomless.

Diagnostically insusceptible

*Nicotiana glutinosa, N. rustica, Ocimum basilicum, Datura stramonium, Cucumis sativus, Cucurbita pepo.*

Maintenance and propagation hosts

*Gomphrena globosa, Nicotiana benthamiana.*

Dicotyledonous Hosts *(**Susceptible** or Insusceptible)*

***Chenopodium quinoa***. *Cucumis sativus. Cucurbita maxima. Datura stramonium.* ***Gomphrena globosa. Nicotiana benthamiana. Nicotiana clevelandii. Nicotiana debneyi.*** *Nicotiana glutinosa.* ***Nicotiana megalosiphon***. *Nicotiana rustica.* ***Nicotiana tabacum***. *Ocimum basilicum.* ***Petunia hybrida***.

**Properties of particles in sap**

TIP: 60 °C. LIV: more than 7 days. DEP: $\log_{10}$ minus 6. Leaf sap contains many virions.

**Purification**

Method: homogenize infected tissues in phosphate buffer (0.07M, pH 7.0) with 0.01M EDTA and 0.1% Na–mercaptoacetate, filter and clarify with 1 vol of 1:1 chloroform–butanol then precipitate with 10% polyethylene glycol and 1% NaCl. Purify by differential and sucrose gradient centrifugation.

**Particle morphology**

Virions quasi bacilliform; 37, 43, 48, and 55 nm in length; 26 nm wide.

**Physical properties**

Two sedimenting components in purified preparations; sedimentation coefficient of the fastest 82 *S*; of the other(s) 77 *S*. Density 1.36 g $cm^{-3}$ in CsCl. $A_{260}/A_{280}$ ratio 1.58.

**Biochemical properties**

Virions contain *c.* 19 % nucleic acid; 81 % protein.

*Genome* consists of RNA; single–stranded; linear; of four parts; largest (or only) genome part 3.3 kb; the 2nd largest 2.8 kb; the 3rd largest 2.45 kb; the 4th largest 2.1 kb.

*Sequence database accession code(s):*

X76993 Em(40)_vi:OLV212 Gb(84)_vi:OLV212 Olive latent virus 2 genomic RNA-3. 2/94 2,438bp.

X77115 Em(40)_vi:OLV2RNA4 Gb(84)_vi:OLV2RNA4 Olive latent virus 2 genomic RNA-4. 2/94 2,078bp. 2 sequences.

*Virion protein(s)* one; $M_r$ 24000; coat protein.

**Cytopathology**

Virions found in leaves and mesophyll; in cytoplasm. Inclusions absent from infected cells. Other cellular changes: major cell organelles, except for nuclei, are affected, there are more cytoplasmic membranes and tonoplast-associated membranous vesicles.

**Taxonomy and Relationships**

Possible ourmiavirus.

Olive latent 2 seems not to be an ilarvirus or related to alfalfa mosaic virus. It shares most properties with pelargonium zonate spot, epirus cherry, cassava Ivorian bacilliform and melon Ourmia viruses, and hence is probably an ourmiavirus.

**References**

Castellano, M.A., Di Franco, A. and Martelli, G.P. (1987). *J. Submicrosc. Cytol.* **19**: 495.

Grieco, F., Martelli, G.P., Savino, V. and Piazzolla, P. (1992). *Riv. Pat. Veg.*,S.V. **2**: 125.

Savino, V., Piazzolla, P., Di Franco, A. and Martelli, G.P. (1984). *5th Congr. Un. Phytopath. Medit. Cairo, Egypt.*

# Onion mite-borne latent (?) *potexvirus*

**Data collated by P. van Dijk, 1991.**

*Strains:* garlic strain.

First reported in *Allium cepa* var. *ascalonicum*; from the Netherlands; by Van Dijk *et al.* (1991).

**Natural host range and symptoms**

Symptoms persist, or none. Symptoms slight mottling or diffuse yellow stripes.

*Allium cepa, A. cepa* var. *ascalonicum* — usually symptomless, sometimes slight mottle. More virulent isolates induce diffuse yellow striping.

**Transmission**

Transmitted by a vector; a mite; *Aceria tulipae*; Eriophyidae. Not transmitted by *Acyrthosiphon pisum, Myzus (Sciamyzus) ascalonicus, M. persicae*. Virus transmitted by mechanical inoculation; not transmitted by seed.

**Ecology and control**

Studies reported by Van Dijk (1991): vegetative propagation of the most important hosts, latent infection and omnipresence and abundance of the vector imply that the virus is readily spread and that it is difficult to control. For information on abundance of *A. tulipae* on onion in the former U.S.S.R., see Cheremushkina (1982).

**Geographical distribution**

Spreads in France, the Netherlands, Spain, and the former USSR.

**Experimental host range**

Few (<3) families susceptible. Experimentally infected plants mostly show symptomless or show chlorotic local lesions.

Diagnostically susceptible

*Allium cepa, A. cepa* var. *ascalonicum, A. ampeloprasum* var. *porrum, A. chinense, A. vineale* — symptomless systemic infection.

*Atriplex hortensis, Chenopodium murale* — chlorotic local lesions.

*Nicotiana occidentalis* — symptomless systemic infection.

Diagnostically insusceptible

*Allium fistulosum, A. schoenoprasum, A. tuberosum, Amaranthus bicolor, Chenopodium quinoa.*

Maintenance and propagation hosts

*Allium cepa, A. cepa* var. *ascalonicum, A. vineale, Chenopodium murale.*

Assay hosts (**L**ocal lesions or **W**hole plants)

*Allium vineale* (W), *Atriplex hortensis* (L), *Chenopodium murale* (L).

Dicotyledonous Hosts (***Susceptible*** *or Insusceptible)*

*Amaranthus caudatus*. ***Atriplex hortensis***. *Celosia argentea*. ***Chenopodium amaranticolor***. *Chenopodium foetidum*. ***Chenopodium murale***. *Chenopodium quinoa*. ***Gomphrena globosa***. *Nicotiana benthamiana*. ***Nicotiana occidentalis***. *Spinacia oleracea*. *Tetragonia expansa*. *Vicia faba*.

Monocotyledonous Hosts (***Susceptible*** *or Insusceptible)*

***Allium ascalonicum***. ***Allium cepa***. *Allium fistulosum*. ***Allium porrum***. *Lilium formosanum*.

Additional hosts

***Allium ampeloprasum*** var. ***porrum, A. chinense, A. vineale***, *A. schoenoprasum, A. tuberosum, Amaranthus bicolor, Chlorophytum comosum* and *Nicotiana hesperis* (**I**).

Sources of host-range data

Van Dijk *et al.* (1991).

**Properties of particles in sap**

Leaf sap contains many virions.

**Particle morphology**

Virions filamentous; not enveloped; usually flexuous; with a clear modal length; of 775 nm. Axial canal obscure. Basic helix obvious.

*Sequence database accession code(s):*

D11159 Em(40)_vi:GVCCP1 Gb(84)_vi:GVCCP Garlic virus C gene for coat protein, complete cds. 10/92 1,372bp.

D49443 Gb(89)_vi:Gmvcp Garlic mite-borne mosaic virus gene for coat protein. 3/95 2,518bp.

L38892 Em(43)_vi:Jgvfra Gb(89)_vi:Jgvfra Mite-Born Filamentous Virus (UK garlic isolate) genomic RNA, 3' clip. 2/95 2,756bp.

D11157 Em(40)_vi:GVACP1 Gb(84)_vi:GVACP Garlic virus A gene for coat protein, complete cds. 10/92 2,962bp.

D11158 Em(40)_vi:GVBCP1 Gb(84)_vi:GVBCP Garlic virus B gene for coat protein, complete cds. 10/92 3,047bp.

D11160 Em(40)_vi:GVDCP1 Gb(84)_vi:GVDCP Garlic virus D gene for coat protein, complete cds. 10/92 2,071bp.

D28590 Em(43)_vi:Gv2cpa Gb(89)_vi:Gv2cpa Garlic virus 2 RNA for polyprotein precursor, nuclear inclusion protein b and coat protein. 1

D28591 Em(43)_vi:Gv1cpb Gb(89)_vi:Gv1cpb Garlic virus 1 RNA for ORF, coat protein and 3'ORF for a 14.2kDa putative protein. 12/94.

**Replication**

Replication does not depend on a helper virus.

**Cytopathology**

Virions found in all parts of the host plant. Inclusions present in infected cells; are unusual in shape; are granular inclusion bodies resembling those of potyviruses.

**Taxonomy and Relationships**

Possible *Potexvirus*.

Virus(es) with serologically related virions: shallot mite-borne latent virus, but distantly.

The garlic strain differs from the type strain by not infecting onion, shallot and rakkyo, and in infecting garlic and sand leek. The garlic strain is also found in other countries including Chile, China, Iran, Japan, Morocco, Phillipines and Syria. Its virion length is 716 nm. Best tests for diagnosis: induction of local lesions in *Chenopodium murale* and not in *C. quinoa*; virions in the electron microscope are clearly more flexuous than those of poty and carlaviruses; no decoration of virions with antisera to poty or carlaviruses (if raised to uncontaminated virus cultures).

**Comments**

The onion mosaic virus reported from the former Russia (Cheremushkina, 1982; Razvjazkina, 1971) was probably a mixture of onion mite-borne latent virus, shallot mite-borne latent virus and onion yellow dwarf virus (Van Dijk *et al.*, 1991). Similarly the garlic yellow streak virus (Mohamed and Young, 1981) of Australasia is also probably a complex of these viruses. Shallot X virus is probably also one or more components of this complex (Van Dijk and van der Vlugt, 1994). An antiserum to a German garlic latent virus contained antibodies to the garlic strain of onion mite-borne latent virus (Van Dijk *et al.*, 1991).

**References**

Cheremushkina, N.P. (1982). *Trudy po Selektsii i Semenovodstvu Ovoshchnyak Kultur* **15**: 74.

Mohamed, N.A. and Young, B.R. (1981). *Ann. appl. Biol.* **97**: 65.

Razvjazkina, G.M. (1971). *Tag. Akad. Landw. D.D.R* **115**: 69.

Van Dijk, P. and van der Vlugt, R.A.A. (1994). *Eur. J. Pl. Path.* **100**: 269.

Van Dijk, P., Verbeek, M. and Bos, L. (1991). *Neth. J. Pl. Path.* **97**: 381.

# Onion yellow dwarf *potyvirus*

**Data collated by L. Bos, 1985.**

*Synonyms:* allium virus 1. *Standard acronym:* OYDV.

First reported in *Allium cepa*; from Iowa, U.S.A. and Wageningen, the Netherlands; by Melhus *et al.* (1929); Bos *et al.* (1978).

**Natural host range and symptoms**

Symptoms persist.
*Allium cepa* — leaf striping and curling, stunting ("yellow dwarf").
*Allium cepa* var. *ascalonicum* — leaf striping and curling, stunting.
*Allium* (ornamental spp.) — symptoms variable.

**Transmission**

Transmitted by a vector; an insect; *Myzus persicae* and several other species; Aphididae. Transmitted in a non–persistent manner. Virus transmitted by mechanical inoculation; not transmitted by contact between plants; not transmitted by seed; not transmitted by pollen.

**Geographical distribution**

Probably distributed worldwide.

**Experimental host range**

Few (<3) families susceptible.

Diagnostically susceptible

*Allium cepa* — systemic leaf striping and curling.

Diagnostically insusceptible

*Allium porrum* (syn. *A. ampeloprasum*), *Chenopodium amaranticolor, Chenopodium quinoa, Nicotiana clevelandii.*

Maintenance and propagation hosts

*Allium cepa.*

Assay hosts (**L**ocal lesions or **W**hole plants)

*Allium cepa* (W).

Dicotyledonous Hosts *(**Susceptible** or Insusceptible)*

*Chenopodium amaranticolor. Chenopodium quinoa. Nicotiana clevelandii. Nicotiana megalosiphon.*

Monocotyledonous Hosts *(**Susceptible** or Insusceptible)*

***Allium ascalonicum. Allium cepa.***

Comments on host-range

Earlier reports that onion yellow dwarf virus infects Chinese sacred lily (*Narcissus tazetta*) and jonquil (*Narcissus jonquilla*) (Anon., 1932; Henderson, 1935) have not been confirmed (Bos, 1982).

Sources of host-range data

Bos *et al.* (1978).

**Properties of particles in sap**

TIP: 60–65 °C. LIV: 2–3 days. DEP: $\log_{10}$ minus 3–4. Infectivity of sap not changed by treatment with di–ethyl ether. Leaf sap contains few virions.

**Purification**

Method: Huttinga (1975).

**Particle morphology**

Virions filamentous; not enveloped; usually flexuous; with no clear modal length; of 772 and 823 nm (Schmidt and Schmelzer, 1978)). Axial canal obscure. Basic helix obscure.

**Physical properties**

One sedimenting component in purified preparations; sedimentation coefficient 143 *S*. Density 1.306 g $cm^{-3}$ in CsCl.

*Sequence database accession code(s):*

L28079 Em(40)_vi:PTWPARD Gb(84)_vi:PTWPARD Shallot potyvirus (probably onion yellow dwarf virus) protease gene, partial cds, RNA-dep polmerase 1 sequence.

*Virion protein(s)* one; $M_r$ 30000. Method of preparation: Huttinga and Mosch (1974).

**Replication**

Replication does not depend on a helper virus.

**Cytopathology**

Virions found in all parts of the host plant; in cytoplasm. Inclusions present in infected cells; are amorphous X–bodies (and granular).

**Taxonomy and Relationships**

*Potyvirus: Potyviridae.*

Virus(es) with serologically related virions: leek yellow stripe virus, but distantly.

Onion yellow dwarf virus infects poorly, or with difficulty, *Allium porrum* (leek); leek yellow stripe virus does not infect or infects onion (*Allium cepa*) and shallot (*Allium cepa* var. *ascalonicum*) with difficulty.

**References**

Anon. (1932). *Rep. Iowa Agric. Exp. Stn* **1932**: 32.

Bos, L. (1976). *CMI/AAB Descr. Pl. Viruses* No. 158, 4 pp.

Bos, L., Huijberts, N., Huttinga, H.L. and Maat, D.Z. (1978). *Neth. J. Pl. Path.* **84**: 105.

Bos, L. (1982). *Acta Hort.* **12**: 11.

Henderson, R.G. (1935). *Res. Bull. Iowa Agric. Exp. Stn* **188**: 209.

Huttinga, H. (1975). *Neth. J. Pl. Path.* **81**: 81.

Huttinga, H. and Mosch, W.H.M. (1974). *Neth. J. Pl. Path.* **80**: 19.

Melhus, I.E., Reddy, C.S., Henderson, W.J. and Vestal, E.F. (1929). *Phytopathology* **19**: 73.

Schmidt, H.E. and Schmelzer, K. (1964). *Phytopath. Z.* **50**: 19.

# Ononis yellow mosaic *tymovirus*

**Data collated by A.J. Gibbs, 1982. Revised 1985.**

*Standard acronym:* OYMV.

First reported in *Ononis repens*; from the U.K; by Gibbs *et al.* (1966).

**Natural host range and symptoms**

Symptoms persist.
*Ononis repens* — systemic yellow mosaic, mottling.

**Transmission**

Transmitted by means not involving a vector. Virus transmitted by mechanical inoculation; not transmitted by grafting; not transmitted by seed; not transmitted by pollen.

**Geographical distribution**

Spreads in the UK.

**Experimental host range**

Several (3–9) families susceptible.

Diagnostically susceptible

*Pisum sativum* — system mosaic, stunting.
*Trifolium incarnatum* — systemic mosaic.
*Datura stramonium* — chlorotic local lesions, not systemic.
*Nicotiana clevelandii* — chlorotic local lesions, then systemic mottle.

Diagnostically insusceptible

*Chenopodium quinoa, Phaseolus vulgaris, Vicia faba, Nicotiana glutinosa.*

Maintenance and propagation hosts

*Pisum sativum.*

Assay hosts (**L**ocal lesions or **W**hole plants)

*Pisum sativum* (W), *Trifolium incarnatum* (W). No local lesion host known.

Dicotyledonous Hosts *(**Susceptible** or Insusceptible)*

*Catharanthus roseus. Chenopodium amaranticolor. Chenopodium foetidum. Chenopodium quinoa. Cucumis sativus. Cucurbita pepo.* ***Datura stramonium****. Daucus carota.* ***Glycine max****. Gomphrena globosa.* ***Lycopersicon esculentum*** (cv. Kondine Red). *Medicago sativa.* ***Nicotiana clevelandii****. Nicotiana debneyi. Nicotiana glutinosa. Nicotiana rustica. Nicotiana tabacum. Pastinaca sativa. Petunia hybrida. Phaseolus vulgaris.* ***Pisum sativum****.* ***Trifolium incarnatum****.* ***Trifolium repens****. Vicia faba. Vigna unguiculata.*

Monocotyledonous Hosts *(**Susceptible** or Insusceptible)*

*Hordeum vulgare. Triticum aestivum.*

Sources of host-range data

Gibbs *et al.* (1966).

**Properties of particles in sap**

TIP: 65–70 °C. LIV: 14 days. DEP: $\log_{10}$ minus 3–5. Infectivity of sap not changed by treatment with di–ethyl ether. Leaf sap contains many virions.

**Purification**

Method: Gibbs *et al.* (1966).

**Particle morphology**

Virions isometric; not enveloped; 25–30 nm in diameter; rounded in profile; with a conspicuous capsomere arrangement.

**Physical properties**

Two sedimenting components in purified preparations; sedimentation coefficient of the fastest 114 *S*; of the other(s) 56 *S*. Isoelectric point pH 3.84.

**Biochemical properties**

Virions contain 34 % nucleic acid; 66 % protein; 0 % lipid.

*Genome* consists of RNA; single–stranded. Base composition 15.6 % G; 21 % A; 34.2 % C; 29.4 % U. Nucleotide sequence references: Ding *et al.* (1990).

*Sequence database accession code(s):*

J04375 Em(40)_vi:MTYCG Gb(84)_vi:OYMCG Ononis yellow mosaic virus complete genome. 6/90 6,211bp.

M15287 Em(40)_vi:TYMTY3OY Gb(84)_vi:OYM3OYMV Ononis yellow mosaic virus 3' end which can form tRNA-like structure. 11/90 131bp.

M58312 Em(40)_vi:OYMRRLS Gb(84)_vi:OYMRRLS Ononis yellow mosaic virus 3' terminus tRNA-like structure. 8/91 84bp. 3 sequences.

*Virion protein(s)* one.

**Replication**

Replication does not depend on a helper virus.

**Cytopathology**

Virions found in all parts of the host plant; in cytoplasm.

**Taxonomy and Relationships**

*Tymovirus.*

Virus(es) with serologically related virions: scrophularia mottle is closely related; belladonna mottle and dulcamara mottle more distantly related; eggplant mosaic, Andean potato latent, wild cucumber mosaic and clitoria yellow vein viruses are very distantly related.

**References**

Ding, S., Keese, P. and Gibbs, A.J. (1990). *J. gen. Virol.* **71**: 925.

Gibbs, A.J., Hecht-Poinar, E., Woods, R.D. and McKee, R.K. (1966). *Ann. appl. Biol.* **58**: 231.

# Opuntia Sammons' *tobamovirus*

**Data collated by C. Wetter, 1989.**

*Synonyms:* Sammons' opuntia virus. *Standard acronym:* SOV.

First reported in *Opuntia engelmanii*; from Arizona, U.S.A.; by Sammons and Chessin (1961).

**Natural host range and symptoms**

*Opuntia engelmanii* and some other species of *Opuntia* — symptomless.

**Transmission**

Virus transmitted by mechanical inoculation.

**Geographical distribution**

Spreads in the USA.

**Experimental host range**

Few (<3) families susceptible.

Diagnostically susceptible

*Chenopodium quinoa, C, murale, C. amaranticolor* — local lesions; not systemic.

Maintenance and propagation hosts

*Chenopodium quinoa, C. murale, C. amaranticolor.*

Assay hosts (**L**ocal lesions or **W**hole plants)

*Chenopodium quinoa* (L), *C. murale* (L), *C. amaranticolor* (L).

Dicotyledonous Hosts *(**Susceptible** or Insusceptible)*

***Chenopodium amaranticolor*. *Chenopodium murale*. *Chenopodium quinoa*.**

Sources of host-range data

Wetter and Paul (1967).

**Properties of particles in sap**

Leaf sap contains many virions.

**Purification**

Method: Wetter: clarify with chloroform—butanol (1:1 v/v), precipitate with 4% polyethylene glycol and 4% NaCl, resuspend and subject to several cycles of differential centrifugation.

**Particle morphology**

Virions rod–shaped; not enveloped; usually straight; with a clear modal length; of 317 nm; 18 nm wide. Axial canal obvious. Basic helix obvious.

**Physical properties**

One sedimenting component in purified preparations; sedimentation coefficient 196 *S*. $A_{260}/A_{280}$ ratio 1.33.

**Taxonomy and Relationships**

*Tobamovirus.*

Virus(es) with serologically related virions: odontoglossum ringspot, tomato mosaic, tobacco mosaic, ribgrass mosaic, tobacco mild green mosaic, pepper mild mottle, sunn-hemp mosaic, frangipani mosaic and cucumber green mottle mosaic viruses.

**References**

Brandes, J. and Chessin, M. (1965). *Virology* **25**: 673.

Sammons, I.M. and Chessin, M. (1961). *Nature, Lond.* **191**: 517.

Wetter, C. and Paul, H.L. (1967). *Phytopath. Z.* **60**: 92.

# Orchid fleck (?) rhabdovirus

**Data collated by D.-E. Lesemann, 1986.**

*Synonyms:* dendrobium leaf streak virus, dendrobium virus, probably laelia red leafspot, short orchid rhabdovirus, orchid rhabdovirus, phalaenopsis chlorotic spot virus, phalaenopsis hybrid virus, phalaenopsis virus. *Standard acronym:* OFV.

First reported in *Cymbidium* spp.; from Japan; by Doi *et al.* (1969); Chang *et al.* (1973).

**Natural host range and symptoms**

Symptoms persist.

*Cymbidium alexanderi* — chlorotic or necrotic flecks.

*Phalaenopsis* spp. and hybrids — chlorotic flecks.

*Renanthera* spp. — chlorotic flecks, then sunken gray necrotic streaks with sharp margins and of variable size.

*Coelogyne* spp., *Dendrobium* spp. (and hybrids), *Miltonia* spp., *Odontoglossum* spp., *Oncidium* spp., and *Paphiopedilum* spp. — chlorotic areas often with necrotic centres or rings.

*Vanda* spp. — irregular chlorotic areas.

**Transmission**

Virus transmitted by mechanical inoculation; not transmitted by seed (some isolates infect better above 30°C).

**Geographical distribution**

Spreads in Australia, Brazil, Germany, Japan, and the USA (Lesemann and Doraiswamy, 1975; Kitajima *et al.*, 1974). Found, but with no evidence of spread, in New Zealand.

**Experimental host range**

Several (3–9) families susceptible.

Diagnostically susceptible

*Dendrobium nobile, Cymbidium alexanderi*—systemic chlorotic or necrotic spots or flecks.

*Nicotiana glutinosa, N. tabacum* cvs Bright Yellows, White Burley, Xanthi-*nc* and KY-57 — chlorotic or necrotic local lesions, no systemic symptoms.

*Chenopodium quinoa* — chlorotic local lesions.

*Vicia faba* — brown necrotic local lesions.

Maintenance and propagation hosts

*Cymbidium alexanderi.*

Assay hosts (**L**ocal lesions or **W**hole plants)

*Nicotiana glutinosa* (L), *Nicotiana tabacum* (L), *Chenopodium quinoa* (L).

Dicotyledonous Hosts *(**Susceptible** or Insusceptible)*

***Chenopodium amaranticolor. Chenopodium quinoa. Nicotiana clevelandii. Nicotiana glutinosa. Nicotiana tabacum. Vicia faba.***

Monocotyledonous Hosts *(**Susceptible** or Insusceptible)*

***Cymbidium alexanderi. Dendrobium nobile.***

**Properties of particles in sap**

LIV: 1 days (at 6°C). DEP: $\log_{10}$ minus 2. Leaf sap contains few virions, or contains many virions. Electron microscopy: for negative staining, mount sap in 2% AM pH 6.5. Virions appear smaller (32–35 × 100–140 nm) in leaf sections.

**Purification**

Method: Chang *et al.* (1976).

**Particle morphology**

Virions rhabdo– or bullet–shaped; not enveloped (or rarely enveloped); with a clear modal length; of 105–150 nm; 40–50 nm wide. Axial canal obvious; *c.* 15 nm in diameter. Basic helix obvious; pitch of basic helix 4.5 nm (Chang *et al.*, 1976).

**Replication**

Replication does not depend on a helper virus.

**Cytopathology**

Virions found in mesophyll; in cytoplasm and in nuclei (where virions are commonly found to attach by one end to the inner nuclear membrane (Chang *et al.*, 1976) or endoplasmic reticulum). Inclusions present in infected cells; are viroplasms and crystals in the nucleus (Chang *et al.*, 1976; Lesemann and Doraiswamy, 1975); they contain virions. Other cellular changes: virions attached to the periphery of the nucleus.

**Taxonomy and Relationships**

Unassigned rhabdovirus.

The virions of orchid fleck virus resemble those of plant rhabdoviruses, but are smaller, lack an outer envelope, and are still infective after treatment with detergents (*e.g.* Triton X-100), thus they may be equivalent to the internal component of rhabdovirus virions. In virion morphology and intracellular effects, orchid fleck virus closely resembles coffee ringspot (Kitajima and Costa, 1972) and citrus leprosis viruses (Kitajima *et al.*, 1972). Best tests for diagnosis: Electron microscopy will distinguish orchid fleck virus from cymbidium mosaic, odontoglossum ringspot and other orchid viruses.

**References**


Chang, M.U., Arai, K., Doi, Y. and Yora, K. (1973). *Ann. Phytopath. Soc. Japan* **39**: 171.

Chang, M.U., Arai, K., Doi, Y. and Yora, K. (1976). *Ann. Phytopath. Soc. Japan* **42**: 156.

Doi, Y., Chang, M.U. and Yora, K. (1977). *CMI/AAB Descr. Pl. Viruses* No. 183, 3 pp.

Doi, Y., Toriyama, S., Yora, K. and Asuyama, H. (1969). *Ann. Phytopath. Soc. Japan* **35**: 388.

Kitajima, E.W. and Costa, A.S. (1972). *Cienc. Cult.* **24**: 542.

Kitajima, E.W., Müller, G.W., Costa, A.S. and Yuki, W. (1972). *Virology* **50**: 254.

Kitajima, E.W., Blumenschein, A. and Costa, A.S. (1974). *Phytopath. Z.* **81**: 280.

Lesemann, D.-E. and Begtrup, J. (1971). *Phytopath. Z.* **71**: 257.
Lesemann, D.-E. and Doraiswamy, S. (1975). *Phytopath. Z.* **83**: 27.
Petzold, H. (1971). *Phytopath. Z.* **70**: 45.

## Ornithogalum mosaic *potyvirus*

**Data collated by J.T. Burger, 1991.**

*Standard acronym:* OrMV.

First reported in *Ornithogalum thyrsoides*; from Oklahoma, U.S.A.; by Smith and Brierley (1944*a*).

**Natural host range and symptoms**

Symptoms persist. Symptoms mottling, mosaic and sunken areas.

*Ornithogalum* spp. and hybrids — mottling of leaves and stems, flowers deformed.

*Lachenalia* spp. and hybrids — mosaic or mottling of leaves and flower stems; flowers deformed.

**Transmission**

Transmitted by a vector; an insect; *Myzus persicae*; Aphididae. Transmitted in a non–persistent manner. Virus transmitted by mechanical inoculation.

**Geographical distribution**

Spreads in the Netherlands, South Africa, and the USA.

**Experimental host range**

Few (<3) families susceptible. Experimentally infected plants mostly show mottling, mosaic and sunken areas in leaves.

Diagnostically susceptible

*Chenopodium quinoa, Tetragonia tetragonioides, Nicotiana clevelandii* (Derks *et al.*, 1983).

Diagnostically insusceptible

*Nicotiana* spp., *Pisum sativum, Petunia hybrida, Phaseolus vulgaris, Chenopodium* spp.

Maintenance and propagation hosts

*Ornithogalum* spp. and hybrids, *Lachenalia* spp. and hybrids.

Assay hosts (**L**ocal lesions or **W**hole plants)

*Ornithogalum* spp. (W), *Lachenalia* spp. (W).

Dicotyledonous Hosts *(**Susceptible** or Insusceptible)*

*Chenopodium album. Chenopodium amaranticolor. Chenopodium capitatum. Chenopodium foetidum. Chenopodium murale. Chenopodium quinoa. Gomphrena globosa. Nicotiana benthamiana. Nicotiana clevelandii. Nicotiana tabacum. Petunia hybrida. Phaseolus vulgaris. Pisum sativum. Tetragonia expansa.*

Monocotyledonous Hosts *(**Susceptible** or Insusceptible)*

*Iris germanica. Lilium longiflorum. Narcissus jonquilla. Nerine bowdenii.*

Sources of host-range data

Derks and Vink-van den Abeele (1980); Dercks *et al.* (1983); Burger and Von Wechmar (1989).

**Properties of particles in sap**

Leaf sap contains few virions. Electron microscopy: elimination of mucilage in plant sap before electron microscopy.

**Purification**

Method: Burger and Von Wechmar (1989).

**Particle morphology**

Virions filamentous; not enveloped; usually flexuous; with a clear modal length; of 740 nm; 12 nm wide. Axial canal obscure. Basic helix obscure.

**Physical properties**

One sedimenting component in purified preparations.

**Biochemical properties**

Virions contain 5 % nucleic acid; 95 % protein.

*Genome* consists of RNA; single–stranded; linear; unipartite; largest (or only) genome part ± 10 kb. Genomic nucleic acid isolated by Burger and Von Wechmar (1989). Poly A region present; the genome has a polyadenylated 3'–terminal tail. Nucleotide sequence references: Accession number: D00615.

*Sequence database accession code(s):*

D00615 Em(40)_vi:OMNIAB Gb(84)_vi:OMVNIAB Ornithogalum mosaic virus genomic RNA for nuclear inclusion protein (NIa, NIb) and coat protein 1 sequence.

*Features of the genome:* single ± 10 kb open reading frame. Sequence of CP, NIb and NIa has been determined.

*Virion protein(s)* one; $M_r$ 30000; it is coat protein and its function is structural. Method of preparation: Burger and Von Wechmar (1989). Amino acid composition: Burger *et al.* (1990).

**Cytopathology**

Virions found in leaves and probably all parts of the host plant.

**Taxonomy and Relationships**

*Potyvirus: Potyviridae.*

Virus(es) with serologically related virions: bean yellow mosaic, potato Y, hyacinth mosaic, zucchini yellow mosaic and passionfruit woodiness viruses.

There are strong sequence homologies with most sequenced potyviruses (Burger *et al.*, 1990).

**References**


Burger, J.T. and Von Wechmar, M.B. (1988). *Acta Hort.* **234**: 31.

Burger, J.T. and Von Wechmar, M.B. (1989). *Phytopathology* **79**: 385.

Burger, J.T., Brand, R.J. and Rybicki, E.P. (1990). *J. gen. Virol.* **71**: 2527.

Derks, A.F.L.M. and Vink-van den Abeele, J.L. (1980). *Acta Hort.* **109**: 495.

Derks, A.F.L.M., Vink-van den Abeele, J.L. and Hollinger Th. C. (1983). *Jversl. Lab. Bloemboll.* Lisse, **1982**: 61.

Smith, F.F. and Brierley, P. (1944*a*). *Phytopathology* **34**: 497.

Smith, F.F. and Brierley, P. (1944*b*). *Phytopathology* **34**: 593.

# Palm mosaic (?) potyvirus

**Data collated by A.A. Brunt and S. Molyneux, 1987.**

*Standard acronym:* PalMV.

First reported in *Washingtonia robusta*; from California, U.S.A.; by Mayhew and Tidwell (1978).

**Natural host range and symptoms**

Symptoms persist.
*Washingtonia robusta*—mosaic, ringspot and line patterns, stunting.

**Transmission**

Virus not transmitted by mechanical inoculation.

**Geographical distribution**

Spreads in the USA.

**Particle morphology**

Virions filamentous; usually flexuous; with a clear modal length; of 686 nm; 13 nm wide. Axial canal obscure. Basic helix obscure.

**Cytopathology**

Virions found in all parts of the host plant; in cytoplasm. Inclusions present in infected cells; are pinwheels; they do not contain virions.

**Taxonomy and Relationships**

Unassigned potyvirus.

**References**

Mayhew, D.E. and Tidwell (1978). *Pl. Dis. Reptr* **62**: 803.

# Pangola stunt *fijivirus*

**Data collated by R.G. Milne, 1984.**

*Standard acronym:* PaSV.

First reported in *Digitaria decumbens*; from Surinam; by Dirven and van Hoof (1960).

**Natural host range and symptoms**

Symptoms persist.
*Digitaria decumbens* — chlorosis and reddening, leaves and flower heads malformed, curled and stunted, small veins swollen.

**Transmission**

Transmitted by a vector; an insect; *Sogatella furcifera*; Delphacidae. Transmitted in a persistent manner. Virus retained when the vector moults; multiplies in the vector; not transmitted congenitally to the progeny of the vector; not transmitted by mechanical inoculation; not transmitted by grafting; not transmitted by contact between plants; not transmitted by seed; not transmitted by pollen.

**Geographical distribution**

Spreads in Australia, Brazil, Fiji, Guyana, Peru, and Taiwan.

**Experimental host range**

Few (<3) families susceptible.

Diagnostically susceptible

Almost all *Digitaria* spp. are susceptible. *D. decumbens* PI 111110, *D. longiflora* PI 299642, *D. pentzii* PI 299769, *D. setivalvola* PI 299795, *D. valida* PI 299875 — stunting, yellowing and twisting of leaves and inflorescences, small veins swollen, excessive tillering.

Maintenance and propagation hosts

*Digitaria decumbens* and other *Digitaria* spp.

Assay hosts (**L**ocal lesions or **W**hole plants)

*Digitaria decumbens* (W).

Monocotyledonous Hosts *(**Susceptible** or Insusceptible)*

***Digitaria decumbens*.**

Sources of host-range data

Hunkar *et al.* (1974); Milne and Lovisolo (1977).

**Properties of particles in sap**

Leaf sap contains few virions (when prepared from whole leaves), or contains many virions (when prepared only from enations). Electron microscopy: to obtain complete or subvirions with intact spikes, use 2% aqueous UA not neutral PTA.

**Purification**

Method: Boccardo *et al.* (1979).

**Particle morphology**

Virions isometric; not enveloped; complete virions including spikes 70 nm in diameter (when negative stained and in thin section); rounded in profile, or angular in profile; with a conspicuous capsomere arrangement.

**Physical properties**

Two sedimenting components in purified preparations (virions and spiked subvirions).

**Biochemical properties**

*Genome* consists of RNA; double–stranded; linear; of 10 parts (called S1–10); largest (or only) genome part the largest 3.405 kb; the 2nd largest 3.385 kb; the 3rd largest 3.385 kb; the 4th largest 2.6 kb; the 5th largest 2.55 kb; the 6th largest and other parts 2.49 kb (pairs; also 2.190, 1.735, 1.728 and 1.720 kb pairs). Genomic nucleic acid isolated by Boccardo *et al.* (1979). Additional factor required for infectivity (only complete virions are infective).

*Non–genomic nucleic acid* not found in the virions.

*Virion protein(s)* virion proteins probably similar in size to those of maize rough dwarf virus.

**Replication**

Genome replicates in cytoplasmic viroplasms. Coat protein *m*RNA translated in the cytoplasm. Replication does not depend on a helper virus.

**Cytopathology**

Virions found in phloem and phloem derived enations; in cytoplasm. Inclusions present in infected cells; are crystals in the cytoplasm, viroplasms, and unusual in shape; tubular structures 100 nm in diameter; they contain virions. Other cellular changes: thickening of cell walls and cell enlargement.

**Taxonomy and Relationships**

*Fijivirus: Reoviridae.*

Virus(es) with serologically related virions: maize rough dwarf and rice black streaked dwarf viruses and are closely related. Virus(es) with serologically unrelated virions: oat sterile dwarf, sugarcane Fiji disease, rice dwarf, rice gall dwarf, rice ragged stunt and clover wound tumor viruses.

**References**

Boccardo, G. and Milne, R.G. (1981). *J. virol. Meth.* **3**: 109.
Boccardo, G. and Milne, R.G. (1984). *CMI/AAB Descr. Pl. Viruses* No. 294, 7 pp.
Boccardo, G., Milne, R.G. and Luisoni, E. (1979). *J. gen. Virol.* **45**: 659.
Conti, M. (1984). *Current Topics in Vector Research*; ed. K.F. Harris. Praeger Scientific, New York.
Dirven, J.G.P. and van Hoof, E.M. (1960). *Tijdsch. PlZiekt.* **66**: 344.
Giannotti, J. and Milne, R.G. (1977). *Virology* **80**: 347.
Hunkar, A.E.S., Hung, A.T.A., Dulder, I.G., Schank, S.C., Holder, N. and Edwards, C. (1974). *Trop. Agric.* **52**: 75.
Milne, R.G. (1977). *CMI/AAB Descr. Pl. Viruses* No. 175, 4 pp.
Milne, R.G. and Lovisolo, O. (1977). *Adv. Virus Res.* **21**: 267.
Milne, R.G. and Luisoni, E. (1977). *Virology* **80**: 12.

# Panicum mosaic satellivirus

**Data collated by A.A. Brunt, 1991.**

*Synonyms:* St. Augustine decline satellite virus.

First reported in *Panicum virgatum*; from the U.S.A.; by Niblett and Paulsen (1975).

**Natural host range and symptoms**

*Panicum virgatum* — unknown as found only with panicum mosaic virus.

**Transmission**

Virus not transmitted by mechanical inoculation.

**Geographical distribution**

Spreads in the USA (Kansas).

**Physical properties**

One sedimenting component in purified preparations; sedimentation coefficient 42 *S*. $A_{260}/A_{280}$ ratio 1.53.

**Biochemical properties**

*Genome* consists of DNA; single–stranded.

*Sequence database accession code(s):*

M17182 Em(40)_vi:SAPCAP Gb(84)_vi:SAPCAP Satellite panicum mosaic virus capsid protein RNA, complete cds. 7/89 826bp.

L10083 Em(40)_vi:SSACAPPRO Gb(84)_vi:SSACAPPRO Satellite St. Augustine decline virus capsid protein gene, complete cds. 2/93 824bp. 2 sequences.

**Taxonomy and Relationships**

Satellivirus.

**References**

Niblett, C.L. and Paulsen, A.Q. (1975). *Phytopathology* **65**: 1157.

# Panicum mosaic (?) *sobemovirus*

**Data collated by A.A. Brunt, A.O. Jackson and C.L. Niblett, 1988, Updated by A.A. Brunt, 1992.**

*Synonyms:* St. Augustine decline virus. *Standard acronym:* PMV. *Strains:* Type strain (TS) and St. Augustine decline virus (SAD), probably molinia streak virus (Paul *et al.*, 1980; Hull, 1988).

First reported in *Panicum virgatum*; from Kansas, U.S.A.; by Sill and Pickett (1957).

**Natural host range and symptoms**

Symptoms persist.
*Panicum virgatum* — faint chlorosis.
*Eremochloa oplicuroides* — chlorotic leaf spots and blotches.
*Muhlenbergia schreberi* — leaf chlorosis.

**Transmission**

Transmitted on cutting shears. Virus transmitted by mechanical inoculation; transmitted by seed (SAD), or not transmitted by seed (TS).

**Geographical distribution**

Spreads in Mexico and the USA.

**Experimental host range**

Few (<3) families susceptible.

Diagnostically susceptible

*Zea mays* inbred lines Ohio 28 and N28 — mild systemic mosaic (type strain only).
*Stenotaphrum secundatum* — systemic chlorotic mottling (St. Augustine decline strain only).
*Avena sativa, Secale cereale, Hordeum vulgare, Triticum aestivum, Sorghum bicolor* — chlorotic local lesions.

Diagnostically insusceptible

*Nicotiana tabacum, Phaseolus vulgaris, Cucumis sativus, Chenopodium quinoa, Tetragonia expansa.*

Maintenance and propagation hosts

*Zea mays* inbred line N28 (type strain), *Stenotaphrum secundatum* (St. Augustine decline strain).

Assay hosts (**L**ocal lesions or **W**hole plants)

*Panicum miliaceum* (W), *Setaria italica* (W).

Dicotyledonous Hosts *(**Susceptible** or Insusceptible)*

*Apium graveolens. Beta vulgaris. Brassica oleracea* ssp. *capitata. Capsicum frutescens. Cassia occidentalis. Catharanthus roseus. Chenopodium amaranticolor. Chenopodium ambrosioides. Chenopodium capitatum. Chenopodium foetidum. Chenopodium murale. Chenopodium quinoa. Cucumis melo. Cucumis sativus. Cucurbita maxima. Cucurbita pepo. Cyamopsis tetragonoloba. Datura metel. Datura stramonium. Daucus carota. Glycine max. Gomphrena globosa. Helianthus annuus. Lactuca sativa. Lycopersicon esculentum. Medicago sativa. Momordica balsamina. Nicotiana debneyi. Nicotiana glutinosa. Nicotiana megalosiphon. Nicotiana sylvestris. Nicotiana tabacum. Petunia hybrida. Phaseolus vulgaris. Physalis floridana. Pisum sativum. Solanum* melongena. *Spinacia oleracea. Tetragonia expansa. Tithonia speciosa. Torenia fournieri. Vigna radiata. Vigna unguiculata. Vigna unguiculata* ssp. *cylindrica. Zinnia elegans.*

Monocotyledonous Hosts ***(Susceptible** or Insusceptible)*

***Avena sativa. Hordeum vulgare. Panicum dichotomiflorum. Panicum miliaceum. Secale cereale. Setaria italica. Sorghum bicolor. Triticum aestivum. Triticum durum. Zea mays.***

Sources of host-range data

McCoy *et al.* (1969); Niblett and Paulsen (1975).

**Properties of particles in sap**

TIP: 85 °C (TS), or 60 °C (SAD). DEP: $\log_{10}$ minus 5 (TS), or 4 (SAD). Leaf sap contains many virions (*c.*0.5 mg/g fresh weight of tissue).

**Purification**

Method: Niblett and Paulsen (1975).

**Particle morphology**

Virions isometric; 25–30 nm in diameter; angular in profile; with a conspicuous capsomere arrangement.

**Physical properties**

One sedimenting component in purified preparations; sedimentation coefficient 109 *S* (TS); of the other(s) 102 *S* (SAD).

**Biochemical properties**

Virions contain 28 % nucleic acid; 72 % protein.

*Genome* consists of RNA; single–stranded; linear; unipartite; largest (or only) genome part 4.2 kb.

*Sequence database accession code(s):* Em_vi:SAPCAP — capsid protein gene, 826bp.

*Non–genomic nucleic acid* not found in the virions. Sub–genomic *m*RNA found in infected cells. 3 virus specified dsRNA species found in infected cells. Size of largest virus specified dsRNA 4.6 kbp; 2nd largest 3 kbp; 3rd largest 1.3 kbp.

*Virion protein(s)* one; $M_r$ 29000; coat protein.

**Replication**

Acts as helper for a satellite virus (with which it is often found. This satellite virus has virions 15–18 nm in diameter which sediment as one component of 42 *S* and contain two ssRNAs (14 *S* and 34 *S*) and a single polypeptide ($M_r$ 15500)).

**Cytopathology**

Virions found in leaves; in cytoplasm. Inclusions present in infected cells; are crystals in the cytoplasm; they contain virions. Other cellular changes: disruption of mitochondria and tonoplast.

**Taxonomy and Relationships**

*Sobemovirus.*

Virus(es) with serologically related virions: cocksfoot mild mosaic (syn. phleum mottle) and panicum mosaic (syn. molinia streak) viruses. Virus(es) with serologically unrelated virions: cocksfoot mild mottle virus.

**References**

Haygood, R.A. and Barnett, O.W. (1992). *Plant Dis.* **76**: 46.

Holcomb, G.E. (1974). *Proc. Am. Phytopath. Soc.* **1**: 21.

Holcomb, G.E., Derrick, K.S., Carver, R.B. and Toler, R.W. (1972). *Pl. Dis. Reptr* **56**: 69.
Hull, R. (1988). In: *The Plant Viruses*, Vol. 3, Polyhedral Virions with Monopartite RNA Genomes, p. 113; ed. R. Koenig. Plenum Press, New York.
Niblett, C.S. and Paulsen, A.Q. (1975). *Phytopathology* **65**: 1157.
Niblett, C.L., Paulsen, A.Q. and Toler, R.W. (1977). *CMI/AAB Descr. Pl. Viruses* No. 177, 4 pp.
Paul, H.L., Querfurth, G. and Huth, W. (1980). *J. gen. Virol.* **47**: 67.
Sill, W.H. and Pickett, R.C. (1957). *Pl. Dis. Reptr* **41**: 241.
Sill, W.H. and Talens, L.T. (1962). *Pl. Dis. Reptr* **46**: 780.
Toler, R.W., McCoy, N.L. and Amadok, J. (1969). *Pl. Dis. Reptr* **53**: 955.

# Panicum streak monogeminivirus

**Data collated by A.A. Brunt, 1995.**

*Synonyms:* Panicum strain of maize streak virus (Bock *et al.*, 1974; Pinner and Markham, 1990).

First reported in *Panicum maximum*; from Kenya; by Storey and McLean (1939) — see Bock *et al.* (1974).

### Natural host range and symptoms

*Panicum maximum* — chlorotic leaf streaking.

### Transmission

Transmitted by a vector; an insect; *Cicadulina mbila*; Cicadellidae. Virus not transmitted by mechanical inoculation.

### Geographical distribution

Spreads in the African region; Kenya, South Africa, and Uganda.

Monocotyledonous Hosts *(**Susceptible** or Insusceptible)*

***Panicum maximum***. *Saccharum officinarum*. *Zea mays*.

### Purification

Method: Briddon *et al.* (1992).

### Particle morphology

Virions geminate; 20 nm in diameter; 30 nm in length.

### Biochemical properties

*Genome* consists of DNA; single–stranded.

*Sequence database accession code(s):*

X60168 Em(40)_vi:PSGIITDNA Gb(84)_vi:PSGIITDNA Panicum streak geminivirus infectious insect_transmissible DNA. 6/92 2,700bp.

L39638 Em(43)_vi:Parpakcg Gb(89)_vi:Pakcg Panicum streak virus (clone pKom100) complete genome. 3/95 2,705bp.

*Features of the genome:* 5 ORFs.

### Taxonomy and Relationships

Monogeminivirus: *Geminiviridae*.

Virus(es) with serologically related virions: Maize streak and sugarcane streak viruses.

The coat protein has clear sequence homologies (46–82% identity) with those of other monogeminiviruses.

**References**

Bock, K.R., Guthrie, E.J. and Woods, R.D. (1974). *Ann. appl. Biol.* **77**: 289.

Briddon, R.W., Luness, P. Chamberlin, L.C.L., Pinner, M.S., Brandish, H. and Markham, P.G. (1992). *J. gen. Virol.* **73**: 1041.

Pinner, M.S. and Markham, P.G. (1990). *J. gen. Virol.* **71**: 1635.

Pinner, M.S., Markham, P.G., Markham, R.H. and Dekker, E.L. (1988). *Pl. Path.* **37**: 74.

Storey, H.H. and McLean, A.P.D. (1930). *Ann. appl. Biol.* **17**: 691.

# Papaya mosaic *potexvirus*

**Data collated by C. Büchen-Osmond, 1987 and E. Hiebert, 1988.**

*Synonyms:* babaco yellow mosaic virus (Barba and Giordano, 1989), boussingaultia mosaic virus (Beczner and Vassanyi, 1980; Phillips *et al.*, 1985), papaw mild mosaic virus. *Standard acronym:* PapMV. *Strains:* Argentine plantago virus (Gracia *et al.*, 1983; Milne, 1988), plantago severe mottle virus (Gracia *et al.*, 1983; Milne, 1988; Rowhani and Peterson, 1980).

First reported in *Carica papaya*; by Conover (1962; 1964*b*); de Bokx (1965).

**Natural host range and symptoms**

Symptoms persist.
*Carica papaya* — mosaic and stunting.
*Ullucus tuberosus* — symptomless infection.

**Transmission**

Transmitted by means not involving a vector (Conover, 1964*b*; Zettler *et al.*, 1968). Virus transmitted by mechanical inoculation; not transmitted by seed.

**Geographical distribution**

Spreads in Bolivia, Peru, the USA, and Venezuela.

**Experimental host range**

Several (3–9) families susceptible.

Diagnostically susceptible

*Carica papaya* — systemic vein-clearing and downward curling of leaves, then mottle or mosaic.

*Gomphrena globosa* — chlorotic local lesions, becoming necrotic with red margins; not systemic.

*Chenopodium amaranticolor* — chlorotic local lesions, not systemic.

*Cassia occidentalis* — necrotic local lesions, not systemic.

Maintenance and propagation hosts

*Carica papaya, Anthirrhinum majus.*

Assay hosts (**L**ocal lesions or **W**hole plants)

*Gomphrena globosa* (L).

Dicotyledonous Hosts *(**Susceptible** or Insusceptible)*

***Antirrhinum majus. Carica papaya. Cassia occidentalis. Chenopodium amaranticolor. Gomphrena globosa.***

Sources of host-range data

Conover (1962; 1964*b*); Cook and Zettler (1970).

**Properties of particles in sap**

TIP: 73–76 °C (in papaya sap; Conover, 1964*b*). LIV: 180 days. DEP: $\log_{10}$ minus 4. Leaf sap contains many virions.

**Purification**

Method: Hiebert (1970); Koenig *et al.* (1970).

**Particle morphology**

Virions filamentous; not enveloped; usually flexuous; with a clear modal length; of 530 nm.

**Physical properties**

One sedimenting component in purified preparations; sedimentation coefficient 118.7 *S*. Isoelectric point pH 5.3.

**Biochemical properties**

Virions contain 7 % nucleic acid; 93 % protein.

*Genome* consists of RNA; single–stranded; linear; unipartite; largest (or only) genome part 6.656 kb. Genomic nucleic acid isolated by Koenig (1971). Base composition 20.7 % G; 33.8 % A; 23.4 % C; 22.1 % U. 5' terminus of RNA has an unusual type (Abouhaidar and Bancroft, 1978). Infectivity retained when deproteinised with phenol or detergent. Poly A region present; a poly A tail (Abouhaidar, 1988). Additional factor not required for infectivity.

*Sequence database accession code(s):*

D00240 Em(40)_vi:PMPCP Gb(84)_vi:PMPCP Papaya mosaic potexvirus (PMV), capsid protein gene, complete cds. 1/93 900bp.

D00580 Em(40)_vi:URNA Papaya mosaic virus genome, complete sequence. 1/93 6,656bp.

D13957 Em(40)_vi:PMPRNA Gb(84)_vi:PMPRNA Papaya mosaic virus genome, complete sequence. 1/93 6,656bp

M35054 Em(40)_vi:PMP5END Gb(84)_vi:PMP5END Papaya mosaic virus, 5' end. 1/93 139bp.

M75794 Em(40)_est:CE02C6 C.elegans cDNA clone cm02c6 encoding hypothetical 176K protein — Papaya mosaic virus homologo 5 sequences.

Sub–genomic *m*RNA found in infected cells; coat protein (Mackie and Bancroft, 1986).

*Virion protein(s)* one; $M_r$ 22930. Method of preparation: Koenig *et al.* (1970). Amino acid sequence: Abouhaidar (1988).

**Replication**

Replication does not depend on a helper virus.

**Cytopathology**

Virions found in all parts of the host plant; in cytoplasm. Inclusions present in infected cells (Zettler *et al.*, 1968; Purcifull *et al.*, 1966; Christie and Edwardson, 1977); virions aggregated in banded inclusion bodies they contain virions (Christie and Edwardson, 1977).

**Taxonomy and Relationships**

*Potexvirus.*

Virus(es) with serologically related virions: clover yellow mosaic virus, but distantly. Virus(es) with serologically unrelated virions: potato X virus (Koenig and Lesemann, 1978; Lesemann and Koenig, 1977).

Best tests for diagnosis: several serious mosaic diseases limit papaya production in various parts of the world, including India, Puerto Rico, South America, Hawaii and Florida. Some are, like papaya mosaic, sap-transmissible, but differ in being aphid-borne and restricted in host range to papaya and cucurbits (Capoor and Varma, 1958; Conover, 1964*a*; Ishii and Holtzmann, 1963; Herold and Weibel, 1962); the best known is papaya ringspot, a typical potyvirus (Purcifull *et al.*, 1984; Zettler *et al.*, 1968).

**References**


Abouhaidar, M. (1988). *J. gen. Virol.* **69**: 219.

Abouhaidar, M. and Bancroft, J.B. (1978). *J. gen. Virol.* **39**: 559.

Barba, M. and Giordano, P. (1989). *Inform. Fitopatol.* **39**: 58.

Beczner, I. and Vassanyi, R. (1980). *Tag. Ber. Akad. Land. Wiss, DDR, Berlin* **184**: 65.

Brunt, A.A., Phillips, S., Jones, R.A.C. and Kenten, R.H. (1982). *Ann. appl. Biol.* **101**: 65.

Capoor, S.P. and Varma, P.M. (1958). *Indian J. agric. Sci.* **28**: 225.

Christie, R.G. and Edwardson, J.R. (1977). *Fla Agric. Exp. Stn Monog.* No. 9, 150 pp.

Conover, R.A. (1962). *Phytopathology* **52**: 6.

Conover, R.A. (1964*a*). *Proc. Fla Stn hort. Soc.* **77**: 440.

Conover, R.A. (1964*b*). *Proc. Fla Stn hort. Soc.* **77**: 444.

Cook, A.A. and Zettler, F.W. (1970). *Pl. Dis. Reptr* **54**: 893.

de Bokx, J.A. (1965). *Pl. Dis. Reptr* **49**: 742.

Gracia, O., Koenig, R. and Lesemann, D.-E. (1983). *Phytopathology* **73**: 1488.

Herold, F. and Weibel, J. (1962). *Virology* **18**: 302.

Hiebert, E. (1970). *Phytopathology* **60**: 1295.

Ishii, M. and Holtzmann, O.V. (1963). *Pl. Dis. Reptr* **47**: 947.

Koenig, R. (1971). *J. gen. Virol.* **10**: 11.

Koenig, R. and Lesemann, D.E. (1978). *CMI/AAB Descr. Pl. Viruses* No. 200, 5 pp.

Koenig, R., Stegemann, H., Francksen, H. and Paul, H.L. (1970). *Biochim. biophys. Acta* **207**: 184.

Lesemann, D.-E. and Koenig, R. (1977). In: *The Atlas of Insect Plant Viruses*, p. 331; eds R.I.B. Francki, R.G. Milne and T. Hatta. C.R.C. Press Inc., Boca Raton, U.S.A.

Mackie, G.A. and Bancroft, J.B. (1986). *Virology* **153**: 215.

Milne, R.G. (1988). In: *The Plant Viruses*, Vol. 4; The Filamentous Plant Viruses, p. 3; ed. R.G. Milne. Plenum Press, New York.

Phillips, S., Brunt, A.A. and Beczner, L. (1985). *Acta Hort.* **164**: 379.

Purcifull, D.E. and Hiebert, E. (1971). *CMI/AAB Descr. Pl. Viruses* No. 56, 4 pp.

Purcifull, D.E., Edwardson, J.R. and Christie, R.G. (1966). *Virology* **29**: 276.

Purcifull, D.E., Edwardson, J.R. and Hiebert, E. (1984). *CMI/AAB Descr. Pl. Viruses* No. 293, 7 pp.

Rowhani, A. and Peterson, J.F. (1980). *Can. J. Pl. Path.* **2**: 12.

Sit, T.L., Abouhaidar, M.G. and Holy, S. (1989). *J. gen. Virol.* **70**: 2325.

Story, G.E. and Halliwell, R.S. (1969). *Phytopathology* **59**: 1336.

Zettler, F.W., Edwardson, J.R. and Purcifull, D.E. (1968). *Phytopathology* **58**: 332.

# Papaya ringspot *potyvirus*

**Data collated by D.E. Purcifull, J.R. Edwardson, E. Hiebert and D. Gonsalves, 1986.**

*Synonyms:* papaya distortion mosaic virus, papaya leaf distortion virus, papaw distortion ringspot virus, papaw mosaic virus, watermelon mosaic virus 1. *Standard acronym:* PRSV. *Strains:* type P, Guadeloupe papaya ringspot, papaya ringspot — type W has been shown to be watermelon mosaic virus 1 potyvirus.

First reported in *Carica papaya*; by Jensen (1949*a*).

**Natural host range and symptoms**

Symptoms persist.

*Carica papaya* — mottling and malformation of leaves, ringspots and streaking on fruit, stems and petioles, plants stunted with less fruit (Conover, 1964*a*; Jensen, 1949*a*). Symptoms severe in cool weather (Jensen, 1949*a*; Conover, 1964*a*; Lima and Gomes, 1975).

**Transmission**

Transmitted by a vector; an insect; *Myzus persicae* (Jensen, 1949*b*; Conover, 1964*a*; Zettler *et al.*, 1968) and *Aphis gossypii*; Aphididae. Transmitted in a non–persistent manner. Amorphous inclusion protein (AI) is required as the helper component for transmission. Virus transmitted by mechanical inoculation; not transmitted by seed.

**Geographical distribution**

Spreads in the Middle East and the South and Central American region; China, France, Germany, India, Italy, Mexico, Taiwan, and the USA.

**Experimental host range**

Few (<3) families susceptible.

Diagnostically susceptible

*Carica papaya* — symptoms variable; vein clearing then mottled and malformed leaves with smaller lobes (Conover, 1964*a*, 1964*b*).

*Cucurbita pepo* — some isolates give mosaic and leaf malformation but others induce mild mottle (Gonsalves and Ishii, 1980).

*Cucumis metuliferus* cv. Accession 2459 — systemic mottling or mosaic.

Diagnostically insusceptible

*Cucumis metuliferus* plant introduction PI 292190, *Nicotiana benthamiana*.

Maintenance and propagation hosts

*Carica papaya, Cucurbita pepo* and *Cucumis metuliferus* cv. Accession 2459.

Assay hosts (**L**ocal lesions or **W**hole plants)

*Chenopodium amaranticolor* (L), *Chenopodium quinoa* (L), *Cucurbita pepo* (W).

Dicotyledonous Hosts *(**Susceptible** or Insusceptible)*

*Carica papaya*. ***Chenopodium amaranticolor***. ***Chenopodium quinoa***. ***Cucumis melo***. *Cucumis metuliferus*. ***Cucumis sativus***. ***Cucurbita maxima***. ***Cucurbita moschata***. *Myosotis sylvatica*.

Sources of host-range data

Jensen (1949*a*); Conover (1964*a*); Wang *et al.* (1978); Milne and Grogan (1969); Purcifull and Hiebert (1979); Yeh *et al.* (1984).

**Properties of particles in sap**

TIP: 54–60 °C. LIV: 0.3 days. DEP: $\log_{10}$ minus 3. Leaf sap contains many virions.

**Purification**

Method: Gonsalves and Ishii (1980), see CMI/AAB description No. 292.

**Particle morphology**

Virions filamentous; not enveloped; usually flexuous; with a clear modal length; of 760–800 nm; 12 nm wide.

**Physical properties**

One sedimenting component in purified preparations. Density 1.32 g $cm^{-3}$ in CsCl.

**Biochemical properties**

Virions contain 5.5 % nucleic acid; 94.5 % protein.

*Genome* consists of RNA; single–stranded; linear; unipartite; largest (or only) genome part 12 kb. Genomic nucleic acid isolated by Brakke and van Pelt (1970); de Mejia (1984); Hiebert *et al.* (1984); de Mejia *et al.* (1984). 5' terminus of RNA probably a genome-linked protein (VPg).

*Sequence database accession code(s):*

D00594 Em(40)_vi:PRVW Gb(84)_vi:PAYW Papaya ringspot virus nuclear inclusion protein (NIb protein) gene, coat protein gene and 3'

D00595 Em(40)_vi:PRVP Gb(84)_vi:PAYP Papaya ringspot virus nuclear inclusion protein (NIb protein) gene, coat protein gene and 3'

S46722 Em(40)_un:S46722 Gb(84)_vi:S46722 polyprotein papaya ringspot virus PRSV, HA, Hawaii isolate, Genomic RNA Complete, 10326 nt.

S89893 Em(40)_vi:S89893 Gb(84)_vi:S89893 3' region: coat protein papaya ringspot virus PRSV, type W Australian isolate, Cucurbita pep

U14736 Gb(84)n:PRU14736 Papaya ringspot virus isolate Bridgeman Downs coat protein mRNA, partial cds. 9/94 842bp

U14737 Gb(84)n:PRU14737 Papaya ringspot virus isolate Bundaberg coat protein mRNA, partial cds. 9/94 842bp.

U14738 Gb(84)n:PRU14738 Papaya ringspot virus isolate coat protein mRNA, partial cds. 9/94 842bp.

U14739 Gb(84)n:PRU14739 Papaya ringspot virus isolate Gatton coat protein mRNA, partial cds. 9/94 842bp

U14740 Gb(84)n:PRU14740 Papaya ringspot virus isolate Wellington Point coat protein mRNA, partial cds. 9/94 842bp.

U14741 Gb(84)n:PRU14741 Papaya ringspot virus isolate Sri Lanka coat protein mRNA, partial cds. 9/94 839bp

U14742 Gb(84)n:PRU14742 Papaya ringspot virus isolate Vietnam coat protein mRNA, partial cds. 9/94 848bp.

U14743 Gb(84)n:PRU14743 Papaya ringspot virus isolate Thailand coat protein mRNA, partial cds. 9/94 839bp.

U14744 Gb(84)n:PRU14744 Papaya ringspot virus isolate Northern territory coat protein mRNA, partial cds. 9/94 842bp.

X67672 Em(40)_vi:PRVNIBCP Gb(84)_vi:PRVNIBCP Papaya ringspot Virus genomic RNA for nuclear inclusion protein b and coat protein. 6/93 2,56

X67673 Em(40)_vi:PRVCG Gb(84)_vi:PRVCG Papaya ringspot Virus complete genomic RNA. 12/92 10,326bp.

D50591 Em(43)_vi:Paycp Gb(89)_vi:Paycp Papaya ringspot virus gene for coat protein. 5/95 1,125bp.

X78557 Em(43)_vi:Prvniab Gb(89)_vi:Prvniab Papaya ringspot virus NIa gene, NiB gene, CP gene. 11/94 2,960bp.

*Virion protein(s)* one; $M_r$ 36000–36500 (two smaller proteins with a $M_r$ of 31000–34000 and 26000–27000 have been reported. Preparations stored for 2 months contained primarily the smaller form, presumably the smaller form arises by proteolytic degradation). Method of preparation: Purcifull and Hiebert (1979); Gonsalves and Ishii (1980); Baum (1980).

*Virus–coded non–virion proteins* isolated (Yeh (1984), for cylindrical (pinwheel) inclusions (CI) and by de Mejia (1984), for amorphous inclusion body (AI)); two proteins found. $M_r$ of the largest 69000–70000; CI non virion protein. $M_r$ of 2nd largest 51000; AI, helper protein in aphid transmission.

**Replication**

Replication does not depend on a helper virus.

**Cytopathology**

Virions found in all parts of the host plant; in cytoplasm and in cell vacuoles. Inclusions present in infected cells; are amorphous X–bodies and pinwheels (*i.e.* scrolls (Edwardson's type 1 inclusions)); they do not contain virions.

**Taxonomy and Relationships**

*Potyvirus: Potyviridae.*

Virus(es) with serologically related virions: the Guadeloupe isolate of PRV (Quiot-Douine *et al.*, 1986) is very closely related, but distinct. PRV antisera react with the D-protein of tobacco etch virus (Shepard *et al.*, 1974) and the virion proteins and CI proteins of a potyvirus isolate from cucurbits in Morocco (Fischer and Lockhart, 1974).

PRV — type P infects papaya and cucurbits, but watermelon mosaic virus 1 (PRV-W) infects cucurbits, but not papaya. Analysis of the coat protein genes of isolates from Australia, Thailand and the U.S.A. indicates that, in each country, papaya ringspot — P isolates have evolved independently from local watermelon mosaic virus 1 isolates (Bateson *et al.*, 1994).

**References**


Bateson, M.F., Henderson, J., Chaleeprom, W., Gibbs, A.J. and Dale, J.L. (1994). *J. gen. Virol.* **75**: 3547.

Baum, R.H. (1980). *Ph.D. Thesis*, Univ. of Florida, 99 pp.

Baum, R.H. and Purcifull, D.E. (1981). *Phytopathology* **71**: 202.

Brakke, M.K. and van Pelt, N. (1970). *Virology* **42**: 699.

Conover, R.A. (1964*a*). *Proc. Fla St. hort. Soc.* **77**: 440.

Conover, R.A. (1964*b*). *Proc. Fla St. hort. Soc.* **77**: 444.

de Mejia, M.V.G. (1984). *Ph.D. Thesis*, Univ. of Florida, 52 pp.

de Mejia, M.V.G., Hiebert, E. and Purcifull, D.E. (1984). *Phytopathology* **74**: 1015.

de Mejia, M.V.G., Hiebert, E. and Purcifull, D.E. (1985). *Virology* **142**: 24,34.

Edwardson, J.R. (1974). *Fla Agric. Exp. Stn Monog.* No. 4, 398 pp.

Fischer, H.U. and Lockhart, B.E. (1974). *Pl. Dis. Reptr* **58**: 143.

Gonsalves, D. and Ishii, M. (1980). *Phytopathology* **70**: 1028.

Hiebert, E., Thornbury, D.W. and Pirone, T.P. (1984). *Virology* **135**: 1.

Jensen, D.D. (1949*a*). *Phytopathology* **39**: 191.

Jensen, D.D. (1949*b*). *Phytopathology* **39**: 212.

Karl, E. and Schmelzer, K. (1971). *Arch. PflSchutz.* **7**: 3.

Lima, J.A.A. and Gomes (1975). *Fitossanidade, Fortaleza* **1**: 56.

Lindner, R.C., Jensen, D.D. and Ikeda, W. (1945). *Haw. Farm and Home* **8**: 10.

Makkouk, K.M. and Lesemann, D.E. (1980). *Plant Dis.* **64**: 789.

Milne, K.S. and Grogan, R.G. (1969). *Phytopathology* **59**: 809.
Purcifull, D.E. and Hiebert, E. (1979). *Phytopathology* **69**: 112.
Purcifull, D.E., Hiebert, E. and Edwardson, J.R. (1984). *CMI/AAB Descr. Pl. Viruses* No. 293, 7 pp.
Purcifull, D., Edwardson, J. and Gonsalves, D. (1984). *CMI/AAB Descr. Pl. Viruses* No. 292, 8 pp.
Quiot-Douine, L., Purcifull, D.E., Hiebert, E. and de Mejia, M.V.G. (1986). *Phytopathology* **76**: 346.
Abu-Samah, N. (1982). *Ph.D. Thesis*, Univ. Adelaide, 109 pp.
Shepard, J.E., Secor, G.A. and Purcifull, D.E. (1974). *Virology* **58**: 464.
Wan, S.H. and Conover, R.A. (1983). *Plant Dis.* **67**: 353.
Wang, H.L., Wang, C.C., Chiu, R.J. and Sun, M.H. (1978). *F.A.O. Pl. Prot. Bull.* **20**: 133.
Yeh, S.D. (1984). *Ph.D. Thesis*, Cornell Univ., 98 pp.
Yeh, S.D., Gonsalves, D. and Provvidenti, R. (1984). *Phytopathology* **74**: 1081.
Zettler, F.W., Edwardson, J.R. and Purcifull, D.E. (1968). *Phytopathology* **58**: 332.

# Paprika mild mottle *tobamovirus*

**Data collated by J.R. Diaz-Ruiz, I. Garcia Luque and M.T. Serra, 1992.**

*Synonyms:* tobacco mosaic virus — P11 pepper isolate. *Standard acronym:* PaMMV.

First reported in *Capsicum annuum*; from the Netherlands; by Rast (1979).

### Natural host range and symptoms

Symptoms persist.
*Capsicum annuum* — mild mosaic.

### Transmission

Virus transmitted by mechanical inoculation; possibly transmitted by seed.

### Geographical distribution

Spreads in the Eurasian region; the Netherlands.

### Experimental host range

Few (<3) families susceptible. Experimentally infected plants mostly show small chlorotic or necrotic local lesions, mosaic and stunting.

Diagnostically susceptible

*Capsicum* spp. with either $L^+L^+$ or $L^1L^1$ genotypes — mild mottle or mosaic.
*Nicotiana benthamiana* — mosaic, stunting.
*Nicotiana clevelandii* — mosaic, malformation.
*Solanum pseudocapsicastrum* — mosaic.

Diagnostically insusceptible

*Lycopersicon esculentum.*

Maintenance and propagation hosts

*Capsicum annuum, Nicotiana benthamiana* and *N. clevelandii.*

Assay hosts (**L**ocal lesions or **W**hole plants)

*Capsicum annuum* (W), *C. frutescens* (L), *Nicotiana clevelandii* (W), *N. sylvestris* (L), *N. tabacum* cv Xanthi (L).

Dicotyledonous Hosts *(**Susceptible** or Insusceptible)*

***Capsicum annuum**. **Capsicum frutescens**. Lycopersicon esculentum. **Nicotiana clevelandii**. **Nicotiana glutinosa**. **Nicotiana sylvestris**. **Nicotiana tabacum**. **Petunia hybrida**. **Solanum melongena**. **Trifolium subterraneum**.*

Sources of host-range data

Tobias *et al.* (1982).

**Properties of particles in sap**

Leaf sap contains many virions.

**Particle morphology**

Virions rod–shaped; not enveloped; usually straight; with a clear modal length.

**Biochemical properties**

Virions contain 5 % nucleic acid; 95 % protein.

*Genome* consists of RNA; single–stranded; linear; unipartite; largest (or only) genome part 6.5 kb. Genomic nucleic acid isolated by Alonso *et al.* (1991). Infectivity retained when deproteinised with phenol or detergent. Additional factor not required for infectivity.

*Sequence database accession code(s):*

X72586 Gb(84)_vi:PMMVCP Paprika mild mottle virus (Pepper strain P11) coat protein gene and 3' non-coding region. 7/9.

*Virion protein(s)* one; $M_r$ 17706; coat protein.

**Replication**

Replication does not depend on a helper virus.

**Cytopathology**

Virions found in leaves; in cytoplasm.

**Taxonomy and Relationships**

*Tobamovirus*.

Virus(es) with serologically unrelated virions: tobacco and tomato mosaic tobamoviruses.

**References**

Alonso, E., Garcia-Luque, I., de la Cruz, A., Wicke, B., Avila-Ricón, M.J., Serra, M.T., Castresana, C. and Diaz-Ruiz, J.R. (1991). *J. gen. Virol.* **72**: 2875.

Garcia-Luque, I., Serra, M.T., Alonso, E., Wicke, B., Ferrero, M.L. and Diaz-Ruiz, J.R. (1990). *J. Phytopath.* **129**:1.

Rast, A.Th.B. (1976). *Med. Fac. Landb. Rijksuniv. Ghent* **44**: 616.

Tobias, I., Rast, A.Th.B. and Maat, D.Z. (1982). *Neth. J. Pl. Path.* **88**: 257.

# Parietaria mottle *ilarvirus*

**Data collated by P. Caciagli, 1991.**

*Standard acronym:* PMoV.

First reported in *Parietaria officinalis*; from Torino, Italy; by Caciagli *et al.* (1989).

**Natural host range and symptoms**

Symptoms vary seasonally. Symptoms mosaic or mottling.

*Parietaria officinalis*—bright yellow mosaic or mottling.

**Transmission**

Virus transmitted by mechanical inoculation; transmitted by grafting.

**Geographical distribution**

Spreads in Italy.

**Experimental host range**

Several (3–9) families susceptible. Experimentally infected plants mostly show chlorotic local lesions and then mottle or leafroll in species of the Chenopodiaceae and necrotic local lesions and necrotic patching in species of Solanaceae.

Diagnostically susceptible

*Chenopodium album* — local lesions and systemic mottle.
*Chenopodium amaranticolor* — local lesions and systemic leafroll.
*Chenopodium quinoa, Spinacia oleracea* — systemic mottle and leaf malformation.
*Lycopersicon esculentum* — mosaic, necrosis.
*Nicotiana clevelandii* — chlorosis, rosetting and mottle.
*Nicotiana megalosiphon* — leaf malformation.
*Nicotiana tabacum* cv. Xanthi — necrotic etching; systemic.
*Nicotiana tabacum* cv. White Burley — local lesions.
*Ocimum basilicum* — necrotic etching.
*Tetragonia expansa* — latent.

Diagnostically insusceptible

*Amaranthus caudatus, Cucumis sativus, Cyamopsis tetragonoloba, Momordica balsamina, Nicotiana glutinosa.*

Maintenance and propagation hosts

*Chenopodium amaranticolor, C. quinoa.*

Assay hosts (**L**ocal lesions or **W**hole plants)

*Chenopodium murale* (L), *C. quinoa* (W), *Phaseolus vulgaris* (L), *Vigna unguiculata* cv. Black (W).

Dicotyledonous Hosts *(**Susceptible** or Insusceptible)*

*Amaranthus caudatus. Beta vulgaris. Capsicum annuum. Cassia occidentalis. Catharanthus roseus.* ***Chenopodium album. Chenopodium amaranticolor. Chenopodium murale. Chenopodium quinoa.*** *Cucumis sativus. Cucurbita pepo. Cyamopsis tetragonoloba. Gomphrena globosa.* ***Lycopersicon esculentum.*** *Momordica balsamina.* ***Nicotiana clevelandii.*** *Nicotiana glutinosa.* ***Nicotiana megalosiphon. Nicotiana tabacum***, or *Nicotiana tabacum.* ***Ocimum basilicum.*** *Petroselinum crispum.* ***Petunia hybrida. Phaseolus vulgaris.*** *Physalis floridana. Saponaria vaccaria.* ***Spinacia oleracea. Tetragonia expansa. Vigna unguiculata*** ssp. ***sesquipedalis.*** *Zinnia elegans.*

Sources of host-range data

Caciagli *et al.* (1989).

**Properties of particles in sap**

TIP: 50 °C. LIV: 1.5 days. DEP: $\log_{10}$ minus 3. Electron microscopy: purification.

**Purification**

Method: Caciagli *et al.* (1989).

**Particle morphology**

Virions isometric; not enveloped; 24 nm in diameter, or 29 nm in diameter, or 36 nm in diameter; rounded in profile; without a conspicuous capsomere arrangement.

**Physical properties**

One sedimenting component in purified preparations (in CsCl), or three sedimenting components in purified preparations (in sucrose). Density 1.354 g $cm^{-3}$ in CsCl. $A_{260}/A_{280}$ ratio 1.63.

**Biochemical properties**

*Genome* consists of RNA; single–stranded; linear; of four parts; largest (or only) genome part the largest 4.3 kb; the 2nd largest 3.7 kb; the 3rd largest 2.7 kb; the 4th largest 1.1 kb. Genomic nucleic acid isolated by Boccardo and Milne (1975); Caciagli *et al.* (1989). Infectivity retained when deproteinised with proteases; retained when deproteinised with phenol or detergent.

*Virion protein(s)* one; $M_r$ 24300. Method of preparation: Francki *et al.* (1980).

**Replication**

Replication does not depend on a helper virus.

**Cytopathology**

Inclusions absent from infected cells.

**Taxonomy and Relationships**

*Ilarvirus: Bromoviridae* (subgroup 10).

Virus(es) with serologically unrelated virions: apple mosaic, hydrangea mosaic, prune dwarf, spinach latent, tobacco streak and Tulare apple mosaic viruses.

**References**

Boccardo, G. and Milne, R.G. (1975). *Virology* **68**: 79.

Caciagli, P., Boccardo, G. and Lovisolo, O. (1989). *Pl. Path.* **38**: 577.

Francki, R.I.B., Hatta, T., Boccardo, G. and Randles, J.W. (1980). *Virology* **101**: 233.

# Parsley 5 (?) *potexvirus*

**Data collated by A.A. Brunt, 1991.**

*Standard acronym:* PaV-5.

First reported in *Petroselinum crispum*; from the U.K.; by Frowd and Tomlinson (1972).

**Natural host range and symptoms**

*Petroselinum crispum* — no conspicuous symptoms.

**Transmission**

Transmitted by a vector; an insect; *Cavariella aegopodii*; Aphididae. Transmitted in a non–persistent manner. Virus not transmitted by mechanical inoculation.

**Geographical distribution**

Found, but with no evidence of spread, in the U.K.

**Particle morphology**

Virions filamentous; usually flexuous; with a clear modal length; of 479 nm; 15 nm wide.

**Taxonomy and Relationships**

Possible *Potexvirus*.

Virus(es) with serologically related virions: potato X and parsnip 3 viruses.

**References**

Frowd, J.A. and Tomlinson, J.A. (1972). *Ann. appl. Biol.* **72**: 177.

# Parsley green mottle (?) potyvirus

**Data collated by A.A. Brunt, 1992.**

First reported in *Petroselinum crispum*; from Italy; by Angelis and Quacquarelli (1974).

**Natural host range and symptoms**

Symptoms persist.

*Petroselinum crispum* — green mottling of leaves; the virus and a strain of chicory yellow mottle nepovirus in complex together cause parsley bushy stunt disease.

**Transmission**

Transmitted by a vector; an insect; *Acyrthosiphon pisum, Aphis spiraecola, Myzus persicae*; Aphididae. Transmitted in a non–persistent manner. Virus transmitted by mechanical inoculation; not transmitted by seed.

**Geographical distribution**

Spreads in Italy.

**Experimental host range**

Few (<3) families susceptible.

Diagnostically susceptible

*Apium graveolens* cv. Doré Barbier — conspicuous chlorosis in systemically infected leaves (infection is symptomless in some other cultivars).

*Ammi majus* — transient systemic leaf chlorosis.

*Chenopodium amaranticolor, C. quinoa* — local lesions, general chlorosis and/or vein yellowing.

*Petroselinum crispum* — mosaic, reduced laminae and distortion of systemicallly infected leaves.

*Pimpinella anisum* — symptomless systemic infection.

Diagnostically insusceptible

*Beta vulgaris.*

Maintenance and propagation hosts

*Petroselinum sativum, Apium graveolens* cv. Doré Barbier.

Assay hosts (**L**ocal lesions or **W**hole plants)

*Petroselinum sativum* (W), *Apium graveolens* (W).

Dicotyledonous Hosts *(**Susceptible** or Insusceptible)*

***Apium graveolens. Chenopodium amaranticolor. Chenopodium quinoa. Petroselinum crispum.***

**Properties of particles in sap**

TIP: 53 °C. LIV: 6 days. DEP: $\log_{10}$ minus 3. Leaf sap contains few virions.

**Purification**

Method: Angelis and Quacquarelli (1974).

**Particle morphology**

Virions filamentous; usually flexuous; with a clear modal length; of 758 nm; 12–15 nm wide. Axial canal obscure. Basic helix obscure.

**Taxonomy and Relationships**

Unassigned potyvirus.
Virus(es) with serologically unrelated virions: celery mosaic potyvirus.

**References**

Angelis, A. and Quacquarelli, A. (1974). *Phytopath. Medit.* **13**: 1.

# Parsley (?) rhabdovirus

**Data collated by A.A. Brunt, 1991.**

*Standard acronym:* PaV.

First reported in *Petroselinum crispum* plants also containing carrot mottle virus; from the U.K.; by Tomlinson and Webb (1974).

**Natural host range and symptoms**

*Petroselinum crispum* — symptoms unknown as the virus has only been found in mixed infections.

**Particle morphology**

Virions rhabdo– or bullet–shaped; with a clear modal length; of 214 nm; 87 nm wide.

**Taxonomy and Relationships**

Possible unassigned rhabdovirus.

**References**

Tomlinson, J.A. and Webb, M.J.W. (1974). *Rep. Nat. Veg. Res. Stn.* 1973, p. 101.

# Parsnip 3 (?) *potexvirus*

**Data collated by A.A. Brunt, 1991.**

*Standard acronym:* ParV-3.

First reported in *Pastinaca sativa*; from the U.K.; by Garrett and Tomlinson (1966).

**Natural host range and symptoms**

*Pastinaca sativa* — black tip necrosis in young leaves.

**Geographical distribution**

Found, but with no evidence of spread, in the U.K.

**Particle morphology**

Virions filamentous; not enveloped; usually flexuous; with a clear modal length; of *c.* 500 nm.

**Taxonomy and Relationships**

Possible *Potexvirus*.
Virus(es) with serologically related virions: potato X virus.

**References**

Garrett, R.G. and Tomlinson, J.A. (1966). *Rep. Nat. Veg. Res. Stn.* 1965, p. 74.

# Parsnip leafcurl virus

**Data collated by P. van Dijk, 1991.**

*Synonyms:* hogweed 4 virus (Bem and Murant, 1979; Van Dijk and Bos, 1989).

First reported in *Heracleum sphondylium*; from the U.K.; by Bem and Murant (1979).

**Natural host range and symptoms**

Symptoms chlorosis and curling of older leaves.
*Pastinaca sativa* — leaf chlorosis and curling.
*Heracleum sphondylium.*

**Transmission**

Virus transmitted by mechanical inoculation; not transmitted by seed.

**Geographical distribution**

Spreads in the Netherlands. Found, but with no evidence of spread, in the U.K.

**Experimental host range**

Several (3–9) families susceptible. Experimentally infected plants mostly show chlorotic to necrotic local lesions.

Diagnostically susceptible

*Ammi majus* — local and systemic necrosis and chlorosis.

*Chenopodium quinoa* — chlorotic local lesions; systemic chlorotic spotting and malformation.

*Chenopodium amaranticolor, Cucumis sativus* — chlorotic local lesions; not systemic.

*Montia perfoliata, Spinacia oleracea* — latent.

Diagnostically insusceptible

*Cyamopsis tetragonoloba, Phaseolus vulgaris, Nicotiana clevelandii, N. tabacum.*

Maintenance and propagation hosts

*Chenopodium quinoa.*

Assay hosts (**L**ocal lesions or **W**hole plants)

*Chenopodium quinoa* (L and W).

Dicotyledonous Hosts *(**Susceptible** or Insusceptible)*

***Anthriscus cerefolium***. *Antirrhinum majus*. ***Apium graveolens***. ***Beta vulgaris***. *Brassica oleracea* ssp. *capitata*. ***Chenopodium album***. ***Chenopodium amaranticolor***. ***Chenopodium capitatum***. ***Chenopodium foetidum***. ***Chenopodium foliosum***. ***Chenopodium murale***. ***Chenopodium quinoa***. ***Coriandrum sativum***. *Cucumis sativus*. *Cyamopsis tetragonoloba*. *Datura stramonium*. ***Gomphrena globosa***. ***Heracleum sphondylium***. *Lycopersicon esculentum*. *Nicotiana benthamiana*. *Nicotiana clevelandii*. *Nicotiana glutinosa*. *Nicotiana tabacum*. ***Pastinaca sativa***. ***Petroselinum crispum***. *Phaseolus vulgaris*. *Pisum sativum*. ***Spinacia oleracea***. ***Tetragonia expansa***. ***Trifolium incarnatum***. *Vicia faba*. *Zinnia elegans*.

Additional hosts

***Ammi majus*** and ***Montia perfoliata***.

Sources of host-range data

Bem and Murant (1979); Van Dijk and Bos (1989).

**Properties of particles in sap**

LIV: 0.18–0.75 days (3–18 hours). Leaf sap contains few virions.

**Particle morphology**

Virions isometric.

**Taxonomy and Relationships**

Best tests for diagnosis: *Chenopodium quinoa* infected systemically, *C. amaranticolor* chlorotic local lesions only, *Cyamopsis tetragonoloba* and *Nicotiana clevelandii* not infected.

**References**

Bem, F. and Murant, A.F. (1979). *Ann. appl. Biol.* **92**: 237.

Van Dijk, P. and Bos, L. (1989). *Neth. J. Pl. Path.* **95(2)**: 34.

# Parsnip mosaic *potyvirus*

**Data collated by A.A. Brunt, 1991.**

*Standard acronym:* ParMV.

First reported in *Pastinaca sativa*; from the U.K. in Scotland; by Murant and Munthe (1967).

**Natural host range and symptoms**

Symptoms disappear soon after infection.

*Pastinaca sativa* — mild mosaic.

**Transmission**

Transmitted by a vector; an insect; *Carvariella aegopodii, C. theobaldi, Myzus persicae*; Aphididae. Transmitted in a non–persistent manner. Virus transmitted by mechanical inoculation.

**Geographical distribution**

Spreads in the UK.

**Experimental host range**

Several (3–9) families susceptible. Experimentally infected plants mostly show veinal necrosis and local lesions.

Diagnostically susceptible

*Anthriscus cerefolium, Bupleurum rotundifolium, Coriandrum sativum, Daucus carota, Thaspium aureum, Trachymene pilosa* — systemic veinal necrosis.

*Chenopodium amaranticolor, C. quinoa, Gomphrena globosa, Spinacia oleracea* — local lesions; no systemic infection.

Diagnostically insusceptible

*Beta vulgaris, Brassica pekinensis, Cucumis melo, Datura stramonium, Nicotiana clevelandii.*

Maintenance and propagation hosts

*Chenopodium quinoa, Coriandrum sativum.*

Assay hosts (**L**ocal lesions or **W**hole plants)

*Chenopodium amaranticolor* (L), *C. quinoa* (L).

Dicotyledonous Hosts *(**Susceptible** or Insusceptible)*

***Anthriscus cerefolium***. *Beta vulgaris. Brassica campestris* ssp. *pekinensis. Catharanthus roseus.* ***Chenopodium amaranticolor***. ***Chenopodium ambrosioides***. ***Chenopodium capitatum***. ***Chenopodium foetidum***. ***Chenopodium murale***. ***Chenopodium quinoa***. ***Coriandrum sativum***. *Cucumis melo. Cucumis sativus. Datura stramonium.* ***Gomphrena globosa***. *Hyoscyamus niger. Nicandra physaloides. Nicotiana clevelandii. Nicotiana debneyi. Nicotiana glutinosa. Nicotiana rustica. Nicotiana tabacum. Petroselinum crispum. Petunia hybrida. Phaseolus vulgaris. Physalis floridana. Tetragonia expansa. Trifolium pratense. Verbesina encelioides. Vigna unguiculata.*

**Properties of particles in sap**

TIP: 55–58 °C. LIV: 7–10 days (at 18°C). DEP: $\log_{10}$ minus 3–4.

**Particle morphology**

Virions filamentous; usually flexuous; with a clear modal length; of 736 nm; 14 nm wide.

**Physical properties**

One sedimenting component in purified preparations; sedimentation coefficient 149 *S*.

**Cytopathology**

Virions found in leaves and mesophyll; in cytoplasm. Inclusions present in infected cells; are pinwheels.

**Taxonomy and Relationships**

*Potyvirus: Potyviridae.*

Virus(es) with serologically unrelated virions: bean yellow mosaic, clover yellow vein, cocksfoot streak, henbane mosaic, iris mild mosaic, lettuce mosaic, pea mosaic, turnip mosaic, tulip breaking, potato Y and celery mosaic viruses.

**References**

Murant, A.F. (1970). *CMI/AAB Descr. Pl. Viruses* No. 91, 4 pp.

Murant, A.F. and Goold, R.A. (1968). *Ann. appl. Biol.* **62**: 123.

Murant, A.F. and Munthe, T. (1967). *Rep. Scottish Hort. Res. Inst.* 1966, p. 64.

Murant, A.F., Munthi, T. and Goold, R.A. (1970). *Ann. appl. Biol.* **65**: 127.

# Parsnip yellow fleck *sequivirus*

**Data collated by P. Jeyanandarajah, 1992.**

*Standard acronym:* PYFV. *Strains:* serotypes P121 from *Pastinaca sativa* and A421 from *Anthriscus cerefolium.*

First reported in *Pastinaca sativum*; from the U.K; by Murant and Goold (1968).

**Natural host range and symptoms**

Symptoms persist. Symptoms yellow flecks, mosaic, necrosis.

*Pastinaca sativa, Heracleum spondylium* — chlorotic vein banding, yellow flecks and leaf mosaic.

*Anthriscus sylvestris, Apium graveolens* — no conspicuous symptoms.

*Aethusa cynapium, Anethum graveolens, Anthriscus cerefolium, Chaerophyllum temulum, Coriandrum sativum, Daucus carota, Oenanthe aquatica, Torilis japonica* — leaf yellowing and dieback (*Anthriscus* serotype).

**Transmission**

Transmitted by a vector; an insect; *Cavariella aegopodii, C. pastinacae*; Aphididae. Not transmitted by *Cavariella theobaldii*. Transmitted in a semi–persistent manner. Virus lost by the vector when it moults; not transmitted congenitally to the progeny of the vector; requires, for vector transmission, a helper virus (anthriscus yellows waikavirus); transmitted by mechanical inoculation; not transmitted by seed; not transmitted by pollen.

**Geographical distribution**

Spreads in Germany and the UK.

**Experimental host range**

Several (3–9) families susceptible. Experimentally infected plants mostly show yellow flecks, mosaic, distortion of leaves, veinal necrosis, local chlorotic or necrotic spots or rings.

Diagnostically susceptible

*Anthriscus cerefolium, Coriandrum sativum, Spinacia oleracea, Nicotiana clevelandii, Petroselinum crispum, Chenopodium quinoa, Gomphrena globosa, Montia perfoliata, Chenopodium amaranticolor* — yellow flecks, necrosis, chlorotic and necrotic local lesions.

*Nicotiana benthamiana* — chlorotic local lesions; systemic vein yellowing, chlorosis and tip necrosis (*Anthriscus* serotype).

Diagnostically insusceptible

*Brassica pekinensis, Foeniculum vulgare, Phaseolus vulgaris, Verbesina encelioides, Vigna unguiculata.*

Maintenance and propagation hosts

*Spinacia oleracea* and, for some isolates, *Anthriscus cerefolium.*

Assay hosts (**L**ocal lesions or **W**hole plants)

*Anthriscus cerefolium* (W), *Chenopodium quinoa* (L).

Dicotyledonous Hosts *(**Susceptible** or Insusceptible)*

***Anthriscus cerefolium***. ***Beta vulgaris***. *Brassica campestris* ssp. *pekinensis*. ***Chenopodium amaranticolor***. ***Chenopodium ambrosioides***. ***Chenopodium capitatum***. ***Chenopodium murale***. ***Chenopodium quinoa***. ***Coriandrum sativum***. ***Cucumis sativus***. ***Datura stramonium***. ***Daucus carota***. ***Gomphrena globosa***. ***Hyoscyamus niger***. ***Lycopersicon esculentum***. ***Nicotiana benthamiana***. ***Nicotiana clevelandii***. ***Nicotiana glutinosa***. ***Nicotiana rustica***. ***Nicotiana tabacum***. ***Pastinaca sativa***. ***Petroselinum crispum***. ***Petunia hybrida***. *Phaseolus vulgaris*. ***Spinacia oleracea***. ***Tetragonia expansa***. *Verbesina encelioides*. *Vigna unguiculata*.

Sources of host-range data

Murant and Goold (1968); Hemida and Murant (1989).

**Properties of particles in sap**

TIP: 57.5–65 °C. LIV: 4–7 days (at room temperature). DEP: $\log_{10}$ minus 3–4.

**Purification**

Method: Hemida and Murant (1989).

**Particle morphology**

Virions isometric; *c.* 31 nm in diameter.

**Physical properties**

Two sedimenting components in purified preparations; sedimentation coefficient 153 *S* (B); of the other(s) 60 *S* (T). Density 1.297 g $cm^{-3}$ in CsCl (T), or 1.49 g $cm^{-3}$ in CsCl (B). $A_{260}/A_{280}$ ratio 1.59 (B), or 1.46 (T).

**Biochemical properties**

Virions contain 40 % nucleic acid.

*Genome* consists of RNA; single–stranded; unipartite; largest (or only) genome part 9.871 kb.

*Sequence database accession code(s):*

D14066 Em(40)_vi:PYFPOLYP Gb(84)_vi:PYFPOLYP Parsnip yellow fleck virus (PYFV) RNA encoding polyprotein, complete cds. 2/93 9,871bp. 1 sequence.

*Virion protein(s)* three; $M_r$ 30900; it is coat protein. $M_r$ of 2nd largest 25500. $M_r$ of 3rd largest 22700.

**Cytopathology**

Virions found in mesophyll. Inclusions present in infected cells; are membranous bodies; they contain virions.

**Taxonomy and Relationships**

*Sequivirus: Sequiviridae* (the type species).

Differ serologically and in host range.

**References**


Elnager, S. (1974). *Ph. D. Thesis*, 168 pp. University of Dundee, U.K.

Elnager, S. and Murant, A.F. (1976). *Ann. appl. Biol.* **84**: 153.

Elnager, S. and Murant, A.F. (1976). *Ann. appl. Biol.* **84**: 169.

Hemida, S.K. and Murant, A.F. (1989). *Ann. appl. Biol.* **114**: 87.

Hemida, S.K. and Murant, A.F. (1989). *Ann. appl. Biol.* **114**: 101.

Lennon, A.M. (1984). *Ph. D. Thesis*, 300 pp. University of Manchester, U.K.

Murant, A.F. (1974). *CMI/AAB Descr. Pl. Viruses* No. 129, 4 pp.

Murant, A.F. (1988). In: *The Plant Viruses*; Vol. 3, Polyhedral Virions with Monopartite Genomes. pp. 273–288, ed. Koenig, R. Plenum Press, New York.

Murant, A.F. (1993). *Arch. Virol.* **131**: 441.

Murant, A.F. and Goold, R.A. (1968). *Ann. appl. Biol.* **62**: 123.

Reavy, B., Mayo, M.A., Turnbull-Ross, A.D. and Murant, A.F. (1993). *Arch. Virol.* **131**: 441.

Singh, H. (1980). *Ph. D. Thesis*. University of Manchester, U.K.

Tomlinson, J.A. and Carter (1970). *Rep. Nat. Veg. Res. Stn.* 1969, p. 110.

Turnbull-Ross, A.D., Reavy, B., Mayo, M.A. and Murant, A.F. (1992). *J. gen. Virol.* **173**: 3203.

Turnbull-Ross, A.D., Mayo, M.A., Reavy, B. and Murant, A.F. (1993). *J. gen. Virol.* **74**: 555.

Van Dijk, P. and Bos, L. (1985). *Neth. J. Pl. Path.* **91**: 169.

Waterhouse, P.M. (1981). *Ph. D. Thesis*, 244 pp. University of Dundee, U.K.

# Paspalum striate mosaic monogeminivirus

**Data collated by R.S. Greber, 1987.**

*Synonyms:* paspalum geminivirus. *Standard acronym:* PSMV. *Strains:* Paspalum (P) and Bromus (BC), also bromus striate virus.

First reported in *Paspalum dilatatum*; by Slykhuis (1962).

**Natural host range and symptoms**

Symptoms persist.
*Paspalum conjugatum, P. dilatatum* — striations, then blotchy mottle.
*P. scrobiculatum* — fine striations and notching of leaf margins, then blotchy mottle.
*Bromus catharticus* — fine chlorotic striations.

**Transmission**

Transmitted by a vector; an insect; *Nesoclutha pallida (obscura)*; Cicadellidae. Transmitted in a semi–persistent manner. Virus retained when the vector moults; does not multiply in the vector; not transmitted congenitally to the progeny of the vector; transmitted by mechanical inoculation (only to the vector); not transmitted by contact between plants.

**Geographical distribution**

Spreads in Australia (New South Wales and Queensland).

**Experimental host range**

Few (<3) families susceptible.

Diagnostically susceptible

*Avena sativa* — striations, stunting, blotchy mottle.
*Chloris gayana* — stunting, leaf margin notching.
*Dactyloctenium aegypticum* — striations, curling.
*Hordeum vulgare* — striations, stunting, blotchy mottle.
*Leptochloa filiformis* — severe stunting, curling.
*Paspalum dilatatum, P. conjugatum, P. scrobiculatum* — striations, mottling.
*Triticum aestivum* — stunting, leaf notching, blotchy mottle.
*Zea mays* (sweet corn) — pale striations.

Diagnostically insusceptible

*Bromus catharticus, Microlaena stipoides.*

Maintenance and propagation hosts

*Triticum aestivum* cv. Gatcher, for propagation, *Paspalum dilatatum* for maintenance.

Assay hosts (**L**ocal lesions or **W**hole plants)

*Triticum aestivum* cv. Gatcher (W), *Paspalum dilatatum* (W).

Monocotyledonous Hosts *(**Susceptible** or Insusceptible)*

***Avena sativa*. *Chloris gayana*. *Hordeum vulgare*. *Leptochloa filiformis*. *Paspalum dilatatum*. *Triticum aestivum*. *Zea mays*.**

**Particle morphology**

Virions geminate; 18 nm in diameter; 30 nm in length; angular in profile; without a conspicuous capsomere arrangement.

**Biochemical properties**

*Genome* consists of DNA; single–stranded; circular.

**Cytopathology**

Virions found in mesophyll and vascular parenchyma; in nuclei. Inclusions present in infected cells; are crystals in the nucleus; they contain virions.

**Taxonomy and Relationships**

Monogeminivirus: *Geminiviridae*.

Virus(es) with serologically unrelated virions: chloris striate mosaic, digitaria streak and maize streak viruses.

Best tests for diagnosis: Readily infects *Paspalum dilatatum* and is vector specific.

**References**

Francki, R.I.B., Milne, R.G. and Hatta, T. (1985). In: *Atlas of Plant Viruses* Vol. I, 222 pp. C.R.C. Press Inc, Boca Raton.

Greber, R.S. (1984). In: *Current Topics in Vector Research*, Vol. 2, p. 141; ed. K.F. Harris. Praeger Scientific, New York.

Grylls, N.E. (1963). *Aust. J. agric. Res.* **14**: 143.

Grylls, N.E. (1979). In: *Leafhopper, Vectors and Plant Disease Agents*, p. 179; ed. K. Maramorosch. Academic Press, New York.

Piner, M.S., Markham, P.G., Rybicki, E. and Greber, R.S. (1992). *Plant Path.* **41**: 618.

Slykhuis, J.T. (1962). *F.A.O. Pl. Prot. Bull.* **10**: 1.

# Passiflora latent *carlavirus*

**Data collated by A.A. Brunt and S. Molyneux, 1986.**

*Standard acronym:* PLV.

First reported in *Passiflora caerulea*; from the Federal Republic of Germany; by Schnepf and Brandes (1961).

**Natural host range and symptoms**

*Passiflora caerulea* — no symptoms.

**Transmission**

Virus transmitted by mechanical inoculation; transmitted by grafting.

**Geographical distribution**

Found, but with no evidence of spread, in the Federal Republic of Germany.

Diagnostically susceptible

*Chenopodium quinoa, C. album* — systemic leaf chlorosis.

Diagnostically insusceptible

*Nicotiana tabacum, Nicotiana glutinosa, Gomphrena globosa, Phaseolus vulgaris, Cucumis sativus.*

Maintenance and propagation hosts

*Chenopodium quinoa.*

Assay hosts (**L**ocal lesions or **W**hole plants)

*Chenopodium quinoa* (W).

Dicotyledonous Hosts (***Susceptible*** *or Insusceptible)*

*Capsicum annuum. Chenopodium amaranticolor. Cucumis sativus. Cyamopsis tetragonoloba. Datura metel. Datura tatula. Dianthus barbatus. Gomphrena globosa.*

*Lathyrus odoratus. Lycopersicon esculentum. Melilotus alba. Nicotiana debneyi. Nicotiana glutinosa. Nicotiana tabacum. Petunia hybrida. Phaseolus vulgaris. Pisum sativum. Solanum tuberosum. Vicia faba. Vigna unguiculata.*

**Properties of particles in sap**

TIP: 65 °C (in sap from infected *Chenopodium quinoa*). Leaf sap contains few virions.

**Purification**

Method: Brandes and Wetter (1964).

**Particle morphology**

Virions filamentous; not enveloped; usually straight; with a clear modal length; of 648 nm. Axial canal obscure. Basic helix obscure.

**Taxonomy and Relationships**

*Carlavirus.*

Virus(es) with serologically related virions: potato S, potato M, carnation latent, chrysanthemum B and cactus 2 viruses.

**References**

Brandes, J. and Wetter, C. (1964). *Phytopath. Z.* **49**: 61.

Schnepf, S. and Brandes, J. (1961). *Phytopath. Z.* **43**: 102.

# Passiflora ringspot *potyvirus*

**Data collated by J.C. Thouvenel, 1984.**

*Synonyms:* passionfruit ringspot virus.

First reported in *Passiflora edulis* f. *flavicarpa*; from the Ivory Coast; by de Wijs (1974*a*).

**Natural host range and symptoms**

Symptoms persist.

*Adenia lobata* — mosaic.

*Passiflora edulis* — mosaic, malformed leaves, ringspots.

*P. foetida* — mosaic, mottle, vein yellowing.

**Transmission**

Transmitted by a vector; an insect; *Aphis gossypii, A. spiraecola*; Aphididae. Transmitted in a non–persistent manner. Virus transmitted by mechanical inoculation; not transmitted by seed.

**Geographical distribution**

Spreads in Cote d'Ivoire.

**Experimental host range**

Several (3–9) families susceptible.

Diagnostically susceptible

*Chenopodium amaranticolor* — faint chlorotic local lesions; not systemic.

*Cassia occidentalis* — systemic ringspotting.

*Phaseolus vulgaris* cv. Triomphe de Farcy — chlorotic spots then systemic vein necrosis.

*Passiflora edulis* f. *flavicarpa* — systemic vein yellowing, epinasty, mottling, spotting, ringspotting.

*P. foetida* — vein yellowing, then spotting and mosaic.
*P. quadrangularis* — death of young plants, older plants not infected.

Diagnostically insusceptible

*Cucumis sativus, Datura stramonium, Nicotiana glutinosa, Physalis floridana.*

Maintenance and propagation hosts

*Adenia lobata, Passiflora edulis.*

Assay hosts (**L**ocal lesions or **W**hole plants)

*Chenopodium amaranticolor* (L), *Phaseolus vulgaris* cv. Triomphe de Farcy (L), *Cassia occidentalis* (W), *Passiflora edulis* (W), *P. foetida* (W).

Dicotyledonous Hosts ***(Susceptible** or Insusceptible)*

***Cassia occidentalis. Chenopodium album. Chenopodium amaranticolor. Crotalaria retusa.*** *Datura stramonium. Glycine max.* ***Lathyrus odoratus.*** *Lycopersicon esculentum.* ***Nicotiana benthamiana.*** *Nicotiana clevelandii. Nicotiana glutinosa.* ***Nicotiana megalosiphon.*** *Nicotiana tabacum. Phaseolus lunatus.* ***Phaseolus vulgaris.*** *Physalis floridana.* ***Tephrosia vogelii.*** *Vigna unguiculata* ssp. *cylindrica.*

Sources of host-range data

De Wijs (1974*a*).

**Properties of particles in sap**

LIV: 12 days. DEP: $\log_{10}$ minus 3–4. Infectivity of sap increased by treatment with di–ethyl ether. Leaf sap contains few virions.

**Purification**

Method: De Wijs (1974*a*).

**Particle morphology**

Virions filamentous; not enveloped; usually flexuous; with a clear modal length; of 810 nm; 15 nm wide. Axial canal obscure. Basic helix obscure.

**Physical properties**

One sedimenting component in purified preparations.

**Biochemical properties**

Virions contain 6 % nucleic acid; 94 % protein; 0 % lipid.
*Virion protein(s)* one; $M_r$ 33000.

**Taxonomy and Relationships**

*Potyvirus: Potyviridae.*
Virus(es) with serologically related virions: passionfruit woodiness virus.

**References**

De Wijs, J.J. (1974*a*). *Ann. appl. Biol.* **77**: 33.
De Wijs, J.J. (1974*b*). *Neth. J. Pl. Path.* **80**: 133.
De Wijs, J.J. (1975). *Neth. J. Pl. Path.* **81**: 144.
De Wijs, J.J. and Mobach, J.D. (1975). *Neth. J. Pl. Path.* **81**: 152.

# Passiflora South African *potyvirus*

*Synonyms:* possibly cowpea Morrocan aphid-borne mosaic potyvirus.
*Sequence database accession code(s):*

D10053 Em(40)_vi:SAPCPNIB Gb(84)_vi:SAPCPNIB South African passiflora virus, coat protein and nuclear inclusion B genes. 4/92 1,418bp.

S51666 Gb(84)_vi:S51666 polyprotein South African passiflora virus SAPV, Genomic RNA, 1436 nt. 3/93 1,436bp. 2 sequences.

**Taxonomy and Relationships**

*Potyvirus: Potyviridae.*

# Passionfruit (?) *nucleorhabdovirus*

**Data collated by R.D. Pares and A.B. Martin, 1986.**

First reported in *Passiflora edulis*; from Sydney, Australia; by Pares *et al.* (1983).

**Natural host range and symptoms**

*Passiflora edulis* — very severe mosaic, rugosity and leaf malformation in mixed infections with passionfruit woodiness virus.

**Transmission**

Virus transmitted by grafting.

**Ecology and control**

Studies reported by Pares and Martin (1984).

**Geographical distribution**

Spreads in Australia (New South Wales in garden plants, but not in passionfruit crops).

Diagnostically susceptible

*Passiflora edulis* — graft-transmission to seedlings.

**Properties of particles in sap**

Electron microscopy: stain with 2% AM.

**Particle morphology**

Virions rhabdo– or bullet–shaped; enveloped; with a clear modal length; of 250 nm; 68 nm wide. Axial canal obscure.

**Cytopathology**

Virions found in mesophyll; in cytoplasm, in nuclei, and in the perinuclear space.

**Taxonomy and Relationships**

Possible *Nucleorhabdovirus: Rhabdoviridae.*

It is not known if this is related to Brazilian passionfruit vein-clearing rhabdovirus.

**References**

Pares, R.D. and Martin, A.B. (1984). *N.S.W. Dep. Agric., Biology Branch Pl. Dis. Surv.* 1982–1983, p. 22.

Pares, R.D., Martin, A.B. and Morrison, W. (1983). *Australas. Pl. Path.* **12**: 51.

# Passionfruit Sri Lankan mottle (?) potyvirus

**Data collated by S.L. Bhattiprolu and A.A. Brunt, 1993.**

*Standard acronym:* SLPMV.

First reported in *Passiflora edulis* cv. *flavicarpa*; from Sri Lanka; by Dassanayake and Hicks (1992).

**Natural host range and symptoms**

Symptoms persist. Symptoms mottling.
*Passiflora edulis* cv. *flavicarpa* — leaf mottling.

**Transmission**

Transmitted by a vector; an insect; *Aphis craccivora, A. gossypii, A. spiraecola. Myzus persicae*; Aphididae. Transmitted in a non–persistent manner. Virus lost by the vector when it moults; does not multiply in the vector; not transmitted congenitally to the progeny of the vector; transmitted by mechanical inoculation; transmitted by grafting.

**Geographical distribution**

Found, but with no evidence of spread, in Sri Lanka.

**Experimental host range**

Several (3–9) families susceptible. Experimentally infected plants mostly show chlorotic/necrotic lesions, chlorotic mottle/mosaic, leaf distortion, vein yellowing.

Diagnostically susceptible

*Cassia occidentalis* — systemic chlorotic mottle, mosaic and vein yellowing.

*Chenopodium album, C. amaranticolor, C. foetidum* — chlorotic local lesions only.

*Chenopodium murale, C. quinoa* — chlorotic mottle and mosaic of inoculated leaves only.

*Gomphrena globosa. Phaseolus vulgaris, Vigna unguiculata* — necrotic local lesions only.

*Passiflora edulis* cvs *edulis* and *flavicarpa* — systemic chlorotic mottle/mosaic with distortion (vein yellowing also in *flavicarpa*).

*P. mollissima* — chlorotic mottle/mosaic and flecking, flower break.

Diagnostically insusceptible

*Datura stramonium, Glycine max, Nicotiana debneyi, N. tabacum, Passiflora suberosa, Pisum sativum.*

Maintenance and propagation hosts

*Chenopodium quinoa, Passiflora edulis* cv. *flavicarpa, P. foetida.*

Assay hosts (**L**ocal lesions or **W**hole plants)

*Chenopodium amaranticolor* (L), *Passiflora foetida* (W).

Dicotyledonous Hosts *(**Susceptible** or Insusceptible)*

*Arachis hypogaea. Beta vulgaris.* ***Cassia occidentalis. Cassia tora. Chenopodium album. Chenopodium amaranticolor. Chenopodium foetidum. Chenopodium murale. Chenopodium quinoa.*** *Cucumis sativus. Datura stramonium. Glycine max.* ***Gomphrena globosa.*** *Lycopersicon esculentum.* ***Nicotiana clevelandii.*** *Nicotiana debneyi. Nicotiana tabacum.* ***Passiflora edulis.*** *Passiflora suberosa.* ***Petunia hybrida. Phaseolus vulgaris.*** *Pisum sativum. Spinacia oleracea. Vicia faba.*

Sources of host-range data

Dassanayake and Hicks (1992).

**Properties of particles in sap**

TIP: 70–75 °C. LIV: 6–7 days. DEP: $\log_{10}$ minus 5–6. Leaf sap contains many virions.

**Purification**

Method: Dassanayake and Hicks (1992).

**Particle morphology**

Virions filamentous; not enveloped; usually flexuous; with a clear modal length; of 841 nm (+/- 4.6nm); 13.1 nm wide (+/- 0.4nm).

**Physical properties**

One sedimenting component in purified preparations. $A_{260}/A_{280}$ ratio 1.33.

1 virus specified dsRNA species found in infected cells. Size of largest virus specified dsRNA 11.2 kbp.

*Virion protein(s)* one; $M_r$ 33200; coat. Method of preparation: Dassanayake and Hicks (1992).

**Cytopathology**

Virions found in leaves and epidermis; in cytoplasm. Inclusions present in infected cells; are crystals in the cytoplasm and pinwheels; they do not contain virions.

**Taxonomy and Relationships**

Possible unassigned potyvirus.

Virus(es) with serologically related virions: passionfruit ringspot, passionfruit woodiness, potato Y and watermelon mosaic 2 viruses. Virus(es) with serologically unrelated virions: bean common mosaic.

**References**

Dassanayake, E.M. and Hicks, R.G.T. (1992). *Ann. appl. Biol.* **120**: 459.

# Passionfruit vein-clearing (?) rhabdovirus

**Data collated by A.A. Brunt, 1990.**

First reported in *Passiflora edulis*; from Brazil; by Kitajima and Crestani (1985).

**Natural host range and symptoms**

*Passiflora edulis* — vein-clearing and chlorotic vein-banding.

**Geographical distribution**

Found, but with no evidence of spread, in Brazil.

**Taxonomy and Relationships**

Possible unassigned rhabdovirus.

It is not known if this is related to Australian passionfruit nucleorhabdovirus.

**References**

Kitajima, E.W. and Crestani, O.A. (1985). *Fitopatol. Bras.* **10**: 681.

# Passionfruit woodiness *potyvirus*

**Data collated by J.L. Dale, 1980.**

*Synonyms:* passiflora chlorotic spot virus (Van Velsen, 1961), passiflora mosaic virus, probably passionfruit mosaic virus (Del Rosario *et al.*, 1964; Martini, 1962; Ong and Ting, 1973). *Standard acronym:* PWV.

First reported in *Passiflora edulis*; from Australia; by Taylor *et al.* (1964).

**Natural host range and symptoms**

Symptoms persist.

*Passiflora edulis, P. suberosa, P. subpeltata, P. aurantia, P. foetida, Arachis hypogaea, Centrosema pubescens, Glycine max, Phaseolus atropurpureus, P. vulgaris* — necrotic and chlorotic local lesions or streaks, mottling, rugosity.

**Transmission**

Transmitted by a vector; an insect; *Aphis gossypii, Myzus persicae*; Aphididae. Transmitted in a non–persistent manner. Virus transmitted by mechanical inoculation.

**Geographical distribution**

Spreads in the South and Central American region; Australia and Brazil. Found, but with no evidence of spread, in New Zealand.

**Experimental host range**

Many (>9) families susceptible.

Diagnostically susceptible

*Macrotyloma uniflorum* — large reddish-brown local lesions; systemic necrotic spots, vein necrosis in upper leaves, stem and petiole necrosis, premature death.

*Macroptilium lathyroides* — necrotic or chlorotic local lesions; systemic vein necrosis in upper leaves, leaf fall, bud blight.

*Phaseolus vulgaris* cv. Bountiful — necrotic or chlorotic local lesions, then systemic mosaic and necrosis along veins.

*Passiflora edulis, P. suberosa* — systemic vein clearing, rugosity, mottle.

Maintenance and propagation hosts

*Phaseolus vulgaris* cvs Bountiful, Climbing Blue Lake, *Nicotiana clevelandii*.

Assay hosts (**L**ocal lesions or **W**hole plants)

*Macrotyloma uniflorum* (L), *Macroptilium lathyroides* (L), *Phaseolus vulgaris* cv. Bountiful, Climbing Blue Lake (L), *Passiflora edulis* (W).

Dicotyledonous Hosts *(**Susceptible** or Insusceptible)*

***Arachis hypogaea*. *Chenopodium album*. *Chenopodium amaranticolor*. *Cucumis sativus***. *Datura stramonium*. ***Glycine max*. *Gomphrena globosa*. *Lablab purpureus*. *Lathyrus odoratus***. *Lycopersicon esculentum*. ***Macroptilium lathyroides*. *Macrotyloma uniflorum*. *Nicotiana clevelandii***. *Nicotiana glutinosa*. ***Nicotiana tabacum*. *Petunia hybrida*. *Phaseolus vulgaris*. *Pisum sativum*. *Sesbania exaltata***. *Trifolium repens*. ***Trifolium subterraneum*. *Vicia faba***. *Vigna unguiculata*.

Sources of host-range data

Teakle and Wildermuth (1967); Edwardson (1974).

**Properties of particles in sap**

TIP: 55–60 °C. LIV: 3–4 days. DEP: $\log_{10}$ minus 4–5. Leaf sap contains few virions.

**Purification**

Method: Moghal and Francki (1974).

**Particle morphology**

Virions filamentous; not enveloped; usually flexuous; with a clear modal length; of 745 nm; 12 nm wide. Axial canal obscure. Basic helix obscure.

*Virion protein(s)* one; $M_r$ 33000. Method of preparation: Dale: 1% SDS or 0.1% mercaptoethanol at 100°C for 1 minute.

**Cytopathology**

Virions found in mesophyll and vascular parenchyma; in cytoplasm. Inclusions present in infected cells; are pinwheels; they do not contain virions.

**Taxonomy and Relationships**

*Potyvirus: Potyviridae.*

Virus(es) with serologically related virions: passionflora ringspot, bean common mosaic and potato Y viruses.

**References**

Del Rosario *et al.* (1964). *Philipp. Agric.* **48**: 95. See *Rev. appl. Mycol.* **44**: 2754.
Edwardson, J.R. (1974). *Fla Agric. Exp. Stn Monog.* No. 5, p. 75.
Martini, C. (1962). *Ann. appl. Biol.* **50**: 163.
McKnight, T. (1953). *Qld. J. agric. Sci.* **10**: 4.
Moghal, S.M. and Francki, R.I.B. (1974). *Aust. J. biol. Sci.* **27**: 341.
Moghal, S.M. and Francki, R.I.B. (1976). *Virology* **73**: 350.
Ong, C.A. and Ting (1973). *MARDI Res. Bull.* **1**: 33. See *Rev. Pl. Path.* **53**: 1031.
Taylor, R.H. and Greber, R.S. (1973). *CMI/AAB Descr. Pl. Viruses* No. 122, 4 pp.
Taylor, R.H. and Kimble, K.A. (1964). *Aust. J. agric. Res.* **15**: 560.
Teakle, D.S. and Wildermuth, G.B. (1967). *Qld. J. agric. anim. Sci.* **24**: 173.
Van Velsen, R.J. (1961). *Papua New Guinea Agric. J.* **13**: 160. See *Rev. appl. Mycol.* **41**: 163.

# Passionfruit yellow mosaic *tymovirus*

**Data collated by E.W. Kitajima, 1985.**

*Standard acronym:* PFYMV.

First reported in *Passiflora edulis, f. flavicarpa*; from Papucaia, Brazil; by Kitajima *et al.* (1984).

**Natural host range and symptoms**

Symptoms persist.

*Passiflora edulis f. flavicarpa* — yellow mosaic, vein clearing.

**Transmission**

Transmitted by a vector; an insect; *Diabrotica speciosa* (experimental transmission); Coleoptera. Transmitted in a non–persistent manner. Virus transmitted by mechanical inoculation; transmitted by grafting; not transmitted by seed.

**Geographical distribution**

Spreads in Brazil.

**Experimental host range**

Few (<3) families susceptible.

Diagnostically susceptible

Several *Passiflora* spp., especially *Passiflora edulis* f. *flavicarpa*.

Maintenance and propagation hosts

*Passiflora edulis* f. *flavicarpa*.

Assay hosts (**L**ocal lesions or **W**hole plants)

*Passiflora edulis f. flavicarpa* (W). No local lesion host.

Dicotyledonous Hosts *(**Susceptible** or Insusceptible)*

***Passiflora edulis***.

Additional hosts

Limited to *Passiflora* spp.

Sources of host-range data

Crestani (1984).

**Properties of particles in sap**

TIP: 55–60 °C. LIV: 8 days. DEP: $\log_{10}$ minus 5. Infectivity of sap not changed by treatment with di–ethyl ether. Leaf sap contains many virions.

**Purification**

Method: Kitajima, adapted from Bozarth *et al.* (1977).

**Particle morphology**

Virions isometric; not enveloped; 28 nm in diameter; rounded in profile, or angular in profile; without a conspicuous capsomere arrangement.

**Physical properties**

Two sedimenting components in purified preparations; sedimentation coefficient of the fastest 125 *S*; of the other(s) 62 *S*.

**Biochemical properties**

Virions contain 25 % nucleic acid (or less).

*Genome* consists of RNA; single–stranded; linear; unipartite; largest (or only) genome part 2 kb. Genomic nucleic acid isolated by Aviv and Leder (1972).

*Virion protein(s)* one; $M_r$ 21000; coat protein. Method of preparation: Laemmli (1970).

**Replication**

Genome replicates perhaps in association with chloroplasts (in the peripheral vesicles). Coat protein *m*RNA translated in the cytoplasm, possibly. Replication does not depend on a helper virus.

**Cytopathology**

Virions found in all parts of the host plant; in cytoplasm and in cell vacuoles. Inclusions absent from infected cells. Other cellular changes: peripheral vesicles in the chloroplasts, which are characteristic for tymoviruses.

**Taxonomy and Relationships**

*Tymovirus*.

Virus(es) with serologically related virions: okra mosaic, kennedya mosaic, clitoria yellow mosaic, cocoa yellow mosaic and turnip yellow mosaic viruses.

**References**

Aviv, H. and Leder, P. (1972). *Proc. Natl. Acad. Sci. U.S.A.* **69**: 1408.

Bozarth, R.F., Lana, A.O., Koenig, R. and Reese, J. (1977). *Phytopathology* **67**: 735.
Crestani, O.A. (1984). *M.Sc. Thesis*, Univ. Brasilia.
Crestani, O.A., Kitajima, E.W., Lin, M.T. and Marinho, V.L.A. (1986). *Phytopathology* **76**: 951.
Crestani, O.A., Kitajima, E.W., Lin, M.T., Marinho, V.L.A. and Pimentel, J.P. (1984). *Fitopatol. Bras.* **9**: 394.
Kitajima, E.W., Ribeiro, R.L.D., Lin, M.T., Ribeiro, M.I.S.D., Costa C.L. and Pimentel J.P. (1984). *Fitopatol. Bras.* **9**: 607.
Laemmli, U.K. (1970). *Nature, Lond.* **227**: 680.

# Patchouli mosaic *potyvirus*

**Data collated by E.W. Kitajima, 1991.**

First reported in *Pogostemon patchouli*; from Campinas, S.P. in Brazil; by Gama (1979).

**Natural host range and symptoms**

Symptoms persist.
*Pogostemon patchouli.*

**Transmission**

Transmitted by a vector (probably); an insect (probably); *Myzus persicae*; Aphididae (probably). Transmitted in a non–persistent manner. Virus transmitted by mechanical inoculation.

**Experimental host range**

Several (3–9) families susceptible. Experimentally infected plants mostly show local lesions, then systemic chlorotic spots or mosaic.

Diagnostically susceptible

*Chenopodium amaranticolor* — local lesion with reddish halo.
*Zinnia elegans* — mosaic.
*Nicotiana tabacum* systemic chlorotic spots.
*Bidens pilosa, Gomphrena globosa* — symptomless.

Diagnostically insusceptible

*Crotalaria juncea, Capsicum annuum, Datura stramonium, Phaseolus vulgaris, Petunia hybrida.*

Maintenance and propagation hosts

*Zinnia elegans.*

Assay hosts (**L**ocal lesions or **W**hole plants)

*Chenopodium amaranticolor* (L), *Zinnia elegans, Nicotiana tabacum* (W).

Dicotyledonous Hosts *(**Susceptible** or Insusceptible)*

*Apium graveolens. Beta vulgaris. Brassica campestris* ssp. *rapa. Brassica oleracea* ssp. *capitata. Capsicum annuum.* ***Chenopodium amaranticolor. Chenopodium murale. Chenopodium quinoa.*** *Coriandrum sativum. Dahlia pinnata. Datura stramonium. Daucus carota. Galinsoga parviflora. Glycine max.* ***Gomphrena globosa. Helianthus annuus.*** *Hibiscus esculentus. Lactuca sativa. Lycopersicon esculentum. Nicandra physaloides.* ***Nicotiana debneyi. Nicotiana glutinosa. Nicotiana sylvestris. Nicotiana tabacum.*** *Petunia hybrida. Phaseolus vulgaris. Physalis floridana. Raphanus sativus.*

*Solanum melongena*. ***Spinacia oleracea***. *Vigna unguiculata* ssp. *cylindrica*. ***Zinnia elegans***.

Sources of host-range data

Gama (1979).

**Properties of particles in sap**

TIP: 55–60 °C. LIV: 16 days. DEP: $\log_{10}$ minus 3–4. Leaf sap contains few virions.

**Particle morphology**

Virions filamentous; not enveloped; usually flexuous; with a clear modal length; of 720 nm; 15 nm wide. Axial canal obscure. Basic helix obscure.

**Cytopathology**

Virions found in leaves, mesophyll and the epidermis; in cytoplasm. Inclusions present in infected cells; are pinwheels.

**Taxonomy and Relationships**

*Potyvirus: Potyviridae.*

Best tests for diagnosis: *Chenopodium amaranticolor* develop local lesions; *Nicotiana tabacum* local and systemic chlorotic spots; and *Zinnia elegans* systemic mosaic.

**References**

Gama, M.I.C.S. (1979). *M.Sc. Dissertation*, Univ. Brasilia.

Gama, M.I.C.S., Lin, M.T., Costa, C.L., Pinto, A.D. and Kitajima, E.W. (1979). *Fitopatol. Bras.* **4**: 113.

# Patchouli mottle (?) *potyvirus*

**Data collated by A.A. Brunt, 1995.**

First reported in *Pogostemon patchouli*; from Japan; by Natsuaki *et al.*, 1994.

**Natural host range and symptoms**

*Pogostemon patchouli* — symptomless or mild leaf mottling.

**Geographical distribution**

Spreads in the Eastern Asian region; India and Japan.

**Experimental host range**

Several (3–9) families susceptible.

Diagnostically susceptible

*Chenopodium amaranticolor, Gomphrena globosa* — chlorotic local lesions, not systemic.

*Chenopodium quinoa, Tetragonia expansa* — local and systemic chlorotic spotting.

*Sesamum indicum* — local and systemic mottle, systemic necrosis.

Diagnostically insusceptible

*Cucumis sativus, Cucurbita pepo, Lycopersicon esculentum, Nicotiana clevelandii, N. glutinosa, N. tabacum, Petunia hybrida, Phaseolus vulgaris* and *Zinnia elegans*.

Maintenance and propagation hosts

*Chenopodium quinoa*.

Assay hosts (**L**ocal lesions or **W**hole plants)

*Chenopodium amaranticolor* (L).

Dicotyledonous Hosts *(**Susceptible** or Insusceptible)*

*Brassica campestris* ssp. *pekinensis*. *Catharanthus roseus*. ***Chenopodium amaranticolor***. ***Chenopodium quinoa***. *Cucumis sativus*. *Cucurbita pepo*. ***Gomphrena globosa***. *Lactuca sativa*. *Lycopersicon esculentum*. *Nicotiana clevelandii*. *Nicotiana glutinosa*. *Nicotiana tabacum*. *Petunia hybrida*. ***Phaseolus vulgaris***. ***Sesamum indicum***. ***Tetragonia expansa***. *Vigna unguiculata* ssp. *sesquipedalis*. *Zinnia elegans*.

Sources of host-range data

Natsuaki *et al.* (1994).

**Properties of particles in sap**

TIP: 60–65 °C. LIV: 3 days (at 20°C). DEP: $\log_{10}$ minus 1.

**Particle morphology**

Virions filamentous; usually flexuous; with a clear modal length; of 760 nm.

**Cytopathology**

Inclusions are pinwheels; they do not contain virions.

**Taxonomy and Relationships**

Possible *Potyvirus: Potyviridae*.

Virus(es) with serologically related virions: turnip mosaic virus. Virus(es) with serologically unrelated virions: lettuce mosaic, pea seed-borne mosaic, watermelon mosaic-2 and zucchini yellow mosaic viruses.

**Comments**

From its properties Natsuaki *et al.* (1994) considered this virus to be a distinct potyvirus, although possibly similar to patchouli mosaic viruses from India (Rao and Nagar, 1986) and Brazil (Gama, 1979).

**References**

Gama, M.I.C.S. (1979). M.S.Thesis. Univ. Brasilia.

Natsuaki, K.T., Tomaru, K., Ushiku, S., Ichikawa, Y., Sugimura, Y., Natsuaki, T., Okuda, S. and Teranaka, M. (1994). *Plant Dis.* **78**: 1094.

Rao, B.L.S. and Nagar, R.S.M. (1986). *Phytopathology* **76**: 1109.

# Patchouli mottle (?) rhabdovirus

**Data collated by A.A. Brunt, 1989.**

*Standard acronym:* PatMV.

First reported in *Pogostemon patchouli*; from Brazil; by Kitajima and Costa (1989).

**Natural host range and symptoms**

*Pogostemon patchouli* — leaf mottling.

**Geographical distribution**

Spreads in Brazil.

**Taxonomy and Relationships**

Possible unassigned rhabdovirus.

**References**

Kitajima, E.W. and Costa, A.S. (1979). *Fitopatol. Bras.* **4**: 55.

# Pea early browning *tobravirus*

**Data collated by B. Gerhardson and K. Ryden, 1982. Revised 1983.**

*Synonyms:* broad bean yellow band virus (Russo *et al.*, 1984; Robinson and Harrison, 1985). *Standard acronym:* PEBV.

First reported in *Pisum sativum*; from The Netherlands; by Bos and van der Want (1962).

**Natural host range and symptoms**

Symptoms persist.

*Pisum sativum, Phaseolus vulgaris* — systemic necrotic spots, streaks and mosaic, leaves malformed, plants stunted.

**Transmission**

Transmitted by a vector; a nematode; *Trichodorus teres, T. pachydermus* in The Netherlands; *T. anenome, T. primitivus, T. viruliferus* in U.K; Trichodoridae. Virus transmitted by mechanical inoculation; transmitted by seed (37% in *Pisum sativum* cv. Rondo).

**Geographical distribution**

Spreads in Belgium, Italy, Morocco, the Netherlands, Sweden, and the UK.

**Experimental host range**

Several (3–9) families susceptible.

Diagnostically susceptible

*Gomphrena globosa* — small white local lesions, not systemic.

*Chenopodium amaranticolor* — small yellow local lesions, not systemic.

*C. quinoa* — necrotic and chlorotic local lesions, sometimes systemic causing chlorotic spots and malformation.

*Cucumis sativus* — necrotic local lesions, not systemic.

*Phaseolus vulgaris* — necrotic local lesions or rings, then systemic mosaic and leaf malformation.

*Pisum sativum* — diffuse chlorotic local lesions, then systemic latent infection or necrotic spots on stems and leaf veins.

*Nicotiana clevelandii* — diffuse local lesions, then faint systemic mosaic.

*N. glutinosa* — diffuse local lesions, then occasional systemic mosaic and chlorotic line patterns.

*N. tabacum* — latent systemic infection.

*Tetragonia expansa* — local and systemic necrotic concentric rings.

Maintenance and propagation hosts

*Phaseolus vulgaris, Pisum sativum, Nicotiana clevelandii.*

Assay hosts (**L**ocal lesions or **W**hole plants)

*Chenopodium amaranticolor, C. quinoa, Phaseolus vulgaris, Tetragonia expansa.*

Dicotyledonous Hosts (***Susceptible*** *or Insusceptible)*

*Beta vulgaris. Brassica campestris* ssp. *rapa.* ***Chenopodium amaranticolor. Chenopodium murale. Chenopodium quinoa. Cucumis sativus. Datura stramonium. Glycine max. Gomphrena globosa.*** *Lactuca sativa.* ***Lathyrus odoratus.*** *Lycopersicon* esculentum. *Medicago sativa*. ***Nicotiana clevelandii***. *Nicotiana glutinosa*. ***Nicotiana tabacum***. ***Petunia hybrida***. ***Phaseolus vulgaris***. ***Pisum sativum***. *Solanum tuberosum*. ***Spinacia oleracea***. ***Tetragonia expansa***. ***Trifolium hybridum***. *Trifolium incarnatum*. *Trifolium pratense*. *Trifolium repens*. ***Tropaeolum majus***. *Vicia faba*. ***Vigna unguiculata***.

Monocotyledonous Hosts (***Susceptible*** *or Insusceptible*)

*Allium cepa*. *Avena sativa*. *Hordeum vulgare*. *Triticum aestivum*.

Sources of host-range data

Hampton *et al.* (1978); Gerhardson *et al.* (1979).

**Properties of particles in sap**

TIP: 74–78 °C. LIV: 183 days (or more). DEP: $\log_{10}$ minus 5. Leaf sap contains few virions.

**Purification**

Method: Cooper and Mayo (1972); frozen sap.

**Particle morphology**

Virions rod–shaped; not enveloped; usually straight; with a clear modal length; of 215 nm, or 207 nm, or 105 nm; 21 nm wide. Axial canal obvious; 4 nm in diameter. Basic helix obvious; pitch of basic helix 2.5 nm.

**Physical properties**

Two sedimenting components in purified preparations; sedimentation coefficient of the fastest 286 *S*; of the other(s) 210 *S*. Density 1.31 g $cm^{-3}$ in CsCl.

**Biochemical properties**

*Genome* consists of RNA; single–stranded; linear; of two parts; largest (or only) genome part the larger 7.9 kb; the 2nd largest 4.1 kb. Nucleotide sequence references: Macfarlanbe *et al.* (1989).

*Sequence database accession code(s):*

M90705 Em(40)_vi:PEB54KA Gb(84)_vi:PEB54KA Pea early browning virus 54k gene, complete cds. 4/92 1,480bp.

X14006 Em(40)_vi:PEBVRNA1 Gb(84)_vi:PEBVRNA1 Pea early browning virus (PEBV) genomic RNA1 for 141K, 201K, 30K and 12K proteins. 9/93 7,073

X15883 Em(40)_vi:TOPEBVR2 Gb(84)_vi:TOPEBVR2 Pea early browning virus RNA 2 for coat protein and putative 29.6 kDa protein. 9/93 2,356bp.

X51828 Em(40)_vi:TOPEBRN2 Gb(84)_vi:TOPEBRN2 Pea early browning virus complete RNA2 genome encoding a coat protein and two unidentified re

X78455 Em(40)_vi:PEBVCOPR Gb(84)_vi:PEBVCOPR Pea early browning virus (isolate TPA56) RNA2 complete sequence, encoding coat protein, 9kDa 5 sequences.

*Virion protein(s)* one; $M_r$ 24000.

**Taxonomy and Relationships**

*Tobravirus*.

Broad bean yellow band tobravirus is a serotype of PEBV (Robinson and Harrison, 1985). It causes yellow vein banding, ringspotting and line patterns in *Vicia faba* in southern Italy. Its RNA-1 has a closely similar sequence to that of PEBV, although their RNA-2s differ. Pseudo-recombinants of BBYBV and PEBV are viable (B.D. Harrison, pers. comm.).

**References**

Bos, L. and van der Want, J.P.H. (1962). *Tijdschr. PltZiekt.* **68**: 368.

Cooper, J.I. and Mayo, M.A. (1972). *J. gen. Virol.* **16**: 285.
Gerhardson, B. and Ryden, K. (1979). *Phytopath. Z.* **95**: 93.
Gibbs, A.J. and Harrison, B.D. (1964). *Ann. appl. Biol.* **54**: 1.
Hampton, R., Beczner, L., Hagedorn, D., Bos, L., Inouye, T., Barnett, O., Musil, M. and Meiners, J. (1978). *Phytopathology* **68**: 989.
Harrison, B.D. (1973). *CMI/AAB Descr. Pl. Viruses* No. 120, 4 pp.
Macfarlane, S.A., Taylor, S.C., King, D.I., Hughes, G. and Davies, J.W. (1989). *Nucl. Acids Res.* **17**: 2245.
Robinson, D.J. and Harrison, B.D. (1985). *J. gen. Virol.* **66**: 2003
Russo, M., Gallitelli, D., Vovlas, C. and Savino, V. (1984). *Ann. appl. Biol.* **105**: 223.

# Pea enation mosaic *enamovirus*

**Data collated by A.J. Cockbain, 1980. Revised 1983.**

*Synonyms:* enation pea mosaic virus. *Standard acronym:* PEMV.

First reported in *Vicia faba*; from New York, U.S.A; by Osborn (1935).

### Natural host range and symptoms

Symptoms persist.

*Cicer arietinum, Lathyrus odoratus, Lens esculenta, Medicago arabica, Pisum sativum, Trifolium incarnatum, Vicia faba, Vicia sativa* — hyaline local lesions with enations, mosaic, puckering and stunting.

### Transmission

Transmitted by a vector; an insect; *Acyrthosiphon pisum, Myzus persicae*; Aphididae. Transmitted in a persistent manner. Virus retained when the vector moults; does not multiply in the vector; not transmitted congenitally to the progeny of the vector; can help the vector transmission of another virus (bean yellow vein banding umbravirus); transmitted by mechanical inoculation; probably transmitted by seed.

### Geographical distribution

Spreads in Canada, China, Iran, the UK, and the USA.

### Experimental host range

Several (3–9) families susceptible.

Diagnostically susceptible

*Chenopodium album, C. amaranticolor* — chlorotic local lesions.

*Lathyrus odoratus, Medicago arabica, Pisum sativum, Trifolium incarnatum, Vicia faba* — systemic mosaic, hyaline lesions with enations, puckering.

*Nicotiana clevelandii* — systemic mosaic and puckering.

Maintenance and propagation hosts

*Pisum sativum* and *Vicia faba*, which is less satisfactory.

Assay hosts (**L**ocal lesions or **W**hole plants)

*Chenopodium album* (L), *C. amaranticolor* (L), *C. quinoa* (L).

Dicotyledonous Hosts *(**Susceptible** or Insusceptible)*

***Chenopodium album*. *Chenopodium amaranticolor*. *Chenopodium quinoa*. *Cicer arietinum*.** *Cyamopsis tetragonoloba. Gomphrena globosa.* ***Lathyrus odoratus. Lespedeza stipulacea*.** *Lotus corniculatus.* ***Lupinus albus. Lupinus angustifolius*.** *Medicago hispida.* ***Medicago sativa. Melilotus alba. Nicotiana clevelandii*.** *Nicotiana* rustica. ***Phaseolus vulgaris***. ***Pisum sativum***. *Sesbania exaltata*. *Trifolium dubium*. ***Trifolium hybridum***. ***Trifolium incarnatum***. *Trifolium pratense*. ***Trifolium repens***. ***Trifolium subterraneum***. ***Vicia faba***. ***Vicia sativa***. ***Vicia villosa***.

Sources of host-range data

McEwen and Schroeder (1956); Hagedorn *et al.* (1964); Thottappilly (1972); Cockbain and Gibbs (1973); Hampton *et al.* (1978).

**Properties of particles in sap**

TIP: 55–65 °C. LIV: 4 days. DEP: $\log_{10}$ minus 4. Infectivity of sap not changed by treatment with di–ethyl ether. Leaf sap contains many virions. Electron microscopy: aldehyde fixation necessary for slower sedimenting virions.

**Purification**

Method: Hull and Lane (1973); Mahmood and Peters (1973).

**Particle morphology**

Virions isometric; not enveloped; 25 and 28 nm in diameter (T and B, respectively); rounded in profile, or angular in profile (some hexagonal, others pentagonal or ovoid); without a conspicuous capsomere arrangement.

**Physical properties**

Two sedimenting components in purified preparations; sedimentation coefficient 112 *S* (B); of the other(s) 99 *S* (T). Density 1.436 g $cm^{-3}$ in CsCl (for B component; top component degraded in CsCl). Density 1.36 in $D_2O$ and 1.380 in $Cs_2SO_4$ for both components. Isoelectric point pH 5–6. $A_{260}/A_{280}$ ratio 1.63 (unfractionated prparations).

**Biochemical properties**

Virions contain 28 % nucleic acid; 72 % protein; 0 % lipid.

*Genome* consists of RNA; single–stranded; linear; of two parts; largest (or only) genome part the larger, RNA-1 5.71 kb; the 2nd largest RNA-2 4.25 kb (see Taxonomy and Relationships — below). Base composition 26.6 % G; 24.1 % A; 24.5 % C; 24.8 % U. 5' terminus of RNA has a VPg. Infectivity lost when deproteinised with proteases. Poly A region absent.

*Sequence database accession code(s):*

L04573 Em(40)_vi:PEQPOLYZ Gb(84)_vi:PEQPOLYZ Pea enation mosaic virus ORF 1, complete cds; ORF 2, complete cds; ORF 3, 3' end cds; coat pr

S53233 Em(40)_vi:S53233 33 kda protein...27 kda protein (RNA 2) pea enation mosaic virus PEMV, WSG, Genomic RNA Comp

U03563 Em(40)_vi:PE03563 Gb(84)_vi:PEU03563 Pea enation mosaic virus 65kDa viral replicase gene, complete cds. 6/94 4,253bp.

U03564 Em(40)_vi:PE03564 Gb(84)_vi:PEU03564 Pea enation mosaic virus WSG satellite RNA. 5/94 717bp.

Z48507 Em(43)_vi:Pemvcp Gb(89)_vi:Pemvcp Pea enation mosaic virus RNA for coat protein. 2/95 1,482bp.

*Features of the genome:* Five ORFs (RNA-1). The coat protein and the 54kDa protein are probably translated from a subgenomic *m*RNA. RNA-2 also has 5 ORFs.

*Non–genomic nucleic acid* found in the virions; is satellite RNA (probably); of 0.72 kb.

*Virion protein(s)* three; $M_r$ of the largest, but minor, protein 54000; facilitates aphid transmission; it comprises the coat protein with a 'read-through' of a 33kDa protein encoded by the 3'terminal gene. $M_r$ of 2nd largest major 21000; coat protein. $M_r$ of 3rd largest 17500; genome-linked protein (VPg).

## Cytopathology

Virions found in leaves and phloem; in nuclei, in cytoplasm, and in cell vacuoles. Inclusions present in infected cells; are unusual in shape; aggregates in the cytoplasm and nucleus; they contain virions. Other cellular changes: phloem necrosis with dark staining spherical bodies in sieve tubes.

## Taxonomy and Relationships

*Enamovirus* (the type species).

This virus is probably a complex that has arisen by 'symbiogenesis' (Gibbs, 1995). Tests with transcripts of cDNAs encoding the two genomic components (Demler *et al.*, 1993, 1994) show that RNA-1 can replicate alone in protoplasts, but not in intact plants, whereas RNA-2 can replicate alone in both protoplasts and plants. Thus RNA-2 complements RNA-1 allowing it to be transmitted in sap, and to spread systemically, whereas RNA-1 directs virion protein synthesis and assembly into virions.

## References


Adam, G., Sander, E. and Shepherd, R.J. (1979). *Virology* **92**: 1.

Clark, R.G. and Bath, J.E. (1977). *Phytopathology* **67**: 1035.

Cockbain, A.J. and Gibbs, A.J. (1973). *Ann. appl. Biol.* **73**: 177.

Demler, S.A. and de Zoeten, G.A. (1989). *J. gen. Virol.* **70**: 1075.

Demler, S.A. and de Zoeten, G.A. (1991). *J. gen. Virol.* **72**: 1819.

Demler, S.A., Rucker, D.G. and de Zoeten, G.A. (1993). *J. gen. Virol.* **74**: 1.

Demler, S.A., Borkhrenious, O.N. Rucker, D.G. and de Zoeten, G.A. (1994). *J. gen. Virol.* **75**: 997.

De Zoeten, G.A., Powell, C.A., Gaard, G. and German, T.L. (1976). *Virology* **70**: 459.

Gabriel, C.J. and de Zoeten, G.A. (1984). *Virology* **139**: 223.

German, T.L. and de Zoeten, G.A. (1975). *Virology* **66**: 172.

German, T.L., de Zoeten, G.A. and Hall, T.C. (1978). *Intervirology* **9**: 226.

Gibbs, M.J. (1995). *D. Phil. thesis*, University of Oxford.

Gonsalves, D. and Shepherd, R.J. (1972). *Virology* **48**: 709.

Harris, K.F., Bath, J.E., Thottapilly, G. and Hooper, G.R. (1975). *Virology* **65**: 148.

Hagedorn, D.J., Layne, R.E.C. and Ruppel, E.G. (1964). *Phytopathology* **54**: 843.

Hampton, R., Beczner, L., Hagedorn, D., Bos, L., Inouye, T., Barnett, O., Musil, M. and Meiners, J. (1978). *Phytopathology* **68**: 989.

Hull, R. (1977). *J. gen. Virol.* **34**: 183.

Hull, R. and Lane, L.C. (1973). *Virology* **55**: 1.

Hull, R. (1977). In: *Aphids as Virus Vectors*, p. 137; ed. K. F. Harris and K. Maramorosch. Academic Press, New York.

Mahmood, K. and Peters, D. (1973). *Neth. J. Pl. Path.* **79**: 138.

McEwen, F.L. and Schroeder, W.T. (1956). *Pl. Dis. Reptr* **40**: 11.

Osborn, H.T. (1935). *Phytopathology* **25**: 160.

Peters, D. (1982). *CMI/AAB Descr. Pl. Viruses* No. 257, 5 pp.

Powell, C.A., de Zoeten, G.A. and Gaard, G. (1977). *Virology* **78**: 135.

Shepherd, R.J. (1970). *CMI/AAB Descr. Pl. Viruses* No. 25, 4 pp.

Thottappilly, G. (1972). *Z. PflKrankh. PflPath. PflSchutz.* **79**: 686.

# Pea mild mosaic *comovirus*

**Data collated by A.A. Brunt, 1991.**

*Standard acronym:* PMiMV.

First reported in *Pisum sativum*; from New Zealand; by Clark (1972).

**Natural host range and symptoms**

*Pisum sativum* — inconspicuous leaf chlorosis or symptomless infection.

**Transmission**

Virus transmitted by mechanical inoculation; transmitted by seed.

**Geographical distribution**

Spreads in New Zealand.

**Experimental host range**

Few (<3) families susceptible.

Diagnostically susceptible

*Phaseolus vulgaris* — small necrotic local lesions in some cultivars or large chlorotic local lesions; no systemic infection. Cultivar Pinto is immune.

*Pisum sativum* — systemic vein-clearing, vein-etching or necrosis, leaf-rolling and, sometimes, necrosis after 10–15 days; plants then recover.

*Vicia faba* — light green mosaic in systemically infected leaves.

Diagnostically insusceptible

*Medicago sativa, Nicotiana tabacum, Trifolium repens, T. pratense, Vigna unguiculata.*

Maintenance and propagation hosts

*Pisum sativum, Vicia faba.*

Assay hosts (**L**ocal lesions or **W**hole plants)

*Phaseolus vulgaris* cv. Bountiful (L), *Pisum sativum* (W).

Dicotyledonous Hosts *(**Susceptible** or Insusceptible)*

*Chenopodium quinoa.* ***Cicer arietinum***. *Cucumis sativus. Datura stramonium.* ***Glycine max***. *Gomphrena globosa.* ***Lathyrus odoratus***. *Lycopersicon esculentum.* ***Medicago hispida***. *Medicago sativa.* ***Melilotus alba***. *Nicotiana glutinosa. Nicotiana tabacum.* ***Trifolium incarnatum***. *Trifolium pratense. Trifolium repens.* ***Trifolium subterraneum***. ***Vicia faba***. *Vigna unguiculata. Zinnia elegans.*

Sources of host-range data

Clark (1972).

**Properties of particles in sap**

TIP: 68 °C. LIV: more than 21 days (at 18°C). DEP: $\log_{10}$ minus 6. Leaf sap contains few virions.

**Purification**

Method: Clark (1972).

**Particle morphology**

Virions isometric; not enveloped; 27 nm in diameter; angular in profile; with a conspicuous capsomere arrangement.

**Physical properties**

Three sedimenting components in purified preparations; sedimentation coefficient of the fastest 119 *S* (B); of the other(s) 94 *S* (M), or 55 *S* (T). $A_{260}/A_{280}$ ratio 1.79 (B), or 1.67 (M), or 0.86 (T).

**Biochemical properties**

*Genome* consists of RNA; single–stranded; of two parts; largest (or only) genome part the larger 8.5 kb; the 2nd largest 5 kb. Base composition 25 % G; 29 % A; 17 % C; 29 % U.

**Taxonomy and Relationships**

*Comovirus: Comoviridae.*

Virus(es) with serologically related virions: cowpea severe mosaic virus. Virus(es) with serologically unrelated virions: bean pod mottle, broad bean stain, cowpea mosaic and red clover mottle viruses.

**References**

Clark, M.F. (1972). *N.Z. Jl agric. Res.* **15**: 846.

# Pea mosaic *potyvirus*

**Data collated by J. Bowyer, 1981.**

*Synonyms:* common pea mosaic virus.

First reported in *Pisum sativum*; from Wisconsin, U.S.A.; by Doolittle and Jones (1925).

**Natural host range and symptoms**

Symptoms persist.

*Pisum sativum, Pisum arvense, Trifolium pratense, Vicia faba, Trifolium incarnatum, Lathyrus odoratus* — marble or yellow mosaic, stunting, delayed unfolding of leaflets.

**Transmission**

Transmitted by a vector; an insect; 15 spp. including *Myzus persicae*; Aphididae. Transmitted in a non–persistent manner. Virus transmitted by mechanical inoculation; not transmitted by seed.

**Geographical distribution**

Probably distributed worldwide. Spreads in China.

**Experimental host range**

Several (3–9) families susceptible.

Diagnostically susceptible

*Chenopodium amaranticolor* — chlorotic local lesions, then systemic mosaic, leaf malformation, apical necrosis.

*C. quinoa* — chlorotic local lesions.

*Lathyrus odoratus, Lupinus angustifolius, Pisum arvense, P. sativum, Trifolium pratense* — systemic mosaic.

*T. subterraneum* — systemic mosaic, vein banding, stunting.

*Vicia faba* — systemic marble mosaic.

Diagnostically insusceptible

Most *Phaseolus vulgaris* cvs.

Maintenance and propagation hosts

*Pisum arvense, P. sativum, Vicia faba.*

Assay hosts (**L**ocal lesions or **W**hole plants)

*Chenopodium amaranticolor, C. quinoa.*

Dicotyledonous Hosts *(**Susceptible** or Insusceptible)*

***Chenopodium amaranticolor. Chenopodium quinoa. Gomphrena globosa.*** *Lactuca sativa.* ***Lathyrus odoratus. Lupinus angustifolius. Nicotiana clevelandii. Nicotiana debneyi. Nicotiana tabacum. Phaseolus vulgaris. Pisum sativum.*** *Spinacia oleracea.* ***Tetragonia expansa. Trifolium incarnatum. Trifolium pratense.*** *Trifolium repens.* ***Trifolium subterraneum. Vicia faba. Vicia sativa.*** *Vigna unguiculata.*

Sources of host-range data

Bos (1970); Bos *et al.* (1974); Jones *et al.* (1978).

**Properties of particles in sap**

TIP: 55–60 °C. LIV: 1–2 days. DEP: $\log_{10}$ minus 3–4. Infectivity of sap not changed by treatment with di–ethyl ether. Leaf sap contains few virions.

**Purification**

Method: Moghal and Francki (1976).

**Particle morphology**

Virions filamentous; not enveloped; usually flexuous; with a clear modal length; of 785–800 nm; 13 nm wide. Axial canal obscure. Basic helix obscure; pitch of basic helix 3.4 nm.

**Physical properties**

One sedimenting component in purified preparations; sedimentation coefficient 140 *S*. Density 1.321 g $cm^{-3}$ in CsCl.

**Biochemical properties**

Virions contain 5.5–6 % nucleic acid; 94–94.5 % protein; 0 % lipid.

*Genome* consists of RNA; single–stranded; linear; unipartite.

*Sequence database accession code(s):*

S71232 Gb(84)n:S71232 3' end: polyprotein pea mosaic potyvirus PMV-I, I, Genomic RNA, 1035 nt. 9/94 1,035bp.

*Virion protein(s)* one; $M_r$ 33000.

**Replication**

Replication does not depend on a helper virus.

**Cytopathology**

Virions found in mesophyll and epidermis; in cytoplasm. Inclusions present in infected cells; are crystals in the cytoplasm, or crystals in the nucleus (associated with the nucleolus, 2–3 μ), or unusual in shape; and granular bodies, 10–30 μ; they contain virions (in the cytoplasm only). Other cellular changes: chloroplast damage.

**Taxonomy and Relationships**

*Potyvirus: Potyviridae.*

Virus(es) with serologically related virions: bean yellow mosaic virus is closely related; bean common mosaic and clover yellow vein viruses are distantly related.

**Comments**

Synonym of bean yellow mosaic potyvirus.

**References**

Bos, L. (1970). *Neth. J. Pl. Path.* **76**: 8.
Bos, L. (1970). *CMI/AAB Descr. Pl. Viruses* No. 40, 4 pp.
Bos, L., Kowalska, C. and Maat, D.Z. (1974). *Neth. J. Pl. Path.* **80**: 173.
Christie, S.R. and Edwardson, J.R. (1977). *Fla Agric. Exp. Stn Monog.* No. 9.
Doolittle, S.P. and Jones, F.R. (1925). *Phytopathology* **15**: 763.
Hagedorn, D.J. (1974). *Ann. Phytop. Soc. Monog.* No. 9.
Huttinga, H. (1975). *Neth. J. Pl. Path.* **81**: 58.
Huttinga, H. and Mosch, W.H.M. (1974). *Neth. J. Pl. Path.* **80**: 19.
Jones *et al.* (1978). *Phytopathology* **68**: 831.
Moghal, S.M. and Francki, R.I.B. (1976). *Virology* **73**: 350.
Randles, J.W., Davies, C., Gibbs, A.J. and Hatta, T. (1980). *Aust. J. biol. Sci.* **33**: 125.
Schroeder, W.T. and Provvidenti, R. (1966). *Pl. Dis. Reptr* **50**: 337.
Taylor, R.H. and Smith, P.R. (1968). *Aust. J. biol. Sci.* **21**: 429.

# Pea seed-borne mosaic *potyvirus*

**Data collated by P. Fry, 1980. Revised 1984.**

*Synonyms:* pea fizzle top virus, pea leaf rolling virus, pea leaf roll mosaic virus, pea leaf rolling mosaic virus. *Standard acronym:* PSbMV.

First reported in *Pisum sativum*; by Musil (1966).

**Natural host range and symptoms**

Symptoms disappear soon after infection.

*Pisum sativum* — transitory vein clearing, rosetting of stem and branches, leaves dark green and leaflets folded adaxially, flowers malformed and often sterile, pods small, few misshapen seeds.

**Transmission**

Transmitted by a vector; an insect; *Acyrthosiphon pisum, Aphis craccivora, A. fabae, Dactynotus escalanti, Macrosiphum crataegarius, Rhopalosiphum padi*; Aphididae. Transmitted in a non–persistent manner. Virus transmitted by mechanical inoculation; transmitted by seed (up to 100% in *Pisum sativum*, especially when testa is split, but little seed transmission in *Vicia faba, V. narbonensis* and *V. pannonica*).

**Geographical distribution**

Spreads in the Central Asian region, the Eastern Asian region, and the North American region; Australia and the UK. Found, but with no evidence of spread, in New Zealand.

**Experimental host range**

Many (>9) families susceptible.

Diagnostically susceptible

*Chenopodium amaranticolor* — necrotic local lesions.
*C. quinoa* — chlorotic local lesions.
*Pisum sativum* — systemic stunting, vein clearing, leaf rolling.
*Vicia faba* — systemic vein chlorosis.

Maintenance and propagation hosts

*Lathyrus odoratus, Pisum sativum, Vicia faba.*

Assay hosts (**L**ocal lesions or **W**hole plants)

*Chenopodium amaranticolor, C. quinoa.*

Dicotyledonous Hosts *(**Susceptible** or Insusceptible)*

*Amaranthus retroflexus. Arachis hypogaea. Atriplex hortensis.* ***Beta vulgaris*** (sugar beet, table beet, var. *cicla*, 4 other cvs (**I**)). ***Brassica campestris*** ssp. ***pekinensis***. *Brassica campestris* ssp. *rapa*. Var. *gemmifera*, var. *viridis* ***Brassica oleracea*** ssp. ***capitata***. ***Capsella bursa–pastoris***. *Capsicum annuum*. ***Catharanthus roseus***. Var. *cristata Celosia argentea*. ***Chenopodium album***. ***Chenopodium amaranticolor***. ***Chenopodium capitatum***. ***Chenopodium foetidum***. *Chenopodium hybridum*. ***Chenopodium murale***. ***Chenopodium quinoa***. *Cucumis melo. Cucumis sativus.* ***Cucurbita maxima***. *Datura stramonium. Daucus carota. Dianthus caryophyllus. Glycine max.* ***Gomphrena globosa***. *Helianthus annuus. Hyoscyamus niger. Lactuca sativa.* ***Lathyrus odoratus***. ***Lens esculenta***. *Lycopersicon esculentum.* ***Medicago sativa***. *Melilotus alba. Nicandra physaloides. Nicotiana clevelandii. Nicotiana debneyi. Nicotiana glutinosa. Nicotiana tabacum.* ***Petunia hybrida***. *Phaseolus vulgaris. Physalis floridana.* ***Pisum sativum***. *Raphanus sativus.* ***Senecio vulgaris***. *Solanum melongena. Solanum nigrum. Solanum tuberosum. Spinacia oleracea.* ***Tetragonia expansa***. *Trifolium pratense. Trifolium repens.* ***Vicia faba***. ***Vicia sativa***. ***Vicia villosa***. *Vigna unguiculata.* ***Zinnia elegans***.

Monocotyledonous Hosts *(**Susceptible** or Insusceptible)*

*Avena sativa. Hordeum vulgare. Secale cereale. Triticum aestivum. Zea mays.*

Sources of host-range data

Hampton (1975).

**Properties of particles in sap**

TIP: 55 °C. LIV: 1 days (in leaves), or 4 days (in roots). DEP: $\log_{10}$ minus 3–4. Leaf sap contains few virions. Electron microscopy: fix with 3.5% gluraraldehyde.

**Purification**

Method: Knesek *et al.* (1974).

**Particle morphology**

Virions filamentous; not enveloped; usually flexuous; with a clear modal length; of 770 nm; 12 nm wide. Axial canal obscure. Basic helix obscure; pitch of basic helix 3.4 nm.

**Physical properties**

One sedimenting component in purified preparations; sedimentation coefficient 154 *S*. Density 1.329 g $cm^{-3}$ in CsCl.

**Biochemical properties**

Virions contain 5.3 % nucleic acid; 94 % protein; 0 % lipid.

*Genome* consists of RNA; single–stranded. Base composition 22.8 % G; 44 % A; 17.6 % C; 15.6 % U.

*Sequence database accession code(s):*

D10453 Em(40)_vi:PSMCP Gb(84)_vi:PSMCP Pea seed-borne mosaic virus (PSbMV) genomic RNA for coat protein, partial sequence. 7/92 1,34

D10930 Em(40)_vi:PSMCOMP1 Gb(84)_vi:PSMCOMP Pea seed-borne mosaic virus mRNA, complete cds. 4/93 9,924bp.

U16215 Em(43)_vi:Ps16215 Gb(89)_vi:Psu16215 Pea seed-borne mosaic virus coat protein gene, partial cds. 11/94 879bp.

Z48508 Em(43)_vi:Psbmvcan Gb(89)_vi:Psbmvcan Pea seed-borne mosaic virus RNA for coat protein and polymerase (partial). 2/95 1,273bp.
Z48509 Em(43)_vi:Psbmvger Gb(89)_vi:Psbmvger Pea seed-borne mosaic virus RNA for coat protein and polymerase (partial). 2/95 1,273bp.
*Virion protein(s)* one; $M_r$ 34000.

**Cytopathology**

Virions found in roots in cortical parenchyma and in mesophyll; in cytoplasm. Inclusions present in infected cells; are pinwheels and unusual in shape; aggregates in tonoplasts; they contain virions (in the aggregates).

**Taxonomy and Relationships**

*Potyvirus: Potyviridae.*
Virus(es) with serologically related virions: bean yellow mosaic virus.

**References**

Aapola, A.A., Knesek, J.E. and Mink, G.I. (1974). *Phytopathology* **64**: 1003.
Hampton, R. and Mink, G.I. (1975). *CMI/AAB Descr. Pl. Viruses* No. 146, 4 pp.
Hampton, R.O., Knesek, J.E. and Mink, G.I. (1974). *Phytopathology* **64**: 1358.
Knesek, J.E., Mink, G.I. and Hampton, R.O. (1974). *Phytopathology* **64**: 1076.
Musil, M. (1966). *Biol. Bratisl.* **21**: 133.
Inouye, T. (1967). *Ann. Phytopath. Soc. Japan* **33**: 38.

# Pea stem necrosis virus

**Data collated by K.T. Natsuaki, 1984.**

First reported in *Pisum sativum*; from Wakayama, Japan; by Nakano *et al.* (1976).

**Natural host range and symptoms**

Symptoms persist.
*Pisum sativum* — yellowing, leaf and stem necrosis.

**Transmission**

Transmitted by a vector; a fungus; *Olpidium* spp; Plasmodiophorales. Virus transmitted by mechanical inoculation; not transmitted by contact between plants; transmitted by seed.

**Geographical distribution**

Spreads in Japan.

**Experimental host range**

Several (3–9) families susceptible.

Diagnostically susceptible

*Gomphrena globosa, Chenopodium amaranticolor, Lathyrus odoratus, Phaseolus vulgaris* — local lesions.
*Pisum sativum* — systemic yellowing, necrosis.
*Sesamum indicum* — local lesions.

Maintenance and propagation hosts

*Pisum sativum.*

Assay hosts (**L**ocal lesions or **W**hole plants)

*Chenopodium amaranticolor* (L), *Pisum sativum* (W).

Dicotyledonous Hosts *(**Susceptible** or Insusceptible)*

***Chenopodium amaranticolor*. *Gomphrena globosa*.** *Lathyrus odoratus*. ***Phaseolus vulgaris*. *Pisum sativum*.**

Sources of host-range data

Nakano *et al.* (1976).

**Properties of particles in sap**

TIP: 70–75 °C. LIV: 2 days. Leaf sap contains many virions.

**Purification**

Method: Inouye *et al.* (1977).

**Particle morphology**

Virions isometric; not enveloped; 34 nm in diameter; rounded in profile; without a conspicuous capsomere arrangement.

**Physical properties**

One sedimenting component in purified preparations; sedimentation coefficient 118 *S*.

**Biochemical properties**

Virions contain 15 % nucleic acid; 85 % protein.

*Genome* consists of RNA; single–stranded; unipartite; largest (or only) genome part 1.4 kb.

*Virion protein(s)* one; $M_r$ 35000.

**Replication**

Genome replicates in cytoplasm. Replication does not depend on a helper virus.

**Cytopathology**

Virions found in all parts of the host plant; in cytoplasm and in cell vacuoles. Inclusions absent from infected cells. Other cellular changes: vesicles in the cytoplasm.

**Taxonomy and Relationships**

Virus(es) with serologically unrelated virions: nepoviruses, tobacco necrosis, cucumber necrosis, satsuma dwarf and melon necrotic spot viruses.

**References**

Inouye *et al.* (1977). *Ann. Phytopath. Soc. Japan* **43**: 371.
Nakano *et al.* (1976). *Ann. Phytopath. Soc. Japan* **42**: 82.
Osaki and Inouye (1982). *Ann. Phytopath. Soc. Japan* **48**: 113.

# Pea streak *carlavirus*

**Data collated by R.O. Hampton, 1981. Revised 1983.**

*Synonyms:* alfalfa latent virus, Steinklee Virus, Wisconsin pea streak virus, probably pea streak New Zealand virus (Bos, 1964; Chamberlain, 1954). *Standard acronym:* PeSV.

First reported in *Pisum sativum*; from the U.S.A; by Hagedorn and Walker (1949).

**Natural host range and symptoms**

Symptoms persist, or disappear soon after infection (in the glasshouse).
*Medicago sativa* — symptomless.
*Pisum sativum* — necrotic lesions or streaks.

**Transmission**

Transmitted by a vector; an insect; *Acyrthosiphon pisum*; Aphididae. Transmitted in a non–persistent manner. Virus transmitted by mechanical inoculation; not transmitted by contact between plants; not transmitted by seed; not transmitted by pollen.

**Geographical distribution**

Spreads in the Eurasian region; Canada and the USA.

**Experimental host range**

Several (3–9) families susceptible.

Diagnostically susceptible

*Pisum sativum* — inoculated leaves symptomless, systemic necrosis, vein banding, reddening.

*Vicia faba* cv. Bell — necrotic local lesions, then systemic necrosis, necrotic spots and streaks.

*Cassia occidentalis* — necrotic or chlorotic local lesions only.

Maintenance and propagation hosts

*Pisum sativum*, *Vicia faba*.

Assay hosts (**L**ocal lesions or **W**hole plants)

*Pisum sativum, Vicia faba.*

Dicotyledonous Hosts *(**Susceptible** or Insusceptible)*

***Antirrhinum majus***. *Cajanus cajan*. *Capsicum frutescens*. ***Chenopodium amaranticolor***. *Chenopodium hybridum*. ***Chenopodium quinoa***. ***Cicer arietinum***. ***Crotalaria spectabilis***. ***Cucumis melo***. ***Cucumis sativus***. ***Cucurbita pepo***. ***Cyamopsis tetragonoloba***. ***Datura stramonium***. ***Fagopyrum esculentum***. ***Glycine max***. ***Gomphrena globosa***. ***Lathyrus odoratus***. ***Lens esculenta***. ***Lupinus albus***. ***Lupinus angustifolius***. ***Lycopersicon esculentum***. ***Medicago sativa***. ***Melilotus alba***. ***Melilotus officinalis***. ***Nicotiana tabacum***. *Petunia hybrida*. *Phaseolus lunatus*. ***Phaseolus vulgaris***. ***Pisum sativum***. *Solanum melongena*. *Solanum tuberosum*. ***Spinacia oleracea***. ***Trifolium hybridum***. ***Trifolium incarnatum***. ***Trifolium pratense***. ***Trifolium repens***. ***Trifolium subterraneum***. ***Vicia faba***. ***Vicia sativa***. ***Vicia villosa***. *Vigna angularis*. *Vigna radiata*. ***Vigna unguiculata***. *Vigna unguiculata* ssp. *sesquipedalis*. ***Zinnia elegans***.

Comments on host-range

The isolate orginally described as alfalfa latent virus induces milder symptoms and infects fewer species.

Sources of host-range data

Bos (1973); Edwardson and Christie (1986); Hampton *et al.* (1978); Thornberry (1966).

**Properties of particles in sap**

TIP: 70 °C. LIV: 7 days. DEP: $\log_{10}$ minus 6. Leaf sap contains many virions.

**Purification**

Method: Veerisetty and Brakke (1978).

**Particle morphology**

Virions filamentous; not enveloped; usually flexuous; with a clear modal length; of 630 nm; 13.5 nm wide. Axial canal obscure. Basic helix obscure.

**Physical properties**

One sedimenting component in purified preparations; sedimentation coefficient 160 *S*.

**Biochemical properties**

Virions contain 5 % nucleic acid; 95 % protein.

*Genome* consists of RNA; single–stranded; unipartite; largest (or only) genome part 7 kb. Infectivity retained when deproteinised with phenol or detergent.

*Non–genomic nucleic acid* not found in the virions. Sub–genomic *m*RNA not found in infected cells.

*Virion protein(s)* two; $M_r$ of the larger 29000. $M_r$ of 2nd largest 24000.

**Cytopathology**

Virions found in mesophyll; in cytoplasm. Inclusions present in infected cells; are unusual in shape; vacuolated, paracrystalline and banded bodies; they contain virions. Other cellular changes: cells and their organelles deteriorate and endoplasmic reticulum increases.

**Taxonomy and Relationships**

*Carlavirus*.

Virus(es) with serologically unrelated virions: red clover vein mosaic, potato M, potato S, hop mosaic, hop latent and American hop latent viruses.

**References**

Bos, L. (1964). *Neth. J. Pl. Path.* **70**: 161.
Bos, L. (1973). *CMI/AAB Descr. Pl. Viruses* No. 112, 4 pp.
Chamberlain, E.E. (1954). *N.Z. Dep. Sci. Ind. Res. Bull.* **108**: 143.
Edwardson, J.R. and Christie, R.G. (1986). *Florida Agric. Exp. Sta. Monogr. Series* **14**.
Hagedorn, D. and Walker, J.C. (1949). *Phytopathology* **39**: 837.
Hampton, R., Beczner, L., Hagedorn, D., Bos, L., Inouye, T., Barnett, O., Musil, M. and Meiners, J. (1978). *Phytopathology* **68**: 989.
Thornberry, H.H. (1966). In: *Index of Plant Virus Diseases* U.S. Dep. Agric. Hdbk No. 307, p. 272.
Veerisetty, V. and Brakke, M.K. (1977*a*). *Phytopathology* **67**: 1202.
Veerisetty, V. and Brakke, M.K. (1977*b*). *Virology* **83**: 226.
Veerisetty, V. and Brakke, M.K. (1978). *Phytopathology* **68**: 59.
Veerisetty, V. (1979). *CMI/AAB Descr. Pl. Viruses* No. 211, 3 pp.

# Peach enation (?) *nepovirus*

**Data collated by A.A. Brunt, 1991.**

First reported in *Prunus persica*; from Japan; by Kishi *et al.* (1973).

**Natural host range and symptoms**

Symptoms persist.

*Prunus persica* — small enations on abaxial leaf surfaces; shortening of internodes.

**Transmission**

Virus transmitted by mechanical inoculation; transmitted by grafting.

**Geographical distribution**

Spreads in Japan.

**Experimental host range**

Several (3–9) families susceptible.

Diagnostically susceptible

*Celosia cristata, Chenopodium quinoa* — local lesions; systemic leaf mottling.

*Chenopodium album, C. amaranticolor* — systemic leaf mottling.

*Gomphrena globosa, Nicotiana glutinosa, N. megalosiphon, N. tabacum, Physalis floridana* — symptomless systemic infection.

*Petunia hybrida* — local lesions; symptomless systemic infection.

Diagnostically insusceptible

*Brassica rapa, Cucumis sativus, Datura stramonium, Phaseolus vulgaris, Pisum sativum.*

Maintenance and propagation hosts

*Chenopodium album, C. amaranticolor, C. quinoa.*

Assay hosts (**L**ocal lesions or **W**hole plants)

*Chenopodium quinoa* (L), *Petunia hybrida* (L).

Dicotyledonous Hosts *(**Susceptible** or Insusceptible)*

*Brassica campestris* ssp. *rapa*. *Capsicum annuum*. ***Catharanthus roseus***. ***Celosia cristata***. ***Chenopodium album***. ***Chenopodium amaranticolor***. ***Chenopodium quinoa***. *Cucumis sativus*. *Cucurbita maxima*. *Datura stramonium*. ***Gomphrena globosa***. *Lycopersicon esculentum*. ***Nicotiana glutinosa***. ***Nicotiana megalosiphon***. ***Nicotiana tabacum***. ***Petunia hybrida***. *Phaseolus vulgaris*. ***Physalis floridana***. *Pisum sativum*. ***Prunus persica***. ***Raphanus sativus***. ***Sesamum indicum***. ***Tetragonia expansa***. ***Torenia fournieri***. ***Vigna unguiculata*** ssp. ***sesquipedalis***. ***Zinnia elegans***.

Sources of host-range data

Kishi *et al.* (1973).

**Properties of particles in sap**

TIP: 60 °C. LIV: 40 days. DEP: $\log_{10}$ minus 5.

**Particle morphology**

Virions isometric; not enveloped; 33 nm in diameter; angular in profile.

**Taxonomy and Relationships**

Possible *Nepovirus: Comoviridae.*

Virus(es) with serologically related virions: arabis mosaic, cherry leaf roll, raspberry ringspot, strawberry latent ringspot and tomato black ring viruses.

**References**

Kishi, K., Abiko, K. and Takanashi, K. (1973). *Ann. Phytopath. Soc. Japan* **39**: 373.

# Peach rosette mosaic *nepovirus*

**Data collated by D.C Ramsdell, 1987.**

*Standard acronym:* PRMV. *Strains:* grape isolate, peach isolate.

First reported in *Prunus persica, Vitis labrusca*; from North America; by Catson (1933).

**Natural host range and symptoms**

Symptoms persist.

*Prunus persica* — shoot terminals rosetted, mosaic, stunting.

*Vitis* spp. — short, crooked cane internodes; leaf malformation, basal stems flat.

*Rumex crispus, Solanum carolinensis, Taraxacum officinale* — no symptoms (Ramsdell and Gillet, 1981).

**Transmission**

Transmitted by a vector; a nematode; *Xiphinema americanum, Longidorus diadecturus*; Dorylamidae. Virus transmitted by mechanical inoculation; transmitted by grafting; not transmitted by contact between plants; transmitted by seed.

**Ecology and control**

Studies reported by Ramsdell and Myers (1974); Ramsdell and Myers (1978); Ramsdell *et al.* (1983).

**Geographical distribution**

Spreads in Canada and the USA (Ontario and Michigan, respectively).

**Experimental host range**

Few (<3) families susceptible.

Diagnostically susceptible

*Chenopodium quinoa* — chlorotic local lesions, tip necrosis and epinasty.

*C. amaranticolor* — faint systemic mottle.

Maintenance and propagation hosts

*Chenopodium quinoa, C. amaranticolor*.

Assay hosts (**L**ocal lesions or **W**hole plants)

*Chenopodium quinoa* (L), *C. amaranticolor* (W).

Dicotyledonous Hosts ***(Susceptible** or Insusceptible)*

***Chenopodium amaranticolor*. *Gomphrena globosa*. *Nicotiana tabacum*.**

Sources of host-range data

Dias (1975).

**Properties of particles in sap**

TIP: 58–68 °C (in *Chenopodium quinoa* sap for 10 minutes). LIV: 15–25 days (at room temperature). DEP: $\log_{10}$ minus 3–5. Leaf sap contains few virions.

**Purification**

Method: Dias (1975); Dias *et al.* (1980).

**Particle morphology**

Virions isometric; not enveloped; 28 nm in diameter; angular in profile.

**Physical properties**

Three sedimenting components in purified preparations; sedimentation coefficient of the fastest 134 *S* (B); of the other(s) 115 *S* (M), or 52 *S* (T). Density 1.51 g $cm^{-3}$ in CsCl (B), or 1.47 g $cm^{-3}$ in CsCl (M; not fixed with aldehydes).

**Biochemical properties**

Virions contain 41 % nucleic acid.

*Genome* consists of RNA; single–stranded; of two parts; largest (or only) genome part the larger 7 kb; the 2nd largest 6.2 kb. Genomic nucleic acid isolated by Allen and Dias (1976). Base composition 23.7 % G; 25.2 % A; 20.9 % C; 30.4 % U.

*Virion protein(s)* one; $M_r$ 57000.

**Replication**

Replication does not depend on a helper virus.

**Taxonomy and Relationships**

*Nepovirus: Comoviridae.*

Virus(es) with serologically unrelated virions: many other nepoviruses.

Best tests for diagnosis: sap inoculate to *Chenopodium quinoa*, from grape in 2% nicotine followed by gel diffusion or ELISA serological tests.

**References**

Catson (1933). *Q. Bull. Mich. State Univ. agric. Exp. Stn* **16**: 79.
Dias, H.F. (1975). *CMI/AAB Descr. Pl. Viruses* No. 150, 4 pp.
Ramsdell, D.C. and Myers, R.L. (1974). *Phytopathology* **64**: 1174.
Ramsdell, D.C. and Myers, R.L. (1978). *Phytopathology* **68**: 447.
Ramsdell, D.C., Andrews, R.W., Gillet, J.M. and Morris, C.E. (1979). *Pl. Dis. Reptr* **63**: 74.

# Peach yellow leaf (?) *closterovirus*

**Data collated by A.A. Brunt, 1992.**

First reported in *Prunus persica*; from Japan; by Namba *et al.* (1980).

**Natural host range and symptoms**

*Prunus persica* — conspicuous leaf chlorosis.

**Particle morphology**

Virions filamentous; usually flexuous; with a clear modal length; of 1500 nm; 11 nm wide.

**Taxonomy and Relationships**

Possible *Closterovirus.*

**References**

Namba, N., Yamashita, S., Doi, Y. and Okayoshi, K. (1980). *Ann. Phytopath. Soc. Japan* **46**: 59.

# Peanut chlorotic streak *caulimovirus*

**Data collated by D.V.R. Reddy, 1984. Revised 1989.**

*Standard acronym:* PCSV.

First reported in *Arachis hypogaea*; from Anantapoor, India; by Reddy *et al.* (1978); Ghanekar *et al.* (1978).

**Natural host range and symptoms**

Symptoms persist.

*Arachis hypogaea* — chlorotic vein streaking and stunting. Symptoms slow to appear.

**Transmission**

Transmitted by a vector; an insect; *Aphis craccivora*; Aphididae. Virus transmitted by mechanical inoculation; transmitted by grafting; transmitted by contact between plants; not transmitted by seed.

**Geographical distribution**

Spreads in India.

**Experimental host range**

Several (3–9) families susceptible.

Diagnostically susceptible

*Canavalia ensiformis* — local and systemic chlorotic spotting.
*Cyamopsis tetragonoloba* — necrotic local lesions; systemic vein-clearing.
*Glycine max, Petunia hybrida* — systemic mosaic.
*Phaseolus vulgaris* cv. Top Crop — local lesions, systemic mosaic.
*Vicia faba* — chlorotic or necrotic local lesions; not systemic.
*Datura stramonium, Nicotiana rustica* — chlorotic local lesions; systemic mosaic.
*Nicotiana glutinosa* — systemic chlorotic spots.
*Petunia hybrida* — systemic mosaic.
*Vigna unguiculata* — chlorotic local lesions with necrotic centre; not systemic.
*Spinacia oleracea* — chlorotic local lesions; not systemic.

Diagnostically insusceptible

*Chenopodium amaranticolor, C. quinoa, Brassica oleracea* var. *botrytis, Cassia obtusifolia, Cajanus cajan, Trifolium incarnatum, Raphanus sativus, Gomphrena globosa, Pisum sativum, Phaseolus vulgaris* cv. Kintoki.

Maintenance and propagation hosts

*Nicotiana benthamiana, N. edwardsonii, N. glutinosa, N. rustica, N. clevelandii, Petunia hybrida.*

Assay hosts (**L**ocal lesions or **W**hole plants)

*Datura stramonium* (L,W), *Spinacia oleracea* (L), *Vigna unguiculata* (L), *Phaseolus vulgaris* cv. Top Crop (L,W).

Dicotyledonous Hosts *(**Susceptible** or Insusceptible)*

***Arachis hypogaea***. *Brassica campestris*. *Brassica juncea*. *Brassica oleracea* ssp. *botrytis*. *Cajanus cajan*. *Chenopodium amaranticolor*. *Chenopodium quinoa*. ***Cyamopsis tetragonoloba***. ***Datura stramonium***. ***Glycine max***. *Gomphrena globosa*. ***Nicandra physaloides***. ***Nicotiana benthamiana***. ***Nicotiana clevelandii***. ***Nicotiana* × *edwardsonii***. ***Nicotiana glutinosa***. ***Nicotiana rustica***. ***Nicotiana tabacum***. ***Petunia hybrida***. ***Phaseolus vulgaris***. *Pisum sativum*. *Raphanus sativus*. ***Spinacia oleracea***. *Trifolium incarnatum*. *Vicia faba*. ***Vigna radiata***. ***Vigna unguiculata***.

**Properties of particles in sap**

TIP: 80–85 °C. LIV: 3 days (at 4°C). DEP: $\log_{10}$ minus 3–4. Leaf sap contains few virions.

**Particle morphology**

Virions isometric; not enveloped; 45–50 nm in diameter; rounded in profile; with a conspicuous capsomere arrangement.

**Physical properties**

One sedimenting component in purified preparations. Density 1.337 g $cm^{-3}$ in CsCl.

**Biochemical properties**

Virions contain 16.1 % nucleic acid; 85.9 % protein; 0 % lipid.

*Genome* consists of DNA; double–stranded; of two parts; largest (or only) genome part 7.6 kb (pairs). Genomic nucleic acid isolated by Gardner and Shepherd (1980). Infectivity retained when deproteinised with phenol or detergent.

*Sequence database accession code(s):*

U13988 Gb(84)n:PCU13988 Peanut chlorotic streak caulimovirus, complete genome. 10/94 8,174bp.

*Virion protein(s)* three; $M_r$ of the largest 75000. $M_r$ of 2nd largest 44000. $M_r$ of 3rd largest 33000.

**Replication**

Replication does not depend on a helper virus.

**Cytopathology**

Virions found in leaves and mesophyll; in cytoplasm. Inclusions present in infected cells; are unusual in shape; a matrix; they contain virions.

**Taxonomy and Relationships**

*Caulimovirus.*

Virus(es) with serologically unrelated virions: cauliflower mosaic, figwort mosaic, soybean chlorotic mottle and dahlia mosaic viruses.

**References**

Gardner, R.C. and Shepherd, R.J. (1980). *Virology* **106**: 159.

Ghanekar, A.M., Reddy, D.V.R. and Iizuka, N. (1978). *Indian Phytopath.* **31**: 118.

# Peanut clump *furovirus*

**Data collated by J.-C. Thouvenel, 1982. Revised 1984.**

*Standard acronym:* IPCV. *Strains:* Indian peanut clump virus.

First reported in *Arachis hypogaea*; from Upper Volta, West Africa; by Thouvenel *et al.* (1974).

**Natural host range and symptoms**

Symptoms persist.

*Arachis hypogaea* — stunting, mottle, mosaic, chlorotic ringspots.

*Sorghum arundinaceum, Sorghum bicolor, Setaria italica* — symptomless.

**Transmission**

Transmitted by a vector; a fungus; *Polymyxa graminis*. pH of soil affects transmission by the vector; Plasmodiophorales. Virus transmitted by mechanical inoculation; transmitted by grafting; transmitted by seed (5–11% in *Arachis hypogaea* and 10% in *Setaria*).

**Geographical distribution**

Spreads in Burkina Faso, Cote d'Ivoire, India, and Senegal.

**Experimental host range**

Several (3–9) families susceptible.

Diagnostically susceptible

*Chenopodium amaranticolor* — chlorotic local lesions, ringspots and line patterns along veins, not systemic.

*Triticum aestivum, Arachis hypogaea* — systemic mosaic.

*Phaseolus vulgaris, Vigna unguiculata, V. mungo, Torenia fournieri* — systemic ringspots and/or mosaic.

*Nicotiana benthamiana* — systemic ringspots and leaf malformation.

*N. glutinosa, Tetragonia expansa* — systemic ringspots.

Maintenance and propagation hosts

*Phaseolus vulgaris, Nicotiana benthamiana.*

Assay hosts (**L**ocal lesions or **W**hole plants)

*Chenopodium quinoa* (L), *C. amaranticolor* (L), *Torenia fournieri* (L), *Nicotiana benthamiana* (L) and *Triticum aestivum* (W).

Dicotyledonous Hosts *(**Susceptible** or Insusceptible)*

***Antirrhinum majus. Arachis hypogaea. Beta vulgaris. Capsicum frutescens. Celosia argentea. Chenopodium amaranticolor. Chenopodium quinoa. Cucumis sativus. Cucurbita pepo. Glycine max. Gomphrena globosa. Lathyrus odoratus. Lycopersicon esculentum. Melilotus alba. Nicotiana benthamiana. Nicotiana clevelandii. Nicotiana × edwardsonii. Nicotiana glutinosa. Nicotiana megalosiphon. Nicotiana tabacum. Petunia hybrida. Phaseolus vulgaris. Physalis floridana. Spinacia oleracea. Tetragonia expansa.*** *Torenia fournieri.* ***Vigna unguiculata. Vigna unguiculata*** ssp. ***sesquipedalis.***

Monocotyledonous Hosts *(**Susceptible** or Insusceptible)*

***Triticum aestivum.***

Sources of host-range data

Thouvenel *et al.* (1981).

**Properties of particles in sap**

TIP: 64 °C. LIV: 28 days. DEP: $\log_{10}$ minus 4. Infectivity of sap not changed by treatment with di–ethyl ether. Leaf sap contains few virions.

**Purification**

Method: Thouvenel *et al.* (1978).

**Particle morphology**

Virions rod–shaped; not enveloped; usually straight; with no clear modal length; of 190 and 245 nm; 21 nm wide. Axial canal obvious. Basic helix obvious; pitch of basic helix 2.6 nm.

**Physical properties**

Two sedimenting components in purified preparations; sedimentation coefficient of the fastest 224 *S*; of the other(s) 183 *S*. Density 1.32 g $cm^{-3}$ in CsCl. Isoelectric point pH 6.45.

**Biochemical properties**

Virions contain 4 % nucleic acid; 96 % protein; 0 % lipid.

*Genome* consists of RNA; single–stranded; of two parts; largest (or only) genome part the larger 6.8 kb; the 2nd largest 5.3 kb. Infectivity retained when deproteinised with phenol or detergent.

*Sequence database accession code(s):*

L07269 Em(40)_vi:PNWRNAII Gb(84)_vi:PNWRNAII Peanut clump virus RNA 2 sequence, coat protein P23 and four hypothetical proteins (P39, P51,

S62815 Em(40)_vi:S62815 3' proximal orf (3' end, RNA 1) peanut clump virus PCV, Genomic RNA, 316 nt. 1/94 316bp.

X76658 Em(40)_vi:IPCVCP Gb(84)_vi:IPCVCP Indian peanut clump virus gene for coat protein. 5/94 1,129bp.

X78602 Em(43)_vi:Pcvrna1 Gb(89)_vi:Pcvrna1 Peanut clump virus genomic RNA 1. 11/94 5,897bp.

*Virion protein(s)* one; $M_r$ 23000. Method of preparation: Wu and Bruening (1971).

### Cytopathology

Virions found in all parts of the host plant; in cytoplasm (near the nucleus and along the plasmalemma). Inclusions absent from infected cells.

### Taxonomy and Relationships

*Furovirus.*

Virus(es) with serologically related virions: beet necrotic yellow vein, potato mop top and soil-borne wheat mosaic. Virus(es) with serologically unrelated virions: Indian peanut clump virus nicotiana velutina mosaic, tobacco mosaic and barley stripe mosaic viruses.

Indian peanut clump furovirus shows no close serological relationship but has much nucleotide homology.

### References


Dollet, M., Fauquet, C. and Thouvenel, J.-C. (1976). *Pl. Dis. Reptr* **60**: 1076.

Mayo, M.A. and Reddy, D.V.R. (1985). *J. gen. Virol.* **66**: 1347.

Reddy, D.V.R., Robinson, D.J., Roberts, I.M. and Harrison, B.D. (1985). *J. gen. Virol.* **66**: 2011.

Reddy, D.V.R., Rajeshwari, R., Iizuka, N., Lesemann, D.-E., Nolt, B.L. and Goto, L. (1983). *Ann. appl. Biol.* **102**: 305.

Thouvenel, J.-C. and Fauquet, C. (1981). *CMI/AAB Descr. Pl. Viruses* No. 235, 4 pp.

Thouvenel, J.-C. and Fauquet, C. (1981). *Ann. appl. Biol.* **97**: 99.

Thouvenel, J.-C., Germani, G. and Pfeiffer, P. (1974). *C. r. hebd. Seanc. Acad. Sci. Paris* **278**: 2847.

Thouvenel, J.-C., Dollet, M. and Fauquet, C. (1976). *Ann. appl. Biol.* **84**: 311.

Thouvenel, J.-C., Fauquet, C. and Dollet, M. (1978). *Annls. Phytopath.* **10**: 93.

Wu, G.J. and Bruening, G. (1971). *Virology* **46**: 596.


## Peanut green mosaic *potyvirus*

**Data collated by N. Iizuka, 1982.**

*Synonyms:* peanut chlorotic mottle virus.

First reported in *Arachis hypogaea*; from India; by Sreenivasulu *et al.* (1981).

### Natural host range and symptoms

Symptoms persist.

*Arachis hypogaea* — mosaic, sometimes chlorotic vein banding, ringspots and/or oak leaf patterns.

### Transmission

Transmitted by a vector; an insect; *Aphis gossypii, Myzus persicae*; Aphididae. Transmitted in a non–persistent manner. Virus transmitted by mechanical inoculation; not transmitted by seed.

**Geographical distribution**

Spreads in India.

**Experimental host range**

Several (3–9) families susceptible.

Diagnostically susceptible

*Chenopodium amaranticolor, C. quinoa* — chlorotic or necrotic local lesions; not systemic.

*Arachis hypogaea, Cassia occidentalis* — severe systemic mosaic.

*Phaseolus vulgaris* cvs Bountiful, Kintoki — necrotic local lesions; systemic in cv. Kintoki.

*Nicotiana benthamiana, Petunia hybrida* — systemic mosaic.

*Tetragonia expansa* — chlorotic local lesions; not systemic.

Diagnostically insusceptible

*Glycine max, Phaseolus vulgaris* cvs Top Crop, Pinto, *Pisum sativum.*

Maintenance and propagation hosts

*Arachis hypogaea, Cassia occidentalis, Nicotiana benthamiana, N. clevelandii, Petunia hybrida.*

Assay hosts (**L**ocal lesions or **W**hole plants)

*Chenopodium quinoa* (L), *Cassia obtusifolia* (L), *Phaseolus vulgaris* cv. Indian (L).

Dicotyledonous Hosts *(**Susceptible** or Insusceptible)*

***Arachis hypogaea***. *Beta vulgaris*. *Cajanus cajan*. *Capsicum annuum*. ***Cassia occidentalis***. ***Chenopodium amaranticolor***. ***Chenopodium quinoa***. *Cucumis sativus*. ***Cyamopsis tetragonoloba***. *Datura stramonium*. *Glycine max*. *Gomphrena globosa*. *Lycopersicon esculentum*. ***Nicotiana benthamiana***. *Nicotiana clevelandii*. *Nicotiana* × *edwardsonii*. *Nicotiana glutinosa*. *Nicotiana rustica*. *Nicotiana tabacum*. *Ocimum basilicum*. ***Phaseolus vulgaris***. *Physalis floridana*. *Pisum sativum*. *Raphanus sativus*. ***Sesamum indicum***. *Solanum melongena*. ***Tetragonia expansa***. *Vicia faba*. *Vigna radiata*. *Vigna unguiculata*. *Zinnia elegans*.

Sources of host-range data

Sreenivasulu *et al.* (1981).

**Properties of particles in sap**

TIP: 55–60 °C. LIV: 3–4 days. DEP: $\log_{10}$ minus 3–4. Leaf sap contains few virions.

**Purification**

Method: Sreenivasulu *et al.* (1981).

**Particle morphology**

Virions filamentous; not enveloped; usually flexuous; with a clear modal length; of 750 nm.

**Physical properties**

One sedimenting component in purified preparations; sedimentation coefficient 171 *S*.

**Biochemical properties**

Virions contain 5 % nucleic acid; 95 % protein.

*Genome* consists of RNA; single–stranded; unipartite; largest (or only) genome part 10.8 kb.

*Virion protein(s)* one; $M_r$ 34500.

**Cytopathology**

Virions found in all parts of the host plant; in cytoplasm. Inclusions present in infected cells; are unusual in shape; cylindrical.

**Taxonomy and Relationships**

*Potyvirus: Potyviridae.*

Virus(es) with serologically unrelated virions: clover yellow vein, soybean mosaic, bean yellow mosaic, blackeye cowpea mosaic and peanut mottle viruses.

**References**

Iizuka, N. (1979). *Proc. Symp. Legumes in Tropics* Univ. Agric. Malaysia.

Sreenivasulu, P., Iizuka, N., Rajeshwari, R., Reddy, D.V.R. and Nayudu, M.V. (1981). *Ann. appl. Biol.* **98**: 255.

# Peanut mottle *potyvirus*

**Data collated by K. Bock, 1980. Revised 1987 by J.R. Edwardson. Updated 1991 by A.A. Brunt.**

*Synonyms:* groundnut mottle virus, peanut mild mosaic virus, peanut severe mosaic virus.
*Standard acronym:* PeMoV.

First reported in *Arachis hypogaea* and *Glycine max*; from Georgia, U.S.A.; Queensland, Australia; west Kenya and Uganda; by Kuhn (1965); Behncken (1970).

**Natural host range and symptoms**

Symptoms persist.

*Arachis hypogaea* — East African isolates, often symptomless or mild mottle. Strains in the USA cause mottling with necrosis and mosaic.

*Arachis pintoi* — chlorotic ringspots in leaves.

*Cassia bicapsularis* (east Africa) and *C. leptocarpa, C. occidentalis, C. tora* (U.S.A.) — mild dark green mottling.

*Glycine max, Phaseolus vulgaris, Pisum sativum* — leaf mottling and necrosis.

*Stylosanthes* sp. — mild leaf chlorosis and stunting.

**Transmission**

Transmitted by a vector; an insect; *Aphis craccivora, A. gossypii, Hyperomyzus lactucae, Myzus persicae, Rhopalosiphum padi*; Aphididae. Transmitted in a non–persistent manner (*A. craccivora* can remain infective for 2 hours and *M. persicae* for 12 hours after acquisition (Paguio and Kuhn, 1974)). Virus transmitted by mechanical inoculation; transmitted by seed (0.02–2% in *Arachis hypogaea*; to 1% in *Phaseolus vulgaris* and to 0.008% in *Vigna unguiculata* (Demski *et al.*, 1983), but not in *Glycine max, Pisum sativum, Cassia obtusifolia*).

**Ecology and control**

Studies reported by Paguio and Kuhn (1974); Kuhn and Demski (1975; 1977); Boerma and Kuhn (1976); Shipe *et al.* (1979).

**Geographical distribution**

Spreads in the African region, the Eastern Asian region, and the South and Central American region; Australia (the north east), Colombia, India (possibly), Japan, Malaysia, the Philippines, Taiwan, the USA (in the south east).

**Experimental host range**

Few (<3) families susceptible.

Diagnostically susceptible

*Arachis hypogaea* — necrotic local lesions, then systemic mottling, mosaic, necrosis.
*Cassia occidentalis* — systemic mosaic; not infected by severe strain.
*Glycine max* — systemic dark green mottling, crinkling, blistering and malformation.
*Phaseolus vulgaris* — chlorotic local lesions; only some strains systemic.
*Nicotiana clevelandii* — systemic mottling.

Maintenance and propagation hosts

*Pisum sativum, Vigna unguiculata, Nicotiana clevelandii* (*Phaseolus lunatus* strain only).

Assay hosts (**L**ocal lesions or **W**hole plants)

*Chenopodium amaranticolor* (*Cassia* and *Voandzeia* isolates only), *Phaseolus vulgaris* (L most isolates).

Dicotyledonous Hosts *(**Susceptible** or Insusceptible)*

*Amaranthus retroflexus*. Cvs Natal Common, Argentine ***Arachis hypogaea***. *Cajanus cajan*. *Capsicum annuum*. *Capsicum frutescens*. *Carica papaya*. ***Cassia bicapsularis*** (host adapted strain in Africa). ***Cassia obtusifolia***. *Cassia occidentalis*. *Cassia tora*. *Catharanthus roseus*. *Chenopodium album*. *Chenopodium amaranticolor*. *Chenopodium quinoa*. ***Citrullus lanatus***, or *Citrullus lanatus*. *Crotalaria spectabilis*. *Cucumis melo*. ***Cucumis sativus***, or *Cucumis sativus*. *Cucurbita pepo*. ***Cyamopsis tetragonoloba***. *Cyphomandra betacea*. *Datura stramonium*. *Fagopyrum esculentum*. ***Glycine max***. ***Gomphrena globosa***, or *Gomphrena globosa*. *Helianthus annuus*. ***Lathyrus odoratus***, or *Lathyrus odoratus*. *Lens esculenta*. ***Lupinus albus***. *Lupinus angustifolius*. *Lycopersicon esculentum*. ***Macroptilium lathyroides***. *Medicago sativa*. *Melilotus alba*. *Melilotus officinalis*. *Nicandra physaloides*. ***Nicotiana clevelandii***. *Nicotiana glutinosa*. *Nicotiana rustica*. *Nicotiana tabacum*. *Passiflora edulis*. *Petunia hybrida*. *Phaseolus acutifolius*. ***Phaseolus lathyroides***. *Phaseolus lunatus*. *Phaseolus vulgaris*. *Phytolacca americana*. ***Pisum sativum***. *Raphanus sativus*. ***Sesamum indicum***. ***Spinacia oleracea***. *Trifolium hybridum*. ***Trifolium incarnatum***, or *Trifolium incarnatum*. *Trifolium pratense*. *Trifolium repens*. *Trifolium subterraneum*. ***Vicia faba***, or *Vicia faba*. *Vicia villosa*. ***Vigna unguiculata***. ***Vigna unguiculata*** ssp. ***cylindrica***. ***Vigna unguiculata*** ssp. ***sesquipedalis***. ***Voandzeia subterranea***. *Zinnia elegans*.

Additional hosts

***Glycine tabacina***.

Sources of host-range data

Bock (1973); Bock *et al.* (1975, 1978); Choopanya (1983); Demski *et al.* (1983); Inouye (1969); Roechan *et al.* (1978).

**Properties of particles in sap**

TIP: 54–65 °C (but varies with isolates, and host adapted variants). LIV: 0.5–6 days, or 8 days. DEP: $\log_{10}$ minus 2–6. Infectivity of sap not changed by treatment with di–ethyl ether. Leaf sap contains few virions (probably).

**Purification**

Method: Sun *et al.* (1972); Bock *et al.* (1975, 1978).

**Particle morphology**

Virions filamentous; not enveloped; usually flexuous; with a clear modal length; of 740–750 nm (but may range between 704–984 nm). Axial canal obscure. Basic helix obscure.

**Physical properties**

One sedimenting component in purified preparations; sedimentation coefficient 151 *S* (Rajeshwari *et al.*, 1983). $A_{260}/A_{280}$ ratio 2.6 (specific absorbance at 260 nm).

**Biochemical properties**

Virions contain *c.* 6 % nucleic acid (Bock *et al.*, 1975).

*Genome* consists of RNA; single–stranded; unipartite; largest (or only) genome part 9.8 kb.

*Sequence database accession code(s):*

L32956 Em(40)_vi:PTWCOATA Gb(84)_vi:PTWCOATA Peanut mottle virus coat protein mRNA. 5/94 1,123bp.

L32957 Em(40)_vi:PTWCOATB Gb(84)_vi:PTWCOATB Peanut mottle virus coat protein mRNA. 5/94 1,122bp.

L32958 Em(40)_vi:PTWCOATC Gb(84)_vi:PTWCOATC Peanut mottle virus coat protein mRNA. 5/94 1,123bp.

L32959 Em(40)_vi:PTWCOATD Gb(84)_vi:PTWCOATD Peanut mottle virus coat protein mRNA. 5/94 1,123bp

L32960 Em(40)_vi:PTWCOATE Gb(84)_vi:PTWCOATE Peanut mottle virus coat protein mRNA. 5/94 1,123bp.

X73422 Em(40)_vi:PMPCPR Gb(84)_vi:PMPCPR Peanut Mottle Potyvirus genomic RNA for coat protein. 5/94 1,244bp. 6 sequences.

*Virion protein(s)* two; $M_r$ of the larger 54100. $M_r$ of 2nd largest 39900; coat protein subunits. Method of preparation: Barton (1978).

**Cytopathology**

Virions found in all parts of the host plant; in cytoplasm. Inclusions present in infected cells; are crystals in the nucleus (Xu, 1984), or unusual in shape; however, most isolates induce pinwheels, bundles, laminated aggregates and scrolls (Christie and Edwardson, 1977; Inouye, 1973; Herold and Munz, 1969; Choopanya, 1983); they do not contain virions.

**Taxonomy and Relationships**

*Potyvirus: Potyviridae.*

Virus(es) with serologically related virions: amaranthus leaf mottle (Rajeshwari *et al.*, 1983), clover yellow vein (Rajeshwari *et al.*, 1983), Colombian datura (Meyer, 1982), cowpea aphid-borne mosaic (Rajeshwari *et al.*, 1983), potato A (Meyer, 1982), potato Y (Meyer, 1982), soybean mosaic (Rajeshwari *et al.*, 1983) and tobacco etch viruses. Virus(es) with serologically unrelated virions: bean common mosaic (Meyer, 1982), bean yellow mosaic, celery mosaic, iris mild mosaic, sugarcane mosaic and tobacco vein mottling viruses.

**References**

Behncken, G.M. (1970). *Aust. J. agric. Res.* **21**: 465.

Bock, K.R. (1973). *Ann. appl. Biol.* **74**: 171.

Bock, K.R. and Kuhn, C.W. (1975). *CMI/AAB Descr. Pl. Viruses* No. 141, 4 pp.

Bock, K.R., Guthrie, E.J. and Meredith, G. (1978). *Ann. appl. Biol.* **89**: 423.

Boerma, H.R. and Kuhn, C.W. (1976). *Crop Sci.* **16**: 533.

Choopanya, D. (1983). *Abstr. 4th. Int. Cong. Pl. Path., Melbourne.* 448.

Demski, J.W., Alexander, A.T., Stefani, M.A. and Kuhn, C.W. (1983). *Plant Dis.* **67**: 267.
Edwardson, J.R. and Christie, R.G. (1986). *Fla Agric. Exp. Stn Monog.* No. 14, p. 431.
Herold, F. and Munz, K. (1969). *Phytopathology* **59**: 663.
Inouye, T. (1969). *Nogaku Kenkyu* **52**: 159.
Inouye, T. (1973). *Nogaku Kenkyu* **54**: 155.
Kuhn, C.W. (1965). *Phytopathology* **55**: 880.
Kuhn, C.W. and Demski, J.W. (1975). *Georgia Agric. Exp. Stn Res. Rep.* **213**: 1.
Meyer, S. (1982). *Phytopath. Z.* **105**: 271.
Morales, F.J., Castano, M., Velasco, A.C., Arroyave, J. and Zettler, F.W. (1991). *Plant Dis.* **75**: 1090.
Paguio, O.R. and Kuhn, C.W. (1974). *Proc. Am. Phytopath. Soc.* **1**: 165.
Paguio, O.R. and Kuhn, C.W. (1973). *Phytopathology* **63**: 976.
Paguio, O.R. and Kuhn, C.W. (1974). *Phytopathology* **64**: 60.
Rajeshwari, R., Iizuka, N., Nolt, B.L. and Reddy, D.V.R. (1983). *Pl. Path.* **32**: 197.
Shipe, E.R., Buss, G.R. and Tolin, S.A. (1979). *Crop Sci.* **19**: 656.
Sun, M.K.C. and Hebert, T.T. (1972). *Phytopathology* **62**: 832.
Tolin, S.A. and Ford, R.H. (1983). *Phytopathology* **73**: 899.
Xu, Z.G. (1984). *M.Sc. Thesis*, Univ. Florida, Gainesville, Florida, U.S.A.

# Peanut stunt *cucumovirus*

**Data collated by G.I. Mink, 1980. Revised by P.L. Guy, 1990.**

*Synonyms:* robinia mosaic virus (Richter *et al.*, 1979; Schmelzer, 1971; Zheng and Kung, 1984), black locust true mosaic virus, clover blotch virus (Richter *et al.*, 1979). *Standard acronym:* PSV. *Strains:* peanut common mosaic virus.

First reported in *Arachis hypogaea*; from Virginia, U.S.A; by Miller *et al.* (1966).

**Natural host range and symptoms**

Symptoms persist.

*Robinia pseudo-acacia* — mosaic and leaf malformation.

*Trifolium repens, T. pratense, T. vesiculosum. T, incarnatum, T. subterraneum, Medicago sativa* — chlorotic mottling.

*Phaseolus* spp. — chlorotic local lesions and mottling, then mosaic, elongated misshapen trifoliate leaves in cv. Bountiful.

*Arachis hypogaea* — stunting, mild mottle, pod malformation.

*Vicia* sp. — systemic chlorotic mottle.

*Glycine max, Pisum sativum, Vigna angularis, Coronilla varia, Tephrosia* sp., *Nicotiana tabacum, Apium graveolens, Lupinus luteus* — leaf chlorosis.

**Transmission**

Transmitted by a vector; an insect; *Aphis craccivora, A. spiraecola* and *Myzus persicae*, but not *Aphis gossypii*; Aphididae. Transmitted in a non–persistent manner. Virus transmitted by mechanical inoculation; transmitted by seed (0.1% in *Arachis hypogaea*).

**Geographical distribution**

Spreads in France, Japan, Korea D.P.R. (North), Korea Republic, Morocco, Poland, Spain, and the USA.

**Experimental host range**

Many (>9) families susceptible.

Diagnostically susceptible

*Chenopodium amaranticolor, C. quinoa* — chlorotic local lesions; systemic spotting.
*Spinacia oleracea, Tetragonia expansa* — systemic mottling.
*Glycine max* — systemic chlorotic mottling.
*Phaseolus vulgaris* — chlorotic or necrotic local lesions, systemic chlorotic mottling or mosaic.
*P. vulgaris* cv. Bountiful — elongated mis-shapen trifoliate leaves.
*Pisum sativum* — systemic chlorotic mottling, stunting.
*Vigna unguiculata* — chlorotic local lesions; systemic vein clearing and severe epinasty.
*Datura stramonium* — systemic chlorotic mottle with islands of green tissue.
*Nicotiana tabacum* — light green and yellow rings, then systemic chlorotic areas.
*Tetragonia expansa* — systemic mottle.

Maintenance and propagation hosts

*Phaseolus vulgaris, Pisum sativum, Vigna unguiculata, Nicotiana tabacum.*

Assay hosts (**L**ocal lesions or **W**hole plants)

*Chenopodium quinoa* (L), *Phaseolus vulgaris* (L), *Vigna unguiculata* (L, W), *Datura stramonium* (W), *Nicotiana tabacum* (W).

Dicotyledonous Hosts *(**Susceptible** or Insusceptible)*

***Amaranthus caudatus***. *Antirrhinum majus*. ***Apium graveolens***. ***Arachis hypogaea***. ***Beta vulgaris***. ***Brassica campestris*** ssp. ***rapa***. ***Capsicum annuum***. ***Capsicum frutescens***. ***Cassia occidentalis***. ***Catharanthus roseus***. ***Chenopodium amaranticolor***. *Chenopodium capitatum*. ***Chenopodium foetidum***. ***Chenopodium murale***. ***Chenopodium quinoa***. ***Cicer arietinum***. ***Citrullus lanatus***. ***Crotalaria spectabilis***. ***Cucumis melo***. ***Cucumis sativus***. ***Cucurbita maxima***. ***Cucurbita moschata***. ***Cucurbita pepo***. ***Cyamopsis tetragonoloba***. ***Datura metel***. ***Datura stramonium***. ***Dianthus barbatus***. ***Glycine max***, or *Glycine max*. ***Gomphrena globosa***. ***Helianthus annuus***, or *Helianthus annuus*. ***Lens esculenta***. ***Lespedeza stipulacea***. ***Lobelia erinus***. ***Lupinus albus***. ***Lupinus angustifolius***. ***Lycopersicon esculentum***. ***Medicago sativa***. ***Melilotus alba***. *Momordica balsamina*. ***Nicotiana clevelandii***. ***Nicotiana glutinosa***. ***Nicotiana rustica***. ***Nicotiana tabacum***. ***Ocimum basilicum***. ***Petunia hybrida***. ***Phaseolus acutifolius***. ***Phaseolus lunatus***. ***Phaseolus vulgaris***. ***Physalis floridana***. ***Physalis peruviana***. ***Pisum sativum***. ***Sesamum indicum***. ***Sesbania exaltata***. ***Solanum melongena***. *Solanum tuberosum*. ***Spinacia oleracea***. ***Tephrosia vogelii***. ***Tetragonia expansa***. ***Torenia fournieri***. ***Trifolium dubium***. ***Trifolium hybridum***. ***Trifolium incarnatum***. ***Trifolium pratense***. ***Trifolium repens***. ***Trifolium subterraneum***. ***Vicia faba***. ***Vicia sativa***. ***Vicia villosa***. ***Vigna radiata***. ***Vigna unguiculata***. ***Vigna unguiculata*** ssp. ***sesquipedalis***. ***Zinnia elegans***.

Sources of host-range data

Mink *et al.* (1969); Waterworth *et al.* (1973); Milbrath and Tolin (1977).

**Properties of particles in sap**

TIP: 50–55 °C. LIV: 1 days (and several days for P strain). DEP: $\log_{10}$ minus 3–4. Leaf sap contains many virions. Electron microscopy: aldehyde fixation necessary.

**Purification**

Method: Mink *et al.* (1969); Waterworth *et al.* (1973).

**Particle morphology**

Virions isometric; not enveloped; 30 nm in diameter; rounded in profile; with a conspicuous capsomere arrangement.

**Physical properties**

One sedimenting component in purified preparations (strain W), or four sedimenting components in purified preparations (strains B and P); sedimentation coefficient 152 *S* (strain P); of the other(s) 135 *S* (strain P), or 98 *S* (strain W), or 90 *S*, or 37 *S*, or 59 *S*, or 75 *S* (strain B). Density 1.3573–1.3645 g $cm^{-3}$ in CsCl (fixed with formaldehyde). Isoelectric point pH 6.05. $A_{260}/A_{280}$ ratio 1.6.

**Biochemical properties**

Virions contain 16 % nucleic acid; 84 % protein; 0 % lipid.

*Genome* consists of RNA; single–stranded; of three parts; largest (or only) genome part the largest 3.355 kb (RNA-1); the 2nd largest 2.946 kb (RNA-2); the 3rd largest 2.186 kb (RNA-3). Base composition 24 % G; 26 % A; 21 % C; 29 % U.

*Sequence database accession code(s):*

D00668 Em(40)_vi:PSVRNA3 Gb(84)_vi:PSVRNA3 Peanut stunt virus RNA 3 for 3a protein and coat protein. 6/92 2,186bp.

D11126 Em(40)_vi:PSVJ1A Gb(84)_vi:PSVJ1A Peanut stunt virus RNA 1 complete genome. 8/92 3,355bp.

D11127 Em(40)_vi:PSVJ2A Gb(84)_vi:PSVJ2A Peanut stunt virus RNA 2 complete genome. 5/92 2,946bp

K03110 Em(40)_vi:CUPSVRN5 Gb(84)_vi:PSVPARNA5 Peanut stunt virus-associated RNA5 (PARNA 5). 4/90 393bp.

X56544 Em(40)_vi:PSVCP Gb(84)_vi:PSVCP Peanut stunt virus coat protein mRNA. 8/91 986bp.

U15728 Em(43)_vi:Ps15728 Gb(89)_vi:Psu15728 Peanut stunt virus RNA 1 complete sequence containing 1a protein mRNA, complete cds. 5/95 3,3

U15729 Em(43)_vi:Ps15729 Gb(89)_vi:Psu15729 Peanut stunt virus RNA 2 complete sequence containing 2a protein mRNA, complete cds. 5/95 2,9

U15730 Em(43)_vi:Ps15730 Gb(89)_vi:Psu15730 Peanut stunt virus RNA 3 complete sequence containing 3a movement protein mRNA, complete cds.

*Non–genomic nucleic acid* found in the virions; is subgenomic *m*RNA, or satellite RNA; the former encodes the coat protein, is named RNA-4 and is of 0.986 kb, and the latter PARNA-5, a satellite RNA of 197 or 620 nucleotides.

Amino acid composition: Mink (1972); Kaper and Waterworth (1981).

**Cytopathology**

Virions found in all parts of the host plant; in cytoplasm and in cell vacuoles. Inclusions present in infected cells; are crystals in the cytoplasm.

**Taxonomy and Relationships**

*Cucumovirus: Bromoviridae.*

Virus(es) with serologically related virions: cucumber mosaic virus.

**References**


Douine, L. and Devergne, J.C. (1978). *Annls. Phytopath.* **10**: 79.

Echandi, E. and Hebert, T.T. (1971). *Phytopathology* **61**: 328.

Fischer, H.U. and Lockhart, B.E.L. (1978). *Phytopathology* **68**: 289.

Kaper, J.M. and Waterworth, H.E. (1981). In: *Handbook of Plant Virus Infections and Comparative Diagnosis,* p. 257; ed. E. Kurstak. Elsevier/North-Holland Biomedical Press, Amsterdam.

Karasawa, A., Nakaho, K., Kakutani, T., Minobe, Y. and Ehara, Y. (1991). *Virology* **185**: 464.
Kim, J.S., Lee, S.H. and Lee, M.W. (1988). *Korean J. Pl. Path* **4**: 88.
Milbrath, G.M. and Tolin, S.A. (1977). *Pl. Dis. Reptr* **61**: 637.
Miller, L.I. and Troutman, J.L. (1966). *Pl. Dis. Reptr* **50**: 139.
Mink, G.I. (1972). *CMI/AAB Descr. Pl. Viruses* No. 92, 4 pp.
Mink, G.I., Silbernagel, M.J. and Saksena, K.N. (1969). *Phytopathology* **59**: 1625.
Richter, J., Proll, E. and Musil, M. (1979). *Acta Virol.* **23**: 489.
Schmelzer, K. (1971). *CMI/AAB Descr. Pl. Viruses* No. 65, 3 pp.
Waterworth, H.E., Monroe, R.L. and Kahn, R.P. (1973). *Phytopathology* **63**: 93.
Xu, Z., Barnett, O.W. and Gibson, P.B. (1986). *Phytopathology* **76**: 390.
Zheng Q.-X. and Kung T.-H. (1984). *Acta Biochim. Biophys. Sin.* **16**: 540.

# Peanut top paralysis (?) potyvirus

**Data collated by A.A. Brunt, 1994.**

First reported in *Arachis hypogaea* (wild hybrid); from Oklahoma, U.S.A.; by Wagih *et al.* (1994).

**Natural host range and symptoms**

Symptoms *Arachis hypogaea* (wild hybrid) — few chlorotic rings, green blotching, smaller leaf laminae.

**Transmission**

Virus transmitted by mechanical inoculation.

**Geographical distribution**

Found, but with no evidence of spread, in Oklahoma, U.S.A.

**Experimental host range**

Few (<3) families susceptible.

Diagnostically susceptible

*Arachis hypogaea* cvs Pronto and Tamnut 74 — severe leaf chlorosis and distortion, plants stunted.

*Glycine max* — systemic leaf chlorosis (some cultivars).

*Phaseolus vulgaris* cv. Top Crop, *Chenopodium amaranticolor* — local necrotic lesions.

Diagnostically insusceptible

*Cucumis sativus, Datura stramonium, Petunia hybrida* and *Vigna unguiculata.*

Maintenance and propagation hosts

*Arachis hypogaea* cv. Tamnut 74.

Assay hosts (**L**ocal lesions or **W**hole plants)

*Phaseolus vulgaris* cv. Top Crop (W), *Chenopodium amaranticolor* (L).

Dicotyledonous Hosts *(**Susceptible** or Insusceptible)*

***Arachis hypogaea***. ***Chenopodium amaranticolor***. *Cucumis sativus*. *Datura stramonium*. ***Glycine max***. *Petunia hybrida*. ***Phaseolus vulgaris***. *Vigna unguiculata*.

**Properties of particles in sap**

TIP: 55–60 °C. LIV: 6 days (at 23°C). DEP: $\log_{10}$ minus 6.

**Purification**

Method: Wagih *et al.* (1994).

**Particle morphology**

Virions filamentous; with a clear modal length; of 750–850 nm. Axial canal obscure. *Virion protein(s)* one; $M_r$ 30000; coat protein.

**Taxonomy and Relationships**

Possible unassigned potyvirus.

Virus(es) with serologically unrelated virions: bean yellow mosaic, bean common mosaic (peanut stripe strain), potato Y, papaya ringspot, watermelon mosaic 2 and zucchini yellow mosaic potyviruses.

**References**

Wagih, E.E., Melouk, H.A. and Sherwood, J.L. (1994). *J. Phytopathology* **141**: 302.

# Peanut veinal chlorosis (?) rhabdovirus

**Data collated by A.A. Brunt, 1989.**

First reported in *Arachis hypogaea*; from India; by Anon. (1986).

**Natural host range and symptoms**

*Arachis hypogaea* — veinal chlorosis.

**Geographical distribution**

Found, but with no evidence of spread, in India.

**Taxonomy and Relationships**

Unassigned rhabdovirus.

**References**

Anon. (1986). *Rep. ICRISAT* 1985, p. 226.

Naidu, A., Manohar, S.K., Reddy, D.V.R. and Reddy, A.S. (1989). *Pl. Path.* **38**: 623.

# Peanut yellow spot *tospovirus*

**Data collated by D.V.R. Reddy, 1989.**

First reported in *Arachis hypogaea*; from India; by Anon. (1980).

**Natural host range and symptoms**

Symptoms persist.

*Arachis hypogaea* — small chlorotic local lesions, which coalesce then become necrotic.

**Transmission**

Transmitted by a vector; an insect; *Scirtothrips dorsalis*; Thysanoptera. Transmitted in a semi–persistent manner. Virus retained when the vector moults; not transmitted congenitally to the progeny of the vector; transmitted by mechanical inoculation; transmitted by grafting; not transmitted by contact between plants; not transmitted by seed.

**Geographical distribution**

Spreads in India and Thailand.

**Experimental host range**

Several (3–9) families susceptible.

Diagnostically susceptible

*Vigna unguiculata* — large chlorotic local lesions becoming necrotic.
*Petunia hybrida* — necrotic local lesions.
*Chenopodium quinoa* — chlorotic local lesions.
*Datura stramonium* — chlorotic local lesions with necrotic rings.

Diagnostically insusceptible

*Lycopersicon esculentum, Nicandra physalodes.*

Maintenance and propagation hosts

*Arachis hypogaea, Vigna mungo.*

Assay hosts (**L**ocal lesions or **W**hole plants)

*Vigna unguiculata* (L), *Petunia hybrida* (L), *Chenopodium quinoa* (L), *Datura stramonium* (L).

Dicotyledonous Hosts *(**Susceptible** or Insusceptible)*

***Chenopodium amaranticolor*. *Chenopodium quinoa*. *Datura stramonium*. *Glycine max*.** *Lycopersicon esculentum. Nicandra physaloides.* ***Nicotiana benthamiana*.** *Nicotiana rustica.* ***Petunia hybrida*. *Phaseolus vulgaris*. *Pisum sativum*. *Vigna radiata*. *Vigna unguiculata*.**

**Properties of particles in sap**

TIP: 40–45 °C. LIV: less than 1 days (*c.* 4 hours at 25–30°C). DEP: $\log_{10}$ minus 2. Infectivity of sap decreased by treatment with di–ethyl ether. Leaf sap contains few virions. Electron microscopy: fix in 1.5% glutaraldehyde.

**Purification**

Method: Reddy: from *Arachis hypogaea* tissue; grind in 20mM phosphate buffer, pH 7.5, clarify. Add NaCl (0.2 M), PEG (2.1); centrifuge 10000 G for 15 minutes; resuspend in 10mM phosphate (pH 7.5) and centrifuge in sucrose gradients.

**Particle morphology**

Virions unusually shaped; quasi-isometric; enveloped; 70–90 nm in diameter; without a conspicuous capsomere arrangement.

**Physical properties**

One sedimenting component in purified preparations.

**Biochemical properties**

*Genome* consists of RNA; single–stranded; linear; of three parts; largest (or only) genome part the largest 8.8 kb; the 2nd largest 5.4 kb; the 3rd largest 3.5 kb. Genomic nucleic acid isolated by Reddy; extract twice in phenol-cresol chloroform mixture. Infectivity lost when deproteinised with proteases; lost when deproteinised with phenol or detergent. Additional factor not required for infectivity.

*Non–genomic nucleic acid* not found in the virions.

*Virion protein(s)* four; $M_r$ of the largest 72000. $M_r$ of 2nd largest 64000. $M_r$ of 3rd largest 62000. $M_r$ of 4th largest 31000. Virion proteins glycosylated. Virion proteins phosphorylated.

**Replication**

Genome replicates in cytoplasm. Replication does not depend on a helper virus.

**Cytopathology**

Virions found in leaves and mesophyll; in cytoplasm; associated with endoplasmic reticulum. Inclusions absent from infected cells.

**Taxonomy and Relationships**

*Tospovirus: Bunyaviridae.*

Virus(es) with serologically unrelated virions: tomato spotted wilt virus.

**References**

Anon. (1980). *Rep. ICRISAT for 1979.*

Anon. (1985). *Rep. ICRISAT* 1984.

Anon. (1986). *Rep. ICRISAT* 1985, p. 226.

Anon. (1987). *Rep. ICRISAT* 1986, p. 227.

Anon. (1988). *Rep. ICRISAT* 1987, p. 235.

Reddy, D.V.R., Sudarshana, M.R., Ratna, A.S., Reddy, A.S., Kiran-Kumar, I. and Murthy, A.K. (1991). *Proc. Tomato Spotted Wilt Virus Wksh. 17–19 April 1991*, Beltsville, Maryland, U.S.A., p. 77.

Wongkaew, S. (1983). *Groundnut Improvement Project, Khon Kaen University*, Progress Report 1983.

# Pecteilis mosaic (?) potyvirus

**Data collated by A.A. Brunt.**

*Standard acronym:* PcMV.

First reported in *Pecteilis* sp.; from Japan; by Yora *et al.* (1983).

**Taxonomy and Relationships**

Possible unassigned potyvirus.

**References**

Yora, K., Saito, Y., Doi, Y., Inouye, T. and Tomaru, K. (1983). In: *Handbook of Plant Viruses*. Asakura Shoten, Tokyo.

# Pelargonium flower break *carmovirus*

**Data collated by I. Bouwen, 1991.**

*Standard acronym:* PFBV.

First reported in *Pelargonium zonale*; from the U.K.; by Stone and Hollings (1971; 1973*a* and *b*).

**Natural host range and symptoms**

Symptoms vary seasonally, or none.

*Pelargonium peltatum* — symptomless.

*P. zonale* — usually symptomless. Some cultivars show white flower streaking, chlorotic spotting of young leaves and stunting.

**Transmission**

Transmitted by means not involving a vector; vegetative propagation and mechanical transmission. Not transmitted by *Cuscuta campestris, Myzus persicae*. Virus transmitted by mechanical inoculation; transmitted by grafting; probably transmitted by contact between plants; not transmitted by seed.

**Ecology and control**

Studies reported by Hollings and Stone (1974), *Pelargonium* can be freed from pelargonium flower break virus by growing for 4 weeks at 30–35°C or 4–8 weeks at 34–37°C (16 hours) and 20–25°C (8 hours) then rooting small cuttings from the stem tips, but few plants survive.

**Geographical distribution**

Spreads in Denmark, Germany, the Netherlands, and the UK. Found, but with no evidence of spread, in the U.S.A.

**Experimental host range**

Several (3–9) families susceptible. Experimentally infected plants mostly show no symptoms, but some *Pelargonium zonale* cultivars show flower breaking.

Diagnostically susceptible

*Chenopodium quinoa* — numerous chlorotic and necrotic local lesions; occasional systemic infection, with leaf mottling.

*Gomphrena globosa* — few, brown necrotic local lesions (or symptomless if inoculum is dilute), not systemic.

*Nicotiana clevelandii* — diffuse chlorotic, occasionally necrotic, local lesions, not systemic.

*Nicotiana megalosiphon* — local and systemic symptomless infection.

*Tetragonia expansa* — chlorotic local lesions not systemic.

Diagnostically insusceptible

*Cucumis sativus, Nicotiana glutinosa, N. tabacum, Phaseolus vulgaris, Pisum sativum.*

Maintenance and propagation hosts

*Nicotiana clevelandii, N. megalosiphon.*

Assay hosts (**L**ocal lesions or **W**hole plants)

*Chenopodium quinoa* (L).

Dicotyledonous Hosts *(**Susceptible** or Insusceptible)*

*Amaranthus caudatus. Antirrhinum majus. Apium graveolens. Arachis hypogaea.* ***Atriplex hortensis****. Bellis perennis. Beta vulgaris. Brassica campestris* ssp. *pekinensis. Brassica juncea. Brassica oleracea* ssp. *capitata. Calendula officinalis. Capsicum frutescens.* ***Celosia argentea****. Cheiranthus cheiri.* ***Chenopodium amaranticolor****.* ***Chenopodium capitatum****.* ***Chenopodium murale****.* ***Chenopodium quinoa****. Chrysanthemum morifolium. Cucumis sativus. Cucurbita maxima. Datura stramonium. Dianthus barbatus. Dianthus caryophyllus. Emilia sagittata. Fagopyrum esculentum. Glycine max.* ***Gomphrena globosa****. Helianthus annuus. Hibiscus esculentus. Lactuca sativa. Lathyrus odoratus. Lycopersicon esculentum. Montia perfoliata. Myosotis sylvatica. Nicandra physaloides.* ***Nicotiana clevelandii****. Nicotiana debneyi. Nicotiana glutinosa.* ***Nicotiana megalosiphon****. Nicotiana rustica. Nicotiana tabacum.* ***Ocimum basilicum****. Petunia hybrida. Phaseolus vulgaris. Phlox drummondii. Physalis floridana. Phytolacca americana. Pisum sativum. Primula malacoides.* ***Saponaria vaccaria****. Senecio vulgaris.* ***Silene armeria****. Solanum melongena. Solanum nigrum. Solanum tuberosum.* ***Spinacia oleracea****. Stellaria media.* ***Tetragonia expansa****. Trifolium incarnatum. Tropaeolum majus. Verbesina encelioides. Vicia faba. Vigna unguiculata. Zinnia elegans*

Monocotyledonous Hosts *(**Susceptible** or Insusceptible)*

*Avena sativa. Zea mays.*

Sources of host-range data

Hollings and Stone (1974); Stone and Hollings (1973*a*).

**Properties of particles in sap**

TIP: 85–90 °C (in sap of *Chenopodium quinoa* and *Nicotiana clevelandii*), or 60 °C (in *Tetragonia expansa* sap). LIV: 21 days (at 20°C), or 189 days (at 2°C in sap of *Chenopodium quinoa* and *Nicotiana clevelandii*), or 32 days (at 20°C in sap of *Tetragonia expansa*). DEP: $\log_{10}$ minus 4 (in sap of *Pelargonium*), or 6 (in sap of *C. quinoa, N. clevelandii* and *Tetragonia expansa*). Leaf sap contains many virions.

**Particle morphology**

Virions isometric; 30 nm in diameter.

**Physical properties**

One sedimenting component in purified preparations; sedimentation coefficient 125 *S*. $A_{260}/A_{280}$ ratio 1.64.

**Biochemical properties**

*Genome* consists of RNA; single–stranded; unipartite; largest (or only) genome part 4 kb.

*Sequence database accession code(s):*

L34289 Gb(84)_vi:CVIRDRP Pelargonium flower-break carmovirus (clone pPFB26) RNA polymerase (RdRp) mRNA fragment. 7/94.

Z28395 Em(43)_vi:Pfbvrdrp Gb(89)_vi:Pfbvrdrp Pelargonium flower break virus RNA (genomic) for RNA-dependent RNA polymerase. 1/95 408bp.

Sub–genomic *m*RNA found in infected cells; 1.5 and 1.7 kb (Morris and Carrington, 1988). 3 virus specified dsRNA species found in infected cells. Size of largest virus specified dsRNA 4 kbp; 2nd largest 1.5 kbp; 3rd largest 1.7 kbp.

*Virion protein(s)* one (plus a fragment); $M_r$ 41200 (plus 35000 — Morris and Carrington, 1988; Stone and Hollings, 1973*a*; Hollings and Stone, 1974).

**Cytopathology**

Virions found in all parts of the host plant. Inclusions absent from infected cells.

**Taxonomy and Relationships**

*Carmovirus: Tombusviridae.*

Virus(es) with serologically unrelated virions: bean mild mosaic, blackgram mottle, carnation mottle, cowpea mottle, cucumber leaf spot, cucumber necrosis, cucumber soil-borne, elderberry latent, galinsoga mosaic, glycine mottle, hibiscus chlorotic ringspot, melon necrotic spot, narcissus tip necrosis, pelargonium line pattern, plantain 6, saguaro cactus, tephrosia symptomless and turnip crinkle viruses.

No serological relationship found to any of 42 other isometric viruses.

**References**


Anon. (1988). In: *Tips on Growing Zonal Geraniums*, p 43; ed. H.K. Tayama, OCES, Ohio State University, Bulletin FP–765 Agdex 281/15.

Hollings, M. and Stone, O.M. (1974). *CMI/AAB Descr. Pl. Viruses* No. 130, 4 pp.

Morris, T.J. and Carrington, J.C. (1988). In: *The Plant Viruses*, Vol. 3, Polyhedral Virions with Monopartite RNA Genomes, p. 73; ed. R. Koenig. Plenum Press, New York.

Paludan, N. (1976). *Acta Hort.* **59**: 119.

Paludan, N. and Begtrup, J. (1987*a*). *Tidsskr. PlAvl.* **91**: 183.

Paludan, N. and Begtrup, J. (1987*b*). *Gron Viden Havebrug* nr. 8, 4 pp.

Stone, O.M. (1974). *Acta Hort.* **36**: 113.

Stone, O.M. (1980). *Acta Hort.* **110**: 177.
Stone, O.M. and Hollings, M. (1971). *Rep. Glasshouse Crops Res. Inst.* 1970, p. 153.
Stone, O.M. and Hollings, M. (1973*a*). *Ann. appl. Biol.* **75:** 15.
Stone, O.M. and Hollings, M. (1973*b*). *Rep. Glasshouse Crops Res. Inst.* 1972, p. 104.
Wiedemann, W. and Oertel, C. (1988). *NachrBl. PflSchutz.* **12**: 42.

# Pelargonium leaf curl *tombusvirus*

**Data collated by A.A. Brunt, 1995.**

*Synonyms:* geranium crinkle virus (Pethybridge and Smith, 1952), tomato bushy stunt virus — pelargonium strain (Martelli *et al.*, 1971).

First reported in *Pelargonium zonale*; from Germany; by Pape (1927).

**Natural host range and symptoms**

*Pelargonium zonale* — pale chlorotic spots, which later become stellate and necrotic, leaf crinkling and plant degeneration.

**Transmission**

Virus transmitted by mechanical inoculation; transmitted by grafting; not transmitted by seed.

**Ecology and control**

Studies reported by Hollings (1962); Hollings and Stone (1965).

**Geographical distribution**

Spreads in the Eurasian region, the Mediterranean region, and the North American region.

Diagnostically susceptible

*Chenopodium amaranticolor, Datura stramonium, Phaseolus vulgaris* — chlorotic and/or necrotic local lesions; systemic chlorosis and necrosis.

*Nicotiana clevelandii* — local chlorotic lesions and rings; severe systemic chlorotic and semi-necrotic spots and rings.

Maintenance and propagation hosts

*Datura stramonium, Nicotiana clevelandii.*

Assay hosts (**L**ocal lesions or **W**hole plants)

*Chenopodium amaranticolor* (L), *Datura stramonium* (L), *Phaseolus vulgaris* (L).

**Properties of particles in sap**

TIP: 90 °C. LIV: 21 days. DEP: $\log_{10}$ minus 4.

**Purification**

Method: Gallitelli *et al.* (1985).

**Particle morphology**

Virions isometric; not enveloped; 28 nm in diameter.

**Physical properties**

One sedimenting component in purified preparations; sedimentation coefficient 136 *S*. Density 1.34 g $cm^{-3}$ in CsCl.

**Biochemical properties**

Virions contain 17 % nucleic acid; 83 % protein; 0 % lipid.

*Genome* consists of RNA; single–stranded; unipartite; largest (or only) genome part 4.7 kb.

*Virion protein(s)* one; $M_r$ 41000; coat protein.

**Cytopathology**

Virions found in cytoplasm, in nuclei, in nucleoli, in the perinuclear space, in chloroplasts, and in cell vacuoles. Inclusions present in infected cells; are crystals in the cytoplasm; they do not contain virions. Other cellular changes: nucleoli enlarged, chromatin depleted.

**Taxonomy and Relationships**

*Tombusvirus: Tombusviridae.*

Virus(es) with serologically related virions: carnation Italian ringspot and tomato bushy stunt viruses.

The genome of the virus has 24% sequence identity with that of tomato bushy stunt virus.

**References**


Gallitelli, D., Hull, R. and Koenig, R. (1985). *J. gen. Virol.* **66**: 1523.

Hollings, M. (1962). *Ann. appl. Biol.* **50**: 189.

Hollings, M. and Stone, O.M. (1965). *Ann. appl. Biol.* **56**: 87.

Koenig, R. and Gibbs, A.J. (1986). *J. gen. Virol.* **67**: 75.

Martelli, G.P. and Castellano, M.A. (1969). *Virology* **39**: 610.

Martelli, G.P. and Russo, M. (1972). *J. gen. Virol.* **15**: 193.

Martelli, G.P., Quacquarelli, A. and Russo, M. (1971). *CMI/AAB Descr. Pl. Viruses* No. **69**, 4 pp.

Martelli, G.P., Gallitelli, D. and Russo, M. (1988). In *The Plant Viruses; 3, Polyhedral Virions with Monopartite Genomes*, pp. 13–72. Plenum Press, New York.

Pape, H. (1927). *Gartenwelt* **31**: 329.

Pethybridge, G.H. and Smith, K.M. (1932). *Gard. Chron.* **92**: 378.

Wetter, C. and Luisoni, E. (1969). *Phytopath. Z.* **65**: 231.


# Pelargonium line pattern (?) *carmovirus*

**Data collated by I. Bouwen and D.Z. Maat, 1991.**

*Synonyms:* pelargonium ring pattern virus. *Standard acronym:* PLPV.

First reported in *Pelargonium zonale*; from Great Britain; by Stone and Hollings (1971 and 1976).

**Natural host range and symptoms**

Symptoms vary seasonally, or none.

*Pelargonium zonale* — mainly symptomless but, during winter, may have yellow/green spots and line patterns.

*Pelargonium peltatum* — no symptoms.

**Transmission**

Not transmitted by *Myzus persicae*. Virus transmitted by mechanical inoculation; transmitted by grafting; transmitted by contact between plants; not transmitted by seed.

**Geographical distribution**

Spreads in Belgium, Denmark, France, the Netherlands, the UK, and the former Yugoslavia. Found, but with no evidence of spread, in the U.S.A. and Germany.

**Experimental host range**

Several (3–9) families susceptible. Experimentally infected plants mostly show symptoms 18 months after inoculation and usually during winter, in *Pelargonium zonale*, as veinal flecks and line pattern.

Diagnostically susceptible

*Chenopodium amaranticolor* — small chlorotic local lesions with whitish necrotic centres, not systemic.

*Chenopodium quinoa* — chlorotic spots, most becoming necrotic, occasional systemic chlorotic spots.

*Nicotiana clevelandii* — faint diffuse spots, occasionally symptomless, systemic mild chlorotic mottle, vein clearing and mild rugosity or symptomless.

*Pisum sativum* — grey-brown necrotic spots, not systemic.

*Vicia faba* — dark green local lesions with necrotic centre, chlorosis, usually symptomless systemic infection.

Diagnostically insusceptible

*Amaranthus caudatus, Chrysanthemum* sp. *Cucumis sativus, Dianthus caryophyllus.*

Maintenance and propagation hosts

*Nicotiana clevelandii.*

Assay hosts (**L**ocal lesions or **W**hole plants)

*Chenopodium quinoa* (L), *Nicotiana clevelandii* (W), *Vicia faba* (L).

Dicotyledonous Hosts *(**Susceptible** or Insusceptible)*

*Amaranthus caudatus*. ***Atriplex hortensis***. ***Beta vulgaris***. ***Celosia argentea***. ***Chenopodium album***. ***Chenopodium amaranticolor***. ***Chenopodium capitatum***. ***Chenopodium murale***. ***Chenopodium quinoa***. *Chrysanthemum morifolium*. *Cucumis sativus*. ***Datura stramonium***. ***Dianthus barbatus***. *Dianthus caryophyllus*. ***Glycine max***. ***Gomphrena globosa***, or *Gomphrena globosa*. ***Lathyrus odoratus***. ***Lycopersicon esculentum***. ***Nicotiana bigelovii***. ***Nicotiana clevelandii***. ***Nicotiana debneyi***. ***Nicotiana glutinosa***. ***Nicotiana megalosiphon***. ***Nicotiana rustica***. ***Nicotiana tabacum***. ***Ocimum basilicum***. ***Petunia hybrida***. ***Phaseolus vulgaris***. ***Pisum sativum***. ***Silene armeria***. ***Solanum melongena***. ***Solanum nigrum***. ***Solanum tuberosum***. ***Spinacia oleracea***. ***Tetragonia expansa***. ***Torenia fournieri***. ***Trifolium incarnatum***. ***Vicia faba***. ***Vigna unguiculata***.

Sources of host-range data

Plese and Stefanac (1980); Stone (1974); Stone and Hollings (1971 and 1977); Welvaert *et al.* (1982).

**Properties of particles in sap**

TIP: 85 °C. LIV: 18–23 days. DEP: $\log_{10}$ minus 5–6. Leaf sap contains many virions.

**Purification**

Method: Maat: harvest leaves of *Nicotiana clevelandii* 10 days after inoculation. Homogenize leaves at 4°C with 2 ml/g of 20mM sodium phosphate, (pH 6), 20mM sodium thioglycolate and 20mM sodium diethyldithiocarbamate, then with 0.16 vol of 50:50 diethyl ether: carbon tetrachloride. Homogenate is clarified, then differentially centrifuged (85,600 g for 2 hours and 8–10000g for 10 minutes) in 20mM sodium phosphate buffer, and 2 cycles of sucrose-gradient centrifugation.

**Particle morphology**

Virions isometric; *c.* 30 nm in diameter.

**Physical properties**

One sedimenting component in purified preparations; sedimentation coefficient 115 *S*.

**Biochemical properties**

*Genome* consists of RNA; single–stranded.

*Sequence database accession code(s):*

S58174 Em(40)_vi:S58174 Gb(84)_vi:S58174 putative RNA polymerase, coat protein pelargonium leaf curl virus PLCV, Genomic RNA, 1787 nt 1 sequence.

*Virion protein(s)* one; $M_r$ 41400; capsid polypeptide.

**Cytopathology**

Virions found in leaves, mesophyll and the epidermis. In *Pelargonium peltatum* the virus has been found in the leaves, petioles, stem, nodes and complete flowers; in cytoplasm and in cell vacuoles. Inclusions present in infected cells; are unusual in shape; in *Chenopodium quinoa* and *Vicia faba* the inclusions are described as delicate granular cytoplasmic bodies of characteristic star-like or amoeboid appearance; they contain virions. Other cellular changes: are described by Stefanac *et al.* (1981 and 1982).

**Taxonomy and Relationships**

Possible *Carmovirus: Tombusviridae.*

Virus(es) with serologically unrelated virions: carnation mottle, bean mild mosaic, blackgram mottle, cowpea mottle, cucumber leaf spot, cucumber necrosis, cucumber soil-borne, elderberry latent, galinsoga mosaic, glycine mottle, hibiscus chlorotic ringspot, melon necrotic spot, narcissus tip necrosis, pelargonium flower break, plantain 6, saguaro cactus, tephrosia symptomless and turnip crinkle viruses.

Best tests for diagnosis: pelargonium line pattern virus is often associated with pelargonium flower break virus. It can be distinguished by ELISA or by mechanical inoculation on *Vicia faba*. Pelargonium line pattern virus produces brown local lesions on inoculated leaves and pelargonium flower break does not infect.

**References**


Anon. (1988). In: *Tips on Growing Zonal Geraniums*, pp. 43–45; ed. H.K. Tayama, OCES, Ohio State University, Bulletin FP–765 Agdex 281/15.

Hoof, H.A., van Maat, D.Z. and Vink, J. (1973). *Vakblad voor de Bloemisterij* **28**: 9.

Morris, T.J. and Carrington, J.C. (1988). In: *The Plant Viruses*, Vol. 3, Polyhedral Virions with Monopartite RNA Genomes, p. 73; ed. R. Koenig. Plenum Press, New York.

Paludan, N. and Begtrup, J. (1987). *Tidsskr. PlAvl.* **91**: 183.

Plese, N. and Stefanac, Z. (1980). *Acta Hort.* **110**: 183.

Stefanac, Z., Plese, N. and Wrischer, M. (1981). *Abstr. 5th Int. Congr. Virol., Strasbourg*, p. 239.

Stefanac, Z., Plese, N. and Wrischer, M. (1982). *Phytopath. Z.* **105**: 288.

Stone, O.M. (1974). *Acta Hort.* **36**: 113.

Stone, O.M. (1980). *Acta Hort.* **110**: 177.

Stone, O.M. and Hollings, M. (1971). *Rep. Glasshouse Crops Res. Inst.* 1970, p. 153.

Stone, O.M. and Hollings, M. (1976). *Rep. Glasshouse Crops Res. Inst.* 1975, p. 119.
Stone, O.M. and Hollings, M. (1977). *Rep. Glasshouse Crops Res. Inst.* 1976, p. 121.
Welvaert, W. and Samyn, G. (1985). *Acta Hort* **164**: 341.
Welvaert, W. and Samyn, G. (1988). *Acta Hort.* **234**: 165.
Welvaert, W., Samyn, G. and De Simpelaere, P. (1982). *Med. Fac. Landb. Rijksuniv. Gent.* **47**: 1033.

## Pelargonium vein clearing (?) *cytorhabdovirus*

**Data collated by A.A. Brunt, 1991.**

*Standard acronym:* PelVCV.

First reported in *Pelargonium peltatum*; from southern Italy; by Russo *et al.* (1979).

**Natural host range and symptoms**

*Pelargonium peltatum, P. zonale* — mild vein clearing.

**Transmission**

Virus transmitted by mechanical inoculation.

**Geographical distribution**

Spreads in Italy.

**Experimental host range**

Few (<3) families susceptible.

Diagnostically susceptible

*Nicotiana benthamiana, N. clevelandii, N. glutinosa, N. rustica, N. tabacum, N. debneyi, N. megalosiphon, Capsicum annuum, Petunia hybrida* — chlorotic or necrotic local lesions in older plants, but systemic vein clearing, chlorotic vein banding and leaf malformation of seedlings inoculated 4 leaf-stage.

*Gomphrena globosa* — small red local lesions; not systemic.

Diagnostically insusceptible

*Chenopodium amaranticolor, C. quinoa, Cucumis sativus, Phaseolus vulgaris, Beta vulgaris.*

Maintenance and propagation hosts

*Nicotiana benthamiana, N. glutinosa.*

Assay hosts (**L**ocal lesions or **W**hole plants)

*Nicotiana benthamiana* (W), *N. glutinosa* (W).

**Properties of particles in sap**

TIP: 60 °C. LIV: 2 days (at 22–24°C). DEP: $\log_{10}$ minus 6.

**Particle morphology**

Virions rhabdo– or bullet–shaped; of 250 nm (in sections), or 175 nm (purified virions); 70 nm wide (in sections), or 90 nm wide (purified virions). Axial canal obvious; 30–35 nm in diameter. Basic helix obvious.

**Cytopathology**

Virions found in leaves, mesophyll, vascular parenchyma and the phloem; in cytoplasm and in the perinuclear space.

**Taxonomy and Relationships**

Possible *Cytorhabdovirus: Rhabdoviridae.*

**References**

Di Franco, A., Russo, M. and Martelli, G.P. (1979). *Phytopathol. Medit.* **18**: 41.
Quacquarelli, A. and Gallitelli, D. (1979). *Phytopathol. Medit.* **18**: 61.
Russo, M., Castellano, M.A. and Martelli, G.P. (1979). *Phytopath. Z.* **96**: 122.

# Pelargonium zonate spot ourmiavirus

**Data collated by D. Gallitelli, 1987.**

*Standard acronym:* PZSV.

First reported in *Lycopersicon esculentum*; from Apulia, Southern Italy; by Martelli and Cirulli (1969).

**Natural host range and symptoms**

Symptoms disappear soon after infection.
*Lycopersicon esculentum* — concentric yellow bands, then necrotic stunting.
*Cynara scolymus* — foliar malformations.
*Chrysanthemum segetum* — yellow mottling.
*Diplotaxis erucoides* — symptomless.
*Pelargonium zonale* — concentric yellow bands.

**Transmission**

Virus transmitted by mechanical inoculation; transmitted by grafting; transmitted by seed; transmitted by pollen to the seed.

**Ecology and control**

Studies reported by Gallitelli (1982); Quacquarelli *et al.* (1969); Vovlas *et al.* (1986).

**Geographical distribution**

Spreads in Italy (Apulia).

**Experimental host range**

Several (3–9) families susceptible.

Diagnostically susceptible

*Cucumis sativus* — chlorotic local lesions.
*Nicotiana glutinosa* — chlorotic or necrotic local lesions, systemic mosaic and marginal necrosis.
*Nicotiana tabacum* cv. Xanthi — chlorotic rings, systemic mottling.
*Cucurbita pepo* — chlorotic or necrotic local lesions, death.
*Phaseolus vulgaris* — chlorotic or necrotic local lesions.
*Lycopersicon esculentum* — concentric yellow bands, then necrosis.

Diagnostically insusceptible

*Callistephus chinensis, Datura stramonium, Nicotiana benthamiana, Ocimum basilicum, Pisum sativum.*

Maintenance and propagation hosts

*Cucumis sativus, Lycopersicon esculentum, Nicotiana glutinosa, N. tabacum* cv. Xanthi.

Assay hosts (**L**ocal lesions or **W**hole plants)

*Chenopodium quinoa* (L), *C. amaranticolor* (L), *Phaseolus aureus* (L), *Cucumis sativus* (W), *Nicotiana glutinosa* (W).

Dicotyledonous Hosts *(**Susceptible** or Insusceptible)*

*Catharanthus roseus*. ***Celosia cristata***. ***Chenopodium amaranticolor***. ***Chenopodium quinoa***. ***Cucumis sativus***. ***Cucurbita pepo***. *Datura stramonium*. ***Glycine max***. ***Gomphrena globosa***. ***Lycopersicon esculentum***. *Nicotiana benthamiana*. ***Nicotiana clevelandii***. ***Nicotiana glutinosa***. ***Nicotiana tabacum***. *Ocimum basilicum*. ***Phaseolus vulgaris***. *Pisum sativum*. *Solanum melongena*. ***Vicia faba***. ***Vigna unguiculata***. *Zinnia elegans*.

Additional hosts

***Phaseolus aureus***.

Sources of host-range data

Quacquarelli *et al.* (1979); Gallitelli (1982); Vovlas *et al.* (1986).

**Properties of particles in sap**

TIP: 40 °C. LIV: 0.3 days. DEP: $\log_{10}$ minus 2. Leaf sap contains many virions. Electron microscopy: fix with formaldehyde before mounting in KPT.

**Purification**

Method: Gallitelli; homogenise *Nicotiana glutinosa* leaves in 20mM sodium phosphate containing 20 mM sodium thioglycollate and 20 mM DIECA, clarify by adjusting pH to 4.8, and precipitate virions with PEG.

**Particle morphology**

Virions bacilliform; not enveloped; 35 nm in length; 25 nm wide.

**Physical properties**

Three sedimenting components in purified preparations; sedimentation coefficient of the fastest 118 *S* (BV, which is probably an aggregate of TV); of the other(s) 80 *S* (TV), or 90 *S* (MV). Density 1.286 g $cm^{-3}$ in CsCl ($Cs_2SO_4$), or 1.346 g $cm^{-3}$ in CsCl (CsCl; formaldehyde-stabilised virions).

**Biochemical properties**

Virions contain 18 % nucleic acid (TV), or 18 % nucleic acid (MV), or 18 % nucleic acid (BV).

*Genome* consists of RNA; single–stranded; of two parts; largest (or only) genome part the larger 4.4 kb; the 2nd largest 3.3 kb. Genomic nucleic acid isolated by Gallitelli; use phenol-SDS. Infectivity retained when deproteinised with phenol or detergent. Additional factor not required for infectivity.

*Non–genomic nucleic acid* found in the virions; is of unknown origin and function; perhaps a satellite-like RNA is associated with some strains.

*Virion protein(s)* one; $M_r$ 23000. Method of preparation: Gallitelli (1982).

**Replication**

Replication does not depend on a helper virus.

**Cytopathology**

Virions found in leaves, mesophyll, epidermis and phloem; in cytoplasm, in nuclei, and in nucleoli. Other cellular changes: damage of chloroplasts and vesicles containing fibrillar material.

**Taxonomy and Relationships**

Possible ourmiavirus.

Virus(es) with serologically unrelated virions: alfalfa mosaic, apple mosaic, arabis mosaic, artichoke Italian latent, artichoke mottled crinkle, artichoke vein banding, broad bean wilt, brome mosaic, carnation ringspot, cherry leaf roll, cherry rasp leaf, chicory yellow mottle, cacao necrosis, cucumber mosaic, grapevine Bulgarian latent, grapevine chrome mosaic, grapevine fanleaf, myrobalan latent ringspot, pelargonium leaf curl, prune dwarf, prunus necrotic ringspot, radish mosaic, raspberry bushy dwarf, raspberry ringspot, southern bean mosaic, sowbane mosaic, tobacco necrosis, tobacco ringspot, tobacco streak, tomato black ring, tomato ringspot, Tulare apple mosaic and turnip yellow mosaic viruses.

The virus shares properties with epirus cherry, cassava Ivorian bacilliform, olive latent 2 and melon Ourmia viruses. Best tests for diagnosis: serology and host range.

**References**

Castellano, M.A. and Martelli, G.P. (1981). *Phytopathol. Medit.* **20**: 64.

Gallitelli, D. (1982). *Ann. appl. biol.* **100**: 457.

Gallitelli, D., Martelli, G.P. and Quacquarelli, A. (1983). *CMI/AAB Descr. Pl. Viruses* No. 272, 4 pp.

Martelli, G.P. and Cirulli, M. (1969). *Phytopathol. Medit.* **8**: 154.

Quacquarelli, A. and Gallitelli, D. (1979). *Phytopathol. Medit.* **18**: 61.

Vovlas, C., Gallitelli, D. and Difranco, A. (1986). *Inform. Fitopatol.* **36**: 9.

# Pepino mosaic *potexvirus*

**Data collated by A.A. Brunt, 1989.**

*Standard acronym:* PepMV.

First reported in *Solanum muricatum*; from Peru; by Jones *et al.* (1980).

**Natural host range and symptoms**

Symptoms persist.

*Solanum muricatum* — mosaic, dark green enations on the abaxial surfaces.

**Transmission**

Transmitted by a vector.

**Geographical distribution**

Spreads in Peru.

**Experimental host range**

Several (3–9) families susceptible.

Diagnostically susceptible

*Nicotiana glutinosa* — systemic mosaic.

*N. debneyi* — systemic chlorosis, necrotic spots, rings and lines.

*Solanum muricatum, Datura metel* and *D. stramonium* — systemic chlorosis.

*S. tuberosum* cvs Merpata and Revolucion — severe systemic necrosis.

*Nicandra physalodes* — irregular systemic necrosis.

Diagnostically insusceptible

*Amaranthus caudatus, Gomphrena globosa, Chenopodium quinoa, Phaseolus vulgaris, Petunia hybrida.*

Maintenance and propagation hosts

*Nicotiana glutinosa, N. debneyi, Datura metel, D. stramonium, Solanum muricatum.*

Assay hosts (**L**ocal lesions or **W**hole plants)

*Nicotiana glutinosa* (W), *N. debneyi* (W).

Dicotyledonous Hosts *(**Susceptible** or Insusceptible)*

*Amaranthus caudatus. Anthriscus cerefolium. Brassica campestris* ssp. *pekinensis. Chenopodium amaranticolor. Chenopodium murale. Chenopodium quinoa. Coriandrum sativum.* ***Cucumis sativus.*** *Cyamopsis tetragonoloba.* ***Datura metel. Datura stramonium.*** *Gomphrena globosa.* ***Lycopersicon chilense. Lycopersicon esculentum. Lycopersicon pimpinellifolium.*** *Montia perfoliata.* ***Nicandra physaloides. Nicotiana bigelovii. Nicotiana clevelandii. Nicotiana debneyi. Nicotiana glutinosa. Nicotiana occidentalis. Nicotiana rustica. Nicotiana tabacum.*** *Petunia hybrida. Phaseolus vulgaris.* ***Physalis floridana. Physalis peruviana. Solanum demissum. Solanum demissum × tuberosum. Solanum tuberosum. Tetragonia expansa.*** *Vigna unguiculata. Vigna unguiculata* ssp. *cylindrica.*

Sources of host-range data

Jones *et al.* (1980).

**Properties of particles in sap**

TIP: in *Nicotiana glutinosa* sap 65–70 °C. LIV: more than 90 days. DEP: $\log_{10}$ minus 5–6. Leaf sap contains many virions.

**Purification**

Method: Jones *et al.* (1980).

**Particle morphology**

Virions filamentous; not enveloped; usually flexuous; with a clear modal length; of 508 nm; 11 nm wide. Axial canal obscure. Basic helix obscure.

*Virion protein(s)* one; $M_r$ 26000; coat protein. Method of preparation: Jones *et al.* (1980).

**Cytopathology**

Virions found in leaves; in cytoplasm. Inclusions present in infected cells; they contain virions.

**Taxonomy and Relationships**

*Potexvirus.*

Virus(es) with serologically related virions: narcissus mosaic and cactus X viruses. Virus(es) with serologically unrelated virions: clover yellow mosaic, cymbidium mosaic, hydrangea ringspot, nerine X, papaya mosaic, potato aucuba mosaic, white clover mosaic and potato X viruses.

**Comments**

The virus frequently occurs together with a strain of potato S virus.

**References**

Jones, R.A.C., Koenig, R. and Lesemann, D.-E. (1980). *Ann. appl. Biol.* **94**: 61.

# Pepper hausteco bigeminivirus

**Data collated by A.A. Brunt, 1995.**

First reported in *Capsicum annuum*; from Mexico; by Garzon-Tiznado *et al.* (1993).

**Natural host range and symptoms**

*Capsicum annuum.*

**Transmission**

Transmitted by a vector; an insect; *Bemisia tabaci*; Aleyrodidae. Virus not transmitted by mechanical inoculation.

**Geographical distribution**

Spreads in Mexico.

**Purification**

Method: Garzon-Tiznado *et al.* (1993).

**Particle morphology**

Virions geminate.

**Biochemical properties**

*Genome* consists of DNA; single–stranded; circular; of two parts; largest (or only) genome part 2.631 kb; the 2nd largest 2.59 kb.

*Sequence database accession code(s):*

X70418 Em(40)_vi:PHVCGAA Gb(84)_vi:PHVCGAA Pepper Huasteco Virus complete genome, component A. 11/93 2,631bp.

X70419 Em(40)_vi:PHVCGAB Gb(84)_vi:PHVCGAB Pepper Huasteco Virus (PHV) complete genome, component B. 11/93 2,589bp. 2 sequences.

**Taxonomy and Relationships**

Bigeminivirus: *Geminiviridae.*

**References**

Garzon-Tiznado, J.A., Torres-Pacheco, I., Ascencio-Ibanez, J.T., Herrera-Estrella, L. and Rivera-Bustamante, R.F. (1993). *Phytopathology* **83**: 514.

Torres-Pacheco, I., Garzon-Tiznado, J.A., Herrera-Estrella, L. and Rivera-Bustamante, R.F. (1993). *J. gen. Virol.* **74**: 2225.

# Pepper Indian mottle *potyvirus*

**Data collated by V.R. Mali, 1990.**

*Synonyms:* chilli mottle virus, Indian pepper mottle virus. *Standard acronym:* IPMV.

First reported in *Capsicum annuum*; from the Punjab, India; by Sandhu and Chohan (1979).

**Natural host range and symptoms**

Symptoms persist.

*Capsicum annuum* — systemic mottling and leaf malformation, including narrowing and split tips.

**Transmission**

Transmitted by a vector; an insect; *Myzus persicae, Aphis gossypii, A. craccivora*; Aphididae. Transmitted in a non–persistent manner. Virus transmitted by mechanical inoculation; transmitted by grafting.

**Ecology and control**

Studies reported by Sandhu and Chohan (1979).

**Geographical distribution**

Spreads in India.

**Experimental host range**

Several (3–9) families susceptible. Experimentally infected plants mostly show systemic mosaics, mottles, ringspots or necrosis showing first on youngest leaves usually as vein clearing.

Diagnostically susceptible

*Capsicum annuum* — systemic mottle.
*Capsicum frutescens* cv. Tabasco — necrotic local lesions, then systemic necrosis.
*Physalis floridana* — systemic mottle.
*Nicotiana glutinosa, N. tabacum* cv. Turkish — systemic mottle.
*Chenopodium amaranticolor* — necrotic local lesions.

Diagnostically insusceptible

*Chenopodium album, Gomphrena globosa, Datura stramonium, Nicotiana rustica, Petunia hybrida.*

Maintenance and propagation hosts

*Cucurbita pepo.*

Assay hosts (**L**ocal lesions or **W**hole plants)

*Chenopodium amaranticolor.*

Dicotyledonous Hosts *(**Susceptible** or Insusceptible)*

*Antirrhinum majus*. ***Capsicum annuum***. ***Capsicum frutescens***. *Chenopodium album*. ***Chenopodium amaranticolor***. *Citrullus lanatus*. *Cucumis melo*. *Cucumis sativus*. *Cucurbita moschata*. ***Cucurbita pepo***. *Dahlia pinnata*. *Datura stramonium*. *Glycine max*. *Gomphrena globosa*. *Luffa acutangula*. ***Lycopersicon esculentum***. ***Nicotiana glutinosa***. *Nicotiana rustica*. ***Nicotiana tabacum***. *Petunia hybrida*. *Phaseolus vulgaris*. *Phlox drummondii*. ***Physalis floridana***. *Pisum sativum*. ***Solanum melongena***. ***Solanum nigrum***. *Vigna radiata*. *Vigna unguiculata*.

Monocotyledonous Hosts *(**Susceptible** or Insusceptible)*

*Triticum aestivum*. *Zea mays*.

Additional hosts

*Abelmoschus esculentus, Benincasa hispida, Citrullus vulgaris* var. *fistulosis, Cucumis melo* var. *momordica, Lagenaria cineraria, Luffa cylindrica, Momordica charantia* and *Vigna mungo*.

Sources of host-range data

Sandhu and Chonan (1979).

**Properties of particles in sap**

TIP: 50–55 °C. LIV: 3 days. DEP: $\log_{10}$ minus 2.

**Taxonomy and Relationships**

*Potyvirus: Potyviridae.*

Virus(es) with serologically related virions: pepper mottle and potato Y viruses. Virus(es) with serologically unrelated virions: tobacco mosaic, tobacco etch, potato X and cucumber mosaic viruses.

Best tests for diagnosis: Indian pepper mottle virus (PeMV-I) can be distinguished from tobacco mosaic (TMV), cucumber mosaic (CMV) and potato X virus (PVX) viruses which naturally infect chilli (*Capsicum annuum*) in India by host range, serology and virion morphology. In India chilli (*Capsicum annuum*) is also infected by potyviruses such as potato Y virus (PVY; Jayarajan and Ramakrishna, 1969; Joshi and Bhargava, 1962; Khatri and Sekhon, 1984) and pepper veinal mottle (PVMV; Bidhari and Reddy, 1981; Prasadrao, 1976). PeMV-I, as reported by Sandhu and Chohan (1979) from the Punjab, can be distinguished from the potyviruses by the reactions of six pepper cultivars (Avelar, Agronomico-S, Yolo Y, Early Calwonder, Fla-23–1–7 and Tabasco) as recommended by Zitter (1973); its host range is similar to that reported by Purcifull *et al.* (1975).

**References**


Bidhari and Reddy (1981). *3rd Int. Symp. Pl. Path. New Delhi* session XXII, p. 129.
Jayarajan and Ramakrishna (1969). *Madras Agric. J.* **56**: 761.
Joshi, B.S. and Bhargava, K.S. (1962). *Indian J. Microbiol.* **2**: 29.
Khatri and Sekhon (1984). *Indian J. Mycol. Pl. Path.* **4**: 121.
Nelson, M.R., Wheeler, R.E. and Zitter, T.A. (1982). *CMI/AAB Descr. Pl. Viruses* No. 253, 4 pp.
Prasadrao, R.D.V. (1976). *Ph.D. Thesis*, U.A.S., Hebbal, Bangalore, India.
Purcifull, D.E., Zitter, T.A. and Hiebert, E. (1975). *Phytopathology* **65**: 559.
Sandhu, K.S. and Chohan, J.S. (1979). *Indian J. Mycol. Pl. Path.* **9**: 177.
Zitter, T.A. (1973). *Pl. Dis. Reptr* **57**: 991.


# Pepper mild mosaic (?) potyvirus

**Data collated by A.A. Brunt, 1989.**

*Standard acronym:* PMMV.

First reported in *Capsicum annuum*; from Venezuela; by Debrot *et al.* (1980).

**Natural host range and symptoms**

*Capsicum annuum* — chlorotic vein-banding and mosaic leaf symptoms in cultivars resistant to potato Y and tobacco etch potyviruses in Venezuela (Debrot *et al.*, 1980).

**Geographical distribution**

Spreads in Venezuela.

Diagnostically susceptible

*Capsicum annuum, Nicotiana glutinosa, N. tabacum* — systemic leaf chlorosis.

Assay hosts (**L**ocal lesions or **W**hole plants)

*Nicotiana glutinosa* (W) (Ladera *et al.*, 1982).

**Purification**

Method: Ladera *et al.* (1982).

**Particle morphology**

Virions filamentous; with a clear modal length; of *c.* 750 nm.

**Physical properties**

Density 1.325 g $cm^{-3}$ in CsCl.

**Biochemical properties**

*Genome* consists of RNA; single–stranded.

*Virion protein(s)* one; $M_r$ 29000; capsid protein.

**Cytopathology**

Inclusions present in infected cells; are pinwheels.

**Taxonomy and Relationships**

Unassigned potyvirus.

Virus(es) with serologically unrelated virions: potato Y and tobacco etch viruses (Salas *et al.*, 1972).

**References**

Debrot, E.A., Lastra, R. and Ladera, P. (1980). *Trop. Agron.*

Ladera, P., Lastra, R. and Debrot, E.A. (1982). *Phytopath. Z.* **104**: 97.

Salas, G., Arcia, A. and Malaguti, G. (1972). *Trop. Agron.* **12**: 45.

# Pepper mild mottle *tobamovirus*

**Data collated by C. Wetter, 1987.**

*Synonyms:* Samsun latent strain of tobacco mosaic virus, pepper mosaic virus (Cohen and Ben-Josef, 1989; Tanzi *et al.*, 1988). *Standard acronym:* PMMoV. *Strains:* capsicum mosaic virus; Samsun latent virus, pepper unusual strain, also isolate P8 of tobacco mosaic virus.

First reported in *Capsicum annuum* L.; from South Carolina, U.S.A.; by McKinney (1952).

**Natural host range and symptoms**

Symptoms persist, or vary seasonally.

*Capsicum* spp. — mild chlorosis and stunting, especially if plants infected when young. Fruits are small, malformed, mottled and some have necrotic depressions.

**Transmission**

Transmitted by means not involving a vector. Virus transmitted by mechanical inoculation; transmitted by grafting; transmitted by contact between plants; transmitted by seed; not transmitted by pollen.

**Geographical distribution**

Spreads in Argentina, Australia, Denmark, France, Hungary, Iceland, Italy, Japan, the Netherlands, Spain, the UK, and the USA.

**Experimental host range**

Several (3–9) families susceptible.

Diagnostically susceptible

*Datura metel* — local necrosis and abscission.

*Datura stramonium* — small necrotic local lesions; not systemic.

*Chenopodium amaranticolor, C. quinoa* — chlorotic local lesions; not systemic.

*Capsicum chacoense, C. praetermissum* — severe systemic necrosis.

*Nicotiana glutinosa, N. sylvestris, N. tabacum* cvs White Burley and Xanthi-*nc.* — small necrotic local lesions; not systemic.

Diagnostically insusceptible

*Cucumis sativus, Gomphrena globosa, Nicotiana rustica.* Many other species in the Solanaceae are susceptible, but not tomato or *Nicotiana glauca.*

Maintenance and propagation hosts

*Capsicum frutescens* and cultivars of *Capsicum annuum* that are immune to tobacco mosaic virus.

Assay hosts (**L**ocal lesions or **W**hole plants)

*Nicotiana glutinosa* (L), *N. sylvestris* (L), *Datura stramonium* (L), *Chenopodium quinoa* (L), *C. amaranticolor* (L).

Dicotyledonous Hosts (***Susceptible*** *or Insusceptible)*

***Capsicum annuum***, or *Capsicum annuum.* ***Capsicum frutescens. Chenopodium amaranticolor. Chenopodium quinoa.*** *Cucumis sativus.* ***Datura metel. Datura stramonium.*** *Gomphrena globosa. Lycopersicon esculentum.* ***Nicotiana clevelandii. Nicotiana debneyi. Nicotiana glutinosa.*** *Nicotiana rustica.* ***Nicotiana sylvestris. Nicotiana tabacum. Ocimum basilicum. Petunia hybrida. Physalis floridana.*** *Vigna unguiculata.*

Additional hosts

***Capsicum baccatum, C. cardenasei, C. chacoense, C. chinense, C. eximum, C. microcarpum, C. praetermissum*** and ***C. pubescens.***

Sources of host-range data

Wetter *et al.* (1984); Wetter and Conti (1987).

## Properties of particles in sap

TIP: 95 °C. LIV: more than 30 days (*i.e* several months). DEP: $\log_{10}$ minus 6–8 (Wetter and Conti, 1984). Infectivity of sap not changed by treatment with di–ethyl ether. Leaf sap contains many virions.

## Purification

Method: Wetter *et al.* (1984).

## Particle morphology

Virions rigid rod–shaped; not enveloped; usually straight; with a clear modal length; of 312 nm; 18 nm wide. Axial canal obvious; 4 nm in diameter. Basic helix obvious; pitch of basic helix *c.* 2.3 nm.

## Physical properties

One sedimenting component in purified preparations. Isoelectric point pH 3.38–3.71.

*Sequence database accession code(s):*

M81413 Em(40)_vi:PPPSLF Gb(84)_vi:PPPSLF Nucleotide sequence of the genomic RNA of the tobamovirus Pepper mild mottle virus, strain S, 1 sequence.

*Virion protein(s)* one. Method of preparation: Fraenkel-Conrat (1957). Amino acid composition: Wetter *et al.* (1984); Creaser *et al.* (1989).

## Replication

Genome replicates in cytoplasm. Replication does not depend on a helper virus.

## Cytopathology

Virions found in mesophyll and phloem; in cytoplasm. Inclusions present in infected cells; are crystals in the cytoplasm and unusual in shape; aggregates layered in angles; they contain virions.

**Taxonomy and Relationships**

*Tobamovirus.*

Virus(es) with serologically related virions: tobacco mosaic, tobacco mild green mosaic, tomato mosaic, odontoglossum ringspot, ribgrass mosaic, sunn-hemp mosaic and cucumber green mottle viruses.

**References**

Cohen, S. and Ben-Josef, R. (1989). *Hassadeh* **70**: 394.
Creaser, W., Gibbs, A.J. and Pares, R.D. (1987). *Australas. Pl. Path.* **16**: 85.
Marte, M. and Wetter, C. (1986). *Z. PflKrankh. PflPath. PflSchutz.* **93**: 37.
McKinney, H.H. (1952). *Pl. Dis. Reptr* **36**: 184.
Paludan, N. (1982). *Acta Hort.* **127**: 65.
Pares, R.D. (1985). *Ann. appl. Biol.* **106**: 469.
Tanzi, M., Betti, L., Bertaccini, A. and Canova, A. (1988). *Phytopath. Medit.* **27**: 28.
Tobias, J., Rast, A.Th.B. and Maat, D.Z. (1982). *Neth. J. Pl. Path.* **88**: 257.
Wetter, C. (1984). *Plant Dis.* **68**: 597.
Wetter, C. and Conti, M. (1987). *AAB Descr. Pl. Viruses* No. 330.
Wetter, C., Conti, M., Altschuh, D., Tabillion, R. and van Regenmortel, M.H.V. (1984). *Phytopathology* **74**: 405.

# Pepper mild tigré (?) bigeminivirus

**Data collated by J.K. Brown, 1989.**

*Synonyms:* tigre disease geminivirus. *Standard acronym:* PepMTV.

First reported in *Capsicum annuum*; from Mexico; by Brown *et al.* (1989).

**Natural host range and symptoms**

Symptoms persist and vary seasonally.

*Capsicum annuum* cvs Jalapeno, Serrano — mild yellow leaf curling and mosaic.

**Transmission**

Transmitted by a vector; an insect; *Bemisia tabaci*; Aleyrodidae. Transmitted in a persistent manner. Virus not transmitted by mechanical inoculation; transmitted by grafting; not transmitted by contact between plants; not transmitted by seed.

**Geographical distribution**

Spreads in Mexico and the USA (Texas).

**Experimental host range**

Few (<3) families susceptible.

Diagnostically susceptible

*Capsicum annuum* cvs Anahiem, Serrano and Tabasco — systemic yellow leaf curling.

*Lycopersicon esculentum* cv. Pole Boy — systemic yellow mosaic or leaf curling.

Diagnostically insusceptible

*Cucumis melo* cv. Topmark, *C. sativus* cv. Straight eight, *Gossypium hirsutum* cv. DP90, *Malva parviflora, Glycine max* cv. Frost beater.

Maintenance and propagation hosts

*Lycopersicon esculentum, Capsicum annuum.*

Assay hosts (Local lesions or Whole plants)

*Capsicum annuum* (W), *Lycopersicon esculentum* (W).

Dicotyledonous Hosts (***Susceptible*** *or Insusceptible)*

***Capsicum annuum. Capsicum frutescens.*** *Cassia obtusifolia. Chenopodium album. Citrullus lanatus. Cucumis melo. Cucumis sativus. Cucurbita maxima. Glycine max. Gossypium hirsutum. Helianthus annuus. Lactuca sativa.* ***Lycopersicon esculentum.*** *Solanum melongena. Vigna unguiculata. Zinnia elegans.*

Sources of host-range data

Brown *et al.* (1989).

**Properties of particles in sap**

Leaf sap contains few virions. Electron microscopy: fix with buffered (pH 7.0) 0.25% glutaraldehyde.

**Particle morphology**

Virions geminate; 20 nm in diameter; dimers 30 nm in length; without a conspicuous capsomere arrangement.

**Biochemical properties**

*Genome* consists of DNA; single–stranded.

*Sequence database accession code(s):*

L27270 Em(40)_vi:GEPEMAC1A Gb(84)_vi:PEMAC1A Pepper geminivirus replication-associated protein (AC1) gene, 3' end. 1/94 299bp

L27271 Em(40)_vi:GEPEMBC1A Gb(84)_vi:PEMBC1A Pepper geminivirus movement protein (BC1) gene, partial cds. 1/94 222bp. 2 sequences.

**Taxonomy and Relationships**

Possible bigeminivirus: *Geminiviridae.*

**Comments**

Alone, the virus induces mild leaf symptoms in pepper, but peppers also containing chino del tomate virus and another uncharacterised whitefly transmitted virus have the tigré disease.

**References**

Brown, J.K. and Nelson, M.R. (1989). *Phytopathology* **79**: 908.

Brown, J.K., Campodonico O.P. and Nelson, M.R. (1989). *Plant Dis.* **73**: 610.

# Pepper Moroccan *tombusvirus*

**Data collated by D. Gallitelli, 1991.**

*Synonyms:* Moroccan pepper virus. *Standard acronym:* MPV.

First reported in *Capsicum annuum*; from Morocco; by Fischer and Lockhart (1977).

**Natural host range and symptoms**

Symptoms persist and vary cyclically over a few weeks. Symptoms stunting, leaf malformations, lethal systemic necrosis.

*Capsicum annuum* — stunting, deformation, leaf and flower abscission, lethal systemic necrosis.

*Lycopersicon esculentum* — leaf mottling and malformation, stunting.

*Pelargonium zonale* — leaf malformation, stunting.

**Transmission**

Transmitted by means not involving a vector. Virus transmitted by mechanical inoculation; transmitted by grafting.

**Geographical distribution**

Spreads in Morocco. Found, but with no evidence of spread, in Germany.

**Experimental host range**

Several (3–9) families susceptible. Experimentally infected plants mostly show necrotic local lesions, systemic necrosis, stunting.

Diagnostically susceptible

*Chenopodium quinoa, C. murale, Cucumis sativus, Cucurbita pepo, Nicotiana tabacum, N. glutinosa, Petunia hybrida, Phaseolus vulgaris, Vicia faba* — necrotic local lesions.

*Capsicum annuum, Nicotiana benthamiana, N. clevelandii, Spinacia oleracea* — necrotic local lesions; systemic necrosis.

Maintenance and propagation hosts

*Nicotiana benthamiana, N. clevelandii.*

Assay hosts (**L**ocal lesions or **W**hole plants)

*Capsicum annuum* (W), *Datura stramonium* (L), *Nicotiana benthamiana* (W), *Nicotiana glutinosa* (L), *Spinacia oleracea* (W).

Dicotyledonous Hosts *(**Susceptible** or Insusceptible)*

***Antirrhinum majus. Capsicum annuum. Chenopodium amaranticolor. Chenopodium foetidum. Chenopodium murale. Chenopodium quinoa. Cucumis melo. Cucumis sativus. Cucurbita pepo. Datura metel. Datura stramonium. Gomphrena globosa. Lupinus albus. Nicotiana benthamiana. Nicotiana clevelandii. Nicotiana debneyi. Nicotiana megalosiphon. Nicotiana tabacum. Ocimum basilicum. Petunia hybrida. Phaseolus vulgaris. Physalis floridana. Pisum sativum. Solanum melongena. Spinacia oleracea. Vicia faba. Vigna unguiculata*** ssp. ***sesquipedalis. Zinnia elegans.***

Sources of host-range data

Fischer and Lockhart (1977); Vetten and Koenig (1983).

**Properties of particles in sap**

TIP: 80–85 °C. LIV: 49–56 days. DEP: $\log_{10}$ minus 6–7. Leaf sap contains many virions.

**Purification**

Method: Gallitelli and Russo (1987).

**Particle morphology**

Virions isometric; not enveloped; 30 nm in diameter; neither clearly angular or nor rounded in profile; without a conspicuous capsomere arrangement.

**Physical properties**

One sedimenting component in purified preparations; sedimentation coefficient 140 *S*. Density 1.36 g $cm^{-3}$ in CsCl.

**Biochemical properties**

Virions contain 18 % nucleic acid; 82 % protein.

*Genome* consists of RNA; single–stranded; linear; unipartite; largest (or only) genome part 4.7 kb. Genomic nucleic acid isolated by Gallitelli and Russo (1987).

Infectivity retained when deproteinised with phenol or detergent. Poly A region absent. Additional factor not required for infectivity.

*Non–genomic nucleic acid* not found in the virions.

*Virion protein(s)* one; $M_r$ 43000; coat protein. Method of preparation: Laemmli (1970).

**Cytopathology**

Virions found in leaves, roots and mesophyll; in cytoplasm and in cell vacuoles. Inclusions present in infected cells; are membranous bodies; they do not contain virions. Other cellular changes: see Martelli *et al.* (1988).

**Taxonomy and Relationships**

*Tombusvirus: Tombusviridae.*

Virus(es) with serologically related virions: artichoke mottled crinkle, eggplant mottled crinkle, grapevine Algerian latent, pelargonium leaf curl, petunia asteroid mosaic and tomato bushy stunt viruses. Virus(es) with serologically unrelated virions: carnation Italian ringspot, cymbidium ringspot and Neckar River tombusviruses.

Dot blot hybridization tests indicate that the virus has 10% nucleotide homology with tomato bushy stunt tombusvirus.

**References**

Fischer, H.V. and Lockhart, B.E. (1977). *Phytopathology* **67**: 1352.

Gallitelli, D. and Russo, M. (1987). *J. Phytopath.* **119**: 106.

Laemmli, U.K. (1970). *Nature, Lond.* **227**: 680.

Martelli, G.P., Gallitelli, D. and Russo, M. (1988). In: *The Plant Viruses*, Vol. 3, Polyhedral Virions with Monopartite RNA Genomes, p. 13; ed. R. Koenig. Plenum Press, New York.

Martelli, G.P., Russo, M. and Gallitelli, D. (1988). *AAB Descr. Pl. Viruses* No. 352. 8pp.

Vetten, H.J. and Koenig, R. (1985). *Phytopath. Z.* **108**: 215.

# Pepper mottle *potyvirus*

**Data collated by T.A. Zitter and V.R. Mali, 1987.**

*Synonyms:* chilli mottle virus. *Standard acronym:* PepMoV.

First reported in *Capsicum annuum*; from Palm Beach County, Delray Beach, Florida, U.S.A.; by Zitter (1972*a* and *b*).

**Natural host range and symptoms**

Symptoms persist and vary seasonally.

*Capsicum annuum* — mottling and malformed leaves.

*Datura meteloides* — mild mottle.

*Solanum* sp. — systemic mild mottle.

**Transmission**

Transmitted by a vector; an insect; *Aphis gossypii, A. craccivora, Myzus persicae*; Aphididae. Transmitted in a non–persistent manner. Virus transmitted by mechanical inoculation; transmitted by grafting; not transmitted by contact between plants; not transmitted by seed.

**Ecology and control**

Studies reported by Sandhu and Chohan (1979); Zitter (1972*a*); Zitter and Cook (1973); Purcifull *et al.* (1975); Zitter and Ozaki (1978*a* and *b*).

**Geographical distribution**

Spreads in El Salvador, India, and the USA (Florida, New Mexico, Texas, Arizona and California).

**Experimental host range**

Few (<3) families susceptible.

Diagnostically susceptible

*Capsicum annuum* — severe systemic leaf mottling in all susceptible cultivars.

*Capsicum frutescens* — necrotic local lesions; systemic leaf necrosis and death of young Tabasco plants.

*Datura meteloides, Lycopersicon esculentum* — mild systemic mottling.

*Nicotiana glutinosa, N. tabacum, Physalis floridana* — systemic leaf mottling.

*Chenopodium amaranticolor* — necrotic local lesions; not systemic.

Diagnostically insusceptible

*Capsicum frutescens* cv. Greenleaf Tabasco, *Datura metel, Datura stramonium, Nicotiana tabacum* cv. 'V-20', *Solanum tuberosum* cv. 'Russet Burbank', *Nicotiana rustica, Petunia hybrida, Chenopodium album, Gomphrena globosa.*

Maintenance and propagation hosts

*Capsicum annuum* cv. Early Calwonder, *Nicotiana benthamiana, N. glutinosa, N. glutinosa x. N. clevelandii* hybrid, *N. tabacum* Turkish NN, Havana 425, Xanthi-*nc*, White Burley, *Cucurbita pepo.*

Assay hosts (**L**ocal lesions or **W**hole plants)

*Capsicum annuum* (W), *C. frutescens* Tabasco (L) and systemic mottle and death of young seedlings (W), *Chenopodium amaranticolor* (L).

Dicotyledonous Hosts *(**Susceptible** or Insusceptible)*

*Antirrhinum majus. Beta vulgaris. Brassica campestris* ssp. *pekinensis.* ***Capsicum annuum*****.** ***Capsicum frutescens*****.** *Cassia tora. Chenopodium album.* ***Chenopodium amaranticolor*****.** *Cichorium endiva. Citrullus lanatus. Cucumis melo. Cucumis sativus. Cucurbita moschata. Cucurbita pepo. Datura metel. Datura stramonium. Glycine max. Gomphrena globosa. Hibiscus esculentus. Luffa acutangula.* ***Lycopersicon esculentum*****.** ***Nicotiana benthamiana*****.** ***Nicotiana clevelandii*****.** ***Nicotiana* × *edwardsonii*****.** ***Nicotiana glutinosa*****.** ***Nicotiana glutinosa* × *clevelandii*****.** *Nicotiana rustica.* ***Nicotiana tabacum*****.** *Petunia hybrida. Phaseolus vulgaris. Phlox drummondii.* ***Physalis floridana*****.** *Phytolacca americana. Pisum sativum.* ***Solanum nigrum*****.** *Vigna radiata. Vigna unguiculata.*

Monocotyledonous Hosts *(**Susceptible** or Insusceptible)*

*Triticum aestivum. Zea mays.*

Sources of host-range data

Nelson and Wheeler (1978); Nelson *et al.* (1982); Purcifull *et al.* (1975); Sandhu and Chohan (1979).

**Properties of particles in sap**

TIP: 50–55 °C. LIV: 3 days. DEP: $\log_{10}$ minus 1–2. Leaf sap contains few virions.

**Purification**

Method: Purcifull *et al.* (1975).

**Particle morphology**

Virions filamentous; not enveloped; usually flexuous; with a clear modal length; of 737 nm.

**Physical properties**

One sedimenting component in purified preparations.

**Biochemical properties**

*Genome* consists of RNA; single–stranded; linear; unipartite.

*Sequence database accession code(s):*

M11598 Em(40)_vi:PMV38KPA Gb(84)_vi:PMV38KPA Pepper mottle virus (PeMV) capsid protein RNA, complete cds. 7/89 1,480bp.

M87827 Em(40)_vi:PMVCPG Gb(84)_vi:PMVCPG Pepper Mottle virus coat protein gene fragment. 3/92 1,009bp.

M96425 Em(40)_vi:PMVCG Gb(84)_vi:PMVCG Pepper mottle virus complete genome. 1/93 9,640bp.

X72587 Gb(84)_vi:PMMVICP Pepper mild mottle virus (Italian strain) coat protein gene and 3' non-coding region. 7/94 69.

*Virion protein(s)* two (Hiebert and McDonald, 1973).

**Replication**

Replication does not depend on a helper virus.

**Cytopathology**

Virions found in leaves and epidermis; in cytoplasm. Inclusions present in infected cells; are pinwheels and unusual in shape; with striated lamellae.

**Taxonomy and Relationships**

*Potyvirus: Potyviridae.*

Virus(es) with serologically related virions: potato Y, tobacco etch and pepper veinal mottle viruses, but distantly. Peptide analyses show that the virus has close affinities with potato Y virus (D. Shukla, unpublished information). Virus(es) with serologically unrelated virions: bidens mottle, lettuce mosaic and turnip mosaic viruses.

Best tests for diagnosis: see Nelson *et al.* (1982).

**References**

Hiebert, E. and McDonald, J.G. (1973). *Virology* **56**: 349.

Nelson, M.R. and Wheeler, R.E. (1978). *Phytopathology* **68**: 979.

Nelson, M.R., Wheeler, R.E. and Zitter, T.A. (1982). *CMI/AAB Descr. Pl. Viruses* No. 253, 4 pp.

Purcifull, D.E., Zitter T.A. and Hiebert, E. (1975). *Phytopathology* **65**: 559.

Sandhu, K.S. and Chohan, J.S. (1979). *Indian J. Mycol. & Pl. Pathol.* **9**: 177.

Zitter, T.A. (1972*a*). *Pl. Dis. Reptr* **56**: 586.

Zitter, T.A. (1972*b*). *Pl. Dis. Reptr* **56**: 731.

Zitter, T.A. (1975). *Phytopathology* **65**: 110.

Zitter, T.A. and Cook, A.A. (1973). *Phytopathology* **63**: 1211.

Zitter, T.A. and Ozaki, H.Y. (1978*a*). *Phytopathology News* **12(99)**: 227.

Zitter, T.A. and Ozaki, H.Y. (1978*b*). *Proc. Fla State Hort. Soc.* **91**: 287.

# Pepper ringspot *tobravirus*

**Data collated by D.J. Robinson, 1987.**

*Synonyms:* artichoke yellow band virus (Chagas, Flores and Caner, 1970; Harrison and Robinson, 1986), tobacco rattle virus — CAM, tobacco rattle virus — serotype III. *Standard acronym:* PepRSV.

First reported in *Lycopersicon esculentum*; from Brazil; by Silberschmidt (1962–63).

**Natural host range and symptoms**

Symptoms persist.
*Capsicum annuum* — ringspots.
*Cynara scolymus* — yellow banding.
*Lycopersicon esculentum* — ringspots and yellow banding.
*Bidens pilosa.*

**Transmission**

Virus transmitted by mechanical inoculation; transmitted by grafting; not transmitted by contact between plants.

**Ecology and control**

Studies reported by reviewed by Harrison (1977).

**Geographical distribution**

Spreads in the South and Central American region; Brazil.

**Experimental host range**

Many (>9) families susceptible.

Diagnostically susceptible

*Chenopodium amaranticolor* — necrotic local lesions, not systemic.
*Cucumis sativus* — chlorotic or necrotic local lesions, not systemic.
*Nicotiana tabacum* cv. Samsun NN — necrotic rings and lines, systemic.
*N. tabacum* cv. White Burley — symptomless, systemic.
*Phaseolus vulgaris* — pinpoint necrotic local lesions, not systemic.
*Pisum sativum* and *Vicia faba* — small necrotic local lesions, not systemic.

Maintenance and propagation hosts

*Nicotiana clevelandii, N. tabacum* cv. White Burley.

Assay hosts (**L**ocal lesions or **W**hole plants)

*Chenopodium amaranticolor* (L), *Phaseolus vulgaris* (L).

Dicotyledonous Hosts *(**Susceptible** or Insusceptible)*

***Beta vulgaris. Capsicum annuum. Catharanthus roseus. Chenopodium album. Chenopodium amaranticolor. Chenopodium murale. Chenopodium quinoa. Cucumis sativus. Datura stramonium. Gomphrena globosa. Lycopersicon esculentum. Nicotiana benthamiana. Nicotiana clevelandii. Nicotiana debneyi. Nicotiana glutinosa. Nicotiana megalosiphon. Nicotiana rustica. Nicotiana tabacum. Ocimum basilicum. Petunia hybrida. Phaseolus vulgaris. Pisum sativum. Spinacia oleracea. Tetragonia expansa. Vicia faba.***

**Properties of particles in sap**

TIP: 80–85 °C. LIV: more than 42 days. DEP: $\log_{10}$ minus 5–6. Infectivity of sap not changed by treatment with di–ethyl ether. Leaf sap contains few virions. Electron microscopy: preservation in uranyl formate/NaOH. Fine detail in UA.

**Purification**

Method: Harrison and Nixon (1959).

**Particle morphology**

Virions rod–shaped; not enveloped; usually straight; with a clear modal length; of 52 and 197 nm; 22 nm wide. Axial canal obvious; 4 nm in diameter. Basic helix obvious; pitch of basic helix 2.5 nm.

**Physical properties**

Two sedimenting components in purified preparations; sedimentation coefficient of the fastest (long) virions 305 *S*; of the other(s) 163 *S* (short virions). Density 1.306 g $cm^{-3}$ in CsCl (unfixed). Density 1.27 g $cm^{-3}$ in sucrose. $A_{260}/A_{280}$ ratio 1.15.

**Biochemical properties**

Virions contain 5 % nucleic acid; 95 % protein; 0 % lipid.

*Genome* consists of RNA; single–stranded; linear; of two parts; largest (or only) genome part the larger 7.5 kb; the 2nd largest 2.1 kb. Genomic nucleic acid isolated by Harrison and Nixon (1959); Darby and Minson, (1972); Cooper and Mayo, (1972). Base composition 25.5 % G (RNA-1), or 24.3 % G (RNA-2); 29.5 % A (RNA-1), or 26.1 % A (RNA-2); 16.5 % C (RNA-1), or 17.6 % C (RNA-2); 28.5 % U (RNA-1), or 32 % U (RNA-2). 5' terminus of RNA has a methylated nucleotide cap (reported only on RNA-1, but probably on both). Infectivity retained when deproteinised with proteases; retained when deproteinised with phenol or detergent. Poly A region absent. Additional factor not required for infectivity. Genome has no *t*RNA–like activity. Nucleotide sequence references: Bergh *et al.* (1985).

*Sequence database accession code(s):*

L23972 Em(40)_vi:PPNRNA Gb(84)_vi:PPNRNA Pepper ringspot viral 12k, 29k 136k and 195k protein RNA, complete cds. 11/93 6,828bp. 1 sequence.

*Non–genomic nucleic acid* found in the virions; is subgenomic *m*RNA. Sub–genomic *m*RNA found in infected cells.

*Virion protein(s)* one; $M_r$ 22000; coat protein. Method of preparation: Semancik (1970); Ghabrial and Lister (1973).

*Virus–coded non–virion proteins* identified by genomic sequence analysis (Bergh *et al.* (1985)); three proteins found. $M_r$ of the largest 170000. $M_r$ of 2nd largest 140000. $M_r$ of 3rd 30000.

**Replication**

Genome replicates in cytoplasm. Coat protein *m*RNA translated in the cytoplasm. Replication does not depend on a helper virus.

**Cytopathology**

Virions found in all parts of the host plant, except xylem; in cytoplasm (mostly near mitochondria, but not in them). Inclusions present in infected cells (by NM forms); are amorphous X–bodies.

**Taxonomy and Relationships**

*Tobravirus.*

Virus(es) with serologically related virions: tobacco rattle and pea early-browning viruses, but distantly.

Although there is distant sequence homology with other tobraviruses, these are not usually detected in hybridization tests.

**Comments**

For many years pepper ringspot virus was thought to be a serotype of tobacco rattle virus, and some research work did not distinguish between them.

**References**

Bergh, S.T., Koziel, M.G., Huang, S.C., Thomas, R.A., Gilley, D.P. and Siegel, A. (1985). *Nucl. Acids Res.* **13**: 8507.
Chagas, C.M., Flores, E. and Caner, J. (1970). *Biologico* **35**: 271.
Cooper, J.I. and Mayo, M.A. (1972). *J. gen. Virol.* **16**: 285.
Darby, G. and Minson, A.C. (1972). *J. gen. Virol.* **14**: 199.
Ghabrial, S.A. and Lister, R.M. (1973). *Virology* **52**: 1.
Harrison, B.D. (1970). *CMI/AAB Descr. Pl. Viruses* No. 12, 4 pp.
Harrison, B.D. (1977). *Ann. Rev. Phytopath.* **15**: 331.
Harrison, B.D. and Nixon, H.L. (1959). *J. gen. Microbiol.* **21**: 591.
Harrison, B.D. and Robinson, D.J. (1986). In: *The Plant Viruses* Vol 2, The Rod-Shaped Viruses, p. 339; eds M.H.V. van Regenmortel and H. Fraenkel-Conrat. Plenum Press, New York.
Semancik, J.S. (1970). *Virology* **40**: 618.
Silberschmidt, K. (1962–63). *Phytopath. Z.* **46**: 209.

# Pepper severe mosaic *potyvirus*

**Data collated by A.A. Brunt, 1990.**

*Standard acronym:* PeSMV. *Strains:* PSMV/type, $PSMV_{20}$, $PSMV_{127}$ and $PSMV_{151}$.

First reported in *Capsicum annuum*; from San Juan, Argentina; by Feldman and Gracia (1977).

**Natural host range and symptoms**

Symptoms persist.

*Capsicum annuum* — necrotic spots and streaks in leaves and stems, leaf abscission, then chlorotic leaves. Fruit yield greatly decreased.

**Transmission**

Transmitted by a vector; an insect; *Aphis craccivora, Macrosiphum euphorbiae* and, except for type strain, *Myzus persicae*; Aphididae. Transmitted in a non–persistent manner. Virus transmitted by mechanical inoculation; transmitted by grafting.

**Ecology and control**

Studies reported by Feldman and Gracia (1977).

**Geographical distribution**

Spreads in Argentina.

**Experimental host range**

Several (3–9) families susceptible.

Diagnostically susceptible

*Chenopodium amaranticolor, C. quinoa* — local lesions; not systemic.

*Nicandra physalodes, Nicotiana debneyi, N. rustica, N. tabacum* — local lesions, followed by systemic vein-clearing mosaic and/or chlorotic rings.

*Capsicum annuum, Nicotiana clevelandii, N. glutinosa, N. sylvestris* — systemic leaf mottling and malformation.

Diagnostically insusceptible

*Datura stramonium, Solanum tuberosum, Gomphrena globosa, Cucumis sativus, Phaseolus vulgaris.*

Maintenance and propagation hosts

*Nicotiana tabacum, N. sylvestris.*

Assay hosts (Local lesions or Whole plants)

*Chenopodium amaranticolor* (L), *Chenopodium quinoa* (L), *Nicotiana tabacum* (L,W), *Nicotiana debneyi* (L,W), *Nicandra physalodes* (W).

Dicotyledonous Hosts *(**Susceptible** or Insusceptible)*

***Capsicum annuum**. **Capsicum frutescens**. **Chenopodium amaranticolor**. **Chenopodium quinoa**. Cucumis melo. Cucumis sativus. Cyamopsis tetragonoloba. **Datura metel**. Datura stramonium. Gomphrena globosa. **Lycopersicon esculentum**. **Lycopersicon pimpinellifolium**. **Nicandra physaloides**. **Nicotiana clevelandii**. **Nicotiana debneyi**. **Nicotiana glutinosa**. **Nicotiana rustica**. **Nicotiana sylvestris**. **Nicotiana tabacum**. **Petunia hybrida**. Phaseolus vulgaris. **Physalis floridana**. **Physalis peruviana**. Pisum sativum. **Solanum nigrum**. Solanum tuberosum. Vicia faba. Vigna unguiculata.*

Sources of host-range data

Feldman and Gracia (1977).

**Properties of particles in sap**

TIP: 85–90 °C. LIV: 32–64 days. DEP: $\log_{10}$ minus 6–7 (values for type strain, but some strains less stable). Leaf sap contains few virions.

**Purification**

Method: Feldman and Gracia (1977).

**Particle morphology**

Virions filamentous; not enveloped; usually flexuous; with a clear modal length; of 761 nm; 13 nm wide. Axial canal obscure. Basic helix obscure.

*Sequence database accession code(s):*

X66027 Em(40)_vi:PSMVCP Gb(84)_vi:PSMVCP Pepper Severe Mosaic Virus genomic RNA for coat protein. 6/92 1,329bp. 1 sequence.

**Cytopathology**

Inclusions present in infected cells; are pinwheels; they do not contain virions.

**Taxonomy and Relationships**

*Potyvirus: Potyviridae.*

Virus(es) with serologically related virions: tobacco etch and potato Y viruses, but very distantly. Virus(es) with serologically unrelated virions: pepper mottle, pepper veinal mottle and turnip mosaic viruses.

**References**

Feldman, J.M. and Gracia, O. (1977). *Phytopath. Z.* **89**: 146.

# Pepper Texas bigeminivirus

**Data collated by D.C. Stenger, 1991.**

*Synonyms:* Texas pepper virus. *Standard acronym:* TPV. *Strains:* distortion strain, yellow mosaic strain.

First reported in *Capsicum annuum*; from Texas, U.S.A.; by Stenger *et al.* (1990).

**Natural host range and symptoms**

Symptoms persist. Symptoms leaf curl, malformation and stunting or mosaic.

*Capsicum annuum* — leaf curling, malformation, vein clearing and stunting (distortion strain); mosaic, vein clearing and stunting (mosaic strain).

**Transmission**

Transmitted by a vector; an insect; *Bemisia tabaci*; Aleyrodidae. Transmitted in a persistent manner. Virus transmitted by mechanical inoculation; transmitted by grafting; not transmitted by contact between plants; not transmitted by seed; not transmitted by pollen.

**Geographical distribution**

Spreads in Mexico and the USA.

**Experimental host range**

Few (<3) families susceptible. Experimentally infected plants mostly show leaf curl, malformation, stunting, vein malformation, mosaic.

Diagnostically susceptible

*Capsicum annuum* — leaf curl, malformation, stunting, vein malformation.
*Datura stramonium, Lycopersicon esculentum* — mosaic, vein clearing, stunting.
*Nicotiana benthamiana* — mosaic, vein clearing, leaf curl, stunting.

Diagnostically insusceptible

*Chenopodium quinoa, Cucumis melo, Malva parviflora, Phaseolus vulgaris.*

Maintenance and propagation hosts

*Nicotiana benthamiana.*

Assay hosts (**L**ocal lesions or **W**hole plants)

*Nicotiana benthamiana* (W), *Capsicum annuum* (W).

Dicotyledonous Hosts *(**Susceptible** or Insusceptible)*

***Capsicum annuum***. *Chenopodium quinoa*. *Citrullus lanatus*. *Cucumis melo*. *Cucurbita maxima*. ***Datura stramonium***. *Gomphrena globosa*. ***Lycopersicon esculentum***. ***Nicotiana benthamiana***. ***Nicotiana clevelandii***. ***Nicotiana rustica***. ***Nicotiana tabacum***. *Phaseolus vulgaris*.

Comments on host-range

Restricted to Solanaceae.

Sources of host-range data

Stenger *et al.* (1990).

**Properties of particles in sap**

Leaf sap contains few virions. Electron microscopy: stain with UA.

**Purification**

Method: Stein *et al.* (1983); Stenger *et al.* (1990).

**Particle morphology**

Virions geminate (isometric virions also present); not enveloped; 18 nm in diameter; dimers 40 nm in length; without a conspicuous capsomere arrangement.

**Physical properties**

Two sedimenting components in purified preparations; sedimentation coefficient of the fastest 88 *S*; of the other(s) 50 *S*.

**Biochemical properties**

*Genome* consists of DNA; single–stranded; circular; of two parts; largest (or only) genome part the larger 2.6 kb; the 2nd largest 2.6 kb. Genomic nucleic acid isolated by : extracting in phenol/ chloroform, precipitate with ethanol. Additional factor not required for infectivity.

*Non–genomic nucleic acid* not found in the virions.

**Replication**

Genome replicates in nuclei. Replication does not depend on a helper virus.

**Cytopathology**

Virions found in leaves and roots; in nuclei.

**Taxonomy and Relationships**

Bigeminivirus: *Geminiviridae.*

Texas pepper geminivirus DNA A has homology with tomato golden mosaic virus DNA A. Best tests for diagnosis: DNA A cross hybridizes with the A DNAs of other bipartite geminiviruses. DNA B does not hybridize with other geminivirus DNAs.

**References**

Stein, V.E., Coutts, R.H.A. and Buck, K.W. (1983). *J. gen. Virol.* **64**: 2493.

Stenger, D.C., Duffus, J.E. and Villalon, B. (1990). *Phytopathology* **80**: 704.

# Pepper veinal mottle *potyvirus*

**Data collated by A.A. Brunt, 1985.**

*Standard acronym:* PVMV. *Strains:* isolates differ in host range and pathogenicity.

First reported in *Capsicum annuum, C. frutescens, Petunia hybrida*; from Tafo, Ghana; by Brunt and Kenten (1971).

**Natural host range and symptoms**

Symptoms persist.

*Capsicum annuum, C. frutescens* — mottle, leaf shape malformation.

*Datura stramonium* — mosaic, fruit malformation, with spine suppression.

*Lycopersicon esculentum* — mosaic and necrotic or chlorotic spots and streaks.

*Petunia hybrida* — mottle and mosaic.

*Datura metel, Nicotiana tabacum, Solanum integrifolium, S. melongena, S. nigrum* — mosaic.

**Transmission**

Transmitted by a vector; an insect; *Aphis gossypii, A. spiraecola, Myzus persicae, Toxoptera citricidus*; Aphididae. Transmitted in a non–persistent manner. Virus transmitted by mechanical inoculation; transmitted by grafting; not transmitted by contact between plants.

**Geographical distribution**

Spreads in Cote d'Ivoire, Ghana, Kenya, Nigeria, and South Africa.

**Experimental host range**

Several (3–9) families susceptible.

Diagnostically susceptible

*Chenopodium amaranticolor, C. foetidum* — local chlorotic or necrotic ringspots.

*Capsicum annuum, C. frutescens* — systemic mosaic, mottle and leaf shape malformation.

*Nicotiana clevelandii, N. megalosiphon, N. occidentalis, Physalis angulata, P. floridana* — systemic mosaic. *N. tabacum* cv. Dutch A — local necrotic or chlorotic ringspots.

Diagnostically insusceptible

*Gomphrena globosa, Beta vulgaris, Brassica pekinensis, Cucumis sativus, Nicotiana sylvestris.*

Maintenance and propagation hosts

*Nicotiana benthamiana, N. clevelandii, N. megalosiphon, N. occidentalis, Petunia hybrida.*

Assay hosts (**L**ocal lesions or **W**hole plants)

*Chenopodium amaranticolor* (L), *C. quinoa* (L), *C. foetidum* (L) at 20–24 °C.

Dicotyledonous Hosts *(**Susceptible** or Insusceptible)*

***Ageratum conyzoides***. *Amaranthus caudatus*. ***Brassica campestris*** ssp. ***pekinensis***. ***Brassica juncea***. ***Brassica oleracea*** ssp. ***capitata***. ***Calendula officinalis***. ***Capsicum annuum***. ***Capsicum frutescens***. ***Catharanthus roseus***. ***Celosia argentea***. *Chenopodium amaranticolor*. *Chenopodium capitatum*. *Chenopodium foetidum*. *Chenopodium murale*. *Chenopodium quinoa*. ***Citrus paradisi***. ***Crotalaria spectabilis***. ***Cucumis melo***. ***Cucumis sativus***. ***Datura metel***. ***Datura stramonium***. ***Emilia sagittata***. ***Glycine max***. ***Gomphrena globosa***. ***Helianthus annuus***. *Hyoscyamus niger*. ***Lactuca sativa***. ***Lycopersicon esculentum***. *Lycopersicon pimpinellifolium*. *Nicandra physaloides*. *Nicotiana benthamiana*. *Nicotiana clevelandii*. *Nicotiana glutinosa*. *Nicotiana megalosiphon*. *Nicotiana rustica*. ***Nicotiana sylvestris***. ***Nicotiana tabacum***. ***Ocimum basilicum***. ***Petunia hybrida***. ***Phaseolus vulgaris***. *Physalis floridana*. *Physalis peruviana*. ***Phytolacca americana***. ***Solanum melongena***. ***Solanum nigrum***. ***Solanum tuberosum***. *Tetragonia expansa*. ***Trifolium incarnatum***. ***Verbesina encelioides***. ***Vicia faba***. ***Zinnia elegans***.

Monocotyledonous Hosts *(**Susceptible** or Insusceptible)*

***Allium cepa***. ***Zea mays***.

Sources of host-range data

Brunt and Kenten (1971); de Wijs (1973); Ladipo and Roberts (1977); Brunt *et al.* (1978).

**Properties of particles in sap**

TIP: 55–60 °C. LIV: 7–8 days. DEP: $\log_{10}$ minus 3–4. Infectivity of sap not changed by treatment with di–ethyl ether. Leaf sap contains few virions.

**Purification**

Method: Brunt and Kenten (1971).

**Particle morphology**

Virions filamentous; not enveloped; usually flexuous; with a clear modal length; of 770 nm; 12 nm wide. Axial canal obscure.

**Physical properties**

One sedimenting component in purified preparations; sedimentation coefficient 155 *S*.

**Biochemical properties**

Virions contain 6 % nucleic acid; 94 % protein.

*Genome* consists of RNA; single–stranded. Base composition 24 % G; 23 % A; 27 % C; 26 % U.

*Virion protein(s)* one; $M_r$ 32000–33000.

**Cytopathology**

Virions found in all parts of the host plant; in cytoplasm. Inclusions present in infected cells; are pinwheels; they do not contain virions.

**Taxonomy and Relationships**

*Potyvirus: Potyviridae.*

Virus(es) with serologically related virions: bean yellow mosaic, clover yellow vein, Colombian datura virus, datura shoestring, potato Y, tobacco etch and watermelon mosaic viruses. Virus(es) with serologically unrelated virions: henbane mosaic, iris mild mosaic, hippeastrum mosaic, lettuce mosaic, pea seed-borne mosaic and turnip mosaic viruses.

**References**

Brunt, A.A. and Kenten, R.H. (1971). *Ann. appl. Biol.* **69**: 235.
Brunt, A.A. and Kenten, R.H. (1972). *CMI/AAB Descr. Pl. Viruses* No. 104, 4 pp.
Brunt, A.A., Kenten, R.H. and Phillips, S. (1978). *Ann. appl. Biol.* **88**: 115.
Christie, R.G. and Edwardson, J.R. (1977). *Fla Agric. Exp. Stn Monog.* No. 9.
de Wijs, J.J. (1973). *Neth. J. Pl. Path.* **79**: 189.
Igwegbe, E.C.K. and Waterworth, H.E. (1982). *Phytopath. Z.* **103**: 9.
Ladipo, J.L. and Roberts, I.M. (1977). *Ann. appl. Biol.* **87**: 133.
Ladipo, J.L. and Roberts, I.M. (1979). *Pl. Dis. Reptr* **63**: 161.
Lana, A.O., Gilmer, R.M., Wilson, G.F. and Shoyinka, S.A. (1975). *Niger. J. Pl. Prot.* **1**: 76.

# Perilla mottle (?) potyvirus

**Data collated by A.A. Brunt, 1991.**

*Standard acronym:* PerMV.

First reported in *Perilla frutescens*; by Lee *et al.* (1980).

**Natural host range and symptoms**

Symptoms persist.

*Perilla frutescens* — leaf mottling.

**Transmission**

Transmitted by a vector; an insect; *Myzus persicae*; Aphididae. Transmitted in a non–persistent manner. Virus transmitted by mechanical inoculation.

**Geographical distribution**

Spreads in Japan.

**Experimental host range**

Several (3–9) families susceptible.

Diagnostically susceptible

*Chenopodium amaranticolor, C. quinoa* — local lesions.

*Perilla frutescens* — systemic leaf mottling.

Diagnostically insusceptible

*Cucumis sativus, Datura stramonium, Nicotiana tabacum, Phaseolus vulgaris, Vigna unguiculata.*

Maintenance and propagation hosts

*Perilla frutescens.*

Assay hosts (**L**ocal lesions or **W**hole plants)

*Chenopodium amaranticolor* (L), *C. quinoa* (L), *Perilla frutescens* (W).

Dicotyledonous Hosts *(**Susceptible** or Insusceptible)*

*Arachis hypogaea. Beta vulgaris. Brassica campestris* ssp. *napus. Brassica oleracea* ssp. *capitata. Capsicum annuum.* ***Chenopodium amaranticolor****. Chenopodium murale.* ***Chenopodium quinoa****. Cucumis sativus. Datura metel. Datura stramonium.* ***Gomphrena globosa****. Lactuca sativa. Lycopersicon esculentum. Nicotiana glutinosa. Nicotiana tabacum. Petunia hybrida. Phaseolus vulgaris. Physalis floridana. Pisum sativum. Salvia splendens. Sesamum indicum. Spinacia oleracea. Tetragonia expansa.* ***Vicia faba****. Vigna angularis. Vigna unguiculata.*

**Particle morphology**

Virions filamentous; not enveloped; usually flexuous; with a clear modal length; of 760 nm. Basic helix obvious; pitch of basic helix 3.4 nm.

**Cytopathology**

Inclusions present in infected cells; are pinwheels; they do not contain virions.

**Taxonomy and Relationships**

Unassigned potyvirus.

**References**

Lee, J.T., Yamashita, S., Doi, Y. and Yora, K. (1980). *Ann. Phytopath. Soc. Japan* **46**: 672.

# Petunia asteroid mosaic *tombusvirus*

**Data collated by A.A. Brunt, 1995.**

*Synonyms:* tomato bushy stunt virus — petunia strain (Martelli *et al.*, 1971). *Standard acronym:* PeAMV.

First reported in *Petunia hybrida*; from Italy; by Lovisolo (1957).

**Natural host range and symptoms**

*Petunia hybrida* — yellow stellate spots in leaves, leaf distortion and flower colour breaking.

*Humulus lupulus* — yellow mottling and leaf distortion.

*Ligustrum vulgare, Cornus sanguinea* — chlorotic leaf spotting.

*Prunus avium* — leaf veinal necrosis and distortion, leaves bunched, shoots stunted with bark splitting, fruits with sunken pits.

*Prunus domestica* — pox-like markings on fruits.

*Spinacia oleracea* — Vein clearing, leaf mottling and distortion and stunting.

*Vitis vinifera* — symptoms unknown as the virus only found in mixtures with others.

**Transmission**

Transmitted by means not involving a vector; probably abiotically in soil. Virus transmitted by mechanical inoculation; transmitted by grafting.

**Geographical distribution**

Spreads in the Eurasian region; Canada, the former Czechoslovakia, Germany, Switzerland, the UK, and the former Yugoslavia.

**Experimental host range**

Several (3–9) families susceptible. Experimentally infected plants mostly show local lesions or necrotic lesions.

Diagnostically susceptible

Details to follow.

**Purification**

Method: Ambrosino *et al.* (1967).

**Particle morphology**

Virions isometric; 30 nm in diameter.

**Physical properties**

One sedimenting component in purified preparations; sedimentation coefficient 134 *S*. Density 1.35 g $cm^{-3}$ in CsCl (unfixed).

**Biochemical properties**

*Genome* consists of RNA; single–stranded; unipartite; largest (or only) genome part 4.7 kb. Genomic nucleic acid isolated by Ambrosino *et al.* (1967). Base composition 28 % G; 27 % A; 22 % C; 23 % U.

*Virion protein(s)* one; $M_r$ 41000; coat protein.

**Taxonomy and Relationships**

*Tombusvirus: Tombusviridae.*

Virus(es) with serologically related virions: Tomato bushy stunt, carnation Italian ringspot, pelargonium leaf curl viruses.

**Comments**

The virus was previously identified in some European countries and Canada as tomato bushy stunt virus.

**References**

Albrechtova, L., Chod, J. and Zimandl, B. (1975). *Phytopath. Z.* **82**: 25.

Allen, W.R. (1968). *Can. J. Bot.* **45**: 229.

Allen, W.R. and Davidson, T.R. (1967). *Can. J. Bot.* **45**: 2375.

Ambrosino, C., Appiano, A., Rialda, G., Papa, G., Redolfi, P. and Carrara, M. (1967). *Alti Acc. Sci. Torino* **101**: 301.

Eric, Z., Stefanac, Z. and Plavsic, B. (1986). *Acta bot. Croat.* **45**: 7.

Gallitelli, D., Hull, R. and Koenig, R. (1985). *J. gen. Virol.* **66**: 1523.

Hollings, M. and Stone, O.M. (1965). *Ann. appl. Biol.* **56**: 87.

Kegler, G. and Kegler, H. (1980). *Tag. Ber. Akad. Landw. Wiss. DDR* **184**: 297.

Koenig, R. and Kunze, L. (1982). *Phytopath. Z.* **103**: 361.

Lovisolo, O. (1957). *Bull. Staz. Pat. Veg. Roma* **14**: 103.

Lovisolo, O., Bode, O. and Völk, J. (1965). *Phytopath. Z.* **53**: 323.

Martelli, G.P., Quacquarelli, A. and Russo, M. (1971). *CMI/AAB Descr. Pl. Viruses* No. 69, 4pp.

Novak, J.B. and Lanzova, J. (1976). *Biol. Plant* **18**: 152.

Novak, J.B. and Lanzova, J. (1977). *Biol. Plant* **19**: 234.

# Petunia vein clearing (?) *caulimovirus*

**Data collated by D.-E. Lesemann, 1991.**

*Standard acronym:* PVCV.

First reported in *Petunia Juss. x hybrida hort.* cv. Himmelsroeschen; from Germany; by Lesemann and Casper (1973).

## Natural host range and symptoms

Symptoms persist and vary seasonally. Symptoms chlorotic vein clearing, leaf malformation.

*Petunia Juss. x hybrida hort.* cv. Himmelsroeschen — chlorotic clearing and banding of leaf veins, leaf malformation, stunting of sprouts.

## Transmission

Virus not transmitted by mechanical inoculation; transmitted by grafting.

## Geographical distribution

Found, but with no evidence of spread, in Germany.

## Experimental host range

Few (<3) families susceptible. Experimentally infected plants mostly show vein clearing, leaf malformation.

Diagnostically susceptible

*Petunia* spp. — vein clearing, chlorotic vein banding and leaf distortion.

Maintenance and propagation hosts

*Petunia* spp.

Dicotyledonous Hosts *(**Susceptible** or Insusceptible)*

***Petunia hybrida**.*

## Properties of particles in sap

Leaf sap contains few virions.

## Particle morphology

Virions isometric; not enveloped; 43–46 nm in diameter; rounded in profile; without a conspicuous capsomere arrangement.

## Cytopathology

Virions found in leaves, mesophyll, epidermis, vascular parenchyma, phloem and the companion cells; in cytoplasm and in nuclei. Inclusions present in infected cells; are viroplasms and unusual in shape; proliferation of endoplasmic reticulum; they do not contain virions. Other cellular changes: proliferation of endoplasmic reticulum into multilayers, surrounding the virion-containing regions which also contain the dark staining, viroplasm like rounded and vacuolated inclusions.

## Taxonomy and Relationships

Tentatively *Caulimovirus*.

Virus(es) with serologically unrelated virions: cauliflower mosaic and dahlia mosaic viruses.

## References

Casper, R., Lesemann, D.-E. and Fischer, D. (1973). *Proc. 7th Conf. Czechoslovak Pl. Virologists, High Tatras 1971*, Slovak Acad. Sci., Bratislava, p. 247.

Lesemann, D.-E. and Casper, R. (1973). *Phytopathology* **63**: 1118.

# Phleum green stripe (?) *tenuivirus*

**Data collated by K. Bremer, 1991.**

First reported in *Phleum* sp.; from Finland; by Heikinheimo and Raatikainen (1976).

**Natural host range and symptoms**

*Phleum* spp. — green stripes in leaves.

**Transmission**

Transmitted by a vector; an insect; *Megadelphax sordidula, Dicranotropis hamata*; Delphacidae. Principal natural vector(s): *Megadelphax sordidula.* Transmitted in a persistent manner. Virus retained when the vector moults; not transmitted by mechanical inoculation; not transmitted by contact between plants.

**Ecology and control**

Studies reported by Heikinheimo and Raatikainen (1976): crop rotation and plough early in autumn.

**Geographical distribution**

Spreads in Finland.

**Experimental host range**

Experimentally infected plants mostly show light green or yellow leaf stripes, seed wrinkling and stunting.

Diagnostically susceptible

*Phleum pratense, Avena sativa, Triticum aestivum, Hordeum vulgare* — light green or yellow leaf stripes, seed wrinkling and stunting.

Diagnostically insusceptible

*Dactylis glomerata, Festuca pratensis, F. rubra, Lolium multiflorum.*

Maintenance and propagation hosts

*Phleum pratense, Avena* spp.

Monocotyledonous Hosts *(**Susceptible** or Insusceptible)*

***Avena sativa***. *Dactylis glomerata*. *Festuca pratensis*. ***Hordeum vulgare***. *Lolium multiflorum*. ***Triticum aestivum***.

**Taxonomy and Relationships**

Possible *Tenuivirus.*

**References**

Gingery, R.E. (1988). In: *The Plant Viruses*, Vol. 4, *The Filamentous Viruses*, p.303. ed. R.G. Milne. Plenum Press, New York.

Heikinheimo, O. and Raatikainen, M. (1976). *Ann. Agric. Fenn.* **15**: 34..

# Physalis mild chlorosis (?) *luteovirus*

**Data collated by A.A. Brunt, 1991.**

*Synonyms:* turnip latent virus disease complex (part; MacKinnon, 1956). *Standard acronym:* PhyMCV.

First reported in *Physalis floridana*; from Canada; by MacKinnon and Lawson (1966).

**Natural host range and symptoms**

*Brassica campestris* var. *rapa* — symptomless infection, even in complex with physalis blotch virus.

**Transmission**

Transmitted by a vector; an insect; *Acyrthosiphon (Aulocorthum) solani, Myzus persicae*; Aphididae. Not transmitted by *Aphis nasturtii, Macrosiphum euphorbiae, Myzus (Sciamyzus) ascalonicus*. Transmitted in a persistent manner.

**Geographical distribution**

Spreads in Canada.

**Experimental host range**

Few (<3) families susceptible.

Diagnostically susceptible

*Physalis floridana* — systemic mild leaf chlorosis; conspicuous chlorotic vein-netting leaf symptoms when present in complex with physalis vein blotch virus.

*Raphanus sativus, Nicandra physaloides* — slight chlorosis of systemically infected leaves.

*Gomphrena globosa, Beta vulgaris, Spinacia oleracea, Brassica napus, B. rapa, Datura bernhardii* — symptomless systemic infection.

Diagnostically insusceptible

*Chenopodium album, Capsicum annuum, Datura metel, D. stramonium, Petunia hybrida* and *Solanum tuberosum*.

Assay hosts (**L**ocal lesions or **W**hole plants)

*Physalis floridana* (W).

Dicotyledonous Hosts *(**Susceptible** or Insusceptible)*

***Brassica campestris*** ssp. ***napus***. ***Brassica campestris*** ssp. ***rapa***. ***Gomphrena globosa***. ***Nicandra physaloides***. ***Physalis floridana***. ***Raphanus sativus***.

**Taxonomy and Relationships**

Possible *Luteovirus*.

Physalis mild chlorosis (?) *luteovirus* is possibly a strain of beet western yellows luteovirus.

**References**

MacKinnon, J.P. (1956). *Can. J. Bot.* **34**: 131.
MacKinnon, J.P. (1965). *Can. J. Bot.* **43**: 509.
MacKinnon, J.P. and Lawson, F.L. (1966). *Can. J. Bot.* **44**: 795.
Waterhouse, P.M., Gildow, F.E. and Johnstone, G.R. (1988). *AAB Descr. Pl. Viruses* No. 339, 9 pp.

# Physalis mosaic *tymovirus*

**Data collated by A.J. Gibbs, 1991.**

*Synonyms:* belladonna mottle virus — US, belladonna mottle virus — Iowa. *Standard acronym:* PhyMV.

First reported in *Physalis subglabrata*; from U.S.A. (Urbana, Illinois); by Peters and Derks (1974).

**Natural host range and symptoms**

Symptoms persist. Symptoms mosaics and mottles.
*Physalis subglabrata, P. heterophylla* — systemic mosaic or mottle.

**Geographical distribution**

Spreads in the USA.

**Experimental host range**

Several (3–9) families susceptible. Experimentally infected plants mostly show necrotic or chlorotic local lesions, systemic mottle or mosaic.

Diagnostically susceptible

*Capsicum annuum* — chlorotic local lesions, systemic mottle.
*Datura metel, D. stramonium* — necrotic local lesions, systemic mosaic.
*Nicandra physaloides* — systemic mosaic.
*Nicotiana clevelandii, N. glutinosa, N. tabacum* cv. White Burley — necrotic local lesions.
*Chenopodium foetidum* — systemic chlorosis.
*Sonchus oleracea* — symptomless infection.

Diagnostically insusceptible

*Atropa belladonna, Brassica campestris* spp. *pekinensis, Phaseolus vulgaris, Pisum sativum, Solanum melongena.*

Maintenance and propagation hosts

*Datura stramonium, Nicotiana clevelandii, N. glutinosa.*

Assay hosts (**L**ocal lesions or **W**hole plants)

*Datura stramonium* (W), *Nicotiana clevelandii* (W).

Dicotyledonous Hosts ***(Susceptible** or Insusceptible)*

*Beta vulgaris. Brassica campestris* ssp. *pekinensis.* ***Capsicum annuum.*** *Catharanthus roseus. Celosia argentea. Chenopodium album.* ***Chenopodium foetidum.*** *Chenopodium hybridum. Chenopodium murale.* ***Chenopodium quinoa***, or *Chenopodium quinoa. Cucumis melo. Cucumis sativus.* ***Datura metel. Datura stramonium.*** *Daucus carota. Dianthus caryophyllus. Glycine max. Gomphrena globosa.* ***Helianthus annuus.*** *Lactuca sativa.* ***Lycopersicon esculentum.*** *Medicago sativa.* ***Nicandra physaloides. Nicotiana benthamiana. Nicotiana clevelandii. Nicotiana debneyi. Nicotiana glutinosa. Nicotiana megalosiphon. Nicotiana rustica. Nicotiana sylvestris. Nicotiana tabacum. Petunia hybrida.*** *Phaseolus vulgaris.* ***Physalis floridana. Physalis peruviana.*** *Pisum sativum. Raphanus sativus. Solanum chacoense. Solanum demissum. Solanum melongena.* ***Solanum nigrum***, or *Solanum nigrum. Solanum tuberosum.* ***Sonchus oleraceus.*** *Trifolium incarnatum. Trifolium repens. Vicia faba. Vigna unguiculata. Zinnia elegans.*

Monocotyledonous Hosts ***(Susceptible** or Insusceptible)*

*Hordeum vulgare.* ***Triticum aestivum.*** *Zea mays.*

Sources of host-range data

Peters and Derks (1974).

**Properties of particles in sap**

LIV: 50 days. DEP: $\log_{10}$ minus 8. Infectivity of sap not changed by treatment with di–ethyl ether. Leaf sap contains many virions.

**Purification**

Method: Peters and Derks (1974).

**Particle morphology**

Virions isometric; not enveloped; 27–29 nm in diameter; rounded in profile; with a conspicuous capsomere arrangement.

**Physical properties**

Two sedimenting components in purified preparations; sedimentation coefficient of the fastest 112–114 *S*; of the other(s) 50–54 *S*. Density 1.42 g $cm^{-3}$ in CsCl (B), or 1.27 g $cm^{-3}$ in CsCl (T). $A_{260}/A_{280}$ ratio 1.63 (B), or 1.06 (T).

**Biochemical properties**

Virions contain 38 % nucleic acid; 62 % protein.

*Genome* consists of RNA; single–stranded; linear; unipartite. Base composition 14.4 % G; 22.9 % A; 37.2 % C; 25.5 % U. Genome has *t*RNA–like activity. Genome accepts valine.

*Sequence database accession code(s):*

S97776 Em(40)_vi:S97776 Gb(84)_vi:S97776 replicase protein, coat protein (3' region) Physalis mottle virus PhMV.

*Virion protein(s)* one; $M_r$ 20000; coat protein. Amino acid sequence: Jacob *et al.* (1991).

**Replication**

Replication does not depend on a helper virus.

**Cytopathology**

Virions found in all parts of the host plant.

**Taxonomy and Relationships**

*Tymovirus*.

Virus(es) with serologically related virions: belladonna mottle and dulcamara mottle viruses.

Most of the plant species inoculated with physalis mosaic virus or with belladonna mottle virus (I) isolate reacted similarly however *Chenopodium quinoa* and *Solanum nigrum* were infected by the latter and not the former. Slight cross protection with Andean potato latent and dulcamara mottle viruses.

**References**


Jacob, A.N.K., Murthy, M.R.N. and Savithri, H.S. (1991). *J. gen. Virol.* in press.

Moline, H.E. and Fries, R.E. (1974). *Phytopathology* **64**: 44.

Peters, D. and Derks, A.F.L.M. (1974). *Neth. J. Pl. Path.* **80**: 124.

Peter, R., Peter, C., Dupin, A. and Witz, J. (1989). *C.r. Acad. Sci. Paris* **309** Serie III, p. 599.

Savithri, H.S., Monski, S.K., Suryanarayana, S., Divikar, S. and Murthy, M.R.N. (1987). *J. gen. Virol.* **68**: 1533.


# Physalis vein blotch (?) *luteovirus*

**Data collated by A.A. Brunt, 1991.**

*Synonyms:* turnip latent virus disease complex (part; MacKinnon, 1956). *Standard acronym:* PhyVBV.

First reported in *Physalis floridana*; from Canada; by MacKinnon and Lawson (1966).

**Natural host range and symptoms**

*Brassica campestris* var. *rapa* — symptomless infection, even in complex with physalis mild chlorosis virus.

**Transmission**

Transmitted by a vector; an insect; *Myzus persicae*; Aphididae. Transmitted in a persistent manner.

**Experimental host range**

Few (<3) families susceptible.

Diagnostically susceptible

*Physalis floridana* — usually symptomless but co-infection with this virus and physalis mild chlorosis (?) *luteovirus* induce chlorotic vein-netting.

Diagnostically insusceptible

*Chenopodium album, Capsicum annuum, Datura metel, D. stramonium, Petunia hybrida* and *Solanum tuberosum.*

Assay hosts (**L**ocal lesions or **W**hole plants)

*Physalis floridana* (W).

**Taxonomy and Relationships**

Possible *Luteovirus.*

**References**

MacKinnon, J.P. (1956). *Can. J. Bot.* **34**: 131.
MacKinnon, J.P. (1965). *Can. J. Bot.* **43**: 509.
MacKinnon, J.P. and Lawson, F.L. (1966). *Can. J. Bot.* **44**: 795.
Waterhouse, P.M., Gildow, F.E. and Johnstone, G.R. (1988). *AAB Descr. Pl. Viruses* No. 339, 9 pp.

# Pigeonpea proliferation (?) rhabdovirus

**Data collated by Y.L. Nene, 1981. Revised 1983.**

*Synonyms:* pigeonpea witches' broom virus, pigeonpea rosette virus. *Standard acronym:* PPPV.

First reported in *Cajanus cajan*; from Dominican Republic; by Hirumi *et al.* (1973).

**Natural host range and symptoms**

Symptoms persist.

*Cajanus cajan* — stunting, bushiness, although leaf size normal, base of midrib of young leaflets chlorotic, rough brown surface of branches.

**Transmission**

Transmitted by a vector; an insect; possibly the leafhopper, *Empoasca papayae*; Cicadellidae. Virus not transmitted by mechanical inoculation; not transmitted by contact between plants; not transmitted by seed; not transmitted by pollen.

**Geographical distribution**

Spreads in the Dominican Republic, Haiti, Jamaica, Papua New Guinea, Puerto Rico, Taiwan, and Trinidad and Tobago.

Dicotyledonous Hosts *(**Susceptible** or Insusceptible)*

***Cajanus cajan.***

**Particle morphology**

Virions rhabdo– or bullet–shaped (MLOs also found).

**Taxonomy and Relationships**

Possible unassigned rhabdovirus.

**References**

Hirumi, H., Maramorosch, K. and Hichez, E. (1973). *Phytopathology* **63**: 202.
Maramorosch, K., Hirumi, H. and Kimura, M. (1974*a*). *FAO Pl. Prot. Bull.* **22**: 32.
Maramorosch, K., Hirumi, H., Kimura, M., Bird, J. and Vakili, N.G. (1974*b*). *Phytopathology* **64**: 582.
Maramorosch, K., Kimura, M. and Nene, Y.L. (1976). *FAO Pl. Prot. Bull.* **24**: 33.

# Pigeonpea sterility mosaic (?) virus

**Natural host range and symptoms**

*Cajanus cajan.*

**Comments**

The pathogen causing the sterility mosaic disease of pigeonpea *Cajanus cajan* in India is transmissible by mites (*Aceria cajani*) and by grafting, and has long been considered to be virus-induced (Capoor, 1952). However, as no virus has been isolated from, or detected in, affected plants during recent intensive investigations, the pathogen is still unknown (Wendy Sachs, personal communication).

**References**

Capoor, S.P. (1952). *Ind. J. agric. Sci.* **22**: 274.

# Pineapple chlorotic leaf streak (?) *nucleorhabdovirus*

**Data collated by A.A. Brunt, 1987.**

*Standard acronym:* PCLSV.

First reported in *Ananas comosus*; from Brazil; by Kitajima *et al.* (1975).

**Natural host range and symptoms**

*Ananas comosus* — conspicuous chlorotic leaf streaks.

**Geographical distribution**

Spreads in Brazil.

**Particle morphology**

Virions bacilliform, or rhabdo– or bullet–shaped; with a clear modal length; of 200–250 nm (when measured in leaf sections); 60–70 nm wide. Axial canal obvious.

**Cytopathology**

Virions found in mesophyll, epidermis and phloem occasionally; in the perinuclear space.

**Taxonomy and Relationships**

Possible *Nucleorhabdovirus: Rhabdoviridae.*

**References**

Kitajima, E.W., Giacomelli, E.J., Costa, A.S., Costa, C.L. and Cupertino, F.P. (1975). *Phytopath. Z.* **82**: 83.

# Pineapple wilt-associated (?) *closterovirus*

**Data collated by A.A. Brunt and U.B. Gunasinghe, 1988. Updated 1991.**

*Synonyms:* pineapple mealybug wilt-associated (?) *closterovirus* (Gunasinghe *et al.*, 1989; Ullman *et al.*, 1989). *Standard acronym:* PMWaV.

First reported in *Ananas comosus* cv. Smooth Cayenne; from Hawaii; by Carter (1933).

**Natural host range and symptoms**

Symptoms persist, vary seasonally, and disappear soon after infection.

*Ananas comosus* — yellowing and flaccidity of young leaves, necrosis of leaf tips; plants wilt and may die.

**Transmission**

Transmitted by a vector; an insect; *Pseudococcus brevipes*; Pseudococcidae. Transmitted in a semi–persistent manner (possibly). Virus not transmitted by mechanical inoculation.

**Ecology and control**

Studies reported by Gunasinghe and German (1986; 1988).

**Geographical distribution**

Spreads in the USA (Hawaii).

**Experimental host range**

Few (<3) families susceptible.

Diagnostically susceptible

*Ananas comosus* — young leaves become yellow and flacid with brown and necrotic tips. Plants wilt.

Maintenance and propagation hosts

*Ananas comosus.*

Assay hosts (**L**ocal lesions or **W**hole plants)

*Ananas comosus* (W).

**Properties of particles in sap**

Leaf sap contains few virions.

**Purification**

Method: Gunasinghe and German (1977). Extract tissue in 0.5M Tris, followed by PEG precipitation and differential centrifugation.

**Particle morphology**

Virions filamentous; not enveloped; usually flexuous; with a clear modal length; of 1200–1500 nm; 12 nm wide. Axial canal obscure.

**Physical properties**

One sedimenting component in purified preparations. Density 1.2 g $cm^{-3}$ in caesium sulphate.

**Biochemical properties**

*Genome* consists of RNA; single–stranded; linear; unipartite; largest (or only) genome part 7.1 kb.

*Virion protein(s)* one; $M_r$ 23000.

**Cytopathology**

Virions found in leaves and phloem.

**Taxonomy and Relationships**

Possible *Closterovirus*.

Best tests for diagnosis: detection of dsRNA can be used for diagnosis (Gunasinghe and German, 1986; 1988).

**Comments**

The virus can be eliminated from planting stock by hot water treatment of "crowns" for up to 120 minutes at 50°C.

**References**

Carter, W.C. (1933). *Phytopathology* **23**: 207.
Gunasinghe, U.B. and German, T. (1986). *Phytopathology* **76**: 1073.
Gunasinghe, U.B. and German, T. (1987). *Phytopathology* **77**: 1776.
Gunasinghe, U.B. and German, T. (1988). *Phytopathology* **78**: 1584.
Gunasinghe, U.B. and German, T.L. (1989). *Phytopathology* **79**: 1337.
Ullman, D.E., German, T.L., Gunasinghe, U.B. and Ebesu, R.H. (1989). *Phytopathology* **79**: 1341.
Ullman, D.E., German, T.L., McIntosh, C.E. and Williams, D.D.F. (1991). *Plant Dis.* **75**: 859.

# Pisum (?) *cytorhabdovirus*

**Data collated by A.A. Brunt, 1991.**

*Standard acronym:* PisV.

First reported in *Pisum sativum*; from Brazil; by Caner *et al.* (1976).

**Natural host range and symptoms**

*Pisum sativum* — no conspicuous symptoms.

**Transmission**

Virus transmitted by mechanical inoculation.

**Geographical distribution**

Spreads in Brazil.

Dicotyledonous Hosts *(**Susceptible** or Insusceptible)*

***Pisum sativum*.**

**Particle morphology**

Virions rhabdo– or bullet–shaped; enveloped; with a clear modal length; of 240 nm (in sections); 45 nm wide (in sections).

**Cytopathology**

Virions found in all parts of the host plant; in cytoplasm.

**Taxonomy and Relationships**

Probable *Cytorhabdovirus: Rhabdoviridae.*

**References**

Caner, J., July, J.R. and Vincente, M. (1976). *Summa Phytopath.* **2**: 264.

# Pittosporum vein yellowing *nucleorhabdovirus*

**Data collated by A.A. Brunt and G.L. Rana, 1991.**

*Synonyms:* pittosporum vein clearing virus (Plásic *et al.*, 1976, 1978; Rana and Di Franco, 1979, 1980), tomato vein yellowing virus, eggplant mottled dwarf virus.

First reported in *Pittosporum tobira*; from Liguria, Italy; by Corte (1957).

**Natural host range and symptoms**

Symptoms persist and vary seasonally. Symptoms stunting, bushy appearance, vein yellowing and mottling, fruit malformation and size reduction.

*Pittosporum tobira* — clearing and chlorotic banding of major leaf veins.

*Lycopersicon esculentum* — stunting, vein yellowing and curling of leaves, yellow spots on fruit.

*Solanum melongena* — stunting, vein yellowing, leaf mottling and crinkling, fruit malformation and size reductions.

*Hibiscus rosa-sinensis, Lonicera* sp. — vein yellowing.

*Solanum sodomaeum* — severe mosaic.

**Transmission**

Not transmitted by *Myzus persicae, Aphis fabae, Aphis spiraecola, Toxoptera aurantium, Aphis gossypii, Macrosiphum* spp. (Rana and Di Franco, 1979; El Maataoui *et al.*, 1985). Virus transmitted by mechanical inoculation; transmitted by grafting; transmitted by contact between plants; not transmitted by seed; transmitted by pollen to the seed.

**Geographical distribution**

Spreads in Italy. Found, but with no evidence of spread, in Yugoslavia and Libya.

**Experimental host range**

Few (<3) families susceptible. Experimentally infected plants mostly show chlorotic local lesions, followed by systemic vein-clearing, leaf curling, veinal necrosis, leaf blade, malformation, mottle, top necrosis.

Diagnostically susceptible

*Capsicum annuum* — systemic vein-clearing and leaf curling.

*Chenopodium amaranticolor* — systemic vein clearing and leaf malformations.

*Gomphrena globosa, Datura stramonium* — chlorotic or necrotic local lesions; not systemic.

*Nicotiana* spp. — chlorotic local lesions; systemic vein-clearing and often veinal necrosis and death of plants.

*Lycopersicon esculentum* — systemic vein clearing and leaf curling.

Diagnostically insusceptible

*Anagyris foetida, Cucumis sativus, Cucurbita pepo, Celosia cristata, Phaseolus aureus, Phaseolus vulgaris, Solanum tuberosum.*

Maintenance and propagation hosts

*Lycopersicon esculentum, Nicotiana alata, N. benthamiana, N. clevelandii, N. debneyi, N. glutinosa, N. rustica, N. tabacum* cv. White Burley.

Assay hosts (**L**ocal lesions or **W**hole plants)

*Gomphrena globosa* (L), *Nicotiana glutinosa* (L and W), *Nicotiana benthamiana* (L and W), *N. rustica* (L and W), *N. alata* (W), *N. clevelandii* (W), *Solanum melongena* (W).

Dicotyledonous Hosts *(**Susceptible** or Insusceptible)*

***Capsicum annuum***. *Celosia cristata*. ***Chenopodium amaranticolor***. ***Chenopodium quinoa***. *Cucumis sativus*. *Cucurbita pepo*. ***Datura metel***. ***Datura stramonium***. ***Gomphrena globosa***. ***Lycopersicon esculentum***. ***Lycopersicon pimpinellifolium***. ***Nicotiana benthamiana***. ***Nicotiana clevelandii***. ***Nicotiana debneyi***. ***Nicotiana glutinosa***. ***Nicotiana megalosiphon***. ***Nicotiana occidentalis***. ***Nicotiana rustica***. ***Nicotiana tabacum***. ***Petunia hybrida***. *Phaseolus vulgaris*. ***Physalis floridana***. ***Solanum melongena***. ***Solanum nigrum***. *Solanum tuberosum*.

Additional hosts

Additional Hosts: *Nicotiana cavicola* (**S**); *Nicotiana langsdorffii* (**S**); *Nicotiana glauca* (**S**); *Solanum sodomaeum* (**S**); *Physalis ixocarpa* (**S**).

Sources of host-range data

Rana and Di Franco (1979); El Maataoui *et al.* (1985); Camele *et al.* (1991).

**Properties of particles in sap**

TIP: 50–55 °C. LIV: 0.12 days (2 hours at about 18°C). DEP: $\log_{10}$ minus 2–3. Leaf sap contains few virions.

**Purification**

Method: at present, no good purification method is known. However, good results were obtained for the TVYV and EMDV isolates using the procedures described by El Maataoui *et al.* (1985). and Adam *et al.* (1987).

**Particle morphology**

Virions rhabdo– or bullet–shaped; enveloped; usually straight; with a clear modal length; of 95–310 nm; 80–96 nm wide. Axial canal obvious; 30 nm in diameter. Basic helix obvious; pitch of basic helix 4–5 nm.

*Virion protein(s)* four; G protein. N protein. $M_1$ protein. $M_2$ protein. Method of preparation: Laemmli (1970); Adam *et al.* (1987). Amino acid sequence: Adam *et al.* (1987). Virion proteins glycosylated; are and it is the G protein.

**Replication**

Replication does not depend on a helper virus.

**Cytopathology**

Virions found in leaves, vascular parenchyma and flowers of the eggplant isolate (Russo and Martelli, 1972); in cytoplasm and in the perinuclear space. Inclusions absent from infected cells. Other cellular changes: less dense nuclei and nucleoli. Perinuclear gaps between the lamellae of the nuclear envelope. Cell wall protrusions contain unusually long rhabdovirus virions.

**Taxonomy and Relationships**

*Nucleorhabdovirus: Rhabdoviridae*.

Virus(es) with serologically related virions: tomato vein yellowing and eggplant mottled dwarf viruses (Adam *et al.*, 1987; Martelli and Cherif, 1987; Camele *et al.*, 1991). Virus(es) with serologically unrelated virions: cynara rhabdovirus.

Pittosporum vein clearing rhabdovirus is very closely related, if not identical, to a Moroccan isolate of TVYV (=EMDV) (Camele *et al.*, 1991). Best tests for diagnosis:

1) sap inoculation of *Nicotiana benthamiana* using fresh young infected leaves of *Pittosporum tobira* as inoculum (Rana and Di Franco, 1979).

2) Inoculate *Cucurbita pepo* to distinguish from cucumber mosaic virus which also infects *Pittosporum* (Rana, 1980).

3) Gel diffusion tests using sap of infected *Nicotiana benthamiana* (Camele *et al.*, 1991; Adam *et al.*, 1987).

**References**

Adam, G., Gnagas, C.M. and Lesemann, D.-E. (1987). *Phytopathology* **120**: 31.

Camele, I., Nuzzaci, M., Rana, G.L., Lafortezza, R., De Stradis, A., Pace, G. and Palumbo, M. (1991). *Inform. Pat. Agrario.*

Corte, A. (1957). *Riv. Ortoflorofruitticol. Ital.* **41**: 97.

Di Franco, A., Russo, M. and Martelli, G.P. (1980). *J. gen. Virol.* **49**: 209.

El Maataoui, M. and Lockhart, B.E.L. (1982). *Phytopathology* **72**: 478.

El Maataoui, M., Lockhart, B.E.L. and Lesemann, D.-E. (1985). *Phytopathology* **75**: 109.

Horvath, J. (1973). *Acta bot. Croat.* **32**: 25.

Laemmli, U.K. (1970). *Nature, Lond.* **227**: 680.

Lockhart, B.E.L. (1987). *Plant Dis.* **71**: 731.

Martelli, G.P. and Cirulli, M. (1969). *Annls. Phytopath.* **1**: 393.

Martelli, G.P. and Cherif, C. (1987). *J. Phytopath.* **119**: 32.

Miličič, D. (1977). *Proc. 8th Conf. Czech. Pl. Virol.*, Bratislava.

Plavsic, B., Corte, A. and Miličič, D. (1976). *Phytopathol. Medit.* **15**: 115.

Plavsic, B., Miličič, D. and Eric, Z. (1976). *Phytopath. Z.* **86**: 225.

Plavsic, B., Miličič, D. and Eric, Z. (1978). *Phytopath. Z.* **91**: 67.

Rana, G.L. (1980). *Inf. tore Fitopatol.* **30(4)**: 7.

Rana, G.L. and Di Franco, A. (1979). *Phytopathol. Medit.* **18**: 48.

Rana, G.L. and Di Franco, A. (1980). *Acta Hort.* **110**: 191.

Russo, M. and Martelli, G.P. (1972). *Phytopathol. Medit.* **11**: 136.

# Plantago 4 (?) *caulimovirus*

**Data collated by A.A. Brunt, 1991.**

*Standard acronym:* PlV-4.

First reported in *Plantago lanceolata, P. major*; from the U.K.; by Hammond (1981).

**Natural host range and symptoms**

*Plantago lanceolata, P. major* — often symptomless.

**Transmission**

Is sap transmissible.

**Geographical distribution**

Spreads in the UK.

Diagnostically susceptible

*Brassica oleracea* var. *botrytis*—symptomless systemic infection.

**Particle morphology**

Virions isometric; 50 nm in diameter.

**Physical properties**

One sedimenting component in purified preparations; sedimentation coefficient *c.* 208 *S*.

**Cytopathology**

Inclusions present in infected cells; they contain virions.

**Taxonomy and Relationships**

Possible *Caulimovirus*.

**References**

Hammond, J. (1981). *Pl. Path.* **30**: 237.
Hammond, J. (1988). *Adv. Virus Res.* **27**: 104.

# Plantago asiatica mosaic *potexvirus*

**Data collated by A.A. Brunt, 1994.**

First reported in *Plantago asiatica*; from Russia; by Kostin and Volkov (1976).

**Natural host range and symptoms**

Symptoms mosaic.
*Plantago asiatica* — mosaic.

**Transmission**

Virus transmitted by mechanical inoculation.

**Geographical distribution**

Spreads in the Central Asian region; the former USSR.

**Experimental host range**

Many (>9) families susceptible. Experimentally infected plants mostly show mosaic.

Diagnostically susceptible

*Plantago asiatica* — mosaic.

**Purification**

Method: Solovyev *et al.* (1994).

**Particle morphology**

Virions filamentous; not enveloped; usually flexuous; with a clear modal length; of 490–530 nm; 11 nm wide. Axial canal obscure.

**Biochemical properties**

Virions contain 6 % nucleic acid; 94 % protein.

*Genome* consists of RNA; single–stranded; unipartite; largest (or only) genome part 6.128 kb. Poly A region present; 3' terminus.

*Sequence database accession code(s):*

S61526 Em(40)_vi:S61526 Gb(84)_vi:S61526 triple block genes: putative helicase...13 kda putative membrane-associated protein Plantago

Z21647 Em(40)_vi:PARNASQ Gb(84)_vi:PARNASQ P.asiatica mosaic virus genomic RNA. 2/94 6,128bp. 2 sequences.

*Non–genomic nucleic acid* not found in the virions. Sub–genomic *m*RNA found in infected cells; mRNA for coat protein, 0.9kb.

*Virion protein(s)* one; $M_r$ 22000; coat. Method of preparation: Solovyev *et al.*, 1994. Virion proteins not glycosylated.

*Virus–coded non–virion proteins* isolated; one protein found (170 kDa).

**Taxonomy and Relationships**

*Potexvirus.*

**References**


Kostin, V.D. and Volkov, Y.G. (1976). *Virusnye Bolezni Rastenij Dalnego Vostoka* **25**: 205.

Minskaya, L.A. (1979). *Virozy Rastenij* **54**: 101.

Minskaya, L.A., Novikov, V.K. and Kostin, V.D. (1977). *Virusnye Bolezni Rastenij Dalnego Vostoka* **48** 61.

Solovyev, A.G., Novikov, V.K., Merits, A., Savenkov, E.I., Zelenina, D.A., Tyulkina, L.G. and Morozov, S.Y. (1994). *J. gen. Virol.* **75**: 259.


# Plantago mottle *tymovirus*

**Data collated by R. Provvidenti, 1981. Revised 1983.**

*Standard acronym:* PlMV.

First reported in *Plantago major*; from the U.S.A; by Granett (1973).

**Natural host range and symptoms**

Symptoms persist.
*Plantago major* — mottle.

**Transmission**

Virus transmitted by mechanical inoculation; not transmitted by seed.

**Geographical distribution**

Spreads in the USA (in New York State).

**Experimental host range**

Several (3–9) families susceptible.

Diagnostically susceptible

*Glycine max, Pisum sativum* — systemic mottle.
*Proboscidea jussieu* — necrotic local lesions, systemic mottle.
*Plantago major* — systemic mottle.
*Antirrhinum majus* — necrotic ringspots not systemic.
*Nicotiana clevelandii* — systemic mottle.
*Petunia hybrida, Tetragonia expansa* — systemic mottle.

Maintenance and propagation hosts

*Pisum sativum.*

Assay hosts (**L**ocal lesions or **W**hole plants)

*Plantago major, Nicotiana clevelandii, Antirrhinum majus.*

Dicotyledonous Hosts *(**Susceptible** or Insusceptible)*

***Antirrhinum majus***. *Brassica campestris* ssp. *pekinensis. Brassica oleracea* ssp. *botrytis. Brassica oleracea* ssp. *capitata. Capsicum frutescens. Catharanthus roseus.*

*Chenopodium amaranticolor. Chenopodium quinoa. Cucumis sativus. Cucurbita moschata. Cucurbita pepo. Datura stramonium.* ***Glycine max****. Gomphrena globosa. Lactuca sativa. Lupinus albus. Lycopersicon esculentum.* ***Nicotiana clevelandii****. Nicotiana sylvestris. Nicotiana tabacum.* ***Petunia hybrida****. Phaseolus lunatus. Phaseolus vulgaris.* ***Pisum sativum****. Spinacia oleracea.* ***Tetragonia expansa****. Trifolium pratense. Trifolium repens. Vicia faba. Vigna unguiculata. Zinnia elegans.*

Sources of host-range data

Granett (1973).

**Properties of particles in sap**

TIP: 65 °C. DEP: $\log_{10}$ minus 7–8.

**Purification**

Method: Granett (1973).

**Particle morphology**

Virions isometric; not enveloped; 26.7 nm in diameter; with a conspicuous capsomere arrangement.

**Physical properties**

$A_{260}/A_{280}$ ratio 8.51 (specific absorbance at 260 nm).

**Biochemical properties**

*Genome* consists of RNA; single–stranded. Base composition 17 % G; 21 % A; 34 % C; 28 % U.

**Cytopathology**

Virions found in all parts of the host plant; in cytoplasm.

**Taxonomy and Relationships**

*Tymovirus.*

Virus(es) with serologically related virions: Andean potato latent, dulcamara mottle, belladonna mottle, ononis yellow mosaic and turnip yellow mosaic viruses.

**References**

Granett, A.L. (1973). *Phytopathology* **73**: 1313.
Provvidenti, R. (1979). *J. Hered.* **70**: 350.
Provvidenti, R. and Granett, A.L. (1976). *Ann. appl. Biol.* **82**: 85.

# Plantain 6 (?) *carmovirus*

**Data collated by A.A. Brunt, 1991.**

*Standard acronym:* PlV-6.

First reported in *Plantago lanceolata*; from Cambridge, U.K.; by Hammond (1981).

**Natural host range and symptoms**

*Plantago lanceolata* — symptomless.

**Transmission**

Virus transmitted by mechanical inoculation.

**Geographical distribution**

Found, but with no evidence of spread, in the U.K.

**Experimental host range**

Several (3–9) families susceptible.

Diagnostically susceptible

*Nicotiana benthamiana, N. clevelandii* — chlorotic local lesions; systemic vein-yellowing, leaf distortion and stunting of plants.

*Apium graveolens, Phaseolus vulgaris, Plantago lanceolata* — symptomless.

Maintenance and propagation hosts

*Nicotiana benthamiana, N. clevelandii.*

Assay hosts (**L**ocal lesions or **W**hole plants)

*Nicotiana benthamiana* (L and W), *N. clevelandii* (L and W).

Dicotyledonous Hosts *(**Susceptible** or Insusceptible)*

***Apium graveolens*. *Nicotiana benthamiana*. *Nicotiana clevelandii*. *Phaseolus vulgaris*.**

**Purification**

Method: Hammond (1981).

**Particle morphology**

Virions isometric; 30 nm in diameter.

**Physical properties**

One sedimenting component in purified preparations; sedimentation coefficient 143 *S*. Density 1.347 g $cm^{-3}$ in CsCl.

**Biochemical properties**

*Genome* consists of RNA; single–stranded; largest (or only) genome part *c*. 4.8 kb.

*Virion protein(s)* 45000.

**Taxonomy and Relationships**

Possible *Carmovirus: Tombusviridae.*

**References**

Hammond, J. (1981). *Pl. Path.* **30**: 237.

# Plantain 7 (?) potyvirus

**Data collated by A.A. Brunt, 1991.**

*Standard acronym:* PV-7.

First reported in *Plantago lanceolata*; from U.K.; by Hammond (1981).

**Natural host range and symptoms**

*Plantago lanceolata* — symptomatology not described (virus found only in complex with others).

**Transmission**

Transmitted by a vector; an insect; *Myzus* sp.; Aphididae.

**Geographical distribution**

Found, but with no evidence of spread, in the U.K. (Cambridge).

Diagnostically susceptible

*Nicotiana benthamiana* — faint leaf chlorosis.

Dicotyledonous Hosts *(**Susceptible** or Insusceptible)*

***Nicotiana benthamiana***.

**Particle morphology**

Virions filamentous; usually flexuous; of 720–750 nm.

**Taxonomy and Relationships**

Unassigned potyvirus.

**References**

Hammond, J. (1981). *Pl. Path.* **30**: 237.

## Plantain 8 (?) *carlavirus*

**Data collated by J. Hammond, 1991.**

*Standard acronym:* PlV-8.

First reported in *Plantago lanceolata*; from the U.K.; by Hammond (1981).

**Natural host range and symptoms**

Symptoms none.

*Plantago lanceolata* — occurs only with other viruses, but alone possibly causes no conspicuous symptoms.

**Transmission**

Presumably transmitted by aphids. Virus transmitted by mechanical inoculation.

**Geographical distribution**

Spreads in the UK. Found, but with no evidence of spread, in England (Cambridge).

Diagnostically susceptible

*Cucumis sativus, Plantago lanceolata* — symptomless.

Dicotyledonous Hosts *(**Susceptible** or Insusceptible)*

***Cucumis sativus***.

Sources of host-range data

Hammond (1980).

**Properties of particles in sap**

Leaf sap contains few virions.

**Particle morphology**

Virions filamentous; not enveloped; usually flexuous; with a clear modal length; of 600–640 nm; 13 nm wide. Axial canal obscure. Basic helix obscure.

**Taxonomy and Relationships**

Possible *Carlavirus*.

**References**

Hammond, J. (1980). *Ph.D. Thesis*, University East Anglia, U.K..
Hammond, J. (1981). *Pl. Path.* **30**: 237.

# Plantain mottle (?) rhabdovirus

**Data collated by A.A. Brunt, 1991.**

*Synonyms:* plantain mottle virus (Smith, 1972). *Standard acronym:* PlMV.

First reported in *Plantago lanceolata*; from the U.K; by Hitchborn *et al.* (1966).

**Natural host range and symptoms**

Symptoms persist.

*Plantago lanceolata* — mottling of younger leaves and necrotic spotting of older leaves.

**Geographical distribution**

Spreads in the UK.

**Particle morphology**

Virions rhabdo– or bullet–shaped; with a clear modal length; of 330 nm (in sap); 63 nm wide (in sap).

**Taxonomy and Relationships**

Possible unassigned rhabdovirus.

**References**

Hitchborn, J.H., Hills, G.J. and Hull, R. (1966). *Virology* **28**: 768.

Smith, K.M. (1972). In: *A Textbook of Plant Viruses*, p.388, 3rd edition. London, Longman.

# Plantain X *potexvirus*

**Data collated by J. Hammond, 1987.**

*Standard acronym:* PlVX.

First reported in *Plantago lanceolata*; from Great Britain; by Hammond and Hull (1981).

**Natural host range and symptoms**

*Plantago lanceolata* — no obvious symptoms.

**Transmission**

Virus transmitted by mechanical inoculation.

**Ecology and control**

Studies reported by Hammond (1982).

**Geographical distribution**

Spreads in the UK.

**Experimental host range**

Several (3–9) families susceptible.

Diagnostically susceptible

*Nicotiana clevelandii* — systemic necrotic flecking when grown in bright light.

*N. edwardsonii* — necrotic flecking of inoculated and systemically infected leaves (white ringspots at 32°C).

*Plantago rugelii* — spreading local and systemic necrotic spots.

Diagnostically insusceptible

*Tetragonia expansa, Montia perfoliata, Brassica rapa, Cucumis sativus Gomphrena globosa.*

Maintenance and propagation hosts

*Nicotiana clevelandii, N. edwardsonii.*

Assay hosts (**L**ocal lesions or **W**hole plants)

*Nicotiana edwardsonii* (W) and *Plantago rugelii* (L).

Dicotyledonous Hosts *(**Susceptible** or Insusceptible)*

***Amaranthus caudatus***. *Brassica campestris* ssp. *chinensis*. *Brassica campestris* ssp. *pekinensis*. *Brassica campestris* ssp. *rapa*. ***Chenopodium amaranticolor***. *Chenopodium hybridum*. *Chenopodium quinoa*. *Cucumis sativus*. ***Datura metel***. ***Datura stramonium***. *Gomphrena globosa*. ***Hyoscyamus niger***. ***Lycopersicon esculentum***. *Momordica balsamina*. *Montia perfoliata*. ***Nicotiana benthamiana***. ***Nicotiana clevelandii***. ***Nicotiana* × *edwardsonii***. *Nicotiana glutinosa*. ***Nicotiana rustica***. ***Nicotiana tabacum***. ***Petunia hybrida***. ***Phaseolus vulgaris***. *Pisum sativum*. ***Plantago rugelii***. *Tetragonia expansa*. *Vicia faba*. ***Vigna unguiculata***. ***Zinnia elegans***.

Sources of host-range data

Hammond and Hull (1981).

**Properties of particles in sap**

TIP: 60–70 °C. Leaf sap contains many virions. Electron microscopy: fixation not necessary. Best detail with UA and UF.

**Purification**

Method: Hammond and Hull (1981).

**Particle morphology**

Virions filamentous; sometimes have a loose external helical structure in negatively stained epidermal dips (Hammond and Hull, 1981); not enveloped; usually flexuous; with a clear modal length; of 570–580 nm; 12 nm wide. Axial canal obscure. Basic helix obvious (in UA or UF); pitch of basic helix 3.4 nm.

**Physical properties**

One sedimenting component in purified preparations; sedimentation coefficient 119 *S*. Density 1.31 g $cm^{-3}$ in CsCl. Density 1.259 in $Cs_2SO_4$. $A_{260}/A_{280}$ ratio 1.19 (partially corrected).

**Biochemical properties**

Virions 94–95 % protein; 0 % lipid.

*Genome* consists of RNA; single–stranded; linear; unipartite; largest (or only) genome part 6.5 kb. Genomic nucleic acid isolated by Wilcockson and Hull; the SDS — sodium perchlorate method; or the phenol-chloroform method is probably equally as efficient.

*Virion protein(s)* one; $M_r$ 26000; coat protein. Method of preparation: Francki and McLean (1968). Amino acid composition: Hammond and Hull (1981).

**Replication**

Replication does not depend on a helper virus.

**Cytopathology**

Virions found in all parts of the host plant; in cytoplasm. Inclusions present in infected cells; are unusual in shape; whorl-like arrays; they contain virions.

**Taxonomy and Relationships**

*Potexvirus.*

Virus(es) with serologically unrelated virions: cactus X, clover yellow mosaic, cymbidium mosaic, narcissus mosaic, papaya mosaic (plantago severe mottle strain), potato X and white clover mosaic viruses.

No significant sequence homologies were found between plantago X, potato X and papaya mosaic viruses by *c*DNA/RNA hybridization. Best tests for diagnosis: Plantain X virus does not infect *Gomphrena globosa*, whereas the plantago severe mottle strain of papaya mosaic virus does. The latter also does not infect *Plantago lanceolata* and *Nicotiana* spp.

**References**

Gracia, O., Koenig, R. and Lesemann, D.-E. (1983). *Phytopathology* **73**: 1488.
Hammond, J. (1977). *Rep. John Innes Inst.* 1976, p.109.
Hammond, J. (1983). *Adv. Virus Res.* **27**: 103.
Hammond, J. and Hull, R. (1981). *J. gen. Virol.* **54**: 75.
Hammond, J. and Hull, R. (1983). *CMI/AAB Descr. Pl. Viruses* No. 266, 4 pp.
Rowhani, A. and Petersen, J.F. (1980). *Can. J. Pl. Path.* **2**: 12.
Wilcockson, J. and Hull, R. (1974). *J. gen. Virol.* **23**: 107.

# Pleioblastus chino (?) potyvirus

**Data collated by A.A. Brunt, 1991.**

*Standard acronym:* PleMV.

First reported in *Pleioblastus chino*; from Japan; by Toriyama and Yora (1972).

**Natural host range and symptoms**

*Pleioblastus chino.*

**Geographical distribution**

Spreads in Japan.

**Particle morphology**

Virions filamentous; not enveloped; usually flexuous; with a clear modal length; of 750 nm.

**Taxonomy and Relationships**

Unassigned potyvirus.

**References**

Toriyama, S. and Yora, K. (1972). In: *Virus Diseases of Wild Grasses and Cereal Crops in Japan.* University of Tokyo Press, Tokyo, Japan.

# Plum American line pattern *ilarvirus*

**Data collated by R.W. Fulton, 1985.**

*Synonyms:* plum line pattern virus, peach line pattern virosis virus, prunus virus 10, American plum line pattern virus. *Standard acronym:* APLPV.

First reported in *Prunus americana x P. salicina* (Shiro plum); from Wenatchee, Washington, U.S.A.; by Kirkpatrick *et al.* (1964).

**Natural host range and symptoms**

Symptoms vary seasonally.

*Prunus domestica, P. serrulata, P. cerasus, P. cerasifera, P. americana x P. salicina, P. persica* — line patterns and chlorotic rings.

**Geographical distribution**

Spreads in the North American region; Canada and the USA (CABI/EPPO data).

**Experimental host range**

Many (>9) families susceptible. Experimentally infected plants mostly show chlorotic or necrotic local lesions; systemic veinbanding, mottles, ringspots.

Diagnostically susceptible

*Vigna unguiculata* ssp. *cylindrica* — tiny red local lesions; systemic red vein necrosis.

*Cucumis sativus, Chenopodium quinoa* — small local lesions, not systemic.

*Nicotiana megalosiphon, N. occidentalis* — chlorotic or diffuse necrotic local lesions, systemic ringspots or mottles.

*Physalis floridana* — inconspicuous chlorotic local lesions, systemic chlorotic veinbanding ringspots, or oak-leaf.

*Petunia hybrida* — chlorotic or necrotic local lesions, systemic necrosis.

Most species of the *Leguminosae* show local necrosis, but many are not very susceptible.

Some isolates infect *Gomphrena globosa* and *Lactuca sativa.*

Diagnostically insusceptible

*Antirrhinum majus, Gomphrena globosa, Lactuca sativa, Torenia fourneri, Tithonia speciosa..*

Maintenance and propagation hosts

*Nicotiana megalosiphon, N. occidentalis, Vigna unguiculata* ssp. *cylindrica, Catharanthus roseus, Prunus domestica.*

Assay hosts (**L**ocal lesions or **W**hole plants)

*Vigna unguiculata* ssp. *cylindrica,* (L), *Nicotiana megalosiphon,* (W).

Dicotyledonous Hosts ***(Susceptible** or Insusceptible)*

*Antirrhinum majus.* ***Catharanthus roseus. Celosia argentea.*** *Chenopodium amaranticolor.* ***Chenopodium murale. Chenopodium quinoa. Crotalaria spectabilis. Cucumis sativus.*** *Cyamopsis tetragonoloba.* ***Datura metel. Datura stramonium.*** *Gomphrena globosa. Helianthus annuus. Lactuca sativa. Momordica balsamina.* ***Nicandra physaloides. Nicotiana benthamiana. Nicotiana clevelandii. Nicotiana debneyi. Nicotiana megalosiphon. Nicotiana sylvestris. Nicotiana tabacum. Petunia hybrida. Phaseolus vulgaris.*** *Phlox drummondii. Physalis floridana. Physalis peruviana. Phytolacca americana.* ***Sesbania exaltata.*** *Tithonia speciosa. Torenia fournieri.* ***Vigna unguiculata. Vigna unguiculata*** ssp. ***cylindrica. Zinnia elegans.***

Sources of host-range data

Paulsen and Fulton (1968); Kirkpatrick *et al.* (1964).

**Properties of particles in sap**

TIP: 66 °C. LIV: 0.2 days (4 hours in sap stabilized with 2-mercaptoethanol. In undiluted sap, infectivity is lost within seconds). DEP: $\log_{10}$ minus 6–7. Electron microscopy: treatment with 1% glutaraldehyde of virions suspended in 0.03 M EDTA.

**Purification**

Method: Fulton (1982); Paulson and Fulton (1969).

**Particle morphology**

Virions isometric; not enveloped; 26, 28, 31, and 33 nm in diameter; rounded in profile; without a conspicuous capsomere arrangement.

**Physical properties**

Four sedimenting components in purified preparations; sedimentation coefficient of the fastest 125 *S*; of the other(s) 95, 100, and 114 *S* (all four are needed for infection). $A_{260}/A_{280}$ ratio 1.67 (unseparated components).

**Biochemical properties**

Virions contain 14 % nucleic acid; 86 % protein; 0 % lipid.

*Genome* probably consists of RNA; single–stranded; probably of three parts.

**Replication**

Replication does not depend on a helper virus.

**Cytopathology**

Virions found in all parts of the host plant.

**Taxonomy and Relationships**

*Ilarvirus: Bromoviridae* (subgroup 5).

Virus(es) with serologically unrelated virions: prunus necrotic ringspot, prune dwarf, tobacco streak, asparagus 2, citrus leaf rugose, citrus variegation, elm mottle, Tulare apple mosaic, apple mosaic, lilac ring mottle and spinach latent viruses.

Best tests for diagnosis: American plum line pattern virus is one of three different viruses that may cause line pattern symptoms in plum and other *Prunus* spp. The other two are apple mosaic virus and some strains of prunus necrotic ringspot virus, which cause line pattern symptoms worldwide. Serological tests are required for unequivocal identification.

**References**


Cation, D. (1941). *Phytopathology* **31**: 1004.

Fulton, R.W. (1982). *Phytopathology* **72**: 1345.

Fulton, R.W. (1984). *CMI/AAB Descr. Pl. Viruses* No. 280, 4 pp.

Kirkpatrick, H.C., Cheney, P.W. and Linder, R.C. (1964). *Pl. Dis. Reptr* **48**: 616.

Kirkpatrick, H.C. and Fulton, R.W. (1976). In: *Virus Diseases and Non Infectious Disorders of Stone Fruits in North America.* U.S. Dep. Agric. Hdbk No. 437, 166 pp.

Paulsen, A.Q. and Fulton, R.W. (1968). *Phytopathology* **58**: 766.

Paulsen, A.Q. and Fulton, R.W. (1969). *Ann. appl. Biol.* **63**: 233.

# Plum pox *potyvirus*

**Data collated by H. Kegler and D. Sutić, 1991.**

*Synonyms:* prunus virus 7, Scharka-Virus, Sarka virus. *Standard acronym:* PPV.

First reported in *Prunus domestica*; from Bulgaria; by Atanasoff (1932).

## Natural host range and symptoms

Symptoms persist. Symptoms leaf mottling and fruit malformation.

*Prunus armeniaca* — fruits pale rings and deformation; stones pale rings.

*P. cerasifera* — latent, occasional mild pale green mottling.

*P. domestica* — pale green mottling, light green or yellow rings, spots or lines, some brown necrotic spots; fruits blue, shallow depressions or ringlike discolouration, deeply engraved rings and spots of the skin, flesh turns brown and gum like, fruit drop; shoots split and die back.

*P. glandulosa.*

*P. persica* — vein clearing, chlorotic blotching and malformation of leaves; fruits develop pale rings and diffuse bands on the skin, some malformation.

*P. insititia, P. spinosa, P. salicina* — symptomless.

## Transmission

Transmitted by a vector; an insect; *Aphis craccivora, A. spiraecola, Brachycaudus helichrysi, B. cardui, Myzus persicae, M. varians, Myzus (Phorodon) humuli*; Aphididae. Principal natural vector(s): *Brachycaudus helichrysi, Myzus persicae, Myzus (Phorodon) humuli*. Not transmitted by *Aphophora alni, Edwardsonia plebeja, Hyalopterus pruni, Lecanium corni, Lygus pratensis, Macrosteles laevis, Myzus cerasi, Olianus quinquecostatus, Paratetranychus pilosus, Typhlocyba rosae*. Transmitted in a non–persistent manner. Virus transmitted by mechanical inoculation; transmitted by grafting; not transmitted by contact between plants; transmitted by seed (in apricots (Nemet and Kolber, 1983)); not transmitted by pollen.

## Ecology and control

Studies reported by Resistant cultivars: *Prunus domestica* Abundance, Anna Spáth, bessarabische Zwetsche, Bluefre, Büler Frühzwetsche, Burbank, Ćcanska lepotica, Ćaćanska najbolja, Ćaćanska, Carpatin Czar, Czernowitzer, Chrudimer, Ersinger, Fecunda, Graf Althans, Grose Grüne Reneklode (distinct sources), Herman Jalomita, Kirkes Pflaume, Montfort, Nancymirabelle, Ontariopflaume, Opal, Oullins, Pamjatj vavilova, Pitestean, President, Ruth Gerstetter, Schöne von Löwen, Stanley, St. Hubertus, Tuleu gras, Valor, Valjevka, Vengerka posdnjaja, Zucker pflaume. *Prunus armeniaca* Stella, Early Orange, Harcot, Nanno, Badami. *Prunus persica var. nectarina* Crimson gold, Fanthasia, May Grand. *Prunus persica* TS–1–TS10, Andross, Coronado, June Berta, Ranger, Rio Oso Gem. Krczal and Kunze (1976); Marenaud *et al.* (1976); Jordović (1967); Minoiu (1973); Verderevskaja *et al.* (1983).

## Geographical distribution

Probably distributed worldwide. Spreads in the African region, the Eurasian region, and the Mediterranean region; Albania, Austria, Bulgaria, Cyprus, the former Czechoslovakia, Egypt, France, Germany, Greece, Hungary, Italy, Luxembourg, Poland, Portugal, Romania, Spain, Syria, Turkey, the UK, the former USSR, the former Yugoslavia.

## Experimental host range

Many (>9) families susceptible. Experimentally infected plants mostly show chlorotic or necrotic local lesions, mild pale green leaf mottling and leaf malformation.

Diagnostically susceptible

*Chenopodium foetidum* — yellow spots, some with necrotic centres or necrotic spots; not systemic.

*Nicandra physaloides* — black-brown necrotic local lesions.

*Prunus domestica* cvs Italian Prune and Pożegaća — chlorotic rings, bands and spots.

*P. japonica* — diffused chlorotic spots.

*P. maritima* — chlorotic spots, vein necrosis.

*P. sibirica* — green spots and faint line pattern.

*P. tomentosa* — epinasty and malformation of young leaves, chlorotic spots, necrotic spots of older leaves.

*Sorbus domestica* — yellow spots, leaf chlorosis.

Diagnostically insusceptible

*Chenopodium amaranticolor, Crataegus monogyna, Malus domestica, Prunus avium, Pyrus communis.*

Maintenance and propagation hosts

*Nicotiana benthamiana, N. clevelandii, N. occidentalis, Stellaria media.*

Assay hosts (**L**ocal lesions or **W**hole plants)

*Chenopodium foetidum* (L), *Nicandra physalodes* (L), *Nicotiana acuminata* (W), *N. benthamiana* (W), *N. clevelandii* (W), *Ranunculus arvensis* (L).

Dicotyledonous Hosts *(**Susceptible** or Insusceptible)*

***Capsella bursa–pastoris*****.** ***Celosia argentea*****.** *Chenopodium amaranticolor.* ***Chenopodium ambrosioides*****.** ***Chenopodium capitatum*****.** ***Chenopodium foetidum*****.** ***Chenopodium foliosum*****.** ***Chenopodium murale*****.** ***Chenopodium quinoa*****.** *Cucumis sativus. Cucurbita pepo.* ***Cyamopsis tetragonoloba*****.** *Datura stramonium.* ***Emilia sagittata*****.** ***Gomphrena globosa*****.** ***Humulus lupulus*****.** ***Hyoscyamus niger*****.** ***Lupinus albus*****.** ***Lycopersicon esculentum*****.** *Malus sylvestris.* ***Melilotus alba*****.** ***Melilotus officinalis*****.** ***Nicandra physaloides*****.** ***Nicotiana benthamiana*****.** ***Nicotiana bigelovii*****.** ***Nicotiana clevelandii*****.** ***Nicotiana debneyi*****.** ***Nicotiana glutinosa*****.** ***Nicotiana megalosiphon*****.** ***Nicotiana occidentalis*****.** ***Nicotiana rustica*****.** ***Nicotiana sylvestris*****.** ***Nicotiana tabacum*****.** ***Petunia hybrida*****.** ***Physalis floridana*****.** ***Physalis peruviana*****.** ***Pisum sativum*****.** ***Prunus armeniaca*****.** ***Prunus cerasifera*****.** ***Prunus domestica*****.** ***Prunus mahaleb*****.** ***Prunus persica*****.** *Prunus serrulata.* ***Ranunculus arvensis*****.** ***Ranunculus sardous*****.** ***Senecio vulgaris*****.** ***Sesbania exaltata*****.** *Solanum nodiflorum. Solanum rostratum.* ***Stellaria media*****.** ***Trifolium pratense*****.** ***Trifolium repens*****.** ***Vicia sativa*****.** ***Vicia villosa*****.** ***Zinnia elegans*****.**

Sources of host-range data

Kassanis and Sutić (1965); Kegler (1977); Kegler and Proll (1989); Kröll (1975); Németh (1963 and 1986); Van Oosten (1970); Sutić (1961 and 1983); Zawadska and Smolarz (1978).

**Properties of particles in sap**

TIP: 52–58 °C. LIV: 3–4 days. DEP: $\log_{10}$ minus 8 ($\times 10^{-3}$). Infectivity of sap increased by treatment with di–ethyl ether, or decreased by treatment with di–ethyl ether. Leaf sap contains few virions, or contains many virions. Electron microscopy: see Bovey (1973).

**Purification**

Method: Ranković and Jordović (1970); Schade (1969); Van Oosten (1972); Sutić *et al.* (1981).

**Particle morphology**

Virions filamentous; usually flexuous; with a clear modal length; of 660–770 nm; 12.5–20 nm wide.

**Physical properties**

Sedimentation coefficient 180 *S*. $A_{260}/A_{280}$ ratio 0.77–0.87.

**Biochemical properties**

Virions contain 7 % nucleic acid.

*Genome* consists of RNA; single–stranded; linear. Additional factor not required for infectivity.

*Sequence database accession code(s):*

D00298 Em(40)_vi:PPVCP Gb(84)_vi:PPVCP Plum pox potyvirus (PPV), capsid protein gene (3' region of PPV RNA). 4/90 1,232bp.

D00424 Em(40)_vi:PPV Plum pox virus, complete genome. 4/90 9,741bp.

D13751 Em(40)_vi:PPV1 Gb(84)_vi:PPV Plum pox virus, complete genome. 11/92 9,741bp.

M21847 Em(40)_vi:PPVCPOL Gb(84)_vi:PPVCPOL Plum pox potyvirus (PPV) capsid protein and RNA polymerase genes, complete cds. 3/91 2,854bp.

M26965 Em(40)_vi:PPVNIAPRO Gb(84)_vi:PPVNIAPRO Plum pox potyvirus ORF containing NI-a-like protein mRNA, partial cds. 4/92 1,695bp.

M55164 Em(40)_sy:AGPPVRC Gb(84)_sy:SYNPPVRC Synthetic bacteriophage T7/Plum pox potyvirus recombined polymerase gene, 5' flank. 7/91 57bp.

M92280 Em(40)_vi:PPVPOLPRO Gb(84)_vi:PPVPOLPRO Plum pox potyvirus polyprotein, complete cds. 5/93 9,786bp.

S57404 Em(40)_vi:S57404 Gb(84)_vi:S57404 polyprotein (3' region) plum pox virus PPV, o6, mRNA Partial, 1768 nt. 1/94 1,768bp.

S57405 Em(40)_vi:S57405 Gb(84)_vi:S57405 polyprotein (3' region) plum pox virus PPV, PS, mRNA Partial, 1768 nt. 1/94 1,768bp

X16415 Em(40)_vi:PPVCG Gb(84)_vi:PPVCG Plum pox virus RNA genome. 9/93 9,787bp.

X56258 Em(40)_vi:PPVPGP Gb(84)_vi:PPVPGP Plum Pox Virus partial genome polyprotein mRNA for CI, 6K, NIa, NIb and coat protein genes. 8

X57975 Em(40)_vi:PPCOAPROA Gb(84)_vi:PPCOAPROA Plum pox potyvirus mRNA for coat protein. 11/92 1,270bp.

X57976 Em(40)_vi:PPCOAPROB Gb(84)_vi:PPCOAPROB Plum pox potyvirus mRNA for coat protein. 11/92 1,270bp.

X81073 Gb(84)n:PPSHCP Plum pox potyvirus gene for coat protein (SH). 8/94 1,228bp

X81074 Gb(84)n:PPNLCP Plum pox potyvirus gene for coat protein (NL). 8/94 1,270bp.

X81075 Gb(84)n:PPNEBCP Plum pox potyvirus gene for coat protein (NEB). 8/94 1,270bp.

X81076 Gb(84)n:PPHBCP Plum pox potyvirus gene for coat protein (HB). 8/94 1,270bp.

X81077 Gb(84)n:PPOB1CP Plum pox potyvirus gene for coat protein (OB1). 8/94 1,270bp.

X81078 Gb(84)n:PPALCP Plum pox potyvirus gene for coat protein (AL). 8/94 1,270bp.

X81079 Gb(84)n:PPSLCP Plum pox potyvirus gene for coat protein (SL). 8/94 1,270bp.

X81080 Gb(84)n:PPGSPCP Plum pox potyvirus gene for coat protein (GSP). 8/94 1,270bp

X81081 Gb(84)n:PPLIHCP Plum pox potyvirus gene for coat protein (LI/H). 8/94 1,270bp.

X81082 Gb(84)n:PPCGCP Plum pox potyvirus gene for coat protein (CG). 8/94 1,270bp.

X81083 Gb(84)n:PPPVCG Plum pox potyvirus complete genomic RNA. 8/94 9,786bp.

X81084 Gb(84)n:PPDOH1CP Plum pox potyvirus gene for coat protein (DOH1). 8/94 1,270bp.

L42470 Em(43)_vi:Ppvp1hc Gb(89)_vi:Ppvp1hc Plum pox virus P1 and helper component (HC) gene, partial cds. 5/95 261bp.

L42471 Em(43)_vi:Ppvhcg Gb(89)_vi:Ppvhcg Plum pox virus helper component (HC) gene, partial cds. 5/95 180bp.

L42472 Em(43)_vi:Ppvnib Gb(89)_vi:Ppvnib Plum pox virus coat protein (CP) and NIb gene, partial cds. 5/95 366bp.

L42473 Em(43)_vi:Ppvniba Gb(89)_vi:Ppvniba Plum pox virus coat protein (CP) and NIb gene, partial cds. 5/95 411bp.

S73776 Gb(89)_un:S73776 CP=coat protein (plum pox virus PPV-S, Skierniewice isolate, Genomic RNA, 993 nt). 3/95 993bp.

*Virion protein(s)* three; $M_r$ of the largest 43500. $M_r$ of 2nd largest 29000. $M_r$ of 3rd largest 27000. Amino acid sequence: Kerlan and Dunez (1976); Ranković (1976).

**Replication**

Genome replicates probably in cytoplasm. Replication does not depend on a helper virus.

**Cytopathology**

Virions found in leaves, roots, mesophyll, xylem, flowers, buds, fruits, bark and young shoots; in cytoplasm. Inclusions present in infected cells; are amorphous X–bodies, pinwheels, and unusual in shape; needle shaped (Pleśe *et al.*, 1969) and are protein paracrystals (Bovey, 1971); they do not contain virions. Other cellular changes: firstly paracrystals appear, then they become surrounded polysomal membranes. Evaginations of external nuclear envelope (Bovey, 1971). Chloroplasts may become invaginated. Increased nucleoprotein synthesis and numbers of ribosomes (Ranković, 1974).

**Taxonomy and Relationships**

*Potyvirus: Potyviridae.*

Virus(es) with serologically related virions: potato Y virus, but distantly (Grüntzig and Fuchs, 1979).

Various strains have been distinguished by symptoms in *Chenopodium foetidum*: yellow strain, intermediate strain, necrotic strain (Sutić *et al.*, 1971). Pathotypes are distinguished by their ability to infect systemically resistant genotype K4: CG strain, D1 strain (Kegler *et al.*, 1991). In France, the south-western ecotype is more virulent than the south-eastern one (Marenaud *et al.*, 1976). Two serotypes closely related have been reported (Kerlan and Dunez, 1979). Best tests for diagnosis: serological tests, local lesion hosts and Petri dish test. Plums are commonly also infected with prunus necrotic ringspot (PNRV), prune dwarf (PDV), apple chlorotic leaf spot (ACLSV) and plum line pattern virus (PLPV). The following differential host reactions allow the mixtures to be sorted: + = symptoms appear — = latent 0 = no infectable host

*Chenopodium foetidum*: PPV +, PNRV 0, PDV 0, ACLSV +, PLPV -

*C. quinoa*: PPV 0, PNRV +, PDV 0, ACLSV +, PLPV +

*Celosia argentea*: PPV 0, PNRV 0, PDV 0, ACLSV +, PLPV 0
*Cucumis sativus*: PPV 0, PNRV +, PDV +, ACLSV 0, PLPV +
*Sesbania exaltata*: PPV 0, PNRV -, PDV +, PLPV —
*Prunus serrulata*: PPV 0, PNRV +, PDV +, ACLSV -, PLPV — .

**References**


Adams, A.N. (1978). *Ann. appl. Biol.* **90**: 125.
Atanasoff, D. (1932). *Jahrb. Univ. Sofia Agr. Fak.* **11**: 49.
Baumann, G. (1968). *Tagungsber. DAL DDR, Berlin* **97**: 211.
Bovey, R. (1971). *Annls. Phytopath.* **225**.
Bovey, R. (1973). *Plant Virology, Prague* **91**, p. 125.
Casper, R. (1975). *Acta Hort.* **44**: 171.
Casper, R. and Meyer, S. (1981). *Nach. dtsch. Pflanzenschutzd.* **33**: 49.
Christoff, A. (1958). *Phytopath. Z.* **31**: 381.
Coman, T. and Çociui, U. (1976). *Bull. d'Inform. Sharka* **2**: 15.
Dunez, J. (1977). *Ann. phytopath.* **9**: 219.
Dunez, J. and Sutić, D. (1988). In: *European Handbook of Plant Diseases*; eds I.M. Smith, J. Dunez, D.H. Phillips, R.A. Lelliot, S.A. Archers, p. 583. Blackwell Scientific Publications.
Grüntzig, M. and Fuchs, E. (1979). *Arch. Phytopathol. Pflanzenschutz, Berlin* **75**: 153.
Hamdorf, G. (1975). *Acta Hort.* **44**: 155.
Jordović, M. (1963). *Zast. Bilja* **18**; 143.
Jordović, M., Festić, H. and Ranković, M. (1971). *Annls. Phytopath.* **179**.
Kegler, H. (1977). In: *Pflanzliche Virologie* pp 315, Klinkowski, M. *et al.*, Akademie-Verlag, Berlin.
Kegler, H. and Proll, E. (1989). *Arch. Phytopathol. Pflanzenschutz* **25**: 435.
Kegler, H. and Schade, C. (1971). *CMI/AAB Descr. Pl. Viruses* No. 70.
Kegler, H., Grüntzig, M. and Schimanski, H.-H. (1991). *NachBr. dtsch. Pflanzen.* **43**: 102.
Kerlan, C. and Dunez, J. (1976). *Annls. Phytopath.* **11**: 241.
Kerlan, C., Mille, B. and Dunez, J. (1981). *Phytopathology* **71**: 400.
Krczal, H. and Kunze, L. (1976). *Acta Hort.* **67**: 171.
Kröll, J. (1975). *Zbl. Bakt. Infekt. Kr. Hyg. II* **130**;; 219.
Macovei, A. (1970). *ISt. si Cerc. de Biologie, S.B. Tom.* **22**: 325.
Macovei, A., Sutić, D. and Savulescu, A. (1971). *Annls. Phytopath. H.S.* **171**.
Marenaud, C. (1972). *C.R. 3rd Long. Union Phytopathol. Medit., Oeiras* **475**.
Marenaud, C. and Mazy, K. (1976). *Phytoma* **280**: 20.
Minoiu, N. (1973). *Anal. Inst. Ceretai Pentru. Prot. Plantelor* **9**: 49.
Németh, M. (1963). *Phytopath. Medit.* **2**: 162.
Németh, M. (1986). In: *Virus, Mycoplasma and Rickettsia Diseases of Fruit Trees*, Akademia Kiado, Budapest.
Németh, M. and Kölber, M. (1983). *Acta Hort.* **130**: 293.
Noel, M.C., Kerlan, C., Garnier, M. and Dunez, J. (1978). *Ann. phytopath.* **10**: 381.
Van Oosten, H.J. (1970). *Neth. J. Pl. Path.* **76**: 252.
Van Oosten, H.J. (1972). *Neth. J. Pl. Path.* **78**: 33.
Plése, N., Rilović, M. and Wrischer, M. (1969). *Zast. Bilja* **104**: 143.
Ranković, M. (1974). *Ph.D. Thesis.*
Ranković, M. and Jordović, M. (1970). *Zast. Bilja* **21**: 195.
Savulescu, A. and Macovei, A. (1965). *Zast. Bilja* **16**: 357.
Schade, C. (1969). *Zbl. Bakt. ParasitKde. Infekt. Kr. Hyg. II* **123**: 299.

Schade, C. (1973). *Plant Virology, Prague*, p. 291.
Schuch, K. (1962). *Z. PflKrankh. PflPath. PflSchutz.* **69**: 137.
Sutić, D. (1961). *T. planteavl.* **65**: 138.
Sutić, D. (1963). *Phytopathol. Medit.* **2**: 171.
Sutić, D. (1965). *Zast. Bilja* **16**: 347.
Sutić, D. (1968). *Acta Hort. Bucharest* **2**: 467.
Sutić, D. (1971). *Ann. Phytopath.* 161.
Sutić, D. (1983). *Viroze bijjaka, Nolit, Beograd.*
Sutić, D. and Ranković, M. (1981). *Agronomie* **1**: 617.
Sutić, D. and Ranković, M. (1983). *Zast. Bilja* **34**: 241.
Sutić, D., Babović, M. and Marković, S. (1976). *Acta Hort.* **67**: 165.
Sutić, D., Jordović, M., Ranković, M. and Festić, H. (1971). *Annls. Phytopath.* **1**: 185.
Sutić, D., Tosić, M. and Virijević, S. (1981). *IV Kongres Mikrobiol. jugoslavije*, 226.
Syrgianidis, G.D. (1980). *Acta Phytopath. Acad. Sci. Hung.* **15**; 85.
Szirmai, J. (1961). *T. Planteavl.* **220**.

# Poa semilatent *hordeivirus*

**Data collated by A.A. Brunt, 1992.**

*Standard acronym:* PSLV. *Strains:* mild and severe strains.

First reported in *Elymus trachycaulum, Poa palustris*; from Canada; by Slykhuis (1972).

**Natural host range and symptoms**

Symptoms disappear soon after infection.

*Elymus trachycaulum, Poa palustris* — probably symptomless infection or faint transient leaf chlorosis (but often occurs together with hordeum mosaic and/or oat necrotic mottle viruses in plants with conspicuous leaf chlorosis).

**Transmission**

Virus transmitted by mechanical inoculation; not transmitted by seed.

**Geographical distribution**

Spreads in Canada (Alberta).

Diagnostically susceptible

*Avena sativa, Triticum aestivum* — chlorotic leaf mottling and plant death.

Diagnostically insusceptible

*Agropyron repens, Agrostis palustris, Bromus inermis, Lolium perenne.*

Maintenance and propagation hosts

*Avena sativa, Poa palustris.*

Assay hosts (**L**ocal lesions or **W**hole plants)

*Avena sativa* (W), *Triticum aestivum* (W).

Monocotyledonous Hosts *(**Susceptible** or Insusceptible)*

*Agropyron repens. Agrostis alba. Agrostis palustris. Agrostis tenuis.* ***Avena sativa.*** *Bromus inermis. Elytrigia intermedium.* ***Hordeum vulgare. Lolium multiflorum.*** *Lolium perenne.* ***Phleum pratense. Poa compressa. Secale cereale. Triticum aestivum. Triticum durum. Zea mays.***

Sources of host-range data

Slykhuis (1972).

Sources of host-range data

Slykhuis (1972).

**Properties of particles in sap**

TIP: 67 °C. LIV: 120 days. DEP: $\log_{10}$ minus 3. Leaf sap contains few virions.

**Purification**

Method: Lane (1974).

**Particle morphology**

Virions rod–shaped; not enveloped; usually straight; with a clear modal length; of 162 nm; 25 nm wide. Axial canal obvious. Basic helix obvious.

**Biochemical properties**

*Genome* consists of RNA; single–stranded; of four parts; largest (or only) genome part 4.5 kb; the 2nd largest 4 kb; the 3rd largest 3.6 kb; the 4th largest 3.5 kb. Genomic nucleic acid isolated by Lane (1974).

*Sequence database accession code(s):*

M81486 Em(40)_vi:PSZBETACD Gb(84)_vi:PSZBETACD Poa semilatent virus beta genomic segment, 3' proximal end. 12/92 795bp.

M81487 Em(40)_vi:PSZGAMRNA Gb(84)_vi:PSZGAMRNAP Poa semilatent virus gamma genomic segment, 3' terminal end. 12/92 1,519bp. 2 sequences.

*Virion protein(s)* one; $M_r$ 23000; and it is coat protein.

**Taxonomy and Relationships**

*Hordeivirus.*

Virus(es) with serologically related virions: barley stripe mosaic and lychnis ringspot viruses.

Several that differ in pathogenicity have been described. The genome of the virus does not cross-hybridize with those of barley stripe mosaic or lychnis ringspot viruses using cDNAs transcribed from them.

**References**

Carroll, T.W. (1986). In: *The Plant Viruses*, Vol. 2, The Rod-Shaped Plant Viruses, p. 388; eds M.H.V. van Regenmortel and H. Fraenkel-Conrat. Plenum Press, New York.

Hunter, B.G., Heaton, L.A., Bracker, C.E. and Jackson, A.O. (1986). *Phytopathology* **76**: 322.

Hunter, B.G., Smith, J., Fattouh, F. and Jackson, A.O. (1989). *Intervirology* **30**: 18.

Jackson, A.O., Hunter, B.G. and Gustafson, G.D. (1989). *Ann. Rev. Phytopath.* **27**: 95.

Lane, L.C. (1974). *Virology* **58**: 323.

Pólak, Z. and Slykhuis, J. (1972). *Can. J. Bot.* **50**: 263.

Slykhuis, J. (1972). *Phytopathology* **62**: 508.

# Poinsettia (?) *alphacryptovirus*

**Data collated by A.J. Gibbs, 1990.**

*Standard acronym:* PnCV.

First reported in *Euphorbia pulcherrima*; from Germany; by Koenig and Lesemann (1980).

**Natural host range and symptoms**

*Euphorbia pulcherrima* — no symptoms.

**Ecology and control**

Studies reported by Preil *et al.* (1982); Paludan *et al.* (1982).

**Geographical distribution**

Found, but with no evidence of spread, in Germany.

**Particle morphology**

Virions isometric; not enveloped; 28 nm in diameter (may be of two sizes 26 nm and 31 nm diameter); rounded in profile; without a conspicuous capsomere arrangement.

**Physical properties**

One sedimenting component in purified preparations; sedimentation coefficient 120 *S*.

**Taxonomy and Relationships**

Possible *Alphacryptovirus: Partitiviridae*.

As plants may be freed of virus by meristem or cell culture, it may not be a typical cryptovirus (Boccardo *et al.*, 1987).

**References**

Boccardo, G., Lisa, V., Luisoni, E. and Milne, R.G. (1987). *Adv. Virus Res.* **32**: 171.

Koenig, R. and Lesemann, D.-E. (1980). *Plant Dis.* **64**: 782.

Paludan, N. and Begtrup, J. (1986). *Tidsskr. PlAvl.* **86**: 283.

Preil, W., Koenig, R., Englehardt, M. and Meier-Dinkel, A. (1982). *Phytopath. Z.* **105**: 193.

# Poinsettia mosaic (?) *tymovirus*

**Data collated by P.L. Guy, 1985.**

*Standard acronym:* PnMV.

First reported in *Euphorbia pulcherrima*; from Wisconsin, U.S.A.; by Fulton and Fulton (1980).

**Natural host range and symptoms**

Symptoms vary seasonally, or disappear soon after infection.

*Euphorbia pulcherrima, E. fulgens* — systemic mosaic and malformation. Bracteoles malformed and may not develop their full cream or red colouration.

**Transmission**

Virus transmitted by mechanical inoculation; transmitted by grafting; not transmitted by contact between plants; not transmitted by seed; not transmitted by pollen.

**Geographical distribution**

Spreads in Australia, Canada, Germany, the UK, and the USA.

**Experimental host range**

Few (<3) families susceptible.

Diagnostically susceptible

*Euphorbia cyathophora* (syn. *E. heterophylla*) — systemic leaf mosaic.

*Nicotiana benthamiana* — patches of irregular chlorosis in systemically infected leaves.

Diagnostically insusceptible

*Brassica pekinensis, Chenopodium amaranticolor, Cucumis sativus, Gomphrena globosa, Phaseolus vulgaris.*

Maintenance and propagation hosts

*Nicotiana benthamiana, Euphorbia cyathophora.*

Assay hosts (**L**ocal lesions or **W**hole plants)

*N. benthamiana* (W), *E. cyathophora* (W).

Dicotyledonous Hosts *(**Susceptible** or Insusceptible)*

*Apium graveolens. Beta patellaris. Beta vulgaris. Brassica campestris* ssp. *pekinensis. Cassia occidentalis. Catharanthus roseus. Celosia argentea. Chenopodium album. Chenopodium amaranticolor. Chenopodium murale. Chenopodium quinoa. Cucumis melo. Cucumis sativus. Cyamopsis tetragonoloba. Dahlia pinnata.* ***Datura stramonium. Euphorbia marginata.*** *Gomphrena globosa. Gossypium hirsutum. Lactuca sativa. Linum usitatissimum. Momordica balsamina.* ***Nicotiana benthamiana. Nicotiana clevelandii. Nicotiana glutinosa. Nicotiana rustica. Nicotiana tabacum.*** *Papaver nudicaule. Pastinaca sativa. Phaseolus vulgaris. Phlox drummondii. Ricinus communis. Tetragonia expansa. Torenia fournieri. Trifolium incarnatum. Trifolium repens. Vigna unguiculata* ssp. *cylindrica. Vigna unguiculata* ssp. *sesquipedalis. Zinnia elegans.*

Monocotyledonous Hosts *(**Susceptible** or Insusceptible)*

*Lilium formosanum.*

Sources of host-range data

Fulton and Fulton (1980); Lesemann *et al.* (1983); Meyer *et al.* (1986); Paludan and Begtrup (1986).

**Properties of particles in sap**

TIP: 60–65 °C. LIV: 8–10 days. DEP: $\log_{10}$ minus 4 (and greater). Leaf sap contains many virions.

**Purification**

Method: Lesemann *et al.* (1983); Koenig *et al.* (1986).

**Particle morphology**

Virions isometric; not enveloped; 26–29 nm in diameter; rounded in profile; without a conspicuous capsomere arrangement.

**Physical properties**

Two sedimenting components in purified preparations; sedimentation coefficient of the fastest 112–117 *S*; of the other(s) 50–55 *S*. Density 1.41 g $cm^{-3}$ in CsCl (B), or 1.29 g $cm^{-3}$ in CsCl (T; when fixed in 9% formaldehyde). Isoelectric point pH 7. $A_{260}/A_{280}$ ratio 1.8 (B), or 1 (T).

**Biochemical properties**

Virions contain 37 % nucleic acid; 63 % protein.

*Genome* consists of RNA; single–stranded; linear; unipartite; largest (or only) genome part 2–2.3 kb. Genomic nucleic acid isolated by Pfannenstiel *et al.* (1982); Lesemann *et al.* (1983). Base composition 20.5 % G; 24.1 % A; 31.3 % C; 24.1 % U.

$M_r$ 21700. Method of preparation: Pfannenstiel *et al.* (1982); Lesemann *et al.* (1983). Amino acid composition: Pfannenstiel *et al.* (1982).

**Replication**

Genome replicates probably in association with chloroplasts (in vesicles). Replication does not depend on a helper virus.

**Cytopathology**

Virions found in all parts of the host plant; in cytoplasm, in nucleoli, and in chloroplasts; the vacuole. However, only empty components (T) are found in the nucleus. Other cellular changes: vesicle formation by invagination of the inner chloroplast membrane only (in contrast to definitive tymoviruses in which the invaginations have a double membrane).

**Taxonomy and Relationships**

Possible *Tymovirus*.

Virus(es) with serologically unrelated virions: 11 tymoviruses.

**References**

Fulton, R.W. and Fulton, J.L. (1980). *Phytopathology* **70**: 321.

Koenig, R. and Lesemann, D.-E. (1980). *Plant Dis.* **65**: 781.

Koenig, R., Lesemann, D.-E. and Fulton, R.W. (1986). *AAB Descr. Pl. Viruses* No. 311, 4 pp.

Lesemann, D.-E., Koenig, R., Huth, W., Brunt, A.A., Phillips, S. and Barton, R.J. (1983). *Phytopath. Z.* **107**: 250.

Meyer, S., Lesemann, D.-E. and Koenig, R. (1986). *NachrBl. dtsch. PflSchutz., Stuttg.* **38**: 163.

Paludan, N. and Begtrup, J. (1986). *Tidsskr. PlAvl* **90**: 283.

Pfannenstiel, M.A., Mintz, K.P. and Fulton, R.W. (1982). *Phytopathology* **72**: 252.

## Pokeweed mosaic *potyvirus*

**Data collated by A.A. Brunt and P. Jeyanandarajah, 1991.**

*Standard acronym:* PkMV.

First reported in *Phytolacca americana*; from the U.S.A.; by Woods (1902).

**Natural host range and symptoms**

Symptoms persist.

*Phytolacca americana* — chlorotic leaf mottling.

**Transmission**

Transmitted by a vector; an insect; *Myzus persicae*; Aphididae. Transmitted in a non–persistent manner. Virus transmitted by mechanical inoculation; transmitted by grafting; not transmitted by seed.

**Geographical distribution**

Spreads in the USA (central, southern and eastern states).

**Experimental host range**

Few (<3) families susceptible.

Diagnostically susceptible

*Phytolacca americana* — systemic chlorotic mottling and malformation.

*Chenopodium quinoa* — chlorotic local lesions; not systemic.

*Gomphrena globosa* — chlorosis and necrotic spotting of systemically infected leaves.

Maintenance and propagation hosts

*Phytolacca americana.*

Assay hosts (**L**ocal lesions or **W**hole plants)

*Chenopodium quinoa* (L).

Dicotyledonous Hosts *(**Susceptible** or Insusceptible)*

*Antirrhinum majus. Beta vulgaris. Brassica oleracea* ssp. *capitata. Capsicum frutescens. Celosia argentea. Chenopodium amaranticolor. Chenopodium capitatum.* ***Chenopodium quinoa****. Citrullus lanatus. Cucumis sativus. Cucurbita pepo. Dahlia pinnata. Datura stramonium.* ***Gomphrena globosa****. Lycopersicon esculentum. Nicotiana clevelandii. Nicotiana glutinosa. Nicotiana tabacum. Phaseolus vulgaris.* ***Phytolacca americana****. Pisum sativum. Solanum melongena. Spinacia oleracea. Vigna unguiculata. Zinnia elegans.*

**Properties of particles in sap**

TIP: 70–75 °C. LIV: 10 days (at 25°C). DEP: $\log_{10}$ minus 5–6.

**Particle morphology**

Virions filamentous; not enveloped; usually flexuous; with a clear modal length; of 776 nm; 12–13 nm wide. Axial canal obscure. Basic helix obscure.

**Cytopathology**

Inclusions present in infected cells; are pinwheels; they do not contain virions.

**Taxonomy and Relationships**

*Potyvirus: Potyviridae.*

Virus(es) with serologically related virions: henbane mosaic, potato Y and tobacco etch viruses.

**References**

Shepherd, R.J., Fulton, J.P. and Wakeman, R.J. (1969). *Phytopathology* **59**: 219.

Woods, A.F. (1902). *Bull. Bur. Pl. Ind. US Dep. Agric.* **18**: 24.

# Poplar decline (?) potyvirus

**Data collated by A.A. Brunt, 1991.**

*Synonyms:* poplar potyvirus.

First reported in *Populus x euramericana, P. tremuloides*; from the U.S.A.; by Martin *et al.* (1982).

**Natural host range and symptoms**

Symptoms persist.

*Populus euramericana, P. grandidenta, P. tremuloides* — chlorotic and/or necrotic leafspots in spring, bronzing and necrosis of leaves later and, ultimately, death of branches and then trees.

**Transmission**

Virus transmitted by mechanical inoculation.

**Geographical distribution**

Spreads in the USA.

**Experimental host range**

Few (<3) families susceptible.

Diagnostically susceptible

*Phaseolus vulgaris, P. lunatus, Vigna unguiculata* — local chlorotic or necrotic spots; systemic mottling and necrosis.

*Nicotiana clevelandii* — mild systemic leaf chlorosis.

*Chenopodium quinoa* — chloroitc local lesions; no systemic infection.

Diagnostically insusceptible

*Cucumis sativus, Nicotiana glutinosa, N. tabacum, Petunia hybrida, Pisum sativum.*

Maintenance and propagation hosts

*Nicotiana clevelandii, Phaseolus vulgaris, P. lunatus, Vigna unguiculata.*

Assay hosts (**L**ocal lesions or **W**hole plants)

*Chenopodium quinoa* (L), *Phaseolus vulgaris* (W), *Vigna unguiculata* (W).

Dicotyledonous Hosts *(**Susceptible** or Insusceptible)*

*Capsicum annuum. Catharanthus roseus. Chenopodium amaranticolor.* ***Chenopodium quinoa***. *Cucumis sativus. Cucurbita pepo. Lactuca sativa. Lycopersicon esculentum. Medicago sativa.* ***Nicotiana clevelandii***. *Nicotiana glutinosa. Nicotiana megalosiphon. Nicotiana rustica. Nicotiana tabacum. Petunia hybrida.* ***Phaseolus lunatus***. ***Phaseolus vulgaris***. *Physalis floridana. Pisum sativum. Raphanus sativus.* ***Vigna unguiculata***. *Zinnia elegans.*

Sources of host-range data

Martin *et al.* (1982).

**Properties of particles in sap**

TIP: 60 °C. LIV: 1.2 days (at 22°C). DEP: $\log_{10}$ minus 5. Leaf sap contains few virions.

**Particle morphology**

Virions filamentous; not enveloped; usually flexuous; with a clear modal length; of 805 nm; 12.6 nm wide. Axial canal obscure.

**Cytopathology**

Inclusions present in infected cells; are pinwheels; they do not contain virions.

**Taxonomy and Relationships**

Unassigned potyvirus.

**References**

Martin, R.R., Berbee, J.G. and Omuema, J.O. (1982). *Phytopathology* **72**: 1158.

# Poplar mosaic *carlavirus*

**Data collated by F.A. van der Meer; J.I. Cooper, 1987.**

*Synonyms:* Canadian poplar mosaic virus (Atanasoff, 1935), poplar latent virus (Matthews, 1979). *Standard acronym:* PopMV.

First reported in *Populus balsamifera*; from Bulgaria; by Atanasoff (1935).

**Natural host range and symptoms**

Symptoms persist (but the virus is often irregularly distributed in trees; Van der Meer *et al.*, 1980; Cooper and Edwards, 1981).

*Populus maximowiczii* — no symptoms (Cooper *et al.*, 1986).

*P. balsamifera, P. nigra, P. deltoides* subspecies angulata, monilifera and missouriensis, *P. trichocarpa, P. candicans, P. euramericana* (= *P. x canadensis*; Berg, 1964), *P. x interamericana* (= *P. trichocarpa* x *P. deltoides*; Van der Meer, 1981) — yellow asteroid spotting of the smaller leaf veins. Very sensitive cultivars may develop necrosis in bark, petioles and leaf veins, growth of poplars may be severely decreased (Berg, 1964; Castellani, 1966; Biddle and Tinsley, 1971; Van der Meer, 1981) but more than one virus may be involved.

**Transmission**

Virus transmitted by mechanical inoculation (Berg, 1962); transmitted by grafting (Corte, 1959; Blattny *et al.*, 1962); not transmitted by seed (Cooper and Edwards, 1981); transmitted by pollen to the pollinated plant (Cooper, 1976).

**Ecology and control**

Studies reported by Cooper (1969); Cooper and Edwards (1981); Van der Meer (1979; 1981); Van der Meer *et al.* (1980*a*). Poplar can be freed from PMV by heat treatment; after 4–10 weeks at 37–39°C (Sweet, 1976; Van der Meer, 1981), one cm long tip cuttings are usually virus-free.

**Geographical distribution**

Spreads in the Netherlands and the UK (Van der Meer, 1979 and 1981; Cooper and Edwards, 1981). Found, but with no evidence of spread, in Bulgaria (Atanasoff, 1935), Canada (Navratil and Boyer, 1968), China (Xiung Yuying *et al.*, 1984), former Czechoslovakia (Blattny *et al.*, 1962), Denmark (Kristensen, 1960), Poland (Benben, 1957), Italy (Corte, 1959), Georgia CIS (Navratil, 1965), Germany (Brandes, 1963; Schmelzer, 1964), Korea (Yong Jooh *et al.*, 1980), U.K (Biddle and Tinsley, 1971*a*) and former Yugoslavia (Perisic, 1951). Also in Austria, Australia, Belgium, France, Hungary, Ireland, Japan, Luxembourg, Spain, Switzerland, Turkey, Tanzania, the U.S.A. and Venezuela.

**Experimental host range**

Many (>9) families susceptible.

Diagnostically susceptible

*Nicotiana megalosiphon* — chlorotic or necrotic local lesions and systemic mosaic.
*Nicotiana glutinosa* — chlorotic local lesions and systemic vein chlorosis.
*Vigna unguiculata* — red or brown local lesions.
*Phaseolus vulgaris* — necrotic local lesions (Van der Meer *et al.*, 1980; 1980*a*).

Diagnostically insusceptible

*Nicotiana tabacum* White Burley, *Gomphrena globosa* (Luisoni *et al.*, 1976).

Maintenance and propagation hosts

*Nicotiana megalosiphon, N. clevelandii.*

Assay hosts (Local lesions or Whole plants)

*Nicotiana megalosiphon* (L and W) (Van der Meer *et al.*, 1980*a*).

Dicotyledonous Hosts *(**Susceptible** or Insusceptible)*

*Amaranthus caudatus.* ***Antirrhinum majus.*** ***Arachis hypogaea.*** *Beta vulgaris. Brassica campestris* ssp. *chinensis. Brassica oleracea* ssp. *botrytis. Brassica oleracea* ssp. *capitata. Capsicum annuum.* ***Catharanthus roseus.*** *Celosia argentea. Chenopodium album. Chenopodium amaranticolor.* ***Chenopodium foetidum.*** *Chenopodium foliosum.* ***Chenopodium murale. Chenopodium quinoa.*** *Crotalaria spectabilis.* ***Cucumis sativus. Cucurbita maxima. Cucurbita pepo. Datura stramonium. Datura tatula. Daucus carota.*** *Dianthus barbatus.* ***Glycine max.*** *Gomphrena globosa. Helianthus annuus.* ***Lathyrus odoratus. Lavatera trimestris.*** *Lupinus albus. Lycopersicon esculentum.* ***Nicotiana clevelandii. Nicotiana × edwardsonii. Nicotiana debneyi. Nicotiana glutinosa. Nicotiana glutinosa × clevelandii. Nicotiana megalosiphon. Nicotiana occidentalis. Nicotiana rustica. Nicotiana tabacum. Ocimum basilicum. Petunia hybrida. Phaseolus vulgaris. Phlox drummondii. Physalis floridana. Pisum sativum.*** *Senecio vulgaris. Spinacia oleracea.* ***Tetragonia expansa.*** *Trifolium repens.* ***Tropaeolum majus.*** *Verbesina encelioides.* ***Vicia faba. Vigna unguiculata. Zinnia elegans.***

Additional hosts

***Lonicera periclymenum.***

Comments on host-range

Infectious inoculum from poplar does not induce symptoms in *Chenopodium quinoa*. However, after passage in *Nicotiana megalosiphon*, the virus causes lesions in *Chenopodium quinoa* and becomes adapted to *Chenopodium quinoa* and then induces numerous local lesions and systemic mosaic in it.

Sources of host-range data

Berg (1964); Schmelzer (1966); Luisoni *et al.* (1976); Van der Meer *et al.* (1980); Xiang Yuying *et al.* (1984). There is some discrepancy between the results in these five references.

**Properties of particles in sap**

TIP: 74 °C (with sap from *Nicotiana megalosiphon*). LIV: 2 days. Leaf sap contains many virions.

**Purification**

Method: Van der Meer *et al.* (1980; 1980*a*).

**Particle morphology**

Virions filamentous; not enveloped; usually flexuous (but only slightly); with a clear modal length; of 675 nm (Biddle and Tinsley, 1971); 13 nm wide (Biddle and Tinsley, 1976). Axial canal obvious (but only partially and is inconsistent (Boccardo and Milne, 1976)); *c.* 2 nm in diameter. Basic helix obvious, or obscure; pitch of basic helix 3.5 nm.

**Physical properties**

One sedimenting component in purified preparations; sedimentation coefficient 165 *S* (Berg, 1964).

**Biochemical properties**

*Genome* consists of RNA; single–stranded; unipartite; largest (or only) genome part 6.48 kb. Genomic nucleic acid isolated by Boccardo and Milne (1976). Infectivity retained when deproteinised with phenol or detergent.

*Sequence database accession code(s):*

D13364 Em(40)_vi:MVWTGBCPG Gb(84)_vi:MVWTGBCPG Poplar mosaic virus (PMV) genes for coat protein, helicase and other proteins, complete cds.

X65102 Em(40)_vi:PMVCOATP Gb(84)_vi:PMVCOATP Poplar mosaic Virus genomic RNA for coat protein and putative 14 kDa protein. 7/93 1,328bp. 2 sequences.

*Virion protein(s)* one (plus a fragment); $M_r$ 40000 (plus 30000); coat protein. Method of preparation: Boccardo and Milne (1976).

**Replication**

Replication does not depend on a helper virus.

**Cytopathology**

Virions found in mesophyll, vascular parenchyma and palisade cells; in cytoplasm; plasmodesmata occasionally, but were virion-like particles. Inclusions absent from infected cells (Boccardo and Milne, 1976; Atkinson and Cooper, 1976; Brunt *et al.*, 1976). Other cellular changes: cell wall overgrowths in poplar caused by poplar mosaic virus (Atkinson and Cooper, 1976), whereas cell wall structure in *Nicotiana megalosiphon* was not altered by the virus (Boccardo and Milne, 1976). Accumulation of endoplasmic reticulum occured in both infected hosts. Loose wefts of virions have been noted (Atkinson and Cooper, 1976).

**Taxonomy and Relationships**

*Carlavirus*.

Virus(es) with serologically related virions: honeysuckle latent virus (Van der Meer *et al.*, 1980; Brunt *et al.*, 1980). Virus(es) with serologically unrelated virions: pea streak, chrysanthemum B, potato S, red clover vein mosaic (Berg, 1964), carnation latent (Berg, 1964; Luisoni *et al.*, 1976), potato M (Berg, 1964; Luisoni *et al.*, 1976) and shallot latent (Bos *et al.*, 1978) viruses.

Poplar mosaic virus induces necrotic local lesions in inoculated leaves of *Phaseolus vulgaris* and *Vigna unguiculata*, but neither are susceptible to the serologically related honeysuckle latent virus (Van der Meer *et al.*, 1980). Although poplar mosaic virus is a carlavirus, its vector is unknown (Berg, 1964; Schmelzer, 1966; Boccardo *et al.*, 1973; Cooper and Edwards, 1981). Best tests for diagnosis: poplar mosaic virus usually occurs alone in poplars. Potyvirus-like virions have been found in poplars in the U.S.A. (Martin *et al.*, 1982), that can be distinguished by virion morphology and by serological tests.

**References**


Atanasoff, D. (1935). *Phytopath. Z.* **8**: 197.

Atkinson, M.A. and Cooper, J.I. (1976). *Ann. appl. Biol.* **83**: 395.

Benben (1957). *Roczn. Nauk les.* **19**: 63 (as cited by Cooper, 1969).

Berg, T.M. (1962). *Nature, Lond.* **194**: 1302.

Berg, T.M. (1964). *Meded. Landb. Wageningen* **64–11**: 1.

Biddle, P.G. and Tinsley, T.W. (1971). *CMI/AAB Descr. Pl. Viruses* No. 75, 3 pp.

Biddle, P.G. and Tinsley, T.W. (1971*a*). *New Phytol.* **70**: 61.

Biddle, P.G. and Tinsley, T.W. (1971*b*). *New Phytol.* **70**: 67.

Blattny, Pozdena, Svobodova, Brčak, Bojansky, Leontovyc and Prochazkova (1962/B). *Acta Musei nationalis Pragae XVIII* **2**: 47.

Boccardo, G., Luisoni, E. and Lovisolo, O. (1973). *Abstr. 2nd. Int. Congr. Pl. Path.*, No. 0928.

Boccardo, G. and Milne, R.G. (1976). *Phytopath. Z.* **87**: 120.

Bos, L., Huttinga, H. and Maat, D.Z. (1978). *Neth. J. Pl. Path.* **84**: 227.

Brandes, J. (1963). *Phytopath. Z.* **47**: 84.

Brunt, A.A., Stace-Smith, R. and Leung, E. (1976). *Intervirology* **7**: 303.

Brunt, A.A., Phillips, S. and Thomas (1980). *Acta Hort.* **110**: 205.

Castellani (1966). In: *Breeding Pest Resistant Trees, NATO and NSF Symp.* 89–96. Pergamon Press.

Cooper, J.I. (1976). *Ann. Rep. Commonw. For. Inst.* **26**, 1975, Oxford (as cited by Cooper, 1979).

Cooper, J.I. (1979). In: *Virus Diseases of Trees and Shrubs* p. 46. Inst. Terr. Ecol., Oxford.

Cooper, J.I. and Edwards, M.L. (1981). *Ann. appl. Biol.* **99**: 53.

Cooper, J.I., Edwards, M.L. and Siwecki, R. (1986). *Eur. J. For. Path.* **16**: 116.

Corte (1959). *Atti Ist, bot. Univ. Lab. Crittogam. Pavia* **17**: 1.

Koenig, R. (1982). *CMI/AAB Descr. Pl. Viruses* No. 259, 4 pp.

Kristensen (1960). *Dansk Skovforen. Tidsskr.* **45**: 155.

Luisoni, E., Boccardo, G. and Milne, R.G. (1976). *Phytopath. Z.* **85**: 65.

Martin, R.R., Berbee, J.G. and Omuema, J.O. (1982). *Phytopathology* **72**: 1158.

Navratil, S. (1965). *Zast. Bilja* **16**: 517 (as cited by Cooper, 1969).

Navratil, S. and Boyer (1968). *Can. J. Bot.* **46**: 722.

Perisic (1951). *Sumarstvo* **4**: 522 (as cited by Cooper, 1969).

Schmelzer, K. (1964). *Arch. Forstw.* **13**: 787.

Schmelzer, K. (1966). *Phytopath. Z.* **55**: 31.

Sweet, J. (1976). *Acta Hort.* **59**: 83.

Van der Meer, F.A. (1979). In: *Proceedings of the IUFRO meeting concerning Poplars in France and Belgium*, 17–22 September 1979, of working parties S2–02–10 and S2–03–07, with participation of the International Energy Agency Forest Biomass (Planning Group B): 104–111. Ed. Dorschkamp Res. Inst. Forestry and Landscape Planning, Wageningen.

Van der Meer, F.A. (1981). *Populier* **18**: 51.

Van der Meer, F.A., Maat, D.Z. and Vink, J. (1980). *Neth. J. Pl. Path.* **86**: 69.

Van der Meer, F.A., Maat, D.Z. and Vink, J. (1980*a*). *Neth. J. Pl. Path.* **86**: 99.

Xiang Y., Xi Z.X. and Zhang H.L. (1984). *Scientia Silvae Sinicae* **20**: No. 4.

Yong Jooh, L., Chang Keun, Y., Woon Hong, Y., Jae Woo, H. and Sang Yung, S. (1980). *J. Korean Forestry Society* **49**: 32.

Zhang Z.-H. (1984). *Journal of the South-Western Forestry College* No. 1.


# Poplar vein yellowing (?) *nucleorhabdovirus*

**Data collated by A.A. Brunt, 1991.**

First reported in *Populus balsamifera*; from Canada; by Navratil (1981).

### Natural host range and symptoms

*Populus balsamifera* — clearing and yellowing of leaf veins.

### Geographical distribution

Spreads in Canada.

### Particle morphology

Virions rhabdo– or bullet–shaped; with a clear modal length; of 286 nm (*in vivo*), or 290 nm (in sap); 60 nm wide (*in vivo*), or 77 nm wide (in sap).

### Cytopathology

Virions found in cytoplasm, in nuclei, and in the perinuclear space.

### Taxonomy and Relationships

Possible *Nucleorhabdovirus: Rhabdoviridae.*

**References**

Navratil, S. (1981). *Phytopathology* **71**: 245.

# Potato 14R (?) *tobamovirus*

**Data collated by A.A. Brunt and S. Molyneux, 1986.**

First reported in *Solanum tuberosum spp. andigena x spp. tuberosum*; from Peru; by Salazar (1977).

**Natural host range and symptoms**

*Solanum tuberosum ssp. andigena x ssp. tuberosum* — yellowish mosaic on leaves.

**Geographical distribution**

Spreads in Peru.

Diagnostically susceptible

*Chenopodium amaranticolor, C. quinoa* — systemic necrotic rings and line patterns.
*Chenopodium ambrosioides* — faint systemic mosaic.
*Nicotiana debneyi, N. tabacum* cv. Turkish — necrotic rings.
*Nicotiana benthamiana, N. glutinosa, N. occidentalis* — necrotic local lesions.

Diagnostically insusceptible

*Trifolium incarnatum, Capsicum annuum, Ocimum basilicum.*

Assay hosts (**L**ocal lesions or **W**hole plants)

*Chenopodium amaranticolor* (L) *Nicotiana glutinosa* (L).

Dicotyledonous Hosts *(**Susceptible** or Insusceptible)*

***Amaranthus caudatus***. ***Beta vulgaris***. *Brassica campestris* ssp. *rapa*. *Capsicum annuum*. ***Catharanthus roseus***. ***Chenopodium album***. ***Chenopodium amaranticolor***. ***Chenopodium ambrosioides***. ***Chenopodium capitatum***. ***Chenopodium foliosum***. ***Chenopodium murale***. ***Chenopodium quinoa***. ***Cucumis sativus***. ***Cyamopsis tetragonoloba***. ***Datura stramonium***. ***Datura tatula***. ***Gomphrena globosa***. *Lactuca sativa*. ***Lycopersicon chilense***. ***Lycopersicon esculentum***. ***Nicotiana benthamiana***. ***Nicotiana clevelandii***. ***Nicotiana debneyi***. ***Nicotiana glutinosa***. ***Nicotiana occidentalis***. ***Nicotiana rustica***. ***Nicotiana tabacum***. *Ocimum basilicum*. ***Petunia hybrida***. ***Phaseolus vulgaris***. ***Physalis floridana***. ***Pisum sativum***. ***Sesbania exaltata***. ***Solanum melongena***. ***Solanum tuberosum***. ***Spinacia oleracea***. ***Tetragonia expansa***. *Trifolium incarnatum*. *Trifolium repens*. ***Vicia faba***. ***Vigna unguiculata***. *Vigna unguiculata* ssp. *cylindrica*. *Vigna unguiculata* ssp. *sesquipedalis*.

Sources of host-range data

Salazar (1977).

**Properties of particles in sap**

TIP: 95 °C. LIV: more than 64 days. DEP: $\log_{10}$ minus 6–7.

**Purification**

Method: Salazar (1978).

**Particle morphology**

Virions rod–shaped; with a clear modal length; of 300 nm; 16.4–18.2 nm wide. Axial canal obvious.

**Physical properties**

One sedimenting component in purified preparations; sedimentation coefficient 180 *S*.

**Biochemical properties**

*Genome* consists of RNA; single–stranded; unipartite; largest (or only) genome part 6.5 kb.

*Virion protein(s)* one; $M_r$ 19400; coat protein.

**Taxonomy and Relationships**

Possible *Tobamovirus*.

Virus(es) with serologically unrelated virions: tobacco mosaic and sunn-hemp mosaic viruses.

**References**

Salazar, L.F. (1977). *Ph.D. Thesis*, Dundee University, Dundee, Scotland.

# Potato A *potyvirus*

**Data collated by C. Büchen-Osmond, 1987.**

*Synonyms:* potato mild mosaic virus, potato virus P, *Solanum* virus 3. *Standard acronym:* PVA.

First reported in *Solanum tuberosum*; from Eire; by Murphy and McKay (1932).

**Natural host range and symptoms**

Symptoms persist.

*Solanum tuberosum* — mild mosaic, leaf surface rough with wavy margins, or symptomless.
Potatoes infected with potato virus A in combination with potato virus X and Y show crinkle symptoms.

**Transmission**

Transmitted by a vector; an insect; *Aphis frangulae, A. nasturtii, Myzus persicae*; Aphididae. Transmitted in a non–persistent manner. Virus does not require a helper virus for vector transmission and can help the vector transmission of another virus (potato aucuba mosaic potexvirus; Clinch *et al.*, 1936); transmitted by mechanical inoculation; not transmitted by seed.

**Geographical distribution**

Probably distributed worldwide (in most potato growing areas).

**Experimental host range**

Few (<3) families susceptible.

Diagnostically susceptible

*Nicotiana tabacum* cv. Samsun — systemic vein-clearing and diffuse mottling.

*N. tabacum* cv. White Burley — systemic vein-clearing and dark green vein-banding.

*Nicandra physalodes* — systemic slight vein-clearing and mottle to severe necrosis, rugosity and stunting.

*Lycopersicon pimpinellifolium* — systemic necrosis and death. (MacLachlan *et al.*, 1953).

*Solanum demissum x S. tuberosum* cv. Aquila (=A6 hybrid) and *S. demissum* SdA — many local lesions.

Maintenance and propagation hosts

*Nicotiana tabacum* cv. Samsun.

Assay hosts (**L**ocal lesions or **W**hole plants)

*Solanum demissum x S. tuberosum* cv. Aquila (L) (Köhler, 1953; Bartels, 1970), *Solanum demissum* SdA (L), *Nicandra physalodes* (W) is useful for aphid transmission tests with *Myzus persicae*.

Dicotyledonous Hosts *(**Susceptible** or Insusceptible)*

***Lycopersicon pimpinellifolium. Nicandra physaloides. Nicotiana tabacum. Solanum demissum. Solanum demissum × tuberosum. Solanum tuberosum.***

**Properties of particles in sap**

TIP: 44–52 °C. LIV: 0.5–0.75 days (12–18 hours MacLachlan *et al.*, 1953). DEP: $\log_{10}$ minus 1.

**Purification**

Method: Bartels (1971).

**Particle morphology**

Virions filamentous; not enveloped; usually flexuous; with a clear modal length; of 730 nm; 11 nm wide (Brandes and Paul, 1957).

**Biochemical properties**

*Genome* consists of RNA; single–stranded.
*Sequence database accession code(s):*
S51667 Em(40)_un:S51667 Gb(84)_vi:S51667 polyprotein potato virus A PVA, Genomic RNA, 2597 nt. 3/93 2,597bp.
X59985 Em(40)_vi:PVACOATM Gb(84)_vi:PVACOATM Potato virus A RNA 3'-terminal for coat protein and polymerase. 6/93 1,383bp
Z21670 Em(40)_vi:PVCGA Gb(84)_vi:PVCGA Potato virus A complete genome. 2/94 9,585bp.
Z49087 Em(43)_vi:Pvaunibcp Gb(89)_vi:Pvaunibcp Potato virus A (USA) mRNA for nuclear inclusion b (partial) and coat protein. 4/95 1,108bp.
Z49088 Em(43)_vi:Pvamnibcp Gb(89)_vi:Pvamnibcp Potato virus A (Maine) mRNA for nuclear inclusion b (partial) and coat protein. 4/95 1,110bp.

**Replication**

Replication does not depend on a helper virus.

**Cytopathology**

Virions found in all parts of the host plant; in cytoplasm. Inclusions present in infected cells.

**Taxonomy and Relationships**

*Potyvirus: Potyviridae.*

Virus(es) with serologically related virions: potato Y (common and tobacco veinal necrosis strain), henbane mosaic and tobacco etch viruses.

Best tests for diagnosis: Potato may be infected by potato A, potato Y and other potyviruses. They are more easily distinguished serologically although only potato virus A induces local lesions in *Solanum demissum* SdA.

**References**

Bartels, R. (1970). *Potato Res.* **13**: 119.
Bartels, R. (1971). *CMI/AAB Descr. Pl. Viruses* No. 54, 4 pp.

Brandes, J. and Paul, H.L. (1957). *Arch. Microbiol.* **26**: 358.
Clinch, P., Loughnane, J.B. and Murphy, D.A. (1936). *Scient. Proc. R. Dubl. Soc.* **21**: 431.
Köhler, E. (1953). *Züchter* **23**: 173.
MacLachlan, D.S., Larson, R.H. and Walker, J.C. (1953). *Res. Bull. agric. Exp. Stn Univ. Wis.*, 180, 36 pp.
Murphy, P.A. and McKay, R. (1932). *Scient. Proc. R. Dubl. Soc.* **20**: 347.

# Potato Andean latent *tymovirus*

**Data collated by R. Koenig, 1985. Revised 1989.**

*Synonyms:* Andean potato latent virus. *Standard acronym:* APLV. *Strains:* isolates differ in serological and electrophoretic properties and symptoms induced (Koenig *et al.*, 1979).

First reported in *Solanum tuberosum* (groups Phureja, Chaucha and Andigena); from material collected in tropical Andes.

**Natural host range and symptoms**

Symptoms persist.
*Solanum tuberosum* — symptomless or sometimes mosaic.

**Transmission**

Transmitted by a vector and means not involving a vector; an insect; *Epithrix* sp; Coleoptera. Transmitted in a non–persistent manner. Virus transmitted by mechanical inoculation; transmitted by grafting; transmitted by contact between plants; transmitted by seed (less than 1% in *Solanum tuberosum*).

**Geographical distribution**

Spreads in the South and Central American region; Bolivia, Chile, Colombia, Ecuador, and Peru (CABI/EPPO data).

**Experimental host range**

Several (3–9) families susceptible.

Diagnostically susceptible

*Chenopodium quinoa* — necrotic spots; systemic chlorosis with some strains.
*Nicotiana bigelovii, N. clevelandii, N. debneyi, N. megalosiphon* — severe mosaic.

Diagnostically insusceptible

*Phaseolus vulgaris, Vigna unguiculata.*

Maintenance and propagation hosts

*Nicotiana bigelovii, N. clevelandii.*

Assay hosts (**L**ocal lesions or **W**hole plants)

*Nicotiana bigelovii* (W), *N. clevelandii* (W).

Dicotyledonous Hosts ***(Susceptible** or Insusceptible)*

***Amaranthus caudatus***. *Brassica campestris* ssp. *chinensis*. ***Chenopodium amaranticolor***. ***Chenopodium quinoa***. ***Cucumis sativus***. ***Datura stramonium***. ***Gomphrena globosa***. ***Hyoscyamus niger***. *Lablab purpureus*. ***Lycopersicon esculentum***. ***Lycopersicon pimpinellifolium***. ***Nicandra physaloides***. ***Nicotiana benthamiana***. ***Nicotiana bigelovii***. ***Nicotiana clevelandii***. ***Nicotiana debneyi***. ***Nicotiana glutinosa***. ***Nicotiana megalosiphon***. ***Nicotiana rustica***. ***Nicotiana sylvestris***. ***Petunia hybrida***. *Phaseolus vulgaris*. *Physalis floridana*. *Physalis peruviana*. *Pisum sativum*. ***Solanum***

chacoense. ***Solanum demissum*****.** ***Solanum melongena*****.** ***Solanum tuberosum*****.** *Trifolium incarnatum. Vicia faba. Vigna unguiculata. Vigna unguiculata* ssp. *cylindrica.*

Monocotyledonous Hosts *(**Susceptible** or Insusceptible)*

*Hordeum vulgare. Triticum aestivum.*

Sources of host-range data

Fribourg *et al.* (1977); Gibbs *et al.* (1966).

**Properties of particles in sap**

TIP: 65–80 °C. LIV: 21 days. DEP: $\log_{10}$ minus 6. Infectivity of sap not changed by treatment with di–ethyl ether. Leaf sap contains many virions.

**Purification**

Method: Dunn and Hitchborn (1965).

**Particle morphology**

Virions isometric; not enveloped; 30 nm in diameter; rounded in profile; with a conspicuous capsomere arrangement.

**Physical properties**

Two sedimenting components in purified preparations; sedimentation coefficient of the fastest 115 *S*; of the other(s) 54 *S* (T). Density 1.26 and 1.41 g $cm^{-3}$ in CsCl (unfixed). Isoelectric point pH of different strains differs. $A_{260}/A_{280}$ ratio 0.81 (T), or 1.51 (B).

**Biochemical properties**

Virions contain 36 % nucleic acid (B); 64 % protein (B).

*Genome* consists of RNA; single–stranded; linear; unipartite; largest (or only) genome part 6.5 kb. Base composition 16 % G; 22 % A; 38 % C; 24 % U. 5' terminus of RNA has a methylated nucleotide cap. Infectivity retained when deproteinised with phenol or detergent. Genome has *t*RNA–like activity. Genome accepts possibly valine.

*Sequence database accession code(s):*

M15284 Em(40)_vi:TYMTY3AP Gb(84)_vi:APV3APLV Andean potato latent virus 3' end which can form tRNA-like structure. 11/90 96bp.

M58313 Em(40)_vi:APVRRLS Gb(84)_vi:APVRRLS Andean potato latent virus (APLV) 3' terminus tRNA-like structure. 8/91 82bp. 2 sequences.

*Virion protein(s)* one; $M_r$ 20000. Amino acid composition: Paul *et al.* (1980).

**Replication**

Genome replicates probably in association with chloroplasts (in virus-induced vesicles). Replication does not depend on a helper virus.

**Cytopathology**

Virions found in all parts of the host plant; in cytoplasm, in nuclei, and in cell vacuoles. Inclusions present in infected cells; are crystals in the cytoplasm and crystals in the nucleus; they contain virions. Other cellular changes: fragmentation of mitochondria.

**Taxonomy and Relationships**

*Tymovirus.*

Virus(es) with serologically related virions: eggplant mosaic virus is closely related; belladonna mottle and dulcamara mottle viruses less so; and most other tymoviruses are distantly related.

**References**

Dunn, D.B. and Hitchborn, J.H. (1965). *Virology* **25**: 171.

Fribourg, C.E., Jones, R.A.C. and Koenig, R. (1977). *Ann. appl. Biol.* **86**: 373.
Gibbs, A.J. and Harrison, B.D. (1973). *CMI/AAB Descr. Pl. Viruses* No. 124, 4 pp.
Gibbs, A.J., Hecht-Poinar, E., Woods, R.D. and McKee, R.K. (1966). *J. gen. Microbiol.* **44**: 177.
Koenig, R., Francksen, H. and Stegemann, H. (1981). *Phytopath. Z.* **100**: 347.
Koenig, R., Fribourg, C.E. and Jones, R.A.C. (1979). *Phytopathology* **69**: 748.
Lesemann, D.-E., Bozarth, R.F and Koenig, R. (1980). *J. gen. Virol.* **48**: 257.
Osorio-Keese, M., Keese, P. and Gibbs, A.J. (1989). *Virology* **172**: 547.
Paul, H.L., Gibbs, A.J. and Wittmann-Liebold, B. (1980). *Intervirology* **13**: 99.

# Potato Andean mottle *comovirus*

**Data collated by R. Koenig, 1984. Revised 1989.**

*Synonyms:* Andean potato mottle virus. *Standard acronym:* APMoV. *Strains:* three recognised (Avila *et al.*, 1984).

First reported in *Solanum tuberosum*; from La Molina, Peru; by Fribourg *et al.* (1975).

### Natural host range and symptoms

Symptoms persist.

*Solanum tuberosum* (unknown cultivar) — yellow local lesions or blotches, mild or severe mottle some leaf malformation.

### Transmission

Not transmitted by *Epitrix* sp. Virus transmitted by mechanical inoculation; transmitted by grafting; transmitted by contact between plants; not transmitted by seed.

### Geographical distribution

Spreads in the South and Central American region; Brazil, Chile, Ecuador, and Peru (CABI/EPPO data).

### Experimental host range

Several (3–9) families susceptible.

Diagnostically susceptible

*Nicotiana bigelovii* — systemic necrotic spotting.
*N. clevelandii, N. glutinosa, Physalis floridana* — mild systemic mosaic.
*Lycopersicon esculentum* — mild systemic mosaic or symptomless.

Diagnostically insusceptible

*Gomphrena globosa, Chenopodium amaranticolor, C. quinoa, Cucumis sativus, Vigna unguiculata.*

Maintenance and propagation hosts

*Nicotiana benthamiana, N. clevelandii.*

Assay hosts (**L**ocal lesions or **W**hole plants)

No local lesion host known. *Nicotiana bigelovii* (W), *N. clevelandii* (W), *N. glutinosa* (W), *Physalis floridana* (W).

Dicotyledonous Hosts *(**Susceptible** or Insusceptible)*

*Amaranthus caudatus. Chenopodium amaranticolor. Chenopodium quinoa. Cucumis sativus.* ***Datura stramonium****. Dolichos biflorus. Gomphrena globosa. Lablab purpureus.* ***Lycopersicon chilense. Lycopersicon esculentum. Lycopersicon pimpinellifolium.*** *Macrotyloma uniflorum.* ***Nicandra physaloides. Nicotiana benthamiana. Nicotiana*** bigelovii. ***Nicotiana clevelandii. Nicotiana debneyi. Nicotiana glutinosa. Nicotiana megalosiphon. Nicotiana occidentalis. Nicotiana rustica. Nicotiana tabacum.*** *Phaseolus vulgaris.* ***Physalis floridana. Physalis peruviana. Solanum demissum. Solanum demissum × tuberosum. Solanum tuberosum.*** *Vigna unguiculata. Vigna unguiculata* ssp. *cylindrica.*

Sources of host-range data

Fribourg *et al.* (1977); Avila *et al.* (1984).

**Properties of particles in sap**

TIP: 65–70 °C. LIV: 28 days. DEP: $\log_{10}$ minus 6. Infectivity of sap not changed by treatment with di–ethyl ether. Leaf sap contains many virions.

**Purification**

Method: Dunn and Hitchborn (1965).

**Particle morphology**

Virions isometric; not enveloped; 28 nm in diameter; rounded in profile, or angular in profile.

**Physical properties**

Three sedimenting components in purified preparations; sedimentation coefficient of the fastest 112 *S* (B); of the other(s) 93 *S* (M), or 53 *S* (T). Density 1.41 g $cm^{-3}$ in CsCl (M), or 1.46 g $cm^{-3}$ in CsCl (B).

**Biochemical properties**

Virions contain 34 % nucleic acid (B), or 27 % nucleic acid (M), or 0 % nucleic acid (T); 66 % protein (B), or 73 % protein (M), or 100 % protein (T); 0 % lipid.

*Genome* consists of RNA; single–stranded; linear; of two parts; largest (or only) genome part the larger 6.5 kb; the 2nd largest 4.6 kb. Genomic nucleic acid isolated by Salazar and Harrison (1978).

*Sequence database accession code(s):* andean potato mottle comovirus

L16239 Em(40)_vi:COMADEPP Gb(84)_vi:ADEPP Andean potato mottle virus polyprotein (which includes small and large coat proteins and a po

M81775 Em(40)_vi:COMADECP Gb(84)_vi:ADECP Andean potato mottle virus coat protein (VP22) mRNA, 3' end. 12/92 1,096bp.

M83536 Em(40)_vi:COMADEVP4 Gb(84)_vi:ADEVP42A Andean potato mottle virus capsid protein (VP42) gene, 5' end. 7/92 1,161bp.

M84483 Em(40)_vi:COMADERN1 Gb(84)_vi:ADERNAPOL Andean potato mottle virus RNA-dependent RNA polymerase. 9/92 2,265bp.

M84806 Em(40)_vi:COMADERNA Gb(84)_vi:ADERNAP Andean potato mottle virus RNA polymerase mRNA, 3' end. 5/93 2,361bp. 5 sequences.

*Virion protein(s)* two; $M_r$ of the larger 41000. $M_r$ of 2nd largest 21000.

**Replication**

Replication does not depend on a helper virus.

**Cytopathology**

Virions found in all parts of the host plant.

**Taxonomy and Relationships**

*Comovirus: Comoviridae.*

Virus(es) with serologically related virions: cowpea mosaic, cowpea severe mosaic, bean pod mottle virus and quail pea mosaic viruses.

**References**

Avila, A.C., Salazar, L.F., Ortega, M. and Daniels, J. (1984). *Plant Dis.* **68**: 997.
Dunn, D.B. and Hitchborn, J.H. (1965). *Virology* **25**: 171.
Fribourg, C.E., Hinostroza de Lekeu, A.M. and Jones, R.A.C. (1975). *Proc. 6th Triennial Conf. Eur. Assoc. for Potato Res.*, p. 103.
Fribourg, C.E., Jones, R.A.C. and Koenig, R. (1977). *Phytopathology* **67**: 969.
Fribourg, C.E., Jones, R.A.C. and Koenig, R. (1979). *CMI/AAB Descr. Pl. Viruses* No. 203, 4 pp.
Koenig, R. and Bode, O. (1978). *Phytopath. Z.* **92**: 275.
Salazar, L.F. and Harrison, B.D. (1978). *J. gen. Virol.* **39**: 171.

# Potato aucuba mosaic *potexvirus*

**Data collated by C. Büchen-Osmond and D.A. Govier, 1987.**

*Synonyms:* potato G virus, potato virus F, tuber blotch virus. *Standard acronym:* PAMV. *Strains:* potato virus F, synonym of tuber blotch virus (Clinch *et al.*, 1936).

First reported in *Solanum tuberosum*; from the Netherlands and U.K.; by Quanjer (1921); Clinch *et al.* (1936).

**Natural host range and symptoms**

Symptoms persist.

*Solanum tuberosum* — chlorotic local lesions on older leaves, necrotic on younger, systemic (top) necrosis (Clinch, 1941). Stored tubers develop necrosis in the cortex and pith with irregular brown or sunken patches on the surface.

**Transmission**

Transmitted by a vector; an insect; *Myzus persicae*; Aphididae. Transmitted in a non–persistent manner. Virus requires, for vector transmission, a helper virus (any potyvirus, such as potato viruses A and Y; Kassanis and Govier, 1971*a;b*); transmitted by mechanical inoculation.

**Geographical distribution**

Probably distributed worldwide (but uncommon).

**Experimental host range**

Few (<3) families susceptible.

Diagnostically susceptible

*Capsicum annuum, C. frutescens* — necrotic local lesions then epinasty, systemic mosaic or necrosis, young plants killed.

*Nicotiana glutinosa* — light green mottle with dark green vein-banding.

*Lycopersicon esculentum* — small chlorotic local lesions.

*Nicotiana tabacum* cvs White Burley and Xanthi-*nc*, *Datura stramonium* and *Petunia* sp. systemic infection but usually symptomless.

Maintenance and propagation hosts

*Nicotiana tabacum* cv. Xanthi-*nc*.

Assay hosts (**L**ocal lesions or **W**hole plants)

*Capsicum annuum* cv. Long Red (L/W), *C. frutescens* cv. Early Calwonder (L/W).

Dicotyledonous Hosts *(**Susceptible** or Insusceptible)*

***Capsicum annuum. Capsicum frutescens. Datura stramonium. Lycopersicon esculentum. Nicotiana glutinosa. Nicotiana tabacum. Petunia hybrida. Solanum tuberosum.***

Monocotyledonous Hosts *(**Susceptible** or Insusceptible)*

*Tinantia erecta.*

**Properties of particles in sap**

TIP: 65–70 °C. LIV: 30–60 days (at 18–20°C; Kollmer and Larson, 1960). DEP: $\log_{10}$ minus 5–6.

**Purification**

Method: Govier and Kassanis (1972).

**Particle morphology**

Virions filamentous; not enveloped; usually flexuous; with a clear modal length; of 580 nm; 11 nm wide.

**Physical properties**

One sedimenting component in purified preparations; sedimentation coefficient 130 *S*.

**Biochemical properties**

Virions contain 5 % nucleic acid; 95 % protein; 0 % lipid.

*Genome* consists of RNA; single–stranded. Base composition 25 % G; 30 % A; 24 % C; 21 % U.

*Sequence database accession code(s):*

S73580 Gb(89)_un:S73580 putative polymerase. capsid protein potato aucuba mosaic potexvirus PAMV. Genomic RNA Complete.

*Virion protein(s)* one; $M_r$ 26000.

**Cytopathology**

Virions found in all parts of the host plant; in cytoplasm (sometimes in dense aggregates).

**Taxonomy and Relationships**

*Potexvirus.*

Best tests for diagnosis: The symptoms in potato resemble those of potato mop-top virus (PMTV); both cause necrosis on the tuber surface, but those of PAMV are less clearly patterned than the raised necrotic rings caused by first-year infection with PMTV; however, PAMV sometimes causes necrotic rings. PAMV, but not PMTV, infects *Capsicum annuum* systemically, they have virions of different sizes (Harrison and Jones, 1970).

**References**

Clinch, P.J. (1941). *Scient. Proc. R. Dubl. Soc. N. S.* **22**: 435.

Clinch, P.J., Loughnane, J.B. and Murphy, P.A. (1936). *Scient. Proc. R. Dubl. Sco. N. S.* **21**: 431.

Harrison, B.D. and Jones, R.A.C. (1970). *Ann. appl. Biol.* **65**: 393.

Kassanis, B. and Govier, D.A. (1971*a*). *J. gen. Virol.* **10**: 99.

Kassanis, B. and Govier, D.A. (1971*b*). *J. gen. Virol.* **13**: 221013.

Kassanis, B. and Govier, D.A. (1972). *CMI/AAB Descr. Pl. Viruses* No. 98, 4 pp.

Kollmer, G.F. and Larson, R.H. (1960). *Res. Bull. agric. Exp. Stn Univ. Wis.* 223.

Quanjer, H.M. (1921). *R. hort. Soc., Lond. Rep. int. Potato Conf.* 127.

# Potato black ringspot *nepovirus*

**Data collated by L.F. Salazar, 1987.**

*Synonyms:* potato calico strain of tobacco ringspot virus (Salazar and Harrison, 1978), Andean potato calico strain of tobacco ringspot virus (Fribourg, 1977). *Standard acronym:* PBRSV. *Strains:* potato calico strain.

First reported in *Solanum tuberosum ssp. tuberosum x Solanum tuberosum ssp. andigena*; from potatoes from Huasa Huasi, Peru; by Salazar and Harrison (1977).

**Natural host range and symptoms**

Symptoms persist.

*Solanum tuberosum* — necrotic ringspots. Potato calico strain causes general leaf chlorosis in some cultivars, especially cv. Ticahuasi.

**Transmission**

Virus transmitted by mechanical inoculation; transmitted by grafting; not transmitted by contact between plants; not transmitted by seed.

**Geographical distribution**

Spreads in the South and Central American region; Peru.

**Experimental host range**

Many (>9) families susceptible.

Diagnostically susceptible

*Beta vulgaris* — necrotic spots and rings; systemic chlorotic or necrotic rings.

*Lycopersicon esculentum*, Kondine Red — systemic necrotic rings spots and/or line patterns.

*Cyamopsis tetragonoloba* — general systemic necrosis.

*Chenopodium amaranticolor, C. quinoa* — necrotic local lesions; systemic necrosis.

*Nicotiana tabacum* — systemic necrotic ringspots and line patterns.

*Solanum tuberosum* ssp. *tuberosum* — necrotic spots; systemic necrotic ringspots.

*Catharanthus roseus* — chlorotic spots; systemic necrosis.

Maintenance and propagation hosts

*Nicotiana clevelandii, N. tabacum* cv. Samsun, *N. benthamiana.*

Assay hosts (**L**ocal lesions or **W**hole plants)

*Chenopodium amaranticolor* (L), *Chenopodium quinoa* (L) and *Nicotiana clevelandii* (S).

Dicotyledonous Hosts *(**Susceptible** or Insusceptible)*

***Amaranthus caudatus. Apium graveolens. Beta vulgaris. Catharanthus roseus. Chenopodium album. Chenopodium amaranticolor. Chenopodium capitatum. Chenopodium foetidum. Chenopodium foliosum. Chenopodium murale. Chenopodium quinoa. Cucumis sativus. Cyamopsis tetragonoloba. Datura stramonium. Datura tatula. Daucus carota. Dianthus barbatus. Lactuca sativa. Lycopersicon esculentum. Nicandra physaloides. Nicotiana benthamiana. Nicotiana bigelovii. Nicotiana clevelandii. Nicotiana debneyi. Nicotiana glutinosa. Nicotiana rustica. Nicotiana tabacum. Petunia hybrida. Phaseolus vulgaris. Physalis floridana. Physalis peruviana. Pisum sativum. Solanum tuberosum. Stellaria media. Tetragonia expansa. Vigna unguiculata.***

Monocotyledonous Hosts *(**Susceptible** or Insusceptible)*

*Zea mays.*

Sources of host-range data

Salazar and Harrison (1978).

**Properties of particles in sap**

TIP: 58–60 °C (in *Nicotiana clevelandii* sap). LIV: 9–10 days. DEP: $\log_{10}$ minus 4. Infectivity of sap not changed by treatment with di–ethyl ether. Leaf sap contains many virions. Electron microscopy: it is better to mount virions in 1% UA.

**Purification**

Method: Salazar and Harrison (1978).

**Particle morphology**

Virions isometric; not enveloped; 25 nm in diameter; angular in profile; without a conspicuous capsomere arrangement.

**Physical properties**

Three sedimenting components in purified preparations; sedimentation coefficient of the fastest 117 *S* (B); of the other(s) 84 *S* (M), or 49 *S* (T). $A_{260}/A_{280}$ ratio 1.97 (B), or 1.78 (M), or 1.03 (T).

**Biochemical properties**

Virions contain 41 % nucleic acid (B), or 28 % nucleic acid (M), or 0 % nucleic acid (T); 59 % protein (B), or 72 % protein (M), or 100 % protein (T).

*Genome* consists of RNA; single–stranded; of two parts; largest (or only) genome part the larger 7 kb; the 2nd largest 4.2 kb. Genomic nucleic acid isolated by Salazar and Harrison (1978). Infectivity lost when deproteinised with proteases; retained when deproteinised with phenol or detergent.

*Non–genomic nucleic acid* not found in the virions.

*Virion protein(s)* one; $M_r$ 59000. Method of preparation: Salazar and Harrison (1978).

**Cytopathology**

Virions found in mesophyll; in cytoplasm. Inclusions absent from infected cells.

**Taxonomy and Relationships**

*Nepovirus: Comoviridae.*

Virus(es) with serologically related virions: tobacco ringspot and eucharis mottle viruses. Virus(es) with serologically unrelated virions: arabis mosaic, artichoke Italian latent, cacao necrosis, cherry leaf roll, grapevine fanleaf, myrobalan latent ringspot, raspberry ringspot, strawberry latent ringspot, tomato black ring and tomato ringspot viruses. Although the virus is serologically distantly related to tobacco ringspot and eucharis mottle viruses, the viruses do not cross-protect in plants and do not form recombinants; potato black ringspot virus is, therefore, considered to be a distinct virus.

Best tests for diagnosis: serology is the only way to distinguish this virus from others with which it is often naturally associated.

**References**

Fribourg, C.E. (1977). *Phytopathology* **67**: 174.

Salazar, L.F. and Harrison, B.D. (1977). *Nature, Lond* **265**: 337.

Salazar, L.F. and Harrison, B.D. (1978*b*). *Ann. appl. Biol.* **90**: 387.

Salazar, L.F. and Harrison, B.D. (1978). *Ann. appl. Biol.* **90**: 375.

Salazar, L.F. and Harrison, B.D. (1979). *CMI/AAB Descr. Pl. Viruses* No. 206, 4 pp.

# Potato leafroll *luteovirus*

**Data collated by J.E. Thomas, 1987.**

*Synonyms:* potato phloem necrosis virus (Quanjer, 1913). *Standard acronym:* PLRV. *Strains:* tobacco yellow top virus, tomato yellow top strain (Sutton, 1955; Thomas, 1984), capsicum yellows virus (Gunn and Pares, 1990).

First reported in *Solanum tuberosum* ssp. *tuberosum*; from the Netherlands; by Quanjer *et al.* (1916); see Peters (1967).

## Natural host range and symptoms

Symptoms persist.

*Solanum tuberosum* ssp. *tuberosum* — pallor or reddening of the tip leaves, which roll and become erect. Plants grown from infected tubers are stunted, leaflets upwardly rolled, oldest leaves first.

*Solanum tuberosum* ssp. *andigena* in South America — stunting and marginal yellowing of tip leaves (Rodriguez and Jones, 1978).

*Lycopersicon esculentum* — stunting of plants, marginal yellowing and curling of leaflets, and death of flower buds (Braithwaite and Blake, 1961).

## Transmission

Transmitted by a vector; an insect; *Myzus persicae* is the most efficient and important; *Macrosiphum euphorbiae* transmits potato strains less effectively but is a good vector of Australian tomato yellow top isolates; Aphididae. Transmitted in a persistent manner. Virus retained when the vector moults; does not multiply in the vector (although Weidemann (1982) found that virion antigen accumulates in the nuclei of cells of the midgut and principal salivary glands of *Myzus persicae* 1–2 days after virus acquisition and suggested that the virus multiplies in these cells); not transmitted by mechanical inoculation; transmitted by grafting; not transmitted by seed; not transmitted by pollen.

## Geographical distribution

Probably distributed worldwide (in potato growing areas. Tomato yellow top diseases are reported from North America (Alstatt and Ivanoff, 1945), South America (Costa, 1949) and Australasia (Sutton, 1955), but not all have been shown to be caused by strains of potato leafroll virus). Spreads in China.

## Experimental host range

Several (3–9) families susceptible.

Diagnostically susceptible

*Datura stramonium* — systemic interveinal yellowing.

*Physalis floridana* — systemic interveinal chlorosis, older leaves slightly rolled, plant stunted.

*Solanum tuberosum* ssp. *tuberosum* — stunted and leaves rolled.

Diagnostically insusceptible

*Brassica campestris* spp. *pekinensis, Raphanus sativus, Vicia faba.*

Maintenance and propagation hosts

*Physalis floridana, Solanum tuberosum.*

Assay hosts (**L**ocal lesions or **W**hole plants)

*Physalis floridana* (W).

Dicotyledonous Hosts *(**Susceptible** or Insusceptible)*

***Amaranthus caudatus***. *Brassica campestris* ssp. *pekinensis*. *Brassica oleracea* ssp. *capitata*. ***Capsella bursa–pastoris***. ***Celosia argentea***. ***Datura stramonium***. ***Gomphrena globosa***. ***Lycopersicon esculentum***. ***Montia perfoliata***. ***Nicotiana clevelandii***. ***Physalis floridana***. *Pisum sativum*. *Senecio vulgaris*. ***Solanum tuberosum***. *Vicia faba*.

Sources of host-range data

Natti *et al.* (1953); Thomas (1984); Tamada *et al.* (1984).

**Properties of particles in sap**

TIP: 70–80 °C (in *Physalis floridana* sap). LIV: 5–10 days (Murayama and Kojima, 1965). DEP: $\log_{10}$ minus 4. Infectivity of sap not changed by treatment with di–ethyl ether. Leaf sap contains few virions.

**Purification**

Method: Takanami and Kubo (1979*a*).

**Particle morphology**

Virions isometric; not enveloped; 24 nm in diameter (Peters, 1967; Takanami and Kubo, 1979*a*); angular in profile; without a conspicuous capsomere arrangement.

**Physical properties**

One sedimenting component in purified preparations; sedimentation coefficient 115 *S*. Density 1.39 g $cm^{-3}$ in CsCl (Rowhani and Stace-Smith, 1979). Density 1.34 g/$cm^3$ in $Cs_2SO_4$ (Thomas, 1984).

**Biochemical properties**

Virions contain 30 % nucleic acid; 70 % protein; 0 % lipid.

*Genome* consists of RNA; single–stranded; linear; unipartite; largest (or only) genome part 5.88–5.99 kb. Genomic nucleic acid isolated by Rowhani and Stace-Smith (1979); Takanami and Kubo (1979*b*). Infectivity retained when deproteinised with proteases; retained when deproteinised with phenol or detergent. Poly A region absent (Mayo *et al.*, 1982).

*Sequence database accession code(s):*

A07935 Em(40)_vi:07935 Gb(84)_pat:A07935 Potato leafroll virus DNA for coat protein. 8/93 1,000bp.

A07940 Em(40)_vi:A07940 Gb(84)_pat:A07940 Potato leafroll virus DNA for coat protein. 8/93 1,000bp.

A07941 Em(40)_vi:A07941 Gb(84)_pat:A07941 Potato leafroll virus DNA for coat protein. 8/93 1,000bp.

A07943 Em(40)_vi:A07943 Gb(84)_pat:A07943 Potato leafroll virus DNA for coat protein. 8/93 2,154bp

D00336 Em(40)_vi:PLV Potato leafroll virus, 3'end of virus genome. 11/92 141bp.

D00436 Em(40)_vi:PLVCP Potato leafroll virus 23K ORF (putative coat protein gene) and 17K ORF. 3/91 693bp.

D00530 Em(40)_vi:PLVGR Gb(84)_vi:PLVGR Potato leafroll luteovirus genomic RNA. 10/92 5,987bp.

D00733 Em(40)_vi:PLVRA Potato leafroll luteovirus (Australian isolate) genomic RNA, complete sequence. 12/92 5,882bp.

D00734 Em(40)_vi:PLVRC Potato leafroll luteovirus (Canadian isolate) genomic RNA, complete sequence. 12/92 5,883bp.

D13746 Em(40)_vi:PLV3E Gb(84)_vi:PLV3E Potato leafroll virus, 3'end of virus genome. 12/92 141bp.

D13753 Em(40)_vi:PLVCP1 Gb(84)_vi:PLVCP Potato leafroll virus 23K ORF (putative coat protein gene) and 17K ORF. 11/92 693bp.

D13953 Em(40)_vi:PLVRA1 Gb(84)_vi:PLVRA Potato leafroll luteovirus (Australian isolate) genomic RNA, complete sequence. 12/92 5,882bp.

D13954 Em(40)_vi:PLVRC1 Gb(84)_vi:PLVRC Potato leafroll luteovirus (Canadian isolate) genomic RNA, complete sequence. 12/92 5,883bp.

M89926 Em(40)_vi:PLVCOAPRO Gb(84)_vi:PLVCOAPRO Potato leafroll virus coat protein and 17kDa protein RNA, complete cds's. 3/92 1,127bp.

X13906 Em(40)_vi:PLRVCOAT Gb(84)_vi:PLRVCOAT Potato leafroll virus coat protein gene. 9/93 627bp.

X14600 Em(40)_vi:PLLGRNA Gb(84)_vi:PLLGRNA Potato leafroll luteovirus genomic RNA. 9/93 5,987bp

X15501 Em(40)_vi:LUPLRPRM Gb(84)_vi:LUPLRPRM Potato leafroll virus genomic RNA for potential promoter. 3/92 119bp.

X74789 Gb(84)n:PLVSQRN Potato leafroll virus RNA sequence. 7/94 5,882bp.

X77321 Em(40)_vi:PLLVCPG1 Gb(84)_vi:PLLVCPG1 Potato leafroll luteovirus (V) gene for coat protein. 2/94 2,154bp.

X77322 Em(40)_vi:PLLVCPG2 Gb(84)_vi:PLLVCPG2 Potato leafroll virus (PLRV-11) gene for coat protein. 2/94 2,154bp.

X77323 Em(40)_vi:PLLVCPG4 Gb(84)_vi:PLLVCPG4 Potato leafroll luteovirus (PLRV-15) gene for coat protein. 2/94 2,154bp.

X77324 Em(40)_vi:PLLVCPG3 Gb(84)_vi:PLLVCPG3 Potato leafroll luteovirus (PLRV-30) gene for coat protein. 2/94 2,154bp.

X77325 Em(40)_vi:PLLVCPG5 Gb(84)_vi:PLLVCPG5 Potato leafroll luteovirus (PLRV-V4) gene for coat protein. 2/94 2,154bp.

X77326 Em(40)_vi:PLLVCPG6 Gb(84)_vi:PLLVCPG6 Potato leafroll luteovirus (PLRV-V31) gene for coat protein. 2/94 2,154bp.

Y07496 Em(40)_vi:PLRVXX Gb(84)_vi:PLRVXX Potato leafroll virus genomic RNA. 9/93 5,882bp.

A25478 Em(43)_sy:A25478 Potato leafroll virus DNA from pCPL1 plasmid. 4/95 1,779bp.

*Non–genomic nucleic acid* found in the virions; is subgenomic *m*RNA (2.66 kb; Barker *et al.*, 1984).

*Virion protein(s)* two; $M_r$ of the larger 26000; coat protein. $M_r$ of 2nd largest 7000; a genome-linked protein (VPg). Method of preparation: Rowhani and Stace-Smith (1979).

*Virus–coded non–virion proteins* isolated (Mayo *et al.* (1982); Mayo and Barker (1984)); three proteins found. $M_r$ of the largest 125000 (Mayo *et al.*, 1982). $M_r$ of 2nd largest 71000 (Mayo *et al.*, 1982). $M_r$ of 3rd 29000 (Mayo and Barker, 1984).

**Replication**

Replication does not depend on a helper virus.

**Cytopathology**

Virions found in phloem and companion cells; in cytoplasm and in cell vacuoles; (Arai *et al.*, 1969; Kojima *et al.*, 1969). Vesicles occur in the cytoplasm attached to the nucleus and fused with the nuclear envelope (Shepardson *et al.*, 1980). Inclusions present in infected cells; are crystals in the cytoplasm; they contain virions. Other cellular changes: thickening of walls in primary phloem cells of stems and petioles, and callose accumulated in some sieve tubes of tubers. The presence of callose is the basis of various staining tests (*e.g.* with 1% Resorcin Blue), used before serological methods were developed (de Bokx, 1967).

**Taxonomy and Relationships**

*Luteovirus.*

Virus(es) with serologically related virions: tobacco necrotic dwarf (SDI 1), beet western yellows/beet mild yellowing (SDI 2–4), bean leafroll, subterranean clover red leaf and barley yellow dwarf viruses (Kubo and Takanami, 1978; Roberts *et al.*, 1980; Kubo, 1981; Tamada *et al.*, 1984; Thomas, 1984).

**References**


Altstatt, G.E. and Ivanoff, S.S. (1945). *Pl. Dis. Reptr* **29**: 29.

Arai, K., Doi, Y., Yora, K. and Asuyama, H. (1969). *Ann. Phytopath. Soc. Japan* **35**: 10.

Barker, H., Mayo, M.A. and Robinson, D.J. (1984). *Rep. Scottish Crop Res. Inst.* 1983, p. 194.

Braithwaite, B.M. and Blake, C.D. (1961). *Aust. J. agric. Res.* **12**: 1100.

Casper, R. (1977). *Phytopath. Z.* **90**: 364.

Clarke, R.G., Converse, R.H. and Kojima, M. (1980). *Plant Dis.* **64**: 43.

Costa, A.S. (1949). *Biologico* **15**: 79.

de Bokx, J.A. (1967). *Eur. Potato J.* **10**: 221.

Gugerli, P. (1979). *Phytopath. Z.* **96**: 97.

Gunn, L.V. and Pares, R.D. (1990). *J. Phytopath.* **129**: 210.

Harrison, B.D. (1984). *CMI/AAB Descr. Pl. Viruses* No. 291, 4 pp.

Kojima, M., Shikata, E., Sugawara, M. and Murayama, D. (1969). *Virology* **39**: 162.

Kubo, S. (1981). *CMI/AAB Descr. Pl. Viruses* No. 234, 4 pp.

Kubo, S. and Takanami, Y. (1978). *Ann. Phytopath. Soc. Japan* **44**: 398.

Maat, D.Z. and de Bokx, J.A. (1978). *Neth. J. Pl. Path.* **84**: 149.

Mayo, M.A. and Barker, H. (1983). *J. gen. Virol.* **64**: 1775.

Mayo, M.A. and Barker, H. (1984). *Rep. Scottish Crop Res. Inst.* 1983, p. 186.

Mayo, M.A., Barker, H., Robinson, D.J., Tamada, T. and Harrison, B.D. (1982). *J. gen. Virol.* **59**: 163.

Mayo, M.A., Robinson, D.J., Jolly, C.A. and Hyman, L. (1989). *J. gen. Virol.* **70**: 1037.

Mehrad, M., Lapierre, H. and Maury, Y. (1978). *C. r. hebd. Seanc. Acad. Sci., Paris D* **286**: 1179.

Murayama, D. and Kojima, M. (1965). *Ann. Phytopath. Soc. Japan* **30**: 209.

Natti, J.J., Kirkpatrick, H.C. and Ross, A.F. (1953). *Am. Potato J.* **30**: 55.

Peters, D. (1967). *Virology* **31**: 46.

Peters, D. (1970). *CMI/AAB Descr. Pl. Viruses* No. 36, 3 pp.

Prufer, D., Tacke, E., Schmitz, J., Kull, B., Kaufmann, A. and Rhodhe, W. (1992). *EMBO J.* **11**: 1111.

Quanjer, H.M. (1916). *Meded. Landb. Wageningen* **6**: 41.

Quanjer, H.M., van der Lek, H.A.A. and Oortwijn Botjes, J. (1913). *Meded. Landb. Wageningen* **10**: 1.

Roberts, I.M. and Harrison, B.D. (1979). *Ann. appl. Biol.* **93**: 289.

Roberts, I.M., Tamada, T. and Harrison, B.D. (1980). *J. gen. Virol.* **47**: 209.

Rodriguez, A. and Jones, R.A.C. (1978). *Phytopathology* **68**: 39.

Rowhani, A. and Stace-Smith, R. (1979). *Virology* **98**: 45.

Shepardson, S., Esau, K. and McCrum, R. (1980). *Virology* **105**: 379.

Sutton, W.S. (1955). *Agric. Gaz. N.S.W.* **66**: 655.

Takanami, T. and Kubo, S. (1979*a*). *J. gen. Virol.* **44**: 153.

Takanami, T. and Kubo, S. (1979*b*). *J. gen. Virol.* **44**: 853.

Tamada, T. and Harrison, B.D. (1980*a*). *Ann. appl. Biol.* **95**: 209.

Tamada, T. and Harrison, B.D. (1980*b*). *Ann. appl. Biol.* **96**: 67.
Tamada, T. and Harrison, B.D. (1981). *Ann. appl. Biol.* **98**: 261.
Tamada, T., Harrison, B.D. and Roberts, I.M. (1984). *Ann. appl. Biol.* **104**: 107.
Thomas, J.E. (1984). *Ann. appl. Biol.* **104**: 79.
Van Balen, E. (1982). *Neth. J. Pl. Path.* **88**: 33.
Weidemann, H.L. (1982). *Z. angew. Ent.* **94**: 321.

# Potato M *carlavirus*

**Data collated by J.A. de Bokx, 1987.**

*Synonyms:* Kartoffel-K-Virus, Kartoffel-Rollmosaik-Virus, potato paracrinkle virus (Bagnall *et al.*, 1959; Kassanis, 1961; 1963; Salaman and Le Pelly, 1930), potato virus E. *Standard acronym:* PVM. *Strains:* potato leafrolling mosaic virus, potato American interveinal mosaic virus (Schultz and Folsom, 1923; Bagnall *et al.*, 1956); Dutch isolates (Rozendaal and van Slogteren, 1958); D1102 and Fortuna isolates (Köhler, 1953; Wetter and Brandes, 1956).

First reported in *Solanum tuberosum*; from the U.S.A.; by Schultz and Folsom (1923); Bagnall *et al.* (1956).

### Natural host range and symptoms

Symptoms persist.

*Solanum tuberosum* — symptoms range from very slight (*e.g.* in cv. King Edward) to severe (*e.g.* in cv. Arran Victory). Causes mottles, mosaic, crinkling and abaxial rolling of leaves, and stunting of shoots.

### Transmission

Transmitted by a vector; an insect; *Myzus persicae* (Wetter and Völk, 1960) but less efficiently by *Aphis frangulae, A. nasturtii* and *Macrosiphum euphorbiae* (Bode and Weidemann, 1970); some isolates not aphid-transmitted (Kassanis, 1961); Aphididae. Transmitted in a non–persistent manner. Virus transmitted by mechanical inoculation; transmitted by grafting; not transmitted by seed; not transmitted by pollen.

### Ecology and control

Studies reported by Kassanis (1957). King Edward potato has been freed from potato M, S and other viruses by apical meristem culture.

### Geographical distribution

Probably distributed worldwide (in potato cultivars).

### Experimental host range

Few (<3) families susceptible.

Diagnostically susceptible

*Datura metel* — chlorotic or necrotic local lesions then systemic rugose chlorotic mottle, leaves abscissed, plants stunted and may die.

*Gomphrena globosa* — chlorotic spots with reddish borders; not systemic.

*Lycopersicon esculentum* — symptomless systemic infection.

*Nicotiana debneyi* — irregular brown necrotic ring-like local lesions; not systemic.

*Solanum rostratum* — systemic necrotic streaking of stem, petioles and leaf-veins.

Diagnostically insusceptible

*Tinantia erecta.*

Maintenance and propagation hosts

*Lycopersicon esculentum, Solanum tuberosum* cv. Saco.

Assay hosts (**L**ocal lesions or **W**hole plants)

*Datura metel* (L), *Gomphrena globosa* (L), *Phaseolus vulgaris* cv. Red Kidney bean (L) (Hiruki, 1970).

Dicotyledonous Hosts *(**Susceptible** or Insusceptible)*

***Beta vulgaris*. *Chenopodium album*. *Chenopodium amaranticolor*. *Chenopodium quinoa*. *Cyamopsis tetragonoloba*. *Datura metel*. *Datura stramonium*. *Dianthus barbatus*. *Dianthus caryophyllus*. *Gomphrena globosa*. *Lycopersicon esculentum*. *Nicotiana debneyi*. *Nicotiana occidentalis*. *Phaseolus vulgaris*. *Solanum demissum*. *Solanum demissum* × *tuberosum*. *Solanum melongena*. *Solanum rostratum*. *Solanum tuberosum*. *Vigna unguiculata*.**

Monocotyledonous Hosts *(**Susceptible** or Insusceptible)*

*Tinantia erecta.*

Sources of host-range data

De Bokx (1984); Vulich and Hunnius (1967); Kowalska and Wás (1976).

**Properties of particles in sap**

TIP: 65–71 °C. LIV: 5–7 days. DEP: $\log_{10}$ minus 2–3.

**Purification**

Method: Wetter (1960); Proll and Richter (1979).

**Particle morphology**

Virions filamentous; not enveloped; usually straight (to slightly curved); with a clear modal length; of 650 nm; 12 nm wide (Brandes *et al.*, 1959). Basic helix obscure.

**Biochemical properties**

Virions contain 6 % nucleic acid.

*Genome* consists of RNA; single–stranded; unipartite; largest (or only) genome part 8.535 kb. Genomic nucleic acid isolated by Proll *et al.* (1981); Tavantzis (1984). Nucleotide sequence references: Rupasov (1989).

*Sequence database accession code(s):*

D00515 Em(40)_vi:PVM3TR Potato virus M genomic RNA, 3'-terminal region. 2/93 3,164bp.

D14449 Em(40)_vi:PTMPVMCG Gb(84)_vi:PTMPVMCG Potato virus M genomic RNA, complete sequence. 2/93 8,535bp.

X53062 Em(40)_vi:POPVMCG Gb(84)_vi:POPVMCG Potato virus M complete genome. 9/93 8,535bp.

X57440 Em(40)_vi:PVMCP Gb(84)_vi:PVMCP Potato Virus M genes for capsid protein CP and 12 kDa protein. 6/92 1,400bp. 4 sequences.

*Features of the genome:* the sequence of the 3' proximal 2630 nucleotides of the genomic RNA has been determined; five ORF's have been recognised which encode polypeptides of *c.* 11000, 34000, 25000, 12000 and 7000.

**Replication**

Replication does not depend on a helper virus.

**Cytopathology**

Virions found in all parts of the host plant; in cytoplasm. Inclusions present in infected cells; are amorphous X–bodies.

**Taxonomy and Relationships**

*Carlavirus.*

Virus(es) with serologically related virions: carnation latent, potato S, chrysanthemum B, passiflora latent, cactus 2 and red clover vein mosaic viruses, but distantly.

Best tests for diagnosis: potatoes are often infected with both potato viruses M and S; PVM can be freed from PVS by inoculating tomato, which is immune to potato virus S, or the potato cultivar Saco, which is very resistant to potato virus S and X.

**References**

Bagnall, R.H., Larson R.H. and Walker, J.C. (1956). *Res. Bull. agric. Exp. Stn. Univ. Wis.* **198**, 45 pp.
Bagnall, R.H., Wetter, C. and Larson, R.H. (1959). *Phytopathology* **49**: 435.
Bode, O. and Weidemann, H.L. (1970). *Proc. 4th trienn. Conf. Eur. Ass. Potato Res. Brest 1969*, p. 224.
Brandes, J., Wetter, C., Bagnall, R.H. and Larson, R.H. (1959). *Phytopathology* **49**: 443.
de Bokx, J.A. (1984). *Potato Res.* **27**: 99.
de Bokx, J.A., Piron, P.J.M. and Cother, E. (1980). *Neth. J. Pl. Path.* **86**: 285.
Hiruki, C. (1970). *Phytopathology* **60**: 739.
Kassanis, B. (1957). *Ann. appl. Biol.* **45**: 422.
Kassanis, B. (1960). *Nature, Lond.* **188**: 688.
Kassanis, B. (1961). *Eur. Potato J.* **4**: 13.
Köhler, E. (1953). *Ber. dt. bot. Ges.* **66**: 63.
Kowalska, A. and Wás, M. (1976). *Potato Res.* **19**: 131.
Proll, E., Leiser, R.M. and Östermann, W.D. (1981). *Potato Res.* **24**: 1.
Proll, E. and Richter, J. (1979). *Arch. Phytopathol. PflSchutz, Berlin* **15**: 233.
Rozendaal, A. and van Slogteren, D.H.M. (1958). *Proc. 3rd Conf. Potato Virus Diseases, Lisse-Wageningen 1957*, p. 20.
Rupasov, V.V., Morozov, S.Y., Kanyuka, K.V. and Zavriev, S.K. (1989). *J. gen. Virol.* **70**: 1861.
Salaman, R.N. and Le Pelley, E.H. (1930). *Proc. R. Soc.* **106**: 140.
Schultz, E.S. and Folsom, D. (1923). *J. agric. Res.* **25**: 43.
Tavantzis, S.M. (1984). *Virology* **133**: 427.
Vulich, M. and Hunnius, W. (1967). *Phytopath. Z.* **59**: 225.
Wetter, C. (1960). *Arch. Microbiol.* **37**: 278.
Wetter, C. (1967). *Z. Naturf.* **22**: 1008.
Wetter, C. (1972). *CMI/AAB Descr. Pl. Viruses* No. 87, 4 pp.
Wetter, C. and Brandes, J. (1956). *Phytopath. Z.* **26**: 81006.
Wetter, C. and Völker, J. (1960). *Eur. Potato J.* **3**: 158.

# Potato mop-top *furovirus*

**Data collated by J.I. Cooper, 1987.**

*Standard acronym:* PMTV.

First reported in *Solanum tuberosum*; from Northern Ireland, U.K.; by Calvert and Harrison (1966).

**Natural host range and symptoms**

Symptoms vary seasonally.

*Solanum tuberosum* — dwarfing (mopping); chlorotic and necrotic chevrons and blotching, tuber cracking and necrotic conchoidal layers in tubers especially in cool weather (15°C).

Chenopodiaceae, Solanaceae — weeds species are susceptible and commonly infected in the Andean region of South America, but not elsewhere.

**Transmission**

Transmitted by a vector; a fungus; potato powdery scab fungus *Spongospora subterranea*. Virus is retained in spore balls, but virus-free *Spongospora* has not been experimentally infected; Jones, 1988; Plasmodiophorales. Virus transmitted by mechanical inoculation; transmitted by grafting.

**Ecology and control**

Studies reported by Cooper *et al.* (1973); Cooper *et al.* (1976); Harrison *et al.* (1970); Jones (1981); Jones *et al.* (1972). Virus does not spread to all tubers of infected plants, thus the virus is maintained by the vector.

**Geographical distribution**

Spreads in the Eurasian region and the South and Central American region (in the Andean region); Israel, Japan, and the UK. Found, but with no evidence of spread, in Chile and Taiwan.

**Experimental host range**

Several (3–9) families susceptible.

Diagnostically susceptible

*Nicotiana debneyi* — systemic necrotic oak leaf patterns.
*Chenopodium amaranticolor* — concentric brown local lesions.

Diagnostically insusceptible

*Lycium halimifolium, Chenopodium capitatum, Solanum dulcamara.*

Maintenance and propagation hosts

*Nicotiana benthamiana, Nicotiana debneyi, Nicotiana clevelandii.*

Assay hosts (**L**ocal lesions or **W**hole plants)

*Chenopodium amaranticolor* (L), *Nicotiana debneyi* (W), *N. benthamiana* (W).

Dicotyledonous Hosts *(**Susceptible** or Insusceptible)*

*Amaranthus caudatus*. *Antirrhinum majus*. ***Atriplex hortensis***. ***Beta vulgaris***. ***Capsicum annuum***. *Celosia argentea*. *Celosia cristata*. ***Chenopodium album***. ***Chenopodium amaranticolor***. *Chenopodium capitatum*. ***Chenopodium foliosum***. ***Chenopodium quinoa***. ***Datura ferox***. ***Datura stramonium***. *Dianthus caryophyllus*. *Gomphrena globosa*. *Gypsophila elegans*. ***Hyoscyamus niger***. *Lobelia erinus*. ***Lycopersicon esculentum***. ***Nicandra physaloides***. ***Nicotiana clevelandii***. ***Nicotiana debneyi***. ***Nicotiana glutinosa***. ***Nicotiana rustica***. ***Nicotiana tabacum***. ***Petunia hybrida***. *Phaseolus vulgaris*. ***Physalis floridana***. ***Physalis peruviana***. *Primula malacoides*. ***Solanum nigrum***. ***Solanum tuberosum***. ***Tetragonia expansa***. *Torenia fournieri*. *Verbesina encelioides*. *Vigna unguiculata*. *Vigna unguiculata* ssp. *cylindrica*.

Sources of host-range data

Jones *et al.* (1972); Harrison *et al.* (1970).

**Properties of particles in sap**

TIP: 75–80 °C. LIV: up to 200 days. DEP: $\log_{10}$ minus 3. Leaf sap contains few virions.

**Purification**

Method: Kassanis *et al.* (1972); Koenig *et al.* (1984).

**Particle morphology**

Virions rod–shaped; not enveloped; usually straight; with no clear modal length; of 100–150 nm, or 250–300 nm; 18–20 nm wide. Axial canal obvious. Basic helix obvious (at the ends).

**Physical properties**

Possibly three sedimenting components in purified preparations (broadbands); sedimentation coefficient 236 *S* (longest most specific infectivity); of the other(s) 126 *S*, or 171 *S*.

**Biochemical properties**

*Genome* consists of RNA; single–stranded; linear.

*Sequence database accession code(s):*

D30753 Em(43)_vi:Pxmtodd Gb(89)_vi:Pxmtodd Potato mop-top virus triple gene for 51K protein, 13K protein, 21K protein and 8K protein.

*Virion protein(s)* one; $M_r$ 18500–20000.

**Cytopathology**

Virions found in leaves; in cytoplasm. Inclusions present in infected cells; are viroplasms and unusual in shape; and small loose sheafs; they contain virions. Other cellular changes: necrosis in arcs or lines in potato tubers.

**Taxonomy and Relationships**

*Furovirus.*

Virus(es) with serologically related virions: soil borne wheat mosaic virus, and also tobacco mosaic virus, but very distantly. Virus(es) with serologically unrelated virions: peanut clump, sugarbeet necrotic yellow vein, hypochoeris mosaic and oat golden stripe viruses.

There is no homology (using *c*DNA or RNA hybridization) between potato mop top and peanut clump, soil borne wheat mosaic and beet necrotic yellow vein viruses. Best tests for diagnosis: virion shape and size. Symptoms in *Chenopodium amaranticolor*, inoculated leaves.

**References**

Calvert, E.L. and Harrison, B.D. (1966). *Pl. Path.* **15**: 134.

Cooper, J.I. and Harrison, B.D. (1973). *Pl. Path.* **22**: 73.

Cooper, J.I., Jones, R.A.C. and Harrison, B.D. (1976). *Ann. appl. Biol.* **83**: 215.

Harrison, B.D. and Jones, R.A.C. (1970). *Ann. appl. Biol.* **65**: 395.

Harrison, B.D. and Jones, R.A.C. (1970). *Ann. appl. Biol.* **67**: 377.

Jones, R.A.C. (1981). In: *Pests, Pathogens and Vegetation* p. 89; ed. J.M. Thresh. Pitman, London.

Jones, R.A.C. (1988). In: *Viruses with Fungal Vectors* p. 255; eds J.I. Cooper and M. Asher. Association of Applied Biologists, Wellesbourne, U.K.

Jones, R.A.C. and Harrison, B.D. (1972). *Ann. appl. Biol.* **71**: 47.

Kassanis, B., Woods, R.D. and White, R.F. (1972). *J. gen. Virol.* **14**: 123.

Koenig, R., Lesemann, D.E. and Burgermeister, W. (1984). *Phytopath. Z.* **111**: 244.

# Potato S *carlavirus*

**Data collated by C. Büchen-Osmond, 1987.**

*Synonyms:* pepino latent virus (Dolby and Jones, 1988; Jones *et al.*, 1980). *Standard acronym:* PVS.

First reported in *Solanum tuberosum*; from the Netherlands; by de Bruyn Ouboter (1952); Rozendaal (1952).

## Natural host range and symptoms

Symptoms persist.

*Solanum tuberosum* — causes few or no symptoms, but decreases yield of potato tubers by up to 20%.

## Transmission

Transmitted by a vector; an insect; *Myzus persicae*. Isolates differ in their transmissibility by *Myzus persicae* (Bode and Weidemann, 1970); an isolate from the U.S. seedling 41956 could not be transmitted under conditions in which the related potato virus M was transmitted (Wetter and Völker, 1960); Aphididae. Transmitted in a non–persistent manner. Virus transmitted by mechanical inoculation; not transmitted by seed.

## Ecology and control

Studies reported by Kassanis (1957). Infected potato, cv. King Edward has been freed from potato M, S and other viruses by apical meristem culture.

## Geographical distribution

Probably distributed worldwide (in cultivated varieties of potato).

## Experimental host range

Few (<3) families susceptible.

Diagnostically susceptible

*Chenopodium amaranticolor, C. quinoa, C. album* — chlorotic local lesions which, in older leaves, may develop a green halo; not systemic.

*Solanum rostratum* — small necrotic local lesions, systemic necrotic spots.

*Nicotiana debneyi* — symptomless in inoculated leaves; systemic veinclearing and mottling and necrosis.

*Cyamopsis tetragonoloba* — small brown necrotic local lesions in cotyledons, not systemic.

Diagnostically insusceptible

*Tinantia erecta.*

Maintenance and propagation hosts

*Nicotiana clevelandii.*

Assay hosts (**L**ocal lesions or **W**hole plants)

*Chenopodium amaranticolor* (L); *Chenopodium quinoa* (L); *Chenopodium album* (L).

Dicotyledonous Hosts *(**Susceptible** or Insusceptible)*

***Chenopodium album*. *Chenopodium amaranticolor*. *Chenopodium quinoa*. *Cyamopsis tetragonoloba*. *Nicotiana clevelandii*. *Nicotiana debneyi*. *Solanum rostratum*. *Solanum tuberosum*.**

Monocotyledonous Hosts *(**Susceptible** or Insusceptible)*

*Tinantia erecta.*

**Properties of particles in sap**

TIP: 55–60 °C. LIV: 3–4 days. DEP: $\log_{10}$ minus 2–3.

**Purification**

Method: Wetter (1960).

**Particle morphology**

Virions filamentous; not enveloped; usually straight (to slightly curved); with a clear modal length; of 650 nm; 12 nm wide.

**Biochemical properties**

Genome unipartite; largest (or only) genome part 7.5 kb. Nucleotide sequence references: Mackenzie *et al.* (1989).

*Sequence database accession code(s):*

D00461 Em(40)_vi:PVS Gb(84)_vi:PVS Potato virus S genomic RNA, 3'-terminal region. 4/90 3,552bp.

S45593 Em(40)_vi:S45593 Gb(84)_vi:S45593 coat protein, 11 kda protein potato virus S PVS, Genomic RNA, 1346 nt. 1/94 1,346bp.

R28712 Em(43)_est:St712 Gb(89)_est:R28712 g63 Solanum tuberosum cDNA clone g63 5' END similar to 11 kD coat protein of potato virus S.

R28713 Em(43)_est:St713 Gb(89)_est:R28713 g64 Solanum tuberosum cDNA clone g64 5' END similar to potato virus S genomic RNA, 3' terminal sequences.

*Features of the genome:* the sequence of the 3' 3553 nucleotides of the genomic RNA has been determined; six ORF's recognised which encode for polypeptides, respectively of $M_r$ 10734, 32515, 7222, 11803, 25092 and 41052.

*Virion protein(s)* one.

**Replication**

Replication does not depend on a helper virus.

**Taxonomy and Relationships**

*Carlavirus.*

Virus(es) with serologically related virions: carnation latent, potato M, chrysanthemum B, passiflora latent, cactus 2 and red clover vein mosaic viruses, but distantly.

Best tests for diagnosis: In some potato cultivars potato S virus is found with the related potato M carlavirus. Potato S virus may be obtained from the mixture by inoculating *Nicotiana debneyi* which is systemically infected by potato S virus, but only locally by potato M virus.

**References**


Bode, O. and Weidemann, H.L. (1970). *Proc. 4th Trienn. Conf. Eur. Ass. Potato Res. Brest, 1969*, p. 224.

de Bruyn Ouboter, M.P. (1952). *Proc. 1st Conf. Potato Virus Diseases, Lisse-Wageningen 1951*, p. 83.

Dolby, C.A. and Jones, R.A.C. (1988). *Ann. appl. Biol.* **112**: 231.

Foster, G.D. and Mills, P.R. (1990). *J. gen. Virol.* **71**: 1247.

Jones, R.A.C., Koenig, R. and Lesemann, D.-E. (1980). *Ann. appl. Biol.* **94**: 61.

Kassanis, B. (1957). *Ann. appl. Biol.* **45**: 42.

Mackenzie, D.J., Tremaine, J.H. and Stace-Smith, R. (1989). *J. gen. Virol.* **70**: 1053.

Rozendaal, A. (1952). *Meded. Ned. alg. KeurDienst LandbZaken Aardappelpootg.* **8**: 94.
Wetter, C. (1960). *Arch. Microbiol.* **37**: 278.
Wetter, C. (1971). *CMI/AAB Descr. Pl. Viruses* No. 60, 3 pp.
Wetter, C. and Völker, J. (1960). *Eur. Potato J.* **3**: 158.

# Potato T *trichovirus*

**Data collated by A.A. Brunt and L.F. Salazar, 1987.**

*Standard acronym:* PVT.

First reported in *Solanum tuberosum*; from Peru; by Salazar and Harrison (1977).

**Natural host range and symptoms**

Symptoms disappear soon after infection.

*Solanum tuberosum* — usually symptomless but occasionally induces mild leaf mottling.

**Transmission**

Virus transmitted by mechanical inoculation; transmitted by grafting; not transmitted by contact between plants; transmitted by seed (10–72% in solanaceous hosts); transmitted by pollen to the seed.

**Ecology and control**

Studies reported by Salazar and Harrison (1978*a*).

**Geographical distribution**

Spreads in the South and Central American region; Bolivia and Peru.

**Experimental host range**

Several (3–9) families susceptible.

Diagnostically susceptible

*Chenopodium amaranticolor* — systemic leaf necrosis.
*C. quinoa* — systemic leaf mosaic and tip necrosis.
*Datura stramonium* — mild systemic leaf chlorosis.
*Nicotiana debneyi* — leaf mottling and systemic necrosis.
*Phaseolus vulgaris* — necrotic local lesions and systemic necrosis.

Maintenance and propagation hosts

*Chenopodium quinoa.*

Assay hosts (**L**ocal lesions or **W**hole plants)

*Phaseolus vulgaris* (L), *Chenopodium amaranticolor* (W).

Dicotyledonous Hosts *(**Susceptible** or Insusceptible)*

*Amaranthus caudatus. Apium graveolens.* ***Beta vulgaris****. Brassica campestris.* ***Catharanthus roseus. Chenopodium album. Chenopodium amaranticolor. Chenopodium ambrosioides. Chenopodium foetidum. Chenopodium foliosum. Chenopodium murale. Chenopodium quinoa****. Cucumis sativus.* ***Cyamopsis tetragonoloba. Datura metel. Datura stramonium. Datura tatula****. Dianthus barbatus.* ***Gomphrena globosa. Hyoscyamus niger****. Lactuca sativa.* ***Lycopersicon chilense****. Lycopersicon esculentum.* ***Nicandra physaloides. Nicotiana benthamiana. Nicotiana clevelandii. Nicotiana debneyi. Nicotiana glutinosa. Nicotiana tabacum****. Petunia hybrida.* ***Phaseolus vulgaris. Physalis floridana. Pisum sativum. Solanum rostratum.***

***Spinacia oleracea***. ***Stellaria media***. *Tetragonia expansa*. *Trifolium pratense*. ***Vicia faba***. ***Vigna unguiculata***. ***Vigna unguiculata*** ssp. ***sesquipedalis***.

Monocotyledonous Hosts (***Susceptible*** *or Insusceptible)*

*Hordeum vulgare*. *Triticum aestivum*. *Zea mays*.

Sources of host-range data

Salazar and Harrison (1978*a*).

**Properties of particles in sap**

TIP: 65 °C (in *Chenopodium quinoa* sap). LIV: 2–4 days (at 20°C). DEP: $\log_{10}$ minus 5. Leaf sap contains few virions.

**Purification**

Method: Salazar and Harrison (1978*b*).

**Particle morphology**

Virions filamentous; not enveloped; usually flexuous; with a clear modal length; of 637 nm; 12 nm wide. Axial canal obscure. Basic helix obvious (when virions mounted in UA); pitch of basic helix 3.4 nm.

**Physical properties**

One sedimenting component in purified preparations; sedimentation coefficient 99 *S*.

**Biochemical properties**

Virions contain 95 % nucleic acid; 5 % protein.

*Genome* consists of RNA; single–stranded; linear; unipartite; largest (or only) genome part 7.5 kb. Poly A region present; at the 3'terminus.

*Sequence database accession code(s):*

D10172 Gb(84)n:PVTGR Potato virus T genomic RNA, 3'-terminal region. 8/94 2,392bp.

*Virion protein(s)* one; $M_r$ 27000; coat protein.

**Taxonomy and Relationships**

*Trichovirus*.

Virus(es) with serologically unrelated virions: apple chlorotic leafspot, potato M, potato S, potato X and potato Y viruses.

**References**

Ochi, M., Kashiwazaki, S., Hiratsuka, K., Namba, S. and Tsuchizaki, T. (1992). *Ann. Phytopath. Soc. Japan* **58**: 416.

Salazar, L.F. and Harrison, B.D. (1977). *Nature, Lond.* **265**: 337.

Salazar, L.F. and Harrison, B.D. (1978*a*). *Ann. appl. Biol.* **89**: 223.

Salazar, L.F. and Harrison, B.D. (1978*b*). *CMI/AAB Descr. Pl. Viruses* No. 187, 4 pp.

# Potato U *nepovirus*

**Data collated by R.A.C. Jones, 1989.**

*Standard acronym:* PVU.

First reported in *Solanum tuberosum*; from the Comas Valley, central Peru, at an altitude of 3600 metres; by Jones *et al.* (1983).

**Natural host range and symptoms**

*Solanum tuberosum* — yellow leaf markings (calico); it is difficult to infect potato, even by grafting, so it may normally infect another unidentified plant species, or may mostly infect potato roots and rarely infect systemically.

**Transmission**

Transmitted by a vector; a nematode; *Longidorus* sp. (P. Jatala and C.E. Fribourg, unpublished); Dorylamidae. Virus transmitted by mechanical inoculation; transmitted by grafting; not transmitted by contact between plants; transmitted by seed (in *Nicotiana debneyi, N. tabacum, Chenopodium amaranticolor* and *C. quinoa*. It is readily seed borne in some test plant species and, therefore, seed transmission may be important for natural spread).

**Ecology and control**

Studies reported by Jones *et al.* (1983).

**Geographical distribution**

Spreads in Peru.

**Experimental host range**

Many (>9) families susceptible.

Diagnostically susceptible

*Chenopodium quinoa* — chlorotic or necrotic spots, then systemic chlorotic mottle, leaf malformation and necrosis.

*C. murale* — necrotic spots, systemic necrosis.

*C. amaranticolor* — chlorotic and necrotic spots, systemic mottle and leaf malformation, later recovery.

*Nicotiana tabacum* — systemic chlorotic ringspots and line patterns, recovery.

*Beta vulgaris* (red beet) — red ringspots, then systemic red vein banding, plus red rings, spots and blotches.

*Cucumis sativus* — chlorotic blotches, systemic vein clearing, chlorotic blotching and mosaic, later recovery.

Diagnostically insusceptible

*Brassica campestris* ssp. *pekinensis, Phaseolus vulgaris, Capsicum sinensis*.

Maintenance and propagation hosts

*Nicotiana clevelandii, N. tabacum*.

Assay hosts (**L**ocal lesions or **W**hole plants)

*Chenopodium murale* (L and W), *C. amaranticolor* (W), *C. quinoa* (W).

Dicotyledonous Hosts *(**Susceptible** or Insusceptible)*

***Beta vulgaris***. *Brassica campestris* ssp. *pekinensis*. ***Chenopodium album***. ***Chenopodium amaranticolor***. ***Chenopodium capitatum***. ***Chenopodium foliosum***. ***Chenopodium murale***. ***Chenopodium quinoa***. ***Cucumis sativus***. ***Datura metel***. ***Datura stramonium***. ***Gomphrena globosa***. ***Hyoscyamus niger***. ***Lycopersicon chilense***. ***Lycopersicon esculentum***. ***Lycopersicon pimpinellifolium***. ***Montia perfoliata***. ***Nicandra physaloides***. ***Nicotiana bigelovii***. ***Nicotiana clevelandii***. ***Nicotiana debneyi***. ***Nicotiana glutinosa***. ***Nicotiana occidentalis***. ***Nicotiana rustica***. ***Nicotiana tabacum***. ***Petunia hybrida***. *Phaseolus vulgaris*. ***Physalis floridana***. ***Physalis peruviana***. ***Solanum chacoense***. ***Solanum demissum***. ***Solanum demissum* × *tuberosum***. ***Solanum tuberosum***.

***Spinacia oleracea. Tetragonia expansa. Vigna unguiculata. Vigna unguiculata*** ssp. ***cylindrica***.

Additional hosts

*Capsicum sinensis*, ***Chenopodium ficifolium, C. glaucum, C. berthaultii, Solanum chancayense, S. goniocalyx, S. megistacrolobum, S. microdontum, S. mochisense, S. stoloniferum*** and ***S. villosum***.

Sources of host-range data

Jones *et al.* (1983).

**Properties of particles in sap**

TIP: 60–65 °C. LIV: 7–8 days. DEP: $\log_{10}$ minus 3–4. Leaf sap contains few virions.

**Purification**

Method: Jones *et al.* (1983); Fribourg *et al.* (1978).

**Particle morphology**

Virions isometric; not enveloped; *c.* 28 nm in diameter; angular in profile; without a conspicuous capsomere arrangement.

**Physical properties**

Four sedimenting components in purified preparations (three major but one subdivides into two); sedimentation coefficient of the fastest 135 *S* (B); of the other(s) 117 *S* (M), or 55 *S* (Tb), or 40 *S* (Ta).

*Virion protein(s)* one; $M_r$ 58000; coat protein. Method of preparation: Koenig *et al.* (1970).

**Taxonomy and Relationships**

*Nepovirus: Comoviridae.*

Virus(es) with serologically unrelated virions: arracacha A, arabis mosaic, artichoke Italian latent, artichoke vein banding, grapevine chrome mosaic, grapevine fan leaf, hibiscus latent ringspot, myrobalan latent ringspot, olive ringspot, potato black ringspot, raspberry ringspot, strawberry latent ringspot, tobacco ringspot (eucharis mottle strain), tomato black ring (beet, bouquet and potato pseudo aucuba strains) and tomato ringspot viruses.

Best tests for diagnosis: to distinguish from other potato viruses use gel diffusion and ISEM.

**References**

Fribourg, C.E., Jones, R.A.C. and Koenig, R. (1978). *Phytopathology* **67**: 969.

Jones, R.A.C., Fribourg, C.E. and Koenig, R. (1983). *Phytopathology* **73**: 195.

Koenig, R., Stegemann, H., Francksen, H. and Paul, H.L. (1970). *Biochim. biophys. Acta* **207**: 184.

# Potato V *potyvirus*

**Data collated by C.E. Fribourg, 1984.**

*Standard acronym:* PVV. *Strains:* PVY-C Gl (Rozendaal *et al.*, 1971), PVY-C AB (Calvert *et al.*, 1980), UF (Fribourg and Nakashima, 1983).

First reported in *Solanum tuberosum*; from the Netherlands; by Rozendaal *et al.* (1971).

**Natural host range and symptoms**

Symptoms persist.

*Solanum tuberosum* — mild or severe mosaic, premature senescence of lower leaves, green spots or rings, partial vein necrosis of leaf undersides.

**Transmission**

Transmitted by a vector; an insect; *Brachycaudus helichrysi, Macrosiphum euphorbiae, Myzus persicae*; Aphididae. Transmitted in a non–persistent manner. Virus transmitted by mechanical inoculation; transmitted by grafting; not transmitted by contact between plants; not transmitted by seed.

**Geographical distribution**

Spreads in France, Ireland, the Netherlands, Peru, and the UK.

**Experimental host range**

Few (<3) families susceptible.

Diagnostically susceptible

*Nicotiana glutinosa* and *N. occidentalis* — systemic mosaic and chlorotic vein banding.

*N. debneyi* — local and systemic chlorotic spotting.

*Solanum demissum* PI 230579 — necrotic spots; systemic veinal necrosis.

*S. tuberosum* clone 'A6' — systemic veinal necrosis.

Diagnostically insusceptible

*Capsicum annuum, Solanum stoloniferum* PI 230557, *S. demissum* PI 175404.

Maintenance and propagation hosts

*Nicotiana clevelandii, N. glutinosa, N. occidentalis.*

Assay hosts (**L**ocal lesions or **W**hole plants)

*Nicotiana occidentalis* (W), *Solanum demissum* PI 230579 (L).

Dicotyledonous Hosts *(**Susceptible** or Insusceptible)*

*Amaranthus caudatus. Cajanus cajan. Capsicum annuum. Chenopodium amaranticolor. Chenopodium murale. Chenopodium quinoa. Clitoria ternatea. Cucumis melo. Cucumis sativus. Cyamopsis tetragonoloba.* ***Datura metel****. Datura stramonium. Gomphrena globosa.* ***Lycopersicon esculentum****. Lycopersicon pimpinellifolium. Matthiola incana.* ***Nicotiana bigelovii****.* ***Nicotiana clevelandii****.* ***Nicotiana debneyi****.* ***Nicotiana glutinosa****.* ***Nicotiana tabacum****. Phaseolus vulgaris.* ***Physalis floridana****. Sesamum indicum.* ***Solanum demissum***, or *Solanum demissum. Solanum demissum* × *tuberosum.* ***Solanum tuberosum****. Vigna unguiculata. Vigna unguiculata* ssp. *cylindrica.*

Sources of host-range data

Calvert *et al.* (1980); Fribourg and Nakashima (1984).

**Properties of particles in sap**

TIP: 55–60 °C. LIV: 21 days. DEP: $\log_{10}$ minus 5–6. Leaf sap contains few virions.

**Purification**

Method: Fribourg and Nakashima (1984).

**Particle morphology**

Virions filamentous; not enveloped; usually flexuous; with a clear modal length; of 760 nm. Axial canal obscure. Basic helix obscure.

**Physical properties**

One sedimenting component in purified preparations.

*Sequence database accession code(s):*
X61279 Gb(89)n:Pvvcp Potato Virus V gene for coat protein. 6/95 2,500bp.

**Replication**

Replication does not depend on a helper virus.

**Cytopathology**

Virions found in all parts of the host plant; in cytoplasm. Inclusions present in infected cells; are pinwheels; they do not contain virions.

**Taxonomy and Relationships**

*Potyvirus: Potyviridae.*

Virus(es) with serologically related virions: potato A and wild potato mosaic viruses; potato virus Y (strains O, N and C) is very distantly related.

**References**

Calvert, E.L., Cooper, P. and McClure, J. (1980). *Rec. Agric. Res.* **28**: 63.
Fribourg, C.E. and Nakashima, J. (1983). In: *Research for the Potato in the Year 2000*; ed. W.J. Hooker. CIP, Peru.
Fribourg, C.E. and Nakashima, J. (1984). *Phytopathology* **74**: 1363.
Gugerli, P. and Fries, P. (1983). *J. gen. Virol.* **64**: 2471.
Jones, R.A.C. and Fribourg, C.E. (1986). *AAB Descr. Pl. Viruses* No. 316, 4 pp.
Rozendaal, A., van Binsbergen, J., Anema, B., van Slogteren, D.H.M. and Bunt, M.H. (1971). *Potato Res.* 14: 241.

# Potato X *potexvirus*

**Data collated by C. Büchen-Osmond, 1987.**

*Synonyms:* potato latent virus, potato mild mosaic virus. *Standard acronym:* PVX. *Strains:* Indian turnip strain; Samad *et al.*, 1991.

First reported in *Solanum tuberosum*; from the U.K.; by Smith (1931).

**Natural host range and symptoms**

Symptoms persist.

*Solanum tuberosum* — symptoms vary, some strains symptomless, others induce necrotic streaks.

*Brassica rapa* — mild mosaic mottling and distortion of leaves, plant stunting.

**Transmission**

Transmitted by means not involving a vector. Virus transmitted by mechanical inoculation; transmitted by contact between plants; not transmitted by seed; not transmitted by pollen.

**Geographical distribution**

Probably distributed worldwide (in potato growing areas). Spreads in India.

**Experimental host range**

Few (<3) families susceptible.

Diagnostically susceptible

*Datura stramonium* — systemic chlorotic rings, then mosaic and mottling.
*Nicotiana tabacum* — systemic ringspot or mottle.

Some strains symptomless at high glasshouse temperatures.

Maintenance and propagation hosts

*Nicotiana tabacum.*

Assay hosts (**L**ocal lesions or **W**hole plants)

*Gomphrena globosa* (L).

Dicotyledonous Hosts *(**Susceptible** or Insusceptible)*

***Datura stramonium. Gomphrena globosa. Nicotiana tabacum. Solanum tuberosum.***

**Properties of particles in sap**

TIP: 68–76 °C. LIV: 40–60 days. DEP: $\log_{10}$ minus 5–6. Leaf sap contains many virions.

**Purification**

Method: Francki and McLean (1968).

**Particle morphology**

Virions filamentous; not enveloped; usually flexuous; with a clear modal length; of 515 nm; 13 nm wide (Brandes, 1964). Axial canal obscure; 3.4 nm in diameter. Basic helix obvious; pitch of basic helix 3.4 nm (Varma *et al.*, 1968).

**Physical properties**

One sedimenting component in purified preparations; sedimentation coefficient 117.7 *S*. Isoelectric point pH 4.4.

**Biochemical properties**

Virions contain 6 % nucleic acid (Knight, 1963); 94 % protein; 0 % lipid.

*Genome* consists of RNA; single–stranded; linear; unipartite; largest (or only) genome part 6.435 kb. Base composition 22 % G; 32 % A; 24 % C; 22 % U. 5' terminus of RNA has a methylated nucleotide cap. Infectivity retained when deproteinised with proteases; retained when deproteinised with phenol or detergent.

*Sequence database accession code(s):*

D00344 Em(40)_vi:PVXX3 Gb(84)_vi:PVXX3 Potato virus X genomic RNA. 4/92 6,435bp.

M31541 Em(40)_vi:PVXKPA Gb(84)_vi:PVXKPA Potato virus X 166K RNA, complete cds; 24K RNA, complete cds; 12K RNA, complete cds; 8K RNA,

M38480 Em(40)_vi:PVXGEN Gb(84)_vi:PVXGEN Potato virus X complete genome. 3/92 6,436bp.

M38655 Em(40)_vi:PVXCT23 Gb(84)_vi:PVXCT23 Potato X virus coat protein gene, complete cds. 10/90 861bp.

M63141 Em(40)_vi:PVXCGA1 Gb(84)_vi:PVXCGA Potato virus X complete genome. 2/92 6,432bp.

M72416 Em(40)_vi:PVXCG Gb(84)_vi:PVXCG Potato virus X complete genome. 10/93 7,568bp.

M95516 Em(40)_vi:PVXCOPR Gb(84)_vi:PVXCOPR Potato virus X mRNA, complete cds. 6/92 6,435bp.

S41511 Em(40)_un:S41511 coat protein potato virus X, PVXHB, Genomic RNA, 783 nt. 2/93 783bp.

X05198 Em(40)_vi:POPVX3 Gb(84)_vi:POPVX3 Potato virus X genome. 10/93 6,435bp.

X12804 Em(40)_vi:PVXCOAT Gb(84)_vi:PVXCOAT Potato Virus X (PVX) mRNA for viral coat protein. 9/93 1,106bp.

X55802 Em(40)_vi:POPOVX Gb(84)_vi:POPOVX Potato Virus X complete genomic RNA. 4/92 6,432bp.

X62442 Em(40)_vi:PVXANPII Gb(84)_vi:PVXANPII Potato virus X artificial neomycin phosphotransferase II mRNA 5'-flank (SEQ1). 9/92 115bp.

X62443 Em(40)_vi:PVXANPTII Gb(84)_vi:PVXANPTII Potato virus X artificial neomycin phosphotransferase II mRNA 5'-flank (SEQ2). 9/92 88bp.

X62444 Em(40)_vi:PVXANPT Gb(84)_vi:PVXANPT Potato virus X artificial neomycin phosphotransferase II mRNA 5'-flank (SEQ3). 9/92 62bp.

X62445 Em(40)_vi:PVXANPTR Gb(84)_vi:PVXANPTR Potato virus X artificial neomycin phosphotransferase II mRNA 5'-flank (SEQ4). 9/92 92bp.

X65015 Em(40)_vi:PVXCP Gb(84)_vi:PVXCP Potato Virus X genomic RNA for coat protein. 6/92 789bp

X72214 Em(40)_vi:PVXHBRNA Gb(84)_vi:PVXHBRNA Potato Virus X strain HB RNA. 11/93 6,432bp.

Z23256 Em(40)_vi:PVCOAPRA Gb(84)_vi:PVCOAPRA Potato virus X coat protein RNA and 3 ORF's, complete CDS's. 12/93 6,433bp.

Z29333 Em(40)_vi:PVX8KCP1 Gb(84)_vi:PVX8KCP1 Potato virus X (P551) mRNA for 8K protein and coat protein. 1/94 877bp.

Z29334 Em(40)_vi:PVX8KCP2 Gb(84)_vi:PVX8KCP2 Potato virus X (CP2) mRNA for 8K protein and coat protein. 1/94 874bp.

Z29335 Em(40)_vi:PVX8KCP3 Gb(84)_vi:PVX8KCP3 Potato virus X (S6111) mRNA for 8K protein and coat protein. 1/94 873bp. Z34261 Gb(84)_vi:PVXCPG Potato virus X (MS) coat protein gene. 6/94 711bp.

M27662 Em(43)_vi:Pvxorf Gb(89)_vi:Pvxorf Potatovirus X (clone pX 1) gene fragment, encoding a partial ORF. 1/95 502bp.

U19790 Gb(89)n:Pvu19790 Potato virus X coat protein mRNA, complete cds. 6/95 789bp.

*Non–genomic nucleic acid* not found in the virions.

*Virion protein(s)* one; $M_r$ 30000; coat protein. Amino acid composition: Shaw and Larson (1962); Shaw *et al.* (1962); Miki and Knight (1968).

**Replication**

Replication does not depend on a helper virus.

**Cytopathology**

Virions found in mesophyll and less frequently in the epidermis; in cytoplasm. Inclusions mainly near the nucleus present in infected cells; are amorphous X–bodies (Kozar and Sheludko, 1969); they contain virions.

**Taxonomy and Relationships**

*Potexvirus.*

Virus(es) with serologically related virions: white clover mosaic, hydrangea ringspot, cactus X and clover yellow mosaic viruses, but distantly.

**References**

Bercks, R, (1970). *CMI/AAB Descr. Pl. Viruses* No. 4, 4 pp.

Brandes, J. (1964). *Mitt. biol. BundesAnst. Land. Forstw.* **110**: 130.

Dolya, V.V., Grama, D.P., Morozov, S.Y. and Atabekov, J.G. (1987). *FEBS Lett.* **214**: 308.

Francki, R.I.B. and McLean, G. (1968). *Aust. J. biol. Sci.* **21**: 1311.

Hussman, M.J., Linthorst, H.J.M., Bol. J.F. and Cornellisen, B.J.C. (1988). *J. gen. Virol.* **69**: 1789.

Knight, C.A. (1963). *Protoplasmatologia* **4**: 177.

Kozar, I.E. and Sheludko, Y.M. (1969). *Virology* **38**: 220.
Miki, T. and Knight, C.A. (1968). *Virology* **36**: 168.
Morozov, S.Y., Miroshnichenko, N.A., Solovyev, A.G., Fedorkin, O.N., Zelenina, D.A., Lukasheva, L.I., Karasev, A.V., Dolya, V.V. and Atabekov, J.G. (1991). *J. gen. Virol.* **72**: 2039.
Paul, H.L. (1959). *Arch. Microbiol.* **32**: 416.
Samad, A., Naqvi, Q.A., Garg, I.D. and Thouvenel, J.-C. (1991). *J. Phytopath.* **133**: 65.
Shaw, J.G. and Larson, R.H. (1962). *Phytopathology* **52**: 170.
Shaw, J.G., Reichmann, M.E. and Hatt, D.L. (1962). *Virology* **18**: 79.
Skryabin, K.G., Morozov, S.Y., Kraev, A.S., Rozanov, M.V., Chernov, B.K., Lukasheva, L.I. and Atabekov, J.G. (1988). *FEBS Lett.* **240**: 33.
Smith, K.M. (1931). *Proc. R. Soc.* **109**: 251.
Varma, A., Gibbs, A.J., Woods, R.D. and Finch, J.T. (1968). *J. gen. Virol.* **2**: 107.
Tomashevskaya, O.L., Solvyev, A.G., Karpova, O.V., Fedorkin, O.N., Rodionova, N.P., Morozov, S.Y. and Atabekov, J.G. (1993). *J. gen. Virol.* **74**: 2717.
Wilson, H.R. and Tollin, P. (1969). *J. gen. Virol.* **5**: 151.
Wodnar-Filipowicz, A., Skizeczkowski, L.J. and Filipowicz, W. (1990). *FEBS Lett.* **109**: 151.

# Potato Y *potyvirus*

**Data collated by C. Büchen-Osmond, 1987.**

*Synonyms:* brinjal mosaic virus (Sastry, 1982; Varma, 1988), datura 437 virus, potato acropetal necrosis virus, potato severe mosaic virus, tobacco vein-banding virus (Fang *et al.*, 1985). *Standard acronym:* PVY. *Strains:* potato C virus, tobacco veinal necrosis virus potato virus $Y^o$ group (common strain), potato virus $Y^N$ group (tobacco veinal necrosis strain), potato virus $Y^C$ group (stipple streak strain, including potato virus C).

First reported in *Solanum tuberosum*; from the U.K.; by Smith (1931).

## Natural host range and symptoms

Symptoms persist.

*Solanum tuberosum* — mild to severe leaf mottling, or streak or 'leaf-drop streak' with vein necrosis ('stipple-streak') (Beemster and Rozendaal, 1972).

*Capsicum* spp. — mild leaf mottling, but severe in complex with other viruses.

*Nicotiana* spp. — mild mottle or veinal necrosis.

*Lycopersicon esculentum* — mild leaf mottling but severe in mixed infections (Edwardson, 1974*a*).

The differences between the primary and secondary symptoms induced by potato virus Y are often uncertain because of the diversity of potato cultivars and virus strains, and effect of climatic conditions (de Bokx and Pirone, 1977).

## Transmission

Transmitted by a vector; an insect; *Myzus persicae* is the most efficient vector; others are *Aphis fabae, Macrosiphum euphorbiae, Myzus (Nectarosiphon) certus, Myzus (Phorodon) humuli* and *Rhopalosiphum insertum* (Kennedy *et al.*, 1962; Van Hoof, 1980); Aphididae. Transmitted in a non–persistent manner. Virus does not require a helper virus for vector transmission and can help the vector transmission of another virus (potato aucuba mosaic potexvirus); transmitted by mechanical inoculation; transmitted by grafting.

**Ecology and control**

Studies reported by Bartels (1970), in certification schemes. In temperate areas, perennials rarely act as virus reservoirs in nature. Potato plants as "ground-keepers" are a reservoir host (Thresh, 1980). In tropical and subtropical areas, weeds such as *Solanum atropurpureum* (Chagas *et al.*, 1977) and other *Solanum* spp. (Edwardson, 1974*a*) may act as important virus sources.

Use of insecticides to control virus spread through vectors have been ineffective. The main control methods are (1) avoidance of infection, *i.e.* growing crops when vectors are absent or numbers are low; (2) not growing crops near established crops of the same species; (3) destroying haulms of seed-potato crops before maturity to restrict virus spread at the end of the growing season; (4) spraying with mineral oils to reduce frequency of transmission (Bradley, Moore and Pond, 1966; Vanderveken, 1977); (5) breeding for resistance, if sources of durable resistance can be obtained; (6) use of reflective surface and sticky yellow sheets which can reduce virus spread (Loebenstein and Raccah, 1980).

**Geographical distribution**

Probably distributed worldwide (in potato growing areas and in outdoor crops of pepper, tobacco and tomato in warmer countries.

$Y^o$ strains are spread worldwide; $Y^N$ strains occur in Europe including the U.S.S.R., parts of Africa and South America; $Y^C$ strains including potato virus C, are probably present in Australia, India and in some parts of the U.K. and continental Europe).

**Experimental host range**

Several (3–9) families susceptible.

Diagnostically susceptible

*Nicotiana glutinosa* — mild to severe systemic mottling.

*N. tabacum* — systemic vein-clearing and leaf mottling.

*Solanum tuberosum* — $Y^C$ strain local lesions in cv. Duke of York (Eersteling); $Y^N$ and $Y^o$ strains no local lesions.

*Tinantia erecta* — severe systemic mottling.

Diagnostically insusceptible

*Datura stramonium, Solanum demissum* 'A'.

Maintenance and propagation hosts

*Nicotiana tabacum* cv. Samsun NN.

Assay hosts (Local lesions or Whole plants)

*Chenopodium amaranticolor* (L); *C. quinoa* (L); *Physalis floridana* (L); *Solanum tuberosum* cvs Duke of York, Saco (L); *Lycium* spp. (L); *S. chacoense* ($TE_1$) (L), *S. demissum* 'Y' (L); *S. demissum x S. tuberosum* 'A6' (L).

Dicotyledonous Hosts *(**Susceptible** or Insusceptible)*

***Capsicum annuum***. ***Capsicum frutescens***. ***Chenopodium amaranticolor***. ***Chenopodium quinoa***. *Datura stramonium*. ***Lycopersicon esculentum***. ***Nicotiana glutinosa***. ***Nicotiana tabacum***. ***Physalis floridana***. ***Solanum demissum***, or *Solanum demissum*. ***Solanum demissum*** × ***tuberosum***. ***Solanum tuberosum***.

Monocotyledonous Hosts *(**Susceptible** or Insusceptible)*

***Tinantia erecta***.

Sources of host-range data

Edwardson (1974*b*); Horváth (1979).

**Properties of particles in sap**

TIP: 50–62 °C. LIV: 7–50 days. DEP: $\log_{10}$ minus 2–6. Infectivity of sap not changed by treatment with di–ethyl ether. Leaf sap contains few virions.

**Purification**

Method: Damirdagh and Shepherd (1970); Huttinga (1973); McDonald *et al.* (1976); Moghal and Francki (1976); Leiser and Richter (1978).

**Particle morphology**

Virions filamentous; not enveloped; usually flexuous; with a clear modal length; of 684 nm (from purified preparations (Delgado-Sanchez and Grogan, 1966)), or 730 nm; 11 nm wide. Axial canal obscure; 2–3 nm in diameter. Basic helix obscure; pitch of basic helix 3.3 nm (Varma *et al.*, 1968).

**Physical properties**

One sedimenting component in purified preparations; sedimentation coefficient 145 *S* (Huttinga, 1975). Density 1.323 g $cm^{-3}$ in CsCl (strain $Y^{o}$), or 1.326 g $cm^{-3}$ in CsCl (strain $Y^{N}$ (Huttinga, 1975)). $A_{260}/A_{280}$ ratio 2.3 (corrected for light-scattering (Leiser and Richter, 1978)), or 2.9 (uncorrected (Stace-Smith and Tremaine, 1970; Leiser and Richter, 1978)).

**Biochemical properties**

Virions contain 5.4–6.4 % nucleic acid (Stace-Smith and Tremaine, 1970; Leiser and Richter, 1978); 93.6–94.6 % protein; 0 % lipid.

*Genome* consists of RNA; single–stranded; linear; unipartite; largest (or only) genome part 10.4 kb (Makkouk and Gumpf, 1974; Hinostroza-Orihuela, 1975). 5' terminus of RNA has a VPg. Infectivity retained when deproteinised with phenol or detergent. Poly A region present; at 3' end, but not essential for infectivity.

*Sequence database accession code(s):*

A08776 Em(40)_vi:A08776 Gb(84)_pat:A08776 Potato virus Y complete genome RNA. 11/93 9,705bp.

D00441 Em(40)_vi:PVY Gb(84)_vi:PVYAAA Potato virus Y (N strain) genomic RNA, complete. 1/94 9,704bp

D00488 Em(40)_vi:PVY5E1 Gb(84)_vi:PVYE1 Potato virus Y (PVY) genome, 5' end. 3/91 173bp.

D00489 Em(40)_vi:PVY3E2 Gb(84)_vi:PVYE2 Potato virus Y (PVY) genome, 3' end. 3/91 276bp.

D12539 Em(40)_vi:PVYOCRNA Gb(84)_vi:PVYOCRNA Potato virus Y genome, 3' half sequence. 9/92 4,998bp.

D12570 Em(40)_vi:PVYCP Gb(84)_vi:PVYCP Potato virus Y (PVY-T) coat protein gene. 10/92 1,268bp

M22470 Em(40)_vi:PVYCPA Gb(84)_vi:PVYCPA Potato virus Y coat protein gene, 3' end. 7/90 1,122bp.

M37180 Em(40)_vi:PVYHEL01 Gb(84)_vi:PVYHELCOA Potato virus Y helper component mRNA, 5' end. 9/90 2,753bp.

M38377 Em(40)_vi:PVYHELCO Gb(84)_vi:PVYHELCO Potato virus Y helper component mRNA, 5' end. 9/90 2,748bp.

M81435 Em(40)_vi:PVYPOLYPR Gb(84)_vi:PVYPOLYPR Potato virus Y polyprotein mRNA, 3' end. 12/91 1,227bp.

M95491 Em(40)_vi:PVYPOLYP Gb(84)_vi:PVYPOLYP Potato virus Y polyprotein gene, complete cds. 5/93 9,703bp

S51663 Em(40)_un:S51663 Gb(84)_vi:S51663 cytoplasmic inclusion protein potato virus Y PVY, ordinary strain PVY-O, Genomic RNA, 1902 n

S51664 Em(40)_un:S51664 Gb(84)_vi:S51664 cytoplasmic inclusion protein potato virus Y PVY, necrotic strain PVY-T13, Genomic RNA, 1902

S63507 Em(40)_vi:S63507 Gb(84)_vi:S63507 polyprotein precursor (5'region) potato virus Y PVY, mRNA Partial, 360 nt. 12/93 360bp.

S70722 Gb(84)n:S70722 NIb=nuclear inclusion b potato virus Y PVY, O, Genomic RNA, 1560 nt. 9/94 1,560bp.

U06789 Em(40)_vi:PV06789 Gb(84)_vi:PVU06789 Potato virus Y-VN genome, 3' untranslated region. 3/94 2,541bp.

U09508 Em(40)_vi:PV09508 Gb(84)_vi:PVU09508 Potato virus Y necrotic strain polyprotein (NIa, NIb, capsid protein) gene, partial cds. 5/94

U10378 Gb(84)_vi:PVU10378 Potato virus Y capsid protein mRNA, partial cds. 6/94 1,231bp.

X12456 Em(40)_vi:PVYNXX Gb(84)_vi:PVYNXX Potatovirus Y strain N genomic RNA. 1/94 9,704bp.

X14136 Em(40)_vi:PVYCOAT Gb(84)_vi:PVYCOAT Potato virus Y (PVY) mRNA for viral coat protein. 9/93 1,182bp

X54058 Em(40)_vi:POCP3 Gb(84)_vi:POCP3 Potato virus Y gene for coat protein, 3' end. 1/91 985bp.

X54611 Em(40)_vi:PVYCOATP Gb(84)_vi:PVYCOATP Potato virus Y RNA for coat protein. 1/92 801bp.

X54636 Em(40)_vi:PVYCAPSI Gb(84)_vi:PVYCAPSI Potato virus RNA for capsid protein and virus-specific polymerase (putative nuclear inclusion

X68221 Em(40)_vi:PVYCAPA Gb(84)_vi:PVYCAPA Potato Virus Y (Chilean) genomic RNA of Capsid protein cistron. 9/93 804bp.

X68222 Em(40)_vi:PVYCAPB Gb(84)_vi:PVYCAPB Potato Virus Y (Potato US) genomic RNA of Capsid protein cistron. 9/93 804bp.

X68223 Em(40)_vi:PVYCAPC Gb(84)_vi:PVYCAPC Potato Virus Y (Europe- H) genomic RNA of Capsid protein cistron. 9/93 804bp.

X68224 Em(40)_vi:PVYCAPD Gb(84)_vi:PVYCAPD Potato Virus Y (NsNr) genomic RNA of Capsid protein cistron. 9/93 804bp.

X68225 Em(40)_vi:PVYCAPE Gb(84)_vi:PVYCAPE Potato Virus Y (MsNr) genomic RNA of Capsid protein cistron. 9/93 804bp.

X68226 Em(40)_vi:PVYCAPF Gb(84)_vi:PVYCAPF Potato Virus Y (O) genomic RNA of Capsid protein cistron. 9/93 804bp.

Z29526 Em(40)_vi:PVYCI6K2 Gb(84)_vi:PVYCI6K2 Potato virus Y (strain O) CI, 6K2, VPg, NIa-Pro, NIa gene. 1/94 1,784bp.

D17386 Em(43)_vi:Pvyia Gb(89)_vi:Pvyia Potato virus Y immunoglobulin gene for monoclonal antibody and L chain. 3/95 927bp.

D17387 Em(43)_vi:Pvyib Gb(89)_vi:Pvyib Potato virus Y immunoglobulin gene for monoclonal antibody and H chain. 3/95 1,488bp.

S74810 Gb(89)_un:S74810 CP=coat protein (potato virus Y, PVY-36, Genomic RNA, 804 nt). 5/95 804bp.

S74813 Gb(89)_un:S74813 CP=coat protein (potato virus Y, PVY-T13, Genomic RNA, 804 nt). 5/95 804bp.

U09509 Em(43)_vi:Pv09509 Gb(89)_vi:Pvu09509 Potato virus Y common strain polyprotein (NIb and capsid protein) gene, partial cds. 12/94 2,

U25672 Gb(89)n:Pvu25672 Potato virus Y coat protein gene, complete cds. 6/95 821bp.

X79305 Em(43)_vi:Pvygcp Gb(89)_vi:Pvygcp Potato Virus Y genomic sequence for coat protein. 1/95 801bp.

X82848 Em(43)_vi:Pvyp1 Gb(89)_vi:Pvyp1 Potato virus Y P1 gene. 5/95 828bp.

*Non–genomic nucleic acid* not found in the virions.

*Virion protein(s)* one; $M_r$ 34000; coat protein. Method of preparation: Hiebert and McDonald (1973); Huttinga and Mosch (1974); Moghal and Francki (1976). Amino acid composition: Stace-Smith and Tremaine (1970); Miki and Oshima (1972); Makkouk and Gumpf (1975); Moghal and Francki (1976).

*Virus–coded non–virion proteins* isolated (Hiebert and McDonald (1973)); one protein found. $M_r$ 67000; cylindrical inclusion protein (Purcifull *et al.*, 1973).

**Replication**

Replication does not depend on a helper virus.

**Cytopathology**

Virions found in epidermis; in cytoplasm and in cell vacuoles. Inclusions present in infected cells; are crystals in the nucleus (when infected by certain strains (Kitajima *et al.*, 1968)), or pinwheels (especially in epidermal tissue in the cytoplasm (Christie and Edwardson, 1977)). Other cellular changes: mitochondria often being surrounded by filaments with a diameter of 9–10 nm but of indeterminate length (Borges and David Ferreira, 1968) when infected with strain $Y^N$ in *Datura metel.*

**Taxonomy and Relationships**

*Potyvirus: Potyviridae.*

Virus(es) with serologically related virions: tobacco etch, henbane mosaic, potato A, pepper veinal mottle and bidens mottle viruses, but distantly (Bartels, 1964; Brunt *et al.*, 1978; Purcifull, 1976).

Best tests for diagnosis: *Tinantia erecta* is immune to potato aucuba mosaic, potato M and S viruses and hence may be used to separate potato Y virus from these viruses (Horváth, 1979). most strains of potato virus Y induce no local lesions in *Solanum demissum* 'A', which gives necrotic local lesions with potato virus A (Cockerham, 1958). Potato cultivars resistant to potato X virus (*e.g.* cv. Saco), can be used to separate potato Y virus from potato X virus. Detached leaves of *Solanum demissum x. S. tuberosum* 'A6' are commonly used to detect potato Y virus, but this host does not distinguish between several viruses which infect it (Bartels, 1970).

**References**


Bartels, R. (1964). *Phytopath. Z.* **49**: 257.

Bartels, R. (1970). *Potato Res.* **13**: 119.

Beemster, A.B.R. and Rozendaal, A. (1972). In: *Viruses of Potatoes and Seed-Potato Production*, p. 115; ed. J.A. de Bokx, Wageningen, PUDOC.

Borges, M. and David-Ferreira, J.F. (1968). *Revta Biol., Lisbon* **6**: 421.

Bradley, R.H.E., Moore, C.A. and Pond, D.D. (1966). *Nature, Lond.* **209**: 1370.

Brunt, A.A., Kenten, R.H. and Phillips, S. (1978). *Ann. appl. Biol.* **88**: 115.

Chagas, C.M., Vincente, M., Alba, A.P.C. and July, J.R. (1977). *Phytopath. Z.* **90**: 147.

Cockerham, G. (1958). *Proc. 3rd Conf. Potato Virus Diseases, Lisse-Wageningen*, p. 199.

Christie, R.G. and Edwardson, J.R. (1977). *Fla Agric. Exp. Stn Monog.* No. 9, 150 pp.

Damirdagh, I.S. and Shepherd, R.J. (1970). *Phytopathology* **60**: 132.

de Bokx, J.A. (1981). *CMI/AAB Descr. Pl. Viruses* No. 242, 6 pp.

de Bokx, J.A. and Pirone, P.G.M. (1978). *Abstr. 7th Trienn. Conf. Eur. Ass. Potato Res.*, p. 244.
Delgado-Sanchez, S. and Grogan, R.G. (1966). *Phytopathology* **56**: 1397.
Delgado-Sanchez, S. and Grogan, R.G. (1970). *CMI/AAB Descr. Pl. Viruses* No. 37, 4 pp.
Edwardson, J.R. (1974*a*). *Fla Agric. Exp. Stn Monog.* No. 4, 398 pp.
Edwardson, J.R. (1974*b*). *Fla Agric. Exp. Stn Monog.* No. 5, p. 95.
Fang, H.S., Nell, H.H. and Chou, T.G. (1985). *Bull. Tobacco Res. Inst. Taiwan Tobacco and Wine Monopoly Bureau.*
Govier, D.A. and Kassanis, B. (1974). *Virology* **61**: 420.
Gugerli, P. and Gehringer, W. (1980). *Potato Res.* **23**: 353.
Hiebert, E. and McDonald, J.G. (1973). *Virology* **56**: 349.
Hinostroza-Orihuela, A.M. (1975). *Virology* **67**: 276.
Horváth, J. (1979). *Acta phytopath. Acad. Sci. hung.* **14**: 157.
Huttinga, H. (1973). *Neth. J. Pl. Path.* **79**: 125.
Huttinga, H. (1975). *Neth. J. Pl. Path.* **81**: 58.
Huttinga, H. and Mosch, W.H.M. (1975). *Neth. J. Pl. Path.* **80**: 19.
Kennedy, J.S., Day, M.F. and Eastop, V.F. (1962). *A Conspectus of Aphids as Vectors of Plant Viruses*, 114 pp. Comm. Inst. Ent., London.
Kitajima, E.W., Camargo, I.J.B. and Costa, A.S. (1968). *J. Electr. Microsc.* **17**: 144.
Leiser, R.M. and Richter, J. (1979). *Arch. Phytopath. PflSchutz.* **14**: 337.
Loebenstein, G. and Raccah, B. (1980). *Phytoparasitica* **8**: 221.
Makkouk, K.M. and Gumpf, D.J. (1974). *Phytopathology* **64**: 1115.
Makkouk, K.M. and Gumpf, D.J. (1975). *Virology* **63**: 336.
McDonald, J.G., Beveridge, T.J. and Bancroft, J.B. (1976). *Virology* **69**: 327.
Moghal, S.M. and Francki, R.I.B. (1976). *Virology* **73**: 350.
Purcifull, D.E. (1976). *CMI/AAB Descr. Pl. Viruses* No. 161, 4 pp.
Purcifull, D.E. and Batchelor, D.L. (1977). *Bull. Fla agric. Exp. Stn* **788**: 39 pp.
Purcifull, D.E. and Gooding, G.V. (1970). *Phytopathology* **60**: 1036.
Purcifull, D.E., Hiebert, E. and McDonald, J.G. (1973). *Virology* **55**: 275.
Rubio-Huertos, M. (1972). In: *Principles and Techniques in Plant Virology*, 300 pp; ed. C.J. Kado and H.O. Agrawal. Van Nostrand Reinhold Company, New York.
Sastry, K.S. (1982). *Curr. Sci.* **51**: 568.
Smith, K.M. (1931). *Proc. R. Soc. B* **109**: 251.
Stace-Smith, R. and Tremaine, J.H. (1970). *Phytopathology* **60**: 1785.
Thresh, J.M. (1980). *Appl. Biol.* **5**: 1.
Vanderveken, J. (1977). In: *Aphids as Virus Vectors*, 559 pp; ed. K.F. Harris and K. Maramorosch. Academic Press, New York.
Van Hoof, H.A. (1980). *Neth. J. Pl. Path.* **86**: 159.
Varma, A. (1988). In: *The Plant Viruses* Vol. 4, The Filamentous Plant Viruses, p. 371; ed. R.G. Milne. Plenum Press, New York.
Varma, A., Gibbs, A.J., Woods, R.D. and Finch, J.T. (1968). *J. gen. Virol.* **2**: 107.

# Potato yellow dwarf *nucleorhabdovirus*

**Data collated by G. Adam, 1991.**

*Standard acronym:* PYDV.

First reported in *Solanum tuberosum*; from the U.S.A.; by Barrus and Chupp (1922).

**Natural host range and symptoms**

*Chrysanthemum leucanthemum, Solanum tuberosum, Trifolium incarnatum.*

**Transmission**

Transmitted by a vector; an insect; *Agallia constricta, A. quadripunctata, Aceratagallia sanguinolenta*; Cicadellidae. Transmitted in a persistent manner. Virus retained when the vector moults; multiplies in the vector; transmitted by mechanical inoculation; not transmitted by seed.

**Geographical distribution**

Spreads in the North American region; Canada and the USA.

**Experimental host range**

Several (3–9) families susceptible. Experimentally infected plants mostly show vein clearing, stunting, leaf malformation, mosaic, chlorotic local lesions.

Diagnostically susceptible

*Nicotiana glutinosa, N. rustica* — chlorotic local lesions, vein clearing, leaf malformation, mosaic.

*Trifolium incarnatum* — vein clearing, death.

Diagnostically insusceptible

*Medicago sativa* cv. Grimm.

Maintenance and propagation hosts

*Nicotiana glutinosa, N. rustica.*

Assay hosts (**L**ocal lesions or **W**hole plants)

*Nicotiana rustica* (L).

Dicotyledonous Hosts *(**Susceptible** or Insusceptible)*

***Nicotiana* × *edwardsonii*. *Nicotiana glutinosa*. *Nicotiana rustica*. *Nicotiana tabacum*. *Solanum tuberosum*. *Trifolium incarnatum*.**

Sources of host-range data

Younkin (1942).

**Properties of particles in sap**

TIP: 50 °C. LIV: 0.25 days (few hours). DEP: $\log_{10}$ minus 3. Infectivity of sap decreased by treatment with di–ethyl ether. Leaf sap contains many virions. Electron microscopy: use ammonium molybdate as the stain.

**Particle morphology**

Virions rhabdo– or bullet–shaped; enveloped; usually straight; with a clear modal length; of 380 nm; 75 nm wide.

**Physical properties**

One sedimenting component in purified preparations; sedimentation coefficient 800 *S*. Density 1.2 g $cm^{-3}$ in sucrose.

**Biochemical properties**

Virions contain 2 % nucleic acid; 78 % protein; 20 % lipid.

*Genome* consists of RNA; single–stranded; linear; unipartite; largest (or only) genome part 12.6 kb. Infectivity lost when deproteinised with proteases; lost when deproteinised with phenol or detergent. Poly A region absent.

*Virion protein(s)* five; $M_r$ of the largest 180000; polymerase. $M_r$ of 2nd largest 80000; spike protein. $M_r$ of 3rd largest 55000; nucleoprotein. $M_r$ of 4th largest 31; matrix protein. $M_r$ of 5th largest 26000. Virion proteins glycosylated; are and it is the G protein.

**Replication**

Replication does not depend on a helper virus.

**Cytopathology**

Virions found in the perinuclear space. Inclusions leaves, roots, mesophyll and phloem.

**Taxonomy and Relationships**

*Nucleorhabdovirus: Rhabdoviridae* (the type species).

Two serotypes, which differ in their proteins and in vector specificity, are known.

**References**


Adam, G. (1984). *Vectors in Virus Biology* **12**: 37.

Adam, G. and Gaedigk, K. (1986). *J. gen. Virol.* **67**; 2775.

Adam, G., Gaedigk, K. and Mundry, K.W. (1983). *Z. PflKrankh. PflPath. PflSchutz.* **90**: 28.

Ahmed, M.E., Black, L.M., Perkins, E.G., Walker, B.L. and Kummerow, F.A. (1964). *Biochem. Biophys. Res. Comm.* **17**: 103.

Barrus and Chupp (1922). *Phytopathology* **12**: 123.

Black, L.M. (1941). *Am. Potato J.* **18**: 231.

Black, L.M. (1965). *Virology* **27**: 446.

Black, L.M. (1979). *Adv. Virus Res.* **25**: 191.

Black, L.M. (1984). *Curr. Topics in Vector Res.* **2**: 1.

Brakke, M.K. (1956). *Virology* **2**: 463.

Chiu, R.J., Liu, H.Y., MacLeod, R. and Black, L.M. (1970). *Virology* **40**: 387.

Dale, J.L. and Peters, D. (1981). *Intervirology* **16**: 86.

Falk, B.W. and Tsai, J.H. (1983). *Phytopathology* **73**: 1536.

Falk, B.W. and Weathers, L.G. (1983). *Phytopathology* **73**: 81.

Falk, B.W., Weathers, L.G. and Greer, F.C. (1981). *Plant Dis.* **65**: 81.

Gaedigk, K., Adam, G. and Mundry, K.-W. (1986). *J. gen. Virol.* **67**: 2763.

Hiebert, E. and Charudattan, R. (1984). *Phytopathology* **74**: 642.

Hsu, H.T. (1978). *Virology* **84**: 9.

Hsu, H.T. and Black, L.M. (1973). *Virology* **52**: 187.

Hsu, H.T. and Black, L.M. (1973). *Virology* **52**: 284.

Hsu, H.T. and Black, L.M. (1973). *Phytopathology* **63**: 692.

Hsu, H.T. and Black, L.M. (1974). *Virology* **59**: 331.

Hsu, H.T., McBeath, J.H. and Black, L.M. (1977). *Virology* **81**: 257.

Hsu, H.T., Nuss, D.L. and Adam, G. (1983). *Curr. Topics Vector Res.* **1** 189.

Jackson, A.O. and Christie, S.R. (1977). *Virology* **77**: 344.

Knudson, D.L. and MacLeod, R. (1972). *Virology* **47**: 285.

Liu, H.Y. and Black, L.M. (1978). *Phytopathology* **68**: ;1243.

MacLeod, R. (1968). *Virology* **34**: 771.

Nagaraj, A.N. and Black, L.M. (1962). *Virology* **16**: 152.

Reeder, Knudson, D.L. and MacLeod, R. (1972). *Virology* **50**: 301.
Sinha, R.C. (1965). *Virology* **27**; 118.
Whitcomb, R.F. (1965). *Phytopathology* **55**: 746.
Younkins (1942). *Am. Potato J.* **19**: 6.

# Potato yellow mosaic bigeminivirus

**Data collated by A.A. Brunt, 1987.**

*Standard acronym:* PYMV.

First reported in *Solanum tuberosum* sp. *tuberosum*; from the U.K. isolated from material from Venezuela; by Roberts *et al.* (1986).

## Natural host range and symptoms

Symptoms persist.

*Solanum tuberosum* subsp. *tuberosum* — bright yellow mosaic, leaf malformation, stunting.

## Transmission

Transmitted by a vector; an insect; *Bemisia tabaci*; Aleyrodidae. Virus transmitted by mechanical inoculation; transmitted by grafting.

## Geographical distribution

Spreads in Venezuela.

## Experimental host range

Several (3–9) families susceptible.

Diagnostically susceptible

*Lycopersicon esculentum, Nicotiana* spp., *Petunia hybrida* — systemic leaf chlorosis.

Dicotyledonous Hosts *(**Susceptible** or Insusceptible)*

***Lycopersicon esculentum. Petunia hybrida. Solanum tuberosum.***

## Particle morphology

Virions geminate.

## Biochemical properties

*Genome* consists of DNA; single–stranded; circular.

*Sequence database accession code(s):*

D00940 Em(40)_vi:UVA Gb(84)_vi:PYVVA Potato yellow mosaic virus (PYMV) DNA A, complete sequence. 6/94 2,593bp.

D00941 Em(40)_un:UVB Gb(84)_vi:PYVVB Potato yellow mosaic virus (PYMV) DNA B, complete sequence. 6/94 2,547bp. 2 sequences.

## Taxonomy and Relationships

Bigeminivirus: *Geminiviridae.*

## References

Roberts, E.J.F., Buck, K.W. and Coutts, R.H.A. (1986). *Plant Dis.* **70**: 603.

# Primula mosaic *potyvirus*

**Data collated by A.A. Brunt and P. Jeyanandarajah, 1992.**

*Standard acronym:* PrMV.

First reported in *Primula obconica*; from Italy; by Lisa and Lovisolo (1976).

**Natural host range and symptoms**

Symptoms persist.
*Primula obconica* — leaf chlorosis and flower breaking.

**Transmission**

Transmitted by a vector; an insect; *Myzus persicae*; Aphididae. Transmitted in a non–persistent manner. Virus transmitted by mechanical inoculation.

**Geographical distribution**

Spreads in Italy.

**Experimental host range**

Many (>9) families susceptible.

Diagnostically susceptible

*Apium graveolens, Coriandrum sativum, Gomphrena globosa, Nicotiana clevelandii, Trigonella foenum-graecum, Zinnia elegans* — chlorotic local lesions; systemic mosaic.

*Chenopodium amaranticolor, C. murale, C. quinoa, Tetragonia expansa* — chlorotic local lesions; no systemic infection.

Diagnostically insusceptible

*Amaranthus caudatus, Beta vulgaris, Datura stramonium, Nicotiana rustica, Spinacia oleracea.*

Maintenance and propagation hosts

*Nicotiana clevelandii.*

Assay hosts (**L**ocal lesions or **W**hole plants)

*Chenopodium amaranticolor* (L), *C, murale* (L), *C. quinoa* (L), *Nicotiana clevelandii* (W).

Dicotyledonous Hosts *(**Susceptible** or Insusceptible)*

*Amaranthus caudatus*. ***Apium graveolens***. *Beta vulgaris*. *Brassica campestris* ssp. *napus*. ***Brassica campestris*** ssp. ***pekinensis***. *Brassica campestris* ssp. *rapa*. ***Brassica oleracea*** ssp. ***botrytis***. *Capsicum annuum*. *Celosia argentea*. ***Chenopodium album***. ***Chenopodium amaranticolor***. ***Chenopodium ambrosioides***. ***Chenopodium foetidum***. *Chenopodium foliosum*. ***Chenopodium murale***. ***Chenopodium quinoa***. ***Coriandrum sativum***. ***Cucumis sativus***. *Datura stramonium*. ***Gomphrena globosa***. *Lactuca sativa*. ***Nicotiana clevelandii***. ***Nicotiana debneyi***. ***Nicotiana glutinosa***. ***Nicotiana megalosiphon***. *Nicotiana rustica*. ***Nicotiana tabacum***. *Ocimum basilicum*. ***Petunia hybrida***. *Phaseolus vulgaris*. ***Physalis floridana***. *Spinacia oleracea*. ***Tetragonia expansa***. *Vigna unguiculata*. ***Zinnia elegans***.

**Properties of particles in sap**

TIP: 60 °C. LIV: 16 days. DEP: $\log_{10}$ minus 4.

**Particle morphology**

Virions filamentous; usually flexuous; with a clear modal length; of 760 nm.

**Cytopathology**

Virions found in leaves and mesophyll; in cytoplasm. Inclusions present in infected cells; are pinwheels; they do not contain virions.

**Taxonomy and Relationships**

*Potyvirus: Potyviridae.*

Virus(es) with serologically related virions: bean yellow mosaic and lettuce mosaic viruses. Virus(es) with serologically unrelated virions: iris severe mosaic, carnation vein mottle, henbane mosaic and potato Y viruses.

**Comments**

It is not known whether a primula mosaic virus in the U.S.A. (Tompkins and Middleton, 1941) has any relationship to that described here. The closeness of the serological relationship between primula mosaic and lettuce mosaic viruses (SDI 2–3) suggests that the former could be regarded as a strain of lettuce mosaic potyvirus.

**References**

Edwardson, J.R. and Christie, R.G. (1991). In: *The Potyvirus Group*, Vol. III, p. 809.
Lisa, V. and Lovisolo, O. (1976). *Acta Hort.* **59**: 167.
Tompkins, C.M. and Middleton, J.T. (1941). *J. agric. Res.* **63**: 671.

# Primula mottle (?) potyvirus

**Data collated by V.R. Mali, 1987.**

*Standard acronym:* PrMoV.

First reported in *Primula malacoides*; from Delhi, India; by Singh *et al.* (1970).

**Natural host range and symptoms**

Symptoms persist.

*Primula malacoides* — mild mosaic of green and chlorotic areas, flower colour breaking and stunting.

**Transmission**

Transmitted by a vector; an insect; *Aphis gossypii, Myzus persicae*; Aphididae. Transmitted in a non–persistent manner. Virus transmitted by mechanical inoculation.

**Ecology and control**

Studies reported by Singh *et al.* (1970); Mali (1985).

**Geographical distribution**

Spreads in India (in Delhi).

**Experimental host range**

Few (<3) families susceptible.

Diagnostically susceptible

*Chenopodium amaranticolor* — chlorotic local lesions.

*C. album* — necrotic local lesions.

*Datura stramonium* — systemic mild mottle, young leaves filiform, malformed and with enations.

*Gomphrena globosa, Nicotiana glutinosa, N. rustica, N. tabacum, Petunia alba* — systemic mild mottle, vein-clearing.

Diagnostically insusceptible

*Chenopodium quinoa, Datura metel, Nicandra physalodes, Solanum nigrum* and *Zinnia elegans.*

Maintenance and propagation hosts

*Primula malacoides, Nicotiana glutinosa.*

Assay hosts (**L**ocal lesions or **W**hole plants)

*Chenopodium amaranticolor* (L), *C. album* (L).

Dicotyledonous Hosts *(**Susceptible** or Insusceptible)*

*Amaranthus caudatus. Antirrhinum majus. Brassica oleracea* ssp. *botrytis. Brassica oleracea* ssp. *capitata. Capsicum annuum.* ***Chenopodium album. Chenopodium amaranticolor.*** *Chenopodium ambrosioides. Chenopodium murale. Chenopodium quinoa. Chrysanthemum morifolium. Cucumis sativus. Cucurbita pepo. Cyamopsis tetragonoloba. Datura metel.* ***Datura stramonium.*** *Dianthus caryophyllus.* ***Gomphrena globosa.*** *Lablab purpureus. Lens esculenta. Luffa acutangula. Nicandra physaloides.* ***Nicotiana glutinosa. Nicotiana rustica. Nicotiana tabacum.*** *Phlox drummondii. Pisum sativum. Solanum melongena. Solanum nigrum. Spinacia oleracea. Tropaeolum majus. Vigna radiata. Vigna unguiculata. Zinnia elegans.*

Sources of host-range data

Singh *et al.* (1970).

**Properties of particles in sap**

TIP: 60–65 °C. LIV: 2 days. DEP: $\log_{10}$ minus 1. Leaf sap contains few virions.

**Particle morphology**

Virions filamentous; not enveloped; usually flexuous; with a clear modal length; of 720–745 nm.

**Taxonomy and Relationships**

Unassigned potyvirus.

PMtV shows some similarities to tobacco etch (TEV) and datura enation mosaic viruses (DEMV). Best tests for diagnosis: *Primula* mosaic caused by alfalfa mosaic virus (AMV) in India (Nagaich and Giri, 1968) can be distinguished from primula mottle virus (PMtV) by its host range, virion morphology and by serology.

**References**

Mali, V.R. (1985). *Int J. Tropical Plant Diseases* **3**: 93.

Singh, S., Sachchidananda, J., Prakesh, N. and Varma, A. (1970). *Indian Phytopath.* **23**: 148.

Verma, G.S. and Verma, H.N. (1963). *Indian Phytopath.* **16**: 368.

## Prune dwarf *ilarvirus*

**Data collated by R.W. Fulton, 1985.**

*Synonyms:* cherry chlorotic ringspot virus, peach stunt virus, sour cherry yellows virus. *Standard acronym:* PDV.

First reported in *Prunus domestica*; from the U.S.A.; by Thomas and Hildebrand (1936).

**Natural host range and symptoms**

Symptoms persist.

*Prunus domestica* — leathery, strap-like leaves.

*Prunus cerasus* — leaf yellowing and abscission.
*Prunus persica* — stunting.

**Transmission**

Transmitted by means not involving a vector. Virus transmitted by mechanical inoculation; transmitted by grafting; not transmitted by contact between plants; transmitted by seed (to 10% in *Prunus cerasus*); transmitted by pollen to the seed and transmitted by pollen to the pollinated plant.

**Geographical distribution**

Probably distributed worldwide.

**Experimental host range**

Many (>9) families susceptible.

Diagnostically susceptible

*Cucumis sativus* — small chlorotic local lesions; systemic mosaic.
*Cucurbita maxima* — interveinal chlorosis.
*Sesbania exaltata* — small dark local lesions in cotyledons; not systemic.
*Crotalaria spectabilis* — small dark local lesions.
*Momordica balsamina* — chlorotic local lesions; systemic mottle.
*Tithonia speciosa* — prominent chlorotic lines and rings.
*Phlox drummondii* — systemic mottle.
*Thunbergia alata* — systemic chlorotic rings.
*Melilotus officinalis* — systemic mottle.

Diagnostically insusceptible

*Gomphrena globosa, Chenopodium amaranticolor, Chenopodium quinoa, Physalis floridana, Cassia tora.*

Maintenance and propagation hosts

*Cucurbita maxima* cv. Buttercup, *Nicotiana occidentalis.*

Assay hosts (**L**ocal lesions or **W**hole plants)

*Sesbania exaltata* (L), *Crotalaria spectabilis* (L), *Cucumis sativus* (W).

Dicotyledonous Hosts *(**Susceptible** or Insusceptible)*

***Antirrhinum majus***. *Cassia tora*. ***Catharanthus roseus***. *Chenopodium amaranticolor*. *Chenopodium quinoa*. ***Crotalaria spectabilis***. ***Cucumis melo***. ***Cucumis sativus***. ***Cucurbita maxima***. ***Cucurbita moschata***. ***Cucurbita pepo***. ***Cyamopsis tetragonoloba***. ***Datura stramonium***. *Gomphrena globosa*. ***Gossypium hirsutum***. ***Helianthus annuus***. ***Ipomoea nil***. ***Lactuca sativa***. ***Melilotus alba***. ***Momordica balsamina***. ***Nicotiana benthamiana***. ***Nicotiana clevelandii***. ***Nicotiana debneyi***. ***Nicotiana glutinosa***. *Nicotiana megalosiphon*. ***Nicotiana rustica***. ***Nicotiana sylvestris***. ***Nicotiana tabacum***. ***Petunia hybrida***. *Physalis floridana*. ***Sesbania exaltata***. ***Tithonia speciosa***. ***Trifolium repens***. ***Zinnia elegans***.

Comments on host-range

Some isolates differ in host range.

Sources of host-range data

Fulton (1957).

**Properties of particles in sap**

TIP: 45–54 °C (in extracts with antioxidant). LIV: 0.5–0.75 days (12–18 hours). DEP: $\log_{10}$ minus 2–3. Electron microscopy: Fixation in 0.5% acrolein plus 0.5% glutaraldehyde.

**Particle morphology**

Virions isometric and bacilliform; not enveloped; 19–20 nm in diameter; 20 nm in length, or 23 nm in length, or 26 nm in length, or 38 nm in length (some 20×73 nm); rounded in profile; without a conspicuous capsomere arrangement.

**Physical properties**

Five sedimenting components in purified preparations (Halk and Fulton, 1978); sedimentation coefficient of the fastest 113 *S*; of the other(s) 75 *S*, or 81 *S*, or 85 *S*, or 98 *S* (the 98 and 113 *S* virions contain the genome). Density 1.33 g $cm^{-3}$ in CsCl (when fixed in formaldehyde). $A_{260}/A_{280}$ ratio 1.56 (for each).

**Biochemical properties**

Virions contain 14 % nucleic acid; 86 % protein; 0 % lipid.

*Genome* consists of RNA; single–stranded. Genomic nucleic acid isolated by Halk and Fulton (1978).

*Sequence database accession code(s):*

L28145 Em(40)_vi:PDVMOVCAP Gb(84)_vi:PDVMOVCAP Prune dwarf virus movement protein, complete cds; coat protein, complete cds. 5/94 2,129bp. 1 sequence.

*Virion protein(s)* probably one.

**Replication**

Replication does not depend on a helper virus.

**Cytopathology**

Virions found in all parts of the host plant.

**Taxonomy and Relationships**

*Ilarvirus: Bromoviridae* (subgroup 4).

Virus(es) with serologically unrelated virions: apple mosaic, prunus necrotic ringspot, tobacco streak, asparagus 2, citrus leaf rugose, citrus variegation, elm mottle and Tulare apple mosaic viruses.

**References**

Cation, D. (1949). *Phytopathology* **39**: 37.
Cropley, R., Gilmer, R.M. and Posnette, A.F. (1964). *Ann. appl. Biol.* **53**: 325.
Fulton, R.W. (1957). *Phytopathology* **47**: 215.
Fulton, R.W. (1957). *Phytopathology* **47**: 683.
Fulton, R.W. (1958). *Virology* **6**: 499.
Fulton, R.W. (1970). *CMI/AAB Descr. Pl. Viruses* No. 19, 3 pp.
George, J.A. and Davidson, T.R. (1963). *Can. J. Pl. Sci.* **43**: 276.
Halk, E.L. and Fulton, R.W. (1978). *Virology* **91**: 434.
Thomas, H.E. and Hildebrand, E.M. (1936). *Phytopathology* **26**: 1145.

# Prunus necrotic ringspot *ilarvirus*

**Data collated by R.W. Fulton, 1985.**

*Synonyms:* European plum line pattern virus, hop B virus, hop C virus, peach ringspot virus, plum line pattern virus, prunus ringspot virus, red currant necrotic ringspot virus, rose chlorotic mottle virus, rose line pattern virus, rose vein banding virus, rose yellow vein mosaic virus (Sastry, 1966), sour cherry necrotic ringspot virus. *Standard acronym:* PNRSV. *Strains:* cherry rugose mosaic virus, Danish plum line pattern virus and North American plum line pattern ilarvirus.

First reported in *Prunus persica*; from the U.S.A.; by Cochran and Hutchins (1941).

**Natural host range and symptoms**

Symptoms persist (in some hosts), or disappear soon after infection (in some hosts).
*Prunus cerasus* — dark necrotic lines and rings, shotholes, recovery.
*Prunus persica* — brown lines and rings, recovery.
*Rosa* (rose) — chlorotic lines and rings, oak leaf patterns, no recovery.
*Prunus* (plum) — chlorotic lines and rings only some strains.
*Prunus* (almond) — bright mosaic.
*Humulus* (hops) — no symptoms.
*Cucumis sativus* — chlorotic local lesions, systemic necrosis and severe stunting.

**Transmission**

Transmitted by means not involving a vector. Virus transmitted by mechanical inoculation; transmitted by grafting; not transmitted by contact between plants; transmitted by seed (to over 80% in *Prunus pennsylvanica* but much less in peach); transmitted by pollen to the seed and transmitted by pollen to the pollinated plant.

**Geographical distribution**

Probably distributed worldwide.

**Experimental host range**

Many (>9) families susceptible.

Diagnostically susceptible

*Cucumis sativus* — large yellow chlorotic local lesions in cotyledons, growing point stunted.
*Momordica balsamina* — necrotic local lesions.
*Cyamopsis tetragonoloba* — dark necrotic local lesions.
*Chenopodium quinoa* — systemic mottle.
*Prunus serrulata* — local necrosis and gumming at site of grafts.
*Lupinus albus* — systemic mottle.
*Helianthus annuus* — systemic chlorotic lines and rings.
*Cassia tora* — local and systemic ringspotting.

Diagnostically insusceptible

*Tithonia speciosa, Lagenaria siceraria, Nepeta cataria, Crotalaria spectabilis, Plantago virginica.*

Maintenance and propagation hosts

*Cucumis sativus, Catharanthus roseus.*

Assay hosts (**L**ocal lesions or **W**hole plants)

*Momordica balsamina* (L), *Cucumis sativus* (W), *Cyamopsis tetragonoloba* (L).

Dicotyledonous Hosts *(**Susceptible** or Insusceptible)*

***Antirrhinum majus. Cassia tora. Catharanthus roseus. Chenopodium album. Chenopodium quinoa.*** *Crotalaria spectabilis. Cucumis melo.* ***Cucumis sativus. Cucurbita maxima. Cucurbita moschata. Cucurbita pepo. Cyamopsis tetragonoloba. Gomphrena globosa. Gossypium hirsutum. Lactuca sativa. Lupinus albus. Melilotus alba. Momordica balsamina. Nicotiana benthamiana.*** *Nicotiana tabacum.* ***Petunia hybrida. Phaseolus vulgaris. Pisum sativum. Sesbania exaltata. Trifolium repens. Vigna unguiculata. Zinnia elegans.***

Additional hosts

Isolates may vary in host range in regard to certain species.

Sources of host-range data

Fulton (1957).

**Properties of particles in sap**

TIP: 55–62 °C. LIV: 0.4–0.75 days (6–18 hours). DEP: $\log_{10}$ minus 2–3 (in presence of antioxidant). Electron microscopy: fix in 1% glutaraldehyde.

**Purification**

Method: Fulton (1968).

**Particle morphology**

Virions isometric; not enveloped; 23, 25, and 27 nm in diameter; rounded in profile; without a conspicuous capsomere arrangement.

**Physical properties**

Three sedimenting components in purified preparations; sedimentation coefficient of the fastest 95 *S* (B); of the other(s) 72 *S* (T), or 90 *S* (M; the B and M components required for infection). Density 1.35 g $cm^{-3}$ in CsCl (unfixed). $A_{260}/A_{280}$ ratio 1.56 (B).

**Biochemical properties**

Virions contain 16 % nucleic acid; 84 % protein; 0 % lipid.

*Genome* consists of RNA; single–stranded; of three parts; largest (or only) genome part the largest 3.662 kb; the 2nd largest 2.507 kb; the 3rd largest 1.887 kb. Genomic nucleic acid isolated by Loesch and Fulton (1975). Base composition 27 % G; 25 % A; 21 % C; 27 % U. Additional factor required for infectivity.

*Sequence database accession code(s):*

L38823 Em(43)_vi:Pvnmopr Gb(89)_vi:Pvnmopr Prunus necrotic ringspot virus movement protein gene, complete cds and coat protein gene.

*Non–genomic nucleic acid* found in the virions; is subgenomic *m*RNA; for the coat protein. Sub–genomic *m*RNA found in infected cells.

$M_r$ 25000. Method of preparation: Gonsalves and Fulton (1977).

**Replication**

Replication does not depend on a helper virus.

**Taxonomy and Relationships**

*Ilarvirus: Bromoviridae* (subgroup 4).

Virus(es) with serologically related virions: virions of prunus necrotic ringspot virus share some serological determinants with those of apple mosaic. Virus(es) with serologically unrelated virions: tobacco streak, asparagus 2, citrus leaf rugose, citrus variegation, elm mottle, Tulare apple mosaic and prune dwarf viruses.

**References**

Barbara, D.J., Clark, M.F., Thresh, J.M. and Casper, R. (1979). *Ann. appl. Biol.* **90**: 395.
Cation, D. (1949). *Phytopathology* **39**: 37.
Cochran, L.C. and Hutchins, L.M. (1941). *Phytopathology* **31**B 860.
Cropley, R., Gilmer, R.M. and Posnette, A.F. (1964). *Ann. appl. Biol.* **53**: 325.
Fulton, R.W. (1957). *Phytopathology* **47**: 215.
Fulton, R.W. (1957). *Phytopathology* **47**: 683.
Fulton, R.W. (1958). *Virology* **6**: 499.
Fulton, R.W. (1968). *Phytopathology* **58**: 635.
Fulton, R.W. (1970). *CMI/AAB Descr. Pl. Viruses* No. 5, 4 pp.
George, J.A. and Davidson, T.R. (1963). *Can. J. Pl. Sci.* **43**: 276.
Gonsalves, D. and Fulton, R.W. (1977). *Virology* **81**: 398.
Loesch, L.S. and Fulton, R.W. (1975). *Virology* **68**: 71.
Moore, J.D., Boyle, J.S. and Keitt, G.W. (1948). *Science* **108**: 623.
Sastry, K.S. (1966). *Indian Phytopath.* **19**: 316.

# Prunus S (?) *carlavirus*

*Standard acronym:* PruVS.

First reported in *Prunus* sp.; from Japan; by Y. Doi (in Hull *et al.*, 1989).

**Taxonomy and Relationships**

Possible *Carlavirus*.

**References**

Hull, R., Brown, F. and Payne, C. (1989). *Directory and Dictionary of Animal, Bacterial and Plant Viruses*, p. 76. MacMillan Reference Books, London.

# Purple granadilla mosaic virus

**Data collated by E.W. Kitajima, 1987.**

First reported in *Passiflora edulis* cv. Sims.; from Cotia, SP, Brazil.

**Natural host range and symptoms**

Symptoms vary cyclically over a few weeks.
*Passiflora edulis* cv. Sims.

**Transmission**

Transmitted by a vector; an insect; *Diabrotica speciosa* Germ.; Coleoptera. Transmitted in a semi–persistent manner. Virus does not multiply in the vector; transmitted by mechanical inoculation; transmitted by grafting.

**Geographical distribution**

Found, but with no evidence of spread, in Brazil.

**Experimental host range**

Few (<3) families susceptible.

Diagnostically susceptible

*Passiflora serrato digitata*, *P. coerulea*, *P. maliformis*, *P. giberth*, *P. alata*, *P. edulis* f. flavicarpa.

Diagnostically insusceptible

*Gomphrena globosa*, *Chenopodium quinoa*, *Nicotiana tabacum*, *Datura stramonium*, *Phaseolus vulgaris* 'marteiga'.

Maintenance and propagation hosts

*Passiflora edulis* cv. Sims.

Assay hosts (**L**ocal lesions or **W**hole plants)

*Passiflora edulis* cv. Sims. (W).

Dicotyledonous Hosts *(**Susceptible** or Insusceptible)*

*Ageratum conyzoides. Beta vulgaris. Brassica campestris. Brassica oleracea* ssp. *botrytis. Brassica oleracea* ssp. *capitata. Capsicum annuum. Celosia argentea. Chenopodium amaranticolor. Chenopodium murale. Chenopodium quinoa. Chrysanthemum morifolium. Cichorium endiva. Citrullus lanatus. Citrus limon. Coriandrum sativum. Crotalaria spectabilis. Cucumis melo. Cucumis sativus. Cucurbita maxima. Cucurbita moschata. Dahlia pinnata. Datura stramonium. Daucus carota. Desmodium tortuosum. Dianthus caryophyllus. Emilia sagittata. Eucalyptus marginata. Euphorbia marginata. Glycine max. Gomphrena globosa. Gossypium hirsutum. Helianthus annuus. Hibiscus esculentus. Ipomoea setosa. Lablab purpureus. Lactuca sativa. Lycopersicon esculentum. Macroptilium lathyroides. Myosotis sylvatica. Nicandra physaloides. Nicotiana debneyi. Nicotiana glutinosa. Nicotiana rustica. Nicotiana tabacum. Passiflora edulis. Petroselinum crispum. Petunia hybrida. Phaseolus lunatus. Phaseolus vulgaris. Phlox drummondii. Physalis floridana. Pisum sativum. Raphanus sativus. Solanum melongena. Solanum tuberosum. Sonchus oleraceus. Tetragonia expansa. Vicia faba. Zinnia elegans.*

Monocotyledonous Hosts *(**Susceptible** or Insusceptible)*

*Allium cepa. Oryza sativa. Saccharum officinarum. Secale cereale. Triticum aestivum. Zea mays.*

Additional hosts

Additional Hosts: *Brassica rapa* (**I**); *Passiflora cerulea* (**S**); *Passiflora edulis* f-flavicarpa (**S**).

Sources of host-range data

Chagas *et al.* (1984); Oliveira (1986).

**Properties of particles in sap**

TIP: 60–65 °C. LIV: 7 days. DEP: $\log_{10}$ minus 5. Leaf sap contains many virions.

**Particle morphology**

Virions isometric; not enveloped; 30 nm in diameter; angular in profile; without a conspicuous capsomere arrangement.

**Physical properties**

One sedimenting component in purified preparations; sedimentation coefficient 129 *S*.

**Biochemical properties**

Virions contain 15 % nucleic acid.
*Genome* consists of RNA; single–stranded.
*Virion protein(s)* one; $M_r$ 35500; coat protein.

**Cytopathology**

Virions found in all parts of the host plant; in cytoplasm and in cell vacuoles. Inclusions absent from infected cells. Other cellular changes: filamentous material within the cristae of mitochondria.

**Taxonomy and Relationships**

Virus(es) with serologically unrelated virions: tymoviruses, comoviruses, cucumoviruses, sobemoviruses, ilarviruses, mimosa mosaic virus and maize rayado fino virus.

Best tests for diagnosis: serology, symptoms in *Passiflora* spp. and host range. (*N.B.* CMV infects species outside *Passifloraceae*). Passionfruit yellow mosaic tymovirus produces bright yellow mosaic in most *Passiflora* spp., but is not serologically related.

**References**

Chagas, C.M., Joazeiro, P.P., Kudamatsu, M. and Vega, J. (1984). *Fitopatol. Bras.* **9**: 241.

Oliveira, C.R.B. (1986). *M.Sc. Thesis*, Univ. Brasilia.

# Quail pea mosaic *comovirus*

**Data collated by K. Boswell, 1985.**

*Synonyms:* bean curly dwarf mosaic virus (Moore and Scott, 1981). *Standard acronym:* QPMV. *Strains:* type, bean curly dwarf mosaic.

First reported in *Strophostyles helvola*; from Arkansas, U.S.A; by Moore (1973).

**Natural host range and symptoms**

*Strophostyles helvola, Glycine max, Phaseolus vulgaris* — mosaic.

**Transmission**

Transmitted by a vector; an insect; *Ceratoma trifurcata.*; Coleoptera. Virus transmitted by mechanical inoculation; not transmitted by seed (type strain not transmitted through seed of *Phaseolus vulgaris* or *P. acutifolius*; Meiners *et al.*, 1977).

**Geographical distribution**

Spreads in Costa Rica and El Salvador (in *Phaseolus vulgaris*), the USA (in *Strophostyles helvola* in Arkansas).

**Experimental host range**

Many (>9) families susceptible.

Diagnostically susceptible

*Gomphrena globosa* — not infected by type strain; systemic chlorosis (bean curly dwarf strain).

*Ipomoea nil* cv.-Scarlet O'Hara — necrotic local lesions; not systemic.

*Vigna unguiculata* cv. Monarch — chlorotic local lesions but not systemic.

Maintenance and propagation hosts

*Phaseolus vulgaris* cvs Black Valentine, Cherokee Wax, *Glycine max* cv. Kanrich.

Assay hosts (Local lesions or Whole plants)

*Phaseolus vulgaris* cv. Pinto (L). *Glycine max* cv. Lee (W) for vector transmission.

Dicotyledonous Hosts (***Susceptible*** *or Insusceptible)*

*Arachis hypogaea*. ***Cajanus cajan***. ***Chenopodium quinoa***. *Cicer arietinum*. ***Glycine max***. ***Gomphrena globosa***, or *Gomphrena globosa*. ***Ipomoea nil***. *Lablab purpureus*, or ***Lablab purpureus***. ***Lens esculenta***. ***Macroptilium lathyroides***. *Macrotyloma uniflorum*. ***Phaseolus lunatus***, or *Phaseolus lunatus*. ***Phaseolus vulgaris***. ***Pisum sativum***. ***Sesbania exaltata***. *Tephrosia vogelii*. ***Vicia faba***. ***Vigna radiata***. ***Vigna unguiculata***, or *Vigna unguiculata*.

Sources of host-range data

Meiners *et al.* (1977).

**Properties of particles in sap**

TIP: 55–60 °C. LIV: 5–6 days. DEP: $\log_{10}$ minus 3–4.

**Purification**

Method: Moore and Scott (1981).

**Particle morphology**

Virions isometric; not enveloped; 30 nm in diameter.

**Physical properties**

Three sedimenting components in purified preparations; sedimentation coefficient of the fastest 112 *S* (B); of the other(s) 93 *S* (M), or 57 *S* (T).

**Biochemical properties**

*Genome* consists of RNA; single–stranded; of two parts.

*Virion protein(s)* two; $M_r$ of the larger 40000. $M_r$ of 2nd largest 22000.

**Taxonomy and Relationships**

*Comovirus: Comoviridae.*

Virus(es) with serologically related virions: members of each of the other four serogroups (Fulton and Scott, 1979).

The bean curly dwarf strain differs from the type strain in several respects. It is transmitted by *Diabrotica undecimpunctata, Epilachna varivestis*, infects *Gomphrena globosa* systemically, induces chlorotic local lesions in *Lablab purpureus*, and diffuse local lesions in *Phaseolus vulgaris* cvs Pinto, Top Crop. Its best local lesion host is *Lablab purpureus*.

**References**

Fulton, J.P. and Scott, H.A. (1979). *Phytopathology* **69**: 305.

Meiners, J.P., Waterworth, H.E., Lawson, R.H. and Smith, F.F. (1977). *Phytopathology* **67**: 163.

Moore, B.J. (1973). *Pl. Dis. Reptr* **57**: 311.

Moore, B.J. and Scott, H.A. (1981). *CMI/AAB Descr. Pl. Viruses* No. 238, 4 pp.

# Radish mosaic *comovirus*

**Data collated by R.N. Campbell, 1988.**

*Synonyms:* radish enation mosaic virus. *Standard acronym:* RaMV. *Strains:* type (North American), European (HZ), Japanese.

First reported in *Raphanus sativus*; from California, U.S.A.; by Tompkins (1939); Severin and Tompkins (1950); Campbell (1964).

**Natural host range and symptoms**

Symptoms persist.

*Brassica campestris* — line and ring patterns, mosaic, necrosis of veins and petioles, rosetting of leaves.

*B. napus* — mild mosaic, and leaf deformation.

*B. oleracea* — variable; from symptomless to severe systemic chlorosis and necrosis.

*Raphanus sativus* — mosaic, occasionally enations.

*Eruca sativa* — severe crinkle, chlorosis, stunting, some necrosis.

**Transmission**

Transmitted by a vector; an insect; *Phyllotreta* spp., *Epitrix hirtipennis* and *Diabrotica undecimpunctata*; Coleoptera. Virus transmitted by mechanical inoculation; transmitted by grafting; not transmitted by seed.

**Ecology and control**

Studies reported by Mamula and Juretic (1985); Koenig and Fischer (1981).

**Geographical distribution**

Spreads in Austria, Germany, Hungary, Italy, Japan, Morocco, the USA, the former USSR, and the former Yugoslavia.

**Experimental host range**

Several (3–9) families susceptible.

Diagnostically susceptible

*Brassica campestris* (tendergreen mustard) — necrotic local lesions; systemic mottling.

*B. campestris* (turnip) — chlorotic or necrotic local lesions; systemic mosaic or line patterns.

*Chenopodium amaranticolor* and other *Chenopodium* spp. — local lesions; not systemic.

*Nicotiana tabacum* cv. Turkish — necrotic rings; not systemic.

Diagnostically insusceptible

*Nicotiana glutinosa, N. tabacum* Havana 425.

Maintenance and propagation hosts

*Brassica campestris.*

Assay hosts (**L**ocal lesions or **W**hole plants)

*Brassica campestris* (W), *Chenopodium amaranticolor* (L), *C. quinoa* (L), *C. murale* (L), *Raphanus sativus* (W).

Dicotyledonous Hosts *(**Susceptible** or Insusceptible)*

***Beta vulgaris***, or *Beta vulgaris*. ***Brassica campestris***. ***Brassica campestris*** ssp. ***chinensis***. ***Brassica campestris*** ssp. ***napus***. ***Brassica campestris*** ssp. ***pekinensis***. ***Brassica campestris*** ssp. ***rapa***. ***Brassica juncea***. ***Brassica oleracea*** ssp. ***botrytis***. ***Brassica oleracea*** ssp. ***capitata***. ***Chenopodium album***. ***Chenopodium amaranticolor***. ***Chenopodium foetidum***. ***Chenopodium foliosum***. ***Chenopodium murale***. ***Chenopodium quinoa***. ***Cucumis sativus***, or *Cucumis sativus*. ***Cucurbita maxima***. *Cucurbita pepo*. ***Datura stramonium***, or *Datura stramonium*. ***Gomphrena globosa***. *Lactuca sativa*. ***Nicotiana clevelandii***. ***Nicotiana megalosiphon***. ***Nicotiana tabacum***, or *Nicotiana tabacum*. *Petunia hybrida*. *Phaseolus vulgaris*. ***Raphanus sativus***. ***Sinapis alba***. ***Spinacia oleracea***. *Tetragonia expansa*. *Vicia faba*. *Vigna radiata*. *Vigna unguiculata*.

Sources of host-range data

Campbell (1964); Stefanac and Mamula (1972).

**Properties of particles in sap**

TIP: 65–70 °C. LIV: 14–21 days. DEP: $\log_{10}$ minus 4. Leaf sap contains many virions.

**Purification**

Method: Stefanac and Mamula (1972).

**Particle morphology**

Virions isometric; not enveloped; about 30 nm in diameter (28 nm side-to-side, 33 nm corner-to-corner; Milne *et al.*, 1980); angular in profile; with a conspicuous capsomere arrangement.

**Physical properties**

Three sedimenting components in purified preparations; sedimentation coefficient of the fastest 128 *S* (B); of the other(s) 103 *S* (M), or 57 *S* (T). Isoelectric point pH 4.3 (fast), or 5.3 (slow). $A_{260}/A_{280}$ ratio 1.65 (M), or 1.78 (B).

**Biochemical properties**

Virions contain 35 % nucleic acid (B), or 26 % nucleic acid (M), or 0 % nucleic acid (T); 65 % protein (B), or 74 % protein (M), or 100 % protein (T); 0 % lipid (all components).

*Genome* consists of RNA; single–stranded; of two parts; largest (or only) genome part the larger 6.2 kb; the 2nd largest 3.6 kb. Genomic nucleic acid isolated by Fukunaga and Furusawa (1981). Base composition 22 % G (M), or 25 % G (B); 29 % A (M), or 30 % A (B); 19 % C (M), or 18 % C (B); 30 % U (M), or 27 % U (B).

*Virion protein(s)* two (plus one or more hydrolysis products, in some isolates); $M_r$ of the larger 40000. $M_r$ of 2nd largest 22000–24000. Method of preparation: Koenig and Fischer (1981).

**Replication**

Replication does not depend on a helper virus.

**Cytopathology**

Virions found in all parts of the host plant; in cytoplasm and in cell vacuoles; in membrane bound layers considered to be tonoplastic channels (Honda and Matsui, 1972). Inclusions present in infected cells; are unusual in shape; vesiculated inclusion bodies (Milne *et al.*, 1980; Stefanac and Ljubesic, 1971); they contain virions. Other cellular changes: enations on chronically infected radish (Chinese white winter types).

**Taxonomy and Relationships**

*Comovirus: Comoviridae.*

Virus(es) with serologically related virions: cowpea severe mosaic, bean pod mottle, squash mosaic and broad bean stain viruses.

Radish mosaic virus has 2 serological subgroups: i). isolates from California and Japan; ii). European isolates including the HZ (Yugoslavian strain). Best tests for diagnosis: serological testing by gel diffusion is most reliable.

**Comments**

An aphid-borne virus designated radish mosaic virus in India (Naqvi *et al.*, 1986; Ahlawat and Chenulu, 1982) differs from radish mosaic comovirus.

**References**

Ahlawat, Y.S. and Chenulu, V.V. (1982). *Indian Phytopath.* **35**: 633.
Campbell, R.N. (1964). *Phytopathology* **54**: 1418.
Campbell, R.N. (1973). *CMI/AAB Descr. Pl. Viruses* No. 121, 4 pp.
Fukunaga, K. and Furusawa, I. (1981). *Ann. Phytopath. Soc. Japan* **47**: 534.
Honda, Y. and Matsui, C. (1972). *Phytopathology* **62**: 448.
Koenig, R. and Fischer, H.U. (1981). *Plant Dis.* **65**: 758.
Mamula, D. and Juretič, N. (1985). *Phyton (Houn, Austria)* **25**: 241.
Milne, R.G., Masenga, V., Lenzi, R. and Lovisolo, O. (1980). *Phytopathol. Medit.* **19**: 145.
Naqvi, Q.A., Khan, J.A. and Mahmood, K. (1986). *J. Indian Bot. Soc.* **65**: 136.
Severin, H.H.P. and Tompkins, C.M. (1950). *Hilgardia* **20**: 191.
Stefanac, Z. and Ljubesic, N. (1971). *J. gen. Virol.* **13**: 51.
Stefanac, Z. and Mamula, D. (1972). *Ann appl. Biol.* **69**: 229.
Tompkins, C.M. (1939). *J. Agric. Res.* **58**: 119.

# Radish yellow edge *alphacryptovirus*

**Data collated by T. Natsuaki, 1987.**

*Standard acronym:* RYEV.

First reported in *Raphanus sativus*; from Japan; by Natsuaki *et al.* (1979).

**Natural host range and symptoms**

Symptoms disappear soon after infection.
*Raphanus sativus* — usually no symptoms, but rarely with yellowed leaf margins.

**Transmission**

Virus not transmitted by mechanical inoculation; not transmitted by grafting; not transmitted by contact between plants; transmitted by seed; transmitted by pollen to the seed.

**Ecology and control**

Studies reported by Natsuaki *et al.* (1979, 1983).

**Geographical distribution**

Found, but with no evidence of spread, in Australia, China, Italy, the U.K. and the U.S.A.

**Experimental host range**

Few (<3) families susceptible.

Diagnostically insusceptible

*Chenopodium amaranticolor, Nicotiana tabacum, Cucumis sativus, Vigna sesquipedalis* and *Brassica campestris* ssp. *rapa*.

Maintenance and propagation hosts

*Raphanus sativus.*

Dicotyledonous Hosts *(**Susceptible** or Insusceptible)*

***Raphanus sativus*.**

Sources of host-range data

Natsuaki *et al.* (1979); Natsuaki (1985).

**Properties of particles in sap**

Leaf sap contains few virions.

**Purification**

Method: Natsuaki (1983).

**Particle morphology**

Virions isometric; not enveloped; 30 nm in diameter; rounded in profile; without a conspicuous capsomere arrangement.

**Physical properties**

One sedimenting component in purified preparations; sedimentation coefficient 118 *S*. Density 1.37 g $cm^{-3}$ in CsCl.

**Biochemical properties**

*Genome* consists of RNA; double–stranded; linear; of three parts; largest (or only) genome part the largest 1.3 kb; the 2nd largest 1.25 kb; the 3rd largest 1.21 kb. Genomic nucleic acid isolated by Natsuaki *et al.* (1983).

*Virion protein(s)* two; $M_r$ of the larger 63000. $M_r$ of 2nd largest 61000. Method of preparation: Natsuaki *et al.* (1983).

**Cytopathology**

Virions found in all parts of the host plant; in cytoplasm. Inclusions present in infected cells; are viroplasms (but are rarely observed); they contain virions.

**Taxonomy and Relationships**

*Alphacryptovirus: Partitiviridae.*

Virus(es) with serologically unrelated virions: all other alphacryptovirus species tested.

Best tests for diagnosis: This virus is readily distinguished from other viruses infecting crucifers by its possession of dsRNA. ISEM and extracting dsRNA directly from leaves are used for diagnosis.

**References**

Natsuaki, T., Yamashita, S., Doi, Y. and Yora, K. (1979). *Ann. Phytopath. Soc. Japan* **45**: 313.

Natsuaki, T., Okuda, S., Teranaka, M., Yamashita, S. and Doi, Y. (1983). *Ann. Phytopath. Soc. Japan* **49**: 593.

# Ranunculus mottle *potyvirus*

**Data collated by P. Jeyanandarajah, 1991.**

*Synonyms:* ranunculus mosaic virus (Raabe and Gold, 1957). *Standard acronym:* RanMV.

First reported in *Ranunculus asiaticus*; from the U.S.A.; by Laird and Dickson (1965).

**Natural host range and symptoms**

Symptoms persist.

*Ranunculus asiaticus* — foliar mottling and distortion.

**Transmission**

Transmitted by a vector; an insect; *Myzus persicae*; Aphididae. Transmitted in a non–persistent manner. Virus transmitted by mechanical inoculation; not transmitted by seed.

**Geographical distribution**

Spreads in the USA.

**Experimental host range**

Few (<3) families susceptible.

Diagnostically susceptible

*Consolidia ambigua*—conspicuous foliar mosaic.

Diagnostically insusceptible

*Anemone coronaria, Aquilegia caerulea, Clematis occidentalis, Thalictrum dipterocarpum.*

Dicotyledonous Hosts *(**Susceptible** or Insusceptible)*

*Brassica campestris* ssp. *rapa. Capsicum annuum. Capsicum frutescens. Cucurbita pepo. Lycopersicon esculentum. Matthiola incana. Nicotiana benthamiana. Nicotiana* × *edwardsonii. Petunia hybrida. Phaseolus vulgaris. Pisum sativum. Solanum melongena.*

**Particle morphology**

Virions filamentous; usually flexuous; of 707 nm.

**Cytopathology**

Virions found in leaves. Inclusions present in infected cells; are pinwheels.

**Taxonomy and Relationships**

*Potyvirus: Potyviridae.*

Virus(es) with serologically related virions: watermelon mosaic 2 virus. Virus(es) with serologically unrelated virions: bean yellow mosaic, bidens mottle, blackeye cowpea mosaic, clover yellow vein, dasheen mosaic, papaya ringspot, peanut mottle, peanut stripe, pepper mottle, potato Y, tobacco etch, turnip mosaic and zucchini yellow mosaic viruses.

**References**

Elliott, M.S., Zettler, F.W., Gallegati, M. and Ko, N.J. (1988). *Acta Hort.* **234**: 39.
Laird, E.F. and Dickson, R.C. (1965). *Pl. Dis. Reptr* **49**: 449.
Laird, E.F. and Dickson, R.C. (1967). *J. econ. Ent.* **60**: 10.
Raabe, R.D. and Gold, A.H. (1957). *Phytopathology* **47**: 28.

# Ranunculus repens symptomless (?) rhabdovirus

**Data collated by A.A. Brunt, 1991.**

*Standard acronym:* RamRSV.

First reported in *Ranunculus repens*; from Italy; by Amici *et al.* (1978).

**Natural host range and symptoms**

*Ranunculus repens* — symptomless infection.

**Particle morphology**

Virions rhabdo– or bullet–shaped.

**Taxonomy and Relationships**

Possible unassigned rhabdovirus.

**References**

Amici, A., Faoro, F. and Tornahgi, R. (1978). *Riv. Patol. Veg.* **14**: 85.

# Raphanus (?) *cytorhabdovirus*

**Data collated by A.A. Brunt, 1991.**

*Standard acronym:* RaV.

First reported in *Raphanus* sp.; from Brazil; by Kitajima and Costa (1979).

**Natural host range and symptoms**

*Raphanus* sp. — no conspicuous symptoms.

**Transmission**

Virus transmitted by mechanical inoculation.

**Geographical distribution**

Spreads in Brazil.

**Particle morphology**

Virions rhabdo– or bullet–shaped; enveloped; with a clear modal length; of 250–300 nm (in sections); 50–70 nm wide (in sections).

**Cytopathology**

Virions found in all parts of the host plant; in cytoplasm.

**Taxonomy and Relationships**

Possible *Cytorhabdovirus: Rhabdoviridae.*

**References**

Kitajima, E.W. and Costa, A.S. (1979). *Fitopatol. Bras.* **4**: 55.

# Raspberry bushy dwarf *idaeovirus*

**Data collated by C. Büchen-Osmond and A.F. Murant, 1987.**

*Synonyms:* loganberry degeneration virus (Barnett and Murant, 1970; Legg, 1960), raspberry yellows virus (Cadman, 1952; Jones *et al.*, 1982), possibly raspberry line-pattern virus (Basak, 1971; Jones and Murant, 1972). *Standard acronym:* RBDV.

First reported in *Rubus idaeus*; from Scotland; by Cadman (1961).

**Natural host range and symptoms**

Symptoms none.

*Rubus idaeus* (raspberry) cvs Lloyd George, Canby, Carnival, Creston, Newburgh and Norfolk Giant — symptomless, but may have high proportion of aborted drupelets.

*R. loganbaccus* (loganberry), *R. ursinus* (boysenberry), *R. occidentalis* (black raspberry) cvs Munger, New Logan and Plum Farmer — symptoms uncertain. This is desirable because, although the virus causes mild symptoms, it may affect fruit quality, especially in mixed infections.

**Transmission**

Transmitted by means not involving a vector (none of the insects common on raspberry in Scotland transmitted the virus (Cadman, 1970)). Virus transmitted by mechanical inoculation; transmitted by grafting; transmitted by seed (up to 77%; Cadman, 1965; Converse, 1973; Murant *et al.*, 1974); possibly transmitted by pollen to the pollinated plant, or transmitted by pollen to the seed.

**Ecology and control**

Studies reported by Murant *et al.* (1974) and Converse (1973). The virus may be eliminated from stocks of *Rubus idaeus* (Murant *et al.*, 1974) and from *R. occidentalis* (Converse, 1973) by heat therapy. Immune cultivars are not infected via pollen.

**Geographical distribution**

Probably distributed worldwide (and probably occurs wherever susceptible *Rubus* cultivars are grown). Spreads in the Eurasian region and the North American region; Australia, New Zealand, and the former USSR.

**Experimental host range**

Many (>9) families susceptible.

Diagnostically susceptible

*Chenopodium amaranticolor* — transient chlorotic local lesions, systemic chlorotic rings and line patterns.

*Chenopodium murale* — sunken necrotic rings; not systemic.

*Chenopodium quinoa* — transient local lesions, systemic chlorotic spots and then mosaic and ring patterns.

*Phaseolus vulgaris* cv. The Prince — small brown local lesions in winter; no systemic infection.

*Nicotiana clevelandii* — infected systemically without symptoms.

Maintenance and propagation hosts

*Chenopodium quinoa.*

Assay hosts (**L**ocal lesions or **W**hole plants)

*Phaseolus vulgaris* cv. The Prince (L), *Chenopodium murale* (L).

Dicotyledonous Hosts *(**Susceptible** or Insusceptible)*

***Chenopodium amaranticolor*. *Chenopodium murale*. *Chenopodium quinoa*. *Fragaria vesca*. *Nicotiana clevelandii*. *Phaseolus vulgaris*. *Rubus idaeus*. *Rubus occidentalis*.**

Sources of host-range data

Barnett and Murant, 1970; 1971.

**Properties of particles in sap**

TIP: 65 °C. LIV: 4 days (at 20°C). DEP: $\log_{10}$ minus 4 (an isolate from *Rubus occidentalis* had a dilution end-point of only $10^{-2}$ in *Chenopodium quinoa* sap and lost infectivity after 2–3 hours at room temperature; Jones and Murant, 1972*a*; Murant and Jones, 1976). Electron microscopy: virions disrupt in PTA but not in UA or UF (Barnett and Murant, 1970).

**Purification**

Method: Murant (1976).

**Particle morphology**

Virions isometric; not enveloped; 33 nm in diameter; angular in profile; without a conspicuous capsomere arrangement.

**Physical properties**

One sedimenting component in purified preparations, or three sedimenting components in purified preparations (in a rather broad zone); sedimentation coefficient 115 *S*. Density 1.37 g $cm^{-3}$ in CsCl (fixed in formaldehyde).

**Biochemical properties**

Virions contain 24 % nucleic acid; 76 % protein.

*Genome* consists of RNA; single–stranded; linear; of three parts; largest (or only) genome part the largest 5.4 kb; the 2nd largest 2.2 kb; the 3rd largest 1 kb. Poly A region absent.

*Sequence database accession code(s):*

D01052 Em(40)_vi:RUDMRCP Gb(84)_vi:RUDMRCP Raspberry bushy dwarf virus (RBDV) mRNA for coat protein. 11/92 946bp.

S51557 Em(40)_un:S51557 Gb(84)_vi:S51557 RNA-1 segment: non-structural polyprotein raspberry bushy dwarf virus RBDV, R15, Genomic RNA 2 sequences.

*Features of the genome:* RNA-1 has one ORF, that encodes a protein of $M_r$ 188000, and is probably a polymerase. RNA-2 has 2 ORFs, one encodes a protein of $M_r$ 39000, that is probably a movement protein, the other the coat protein.

*Non–genomic nucleic acid* found in the virions; is subgenomic *m*RNA (RNA-3).

*Virion protein(s)* one (probably); $M_r$ 29000; coat protein. Method of preparation: Murant (1975).

**Replication**

Replication does not depend on a helper virus.

**Taxonomy and Relationships**

*Idaeovirus* (the type, and only, species of the genus).

Virus(es) with serologically unrelated virions: twenty-four isometric viruses, including seven ilarviruses (Barnett and Murant, 1970).

Earlier reports (Cadman, 1963; 1970) that raspberry bushy dwarf virus is related to apple chlorotic leaf spot trichovirus have not been confirmed (Murant, 1976). Best tests for diagnosis: Raspberry bushy dwarf virus (RBDV) is one of several viruses isolated from Lloyd George raspberry with 'bushy dwarf' (Cadman and Harris, 1951) or 'symptomless decline' (Cadman, 1952). Plants infected with RBDV alone produce few, stunted canes and are late fruiting (Barnett and Murant, 1970), but so do plants infected with mixtures of RBDV and aphid-borne black raspberry necrosis virus (Jones, 1976). RBDV is readily distinguished from most nepoviruses from *Rubus*, except strawberry latent ringspot virus, by the symptoms it induces in *Chenopodium quinoa*; and their virions differ. The symptoms caused in *C. quinoa* also distinguish it from black raspberry necrosis virus (Jones and Murant, 1972*b*); which is transmitted by sap inoculation with difficulty, has isometric virions, but occurs in very small concentrations in sap. The black raspberry latent strain of tobacco streak virus (Lister and Converse, 1972) has virions like those of RBDV and is transmitted by pollen but causes severe systemic necrosis in *C. quinoa* and *Phaseolus vulgaris*.

**References**


Barnett, O.W. and Murant, A.F. (1970). *Ann. appl. Biol.* **65**: 435.

Basak, W. (1971). *Bull. Acad. Polmarse Sci., Serie sci. biol. C.I.V.* **19**: 681.

Cadman, C.H. (1952). *Ann. appl. Biol.* **39**: 495.

Cadman, C.H. (1961). *Hort. Res.* **1**: 47.

Cadman, C.H. (1963). *Pl. Dis. Reptr* **47**: 459.

Cadman, C.H. (1965). *Pl. Dis. Reptr* **49**: 230.

Cadman, C.H. (1970). In: *Virus Diseases of Small Fruit and Grapevines*, p. 149. University of California Press, Berkeley.

Cadman, C.H. and Harris, R.V. (1951). *Rep. E. Malling Res. Stn.* 1950, p.127.

Converse, R.H. (1973). *Phytopathology* **63**: 780.

Jones, A.T. (1976). *Rep. Scottish Hort. Res. Inst.* 1975, p. 73.
Jones, A.T. and Murant, A.F. (1972*a*). *Rep. Scottish Hort. Res. Inst.* 1971, p. 60.
Jones, A.T. and Murant, A.F. (1972*b*). *Pl. Path.* **21**: 166.
Jones, A.T., Murant, A.F., Jennings, D.L. and Wood, G.A. (1982). *Ann. appl. Biol.* **100**: 135.
Legg, J.T. (1960). *Rep. E. Malling Res. Stn.* 1959, 102.
Lister, R.M. and Converse, R.H. (1972). *CMI/AAB Descr. Pl. Viruses* No.166, 4 pp.
Mayo, M.A., Jolly, C.A., Muran, A.F. and Raschke, J.H. (1991). *J. gen. Virol.* **72**: 469.
Murant, A.F. (1976). *CMI/AAB Descr. Pl. Viruses* No. 165, 4 pp.
Murant, A.F. (1987). In: *Virus Diseases of Small Fruits*, U.S. Dep. Agric Hdbk No. 631, p. 229.
Murant, A.F., Chambers, J. and Jones, A.T. (1974). *Ann. appl. Biol.* **77**: 271.
Murant, A.F. and Jones, A.T. (1976). *Acta Hort.* **66**: 47.
Murant, A.F., Mayo, M.A. and Raschke, J.H. (1986). *Acta Hort.* **186**: 23.
Natsuaki, T., Mayo, M.A., Jolly, C.A. and Murant, A.F. (1991). *J. gen. Virol.* **72**: 2183.
Ziegler, A., Natsuaki, T., Mayo, M.A., Jolly, C.A. and Murant, A.F. (1992). *J. gen. Virol.* **73**: 3213.

## Raspberry leaf curl (?) *luteovirus*

**Data collated by R. Stace-Smith, 1991.**

*Standard acronym:* RLCV. *Strains:* alpha, beta.

First reported in *Rubus idaeus*; from Ontario, Canada and New York, U.S.A.; by Rankin and Hockey (1922).

### Natural host range and symptoms

Symptoms persist. Symptoms leaf curling, shoot proliferation, stunting, fruits are small and crumbly.

*Rubus idaeus, R. occidentalis, R. fruticosus, R. x neglectus* — leaf curling, proliferation of shoots and stunting.

### Transmission

Transmitted by a vector; an insect; *Aphis rubicola*; Aphididae. Not transmitted by *Amphorophora agathonica*. Transmitted in a persistent manner. Virus retained when the vector moults; not transmitted congenitally to the progeny of the vector; not transmitted by mechanical inoculation; transmitted by grafting; not transmitted by contact between plants; not transmitted by seed; not transmitted by pollen.

### Ecology and control

Studies reported by Stace-Smith (1962).

### Geographical distribution

Spreads in the North American region; Canada and the USA.

### Experimental host range

Few (<3) families susceptible. Experimentally infected plants mostly show downcurling of leaves, yellowing and stunting.

Diagnostically susceptible

*Rubus idaeus, R. albescens, R. henryii, Fragaria vesca* — downcurling of leaves and leaf yellowing.

Diagnostically insusceptible

*Chenopodium quinoa, Nicotiana tabacum, Cucumis sativus, Phaseolus vulgaris, Vigna unguiculata* syn. *sinensis.*

Maintenance and propagation hosts

*Rubus idaeus.*

Assay hosts (**L**ocal lesions or **W**hole plants)

*Rubus idaeus* (W), *R. henryii* (W), *R. phoenicolasius* (W).

Dicotyledonous Hosts *(**Susceptible** or Insusceptible)*

***Fragaria vesca. Rubus idaeus.***

**Taxonomy and Relationships**

Possible *Luteovirus.*

**References**

Rankin, W.H. and Hockey, J.F. (1922). *Phytopathology* **12**: 253.
Stace-Smith, R. (1962). *Can. J. Bot.* **40**: 651.

## Raspberry ringspot *nepovirus*

**Data collated by A.A. Brunt, 1992.**

*Synonyms:* raspberry Scottish leaf curl virus, red currant ringspot virus. *Standard acronym:* RpRSV. *Strains:* type strain, Lloyd George yellow blotch strain, English strain.

First reported in *Rubus idaeus*; from Scotland; by Cadman (1956).

**Natural host range and symptoms**

*Rubus idaeus* cv. Norfolk Giant — leaf curling, dieback and death (cv. Lloyd George is immune), but less conspicuous symptoms in other cultivars.

*Fragaria vesca* —dwarfing and eventually death.

*Narcissus pseudonarcissus* — symptomless infection.

The virus may infect species of wild and cultivated dicotyledonous and monocotyledonous plants.

**Transmission**

Transmitted by a vector; a nematode; *Longidorus elongatus, L. macrosoma*; Dorylamidae. Virus transmitted by mechanical inoculation; transmitted by seed; transmitted by pollen to the seed.

**Ecology and control**

Studies reported by Cadman (1956); Murant *et al.* (1968); Kleinhempel (1970); Murant (1970); Bovey *et al.* (1972); Jennings (1964).

**Geographical distribution**

Spreads in the Eurasian region and the Mediterranean region; Austria, Belgium, Bulgaria, the former Czechoslovakia, Finland, France, Germany, Greece, Hungary, Ireland, Luxembourg, the Netherlands, Norway, Poland, Spain, Switzerland, Turkey, the UK, the USA, the former USSR, the former Yugoslavia. Found, but with no evidence of spread, in Denmark.

**Experimental host range**

Many (>9) families susceptible.

Diagnostically susceptible

*Chenopodium amaranticolor* — chlorotic or necrotic local lesions; no systemic infection.

*C. quinoa* — chlorotic or necrotic local lesions; systemic mottling and apical necrosis.

*Nicotiana rustica* — local lesions or rings; systemic ring and line patterns but systemic infection later often symptomless.

*Phaseolus vulgaris* — necrotic local lesions (in winter) and chlorotic local lesions (in summer); systemic mottling, distortion and necrosis.

*Petunia hybrida* — local chlorotic lesions; systemic chlorosis, line patterns and/or necrosis.

Maintenance and propagation hosts

*Nicotiana clevelandii, N. rustica, Petunia hybrida.*

Assay hosts (**L**ocal lesions or **W**hole plants)

*Chenopodium amaranticolor* (L), *C. quinoa* (L/W), *Spinacia oleracea* (W, especially useful as a bait plant for vector transmission tests).

Dicotyledonous Hosts *(**Susceptible** or Insusceptible)*

***Beta vulgaris*. *Chenopodium amaranticolor*. *Cucumis sativus*. *Datura stramonium*. *Lycopersicon esculentum*. *Nicotiana rustica*. *Nicotiana tabacum*. *Petunia hybrida*. *Phaseolus vulgaris*. *Spinacia oleracea*. *Vigna unguiculata*.**

**Properties of particles in sap**

TIP: 70 °C. LIV: 21 days. DEP: $\log_{10}$ minus 4.

**Particle morphology**

Virions isometric; 30 nm in diameter; angular in profile.

**Physical properties**

Three sedimenting components in purified preparations; sedimentation coefficient 125 *S* (B); of the other(s) 91 *S* (M), or 50 *S* (T). $A_{260}/A_{280}$ ratio 1.78 (B), or 1.62 (M).

**Biochemical properties**

Virions contain 43 % nucleic acid (B), or 29 % nucleic acid (M), or 0 % nucleic acid (T); 57 % protein (B), or 71 % protein (M), or 100 % protein (T).

*Genome* consists of RNA; single–stranded; of two parts; largest (or only) genome part 8.485 kb; the 2nd largest 4.242 kb. Nucleotide sequence references: Mayo *et al.* (1992).

*Sequence database accession code(s):*

S46011 Em(40)_un:S46011 Gb(84)_vi:S46011 polyprotein raspberry ringspot virus RRV, Genomic RNA Complete, 3928 nt. 2/93 3,928bp. 1 sequence.

*Virion protein(s)* one; $M_r$ 54000; coat protein.

**Cytopathology**

Virions found in leaves and mesophyll; in cytoplasm.

**Taxonomy and Relationships**

*Nepovirus: Comoviridae.*

Virus(es) with serologically unrelated virions: arabis mosaic, strawberry latent ringspot, sweet potato ringspot, tobacco ringspot, tomato black ring and tomato ringspot viruses.

Many variants occur. Cultivars differ in susceptibility; Norfolk Giant is very susceptible to the type strain, but Lloyd George is immune to it.

**References**

Bovey, R. *et al.* (1972). In: *La Défense des Plantes Cultivées* 6th Edition. Payot, Switzerland.

Cadman, C.H. (1956). *J. Hort. Sci.* **31**: 111.

Cadman, C.H. (1960). *Virology* **11**: 653.

Cropley, R. (1961). *Ann. appl. Biol.* **49**: 530.

Hanada, K. and Harrison, B.D. (1977). *Ann. appl. Biol.* **85**: 79.

Harrison, B.D. (1958). *Ann. appl. Biol.* **46**: 571.

Harrison, B.D. (1961). *Tijdschr. PlZiekt.* **67**: 562.

Harrison, B.D. (1964). *Virology* **22**: 544.

Kleinhempel, H. (1970). *Archiv. Gartenbau.* **18**: 267.

Lister, R.M. and Murant, A.F. (1967). *Ann. appl. Biol.* **59**: 49.

Maat, D.Z., van der Meer, F.A. and Pfaeltzer (1962). *Tijdschr. PlZiekt.* **68**: 120.

Mayo, M.A., Acosta, O., Blok, V.C., Wardell, I.M., Manoukian, A., Jolly, C.A. and Robinson, D.J. (1992). *Rep. Scottish Crop Res. Inst.* 1991, p. 84.

Murant, A.F. (1970). In: *Virus Diseases of Small Fruits and Grapevines*, pp. 132–148; ed. N.W. Frazier, University of California, Berkeley, California, U.S.A.

Murant, A.F. (1978). *CMI/AAB Descr. Pl. Viruses* No. 198, 5 pp.

Murant, A.F. (1981). In: *Handbook of Plant Virus Infections: Comparative Diagnosis*, pp. 197–238; ed. E. Kurstak, Elsevier/North Holland Press, Amsterdam.

Murant, A.F. and Taylor, M. (1978). *J. gen. Virol.* **41**: 53.

Murant, A.F. and Lister, R.M. (1967). *Ann. appl. Biol.* **59**: 63.

Murant, A.F., Taylor, C.E. and Chambers, J. (1968). *Ann. appl. Biol.* **61**: 175.

# Raspberry vein chlorosis (?) *nucleorhabdovirus*

**Data collated by R. Stace-Smith, 1987.**

*Synonyms:* raspberry chlorotic virus. *Standard acronym:* RVCV.

First reported in *Rubus idaeus*; from Scotland, U.K.; by Cadman (1952).

**Natural host range and symptoms**

Symptoms vary seasonally.

*Rubus idaeus.*

**Transmission**

Transmitted by a vector; an insect; *Aphis idaei*; Aphididae. Transmitted in a persistent manner. Virus not transmitted by mechanical inoculation; transmitted by grafting; not transmitted by contact between plants; not transmitted by seed; not transmitted by pollen.

**Ecology and control**

Studies reported by Cadman and Stace-Smith (1970); Jones *et al.* (1987).

**Geographical distribution**

Probably distributed worldwide (wherever *Rubus* is grown). Spreads in Canada and New Zealand. Found, but with no evidence of spread, in Canada and New Zealand.

**Experimental host range**

Few (<3) families susceptible.

Diagnostically susceptible

*Fragaria vesca.*

Maintenance and propagation hosts

*Rubus idaeus.*

Assay hosts (**L**ocal lesions or **W**hole plants)

*Rubus idaeus* (W), *Fragaria vesca* (W).

Dicotyledonous Hosts *(**Susceptible** or Insusceptible)*

***Fragaria vesca. Rubus idaeus.*** *Rubus occidentalis.*

Sources of host-range data

Jones *et al.* (1977).

**Properties of particles in sap**

Leaf sap contains few virions.

**Particle morphology**

Virions rhabdo– or bullet–shaped; with a clear modal length; of 500 nm; 70–80 nm wide. Axial canal obscure. Basic helix obvious; pitch of basic helix 4–5 nm.

**Cytopathology**

Virions found in vascular parenchyma; in cytoplasm and in the perinuclear space. Inclusions absent from infected cells.

**Taxonomy and Relationships**

Possible *Nucleorhabdovirus: Rhabdoviridae.*

Best tests for diagnosis: transmission by aphid vector to virus-free raspberry clones.

**References**


Cadman, C.H. (1952). *Ann. appl. Biol.* **39**: 61.

Cadman, C.H. and Stace-Smith, R. (1970). In: *Virus Diseases of Small Fruits and Grapevines* p. 119; ed. N.W. Frazier. University California, Berkeley.

Jones, A.T., Murant, A.F. and Stace-Smith, R. (1987). In: *Virus Diseases of Small Fruits* p. 194; ed. R.H. Converse. USDA Agric. Handbook 631.

Jones, A.T., Murant, A.F. and Stace-Smith, R. (1977). *CMI/AAB Descr. Pl. Viruses* No. 174, 4 pp.


# Red clover 2 *betacryptovirus*

**Data collated by A.A. Brunt, 1995.**

*Synonyms:* clover red cryptic 2 virus. *Standard acronym:* RCCV-2.

First reported in *Trifolium pratense*; from Italy; by Boccardo *et al.* (1983).

**Natural host range and symptoms**

*Trifolium pratense* — symptomless.

**Transmission**

Virus not transmitted by mechanical inoculation; not transmitted by grafting; transmitted by seed.

**Purification**

Method: Boccardo *et al.* (1983).

**Particle morphology**

Virions isometric; 38 nm in diameter.

**Biochemical properties**

*Genome* consists of RNA; double–stranded.

**Taxonomy and Relationships**

*Betacryptovirus: Partitiviridae.*

**References**

Boccardo, G., Lisa, V. and Milne, R.G. (1983). In: *Double-stranded RNA Viruses* p. 425; eds R.W. Compans and D.H.L. Bishop. Elsevier, New York.

Boccardo, G., Lisa, V., Luisoni, E. and Milne, R.G. (1987). *Adv. Virus Res.* **32**: 171.

# Red clover mosaic (?) *nucleorhabdovirus*

**Data collated by A.A. Brunt, 1991.**

*Synonyms:* clover red mosaic virus. *Standard acronym:* RClMV.

First reported in *Trifolium incarnatum*; from Spain; by Vela and Rubio-Huertos, 1974.

**Natural host range and symptoms**

*Trifolium incarnatum* — leaf mosaic.

**Geographical distribution**

Spreads in Spain.

Dicotyledonous Hosts *(**Susceptible** or Insusceptible)*

***Trifolium incarnatum*.**

**Particle morphology**

Virions rhabdo– or bullet–shaped; enveloped; of 300 nm; 65 nm wide.

**Cytopathology**

Virions found in leaves; in nuclei.

**Taxonomy and Relationships**

Possible *Nucleorhabdovirus: Rhabdoviridae.*

**References**

Vela, A. and Rubio-Huertos, M. (1974). *Phytopath. Z.* **79**: 343.

# Red clover mottle *comovirus*

**Data collated by K. Marcinka, 1980. Revised 1983.**

*Synonyms:* clover red mottle virus. *Standard acronym:* RCMV.

First reported in *Trifolium pratense*; from U.K; by Sinha (1960).

**Natural host range and symptoms**

Symptoms persist.
*Trifolium pratense* — mosaic, stunting.

**Transmission**

Transmitted by a vector; an insect; *Apion africans, A. varipes*; Coleoptera. Virus transmitted by mechanical inoculation.

**Geographical distribution**

Spreads in the Eurasian region.

**Experimental host range**

Several (3–9) families susceptible.

Diagnostically susceptible

*Gomphrena globosa* and *Vicia faba* — necrotic local lesions.
*Chenopodium amaranticolor* and
*C. quinoa* — chlorotic local lesions becoming necrotic.
*Glycine max* — systemic mosaic, dwarfing.
*Lathyrus odoratus* — latent infection.
*Phaseolus vulgaris* — necrotic local lesions (14/82 cultivars).
*Pisum sativum* — chlorosis, stunting.
*Trifolium pratense* — mosaic, dwarfing.
*Vicia sativa* — necrotic and chlorotic local lesions, leaf abscission.

Diagnostically insusceptible

*Cucumis sativus*, all *Nicotiana* spp., *Datura stramonium, Nicandra physalodes, Petunia hybrida.*

Maintenance and propagation hosts

*Phaseolus vulgaris, Pisum sativum, Vicia faba, Vigna unguiculata, Trifolium pratense, T. hybridum.*

Assay hosts (**L**ocal lesions or **W**hole plants)

*Gomphrena globosa (L), Chenopodium amaranticolor (L), C. quinoa (L), Phaseolus vulgaris (L), Trifolium incarnatum(W).*

Dicotyledonous Hosts *(**Susceptible** or Insusceptible)*

*Brassica campestris* ssp. *rapa*. ***Chenopodium amaranticolor***. ***Chenopodium quinoa***. *Cucumis sativus*. *Datura stramonium*. ***Glycine max***. ***Gomphrena globosa***. ***Lathyrus odoratus***. ***Medicago sativa***. ***Melilotus alba***. *Nicandra physaloides*. *Nicotiana clevelandii*. *Nicotiana glutinosa*. *Nicotiana tabacum*. *Petunia hybrida*. ***Phaseolus vulgaris***. ***Pisum sativum***. *Trifolium incarnatum*. ***Trifolium pratense***. ***Trifolium repens***. ***Trifolium subterraneum***. ***Vicia faba***. ***Vicia sativa***. ***Vigna unguiculata***.

Sources of host-range data

Hampton *et al.* (1978).

**Properties of particles in sap**

TIP: 55–60 °C. LIV: 14 days. DEP: $\log_{10}$ minus 4. Infectivity of sap decreased by treatment with di–ethyl ether. Leaf sap contains many virions.

**Purification**

Method: Marcinka (1971) from *Pisum sativum.*

**Particle morphology**

Virions isometric; not enveloped; 28 nm in diameter; angular in profile; without a conspicuous capsomere arrangement.

**Physical properties**

Three sedimenting components in purified preparations; sedimentation coefficient of the fastest 122 *S* (B); of the other(s) 96 *S* (M), or 60 *S* (T). Isoelectric point pH 5.95 and 6.4.

**Biochemical properties**

Virions contain 36 % nucleic acid (B), or 25 % nucleic acid (M), or 0 % nucleic acid (T); 64 % protein (B), or 75 % protein (M), or 100 % protein (T); 0 % lipid.

*Genome* consists of RNA; single–stranded; linear; of two parts; largest (or only) genome part 6.03 kb (RNA–1); the 2nd largest 3.54 kb (RNA-2). Base composition 20.3 % G; 29 % A; 20.4 % C; 30.2 % U. 5' terminus of RNA has a VPg. Infectivity retained when deproteinised with proteases; retained when deproteinised with phenol or detergent. Poly A region present.

*Sequence database accession code(s):*

D00657 Em(40)_vi:COMRCM24 Gb(84)_vi:RCM24KP Red clover mottle virus RNA encoding 24K protease. 3/91 771bp.

M14913 Em(40)_vi:COMRCMMI Gb(84)_vi:RCMMIDCA Red clover mottle virus (RCMV) middle component (M) gene, complete cds. 7/89 3,543bp.

X64886 Em(40)_vi:RCMVBCRNA Gb(84)_vi:RCMVBCRNA Red Clover Mottle Virus genomic RNA sequence for bottom component RNA. 7/93 6,033bp. 3 sequences.

*Non–genomic nucleic acid* not found in the virions.

*Virion protein(s)* three; $M_r$ of the largest 40000. $M_r$ of 2nd largest 22200. $M_r$ of 3rd largest 18300. Method of preparation: Marcinka: heating 100°C for 5–10 minutes in presence of 1% SDS and 1% 2-mercaptoethanol.

**Replication**

Replication does not depend on a helper virus.

**Cytopathology**

Virions found in roots and all parts of the host plant; in cytoplasm. Other cellular changes: necrosis of tissues responding hypersensitively to infection.

**Taxonomy and Relationships**

*Comovirus: Comoviridae.*

Virus(es) with serologically related virions: most comoviruses.

**References**

Bos, L. and Maat, D.Z. (1965). *Neth. J. Pl. Path.* **71**: 8.

Hampton, R., Beczner, L., Hagedorn, D., Bos, L., Inouye, T., Barnett, O., Musil, M. and Meiners, J. (1978). *Phytopathology* **68**: 989.

Marcinka, K. (1971). *Acta virol.* **15**: 316.

Musil *et al.* (1969). *Biol. Brat.* **24**: 23.

Oxelfelt, P. (1976). *Virology* **74**: 73.
Sinha, R.C. (1960). *Ann. appl. Biol.* **48**: 742.
Valenta, V. and Marcinka, K. (1968). *Act. Virol.* **12**: 288.
Valenta, V. and Marcinka, K. (1971). *CMI/AAB Descr. Pl. Viruses* No. 74, 3 pp.

# Red clover necrotic mosaic *dianthovirus*

**Data collated by M. Hollings, 1980.**

*Synonyms:* clover primary leaf necrosis virus (Ragetli and Elder, 1977), clover red necrotic mosaic virus. *Standard acronym:* RCNMV. *Strains:* isolates differing in their serological types reported from England, Scotland, Sweden and the former Czechoslovakia.

First reported in *Trifolium pratense*; from the former Czechoslovakia; by Musil (1969*a*); Musil *et al.* (1967); Gould *et al.* (1981).

## Natural host range and symptoms

Symptoms persist.
*Trifolium pratense* — necrotic local lesions, mosaics, leaf deformation.
*Medicago sativa, Melilotus officinalis, Trifolium repens.*

## Transmission

Transmitted by a vector; Experimentally transmitted by the chytrid *Olpidium radicale.* Virus does not require a helper virus for vector transmission; transmitted by mechanical inoculation; not transmitted by seed.

## Geographical distribution

Spreads in the former Czechoslovakia, Sweden, and the UK. Found, but with no evidence of spread, in Australia, Poland, Canada and New Zealand.

## Experimental host range

Many (>9) families susceptible.

Diagnostically susceptible

*Gomphrena globosa* — necrotic local lesions, not systemic.

*Chenopodium capitatum* — necrotic local lesions, systemic chlorosis, mottle, rosetting, stunting and leaf deformation.

*Chenopodium quinoa, Ocimum basilicum* — necrotic local lesions, not systemic.

*Cucumis sativus* cv. Butcher's Disease Resistor — necrotic local lesions in cotyledons; not systemic.

*Phaseolus vulgaris* cv. Prince — chlorotic/necrotic local lesions, systemic necrotic streaks, stunting, leaf deformation.

*Vigna unguiculata* cv. Blackeye — necrotic local lesions, systemic mottle.

*Nicotiana clevelandii* — necrotic local lesions, systemic chlorotic rings, mottle, stunting, severe puckering and leaf deformation.

*N. tabacum* cv. White Burley — chlorotic or necrotic local lesions; not systemic.

*Tetragonia expansa* — necrotic local lesions, not systemic.

Maintenance and propagation hosts

*Phaseolus vulgaris, Nicotiana clevelandii.*

Assay hosts (**L**ocal lesions or **W**hole plants)

*Cucumis sativus* (L), *Chenopodium quinoa* (L), *Ocimum basilicum* (L), *Phaseolus vulgaris* (L,W), *Vigna unguiculata* (L,W), *Nicotiana clevelandii* (L,W).

Dicotyledonous Hosts *(**Susceptible** or Insusceptible)*

*Ageratum conyzoides*. ***Amaranthus caudatus***. ***Antirrhinum majus***. ***Apium graveolens***. ***Arachis hypogaea***. ***Atriplex hortensis***. *Bellis perennis*. ***Beta vulgaris***. *Brassica campestris* ssp. *napus*. *Brassica campestris* ssp. *pekinensis*. *Brassica campestris* ssp. *rapa*. *Brassica juncea*. *Brassica oleracea* ssp. *botrytis*. *Brassica oleracea* ssp. *capitata*. *Capsicum annuum*. *Capsicum frutescens*. *Catharanthus roseus*. ***Celosia argentea***. *Cheiranthus cheiri*. ***Chenopodium amaranticolor***. *Chenopodium capitatum*. ***Chenopodium murale***. ***Chenopodium quinoa***. *Chrysanthemum morifolium*. ***Cucumis sativus***. ***Datura stramonium***. *Dianthus barbatus*. *Dianthus caryophyllus*. *Emilia sagittata*. ***Fagopyrum esculentum***. ***Glycine max***. ***Gomphrena globosa***. ***Helianthus annuus***. *Hibiscus esculentus*. ***Hyoscyamus niger***. *Ipomoea nil*. *Ipomoea setosa*. ***Lactuca sativa***. ***Lathyrus odoratus***. *Lupinus angustifolius*. ***Lycopersicon esculentum***. ***Nicandra physaloides***. ***Nicotiana benthamiana***. ***Nicotiana clevelandii***. ***Nicotiana × edwardsonii***. ***Nicotiana debneyi***. ***Nicotiana glutinosa***. ***Nicotiana megalosiphon***. ***Nicotiana rustica***. ***Nicotiana tabacum***. ***Ocimum basilicum***. *Petunia hybrida*. ***Phaseolus vulgaris***. *Physalis floridana*. ***Phytolacca americana***. ***Pisum sativum***. ***Rumex acetosa***. ***Saponaria vaccaria***. *Senecio vulgaris*. ***Silene armeria***. ***Solanum demissum***. *Solanum melongena*. *Solanum nigrum*. *Solanum tuberosum*. ***Spinacia oleracea***. *Stellaria media*. ***Tetragonia expansa***. *Torenia fournieri*. ***Trifolium incarnatum***. ***Trifolium pratense***. ***Tropaeolum majus***. *Verbesina encelioides*. ***Vicia faba***. ***Vigna unguiculata***. *Zinnia elegans*.

Sources of host-range data

Hollings *et al.* (1977*b*); Hollings unpublished data.

**Properties of particles in sap**

TIP: 85–90 °C. LIV: 63–70 days. DEP: $\log_{10}$ minus 5. Infectivity of sap decreased by treatment with di–ethyl ether. Leaf sap contains many virions. Electron microscopy: virions swell in UA.

**Purification**

Method: Hollings and Stone (1974); Hollings *et al.* (1977*b*).

**Particle morphology**

Virions isometric; not enveloped; 31–34 nm in diameter; rounded in profile; without a conspicuous capsomere arrangement.

**Physical properties**

One sedimenting component in purified preparations; sedimentation coefficient 135 *S*. Density 1.363 g $cm^{-3}$ in CsCl (1.37 when fixed with aldehydes).

**Biochemical properties**

Virions contain 23 % nucleic acid; 77 % protein; 0 % lipid.

*Genome* consists of RNA; single–stranded; of two parts; largest (or only) genome part the larger 3.89 kb (RNA-1); the 2nd largest 1.448 kb (RNA-2). Infectivity retained when deproteinised with phenol or detergent.

*Sequence database accession code(s):*

J04357 Em(40)_vi:RNMRS1 Gb(84)_vi:RNMRS1 Red clover necrotic mosaic virus RNA-1, complete sequence. 3/91 3,890bp.

X08021 Em(40)_vi:RCNMV2 Gb(84)_vi:RCNMV2 Red clover necrotic mosaic virus (RCNMV) RNA2. 9/93 1,448bp. 2 sequences.

*Virion protein(s)* one; $M_r$ 39500, or 40800 (in 2 strains).

**Replication**

Replication does not depend on a helper virus.

### Cytopathology

Virions found in leaves, roots, stems and flowers. Inclusions present in infected cells; are unusual in shape; amoeboid.

### Taxonomy and Relationships

*Dianthovirus*.

Virus(es) with serologically unrelated virions: 60 strains of 45 isometric viruses.

Some isolates induce different symptoms, for example with those from England and Scotland are different. *Tetragonia expansa* is infected systemically only by strain TpM-34.

### References

Bowen, R. and Plumb, R.T. (1979). *Ann. appl. Biol.* **91**: 227.
Gerhardson, B. and Lindsten, K. (1973). *Phytopath. Z.* **76**: 67.
Gould, A.R., Francki, R.I.B., Hatta, T. and Hollings, M. (1981). *Virology* **108**: 499.
Hiruki, C. (1987). *Adv. Virus Res.* **33**: 251.
Hollings, M. and Stone, O.M. (1974). *Rep. Glasshouse Crops Res. Inst.* 1973, p. 122.
Hollings, M. and Stone, O.M. (1977). *CMI/AAB Descr. Pl. Viruses* No. 181, 4 pp.
Hollings, M., Stone, O.M. and Pawley (1977*b*). *Rep. Glasshouse Crops Res. Inst.* 1976, p. 129.
Hollings, M. and Stone, O.M. (1980). *Rep. Glasshouse Crops Res. Inst.* 1979, p. 160.
Lommel, S.A., Weston-Fina, M., Xiong, Z. and Lomonossoff, G.P. (1988). *Nucl. Acids Res.* **16**: 8587.
MacFarlane, I. (1982). *Rep. Rothamsted Exp. Stn.* 1982, p. 190.
Marcinka, K., Musil, M. and Mrena (1969). *Acta. Virol. Prague* **13**: 247.
Musil, M. (1969*a*). *Acta. Virol. Prague* **13**: 226.
Musil, M. (1969*b*). *Biol. Brat.* **24**: 33.
Musil, M. and Matisova, J. (1967). *Ochr. Rost.* **3**: 225.
Osman, T.A.M. and Buck, K.W. (1987). *J. gen. Virol.* **68**: 289.
Osman, T.A.M. and Buck, K.W. (1990). *J. gen. Virol.* **71**: 945.
Osman, T.A.M., Miller, S.J., Marriott, A.C. and Buck, K.W. (1991). *J. gen. Virol.* **72**: 213.
Ragetli, H.W.J. and Elder, M. (1977). *Can. J. Bot.* **55**: 2122.
Xiong, Z. and Lommel, S.A. (1989). *Virology* **171**: 543.

## Red clover vein mosaic *carlavirus*

**Data collated by D.J. Hagedorn, 1981. Revised 1983.**

*Synonyms:* pea stunt virus, vein mosaic virus of red clover. *Standard acronym:* RCVMV.

First reported in *Trifolium pratense*; from the U.S.A.; by Osborne (1937).

### Natural host range and symptoms

Symptoms persist.

*Trifolium pratense* — mosaic, vein yellowing.

### Transmission

Transmitted by a vector; an insect; *Acyrthosiphon pisum, Myzus persicae, Therioaphis (Pterocallidium) maculatus, Cavariella aegopodii, C. theobaldi*; Aphididae. Transmitted in a non–persistent manner. Virus transmitted by mechanical inoculation; transmitted by seed (in *Trifolium pratense* and *Vicia faba*); not transmitted by pollen.

**Geographical distribution**

Spreads in the North American region; Germany and the Netherlands. Found, but with no evidence of spread, in the U.K., Sweden and Switzerland.

**Experimental host range**

Several (3–9) families susceptible.

Diagnostically susceptible

*Medicago sativa, Melilotus alba, Phaseolus vulgaris, Pisum sativum* cv. Perfection, *Trifolium incarnatum, T. pratense, T. repens, Vicia sativa* — systemic mosaic, yellowing and leaf deformation.

Diagnostically insusceptible

*Medicago sativa.*

Maintenance and propagation hosts

*Melilotus alba, Phaseolus vulgaris, Pisum sativum* cv. Perfection, *Trifolium incarnatum, T. pratense.*

Assay hosts (**L**ocal lesions or **W**hole plants)

*Melilotus alba* (W), *Phaseolus vulgaris* (W), *Pisum sativum* cv. Perfection (W), *Trifolium incarnatum* (W), *T. pratense* (W).

Dicotyledonous Hosts *(**Susceptible** or Insusceptible)*

***Antirrhinum majus***. *Capsicum frutescens*. ***Chenopodium amaranticolor***. ***Chenopodium quinoa***. *Cucumis sativus. Datura stramonium. Glycine max. Gomphrena globosa*. ***Lathyrus odoratus***. *Lotus corniculatus. Lupinus angustifolius. Lycopersicon esculentum. Medicago sativa*. ***Melilotus alba***. *Nicotiana glutinosa. Nicotiana rustica. Nicotiana sylvestris*. ***Nicotiana tabacum***. ***Phaseolus vulgaris***. ***Pisum sativum***. *Solanum tuberosum*. ***Trifolium hybridum***. ***Trifolium incarnatum***. ***Trifolium pratense***. ***Trifolium repens***. ***Vicia faba***. ***Vicia sativa***. *Vicia villosa. Vigna radiata. Vigna unguiculata.*

Sources of host-range data

Hagedorn and Hanson (1951); Sander (1959); Thornberry (1966); Varma (1970); Hampton *et al.* (1978).

**Properties of particles in sap**

TIP: 60 °C. LIV: 2–3 days. DEP: $\log_{10}$ minus 3. Leaf sap contains many virions.

**Purification**

Method: Varma *et al.* (1970).

**Particle morphology**

Virions filamentous; not enveloped; usually straight; with a clear modal length; of 600–700 nm; 12 nm wide. Axial canal obvious; 3.5 nm in diameter. Basic helix obvious; pitch of basic helix 3.4 nm.

**Physical properties**

One sedimenting component in purified preparations; sedimentation coefficient 160 *S*. Isoelectric point pH 4.5.

**Biochemical properties**

Virions contain 6.25 % nucleic acid; 93.75 % protein; 0 % lipid.

*Genome* consists of RNA; single–stranded; unipartite; largest (or only) genome part 7.5 kb. Base composition 31 % G; 24 % A; 23 % C; 22 % U.

**Cytopathology**

Virions found in leaves. Inclusions present in infected cells (*Pisum sativum* and in leaf hairs of *Trifolium pratense*); are amorphous X–bodies and unusual in shape; and crystalline; bundles of virions and rings of coiled virions in ultrathin sections; they contain virions.

**Taxonomy and Relationships**

*Carlavirus.*

Virus(es) with serologically related virions: pea streak virus.

**References**

Hagedorn, D.J. (1958). *Tijdschr. PlZiekt.* **64**: 263.
Hagedorn, D.J. (1974). *Ann. Phytop. Soc. Monog.* No. 9.
Hagedorn, D.J. and Hanson (1951). *Phytopathology* **41**: 813.
Hagedorn, D.J. and Walker, J.C. (1949). *J. Agric. Res.* **78**: 617.
Hagedorn, D.J., Bos, L. and van der Want, J.P.H. (1959). *Tijdschr. PlZiekt.* **65**: 13.
Hampton, R., Beczner, L., Hagedorn, D., Bos, L., Inouye, T., Barnett, O., Musil, M. and Meiners, J. (1978). *Phytopathology* **68**: 989.
Osborne, H.T. (1937). *Phytopathology* **27**: 1051.
Sander, E. (1959). *Phytopathology* **49**: 748.
Thornberry, H.H. (1966). In: *Index of Plant Virus Diseases* U.S. Dep. Agric. Hdbk No. 307, p. 294.
Varma, A. (1970). *CMI/AAB Descr. Pl. Viruses* No. 22, 4 pp.
Varma, A., Gibbs, A.J. and Woods, R.D. (1970). *J. gen. Virol.* **8**: 21.

# Red pepper 1 (?) *alphacryptovirus*

**Data collated by A.A. Brunt, 1991.**

*Synonyms:* pepper red cryptic 1 virus. *Standard acronym:* RPCV-1.

First reported in *Capsicum annuum, C. frutescens*; from Japan; by Natsuaki *et al.* in Boccardo *et al.* (1983) and Milne and Natsuaki (1994).

**Natural host range and symptoms**

*Capsicum annuum, C. frutescens* — symptomless.

**Transmission**

Virus not transmitted by mechanical inoculation; transmitted by seed.

**Geographical distribution**

Found, but with no evidence of spread, in Japan.

**Particle morphology**

Virions isometric; not enveloped; 30 nm in diameter; rounded in profile.

**Biochemical properties**

*Genome* consists of RNA; double–stranded.

**Taxonomy and Relationships**

Possible *Alphacryptovirus: Partitiviridae.*

**References**

Boccardo, G., Lisa, V. and Milne, R.G. (1983). In: *Double-Stranded RNA Viruses*, p. 425, eds R.W. Compans and D.H.L. Bishop. Elsevier, New York.

Boccardo, G., Milne, R.G., Luisoni, E., Lisa, V. and Accotto, G.P. (1985). *Virology* **147**: 29.
Boccardo, G., Lisa, V., Luisoni, E. and Milne, R.G. (1987). *Adv. Virus Res.* **32**: 171.
Milne, R.G. and Natsuaki, T (1994). In: *Pathogenesis and Host Specificity in Plant Diseases*, Vol. 3, ed. R.P Singh. Elsevier Science, Oxford.

# Red pepper 2 (?) *alphacryptovirus*

**Data collated by A.A. Brunt, 1991.**

*Synonyms:* pepper red cryptic 2 virus. *Standard acronym:* RPCV-2.

First reported in *Capsicum annuum, C. frutescens*; from Japan; by Natsuaki in Boccardo *et al.* (1983) and Milne and Natsuaki (1994).

**Natural host range and symptoms**

*Capsicum annuum, C. frutescens* — symptomless.

**Transmission**

Virus not transmitted by mechanical inoculation; transmitted by seed.

**Geographical distribution**

Found, but with no evidence of spread, in Japan.

**Particle morphology**

Virions isometric; not enveloped; 30 nm in diameter; rounded in profile.

**Biochemical properties**

*Genome* consists of RNA; double–stranded.

**Taxonomy and Relationships**

Possible *Alphacryptovirus: Partitiviridae.*

**References**

Boccardo, G., Lisa, V. and Milne, R.G. (1983). In: *Double-Stranded RNA Viruses.* p. 425, eds R.W. Compans and D.H.L. Bishop. Elsevier, New York.
Milne, R.G. and Natsuaki, T (1994). In: *Pathogenesis and Host Specificity in Plant Diseases*, Vol. 3, ed. R.P Singh. Elsevier Science, Oxford.

# Rehmannia X (?) *potexvirus*

**Data collated by A.A. Brunt, 1994.**

First reported in *Rehmannia* sp.; from Japan; by Y. Doi in Hull *et al.*, 1989.

**Particle morphology**

Virions filamentous; not enveloped; usually flexuous; with a clear modal length; of 520 nm.

**Taxonomy and Relationships**

Possible *Potexvirus.*

**References**

Hull, R., Brown, F. and Payne, C. (1989). *Directory and Dictionary of Animal. Bacterial and Plant Viruses.* MacMillan Reference Books, London.
Yora, K., Saito, Y., Doi, Y., Inouye, T. and Tomaru, K. (1983). In: *Handbook of Plant Viruses.* Asakura Shoten, Tokyo.

# Rhododendron necrotic ringspot (?) *potexvirus*

**Data collated by A.A. Brunt, 1991.**

*Standard acronym:* RoNRSV.

First reported in *Rhododendron* sp. and *Kalmia latifolia*; from Oregon, U.S.A.; by Coyier *et al.* (1976).

**Natural host range and symptoms**

Symptoms persist.
*Rhododendron* spp., *Kalmia latifolia* — concentric necrotic rings in leaves.

**Transmission**

Virus not transmitted by mechanical inoculation; transmitted by grafting.

**Geographical distribution**

Spreads in Canada, the USA, and the UK (probably).

**Properties of particles in sap**

Leaf sap contains few virions.

**Particle morphology**

Virions filamentous; not enveloped; usually flexuous; with a clear modal length; of 504 nm; 13 nm wide.

**Cytopathology**

Virions found in leaves and mesophyll; in cytoplasm. Inclusions present in infected cells; aggregated they contain virions.

**Taxonomy and Relationships**

Possible *Potexvirus*.

**References**

Coyier, D.L., Stace-Smith, R., Allen, T.C. and Leung, E. (1976). *Acta Hort.* **59**: 93.
Coyier, D.L., Stace-Smith, R., Allen, T.C. and Leung, E. (1977). *Phytopathology* **67**: 1090.
Pearse, A.G.E. (1968). *Gard. Chron.* **164**: 5.

# Rhubarb 1 (?) *potexvirus*

*Standard acronym:* RV-1.

**Taxonomy and Relationships**

Possible *Potexvirus*.

# Rhubarb temperate (?) *alphacryptovirus*

**Data collated by A.A. Brunt, 1991.**

*Standard acronym:* RTV.

First reported in *Rheum rhaponticum*; from Japan; by Natsuaki *et al.* (1983).

**Natural host range and symptoms**

*Rheum rhaponticum* — symptomless.

**Transmission**

Virus not transmitted by mechanical inoculation; transmitted by seed.

**Geographical distribution**

Found, but with no evidence of spread, in Japan.

**Particle morphology**

Virions isometric; not enveloped; 30 nm in diameter.

**Taxonomy and Relationships**

Possible *Alphacryptovirus: Partitiviridae.*

**References**

Boccardo, G., Lisa, V., Luisoni, E. and Milne, R.G. (1987). *Adv. Virus Res.* **32**: 171.

Natsuaki, T., Yamashita, S., Doi, Y., Okuda, S. and Teranaka, M. (1983). *Ann. Phytopath. Soc. Japan* **49**: 593.

Natsuaki, T., Natsuaki, K.T., Okuda, S., Teranaka, M., Milne, R.G., Boccardo, G. and Luisoni, E. (1986). *Intervirology* **25**: 69.

# Rhynchosia mosaic bigeminivirus

**Data collated by A.A. Brunt, 1990.**

*Standard acronym:* RhMV.

First reported in *Rhynchosia minima*; from Puerto Rico; by Bird (1962).

**Natural host range and symptoms**

*Rhynchosia minima* — leaf chlorosis and curling.

**Transmission**

Transmitted by a vector; an insect; *Bemisia tabaci*; Aleyrodidae. Virus transmitted by mechanical inoculation (with difficulty from *Phaseolus vulgaris* to *Glycine max* (Jilke and Utermohlen, 1986)).

Diagnostically susceptible

*Phaseolus vulgaris, P. lathyroides, Canavalia ensiformis, Cajanus indicus, Nicotiana glutinosa, N. tabacum, Abelmoschus esculentus, Gossypium hirsutum* — leaf curling and chlorosis.

*Glycine max* cv. Williams — yellow-green mosaic and stunting.

Dicotyledonous Hosts *(**Susceptible** or Insusceptible)*

***Gossypium hirsutum. Nicotiana glutinosa. Nicotiana tabacum. Phaseolus lathyroides. Phaseolus vulgaris.***

**Biochemical properties**

*Genome* consists of DNA; single–stranded.

**Taxonomy and Relationships**

Bigeminivirus: *Geminiviridae.*

**References**

Bird, J. (1962). *Phytopathology* **52**: 286.

Bird, J. and Maramorosch, K. (1978). *Adv. Virus Res.* **22**: 55.

Bird, J. and Sanchez, J. (1971). *J. agric. Univ. Puerto Rico* **55**: 461.

Bird, J., Sanchez, J., Rodriguez, R.L. and Julia, F.J. (1975). In: *Tropical Diseases of Legumes*, p. 3; ed. J. Bird and K. Maramorosch. Academic Press, New York.

Jilke, J.M. and Utermohlen, J.G. (1986). *Phytopathology* **76**: 1073.

# Ribgrass mosaic *tobamovirus*

**Data collated by N. Oshima, 1986.**

*Synonyms:* Holmes' ribgrass virus, plantago mosaic virus, ribgrass strain of tobacco mosaic virus, rai mosaic virus, youcai mosaic virus. *Standard acronym:* RMV. *Strains:* primula isolate, RMV-K, TMV-C, Wasabi isolate.

First reported in *Plantago lanceolata*; from New Jersey, U.S.A.; by Holmes (1941).

**Natural host range and symptoms**

Symptoms persist.
*Plantago lanceolata, P. major* — chlorotic mottling and rings.
*Digitalis lanata* — mosaic.
*Rorippa amphibia* — mild mottling and necrotic lines.
*Rorippa sylvestris* — symptomless carrier.
*Eutrema wasabi* — mottling.
*Sisymbrium loeselii* — yellow leaf edges and vein clearing.
*Lychnis alba* — chlorotic ringspots and lines.
*Nicotiana tabacum* — stunting and necrosis.

**Transmission**

Transmitted by means not involving a vector. Virus transmitted by mechanical inoculation; transmitted by grafting; transmitted by contact between plants.

**Geographical distribution**

Probably distributed worldwide.

**Experimental host range**

Many (>9) families susceptible.

Diagnostically susceptible

*Nicotiana tabacum* cv. Xanthi-*nc* — brown necrotic local lesions; not systemic.
*Nicotiana tabacum* cv. Samsun — necrotic concentric rings and line patterns.
*Nicotiana glutinosa, N. sylvestris, Petunia hybrida* — brown necrotic local lesions; not systemic.
*Chenopodium amaranticolor* — small necrotic local lesions, occasional systemic necrotic spotting.
*Cyphomandra betacea* — chlorotic local lesions; systemic mosaic and deformation.
*Momordica charantia* — necrotic local lesions; not systemic.
*Brassica rapa, B. nigra* — systemic mosaic symptoms or symptomless carrier.

Diagnostically insusceptible

*Phaseolus vulgaris.*

Maintenance and propagation hosts

*Nicotiana tabacum* cv. Samsun.

Assay hosts (**L**ocal lesions or **W**hole plants)

*Nicotiana tabacum* cv. Xanthi-*nc* (L). *N. glutinosa* (L); *N. sylvestris* (L); *N. tabacum* cv. Samsun (W); *Chenopodium amaranticolor* (L).

Dicotyledonous Hosts *(**Susceptible** or Insusceptible)*

***Amaranthus caudatus*. *Amaranthus retroflexus*. *Beta vulgaris*. *Brassica campestris*. *Brassica campestris*** ssp. ***napus*. *Brassica campestris*** ssp. ***pekinensis*. *Brassica campestris*** ssp. ***rapa*. *Brassica juncea*. *Brassica oleracea*** ssp. ***botrytis*. *Brassica oleracea*** ssp. ***capitata*. *Capsicum annuum*. *Capsicum frutescens*. *Chenopodium album*. *Chenopodium amaranticolor*. *Chenopodium hybridum*. *Chenopodium murale*. *Chenopodium quinoa*. *Cichorium endiva*. *Citrullus lanatus*.** *Cucumis melo*. ***Cucumis sativus*. *Datura metel*. *Datura stramonium*. *Datura tatula*. *Gomphrena globosa*. *Hyoscyamus niger*. *Lactuca sativa*. *Lupinus albus*. *Lycopersicon esculentum*. *Lycopersicon pimpinellifolium*. *Nicandra physaloides*. *Nicotiana clevelandii*. *Nicotiana debneyi*. *Nicotiana glutinosa*. *Nicotiana megalosiphon*. *Nicotiana rustica*. *Nicotiana sylvestris*. *Nicotiana tabacum*. *Ocimum basilicum*. *Petunia hybrida*.** *Phaseolus vulgaris*. ***Phlox drummondii*. *Physalis peruviana*. *Pisum sativum*. *Raphanus sativus*. *Solanum melongena*. *Solanum nigrum*. *Spinacia oleracea*. *Stellaria media*. *Tetragonia expansa*. *Tropaeolum majus*. *Vicia faba*.** *Vicia sativa*. ***Vigna unguiculata*** ssp. ***sesquipedalis*. *Zinnia elegans*.**

Monocotyledonous Hosts *(**Susceptible** or Insusceptible)*

***Hordeum vulgare*. *Triticum aestivum*.**

Sources of host-range data

Oshima *et al.*, (1974); Schumann (1963); Kovachevsky (1963); Fujisawa *et al.* (1982).

**Properties of particles in sap**

TIP: 93 °C. LIV: 160 days (or more at 20–25°C). DEP: $\log_{10}$ minus 6–7. Leaf sap contains many virions.

**Purification**

Method: Siegel and Wildman (1954).

**Particle morphology**

Virions rod–shaped; not enveloped; usually straight; with a clear modal length; of 300 nm; 18 nm wide. Axial canal obvious; *c.* 4 nm in diameter. Basic helix obvious; pitch of basic helix 2.3 nm.

**Physical properties**

One sedimenting component in purified preparations; sedimentation coefficient 180–200 *S*. Density *c.* 1.37 g $cm^{-3}$ in CsCl. Isoelectric point pH 4.49.

**Biochemical properties**

Virions contain 5 % nucleic acid; 95 % protein; 0 % lipid.

*Genome* consists of RNA; single–stranded; linear; unipartite; largest (or only) genome part 6.5 kb. Genomic nucleic acid isolated by Fraenkel-Conrat *et al.* (1961). Base composition 25.8 % G; 29.3 % A; 18 % C; 27 % U. 5' terminus of RNA has a methylated nucleotide cap ($m^7$GpppX). Infectivity decreased when deproteinised with proteases; retained when deproteinised with phenol or detergent.

*Non–genomic nucleic acid* not found in the virions. Sub–genomic *m*RNA possibly found in infected cells.

*Virion protein(s)* one; $M_r$ 17500. Method of preparation: Fraenkel-Conrat (1957).

**Replication**

Coat protein *m*RNA translated in the cytoplasm, possibly. Replication does not depend on a helper virus.

**Cytopathology**

Virions found in all parts of the host plant; in cytoplasm and in cell vacuoles. Inclusions present in infected cells; are crystals in the cytoplasm and amorphous X–bodies; they contain virions.

**Taxonomy and Relationships**

*Tobamovirus.*

Virus(es) with serologically related virions: all tobamoviruses to a greater or lesser extent, and this is correlated with their virion protein sequence similarities (Gibbs, 1986). Distant serological relationship with broad bean necrosis, potato mop—top and soil—borne wheat mosaic (Van Regenmortel, 1982).

**References**


Chessin, M., Zaitlin, M. and Solberg, R.A. (1967). *Phytopathology* **57**: 452.

Fraenkel-Conrat, H. (1957). *Virology* **4**: 1.

Fraenkel-Conrat, H., Singer, B. and Tsugita, A. (1961). *Virology* **14**: 54.

Fujisawa, K., Tsuchizaki, T. and Iizuka, N. (1982). *Ann. Phytopath. Soc. Japan* **48**: 592.

Gibbs, A.J. (1986). In: *The Plant Viruses*, Vol. 2, The Rod-Shaped Plant Viruses, p. 167; eds M.H.V. van Regenmortel and H. Fraenkel-Conrat. Plenum Press, New York.

Goto, T. and Oshima, N. (1962). *Ann. Phytopath. Soc. Japan* **27**: 109.

Holmes, F.O. (1941). *Phytopathology* **31**: 1089.

Juretič, N., Miličič, D. and Mamula, D. (1973). *Acta bot. Croat.* **32**: 29.

Kovachevsky, I.C. (1963). *Phytopath. Z.* **49**: 127.

Kovachevsky, I.C. (1969). *Proc. 7th Conf. Czechoslov. Pl. Virol., Olomouc, 1967*, p. 250.

Oshima, N. and Harrison, B.D. (1975). *CMI/AAB Descr. Pl. Viruses* No. 152, 4 pp.

Oshima, N., Ohashi, Y. and Ume K. (1974). *Ann. Phytopath. Soc. Japan* **40**: 241.

Polak, Z. (1962). *Proc. 5th Conf. Czechoslov. Pl. Virol., Prague, 1962*, p. 168.

Schumann, K. (1963). *Phytopath. Z.* **48**: 1.

Siegel, A. and Wildmann, S.G. (1954). *Phytopathology* **54**: 277.

Tochihara, H., Komuro, Y. and Kobari, Y. (1964). *Proc. Kanto Pl. Prot. Soc.* **11**: 46.

Van Regenmortel, M.H.V. (1982). In: *Serology and Immunochemistry of Plant Viruses*, p. 147 and p. 166. Academic Press, New York.

Zhu, P.K., Lu, M.K., Wang, M.Q., Xia, Y.N., Sun, X.G., Jin, C.X. and Shen, Q. (1984). *Acta phytopath. Sin.* **14**: 59.


# Rice black-streaked dwarf *fijivirus*

**Data collated by E. Shikata, 1987; Li Jing, 1987.**

*Synonyms:* rice black streak virus. *Standard acronym:* RBSDV.

First reported in *Oryza sativa* var. *japonica*; from Japan; by Kuribayashi and Shinkai (1952).

**Natural host range and symptoms**

Symptoms persist.

*Oryza sativa, Zea mays, Triticum sativa, Avena sativa, Hordeum vulgare* — stunted, deformed leaves with white waxy or black-streaked swellings along the veins.

**Transmission**

Transmitted by a vector; an insect; *Laodelphax striatellus* (Kuribayashi and Shinkai, 1952; Shinkai, 1962), *Unkanodes sapporona* (Shinkai, 1966) and *U. albifascia* (Shinkai, 1967); Delphacidae. Transmitted in a persistent manner. Virus retained when the vector moults; multiplies in the vector; not transmitted congenitally to the progeny of the vector; not transmitted by mechanical inoculation; not transmitted by seed.

**Geographical distribution**

Spreads in China, Japan, Korea D.P.R. (North), and Korea Republic.

**Experimental host range**

Few (<3) families susceptible.

Diagnostically susceptible

*Oryza sativa* var. *japonica* — stunting, darkening of leaves, twisting of tips of young leaves, white waxy swellings along veins on the abaxial leaf surface, later becoming brown and forming black-streaked tumours.

*Zea mays* — dwarfening, darkening of leaves, white streaks along the veins and white waxy swollen veins on the abaxial surface of leaves.

*Triticum sativum, Avena sativa, Hordeum vulgare* — severe stunting, twisting of leaves, sometimes with waxy swollen veins on the abaxial surface of leaves and culms.

Maintenance and propagation hosts

*Oryza sativa, Triticum sativum, Avena sativa* and *Hordeum vulgare*.

Assay hosts (**L**ocal lesions or **W**hole plants)

*Zea mays* cv. Golden Cross Bantam (W) (*Zea mays* useless as a virus source for insect transmission).

Monocotyledonous Hosts *(**Susceptible** or Insusceptible)*

***Avena sativa*. *Hordeum vulgare*. *Oryza sativa*. *Triticum aestivum*. *Zea mays*.**

Sources of host-range data

Shinkai (1962); Morinaka and Sakurai (1968).

**Properties of particles in sap**

TIP: 60 °C. LIV: 6 days, or 7 days. DEP: $\log_{10}$ minus 5 (in rice plants), or 6 (in insects). Leaf sap contains many virions. Electron microscopy: leaves fixed in 2% osmium tetroxide or 2% paraformaldehyde before extraction yield the large virions of 75–80 nm with the outer projections attached.

**Purification**

Method: Kitagawa and Shikata (1969); Kawano *et al.* (1984).

**Particle morphology**

Virions isometric; not enveloped; 75–80 nm in diameter (in cells of infected plants and insects), or 55 nm in diameter (B-spiked subvirions), or 70 nm in diameter (intact virions); angular in profile; without a conspicuous capsomere arrangement.

**Biochemical properties**

*Genome* consists of RNA; double–stranded; of 10 parts (named S1–10); largest (or only) genome part 3.998 kb; the 2nd largest 3.385 kb; the 3rd largest 3.31 kb; the 4th largest 3.31 kb; the 5th largest 3.1 kb; the 6th largest and other parts 2.45 kb (pairs; also 2.198, 1.830, 1.826 and 1.810 kb pairs).

*Sequence database accession code(s):*

D00606 Em(40)_vi:RBDSEG10 Gb(84)_vi:RBDSEG10 Rice black-streaked dwarf virus genome segment 10. 11/90 1,801bp.
L36524 Gb(84)n:RBDGS3A Rice black-streaked dwarf virus genome segment 3 fragment. 10/94 939bp.
S63914 Em(40)_vi:S63914 Gb(84)_vi:S63914 orf (segment 8) rice black-streaked dwarf virus RBSDV, Genomic RNA, 1927 nt. 1/94 1,927bp.
S63917 Em(40)_vi:S63917 Gb(84)_vi:S63917 orf1, orf2 (segment 7) rice black-streaked dwarf virus RBSDV, Genomic RNA, 2193 nt. 1/94 2.

#### Cytopathology

Virions found in phloem; in cytoplasm. Inclusions present in infected cells; are viroplasms, or crystals in the nucleus; they contain virions. Other cellular changes: hyperplasia and hypertrophy of phloem cells.

### Taxonomy and Relationships

*Fijivirus: Reoviridae.*

Reported to be serologically closely related to maize rough dwarf fijivirus (Luisoni *et al.*, 1973); their relationship is so close that they may be the same species.

### References


Kashiwagi, Y. (1966). *Ann. Phytopath. Soc. Japan* **32**: 168.
Kawano, S., Uyeda, I. and Shikata, E. (1984). *J. Fac. Agric. Hokkaido Univ.* **61**: 408.
Kitagawa, Y. and Shikata, E. (1969). *Mem. Fac. Agric. Hokkaido Univ.* **6**: 446.
Kuribayashi, K. and Shinkai, A. (1952). *Ann. Phytopath. Soc. Japan* **16**: 41.
Luisoni, E., Lovisolo, O., Kitagawa, Y. and Shikata, E. (1973). *Virology* **52**: 281.
Morinaka, T. and Sakurai, Y. (1968). *Bull. Chugoku agric. exp. Stn. Ser. E* **2**: 1.
Shikata, E. (1974). *CMI/AAB Descr. Pl. Viruses* No. 135, 4 pp.
Shinkai, A. (1962). *Bull. nat. Inst. agric. Sci., Tokyo Ser. C.*, p. 223.
Shinkai, A. (1966). *Ann. Phytopath. Soc. Japan* **32**: 317.
Shinkai, A. (1967). *Ann. Phytopath. Soc. Japan* **35**: 318.


## Rice bunchy stunt (?) *phytoreovirus*

**Data collated by A.A. Brunt and H. Hibino, 1988.**

First reported in *Oryza sativa*; from southern China; by Chen *et al.* (1978).

### Natural host range and symptoms

Symptoms persist.

*Oryza sativa* — most cultivars have short narrow mottled leaves, are severely stunted with increased tillering; intolerant cultivars also produce nodal branches and bunches of short narrow leaves which have the appearance of birds nests.

### Transmission

Transmitted by a vector; an insect; *Nephotettix virescens, N. cincticeps*; Cicadellidae. Transmitted in a persistent manner. Virus multiplies in the vector; not transmitted congenitally to the progeny of the vector; not transmitted by mechanical inoculation; not transmitted by contact between plants; not transmitted by seed.

### Ecology and control

Studies reported by Chen *et al.* (1978); Xie *et al.* (1979, 1980); Xie and Lin (1980, 1982). The virus occurs sporadically in rice fields; average incidence differs according to year, location and cultivar, it was less than 1% in 1974, 1975, 1980 and 1981, over 3% from 1976–1979 and highest (3.2%) in 1977, but has declined since 1981. Infected rice plants can over-winter and produce new leaves from which non-viruliferous leafhoppers can acquire virus. Viruliferous nymph and adult leafhoppers can also over-winter.

**Geographical distribution**

Spreads in China.

Diagnostically susceptible

*Oryza sativa* — symptoms as in naturally infected plants.

Maintenance and propagation hosts

*Oryza sativa*.

Assay hosts (**L**ocal lesions or **W**hole plants)

*Oryza sativa* (W).

**Properties of particles in sap**

TIP: 60 °C (in rice sap; determined by micro injection of vectors). LIV: 4 days (at 2°C).

**Particle morphology**

Virions isometric; not enveloped; about 60 nm in diameter.

**Cytopathology**

Virions found in cells of leaves and roots.

**Taxonomy and Relationships**

Possible *Phytoreovirus: Reoviridae*.

Virus(es) with serologically unrelated virions: rice dwarf virus.

No cross protection in rice plants previously infected with rice dwarf or rice ragged stunt viruses.

**References**


Chen, S.H., Lin, H.Y., Xi, L.H. and Hu, F.P. (1978). *Scientia Agric. Sinica* **3**: 79.
Xie, L.H., Chen, Z.X. and Lin, Q.Y. (1979). *Acta Phytopath. Sin.* **9**: 93.
Xie, L.H. and Lin, J.Y. (1980). *Kexue Tongbao* **25**: 786.
Xie, L.H. and Lin, J.Y. (1982). *Int. Rice Res. Newsl.* **7**: 6.
Xie, L.H. and Lin, H.Y. (1983). *Acta Phytopath. Sin.* **13**: 15.
Xie, L.H., Lin, Q.Y. and Zhu, Q.L. (1982). *Acta Phytopath. Sinica* **12**: 16.


# Rice dwarf *phytoreovirus*

**Data collated by A.J. Gibbs, 1990; H. Hibino, 1992.**

*Standard acronym:* RDV. *Strains:* severe strain and a vector non-transmissible strain.

First reported in *Oryza sativa*; from Shiga, Japan; by Takada (1895 and 1896).

**Natural host range and symptoms**

Symptoms persist. Symptoms stunting, fine chlorotic specks on leaves.

*Echinochloa crus-galli, Paspalum thumbergii, Oryza sativa, Alopecurus aequalis* — stunting, white chlorotic specks on leaves.

**Transmission**

Transmitted by a vector; an insect; *Nephotettix cincticeps, N. apicalis, N. nigropictus, N. virescens, Recilia dorsalis*; Cicadellidae. Principal natural vector(s): *Nephotettix cincticeps*. Transmitted in a persistent manner. Virus retained when the vector moults;

multiplies in the vector; transmitted congenitally to the progeny of the vector; not transmitted by mechanical inoculation; not transmitted by contact between plants; not transmitted by seed; not transmitted by pollen.

**Ecology and control**

Studies reported by Miyai *et al.* (1986).

**Geographical distribution**

Spreads in China, Japan, Korea D.P.R. (North), Korea Republic, and Nepal.

**Experimental host range**

Few (<3) families susceptible. Experimentally infected plants mostly show stunting, white specks on leaves, darker leaf colour.

Diagnostically susceptible

*Alopecurus japonicus, A. aequalis, Avena sativa, Echinochloa crus-galli, Hordeum sativum, Oryza cubensis, O. sativa, Panicum miliaceum, Poa annua, Secale cereale, Triticum aestivum* — stunting and systemic chlorotic specks.

Diagnostically insusceptible

*Andropogon sorghum, Setaria italica, Zea mays.*

Maintenance and propagation hosts

*Alopecurus japonicus, Oryza sativa.*

Assay hosts (**L**ocal lesions or **W**hole plants)

*Oryza sativa* (W).

Monocotyledonous Hosts *(**Susceptible** or Insusceptible)*

***Avena sativa*. *Echinochloa crus–galli*. *Hordeum vulgare*. *Oryza sativa*. *Panicum miliaceum*. *Poa annua*. *Secale cereale*.** *Setaria italica*. ***Triticum aestivum***. *Zea mays*.

Sources of host-range data

Fukushi (1934); Shinkai (1962).

**Properties of particles in sap**

TIP: 45–50 °C. LIV: 10 days (at 4°C). DEP: $\log_{10}$ minus 3–4 (in sap from rice leaves). Infectivity of sap not changed by treatment with di–ethyl ether. Leaf sap contains many virions.

**Purification**

Method: Iida *et al.* (1972).

**Particle morphology**

Virions isometric; not enveloped; 70 nm in diameter; angular in profile; with a conspicuous capsomere arrangement.

**Physical properties**

Two sedimenting components in purified preparations; sedimentation coefficient 510 *S*. Density 1.26 g $cm^{-3}$ in CsCl (T), or 1.38 g $cm^{-3}$ in CsCl (M1), or 1.43 g $cm^{-3}$ in CsCl (M2), or 1.46 g $cm^{-3}$ in CsCl (B). $A_{260}/A_{280}$ ratio 1.4.

**Biochemical properties**

Virions contain 20 % nucleic acid; 80 % protein.

*Genome* consists of RNA; double–stranded; linear; of 12 parts (named S1–10); largest (or only) genome part 4.354 kb (pairs); the 2nd largest 3.385 kb (pairs); the 3rd largest 3.195 kb (pairs); the 4th largest 2.468 kb (pairs); the 5th largest 2.57 kb (pairs);

the 6th largest and other parts 1.699 kb (pairs; also 1.696, 1.424, 1.305, 1.319, 0.887 and 0.828 kb pairs). Genomic nucleic acid isolated by Kimura *et al.* (1987). Base composition 21.8–22.8 % G; 27.8–28.4 % A; 21.4–21.6 % C; 28–28.2 % U. Base composition of ds genome 42.3–48.1 % (G + C). 5' terminus of RNA has an unusual type; shows the sequence of GGCA or GGUA. Infectivity lost when deproteinised with phenol or detergent. Poly A region absent.

*Sequence database accession code(s):*

D00241 Em(40)_vi:RDVS10 Gb(84)_vi:RDVS10 Rice dwarf virus segment S10 mRNA, encoding an ORF, complete cds. 7/90 1,319bp.

D00465 Em(40)_vi:RDVSEG9 Gb(84)_vi:RDVSEG9 Rice dwarf virus segment 9 RNA, complete sequence. 3/91 1,305bp.

D00473 Em(40)_vi:RDVS10A Rice dwarf virus genome segment 10, complete sequence. 7/90 1,321bp.

D00536 Em(40)_vi:RDVS8 Gb(84)_vi:RDVS8 Rice dwarf virus outer capsid protein RNA (S8), complete cds. 3/91 1,424bp.

D00607 Em(40)_vi:RDVSEG3 Gb(84)_vi:RDVSEG3 Rice dwarf virus genome segment 3, complete sequence. 11/90 3,195bp.

D00608 Em(40)_vi:RDVSEG4 Gb(84)_vi:RDVSEG4 Rice dwarf virus genome segment 4, complete sequence. 3/91 2,468bp.

D00639 Em(40)_vi:RDVS7 Gb(84)_vi:RDVS7 Rice dwarf virus genome segment 7, complete sequence. 3/91 1,696bp.

D00693 Em(40)_vi:RDVS3CP Gb(84)_vi:RDVS3CP Rice dwarf virus genome segment S3 encoding 114K core protein. 2/92 3,130bp.

D10218 Em(40)_vi:RDVSEG7RD Gb(84)_vi:RDVSEG7RD Rice dwarf virus segment 7, complete sequence. 3/94 1,696bp.

D10219 Em(40)_vi:RDVSEG8RD Gb(84)_vi:RDVSEG8RD Rice dwarf virus segment 8, complete sequence. 3/94 1,427bp.

D10220 Em(40)_vi:RDVSEG9RD Gb(84)_vi:RDVSEG9RD Rice dwarf virus segment 9, complete sequence. 3/94 1,305bp.

D10221 Em(40)_vi:RDVSEG10R Gb(84)_vi:RDVSEG10RD Rice dwarf virus segment 10, complete sequence. 3/94 1,321bp.

D10222 Em(40)_vi:RDVSEG1H Gb(84)_vi:RDVSEG1H Rice dwarf virus segment 1, RNA-dependent RNA polymerase. 12/92 4,423bp.

D10249 Em(40)_vi:RDVS11 Gb(84)_vi:RDVS11 Rice dwarf *phytoreovirus* segment S11 mRNA. 2/93 1,067bp.

D11369 Em(40)_vi:RDVGS12 Gb(84)_vi:RDVGS12 Rice dwarf virus genome segment 12, complete cds. 10/92 1,066bp.

D13404 Em(40)_vi:RDVS9 Gb(84)_vi:RDVS9 Rice dwarf virus segment 9 RNA. 1/93 1,305bp.

D13773 Em(40)_vi:RDVSS8 Gb(84)_vi:RDVSS8 Rice dwarf virus RNA for outer capsid protein. 4/94 1,427bp. D90033 Em(40)_vi:RDVAS5 Gb(84)_vi:RDVAS5 Rice dwarf virus A segment 5, complete. 5/92 2,570bp. D90198 Em(40)_vi:RDVS1 Gb(84)_vi:RDVSEG1A Rice dwarf virus segment 1, RNA-dependent RNA polymerase mRNA. 11/92 4,423bp. D90199 Em(40)_vi:US11 Rice dwarf phytoreovirus segment 11 mRNA. 8/92 1,067bp. D90200 Em(40)_vi:RDVS12 Gb(84)_vi:RDVS12 Rice dwarf phytoreovirus segment 12 mRNA. 2/93 1,066bp.

M35118 Em(40)_vi:RDVS1001 Gb(84)_vi:RDVS10A Rice dwarf virus segment 10 RNA, encoding an ORF, complete cds. 2/91 1,321bp.

M91653 Em(40)_vi:RDVASEG6 Gb(84)_vi:RDVASEG6 Rice dwarf virus segment 6, complete. 3/93 1,699bp.

S77255 Em(40)_vi:S77255 Gb(84)_vi:S77255 (3' terminus) rice dwarf virus RDV, Genomic RNA, 10 nt. 12/93 10bp.
S77256 Em(40)_vi:S77256 Gb(84)_vi:S77256 (3' terminus) rice dwarf virus RDV, Genomic RNA, 10 nt. 12/93 10bp.
S77257 Em(40)_vi:S77257 Gb(84)_vi:S77257 (3' terminus) rice dwarf virus RDV, Genomic RNA, 10 nt. 12/93 10bp.
S77258 Em(40)_vi:S77258 Gb(84)_vi:S77258 (3' terminus) rice dwarf virus RDV, Genomic RNA, 10 nt. 12/93 10bp.
S77259 Em(40)_vi:S77259 Gb(84)_vi:S77259 (3' terminus) rice dwarf virus RDV, Genomic RNA, 10 nt. 12/93 10bp.
S77260 Em(40)_vi:S77260 Gb(84)_vi:S77260 (3' terminus) rice dwarf virus RDV, Genomic RNA, 10 nt. 12/93 10bp.
S77261 Em(40)_vi:S77261 Gb(84)_vi:S77261 (3' terminus) rice dwarf virus RDV, Genomic RNA, 10 nt. 12/93 10bp.
S77262 Em(40)_vi:S77262 Gb(84)_vi:S77262 (3' terminus) rice dwarf virus RDV, Genomic RNA, 10 nt. 12/93 10bp.
S77263 Em(40)_vi:S77263 Gb(84)_vi:S77263 (3' terminus) rice dwarf virus RDV, Genomic RNA, 10 nt. 12/93 10bp.
S77264 Em(40)_vi:S77264 Gb(84)_vi:S77264 (3' terminus) rice dwarf virus RDV, Genomic RNA, 10 nt. 12/93 10bp.
S77265 Em(40)_vi:S77265 Gb(84)_vi:S77265 (3' terminus) rice dwarf virus RDV, Genomic RNA, 10 nt. 12/93 10bp.
S77266 Em(40)_vi:S77266 Gb(84)_vi:S77266 (3' terminus) rice dwarf virus RDV, Genomic RNA, 10 nt. 12/93 10bp.
S77279 Em(40)_vi:S77279 Gb(84)_vi:S77279 (5' terminus) rice dwarf virus RDV, Genomic RNA, 10 nt. 12/93 10bp.
S77280 Em(40)_vi:S77280 Gb(84)_vi:S77280 (5' terminus) rice dwarf virus RDV, Genomic RNA, 10 nt. 12/93 10bp.
S77281 Em(40)_vi:S77281 Gb(84)_vi:S77281 (5' terminus) rice dwarf virus RDV, Genomic RNA, 10 nt. 12/93 10bp.
S77282 Em(40)_vi:S77282 Gb(84)_vi:S77282 (5' terminus) rice dwarf virus RDV, Genomic RNA, 10 nt. 12/93 10bp.
S77283 Em(40)_vi:S77283 Gb(84)_vi:S77283 (5' terminus) rice dwarf virus RDV, Genomic RNA, 10 nt. 12/93 10bp.
S77284 Em(40)_vi:S77284 Gb(84)_vi:S77284 (5' terminus) rice dwarf virus RDV, Genomic RNA, 10 nt. 12/93 10bp.
S77285 Em(40)_vi:S77285 Gb(84)_vi:S77285 (5' terminus) rice dwarf virus RDV, Genomic RNA, 10 nt. 12/93 10bp.
S77286 Em(40)_vi:S77286 Gb(84)_vi:S77286 (5' terminus) rice dwarf virus RDV, Genomic RNA, 10 nt. 12/93 10bp.
S77287 Em(40)_vi:S77287 Gb(84)_vi:S77287 (5' terminus) rice dwarf virus RDV, Genomic RNA, 10 nt. 12/93 10bp.
S77288 Em(40)_vi:S77288 Gb(84)_vi:S77288 (5' terminus) rice dwarf virus RDV, Genomic RNA, 10 nt. 12/93 10bp.
S77289 Em(40)_vi:S77289 Gb(84)_vi:S77289 (5' terminus) rice dwarf virus RDV, Genomic RNA, 10 nt. 12/93 10bp.
S77290 Em(40)_vi:S77290 Gb(84)_vi:S77290 (5' terminus) rice dwarf virus RDV, Genomic RNA, 10 nt. 12/93 10bp.
X16017 Em(40)_vi:RDVSEG5A Gb(84)_vi:RDVSEG5A Rice Dwarf Virus RNA for genomic segment 5. 9/93 2,552bp.

X54620 Em(40)_vi:RDVSEG3A Gb(84)_vi:RDVSEG3A Rice dwarf virus S3 mRNA for segment 3. 9/93 3,195bp.
X54622 Em(40)_vi:RDVSEG4A Gb(84)_vi:RDVSEG4A Rice dwarf virus S4 mRNA for segment 4. 9/93 2,468bp.
S72083 Gb(89)_vi:S72083 P12...P120Pa (segment 12) rice dwarf virus RDV, RDV-H, Genomic RNA, 3 genes, 1066 nt). 10/94
S72085 Gb(89)_vi:S72085 P12...P120Pb (segment 12) rice dwarf virus RDV, RDV-AKITA, Genomic RNA, 3 genes, 1066 nt). 1
S74804 Gb(89)_vi:S74804 polyprotein (S2, segment 2) (rice dwarf phytoreovirus, RDV-H, Genomic RNA, 3512 nt). 5/95 3,5
S74805 Gb(89)_vi:S74805 polyprotein (S2, segment 2) (rice dwarf phytoreovirus, RDV-Akita, Genomic RNA, 3512 nt). 5/95
U25671 Gb(89)n:Rdu25671 Rice dwarf virus segment 10 nonstructural protein gene, complete cds. 6/95 1,164bp.

*Features of the genome:* s3 has an ORF which encodes 114000 polypeptide; s4 ORF encodes 79800 polypeptide; s5 ORF encodes 90500 polypeptide; s6 ORF encodes 57400 polypeptide; s7 ORF encodes 55300 polypeptide; s8 ORF encodes 46400 polypeptide; s9 ORF encodes 38900 polypeptide; s10 ORF encodes 39100 or 39200 polypeptide; s11 ORF encodes 20800 polypeptide.

Sub–genomic *m*RNA not found in infected cells.

*Virion protein(s)* six; $M_r$ 210000. $M_r$ of 2nd largest 185000. $M_r$ of 3rd largest 165000. $M_r$ of 4th largest 114000; major core protein. $M_r$ of 5th largest 55000; core protein. $M_r$ of 6th largest and smaller, 46000; outer capsid protein.

**Replication**

Genome replicates in cytoplasm. Replication does not depend on a helper virus.

**Cytopathology**

Virions found in leaves, mesophyll, epidermis, vascular parenchyma and all parts of the host plant; in cytoplasm. Inclusions present in infected cells; are viroplasms; they contain virions.

**Taxonomy and Relationships**

*Phytoreovirus: Reoviridae.*

Virus(es) with serologically related virions: rice gall dwarf phytoreovirus. Virus(es) with serologically unrelated virions: clover wound tumour phytoreovirus.

Each segment has a high homology with corresponding segments of clover wound tumour and rice gall dwarf phytoreoviruses. Best tests for diagnosis: distinguish by characteristic symptoms and serological tests.

**References**


Boccardo, G. and Milne, R.G. (1984). *CMI/AAB Descr. Pl. Viruses* No. 294, 7 pp.
Fujii-Kawata, I. and Miura, K. (1970). *J. mol. Biol.* **51**: 247.
Fukushi, T. and Shikata, E. (1963). *Virology* **21**: 500.
Fukushi, T. (1934). *J. Fac. Agric. Hokkaido Imp. Univ.* **39**: 41.
Fukushi, T. (1937). *Proc. imp. Acad. Japan* **13**: 328.
Iida, T.T., Shinkai, A. and Kimura, I. (1972). *CMI/AAB Descr. Pl. Viruses* No. 102.
Kimura, I., Kodama, T. and Suzuki, N. (1968). *Ann. Phytopath. Soc. Japan* **34**: 206.
Kimura, I., Minobe, Y. and Omura, T. (1987). *J. gen. Virol.* **68**: 3211.
Miyai, S., Kinitani, K. and Nakasuji, F. (1986). In: *Plant Virus Epidemics: Monitoring, Modelling and Predicting Outbreaks*. Eds G.D. McLean, R.G. Garrett and W.G. Ruesnik. Academic Press, London.

Nakata, M., Fukunaga, K. and Suzuki, N. (1978). *Ann. Phytopath. Soc. Japan* **44**: 288.
Nasu, S. (1963). *Bull. Kyushu agric. exp. Stn* **8**: 153.
Nasu, S. (1965). *Jap. J. appl. Ent. Zool.* **9**: 225.
Omura, T., Inoue, H., Pradhan, R.B. and Thapa, B.J. (1986). *Tech. Bull. Trop. agric. Res. Center, Japan* **21**: 78.
Omura, T., Minobe, Y. and Tsuchizaki, T. (1988). *J. gen. Virol.* **69**: 227.
Shikata, E. (1966). *J. Fac. Agric. Hokkaido Univ.* **55**: 1.
Shinkai, A. (1962). *Bull. Nat. Inst. Agric. Sci. (Japan), Ser. C* **14**: 1.
Takada, K. (1895). *J. Jap. Agric. Soc.* **171**: 1.
Takada, K. (1896). *J. Jap. Agric. Soc.* **172**: 13.
Uyeda, I. and Shikata, E. (1982). *Ann. Phytopath. Soc. Japan* **48**: 295.
Uyeda, I., Kudo, H., Takahashi, T., Sano, T., Oshima, K., Matsumara, T. and Shikata, E. (1989). *J. gen. Virol.* **70**: 1297.

## Rice gall dwarf *phytoreovirus*

**Data collated by T. Omura, 1984. Revised 1989.**

*Standard acronym:* RGDV.

First reported in *Oryza sativa*; from Thailand; by Omura *et al.* (1982).

**Natural host range and symptoms**

Symptoms persist.
*Oryza sativa* — stunting, gall formation and dark green leaves.

**Transmission**

Transmitted by a vector; an insect; *Nephotettix nigropictus, N. cincticeps, N. malayanus, N. virescens, Recilia dorsalis*; Cicadellidae. Transmitted in a persistent manner. Virus retained when the vector moults; multiplies in the vector; transmitted congenitally to the progeny of the vector; not transmitted by mechanical inoculation; not transmitted by contact between plants; not transmitted by seed; not transmitted by pollen.

**Geographical distribution**

Spreads in China. Found, but with no evidence of spread, in Malaysia and Thailand.

**Experimental host range**

Few (<3) families susceptible.

Diagnostically susceptible

*Avena sativa, Hordeum vulgare, Lolium multiflorum, Secale cereale, Triticum aestivum* — stunting, gall formation and leaf discoloration.

Maintenance and propagation hosts

*Oryza sativa.*

Assay hosts (**L**ocal lesions or **W**hole plants)

*Avena sativa* (W), *Hordeum vulgare* (W), *Oryza sativa* (W), *Secale cereale* (W), *Triticum aestivum* (W).

Monocotyledonous Hosts *(**Susceptible** or Insusceptible)*

***Avena sativa*. *Hordeum vulgare*. *Oryza sativa*.** *Secale cereale*. ***Triticum aestivum*.**

Sources of host-range data

Putta *et al.* (1982).

**Properties of particles in sap**

Leaf sap contains few virions.

**Purification**

Method: Omura *et al.* (1982).

**Particle morphology**

Virions isometric; 65 nm in diameter; angular in profile; with a conspicuous capsomere arrangement.

**Physical properties**

Two sedimenting components in purified preparations. Density 1.406 g $cm^{-3}$ in CsCl. $A_{260}/A_{280}$ ratio 1.42 (uncorrected for scattering).

**Biochemical properties**

*Genome* consists of RNA; double–stranded; of 12 parts (named S1–12); largest (or only) genome part the largest 4.423 kb (pairs); the 2nd largest 3.385 kb (pairs); the 3rd largest 3.205 kb (pairs); the 4th largest 2.5 kb (pairs); the 5th largest 2.47 kb (pairs); the 6th largest and other parts 1.72 kb (pairs; also 1.720, 1.578, 1.202, 1.198, 1.190 and 0.600 kb pairs). Genomic nucleic acid isolated by Hibi *et al.* (1984).

*Sequence database accession code(s):*

D01047 Em(40)_vi:RGDS9 Gb(84)_vi:RGDS9 Rice gall dwarf virus genome segment S9, complete sequence. 2/94 1,202bp.

D13410 Em(40)_vi:RGD47K Gb(84)_vi:RGD47K Rice gall dwarf virus segment 8 RNA 47K outer capsid protein gene. 12/92 1,578bp.

D13411 Em(40)_vi:RGDS10 Gb(84)_vi:RGDS10 Rice gall dwarf virus segment 10 RNA. 12/92 1,198bp.

D13774 Em(40)_vi:RGDVS3CP Gb(84)_vi:RGDVS3CP Rice gall dwarf virus RNA for 116K protein. 4/94 3,224bp.

S77267 Em(40)_vi:S77267 Gb(84)_vi:S77267 (3' terminus) rice gall dwarf virus RGDV, Genomic RNA, 10 nt. 12/93 10bp

S77268 Em(40)_vi:S77268 Gb(84)_vi:S77268 (3' terminus) rice gall dwarf virus RGDV, Genomic RNA, 10 nt. 12/93 10bp.

S77269 Em(40)_vi:S77269 Gb(84)_vi:S77269 (3' terminus) rice gall dwarf virus RGDV, Genomic RNA, 10 nt. 12/93 10bp.

S77270 Em(40)_vi:S77270 Gb(84)_vi:S77270 (3' terminus) rice gall dwarf virus RGDV, Genomic RNA, 10 nt. 12/93 10bp.

S77271 Em(40)_vi:S77271 Gb(84)_vi:S77271 (3' terminus) rice gall dwarf virus RGDV, Genomic RNA, 10 nt. 12/93 10bp.

S77272 Em(40)_vi:S77272 Gb(84)_vi:S77272 (3' terminus) rice gall dwarf virus RGDV, Genomic RNA, 10 nt. 12/93 10bp.

S77273 Em(40)_vi:S77273 Gb(84)_vi:S77273 (3' terminus) rice gall dwarf virus RGDV, Genomic RNA, 10 nt. 12/93 10bp.

S77274 Em(40)_vi:S77274 Gb(84)_vi:S77274 (3' terminus) rice gall dwarf virus RGDV, Genomic RNA, 10 nt. 12/93 10bp.

S77275 Em(40)_vi:S77275 Gb(84)_vi:S77275 (3' terminus) rice gall dwarf virus RGDV, Genomic RNA, 10 nt. 12/93 10bp.

S77276 Em(40)_vi:S77276 Gb(84)_vi:S77276 (3' terminus) rice gall dwarf virus RGDV, Genomic RNA, 10 nt. 12/93 10bp.

S77277 Em(40)_vi:S77277 Gb(84)_vi:S77277 (3' terminus) rice gall dwarf virus RGDV, Genomic RNA, 10 nt. 12/93 10bp.

S77278 Em(40)_vi:S77278 Gb(84)_vi:S77278 (3' terminus) rice gall dwarf virus RGDV, Genomic RNA, 10 nt. 12/93 10bp.
S77291 Em(40)_vi:S77291 Gb(84)_vi:S77291 (5' terminus) rice gall dwarf virus RGDV, Genomic RNA, 10 nt. 12/93 10bp.
S77292 Em(40)_vi:S77292 Gb(84)_vi:S77292 (5' terminus) rice gall dwarf virus RGDV, Genomic RNA, 10 nt. 12/93 10bp.
S77293 Em(40)_vi:S77293 Gb(84)_vi:S77293 (5' terminus) rice gall dwarf virus RGDV, Genomic RNA, 10 nt. 12/93 10bp.
S77294 Em(40)_vi:S77294 Gb(84)_vi:S77294 (5' terminus) rice gall dwarf virus RGDV, Genomic RNA, 10 nt. 12/93 10bp.
S77295 Em(40)_vi:S77295 Gb(84)_vi:S77295 (5' terminus) rice gall dwarf virus RGDV, Genomic RNA, 10 nt. 12/93 10bp.
S77296 Em(40)_vi:S77296 Gb(84)_vi:S77296 (5' terminus) rice gall dwarf virus RGDV, Genomic RNA, 10 nt. 12/93 10bp.
S77297 Em(40)_vi:S77297 Gb(84)_vi:S77297 (5' terminus) rice gall dwarf virus RGDV, Genomic RNA, 10 nt. 12/93 10bp.
S77298 Em(40)_vi:S77298 Gb(84)_vi:S77298 (5' terminus) rice gall dwarf virus RGDV, Genomic RNA, 10 nt. 12/93 10bp.
S77299 Em(40)_vi:S77299 Gb(84)_vi:S77299 (5' terminus) rice gall dwarf virus RGDV, Genomic RNA, 10 nt. 12/93 10bp.
S77300 Em(40)_vi:S77300 Gb(84)_vi:S77300 (5' terminus) rice gall dwarf virus RGDV, Genomic RNA, 10 nt. 12/93 10bp.
S77301 Em(40)_vi:S77301 Gb(84)_vi:S77301 (5' terminus) rice gall dwarf virus RGDV, Genomic RNA, 10 nt. 12/93 10bp.
S77302 Em(40)_vi:S77302 Gb(84)_vi:S77302 (5' terminus) rice gall dwarf virus RGDV, Genomic RNA, 10 nt. 12/93 10bp. 28 sequences.

*Virion protein(s)* 7; $M_r$ of the largest 183000. $M_r$ of 2nd largest 165000. $M_r$ of 3rd largest 150000. $M_r$ of 4th largest 143000. $M_r$ of 5th largest 120000. $M_r$ of 6th largest and smaller, 45000 and 56000. Method of preparation: Omura (1985).

**Cytopathology**

Virions found in phloem; in cytoplasm and in cell vacuoles. Inclusions present in infected cells; are viroplasms; they do not contain virions (but incomplete virion-like particles are frequently observed).

**Taxonomy and Relationships**

*Phytoreovirus: Reoviridae.*

Virus(es) with serologically unrelated virions: rice dwarf and clover wound tumour viruses. The antiserum against SDS-dissociated rice dwarf virus cross-reacted with one peptide component of rice gall dwarf virus when tested by immunoblotting.

**References**


Hibi, T., Omura, T. and Saito, Y. (1984). *J. gen. Virol.* **65**: 1585.
Matsuoka, M., Minobe, Y. and Omura, T. (1985). *Phytopathology* **75**: 1125.
Omura, T. (1985). *J. gen. Virol.* **66**: 811.
Omura, T. and Inoue, H. (1985). *AAB Descr. Pl. Viruses* No. 296, 4 pp.
Omura, T., Morinaka, T., Inoue, H. and Saito, Y. (1982). *Phytopathology* **72**: 1246.
Omura, T., Hibino, H., Usugi, T., Inoue, H., Morinaka, T., Tsurumachi, S., Ong, C.A., Putta, M., Tsuchizaki, T. and Saito, Y. (1984). *Plant Dis.* **68**: 374.
Omura, T., Hibino, H., Inoue, H. and Tsuchizaki, T. (1985). *J. gen. Virol.* **66**: 2581.
Omura, T., Inoue, H., Morinaka, T., Saito, Y., Chettanachit, D., Putta, M., Parejarearm, A. and Disthaporn, S. (1980). *Plant Dis.* **64**: 795.

Omura, T., Kimura, I., Tsuchizaki, T. and Saito, Y. (1988). *J. gen. Virol.* **69**: 429.
Putta, M., Chettanachit, D., Omura, T., Inoue, H., Morinaka, T., Honda, Y., Saito, Y. and Disthaporn, S. (1982). *Int. Rice Res. Newsl.* **7**: 13.

# Rice grassy stunt *tenuivirus*

**Data collated by H. Hibino, 1985. Revised 1989.**

*Synonyms:* rice rosette Philippines virus, probably rice rosette virus (Bergonia *et al.*, 1966; Gingery, 1988; Ling, 1972). *Standard acronym:* RGSV. *Strains:* GSV1 and GSV2 in the Philippines, wilted stunt virus (GSW), grassy stunt B (GSV), grassy stunt Y (GSY) in Taiwan. GSV 2-like strains in India, Indonesia and Thailand.

First reported in *Oryza sativa*; from the Philippines; by Ribera *et al.* (1966).

**Natural host range and symptoms**

Symptoms persist.

*Oryza sativa* — severe stunting, excessive tillering, erect growth habit, short, narrow pale green leaves.

**Transmission**

Transmitted by a vector; an insect; *Nilaparvata bakeri, N. lugens, N. muiri*; Delphacidae. Transmitted in a persistent manner. Virus retained when the vector moults; multiplies in the vector; not transmitted congenitally to the progeny of the vector; not transmitted by mechanical inoculation; not transmitted by contact between plants; not transmitted by seed; not transmitted by pollen.

**Ecology and control**

Studies reported by Khush and Ling (1974); Ling *et al.* (1976); Chen (1984).

**Geographical distribution**

Spreads in China, India, Indonesia, Japan, Malaysia, the Philippines, Sri Lanka, Taiwan, and Thailand.

**Experimental host range**

Few (<3) families susceptible.

Diagnostically susceptible

*Oryza sativa* cv. Shan San Sa San — striping of leaves.

*Oryza sativa* cv. Taichung Native 1 — excess tillering, short, narrow pale green leaves.

Maintenance and propagation hosts

*Oryza sativa*.

Assay hosts (**L**ocal lesions or **W**hole plants)

*Oryza sativa* (W).

Monocotyledonous Hosts *(**Susceptible** or Insusceptible)*

***Echinochloa colonum*. *Oryza punctata*. *Oryza sativa*.**

Additional hosts

***Oryza* spp., *Cynodon dactylon, Cyperus rotundus, Leersia hexandra, Monochoria vaginalis*.**

Sources of host-range data

Ling (1979); Anjaneyulu *et al.* (1988).

**Properties of particles in sap**

LIV: 0.5 days (12 hours). DEP: $\log_{10}$ minus 2. Electron microscopy: virions difficult to detect in sap.

**Purification**

Method: Hibino *et al.* (1985*a*; 1985*b*).

**Particle morphology**

Virions filamentous (often circular); not enveloped; of 950–1350 nm; 6 nm wide.

**Physical properties**

Three sedimenting components in purified preparations.

**Biochemical properties**

*Genome* consists of RNA; single–stranded; linear; of four parts; largest (or only) genome part 4.39 kb; the 2nd largest 3.94 kb; the 3rd largest 3.64 kb; the 4th largest 3.49 kb.

*Virion protein(s)* one; $M_r$ 31000. Method of preparation: Hibino *et al.* (1985*a*).

**Cytopathology**

Inclusions present in infected cells; are crystals in the cytoplasm, crystals in the nucleus, and unusual in shape; bundles of fibrillar structures. Other cellular changes: occurrence of small spherical virions in sieve tubes and phloem cells (Pellegrini and Bassi, 1978, Shikata *et al.*, 1980).

**Taxonomy and Relationships**

*Tenuivirus*.

Virus(es) with serologically related virions: rice stripe virus. Virus(es) with serologically unrelated virions: maize stripe and rice hoja blanca viruses.

Best tests for diagnosis: use combination of symptomatology, transmission tests using *Nilaparvata lugens* and serology.

**References**

Anjaneyulu, A., Daquioag, R.D., Mesina, M. E., Hibino, H., Lubigan. R.T. and Moody, K. (1988). *Int. Rice Res. Newsl.* **13**: 37.

Bergonia, Capule, Novera and Calica (1966). *Philipp. J. Pl. Ind.* **31**: 47.

Chen, C.C. (1984). *Plant Prot. Bull.* Taiwan **26**: 315.

Chen, C.C. and Chiu, R.J. (1982). *Plant Dis.* **66**: 15.

Gingery, R.E. (1988). In: *The Plant Viruses* Vol. 4, The Filamentous Plant Viruses, p. 302; ed. R.G. Milne. Plenum Press, New York.

Hibino, H. (1986). *AAB Descr. Pl. Viruses* No. 320, 5 pp.

Hibino, H., Iwasaki, M. and Izumi, K. (1982). *Ann. Phytopath. Soc. Japan* **48**: 388.

Hibino, H., Usugi, T. and Tsuchizaki, T. (1983*a*). *Int. Rice Res. Newsl.* **8**: 11.

Hibino, H., Usugi, T., Omura, T. and Shohara, K. (1983*b*). *Int. Rice Res. Newsl.* **8**: 9.

Hibino, H., Cabauatan, P.Q., Omura, T. and Tsuchizaki, T. (1985*a*). *Plant Dis.* **69**: 538.

Hibino, H., Usugi, T., Omura, T., Tsuchizaki, T., Shohara, K. and Iwasaki, M. (1985*b*). *Phytopathology* **75**: 894.

Khush, G.S. and Ling, K.C. (1974). *J. Hered.* **65**: 134.

Ling, K.C. (1972). *Rice Virus Diseases*. Los Banos, Philippines, Int. Rice Res. Inst.

Ling, K.C. (1979). Int. Rice Res. Inst., Los Banos, Philippines.
Ling, K.C., Aquieno, V.M. and Lee, S.H. (1976). *Plant Dis.* **54**: 565.
Omura, T., Hibino, H., Usugi, T., Inoue, H., Morinaka, T., Tsurumachi, S., Ong, C.A., Putta, M., Tsuchizaki, T. and Saito, Y. (1984). *Plant Dis.* **68**: 374.
Pellegrini, S. and Bassi, M. (1978). *Phytopath. Z.* **92**: 247.
Ribera, C.T., Ou, S.H. and Iida, T.T. (1966). *Pl. Dis. Reptr* **50**: 453.
Shikata, E., Senboku, T. and Ishimizu, T. (1980). *Proc. Japan Acad. 56, Ser.* **B**: 89.
Toriyama, S. (1985). *Ann. Phytopath. Soc. Japan* **51**: 59.

# Rice hoja blanca *tenuivirus*

**Data collated by F.J. Morales, 1984. Revised 1989.**

*Synonyms:* echinochloa hoja blanca virus (Falk *et al.*, 1987). *Standard acronym:* RHBV.
*Strains:* echinochloa strain (Morales and Niessen, 1985).

First reported in *Oryza sativa*; from Colombia; by Garcés (1940); Morales and Niessen (1983).

### Natural host range and symptoms

Symptoms persist.

*Oryza sativa*, wild *Oryza* spp. — systemic necrosis, systemic chlorotic striping in early infections.

### Transmission

Transmitted by a vector; an insect; *Sogatodes orizicola*; Delphacidae. Transmitted in a persistent manner. Virus retained when the vector moults; multiplies in the vector; transmitted congenitally to the progeny of the vector; not transmitted by mechanical inoculation; not transmitted by contact between plants; not transmitted by seed.

### Ecology and control

Studies reported by Everett and Lamey (1969); Lamey (1967).

### Geographical distribution

Spreads in the South and Central American region; Colombia, Costa Rica, Cuba, the Dominican Republic, Ecuador, El Salvador, Honduras, Nicaragua, Panama, Peru, and Venezuela (and elsewhere in the Caribbean). Found, but with no evidence of spread, in the U.S.A.

### Experimental host range

Few (<3) families susceptible.

Diagnostically susceptible

*Avena sativa, Digitaria* spp., *Hordeum vulgare, Leptochloa* spp. *Oryza sativa* cv Bluebonnet-50, *Secale cereale, Triticum aestivum* — chlorotic stripes.

Diagnostically insusceptible

Host range restricted to Gramineae.

Maintenance and propagation hosts

Purified from susceptible *Oryza sativa* cultivars only.

Assay hosts (**L**ocal lesions or **W**hole plants)

Susceptible *Oryza sativa* cultivars (W).

Monocotyledonous Hosts *(**Susceptible** or Insusceptible)*

***Avena sativa. Hordeum vulgare. Oryza sativa. Secale cereale. Triticum aestivum.***

Sources of host-range data

Lamey *et al.* (1964).

**Properties of particles in sap**

Leaf sap contains few virions.

**Purification**

Method: Morales and Niessen (1983).

**Particle morphology**

Virions filamentous; not enveloped; with no clear modal length; 3 nm wide.

**Physical properties**

One sedimenting component in purified preparations to two sedimenting components in purified preparations (in caesium salts), or one sedimenting component in purified preparations to three sedimenting components in purified preparations (in sucrose). Density 1.288 g $cm^{-3}$ in CsCl. Isoelectric point pH 4.5.

**Biochemical properties**

Virions contain 12 % nucleic acid (determined from $A_{260/280}$).

*Genome* consists of RNA; single–stranded; linear; of four parts; largest (or only) genome part 9.8 kb; the 2nd largest 3.5 kb; the 3rd largest 2.3 kb; the 4th largest 1.99 kb.

*Sequence database accession code(s):*

L07940 Em(40)_vi:HBVMNS3NP Gb(84)_vi:HBVMNS3NP Rice hoja blanca virus non-structural protein 3 (NS-3) and nucleoprotein (N-protein) genes, c

L14952 Em(40)_vi:HBVNS4VC Gb(84)_vi:HBVNS4VC Rice hoja blanca virus nonstructural protein (NS4) and (NS4vc) genes, complete cds's. 3/94 1.

L39989 Em(43)_vi:Hbvpv2a Gb(89)_vi:Hbvpv2a Rice hoja blanca virus membrane associated protein (pv2) gene, complete cds. 3/95 988bp.

*Virion protein(s)* one; $M_r$ 34000.

*Virus–coded non–virion proteins* isolated (Gingery *et al.* (1981)); one protein found. $M_r$ 17500.

**Cytopathology**

Virions found in all parts of the host plant; in cytoplasm. Inclusions present in infected cells; are amorphous X–bodies; they contain virions. Other cellular changes: disorganisation of chloroplasts, mitochondria and nuclei.

**Taxonomy and Relationships**

*Tenuivirus.*

Virus(es) with serologically related virions: echinochloa hoja blanca virus. Virus(es) with serologically unrelated virions: maize stripe and rice stripe viruses.

Best tests for diagnosis: rice hoja blanca is the only virus known to affect rice in the Americas.

**References**

Everett, T.R. and Lamey, H.A. (1969). In: *Viruses, Vectors, and Vegetation*; ed. K. Maramorosch. Interscience, New York.

Flak, B.W., Morales, F.J., Tsai, J.H. and Niessen, A. (1987). *Phytopathology* **77**: 196.

Garcés (1940). In: *Projects in Plant Pathology at the Palmira Experimental Station*, unnumbered report to the Ministry of Economy, Bogota.

Gingery, R.E., Nault, L.R. and Bradfute, O.E. (1981). *Virology* **112**: 99.
Lamey, H.A. (1967). In: *The Virus Diseases of the Rice Plant*, 354 pp. John Hopkins, Baltimore.
Lamey, H.A., McMillan, W.W. and McGuire, J.U. (1964). *Phytopathology* **54**: 536.
Morales, F.J. and Niessen, A.I. (1983). *Phytopathology* **73**: 971.
Morales, F.J. and Niessen, A.I. (1985). *AAB Descr. Pl. Viruses* No. 299, 4 pp.
Ramirez, B.C., Macaya, G., Calvert, L.A. and Haenni, A.-L. (1992). *J. gen. Virol.* **73**: 1457.
Ramirez, B.C., Lozano, I., Constantino, L.-M., Haenni, A.-L. and Calvert, L.A. (1993). *J. gen. Virol.* **74**: 2463.

## Rice necrosis mosaic *bymovirus*

**Data collated by T. Inouye, 1989.**

*Standard acronym:* RNMV.

First reported in *Oryza sativa*; from Okayama, Japan; by Fujii *et al.* (1966; 1967).

**Natural host range and symptoms**

Symptoms persist.
*Oryza sativa* — spindle shaped yellow flecks and streaks in older leaves.

**Transmission**

Transmitted by a vector; a fungus; *Polymyxa graminis*; Plasmodiophorales. Virus transmitted by mechanical inoculation; not transmitted by contact between plants; not transmitted by seed.

**Geographical distribution**

Spreads in Japan.

**Experimental host range**

Few (<3) families susceptible.

Diagnostically susceptible
*Oryza sativa* — spindle shaped yellow flecks and streaks.

Diagnostically insusceptible
*Avena sativa, Hordeum vulgare, Triticum aestivum, Zea mays.*

Maintenance and propagation hosts
*Oryza sativa.*

Assay hosts (**L**ocal lesions or **W**hole plants)
*Oryza sativa* (W).

Monocotyledonous Hosts *(**Susceptible** or Insusceptible)*
*Avena sativa. Hordeum vulgare.* ***Oryza sativa****. Triticum aestivum. Zea mays.*

Sources of host-range data
Inouye and Fujii (1977).

**Properties of particles in sap**

TIP: 60–65 °C. LIV: 7–14 days. DEP: $\log_{10}$ minus 3–4. Leaf sap contains few virions.

**Purification**

Method: Fujii (1978).

**Particle morphology**

Virions filamentous; not enveloped; usually flexuous; with clear modal lengths (two); of 257 and 550 nm; 13–14 nm wide. Axial canal obscure. Basic helix obscure.

**Cytopathology**

Virions found in all parts of the host plant; in cytoplasm. Inclusions present in infected cells; are amorphous X–bodies, membranous bodies, and pinwheels; they do not contain virions (but virions are often associated with inclusions).

**Taxonomy and Relationships**

*Bymovirus: Potyviridae.*

Virus(es) with serologically related virions: barley yellow mosaic and wheat yellow mosaic viruses.

**Comments**

Data on rice necrosis mosaic virus (Ghosh, 1979; 1981) in India has not been included, as its relationship to this virus is not known.

**References**

Fujii, S. (1978). *Special Bull. Okayama Prefectural Agric. Exp. Stn* 69.

Fujii, S., Okamoto, Y., Idei, T. and Inouye, T. (1966). *Ann. Phytopath. Soc. Japan* **32**: 82.

Fujii, S., Okamoto, Y., Idei, T., Shiomi, M., Inouye, T., Inouye, N., Mitsuhata, M. and Asatani, K. (1967). *Ann. Phytopath. Soc. Japan* **33**: 105.

Ghosh, S.K. (1981). *Plant Dis.* **65**: 602.

Inouye, T. and Fujii, S. (1977). *CMI/AAB Descr. Pl. Viruses* No. 172, 4 pp.

Usugi, T. and Saito, Y. (1976). *Ann. Phytopath. Soc. Japan* **42**: 12.

# Rice ragged stunt *oryzavirus*

**Data collated by G. Boccardo, 1984.**

*Synonyms:* rice infectious gall virus. *Standard acronym:* RSSV.

First reported in *Oryza sativa*; from Indonesia and the Philippines; by Hibino *et al.* (1977); Ling (1977); Shikata *et al.* (1977).

**Natural host range and symptoms**

Symptoms persist.

*Oryza latifolia, O. nivara, O. sativa* — stunting, enations on veins of leaves and leaf sheaths, ragged leaves, flower suppression.

**Transmission**

Transmitted by a vector; an insect; *Nilaparvata lugens*; Delphacidae. Transmitted in a persistent manner. Virus retained when the vector moults; multiplies in the vector; not transmitted congenitally to the progeny of the vector; does not require a helper virus for vector transmission; not transmitted by mechanical inoculation; not transmitted by contact between plants; not transmitted by seed; not transmitted by pollen.

**Geographical distribution**

Spreads in Bangladesh, China, India, Indonesia, Japan, Malaysia, the Philippines, Sri Lanka, Taiwan, and Thailand.

**Experimental host range**

Few (<3) families susceptible.

Diagnostically susceptible

*Oryza sativa* cv. Taichung Native (TN-1) — pale spindle-shaped enations on abaxial surfaces of leaves and leaf sheaths, twisted and ragged leaves, dark green colour, suppressed flowering, infected plants remain green and appear grassy after healthy plants have senesced.

Diagnostically insusceptible

*Triticum aestivum, Saccharum* spp.

Maintenance and propagation hosts

*Oryza sativa* cv. Taichung Native (TN-1).

Assay hosts (**L**ocal lesions or **W**hole plants)

*Oryza sativa* (W).

Monocotyledonous Hosts ***(Susceptible** or Insusceptible)*

***Hordeum vulgare***. ***Oryza sativa***. *Saccharum officinarum*. *Triticum aestivum*. ***Zea mays***.

Sources of host-range data

Hibino (1979); Shikata *et al.* (1979).

**Properties of particles in sap**

Leaf sap contains few virions (whole leaf sap), or contains many virions (in sap from enations).

**Purification**

Method: Milne *et al.* (1982).

**Particle morphology**

Virions isometric; not enveloped; 65 nm in diameter; angular in profile; without a conspicuous capsomere arrangement (but with 12 "B" type spikes 8–10 nm long).

**Physical properties**

One sedimenting component in purified preparations.

**Biochemical properties**

*Genome* consists of RNA; double–stranded; of 10 parts (named S1–10); largest (or only) genome part the largest 3.9 kb (pairs); the 2nd largest 3.9 kb (pairs); the 3rd largest 3.8 kb (pairs); the 4th largest 3.8 kb (pairs); the 5th largest 2.75 kb (pairs); the 6th largest and other parts 2.3 kb (pairs; also 1.95, 1.90, 1.20 and 1.16 kb pairs). Additional factor required for infectivity; as only the whole virion is infective.

*Non–genomic nucleic acid* not found in the virions.

*Virion protein(s)* 8; $M_r$ 125000. $M_r$ of 2nd largest 97000. $M_r$ of 3rd largest 66000. $M_r$ of 4th largest 64000. $M_r$ of 5th largest 48000. $M_r$ of 6th largest and smaller, 43000 (also 36000 and 32000).

**Replication**

Genome replicates in cytoplasmic viroplasms. Coat protein *m*RNA translated in the cytoplasm. Replication does not depend on a helper virus.

**Cytopathology**

Virions found in phloem enations; in cytoplasm. Inclusions present in infected cells; are viroplasms; they contain virions. Other cellular changes: cell enlargement and sometimes cell wall proliferation.

**Taxonomy and Relationships**

*Oryzavirus: Reoviridae* (the type species).

Virus(es) with serologically unrelated virions: sugarcane Fiji disease, maize rough dwarf, pangola stunt, rice black-streaked dwarf, oat sterile dwarf, rice dwarf, rice gall dwarf and clover wound tumor viruses.

**References**


Boccardo, G. and Milne, R.G. (1980). *Intervirology* **14**: 57.

Boccardo, G. and Milne, R.G. (1984). *CMI/AAB Descr. Pl. Viruses* No. 294, 7 pp.

Chen, C.C., Hsu, Y.H., Chen, M.J., and Chin, R.J. (1989). *Intervirology* **30**: 278.

Hibino, H. (1979). *Rev. Pl. Protect. Res.* **12**: 98.

Hibino, H. and Kimura, I. (1982). *Phytopathology* **72**: 656.

Hibino, H., Roechan, M., Sudarisman, S. and Tantera, D.M. (1977). *Contr. Centr. Res. Inst. Agric. Bogor* No. 35: 15.

Hibino, H., Saleh, N. and Roechan, M. (1979). *Ann. Phytopath. Soc. Japan* **45**: 228.

Lee, S.Y., Uyeda, I. and Shikata, E. (1987). *intervirology* **27**: 189.

Ling, K.C. (1977). *Int. Rice Res. Newsl.* **2**: 6.

Ling, K.C., Tiongco, E.R., Aguiero, V.M. and Cabauatan, P.Q. (1978). *Int. Rice Res. Paper Series 16*, 25 pp.

Luisoni, E., Milne, R.G. and Roggero, P. (1982). *Plant Dis.* **66**: 929.

Milne, R.G. (1980). *Intervirology* **14**: 331.

Milne, R.G., Luisoni, E. and Ling, K.C. (1979). *Pl. Dis. Reptr* **63**: 445.

Milne, R.G., Boccardo, G. and Ling, K.C. (1982). *CMI/AAB Descr. Pl. Viruses* No. 248, 5 pp.

Shikata, E., Senboku, T., Kamjaipai, K., Chou, T., Tiongco, E.R. and Ling, K.C. (1979). *Ann. Phytopath. Soc. Japan* **45**: 436.

Shikata, E., Leelopang, Tiongo, E.R. and Ling, K.C. (1977). *Int. Rice Res. Newsl.* **2**: 7.

Yan, J., Kudo, H., Uyeda, I., Lee, S.Y. and Shikata, E. (1992). *J. gen. Virol.* **73**: 785.


# Rice stripe necrosis (?) *furovirus*

**Data collated by A.A. Brunt, 1987 and C. Fauquet, 1987.**

*Standard acronym:* RSNV.

First reported in *Oryza sativa*; from Ivory Coast; by Louvel and Bidaux (1977).

**Natural host range and symptoms**

Symptoms persist.

*Oryza sativa* — chlorotic striping and necrosis of leaves, stunting and reduced tillering.

**Transmission**

Transmitted by a vector; a fungus; *Polymyxa graminis*. Virus transmitted by mechanical inoculation; not transmitted by seed.

**Geographical distribution**

Spreads in Cote d'Ivoire.

**Experimental host range**

Few (<3) families susceptible.

Diagnostically susceptible

*Chenopodium amaranticolor* — chlorotic local lesions.
*Nicotiana benthamiana* — necrotic ringspots.

Assay hosts (Local lesions or Whole plants)

*Chenopodium amaranticolor* (L).

Dicotyledonous Hosts *(**Susceptible** or Insusceptible)*

***Chenopodium amaranticolor*. *Nicotiana benthamiana*.**

Monocotyledonous Hosts *(**Susceptible** or Insusceptible)*

***Oryza sativa*.**

Sources of host-range data

Fauquet and Thouvenel (1983).

**Properties of particles in sap**

TIP: more than 50 °C.

**Purification**

Method: Fauquet and Thouvenel (1983).

**Particle morphology**

Virions rod–shaped; with a clear modal length; of 110–160 nm, or 270–380 nm; 20 nm wide.

**Physical properties**

Three sedimenting components in purified preparations.

**Taxonomy and Relationships**

Possible *Furovirus*.

Virus(es) with serologically unrelated virions: beet necrotic yellow vein, nicotiana velutina mosaic and peanut clump viruses.

**References**

Fauquet, C. and Thouvenel, J.C. (1983). *C. R. Acad. Sci. Paris* **296**: 575.
Louvel, D. and Bidaux, J.M. (1977). *Agron. Trop., Nogent* **32**: 257.

## Rice stripe *tenuivirus*

**Data collated by S. Toriyama, 1984. Revised 1989.**

*Synonyms:* ine shima-ha gare uirusu. *Standard acronym:* RSV.

First reported in *Oryza sativa*; from Japan; by Koganezawa *et al.*, 1975.

**Natural host range and symptoms**

Symptoms persist.

*Avena sativa, Digitaria adscendens, D. violascens, Eragrostis multiflorum, Hordeum vulgare, Oryza sativa, Setaria italica, S. viridis, Triticum aestivum, Zea mays* — systemic necrotic or chlorotic spots and streaking.

**Transmission**

Transmitted by a vector; an insect; *Laodelphax striatellus, Terthron albovittatum, Unkanodes sapporona, U. albifascia*; Delphacidae. Transmitted in a persistent manner. Virus retained when the vector moults; multiplies in the vector; transmitted congenitally to the progeny of the vector; transmitted by mechanical inoculation (but only with great difficulty); not transmitted by contact between plants; not transmitted by seed.

**Geographical distribution**

Spreads in the Eastern Asian region; China, Japan, Taiwan, and the former USSR.

**Experimental host range**

Few (<3) families susceptible.

Diagnostically susceptible

*Oryza sativa* (Japonica type), *Triticum aestivum, Zea mays* — chlorotic stripe and pale streaking.

Diagnostically insusceptible

*Anthoxanthum odoratum, Elymus tsukushiense* var. *transiens, Festuca arundinacea, Phalaris arundinacea, Poa pratensis.*

Maintenance and propagation hosts

*Oryza sativa* Japonica type varieties, *Triticum aestivum, Zea mays.*

Assay hosts (**L**ocal lesions or **W**hole plants)

*Oryza sativa* japonica type cvs Norin 8, Kinmaze, *Triticum aestivum, Zea mays* cv. Golden Cross Bantam, all (W).

Monocotyledonous Hosts *(**Susceptible** or Insusceptible)*

*Anthoxanthum odoratum*. ***Avena sativa***. ***Cynodon dactylon***. ***Dactylis glomerata***. *Echinochloa crus–galli*. ***Hordeum vulgare***. ***Lolium perenne***. ***Oryza sativa***. ***Phleum pratense***. ***Secale cereale***. ***Setaria italica***. ***Triticum aestivum***. ***Zea mays***.

Sources of host-range data

Iida (1969); Toriyama (1983*a*).

**Properties of particles in sap**

TIP: 50–55 °C. LIV: 4 days in extracts of viruliferous insects kept at 4°C, 8–12 months in viruliferous insects and diseased rice plants at -20°C and 1–2 months in purified preparations at -20°C. DEP: $\log_{10}$ minus 3–4 (from plants, $10^{-4}$ to $10^{-5}$ from viruliferous insects). Leaf sap contains many virions.

**Purification**

Method: Toriyama (1982*a*).

**Particle morphology**

Virions filamentous; not enveloped; usually flexuous; 8 nm wide. Axial canal obvious (canal- like structures are seen in some virions). Basic helix obvious.

**Physical properties**

Four sedimenting components in purified preparations (T, $M_1$, $M_2$, B and nB); sedimentation coefficient 98 *S* (nB); of the other(s) 68 *S* (M), or 80 *S* (B). Density 1.282 g $cm^{-3}$ in CsCl (not fixed with aldehydes). Isoelectric point pH 4.5.

**Biochemical properties**

Virions contain 12 % nucleic acid; 88 % protein.

*Genome* consists of RNA; single–stranded; linear; of four parts; largest (or only) genome part 9 kb; the 2nd largest 3.5 kb; the 3rd largest 2.48 kb; the 4th largest 2.14 kb. Genomic nucleic acid isolated by Toriyama (1982*a*, 1982*b*, 1989).

*Sequence database accession code(s):*

D01039 Em(40)_vi:RCSSEG4 Gb(84)_vi:RISSEG4 Rice stripe virus genome, segment4, complete sequence. 12/92 2,137bp.

D01094 Em(40)_vi:RCSRSVAS3 Gb(84)_vi:RSTRSVAS3 Rice stripe virus (RSV) ambisense genes in segment 3. 8/92 2,475bp.

D10979 Em(40)_vi:RCSRNA41 Gb(84)_vi:RCSRNA4 Rice stripe virus RNA 4, complete sequence. 9/92 2,157bp

D13787 Gb(84)n:RCSRNA2 Rice stripe virus segment 2 RNA for ORF 1,2. 9/94 3,514bp.

S58538 Em(40)_vi:S58538 Gb(84)_vi:S58538 22.7 kda protein, 94 kda putative membrane protein (RNA segment 2) rice stripe virus RSV, Ge

X53563 Em(40)_vi:RSVRNA3 Gb(84)_vi:RSVRNA3 Rice stripe virus RNA3 genomic RNA for coat protein (two ORF's). 9/91 2,504bp.

D13176 Em(43)_vi:Rcsrna Gb(89)_st:Rcsrna Rice stripe virus RNA. 9/94 3,514bp.

D31879 Em(43)_vi:Rcsrnap Gb(89)_vi:Rcsrnap Rice stripe virus RNA for polymerase. 1/95 8,970bp.

*Features of the genome:* Purified rice stripe virus preparation has a RNA-dependent RNA polymerase.

*Virion protein(s)* two; $M_r$ of the larger 32000. $M_r$ of 2nd largest 230000. Amino acid composition: Toriyama (1983).

**Cytopathology**

Virions found in epidermis; in nuclei (rarely), or in cytoplasm. Inclusions present in infected cells; are crystals in the cytoplasm and amorphous X–bodies; they do not contain virions.

**Taxonomy and Relationships**

*Tenuivirus.*

Virus(es) with serologically related virions: maize stripe tenuivirus, also possibly rice grassy stunt tenuivirus.

**References**


Iida, T.T. (1969). In: *The Virus Diseases of the Rice Plant*, p. 3. John Hopkins, Baltimore.

Kakutani, T., Hayano, Y., Hayashi, T. and Minobe, Y. (1990). *J. gen. Virol.* **71**: 1427.

Kakutani, T., Hayano, Y., Hayashi, T. and Minobe, Y. (1991). *J. gen. Virol.* **72**: 465.

Kishimoto, R. (1967). *Virology* **32**: 144.

Kiso, A. and Yamamoto, T. (1973). *Rev. Pl. Prot. Res.* **6**: 75.

Koganezawa, H., Doi, Y. and Yora, K. (1975). *Ann. Phytopath. Soc. Japan* **41**: 148.

Kuribayashi, K. (1931). *Bull. Nagano Agr. Exp. Stn* **2**: 45.

Shinkai, A. (1962). *Bull. Natl. Inst. Agr. Sci.* **14**: 1.

Takahashi, M., Toriyama, S., Kikuchi, Y., Hayakawa, T. and Ishihama, A. (1990). *J. gen. Virol.* **71**: 2817.

Takahashi, M., Toriyama, S., Hamamatsu, C. and Ishihama, A. (1993). *J. gen. Virol.* **74**: 769.

Toriyama, S. (1982*a*). *J. gen. Virol.* **61**: 187.

Toriyama, S. (1982*b*). *Ann. Phytopath. Soc. Japan* **48**: 482.

Toriyama, S. (1983*a*). *CMI/AAB Descr. Pl. Viruses* No. 269, 5 pp.

Toriyama, S. (1987). *J. gen. Virol.* **68**: 925.

Toriyama, S. (1986). *J. gen. Virol.* **67**: 1247.
Toriyama, S. (1986). *Microbiol. Sciences* **3**: 347.
Toriyama, S. and Watanabe, Y. (1989). *J. gen. Virol.* **70**: 505.
Zhu, Y., Hayakawa, T., Toriyama, S. and Takahashi, M. (1992). *J. gen. Virol.* **72**: 763.
Zhu, Y., Hayakawa, T. and Toriyama, S. (1992). *J. gen. Virol.* **73**: 1309.

# Rice transitory yellowing (?) *nucleorhabdovirus*

**Data collated by E. Shikata, 1989.**

*Standard acronym:* RTYV.

First reported in *Oryza sativa*; from Taiwan; by Chiu *et al.* (1965).

**Natural host range and symptoms**

Symptoms disappear soon after infection.

*Oryza sativa* — japonica types are more susceptible than indica types. Yellow or orange-buff colouring of oldest leaves, some rusty flecking. No systemic symptoms.

**Transmission**

Transmitted by a vector; an insect; *Nephotettix nigropictus, N. cincticeps, N. impictus*; Cicadellidae. Transmitted in a persistent manner. Virus retained when the vector moults; multiplies in the vector; not transmitted congenitally to the progeny of the vector; not transmitted by mechanical inoculation; not transmitted by contact between plants.

**Geographical distribution**

Spreads in Taiwan. Found, but with no evidence of spread, in Okinawa, Japan.

**Experimental host range**

Few (<3) families susceptible.

Diagnostically susceptible

*Oryza sativa* japonica type — yellow or orange-buff colouring of oldest leaves.

Maintenance and propagation hosts

*Oryza sativa* japonica type.

Assay hosts (**L**ocal lesions or **W**hole plants)

*Oryza sativa* japonica type (W).

Monocotyledonous Hosts *(**Susceptible** or Insusceptible)*

*Echinochloa colonum. Echinochloa crus–galli.* ***Oryza sativa*.**

Sources of host-range data

Chiu *et al.* (1968).

**Properties of particles in sap**

LIV: 11 days (at 0°C to 2°C). DEP: $\log_{10}$ minus 5.

**Purification**

Method: Inoue *et al.* (1986).

**Particle morphology**

Virions rhabdo– or bullet–shaped; of 129 nm (120–140 nm); 96 nm wide.

**Cytopathology**

Virions found in leaves, phloem and companion cells; in the perinuclear space.

**Taxonomy and Relationships**

Possible *Nucleorhabdovirus: Rhabdoviridae*.

**References**

Chiu, R.J., Jean, J.H., Chen, M.H. and Lo, T.C. (1968). *Phytopathology* **58**: 740.

Chiu, R.J., Lo, T.C., Pi, C.L. and Chen, M.H. (1965). *Bot. Bull. Acad. Sin., Taipei* **6**: 1.

Inoue *et al.* (1986). *Tech. Bull. Trop. Agr. Res. Center, Japan* **21**: 65.

Shikata, E. (1972). *CMI/AAB Descr. Pl. Viruses* No. 100, 3 pp.

## Rice tungro bacilliform *badnavirus*

**Data collated by H. Hibino, 1987. Updated by A.A. Brunt, 1991.**

*Synonyms:* rice yellow orange leaf virus. *Standard acronym:* RTBV.

First reported in *Oryza sativa*; from Indonesia; by Rivera and Ou (1965).

**Natural host range and symptoms**

Symptoms persist.

*Oryza sativa* — leaves yellow-orange, stunting.

**Transmission**

Transmitted by a vector; an insect; especially *Nephotettix virescens*, but also *N. nigropictus, N. cincticeps, N. malayanus, N. parvus, Recilia dorsalis*; Cicadellidae. Transmitted in a semi–persistent manner. Virus lost by the vector when it moults; does not multiply in the vector; not transmitted congenitally to the progeny of the vector; requires, for vector transmission, a helper virus (rice tungro spherical *waikavirus*); not transmitted by mechanical inoculation; not transmitted by contact between plants; not transmitted by seed; not transmitted by pollen.

**Ecology and control**

Studies reported by Cabunagan *et al.* (1987); Hibino *et al.* (1987); Hibino *et al.* (1988); Tiongco *et al.* (1988); Khan *et al.* (1991).

**Geographical distribution**

Spreads in the Eastern Asian region; China.

**Experimental host range**

Few (<3) families susceptible.

Diagnostically susceptible

*Oryza sativa* — leaves yellow-orange, stunting.

Maintenance and propagation hosts

*Oryza sativa*.

Assay hosts (**L**ocal lesions or **W**hole plants)

*Oryza sativa* (W), *Oryza australiensis* (W), *Oryza longistaminata* (W), *Oryza barthii* (W), *Oryza latifolia* (W), *Oryza nivara* (W), *Oryza glaberima* (W), *Oryza perennis* (W), *Eleusine indica* (W), *Cyperus rotundus* (W), *Fimbristylis miliacea* (W), *Leersia hexandria* (W), *Monochoria vaginalis* (W).

Monocotyledonous Hosts *(**Susceptible** or Insusceptible)*

***Oryza sativa*.**

Sources of host-range data

Anjaneyulu *et al.* (1988).

**Properties of particles in sap**

Leaf sap contains few virions. Electron microscopy: use PTA or UA, fixation not required.

**Purification**

Method: Omura *et al.* (1983); Cabauatan and Hibino (1988).

**Particle morphology**

Virions bacilliform; not enveloped; 110–400 nm in length; 30–35 nm wide.

**Physical properties**

One sedimenting component in purified preparations. Density 1.312 g $cm^{-3}$ in CsCl.

**Biochemical properties**

*Genome* consists of DNA; double–stranded; circular; unipartite; largest (or only) genome part 8.3 kb. Genomic nucleic acid isolated by Hibino *et al.* (1991).

*Sequence database accession code(s):*

D10774 Em(40)_vi:RTBWG Gb(84)_vi:RTBWG Rice tungro bacilliform virus whole genome. 9/92 8,002bp.

M65026 Em(40)_vi:LERTUORFS Gb(84)_vi:RTUORFS Rice tungro bacilliform open reading frames 1, 2, 3, and 4 genomic RNA. 12/92 8,000bp.

M95497 Em(40)_vi:LERTUPOLY Gb(84)_vi:RTUPOLYP Rice tungro virus polyprotein gene, complete cds. 1/93 12,484bp.

X57924 Em(40)_vi:RTBVPHIL Gb(84)_vi:RTBVPHIL Rice tungro bacilliform virus (infectious clone) ORF's P24, P12, P194 & P46. 6/91 8,002bp. 4 sequences.

*Virion protein(s)* one; $M_r$ 32000; coat protein.

**Cytopathology**

Virions found in leaves, roots, vascular parenchyma and phloem; in cytoplasm. Inclusions absent from infected cells. Other cellular changes: starch accumulation.

**Taxonomy and Relationships**

*Badnavirus.*

Best tests for diagnosis: ELISA or latex serology, transmission test using *Nephotettix virescens*, symptomatology.

**Comments**

Rice "tungro" disease is caused by a complex of rice tungro badnavirus and rice tungro spherical *waikavirus*; the badnavirus depends on the *waikavirus* for its transmission by leafhoppers, and causes the tungro symptoms, whereas the *waikavirus* alone causes mild stunting and enhances the tungro symptoms.

**References**

Anjaneyulu, A., Daquioag, R.D., Mesina, M. E., Hibino, H., Lubigan, R.T. and Moody, K. (1988). *Int. Rice Res. Newsl.* **13**: 30.

Bajet, N.B., Daquioag, R.D. and Hibino, H. (1985). *J. Pl. Prot. Tropics* **2**: 125.

Bao, Y. and Hull, R. (1992). *J. gen. Virol.* **73**: 1297.

Cabauatan, P.Q. and Hibino, H. (1985). *Philipp. Phytopathol.* **21**: 103.

Cabauatan, L.Q. and Hibino, H. (1988). *Plant Dis.* **72**: 525.

Cabunagan, R.C., Flores, Z.M., Hibino, H., Elazegui, F. and Mew, T.W. (1987). *Int. Rice Res. Newsl.* **12**: 25.
Favali, M.A., Pellegrini, S. and Bassi, M. (1975). *Virology* **66**: 793.
Hay, M., Jones, M.C., Blakebrough, M.L., Dasgupta, I., Davies, J.W. and Hull, R. (1991). *Nucl. Acids Res.* **19**: 2615.
Hibino, H. (1983). *Plant Dis.* **17**: 774.
Hibino, H. (1983). *Ann. Phytopath. Soc. Japan* **49**: 545.
Hibino, H. and Cabauatan, P.Q. (1986). *Phytopathology* **77**: 473.
Hibino, H. and Cabunagan, R.C. (1986). *Tropical Agric. Res. Series* **19**: 173.
Hibino, H., Roechan, M. and Sudarisman, S. (1978). *Phytopathology* **68**: 1412.
Hibino, H., Saleh, N. and Roechan, M. (1979). *Phytopathology* **69**: 1266.
Hibino, H., Tiongco, E.R., Cabunagan, R.C. and Flores, Z.M. (1987). *Phytopathology* **77**: 871.
Hibino, H., Ishikawa, K., Omura, T., Cabauatan, P.Q. and Koganezawa, H. (1991). *Phytopathology* **81**: 1130.
Khan, M.A., Hibino, H., Aguiero, V.M., Daquioag, R.D. and Opina, O.S. (1919). *Plant Dis.* **75**: 926.
Ling, K.C. (1966). *Phytopathology* **60**: 795.
Omura, T., Saito, Y., Usugi, T. and Hibino, H. (1982). *Ann. Phytopath. Soc. Japan* **49**: 73.
Omura, T., Hibino, H. and Usugi, T. (1984). *Plant Dis.* **68**: 374.
Qu, R., Bhattacharya, M., Laco, G.S., Subba Rao, B.L., Kaniewska, M.B., Elmer, J.S., Rochester, D.E., Smith, C.E. and Beachy, R.N. (1991). *Virology* **185**: 354.
Rivera, C.T. and Ou, S.H. (1965). *Pl. Dis. Reptr* **49**: 127.
Saito, Y. (1977). *Tropical Agric. Res. Series* **10**: 129.
Saito, Y., Roechan, M., Tantera, D.M. and Iwaki, M. (1975). *Phytopathology* **65**: 793.
Tiongco, E.R., Cabunagan, R.C., Flores, Z.M. and Hibino, H. (1988). *Int. Rice Res. Newsl.* **13**: 10.

# Rice tungro spherical *waikavirus*

**Data collated by H. Hibino, 1987. Updated by A.A. Brunt, 1991.**

*Synonyms:* rice leaf yellowing virus, rice penyakit habeng virus, rice penyakit mentek virus, rice waika virus, rice yellow leaf virus. *Standard acronym:* RTSV.

First reported in *Oryza sativa*; from the Philippines; by Gálvez (1968).

### Natural host range and symptoms

Symptoms persist.

*Oryza sativa* — slight stunting (but when in complex exacerbates the symptoms induced by rice tungro badnavirus).

### Transmission

Transmitted by a vector; an insect; especially *Nephotettix virescens*; Cicadellidae. Transmitted in a semi–persistent manner. Virus lost by the vector when it moults; does not multiply in the vector; not transmitted congenitally to the progeny of the vector; can help the vector transmission of another virus (rice tungro bacilliform badnavirus); not transmitted by mechanical inoculation; not transmitted by contact between plants; not transmitted by seed; not transmitted by pollen.

**Ecology and control**

Studies reported by Furuta (1977); Bajet *et al.* (1986); Hibino *et al.* (1987); Khan *et al.* (1991).

**Geographical distribution**

Spreads in the Eastern Asian region; China and Japan.

**Experimental host range**

Few (<3) families susceptible.

Diagnostically susceptible

*Oryza sativa* — mild stunting, grain discoloration.

Maintenance and propagation hosts

*Oryza sativa.*

Assay hosts (**L**ocal lesions or **W**hole plants)

*Oryza sativa* (W).

Monocotyledonous Hosts *(**Susceptible** or Insusceptible)*

***Cynodon dactylon*. *Echinochloa colonum*. *Oryza sativa*.**

Additional hosts

***Eleusine indica, Cyperus rotundus, Fimbristylis miliaceae, Cyperus difformis, Cynodon dactylon, Leersia hexandra, Digitaria ciliaris, Brachiaria mutica, Imperata cylindrica, Cyperus brevifolius, Eragrostis tenella, Monochoria vaginalis, Axonopus compresses, Oryza australiensis, O. longistaminata, O. barthii, O. latifolia, O. glaberima, O. nivara*** and ***O. perennis***.

Sources of host-range data

Anjaneyulu *et al.* (1988).

**Properties of particles in sap**

Leaf sap contains many virions. Electron microscopy: use PTA or UA.

**Purification**

Method: Omura *et al.* (1983); Cabauatan and Hibino (1988).

**Particle morphology**

Virions isometric; not enveloped; 30 nm in diameter; rounded in profile; with a conspicuous capsomere arrangement.

**Physical properties**

One sedimenting component in purified preparations; sedimentation coefficient 175 *S* (±5). Density 1.551 g $cm^{-3}$ in CsCl.

**Biochemical properties**

*Genome* consists of RNA; single–stranded; unipartite; largest (or only) genome part 10.422 kb. Genomic nucleic acid isolated by Hibino *et al.* (1991).

*Sequence database accession code(s):*

S65252 Em(40)_vi:S65252 Gb(84)_vi:S65252 polyprotein rice tungro spherical virus RTSV, Genomic RNA, 2823 nt. 1/94 2,823bp. 1 sequence.

*Virion protein(s)* one; $M_r$ 23000; and it is coat protein. $M_r$ of 2nd largest 24000; coat protein.

**Cytopathology**

Virions found in leaves, roots, vascular parenchyma and phloem; in cytoplasm. Inclusions present in infected cells; are amorphous X–bodies, viroplasms, and unusual in shape; lattice or tubular structures; they contain virions (in viroplasms), or they do not contain virions (in lattice structures). Other cellular changes: some starch accumulation.

**Taxonomy and Relationships**

*Waikavirus: Sequiviridae* (the type species).

Virus(es) with serologically unrelated virions: maize chlorotic dwarf virus.

Best tests for diagnosis: use electron microscopy, ELISA or latex agglutination serology.

**Comments**

Rice tungro disease is caused by a complex of rice tungro badnavirus and rice tungro spherical *waikavirus*; the badnavirus depends on the *waikavirus* for its transmission by leafhoppers, and causes the tungro symptoms, whereas the *waikavirus* alone causes mild stunting and enhances the tungro symptoms.

**References**


Anjaneyulu, A., Daquioag, R.D., Mesina, Ma.E., Hibino, H., Lubigan, R.T. and Moody, K. (1988). *Int. Rice Res. Newsl.* **13**: 30.

Bajet, N.B., Daquioag, R.D. and Hibino, H. (1985). *J. Pl. Prot. Tropics* **2**: 125.

Bajet, N.B., Aguiero, V.M., Daquioag, R.D., Jonson, G.B., Cabunagan, R.C., Mesina, E.M. and Hibino, H. (1986). *Plant Dis.* **70**: 971.

Cabauatan, P.Q. and Hibino, H. (1985). *Philipp. Phytopathol.* **21**: 103.

Cabauatan, P.Q. and Hibino, H. (1988). *Plant Dis.* **72**: 526.

Favali, M.A., Pellegrini, S. and Bassi, M. (1975). *Virology* **66**: 502.

Furata, T. (1977). *Rev. Pl. Prot. Res.* **10**: 70.

Gálvez, G.E. (1968). *Virology* **35**: 418.

Gálvez, G.E. (1971). *CMI/AAB Descr. Pl. Viruses* No. 67, 3 pp.

Hibino, H. (1983). *Plant Dis.* **17**: 774.

Hibino, H. (1983). *Ann. Phytopath. Soc. Japan* **49**: 545.

Hibino, H. and Cabauatan, P.Q. (1986). *Phytopathology* **77**: 473.

Hibino, H. and Cabunagan, R.C. (1986). *Tropical Agric. Res. Series* **19**: 173.

Hibino, H., Ishikawa, K., Omura, T., Cabauatan, P.Q. and Kaganezawa, H. (1991). *Phytopathology* **81**: 1130.

Hibino, H., Roechan, S. and Sudarisman, S. (1978). *Phytopathology* **68**: 1412.

Hibino, H., Saleh, N. and Roechan, M. (1979). *Phytopathology* **69**: 1266. 1346.39

Hibino, H., Tiongco, E.R., Cabunagan, K.C. and Flores, Z.M. (1987). *Phytopathology* **77**: 871.

Khan, M.A., Hibino, H., Aguiero, V.M., Daquioag, R.D. and Opina, O.S. (1991). *Plant Dis.* **75**: 926.

Ling, K.C. (1966). *Phytopathology* **60**: 795.

Murant, A.F. (1993). *Arch. Virol.* **131**: 441.

Omura, T., Saito, Y., Usugi, T. and Hibino, H. (1982). *Ann. Phytopath. Soc. Japan* **49**: 73.

Omura, T., Hibino, H. and Usugi, T. (1984). *Plant Dis.* **68**: 374.

Rivera, C.T. and Ou, S.H. (1965). *Pl. Dis. Reptr* **49**: 127.

Saito, Y. (1977). *Trop. Agric. Res. Series* **10**: 129.

Shen, P., Kaniewska, M.B., Smith, C. and Beachy, R.N. (1993). *Virology* **193**: 621.

Yamashita, S., Doi, Y. and Yora, K. (1977). *Ann. Phytopath. Soc. Japan* **43**: 278.

Zhang, S., Jones, M.C., Barker, P., Davies, J.W. and Hull, R. (1993). *Virus Genes* **7**: 121.

# Rice yellow mottle *sobemovirus*

**Data collated by A.A. Brunt and W. Bakker, 1987.**

*Standard acronym:* RYMV.

First reported in *Oryza sativa*; from Kisumu, Western Kenya; by Bakker (1970).

**Natural host range and symptoms**

Symptoms persist.
*Oryza sativa* — yellow or orange leaves, stunting, reduced tillering and sterility.
*Oryza longistaminata* — chlorosis and mottling.

**Transmission**

Transmitted by a vector; an insect; *Sesselia pusilla, Chaetocnema pulla, Trichispa serica* and *Dicladispa viridicyanea*; Coleoptera. Transmitted in a semi–persistent manner. Virus lost by the vector when it moults; does not multiply in the vector; not transmitted congenitally to the progeny of the vector; does not require a helper virus for vector transmission; transmitted by mechanical inoculation; not transmitted by seed.

**Ecology and control**

Studies reported by Bakker (1970, 1971, 1974); Fauquet and Thouvenel (1977); Rossel *et al.* (1982); John *et al.* (1984).

**Geographical distribution**

Spreads in Burkina Faso, Cote d'Ivoire, Kenya, Liberia, Mali, Niger, Nigeria, Sierra Leone, and Tanzania (and, possibly, Zanzibar).

**Experimental host range**

Few (<3) families susceptible.

Diagnostically susceptible

*Oryza sativa* — chlorotic dotting of leaf bases, elongating parallel to veins to produce mottling, stunting.
*Oryza barthii* — yellow spots and vein streaking.
*Oryza punctata* — systemic chlorotic dots.
*Dinebra retroflexa* — mild systemic mottling.
*Phleum arenarium* — systemic mottling and leaf death.

Maintenance and propagation hosts

*Oryza sativa* cv. Sindano.

Assay hosts (**L**ocal lesions or **W**hole plants)

*Oryza sativa* cvs Sindano and IR 20 (W).

Sources of host-range data

Bakker (1970, 1974).

**Properties of particles in sap**

TIP: 65–70 °C (in sap from infected rice). LIV: 99 days (at 20°C), or 260 days (at 4°C). DEP: $\log_{10}$ minus 6–9. Infectivity of sap not changed by treatment with di–ethyl ether. Leaf sap contains many virions.

**Purification**

Method: Bakker (1975); Fauquet and Thouvenel (1977).

**Particle morphology**

Virions isometric; not enveloped; 25 nm in diameter; angular in profile; with a conspicuous capsomere arrangement.

**Physical properties**

One sedimenting component in purified preparations; sedimentation coefficient 116 *S*. Density 1.359 g $cm^{-3}$ in CsCl. Isoelectric point pH 6. $A_{260}/A_{280}$ ratio 1.46–1.54.

**Biochemical properties**

Virions contain 23 % nucleic acid; 77 % protein; 0 % lipid.

*Genome* consists of RNA; single–stranded; unipartite; largest (or only) genome part 4.2 kb. Base composition 29 % G; 21 % A; 25 % C; 25 % U.

*Sequence database accession code(s):*

L20893 Em(40)_vi:RYVCGEN Gb(84)_vi:RYVCGEN Rice yellow mottle virus complete genome. 2/94 4,450bp.

U23142 Em(43)_vi:Ry23142 Gb(89)_vi:Ryu23142 Rice yellow mottle virus from Nigeria, complete genome. 4/95 4,451bp.

**Cytopathology**

Virions found in mesophyll and epidermis; in cytoplasm. Inclusions absent from infected cells.

**Taxonomy and Relationships**

*Sobemovirus*.

Virus(es) with serologically unrelated virions: cocksfoot mottle and cocksfoot mild mosaic (syn. phleum mottle) viruses.

**References**

Bakker, W. (1970). *Neth. J. Pl. Path* **76**: 53.

Bakker, W. (1971). *Neth J. Pl. Path.* **77**: 201.

Bakker, W. (1974). *Agric. Res. Reports* **829**: 152 pp.

Bakker, W. (1975). *CMI/AAB Descr. Pl. Viruses* No. 149, 4pp.

Fauquet, C. and Thouvenel, J.C. (1977). *Pl. Dis. Reptr* **61**: 443.

John, V.T., Thottappilly, G. and Awoderu, V.A. (1984). *FAO Pl. Prot. Bull.* **32**: 86.

Rossel, H.W., Thottappilly, G. and Buddenhagen, I.W. (1982). *FAO Pl. Prot. Bull.* **30**: 137.

# Rice yellow stunt virus

*Sequence database accession code(s):*

X75534 Em(40)_vi:RYSVNP Gb(84)_vi:RYSVNP Rice yellow stunt virus mRNA for nucleocapsid protein. 11/93 1,920bp. 1 sequence.

# Rose (?) *tobamovirus*

**Data collated by R.G.T. Hicks, 1991.**

*Synonyms:* rose colour break virus (Hunter, 1966; Farrar and Frost, 1972).

First reported in *Rosa* spp.; from Essex, U.K.; by Hicks and Frost (1984).

**Natural host range and symptoms**

Symptoms persist. Symptoms flower break.

*Rosa* spp. — deformed flecked and streaked petals, but roses infected with this tobamovirus, by sap or grafting, show no flower symptoms so the cause of the flower break disease is unknown.

**Transmission**

Disseminated by vegetative propagation. Virus transmitted by mechanical inoculation; transmitted by grafting (with difficulty); not transmitted by seed.

**Geographical distribution**

Spreads in the UK.

**Experimental host range**

Several (3–9) families susceptible. Experimentally infected plants mostly show local lesions, usually necrotic; systemic mosaics or mottles, or symptomless.

Diagnostically susceptible

*Chenopodium amaranticolor* — chlorotic local lesions; systemic chlorotic lines and spots.

*Cucumis sativus* cvs Lemon, BDR — starch lesions in cotyledons; not systemic.

*Datura stramonium* — chocolate local lesions; not systemic.

*Nicotiana debneyi* — chlorotic mottle; systemic necrotic local lesions or mosaic.

*Nicotiana glutinosa* — necrotic local lesions; systemic mosaic and leaf deformation.

*Nicotiana megalosiphon, N. tabacum* — cv. White Burley — necrotic local lesions; symptomless systemic.

*Petunia hybrida* cv. Rose of Heaven — necrotic local lesions; systemic mottle or flower break or symptomless.

Diagnostically insusceptible

*Pisum sativum* cv. Onward, *Vicia faba, Catharanthus roseus* cv. Little Pinkie.

Maintenance and propagation hosts

*Nicotiana clevelandii, N. tabacum* cv. White Burley, *Petunia hybrida* cv. Rose of Heaven.

Assay hosts (**L**ocal lesions or **W**hole plants)

*Chenopodium amaranticolor* (L), *Cucumis sativus* (L).

Dicotyledonous Hosts ***(Susceptible** or Insusceptible)*

***Chenopodium amaranticolor***. *Chenopodium foetidum*. ***Chenopodium murale***. ***Chenopodium quinoa***. ***Cucumis sativus***. ***Cucurbita moschata***. ***Datura stramonium***. ***Gomphrena globosa***. ***Lycopersicon esculentum***. ***Nicotiana clevelandii***. ***Nicotiana debneyi***. ***Nicotiana glutinosa***. ***Nicotiana rustica***. *Nicotiana sylvestris*. ***Petunia hybrida***. ***Phaseolus vulgaris***. *Pisum sativum*. ***Spinacia oleracea***. *Vicia faba*. ***Vigna unguiculata***.

Sources of host-range data

Hicks (1979); Hicks and Frost (1984).

**Properties of particles in sap**

TIP: 95 °C. LIV: 730 days. DEP: $\log_{10}$ minus 6–7. Leaf sap contains few virions.

**Purification**

Method: Hicks and Frost (1984).

**Particle morphology**

Virions rod–shaped; not enveloped; usually straight; with no clear modal length; of 310–320 nm; 15 nm wide. Axial canal obvious; 2.5 nm in diameter. Basic helix obscure.

**Physical properties**

$A_{260}/A_{280}$ ratio 1.2.

**Biochemical properties**

Virions contain 5 % nucleic acid; 95 % protein; 0 % lipid.

*Genome* consists of RNA; single–stranded.

**Cytopathology**

Virions found in leaves.

**Taxonomy and Relationships**

Possible *Tobamovirus*.

Virus(es) with serologically related virions: the ash isolate of tobacco mosaic virus. Virus(es) with serologically unrelated virions: sunn-hemp mosaic virus.

Best tests for diagnosis: *Nicotiana tabacum* produces no systemic symptoms. *Petunia hybrida* shows flower break at 14 or 17°C, but not 22°C or higher.

**References**

Farrar, E.H. and Frost, R.R. (1972). *Pl. Path.* **21**: 97.
Hicks, R.G.T. (1979). *Ph.D. Thesis* University of Manchester, U.K.
Hicks, R.G.T. and Frost, R.R. (1984). *Pl. Path.* **33**: 581.
Hunter, J.A. (1966). *N.Z. Jl agric. Res.* **9**: 1070.

# Rottboellia yellow mottle *sobemovirus*

**Data collated by S.L. Bhattiprolu, 1993.**

First reported in *Rottboellia cochinchinensis* syn. *R. exaltata*; from Nigeria; by Thottapilly *et al.* (1992).

**Natural host range and symptoms**

Symptoms persist. Symptoms chlorotic mottling and severe stunting.

*Rottboellia cochinchinensis* — chlorotic leaf mottling.

**Transmission**

Virus transmitted by mechanical inoculation.

**Geographical distribution**

Spreads in Nigeria.

**Experimental host range**

Few (<3) families susceptible. Experimentally infected plants mostly show systemic chlorotic mottle and stunting.

Diagnostically susceptible

*Rottboellia cochinchinensis* — systemic chlorotic mottle.

*Zea mays* — chlorotic spots, systemic mottle and stunting.

Diagnostically insusceptible

*Chenopodium amaranticolor, Datura stramonium, Glycine max, Oryza sativa, Phaseolus vulgaris, Triticum aestivum.*

Maintenance and propagation hosts

*Rottboellia cochinchinensis.*

Assay hosts (**L**ocal lesions or **W**hole plants)

*Rottboellia cochinchinensis* (W), *Zea mays* (W).

Dicotyledonous Hosts *(**Susceptible** or Insusceptible)*

*Beta vulgaris. Capsicum annuum. Capsicum frutescens. Chenopodium amaranticolor. Chenopodium foliosum. Chenopodium hybridum. Cucumis melo. Cucumis sativus. Cucurbita maxima. Cyamopsis tetragonoloba. Datura stramonium. Glycine max. Gomphrena globosa. Helianthus annuus. Lycopersicon esculentum. Nicotiana benthamiana. Nicotiana clevelandii. Nicotiana glutinosa. Nicotiana tabacum. Petunia hybrida. Phaseolus lunatus. Phaseolus vulgaris. Pisum sativum. Vigna unguiculata.*

Monocotyledonous Hosts *(**Susceptible** or Insusceptible)*

*Avena sativa: Hordeum vulgare. Oryza sativa. Phalaris arundinacea. Phleum pratense. Secale cereale. Triticum aestivum. Triticum durum.* ***Zea mays****.*

Additional hosts

*Bromus commutalis* (insusceptible), *B. sterilis* (insusceptible).

Sources of host-range data

Thottapilly *et al.* (1992).

**Purification**

Method: Thottapilly *et al.*, 1992.

**Particle morphology**

Virions isometric; not enveloped; 29 nm in diameter; angular in profile.

**Physical properties**

One sedimenting component in purified preparations; sedimentation coefficient 114 *S* (decreasing to *c.* 87S after EDTA treatment). Density 1.379 g $cm^{-3}$ in CsCl. Density two components in caesium sulphate; 1.29–1.31 and 1.32–1.33 g $cm^{-3}$. $A_{260}/A_{280}$ ratio 1.51.

**Biochemical properties**

*Genome* consists of RNA; single–stranded; linear; unipartite; largest (or only) genome part 4.4 kb. Genomic nucleic acid isolated by Thottapilly *et al.*, 1992. Infectivity lost when deproteinised with proteases. Additional factor required for infectivity.

*Non–genomic nucleic acid* not found in the virions.

*Virion protein(s)* one; $M_r$ 27000; coat. Method of preparation: Thottapilly *et al.*, 1992.

**Cytopathology**

Virions found in leaves; in cytoplasm and in cell vacuoles. Inclusions present in infected cells; are crystals in the cytoplasm (virion aggregates); they contain virions.

**Taxonomy and Relationships**

*Sobemovirus.*

Virus(es) with serologically unrelated virions: cocksfoot mild mosaic, cocksfoot mottle, cynosurus mottle, lucerne transient streak, panicum mosaic, rice yellow mottle and sowbane mosaic viruses.

**References**

Thottapilly, G., van Lent, J.W.M., Rossel, H.W. and Sehgal, O.P. (1992). *Ann. appl. Biol.* **120**: 405.

## Rubus Chinese seed-borne (?) *nepovirus*

**Data collated by D.J. Barbara, 1986.**

*Standard acronym:* RCSV.

First reported in *Rubus* spp.; from the U.K. (in seed imported from China); by Barbara *et al.* (1985).

**Natural host range and symptoms**

*Rubus* spp. — infected seedlings show no symptoms.

**Transmission**

As the virus is probably a nepovirus, it is likely to have a nematode vector. Virus transmitted by mechanical inoculation; transmitted by seed.

**Geographical distribution**

Spreads in China. Found, but with no evidence of spread, in the U.K. from seed imported from China.

**Experimental host range**

Several (3–9) families susceptible.

Diagnostically susceptible

*Chenopodium quinoa* — chlorotic local lesions, systemic chlorosis, deformation, necrotic flecking.

*Cucumis sativus* — chlorotic local lesions, systemic chlorotic flecking or chlorosis.

Maintenance and propagation hosts

*Nicotiana clevelandii, Chenopodium quinoa.*

Assay hosts (**L**ocal lesions or **W**hole plants)

*Chenopodium quinoa* (L).

Dicotyledonous Hosts *(**Susceptible** or Insusceptible)*

*Alyssum saxatile* (**I**) ***Antirrhinum majus*** (*Aubrieta deltoides* (**I**)). ***Beta vulgaris***. *Brassica oleracea* ssp. *botrytis*. ***Chenopodium album***. ***Chenopodium amaranticolor***. ***Chenopodium foetidum***. ***Chenopodium foliosum***. ***Chenopodium murale***. ***Chenopodium quinoa***. ***Cucumis sativus***. *Cucurbita pepo* (*Datura bernhardii* (**S**)). ***Datura stramonium***. *Helianthus annuus*. ***Lactuca sativa*** (*Lupinus polyphyllus* (**I**)). ***Lycopersicon esculentum***. *Medicago sativa*. ***Nicotiana benthamiana***. ***Nicotiana bigelovii***. ***Nicotiana clevelandii***. ***Nicotiana debneyi***. ***Nicotiana glutinosa***. ***Nicotiana rustica***. ***Nicotiana tabacum***. ***Petunia hybrida***. *Phaseolus vulgaris*. ***Pisum sativum***. *Raphanus sativus*. *Rubus idaeus*. *Solanum melongena*. *Trifolium hybridum*. *Trifolium pratense*. *Trifolium repens*. *Vicia faba*. *Vigna unguiculata*.

Sources of host-range data

Barbara *et al.* (1985).

**Properties of particles in sap**

LIV: 1–2 days. DEP: $\log_{10}$ minus 4–5. Leaf sap contains few virions. Electron microscopy: stable in 2% Na-PTA, pH 6.5, partially disrupted in 2% UA.

**Purification**

Method: Barbara *et al.* (1985).

**Particle morphology**

Virions isometric; not enveloped; 30 nm in diameter (some virions are penetrated by negative stain and these are 8–10% smaller in some preparations); angular in profile; without a conspicuous capsomere arrangement.

**Physical properties**

Two sedimenting components in purified preparations (in sucrose density gradients), or three sedimenting components in purified preparations (in CsCl). Density 1.461 g $cm^{-3}$ in CsCl, or 1.452 g $cm^{-3}$ in CsCl (not fixed with aldehydes). $A_{260}/A_{280}$ ratio 1.71 (B), or 0.87 (T).

**Biochemical properties**

Virions contain 37 % nucleic acid (B), or 36 % nucleic acid (M), or 0 % nucleic acid (T); 63 % protein (B), or 64 % protein (T), or 100 % protein (T).

*Genome* consists of RNA; single–stranded; possibly linear; of two parts. Genomic nucleic acid isolated by Barbara *et al.* (1985).

*Non–genomic nucleic acid* not found in the virions.

*Virion protein(s)* two; $M_r$ of the larger 47000; probably structural protein. $M_r$ of 2nd largest 25200; structural protein. Method of preparation: Barbara *et al.* (1985).

**Replication**

Replication does not depend on a helper virus.

**Cytopathology**

Virions found in mesophyll and epidermis (so far only these have been tested); in cytoplasm. Inclusions present in infected cells; are membranous bodies and unusual in shape; and tubular structures within the outer membrane; they contain virions.

**Taxonomy and Relationships**

Possible *Nepovirus: Comoviridae*.

Virus(es) with serologically related virions: strawberry latent ringspot virus. Virus(es) with serologically unrelated virions: arabis mosaic, cherry leaf roll, elm mosaic, myrobalan latent ringspot, satsuma dwarf, tobacco ringspot, tomato black ring, and tomato ringspot viruses.

Best tests for diagnosis: use serology — ELISA.

**References**

Barbara, D.J., Ashby, S.C. and McNamara, D.G. (1985). *Ann. appl. Biol.* **107**: 45.

# Rubus yellow net (?) *badnavirus*

**Data collated by A.T. Jones, 1987. Checked 1989.**

*Synonyms:* probably raspberry yellow mosaic virus. *Standard acronym:* RYNV.

First reported in *Rubus procerus*; from Canada; by Stace-Smith (1955).

**Natural host range and symptoms**

Symptoms persist.

*Rubus idaeus* var. *strigosus* (red raspberry) — chlorotic netting of leaf veins.

*R. procerus* (Himalayan blackberry) — chlorotic netting or yellow chlorosis, or no symptoms.

*R. occidentalis* (black raspberry) — chlorotic netting of leaf veins. *N.B.* the virus usually occurs mixed with other aphid-borne viruses, and these induce mosaic diseases of red and black raspberry.

**Transmission**

Transmitted by a vector; an insect; Aphididae. Transmitted in a semi–persistent manner. Virus lost by the vector when it moults; not transmitted congenitally to the progeny of the vector; not transmitted by mechanical inoculation; transmitted by grafting; not transmitted by contact between plants; not transmitted by pollen.

**Ecology and control**

Studies reported by Stace-Smith and Jones (1978).

**Geographical distribution**

Spreads in the Eurasian region and the North American region.

**Experimental host range**

Few (<3) families susceptible.

Diagnostically susceptible

*Rubus idaeus, R. occidentalis, R. procerus* — vein netting.

*Fragaria vesca* — downcurling of leaves, then necrosis of petiole, wilting and plant death.

*Rubus macraei* — vein netting.

Maintenance and propagation hosts

*Rubus macraei, R. idaeus, R. occidentalis.*

Dicotyledonous Hosts *(**Susceptible** or Insusceptible)*

***Fragaria vesca*. *Rubus idaeus*. *Rubus occidentalis*.**

Sources of host-range data

Stace-Smith and Jones (1978); Jones and Badenoch (1982).

**Properties of particles in sap**

Leaf sap contains few virions. Electron microscopy: 2% AM (pH 6.5).

**Particle morphology**

Virions bacilliform; not enveloped; 80–150 nm in length; 25–31 nm wide.

**Replication**

Replication does not depend on a helper virus.

**Cytopathology**

Virions found in leaves, mesophyll, epidermis and vascular parenchyma; in cytoplasm. Inclusions absent from infected cells. Other cellular changes: cell wall outgrowths and parallel arrangements of long tubules.

**Taxonomy and Relationships**

Probably *Badnavirus*.

**References**

Stace-Smith, R. (1955). *Can J. Bot.* **33**: 269.

Stace-Smith, R. and Jones, A.T. (1978). *CMI/AAB Descr. Pl. Viruses* No. 188, 4 pp.

Jones, A.T. and Badenoch, J. (1982). *Acta Hort.* **129**: 49.

## Ryegrass *alphacryptovirus*

**Data collated by P.L. Guy, 1987.**

*Synonyms:* ryegrass spherical cryptic virus. *Standard acronym:* RGCV.

First reported in *Lolium multiflorum*, (unknown cultivar) which may have also contained ryegrass mosaic virus; from the U.K.; by Plumb (1973).

### Natural host range and symptoms

*Lolium perenne, Lolium multiflorum, Lolium multiflorum x perenne* — symptomless but may enhance symptoms of ryegrass mosaic virus in some cultivars.

### Transmission

Virus not transmitted by mechanical inoculation; not transmitted by contact between plants; transmitted by seed; transmitted by pollen to the seed.

### Geographical distribution

Spreads in Australia, Germany, Italy, and the UK.

### Experimental host range

Few (<3) families susceptible.

Monocotyledonous Hosts *(**Susceptible** or Insusceptible)*

***Lolium multiflorum. Lolium perenne.***

### Properties of particles in sap

Leaf sap contains few virions (or very few virions; more virions in plants also infected with ryegrass mosaic virus). Electron microscopy: repeated testing with ISEM necessary to ensure that preparation is free of other viruses.

### Purification

Method: extract sap in distilled water, heat 40°C for 10 minutes, then purify through differential centrifugation.

### Particle morphology

Virions isometric; not enveloped; 29–30 nm in diameter; rounded in profile; without a conspicuous capsomere arrangement.

### Biochemical properties

*Genome* consists of RNA; double–stranded.

### Cytopathology

Virions found in leaves; in cytoplasm and in nuclei.

### Taxonomy and Relationships

*Alphacryptovirus: Partitiviridae.*

Virus(es) with serologically unrelated virions: beet cryptic and carnation cryptic viruses.

### References

Luisoni, E. and Milne, R.G. (1980). *Proc. 3rd Conf. on Virus Diseases of Gramineae in Europe*, p. 73.

Plumb, R.T. and Lennon, A.M. (1980). *Proc. 3rd Conf. on Virus Diseases of Gramineae in Europe*, p. 75.

# Ryegrass bacilliform (?) *nucleorhabdovirus*

**Data collated by A.A. Brunt, 1991.**

First reported in *Lolium perenne, L. multiflorum x L. perenne*; from the U.K.; by Plumb and James (1975).

**Natural host range and symptoms**

*Lolium* spp. — probably symptomless, but may intensify the symptoms of ryegrass mosaic rymovirus.

**Particle morphology**

Virions rhabdo– or bullet–shaped; of 288 nm (in vivo), or 210 nm (for damaged virions found in negatively stained sap); 68 nm wide.

**Cytopathology**

Virions found in mesophyll within membrane bound vesicles; in the perinuclear space.

**Taxonomy and Relationships**

Possible *Nucleorhabdovirus: Rhabdoviridae.*

**References**

Plumb, R.T. and James, M. (1975). *Ann. appl. Biol.* **80**: 181.

# Ryegrass mosaic *rymovirus*

**Data collated by R.T. Plumb, 1980. Revised 1984.**

*Synonyms:* ryegrass streak mosaic virus. *Standard acronym:* RGMV.

First reported in *Lolium multiflorum*; from Washington state U.S.A; by Bruehl *et al.* (1957).

**Natural host range and symptoms**

Symptoms persist.

*Lolium* spp. in Europe, *Lolium* spp. and *Dactylis glomerata* in North America — chlorotic flecking, mosaic, necrosis with some isolates, stunting and fewer tillers. Symptoms usually more obvious in *L. multiflorum* and *L. multiflorum x, L. perenne* than in *L. perenne*. Necrosis rare in *L. perenne*.

**Transmission**

Transmitted by a vector; a mite; *Abacarus hystrix*; Eriophyidae. Virus transmitted by mechanical inoculation; not transmitted by seed; not transmitted by pollen.

**Geographical distribution**

Spreads in the Eurasian region and the North American region; Australia.

**Experimental host range**

Few (<3) families susceptible.

Diagnostically susceptible

*Avena sativa, Lolium multiflorum* — systemic mosaic, necrotic flecking and striping.
*L. perenne* — systemic mosaic.

Monocotyledonous Hosts *(**Susceptible** or Insusceptible)*

*Avena sativa*. ***Bromus mollis***. ***Dactylis glomerata***. ***Festuca pratensis***. ***Lolium perenne***. ***Oryza sativa***. ***Triticum aestivum***.

Sources of host-range data

Mulligan (1960).

**Properties of particles in sap**

TIP: 60 °C. LIV: 1 days. DEP: $log_{10}$ minus 3. Leaf sap contains few virions.

**Purification**

Method: Paliwal *et al.* (1976).

**Particle morphology**

Virions filamentous; not enveloped; usually flexuous; with a clear modal length; of 700 nm; 15 nm wide. Axial canal obscure. Basic helix obscure.

**Physical properties**

One sedimenting component in purified preparations; sedimentation coefficient 166 *S*. Density 1.325 g $cm^{-3}$ in CsCl (at 25°C), or 1.3049 g $cm^{-3}$ in CsCl (when fixed with glutaraldehyde).

**Biochemical properties**

Virions contain 5.3 % nucleic acid; 0 % lipid.

*Genome* consists of RNA; single–stranded; unipartite; largest (or only) genome part 9–10 kb. Base composition 31 % G; 23 % A; 24 % C; 22 % U. Infectivity retained when deproteinised with phenol or detergent.

*Virion protein(s)* one; $M_r$ 29200. Method of preparation: Paliwal *et al.* (1976).

**Replication**

Replication does not depend on a helper virus.

**Cytopathology**

Virions found in mesophyll; in cytoplasm. Inclusions present in infected cells; are pinwheels and unusual in shape; and cylindrical structures; they do not contain virions. Other cellular changes: vacuolation of cytoplasm.

**Taxonomy and Relationships**

*Rymovirus: Potyviridae* (the type species).

Virus(es) with serologically related virions: isolates of the virus from Europe and North America are closely related. Virus(es) with serologically unrelated virions: agropyron mosaic, spartina mottle, hordeum mosaic, wheat streak mosaic and oat necrotic mottle viruses.

**References**

Bruehl, G.W., Toko, H. and McKinney, H.H. (1957). *Phytopathology* **47**: 517.
Gibson, R.W. and Heard, A.J. (1979). *Ann. appl. Biol.* **91**: 101.
Mulligan, T.E. (1960). *Ann. appl. Biol.* **48**: 575.
Paliwal, Y.C. and Tremaine, J.H. (1976). *Phytopathology* **66**: 406.
Plumb, R.T. and James, M. (1973). *J. gen. Virol.* **18**: 409.
Slykhuis, J.T. (1958). *FAO Pl. Prot. Bull.* **6**: 129.
Slykhuis, J.T. and Paliwal, Y.C. (1972). *CMI/AAB Descr. Pl. Viruses* No. 86, 4 pp.
Slykhuis, J.T., Watson, M.A. and Mulligan, T.E. (1956). *Rep. Rothamsted Exp. Stn.* 1956, p. 104.

# Ryegrass mottle (?) *sobemovirus*

**Data collated by A.A. Brunt, 1991.**

First reported in *Lolium multiflorum* and *Dactylis glomerata*; from Japan; by Toriyama *et al.* (1983).

**Natural host range and symptoms**

Symptoms persist.

*Dactylis glomerata, Lolium multiflorum* — leaf mottling and necrosis.

**Geographical distribution**

Spreads in Japan.

**Experimental host range**

Few (<3) families susceptible.

Diagnostically susceptible

*Dactylis glomerata, Festuca rubra, Avena sativa, Hordeum vulgare, Lolium multiflorum, L. perenne, Setaria italica, Triticum aestivum, Secale cereale* — leaf mottling and, with some species, leaf necrosis.

Maintenance and propagation hosts

*Lolium multiflorum, Triticum aestivum.*

Assay hosts (**L**ocal lesions or **W**hole plants)

*Lolium multiflorum* (W), *Triticum aestivum* (W).

Monocotyledonous Hosts ***(Susceptible** or Insusceptible)*

***Avena sativa. Dactylis glomerata. Hordeum vulgare. Lolium multiflorum. Lolium perenne. Secale cereale. Setaria italica. Triticum aestivum.***

Sources of host-range data

Toriyama *et al.* (1983).

**Particle morphology**

Virions isometric; not enveloped; 28 nm in diameter.

**Physical properties**

One sedimenting component in purified preparations; sedimentation coefficient 108 *S*. Density 1.366 g $cm^{-3}$ in CsCl.

**Biochemical properties**

Virions contain 23 % nucleic acid; 77 % protein; 0 % lipid.

*Genome* consists of RNA; single–stranded; unipartite; largest (or only) genome part 4.2 kb.

*Virion protein(s)* one; $M_r$ 26000; coat protein.

**Taxonomy and Relationships**

Possible *Sobemovirus.*

Virus(es) with serologically unrelated virions: cocksfoot mild mosaic, cocksfoot mottle and cynosurus mottle viruses.

**References**

Toriyama, S., Mikoshiba, Y. and Doi, Y. (1983). *Ann. Phytopath. Soc. Japan* **49**: 610.

# Saguaro cactus (?) *carmovirus*

**Data collated by J.H. Tremaine, 1988.**

*Synonyms:* cactus Saguaro virus. *Standard acronym:* SCV.

First reported in *Carnegiea gigantea*; from Arizona, U.S.A.; by Milbrath and Nelson (1972).

**Natural host range and symptoms**

Symptoms none.
*Carnegiea gigantea, Chamaecereus sylvestrii* — symptomless.

**Transmission**

Virus transmitted by mechanical inoculation.

**Geographical distribution**

Found, but with no evidence of spread, in Arizona, U.S.A. or Germany.

Diagnostically susceptible

*Chenopodium capitatum* — systemic vein-clearing and necrosis.
*Chenopodium amaranticolor, C. quinoa* — local lesions, not systemic.
*Gomphrena globosa* — local lesions, systemic spots.

Maintenance and propagation hosts

*Chenopodium capitatum.*

Assay hosts (**L**ocal lesions or **W**hole plants)

*Chenopodium amaranticolor* (L), *C. quinoa* (L).

**Properties of particles in sap**

TIP: 55–65 °C. LIV: 34 days. DEP: $\log_{10}$ minus 4. Leaf sap contains many virions.

**Purification**

Method: Nelson and Tremaine (1975).

**Particle morphology**

Virions isometric; not enveloped; 32 nm in diameter; rounded in profile; without a conspicuous capsomere arrangement.

**Physical properties**

One sedimenting component in purified preparations; sedimentation coefficient 118 *S*. Density 1.33 g $cm^{-3}$ in CsCl. Isoelectric point pH 3.5, 5, and 5.5.

**Biochemical properties**

Virions contain 17 % nucleic acid; 83 % protein; 0 % lipid.

*Genome* consists of RNA; single–stranded; linear; unipartite; largest (or only) genome part 4.2 kb. Base composition 29 % G; 24 % A; 21 % C; 27 % U.

*Non–genomic nucleic acid* found in the virions; is subgenomic *m*RNA.

*Virion protein(s)* one (and degraded protein and dimers 5%); $M_r$ 38900; coat protein. Method of preparation: Nelson and Tremaine (1975). Amino acid composition: Nelson and Tremaine (1975).

**Replication**

Replication does not depend on a helper virus.

**Cytopathology**

Virions found in cactus buds, flowers and fruit.

**Taxonomy and Relationships**

Possible *Carmovirus: Tombusviridae.*

Virus(es) with serologically unrelated virions: carnation mottle and turnip crinkle viruses.

Like carmoviruses it produces dsRNAs, and of a similar size (1.7 and 1.5 kbp) (Morris and Carrington, 1985).

**References**

Milbrath, G.M. and Nelson, M.R. (1972). *Phytopathology* **62**: 739.

Morris, T.J. and Carrington, J.C. (1985). In: *The Plant Viruses*, Vol. 3, Polyhedral Virions with Monopartite RNA Genomes, p. 73 ed. R. Koenig. Plenum Press, New York.

Nelson, M.R. and Tremaine, J.H. (1975). *Virology* **65**: 309.

Nelson, M.R., Yoshimura, M.A. and Tremaine, J.H. (1975). *CMI/AAB Descr. Pl. Viruses* No. 148, 4 pp.

# Saintpaulia leaf necrosis (?) rhabdovirus

**Data collated by A.A. Brunt, 1991.**

*Standard acronym:* SLNV.

First reported in *Saintpaulia ionantha*; from the former Czechoslovakia; by Ciampor and Dokoupil (1974).

**Natural host range and symptoms**

*Saintpaulia ionantha* — leaf necrosis.

**Particle morphology**

Virions rhabdo– or bullet–shaped; enveloped; with a clear modal length; of 200–220 nm; 60–65 nm wide.

**Taxonomy and Relationships**

Possible unassigned rhabdovirus.

**References**

Ciampor, F. and Dokoupil, N. (1974). *Acta Virol.* **18**: 355.

# Sambucus vein clearing (?) rhabdovirus

**Data collated by A.A. Brunt, 1991.**

*Synonyms:* elderberry vein clearing virus. *Standard acronym:* SVCV.

First reported in *Sambucus* sp.; from Japan; by Kusonoki *et al.* (1977).

**Natural host range and symptoms**

*Sambucus* sp. — clearing of leaf veins.

Diagnostically susceptible

*Sambucus* sp. — vein clearing.

**Particle morphology**

Virions rhabdo– or bullet–shaped; enveloped; with a clear modal length; of 275 nm (in sections); 80 nm wide (in sections).

**Taxonomy and Relationships**

Possible unassigned rhabdovirus.

**References**

Kusonoki, M., Chang, M.U., Arai, K., Doi, Y. and Yora, K. (1977). *Ann. Phytopath. Soc. Japan* **43**: 125.

# Santosai temperate (?) *alphacryptovirus*

**Data collated by A.A. Brunt, 1991.**

*Standard acronym:* STV.

First reported in *Brassica rapa var. amplexicaulis subvar. dentata*; from Japan; by Natsuaki *et al.* (1983).

**Natural host range and symptoms**

*Brassica rapa var. amplexicaulis subvar. dentata* — symptomless infection.

**Transmission**

Virus not transmitted by mechanical inoculation; transmitted by seed.

**Geographical distribution**

Found, but with no evidence of spread, in Japan.

**Particle morphology**

Virions isometric; not enveloped; 30 nm in diameter.

**Taxonomy and Relationships**

Possible *Alphacryptovirus: Partitiviridae.*

**References**

Boccardo, G., Lisa, V., Luisoni, E. and Milne, R.G. (1987). *Adv. Virus Res.* **32**: 171.

Natsuaki, T., Yamashita, S., Doi, Y., Okuda, S. and Teranaka, M. (1983). *Ann. Phytopath. Soc. Japan* **49**: 593.

Natsuaki, T., Natsuaki, K.T., Okuda, S., Teranaka, M., Milne, R.G., Boccardo, G. and Luisoni, E. (1986). *Intervirology* **25**: 69.

# Sarracenia purpurea (?) rhabdovirus

**Data collated by A.A. Brunt, 1991.**

*Standard acronym:* SPV.

First reported in *Sarracenia purpurea*; from Germany; by Barckhaus and Weinert (1975).

**Natural host range and symptoms**

*Sarracenia purpurea* — symptoms unrecorded.

**Particle morphology**

Virions rhabdo– or bullet–shaped.

**Taxonomy and Relationships**

Possible unassigned rhabdovirus.

**References**

Barckhaus, R.H. and Weinert, H. (1975). *Protoplasma* **84**: 101.

## Satsuma dwarf (?) *nepovirus*

**Data collated by A.A. Brunt and Y. Saito, 1987.**

*Standard acronym:* SDV. *Strains:* citrus mosaic, natsudaidai dwarf and navel orange infectious mottling.

First reported in *Citrus unshia* (Satsuma); from Japan; by Yamada and Sawamura (1950).

**Natural host range and symptoms**

Symptoms persist.
*Citrus unshiu* — stunting, leaf malformation and rosetting.

**Transmission**

Virus transmitted by mechanical inoculation; transmitted by grafting; not transmitted by contact between plants; transmitted by seed (8.6% transmission in *Phaseolus vulgaris*).

**Ecology and control**

Studies reported by Usugi and Saito (1979).

**Geographical distribution**

Spreads in Japan and Turkey.

**Experimental host range**

Several (3–9) families susceptible.

Diagnostically susceptible

*Chenopodium quinoa* — chlorotic and necrotic local lesions; not systemic.
*Physalis floridana* — systemic mottling and leaf malformation.
*Sesamum indicum* — local lesions; systemic vein clearing, necrosis and leaf deformation.
*Vigna unguiculata* — local lesions; vein clearing and mottling, necrotic streaking of petioles and stems.

Maintenance and propagation hosts

*Physalis floridana*.

Assay hosts (**L**ocal lesions or **W**hole plants)

*Chenopodium quinoa* (L).

Sources of host-range data

Tanaka and Nakanishi (1972); Usugi and Saito (1977).

**Properties of particles in sap**

TIP: 50–55 °C (in *Physalis floridana* sap). LIV: 8–12 days (at room temperature). DEP: $\log_{10}$ minus 3–4. Leaf sap contains few virions.

**Purification**

Method: Usugi and Saito (1979).

**Particle morphology**

Virions isometric; not enveloped; 26 nm in diameter; angular in profile; without a conspicuous capsomere arrangement.

**Physical properties**

Three sedimenting components in purified preparations; sedimentation coefficient of the fastest 129 *S* (B); of the other(s) 119 *S* (M). Density 1.46 g $cm^{-3}$ in CsCl (B), or 1.43 g $cm^{-3}$ in CsCl (M).

**Biochemical properties**

Virions contain 37 % nucleic acid (B), or 33 % nucleic acid (M), or 0 % nucleic acid (T); 63 % protein (B), or 67 % protein (M), or 100 % protein (T).

*Genome* consists of RNA; single–stranded; of two parts; largest (or only) genome part the larger 5.3 kb; the 2nd largest 4.8 kb.

*Virion protein(s)* two; $M_r$ of the larger 42000; coat protein. $M_r$ of 2nd largest 21000; coat protein. Method of preparation: Usugi and Saito (1979).

**Cytopathology**

Virions found in mesophyll and epidermis; in cytoplasm. Inclusions present in infected cells; are membranous bodies (vesicular). Other cellular changes: tubular outgrowths of cell walls containing a single row of virions.

**Taxonomy and Relationships**

Possible *Nepovirus: Comoviridae.*

Virus(es) with serologically unrelated virions: cherry leaf roll, citrus leaf rugose, grapevine Bulgarian latent, mulberry ringspot, raspberry ringspot, strawberry latent ringspot, tobacco ringspot, tomato black ring and tomato ringspot viruses.

**References**

Tanaka and Nakanishi (1972). *Ann. Phytopath. Soc. Japan* **38**: 212.
Usugi, T. and Saito, Y. (1977). *Ann. Phytopath. Soc. Japan* **43**: 137.
Usugi, T. and Saito, Y. (1979). *CMI/AAB Descr. Pl. Viruses* No. 208, 4 pp.
Yamada, M. (1950). *Bull. Hort. Div. Tokai-Kirki Agric. Exp. Stn* **1**: 61.

# Schefflera ringspot *badnavirus*

**Data collated by A.A. Brunt, 1992.**

*Standard acronym:* SRV.

First reported in *Brassica actinophylla*; from the U.S.A.; by Lockhart.

**Natural host range and symptoms**

*Brassica actinophylla* — spreading chlorotic ringspots.

**Geographical distribution**

Spreads in the USA.

**Particle morphology**

Virions bacilliform; not enveloped; 120–150 nm in length; 25–30 nm wide.

**Taxonomy and Relationships**

*Badnavirus.*

**References**

Lockhart, B.E. and Hull, R. (1991). In: *Classification and Nomenclature of Viruses; Report of the ICTV.* Eds R.I.B. Francki, C.M. Fauquet, D.L. Knudson and F. Brown. *Arch. Virol.*, Suppl. 2, p. 153.

# Scrophularia mottle *tymovirus*

**Data collated by A.J. Gibbs, 1991.**

*Synonyms:* scrophularia-Scheckungsvirus. *Standard acronym:* ScrMV.

First reported in *Scrophularia nodosa*; from Germany; by Hein (1959).

**Natural host range and symptoms**

Symptoms persist. Symptoms mottling.
*Scrophularia nodosa* — mottling and vein clearing.

**Transmission**

Transmitted by a vector; an insect; *Cionus tuberculosis, C. scrophularia, C. hortulanus, C. alauda*; Coleoptera. Virus transmitted by mechanical inoculation.

**Geographical distribution**

Spreads in Germany.

**Experimental host range**

Several (3–9) families susceptible. Experimentally infected plants mostly show local lesions, systemic mottle or mosaic.

Diagnostically susceptible

*Anthirrhinum majus* — systemic green spots.
*Chenopodium quinoa* — chlorotic local lesions, systemic mosaic.
*Datura stramonium* — yellow spreading local lesions, systemic spotting and mosaic.
*Vicia faba* — chlorotic local lesions.

Diagnostically insusceptible

*Arachis hypogaea, Brassica campestris* spp. *pekinensis, Cucumis sativus, Petunia hybrida.*

Maintenance and propagation hosts

*Anthirrhinum majus, Datura stramonium.*

Assay hosts (**L**ocal lesions or **W**hole plants)

*Vicia faba* (L).

Dicotyledonous Hosts *(**Susceptible** or Insusceptible)*

***Antirrhinum majus***. *Arachis hypogaea. Bellis perennis. Beta vulgaris. Brassica campestris* ssp. *pekinensis*. ***Catalpa bignonioides***. ***Catharanthus roseus***. ***Chenopodium quinoa***. *Cucumis sativus*. ***Datura stramonium***. *Daucus carota. Delphinium hybridum. Eucalyptus cloesiana. Euphorbia marginata. Fagopyrum esculentum. Fragaria vesca. Gossypium hirsutum. Gypsophila elegans. Helianthus annuus. Lactuca sativa. Lobelia erinus. Myosotis sylvatica. Papaver nudicaule. Pastinaca sativa. Petroselinum crispum. Petunia hybrida*. ***Pisum sativum***. *Podolepis robusta. Salvia splendens. Sinningia speciosa. Tropaeolum majus*. ***Vicia faba***. *Vigna radiata. Viola cornuta.*

Additional hosts

*Cassia artemisiodes, Ipomoea purpurea, Leptosiphon* sp., ***Valerianella*** sp.

Sources of host-range data

Bercks (1973); Hein (1959); Guy *et al.* (1984).

**Properties of particles in sap**

TIP: 90 °C. LIV: 30 days. DEP: $\log_{10}$ minus 5–6. Infectivity of sap not changed by treatment with di–ethyl ether. Leaf sap contains many virions.

**Purification**

Method: Bercks (1973).

**Particle morphology**

Virions isometric; not enveloped; 26 nm in diameter; rounded in profile; with a conspicuous capsomere arrangement.

**Physical properties**

Two sedimenting components in purified preparations; sedimentation coefficient of the fastest 116 *S*; of the other(s) 54 *S*. Density 1.351 and 1.449 g $cm^{-3}$ in CsCl. Isoelectric point pH 7.

**Biochemical properties**

Virions contain 37 % nucleic acid; 63 % protein; 0 % lipid.

*Genome* consists of RNA; single–stranded; linear; unipartite; largest (or only) genome part 6 kb. Base composition 15.9 % G; 21.3 % A; 33.7 % C; 29.3 % U.

*Virion protein(s)* one; $M_r$ 21600; coat protein. Amino acid composition: Bercks (1973).

**Replication**

Replication does not depend on a helper virus.

**Cytopathology**

Virions found in all parts of the host plant.

**Taxonomy and Relationships**

*Tymovirus*.

Virus(es) with serologically related virions: ononis yellow mosaic virus is the closest related, but also Andean potato latent, belladonna mottle, dulcamara mottle, eggplant mosaic and turnip yellow mosaic viruses. A virus from *Anagyris foetida* in Italy is closely related (Rana *et al.*, 1988).

Best tests for diagnosis: virions serologically close to ononis yellow mosaic virus but the virions of ononis yellow mosaic virus have a lesser IEP.

**References**

Bercks, R. (1973). *CMI/AAB Descr. Pl. Viruses* No. 113.

Bercks, R., Huth, W., Koenig, R., Lesemann, D.-E., Paul, H.L. and Querfurth, G. (1971). *Phytopath. Z.* **71**: 341.

Guy, P.L., Dale, J.L., Adena, M.A. and Gibbs, A.J. (1984). *Pl. Path.* **33**: 337.

Hein, A. (1959). *Phytopath. Z.* **36**: 290.

Rana, G.L., Castellano, M.A. and Koenig, R. (1988). *J. Phytopath.* **121**: 239.

# Serrano golden mosaic bigeminivirus

**Data collated by J.K. Brown, 1991.**

*Standard acronym:* SGMV.

First reported in *Capsicum annuum* cv. Serrano; from Sinaloa, Mexico; by Brown and Poulos (1990).

**Natural host range and symptoms**

Symptoms vary cyclically over a few weeks. Symptoms golden mosaic.

*Capsicum annuum* cvs Serrano, Bell, Anaheim — golden mosaic; green mosaic, stunting and fruit deformation.

*Lycopersicon esculentum* — interveinal chlorosis; sometimes necrosis of blossom end of fruit and deformation.

**Transmission**

Transmitted by a vector; an insect; *Bemisia tabaci*; Aleyrodidae. Not transmitted by *Trialeurodes vaporariorum*. Transmitted in a semi–persistent manner, or in a persistent manner. Virus lost by the vector when it moults; transmitted by mechanical inoculation (with difficulty); transmitted by grafting; not transmitted by contact between plants; not transmitted by seed.

**Geographical distribution**

Spreads in Mexico (Sinaloa and, possibly, Sonora), the USA (in greenhouse grown tomatoes in Arizona).

**Experimental host range**

Few (<3) families susceptible. Experimentally infected plants mostly show golden mosaic.

Diagnostically susceptible

*Capsicum frutescens, C. annuum*—golden mosaic.
*Lycopersicon esculentum, Datura stramonium* — yellow mosaic.

Diagnostically insusceptible

*Cucumis melo, Cucurbita maxima, C. pepo, Lactuca sativa, Gossypium hirsutum.*

Maintenance and propagation hosts

*Capsicum annuum, Lycopersicon esculentum.*

Assay hosts (**L**ocal lesions or **W**hole plants)

*Capsicum annuum* (W).

Dicotyledonous Hosts *(**Susceptible** or Insusceptible)*

***Capsicum annuum***. ***Capsicum frutescens***. *Cucurbita maxima*. *Cucurbita pepo*. ***Datura stramonium***. *Gossypium hirsutum*. *Lactuca sativa*. ***Lycopersicon esculentum***.

**Properties of particles in sap**

Leaf sap contains few virions.

**Purification**

Method: similar to that for other geminiviruses, but variable concentrations in pepper add difficulty.

**Particle morphology**

Virions geminate; 20 nm in diameter; dimers 30 nm in length.

**Physical properties**

$A_{260}/A_{280}$ ratio 1.37.

**Biochemical properties**

Virions contain 20 % nucleic acid; 80 % protein.

*Genome* consists of DNA; single–stranded; circular; of two parts; largest (or only) genome part 2.6 kb; the 2nd largest 2.6 kb. Genomic nucleic acid isolated by Brown: incubate 1% SDS at 42°C; extract with phenol chloroform, ethanol precipitate. Infectivity retained when deproteinised with proteases; retained when deproteinised with phenol or detergent.

**Taxonomy and Relationships**

Bigeminivirus: *Geminiviridae.*

Best tests for diagnosis: distinguish by transmission with *Bemisia tabaci* to susceptible pepper.

**References**

Brown, J.K. and Poulos, B.T. (1990). *Plant Dis.* **74**: 720.

# Shallot latent *carlavirus*

**Data collated by L. Bos, 1987.**

*Synonyms:* garlic laent virus (Van Dijk, 1993). *Standard acronym:* SLV. *Strains:* type, garlic and Asian shallot strains recognized (Van Dijk, 1993).

First reported in *Allium cepa* var. *ascalonicum*; from the Netherlands; by Bos (1972).

**Natural host range and symptoms**

*Allium ampeloprasum* var. *porrum* — mild chlorotic streaking.
*Allium cepa* var. *cepa, A. sativum, A. cepa* var. *ascalonicum* — symptomless.

**Transmission**

Transmitted by a vector; an insect; *Myzus (Sciamyzus) ascolonicus* and, perhaps, *Aphis fabae*; Aphididae. Transmitted in a non–persistent manner. Virus transmitted by mechanical inoculation.

**Ecology and control**

Studies reported by Bos *et al.* (1978); Paludan (1980).

**Geographical distribution**

Probably distributed worldwide (all shallots tested have been found to be infected). Spreads in Belgium, Denmark, France, Germany, Japan, and the UK.

**Experimental host range**

Few (<3) families susceptible.

Diagnostically susceptible

*Chenopodium album, C. amaranticolor, C. quinoa* — small dry local lesions in yellowing leaves, often developing green rings.

Diagnostically insusceptible

*Gomphrena globosa, Nicotiana clevelandii, N. tabacum, Petunia hybrida, Phaseolus vulgaris, Pisum sativum.*

Maintenance and propagation hosts

*Allium porrum, A. fistulosum* ('filter host' to separate the leek yellow stripe virus).

Assay hosts (**L**ocal lesions or **W**hole plants)

*Chenopodium amaranticolor* (L), *C. quinoa* (L).

Dicotyledonous Hosts *(**Susceptible** or Insusceptible)*

***Chenopodium album*. *Chenopodium amaranticolor*. *Chenopodium quinoa*.** *Cichorium endiva. Gomphrena globosa. Lycopersicon esculentum. Nicotiana clevelandii. Nicotiana glutinosa. Nicotiana tabacum. Petunia hybrida. Phaseolus vulgaris. Pisum sativum. Tetragonia expansa. Vicia faba. Vigna unguiculata.*

Monocotyledonous Hosts *(**Susceptible** or Insusceptible)*

***Allium ascalonicum**. **Allium cepa**. **Allium fistulosum**. **Allium porrum**. Lilium formosanum. Narcissus pseudonarcissus. Zea mays.*

Additional hosts

Additional Hosts: *Chrysanthemum sp.* (**I**); *Crotalaria sp.* (**I**); *Tulipa sp.* (**I**).

Sources of host-range data

Bos *et al.* (1978).

**Properties of particles in sap**

TIP: about 80 °C. LIV: 8–11 days. DEP: $\log_{10}$ minus 4–5. Leaf sap contains many virions. Electron microscopy: virions best seen in PTA, often aggregated end-to-end.

**Purification**

Method: Bos *et al.* (1978); molecular sieving through Sephadex G-200, then equilibrium density gradient centrifuging in CsCl (Bos *et al.*, 1978).

**Particle morphology**

Virions filamentous; not enveloped; usually straight (or slightly curved); with a clear modal length; of 650 nm. Axial canal obscure. Basic helix obscure.

**Physical properties**

One sedimenting component in purified preparations; sedimentation coefficient 147.5 *S*. Density 1.313 g $cm^{-3}$ in CsCl.

*Virion protein(s)* one; $M_r$ 23200; coat protein. Method of preparation: Huttinga (1975).

**Cytopathology**

Virions found in leaves; in cytoplasm (Paludan, 1980). Inclusions absent from infected cells.

**Taxonomy and Relationships**

*Carlavirus*.

Virus(es) with serologically related virions: narcissus latent and lonicera latent viruses. Virus(es) with serologically unrelated virions: chrysanthemum B, nerine latent, pea streak, poplar mosaic, potato S, potato M, red clover vein mosaic, lily symptomless and carnation latent viruses.

Best tests for diagnosis: ISEM with antiserum or shallot latent virus, rapid formation of local lesions on *Chenopodium amaranticolor* and *C. quinoa* and local lesions more numerous and smaller than those of leek yellow stripe virus.

**References**

Bos, L. (1972). *Gewasbesch.* **3**:81.

Bos, L. (1982). *CMI/AAB Descr. Pl. Viruses* No. 250, 4 pp.

Bos, L., Huttinga, H. and Maat, D.Z. (1975). *Neth. J. Pl. Path.* **84**: 227.

Huttinga, H. (1975). *Neth. J. Pl. Path.* **81**: 81.

Paludan, N. (1980). *Tidsskr. PlAvl.* **84**: 371.

Van Dijk, P. (1993). *Neth. J. Pl. Path.* **99**: 233.

# Shallot mite-borne latent (?) *potexvirus*

**Data collated by P. van Dijk, 1991.**

First reported in *Allium cepa* var. *ascalonicum*; from the Netherlands (in shallots imported from Asia); by Van Dijk *et al.* (1991).

## Natural host range and symptoms

Symptoms none.

*Allium cepa* var. *ascalonicum* — symptomless infection or sometimes slight mottle.

## Transmission

Transmitted by a vector; a mite; *Aceria tulipae*; Eriophyidae. Not transmitted by aphids; *Acyrthosiphon pisum, Myzus (Sciamyzus) ascolonicus.* Virus does not require a helper virus for vector transmission; transmitted by mechanical inoculation.

## Ecology and control

Studies reported by Van Dijk *et al.* (1991): vegetative propagation, latent infection, omnipresence and abundance of the vector imply that the virus can be readily spread and is difficult to control.

## Geographical distribution

Spreads in China, the Netherlands, Thailand, and the former USSR.

## Experimental host range

Few (<3) families susceptible. Experimentally infected plants mostly show mostly symptomless or show chlorotic local lesions.

Diagnostically susceptible

*Allium cepa, A. cepa* var. *ascalonicum, A. vineale* — symptomless infection or slight mottle.

Diagnostically insusceptible

*Allium ampeloprasum* var. *porrum, A. sativum, A. fistulosum, Chenopodium murale.*

Maintenance and propagation hosts

*Allium cepa, A. cepa* var. *ascalonicum.*

Assay hosts (**L**ocal lesions or **W**hole plants)

*Allium cepa, A. cepa* var. *ascalonicum* (W).

Dicotyledonous Hosts *(**Susceptible** or Insusceptible)*

*Celosia argentea. Chenopodium amaranticolor. Chenopodium murale. Chenopodium quinoa. Nicotiana benthamiana. Nicotiana occidentalis.*

Additional hosts

***Allium cepa* var. *ascalonicum, Allium cepa***, *Allium sativum, Allium fistulosum,* ***Allium vineale, Allium porrum, Chlorophytum comosum, Nicotiana hesperis*** and *Vicia faba.*

Sources of host-range data

Van Dijk *et al.* (1991).

## Properties of particles in sap

Leaf sap contains many virions.

**Particle morphology**

Virions filamentous; not enveloped; usually flexuous; with a clear modal length; of 767 nm. Axial canal obscure.

*Sequence database accession code(s):*

M97264 Em(40)_vi:SLVRNAGEN Gb(84)_vi:SLVRNAGENM Shallot virus X DNA sequence, open reading frames 1 ( RNA replicase) through 5 (coat protein) 1 sequence.

**Replication**

Replication does not depend on a helper virus.

**Cytopathology**

Virions found in all parts of the host plant. Inclusions present in infected cells; are unusual in shape.

**Taxonomy and Relationships**

Possible *Potexvirus.*

Virus(es) with serologically related virions: onion mite-borne latent virus is distantly related to shallot mite-borne latent virus, but distantly. Shallot virus X is probably onion mite-borne latent virus, and/or shallot mite-borne latent virus (Van Dijk and van der Vlugt, 1994). Virus(es) with serologically unrelated virions: no serological relationship with potyviruses, when tested in ISEM with polyclonal antibodies raised against whole potyvirus virions.

Best tests for diagnosis: no local lesions in *Chenopodium* spp., especially *Chenopodium murale*; virions in the electron microscope are clearly more flexuous than those of poty and carlaviruses; no decoration of virions with antisera to poty or carlaviruses (if raised to uncontaminated virus cultures).

**Comments**

The onion mosaic virus reported from the former U.S.S.R. (Cheremushkina, 1982; Razvjaskina, 1971), was probably a mixture of onion mite-borne latent, shallot mite-borne latent and onion yellow dwarf viruses (Van Dijk *et al.*, 1991).

**References**

Cheremushkina, N.P. (1982). *Trudy po Selektsii i Semenovodstvu Ovoshchnyak Kultur* **15**: 74.

Razvjazkina, G.M. (1971). *Tag. Akad. Landw. D.D.R.* **115**: 69.

Van Dijk, P., Verbeek, M. and Bos, L. (1991). *Neth. J. Pl. Path.* **97**: 381.

Van Dijk, P. and van der Vlugt, R.A.A. (1994). *Eur. J. Pl. Path.* **100**: 269.

## Shallot yellow stripe (?) *potyvirus*

**Data collated by A.A. Brunt, 1995.**

*Strains:* Common and virulent (Van Dijk, 1993).

First reported in *Allium cepa* var. *ascolonicum* (shallot); from the Netherlands from bulbs from Asia; by Van Dijk (1993).

**Natural host range and symptoms**

*Allium cepa* var. *ascolonicum* — mild striping of young leaves.

**Transmission**

Transmitted by a vector; an insect; *Acyrthosiphon pisum, Myzus cymbalariae*; Aphididae. Virus transmitted by mechanical inoculation.

**Geographical distribution**

Spreads in China, Indonesia, and Thailand.

**Experimental host range**

Few (<3) families susceptible.

Diagnostically susceptible

*Allium cepa* var. *cepa* cv. Stuttgarter Riesen — systemic leaf chlorosis, necrosis and malformation, plant stunting and, occasionally, death (virulent strain), or systemic chlorotic striping (common strain).

*Allium sativum, Lilium formosanum* — systemic yellow leaf striping and etching (virulent strain).

Diagnostically insusceptible

*Allium ampeloprasum* var. *porrum* (leek), *A. fistulosum* (Welsh onion), *Chenopodium amaranticolor, C. murale, C. quinoa, Nicotiana benthamiana, N. occidentalis, Vicia faba.*

Maintenance and propagation hosts

*Allium cepa* var. *cepa.*

Assay hosts (**L**ocal lesions or **W**hole plants)

*Allium cepa* var. *cepa* (W).

Dicotyledonous Hosts *(**Susceptible** or Insusceptible)*

*Celosia argentea. Chenopodium amaranticolor. Chenopodium murale. Chenopodium quinoa. Nicotiana benthamiana. Nicotiana occidentalis. Spinacia oleracea. Vicia faba.*

Monocotyledonous Hosts *(**Susceptible** or Insusceptible)*

***Allium ascalonicum**. **Allium cepa**. Allium fistulosum. Allium porrum.*

**Particle morphology**

Virions filamentous; usually flexuous; of 750 nm.

**Cytopathology**

Inclusions present in infected cells; are viroplasms.

**Taxonomy and Relationships**

Possible *Potyvirus: Potyviridae.*

Virus(es) with serologically unrelated virions: leek yellow stripe and onion yellow dwarf viruses.

**References**

Van Dijk, P. (1993). *Neth. J. Pl. Path.* **99**: Suppl.2, 48pp.

# Shamrock chlorotic ringspot (?) potyvirus

**Data collated by A.A. Brunt, 1991.**

First reported in *Oxalis regnellii*; from the U.S.A.; by Coyier (1981).

**Natural host range and symptoms**

*Oxalis regnellii* — chlorotic ringspots in leaves, subsequent decline and death of plants.

**Transmission**

Transmitted by a vector; an insect; *Aulocorthum circumflexum*; Aphididae. Virus transmitted by mechanical inoculation.

**Geographical distribution**

Spreads in the USA.

**Particle morphology**

Virions filamentous; usually flexuous; with a clear modal length; of 850 nm.

**Taxonomy and Relationships**

Unassigned potyvirus.

**References**

Coyier, D.L. (1981). *Plant Dis.* **65**: 275.

# Sieg River (?) *potexvirus*

**Data collated by R. Koenig, 1990.**

First reported in river water; from the river Sieg, Germany; by Koenig and Lesemann (1985).

**Transmission**

Probably spread by water. Virus transmitted by mechanical inoculation.

**Geographical distribution**

Spreads in Germany.

**Experimental host range**

Few (<3) families susceptible. Experimentally infected plants mostly show systemic mottle.

Diagnostically susceptible

*Chenopodium quinoa, C. amaranticolor, C. capitatum, C. foliosum, C. murale, C. schraderianum* — systemic mottle.

*Gomphrena globosa* — necrotic local lesions.

Diagnostically insusceptible

*Nicotiana* spp., *Cucumis sativus, Datura stramonium, Vicia faba, Anthirrhinum majus.*

Maintenance and propagation hosts

*Chenopodium quinoa.*

Assay hosts (**L**ocal lesions or **W**hole plants)

*Gomphrena globosa* (L), *Chenopodium quinoa* (W).

Dicotyledonous Hosts *(**Susceptible** or Insusceptible)*

*Antirrhinum majus*. *Capsicum annuum*. ***Chenopodium amaranticolor***. ***Chenopodium capitatum***. ***Chenopodium foliosum***. ***Chenopodium murale***. ***Chenopodium quinoa***. *Cucumis sativus*. *Datura metel*. *Datura stramonium*. ***Gomphrena globosa***. *Lycopersicon esculentum*. *Nicandra physaloides*. *Nicotiana benthamiana*. *Nicotiana clevelandii*. *Nicotiana debneyi*. *Nicotiana glutinosa*. *Nicotiana tabacum*. *Vicia faba*.

**Purification**

Method: Koenig and Lesemann (1983 and 1985).

**Particle morphology**

Virions filamentous; not enveloped; usually flexuous; with a clear modal length; of 466 nm; 13 nm wide.

**Physical properties**

One sedimenting component in purified preparations.

**Biochemical properties**

*Genome* consists of RNA; single–stranded; linear; unipartite.
*Virion protein(s)* one; $M_r$ 22800.

**Cytopathology**

Virions found in leaves, mesophyll and the epidermis; in cytoplasm. Inclusions present in infected cells; are crystals in the cytoplasm and unusual in shape; paracrystalline aggregates; they contain virions.

**Taxonomy and Relationships**

Possible *Potexvirus*.

Virus(es) with serologically related virions: plantago severe mosaic and viola mottle viruses. Virus(es) with serologically unrelated virions: potato X, cactus X, cymbidium mosaic, papaya mosaic and white clover mosaic viruses.

**References**

Koenig, R. and Lesemann, D.-E. (1983). *CMI/AAB Descr. Pl. Viruses* No. 265.
Koenig, R. and Lesemann, D.-E. (1985). *Phytopath. Z.* **112**: 105.

# Silene X (?) *potexvirus*

**Data collated by R.I. Hamilton, 1991.**

First reported in *Silene pratensis*; from British Colombia, Canada; by Hamilton (1984).

**Natural host range and symptoms**

Symptoms persist.
*Silene pratensis* — usually symptomless, sometimes a slight mosaic, stunting.

**Transmission**

Infected plants occur in orchards and on roadsides which are frequently mown. Virus transmitted by mechanical inoculation; not transmitted by seed.

**Geographical distribution**

Spreads in Canada.

**Experimental host range**

Several (3–9) families susceptible. Experimentally infected plants mostly show usually symptomless or systemic mosaic.

Diagnostically susceptible

*Chenopodium amaranticolor, C. quinoa* — chlorotic local lesions; systemic mosaic.
*Nicotiana benthamiana, N. clevelandii* — necrotic local lesions; systemic mosaic.
*Vigna unguiculata* — chlorotic rings; systemic mosaic.

Diagnostically insusceptible

*Narcissus* spp. *Solanum tuberosum* (5 cultivars).

Maintenance and propagation hosts

*Chenopodium quinoa*.

Assay hosts (**L**ocal lesions or **W**hole plants)

*Chenopodium amaranticolor* (W), *C. quinoa* (W), *Gomphrena globosa* (L), *Phaseolus vulgaris* (L), *Pisum sativum* (L).

Dicotyledonous Hosts *(**Susceptible** or Insusceptible)*

***Chenopodium amaranticolor*. *Chenopodium quinoa*. *Cucumis sativus*. *Dianthus barbatus*. *Gomphrena globosa*. *Nicotiana benthamiana*. *Nicotiana clevelandii*. *Nicotiana tabacum*. *Phaseolus vulgaris*. *Pisum sativum*. *Stellaria media*. *Vigna unguiculata*.**

Additional hosts

***Spergula arvensis*.**

Sources of host-range data

Wakarchuk and Hamilton (1989).

**Properties of particles in sap**

Leaf sap contains few virions. Electron microscopy: aldehyde fixation.

**Purification**

Method: Brunt (1966).

**Particle morphology**

Virions filamentous; not enveloped; usually flexuous; with a clear modal length; of 546 nm; 13 nm wide. Axial canal obscure. Basic helix obscure.

**Physical properties**

Density 1.32 g $cm^{-3}$ in CsCl (±0.3). $A_{260}/A_{280}$ ratio 1.18.

**Biochemical properties**

*Genome* consists of RNA; single–stranded; linear; unipartite; largest (or only) genome part 6.9 kb. Genomic nucleic acid isolated by Hamilton: phenol extraction of purified virus in 100 mm $NH_4CO_3$ (pH 9.0), 10 mm EDTA and 1% SDS; precipitation of RNA in EtOH. Infectivity retained when deproteinised with phenol or detergent. Poly A region present. Additional factor not required for infectivity.

*Non–genomic nucleic acid* found in the virions; is subgenomic *m*RNA. 4 virus specified dsRNA species found in infected cells. Size of largest virus specified dsRNA 17.1 kbp; 2nd largest 7 kbp; 3rd largest 5.7 kbp; 4th largest 4.6 kbp.

*Virion protein(s)* one; $M_r$ 29000; capsid protein. Method of preparation: Laemmli (1970).

**Replication**

Replication does not depend on a helper virus.

**Cytopathology**

Virions found in leaves and mesophyll; in cytoplasm and in cell vacuoles. Inclusions absent from infected cells.

**Taxonomy and Relationships**

Possible *Potexvirus*.

Virus(es) with serologically related virions: Argentine plantago, clover yellow mosaic, cymbidium mosaic, daphne X, hydrangea ringspot, narcissus mosaic, viola mottle and white clover mosaic viruses.

Best tests for diagnosis: silene X virus often occurs mixed with ribgrass mosaic virus in *Silene pratensis*. It rapidly infects *Chenopodium quinoa* systemically, whereas ribgrass mosaic virus does it more slowly.

**References**

Brunt, A.A. (1966). *Ann. appl. Biol.* **58**: 13.
Laemmli, U.K. (1970). *Nature, Lond.* **227**: 680.
Wakarchuk, D.A. and Hamilton, R.I. (1989). *Intervirology* **30**: 330.

# Sint-Jan's onion latent (?) *carlavirus*

**Data collated by A.A. Brunt, 1995.**

*Synonyms:* Welsh onion latent virus (Van Dijk and Sutarya, 1992; Van Dijk, 1993).

First reported in *Allium* sp. (unidentified); from the Netherlands; by Van Dijk (1993).

**Natural host range and symptoms**

*Allium* sp. unident. (Sint-Jan's onion), *A. ampeloprasum* var. *sectivum* (pearl onion), *A. cepa* var. *ascolonicum* (shallot), *A. cepa* var. *perutile* (ever-ready onion), *A. chinense* (rakkyo), *A. fistulosum* (Welsh onion) — symptomless.

**Transmission**

Transmitted by a vector; an insect; *Myzus (Sciamyzus) ascalonicus*; Aphididae. Virus transmitted by mechanical inoculation.

**Geographical distribution**

Found, but with no evidence of spread, in Indonesia, The Netherlands.

**Experimental host range**

Few (<3) families susceptible. Experimentally infected plants mostly show No symptoms.

Diagnostically susceptible

*Allium cepa* var. *cepa*, *A. vineale* (crow garlic), *A. cepa* var. *ascolonicum*, *A. fistulosum* (Welsh onion) — symptomless infection.

Diagnostically insusceptible

*Allium sativum, Chlorophytum comosum, Lilium formosanum.*

Maintenance and propagation hosts

*Allium cepa* var. *cepa*, *A. vineale.*

Assay hosts (**L**ocal lesions or **W**hole plants)

*Allium cepa* var. *cepa* (W).

Monocotyledonous Hosts *(**Susceptible** or Insusceptible)*

***Allium ascalonicum**. **Allium cepa**. **Allium fistulosum**. Lilium formosanum.*

**Particle morphology**

Virions filamentous; with a clear modal length; of 650 nm.

**Taxonomy and Relationships**

Possible *Carlavirus*.

Virus(es) with serologically related virions: Shallot latent and garlic common latent carlaviruses.

**Comments**

The virus is also considered to be a serological variant of shallot latent virus (Barg *et al.* 1995).

**References**

Barg, E., Lesemann, D.-E., Vetten, H.J. and Maiss, E. (1995). *Proc. 8th Conf. Virus Dis. Vegetables*, p.26.
Van Dijk, P. (1993). *Neth. J. Pl. Path.* **93**: 233.
Van Dijk, P. and Sutarya, R. (1992). *Internal Doc. DLO, Wageningen*, pp.70.

# Sitke waterborne (?) *tombusvirus*

**Data collated by A.A.Brunt, 1994.**

*Standard acronym:* SWBV.

First reported in water from River Sitke; from Germany; by Yi *et al.* (1992).

**Geographical distribution**

Found, but with no evidence of spread, in plants in Germany (River Sitke at Braunschweig and Neustadt).

**Experimental host range**

Few (<3) families susceptible.

Diagnostically susceptible

*Chenopodium amaranticolor, C. quinoa, Nicotiana benthamiana, N. clevelandii.*

Maintenance and propagation hosts

*Chenopodium quinoa.*

Dicotyledonous Hosts *(**Susceptible** or Insusceptible)*

***Chenopodium amaranticolor***. ***Chenopodium quinoa***. ***Nicotiana benthamiana***. ***Nicotiana clevelandii***.

**Purification**

Method: Gallitelli *et al.* (1985).

**Particle morphology**

Virions isometric; not enveloped; 33 nm in diameter.

**Biochemical properties**

*Genome* consists of RNA; single–stranded; unipartite; largest (or only) genome part 4.8 kb.

*Virion protein(s)* one; $M_r$ 35800; coat protein.

**Cytopathology**

Virions found in cytoplasm. Other cellular changes: few peroxisomal vesicles induced; endoplasmic reticulum adjacent to peroxisomes.

**Taxonomy and Relationships**

Possible *Tombusvirus: Tombusviridae*.

Virus(es) with serologically unrelated virions: Artichoke mottled crinkle, carnation Italian ringspot, cymbidium ringspot, cucumber necrosis, eggplant mottled crinkle, grapevine Algerian latent, Lato River, pepper Moroccan, Neckar River, petunia asteroid mosaic, pelargonium leaf curl and tomato bushy stunt tombusviruses, and 15 carmoviruses.

In Northern blot hybridization tests, the genomic RNA showed some homology with the genomic RNAs of grapevine Algerian latent, petunia asteroid mosaic and cymbidium ringspot tombusviruses.

**References**

Gallitelli, D., Hull, R. and Koenig, R. (1985). *J. gen Virol.* **66**: 1523.

Yi, L., Lesemann, D.-E., Koenig, R., Rüdel, M. and Pfeilstetter, E. (1992). *J. Phytopathology* **134**: 121.

# Smithiana (?) *potexvirus*

**Data collated by A.A. Brunt, 1991.**

*Standard acronym:* SmiV.

First reported in *Smithiana* spp.; from the U.S.A.; by Zettler and Nagel (1983).

**Natural host range and symptoms**

*Smithiantha* spp. — no obvious symptoms.

**Particle morphology**

Virions filamentous; usually flexuous; with a clear modal length; of 486 nm.

**Taxonomy and Relationships**

Possible *Potexvirus*.

Virus(es) with serologically unrelated virions: clover yellow mosaic, cymbidium mosaic, foxtail mosaic, nandina mosaic, papaya mosaic and potato X viruses.

**References**

Zettler, F.W. and Nagel, J. (1983). *Plant Dis.* **67**: 1123.

# Solanum apical leaf curling (?) bigeminivirus

**Data collated by L.F. Salazar, 1991.**

*Standard acronym:* SALCV.

First reported in *Solanum tuberosum*; from San Ramon, Junin, Peru; by Hooker and Salazar (1983).

**Natural host range and symptoms**

Symptoms persist. Symptoms curling, dwarfing and chlorosis of tip leaves.

*Solanum tuberosum* — upward rolling, narrowing and dwarfing of tip leaflets. Red or purple discoloured leaflet bases. Tubers do not sprout or produce short, thin sprouts.

*Solanum basendopogon, S. basendopogon f. obtusum* — mild chlorosis of apical leaves.

*Nicandra physalodes* — chlorosis.

*Physalis peruviana* — stunting and chlorosis with green leaf veins.

*Datura tatula* — vein yellowing, tip leaves curled downward and dwarfed. Severe chlorosis in older infections.

**Transmission**

Probably transmitted by leafhoppers. Not transmitted by *Myzus persicae*. Virus not transmitted by mechanical inoculation; transmitted by grafting; not transmitted by contact between plants; not transmitted by seed; not transmitted by pollen.

**Geographical distribution**

Spreads in Peru.

**Experimental host range**

Few (<3) families susceptible. Experimentally infected plants mostly show curling, dwarfing and chlorosis.

Diagnostically susceptible

*Datura tatula, D. stramonium, Nicotiana benthamiana* — curling, dwarfing and chlorosis of apical leaves.

*Solanum nigrum* — chlorosis.

*Lycopersicon esculentum, Nicandra physalodes* — symptomless.

Maintenance and propagation hosts

*Nicotiana benthamiana, Datura tatula, D. stramonium, Lycopersicon esculentum.*

Assay hosts (**L**ocal lesions or **W**hole plants)

*Datura tatula* (W), *Datura stramonium* (W).

Dicotyledonous Hosts *(**Susceptible** or Insusceptible)*

***Lycopersicon esculentum. Nicotiana benthamiana. Solanum nigrum.***

**Properties of particles in sap**

Leaf sap contains few virions.

**Purification**

Method: Hooker and Salazar (1983).

**Particle morphology**

Virions geminate, or unusually shaped; trisegmented virions; not enveloped; 17 nm in diameter; up to 52 nm in length; without a conspicuous capsomere arrangement.

**Physical properties**

Three sedimenting components in purified preparations (usually, but 1 or 2 components seen in some preparations).

**Biochemical properties**

*Genome* consists of DNA; single–stranded; circular.

**Cytopathology**

Virions found in leaves, phloem and companion cells.

**Taxonomy and Relationships**

Possible bigeminivirus: *Geminiviridae*.

Best tests for diagnosis: serology (ELISA).

**References**

Hooker, W.J. and Salazar, L.F. (1983). *Ann. appl. Biol.* **103**: 449.

Hooker, W.J., Salazar, L.F. and Brown, C.R. (1985). *Am. Potato J.* **62**: 263.

# Solanum nodiflorum mottle *sobemovirus*

**Data collated by R.S. Greber and J.W. Randles, 1986.**

*Synonyms:* solanum nigrum mosaic virus, solanum nodiflorum virus. *Standard acronym:* SNMV.

First reported in *Solanum nodiflorum*, an endemic species (Henderson, 1974); from eastern Australia; by Greber (1965).

**Natural host range and symptoms**

Symptoms persist.

*Solanum nodiflorum, S. nitidibaccatum, S. nigrum* — leaf mottling and rugosity.

**Transmission**

Transmitted by a vector; an insect; *Epilachna vigintioctopunctata pardalis* (syn. *E. sparsa*), *E. vigintisexpunctata vigintisexpunctata* (syn. *E. doryca australica*) and *E. guttatopustulata*; Coleoptera. Virus not transmitted by seed.

**Geographical distribution**

Spreads in Australia (the east).

**Experimental host range**

Few (<3) families susceptible.

Diagnostically susceptible

*Solanum nodiflorum* — systemic leaf mottling and puckering.

*Nicotiana clevelandii* — chlorotic local lesions; systemic vein chlorosis, mottling and puckering.

*Nicotiana debneyi* — necrotic local lesions and small chlorotic ring local lesions; not systemic.

Maintenance and propagation hosts

*Nicotiana clevelandii* — for purification. *Solanum nodiflorum* — for culture and vector maintenance.

Assay hosts (**L**ocal lesions or **W**hole plants)

*Nicotiana debneyi* (L).

Dicotyledonous Hosts *(**Susceptible** or Insusceptible)*

***Chenopodium quinoa*. *Datura stramonium*. *Gomphrena globosa*. *Nicandra physaloides*. *Nicotiana benthamiana*. *Nicotiana clevelandii*. *Nicotiana debneyi*. *Nicotiana glutinosa*. *Nicotiana tabacum*. *Petunia hybrida*. *Solanum melongena*. *Solanum nigrum*. *Solanum nodiflorum*. *Solanum rostratum*. *Tetragonia expansa*.**

Sources of host-range data

Greber (1981); Jones and Mayo (1984).

**Properties of particles in sap**

TIP: 90 °C. LIV: 365 days (at 2°C and 7 years at -20°C). DEP: $\log_{10}$ minus 5. Infectivity of sap not changed by treatment with di–ethyl ether. Leaf sap contains many virions.

**Purification**

Method: Greber (1981); Hollings *et al.* (1979); Kiberstis and Zimmern (1984).

**Particle morphology**

Virions isometric; not enveloped; 28–30 nm in diameter; angular in profile.

**Physical properties**

One sedimenting component in purified preparations; sedimentation coefficient 114 *S* (Hollings *et al.*, 1979). Density 1.35 g $cm^{-3}$ in CsCl.

**Biochemical properties**

*Genome* consists of RNA; single–stranded; linear; unipartite; largest (or only) genome part 4.2 kb. Genomic nucleic acid isolated by Gould and Hatta (1981).

*Sequence database accession code(s):*

J02386 Em(40)_vi:NOR2 Gb(84)_vi:MSNR2 solanum nodiflorum mottle virus (circular viroid-like) RNA2. 4/90 377bp. 1 sequence.

*Non–genomic nucleic acid* found in the virions; is satellite RNA (of 2 species, (RNA-2 and RNA-3), which are similar in size (*c.* 0.37 kb) but one is circular (RNA-2) the other linear (Gould *et al.*, 1981; Jones and Mayo, 1984)).

*Virion protein(s)* one; $M_r$ 31000; coat protein (Hollings *et al.*, 1979; Greber, 1981; Chu and Francki, 1983).

**Replication**

Replication does not depend on a helper virus. Acts as helper for a satellite RNA (Some of the isolates of the virus have no satellite RNA; Jones and Mayo, 1984).

**Cytopathology**

Virions found in all parts of the host plant; in cytoplasm and in nuclei. Inclusions present in infected cells; are crystals in the cytoplasm, membranous bodies, and unusual in shape; long tubules or rod structures.

**Taxonomy and Relationships**

*Sobemovirus.*

Virus(es) with serologically related virions: velvet tobacco mottle virus (Randles *et al.*, 1981; Chu and Francki, 1983).

No serological relationship was found to 53 other viruses with isometric virions (Hollings *et al.*, 1979).

**References**

Chu, P. and Francki, R.I.B. (1983). *Virology* **129**: 350.

Gould, A.R. and Hatta, T. (1981). *Virology* **109**: 137.

Gould, A.R., Francki, R.I.B. and Randles, J.W. (1981). *Virology* **110**: 420.

Greber, R.S. (1965). *Aust. Plant Dis. Rec.* **17**: 18.

Greber, R.S. (1973). *Aust. Pl. Path. Soc. Newsl.* **2**: 3.

Greber, R.S. (1981). *Aust. J. biol. Sci.* **34**: 369.

Greber, R.S. and Randles, J.W. (1986). *AAB Descr. Pl. Viruses* No. 318, 5 pp.

Henderson, J. (1974). *Qld. Herb. Contrib.* **16**: 78 pp.

Hollings, M., Stone, O.M., Barton, R.J. and Greber, R.S. (1979). *Rep. Glasshouse Crops Res. Inst.* 1978, p. 150.

Jones, A.T. and Mayo, M.A. (1984). *J. gen. Virol.* **65**: 1713.

Kiberstis, P.A. and Zimmern, D. (1984). *Nucl. Acids Res.* **12**: 933.

Kiberstis, P.A., Haseloff, J. and Zimmern, D. (1985). *EMBO J.* **4**: 1984.

Randles, J.W., Davis, C., Hatta, T., Gould, A.R. and Francki, R.I.B. (1981). *Virology* **108**: 111.

# Solanum yellows *luteovirus*

**Data collated by A.A. Brunt, 1989.**

*Standard acronym:* SYV.

First reported in *Solanum tuberosum*; from the U.S.A.; by Milbrath and Duffus (1978).

**Natural host range and symptoms**

*Solanum tuberosum.*

**Taxonomy and Relationships**

*Luteovirus.*

**References**

Casper, R. (1988). In: *The Plant Viruses*, Vol. 3, Polyhedral Virions with Monopartite RNA Genomes, p. 244; ed. R. Koenig. Plenum Press, New York.

Milbrath, G.M. and Duffus, J.E. (1978). *Phytopath. News* **12**: 170.

Waterhouse, P.M., Gildow, F.E. and Johnstone, G.R. (1988). *AAB Descr. Pl. Viruses* No. 339, 9 pp.

# Sonchus *cytorhabdovirus*

**Data collated by A.A. Brunt, 1991.**

*Standard acronym:* SonV.

First reported in *Sonchus oleraceus*; from Argentina; by Vega *et al.* (1976).

**Natural host range and symptoms**

Symptoms persist.

*Sonchus oleraceus* — conspicuous leaf mottling and, sometimes, chlorotic lines, spots and rings.

**Transmission**

Virus transmitted by mechanical inoculation.

**Geographical distribution**

Spreads in Argentina.

**Experimental host range**

Several (3–9) families susceptible.

Diagnostically susceptible

*Chenopodium quinoa, Gomphrena globosa* — local lesions; no systemic infection.

*Datura stramonium, Lactuca sativa, L. serriola, Nicotiana sylvestris, N. tabacum, Petunia hybrida, Physalis floridana* — systemic leaf chlorosis.

*Datura metel, Matthiola incarna* — symptomless systemic infection.

*Nicotiana clevelandii, N. glutinosa* — chlorotic local lesions; systemic vein-clearing, chlorotic spots and rings, chlorosis and slight distortion.

Diagnostically insusceptible

*Beta vulgaris, Chenopodium amaranticolor, Calendula officinalis, Callistephus chinensis, Zinnia elegans.*

Maintenance and propagation hosts

*Nicotiana clevelandii, N. glutinosa.*

Assay hosts (**L**ocal lesions or **W**hole plants)

*Chenopodium quinoa* (L), *Nicotiana glutinosa* (W).

Dicotyledonous Hosts (***Susceptible*** *or Insusceptible)*

*Antirrhinum majus*. *Beta vulgaris*. *Brassica campestris* ssp. *rapa*. *Brassica oleracea* ssp. *botrytis*. *Calendula officinalis*. *Capsella bursa–pastoris*. *Capsicum annuum*. *Capsicum frutescens*. *Chenopodium amaranticolor*. ***Chenopodium quinoa***. *Cucumis melo*. *Cucumis sativus*. ***Datura metel***. ***Datura stramonium***. *Glycine max*. ***Gomphrena globosa***. ***Lactuca sativa***. *Lycopersicon esculentum*. ***Matthiola incana***. ***Nicotiana clevelandii***. *Nicotiana debneyi*. ***Nicotiana glutinosa***. *Nicotiana rustica*. ***Nicotiana sylvestris***. ***Nicotiana tabacum***. ***Petunia hybrida***. *Phaseolus vulgaris*. ***Physalis floridana***. *Physalis peruviana*. *Pisum sativum*. *Solanum nigrum*. ***Sonchus oleraceus***. *Spinacia oleracea*. *Vicia faba*. *Vigna unguiculata*. *Zinnia elegans*.

Sources of host-range data

Vega *et al.* (1976).

**Properties of particles in sap**

TIP: 50 °C. LIV: 0.33 days (at *c.* 18°C). DEP: $\log_{10}$ minus 4. Leaf sap contains few virions.

**Particle morphology**

Virions rhabdo– or bullet–shaped; enveloped; usually straight; with a clear modal length; of 270–350 nm; 45–57 nm wide.

**Cytopathology**

Virions found in leaves and mesophyll; in cytoplasm (within cisternae of endoplasmic reticulum).

**Taxonomy and Relationships**

*Cytorhabdovirus: Rhabdoviridae.*

**References**

Vega, J., Gracia, O., Rubio-Huertos, M. and Feldman, J.M. (1976). *Phytopath. Z.* **85**: 7.

## Sonchus mosaic (?) *potexvirus*

**Data collated by A.A. Brunt, 1992.**

First reported in *Sonchus asper*; from India; by Quereshi *et al.* (1979).

**Natural host range and symptoms**

*Sonchus asper* — leaf chlorosis.

**Transmission**

Transmitted by a vector; an insect; *Myzus persicae*; Aphididae. Virus transmitted by mechanical inoculation.

**Geographical distribution**

Spreads in India (Uttar Pradesh).

**Particle morphology**

Virions filamentous; usually flexuous; with a clear modal length; of 472 nm; 13–14 nm wide. Axial canal obscure.

**Physical properties**

One sedimenting component in purified preparations; sedimentation coefficient 120 *S*.

**Taxonomy and Relationships**

Possible *Potexvirus*.

**References**

Naqvi, Q.A., Quereshi, S. and Mahmood, K. (1985). *Indian J. Phytopath.* **3**: 125.
Quereshi, S., Naqvi, Q.A. and Mahmood, K. (1979). *Indian Phytopath.* **32**: 264.

# Sonchus yellow net *nucleorhabdovirus*

**Data collated by J.D. Wagner, 1991.**

*Standard acronym:* SYNV.

First reported in *Sonchus oleraceus*; from Florida, U.S.A.; by Christie *et al.* (1974).

**Natural host range and symptoms**

Symptoms persist. Symptoms vein clearing, yellow netting, stunting.
*Bidens pilosa* — symptomless; mixed infection with bidens mottle gives necrotic etching.
*Lactuca sativa* — bright yellow interveinal spotting of older leaves.
*Sonchus oleraceus* — chlorotic local lesions, vein clearing, yellow patches between veins.

**Transmission**

Transmitted by a vector; an insect; *Aphis coreopsidis*; Aphididae. Not transmitted by *Dactynotus* sp., *Hyperomyzus lactucae*. Transmitted in a persistent manner. Virus multiplies in the vector; transmitted by mechanical inoculation.

**Geographical distribution**

Spreads in the USA.

**Experimental host range**

Few (<3) families susceptible. Experimentally infected plants mostly show vein clearing, chlorotic local lesions, malformation or cupping, symptoms often disappear after 4–6 weeks.

Diagnostically susceptible

*Bidens pilosa* — symptomless; mixed infection with bidens mottle gives necrotic etching.
*Chenopodium quinoa* — local chlorotic local lesions, not systemic.
*Lactuca sativa* — bright yellow interveinal spotting of older leaves.
*Nicotiana benthamiana* — severe systemic vein clearing.
*N. clevelandii, N. debneyi, N. x edwardsonii, N. glutinosa* — yellow netting, leaf cupping.
*Sonchus oleraceus* — chlorotic local lesions, vein clearing, yellow patches between veins.
*Zinnia elegans* — transient systemic vein clearing.

Diagnostically insusceptible

*Capsicum frutescens, Chenopodium amaranticolor, Datura stramonium, Gomphrena globosa, Nicotiana tabacum.*

Maintenance and propagation hosts

*Nicotiana x edwardsonii.*

Assay hosts (**L**ocal lesions or **W**hole plants)

*Chenopodium quinoa* (L).

Dicotyledonous Hosts (***Susceptible*** *or Insusceptible)*

*Capsicum frutescens. Chenopodium amaranticolor.* ***Chenopodium quinoa.*** *Datura stramonium. Gomphrena globosa.* ***Lactuca sativa. Nicotiana benthamiana.*** *Nicotiana clevelandii.* ***Nicotiana × edwardsonii. Nicotiana debneyi. Nicotiana glutinosa.*** *Nicotiana rustica. Nicotiana tabacum. Solanum nigrum.* ***Sonchus oleraceus.*** *Vigna radiata.* ***Zinnia elegans.***

Sources of host-range data

Christie *et al.* (1974); Jackson and Christie (1974); Falk *et al.* (1986).

**Properties of particles in sap**

LIV: 0.33 days (less than 8 hours at 20°C). DEP: $\log_{10}$ minus 3–4. Leaf sap contains few virions. Electron microscopy: Christie *et al.* (1974) Jackson and Christie (1974).

**Particle morphology**

Virions rhabdo– or bullet–shaped; enveloped; usually straight; with a clear modal length; of 248 nm; 94 nm wide. Axial canal obvious. Basic helix obvious; pitch of basic helix 4.1 nm.

**Physical properties**

One sedimenting component in purified preparations; sedimentation coefficient 1044 *S*. Density 1.183 g $cm^{-3}$ in sucrose.

**Biochemical properties**

Virions contain 1 % nucleic acid; 80 % protein; 15–20 % lipid. Also carbohydrate.

*Genome* consists of RNA; single–stranded; linear; unipartite; largest (or only) genome part 13.6 kb. Genomic nucleic acid isolated by Jackson and Christie (1974). Infectivity lost when deproteinised with proteases; lost when deproteinised with phenol or detergent. Poly A region absent. Additional factor required for infectivity; viral proteins required for infectivity. Genome has no *t*RNA–like activity. Nucleotide sequence references: GenBank Accession Numbers: SYNV (+) strand leader RNA M13950, M2 M23023, N M17210, M1 M35689, G M73626.

*Sequence database accession code(s):*

L16768 Em(40)_vi:SYETRAILE Gb(84)_vi:SYETRAILER Sonchus yellow net virus 5' trailer region sequence. 5/93 160bp.

L32603 Em(40)_vi:SYENMGL Gb(84)_vi:SYENMGL Sonchus yellow net virus complete nucleotide sequence. 5/94 13,720bp.

L32604 Em(40)_vi:SYESC4X Gb(84)_vi:SYESC4X Sonchus yellow net virus sc4 mRNA, complete cds. 5/94 1,196bp.

M13950 Em(40)_vi:SYERNA Gb(84)_vi:SYERNA Sonchus yellow net virus leader RNA. 4/90 205bp.

M17210 Em(40)_vi:SYENCP Gb(84)_vi:SYENCP Sonchus yellow net virus nucleocapsid protein mRNA, complete cds. 7/89 1,546bp.

M23023 Gb(84)_vi:SYESYNVM2 Sonchus yello net virus SYNV M2 protein gene, complete CDS. 6/89 1,137bp.

M35689 Em(40)_vi:SYESC6 Gb(84)_vi:SYESC6 Sonchus yellow net virus M1 protein gene, complete cds. 12/90 1,071bp.

M73626 Em(40)_vi:SYESYNVG Gb(84)_vi:SYESYNVG Sonchus yellow net virus SYNV glycoprotein gene, complete CDS. 12/91 2,045bp.

M87829 Em(40)_vi:SYEORF Gb(84)_vi:SYEORF Sonchus yellow net virus L protein gene mRNA, complete cds. 5/94 6,401bp. 9 sequences.

*Features of the genome:* 6 large ORF's; conserved intergenic regions which show homology to other rhabdoviruses.

*Non–genomic nucleic acid* not found in the virions. Sub–genomic *m*RNA found in infected cells; all six (for each viral protein) and for the plus-strand leader.

*Virion protein(s)* five; $M_r$ of the largest 241000; L and is core polymerase. $M_r$ of 2nd largest 70000; G and is glycoprotein. $M_r$ of 3rd largest 54000; N, nucleocapsid protein. $M_r$ of 4th largest 38000; M2, matrix protein. $M_r$ of 5th largest 32000; M1, matrix protein. Method of preparation: Jackson (1978). Amino acid sequence: GenBank Accession Numbers: SYNV (+) strand leader RNA M13950, M2 M23023, N M17210, M1 M35689, G M73626. Virion proteins glycosylated. Are G proteins.

And it is NV non-virion protein.

### Replication

Genome replicates probably in nuclei. Replication does not depend on a helper virus.

### Cytopathology

Virions found in leaves, roots and all parts of the host plant; in the perinuclear space. Inclusions present in infected cells; are unusual in shape; are amorphous in the nucleus; they do not contain virions. Other cellular changes: enlarged nucleus (Christie and Edwardson, 1978).

### Taxonomy and Relationships

*Nucleorhabdovirus: Rhabdoviridae.*

Virus(es) with serologically unrelated virions: sonchus yellow vein virus.

Best tests for diagnosis: leaf dips show bacilliform virions.

### References

Christie, S.R. and Edwardson, J.R. (1978). *Ann. Rev. Phytopath.* **16**: 31.

Christie, S.R., Christie, R.G. and Edwardson, J.R. (1974). *Phytopathology* **64**: 840.

Falk, B.W., Purcifull, D.E. and Christie, S.R. (1986). *Plant Dis.* **70**: 591.

Heaton, L.A., Hillman, B.I., Hunter, B.G., Zuidema, D. and Jackson, A.O. (1989). *Proc. Natl. Acad. Sci., U.S.A.* **86**: 8665.

Jackson, A.O. (1978). *Virology* **87**: 172.

Jackson, A.O. and Christie, S.R. (1974). *Virology* **77**; 344.

Jackson, A.O., Francki, R.I.B. and Zuidema, D. (1987). In: *The Rhabdoviruses*, p. 427; ed. R. Wagner. Plenum Press, New York.

## Sorghum chlorotic spot *furovirus*

**Data collated by T.L. Kendall and W.G. Langenberg, 1990.**

*Standard acronym:* SgCSV.

First reported in *Sorghum bicolor*; from Kansas, U.S.A.

### Natural host range and symptoms

*Sorghum bicolor* — elongated, concentric chlorotic spots.

### Transmission

A fungal vector is suspected. Virus transmitted by mechanical inoculation.

**Geographical distribution**

Found, but with no evidence of spread, in Kansas, but not reported from other states in the U.S.A.

Diagnostically susceptible

*Chenopodium amaranticolor* — chlorotic local lesions.
*C. quinoa* — ringspots.
*Zea mays* — elongated chlorotic spots.
*Nicotiana clevelandii* — yellowing and necrotic local lesions.

Maintenance and propagation hosts

*Zea mays*.

**Properties of particles in sap**

Leaf sap contains many virions.

**Purification**

Method: Kendall *et al.* (1988).

**Particle morphology**

Virions rod–shaped; usually straight; with a clear modal length; of 140 and 260 nm; 20 nm wide. Axial canal obvious.

**Biochemical properties**

*Genome* consists of RNA; single–stranded; of two parts; largest (or only) genome part 6.2 kb; the 2nd largest 3.5 kb. Poly A region absent.

*Non–genomic nucleic acid* not found in the virions.

*Virion protein(s)* one; $M_r$ 20500; capsid protein.

**Taxonomy and Relationships**

*Furovirus*.

Virus(es) with serologically related virions: wheat soil borne mosaic virus.

**References**

Kendall, T.L. and Lommel, S.A. (1988). In: *Viruses with Fungal Vectors*; eds Cooper and Asher. Association of Applied Biologists.

Kendall, T.L., Langenberg, W.G. and Lommel, S.A. (1988). *J. gen. Virol.* **69**: 2335.

Langenberg, W.G., Lommel, S.A. and Purcifull, D.E. (1989). *J. Ultrastruct. Mol. Struct. Res.* **102**: 47.

# Sorghum mosaic *potyvirus*

**Data collated by A.J. Gibbs, 1990.**

*Standard acronym:* SrMV. *Strains:* SCMV–H (Florida) (Abbot and Tibbet, 1966), SCMV–I, SCMV–M (Koike and Gillespie, 1976), sorghum mosaic virus — H (Florida), sugarcane mosaic virus — I, sugarcane mosaic virus — M.

First reported in *Saccharum officinarum*; from the U.S.A.; by Abbott and Tippet (1966).

**Natural host range and symptoms**

Symptoms persist.
*Saccharum officinarum* — mosaic.
*Sorghum bicolor* — mosaic, red-leaf.

**Transmission**

Transmitted by a vector; an insect; Aphididae. Transmitted in a non–persistent manner.

**Experimental host range**

Few (<3) families susceptible.

Monocotyledonous Hosts (***Susceptible*** *or Insusceptible)*

***Sorghum bicolor*.**

**Properties of particles in sap**

Leaf sap contains few virions.

**Particle morphology**

Virions filamentous; not enveloped; usually flexuous; with a clear modal length; of 702–714 nm. Axial canal obscure.

*Sequence database accession code(s):*

U07219 Em(40)_vi:SM07219 Gb(84)_vi:SMU07219 Sorghum mosaic virus strain Sugarcane mosaic virus H polyprotein (nuclear inclusion II protei 1 sequence).

**Cytopathology**

Virions found in all parts of the host plant; in cytoplasm. Inclusions present in infected cells; are pinwheels and unusual in shape; bundles, scrolls, tubes and laminated aggregates; group I inclusions (Edwardson, 1974).

**Taxonomy and Relationships**

*Potyvirus: Potyviridae.*

This virus is one of a complex of potyviruses infecting tropical grasses. These have been shown by immunoblot and virion protein sequencing to fall into four species; maize dwarf mosaic, sugarcane mosaic, sorghum mosaic and Johnson grass mosaic. For details also check descriptions of the other three species.

**References**


Abbot, E.V. and Tippet, R.L. (1966). *U.S. Dep. Agric. Res. Serv. Tech. Bull.* 1340.

Koike, H. and Gillespie, A.G. (1976). *Pl. Dis. Reptr* **60**: 50.

Edwardson, J.R. (1974). *Fla Agric. Exp. Stn Monog. No. 4.*

Giorda. L.M., Toler, R.W. and Miller, F.R. (1986). *Plant Dis.* **70**: 624.

Shukla, D.D., Tosic, M., Jilka, J., Ford, R.E., Toler, R.W. and Langham, M.A.C. (1989). *Phytopathology* **79**: 223.


# Sorghum stunt mosaic (?) *nucleorhabdovirus*

**Data collated by A.A. Brunt and D.E. Mayhew, 1988.**

First reported in *Sorghum vulgare*; from California, U.S.A.; by Mayhew and Flock (1981).

**Natural host range and symptoms**

Symptoms persist.

*Sorghum vulgare* — chlorotic and necrotic mottling and streaking, severe stunting and fewer seeds.

**Transmission**

Transmitted by a vector; an insect; *Graminella sonora*; Delphacidae. Virus not transmitted by mechanical inoculation.

**Ecology and control**

Studies reported by there may be different optimum temperatures for virus transmission and symptom expression. Temperatures above 38°C kill viruliferous leafhoppers, but induce severe symptoms. Below 27°C, the vector thrives, but no virus can be detected in the hosts (Mayhew and Flock, 1981).

**Geographical distribution**

Spreads in the USA.

**Experimental host range**

Few (<3) families susceptible.

Diagnostically susceptible

*Sorghum vulgare, Zea mays, Triticum aestivum* — systemic chlorotic and necrotic mottling and streaking, severe stunting and fewer seeds.

Diagnostically insusceptible

*Hordeum vulgare, Avena sativa.*

Maintenance and propagation hosts

*Sorghum vulgare, Zea mays.*

Assay hosts (**L**ocal lesions or **W**hole plants)

*Sorghum vulgare* (W), *Zea mays* (W).

Monocotyledonous Hosts *(**Susceptible** or Insusceptible)*

*Avena sativa. Hordeum vulgare.* ***Triticum aestivum*. *Zea mays*.**

Additional hosts

***Sorghum vulgare*.**

Sources of host-range data

Mayhew and Flock (1981).

**Properties of particles in sap**

Leaf sap contains few virions.

**Particle morphology**

Virions rhabdo– or bullet–shaped; enveloped; with a clear modal length; of 220 nm; 68 nm wide (in sections). Basic helix obvious.

**Cytopathology**

Virions found in leaves and mesophyll; in the perinuclear space.

**Taxonomy and Relationships**

Possible *Nucleorhabdovirus: Rhabdoviridae*.

Virus(es) with serologically unrelated virions: maize mosaic rhabdovirus.

**References**

Mayhew, D.E. and Flock, R.A. (1981). *Plant Dis.* **65**: 84.

# Soursop yellow blotch (?) *cytorhabdovirus*

**Data collated by A.A. Brunt, 1994.**

*Standard acronym:* SYBV.

First reported in *Annona muricata*; from Ceara, Brazil; by Kitajima *et al.* (1993).

**Natural host range and symptoms**

*Annona muricata* (soursop) — vein-clearing, yellow blotch and leaf distortion, stunting.

**Transmission**

Virus transmitted by mechanical inoculation; transmitted by grafting; not transmitted by seed.

**Geographical distribution**

Spreads in the South and Central American region; Brazil.

**Experimental host range**

Few (<3) families susceptible.

Diagnostically susceptible

*Annona muricata* — vein-clearing, yellow blotching and leaf distortion.

*Annona squamosa, Rollinia deliciosa* (biriba) — leaves near graft site show vein-clearing.

Diagnostically insusceptible

*Brassica oleracea* cv. *acephala, Capsicum annuum, Chenopodium amaranticolor, C. quinoa, Datura stramonium, Gomphrena globosa, Lactuca sativa, Lycopersicon esculentum, Nicandra physaloides, Nicotiana benthamiana, N. glutinosa, N. rustica, N. tabacum* and *Vigna unguiculata.*

Maintenance and propagation hosts

*Annona muricata.*

Assay hosts (**L**ocal lesions or **W**hole plants)

*Annona muricata* (W).

Dicotyledonous Hosts *(**Susceptible** or Insusceptible)*

*Capsicum annuum. Chenopodium amaranticolor. Chenopodium quinoa. Datura stramonium. Gomphrena globosa. Lactuca sativa. Lycopersicon esculentum. Nicandra physaloides. Nicotiana benthamiana. Nicotiana glutinosa. Nicotiana rustica. Nicotiana tabacum. Vigna unguiculata.*

**Particle morphology**

Virions rhabdo– or bullet–shaped; of 250–300 nm; 60–70 nm wide.

**Cytopathology**

Virions found in cytoplasm (mostly *c.*400 x 60 nm.); in cisternae of endoplasmic reticulum, but not perinuclear space.

**Taxonomy and Relationships**

Probably *Cytorhabdovirus: Rhabdoviridae.*

Virus(es) with serologically unrelated virions: lettuce necrotic yellows, sowthistle yellow vein and maize mosaic viruses.

**References**

Kitajima, E.W., Martins, C.R.F. and Santos, A.A. (1993). *Plant Dis.* **77**: 276.

# Southern potato latent (?) *carlavirus*

**Data collated by A.A. Brunt, 1995.**

First reported in *Solanum tuberosum*; from Japan; by Yora *et al.*, 1983.

**Taxonomy and Relationships**

Possible *Carlavirus*.

**References**

Yora, K., Saito, Y., Doi, Y., Inouye, T. and Tomaru,K. (1983). In: *Handbook of Plant Viruses*. Asakura Shoten, Tokyo.

# Sowbane mosaic *sobemovirus*

**Data collated by D.S. Teakle, 1985.**

*Synonyms:* apple latent virus 2 (Bancroft and Tolin, 1967; Kirkpatrick *et al.*, 1965), chenopodium mosaic virus (Bennett and Costa, 1961; Dias and Waterworth, 1967; Silva *et al.*, 1958; Vuittenez, 1962), chenopodium seed-borne mosaic virus (Dias and Waterworth, 1967), chenopodium star mottle virus (Bancroft and Tolin, 1967). *Standard acronym:* SoMV.

First reported in *Chenopodium murale*; from Riverside, California, U.S.A.; by Bennett and Costa (1961).

**Natural host range and symptoms**

Symptoms persist.
*Chenopodium* spp. — systemic chlorotic mottling.
*Vitis* sp., *Prunus domestica* — latent infection, but rarely infected.
*Atriplex suberecta* — stunting and leaf deformation.
*Chenopodium murale* — symptomless.

**Transmission**

Transmitted by probably a vector; an insect; *Myzus persicae*, *Circulifer tenellus*, *Liriomyza langei*, *Halticus citri* (Bennett and Costa, 1961). Probably none of these transmit frequently or specifically; Aphididae, or Cicadellidae, or Coleoptera (and Diptera). Not transmitted by *Myzus persicae*; Dias and Waterworth (1967). Transmitted in a non–persistent manner (Bennett and Costa, 1961). Virus transmitted by mechanical inoculation; transmitted by grafting; transmitted by seed (up to 60% in *Chenopodium quinoa*; Dias and Waterworth, 1967, and 20–70% in *Chenopodium murale*; Bennett and Costa, 1961); transmitted by pollen to the seed and transmitted by pollen to the pollinated plant (pollen contains much virus, Francki and Miles, 1985).

**Geographical distribution**

Spreads in the South and Central American region; Australia, Bulgaria, Canada, the former Czechoslovakia, Italy, Japan, the USA, and the former Yugoslavia.

**Experimental host range**

Several (3–9) families susceptible.

Diagnostically susceptible

*Chenopodium amaranticolor, C. quinoa, C. murale* — chlorotic local lesions, systemic yellow flecking, star shaped patterns, leaf deformation.

*Beta vulgaris* — mild systemic mosaic.

Only species of Chenopodiaceae show symptoms, but the virus locally infects plant species of several families including Solanaceae and Fabaceae (Dias and Waterworth, 1967).

Maintenance and propagation hosts

*Chenopodium amaranticolor; C. quinoa; C. murale* — for maintenance. *Spinacia oleracea* for purification.

Assay hosts (**L**ocal lesions or **W**hole plants)

*Chenopodium amaranticolor* (L); *C. quinoa* (L).

Dicotyledonous Hosts *(**Susceptible** or Insusceptible)*

***Atriplex hortensis. Beta macrocarpa. Beta vulgaris. Chenopodium album. Chenopodium amaranticolor. Chenopodium capitatum. Chenopodium murale. Chenopodium quinoa. Cucumis sativus. Cucurbita maxima. Datura metel. Datura stramonium. Gomphrena globosa. Nicotiana clevelandii.*** *Nicotiana glutinosa.* ***Nicotiana tabacum. Phaseolus lunatus. Phaseolus vulgaris. Spinacia oleracea. Vigna unguiculata.***

**Properties of particles in sap**

TIP: usually 86–96 °C (Dias and Waterworth, 1967; Kado, 1971). LIV: 60 days (at 23°C; Kado, 1971). DEP: $\log_{10}$ minus 7–9 (or more; Kado, 1971; Teakle, 1968). Leaf sap contains many virions (Teakle, 1968). Electron microscopy: virions of some isolates may be more stable in PTA at pH 4 than at pH 7 (Finlay and Teakle, 1969; Guy, 1982).

**Purification**

Method: Steere (1956); Kado (1967); Guy (1982) used the method of Steere (1956), but resuspended virions in 10 mM sodium acetate, 1 mM magnesium acetate, pH 6.

**Particle morphology**

Virions isometric; not enveloped; 26–28 nm in diameter; rounded in profile, or angular in profile; without a conspicuous capsomere arrangement.

**Physical properties**

One sedimenting component in purified preparations; sedimentation coefficient 104–107 *S* (Kado, 1971; Guy, 1982). Isoelectric point pH 4.4 (Kado, 1971).

**Biochemical properties**

Virions contain 20 % nucleic acid; 80 % protein; 0 % lipid.

*Genome* consists of RNA; single–stranded; unipartite; largest (or only) genome part 4.2 kb. Genomic nucleic acid isolated by Tremaine and Ronald (1977). Base composition 26 % G; 23.2 % A; 27.2 % C; 23.2 % U. Infectivity retained when deproteinised with phenol or detergent (but infectivity is reduced; Saksena and Mink, 1970).

*Virion protein(s)* one; $M_r$ 31000 (Kado, 1967), or 32000 (Guy, 1982). Amino acid composition: Kado, 1967.

**Cytopathology**

Virions found in mesophyll and xylem, but never in phloem (Lombardo *et al.*, 1971); in cytoplasm and in cell vacuoles. Inclusions present in infected cells; are crystals in the cytoplasm; they contain virions. Other cellular changes: proliferation of tonoplasts with concentric membranes bulging into the vacuole (Lombardo *et al.*, 1971).

**Taxonomy and Relationships**

*Sobemovirus.*

Virus(es) with serologically unrelated virions: cocksfoot mottle, lucerne transient streak, rice yellow mottle, southern bean mosaic and turnip rosette viruses.

**References**


Bancroft, J.B. and Tolin, S.A. (1967). *Phytopathology* **57**: 639.
Bennett, C.W. and Costa, A.S. (1961). *Phytopathology* **51**: 546.
Dias, H.J. and Waterworth, H.E. (1967). *Can. J. Bot.* **45**: 1285.
Finlay, J.R. and Teakle, D.S. (1969). *J. gen. Virol.* **5**: 93.
Francki, R.I.B. and Miles, R. (1985). *Pl. Path.* **34**: 11.
Guy, P. (1982). *Australas. Pl. Path.* **11**: 51.
Kado, C.J. (1967). *Virology* **31**: 217.
Kado, C.I. (1971). *CMI/AAB Descr. Pl. Viruses* No. 64, 4 pp.
Kirkpatrick, H.C., Mink, G.I. and Lindner, R.C. (1965). *Phytopathology* **55**: 286.
Lombardo, G., Simonetta, M. and Bassi, M. (1971). *Cariologia* **24**: 323.
Milne, R.G. (1967). *Virology* **32**: 58.
Paul, H.L. and Huth, W. (1970). *Phytopath. Z.* **69**: 1.
Saksena, K.N. and Mink, G.I. (1970). *Virology* **40**: 540.
Silva, D.M., Meneghini, M. and Santos, H.L.S. (1958). *Bragantia* **17**: 167.
Steere, R.L. (1956). *Phytopathology* **46**: 6.
Teakle, D.S. (1968). *Aust. J. biol. Sci.* **21**: 649.
Tremaine, J.H. and Ronald, W.P. (1977). *Can. J. Bot.* **55**: 2274.
Vuittenez, A. (1962). *Etudes Virologie Appliquee (Versailles)* **3**: 15.


# Sowthistle yellow vein *nucleorhabdovirus*

**Data collated by D.C. Stenger, 1991.**

*Standard acronym:* SYVV.

First reported in *Sonchus oleraceus*; from the U.S.A.

**Transmission**

Transmitted by a vector; an insect; *Hyperomyzus lactucae*; Aphididae. Principal natural vector(s): *Hyperomyzus lactucae*. Not transmitted by *Chaetosiphon jacobi*. Transmitted in a persistent manner. Virus retained when the vector moults; transmitted congenitally to the progeny of the vector; not transmitted by mechanical inoculation; probably transmitted by grafting; not transmitted by contact between plants; not transmitted by seed; not transmitted by pollen.

**Experimental host range**

Few (<3) families susceptible.

Dicotyledonous Hosts *(**Susceptible** or Insusceptible)*

***Lactuca sativa*. *Sonchus oleraceus*.**

Sources of host-range data

Peters (1971).

**Properties of particles in sap**

Electron microscopy: to retain membrane, fix with glutaraldehyde and stain with neutral PTA, otherwise 'bullets' are found.

**Purification**

Method: Jackson and Christie (1977).

**Particle morphology**

Virions rhabdo– or bullet–shaped; enveloped; usually straight; with a clear modal length; of 230 nm; 100 nm wide. Axial canal obscure. Basic helix obvious; pitch of basic helix 4.4 nm.

**Physical properties**

One sedimenting component in purified preparations; sedimentation coefficient 1000 *S*. Density *c*. 1.18 g $cm^{-3}$ in sucrose.

**Biochemical properties**

Virions contain 1 % nucleic acid; 80 % protein; 19 % lipid (probably).

*Genome* consists of RNA; single–stranded; linear; unipartite; largest (or only) genome part 13–14 kb. Genomic nucleic acid isolated by Stenger *et al.* (1988). Infectivity lost when deproteinised with proteases (probably); lost when deproteinised with phenol or detergent (probably).

*Features of the genome:* genome encodes at least 4 size classes of polyadenylated mRNA species.

*Non–genomic nucleic acid* not found in the virions. Sub–genomic *m*RNA not found in infected cells.

*Virion protein(s)* five; $M_r$ of the largest 200; L and it is transcriptase. $M_r$ of 2nd largest 78; G and appear as spikes. $M_r$ of 3rd largest 56; N, nucleocapsid. $M_r$ of 4th largest 44; M1. $M_r$ of 5th largest 37; M2. Method of preparation: Dale and Peters (1981). Virion proteins glycosylated; are and it is the G protein.

**Replication**

Replication does not depend on a helper virus.

**Cytopathology**

Virions found in leaves, roots, phloem and companion cells; in cytoplasm, in nuclei, and in the perinuclear space. Inclusions present in infected cells; are viroplasms and unusual in shape; and many bacilliform virions found in the perinuclear space; they contain virions.

**Taxonomy and Relationships**

*Nucleorhabdovirus: Rhabdoviridae*.

Virus(es) with serologically unrelated virions: sonchus yellow net virus.

Best tests for diagnosis: 32p probes using cloned DNA fragments in Northern or dot blots. Bacilliform virions in *Sonchus oleraceus* transmitted persistently with a long latent period by *Hyperomyzus lactucae*.

**References**

Dale, L.J. and Peters, D. (1981). *Intervirology* **16**: 86.

Jackson, A.O., Francki, R.I.B. and Zuideman, D. (1987). In: *The Rhabdoviruses*; ed. R.R. Wagner. Plenum Press, New York.

Peters, D. (1971). *CMI/AAB Descr. Pl. Viruses* No. 62, 3 pp.

Stenger, D.C., Richardson, J., Sylvester, E.S., Jackson, A.O. and Morris, T.J. (1988). *Phytopathology* **78**: 1473.

# Soybean chlorotic mottle *caulimovirus*

**Data collated by Mitsuro Kameya-Iwaki, 1986.**

*Standard acronym:* SbCMV.

First reported in *Glycine max*; from Aichi Prefecture, Japan; by Iwaki *et al.* (1984).

**Natural host range and symptoms**

Symptoms persist.
*Glycine max* — systemic mosaic, stunting.
*Dolichos lablab* — systemic vein-clearing.
*Phaseolus vulgaris* — chlorotic local lesions. systemic vein clearing, mottling, leaf curling.
*Vigna unguiculata* — systemic infection, no symptoms.

**Transmission**

Virus transmitted by mechanical inoculation; not transmitted by seed.

**Geographical distribution**

Spreads in Japan.

**Experimental host range**

Few (<3) families susceptible.

Diagnostically susceptible

*Glycine max* — systemic vein-clearing, followed by leaf mottling and stunting.
*Phaseolus vulgaris* — chlorotic local lesions; systemic vein-clearing, mottling and leaf curling.
*Dolichos lablab* — systemic vein-clearing.
*Vigna unguiculata* cv. Blackeye — symptomless systemic infection.

Diagnostically insusceptible

*Chenopodium amaranticolor, Nicotiana tabacum, Vicia faba, Vigna sesquipedalis.*

Maintenance and propagation hosts

*Glycine max, Phaseolus vulgaris.*

Assay hosts (**L**ocal lesions or **W**hole plants)

*Glycine max* (W) *Phaseolus vulgaris* (L, W).

Dicotyledonous Hosts *(**Susceptible** or Insusceptible)*

*Arachis hypogaea. Beta vulgaris. Brassica campestris* ssp. *rapa. Brassica oleracea* ssp. *capitata. Cajanus cajan. Capsicum annuum. Cassia tora. Chenopodium amaranticolor. Chenopodium quinoa. Cucumis sativus. Cucurbita moschata. Datura stramonium. Dianthus caryophyllus.* ***Glycine max****. Gomphrena globosa. Hibiscus esculentus. Lactuca sativa. Lycopersicon esculentum. Nicotiana clevelandii. Nicotiana debneyi. Nicotiana glutinosa. Nicotiana tabacum. Petunia hybrida.* ***Phaseolus vulgaris****. Pisum sativum. Sesamum indicum. Spinacia oleracea. Tetragonia expansa. Trifolium pratense. Trifolium repens. Vicia faba. Vigna angularis. Vigna radiata.* ***Vigna unguiculata****. Vigna unguiculata* ssp. *sesquipedalis. Zinnia elegans.*

Additional hosts

***Dolichos lablab****.*

Sources of host-range data

Iwaki *et al.* (1984).

**Properties of particles in sap**

TIP: 85–90 °C. LIV: 1 days. DEP: $\log_{10}$ minus 3–4. Leaf sap contains few virions.

**Purification**

Method: Iwaki *et al.* (1984).

**Particle morphology**

Virions isometric; not enveloped; 50 nm in diameter; rounded in profile; with a conspicuous capsomere arrangement.

**Physical properties**

One sedimenting component in purified preparations.

**Biochemical properties**

*Genome* consists of DNA; double–stranded; circular; unipartite; largest (or only) genome part 8.175 kb (pairs). Genomic nucleic acid isolated by Hibi *et al.* (1986). Base composition of ds genome 28 % (G + C). Infectivity decreased when deproteinised with proteases; retained when deproteinised with phenol or detergent. Nucleotide sequence references: Verver *et al.* (1986).

*Sequence database accession code(s):*

X15828 Em(40)_vi:CASCMVX Gb(84)_vi:CASCMVX Soybean chlorotic mottle virus complete genomic sequence. 9/93 8,175bp. 1 sequence.

**Cytopathology**

Virions found in all parts of the host plant; in cytoplasm. Inclusions present in infected cells; are viroplasms; they contain virions.

**Taxonomy and Relationships**

*Caulimovirus.*

**References**

Hibi, T. and Kameya-Iwaki, M. (1988). *AAB Descr. Pl. Viruses* No. 331, 4 pp.

Hibi, T., Iwaki, M., Saito, Y., Verver, J. and Goldbach, R. (1986). *Ann. Phytopath. Soc. Japan* **52** (**5**): 785.

Iwaki, M., Isogawa, Y., Tsuzuki, H. and Honda, Y. (1984). *Plant Dis.* **68**: 1009.

Verver, J., Schijns, P., Hibi, T. and Goldbach, R. (1986). *J. gen. Virol.* **68**: 159.

# Soybean crinkle leaf (?) bigeminivirus

**Data collated by M. Iwaki, 1984. Revised 1989.**

*Synonyms:* soybean crinkle virus. *Standard acronym:* SCLV.

First reported in *Glycine max*; from Phitsanulok, Thailand; by Iwaki *et al.* (1983).

**Natural host range and symptoms**

Symptoms persist.

*Glycine max* — leaf deformation and leaf vein enations.

**Transmission**

Transmitted by a vector; an insect; *Bemisia tabaci*; Aleyrodidae. Transmitted in a persistent manner. Virus not transmitted by mechanical inoculation; transmitted by grafting; not transmitted by seed.

**Geographical distribution**

Spreads in Thailand.

**Experimental host range**

Several (3–9) families susceptible. Experimentally infected plants mostly show vein-clearing, leaf mottling, vein-enations.

Diagnostically susceptible

*Cassia tora, Datura stramonium, Glycine max, Lycopersicon esculentum, Nicotiana clevelandii, N. debneyi, N. glutinosa, N. tabacum, Petunia hybrida, Phaseolus vulgaris cv. Top Crop, Zinnia elegans* — systemic vein-clearing followed by leaf curling and crinkling.

Diagnostically insusceptible

*Arachis hypogaea, Brassica rapa, Chenopodium quinoa, Cucumis sativus, Phaseolus lunatus, Vigna sesquipedalis, V. unguiculata.*

Maintenance and propagation hosts

*Glycine max.*

Assay hosts (**L**ocal lesions or **W**hole plants)

*Nicotiana tabacum* (W), *Lycopersicon esculentum* (W), *Petunia hybrida* (W), *Datura stramonium* (W), *Phaseolus vulgaris* (W).

Dicotyledonous Hosts *(**Susceptible** or Insusceptible)*

*Arachis hypogaea. Brassica campestris* ssp. *rapa. Cajanus cajan. Capsicum annuum.* ***Cassia tora***. *Chenopodium amaranticolor. Chenopodium quinoa. Cucumis sativus.* ***Datura stramonium***. ***Glycine max***. *Gomphrena globosa. Gossypium hirsutum. Hibiscus esculentus. Lablab purpureus.* ***Lycopersicon esculentum***. ***Nicotiana clevelandii***. ***Nicotiana debneyi***. ***Nicotiana glutinosa***. ***Nicotiana tabacum***. ***Petunia hybrida***. *Phaseolus lunatus.* ***Phaseolus vulgaris***, or *Phaseolus vulgaris. Sesamum indicum. Solanum melongena. Tetragonia expansa. Trifolium pratense. Trifolium repens. Vigna radiata. Vigna unguiculata. Vigna unguiculata* ssp. *sesquipedalis*. ***Zinnia elegans***.

Sources of host-range data

Iwaki *et al.* (1983).

**Properties of particles in sap**

Leaf sap contains few virions.

**Particle morphology**

Virions possibly geminate; not enveloped.

**Cytopathology**

Virions found in phloem; in nuclei. Inclusions present in infected cells; are crystals in the nucleus and amorphous X–bodies; they contain virions.

**Taxonomy and Relationships**

Possible bigeminivirus: *Geminiviridae*.

**References**

Iwaki, M., Thongmeearkom, P., Honda, Y. and Deema, N (1983). *Plant Dis.* 67: 546.

Iwaki, M., Thongmeearkom, P., Honda, Y., Sarindu, N., Deema, N. and Surin, P. (1986). *Tech. Bull. Trop. Agric. Res. Center, Japan*, p. 132.

# Soybean dwarf *luteovirus*

**Data collated by T. Tamada, 1980. Revised by G.R. Johnstone, R.R. Martin and P. Waterhouse, 1989.**

*Synonyms:* subterranean clover red leaf luteovirus. *Standard acronym:* SbDV. *Strains:* soybean dwarf virus — dwarfing strain (SDV–D), soybean dwarf — leaf yellowing strain (SDV–Y).

First reported in *Glycine max*; from Hokkaido, Japan; by Tamada *et al.* (1969); Tamada (1970).

**Natural host range and symptoms**

Symptoms persist.

*Glycine max, Trifolium pratense* — puckering, rugosity and yellowing, stunting.

*Pisum sativum* — mild yellowing of older leaves.

*Phaseolus vulgaris* — yellowing of leaves and stunting.

*Gomphrena globosa, Erodium moschatum, Trifolium subterraneum, Vicia sativa, Calandrinia caulescens* — leaf reddening.

*Beta vulgaris, Lupinus albus, Lupinus angustifolius, Medicago polymorpha* — leaf yellowing.

*Trifolium repens, T. pratense* — no symptoms.

*Vicia faba, Vicia articulata* — leaf yellowing, rolling.

**Transmission**

Transmitted by a vector; an insect; *Acyrthosiphon (Aulacorthum) solani* (some isolates also by *Acyrthosiphon pisum*); Aphididae. Transmitted in a persistent manner. Virus retained when the vector moults; does not multiply in the vector; not transmitted congenitally to the progeny of the vector; does not require a helper virus for vector transmission; not transmitted by mechanical inoculation; not transmitted by contact between plants; not transmitted by seed.

**Ecology and control**

Studies reported by Ashby *et al.* (1979); Johnstone (1978); Tamada (1975).

**Geographical distribution**

Spreads in Australia, Japan, New Zealand, and the USA (in California).

**Experimental host range**

Several (3–9) families susceptible.

Diagnostically susceptible

*Astragalus sinicus* — systemic leaf yellowing or reddening.

*Glycine max* — systemic puckering, rugosity and yellowing of leaves, stunting.

*Trifolium incarnatum, T. subterraneum* — systemic leaf reddening.

*Vicia faba* — systemic chlorosis and yellowing.

Maintenance and propagation hosts

*Glycine max, Pisum sativum.*

Assay hosts (**L**ocal lesions or **W**hole plants)

*Glycine max* (W), *Pisum sativum* (W), *Trifolium subterraneum* (W).

Dicotyledonous Hosts *(**Susceptible** or Insusceptible)*

***Arachis hypogaea. Astragalus sinicus. Beta vulgaris.*** *Brassica campestris* ssp. *napus. Capsella bursa–pastoris. Cicer arietinum. Datura stramonium.* ***Glycine max.***

***Gomphrena globosa***. *Lablab purpureus*. ***Lactuca sativa***. *Lathyrus odoratus*. ***Lens esculenta***. *Linum usitatissimum*. *Lotus corniculatus*. ***Lupinus albus***. ***Lupinus angustifolius***. ***Lupinus cosentinii***. ***Medicago hispida***. *Medicago sativa*. *Melilotus alba*. ***Phaseolus vulgaris***. ***Pisum sativum***. *Stellaria media*. ***Tetragonia expansa***. ***Trifolium dubium***. ***Trifolium hybridum***. ***Trifolium incarnatum***. ***Trifolium pratense***. ***Trifolium repens***. ***Trifolium subterraneum***. ***Vicia faba***. ***Vicia sativa***. *Vicia villosa*. *Vigna angularis*. *Vigna radiata*. *Vigna unguiculata*. *Vigna unguiculata* ssp. *sesquipedalis*.

Sources of host-range data

Tamada (1973, 1975); Johnstone *et al.* (1984).

**Properties of particles in sap**

TIP: 45–50 °C (when tested by insect injection or membrane feeding). LIV: more than 14 days. DEP: $\log_{10}$ minus 2–3. Infectivity of sap decreased by treatment with di–ethyl ether. Leaf sap contains few virions. Electron microscopy: not necessary to fix the virions.

**Purification**

Method: Tamada *et al.* (1969); Kojima and Tamada (1976); Tamada (1973); Ashby and Kyriakou (1982); Johnstone *et al.* (1982); Waterhouse and Helms (1985).

**Particle morphology**

Virions isometric; not enveloped; 25–27 nm in diameter; rounded in profile; without a conspicuous capsomere arrangement.

**Physical properties**

One sedimenting component in purified preparations; sedimentation coefficient 114 *S*.

**Biochemical properties**

Virions contain 30 % nucleic acid; 70 % protein; 0 % lipid.

*Genome* consists of RNA; single–stranded; linear; unipartite; largest (or only) genome part 5.861 kb. Genomic nucleic acid isolated by Waterhouse *et al.* (1987). Poly A region absent.

*Sequence database accession code(s):*

L20835 Em(40)_vi:SDVCOATX Gb(84)_vi:SDVCOATX Soybean dwarf virus coat protein gene and 21 kDa ORF, complete cds. 3/94 612bp.

L24049 Em(40)_vi:SDVRNAP Gb(84)_vi:SDVRNAP Soybean dwarf virus RNA-dependent RNA polymerase, coat protein, genome-linked protein genes,

X15502 Em(40)_vi:LUSDPRM Gb(84)_vi:LUSDPRM Soybean dwarf virus genomic RNA for potential promoter. 3/92 122bp. 3 sequences.

*Virion protein(s)* two; $M_r$ of the major 24000; coat protein. $M_r$ of 2nd largest 55000; possibly a read through protein from the coat protein. Method of preparation: Ashby and Kyriakou (1982).

**Cytopathology**

Virions found in phloem; in cytoplasm and in cell vacuoles. Inclusions absent from infected cells. Other cellular changes: phloem necrosis, starch accumulates in chloroplasts of epidermis and mesophyll cells.

**Taxonomy and Relationships**

*Luteovirus*.

Virus(es) with serologically related virions: beet western yellows, barley yellow dwarf (strains RPV and MAV), bean leaf roll, potato leafroll and tobacco necrotic dwarf viruses. Subterranean clover red leaf virus is so closely related that it is often considered to be the same species.

Best tests for diagnosis: ELISA tests using monoclonal antibodies.

**Comments**

A severe stunting disease of soybean (*Glycine max*), African soybean dwarf, was first noted in 1975 in Nigeria (Rossel and Thottappilly, 1983). The causal virus was transmitted in the persistent manner by *Bemisia tabaci*. It resembles that caused by soybean dwarf luteovirus, which is aphid-borne. Some soybean genotypes (*e.g.* cvs Improved Pelican, Indo 100 and Bossier) are very resistant to the Nigerian virus (Rossel and Thottappilly, 1985).

**References**


Ashby, J.W. and Kyriakou, A. (1982). *N.Z. Jl agric. Res.* **25**: 607.

Ashby, J.W., Teh, P.B. and Close, R.C. (1979). *N.Z. Jl agric. Res.* **22**: 361.

D'Arcy, C.J., Torrance, L. and Martin, R.R. (1989). *Phytopathology* **79**: 869.

Johnstone, G.R. (1978). *Aust. J. agric. Res.* **29**: 1003.

Johnstone, G.R., Duffus, J.E., Munro, D. and Ashby, J.W. (1982). *Aust. J. agric. Res.* **33**: 697.

Johnstone, G.R., Ashby, J.W., Gibbs, A.J., Duffus, J.E., Thottappilly, G. and Fletcher, J.D. (1984). *Neth. J. Pl. Path.* **90**: 225.

Kojima, M. and Tamada, T. (1976). *Phytopath. Z.* **85(3)**: 347.

Rathjen, J.P., Karageorgos, L.E., Habili, N., Waterhouse, P.M. and Symons, R.H. (1994). *Virology* **198**: 671.

Rossel, H.W. and Thottappilly, G. (1983). *I.I.T.A. Res. Briefs* **3**: 3.

Rossel, H.W. and Thottappilly, G. (1985). *Virus Diseases of Important Food Crops in Tropical Africa*. Ibadan: I.I.T.A.

Tamada, T. (1970). *Ann. Phytopath. Soc. Japan* **36**: 266.

Tamada, T. (1973). *Ann. Phytopath. Soc. Japan* **39**: 27.

Tamada, T. (1975). *Rep. Hokkaido Pref. Agric. Exp. Stn* **24**.

Tamada, T. (1975). *Rep. Hok. Pref. Agric. Exp. Stn* **25**: pp.144.

Tamada, T. and Kojima, M. (1977). *CMI/AAB Descr. Pl. Viruses* No. 179, 4 pp.

Tamada, T., Goto, T., Chiba, I. and Suwa, T. (1969). *Ann. Phytopath. Soc. Japan* **35**: 282.


# Soybean Indonesian dwarf *luteovirus*

**Data collated by M. Iwaki, 1983.**

*Synonyms:* Indonesian soybean dwarf virus. *Standard acronym:* ISDV.

First reported in *Glycine max*; from Bogor, Indonesia; by Iwaki *et al.* (1980).

**Natural host range and symptoms**

Symptoms persist.

*Glycine max* — stunting, mottling and leaf malformation.

**Transmission**

Transmitted by a vector; an insect; *Aphis glycines*; Aphididae. Transmitted in a persistent manner. Virus retained when the vector moults; does not multiply in the vector; not transmitted by mechanical inoculation; not transmitted by contact between plants; not transmitted by seed; not transmitted by pollen.

**Geographical distribution**

Spreads in Indonesia and Thailand.

**Experimental host range**

Few (<3) families susceptible.

Diagnostically susceptible

*Glycine max* — stunting, mottling, leaf shape malformation.

Maintenance and propagation hosts

*Glycine max*.

Assay hosts (**L**ocal lesions or **W**hole plants)

*Glycine max* (W).

Dicotyledonous Hosts (***Susceptible*** *or Insusceptible)*

*Arachis hypogaea. Astragalus sinicus. Brassica campestris* ssp. *rapa. Chenopodium amaranticolor. Cucumis sativus. Cucurbita pepo.* ***Glycine max****. Lathyrus odoratus. Nicotiana tabacum. Petunia hybrida. Phaseolus vulgaris. Pisum sativum. Sesamum indicum. Trifolium hybridum. Trifolium incarnatum. Trifolium pratense. Trifolium repens. Vicia faba. Vigna radiata. Vigna unguiculata. Vigna unguiculata* ssp. *sesquipedalis*.

Sources of host-range data

Iwaki *et al.* (1980).

**Properties of particles in sap**

Leaf sap contains few virions.

**Purification**

Method: Takanami and Kubo (1979).

**Particle morphology**

Virions isometric; not enveloped; 26 nm in diameter; rounded in profile; without a conspicuous capsomere arrangement.

**Physical properties**

One sedimenting component in purified preparations.

**Cytopathology**

Virions found in phloem.

**Taxonomy and Relationships**

*Luteovirus*.

Virus(es) with serologically unrelated virions: soybean dwarf and beet western yellows luteoviruses (Ashby and Johnstone, 1985).

**References**


Ashby, J.W. and Johnstone, G.R. (1985). *Australas. Pl. Path.* **14**: 2.

Iwaki, M., Roechan, M., Hibino, H., Tochihara, H. and Tantera, D.M. (1980). *Plant Dis.* **64**: 1027.

Takanami, Y. and Kubo, S. (1979). *J. gen. Virol.* **44**: 153.

# Soybean mild mosaic virus

**Data collated by T. Inouye, 1981. Revised 1989.**

First reported in *Glycine max*; from Japan; by Takahashi *et al.* (1974).

**Natural host range and symptoms**

Symptoms persist.

*Glycine max* — chlorotic spots, mild mottling.

**Transmission**

Transmitted by a vector; an insect; *Myzus persicae*; Aphididae. Transmitted in a non–persistent manner. Virus transmitted by mechanical inoculation; transmitted by seed (22–70% in *Glycine max* cultivars).

**Geographical distribution**

Spreads in Japan.

**Experimental host range**

Several (3–9) families susceptible.

Diagnostically susceptible

*Chenopodium amaranticolor* — chlorotic local lesions; systemic mosaic.

*Spinacia oleracea* — necrotic and chlorotic local lesions; systemic chlorotic spotting and vein-clearing.

*Cucumis sativus* — chlorotic local lesions or latent.

*Glycine max, Pisum sativum* — systemic mosaic.

*Vigna radiata, V. unguiculata* — necrotic local lesions; systemic vein-clearing and necrotic spotting.

*Nicotiana glutinosa, N. tabacum* — systemic mosaic.

Diagnostically insusceptible

*Astragalus sinicus, Datura stramonium, Tetragonia expansa.*

Maintenance and propagation hosts

*Spinacia oleracea, Glycine max, Lupinus albus, Macroptilium lathyroides, Vigna unguiculata.*

Assay hosts (**L**ocal lesions or **W**hole plants)

*Phaseolus lunatus* cv. Henderson Bush (L), *P. vulgaris* cv. Top Crop (L), *Vigna radiata*, (L).

Dicotyledonous Hosts *(**Susceptible** or Insusceptible)*

***Antirrhinum majus***. *Apium graveolens*. *Cassia tora*. ***Chenopodium album***. *Chenopodium amaranticolor*. ***Chenopodium quinoa***. ***Cucumis sativus***. *Cucurbita maxima*. ***Cyamopsis tetragonoloba***. *Datura metel*. *Datura stramonium*. ***Glycine max***. ***Gomphrena globosa***. *Ipomoea nil*. *Lablab purpureus*. ***Lathyrus odoratus***. ***Lupinus albus***. ***Lycopersicon esculentum***. ***Macroptilium lathyroides***. ***Macrotyloma uniflorum***. ***Nicotiana clevelandii***. *Nicotiana glutinosa*. ***Nicotiana tabacum***. ***Petunia hybrida***. ***Phaseolus lunatus***. ***Phaseolus vulgaris***, or *Phaseolus vulgaris*. ***Physalis floridana***. ***Pisum sativum***. ***Sesamum indicum***. *Spinacia oleracea*. ***Vicia faba***. ***Vicia sativa***. ***Vicia villosa***. *Vigna angularis*. ***Vigna radiata***. ***Vigna unguiculata***. ***Vigna unguiculata*** ssp. ***sesquipedalis***. *Zinnia elegans*.

Sources of host-range data

Takahashi *et al.* (1974, 1980); Hampton *et al.* (1978).

**Properties of particles in sap**

TIP: 60–65 °C. LIV: 3–4 days. DEP: $\log_{10}$ minus 4–5. Leaf sap contains few virions.

**Particle morphology**

Virions isometric; not enveloped; 26–27 nm in diameter; angular in profile; without a conspicuous capsomere arrangement.

**Cytopathology**

Virions found in leaves. Inclusions absent from infected cells.

**Taxonomy and Relationships**

Virus(es) with serologically unrelated virions: soybean stunt, alfalfa mosaic, cucumber mosaic, mulberry ringspot, broad bean mottle and tomato black ring viruses.

**References**

Hampton, R., Beczner, L., Hagedorn, D., Bos, L., Inouye, T., Barnett, O.W., Musil, M. and Meiners, J. (1978). *Phytopathology* **68**: 989.

Kolte, S.J. and Nene, Y.L. (1972). *Indian Phytopath.* **25**: 401.

Takahashi, K., Tanaka, T. and Tsuda, Y. (1974). *Ann. Phytopath. Soc. Japan* **40**: 103.

Takahashi, K., Tanaka, T., Iida, W. and Tsuda, Y. (1980). *Tohoku Agric. Exp. Stn Bull.* **62**: 1.

# Soybean mosaic *potyvirus*

**Data collated by R. Goodman, 1980. Revised 1987 by J.R. Edwardson and L. Bos.**

*Standard acronym:* SMV.

First reported in *Glycine max*; from Connecticut, U.S.A.; by Clinton (1915); Gardener and Kendrick (1921).

**Natural host range and symptoms**

Symptoms persist, or disappear soon after infection (especially at high temperatures).

*Glycine max* — Rugosity, dark green vein banding and light green interveinal areas, stunting, leaf curling and seed coat mottling, male sterility, flower deformation, less pubescent, necrosis, sometimes necrotic local lesions, systemic necrosis and bud blight.

**Transmission**

Transmitted by normally a vector; an insect; 16 species, including *Acyrthosiphon pisum, Aphis fabae, Myzus persicae* (Edwardson and Christie, 1986); Aphididae. Transmitted in a non–persistent manner. Virus does not require a helper virus for vector transmission (One strain is not aphid transmitted (Ross, 1975; Cho and Goodman, 1981)); transmitted by mechanical inoculation; up to 30% or higher transmitted by seed (not in *Glycine max* cvs Kawanggyo, Hill or Bienville (Cho *et al.*, 1977; Ross, 1968; Porto and Hagedorn, 1975)); transmitted by pollen to the seed, or transmitted by pollen to the pollinated plant.

**Ecology and control**

Studies reported by Ross (1968, 1977); Quiniones *et al.* (1971); Sun *et al.* (1977); Cho and Goodman (1979); Kiihl and Hartwig (1979); Dhingra and Chenulu (1980); Kwon and Oh (1980); Irwin and Schultz (1981); Polsten and Goodman (1981).

**Geographical distribution**

Probably distributed worldwide.

**Experimental host range**

Few (<3) families susceptible.

Diagnostically susceptible

*Chenopodium album, C. quinoa* — chlorotic local lesions.
*Lablab purpureus* — necrotic local lesion.
*Macroptilium lathyroides* — systemic mosaic.
*Phaseolus vulgaris* — systemic mosaic with some strains in some cultivars, but often latent or no infection.
*P. vulgaris* cv. Top Crop — necrotic local lesions at 30°C in detached leaves.

Diagnostically insusceptible

*Cucumis sativus, Vicia faba, Petunia hybrida.*

Maintenance and propagation hosts

*Glycine max.*

Assay hosts (Local lesions or Whole plants)

*Cyamopsis tetragonoloba* (L), *Lablab purpureus* (L), *Macrotyloma uniflorum* (W), *Glycine max* (L/W), *Phaseolus vulgaris* (L).

Dicotyledonous Hosts *(**Susceptible** or Insusceptible)*

***Antirrhinum majus***. *Beta vulgaris*. *Brassica campestris*. *Cajanus cajan*. *Capsicum frutescens*. ***Cassia occidentalis***. *Cassia tora*. *Catharanthus roseus*. ***Chenopodium album***. *Chenopodium amaranticolor*. ***Chenopodium quinoa***. ***Crotalaria spectabilis***. *Cucumis sativus*. *Cucurbita maxima*. *Cucurbita pepo*. ***Cyamopsis tetragonoloba***. *Datura metel*. *Datura stramonium*. *Dianthus barbatus*. ***Dolichos biflorus***. ***Glycine max***. *Gomphrena globosa*. *Gossypium hirsutum*. *Helianthus annuus*. *Lablab purpureus*. *Lactuca sativa*. ***Lespedeza stipulacea***. *Lotus corniculatus*. ***Lupinus albus***. *Lupinus angustifolius*. *Lycopersicon esculentum*. *Macroptilium lathyroides*. *Macrotyloma uniflorum*. *Medicago sativa*. *Melilotus alba*. *Melilotus officinalis*. *Mimosa pudica*. ***Nicandra physaloides***. ***Nicotiana benthamiana***. *Nicotiana glutinosa*. ***Nicotiana tabacum***, or *Nicotiana tabacum*. ***Petunia hybrida***, or *Petunia hybrida*. ***Phaseolus lathyroides***. *Phaseolus lunatus*, or ***Phaseolus lunatus***. ***Phaseolus vulgaris***. *Physalis floridana*. ***Pisum sativum***. *Raphanus sativus*. ***Sesbania exaltata***. *Solanum nigrum*. *Spinacia oleracea*. *Trifolium hybridum*. *Trifolium incarnatum*. *Trifolium pratense*. *Trifolium repens*. *Tropaeolum majus*. ***Vicia faba***, or *Vicia faba*. *Vigna radiata*. ***Vigna unguiculata***. *Vigna unguiculata* ssp. *sesquipedalis*. *Zinnia elegans*.

Monocotyledonous Hosts *(**Susceptible** or Insusceptible)*

*Avena sativa*. *Hordeum vulgare*. *Triticum aestivum*. *Zea mays*.

Additional hosts

***Glycine usuriensis, Glycine wightii*** and ***Lespedeza striata***.

Sources of host-range data

Conover (1948); Quantz (1961); Galvez (1963); Walters (1963); Edwardson (1974); Cho *et al.* (1977); Christie and Crawford (1978); Hampton *et al.* (1978).

**Properties of particles in sap**

TIP: usually 50 °C, or 65–70 °C (with some isolates). LIV: 1–4 days. DEP: $\log_{10}$ minus 3–4. Leaf sap contains few virions.

**Purification**

Method: Ross (1967); Hill and Benner (1980).

**Particle morphology**

Virions filamentous; not enveloped; usually flexuous; with a clear modal length; of 650–700 nm, or 760 nm; 15–18 nm wide. Axial canal obscure. Basic helix obscure.

**Physical properties**

One sedimenting component in purified preparations (but tend to aggregate).

**Biochemical properties**

Virions contain 5.3 % nucleic acid; 94.7 % protein; 0 % lipid.

*Genome* consists of RNA; single–stranded; linear; unipartite; largest (or only) genome part 10.4 kb. Genomic nucleic acid isolated by Hill and Benner (1978, 1980*a*). Base composition 24.3 % G; 29.9 % A; 14.9 % C; 30.9 % U. Infectivity retained when deproteinised with phenol or detergent.

*Sequence database accession code(s):*

D00507 Em(40)_vi:MSVCP Gb(84)_vi:SMVCPA Soybean mosaic virus, coat protein gene, complete cds. 7/93 1,168bp.

D00717 Em(40)_vi:SMVNIA Gb(84)_vi:SMVNIA Soybean mosaic virus genome, CI protein, 3' end, NIa proteinase, complete cds, NIb protein, 5

S42280 Em(40)_un:S42280 Gb(84)_vi:S42280 SMV genome soybean mosaic virus SMV, strain G2, Genomic RNA Complete, 9588 nt. 2/93 9,588bp.

X63771 Em(40)_vi:SMVCP Gb(84)_vi:X63771 Soybean Mosaic Virus gene for coat protein. 1/92 807bp.

S73815 Em(43)_vi:S73815 Gb(89)_vi:S73815 CP=coat protein (soybean mosaic virus SMV, Beijing isolate, Genomic, 1055 nt). 5/95 1,055bp.

U25673 Gb(89)n:Smu25673 Soybean mosaic virus coat protein gene, complete cds. 6/95 807bp.

*Non–genomic nucleic acid* not found in the virions.

*Virion protein(s)* one; $M_r$ 28300. Method of preparation: Hill and Benner (1980*b*).

*Virus–coded non–virion proteins* isolated (Vance and Beachy (1984)); two proteins found.

**Replication**

Replication does not depend on a helper virus.

**Cytopathology**

Virions found in roots cortex, epidermis, phloem, testa and embryos; in cytoplasm (sometimes as cylindrical aggregations (Tu, 1975)). Inclusions present in infected cells; are cylindrical pinwheels and unusual in shape; , bundles, scrolls and laminated aggregates and are pinwheels, bundles, scrolls and short curved laminated aggregates; they do not contain virions. Other cellular changes: the production of fewer starch grains, but more vesicles in chloroplasts.

**Taxonomy and Relationships**

*Potyvirus: Potyviridae.*

Virus(es) with serologically related virions: bean yellow mosaic, bean common mosaic, clover yellow vein and watermelon mosaic II viruses. Virus(es) with serologically unrelated virions: bean yellow mosaic strains 240–1 and OH-S, blackeye cowpea mosaic, gloriosa stripe and peanut mottle viruses.

Peptide analysis and immunoblot tests show that this virus is very closely related to blackeye cowpea mosaic potyvirus (D. Shukla, personal communication). Best tests for diagnosis: Soybean mosaic virus (SMV) does not infect *Phaseolus vulgaris* systemically, except cv. Double White Princess. Bean common mosaic (BCMV) and bean yellow mosaic (BYMV) viruses infect *Phaseolus vulgaris* systemically. BCMV does not infect *Glycine max* systemically and BYMV infects *Vicia faba* systemically.

**References**


Bos, L. (1972). *CMI/AAB Descr. Pl. Viruses* No. 93, 4 pp.
Cho, E.K. and Goodman, R.M. (1979). *Phytopathology* **69**: 467.
Cho, E.K. and Goodman, R.M. (1981). *Phytopathology* **71**: 208.
Clinton, G.P. (1915). *Rep. Connecticut Agric. Exp. Stn 1915*, p. 421.
Dhingra, K.L. and Chenulu, V.V. (1980). *Indian Phytopath.* **33**: 586.
Edwardson, J.R. (1974). *Fla Agric. Exp. Stn Monog.* No. 5, p. 103.
Edwardson, J.R. and Christie, R.G. (1986). *Fla Agric. Exp. Stn Monog.* No. 14, p. 432.
Galvez, G.E. (1963). *Phytopathology* **53**: 388.
Gardener, M.W. and Kendrick, J.B. (1921). *J. Agric. Res.* **22**: 111.
Hampton, R., Beczner, L., Hagedorn, D., Bos, L., Inouye, T., Barnett, O.W., Musil, M. and Meiners, J. (1978). *Phytopathology* **68**: 989.
Hill, J.H. and Benner, H.I. (1978). *Phytopath. News* **12**: 150.
Hill, J.H. and Benner, H.I. (1980*a*). *Phytopathology* **70**: 236.
Hill, J.H. and Benner, H.I. (1980*b*). *Phytopath. Z.* **97**: 272.
Irwin, M.E. and Goodman, R.M. (1980). In: *Ecological and Epidemiological Aspects*, Spread of Plant Virus Diseases. ed. K.F. Harris. Academic Press, New York.
Irwin, M.E and Schultz, G.A. (1981). *FAO Pl. Prot. Bull.* **29**: 41.
Kiihl, R.A.S. and Hartwig, E.E. (1979). *Crop Sci.* **19**: 372.
Kwon, S.H. and Oh, J.H. (1980). *Crop Sci.* **20**: 403.
Lucas, B.S. and Hill, J.H. (1980). *Phytopath. Z.* **99**: 47.
McLaughlin, M.R., Bryant, G.R., Hill, J.H., Benner, H.I. and Duvant, D.P. (1980). *Phytopathology* **70**: 831.
Quantz, L. (1916). *Phytopath. Z.* **43**: 79.
Quiniones, S.S., Dunleavy, J.M. and Fischer, J.W. (1971). *Crop Sci.* **11**: 662.
Ross, J.P. (1967). *Phytopathology* **57**: 465.
Ross, J.P. (1968). *Pl. Dis. Reptr* **52**: 344.
Ross, J.P. (1975). *Pl. Dis. Reptr* **59**: 806.
Ross, J.P. (1977). *Crop Sci.* **17**: 869.
Soong, M.M.C (1977). *Ph.D. Thesis*, Univ. Illinois, Urbana-Champaign, 128 pp.
Sun, M., Hu, C.H., Lai, P.L. and Ng, O.D. (1977). *Abstr. Am. Phytopath. Soc. 69th Ann. Meet.* No. 44.
Tu, J.C. (1975). *Microbios* **14**: 151.
Vance, V.B. and Beachy, R.N. (1984). *Virology* **132**: 271.
Walters, H.J. (1963). *Pl. Dis. Reptr* **47**: 726.


# Soybean (?) rhabdovirus

**Data collated by A.A. Brunt, 1994.**

First reported in *Glycine max*; from South Africa; by Pietersen, 1993.

**Natural host range and symptoms**

*Glycine max* — mosaic.

**Transmission**

Virus not transmitted by mechanical inoculation.

**Geographical distribution**

Spreads in the African region; South Africa.

**Particle morphology**

Virions rhabdo– or bullet–shaped; enveloped; of 290–440 nm; 79 nm wide.

**Taxonomy and Relationships**

Unassigned rhabdovirus.

Virus(es) with serologically unrelated virions: thirteen other rhabdoviruses.

**References**

Pietersen, G. (1993). *Abstr. Int. Cong. Pl. Path., Montreal*, 1993, p.306.

# Soybean spherical virus

**Data collated by J.L. Dale, 1987.**

First reported in *Glycine max*; from Warwick, Queensland, Australia; by Behncken (1972).

**Natural host range and symptoms**

Symptoms persist.

*Glycine max* — leaf mottling.

**Geographical distribution**

Spreads in Australia.

**Experimental host range**

Few (<3) families susceptible.

Diagnostically susceptible

*Glycine max* — systemic 'spot' mottle.

*Chenopodium quinoa, Phaseolus vulgaris* cv. Bountiful — chlorotic local lesions, systemic chlorotic mottle.

*Chenopodium amaranticolor* — necrotic local lesions; systemic chlorotic spots.

*Vigna unguiculata* cv. Blackeye — systemic green mottle.

*Cyamopsis tetragonoloba* — systemic necrotic rings.

Diagnostically insusceptible

*Nicotiana tabacum, N. clevelandii, N. glutinosa, Vicia faba.*

Maintenance and propagation hosts

*Glycine max, Chenopodium quinoa.*

Assay hosts (**L**ocal lesions or **W**hole plants)

*Chenopodium quinoa* (L).

Dicotyledonous Hosts (***Susceptible*** *or Insusceptible)*

*Beta vulgaris*. *Brassica campestris* ssp. *rapa*. *Brassica oleracea* ssp. *capitata*. ***Cassia occidentalis***. ***Chenopodium amaranticolor***. ***Chenopodium quinoa***. *Cucumis sativus*. ***Cyamopsis tetragonoloba***. *Galinsoga parviflora*. ***Glycine max***. ***Gomphrena globosa***. *Lablab purpureus*. ***Lathyrus odoratus***. ***Lupinus angustifolius***. *Lycopersicon esculentum*. ***Macroptilium lathyroides***. *Medicago sativa*. *Nicotiana clevelandii*. *Nicotiana glutinosa*. *Nicotiana tabacum*. *Petunia hybrida*. ***Phaseolus lathyroides***. ***Phaseolus lunatus***. ***Phaseolus vulgaris***. *Solanum nigrum*. *Vicia faba*. ***Vigna unguiculata***.

Sources of host-range data

Behncken (1972).

**Properties of particles in sap**

TIP: 65–70 °C. LIV: more than 42 days. DEP: $\log_{10}$ minus 4–5.

**Particle morphology**

Virions isometric; not enveloped; 25 nm in diameter; angular in profile; without a conspicuous capsomere arrangement.

**Physical properties**

Probably two sedimenting components in purified preparations.

**Cytopathology**

Virions found in leaves.

**Taxonomy and Relationships**

Virus(es) with serologically unrelated virions: galinsoga mosaic, tobacco ringspot, tomato ringspot, cowpea chlorotic mottle, bean pod mottle, broad bean wilt, arabis mosaic, cucumber mosaic, tobacco necrosis, southern bean mosaic, squash mosaic, sowbane mosaic, tomato bushy stunt, raspberry ringspot, cymbidium ringspot and tobacco streak viruses.

**References**

Behncken, G.M. (1972). *Aust. Pl. Path. Soc. Newsl.* **1**: 17.

# Soybean yellow vein virus

**Data collated by A.A. Brunt and T. Senboku, 1987.**

First reported in *Glycine max*; from Thailand; by Senboku *et al.* (1986).

**Natural host range and symptoms**

Symptoms persist.

*Glycine max* — yellow vein banding or mottling.

**Transmission**

Transmitted by means not involving a vector. Virus transmitted by mechanical inoculation; transmitted by grafting.

**Geographical distribution**

Spreads in Thailand.

**Experimental host range**

Few (<3) families susceptible.

Diagnostically susceptible

*Chenopodium amaranticolor, C. quinoa* — chlorotic local lesions; not systemic.

*Glycine max* — systemic yellow vein symptoms.

Diagnostically insusceptible

*Cucumis sativus, Nicotiana tabacum, Vigna mungo, V. radiata, V. sesquipedalis.*

Maintenance and propagation hosts

*Glycine max.*

Assay hosts (Local lesions or Whole plants)

*Chenopodium amaranticolor* (L).

**Properties of particles in sap**

TIP: 35–40 °C (in *Glycine max* leaf sap). LIV: 0.125–0.25 days (at 4°C). DEP: $\log_{10}$ minus 3–4.

**Particle morphology**

Virions rod–shaped, or filamentous; usually straight; with a clear modal length; of 500–550 nm; 15–20 nm wide. Axial canal obscure. Basic helix obscure.

**References**

Senboku, T., Thongmeearkom, P., Kittipakorn, K., Kiratiya-Angul, S., Srithongchai, W. and Deema, N. (1986). *Tech. Bull. Trop. Agric. Res. Center, Japan* **21**: 121.

# Soybean Z (?) potyvirus

**Data collated by J.L. Dale, 1987.**

First reported in *Glycine max*; from northern New South Wales, Australia; by Dale *et al.* (1975).

**Natural host range and symptoms**

Symptoms persist.

*Glycine max* — mosaic.

**Transmission**

Virus transmitted by mechanical inoculation.

**Geographical distribution**

Spreads in Australia.

**Experimental host range**

Few (<3) families susceptible.

Diagnostically susceptible

*Glycine max* — mosaic.

*Phaseolus vulgaris* cv. Bountiful — green local lesions when inoculated leaves senesce.

*Lathyrus odoratus, Phaseolus lathyroides* — symptomless systemic infection.

Diagnostically insusceptible

*Chenopodium amaranticolor, Chenopodium album, Pisum sativum, Vicia faba, Datura stramonium.*

Maintenance and propagation hosts

*Glycine max.*

Assay hosts (Local lesions or Whole plants)

*Phaseolus vulgaris* (L), *Glycine max* (W).

Dicotyledonous Hosts *(**Susceptible** or Insusceptible)*

*Chenopodium album. Chenopodium amaranticolor. Cucumis sativus. Datura stramonium. Dolichos biflorus.* ***Glycine max***. ***Gomphrena globosa***. *Lablab purpureus.* ***Lathyrus odoratus***. *Lycopersicon esculentum. Medicago sativa.* ***Phaseolus lathyroides***.

***Phaseolus vulgaris***. *Pisum sativum. Trifolium pratense. Trifolium repens. Trifolium subterraneum. Vicia faba. Vigna unguiculata.*

Sources of host-range data

Dale and Gibbs (1975).

**Properties of particles in sap**

TIP: 60–64 °C. LIV: more than 10 days. DEP: $\log_{10}$ minus 3–4. Leaf sap contains few virions.

**Purification**

Method: Dale and Gibbs (1975).

**Particle morphology**

Virions filamentous; not enveloped; usually flexuous; with a clear modal length; of 610 nm. Axial canal obscure.

**Physical properties**

One sedimenting component in purified preparations.

**Biochemical properties**

Virions contain 6 % nucleic acid; 94 % protein.

*Genome* consists of RNA; single–stranded. Base composition 23.3 % G; 25.5 % A; 24.8 % C; 26.3 % U.

*Virion protein(s)* one; $M_r$ 27400; coat protein. Method of preparation: Dale and Gibbs (1975). Amino acid composition: Dale and Gibbs (1975).

**Cytopathology**

Virions found in leaves; in cytoplasm. Inclusions present in infected cells; are amorphous X–bodies; they contain virions.

**Taxonomy and Relationships**

Unassigned potyvirus.

Best tests for diagnosis: this virus could be confused with soybean mosaic virus, but can be distinguished by serological tests and is not transmitted by aphids.

**References**

Dale, J.L. and Gibbs, A.J. (1975). *Intervirology* **6**: 325.

# Spartina mottle (?) *rymovirus*

**Data collated by P. Jones, 1990.**

*Standard acronym:* SpMV.

First reported in *Spartina anglica*; from the U.K.; by Jones (1980).

**Natural host range and symptoms**

Symptoms persist. Symptoms leaf mottle.

*Spartina anglica, S. maritima, S. ×. townsendii, S. glabra, S. alternifolia, S. sp. 2n=76, S. sp. 2n=92* — mottling.

**Transmission**

Transmitted by means not involving a vector. Virus transmitted by mechanical inoculation.

**Ecology and control**

Studies reported by Jones (1980).

**Geographical distribution**

Spreads in France and the UK.

**Experimental host range**

Few (<3) families susceptible. Experimentally infected plants mostly show green-yellow leaf mottling.

Diagnostically susceptible

*Spartina* spp. — mottling.

Diagnostically insusceptible

*Chenopodium amaranticolor, Avena sativa, Agropyron repens, Nicotiana glutinosa, Phleum pratense.*

Maintenance and propagation hosts

*Spartina anglica.*

Assay hosts (**L**ocal lesions or **W**hole plants)

*Spartina anglica* (W).

Dicotyledonous Hosts *(**Susceptible** or Insusceptible)*

*Beta vulgaris. Chenopodium amaranticolor. Cucumis sativus. Nicotiana clevelandii. Nicotiana glutinosa. Nicotiana tabacum. Petunia hybrida. Phaseolus vulgaris.*

Monocotyledonous Hosts *(**Susceptible** or Insusceptible)*

*Agropyron repens. Avena sativa. Chloris gayana. Cynodon dactylon. Dactylis glomerata. Hordeum vulgare. Lolium multiflorum. Oryza sativa. Phleum pratense. Setaria italica. Triticum aestivum. Zea mays.*

Sources of host-range data

Jones (1980).

**Properties of particles in sap**

Leaf sap contains few virions.

**Purification**

Method: Jones (1980).

**Particle morphology**

Virions filamentous; not enveloped; usually flexuous; with a clear modal length; of 725 nm; 12 nm wide. Axial canal obscure. Basic helix obscure.

**Physical properties**

One sedimenting component in purified preparations; sedimentation coefficient 150 *S*.

**Cytopathology**

Virions found in leaves, roots and mesophyll; in cytoplasm. Inclusions present in infected cells; are pinwheels; and are laminar inclusions.

**Taxonomy and Relationships**

Possible *Rymovirus: Potyviridae.*

Virus(es) with serologically related virions: agropyron mosaic virus. Virus(es) with serologically unrelated virions: sugarcane mosaic, ryegrass mosaic and oat necrotic mottle viruses.

**References**

Jones, P. (1980). *Ann. appl. Biol.* **94**: 77.

# Spinach latent *ilarvirus*

**Data collated by L. Bos, 1987.**

*Synonyms:* GE 36 virus (Van der Meer, 1968; Maat and Vink, 1971). *Standard acronym:* SpLV.

First reported in *Spinacia oleracea*; from plants grown from seed of spinach (*Spinacia oleracea*) from several countries; by Bos *et al.* (1980).

**Natural host range and symptoms**

Symptoms disappear soon after infection.
*Spinacia oleracea* — symptomless or transient faint mosaic and vein clearing.

**Transmission**

Transmitted by means not involving a vector. Virus transmitted by mechanical inoculation; transmitted by seed; transmitted by pollen to the seed and transmitted by pollen to the pollinated plant (Štefanac and Wrischer, 1983).

**Ecology and control**

Studies reported by Bos *et al.* (1980).

**Geographical distribution**

Spreads in the Netherlands, the USA, and the former Yugoslavia.

**Experimental host range**

Many (>9) families susceptible.

Diagnostically susceptible

*Beta vulgaris* — small chlorotic or etched rings.
*Celosia cristata* — symptomless.
*Chenopodium amaranticolor* — pin-point desiccated chlorotic local lesions.
*Chenopodium quinoa* — diffuse chlorotic local lesions then systemic leaf rolling, chlorotic vein banding, mottling, stunting and tip blight. leaves.
*Cucumis sativus* cv. Gele Tros — transient chlorotic local lesions.
*Phaseolus vulgaris* cv. Bataaf — necrotic local lesions.
*Nicotiana clevelandii, N. rustica, N. glutinosa, N. tabacum, Tetragonia expansa, Gomphrena globosa* — symptomless systemic infection.

Diagnostically insusceptible

*Apium graveolens, Cyamopsis tetragonoloba, Datura stramonium, Lactuca sativa, Lycopersicon esculentum, Pisum sativum.*

Maintenance and propagation hosts

*Chenopodium quinoa, Nicotiana spp.*

Assay hosts (**L**ocal lesions or **W**hole plants)

*Chenopodium amaranticolor* (L), *C. quinoa* (L) and (W), *Phaseolus vulgaris* (L).

Dicotyledonous Hosts *(**Susceptible** or Insusceptible)*

***Amaranthus retroflexus***. ***Antirrhinum majus***. *Apium graveolens*. ***Beta macrocarpa***. ***Beta vulgaris***. *Brassica campestris* ssp. *chinensis*. *Brassica campestris* ssp. *rapa*. *Catharanthus roseus*. ***Celosia argentea***. ***Celosia cristata***. ***Chenopodium amaranticolor***. ***Chenopodium quinoa***. ***Cucumis sativus***. *Cyamopsis tetragonoloba*. *Datura stramonium*. ***Gomphrena globosa***. *Hyoscyamus niger*. *Lactuca sativa*. *Lycopersicon esculentum*. *Medicago sativa*. ***Nicotiana bigelovii***. ***Nicotiana clevelandii***. ***Nicotiana glutinosa***. ***Nicotiana megalosiphon***. ***Nicotiana rustica***. ***Nicotiana tabacum***. *Ocimum basilicum*. *Petroselinum crispum*. ***Petunia hybrida***. ***Phaseolus vulgaris***. ***Phlox drummondii***. *Physalis floridana*. *Pisum sativum*. *Sinapis alba*. ***Spinacia oleracea***. ***Tetragonia expansa***. *Trifolium incarnatum*. *Trifolium pratense*. *Trifolium repens*. *Vicia faba*. *Vigna unguiculata*. *Zinnia elegans*.

Comments on host-range

Infection in most species is symptomless.

Sources of host-range data

Bos *et al.* (1980).

**Properties of particles in sap**

TIP: 60–65 °C. LIV: 6–8 days. DEP: $\log_{10}$ minus 3–4. Electron microscopy: virions are labile and difficult to locate without fixation.

**Purification**

Method: Bos *et al.* (1980).

**Particle morphology**

Virions isometric; not enveloped; 27 nm in diameter; rounded in profile; without a conspicuous capsomere arrangement.

**Physical properties**

Three sedimenting components in purified preparations; sedimentation coefficient of the fastest 103 *S* (B); of the other(s) 87 and 98 *S* (T and (M), respectively). Density 1.269 in $Cs_2SO_4$.

**Biochemical properties**

*Genome* consists of RNA; single–stranded; linear; of five parts; largest (or only) genome part the largest 3.662 kb; the 2nd largest 3.324 kb; the 3rd largest 2.563 kb; the 4th largest 0.986 kb; the 5th largest 0.76 kb. Genomic nucleic acid isolated by Bos *et al.* (1980). Additional factor required for infectivity; one or two of the two smallest RNA segments or coat protein required for infectivity of the three largest segments.

*Virion protein(s)* one; $M_r$ 20000. Method of preparation: Bos *et al.* (1980).

**Replication**

Replication does not depend on a helper virus.

**Cytopathology**

Virions found in leaves and mesophyll. Inclusions absent from infected cells. Other cellular changes: groups of vesicles surrounded by a unit membrane in cytoplasm and small vacuoles of leaf parenchyma cells (Stefanac and Wrischer, 1983).

**Taxonomy and Relationships**

*Ilarvirus: Bromoviridae* (subgroup 6).

Virus(es) with serologically unrelated virions: apple mosaic, elm mottle, lilac ring mottle, prunus necrotic ringspot and tobacco streak viruses (but ilarviruses are poorly immunogenic).

Best tests for diagnosis: Cucumber mosaic virus induces characteristic local lesions in *Chenopodium quinoa* and infects cucumber and *Nicotiana spp.* systemically. Cucumber mosaic virus and tobacco rattle virus rarely infect *Chenopodium quinoa* systemically. Spinach latent virus induces characteristic local and systemic reactions in *Chenopodium quinoa*.

**References**

Bos, L. (1984). *CMI/AAB Descr. Pl. Viruses* No. 281, 4 pp.
Bos, L., Huttinga, H. and Maat, D.Z. (1980). *Neth. J. Pl. Path.* **86**: 79.
Maat, D.Z. and Vink, J. (1971). *Neth. J. Pl. Path.* **77**: 93.
Štefanac, Z. (1978). *Acta bot. Croat.* **37**: 39.
Štefanac, Z. and Wrischer, M. (1983). *Acta bot. Croat.* **42**: 1.
Van der Meer, F.A. (1968). *TagBer. dt. Akad. Landw. Berlin* **97**: 27.

# Spinach temperate *alphacryptovirus*

**Data collated by A.A. Brunt, 1991.**

*Standard acronym:* SpTV.

First reported in *Spinacia oleracea*; from Japan; by Natsuaki *et al.* (1983).

**Natural host range and symptoms**

*Spinacia oleracea* — symptomless infection.

**Transmission**

Virus not transmitted by mechanical inoculation; transmitted by seed.

**Geographical distribution**

Found, but with no evidence of spread, in Japan.

**Particle morphology**

Virions isometric; not enveloped; 30 nm in diameter; rounded in profile; with a conspicuous capsomere arrangement.

**Biochemical properties**

*Genome* consists of RNA; double-stranded.

**Taxonomy and Relationships**

*Alphacryptovirus: Partitiviridae.*

**References**

Boccardo, G., Lisa, V., Luisoni, E. and Milne, R.G. (1987). *Adv. Virus Res.* **32**: 171.
Natsuaki, T., Yamashita, S., Doi, Y., Okuda, S. and Teranaka, M. (1983). *Ann. Phytopath. Soc. Japan* **49**: 709.

# Spring beauty latent *bromovirus*

**Data collated by R.A. Valverde, 1991.**

*Standard acronym:* SBLV.

First reported in *Claytonia virginica*; from Arkansas, U.S.A.; by Valverde (1985).

**Natural host range and symptoms**

Symptoms none.
*Claytonia virginica* — symptomless.

**Transmission**

Possibly on lawn mowers and tractors in lawns and public areas. Virus transmitted by mechanical inoculation; transmitted by grafting; possibly transmitted by contact between plants; not transmitted by seed.

**Geographical distribution**

Spreads in the USA.

**Experimental host range**

Several (3–9) families susceptible. Experimentally infected plants mostly show necrotic local lesions, systemic mottle and necrosis.

Diagnostically susceptible

*Chenopodium quinoa, C. album, C. amaranticolor, Cucurbita pepo* cv. Small Sugar, *Datura stramonium, Glycine max* cvs Lee and Bragg, *Phaseolus aureus, P. vulgaris* cvs Pinto, Bountiful, Black Valentine, *Vigna unguiculata* ssp. *unguiculata* Monarch — necrotic local lesions.
*Cucumis sativus* cv. Model — chlorotic local lesions.
*Commelina diffusa* — systemic mottle.
*Gomphrena globosa* — necrotic local lesions, systemic mottle and necrosis.
*Nicotiana megalosiphon, N. clevelandii* — necrotic local lesions, systemic necrosis.
*Nicotiana debneyi* — necrotic local lesions, systemic mottle.
*Pisum sativum* cv. little Marvel — systemic mottle and necrosis.
*Vicia faba* — systemic necrosis.
*Catharanthus roseus* — systemic necrosis and mottle.

Diagnostically insusceptible

*Avena sativa, Capsicum annuum, Lycopersicon esculentum, Nicotiana tabacum* cv. Havana 38, *Zea mays*.

Maintenance and propagation hosts

*Commelina diffusa, Gomphrena globosa, Nicotiana clevelandii, Pisum sativum, Catharanthus roseus*.

Assay hosts (**L**ocal lesions or **W**hole plants)

*Chenopodium quinoa* (L), *Datura stramonium* (L), *Gomphrena globosa* (L and W), *Pisum sativum* (W), *Vicia faba* (W).

Dicotyledonous Hosts ***(Susceptible** or Insusceptible)*

*Capsicum annuum*. ***Chenopodium amaranticolor***. ***Chenopodium quinoa***. *Cucumis melo*. ***Cucumis sativus***. ***Cucurbita pepo***. ***Datura stramonium***. ***Glycine max***. ***Gomphrena globosa***. *Lycopersicon esculentum*. ***Nicotiana clevelandii***. ***Nicotiana debneyi***. ***Nicotiana glutinosa***. *Nicotiana tabacum*. ***Phaseolus vulgaris***. ***Pisum sativum***. ***Vicia faba***. ***Vigna unguiculata***. *Zinnia elegans*.

Monocotyledonous Hosts *(**Susceptible** or Insusceptible)*

*Triticum aestivum. Zea mays.*

Additional hosts

Additional Hosts: *Portulaca oleracea* (**I**).

Sources of host-range data

Valverde (1985).

**Properties of particles in sap**

Leaf sap contains many virions.

**Purification**

Method: Valverde (1985).

**Particle morphology**

Virions isometric; not enveloped; 28 nm in diameter; rounded in profile; without a conspicuous capsomere arrangement.

**Physical properties**

One sedimenting component in purified preparations; sedimentation coefficient 90 *S*.

**Biochemical properties**

*Genome* consists of RNA; single–stranded; linear; of three parts; largest (or only) genome part the largest 3.9 kb; the 2nd largest 3.4 kb; the 3rd largest 2.6 kb. Genomic nucleic acid isolated by Valverde (1985). Additional factor not required for infectivity.

*Non–genomic nucleic acid* found in the virions; is subgenomic *m*RNA. 5 virus specified dsRNA species found in infected cells. Size of largest virus specified dsRNA 2.2 kbp; 2nd largest 2 kbp; 3rd largest 1.4 kbp; 4th largest 1.2 kbp; 5th largest 0.5 kbp.

*Virion protein(s)* one; $M_r$ 23000; and it is coat protein. Method of preparation: Valverde (1985).

**Replication**

Replication does not depend on a helper virus.

**Cytopathology**

Virions found in all parts of the host plant; in cytoplasm.

**Taxonomy and Relationships**

*Bromovirus: Bromoviridae.*

Virus(es) with serologically related virions: broad bean mottle, brome mosaic, cowpea chlorotic mottle and melandrium yellow fleck viruses.

Best tests for diagnosis: serology and double diffusion tests.

**References**

Valverde, R.A. (1985). *Phytopathology* **75**: 395.

Valverde, R.A. and Glascock, C.B. (1991). *Phytopathology* **81**: 401.

# Squash leaf curl bigeminivirus

**Data collated by A.A. Brunt, S. Molyneux and S. Cohen, 1986.**

*Synonyms:* probably muskmelon necrotic mosaic virus. *Standard acronym:* SLCV.

First reported in *Cucurbita maxima, Cucurbita moschata, Cucurbita pepo, Phaseolus vulgaris*; from the south western U.S.A.; by Cohen *et al.* (1983).

## Natural host range and symptoms

Symptoms persist.
*Cucurbita pepo* — severe stunting and leaf curl.
*Cucurbita moschata, C. pepo, Phaseolus vulgaris.*

## Transmission

Transmitted by a vector; an insect; *Bemisia tabaci*; Aleyrodidae. Transmitted in a persistent manner. Virus not transmitted by mechanical inoculation.

## Ecology and control

Studies reported by Cohen *et al.* (1983).

## Geographical distribution

Spreads in the USA (California).

Diagnostically susceptible

*Cucurbita pepo, Cucurbita maxima, Cucurbita moschata* — severe stunting and leaf curling.
*Phaseolus vulgaris* — green mosaic leaf symptoms and deformation.

Diagnostically insusceptible

*Beta vulgaris, Cucumis melo, Cucumis sativus, Lycopersicon esculentum, Pisum sativum.*

Maintenance and propagation hosts

*Cucurbita pepo.*

Assay hosts (**L**ocal lesions or **W**hole plants)

*Cucurbita maxima* (W), *Cucurbita moschata* (W), *Cucurbita pepo* (W), *Phaseolus vulgaris* (W).

Dicotyledonous Hosts *(**Susceptible** or Insusceptible)*

*Beta macrocarpa. Beta vulgaris. Brassica oleracea* ssp. *botrytis. Capsella bursa–pastoris. Carthamus tinctorius. Chenopodium murale. Cichorium endiva. Cucumis melo. Cucumis sativus.* ***Cucurbita maxima. Cucurbita moschata. Cucurbita pepo.*** *Daucus carota. Gossypium hirsutum. Helianthus annuus. Hibiscus esculentus. Lactuca sativa. Lycopersicon esculentum. Medicago sativa.* ***Phaseolus vulgaris.*** *Pisum sativum. Sonchus oleraceus. Spinacia oleracea. Trifolium subterraneum. Zinnia elegans.*

Additional hosts

*Citrullus vulgaris.*

## Purification

Method: Cohen *et al.* (1983).

## Particle morphology

Virions geminate; 22 nm in diameter.
*Sequence database accession code(s):*

L20240 Em(40)_vi:CPBR1A Gb(84)_vi:SQLBR1A Squash leaf curl virus movement protein (BR1) gene, complete cds. 11/93 765bp.
L27272 Em(40)_vi:GESQLAC1A Gb(84)_vi:SQLAC1A Squash leaf curl virus replication-associated protein (AC1) gene, partial cds. 1/94 279bp.
L27273 Em(40)_vi:GESQLBC1A Gb(84)_vi:SQLBC1A Squash leaf curl virus movement protein (BC1) gene, partial cds. 1/94 141bp.
M38182 Em(40)_vi:GESQLCVE Gb(84)_vi:SQLCVE Squash leaf curl virus circular viral B component DNA. 6/92 2,607bp.
M38183 Em(40)_vi:GESQLCVEA Gb(84)_vi:SQLCVEA Squash leaf curl virus circular viral A component DNA. 6/92 2,634bp.
M63155 Em(40)_vi:GESQLC01 Gb(84)_vi:SQLCVRAA Squash leaf curl virus A component DNA. 8/93 212bp.
M63156 Em(40)_vi:GESQLCVR Gb(84)_vi:SQLCVRAB Squash leaf curl virus B component DNA. 3/91 214bp.
M63157 Em(40)_vi:GESQLC03 Gb(84)_vi:SQLCVEAA Squash leaf curl virus A component DNA. 8/93 228bp.
M63158 Em(40)_vi:GESQLC02 Gb(84)_vi:SQLCVEAB Squash leaf curl virus B component DNA. 3/91 229bp. 9 sequences.

**Taxonomy and Relationships**

Bigeminivirus: *Geminiviridae.*

Virus(es) with serologically related virions: African cassava mosaic virus. Virus(es) with serologically unrelated virions: bean golden mosaic, beet curly top and tobacco yellow dwarf viruses.

**References**

Cohen, S., Duffus, J.E., Larsen, R.C., Liu, H.Y. and Flock, R.A. (1983). *Phytopathology* **73**: 1669.

# Squash mosaic *comovirus*

**Data collated by R.N. Campbell, 1985.**

*Synonyms:* cucurbit ring mosaic virus, muskmelon mosaic virus, pumpkin mosaic virus (Stoner, 1963). *Standard acronym:* SqMV. *Strains:* Arizona watermelon stunt virus; 2 serological groups, named I and II, have been recognized.

First reported in *Cucurbita pepo*; from California, U.S.A.; by Freitag (1956).

**Natural host range and symptoms**

*Cucumis melo, C. sativus, Cucurbita pepo, C. moschata, C. maxima* — systemic mosaic, ringspots and leaf deformation.

*Ecballium elaterium* — mild yellow mosaic (reported from Israel).

*Citrullus lanatus* — infected by some isolates.

*Chenopodium album* — one report from Morocco.

**Transmission**

Transmitted by a vector; an insect; *Acalymma trivittata; A. thiemei thiemei; Diabrotica undecimpunctata undecimpunctata; D. bivittula; Epilachna chrysomelina*; *E. paenulata*; Coleoptera. Transmitted in a non–persistent manner. Virus transmitted by mechanical inoculation; transmitted by seed (*c.* 10% in *Cucumis melo* and up to 35% in *Cucurbita pepo*); not transmitted by pollen (Alvarez and Campbell, 1978; Nolan and Campbell, 1984).

**Geographical distribution**

Probably distributed worldwide (presumably *via* imported seeds). Spreads in China. Found, but with no evidence of spread, in New Zealand.

**Experimental host range**

Few (<3) families susceptible.

Diagnostically susceptible

*Cucumis melo, Cucurbita pepo* — systemic mosaic, often with ringspots, leaf deformation.

*Cucumis metuliferus* — chlorotic local lesions.

Diagnostically insusceptible

*Chenopodium amaranticolor, Datura stramonium, Nicotiana glutinosa, N. tabacum* cv. Turkish, Havana, *Citrullus lanatus* (resistant to all but the watermelon strain of SqMV).

Maintenance and propagation hosts

*Cucurbita pepo, Cucumis melo.*

Assay hosts (**L**ocal lesions or **W**hole plants)

*Cucurbita pepo* (W), *Cucumis metuliferus* — some collections *e.g.* P.I. 202681 or P.I. 292190 (L).

Dicotyledonous Hosts *(**Susceptible** or Insusceptible)*

*Amaranthus retroflexus. Antirrhinum majus. Apium graveolens. Arachis hypogaea. Beta vulgaris* (Lockhart (1982) found *Beta vulgaris* susceptible in Morocco). *Brassica campestris* ssp. *rapa. Brassica oleracea* ssp. *botrytis. Brassica oleracea* ssp. *capitata. Calendula officinalis. Capsicum frutescens. Catharanthus roseus. Celosia argentea. Cheiranthus cheiri.* ***Chenopodium album. Chenopodium quinoa*** (in Morocco (Lockhart, 1982)). *Citrullus lanatus*, or ***Citrullus lanatus*** (when infected with Arizona strain). ***Cucumis melo. Cucumis sativus. Cucurbita maxima. Cucurbita moschata. Cucurbita pepo.*** *Datura stramonium. Daucus carota. Delphinium hybridum. Dianthus caryophyllus. Glycine max. Gossypium hirsutum. Helianthus annuus. Hibiscus esculentus. Ipomoea nil. Lactuca sativa. Lycopersicon esculentum. Matthiola incana. Medicago hispida. Medicago sativa. Melilotus alba. Nicotiana glutinosa. Nicotiana rustica. Nicotiana tabacum. Pastinaca sativa. Petroselinum crispum. Petunia hybrida. Phaseolus lunatus. Phaseolus vulgaris. Phlox drummondii. Pisum sativum*, or ***Pisum sativum*** (in Morocco (Lockhart, 1982)). *Raphanus sativus. Ricinus communis. Rumex acetosa. Solanum tuberosum. Spinacia oleracea. Trifolium incarnatum. Trifolium pratense. Trifolium repens. Tropaeolum majus. Vicia faba. Vigna unguiculata. Vigna unguiculata* ssp. *sesquipedalis. Zinnia elegans.*

Monocotyledonous Hosts *(**Susceptible** or Insusceptible)*

*Zea mays.*

Sources of host-range data

Freitag (1956); Lockhart (1982).

**Properties of particles in sap**

TIP: 70–80 °C. LIV: 30 days. DEP: $\log_{10}$ minus 4–6.

**Purification**

Method: Alvarez and Campbell (1978).

## Particle morphology

Virions isometric; not enveloped; 28 nm in diameter; angular in profile; without a conspicuous capsomere arrangement.

## Physical properties

Three sedimenting components in purified preparations; sedimentation coefficient of the fastest 118 *S* (B); of the other(s) 95 *S* (M), or 57 *S* (T). Isoelectric point pH 4.56 (Rice *et al.*, 1955).

## Biochemical properties

Virions contain 35 % nucleic acid (B), or 27 % nucleic acid (M), or 0 % nucleic acid (T); 65 % protein (B), or 73 % protein (M), or 100 % protein (T).

*Genome* consists of RNA; single–stranded; linear; of two parts; largest (or only) genome part the larger 6.8 kb; the 2nd largest 4.5 kb (Mazzone *et al.*, 1962). Genomic nucleic acid isolated by Hiebert and Purcifull (1981). Base composition 22.5 % G; 31.6 % A; 16.2 % C; 29.5 % U. 5' terminus of RNA has a VPg. Poly A region present; Daubert and Bruening, 1979 at the 3' end.

*Sequence database accession code(s):*

M96148 Em(40)_vi:COMSQVCAP Gb(84)_vi:SQVCAPPRO Squash Moasic Virus 42 kDa capsid protein and 22 kDa capsid protein. 7/93 2,536bp. 1 sequence.

*Non–genomic nucleic acid* not found in the virions.

*Virion protein(s)* three; $M_r$ of the largest 42000; coat protein. $M_r$ of 2nd largest 22000; coat protein. $M_r$ of 3rd largest 5000; genome-linked protein (VPg) — data given are based on analogy to other comoviruses, see Daubert and Bruening 1979 and 1984. Method of preparation: Hiebert and Purcifull (1981). Amino acid composition: Mazzone *et al.* (1962).

*Virus–coded non–virion proteins* isolated (Hiebert and Purcifull (1981)); six proteins found. $M_r$ of the largest 190000. $M_r$ of 2nd largest 51000. $M_r$ of 3rd 32000 (all translations of B–RNA). $M_r$ of 4th 112000. $M_r$ of 5th and smaller 105000, or 90000, or 64000, or 40000, or 22000 (all translations of M-RNA, except the last value).

## Replication

Replication does not depend on a helper virus.

## Cytopathology

Virions found in all parts of the host plant.

## Taxonomy and Relationships

*Comovirus: Comoviridae.*

Virus(es) with serologically related virions: cowpea severe mosaic (Arkansas strain), radish mosaic, bean pod mottle, glycine mosaic, quail pea mosaic and broad bean stain viruses. Virus(es) with serologically unrelated virions: Andean potato mottle virus.

## References


Alvarez, M. and Campbell, R.N. (1978). *Phytopathology* **68**: 257.

Bruening, G. (1978). *CMI/AAB Descr. Pl. Viruses* No. 199, 5 pp.

Campbell, R.N. (1971). *CMI/AAB Descr. Pl. Viruses* No. 43, 4 pp.

Daubert, S.D. and Bruening, G. (1979). *Virology* **98**: 246.

Daubert, S.D. and Bruening, G. (1984). In: *Methods in Virology*, Vol. 8, pp. 364–365; eds. K. Maramorosch and H. Koprowski. Academic Press, New York.

Freitag, J.H. (1956). *Phytopathology* **46**: 73.

Hiebert, E. and Purcifull, D.E. (1981). *Virology* **113**: 630.

Lindberg, G.D., Hall, D.H. and Walker, J.C. (1956). *Phytopathology* **46**: 489.

Lockhart, B.E.L. (1982). *Plant Dis.* **66**: 1191.
Mazzone, H.M., Incardona, N.L. and Kaesberg, P. (1962). *Biochim. biophys. Acta* **55**: 164.
Nolan, P.A. and Campbell, R.N. (1984). *Plant Dis.* **68**: 971.
Rice, R.V., Lindberg, G.D., Kaesberg, P., Walker, J.C. and Stahmann, S.H. (1955). *Phytopathology* **45**: 145.
Stoner, W.N. (1963). *Phytopathology* **53**: 890.

# Statice Y *potyvirus*

**Data collated by D.-E. Lesemann, 1991.**

*Standard acronym:* SVY.

First reported in *Limonium sinuatum*; from southern Germany; by Lesemann *et al.* (1979).

**Natural host range and symptoms**

Symptoms persist. Symptoms light green mosaic, stunting, narrow leaves, deformation of leaves.

*Limonium sinuatum.*

**Transmission**

Virus transmitted by mechanical inoculation.

**Geographical distribution**

Found, but with no evidence of spread, in southern Germany.

**Experimental host range**

Several (3–9) families susceptible. Experimentally infected plants mostly show local lesions, mosaic.

Diagnostically susceptible

*Chenopodium quinoa* — systemic, chlorotic, later necrotic spots.
*C. amaranticolor, C. murale, C. foliosum* — local lesions.
*Gomphrena globosa, Amaranthus retroflexus* — symptomless.
*Lathyrus odoratus* — light mosaic.
*Nicotiana clevelandii* — faint mottle.
*Phaseolus vulgaris* — yellow mosaic.
*Vicia faba* — mosaic.

Diagnostically insusceptible

*Beta vulgaris, Nicotiana glutinosa, N. tabacum* cv. Samsun, *Pisum sativum* cv. Kelvex, *Zinnia elegans*.

Maintenance and propagation hosts

*Chenopodium quinoa.*

Assay hosts (**L**ocal lesions or **W**hole plants)

*Chenopodium quinoa* (W), *C. amaranticolor* (L), *Phaseolus vulgaris* (W).

Dicotyledonous Hosts *(**Susceptible** or Insusceptible)*

***Amaranthus retroflexus***. *Apium graveolens*. *Beta vulgaris*. ***Chenopodium amaranticolor***. ***Chenopodium ambrosioides***. ***Chenopodium capitatum***. ***Chenopodium foliosum***. ***Chenopodium murale***. ***Chenopodium quinoa***. *Cucumis sativus*. ***Gomphrena globosa***. ***Lathyrus odoratus***. ***Nicotiana clevelandii***. *Nicotiana glutinosa*. *Nicotiana* tabacum. *Petunia hybrida*. ***Phaseolus vulgaris***. *Pisum sativum*. *Raphanus sativus*. *Sinapis alba*. ***Spinacia oleracea***. ***Vicia faba***. *Zinnia elegans*.

**Properties of particles in sap**

TIP: 54–56 °C. LIV: 15 days. DEP: $\log_{10}$ minus 5–6. Leaf sap contains few virions. Electron microscopy: glutaraldehyde fixation reduces virion breakage.

**Purification**

Method: Lesemann *et al.* (1979).

**Particle morphology**

Virions filamentous; not enveloped; usually flexuous; with a clear modal length; of 748 nm. Axial canal obscure. Basic helix obscure.

*Virion protein(s)* one; $M_r$ 31000; coat protein. Method of preparation: Lesemann *et al.* (1979): SDS — polyacrylamide electrophoresis.

**Cytopathology**

Virions found in leaves, mesophyll and the epidermis; in cytoplasm. Inclusions present in infected cells; are crystals in the cytoplasm, crystals in the nucleus, and pinwheels; and are hypertrophied mitochondria; they do not contain virions. Other cellular changes: pinwheels of laminated aggregate type, small electron dense crystals in nucleolus and cytoplasm and hypertrophied mitochondria (Lesemann *et al.*, 1979).

**Taxonomy and Relationships**

*Potyvirus: Potyviridae.*

Virus(es) with serologically related virions: bean yellow mosaic, bean common mosaic, clover yellow vein, gloriosa stripe, pea seed borne mosaic and potato Y viruses. Virus(es) with serologically unrelated virions: lettuce mosaic virus.

Best tests for diagnosis: Systemically infects *Chenopodium quinoa* and fails to infect *Pisum sativum* unlike bean yellow mosaic virus. Little or no reaction with antisera to clover yellow virus (CYVV) and fails to produce local lesions on cucumber like CYVV.

**References**

Lesemann, D.-E., Koenig, R. and Hein, H. (1979). *Phytopath. Z.* **95**: 128.

# Strawberry crinkle *cytorhabdovirus*

**Data collated by R.H. Converse, 1985 and J.A. Cooper, 1985.**

*Standard acronym:* SCrV. *Strains:* strawberry latent A (mild form); strawberry latent A and latent B; strawberry vein chlorosis.

First reported in *Fragaria vesca* cv. Marshall; from Oregon, U.S.A.; by Zeller and Vaughan (1932).

**Natural host range and symptoms**

Symptoms persist.

*Fragaria vesca* — chlorotic and necrotic spotting of veins, epinasty, deformation, local lesions in petioles and stolons, streaking and deformation of petals.

*Fragaria virginiana* — leaflets malformed, vigour reduced.

*Fragaria chiloensis*, *F. chiloensis* var *ananassa* (cultivated strawberry) — usually symptomless.

**Transmission**

Transmitted by a vector; an insect; *Chaetosiphon jacobi, Chaetosiphon fragaraefolii*; Aphididae. Transmitted in a persistent manner. Virus retained when the vector moults; multiplies in the vector; not transmitted congenitally to the progeny of the vector; transmitted by mechanical inoculation (by injection of haemolymph from aphid to aphid); transmitted by grafting; not transmitted by contact between plants; not transmitted by seed; not transmitted by pollen.

**Geographical distribution**

Spreads in the African region, the Eurasian region, the North American region, and the South and Central American region; Australia, Belgium, Bulgaria, Canada, Chile, China, the former Czechoslovakia, France, Germany, Israel, Italy, Japan, the Netherlands, New Zealand, Poland, South Africa, the UK, the USA, the former Yugoslavia.

**Experimental host range**

Few (<3) families susceptible.

Diagnostically susceptible

*Fragaria vesca* — chlorotic and necrotic spotting of veins, epinasty, deformation, local lesions in petioles and stolons, streaking and deformation of petals.

*Fragaria virginiana* — leaflets malformed, vigour reduced.

Maintenance and propagation hosts

*Fragaria chiloensis, F. chiloensis var. ananassa, F. vesca, F. virginiana.*

Assay hosts (**L**ocal lesions or **W**hole plants)

*Fragaria vesca* var. *semperflorens* (W).

Dicotyledonous Hosts *(**Susceptible** or Insusceptible)*

***Fragaria vesca.***

Additional hosts

***Sanguisorba minor.***

**Properties of particles in sap**

LIV: 0.1–0.2 days (the virus remains infectious for at least 1 hour in refrigerated extracts made of infective aphids. Infectious virus can be recovered from aphids frozen at -54°C for two years). Leaf sap contains few virions.

**Particle morphology**

Virions rhabdo– or bullet–shaped (ranging between 190–380 nm); enveloped (less than 80 nm and uncoated); with no clear modal length; 42 nm wide (±3 nm).

**Cytopathology**

Virions found in epidermis and in parenchyma cells near vascular bundles and sieve tubes; in cytoplasm; sometimes enclosed in membrane bound cisternae.

**Taxonomy and Relationships**

*Cytorhabdovirus: Rhabdoviridae.*

**References**

Frazier, N.W. and Mellor, F.C. (1970). In: *Virus Diseases of Small Fruits and Grapevines*, pp. 18–23; ed. Frazier, N.W.. Univ. Calif., Div. Agric. Sci., Berkeley.

Frazier, N.W., Sylvester, E.S. and Richardson, J. (1987). In: *Virus Diseases of Small Fruits*; ed. R.H. Converse. U.S. Dep. Agric. Hdbk No. 631, p. 20.

Richardson, J., Frazier, N.W. and Sylvester, E.S. (1972). *Phytopathology* **62**: 491.
Sylvester, E.S., Frazier, N.W. and Richardson, J. (1976). *CMI/AAB Descr. Pl. Viruses* No. 163, 4 pp.
Zeller, S.M. and Vaughan, E.K. (1932). *Phytopathology* **22**: 709.

# Strawberry latent C (?) rhabdovirus

**Data collated by A.A. Brunt, 1993.**

First reported in *Fragaria* spp.; from Japan and North America; by Yoshikawa *et al.* (1986).

**Natural host range and symptoms**

*Fragaria* spp. — most cultivars symptomless.

**Transmission**

Transmitted by a vector; an insect; *Chaetosiphon fragaefolii*; Aphididae. Virus transmitted by grafting.

**Geographical distribution**

Spreads in the Eastern Asian region and the North American region; Canada, Japan, and the USA.

**Experimental host range**

Few (<3) families susceptible.

Diagnostically susceptible

*Fragaria vesca* — symptomless in fection of clones UC4. Severe leaf epinasty of young leaves of clone UC5, mottling and distortion of leaves produced subsequently, plants stunted.

Maintenance and propagation hosts

*Fragaria vesca* clone UC5.

Assay hosts (**L**ocal lesions or **W**hole plants)

*Fragaria vesca* clone UC5 (W).

Dicotyledonous Hosts *(**Susceptible** or Insusceptible)*

***Fragaria vesca***.

**Particle morphology**

Virions rhabdo– or bullet–shaped; enveloped; of 190–380 nm; 68 nm wide (from sap).

**Taxonomy and Relationships**

Unassigned rhabdovirus.

**References**

Yoshikawa, N., Inouye, T. and Converse, R.H. (1986). *Ann. Phytopath. Soc. Japan* **52**: 437.

# Strawberry latent ringspot (?) *nepovirus*

**Data collated by J. Cooper, 1986.**

*Synonyms:* aesculus line pattern virus (Schmelzer and Schmidt, 1968), rhubarb virus 5. *Standard acronym:* SLRSV.

First reported in *Fragaria vesca* and *Rubus idaeus*; from Dundee, Scotland; by Lister (1964).

## Natural host range and symptoms

*Fragaria vesca* (strawberry) — mottle and death.
*Apium graveolens* — strap leaf.
*Robinia pseudoacacia* — mosaic.
*Rosa* spp. — chlorotic ringspots and stunting.
*Euonymus europaeus* — yellow mottle.
*Aesculus carnea* — line pattern.
*Rubus idaeus* — stunting. Other natural hosts include:
*Rubus fruticosus.*
*Ribes nigrum.*
*Ribes rubrum.*
*Asparagus officinalis.*
*Vitis vinifera.*
*Capsella bursa-pastoris, Delphinium* spp., *Lamium amplexicaule, Narcissus* spp., *Prunus domestica, P. persica, Rheum rhaponticum, Sambucus nigra, Senecio vulgaris, Stellaria media, Taraxacum officinale, Trifolium repens* and *Urtica dioica.*

## Transmission

Transmitted by a vector; a nematode; *Xiphinema diversicaudatum*; Dorylamidae. Virus transmitted by mechanical inoculation; transmitted by grafting; transmitted by seed (often more than 70% in *Capsella bursa-pastoris, Mentha arvensis, Lamium amplexicaule, Rubus idaeus, Senecio vulgaris, Stellaria media, Chenopodium quinoa, Pastinaca sativa, Petroselinum crispum* and *Apium graveolens*).

## Geographical distribution

Spreads in Australia, Belgium, Canada, Finland, France, Germany, Ireland, Israel, Italy, Luxembourg, the Netherlands, New Zealand, Poland, Portugal, Romania, Spain, Switzerland, Turkey, the UK, the USA, the former Yugoslavia. Found, but with no evidence of spread, in Canada and New Zealand.

## Experimental host range

Many (>9) families susceptible.

Diagnostically susceptible

*Chenopodium amaranticolor, C. murale, C. quinoa* — chlorotic or necrotic local lesions, systemic chlorosis and deformation, necrosis or faint chlorotic mottle.

*Cucumis sativus* — chlorotic local lesions or none; systemic interveinal chlorosis or necrosis; recovery in summer, in winter symptoms may persist. Some isolates induce enations.

Diagnostically insusceptible

*Nicotiana rustica, Nicotiana tabacum, Petunia hybrida.*

Maintenance and propagation hosts

*Cucumis sativus.*

Assay hosts (**L**ocal lesions or **W**hole plants)

*Chenopodium murale* (L), *Chenopodium amaranticolor* (L).

Dicotyledonous Hosts *(**Susceptible** or Insusceptible)*

*Amaranthus caudatus*. ***Amaranthus retroflexus***. ***Antirrhinum majus***. ***Apium graveolens***. ***Arachis hypogaea***. ***Atriplex hortensis***. ***Beta vulgaris***. ***Brassica campestris*** ssp. ***rapa***. ***Brassica oleracea*** ssp. ***botrytis***. ***Calendula officinalis***. ***Capsella bursa–pastoris***. ***Celosia argentea***. ***Chenopodium album***. ***Chenopodium amaranticolor***. ***Chenopodium ambrosioides***. ***Chenopodium capitatum***. ***Chenopodium foetidum***. ***Chenopodium foliosum***. ***Chenopodium hybridum***. ***Chenopodium murale***. ***Chenopodium quinoa***. ***Citrullus lanatus***. ***Citrus aurantium***. ***Coriandrum sativum***. ***Cucumis melo***. ***Cucumis sativus***. ***Cucurbita maxima***. ***Cucurbita pepo***. ***Datura stramonium***. ***Daucus carota***. ***Delphinium hybridum***. ***Dianthus barbatus***. ***Emilia sagittata***. ***Fragaria vesca***. ***Gomphrena globosa***. ***Hyoscyamus niger***. ***Lactuca sativa***. ***Lavatera trimestris***. ***Lobelia erinus***. ***Lycopersicon esculentum***. *Myosotis sylvatica*. ***Nicandra physaloides***. ***Nicotiana clevelandii***. ***Nicotiana glutinosa***. ***Nicotiana megalosiphon***. ***Nicotiana occidentalis***. *Nicotiana rustica*. ***Nicotiana sylvestris***. ***Nicotiana tabacum***. ***Ocimum basilicum***. ***Pastinaca sativa***. ***Petroselinum crispum***. *Petunia hybrida*. ***Phaseolus vulgaris***. ***Phlox drummondii***. ***Physalis floridana***. ***Physalis peruviana***. ***Pisum sativum***. ***Prunus domestica***. ***Prunus persica***. ***Rubus idaeus***. ***Senecio vulgaris***. ***Solanum nodiflorum***. ***Solanum tuberosum***. ***Spinacia oleracea***. ***Stellaria media***. ***Tetragonia expansa***. ***Torenia fournieri***. ***Trifolium incarnatum***. ***Trifolium pratense***. ***Trifolium repens***. ***Tropaeolum majus***. ***Verbesina encelioides***. ***Vicia faba***. ***Vigna unguiculata***. ***Vitis vinifera***. ***Zinnia elegans***.

Monocotyledonous Hosts *(**Susceptible** or Insusceptible)*

***Allium porrum***. ***Hordeum vulgare***. ***Lolium multiflorum***. ***Narcissus jonquilla***. ***Poa annua***.

Sources of host-range data

Schmelzer (1969); Murant (1974).

**Properties of particles in sap**

TIP: 52–58 °C (in *Chenopodium quinoa*). LIV: 12–50 days. DEP: $\log_{10}$ minus 3–5. Infectivity of sap not changed by treatment with di–ethyl ether.

**Purification**

Method: Lister (1964); Richard and Proll (1970); Allen *et al.* (1970).

**Particle morphology**

Virions isometric; 30 nm in diameter; angular in profile.

**Physical properties**

Two sedimenting components in purified preparations; sedimentation coefficient of the fastest 126–134 *S* (B); of the other(s) 50–58 *S* (T). Density 1.46 g $cm^{-3}$ in CsCl (B).

**Biochemical properties**

Virions contain 38 % nucleic acid (B); 62 % protein (T).

*Genome* consists of RNA; single–stranded; of two parts (Mayo *et al.*, 1974); largest (or only) genome part 7.8 kb; the 2nd largest 4.8 kb. Genomic nucleic acid isolated by Mayo *et al.* (1974).

*Sequence database accession code(s):*

X77466 Em(40)_vi:SLRVRNA2 Gb(84)_vi:SLRVRNA2 Strawberry latent ringspot virus genomic RNA-2. 4/94 3,824bp.

X75165 Em(43)_vi:Slrvcp Gb(89)_vi:Slrvcp Strawberry latent ringspot virus mRNA for 43K/27K coat proteins. 2/95 2,424bp.

*Virion protein(s)* two; $M_r$ of the larger 44000. $M_r$ of 2nd largest 29000.

**Cytopathology**

Virions found in mesophyll and epidermis; in the perinuclear space. Inclusions present in infected cells; are membranous bodies and unusual in shape; and tubules; they contain virions (and empty shells).

**Taxonomy and Relationships**

Possible *Nepovirus: Comoviridae.*

Virus(es) with serologically unrelated virions: arabis mosaic, cherry leafroll, raspberry ringspot, tobacco ringspot, tomato black ring and tomato ringspot viruses.

**References**

Allen, W.R., Davidson, T.R. and Briscoe, M.R. (1970). *Phytopathology* **60**: 1262.

Lister, R.M. (1964). *Ann. appl. Biol.* **54**: 167.

Mayo, M.A., Murant, A.F., Harrison, B.D. and Goold, R.A. (1974). *J. gen. Virol.* **24**: 29.

Murant, A.F. (1974). *CMI/AAB Descr. Pl. Viruses* No. 126, 4 pp.

Murant, A.F. and Lister, R.M. (1987). In: *Virus Diseases of Small Fruits*; ed. R.H. Converse. U.S. Dep. Agric. Hdbk No. 631, p. 49.

Richter, J. and Proll, E. (1970). *Acta phytopath. Acad. Sci. hung.* **5**: 151.

Schmelzer, K. (1969). *Phytopath. Z.* **66**: 1.

Schmelzer, K. and Schmidt, H.E. (1968). *Phytopath. Z.* **62**: 105.

Walkey, D.G.A. and Webb, M.J.W. (1970). *J. gen. Virol.* **7**: 159.

# Strawberry latent ringspot satellite RNA

*Sequence database accession code(s):*

X69826 Em(40)_vi:SLRVSAT Gb(84)_vi:SLRVSAT Strawberry latent ringspot virus satellite RNA encoding 36 kDa protein. 6/93 1,118bp.

**Taxonomy and Relationships**

Satellite RNA.

# Strawberry mild yellow edge-associated (?) *potexvirus*

**Data collated by W. Jelkmann, 1991.**

*Standard acronym:* SMYEaV.

First reported in *Fragaria vesca*; from California, U.S.A.; by Horne (1922).

**Natural host range and symptoms**

*Fragaria vesca* — chlorotic flecks in young leaves, gradual loss of vigour, chlorotic mottling, interveinal necrosis of older leaves.

*Rubus rosifolius, Chenopodium quinoa* — symptomless.

**Transmission**

Transmitted by a vector; an insect; *Chaetosiphon fragaraefolii*; Aphididae. Transmitted in a non–persistent manner. Virus possibly requires, for vector transmission, a helper virus (strawberry mild yellow edge luteovirus); transmitted by mechanical inoculation; transmitted by grafting.

**Ecology and control**

Studies reported by Converse *et al.* (1987).

**Geographical distribution**

Probably distributed worldwide. Spreads in Australia, Belgium, Bulgaria, Canada, Chile, China, the former Czechoslovakia, France, Germany, Ireland, Israel, Italy, Japan, Luxembourg, New Zealand, Paraguay, South Africa, Switzerland, the UK, the USA.

Assay hosts (Local lesions or Whole plants)

*Fragaria vesca* UC4 (W), UC5 (W), EMC (W), *F. vesca* var. semperfloreus 'Alpine' (W).

Dicotyledonous Hosts *(**Susceptible** or Insusceptible)*

***Chenopodium murale. Chenopodium quinoa.***

**Properties of particles in sap**

Leaf sap contains few virions.

**Particle morphology**

Virions filamentous; usually flexuous; with a clear modal length; of 482 nm; 13 nm wide.

**Biochemical properties**

*Genome* consists of RNA; single–stranded; linear; unipartite; largest (or only) genome part 5.966 kb. Poly A region present. Nucleotide sequence references: Jelkmann *et al.* (1991).

*Sequence database accession code(s):*

D12515 Em(40)_vi:SYVCPH Gb(84)_vi:SYVCPH Strawberry mild yellow edge-associated virus (SMYEAV), 3'terminus of the viral genome encodin

D12517 Em(40)_vi:SYVCPE Gb(84)_vi:SYVCPE Strawberry mild yellow edge-associated virus (SMYEAV) genomic RNA encoding coat protein and 5 2 sequences.

4 virus specified dsRNA species found in infected cells. Size of largest virus specified dsRNA 6 kbp; 2nd largest 4.2 kbp; 3rd largest 1.7 kbp; 4th largest 0.5 kbp.

*Virion protein(s)* one; $M_r$ 25714. Method of preparation: Jelkmann *et al.* (1990). Amino acid sequence: Jelkmann *et al.* (1991). Amino acid composition: Jelkmann *et al.* (1991).

*Virus–coded non–virion proteins* identified by genomic sequence analysis; five proteins found (ORFs in genome in addition to coat protein). $M_r$ of the largest 149423; probably replicase. $M_r$ of 2nd largest 25344. $M_r$ of 3rd 11576. $M_r$ of 4th 11216. $M_r$ of 5th and smaller 8079.

**Cytopathology**

Virions found in the companion cells.

**Taxonomy and Relationships**

Possible *Potexvirus*.

Virus(es) with serologically related virions: chicory X, cassava Colombian common mosaic, papaya mosaic, Sieg river potexviruses and, but distantly, wineberry latent virus. Virus(es) with serologically unrelated virions: plantago Argentinian, cactus X, cassava common mosaic (WF), clover yellow mosaic, cymbidium mosaic, daphne X, hydrangea ringspot, lily X, lychnis potex, narcissus mosaic, pepino mosaic, plantago X, potato aucuba mosaic, potato X, Tamus potex, tulip X, viola mottle and white clover mosaic viruses.

The sequence of the coat protein of SMYEAV, like that of other potexviruses, has significant homologies with the protein sequences of carlaviruses. Best tests for diagnosis: ISEM useful, antisera prepared from coat protein expressed as a fusion protein in *E. coli*.

**References**


Converse, R.H., Martin, R.R. and Spiegel, S. (1987). In: *Virus Diseases of Small Fruits*; ed. R.H. Converse. U.S. Dep. Agric. Hdbk No. 631, p. 25.

Horne, W.T. (1922). *Rep. Cal. Exp. Stn.* 1921–1922, p. 122.

Jelkmann, W., Martin, R.R. and Maiss, E. (1989). *Phytopathology* **79**: 1250.

Jelkmann, W., Martin, R.R., Lesemann, D.-E., Vetten, H.J. and Skelton, F. (1990). *J. gen. Virol.* **71**: 1251.

Jelkmann, W., Maiss, E. and Martin, R.R. (1990). *Abstr. 8th Int. Congr. Virol.*, p. 1445.

Jelkmann, W., Maiss, E. and Martin, R.R. (1991). *J. gen. Virol.* submitted.

Martin, R.R., Jelkmann, W., Spiegel, S. and Converse, R.H. (1989). *Acta Hort.* **236**: 111.


## Strawberry mild yellow edge *luteovirus*

**Data collated by R.H. Converse, 1985 and J.A. Cooper, 1985.**

*Synonyms:* soybean dwarf luteovirus.

First reported in *Fragaria vesca*; from California, U.S.A. and England; by Horne (1922); Harris (1933).

**Natural host range and symptoms**

*Fragaria chiloensis, F. chiloensis* var. *ananassa* — usually symptomless.

*F. vesca* — leaflets cupped, chlorotic margins, vigour reduced. An isolate from the Pacific Coast of North America causes chlorotic vein netting and necrosis of youngest leaves.

*F. virginiana* — leaflets cupped, vigour reduced.

**Transmission**

Transmitted by a vector; an insect; *Chaetosiphon fragaraefolii, C. thomasi, C. thomasi jacobi*; Aphididae. Transmitted in a persistent manner. Virus can help the vector transmission of another virus (strawberry mild yellow edge-associated potexvirus); not transmitted by mechanical inoculation; transmitted by grafting; not transmitted by contact between plants (of *Fragaria vesca* clone UC–6); not transmitted by seed; not transmitted by pollen.

**Geographical distribution**

Probably distributed worldwide. Spreads in the Eastern Asian region, the Eurasian region, the Mediterranean region, the North American region, and the Pacific region.

**Experimental host range**

Few (<3) families susceptible.

Diagnostically susceptible

*Fragaria vesca* — leaflets cupped, chlorotic margins, vigour reduced; chlorotic vein netting and necrosis of young leaves with Pacific coast strain.

*F. virginiana* — leaflets cupped, vigour reduced.

*F. ovalis* — symptomless carrier.

Diagnostically insusceptible

*Fragaria vesca* strain UC–6 is insusceptible or only very mildly infected when graft inoculated.

Maintenance and propagation hosts

*Fragaria chiloensis, F. chiloensis* var *F. vesca, F. virginiana.*

Assay hosts (**L**ocal lesions or **W**hole plants)

*Fragaria vesca* (W).

Dicotyledonous Hosts *(**Susceptible** or Insusceptible)*

***Citrus aurantium*. *Fragaria vesca*.**

Additional hosts

***Duchesnea indica*** and ***Sanguisorba minor***.

**Properties of particles in sap**

TIP: cannot be directly measured but is 45–50 °C (for aphid acquisition from living leaves). Leaf sap contains few virions.

**Purification**

Method: Martin and Converse (1982; 1985).

**Particle morphology**

Virions isometric; not enveloped; 23–28 nm in diameter; angular in profile; without a conspicuous capsomere arrangement.

**Physical properties**

One sedimenting component in purified preparations.

**Cytopathology**

Virions found in roots (Yosuikewa, 1985).

**Taxonomy and Relationships**

*Luteovirus.*

Virus(es) with serologically related virions: soybean dwarf virus, closely; beet western yellows virus. Virus(es) with serologically unrelated virions: potato leafroll, bean leaf roll and tobacco necrotic dwarf viruses. The relationship to soybean dwarf virus is so close that they are probably the same species.

**References**

Converse, R.H., Martin, R.R. and Spiegel, S. (1987). In: *Virus Diseases of Small Fruits*; ed. R.H. Converse. U.S. Dep. Agric. Hdbk No. 631, p. 25.

Harris, R.V. (1933). *J. Pomol.* **11**: 56.

Harris, R.V. and King, M.E. (1940). *Rep. East Malling Res. Stn.* 1939, p. 66.

Horne, W.T. (1922). *Rep. Calif. Agric. Exp. Stn.* 1921–1922, p. 122.

Martin, R.R. and Converse, R.H. (1982). *Acta Hort.* **129**: 75.

Martin, R.R. and Converse, R.H. (1985). *Phytopath. Z.* **114**: 21.

Mellor, F.C. and Frazier, N.W. (1970). In: *Virus Diseases of Small Fruits and Grapevines*, p. 14; ed. N.W. Frazier. Univ. Calif. Div. Agric. Sci., Berkeley.

Yankulova, M. and Schmelzer, K. (1974). *Virusni bolesti po rastenyala*, pp.33–42.

Yoshikawa, N., Ohki, S.T., Kobatake, H., Osaki, T. and Inouye, T. (1984). *Ann. Phytopath. Soc. Japan* **50**: 659.

Yoshikawa, N. and Inouye, T. (1986). *Ann. Phytopath. Soc. Japan* **52**: 643.

# Strawberry mottle virus

**Data collated by N. Yoshikawa, 1991.**

*Synonyms:* strawberry mild crinkle virus. *Strains:* many strains of differing severity have been described.

First reported in *Fragaria ananassa*; from England; by Harris (1938).

**Natural host range and symptoms**

Symptoms vary seasonally, or none (reduced vigour and yield).
*Fragaria ananassa.*

**Transmission**

Transmitted by a vector; an insect; *Chaetosiphon fragaraefolii, C. thomasi, C. minor, C. jacobi, Aphis gossypii*; Aphididae. Transmitted in a semi–persistent manner. Virus lost by the vector when it moults; does not multiply in the vector; not transmitted congenitally to the progeny of the vector; transmitted by mechanical inoculation; transmitted by grafting; not transmitted by contact between plants; not transmitted by seed.

**Geographical distribution**

Probably distributed worldwide (where strawberries are grown).

**Experimental host range**

Several (3–9) families susceptible. Experimentally infected plants mostly show *Fragaria vesca* shows vein clearing, mottling and dwarfing.

Diagnostically susceptible

*Fragaria ananassa, Gomphrena globosa, Potentilla canadensis, P. sterilis* — symptomless.
*Fragaria virginiana* — dwarfing of leaves.
*F. vesca* — mottle, vein clearing and stunting.
*Potentilla adscherica* — vein chlorosis.
*Chenopodium quinoa* — chlorotic local lesions and systemic mottling.
*C. amaranticolor* — vein clearing and necrotic spots.
*Cyamopsis tetragonoloba* — chlorotic local lesions, systemic vein clearing.

Maintenance and propagation hosts

*Fragaria ananassa, F. virginiana.*

Assay hosts (**L**ocal lesions or **W**hole plants)

*Fragaria vesca* (W).

Dicotyledonous Hosts *(**Susceptible** or Insusceptible)*

***Chenopodium amaranticolor*. *Cyamopsis tetragonoloba*. *Fragaria chiloensis*. *Fragaria virginiana*.**

Sources of host-range data

Frazier (1968); Converse (1987).

**Purification**

Method: Yoshikawa and Converse (1991).

**Particle morphology**

Virions isometric; not enveloped; 37 nm in diameter; angular in profile; without a conspicuous capsomere arrangement.

**Physical properties**

One sedimenting component in purified preparations. Density 1.42 g $cm^{-3}$ in CsCl.

**Biochemical properties**

*Genome* consists of RNA; single–stranded (possibly); unipartite; largest (or only) genome part 6.6 kb.

*Virion protein(s)* one; $M_r$ 26000; coat protein.

**Cytopathology**

Virions found in leaves, mesophyll, epidermis and the phloem; in cytoplasm. Inclusions absent from infected cells.

**References**


Frazier, N.W. (1968). *Pl. Dis. Reptr* **52**: 64.

Frazier, N.W. (1974). *Pl. Dis. Reptr* **58**: 28.

Frazier, N.W. and Sylvester, E.S. (1960). *Virology* **12**: 233.

Kitajima, E.W., Betti, J.A. and Costa, A.S. (1971). *Cienc. Cult.* **23**: 649.

Mellor, F.C. and Krczal, H. (1987). In: *Viruses Diseases of Small Fruits*; ed. R.H. Converse. U.S. Dep. Agric. Hdbk No. 631, p, 10.

Yoshikawa, N. and Converse, R.H. (1991). *Ann. appl. Biol.*.


# Strawberry pallidosis (?) virus

**Data collated by N. Yoshikawa, 1991.**

First reported in *Fragaria ananassa*; from the U.S.A and Australia; by Frazier and Stubbs (1957).

**Natural host range and symptoms**

Symptoms none.

*Fragaria ananassa.*

**Transmission**

Virus not transmitted by mechanical inoculation; transmitted by grafting; not transmitted by contact between plants.

**Ecology and control**

Studies reported by Mullin *et al.* (1975).

**Geographical distribution**

Spreads in Australia, Canada, and the USA.

**Experimental host range**

Few (<3) families susceptible. Experimentally infected plants mostly show leaf deformation and chlorosis and dwarfing of the plant.

Diagnostically susceptible

*Fragaria ananassa, F. vesca* — symptomless.

*Fragaria virginiana* — leaf deformation, chlorosis and dwarfing.

Maintenance and propagation hosts

*Fragaria ananassa, F. vesca.*

Assay hosts (**L**ocal lesions or **W**hole plants)

*Fragaria virginiana.*

Dicotyledonous Hosts *(**Susceptible** or Insusceptible)*

***Fragaria chiloensis*. *Fragaria virginiana*.**

3–14 virus specified dsRNA species found in infected cells (depending on isolates). Size of largest virus specified dsRNA 4.3–5.2 kbp.

**References**

Frazier, N.W. and Stubbs, L.L. (1969). *Pl. Dis. Reptr* **53**: 524.
Mullin, R.H., Frazier, N.W. and Schlegel, D.E. (1975). *Pl. Dis. Reptr* **59**: 268.
Yoshikawa, N. and Converse, R.H. (1990). *Phytopathology* **80**: 543.

# Strawberry pseudo mild yellow edge *carlavirus*

**Data collated by N. Yoshikawa, 1991.**

*Standard acronym:* SPMYEV.

First reported in *Fragaria virginiana*; from the U.S.A.; by Frazier (1966).

**Natural host range and symptoms**

Symptoms none.

*Fragaria virginiana, F. ananassa* — symptomless.

**Transmission**

Transmitted by a vector; an insect; *Chaetosiphon thomasi jacobi, C. fragaraefolii, Aphis gossypii*; Aphididae. Principal natural vector(s): *Chaetosiphon fragaraefolii, Aphis gossypii*. Transmitted in a semi–persistent manner. Virus lost by the vector when it moults; does not multiply in the vector; not transmitted congenitally to the progeny of the vector; not transmitted by mechanical inoculation; transmitted by grafting; not transmitted by contact between plants; not transmitted by seed.

**Geographical distribution**

Spreads in Japan and the USA.

**Experimental host range**

Few (<3) families susceptible. Experimentally infected plants mostly show yellow to reddish colouration on lower leaves, followed by necrosis and leaf roll.

Diagnostically susceptible

*Fragaria ananassa* — symptomless.

*Fragaria vesca* — chlorotic mottle and reddening on older leaves, then necrosis and leaf fall.

*Fragaria virginiana* — yellow to red colouration of old leaves.

Maintenance and propagation hosts

*Fragaria ananassa, F. virginiana, F. vesca.*

Assay hosts (**L**ocal lesions or **W**hole plants)

*Fragaria vesca* (W).

Dicotyledonous Hosts *(**Susceptible** or Insusceptible)*

***Fragaria chiloensis*. *Fragaria virginiana*.**

Sources of host-range data

Frazier (1966); Yoshikawa and Inouye (1986).

**Properties of particles in sap**

Leaf sap contains few virions.

**Purification**

Method: Yoshikawa and Inouye (1986).

**Particle morphology**

Virions filamentous; not enveloped; usually straight; with a clear modal length; of 625 nm; 12 nm wide. Axial canal obscure. Basic helix obscure.

**Physical properties**

One sedimenting component in purified preparations. Density 1.32 g $cm^{-3}$ in CsCl. $A_{260}/A_{280}$ ratio 1.19.

**Biochemical properties**

Virions contain 5 % nucleic acid.

*Genome* consists of RNA; single–stranded; linear (possibly); unipartite; largest (or only) genome part 8 kb. Genomic nucleic acid isolated by Yoshikawa and Inouye (1986).

*Non–genomic nucleic acid* not found in the virions.

*Virion protein(s)* one; $M_r$ 33500; coat protein.

**Cytopathology**

Virions found in leaves, mesophyll, epidermis, vascular parenchyma and companion cells; in cytoplasm. Inclusions absent from infected cells.

**Taxonomy and Relationships**

*Carlavirus*.

Virus(es) with serologically related virions: carnation latent virus.

Best tests for diagnosis: dot immunobinding assay.

**References**

Frazier, N.W. (1966). *Phytopathology* **56**: 571.

Yoshikawa, N. and Inouye, T. (1986). *Ann. Phytopath. Soc. Japan* **52**: 643.

Yoshikawa, N., Poolpol. P. and Inouye, T. (1986). *Ann. Phytopath. Soc. Japan* **52**: 728.

# Strawberry vein banding (?) *caulimovirus*

**Data collated by T.J. Morris and J.A. Cooper, 1985.**

*Synonyms:* yellow vein banding virus, chiloensis vein banding virus, Eastern veinbanding virus. *Standard acronym:* SVBV. *Strains:* possibly Erdbeer-Nekrose virus (Schöniger, 1958; Domes, 1957).

First reported in *Fragaria vesca*; from Great Britain from plants of cv. 'Fairfax' imported from the U.S.A. and from Brazil; by Prentice (1952); Kitajima *et al.* (1973).

**Natural host range and symptoms**

Symptoms persist (but symptoms fluctuate in severity).

*Fragaria vesca, F. virginiana, F. chiloensis and F. ananassa* — yellow vein banding, discontinuous banding, streaking and spotting of older leaves, twisting of leaflets.

**Transmission**

Transmitted by a vector; an insect; *Chaetosiphon jacobi, C. fragaraefolii, C. thomasi*; Aphididae. Transmitted in a semi–persistent manner. Virus does not multiply in the vector; not transmitted congenitally to the progeny of the vector; not transmitted by mechanical inoculation; transmitted by grafting; not transmitted by contact between plants; not transmitted by seed; not transmitted by pollen.

**Geographical distribution**

Spreads in the African region; Australia, Brazil, Canada, Chile, China, the former Czechoslovakia, Germany, Hungary, Ireland, Italy, Japan, the USA, and the former USSR (Schöniger, 1958).

**Experimental host range**

Few (<3) families susceptible.

Diagnostically susceptible

*Fragaria vesca, F. virginiana, F. chiloensis and F. ananassa* — yellow vein banding, discontinuous banding, streaking and spotting of older leaves, twisting of leaflets.

Maintenance and propagation hosts

Most species and clones of *Fragaria vesca, F. virginiana, F. chiloensis virginiana and F. ananassa.*

Assay hosts (**L**ocal lesions or **W**hole plants)

*Fragaria vesca* (W) *F. virginiana* (W).

Dicotyledonous Hosts *(**Susceptible** or Insusceptible)*

***Fragaria vesca.***

Additional hosts

***Sanguisorba minor*** (by aphids and grafting).

**Properties of particles in sap**

Leaf sap contains few virions. Electron microscopy: purification, 1% formaldehyde treatment, stain with UA. For ISEM use cauliflower mosaic virus antiserum to trap virions.

**Purification**

Method: Morris *et al.* (1980).

**Particle morphology**

Virions isometric; not enveloped; 43 nm in diameter (± 4nm); rounded in profile.

**Physical properties**

One sedimenting component in purified preparations; sedimentation coefficient 200 *S* (±10).

**Biochemical properties**

*Genome* consists of DNA; double–stranded; circular; unipartite; largest (or only) genome part 7.8 kb (pairs).

**Replication**

Genome replicates in cytoplasmic inclusions.

**Cytopathology**

Virions found in leaves, mesophyll, epidermis and vascular parenchyma particularly in adjoining xylem vessels; in cytoplasm (where unique spike structures are observed when stained with UA). Inclusions present in infected cells; are viroplasms; they contain virions.

**Taxonomy and Relationships**

Possible *Caulimovirus.*

Virus(es) with serologically related virions: cauliflower mosaic virus. Virus(es) with serologically unrelated virions: carnation etched ring virus.

The strawberry vein-banding virus genome has no significant homology in hybridization tests with that of cauliflower mosaic virus.

**References**


Domes, R. (1957). *Phytopath. Z.* **31**: 113.

Frazier, N.W. and Converse, R.H. (1980). *CMI/AAB Descr. Pl. Viruses* No. 219, 4 pp.

Frazier, N.W. and Morris, T.J. (1985). In: *Virus Diseases of Small Fruits*; ed. R.H. Converse. U.S. Dep. Agric. Hdbk No. 631, p. 16.

Kitajima, E.W., Betti, J.A. and Costa, A.S. (1973). *J. gen. Virol.* **20**: 117.

Miller, P.W. and Frazier, N.W. (1970). In: *Virus Diseases of Small Fruits and Grapevines* pp. 8–10; ed. Frazier. Univ. Calif. Div. Agric. Sci., Berkeley.

Morris, T.J., Mullin, R.H., Schlegel, D.E., Cole, A. and Alosi, M.C. (1980). *Phytopathology* **70**: 156.

Prentice, I.W. (1952). *Ann. appl. Biol.* **39**: 487.

Schöniger, G. (1958). *Phytopath. Z.* **32**: 325.


# Subterranean clover mottle *sobemovirus*

**Data collated by R.I.B. Francki, 1983. Revised 1987.**

*Synonyms:* subterranean clover virus. *Standard acronym:* SCMoV.

First reported in *Trifolium subterraneum*; from Perth, Australia; by Francki *et al.* (1983).

**Natural host range and symptoms**

Symptoms persist.

*Trifolium subterraneum* — systemic mosaic, stunting.

**Transmission**

Virus transmitted by mechanical inoculation; transmitted by seed.

**Geographical distribution**

Spreads in Australia.

**Experimental host range**

Few (<3) families susceptible.

Diagnostically susceptible

*Pisum sativum* — necrotic local lesions; No systemic infection.

*Trifolium subterraneum* — systemic mosaic, stunting.

Diagnostically insusceptible

*Chenopodium amaranticolor, C. quinoa, Nicotiana clevelandii.*

Maintenance and propagation hosts

*Trifolium subterraneum.*

Assay hosts (**L**ocal lesions or **W**hole plants)

*Pisum sativum* (L).

Dicotyledonous Hosts *(**Susceptible** or Insusceptible)*

*Chenopodium amaranticolor. Chenopodium quinoa. Nicotiana clevelandii.* ***Pisum sativum. Trifolium subterraneum.***

Sources of host-range data

Francki *et al.* (1983).

**Purification**

Method: Francki *et al.* (1983).

**Particle morphology**

Virions isometric; not enveloped; 30 nm in diameter; angular in profile; without a conspicuous capsomere arrangement.

**Physical properties**

One sedimenting component in purified preparations; sedimentation coefficient 115 *S*.

**Biochemical properties**

Virions contain *c.* 20 % nucleic acid; *c.* 80 % protein; 0 % lipid.

*Genome* consists of RNA; single–stranded; unipartite; largest (or only) genome part 4.2 kb.

*Sequence database accession code(s):*

M33000 Em(40)_vi:SUVSATA Gb(84)_vi:SUVSATA Subterranean clover mottle virus satellite RNA (virusoid) sequence. 7/90 332bp.

M33001 Em(40)_vi:SUVSATB Gb(84)_vi:SUVSATB Subterranean clover mottle virus satellite RNA (virusoid) sequence. 7/90 388bp. 2 sequences.

*Non–genomic nucleic acid* found in the virions; is of an unusual type of structure; satellite RNA.

*Virion protein(s)* one; $M_r$ 29000; coat protein.

**Cytopathology**

Virions found in leaves and all parts of the host plant; in nuclei. Inclusions present in infected cells; are unusual in shape; aggregates; they contain virions.

**Taxonomy and Relationships**

*Sobemovirus.*

Virus(es) with serologically related virions: lucerne transient streak virus, but distantly. Virus(es) with serologically unrelated virions: velvet tobacco mottle, solanum nodiflorum mottle, southern bean mosaic and sowbane mosaic viruses.

**References**

Francki, R.I.B., Randles, J.W. and Graddon, D. (1987). *CMI/AAB Descr. Pl. Viruses* No. 329.

Francki, R.I.B., Randles, J.W., Hatta, T., Davies, C., Chu, P.W.G. and McLean, G.D. (1983). *Pl. Path.* **32**: 47.

Graddon, D. and Randles, J.W. (1986). *J. virol. Meth.* **13**: 63.

# Subterranean clover red leaf *luteovirus*

**Data collated by G.R. Johnstone, P.M. Waterhouse, 1980; J.W. Ashby 1980. Revised 1984.**

*Synonyms:* clover subterranean red leaf virus, soybean dwarf luteovirus (Ashby and Kyriakou, 1982; Johnstone and McLean, 1987).

First reported in *Trifolium subterraneum*; from Northern Victoria, Australia; by Anon. (1968).

**Natural host range and symptoms**

Symptoms persist.

*Erodium moschatum, Trifolium repens, T. subterraneum, Pisum sativum, Vicia faba, Lupinus spp., Glycine max, Lens esculenta* — mild yellowing, stunting and reddening.

**Transmission**

Transmitted by a vector; an insect; *Acyrthosiphon (Aulacorthum) solani, Neomyzus circumflexus, Acyrthosiphon pisum, Macrosiphum euphorbiae* (in New Zealand); Aphididae. Transmitted in a persistent manner. Virus retained when the vector moults; does not multiply in the vector; not transmitted congenitally to the progeny of the vector; not transmitted by mechanical inoculation; not transmitted by contact between plants; not transmitted by seed; not transmitted by pollen.

**Geographical distribution**

Spreads in Australia, New Zealand, and the USA (in California).

**Experimental host range**

Several (3–9) families susceptible.

Diagnostically susceptible

*Gomphrena globosa* — slight leaf reddening.
*Beta vulgaris* — systemic yellowing.
*Erodium cicutarium* — systemic reddening.
*Glycine max* cv. Shirotsuronoko — interveinal yellowing, leaves rolled down.
*Lupinus cosentinii* — systemic yellowing and reddening.
*Phaseolus vulgaris* cvs Galatin 50, Tendergreen, Top Crop — interveinal chlorosis, leaf puckering, leaves rolled down.
*Pisum sativum* cvs Laxton superb, Onyx, Puget — general yellowing, leaves brittle, leaves rolled.
*Trifolium hybridum* and *T. repens* — systemic infection but few or no symptoms.
*T. subterraneum* — leaf reddening.

Diagnostically insusceptible

*Brassica rapus, Capsella bursa-pastoris, Arachis hypogaea, Lathyrus odoratus, Trifolium pratense.*

Maintenance and propagation hosts

*Pisum sativum* cvs Onyx, Puget.

Assay hosts (**L**ocal lesions or **W**hole plants)

*Pisum sativum* cvs Onyx, Puget, *Trifolium subterraneum* cvs Bacchus Marsh, Geraldton, Mt Barker (W).

Dicotyledonous Hosts *(**Susceptible** or Insusceptible)*

*Arachis hypogaea.* ***Astragalus sinicus.*** ***Beta vulgaris.*** *Brassica campestris* ssp. *napus. Capsella bursa–pastoris. Chenopodium amaranticolor. Chenopodium quinoa. Cicer arietinum.* ***Glycine max.*** ***Gomphrena globosa.*** ***Lactuca sativa.*** *Lathyrus odoratus.* ***Lens esculenta.*** *Linum usitatissimum.* ***Lupinus albus.*** ***Lupinus cosentinii.*** ***Medicago hispida.*** *Medicago sativa. Melilotus alba.* ***Phaseolus vulgaris.*** ***Pisum sativum.*** ***Senecio*** vulgaris. *Stellaria media. Trifolium hybridum. Trifolium incarnatum. Trifolium pratense.* ***Trifolium repens. Trifolium subterraneum. Vicia faba. Vicia sativa.*** *Vigna unguiculata.* ***Zinnia elegans.***

Sources of host-range data

Kellock (1971); Kellock (unpublished data).

**Properties of particles in sap**

TIP: 55–60 °C. LIV: 14 days. DEP: $\log_{10}$ minus 2. Leaf sap contains few virions.

**Purification**

Method: Waterhouse and Helms (1984); Ashby and Kyriakou (1982). Add 1–1.5% cellulase when using standard luteovirus purification method.

**Particle morphology**

Virions isometric; not enveloped; 27 nm in diameter, or 30 nm in diameter; angular in profile; without a conspicuous capsomere arrangement.

**Physical properties**

One sedimenting component in purified preparations (dimers often found); sedimentation coefficient 114 *S*. Density 1.39 g $cm^{-3}$ in CsCl.

**Biochemical properties**

Virions contain 28–30 % nucleic acid; 70–72 % protein; 0 % lipid.

*Genome* consists of RNA; single–stranded; linear; unipartite; largest (or only) genome part 5.86 kb.

*Virion protein(s)* one; $M_r$ 22600, or 25000. Method of preparation: Ashby and Kyriakou (1982). Amino acid composition: Waterhouse and Kortt in Waterhouse *et al.* (1987).

**Replication**

Replication does not depend on a helper virus.

**Cytopathology**

Virions found in phloem and companion cells in transfer cells; in cytoplasm, in nuclei, and in cell vacuoles. Inclusions absent from infected cells. Other cellular changes: small vesicles with stranded material.

**Taxonomy and Relationships**

*Luteovirus.*

Virus(es) with serologically related virions: soybean dwarf virus, so closely that it is probably the same species; bean leaf roll, beet western yellows and potato leafroll more distantly.

**References**

Anon. (1968). *J. Dep. Agric. Vic.* **66**: 182.

Ashby, J.W. and Kyriakou, A. (1982). *N.Z. Jl agric. Res.* **25**: 607.

Ashby, J.W., Teh, P.B. and Close, R.C. (1979). *N.Z. Jl agric. Res.* **22**: 361.

Jayasena, K.W., Hatta, T., Francki, R.I.B. and Randles, J.W. (1981). *J. gen. Virol.* **57**: 205.

Johnstone, G.R. (1978). *Aust. J. agric. Res.* **29**: 1003.

Johnstone, G.R. and Rapley, P.E.L. (1979). *Ann. appl. Biol.* **91**: 345.

Johnstone, G.R. and McLean, G.D. (1987). *Ann. appl. Biol.* **110**: 421.

Johnstone, G.R., Duffus, J.E., Munro, D. and Ashby, J.W. (1982). *Aust. J. agric. Res.* **33**: 697.

Kellock, A.W. (1971). *Aust. J. agric. Res.* **22**: 615.
Waterhouse, P.M. and Helms, K. (1984). *J. Virol. Meth.* **8**: 321.
Waterhouse, P.M., Gildow, F.E. and Johnstone, G.R. (1987). *AAB Descr. Pl. Viruses* No. 339.
Wilson, J. and Close, R.C. (1973). *N.Z. Jl agric. Res.* 16: 305.

# Subterranean clover stunt nanavirus

**Data collated by G.R. Johnstone, 1983. Revised by P.W.G. Chu, 1987.**

*Synonyms:* clover stunt virus, clover subterranean stunt virus. *Standard acronym:* SCSV. *Strains:* Dickson isolate; Clandulla isolate; Ginnindera isolate a severe strain and Wattamondara isolate.

First reported in *Trifolium subterraneum*; from New South Wales, Australia; by Grylls and Butler (1956).

**Natural host range and symptoms**

Symptoms persist.

*Trifolium subterraneum, T. cernuum* — mild or severe stunting, marginal chlorosis and puckering or cupping of leaflets, older leaves purple or crimson.

*Medicago lupulina, M. hispida* var. *denticulata, M. minima, Trifolium repens, T. glomeratum, Wisteria sinensis* — typical stunting, shortening of petioles and internodes; chlorosis of young leaves.

*Phaseolus vulgaris* — severe epinasty and leaves thickened, smaller and chlorotic; severe stunting and terminal necrosis, few or no seeds.

*Pisum sativum, Vicia faba* — leaf size reduced, petioles and internodes shortened, leaf chlorosis and rolling.

*Trifolium dubium, T. pratense, Medicago arabica* — typical stunting and yellowing on new growths.

**Transmission**

Transmitted by a vector; an insect; *Aphis craccivora, A. gossypii, Myzus persicae, Macrosiphum euphorbiae*; Aphididae. Transmitted in a persistent manner. Virus retained when the vector moults; does not multiply in the vector; not transmitted congenitally to the progeny of the vector; not transmitted by mechanical inoculation; transmitted by grafting; not transmitted by contact between plants; not transmitted by seed; not transmitted by pollen.

**Ecology and control**

Studies reported by Grylls and Butler (1956; 1959); O'Loughlin (1958); Grylls and Peak (1960; 1969); Peak *et al.* (1963); Gutierrez *et al.* (1971).

**Geographical distribution**

Spreads in Australia.

**Experimental host range**

Few (<3) families susceptible.

Diagnostically susceptible

*Trifolium subterraneum* cv. Mt Barker — severe stunting, rosetting, puckering and marginal chlorosis, reddening of older leaves.

*Medicago hispida* var. *denticulata* — severe stunting, rosetting, puckering and marginal chlorosis, reddening of older leaves.

*Phaseolus lathyroides* — severe stunting and small leaves, then marginal chlorosis and puckering of terminal leaves.

*Trigonella procumbens* — extreme stunting and puckering of leaflets.

*Astragalus sinicus* — stunting, leaf yellowing, size reduction and rolling down.

Diagnostically insusceptible

*Gomphrena globosa, Capsella bursa-pastoris, Erodium moschatum.*

Maintenance and propagation hosts

*Trifolium subterraneum* cv. Mt Baker, *Pisum sativum* cv. Greenfeast — propagation only.

Assay hosts (**L**ocal lesions or **W**hole plants)

*Trifolium subterraneum* (W).

Dicotyledonous Hosts *(**Susceptible** or Insusceptible)*

***Arachis hypogaea***, or *Arachis hypogaea*. ***Astragalus sinicus***. *Beta vulgaris*. *Brassica campestris* ssp. *napus*. *Capsella bursa–pastoris*. *Cicer arietinum*. ***Crotalaria spectabilis***. *Datura stramonium*. *Glycine max*. *Gomphrena globosa*. *Lactuca sativa*. *Lathyrus odoratus*. ***Lens esculenta***. *Linum usitatissimum*. *Lupinus albus*. ***Macrotyloma uniflorum***. ***Medicago hispida***. ***Medicago sativa***, or *Medicago sativa*. ***Melilotus alba***, or *Melilotus alba*. ***Phaseolus lathyroides***. ***Phaseolus vulgaris***. ***Pisum sativum***. *Senecio vulgaris*. *Stellaria media*. *Tetragonia expansa*. ***Trifolium dubium***. *Trifolium hybridum*. ***Trifolium incarnatum***. *Trifolium pratense*. ***Trifolium repens***, or *Trifolium repens*. ***Trifolium subterraneum***. ***Vicia faba***. ***Vicia sativa***. ***Vigna unguiculata***. *Vitis rupestris*.

Additional hosts

*Crotalaria intermedia* (**S**); *Desmodium intortum* (**S**); *Dolichos lablab*; *Medicago arabica* (**S**); *Medicago polymorpha* (**S**); *Medicago lupulina* (**S**); *Medicago minima* (**S**); *Medicago scutellata* (**S**); *Medicago orbicularis* (**S**); *Melilotus indica* (**S**); *Onobrychis viciifolia* (**S**); *Onobrychis biebersteinii* (**S**); *Trifolium cernuum* (**S**); *Trifolium glomeratum* (**S**); *Trifolium pannonicum* (**S**); *Trigonella procumbens* (**S**); *Wisteria sinensis* (**S**). Host range data variable, probably resulting from strain differences or contaminants. All varieties of *Trifolium subterraneum* appear to be susceptible as well as all varieties of *Phaseolus vulgaris*. Many species of *Medicago* and *Trifolium* are susceptible.

Sources of host-range data

Grylls and Butler (1959); O'Loughlin (1958); Gutierrez *et al.* (1971); Smith (1966).

**Properties of particles in sap**

Leaf sap contains many virions (when tested by ISEM). Electron microscopy: virus preparations are stable in AM, pH 6.5.

**Purification**

Method: Chu and Helms (1987). Use of cellulases helps purify virions.

**Particle morphology**

Virions isometric; not enveloped; 17–19 nm in diameter; angular in profile; without a conspicuous capsomere arrangement.

**Physical properties**

One sedimenting component in purified preparations. Density 1.34 g $cm^{-3}$ in CsCl (unfixed). $A_{260}/A_{280}$ ratio 1.35.

**Biochemical properties**

Virions contain 16–17 % nucleic acid; 83–84 % protein.

*Genome* consists of DNA; single–stranded; circular; of 7 parts (or more); largest (or only) genome part the largest 0.89 kb; the 2nd largest 0.88 kb; the 3rd largest 0.88 kb; the 4th largest 0.87 kb. Genomic nucleic acid isolated by Chu and Helms (1988): resuspend virus purified by sucrose density gradient in 50 mM tris–HCl buffer, pH 7.6, 1 mM EDTA, 100 mM Na–acetate, and extract with phenol and 2% SDS.

*Sequence database accession code(s):*

S57687 Em(40)_vi:S57687 Gb(84)_vi:S57687 coat protein (component 3) subterranean clover stunt virus SCSV, F isolate, Genomic, 48 nt.

U16730 Em(43)_vi:Sc16730 Gb(89)_vi:Scu16730 Subterranean clover stunt virus (SCSV1) gene, complete cds. 12/94 1,001bp.

U16731 Em(43)_vi:Sc16731 Gb(89)_vi:Scu16731 Subterranean clover stunt virus possible replication asociated protein (SCSV2) gene, complete

U16732 Em(43)_vi:Sc16732 Gb(89)_vi:Scu16732 Subterranean clover stunt virus (SCSV3) gene, complete cds. 12/94 991bp.

U16733 Em(43)_vi:Sc16733 Gb(89)_vi:Scu16733 Subterranean clover stunt virus (SCSV4) gene, complete cds. 12/94 1,002bp.

U16734 Em(43)_vi:Sc16734 Gb(89)_vi:Scu16734 Subterranean clover stunt virus coat protein (SCSV5) gene, complete cds. 12/94 998bp.

U16735 Em(43)_vi:Sc16735 Gb(89)_vi:Scu16735 Subterranean clover stunt virus possible replication associated protein (SCSV6) gene, complet

U16736 Em(43)_vi:Sc16736 Gb(89)_vi:Scu16736 Subterranean clover stunt virus (SCSV7) gene, complete cds. 12/94 988bp.

*Virion protein(s)* one; $M_r$ 19000; coat protein.

**Cytopathology**

Virions found in phloem. Inclusions absent from infected cells. Other cellular changes: phloem necrosis.

**Taxonomy and Relationships**

Nanavirus (the type species).

Best tests for diagnosis: often occurs naturally with bean yellow mosaic and clover yellow mosaic viruses; ELISA tests are best to distinguish these viruses.

**References**

Chu, P.W.G. and Helms, K. (1988). *Virology* **167**: 38.
Grylls, N.E. and Butler, F.C. (1956). *J. Aust. Inst. agric. Sci.* **22**: 73.
Grylls, N.E. and Butler, F.C. (1959). *Aust. J. agric. Res.* **10**: 145.
Grylls, N.E. and Peak, J.W. (1960). *Aust. J. agric. Res.* **11**: 723.
Grylls, N.E. and Peak, J.W. (1969). *Aust. J. agric. Res.* **20**: 37.
Gutierrez, A.P., Morgan, D.J. and Havenstein, D.E. (1971). *J. appl. Ecol.* **8**: 69.
O'Loughlin, G.T. (1958). *J. Agric. Vic.* **56**: 385.
Peak, J.W., Morley, F.H.W. and Grylls, N.E. (1963). *Aust. J. agric. Res.* **14**: 639.
Smith, P.R. (1966). *Aust. J. agric. Res.* **17**: 875.

# Sugarcane bacilliform *badnavirus*

**Data collated by A.A. Brunt and B.E.L. Lockhart, 1988.**

*Standard acronym:* SCBV.

First reported in *Saccharum officinarum*; from north west Morocco; by Lockhart and Autry (1988).

**Natural host range and symptoms**

*Saccharum officinarum* — no conspicuous symptoms.

**Geographical distribution**

Spreads in Cuba, Morocco, and the USA (Florida and Hawaii).

**Experimental host range**

Few (<3) families susceptible.

Diagnostically susceptible

*Saccharum officinarum* (clone CP 44–101) — symptomless systemic infection.

Diagnostically insusceptible

*Chenopodium quinoa, Nicotiana benthamiana, N. glutinosa, Sorghum halepense, Zea mays.*

Maintenance and propagation hosts

*Saccharum officinarum.*

Assay hosts (**L**ocal lesions or **W**hole plants)

*Saccharum officinarum* (clone CP 44–101) (W).

Dicotyledonous Hosts *(**Susceptible** or Insusceptible)*

*Chenopodium quinoa. Nicotiana benthamiana. Nicotiana glutinosa.*

Monocotyledonous Hosts *(**Susceptible** or Insusceptible)*

*Avena sativa. Hordeum vulgare. Triticum aestivum. Zea mays.*

Sources of host-range data

Lockhart and Autry (1988).

**Purification**

Method: Lockhart and Autry (1988).

**Particle morphology**

Virions bacilliform; not enveloped; 131 nm in length; 31 nm wide.

*Sequence database accession code(s):*

M89923 Em(40)_vi:SBVORFS Gb(84)_vi:SBVORFS Sugarcane bacilliform virus ORF 1,2,and 3 DNA, complete cds. 3/93 7,568bp. 1 sequence.

**Taxonomy and Relationships**

*Badnavirus.*

Virus(es) with serologically related virions: banana streak badnavirus. Virus(es) with serologically unrelated virions: cacao swollen shoot, commelina yellow mottle virus, kalanchoe top spotting, yam bacilliform, rice tungro and canna yellow mottle 'commelina yellow mottle viruses'.

The virus is serologically closely related to banana streak badnavirus, but is probably distinct.

**References**

Bouhida, M., Lockhart, B.E.L. and Olszewski, N.E. (1993). *J. gen. Virol.* **74**: 15.
Lockhart, B.E.L. and Autry, L.J.C. (1988). *Plant Dis.* **72**: 230.

# Sugarcane Fiji disease *fijivirus*

**Data collated by T. Hatta, 1980.**

*Standard acronym:* FDV.

First reported in *Saccharum* spp.; from Australia and the Philippines; by Anon. (1886); North (1915); Ocfemia (1923); Bell (1933).

### Natural host range and symptoms

Symptoms persist.

*Saccharum* spp. — leaf deformation and leaf galls.

### Transmission

Transmitted by a vector; an insect; *Perkinsiella saccharicida, P. vitiensis* and *P. vastatrix*; Delphacidae. Transmitted in a persistent manner. Virus retained when the vector moults; multiplies in the vector; transmitted congenitally to the progeny of the vector; not transmitted by mechanical inoculation.

### Geographical distribution

Spreads in the African region and the Pacific region (New Britain and the New Hebrides); Australia, Fiji, Papua New Guinea, the Philippines, Thailand, and Western Samoa. Found, but with no evidence of spread, in the U.S.A.

### Experimental host range

Few (<3) families susceptible.

Diagnostically susceptible

*Saccharum* spp., *Sorghum* spp., *Zea mays* — enations on deformed leaves and leaf sheaths.

Monocotyledonous Hosts ***(Susceptible** or Insusceptible)*

***Saccharum officinarum*. *Sorghum bicolor*. *Zea mays*.**

Sources of host-range data

Hutchinson *et al.* (1972).

### Properties of particles in sap

Leaf sap galls contains many virions, or contains few virions (in leaves). Electron microscopy: stain with UA.

### Purification

Method: Van der Lubbe *et al.* (1979).

### Particle morphology

Virions isometric; 71 nm in diameter (intact virions), or 54 nm in diameter (subvirions); rounded in profile; without a conspicuous capsomere arrangement.

### Physical properties

One sedimenting component in purified preparations.

### Biochemical properties

Virions 0 % lipid.

*Genome* consists of RNA; double–stranded; linear; of 10 parts (named S1–10); largest (or only) genome part 3.998 kb (pairs); the 2nd largest 3.385 kb (pairs); the 3rd largest 3.38 kb (pairs); the 4th largest 3.38 kb (pairs); the 5th largest 3.1 kb (pairs); the 6th largest and other parts 2.5 kb (pairs; also 2.190, 1.732, 1.724 and 1.720 kb pairs).

*Virion protein(s)* at least three (in the core); $M_r$ 145000–150000. $M_r$ of 2nd largest 125000–129000. $M_r$ of 3rd largest 36000–39000. Method of preparation: Van der Lubbe *et al.* (1979).

**Cytopathology**

Virions found in gall tissue; in cytoplasm and in cell vacuoles. Inclusions present in infected cells; are unusual in shape; round, containing strands and amorphous material; they contain virions. Other cellular changes: proliferations in both phloem and xylem.

**Taxonomy and Relationships**

*Fijivirus: Reoviridae* (the type species).

**References**

Bell, A.F. (1933). *Qld Bur. Sug. Exp. Stn. Div. Soils* **8**: 1.
Hatta, T. and Francki, R.I.B. (1976). *Physiol. Pl. Path.* **9**: 321.
Hatta, T. and Francki, R.I.B. (1979). *Virology* **76**: 797.
Hutchinson, P.B. and Francki, R.I.B. (1973). *CMI/AAB Descr. Pl. Viruses* No. 119, 3 pp.
Hutchinson, P.B., Forteath, G.N.R. and Osborn, A.W. (1972). *Sugarcane Path. Newsl.* **9**: 12.
Ikegami, M. and Francki, R.I.B. (1974). *Virology* **64**: 161.
Kunkel, L.O. (1924). *Bull. Hawaiian Sugar Plrs' Assoc. Exp. Stn Bot. Ser.* **3**: 91.
Lyon, L. (1910). *Hawaiian Plrs' Rec.* **3**: 200.
North, D.S. (1915). *Plant Rec.* **12**:.
North, D.S. (1923). *Aust. Sug. J.* **14**: 1.
North, D.S. (1923). *Aust. Sug. J.* **15**: 1.
Ocfemia, G.O. (1933). *Univ. Philipp. Nat. Appl. Sci. Bull.* **3**: 277.
Ocfemia, G.O. (1934). *Am. J. Bot.* **21**: 113.
Reddy, D.V.R., Boccardo, G., Outridge, R., Teakle, D.S. and Black, L.M. (1975). *Virology* **63**: 287.
Van der Lubbe, J.L.M., Hatta, T. and Francki, R.I.B. (1979). *Virology* **95**: 405.

# Sugarcane mosaic *potyvirus*

**Data collated by D.M. Persley, 1980.**

*Synonyms:* grass mosaic virus, maize dwarf mosaic virus strain B (MacKenzie *et al.*, 1966; Louie and Knoke, 1975; Shukla *et al.*, 1989), sorghum red stripe virus. *Standard acronym:* SCMV. *Strains:* European maize dwarf virus, MDMV–B, SCMV–A, SCMV–B, SCMV–D, SCMV–E, SCMV–SC, SCMV–Sabi, sorghum concentric ringspot virus, possibly also abaca mosaic potyvirus.

First reported in *Saccharum* spp.; by Brandes (1919); Gold and Martin (1955).

**Natural host range and symptoms**

Symptoms persist.

*Saccharum* spp., *Sorghum bicolor*, *Panicum* spp., *Eleusine* spp., *Setaria* spp., *Zea mays* — mosaic and/or ringspots, stunting.

*Sorghum halepense* (particular strains) — mosaic and/or necrosis.

**Transmission**

Transmitted by a vector; an insect; *Dactynotus ambrosiae, Hysteroneura setariae, Rhopalosiphum maidis, Toxoptera graminum*; Aphididae. Transmitted in a non–persistent manner. Virus transmitted by mechanical inoculation; transmitted by grafting; transmitted by seed; not transmitted by pollen.

**Geographical distribution**

Probably distributed worldwide. Spreads in Australia.

**Experimental host range**

Several (3–9) families susceptible.

Diagnostically susceptible

*Saccharum* sp. — systemic mosaic.
*Sorghum bicolor* — necrotic local lesion, then systemic mosaic, necrosis.
*S. halepense* — systemic mosaic.
*Zea mays* — systemic mosaic, ringspots.
Most dicotyledonous plants.

Maintenance and propagation hosts

*Zea mays* (sweet corn).

Assay hosts (**L**ocal lesions or **W**hole plants)

*Sorghum* genotypes (L,W).

Dicotyledonous Hosts *(**Susceptible** or Insusceptible)*

***Phaseolus vulgaris*** (symptomless local infection on cv. Bountiful with some strains).

Monocotyledonous Hosts *(**Susceptible** or Insusceptible)*

***Agropyron repens***. *Agrostis tenuis*. ***Anthoxanthum odoratum***. ***Arundinaria amabilis***. ***Arundo donax***. *Avena sativa*. *Bromus mollis*. *Cortaderia selloana*. ***Cynodon dactylon***. *Dactylis glomerata*. ***Digitaria decumbens***. ***Echinochloa crus–galli***. ***Eleusine coracana***. ***Eragrostis cilianensis***. *Festuca pratensis*. ***Hordeum vulgare***. ***Imperata cylindrica***. *Isachne globosa*. ***Lolium perenne***. *Lolium temulentum*. ***Oplismenus compositae***. ***Oryza sativa***. ***Panicum capillare***. ***Paspalum dilatatum***. ***Phalaris paradoxa***. *Phleum pratense*. ***Phragmites australis***. *Poa pratensis*. ***Saccharum officinarum***. ***Sacciolepis indica***. *Secale cereale*. ***Setaria italica***. ***Setaria viridis***. ***Sorghum bicolor***. *Triticum aestivum*. ***Zea mays***.

Sources of host-range data

Ford *et al.* (1972); Tosic *et al.* (1972); Penrose (1974); Rosenkranz (1978; 1987).

**Properties of particles in sap**

TIP: 50–60 °C. LIV: 2–4 days. DEP: $\log_{10}$ minus 2–4. Infectivity of sap not changed by treatment with di–ethyl ether. Leaf sap contains few virions.

**Purification**

Method: Moghal and Francki (1976); Shukla *et al.* (1989*a*); Bond and Pirone (1971).

**Particle morphology**

Virions filamentous; not enveloped; usually flexuous; with a clear modal length; of 730–755 nm; 13 nm wide. Axial canal obscure. Basic helix obscure.

**Physical properties**

One sedimenting component in purified preparations; sedimentation coefficient 148–176 *S*. Density 1.3327 g $cm^{-3}$ in CsCl.

**Biochemical properties**

Virions contain 5.5–6 % nucleic acid; 94–94.5 % protein.

*Genome* consists of RNA; single–stranded; unipartite; largest (or only) genome part 9.5 kb.

*Sequence database accession code(s):*

D00094 Gb(84)_vi:MSCVCP Sugarcane mosaic virus (SCMV), the 3'terminal 1782 nucleotides of the genome, encoding the vi

D00948 Em(40)_vi:MSCCP Gb(84)_vi:MSCCP Sugarcane mosaic virus mRNA for coat protein, 3'-terminal region. 8/92 1,330bp.

X05040 Em(40)_vi:POSCMVCA Gb(84)_vi:POSCMVCA Sugar cane mosaic virus RNA for capsid/nuclear inclusion protein (3' terminal genome). 7/89 1 3 sequences.

*Virion protein(s)* one; $M_r$ 34998 (328 residues (MDMV–B)). Method of preparation: Francki and McLean (1968).

**Cytopathology**

Virions found in mesophyll; in cytoplasm and in Golgi apparatus (bundles, scrolls, tubes and laminated aggregates, namely inclusion group III (Edwardson, 1974)). Inclusions present in infected cells; are pinwheels; they do not contain virions. Other cellular changes: alteration of chloroplasts organisation (*i.e.* small grana with few thylakoids and uniform distribution of intergrana (stromal) lamellae).

**Taxonomy and Relationships**

*Potyvirus: Potyviridae.*

Virus(es) with serologically related virions: using electro-blot immunoassay with cross-absorbed polyclonal antibodies against the N-terminus of the virion proteins that the potyviruses infecting sugarcane, maize and sorghum are of four distinct potyviruses: 1) Johnson grass mosaic potyvirus — SCMV–JG and maize dwarf mosaic virus strain O. 2) Maize dwarf mosaic potyvirus — MDMV–A, MDMV–D, MDMV–E and MDMV–F. 3) Sugarcane mosaic potyvirus — MDMV–B, SCMV–A, SCMV–B, SCMV–D, SCMV–E, SCMV–Sc, SCMV–Bc and SCMV–Sabi. 4) Sorghum mosaic potyvirus — SCMV–H, SCMV–I and SCMV–M (Shukla *et al.*, 1989).

**References**


Bond, W.P. and Pirone, T.P. (1971). *Phytopath. Z.* **71**: 56.

Brandes, J. (1919). *Tech. Bull. U.S. Dep. Agric.* **829**.

Edwardson, J.R. (1974). *Fla Agric. Exp. Stn Monog. No. 4.*

Ford, R.E. and Tosic, M. (1972). *Phytopath. Z.* **75**: 315.

Francki, R.I.B. and McLean, G.D. (1968). *Aust. J. biol. Sci.* **21**: 1311.

Gold, A.H. and Martin, J.P. (1955). *Phytopathology* **45**: 694.

Hill, J.H. and Benner, H.I. (1976). *Virology* **75**: 419.

Hill, J.H., Ford, R.E. and Benner, H.I. (1973). *J. gen. Virol.* **20**: 327.

Louie, R. and Knoke, J.K. (1975). *Pl. Dis. Reptr* **59**: 518.

MacKenzie, D.R., Wernham, C.C. and Ford, R.E. (1966). *Pl. Dis. Reptr* **50**: 814.

Moghal, S.M. and Francki, R.I.B. (1976). *Virology* **73**: 350.

Osborn, H.T. (1935). *Phytopathology* **25**: 160.

Penrose, L.J. (1974). *Aust. J. agric. Res.* **25**: 99.

Pirone, T.P. (1972). *CMI/AAB Descr. Pl. Viruses* No. 88, 4 pp.

Rosenkranz, E. (1978). *Phytopathology* **68**: 176.

Rosenkranz, E. (1980). *Phytopathology* **70**: 1056.

Rosenkranz, E. (1983). *Phytopathology* **73**: 1314.

Rosenkranz, E. (1987). *Phytopathology* **77**: 598.

Shepherd, R.J. (1965). *Phytopathology* **55**: 1250.

Shukla, D.D., Jilka, J., Tosic, M. and Ford, R.E. (1989*a*). *J. gen. Virol.* **70**: 13.

Shukla, D.D., Tosic, M., Jilka, J., Ford, R.E., Toler, R.W. and Langham, M.A.C. (1989*b*). *Phytopathology* **79**: 223.
Tosic, M. and Ford, R.E. (1972). *Phytopathology* **62**: 1466.
Tosic, M., Ford, R.E., Moline, H.E. and Mayhew, D.E. (1974). *Phytopathology* **64**: 312.

# Sugarcane streak monogeminivirus

**Data collated by F. Hughes, 1991.**

*Standard acronym:* SSV. *Strains:* Natal (type) and Mauritius.

First reported in *Saccharum officinarum*; from Natal, South Africa; by Storey (1925).

**Natural host range and symptoms**

Symptoms persist. Symptoms chlorotic spots and narrow chlorotic streaks.

*Cenchrus echinatus*, *Saccharum officinarum*—narrow, elongated chlorotic spots and streaks.

**Transmission**

Transmitted by a vector; an insect; *Cicadulina mbila, C. bipunctata bipunctata. C. bipunctella-zeae*; Cicadellidae. Principal natural vector(s): *Cicadulina mbila, C. bipunctata bipunctata, C. bipunctella-zeae.* Transmitted in a non–persistent manner, or in a semi–persistent manner. Virus retained when the vector moults; does not multiply in the vector; not transmitted congenitally to the progeny of the vector; does not require a helper virus for vector transmission; not transmitted by mechanical inoculation; not transmitted by grafting; not transmitted by contact between plants; not transmitted by seed; not transmitted by pollen.

**Geographical distribution**

Spreads in Benin, Cape Verde, Cote d'Ivoire, Egypt, India, Kenya, Malawi, Mauritius, Mozambique, Pakistan, Reunion, Sudan, Uganda, and Zimbabwe.

**Experimental host range**

Several (3–9) families susceptible. Experimentally infected plants mostly show mild chlorotic streaks and spots.

Diagnostically susceptible

*Saccharum officinarum* — yellow and white streaks.
*Cenchrus echinatus.*
*Eleusine indica* — mild permanent leaf streaks.
*Zea mays* — mild, short, pale streaks.

Diagnostically insusceptible

*Digitaria horizontalis.*

Maintenance and propagation hosts

*Saccharum officinarum, Zea mays.*

Monocotyledonous Hosts *(**Susceptible** or Insusceptible)*

***Zea mays.***

Sources of host-range data

McClean (1947); Pinner *et al.* (1988).

**Properties of particles in sap**

LIV: 7 days. Leaf sap contains many virions.

**Purification**

Method: Bock and Bailey (1989); Pinner *et al.* (1988).

**Particle morphology**

Virions geminate; not enveloped; 20 nm in diameter; 30 nm in length.

**Physical properties**

Sedimentation coefficient 76 *S*.

**Biochemical properties**

Virions 0 % lipid.

*Genome* consists of DNA; single–stranded; circular; unipartite; largest (or only) genome part 2.76 kb. Genomic nucleic acid isolated by Hayes *et al.* (1988). Base composition 27.7 % G; 21.8 % A; 23.1 % C; 28.1 % T. Nucleotide sequence references: GENBANK database accession number M33829.

*Sequence database accession code(s):*

D00597 Em(40)_vi:SSVL1 Gb(84)_vi:SSVL1 Sugarcane streak virus, L1'' gene encoding replication associated protein. 3/91 254bp.

M82918 Em(40)_vi:GESSV4ORF Gb(84)_vi:SSV4ORFS Sugarcane streak virus (strain Natal) DNA. 2/94 2,758bp.

S64567 Em(40)_vi:S64567 Gb(84)_un:S64567 11.16 kda orf(V1)...34.9 kda orf(C1) sugarcane streak virus SSV, Genomic Complete, 4 genes, 3 sequences.

*Features of the genome:* 2 virion sense ORFs encoding proteins of 11,000 and 27,000 Mr; the smaller ORF has homology with V1 of MSV (transport protein), and the larger with V2 of MSV (capsid protein). There are two ORFs in the complementary strand; C1 encoding a protein of 34,940 Mr and C2 of 17,000 Mr, these, like other monogeminiviruses, may be spliced to give c', which encodes the replicase protein of 41,000 Mr. There is a TATA box upstream of V1 and upstream of C1, and a AATAA consensus polyadenylation signal downstream of C2 and of V2. Sequence with homology to the core upstream activator sequence of MSV is upstream of C1. There are two intergenic regions. The large one is between the starts of C1 and V1 and contains promoter sequences and a conserved stem and loop structure with the TAATATTAC sequence common to all geminivirus genomes. The small intergenic region lies between the termini of V2 and C2 and contains AATAA sequences and a possible binding site of about 72 nucleotides in the genomic strand for an endogenous DNA primer molecule.

*Non–genomic nucleic acid* not found in the virions.

*Virion protein(s)* one; $M_r$ 28000; capsid protein.

**Replication**

Genome replicates in nuclei. Replication does not depend on a helper virus.

**Cytopathology**

Virions found in leaves, mesophyll, meristem and phloem; in nuclei. Inclusions present in infected cells; are unusual in shape; non-crystalline.

**Taxonomy and Relationships**

Monogeminivirus: *Geminiviridae*.

Virus(es) with serologically related virions: digitaria streak and maize streak viruses.

The Mauritius strain is milder and induces finer streak symptoms. Until recently sugarcane streak geminivirus was considered to be a strain of maize streak virus. Best tests for diagnosis: serology (Pinner and Markham, 1990; Peterschmitt *et al.*, 1991), transmission to maize and sugarcane and symptoms (Bock and Bailey, 1989).

**References**

Bock, K.R. and Bailey, R.A. (1989). In: *Diseases of Sugarcane; Major Diseases*, p. 322; eds C. Ricaud, B.T. Egan, A.G. Gillespie Jr. and G.C. Hughes. Elsevier/North Holland, Amsterdam, Oxford, New York, Tokyo.

Hatta, T. and Francki, R.I.B. (1979). *Virology* **92**: 428.

Hayes, R.J., MacDonald, H., Coutts, R.H.A. and Buck, K.W. (1988). *J. gen. Virol.* **69**: 1345.

Hughes, F.L., Rybicki, E.P., Kirby, R. and Von Wechmar, M.B. (1990). *Intervirology* **32**: 19.

McClean, A.P.D. (1947). *S.A. Dep. Agric. Sci. Bull.* **265**: 33 pp.

Peterschmitt, M., Reynaud, B., Sommermeyer, G. and Baudin, P. (1991). *Plant Dis.* **75**: 27.

Pinner, M.S., Markham, P.G., Markham, R.H. and Dekker, E.L. (1988). *Pl. Path.* **37**: 74.

Storey, H.H (1925). *S. Afr. Sugar Technol. Assoc.* **4**: 69.

# Sunflower crinkle (?) *umbravirus*

**Data collated by A.A. Brunt, 1995.**

*Synonyms:* sunflower rugose mosaic virus. *Standard acronym:* SuCV.

First reported in *Helianthus annuus*; from Kenya; by Martens *et al.* (1970).

**Geographical distribution**

Spreads in Kenya and Tanzania.

**Taxonomy and Relationships**

Possible *Umbravirus*.

**References**

Martens, J.W., Ravagan, M. and McDonald, W.C. (1970). *E.A. Agric. Forest. J.* **35**: 389.

Murant, A.F., Robinson, D.J. and Gibbs, M.J. (1995). *Arch. Virol.* Suppl. **10**: 388.

Singh, J.P. (1979). *Sunflower Newsl.* **3**: 13.

Theuri, J.M., Bock, K.R. and Woods, R.D. (1987). *Trop. Pest Manag.* **33**: 202.

# Sunflower mosaic (?) potyvirus

**Data collated by A.A. Brunt, 1988.**

*Synonyms:* helianthus mosaic virus, sunflower potyvirus. *Standard acronym:* SuMV.

First reported in *Helianthus annuus*; from Texas, U.S.A.; by Arnott and Smith (1967).

**Natural host range and symptoms**

Symptoms persist.

*Helianthus annuus* — mild systemic mosaic and mottling, mosaic then severe necrosis of leaves and stems.

**Geographical distribution**

Spreads in China and the USA (Texas).

**Experimental host range**

Few (<3) families susceptible.

Diagnostically susceptible

*Helianthus annuus* — systemic leaf mottling.

Maintenance and propagation hosts

*Helianthus annuus.*

Dicotyledonous Hosts *(**Susceptible** or Insusceptible)*

***Helianthus annuus.***

**Particle morphology**

Virions filamentous; not enveloped; usually flexuous.

**Cytopathology**

Virions found in leaves and mesophyll; in cytoplasm. Inclusions present in infected cells; are pinwheels; they do not contain virions.

**Taxonomy and Relationships**

Unassigned potyvirus.

**References**

Arnott, H.J. and Smith, K.M. (1967). *J. Ultrastruct. Res.* **19**: 173.

## Sunflower ringspot (?) *ilarvirus*

**Data collated by J.L. Dale, 1986.**

First reported in *Helianthus annuus*; from Australia; by J.L. Dale; unpublished information.

**Natural host range and symptoms**

Symptoms disappear soon after infection.

*Glycine max, Helianthus annuus, Lycopersicon esculentum* — mosaic, ringspots and/or line patterns.

**Transmission**

Virus transmitted by mechanical inoculation; transmitted by grafting; probably transmitted by seed.

**Geographical distribution**

Spreads in Australia.

**Experimental host range**

Several (3–9) families susceptible.

Diagnostically susceptible

*Pisum sativum* — systemic chlorosis.

*Chenopodium amaranticolor* — local lesions.

Maintenance and propagation hosts

*Pisum sativum.*

Assay hosts (**L**ocal lesions or **W**hole plants)

*Chenopodium amaranticolor* (L).

Dicotyledonous Hosts (***Susceptible*** *or Insusceptible)*

***Capsella bursa–pastoris****.* ***Chenopodium amaranticolor****.* ***Chenopodium quinoa****.* ***Cucurbita maxima****.* ***Galinsoga parviflora****.* ***Glycine max****.* ***Gomphrena globosa****.* ***Helianthus annuus****.* ***Lathyrus odoratus****.* ***Lens esculenta****.* ***Lycopersicon esculentum****.* ***Macrotyloma uniflorum****.* *Nicandra physaloides.* ***Nicotiana clevelandii****.* ***Nicotiana glutinosa****.* ***Petunia hybrida****.* ***Phaseolus vulgaris****.* ***Physalis floridana****.* ***Pisum sativum****.* ***Spinacia oleracea****.* ***Stellaria media****.* ***Trifolium incarnatum****.* ***Vicia faba****.* ***Vigna unguiculata****.*

**Properties of particles in sap**

TIP: 55–60 °C. LIV: 1–2 days. DEP: $\log_{10}$ minus 4–5. Leaf sap contains few virions.

**Particle morphology**

Virions isometric; 28 nm in diameter; rounded in profile; without a conspicuous capsomere arrangement.

**Physical properties**

One sedimenting component in purified preparations.

**Biochemical properties**

*Genome* consists of RNA; single–stranded; probably linear; of four parts, or of five parts; largest (or only) genome part the largest 3.38 kb; the 2nd largest 2.648 kb; the 3rd largest 2.507 kb; the 4th largest 1.409 kb; the 5th largest 0.817 kb. Infectivity retained when deproteinised with phenol or detergent.

*Non–genomic nucleic acid* found in the virions; is subgenomic *m*RNA.

*Virion protein(s)* one; $M_r$ 24000. Amino acid composition: J.L. Dale (unpublished data).

**Replication**

Replication does not depend on a helper virus.

**Cytopathology**

Virions found in mesophyll; in cytoplasm. Inclusions absent from infected cells.

**Taxonomy and Relationships**

Probably *Ilarvirus: Bromoviridae.*

Virus(es) with serologically unrelated virions: prunus necrotic ringspot, cucumber mosaic, alfalfa mosaic, prune dwarf and tobacco streak viruses.

# Sunflower yellow blotch (?) *umbravirus*

**Data collated by A.A. Brunt, 1995.**

*Synonyms:* sunflower yellow ringspot virus. *Standard acronym:* SuYBV.

First reported in *Helianthus annuus*; from Kenya; by Martens *et al.* (1970).

**Geographical distribution**

Spreads in Kenya and Tanzania.

**Taxonomy and Relationships**

Possible *Umbravirus.*

**References**

Martens, J.W., Ravagan, M. and McDonald, W.C. (1970). *E.A. Agric. Forest. J.* **35**: 389.

Murant, A.F., Robinson, D.J. and Gibbs, M.J. (1995). *Arch. Virol.* Suppl. **10**: 388.
Singh, J.P. (1979). *Sunflower Newsl.* **3**: 13.
Theuri, J.M., Bock, K.R. and Woods, R.D. (1987). *Trop. Pest Manag.* **33**: 202.

# Sunn-hemp mosaic *tobamovirus*

**Data collated by A.J. Gibbs, 1983. Revised 1985.**

*Synonyms:* bean strain of tobacco mosaic virus, cowpea strain of tobacco mosaic virus, cowpea chlorotic spot virus (Sharma and Varma, 1975; Varma, 1986), cowpea mosaic virus (Lister and Thresh, 1955; Kassanis and Varma, 1975), cowpea yellow mosaic virus, Crotalaria mucronata mosaic virus, dolichos enation mosaic virus (Capoor and Varma, 1948; Kassanis and Varma, 1975), sunn-hemp rosette virus (Varma, 1986; Verma and Awasthi, 1976; 1978). *Standard acronym:* SHMV.

First reported in *Crotalaria juncea*; from India; by Capoor and Varma (1948).

**Natural host range and symptoms**

Symptoms persist.

*Crotalaria juncea, Lablab purpureus, Vigna unguiculata, Mucuna aterrima* — systemic mosaic, mottling, stunting.

**Transmission**

Transmitted by means not involving a vector. Virus transmitted by mechanical inoculation; transmitted by contact between plants; probably transmitted by seed; not transmitted by pollen.

**Geographical distribution**

Spreads in in the tropics.

**Experimental host range**

Several (3–9) families susceptible.

Diagnostically susceptible

*Cajanus cajan* — systemic leaf mosaic and mottling.
*Crotalaria juncea* — systemic vein yellowing, mosaic, mottling and stunting.
*Cyamopsis tetragonoloba* — necrotic local lesions.
*Phaseolus vulgaris* — necrotic or chlorotic local lesions; systemic mosaic and vein yellowing.
*Vigna unguiculata* — systemic mosaic, mottling and vein yellowing.
*Nicotiana glutinosa* — necrotic local lesions; not systemic.
*Petunia hybrida, Solanum melongena, Tropaeolum majus* — few or no systemic symptoms.

Diagnostically insusceptible

*Catharanthus roseus, Cucumis sativus, Lablab purpureus, Vicia faba* (with most, but not all, isolates).

Maintenance and propagation hosts

*Phaseolus vulgaris.*

Assay hosts (**L**ocal lesions or **W**hole plants)

*Nicotiana glutinosa* (L), *N. tabacum* cv. Xanthi-*nc* (L), *Phaseolus vulgaris* (L), most cultivars.

Dicotyledonous Hosts *(**Susceptible** or Insusceptible)*

***Beta vulgaris**. **Cajanus cajan**. **Capsicum frutescens**. **Cassia tora**. Catharanthus roseus. **Crotalaria retusa**. **Crotalaria spectabilis**. Cucumis sativus. **Cyamopsis tetragonoloba**. **Datura stramonium**. **Datura tatula**. **Gomphrena globosa**. Hibiscus esculentus. Lablab purpureus. Macrotyloma uniflorum. **Nicandra physaloides**. **Nicotiana glutinosa**. **Nicotiana rustica**. **Nicotiana sylvestris**. **Nicotiana tabacum**. **Petunia hybrida**. **Phaseolus lunatus**. **Phaseolus vulgaris**. **Physalis peruviana**. **Pisum sativum**. Rumex acetosa. **Solanum melongena**. **Solanum nigrum**. **Spinacia oleracea**. **Tropaeolum majus**. Vicia faba. **Vigna unguiculata**. **Vigna unguiculata*** ssp. ***cylindrica**. **Vigna unguiculata*** ssp. ***sesquipedalis**. Zinnia elegans.*

Sources of host-range data

Capoor (1962).

**Properties of particles in sap**

TIP: 90–95 °C. LIV: 3000 days (or longer). DEP: $\log_{10}$ minus 6–7. Infectivity of sap not changed by treatment with di–ethyl ether. Leaf sap contains many virions.

**Purification**

Method: Kassanis and Varma (1975).

**Particle morphology**

Virions rod–shaped; not enveloped; usually straight; with a clear modal length; of 300 nm (also *c.* 25 nm); 17 nm wide. Axial canal obscure; 4 nm in diameter. Basic helix obvious; pitch of basic helix 2.3 nm.

**Physical properties**

Three sedimenting components in purified preparations; sedimentation coefficient of the fastest 187 *S*; of the other(s) 35 *S*, or 75 *S*. Density 1.318 g $cm^{-3}$ in CsCl.

**Biochemical properties**

Virions contain 5 % nucleic acid; 95 % protein; 0 % lipid.

*Genome* consists of RNA; single–stranded; linear; unipartite; largest (or only) genome part 6.5 kb. 5' terminus of RNA has a methylated nucleotide cap. Infectivity retained when deproteinised with proteases; retained when deproteinised with phenol or detergent. Poly A region absent. Genome has *t*RNA–like activity. Genome accepts valine.

*Sequence database accession code(s):* Em_vi:TOC30KCP — 30k and coat protein genes, 1,800bp.

*Non–genomic nucleic acid* found in the virions (short virions); is subgenomic *m*RNA (coat protein). Sub–genomic *m*RNA found in infected cells; found in short virions.

*Virion protein(s)* one; $M_r$ 18062. Amino acid composition: Rees and Short (1975).

**Replication**

Replication does not depend on a helper virus.

**Cytopathology**

Virions found in all parts of the host plant. Inclusions present in infected cells; are amorphous X–bodies.

**Taxonomy and Relationships**

*Tobamovirus.*

Virus(es) with serologically related virions: most tobamoviruses.

**References**

Capoor, S.P. and Varma, S.P. (1948). *Curr. Sci.* **17**: 57.

Capoor, S.P. (1962). *Phytopathology* **52**: 393.
Kassanis, B. and McCarthy, D. (1967). *J. gen. Virol.* **1**: 425.
Kassanis, B. and Varma, A. (1975). *CMI/AAB Descr. Pl. Viruses* No. 153, 4 pp.
Lister, R.M. and Thresh, J.M. (1955). *Nature, Lond.* **175**: 1047
Rees, M.N. and Short, M.N. (1975). *Biochim. biophys. Acta* **393**: 15.
Sharma, S.R. and Varma, A. (1975). *Indian Phytopath.* **28**: 292.
Varma, A. (1986). In: *The Plant Viruses*, Vol. 2, The Rod-Shaped Plant Viruses, p. 249; eds M.H.V. van Regenmortel and H. Fraenkel-Conrat. Plenum Press, New York.
Verma, H.N. and Awasthi, L.P. (1976). *Curr. Sci.* **45**: 642.
Verma, H.N. and Awasthi, L.P. (1978). *Phytopath. Z.* **92**: 83.

## Sweet clover latent (?) *nucleorhabdovirus*

**Data collated by G.M. Milbrath, 1982. Revised 1984.**

*Synonyms:* melilotus latent virus.

First reported in *Melilotus alba, M. officinalis*; from Chicago, U.S.A; by Kitajima *et al.* (1969).

**Natural host range and symptoms**

*Melilotus alba, M. officinalis* — no distinct symptoms, but often present with bean yellow mosaic potyvirus in plants with leaf chlorosis.

**Transmission**

Virus transmitted by grafting.

**Geographical distribution**

Spreads in the USA. Found, but with no evidence of spread, in Spain, the Netherlands and the U.S.A.

**Experimental host range**

Few (<3) families susceptible.

Diagnostically susceptible

*Melilotus alba, M. officinalis* — symptomless infection.

Maintenance and propagation hosts

*Melilotus alba, M. officinalis*.

Assay hosts (**L**ocal lesions or **W**hole plants)

*Melilotus alba* (W).

Dicotyledonous Hosts *(**Susceptible** or Insusceptible)*

***Melilotus alba***.

**Properties of particles in sap**

Leaf sap contains few virions.

**Particle morphology**

Virions rhabdo– or bullet–shaped; enveloped; of 250–300 nm; 80–100 nm wide.

**Cytopathology**

Virions found in vascular parenchyma in sieve tubes and young tracheids; in mitochondria and in the perinuclear space. Inclusions absent from infected cells.

**Taxonomy and Relationships**

Possible *Nucleorhabdovirus: Rhabdoviridae.*

**References**

Kitajima, E.W., Lauritis, J.A. and Swift, H. (1969). *J. Ultrastruct. Res.* **29**: 141.
Knudson, D.L. (1973). *J. gen. Virol.* **20**: 105.

# Sweet clover necrotic mosaic *dianthovirus*

**Data collated by K. Boswell, 1985.**

*Synonyms:* clover sweet necrotic mosaic virus. *Standard acronym:* SCNMV.

First reported in *Melilotus officinalis*; from Athabasca, Alberta, Canada; by Hiruki *et al.* (1981).

**Natural host range and symptoms**

Symptoms persist.

*Melilotus alba, M. officinalis* — chlorotic ringspots, then mottle, mosaic, leaf malformation, veinal necrosis especially during winter.

**Transmission**

Virus transmitted by mechanical inoculation.

**Geographical distribution**

Spreads in Canada.

**Experimental host range**

Several (3–9) families susceptible.

Diagnostically susceptible

*Chenopodium amaranticolor, C. quinoa, Gomphrena globosa, Vicia faba* — necrotic local lesions, not systemic.

*Nicotiana clevelandii, N. glutinosa* — systemic mosaic.

*Phaseolus vulgaris* cv Red Kidney — numerous chlorotic local lesions, irregular systemic necrosis and distortion.

*Tetragonia expansa* — systemic mosaic.

Diagnostically insusceptible

*Petunia hybrida.*

Maintenance and propagation hosts

*Phaseolus vulgaris* cv. Red Kidney.

Assay hosts (**L**ocal lesions or **W**hole plants)

*Chenopodium amaranticolor (L), Phaseolus vulgaris* cv. Red Kidney (W).

Dicotyledonous Hosts ***(Susceptible** or Insusceptible)*

*Beta vulgaris.* ***Chenopodium amaranticolor. Chenopodium quinoa. Cucumis sativus.*** *Datura stramonium.* ***Gomphrena globosa.*** *Lathyrus odoratus. Lycopersicon esculentum. Medicago sativa.* ***Melilotus alba. Nicotiana clevelandii. Nicotiana glutinosa. Nicotiana tabacum.*** *Petroselinum crispum. Petunia hybrida.* ***Phaseolus vulgaris. Pisum sativum. Tetragonia expansa.*** *Trifolium hybridum.* ***Trifolium pratense.*** *Trifolium repens.* ***Vicia faba. Vigna unguiculata.*** *Zinnia elegans.*

Sources of host-range data

Hiruki *et al.* (1984).

**Properties of particles in sap**

TIP: 90 °C. DEP: $\log_{10}$ minus 6–7.

**Purification**

Method: Hiruki *et al.* (1984).

**Particle morphology**

Virions isometric; not enveloped; 35 nm in diameter; angular in profile.

**Physical properties**

One sedimenting component in purified preparations; sedimentation coefficient 126 *S*. $A_{260}/A_{280}$ ratio 1.8.

**Biochemical properties**

Virions contain 20 % nucleic acid; 80 % protein; 0 % lipid.

*Genome* consists of RNA; single–stranded; of two parts; largest (or only) genome part the larger 3.75 kb; the 2nd largest 1.5 kb.

*Sequence database accession code(s):*

L07884 Em(40)_vi:SCWFUSPRO Gb(84)_vi:SCWFUSPRO Sweet clover necrotic mosaic virus unidentified genes, three complete cds's including fusion

S46027 Em(40)_un:S46027 Gb(84)_vi:S46027 SC38P2=36.5 kda protein sweet clover necrotic mosaic virus SCNMV, strain 38, Genomic RNA Com

S46028 Em(40)_un:S46028 Gb(84)_vi:S46028 SC59P2=36.5 kda protein sweet clover necrotic mosaic virus SCNMV, strain 59, Genomic RNA Com 3 sequences.

*Virion protein(s)* one; $M_r$ 38000; coat protein.

**Cytopathology**

Virions found in all parts of the host plant; in cytoplasm. Inclusions present in infected cells; are amorphous X–bodies.

**Taxonomy and Relationships**

*Dianthovirus*.

Virus(es) with serologically related virions: red clover necrotic mosaic virus (strains SW and E). Virus(es) with serologically unrelated virions: carnation ringspot viruses (distantly). They produce viable recombinants, but are considered to be different species (Hiruki, 1986).

**References**


Hiruki, C. (1986). *AAB/CMI Descr. Pl. Viruses*, No. 321, 4pp.

Hiruki, C., Okuno, T., Rao, D.V. and Chen, M.H. (1981). *Abstr. 5th Int. Cong. Virol. Strasbourg, France*, W22/08, p.235.

Hiruki, C., Rao, D.V., Chen, M.H., Okuno, T. and Figueiredo, G. (1984). *Phytopathology* **74**: 482.

Hiruki, C., Rao, D.V., Furuya, Y. and Figueiredo, G. (1983). *J. gen. Virol.*

Okuna, T., Hiruki, C., Rao, D.V. and Figueiredo, G. (1983). *J. gen. Virol.* **64**: 1907.

Rao, D.V. and Hiruki, C. (1985a). *Plant Dis.* **69**: 610.

Rao, D.V. and Hiruki, C. (1985b). *Plant Dis.* **69**: 959.

# Sweet potato *caulimovirus*

**Data collated by A.A Brunt, 1986; F. Nienhaus, 1987.**

First reported in *Ipomoea batatas*; from the U.K., in plants imported from Puerto Rico; by Atkey and Brunt (1987).

**Natural host range and symptoms**

*Ipomoea batatas* — no conspicuous symptoms.

**Transmission**

Virus not transmitted by mechanical inoculation; transmitted by grafting; not transmitted by contact between plants; not transmitted by seed.

**Geographical distribution**

Found, but with no evidence of spread, in the U.S.A., Puerto Rico, New Zealand, Kenya, Papua New Guinea, Solomon Islands, Australia, Tonga and Madeira.

Diagnostically susceptible

*Ipomoea setosa* — chlorotic veinal flecks and/or circular interveinal chlorotic spots.

Diagnostically insusceptible

*Nicotiana clevelandii, N. benthamiana, Chenopodium quinoa, Gomphrena globosa, Phaseolus vulgaris.*

Maintenance and propagation hosts

*Ipomoea setosa.*

**Properties of particles in sap**

Leaf sap contains few virions.

**Particle morphology**

Virions isometric; 50 nm in diameter; rounded in profile; without a conspicuous capsomere arrangement.

**Biochemical properties**

*Genome* consists of DNA; double–stranded.

*Virion protein(s)* one (major); $M_r$ 42000–44000.

**Cytopathology**

Virions found in vascular parenchyma; in cytoplasm. Inclusions present in infected cells; are unusual in shape; ovoid and up to 4 µm in diameter with a large central lacuna and usually several smaller peripheral lacunae; they do not contain virions. Other cellular changes: vascular parenchyma cells, which protrude into and sometimes completely occlude, adjacent xylem vessels.

**Taxonomy and Relationships**

Probable *Caulimovirus*.

Virus(es) with serologically unrelated virions: cauliflower mosaic, dahlia mosaic, carnation etched ring and soybean chlorotic mottle viruses.

**References**

Atkey, P.T. and Brunt, A.A. (1987). *J. Phytopath.* **118**: 370.

# Sweet potato feathery mottle *potyvirus*

**Data collated by R.N. Campbell, 1984; F. Nienhaus, 1987. Revised 1989 by J.W. Moyer.**

*Synonyms:* sweet potato chlorotic leaf spot virus, sweet potato internal cork virus, sweet potato russet crack virus, sweet potato virus A (Hollings *et al.*, 1976; Sheffield, 1957). *Standard acronym:* SPFMV. *Strains:* russet crack strain and sweet potato vein-clearing virus.

First reported in *Ipomoea batatas*; from Beltsville, Maryland, U.S.A; by Doolittle and Harter (1945).

**Natural host range and symptoms**

Symptoms disappear soon after infection (except for russet crack strain symptoms which persist on roots).

*Ipomoea* spp. — vein clearing, vein feathering, chlorotic spots, in cv. Jersey also causes root necrosis if infected by russet crack strain.

**Transmission**

Transmitted by a vector; an insect; *Aphis gossypii, A. craccivora, Lipaphis erysimi, Myzus persicae*; Aphididae. Transmitted in a non–persistent manner. Virus transmitted by mechanical inoculation; transmitted by grafting; not transmitted by contact between plants; not transmitted by seed; not transmitted by pollen.

**Geographical distribution**

Probably distributed worldwide. Found, but with no evidence of spread, in Louisiana, U.S.A.

**Experimental host range**

Few (<3) families susceptible.

Diagnostically susceptible

*Ipomoea incarnata, I. nil* — systemic vein clearing, vein banding, ringspots.

*I. setosa* — vein clearing followed by remission.

Maintenance and propagation hosts

*Ipomoea batatas, I. incarnata, I. nil, I. purpurea, I. setosa, Nicotiana benthamiana* (some strains).

Assay hosts (**L**ocal lesions or **W**hole plants)

*Ipomoea setosa* (W), *I. nil* (W), *Chenopodium amaranticolor* (L) for some isolates in certain environments.

Dicotyledonous Hosts *(**Susceptible** or Insusceptible)*

*Beta vulgaris. Brassica campestris. Brassica juncea. Chenopodium amaranticolor. Chenopodium capitatum. Chenopodium quinoa. Cucumis sativus. Cucurbita pepo. Datura stramonium. Gomphrena globosa. Lycopersicon esculentum.* ***Nicotiana benthamiana****. Nicotiana glutinosa. Nicotiana tabacum. Phaseolus vulgaris. Raphanus sativus. Vigna unguiculata.*

Sources of host-range data

Campbell *et al.* (1974); Moyer and Kennedy (1978).

**Properties of particles in sap**

TIP: 60–65 °C. LIV: 0.3–0.5 days (7–12 hours). DEP: $\log_{10}$ minus 3–4. Leaf sap contains few virions.

**Purification**

Method: Moyer and Kennedy (1978); Cali and Moyer (1981); Cadena-Hinojosa and Campbell (1981*a*); Loebenstein *et al.* (1988).

**Particle morphology**

Virions filamentous; not enveloped; usually flexuous; with a clear modal length; of 830–850 nm. Axial canal obscure. Basic helix obscure.

**Physical properties**

One sedimenting component in purified preparations.

**Biochemical properties**

*Genome* consists of RNA; single–stranded; linear; unipartite; largest (or only) genome part 11.6 kb. Genomic nucleic acid isolated by Moyer and Cali (1985). Poly A region present. Additional factor not required for infectivity.

*Sequence database accession code(s):*

D16664 Em(40)_vi:SPFNIBCP Gb(84)_vi:SPFNIBCP Sweet potato feathery mottle virus 3' terminal region. 4/94 2,290bp.

S43450 Em(40)_un:S43450 Gb(84)_vi:S43450 capsid protein sweetpotato feathery mottle virus SPFMV, RC, mRNA Partial, 1196 nt. 2/93 1,1

S43451 Em(40)_un:S43451 Gb(84)_vi:S43451 capsid protein sweetpotato feathery mottle virus SPFMV, C, mRNA Partial, 1196 nt. 2/93 1,19

S69825 Gb(84)n:S69825 NIb-CP region: polyprotein...CP=coat/capsid protein sweet potato feathery mottle virus SPFMV

S69826 Gb(84)n:S69826 NIa-NIb region: polyprotein...NIb=RNA polymerase/nuclear inclusion b protein.

*Virion protein(s)* one; $M_r$ 38000. Method of preparation: Moyer and Cali (1985).

**Cytopathology**

Virions found in all parts of the host plant; in cytoplasm. Inclusions present in infected cells; are pinwheels; they do not contain virions (Lawson *et al.*, 1971; Nome *et al.*, 1974).

**Taxonomy and Relationships**

*Potyvirus: Potyviridae.*

Best tests for diagnosis: graft sweet potato to *Ipomoea setosa* and check it serologically.

**References**


Cadena-Hinojosa, M.A. and Campbell, R.N. (1981*a*). *Phytopathology* **71**: 1086.

Cadena-Hinojosa, M.A. and Campbell, R.N. (1981*b*). *Plant Dis.* **65**: 412.

Cali, B.B. and Moyer, J.W. (1981). *Phytopathology* **71**: 302.

Campbell, R.N., Hall, D.H. and Mielinis, N.M. (1974). *Phytopathology* **64**: 210.

Clark, C.A., Derrick, K.S., Pace, C.S. and Watson, B. (1986). *Plant Dis.* **70**: 931.

Cohen, J., Salomon, R. and Loebenstein, G. (1988). *Phytopathology* **78**: 809.

Doolittle, S.P. and Harter, L.L. (1945). *Phytopathology* **35**: 695.

Esbenshade, P.R. and Moyer, J.W. (1982). *Plant Dis.* **66**: 911.

Hollings, M., Stone, O.M. and Bock, K.R. (1976). *Ann. appl. Biol.* **82**: 511.

Kennedy, G.G. and Moyer, J.W. (1982). *J. econ. Ent.* **75**: 130.

Lawson, R.H., Hearon, S.S. and Smith, F.F. (1971). *Virology* **46**: 453.

McLean, D.L. (1959). *J. econ. Ent* **52**: 1057.

Moyer, J.W. and Cali, B.B. (1985). *J. gen. Virol.* **66**: 1185.

Moyer, J.W. and Kennedy, G.G. (1978). *Phytopathology* **68**: 998.

Nome, S.F., Shalla, T.A. and Petersen, L.J. (1974). *Phytopath. Z.* **79**: 169.
Sheffield, F.M.L. (1957). *Phytopathology* **47**: 582.

# Sweet potato G *potyvirus*

First reported in *Ipomoea batatas*; by Colinet *et al.* (1994).

*Sequence database accession code(s):*

X76944 Em(40)_vi:SPGVCPG Gb(84)_vi:SPGVCPG Sweet potato G virus coat protein gene. 1/94 1,288bp. 1 sequence.

**Taxonomy and Relationships**

*Potyvirus: Potyviridae.*

The sequence of the 3'terminal region is more than 80% identical to that of sweet potato feathery mottle potyvirus.

**References**

Colinet, D., Kummert, J. and Lepoivre, P. (1994). *Arch. Virol.* **139**: 327.

# Sweet potato latent (?) potyvirus

**Data collated by A.A. Brunt, S. Molyneux, 1986, F. Nienhaus, 1987.**

*Synonyms:* sweet potato virus N. *Standard acronym:* SwPLV.

First reported in *Ipomoea batatas* cv. Tainung 63; from Taiwan; by Liao *et al.* (1979).

**Natural host range and symptoms**

Symptoms disappear soon after infection.

*Ipomoea batatas* — most cultivars infected symptomlessly.

**Transmission**

Virus transmitted by mechanical inoculation; transmitted by grafting; not transmitted by contact between plants; not transmitted by seed.

**Ecology and control**

Studies reported by Liao *et al.* (1979); Liao *et al.* (1982).

**Geographical distribution**

Spreads in Taiwan.

**Experimental host range**

Several (3–9) families susceptible.

Diagnostically susceptible

*Nicotiana benthamiana* — systemic mosaic and stunting.

*Chenopodium quinoa, C. amaranticolor* — brown necrotic local lesions; no systemic infection.

*Ipomoea nil* — systemic mottle on *c.* 6–8 leaves but symptomless on subsequent leaves.

*Nicotiana tabacum cv. Dutch 'A'* — chlorotic rings.

*Nicotiana megalosiphon* — systemic deformation and mottling.

*Nicotiana clevelandii* — systemic pin-prick chlorotic local lesions.

Diagnostically insusceptible

*Nicotiana glutinosa, N. debneyi, Chenopodium murale, C. capitatum.*

Maintenance and propagation hosts

*Nicotiana benthamiana, N. clevelandii.*

Assay hosts (**L**ocal lesions or **W**hole plants)

*Chenopodium quinoa* (L), *Chenopodium amaranticolor* (L), *Nicotiana benthamiana* (W).

Dicotyledonous Hosts ***(Susceptible** or Insusceptible)*

*Brassica campestris* ssp. *pekinensis. Brassica campestris* ssp. *rapa. Calendula officinalis. Capsicum annuum. Celosia cristata.* ***Chenopodium amaranticolor****. Chenopodium capitatum.* ***Chenopodium murale****.* ***Chenopodium quinoa****. Cucumis metuliferus. Cucumis sativus. Glycine max. Gomphrena globosa.* ***Ipomoea nil****.* ***Ipomoea setosa****. Lycopersicon esculentum. Momordica balsamina.* ***Nicotiana benthamiana****.* ***Nicotiana clevelandii****.* ***Nicotiana debneyi****. Nicotiana glutinosa.* ***Nicotiana megalosiphon****. Nicotiana rustica. Nicotiana sylvestris.* ***Nicotiana tabacum****. Petunia hybrida. Phaseolus vulgaris. Physalis floridana. Pisum sativum. Raphanus sativus. Spinacia oleracea. Tetragonia expansa. Trifolium incarnatum. Trifolium pratense. Trifolium repens. Vigna unguiculata.*

Monocotyledonous Hosts ***(Susceptible** or Insusceptible)*

*Avena byzantina. Avena sativa. Hordeum vulgare. Sorghum bicolor. Triticum aestivum.*

Sources of host-range data

Liao *et al.* (1979).

**Properties of particles in sap**

TIP: 60–65 °C (in *Nicotiana benthamiana* sap). LIV: 0.6–1 days (at 25°C). DEP: $\log_{10}$ minus 2–3. Leaf sap contains few virions.

**Purification**

Method: Liao *et al.* (1979).

**Particle morphology**

Virions filamentous; usually flexuous; with a clear modal length; of 700–750 nm. Axial canal obscure. Basic helix obvious.

*Sequence database accession code(s):*

Z26135 Em(40)_vi:SPLVCCP Gb(84)_vi:SPLVCCP Sweet potato latent virus (China) coat protein gene (partial). 9/93 423bp.

Z26136 Em(40)_vi:SPLVTCP Gb(84)_vi:SPLVTCP Sweet potato latent virus (Taiwan) coat protein gene (partial). 9/93 426bp.

X84011 Em(43)_vi:Splvchicp Gb(89)_vi:Splvchicp Sweet potato latent virus mRNA for coat protein (Chinese). 1/95 1,072bp.

X84012 Em(43)_vi:Splvtwacp Gb(89)_vi:Splvtwacp Sweet potato latent virus mRNA for coat protein (Taiwan). 1/95 1,076bp.

*Virion protein(s)* one; $M_r$ 36000. Method of preparation: Brunt: proteins identified by SDS polyacrylamide gel electrophoresis.

**Replication**

Replication does not depend on a helper virus.

**Cytopathology**

Virions found in all parts of the host plant. Inclusions present in infected cells; are crystals in the nucleus and pinwheels.

**Taxonomy and Relationships**

Unassigned potyvirus.

Virus(es) with serologically unrelated virions: sweet potato feathery mottle virus and several other potyviruses.

**References**

Liao, C.H., Chien, K., Chung, M.L., Chiu, R.J. and Han, Y.H. (1979). *J. agric. Res. China* **28**: 127.

Liao, C.H., Tsay, H.S. and Lu, Y.C. (1982). *J. agric. Res. China* **31**: 239.

# Sweet potato leaf curl (?) *badnavirus*

**Data collated by A.A. Brunt and S. Green, 1988.**

First reported in *Ipomoea batatas*; from Taiwan; by Liao *et al.* (1979).

**Natural host range and symptoms**

Symptoms vary seasonally.

*Ipomoea batatas* — upward curling of leaves and vein swelling on young plants, but later few or no symptoms.

**Transmission**

Transmitted by a vector; an insect; *Bemisia tabaci*; Aleyrodidae. Transmitted in a persistent manner. Virus not transmitted by mechanical inoculation; transmitted by grafting; not transmitted by contact between plants; not transmitted by seed.

**Geographical distribution**

Spreads in Japan and Taiwan.

**Experimental host range**

Few (<3) families susceptible.

Diagnostically susceptible

*Ipomoea batatas* cv. Tainung 63, *Ipomoea nil* — curling of young leaves.

Dicotyledonous Hosts *(**Susceptible** or Insusceptible)*

***Ipomoea nil*.**

Sources of host-range data

Chung *et al.* (1985).

**Particle morphology**

Virions bacilliform; not enveloped.

**Cytopathology**

Virions found in phloem; in cytoplasm.

**Taxonomy and Relationships**

Possible *Badnavirus*.

**References**

Chung, M.L., Liao, C.H., Chen, M.J. and Chiu, R.J. (1985). *Pl. Prot. Bull., Taiwan* **27**: 333.

Liao, C.H., Chien, I.C., Chung, M.L., Chiu, R.J. and Han, Y.H. (1979). *J. agric. Res. China* **28**: 121.

Yamashita, S., Doi, Y. and Shin, K.A. (1984). *Ann. Phytopath. Soc. Japan.* **50**: 438.

# Sweet potato mild mottle ipomovirus

**Data collated by A.A. Brunt, S. Molyneux, 1986; F. Nienhaus, 1987.**

*Synonyms:* sweet potato T virus, sweet potato B virus (Hollings *et al.*, 1976a; Sheffield, 1957). *Standard acronym:* SPMMV.

First reported in *Ipomoea batatas*; from Kenya, Uganda and Tanzania; by Sheffield, 1957; Hollings *et al.* (1971).

**Natural host range and symptoms**

Symptoms persist.

*Ipomoea batatas* — mild leaf mottling and stunting in intolerant cultivars, but symptomless infection in others. Some cultivars are immune.

**Transmission**

Transmitted by a vector; an insect; *Bemisia tabaci*; Aleyrodidae. Transmitted in a persistent manner. Virus transmitted by mechanical inoculation; transmitted by grafting; not transmitted by contact between plants; not transmitted by seed.

**Ecology and control**

Studies reported by Hollings *et al.* (1976*a*); Wambugu (1990).

**Geographical distribution**

Spreads in Burundi, Kenya, Tanzania, and Uganda.

**Experimental host range**

Many (>9) families susceptible.

Diagnostically susceptible

*Nicotiana glutinosa, N. tabacum* — vein clearing, leaf curling and deformation.
*Chenopodium quinoa* — local lesions; not systemic.

Maintenance and propagation hosts

*Nicotiana glutinosa, N. clevelandii, N. tabacum.*

Assay hosts (**L**ocal lesions or **W**hole plants)

*Chenopodium quinoa* (L).

Dicotyledonous Hosts *(**Susceptible** or Insusceptible)*

*Ageratum conyzoides*. *Amaranthus caudatus*. *Antirrhinum majus*. *Apium graveolens*. *Arachis hypogaea*. ***Atriplex hortensis***. *Bellis perennis*. ***Beta vulgaris***. *Brassica campestris* ssp. *pekinensis*. *Brassica campestris* ssp. *rapa*. *Brassica juncea*. *Brassica oleracea* ssp. *capitata*. *Calendula officinalis*. *Capsicum frutescens*. *Catharanthus roseus*. ***Celosia argentea***. *Cheiranthus cheiri*. *Chenopodium capitatum*. ***Chenopodium murale***. *Chrysanthemum morifolium*. *Cucumis sativus*. *Cucurbita maxima*. *Cucurbita moschata*. ***Datura stramonium***. *Dianthus barbatus*. *Dianthus caryophyllus*. *Emilia sagittata*. *Fagopyrum esculentum*. *Glycine max*. ***Gomphrena globosa***. *Helianthus annuus*. *Hibiscus esculentus*. ***Hyoscyamus niger***. *Lactuca sativa*. ***Lathyrus odoratus***. ***Lycopersicon esculentum***. ***Montia perfoliata***. *Myosotis sylvatica*. ***Nicandra physaloides***. ***Nicotiana benthamiana***. ***Nicotiana debneyi***. ***Nicotiana glutinosa × clevelandii***. ***Nicotiana megalosiphon***. ***Nicotiana rustica***. *Ocimum basilicum*. ***Petunia hybrida***. *Phaseolus vulgaris*. ***Phlox drummondii***. *Physalis floridana*. *Phytolacca americana*. *Pisum sativum*. ***Primula malacoides***. ***Saponaria vaccaria***. *Senecio vulgaris*. ***Silene armeria***. *Solanum melongena*. ***Solanum nigrum***. *Solanum tuberosum*. ***Spinacia oleracea***. *Stellaria media*.

***Tetragonia expansa*. *Torenia fournieri*. *Trifolium incarnatum*.** *Tropaeolum majus*. *Verbesina encelioides*. *Vicia faba*. *Vigna unguiculata*. ***Zinnia elegans*.**

Monocotyledonous Hosts (***Susceptible*** *or Insusceptible)*

*Triticum aestivum*. *Zea mays*.

Sources of host-range data

Hollings *et al.* (1976).

**Properties of particles in sap**

TIP: 60 °C (in *Nicotiana clevelandii* sap). LIV: 3–7 days (at 18°C), or 42–49 days (at 2°C). DEP: $\log_{10}$ minus 2–3.

**Purification**

Method: Hollings *et al.* (1976 *a,b*).

**Particle morphology**

Virions filamentous; of 800–950 nm.

**Physical properties**

One sedimenting component in purified preparations; sedimentation coefficient 155 *S*.

**Biochemical properties**

Virions contain 5 % nucleic acid; 95 % protein.

*Genome* consists of RNA; single-stranded.

*Virion protein(s)* one; $M_r$ 37700; coat protein.

**Cytopathology**

Virions found in mesophyll and epidermis; in cytoplasm. Inclusions present in infected cells; are pinwheels; they do not contain virions.

**Taxonomy and Relationships**

Ipomovirus: *Potyviridae*.

Virus(es) with serologically unrelated virions: bean yellow mosaic, carnation vein mottle, clover yellow vein, henbane mosaic, hippeastrum mosaic, iris mild mosaic, pepper veinal mottle, potato Y, sweet potato feathery mottle, tobacco etch, turnip mosaic, watermelon mosaic 2 and narcissus yellow stripe viruses.

**References**

Hollings, M., Stone, O.M. and Bock, K.R. (1971). *Rep. Glasshouse Crops Res. Inst.* 1970, p. 155.

Hollings, M., Stone, O.M. and Bock, K.R. (1976*a*). *Ann. appl. Biol.* **82**: 511.

Hollings, M., Stone, O.M. and Bock, K.R. (1976*b*). *CMI/AAB Descr. Pl. Viruses* No. 162, 4 pp.

Sheffield, F.M.L. (1957). *Phytopathology* **47**: 582.

Wambugu, F. (1990). Ph.D. Thesis, pp. 271. University of Bath, U.K.

# Sweet potato (?) *phytoreovirus*

**Data collated by A.A. Brunt, 1990.**

First reported in *Ipomoea aquatica*; from material from Thailand, tested in the U.K.; by A.A. Brunt; unpublished information.

**Natural host range and symptoms**

Symptoms mild leaf chlorosis.

**Geographical distribution**

Found, but with no evidence of spread, in isolated in the U.K. from plants from Thailand.

Dicotyledonous Hosts (***Susceptible*** *or Insusceptible)*

*Ipomoea setosa.*

**Taxonomy and Relationships**

Possible *Phytoreovirus: Reoviridae.*

# Sweet potato ringspot (?) *nepovirus*

**Data collated by J.D. Brown and A.A. Brunt, 1988.**

First reported in *Ipomoea batatas* cv. Wanmun; from the U.K. in a sweet potato imported from Papua New Guinea; by Brown *et al.* (1988).

**Natural host range and symptoms**

Symptoms disappear soon after infection.
*Ipomoea batatas* — occasional chlorotic ringspots, then symptomless.

**Transmission**

Virus transmitted by mechanical inoculation; transmitted by grafting; not transmitted by contact between plants; transmitted by seed.

**Geographical distribution**

Spreads in Papua New Guinea.

**Experimental host range**

Several (3–9) families susceptible.

Diagnostically susceptible

*Ipomoea batatas* cvs Jasper and Rose Centennial — chlorotic ringspots (after graft inoculation), then symptomless.
*Ipomoea setosa* — faint systemic chlorotic leaf mottling.
*Nicotiana megalosiphon* — necrotic rings, systemic leaf mottling.
*Nicotiana benthamiana* — systemic leaf deformation.
*Glycine max* — systemic leaf chlorosis and stunting.
*Chenopodium quinoa, C. murale* — transient systemic chlorosis.

Diagnostically insusceptible

*Nicotiana debneyi, Trifolium incarnatum, Vicia faba, Pisum sativum, Brassica pekinensis.*

Maintenance and propagation hosts

*Nicotiana megalosiphon, N. benthamiana, Ipomoea setosa.*

Assay hosts (**L**ocal lesions or **W**hole plants)

*Nicotiana megalosiphon* (W), *Chenopodium murale* (W), *C. quinoa* (W).

Dicotyledonous Hosts (***Susceptible*** *or Insusceptible)*

*Brassica campestris* ssp. *pekinensis.* ***Capsicum annuum. Chenopodium amaranticolor. Chenopodium capitatum. Chenopodium murale. Chenopodium quinoa.***

***Cucumis sativus*. *Datura stramonium*. *Glycine max*. *Gomphrena globosa*. *Hibiscus esculentus*. *Ipomoea nil*. *Ipomoea setosa*. *Nicotiana benthamiana*. *Nicotiana clevelandii*.** *Nicotiana debneyi*. ***Nicotiana glutinosa*. *Nicotiana megalosiphon*. *Nicotiana tabacum*. *Phaseolus vulgaris*.** *Pisum sativum*. ***Tetragonia expansa*.** *Trifolium incarnatum*. ***Vigna unguiculata*.**

Sources of host-range data

Brunt and Brown (unpublished information).

**Properties of particles in sap**

Leaf sap contains few virions.

**Purification**

Method: Brunt and Brown (unpublished information).

**Particle morphology**

Virions isometric; not enveloped; 28 nm in diameter; angular in profile; with a conspicuous capsomere arrangement.

**Physical properties**

Sedimentation coefficient 130 *S* (B); of the other(s) 87 *S* (M), or 47 *S* (T).

**Biochemical properties**

*Genome* consists of RNA; single–stranded; linear; of two parts; largest (or only) genome part the larger 8.4 kb; the 2nd largest 6.7 kb. Genomic nucleic acid isolated by Brunt *et al.* (1989).

*Virion protein(s)* one; $M_r$ 56500; coat protein. Method of preparation: Brunt: dissociation in 0.625 M Tris-HCl pH 6.8 containing 2% SDS and 5% 2-mercaptoethanol for 3 minutes at 100°C.

**Cytopathology**

Virions found in leaves and roots.

**Taxonomy and Relationships**

Possible *Nepovirus: Comoviridae*.

Virus(es) with serologically unrelated virions: arabis mosaic, cassava green mottle, crimson clover latent, hibiscus latent ringspot, cacao necrosis, grapevine chrome mosaic, cherry leaf roll, peach rosette mosaic, raspberry ringspot, strawberry latent ringspot, tomato black ring, tomato ringspot and tobacco ringspot viruses.

**References**

Brown, J.D., Brunt, A.A. and Hugo, S.A. (1988). *Rep. Glasshouse Crops Res. Inst. for 1986–1987*, p. 104.

# Sweet potato sunken vein (?) *closterovirus*

**Data collated by A.A. Brunt and G. Loebenstein, 1995.**

*Synonyms:* possibly sweet potato whitefly transmitted agent (Schaefers and Terry, 1976; Rossel and Thottapilly, 1985).

First reported in *Ipomoea batatas*; from Israel; by Cohen *et al.* (1992).

**Natural host range and symptoms**

*Ipomoea batatas* — symptomless or mild vein yellowing, some sunken secondary veins on adaxial leaf surfaces and swollen veins on abaxial surfaces. Very severe diseases occur when the virus is in mixed infections with sweet potato feathery mottle virus.

**Transmission**

Transmitted by a vector; an insect; *Bemisia tabaci*; Aleyrodidae. Transmitted in a semi–persistent manner. Virus not transmitted by mechanical inoculation; transmitted by grafting; not transmitted by contact between plants.

**Geographical distribution**

Spreads in Israel and Nigeria (possibly).

**Experimental host range**

Few (<3) families susceptible.

Diagnostically susceptible

*Ipomoea hederacea, I. setosa, I. trichocarpa, I. trifida* — leaf rolling and cupping.

Diagnostically insusceptible

*Brassica campestris* ssp. *pekinensis, Chenopodium murale, C. quinoa, Cucumis sativus, Datura stramonium, Petunia hybrida, Physalis floridana.*

Dicotyledonous Hosts *(**Susceptible** or Insusceptible)*

*Brassica campestris* ssp. *pekinensis. Chenopodium murale. Chenopodium quinoa. Cucumis sativus. Datura stramonium. Daucus carota. Helianthus annuus.* ***Ipomoea nil. Ipomoea setosa****. Lactuca sativa. Lycopersicon esculentum.* ***Nicotiana benthamiana. Nicotiana clevelandii****. Petunia hybrida. Physalis floridana. Raphanus sativus. Zinnia elegans.*

**Purification**

Method: Cohen *et al.* (1992).

**Particle morphology**

Virions filamentous; usually flexuous; of 850 nm; 12 nm wide.

*Sequence database accession code(s):*

X80995 Em(43)_vi:Spsvvcp Gb(89)_vi:Spsvvcp Sweet potato sunken vein virus genomic RNA for coat protein. 4/95 1,182bp.

3 virus specified dsRNA species found in infected cells. Size of largest virus specified dsRNA 10.5 kbp; 2nd largest 7 kbp; 3rd largest 5 kbp.

*Virion protein(s)* one; $M_r$ 34000; coat protein.

**Cytopathology**

Other cellular changes: phloem cells have fibril-containing vesicles.

**Taxonomy and Relationships**

Possible *Closterovirus.*

Virus(es) with serologically unrelated virions: lettuce infectious yellows virus.

**Comments**

The whitefly-transmitted component of the sweet potato virus disease complex occurring in Nigeria has properties very similar to those of sweet potato sunken vein virus; the genome of the former has also been shown to be of 9 kb.

## References

Cohen, J. and Loebenstein, G. (1991) *Plant Dis.* **75**: 291.
Cohen, J., Frank, A., Vetten, H.J., Lesemann, D.-E. and Loebenstein, G. (1992). *Ann. appl. Biol.* **121**: 257.
Rossel, H.W. and Thottappilly, G. (1985). *Virus Diseases of Important Food Crops in Tropical Africa*, pp. 10–13. IITA, Ibadan, Nigeria.
Schaefers, G.A. and Terry, E.R. (1976). *Phytopathology* **66**: 642.
Winter, S., Purac, A., Leggett, F., Frison, E.A., Rossel, H.W. and Hamilton, R.I. (1992). *Phytopathology* **82**: 869.

# Sweet potato vein mosaic (?) potyvirus

**Data collated by F. Nienhaus, 1987.**

*Standard acronym:* SPVMV.

First reported in *Ipomoea batatas*; from Argentina; by Nome (1973).

## Natural host range and symptoms

*Ipomoea batatas* — vein-clearing, mosaic, twisting of leaves and stunting.

## Transmission

Transmitted by a vector; an insect; *Myzus persicae*; Aphididae. Transmitted in a non–persistent manner. Virus transmitted by mechanical inoculation.

## Geographical distribution

Spreads in Argentina.

## Experimental host range

Several (3–9) families susceptible.

Diagnostically susceptible

*Ipomoea batatas* and other *Ipomoea* spp. — systemic vein-clearing and mosaic.

Dicotyledonous Hosts *(**Susceptible** or Insusceptible)*

***Ipomoea nil. Ipomoea setosa.***

## Particle morphology

Virions filamentous; not enveloped; usually flexuous; with a clear modal length; of 761 nm.

## Cytopathology

Virions found in all parts of the host plant; in cytoplasm. Inclusions present in infected cells; are pinwheels and unusual in shape; scrolls and laminated aggregates; they do not contain virions.

## Taxonomy and Relationships

Probably *Potyvirus: Potyviridae.*

## References

Nome, S.F. (1973). *Phytopath. Z.* **77**: 44.
Nome, S.F., Shalla, T.A. and Petersen, L.J. (1974). *Phytopath. Z.* **79**: 169.

# Sweet potato yellow dwarf (?) *ipomovirus*

**Data collated by A.A. Brunt, 1988.**

*Synonyms:* whitefly transmissible sweet potato virus. *Standard acronym:* SPYDV.

First reported in *Ipomoea batatas*; from Taiwan; by Liao *et al.* (1979).

### Natural host range and symptoms

Symptoms persist.

*Ipomoea batatas*—stunting, chlorosis (chlorophyll content decreased by 50%), poor roots and tubers.

### Transmission

Transmitted by a vector; an insect; *Bemisia tabaci*; Aleyrodidae. Transmitted in a persistent manner. Virus transmitted by mechanical inoculation; transmitted by grafting.

### Geographical distribution

Spreads in Taiwan.

### Experimental host range

Several (3–9) families susceptible.

Diagnostically susceptible

*Ipomoea batatas* cv. Tainung 63. — severe systemic leaf chlorosis and stunting.

*Chenopodium* spp., *Gomphrena globosa, Datura stramonium, Cassia occidentalis, Sesamum indicum.*

Dicotyledonous Hosts *(**Susceptible** or Insusceptible)*

***Cassia occidentalis*. *Datura stramonium*. *Gomphrena globosa*. *Sesamum indicum*.**

Additional hosts

***Sesamum orientale*.**

Sources of host-range data

Chung *et al.* (1986).

### Properties of particles in sap

TIP: 85–90 °C (in sweet potato sap). LIV: more than 7 days. DEP: $\log_{10}$ minus 6–7.

### Particle morphology

Virions filamentous; not enveloped; usually flexuous; with a clear modal length; of 750 nm.

### Physical properties

One sedimenting component in purified preparations.

*Virion protein(s)* one; $M_r$ 33000; coat protein.

### Cytopathology

Virions found in leaves; in cytoplasm. Inclusions present in infected cells; are pinwheels.

### Taxonomy and Relationships

Possible ipomovirus: *Potyviridae*.

### References

Chung, M.L., Hsu, Y.H., Chen, M.J. and Chiu, R.J. (1986). In: *Pl. Virus Diseases of Hort. Crops in the Tropics and Sub-tropics*, No. 33, p. 84. FFTC Book Series.

Liao, C.H., Chien, I.C., Chung, M.L., Chin, R.J. and Han, Y.H. (1979). *J. agric. Res. China* **28**: 121.

## Swordbean distortion mosaic *potyvirus*

**Data collated by V.R. Mali, 1987. Revised 1989.**

*Synonyms:* swordbean mosaic virus. *Standard acronym:* SBDMV.

First reported in *Canavalia ensiformis*; from Parbhani, Maharashtra State, India; by Mali (1979).

**Natural host range and symptoms**

*Canavalia ensiformis, C. virosa, C. ensiformis x C. virosa* — mosaic, leaf deformation, shoe-stringing, small deformed pods, stunting.

*Canavalia gladiata* — mosaic, mottling and slight leaf deformation.

**Transmission**

Transmitted by a vector; an insect; *Aphis gossypii, Dactynotus sonchi, Myzus persicae, Rhopalosiphum maidis*; Aphididae. Transmitted in a non–persistent manner. Virus transmitted by mechanical inoculation; transmitted by grafting; not transmitted by contact between plants; not transmitted by seed.

**Ecology and control**

Studies reported by Mali (1979, 1985); Mali *et al.* (1983); Raut (1981).

**Geographical distribution**

Spreads in India (Maharashtra and Karnataka States).

**Experimental host range**

Several (3–9) families susceptible.

Diagnostically susceptible

*Canavalia gladiata* — veinal necrosis; systemic mosaic, mottling, veinal necrosis and slight leaf deformation.

*Canavalia virosa* — veinal chlorosis; mosaic and systemic leaf deformation.

*Phaseolus vulgaris, Glycine max* — chlorotic or necrotic local lesions, veinal necrosis; systemic mosaic mottling and slight puckering.

*Cicer arietinum, Trigonella foenum graecum* — systemic mosaic and leaf deformation.

*Vigna unguiculata* — systemic mosaic.

*Chenopodium album* — faint chlorotic local lesions; not systemic.

Diagnostically insusceptible

*Chenopodium amaranticolor*, *C. quinoa*, *Arachis hypogaea*, *Nicotiana glutinosa*, *Phaseolus vulgaris* cv. Top Crop, *Solanum nigrum*.

Maintenance and propagation hosts

*Canavalia ensiformis*, *Phaseolus vulgaris*, *Cicer arietinum*.

Assay hosts (**L**ocal lesions or **W**hole plants)

*Phaseolus vulgaris* (W), *Cicer arietinum* (W).

Dicotyledonous Hosts *(**Susceptible** or Insusceptible)*

*Arachis hypogaea. Beta vulgaris.* ***Cajanus cajan****. Capsicum annuum. Carthamus tinctorius.* ***Chenopodium album****. Chenopodium amaranticolor. Chenopodium murale.*

*Chenopodium quinoa*. ***Cicer arietinum***. *Cucumis sativus*. ***Cyamopsis tetragonoloba***. ***Glycine max***. *Helianthus annuus*. *Lablab purpureus*. *Lactuca sativa*. *Lens esculenta*. *Lycopersicon esculentum*. *Medicago sativa*. *Nicotiana glutinosa*. *Nicotiana rustica*. *Nicotiana tabacum*. *Ocimum basilicum*. *Petunia hybrida*. ***Phaseolus lunatus***. ***Phaseolus vulgaris***. ***Pisum sativum***. *Sesamum indicum*. *Solanum melongena*. *Solanum nigrum*. *Spinacia oleracea*. *Vicia faba*. ***Vigna radiata***. ***Vigna unguiculata***. ***Vigna unguiculata*** ssp. ***cylindrica***. ***Vigna unguiculata*** ssp. ***sesquipedalis***.

Additional hosts

***Canavalia gladiata, Canavalia virosa*** and ***Canavalia virosa*** x ***C. ensiformis***.

Sources of host-range data

Mali (1979, 1985); Mali *et al.* (1985); Raut (1981).

**Properties of particles in sap**

TIP: 65–70 °C. LIV: 2 days. DEP: $\log_{10}$ minus 5–6. Leaf sap contains few virions.

**Purification**

Method: Mali (1985).

**Particle morphology**

Virions filamentous; not enveloped; usually flexuous; with a clear modal length; of 790 nm; 11 nm wide. Axial canal obscure.

**Cytopathology**

Inclusions cytoplasmic cylindrical present in infected cells; are pinwheels and unusual in shape; scrolls, resembling those of sub-division I type (Edwardson, 1974).

**Taxonomy and Relationships**

*Potyvirus: Potyviridae.*

Virus(es) with serologically related virions: bean common mosaic, bean yellow mosaic, peanut mottle, soybean mosaic and wisteria vein mosaic viruses. Virus(es) with serologically unrelated virions: blackeye cowpea mosaic, celery mosaic, datura distortion mosaic, henbane mosaic, papaya ringspot, potato Y and watermelon mosaic 2 viruses.

Although serologically related to five potyviruses, swordbean distortion mosaic virus is distinguished from them by its host range and symptomatology. Best tests for diagnosis: insusceptibility of *Chenopodium amaranticolor* and *C. quinoa* aids the distinction of the swordbean distortion mosaic virus from a similar virus (probably caused by a strain of bean yellow mosaic virus) infecting *Canavalia ensiformis* in India.

**References**

Edwardson, J.R. (1974). *Fla Agric. Exp. Stn Monog.* No. 4, p. 1.

Hollings, M. and Brunt, A.A. (1981). *CMI/AAB Descr. Pl. Viruses* No. 245, 7 pp.

Mali, V.R. (1979). *Curr. Sci.* **48**: 162.

Mali, V.R. (1985). *Int. J. Trop. Pl. Dis.* **3**: 93.

Mali, V.R., Nirmal, D.D., Mundhe, G.E., Vyanjane, N.T. and Raut, K.G. (1983). *Abstr. Papers Virol. Conf. Publ. HAU, Hissar* p. 41.

Raut, K.G. (1981). *M.Sc. Thesis*, M.A.U., Parbhani, M.S., India.

Rodriguez, R.L., Bird, J., Monllor, A.C., Waterworth, H.E., Kimura, M. and Maramorosch, K. (1975). In: *Tropical Diseases of Legumes* p. 91; eds J. Bird and K. Maramorosch. Academic Press, New York.

Silberschmidt, K. and Nobrega, N.R. (1943). *O. Biologica* **8**: 129.

# Tamarillo mosaic *potyvirus*

**Data collated by D.W. Mossop, 1988.**

*Synonyms:* tomato tree virus, Cyphomandra virus. *Standard acronym:* TamMV.

First reported in *Cyphomandra betacea*; from New Zealand; by Chamberlain (1954).

## Natural host range and symptoms

Symptoms vary seasonally.
*Cyphomandra betacea* — vein-banding and mosaic, fruit blotching.
*Solanum nigrum* — usually symptomless, sometimes leaf blotching.

## Transmission

Transmitted by a vector; an insect; *Myzus persicae*; Aphididae. Transmitted in a non–persistent manner. Virus transmitted by mechanical inoculation; transmitted by grafting; not transmitted by contact between plants; not transmitted by seed; not transmitted by pollen.

## Ecology and control

Studies reported by Mossop (1977).

## Geographical distribution

Spreads in New Zealand.

## Experimental host range

Few (<3) families susceptible.

Diagnostically susceptible

*Nicotiana glutinosa* — systemic necrosis and, usually, plant death.
*N. clevelandii, N. rustica* and *N. tabacum* — systemic vein-clearing and mosaic.

Maintenance and propagation hosts

*Nicotiana clevelandii, N. glutinosa.*

Assay hosts (**L**ocal lesions or **W**hole plants)

*Nicotiana glutinosa* (W).

Dicotyledonous Hosts *(**Susceptible** or Insusceptible)*

*Beta vulgaris. Brassica campestris* ssp. *pekinensis. Calendula officinalis. Capsicum annuum. Chenopodium amaranticolor. Chenopodium quinoa. Cucumis sativus.* ***Cyphomandra betacea****. Datura stramonium. Ipomoea setosa. Lactuca sativa. Lupinus angustifolius.* ***Lycopersicon esculentum. Nicotiana clevelandii. Nicotiana × edwardsonii. Nicotiana glutinosa. Nicotiana rustica****. Nicotiana sylvestris.* ***Nicotiana tabacum. Petunia hybrida****. Phaseolus vulgaris. Pisum sativum. Solanum melongena.* ***Solanum nigrum****. Solanum tuberosum. Spinacia oleracea. Trifolium repens. Vicia faba. Zinnia elegans.*

Monocotyledonous Hosts *(**Susceptible** or Insusceptible)*

*Zea mays.*

Sources of host-range data

Mossop (1977).

## Properties of particles in sap

TIP: 55–60 °C. LIV: 2 days. DEP: $\log_{10}$ minus 3–4. Leaf sap contains few virions.

**Purification**

Method: Mossop (1977).

**Particle morphology**

Virions filamentous; not enveloped; usually flexuous; with a clear modal length; of 745 nm; 12 nm wide. Axial canal obscure. Basic helix obscure.

**Physical properties**

One sedimenting component in purified preparations; sedimentation coefficient 148 *S* (±3).

**Biochemical properties**

Virions contain 6 % nucleic acid (based on $A_{260}/A_{280}$ values).

*Genome* consists of RNA; single–stranded; linear; unipartite. Infectivity retained when deproteinised with phenol or detergent. Poly A region present; probably at the 3' end. Additional factor not required for infectivity.

*Sequence database accession code(s):*

X54804 Em(40)_vi:TAMVCPG Gb(84)_vi:TAMVCPG Tamarillo mosaic virus gene for the coat protein. 4/91 1,066bp 1 sequence.

*Virion protein(s)* one; $M_r$ 30000; coat protein.

**Replication**

Replication does not depend on a helper virus.

**Cytopathology**

Virions found in all parts of the host plant; in cytoplasm. Inclusions present in infected cells; are pinwheels; they do not contain virions.

**Taxonomy and Relationships**

*Potyvirus: Potyviridae.*

Virus(es) with serologically unrelated virions: potato Y, turnip mosaic, bean yellow mosaic, lettuce mosaic, passionfruit woodiness and tobacco etch viruses.

Originally incorrectly identified as potato virus Y in tree tomato, furthermore PVY does not protect tamarillo against tamarillo mosaic potyvirus. Best tests for diagnosis: tamarillo mosaic virus is often in mixed infections with potato aucuba mosaic virus, which depends on TaMV for aphid transmission. PaMV causes severe necrosis of *Capsicum frutescens*, but no host species is known that separates them.

**References**


Chamberlain, E.E. (1954). *Plant Viruses in New Zealand. Bulletin DSIR*, No.108, p.129.

Mossop, D.W. (1977). *N. Z. Jl agric. Res.* **20**: 535.

McLean, G.D. and Mossop, D.W. (1988). In: *The Plant Viruses*; Vol. 4, *The Filamentous Viruses*, p. 401; ed R.G. Milne. Plenum Press, New York.


# Tamus latent (?) *potexvirus*

**Data collated by A.A. Brunt and D.-E. Lesemann, 1991.**

First reported in *Tamus communis*; from south west England; by Vetten *et al.* (1987).

**Natural host range and symptoms**

*Tamus communis* — symptomless.

**Transmission**

Virus transmitted by mechanical inoculation.

**Geographical distribution**

Found, but with no evidence of spread, in the U.K.

**Experimental host range**

Several (3–9) families susceptible. Experimentally infected plants mostly show local lesions, then systemic vein chlorosis or necrosis.

Diagnostically susceptible

*Nicotiana benthamiana, N. megalosiphon* — systemic chlorotic and necrotic vein banding.

*Chenopodium amaranticolor, C. murale, C. quinoa, Gomphrena globosa, Tetragonia expansa* — local chlorotic or necrotic lesions,

*Datura stramonium, Lycopersicon esculentum, Nicotiana clevelandii* — symptomless systemic infection.

Diagnostically insusceptible

*Cucumis sativus, Nicotiana tabacum, Petunia hybrida, Phaseolus vulgaris, Physalis floridana.*

Maintenance and propagation hosts

*Nicotiana benthamiana, N. megalosiphon.*

Assay hosts (**L**ocal lesions or **W**hole plants)

*Chenopodium amaranticolor* (L), *C. quinoa* (L).

Dicotyledonous Hosts ***(Susceptible** or Insusceptible)*

***Antirrhinum majus*.** ***Chenopodium amaranticolor*.** ***Chenopodium murale*.** ***Chenopodium quinoa*.** *Cucumis sativus.* ***Datura metel*.** ***Datura stramonium*.** ***Gomphrena globosa*.** ***Lycopersicon esculentum*.** ***Nicotiana benthamiana*.** ***Nicotiana clevelandii*.** ***Nicotiana debneyi*.** ***Nicotiana glutinosa*.** ***Nicotiana megalosiphon*.** *Nicotiana tabacum. Petunia hybrida. Phaseolus vulgaris. Physalis floridana.* ***Solanum demissum*.** ***Tetragonia expansa*.**

**Properties of particles in sap**

Leaf sap contains many virions.

**Purification**

Method: Vetten *et al.* (1987).

**Particle morphology**

Virions filamentous; not enveloped; usually flexuous; with a clear modal length; of 553 nm. Axial canal obscure. Basic helix obvious.

**Physical properties**

$A_{260}/A_{280}$ ratio 1.26.

*Virion protein(s)* one; $M_r$ 27700; and it is coat protein. Method of preparation: Vetten *et al.* (1987).

**Replication**

Replication does not depend on a helper virus.

**Cytopathology**

Virions found in leaves, mesophyll and all parts of the host plant; in cytoplasm. Inclusions present in infected cells; are unusual in shape; cytoplasmic masses of filamentous virions mainly parallel or in whorls; they contain virions.

**Taxonomy and Relationships**

Possible *Potexvirus*.

Virus(es) with serologically related virions: papaya mosaic (Argentine plantago and boussingaultia mosaic strains), hydrangea ringspot and potato X viruses. Virus(es) with serologically unrelated virions: Brazilian cassava common mosaic, cymbidium mosaic, dioscorea latent, lily X, narcissus mosaic, potato aucuba mosaic, potato X and white clover mosaic potexviruses.

Best tests for diagnosis: In *Tamus communis* in England, potyviruses related to yam mosaic virus are common, but may be distinguished from this virus by serology and virion morphology.

**References**

Vetten, H.J., Lesemann, D.-E. and Allen, D.J. (1987). *Pl. Path.* **36**: 492.

# Teasel mosaic (?) potyvirus

**Data collated by A.A. Brunt and P. Jeyanandarajah, 1991.**

*Standard acronym:* TeaMV.

First reported in *Dipsacus fullonum*; from California, U.S.A.; by Stoner (1951).

**Natural host range and symptoms**

Symptoms persist.

*Dipsacus fullonum* — vein clearing, chlorotic mottling and malformed filiform leaves.

**Transmission**

Transmitted by a vector; an insect; *Macrosiphum rosae, Myzus persicae*; Aphididae. Not transmitted by *Rhopalosiphum padi, Macrosiphum avenae.* Transmitted in a non–persistent manner. Virus not transmitted by mechanical inoculation.

**Geographical distribution**

Spreads in the USA (California).

**Experimental host range**

Few (<3) families susceptible. Experimentally infected plants mostly show leaf mottling and malformation.

Diagnostically susceptible

*Dipsacus fullonum, Scabiosa atropurpurea* — systemic vein clearing, then mottling and malformed leaves.

Diagnostically insusceptible

*Cucurbita pepo, Beta vulgaris, Nicotiana tabacum, Cucumis sativus.*

Maintenance and propagation hosts

*Dipsacus fullonum, Scabiosa atropurpurea.*

Assay hosts (**L**ocal lesions or **W**hole plants)

*Dipsacus fullonum* (W).

Dicotyledonous Hosts (***Susceptible*** *or Insusceptible)*

*Apium graveolens. Beta vulgaris. Cucumis sativus. Cucurbita pepo. Nicotiana tabacum.*

Sources of host-range data

Stoner (1951).

**Particle morphology**

Virions possibly filamentous; of 750 nm.

**Taxonomy and Relationships**

Unassigned potyvirus.

**References**

Stoner, W.N. (1951). *Phytopathology* **41**: 91.

# Telfairia mosaic *potyvirus*

**Data collated by A.A. Brunt, 1987.**

*Standard acronym:* TeMV.

First reported in *Telfairia occidentalis*; from Nigeria; by Nwauzo and Brown (1975).

**Natural host range and symptoms**

Symptoms persist.

*Telfairia occidentalis* — stunting, chlorosis of leaves, malformation of fruits.

**Transmission**

Transmitted by a vector; an insect; *Aphis spiraecola*; Aphididae. Transmitted in a non–persistent manner. Virus transmitted by mechanical inoculation; transmitted by grafting; not transmitted by contact between plants; transmitted by seed (Anno-Nyako, 1988); not transmitted by pollen.

**Ecology and control**

Studies reported by Atiri and Varma (1983).

**Geographical distribution**

Spreads in Nigeria.

**Experimental host range**

Several (3–9) families susceptible.

Diagnostically susceptible

*Citrullus vulgaris, Cucumis melo, Cucumis sativus, Cucurbita pepo* — systemic leaf chlorosis.

*Chenopodium amaranticolor, C. quinoa, C. murale, Phaseolus vulgaris* — chlorotic or necrotic local lesions; not systemic.

*Nicotiana benthamiana, N. clevelandii* — systemic chlorosis.

Diagnostically insusceptible

*Nicotiana rustica, N. glutinosa, Datura metel, Brassica pekinensis, Ocimum basilicum.*

Maintenance and propagation hosts

*Nicotiana benthamiana, N. clevelandii.*

Assay hosts (**L**ocal lesions or **W**hole plants)

*Chenopodium quinoa* (L), *Phaseolus vulgaris* (L).

Dicotyledonous Hosts *(**Susceptible** or Insusceptible)*

***Amaranthus caudatus***. *Brassica campestris* ssp. *pekinensis*. ***Celosia argentea***. ***Chenopodium amaranticolor***. ***Chenopodium capitatum***. ***Chenopodium murale***. ***Chenopodium quinoa***. ***Cucumis melo***. ***Cucumis sativus***. ***Cucurbita pepo***. *Cyamopsis tetragonoloba*. *Datura metel*. ***Datura stramonium***. ***Datura tatula***. ***Gomphrena globosa***. ***Nicotiana benthamiana***. ***Nicotiana clevelandii***. *Nicotiana glutinosa*. ***Nicotiana megalosiphon***. *Nicotiana rustica*. ***Nicotiana sylvestris***. ***Nicotiana tabacum***. *Ocimum basilicum*. ***Phaseolus lathyroides***. ***Phaseolus vulgaris***. ***Physalis floridana***. ***Tetragonia expansa***. ***Trifolium incarnatum***. ***Vicia faba***. *Vigna unguiculata*.

**Properties of particles in sap**

TIP: 65–70 °C (in *Telfairia occidentalis* sap). LIV: 1 days (or less at about 25°C). DEP: $\log_{10}$ minus 3–4.

**Purification**

Method: Shoyinka *et al.* (1987).

**Particle morphology**

Virions filamentous; usually flexuous; with a clear modal length; of 806 nm; 13 nm wide. Basic helix obscure.

**Physical properties**

One sedimenting component in purified preparations; sedimentation coefficient 145 *S*. Density 1.32 g $cm^{-3}$ in CsCl.

**Biochemical properties**

Virions contain *c.* 5 % nucleic acid; 95 % protein.

*Genome* consists of RNA; single–stranded.

*Virion protein(s)* one; $M_r$ 32000; coat protein.

**Cytopathology**

Virions found in leaves and roots; in cytoplasm. Inclusions present in infected cells; are pinwheels (often seen as scrolls); they do not contain virions. Other cellular changes: hypertrophy of elements of endoplasmic reticulum with vesicles containing finely stranded material.

**Taxonomy and Relationships**

*Potyvirus: Potyviridae.*

Virus(es) with serologically related virions: bean yellow mosaic, watermelon mosaic 2 and zucchini yellow mosaic viruses. Virus(es) with serologically unrelated virions: alstroemeria mosaic, arracacha Y, bean common mosaic, bidens mottle, blackeye cowpea mosaic, commelina mosaic, clover yellow vein, datura shoestring, freesia mosaic, henbane mosaic, hippeastrum mosaic, iris mild mosaic, iris severe mosaic, lettuce mosaic, nerine yellow stripe, papaya ringspot, passiflora woodiness, peanut mottle, pepper veinal mottle, potato Y, sugarbeet mosaic, sweet potato feathery mottle, tobacco etch, turnip mosaic and ullucus mosaic viruses.

**References**

Anno-Nyako, F.O. (1988). *J. Phytopath.* **121**: 85.

Atiri, G.I. and Varma, A. (1983). *Trop. Agric., Trin.* **60**: 95.

Nwauzo, E.E. and Brown, W.M. (1975). *Pl. Dis. Reptr* **50**: 430.

Shoyinka, S.A., Brunt, A.A., Phillips, S., Lesemann, D.-E., Thottappilly, G. and Lastra, R. (1987). *J. Phytopath.* **119**: 13.

## Tephrosia symptomless (?) *carmovirus*

**Data collated by K.R. Bock, 1982. Revised 1983.**

*Standard acronym:* TeSV.

First reported in *Tephrosia villosa*; from Mombasa, Kenya; by Bock *et al.* (1981).

**Natural host range and symptoms**

*Tephrosia noctiflora, T. villosa* — symptomless.

**Transmission**

Virus transmitted by mechanical inoculation.

**Geographical distribution**

Spreads in Kenya.

**Experimental host range**

Few (<3) families susceptible.

Diagnostically susceptible

*Canavalia ensiformis* — necrotic local lesions; systemic mosaic.
*Glycine max* — systemic mosaic.
*Phaseolus acutifolius, P. vulgaris* — necrotic or chlorotic local lesions; not systemic.

Diagnostically insusceptible

*Gomphrena globosa, Chenopodium amaranticolor, C. quinoa, Nicotiana clevelandii, N. tabacum.*

Maintenance and propagation hosts

*Glycine max.*

Assay hosts (**L**ocal lesions or **W**hole plants)

*Glycine max* (W).

Dicotyledonous Hosts *(**Susceptible** or Insusceptible)*

*Amaranthus caudatus. Arachis hypogaea. Atriplex hortensis. Bauhinia purpurea. Beta vulgaris. Cajanus cajan. Calendula officinalis. Capsicum frutescens. Cassia occidentalis. Chenopodium amaranticolor. Chenopodium quinoa. Cicer arietinum. Citrullus lanatus. Clitoria ternatea. Cucumis melo. Cucumis sativus. Cucurbita maxima. Cucurbita pepo. Cyamopsis tetragonoloba. Datura ferox. Datura metel. Datura stramonium.* ***Glycine max****. Gomphrena globosa. Lablab purpureus. Macroptilium lathyroides. Nicotiana clevelandii. Nicotiana glutinosa. Nicotiana rustica. Nicotiana tabacum. Ocimum basilicum. Phaseolus lunatus.* ***Phaseolus vulgaris****. Physalis floridana. Pisum sativum. Solanum melongena. Solanum nigrum. Tetragonia expansa. Trifolium subterraneum. Vicia faba. Vigna angularis. Vigna unguiculata. Zinnia elegans.*

Sources of host-range data

Bock *et al.* (1981).

**Properties of particles in sap**

TIP: 85–90 °C. LIV: 23 days. DEP: $\log_{10}$ minus 6–7. Leaf sap contains many virions.

**Purification**

Method: Bock *et al.* (1981).

**Particle morphology**

Virions isometric; not enveloped; 33 nm in diameter; rounded in profile; with a conspicuous capsomere arrangement.

**Physical properties**

One sedimenting component in purified preparations; sedimentation coefficient 126–127 *S*. Density 1.361 g $cm^{-3}$ in CsCl.

**Biochemical properties**

Virions contain 20 % nucleic acid (estimated from bouyant density in CsCl and $A_{260}/A_{280}$); 80 % protein.

*Genome* consists of RNA; single–stranded; unipartite; largest (or only) genome part 4.5–4.7 kb.

*Virion protein(s)* one; $M_r$ 42000; coat protein. Method of preparation: Bock *et al.* (1981).

**Taxonomy and Relationships**

Possible *Carmovirus: Tombusviridae.*

Virus(es) with serologically unrelated virions: 44 other viruses, including eight tombus- or carmoviruses (Bock *et al.*, 1981).

**References**

Bock, K.R. (1982). *CMI/AAB Descr. Pl. Viruses* No. 256, 3 pp.

Bock, K.R., Guthrie, E.J. and Figueiredo, G. (1981). *Ann. appl. Biol.* **97**: 277.

# Thimbleberry ringspot virus

**Data collated by R. Stace-Smith, 1991.**

First reported in *Rubus parviflorus*; from British Columbia, Canada; by Stace-Smith (1958).

**Natural host range and symptoms**

Symptoms persist.

*Rubus parviflorus* — irregular leaf ringspotting and oak leaf patterns.

**Transmission**

Transmitted by a vector; an insect; *Macrosiphum (Oestlundia) davidsonii, M.(O.) maxima* and *Amphorophora parviflori*; Aphididae. Not transmitted by *Amphorophora agathonica.* Virus not transmitted by mechanical inoculation; transmitted by grafting; not transmitted by contact between plants; not transmitted by seed; not transmitted by pollen.

**Geographical distribution**

Spreads in Canada.

**Experimental host range**

Few (<3) families susceptible. Experimentally infected plants mostly show ringspots and oak leaf markings.

Diagnostically susceptible

*Rubus parviflorus* — ringspot and oak leaf markings.

*Rubus idaeus* — faint chlorotic markings.

*R. occidentalis* — diffuse mottle.
*R. henryi* — chlorotic leaf spots.

Diagnostically insusceptible

All herbaceous hosts tested were insusceptible including *Chenopodium quinoa, Cucumis sativus, Nicotiana tabacum, Vigna unguiculata* syn. *sinensis, Phaseolus vulgaris.*

Maintenance and propagation hosts

*Rubus parviflorus.*

Assay hosts (**L**ocal lesions or **W**hole plants)

*Rubus parviflorus* (W), *Rubus occidentalis* (W).

Dicotyledonous Hosts *(**Susceptible** or Insusceptible)*

***Rubus idaeus.***

Sources of host-range data

Stace-Smith (1958).

**Particle morphology**

Virions isometric; rounded in profile; with a conspicuous capsomere arrangement.

**References**

Stace-Smith, R. (1958). *Can. J. Bot.* **36**: 385.
Stace-Smith, R. (1987). In: *Virus Diseases of Small Fruits*; ed. R.H. Converse, U.S. Dep. Agric. Hdbk No.631, p. 277.

# Thistle mottle *caulimovirus*

**Data collated by R. Hull, 1987.**

*Standard acronym:* ThMoV.

First reported in *Cirsium arvense*; from Norwich, U.K.; by Donson and Hull (1983).

**Natural host range and symptoms**

Symptoms persist.
*Cirsium arvense* — slight mottle.

**Geographical distribution**

Found, but with no evidence of spread, in the U.K.

Maintenance and propagation hosts

*Cirsium arvense.*

**Properties of particles in sap**

Leaf sap contains few virions.

**Purification**

Method: Donson and Hull (1983).

**Particle morphology**

Virions isometric; not enveloped; 50 nm in diameter; rounded in profile; without a conspicuous capsomere arrangement.

### Physical properties

One sedimenting component in purified preparations; sedimentation coefficient 210 *S*. Density 1.36 g $cm^{-3}$ in CsCl.

### Biochemical properties

*Genome* consists of DNA; double–stranded; circular; unipartite; largest (or only) genome part 7.8 kb (pairs). Genomic nucleic acid isolated by Donson and Hull (1983).

*Non–genomic nucleic acid* not found in the virions.

### Replication

Genome replicates presumably in nuclei and in cytoplasm (DNA to RNA in the former, RNA to DNA in the latter). Replication does not depend on a helper virus.

### Cytopathology

Virions found in mesophyll; in cytoplasm. Inclusions present in infected cells; are viroplasms; they contain virions.

### Taxonomy and Relationships

*Caulimovirus.*

### References

Donson, J. and Hull, R. (1983). *J. gen. Virol.* **64**: 2281.

# Thyme (?) *nucleorhabdovirus*

**Data collated by A.A. Brunt, 1991.**

First reported in *Thymus x citriodorus*; from the U.K.; by Schultz *et al.* (1975).

### Natural host range and symptoms

*Thymus x citriodorus* — leaf chlorosis.

### Transmission

Not transmitted by *Aphis serphylli*. Virus not transmitted by mechanical inoculation.

### Particle morphology

Virions rhabdo– or bullet–shaped; with a clear modal length; of 219 nm; 72 nm wide.

### Cytopathology

Virions found in phloem parenchyma cells; in the perinuclear space and in unusual sites; nucleoplasm.

### Taxonomy and Relationships

Possible *Nucleorhabdovirus: Rhabdoviridae.*

### References

Schultz, M.G., Harrap, K.A. and Land, J.B. (1975). *Ann. appl. Biol.* **80**: 251.

# Tobacco etch *potyvirus*

**Data collated by D.E. Purcifull, 1981.**

*Synonyms:* datura Z virus, tobacco severe etch virus, tomato etch virus. *Standard acronym:* TEV.

First reported in *Datura* spp.; from the U.S.A; by Blakeslee (1921).

**Natural host range and symptoms**

Symptoms persist (may be strain-dependant).

*Cassia obtusifolia* — leaf mottling, malformation and necrosis, stunting and reduced seed yield.

*Capsicum annuum* — mottling, mosaic and leaf malformation, fruit malformation, stunting.

*Datura stramonium* — leaf mottling, vein banding and malformation, suppression of fruit spines.

*Lycopersicon esculentum* — leaf mottling and distortion, severe yield reductions.

*Nicotiana tabacum* — mottling and necrotic leaf etching.

*Physalis* spp. — leaf chlorosis.

**Transmission**

Transmitted by a vector; an insect; more than 10 spp., especially *Myzus persicae*; Aphididae. Transmitted in a non–persistent manner. Virus does not require a helper virus for vector transmission (Simons, 1976; Pirone, 1979); transmitted by mechanical inoculation; not transmitted by seed.

**Geographical distribution**

Spreads in the Eastern Asian region and the South and Central American region; Canada, China, Mexico, Puerto Rico, Venezuela, and the USA.

**Experimental host range**

Many (>9) families susceptible.

Diagnostically susceptible

*Cassia obtusifolia* — necrotic local lesions; systemic mosaic and leaf malformation.

*Capsicum annuum* — systemic mosaic, rugosity and leaf malformation.

*C. frutescens* — root necrosis, wilting and death.

*Chenopodium album, C. amaranticolor* and *C. quinoa* — local chlorotic or necrotic lesions; not systemic.

*Datura stramonium* — systemic mosaic, ringspots, rugosity and leaf malformation.

*Nicotiana tabacum* — chlorotic local lesions; systemic mosaic, ringspots, leaf malformation and necrotic etching.

Maintenance and propagation hosts

*Capsicum annuum, Datura stramonium, Nicotiana tabacum.*

Assay hosts (**L**ocal lesions or **W**hole plants)

*Chenopodium amaranticolor* (L) *C. quinoa* (L).

Dicotyledonous Hosts *(**Susceptible** or Insusceptible)*

*Ageratum conyzoides*. ***Amaranthus caudatus***. *Antirrhinum majus*. *Apium graveolens*. *Bellis perennis*. ***Beta vulgaris***. *Brassica campestris* ssp. *pekinensis*. *Brassica campestris* ssp. *rapa*. *Calendula officinalis*. *Capsella bursa–pastoris*. ***Capsicum frutescens***. ***Cassia tora***. *Catharanthus roseus*. ***Celosia argentea***. *Cheiranthus cheiri*. *Chenopodium album*. ***Chenopodium amaranticolor***. ***Chenopodium foetidum***. ***Chenopodium quinoa***. *Cichorium endiva*. *Cucumis melo*. *Cucumis sativus*. *Cucurbita maxima*. *Cucurbita pepo*. *Dahlia pinnata*. ***Datura ferox***. ***Datura metel***. ***Datura stramonium***. *Daucus carota*. *Dianthus barbatus*. *Euphorbia marginata*. *Fagopyrum esculentum*. *Glycine max*. ***Gomphrena globosa***. *Gossypium hirsutum*. ***Gypsophila elegans***. *Hibiscus esculentus*. ***Hyoscyamus niger***. *Ipomoea nil*. *Ipomoea setosa*. *Lablab purpureus*. *Lens esculenta*.

*Lobelia erinus. Lycopersicon esculentum. Lycopersicon pimpinellifolium. Matthiola incana. Medicago hispida. Medicago sativa.* ***Melilotus alba.*** *Myosotis sylvatica.* ***Nicandra physaloides. Nicotiana bigelovii. Nicotiana clevelandii. Nicotiana × edwardsonii. Nicotiana glutinosa. Nicotiana rustica. Nicotiana sylvestris. Nicotiana tabacum.*** *Papaver nudicaule. Pastinaca sativa.* ***Petunia hybrida.*** *Phaseolus vulgaris.* ***Physalis floridana. Physalis peruviana.*** *Pisum sativum. Raphanus sativus. Ricinus communis. Salvia splendens.* ***Senecio vulgaris.*** *Sinningia speciosa.* ***Solanum melongena. Solanum nigrum.*** *Solanum nodiflorum.* ***Solanum tuberosum.*** *Spinacia oleracea. Stellaria media.* ***Tetragonia expansa. Torenia fournieri.*** *Trifolium incarnatum. Trifolium pratense. Trifolium repens. Tropaeolum majus. Vicia faba. Vicia sativa. Vigna radiata. Vigna unguiculata.* ***Zinnia elegans.***

Monocotyledonous Hosts *(**Susceptible** or Insusceptible)*

*Allium cepa. Avena sativa. Hordeum vulgare. Secale cereale. Triticum aestivum. Zea mays.*

Sources of host-range data

Holmes (1946); Schmelzer (1967); Edwardson (1974).

**Properties of particles in sap**

TIP: 55 °C. LIV: 5–10 days. DEP: $\log_{10}$ minus 4. Leaf sap contains few virions.

**Purification**

Method: Dougherty and Hiebert (1980).

**Particle morphology**

Virions filamentous; not enveloped; usually flexuous; with a clear modal length; of 730–750 nm; 12–13 nm wide. Basic helix obvious (in UF).

**Physical properties**

One sedimenting component in purified preparations; sedimentation coefficient 154 *S*. Density 1.33 g $cm^{-3}$ in CsCl.

**Biochemical properties**

Virions contain 5 % nucleic acid; 95 % protein; 0 % lipid.

*Genome* consists of RNA; single–stranded; unipartite; largest (or only) genome part 9.494 kb. Base composition 23 % G; 30 % A; 20 % C; 27 % U. 5' terminus of RNA has a VPg. Infectivity retained when deproteinised with proteases, or decreased when deproteinised with proteases. Poly A region present; at the 3' terminus.

*Sequence database accession code(s):*

M11216 Em(40)_vi:TEVCAPG Gb(84)_vi:TEVCAPG Tobacco etch virus capsid protein gene. 4/90 2,324bp.

M11458 Em(40)_vi:TECGHAT Gb(84)_vi:TEVCGHAT Tobacco etch virus (highly aphid transmissible (HAT)) complete genome. 4/90 9,494bp.

M15239 Em(40)_vi:TEVGEN Gb(84)_vi:TEVGEN Tobacco etch virus RNA, complete genome. 7/89 9,497bp.

L38714 Em(43)_vi:Tevnwp Gb(89)_vi:Tevnwp Tobacco etch virus P1, Helper Component and Proteinase, P3, Cylindric Inclusion, Nuclear Inclusion.

Sub–genomic *m*RNA not found in infected cells.

*Virion protein(s)* one; $M_r$ 30000–32000; coat protein. Method of preparation: Hiebert and McDonald (1973); Hiebert *et al.* (1979).

**Replication**

Replication does not depend on a helper virus.

**Cytopathology**

Virions found in mesophyll and epidermis; in cytoplasm; plasmodesmata. Inclusions present in infected cells; are crystals in the nucleus and pinwheels; they do not contain virions.

**Taxonomy and Relationships**

*Potyvirus: Potyviridae.*

Virus(es) with serologically related virions: potato Y and 13 other viruses (Shepard *et al.*, 1974).

**References**


Blakeslee, A.F. (1921). *J. Genet.* **11**: 17.
Christie, R.G. and Edwardson, J.R. (1977). *Fla Agric. Exp. Stn Monog.* No. 9, p. 55.
Damirdagh, I.S. and Shepherd, R.J. (1970). *Virology* **40**: 84.
Dougherty, W.G. and Hiebert, E. (1980). *Virology* **101**: 466.
Dougherty, W.G. and Hiebert, E. (1980*a*). *Virology* **104**: 183.
Dougherty, W.G. and Hiebert, E. (1980*b*). *Virology* **104**: 174.
Edwardson, J.R. (1968). *Virology* **34**: 250.
Edwardson, J.R. (1974). *Fla Agric. Exp. Stn Monog.* No. 4, p. 398.
Hari, V. (1981). *Virology* **112**: 391.
Hiebert, E. and McDonald, J.G. (1973). *Virology* **56**: 349.
Hiebert, E., Tremaine, J.H. and Ronald, W. (1979). *Phytopathology* **69**: 1031.
Hill, J.H. and Benner, H.I. (1976). *Virology* **75**: 419.
Holmes, F.O. (1946). *Phytopathology* **36**: 643.
Jones, A.T. and Barker, H. (1976). *Ann. appl. Biol.* **83**: 231.
Knuhtsen, H., Hiebert, E. and Purcifull, D.E. (1974). *Virology* **61**: 200.
Pirone, T.P. (1979). *Phytopathology* **69**: 531.
Purcifull, D.E. (1973). *Virology* **55**: 275.
Purcifull, D.E. and Batchelor, D.L. (1977). *Bull. Univ. Fla Agric. Exp. Stn* **788**: 39 pp.
Purcifull, D.E. and Hiebert, E. (1982). *CMI/AAB Descr. Pl. Viruses* No. 258, 7 pp.
Shepard, J.E., Secor, G.A. and Purcifull, D.E. (1974). *Virology* **58**: 464.
Shepherd, R.J. and Purcifull, D.E. (1972). *CMI/AAB Descr. Pl. Viruses* No. 55, 4 pp.
Schmelzer, K. (1967). *Phytopath. Z.* **60**: 301.
Simons, J.N. (1976). *Phytopathology* **66**: 652.


# Tobacco leaf curl bigeminivirus

**Data collated by T. Osaki, 1984. Revised 1989.**

*Synonyms:* eggplant yellow mosaic virus, eupatorium yellow vein virus, honeysuckle yellow vein mosaic virus (Osaki *et al.*, 1979), tobacco cabbaging virus, tobacco curly leaf virus, tobacco frenching virus, tobacco leaf curl virus 1, tomato yellow dwarf virus, zinnia leaf curl virus. *Standard acronym:* TLCV.

First reported in *Nicotiana tabacum*; from Amani, Tanzania; by Storey (1931).

**Natural host range and symptoms**

Symptoms persist.

*Ageratum conyzoides, Zinnia elegans, Capsicum annuum, Datura stramonium, Nicotiana tabacum* — leaf curling.

*Lycopersicon esculentum* — leaf curling, yellow dwarfing.

*Lonicera japonica* — vein yellowing.

**Transmission**

Transmitted by a vector; an insect; *Bemisia tabaci*; Aleyrodidae. Transmitted in a persistent manner. Virus lost by the vector when it moults; not transmitted congenitally to the progeny of the vector; not transmitted by mechanical inoculation; transmitted by grafting; not transmitted by contact between plants; not transmitted by seed.

**Geographical distribution**

Probably distributed worldwide (in tropical and subtropal regions).

**Experimental host range**

Several (3–9) families susceptible.

Diagnostically susceptible

*Datura stramonium, Nicotiana glutinosa* — systemic leaf curling.
*Lonicera japonica* — systemic vein yellowing.
*Lycopersicon esculentum* — systemic yellow dwarfing, leaf curling.

Diagnostically insusceptible

*Cassia tora, Phaseolus vulgaris, Vicia faba, Vigna unguiculata* ssp. *sesquipedalis, Solanum nodiflorum.*

Maintenance and propagation hosts

*Datura stramonium, Lycopersicon esculentum, Nicotiana tabacum, N. glutinosa.*

Assay hosts (**L**ocal lesions or **W**hole plants)

*Datura stramonium* (W), *Lycopersicon esculentum* (W), *Nicotiana glutinosa* (W).

Dicotyledonous Hosts *(**Susceptible** or Insusceptible)*

***Ageratum conyzoides***. *Beta vulgaris*. *Brassica campestris* ssp. *napus*. *Brassica campestris* ssp. *rapa*. ***Capsicum annuum***. *Cassia tora*. *Catharanthus roseus*. *Chenopodium amaranticolor*. *Chenopodium quinoa*. *Cucumis sativus*. *Cucurbita maxima*. *Cucurbita moschata*. *Cucurbita pepo*. *Dahlia pinnata*. ***Datura stramonium***. *Glycine max*. *Gomphrena globosa*. *Gossypium hirsutum*. *Lablab purpureus*. ***Lycopersicon esculentum***. ***Nicotiana clevelandii***. ***Nicotiana debneyi***. ***Nicotiana glutinosa***. *Nicotiana rustica*. *Nicotiana sylvestris*. ***Nicotiana tabacum***. ***Petunia hybrida***. *Phaseolus vulgaris*. ***Physalis floridana***. *Raphanus sativus*. ***Solanum melongena***. *Solanum nodiflorum*. *Vicia faba*. *Vigna unguiculata* ssp. *sesquipedalis*. ***Zinnia elegans***.

Sources of host-range data

Osaki and Inouye (1981).

**Properties of particles in sap**

Electron microscopy: fix with 2% glutaraldehyde.

**Purification**

Method: Osaki and Inouye (1978).

**Particle morphology**

Virions geminate; not enveloped; 18 nm in diameter; dimers 30 nm in length.

**Biochemical properties**

*Genome* consists of DNA; single–stranded; circular; of two parts; largest (or only) genome part 2.7 kb; the 2nd largest 2.7 kb.

**Cytopathology**

Virions found in phloem; in nuclei. Inclusions absent from infected cells.

**Taxonomy and Relationships**

Bigeminivirus: *Geminiviridae.*

Virus(es) with serologically unrelated virions: bean golden mosaic and mung bean yellow mosaic viruses.

**References**

Ikegami, M., Osaki, T. and Inouye, T. (1987). *Ann. Phytopath. Soc. Japan* **53**: 269.
Osaki, T. and Inouye, T. (1978). *Ann. Phytopath. Soc. Japan* **44**: 167.
Osaki, T. and Inouye, T. (1979). *Ann. Phytopath. Soc. Japan* **45**: 62.
Osaki, T. and Inouye, T. (1981). *CMI/AAB Descr. Pl. Viruses* No. 232, 4 pp.
Osaki, T., Kobatake, H. and Inouye, T. (1979). *Ann. Phytopath. Soc. Japan* **45**: 62.
Pruthi, H.S. and Samuel, C.K. (1937). *Indian J. agric. Sci.* **7**: 659.
Pruthi, H.S. and Samuel, C.K. (1942). *Indian J. agric. Sci.* **12**: 37.
Storey, H.H. (1931). *Nature, Lond.* **128**: 187.
Vasudeva, R.S. and Samraj, J. (1948). *Phytopathology* **38**: 364.
Yassin, A.M. and Nour, M.A. (1965). *Ann. appl. Biol.* **56**: 207.

# Tobacco mild green mosaic *tobamovirus*

**Data collated by A.J. Gibbs, 1988.**

*Synonyms:* green-tomato atypical mosaic virus (Wetter, 1984; 1986), mild strain of tobacco mosaic virus (Johnson, 1947; Wetter, 1984), para—tobacco mosaic virus (Köhler and Panjan, 1943; Wetter, 1984), tobacco mosaic virus — South Carolina mild mottling strain (McKinney, 1952; Wetter, 1984), tobacco mosaic virus strain U2 (Johnson, 1947; Siegel and Wildman, 1954; Wetter, 1984), tobacco mosaic virus strain U5, tomato atypical mosaic green mottling strain (Knight *et al.*, 1962; Miller and Thornberry, 1958).
*Standard acronym:* TMGMV.

First reported in *Nicotiana glauca*; from the Canary islands; by McKinney (1929).

**Natural host range and symptoms**

Symptoms persist.
*Nicotiana glauca* — bright yellow mosaic, mottling, ringspots and cupping; stunting.
*N. tabacum* cv. Samsun — mild green mosaic.
*Capsicum annuum* — severe necrotic mosaic; plants often killed.
*Eryngium aquaticum, E. planum* — systemic yellow flecking.

**Transmission**

Transmitted by means not involving a vector. Virus transmitted by mechanical inoculation; transmitted by grafting; not transmitted by seed.

**Ecology and control**

Studies reported by McKinney (1929); Randles *et al.* (1981); Wetter (1980).

**Geographical distribution**

Probably distributed worldwide (in *Nicotiana glauca* in Mediterranean climatic regions).

**Experimental host range**

Several (3–9) families susceptible.

Diagnostically susceptible

*Nicotiana glutinosa, N. sylvestris, N. tabacum* cvs Xanthi-*nc*, White Burley — systemic yellow mosaic, ringspots.

*Datura stramonium, Chenopodium amaranticolor* — local lesions; not systemic.
*Eryngium aquaticum, E. planum* — systemic yellow flecks.
*Lycopersicon esculentum* cv. Rutgers — yellow mosaic with Australian isolate.

Diagnostically insusceptible

*Lycopersicon esculentum*, Canary Island and American cvs to Canary Island isolates (McKinney, 1929) *L. esculentum* cv. Marmande (Wetter *et al.*, 1984).

Maintenance and propagation hosts

*Nicotiana glauca, N. tabacum* Samsun, *Eryngium aquaticum* (maintenance), *Nicotiana clevelandii.*

Assay hosts (**L**ocal lesions or **W**hole plants)

*Nicotiana glutinosa* (L), *Datura stramonium* (L).

Dicotyledonous Hosts *(**Susceptible** or Insusceptible)*

***Capsicum annuum*. *Chenopodium amaranticolor*. *Chenopodium quinoa*. *Datura stramonium*. *Lycopersicon esculentum***, or *Lycopersicon esculentum*. ***Nicotiana clevelandii*. *Nicotiana glutinosa*. *Nicotiana glutinosa* × *clevelandii*. *Nicotiana sylvestris*. *Nicotiana tabacum*. *Ocimum basilicum*. *Petunia hybrida*.**

Sources of host-range data

Wetter (1980; 1984); Wetter *et al.* (1984).

**Properties of particles in sap**

TIP: 85–90 °C. LIV: more than 35000 days (i.e. more than 90 years). Infectivity of sap not changed by treatment with di–ethyl ether. Leaf sap contains many virions.

**Purification**

Method: Boedtker and Simmons (1958).

**Particle morphology**

Virions rod–shaped; not enveloped; usually straight; with a clear modal length; of 308 nm; 18 nm wide. Axial canal obvious. Basic helix obvious.

**Physical properties**

One sedimenting component in purified preparations; sedimentation coefficient 190 *S*. Density 1.307 g $cm^{-3}$ in CsCl. Isoelectric point pH 3.16.

**Biochemical properties**

Virions contain 5 % nucleic acid; 95 % protein; 0 % lipid.

*Genome* consists of RNA; single–stranded; linear; unipartite; largest (or only) genome part 6.355 kb. Additional factor not required for infectivity.

*Sequence database accession code(s):*

M34077 Em(40)_vi:TMGCG Gb(84)_vi:TMGCG Tobacco mild green mosaic virus complete genome. 1/91 6,355bp

M34236 Em(40)_vi:TMGMPCPA Gb(84)_vi:TMGMPCPA Tobacco mild green mosaic virus movement and coat protein genes, complete cds. 2/91 1,413bp.

X81172 Gb(84)n:TMGMVAUS6 Tobacco mild green mosaic Virus AUS6 genomic RNA. 8/94 414bp

X81173 Gb(84)n:TMGVCA7 Tobacco mild green mosaic Virus CA7 genomic RNA. 8/94 414bp.

X81174 Gb(84)n:TMGVCRA Tobacco mild green mosaic Virus CR-A genomic RNA. 8/94 414bp.

X81175 Gb(84)n:TMGVCRB Tobacco mild green mosaic Virus CR-B genomic RNA. 8/94 414bp.

X81176 Gb(84)n:TMGVE5 Tobacco mild green mosaic Virus E5 genomic RNA. 8/94 414bp.

X81177 Gb(84)n:TMGVEGYPT Tobacco mild green mosaic Virus EGYPT genomic RNA. 8/94 414bp.

X81178 Gb(84)n:TMGVGAE21 Tobacco mild green mosaic Virus GAE2.1 genomic RNA. 8/94 414bp.

X81179 Gb(84)n:TMGVMV1 Tobacco mild green mosaic Virus MV1 genomic RNA. 8/94 414bp

X81180 Gb(84)n:TGMVNSW38 Tobacco mild green mosaic Virus NSW/38 genomic RNA. 8/94 414bp.

X81181 Gb(84)n:TMGMVU2D Tobacco mild green mosaic Virus U2-D genomic RNA. 8/94 414bp.

Sub–genomic *m*RNA found in infected cells.

*Virion protein(s)* one. Amino acid sequence: Altschuh *et al.* (1981).

**Replication**

Replication does not depend on a helper virus.

**Cytopathology**

Virions found in all parts of the host plant (Solberg and Bald, 1963); in cytoplasm, in nuclei, and in chloroplasts. Inclusions present in infected cells; are crystals in the cytoplasm and unusual in shape; angled-layer aggregates and paracrystals; they contain virions.

**Taxonomy and Relationships**

*Tobamovirus.*

Virus(es) with serologically related virions: tobacco mosaic, tomato mosaic and odontoglossum ringspot viruses are closely related and cucumber green mottle mosaic and sunn-hemp mosaic tobamoviruses are distantly related.

The virion protein of tobacco mild green mosaic virus has 67–68% amino acid homology with those of tobacco mosaic and tomato mosaic, 60% with odontoglossum ringspot and less than 45% with those of other tobamoviruses; thus it is most closely related to other 'solanaceous tobamoviruses' (Gibbs, 1986). Nucleic acid hybridization tests confirm these results (Palukaitis and Symons, 1980). Best tests for diagnosis: tobacco mild green mosaic virus infects *Eryngium aquaticum* and *E. planum* systemically, whereas tobacco mosaic and pepper mild mottle viruses infect only inoculated leaves.

**References**

Bald, J.G. and Goodchild, D.J. (1960). *Phytopathology* **50**: 497.

Boedtker, H. and Simmons, N.S. (1958). *J. Am. chem. Soc.* **80**: 2550.

Gibbs, A.J. (1986). In: *The Plant Viruses*; Vol. 2, *The Rod-Shaped Plant Viruses*; eds M.H.V. van Regenmortel and H. Fraenkel-Conrat. Plenum Press, New York.

Johnson, J. (1947). *Phytopathology* **37**: 822.

Knight, C.A., Silva, D.M., Dahl, D. and Tsugita, A. (1962). *Virology* **16**: 236.

Köhler, E. and Panjan, M. (1943). *Ber. Deutsch. bot. Ges.* **61**: 175.

McKinney, H.H. (1929). *J. Agric. Res.* **39**: 557.

McKinney, H.H. (1952). *Pl. Dis. Reptr* **36**: 184.

Miller, P.M. and Thornberry, H.H. (1958). *Phytopathology* **48**: 665.

Palukaitis, P. and Symons, R.H. (1980). *Virology* **107**: 354.

Randles, J., Palukaitis, P. and Davies, C. (1981). *Ann. appl. Biol.* **98**: 109.

Solis, I. and Garcia-Arenal, F. (1990). *Virology* **177**: 553.

Siegel, A. and Wildman, S.G. (1954). *Phytopathology* **44**: 277.
Van de Walle, M.J. and Siegel, A. (1982). *Phytopathology* **72**: 390.
Wetter, C. (1980). *Z. PflKrankh. PflPath. PflSchutz.* **87**: 150.
Wetter, C. (1984). *Phytopathology* **74**: 1308.
Wetter, C. (1986). In: *Plant Viruses*; Vol. 2, *The Rod-Shaped Plant Viruses*; eds M.H.V. van Regenmortel and H. Fraenkel-Conrat. Plenum Press, New York.
Wetter, C., Conti, M., Altschuh, D., Tabillion, R. and van Regenmortel, M.H.V. (1984). *Phytopathology* **74**: 405.

# Tobacco mosaic satellivirus

**Data collated by G. Kurath and J.A. Dodds, 1992.**

*Synonyms:* satellite tobacco mosaic virus. *Strains:* known to exist but none yet described in the literature.

First reported in *Nicotiana glauca*; from southern California, U.S.A.; by Valverde and Dodds (1986).

**Natural host range and symptoms**

*Nicotiana glauca* — symptomless; symptoms may be caused by mixed infection with cucumber mosaic virus or potyviruses.

**Transmission**

Virus transmitted by mechanical inoculation; not transmitted by seed.

**Geographical distribution**

Spreads in the USA.

**Experimental host range**

Several (3–9) families susceptible.

Diagnostically susceptible

*Nicotiana tabacum, N. glauca, N. benthamiana, N. rustica, N. sylvestris, Lycopersicon esculentum, Physalis floridana, Datura metaloides, Catharanthus roseus* — symptoms of the helper virus are unaltered.

*Capsicum annuum* — modifies helper virus symptoms by increased chlorosis and transient amelioration of leaf blister.

Diagnostically insusceptible

Only those that do not support the helper virus.

Maintenance and propagation hosts

*Nicotiana tabacum* cv. Xanthi for systemic infections, *N. tabacum* cv. Xanthi-*nc* for local lesions passage.

Assay hosts (**L**ocal lesions or **W**hole plants)

*Capsicum annuum* (W).

Dicotyledonous Hosts *(**Susceptible** or Insusceptible)*

***Capsicum annuum. Catharanthus roseus. Chenopodium amaranticolor. Chenopodium quinoa. Cucurbita pepo. Datura metel. Datura stramonium. Gomphrena globosa. Lycopersicon esculentum. Nicotiana benthamiana. Nicotiana clevelandii. Nicotiana glutinosa. Nicotiana rustica. Nicotiana sylvestris. Nicotiana tabacum. Physalis floridana.***

Sources of host-range data

Valverde and Dodds (1986); Valverde *et al.* (1991).

**Purification**

Method: Valverde *et al.* (1991).

**Particle morphology**

Virions isometric; not enveloped; less than 17 nm in diameter; neither clearly angular or nor rounded in profile; without a conspicuous capsomere arrangement.

**Physical properties**

One sedimenting component in purified preparations; sedimentation coefficient 42 *S*.

**Biochemical properties**

Virions contain 25 % nucleic acid; 75 % protein.

*Genome* consists of RNA; single–stranded; linear; unipartite; largest (or only) genome part 1.059 kb. Genomic nucleic acid isolated by proteinase K treatment and phenol:chloroform extraction. Base composition 24.4 % G; 26.1 % A; 21.3 % C; 28.2 % U. 5' terminus of RNA has an unusual type; is non-phosphorylated. Infectivity retained when deproteinised with proteases; retained when deproteinised with phenol or detergent. Poly A region absent. Additional factor not required for infectivity. Nucleotide sequence references: Mirkov *et al.* (1989).

*Sequence database accession code(s):* Genbank M25782.

*Features of the genome:* ORF 1, nt 53–229, encodes a 6.8kDa protein of unknown function. This ORF has been translated *in vitro* but its role *in vivo* has not been ascertained. ORF 2, nt 163–642, encodes the virion coat protein both *in vitro* and *in vivo*. The 3' terminal 240 nt share 65% overall sequence similarity with the 3' terminus of TMV–U1. This includes two regions of near identity which are 40 and 50 nt long.

The sequence of tobacco mosaic satellite virus genomic RNA indicates that, probably, much of it is base paired.

*Non–genomic nucleic acid* not found in the virions. Sub–genomic *m*RNA not found in infected cells. 1 virus specified dsRNA species found in infected cells. Size of largest virus specified dsRNA 1.1 kbp.

*Virion protein(s)* one; $M_r$ 17500; coat protein. Method of preparation: Valverde and Dodds (1987).

*Virus–coded non–virion proteins* isolated (Mirkov *et al.* (1989)); one protein found (*in vitro*). $M_r$ 68000.

**Replication**

Genome replicates in cytoplasmic viroplasms. Replication depends on co–infection with a helper virus; but the specific basis of dependence is not known.

**Cytopathology**

Virions found in leaves, roots, mesophyll and epidermis; in cytoplasm. Inclusions present in infected cells; are crystals in the cytoplasm, viroplasms, and unusual in shape; unit membranes with vesicles containing fibrillar material surround viral crystals; proteinaceous bodies are contained within this inclusion; they contain virions.

**Taxonomy and Relationships**

Satellivirus.

Virus(es) with serologically unrelated virions: tobacco necrosis satellite, panicum mosaic satellite, and maize white line mosaic satellite viruses.

This satellite virus is found naturally associated with tobacco mild green mosaic tobamovirus, but can be experimentally coinfect with any of tobamoviruses. Best tests for diagnosis: tobacco mosaic satellite virus causes no symptom modification in any host plants except *Capsicum annuum*. Therefore the quickest way to identify it is by Ouchterlony diffusion tests with tobacco mosaic satellite virus specific antiserum available from J.A. Dodds.

**Comments**

Tobacco mosaic satellite virus crystallizes very readily (Koszelak *et al.*, 1989). The virion structure has recently been resolved to the 0.3nm level (A. McPherson, personal communication).

**References**


Dodds, J.A. (1991). *Can. J. Pl. Pathol.* **13**: 192.

Kim, K.S., Valverde, R.A. and Dodds, J.A. (1989). *J. Ultrastruct. Mol. Struct. Res.* **102**: 196.

Koszelak, S., Dodds, J.A. and McPherson, A. (1989). *J. mol. Biol.* **209**: 323.

Mirkov, T.E., Mathews, D.M., DuPlessis, D.H. and Dodds, J.A. (1989). *Virology* **170**: 139.

Mirkov, T.E., Kurath, G., Mathews, D.M., Elliot, K., Dodds, J.A. and Fitzmaurice, L. (1990). *Virology* **179**: 395.

Valverde, R.A. and Dodds, J.A. (1986). *J. gen. Virol.* **67**: 1875.

Valverde, R.A. and Dodds, J.A. (1987). *J. gen. Virol.* **68**: 965.

Valverde, R.A., Heick, J.A. and Dodds, J.A. (1991). *Phytopathology* **81**: 99.


# Tobacco mosaic *tobamovirus*

**Data collated by M. Zaitlin, 1984.**

*Synonyms:* virus mosaique de tabac, tobacco mosaic virus — U1, tobacco mosaic virus — type, tobacco mosaic virus — vulgare, tobacco mosaic virus common strain. *Standard acronym:* TMV. *Strains:* Japanese common strain, masked strain M and many others.

First reported in *Nicotiana tabacum*; from Russia and the U.S.A.; by Mayer (1886); Iwanowski (1892); Allard (1914); Stanley (1935).

**Natural host range and symptoms**

Symptoms persist.

*Nicotiana tabacum* — leaf mosaic, severe crop losses.

Also found in many other plant species.

**Transmission**

Transmitted by means not involving a vector. Virus transmitted by mechanical inoculation; transmitted by grafting; transmitted by contact between plants; transmitted by seed (occasionally transmitted through the testa, but not through the embryo); not transmitted by pollen.

**Geographical distribution**

Probably distributed worldwide.

**Experimental host range**

Many (>9) families susceptible.

Diagnostically susceptible

*Nicotiana tabacum* (various cultivars) — systemic mosaic.

*Chenopodium quinoa, Nicotiana glutinosa and N. tabacum* cv. Xanthi-*nc* — local lesions.

Maintenance and propagation hosts

*Nicotiana tabacum* cv. Turkish Samsun.

Assay hosts (**L**ocal lesions or **W**hole plants)

*Chenopodium quinoa* (L), *Nicotiana glutinosa* (L), *N. tabacum* cvs Samsun NN or Xanthi-*nc* (L).

Dicotyledonous Hosts (***Susceptible*** *or Insusceptible)*

*Apium graveolens. Bellis perennis.* ***Beta vulgaris****. Brassica campestris* ssp. *pekinensis. Capsella bursa–pastoris.* ***Capsicum frutescens****. Cheiranthus cheiri.* ***Chenopodium amaranticolor****.* ***Chenopodium hybridum****.* ***Chenopodium quinoa****. Chrysanthemum morifolium.* ***Cucumis melo****.* ***Cucumis sativus****.* ***Cucurbita pepo****. Dahlia pinnata.* ***Datura stramonium****. Daucus carota. Dianthus barbatus. Gossypium hirsutum. Helianthus annuus. Ipomoea nil. Ipomoea setosa.* ***Lactuca sativa****.* ***Lycopersicon esculentum****.* ***Lycopersicon pimpinellifolium****. Matthiola incana.* ***Nicotiana benthamiana****.* ***Nicotiana bigelovii****.* ***Nicotiana clevelandii****.* ***Nicotiana debneyi****.* ***Nicotiana glutinosa****. Nicotiana megalosiphon.* ***Nicotiana rustica****.* ***Nicotiana sylvestris****.* ***Nicotiana tabacum****.* ***Papaver nudicaule****. Pastinaca sativa. Petroselinum crispum. Petunia hybrida.* ***Phaseolus vulgaris****.* ***Physalis floridana****.* ***Physalis peruviana****. Pisum sativum. Raphanus sativus. Salvia splendens. Sinningia speciosa. Solanum melongena. Solanum nigrum.* ***Solanum tuberosum****. Spinacia oleracea. Stellaria media. Zinnia elegans.*

Monocotyledonous Hosts (***Susceptible*** *or Insusceptible)*

*Avena sativa. Hordeum vulgare. Secale cereale. Triticum aestivum. Zea mays.*

Sources of host-range data

Holmes (1946); Cheo and Gerard (1971).

**Properties of particles in sap**

LIV: 3000 days. Infectivity of sap not changed by treatment with di–ethyl ether. Leaf sap contains many virions.

**Purification**

Method: Gooding and Hebert (1967); Asselin and Zaitlin (1978).

**Particle morphology**

Virions rod–shaped; not enveloped; usually straight; with a clear modal length; of 300 nm; 18 nm wide. Axial canal obvious; 2 nm in diameter. Basic helix obvious; pitch of basic helix 2.3 nm.

**Physical properties**

One sedimenting component in purified preparations (the main component is of intact virions, but shorter rods are found, and two of these contain encapsidated sub-genomic RNAs); sedimentation coefficient 194 *S*. Density 1.325 g $cm^{-3}$ in CsCl. Isoelectric point pH 3.5.

**Biochemical properties**

Virions contain 5 % nucleic acid; 95 % protein; 0 % lipid. Also include a trace amount of metal ions.

*Genome* consists of RNA; single–stranded; linear; unipartite; largest (or only) genome part 6.395 kb. Genomic nucleic acid isolated by Mandeles and Bruening (1968). Base composition 25.3 % G; 29.8 % A; 18.5 % C; 26.3 % U. 5' terminus of RNA has a methylated nucleotide cap. Infectivity retained when deproteinised with proteases; retained when deproteinised with phenol or detergent. Poly A region absent (at 3' terminus). Additional factor not required for infectivity. Genome has *t*RNA–like activity. Genome accepts histidine. Nucleotide sequence references: Goelet *et al.*, (1982).

*Sequence database accession code(s):*

D13367 Gb(84)_vi:MTVCP Tobacco mosaic virus CP gene. 7/94 474bp.

D13438 Em(40)_vi:MTVGRNA Gb(84)_vi:MTVGRNA Tobacco mosaic virus genomic RNA. 12/93 6,507bp.

D17458 Em(40)_vi:MTV30KP Gb(84)_vi:MTV30KP Tobacco mosaic virus (TMV) RNA for 30K protein, complete cds. 3/94 795bp. D38444 Gb(84)n:MTVCG Tobacco mosaic virus RNA. 10/94 6,303bp.

J02412 Em(40)_vi:TO30KOM Gb(84)_vi:MTV30KOM tobacco mosaic virus 30k protein gene. 4/90 961bp.

J02413 Em(40)_vi:TOC30KCP Gb(84)_vi:MTV30KCP tobacco mosaic virus(cowpea strain) 30k & coat protein genes. 4/90 1,800bp.

J02414 Em(40)_vi:TOL30KCP Tobacco mosaic virus 3' end (30k & coat protein genes). 6/85 1,614bp.

J02415 Gb(84)_vi:MTVVCG Tobacco mosaic virus (strain vulgare), complete genome. 9/88 6,395bp.

L11665 Em(40)_vi:MTVNGHYPE Gb(84)_vi:MTVNGHYPER Tobacco mosaic virus RNA. 8/93 6,506bp

L35073 Gb(84)n:MTVCOATPRA Tobacco mosaic virus coat protein, complete cds. 8/94 678bp.

L35074 Gb(84)n:MTVCOATPRO Tobacco mosaic virus coat protein, complete cds. 8/94 680bp.

M19101 Em(40)_sy:AGVCHY Gb(84)_sy:SYNRMTVCHY Tobacco mosaic virus/calf chymosin recombined mRNA, promoter and 5' end. 7/89 101bp.

M19102 Em(40)_sy:AGVLSZ Gb(84)_sy:SYNRMTVLSZ Tobacco mosaic virus/chicken lysozyme recombined mRNA, promotor and 5' end. 7/89 107bp

M19103 Em(40)_sy:AGVCAT Gb(84)_sy:SYNRMTVCAT Tobacco mosaic virus/plasmid pJII2 chloramphenicol transferase recombined mRNA, leader and 5'

M19104 Em(40)_sy:AGVNTP Gb(84)_sy:SYNRMTVNTP Tobacco mosaic virus/plasmid pJII3 neomycin phosphotransferase II recombined mRNA, leader and

M19105 Em(40)_sy:AGVGUSA Gb(84)_sy:SYNVGUSA Tobacco mosaic virus/plasmid pJII119 beta-glucuronidase recombined mRNA, leader and 5' end. 7

M19106 Em(40)_sy:AGVGUSB Gb(84)_sy:SYNVGUSB Tobacco mosaic virus/plasmid pJII139 beta-glucuronidase recombined mRNA, leader and 5' end. 7

M22483 Em(40)_vi:TOBMTVCO Tobacco mosaic virus coat protein, 3' end. 4/90 229bp

M24809 Em(40)_vi:TOBMTVGT Gb(84)_vi:MTVGTAMV tobacco mosaic virus RNA, 3' end. 2/90 72bp.

M24955 Em(40)_vi:MTVU1RAA Gb(84)_vi:MTVU1RAA Tobacco mosaic virus (U1) omega RNA. 9/90 70bp.

M24992 Gb(84)_vi:MTVU2RAA Tobacco mosaic virus (U2) omega RNA. 9/89 93bp.

V01405 Em(40)_vi:TOTMV1 Gb(84)_vi:TOTMV1 Tobacco mosaic virus (TMV) RNA 5' coding region (nucleotides 69 to 236). 7/89 168bp.

V01407 Em(40)_vi:TOTMV3 Gb(84)_vi:TOTMV3 Two tobacco mosaic virus genes (viral transport and coat protein). 9/93 961bp.

V01408 Em(40)_vi:TOTMV4 Gb(84)_vi:TOTMV4 Tobacco mosaic virus genome (variant 1). 7/83 6,395bp

V01409 Em(40)_vi:TOTMV5 Gb(84)_vi:TOTMV5 Tobacco mosaic virus genome (variant 2). 7/83 6,398bp.

X00052 Em(40)_vi:TOTMV6 Gb(84)_vi:TOTMV6 Tobacco mosaic virus (TMV) common strain OM. 5'-terminal region. 6/85 275bp.

X00053 Em(40)_vi:TOTMV7 Gb(84)_vi:TOTMV7 Tobacco mosaic virus tomato (TMV) strain L. 5'-terminal region. 6/85 278bp.

X02144 Em(40)_vi:TOTMV8 Gb(84)_vi:TOTMV8 Tobacco mosaic virus tomato strain (L) genome. 9/93 6,384bp.

X66047 Em(40)_vi:TMV54KDA Gb(84)_vi:TMV54KDA Tobacco Mosaic Virus RNA for 54 kDa protein. 6/92 1,566bp.

X68110 Em(40)_vi:TMVCG Gb(84)_vi:TMVCG Tobacco mosaic virus, complete genome. 10/92 6,395bp.

X70882 Em(40)_vi:TMVPM2CP Gb(84)_vi:TMVPM2CP Tobacco mosaic virus PM2 mRNA for capsid protein. 7/93 765bp

X70883 Em(40)_vi:TMVDT1CP Gb(84)_vi:TMVDT1CP Tobacco mosaic virus DT1 mRNA for capsid protein. 7/93 765bp.

X70884 Em(40)_vi:TMVDT2CP Gb(84)_vi:TMVDT2CP Tobacco mosaic virus DT2 mRNA for capsid protein. 7/93 763bp.

X70885 Em(40)_vi:TMVDT1GCP Gb(84)_vi:TMVDT1GCP Tobacco mosaic virus DT1G mRNA for capsid protein. 7/93 763bp.

Z29370 Em(40)_vi:TMVRPTPCP Gb(84)_vi:TMVRPTPCP Tobacco mosaic virus (Crucifer) genomic RNA for RNA-dependent RNA polymerase; 122K protein.

M25782 Em(43)_vi:Sllcp Gb(89)_vi:Sllcp Satellite tobacco mosaic virus coat protein RNA, complete cds. 11/94 1,058bp.

*Non–genomic nucleic acid* found in the virions; is subgenomic *m*RNA and host genomic RNA. Sub–genomic *m*RNA found in infected cells.

*Virion protein(s)* two; $M_r$ of the larger 26500; protein of unknown function; about 1 copy per virion. $M_r$ of 2nd largest 17500; coat protein. Method of preparation: Fraenkel-Conrat (1957); Collmer *et al.* (1983). Amino acid sequence: Goelet *et al.*, (1982).

*Virus–coded non–virion proteins* identified by genomic sequence analysis; four proteins found. $M_r$ of the largest 183000; possibly a replicase protein. $M_r$ of 2nd largest 126000; possibly a helicase protein. $M_r$ of 3rd 54000. $M_r$ of 4th 30000.

**Replication**

Genome replicates possibly on membranous structures in the cytoplasm: the protein synthesized by cytoplasmic ribosomes, the assembly site for virions not known. Coat protein *m*RNA translated in the cytoplasm. Replication does not depend on a helper virus.

**Cytopathology**

Virions found in all parts of the host plant; in cytoplasm and in nuclei; and possibly chloroplasts. Inclusions present in infected cells; are crystals in the cytoplasm and amorphous X–bodies; they contain virions. Other cellular changes: chloroplasts disrupted in systemic infections.

**Taxonomy and Relationships**

*Tobamovirus* (the type species).

Virus(es) with serologically related virions: all species of the genus related to different extents (Van Regenmortel, 1970).

**References**

Allard, H.A. (1914). *Bull. U.S. Dep. Agric.* **40**: 33 pp.

Asselin, A. and Zaitlin, M. (1978). *Virology* **91**: 173.

Butler, P.J.G. (1984). *J. gen. Virol.* **65**: 253.

Cheo, P.C. and Gerard, J.S. (1971). *Phytopathology* **61**: 1010.

Clark, M.F. and Bar-Joseph, M. (1984). *Methods in Virol.* VII, p. 51.

Collmer, C.H., Vogt, V.M. and Zaitlin, M. (1983). *Virology* **126**: 429.

Fraenkel-Conrat, H. (1957). *Virology* **4**: 1.

Garcia-Arenal, F., Palukaitis, F. and Zaitlin, M. (1984). *Virology* **132**: 131.

Gibbs, A.J. (1977). *CMI/AAB Descr. Pl. Viruses* No. 184, 6 pp.

Goelet, P., Lomonossoff, G.P., Butler, P.J.G., Akam, M.E., Gait, M.J. and Karn, J. (1982). *Proc. Natl. Acad. Sci., U.S.A.* **79**: 5818.

Gooding, J.V. and Hebert, T.T. (1967). *Phytopathology* **57**: 1285.

Haseloff, J., Goelet, P., Zimmern, D., Ahlquist, P., Dasgupta, R. and Kaesberg, P. (1984). *Proc. Natl. Acad. Sci., U.S.A.* **81**: 4358.

Hills, J.G., Plaskitt, K.A., Young, M.D., Dunigan, D.D., Watts, J.W., Wilson, T.M.A. and Zaitlin, M. (1987). *Virology* **161**: 488.

Hirth, L. and Richards, K.E. (1981). *Adv. Virus Res.* **26**: 145.

Holmes, F.O. (1946). *Phytopathology* **36**: 643.

Iwanovski, D. (1892). *Izv. imp. Akad. Nauk.* **35**: 67.

Mandeles, S. and Brüening, G. (1968). *Biochem. Prep.* **12**: 111.

Mayer, A. (1886). *Landw. VersStnen* **32**: 451.

Milne, R.G. (1984). *Methods in Virol.* VII, p. 88.

Rezaian, M.A., Williams, R.H.V., Gordon, K.H.J., Gould, A.R. and Symons, R.H. (1984). *Eur. J. Biochem.* **143**: 277.

Stanley, W.M. (1935). *Science* **81**: 644.

Tsugita, A., Gish, D.T., Yopung, J., Fraenkel-Conrat, H., Knight, C.A. and Stanley, W.M. (1960). *Proc. Natl. Acad. Sci. U.S.A.* **46**: 1463.

Van Regenmortel, M.H.V. (1970). *Virology* **31**: 467.

Van Regenmortel, M.H.V. (1981). In: *Handbook of Plant Virus Infections and Comparative Diagnosis*, p. 541; ed. E. Kurstak. Elsevier/North Holland Biomedical Press, Amsterdam.

Zaitlin, M. and Israel, H.W. (1975). *CMI/AAB Descr. Pl. Viruses* No. 151, 5 pp.

# Tobacco mottle *umbravirus*

**Data collated by A.A. Brunt, 1987.**

*Standard acronym:* TMoV.

First reported in *Nicotiana tabacum*; from Zimbabwe; by Smith (1946*a*).

**Natural host range and symptoms**

Symptoms persist.

*Nicotiana tabacum* — vein clearing and dark green leaf mottling. Mixed infections with tobacco vein distorting luteovirus give the rosette disease.

**Transmission**

Transmitted by a vector; an insect; *Myzus persicae, Acyrthosiphon (Aulacorthum) solani*; Aphididae. Virus requires, for vector transmission, a helper virus (tobacco vein-distorting luteovirus); transmitted by mechanical inoculation.

**Ecology and control**

Studies reported by Smith (1946*a,b*).

**Geographical distribution**

Spreads in Zimbabwe.

**Experimental host range**

Few (<3) families susceptible.

Diagnostically susceptible

*Nicotiana tabacum, N. glutinosa* — dark green leaf mottling.
*Lycopersicon esculentum* — systemic transient leaf mottling.
*Datura stramonium* — systemic leaf mottling and malformation.

Maintenance and propagation hosts

*Nicotiana tabacum, N. glutinosa, Lycopersicon esculentum.*

Assay hosts (**L**ocal lesions or **W**hole plants)

*Nicotiana tabacum* (W).

Dicotyledonous Hosts *(**Susceptible** or Insusceptible)*

***Datura stramonium. Lycopersicon esculentum. Nicotiana glutinosa. Nicotiana tabacum.***

**Properties of particles in sap**

TIP: 55 °C (in tobacco sap). LIV: 2 days. DEP: $\log_{10}$ minus 2–3.

**Taxonomy and Relationships**

*Umbravirus.*

**Comments**

Tobacco mottle was probably the first virus found to require an assistor virus (tobacco vein distorting luteovirus) for its transmission by aphids (Smith 1946*a,b*). Few properties of the virus are known, but it may resemble groundnut rosette and carrot mottle 'umbraviruses'.

**References**

Smith, K.M. (1946*a*). *Parasitol.* **37**: 21.
Smith, K.M. (1946*b*). *Parasitol.* **37**: 131.

# Tobacco necrosis *necrovirus*

**Data collated by D.S. Teakle, 1982. Revised 1985. Updated 1990 by A.A. Brunt.**

*Synonyms:* tulip Augusta disease virus, bean stipple streak virus, chenopodium necrosis *necrovirus* (Tomlinson *et al.*, 1983), cucumber systemic necrosis virus, strawberry necrotic rosette virus (Faccioli, 1970; 1974). *Standard acronym:* TNV. *Strains:* many reported, including A, B, C, D E, and S (Babos and Kassanis, 1963), cucumber necrosis (Van Koot and van Dorst, 1955), stipple streak (Van der Want, 1948), Ac 36, Ac 38 and Ac 43 (Uyemoto *et al.*, 1968), Urbana (Lesnaw and Reichmann, 1969), chenopodium necrosis (Tomlinson *et al.*, 1983).

First reported in *Nicotiana tabacum*; from Cambridge, U.K; by Smith and Bald (1935).

**Natural host range and symptoms**

Symptoms persist.

*Chenopodium amaranticolor, C. quinoa* — local lesions (some strains), systemic necrosis (Chenopodium necrosis strain only).

*Cucumis sativus* — systemic necrosis of leaves and stems (cucumber necrosis strain only).

*Nicotiana clevelandii, N. tabacum* — necrotic local lesions.

*Phaseolus vulgaris* — discrete local lesions only (D strains), local lesions that spread along minor veins (A strains), systemic infection (stipple streak strain only).

*Tulipa gesneriana* — necrosis of leaves and stems (Augusta disease).

**Transmission**

Transmitted by a vector; a fungus; *Olpidium brassicae*; Chytridiales. Virus transmitted by mechanical inoculation; not transmitted by seed; not transmitted by pollen.

**Geographical distribution**

Probably distributed worldwide.

**Experimental host range**

Many (>9) families susceptible.

Diagnostically susceptible

*Chenopodium amaranticolor, C. quinoa, Lactuca sativa, Cucumis sativus, Phaseolus vulgaris, Vigna radiata, V. unguiculata, Nicotiana clevelandii, N. tabacum* — necrotic local lesions, usually no systemic spread (except the bean stipple streak strain in *P. vulgaris*).

Maintenance and propagation hosts

*Phaseolus vulgaris, Vigna unguiculata, Nicotiana clevelandii, N. tabacum.*

Assay hosts (**L**ocal lesions or **W**hole plants)

*Chenopodium amaranticolor* (L), *Phaseolus vulgaris* (L), *Vigna radiata* (L), *V. unguiculata* (L).

Dicotyledonous Hosts *(**Susceptible** or Insusceptible)*

***Amaranthus caudatus*. *Antirrhinum majus*. *Arachis hypogaea*. *Beta vulgaris*. *Brassica campestris*** ssp. ***rapa*. *Brassica oleracea*** ssp. ***capitata***. *Calendula officinalis*. ***Capsicum frutescens*. *Catharanthus roseus*. *Chenopodium album*. *Chenopodium amaranticolor*. *Cichorium endiva*. *Clarkia amoena*. *Cucumis melo*. *Cucumis sativus*. *Cucurbita maxima*. *Cucurbita pepo*. *Cyamopsis tetragonoloba*. *Datura stramonium***. *Daucus carota*. *Dianthus barbatus*. *Euphorbia marginata*. ***Fragaria vesca*. *Glycine max*. *Gomphrena globosa*. *Gypsophila elegans*. *Helianthus annuus*. *Hyoscyamus niger***. *Ipomoea setosa*. ***Lactuca sativa*. *Lathyrus odoratus*. *Lycopersicon esculentum***. *Medicago sativa*. *Myosotis sylvatica*. ***Nicandra physaloides*. *Nicotiana glutinosa*. *Nicotiana rustica*. *Nicotiana sylvestris*. *Nicotiana tabacum*. *Phaseolus vulgaris***. *Physalis peruviana*. ***Pisum sativum***. *Raphanus sativus*. ***Ricinus communis*. *Sinningia speciosa***. *Solanum melongena*. ***Solanum nodiflorum*. *Sonchus oleraceus*. *Spinacia oleracea*. *Tetragonia expansa*. *Trifolium incarnatum*. *Trifolium pratense*. *Trifolium repens***. *Tropaeolum majus*. *Vicia faba*. ***Vigna radiata*. *Vigna unguiculata*. *Zinnia elegans***.

Monocotyledonous Hosts *(**Susceptible** or Insusceptible)*

***Avena sativa*. *Hordeum vulgare***. *Secale cereale*. ***Triticum aestivum*. *Zea mays***.

Sources of host-range data

Price (1940); Thornberry (1966).

**Properties of particles in sap**

TIP: 85–95 °C. LIV: 7–28 days. DEP: $\log_{10}$ minus 4. Leaf sap contains few virions. Electron microscopy: aldehyde fixation necessary for some strains (Finlay and Teakle, 1969).

**Purification**

Method: Finlay and Teakle (1969); Kassanis (1970).

**Particle morphology**

Virions isometric; not enveloped; 26 nm in diameter; rounded in profile; without a conspicuous capsomere arrangement.

**Physical properties**

One sedimenting component in purified preparations; sedimentation coefficient 118 *S*. Density 1.399 g $cm^{-3}$ in CsCl. Isoelectric point pH 4.5 (B strain).

**Biochemical properties**

Virions contain 19 % nucleic acid; 81 % protein; 0 % lipid.

*Genome* consists of RNA; single–stranded; linear; unipartite; largest (or only) genome part 3.759 kb. Base composition 24 % G; 27.5 % A; 22.5 % C; 25.5 % U. 5' terminus of RNA has an unusual type (uncapped). Infectivity retained when deproteinised with phenol or detergent. Poly A region at 3' terminus absent.

*Sequence database accession code(s):*

D00942 Em(40)_vi:TNCDNSC Gb(84)_vi:TNCDNSC Tobacco necrosis virus genome, complete sequence. 1/94 3,759bp.

L04261 Em(40)_vi:TNCPOLCOA Gb(84)_vi:TNCPOLCOAT Tobacco necrosis virus (Nebraska) RNA-dependent RNA polymerase gene, 3' end p10.7, p6.2 and c

M33002 Em(40)_vi:TNCPOLY Gb(84)_vi:TNCPOLY Tobacco necrosis virus RNA dependent RNA polymerase and coat protein genes, complete cds. 7/9

X58455 Gb(84)_vi:TNVARNA Tobacco necrosis virus type A RNA. 6/94 2,236bp.

*Features of the genome:* 5 ORFs encoding proteins of 23, 82, 7.9, 6.2 and 30 kDa with a sixth encoding a 6.7 kDa protein found only in strain A. The 82 kDa protein is probably a polymerase, and the 30 kDa the coat protein.

*Non–genomic nucleic acid* not found in the virions. Sub–genomic *m*RNA found in infected cells; Two subgenomic RNAs (1.6 and 1.3 kb); the smaller encodes the coat protein, and the larger the proteins of ORF 3 and, possibly, ORF 4.

*Virion protein(s)* one; $M_r$ 30000 (268–275 amino acid residues); coat protein. Method of preparation: Lesnaw and Reichmann (1969); Uyemoto and Grogan (1969).

**Replication**

Replication does not depend on a helper virus.

**Cytopathology**

Virions found in roots; in cytoplasm. Inclusions present in infected cells; are crystals in the cytoplasm, or unusual in shape; string-like arrays and aggregates in the nucleus; they contain virions (Edwardson and Christie, 1986).

**Taxonomy and Relationships**

*Necrovirus* (the type species).

**References**

Adam, G., Winter, S. and Lesemann, D.-E. (1990). *Ann. appl. Biol.* **116**: 523.

Babos, P. and Kassanis, B. (1962). *Virology* **18**: 206.

Babos, P. and Kassanis, B. (1963). *J. gen. Microbiol.* **32**: 135.
Bawden, F.C. and van der Want, J.P.H. (1949). *Tijdschr. PlZiekt.* **55**: 142.
Coutts, R.H.A., Rigden, J.E., Slabar, A.R., Lomonossoff, G.P. and Wise, P.J. (1991). *J. gen. Virol.* **72**: 1521.
Edwardson, J.R. and Christie, R.G. (1986). *Fla Agric. Exp. Stn Monog.* No. 14, p. 525.
Faccioli, G. (1970). *Proc. 4th Nat. Strawberry Conf., Cesena, Italy*, p. 351.
Faccioli, G. (1974). *Phytopathol. Medit.* **13**: 71.
Finlay, J.R. and Teakle, D.S. (1969). *J. gen. Virol.* **5**: 93.
Gama, M.I.C.S., Kitajima, E.W. and Lin, M.T. (1982). *Phytopathology* **72**: 529.
Kassanis, B. (1970). *CMI/AAB Descr. Pl. Viruses* No. 14, 4 pp.
Kassanis, B. (1977). In: *Atlas of Insect and Plant Viruses*, p. 281; ed K. Maramorosch. Academic Press, New York.
Kassanis, B. and Macfarlane, I. (1965). *Virology* **26**: 603.
Lesnaw, J.A. and Reichmann, M.E. (1969). *Virology* **39**: 738.
Mali, V.R. (1976). *Indian Phytopath.* **29**: 266.
Meulewaeter, F., Seurinck, J. and van Emmelo, J. (1990). *Virology* **177**: 699.
Price, W.C. (1940). *Am. J. Bot.* **27**: 530.
Smith, K.M. and Bald, J.G. (1935). *Parasitol.* **27**: 231.
Stussi-Garaud, C., Lemins, J. and Fraenkel-Conrat, H. (1977). *Virology* **81**: 224.
Teakle, D.S. (1962). *Phytopathology* **52**: 1037.
Temmink, J.H.M., Campbell, R.N. and Smith, P.R. (1970). *J. gen. Virol.* **9**: 201.
Thornberry, H.H. (1966). In: *Index of Plant Virus Diseases.* U.S. Dep. Agric. Hdbk No. 307, p. 312.
Tomlinson, J.A., Faithfull, E.M., Webb, M.J.W., Fraser, R.S.S. and Seeley, N.D. (1983). *Ann. appl. Biol.* **102**: 135.
Uyemoto, J.K. (1981). in *Handbook of Plant Infections and Comparative Diagnosis.* p.123, ed. E. Kurstak. Elsevier/North Holland Biomedical Press, Amsterdam.
Uyemoto, J.K. and Grogan, R.G. (1969). *Virology* **39**: 79
Uyemoto, J.K., Grogan, R.G. and Wakeman, J.R. (1968). *Virology* **34**: 410.
Van Koot, Y. and van Dorst, H.J.M. (1955). *Tijdschr. PlZiekt.* **61**: 163.
Van der Want, J.P.H. (1948). *Tijdschr. PlZiekt.* **54**: 85.

# Tobacco necrosis satellivirus

**Data collated by A.A. Brunt, 1991.**

*Synonyms:* satellite virus. *Standard acronym:* STNV. *Strains:* $SV_1$, $SV_2$ and $SV_3$ (England) and $SV_A$, $SV_B$ and $SV_C$ (U.S.A.).

First reported in *Nicotiana tabacum*; from the U.K.; by Kassanis and Nixon, 1960.

### Natural host range and symptoms

*Phaseolus vulgaris, Tulipa gesneriana* — symptoms unknown. It only infects when with its natural helper, tobacco necrosis necrovirus, and makes its lesions smaller.

### Transmission

Transmitted by a vector; a fungus; *Olpidium brassicae*; Chytridiales. Virus transmitted by mechanical inoculation; not transmitted by seed.

### Geographical distribution

Spreads in the Eurasian region; Canada and the USA.

Diagnostically susceptible

*Phaseolus vulgaris, Nicotiana tabacum* — inocula containing both tobacco necrosis and its satellite virus induce smaller local lesions than the former alone.

Maintenance and propagation hosts

*Nicotiana clevelandii, N. tabacum, Phaseolus vulgaris* (but dependent upon strain of tobacco necrosis virus in inoculum).

Assay hosts (**L**ocal lesions or **W**hole plants)

*Phaseolus vulgaris* (L), *Nicotiana tabacum* (W).

Dicotyledonous Hosts *(**Susceptible** or Insusceptible)*

***Nicotiana clevelandii*. *Nicotiana tabacum*. *Phaseolus vulgaris*.**

Monocotyledonous Hosts *(**Susceptible** or Insusceptible)*

***Tulipa gesneriana*.**

**Properties of particles in sap**

TIP: 95 °C.

**Purification**

Method: Kassanis (1970).

**Particle morphology**

Virions isometric; not enveloped; 17 nm in diameter; rounded in profile; with a conspicuous capsomere arrangement.

**Physical properties**

One sedimenting component in purified preparations; sedimentation coefficient 50 *S*. Isoelectric point pH 7. $A_{260}/A_{280}$ ratio 1.65.

**Biochemical properties**

Virions contain 20 % nucleic acid; 80 % protein.

*Genome* consists of RNA; single–stranded; unipartite; largest (or only) genome part 1.239 kb. Base composition 22–24 % G; 28–29 % A; 20–22 % C; 25–29 % U.

*Sequence database accession code(s):*

J02399 Gb(84)_vi:TNSCG Satellite tobacco necrosis virus, complete genome. 3/89 1,239bp

L06057 Em(40)_vi:TNSUTR Gb(84)_vi:TNSUTR Satellite tobacco necrosis virus RNA sequence. 11/92 1,239bp.

M16603 Em(40)_sy:CVGNL1 Gb(84)_sy:SYNTNSGNL1 Satellite tobacco necrosis virus genome 5' end with plasmid linker pSTNV-403. 7/89 80bp.

M16604 Em(40)_sy:CVGNL2 Gb(84)_sy:SYNTNSGNL2 Satellite tobacco necrosis virus genome 3' end with plasmid linker pSTNV-403. 7/89 64bp.

M20900 Em(40)_vi:TNSFRNA Gb(84)_vi:TNSFRNA Satellite tobacco necrosis virus (STNV) folded RNA region with an unidentified ORF, 5' end. 7

M64479 Em(40)_vi:TNSCP Gb(84)_vi:TNSCP Satellite of tobacco necrosis virus coat protein gene, complete cds. 12/91 1,245bp.

V01468 Em(40)_vi:XXSTNV Gb(84)_vi:XXSTNV Satellite tobacco necrosis virus genome including reading frame for the viral coat protein.

*Virion protein(s)* one; $M_r$ 23000; coat protein. Method of preparation: Rees *et al.* (1970); Uyemoto and Grogan (1969).

**Cytopathology**

Virions found in all parts of the host plant; in cytoplasm. Inclusions absent from infected cells.

**Taxonomy and Relationships**

Satellivirus (type species).

**Comments**

Type strain of above group.

**References**

Kassanis, B. (1962). *J. gen. Microbiol.* **27**: 477.
Kassanis, B. (1970). *CMI/AAB Descr. Pl. Viruses* No. 15, 4 pp.
Kassanis, B. and Kleczkowski, A. (1965). *Nature, Lond.* **205**: 310.
Kassanis, B. and MacFarlane, I. (1968). *J. gen. Virol.* **3**: 277.
Kassanis, B. and Nixon, H.L. (1960). *Nature, Lond.* **187**: 713.
Kassanis, B. and Nixon, H.L. (1961). *J. gen. Microbiol.* **25**: 459.
Kassanis, B. and Phillips, M.P. (1970). *J. gen. Virol.* **9**: 119.
Kassanis, B., Vince, D.A. and Woods, R.D. (1970). *J. gen. Virol.* **7**: 143.
Kassanis, B. and Woods, R.D. (1968). *J. gen. Virol.* **2**: 395.
Lesnaw, J.A. and Reichmann, M.E. (1969). *Virology* **38**: 368.
Lesnaw, J.A. and Reichmann, M.E. (1969). *Virology* **39**: 729.
Rees, M., Short, M.N. and Kassanis, B. (1970). *Virology* **40**: 448.
Uyemoto, J.K. and Grogan, R.G. (1969). *Virology* **39**: 79.
Uyemoto, J.K., Grogan, R.G. and Wakeman, J.R. (1968). *Virology* **34**: 410.

# Tobacco necrotic dwarf *luteovirus*

**Data collated by S. Kubo, 1984. Revised 1989.**

*Standard acronym:* TNDV.

First reported in *Nicotiana tabacum*; from Tochigi Prefecture, Japan; by Kubo and Takanami (1977).

**Natural host range and symptoms**

Symptoms persist.
*Spinacia oleracea* — leaf vein yellowing.
*Nicotiana tabacum* — interveinal leaf chlorosis, veinal necrosis, severe stunting.
*Capsella bursa-pastoris* — interveinal reddening, stunting.

**Transmission**

Transmitted by a vector; an insect; *Myzus persicae*; Aphididae. Transmitted in a persistent manner. Virus retained when the vector moults; does not multiply in the vector; not transmitted congenitally to the progeny of the vector; transmitted by grafting; not transmitted by contact between plants; not transmitted by seed; not transmitted by pollen.

**Geographical distribution**

Spreads in Japan.

**Experimental host range**

Few (<3) families susceptible.

Diagnostically susceptible

*Capsella bursa-pastoris, Datura stramonium, Nicotiana sylvestris, N. tabacum* — systemic interveinal chlorosis, veinal necrosis, stunting.

*Physalis floridana* — systemic interveinal chlorosis, leaf rolling.

*Montia perfoliata* — mild systemic interveinal chlorosis.

Diagnostically insusceptible

*Beta vulgaris, Brassica rapa, Astragalus sinicus, Pisum sativum, Lactuca sativa.*

Maintenance and propagation hosts

*Physalis floridana.*

Assay hosts (Local lesions or Whole plants)

*Nicotiana sylvestris* (W).

Dicotyledonous Hosts ***(Susceptible** or Insusceptible)*

*Astragalus sinicus. Beta vulgaris. Brassica campestris* ssp. *pekinensis. Brassica campestris* ssp. *rapa. Brassica oleracea* ssp. *capitata.* ***Capsella bursa–pastoris****. Chenopodium amaranticolor. Chenopodium quinoa. Cucumis sativus.* ***Datura metel. Datura stramonium****. Glycine max.* ***Gomphrena globosa****. Lactuca sativa.* ***Lycopersicon esculentum. Montia perfoliata. Nicotiana clevelandii. Nicotiana glutinosa. Nicotiana rustica. Nicotiana sylvestris. Nicotiana tabacum****. Petunia hybrida.* ***Physalis floridana. Physalis peruviana****. Pisum sativum. Raphanus sativus. Senecio vulgaris. Solanum melongena.* ***Solanum nigrum. Solanum nodiflorum. Solanum tuberosum. Spinacia oleracea****. Stellaria media. Tetragonia expansa. Vicia faba. Vigna unguiculata* ssp. *sesquipedalis. Zinnia elegans.*

Sources of host-range data

Kubo (1981).

**Properties of particles in sap**

TIP: 80 °C (assayed using protoplasts). LIV: more than 6 months at 4°C. Leaf sap contains few virions.

**Purification**

Method: Takanami and Kubo (1979).

**Particle morphology**

Virions isometric; not enveloped; 25 nm in diameter; angular in profile; without a conspicuous capsomere arrangement.

**Physical properties**

One sedimenting component in purified preparations; sedimentation coefficient 115 *S*. Density virions disrupted in caesium salts. Isoelectric point pH 5.3.

**Biochemical properties**

Virions contain 30 % nucleic acid; 70 % protein; 0 % lipid.

*Genome* consists of RNA; single–stranded; linear; unipartite; largest (or only) genome part 6.1 kb.

*Non–genomic nucleic acid* found in the virions; is satellite RNA.

*Virion protein(s)* one; $M_r$ 25700; coat protein. Amino acid composition: Kubo (1981).

**Replication**

Replication does not depend on a helper virus.

**Cytopathology**

Virions found in phloem and companion cells; in cytoplasm, in nuclei, and in cell vacuoles. Inclusions absent from infected cells. Other cellular changes: a fibrous appearance in cytoplasm.

**Taxonomy and Relationships**

*Luteovirus.*

Virus(es) with serologically related virions: potato leafroll virus is closely related, but distinct; carrot red leaf and soybean dwarf viruses are distantly related.

**References**

Kubo, S. (1981). *CMI/AAB Descr. Pl. Viruses* No. 234, 4 pp.

Kubo, S. and Takanami, Y. (1977). *Ann. Phytopath. Soc. Japan* **43**: 76.

Takanami, Y. and Kubo, S. (1979). *J. gen. Virol.* **44**: 153.

Takanami, Y. and Kubo, S. (1979). *J. gen. Virol.* **44**: 853.

# Tobacco rattle *tobravirus*

**Data collated by B.D. Harrison and D.J. Robinson, 1984.**

*Synonyms:* aster ringspot virus, belladonna mosaic virus, paeony mosaic virus (Brierley, 1944), paeony ringspot virus (Chang *et al.*, 1976; Jones and Young, 1978; Martin, 1929), potato corky ringspot virus, potato stem mottle virus, stengelbonk virus (Rozendaal and van der Want, 1948), ratel virus, Tabakmauche Virus, Tabakstreifen- und Kraüselkrankheit Virus, spinach yellow mottle virus, tulip white streak virus (Smith, 1950). *Standard acronym:* TRV. *Strains:* fraxinus tobravirus (Cooper *et al.*, 1983), Oregon yellow virus, PRN (Scottish type strain), spinach yellow mottle. NM isolates, which lack the smaller part of the genome and do not produce nucleoprotein, can be prepared from all strains.

First reported in *Nicotiana tabacum*; from Germany; by Böning (1931).

**Natural host range and symptoms**

Symptoms persist (some strains), or vary seasonally, or disappear soon after infection.

*Stellaria media, Viola arvensis* — few symptoms, mottling.

*Beta vulgaris, Spinacia oleracea* — chlorotic or necrotic local lesions; systemic mottle.

*Capsicum annuum* — ringspots or line patterns.

*Solanum tuberosum* — necrotic local lesions; systemic chlorotic or necrotic spots and streaks.

*Nicotiana tabacum* — necrotic local lesions, necrotic or chlorotic ringspots; systemic necrosis, ringspots or line patterns, mottling.

*Narcissus pseudonarcissus, Tulipa* sp., *Hyacinthus* sp. — mottling.

**Transmission**

Transmitted by a vector; a nematode; *Paratrichodorus allius, P. anemones, P. christiei, P. nanus, P. pachydermus, P. teres, Trichodorus minor, T. primitivus, T. viruliferus*; Trichodoridae. Virus transmitted by mechanical inoculation; transmitted by grafting; not transmitted by contact between plants; transmitted by seed (up to 40% in *Viola arvensis* or only to 1% in *Capsella bursa-pastoris*).

**Geographical distribution**

Spreads in the Eurasian region, the North American region, and the South and Central American region; China, Japan, and the former USSR. Found, but with no evidence of spread, in New Zealand and Australia.

**Experimental host range**

Many (>9) families susceptible.

Diagnostically susceptible

*Chenopodium amaranticolor, C. quinoa, Cucumis sativus* — necrotic and chlorotic local lesions; no systemic infection.

*Phaseolus vulgaris* cv. The Prince, *Pisum sativum, Vicia faba* — necrotic local lesions; no systemic infection.

*Nicotiana clevelandii* — necrotic and chlorotic local lesions; few systemic symptoms, mottle, necrosis.

*N. tabacum* cv. Samsun NN — necrotic local lesions and ringspots; systemic necrosis, ringspots, mottle.

Diagnostically insusceptible

Probably *Hordeum vulgare.*

Maintenance and propagation hosts

*Nicotiana clevelandii, N. tabacum.*

Assay hosts (**L**ocal lesions or **W**hole plants)

*Chenopodium amaranticolor* (L), *Phaseolus vulgaris* (L), *Nicotiana tabacum* for some strains (L).

Dicotyledonous Hosts *(**Susceptible** or Insusceptible)*

***Amaranthus caudatus***. ***Amaranthus retroflexus***. ***Antirrhinum majus***. ***Arachis hypogaea***. ***Bellis perennis***. ***Beta vulgaris***. ***Brassica campestris***. ***Brassica campestris*** ssp. ***napus***. ***Brassica campestris*** ssp. ***pekinensis***. ***Brassica juncea***. ***Calendula officinalis***. ***Capsella bursa–pastoris***. ***Capsicum annuum***. ***Catharanthus roseus***. ***Cheiranthus cheiri***. ***Chenopodium album***. ***Chenopodium amaranticolor***. ***Chenopodium foetidum***. ***Chenopodium quinoa***. *Citrus medica*. ***Coriandrum sativum***. ***Cucumis melo***. ***Cucumis sativus***. *Cucurbita pepo*. *Datura ferox*. *Datura metel*. *Datura stramonium*. *Daucus carota*. *Dianthus barbatus*. *Dianthus caryophyllus*. *Fragaria vesca*. ***Glycine max***. ***Gomphrena globosa***. ***Gypsophila elegans***. ***Helianthus annuus***. ***Hyoscyamus niger***. ***Lactuca sativa***. ***Lathyrus odoratus***. ***Linum usitatissimum***. ***Lobelia erinus***. ***Lupinus mutabilis***. ***Lycopersicon esculentum***. ***Lycopersicon pimpinellifolium***. *Matthiola incana*. ***Melilotus alba***. ***Momordica balsamina***. ***Myosotis sylvatica***. ***Nicandra physaloides***. ***Nicotiana benthamiana***. ***Nicotiana clevelandii***. ***Nicotiana × edwardsonii***. ***Nicotiana glutinosa***. ***Nicotiana rustica***. ***Nicotiana sylvestris***. ***Nicotiana tabacum***. ***Ocimum basilicum***. ***Petunia hybrida***. ***Phaseolus vulgaris***. *Phlox drummondii*. *Physalis floridana*. ***Phytolacca americana***. ***Pisum sativum***. ***Raphanus sativus***. ***Ricinus communis***. *Rumex acetosa*. ***Salvia splendens***. ***Senecio vulgaris***. *Sinningia speciosa*. ***Solanum melongena***. ***Solanum nigrum***. ***Solanum tuberosum***. ***Spinacia oleracea***. ***Stellaria media***. ***Trifolium pratense***. ***Trifolium repens***. ***Tropaeolum majus***. ***Vicia faba***. ***Vicia villosa***. *Zinnia elegans*.

Monocotyledonous Hosts *(**Susceptible** or Insusceptible)*

***Allium cepa***. ***Avena sativa***. *Hordeum vulgare*. ***Narcissus pseudonarcissus***. *Secale cereale*. *Triticum aestivum*. *Zea mays*.

Sources of host-range data

Schmelzer (1957); Uschdraweit and Valentin (1956); Noordam (1956).

**Properties of particles in sap**

TIP: 80–85 °C. LIV: 40–50 days. DEP: $\log_{10}$ minus 6. Infectivity of sap not changed by treatment with di–ethyl ether. Leaf sap contains few virions.

**Purification**

Method: Harrison and Nixon (1959); Lister and Bracker (1969).

**Particle morphology**

Virions rod–shaped; not enveloped; usually straight; with a clear modal length; of 46–114 nm (T), or 180–197 nm (B); 22 nm wide. Axial canal obvious; 4 nm in diameter. Basic helix obvious; pitch of basic helix 2.5 nm.

**Physical properties**

Two sedimenting components in purified preparations; sedimentation coefficient B 300 *S*. Density 1.306–1.324 g $cm^{-3}$ in CsCl (not fixed with aldehydes). $A_{260}/A_{280}$ ratio 1.1–1.15.

**Biochemical properties**

Virions contain 5 % nucleic acid; 95 % protein; 0 % lipid.

*Genome* consists of RNA; single–stranded; linear; of two parts; largest (or only) genome part the larger 7.272 kb (RNA-1 from B component virions); the 2nd largest 1.818–4.242 kb (RNA-2 from T component virions). Base composition 24–25 % G; 26–29 % A; 17–18 % C; 29–32 % U. 5' terminus of RNA has a methylated nucleotide cap (in RNA-2; not yet found on RNA-1). Infectivity retained when deproteinised with proteases; retained when deproteinised with phenol or detergent. Poly A region absent. Additional factor not required for infectivity. Nucleotide sequence references: Minson and Darby (1973); Abou Harter and Hirth (1977).

*Sequence database accession code(s):*

D00155 Gb(84)_vi:MTRRNA1 Tobacco rattle virus (TRV), RNA-1, complete cds. 1/92 6,791bp

J04347 Em(40)_vi:TOBMTRCP Gb(84)_vi:MTRCPSA Tobacco Rattle virus coat protein, 16K protein, and 13K protein genes, complete cds. 3/92 2,1

X03241 Em(40)_vi:TOBTRVR2 Gb(84)_vi:TOBTRVR2 Tobacco rattle virus RNA-2 (CAM strain) for capsid protein. 11/93 1,799bp.

X03685 Em(40)_vi:TOTRVRG1 Gb(84)_vi:TOTRVRG1 Tobacco rattle virus strain PSG RNA1 3' terminal sequence (2 kb). 9/93 2,077bp.

X03686 Em(40)_vi:TOTRVRG2 Gb(84)_vi:TOTRVRG2 Tobacco rattle virus strain PSG RNA2. 9/93 1,905bp.

X03955 Em(40)_vi:TOTRVRN2 Gb(84)_vi:TOTRVRN2 Tobacco rattle virus (TRV) strain TCM RNA2 sequence. 9/93 3,389bp.

X06172 Em(40)_vi:TOTRRNA1 Gb(84)_vi:TOTRRNA1 Tobacco Rattle virus (TRV) RNA-1 complete sequence. 9/93 6,791bp.

S72875 Gb(89)_un:S72876s2 (RNA-2 5' and 3' regions) (tobacco rattle virus TRV, TCM, I6, Genomic RNA, 565 nt, segment 2

S72876 Gb(89)_un:S72876s1 (RNA-2 5' and 3' regions) (tobacco rattle virus TRV, TCM, I6, Genomic RNA, 369 nt, segment 1

Z36974 Em(43)_vi:Trvrna2cp Gb(89)_vi:Trvrna2cp Tobacco rattle virus genes for coat protein, 28.7 kDa & 32.8 kDa proteins. 3/95 3,856bp.

Z36983 Em(43)_vi:Trvpprncp Gb(89)_vi:Trvpprncp Tobacco rattle virus coat protein gene. 9/94 627bp.

*Non–genomic nucleic acid* found in the virions; is subgenomic *m*RNA (in 3 parts namely RNA-3, RNA-4 and RNA-5). Sub–genomic *m*RNA found in infected cells.

*Virion protein(s)* one; $M_r$ 21000–23000; coat protein. Method of preparation: Semancik (1970); Ghabrial and Lister (1973). Amino acid sequence: Cornelissen *et al.* (1986). Amino acid composition: Offord and Harris (1965); Offord (1966); Semancik (1966); Miki and Okada (1970); Ghabrial and Lister (1973).

*Virus–coded non–virion proteins* isolated; two proteins found. $M_r$ of the largest 187000. $M_r$ of 2nd largest 142000.

**Replication**

Coat protein *m*RNA translated in the cytoplasm. Replication does not depend on a helper virus.

**Cytopathology**

Virions found in all parts of the host plant; in cytoplasm (where they stack on the surface of the mitochondria). Inclusions present in infected cells (some strains); are unusual in shape; consisting of clusters of abnormal mitochondria; they contain virions. Other cellular changes: chloroplast break down.

**Taxonomy and Relationships**

*Tobravirus*.

Virus(es) with serologically related virions: some isolates of pea early browning are distantly related.

The genomes of different strains differ in size, especially their RNA-2s.

**References**

Abou Harter, M. and Hirth, L. (1977). *Virology* **76**: 173.
Böning, K. (1931). *Z. ParasitKde.* **3**: 103.
Brierley, P. (1944). *Pl. Dis. Reptr* **150 Suppl.**: 410.
Chang, M.U., Doi, Y. and Yora, K. (1976). *Ann. Phytopath. Soc. Japan* **42**: 325.
Cooper, J.I., Edwards, M.L., Arnold, M.K. and Massalski, P.R. (1983). *Pl. Path.* **32**: 469.
Cornelissen, B.J.C., Linthorst, H.J.M., Brederode F.Th. and Bol, J.F. (1986). *Nucl. Acids Res.* **14**: 2157.
Ghabrial, S.A. and Lister, R.M. (1973). *Virology* 52: 1.
Gugerli, P., Darby, G. and Minson, A.C. (1978). *J. gen. Virol.* **38**: 273.
Harrison, B.D. (1970). *CMI/AAB Descr. Pl. Viruses* No. 12, 4 pp.
Harrison, B.D. and Nixon, H.L. (1959). *J. gen. Microbiol.* **21**: 569.
Harrison, B.D. and Robinson, D.J. (1978). *Adv. Virus Res.* **23**: 25
Harrison, B.D. and Robinson, D.J. (1981). In: *Handbook of Plant Virus Infections*, p. 515; ed. E. Kurstak. Elsevier/North Holland Biomedical Press, Amsterdam.
Jones, A.T. and Young, B.R. (1978). *Pl. Dis. Reptr* **62**: 925.
Lister, R.M. and Bracker, C.E. (1969). *Virology* **37**: 262.
Martin, G.H. (1929). *Pl. Dis. Reptr Suppl.* **73**: 390.
Miki, T. and Okada, Y. (1970). *Virology* **42**: 993.
Minson, T. and Darby, G. (1973). *J. mol. Biol.* **77**: 337.
Noordam, D. (1956). *Tijdschr. Plziekt.* **62**: 219.
Offord, R.E. and Harris, J.I. (1965). *Proc. FEBS Meeting*, p. 216.
Offord, R.E. (1966). *Ph.D. Thesis*, University of Cambridge, England.
Robinson, D.J. (1977). *J. gen. Virol.* **35**: 37.
Rozendaal, A. and van der Want, J.P.H. (1948). *Tijdschr. PlZiekt.* **54**: 113.
Schmelzer, K. (1957). *Phytopath. Z.* **30**: 281.
Semancik, J.S. (1966). *Phytopathology* **56**: 1190.

Semancik, J.S. (1970). *Virology* **40**: 618.
Smith, K.M. (1950). *J. R. hort. Sci.* **75**: 350.
Uschdraweit, H.A. and Valentin, H. (1956). *NachrBl. dtsch. Pflschutz.*, Braunschweig **8**: 132.
Van Hoof, H.A. (1968). *Nematologica* **14**: 20.

# Tobacco ringspot *nepovirus*

**Data collated by P.G.W. Chu, 1980. Revised 1984.**

*Synonyms:* anemone necrosis virus (Hollings and Stone, 1963; Hollings, 1965), blueberry necrotic ringspot virus (Lister *et al.*, 1963; Varney and Raniere, 1960), tobacco ringspot virus 1. *Standard acronym:* TRSV. *Strains:* anemone necrosis strain, gladiolus strain, North American strains WS, ST, NC–39 (these contain satellite RNA), tobacco green ringspot strain, tobacco yellow ringspot strain; possibly potato black ringspot virus (Salazar and Harrison, 1977; 1978a and b; 1979) and also ash ringspot virus (Hibben and Bozarth, 1972).

First reported in *Nicotiana tabacum*; by Fromme *et al.* (1927).

### Natural host range and symptoms

Symptoms disappear soon after infection.

*Cucumis sativus, Glycine max, Lycopersicon esculentum, Nicotiana tabacum* — necrotic spots, mottling, chlorotic ringspots and vein banding.

### Transmission

Transmitted by a vector; a nematode (and also, non-specifically by insects and mites — *Aphis gossypii, Myzus persicae, Melanopus* sp. *Epitrix hirtipennis, Thrips tabaci* (possibly nymphs) and *Tetranychus* sp.); *Xiphinema americanum*; Dorylamidae. Virus lost by the vector when it moults; does not multiply in the vector; not transmitted congenitally to the progeny of the vector; does not require a helper virus for vector transmission; transmitted by mechanical inoculation; not transmitted by contact between plants; transmitted by seed; transmitted by pollen to the seed.

### Geographical distribution

Spreads in the North American region; China. Found, but with no evidence of spread, in Australia, the U.K., Germany and New Zealand.

### Experimental host range

Many (>9) families susceptible.

Diagnostically susceptible

*Chenopodium amaranticolor, C. quinoa* — necrotic or chlorotic local lesions.

*Cucumis sativus* — chlorotic local lesions; then systemic ringspots, and leaves malformed.

*Phaseolus vulgaris* — necrotic local lesions; systemic ringspots.

*Nicotiana clevelandii, Vigna unguiculata* — necrotic local lesions; systemic leaf shape malformation.

*N. glutinosa, N. tabacum* — necrotic local lesions; systemic ringspots.

Diagnostically insusceptible

*Brassica rapa, Abelmoschus manihot, Reseda odorata, Asperula odorata, Secale cereale.*

Maintenance and propagation hosts

*Cucumis sativus, Phaseolus vulgaris, Nicotiana tabacum.*

Assay hosts (**L**ocal lesions or **W**hole plants)

*Cassia occidentalis* (L), *Vigna unguiculata* (L), *Nicotiana clevelandii* (L), *N. tabacum* (L,W).

Dicotyledonous Hosts ***(Susceptible** or Insusceptible)*

***Ageratum conyzoides. Amaranthus caudatus. Antirrhinum majus.*** And var. *dulce* ***Apium graveolens. Atriplex hortensis. Bellis perennis.*** And var. *cicla* ***Beta vulgaris.*** Cv. Wong Bok ***Brassica campestris*** ssp. ***pekinensis.*** *Brassica campestris* ssp. *rapa.* ***Calendula officinalis. Capsella bursa–pastoris. Capsicum frutescens. Catharanthus roseus. Celosia argentea. Cheiranthus cheiri. Chenopodium album. Chenopodium amaranticolor. Chenopodium quinoa. Cichorium endiva. Clarkia amoena. Crotalaria spectabilis. Cucumis melo.*** Cv. Davis Perfect ***Cucumis sativus. Cucurbita maxima. Cucurbita moschata.*** And vars. *medullosa, melopepo, orifera* ***Cucurbita pepo. Cyamopsis tetragonoloba. Datura metel. Datura stramonium. Daucus carota*** (unconfirmed). ***Dianthus barbatus*** (latent). ***Fagopyrum esculentum. Glycine max*** (cv. Lincoln). ***Gomphrena globosa. Gypsophila elegans.*** *Helianthus annuus.* ***Hibiscus esculentus. Hyoscyamus niger. Lablab purpureus. Lactuca sativa. Lathyrus odoratus.*** *Linum usitatissimum. Lupinus albus.* ***Lycopersicon esculentum. Macroptilium lathyroides. Medicago sativa. Melilotus alba. Myosotis sylvatica. Nicandra physaloides. Nicotiana bigelovii. Nicotiana clevelandii.*** *Nicotiana glutinosa.* ***Nicotiana rustica. Nicotiana sylvestris. Nicotiana tabacum. Pastinaca sativa. Petunia hybrida. Phaseolus lunatus. Phaseolus vulgaris. Phlox drummondii. Physalis floridana. Physalis peruviana. Phytolacca americana. Pisum sativum. Ricinus communis. Salvia splendens. Senecio vulgaris. Sesamum indicum.*** *Sinningia speciosa.* ***Solanum melongena. Solanum nigrum. Solanum nodiflorum. Solanum tuberosum. Sonchus oleraceus. Spinacia oleracea. Stellaria media. Tetragonia expansa. Torenia fournieri.*** *Trifolium hybridum. Trifolium incarnatum. Trifolium pratense. Trifolium repens.* ***Tropaeolum majus. Vicia faba. Vicia sativa.*** *Vicia villosa.* ***Vigna angularis. Vigna radiata. Vigna unguiculata. Vigna unguiculata*** ssp. ***cylindrica. Vigna unguiculata*** ssp. ***sesquipedalis. Viola cornuta. Zinnia elegans.***

Monocotyledonous Hosts ***(Susceptible** or Insusceptible)*

*Allium cepa.* ***Iris germanica. Lilium longiflorum.*** *Secale cereale. Triticum aestivum.* ***Zea mays.***

Sources of host-range data

Wingard (1928); Hill (1937); Price Francki (1965); Thornberry (1966); Bowyer and Atherton (1972).

**Properties of particles in sap**

TIP: 55–65 °C. LIV: 22 days. DEP: $\log_{10}$ minus 4. Infectivity of sap not changed by treatment with di–ethyl ether. Leaf sap contains many virions.

**Purification**

Method: Rezaian *et al.* (1973).

**Particle morphology**

Virions isometric; not enveloped; 25–29 nm in diameter (virions may swell at pH 8.0); angular in profile; without a conspicuous capsomere arrangement.

### Physical properties

Three sedimenting components in purified preparations; sedimentation coefficient of the fastest 126 *S* (B); of the other(s) 91 *S* (M), or 53 *S* (T). Density 1.52 g $cm^{-3}$ in CsCl ($B_2$), or 1.51 g $cm^{-3}$ in CsCl ($B_1$), or 1.42 g $cm^{-3}$ in CsCl (M). Isoelectric point pH 4.2–6.1. $A_{260}/A_{280}$ ratio 1.88 (B), or 1.37 (M), or 0.91 (T).

### Biochemical properties

Virions contain 40 % nucleic acid (B), or 28 % nucleic acid (M), or 0 % nucleic acid (T); 60 % protein (B), or 72 % protein (M), or 100 % protein (T); 0 % lipid.

*Genome* consists of RNA; single–stranded; linear; of two parts; largest (or only) genome part the larger 6.8 kb; the 2nd largest 4.364 kb. Genomic nucleic acid isolated by Murant *et al.* (1972); Rezaian and Francki (1974). Base composition 24.7 % G; 23.1 % A; 22.4 % C; 29.8 % U. 5' terminus of RNA has a VPg. Infectivity lost when deproteinised with proteases; retained when deproteinised with phenol or detergent. Poly A region present. Genome has no *t*RNA–like activity. Nucleotide sequence references: Rezaian and Francki (1974).

*Sequence database accession code(s):*

L09205 Em(40)_vi:TRSCAPSID Gb(84)_vi:TRSCAPSID Tobacco ringspot virus capsid protein gene, complete cds. 4/94 2,022bp.

M17439 Em(40)_vi:TRSSTO01 Gb(84)_vi:TRSSTOBAA Tobacco ringspot virus (TobRV) mRNA. 12/90 360bp.

A13898 Em(43)_pat:A13898 Gb(89)_pat:A13898 Tobacco ringspot virus consensus self-cleavage region. 10/94 52bp.

A13899 Em(43)_vi:A13899 Gb(89)_pat:A13899 Tobacco ringspot virus mutant D–51 self-cleavage region. 10/94 63bp.

*Non–genomic nucleic acid* found in the virions; is satellite RNA (of 90–120 Kd).

*Virion protein(s)* two; $M_r$ 58000; coat protein. $M_r$ of 2nd largest 4000; VPg. Method of preparation: Chu and Francki (1979). Amino acid composition: Chu and Francki (1979).

### Replication

Genome replicates in cytoplasm. Coat protein *m*RNA translated in the cytoplasm. Replication does not depend on a helper virus.

### Cytopathology

Virions found in all parts of the host plant; in cell vacuoles (tubules, and plasmodesmata). Inclusions present in infected cells; are unusual in shape; irregular, crystalline arrays, that may be vacuolated; they contain virions. Other cellular changes: the excess development of Golgi and rough endoplasmic reticulum.

### Taxonomy and Relationships

*Nepovirus: Comoviridae.*

Virus(es) with serologically related virions: potato black ringspot virus. Virus(es) with serologically unrelated virions: several other nepoviruses, such as tomato ringspot nepovirus.

### References

Bowyer, J.W. and Atherton, J.G. (1972). *Phytopathology* **61**: 1451.

Chu, P.W.G. and Francki, R.I.B. (1979). *Virology* **93**: 398.

Chu, P.W.G., Boccardo, G. and Francki, R.I.B. (1981). *Virology* **109**: 428.

Dodd, S.M. and Robinson, D.J. (1984). *J. gen. Virol.* **65**: 1731.

Francki, R.I.B. and Hatta, T. (1977). In: *The Atlas of Insect and Plant Viruses*, p. 221; ed. K. Maramorosch. Academic Press, New York.

Fromme, F.D., Wingard, S.A. and Priode, C.N.(1927). *Phytopathology* **17**: 321.
Helson, G.A.H. (1950). *Aust. J. agric. Res.* **1**: 144.
Hibben, C.R. and Bozarth, R.F. (1972). *Phytopathology* **62**: 1023.
Hill, A.V. (1937). *J. Coun. Scient. ind. Res. Aust.* **10**: 228.
Hollings, M. (1965). *Ann. appl. Biol.* **55**: 447.
Hollings, M. and Stone, O.M. (1963). *Rep. Glasshouse Crops Res. Inst. 1962*, p. 89.
Kahn, R.P., Scott, H.A. and Monroe, R.L. (1962). *Phytopathology* **52**: 1211.
Lister, R.M., Raniere, L.C. and Varney, E.H. (1963). *Phytopathology* **53**: 1031.
Murant, A.F., Mayo, M.A., Harrison, B.D. and Gould, R.A. (1972). *J. gen. Virol.* **16**: 327.
Price, W.C. (1940). *Am. J. Bot.* **27**: 530
Randles, J.W. and Francki, R.I.B. (1965). *Aust. J. biol. Sci.* **18**: 979.
Rezaian, M.A. and Francki, R.I.B. (1973). *Virology* **56**: 238.
Rezaian, M.A. and Francki, R.I.B. (1974). *Virology* **59**: 275.
Rezaian, M.A. and Jackson, A.O. (1981). *Virology* **114**: 534.
Rezaian, M.A, Francki, R.I.B., Chu, P.W.G. and Hatta, T. (1976). *Virology* **74**: 481.
Salazar, L.F. and Harrison, B.D. (1977). *Nature, Lond* **265**: 337.
Salazar, L.F. and Harrison, B.D. (1978*b*). *Ann. appl. Biol.* **90**: 387.
Salazar, L.F. and Harrison, B.D. (1978). *Ann. appl. Biol.* **90**: 375.
Salazar, L.F. and Harrison, B.D. (1979). *CMI/AAB Descr. Pl. Viruses* No. 206, 4 pp.
Schneider, I.R., Hull, R. and Markham, R. (1972). *Virology* **47**: 320.
Stace-Smith, R. (1970). *CMI/AAB Descr. Pl. Viruses* No. 17, 4 pp.
Stace-Smith, R. (1985). *CMI/AAB Descr. Pl. Viruses* No. 309, 6 pp.
Thornberry, H.H. (1966). In: *Index of Plant Virus Diseases.* U.S. Dept Agric. Hdbk No. 307, p.313.
Varney, E.H. and Raniere, L.C. (1960). *Phytopathology* **50**: 24.
Wingard, S.A. (1928). *J. Agric. Res.* **37**: 127.

# Tobacco ringspot virus satellite RNA

*Sequence database accession code(s):*

M14879 Em(40)_vi:TRSSATA Gb(84)_vi:TRSSATA Tobacco ringspot virus satellite RNA. 7/89 359bp.

M31515 Em(40)_vi:TRSSTOBA Gb(84)_vi:TRSSTOBA Tobacco ringspot virus (STobRV) autolytic junction. 12/90 71bp.

S63883 Em(40)_vi:S63883 Gb(84)_vi:S63883 satellite RNA tobacco ringspot virus satellite RNA sTobRV RNA, D-94, Genomic RNA, 75 nt. 12

S63888 Em(40)_vi:S63888 Gb(84)_vi:S63888 (D80.2)=satellite RNA tobacco ringspot virus satellite RNA sTobRV RNA, D-94, Genomic RNA Mut

S63895 Em(40)_vi:S63895 Gb(84)_vi:S63895 (D80.3)=satellite RNA tobacco ringspot virus satellite RNA sTobRV RNA, D-94, Genomic RNA Mut

S63896 Em(40)_vi:S63896 Gb(84)_vi:S63896 (D80.4)=satellite RNA tobacco ringspot virus satellite RNA sTobRV RNA, D-94, Genomic RNA Mut

S63897 Em(40)_vi:S63897 Gb(84)_vi:S63897 (D83.2)=satellite RNA tobacco ringspot virus satellite RNA sTobRV RNA, D-94, Genomic RNA Mut

S63901 Em(40)_vi:S63901 Gb(84)_vi:S63901 satellite RNA tobacco ringspot virus satellite RNA sTobRV RNA, D-16, Genomic RNA, 77 nt. 12

S63903 Em(40)_vi:S63903 Gb(84)_vi:S63903 (E.8)=satellite RNA tobacco ringspot virus satellite RNA sTobRV RNA, D-16, Genomic RNA Mutan

S63904 Em(40)_vi:S63904 Gb(84)_vi:S63904 (E.59)=satellite RNA tobacco ringspot virus satellite RNA sTobRV RNA, D-16, Genomic RNA Muta
S63907 Em(40)_vi:S63907 Gb(84)_vi:S63907 (J63.B1)=satellite RNA tobacco ringspot virus satellite RNA sTobRV RNA, D-16, Genomic RNA Mu
S63908 Em(40)_vi:S63908 Gb(84)_vi:S63908 (J63.A1)=satellite RNA tobacco ringspot virus satellite RNA sTobRV RNA, D-16, Genomic RNA Mu
S63910 Em(40)_vi:S63910 Gb(84)_un:S63910 satellite RNA tobacco ringspot virus satellite RNA sTobRV RNA, mutant B–15, Genomic RNA, 55
S63911 Em(40)_vi:S63911 Gb(84)_un:S63911 (B.5.15)=satellite RNA tobacco ringspot virus satellite RNA sTobRV RNA, mutant B–15.

**Taxonomy and Relationships**

Satellite RNA.

# Tobacco streak *ilarvirus*

**Data collated by R.W. Fulton, 1984.**

*Synonyms:* asparagus stunt virus (Brunt and Paludan, 1970; Paludan, 1964), datura quercina virus, strawberry necrotic shock virus (Frazier *et al.*, 1962; Stace-Smith and Frazier, 1971). *Standard acronym:* TSV. *Strains:* bean red node strain (Scott *et al.*, 1961; Thomas and Zaumeyér, 1950; Virgin, 1943), black raspberry latent strain (Brunt and Stace-Smith, 1976; Converse and Lister, 1969; Jones and Mayo, 1975).

First reported in *Nicotiana tabacum*; from Wisconsin, U.S.A; by Johnson (1936).

**Natural host range and symptoms**

Symptoms persist (most hosts do not recover).
*Dahlia* spp. — leaf mottling or symptomless.
*Gossypium herbaceum, Melilotus alba, Trifolium pratense* — leaf mottling.
*Phaseolus vulgaris* — reddening of nodes.
*Glycine max* — systemic necrosis, bud blight.
*Rosa setigera* — leaf vein yellowing.
*Nicotiana tabacum* — systemic necrosis, with recovery.
*Asparagus officinalis* — stunting.

**Transmission**

Transmitted by a vector; an insect; *Frankliniella occidentalis* and *Thrips tabaci*; Thysanoptera (possibly by allowing virus from the surface of infected pollen to enter through feeding wounds; Sdoodee and Teakle, 1987). Virus transmitted by mechanical inoculation; transmitted by grafting; not transmitted by contact between plants; transmitted by seed (in *Chenopodium quinoa, Phaseolus vulgaris, Datura stramonium* to different extents); transmitted by pollen to the pollinated plant.

**Geographical distribution**

Probably distributed worldwide. Spreads in the North American region and the Pacific region; Australia, Canada, Peru, and the USA (CABI/EPPO data for black raspberry latent strain).

**Experimental host range**

Many (>9) families susceptible.

Diagnostically susceptible

*Beta patellaris* — necrotic local lesions.

*Cyamopsis tetragonoloba* — small dark local lesions.
*Phaseolus vulgaris* cv. Manteiga — necrotic local lesions.
*Vigna unguiculata* — chlorotic or necrotic local lesions.
*Nicotiana tabacum* cv. Turkish — upper leaves become toothed.

Maintenance and propagation hosts

*Catharanthus roseus, Cucumis sativus, Datura stramonium, Nicotiana tabacum.*

Assay hosts (**L**ocal lesions or **W**hole plants)

*Beta patellaris* (L), *Cyamopsis tetragonoloba* (L), *Macrotyloma uniflorum* (L), *Phaseolus vulgaris* cv. Manteiga (L), *Vigna unguiculata* ssp. *cylindrica*, (L) for most strains.

Dicotyledonous Hosts ***(Susceptible** or Insusceptible)*

***Amaranthus caudatus*. *Amaranthus retroflexus*. *Antirrhinum majus*. *Arachis hypogaea*. *Beta macrocarpa*. *Beta patellaris*. *Beta vulgaris*. *Brassica campestris*.** *Brassica juncea*. ***Capsicum annuum*. *Capsicum frutescens*. *Cassia occidentalis*. *Cassia tora*. *Catharanthus roseus*. *Chenopodium album*. *Chenopodium amaranticolor*. *Chenopodium capitatum*. *Chenopodium foetidum*. *Chenopodium hybridum*. *Chenopodium murale*. *Chenopodium quinoa*.** *Citrullus lanatus*. ***Crotalaria spectabilis*. *Cucumis melo*. *Cucumis sativus*. *Cucurbita maxima*. *Cucurbita moschata*. *Cucurbita pepo*. *Cyamopsis tetragonoloba*. *Dahlia pinnata*.** *Datura ferox*. *Datura metel*. *Datura stramonium*. *Datura tatula*. ***Fragaria vesca*. *Glycine max*. *Gomphrena globosa*. *Gossypium hirsutum*. *Helianthus annuus*. *Hibiscus esculentus*. *Hyoscyamus niger*. *Lactuca sativa*. *Lens esculenta*.** *Lupinus albus*. ***Lycopersicon esculentum*. *Lycopersicon pimpinellifolium*. *Matthiola incana*. *Medicago sativa*. *Melilotus alba*. *Momordica balsamina*. *Nicandra physaloides*. *Nicotiana benthamiana*. *Nicotiana bigelovii*. *Nicotiana clevelandii*. *Nicotiana* × *edwardsonii*. *Nicotiana debneyi*. *Nicotiana glutinosa*. *Nicotiana megalosiphon*.** *Nicotiana rustica*. *Nicotiana sylvestris*. ***Nicotiana tabacum*. *Ocimum basilicum*. *Petunia hybrida*. *Phaseolus vulgaris*. *Phlox drummondii*. *Physalis floridana*. *Physalis peruviana*. *Pisum sativum*. *Salvia splendens*.** *Sesamum indicum*. *Sesbania exaltata*. ***Solanum demissum*. *Solanum demissum* × *tuberosum*. *Solanum melongena*. *Solanum nigrum*. *Solanum nodiflorum*. *Solanum rostratum*. *Solanum tuberosum*. *Spinacia oleracea*. *Stellaria media*. *Tithonia speciosa*. *Torenia fournieri*. *Trifolium incarnatum*. *Trifolium pratense*. *Trifolium repens*.** *Tropaeolum majus*. ***Vicia faba*. *Vigna radiata*. *Vigna unguiculata*. *Vigna unguiculata*** ssp. ***cylindrica*. *Vigna unguiculata*** ssp. ***sesquipedalis*. *Zinnia elegans*.**

Monocotyledonous Hosts ***(Susceptible** or Insusceptible)*

***Zea mays*.**

Sources of host-range data

Berkeley and Phillips (1943); Fulton (1948); Kaiser *et al.* (1982).

**Properties of particles in sap**

TIP: 64 °C (in sap diluted in buffer containing anti-oxidant). LIV: 1.5 days (most infectivity lost within 5 min. in crude sap). DEP: $\log_{10}$ minus 1–4 (but varies over a wide range depending on host species, antioxidants used, *etc.*). Electron microscopy: fixation with 1% glutaraldehyde essential.

**Purification**

Method: Fulton (1967).

**Particle morphology**

Virions isometric; not enveloped; 27, 30, and 35 nm in diameter; rounded in profile; without a conspicuous capsomere arrangement.

**Physical properties**

Three sedimenting components in purified preparations; sedimentation coefficient of the fastest 113 *S* (B); of the other(s) 90 *S* (T), or 98 *S* (M). Density 1.35 g $cm^{-3}$ in CsCl. Isoelectric point pH 4.6. $A_{260}/A_{280}$ ratio 1.56.

**Biochemical properties**

Virions contain 14 % nucleic acid; 86 % protein; 0 % lipid.

*Genome* consists of RNA; single–stranded; of three parts; largest (or only) genome part the largest 2.94 kb; the 2nd largest 2.77 kb; the 3rd largest 2.205 kb. Genomic nucleic acid isolated by Van Vloten-Doting (1975). Base composition 25 % G; 25 % A; 20 % C; 30 % U. 5' terminus of RNA has a methylated nucleotide cap (Van Vloten-Doting, 1981). Infectivity lost when deproteinised with proteases; lost when deproteinised with phenol or detergent. Poly A region present (Van Vloten-Doting, 1981). Additional factor required for infectivity. Genome has no *t*RNA–like activity.

*Sequence database accession code(s):*

J02416 Gb(84)_vi:TSVRNA2 tobacco streak virus rna 2: 3' end. 9/83 180bp

J02417 Gb(84)_vi:TSVRNA34 tobacco streak virus rna 3/4 : 3' end. 9/83 140bp

V00600 Em(40)_vi:ILTSV1 Gb(84)_vi:ILTSV1 Tobacco streak virus RNA 2 (3' end). 7/91 180bp.

V00601 Em(40)_vi:ILTSV2 Gb(84)_vi:ILTSV2 Tobacco streak virus RNA 3, 3' end (identical to the 3' end of RNA 4). 7/91 140bp.

X00435 Em(40)_vi:TOTSV3 Gb(84)_vi:TOTSV3 Complete sequence of tobacco streak virus RNA 3 (TSV RNA 3). 9/93 2,205bp. 5 sequences.

*Non–genomic nucleic acid* found in the virions; is subgenomic *m*RNA; of 0.881 kb. Sub–genomic *m*RNA not found in infected cells.

*Virion protein(s)* one; $M_r$ 28500–30000; coat protein. Method of preparation: Van Vloten-Doting (1975). Amino acid composition: Ghabrial and Lister (1974).

**Replication**

Replication does not depend on a helper virus.

**Cytopathology**

Virions found in all parts of the host plant; in cytoplasm and in nuclei. Inclusions present in infected cells (epidermal or guard cells); are unusual in shape; irregularly shaped; they do not contain virions. Other cellular changes: fibrils in the cytoplasm and nucleus that are associated with virion aggregates and masses of fibrils in the cytoplasm that are not associated with virions (Edwardson and Christie, 1986).

**Taxonomy and Relationships**

*Ilarvirus: Bromoviridae* (subgroup 1 and the type species).

Virus(es) with serologically unrelated virions: apple mosaic, American plum line pattern, asparagus C, citrus leaf rugose, citrus variegation, elm mottle, lilac ring mottle, prune dwarf, prunus necrotic ringspot, spinach latent, Tulare apple mosaic viruses.

**References**

Berkeley, G.H. and Phillips, J.H.H. (1943). *Can. J. Res.* **21**: 181.

Brunt, A.A. and Paludan, N. (1970). *Phytopath. Z.* **69**: 277.

Brunt, A.A. and Stace-Smith, R. (1976). *Acta Hort.* **66**: 71.

Converse, R.H. and Lister, R.M. (1969). *Phytopathology* **59**: 325.

Edwardson, J.R. and Christie, R.G. (1986). *Fla Agric. Exp. Stn Monog.* No. 14, p. 221.
Frazier, N.W., Jorgensen, P.S., Thomas, H.E. and Johnson, H.A. (1962). *Pl. Dis. Reptr* **46**: 547.
Fulton, R.W. (1948). *Phytopathology* **38**: 421.
Fulton, R.W. (1967). *Virology* **32**: 153.
Fulton, R.W. (1970). *Virology* **41**: 288.
Fulton, R.W. (1971). *CMI/AAB Descr. Pl. Viruses* No. 44, 4 pp.
Fulton, R.W. (1985). *CMI/AAB Descr. Pl. Viruses* No. 307, 5 pp.
Ghabrial, S.A. and Lister, R.M. (1974). *Virology* **57**: 1.
Jones, A.T. and Mayo, M.A. (1975). *Ann. appl. Biol.* **79**: 297.
Johnson, J. (1936). *Phytopathology* **26**: 285.
Kaiser, W.J., Wyatt, S.D. and Pesho, G.R. (1982). *Phytopathology* **72**: 1508.
Lister, R.M. and Saksena, K.N. (1976). *Virology* **70**: 440.
Lister, R.M., Ghabrial, S.A. and Saksena, K.N. (1972). *Virology* **49**: 290.
Paludan, N. (1964). *Maanedsovers. PlSygd.* **407**: 11.
Scott, H.A., Vincint, M. and Zaumeyer, W.J. (1961). *Phytopathology* **51**: 755.
Sdoodee, R. and Teakle, D.S. (1987). *Plant Path.* **36**: 377.
Stace-Smith, R. and Frazier, N.W. (1971). *Phytopathology* **61**: 757.
Thomas, H.R. and Zaumeyer, W.J. (1950). *Phytopathology* **40**: 832.
Van Vloten-Doting, L. (1975). *Virology* **65**: 215.
Van Vloten-Doting, L. (1981). *Intervirology* **15**: 198.
Virgin, W.J. (1943). *Phytopathology* **33**: 743.

# Tobacco stunt varicosavirus

**Data collated by S. Kuwata, 1985. Revised 1989.**

*Standard acronym:* TStV.

First reported in *Nicotiana tabacum*; from Hiroshima Prefecture, Japan; by Hidaka (1950).

### Natural host range and symptoms

Symptoms vary seasonally.
*Nicotiana tabacum* — stunting, vein necrosis, necrotic leaf spotting.

### Transmission

Transmitted by a vector; a fungus; *Olpidium brassicae*; Chytridiales. Virus does not require a helper virus for vector transmission; transmitted by mechanical inoculation; transmitted by grafting; not transmitted by contact between plants; not transmitted by seed; not transmitted by pollen.

### Geographical distribution

Spreads in Japan.

### Experimental host range

Several (3–9) families susceptible.

Diagnostically susceptible

*Chenopodium amaranticolor, Cyamopsis tetragonoloba, Nicotiana tabacum* cv. Burley 21 — necrotic leaf spotting.
*Nicotiana tabacum* cv. Xanthi-*nc* — faint leaf spotting.
*Tetragonia expansa* — yellow leaf spotting; not systemic.

Diagnostically insusceptible

*Brassica oleracea, Lactuca sativa, Capsicum annuum.*

Maintenance and propagation hosts

*Nicotiana tabacum* cvs Burley 21, Bright Yellow.

Assay hosts (**L**ocal lesions or **W**hole plants)

*Chenopodium amaranticolor* (L), *Cyamopsis tetragonoloba* (L), *Nicotiana glutinosa* (L,W), *N. tabacum* (L,W), *Tetragonia expansa* (L).

Dicotyledonous Hosts *(**Susceptible** or Insusceptible)*

*Antirrhinum majus. Apium graveolens. Arachis hypogaea.* ***Beta vulgaris.*** *Brassica campestris* ssp. *napus. Brassica oleracea* ssp. *botrytis. Brassica oleracea* ssp. *capitata. Calendula officinalis. Capsicum annuum. Celosia argentea.* ***Chenopodium album. Chenopodium amaranticolor. Chenopodium quinoa. Cucumis sativus.*** *Cucurbita pepo. Cyamopsis tetragonoloba.* ***Datura stramonium. Datura tatula.*** *Daucus carota.* ***Dianthus caryophyllus. Glycine max. Gomphrena globosa.*** *Lactuca sativa.* ***Lycopersicon esculentum.*** *Medicago sativa.* ***Nicotiana benthamiana.*** *Nicotiana bigelovii. Nicotiana clevelandii.* ***Nicotiana debneyi. Nicotiana glutinosa. Nicotiana megalosiphon. Nicotiana rustica. Nicotiana sylvestris. Nicotiana tabacum.*** *Petroselinum crispum.* ***Petunia hybrida.*** *Phaseolus vulgaris.* ***Physalis floridana.*** *Phytolacca americana.* ***Pisum sativum.*** *Raphanus sativus. Solanum demissum.* ***Solanum tuberosum. Spinacia oleracea. Tetragonia expansa.*** *Trifolium hybridum. Trifolium pratense. Vicia faba. Zinnia elegans.*

Monocotyledonous Hosts *(**Susceptible** or Insusceptible)*

*Triticum aestivum. Zea mays.*

Additional hosts

***Catharanthus roseus.***

Sources of host-range data

Hiruki (1975).

**Properties of particles in sap**

TIP: 40–50 °C (sap transmissible only after addition of reducing agents such as 0.01M $Na_2SO_3$). LIV: 1 days. DEP: $\log_{10}$ minus 3–4. Leaf sap contains few virions. Electron microscopy: fixation with glutaraldehyde necessary.

**Purification**

Method: Kuwata and Kubo (1981).

**Particle morphology**

Virions rod–shaped; not enveloped; usually straight; with a clear modal length; of 300 and 340 nm; 18 nm wide. Axial canal obvious; 3.5 nm in diameter. Basic helix obvious; pitch of basic helix 5 nm.

**Physical properties**

Density 1.27 in $Cs_2SO_4$.

**Biochemical properties**

*Genome* consists of RNA; double–stranded; linear; of two parts; largest (or only) genome part the larger 6.35 kb (pairs); the 2nd largest 5.63 kb (pairs).

*Virion protein(s)* one; $M_r$ 48000; coat protein.

**Replication**

Replication does not depend on a helper virus.

**Cytopathology**

Virions found in all parts of the host plant; in cytoplasm. Inclusions absent from infected cells.

**Taxonomy and Relationships**

Varicosavirus.

Virions of the virus are closely serologically related to those of lettuce big vein virus, but as the two differ in host range, they are considered as distinct species, rather than related strains (Kuwata and Kubo, 1986).

**References**

Hidaka, Z. (1950). *Ann. Phytopath. Soc. Japan* **15**: 40.
Hidaka, Z., Hiruki, C., Nakano, K., Shimizu, T. and Uozumi, T. (1956). *Bull. Hatano Tobacco Exp. Stn* **40**: 1.
Hiruki, C. (1965). *Virology* **25**: 288.
Hiruki, C. (1975). *Can. J. Bot.* **53**: 2425.
Kuwata, S. and Kubo, S. (1981). *Ann. Phytopath. Soc. Japan* **47**: 264.
Kuwata, S. and Kubo, S. (1986). *AAB Descr. Pl. Viruses* No. 313, 4 pp.

# Tobacco vein-banding mosaic *potyvirus*

*Sequence database accession code(s):*

L28816 Em(40)_vi:PTWCOPOL Gb(84)_vi:PTWCOPOL Tobacco vein banding mosaic virus nuclear inclusion body and coat protein. 2/94 1,902bp.

X77637 Em(40)_vi:TVBMVCP Gb(84)_vi:TVBMVCP Tobacco vein-banding mosaic virus cp gene. 2/94 1,204bp.

**Taxonomy and Relationships**

*Potyvirus: Potyviridae.*

**References**

Habera, L.F., Berger, P.H., Reddick, B.B. (1994). *Arch. Virol.* **138**: 27.
Huang, C.R., Yeh, S.D., Chiang, J.K., Hung, L.M. and Hu, H.Y. (1994). *Arch. Virol.* **138**: 17.

# Tobacco vein-distorting (?) *luteovirus*

**Data collated by A.A. Brunt, 1987.**

*Standard acronym:* TVDV.

First reported in *Nicotiana tabacum*; from Zimbabwe; by Smith (1946*a*).

**Natural host range and symptoms**

Symptoms persist.

*Nicotiana tabacum* — malformation of leaf veins. Tobacco vein-distorting and tobacco mottle viruses together causes tobacco rosette disease.

**Transmission**

Transmitted by a vector; an insect; *Myzus persicae, Acyrthosiphon (Aulacorthum) solani*; Aphididae. Transmitted in a persistent manner. Virus can help the vector transmission of another virus (tobacco mottle (?) umbravirus); not transmitted by mechanical inoculation; transmitted by grafting; not transmitted by contact between plants.

**Ecology and control**

Studies reported by Smith (1946*a,b*).

**Geographical distribution**

Spreads in Zimbabwe.

**Experimental host range**

Few (<3) families susceptible.

Diagnostically susceptible

*Nicotiana tabacum, N. glutinosa, N. sylvestris, Lycopersicon esculentum, Datura stramonium* — leaf vein malformation.

Maintenance and propagation hosts

*Nicotiana tabacum, N. glutinosa, N. sylvestris.*

Assay hosts (**L**ocal lesions or **W**hole plants)

*Nicotiana tabacum* (W).

Dicotyledonous Hosts *(**Susceptible** or Insusceptible)*

***Datura stramonium. Lycopersicon esculentum. Nicotiana glutinosa. Nicotiana sylvestris. Nicotiana tabacum.***

Sources of host-range data

Smith (1946*a*).

**Taxonomy and Relationships**

Probably *Luteovirus.*

Few of the properties of this virus are known, but some indicate that it is probably a luteovirus (Waterhouse *et al.*, 1988).

**Comments**

This virus was probably the first to be shown to assist the aphid transmission of another virus, tobacco mottle umbravirus (Smith, 1946 *a* and *b*).

**References**

Smith, K.M. (1946*a*). *Parasitol.* **37**: 21.
Smith, K.M. (1946*b*). *Parasitol.* **37**: 131.
Waterhouse, P.M., Gildow, F.E. and Johnstone, G.R. (1988). *AAB Descr. Pl. Viruses*, No.339, 9 pp.

# Tobacco vein mottling *potyvirus*

**Data collated by M.K. Sun, 1985.**

*Standard acronym:* TVMV.

First reported in *Nicotiana tabacum*; from the U.S.A.; by Gooding and Sun (1972).

**Natural host range and symptoms**

Symptoms vary cyclically over a few weeks.

*Rumex* sp., *Nicotiana tabacum, Solanum carolinense* — chlorotic vein banding of leaves.

**Transmission**

Transmitted by a vector; an insect; *Myzus persicae*; Aphididae. Transmitted in a non–persistent manner. Virus transmitted by mechanical inoculation; transmitted by grafting; transmitted by contact between plants; not transmitted by pollen.

**Geographical distribution**

Spreads in the North American region.

**Experimental host range**

Few (<3) families susceptible.

Diagnostically susceptible

*Lycopersicon esculentum, Nicotiana glutinsoa, N. tabacum* — chlorotic vein banding.

Maintenance and propagation hosts

*Nicotiana tabacum.*

Assay hosts (Local lesions or Whole plants)

*Nicotiana tabacum* (W).

Dicotyledonous Hosts *(**Susceptible** or Insusceptible)*

*Capsicum annuum. Capsicum frutescens. Chenopodium amaranticolor. Cucumis sativus.* ***Datura metel****. Datura stramonium. Glycine max. Gomphrena globosa.* ***Lycopersicon esculentum****. Nicotiana clevelandii.* ***Nicotiana glutinosa. Nicotiana tabacum****. Phaseolus vulgaris. Pisum sativum. Solanum tuberosum. Trifolium incarnatum. Vigna unguiculata.*

Sources of host-range data

Sun *et al.* (1974).

**Properties of particles in sap**

TIP: 60–70 °C. LIV: 1–2 days. DEP: $\log_{10}$ minus 2–3. Leaf sap contains many virions.

**Purification**

Method: Sun *et al.* (1974).

**Particle morphology**

Virions filamentous; not enveloped; usually flexuous; with a clear modal length; of 765 nm; 13 nm wide. Basic helix obvious.

**Physical properties**

One sedimenting component in purified preparations.

*Sequence database accession code(s):*

K02991 Em(40)_vi:TVMURF Gb(84)_vi:TVMURF Tobacco vein mottling virus helper component-related P75 gene, partial cds. 4/90 420bp

S68375 Em(40)_vi:S68375 Gb(84)_vi:S68375 coat protein (3' UTR region, direct repeats) tobacco vein mottling virus (TVMV), variant pXB

X04083 Em(40)_vi:POTTVMVX Gb(84)_vi:POTTVMVX Tobacco vein mottling virus (TVMV) RNA genome. 9/93 9,472bp. 3 sequences.

**Replication**

Genome replicates in cytoplasm.

**Cytopathology**

Virions found in all parts of the host plant; in cytoplasm. Inclusions present in infected cells; are pinwheels; they do not contain virions.

**Taxonomy and Relationships**

*Potyvirus: Potyviridae.*

Virus(es) with serologically unrelated virions: bean common mosaic, iris severe mosaic, clover yellow vein, henbane mosaic, peanut mottle, potato Y, soybean mosaic and tobacco etch viruses.

**References**

Gooding, G.V. and Sun, M. (1972). *Phytopathology* **62**: 803.

Sun, M.K.C., Gooding, G.V., Pirone, T.P. and Tolin, S.A. (1974). *Phytopathology* **64**: 1133.

# Tobacco wilt *potyvirus*

**Data collated by A.A. Brunt and A. Varma, 1987.**

*Standard acronym:* TWV.

First reported in *Solanum jasminoides*; from England from infected plants imported from India; by Badami and Kassanis (1959).

**Natural host range and symptoms**

*Solanum jasminoides* — symptoms unknown; naturally infected plants are also infected with datura necrosis virus.

**Transmission**

Transmitted by a vector; an insect; *Myzus persicae*; Aphididae. Transmitted in a non–persistent manner. Virus transmitted by mechanical inoculation.

**Geographical distribution**

Spreads in India.

**Experimental host range**

Several (3–9) families susceptible.

Diagnostically susceptible

*Nicotiana tabacum* — chlorotic local lesions; systemic veinal necrosis.

*Nicotiana glutinosa, Datura tatula, Hyoscyamus niger* — systemic vein banding, mottling, severe malformation.

*Solanum jasminoides* — necrotic local lesions; systemic veinal necrotic spotting.

Diagnostically insusceptible

*Solanum tuberosum, Cucumis sativus, Phaseolus vulgaris.*

Maintenance and propagation hosts

*Nicotiana glutinosa, Datura tatula, Hyoscyamus niger.*

Assay hosts (**L**ocal lesions or **W**hole plants)

*Nicotiana tabacum* (W), *N. glutinosa* (W).

Dicotyledonous Hosts *(**Susceptible** or Insusceptible)*

*Cucumis sativus*. ***Datura tatula***. ***Hyoscyamus niger***. ***Nicotiana glutinosa***. ***Nicotiana tabacum***. *Phaseolus vulgaris*.

**Properties of particles in sap**

TIP: 60–65 °C (in *Nicotiana glutinosa* sap). LIV: 4–8 days. DEP: $\log_{10}$ minus 4.

**Particle morphology**

Virions filamentous; not enveloped; usually flexuous; with a clear modal length; of 750 nm. Axial canal obscure.

**Taxonomy and Relationships**

*Potyvirus: Potyviridae.*

Virus(es) with serologically unrelated virions: potato Y and datura necrosis viruses.

**References**

Badami, R.S. and Kassanis, B. (1959). *Ann. appl. Biol.* **47**: 90.

# Tobacco yellow dwarf monogeminivirus

**Data collated by J.E. Thomas, 1980. Revised 1984.**

*Synonyms:* bean summer death virus. *Standard acronym:* TYDV.

First reported in *Nicotiana tabacum*; from Myrtleford District, Victoria, Australia; by Hill (1937).

**Natural host range and symptoms**

Symptoms persist.

*Nicotiana tabacum* — downcurling of tips and margins of younger leaves, general chlorosis.

*Datura stramonium, D. tatula, Lycopersicon esculentum* — symptomless.

**Transmission**

Transmitted by a vector; an insect; *Orosius argentatus*; Cicadellidae. Transmitted in a persistent manner. Virus retained when the vector moults; not transmitted by mechanical inoculation; transmitted by grafting (between solanaceous hosts); not transmitted by seed.

**Geographical distribution**

Spreads in Australia.

**Experimental host range**

Several (3–9) families susceptible.

Diagnostically susceptible

*Phaseolus vulgaris* cv. Spartan Arrow, *Datura stramonium, Nicotiana tabacum* — systemic leaf yellowing.

Maintenance and propagation hosts

*Datura stramonium, Lycopersicon esculentum* cv. Grosse Lisse, *Nicotiana tabacum.*

Assay hosts (**L**ocal lesions or **W**hole plants)

*Phaseolus vulgaris* cv. Spartan Arrow (W).

Dicotyledonous Hosts *(**Susceptible** or Insusceptible)*

***Beta vulgaris***. *Cucumis sativus*. ***Datura stramonium***. ***Datura tatula***. ***Lycopersicon esculentum***. *Macroptilium lathyroides*. ***Medicago sativa***. *Nicotiana debneyi*. ***Nicotiana glutinosa***. ***Nicotiana rustica***. ***Nicotiana tabacum***. ***Phaseolus vulgaris***. ***Solanum nigrum***. *Solanum tuberosum*. ***Sonchus oleraceus***. ***Trifolium repens***. *Vigna radiata*.

Sources of host-range data

Hill (1950); Helson (1950); Hill and Mandryk (1954); Thornberry (1966); Bowyer and Atherton (1972).

**Properties of particles in sap**

TIP: 50–60 °C (determined by feeding leafhoppers through membranes on virus-containing extracts). LIV: 26 days (in bean tissue stored at –18°C and in purified preparations stored at 4°C). DEP: $\log_{10}$ minus 2–4. Leaf sap contains few virions.

**Purification**

Method: Thomas and Bowyer (1980).

**Particle morphology**

Virions geminate; not enveloped; 18 nm in diameter; 35 nm in length.

**Physical properties**

One sedimenting component in purified preparations; sedimentation coefficient 76 *S*.

**Biochemical properties**

*Genome* consists of DNA; single–stranded; circular; unipartite; largest (or only) genome part 2.8 kb.

*Sequence database accession code(s):*

M81103 Em(40)_vi:GETYDMONO Gb(84)_vi:TYDMONOCOT Tobacco yellow dwarf virus gene, complete cds. 2/94 2,580bp. 1 sequence.

*Virion protein(s)* one; $M_r$ 27500; coat protein. Method of preparation: Thomas and Bowyer (1980).

**Replication**

Replication does not depend on a helper virus.

**Cytopathology**

Virions found in phloem. Inclusions absent from infected cells. Other cellular changes: necrosis of phloem, especially in sieve tubes.

**Taxonomy and Relationships**

Monogeminivirus: *Geminiviridae*.

**References**


Ballantyne, B. (1968). *Agric. Gaz. N.S.W.* **79**: 486.

Bowyer, J.W. and Atherton, J.G. (1972). *Phytopathology* **61**: 1451.

Helson, G.A.H. (1950). *Aust. J. agric. Res.* **1**: 144.

Hill, A.V. (1937). *J. Coun. Scient. Ind. Res. Aust.* **10**: 228.

Hill, A.V. (1950). *Aust. J. agric. Res.* **1**: 141.

Hill, A.V. and Mandryk, M. (1954). *Aust. J. agric. Res.* **5**: 617.

Thomas, J.E. and Bowyer, J.W. (1980). *Phytopathology* **70**: 214.

Thomas, J.E. and Bowyer, J.W. (1984). *CMI/AAB Descr. Pl. Viruses* No. 278, 4 pp.

Thornberry, H.H. (1966). In: *Index of Plant Virus Diseases.* U.S. Dep. Agric. Hdbk No. 307.

# Tobacco yellow net (?) *luteovirus*

**Data collated by A.A. Brunt, 1991.**

*Standard acronym:* TYNV.

First reported in *Nicotiana tabacum*; from the U.S.A.; by Sylvester (1954).

**Natural host range and symptoms**

Symptoms vary seasonally.
*Nicotiana tabacum, Lycopersicon esculentum* — transient chlorotic vein-banding.

**Transmission**

Transmitted by a vector; an insect; *Myzus persicae*; Aphididae. Transmitted in a persistent manner.

**Geographical distribution**

Spreads in the USA (California).

Diagnostically susceptible

*Lycopersicon esculentum, L. pimpinellifolium, Nicotiana tabacum* — transient chlorotic vein-banding.

Diagnostically insusceptible

*Geum chiloense, Beta vulgaris, Nicotiana glutinosa.*

Maintenance and propagation hosts

*Lycopersicon esculentum, Nicotiana tabacum.*

Assay hosts (**L**ocal lesions or **W**hole plants)

*Lycopersicon esculentum* (W), *Nicotiana tabacum* (W).

Dicotyledonous Hosts *(**Susceptible** or Insusceptible)*

*Beta vulgaris*. ***Lycopersicon esculentum***. ***Lycopersicon pimpinellifolium***. *Nicotiana glutinosa*. ***Nicotiana tabacum***.

**Taxonomy and Relationships**

Possible *Luteovirus*.
Host range tests indicate that this virus differs from beet yellow net (?) *luteovirus*.

**References**

Sylvester, E.S. (1954). *Phytopathology* **44**: 219.
Waterhouse, P.M., Gildow, F.E. and Johnstone, G.R. (1988). *AAB Descr. Pl. Viruses* No. 339, 9 pp.

# Tobacco yellow vein assistor (?) *luteovirus*

**Data collated by A.N. Adams, 1987.**

*Standard acronym:* TYVAV. *Strains:* virulent and mild (Adams and Hull, 1972).

First reported in *Nicotiana rustica*; from Malawi; by Adams and Hull (1972).

**Natural host range and symptoms**

*Nicotiana rustica* — only found in natural infections with tobacco yellow vein virus; which together cause tobacco yellow vein disease.

**Transmission**

Transmitted by a vector; an insect; *Myzus persicae*; Aphididae. Transmitted in a persistent manner. Virus retained when the vector moults; can help the vector transmission of another virus (tobacco yellow vein umbravirus); not transmitted by mechanical inoculation.

**Ecology and control**

Studies reported by Adams and Hull (1972).

**Geographical distribution**

Spreads in Malawi.

**Experimental host range**

Few (<3) families susceptible.

Diagnostically susceptible

*Nicotiana tabacum, N. clevelandii, N. glutinosa, N. rustica* — systemic veinal chlorosis and mottling.

*Lycopersicon esculentum* — mottling and rugosity (virulent strain only).

Diagnostically insusceptible

*Chenopodium amaranticolor, C. quinoa, Cucumis sativus, Datura stramonium, Phaseolus vulgaris, Vicia faba.*

Maintenance and propagation hosts

*Nicotiana clevelandii* (W), *N. tabacum* (W).

Dicotyledonous Hosts *(**Susceptible** or Insusceptible)*

*Arachis hypogaea*. ***Nicotiana tabacum***.

Sources of host-range data

Adams and Hull (1972).

**Taxonomy and Relationships**

Probably *Luteovirus*.

Few of the properties of this virus are known; it is probably a luteovirus (Waterhouse *et al.*, 1988).

**References**

Adams, A.N. and Hull, R. (1972). *Ann. appl. Biol.* **71**: 135.

Waterhouse, P.M., Gildow, F.E. and Johnstone, G.R. (1988). *AAB Descr. Pl. Viruses* No. 339, 9pp.

# Tobacco yellow vein (?) *umbravirus*

**Data collated by A.A. Brunt and A.N. Adams, 1987.**

*Standard acronym:* TYVV. *Strains:* mild and virulent strains.

First reported in *Nicotiana tabacum*; from Malawi; by Adams and Hull (1972).

**Natural host range and symptoms**

*Nicotiana rustica, Nicotiana tabacum* — chlorotic vein-banding and leaf malformation.

**Transmission**

Transmitted by a vector; an insect; *Myzus persicae*; Aphididae. Transmitted in a persistent manner. Virus retained when the vector moults; requires, for vector transmission, a helper virus (tobacco yellow vein assistor (?) *luteovirus*); transmitted by mechanical inoculation.

**Ecology and control**

Studies reported by Adams and Hull (1972).

**Geographical distribution**

Spreads in Malawi.

**Experimental host range**

Few (<3) families susceptible.

Diagnostically susceptible

*Nicotiana tabacum, N. clevelandii, N. glutinosa, N. rustica* — systemic chlorotic vein-banding; severe isolates also cause malformation.

*Lycopersicon esculentum, Nicandra physalodes, Petunia hybrida* — light and dark green systemic leaf mottling.

*Arachis hypogaea* — systemic pale green rings and pale green areas along midribs.

*Glycine max* — mild systemic leaf mottling.

Diagnostically insusceptible

*Brassica pekinensis, Chenopodium amaranticolor, C. quinoa, Cucumis sativus, Datura stramonium.*

Maintenance and propagation hosts

*Nicotiana tabacum.*

Assay hosts (**L**ocal lesions or **W**hole plants)

*Nicotiana tabacum* (W), *N. rustica* (W), *N. clevelandii* (W), *N. glutinosa* (W).

Dicotyledonous Hosts *(**Susceptible** or Insusceptible)*

***Arachis hypogaea***. *Brassica campestris* ssp. *pekinensis*. *Chenopodium amaranticolor*. *Chenopodium quinoa*. *Cucumis sativus*. *Datura stramonium*. ***Glycine max***. ***Lycopersicon esculentum***. ***Nicandra physaloides***. ***Nicotiana clevelandii***. ***Nicotiana glutinosa***. ***Nicotiana rustica***. ***Nicotiana tabacum***. *Phaseolus vulgaris*. *Raphanus sativus*. *Vicia faba*. *Vigna unguiculata*.

Sources of host-range data

Adams and Hull (1972).

**Taxonomy and Relationships**

Possible *Umbravirus*.

Does not cross-protect against tobacco mottle and groundnut rosette viruses.

**References**

Adams, A.N. and Hull, R. (1972). *Ann. appl. Biol.* **71**: 135.

# Tomato aspermy *cucumovirus*

**Data collated by J. Hammond and J.M. Kaper, 1986; N. Habili, 1987.**

*Synonyms:* chrysanthemum mild mottle virus, chrysanthemum mosaic virus, chrysanthemum aspermy virus. *Standard acronym:* TAV. *Strains:* type strain, TAV-Can

(Blencowe and Caldwell, 1949; Hollings and Stone, 1969) and TAV-V (Habili and Francki, 1974*a,b*).

First reported in *Chrysanthemum morifolium*; from England; by Ainsworth (1939); Blencowe and Caldwell (1949).

**Natural host range and symptoms**

Symptoms persist (but vary from plant to plant).
*Chrysanthemum morifolium* — severe flower-breaking, dwarfing and malformation.
*Lycopersicon esculentum* — severe leaf malformation and seedless fruit.
*Canna* spp., *Lilium* spp. — mosaic.

**Transmission**

Transmitted by a vector; an insect; *Myzus persicae* and many other species; Aphididae (Kennedy *et al.*, 1962). Transmitted in a non–persistent manner. Virus transmitted by mechanical inoculation; transmitted by grafting; not transmitted by contact between plants; transmitted by seed (transmitted in seeds of *Stellaria media*; Noordam *et al.*, 1965).

**Ecology and control**

Studies reported by Hollings and Kassanis (1957); Brierley and Lorentz (1980); most strains of the virus can be eliminated from *Chrysanthemum* after *c.* 4 weeks at 37°C; or by meristem-tip culture (Dunez and Monsion, 1968). A few heat-tolerant isolates have been found in Britain, able to persist at 37°C, but their other properties were similar to those of typical strains.

**Geographical distribution**

Spreads in the Eurasian region; Australia, Canada, India, Japan, New Zealand, the USA, and the former USSR.

**Experimental host range**

Many (>9) families susceptible.

Diagnostically susceptible

*Nicotiana glutinosa* — chlorotic or ring local lesions, severe blister-mottle with malformation and dwarfing. Leaves become tendril-like and enations form on abaxial leaf surfaces.

*Nicotiana tabacum* — severe mottle, dwarfing and malformation; symptoms less diagnostic than in *N. glutinosa*.

*Nicotiana clevelandii* — severe mottling, dwarfing and malformation.

*Cucumis sativus* — small local chlorotic spots in cotyledons only; type strain does not infect.

*Lycopersicon esculentum* — leaf mottling, malformation, dwarfing, fruits dwarfed, malformed and seedless.

*Phaseolus vulgaris* — small pale necrotic local lesions in winter, not infected by type strain.

*Chenopodium amaranticolor, C. quinoa* — numerous chlorotic or necrotic local lesions; not systemic.

*Vigna unguiculata* — necrotic local lesions; not infected by type strain.

Maintenance and propagation hosts

*Nicotiana glutinosa* — maintenance; *Nicotiana clevelandii* — purification.

Assay hosts (**L**ocal lesions or **W**hole plants)

*Chenopodium amaranticolor* (L); *Chenopodium quinoa* (L); *Vigna unguiculata* (L).

Dicotyledonous Hosts *(**Susceptible** or Insusceptible)*

***Chenopodium amaranticolor. Chenopodium quinoa. Chrysanthemum morifolium. Cucumis sativus***, or *Cucumis sativus. **Lycopersicon esculentum. Nicotiana clevelandii. Nicotiana glutinosa. Nicotiana tabacum. Phaseolus vulgaris***, or *Phaseolus vulgaris. **Stellaria media. Vigna unguiculata***, or *Vigna unguiculata.*

Sources of host-range data

Brierley (1955); Govier (1957); Hollings (1955); Hollings *et al.* (1968); Noordam (1952); Oertel (1967).

**Properties of particles in sap**

TIP: 50–60 °C. LIV: 2–6 days. DEP: $\log_{10}$ minus 4–6. Electron microscopy: best to fix the virions with formaldehyde. Few virions are found in dips, and they are difficult to distinguish from normal cell components. Strains vary in stability in PTA; UA is better.

**Purification**

Method: Marani (1969), a variant of Steere's method), Grogan *et al.* (1963), Hollings *et al.* (1968), Habili and Francki (1974*b*), Lot *et al.* (1972).

**Particle morphology**

Virions isometric; not enveloped; 29 nm in diameter; rounded in profile, or angular in profile (V-strain); without a conspicuous capsomere arrangement.

**Physical properties**

One sedimenting component in purified preparations; sedimentation coefficient 98–100 *S*. Density 1.367 g $cm^{-3}$ in CsCl (labile in CsCl unless fixed; Habili and Francki, 1974*a*). Density 1.304 g/$cm^3$ in $Cs_2SO_4$ (Hull, 1976). Isoelectric point pH 5.6.

**Biochemical properties**

Virions contain 21.2 % nucleic acid; 78.8 % protein; 0 % lipid.

*Genome* consists of RNA; single–stranded; linear; of three parts; largest (or only) genome part the largest 3.41 kb; the 2nd largest 3.074 kb; the 3rd largest 2.214 kb. Genomic nucleic acid isolated by Habili and Francki (1974*a*). Base composition 23.1 % G (TAV-Can), or 23.9 % G (TAV-V); 24.6 % A (TAV-V), or 26.2 % A (TAV-Can); 21.4 % C (TAV-V), or 21.6 % C (TAV-Can); 29 % U (TAV-Can), or 30.1 % U (TAV-V; Stace-Smith and Tremaine, 1973; Habili and Francki, 1974*a*). 5' terminus of RNA has a VPg. Infectivity decreased when deproteinised with proteases; retained when deproteinised with phenol or detergent. Poly A region absent. Genome has *t*RNA–like activity. Genome accepts tyrosine. Nucleotide sequence references: Wilson and Symons (1981); 189 residues from 3' ends of RNAs of two Australian TAV isolates.

*Sequence database accession code(s):*

D01015 Em(40)_vi:TOARNA3 Gb(84)_vi:TOARNA3 Tomato aspermy virus, RNA3 genome segment, complete sequence. 8/92 2,214bp.

D10044 Em(40)_vi:TOAVRNA1 Gb(84)_vi:TOAVRNA1 Tomato aspermy virus (V-TAV) RNA1. 2/92 3,410bp.

D10663 Em(40)_vi:TOARNA2 Gb(84)_vi:TOARNA2 Tomato aspermy virus (TAV) RNA 2, complete sequence. 7/92 3,074bp.

L15335 Em(40)_vi:TOA3APCOA Gb(84)_vi:TOA3APCOAT Tomato aspermy virus 3A protein and coat protein mRNA (RNA3), complete cds. 5/93 2,222bp.

M10342 Em(40)_vi:CUTOARN Gb(84)_vi:TOARNAV4 Tomato aspermy virus (V-TAV) RNA 4, 3' end. 7/89 188bp.

M10343 Em(40)_vi:CUTOARN2 Gb(84)_vi:TOARNAV3 Tomato aspermy virus (V-TAV) RNA 3, 3' end. 7/89 188bp

M10344 Em(40)_vi:CUTOARN3 Gb(84)_vi:TOARNAV1 Tomato aspermy virus (V-TAV) RNA 1, 3' end. 7/89 189bp.

M10345 Em(40)_vi:CUTOARN4 Gb(84)_vi:TOARNAN3 Tomato aspermy virus (N-TAV) RNA 3, 3' end. 7/89 189bp.

M10346 Em(40)_vi:CUTOARNA Gb(84)_vi:TOARNAN1 Tomato aspermy virus (N-TAV) RNA 1, 3' end. 7/89 189bp.

S72468 Gb(89)_un:S72468 capsid protein (RNA 3) (tomato aspermy virus TAV-B, Blencowe, Genomic RNA, 2213 nt). 1/95 2,2.

*Non–genomic nucleic acid* found in the virions; is subgenomic *m*RNA, or satellite RNA; the former encodes the coat protein and is called RNA-4 and is of 1.303 kb.

$M_r$ 24500; coat protein. Method of preparation: Habili and Francki (1974*a*). Amino acid composition: Habili and Francki (1974*a*).

*Virus–coded non–virion proteins* isolated; one protein found. $M_r$ 30000; non structural protein.

**Replication**

Replication does not depend on a helper virus.

**Cytopathology**

Virions found in all parts of the host plant and particularly in the mesophyll; in cytoplasm and in cell vacuoles. Inclusions present in infected cells; are crystals in the cytoplasm; they contain virions.

**Taxonomy and Relationships**

*Cucumovirus: Bromoviridae.*

Virus(es) with serologically related virions: cucumber mosaic and peanut stunt viruses (Waterworth *et al.*, 1973).

Tomato aspermy and cucumber mosaic virus are similar; some isolates of each have serologically related virions and cross-protect plants against each other. Most strains of tomato aspermy virus differ from most of cucumber mosaic virus in that they (1) infect only the cotyledons of cucumber; (2) induce seedlessness of tomato fruit; (3) induce enations to form on the abaxial surfaces of leaves of several *Nicotiana* species; and (4) infect *Chrysanthemum morifolium.*

**References**


Ainsworth, G.C. (1939). *Rep. Exp. Res. Stn Cheshunt, 1938*, p. 60.

Bernal, J.J., Moriones, E. and Garcia-Arenal, F. (1991). *J. gen. Virol.* **72**: 2191.

Blencowe, J.W. and Caldwell, J. (1949). *Ann. appl. Biol.* **36**: 320.

Brierley, P. (1955). *Phytopathology* **45**: 2.

Brierley, P. and Lorentz, P. (1960). *Phytopathology* **50**: 404.

Christie, R.G. and Edwardson, J.R. (1977). *Fla Agric. Exp. Stn Monog.* No. 9, p.89.

Dunez, J. and Monsion, M. (1968). *Annls. Épiphyt.* **19**: 165.

Govier, D.A. (1957). *Ann. appl. Biol.* **45**: 62.

Grogan, R.G., Uyemoto, J.K. and Kimble, K.A. (1963). *Virology* **21**: 36.

Habili, N. and Francki, R.I.B. (1974*a*). *Virology* **57**: 392.

Habili, N. and Francki, R.I.B. (1974*b*). *Virology* **60**: 29.

Hollings, M. (1955). *Ann. appl. Biol.* **43**: 86.

Hollings, M. and Kassanis, B. (1957). *J. R. hort. Soc.* **82**: 339.

Hollings, M. and Stone, O.M. (1969). *Rep. Glasshouse Crops Res. Inst.* 1967, p. 95.

Hollings, M and Stone, O.M. (1971). *CMI/AAB Descr. Pl. Viruses* No. 79, 4 pp.

Hollings, M., Stone, O.M. and Brunt, A.A. (1968). *Rep. Glasshouse Crops Res. Inst.* 1968, p. 95.

Hull, R. (1976). *Virology* **75**: 18.

Kennedy, J.S., Day, M.F. and Eastop, V.F. (1962). *A Conspectus of Aphids as Vectors of Plant Viruses*. Comm. Inst. Ent., London.
Lot, H., Marrou, J., Quiot, J.B. and Esvan, C. (1972). *Annls. Phytopath.* **4**: 25.
Marani, F. (1969). *Phytopathol. Medit.* **8**: 142.
Moriones, E., Roossinck, M. and Garcia-Arenal, F. (1991). *J. gen. Virol.* **72**: 779.
Noordam, D. (1952). *Tijdschr. PlZiekt.* **58**: 121.
Noordam, D., Bijl, M., Overbeek, S.C. and Quiniones, S.S. (1965). *Neth. J. Pl. Path.* **71**: 61.
O'Reilly, D., Thomas, C.J.R. and Coutts, R.H.A. (1991). *J. gen. Virol.* **72**: 1.
Oertel, C. (1967). *Zbl. Bakt. ParasitKde.* Abt. 2, **121**: 276.
Stace-Smith, R. and Tremaine, J.H. (1973). *Virology* **51**: 401.
Waterworth, H.E., Monroe, R.L. and Kahn, R.P. (1973). *Phytopathology* **63**: 93.
Wilson, P.A. and Symons, R.H. (1981). *Virology* **112**: 342.

# Tomato Australian leafcurl bigeminivirus

**Data collated by B.D. Conde and J.E. Thomas, 1987.**

*Synonyms:* tomato transmissible leafroll virus.

First reported in *Lycopersicon esculentum*; from Darwin, Australia; by Aldrick (1971).

## Natural host range and symptoms

Symptoms persist.

*Lycopersicon esculentum* — chlorosis and upward curling of leaflet margins, dark stunting.

## Transmission

Transmitted by a vector; possible an insect; *Bemisia tabaci*; Aleyrodidae. Virus not transmitted by mechanical inoculation; transmitted by grafting.

## Geographical distribution

Spreads in Australia.

## Experimental host range

Few (<3) families susceptible (and limited to graft transmission tests until vector transmission is obtained).

Diagnostically susceptible

*Lycopersicon esculentum* — chlorosis and upward curling of leaflet margins, stunting.

*Datura stramonium* — downward curling of leaf margins, interveinal chlorosis, and chlorosis of minor veins.

*Nicotiana benthamiana* — leaf malformation, vein chlorosis, prominent veins on abaxial leaf and sepal surfaces.

*Nicotiana tabacum* — stunting, chlorosis and upward cupping of leaves.

Diagnostically insusceptible

*Capsicum annuum, Physalis minima, Physalis peruvianum.*

Maintenance and propagation hosts

*Lycopersicon esculentum, Datura stramonium, Nicotiana benthamiana.*

Assay hosts (**L**ocal lesions or **W**hole plants)

*Datura stramonium* (W); *Nicotiana benthamiana* (W).

Dicotyledonous Hosts *(**Susceptible** or Insusceptible)*

*Capsicum annuum*. ***Datura stramonium***. ***Lycopersicon esculentum***. ***Nicandra physaloides***. ***Nicotiana benthamiana***. ***Nicotiana glutinosa***. *Nicotiana rustica*. ***Nicotiana tabacum***. ***Physalis floridana***. *Physalis peruviana*. ***Solanum melongena***. ***Solanum tuberosum***.

Sources of host-range data

B.D. Conde and J.E. Thomas (unpublished data).

**Particle morphology**

Virions geminate; not enveloped.

**Biochemical properties**

*Genome* probably consists of DNA.

*Sequence database accession code(s):*

L11746 Em(40)_vi:GETLCAL1R Gb(84)_vi:TLCAL1REP Tomato leaf curl virus AL1 replication protein gene, complete cds. 3/93 294bp.

L12738 Em(40)_vi:GETLCVSEQ Gb(84)_vi:TLCVSEQ Tomato leaf curl virus coat protein (AR1) gene, partial cds. 4/93 219bp.

L12739 Em(40)_vi:GETLCVSE1 Gb(84)_vi:TLCVSEQA Tomato leaf curl virus origin of replication. 5/93 165bp.

S53251 Em(40)_vi:S53251 Gb(84)_vi:S53251 orf V1...orf C4 tomato leaf curl virus TLCV, Australian isolate, Genomic Complete, 6 genes,

S55329 Em(40)_vi:S55329 Gb(84)_vi:S55329 V1 orf (5' region) tomato leaf curl virus TLCV, Genomic, 71 nt. 12/93 71bp.

S55330 Em(40)_vi:S55330 Gb(84)_vi:S55330 V2 orf, C2 orf (3' regions) tomato leaf curl virus TLCV, Genomic, 51 nt. 12/93 51bp.

S55334 Em(40)_vi:S55334 Gb(84)_vi:S55334 C2 orf (5' region) tomato leaf curl virus TLCV, Genomic, 72 nt. 12/93 72bp.

S55445 Em(40)_vi:S55445 Gb(84)_vi:S55445 C1 orf (5' region) tomato leaf curl virus TLCV, Genomic, 240 nt. 12/93 240bp

U15015 Gb(84)n:TLU15015 Tomato leaf curl virus coat protein AV1, AV2, AV3, replication-associated protein AC1, AC2, A

U15016 Gb(84)n:TLU15016 Tomato leaf curl virus coat protein AV1, AV2, AV3, replication-associated protein AC1, AC2.

**Cytopathology**

Virions found in phloem; in nuclei.

**Taxonomy and Relationships**

Bigeminivirus: *Geminiviridae* (based on virion morphology and *c*-DNA reaction with a probe to African cassava mosaic virus).

Virus(es) with serologically related virions: African cassava mosaic virus.

Genomic DNA reacts with a c-DNA probe to African cassava mosaic virus (D.J. Robinson, unpublished data).

**References**

Aldrick, S. (1970–71). *Plant Path. Ann. Rep., Northern Territory Administration* (unpublished).

Thomas, J.E., Massalski, P.R. and Harrison, B.D. (1986). *J. gen. Virol.* **67**: 2739.

# Tomato black ring *nepovirus*

**Data collated by B.D. Harrison, 1982. Revised 1983.**

*Synonyms:* bean ringspot virus, beet ringspot virus (Harrison, 1958), celery yellow vein virus (Hollings, 1960; 1965), lettuce ringspot virus (Hollings, 1965), potato bouquet virus, potato pseudo-aucuba virus. *Standard acronym:* TBRV.

First reported in *Lycopersicon esculentum*; from the U.K; by Smith (1946).

**Natural host range and symptoms**

Symptoms disappear soon after infection.

*Allium porrum, Beta vulgaris, Lactuca sativa, Phaseolus vulgaris, Robinia pseudo-acacia, Fraxinus* sp., *Rubus* sp., *Lycopersicon esculentum, Solanum tuberosum, Apium graveolens, Narcissus pseudo-narcissus, Tulipa* sp., *Vitis* sp., and many weed and endemic species. — necrotic rings, spots and flecks, systemic chlorotic ringspots, mottle, stunting, leaf malformation, vein yellowing.

**Transmission**

Transmitted by a vector; a nematode; *Longidorus elongatus, L. attenuatus*; Dorylamidae. Virus transmitted by mechanical inoculation; transmitted by grafting; transmitted by seed (10–100% depending on host); transmitted by pollen to the seed, or transmitted by pollen to the pollinated plant.

**Geographical distribution**

Spreads in Brazil, Canada, the former Czechoslovakia, Denmark, Finland, France, Germany, Greece, Hungary, India, Ireland, Italy, Japan, Kenya, the Netherlands, Norway, Poland, Romania, Sweden, Turkey, the UK, the USA, the former USSR, the former Yugoslavia. Found, but with no evidence of spread, in Japan.

**Experimental host range**

Many (>9) families susceptible.

Diagnostically susceptible

*Beta vulgaris, Chenopodium amaranticolor, C. quinoa* — necrotic local lesions, systemic chlorotic mottling, necrosis.

*Cucumis sativus* — necrotic local lesions, systemic mottling, necrosis.

*Phaseolus vulgaris* — chlorotic local lesions, systemic chlorotic mottling, necrosis, malformation.

*Lycopersicon esculentum* — necrotic local lesions, systemic necrotic spotting, streaking, mottling and malformation.

*Nicotiana clevelandii* — necrotic local lesions, systemic necrotic streaking, mottling and stunting.

*N. rustica, N. tabacum* cvs White Burley, Xanthi-*nc* — necrotic and chlorotic local lesions, systemic necrotic spotting and streaking, mottling.

*Petunia hybrida* — chlorotic local lesions, systemic veinal necrosis or chlorosis.

Maintenance and propagation hosts

*Cucumis sativus, Nicotiana clevelandii, N. tabacum.*

Assay hosts (**L**ocal lesions or **W**hole plants)

*Chenopodium amaranticolor, C. quinoa, Nicotiana tabacum.*

Dicotyledonous Hosts *(**Susceptible** or Insusceptible)*

*Ageratum conyzoides*. ***Amaranthus caudatus***. ***Amaranthus retroflexus***. ***Antirrhinum majus***. ***Apium graveolens***. ***Atriplex hortensis***. ***Bellis perennis***. ***Beta*** vulgaris. *Brassica campestris* ssp. *pekinensis*. *Brassica campestris* ssp. *rapa*. ***Calendula officinalis***. ***Capsella bursa–pastoris***. *Capsicum frutescens*. ***Celosia argentea***. *Cheiranthus cheiri*. *Chenopodium album*. *Chenopodium amaranticolor*. ***Chenopodium foetidum***. ***Chenopodium quinoa***. *Chrysanthemum morifolium*. ***Coriandrum sativum***. ***Cucumis melo***. ***Cucumis sativus***. ***Cucurbita pepo*** (var. medullosa). *Dahlia pinnata*. ***Datura stramonium***. ***Daucus carota***. *Dianthus barbatus*. ***Emilia sagittata***. ***Fagopyrum esculentum***. ***Glycine max***. ***Gomphrena globosa***. ***Helianthus annuus***. ***Hyoscyamus niger***. ***Lactuca sativa***. *Lathyrus odoratus*. ***Lens esculenta***. ***Lobelia erinus***. ***Lycopersicon esculentum***. *Lycopersicon pimpinellifolium*. ***Matthiola incana***. *Medicago sativa*. *Melilotus alba*. ***Myosotis sylvatica***. ***Nicandra physaloides***. ***Nicotiana clevelandii***. *Nicotiana glutinosa*. ***Nicotiana rustica***. ***Nicotiana sylvestris***. ***Nicotiana tabacum***. *Pastinaca sativa*. *Petroselinum crispum*. *Petunia hybrida*. ***Phaseolus vulgaris***. ***Physalis floridana***. ***Phytolacca americana***. ***Pisum sativum***. ***Salvia splendens***. ***Senecio vulgaris***. ***Solanum tuberosum***. ***Sonchus oleraceus***. ***Spinacia oleracea***. ***Stellaria media***. ***Tetragonia expansa***. ***Torenia fournieri***. *Trifolium pratense*. ***Trifolium repens***. ***Tropaeolum majus***. ***Vicia faba***. ***Vigna unguiculata***. ***Zinnia elegans***.

Monocotyledonous Hosts *(**Susceptible** or Insusceptible)*

*Avena sativa*. *Chloris gayana*. ***Narcissus pseudonarcissus***.

Sources of host-range data

Schmelzer (1963); Hollings (1965); Thornberry (1966).

**Properties of particles in sap**

TIP: 60–65 °C. LIV: 14–21 days. DEP: $\log_{10}$ minus 3–4. Infectivity of sap not changed by treatment with di–ethyl ether. Leaf sap contains few virions.

**Purification**

Method: Mayo *et al.* (1979).

**Particle morphology**

Virions isometric; not enveloped; 26 nm in diameter; angular in profile; without a conspicuous capsomere arrangement.

**Physical properties**

Three sedimenting components in purified preparations; sedimentation coefficient of the fastest 121 *S* (B). Density 1.5 g $cm^{-3}$ in CsCl (B), or 1.44 g $cm^{-3}$ in CsCl (M), or 1.285 g $cm^{-3}$ in CsCl (T). $A_{260}/A_{280}$ ratio 1.78 (B), or 1.62 (M).

**Biochemical properties**

Virions contain 44 % nucleic acid (B), or 32 % nucleic acid (M), or 0 % nucleic acid (T); 56 % protein (B), or 68 % protein (M), or 100 % protein (T); 0 % lipid.

*Genome* consists of RNA; single–stranded; linear; of two parts; largest (or only) genome part the larger 8.2 kb; the 2nd largest 4.8 kb. 5' terminus of RNA has a VPg. Infectivity lost when deproteinised with proteases; retained when deproteinised with phenol or detergent. Poly A region present; at 3' end and 100 adjacent nucleotides of RNA–2. Nucleotide sequence references: Harrison (unpublished).

*Sequence database accession code(s):*

D00322 Em(40)_vi:TBRRNA1 Gb(84)_vi:TBRRNA1 Tomato black ring virus (TBRV), RNA-1, complete cds. 3/91 7,356bp.

X04062 Em(40)_vi:NETBRVR2 Gb(84)_vi:NETBRVR2 Tomato black ring virus strain S RNA-2. 9/93 4,662bp.

X05304 Em(40)_vi:NETBRVC1 Gb(84)_vi:NETBRVC1 Tomato black ring virus strain C RNA-1 3'terminal region. 7/89 166bp.

X05305 Em(40)_vi:NETBRVC2 Gb(84)_vi:NETBRVC2 Tomato black ring virus strain C RNA-2 3'terminal region. 7/89 151bp.

X05306 Em(40)_vi:NETBRVG2 Gb(84)_vi:NETBRVG2 Tomato black ring virus strain G RNA-2 3'terminal region. 7/89 100bp.

X05307 Em(40)_vi:NETBRVA2 Gb(84)_vi:NETBRVA2 Tomato black ring virus strain A RNA-1 3'terminal region. 7/89 163bp.

X05308 Em(40)_vi:NETBRVA1 Gb(84)_vi:NETBRVA1 Tomato black ring virus strain A RNA-2 3'terminal region. 11/87 127bp.

X80831 Gb(84)n:TBRVEDRN2 Tomato black ring Virus (strain ED) genomic RNA-2. 8/94 4,618bp.

*Non–genomic nucleic acid* found in the virions; is satellite RNA (1.515kb in size; satellite RNAs of TBRV-G and TBRV-S differ in nucleotide sequence).

*Virion protein(s)* two, or one (possibly); $M_r$ of the larger 58000; coat protein. $M_r$ of 2nd largest 4000; VPg.

**Replication**

Replication does not depend on a helper virus.

**Cytopathology**

Virions found in all parts of the host plant and particularly in the mesophyll; in cytoplasm and in cell vacuoles; tubules extending through plasmodesmata, which is possibly a sign of degenerating nuclei. Inclusions present in infected cells; are unusual in shape; vesicles; they contain virions. Other cellular changes: virions in single files and tubules which are surrounded by a single membrane.

**Taxonomy and Relationships**

*Nepovirus: Comoviridae.*

Virus(es) with serologically related virions: cacao necrosis and grapevine chrome mosaic viruses, but distantly.

**References**


Greif, C., Hemmer, O. and Fritsch, C. (1988). *J. gen. Virol.* **69**: 1517.

Harrison, B.D. (1957). *Ann. appl. Biol.* **45**: 462.

Harrison, B.D. (1958). *J. gen. Microbiol.* **18**: 450.

Harrison, B.D., Mowat, W.P. and Taylor, W.P. (1961). *Virology* **14**: 480.

Hollings, M. (1960). *Rep. Glasshouse Crops Res. Inst.* 1959, p. 73.

Hollings, M. (1965). *Ann. appl. Biol.* **55**: 459.

Mayo, M.A., Barker, H. and Harrison, B.D. (1979). *J. gen. Virol.* **43**: 603.

Meyer, M., Hemmer, O., Mayo, M.A. and Fritsch, C. (1986). *J. gen. Virol.* **67**: 1257.

Murant, A.F. (1970). *CMI/AAB Descr. Pl. Viruses* No. 38, 4 pp.

Murant, A.F., Mayo, M.A., Harrison, B.D. and Goold, R.A. (1973). *J. gen. Virol.* **19**: 275.

Murant, A.F., Taylor, M., Duncan, G.H. and Raschke, S.H. (1981). *J. gen. Virol.* **53**: 321.

Najarian, R.C. and Bruening, G. (1980). *Virology* **106**: 301.

Randles, J.W., Harrison, B.D., Murant, A.F. and Mayo, M.A. (1977). *J. gen. Virol.* **36**: 187.

Schmelzer, K.M. (1963). *Phytopath. Z.* **46**: 235.

Smith (1946). *Parasitol.* **37**: 126.

Stace-Smith, R. (1970). *CMI/AAB Descr. Pl. Viruses* No. 18, 4 pp.

Thornberry, H.H. (1966). In: *Index of Plant Virus Diseases*. U.S. Dep. Agric. Hdbk No. 307, p.323.

# Tomato black ring virus satellite RNA

*Sequence database accession code(s):*

D00142 Gb(84)_vi:TBRVRNAC Tomato black ring virus (TBRV) isolate C satellite RNA. 2/91 1,374bp.

D00143 Gb(84)_vi:TBRVRNAE Tomato black ring virus (TBRV) isolate E satellite RNA. 2/91 1,372bp.

D00144 Gb(84)_vi:TBRVRNAL Tomato black ring virus (TBRV) isolate L satellite RNA. 2/91 1,376bp.

X00978 Em(40)_vi:NETBRV01 Gb(84)_vi:NETBRV01 Tomato black ring virus (TBRV) satellite RNA. 9/93 1,375bp.

X05687 Em(40)_vi:NCTBRVLS Gb(84)_vi:NCTBRVLS Tomato black ring virus (TBRV-L) satellite RNA. 7/91 1,376bp.

X05688 Em(40)_vi:NCTBRVES Gb(84)_vi:NCTBRVES Tomato black ring virus (TBRV-E) satellite RNA. 7/91 1,372bp

X05689 Em(40)_vi:NCTBRVCS Gb(84)_vi:NCTBRVCS Tomato black ring virus (TBRV-C) satellite RNA. 7/91 1,374bp.

### Taxonomy and Relationships

Satellite RNA.

# Tomato bushy stunt *tombusvirus*

**Data collated by G.P. Martelli, 1990.**

*Standard acronym:* TBSV. *Strains:* type, BS-3 and others.

First reported in *Lycopersicon esculentum*; from England; by Smith, 1935.

### Natural host range and symptoms

Symptoms persist.

*Lycopersicon esculentum* — bushy growth, fewer smaller fruits with chlorotic blotching, rings and line pattern.

*Capsicum annuum* — stunting, mottling, deformation of the leaves, no fruit.

*Solanum melongena* — stunting, leaf mottling and crinkling; few spotted and deformed fruit.

*Tulipa* spp. — leaf and petal necrosis.

*Tolmiea menziesii* — stunting and leaf mottling.

*Malus* spp., *Pyrus* spp. — no symptoms.

### Transmission

Transmitted by means not involving a vector. Virus transmitted by mechanical inoculation; transmitted by grafting; not transmitted by contact between plants; transmitted by seed (at a low rate); possibly transmitted by pollen to the seed.

### Ecology and control

Studies reported by Cherif and Spire (1983); Kegler and Kegler (1981); Tomlinson and Faithfull (1984).

### Geographical distribution

Spreads in Argentina, Morocco, Tunisia, the UK, and the USA (in California). Found, but with no evidence of spread, in Portugal, France, Italy, Germany and Canada.

**Experimental host range**

Many (>9) families susceptible. Experimentally infected plants mostly show systemic mosaics and mottles.

Diagnostically susceptible

*Gomphrena globosa* — necrotic reddish local lesions, some isolates infect systemically.

*Ocimum basilicum* — necrotic black local lesions with lighter centre; not systemic.

*Chenopodium amaranticolor* — whitish necrotic dots with chlorotic haloes; rarely systemic.

*Chenopodium quinoa* — chlorotic local lesions; rarely systemic.

*Nicotiana clevelandii* — chlorotic or necrotic local lesions; systemic mottle and necrosis.

*Nicotiana glutinosa* — brown necrotic local lesions.

Maintenance and propagation hosts

*Datura stramonium, Nicotiana clevelandii, Gomphrena globosa, Chenopodium amaranticolor, Chenopodium quinoa.*

Assay hosts (**L**ocal lesions or **W**hole plants)

*Gomphrena globosa* (L), *Chenopodium amaranticolor* (L), *Nicotiana glutinosa* (L), *Nicotiana clevelandii* (W), *Datura stramonium* (W).

Dicotyledonous Hosts *(**Susceptible** or Insusceptible)*

***Antirrhinum majus*. *Capsicum annuum*. *Capsicum frutescens*. *Celosia argentea*. *Chenopodium album*. *Chenopodium amaranticolor*. *Chenopodium murale*. *Chenopodium quinoa*. *Cucumis sativus*. *Cucurbita maxima*. *Cucurbita pepo*. *Datura stramonium*. *Dianthus barbatus*. *Gomphrena globosa*. *Gypsophila elegans*. *Hyoscyamus niger*. *Lavatera trimestris*. *Lycopersicon esculentum*.** *Lycopersicon pimpinellifolium*. ***Nicotiana benthamiana*. *Nicotiana bigelovii*. *Nicotiana clevelandii*. *Nicotiana debneyi*. *Nicotiana glutinosa*. *Nicotiana rustica*. *Nicotiana tabacum*. *Ocimum basilicum*. *Petunia hybrida*. *Phaseolus vulgaris*. *Physalis floridana*. *Solanum melongena*. *Solanum nigrum*. *Solanum tuberosum*. *Spinacia oleracea*. *Tetragonia expansa*. *Vicia faba*. *Vigna radiata*. *Vigna unguiculata*. *Zinnia elegans*.**

Comments on host-range

Most hosts are only infected locally.

Sources of host-range data

Schmelzer (1958); Cherif and Spire (1983).

**Properties of particles in sap**

TIP: 80–90 °C. LIV: 130–150 days. DEP: $\log_{10}$ minus often above 6. Leaf sap contains many virions.

**Purification**

Method: Gallitelli *et al.* (1985).

**Particle morphology**

Virions isometric; not enveloped; 30 nm in diameter; rounded in profile; without a conspicuous capsomere arrangement.

**Physical properties**

One sedimenting component in purified preparations; sedimentation coefficient 130 *S*. Density 1.35 g $cm^{-3}$ in CsCl. Isoelectric point pH 4.1. $A_{260}/A_{280}$ ratio 1.64.

**Biochemical properties**

Virions contain 17 % nucleic acid; 83 % protein.

*Genome* consists of RNA; single–stranded; linear; unipartite; largest (or only) genome part 4.7 kb. Genomic nucleic acid isolated by Gallitelli *et al.* (1985). Base composition 28.6 % G; 26.3 % A; 21.2 % C; 26.3 % U. 5' terminus of RNA has probably a methylated nucleotide cap. Infectivity retained when deproteinised with proteases; retained when deproteinised with phenol or detergent. Poly A region absent. Additional factor not required for infectivity. Genome has no *t*RNA–like activity.

*Sequence database accession code(s):*

M18056 Em(40)_vi:TBSVDI Gb(84)_vi:TBSVDI Tomato bushy stunt virus defective interfering virion RNA. 7/89 396bp.

M21958 Em(40)_vi:TBSCPP Gb(84)_vi:TBSACG Tomato bushy stunt virus complete genome. 12/90 4,776bp.

M59041 Em(40)_vi:TBSDIPA Gb(84)_vi:TBSDIPA Tomato bushy stunt virus defective interfering particle. 3/91 591bp.

M59042 Em(40)_vi:TBSDIPB Gb(84)_vi:TBSDIPB Tomato bushy stunt virus defective interfering particle. 3/91 546bp.

M59043 Em(40)_vi:TBSDIPC Gb(84)_vi:TBSDIPC Tomato bushy stunt virus defective interfering particle. 3/91 591bp.

M59044 Em(40)_vi:TBSDIPD Gb(84)_vi:TBSDIPD Tomato bushy stunt virus defective interfering particle. 3/91 595bp.

M59045 Em(40)_vi:TBSDIPE Gb(84)_vi:TBSDIPE Tomato bushy stunt virus defective interfering particle. 3/91 599bp.

M59046 Em(40)_vi:TBSDIPF Gb(84)_vi:TBSDIPF Tomato bushy stunt virus defective interfering particle. 3/91 594bp.

M59047 Em(40)_vi:TBSDIPG Gb(84)_vi:TBSDIPG Tomato bushy stunt virus defective interfering particle. 3/91 620bp.

M59048 Em(40)_vi:TBSDIPH Gb(84)_vi:TBSDIPH Tomato bushy stunt virus defective interfering particle. 3/91 578bp.

M59049 Em(40)_vi:TBSDIPI Gb(84)_vi:TBSDIPI Tomato bushy stunt virus defective interfering particle. 3/91 583bp.

M59050 Em(40)_vi:TBSDIPJ Gb(84)_vi:TBSDIPJ Tomato bushy stunt virus defective interfering particle. 3/91 592bp. 12 sequences.

*Non–genomic nucleic acid* found in the virions; is subgenomic *m*RNA, satellite RNA, and of an unusual type of structure (defective interfering RNA). Sub–genomic *m*RNA found in infected cells. 4 virus specified dsRNA species found in infected cells (plus 1). Size of largest virus specified dsRNA 4.733 kbp (corresponding to the genome); 2nd largest 2.188 kbp (first subgenomic); 3rd largest 0.936 kbp (2nd subgenomic); 4th largest 0.62 kbp (satellite RNA); 5th largest 0.499 kbp (DI RNA).

*Virion protein(s)* one; $M_r$ 42000; coat protein. Method of preparation: Gallitelli and Russo (1987). Amino acid sequence: Hopper *et al.* (1984). Virion proteins not glycosylated. Virion proteins not phosphorylated.

*Virus–coded non–virion proteins* isolated (Martelli: by electrophoresis from *in vitro* translation systems); three proteins found. $M_r$ of the largest 92000. $M_r$ of 2nd largest 33000. $M_r$ of 3rd 22000.

**Replication**

Genome replicates probably in cytoplasm (in association with multivesicular bodies). Coat protein *m*RNA translated in in the cytoplasm probably. Replication does not depend on a helper virus. Acts as helper for a satellite RNA.

**Cytopathology**

Virions found in all parts of the host plant; in cytoplasm, in nuclei, in nucleoli, in mitochondria, and in cell vacuoles. Inclusions present in infected cells; are crystals in the cytoplasm and unusual in shape; multivesicular bodies derived from peroxisomes or mitochondria; they do not contain virions.

**Taxonomy and Relationships**

*Tombusvirus: Tombusviridae* (the type species).

Virus(es) with serologically related virions: artichoke mottled crinkle, pelargonium leaf curl and petunia asteroid mosaic viruses are closely related; Moroccan pepper virus is distantly related; eggplant mottled crinkle, carnation Italian ringspot, Lato River and Neckar River viruses are very distantly related. Virus(es) with serologically unrelated virions: cymbidium ringspot, cucumber necrosis and grapevine Algerian latent viruses.

Sequences homologous with those of other species of genus (especially in certain ORFs). The coat protein gene is, unusually, located upstream of a 3' terminal ORF coding for a non-virion protein. Best tests for diagnosis: mechanically inoculated plants of *Gomphrena globosa* produce characteristic local lesions within 24–36 hours and those of *Ocimum basilicum* in less than a week.

**References**


Cherif, C. and Spire, D. (1983). *Agronomie* **3**: 70.

Gallitelli, D. and Russo, M. (1987). *Phytopath. Z.* **119**: 106.

Gallitelli, D., Hull, R. and Koenig, R. (1985). *J. gen. Virol.* **66**: 1523.

Herne, P.Q., Knorr, D.A., Hillman, B.I. and Morris, T.J. (1990). *Virology* **177**: 141.

Hopper, P., Harrison, S.C. and Sauer, R.T. (1984). *J. mol. Biol.* **177**: 701.

Koenig, R. and Gibbs, A.J. (1986). *J. gen. Virol.* **67**: 75.

Kegler, G. and Kegler, H. (1981). *Arch. Phytopathol. PflSchutz.* **17**: 307.

Martelli, G.P., Gallitelli, D. and Russo, M. (1988). In: *The Plant Viruses*; Vol. 3, *Polyhedral Virions with Monopartite RNA Genomes*, p. 13; ed. R. Koenig. Plenum Press, New York.

Martelli, G.P., Russo, M. and Gallitelli, D. (1989). *AAB Descr. Pl. Viruses* No. 352, 8pp.

Martelli, G.P., Quacquarelli, A. and Russo, M. (1971). *CMI/AAB Descr. Pl. Viruses* No. 69, 4 pp.

Olsen, A.J., Briscogne, G. and Harrison, S.C. (1983). *J. mol. Biol.* **171**: 61.

Schmelzer, K. (1958). *Z. PflKrankh. PflPath. PflSchutz.* **65**: 80.

Smith, K.M. (1935). *Ann. appl. Biol.* **22**: 731.

Tomlinson, J.A. and Faithfull, E.M. (1984). *Ann. appl. Biol.* **104**: 485.


# Tomato chlorotic spot (?) *tospovirus*

**Data collated by A.A. Brunt, 1995.**

First reported in *Lycopersicon esculentum*; from Brazil; by De Avila *et al.* (1990).

*Sequence database accession code(s):*

S54325 Em(40)_un:S54325 Gb(84)_vi:S54325 nucleoprotein tomato chlorotic spot virus, isolate BR-03, Genomic RNA, 929 nt. 3/93 929bp. 1 sequence.

**Taxonomy and Relationships**

Possible *Tospovirus: Bunyaviridae*.

**References**

De Avila, A.C., Huguenot, R., Resende, R. de O., Kitajima, E.W., Goldbach, R.W. and Peters, D. (1990). *J. gen. Virol.* **71**: 2801.

De Avila, A.C., de Haan, P., Smeets, M.L.L., Resende, R. de O., Kitajima, E.W., Goldbach, R.W. and Peters, D. (1992). *Arch. Virol.* **128**: 211.

De Avila, A.C., de Haan, P., Kormelink, R., Resende, R. de O., Goldbach, R.W. and Peters, D. (1993). *J. gen. Virol.* **74**: 153.

# Tomato golden mosaic bigeminivirus

**Data collated by K.W. Buck, 1987.**

*Synonyms:* virus do mosaico dourado do tomateiro, possibly tomato yellow mosaic virus.
*Standard acronym:* TGMV.

First reported in *Lycopersicon esculentum*; from Brazil; by Costa *et al.* (1975).

**Natural host range and symptoms**

*Lycopersicon esculentum* — stunting and severely deformed young leaves and shoots; bright yellow mosaic.

**Transmission**

Transmitted by a vector; an insect; *Bemisia tabaci*; Aleyrodidae. Virus transmitted by mechanical inoculation; transmitted by grafting.

**Ecology and control**

Studies reported by Costa *et al.* (1976).

**Geographical distribution**

Spreads in Brazil.

Diagnostically susceptible

*Nicotiana benthamiana* — faint chlorotic systemic mottling then golden yellow mosaic, curling and malformation.

*Datura stramonium* — yellow local lesions, then a systemic golden yellow mosaic especially around veins.

Diagnostically insusceptible

Virus only known to infect *Solanaceae*; few other species tested.

Maintenance and propagation hosts

*Nicotiana benthamiana, Nicotiana tabacum* cv. Xanthi-*nc*.

Dicotyledonous Hosts *(**Susceptible** or Insusceptible)*

***Datura stramonium*. *Lycopersicon esculentum*. *Nicotiana benthamiana*. *Nicotiana clevelandii*. *Nicotiana debneyi*. *Nicotiana glutinosa*. *Nicotiana tabacum*. *Petunia hybrida*.** *Phaseolus vulgaris*. ***Physalis floridana*.** *Solanum nigrum*. *Solanum tuberosum*.

Additional hosts

*Physalis* sp.; *Solanum pennelli* (**S**).

Species listed as insusceptible have been tested by mechanical inoculation or grafting; infection by whitefly has not been tested.

Sources of host-range data

Costa *et al.* (1975); Costa (1976); Hamilton *et al.* (1981); Matyis *et al.* (1975).

**Properties of particles in sap**

Leaf sap contains few virions.

**Purification**

Method: Stein *et al.* (1983).

**Particle morphology**

Virions geminate; not enveloped; 25 nm in diameter.

**Physical properties**

One sedimenting component in purified preparations. Density 1.34 in $Cs_2SO_4$; (unstable in CsCl).

**Biochemical properties**

*Genome* consists of DNA; single–stranded; circular; of two parts; largest (or only) genome part the larger 2.59 kb; the 2nd largest 2.51 kb. Genomic nucleic acid isolated by Hamilton *et al.* (1981). Base composition 23.4 % G (DNA A), or 21.5 % G (DNA B); 26 % A (DNA A), or 29 % A (DNA B); 19.8 % C (DNA A), or 18.7 % C (DNA B); 30.8 % T (DNA A), or 30.8 % T (DNA B). Infectivity retained when deproteinised with proteases; retained when deproteinised with phenol or detergent. Poly A region absent. Nucleotide sequence references: Hamilton *et al.* (1984).

*Sequence database accession code(s):*

K02029 Em(40)_vi:GETGMVA Gb(84)_vi:MTGA Tomato golden mosaic virus, component A of complete genome. 4/90 2,588bp.

K02030 Em(40)_vi:GETGMVB Gb(84)_vi:MTGB Tomato golden mosaic virus, component B of complete genome. 4/90 2,508bp.

M73794 Em(40)_vi:GEMTGDNAB Gb(84)_vi:MTGDNAB Tomato golden mosaic virus DNA B gene, complete cds. 2/94 2,524bp.

S44878 Em(40)_un:S44878 Gb(84)_vi:S44878 (5'end of AL61 RNA) tomato golden mosaic virus TGMV, mRNA Partial, 50 nt. 2/93 50bp.

S44879 Em(40)_un:S44879 Gb(84)_vi:S44879 (5'end of AL2540 RNA) tomato golden mosaic virus TGMV, mRNA Partial, 50 nt. 2/93 50bp.

S44890 Em(40)_un:S44890 Gb(84)_vi:S44890 (5'end of AL2515 RNA) tomato golden mosaic virus TGMV, mRNA Partial, 50 nt. 2/93 50bp.

S44891 Em(40)_un:S44891 Gb(84)_vi:S44891 (5'end of AL1935 RNA) tomato golden mosaic virus TGMV, mRNA Partial, 50 nt. 2/93 50bp

S44892 Em(40)_un:S44892 Gb(84)_vi:S44892 (5'end of AL1629 RNA) tomato golden mosaic virus TGMV, mRNA Partial, 50 nt. 2/93 50bp.

X04485 Em(40)_vi:GETGMVS1 Gb(84)_vi:GETGMVS1 Tomato golden mosaic virus subgenomic DNA derived from DNA B cccds = covalently closed circul 9 sequences.

*Non–genomic nucleic acid* found in the virions; is of an unusual type of structure; subgenomic DNA; single stranded DNA circles, formed by specific deletions in DNA B (Macdowell *et al.*, 1986).

*Virion protein(s)* one; $M_r$ 28500; coat protein. Method of preparation: Stein (1985). Amino acid sequence: Stein (1985) and Hamilton *et al.* (1984).

**Replication**

Genome replicates in nuclei. Replication does not depend on a helper virus (DNA A will replicate alone and form virions in petunia plants transformed with dimers of DNA A using a Ti plasmid vector without causing symptoms; Rogers *et al.*, 1986; Sunter *et al.*, 1987).

**Cytopathology**

Virions found in phloem; in nuclei.

**Taxonomy and Relationships**

Bigeminivirus: *Geminiviridae.*

Virus(es) with serologically related virions: African cassava mosaic, bean golden mosaic, euphorbia mosaic and squash leaf curl viruses, but distantly. Virus(es) with serologically unrelated virions: wheat dwarf, chloris striate mosaic, maize streak and tobacco yellow dwarf viruses.

Does not form pseudo-recombinants with African cassava mosaic virus. No host shared with bean golden mosaic virus. The DNAs of tomato golden mosaic and euphorbia mosaic viruses will hybridize. Nucleotide sequence of DNA A related to those of wheat dwarf, maize streak and beet curly top viruses (Macdowell *et al.*, 1985; Mullineaux *et al.*, 1985; Stanley *et al.*, 1986). Best tests for diagnosis: monoclonal antibodies may be useful for diagnosis, see Thomas *et al.* (1986).

**References**


Bisaro, O.M., Hamilton, W.D.O., Coutts, R.H.A. and Buck, K.W. (1982). *Nucl. Acids Res.* **10**: 4913.

Buck, K.W. and Coutts, R.H.A. (1985). *CMI/AAB Descr. Pl. Viruses* No. 303, 6 pp.

Costa, A.S. (1976). *Ann. Rev. Phytopath.* **14**: 429.

Costa, A.S., Oliveira, A.R. and Silva, D.M. (1975). *Abstr. 8th Congr. Soc. Bras. Fitopatol. Mossoro, Brazil.*

Hamilton, W.D.O., Sanders, R. and Coutts, R.H.A. (1981). *FEMS Microbiol. Lett.* **11**: 263.

Hamilton, W.D.O., Stein, V.E., Coutts, R.H.A. and Buck, K.W. (1984). *EMBO. J.* **3**: 2197.

Matyis, J.C., Silva, D.M., Oliveira, A.R. and Costa, A.S. (1975). *Summa Phytopath.* **1**: 267.

Macdowell, S.W., Macdonald, H., Hamilton, W.D.O., Coutts, R.H.A. and Buck, K.W. (1985) *EMBO J.* **4**: 2173.

Macdowell, S.W., Coutts, R.H.A. and Buck, K.W. (1986). *Nucl. Acids Res.* **14**: 7967.

Mullineaux, P.M., Donson, J., Stanley, J., Boulton, M.I., Morris-Krsinich, B.A.M., Markham, P.G. and Davies, J.W. (1985). *Pl. Mol. Biol.* **5**: 125.

Rogers, S.G., Bisaro, D.M., Horsch, R.B., Fraley, R.T., Hoffman, N.L., Brand, L., Elmer, J.S. and Lloyd, A.M. (1986). *Cell* **45**: 593.

Stanley, J., Markham, P.G., Callis, R.J. and Pinner, M.S. (1986). *EMBO J.* **5**: 1761.

Stein, V.E. (1985). *Ph.D. Thesis*, University of London, U.K..

Stein, V.E., Coutts, R.H.A. and Buck, K.W. (1983). *J. gen. Virol.* **64**: 2493.

Sunter, G., Gardiner, W.E., Rushing, A.E., Rogers, S.G. and Bisaro, D.M. (1987). *Pl. mol. Biol.* **8**: 477.

Thomas, J.E., Massalski, P.R. and Harrison, B.D. (1986). *J. gen. Virol.* **67**: 2739.


# Tomato Indian leafcurl bigeminivirus

*Sequence database accession code(s):*

X78956 Em(40)_vi:ITLCGCP Gb(84)_vi:ITLCGCP Indian Tomato Leaf Curl Geminivirus coat protein gene. 4/94 690bp. 1 sequence.

**Taxonomy and Relationships**

Bigeminivirus: *Geminiviridae.*

# Tomato infectious chlorosis (?) *closterovirus*

**Data collated by A.A. Brunt, 1995.**

First reported in *Lycopersicon esculentum*; from U.S.A. (California); by Duffus (1995).

**Natural host range and symptoms**

*Lycopersicon esculentum* — interveinal yellowing, necrosis and severe yield loss.

**Transmission**

Transmitted by a vector; an insect; *Trialeurodes vaporariorum* (but not *Bemisia* spp.). Virus not transmitted by mechanical inoculation.

**Geographical distribution**

Spreads in possibly Italy and the USA.

**Experimental host range**

Several (3–9) families susceptible.

Diagnostically susceptible

*Lycopersicon esculentum* — interveinal yellowing, necrosis and stunting.

Dicotyledonous Hosts *(**Susceptible** or Insusceptible)*

***Lactuca sativa***. ***Lycopersicon esculentum***. ***Petunia hybrida***. ***Solanum tuberosum***. ***Zinnia elegans***.

**Purification**

Method: Duffus (1995).

**Particle morphology**

Virions filamentous; usually flexuous; with a clear modal length; of 850–900 nm; 12 nm wide.

**Taxonomy and Relationships**

Possible *Closterovirus*.

Virus(es) with serologically unrelated virions: Beet pseudo yellows, lettuce infectious yellows and cucurbit yellow stunt (?) closteroviruses.

**References**

Dellavalle, G., Caciagli, P., Bosco, D., Lisa, V., d'Aquilio, M., Milne, R.G. and Mascaga, V. (1995). *Proc. 8th Conf. Virus Dis. Veg. Prague* (July 9–15, 1995), p. 35.

Duffus, J. (1995). *Proc. 8th Conf. Virus Dis. Veg., Prague* (July 9–15, 1995), p. 39.

Duffus, J., Lui, H.Y. and Wisler, G.C. (1994). *Phytopathology* **84**: 1072.

# Tomato leaf crumple bigeminivirus

*Standard acronym:* TLCrV.

*Sequence database accession code(s):*

L27267 Em(40)_vi:GETLGAC1A Gb(84)_vi:TLGAC1A Tomato leaf crumple geminivirus replication-associated protein (AC1) gene, partial cds. 1/94

L27268 Em(40)_vi:GETLGBC1B Gb(84)_vi:TLGBC1B Tomato leaf crumple geminivirus movement protein (BC1) gene, partial cds. 1/94 223bp.

L34746 Gb(84)n:TLGNEFA Tomato leaf crumple geminivirus (clone pBMX9B) nuclear export factor (nef) gene, complete cds

L34747 Gb(84)n:TLGCPA Tomato leaf crumple geminivirus (clone pTMX22A) coat protein (cp) gene, complete cds. 9/94 75

L34748 Gb(84)n:TLGREG Tomato leaf crumple geminivirus (clone pBMX5A) regulatory sequence, 5' end. 9/94 161bp.

L34749 Gb(84)n:TLGNEFB Tomato leaf crumple geminivirus (clone pTMX44B) nuclear export factor (nef) gene, complete.

**Taxonomy and Relationships**

Bigeminivirus: *Geminiviridae*.

# Tomato mild mottle (?) potyvirus

**Data collated by A.A. Brunt.**

First reported in *Datura stramonium, Lycopersicon esculentum, Solanum nigrum*; from Yemen Republic; by Walkey *et al.* (1994).

**Natural host range and symptoms**

Symptoms

*Datura stramonium, Lycopersicon esculentum, Solanum nigrum* — faint leaf mottling and plant stunting.

**Transmission**

Transmitted by a vector; an insect; *Myzus persicae*; Aphididae. Transmitted in a non–persistent manner. Virus transmitted by mechanical inoculation.

**Geographical distribution**

Spreads in the Middle East; Yemen.

**Experimental host range**

Few (<3) families susceptible.

Diagnostically susceptible

*Nicotiana clevelandii, N. glutinosa, N. tabacum* cv. White Burley — necrotic local lesions, systemic veinal chlorosis.

*Nicotiana tabacum* cv. Xanthi — chlorotic local lesions, systemic mottling.

*Datura metel, Nicotiana benthamiana, N. occidentalis* — systemic veinal chlorosis or necrosis.

*Lycopersicon esculentum, Nicotiana rustica* — systemic mottling.

Diagnostically insusceptible

*Capsicum annuum, C. frutescens, Chenopodium quinoa, Cucumis sativus, Petunia hybrida, Phaseolus vulgaris.*

Maintenance and propagation hosts

*Nicotiana glutinosa.*

Assay hosts (**L**ocal lesions or **W**hole plants)

*Lycopersicon esculentum* (W), *Nicotiana glutinosa* (W).

Dicotyledonous Hosts *(**Susceptible** or Insusceptible)*

*Capsicum annuum. Capsicum frutescens. Chenopodium quinoa. Cucumis sativus.* ***Datura metel. Lycopersicon esculentum. Nicotiana benthamiana. Nicotiana clevelandii. Nicotiana glutinosa. Nicotiana occidentalis. Nicotiana rustica. Nicotiana tabacum.*** *Phaseolus vulgaris. Sonchus oleraceus.*

**Properties of particles in sap**

TIP: 70–75 °C. LIV: 8–16 days (at 2°C and 18°C). DEP: $\log_{10}$ minus 5.

**Particle morphology**

Virions filamentous; usually flexuous; with a clear modal length; of 719 nm.
*Virion protein(s)* one; $M_r$ 41500; coat protein.

**Cytopathology**

Virions found in cytoplasm. Inclusions present in infected cells; are pinwheels.

**Taxonomy and Relationships**

Possible unassigned potyvirus.

Virus(es) with serologically unrelated virions: Chilli veinal mottle, datura shoestring, pepper mottle, pepper veinal mottle, potato Y and tobacco etch potyviruses.

**References**

Walkey, D.G.A., Spence, N.J., Clay, C.M. and Miller, A. (1994). *Pl. Path.* **43**: 931.

# Tomato mosaic *tobamovirus*

**Data collated by H. Huttinga and A.Th.B. Rast, 1985.**

*Standard acronym:* ToMV. *Strains:* tomato aucuba mosaic, tomato enation mottle, dahlemense and other strains (Hollings and Huttinga, 1976).

First reported in *Lycopersicon esculentum*; from Connecticut, U.S.A.; by Clinton (1909).

**Natural host range and symptoms**

*Lycopersicon esculentum* — mosaic with leaf narrowing in winter.

**Transmission**

Transmitted by means not involving a vector. Virus transmitted by mechanical inoculation; transmitted by grafting; transmitted by contact between plants; transmitted by seed (up to 94% in tomato. The virus is found on the outside of the seed, sometimes in the endosperm, but not in the embryo; infection occurs during transplanting).

**Geographical distribution**

Probably distributed worldwide.

**Experimental host range**

Many (>9) families susceptible.

Diagnostically susceptible

*Solanum giganteum* — systemic leaf mosaic.

*Nicotiana tabacum* cv. White Burley (necrosis line or Dutch A) — local lesions; not systemic.

*Nicotiana rustica, Petunia hybrida* — local lesions; not systemic.

Diagnostically insusceptible

*Cucumis sativus, Cyphomandra betacea* (tree tomato), *Phaseolus vulgaris.*

Maintenance and propagation hosts

*Nicotiana tabacum* cv. Samsun, *Lycopersicon esculentum.*

Assay hosts (Local lesions or Whole plants)

*Nicotiana glutinosa* (L), *N. tabacum* cv. Xanthi-*nc* and Samsun NN (L), *N. clevelandii* (W).

Dicotyledonous Hosts *(**Susceptible** or Insusceptible)*

***Capsicum annuum. Capsicum frutescens. Chenopodium amaranticolor. Chenopodium murale. Chenopodium quinoa.*** *Cucumis melo. Cucumis sativus. Cucurbita pepo.* ***Datura metel. Lycopersicon esculentum. Lycopersicon pimpinellifolium. Nicotiana benthamiana. Nicotiana clevelandii. Nicotiana glutinosa. Nicotiana megalosiphon. Nicotiana rustica. Nicotiana tabacum. Petunia hybrida.*** *Phaseolus vulgaris.* ***Physalis floridana. Physalis peruviana. Solanum melongena. Solanum nigrum.*** *Vigna unguiculata. Vigna unguiculata* ssp. *sesquipedalis.*

**Properties of particles in sap**

TIP: 85–90 °C. LIV: 500 days. DEP: $\log_{10}$ minus 5–7. Infectivity of sap not changed by treatment with di–ethyl ether. Leaf sap contains many virions.

**Purification**

Method: Gooding and Hebert (1967).

**Particle morphology**

Virions rod–shaped; not enveloped; usually straight; with a clear modal length; of 300 nm; 18 nm wide. Axial canal obvious; 4 nm in diameter. Basic helix obvious; pitch of basic helix 2.3 nm.

**Physical properties**

One sedimenting component in purified preparations; sedimentation coefficient 190 *S*. Density 1.33 g $cm^{-3}$ in CsCl. Isoelectric point pH 4.5–4.64 (depending on strain).

**Biochemical properties**

Virions contain 5 % nucleic acid; 95 % protein; 0 % lipid.

*Genome* consists of RNA; single–stranded; linear; unipartite; largest (or only) genome part 6.384 kb. Genomic nucleic acid isolated by Mandeles and Bruening (1968). Base composition 23 % G; 28 % A; 19 % C; 30 % U. Infectivity decreased when deproteinised with proteases (due to action of RNases at the site of infection); retained when deproteinised with phenol or detergent. Additional factor not required for infectivity.

*Sequence database accession code(s):* Em_vi:TOTMV7 — strain L. 5'-terminal region, 278bp; Em_vi:TOTMV8 — strain L genome, 6,384bp.

*Virion protein(s)* one; $M_r$ 17600; coat protein. Method of preparation: Fraenkel-Conrat (1957). Amino acid sequence: Wittmann-Liebold and Wittmann (1967).

**Replication**

Genome replicates probably in cytoplasm. Coat protein *m*RNA translated in the cytoplasm, probably. Replication does not depend on a helper virus.

**Cytopathology**

Virions found in all parts of the host plant; in cytoplasm. Inclusions present in infected cells; are crystals in the cytoplasm and amorphous X–bodies; they contain virions.

**Taxonomy and Relationships**

*Tobamovirus.*

Virus(es) with serologically related virions: other tobamoviruses (Van Regenmortel, 1975).

**References**

Broadbent, L. (1976). *Ann. Rev. Phytopath.* **14**: 75.

Brunt, A.A. (1986). in *The Plant Viruses*; Vol.2, *The Rod-Shaped Viruses*, p.181; eds Van Regenmortel, M.H.V. and Fraenkel-Conrat, H. Plenum Press, New York.
Dekker, E.L., Dore, I., Porta, C. and van Regenmortel, M.H.V. (1987). *Arch. Virol.* **94**: 191.
Fraenkel-Conrat, H. (1957). *Virology* **4**: 1.
Gooding, G.V. and Hebert, T.T. (1967). *Phytopathology* **57**: 1285.
Hollings, M. and Huttinga, H. (1976). *CMI/AAB Descr. Pl. Viruses* No. 156, 6 pp.
Mandeles, S. and Bruening, G. (1968). *Biochem. Prep.* **12**: 111.
Mosch, W.H.M., Huttinga, H. and Rast, A.Th.B. (1977). *Neth. J. Pl. Path.* **79**: 104.
Ohno, T., Aoyagi, M., Yamanashi, Y., Saito, H., Ikawa, S., Neshi, T. and Okada, Y. (1984). *J. Biochem* (Tokyo) **96**: 1915.
Rast, A.Th.B. (1975). *Agric. Research Reports Wageningen*, p. 834.
Wittmann-Liebold, B. and Wittmann, H.G. (1967). *Mol. gen. Genetics* **100**: 358.
Van Regenmortel, M.H.V. (1975). *Virology* **64**: 415.

# Tomato mottle bigeminivirus

**Data collated by A.A. Brunt, 1995.**

*Standard acronym:* ToMoV.

First reported in *Lycopersicon esculentum*; from Florida, U.S.A.; by J.K. Brown (in Polston *et al.*, 1993).

**Natural host range and symptoms**

Symptoms persist.

*Lycopersicon esculentum* — chlorotic mottling of younger leaves, curling and chlorosis of older leaves; plants stunted, especially those infected when young.

**Transmission**

Transmitted by a vector; an insect; *Bemisia tabaci*; Aleyrodidae. Principal natural vector(s): *B. tabaci* — biotype B. Virus not transmitted by mechanical inoculation; transmitted by grafting; not transmitted by seed; Kring *et al.* (1991).

**Geographical distribution**

Spreads in the North American region; the USA.

**Experimental host range**

Few (<3) families susceptible.

Diagnostically susceptible

*L. pimpinellifolium, Nicotiana benthamiana* — systemic mottling and leaf curling; plants stunted.

*Lycopersicon hirsutum, L. peruvianum* — systemic yellow mosaic.

*Physalis ixocarpa* — systemic yellow mosaic, leaf curling; plants stunted.

*Phaseolus vulgaris* — symptomless systemic infection.

Diagnostically insusceptible

*Brassica oleracea, Capsicum annuum, Cucurbita pepo, Datura stramonium.*

Maintenance and propagation hosts

*Lycopersicon esculentum.*

Assay hosts (**L**ocal lesions or **W**hole plants)

*Lycopersicon esculentum* (W).

Dicotyledonous Hosts *(**Susceptible** or Insusceptible)*

*Arachis hypogaea. Brassica oleracea* ssp. *capitata. Capsicum annuum. Cucurbita pepo. Datura stramonium. Glycine max. Gossypium hirsutum.* ***Lycopersicon chilense. Lycopersicon esculentum. Lycopersicon pimpinellifolium.*** *Macroptilium lathyroides. Medicago sativa.* ***Nicotiana benthamiana. Nicotiana × edwardsonii. Nicotiana tabacum. Petunia hybrida.*** *Phaseolus acutifolius.* ***Phaseolus vulgaris.*** *Solanum melongena. Solanum tuberosum. Vigna radiata.*

**Particle morphology**

Virions geminate; 18 nm in diameter; dimers 30 nm in length.

**Biochemical properties**

*Genome* consists of DNA; single–stranded; of two parts.

*Sequence database accession code(s):*

L02618 Em(40)_vi:GETMWBR1 Gb(84)_vi:TMWBR1 Tomato mottle geminivirus (clone pTFB-1) BR1 gene, complete cds. 3/93 771bp.

L14460 Em(40)_vi:GETMWALR Gb(84)_vi:TMWALR Tomato mottle geminivirus complete A component DNA; coat protein (AR1) gene, replicative prot

L14461 Em(40)_vi:GETMWBRL Gb(84)_vi:TMWBRL Tomato mottle geminivirus complete B component DNA; BR1 and BL1 genes, complete cds's. 5/93 2

M90494 Em(40)_vi:GETMVREPO Gb(84)_vi:TMWREPORI Tomato mottle viurs gene, putative origin of replication. 5/92 182bp.

M90495 Em(40)_vi:GETMVREPP Gb(84)_vi:TMWREPPRO Tomato mottle viurs replicative protein (AL1) gene, putative cds. 5/92 1,086bp. 5 sequences.

**Cytopathology**

Virions found in cytoplasm and in nuclei. Inclusions present in infected cells; are unusual in shape; fibrillar rings.

**Taxonomy and Relationships**

Bigeminivirus: *Geminiviridae.*

Its host range and genome nucleotide sequence indicate that the virus is distinct from other New World geminiviruses.

**References**


Abouzid, A., Polston, J.E. and Hiebert, E. (1992). *J. gen. Virol.* **73**: 3225.

Kring, J.B., Shuster, D.J., Price, J.F. and Simone, G.W. (1991). *Plant Dis.* **75**: 1186.

Polston, J.E., Hiebert, E., McGovern, R.J., Stansly, P.A. and Schuster, D.J. (1993). *Plant Dis.* **77**: 1181.

Simone, G.W., Brown, J.K., Hiebert, E. and Cullen, R.E. (1990). *Phytopathology* **80**: 1063.


# Tomato Peru *potyvirus*

**Data collated by E.N. Fernandez-Northcote, 1985.**

*Synonyms:* Peru tomato mosaic virus, Peru tomato virus, Peruvian tomato virus, virus Peruano del tomate. *Standard acronym:* PTV. *Strains:* M; N; $O^c$; $O^m$; $O^s$.

First reported in *Lycopersicon esculentum*; from Trujillo, Peru; by Raymer *et al.* (1972).

**Natural host range and symptoms**

Symptoms persist.

*Capsicum annuum, C. baccatum* var. *pendulum, C. chinense, Physalis peruviana* — mosaic, green vein banding.

*Lycopersicon esculentum* — vein clearing, twisting, mosaic, stem and leaf necrosis by some strains, fruit mottling.

*Solanum nigrum, Nicandra physalodes* — leaf chlorosis.

**Transmission**

Transmitted by a vector; an insect; *Myzus persicae*; Aphididae. Transmitted in a non–persistent manner. Virus transmitted by mechanical inoculation; transmitted by grafting; not transmitted by seed.

**Geographical distribution**

Spreads in Peru.

**Experimental host range**

Few (<3) families susceptible.

Diagnostically susceptible

*Chenopodium amaranticolor, C. quinoa* — chlorotic local lesions; not systemic.

*Lycopersicon esculentum* — systemic leaf chlorosis and/or necrosis, epinasty, apical necrosis and stunting.

*Nicotiana occidentalis* — systemic chlorotic blotching and leaf malformation.

*Nicotiana debneyi* — systemic chlorotic rings and spots.

*N. glutinosa* — systemic mosaic.

Diagnostically insusceptible

*Gomphrena globosa, Capsicum annuum* cv. Agronomico 8, *Datura stramonium, Solanum demissum x S. tuberosum* cv. A6, *S. tuberosum* cv. Katahdin.

Maintenance and propagation hosts

*Lycopersicon esculentum, Nicotiana glutinosa, N. occidentalis, N. tabacum* cvs Burley 21, White Burley.

Assay hosts (**L**ocal lesions or **W**hole plants)

*Chenopodium amaranticolor* (L), *Nicotiana occidentalis* (W), *N. tabacum* cv. Burley 21 (W).

Dicotyledonous Hosts *(**Susceptible** or Insusceptible)*

*Apium graveolens. Beta patellaris. Beta vulgaris. Brassica oleracea* ssp. *capitata.* ***Capsicum annuum. Capsicum frutescens.*** *Cassia occidentalis. Catharanthus roseus. Celosia argentea. Chenopodium album. Chenopodium amaranticolor. Chenopodium quinoa. Citrullus lanatus. Crotalaria spectabilis. Cucumis melo. Cucumis sativus. Cucurbita maxima. Cucurbita pepo.* ***Datura metel.*** *Datura stramonium. Dianthus barbatus. Glycine max. Gomphrena globosa. Gossypium hirsutum. Helianthus annuus. Lactuca sativa.* ***Lycopersicon esculentum. Lycopersicon pimpinellifolium.*** *Momordica balsamina.* ***Nicandra physaloides. Nicotiana bigelovii. Nicotiana clevelandii. Nicotiana debneyi. Nicotiana glutinosa. Nicotiana megalosiphon. Nicotiana rustica. Nicotiana sylvestris. Nicotiana tabacum.*** *Ocimum basilicum.* ***Petunia hybrida.*** *Phaseolus lunatus. Phaseolus vulgaris.* ***Physalis floridana. Physalis peruviana.*** *Pisum sativum. Raphanus sativus. Ricinus communis. Sesamum indicum. Sesbania exaltata. Solanum demissum* × *tuberosum* (cv. A6). ***Solanum nigrum.*** *Solanum tuberosum. Trifolium hybridum. Verbesina encelioides. Vicia faba. Vigna angularis. Vigna radiata. Vigna unguiculata. Vigna unguiculata* ssp. *cylindrica. Vigna unguiculata* ssp. *sesquipedalis. Zinnia elegans.*

Sources of host-range data

Fernandez-Northcote (1978); Fribourg (1979); Fernandez-Northcote and Fulton (1980).

**Properties of particles in sap**

TIP: 55–60 °C. LIV: 10–21 days. DEP: $\log_{10}$ minus 5–6. Leaf sap contains few virions.

**Purification**

Method: Fernandez-Northcote and Fulton (1980).

**Particle morphology**

Virions filamentous; not enveloped; usually flexuous; with a clear modal length; of 741–746 nm; 11–13 nm wide. Axial canal obscure. Basic helix obscure.

**Physical properties**

One sedimenting component in purified preparations.

**Biochemical properties**

Virions contain 7 % nucleic acid; 93 % protein.

**Cytopathology**

Virions found in all parts of the host plant; in cytoplasm. Inclusions present in infected cells; are pinwheels; they do not contain virions.

**Taxonomy and Relationships**

*Potyvirus: Potyviridae.*

Virus(es) with serologically related virions: potato Y, pepper mottle, tobacco etch and bidens mottle viruses. Virus(es) with serologically unrelated virions: bean yellow mosaic, lettuce mosaic, pepper veinal mottle, pepper severe mosaic, potato A, tobacco vein mottling, turnip mosaic and wild potato mosaic viruses.

Best tests for diagnosis: check for pinwheel inclusions to differentiate from tobacco mosaic virus.

**References**

Christie, R.G. (1967). *Virology* **31**: 268.

Raymer, W.B., Kahn, R.P., Hikida, H.R. and Waterworth, H.E. (1972). *Phytopathology* **62**: 784.

Fernandez-Northcote, E.N. (1978). *Ph.D. Thesis*, Univ. Wisconsin, U.S.A.

Fribourg, C.S. (1979). *Phytopathology* **69**: 441.

Fribourg, C.E. and Fernandez-Northcote, E.N. (1982). *CMI/AAB Descr. Pl. Viruses* No. 255, 3 pp.

Fernandez-Northcote, E.N. and Fulton, R.W. (1980). *Phytopathology* **70**: 315.

# Tomato pseudo curly top (?) hybrigeminivirus

**Data collated by A.A. Brunt, 1991.**

*Standard acronym:* TPCTV.

First reported in *Lycopersicon esculentum*; from the U.S.A.; by Giddings *et al.* (1951).

**Natural host range and symptoms**

*Lycopersicon esculentum* — chlorosis (especially at leaf edges), vein-clearing, leaf curling and cupping and shoot proliferation. Plants often severely stunted and set few fruit.

*Ambrosia* sp., *Solanum nigrum* — vein-clearing and leaf rolling.

**Transmission**

Transmitted by a vector; an insect; *Micrutalis malleifera*; Membracidae. Not transmitted by *Acutalis* sp., *Micrutalis clava, Spissistilus festinus*. Transmitted in a semi–persistent manner. Virus retained when the vector moults; not transmitted by mechanical inoculation; not transmitted by seed.

**Ecology and control**

Studies reported by Tsai and Brown (1991).

**Geographical distribution**

Spreads in the USA (Florida).

**Experimental host range**

Several (3–9) families susceptible.

Diagnostically susceptible

*Ambrosia* sp., *Datura stramonium, Lactuca sativa, Lycopersicon esculentum, Nicotiana glutinosa, N. x. edwardsonii, Solanum melongena, Solanum nigrum* — systemic vein-clearing, chlorosis, leaf rolling or cupping and plant stunting.

Diagnostically insusceptible

*Beta vulgaris, Capsicum annuum, Cucumis sativus, Nicotiana tabacum, Solanum melongena.*

Maintenance and propagation hosts

*Lycopersicon esculentum, Nicotiana* x *edwardsonii, Solanum nigrum.*

Assay hosts (**L**ocal lesions or **W**hole plants)

*Lycopersicon esculentum* (W), *Solanum nigrum* (W).

Dicotyledonous Hosts *(**Susceptible** or Insusceptible)*

*Beta vulgaris. Capsicum annuum. Catharanthus roseus. Chenopodium ambrosioides. Cucumis sativus. Cucurbita pepo. Gomphrena globosa.* ***Lactuca sativa. Lycopersicon esculentum. Nicotiana × edwardsonii. Nicotiana glutinosa.*** *Nicotiana tabacum. Petunia hybrida. Phaseolus vulgaris. Physalis floridana.* ***Solanum melongena. Solanum nigrum.*** *Zinnia elegans.*

Sources of host-range data

Simons (1962*a*).

**Particle morphology**

Virions geminate; not enveloped.

**Biochemical properties**

*Genome* consists of DNA; single–stranded; circular.

**Cytopathology**

Virions found in leaves; in nuclei. Inclusions present in infected cells; are viroplasms and unusual in shape; fibrillar rings in nuclei; they contain virions.

**Taxonomy and Relationships**

Possible hybrigeminivirus: *Geminiviridae*.

Virus(es) with serologically related virions: beet curly top hybrigeminivirus.

Reported to have virions that are serologically related to those of beet curly top hybrigeminivirus, however this virus is transmitted by the treehopper *Micrutalis* malleifera (Hemiptera: Membracidae) and fails to infect *Beta vulgaris*, whereas beet curly top hybrigeminivirus is transmitted by the leafhopper *Circulifer tenellus* (Hemiptera: Cicadellidae).

**References**

Bennett, C.W. (1971). *Am. Phytopathol. Soc. Monog.* **7**, 81 pp.

Christie, R.G., Ko, N.-J., Flak, B.W., Hiebert, E., Lastra, R., Bird, J. and Kim, K.S. (1986). *Phytopathology* **76**: 124.

Giddings, N.J., Bennett, C.W. and Harrison, A.L. (1951). *Phytopathology* **41**: 415.

McDaniel, L.L. and Tsai, J.H. (1990). *Plant Dis.* **74**: 17.

Simons, J.N. (1962*a*). *J. econ. Ent.* **55**: 358.

Simons, J.N. (1962*b*). *J. econ. Ent.* **55**: 363.

Simons, J.N. and Coe, D.M. (1958). *Virology* **6**: 43.

Tsai, J.H. and Brown, L.G. (1991). *Fla Dep. Agric. and Consumer Serv., Plant Path. Circ.* No. 344, 2 pp.

Zitter, T.A. and Tsai, J.H. (1981). *Plant Dis.* **65**: 787.

# Tomato ringspot *nepovirus*

**Data collated by R. Stace-Smith, 1981. Revised 1987.**

*Synonyms:* blackberry Himalaya mosaic virus, euonymus ringspot virus (Puffinberger and Corbett, 1985), grape yellow vein virus, grapevine yellow vein virus (Gooding *et al.*, 1967), peach yellow bud mosaic virus (Cadman and Lister, 1961; Thomas and Rawlins, 1950), prune brown line virus, prunus stem-pitting virus, red currant mosaic virus (Hildebrand, 1942), tobacco ringspot virus 2, winter peach mosaic virus. *Standard acronym:* ToRSV. *Strains:* tobacco strain, grape yellow vein strain, apple union necrosis nepovirus, euonymus chlorotic ringspot virus.

First reported in *Nicotiana tabacum*; from the U.S.A; by Price (1936).

**Natural host range and symptoms**

Symptoms disappear soon after infection.

*Pelargonium* spp., *Rubus* spp. and a very wide range of other species — mosaic or ringspot.

*Prunus* spp. — rasp leaf, yellow bud or yellow vein.

*Nicotiana tabacum* — ringspots and chlorosis.

**Transmission**

Transmitted by a vector; a nematode; *Xiphinema americanum (sensu lato)*; Dorylamidae. Virus transmitted by mechanical inoculation; transmitted by grafting; transmitted by seed (of *Rubus idaeus, Nicotiana tabacum, Glycine max, Fragaria x ananassa*); transmitted by pollen to the seed, or transmitted by pollen to the pollinated plant.

**Ecology and control**

Studies reported by Stace-Smith and Ramsdell (1987).

**Geographical distribution**

Spreads in the North American region; Australia, Bulgaria, Canada, Chile, China, Cyprus, Germany, Italy, Japan, Korea Republic, New Zealand, Peru, Puerto Rico, Turkey, the USA, the former USSR, the former Yugoslavia.

**Experimental host range**

Several (3–9) families susceptible.

Diagnostically susceptible

*Chenopodium amaranticolor* and *C. quinoa* — chlorotic local lesions, systemic apical necrosis.

*Cucumis sativus* — necrotic or chlorotic local lesions; systemic mottle.

*Phaseolus vulgaris* — chlorotic local lesions; systemic rugosity, tip necrosis.

*Vigna unguiculata* — necrotic or chlorotic local lesions; systemic tip necrosis.

*Lycopersicon esculentum* — necrotic flecking; systemic mottle, necrosis.

*Nicotiana clevelandii* — necrotic local lesions; systemic chlorosis, necrosis.

*N. tabacum* — necrotic local lesions or ringspot markings; systemic ring or line patterns.

*Petunia hybrida* — necrotic local lesions; tip necrosis.

Maintenance and propagation hosts

*Cucumis sativus, Phaseolus vulgaris, Nicotiana tabacum, Petunia hybrida.*

Assay hosts (**L**ocal lesions or **W**hole plants)

*Chenopodium amaranticolor* (L), *C. quinoa* (L), *Vigna unguiculata* (L), *Nicotiana clevelandii* (L), *N. tabacum* (L).

Dicotyledonous Hosts *(**Susceptible** or Insusceptible)*

***Amaranthus retroflexus*. *Apium graveolens*. *Capsicum frutescens*.** *Catharanthus roseus*. ***Chenopodium album*. *Chenopodium amaranticolor*. *Chenopodium capitatum*. *Chenopodium foetidum*. *Chenopodium hybridum*. *Chenopodium murale*. *Chenopodium quinoa*. *Cucumis sativus*. *Cucurbita pepo*. *Datura stramonium*.** *Daucus carota*. ***Fragaria chiloensis*. *Fragaria vesca*. *Fragaria virginiana*. *Glycine max*. *Gomphrena globosa*.** *Hibiscus esculentus*. *Hyoscyamus niger*. ***Lycopersicon esculentum*. *Lycopersicon pimpinellifolium*. *Malus sylvestris*. *Nicandra physaloides*. *Nicotiana benthamiana*. *Nicotiana bigelovii*. *Nicotiana clevelandii*. *Nicotiana debneyi*.** *Nicotiana glutinosa*. ***Nicotiana megalosiphon*. *Nicotiana occidentalis*. *Nicotiana rustica*. *Nicotiana sylvestris*. *Nicotiana tabacum*. *Petunia hybrida*. *Phaseolus vulgaris*. *Prunus persica*. *Rubus idaeus*. *Rubus occidentalis*. *Solanum melongena*. *Solanum tuberosum*. *Stellaria media*.** *Tetragonia expansa*. ***Trifolium hybridum*. *Trifolium pratense*. *Trifolium repens*.** *Tropaeolum majus*. ***Vigna unguiculata*.** *Zinnia elegans*.

Comments on host-range

In nature, the virus occurs mainly in ornamentals and woody or semi-woody plants.

Sources of host-range data

Stace-Smith (1962); Thornberry (1966); Najarian *et al.* (1980).

**Properties of particles in sap**

TIP: 58 °C. LIV: 21 days. DEP: $\log_{10}$ minus 3. Leaf sap contains many virions.

**Purification**

Method: Stace-Smith (1984).

**Particle morphology**

Virions isometric; not enveloped; 25 nm in diameter; angular in profile; without a conspicuous capsomere arrangement.

**Physical properties**

Three sedimenting components in purified preparations; sedimentation coefficient of the fastest 127 *S* (B); of the other(s) 119 *S* (M), or 53 *S* (T). Density 1.49 and 1.5 g $cm^{-3}$ in CsCl. Isoelectric point pH 5.1. $A_{260}/A_{280}$ ratio 1.8 (M and B), or 0.9 (T).

**Biochemical properties**

Virions contain 44 % nucleic acid (B), or 41 % nucleic acid (M), or 0 % nucleic acid (T); 56 % protein (B), or 59 % protein (M), or 100 % protein (T); 0 % lipid.

*Genome* consists of RNA; single–stranded; of two parts; largest (or only) genome part the larger 8.5 kb; the 2nd largest 7.3 kb. Base composition 26 % G; 23 % A; 22 % C; 29 % U. Infectivity retained when deproteinised with phenol or detergent. Nucleotide sequence references: RNA 2 has been sequenced (Rott *et al.*, 1991).

*Sequence database accession code(s):*

D12477 Em(40)_vi:TOSRNA2 Gb(84)_vi:TOSRNA2 Tomato ringspot virus (TomRSV) RNA-2. 4/94 7,273bp.

L19655 Gb(84)n:TOSRNA1X Tomato ringspot virus polyprotein (RNA-1) gene, complete cds. 8/94 8,114bp.

M27935 Em(40)_vi:TOSNC2 Gb(84)_vi:TOSNC2 Tomato ringspot virus genomic RNA-2 3' non-coding region. 2/90 1,550bp.

M27936 Em(40)_vi:TOSNC1 Gb(84)_vi:TOSNC1 Tomato ringspot virus genomic RNA-1 3' non-coding region. 2/90 1,546bp.

M73822 Em(40)_vi:TOSRNA1A Gb(84)_vi:TOSRNA1A Tomato ringspot virus RNA1 gene, 5' end. 12/91 1,140bp.

*Virion protein(s)* one; $M_r$ 58000; coat protein. Method of preparation: Allen *et al.* (1977).

**Cytopathology**

Virions found in mesophyll; in cytoplasm; plasmodesmata. Inclusions present in infected cells; are unusual in shape; large vacuoles in the cytoplasm (Smith and McWhorter, 1957) or crystals consisting of nucleoprotein (Gordejchuk *et al.*, 1977). Other cellular changes: general tissue disintegration in florets of *Pelargonium x hortorum*.

**Taxonomy and Relationships**

*Nepovirus: Comoviridae.*

Virus(es) with serologically unrelated virions: cherry leaf roll virus, and many other members of the nepovirus group.

Best tests for diagnosis: serological tests are the only reliable ones because symptoms resemble those caused by other nepoviruses.

**References**

Allen, W.R. and Dias, H.F. (1977). *Can. J. Bot.* **55**: 1028.

Cadman, C.H. and Lister, R.M. (1961). *Phytopathology* **51**: 29.

Capoor, S.P. (1952). *Indian J. agric. Sci.* **22**: 274.

Forer, L.B. and Stouffer, R.F. (1982). *Plant Dis.* **66**: 735.

Gooding, G.V., Hewitt, W.B. and Cory, L. (1967). *Phytopathology* **57**: 236.

Gordejchuk, O.G., Krylova, A.V., Krylova, N.V. and Samonia, I.N. (1977). *Zbl. Bakt. ParasitKde.* Abt. 2, **132**: 686.

Hildebrand, E.M. (1942). *Am. J. Bot.* **29**: 362.

Hoy, J.W., Mircetich, S.M. and Lownsbery, B.F. (1984). *Phytopathology* **74**: 332.

Mayo, M.A., Barker, H. and Harrison, B.D. (1982). *J. gen Virol.* **59**: 149.

Murant, A.F., Taylor, M., Duncan, G.H. and Raschke, J.H. (1981). *J. gen Virol.* **53**: 321.

Price, W.C. (1936). *Phytopathology* **26**: 665.
Puffinberger, C.W. and Corbett, M.K. (1985). *Phytopathology* **75**: 423.
Rosenberger, D.A., Harrison, M.B. and Gonsalves, D. (1983). *Plant Dis.* **67**: 356.
Rott, M.E., Rochon, D.M. and Tremaine, J.H. (1988). *J. gen. Virol.* **69**: 745.
Rott, M.E., Tremaine, J.H. and Rochon, D.M. (1991). *J. gen. Virol.* **72**: 1505.
Schneider, I.R., White, R.M. and Civerolo, E.L. (1974). *Virology* **57**: 139.
Smith, F.H. and McWhorter, F.P. (1957). *Am. J. Bot.* **44**: 470.
Stace-Smith, R. (1962). *Can. J. Bot.* **40**: 905.
Stace-Smith, R. (1966). *Virology* **29**: 240.
Stace-Smith, R. (1970). *CMI/AAB Descr. Pl. Viruses* No. 18, 4 pp.
Stace-Smith, R. (1984). *CMI/AAB Descr. Pl. Viruses* No. 290, 6 pp.
Stace-Smith, R. (1984). *Plant Dis.* **68**: 274.
Thomas, H.E. and Rawlins, T.E. (1950). *USDA Agric. Hdbk* No. 10.
Tremaine, J. and Stace-Smith, R. (1968). *Virology* **35**: 102.

# Tomato spotted wilt *tospovirus*

**Data collated by A.J. Gibbs, 1983. Revised 1985; 1990.**

*Synonyms:* dahlia oakleaf virus, dahlia ringspot virus, dahlia yellow ringspot virus (Brunt, 1959), groundnut ringspot virus (Klesser, 1966), mung bean leaf curl virus, pineapple yellow spot virus, watermelon silver mottle virus. *Standard acronym:* TSWV.

First reported in *Lycopersicon esculentum*; by Brittlebank (1919); Samuel *et al.* (1930).

## Natural host range and symptoms

Symptoms persist.

*Ananas comosus, Bidens pilosa, Capsicum annuum, Datura stramonium, Helianthus annuus, Ipomoea congesta, Lactuca sativa, Malva parviflora, Nicandra physalodes, Phaseolus vulgaris, Physalis peruviana, Zinnia elegans, Arachis hypogaea, Canavalia gladiata, C. obtusifolia, C. occidentalis, Crotalaria juncea, Desmodium triflorum, Glycine max, Pisum sativum, Tephrosia purpurea, Vicia faba, Vigna mungo, V. radiata, V. unguiculata, Lycopersicon esculentum, Nicotiana tabacum, Solanum melongena, S. carpsicatrum, S. tuberosum* and many other species — necrotic and chlorotic local lesions, systemic wilting, necrosis, spotting, streaking, mosaic, mottling, leaf shape malformation, vein yellowing, ringspots, line patterns, yellow netting and flower colour-breaking.

## Transmission

Transmitted by a vector; an insect; *Thrips tabaci, T. setosus, T. parmi, Frankliniella schultzei, F. occidentalis, F. fusca* and *Scirtothrips dorsalis*; Thysanoptera. Transmitted in a persistent manner. Virus retained when the vector moults; multiplies in the vector (probably); not transmitted congenitally to the progeny of the vector; transmitted by mechanical inoculation; transmitted by grafting; not transmitted by contact between plants; not transmitted by seed; not transmitted by pollen.

## Ecology and control

Studies reported by Cho *et al.* (1989).

## Geographical distribution

Probably distributed worldwide. Spreads in Afghanistan, Algeria, Argentina, Australia, Austria, Belgium, Bolivia, Brazil, Bulgaria, Canada, Chile, China, Cote d'Ivoire, Cyprus, the former Czechoslovakia, Egypt, France, Germany, Greece, Guyana,

Haiti, India, Ireland, Israel, Italy, Jamaica, Japan, Libya, Madagascar, Malaysia, Malta, Mauritius, Mexico, Nepal, the Netherlands, New Zealand, Niger, Nigeria, Pakistan, Papua New Guinea, Paraguay, Poland, Portugal, Puerto Rico, Reunion, Romania, Senegal, South Africa, Spain, Sri Lanka, Suriname, Sweden, Switzerland, Taiwan, Tanzania, Thailand, Turkey, the UK, the USA, the former USSR, Uganda, Uruguay, the former Yugoslavia, Zaire, Zimbabwe.

**Experimental host range**

Many (>9) families susceptible.

Diagnostically susceptible

*Catharanthus roseus* — local black spots, leaves sometimes becoming yellow and abscissing; systemic mosaic and leaf deformation.

*Cucumis sativus* — chlorotic spots with necrotic centres in cotyledons; not systemic.

*Petunia hybrida* cvs Pink Beauty, Minstrel — necrotic local lesions; not systemic.

*Nicotiana clevelandii, N. glutinosa, N. tabacum* — necrotic local lesions; systemic necrotic patterns and leaf deformation.

*Tropaeolum majus* — symptomless local infection; systemic necrotic spotting and streaking.

Maintenance and propagation hosts

*Gomphrena globosa, Nicotiana glutinosa, N. rustica, Tropaeolum majus.*

Assay hosts (**L**ocal lesions or **W**hole plants)

*Nicotiana clevelandii* (L), *Petunia hybrida* cvs Pink Beauty, Minstrel (L).

Dicotyledonous Hosts ***(Susceptible** or Insusceptible)*

***Amaranthus caudatus*. *Amaranthus retroflexus*** (*Ananas comosis* (**S**)). *Apium graveolens*. ***Arachis hypogaea*. *Brassica oleracea*** ssp. ***botrytis*. *Calendula officinalis*. *Capsella bursa-pastoris*. *Capsicum annuum*. *Cassia occidentalis*. *Cassia tora*. *Catharanthus roseus*. *Cheiranthus cheiri*. *Cichorium endiva***. *Coriandrum sativum*. ***Dahlia pinnata*. *Datura stramonium*. *Glycine max*. *Helianthus annuus*. *Hyoscyamus niger*. *Lactuca sativa*. *Lathyrus odoratus*. *Lupinus mutabilis*. *Lycopersicon esculentum*. *Lycopersicon pimpinellifolium*. *Matthiola incana*. *Nicotiana bigelovii***. *Nicotiana debneyi*. *Nicotiana glutinosa*. ***Nicotiana rustica*. *Nicotiana sylvestris*. *Nicotiana tabacum*. *Papaver nudicaule*. *Petunia hybrida*. *Phaseolus vulgaris*. *Phlox drummondii*. *Physalis peruviana*. *Pisum sativum*. *Solanum melongena*. *Solanum nigrum*. *Solanum nodiflorum*. *Solanum tuberosum*. *Sonchus oleraceus*. *Spinacia oleracea*. *Stellaria media*. *Trifolium subterraneum*. *Tropaeolum majus*. *Vicia faba*. *Vigna radiata*. *Vigna unguiculata*. *Zinnia elegans***.

Monocotyledonous Hosts ***(Susceptible** or Insusceptible)*

***Belamcanda chinensis*. *Hippeastrum hybridum***.

Sources of host-range data

Klinkowski and Uschdraweit (1952); Smith (1957); Best (1968).

**Properties of particles in sap**

TIP: 45 °C. LIV: 0.2 days (5 hours). DEP: $\log_{10}$ minus 3. Infectivity of sap decreased by treatment with di–ethyl ether. Leaf sap contains few virions. Electron microscopy: fix with formaldehyde.

**Purification**

Method: Van den Hurk *et al.* (1977).

**Particle morphology**

Virions isometric; enveloped; 85 nm in diameter; rounded in profile; without a conspicuous capsomere arrangement.

**Physical properties**

One sedimenting component in purified preparations; sedimentation coefficient 550 *S*. Density 1.21 g $cm^{-3}$ in sucrose.

**Biochemical properties**

Virions contain 5 % nucleic acid; 70 % protein; 20 % lipid. Also 5% carbohydrate.

*Genome* consists of RNA; single–stranded; linear; largest (or only) genome part the largest 8.897 kb (L-RNA); the 2nd largest 5.4 kb; the 3rd largest 2.916 kb (S-RNA). Genomic nucleic acid isolated by de Haan (1990). Base composition 16.2 % G; 31.6 % A; 19.3 % C; 32.9 % U. Infectivity lost when deproteinised with proteases; lost when deproteinised with phenol or detergent. Poly A region absent (in *s*RNA). Additional factor required for infectivity. Nucleotide sequence references: Haan *et al.* (1990).

*Sequence database accession code(s):*

D00645 Em(40)_vi:TSWSRS Gb(84)_vi:TSWSRS Tomato spotted wilt virus S RNA segment, complete sequence. 3/91 2,916bp.

D00821 Em(40)_vi:TSWSNS tomato spotted wilt virus, genomic S RNA. 5/92 2,837bp

D10066 Em(40)_vi:TSWLRPOLM Gb(84)_vi:TSWLRPOLM Tomato spotted wilt virus L RNA encoding RNA polymerase, complete cds. 4/92 8,897bp.

D13926 Em(40)_vi:TSWSNS1 Gb(84)_vi:TSWSNS Tomato spotted wilt virus, genomic S RNA. 2/93 2,837bp.

M64305 Em(40)_vi:TSWNSS Gb(84)_vi:TSWNSS Tomato spotted wilt virus nonstructural protein (NS-S) gene, 5' flank. 4/91 123bp.

S48091 Em(40)_un:S48091 Gb(84)_vi:S48091 NSM, glycoprotein precursor (M segment) tomato spotted wilt virus TSWV, Genomic RNA Complete

S58512 Em(40)_vi:S58512 Gb(84)_vi:S58512 33.6 kda protein (RNA seqment M) tomato spotted wilt virus, Genomic RNA, 1120 nt. 11/93 1,1

X61799 Em(40)_vi:TSWVN Gb(84)_vi:TSWVN Tomato spotted wilt virus N mRNA for Nucleocapsid (N) protein. 10/91 777bp.

Z36882 Em(43)_vi:Tswvng Gb(89)_vi:Tswvng Tomato spotted wilt virus (Italy) N gene. 8/94 777bp.

*Features of the genome: s*RNA is an ambisense RNA with two terminal ORFs that are expressed via subgenomic *m*RNAs. that in the genomic strand is of 52.4K and in the complementary strand of 28.8K (the nucleocapsid protein). *s*RNA has long terminal untranslated regions that are of inverted complementary to each other, and also an intergenic stable hairpin structure. Preliminary data of the *m*RNA indicates it is all negative strand.

Sub–genomic *m*RNA found in infected cells (2 *m*RNAs for *s*RNA).

*Virion protein(s)* four; $M_r$ of the largest 120000; L protein. $M_r$ of 2nd largest 78000; G1 membrane glycoprotein. $M_r$ of 3rd largest 58000; G2 membrane glycoprotein. $M_r$ of 4th largest 27000; N protein. Method of preparation: Mohamed *et al.* (1973); Tas *et al.* (1977). Amino acid sequence: Haan *et al.* (1990). Virion proteins glycosylated (G1 and G2).

**Replication**

Replication does not depend on a helper virus.

### Cytopathology

Virions found in all parts of the host plant; in cytoplasm. Inclusions present in infected cells; are unusual in shape; clusters of virions in the cell vacuole, which possibly bud from the cisternae of endoplasmic reticulum.

### Taxonomy and Relationships

*Tospovirus: Bunyaviridae.*

The structure of the genome resembles that of arthropod-borne phleboviruses (Bunyaviridae), but their sequences are not significantly similar A tomato spotted wilt like virus first found in *Impatiens* species, and then considered to be a strain of the virus, is now called Impatiens necrotic spot virus, and is recognised as a distinct member of the tospovirus group; (de Avila *et al.*, 1992).

### References

Best, R.J. (1968). *Adv. Virus Res.* **13**: 65.

Brittlebank, C.C. (1919). *J. Agric. Victoria, Aust.* **17**: 213.

Brunt, A.A. (1959). *Nature, Lond.* **183**: 627.

Cho, J.J., Mau, R.F.L., German, T.L., Hartmann, R.W., Yurdin, L.S., Gonsalves, D. and Provvidenti, R. (1989). *Plant Dis.* **73**: 375.

de Avila, A.C.P., de Haan, P., Kitajima, E.W., Kormelink, R. de O., Resende, R., Goldbach, R.W. and Peters, D. (1992). *J. Phytopath.* **134**: 133.

Francki, R.I.B. and Hatta, T. (1981). In: *Handbook of Plant Virus Infections*, p. 491; ed. E. Kurstak. Elsevier/North Holland Biomedical Press, Amsterdam.

Haan, P. de, Wagemakers, L., Peters, D. and Goldenbach, R. (1990). *J. gen. Virol.* **71**: 1001.

Ie, T.S. (1970). *CMI/AAB Descr. Pl. Viruses* No. 39, 4 pp.

Klesser, P.J. (1966). *S. Afr. agric. Sci.* **9**: 711.

Klinkowski, M. and Uschdraweit, H.A. (1952). *Phytopath. Z.* **19**: 269.

Law, M.D. and Moyer, J.W. (1990). *J. gen. Virol.* **71**: 933.

Law, M.D., Speck, J. and Moyer, J.W. (1991). *J. gen. Virol.* **72**: 2597.

Matthews, R.E.F. (1982). *Intervirology* **17**: 11.

Mohamed, N.A., Randles, J.W. and Francki, R.I.B. (1973). *Virology* **56**: 12.

Samuel, G., Bald, J.G. and Pittman, H.A. (1930). *Bull. Coun. scient. ind. Res. Melbourne* **44**, 65 pp..

Smith, K.M. (1957). *Textbook of Plant Virus Diseases*, second edition. Churchill, London.

Tas, P.W.L., Boerian, M.L. and Peters, D. (1977). *J. gen. Virol.* **36**: 267.

Van den Hurk, J., Tas, P.W.L. and Peters, D. (1977). *J. gen. Virol.* **36**: 81.

# Tomato top necrosis (?) *nepovirus*

**Data collated by R. Stace-Smith, 1986.**

*Standard acronym:* ToTNV.

First reported in *Lycopersicon esculentum*; from Indianna, Missouri and Illinois, U.S.A.; by Imle and Samson (1942).

### Natural host range and symptoms

Symptoms disappear soon after infection.

*Lycopersicon esculentum* — rings and lines on young leaves; stripes on the stem; concentric rings on the fruit.

**Geographical distribution**

Spreads in the USA.

**Experimental host range**

Many (>9) families susceptible.

Diagnostically susceptible

*Chenopodium quinoa* — local lesions and systemic necrosis.
*C. murale* — local lesions only.
*Nicotiana tabacum* — mild ring and line patterns.
*N. affinis* — numerous faint small rings.
*Datura stramonium* — indefinite rings and lines.

Diagnostically insusceptible

*Tetragonia expansa, Beta vulgaris, Zinnia elegans, Brassica rapa, Cucumis sativus.*

Maintenance and propagation hosts

*Chenopodium quinoa, Nicotiana tabacum.*

Assay hosts (**L**ocal lesions or **W**hole plants)

*Chenopodium quinoa* (L), *Nicotiana tabacum* (L), *Lycopersicon esculentum* (W).

Dicotyledonous Hosts *(**Susceptible** or Insusceptible)*

***Amaranthus retroflexus***. *Beta vulgaris*. *Brassica oleracea* ssp. *capitata*. *Chenopodium album*. ***Chenopodium murale***. ***Chenopodium quinoa***. *Cucumis melo*. *Cucumis sativus*. *Cucurbita maxima*. *Cucurbita moschata*. ***Datura stramonium***. ***Datura tatula***. ***Lycopersicon esculentum***. ***Lycopersicon pimpinellifolium***. ***Nicandra physaloides***. *Nicotiana glutinosa*. *Nicotiana sylvestris*. ***Nicotiana tabacum***. ***Petunia hybrida***. ***Solanum melongena***. ***Solanum tuberosum***.

Sources of host-range data

Imle and Samson (1942).

**Properties of particles in sap**

TIP: 58 °C. LIV: 2 days. DEP: $\log_{10}$ minus 1:500.

**Purification**

Method: Bancroft (1968).

**Particle morphology**

Virions isometric; not enveloped; 26 nm in diameter; angular in profile; without a conspicuous capsomere arrangement.

**Physical properties**

Three sedimenting components in purified preparations; sedimentation coefficient of the fastest 126 *S* (B); of the other(s) 52 *S* (T), or 102 *S* (M). $A_{260}/A_{280}$ ratio 1.71 (M), or 1.82 (B).

**Taxonomy and Relationships**

Possible *Nepovirus: Comoviridae.*

Virus(es) with serologically unrelated virions: arabis mosaic, raspberry ringspot, strawberry latent ringspot, tobacco ringspot, tomato black ring and tomato ringspot viruses.

**Comments**

This virus was isolated and studied in the 1960's. Stace-Smith obtained an isolate from Purdue University, where it was last studied. However, the culture was not infective, and no other source is known.

**References**

Bancroft, J.B. (1968). *Phytopathology* **32**: 1037.
Harrison, B.D. and Murant A.F. (1977). *CMI/AAB Descr. Pl. Viruses* No. 185, 4 pp.
Imle, E.P. and Samson, R.W. (1942). *Phytopathology* **58**: 1360.

# Tomato vein clearing *nucleorhabdovirus*

**Data collated by S. Yamashita, 1987.**

*Synonyms:* possibly eggplant mottled dwarf rhabdovirus.

First reported in *Lycopersicon esculentum*; from Tokyo, Japan; by Kano *et al.* (1981*a*).

**Natural host range and symptoms**

Symptoms persist.

*Lycopersicon esculentum* — vein-clearing and vein-yellowing, mottling, leaf crinkle and dwarfing.

**Transmission**

Transmitted by a vector (probably); an insect. Virus transmitted by mechanical inoculation; transmitted by grafting; not transmitted by contact between plants; not transmitted by seed; not transmitted by pollen.

**Ecology and control**

Studies reported by Kano *et al.* (1981*a*); Kano *et al.* (1981*b*); Kano *et al.* (1985).

**Geographical distribution**

Spreads in Japan.

**Experimental host range**

Few (<3) families susceptible.

Diagnostically susceptible

*Lycopersicon esculentum* — systemic local lesions; vein-clearing, vein-yellowing, leaf crinkling.

*Nicotiana glutinosa* — systemic necrotic local lesions, vein-clearing, vein-yellowing.

*N. rustica* — local and systemic necrotic local lesions, vein-yellowing.

*Petunia hybrida* — local and systemic chlorotic or systemic local lesions, vein-clearing, vein-yellowing.

*Datura stramonium* — necrotic local lesions.

*Chenopodium murale, Tetragonia expansa* — chlorotic local lesions.

Diagnostically insusceptible

*Solanum melongena, Chenopodium amaranticolor, Gomphrena globosa, Lactuca sativa, Phaseolus vulgaris.*

Maintenance and propagation hosts

*Lycopersicon esculentum, Nicotiana glutinosa, N. rustica, Petunia hybrida.*

Assay hosts (**L**ocal lesions or **W**hole plants)

*Lycopersicon esculentum* (W), *Nicotiana glutinosa* (L and W), *N. rustica* (L and W), *Chenopodium murale* (L), *Tetragonia expansa* (L).

Dicotyledonous Hosts *(**Susceptible** or Insusceptible)*

*Arachis hypogaea. Beta vulgaris. Brassica campestris* ssp. *chinensis. Brassica campestris* ssp. *napus. Brassica campestris* ssp. *pekinensis. Brassica campestris* ssp. *rapa. Catharanthus roseus. Chenopodium amaranticolor.* ***Chenopodium murale.*** *Chenopodium quinoa. Citrullus lanatus. Cucumis melo. Cucumis sativus. Cucurbita maxima. Cucurbita moschata. Cucurbita pepo.* ***Datura stramonium.*** *Glycine max. Gomphrena globosa. Lactuca sativa.* ***Lycopersicon esculentum. Nicotiana glutinosa. Nicotiana rustica.*** *Nicotiana sylvestris. Nicotiana tabacum.* ***Petunia hybrida.*** *Phaseolus vulgaris. Physalis floridana. Pisum sativum. Solanum melongena. Solanum tuberosum. Spinacia oleracea.* ***Tetragonia expansa.*** *Vicia faba. Vigna unguiculata. Zinnia elegans.*

Monocotyledonous Hosts *(**Susceptible** or Insusceptible)*

*Avena sativa. Hordeum vulgare. Oryza sativa. Triticum aestivum. Zea mays.*

Sources of host-range data

Kano *et al.* (1981*a*); Kano *et al.* (1981*b*); Kano *et al.* (1985).

**Properties of particles in sap**

TIP: 40 °C. LIV: 2 days (4°C), or 0.5–1 days (20°C). DEP: $\log_{10}$ minus 3–4. Infectivity of sap decreased by treatment with di–ethyl ether. Leaf sap contains many virions. Electron microscopy: rapid preparations using PTA after fixing.

**Purification**

Method: Kano *et al.* (1981*a*); Kano *et al.* (1981*b*); Kano *et al.* (1985).

**Particle morphology**

Virions rhabdo– or bullet–shaped; enveloped; with a clear modal length; of 240–250 nm; 86–88 nm wide. Axial canal obvious. Basic helix obvious; pitch of basic helix 4.5 nm.

**Physical properties**

One sedimenting component in purified preparations; sedimentation coefficient 1040 *S*.

**Biochemical properties**

Also are polysaccharides.

*Virion protein(s)* four.

**Replication**

Replication does not depend on a helper virus.

**Cytopathology**

Virions found in all parts of the host plant; in cytoplasm, in nucleoli, and in the perinuclear space; in the endoplasmic reticulum. Inclusions present in infected cells; are viroplasms; they contain virions.

**Taxonomy and Relationships**

*Nucleorhabdovirus: Rhabdoviridae.*

**References**

Kano, R., Yamashita, S., Doi, Y. and Yora, K. (1981*a*). *Ann. Phytopath. Soc. Japan* **47**: 138.

Kano, R., Yamashita, S., Doi, Y. and Yora, K. (1981*b*). *Ann. Phytopath. Soc. Japan* **47**: 411.
Kano, R., Namba, S., Yamashita, S., Doi, Y. and Yora, K. (1985). *Ann. Phytopath. Soc. Japan* **51**: 606.
Yamashita, S. (1983). In: *Handbook of Plant Viruses in Japan* p. 532; eds K. Yora, Y. Saito, Y. Doi, T. Inouye and K. Tomaru. Asakura Shoten, Tokyo, Japan.
Yamashita, S., Maeda, T., Doi, Y. and Yora, K. (1982). *Ann. Phytopath. Soc. Japan* **48**: 395.

# Tomato yellow leaf curl bigeminivirus

**Data collated by A.A. Brunt and S. Cohen, 1988.**

*Synonyms:* tomato leaf curl virus. *Standard acronym:* TYLCV-Is, -Sr, -Th, -Ye.

First reported in *Lycopersicon esculentum*; from Israel; by Cohen and Harpaz (1964).

## Natural host range and symptoms

Symptoms persist.

*Lycopersicon esculentum* — plants stunted with small chlorotic puckered leaves; fruit yield much decreased.

*Cynanchum acutum, Malva parviflora* — symptomless infection.

## Transmission

Transmitted by a vector; an insect; *Bemisia tabaci*; Aleyrodidae. Transmitted in a persistent manner. Virus retained when the vector moults; does not multiply in the vector; not transmitted congenitally to the progeny of the vector; does not require a helper virus for vector transmission; transmitted by mechanical inoculation (poorly); transmitted by grafting; not transmitted by contact between plants.

## Ecology and control

Studies reported by Cohen *et al.* (1988), Ioannou (1987) — epidemiology of tomato yellow leaf curl virus in the Jordan valley and in Cyprus; Sharaf (1986) — control by insecticides; Al-Musa (1982), Cohen and Melamed-Madjar (1978) — control by trap plants and yellow mulches; Pilowsky and Cohen (1974) — resistant cultivars.

## Geographical distribution

Spreads in Cyprus, India, Iraq, Israel, Jordan, Lebanon, Nigeria, Senegal, Sudan, Thailand, Tunisia, and Turkey (Verma *et al.*, 1975; Lana and Wilson, 1976).

## Experimental host range

Several (3–9) families susceptible.

Diagnostically susceptible

*Lycopersicon esculentum* — severe systemic leaf chlorosis and malformation.

*Datura stramonium* — interveinal leaf chlorosis.

*Nicotiana glutinosa, N. benthamiana* — systemic vein clearing, then downward curling of the leaf and its margins.

*Phaseolus vulgaris* — interveinal chlorosis, downward curling of the leaves.

Diagnostically insusceptible

*Nicotiana rustica, Physalis floridana, Vicia faba, Gomphrena globosa, Cucumis sativus.*

Maintenance and propagation hosts

*Lycopersicon esculentum, Datura stramonium.*

Assay hosts (**L**ocal lesions or **W**hole plants)

*Lycopersicon esculentum* (W), *Datura stramonium* (W).

Dicotyledonous Hosts ***(Susceptible** or Insusceptible)*

*Amaranthus retroflexus. Arachis hypogaea. Beta vulgaris. Chenopodium amaranticolor. Cucumis sativus.* ***Datura stramonium****. Gomphrena globosa. Gossypium hirsutum. Hibiscus esculentus.* ***Lens esculenta****.* ***Lycopersicon esculentum****.* ***Lycopersicon pimpinellifolium****. Medicago sativa. Momordica balsamina.* ***Nicotiana glutinosa****. Nicotiana rustica.* ***Nicotiana tabacum****.* ***Phaseolus vulgaris****. Physalis floridana. Pisum sativum. Ricinus communis.* ***Sonchus oleraceus****. Vicia faba. Zinnia elegans.*

Additional hosts

***Cynanchum acutum, Malva nicaensis, Malva parviflora, Hyoscyamus desertorum****, Nerium oleander, Calotropis aegyptia, Capparis aegyptia, Lonicera* sp., *Atriplex* sp., *Xanthium strumarium, Momordica balsamina, Ricinus communis, Prosopis farcata, Althaea* sp., *Hibiscus rosa-sinensis, Lavatera cretica, Plumbago capensis, Polygonum equisetiforme, Portulaca oleracea, Ochradenus baccatus, Lycium* sp., *Solanum incanum, Solanum villosum, Withania somnifera, Tamarix* sp., and *Tribulus* sp.

**Purification**

Method: Czosnek *et al.* (1988).

**Particle morphology**

Virions geminate; 20 nm in diameter; dimers 30 nm in length; angular in profile; without a conspicuous capsomere arrangement.

**Biochemical properties**

*Genome* consists of DNA; single–stranded; circular; of two parts; largest (or only) genome part 2.787 kb; the 2nd largest 2.7 kb.

*Sequence database accession code(s):*

L12219 Em(40)_vi:GETYLREPO Gb(84)_vi:TYLREPORIG Tomato yellow leaf curl virus replication origin DNA. 3/93 312bp.

L27708 Em(40)_vi:GETYLV12C Gb(84)_vi:TYLV12C14X Tomato yellow leaf curl virus (V1-V2, C1-C4) genes, complete cds. 1/94 2,777bp.

M59838 Em(40)_vi:GETYLSEQ Gb(84)_vi:TYLSEQA Tomato yellow leaf curl virus DNA sequence. 3/94 168bp

M59839 Em(40)_vi:GETYLSEQB Gb(84)_vi:TYLSEQB Tomato yellow leaf curl virus DNA sequence. 3/94 167bp.

X15656 Em(40)_vi:TYLCV Gb(84)_vi:TYLCV Tomato yellow leaf curl virus TYLCV virion DNA. 11/91 2,787bp.

X61153 Em(40)_vi:TYLCVORF Gb(84)_vi:TYLCVORF Tomato Yellow Leaf Curl Virus DNA for ORF's V1, V2, C1, C2, C3 & C4. 12/91 2,770bp.

X63016 Gb(84)_vi:TYLCVTHB Tomato Yellow Leaf Curl Virus complete genome (TYLCV-TH B-DNA). 7/94 2,737bp

X76319 Gb(84)n:TYLCV12 Tomato yellow leaf curl virus V1, V2, C1, C2, C3 and C4 genes. 7/94 2,790bp.

X79429 Em(40)_vi:TYLCVAC1 Gb(84)_vi:TYLCVAC1 Tomato yellow leaf curl virus AC1 gene (partial). 5/94 393bp.

Z25751 Em(40)_vi:TYLCVVC Gb(84)_vi:TYLCVVC Tomato yellow leaf curl virus V1, V2, C1, C2, C3 and C4 genes. 4/94 2,777bp.

Z28390 Em(40)_vi:TYLCVVC2 Gb(84)_vi:TYLCVVC2 Tomato yellow leaf curl virus (Sicily) genes for V1, V2, C1, C2, C3, C4. 12/93 2,773bp.

**Cytopathology**

Virions found in leaves and phloem parenchyma; in nuclei. Inclusions present in infected cells; are unusual in shape; spherical (400–3000 nm in diameter); seen in section as fibrillar rings with a large central and several peripheral lacunae. Other cellular changes: the reduction of chromatin and the apparent absence of nucleoli.

**Taxonomy and Relationships**

Bigeminivirus: *Geminiviridae.*

**Comments**

*Bemisia tabaci* limits the transmission of this virus by an unknown mechanism (Cohen 1967, 1969; Cohen and Marco, 1970; Marco *et al.*, 1972; 1975).

**References**


Al-Musa, A. (1982). *Plant Dis.* **66**: 561.

Cherif, C. and Russo, M. (1983). *Phytopath. Z.* **108**: 221.

Cohen, S. (1967). *Virology* **31**: 180.

Cohen, S. (1969). *Virology* **37**: 448.

Cohen, S. and Harpaz, I. (1964). *Ent. exp. appl.* **7**: 155.

Cohen, S. and Marco, S. (1970). *Virology* **40**: 363.

Cohen, S. and Melamed-Madjar, V. (1978). *Bull. ent. Res.* **68**: 465.

Cohen, S. and Nitzany, F.E. (1966). *Phytopathology* **56**: 1127.

Cohen, S., Kern, J., Harpaz, I. and Bar-Joseph, M. (1988). *Phytoparasitica* **16**: 259.

Czosnek, H., Ber, R., Antignus, Y., Cohen, S., Vavot, N. and Zamir, D. (1988). *Phytopathology* **78**: 508.

De Francq, D'Hondt, M. and Russo, M. (1985). *Phytopath. Z.* **112**: 153.

Ioannou, N. (1987). *Pl. Path* **36**: 367.

Lana, A.F. and Wilson, G.F. (1976). *Pl. Dis. Reptr* **60**: 296.

Makkouk, K.M. and Laterrot, H. (1983). In: *Plant Virus Epidemiology*, p. 315; eds R.T. Plumb and J.M. Thresh. Blackwell Scientific Publications, Oxford.

Makkouk, K.M. and Shebab, S. (1978). *Abstr. 3rd Int. Congr. Pl. Path.*, p. 51.

Marco, S., Cohen, S., Harpaz, I. and Birk, Y. (1975). *J. Insect Physiol.* **21**: 1821.

Russo, M., Cohen, S. and Martelli, G.P. (1980). *J. gen. Virol.* **49**: 209.

Sharaf, N. (1986). *Agric., Ecosystems and Environ.* **17**: 111.

Verma, H.N., Srivastava, K.M. and Mathur, A.K. (1975). *Pl. Dis. Reptr* **59**: 494.


# Tomato yellow mosaic bigeminivirus

**Data collated by R. Lastra, 1985. Revised by K.W. Buck, 1989.**

*Synonyms:* virus del mosaico dourado do tomateiro, virus del mosaico amarillo del tomate. *Standard acronym:* ToYMV. *Strains:* tomato yellow leaf curl and tomato yellow mosaic, induce similar symptoms but seem not to be related.

First reported in *Lycopersicon esculentum*; from Brazil; by Matys *et al.* (1975).

**Natural host range and symptoms**

Symptoms persist.

*Lycopersicon hirsutum* — yellow mosaic.

*L. esculentum* — yellow mosaic, stunting, small leaves.

*L. pimpinellifolium, Solanum tuberosum* — mild yellow mosaic.

**Transmission**

Transmitted by a vector; an insect; *Bemisia tabaci*; Aleyrodidae. Transmitted in a persistent manner (for up to 7 days). Virus retained when the vector moults; does not multiply in the vector; not transmitted congenitally to the progeny of the vector; transmitted by mechanical inoculation (poorly when inoculum is from old tissue); transmitted by grafting; not transmitted by contact between plants; not transmitted by seed; not transmitted by pollen.

**Geographical distribution**

Spreads in Brazil and Venezuela.

**Experimental host range**

Few (<3) families susceptible.

Diagnostically susceptible

*Datura stramonium, Nicotiana benthamiana* — systemic yellow mosaic.

*Nicandra physalodes, Petunia hybrida, Physalis peruviana* — systemic mild yellow mosaic.

*N. glutinosa* — systemic yellow spots at base of first infected leaves, then mosaic.

*N. tabacum* cv. Samsun — systemic mosaic and leaf deformation.

Diagnostically insusceptible

*Chenopodium quinoa, Euphorbia prunifolium, Phaseolus vulgaris, Vigna unguiculata, Sida rhombifolia, Datura metel.*

Maintenance and propagation hosts

*Nicotiana benthamiana, N. tabacum* cv. Samsun.

Assay hosts (**L**ocal lesions or **W**hole plants)

*Nicotiana benthamiana* (W), *N. glutinosa* (W).

Dicotyledonous Hosts *(**Susceptible** or Insusceptible)*

*Arachis hypogaea. Brassica oleracea* ssp. *botrytis. Capsicum frutescens. Chenopodium album. Chenopodium amaranticolor. Chenopodium quinoa. Cucumis sativus. Cucurbita pepo. Datura metel.* ***Datura stramonium****. Gossypium hirsutum. Lens esculenta.* ***Lycopersicon esculentum****.* ***Lycopersicon pimpinellifolium****. Momordica balsamina.* ***Nicandra physaloides****.* ***Nicotiana benthamiana****. Nicotiana clevelandii.* ***Nicotiana glutinosa****.* ***Nicotiana tabacum****.* ***Petunia hybrida****. Phaseolus lunatus. Phaseolus vulgaris.* ***Physalis floridana****. Pisum sativum. Raphanus sativus.* ***Solanum melongena****.* ***Solanum tuberosum****. Vigna unguiculata.*

Comments on host-range

Mainly Solanaceae infected.

Sources of host-range data

Uzcátegui and Lastra (1978).

**Properties of particles in sap**

TIP: 40–42 °C. LIV: 0.01 days (15 minutes). DEP: $\log_{10}$ minus 1. Leaf sap contains few virions. Electron microscopy: best stained with UA.

**Purification**

Method: Lastra *et al.* (1981).

**Particle morphology**

Virions geminate; not enveloped; 18–20 nm in diameter; angular in profile; without a conspicuous capsomere arrangement.

**Physical properties**

One sedimenting component in purified preparations; sedimentation coefficient 70 *S* (N.B. unfixed virions disrupt in CsCl).

**Biochemical properties**

Virions contain 20 % nucleic acid; 80 % protein; 0 % lipid.

*Genome* consists of DNA; single–stranded; circular; of two parts; largest (or only) genome part the larger 2.588 kb; the 2nd largest 2.508 kb (Hamilton *et al.*, 1984). Genomic nucleic acid isolated by Hamilton *et al.* (1981). Infectivity retained when deproteinised with phenol or detergent. Additional factor not required for infectivity.

*Non–genomic nucleic acid* not found in the virions.

*Virion protein(s)* one; $M_r$ 28000; coat protein.

**Replication**

Genome replicates in nuclei. Replication does not depend on a helper virus.

**Cytopathology**

Virions found in phloem; in nuclei and in nucleoli. Inclusions present in infected cells; are unusual in shape; clusters of virions; they contain virions. Other cellular changes: fibrillar rings in the nucleus that may be where virions assemble, and are composed of DNA and protein.

**Taxonomy and Relationships**

Bigeminivirus: *Geminiviridae.*

Virus(es) with serologically related virions: African cassava mosaic and beet curly top viruses, but distantly. Virus(es) with serologically unrelated virions: maize streak, wheat dwarf, chloris striate mosaic and tobacco yellow dwarf viruses.

**References**

Anzola, D. and Lastra, R. (1985). *Phytopath. Z.* **112**: 363.

De Uzcátegui, R.C. and Lastra, R. (1978). *Phytopathology* **68**: 985.

Hamilton, W.D.O., Saunders, R.C., Coutts, R.H.A. and Buck, K.W. (1981). *FEMS Microbiol. Lett.* **11**: 263.

Hamilton, W.D.O., Bisaro, D.M. and Buck, K.W. (1982). *Nucl. Acids Res.* **10**: 4901.

Hamilton, W.D.O., Stein, V.E., Coutts, R.H.A. and Buck, K.W. (1984). *EMBO J.* **3**: 2197.

Lastra, R. and Anzola, D. (1978). *Agron. trop.* **28**: 473.

Lastra, R. and Gil, F. (1981). *Phytopathology* **71**: 524.

Lastra, R., de Uzcátegui, R.C. and Gil, F. (1981). *Abstr. Vth Int. Congr. Virol., Strasbourg,* p.206.

Matys, J.C., Silva, D.M., Oliveira, A.R. and Costa, A.S. (1975). *Summa Phytopath.* **1**: 267.

Roberts, I.M., Robinson, D.J. and Harrison, B.D. (1984). *Rep. Scottish Crop Res. Inst., 1985*, p. 179.

# Tradescantia-Zebrina *potyvirus*

**Data collated by A.A. Brunt, 1991.**

*Standard acronym:* TZV.

First reported in *Tradescantia albiflora*; from Minnesota, U.S.A; by Lockhart *et al.* (1981).

**Natural host range and symptoms**

Symptoms persist.

*Tradescantia* and *Zebrina* spp., *Commelina diffusa, Rhoeo discolor* — mosaic and malformation of leaves, stunting.

**Transmission**

Transmitted by a vector; an insect; *Rhopalosiphum padi, Myzus persicae*; Aphididae. Transmitted in a non–persistent manner. Virus transmitted by mechanical inoculation.

**Geographical distribution**

Spreads in the USA.

**Experimental host range**

Few (<3) families susceptible.

Diagnostically susceptible

*Tradescantia albiflora, T. blossfeldiana, T. fluminensis, T. navicularis, Zebrina pendula, Rhoeo discolor* — mosaic and malformation of leaves, stunting.

Diagnostically insusceptible

*Chenopodium quinoa, Chlorophytum comosum, Commelina elegans, Nicotiana clevelandii.*

Maintenance and propagation hosts

*Rhoeo discolor, Tradescantia albiflora, Zebrina pendula.*

Assay hosts (**L**ocal lesions or **W**hole plants)

*Zebrina pendula* (W).

Dicotyledonous Hosts *(**Susceptible** or Insusceptible)*

*Beta vulgaris. Brassica campestris* ssp. *pekinensis. Capsicum annuum. Chenopodium amaranticolor. Chenopodium quinoa. Cucumis metuliferus. Cucurbita pepo. Datura stramonium. Gomphrena globosa. Helianthus annuus. Lactuca sativa. Nicotiana clevelandii. Nicotiana debneyi. Nicotiana glutinosa. Nicotiana tabacum. Petunia hybrida. Phaseolus vulgaris. Physalis floridana. Pisum sativum. Vicia faba. Vigna unguiculata. Zinnia elegans.*

Monocotyledonous Hosts *(**Susceptible** or Insusceptible)*

*Avena sativa. Hordeum vulgare. Secale cereale. Triticum aestivum. Tulipa gesneriana. Zea mays.*

Sources of host-range data

Lockhart *et al.* (1981).

**Properties of particles in sap**

Leaf sap contains few virions.

**Purification**

Method: Lockhart *et al.* (1981).

**Particle morphology**

Virions filamentous; not enveloped; usually flexuous; with a clear modal length; of 754 nm. Axial canal obscure.

**Physical properties**

One sedimenting component in purified preparations.

**Cytopathology**

Inclusions present in infected cells; are pinwheels; they do not contain virions.

**Taxonomy and Relationships**

*Potyvirus: Potyviridae.*

Virus(es) with serologically unrelated virions: bean common mosaic, bean yellow mosaic, bidens mottle, blackeye cowpea mosaic, commelina mosaic, dasheen mosaic, maize dwarf mosaic, papaya ringspot, pepper mottle, potato Y, tobacco etch and watermelon mosaic viruses. It is not known if it is related to potyviruses reported in *Commelina diffusa* in Guadeloupe (Migliori and Lastra, 1978) and in 12 species of Commelinaceae in Florida, U.S.A. (Baker and Zettler, 1986).

**References**

Baker, C. and Zettler, F.W. (1986). *Phytopathology* **76**: 372.
Lockhart, B.E.L., Betzold, J.A. and Pfleger, F.L. (1981). *Phytopathology* **71**: 602.
Migliori, A. and Lastra, R. (1978). *Annls. Phytopath.* **10**: 467.

# Trichosanthes mottle (?) potyvirus

**Data collated by A.A. Brunt, 1994.**

*Standard acronym:* TrMV.

First reported in *Trichosanthes rostrata*; from Japan; by Yonaha *et al.*, 1988.

**Natural host range and symptoms**

*Trichosanthes rostrata* — leaf mottling.

**Transmission**

Transmitted by a vector; an insect; Aphididae; *Aphis gossypii, Myzus persicae.* Transmitted in a non–persistent manner. Virus transmitted by mechanical inoculation.

**Geographical distribution**

Spreads in the Eastern Asian region; Japan.

**Experimental host range**

Few (<3) families susceptible.

Diagnostically susceptible

*Benincasa hispida, Cucumis sativus, Cucurbita maxima X moschata, C. moschata, Luffa cylindrica* and *Momordica charantia* — systemic chlorotic spotting.

*Citrullus lanatus, Cucumis melo, Lagenaria leucantha* and *Trichosanthes anguina* — systemic vein clearing, vein necrosis, mottling and/or distortion.

*Trichosanthes rostrata* — systemic mottling.

Diagnostically insusceptible

*Chenopodium amaranticolor, C. murale, C. quinoa, Datura stramonium, Nicotiana clevelandii, N. glutinosa.*

Maintenance and propagation hosts

*Cucurbita pepo*.

Assay hosts (**L**ocal lesions or **W**hole plants)

*Cucurbita pepo* (W).

Dicotyledonous Hosts *(**Susceptible** or Insusceptible)*

*Chenopodium amaranticolor*. *Chenopodium murale*. *Chenopodium quinoa*. ***Citrullus lanatus***. ***Cucumis melo***. ***Cucumis sativus***. ***Cucurbita moschata***. ***Cucurbita pepo***. *Datura stramonium*. *Nicotiana clevelandii*. *Nicotiana glutinosa*.

**Properties of particles in sap**

TIP: 50 °C. LIV: 5–8 days. DEP: $\log_{10}$ minus 3.

**Purification**

Method: Yonaha *et al.*, 1988.

**Particle morphology**

Virions filamentous; not enveloped; usually flexuous; with a clear modal length; of 750 nm.

**Cytopathology**

Virions found in cytoplasm. Inclusions are pinwheels.

**Taxonomy and Relationships**

Possible unassigned potyvirus.

Virus(es) with serologically related virions: papaya leaf-distortion mosaic. Virus(es) with serologically unrelated virions: papaya ringspot, zucchini yellow mosaic and watermelon mosaic-2 viruses.

**References**

Yonaha, T., Tamori, M., Fujiwara, Y. and Ishimine, M. (1988). *Bull. Coll. Agric. Univ. Ryukus* **35**: 1–15.

# Tropaeolum 1 *potyvirus*

**Data collated by A.A. Brunt, 1990.**

*Synonyms:* isanu mosaic virus. *Standard acronym:* TV-1.

First reported in *Tropaeolum tuberosum*; from material imported to U.K. from Peru; by A.A. Brunt and R.S. Phillips, unpublished information.

**Natural host range and symptoms**

*Tropaeolum tuberosum* — leaf chlorosis.

**Transmission**

Transmitted by a vector; an insect; *Myzus persicae*; Aphididae. Virus transmitted by mechanical inoculation; not transmitted by seed.

Diagnostically susceptible

*Nicotiana benthamiana, N. clevelandii* and *N. megalosiphon* — chlorotic local lesions; systemic leaf chlorosis.

**Particle morphology**

Virions filamentous; with a clear modal length; of *c.* 750 nm.

**Taxonomy and Relationships**

*Potyvirus: Potyviridae.*

Virus(es) with serologically unrelated virions: 17 other potyviruses.

It is not yet known if the virus is related to a potyvirus isolated from *Tropaeolum tuberosum* in Bolivia (Delhey and Monasterios, 1977).

**Comments**

Tropaeolum potyviruses 1 and 2 frequently occur as a mixture, isanu mosaic virus, in *Tropaeolum tuberosum.*

**References**

Delhey, R. and Monasterios, T. (1977). *Z. PflKrankh. PflPath. PflSchutz.* **84**: 224.
Mathu, R.W. (1987). *M.Sc. Thesis*, pp. 91, University of London, U.K.

# Tropaeolum 2 *potyvirus*

**Data collated by A.A. Brunt, 1990.**

*Synonyms:* isanu mosaic virus. *Standard acronym:* TV-2.

First reported in *Tropaeolum tuberosum*; from material imported to U.K. from Peru; by A.A. Brunt and R.S. Phillips, unpublished information.

**Natural host range and symptoms**

*Tropaeolum tuberosum* — leaf chlorosis.

**Transmission**

Transmitted by a vector; an insect; *Myzus persicae*; Aphididae. Virus transmitted by mechanical inoculation; not transmitted by seed.

Diagnostically susceptible

*Nicotiana benthamiana, N. clevelandii* and *N. megalosiphon* — chlorotic local lesions; systemic leaf chlorosis.

*Tropaeolum majus* — systemic leaf chlorosis.

*Datura metel, Nicotiana sylvestris* and *Zinnia elegans* — chlorotic local lesions.

*Nicotiana plumbaginifolia* and *Helianthus annuus* — symptomless systemic infection.

**Particle morphology**

Virions filamentous; with a clear modal length; of *c.* 750 nm.

**Taxonomy and Relationships**

*Potyvirus: Potyviridae.*

Virus(es) with serologically related virions: potato Y potyvirus is distantly related. Virus(es) with serologically unrelated virions: 17 other potyviruses.

It is not yet known if the virus is related to a potyvirus isolated from *Tropaeolum tuberosum* in Bolivia (Delhey and Monasterios, 1977).

**Comments**

Tropaeolum potyviruses 1 and 2 frequently occur as a mixture, isanu mosaic virus, in *Tropaeolum tuberosum.*

**References**

Delhey, R. and Monasterios, T. (1977). *Z. PflKrankh. PflPath. PflSchutz.* **84**: 224.
Mathu, R.W. (1987). *M.Sc. Thesis*, University of London, U.K., pp. 91.

# Tuberose *potyvirus*

**Data collated by M.N. Pearson, 1991.**

First reported in *Polianthes tuberosa*; from Pukekohe, New Zealand; by Pearson and Horner (1986).

**Natural host range and symptoms**

Symptoms persist and vary seasonally. Symptoms leaf mottling.
*Polianthes tuberosa* — irregular foliar mottling.

**Transmission**

Spread in infected vegetative planting material. Virus transmitted by mechanical inoculation.

**Geographical distribution**

Spreads in New Zealand.

**Experimental host range**

Few (<3) families susceptible.

Diagnostically susceptible

*Polianthes tuberosa* — leaf mottling.

Diagnostically insusceptible

*Chenopodium amaranticolor, C. quinoa, Nicotiana clevelandii, N. glutinosa, Phaseolus vulgaris.*

Maintenance and propagation hosts

*Polianthes tuberosa.*

Dicotyledonous Hosts *(**Susceptible** or Insusceptible)*

*Beta vulgaris. Chenopodium amaranticolor. Chenopodium murale. Cucumis sativus. Datura stramonium. Gomphrena globosa. Nicotiana clevelandii. Nicotiana glutinosa. Nicotiana tabacum. Petunia hybrida. Phaseolus vulgaris. Pisum sativum. Spinacia oleracea. Vicia faba. Zinnia elegans.*

Monocotyledonous Hosts *(**Susceptible** or Insusceptible)*

*Zea mays.*

**Properties of particles in sap**

Leaf sap contains many virions. Electron microscopy: negatively-stained virions from cut leaf edges, better than sap extracts.

**Purification**

Method: Horner and Pearson (1988).

**Particle morphology**

Virions filamentous; not enveloped; usually flexuous; with a clear modal length; of 749 nm; 15 nm wide. Axial canal obscure. Basic helix obscure.

**Physical properties**

One sedimenting component in purified preparations. $A_{260}/A_{280}$ ratio 1.038.

**Replication**

Replication does not depend on a helper virus.

**Cytopathology**

Virions found in leaves, mesophyll and meristems; in cytoplasm. Inclusions present in infected cells; are unusual in shape; stacked plates (laminated aggregates) thought to be parts of cylindrical or pinwheel inclusions; they do not contain virions.

**Taxonomy and Relationships**

*Potyvirus: Potyviridae.*

Virus(es) with serologically unrelated virions: tulip breaking virus.

**References**

Horner, M.B. and Pearson, M.N. (1988). *J. Phytopath.* **122**: 261.
Pearson, M.N. and Horner, M.B. (1986). *Australas. Plant Path.* **15**: 39.

# Tulare apple mosaic *ilarvirus*

**Data collated by A.A. Brunt, 1992.**

*Standard acronym:* TAMV.

First reported in *Malus sylvestris*; from the U.S.A.; by Yarwood (1955).

**Natural host range and symptoms**

Symptoms persist.

*Corylus avellana, Malus sylvestris* — conspicuous leaf chlorosis.

**Transmission**

Virus transmitted by mechanical inoculation; not transmitted by seed.

**Geographical distribution**

Found, but with no evidence of spread, in France and the U.S.A. (California).

**Experimental host range**

Several (3–9) families susceptible.

Diagnostically susceptible

*Nicotiana tabacum* — local necrotic lines and rings; systemic necrotic rings and oak-leaf patterns, but later formed leaves are symptomless.

*Phaseolus vulgaris* cv. Bountiful — brown local lesions; no systemic infection.

Maintenance and propagation hosts

*Catharanthus roseus, Nicotiana tabacum.*

Assay hosts (**L**ocal lesions or **W**hole plants)

*Phaseolus vulgaris* cv. Bountiful (L), *Nicotiana tabacum* (W).

Dicotyledonous Hosts *(**Susceptible** or Insusceptible)*

***Catharanthus roseus*. *Nicotiana tabacum*. *Phaseolus vulgaris*.**

**Properties of particles in sap**

TIP: 62 °C (in sap containing reducing agent). LIV: 0.0069 days.

**Purification**

Method: Mink and Bancroft (1962); Mink *et al.* (1963); Fulton (1967).

**Particle morphology**

Virions isometric; not enveloped; 33 nm in diameter; rounded in profile.

**Physical properties**

Three sedimenting components in purified preparations; sedimentation coefficient of the fastest 118 *S*; of the other(s) 93 and 108 *S*. $A_{260}/A_{280}$ ratio 1.36.

**Biochemical properties**

Virions contain 12 % nucleic acid; 88 % protein; 0 % lipid.

*Genome* consists of RNA; single–stranded. Base composition 24 % G; 24 % A; 21 % C; 31 % U.

*Virion protein(s)* one; $M_r$ 19000; coat protein. Amino acid composition: Barnett and Fulton (1969).

**Taxonomy and Relationships**

*Ilarvirus: Bromoviridae* (subgroup 2).

Virus(es) with serologically related virions: citrus leaf rugose virus.

**Comments**

The virus has been isolated from a single apple tree (since removed) in Tulare County, California, and later from hazel in France (Cardin and Marenaud, 1975).

**References**

Barnett, O.W. and Fulton, R.W. (1969). *Virology* **39**: 556.
Cardin, L. and Marénaud, C. (1975). *Annls. Phytopath.* **7**: 159.
Fulton, R.W. (1956). *Phytopathology* **46**: 694.
Fulton, R.W. (1967). *Virology* **32**: 153.
Fulton, R.W. (1971). *CMI/AAB Descr. Pl. Viruses* No. 42, 3 pp.
Gilmer, R.M. (1958). *Phytopathology* **48**: 432.
Lister, R.M. and Saksena, K.N. (1976). *Virology* **70**: 440.
Mink, G.I. (1965). *Virology* **26**: 700.
Mink, G.I. and Bancroft, J.B. (1962). *Nature, Lond.* **194**: 214.
Mink, G.I., Bancroft, J.B. and Nadakavukaren, M.J. (1963). *Phytopathology* **53**: 973.
Uyeda, I. and Mink, G.I. (1983). *Phytopathology* **73**: 47.
Yarwood, C.E. (1955). *Hilgardia* **23**: 613.

# Tulip band-breaking *potyvirus*

**Data collated by A.A.Brunt, 1994.**

*Standard acronym:* TBBV.

First reported in *Tulipa* sp.; from The Netherlands; by Asjes and Segers (1985), Dekker *et al.* (1993).

**Geographical distribution**

Spreads in the Netherlands.

*Sequence database accession code(s):* tulip band-breaking virus

S60805 Em(40)_vi:S60805 Gb(84)_vi:S60805 coat protein tulip band-breaking virus TBBV, Genomic RNA, 277 nt. 1/94 277bp. 1 sequence.

**Taxonomy and Relationships**

*Potyvirus: Potyviridae.*

**References**

Asjes, C.J. and Segers,L.Ch. (1985). *Jvers. Lab. Bloemboll.*, Lisse, 1984, p.81.
Langeveld, S.A., Dore, J.-M., Memelink, J., Derks, A.F.L.M., van der Vlugt, C.I.M., Asjes, C.J. and Bol, J.F. (1991). *J. gen. Virol.* **72**: 1531.

Dekker, E.L., Derks, A.F.L.M., Asjes, C.J., Lemmers, M.E.C., Bol, J.F. and Langeveld, S.A. (1993). *J. gen. Virol.* **74**: 881.

# Tulip breaking *potyvirus*

**Data collated by S. Phillips, 1986.**

*Synonyms:* lily mosaic virus, lily mottle virus, lily streak and tulip mosaic virus. *Standard acronym:* TBV. *Strains:* severe tulip breaking virus; mild tulip breaking virus.

First reported in *Tulipa* spp.; from Surrey, England; by Cayley (1928, 1932).

**Natural host range and symptoms**

Symptoms persist.
*Tulipa* spp. and Darwin hybrids — colour-breaking of flowers and leaf chlorosis.
*Lilium* spp. and Mid-Century hybrids — leaf chlorosis and degeneration.

**Transmission**

Transmitted by a vector; an insect; *Myzus persicae, Aphis gossypii, A. fabae, Macrosiphum euphorbiae, Dysaphis tulipae, Aulocorthum circumflexum*; Aphididae. Transmitted in a non–persistent manner. Virus does not require a helper virus for vector transmission; transmitted by mechanical inoculation; transmitted by grafting; not transmitted by contact between plants; not transmitted by seed; not transmitted by pollen.

**Geographical distribution**

Probably distributed worldwide.

**Experimental host range**

Few (<3) families susceptible.

Diagnostically susceptible

*Lilium formosanum* — mottle, malformed leaves and flowers.
*Tulipa* hybrids — flower colour- breaking.

Diagnostically insusceptible

Not known to infect any herbaceous indicator species. *Narcissus pseudonarcissus, Hyacinthus orientalis* and *Hippeastrum hybridum.*

Maintenance and propagation hosts

*Tulipa* varieties Croisette, Grand Pride and Paljas. *Lilium* varieties Concorde and Sterling Star.

Assay hosts (**L**ocal lesions or **W**hole plants)

None known, but *Lilium formosanum* (W) could be used.

Dicotyledonous Hosts *(**Susceptible** or Insusceptible)*

*Amaranthus retroflexus. Atriplex hortensis. Chenopodium amaranticolor. Chenopodium murale. Chenopodium quinoa. Crotalaria spectabilis. Cucumis sativus. Datura stramonium. Glycine max. Gomphrena globosa. Nicotiana clevelandii. Nicotiana glutinosa. Nicotiana tabacum. Petunia hybrida. Phaseolus vulgaris. Tetragonia expansa. Torenia fournieri. Vicia faba. Vigna unguiculata. Zinnia elegans.*

Monocotyledonous Hosts *(**Susceptible** or Insusceptible)*

*Allium cepa. Avena sativa. Hippeastrum hybridum.* ***Lilium formosanum. Lilium longiflorum.*** *Narcissus pseudonarcissus. Saccharum officinarum. Secale cereale. Zea mays.*

Additional hosts

Tulip breaking virus may also infect *Zygadenus fremontii* — mottling; *Fritillaria pudica* — symptomless; *Calochortus* sp — symptomless. Brierley and Smith (1944) inoculated 79 dicotyledonous species from 66 genera and 22 families with tulip breaking virus, and back tested by inoculation to *Lilium formosanum*. All tests were negative, but the species were not named, and in a further survey, 39 monocotyledous species also proved to be insusceptible. The 3 species listed above plus *Ornithogalum thyrsoides* were found to be susceptible but this has not been confirmed.

Sources of host-range data

Brierley and Smith (1944).

**Properties of particles in sap**

TIP: 65–70 °C. LIV: 4–6 days (in lily sap). DEP: $\log_{10}$ minus 5. Leaf sap contains few virions.

**Purification**

Method: Derks *et al.* (1982).

**Particle morphology**

Virions filamentous; not enveloped; usually flexuous; with a clear modal length; of 750–775 nm; 14 nm wide. Axial canal obscure. Basic helix obscure.

**Physical properties**

One sedimenting component in purified preparations; sedimentation coefficient 153 *S*. $A_{260}/A_{280}$ ratio 1.03–1.21.

**Biochemical properties**

*Genome* consists of RNA; single–stranded; linear. Nucleotide sequence references: Langeveld *et al.*, 1991; 1586 nucleotides of the 3'terminal end of the genome of a lily isolate.

*Sequence database accession code(s):*

S44147 Em(40)_vi:S44147 Gb(84)_vi:S44147 polyprotein precursor tulip breaking virus, lily, Genomic RNA, 1556 nt. 1/94 1,556bp.

S60804 Em(40)_vi:S60804 Gb(84)_vi:S60804 coat protein tulip- breaking virus TBV, Genomic RNA, 277 nt. 1/94 277bp.

S60808 Em(40)_vi:S60808 Gb(84)_vi:S60808 coat protein Rembrandt tulip-breaking virus ReTBV, Genomic RNA, 277 nt. 1/94 277bp.

X63630 Gb(84)_vi:TMVCPRNA Tulip mosaic virus gene for coat protein. 7/94 1,479bp.

S60806 Em(40)_vi:S60806 Gb(84)_vi:S60806 coat protein tulip top-breaking virus TTBV, Genomic RNA, 277 nt. 1/94 277bp.

S60810 Em(40)_vi:S60810 Gb(84)_vi:S60810 coat protein lily mottle virus LiMV, Genomic RNA, 277 nt. 1/94 277bp. 6 sequences.

**Replication**

Replication does not depend on a helper virus.

**Cytopathology**

Virions found in all parts of the host plant; in cytoplasm (scattered, or in bundles, but not in chloroplasts or mitochondria). Inclusions present in infected cells; are crystals in the cytoplasm, or viroplasms (N.B. these differ from the X-bodies produced by lily symptomless virus ; McWhorter, 1940), or pinwheels.

**Taxonomy and Relationships**

*Potyvirus: Potyviridae.*

Virus(es) with serologically related virions: tobacco etch (SDI: 5) and henbane mosaic viruses (SDI: 7). Virus(es) with serologically unrelated virions: alstroemeria mosaic, bean yellow mosaic, clover yellow vein, Columbian datura, hippeastrum mosaic, iris mild mosaic, lettuce mosaic, narcissus yellow stripe, potato A, potato Y or turnip mosaic viruses.

**References**

Brierley, P. and Smith, F.F. (1944). *Phytopathology* **34**: 718.
Cayley, D.M. (1928). *Ann. appl. Biol.* **15**: 529.
Cayley, D.M. (1932). *Ann. appl. Biol.* **19**: 153.
Derks, A.F.L.M., Vink-van den Abeele, J.L. and van Schadewijk, A.R. (1982). *Neth. J. Pl. Path.* **88**: 87.
Langeveld, S.A., Dore, J.-M., Memelink, J., Derks, A.F.L.M., van der Vlugt, C.I.M., Asjes, C.J. and Bol, J.F. (1991). *J. gen. Virol.* **72**: 1531.
McWhorter, F.P. (1940). *Phytopathology* **30**: 788.
Van Slogteren, D.H.M. (1971). *CMI/AAB Descr. Pl. Viruses* No. 71, 4 pp.
Van Slogteren, D.H.M. and de Bruyn Ouboter, M.P. (1941). *Meded. Landb. Wageningen* **45**: 54.
Van Slogteren, D.H.M. and de Vos, N.P. (1966). In: *Viruses of Plants*, pp. 320; eds. A.B.R. Beemster and J. Dijkstra. North Holland Publishing Co., Amsterdam.

# Tulip chlorotic blotch *potyvirus*

**Data collated by W.P. Mowat, 1990.**

*Standard acronym:* TCBV.

First reported in *Tulipa* spp.; from Australia; by Mowat (1985).

**Natural host range and symptoms**

Symptoms persist. Symptoms pale green blotches or streaks in leaves. Colour break symptoms in flowers.

*Tulipa* spp. — blotching or streaking of leaves. Flower break symptoms indistinguishable from those caused by tulip breaking virus.

**Transmission**

Virus transmitted by mechanical inoculation.

**Geographical distribution**

Spreads in Australia.

**Experimental host range**

Several (3–9) families susceptible. Experimentally infected plants mostly show systemic chlorotic mottle and chlorotic local lesions.

Diagnostically susceptible

*Nicotiana benthamiana* — systemic chlorotic mottle.
*Nicotiana clevelandii* — systemic chlorotic mottle, sometimes with veinal necrosis.
*Chenopodium amaranticolor, C. quinoa* — chlorotic local lesions; not systemic.

Diagnostically insusceptible

*Beta vulgaris, Brassica campestris, B. rapa, Cucumis sativus, Lilium formosum.*

Maintenance and propagation hosts

*Nicotiana benthamiana, N. clevelandii.*

Assay hosts (**L**ocal lesions or **W**hole plants)

*Chenopodium amaranticolor* (L).

Dicotyledonous Hosts *(**Susceptible** or Insusceptible)*

*Apium graveolens. Beta vulgaris. Brassica campestris. Brassica campestris* ssp. *rapa.* ***Chenopodium amaranticolor. Chenopodium quinoa.*** *Coriandrum sativum. Cucumis sativus. Datura stramonium. Dianthus barbatus.* ***Gomphrena globosa. Lycopersicon esculentum. Nicotiana benthamiana. Nicotiana clevelandii. Nicotiana debneyi. Nicotiana glutinosa. Nicotiana rustica. Nicotiana tabacum.*** *Ocimum basilicum.* ***Petunia hybrida.*** *Phaseolus vulgaris. Physalis floridana. Pisum sativum.* ***Tetragonia expansa.*** *Trifolium incarnatum. Trifolium pratense. Trifolium repens.* ***Vicia faba.***

Monocotyledonous Hosts *(**Susceptible** or Insusceptible)*

*Poa annua.*

Sources of host-range data

Mowat (1985).

**Properties of particles in sap**

TIP: 50–60 °C. DEP: $\log_{10}$ minus 3. Leaf sap contains few virions (probably).

**Purification**

Method: Mowat (1985).

**Particle morphology**

Virions filamentous; usually flexuous; with a clear modal length; of 720 nm; 12 nm wide. Axial canal obscure. Basic helix obscure.

**Physical properties**

One sedimenting component in purified preparations, or two sedimenting components in purified preparations.

**Cytopathology**

Other cellular changes: interference with pigment formation in petals (Mowat, 1985).

**Taxonomy and Relationships**

*Potyvirus: Potyviridae.*

Virus(es) with serologically related virions: bean yellow mosaic virus. Virus(es) with serologically unrelated virions: tulip breaking, pepper veinal mottle, tobacco etch and potato Y viruses.

Best tests for diagnosis: ability to infect *Chenopodium amaranticolor, C. quinoa* and failure to react with antiserum to tulip breaking virus virions.

**References**

Mowat, W.P. (1985). *Ann. appl. Biol.* **106**: 65.

# Tulip halo necrosis (?) virus

**Data collated by W.P. Mowat, 1990.**

First reported in *Tulipa* spp.; from the U.K.

**Natural host range and symptoms**

Symptoms persist. Symptoms necrotic local lesions in leaves and flower stems; often with translucent border.

*Tulipa* spp. — necrotic local lesions in leaves and flower stems; often with translucent border.

**Transmission**

Not transmitted by *Myzus persicae* (after short or long acquisition feeds). Virus transmitted by mechanical inoculation (inoculum for *Chenopodium quinoa* prepared in 0.07M $PO_4$ buffer pH 8 containing 0.2M 2–mercaptoethanol is much more infective than that prepared in the same buffer at pH 7 without reducing agent).

**Geographical distribution**

Found, but with no evidence of spread, in the U.K.

**Experimental host range**

Several (3–9) families susceptible. Experimentally infected plants mostly show necrotic and chlorotic local lesions and veinal necrosis.

Diagnostically susceptible

*Chenopodium amaranticolor, C. quinoa* — necrotic local lesions; not systemic.

*Phaseolus vulgaris* — brown local lesions, rings.

*Datura stramonium* — necrotic local lesions and multiple green rings; sometimes systemic.

*Nicotiana benthamiana* — chlorotic local lesions, becoming necrotic; necrosis and chlorotic mottle.

*Nicotiana clevelandii* — necrotic and chlorotic local lesions, veinal necrosis and mottle.

*Beta vulgaris* — necrotic or green local lesions.

*Nicotiana debneyi* — necrotic local lesions; systemic.

*Nicotiana tabacum* cv. Xanthi-*nc* — necrotic local lesions; necrotic spots and rings.

Assay hosts (**L**ocal lesions or **W**hole plants)

*Nicotiana clevelandii* (W), *Chenopodium quinoa* (L).

Dicotyledonous Hosts *(**Susceptible** or Insusceptible)*

***Beta vulgaris*. *Chenopodium amaranticolor*. *Chenopodium quinoa*. *Datura stramonium*. *Gomphrena globosa*. *Nicotiana benthamiana*. *Nicotiana clevelandii*. *Nicotiana debneyi*. *Nicotiana glutinosa*. *Nicotiana tabacum*. *Petunia hybrida*. *Phaseolus vulgaris*. *Spinacia oleracea*. *Tetragonia expansa*.**

Sources of host-range data

Mowat (1971).

**Properties of particles in sap**

TIP: 45 °C. LIV: 0.25 days. DEP: $\log_{10}$ minus 6. Infectivity of sap not changed by treatment with di–ethyl ether.

**Purification**

Method: Mowat (1971).

**Taxonomy and Relationships**

Best tests for diagnosis: the virus and cucumber mosaic virus, tobacco necrosis virus, tomato bushy stunt and tulip X viruses all all cause similar symptoms in tulips. These viruses are best distinguished by electron microscopy and serology.

**Comments**

Studies involving centrifugation of sap extracts, and the effect of diethyl ether treatment strongly suggests that THN is caused by a virus, but no virions have been seen. Extracts treated with water saturated phenol are not infective.

**References**

Mowat, W.P. (1970). *Ann. appl. Biol.* **66**: 17.
Mowat, W.P. (1971). *Acta Hort.* **23**: 283.
Mowat, W.P. (1971). *Ann. appl. Biol.* **69**: 147.

# Tulip mosaic virus

*Sequence database accession code(s):*

X63630 Em(43)_vi:Tmvcprna Gb(89)_vi:Tmvcprna Tulip mosaic virus gene for coat protein. 7/94 1,479bp.

# Tulip X *potexvirus*

**Data collated by W.P. Mowat, 1990.**

*Standard acronym:* TVX.

First reported in *Tulipa* sp.; from the U.K.; by Mowat (1982).

**Natural host range and symptoms**

Symptoms persist. Symptoms chlorotic or necrotic grey-brown streaking of leaves; streaks of intensified pigment (or of necrosis) in petals.

*Tulipa* spp. — chlorotic or necrotic grey-brown streaking of leaves; streaks of intensified pigment (or of necrosis) in petals.

**Transmission**

Transmitted by means not involving a vector; manual inoculation of tulips is only occasionally successful, thus spread by contact between plants may be rare. Virus transmitted by mechanical inoculation; not transmitted by contact between plants (probably).

**Geographical distribution**

Found, but with no evidence of spread, in the U.K. and Sweden.

**Experimental host range**

Many (>9) families susceptible. Experimentally infected plants mostly show chlorotic local lesions and chlorotic mottle.

Diagnostically susceptible

*Chenopodium amaranticolor, C. quinoa, Trifolium incarnatum* — chlorotic local lesions; systemic chlorotic mottle.

*Spinacia oleracea, Nicotiana benthamiana, Apium graveolens, Viola odorata* — symptomless systemic infection.

Diagnostically insusceptible

*Brassica campestris, Trifolium pratense, Nicotiana clevelandii, N. tabacum* cv. White Burley.

Maintenance and propagation hosts

*Chenopodium amaranticolor, C. quinoa.*

Assay hosts (**L**ocal lesions or **W**hole plants)

*Chenopodium amaranticolor* (L).

Dicotyledonous Hosts *(**Susceptible** or Insusceptible)*

***Anthriscus cerefolium. Apium graveolens. Beta vulgaris.*** *Brassica campestris. Brassica campestris* ssp. *rapa. Capsicum annuum.* ***Chenopodium amaranticolor. Chenopodium quinoa. Coriandrum sativum. Cucumis sativus.*** *Cyamopsis tetragonoloba. Datura stramonium.* ***Daucus carota. Gomphrena globosa. Heracleum sphondylium.*** *Lactuca sativa.* ***Lycopersicon esculentum. Nicotiana benthamiana.*** *Nicotiana clevelandii. Nicotiana debneyi. Nicotiana glutinosa. Nicotiana rustica. Nicotiana tabacum.* ***Petroselinum crispum.*** *Petunia hybrida. Phaseolus vulgaris. Pisum sativum.* ***Spinacia oleracea. Tetragonia expansa. Trifolium incarnatum.*** *Trifolium pratense. Trifolium repens. Vicia faba.*

Monocotyledonous Hosts *(**Susceptible** or Insusceptible)*

*Lilium formosanum. Setaria italica.*

Additional hosts

***Viola odorata.***

Sources of host-range data

Mowat (1982).

**Properties of particles in sap**

TIP: 60–65 °C. LIV: 30 days. DEP: $\log_{10}$ minus 9. Leaf sap contains many virions.

**Purification**

Method: Mowat (1982).

**Particle morphology**

Virions filamentous; not enveloped; usually flexuous; with a clear modal length; of 495 nm; *c.* 13 nm wide. Axial canal obscure. Basic helix obscure; pitch of basic helix 3.25 nm.

**Physical properties**

Two sedimenting components in purified preparations; sedimentation coefficient of the fastest 118 *S*; of the other(s) 102 *S*.

**Biochemical properties**

Virions contain 8 % nucleic acid; 92 % protein; 0 % lipid.

Genome unipartite; largest (or only) genome part 6.459 kb. Genomic nucleic acid isolated by Mowat (1982): two cycles of precipitation with 8% polyethylene glycol ($M_r$ 6000) in the presence of 0.2M NaCl.

*Virion protein(s)* one; $M_r$ 22500 (sometimes one of 19000 is found).

**Cytopathology**

Virions found in epidermal cells of leaves. Inclusions present in infected cells; are unusual in shape; spindle shaped; they contain virions.

**Taxonomy and Relationships**

*Potexvirus.*

Virus(es) with serologically related virions: hydrangea ringspot and viola mottle viruses. Virus(es) with serologically unrelated virions: bamboo mosaic, cactus X, clover yellow mosaic, foxtail mosaic, lily X, narcissus mosaic, nerine X, plantain X and potato X viruses.

Best tests for diagnosis: symptoms in tulip resemble those caused by other viruses, but systemic line patterns in *Chenopodium amaranticolor* and the failure to infect *Nicotiana clevelandii*, combined with the presence of potexvirus virions, give a reliable diagnosis, especially when confirmed by serological tests.

**References**

Mowat, W.P. (1982). *Ann. appl. Biol.* **101**: 51.

Mowat, W.P. (1984). *CMI/AAB Descr. Pl. Viruses* No. 276, 4pp.

# Turnip crinkle *carmovirus*

**Data collated by A.A. Brunt, 1991.**

*Standard acronym:* TCV.

First reported in *Brassica campestris* ssp. *rapa*; from Scotland; by Broadbent and Blencowe (1955).

**Natural host range and symptoms**

Symptoms persist.

*Brassica campestris* ssp. *rapa* — mottling and leaf distortion; stunting of plants.

*Brassica napo-brassica* — faint leaf chlorosis or symptomless infection.

*Brassica sinapis* — severe leaf distortion.

**Transmission**

Transmitted by a vector; an insect; *Phyllotreta* (nine species) and *Psylloides* (two species). Transmitted in a non–persistent manner. Virus transmitted by mechanical inoculation; transmitted by contact between plants; not transmitted by seed.

**Geographical distribution**

Spreads in the UK and the former Yugoslavia.

**Experimental host range**

Many (>9) families susceptible.

Diagnostically susceptible

*Brassica campestris* ssp. *pekinensis* — local chlorotic lesions; systemic leaf mottling and crinkling.

*Chenopodium amaranticolor, C. quinoa, Datura stramonium* — local chlorotic lesions; no systemic infection.

*Chenopodium murale* — local chlorotic lesions; systemic chlorotic flecking.

*Tetragonia expansa* — local chlorotic lesions; symptomless systemic infection.

Maintenance and propagation hosts

*Brassica juncea, Brassica campestris* ssp. *pekinensis.*

Assay hosts (**L**ocal lesions or **W**hole plants)

*Brassica juncea* (W), *Brassica campestris* ssp. *pekinensis* (W), *Chenopodium amaranticolor* (L), *C. murale* (L), *C. quinoa* (L), *Tetragonia expansa* (L).

Dicotyledonous Hosts *(**Susceptible** or Insusceptible)*

*Amaranthus caudatus. Beta vulgaris.* ***Brassica campestris*** ssp. ***napus. Brassica campestris*** ssp. ***pekinensis. Brassica campestris*** ssp. ***rapa. Brassica oleracea*** ssp. ***botrytis. Brassica oleracea*** ssp. ***capitata. Capsella bursa–pastoris. Catharanthus roseus. Cheiranthus cheiri. Chenopodium album. Chenopodium amaranticolor. Cucumis sativus. Datura stramonium. Gomphrena globosa. Lactuca sativa. Lunaria annua. Lycopersicon esculentum. Nicotiana bigelovii.*** *Nicotiana clevelandii. Nicotiana tabacum. Phaseolus vulgaris.* ***Raphanus sativus. Sinapis alba. Tetragonia expansa.*** *Vicia faba.*

Sources of host-range data

Broadbent and Heathcote (1958).

**Properties of particles in sap**

TIP: 95 °C. LIV: more than 42 days (at 20°C). DEP: $\log_{10}$ minus 6.

**Purification**

Method: Hollings and Stone (1969).

**Particle morphology**

Virions isometric; not enveloped; 28 nm in diameter; rounded in profile; with a conspicuous capsomere arrangement.

**Physical properties**

One sedimenting component in purified preparations; sedimentation coefficient 129 *S*. Density 1.34 g $cm^{-3}$ in CsCl. $A_{260}/A_{280}$ ratio 1.48.

**Biochemical properties**

Virions contain 17 % nucleic acid; 83 % protein; 0 % lipid.

*Genome* consists of RNA; single–stranded; unipartite; largest (or only) genome part 4.2 kb. Poly A region absent. Nucleotide sequence references: Carrington *et al.* (1987).

*Sequence database accession code(s):*

M22445 Em(40)_vi:TCVPORF Gb(84)_vi:TCVPORF Turnip crinkle virus P28, P88, P8 and coat protein genes RNA, complete cds. 11/89 4,050bp

M29290 Em(40)_vi:TCVDIGAA Gb(84)_vi:TCVDIGAA Turnip crinkle virus defective interfering RNA. 5/90 347bp.

X05193 Em(40)_vi:TOTCVCP Gb(84)_vi:TOTCVCP Turnip crinkle virus RNA for coat protein. 9/93 1,450bp.

X12749 Em(40)_vi:TOTCVRF Gb(84)_vi:TOTCVRF Turnip crinkle virus (TCV) avirulent satellite RNA F. 12/89 230bp.

X12750 Em(40)_vi:TOTCVRC Gb(84)_vi:TOTCVRC Turnip crinkle virus (TCV) virulent satellite RNA C. 4/89 355bp. 5 sequences.

*Virion protein(s)* one; $M_r$ 38000; coat protein. Amino acid sequence: Carrington (1986); Carrington *et al.* (1987). Amino acid composition: Butler (1970).

**Cytopathology**

Virions found in all parts of the host plant. Inclusions absent from infected cells.

**Taxonomy and Relationships**

*Carmovirus: Tombusviridae.*

Virus(es) with serologically related virions: galinsoga mosaic virus. Suggested relationship to carnation mottle carmovirus not yet confirmed. Virus(es) with serologically unrelated virions: carnation mottle virus.

The virus showed no RNA sequence homology with galinsoga mosaic, cucumber soil-borne, saguaro cactus or pelargonium flower break viruses.

**References**

Altenbach, S.B. and Howell, S.H. (1982). *Virology* **118**: 128.
Broadbent, L. and Blencowe, J.W. (1955). *Rep. Rothamsted Exp. Stn.* 1954, p. 87.
Broadbent, L. and Heathcote, G.D. (1958). *Ann. appl. Biol.* **46**: 585.
Butler, P.J.G. (1970). *J. mol. Biol.* **52**: 589.
Carrington, J.C. (1986). *Ph.D. Thesis.* University of California, Berkeley, U.S.A.
Carrington, J.C., Heaton, L.A., Zuidema, D., Hillman, B.I. and Morris, T.J. (1989). *Virology* **170**: 219.
Carrington, J.C., Morris, T.J., Stockley, P.G. and Harrison, S.C. (1987). *J. mol. Biol.* **194**: 265.
Dougherty, W.G. and Kaesberg, P. (1981). *Virology* **115**: 45.
Finch, J.T., Klug, A. and Leberman, R. (1970). *J. mol. Biol.* **50**: 215.
Hacker, D.L., Petty, I.R.D., Wei, N. and Morris, T.J. (1992). *Virology* **186**: 1.
Hollings, M. and Stone, O.M. (1969). *Zbl. Bakt. ParasitKde. Abt. 2,* **123**: 237.
Leberman, R. and Finch, J.T. (1970). *J. mol. Biol.* **50**: 209.
McLeod, R. and Markham, R. (1963). *Virology* **19**: 190.
Martin, C. (1958). *Proc. 3rd Conf. Potato Virus Dis., Lisse-Wageningen, 1957*, p. 106.
Symons, R.H., Rees, M.W., Short, M.N. and Markham, R. (1963). *J. mol. Biol.* **6**: 1.
Tremaine, J. (1970). *Virology* **42**: 611.

## Turnip mosaic *potyvirus*

**Data collated by R. Provvidenti, 1981. Revised 1983.**

*Synonyms:* anemone mosaic virus, cabbage A virus (Walker *et al.*, 1945; Yoshii, 1963), cabbage black ring virus (Tompkins *et al.*, 1938), cabbage black ringspot virus (Smith, 1935), cabbage ring necrosis virus (Larson and Walker, 1941), daikon mosaic virus (Yoshii, 1963), horseradish mosaic virus (Pound, 1948; Yoshii, 1963), radish P virus (Tochihara, 1959; Yoshii, 1963), tigridia mosaic virus (Brunt, 1976; Smith and Brierley, 1944). *Standard acronym:* TuMV.

First reported in *Brassica chinensis, B. japonica, B. rapa*; from the U.S.A; by Schultz (1921).

**Natural host range and symptoms**

Symptoms persist.

*Capsella bursa-pastoris, Brassica nigra, Brassica pekinensis, Stellaria media, Trifolium hybridum, Alliaria officinalis, Calanthe* sp., *Hesperis matronalis* — chlorotic local lesions, mosaic, mottling, puckering or rugosity.

*Brassica chinensis. B. rapa, B. japonica.*

**Transmission**

Transmitted by a vector; an insect; 40–50 spp. especially *Myzus persicae* and *Brevicoryne brassicae*; Aphididae. Transmitted in a non–persistent manner. Virus transmitted by mechanical inoculation; not transmitted by seed.

**Geographical distribution**

Probably distributed worldwide. Spreads in Australia, China, and the UK.

**Experimental host range**

Many (>9) families susceptible.

Diagnostically susceptible

*Chenopodium quinoa* — chlorotic local lesions; not systemic.

*Brassica nigra, Brassica oleracea* var. *capitata, Cichorium endiva, Cucurbita pepo, Pisum sativum, Nicotiana glutinosa, Petunia hybrida, Spinacia oleracea* — systemic mosaic, rugosity.

*Brassica pekinensis* — chlorotic local lesions; systemic mosaic, rugosity.

Maintenance and propagation hosts

*Brassica pekinensis, B. perviridis, B. rapa, Nicotiana glutinosa, Petunia hybrida.*

Assay hosts (**L**ocal lesions or **W**hole plants)

*Chenopodium amaranticolor* (L), *C. quinoa* (L), *Nicotiana tabacum* (W).

Dicotyledonous Hosts *(**Susceptible** or Insusceptible)*

***Amaranthus caudatus***. *Antirrhinum majus. Apium graveolens. Bellis perennis.* ***Beta vulgaris. Brassica campestris*** ssp. ***napus. Brassica campestris*** ssp. ***pekinensis. Brassica campestris*** ssp. ***rapa. Brassica juncea. Brassica oleracea*** ssp. ***botrytis. Brassica oleracea*** ssp. ***capitata***. *Cajanus cajan.* ***Calendula officinalis. Capsella bursa-pastoris***. *Capsicum frutescens.* ***Cassia occidentalis. Celosia argentea. Cheiranthus cheiri*** (U.K. isolate), or *Cheiranthus cheiri. Chenopodium album.* ***Chenopodium amaranticolor. Chenopodium murale. Chenopodium quinoa. Cicer arietinum. Cichorium endiva. Clarkia amoena***. *Cucumis sativus.* ***Cucurbita pepo***. *Cyamopsis tetragonoloba.* ***Datura metel. Datura stramonium***. *Dianthus barbatus. Fagopyrum esculentum. Glycine max.* ***Gomphrena globosa***. *Hibiscus esculentus. Lablab purpureus.* ***Lactuca sativa***, or *Lactuca sativa.* ***Lens esculenta***. *Lotus corniculatus.* ***Lupinus albus***. *Lupinus angustifolius. Lycopersicon esculentum.* ***Lycopersicon pimpinellifolium. Matthiola incana***. *Medicago sativa.* ***Melilotus alba. Nicandra physaloides. Nicotiana × edwardsonii. Nicotiana glutinosa. Nicotiana rustica***, or *Nicotiana rustica.* ***Nicotiana sylvestris. Nicotiana tabacum. Papaver nudicaule. Petunia hybrida***. *Phaseolus vulgaris.* ***Physalis floridana. Phytolacca americana. Pisum sativum. Raphanus sativus***, or *Raphanus sativus. Ricinus communis.* ***Senecio vulgaris***. *Solanum nigrum. Solanum tuberosum.* ***Spinacia oleracea. Stellaria media. Tetragonia expansa. Trifolium hybridum. Trifolium incarnatum***. *Trifolium pratense. Trifolium repens. Tropaeolum majus. Vicia faba. Vicia sativa. Vicia villosa. Vigna angularis. Vigna unguiculata. Vigna unguiculata* ssp. *sesquipedalis. Viola cornuta.* ***Zinnia elegans***.

Monocotyledonous Hosts *(**Susceptible** or Insusceptible)*

*Avena sativa. Zea mays.*

Sources of host-range data

Thornberry (1966); Purcifull (1968); Edwardson (1974); Fischer and Lockhart (1976); Provvidenti (1978).

**Properties of particles in sap**

TIP: 62 °C. LIV: 3–4 days. DEP: $\log_{10}$ minus 3–4.

**Purification**

Method: Shepherd and Pound (1960); Tomlinson (1964).

**Particle morphology**

Virions filamentous; not enveloped; usually flexuous; with a clear modal length; of 720 nm.

*Sequence database accession code(s):*

D10601 Em(40)_vi:TMVCAP Gb(84)_vi:TMVCAP Turnip mosaic virus cDNA to genomic RNA for CI, 6K, NIa, NIb and capsid protein. 7/92 5,102bp.

D10927 Em(40)_vi:TMVCAPP1 Gb(84)_vi:TMVCAPP Turnip mosaic virus genomic RNA for Capsid protein, complete cds. 2/93 9,830bp.

L12396 Em(40)_vi:TMVNIAP Gb(84)_vi:TMVNIAP Turnip mosaic virus polyprotein (including nuclear inclusion a and b proteins and core protein

S43452 Em(40)_un:S43452 coat protein turnip mosaic virus TuMV-Jap, mRNA Partial, 1224 nt. 2/93 1,224bp.

X52804 Em(40)_vi:TMVCPP Gb(84)_vi:TMVCPP Turnip mosaic virus gene for coat protein (partial). 9/93 1,448bp.

X65978 Em(40)_vi:TMVCP Gb(84)_vi:TMVCP Turnip mosaic virus gene for coat protein. 5/92 867bp.

X79366 Em(40)_vi:TUMVCP Gb(84)_vi:TUMVCP Turnip Mosaic Virus-Ca coat protein gene, partial. 5/94 297bp

X80158 Gb(84)_vi:TMVNUCIN Turnip Mosaic Virus RNA for nuclear inclusion body protein. 7/94 1,023bp.

X81140 Gb(84)n:TMVCPRN Turnip mosaic virus mRNA for coat protein. 8/94 867bp.

X81141 Gb(84)n:TMVCPMUT Turnip mosaic virus mRNA for coat protein, mutant. 8/94 867bp.

U18654 Gb(89)n:Tmu18654 Turnip mosaic virus nuclear inclusion protein a gene, partial cds. 6/95 729bp.

X83968 Em(43)_vi:Tumvcqs Gb(89)_vi:Tumvcqs Turnip mosaic virus RNA for coat protein. 2/95 1,081bp.

### Cytopathology

Virions found in all parts of the host plant; in cytoplasm. Inclusions cytoplasmic cylindrical present in infected cells; are pinwheels and unusual in shape; bundles, scrolls and short, often curved laminated aggregates; they do not contain virions.

## Taxonomy and Relationships

*Potyvirus: Potyviridae.*

Virus(es) with serologically related virions: maize dwarf mosaic, potato Y and tobacco etch viruses, but distantly. Virus(es) with serologically unrelated virions: bean yellow mosaic, iris severe mosaic, narcissus yellow stripe, peanut mottle and pepper veinal mottle viruses.

## References


Brunt, A.A. (1976). *J. hort. Sci.* **5**: 99.

Edwardson, J.R. (1974). *Fla Agric. Exp. Stn Monog.* No.5.

Fischer, H.U. and Lockhart, B.E.L. (1976). *Pl. Dis. Reptr* **60**: 398.

Larson, R.H. and Walker, J.C. (1941). *J. agric. Res.* **62**: 475.

Pound, G.S. (1948). *J. agric. Res.* **77**: 97.

Provvidenti, R. (1978). *Pl. Dis. Reptr* **62**: 482.

Purcifull, D.E. (1968). *Pl. Dis. Reptr* **52**: 759.

Schultz, E.S. (1921). *J. agric. Res.* **22**: 123.

Shepherd, R.J. and Pound, G.S. (1960). *Phytopathology* **50**: 797.

Smith, F.F. and Brierley, P. (1944). *Phytopathology* **34**: 593.

Smith, K.M. (1935). *Ann. appl. Biol.* **22**: 239.

Thornberry, H.H. (1966). In: *Index of Plant Virus Diseases.* U.S. Dep. Agric. Hdbk No. 307, p.329.

Tochihara, H. (1959). *Ann. Phytopath. Soc. Japan* **24**: 287.
Tomlinson, J.A. (1964). *Ann. appl. Biol.* **53**: 95.
Tomlinson, J.A. (1970). *CMI/AAB Descr. Pl. Viruses* No. 8, 4 pp.
Tompkins, C.M., Gardner, M.W. and Thomas, H.R. (1938). *J. agric. Res.* **57**: 929.
Walker, J.C., Le Bean, F.J. and Pound, G.S. (1945). *J. agric. Res.* **70**: 379.
Yoshii, H. (1963). *Ann. Phytopath. Soc. Japan* **28**: 221.

# Turnip rosette *sobemovirus*

**Data collated by A.A. Brunt, 1995.**

*Standard acronym:* TRoV.

First reported in *Brassica campestris* ssp. *napus* and ssp. *rapa*; from Scotland, U.K.; by Blencowe and Broadbent, 1957.

**Natural host range and symptoms**

*Brassica campestris* ssp. *napus* — chlorotic vein-banding and severe stunting.

*Brassica campestris* ssp. *rapa* — necrosis of veins, leaf distortion and severe stunting ("rosetting").

**Transmission**

Virus transmitted by mechanical inoculation.

**Geographical distribution**

Spreads in the UK.

**Experimental host range**

Several (3–9) families susceptible.

Diagnostically susceptible

*Brassica campestris* ssp. *napus* and ssp. *rapa* — local chlorotic or necrotic lesions; systemic necrosis of veins and petioles, severe stunting.

*Brassica campestris* ssp. *pekinensis* — local necrotic and/or chlorotic lesions; occasional systemic veinal necrosis and leaf distortion.

Diagnostically insusceptible

*Nicotiana tabacum.*

Maintenance and propagation hosts

*Brassica campestris* ssp. *rapa.*

Assay hosts (**L**ocal lesions or **W**hole plants)

*Brassica campestris* ssp. *pekinensis* (L).

Dicotyledonous Hosts *(**Susceptible** or Insusceptible)*

***Bauhinia purpurea**. **Brassica campestris** ssp. **pekinensis**. **Brassica campestris** ssp. **rapa**. **Nicotiana bigelovii**. **Nicotiana clevelandii**. Nicotiana tabacum.*

Sources of host-range data

Hollings and Stone 1969.

**Properties of particles in sap**

TIP: 90 °C. LIV: more than 30 days. DEP: $\log_{10}$ minus 5–6.

**Purification**

Method: Hollings and Stone (1969, 1973).

**Particle morphology**

Virions isometric; not enveloped; 28 nm in diameter.

**Physical properties**

One sedimenting component in purified preparations; sedimentation coefficient 112 *S*. $A_{260}/A_{280}$ ratio 1.56.

**Biochemical properties**

*Genome* consists of RNA; single–stranded. Base composition 25 % G; 26 % A; 22 % C; 27 % U.

*Virion protein(s)* one; $M_r$ 27700; coat protein.

**Cytopathology**

Inclusions absent from infected cells.

**Taxonomy and Relationships**

*Sobemovirus*.

Virus(es) with serologically unrelated virions: 27 viruses with isometric virions.

**References**

Blencowe, J.W. and Broadbent, L. (1957). *Rep. Rothamsted exp. Stn.* 1956, p.105.
Broadbent, L. and Heathcote, G. (1958). *Ann. appl. Biol.* **46**: 585.
Hollings, M. and Stone, O.M. (1969). *Zbl. Bakt. ParasitKde.* Abt 2, **123**: 237.
Hollings, M. and Stone, O.M. (1973). *C.M.I./A.A.B. Descr. Pl. Viruses* No. 125, 3pp.
Lister, R.M. (1958). *Pl. Path.* **7**: 1958.
Martini, C. (1958). *Proc. 3rd Conf. Potato Virus Dis.* Lisse-Wageningen 1957, p. 106.

# Turnip vein-clearing virus

*Sequence database accession code(s):*

L22518 Em(40)_vi:TUVRNAREP Turnip vein- clearing virus RNA replicase component, complete cds. 1/94 3,324bp.

U03387 Em(40)_vi:TV03387 Gb(84)_vi:BRU03387 Turnip vein-clearing virus OSU, complete genome. 1/94 6,313bp 2 sequences.

# Turnip yellow mosaic *tymovirus*

**Data collated by P.L. Guy, 1985; A.J. Gibbs, 1987.**

*Synonyms:* cardamine yellow mosaic virus. *Standard acronym:* TYMV. *Strains:* Type or TYMV-1 isolates include the type (Cambridge), honesty, Rademacher and the closely related Australian isolates Club Lake, Blue Lake, RR, Lake Albina; the cauliflower or TYMV-2 isolates include cauliflower (Northumberland), Rothamsted and Denmark.

First reported in *Brassica pekinensis*; from Europe; by Markham and Smith (1949).

**Natural host range and symptoms**

*Brassica* spp. and many Brassicaceae — vein clearing and mosaic or blotching symptoms.

*Cardamine lilaciana* sward forming variant — vein clearing and mosaic.

**Transmission**

Transmitted by a vector; an insect; *Phyllotreta* and *Psylloides* species (flea beetles) and *Phaedon cochleariae* (the mustard beetle) in Europe and *Pedilophorus* sp in

Australia; Coleoptera. Virus transmitted by mechanical inoculation; transmitted by grafting; transmitted by seed (of *Camelina sativa* (Hein, 1984)); not transmitted by pollen.

**Ecology and control**

Studies reported by Broadbent (1957).

**Geographical distribution**

Spreads in the Eurasian region (in the north west); Australia.

**Experimental host range**

Few (<3) families susceptible.

Diagnostically susceptible

*Brassica pekinensis, Raphanus sativus* — vein clearing and mosaic or blotching.

Diagnostically insusceptible

*Chenopodium quinoa, Nicotiana glutinosa, Pisum sativum.*

Maintenance and propagation hosts

*Brassica pekinensis.*

Assay hosts (**L**ocal lesions or **W**hole plants)

*Brassica pekinensis* (W).

Dicotyledonous Hosts *(**Susceptible** or Insusceptible)*

*Antirrhinum majus. Arachis hypogaea. Bellis perennis. Beta vulgaris.* ***Brassica campestris*****.** ***Brassica campestris*** ssp. ***chinensis*****.** ***Brassica campestris*** ssp. ***pekinensis*****.** ***Brassica juncea*****.** ***Brassica oleracea*** ssp. ***botrytis*****.** ***Capsella bursa–pastoris*****.** *Catalpa bignonioides. Catharanthus roseus.* ***Cheiranthus cheiri*****.** *Chenopodium album. Chenopodium amaranticolor. Chenopodium quinoa. Cicer arietinum. Clarkia amoena.* ***Crambe abyssinica*****.** *Cucumis sativus. Datura stramonium. Daucus carota. Delphinium hybridum. Eucalyptus cloesiana. Euphorbia marginata. Fagopyrum esculentum. Fragaria vesca.* ***Glycine max*****.** *Gomphrena globosa. Gossypium hirsutum. Gypsophila elegans. Helianthus annuus. Hibiscus esculentus. Lablab purpureus. Lobelia erinus.* ***Lunaria annua*****.** *Lupinus albus.* ***Matthiola incana*****.** *Myosotis sylvatica. Nicotiana bigelovii. Nicotiana clevelandii. Nicotiana glutinosa. Nicotiana tabacum. Papaver nudicaule. Pastinaca sativa. Petroselinum crispum.* ***Phaseolus lathyroides*****.** *Phaseolus vulgaris. Pisum sativum. Podolepis robusta.* ***Raphanus sativus*****.** *Salvia splendens.* ***Sinapis alba*****.** *Sinningia speciosa. Solanum melongena. Spinacia oleracea. Tetragonia expansa. Tropaeolum majus. Valeriana officinalis. Vicia faba. Vigna unguiculata. Viola cornuta.*

Additional hosts

***Barbarea vulgaris, Brassica arvensis, Brassica carinata, Brassica napus, Brassica nigra, Brassica rapa, Camelina sativa, Cleome spinosa, Conringia orientalis***, *Crambe arreria, Crambe cordifolia, Crambe hispanica, Crambe tataria* Sebeok, ***Diplotaxis teninfolia, Erysimum cheiranthoides, Lepidium campestre, Neslia paniculata***, *Papaver rhoeas, Physalis aequata, Physalis ixocarpa, Physalis viscosa*, ***Raphanus raphanistrum, Reseda odorata, Sinapis arvensis, Sisymbrium officinale, Thlaspi arvense*** and *Viola tricolor.*

**Properties of particles in sap**

TIP: 60–65 °C. LIV: 10 days (and more). DEP: $\log_{10}$ minus 7. Leaf sap contains many virions.

**Purification**

Method: Gibbs *et al.* (1966); Matthews (1980).

**Particle morphology**

Virions isometric; not enveloped; 28 nm in diameter; rounded in profile; without a conspicuous capsomere arrangement.

**Physical properties**

Two sedimenting components in purified preparations; sedimentation coefficient of the fastest 118 *S* (B); of the other(s) 54 *S* (T). Density 1.28 g $cm^{-3}$ in CsCl (T), or 1.395–1.402 g $cm^{-3}$ in CsCl (B). Isoelectric point pH 3.77–3.97. $A_{260}/A_{280}$ ratio 0.81 (T), or 1.51 (B).

**Biochemical properties**

Virions contain 37 % nucleic acid (B); 63 % protein (B).

*Genome* consists of RNA; single–stranded; linear; unipartite; largest (or only) genome part 6.319 kb. Genomic nucleic acid isolated by Matthews (1980). Base composition 17.6 % G; 23.75 % A; 37.05 % C; 22.15 % U (type strain and Australian isolates). 5' terminus of RNA has an unusual type; a $^{7}$methyl Gppp $^{5}$G... cap. Infectivity retained when deproteinised with phenol or detergent. Genome has *t*RNA–like activity. Genome accepts valine (Pinck *et al.*, 1970).

*Sequence database accession code(s):*

J02418 Gb(84)_vi:MTY1 Turnip yellow mosaic virus genomic (heavy) RNA, 5' end. 8/87 110bp.

J02419 Gb(84)_vi:MTYCOAT Turnip yellow mosaic virus coat protein mRNA. 2/87 694bp.

J04373 Em(40)_vi:MTYCLCGA Gb(84)_vi:MTYCLCGA Turnip yellow mosaic virus Club Lake isolate, complete genome. 5/90 6,319bp.

K00602 Em(40)_vi:MTY2 Gb(84)_vi:MTY2 Turnip yellow mosaic virus genomic (heavy) RNA: 3' end. 6/92 694bp.

M24796 Em(40)_vi:MTYCPCDBL Gb(84)_vi:MTYCPCDBL Turnip yellow mosaic virus coat (Blue Lake isolate) protein mRNA, 3' end. 6/92 166bp

M24797 Em(40)_vi:MTYCPCDCL Gb(84)_vi:MTYCPCDCL Turnip yellow mosaic virus (Club Lake isolate) coat protein mRNA, 3' end. 6/92 169bp.

M24798 Em(40)_vi:MTYCPCDLA Gb(84)_vi:MTYCPCDLA Turnip yellow mosaic virus (Lake Albina isolate) coat protein mRNA, 3' end. 6/92 167bp.

M24799 Em(40)_vi:MTYCPCDRR Gb(84)_vi:MTYCPCDRR Turnip yellow mosaic virus (Mt.Townsend isolate) coat protein mRNA, 3' end. 7/92 196bp.

M24800 Em(40)_vi:MTYCP1HO Gb(84)_vi:MTYCP1HON Turnip yellow mosaic virus (honesty isolate) coat protein mRNA, 3' end. 6/92 178bp.

M24801 Em(40)_vi:MTYCP1RAD Gb(84)_vi:MTYCP1RAD Turnip yellow mosaic virus (Rademacher isolate) coat protein mRNA, 3' end. 7/92 178bp.

M24802 Em(40)_vi:MTYCP1TYP Gb(84)_vi:MTYCP1TYPE Turnip yellow mosaic virus (type isolate) coat protein mRNA, 3' end. 6/92 163bp.

M24803 Em(40)_vi:MTYCP2CAU Gb(84)_vi:MTYCP2CAUL Turnip yellow mosaic virus (cauliflower isolate) coat protein mRNA, 3' end. 6/92 100bp.

M24804 Em(40)_vi:MTYCP2DEN Gb(84)_vi:MTYCP2DEN Turnip yellow mosaic virus (Denmark isolate) coat protein mRNA, 3' end. 6/92 181bp.

M24805 Em(40)_vi:MTYCP2ROT Gb(84)_vi:MTYCP2ROTH Turnip yellow mosaic virus (Rothamsted isolate) coat protein mRNA, 3' end. 6/92 166bp.

M54918 Em(40)_vi:MTYCPZ Gb(84)_vi:MTYCPZ Turnip yellow mosaic virus coat protein gene, 3' end. 8/91 159bp.

M58309 Em(40)_vi:MTYTRST Gb(84)_vi:MTYTRST Turnip yellow mosaic virus (TYMV) 3' terminus tRNA-like structure. 8/91 86bp.

V01418 Em(40)_vi:TYTYM1 Gb(84)_vi:TYTYM1 Turnip yellow mosaic virus mRNA for the coat protein. 9/93 695bp.

V01419 Em(40)_vi:TYTYM2 Gb(84)_vi:TYTYM2 5' end of turnip yellow mosaic virus. 7/89 110bp.

X07441 Em(40)_vi:TYMVXX Gb(84)_vi:TYMVXX Turnip yellow mosaic virus genome. 9/93 6,318bp.

X16378 Em(40)_vi:TYMVCG Gb(84)_vi:TYMVCG Turnip yellow mosaic virus complete RNA genome. 9/93 6,318bp. 20 sequences.

*Non–genomic nucleic acid* found in the virions; is subgenomic *m*RNA (with $M_r$ 180000 coding for coat protein (Ricard *et al.*, 1977; Pleij *et al.*, 1977; Higgins *et al.*, 1978)). Sub–genomic *m*RNA found in infected cells (Hatta and Matthews, 1976).

*Virion protein(s)* one; $M_r$ 20600; coat protein. Method of preparation: Matthews (1980). Amino acid sequence: Peter *et al.* (1972).

## Replication

Genome replicates in association with chloroplasts (in vesicles in the margins of the chloroplasts, that form as an invagination of the outer chloroplast bilayer membrane and open to the cytoplasm). Replication does not depend on a helper virus.

## Cytopathology

Virions found in leaves and all parts of the host plant including the apical meristem but mainly in the leaf lamina; in cytoplasm (in electron dense material close to or replacing the endoplasmic reticulum). Inclusions absent from infected cells. Other cellular changes: the formation of vesicles deriving from the bilayer of the chloroplast membrane and the rounding of chloroplasts.

## Taxonomy and Relationships

*Tymovirus.*

Virus(es) with serologically related virions: type 1 (TYMV-1) strains, which are indistinguishable, but TYMV-2 isolates, which are distinct; kennedya yellow mosaic, dulcamara mottle and scrophularia mottle viruses are more distantly related. Nucleotide sequencing confirms these results and shows that European and Australian TYMV-1 isolates differ by about 5%. TYMV-1 and TYMV-2 genome sequences differ by around 10%.

## References


Broadbent, L. (1957). *Agric. Res. Council Rep* No. 14. Cambridge Univ. Press, Cambridge, UK.

Gibbs, A.J., Hecht-Poinar, E., Woods, R.D. and McKee, R.K. (1966). *J. gen. Microbiol.* **44**: 177.

Guy, P. and Gibbs, A.J. (1981). *Australas. Pl. Path.* **10**: 12.

Guy, P. and Gibbs, A.J. (1985). *Pl. Path.* **34**: 532

Hatta, T. and Matthews, R.E.F. (1976). *Virology* **73**: 1.

Hein, A. (1984). *Z. PflKrankh. PflPath. PflSchutz.* **91**: 549.

Higgins, T.J.V., Whitfield, P.R. and Matthews, R.E.F. (1978). *Virology* **84**: 153.

Markham, R.E.F. and Smith, K.M. (1949). *Parasitol.* **39**: 330.

Matthews, R.E.F. (1970). *CMI/AAB Descr. Pl. Viruses* No. 2, 4 pp.

Matthews, R.E.F. (1980). *CMI/AAB Descr. Pl. Viruses* No. 230, 6 pp.

Peter, R., Stehelin, D., Reinbalt, J., Collot, P. and Duranton, H. (1972). *Virology* **49**: 615.

Pleij, C.W.A., Mellema, J.R., Noort, A. and Bosch, L. (1977). *FEBS Lett.* **80**: 19.

Ricard, B., Barreau, C., Renaudin, H., Moucher, C. and Bové, J.M. (1977). *Virology* **79**: 231.

# Ullucus C *comovirus*

**Data collated by A.A. Brunt, 1986.**

*Standard acronym:* UVC.

First reported in *Ullucus tuberosus*; from plants obtained from South America; by Brunt *et al.* (1982*a*).

**Natural host range and symptoms**

*Ullucus tuberosus* — symptomless infection.

**Transmission**

Virus transmitted by mechanical inoculation; transmitted by grafting; not transmitted by seed.

**Geographical distribution**

Spreads in Bolivia and Peru.

**Experimental host range**

Several (3–9) families susceptible.

Diagnostically susceptible

*Chenopodium amaranticolor, C. quinoa* — chlorotic local lesions and systemic leaf chlorosis.
*Chenopodium murale* — chlorotic local lesions; not systemic.
*Tetragonia expansa* — local lesions; not systemic.
*Nicotiana clevelandii* — symptomless infection of inoculated leaves; not systemic.

Diagnostically insusceptible

*Nicotiana glutinosa, N. tabacum, Gomphrena globosa, Cucumis sativus, Phaseolus vulgaris.*

Maintenance and propagation hosts

*Chenopodium quinoa, Ullucus tuberosus.*

Assay hosts (**L**ocal lesions or **W**hole plants)

*Chenopodium amaranticolor* (W) *Chenopodium quinoa* (W).

Dicotyledonous Hosts *(**Susceptible** or Insusceptible)*

*Amaranthus caudatus. Arachis hypogaea. Brassica campestris* ssp. *pekinensis.* ***Chenopodium amaranticolor. Chenopodium murale. Chenopodium quinoa.*** *Cucumis sativus. Datura metel. Datura stramonium. Gomphrena globosa. Lycopersicon esculentum. Montia perfoliata. Nicandra physaloides.* ***Nicotiana clevelandii.*** *Nicotiana debneyi. Nicotiana glutinosa. Nicotiana megalosiphon. Nicotiana rustica. Nicotiana tabacum. Phaseolus vulgaris. Physalis floridana. Pisum sativum. Solanum tuberosum.* ***Tetragonia expansa.*** *Vicia faba. Vigna unguiculata.*

**Properties of particles in sap**

TIP: 70–75 °C. LIV: 45–112 days. DEP: $\log_{10}$ minus 7–8. Leaf sap contains many virions.

**Purification**

Method: Brunt *et al.* (1982*b*).

**Particle morphology**

Virions isometric; 28 nm in diameter; angular in profile; with a conspicuous capsomere arrangement.

**Physical properties**

Three sedimenting components in purified preparations; sedimentation coefficient of the fastest 116 *S* (B); of the other(s) 51 *S* (T), or 95 *S* (M). Density 1.404 g $cm^{-3}$ in CsCl (M), or 1.463 g $cm^{-3}$ in CsCl (B). $A_{260}/A_{280}$ ratio 1.58 (M), or 1.65 (B; all corrected for light scattering).

**Biochemical properties**

Virions contain 38 % nucleic acid (B), or 30 % nucleic acid (M), or 0 % nucleic acid (T); 62 % protein (B), or 70 % protein (M), or 100 % protein (T).

*Genome* consists of RNA; single–stranded; of two parts; largest (or only) genome part the larger 6.2 kb; the 2nd largest 4.5 kb.

*Virion protein(s)* two; $M_r$ of the larger 45100. $M_r$ of 2nd largest 20700.

**Taxonomy and Relationships**

*Comovirus: Comoviridae.*

Virus(es) with serologically related virions: cowpea mosaic virus is distantly related. Virus(es) with serologically unrelated virions: Andean potato mottle, bean pod mottle, bean rugose mosaic, broad bean stain, cowpea severe mosaic, broad bean true mosaic, glycine mosaic, quail pea mosaic, radish mosaic, red clover mottle and squash mosaic viruses.

**References**

Brunt, A.A., Phillips, S., Jones, R.A.C. and Kenten, R.H. (1982*a*). *Ann. appl. Biol.* **101**: 65.

Brunt, A.A., Barton, R.J., Phillips, S. and Jones, R.A.C. (1982*b*). *Ann. appl. Biol.* **101**: 73.

Brunt, A.A., Phillips, S., Barton, R.J. and Jones, R.A.C. (1986). *CMI/AAB Descr. Pl. Viruses* No. 277, 4 pp.

# Ullucus mild mottle *tobamovirus*

**Data collated by A.A.Brunt and S. Molyneux, 1986.**

*Standard acronym:* UMMV.

First reported in *Ullucus tuberosus*; from in plants from Peru; by Brunt *et al.* (1982).

**Natural host range and symptoms**

Symptoms disappear soon after infection.

*Ullucus tuberosus* — plants also naturally infected with other viruses. Inoculated plants show mottling.

**Transmission**

Transmitted by means not involving a vector. Virus transmitted by mechanical inoculation; transmitted by grafting; transmitted by contact between plants.

**Ecology and control**

Studies reported by Brunt *et al.* (1982).

**Geographical distribution**

Spreads in Peru.

**Experimental host range**

Several (3–9) families susceptible.

Diagnostically susceptible

*Datura metel, Nicotiana clevelandii, N. benthamiana* — systemic chlorosis.

*Nicotiana glutinosa, N. megalosiphon, N. rustica, N. tabacum* — chlorotic or necrotic local lesions; not systemic.

*Chenopodium capitatum* — severe systemic leaf chlorosis.

*Chenopodium amaranticolor, C. quinoa* — chlorotic rings or local lesions; not systemic.

Diagnostically insusceptible

*Capsicum frutescens, Solanum tuberosum, Phaseolus vulgaris, Vicia faba, Vigna unguiculata.*

Maintenance and propagation hosts

*Nicotiana clevelandii* and *N. benthamiana.*

Assay hosts (**L**ocal lesions or **W**hole plants)

*Chenopodium quinoa* (L), *Chenopodium amaranticolor* (L), *Chenopodium murale* (L), *Nicotiana glutinosa* (L).

Dicotyledonous Hosts *(**Susceptible** or Insusceptible)*

***Amaranthus caudatus***. *Arachis hypogaea*. *Brassica campestris* ssp. *pekinensis*. *Capsicum frutescens*. ***Chenopodium amaranticolor***. ***Chenopodium capitatum***. ***Chenopodium murale***. ***Chenopodium quinoa***. ***Cucumis sativus***. ***Datura metel***. ***Datura stramonium***. ***Gomphrena globosa***. ***Lycopersicon esculentum***. ***Montia perfoliata***. ***Nicandra physaloides***. ***Nicotiana benthamiana***. ***Nicotiana clevelandii***. ***Nicotiana debneyi***. ***Nicotiana glutinosa***. ***Nicotiana megalosiphon***. ***Nicotiana rustica***. ***Nicotiana tabacum***. ***Petunia hybrida***. *Phaseolus vulgaris*. ***Physalis floridana***. *Pisum sativum*. *Solanum tuberosum*. ***Tetragonia expansa***. *Trifolium incarnatum*. *Vicia faba*. *Vigna unguiculata*.

Sources of host-range data

Brunt *et al.* (1982); Brunt (1986).

**Properties of particles in sap**

TIP: 90–95 °C. DEP: $\log_{10}$ minus 7. Leaf sap contains many virions.

**Purification**

Method: Brunt (1986).

**Particle morphology**

Virions rod–shaped; not enveloped; usually straight; with a clear modal length; of 300 nm; 18 nm wide. Axial canal obvious. Basic helix obvious; pitch of basic helix 2.4 nm.

**Physical properties**

One sedimenting component in purified preparations; sedimentation coefficient 176 *S*. Density 1.32 g $cm^{-3}$ in CsCl (unfixed).

**Biochemical properties**

Virions contain 5 % nucleic acid; 95 % protein.

*Genome* consists of RNA; single–stranded; unipartite; largest (or only) genome part 6.4 kb.

*Virion protein(s)* one; $M_r$ 21000; coat protein. Amino acid sequence: Van Regenmortel and Brunt, in Brunt (1986).

**Cytopathology**

Virions found in all parts of the host plant; in cytoplasm.

**Taxonomy and Relationships**

*Tobamovirus*.

Virus(es) with serologically related virions: ribgrass mosaic virus. Virus(es) with serologically unrelated virions: cucumber green mottle mosaic, frangipani mosaic, odontoglossum ringspot and sunn-hemp mosaic viruses (Brunt *et al.*, 1982; Duarte, 1986).

**References**

Brunt, A.A., Phillips, S., Jones, R.A.C. and Kenten, R.H. (1982). *Ann. appl. Biol.* **101**: 65.

Brunt, A.A. (1986). In: *The Plant Viruses* Vol. 2, *The Rod-Shaped Viruses*, p. 283; eds M.H.V. van Regenmortel and H. Fraenkel-Conrat. Plenum Press, New York.

Duarte, I.M.N. (1986). *M.Sc. Thesis*, 98 pp. University of Reading, U.K.

# Ullucus mosaic *potyvirus*

**Data collated by S. Phillips, 1986.**

*Standard acronym:* UMV.

First reported in *Ullucus tuberosus*; from Peru and Bolivia; by Brunt *et al.* (1982).

**Natural host range and symptoms**

Symptoms persist.

*Ullucus tuberosus* — systemic leaf mottling and malformation.

**Transmission**

Transmitted by a vector; an insect; *Myzus persicae*; Aphididae. Transmitted in a non–persistent manner. Virus transmitted by mechanical inoculation; not transmitted by contact between plants.

**Geographical distribution**

Spreads in Bolivia and Peru.

**Experimental host range**

Several (3–9) families susceptible.

Diagnostically susceptible

*Nicotiana clevelandii* — spotty systemic chlorosis.
*N. benthamiana* — systemic chlorosis and malformation.
*N. glutinosa x clevelandii* hybrid — systemic chlorotic rings and leaf malformation.
*Physalis floridana* — systemic chlorotic mosaic.
*Chenopodium quinoa, C. amaranticolor* — chlorotic local lesions; not systemic.

Diagnostically insusceptible

*Phaseolus vulgaris, Vigna unguiculata, Trifolium incarnatum, Solanum tuberosum, Capsicum annuum.*

Maintenance and propagation hosts

*Nicotiana clevelandii, Nicotiana benthamiana, Nicotiana clevelandii x glutinosa* hybrid, *Nicotiana debneyi.*

Assay hosts (**L**ocal lesions or **W**hole plants)

*Chenopodium quinoa* (L) *Chenopodium amaranticolor* (L), *Nicotiana clevelandii* (W), and *Nicotiana benthamiana* (W).

Dicotyledonous Hosts (***Susceptible*** *or Insusceptible)*

*Brassica campestris* ssp. *pekinensis*. *Capsicum annuum*. ***Chenopodium amaranticolor***. ***Chenopodium murale***. ***Chenopodium quinoa***. *Cucumis sativus*. *Datura metel*. *Datura stramonium*. ***Gomphrena globosa***. *Ipomoea nil*. *Ipomoea setosa*. *Lycopersicon esculentum*. ***Nicotiana benthamiana***. ***Nicotiana clevelandii***. ***Nicotiana debneyi***. *Nicotiana glutinosa*. ***Nicotiana megalosiphon***. ***Nicotiana rustica***. ***Nicotiana tabacum***. ***Petunia hybrida***. *Phaseolus vulgaris*. ***Physalis floridana***. *Solanum demissum*. *Solanum tuberosum*. ***Tetragonia expansa***. *Trifolium incarnatum*. *Verbesina encelioides*. *Vigna unguiculata*.

Sources of host-range data

Brunt *et al.* (1982).

**Properties of particles in sap**

TIP: 65 °C (in *Nicotiana clevelandii* sap). LIV: 8–16 days. DEP: $\log_{10}$ minus 5. Leaf sap contains few virions.

**Purification**

Method: Brunt *et al.* (1982).

**Particle morphology**

Virions filamentous; not enveloped; usually flexuous; with a clear modal length; of 752 nm; 12 nm wide. Axial canal obscure. Basic helix obscure.

**Physical properties**

One sedimenting component in purified preparations; sedimentation coefficient 150 *S*.

**Biochemical properties**

Virions contain 5 % nucleic acid; 95 % protein.

*Genome* consists of RNA; single–stranded; unipartite; largest (or only) genome part 9 kb.

*Non–genomic nucleic acid* not found in the virions.

$M_r$ 32000; coat protein.

**Replication**

Replication does not depend on a helper virus.

**Cytopathology**

Virions found in all parts of the host plant; in cytoplasm. Inclusions present in infected cells; are pinwheels; they do not contain virions.

**Taxonomy and Relationships**

*Potyvirus: Potyviridae*.

Virus(es) with serologically related virions: alstroemeria mosaic (SDI 8), bidens mottle (SDI 6), dasheen mosaic (SDI 4), iris severe mosaic (SDI 7) and potato Y (SDI 6) viruses. Virus(es) with serologically unrelated virions: asparagus 1, bean yellow mosaic, bean common mosaic, beet mosaic, blackeye cowpea mosaic, carnation vein mosaic, clover yellow vein, datura shoestring, hyacinth mosaic, hippeastrum mosaic, iris mild mosaic, lettuce mosaic, passiflora woodiness, pepper veinal mottle, peanut mottle, sweet potato feathery mottle, narcissus yellow stripe, tobacco etch, tulip breaking, watermelon mosaic and yam mosaic viruses.

**References**

Brunt, A.A., Phillips, S., Jones, R.A.C. and Kenten, R.H. (1982). *Ann. appl. Biol.* **101**: 65.

de Silva, D.P. (1986). *M.Sc. Thesis*, 65 pp. University of Reading, U.K.

# Urd bean leaf crinkle virus

**Data collated by S.P.S. Beniwal and Y.L. Nene, 1983.**

*Synonyms:* blackgram leaf crinkle virus, bean urd leaf crinkle virus.

First reported in *Vigna mungo*; from the Pantnagar District, Nainital, India; by Kolte and Nene (1972).

**Natural host range and symptoms**

Symptoms persist.

*Cajanus cajan, Vigna aconitifolia, V. mungo, V. radiata, V. unguiculata* — leaf puckering and malformation.

**Transmission**

Transmitted by a vector; an insect; *Henosepilachna dodecastigma*; Coleoptera. Transmitted in a non–persistent manner. Virus transmitted by mechanical inoculation; not transmitted by contact between plants; transmitted by seed (18%); not transmitted by pollen.

**Geographical distribution**

Spreads in India.

**Experimental host range**

Few (<3) families susceptible.

Diagnostically susceptible

*Cucumis sativus* cv. National Pickling — systemic mosaic, rugosity, leaf malformation.

*Lagenaria cylindrica* — systemic rugosity, leaf malformation, yellowing.

*Macroptilium lathyroides, Phaseolus vulgaris* cv. Top Crop — symptomless systemic infection.

*Vigna aconitifolia, V. mungo, V. radiata, V. unguiculata* — rugosity, leaf malformation.

Diagnostically insusceptible

*Chenopodium amaranticolor, Datura stramonium.*

Maintenance and propagation hosts

*Cucumis sativus, Lagenaria cylindrica, Vigna aconitifolia, V. mungo, V. unguiculata.*

Assay hosts (**L**ocal lesions or **W**hole plants)

*Cucumis sativus* cv. National Pickling (W), *Lagenaria cylindrica* (W), *Vigna aconitifolia* (W), *V. mungo* (W), *V. unguiculata* (W).

Dicotyledonous Hosts *(**Susceptible** or Insusceptible)*

*Arachis hypogaea. Beta vulgaris.* ***Cajanus cajan.*** *Capsicum annuum. Cassia tora. Chenopodium album. Chenopodium amaranticolor. Chenopodium quinoa. Crotalaria* spectabilis. *Cucumis melo*. ***Cucumis sativus***. *Cucurbita pepo*. *Cyamopsis tetragonoloba*. *Datura stramonium*. *Glycine max*. *Gomphrena globosa*. *Gossypium hirsutum*. *Helianthus annuus*. *Lablab purpureus*. *Lathyrus odoratus*. *Lycopersicon esculentum*. ***Macroptilium lathyroides***. *Medicago sativa*. *Nicotiana glutinosa*. *Nicotiana tabacum*. *Phaseolus lunatus*. ***Phaseolus vulgaris***. *Pisum sativum*. *Sesamum indicum*. *Spinacia oleracea*. *Tetragonia expansa*. *Vicia faba*. ***Vigna radiata***. ***Vigna unguiculata***. *Vigna unguiculata* ssp. *sesquipedalis*.

Sources of host-range data

Kolte and Nene (1975); Bharathan (1980); Beniwal *et al.* (1983).

**Properties of particles in sap**

TIP: 60–70 °C. LIV: 3 days. DEP: $\log_{10}$ minus 4. Infectivity of sap not changed by treatment with di–ethyl ether. Leaf sap contains few virions.

**Purification**

Method: Beniwal and Nene: PEG and column chromatography.

**Particle morphology**

Virions isometric; not enveloped; 25–30 nm in diameter; rounded in profile; without a conspicuous capsomere arrangement.

**Physical properties**

One sedimenting component in purified preparations.

**Biochemical properties**

Virions contain 17 % nucleic acid.

*Genome* consists of RNA; single–stranded; unipartite.

**Replication**

Replication does not depend on a helper virus.

**Cytopathology**

Virions found in leaves; in cytoplasm, in nuclei, and in chloroplasts. Inclusions absent from infected cells. Other cellular changes: hypertrophy of infected cells. Mitochondria become filiform.

**Taxonomy and Relationships**

Virus(es) with serologically unrelated virions: bean pod mottle, bean rugose mosaic, bean yellow stipple, blackgram mottle, cowpea mosaic, cowpea severe mosaic, cowpea chlorotic mottle, southern bean mosaic, squash mosaic, turnip crinkle and turnip yellow mosaic viruses.

**Comments**

In addition to this beetle-transmitted crinkle virus of green gram (Kolte and Nene, 1972), two others causing crinkle diseases of green gram are known. One is transmitted by aphids (Dingra, 1975), and the other by whitefly (Narayanasamy and Jaganathan, 1974).

**References**

Beniwal, S.P.S. and Bharathan, N. (1980). *Indian Phytopath.* **33**: 600.

Beniwal, S.P.S. and Chaubey, S.N. (1979). *Seed Research* **7**: 175.

Beniwal, S.P.S, Kolte, S.J. and Nene, Y.L. (1980). *Ind. J. Mycol. Pl. Path.* **9**: 188.

Beniwal, S.P.S, Bharathan, N. and Chaubey, S.N. (1983). *Indian Phytopath.* **36**: 577.

Bhaktavatsalam, G., Nene, Y.L. and Beniwal, S.P.S (1982). *Indian Phytopath.* **35**: 683.

Bharathan, N. (1980). *M.Sc. Thesis*, G.B. Pant. Univ. Tech., Pantnagar, Nainital, U.P. India.
Dingra, K.L. (1975). *Indian Phytopath.* **28**: 80.
Kolte, S.J. and Nene, Y.L. (1972). *Indian Phytopath.* **25**: 401.
Kolte, S.J. and Nene, Y.L. (1975). *Indian Phytopath.* **28**: 430.
Narayanasamy, P. and Jaganathan, T. (1974). *Madras Agric. J.* **61**: 979.
Nene, Y.L. (1972). *G. B. Pant. Univ. Agric. Tech. Bull.* **4**: 191.

# Vallota mosaic *potyvirus*

**Data collated by A.A. Brunt and P. Jeyanandarajah, 1991.**

*Standard acronym:* ValMV.

First reported in *Vallota speciosa*; from the Netherlands; by Inouye and Hakkaart (1980).

**Natural host range and symptoms**

Symptoms persist.

*Vallota speciosa* — chlorotic leaf striping and flower breaking.

**Transmission**

Transmitted by a vector; an insect; *Myzus persicae*; Aphididae. Transmitted in a non–persistent manner. Virus transmitted by mechanical inoculation.

**Geographical distribution**

Spreads in the Netherlands. Found, but with no evidence of spread, in the U.K.

**Experimental host range**

Several (3–9) families susceptible. Experimentally infected plants mostly show chlorotic local lesions.

Diagnostically susceptible

*Vallota speciosa* — chlorotic leaf striping.

*Chenopodium amaranticolor, C. quinoa, Hyoscyamus niger* — chlorotic local lesions; not systemic.

*Tetragonia expansa* — chlorotic local lesions, green ringspots and/or green spots; not systemic.

*Gomphrena globosa* — red-edged local lesions; not systemic.

*Freesia* spp. — systemic mosaic.

*Nicotiana clevelandii, Spinacia oleracea* — symptomless infection.

Diagnostically insusceptible

*Datura stramonium, Nicotiana glutinosa, N. tabacum, Phaseolus vulgaris, Cucumis sativus.*

Maintenance and propagation hosts

*Vallota speciosa.*

Assay hosts (**L**ocal lesions or **W**hole plants)

*Chenopodium amaranticolor* (L), *C. quinoa* (L), *Hyoscyamus niger* (L), *Tetragonia expansa* (L).

Dicotyledonous Hosts *(**Susceptible** or Insusceptible)*

*Beta vulgaris.* ***Chenopodium amaranticolor.*** ***Chenopodium quinoa.*** *Cucumis sativus. Cucurbita moschata. Cucurbita pepo. Datura stramonium. Glycine max.* ***Gomphrena globosa.*** ***Hyoscyamus niger.*** *Lycopersicon esculentum.* ***Nicotiana*** clevelandii. *Nicotiana glutinosa. Nicotiana rustica. Nicotiana tabacum. Petunia hybrida. Phaseolus vulgaris. Physalis floridana. Pisum sativum. Raphanus sativus. Sesamum indicum.* ***Spinacia oleracea. Tetragonia expansa.*** *Trifolium incarnatum. Vicia faba. Vigna unguiculata* ssp. *sesquipedalis. Zinnia elegans.*

Sources of host-range data

Inouye and Hakkaart (1980).

**Properties of particles in sap**

TIP: 60–65 °C. LIV: 4–8 days. DEP: $\log_{10}$ minus 4–5. Leaf sap contains few virions.

**Particle morphology**

Virions filamentous; not enveloped; usually flexuous; with a clear modal length; of 750 nm; 13 nm wide. Axial canal obscure. Basic helix obscure.

**Cytopathology**

Virions found in leaves and mesophyll; in cytoplasm. Inclusions present in infected cells; are pinwheels; they do not contain virions.

**Taxonomy and Relationships**

*Potyvirus: Potyviridae.*

**References**

Inouye, N. and Hakkaart, F.A. (1980). *Neth. J. Pl. Path.* **86**: 265.

# Vanilla mosaic *potyvirus*

**Data collated by F.W. Zettler and G.C. Wisler, 1986.**

First reported in *Vanilla tahitensis*; from Huahine, Raiatea, Tahaa and Tahiti, French Polynesia; by Wisler *et al.* (1986).

**Natural host range and symptoms**

Symptoms persist.

*Vanilla tahitensis, Vanilla pompona* — leaf mosaic and malformation.

**Transmission**

Transmitted by a vector; an insect; *Myzus persicae*; Aphididae. Transmitted in a non–persistent manner.

**Geographical distribution**

Found, but with no evidence of spread, in French Polynesia on the islands of Huahine, Raiatea, Tahaa and Tahiti.

Diagnostically susceptible

*Vanilla pompona* — no symptoms in leaves inoculated by aphids but then systemic terminal necrosis and mosaic and malformation.

Dicotyledonous Hosts *(**Susceptible** or Insusceptible)*

VanMV.

**Particle morphology**

Virions filamentous; not enveloped; usually flexuous; with a clear modal length; of 767 nm. Axial canal obscure. Basic helix obscure.

### Cytopathology

Virions found in mesophyll; in cytoplasm. Inclusions present in infected cells; are pinwheels and unusual in shape; scrolls and laminated aggregates.

### Taxonomy and Relationships

*Potyvirus: Potyviridae.*

Virus(es) with serologically related virions: bidens mottle and tobacco etch viruses are serologically related to, but distinct from, vanilla mosaic potyvirus in SDS immunodiffusion tests. In tests with dasheen mosaic virus, and its homologous antiserum, bands of precipitate formed that were fully confluent with those of the vanilla potyvirus.

### References

Wisler, G.C., Zettler, F.W. and Mu, L. (1986). *Phytopathology* **76**: 1091.
Wisler, G.C., Zettler, F.W. and Mu, L. (1987). *Amer. Orchid Soc. Bull.* **56**: 381.
Wisler, G.C., Zettler, F.W. and Mu, L. (1987). *Plant Dis.* **71**: 1125.

## Vanilla necrosis *potyvirus*

**Data collated by M.N. Pearson and S.P. Pone, 1989.**

First reported in *Vanilla fragrans*; from Tonga; by Pearson and Pone (1988).

### Natural host range and symptoms

Symptoms persist (but symptoms patchy).

*Vanilla fragrans* — chlorotic blotching or streaking and malformed margins of tip leaves which, as they age, develop brown scab-like and necrotic local lesions, which also form on the vines causing die-back.

### Transmission

Transmitted by a vector; possibly an insect; *Aphis gossypii* and *Myzus persicae*; Aphididae. Transmitted in a non–persistent manner. Virus transmitted by mechanical inoculation; transmitted by grafting.

### Geographical distribution

Spreads in Fiji and Tonga.

### Experimental host range

Several (3–9) families susceptible. Experimentally infected plants mostly show systemic mosaics, mottles, ringspots or necrosis and leaf malformation.

Diagnostically susceptible

*Nicotiana benthamiana* — systemic vein chlorosis and malformed leaves.
*Nicotiana clevelandii* — mild systemic mottling.
*Chenopodium amaranticolor* — necrotic local lesions; not systemic.
*C. quinoa* — diffuse chlorotic local lesions; not systemic.

Maintenance and propagation hosts

*Nicotiana benthamiana, N. clevelandii.*

Assay hosts (**L**ocal lesions or **W**hole plants)

*Chenopodium amaranticolor* (L), *C. quinoa* (L).

Dicotyledonous Hosts *(**Susceptible** or Insusceptible)*

*Cassia occidentalis. Cassia tora.* ***Chenopodium amaranticolor****. Chenopodium murale.* ***Chenopodium quinoa****. Datura stramonium. Gomphrena globosa.* ***Nicotiana*** benthamiana. ***Nicotiana clevelandii***. *Nicotiana glutinosa*. *Nicotiana megalosiphon*. *Nicotiana tabacum*. *Phaseolus vulgaris*. *Vigna unguiculata*.

Monocotyledonous Hosts *(**Susceptible** or Insusceptible)*

*Colocasia esculenta*.

Additional hosts

*Chenopodium amaranticolor, C. quinoa* and *Nicotiana clevelandii* only successfully infected from *Nicotiana benthamiana* but not directly from *Vanilla fragrans*.

Sources of host-range data

Pearson *et al.* (1990).

**Properties of particles in sap**

TIP: 58–60 °C. LIV: 1–2 days. DEP: $\log_{10}$ minus 4–5. Leaf sap contains few virions.

**Purification**

Method: Pearson *et al.* (1990).

**Particle morphology**

Virions filamentous; not enveloped; usually flexuous; with a clear modal length; of 775 nm. Axial canal obscure. Basic helix obscure.

**Physical properties**

One sedimenting component in purified preparations.

*Virion protein(s)* one; $M_r$ 32700; coat protein. Method of preparation: Pearson *et al.* (1990).

**Cytopathology**

Virions found in leaves and mesophyll; in cytoplasm. Inclusions present in infected cells; are pinwheels; they do not contain virions.

**Taxonomy and Relationships**

*Potyvirus: Potyviridae*.

Virus(es) with serologically unrelated virions: alstroemeria mosaic, asparagus I, bean yellow mosaic, commelina Y, dasheen mosaic, hippeastrum mosaic, iris mild mosaic, iris severe mosaic, narcissus late season yellows, narcissus degeneration, narcissus yellow stripe, nerine yellow stripe, potato Y, turnip mosaic, tulip breaking, watermelon mosaic 2, yam mosaic and vanilla mosaic viruses.

**References**

Pearson, M.N. and Pone, S.P. (1988). *Australas. Pl. Path.* **17**: 59.

Pearson, M.N., Brunt, A.A. and Pone, S.P. (1990). *J. Phytopath.* **128**: 46.

# Velvet tobacco mottle *sobemovirus*

**Data collated by J.W. Randles, 1986.**

*Synonyms:* tobacco velvet mottle virus. *Standard acronym:* VTMoV.

First reported in *Nicotiana velutina*; from Cobblers Sandhill and Innamincka, northeastern South Australia; by Randles *et al.* (1981).

**Natural host range and symptoms**

Symptoms persist.

*Nicotiana velutina* — systemic mosaic and leaf malformation.

**Transmission**

Transmitted by a vector; an insect; *Cyrtopeltis nicotianae*: Miridae. Transmitted in a semi–persistent manner ( atypically). Virus does not multiply in the vector; does not require a helper virus for vector transmission (Gibb and Randles, 1988); transmitted by mechanical inoculation; not transmitted by seed.

**Geographical distribution**

Spreads in Australia (Randles *et al.*, 1981; G. Behncken and J.W. Randles, unpublished data; Horton, 1981).

**Experimental host range**

Few (<3) families susceptible.

Diagnostically susceptible

*Nicotiana clevelandii* — chlorotic to necrotic local lesions, then systemic vein yellowing and mosaic.

*Nicotiana glutinosa, N. edwardsonii* — mild systemic mosaic.

*Nicotiana velutina* — systemic vein yellowing and mosaic.

Diagnostically insusceptible

*Solanum nodiflorum.*

Maintenance and propagation hosts

*Nicotiana clevelandii.*

Assay hosts (**L**ocal lesions or **W**hole plants)

*Nicotiana clevelandii* (L).

Dicotyledonous Hosts *(**Susceptible** or Insusceptible)*

***Nicotiana clevelandii***. ***Nicotiana*** × ***edwardsonii***. ***Nicotiana glutinosa***. *Solanum nodiflorum.*

Comments on host-range

An isolate of velvet tobacco mottle virus freed from the satellite RNA did not cause local lesions in *Nicotiana clevelandii* (Francki *et al.*, 1986).

Sources of host-range data

Francki *et al.* (1985).

**Properties of particles in sap**

TIP: 70 °C. Leaf sap contains many virions.

**Purification**

Method: Randles *et al.* (1981).

**Particle morphology**

Virions isometric; not enveloped; *c.* 30 nm in diameter; rounded in profile; without a conspicuous capsomere arrangement.

**Physical properties**

One sedimenting component in purified preparations; sedimentation coefficient 115 *S*. Density 1.37 g $cm^{-3}$ in CsCl.

**Biochemical properties**

Virions contain 20 % nucleic acid; 80 % protein; 0 % lipid.

*Genome* consists of RNA; single–stranded; linear; unipartite; largest (or only) genome part 4.2 kb. Genomic nucleic acid isolated by Gould (1981); Randles *et al.* (1981).

Infectivity retained when deproteinised with proteases; retained when deproteinised with phenol or detergent. Nucleotide sequence references: Randles *et al.* (1981); Gould (1981); Haseloff and Symons (1982); Keese and Symons (1986).

*Non–genomic nucleic acid* found in the virions; is subgenomic *m*RNA (*c.*1.77 kb for RNA1a and *c.*0.71 kb for RNA1b, and RNA2 which is of 366–367 nucleotides and has a circular, viroid-like structure, and RNA3, which is the same size and sequence as RNA2, but linear).

*Virion protein(s)* one (in native), or three (in dissociated preparations); $M_r$ 37000; coat protein. $M_r$ of 2nd largest 33000. $M_r$ of 3rd largest 31500. Method of preparation: Chu and Francki (1983); Randles *et al.* (1981).

**Replication**

Replication does not depend on a helper virus.

**Cytopathology**

Virions found in all parts of the host plant; in cytoplasm, in nuclei, and in cell vacuoles. Inclusions present in infected cells; are membranous bodies.

**Taxonomy and Relationships**

*Sobemovirus.*

Virus(es) with serologically related virions: solanum nodiflorum mottle virus (SDI 1–2). Virus(es) with serologically unrelated virions: lucerne transient streak, subterranean clover mottle, southern bean mosaic, broad bean mottle, carnation ringspot, cymbidium ringspot, galinsoga mosaic, glycinè mottle, narcissus tip necrosis, red clover necrotic mosaic, saguaro cactus, tomato bushy stunt and turnip crinkle viruses (Randles *et al.*, 1981; Francki *et al.*, 1983; Tremaine and Hamilton, 1983).

Velvet tobacco mottle virus (VTMoV) may be distinguished from solanum nodiflorum mottle virus (SNMV) in host tests; *Nicotiana velutina* and *N. glutinosa* are infected systemically by VTMoV, but not by SNMV, whereas *Solanum nodiflorum* is infected by SNMV but not VTMoV (Randles *et al.*, 1981; Greber, 1981). VTMoV is serologically unrelated to two other viruses which have satellite RNAs, namely lucerne transient streak and subterranean clover mottle viruses (Tien *et al.*, 1981; Francki *et al.*, 1983); these satellites differ in size and sequence from that of VTMoV (Keese *et al.*, 1983; Francki *et al.*, 1985), but those of both VTMoV and SNMV replicate in plants infected with lucerne transient streak virus (Jones and Mayo, 1983; 1985). Best tests for diagnosis: VTMoV has virions serologically related to those of SNMV, which has a genome with 20–50% homology (Gould and Hatta, 1981), but it fails to infect *Solanum nodiflorum* and has a smaller satellite (Haseloff and Symons, 1982).

**Comments**

The virus occurs in gut, haemolymph and faeces, but not salivary glands, of its vector.

**References**

Chu, P.W.G. and Francki, R.I.B. (1983). *Virology* **129**. 350.

Francki, R.I.B., Randles, J.W., Hatta, T., Davies, C. and Chu, P.W.G. (1983). *Pl. Path.* **32**: 47.

Francki, R.I.B., Randles, J.W., Chu, P.W.G., Rohozinski, J. and Hatta, T. (1985). In: *Subviral Pathogens of Plants and Animals, Viroids and Prions*, pp. 265–297; ed. K. Maramorosch and J.J. McKelvey. Academic Press, New York.

Francki, R.I.B., Grivell, C.J. and Gibb, K.S. (1986). *Virology* **148**. 381.

Gibb, K.S. and Randles, J.W. (1988). *Ann. appl. Biol.* **112**: 427.

Gibb, K.S. and Randles, J.W. (1990). *Ann. appl. Biol.* **116**: 513.

Greber, R.S. (1981). *Aust. J. biol. Sci.* **34**: 369.

Gould, A.R. (1981). *Virology* **108**: 123.
Haseloff, J. and Symons, R.H. (1982). *Nucl. Acids Res.* **10**: 3681.
Horton, P. (1981). *J. Adelaide Bot. Gard.* **3**: 1.
Jones, A.T. and Mayo, M.A. (1983). *J. gen. Virol.* **64**: 1771.
Jones, A.T. and Mayo, M.A. (1985). *Rep. Scottish Crop Res. Inst.* 1984, p.190.
Keese, P. and Symons, R.H. (1986). In: *Viroids and Viroid-Like Pathogens*, p. 1; ed. J.S. Semancik. CRC Press, Boca Raton, Florida.
Keese, P., Bruening, G. and Symons, R.H. (1983). *FEBS Lett.* **159**: 185.
Randles, J.W. and Francki, R.I.B. (1986). *CMI/AAB Descr. Pl. Viruses* No. 317, 5 pp.
Randles, J.W., Davies, C., Hatta, T. and Francki, R.I.B. (1981). *Virology* **108**: 111.
Tien, P., Davies, C., Hatta, T. and Francki, R.I.B. (1981). *FEBS Lett.* **132**: 353.
Tremaine, J.H. and Hamilton, R.I. (1983). *CMI/AAB Descr. Pl. Viruses* No. 274, 6 pp.

## Velvet tobacco mottle virus viroid-like satellite RNA

*Sequence database accession code(s):*
J02439 Em(40)_vi:VTMOVR2 Gb(84)_vi:MTVR2 Velvet tobacco mottle virus circular viroid-like RNA (RNA 2). 4/90 366bp. 1 sequence.

**Taxonomy and Relationships**

Satellite RNA.

## Vicia *alphacryptovirus*

**Data collated by A.J. Cockbain, 1982. Revised 1983.**

*Standard acronym:* VCV.

First reported in *Vicia faba*; from Hertfordshire, U.K; by Kenten *et al.* (1978).

**Natural host range and symptoms**

*Vicia faba* — symptomless.

**Transmission**

Virus not transmitted by mechanical inoculation; not transmitted by contact between plants; transmitted by seed; transmitted by pollen to the seed.

**Geographical distribution**

Found, but with no evidence of spread, in Germany, the Netherlands and the U.K.

**Experimental host range**

Few (<3) families susceptible.

Maintenance and propagation hosts

*Vicia faba.*

Dicotyledonous Hosts *(**Susceptible** or Insusceptible)*

***Vicia faba. Vigna unguiculata.***

**Properties of particles in sap**

Leaf sap contains few virions.

**Purification**

Method: Cockbain: grind frozen tissue in 0.5M phosphate buffer, pH 8.0 containing 0.01M EDTA and 0.4% 2-mercaptoethanol with chloroform/butanol (1 g leaf: 2 ml buffer: 0.5 ml chloroform: 0.5 ml N-butanol) followed by differential centrifugation.

**Particle morphology**

Virions isometric; not enveloped; 30 nm in diameter; angular in profile; without a conspicuous capsomere arrangement.

**Physical properties**

Density 1.38 and 1.39 g $cm^{-3}$ in CsCl.

**Taxonomy and Relationships**

*Alphacryptovirus: Partitiviridae.*

Virus(es) with serologically unrelated virions: bean leaf roll, broad bean mottle, broad bean true mosaic, broad bean wilt, carnation cryptic, crimson clover latent, pea enation mosaic and southern bean mosaic viruses.

**References**

Kenten, R.H., Cockbain, A.J. and Woods, R.D. (1978). *Rep. Rothamsted Exp. Stn.* 1977. p.222.

# Vigna sinensis mosaic (?) rhabdovirus

**Data collated by A.A. Brunt, 1989.**

*Synonyms:* cowpea mosaic virus.

First reported in *Vigna unguiculata*; from Brazil; by Kitajima and Costa (see Peters, 1981; Francki *et al.*, 1981).

**Natural host range and symptoms**

*Vigna unguiculata.*

**Geographical distribution**

Spreads in Brazil.

**Taxonomy and Relationships**

Unassigned rhabdovirus.

**References**

Francki, R.I.B., Kitajima, E.W. and Peters, D. (1981). In: *Handbook of Plant Virus Infection and Comparative Diagnosis*, ed. E. Kurstak. Elsevier/North Holland Biomedical Press, Amsterdam.

Peters, D. (1981). *CMI/AAB Descr. Pl. Viruses* No. 244, 6pp.

# Viola mottle *potexvirus*

**Data collated by V. Lisa, 1986; G.R. Johnstone, 1987.**

*Standard acronym:* VMV.

First reported in *Viola odorata*; from Liguria, Italy; by Lisa and Dellavalle (1977).

**Natural host range and symptoms**

Symptoms persist.

*Viola odorata* — leaf mottling, white stripes on petals.

**Transmission**

Transmitted by means not involving a vector. Virus transmitted by mechanical inoculation.

**Geographical distribution**

Spreads in Australia, France, and Italy.

**Experimental host range**

Many (>9) families susceptible.

Diagnostically susceptible

*Amaranthus caudatus* — necrotic local lesions; systemic mottle.
*Chenopodium amaranticolor* — chlorotic local lesions; systemic mosaic.
*Gomphrena globosa* — necrotic local lesions; systemic infection usually latent.
*Nicotiana clevelandii* — symptomless local infection, not systemic.

Diagnostically insusceptible

*Datura stramonium, Nicotiana tabacum, Nicotiana glutinosa, Cucumis sativus.*

Maintenance and propagation hosts

*Chenopodium quinoa.*

Assay hosts (**L**ocal lesions or **W**hole plants)

*Chenopodium quinoa* (L), *Gomphrena globosa* (L).

Dicotyledonous Hosts (***Susceptible*** *or Insusceptible)*

***Amaranthus caudatus***. ***Antirrhinum majus***. ***Beta vulgaris***. ***Chenopodium album***. ***Chenopodium amaranticolor***. ***Chenopodium murale***. ***Chenopodium quinoa***. *Cucumis sativus. Datura stramonium.* ***Dianthus barbatus***. ***Gomphrena globosa***. ***Lactuca sativa***. *Lupinus albus.* ***Nicotiana clevelandii***. *Nicotiana glutinosa. Nicotiana tabacum. Ocimum basilicum. Petunia hybrida. Phaseolus vulgaris.* ***Primula malacoides***. ***Solanum melongena***. ***Spinacia oleracea***. ***Tetragonia expansa***. ***Trifolium pratense***. *Vigna unguiculata. Zinnia elegans.*

Additional hosts

***Amaranthus hybridus***, *Callistephus chinensis, Hydrangea macrophylla*, ***Primula malacoides, Primula obconica***, *Tagetes patulus, Trigonella foenum-graecum*, ***Viola tricolor***.

Sources of host-range data

Lisa and Dellavalle (1977).

**Properties of particles in sap**

TIP: 95 °C. LIV: 240 days (at 20°C). DEP: $\log_{10}$ minus 9–10. Leaf sap contains many virions.

**Purification**

Method: Lisa and Dellavalle (1977); Richardson *et al.* (1981).

**Particle morphology**

Virions filamentous; not enveloped; usually flexuous; with a clear modal length; of 478–495 nm; *c.* 13 nm wide. Axial canal faint indication obvious. Basic helix obscure; pitch of basic helix 3.5 nm.

**Physical properties**

One sedimenting component in purified preparations. Density 1.31 g $cm^{-3}$ in CsCl.

**Biochemical properties**

Virions contain 6 % nucleic acid; 94 % protein.

*Genome* consists of RNA; single–stranded; linear; unipartite; largest (or only) genome part 7.2 kb. Genomic nucleic acid isolated by Lisa *et al.* (1982). Base composition 22.8 % G; 26.4 % A; 30.3 % C; 20.5 % U.

*Non–genomic nucleic acid* not found in the virions.

*Virion protein(s)* one; $M_r$ 21473; coat protein. Amino acid sequence: Bendena and Mackie (1986). Amino acid composition: M.N. Short (unpublished data) *cf.* Lisa *et al.* (1982).

*Virus–coded non–virion proteins* isolated (Bendena and Mackie (1986)); one protein found. $M_r$ 160000.

**Replication**

Replication does not depend on a helper virus.

**Cytopathology**

Virions found in mesophyll, epidermis and all parts of the host plant; in cytoplasm. Inclusions absent from infected cells.

**Taxonomy and Relationships**

*Potexvirus.*

Virus(es) with serologically related virions: cactus virus X (strains B5 and 884), and hydrangea ringspot virus (Lisa and Dellavalle, 1977) are distantly related as are foxtail mosaic virus (Short, 1981), nerine virus X (*Agapanthus* isolate; Phillips and Brunt, 1980) and tulip virus X (Mowat, 1979). Virus(es) with serologically unrelated virions: cactus virus X (strains CC10 and K11), clover yellow mosaic, narcissus mosaic, papaya mosaic, pepino mosaic, white clover mosaic, cymbidium mosaic and potato virus X viruses.

Best tests for diagnosis: viola mottle virus is often mixed with cucumber mosaic virus in naturally-infected violets, but these can be easily distinguished by electron microscopy, or better by ISEM. The ability of viola mottle virus to infect *Chenopodium amaranticolor* also helps.

**References**


Bendena, W.G. and Mackie, G.A. (1986). *Virology* **153**: 220.

Lisa, V. and Dellavalle, G. (1977). *Phytopath. Z.* **89**: 82.

Lisa, V., Boccardo, G. and Milne, R.G. (1982). *CMI/AAB Descr. Pl. Viruses* No. 247, 4 pp.

Mowat, W.P. (1979). *Rep. Scottish Hort. Res. Inst.* 1978, p.126.

Phillips, S. and Brunt, A.A. (1980). *Acta Hort.* **110**: 65.

Richardson, J.F., Tollin, P. and Bancroft, J.B. (1981). *Virology* **112**: 34.

Short, N. (1981). *Rep. John Innes Inst.* 1980, p.126.


# Voandzeia distortion mosaic *potyvirus*

**Data collated by A.A. Brunt, 1990.**

First reported in *Voandzeia subterranea*; from the Ivory Coast; by Fauquet and Thouvenel (1987).

**Natural host range and symptoms**

*Voandzeia subterranea* — severe leaf chlorosis, malformation, and stunting.

**Transmission**

Transmitted by a vector; an insect; *Aphis craccivora* and *A. spiraecola*; Aphididae. Transmitted in a non–persistent manner. Virus transmitted by mechanical inoculation; not transmitted by seed.

**Geographical distribution**

Spreads in Cote d'Ivoire.

Diagnostically susceptible

*Voandzeia subterranea* — conspicuous leaf chlorosis and malformation.

**Properties of particles in sap**

TIP: 50–55 °C. LIV: 14 days (at 4°C, or a few hours at 24°C). DEP: $\log_{10}$ minus 2–3. Leaf sap contains few virions.

**Purification**

Method: Fauquet and Thouvenel (1987).

**Particle morphology**

Virions filamentous; not enveloped; usually flexuous; with a clear modal length; of 744 nm; 14 nm wide. Axial canal obscure. Basic helix obscure.

**Physical properties**

$A_{260}/A_{280}$ ratio 1.14.

**Biochemical properties**

Virions contain 5 % nucleic acid; 95 % protein.

*Virion protein(s)* one; $M_r$ 34000; coat protein. Amino acid composition: Fauquet and Thouvenel (1987).

**Taxonomy and Relationships**

*Potyvirus: Potyviridae.*

Virus(es) with serologically related virions: canavalia mosaic, groundnut eyespot, guinea grass mosaic, passionfruit ringpspot, pepper veinal mottle and yam mosaic viruses.

**References**

Fauquet, C. and Thouvenel, J.-C. (1987). In: *Plant Viral Diseases in the Ivory Coast*, p. 173. ORSTOM, Paris.

# Voandzeia necrotic mosaic *tymovirus*

**Data collated by A.A. Brunt and A. Monsarrat, 1988.**

*Standard acronym:* VNMV.

First reported in *Voandzeia subterranea*; from the Ivory Coast; by Fauquet *et al.* (1981).

**Natural host range and symptoms**

Symptoms persist.

*Voandzeia subterranea* — mosaic, necrosis and malformation of leaves, occasionally lethal.

**Transmission**

Possibly transmitted by a vector. Virus transmitted by mechanical inoculation; transmitted by grafting; not transmitted by seed.

**Ecology and control**

Studies reported by Monsarrat *et al.* (1984).

**Geographical distribution**

Spreads in Cote d'Ivoire.

**Experimental host range**

Few (<3) families susceptible.

Diagnostically susceptible

*Voandzeia subterranea* — systemic chlorosis, veinal necrosis and leaf malformation.

*Dolichos lablab, Vigna unguiculata* ssp. *sinensis, Vigna unguiculata* ssp. *unguiculata* — chlorotic local lesions, then systemic leaf chlorosis.

*Chenopodium amaranticolor* — necrotic local lesions; not systemic.

Diagnostically insusceptible

*Arachis hypogaea, Phaseolus vulgaris, Nicotiana benthamiana, Nicotiana glutinosa, Abelomoschus esculentus.*

Maintenance and propagation hosts

*Voandzeia subterranea, Dolichos lablab, Vigna unguiculata* ssp *sinensis, Vigna unguiculata* ssp *unguiculata.*

Assay hosts (**L**ocal lesions or **W**hole plants)

*Chenopodium amaranticolor* (L).

Dicotyledonous Hosts *(**Susceptible** or Insusceptible)*

*Arachis hypogaea.* ***Chenopodium amaranticolor.*** ***Dolichos biflorus.*** *Nicotiana benthamiana. Nicotiana glutinosa. Phaseolus vulgaris.* ***Vigna unguiculata. Vigna unguiculata*** ssp. ***cylindrica. Vigna unguiculata*** ssp. ***sesquipedalis.***

Additional hosts

An unspecified 82 species in Amaranthaceae, Chenopodiaceae, Cruciferae, Leguminosae, Malvaceae, Passifloraceae and Solanaceae are not infected.

Sources of host-range data

Monsarrat *et al.* (1984).

**Properties of particles in sap**

TIP: in *Voandzeia subterranea* sap 70 °C. LIV: 2 days at 10°C and 10 days (at 4°C or -20°C). DEP: $\log_{10}$ minus 2. Leaf sap contains few virions.

**Purification**

Method: Monsarrat *et al.* (1984).

**Particle morphology**

Virions isometric; not enveloped; 28 nm in diameter; angular in profile; without a conspicuous capsomere arrangement.

**Physical properties**

Two sedimenting components in purified preparations; sedimentation coefficient of the fastest 113 *S* (B); of the other(s) 51 *S* (T). Density 1.45 g $cm^{-3}$ in CsCl (B), or 1.29 g $cm^{-3}$ in CsCl (T). Isoelectric point pH 4.65. $A_{260}/A_{280}$ ratio 1.75 (B), or 0.64 (T).

**Biochemical properties**

Virions contain 35 % nucleic acid (B); 65 % protein (B).

*Genome* consists of RNA; single–stranded.

*Virion protein(s)* one; $M_r$ 20000; coat protein.

**Cytopathology**

Virions found in leaves and mesophyll; in cytoplasm. Inclusions present in infected cells; are crystals in the cytoplasm; they contain virions. Other cellular changes: the occurence of peripheral vesicles in chloroplasts bounded by a double membrane.

**Taxonomy and Relationships**

*Tymovirus*.

Virus(es) with serologically related virions: kennedya yellow mosaic, desmodium yellow mottle, okra mosaic and turnip yellow mosaic viruses. Virus(es) with serologically unrelated virions: cacao yellow mosaic and clitoria yellow vein viruses.

**References**

Fauquet, C., Monsarrat, A. and Thouvenel, J.-C. (1981). *Abstr. 5th. Congr. Virol., Strasbourg*, 1981, p. 273.

Fauquet, C., Monsarrat, A. and Thouvenel, J.-C. (1984). *CMI/AAB Descr. Pl. Viruses* No. 279, 4 pp.

Monsarrat, A., Fauquet, C. and Thouvenel, J.-C. (1984). *C.r. hebd. Seanc. Acad. Sci. Paris, Series III* **299**: 53.

# Watercress yellow spot virus

**Data collated by J.A. Walsh, 1991.**

*Synonyms:* virus des taches jaunes du cresson.

First reported in *Rorippa nasturtium-aquaticum*; from France; by Spire (1962).

**Natural host range and symptoms**

Symptoms vary cyclically over a few weeks, vary seasonally, and disappear soon after infection (sometimes). Symptoms chlorotic spotting and blotching.

*Rorippa nasturtium-aquaticum* — chlorotic spotting and blotching, mainly on leaf veins, some leaves malformed and maybe totally chlorotic.

**Transmission**

Transmitted by probably a vector; probably a fungus; probably *Spongospora subterranea* f.sp. *Nasturtium*; Plasmodiophorales. Not transmitted by *Myzus persicae*. Virus transmitted by mechanical inoculation (with great difficulty).

**Ecology and control**

Studies reported by Spire (1962); Walsh *et al.* (1989*a*); Walsh (1990); Walsh and Phelps (1991).

**Geographical distribution**

Spreads in France and the UK (Spire, 1962; Walsh *et al.*, 1989*a*).

**Experimental host range**

Few (<3) families susceptible. Experimentally infected plants mostly show necrotic local lesions, then indistinct chlorotic blotches or ringspots.

Diagnostically susceptible

*Rorippa nasturtium-aquaticum* — spotting and blotching, but less severe than in naturally infected plants.

*Chenopodium quinoa* — necrotic local lesions, not systemic.

*Nicotiana clevelandii* — indistinct chlorotic blotches, sometimes ringspots, not systemic.

*Brassica juncea* cv. Tendergreen — symptomless.

*Nicotiana glutinosa* — ringspots, not systemic.

*Brassica napus* ssp. *oleifera* — cv. Mikado — pinpoint necrotic local lesions.

*Barbarea verna* — symptomless.

*Chenopodium amaranticolor* — necrotic local lesions, not systemic.

Diagnostically insusceptible

*Brassica campestris* ssp. *pekinensis, Lactuca sativa, Brassica campestris*, ssp. *rapa* cv. Just Right, *Nicotiana tabacum* cv. White Burley.

Assay hosts (**L**ocal lesions or **W**hole plants)

*Chenopodium quinoa* (L), *C. amaranticolor* (L), *Nicotiana clevelandii* (L).

Dicotyledonous Hosts ***(Susceptible** or Insusceptible)*

*Brassica campestris* ssp. *pekinensis*. ***Brassica juncea***. ***Chenopodium amaranticolor***. ***Chenopodium quinoa***. ***Nicotiana benthamiana***. ***Nicotiana clevelandii***. ***Nicotiana glutinosa***. *Nicotiana tabacum*.

Sources of host-range data

Walsh *et al.* (1989*b*).

**Properties of particles in sap**

TIP: 65 °C. LIV: less than 1 days. DEP: $\log_{10}$ minus 2–3. Infectivity of sap decreased by treatment with di–ethyl ether. Leaf sap contains few virions. Electron microscopy: ISEM often required to see virions in leaf sap.

**Purification**

Method: Walsh *et al.* (1989*a*).

**Particle morphology**

Virions isometric; not enveloped; 27 nm in diameter (Spire, 1962), or 37–38 nm in diameter (Walsh *et al.*, 1989*a*); rounded in profile; with a conspicuous capsomere arrangement (sometimes).

**Physical properties**

Probably one sedimenting component in purified preparations; sedimentation coefficient 133.8 *S*. Density 1.335 g $cm^{-3}$ in CsCl. $A_{260}/A_{280}$ ratio 1.53–1.66.

**Biochemical properties**

Virions contain 18 % nucleic acid.

*Genome* possibly consists of RNA; single–stranded; possibly linear.

*Virion protein(s)* one; $M_r$ 43500; coat protein.

**Cytopathology**

Virions found in roots and sometimes in leaves.

**Taxonomy and Relationships**

Shares some properties with tombusviruses, including the sedimentation coefficient of the virions and the coat protein molecular weight, but its virions are unstable, and are larger than those of tombusviruses. Best tests for diagnosis: sero-diagnosis by ELISA, ISEM and gel diffusion.

**Comments**

Watercress chlorotic leafspot, although initially attributed to virus infection (Tomlinson and Hunt, 1987), is now considered to be induced by a virus-like agent (Tomlinson, 1988).

**References**

Spire, D. (1962). *Ann. Epiphyt.* **13**: 39.

Tomlinson, J.A. and Hunt, J. (1987). *Ann. appl. Biol.* **110**: 75.

Tomlinson, J.A. (1988). In: *Viruses with Fungal Vectors*; ed. J.I. Cooper and M.J.C. Asher. Wellesbourne, U.K., Association of Applied Biologists.

Walsh, J.A. (1990). *First Symp. of Int. Working Group on Plant Viruses with Fungal Vectors*, German Phytomedical Society. Ulmer, Stuttgart, 173

Walsh, J.A. and Phelps, K. (1991). *Pl. Path.* in press.

Walsh, J.A., Clay, C.M. and Miller, A. (1989*a*). *EPPO Bull.* **19**: 463.

Walsh, J.A., Clay, C.M., Miller, A. and Rowe, J.C. (1989*b*). *Aspects Appl. Biol.* **22**: 101.

# Watermelon chlorotic stunt bigeminivirus

**Data collated by D.G.A. Walkey and P. Jones, 1989.**

*Standard acronym:* WmCSV.

First reported in *Citrullus vulgaris*; from the Peoples Democratic Republic of South Yemen; by Jones *et al.* (1987).

**Natural host range and symptoms**

Symptoms persist.

*Citrullus vulgaris, C. colocynthis* — crinkling and chlorotic mottling of young leaves; severe stunting.

**Transmission**

Transmitted by a vector; an insect; probably *Bemisia tabaci*, but this has still to be confirmed; Aleyrodidae. Virus not transmitted by mechanical inoculation; transmitted by grafting.

**Ecology and control**

Studies reported by Jones *et al.* (1988); Walkey *et al.* (1989).

**Geographical distribution**

Spreads in Saudi Arabia (possibly), Yemen (the north and south).

**Experimental host range**

Few (<3) families susceptible. Experimentally infected plants mostly show systemic mosaic, mottling, ringspotting or necrosis.

Diagnostically susceptible

*Citrullus vulgaris* — systemic chlorotic mottling, crinkling and malformation of tip leaves.

Maintenance and propagation hosts

*Citrullus vulgaris* cv. Sugar Baby, Charleston Grey and other cultivars.

Dicotyledonous Hosts *(**Susceptible** or Insusceptible)*

***Citrullus lanatus***. *Cucumis melo*. *Cucumis sativus*. *Cucurbita pepo*.

Additional hosts

*Citrullus colocynthis* and *Coccinea grandis*.

Comments on host-range

Has only been graft transmitted to *Citrullus vulgaris*.

**Properties of particles in sap**

Leaf sap contains few virions. Electron microscopy: ISEM trapping procedures have so far been essential to see virions.

**Purification**

Method: purification is difficult because of the very small concentrations of virions in plants after symptoms show.

**Particle morphology**

Virions geminate; not enveloped; individual virions of geminate pair are *c.* 19 nm in diameter; 19 nm in length; rounded in profile; without a conspicuous capsomere arrangement.

*Sequence database accession code(s):*

X79430 Em(40)_vi:WCSVAC1 Gb(84)_vi:WCSVAC1 Watermelon chlorotic stunt virus AC1 gene (partial). 5/94 410bp. 1 sequence.

**Taxonomy and Relationships**

Bigeminivirus: *Geminiviridae*.

Virus(es) with serologically related virions: African cassava mosaic and squash leaf curl viruses. Virus(es) with serologically unrelated virions: beet curly top virus.

No homology found in tests with probes to watermelon curly mottle virus.

**References**

Alhubaishi, A.A., Walkey, D.G.A., Webb, M.J.W., Bolland, C.J. and Cook, A.A. (1987). *FAO Pl. Prot. Bull.* **35**: 135.

Jones, P., Satar, M.H.A. and Alkaff, N. (1987). *Rep. Rothamsted Exp. Stn.* 1987, p. 76.

Jones, P., Sattar, M.H.A. and Alkaff, N. (1988). *Asp. Appl. Biol.* **17**: 203.

Walkey, D.G.A., Alhubaishi, A.A. and Webb, M.J.W. (1989). *Trop. Pest Manag.* **36**: 195.

# Watermelon curly mottle bigeminivirus

**Data collated by A.A. Brunt and S. Molyneux, 1986.**

*Standard acronym:* WmCMV.

First reported in *Citrullus lanatus, Cucumis melo* var. *cantalupensis*; from Arizona, U.S.A.; by Brown and Nelson (1984).

**Natural host range and symptoms**

*Citrullus lanatus* and *Cucumis melo* var. *cantalupensis* — leaf chlorosis, vein-banding, curling and stunting.

**Transmission**

Transmitted by a vector; an insect; *Bemisia tabaci*; Aleyrodidae. Transmitted in a persistent manner. Virus lost by the vector when it moults; does not multiply in the vector; not transmitted congenitally to the progeny of the vector; transmitted by mechanical inoculation; transmitted by grafting; not transmitted by contact between plants; not transmitted by seed; not transmitted by pollen.

**Ecology and control**

Studies reported by Brown and Nelson (1986).

**Geographical distribution**

Spreads in the USA (Arizona).

Diagnostically susceptible

*Citrullus lanatus, Cucurbita maxima, C. moschata, C. pepo, Phaseolus vulgaris* cv. Red Kidney — severe systemic leaf mottling and curling.

*Cucumis melo, C. sativus* — mild systemic leaf mottling and leaf curling.

Dicotyledonous Hosts *(**Susceptible** or Insusceptible)*

*Beta vulgaris. Carthamus tinctorius. Chenopodium capitatum. Cichorium endiva.* ***Citrullus lanatus. Cucumis melo. Cucumis sativus. Cucurbita maxima. Cucurbita moschata. Cucurbita pepo.*** *Daucus carota. Gossypium hirsutum. Helianthus annuus. Lactuca sativa. Medicago sativa.* ***Phaseolus vulgaris.*** *Physalis peruviana. Raphanus sativus. Rumex acetosa. Sonchus oleraceus. Spinacia oleracea. Zinnia elegans.*

Monocotyledonous Hosts *(**Susceptible** or Insusceptible)*

*Zea mays.*

Sources of host-range data

Brown and Nelson (1986).

**Properties of particles in sap**

Leaf sap contains few virions.

**Purification**

Method: Brown and Nelson (1986).

**Particle morphology**

Virions geminate; not enveloped.

**Taxonomy and Relationships**

Bigeminivirus: *Geminiviridae.*

**References**

Brown, J.K. and Nelson, M.R. (1984). *Phytopathology* **74**: 1136.
Brown, J.K. and Nelson, M.R. (1986). *Phytopathology* **76**: 236.

# Watermelon Moroccan mosaic (?) potyvirus

**Data collated by A.A.Brunt, 1995.**

*Synonyms:* Moroccan watermelon mosaic virus. *Standard acronym:* MWMV.

First reported in *Citrullus lanatus*; from Morocco; by Fischer and Lockhart (1974).

**Natural host range and symptoms**

*Citrullus lanatus, Cucumis melo, Cucumis sativus, Cucurbita pepo* and *Malva parviflora* — leaf chlorosis.

**Geographical distribution**

Spreads in Morocco.

Dicotyledonous Hosts *(**Susceptible** or Insusceptible)*

***Citrullus lanatus*. *Cucumis melo*. *Cucumis sativus*. *Cucurbita pepo*.**

**Taxonomy and Relationships**

Possible unassigned potyvirus.

**References**

Fischer, H.U. and Lockhart B.E.L. (1974). *Pl. Dis. Reptr.* **58**: 143.

McKern, N.M., Strike, P.M., Barnett, O.W., Ward, C.W. and Shukla, D.D. (1993). *Arch. Virol.* **131**: 467.

# Watermelon mosaic 1 *potyvirus*

**Data collated by D.E. Purcifull, J.R. Edwardson, E. Hiebert and D. Gonsalves, 1986; R.S. Greber, 1986.**

*Synonyms:* papaya ringspot virus type W, watermelon papaya ringspot.

First reported in *Citrullus lanatus*; by Webb (1965).

**Natural host range and symptoms**

Symptoms persist.

*Cucurbita pepo* — mosaic, green blistering, leaves and fruit malformed.

*Citrullus lanatus, Cucumis melo, Melothria pendula, Momordica charantia* — mosaic, mottling, leaves malformed.

**Transmission**

Transmitted by a vector; an insect; *Myzus persicae, Acyrthosiphon (Aulacorthum) solani, Aphis craccivora* and *Macrosiphum euphorbiae* (Karl and Schmelzer, 1971); Aphididae. Transmitted in a non–persistent manner. Virus transmitted by mechanical inoculation; not transmitted by seed.

**Geographical distribution**

Spreads in the Middle East and the South and Central American region; Australia, China, France, Germany, India, Italy, Mexico, and the USA.

**Experimental host range**

Few (<3) families susceptible.

Diagnostically susceptible

*Cucurbita pepo* — mosaic, green blistering, leaves and fruit malformed.

*Cucumis metuliferus* cv. Accession 2459 — systemic mottle or mosaic.

*Luffa acutangula* — systemic chlorotic mottling or spotting.

Diagnostically insusceptible

*Carica papaya, Cucumis metuliferus* plant introduction PI 292190, *Nicotiana benthamiana, Phaseolus vulgaris* cv. Bountiful.

Maintenance and propagation hosts

*Cucurbita maxima, Cucurbita moschata, Cucurbita pepo.*

Assay hosts (**L**ocal lesions or **W**hole plants)

*Chenopodium amaranticolor* (L) but not for all isolates, *Chenopodium quinoa* (L) but not for all isolates, *Cucumis melo* cv. B633 (L), *Cucumis sativus* (W), *Cucurbita pepo* (W).

Dicotyledonous Hosts *(**Susceptible** or Insusceptible)*

***Carica papaya**. **Chenopodium amaranticolor**. **Chenopodium quinoa**. **Cucumis melo**. Cucumis metuliferus. **Cucumis sativus**. **Cucurbita maxima**. **Cucurbita moschata**. Myosotis sylvatica. **Phaseolus lathyroides**.*

Comments on host-range

Many watermelon mosaic virus 2 isolates (Purcifull, Hiebert and Edwardson, 1984) do not infect *Luffa acutangula*, and therefore this species has been used to obtain PRSV isolates free from watermelon mosaic virus 2 (Webb, 1965; Greber, 1978; D. Purcifull, unpublished data).

Sources of host-range data

Wang *et al.* (1978); Adlerz (1972*a*; 1972*b*); Milne and Grogan (1969); Purcifull and Hiebert (1979); Yeh *et al.* (1984).

**Properties of particles in sap**

TIP: 60 °C. LIV: 40–60 days. DEP: $\log_{10}$ minus 4–5. Leaf sap contains many virions.

**Purification**

Method: Gonsalves and Ishii (1980); modification of the method described by Purcifull and Hiebert (1979) for a type W isolate. Both purification methods are outlined in the CMI/AAB description No. 292.

**Particle morphology**

Virions filamentous; not enveloped; usually flexuous; with a clear modal length; of 760–800 nm; 12 nm wide.

**Physical properties**

One sedimenting component in purified preparations. Density 1.32 g $cm^{-3}$ in CsCl.

**Biochemical properties**

Virions contain 5.5 % nucleic acid; 94.5 % protein.

*Genome* consists of RNA; single–stranded; linear; unipartite; largest (or only) genome part 11.4 kb. Genomic nucleic acid isolated by Brakke and van Pelt (1970); de Mejia (1984); Hiebert *et al.* (1984); de Mejia *et al.* (1984). 5' terminus of RNA has a VPg. Poly A region present; at the 3' terminus (Nagel and Hiebert, 1985). The 3' end cloned and expressed in *E. coli*.

*Virion protein(s)* one; $M_r$ 36000–36500 (two smaller proteins with $M_r$s of 31000–34000 and 26000–27000 have been found, and virions stored for 2 months contained primarily the smaller one, which presumably results from proteolytic degradation). Method of preparation: Purcifull and Hiebert (1979), Gonsalves and Ishii (1980), Baum (1980).

*Virus–coded non–virion proteins* isolated (Yeh (1984) for cylindrical (pinwheel) inclusion (CI) protein and by de Mejia (1984) for amorphous inclusion body (AI) protein); two proteins found. $M_r$ of the largest 69000–70000; for CI. $M_r$ of 2nd largest 51000; for AI, a helper protein for aphid transmission.

**Replication**

Replication does not depend on a helper virus.

**Cytopathology**

Virions found in mesophyll, epidermis and phloem; in cytoplasm. Inclusions present in infected cells; are amorphous X–bodies and pinwheels; they do not contain virions.

**Taxonomy and Relationships**

*Potyvirus: Potyviridae.*

Virus(es) with serologically related virions: the Guadeloupe isolate of papaya ringspot is very closely related, but distinct (Quiuot-Douine *et al.*, 1986). Furthermore PRV-W reacts with the D-protein of tobacco etch virus (Shepard *et al.*, 1974) and the virion proteins and CI proteins of a potyvirus isolate from cucurbits in Morocco (Fischer and Lockhart, 1974).

**References**


Abu-Samah, N. (1982). *Ph.D. Thesis*, Univ. Adelaide, 109 pp.

Adlerz, W.C. (1972*a*). *J. econ. Ent.* **65**: 1303.

Adlerz, W.C. (1972*b*). *Pl. Dis. Reptr* **56**: 563.

Baum, R.H. (1980). *Ph.D. Thesis*, Univ. of Florida 99 pp.

Baum, R.H. and Purcifull, D.E. (1981). *Phytopathology* **71**: 202.

Brakke, M.K. and van Pelt, N. (1970). *Virology* **42**: 699.

de Mejia, M.V.G. (1984). *Ph.D. Thesis*, Univ. of Florida 52 pp.

de Mejia, M.V.G., Hiebert, E. and Purcifull, D.E. (1984). *Phytopathology* **74**: 1015.

de Mejia, M.V.G., Hiebert, E. and Purcifull, D.E. (1985*a*). *Virology* **142**: 24.

de Mejia, M.V.G., Hiebert, E. and Purcifull, D.E. (1985*b*). *Virology* **142**: 34.

Edwardson, J.R. and Christie, R.G. (1984). *Fla Agric. Exp. Stn Monog* No. 14, 398 pp.

Fischer, H.U. and Lockhart, B.E. (1974). *Pl. Dis. Reptr* **58**: 143.

Gonsalves, D. and Ishii, M. (1980). *Phytopathology* **70**: 1028.

Greber, R.S. (1978). *Aust. J. agric. Res.* **29**: 1235.

Hiebert, E., Thornbury, D.W. and Pirone, T.P. (1984). *Virology* **135**: 1.

Karl, E. and Schmelzer, K. (1971). *Arch. PflSchutz.* **7**: 3.

Makkouk, K.M. and Lesemann, D.E. (1980). *Plant Dis.* **64**: 789.

Milne, K.S. and Grogan, R.G. (1969). *Phytopathology* **59**: 809.

Nagel, J. and Hiebert, E. (1985). *Virology* **143**: 435.

Purcifull, D.E. (1972). *CMI/AAB Descr. Pl. Viruses* No. 84, 3 pp.

Purcifull, D.E. and Hiebert, E. (1979). *Phytopathology* **69**: 112.

Purcifull, D.E., Hiebert, E. and Edwardson, J.R. (1984). *CMI/AAB Descr. Pl. Viruses* No. 293, 7 pp.

Purcifull, D.E., Edwardson, J.R., Hiebert, E. and Gonsalves, D. (1984). *CMI/AAB Descr. Pl. Viruses* No. 292, 8 pp.

Quiot-Douine, L., Purcifull, D.E., Hiebert, E. and de Mejia, M.V.G. (1986). *Phytopathology* **76**: 346.

Shepard, J.E., Secor, G.A. and Purcifull, D.E. (1974). *Virology* **58**: 464.

Wang, H.L., Wang, C.C., Chiu, R.J. and Sun, M.H. (1978). *Pl. Prot. Bull.* **20**: 133.

Webb, R.E. (1965). *Phytopathology* **55**: 1379.

Yeh, S.D. (1984). *Ph.D. Thesis*, Cornell Univ., 98 pp.

Yeh, S.D., Gonsalves, D. and Provvidenti, R. (1984). *Phytopathology* **74**: 1081.


# Watermelon mosaic 2 *potyvirus*

**Data collated by D. Purcifull, 1981.**

*Synonyms:* marrow mosaic virus (Raychaudhuri and Varma, 1975; Varma, 1988), melon mosaic virus (Iwaki *et al.*, 1984; Komuro, 1962), watermelon mosaic virus. *Standard acronym:* WMV-2. *Strains:* soybean mosaic virus (Frenkel *et al.*, 1989; Yu *et al.*, 1989;

and D.D. Shukla, personal communication), indeed it is proposed that all are a single species and should be called soybean mosaic virus.

First reported in *Citrullus lanatus*; by Webb *et al.* (1965).

**Natural host range and symptoms**

Symptoms persist.

*Cucurbita pepo, Cucumis melo, C. sativus, Citrullus lanatus*, and some legumes — mosaic, mottling, leaf shape malformation.

**Transmission**

Transmitted by a vector; an insect; *Myzus persicae, Aphis craccivora*; at least 29 species of aphids transmit watermelon mosaic 2 potyvirus (Edwardson and Christie, 1986); Aphididae. Virus transmitted by mechanical inoculation; not transmitted by seed.

**Geographical distribution**

Probably distributed worldwide.

**Experimental host range**

Many (>9) families susceptible.

Diagnostically susceptible

*Chenopodium amaranticolor* — chlorotic local lesions; not systemic.
*Cucurbita pepo* — systemic mosaic and occasional leaf malformation.
*Pisum sativum* cv. Alaska — necrotic local lesions; systemic mottling and necrosis.
*Nicotiana benthamiana* — systemic mosaic.

Diagnostically insusceptible

*Pisum sativum* cv. Little Marvel, *Nicotiana glutinosa.*

Maintenance and propagation hosts

*Cucurbita pepo* cvs Small Sugar, Zucchini, *Nicotiana benthamiana.*

Assay hosts (**L**ocal lesions or **W**hole plants)

*Chenopodium amaranticolor, C. quinoa, Hibiscus esculentus, Lavatera trimestris* (L).

Dicotyledonous Hosts *(**Susceptible** or Insusceptible)*

***Amaranthus caudatus***. ***Astragalus sinicus***. ***Atriplex hortensis***. *Beta macrocarpa*. *Beta vulgaris*. ***Capsella bursa-pastoris***. *Capsicum annuum*. ***Cassia occidentalis***. ***Catharanthus roseus***. ***Celosia argentea***. *Cheiranthus cheiri*. ***Chenopodium album***. ***Chenopodium amaranticolor***. ***Chenopodium foetidum***. ***Chenopodium foliosum***. ***Chenopodium hybridum***. ***Chenopodium murale***. ***Chenopodium quinoa***. ***Cichorium endiva***. ***Citrullus lanatus***. ***Coriandrum sativum***. ***Crotalaria spectabilis***. ***Cucumis melo***. ***Cucumis metuliferus***. ***Cucumis sativus***. ***Cucurbita maxima***. ***Cucurbita moschata***. ***Cucurbita okeechobeensis***. ***Cucurbita pepo***. ***Cyamopsis tetragonoloba***. ***Datura stramonium***. ***Euphorbia marginata***. ***Glycine max***. ***Gomphrena globosa***. ***Hibiscus esculentus***. ***Ipomoea nil***. ***Lathyrus odoratus***. ***Lavatera trimestris***. ***Luffa acutangula***. ***Lupinus albus***. ***Lupinus angustifolius***. ***Macroptilium lathyroides***. ***Macrotyloma uniflorum***. ***Medicago hispida***. ***Medicago sativa***. ***Melilotus alba***. ***Melilotus officinalis***. ***Nicotiana benthamiana***. ***Nicotiana clevelandii***. ***Nicotiana glutinosa***. ***Nicotiana megalosiphon***. ***Nicotiana tabacum***. *Ocimum basilicum*. *Petunia hybrida*. ***Phaseolus lathyroides***. ***Phaseolus vulgaris***. ***Pisum sativum***. ***Senecio vulgaris***. ***Sesamum indicum***. ***Sonchus oleraceus***. ***Spinacia oleracea***. ***Tetragonia expansa***. ***Torenia fournieri***. ***Trifolium incarnatum***. ***Trifolium pratense***. *Tropaeolum majus*. ***Valeriana officinalis***.

***Vicia faba*. *Vicia sativa*.** *Vigna angularis*. ***Vigna unguiculata*. *Vigna unguiculata*** ssp. ***cylindrica*. *Zinnia elegans*.**

Sources of host-range data

Edwardson (1974); Horváth *et al.* (1975); Greber (1978); Hampton *et al.* (1978); Horváth (1979); Lecoq *et al.* (1981); Edwardson and Christie (1986).

**Properties of particles in sap**

TIP: 55–65 °C. LIV: 10–50 days. DEP: $\log_{10}$ minus 3–5. Leaf sap contains few virions.

**Purification**

Method: Purcifull *et al.* (1979).

**Particle morphology**

Virions filamentous; not enveloped; usually flexuous; of 730–765 nm. Axial canal obscure. Basic helix obscure.

**Physical properties**

One sedimenting component in purified preparations; sedimentation coefficient 150 *S*. Density 1.32 g $cm^{-3}$ in CsCl.

**Biochemical properties**

Virions contain 5 % nucleic acid; 95 % protein.

*Genome* consists of RNA; single–stranded; linear; unipartite. Genomic nucleic acid isolated by Purcifull *et al.* (1984).

*Sequence database accession code(s):*

D00535 Em(40)_vi:WMVCOAT Gb(84)_vi:WMVCOAT Watermelon mosaic virus 2 coat protein gene. 3/91 1,157bp.

D00592 Em(40)_vi:MTWCP Watermelon mosaic virus 2 49kda protease, nuclear inclusion protein and coat protein genes. 3

D13913 Em(40)_vi:WMVCP Gb(84)_vi:WMVCP Watermelon mosaic virus 2 49kda protease, nuclear inclusion protein and coat protein genes. 1 L22907 Em(40)_vi:WMVCOATA Gb(84)_vi:WMVCOATA Watermelon mosaic virus 2 coat protein gene, complete cds. 5/94 1,656bp. 4 sequences.

*Virion protein(s)* one; $M_r$ 34000.

*Virus–coded non–virion proteins* identified by genomic sequence analysis (Purcifull *et al.* (1984)); seven proteins found. $M_r$ of the largest 58000. $M_r$ of 2nd largest 51000; helper component. $M_r$ of 3rd 50000. $M_r$ of 4th 69000; cylindrical inclusions. $M_r$ of 5th and smaller 49000, or 54000, or 34000, or 69000; protease, polymerase, coat protein, cylindrical inclusions.

**Cytopathology**

Virions found in mesophyll and epidermis; in cytoplasm. Inclusions present in infected cells; are unusual in shape; scrolls and laminated aggregates (Edwardson's Type III inclusions) in the cytoplasm and thin plates, of unknown nature, in the nucleus; they do not contain virions and they contain virions (in cytoplasmic aggregates associated with membranes).

**Taxonomy and Relationships**

*Potyvirus: Potyviridae.*

Virus(es) with serologically related virions: soybean mosaic and blackeye cowpea mosaic viruses, but distantly.

**References**

Broderode, F.Th., Koper-Zwarthoff, C.E. and Bol, J.F. (1980). *Nucl. Acids Res.* **8**: 2213:
Edwardson, J.R. (1974). *Fla Agric. Exp. Stn Monog.* No. 5, p. 131.
Edwardson, J.R. and Christie, R.G. (1986). *Fla Agric. Exp. Stn Monog.* No. 14, p. 454.
Frenkel, M.J., Ward, C.W. and Shukla, D.D. (1989). *J. gen. Virol.* **70**: 2775.
Greber, R.S. (1978). *Aust. J. agric. Res.* **29**: 1235.
Hampton, R., Beczner, L., Hagedorn, D., Bos, L., Inouye, T., Barnett, O.W., Musil, M. and Meiners, J. (1978). *Phytopathology* **68**: 989.
Horváth, J. (1979). *Acta Phytop. Acad. Sci. Hung.* **14**: 157.
Horváth, J., Juretič, N., Besada, W.H. and Kuroli, G. (1975). *Acta Phytopath. Acad. Sci. Hung.* **10**: 93.
Iwaki, M., Honda, Y., Hanada, K., Tochihara, H., Yonaha, T., Hokama, K. and Yokoyama, T. (1984). *Plant Dis.* **68**: 1006.
Komuro, Y. (1962). *Ann. Phytopath. Soc. Japan* **27**: 31.
Lecoq, H., Pitrat, M. and Clement, M. (1981). *Agronomie* **1**: 827.
Milne, K.S. and Grogan, R.G. (1969). *Phytopathology* **59**: 809.
Milne, K.S., Grogan, R.G. and Kimble, K.A. (1969). *Phytopathology* **59**: 819.
Purcifull, D.E. and Hiebert, E. (1979). *Phytopathology* **69**: 112.
Purcifull, D., Hiebert, E. and Edwardson, J.R. (1984). *CMI/AAB Descr. Pl. Viruses* No. 293, 7 pp.
Raychaudhuri, M. and Varma, A. (1975). *Indian J. Ent.* **37**: 247.
Thomas, W. (1971). *N.Z. Jl agric. Res.* **14**: 235.
Van Regenmortel, M.H.V. (1971). *CMI/AAB Descr. Pl. Viruses* No. 63, 4 pp.
Varma, A. (1988). In: *The Plant Viruses* Vol. 4, The Filamentous Plant Viruses, p. 371; ed. R.G. Milne. Plenum Press, New York.
Webb, R.E. and Scott, H.A. (1965). *Phytopathology* **55**: 895.
Yu, M., Frenkel, M.J., McKern, N.M., Shukla, D.D., Strike, P.M. and Ward, C.M. (1989). *Arch. Virol.* **105**: 55.

# Weddel waterborne (?) *carmovirus*

**Data collated by A.A.Brunt, 1994.**

*Standard acronym:* WWBV.

First reported in water from the River Weddel; from Germany; by Yi *et al.* (1992).

**Experimental host range**

Few (<3) families susceptible.

Diagnostically susceptible

*Chenopodium amaranticolor, C. quinoa.*

Diagnostically insusceptible

*Nicotiana benthamiana, N. clevelandii.*

Maintenance and propagation hosts

*Chenopodium quinoa.*

Dicotyledonous Hosts *(**Susceptible** or Insusceptible)*

***Chenopodium amaranticolor*. *Chenopodium quinoa*. *Nicotiana benthamiana*. *Nicotiana clevelandii*.**

**Purification**

Method: Gallitelli *et al.* (1985).

**Particle morphology**

Virions isometric; not enveloped; 33 nm in diameter.

**Biochemical properties**

*Genome* consists of RNA; single–stranded; unipartite; largest (or only) genome part 4.1 kb.

*Virion protein(s)* one; $M_r$ 37100; coat protein.

**Replication**

Genome replicates in cytoplasm.

**Taxonomy and Relationships**

Possible *Carmovirus: Tombusviridae.*

Virus(es) with serologically unrelated virions: Bean mild mosaic, blackgram mottle, carnation mottle, cucumber leaf spot, cucumber soil-borne, elderberry latent, galinsoga mosaic, glycine mottle, hibiscus chlorotic ringspot, melon necrotic spot, narcissus tip necrosis, pelargonium flower break, saguaro cactus, tephrosia symptomless and turnip crinkle carmoviruses and 12 tombusviruses.

In Northern blot hybridization tests, the genomic RNA showed some homology with that of Ahlum waterborne and carnation mottle viruses.

**References**


Gallitelli, D., Hull, R. and Koenig, R. (1985). *J. gen. Virol.* **66**: 1523.

Yi, L., Lesemann, D.-E., Koenig, R., Rüdel, M. and Pfeilstetter, E. (1992). *J. Phytopathology* **134**: 121.


# Welsh onion yellow stripe (?) potyvirus

**Data collated by A.A. Brunt, 1995.**

*Synonyms:* onion yellow dwarf virus — Brazil (Costa *et al.*, 1971), onion yellow dwarf virus — Japan (Fukami *et al.*, 1989; Sako *et al.*, 1991; Yoshino and Yasu, 1965).

First reported in *Allium fistulosum*; from Indonesia; by Van Dijk and Sutarya (1992).

**Natural host range and symptoms**

*Allium chinense* (rakkyo), *A. fistulosum* (Welsh onion) — yellow leaf striping.

**Transmission**

Transmitted by a vector; an insect; *Myzus persicae, Rhopalosiphum maidis, Acyrthosiphon pisum*; Aphididae. Transmitted in a non–persistent manner. Virus transmitted by mechanical inoculation.

**Geographical distribution**

Spreads in Brazil, Indonesia, and Japan.

**Experimental host range**

Few (<3) families susceptible.

Diagnostically susceptible

*Allium fistulosum, A. vineale* (crow garlic) — systemic yellow leaf striping.

*Allium cepa* var. *cepa* (onion) cv. Stuttgarter Riesen — bright systemic leaf striping, leaf distortion and stunting.

*Chenopodium amaranticolor, C. quinoa* — local chlorotic lesions, no systemic infection.

Diagnostically insusceptible

*Allium tuberosum* (Chinese chive), *A. sativum* (garlic), *A. ampeloprasum* var. *porrum* (leek).

Maintenance and propagation hosts

*Allium cepa* var. *cepa* cv. Stuttgarter Riesen.

Assay hosts (Local lesions or Whole plants)

*Allium cepa* var. *cepa* cv. Stuttgarter Riesen.

Dicotyledonous Hosts *(**Susceptible** or Insusceptible)*

***Chenopodium amaranticolor***. ***Chenopodium quinoa***. *Nicotiana benthamiana. Nicotiana occidentalis. Vicia faba.*

Monocotyledonous Hosts *(**Susceptible** or Insusceptible)*

***Allium cepa***. ***Allium fistulosum***. *Allium porrum.*

**Particle morphology**

Virions filamentous; usually flexuous; of 750 nm.

**Taxonomy and Relationships**

Possible *Potyvirus: Potyviridae.*

Virus(es) with serologically related virions: shallot yellow stripe virus. Virus(es) with serologically unrelated virions: leek yellow stripe virus.

**References**

Costa, A.S., Costa, C.L., Nagani, H. and Kitajima, E.W. (1971). *O. Biologica* **37**: 157.

Fukami, M., Natsuaki, K.T. and Tomaru, K. (1989). *Ann. Phytopath. Soc. Japan* **55**: 542.

Sako, I., Nakasone, W., Okada, K., Ohki, S.T., Osaki, T. and Inouye, T. (1991). *Ann. Phytopath. Soc. Japan* **57**: 65.

Van Dijk, P. and Sutarya, R. (1992). *Onion Newsl. Trop.* **4**: 57.

Van Dijk, P. (1993). *Neth. J. Pl. Path.* **99**, Suppl. 2, 48 pp.

Yoshino, M. and Yasu, M. (1965). *Bull. Saitama Pref. Agric. Exp. Stn* **26**, 67pp.

# Wheat American striate mosaic *nucleorhabdovirus*

**Data collated by R.G. Timian, 1987.**

*Synonyms:* oat rhabdovirus, oat striate rhabdovirus, oat striate mosaic virus, American wheat striate mosaic virus, wheat striate mosaic virus. *Standard acronym:* WASMV.

First reported in *Triticum vulgare*; from South Dakota, U.S.A.; by Slykhuis (1953).

**Natural host range and symptoms**

Symptoms persist.

*Triticum dicoccum, T. vulgare, T. durum, Bromus inermis, Eragrostis cilianensis, Panicum capillare* — faint chlorotic streaks along the veins, clearest on the abaxial surfaces of leaves, then necrosis. Brown necrotic spots and streaks develop in some cultivars. Wheat plants die or become stunted with sterile heads.

**Transmission**

Transmitted by a vector; an insect; *Elymana virescens* and *Endria inimica*; Cicadellidae. Transmitted in a persistent manner. Virus lost by the vector when it moults; multiplies in the vector; not transmitted congenitally to the progeny of the vector; not transmitted by mechanical inoculation; not transmitted by contact between plants; not transmitted by seed; not transmitted by pollen.

**Ecology and control**

Studies reported by Slykhuis (1953); Slykhuis (1963).

**Geographical distribution**

Spreads in the North American region; Canada and the USA.

**Experimental host range**

Few (<3) families susceptible. Experimentally infected plants mostly show severe striate mosaic.

Diagnostically susceptible

*Avena sativa, Eragrostis cilianensis, Panicum capillare* and *Triticum durum* — severe chlorotic striping.

*Hordeum vulgare* and *Zea mays* — mild chlorotic striping.

Diagnostically insusceptible

*Triticum vulgare.*

Maintenance and propagation hosts

*Bromus inermis, Triticum durum.*

Assay hosts (**L**ocal lesions or **W**hole plants)

*Triticum durum* (W).

Monocotyledonous Hosts *(**Susceptible** or Insusceptible)*

*Agropyron intermedium. Agropyron repens. Agrostis alba.* ***Avena sativa. Bromus inermis. Bromus secalinus. Bromus tectorum.*** *Dactylis glomerata.* ***Echinochloa crus-galli. Eragrostis cilianensis. Hordeum vulgare. Lolium multiflorum. Lolium perenne. Panicum capillare. Panicum miliaceum.*** *Phleum pratense. Poa compressa. Poa pratensis.* ***Secale cereale. Setaria viridis. Triticum durum. Zea mays.***

Comments on host-range

Few species tested, none of them dicotyledons.

Sources of host-range data

Slykhuis (1953); Slykhuis (1963).

**Properties of particles in sap**

TIP: 55 °C. LIV: 3 days (at 4°C or -10°C). DEP: $\log_{10}$ minus 5. Leaf sap contains many virions. Electron microscopy: fix with gluteraldehyde before mounting in PTA.

**Purification**

Method: Ahmed *et al.* (1970), for infective virus; Thottappilly and Sinha (1973) for antigen.

**Particle morphology**

Virions rhabdo– or bullet–shaped; enveloped; usually straight; with a clear modal length; of 200–250 nm; 75 nm wide. Axial canal obvious; 23 nm in diameter. Basic helix obscure.

**Physical properties**

One sedimenting component in purified preparations, or two sedimenting components in purified preparations; sedimentation coefficient 875 *S* (major component).

**Biochemical properties**

Virions contain 5 % nucleic acid; 68 % protein; 25 % lipid. Also carbohydrate (probably) 2%.

*Genome* consists of RNA; single–stranded; linear; unipartite; largest (or only) genome part 7 kb. Genomic nucleic acid isolated by Sinha *et al.* (1976). Base composition 26.4 % G; 21.3 % A; 22.5 % C; 29.8 % U.

*Virion protein(s)* four; $M_r$ of the largest 145000. $M_r$ of 2nd largest 92000; glycoprotein. $M_r$ of 3rd largest 59000; N protein. $M_r$ of 4th largest 25000. Method of preparation: Trefzger *et al.* (1976). Amino acid composition: Sinha *et al.* (1976).

**Replication**

Genome replicates in nuclei and in cytoplasm (associated with the nuclear membrane and within cytoplasmic cisternae). Replication does not depend on a helper virus.

**Cytopathology**

Virions found in leaves, roots, mesophyll, epidermis, vascular parenchyma, companion cells and all parts of the host plant; in cytoplasm, in nuclei, and in the perinuclear space. Other cellular changes: the plasma membrane separating from the cell wall.

**Taxonomy and Relationships**

*Nucleorhabdovirus: Rhabdoviridae.*

Best tests for diagnosis: Wheat streak mosaic rymovirus is transmitted mechanically and by mites whereas wheat striate mosaic cytorhabdovirus is transmitted by *Endria inimica* but not mechanically.

**References**


Ahmed, M.E., Sinha, R.C. and Hochster, R.M. (1970). *Virology* **41**: 768.

Sinha, R.C. and Behki, R.M. (1972). *CMI/AAB Descr. Pl. Viruses* No. 99, 4 pp.

Sinha, R.C., Harwalkar, V.R. and Behki, R.M. (1976). *Phytopath. Z.* **87**: 314.

Slykhuis, J.T. (1953). *Phytopathology* **43**: 537.

Slykhuis, J.T. (1963). *Can. J. Bot.* **41**: 1171.

Thottappilly, G. and Sinha, R.C. (1973). *Virology* **33**: 312.

Trefzger-Stevens, J. and Lee, P. (1976). *Virology* **78**: 144.


# Wheat chlorotic spot (?) rhabdovirus

**Data collated by A.A. Brunt, 1991.**

First reported in *Triticum aestivum*; from Brazil; by Kitajima *et al.* (1976).

**Natural host range and symptoms**

*Triticum aestivum* — chlorotic spots in leaves.

**Particle morphology**

Virions rhabdo– or bullet–shaped.

**Taxonomy and Relationships**

Possible unassigned rhabdovirus.

**References**

Kitajima, E.W., Cupertino, F.P. and Caetano, V.R. (1976). *Fitopatol. Lima* **11**: 16.

# Wheat dwarf monogeminivirus

**Data collated by K. Lindsten, 1991.**

*Standard acronym:* WDV. *Strains:* barley and wheat strains.

First reported in *Triticum aestivum*; from the former Czechoslovakia; by Vacke (1961).

**Natural host range and symptoms**

Symptoms persist. Symptoms yellowing and dwarfing.
*Avena sativa* — yellowing and dwarfing.
*Hordeum vulgare* — yellowing and dwarfing, poor cropping.
*Lolium multiflorum, Poa annua* — growth reduction.
*Secale cereale, Triticum aestivum* — yellowing and severe dwarfing.

**Transmission**

Transmitted by a vector; an insect; *Psammotettix alienus*; Cicadellidae. Not transmitted by *Javesella pellucida, Laodelphax striatellus, Macrosteles laevis.* Transmitted in a persistent manner. Virus retained when the vector moults; does not multiply in the vector; not transmitted congenitally to the progeny of the vector; not transmitted by mechanical inoculation; not transmitted by contact between plants; not transmitted by seed; not transmitted by pollen.

**Ecology and control**

Studies reported by Lindsten (1980; 1991).

**Geographical distribution**

Spreads in Bulgaria, the former Czechoslovakia, France, Hungary, and the former USSR. Found, but with no evidence of spread, in Sweden.

**Experimental host range**

Few (<3) families susceptible. Experimentally infected plants mostly show yellowing and severe dwarfing.

Diagnostically susceptible

*Avena sativa* — yellowing and dwarfing.
*Bromus secalinus* — yellowing, plants die prematurely.
*Hordeum vulgare* — dwarfing, poor heads and yellowing.
*Lolium perenne* — symptomless.
*L. remotum* — yellowing.
*Secale cereale, Triticum aestivum* — yellowing and severe dwarfing, no crop.

Diagnostically insusceptible

*Dactylis glomerata, Festuca* spp., *Phleum pratense, Poa pratensis, Zea mays.*

Maintenance and propagation hosts

*Lolium multiflorum, Poa annua, Triticum aestivum, Hordeum vulgare* for the barley strain.

Assay hosts (**L**ocal lesions or **W**hole plants)

*Bromus secalinus* (W), *Lagurus ovatus* (W), *Triticum aestivum* (W), *Hordeum vulgare* (W).

Monocotyledonous Hosts *(**Susceptible** or Insusceptible)*

*Agropyron repens.* ***Avena sativa.*** ***Avena strigosa.*** *Bromus inermis. Bromus mollis.* ***Bromus secalinus.*** *Dactylis glomerata. Echinochloa crus–galli. Festuca pratensis.* ***Lagurus ovatus.*** ***Lolium multiflorum.*** ***Lolium perenne.*** ***Lolium temulentum.*** *Phalaris arundinacea. Phleum pratense.* ***Poa annua.*** *Poa pratensis.* ***Secale cereale.*** *Setaria italica. Setaria viridis.* ***Triticum aestivum.*** ***Triticum durum.*** *Zea mays.*

Sources of host-range data

Lindsten (1991); Vacke (1972).

**Properties of particles in sap**

Leaf sap contains few virions.

**Purification**

Method: Bisztray *et al.* (1989); Lindsten *et al.* (1980).

**Particle morphology**

Virions geminate; not enveloped; 18 nm in diameter; 30 nm in length; angular in profile; without a conspicuous capsomere arrangement.

**Physical properties**

One sedimenting component in purified preparations, or two sedimenting components in purified preparations; sedimentation coefficient 70 *S*; of the other(s) 50 *S*. $A_{260}/A_{280}$ ratio 1.3–1.4.

**Biochemical properties**

Virions contain 20 % nucleic acid.

*Genome* consists of DNA; single–stranded; circular; unipartite; largest (or only) genome part 2.749 kb. Infectivity lost when deproteinised with proteases.

*Sequence database accession code(s):*

D00305 Em(40)_vi:GEWDVASD Gb(84)_vi:WDVASD Wheat dwarf virus (WDV), virion-associated small DNA. 4/90 80bp.

X02869 Em(40)_vi:GEWDVXX Gb(84)_vi:GEWDVXX Wheat dwarf virus (WDV) genome. 9/93 2,749bp.

X82104 Em(43)_vi:Wdvgns Gb(89)_vi:Wdvgns Wheat dwarf virus genes for V1, V2, C1–1, C1–2 and Cx. 10/94 2,750bp.

*Features of the genome:* see MacDowell *et al.* (1985); Matzeit *et al.* (1990); Woolston *et al.* (1988).

*Non–genomic nucleic acid* not found in the virions. Sub–genomic *m*RNA not found in infected cells.

*Virion protein(s)* one; $M_r$ 28000.

**Replication**

Genome replicates in nuclei. Replication does not depend on a helper virus.

**Cytopathology**

Virions found in leaves and roots; in nuclei. Inclusions absent from infected cells.

**Taxonomy and Relationships**

Monogeminivirus: *Geminiviridae.*

Virus(es) with serologically unrelated virions: chloris striate mosaic and maize streak viruses.

*Triticum aestivum* (wheat) and *Hordeum vulgare* (barley) react differently to infection with the common and barley strains. The barley strain causes dwarfing, poor heads and yellowing in barley, but which is resistant to wheat strains. Not serologically related to other geminiviruses tested. Best tests for diagnosis: transmission by *Psammotettix alienus* to wheat and barley plants, and serological tests.

**References**

Bisztray, G., Gaborjanyi, D. and Gaborjanyi, R. (1989). *Z. PflKrankh. PflSchutz.* **96**: 449.

Lindsten, K. (1980). *Vaxtskyddsnotiser* **44**: 54.

Lindsten, K. (1985). *Abstr. A.A.B. Virol. Gp., Cambridge* p. 36.

Lindsten, K. (1991). In: *Proc. 6th Conf. on Virus Diseases of Gramineae in Europe, Torino*, June 1991.

Lindsten, K., Vacke, J. and Gerhardson, B. (1970). *Nat. Swedish Inst. Pl. Prot. Contr.* **14**: 281.

Lindsten, K., Lindsten, B., Abdelmoeti, M. and Junti, N. (1980). In: *Proc. 3rd. Conf. on Virus Diseases of Gramineae in Europe, Rothamsted*, May 1980.

MacDowell, S.W., MacDonald, H., Hamilton, W.D.O., Coutts, R.H.A. and Buck, K.W. (1985). *EMBO J.* **4**: 2173.

Matzeit, V., Laufs, J., Kammann, M., Schaefer, S., Schalk, H.-J., Schmidt, B., Wirtz, U., Baker, B. and Gronenborn, B. (1990). *Monograph Series, Plant Biology* **11**: 71.

Tomenius, K. and Oxelfelt, P. (1981). *Phytopath. Z.* **101**: 163.

Vacke, J. (1961). *Biol. Plant. Praha* **3**: 228.

Vacke, J. (1972). *Vedecke Prace Vyzk. Ustavu Rostl. Vyroby, Praha-Ryzyne* **17**: 151.

Woolston, C.J., Barker, R., Gunn, H., Boulton, M.I. and Mullineaux, P.M. (1988). *Pl. mol. Biol.* **11**: 35.

# Wheat European striate mosaic (?) *tenuivirus*

**Data collated by A.A. Brunt, 1992.**

*Synonyms:* European wheat striate mosaic virus, oat striate and red disease virus (Lindsten, 1961). *Standard acronym:* EWSMV.

First reported in *Triticum aestivum*; from Europe; by Slykhuis and Watson (1958).

**Natural host range and symptoms**

*Lolium perenne, Triticum aestivum* — chlorotic spotting and streaking of leaves, general chlorosis, stunting and death of plants.

**Transmission**

Transmitted by a vector; an insect; *Javesella pellucida, J. dubia, J. obscurella*; Delphacidae. Transmitted in a persistent manner. Virus multiplies in the vector; transmitted congenitally to the progeny of the vector; not transmitted by mechanical inoculation.

**Ecology and control**

Studies reported by Ammar (1975); Lindsten (1979).

**Geographical distribution**

Spreads in the former Czechoslovakia, Denmark, Finland, Germany, Poland, Romania, Spain, Sweden, and the UK.

Diagnostically susceptible

*Avena sativa, Hordeum vulgare, Lolium multiflorum, L. perenne, Secale cereale, Triticum aestivum* — chlorotic spotting and fine chlorotic streaking of leaves, general chlorosis.

Maintenance and propagation hosts

*Avena sativa, Hordeum vulgare, Secale cereale, Triticum aestivum.*

Assay hosts (**L**ocal lesions or **W**hole plants)

*Triticum aestivum* (W).

Monocotyledonous Hosts ***(Susceptible** or Insusceptible)*

***Avena sativa. Hordeum vulgare. Lolium multiflorum. Lolium perenne. Secale cereale. Triticum aestivum.***

Sources of host-range data

Slykhuis and Watson (1958); Ikäheimo (1964); Vacke and Pruša (1961).

**Purification**

Method: Serjeant (1967).

**Particle morphology**

Virions filamentous; not enveloped; usually flexuous; with no clear modal length; 3 nm wide.

**Physical properties**

One sedimenting component in purified preparations; sedimentation coefficient less than 178 *S*.

**Cytopathology**

Virions found in leaves; in cytoplasm. Inclusions present in infected cells; are crystals in the cytoplasm; they do not contain virions.

**Taxonomy and Relationships**

Possible *Tenuivirus*.

**References**


Ammar, E.-D. (1974). *Riv. Patol. Veg.* **10**: 143.

Ammar, E.-D. (1975). *Ann. appl. Biol.* **79**: 195.

Ammar, E.-D. (1975). *Ann. appl. Biol.* **79**: 203.

Brčak, J. (1979). In: *Leafhopper Vectors and Plant Disease Agents*, p. 97; eds K. Maramorosch and K.F. Harris. Academic Press, New York.

Gingery, R.E. and Plumb, R. (1988). In: *The Plant Viruses*, Vol. 4, The Filamentous Viruses, p. 297; ed. R.G. Milne. Plenum Press, New York.

Ikäheimo, K. (1964). *Ann. Phytopath. Fenniae* **3**: 133.

Lindsten, K. (1961). *Ann. Roy. agric. College, Sweden* **27**: 199.

Lindsten, K. (1979). In: *Leafhopper Vectors and Plant Disease Agents*, p. 155; eds K. Maramorosch and K.F. Harris. Plenum Press, New York.

Nuorteva, P. (1965). *Zool. Beitr. (N.F.)* **11**: 191.

Serjeant, E.P. (1967). *Ann. appl. Biol.* 59: 39.

Slykhuis, J.T. and Watson, M.A. (1958). *Ann. appl. Biol.* **46**: 542.

Vacke, J. and Pruša, V. (1961). *Biol. Plant* **3**: 277.

Watson, M.A. and Sinha, R.C. (1959). *Virology* **8**: 139.

# Wheat Iranian stripe (?) *tenuivirus*

**Data collated by A.A. Brunt, 1994.**

First reported in *Triticum aestivum*; from Iran; by Heydarnejad and Izadpanah (1989).

**Taxonomy and Relationships**

Possible *Tenuivirus*.

**References**

Heydarnejad, J. and Izadpanah, K. (1989). *Proc. 9th Pl. Prot. Congr. Iran*, p.160. Ferdowsi Univ., Mashad, Iran.

# Wheat soil-borne mosaic *furovirus*

**Data collated by A.J. Gibbs, 1995.**

*Synonyms:* wheat viruses 1 and 2, soil-borne wheat green mosaic virus, wheat mosaic virus. *Strains:* green mosaic, yellow mosaic and rosette strains.

First reported in *Triticum aestivum*; from the U.S.A.; by McKinney (1923).

**Natural host range and symptoms**

Symptoms vary seasonally (brightest in spring).

*Triticum aestivum* (winter wheat) and *Hordeum vulgare* — light green to yellow leaf mosaic, roots stunted, severe rosetting and tillering of susceptible cultivars (Brakke, 1971).

**Transmission**

Transmitted by a vector; a fungus; *Polymyxa graminis*; Estes and Brakke (1966); Plasmodiophorales (Brakke, 1971). Virus transmitted by mechanical inoculation; not transmitted by seed.

**Ecology and control**

Studies reported by Himmel *et al.* (1991). When the virus is transmitted to plants of resistant cultivars by growing them in infested soil, their roots are as susceptible as those of other cultivars, but the virus does not spread systemically (Myers *et al.* 1993).

**Geographical distribution**

Spreads in China, Italy, Japan, and the USA (Cai *et al.* (1983); Canova (1966); Putnam *et al.* (1994); Saito *et al.* (1964)).

**Experimental host range**

Several (3–9) families susceptible.

Diagnostically susceptible

*Triticum aestivum* and other *Triticum* spp. — mosaic and rosetting.

Diagnostically insusceptible

*Agropyron repens, Avena sativa, Bromus inermis, Cucumis sativus, Lycopersicon esculentum, Nicotiana tabacum, Phaseolus vulgaris, Zea mays.*

Maintenance and propagation hosts

*Triticum aestivum* cv. Michigan Amber and others.

Assay hosts (Local lesions or Whole plants)

*Chenopodium amaranticolor* (L), *C. quinoa* (L), *Triticum aestivum* (W).

Dicotyledonous Hosts *(**Susceptible** or Insusceptible)*

***Chenopodium amaranticolor*. *Chenopodium quinoa*.** *Cucumis sativus. Lycopersicon esculentum. Nicotiana tabacum. Phaseolus vulgaris.*

Monocotyledonous Hosts *(**Susceptible** or Insusceptible)*

*Agropyron repens. Avena byzantina. Avena sativa. Bromus inermis.* ***Bromus tectorum*. *Hordeum vulgare*. *Secale cereale*.** *Zea mays.*

Additional hosts

*Bromus commutatus.*

**Properties of particles in sap**

TIP: 60–65 °C. LIV: 4000 days (at least, in dried leaves). DEP: $\log_{10}$ minus 2–3. Leaf sap contains few virions.

**Purification**

Method: Gumpf (1971).

**Particle morphology**

Virions rod–shaped; not enveloped; usually straight; with clear modal lengths; of 82–110 nm (T), or 138–160 nm (M), or 280–300 nm (B); 20 nm wide. Axial canal obvious.

**Physical properties**

Two sedimenting components in purified preparations, or three sedimenting components in purified preparations; sedimentation coefficient 212 *S* (of the fastest); of the other(s) 173 *S*. $A_{260}/A_{280}$ ratio 1.2.

**Biochemical properties**

Virions contain 5 % nucleic acid; 95 % protein; 0 % lipid.

*Genome* consists of RNA; single–stranded; linear; of two parts (Shirako and Brakke, 1984); largest (or only) genome part 6.5 kb; the 2nd largest 3.5 kb.

*Sequence database accession code(s):*

L07937 Em(40)_vi:WMORNA1A Gb(84)_vi:WMORNA1A Soil-borne wheat mosaic virus (SBWMV) complete RNA 1 segment: 150 kDa protein, 209 kDa readth

L07938 Em(40)_vi:WMORNA2A Gb(84)_vi:WMORNA2A Soil-borne wheat mosaic virus (SBWMV) complete RNA 2 segment: 84 kDa readthrough protein, 19

X81639 Gb(84)n:SBWMV19K Soil-borne wheat mosaic virus genomic RNA2. 9/94 3,592bp.

*Non–genomic nucleic acid* found in the virions; deletion variants of the genomic RNAs.

*Virion protein(s)* one; $M_r$ 19700; coat protein.

**Cytopathology**

Virions found in roots and, occasionally, leaves. Inclusions present in infected cells; are crystals in the cytoplasm and amorphous X–bodies.

**Taxonomy and Relationships**

*Furovirus.*

Virus(es) with serologically related virions: tobacco mosaic tobamovirus (Brunt and Shikata, 1986).

Isolates vary greatly in virion lengths, those from Japan are serologically related to those from the U.S.A., but infect *Nicotiana tabacum* and *Zea mays*.

**References**

Brakke, M.K. (1971). *CMI/AAB Descr. Pl. Viruses* No. 77. 4 pp.
Brunt, A.A. and Richards, K.E. (1989). *Adv. Virus Res.* **36**: 1.
Brunt, A.A. and Shikata, E. (1986). In: *The Plant Viruses*, No. 2 *The Rod-Shaped Plant Viruses* eds M.H.V. van Regenmortel and H. Fraenkel-Conrat, p. 305, Plenum Press, New York.
Cai, W., Peng, X. and Mang, K. (1983). *Acta Phytopathol. Sin.* **13**: 7.
Canova, A. (1966). *Phytopathol. Medit.* **5**: 53.
Gumpf, D.J. (1971). *Virology* **43**: 586.
Himmel, P.T., Hewings, A.D. and Glawe, D.A. (1991). *Plant Dis.* **75**: 1008.
McKinney, H.H. (1923). *J. Agric. Res.* **23**: 771.
Myers, L.D., Sherwood, J.L., Siegerist, W.C. and Hunger, R.M. (1993). *Phytopathology* **83**: 548.
Putnam, M.L., Carroll, T.W. and French, R. (1994). *Plant Dis.* **78**: 102.
Saito, Y., Takanashi, K., Iwata, Y. and Okamoto, H. (1964). *Bull. Natl. Inst. Agric. Sci.*, Tokyo, Ser. C **1964**: 1711.
Shirako, Y. and Brakke, M.K. (1984). *J. gen. Virol.* **65**: 119.
Shirako, Y. and Wilson, T.M.A. (1992). *Virology* **195**: 16.

# Wheat spindle streak mosaic *bymovirus*

**Data collated by K.Z. Haufler, 1988.**

*Standard acronym:* WSSMV. *Strains:* wheat yellow mosaic virus (Usugi and Saito, 1979; Wiese, 1987).

First reported in *Triticum aestivum*; from Ontario, Canada; by Slykhuis (1960).

**Natural host range and symptoms**

Symptoms vary seasonally and disappear soon after infection (symptoms develop best at 20°C or less).

*Secale cereale, Triticum aestivum, T. durum* — chlorotic streaking.

**Transmission**

Transmitted by a vector; a fungus; *Polymyxa graminis*; Plasmodiophorales. Virus transmitted by mechanical inoculation (but not readily); not transmitted by grafting; not transmitted by contact between plants; not transmitted by seed; not transmitted by pollen.

**Ecology and control**

Studies reported by Slykhuis (1970); Slykhuis (1974); Wiese *et al.* (1974); Haufler and Fulbright (1986); Van Koevering *et al.* (1987).

**Geographical distribution**

Spreads in France, Germany, India, Italy, Japan, and the USA (possibly also China).

**Experimental host range**

Few (<3) families susceptible (No susceptible dicotyledonous species are known).

Diagnostically susceptible

*Secale cereal, Triticum aestivum, T. durum* — chlorotic streaking.

Maintenance and propagation hosts

*Triticum aestivum.*

Assay hosts (**L**ocal lesions or **W**hole plants)

*Triticum aestivum* (W).

Monocotyledonous Hosts *(**Susceptible** or Insusceptible)*

***Secale cereale. Triticum aestivum. Triticum durum.***

Sources of host-range data

Slykhuis (1961); Slykhuis (1970); Jackson *et al.* (1976); Proeseler and Stanarius (1983); Bays *et al.* (1985).

**Properties of particles in sap**

TIP: 45–50 °C (for 10 minutes). DEP: $\log_{10}$ minus 1–5. Leaf sap contains few virions.

**Purification**

Method: Haufler and Fulbright (1985; unpublished).

**Particle morphology**

Virions filamentous; not enveloped; usually flexuous (but only slightly); with no clear modal length; of 300–2000 nm (modal length probably 700–1000 nm); 16 nm wide. Axial canal obscure. Basic helix obscure.

**Physical properties**

One sedimenting component in purified preparations. Density 1.28 g $cm^{-3}$ in CsCl. Density 1.20–1.21 g $cm^{-3}$ in caesium sulphate.

*Sequence database accession code(s):*

X73883 Gb(84)_vi:WSSMVRNA Wheat spindle streak mosaic virus RNA. 6/94 4,646bp.

*Virion protein(s)* three; $M_r$ of the largest 35000–36000. $M_r$ of 2nd largest 30000; coat protein. $M_r$ of 3rd largest 27000; coat protein.

$M_r$ 69000; cytoplasmic inclusion protein.

**Cytopathology**

Virions found in leaves, roots, mesophyll and vascular parenchyma; in cytoplasm. Inclusions present in infected cells; are pinwheels; they do not contain virions. Other cellular changes: disruption of organelles (chloroplasts, mitochondria swell and then are disrupted).

**Taxonomy and Relationships**

*Bymovirus: Potyviridae.*

Virus(es) with serologically related virions: barley yellow mosaic, oat mosaic and rice necrosis mosaic; wheat spindle streak and wheat yellow mosaic viruses are now considered the one species.

Best tests for diagnosis: Symptoms of the virus appear first in the spring, those of barley yellow dwarf virus and wheat streak mosaic virus appear later, those of wheat soil-borne mosaic virus are also brightest in cool weather, but are brighter.

**Comments**

Although this virus is probably a strain of wheat yellow mosaic bymovirus, it is listed separately here.

**References**

Bays, D.C., Cunfer, B.M. and Demski, J.W. (1985). *Plant Dis.* **69**: 1094.

Haufler, K., Zagula, K.R. and Fulbright, D.W. (1986). *Plant Dis.* **70**: 31.

Haufler, K., Zagula, K.R. and Fulbright, D.W. (1985). *Phytopathology* **75**: 1349.
Jackson, A.O. *et al.* (1976). *Pl. Dis. Reptr* **60**: 202.
Proeseler, G. and Stanarius, A. (1983). *Arch. Phytopathol. Pflanzenschutz, Berlin* **19**: 345.
Slykhuis, J.T. (1960). *Can. Pl. Dis. Surv.* **40**: 43.
Slykhuis, J.T. (1961). *Can. Pl. Dis. Surv.* **41**: 330.
Slykhuis, J.T. (1970). *Phytopathology* **60**: 319.
Slykhuis, J.T. (1974). *Phytopathology* **64**: 554.
Slykhuis, J.T. (1976). *CMI/AAB Descr. Pl. Viruses* No. 167, 3 pp.
Usugi, Y. and Saito, Y. (1976). *Ann. Phytopath. Soc. Japan* **42**: 12.
Van Koevering, M., Haufler, Z., Fulbright, D.W., Isleib, T.C. and Everson, E.H. (1987). *Phytopathology* **77**: 742.
Weise, M.V. (1987). *Compendium of Wheat Diseases* 2nd Ed. Amer. Phytopath. Soc. St Paul, Minnesota.
Wiese, M.V., Ravenscroft, A.V. and Everson, E.H. (1974). *Pl. Dis. Reptr* **58**: 522.

## Wheat streak mosaic *rymovirus*

**Data collated by A.A. Brunt, 1995.**

*Standard acronym:* WSMV. *Strains:* Several variants giving different symptoms have been described.

First reported in *Triticum aestivum*; from the U.S.A.; by McKinney (1937).

**Natural host range and symptoms**

*Echinochloa, Panicum, Setaria* spp. — usually severe mosaic.
*Triticum aestivum* — severe leaf mosaic, necrosis of tillers, stunting and crop loss.
*Zea mays* — mild mosaic (some cultivars only).

**Transmission**

Transmitted by a vector; a mite; *Aceria tulipae*; Eriophyidae. Virus not transmitted congenitally to the progeny of the vector; transmitted by mechanical inoculation; not transmitted by contact between plants; transmitted by seed (very low levels).

**Ecology and control**

Studies reported by Slykhuis (1955); Ashworth and Futrell (1961); Orlob (1966); Nault and Styer (1969).

**Geographical distribution**

Spreads in Canada, Jordan, Romania, and the USA.

**Experimental host range**

Few (<3) families susceptible.

Diagnostically susceptible

*Avena sativa, Hordeum vulgare, Poa compressa, Secale cereale, Triticum aestivum, Zea mays* (some cultivars) — systemic mosaic.

Diagnostically insusceptible

*Agropyron repens, Bromus inermis, Hordeum jubilatum, Saccharum officinarum, Sorghum bicolor.*

Maintenance and propagation hosts

*Triticum aestivum* (especially cvs Michigan, Amber and Kent).

Assay hosts (**L**ocal lesions or **W**hole plants)

*Triticum aestivum* (W).

Monocotyledonous Hosts *(**Susceptible** or Insusceptible)*

*Agropyron repens*. ***Avena sativa***. *Bromus inermis*. ***Hordeum vulgare***. *Saccharum officinarum*. ***Secale cereale***. *Sorghum bicolor*. ***Triticum aestivum***. ***Zea mays***.

Additional hosts

The virus also infects many wild grass species in genera that include *Aegilops, Agropyron, Bromus, Digitaria, Echinochloa, Eragrostis, Hordeum, Lolium, Panicum, Phalaris, Poa* and *Stipa*.

**Properties of particles in sap**

TIP: 54 °C. LIV: 4–8 days (at 20°C, 1–2 months at 2°C, and at least 16 years in dried leaf at 2°C). DEP: $\log_{10}$ minus 3–4.

**Purification**

Method: Brakke and Ball (1968).

**Particle morphology**

Virions filamentous; with a clear modal length; of 700 nm.

**Physical properties**

One sedimenting component in purified preparations; sedimentation coefficient *c.* 165 *S*. $A_{260}/A_{280}$ ratio 1.37.

**Biochemical properties**

*Genome* consists of RNA; single–stranded; unipartite; largest (or only) genome part 8.5 kb. Genomic nucleic acid isolated by Zagula *et al.* (1992). Nucleotide sequence references: Niblett *et al.* (1991); Zagula *et al.* (1992).

*Virion protein(s)* one; $M_r$ 47000; coat protein.

**Cytopathology**

Virions found in cytoplasm. Inclusions present in infected cells; are pinwheels.

**Taxonomy and Relationships**

*Rymovirus: Potyviridae.*

Virus(es) with serologically related virions: ryegrass mosaic, agropyron mosaic and oat necrotic mottle viruses.

**References**


Ashworth, L.J. and Futrell, M.C. (1961). *Pl. Dis. Reptr.* **45**: 220.

Brakke, M.K. (1971). *CMI/AAB Descr. Pl. Viruses* No. 48, 4 pp.

Brakke, M.K. and Ball, E.M. (1968). *Virology* **36**: 152.

Brakke, M.K., Ball, E.M., Hsu, Y.-H. and Langenberg, W.G. (1987). *J. gen. Virol.* **68**: 281.

Brakke, M.K., Skopp, R.N. and Lane, L.C. (1990). *Phytopathology* **80**: 1401.

Hill, J.H., Martinson, C.A. and Russell, W.A. (1974). *Crop Sci.* **67**: 1161.

Langenberg, W.G. (1986) *J. gen. Virol.* **67**: 1161.

Langenberg, W.G. (1988) *Phytopathology* **78**: 589.

Langenberg, W.G. (1989) *J. gen. Virol.* **70**: 969.

Langenberg, W.G. (1991) *J. gen. Virol.* **72**: 493.

Langenberg, W.G. and Schroeder, H.F. (1973) *Virology* **55**: 218.

McKinney, H.H. (1973). *Circ. U.S. Dep. Agric.* **442**, 23 pp.

McMullen, M.D. and Louie, R. (1991). *Phytopathology* **81**: 624.

Nault, I.R. and Styer, W.E. (1969). *Ann. Ent. Soc. Amer.* **62**: 1446.
Niblett, C.L., Zagula, K.R. Calvert, L.A., Kendall, T.L., Stark, D.M., Smith, C.E., Beachy, R.N. and Lommel, S.A. (1991). *J. gen. Virol.* **72**: 499.
Orlob, G.B. (1966). *Phytopath. Z.* **55** 218.
Pfannenstiel, M.A. and Niblett, C.L. (1978). *Phytopathology* **68**: 1204.
Shahwan, J.M. and Hill, J.P. (1984). *Plant Dis.* **68**: 579.
Slykhuis, J.T. (1955). *Phytopathology* **45**: 116.
Slykhuis, J.T. (1955). *Rev. Appl. Mycol.* **46**: 401.
Slykhuis, J.T. and Bell, W. (1955). *Can. J. Bot.* **44**: 1191.
Zagula, K.R., Niblett, C.L., Robertson, N.L., French, R. and Lommel, S.A. (1992). *Arch. Virol.*, Suppl. **5**: 269.

# Wheat yellow leaf *closterovirus*

**Data collated by T. Inouye, 1984.**

*Standard acronym:* WYLV.

First reported in *Hordeum vulgare, Triticum aestivum*; from Japan; by Inouye *et al.* (1973).

**Natural host range and symptoms**

Symptoms persist.
*Avena sativa, Lolium multiflorum* — leaf reddening.
*Hordeum vulgare, Secale cereale, Triticum aestivum* — leaf yellowing.

**Transmission**

Transmitted by a vector; an insect; *Rhopalosiphum maidis, R. padi*; Aphididae. Transmitted in a semi–persistent manner. Virus lost by the vector when it moults; not transmitted by mechanical inoculation; not transmitted by contact between plants; not transmitted by seed.

**Geographical distribution**

Spreads in Italy and Japan.

**Experimental host range**

Few (<3) families susceptible.

Diagnostically susceptible

*Avena sativa, Lolium multiflorum* — leaf reddening.
*Hordeum vulgare, Secale cereale, Triticum aestivum* — leaf yellowing.

Diagnostically insusceptible

*Oryza sativa, Zea mays.*

Maintenance and propagation hosts

*Avena sativa, Triticum aestivum.*

Assay hosts (**L**ocal lesions or **W**hole plants)

*Avena sativa* (W) *Triticum aestivum* (W).

Monocotyledonous Hosts *(**Susceptible** or Insusceptible)*

***Avena sativa***. ***Hordeum vulgare***. ***Lolium perenne***. *Oryza sativa*. ***Secale cereale***. ***Triticum aestivum***. *Zea mays*.

Sources of host-range data

Inouye (1976).

**Properties of particles in sap**

Leaf sap contains few virions.

**Purification**

Method: Inouye (1976).

**Particle morphology**

Virions filamentous; usually flexuous; with no clear modal length; of 1600–1850 nm; 10 nm wide. Axial canal obscure. Basic helix obvious; pitch of basic helix 3.4 nm.

**Cytopathology**

Virions found in phloem; in cytoplasm. Inclusions present in infected cells; are membranous bodies.

**Taxonomy and Relationships**

*Closterovirus.*

Virus(es) with serologically related virions: carnation necrotic fleck and carnation yellow fleck viruses, but distantly. Virus(es) with serologically unrelated virions: burdock yellow and citrus tristeza viruses.

**References**

Amici, A. and Faoro, F. (1980). *Riv. Patologia Veg. S. I*, **16**: 9.

Inouye, T., Mitsuhata, K., Heta, H. and Hiura, U. (1973). *Nogaku Kenkyu* **55**: 1.

# Wheat yellow mosaic *bymovirus*

**Data collated by T. Inouye, 1991.**

*Synonyms:* soil-borne wheat yellow mosaic virus. *Strains:* wheat spindle streak mosaic virus (Weise, 1987).

First reported in *Triticum aestivum*; from Japan; by Sawada (1927).

**Natural host range and symptoms**

Symptoms vary seasonally. Symptoms mosaic.

*Triticum aestivum* — mosaic; chlorotic to necrotic streaks.

**Transmission**

Transmitted by a vector; a fungus; *Polymyxa graminis* probably; Plasmodiophorales. Not transmitted by *Macrosiphum akebiae, Onychiurus matsumotoi, O. watanabei.* Virus transmitted by mechanical inoculation; not transmitted by contact between plants; not transmitted by seed; not transmitted by pollen.

**Ecology and control**

Studies reported by Ikata and Kawai (1940); Saito *et al.* (1964).

**Geographical distribution**

Spreads in Canada, China, France, Japan, Korea D.P.R. (North), and Korea Republic.

**Experimental host range**

Few (<3) families susceptible. Experimentally infected plants mostly show mosaic.

Diagnostically susceptible

*Secale cereale, Triticum aestivum* — mosaic.

Diagnostically insusceptible

*Chenopodium amaranticolor, Cucumis sativus, Hordeum vulgare, Nicotiana tabacum, Zea mays.*

Maintenance and propagation hosts

*Triticum aestivum.*

Assay hosts (**L**ocal lesions or **W**hole plants)

*Triticum aestivum* (W).

Dicotyledonous Hosts (***Susceptible*** *or Insusceptible)*

*Chenopodium amaranticolor. Cucumis sativus. Nicotiana tabacum. Phaseolus vulgaris.*

Monocotyledonous Hosts (***Susceptible*** *or Insusceptible)*

*Agropyron kamoji. Hordeum vulgare. Poa annua.* ***Secale cereale. Triticum aestivum.***

Sources of host-range data

Ikata and Kawai (1940); Saito *et al.* (1968); Inouye (1969).

**Properties of particles in sap**

TIP: less than 50 °C. LIV: 1 days. DEP: $\log_{10}$ minus 2–3. Leaf sap contains few virions.

**Purification**

Method: Usugi and Saito (1976).

**Particle morphology**

Virions filamentous; not enveloped; usually flexuous; with clear modal lengths (two); of 275–300 nm, or 575–600 nm; 13–14 nm wide. Axial canal obscure.

**Physical properties**

Two sedimenting components in purified preparations. Density 1.281 g $cm^{-3}$ in CsCl. $A_{260}/A_{280}$ ratio 1.14.

**Cytopathology**

Virions found in all parts of the host plant; in cytoplasm. Inclusions present in infected cells; are amorphous X–bodies, membranous bodies, and pinwheels; they do not contain virions.

**Taxonomy and Relationships**

*Bymovirus: Potyviridae.*

Virus(es) with serologically related virions: barley yellow mosaic and rice necrosis mosaic viruses.

**References**

Hariri, D., Courtillot, M., Zaoni, P. and Lapierre, H. (1987). *Agronomie* **7**: 789.
Hibino, H., Usugi, T. and Saito, Y. (1981). *Ann. Phytopath. Soc. Japan* **47**: 510.
Ikata, S. and Kawai, I. (1940). *Noji Kairyo Shiryo* **154**: 1.
Inouye, T. (1969). *Nogaku Kenkyu* **53**:; 61.
Lin, M.C. and Ruan, Y.L. (1986). *Acta Phytopath. Sinica* **16**: 73.
Saito, Y., Tsuchizaki, T. and Hibino, H. (1968). *Ann. Phytopath. Soc. Japan* **34**: 347.
Sawada, E. (1927). *Byochugai Zasshi* **14**: 444.
Slykhuis, J.T. (1976). *CMI/AAB Descr. Pl. Viruses* No. 167.
Usugi, T. and Saito, Y. (1976). *Ann. Phytopath. Soc. Japan* **42**: 12.
Usugi, T. and Saito, Y. (1979). *Ann. Phytopath. Soc. Japan* **45**: 397.

Usugi, T., Kuwabara, T. and Tsuchizaki, T. (1984). *Ann. Phytopath. Soc. Japan* **50**: 63.

Weise, M.V. (1987). *Compendium of Wheat Diseases* 2nd Ed. Amer. Phytopath. Soc., St Paul, Minnesota.

# White bryony mosaic (?) *carlavirus*

**Data collated by A.A. Brunt, 1991.**

*Synonyms:* bryony white mosaic virus. *Standard acronym:* WBMV.

First reported in *Bryonia dioica*; by Tomlinson and Carter (1969).

### Natural host range and symptoms

Symptoms persist.

*Bryonia dioica* — mosaic.

### Geographical distribution

Spreads in the UK and France (Spire *et al.*, 1966; Quiot *et al.*, 1979).

### Particle morphology

Virions filamentous; not enveloped; usually straight; with a clear modal length; of 650 nm; 13 nm wide.

### Taxonomy and Relationships

Possible *Carlavirus.*

### Comments

Infected plants often also contain a potyvirus. Four potyviruses, all causing conspicuous symptoms, have been isolated from *Bryonia* spp; bryonia mottle virus occurs in Morocco (Lockhart and Fischer, 1979), and the other three (P1, P2 and P3) were isolated in northern Italy (Milne *et al.*, 1980). The relationships of these potyviruses are unknown.

### References

Hollings, M., Stone, O.M. and Brunt, A.A. (1967). *Rep. Glasshouse Crops Res. Inst.* 1966, p. 102.

Lockhart, B.E.L. and Fischer, H.U. (1979). *Phytopath. Z.* **96**: 244.

Milne, R.G., Masenga, V. and Lovisolo, O. (1980). *Phytopathol. Medit.* **19**: 115.

Quiot, J.B., Douine, L. and Gebre Selassie, K. (1979). *Annls. Phytopath.* **11**: 283.

Spire, D., Bertrand, J. and Ferault, C. (1966). *Ann. Epiphyt.* **17**: 121.

# White bryony (?) potyvirus

**Data collated by A.A. Brunt, 1994.**

*Standard acronym:* WBV.

First reported in *Bryonia cretica*; from Italy; by Milne *et al.* (1980).

### Taxonomy and Relationships

Unassigned potyvirus.

### References

Milne, R.G., Masenga, V. and Lovisolo, O. (1980). *Phytopath. Medit.* **19**: 115.

# White clover 1 *alphacryptovirus*

**Data collated by A.A. Brunt, 1991.**

*Synonyms:* red clover cryptic virus 1. *Standard acronym:* WCCV-1.

First reported in *Trifolium repens*; from Italy; by Boccardo *et al.* (1985).

**Natural host range and symptoms**

*Trifolium repens* — symptomless infection.

**Transmission**

Virus transmitted by mechanical inoculation; not transmitted by grafting; transmitted by seed.

**Geographical distribution**

Found, but with no evidence of spread, in Japan.

Dicotyledonous Hosts *(**Susceptible** or Insusceptible)*

***Trifolium repens*.**

**Particle morphology**

Virions isometric; not enveloped; 30 nm in diameter; rounded in profile.

**Biochemical properties**

*Genome* consists of RNA; double–stranded; of two parts.

**Taxonomy and Relationships**

*Alphacryptovirus: Partitiviridae* (the type species).

**Comments**

White clover temperate virus (Natsuaki *et al.* (1984) is a mixture of white clover cryptic viruses 1, 2 and 3 (Natsuaki *et al.* (1986).

**References**


Boccardo, G., Milne, R.G., Luisoni, E., Lisa, V. and Accotto, G.P. (1985). *Virology* **147**: 29.

Boccardo, G., Lisa, V., Luisoni, E. and Milne, R.G. (1987). *Adv. Virus Res.* **32**: 171.

Natsuaki, K.T., Natsuaki, T., Okuda, S., Teranaka, M., Yamashita, S. and Doi, Y. (1984). *J. agric. Sci., Tokyo Nogyo Daigaku* **29**: 49.

Natsuaki, T., Natsuaki, K.T., Okuda, S., Teranaka, M., Milne, R.G., Boccardo, G. and Luisoni, E. (1986). *Intervirology* **25**: 69.


# White clover 2 *betacryptovirus*

**Data collated by A.A. Brunt, 1991.**

*Standard acronym:* WCCV-2.

First reported in *Trifolium repens*; from Italy; by Boccardo *et al.* (1985).

**Natural host range and symptoms**

*Trifolium repens* — no symptoms.

**Transmission**

Virus not transmitted by mechanical inoculation; not transmitted by grafting; transmitted by seed.

**Geographical distribution**

Found, but with no evidence of spread, in Europe, Japan, New Zealand and the U.S.A.

**Experimental host range**

Few (<3) families susceptible.

Diagnostically susceptible

None known.

Maintenance and propagation hosts

*Trifolium repens* containing no other viruses.

Dicotyledonous Hosts (***Susceptible*** *or Insusceptible)*

***Trifolium repens***.

**Purification**

Method: Boccardo *et al.* (1985).

**Particle morphology**

Virions isometric; not enveloped; 38 nm in diameter; rounded in profile; with a conspicuous capsomere arrangement.

**Physical properties**

One sedimenting component in purified preparations. Density 1.375 g $cm^{-3}$ in CsCl. $A_{260}/A_{280}$ ratio 1.31.

**Biochemical properties**

Virions contain 24 % nucleic acid; 76 % protein.

*Genome* consists of RNA; double–stranded; linear; of two parts; largest (or only) genome part the larger 2.238 kb (pairs); the 2nd largest 2.072 kb (pairs).

**Cytopathology**

Virions found in all parts of the host plant; in cytoplasm. Inclusions absent from infected cells.

**Taxonomy and Relationships**

*Betacryptovirus: Partitiviridae* (the type species).

Virus(es) with serologically related virions: red clover 2 and hop trefoil 2 betacryptoviruses.

Complementary *c*DNA probes to white clover 2 betacryptovirus failed to detect any homology with white clover 1 and 3 alphacryptoviruses.

**Comments**

This virus was originally described as white clover temperate virus (Natsuaki *et al.*, 1984), and was shown later to be a mixture of white clover cryptic viruses 1, 2 and 3 (Natsuaki *et al.*, 1986).

**References**

Boccardo, G., Lisa, V. and Milne, R.G. (1983). In: *Double-Stranded RNA Viruses*, p. 425; eds R.W. Compans and D.H.L. Bishop. Elsevier, New York.

Boccardo, G., Milne, R.G., Luisoni, E., Lisa, V. and Accotto, G.P. (1985). *Virology* **147**: 29.

Boccardo, G., Lisa, V., Luisoni, E. and Milne, R.G. (1987). *Adv. Virus Res.* **32**: 171.

Luisoni, E. and Milne, R.G. (1988). *AAB Descr. Pl. Viruses* No. 332, 4 pp.

Luisoni, E., Milne, R.G., Accotto, G.P. and Boccardo, G. (1987). *Intervirology* **28**: 144.

Natsuaki, K.T., Natsuaki, T., Okuda, S., Teranaka, M., Yamashita, S. and Doi, Y. (1984). *J. agric. Sci., Tokyo Nogyo Daigaku* **29**: 49.

Natsuaki, T., Natsuaki, K.T., Okuda, S., Teranaka, M., Milne, R.G., Boccardo, G. and Luisoni, E. (1986). *Intervirology* **25**: 69.

# White clover 3 *alphacryptovirus*

**Data collated by A.A. Brunt, 1991.**

*Synonyms:* red clover cryptic virus 3. *Standard acronym:* WCCV-3.

First reported in *Trifolium repens*; from Italy; by Boccardo *et al.* (1985).

**Natural host range and symptoms**

*Trifolium repens* — symptomless infection.

**Transmission**

Virus transmitted by mechanical inoculation; not transmitted by grafting; transmitted by seed.

**Geographical distribution**

Found, but with no evidence of spread, in Japan.

Dicotyledonous Hosts ***(Susceptible** or Insusceptible)*

***Trifolium repens**.*

**Particle morphology**

Virions isometric; not enveloped; 30 nm in diameter; rounded in profile.

**Biochemical properties**

*Genome* consists of RNA; double–stranded; of two parts; largest (or only) genome part the larger 3.9 kb (pairs); the 2nd largest 3.2 kb (pairs).

**Taxonomy and Relationships**

*Alphacryptovirus: Partitiviridae.*

**Comments**

White clover temperate virus (Natsuaki *et al.*, 1984) is a mixture of white clover cryptic viruses 1, 2 and 3 (Natsuaki *et al.*, 1986).

**References**

Boccardo, G., Milne, R.G., Luisoni, E., Lisa, V. and Accotto, G.P. (1985). *Virology* **147**: 29.

Boccardo, G., Lisa, V., Luisoni, E. and Milne, R.G. (1987). *Adv. Virus Res.* **32**: 171.

Natsuaki, K.T., Natsuaki, T., Okuda, S., Teranaka, M., Yamashita, S. and Doi, Y. (1984). *J. agric. Sci., Tokyo Nogyo Daigaku* **29**: 49.

Natsuaki, T., Natsuaki, K.T., Okuda, S., Teranaka, M., Milne, R.G., Boccardo, G. and Luisoni, E. (1986). *Intervirology* **25**: 69.

# White clover mosaic *potexvirus*

**Data collated by P.L. Guy, 1980. Revised 1983.**

*Synonyms:* clover mosaic virus, pea wilt virus (Bos *et al.*, 1959; Johnson, 1942). *Standard acronym:* WClMV.

First reported in *Trifolium repens*; from the U.S.A; by Pierce (1935).

**Natural host range and symptoms**

Symptoms vary seasonally.

*Trifolium* spp. — chlorotic mottling, streaking and systemic mosaic.

**Transmission**

Virus transmitted by mechanical inoculation; transmitted by contact between plants; transmitted by seed (to 6% in *Trifolium pratense*).

**Geographical distribution**

Probably distributed worldwide (in temperate regions).

**Experimental host range**

Several (3–9) families susceptible.

Diagnostically susceptible

*Cucumis sativus* — necrotic local lesions.
*Phaseolus vulgaris* — necrotic and chlorotic local lesions, systemic yellowing.
*Pisum sativum* — chlorotic local lesions, leaf shape malformation.
*Trifolium* spp. — systemic mosaic.
*Vicia faba* — necrotic local lesions, systemic mosaic.
*Vigna unguiculata* — necrotic and chlorotic local lesions.

Diagnostically insusceptible

*Chenopodium quinoa.*

Maintenance and propagation hosts

*Phaseolus vulgaris, Pisum sativum.*

Assay hosts (Local lesions or Whole plants)

*Phaseolus vulgaris.*

Dicotyledonous Hosts *(**Susceptible** or Insusceptible)*

*Amaranthus retroflexus*. ***Antirrhinum majus***. *Apium graveolens*. *Beta vulgaris*. *Brassica oleracea* ssp. *capitata*. *Calendula officinalis*. *Capsicum frutescens*. *Chenopodium amaranticolor*. *Chenopodium quinoa*. ***Crotalaria spectabilis***. ***Cucumis sativus***. ***Cucurbita pepo***. ***Cyamopsis tetragonoloba***. ***Datura stramonium***. ***Glycine max***. *Gomphrena globosa*. ***Lycopersicon esculentum***. *Matthiola incana*. ***Medicago sativa***. ***Melilotus alba***. ***Nicotiana debneyi***. *Nicotiana glutinosa*. ***Nicotiana megalosiphon***. ***Nicotiana rustica***. ***Nicotiana sylvestris***. *Nicotiana tabacum*. ***Petunia hybrida***. ***Phaseolus lunatus***. ***Phaseolus vulgaris***. ***Pisum sativum***. *Ricinus communis*. *Solanum melongena*. *Solanum nigrum*. *Tetragonia expansa*. ***Trifolium hybridum***. ***Trifolium incarnatum***. ***Trifolium pratense***. ***Trifolium repens***. ***Tropaeolum majus***. ***Vicia faba***. ***Vigna unguiculata***. *Zinnia elegans*.

Sources of host-range data

Bancroft *et al.* (1960); Beczner and Vassanyi (1981).

**Properties of particles in sap**

TIP: 60 °C. LIV: 10–99 days. DEP: $\log_{10}$ minus 4–5. Infectivity of sap not changed by treatment with di–ethyl ether. Leaf sap contains many virions (especially that from French bean, although often fewer in sap of infected clover).

**Purification**

Method: Francki and McLean (1968).

**Particle morphology**

Virions filamentous; not enveloped; usually flexuous; with a clear modal length; of 480 nm; 13 nm wide. Axial canal obscure; 3.5 nm in diameter. Basic helix obvious; pitch of basic helix 3.4 nm.

**Physical properties**

One sedimenting component in purified preparations; sedimentation coefficient 119 *S*. Isoelectric point pH 4.5.

**Biochemical properties**

Virions contain 6 % nucleic acid; 94 % protein; 0 % lipid.

*Genome* consists of RNA; single–stranded; linear; unipartite; largest (or only) genome part 5.845 kb. Base composition 15.5 % G; 31.8 % A; 26.9 % C; 25.7 % U. 5' terminus of RNA has an unusual type; a nucleotide cap probably of the type $m^7$GpppG. Infectivity retained when deproteinised with phenol or detergent.

*Sequence database accession code(s):*

M18920 Em(40)_vi:WCP3EA Gb(84)_vi:WCP3EA White clover mosaic potexvirus 3'-terminal RNA region encoding ORFs of 17K, 5' end; and of 13

X06728 Em(40)_vi:POWC1MVX Gb(84)_vi:POWC1MVX Potexvirus white clover mosaic virus (WC1MV) RNA genome. 9/93 5,845bp.

X16636 Em(40)_vi:POWCLMV4 Gb(84)_vi:POWCLMV4 White clover mosaic virus total genomic mRNA. 9/93 5,846bp. 3 sequences.

*Virion protein(s)* one; $M_r$ 14000–20000. Method of preparation: Miki *et al.* (1967); Koenig *et al.* (1971).

**Replication**

Replication does not depend on a helper virus.

**Cytopathology**

Virions found in all parts of the host plant; in cytoplasm. Inclusions present in infected cells; are unusual in shape; cytoplasmic banded bodies and few vesiculated aggregates; they contain virions.

**Taxonomy and Relationships**

*Potexvirus.*

Virus(es) with serologically related virions: potato X, cactus X, hydrangea ringspot and clover yellow mosaic viruses, but distantly.

Best tests for diagnosis: the banded body inclusions induced by white clover mosaic virus are disrupted when preparing epidermal strips for examination, but these inclusions distinguish white clover mosaic virus from clover yellow mosaic virus.

**References**


Agrawal, H., Bos, L. and Chessin, M. (1962). *Phytopathology* **62**: 517.

Bancroft, J.B., Tuite, J. and Hissong, G. (1960). *Phytopathology* **50**: 711.

Beczner, L. and Vassanyi, R. (1981). *Acta Phytopath. Acad. Sci. Hung.* **16**: 109.

Bos, L., Delvic, B. and van der Want, J.P.H. (1959). *Tijdschr. PlZiekt.* **65**: 89.

Bucks, R. (1971). *CMI/AAB Descr. Pl. Viruses* No. 41, 4 pp.

Edwardson, J.R. and Christie, R.G. (1986). *Fla Agric. Exp. Stn Monog.* No. 14, p. 350.

Forster, R.L.S., Bevan, M.V., Harrison, S.A. and Gardner, R.C. (1988). *Nucl. Acids Res.* **16**: 291.

Francki, R.I.B. and McLean, G.D. (1968). *Aust. J. biol. Sci.* **21**: 1311.

Fry, P.R., Grogan, R.G. and Lyttleton, J.W. (1960). *Phytopathology* **50**: 175.

Johnson, F. (1942). *Phytopathology* **32**: 103.
Koenig, R. (1971). *J. gen. Virol.* **10**: 111.
Miki, T. and Knight, C.A. (1967). *Virology* **31**: 55.
Pierce, W.H. (1935). *J. Agric. Res.* **51**: 1017.

# White mustard (?) *alphacryptovirus*

**Data collated by A.A. Brunt, 1994.**

First reported in *Sinapis alba*; from Germany; by Kühne *et al.* (1987).

**Natural host range and symptoms**

Symptoms
*Sinapis alba* — symptomless.

**Geographical distribution**

Found, but with no evidence of spread, in Germany.

**Taxonomy and Relationships**

Possible *Alphacryptovirus: Partitiviridae.*

**References**

Kühne, T., Stanarius, A. and Schubert, J. (1987). *Arch. Phytopath. Pflschutz.* **23**: 499.

# Wild cucumber mosaic *tymovirus*

**Data collated by A.A. Brunt and A.J. Gibbs, 1989.**

*Synonyms:* cucumber wild mosaic tymovirus. *Standard acronym:* WCMV.

First reported in *Marah macrocarpa*; from California, U.S.A.; by Freitag (1952).

**Natural host range and symptoms**

Symptoms persist.
*Marah macrocarpus, Marah oreganus, Echinocystis* sp. — mild leaf chlorosis.

**Transmission**

Transmitted by a vector; an insect; *Acalymma trivittata*; Coleoptera. Virus transmitted by mechanical inoculation; transmitted by grafting; not transmitted by contact between plants.

**Geographical distribution**

Spreads in the USA (California and Oregon).

**Experimental host range**

Several (3–9) families susceptible.

Diagnostically susceptible

*Cucurbita pepo, C. maxima, Citrullus lanatus, Cucumis sativus* — chlorotic local lesions and systemic chlorosis.

*Catharanthus roseus* — symptomless systemic infection.

Diagnostically insusceptible

*Nicotiana tabacum, N. glutinosa, Datura stramonium, Vigna unguiculata, Chenopodium amaranticolor.*

Maintenance and propagation hosts

*Cucurbita pepo* (pumpkin and zucchini), *Citrullus lanatus.*

Assay hosts (**L**ocal lesions or **W**hole plants)

*Citrullus vulgaris* (L) for some isolates, *Cucumis melo* (L) for some isolates.

Dicotyledonous Hosts *(**Susceptible** or Insusceptible)*

*Brassica campestris* ssp. *pekinensis.* ***Catharanthus roseus.*** *Chenopodium amaranticolor.* ***Citrullus lanatus. Cucumis melo. Cucumis sativus. Cucurbita maxima. Cucurbita pepo.*** *Datura stramonium. Nicotiana glutinosa. Nicotiana tabacum. Vigna unguiculata.*

Sources of host-range data

Lindberg *et al.* (1956); Milne *et al.* (1969).

**Properties of particles in sap**

TIP: 70 °C. LIV: more than 28 days (at 20°C). DEP: $\log_{10}$ minus 4. Infectivity of sap not changed by treatment with di–ethyl ether. Leaf sap contains many virions. Electron microscopy: stable in all stains.

**Purification**

Method: chloroform/n–butanol clarification, concentrated by PEG then differentail centrifugation.

**Particle morphology**

Virions isometric; not enveloped; 28 nm in diameter; rounded in profile; with a conspicuous capsomere arrangement.

**Physical properties**

Two sedimenting components in purified preparations; sedimentation coefficient of the fastest 119 *S* (possibly); of the other(s) 53 *S* (possibly). Isoelectric point pH 5.2.

**Biochemical properties**

Virions contain 35 % nucleic acid (B); 65 % protein (B).

*Genome* consists of RNA; single–stranded; linear; unipartite; largest (or only) genome part 2 kb. Genomic nucleic acid isolated by : standard methods. Base composition 16.4 % G; 17 % A; 41 % C; 25.6 % U. Infectivity retained when deproteinised with phenol or detergent.

*Virion protein(s)* one; $M_r$ 22200; coat protein.

**Cytopathology**

Virions found in mesophyll and all parts of the host plant; in cytoplasm and in chloroplasts (in chronically infected leaves). Inclusions present in infected cells; are crystals in the cytoplasm; they contain virions. Other cellular changes: development of vesicles in chloroplasts.

**Taxonomy and Relationships**

*Tymovirus.*

Virus(es) with serologically related virions: ononis yellow mosaic and turnip yellow mosaic viruses are most closely related; eggplant mosaic, Andean potato latent, belladona mottle, dulcamara mottle, scrophularia mottle, okra mosaic and desmodium yellow mottle viruses are related. Virus(es) with serologically unrelated virions: kennedya yellow mosaic, clitoria yellow vein and cacao yellow mosaic viruses.

**References**

Allen, T.C. (1972). *Virology* **47**: 467.

Allen, T.C. and Fernald, E.K. (1971). *Pl. Dis. Reptr* **55**: 546.
Brunt, A.A., Kenten, R.H., Gibbs, A.J. and Nixon, H. (1965). *J. gen. Microbiol.* **33**: 81.
Freitag, J.H. (1952). *Phytopathology* **42**: 8.
Koenig, R. (1976). *Virology* **72**: 1.
Lindberg, G.D., Hall, D.H. and Walker, J.C. (1956). *Phytopathology* **46**: 489.
Macleod, R. and Markham, R. (1963). *Virology* **19**: 190.
Milne, K.S., Grogan, R.G. and Kimble, K.A. (1969). *Phytopathology* **59**: 819.
Sinclair, J.B., Geil, P.H. and Kaesberg, P. (1957). *Phytopathology* **47**: 372.
Van Regenmortel, M.H.V. (1972). *CMI/AAB Descr. Pl. Viruses* No. 105, 4 pp.
Walkers, H.J. (1969). *Adv. Virus Res.* **15**: 339.

# Wild potato mosaic *potyvirus*

**Data collated by R.A.C. Jones and C.E. Fribourg, 1989.**

*Synonyms:* potato wild mosaic potyvirus. *Standard acronym:* WPMV.

First reported in *Solanum chancayense*; from Lachay in the Peruvian coastal desert.

## Natural host range and symptoms

Symptoms persist.

*Solanum chancayense* — severe leaf mosaic and deformation.

*Solanum muricatum* (pepino) — chlorotic mosaic, rugosity and enations.

## Transmission

Transmitted by a vector; an insect; *Myzus persicae*; Aphididae. Transmitted in a non–persistent manner. Virus transmitted by mechanical inoculation; not transmitted by contact between plants.

## Ecology and control

Studies reported by Jones and Fribourg (1979); Jones (1981).

## Geographical distribution

Spreads in Peru.

## Experimental host range

Few (<3) families susceptible. Experimentally infected plants mostly show systemic mosaics, mottles, ringspots or necrosis leaf deformation.

Diagnostically susceptible

*Nicotiana clevelandii* — systemic mosaic.

*Nicotiana bigelovii* — blotchy systemic mosaic with twisting and deformation of young leaves.

*Nicotiana occidentalis* — systemic vein clearing, mosaic and leaf curling.

*Nicotiana rustica* — chlorotic blotches with brown necrotic rings; occasional systemic chlorotic leaf blotching.

*Solanum chancayense, S. mochicense* — systemic mosaic and deformation.

Diagnostically insusceptible

*Solanum tuberosum, Datura stramonium, Gomphrena globosa, Chenopodium amaranticolor, Cucumis sativus.*

Maintenance and propagation hosts

*Nicotiana clevelandii, N. bigelovii, N. occidentalis, Solanum chancayense.*

Assay hosts (**L**ocal lesions or **W**hole plants)

*Nicotiana clevelandii* (W) and *N. bigelovii* (W).

Dicotyledonous Hosts *(**Susceptible** or Insusceptible)*

*Amaranthus caudatus. Anthriscus cerefolium. Brassica campestris* ssp. *pekinensis. Chenopodium amaranticolor. Chenopodium murale. Chenopodium quinoa. Clitoria ternatea. Coriandrum sativum. Cucumis sativus. Cucurbita pepo. Cyamopsis tetragonoloba. Datura stramonium. Gomphrena globosa. Lycopersicon chilense.* ***Lycopersicon esculentum****. Lycopersicon pimpinellifolium.* ***Nicandra physaloides. Nicotiana benthamiana. Nicotiana bigelovii. Nicotiana clevelandii. Nicotiana debneyi. Nicotiana glutinosa. Nicotiana occidentalis. Nicotiana rustica. Nicotiana tabacum.*** *Ocimum basilicum. Phaseolus acutifolius. Phaseolus vulgaris.* ***Physalis floridana.*** *Physalis peruviana. Sesamum indicum.* ***Solanum chacoense. Solanum demissum.*** *Solanum demissum* × *tuberosum. Solanum tuberosum. Vigna unguiculata. Vigna unguiculata* ssp. *cylindrica. Zinnia elegans.*

Additional hosts

***Solanum mochicense, Solanum muricatum, Solanum chancayense, Solanum berthaultii, Solanum brachycarpum, Solanum megistacrolobum, Solanum microdontum, Solanum raphanifolium*** and ***Solanum vernei***.

Sources of host-range data

Jones and Fribourg (1974); Jones *et al.* (1980).

**Properties of particles in sap**

TIP: 60–65 °C. LIV: 5–6 days. DEP: $\log_{10}$ minus 3–4. Leaf sap contains few virions. Electron microscopy: neutral PTA stains virions well.

**Purification**

Method: Fribourg (1979).

**Particle morphology**

Virions filamentous; not enveloped; usually flexuous; with a clear modal length; of 735 nm. Axial canal obscure. Basic helix obscure.

**Taxonomy and Relationships**

*Potyvirus: Potyviridae.*

Virus(es) with serologically related virions: Peru tomato, tobacco vein mottling, potato Y, potato A, potato V, pepper veinal mottle, tobacco etch, henbane mosaic and Colombian datura viruses.

Isolates from pepino may be a distinct strain. They cause vein clearing, mosaic, leaf deformation and stunting in *Nicotiana glutinosa* whereas the type isolate fails to infect this species. Virions of wild potato mosaic potyvirus were serologically most closely related to those of Peru tomato virus, tobacco vein mottling virus, potato V virus and potato Y virus. The type isolate of WPMV is readily distinguished from Peru tomato virus, potato V virus and potato Y virus by its narrower host range, distinctive symptoms and, unlike tobacco vein mottling virus, fails to infect *Nicotiana clevelandii* (Fribourg and Nakashima, 1984; Jones and Fribourg, 1979; Jones and Fribourg, 1986). Best tests for diagnosis: No other viruses known in *Solanum chancayense*. Other filamentous viruses in *Solanum muricatum* (pepino) can be distinguished by electron microscopy and inoculation to *Solanum tuberosum* (potato), which wild potato mosaic virus does not infect (Jones *et al.*, 1980; Dolby and Jones, 1988).

**Comments**

This virus was first found in the 'Lomas' vegetation of the Peruvian coastal desert, where, in the winter, low lying clouds moisten the desert. Very few plant species survive in this habitat, *Solanum chancayense* is one; it survives in summer as underground tubers and some are infected with wild potato mosaic potyvirus. Spread from plant to plant is by aphids. WPMV is also found in cultivated pepino in irrigated coastal valleys, far from Lomas vegetation. There it is spread in infected cuttings and from plant to plant by aphids.

**References**

Dolby, C.A. and Jones, R.A.C. (1988). *Ann. appl. Biol.* **112**: 231.
Fribourg, C.E. (1979). *Phytopathology* **69**: 441.
Fribourg, C.E. and Nakashima, J. (1984). *Phytopathology* **74**: 1363.
Jones, R.A.C. (1981). In: *Pests, Pathogens and Vegetation*, p. 89; ed. J.M. Thresh. Pitman, London.
Jones, R.A.C. and Fribourg, C.E. (1979). *Phytopathology* **69**: 446.
Jones, R.A.C. and Fribourg, C.E. (1986). *CMI/AAB Descr. Pl. Viruses* No. 316, 4 pp.
Jones, R.A.C., Koenig, R. and Lesemann, D.-E. (1980). *Ann. appl. Biol.* **94**: 61.

# Wineberry latent virus

**Data collated by A.T. Jones, 1991.**

*Synonyms:* loganberry calico virus. *Standard acronym:* WLV.

First reported in *Rubus phoenicolasius*; from Scotland in imported material; by Jones (1977); Jones *et al.* (1990).

**Natural host range and symptoms**

Symptoms vary seasonally. Symptoms chlorotic or yellow line pattern.

*Rubus phoenicolasius* — symptomless. However, it has been shown to induce calico disease-like symptoms in blackberry.

**Transmission**

Not transmitted by *Macrosiphum euphorbiae*. Virus transmitted by mechanical inoculation; transmitted by grafting; not transmitted by contact between plants; not transmitted by seed; not transmitted by pollen.

**Geographical distribution**

Spreads in the USA. Found, but with no evidence of spread, in the U.K.

**Experimental host range**

Few (<3) families susceptible. Experimentally infected plants mostly show necrotic or red local lesions.

Diagnostically susceptible

*Gomphrena globosa, Chenopodium album, C. amaranticolor, C. ambrosioides, C. murale, C. quinoa* — necrotic local lesions.

Diagnostically insusceptible

*Nicotiana clevelandii, N. debneyi, N. glutinosa, N. tabacum, Phaseolus vulgaris.*

Maintenance and propagation hosts

*Chenopodium amaranticolor, C. quinoa, Gomphrena globosa.*

Assay hosts (Local lesions or Whole plants)

*Chenopodium amaranticolor* (L), *C. quinoa* (L).

Dicotyledonous Hosts *(**Susceptible** or Insusceptible)*

*Apium graveolens*. ***Beta macrocarpa***. ***Chenopodium album***. ***Chenopodium amaranticolor***. ***Chenopodium ambrosioides***. ***Chenopodium foetidum***. ***Chenopodium murale***. ***Chenopodium quinoa***. *Datura stramonium*. ***Gomphrena globosa***. ***Lycopersicon esculentum***. *Nicotiana clevelandii*. *Nicotiana debneyi*. *Nicotiana glutinosa*. *Nicotiana rustica*. *Nicotiana tabacum*. *Petunia hybrida*. *Phaseolus vulgaris*. *Pisum sativum*. ***Spinacia oleracea***. ***Tetragonia expansa***. *Vicia faba*.

Sources of host-range data

Jones (1977).

**Properties of particles in sap**

TIP: less than 70 °C. LIV: less than 8 days (at 18°C). DEP: $\log_{10}$ minus 3–4. Infectivity of sap not changed by treatment with di–ethyl ether. Leaf sap contains few virions.

**Purification**

Method: Jones *et al.* (1990).

**Particle morphology**

Virions filamentous; not enveloped; usually flexuous; with a clear modal length; of 620 nm; 12 nm wide. Axial canal obscure. Basic helix obscure.

**Physical properties**

Sometimes two sedimenting components in purified preparations; sedimentation coefficient 115–125 *S*. Density 1.29 and 1.3 g $cm^{-3}$ in CsCl. Density 1.25 and 1.26 $g/cm^3$ in $Cs_2SO_4$. $A_{260}/A_{280}$ ratio 1.26.

**Biochemical properties**

Virions contain 5 % nucleic acid; 95 % protein.

*Genome* consists of RNA; single–stranded; linear; unipartite. Genomic nucleic acid isolated by Jones: use proteinase K + phenol.

*Non–genomic nucleic acid* not found in the virions. 1 virus specified dsRNA species found in infected cells. Size of largest virus specified dsRNA 17 kbp.

*Virion protein(s)* one; $M_r$ 31000. Method of preparation: Jones *et al.* (1990).

**Replication**

Replication does not depend on a helper virus.

**Cytopathology**

Virions found in leaves and all parts of the host plant.

**Taxonomy and Relationships**

Virus(es) with serologically unrelated virions: 9 carlaviruses.

Best tests for diagnosis: ISEM and mechanical inoculation to *Chenopodium quinoa*.

**References**

Jones, A.T. (1977). *Ann. appl. Biol.* **86**: 199.

Jones, A.T. (1985). *AAB Descr. Pl. Viruses* No. 304.

Jones, A.T., Mitchell, M.J., McGavin, W.J. and Roberts, I.M. (1990). *Ann. appl. Biol.* **117**: 571.

# Winter wheat Russian mosaic (?) *cytorhabdovirus*

**Data collated by A.A. Brunt, 1991.**

*Synonyms:* Russian winter wheat mosaic virus, cereal mosaic virus. *Standard acronym:* WWRMV.

First reported in *Triticum aestivum*; from the former USSR; by Sukhov (1943).

**Natural host range and symptoms**

*Triticum aestivum* — mosaic leaf symptoms.

**Transmission**

Transmitted by a vector; an insect; *Psammotettix striatus, P. alienus*; Delphacidae. Transmitted in a persistent manner (circulative). Virus transmitted congenitally to the progeny of the vector; not transmitted by mechanical inoculation.

**Geographical distribution**

Spreads in the former USSR.

Diagnostically susceptible

*Triticum aestivum* — mosaic leaf symptoms.

Monocotyledonous Hosts *(**Susceptible** or Insusceptible)*

***Triticum aestivum***.

**Particle morphology**

Virions rhabdo– or bullet–shaped; enveloped; with a clear modal length; of 260 nm; 60 nm wide (Razvjazkina and Polyakova, 1967; 1969).

**Cytopathology**

Virions found in leaves; in cytoplasm.

**Taxonomy and Relationships**

Probable *Cytorhabdovirus: Rhabdoviridae*.

**References**

Razvjaskina, G.M. and Poljakova, G.P. (1969). *Proc. 6th Conf. Czech. Pl. Virologists, Prague 1967. Czech. Acad. Sci., Prague*, p. 129.
Sukhov, K.S. (1943). *C.R. (Dokl.) Acad. Sci. URSS* **41**: 73.
Shaskolskaya (1962). *Zool. Zh.* **41**: 717.

# Wisteria vein mosaic *potyvirus*

**Data collated by L. Bos, 1991.**

*Standard acronym:* WVMV.

First reported in *Wisteria floribunda, W. sinensis*; from the Netherlands and Italy; by Bos (1970); Conti and Lovisolo (1970).

**Natural host range and symptoms**

Symptoms persist and vary seasonally. Symptoms irregular vein mosaic.

*Wisteria floribunda, W. sinensis* — yellowing, vein clearing, flecking, mottling or mosaic, some malformation.

**Transmission**

Transmitted by a vector; an insect; *Aphis craccivora, Myzus persicae*; Aphididae. Transmitted in a non–persistent manner. Virus transmitted by mechanical inoculation; not transmitted by seed.

**Geographical distribution**

Spreads in the former Czechoslovakia, Italy, and the Netherlands (and probably several other countries).

**Experimental host range**

Several (3–9) families susceptible. Experimentally infected plants mostly show chlorotic, sometimes necrotic local lesions, sometimes followed by chlorotic mottling to mosaic.

Diagnostically susceptible

*Atriplex hortense, Chenopodium amaranticolor* — chlorotic local lesions.

*Chenopodium quinoa* — chlorotic to yellow local lesions, becoming green.

*Cicer arietinum* — chlorotic local lesions, then necrotic mosaic, stunting and withering.

*Lupinus albus* — chlorotic, later necrotic, local lesions followed by bright mosaic.

*Nicotiana clevelandii* — systemic chlorotic spots and rings, sometimes faint chlorotic local lesions.

*N. tabacum* cv. White Burley — chlorotic local lesions.

*Phaseolus vulgaris* — chlorotic, later necrotic local lesions then, usually, systemic mottling, flecking or stippling.

*Pisum sativum* — interveinal chlorotic spotting or mottling.

*Vicia faba* — chlorotic ring-like local lesions; diffuse chlorotic systemic spots.

Diagnostically insusceptible

*Beta vulgaris, Cucumis sativus* cv. Gele Tros, *Datura stramonium, Nicotiana glutinosa, Vigna unguiculata.*

Maintenance and propagation hosts

*Pisum sativum, N. clevelandii.*

Assay hosts (**L**ocal lesions or **W**hole plants)

*Chenopodium amaranticolor* (L), *C. quinoa* (L and W), *Phaseolus vulgaris* cv. Saxa (L and W).

Dicotyledonous Hosts *(**Susceptible** or Insusceptible)*

*Beta vulgaris. Capsicum annuum. Cassia tora.* ***Chenopodium album*****.** ***Chenopodium amaranticolor*****.** ***Chenopodium murale*****.** ***Chenopodium quinoa*****.** ***Cicer arietinum*****.** ***Crotalaria spectabilis*****.** *Cucumis sativus. Cucurbita pepo. Datura stramonium.* ***Glycine max*****.** ***Gomphrena globosa*****.** ***Lathyrus odoratus*****.** ***Lupinus albus*****.** *Lycopersicon esculentum.* ***Melilotus officinalis*****.** ***Nicotiana clevelandii*****.** *Nicotiana glutinosa. Nicotiana rustica.* ***Nicotiana tabacum*****.** ***Petunia hybrida*****.** ***Phaseolus lunatus*****.** ***Phaseolus vulgaris*****.** *Physalis floridana.* ***Pisum sativum*****.** ***Spinacia oleracea*****.** ***Tetragonia expansa*****.** ***Trifolium incarnatum*****.** *Trifolium pratense. Trifolium repens.* ***Vicia faba*****.** ***Vicia sativa*****.** ***Zinnia elegans*****.**

Sources of host-range data

Bos (1970); Conti and Lovisolo (1970).

**Properties of particles in sap**

TIP: 55 °C, or 60 °C. LIV: 2 days, or 12 days. DEP: $\log_{10}$ minus 3, or 5. Leaf sap contains few virions.

**Purification**

Method: Bos (1970).

**Particle morphology**

Virions filamentous; usually flexuous; with a clear modal length; of 750–775 nm. Axial canal obscure. Basic helix obscure.

**Cytopathology**

Virions found in leaves. Inclusions present in infected cells; are pinwheels and unusual in shape; granular cytoplasmic inclusions; they contain virions.

**Taxonomy and Relationships**

*Potyvirus: Potyviridae.*

Virus(es) with serologically related virions: bean yellow mosaic, bean common mosaic and clover yellow vein viruses.

Best tests for diagnosis: the virions of WVMV are only distantly related serologically to those of bean yellow mosaic virus and clover yellow vein virus when using polyclonal antisera. WVMV differs from bean yellow mosaic virus in not infecting red clover and cowpea and in infecting *Chenopodium quinoa* systemically. It differs from clover yellow vein virus in not infecting cucumber (in which clover yellow vein virus causes local lesions) and in not infecting white clover.

**References**

Bos, L. (1970). *Neth. J. Pl. Path.* **96**: 8.
Brčak, J. (1980). *Biologia Pl.* **22**: 465.
Brčak, J. and Králik, O. (1983). *Zesz. probl. Postep. Nauk roln.* **291**: 69.
Conti, M. and Lovisolo, O. (1969/70). *Riv. Patol. veg. Serie IV* **5**: 115.
de Beni, P.V. (1964). *Ricerca Scientifica* **34**: 21.

# Yam internal brown spot (?) *badnavirus*

**Data collated by A.A. Brunt and I.M. Roberts, 1987.**

*Synonyms:* dioscorea badnavirus.

First reported in *Dioscorea alata*; from Barbados; by Coursey (1967); Jeffers and Headley (1967).

**Natural host range and symptoms**

Symptoms persist.

*Dioscorea alata* cv. White Lisbon — probably causes internal brown spot of tubers, either alone or together with an uncharacterised potyvirus.

**Transmission**

Virus not transmitted by mechanical inoculation.

Diagnostically susceptible

*Dioscorea alata* — internal brown spots in tubers.

**Properties of particles in sap**

Leaf sap contains few virions. Electron microscopy: virions often bind to cell debris.

**Particle morphology**

Virions bacilliform; not enveloped; 130 nm in length; 29 nm wide.

**Cytopathology**

Virions found in clusters in phloem parenchyma; in cytoplasm (close to the nuclei). Inclusions may be present in infected cells (as clusters of virions bound together with amorphous material).

**Taxonomy and Relationships**

Possible *Badnavirus*.

**References**

Coursey, D.G. (1967). *J. Agric. Soc. Trinidad* **67**: 473.
Harrison, B.D. and Roberts, I.M. (1973). *Trop. Agric., Trin.* **50**: 335.
Jeffers, W. de C. and Headley, W. (1967). *Minist. Agric. Barbados Bull.* **46**: 7.

# Yam mosaic *potyvirus*

**Data collated by J.C. Thouvenel, 1984; F. Nienhaus, 1987.**

*Standard acronym:* YMV.

First reported in *Dioscorea cayenensis*; from the Ivory Coast; by Thouvenel and Fauquet (1977).

**Natural host range and symptoms**

Symptoms disappear soon after infection.

*Dioscorea alata, D. cayenensis, D. dumetorum, D. praehensilis, D. preussii, D. rotundata* — mosaic, vein banding.

**Transmission**

Transmitted by a vector; an insect; *Aphis craccivora, A. gossypii, Rhopalosiphum maidis, Toxoptera citricidus*; Aphididae. Transmitted in a non–persistent manner. Virus transmitted by mechanical inoculation; not transmitted by contact between plants; not transmitted by seed (is by vegetative propagation).

**Geographical distribution**

Spreads in the African region (in the west); Guadeloupe and Guyana.

**Experimental host range**

Few (<3) families susceptible.

Diagnostically susceptible

*Nicotiana benthamiana* — faint systemic chlorosis and mosaic.
*Dioscorea cayenensis, D. praehensilis* — systemic mosaic and vein banding.

Diagnostically insusceptible

*Dioscorea composita, D. floribunda.*

Maintenance and propagation hosts

*Nicotiana benthamiana, Dioscorea cayenensis.*

Assay hosts (**L**ocal lesions or **W**hole plants)

*Nicotiana benthamiana* (W), *Dioscorea cayenensis* (W).

Dicotyledonous Hosts *(**Susceptible** or Insusceptible)*

***Nicotiana benthamiana*.**

Sources of host-range data

Thouvenel and Fauquet (1979).

**Properties of particles in sap**

TIP: 55–60 °C. LIV: 0.5–1 days. DEP: $\log_{10}$ minus 2–3. Infectivity of sap not changed by treatment with di–ethyl ether. Leaf sap contains few virions.

**Purification**

Method: Thouvenel and Fauquet (1979).

**Particle morphology**

Virions filamentous; not enveloped; usually flexuous; with a clear modal length; of 785 nm; 15 nm wide. Axial canal obscure. Basic helix obscure.

**Physical properties**

One sedimenting component in purified preparations. Isoelectric point pH 4.3.

**Biochemical properties**

Virions contain 6 % nucleic acid; 94 % protein; 0 % lipid.

*Genome* consists of RNA; single–stranded. Infectivity retained when deproteinised with phenol or detergent.

*Virion protein(s)* one; $M_r$ 34000.

**Cytopathology**

Virions found in all parts of the host plant; in cytoplasm. Inclusions present in infected cells; are pinwheels; they do not contain virions.

**Taxonomy and Relationships**

*Potyvirus: Potyviridae.*

Virus(es) with serologically related virions: groundnut eyespot, Guinea grass mosaic and pepper veinal mottle viruses, but distantly.

**References**

Thouvenel, J.C. and Fauquet, C. (1977). *C. r. Acad. Sci. Paris* **284**: 1947.

Thouvenel, J.C. and Fauquet, C. (1979). *Ann. appl. Biol.* **93**: 279.

Thouvenel, J.-C. and Fauquet, C. (1986). *AAB Descr. Pl. Viruses* No. 314, 4 pp.

## Yucca bacilliform (?) *badnavirus*

**Data collated by A.A. Brunt, 1986; R. Milne, 1987.**

*Standard acronym:* YBV.

First reported in *Yucca elephantipes*; from Italy in plants imported from Guatemala; by Milne *et al.* (1985).

**Natural host range and symptoms**

*Yucca elephantipes* — chlorotic lesions which later become necrotic.

**Particle morphology**

Virions bacilliform; 120–180 nm in length; 30 nm wide.

**Cytopathology**

Virions found in vascular parenchyma; in cytoplasm.

**Taxonomy and Relationships**

Possible *Badnavirus*.

**References**

Milne, R.G., Masenga, V., Accotto, G.P., Scapiri, I. and D'Agliano, G. (1985). *Inform. Fitopatol.* **6**: 43.

# Zinnia mild mottle (?) potyvirus

**Data collated by A.A. Brunt, 1991.**

First reported in *Zinnia elegans*; from India; by Padma *et al.* (1974).

**Natural host range and symptoms**

Symptoms persist.
*Zinnia elegans* — faint leaf mottling.

**Geographical distribution**

Spreads in India.

Dicotyledonous Hosts *(**Susceptible** or Insusceptible)*

***Zinnia elegans***.

**Particle morphology**

Virions filamentous; not enveloped; usually flexuous; with a clear modal length; of 730 nm. Axial canal obscure. Basic helix obscure.

**Taxonomy and Relationships**

Unassigned potyvirus.

**References**

Padma, R., Phatak, H.C., Singh, S. and Verma, V.S. (1974). *Hort. Res.* **14**: 55.

# Zoysia mosaic (?) potyvirus

**Data collated by A.A. Brunt, 1991.**

*Standard acronym:* ZMV.

First reported in *Zoysia* sp.; from Japan; by Toriyama and Yora (1972).

**Natural host range and symptoms**

*Zoysia* sp.

**Geographical distribution**

Spreads in Japan.

**Particle morphology**

Virions filamentous; not enveloped; usually flexuous; with a clear modal length; of 750 nm. Axial canal obscure.

**Taxonomy and Relationships**

Unassigned potyvirus.

**References**

Toriyama, S. and Yora, K. (1972). In: *Virus Diseases of Wild Grasses and Cereal Crops in Japan*. University of Tokyo Press, Tokyo, Japan.

# Zucchini yellow fleck *potyvirus*

**Data collated by C. Vovlas, 1991.**

*Standard acronym:* ZYFV.

First reported in *Cucurbita pepo*; from Apulia, southern Italy.

## Natural host range and symptoms

Symptoms persist.

*Cucurbita pepo* — yellow pin-points on leaves, becoming completely yellow and necrotic, stunting and fruitlessness with severe infection.

## Transmission

Transmitted by a vector; an insect; Aphididae. Transmitted in a non–persistent manner. Virus transmitted by mechanical inoculation; transmitted by contact between plants.

## Geographical distribution

Spreads in Italy (the south).

## Experimental host range

Several (3–9) families susceptible. Experimentally infected plants mostly show yellow local lesions, systemic mottling.

Diagnostically susceptible

*Cucumis sativus* — necrotic local lesions, systemic mottling.
*C. melo, Cucurbita maxima* — systemic yellow mottle.
*Cucurbita pepo* — local lesions, total necrosis.
*Citrullus lanatus* — small local lesions and yellow flecking.
*Lagenaria siceraria* — systemic vein clearing, mottling.

Diagnostically insusceptible

*Chenopodium quinoa, C. amaranticolor, Gomphrena globosa, Vigna sinensis, Phaseolus vulgaris* cvs Bountiful and La Victoire.

Maintenance and propagation hosts

*Cucumis sativus, Cucurbita maxima, Cucumis melo, Lagenaria siceraria.*

Assay hosts (**L**ocal lesions or **W**hole plants)

*Cucurbita pepo* (L), *Cucumis sativus* (W), *Cucumis melo* (W), *Citrullus lanatus* (L).

Dicotyledonous Hosts *(**Susceptible** or Insusceptible)*

*Celosia cristata. Chenopodium amaranticolor. Chenopodium quinoa.* ***Citrullus lanatus. Cucumis melo. Cucumis sativus. Cucurbita maxima. Cucurbita pepo.*** *Gomphrena globosa. Lupinus albus. Nicotiana clevelandii. Phaseolus vulgaris. Vicia faba. Zinnia elegans.*

Additional hosts

*Luffa acutangula, Ocimum basilicum, Nicotiana tabacum* cv. White Burley, *Nicotiana tabacum* cv. Xanthi.

Sources of host-range data

Vovlas *et al.* (1981).

## Properties of particles in sap

TIP: 55–60 °C. LIV: 30 days. DEP: $\log_{10}$ minus 4–5.

**Purification**

Method: systemically infected leaves homogenised in potassium phosphate buffer and carbon tetrachloride with chloroform (1/2 volume each) for clarification. Precipitation with PEG.

**Particle morphology**

Virions filamentous; with a clear modal length; of 750 nm.

**Physical properties**

One sedimenting component in purified preparations. Density 1.30.

**Replication**

Replication does not depend on a helper virus.

**Cytopathology**

Virions found in leaves; in cytoplasm. Inclusions present in infected cells; are crystals in the cytoplasm, membranous bodies, pinwheels, and unusual in shape; amorphous inclusions; they contain virions. Other cellular changes: cytological modifications caused mainly by fibriate bodies and other types of inclusions.

**Taxonomy and Relationships**

*Potyvirus: Potyviridae.*

**References**

Martelli, G.P., Russo, M. and Vovlas, C. (1981). *Phytopathol. Medit.* **20**: 193.
Vovlas, C. (1979). *3rd Conf. of ISHS Working Group on Vegetable Crops*, Bari, Italy.
Volvas, C., Hiebert, E. and Russo, M. (1981). *Phytopathol. Medit.* **20**: 123.

## Zucchini yellow mosaic *potyvirus*

**Data collated by C. Büchen-Osmond and D. Purcifull, 1987.**

*Synonyms:* muskmelon yellow stunt virus (Lecoq *et al.*, 1981; 1983). *Standard acronym:* ZYMV. *Strains:* 22 isolates of the virus have been grouped into three pathotypes by their effect on muskmelon line PI 414723 (Pitrat and Lecoq, 1984; Risser *et al.*, 1981; Lecoq and Pitrat, 1985 and unpublished data). ZYMV isolates differ in the symptoms they cause (Lecoq *et al.*, 1981; Provvidenti *et al.*, 1984), aphid transmissibility (Lecoq, 1986) or virulence towards a resistance gene (Lecoq and Pitrat, 1985).

First reported in *Cucurbita pepo*; from Italy; by Lisa *et al.* (1981).

**Natural host range and symptoms**

Symptoms persist.

*Cucurbita pepo* (zucchini squash), *Cucumis melo* (muskmelon), *Cucumis sativus* (cucumber), and *Citrullus lanatus* (watermelon) — mosaic, yellowing, shoestring leaves, stunting, and fruit and seed deformation.

*Melothria pendula* — mosaic, yellowing.

**Transmission**

Transmitted by a vector; an insect; *Aphis citricola* (Purcifull *et al.*, 1984), *Aphis gossypii; Myzus persicae* (Lecoq *et al.*, 1981; Lisa *et al.*, 1981) and *Macrosiphum euphorbiae* (Lecoq, unpublished data) and *Aphis middletonii, Aphis craccivora, Acrythosiphon pisum, Lipaphis erysimi, Uroleucon* sp. (Adlerz, 1987); Aphididae. Transmitted in a non–persistent manner. Virus transmitted by mechanical inoculation; not transmitted by seed (Lecoq *et al.*, 1981; Dodds *et al.*, 1984).

**Geographical distribution**

Spreads in Algeria, Australia, Egypt, France, Germany, Israel, Italy, Japan, Jordan, Lebanon, Mauritius, Morocco, Spain, Taiwan, Turkey, the UK, and the USA (Zucchini yellow mosaic virus has been isolated from the wild perennial cucurbit *Melothria pendula* in Florida (Adlerz *et al.*, 1983*b*)).

**Experimental host range**

Several (3–9) families susceptible.

Diagnostically susceptible

*Chenopodium amaranticolor, C. quinoa* — chlorotic local lesions; not systemic.

*Cucumis melo* — chlorotic local lesions; systemic vein clearing, yellowing, mosaic, leaf deformation, stunting and occasional necrosis.

*Cucurbita okeechobeensis* — systemic mosaic (N.B. this species is not susceptible to most cucumber mosaic virus isolates).

*Cucurbita pepo* — chlorotic local lesions; systemic vein netting, yellowing, mosaic and leaf deformation, often lethal.

*Gomphrena globosa* — local lesions; not systemic (N.B. this species is not infected by squash mosaic virus).

*Luffa acutangula* — systemic mosaic or latent.

*Ranunculus sardous* — symptomless systemic infection with most isolates (N.B. this species is not infected by papaya ringspot virus-W or watermelon mosaic virus 2).

Diagnostically insusceptible

*Lavatera trimestris* — but watermelon virus 2 induces necrotic local lesions.

Maintenance and propagation hosts

*Cucurbita pepo*.

Assay hosts (**L**ocal lesions or **W**hole plants)

*Chenopodium amaranticolor* (L), *Cucurbita pepo* (W).

Dicotyledonous Hosts *(**Susceptible** or Insusceptible)*

***Chenopodium amaranticolor*. *Chenopodium quinoa*. *Citrullus lanatus*. *Cucumis melo*. *Cucumis metuliferus*. *Cucumis sativus*. *Cucurbita maxima*. *Cucurbita moschata*. *Cucurbita okeechobeensis*. *Cucurbita pepo*. *Gomphrena globosa*.** *Lavatera trimestris*. ***Luffa acutangula*. *Phaseolus vulgaris*. *Pisum sativum*. *Ranunculus sardous*. *Senecio vulgaris*.**

Sources of host-range data

Lecoq *et al.* (1983); Lesemann *et al.* (1983); Provvidenti *et al.* (1984); Lecoq *et al.* (1981); Lisa *et al.* (1981); Purcifull *et al.* (1984).

**Properties of particles in sap**

TIP: 55–60 °C. LIV: 3–5 days (at room temperature). DEP: $\log_{10}$ minus 4–5. Leaf sap contains few virions.

**Purification**

Method: Lisa *et al.* (1981); Lisa and Lecoq (1984); Lesemann *et al.* (1983); Purcifull *et al.* (1984).

**Particle morphology**

Virions filamentous; not enveloped; usually flexuous; with a clear modal length; of 750 nm; 11 nm wide. Axial canal obscure. Basic helix obscure.

**Physical properties**

One sedimenting component in purified preparations. Density 1.323 g $cm^{-3}$ in CsCl (at 10°C).

**Biochemical properties**

Virions contain 4.5–7 % nucleic acid; 93–95.5 % protein.

*Genome* consists of RNA; single–stranded; linear; unipartite; largest (or only) genome part 9 kb.

*Sequence database accession code(s):*

D00593 Em(40)_vi:ZYMCP Zucchini yellow mosaic virus, nuclear inclusion protein and coat protein genes. 2/92 1,963bp.

D00692 Em(40)_vi:ZYMV Gb(84)_vi:ZYMV Zucchini yellow mosaic virus gene, 3' terminal region, coat protein and RNA polymerase. 6/91

D13914 Em(40)_vi:ZYMCP1 Gb(84)_vi:ZYMCP Zucchini yellow mosaic virus, nuclear inclusion protein and coat protein genes. 1/94 1,944bp.

L29569 Em(40)_vi:ZYMP13PRE Gb(84)_vi:ZYMP13PREP Zucchini yellow mosaic virus polyprotein; P1 protease; helper component-protease; P3 protein;

L31350 Em(40)_vi:ZYMPROPOL Gb(84)_vi:ZYMPROPOLR Zucchini yellow mosaic virus polyprotein, complete cds; P1 protease; P2 HC-protease

L35588 Gb(84)n:ZYMP1PROTA Zucchini yellow mosaic virus P1 protease, 5' end. 8/94 918bp.

L35589 Gb(84)n:ZYMP1PROTB Zucchini yellow mosaic virus P1 protease, 5' end. 8/94 897bp.

L35590 Gb(84)n:ZYMP13PROT Zucchini yellow mosaic virus P1 protease, helper component, and P3 protease; 5' end. 8/94 3,5

M35095 Em(40)_vi:ZYMCPA Gb(84)_vi:ZYMCPA Zucchini yellow mosaic virus coat protein (cp) mRNA, 3' end. 7/90 1,374bp

S46009 Em(40)_un:S46009 Gb(84)_vi:S46009 coat protein zucchini yellow mosaic virus ZYMV, aphid-transmissible isolate, Genomic RNA, 12

X62662 Em(40)_vi:ZYMVCP Gb(84)_vi:ZYMVCP Zucchini yellow mosaic virus (ZYMV-S) mRNA for coat protein. 4/94 1,250bp.

X68509 Em(40)_vi:ZYMVSR Gb(84)_vi:ZYMVSR Zucchini Yellow Mosaic Virus-S RNA. 3/93 3,048bp.

X77756 Em(40)_vi:ZYMVHCPRO Gb(84)_vi:ZYMVHCPRO Zucchini Yellow Mosaic Virus (ZYMV) HC-PRO gene. 6/94 1,368bp.

Z22759 Em(40)_vi:ZCHELCOMA Gb(84)_vi:ZCHELCOMA Zucchini yellow mosaic virus helper component RNA. 3/94 1,368bp.

*Virion protein(s)* one; $M_r$ 36000; coat protein. Method of preparation: Lisa *et al.* (1981).

*Virus–coded non–virion proteins* identified by genomic sequence analysis; seven proteins found. $M_r$ of the largest 39000. $M_r$ of 2nd largest 51000; helper component. $M_r$ of 3rd 50000. $M_r$ of 4th 70000; cylindrical inclusion. $M_r$ of 5th and smaller 49000, or 58000, or 34000; protease, polymerase, coat protein.

**Replication**

Replication does not depend on a helper virus.

**Cytopathology**

Virions found in all parts of the host plant; in cytoplasm. Inclusions present in infected cells; are pinwheels (but not laminated aggregates (Lisa *et al.*, 1981)); they do not contain virions (endoplasmic reticulum and vesicles containing fibrillar material accumulate (Lesemann *et al.*, 1983)).

**Taxonomy and Relationships**

*Potyvirus: Potyviridae.*

Virus(es) with serologically related virions: watermelon mosaic virus II, bean yellow mosaic virus (Lecoq *et al.*, 1981; Lisa *et al.*, 1981; Adlerz *et al.*, 1983*a*; Lesemann, 1983; Lesemann *et al.*, 1983; Makkouk and Abbasher, 1983; Purcifull *et al.*, 1984) and amaranthus leaf mottle viruses (Lovisolo and Lisa, 1976; V. Lisa and G. D'Agostino, unpublished data), but distantly.

No serological differences have been detected between Italian and other isolates of the virus (Lecoq *et al.*, 1983; Lesemann *et al.*, 1983; Purcifull *et al.*, 1984; Provvidenti *et al.*, 1984). Different isolates cross-protect in muskmelon (Lecoq *et al.*, 1981). Best tests for diagnosis: Some isolates of ZYMV cause symptoms similar to those attributed to other cucurbit viruses (cucumber mosaic virus, watermelon mosaic virus 1, strain W of papaya ringspot virus, watermelon mosaic virus 2 and squash mosaic virus), thus a diagnosis based on symptoms is uncertain, and serological tests must be used. ZYMV frequently occurs with other viruses in cucurbits, but can be isolated alone by inoculating, for example, *Cucurbita okeechobeensis* and *Ranunculus sardous*. ZYMV can be separated from squash mosaic by aphid transmission or by inoculation to *Gomphrena globosa*. Zucchini yellow fleck potyvirus, found in the the Mediterranean region (Vovlas *et al.*, 1981), is not serologically related to ZYMV (M. Russo, unpublished data).

**References**


Adlerz, W.C. (1987). *Econ. Ent.* **71**: 531.

Adlerz, W.C., Purcifull, D.E., Simone, G.W. and Hiebert, E. (1983*a*). *Abstr. 4th Int. Congr. Pl. Path., Melbourne,* 1983, No. **440**:

Adlerz, W.C., Purcifull, D.E., Simone, G.W. and Hiebert, E. (1983*b*). *Proc. Fla Stn hort. Soc.* **96**: 72.

Dodds, J.A., Lee, J.G., Nameth, S.T. and Laemmlen, F.F. (1984). *Phytopathology* **74**: 221.

Hiebert, E., Thornbury, H.H. and Pirone, T.P. (1984). *Virology* **135**: 1.

Lecoq, H. (1986). *Phytopathology* **76**: 1063.

Lecoq, H. and Pitrat, M. (1985). *Phytopathology* 75: 890.

Lecoq, H., Pitrat, M. and Clement, M. (1981). *Agronomie* **1**: 827.

Lecoq, H., Lisa, V. and Dellavalle, G. (1983). *Plant Dis.* **67**: 824.

Lesemann, D.-E. (1983). *Acta Hort.* **127**: 159.

Lesemann, D.-E., Makkouk, K.M., Koenig, R. and Samman, E.N. (1983). *Phytopath. Z.* **108**: 304.

Lisa, V. and Lecoq, H. (1984). *CMI/AAB Descr. Pl. Viruses* No. 282, 4 pp.

Lisa, V., Boccardo, G., D'Agostino, G., Dellavalle, G. and d'Aquilio, M. (1981). *Phytopathology* **71**: 667.

Lovisolo, O. and Lisa, V. (1976). *Poljopr. znanst. Smotra* **39(49)**: 553.

Makkouk, K.M. and Abbasher, A. (1983). *Abstr. 4th Int. Congr. Pl. Path., Melbourne,* 1983, No. **472**:

Pitrat, M. and Lecoq, H. (1984). *Euphytica* **33**: 57.

Provvidenti, R., Gonsalves, D. and Humaydan, H. (1984). *Plant Dis.* **68**: 443.

Purcifull, D.E., Adlerz, W.C., Simone, G.W., Hiebert, E. and Christie, R.G. (1984). *Plant Dis.* **68**: 230.

Risser, G., Pitrat, M., Lecoq, H. and Rode, J.-C. (1981). *Agronomie* **1**: 835.

Vovlas, C., Hiebert, E. and Russo, M. (1981). *Phytopathol. Medit.* **20**: 123.

# Zygocactus Montana X (?) *potexvirus*

**Data collated by M. Chessin, 1989.**

*Synonyms:* zygocactus virus X, ZyV-58, cactus zygocactus virus.

First reported in *Zygocactus truncatus* (Haworth) Schumann; from Missoula, Montana, U.S.A.; by Chessin and Giri (1975).

**Natural host range and symptoms**

Symptoms persist.
*Zygocactus truncatus* — reddening of pads.

**Transmission**

Virus transmitted by mechanical inoculation.

**Geographical distribution**

Found, but with no evidence of spread, in the U.S.A.

**Experimental host range**

Few (<3) families susceptible. Experimentally infected plants mostly show systemic mosaics, mottles, ringspots or necrosis.

Diagnostically susceptible

*Atriplex hastata* — systemic leaf mottling.
*Chenopodium quinoa* — necrotic local lesions, systemic mosaic.
*Chenopodium polyspermum* — systemic leaf mottling.
*Chenopodium album, C. giganteum, C. urbicum* — chlorotic or necrotic local lesions; not systemic.
*Hablitzia tamnoides* — chlorotic local lesions; systemic green flecking.

Diagnostically insusceptible

*Amaranthus tricolor, Chenopodium capitatum, C. hybridum, C. foliosum, Datura stramonium, Gomphrena globosa, Nicotiana tabacum, N. rustica, N. sylvestris, Phaseolus vulgaris.*

Maintenance and propagation hosts

*Chenopodium quinoa.*

Assay hosts (**L**ocal lesions or **W**hole plants)

*Chenopodium quinoa* (L).

Dicotyledonous Hosts *(**Susceptible** or Insusceptible)*

***Amaranthus caudatus***. ***Chenopodium album***. *Chenopodium capitatum. Chenopodium foliosum. Chenopodium hybridum.* ***Chenopodium quinoa***. *Datura stramonium. Gomphrena globosa. Nicotiana glutinosa. Nicotiana rustica. Nicotiana sylvestris. Nicotiana tabacum. Phaseolus vulgaris.*

Sources of host-range data

Giri (1974).

**Properties of particles in sap**

TIP: 75–80 °C. LIV: 6–7 days. DEP: $\log_{10}$ minus 5–6. Leaf sap contains many virions. Electron microscopy: PTA with 0.1% (w/v) bovine serum albumin.

**Purification**

Method: cycles of differential and density gradient centrifugation in CsCl at *c.* 100000 rpm for 12–18 hours.

**Particle morphology**

Virions filamentous; not enveloped; usually flexuous; with a clear modal length; of 519 nm (±10). Axial canal obscure. Basic helix obscure.

**Physical properties**

Two sedimenting components in purified preparations; sedimentation coefficient of the fastest 120–123 *S* (a component sedimenting at *c.*200*S* probably consists of virion aggregates).

**Cytopathology**

Virions found in mesophyll and epidermis; in cytoplasm. Inclusions present in infected cells; are unusual in shape; spindles; they contain virions.

**Taxonomy and Relationships**

Possible *Potexvirus*.

Virus(es) with serologically related virions: potato X, cactus X and white clover mosaic viruses. Virus(es) with serologically unrelated virions: zygocactus potexvirus (Casper and Brandes, 1969), potato aucuba mosaic, clover yellow mosaic and narcissus mosaic viruses.

**References**

Casper, R. and Brandes, J. (1969). *J. gen. Virol.* **5**: 155.
Chessin, M. and Giri, L. (1975). *Phytopath. Z.* **83**: 40.
Giri, L. (1974). *Ph.D. Thesis*, University of Montana, U.S.A.

# Zygocactus (?) *potexvirus*

**Data collated by A.A. Brunt, 1987 and R. Casper, 1987.**

*Standard acronym:* ZV.

First reported in *Zygocactus x schlumbergera*; from Europe; by Casper and Brandes (1969).

**Natural host range and symptoms**

*Zygocactus x Schlumbergera* — symptomless.

**Transmission**

Virus transmitted by mechanical inoculation.

**Geographical distribution**

Spreads in the Eurasian region (but is probably more widespread).

**Experimental host range**

Few (<3) families susceptible.

Diagnostically susceptible

*Chenopodium quinoa* — slight systemic leaf curling.
*Nicotiana glutinosa* — systemic light green spotting leaves.
*N. clevelandii, Solanum demissum* — symptomless systemic infection.

Maintenance and propagation hosts

*Nicotiana glutinosa, N. clevelandii.*

Assay hosts (**L**ocal lesions or **W**hole plants)

*Nicotiana glutinosa* (W).

Dicotyledonous Hosts *(**Susceptible** or Insusceptible)*

***Chenopodium quinoa. Nicotiana clevelandii. Nicotiana glutinosa. Solanum demissum.***

Sources of host-range data

Casper and Brandes (1969).

**Properties of particles in sap**

TIP: 72–74 °C. Leaf sap contains few virions.

**Particle morphology**

Virions filamentous; not enveloped; usually flexuous; with a clear modal length; of 580 nm. Axial canal obscure. Basic helix obscure.

**Taxonomy and Relationships**

Possible *Potexvirus*.

Virus(es) with serologically unrelated virions: cactus X, potato X, potato aucuba and zygocactus X viruses.

**References**

Casper, R. and Brandes, J. (1969). *J. gen. Virol.* **5**: 155.

# APPENDIX 1. ABBREVIATIONS

| | |
|---|---|
| A | adenine |
| $A_{260/280}$ | ratio of absorption of light of 260 and 280 nm wavelength |
| AM | ammonium molybdate |
| (B) | bottom (fastest sedimenting) component of multicomponent virus |
| C | cytosine |
| c-DNA | complementary deoxyribonucleic acid |
| DAS-ELISA | double antibody sandwich form of ELISA |
| DEP | dilution end point of the infectivity of sap |
| (ds)DNA | (double-stranded) deoxyribonucleic acid |
| (ds)RNA | (double-stranded) ribonucleic acid |
| EDTA | ethylene diamine tetraacetate |
| ELISA | enzyme-linked immuno-sorbent assay |
| G | guanine |
| G+C | guanine plus cytosine |
| ICTV | International Committee on the Taxonomy of Viruses |
| ISEM | immuno-sorbent electron microscopy |
| Kb | kilobases; the size of a ssDNA or ssRNA molecule |
| Kbp | kilobase pairs; the size of a dsDNA or dsRNA molecule |
| kDa | kilodaltons; the size of a protein |
| KPT | potassium phosphotungstate |
| LIV | longevity of the infectivity of sap *in vitro* |
| (M) | middle component of multicomponent virus |
| (MAb) | monoclonal antibody |
| MLO(s) | mycoplasma-like organism(s) or phytoplasma |
| Mr | relative molecular mass |
| mRNA | messenger RNA |
| nm | nanometers |
| nt(s) | nucleotide(s) |
| ORF(s) | open reading frame(s) |
| PEG | polyethylene glycol |
| *S* | sedimentation coefficient; Svedberg units |
| $s_{20°,w}$ | corrected sedimentation coefficient; extrapolated to that in water at 20°C and infinitely diluted |
| SDI | serological differentiation index |
| SDS | sodium dodecyl sulphate |
| ssDNA | single-stranded DNA |
| ssRNA | single-stranded RNA |
| T | thymine |
| (T) | top (slowest sedimenting) component of a virus with more than one type of virion |
| TIP | thermal inactivation point; lowest temperature at which infectivity of sap is destroyed in 10 min |
| U | uracil |
| UA | uranyl acetate |
| UF | uranyl formate |
| Vpg | genome-linked protein |

# APPENDIX 2. CHARACTER LIST

#1. <Data collated by: author, year>/

*Synonyms*

#2. Synonyms:/
#3. Standard acronym:/

*Strains*

#4. Strains: <of the virus>/
#5. Definitive species <of the genus>/
#6. Tentative species <of the genus>/

*Historical details*

#7. First reported in <plant species>/
#8. <First reported> from <locality>/
#9. <Key early references> by/

**Natural host range and symptoms**

#10. <Intensity and duration of> symptoms <shown by naturally infected plants>/
1. persist/
2. vary cyclically over a few weeks/
3. vary seasonally/
4. disappear soon after infection/
5. none/

#11. Symptoms <shown by naturally infected plants are>/
#12. <Natural hosts and symptoms>/

**Transmission and ecology**

#13. <Whether> transmitted by <vector>/
1. a vector/
2. means not involving a vector/

#14. <Comments on transmission by unknown means>/
#15. Transmitted by <vector type>/
1. an insect/
2. a mite/
3. a nematode/
4. a fungus/
5. a vector that is not an insect, mite, nematode or fungus/

#16. Transmitted by <vector species that are insect, mite, nematode or fungus>/
#17. Transmitted by <insect group providing vector species>/
1. Aleyrodidae/
2. Aphididae/
3. Cicadellidae/
4. Delphacidae/
5. Gymnocerata/
6. Psyllidae/
7. Membracidae/
8. Pseudococcidae/
9. Coleoptera/
10. Thysanoptera/
11. an unusual insect group <specify>/

#18. Transmitted by <mite group providing vector species>/
1. Eriophyidae/
2. Tetranychidae/
3. neither Eriophyidae nor Tetranychidae/

#19. Transmitted by <nematode group providing vector species>/
1. Dorylamidae/
2. Trichodoridae/
3. neither Dorylamidae nor Trichodoridae/

#20. Transmitted by <fungus group providing vector species>/
1. Plasmodiophorales/
2. Chytridiales/
3. neither Plasmodiophorales nor Chytridiales/

#21. Principal natural vector(s): <species of insect, mite, nematode or fungus that are principal natural vectors>/

#22. Transmitted by <the following vector species, that are neither insect, mite, nematode nor fungus>/

#23. Not transmitted by <species (and group) of known non–vectors>/

#24. Transmitted <whether in a persistent /non-pers /semi-persistent manner by the vector>/
  1. in a non–persistent manner/
  2. in a semi–persistent manner/
  3. in a persistent manner/

#25. Virus <whether retained or lost when the vector moults>/
  1. retained when the vector moults/
  2. lost by the vector when it moults/

#26. Virus <whether it multiplies in the vector>/
  1. multiplies in the vector/
  2. does not multiply in the vector/

#27. Virus <whether transmitted congenitally from the vector to its progeny>/
  1. transmitted congenitally to the progeny of the vector/
  2. not transmitted congenitally to the progeny of the vector/

#28. Virus <whether it requires, or is required by, another virus for vector transmission>/
  1. requires, for vector transmission, a helper virus <named>/
  2. does not require a helper virus for vector transmission/
  3. can help the vector transmission of another virus <named virus>/

#29. <Details of how the helper virus aids transmission>/

#30. Virus <whether transmitted by mechanical inoculation>/
  1. transmitted by mechanical inoculation/
  2. not transmitted by mechanical inoculation/

#31. Virus <whether transmitted by grafting>/
  1. transmitted by grafting/
  2. not transmitted by grafting/

#32. Virus <whether transmitted by contact between plants>/
  1. transmitted by contact between plants/
  2. not transmitted by contact between plants/

#33. Virus <whether transmitted by seed>/
  1. transmitted by seed/
  2. not transmitted by seed/

#34. Virus <whether transmitted by pollen>/
  1. transmitted by pollen to the seed/
  2. transmitted by pollen to the pollinated plant/
  3. not transmitted by pollen/

#35. <Ecology and control> studies reported by/

**Geographical distribution**

#36. <Whether distributed worldwide or not>/
  1. probably distributed worldwide/
  2. geographical distribution probably restricted/

#37. Spreads in <major geographic regions>/
  1. the African region <and Arabian Peninsula>/
  2. Antarctica <and sub-antarctic islands>/
  3. the Arctic <Greenland and sub-arctic islands>/
  4. the Central Asian region <India, Siberia>/
  5. the Eastern Asian region <China, Indonesia, Japan>/
  6. the Eurasian region/
  7. the Mediterranean region/
  8. the Middle East/
  9. the North American region <Alaska, Canada, Mexico, U.S.A.>/
  10. the Pacific region <Australasia, Pacific Islands>/

11. the South and Central American region/

#38. Spreads in <countries>/

1. Afghanistan/
2. Albania/
3. Algeria/
4. American Samoa/
5. Andorra/
6. Angola/
7. Anguilla Island/
8. Antarctica/
9. Antigua and Barbuda/
10. Argentina/
11. Aruba/
12. Australia/
13. Austria/
14. the Bahamas/
15. Bahrain/
16. Bangladesh/
17. Barbados/
18. Belgium/
19. Belize/
20. Benin/
21. Bermuda/
22. Bhutan/
23. Bolivia/
24. Botswana/
25. Brazil/
26. the British Indian Ocean Territory/
27. the British Virgin Islands/
28. Brunei Darussalam/
29. Bulgaria/
30. Burkina Faso/
31. Burma (Myanmar)/
32. Burundi/
33. Cameroon/
34. Canada/
35. Cape Verde/
36. the Cayman Islands/
37. the Central African Republic/
38. Chad/
39. Chile/
40. China/
41. the Christmas Islands/
42. the Cocos Islands/
43. Colombia/
44. Comoros/
45. the Congo/
46. the Cook Islands/
47. Costa Rica/
48. Cote d'Ivoire/
49. Cuba/
50. Cyprus/
51. the former Czechoslovakia/
52. Democratic Yemen/
53. Denmark/
54. Djibouti/
55. Dominica/
56. the Dominican Republic/
57. East Timor/
58. Ecuador/
59. Egypt/
60. El Salvador/
61. Equatorial Guinea/
62. Ethiopia/
63. the Falkland Islands/
64. the Faroe Islands/
65. Fiji/
66. Finland/
67. France/
68. French Guiana/
69. French Polynesia/
70. Gabon/
71. Gambia/
72. Germany/
73. Ghana/
74. Gibraltar/
75. Greece/
76. Greenland/
77. Grenada/
78. Guadeloupe/
79. Guam/
80. Guatemala/
81. Guinea/
82. Guinea Bissau/
83. Guyana/
84. Haiti/
85. Heard and McDonald Islands/
86. Honduras/
87. Hong Kong/
88. Hungary/
89. Iceland/
90. India/
91. Indonesia/
92. Iran/
93. Iraq/
94. Ireland/
95. Israel/
96. Italy/
97. Jamaica/
98. Japan/

99. Jordan/
100. Kampuchea/
101. Kenya/
102. Kiribati/
103. Korea D.P.R. (North)/
104. Korea Republic/
105. Kuwait/
106. Lao/
107. Lebanon/
108. Lesotho/
109. Liberia/
110. Libya/
111. Lichtenstein/
112. Luxembourg/
113. Macau/
114. Madagascar/
115. Malawi/
116. Malaysia/
117. the Maldives/
118. Mali/
119. Malta/
120. the Marshall Islands/
121. Martinique/
122. Mauritania/
123. Mauritius/
124. Mexico/
125. Micronesia/
126. Monaco/
127. Mongolia/
128. Montserrat/
129. Morocco/
130. Mozambique/
131. Namibia/
132. Nauru/
133. Nepal/
134. the Netherlands/
135. Netherlands Antilles/
136. New Caledonia/
137. New Zealand/
138. Nicaragua/
139. Niger/
140. Nigeria/
141. Niue/
142. Norfolk Island/
143. the Northern Mariana Islands/
144. Norway/
145. Oman/
146. Pakistan/
147. Palau/
148. Panama/
149. Papua New Guinea/
150. Paraguay/
151. Peru/
152. the Philippines/
153. Pitcairn/
154. Poland/
155. Portugal/
156. Puerto Rico/
157. Qatar/
158. Reunion/
159. Romania/
160. Rwanda/
161. San Marino/
162. Sao Tome and Principe/
163. Saudi Arabia/
164. Senegal/
165. the Seychelles/
166. Sierra Leone/
167. Singapore/
168. the Solomon Islands/
169. Somalia/
170. South Africa/
171. Spain/
172. Sri Lanka/
173. St. Helena/
174. St. Kitts and Nevis/
175. St. Lucia/
176. St. Pierre and Miquelon/
177. St. Vincent and Grenadines/
178. Sudan/
179. Suriname/
180. Swaziland/
181. Sweden/
182. Switzerland/
183. Syria/
184. Taiwan/
185. Tanzania/
186. Thailand/
187. Togo/
188. Tokelau/
189. Tonga/
190. Trinidad and Tobago/
191. Tunisia/
192. Turkey/
193. the Turks and Caicos Islands/
194. Tuvalu/
195. the UK/
196. the USA/
197. the former USSR/
198. Uganda/
199. the United Arab Emirates/
200. the United States Virgin

Islands/
201. Uruguay/
202. Vanuatu/
203. Vatican/
204. Venezuela/
205. Vietnam/
206. Wallis and Futuna Islands/
207. Western Sahara/
208. Western Samoa/
209. Yemen/
210. the former Yugoslavia/
211. Zaire/
212. Zambia/
213. Zimbabwe/

#39. Found, but with no evidence of spread, in <countries>/

**Experimental host range**

#40. <Experimental host range wide or narrow>/
1. many (>9) families susceptible/
2. several (3–9) families susceptible/
3. few (<3) families susceptible/

#41. Experimentally infected plants mostly show <symptoms of>/
#42. <Most useful susceptible diagnostic host species and symptoms>/
#43. <Diagnostically insusceptible host species>/
#44. <Useful maintenance and propagation host species>/
#45. <Best assay hosts (Local lesions or Whole plants)>/

***Dicotyledonous hosts***

#46. /
1. ***Acacia baileyana*** <susceptible>/
2. *Acacia baileyana* <insusceptible>/

#47. /
1. ***Adansonia digitata*** <susceptible>/
2. *Adansonia digitata* <insusceptible>/

#48. /
1. ***Ageratum conyzoides*** <susceptible>/
2. *Ageratum conyzoides* <insusceptible>/

#49. /
1. ***Amaranthus caudatus*** <susceptible>/
2. *Amaranthus caudatus* <insusceptible>/

#50. /
1. ***Amaranthus retroflexus*** <susceptible>/
2. *Amaranthus retroflexus* <insusceptible>/

#51. /
1. ***Anthriscus cerefolium*** <susceptible>/
2. *Anthriscus cerefolium* <insusceptible>/

#52. /
1. ***Antirrhinum majus*** <susceptible>/
2. *Antirrhinum majus* <insusceptible>/

#53. /
1. ***Apium graveolens*** <susceptible>/
2. *Apium graveolens* <insusceptible>/

#54. /
1. ***Apium leptophyllum*** <susceptible>/
2. *Apium leptophyllum* <insusceptible>/

#55. /
1. ***Arachis hypogaea*** <susceptible>/
2. *Arachis hypogaea* <insusceptible>/

#56. /
1. ***Astragalus sinicus*** <susceptible>/
2. *Astragalns sinicus* <insusceptible>/

#57. /
1. ***Atriplex hortensis*** <susceptible>/
2. *Atriplex hortensis* <insusceptible>/

#58. /
1. ***Bauhinia carronnii*** <susceptible>/

2. *Bauhinia carronnii* <insusceptible>/

#59. /

1. ***Bauhinia purpurea*** <susceptible>/
2. *Bauhinia purpurea* <insusceptible>/

#60. /

1. ***Bellis perennis*** <susceptible>/
2. *Bellis perennis* <insusceptible>/

#61. /

1. ***Beta macrocarpa*** <susceptible>/
2. *Beta macrocarpa* <insusceptible>/

#62. /

1. ***Beta patellaris*** <susceptible>/
2. *Beta patellaris* <insusceptible>/

#63. /

1. ***Beta vulgaris*** <susceptible>/
2. *Beta vulgaris* <insusceptible>/

#64. /

1. ***Brassica campestris*** <susceptible>/
2. *Brassica campestris* <insusceptible>/

#65. /

1. ***Brassica campestris*** ssp. ***chinensis*** <susceptible>/
2. *Brassica campestris* ssp. *chinensis* <insusceptible>/

#66. /

1. ***Brassica campestris*** ssp. ***napus*** <susceptible>/
2. *Brassica campestris* ssp. *napus* <insusceptible>/

#67. /

1. ***Brassica campestris*** ssp. ***pekinensis*** <susceptible>/
2. *Brassica campestris* ssp. *pekinensis* <insusceptible>/

#68. /

1. ***Brassica campestris*** ssp. ***rapa*** <susceptible>/
2. *Brassica campestris* ssp. *rapa* <insusceptible>/

#69. /

1. ***Brassica juncea*** <susceptible>/
2. *Brassica juncea* <insusceptible>/

#70. /

1. ***Brassica oleracea*** ssp. ***botrytis*** <susceptible>/
2. *Brassica oleracea* ssp. *botrytis* <insusceptible>/

#71. /

1. ***Brassica oleracea*** ssp. ***capitata*** <susceptible>/
2. *Brassica oleracea* ssp. *capitata* <insusceptible>/

#72. /

1. ***Caesalpinia pulcherrima*** <susceptible>/
2. *Caesalpinia pulcherrima* <insusceptible>/

#73. /

1. ***Cajanus cajan*** <susceptible>/
2. *Cajanus cajan* <insusceptible>/

#74. /

1. ***Calendula officinalis*** <susceptible>/
2. *Calendula officinalis* <insusceptible>/

#75. /

1. ***Capsella bursa–pastoris*** <susceptible>/
2. *Capsella bursa–pastoris* <insusceptible>/

#76. /

1. ***Capsicum annuum*** <susceptible>/
2. *Capsicum annuum* <insusceptible>/

#77. /

1. ***Capsicum frutescens*** <susceptible>/
2. *Capsicum frutescens* <insusceptible>/

#78. /

1. ***Carica papaya*** <susceptible>/
2. *Carica papaya* <insusceptible>/

#79. /

1. ***Carnegiea gigantea*** <susceptible>/
2. *Carnegiea gigantea* <insusceptible>/

#80. /

1. ***Carthamus tinctorius*** <susceptible>/
2. *Carthamus tinctorius* <insusceptible>/

#81. /
1. ***Cassia bicapsularis*** <susceptible>/
2. *Cassia bicapsularis* <insusceptible>/

#82. /
1. ***Cassia fistula*** <susceptible>/
2. *Cassia fistula* <insusceptible>/

#83. /
1. ***Cassia obtusifolia*** <susceptible>/
2. *Cassia obtusifolia* <insusceptible>/

#84. /
1. ***Cassia occidentalis*** <susceptible>/
2. *Cassia occidentalis* <insusceptible>/

#85. /
1. ***Cassia tora*** <susceptible>/
2. *Cassia tora* <insusceptible>/

#86. /
1. ***Catalpa bignonioides*** <susceptible>/
2. *Catalpa bignonioides* <insusceptible>/

#87. /
1. ***Catharanthus roseus*** <susceptible>/
2. *Catharanthus roseus* <insusceptible>/

#88. /
1. ***Ceiba pentandra*** <susceptible>/
2. *Ceiba pentandra* <insusceptible>/

#89. /
1. ***Celosia argentea*** <susceptible>/
2. *Celosia argentea* <insusceptible>/

#90. /
1. ***Celosia cristata*** <susceptible>/
2. *Celosia cristata* <insusceptible>/

#91. /
1. ***Chaenomeles japonica*** <susceptible>/
2. *Chaenomeles japonica* <insusceptible>/

#92. /
1. ***Cheiranthus cheiri*** <susceptible>/
2. *Cheiranthus cheiri* <insusceptible>/

#93. /
1. ***Chenopodium album*** <susceptible>/
2. *Chenopodium album* <insusceptible>/

#94. /
1. ***Chenopodium amaranticolor*** <susceptible>/
2. *Chenopodium amaranticolor* <insusceptible>/

#95. /
1. ***Chenopodium ambrosioides*** <susceptible>/
2. *Chenopodium ambrosioides* <insusceptible>/

#96. /
1. ***Chenopodium capitatum*** <susceptible>/
2. *Chenopodium capitatum* <insusceptible>/

#97. /
1. ***Chenopodium foetidum*** <susceptible>/
2. *Chenopodium foetidum* <insusceptible>/

#98. /
1. ***Chenopodium foliosum*** <susceptible>/
2. *Chenopodium foliosum* <insusceptible>/

#99. /
1. ***Chenopodium hybridum*** <susceptible>/
2. *Chenopodium hybridum* <insusceptible>/

#100. /
1. ***Chenopodium murale*** <susceptible>/
2. *Chenopodium murale* <insusceptible>/

#101. /
1. ***Chenopodium quinoa*** <susceptible>/
2. *Chenopodium quinoa* <insusceptible>/

#102. /
1. ***Chrysanthemum morifolium*** <susceptible>/

2. *Chrysanthemum morifolium* <insusceptible>/

#103. /
1. ***Cicer arietinum*** <susceptible>/
2. *Cicer arietinum* <insusceptible>/

#104. /
1. ***Cichorium endiva*** <susceptible>/
2. *Cichorium endiva* <insusceptible>/

#105. /
1. ***Citrullus lanatus*** <susceptible>/
2. *Citrullus lanatus* <insusceptible>/

#106. /
1. ***Citrullus vulgaris*** <susceptible>/
2. *Citrullus vulgaris* <insusceptible>/

#107. /
1. ***Citrus aurantifolia*** <susceptible>/
2. *Citrus aurantifolia* <insusceptible>/

#108. /
1. ***Citrus aurantium*** <susceptible>/
2. *Citrus aurantium* <insusceptible>/

#109. /
1. ***Citrus limon*** <susceptible>/
2. *Citrus limon* <insusceptible>/

#110. /
1. ***Citrus medica*** <susceptible>/
2. *Citrus medica* <insusceptible>/

#111. /
1. ***Citrus paradisi*** <susceptible>/
2. *Citrus paradisi* <insusceptible>/

#112. /
1. ***Citrus sinensis*** <susceptible>/
2. *Citrus sinensis* <insusceptible>/

#113. /
1. ***Clarkia amoena*** <susceptible>/
2. *Clarkia amoena* <insusceptible>/

#114. /
1. ***Clitoria ternatea*** <susceptible>/
2. *Clitoria ternatea* <insusceptible>/

#115. /
1. ***Cola chlamydantha*** <susceptible>/
2. *Cola chlamydantha* <insusceptible>/

#116. /
1. ***Cola gigantea*** <susceptible>/
2. *Cola gigantea* <insusceptible>/

#117. /
1. ***Conium maculatum*** <susceptible>/
2. *Conium maculatum* <insusceptible>/

#118. /
1. ***Corchorus spp.*** <susceptible>/
2. *Corchorus spp.* <insusceptible>/

#119. /
1. ***Coriandrum sativum*** <susceptible>/
2. *Coriandrum sativum* <insusceptible>/

#120. /
1. ***Crambe abyssinica*** <susceptible>/
2. *Crambe abyssinica* <insusceptible>/

#121. /
1. ***Crotalaria retusa*** <susceptible>/
2. *Crotalaria retusa* <insusceptible>/

#122. /
1. ***Crotalaria spectabilis*** <susceptible>/
2. *Crotalaria spectabilis* <insusceptible>/

#123. /
1. ***Cucumis melo*** <susceptible>/
2. *Cucumis melo* <insusceptible>/

#124. /
1. ***Cucumis metuliferus*** <susceptible>/
2. *Cucumis metuliferus* <insusceptible>/

#125. /
1. ***Cucumis sativus*** <susceptible>/
2. *Cucumis sativus* <insusceptible>/

#126. /
1. ***Cucurbita maxima*** <susceptible>/
2. *Cucurbita maxima* <insusceptible>/

#127. /
1. ***Cucurbita moschata*** <susceptible>/
2. *Cucurbita moschata* <insusceptible>/

#128. /
1. ***Cucurbita okeechobeensis*** <susceptible>/
2. *Cucurbita okeechobeensis* <insusceptible>/

#129. /
1. ***Cucurbita pepo*** <susceptible>/
2. *Cucurbita pepo* <insusceptible>/

#130. /
1. ***Cyamopsis tetragonoloba*** <susceptible>/
2. *Cyamopsis tetragonoloba* <insusceptible>/

#131. /
1. ***Cyphomandra betacea*** <susceptible>/
2. *Cyphomandra betacea* <insusceptible>/

#132. /
1. ***Dahlia pinnata*** <susceptible>/
2. *Dahlia pinnata* <insusceptible>/

#133. /
1. ***Dalbergia sissoo*** <susceptible>/
2. *Dalbergia sissoo* <insusceptible>/

#134. /
1. ***Daniellia ogea*** <susceptible>/
2. *Daniellia ogea* <insusceptible>/

#135. /
1. ***Datura ferox*** <susceptible>/
2. *Datura ferox* <insusceptible>/

#136. /
1. ***Datura metel*** <susceptible>/
2. *Datura metel* <insusceptible>/

#137. /
1. ***Datura stramonium*** <susceptible>/
2. *Datura stramonium* <insusceptible>/

#138. /
1. ***Datura tatula*** <susceptible>/
2. *Datura tatula* <insusceptible>/

#139. /
1. ***Daucus carota*** <susceptible>/
2. *Daucus carota* <insusceptible>/

#140. /
1. ***Delphinium hybridum*** <susceptible>/
2. *Delphinium hybridum* <insusceptible>/

#141. /
1. ***Desmodium tortuosum*** <susceptible>/
2. *Desmodium tortuosum* <insusceptible>/

#142. /
1. ***Dianthus barbatus*** <susceptible>/
2. *Dianthus barbatus* <insusceptible>/

#143. /
1. ***Dianthus caryophyllus*** <susceptible>/
2. *Dianthus caryophyllus* <insusceptible>/

#144. /
1. ***Dolichos biflorus*** <susceptible>/
2. *Dolichos biflorus* <insusceptible>/

#145. /
1. ***Emilia sagittata*** <susceptible>/
2. *Emilia sagittata* <insusceptible>/

#146. /
1. ***Eucalyptus chiloensis*** <susceptible>/
2. *Eucalyptus chiloensis* <insusceptible>/

#147. /
1. ***Eucalyptus cloesiana*** <susceptible>/
2. *Eucalyptus cloesiana* <insusceptible>/

#148. /
1. ***Eucalyptus marginata*** <susceptible>/
2. *Eucalyptus marginata* <insusceptible>/

#149. /
1. ***Euphorbia marginata*** <susceptible>/
2. *Euphorbia marginata* <insusceptible>/

#150. /
1. ***Fagopyrum esculentum*** <susceptible>/
2. *Fagopyrum esculentum* <insusceptible>/

#151. /
1. ***Fragaria chiloensis*** <susceptible>/
2. *Fragaria chiloensis* <insusceptible>/

#152. /
1. ***Fragaria vesca*** <susceptible>/
2. *Fragaria vesca* <insusceptible>/

#153. /
1. ***Fragaria virginiana*** <susceptible>/
2. *Fragaria virginiana* <insusceptible>/

#154. /
1. ***Galinsoga parviflora*** <susceptible>/
2. *Galinsoga parviflora* <insusceptible>/

#155. /
1. ***Glycine max*** <susceptible>/
2. *Glycine max* <insusceptible>/

#156. /
1. ***Gomphrena globosa*** <susceptible>/
2. *Gomphrena globosa* <insusceptible>/

#157. /
1. ***Gossypium hirsutum*** <susceptible>/
2. *Gossypium hirsutum* <insusceptible>/

#158. /
1. ***Gynura aurantiaca*** <susceptible>/
2. *Gynura aurantiaca* <insusceptible>/

#159. /
1. ***Gypsophila elegans*** <susceptible>/
2. *Gypsophila elegans* <insusceptible>/

#160. /
1. ***Helianthus annuus*** <susceptible>/
2. *Helianthus annuus* <insusceptible>/

#161. /
1. ***Heracleum sphondylium*** <susceptible>/
2. *Heracleum sphondylium* <insusceptible>/

#162. /
1. ***Hibiscus cannabinus*** <susceptible>/
2. *Hibiscus cannabinus* <insusceptible>/

#163. /
1. ***Hibiscus esculentus*** <susceptible>/
2. *Hibiscus esculentus* <insusceptible>/

#164. /
1. ***Humulus lupulus*** <susceptible>/
2. *Humulus lupulus* <insusceptible>/

#165. /
1. ***Hyoscyamus niger*** <susceptible>/
2. *Hyoscyamus niger* <insusceptible>/

#166. /
1. ***Ipomoea nil*** <susceptible>/
2. *Ipomoea nil* <insusceptible>/

#167. /
1. ***Ipomoea setosa*** <susceptible>/
2. *Ipomoea setosa* <insusceptible>/

#168. /
1. ***Kennedya rubicunda*** <susceptible>/
2. *Kennedya rubicunda* <insusceptible>/

#169. /
1. ***Lablab purpureus*** <susceptible>/
2. *Lablab purpureus* <insusceptible>/

#170. /
1. ***Lactuca sativa*** <susceptible>/
2. *Lactuca sativa* <insusceptible>/

#171. /
1. ***Lathyrus odoratus*** <susceptible>/
2. *Lathyrus odoratus* <insusceptible>/

#172. /
1. ***Lavatera trimestris*** <susceptible>/
2. *Lavatera trimestris* <insusceptible>/

#173. /
1. ***Lens esculenta*** <susceptible>/
2. *Lens esculenta* <insusceptible>/

#174. /
1. ***Lespedeza stipulacea*** <susceptible>/
2. *Lespedeza stipulacea* <insusceptible>/

#175. /
1. ***Linum usitatissimum*** <susceptible>/
2. *Linum usitatissimum* <insusceptible>/

#176. /
1. ***Lobelia erinus*** <susceptible>/
2. *Lobelia erinus* <insusceptible>/

#177. /
1. ***Lotus corniculatus*** <susceptible>/
2. *Lotus corniculatus* <insusceptible>/

#178. /
1. ***Luffa acutangula*** <susceptible>/
2. *Luffa acutangula* <insusceptible>/

#179. /
1. ***Lunaria annua*** <susceptible>/
2. *Lunaria annua* <insusceptible>/

#180. /
1. ***Lupinus albus*** <susceptible>/
2. *Lupinus albus* <insusceptible>/

#181. /
1. ***Lupinus angustifolius*** <susceptible>/
2. *Lupinus angustifolius* <insusceptible>/

#182. /
1. ***Lupinus cosentinii*** <susceptible>/
2. *Lupinus cosentinii* <insusceptible>/

#183. /
1. ***Lupinus mutabilis*** <susceptible>/
2. *Lupinus mutabilis* <insusceptible>/

#184. /
1. ***Lycopersicon chilense*** <susceptible>/
2. *Lycopersicon chilense* <insusceptible>/

#185. /
1. ***Lycopersicon esculentum*** <susceptible>/
2. *Lycopersicon esculentum* <insusceptible>/

#186. /
1. ***Lycopersicon pimpinellifolium*** <susceptible>/
2. *Lycopersicon pimpinellifolium* <insusceptible>/

#187. /
1. ***Maclura pomifera*** <susceptible>/
2. *Maclura pomifera* <insusceptible>/

#188. /
1. ***Macroptilium lathyroides*** <susceptible>/
2. *Macroptilium lathyroides* <insusceptible>/

#189. /
1. ***Macrotyloma uniflorum*** <susceptible>/
2. *Macrotyloma uniflorum* <insusceptible>/

#190. /
1. ***Malus platycarpa*** <susceptible>/
2. *Malus platycarpa* <insusceptible>/

#191. /
1. ***Malus sylvestris*** <susceptible>/
2. *Malus sylvestris* <insusceptible>/

#192. /
1. ***Manihot esculenta*** <susceptible>/
2. *Manihot esculenta* <insusceptible>/

#193. /
1. ***Matthiola incana*** <susceptible>/
2. *Matthiola incana* <insusceptible>/

#194. /
1. ***Medicago hispida*** <susceptible>/
2. *Medicago hispida* <insusceptible>/

#195. /
1. ***Medicago sativa*** <susceptible>/
2. *Medicago sativa* <insusceptible>/

#196. /
1. ***Melandrium album*** <susceptible>/
2. *Melandrium album* <insusceptible>/

#197. /
1. ***Melilotus alba*** <susceptible>/
2. *Melilotus alba* <insusceptible>/

#198. /
1. ***Melilotus officinalis*** <susceptible>/
2. *Melilotus officinalis* <insusceptible>/

#199. /
1. ***Mimosa pudica*** <susceptible>/
2. *Mimosa pudica* <insusceptible>/

#200. /
1. ***Momordica balsamina*** <susceptible>/
2. *Momordica balsamina* <insusceptible>/

#201. /
1. ***Montia perfoliata*** <susceptible>/
2. *Montia perfoliata* <insusceptible>/

#202. /
1. ***Morus alba*** <susceptible>/
2. *Morus alba* <insusceptible>/

#203. /
1. ***Myosotis sylvatica*** <susceptible>/
2. *Myosotis sylvatica* <insusceptible>/

#204. /
1. ***Nicandra physaloides*** <susceptible>/
2. *Nicandra physaloides* <insusceptible>/

#205. /
1. ***Nicotiana benthamiana*** <susceptible>/
2. *Nicotiana benthamiana* <insusceptible>/

#206. /
1. ***Nicotiana bigelovii*** <susceptible>/
2. *Nicotiana bigelovii* <insusceptible>/

#207. /
1. ***Nicotiana clevelandii*** <susceptible>/
2. *Nicotiana clevelandii* <insusceptible>/

#208. /
1. ***Nicotiana* × *edwardsonii*** <susceptible>/
2. *Nicotiana* × *edwardsonii* <insusceptible>/

#209. /
1. ***Nicotiana debneyi*** <susceptible>/
2. *Nicotiana debneyi* <insusceptible>/

#210. /
1. ***Nicotiana glutinosa*** <susceptible>/
2. *Nicotiana glutinosa* <insusceptible>/

#211. /
1. ***Nicotiana glutinosa* × *clevelandii*** <susceptible>/
2. *Nicotiana glutinosa* × *clevelandii* <insusceptible>/

#212. /
1. ***Nicotiana megalosiphon*** <susceptible>/
2. *Nicotiana megalosiphon* <insusceptible>/

#213. /
1. ***Nicotiana occidentalis*** <susceptible>/
2. *Nicotiana occidentalis* <insusceptible>/

#214. /
1. ***Nicotiana rustica*** <susceptible>/
2. *Nicotiana rustica* <insusceptible>/

#215. /
1. ***Nicotiana sylvestris*** <susceptible>/
2. *Nicotiana sylvestris* <insusceptible>/

#216. /
1. ***Nicotiana tabacum*** <susceptible>/
2. *Nicotiana tabacum* <insusceptible>/

#217. /
1. ***Ocimum basilicum*** <susceptible>/
2. *Ocimum basilicum* <insusceptible>/

#218. /
1. ***Opuntia vulgaris*** <susceptible>/
2. *Opuntia vulgaris* <insusceptible>/

#219. /
1. ***Papaver nudicaule*** <susceptible>/
2. *Papaver nudicaule* <insusceptible>/

#220. /
1. ***Passiflora edulis*** <susceptible>/
2. *Passiflora edulis* <insusceptible>/

#221. /
1. ***Passiflora suberosa*** <susceptible>/
2. *Passiflora suberosa* <insusceptible>/

#222. /
1. ***Pastinaca sativa*** <susceptible>/
2. *Pastinaca sativa* <insusceptible>/

#223. /
1. ***Persea americana*** <susceptible>/
2. *Persea americana* <insusceptible>/

#224. /
1. ***Petroselinum crispum*** <susceptible>/
2. *Petroselinum crispum* <insusceptible>/

#225. /
1. ***Petunia hybrida*** <susceptible>/
2. *Petunia hybrida* <insusceptible>/

#226. /
1. ***Phaseolus acutifolius*** <susceptible>/
2. *Phaseolus acutifolius* <insusceptible>/

#227. /
1. ***Phaseolus lathyroides*** <susceptible>/
2. *Phaseolus lathyroides* <insusceptible>/

#228. /
1. ***Phaseolus lunatus*** <susceptible>/
2. *Phaseolus lunatus* <insusceptible>/

#229. /
1. ***Phaseolus vulgaris*** <susceptible>/
2. *Phaseolus vulgaris* <insusceptible>/

#230. /
1. ***Phlox drummondii*** <susceptible>/
2. *Phlox drummondii* <insusceptible>/

#231. /
1. ***Physalis floridana*** <susceptible>/
2. *Physalis floridana* <insusceptible>/

#232. /
1. ***Physalis peruviana*** <susceptible>/
2. *Physalis peruviana* <insusceptible>/

#233. /
1. ***Phytolacca americana*** <susceptible>/
2. *Phytolacca americana* <insusceptible>/

#234. /
1. ***Pisum sativum*** <susceptible>/
2. *Pisum sativum* <insusceptible>/

#235. /
1. ***Plantago rugelii*** <susceptible>/
2. *Plantago rugelii* <insusceptible>/

#236. /
1. ***Podolepis robusta*** <susceptible>/
2. *Podolepis robusta* <insusceptible>/

#237. /
1. ***Poncirus trifoliata*** <susceptible>/
2. *Poncirus trifoliata* <insusceptible>/

#238. /
1. ***Primula malacoides*** <susceptible>/
2. *Primula malacoides* <insusceptible>/

#239. /
1. ***Prunus americana*** × ***salicina*** <susceptible>/
2. *Prunus americana* × *salicina* <insusceptible>/

#240. /
1. ***Prunus armeniaca*** <susceptible>/
2. *Prunus armeniaca* <insusceptible>/

#241. /
1. ***Prunus cerasifera*** <susceptible>/
2. *Prunus cerasifera* <insusceptible>/

#242. /
1. ***Prunus domestica*** <susceptible>/
2. *Prunus domestica* <insusceptible>/

#243. /
1. ***Prunus mahaleb*** <susceptible>/
2. *Prunus mahaleb* <insusceptible>/

#244. /
1. ***Prunus persica*** <susceptible>/
2. *Prunus persica* <insusceptible>/

#245. /
1. ***Prunus serrulata*** <susceptible>/
2. *Prunus serrulata* <insusceptible>/

#246. /
1. ***Ranunculus arvensis*** <susceptible>/
2. *Ranunculus arvensis* <insusceptible>/

#247. /
1. ***Ranunculus sardous*** <susceptible>/
2. *Ranunculus sardous* <insusceptible>/

#248. /
1. ***Raphanus sativus*** <susceptible>/
2. *Raphanus sativus* <insusceptible>/

#249. /
1. ***Ricinus communis*** <susceptible>/
2. *Ricinus communis* <insusceptible>/

#250. /
1. ***Rubus idaeus*** <susceptible>/
2. *Rubus idaeus* <insusceptible>/

#251. /
1. ***Rubus occidentalis*** <susceptible>/
2. *Rubus occidentalis* <insusceptible>/

#252. /
1. ***Rumex acetosa*** <susceptible>/
2. *Rumex acetosa* <insusceptible>/

#253. /
1. ***Salvia splendens*** <susceptible>/
2. *Salvia splendens* <insusceptible>/

#254. /
1. ***Sambucus racemosus*** <susceptible>/
2. *Sambucus racemosus* <insusceptible>/

#255. /
1. ***Saponaria vaccaria*** <susceptible>/
2. *Saponaria vaccaria* <insusceptible>/

#256. /
1. ***Senecio vulgaris*** <susceptible>/
2. *Senecio vulgaris* <insusceptible>/

#257. /
1. ***Sesamum indicum*** <susceptible>/
2. *Sesamum indicum* <insusceptible>/

#258. /
1. ***Sesbania exaltata*** <susceptible>/
2. *Sesbania exaltata* <insusceptible>/

#259. /
1. ***Silene armeria*** <susceptible>/
2. *Silene armeria* <insusceptible>/

#260. /
1. ***Sinapis alba*** <susceptible>/
2. *Sinapis alba* <insusceptible>/

#261. /
1. ***Sinningia speciosa*** <susceptible>/
2. *Sinningia speciosa* <insusceptible>/

#262. /
1. ***Solanum chacoense*** <susceptible>/
2. *Solanum chacoense* <insusceptible>/

#263. /
1. ***Solanum demissum*** <susceptible>/
2. *Solanum demissum* <insusceptible>/

#264. /
1. ***Solanum demissum* × *tuberosum*** <susceptible>/
2. *Solanum demissum* × *tuberosum* <insusceptible>/

#265. /
1. ***Solanum melongena*** <susceptible>/
2. *Solanum melongena* <insusceptible>/

#266. /
1. ***Solanum nigrum*** <susceptible>/
2. *Solanum nigrum* <insusceptible>/

#267. /
1. ***Solanum nodiflorum*** <susceptible>/
2. *Solanum nodiflorum* <insusceptible>/

#268. /
1. ***Solanum rostratum*** <susceptible>/
2. *Solanum rostratum* <insusceptible>/

#269. /
1. ***Solanum tuberosum*** <susceptible>/
2. *Solanum tuberosum* <insusceptible>/

#270. /
1. ***Sonchus oleraceus*** <susceptible>/
2. *Sonchus oleraceus* <insusceptible>/

#271. /
1. ***Sophora microphylla*** <susceptible>/
2. *Sophora microphylla* <insusceptible>/

#272. /
1. ***Spinacia oleracea*** <susceptible>/
2. *Spinacia oleracea* <insusceptible>/

#273. /
1. ***Stellaria media*** <susceptible>/
2. *Stellaria media* <insusceptible>/

#274. /
1. ***Tephrosia villosa*** <susceptible>/
2. *Tephrosia villosa* <insusceptible>/

#275. /
1. ***Tephrosia vogelii*** <susceptible>/
2. *Tephrosia vogelii* <insusceptible>/

#276. /
1. ***Tetragonia expansa*** <susceptible>/
2. *Tetragonia expansa* <insusceptible>/

#277. /
1. ***Theobroma cacao*** <susceptible>/
2. *Theobroma cacao* <insusceptible>/

#278. /
1. ***Tithonia speciosa*** <susceptible>/

2. *Tithonia speciosa* <insusceptible>/

#279. /
1. ***Torenia fournieri*** <susceptible>/
2. *Torenia fournieri* <insusceptible>/

#280. /
1. ***Trifolium dubium*** <susceptible>/
2. *Trifolium dubium* <insusceptible>/

#281. /
1. ***Trifolium hybridum*** <susceptible>/
2. *Trifolium hybridum* <insusceptible>/

#282. /
1. ***Trifolium incarnatum*** <susceptible>/
2. *Trifolium incarnatum* <insusceptible>/

#283. /
1. ***Trifolium pratense*** <susceptible>/
2. *Trifolium pratense* <insusceptible>/

#284. /
1. ***Trifolium repens*** <susceptible>/
2. *Trifolium repens* <insusceptible>/

#285. /
1. ***Trifolium subterraneum*** <susceptible>/
2. *Trifolium subterraneum* <insusceptible>/

#286. /
1. ***Tropaeolum majus*** <susceptible>/
2. *Tropaeolum majus* <insusceptible>/

#287. /
1. ***Valeriana officinalis*** <susceptible>/
2. *Valeriana officinalis* <insusceptible>/

#288. /
1. ***Verbesina encelioides*** <susceptible>/
2. *Verbesina encelioides* <insusceptible>/

#289. /
1. ***Vicia faba*** <susceptible>/
2. *Vicia faba* <insusceptible>/

#290. /
1. ***Vicia sativa*** <susceptible>/
2. *Vicia sativa* <insusceptible>/

#291. /
1. ***Vicia villosa*** <susceptible>/
2. *Vicia villosa* <insusceptible>/

#292. /
1. ***Vigna angularis*** <susceptible>/
2. *Vigna angularis* <insusceptible>/

#293. /
1. ***Vigna radiata*** <susceptible>/
2. *Vigna radiata* <insusceptible>/

#294. /
1. ***Vigna unguiculata*** <susceptible>/
2. *Vigna unguiculata* <insusceptible>/

#295. /
1. ***Vigna unguiculata*** ssp. ***cylindrica*** <susceptible>/
2. *Vigna unguiculata* ssp. *cylindrica* <insusceptible>/

#296. /
1. ***Vigna unguiculata*** ssp. ***sesquipedalis*** <susceptible>/
2. *Vigna unguiculata* ssp. *sesquipedalis* <insusceptible>/

#297. /
1. ***Viola cornuta*** <susceptible>/
2. *Viola cornuta* <insusceptible>/

#298. /
1. ***Vitis labrusca*** <susceptible>/
2. *Vitis labrusca* <insusceptible>/

#299. /
1. ***Vitis rupestris*** <susceptible>/
2. *Vitis rupestris* <insusceptible>/

#300. /
1. ***Vitis vinifera*** <susceptible>/
2. *Vitis vinifera* <insusceptible>/

#301. /
1. ***Voandzeia subterranea*** <susceptible>/
2. *Voandzeia subterranea* <insusceptible>/

#302. /
1. ***Zinnia elegans*** <susceptible>/
2. *Zinnia elegans* <insusceptible>/

***Monocotyledonous hosts***

#303. /
1. ***Agropyron elongatum*** <susceptible>/
2. *Agropyron elongatum* <insusceptible>/

#304. /
1. ***Agropyron intermedium*** <susceptible>/
2. *Agropyron intermedium* <insusceptible>/

#305. /
1. ***Agropyron kamoji*** <susceptible>/
2. *Agropyron kamoji* <insusceptible>/

#306. /
1. ***Agropyron repens*** <susceptible>/
2. *Agropyron repens* <insusceptible>/

#307. /
1. ***Agrostis alba*** <susceptible>/
2. *Agrostis alba* <insusceptible>/

#308. /
1. ***Agrostis palustris*** <susceptible>/
2. *Agrostis palustris* <insusceptible>/

#309. /
1. ***Agrostis tenuis*** <susceptible>/
2. *Agrostis tenuis* <insusceptible>/

#310. /
1. ***Allium ascalonicum*** <susceptible>/
2. *Allium ascalonicum* <insusceptible>/

#311. /
1. ***Allium cepa*** <susceptible>/
2. *Allium cepa* <insusceptible>/

#312. /
1. ***Allium fistulosum*** <susceptible>/
2. *Allium fistulosum* <insusceptible>/

#313. /
1. ***Allium porrum*** <susceptible>/
2. *Allium porrum* <insusceptible>/

#314. /
1. ***Anthoxanthum cristatum*** <susceptible>/
2. *Anthoxanthum cristatum* <insusceptible>/

#315. /
1. ***Anthoxanthum odoratum*** <susceptible>/
2. *Anthoxanthum odoratum* <insusceptible>/

#316. /
1. ***Areca catechu*** <susceptible>/
2. *Areca catechu* <insusceptible>/

#317. /
1. ***Arundinaria amabilis*** <susceptible>/
2. *Arundinaria amabilis* <insusceptible>/

#318. /
1. ***Arundo donax*** <susceptible>/
2. *Arundo donax* <insusceptible>/

#319. /
1. ***Avena byzantina*** <susceptible>/
2. *Avena byzantina* <insusceptible>/

#320. /
1. ***Avena sativa*** <susceptible>/
2. *Avena sativa* <insusceptible>/

#321. /
1. ***Avena strigosa*** <susceptible>/
2. *Avena strigosa* <insusceptible>/

#322. /
1. ***Belamcanda chinensis*** <susceptible>/
2. *Belamcanda chinensis* <insusceptible>/

#323. /
1. ***Bromus inermis*** <susceptible>/
2. *Bromus inermis* <insusceptible>/

#324. /
1. ***Bromus mollis*** <susceptible>/
2. *Bromus mollis* <insusceptible>/

#325. /
1. ***Bromus racemosus*** <susceptible>/
2. *Bromus racemosus* <insusceptible>/

#326. /
1. ***Bromus secalinus*** <susceptible>/
2. *Bromus secalinus* <insusceptible>/

#327. /
1. ***Bromus tectorum*** <susceptible>/
2. *Bromus tectorum* <insusceptible>/

#328. /
1. ***Caladium hortulanum*** <susceptible>/
2. *Caladium hortulanum* <insusceptible>/

#329. /
1. ***Chloris gayana*** <susceptible>/
2. *Chloris gayana* <insusceptible>/

#330. /
1. ***Chrysalidocarpus lutescens*** <susceptible>/
2. *Chrysalidocarpus lutescens* <insusceptible>/

#331. /
1. ***Cocos nucifera*** <susceptible>/
2. *Cocos nucifera* <insusceptible>/

#332. /
1. ***Colocasia esculenta*** <susceptible>/
2. *Colocasia esculenta* <insusceptible>/

#333. /
1. ***Cortaderia selloana*** <susceptible>/
2. *Cortaderia selloana* <insusceptible>/

#334. /
1. ***Corypha elata*** <susceptible>/
2. *Corypha elata* <insusceptible>/

#335. /
1. ***Cymbidium alexanderi*** <susceptible>/
2. *Cymbidium alexanderi* <insusceptible>/

#336. /
1. ***Cynodon dactylon*** <susceptible>/
2. *Cynodon dactylon* <insusceptible>/

#337. /
1. ***Dactylis glomerata*** <susceptible>/
2. *Dactylis glomerata* <insusceptible>/

#338. /
1. ***Dendrobium nobile*** <susceptible>/
2. *Dendrobium nobile* <insusceptible>/

#339. /
1. ***Dieffenbachia picta*** <susceptible>/
2. *Dieffenbachia picta* <insusceptible>/

#340. /
1. ***Digitaria decumbens*** <susceptible>/
2. *Digitaria decumbens* <insusceptible>/

#341. /
1. ***Digitaria longiflora*** <susceptible>/
2. *Digitaria longiflora* <insusceptible>/

#342. /
1. ***Digitaria pentzii*** <susceptible>/
2. *Digitaria pentzii* <insusceptible>/

#343. /
1. ***Digitaria setivalvola*** <susceptible>/
2. *Digitaria setivalvola* <insusceptible>/

#344. /
1. ***Digitaria valida*** <susceptible>/
2. *Digitaria valida* <insusceptible>/

#345. /
1. ***Dinebra retroflexa*** <susceptible>/
2. *Dinebra retroflexa* <insusceptible>/

#346. /
1. ***Echinochloa colonum*** <susceptible>/
2. *Echinochloa colonum* <insusceptible>/

#347. /
1. ***Echinochloa crus–galli*** <susceptible>/

2. *Echinochloa crus–galli* <insusceptible>/

#348. /
1. ***Eleusine coracana*** <susceptible>/
2. *Eleusine coracana* <insusceptible>/

#349. /
1. ***Elymus junceus*** <susceptible>/
2. *Elymus junceus* <insusceptible>/

#350. /
1. ***Eragrostis capillata*** <susceptible>/
2. *Eragrostis capillata* <insusceptible>/

#351. /
1. ***Eragrostis cilianensis*** <susceptible>/
2. *Eragrostis cilianensis* <insusceptible>/

#352. /
1. ***Eragrostis elatior*** <susceptible>/
2. *Eragrostis elatior* <insusceptible>/

#353. /
1. ***Eragrostis rubra*** <susceptible>/
2. *Eragrostis rubra* <insusceptible>/

#354. /
1. ***Festuca pratensis*** <susceptible>/
2. *Festuca pratensis* <insusceptible>/

#355. /
1. ***Hippeastrum hybridum*** <susceptible>/
2. *Hippeastrum hybridum* <insusceptible>/

#356. /
1. ***Hippeastrum jubatum*** <susceptible>/
2. *Hippeastrum jubatum* <insusceptible>/

#357. /
1. ***Hippeastrum murinum*** <susceptible>/
2. *Hippeastrum murinum* <insusceptible>/

#358. /
1. ***Hordeum vulgare*** <susceptible>/
2. *Hordeum vulgare* <insusceptible>/

#359. /
1. ***Imperata cylindrica*** <susceptible>/
2. *Imperata cylindrica* <insusceptible>/

#360. /
1. ***Iris germanica*** <susceptible>/
2. *Iris germanica* <insusceptible>/

#361. /
1. ***Isachne globosa*** <susceptible>/
2. *Isachne globosa* <insusceptible>/

#362. /
1. ***Lagurus ovatus*** <susceptible>/
2. *Lagurus ovatus* <insusceptible>/

#363. /
1. ***Lamarkia aurea*** <susceptible>/
2. *Lamarkia aurea* <insusceptible>/

#364. /
1. ***Leptochloa filiformis*** <susceptible>/
2. *Leptochloa filiformis* <insusceptible>/

#365. /
1. ***Lilium formosanum*** <susceptible>/
2. *Lilium formosanum* <insusceptible>/

#366. /
1. ***Lilium longiflorum*** <susceptible>/
2. *Lilium longiflorum* <insusceptible>/

#367. /
1. ***Lolium multiflorum*** <susceptible>/
2. *Lolium multiflorum* <insusceptible>/

#368. /
1. ***Lolium perenne*** <susceptible>/
2. *Lolium perenne* <insusceptible>/

#369. /
1. ***Lolium temulentum*** <susceptible>/
2. *Lolium temulentum* <insusceptible>/

#370. /
1. ***Lolium persicum*** <susceptible>/
2. *Lolium persicum* <insusceptible>/

#371. /
1. ***Narcissus jonquilla*** <susceptible>/
2. *Narcissus jonquilla* <insusceptible>/

#372. /
1. ***Narcissus pseudonarcissus*** <susceptible>/
2. *Narcissus pseudonarcissus* <insusceptible>/

#373. /
1. ***Nerine bowdenii*** <susceptible>/
2. *Nerine bowdenii* <insusceptible>/

#374. /
1. ***Oplismenus compositae*** <susceptible>/
2. *Oplismenus compositae* <insusceptible>/

#375. /
1. ***Oryza punctata*** <susceptible>/
2. *Oryza punctata* <insusceptible>/

#376. /
1. ***Oryza sativa*** <susceptible>/
2. *Oryza sativa* <insusceptible>/

#377. /
1. ***Oryzopsis miliacea*** <susceptible>/
2. *Oryzopsis miliacea* <insusceptible>/

#378. /
1. ***Panicum capillare*** <susceptible>/
2. *Panicum capillare* <insusceptible>/

#379. /
1. ***Panicum dichotomiflorum*** <susceptible>/
2. *Panicum dichotomiflorum* <insusceptible>/

#380. /
1. ***Panicum maximum*** <susceptible>/
2. *Panicum maximum* <insusceptible>/

#381. /
1. ***Panicum miliaceum*** <susceptible>/
2. *Panicum miliaceum* <insusceptible>/

#382. /
1. ***Paspalum dilatatum*** <susceptible>/
2. *Paspalum dilatatum* <insusceptible>/

#383. /
1. ***Paspalum membranaceum*** <susceptible>/
2. *Paspalum membranaceum* <insusceptible>/

#384. /
1. ***Phalaris arundinacea*** <susceptible>/
2. *Phalaris arundinacea* <insusceptible>/

#385. /
1. ***Phalaris paradoxa*** <susceptible>/
2. *Phalaris paradoxa* <insusceptible>/

#386. /
1. ***Philodendron selloum*** <susceptible>/
2. *Philodendron selloum* <insusceptible>/

#387. /
1. ***Philodendron verrucosum*** <susceptible>/
2. *Philodendron verrucosum* <insusceptible>/

#388. /
1. ***Phleum arenarium*** <susceptible>/
2. *Phleum arenarium* <insusceptible>/

#389. /
1. ***Phleum pratense*** <susceptible>/
2. *Phleum pratense* <insusceptible>/

#390. /
1. ***Phragmites australis*** <susceptible>/
2. *Phragmites australis* <insusceptible>/

#391. /
1. ***Poa annua*** <susceptible>/
2. *Poa annua* <insusceptible>/

#392. /
1. ***Poa compressa*** <susceptible>/
2. *Poa compressa* <insusceptible>/

#393. /
1. ***Poa pratensis*** <susceptible>/
2. *Poa pratensis* <insusceptible>/

#394. /
1. ***Poa trivialis*** <susceptible>/
2. *Poa trivialis* <insusceptible>/

#395. /
1. ***Rottboellia exaltata*** <susceptible>/
2. *Rottboellia exaltata* <insusceptible>/

#396. /
1. ***Roystonea regia*** <susceptible>/
2. *Roystonea regia* <insusceptible>/

#397. /
1. ***Saccharum officinarum*** <susceptible>/
2. *Saccharum officinarum* <insusceptible>/

#398. /
1. ***Sacciolepis indica*** <susceptible>/
2. *Sacciolepis indica* <insusceptible>/

#399. /
1. ***Secale cereale*** <susceptible>/
2. *Secale cereale* <insusceptible>/

#400. /
1. ***Setaria italica*** <susceptible>/
2. *Setaria italica* <insusceptible>/

#401. /
1. ***Setaria macrostachia*** <susceptible>/
2. *Setaria macrostachia* <insusceptible>/

#402. /
1. ***Setaria viridis*** <susceptible>/
2. *Setaria viridis* <insusceptible>/

#403. /
1. ***Sorghum arundinaceum*** <susceptible>/
2. *Sorghum arundinaceum* <insusceptible>/

#404. /
1. ***Sorghum bicolor*** <susceptible>/
2. *Sorghum bicolor* <insusceptible>/

#405. /
1. ***Stenotaphrum secundatum*** <susceptible>/
2. *Stenotaphrum secundatum* <insusceptible>/

#406. /
1. ***Tinantia erecta*** <susceptible>/
2. *Tinantia erecta* <insusceptible>/

#407. /
1. ***Triticum aestivum*** <susceptible>/
2. *Triticum aestivum* <insusceptible>/

#408. /
1. ***Triticum durum*** <susceptible>/
2. *Triticum durum* <insusceptible>/

#409. /
1. ***Tulipa gesneriana*** <susceptible>/
2. *Tulipa gesneriana* <insusceptible>/

#410. /
1. ***Veitchia merrillii*** <susceptible>/
2. *Veitchia merrillii* <insusceptible>/

#411. /
1. ***Xanthosoma caracu*** <susceptible>/
2. *Xanthosoma caracu* <insusceptible>/

#412. /
1. ***Zantedeschia sp.*** <susceptible>/
2. *Zantedeschia sp.* <insusceptible>/

#413. /
1. ***Zea mays*** <susceptible>/
2. *Zea mays* <insusceptible>/

*Additional hosts, comments on host-range, data sources*

#414. <Additional hosts>/
#415. <Comments on host-range>/
#416. <Sources of host-range data: references>/

**Properties of particles in sap**

#417. TIP:/
°C/
#418. LIV:/
days/
#419. DEP: $\log_{10}$ minus/
/
#420. Infectivity of sap <after treatment with di–ethyl ether>/
1. increased by treatment with di–ethyl ether/
2. decreased by treatment with di–ethyl ether/
3. not changed by treatment with di–ethyl ether/

#421. Leaf sap <concentration of virions contained>/
1. contains few virions <fewer than 10 per field at a magnification of 20,000>/
2. contains many virions <more than 10 per field at a magnification of 20,000>/

#422. Electron microscopy: <comments: e.g. special fixation or stains required, differences between negatively stained virions and those in thin sections>/

**Purification**

#423. <References or outline of> method:/

**Particle morphology**

#424. Virions <shape>/
1. isometric <or quasi-isometric>/
2. geminate/
3. bacilliform <non-rhabdovirus like>/
4. rod–shaped/
5. filamentous/
6. rhabdo– or bullet–shaped/
7. unusually shaped <see text character for details>/

#425. Virions <details when unusually shaped>/
#426. Virions <whether enveloped>/
1. enveloped/
2. not enveloped/

#427. Virions <isometric or geminate virions or subvirions, diameter or width>/
nm in diameter/
#428. Virions <geminate or bacilliform, length>/
nm in length/
#429. Virions <isometric, whether angular or rounded in profile>/
1. rounded in profile/
2. angular in profile/
3. neither clearly angular or nor rounded in profile/

#430. Virions <isometric, whether capsomeres of can be seen>/
1. with a conspicuous capsomere arrangement/
2. without a conspicuous capsomere arrangement/

#431. Virions <elongate, whether usually straight or flexuous>/
1. usually straight/
2. usually flexuous/

#432. Virions <elongate, whether having clear modal length(s)>/
1. with a clear modal length/
2. with no clear modal length <range between>/
3. with clear modal lengths/

#433. Virions of <modal length(s)>/
nm/
#434. Virions <rod/filament/bacilliforms etc, width>/
nm wide/
#435. Axial canal <obvious>/
1. obvious/
2. obscure/

#436. Axial canal <diameter>/
nm in diameter/
#437. Basic helix <obvious>/
1. obvious/
2. obscure/

#438. Pitch of basic helix <of elongate virions>/
nm/

**Physical properties**

#439. <Number of sedimenting components in purified preparations>/
1. one sedimenting component in purified preparations/
2. two sedimenting components in purified preparations/
3. three sedimenting components in purified preparations/
4. four sedimenting components in purified preparations/
5. five sedimenting components in purified preparations/
6. six sedimenting components in purified preparations/
7. seven sedimenting components in purified preparations/
8. eight sedimenting components in purified preparations/
9. nine sedimenting components in purified preparations/
10. ten sedimenting components in purified preparations/

#440. Sedimentation coefficient <of fastest or only component>/
*S*/
#441. <Sedimentation coefficient(s)> of the other(s) <slower ones>/
*S*/
#442. Density <in CsCl, state whether fixed with aldehydes>/
g cm$^{-3}$ in CsCl/
#443. Density <in sucrose>/
g cm$^{-3}$ in sucrose/
#444. Density <in solutions other than sucrose or CsCl>/
#445. Isoelectric point pH/
#446. $A_{260}/A_{280}$ ratio <list if separated components>/

**Biochemical properties**

#447. Virions contain <% nucleic acid>/
% nucleic acid/
#448. Virions <% protein>/
% protein/
#449. Virions <% lipid>/
% lipid/
#450. Also <compounds others than nucleic acid, protein and lipid, found in the virions>/

*Genomic nucleic acids*

#451. Genome <DNA/RNA>/
1. consists of DNA/
2. consists of RNA/

#452. <Genome NA strandedness>/
1. single–stranded/
2. double–stranded/

#453. <Genome whether circular or linear>/
1. circular/
2. linear/

#454. Total genome size <in kb or kb pairs>/
kb <state when pairs>/
#455. Genome <number of genome parts>/
1. unipartite/
2. of two parts/
3. of three parts/
4. of four parts/
5. of five parts/
6. of six parts/
7. of 7 parts/
8. of 8 parts/
9. of 9 parts/
10. of 10 parts/
11. of 11 parts/
12. of 12 parts/
13. of 13 parts/
14. of 14 parts/
15. of 15 parts/

#456. Largest (or only) genome part <size>/
kb/
#457. The 2nd largest <genome part, size>/
kb/
#458. The 3rd largest <genome part, size>/
kb/
#459. The 4th largest <genome part, size>/
kb/

#460. The 5th largest <genome part, size>/
kb/
#461. The 6th largest and other <genome> parts <size>/
kb/
#462. Genomic nucleic acid isolated by <references or method>/
#463. Base composition <genome guanine>/
% G/
#464. Base composition <genome adenine>/
% A/
#465. Base composition <genome cytosine>/
% C/
#466. Base composition <genome uracil>/
% U/
#467. Base composition <genome thymine>/
% T/
#468. Base composition of ds genome <(guanine + cytosine)>/
% (G + C)/
#469. 5' terminus of RNA has <specific terminal group>/
1. a methylated nucleotide cap/
2. a VPg/
3. a monophosphate/
4. a diphosphate/
5. an unusual type <see text character>/
#470. 5' terminus of RNA <unusual – not a cap, VPg, mono or di-phosphate>/
#471. Infectivity <of genome, whether retained when deproteinised with proteases>/
1. retained when deproteinised with proteases/
2. decreased when deproteinised with proteases/
3. lost when deproteinised with proteases/
#472. Infectivity <of genome, whether retained when genome deproteinised with phenol or detergent>/
1. retained when deproteinised with phenol or detergent/
2. lost when deproteinised with phenol or detergent/
#473. Poly A region <whether present or absent in the genome>/
1. present/
2. absent/
#474. Poly A region <position>/
#475. Additional factor <whether required for infectivity>/
1. required for infectivity/
2. not required for infectivity/
#476. Additional factor <required for infectivity e.g. coat protein or core>/
#477. Genome <whether it has *t*RNA-like activity>/
1. has *t*RNA–like activity/
2. has no *t*RNA–like activity/
#478. Genome <with *t*RNA-like activity> accepts <amino acid>/
#479. Nucleotide sequence references:/
#480. Sequence database accession code(s):/
#481. Features of the genome: <e.g. promoters, stem and loop structures, open reading frames (ORFs), and the genes they encode, starting at the 5'terminus of the largest *m*RNA>/

*Non-genomic viral nucleic acids*

#482. Non–genomic nucleic acid <whether found in the virions>/
1. found in the virions/
2. not found in the virions/
#483. <Non-genomic nucleic acid> is/
1. subgenomic *m*RNA/
2. host genomic RNA/
3. *t*RNA/
4. *r*RNA/
5. satellite RNA/
6. of unknown origin and function/
7. of an unusual type of structure/
#484. Non–genomic nucleic acid <description of unusual types>/

#485. Sub–genomic *m*RNA <whether found in infected cells>/
1. found in infected cells/
2. not found in infected cells/

#486. <Description of subgenomic mRNA found in infected cells>/

#487. <Number of virus specified dsRNA species found in infected cells>/
virus specified dsRNA species found in infected cells/

#488. Size of largest <or only> virus specified dsRNA/
kbp/

#489. <Size of> 2nd largest <virus specified dsRNA>/
kbp/

#490. <Size of> 3rd largest <virus specified dsRNA>/
kbp/

#491. <Size of> 4th largest <virus specified dsRNA>/
kbp/

#492. <Size of> 5th largest <virus specified dsRNA>/
kbp/

#493. <Size of> 6th largest <virus specified dsRNA>/
kbp/

*Virion proteins*

#494. Virion protein(s) <number>/
1. one/
2. two/
3. three/
4. four/
5. five/
6. six/
7. 7/
8. 8/
9. 9/
10. 10/
11. 11/
12. 12/
13. 13/
14. 14/
15. none/

#495. $M_r$ <of largest (or only) protein in the virion, e.g. 17000>/

#496. <Function of largest or only virion protein, e.g. coat protein>/

#497. $M_r$ of 2nd largest <virion protein>/

#498. <Function of 2nd largest virion protein>/

#499. $M_r$ of 3rd largest <virion protein>/

#500. <Function of 3rd largest virion protein>/

#501. $M_r$ of 4th largest <virion protein>/

#502. <Function of 4th largest virion protein>/

#503. $M_r$ of 5th largest <virion protein>/

#504. <Function of 5th largest virion protein, number>/

#505. $M_r$ of 6th largest and smaller, <virion proteins>/

#506. <Function of 6th largest virion protein>/

#507. Method of preparation: <of virion protein; references>/

#508. Amino acid sequence: <references>/

#509. Amino acid composition: <references>/

#510. Virion proteins <whether glycosylated>/
1. glycosylated/
2. not glycosylated/

#511. <The virion proteins which are glycosylated> are/

#512. Virion proteins <whether phosphorylated>/
1. phosphorylated/
2. not phosphorylated/

#513. <The virion proteins which are phosphorylated> are/

*Non-virion viral proteins*

#514. Virus–coded non–virion proteins <identified>/
1. isolated <methods or references>/
2. identified by genomic sequence analysis <references>/

#515. <Number of virus-coded non-virion proteins found>/
1. one protein found/
2. two proteins found/
3. three proteins found/
4. four proteins found/

5. five proteins found/
6. six proteins found/
7. seven proteins found/
8. eight proteins found/
9. nine proteins found/
10. ten proteins found/

#516. $M_r$ <of largest or only>/
#517. <Comments on largest or only non-virion protein>/
#518. $M_r$ of 2nd largest/
#519. <Comments on 2nd largest non-virion protein>/
#520. $M_r$ of 3rd/
#521. <Comments on 3rd largest non-virion protein>/
#522. $M_r$ of 4th/
#523. <Comments on 4th largest non-virion protein>/
#524. $M_r$ of 5th and smaller/
#525. <Comments on 5th largest non-virion protein>/

**Replication and helper functions**

#526. Genome replicates <in a specific cellular site>/
1. in nuclei/
2. in cytoplasm/
3. in association with chloroplasts/
4. in cytoplasmic viroplasms/

#527. Coat protein *m*RNA translated in <specific cellular site>/
#528. Replication <whether dependent on a helper virus>/
1. depends on co–infection with a helper virus <see text character>/
2. does not depend on a helper virus/

#529. <Details of interaction with a helper virus>/
#530. Acts as helper for/
1. a satellite virus/
2. a satellite RNA/
3. another virus <specify>/

#531. Acts as a helper for <name the other entity>/
#532. Acts as a helper by <the following mechanism>/

**Cytopathology**

#533. Virions found in <specific tissue types, list>/
#534. Virions found <particular organelles>/
1. in cytoplasm/
2. in nuclei/
3. in nucleoli/
4. in the perinuclear space/
5. in chloroplasts/
6. in mitochondria/
7. in Golgi apparatus/
8. in cell vacuoles/
9. in unusual sites <i.e. in sites other than those listed>/

#535. <Details other organelles> virions found in/
#536. Inclusions <found/not found in infected cells>/
1. present in infected cells/
2. absent from infected cells/

#537. Inclusions are <of a particular morphology>/
1. crystals in the cytoplasm/
2. crystals in the nucleus/
3. amorphous X–bodies/
4. membranous bodies/
5. viroplasms/
6. pinwheels/
7. unusual in shape <described in following character>/

#538. <The unusual inclusions, see last character, are>/
#539. <Inclusions, whether containing virions>/
1. they contain virions/
2. they do not contain virions/

#540. Other <structural and> cellular changes:/

**Taxonomy and Relationships**

#541. <Virus taxonomic group>/
1. *Alfamovirus: Bromoviridae*/
2. *Alphacryptovirus: Partitiviridae*/
3. *Badnavirus*/
4. *Betacryptovirus: Partitiviridae*/
5. bigeminivirus: *Geminiviridae*/
6. *Bromovirus: Bromoviridae*/
7. *Bymovirus: Potyviridae*/

8. *Capillovirus*/
9. *Carlavirus*/
10. *Carmovirus: Tombusviridae*/
11. *Caulimovirus*/
12. *Closterovirus*/
13. *Comovirus: Comoviridae*/
14. *Cucumovirus: Bromoviridae*/
15. *Cytorhabdovirus: Rhabdoviridae*/
16. *Dianthovirus*/
17. *Enamovirus*/
18. *Fabavirus: Comoviridae*/
19. *Fijivirus: Reoviridae*/
20. *Furovirus*/
21. *Hordeivirus*/
22. hybrigeminivirus: *Geminiviridae*/
23. *Idaeovirus*/
24. *Ilarvirus: Bromoviridae*/
25. ipomovirus: *Potyviridae*/
26. *Luteovirus*/
27. *Machlomovirus*/
28. macluravirus/
29. *Marafivirus*/
30. monogeminivirus: *Geminiviridae*/
31. nanavirus/
32. *Necrovirus*/
33. *Nepovirus: Comoviridae*/
34. *Nucleorhabdovirus: Rhabdoviridae*/
35. *Oryzavirus: Reoviridae*/
36. ourmiavirus/
37. *Phytoreovirus: Reoviridae*/
38. *Potexvirus*/
39. *Potyvirus: Potyviridae*/
40. *Rymovirus: Potyviridae*/
41. satellite RNA/
42. satellivirus/
43. *Sequivirus: Sequiviridae*/
44. *Sobemovirus*/
45. *Tenuivirus*/
46. *Tobamovirus*/
47. *Tobravirus*/
48. *Tombusvirus: Tombusviridae*/
49. *Tospovirus: Bunyaviridae*/
50. *Trichovirus*/
51. *Tymovirus*/
52. *Umbravirus*/
53. unassigned potyvirus/
54. unassigned rhabdovirus/
55. varicosavirus/
56. *Waikavirus: Sequiviridae*/

#542. Virus(es) with serologically related virions:/

#543. Virus(es) with serologically unrelated virions:/

#544. <Differences between type strain and others>/

#545. <Additional comments on relationships, e.g. known sequence homologies with other viruses>/

#546. Best tests for diagnosis: <hints for diagnosis>/

**Comments**

#547. <General comments>/

**References**

#548. <References to data in this description: list>/

# INDEX OF NAMES, SYNONYMS AND ACRONYMS

anemone mosaic virus = Turnip mosaic *potyvirus*
anemone necrosis virus = Tobacco ringspot *nepovirus*
Anjermozaick virus = Carnation ringspot *dianthovirus*
ANMV = Aquilegia necrotic mosaic *caulimovirus*
**Anthoxanthum latent blanching *hordeivirus***
**Anthoxanthum mosaic potyvirus**
**Anthriscus *carlavirus***
**Anthriscus yellows *waikavirus***
AntMV = Anthoxanthum mosaic potyvirus
AntV = Anthriscus *carlavirus*
apium virus 1 = Celery mosaic *potyvirus*
APLPV = Plum American line pattern *ilarvirus*
APLV = Potato Andean latent *tymovirus*
APMoV = Potato Andean mottle *comovirus*
ApMV = Apple mosaic *ilarvirus*
**Apple chlorotic leaf spot *trichovirus***
apple latent virus 2 = Sowbane mosaic *sobemovirus*
apple latent virus type 1 = Apple chlorotic leaf spot *trichovirus*
**Apple mosaic *ilarvirus***
**Apple necrosis *ilarvirus***
apple spy 227 epinasty and decline virus = Apple stem pitting virus
**Apple stem grooving *capillovirus***
**Apple stem pitting virus**
**Aquilegia necrotic mosaic *caulimovirus***
**Aquilegia potyvirus**
AqV = Aquilegia potyvirus
**Arabis mosaic *nepovirus***
**Arabis mosaic satellite RNA**
arabis mosaic virus = Dogwood mosaic *nepovirus*
**Araujia mosaic *potyvirus***
ArjMV = Araujia mosaic *potyvirus*
ArLV = Artichoke latent *potyvirus*
ArLVM = Artichoke latent M *carlavirus*
ArLVS = Artichoke latent S *carlavirus*
ArMV = Arabis mosaic *nepovirus*
**Arracacha A *nepovirus***
**Arracacha B *nepovirus***
**Arracacha latent *carlavirus***
**Arracacha Y *potyvirus***
arrhenatherum blue dwarf virus = Oat sterile dwarf *fijivirus*
artichoke Californian latent virus = Artichoke latent *potyvirus*
**Artichoke curly dwarf *potexvirus***
**Artichoke Italian latent *nepovirus***
**Artichoke latent M *carlavirus***
**Artichoke latent *potyvirus***
**Artichoke latent S *carlavirus***
**Artichoke mottled crinkle *tombusvirus***
artichoke potyvirus = Artichoke latent *potyvirus*
**Artichoke vein banding *nepovirus***
artichoke yellow band virus = Pepper ringspot *tobravirus*
**Artichoke yellow ringspot *nepovirus***
**Asclepias rhabdovirus**
ASGV = Apple stem grooving *capillovirus*
ash mosaic virus = Cherry leaf roll *nepovirus*
ash ring and line pattern virus = Arabis mosaic *nepovirus*
**Asparagus 1 *potyvirus***
**Asparagus 2 *ilarvirus***
**Asparagus 3 *potexvirus***
asparagus latent virus = Asparagus 2 *ilarvirus*
asparagus stunt virus = Tobacco streak *ilarvirus*
asparagus virus B = Asparagus 1 *potyvirus*
asparagus virus C = Asparagus 2 *ilarvirus*
**Aster chlorotic stunt *carlavirus***
aster ringspot virus = Tobacco rattle *tobravirus*
**Asystasia gangetica mottle potyvirus**
AtBV = Atropa belladonna rhabdovirus
**Atropa belladonna rhabdovirus**
atropa mild mosaic virus = Henbane mosaic *potyvirus*
AuBV = Aucuba ringspot *badnavirus*
**Aucuba ringspot *badnavirus***
*Aureogenus magnivena* virus = Clover wound tumor *phytoreovirus*
Australian wheat striate mosaic virus = Chloris striate mosaic monogeminivirus
AV-1 = Asparagus 1 *potyvirus*
AV-2 = Asparagus 2 *ilarvirus*
AV-3 = Asparagus 3 *potexvirus*

BMV = Brome mosaic *bromovirus*
BMYV = Beet mild yellowing *luteovirus*
BNYV = Broccoli necrotic yellows *cytorhabdovirus*
BNYVV = Beet necrotic yellow vein *furovirus*
bottlegourd Indian mosaic virus = Cucumber green mottle mosaic *tobamovirus*
boussingaultia mosaic virus = Papaya mosaic *potexvirus*
BPMV = Bean pod mottle *comovirus*
BPYV = Beet pseudo-yellows *closterovirus*
**Bramble yellow mosaic potyvirus**
brassica virus 3 = Cauliflower mosaic *caulimovirus*
Brazilian cassava common mosaic virus = Cassava common mosaic *potexvirus*
**Brinjal mild mosaic potyvirus**
brinjal mosaic virus = Potato Y *potyvirus*
BRMV = Bean rugose mosaic *comovirus*
BrmYMV = Bramble yellow mosaic potyvirus
BRNV = Black raspberry necrosis virus
**Broad bean B virus**
broad bean Evesham stain virus = Broad bean stain *comovirus*
broad bean F1 virus = Broad bean stain *comovirus*
**Broad bean mottle *bromovirus***
**Broad bean necrosis *furovirus***
**Broad bean severe chlorosis *closterovirus***
**Broad bean stain *comovirus***
**Broad bean true mosaic *comovirus***
**Broad bean V potyvirus**
**Broad bean wilt *fabavirus***
broad bean yellow band virus = Pea early browning *tobravirus*
**Broad bean yellow ringspot virus**
**Broad bean yellow vein *cytorhabdovirus***
broccoli mosaic virus = Cauliflower mosaic *caulimovirus*
**Broccoli necrotic yellows *cytorhabdovirus***
**Brome mosaic *bromovirus***
**Brome streak mosaic *rymovirus***
brown line disease virus = Apple stem grooving *capillovirus*
BRRV = Blueberry red ringspot *caulimovirus*
BryMV = Bryonia mottle potyvirus
**Bryonia mottle potyvirus**
bryony white mosaic virus = White bryony mosaic *carlavirus*
BSBV = Beet soil-borne *furovirus*
BSMV = Barley stripe mosaic *hordeivirus*
BSSV = Blueberry shoestring *sobemovirus*
BStV = Brome streak mosaic *rymovirus*
BSV = Banana streak *badnavirus*
BtMV = Beet mosaic *potyvirus*
**Burdock mosaic virus**
**Burdock mottle virus**
**Burdock yellow mosaic *potexvirus***
**Burdock yellows *closterovirus***
ButMV = Butterbur mosaic *carlavirus*
**Butterbur mosaic *carlavirus***
**Butterbur *nucleorhabdovirus***
BuYV = Burdock yellows *closterovirus*
BWYV = Beet western yellows *luteovirus*
BYDV-MAV, BYDV-PAV, BYDV-RGV, BYDV-RMV, BYDV-RPV, BYDV-SGV = Barley yellow dwarf *luteovirus*
BYMV = Bean yellow mosaic *potyvirus*
BYNV = Beet yellow net *luteovirus*
BYSMV = Barley yellow striate mosaic *cytorhabdovirus*
BYSV = Beet yellow stunt *closterovirus*
BYV = Beet yellows *closterovirus*
BYVBV = Bean yellow vein banding *umbravirus*
cabbage A virus = Turnip mosaic *potyvirus*
cabbage black ringspot virus = Turnip mosaic *potyvirus*
cabbage black ring virus = Turnip mosaic *potyvirus*
cabbage mosaic virus = Cauliflower mosaic *caulimovirus*
cabbage ring necrosis virus = Turnip mosaic *potyvirus*
cabbage virus B = Cauliflower mosaic *caulimovirus*
cabuya necrotic streak mosaic virus = Furcraea necrotic streak *dianthovirus*

chenopodium necrosis *necrovirus* = Tobacco necrosis *necrovirus*
chenopodium seed-borne mosaic virus = Sowbane mosaic *sobemovirus*
chenopodium star mottle virus = Sowbane mosaic *sobemovirus*
**Cherry A *capillovirus***
cherry chlorotic ringspot virus = Prune dwarf *ilarvirus*
**Cherry leaf roll *nepovirus***
**Cherry mottle leaf *trichovirus***
**Cherry rasp leaf *nepovirus***
**Chickpea bushy dwarf *potyvirus***
**Chickpea chlorotic dwarf monogeminivirus**
**Chickpea distortion mosaic *potyvirus***
**Chickpea filiform potyvirus**
chicory blotch virus = Chicory yellow blotch *carlavirus*
**Chicory X *potexvirus***
**Chicory yellow blotch *carlavirus***
**Chicory yellow mottle *nepovirus***
**Chicory yellow mottle satellite RNA**
chicory yellows virus = Chicory yellow blotch *carlavirus*
chilli mottle virus = Pepper Indian mottle *potyvirus*
chilli mottle virus = Pepper mottle *potyvirus*
**Chilli veinal mottle potyvirus**
chiloensis vein banding virus = Strawberry vein banding *caulimovirus*
**Chinese yam necrotic mosaic *carlavirus***
**Chino del tomaté bigeminivirus**
ChiVMV = Chilli veinal mottle potyvirus
**Chloris striate mosaic monogeminivirus**
**Chondrilla juncea stunting *nucleorhabdovirus***
chrysanthemum aspermy virus = Tomato aspermy *cucumovirus*
**Chrysanthemum B *carlavirus***
chrysanthemum dwarf mottle virus = Chrysanthemum B *carlavirus*
chrysanthemum mild mosaic virus = Chrysanthemum B *carlavirus*
chrysanthemum mild mottle virus = Tomato aspermy *cucumovirus*
chrysanthemum mosaic virus = Tomato aspermy *cucumovirus*
chrysanthemum necrotic mottle virus = Chrysanthemum B *carlavirus*
**Chrysanthemum spot potyvirus**
**Chrysanthemum vein chlorosis rhabdovirus**
chrysanthemum virus Q = Chrysanthemum B *carlavirus*
ChVX = Chicory X *potexvirus*
ChYBV = Chicory yellow blotch *carlavirus*
ChYMV = Chicory yellow mottle *nepovirus*
ChYNMV = Chinese yam necrotic mosaic *carlavirus*
CiLRV = Citrus leaf rugose *ilarvirus*
CiLV = Citrus leprosis rhabdovirus
CIRV = Carnation Italian ringspot *tombusvirus*
citrange stunt virus = Citrus tatter leaf *capillovirus*
citrus crinkly leaf ilarvirus = Citrus leaf rugose *ilarvirus*
**Citrus enation – woody gall *luteovirus***
**Citrus leaf rugose *ilarvirus***
**Citrus leprosis rhabdovirus**
citrus psorosis virus complex = Citrus leaf rugose *ilarvirus*
citrus psorosis virus complex — A and B = Citrus ringspot virus
citrus psorosis virus complex — infectious variegation component = Citrus variegation *ilarvirus*
citrus quick decline virus = Citrus tristeza *closterovirus*
**Citrus ringspot virus**
**Citrus tatter leaf *capillovirus***
**Citrus tristeza *closterovirus***
**Citrus variegation *ilarvirus***
CLCrV = Cotton leaf crumple bigeminivirus
CLCuV = Cotton leaf curl bigeminivirus
ClEV = Clover enation *nucleorhabdovirus*
**Clitoria mosaic *potexvirus***
**Clitoria yellow vein *tymovirus***
clover Alsike vein mosaic virus = Alsike clover vein mosaic virus
clover big vein virus = Clover wound tumor *phytoreovirus*
clover blotch virus = Peanut stunt *cucumovirus*

**Cowpea Moroccan aphid-borne mosaic *potyvirus***
cowpea Morrocan aphid-borne mosaic potyvirus = Passiflora South African *potyvirus*
**Cowpea mosaic *comovirus***
cowpea mosaic virus = Sunn-hemp mosaic *tobamovirus*
cowpea mosaic virus = Vigna sinensis mosaic rhabdovirus
**Cowpea mottle *carmovirus***
cowpea ringspot virus = Cucumber mosaic *cucumovirus*
**Cowpea rugose mosaic *potyvirus***
**Cowpea severe mosaic *comovirus***
cowpea strain of tobacco mosaic virus = Sunn-hemp mosaic *tobamovirus*
**Cowpea stunt *luteovirus***
cowpea yellow mosaic virus = Cowpea mosaic *comovirus*
cowpea yellow mosaic virus = Sunn-hemp mosaic *tobamovirus*
CowPMV = Cow parsnip mosaic *nucleorhabdovirus*
CpBDV = Chickpea bushy dwarf *potyvirus*
CpFV = Chickpea filiform potyvirus
CPGMV = Cowpea golden mosaic bigeminivirus
CPMMV = Cowpea mild mottle *carlavirus*
CPMoV = Cowpea mottle *carmovirus*
CPMV = Cowpea mosaic *comovirus*
CPRMV = Cowpea rugose mosaic *potyvirus*
CPSMV = Cowpea severe mosaic *comovirus*
CraV = Cynara *nucleorhabdovirus*
**Crimson clover latent *nepovirus***
CriMV = Crinum mosaic *potyvirus*
**Crinum mosaic *potyvirus***
CRLV = Cherry rasp leaf *nepovirus*
CRMV = Canary reed mosaic potyvirus
Crotalaria mucronata mosaic virus = Sunn-hemp mosaic *tobamovirus*
**Crotalaria spectabilis yellow mosaic *potexvirus***
croton rhabdovirus = Croton vein yellowing *nucleorhabdovirus*
**Croton vein yellowing *nucleorhabdovirus***
**Croton yellow vein mosaic bigeminivirus**
CRSV = Carnation ringspot *dianthovirus*
CsCMV = Cassava common mosaic *potexvirus*
CSMV = Chloris striate mosaic monogeminivirus
CsSLV = Cassava symptomless rhabdovirus
CSSV = Cacao swollen shoot *badnavirus*
CSV = Cocksfoot streak potyvirus
CsVX = Cassava X *potexvirus*
CTeV-1 = Carrot temperate 1 *alphacryptovirus*
CTeV-2 = Carrot temperate 2 *betacryptovirus*
CTeV-3 = Carrot temperate 3 *alphacryptovirus*
CTeV-4 = Carrot temperate 4 *alphacryptovirus*
CtLtV = Carrot latent *nucleorhabdovirus*
CTLV = Carrot thin leaf *potyvirus*
CTLV = Citrus tatter leaf *capillovirus*
CtMV = Carrot mosaic potyvirus
CtRLV = Carrot red leaf *luteovirus*
CTV = Citrus tristeza *closterovirus*
**Cucumber chlorotic spot *closterovirus***
cucumber fruit streak virus = Cucumber leaf spot *carmovirus*
**Cucumber green mottle mosaic *tobamovirus***
cucumber green mottle mosaic virus = Cucumber green mottle mosaic *tobamovirus*
cucumber green mottle mosaic virus - cucumber strain or strain C = Kyuri green mottle mosaic *tobamovirus*
**Cucumber leaf spot *carmovirus***
**Cucumber mosaic *cucumovirus***
**Cucumber mosaic virus satellite RNA**
**Cucumber necrosis *tombusvirus***
**Cucumber soil-borne *carmovirus***
cucumber systemic necrosis virus = Tobacco necrosis *necrovirus*
**Cucumber toad-skin rhabdovirus**
**Cucumber vein yellowing virus**
cucumber virus 1 = Cucumber mosaic *cucumovirus*
cucumber virus 3 = Cucumber green mottle mosaic *tobamovirus*

**Dendrobium vein necrosis *closterovirus***
dendrobium virus = Orchid fleck rhabdovirus
**Desmodium mosaic *potyvirus***
**Desmodium yellow mottle *tymovirus***
DesMV = Desmodium mosaic *potyvirus*
Digitaria didactyla striate mosaic virus = Datura striate mosaic monogeminivirus
**Digitaria streak monogeminivirus**
**Digitaria striate *cytorhabdovirus***
**Diodia vein chlorosis *closterovirus***
**Dioscorea alata *potyvirus***
**Dioscorea bacilliform *badnavirus***
dioscorea badnavirus = Yam internal brown spot *badnavirus*
**Dioscorea green banding mosaic *potyvirus***
**Dioscorea latent *potexvirus***
**Dioscorea trifida potyvirus**
DiSV = Digitaria striate *cytorhabdovirus*
DLV = Dioscorea latent *potexvirus*
DMMV = Dock mottling mosaic potyvirus
DMV = Dahlia mosaic *caulimovirus*
DNV = Datura necrosis *potyvirus*
dock mosaic virus = Dock mottling mosaic potyvirus
**Dock mottling mosaic potyvirus**
**Dodonaea yellows-associated virus**
**Dogwood mosaic *nepovirus***
dolichos enation mosaic virus = Sunn-hemp mosaic *tobamovirus*
**Dolichos yellow mosaic bigeminivirus**
DoYMV = Dolichos yellow mosaic bigeminivirus
DsMV = Dasheen mosaic *potyvirus*
DSTV = Datura shoestring *potyvirus*
DSV = Digitaria streak monogeminivirus
DTMV = Datura mosaic potyvirus
DTV = Dioscorea trifida potyvirus
dulcamara A virus = Carnation latent *carlavirus*
dulcamara B virus = Carnation latent *carlavirus*
**Dulcamara mottle *tymovirus***
DuMV = Dulcamara mottle *tymovirus*
Dutch plum line pattern virus = Apple mosaic *ilarvirus*
DVCV = Diodia vein chlorosis *closterovirus*
DVNV = Dendrobium vein necrosis *closterovirus*
DVS = Daphne S *carlavirus*
DVX = Daphne X *potexvirus*
DVY = Daphne Y *potyvirus*
DYMV = Desmodium yellow mottle *tymovirus*
DYVV = Datura yellow vein *nucleorhabdovirus*
Eastern veinbanding virus = Strawberry vein banding *caulimovirus*
echinochloa hoja blanca virus = Rice hoja blanca *tenuivirus*
**Echinochloa ragged stunt *oryzavirus***
Echtes Ackerbohnenmosaikvirus = Broad bean true mosaic *comovirus*
EFV = Euonymus fasciation rhabdovirus
**Eggplant green mosaic *potyvirus***
**Eggplant mild mottle *carlavirus***
**Eggplant mosaic *tymovirus***
**Eggplant mottled crinkle *tombusvirus***
**Eggplant mottled dwarf *nucleorhabdovirus***
eggplant mottled dwarf rhabdovirus = Tomato vein clearing *nucleorhabdovirus*
eggplant mottled dwarf virus = Pittosporum vein yellowing *nucleorhabdovirus*
eggplant severe mosaic virus = Eggplant severe mottle potyvirus
**Eggplant severe mottle potyvirus**
eggplant yellow mosaic virus = Tobacco leaf curl bigeminivirus
EGMV = Eggplant green mosaic *potyvirus*
elderberry A virus = Elderberry *carlavirus*
**Elderberry *carlavirus***
**Elderberry latent *carmovirus***
elderberry vein clearing virus = Sambucus vein clearing rhabdovirus
eleusine mosaic virus = Finger millet mosaic *nucleorhabdovirus*
**Elm mottle *ilarvirus***
ELV = Elderberry latent *carmovirus*
EMCV = Eggplant mottled crinkle *tombusvirus*
EMDV = Eggplant mottled dwarf *nucleorhabdovirus*

EMMV = Eggplant mild mottle *carlavirus*
EMoV = Elm mottle *ilarvirus*
EMV = Eggplant mosaic *tymovirus*
enation pea mosaic virus = Pea enation mosaic *enamovirus*
**Endive necrotic mosaic *potyvirus***
EpCV = Epirus cherry ourmiavirus
**Epirus cherry ourmiavirus**
ErLV = Erysimum latent *tymovirus*
ERSV = Echinochloa ragged stunt *oryzavirus*
**Erysimum latent *tymovirus***
ESMV = Eggplant severe mottle potyvirus
**Eucharis mottle *nepovirus***
EuMV = Euphorbia mosaic bigeminivirus
**Euonymus fasciation rhabdovirus**
**Euonymus rhabdovirus**
euonymus ringspot virus = Tomato ringspot *nepovirus*
eupatorium yellow vein virus = Tobacco leaf curl bigeminivirus
**Euphorbia mosaic bigeminivirus**
**Euphorbia ringspot *potyvirus***
European plum line pattern virus = Apple mosaic *ilarvirus*
European plum line pattern virus = Prunus necrotic ringspot *ilarvirus*
European wheat striate mosaic virus = Wheat European striate mosaic *tenuivirus*
EuRV = Euphorbia ringspot *potyvirus*
EV = Elderberry *carlavirus*
EWSMV = Wheat European striate mosaic *tenuivirus*
**Faba bean necrotic yellows nanavirus**
FCV = Fescue *alphacryptovirus*
FDV = Sugarcane Fiji disease *fijivirus*
**Fern potyvirus**
fern tobravirus = Harts tongue fern *tobravirus*
**Fescue *alphacryptovirus***
**Festuca leaf streak *cytorhabdovirus***
**Festuca necrosis *closterovirus***
FicCV = Fig potyvirus
**Fig potyvirus**
**Fig S *carlavirus***
**Figwort mosaic *caulimovirus***
figwort mottle virus = Figwort mosaic *caulimovirus*
**Filaree red leaf *luteovirus***
**Finger millet mosaic *nucleorhabdovirus***
flat apple virus = Cherry rasp leaf *nepovirus*
flax crinkle virus = Oat blue dwarf *marafivirus*
FLRV = Filaree red leaf *luteovirus*
FLSV = Festuca leaf streak *cytorhabdovirus*
FLV = Fuchsia latent *carlavirus*
FMMV = Finger millet mosaic *nucleorhabdovirus*
FMV = Figwort mosaic *caulimovirus*
FNSV = Furcraea necrotic streak *dianthovirus*
FNV = Festuca necrosis *closterovirus*
FoMV = Foxtail mosaic *potexvirus*
forsythia yellow net virus = Arabis mosaic *nepovirus*
**Foxtail mosaic *potexvirus***
**Fragaria chiloensis *ilarvirus***
**Frangipani mosaic *tobamovirus***
**Freesia leaf necrosis varicosavirus**
**Freesia mosaic *potyvirus***
freesia streak virus = Freesia mosaic *potyvirus*
FreMV = Freesia mosaic *potyvirus*
FrMV = Frangipani mosaic *tobamovirus*
**Fuchsia latent *carlavirus***
fuchsia S virus = Fuchsia latent *carlavirus*
**Furcraea necrotic streak *dianthovirus***
**Galinsoga mosaic *carmovirus***
GALV = Grapevine Algerian latent *tombusvirus*
GaMV = Galinsoga mosaic *carmovirus*
**Garland chrysanthemum temperate *alphacryptovirus***
garland chrysanthemum temperate cryptovirus = Garland chrysanthemum temperate *alphacryptovirus*
**Garlic common latent *carlavirus***
**Garlic dwarf *fijivirus***
garlic laent virus = Shallot latent *carlavirus*
garlic latent virus — France = Garlic common latent *carlavirus*
garlic latent virus — Germany = Garlic common latent *carlavirus*
GAV = Grapevine ajinashika disease *luteovirus*

GRV = Groundnut rosette *umbravirus*
GSLV = Guar symptomless *potyvirus*
**Guar symptomless *potyvirus***
**Guar top necrosis virus**
**Guinea grass mosaic *potyvirus***
gynura latent virus = Chrysanthemum B *carlavirus*
**Habenaria mosaic potyvirus**
HaMV = Habenaria mosaic potyvirus
**Harts tongue fern *tobravirus***
hassaku dwarf virus = Citrus tristeza *closterovirus*
hawthorn ring pattern mosaic virus = Apple stem pitting virus
HCRSV = Hibiscus chlorotic ringspot *carmovirus*
HDIV = Datura innoxia Hungarian mosaic potyvirus
HdLV = Hydrangea latent *carlavirus*
HdMV = Hydrangea mosaic *ilarvirus*
**Helenium S *carlavirus***
**Helenium Y *potyvirus***
helianthus mosaic virus = Sunflower mosaic potyvirus
**Helleborus mosaic *carlavirus***
HeMV = Helleborus mosaic *carlavirus*
**Henbane mosaic *potyvirus***
heracleum 6 capillovirus = Carrot yellow leaf *closterovirus*
**Heracleum latent *trichovirus***
HgYMV = Horsegram yellow mosaic bigeminivirus
**Hibiscus chlorotic ringspot *carmovirus***
**Hibiscus latent ringspot *nepovirus***
**Hibiscus yellow mosaic *tobamovirus***
HiMV = Hippeastrum mosaic *potyvirus*
hippeastrum latent virus = Nerine latent *carlavirus*
**Hippeastrum mosaic *potyvirus***
HjV = Humulus japonicus *ilarvirus*
HLRSV = Hibiscus latent ringspot *nepovirus*
HLV = Heracleum latent *trichovirus*
HLYV = Holcus lanatus yellowing rhabdovirus
HMV = Henbane mosaic *potyvirus*
HnLV = Honeysuckle latent *carlavirus*
hogweed 4 virus = Parsnip leafcurl virus
hogweed 6 virus = Carrot yellow leaf *closterovirus*
**Holcus lanatus yellowing rhabdovirus**
**Holcus streak potyvirus**
Holmes' ribgrass virus = Ribgrass mosaic *tobamovirus*
HoMV = Hordeum mosaic *rymovirus*
**Honeysuckle latent *carlavirus***
honeysuckle yellow vein mosaic virus = Tobacco leaf curl bigeminivirus
**Hop American latent *carlavirus***
hop A virus = Apple mosaic *ilarvirus*
hop B virus = Prunus necrotic ringspot *ilarvirus*
hop C virus = Prunus necrotic ringspot *ilarvirus*
**Hop latent *carlavirus***
**Hop mosaic *carlavirus***
**Hop trefoil 1 *alphacryptovirus***
**Hop trefoil 2 *betacryptovirus***
**Hop trefoil 3 *alphacryptovirus***
**Hordeum mosaic *rymovirus***
hordeum nanescens virus = Barley yellow dwarf *luteovirus*
horse chestnut yellow mosaic virus = Apple mosaic *ilarvirus*
**Horsegram yellow mosaic bigeminivirus**
**Horseradish latent *caulimovirus***
horseradish mosaic virus = Turnip mosaic *potyvirus*
HpLV = Hop latent *carlavirus*
HpMV = Hop mosaic *carlavirus*
HRLV = Horseradish latent *caulimovirus*
HRSV = Hydrangea ringspot *potexvirus*
HSV = Holcus streak potyvirus
HTCV-1 = Hop trefoil 1 *alphacryptovirus*
HTCV-2 = Hop trefoil 2 *betacryptovirus*
HTCV-3 = Hop trefoil 3 *alphacryptovirus*
**Humulus japonicus *ilarvirus***
Hungarian chrome mosaic virus = Grapevine chrome mosaic *nepovirus*
HV2 = Heracleum latent *trichovirus*
HVS = Helenium S *carlavirus*
HVY = Helenium Y *potyvirus*
**Hyacinth mosaic *potyvirus***
HyaMV = Hyacinth mosaic *potyvirus*
**Hydrangea latent *carlavirus***
**Hydrangea mosaic *ilarvirus***
hydrangea mosaic virus = Elm mottle *ilarvirus*
**Hydrangea ringspot *potexvirus***
HyMV = Hypochoeris mosaic *furovirus*

hyoscyamus virus I = Henbane mosaic *potyvirus*
hyoscyamus virus III = Henbane mosaic *potyvirus*
**Hypochoeris mosaic *furovirus***
ICMV = Cassava Indian mosaic bigeminivirus
IFMV = Iris fulva mosaic *potyvirus*
IGLSV = Iris germanica leaf stripe rhabdovirus
ILV = Impatiens latent *potexvirus*
IMMV = Iris mild mosaic *potyvirus*
**Impatiens latent *potexvirus***
**Impatiens necrotic spot *tospovirus***
Indian cassava mosaic virus = Cassava Indian mosaic bigeminivirus
Indian pepper mottle virus = Pepper Indian mottle *potyvirus*
Indonesian soybean dwarf virus = Soybean Indonesian dwarf *luteovirus*
ine shima-ha gare uirusu = Rice stripe *tenuivirus*
infectious chlorosis of Malvaceae = Abutilon mosaic bigeminivirus
INSV = Impatiens necrotic spot *tospovirus*
IPCV = Peanut clump *furovirus*
IPMV = Pepper Indian mottle *potyvirus*
Iranian Shiraz maize mosaic virus = Maize Iranian mosaic *nucleorhabdovirus*
irisbontvirus = Narcissus latent macluravirus
**Iris fulva mosaic *potyvirus***
**Iris germanica leaf stripe rhabdovirus**
iris grijs virus = Iris severe mosaic *potyvirus*
**Iris Japanese necrotic ring virus**
iris latent mosaic virus = Iris mild mosaic *potyvirus*
**Iris mild mosaic *potyvirus***
iris mild mosaic virus = Narcissus latent macluravirus
iris mild yellow mosaic virus = Narcissus latent macluravirus
iris mosaic virus = Iris mild mosaic *potyvirus*
**Iris severe mosaic *potyvirus***
iris stripe virus = Iris severe mosaic *potyvirus*
iris yellow mosaic virus = Iris severe mosaic *potyvirus*
**Isachne mosaic potyvirus**
IsaMV = Isachne mosaic potyvirus
isanu mosaic virus = Tropaeolum 1 *potyvirus*
isanu mosaic virus = Tropaeolum 2 *potyvirus*
ISDV = Soybean Indonesian dwarf *luteovirus*
ISMV = Iris severe mosaic *potyvirus*
IVCV = Ivy vein clearing *cytorhabdovirus*
IVLP = Kalanchoe isometric virus
**Ivy vein clearing *cytorhabdovirus***
Japanese CV3 = Kyuri green mottle mosaic *tobamovirus*
jasmine yellow blotch virus = Arabis mosaic *nepovirus*
**Jatropha mosaic bigeminivirus**
JGMV = Johnsongrass mosaic *potyvirus*
JMV = Jatropha mosaic bigeminivirus
**Johnsongrass chlorotic stripe *carmovirus***
**Johnsongrass mosaic *potyvirus***
kalanchoe 'commelina yellow mottle badnavirus' = Kalanchoe top-spotting *badnavirus*
**Kalanchoe isometric virus**
kalanchoe I virus = Kalanchoe latent *carlavirus*
**Kalanchoe latent *carlavirus***
**Kalanchoe mosaic potyvirus**
**Kalanchoe top-spotting *badnavirus***
Kartoffel-K-Virus = Potato M *carlavirus*
Kartoffel-Rollmosaik-Virus = Potato M *carlavirus*
**Kenaf vein-clearing rhabdovirus**
**Kennedya yellow mosaic *tymovirus***
**Kennedya Y *potyvirus***
KGMMV = Kyuri green mottle mosaic *tobamovirus*
KLV = Kalanchoe latent *carlavirus*
KMV = Konjak mosaic *potyvirus*
**Konjak mosaic *potyvirus***
KTSV = Kalanchoe top-spotting *badnavirus*
KVY = Kennedya Y *potyvirus*
KYMV = Kennedya yellow mosaic *tymovirus*
**Kyuri green mottle mosaic *tobamovirus***

melilotus latent virus = Sweet clover latent *nucleorhabdovirus*
**Melilotus mosaic potyvirus**
**Melon leaf curl bigeminivirus**
melon mosaic virus = Watermelon mosaic 2 *potyvirus*
**Melon necrotic spot *carmovirus***
**Melon Ourmia ourmiavirus**
**Melon rugose mosaic *tymovirus***
**Melon variegation *cytorhabdovirus***
**Melon vein-banding mosaic potyvirus**
**Melothria mottle potyvirus**
MeMV = Melilotus mosaic potyvirus
MF virus = Broad bean stain *comovirus*
**Mibuna temperate *alphacryptovirus***
mild strain of tobacco mosaic virus = Tobacco mild green mosaic *tobamovirus*
**Milk vetch dwarf nanavirus**
millet finger mosaic virus = Finger millet mosaic *nucleorhabdovirus*
**Millet red leaf *luteovirus***
mimosa bacilliform virus = Mimosa striped chlorosis *badnavirus*
**Mimosa mosaic virus**
**Mimosa striped chlorosis *badnavirus***
MiMV = Mirabilis mosaic *caulimovirus*
**Mirabilis mosaic *caulimovirus***
**Miscanthus streak monogeminivirus**
MiSV = Miscanthus streak monogeminivirus
MLCV = Melon leaf curl bigeminivirus
MLRSV = Myrobalan latent ringspot *nepovirus*
MLV = Mulberry latent *carlavirus*
MMTV = Mung bean mottle potyvirus
MMV = Maize mosaic *nucleorhabdovirus*
MNSV = Melon necrotic spot *carmovirus*
Moroccan pepper virus = Pepper Moroccan *tombusvirus*
Moroccan watermelon mosaic virus = Watermelon Moroccan mosaic potyvirus
MPV = Pepper Moroccan *tombusvirus*
MRDV = Maize rough dwarf *fijivirus*
MRFV = Maize rayado fino *marafivirus*
MRLV = Millet red leaf *luteovirus*
MRSV = Mulberry ringspot *nepovirus*
MSpV = Maize stripe *tenuivirus*
MSV = Maize streak monogeminivirus
MTV = Mibuna temperate *alphacryptovirus*
**Mulberry latent *carlavirus***
**Mulberry ringspot *nepovirus***
mung bean leaf curl virus = Tomato spotted wilt *tospovirus*
mungbean mosaic virus = Bean common mosaic *potyvirus*
**Mung bean mottle potyvirus**
**Mung bean yellow mosaic bigeminivirus**
muskmelon mosaic virus = Squash mosaic *comovirus*
muskmelon necrotic mosaic virus = Squash leaf curl bigeminivirus
**Muskmelon vein necrosis *carlavirus***
muskmelon yellow stunt virus = Zucchini yellow mosaic *potyvirus*
muskmelon yellows virus = Beet pseudo-yellows *closterovirus*
MuVNV = Muskmelon vein necrosis *carlavirus*
MVBMV = Melon vein-banding mosaic potyvirus
MVCV = Malva vein clearing *potyvirus*
MVDV = Milk vetch dwarf nanavirus
MVNV = Malva veinal necrosis *potexvirus*
MVV = Melon variegation *cytorhabdovirus*
MWLMV = Maize white line mosaic virus
MWMV = Watermelon Moroccan mosaic potyvirus
MYFV = Melandrium yellow fleck *bromovirus*
MYMV = Mung bean yellow mosaic bigeminivirus
**Myrobalan latent ringspot *nepovirus***
NaMV = Nandina mosaic *potexvirus*
**Nandina mosaic *potexvirus***
nandina *potexvirus* = Nandina mosaic *potexvirus*
**Nandina stem pitting *capillovirus***
**Narcissus degeneration potyvirus**
narcissus degeneration potyvirus = Narcissus yellow stripe *potyvirus*
**Narcissus latent macluravirus**
**Narcissus late season yellows potyvirus**
narcissus mild mottle carlavirus = Narcissus latent macluravirus

**Narcissus mosaic *potexvirus***
**Narcissus tip necrosis *carmovirus***
narcissus yellow streak virus = Narcissus yellow stripe *potyvirus*
**Narcissus yellow stripe *potyvirus***
NasMV = Nasturtium mosaic *potyvirus*
**Nasturtium mosaic *potyvirus***
nasturtium ringspot virus = Broad bean wilt *fabavirus*
NCMV = Cereal northern mosaic *cytorhabdovirus*
NDV = Narcissus degeneration potyvirus
**Neckar River *tombusvirus***
NeCMV = Negro coffee mosaic *potexvirus*
**Negro coffee mosaic *potexvirus***
NeLV = Nerine latent *carlavirus*
**Nerine latent *carlavirus***
**Nerine *potyvirus***
**Nerine X *potexvirus***
nerine yellow stripe virus = Nerine *potyvirus*
**Nerine Y potyvirus**
New Zealand hop virus = Hop American latent *carlavirus*
**Nicotiana velutina mosaic *furovirus***
NLSYV = Narcissus late season yellows potyvirus
NLV = Narcissus latent macluravirus
NMV = Narcissus mosaic *potexvirus*
NoMV = Nothoscordum mosaic potyvirus
Norfolk virus = Beet soil-borne *furovirus*
northern cereal mosaic virus = Cereal northern mosaic *cytorhabdovirus*
**Nothoscordum mosaic potyvirus**
NRV = Neckar River *tombusvirus*
NSPV = Nandina stem pitting *capillovirus*
NTNV = Narcissus tip necrosis *carmovirus*
NVMV = Nicotiana velutina mosaic *furovirus*
NV = Nerine *potyvirus*
NVX = Nerine X *potexvirus*
NYSV = Narcissus yellow stripe *potyvirus*
**Oak ringspot virus**
**Oat blue dwarf *marafivirus***
oat chlorotic stripe virus = Oat golden stripe *furovirus*
**Oat chlorotic stunt *tombusvirus***
**Oat golden stripe *furovirus***
**Oat mosaic *bymovirus***
**Oat necrotic mottle *rymovirus***
**Oat pseudorosette *tenuivirus***
oat pupation disease virus = Oat pseudorosette *tenuivirus*
oat red leaf virus = Barley yellow dwarf *luteovirus*
oat rhabdovirus = Wheat American striate mosaic *nucleorhabdovirus*
oat Siberian mosaic virus = Oat pseudorosette *tenuivirus*
**Oat sterile dwarf *fijivirus***
oat striate and red disease virus = Wheat European striate mosaic *tenuivirus*
oat striate mosaic virus = Wheat American striate mosaic *nucleorhabdovirus*
oat striate rhabdovirus = Wheat American striate mosaic *nucleorhabdovirus*
oat stripe mosaic virus = Barley stripe mosaic *hordeivirus*
oat tubular virus = Oat golden stripe *furovirus*
OBDV = Oat blue dwarf *marafivirus*
**Odontoglossum ringspot *tobamovirus***
OFV = Orchid fleck rhabdovirus
OGSV = Oat golden stripe *furovirus*
Ohio corn stunt agent = Maize chlorotic dwarf *waikavirus*
OkMV = Okra mosaic *tymovirus*
**Okra leaf-curl bigeminivirus**
**Okra mosaic *tymovirus***
OLCV = Okra leaf-curl bigeminivirus
**Olive latent 1 *sobemovirus***
**Olive latent 2 ourmiavirus**
**Olive latent ringspot *nepovirus***
OLRSV = Olive latent ringspot *nepovirus*
OLV-1 = Olive latent 1 *sobemovirus*
OLV-2 = Olive latent 2 ourmiavirus
OMV = Oat mosaic *bymovirus*
**Onion mite-borne latent *potexvirus***
**Onion yellow dwarf *potyvirus***
onion yellow dwarf virus — Brazil = Welsh onion yellow stripe potyvirus
onion yellow dwarf virus — Japan = Welsh onion yellow stripe potyvirus
ONMV = Oat necrotic mottle *rymovirus*
**Ononis yellow mosaic *tymovirus***
**Opuntia Sammons' *tobamovirus***

orchard grass mosaic virus = Cocksfoot streak potyvirus

**Orchid fleck rhabdovirus**

orchid mosaic virus = Cymbidium mosaic *potexvirus*

orchid rhabdovirus = Orchid fleck rhabdovirus

OrMV = Ornithogalum mosaic *potyvirus*

**Ornithogalum mosaic *potyvirus***

ORSV = Odontoglossum ringspot *tobamovirus*

OSDV = Oat sterile dwarf *fijivirus*

OuMV = Melon Ourmia ourmiavirus

Ourmia melon virus = Melon Ourmia ourmiavirus

OYDV = Onion yellow dwarf *potyvirus*

OYMV = Ononis yellow mosaic *tymovirus*

paeony mosaic virus = Tobacco rattle *tobravirus*

paeony ringspot virus = Tobacco rattle *tobravirus*

**Palm mosaic potyvirus**

PalMV = Palm mosaic potyvirus

PaMMV = Paprika mild mottle *tobamovirus*

PAMV = Potato aucuba mosaic *potexvirus*

**Pangola stunt *fijivirus***

**Panicum mosaic satellivirus**

**Panicum mosaic *sobemovirus***

Panicum strain of maize streak virus = Panicum streak monogeminivirus

**Panicum streak monogeminivirus**

papaw distortion ringspot virus = Papaya ringspot *potyvirus*

papaw mild mosaic virus = Papaya mosaic *potexvirus*

papaw mosaic virus = Papaya ringspot *potyvirus*

papaya distortion mosaic virus = Papaya ringspot *potyvirus*

papaya leaf distortion virus = Papaya ringspot *potyvirus*

**Papaya mosaic *potexvirus***

**Papaya ringspot *potyvirus***

papaya ringspot virus type W = Watermelon mosaic 1 *potyvirus*

PapMV = Papaya mosaic *potexvirus*

**Paprika mild mottle *tobamovirus***

para—tobacco mosaic virus = Tobacco mild green mosaic *tobamovirus*

**Parietaria mottle *ilarvirus***

ParMV = Parsnip mosaic *potyvirus*

**Parsley 5 *potexvirus***

parsley carrot leaf virus = Chicory yellow mottle *nepovirus*

**Parsley green mottle potyvirus**

**Parsley rhabdovirus**

parsley virus 3 = Broad bean wilt *fabavirus*

**Parsnip 3 *potexvirus***

**Parsnip leafcurl virus**

**Parsnip mosaic *potyvirus***

**Parsnip yellow fleck *sequivirus***

ParV-3 = Parsnip 3 *potexvirus*

paspalum geminivirus = Paspalum striate mosaic monogeminivirus

**Paspalum striate mosaic monogeminivirus**

passiflora chlorotic spot virus = Passionfruit woodiness *potyvirus*

**Passiflora latent *carlavirus***

passiflora mosaic virus = Passionfruit woodiness *potyvirus*

**Passiflora ringspot *potyvirus***

**Passiflora South African *potyvirus***

passiflora South African potyvirus = Cowpea Moroccan aphid-borne mosaic *potyvirus*

passionfruit mosaic virus = Passionfruit woodiness *potyvirus*

**Passionfruit *nucleorhabdovirus***

passionfruit ringspot virus = Passiflora ringspot *potyvirus*

**Passionfruit Sri Lankan mottle potyvirus**

**Passionfruit vein-clearing rhabdovirus**

**Passionfruit woodiness *potyvirus***

**Passionfruit yellow mosaic *tymovirus***

PaSV = Pangola stunt *fijivirus*

**Patchouli mosaic *potyvirus***

**Patchouli mottle *potyvirus***

**Patchouli mottle rhabdovirus**

PatMV = Patchouli mottle rhabdovirus

PaV-5 = Parsley 5 *potexvirus*

PaV = Parsley rhabdovirus

PBRSV = Potato black ringspot *nepovirus*

PCLSV = Pineapple chlorotic leaf streak *nucleorhabdovirus*

PcMV = Pecteilis mosaic potyvirus
PCSV = Peanut chlorotic streak *caulimovirus*
PDV = Prune dwarf *ilarvirus*
**Peach enation *nepovirus***
peach line pattern virosis virus = Plum American line pattern *ilarvirus*
peach ringspot virus = Prunus necrotic ringspot *ilarvirus*
**Peach rosette mosaic *nepovirus***
peach stunt virus = Prune dwarf *ilarvirus*
peach yellow bud mosaic virus = Tomato ringspot *nepovirus*
**Peach yellow leaf *closterovirus***
**Pea early browning *tobravirus***
**Pea enation mosaic *enamovirus***
pea fizzle top virus = Pea seed-borne mosaic *potyvirus*
pea leaf rolling mosaic virus = Pea seed-borne mosaic *potyvirus*
pea leaf rolling virus = Pea seed-borne mosaic *potyvirus*
pea leaf roll mosaic virus = Pea seed-borne mosaic *potyvirus*
pea leaf roll virus = Bean leaf roll *luteovirus*
pea leaf roll virus — New Zealand = Beet western yellows *luteovirus*
**Pea mild mosaic *comovirus***
**Pea mosaic *potyvirus***
pea mottle virus = Clover yellow mosaic *potexvirus*
pea mottle virus = Clover yellow vein *potyvirus*
PeAMV = Petunia asteroid mosaic *tombusvirus*
pea necrosis virus = Clover yellow vein *potyvirus*
peanut chlorotic mottle virus = Peanut green mosaic *potyvirus*
**Peanut chlorotic streak *caulimovirus***
**Peanut clump *furovirus***
**Peanut green mosaic *potyvirus***
peanut mild mosaic virus = Peanut mottle *potyvirus*
**Peanut mottle *potyvirus***
peanut severe mosaic virus = Peanut mottle *potyvirus*
**Peanut stunt *cucumovirus***
**Peanut top paralysis potyvirus**
**Peanut veinal chlorosis rhabdovirus**
peanut yellow mosaic virus = Cucumber mosaic *cucumovirus*
**Peanut yellow spot *tospovirus***
pear necrotic spot virus = Apple stem pitting virus
pear ring pattern mosaic virus = Apple chlorotic leaf spot *trichovirus*
pear stony pit virus = Apple stem pitting virus
pear vein yellows virus = Apple stem pitting virus
**Pea seed-borne mosaic *potyvirus***
pea silky green virus = Bean yellow vein banding *umbravirus*
**Pea stem necrosis virus**
**Pea streak *carlavirus***
pea streak New Zealand virus = Pea streak *carlavirus*
pea stunt virus = Red clover vein mosaic *carlavirus*
pea tip yellowing virus = Bean leaf roll *luteovirus*
pea top necrosis virus = Cucumber mosaic *cucumovirus*
pea top yellows virus = Bean leaf roll *luteovirus*
pea western ringspot virus = Clover yellow vein *potyvirus*
pea western ringspot virus = Cucumber mosaic *cucumovirus*
pea wilt virus = White clover mosaic *potexvirus*
PEBV = Pea early browning *tobravirus*
**Pecteilis mosaic potyvirus**
**Pelargonium flower break *carmovirus***
**Pelargonium leaf curl *tombusvirus***
**Pelargonium line pattern *carmovirus***
pelargonium ring pattern virus = Pelargonium line pattern *carmovirus*
pelargonium ringspot virus = Elderberry latent *carmovirus*
**Pelargonium vein clearing *cytorhabdovirus***
**Pelargonium zonate spot ourmiavirus**
PelVCV = Pelargonium vein clearing *cytorhabdovirus*
PeMoV = Peanut mottle *potyvirus*
PEMV = Pea enation mosaic *enamovirus*
pepino latent virus = Potato S *carlavirus*
**Pepino mosaic *potexvirus***
PepMoV = Pepper mottle *potyvirus*

**Plum pox *potyvirus***
plum pseudopox virus = Apple chlorotic leaf spot *trichovirus*
PlV-4 = Plantago 4 *caulimovirus*
PlV-6 = Plantain 6 *carmovirus*
PlV-8 = Plantain 8 *carlavirus*
PlVX = Plantain X *potexvirus*
PLV = Passiflora latent *carlavirus*
PMiMV = Pea mild mosaic *comovirus*
PMMoV = Pepper mild mottle *tobamovirus*
PMMV = Pepper mild mosaic potyvirus
PMoV = Parietaria mottle *ilarvirus*
PMTV = Potato mop-top *furovirus*
PMV = Panicum mosaic *sobemovirus*
PMWaV = Pineapple wilt-associated *closterovirus*
PnCV = Poinsettia *alphacryptovirus*
PnMV = Poinsettia mosaic *tymovirus*
PNRSV = Prunus necrotic ringspot *ilarvirus*
**Poa semilatent *hordeivirus***
pod mottle virus = Bean pod mottle *comovirus*
**Poinsettia *alphacryptovirus***
**Poinsettia mosaic *tymovirus***
**Pokeweed mosaic *potyvirus***
P.O. pea streak virus = Broad bean wilt *fabavirus*
**Poplar decline potyvirus**
poplar latent virus = Poplar mosaic *carlavirus*
**Poplar mosaic *carlavirus***
poplar potyvirus = Poplar decline potyvirus
**Poplar vein yellowing *nucleorhabdovirus***
PopMV = Poplar mosaic *carlavirus*
**Potato 14R *tobamovirus***
potato acropetal necrosis virus = Potato Y *potyvirus*
**Potato Andean latent *tymovirus***
**Potato Andean mottle *comovirus***
**Potato A *potyvirus***
**Potato aucuba mosaic *potexvirus***
**Potato black ringspot *nepovirus***
potato bouquet virus = Tomato black ring *nepovirus*
potato calico strain of tobacco ringspot virus = Potato black ringspot *nepovirus*
potato calico virus = Alfalfa mosaic *alfamovirus*
potato corky ringspot virus = Tobacco rattle *tobravirus*
potato G virus = Potato aucuba mosaic *potexvirus*
potato latent virus = Potato X *potexvirus*
**Potato leafroll *luteovirus***
**Potato M *carlavirus***
potato mild mosaic virus = Potato A *potyvirus*
potato mild mosaic virus = Potato X *potexvirus*
**Potato mop-top *furovirus***
potato paracrinkle virus = Potato M *carlavirus*
potato phloem necrosis virus = Potato leafroll *luteovirus*
potato pseudo-aucuba virus = Tomato black ring *nepovirus*
**Potato S *carlavirus***
potato severe mosaic virus = Potato Y *potyvirus*
potato stem mottle virus = Tobacco rattle *tobravirus*
**Potato T *trichovirus***
**Potato U *nepovirus***
potato virus E = Potato M *carlavirus*
potato virus F = Potato aucuba mosaic *potexvirus*
potato virus P = Potato A *potyvirus*
**Potato V *potyvirus***
potato wild mosaic potyvirus = Wild potato mosaic *potyvirus*
**Potato X *potexvirus***
**Potato yellow dwarf *nucleorhabdovirus***
**Potato yellow mosaic bigeminivirus**
**Potato Y *potyvirus***
PPPV = Pigeonpea proliferation rhabdovirus
PPV = Plum pox *potyvirus*
**Primula mosaic *potyvirus***
**Primula mottle potyvirus**
PrMoV = Primula mottle potyvirus
PRMV = Peach rosette mosaic *nepovirus*
PrMV = Primula mosaic *potyvirus*
PRSV = Papaya ringspot *potyvirus*
prune brown line virus = Tomato ringspot *nepovirus*
**Prune dwarf *ilarvirus***

*Prunivir cerasi* virus = Cherry mottle leaf *trichovirus*
**Prunus necrotic ringspot *ilarvirus***
prunus ringspot virus = Prunus necrotic ringspot *ilarvirus*
**Prunus S *carlavirus***
prunus stem-pitting virus = Tomato ringspot *nepovirus*
prunus virus 10 = Plum American line pattern *ilarvirus*
Prunus virus 1 = Cherry mottle leaf *trichovirus*
prunus virus 7 = Plum pox *potyvirus*
PruVS = Prunus S *carlavirus*
PSbMV = Pea seed-borne mosaic *potyvirus*
PSLV = Poa semilatent *hordeivirus*
PSMV = Paspalum striate mosaic monogeminivirus
Psophocarpus necrotic mosaic virus = Cowpea mild mottle *carlavirus*
PSV = Peanut stunt *cucumovirus*
PTV = Tomato Peru *potyvirus*
Puerto Rico cowpea mosaic virus = Cowpea severe mosaic *comovirus*
pumpkin mosaic virus = Squash mosaic *comovirus*
**Purple granadilla mosaic virus**
PV-7 = Plantain 7 potyvirus
PVA = Potato A *potyvirus*
PVCV = Petunia vein clearing *caulimovirus*
PVM = Potato M *carlavirus*
PVMV = Pepper veinal mottle *potyvirus*
PVS = Potato S *carlavirus*
PVT = Potato T *trichovirus*
PVU = Potato U *nepovirus*
PVV = Potato V *potyvirus*
PVX = Potato X *potexvirus*
PVY = Potato Y *potyvirus*
PWV = Passionfruit woodiness *potyvirus*
PYDV = Potato yellow dwarf *nucleorhabdovirus*
PYFV = Parsnip yellow fleck *sequivirus*
PYMV = Potato yellow mosaic bigeminivirus
PZSV = Pelargonium zonate spot ourmiavirus
QPMV = Quail pea mosaic *comovirus*
**Quail pea mosaic *comovirus***
quince stunt virus = Apple chlorotic leaf spot *trichovirus*
radish enation mosaic virus = Radish mosaic *comovirus*
**Radish mosaic *comovirus***
radish P virus = Turnip mosaic *potyvirus*
**Radish yellow edge *alphacryptovirus***
radish yellows virus = Beet western yellows *luteovirus*
rai mosaic virus = Ribgrass mosaic *tobamovirus*
RamRSV = Ranunculus repens symptomless rhabdovirus
RaMV = Radish mosaic *comovirus*
RanMV = Ranunculus mottle *potyvirus*
ranunculus mosaic virus = Ranunculus mottle *potyvirus*
**Ranunculus mottle *potyvirus***
**Ranunculus repens symptomless rhabdovirus**
**Raphanus *cytorhabdovirus***
raspberry black necrosis virus = Black raspberry necrosis virus
**Raspberry bushy dwarf *idaeovirus***
raspberry chlorotic virus = Raspberry vein chlorosis *nucleorhabdovirus*
**Raspberry leaf curl *luteovirus***
raspberry line-pattern virus = Raspberry bushy dwarf *idaeovirus*
**Raspberry ringspot *nepovirus***
raspberry Scottish leaf curl virus = Raspberry ringspot *nepovirus*
**Raspberry vein chlorosis *nucleorhabdovirus***
raspberry yellow dwarf virus = Arabis mosaic *nepovirus*
raspberry yellow mosaic virus = Rubus yellow net *badnavirus*
raspberry yellows virus = Raspberry bushy dwarf *idaeovirus*
ratel virus = Tobacco rattle *tobravirus*
RaV = Raphanus *cytorhabdovirus*
RBDV = Raspberry bushy dwarf *idaeovirus*
RBSDV = Rice black-streaked dwarf *fijivirus*
RCCV-2 = Red clover 2 *betacryptovirus*
RClMV = Red clover mosaic *nucleorhabdovirus*
RCMV = Red clover mottle *comovirus*

RCNMV = Red clover necrotic mosaic *dianthovirus*
RCSV = Rubus Chinese seed-borne *nepovirus*
RCVMV = Red clover vein mosaic *carlavirus*
RDV = Rice dwarf *phytoreovirus*
**Red clover 2 *betacryptovirus***
red clover cryptic virus 1 = White clover 1 *alphacryptovirus*
red clover cryptic virus 3 = White clover 3 *alphacryptovirus*
**Red clover mosaic *nucleorhabdovirus***
**Red clover mottle *comovirus***
**Red clover necrotic mosaic *dianthovirus***
**Red clover vein mosaic *carlavirus***
red currant mosaic virus = Tomato ringspot *nepovirus*
red currant necrotic ringspot virus = Prunus necrotic ringspot *ilarvirus*
red currant ringspot virus = Raspberry ringspot *nepovirus*
**Red pepper 1 *alphacryptovirus***
**Red pepper 2 *alphacryptovirus***
**Rehmannia X *potexvirus***
RGCV = Ryegrass *alphacryptovirus*
RGDV = Rice gall dwarf *phytoreovirus*
RGMV = Ryegrass mosaic *rymovirus*
RGSV = Rice grassy stunt *tenuivirus*
Rhabarber-Mosaik-Virus = Arabis mosaic *nepovirus*
RHBV = Rice hoja blanca *tenuivirus*
RhMV = Rhynchosia mosaic bigeminivirus
**Rhododendron necrotic ringspot *potexvirus***
**Rhubarb 1 *potexvirus***
rhubarb mosaic virus = Arabis mosaic *nepovirus*
**Rhubarb temperate *alphacryptovirus***
rhubarb virus 5 = Strawberry latent ringspot *nepovirus*
**Rhynchosia mosaic bigeminivirus**
**Ribgrass mosaic *tobamovirus***
ribgrass strain of tobacco mosaic virus = Ribgrass mosaic *tobamovirus*
**Rice black-streaked dwarf *fijivirus***
rice black streak virus = Rice black-streaked dwarf *fijivirus*
**Rice bunchy stunt *phytoreovirus***
**Rice dwarf *phytoreovirus***
**Rice gall dwarf *phytoreovirus***
rice giallume virus = Barley yellow dwarf *luteovirus*
**Rice grassy stunt *tenuivirus***
**Rice hoja blanca *tenuivirus***
rice infectious gall virus = Rice ragged stunt *oryzavirus*
rice leaf yellowing virus = Rice tungro spherical *waikavirus*
**Rice necrosis mosaic *bymovirus***
rice penyakit habeng virus = Rice tungro spherical *waikavirus*
rice penyakit mentek virus = Rice tungro spherical *waikavirus*
**Rice ragged stunt *oryzavirus***
rice rosette Philippines virus = Rice grassy stunt *tenuivirus*
rice rosette virus = Rice grassy stunt *tenuivirus*
**Rice stripe necrosis *furovirus***
**Rice stripe *tenuivirus***
**Rice transitory yellowing *nucleorhabdovirus***
**Rice tungro bacilliform *badnavirus***
**Rice tungro spherical *waikavirus***
rice waika virus = Rice tungro spherical *waikavirus*
rice yellow leaf virus = Rice tungro spherical *waikavirus*
**Rice yellow mottle *sobemovirus***
rice yellow orange leaf virus = Rice tungro bacilliform *badnavirus*
**Rice yellow stunt virus**
Ringmosaikvirus der Kapuzinerkresse = Broad bean wilt *fabavirus*
RLCV = Raspberry leaf curl *luteovirus*
RMV = Ribgrass mosaic *tobamovirus*
RNMV = Rice necrosis mosaic *bymovirus*
robinia mosaic virus = Peanut stunt *cucumovirus*
RoNRSV = Rhododendron necrotic ringspot *potexvirus*
rose chlorotic mottle virus = Prunus necrotic ringspot *ilarvirus*
rose colour break virus = Rose *tobamovirus*
rose line pattern virus = Prunus necrotic ringspot *ilarvirus*

rose mosaic virus = Apple mosaic *ilarvirus*
**Rose *tobamovirus***
rose vein banding virus = Prunus necrotic ringspot *ilarvirus*
rose yellow vein mosaic virus = Prunus necrotic ringspot *ilarvirus*
**Rottboellia yellow mottle *sobemovirus***
RPCV-1 = Red pepper 1 *alphacryptovirus*
RPCV-2 = Red pepper 2 *alphacryptovirus*
RpRSV = Raspberry ringspot *nepovirus*
RSNV = Rice stripe necrosis *furovirus*
RSSV = Rice ragged stunt *oryzavirus*
RSV = Rice stripe *tenuivirus*
RTBV = Rice tungro bacilliform *badnavirus*
RTSV = Rice tungro spherical *waikavirus*
RTV = Rhubarb temperate *alphacryptovirus*
RTYV = Rice transitory yellowing *nucleorhabdovirus*
**Rubus Chinese seed-borne *nepovirus***
**Rubus yellow net *badnavirus***
Russian winter wheat mosaic virus = Winter wheat Russian mosaic *cytorhabdovirus*
Rübenkräusel-Virus = Beet leaf curl rhabdovirus
RV-1 = Rhubarb 1 *potexvirus*
RVCV = Raspberry vein chlorosis *nucleorhabdovirus*
**Ryegrass *alphacryptovirus***
**Ryegrass bacilliform *nucleorhabdovirus***
ryegrass chlorotic streak virus = Barley yellow dwarf *luteovirus*
**Ryegrass mosaic *rymovirus***
**Ryegrass mottle *sobemovirus***
ryegrass spherical cryptic virus = Ryegrass *alphacryptovirus*
ryegrass streak mosaic virus = Ryegrass mosaic *rymovirus*
ryegrass streak virus = Brome mosaic *bromovirus*
RYEV = Radish yellow edge *alphacryptovirus*
RYMV = Rice yellow mottle *sobemovirus*
RYNV = Rubus yellow net *badnavirus*
**Saguaro cactus *carmovirus***
**Saintpaulia leaf necrosis rhabdovirus**
SALCV = Solanum apical leaf curling bigeminivirus
sambucus ringspot and yellow net virus = Cherry leaf roll *nepovirus*
**Sambucus vein clearing rhabdovirus**
Sammons' opuntia virus = Opuntia Sammons' *tobamovirus*
Samsun latent strain of tobacco mosaic virus = Pepper mild mottle *tobamovirus*
**Santosai temperate *alphacryptovirus***
Sarka virus = Plum pox *potyvirus*
**Sarracenia purpurea rhabdovirus**
satellite tobacco mosaic virus = Tobacco mosaic satellivirus
satellite virus = Tobacco necrosis satellivirus
**Satsuma dwarf *nepovirus***
SbCMV = Soybean chlorotic mottle *caulimovirus*
SBDMV = Swordbean distortion mosaic *potyvirus*
SbDV = Soybean dwarf *luteovirus*
SBLV = Spring beauty latent *bromovirus*
SBMV = Bean southern mosaic *sobemovirus*
SCBV = Sugarcane bacilliform *badnavirus*
Scharka-Virus = Plum pox *potyvirus*
**Schefflera ringspot *badnavirus***
SCLV = Soybean crinkle leaf bigeminivirus
SCMoV = Subterranean clover mottle *sobemovirus*
SCMV = Sugarcane mosaic *potyvirus*
SCNMV = Sweet clover necrotic mosaic *dianthovirus*
ScrMV = Scrophularia mottle *tymovirus*
**Scrophularia mottle *tymovirus***
scrophularia-Scheckungsvirus = Scrophularia mottle *tymovirus*
SCrV = Strawberry crinkle *cytorhabdovirus*
SCSV = Subterranean clover stunt nanavirus
SCV = Saguaro cactus *carmovirus*
SDV = Satsuma dwarf *nepovirus*
**Serrano golden mosaic bigeminivirus**
SgCSV = Sorghum chlorotic spot *furovirus*

**Tomato pseudo curly top hybrigeminivirus**
**Tomato ringspot *nepovirus***
**Tomato spotted wilt *tospovirus***
**Tomato top necrosis *nepovirus***
tomato transmissible leafroll virus = Tomato Australian leafcurl bigeminivirus
tomato tree virus = Tamarillo mosaic *potyvirus*
tomato vein chlorosis virus = Eggplant mottled dwarf *nucleorhabdovirus*
**Tomato vein clearing *nucleorhabdovirus***
tomato vein clearing virus = Eggplant mottled dwarf *nucleorhabdovirus*
tomato vein yellowing virus = Eggplant mottled dwarf *nucleorhabdovirus*
tomato vein yellowing virus = Pittosporum vein yellowing *nucleorhabdovirus*
tomato yellow dwarf virus = Tobacco leaf curl bigeminivirus
**Tomato yellow leaf curl bigeminivirus**
**Tomato yellow mosaic bigeminivirus**
tomato yellow mosaic virus = Tomato golden mosaic bigeminivirus
tomato yellows virus = Beet curly top hybrigeminivirus
tomato yellow vein virus = Eggplant mottled dwarf *nucleorhabdovirus*
ToMoV = Tomato mottle bigeminivirus
ToMV = Tomato mosaic *tobamovirus*
ToRSV = Tomato ringspot *nepovirus*
ToTNV = Tomato top necrosis *nepovirus*
ToYMV = Tomato yellow mosaic bigeminivirus
TPCTV = Tomato pseudo curly top hybrigeminivirus
TPV = Pepper Texas bigeminivirus
**Tradescantia-Zebrina *potyvirus***
trespengrasmosaikvirus = Brome mosaic *bromovirus*
**Trichosanthes mottle potyvirus**
*Trifoliumvirus nervicrassans* virus = Clover wound tumor *phytoreovirus*
TrMV = Trichosanthes mottle potyvirus
**Tropaeolum 1 *potyvirus***
**Tropaeolum 2 *potyvirus***
tropaeolum ringspot virus = Broad bean wilt *fabavirus*
TRoV = Turnip rosette *sobemovirus*
TRSV = Tobacco ringspot *nepovirus*
TRV = Tobacco rattle *tobravirus*
TStV = Tobacco stunt varicosavirus
TSV = Tobacco streak *ilarvirus*
TSWV = Tomato spotted wilt *tospovirus*
tuber blotch virus = Potato aucuba mosaic *potexvirus*
**Tuberose *potyvirus***
**Tulare apple mosaic *ilarvirus***
tulip Augusta disease virus = Tobacco necrosis *necrovirus*
**Tulip band-breaking *potyvirus***
**Tulip breaking *potyvirus***
tulip breaking virus — lily strain = Lily mottle *potyvirus*
**Tulip chlorotic blotch *potyvirus***
**Tulip halo necrosis virus**
**Tulip mosaic virus**
tulip white streak virus = Tobacco rattle *tobravirus*
**Tulip X *potexvirus***
TuMV = Turnip mosaic *potyvirus*
**Turnip crinkle *carmovirus***
turnip latent virus disease complex = Physalis mild chlorosis *luteovirus*
turnip latent virus disease complex = Physalis vein blotch *luteovirus*
turnip mild yellows virus = Beet western yellows *luteovirus*
**Turnip mosaic *potyvirus***
**Turnip rosette *sobemovirus***
**Turnip vein-clearing virus**
**Turnip yellow mosaic *tymovirus***
TV-1 = Tropaeolum 1 *potyvirus*
TV-2 = Tropaeolum 2 *potyvirus*
TVDV = Tobacco vein-distorting *luteovirus*
TVMV = Tobacco vein mottling *potyvirus*
TVX = Tulip X *potexvirus*
TWV = Tobacco wilt *potyvirus*
TYDV = Tobacco yellow dwarf monogeminivirus
TYLCV-Is, -Sr, -Th, -Ye = Tomato yellow leaf curl bigeminivirus
TYMV = Turnip yellow mosaic *tymovirus*
TYNV = Tobacco yellow net *luteovirus*
TYVAV = Tobacco yellow vein assistor *luteovirus*

TYVV = Tobacco yellow vein *umbravirus*
TZV = Tradescantia-Zebrina *potyvirus*
**Ullucus C *comovirus***
**Ullucus mild mottle *tobamovirus***
**Ullucus mosaic *potyvirus***
UMMV = Ullucus mild mottle *tobamovirus*
UMV = Ullucus mosaic *potyvirus*
**Urd bean leaf crinkle virus**
UVC = Ullucus C *comovirus*
**Vallota mosaic *potyvirus***
ValMV = Vallota mosaic *potyvirus*
**Vanilla mosaic *potyvirus***
**Vanilla necrosis *potyvirus***
VCV = Vicia *alphacryptovirus*
vein mosaic virus of red clover = Red clover vein mosaic *carlavirus*
**Velvet tobacco mottle *sobemovirus***
**Velvet tobacco mottle virus viroid-like satellite RNA**
**Vicia *alphacryptovirus***
*Vicia* virus 1 = Broad bean true mosaic *comovirus*
**Vigna sinensis mosaic rhabdovirus**
**Viola mottle *potexvirus***
Virginia Crab stem grooving virus = Apple stem grooving *capillovirus*
virus ampollado del frijol = Bean rugose mosaic *comovirus*
virus de la mosaique de la fève = Broad bean stain *comovirus*
virus de la mosaique du Gombo = Okra mosaic *tymovirus*
virus del enanismo rayado maiz = Maize mosaic *nucleorhabdovirus*
virus del mosaico amarillo del tomate = Tomato yellow mosaic bigeminivirus
virus del mosaico dorado = Bean golden mosaic bigeminivirus
virus del mosaico dourado do tomateiro = Tomato yellow mosaic bigeminivirus
virus del mosaico rugoso de frijol = Bean rugose mosaic *comovirus*
virus del moteado amarillo = Cowpea chlorotic mottle *bromovirus*
virus del nanismo ruvido del mais = Maize rough dwarf *fijivirus*
virus des taches jaunes du cresson = Watercress yellow spot virus
virus do mosaico dourado do tomateiro = Tomato golden mosaic bigeminivirus
virus mosaique de tabac = Tobacco mosaic *tobamovirus*
virus Peruano del tomate = Tomato Peru *potyvirus*
VMV = Viola mottle *potexvirus*
VNMV = Voandzeia necrotic mosaic *tymovirus*
voandezia mosaic virus = Cowpea mild mottle *carlavirus*
**Voandzeia distortion mosaic *potyvirus***
**Voandzeia necrotic mosaic *tymovirus***
VTMoV = Velvet tobacco mottle *sobemovirus*
WASMV = Wheat American striate mosaic *nucleorhabdovirus*
**Watercress yellow spot virus**
**Watermelon chlorotic stunt bigeminivirus**
**Watermelon curly mottle bigeminivirus**
**Watermelon Moroccan mosaic potyvirus**
**Watermelon mosaic 1 *potyvirus***
**Watermelon mosaic 2 *potyvirus***
watermelon mosaic virus 1 = Papaya ringspot *potyvirus*
watermelon mosaic virus = Watermelon mosaic 2 *potyvirus*
watermelon papaya ringspot = Watermelon mosaic 1 *potyvirus*
watermelon silver mottle virus = Tomato spotted wilt *tospovirus*
WBMV = White bryony mosaic *carlavirus*
WBV = White bryony potyvirus
WCCV-1 = White clover 1 *alphacryptovirus*
WCCV-2 = White clover 2 *betacryptovirus*
WCCV-3 = White clover 3 *alphacryptovirus*
WClMV = White clover mosaic *potexvirus*
WCMV = Wild cucumber mosaic *tymovirus*
WDV = Wheat dwarf monogeminivirus
**Weddel waterborne *carmovirus***
weidelgrasmosaikvirus = Brome mosaic *bromovirus*